List of Symbols in General Use in Machine Design

Symbol	Meaning
A	Area; constant
$\mathbf{a}, a$	Acceleration
a	Constant; dimension; addendum
b	Weibull exponent; dedendum; section width; dimension; constant; fatigue-strength exponent.
C	Coefficient; spring index; bearing-load rating; column end-condition constant; gear factor; center distance; specific heat; diametral clearance
c	Clearance; radial clearance; distance from neutral axis in beam; fatigue ductility exponent; coefficient of viscosity
D, d	Diameter
E	Modulus of elasticity; kinetic energy
$\mathbf{F}, F$	Force
F	Face width
f	Frequency; coefficient of friction
G	Shear modulus of elasticity
g	Acceleration due to gravity
H	Hardness number; heat gained or lost; power
h	Section depth; bearing clearance
I	Moment of inertia; second area moment; geometry factor
i	Integer
J	Polar moment of inertia; second polar area moment; geometry factor; mechanical equivalent of heat
j	Integer
K	Stress-concentration factor; wear factor; Wahl correction factor; gear factor; bearing-rating ratio; strength coefficient; stress-intensity factor; bolt-torque coefficient
k	Spring rate; endurance-limit modification factor; radius of gyration
L	Length; life; lead
l	Length
$\mathbf{M}, M$	Moment
m	Mass, margin of safety; slope; contact ratio; module
n	Design factor; strain-hardening exponent; speed in rpm
$\mathbf{P}, P$	Force
P	Diametral pitch; bearing pressure
p	Pressure; linear or circular pitch
Q	First moment of area; flow volume
q	Load intensity; notch sensitivity; arc length
R	Reliability; reduction in area
r	Radius; radial direction indicator; correlation coefficient
S	Strength; bearing characteristic number; scale
s	Sample standard deviation; distance
T	Torque; temperature
t	Thickness; tangential direction; time
U	Energy; velocity; coefficient
V	Shear force; velocity; rotation factor for bearings
W	Weight; load; cold-work factor
X	Radial factor for bearings
x	Rectangular coordinate; distance
Y	Thrust factor for bearings
y	Rectangular coordinate; distance
Z	Section modulus; viscosity
z	Rectangular coordinate; distance; standard statistical variable

Symbol	Meaning
α	Coefficient of thermal expansion; thread angle; axial fatigue-stress correction factor; angle
β	Partial bearing angle
Γ	Pitch angle; gamma function
γ	Shear strain; pitch angle; articulation angle
Δ	Increment or change
δ	Total deformation or elongation
ϵ	Unit engineering strain
ε	True strain; eccentricity ratio
η	Efficiency
η	Factor of safety (distinguished from design factor n)
θ	Twist angle; slope; a Weibull parameter
λ	Lead angle
μ	Population mean; absolute viscosity
ν	Poisson's ratio
ρ	Radius of curvature; density
Σ	Summation sign
$\sigma, \bar{\sigma}$	Normal stress (engineering or nominal)
σ	True stress
$\hat{\sigma}$	Population standard deviation
τ	Shear stress
ϕ	Pressure angle; angle
ψ	Helix angle
ω	Angular velocity

STANDARD HANDBOOK OF MACHINE DESIGN

Joseph E. Shigley Editor in Chief

Late Professor Emeritus
The University of Michigan
Ann Arbor, Michigan

Charles R. Mischke Editor in Chief

Professor Emeritus of Mechanical Engineering
Iowa State University
Ames, Iowa

Second Edition

McGraw-Hill

New York   San Francisco   Washington, D.C.   Auckland   Bogotá
Caracas   Lisbon   London   Madrid   Mexico City   Milan
Montreal   New Delhi   San Juan   Singapore
Sydney   Tokyo   Toronto

Library of Congress Cataloging-in-Publication Data

Standard handbook of machine design / editors in chief, Joseph E.
  Shigley, Charles R. Mischke. — 2nd ed.
        p.   cm.
  Includes index.
  ISBN 0-07-056958-4
  1. Machine design—Handbooks, manuals, etc.   I. Shigley, Joseph
Edward.   II. Mischke, Charles R.
TJ230.S8235   1996
621.8'15—dc20   95-50600
                                                         CIP

McGraw-Hill

A Division of The McGraw·Hill Companies

    2 3 4 5 6 7 8 9 0   DOC/DOC   9 0 1 0 9 8 7 6

ISBN 0-07-056958-4

*The sponsoring editor for this book was Harold Crawford, the editing
supervisor was Bernard Onken, and the production supervisor was
Pamela Pelton. It was set in Times Roman by North Market Street Graphics.*

Printed and bound by R.R. Donnelley & Sons Company.

McGraw-Hill books are available at special quantity discounts to use as premiums and sales promotions, or for use in corporate training programs. For
more information, please write to the Director of Special Sales, McGraw-
Hill, 11 West 19th Street, New York, NY 10011. Or contact your local bookstore.

To the late Joseph Edward Shigley

Joseph Edward Shigley was awarded bachelor degrees in electrical (1931) and mechanical (1932) engineering by Purdue University, and a master of science in engineering mechanics (1946) by The University of Michigan. His career in engineering education began at Clemson College (1936–1956) and continued at The University of Michigan (1956–1978). Upon retirement, he was named Professor Emeritus of Mechanical Engineering by the Regents in recognition of his outstanding achievement and dedicated service.

At the time when Professor Shigley began thinking about his first book on machine design, many designers were unschooled, and textbooks tended to give results with only a brief explanation—they did not offer the reader many tools with which to proceed in other or new directions. Professor Shigley's first book, *Machine Design* (1956), showed his attention to learning and understanding. That milestone book is currently in its fifth edition. Other books followed, among which are *Theory of Machines and Mechanisms* (with John J. Uicker, Jr.), *Mechanical Engineering Design* (with Charles R. Mischke), and *Applied Mechanics of Materials*.

Early in the 1980s, Professor Shigley called Professor Mischke and said, "I've never done a Handbook before; there is no precedent in machine design, and it is time there was one. I propose we do it together. Take a couple of months to consider what ought to be in it, the organization and presentation style. Then we can get together and compare notes."

The result was the first edition of the *Standard Handbook of Machine Design* (1986), which won the Association of American Publishers Award for the best book in engineering and technology published in 1986. Eight *Mechanical Designers Workbooks* followed.

Professor Shigley received recognitions such as the grade of Fellow in the American Society of Mechanical Engineers, from which he also received the Mechanisms Committee Award in 1974, the Worcester Reed Warner Medal in 1977, and the Machine Design Award in 1985. I believe he would have given up all the above rather than give up the effect he had as mentor and tutor to students, and in guiding boys toward manhood as a scoutmaster.

He indeed made a difference.

Charles R. Mischke

CONTENTS

CONTRIBUTORS

Erich K. Bender *Division Vice President,* Bolt, Beranek and Newman Inc., Cambridge, Mass.

R. B. Bhat *Associate Professor,* Department of Mechanical Engineering, Concordia University, Montreal, Quebec, Canada.

John H. Bickford *Retired Vice President, Manager of the Power-Dyne Division,* Raymond Engineering Inc., Middletown, Conn.

Omer W. Blodgett *Design Consultant,* The Lincoln Electric Company, Cleveland, Ohio.

Daniel M. Curtis *Senior Mechanical Engineer,* NKF Engineering, Inc., Arlington, Va.

Daniel E. Czernik *Director of Product Engineering,* Fel-Pro Inc., Skokie, Ill.

Joseph Datsko *Professor of Mechanical Engineering Emeritus,* The University of Michigan, Ann Arbor, Mich.

Raymond J. Drago *Senior Engineer,* Advanced Power Train Technology, Boeing Vertol, Philadelphia, Pa.

K. S. Edwards *Professor of Mechanical Engineering,* The University of Texas at El Paso, Tex.

Rudolph J. Eggert *Associate Professor of Mechanical Engineering,* University of Idaho, Boise, Idaho.

Wolfram Funk *Professor,* Fachbereich Maschinenbau, Fachgebiet Maschinenelemente und Getriebetechnik, Universität der Bundeswehr Hamburg, Hamburg, Federal Republic of Germany.

Richard E. Gustavson *Technical Staff Member,* The Charles Draper Laboratory Inc., Cambridge, Mass.

Jerry Lee Hall *Professor of Mechanical Engineering,* Iowa State University, Ames, Iowa.

Russ Henke Russ Henke Associates, Elm Grove, Wis.

Harry Herman *Professor of Mechanical Engineering,* New Jersey Institute of Technology, Newark, N.J.

R. Bruce Hopkins The Hopkins Engineering Co., Cedar Falls, Iowa.

Robert J. Hotchkiss *Director,* Gear Technology, Gleason Machine Division, Rochester, N.Y.

Robert E. Joerres *Applications Engineering Manager,* Associated Spring, Barnes Group Inc., Bristol, Conn.

Harold L. Johnson *Associate Professor Emeritus,* School of Mechanical Engineering, Georgia Institute of Technology, Atlanta, Ga.

Ray C. Johnson *Higgins Professor of Mechanical Engineering Emeritus,* Worcester Polytechnic Institute, Worcester, Mass.

Theo J. Keith, Jr. *Professor and Chairman of Mechanical Engineering,* University of Toledo, Toledo, Ohio.

Theodore K. Krenzer *Manager,* Gear Theory Department, Gleason Machine Division, Rochester, N.Y.

Karl H. E. Kroemer   *Professor,* Industrial and Systems Engineering Department, Virginia Tech (VPI & SU), Blacksburg, Va.

A. R. Lansdown   *Director,* Swansea Tribology Centre, University of Swansea, United Kingdom.

Kenneth C. Ludema   *Professor of Mechanical Engineering,* Department of Mechanical Engineering and Applied Mechanics, The University of Michigan, Ann Arbor, Mich.

Charles R. Mischke   *Professor of Mechanical Engineering Emeritus,* Iowa State University, Ames, Iowa.

Andrzej A. Olędzki   *Professor Emeritus,* Warsaw Technical University, Warsaw, Poland.

Leo C. Peters   *Professor of Mechanical Engineering,* Iowa State University, Ames, Iowa.

Paul J. Remington   *Principal Engineer,* Bolt, Beranek and Newman, Inc., Cambridge, Mass.

Richard S. Sabo   *Manager,* Educational Services, The Lincoln Electric Company, Cleveland, Ohio.

T. S. Sankar   *Professor and Chairman,* Department of Mechanical Engineering, Concordia University, Montreal, Quebec, Canada.

Howard B. Schwerdlin   *Engineering Manager,* Lovejoy, Inc., Downers Grove, Ill.

Joseph E. Shigley   *Professor Emeritus,* The University of Michigan, Ann Arbor, Mich.

Charles O. Smith   *Consulting Engineer,* Terre Haute, Ind.

L. E. Torfason   *Professor of Mechanical Engineering,* University of New Brunswick, Fredericton, New Brunswick, Canada.

David A. Towers   *Senior Consulting Engineer,* Harris Miller & Hanson Inc., Burlington, Mass.

Eric E. Ungar   *Chief Consulting Engineer,* Bolt, Beranek and Newman, Inc., Cambridge, Mass.

Kenneth J. Waldron   *Professor of Mechanical Engineering,* The Ohio State University, Columbus, Ohio.

Milton G. Wille   *Professor of Mechanical Engineering,* Brigham Young University, Provo, Utah.

John L. Wright   *General Product Manager,* Diamond Chain Company, Indianapolis, Ind.

John R. Zimmerman   *Professor of Mechanical and Aerospace Engineering,* University of Delaware, Newark, Del.

PREFACE TO THE FIRST EDITION

There is no lack of good textbooks dealing with the subject of machine design. These books are directed primarily to the engineering student. Because of this, they contain much theoretical material that is amenable to mathematical analysis. Such topics are preferred by the instructor as well as the student because they appeal to the student's scientific, mathematical, and computer backgrounds; are well-defined topics with a beginning, a middle, and an end; and are easy to use in testing the student's knowledge acquisition. The limited amount of time available for academic studies severely limits the number of topics that can be used as well as their treatment. Since textbooks devoted to mechanical design inevitably reflect this bias, there is great need for a handbook that treats the universe of machine design—not just the readily teachable part.

The beginning designer quickly learns that there is a great deal more to successful design than is presented in textbooks or taught in technical schools or colleges. This handbook connects formal education and the practice of design engineering by including the general knowledge required by every machine designer.

Much of the practicing designer's daily informational needs are satisfied by various pamphlets or brochures, such as those published by the various standards organizations. Other sources include research papers, design magazines, and the various corporate publications concerned with specific products. More often than not, however, a visit to the design library or to the file cabinet will reveal that a specific publication is on loan, lost, or out of date. This handbook is intended to serve such needs quickly and immediately by giving the designer authoritative, up-to-date, understandable, and informative answers to the hundreds of such questions that arise every day in his or her work. Mathematical and statistical formulas and tabulations are available in every design office and, for this reason, are not included in this handbook.

This handbook has been written for working designers, and its place is on the designer's desk—not on the bookshelf. It contains a great many formulas, tables, charts, and graphs, many in condensed form. These are intended to give quick answers to the many questions that seem constantly to arise.

The introduction of new materials, new processes, and new analytical tools and approaches changes the way we design machines. Higher speeds, greater efficiencies, compactness, and safer, lighter-weight, and predictably reliable machines can result if designers keep themselves up to date on technological changes. This book presents machine design as it is practiced today; it is intended to keep the user in touch with the latest aspects of design.

Computer-aided design methods and a host of other machine-computation capabilities of tremendous value to designers have multiplied in the last few years. These have made large and lasting changes in the way we design. This book has been planned and written to make it easy to take advantage of machine-computation facilities of whatever kind may be available. Future developments in computer hardware and software will not render the content of this book obsolete.

This Handbook consists of the writings of 42 different contributors, all well-known experts in their field. We have tried to assemble and to organize the 47 chapters so as to form a unified approach to machine design instead of a collection of unrelated discourses. This has been done by attempting to preserve the same level of mathematical sophistication throughout and by using the same notation wherever possible.

The ultimate responsibility for design decisions rests with the engineer in charge of the design project. Only he or she can judge if the conditions surrounding the application are congruent with the conditions which formed the bases of the presentations in this Handbook, in references, or in any other literature source. In view of the large number of considerations that enter into any design, it is impossible for the editors of this Handbook to assume any responsibility for the manner in which the material presented here is used in design.

We wish to thank all contributors, domestic and foreign, for their patience and understanding in permitting us to fine-tune their manuscripts and for meeting and tolerating our exacting demands. We are also grateful to the many manufacturers who so generously provided us with advice, literature, and photographs. Most of the artwork was competently prepared and supervised by Mr. Gary Roys of Madrid, Iowa, to whom the editors are indebted.

Care has been exercised to avoid error. The editors will appreciate being informed of errors discovered, so that they may be eliminated in subsequent printings.

Joseph E. Shigley
Charles R. Mischke

PREFACE TO THE SECOND EDITION

The introduction of new materials, new processes, and new (or more refined) analytical tools and approaches changes the way in which machines are designed. Complementary to the urge to update and improve, it is useful to look back in order to retain a perspective and appreciate how all this fits into the fabric of machine design methodology. Many of the machine elements we know today were known to the ancients. We have the advantage of improved materials, better manufacturing methods, and finer geometric control, as well as insightful theory and the opportunity to stand on the shoulders of the giants among our predecessors.

Assuring the integrity of a contemplated design, its components, and the aggregate machine or mechanism has always been a problem for the engineer. The methods of record include the following:

- *The Roman method* This method, developed in the Macedonia-Roman period, was to replicate a proven, durable design (with some peripheral improvements). Encyclopedic "books" were compiled for the guidance of designers. In strength-limited designs, the essential thought was, "Don't lean on your element any harder than was done in the durable, extant designs of the past." There are times when contemporary engineers still employ this method.

- *The factor of safety method (of Philon of Byzantium)* In today's terms, one might express this idea as

$$n = \frac{\text{loss-of-function load}}{\text{impressed load}} = \frac{\text{strength}}{\text{stress}}$$

for linear load-stress relations. Alternatively,

$$\text{Allowable load} = \frac{\text{loss-of-function load}}{n}$$

or

$$\text{Allowable stress} = \frac{\text{strength}}{n}$$

for linear load-stress relations. The factor of safety or design factor was experiential and came to consider uncertainty in load as well as in strength.

- *The permissible stress method* Since the concept of stress was introduced by Cauchy in 1822, some engineers have used the idea of permissible stress with load uncertainty considered, and later with the relevant material strength included, as for example in

$$0.40 S_y \leq (\sigma_{\text{all}})_{\text{bending}} \leq 0.60 S_y$$

It is not clear whether the material strength uncertainty is included or not. When the word "allowable" or "permissible" is used among engineers, it is important to clearly define what is, and what is not, included.

- *Allowable stress by design factor* The definition of allowable stress σ_{all} is expressed as

$$\sigma_{all} = \frac{\text{strength}}{n_d^m}$$

where $\sigma = \phi P^m$, i.e., stress is proportional to the mth power of load P. The design factor n_d is experiential and includes load and material strength uncertainty. In gear and cam design, contact stresses are not linear with load. In the form above, if the design factor is 2, then doubling the load creates the loss-of-function load, *whether the stress is bending or hertzian.*

- *Stochastic design factor method* Recognizing that strength **S**, load-induced stress $\boldsymbol{\sigma}$, and the design factor $\mathbf{n}_d$ are stochastic (random) variables, one writes $\mathbf{n}_d = \mathbf{S}/\boldsymbol{\sigma}$. For lognormal strength and lognormal loading, the mean design factor $\bar{n}_d$ is

$$\bar{n}_d = \exp[C_n(z - C_n/2)]$$

where

$$C_n = \sqrt{C_S^2 + C_\sigma^2} = \sqrt{C_S^2 + C_P^2}$$

in which the C's are coefficients of variation of strength, stress, load, or design factor as subscripted. From this point on,

$$\bar{\sigma}_{all} = \frac{\bar{S}}{\bar{n}_d}$$

and one proceeds deterministically using mean values. Note in particular that $\bar{n}_d$ is *quantitatively* experiential from data.

- *The stochastic method* The design factor n_d and the factor n are not used. Distributions are identified, and by simulation procedures the reliability corresponding to a decision set is identified. The computer becomes an important tool.

The practicing designer should be familiar with all of these methods. Although some of them may not be the method of choice under particular circumstances, it is important to understand them all in order to communicate with others and to follow their work.

Developments since the appearance of the first edition of this book in 1986 are reflected in additions to chapters, rewritten chapters, and completely new chapters. More attention is being paid to probabilistic approaches to the design of machinery, and information and methods continue to develop. The reader can appreciate where parts fit in with the historical summary above.

Chapter 2, "Statistical Considerations," has been rewritten to show the relationship between the design factor method and the stochastic methods of design as the inevitable uncertainties are considered. The result on the necessary size of the mean design factor of the interfering of a normal stress with a normal strength is shown. The more useful result on the necessary size of the mean design factor of the interference of a lognormal stress with a lognormal strength is explained. General interference methodology is included, as well as a caution on the nature of numbers, with the reason why significant numbers are rarely useful and possibly harmful.

Chapter 5, "Computer Considerations," has had material added on the important application of computer simulation and, very importantly, on assessing the confidence interval on the result.

Chapter 8, "The Strength of Cold-Worked and Heat-Treated Steels," now includes a Fortran code for cold-work property predictions using the method of Datsko.

Chapter 13, "Strength under Dynamic Conditions," now includes both stochastic and deterministic Marin fatigue reduction factors, and the correlation method for estimating endurance limits in steels. A tabular summary of fatigue equations is also presented in Customary Engineering Units and in SI. The methods for estimating the strength amplitude component and its coefficient of variation are shown for distortion energy–Gerber, ASME-elliptic, and Smith-Dolan fatigue loci. A section on complicated stress variation patterns has also been added.

Chapter 37, "Shafts," has been completely rewritten to show the interplay between deflections (including shear deflections in short shafts) and stress-strength considerations. The fatigue failure loci featured are those that cannot be statistically rejected. These are the distortion energy–Gerber and the ASME-elliptic loci. A section on estimating the first critical speed using Rayleigh's equation has been added.

A new Chapter 9, "Usability," has been added, recognizing that human capabilities and limitations are an integral part of designing tools and machines, and that a practicing machine designer has need for a handy reference. The four essential steps to assure that the product or system fits the operator are enumerated, and sources of available anthropometric information are given. Some basic information is included and references identified.

A new Chapter 10, "Safety," has been added. The ASME Code of Ethics states, "Engineers shall hold paramount the safety, health and welfare of the public in the performance of their professional duties." This chapter identifies the why of safety; what safety is; the nature of hazard, risk, and danger; and the designer's obligation. Human beings interact with all products in the processes of designing, manufacturing, and maintaining them. This ties in with Chaps. 1 and 9.

A new Chapter 11, "Minimizing Engineering Effort," addresses a topic that engineers are reluctant to discuss, namely, "How do you know you are right?" Intertwined with this is the matter of checking and its effectiveness.

The chapter "Gaskets and Seals" has been partitioned into "Gaskets" and "Seals" for this edition. New contributors present the first edition topics of "Power Screws" and "Chain Drives." As in the previous edition, the Handbook continues to be written to take easy advantage of whatever kind of machine-computation facilities may be available. Future developments in hardware and software will not render the contents of this book obsolete.

This edition contains the work of 41 different contributors, all well-known experts in their fields. There are now 50 chapters, assembled and organized to form a coherent approach to machine design.

The ultimate responsibility for design decisions rests with the engineer in charge of the design project. Only he or she can judge if the conditions surrounding the applications are congruent with the circumstances which formed the bases of the presentations in this Handbook, in references, or in any other literature source. In view of the large number of conditions that enter into any design, it is impossible for the editors of this Handbook to assume any responsibility for the manner in which the material presented here is used in design.

We wish to thank all the contributors, domestic and foreign, for their patience and understanding in letting us fine-tune their manuscripts, and for meeting and tolerating our exacting demands. We are also grateful to the many manufacturers who so generously provided us with advice, literature, and photographs. The new artwork for this edition was competently prepared by Ms. Lynn Ekblad, Graphic Designer, of Ames, Iowa.

Care has been taken to avoid error. The editors would appreciate being informed of any errors discovered, so that they may be eliminated in subsequent printings.

Joseph E. Shigley
Charles R. Mischke

CHAPTER 1
INTRODUCTION: STANDARDS, CODES, REGULATIONS

Leo C. Peters, Ph.D., P.E.
Professor of Mechanical Engineering
Iowa State University
Ames, Iowa

R. Bruce Hopkins, Ph.D., P.E.
The Hopkins Engineering Co., P.C.
Cedar Falls, Iowa

1.1   THE DESIGNER AND THE DESIGNER'S PROBLEMS

1.1.1   Design and the Designer

Design and engineering, although sometimes viewed as distinct, are two facets of the same profession. Krick [1.1] states that engineering is a profession concerned primarily with the application of a certain body of knowledge, set of skills, and point of view in the creation of devices, structures, and processes used to transform resources to forms which satisfy the needs of society.

Design is the activity in which engineers accomplish the preceding task, usually by responding to a design imperative for the required task. The design imperative is the result of a problem definition and has the following general form [1.2]: "Design (subject to certain problem-solving constraints) a component, system or process that will perform a specified task (subject to certain solution constraints) optimally."

The end result of the engineering design process is a specification set from which a machine, process, or system may be built and operated to meet the original need.

The designer's task is then to create this specification set for the manufacture, assembly, testing, installation, operation, repair, and use of a solution to a problem. Although primarily decision making and problem solving, the task is a complex activity requiring special knowledge and abilities. A designer cannot effectively operate in a vacuum, but must know, or be able to discover, information affecting the design, such as the state of the art, the custom of the industry, governmental regulations, standards, good engineering practice, user expectations, legal considerations (such as product liability), and legal design requirements.

In addition, an effective designer possesses the ability to make decisions; to innovate solutions to engineering problems; to exhibit knowledge of other technologies and the economics involved; to judge, promote, negotiate, and trade off; and finally, to sell an acceptable problem solution which meets the imposed constraints.

The designer must also be an effective communicator, not only with design supervisors and peers, but also with the public, as represented by federal, state, and local governments, the courts, and the news media.

Most of the time design proceeds by evolution rather than revolution. Thus many of the requirements may have already been met by contributions of others, and most of the time the engineer has to work on only a small portion of the design, requiring only some of the requisites previously identified.

1.1.2   Design Criteria

Although the general criteria used by a designer are many, the following list addresses almost all concerns:

- Function
- Safety
- Reliability
- Cost
- Manufacturability
- Marketability

The inclusion of safety and reliability at or near the level of importance of function is a recent development that has resulted from governmental regulation, expansion in the numbers of standards created, and development of product liability law, all of which occurred in the late 1960s and early 1970s.

Although cost is explicitly fourth on the list, its consideration permeates all the criteria just listed and is part of all design decisions.

As taught and practiced in the past, design criteria emphasized function, cost, manufacturability, and marketability. Reliability was generally included as a part of functional considerations. If product safety was included, it was somewhere in the function-cost considerations.

Design critiques were accomplished at in-house policy committee meetings or their equivalent involving design engineers, a production representative, a materials representative, and possibly representatives of marketing and service.

In the current design climate, the traditional design criteria are still valid; however, the additional constraints of governmental regulations, standards, and society's desire for safety, as exemplified in product liability litigation, have to be included in the design process. In addition, engineers must now be prepared to have their designs evaluated by nondesigners or nontechnical people. This evaluation will not be in the inner confines of a design department by peers or supervisors, as in the past, but may be in a courtroom by a jury of nontechnical people and attorneys who have an ulterior motive for their approach or in the public arena.

Since such a design evaluation is generally a result of an incident which caused damage or injury, to mitigate the nontechnical evaluation, current design procedures should emphasize the following factors in addition to traditional design criteria:

1. *Safety* This is associated with all modes of product usage. In providing for safety, the priorities in design are first, if at all possible, to design the hazards out of the product. If this cannot be done, then shielding and guarding should be provided so that operators and bystanders cannot be exposed to the hazard. Otherwise, if a risk-benefit analysis shows that production and sale of the machine are still justified (and only as a last resort), effective warning should be given against the hazard present. Even though warnings are the least expensive and easiest way to handle hazards in the design process, there has never been a warning that physically prevented an accident in progress. Warnings require human action or intervention. If warnings are required, excellent reference sources are publications of the National Safety Council in Chicago and a notebook entitled *Machinery Product Safety Signs and Labels* [1.78].

2. *Failure analysis* If failure cannot be prevented, it is necessary that it be foreseen and its consequences controlled.

3. *Documentation* Associated with the evolution of the design, documentation is developed so that it can satisfy the involved nontechnical public as to the rationale behind the design and the decisions and tradeoffs that were made.

The designer is in a new mode which places safety on the same level of importance in design considerations as the function or the ability of the design to perform as intended.

Arguments may be made that cost considerations are the most important. This is true only if the cost of the design includes the costs of anticipated litigation. These costs include product liability insurance premiums; direct out-of-pocket costs of investigating and defending claims; and indirect costs in the loss of otherwise productive time used in reviewing the design involved, in finding information for interrogatories, in being deposed, and in developing defense testimony and exhibits. If a lawsuit is lost, the amount of the verdict and the probable increase in product liability insurance premiums must also be included.

No longer can product liability be considered after the design is on the market and the first lawsuit is filed. Product liability considerations must be an integral part of the entire design process throughout the function, safety, cost, manufacturing, and marketing phases.

Additional criteria, considerations, and procedures should be included in programs to address specifically the product safety, failure, or malfunction problems which have contributed significantly to the existing product liability situation. Some of the important considerations and procedures are

1. Development and utilization of a design review system specifically emphasizing failure analysis, safety considerations, and compliance with standards and governmental regulations

2. Development of a list of modes of operation and examination of the product utilization in each mode

3. Identification of the environments of usage for the product, including expected uses, foreseeable misuses, and intended uses

4. Utilization of specific design theories emphasizing failure or malfunction analysis and safety considerations in each mode of operation

Design reviews have been used extensively for improving product performance, reducing cost, and improving manufacturability. In the current product liability climate, it is very important to include, and document in the review, specific failure analysis and safety emphases as well as to check compliance with standards and governmental regulations.

An important consideration in the design review process is to have it conducted by personnel who were not involved in the original design work, so that a fresh, disinterested, competent outlook and approach can be applied in the review.

1.1.3   Influences on the Designer

While attempting to meet the general criteria discussed earlier, the designer's work and the results are affected by both internal and external influences. The external influences, shown in Fig. 1.1, reflect the desires of society as represented by economics, governmental regulations, standards, legal requirements, and ethics, as well as the items shown as human taste.

The other broad area of external influences reflects what is known and available for use in a design problem. The designer is limited by human knowledge, human skills, and, again, economics as to what can be made.

Another important external influence on the designer and the design is legal in nature. The designer is directly influenced by the in-house legal staff or outside attorney retained for legal advice on patents, product liability, and other legal matters and also is affected by product liability suits against the product being designed or similar products.

Internal influences also affect the design. Figure 1.2 identifies some of these. They are a result of the designer's environment while maturing, education, life experiences, moral and ethical codes, personality, and personal needs. These personal or internal influences help shape the engineer's philosophy of design as well as the approach and execution. Individual designs will vary depending on the most important local influences at any given time.

1.1.4   Design Procedure

The general procedure for design is widely available in the literature (see Refs. [1.3] to [1.12]). The following procedure is representative of those found in the literature and is discussed extensively by Hill [1.3]:

1. Identification of need
2. Problem statement or definition of goal
3. Research
4. Development of specifications
5. Generation of ideas
6. Creation of concepts based on the ideas

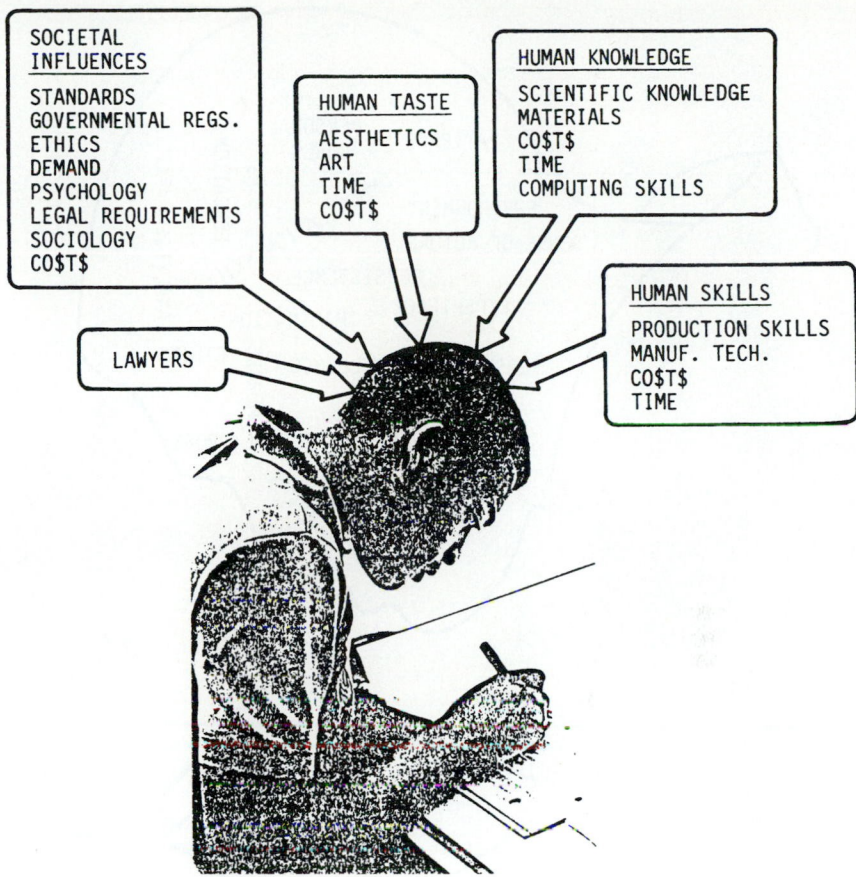

SOCIETAL
INFLUENCES

STANDARDS
GOVERNMENTAL REGS.
ETHICS
DEMAND
PSYCHOLOGY
LEGAL REQUIREMENTS
SOCIOLOGY
COT

HUMAN TASTE

AESTHETICS
ART
TIME
COT

HUMAN KNOWLEDGE

SCIENTIFIC KNOWLEDGE
MATERIALS
COT
TIME
COMPUTING SKILLS

HUMAN SKILLS

PRODUCTION SKILLS
MANUF. TECH.
COT
TIME

LAWYERS

FIGURE 1.1   External influences on the engineering designer.

7. Analysis of alternative concepts
8. Prototype and laboratory testing
9. Selection and specification of best concept
10. Production
11. Marketing
12. Usage (maintenance and repair)

The flowchart in Fig. 1.3 (taken from Ref. [1.13]) illustrates the design process. Note that although not all feedback paths are shown, each step in the process can result in arresting progress and reverting to a prior step, emphasizing that product design is an iterative process.

Much of the design work done is in a small part of one of the feedback or feed-forward portions of the chart and thus is evolutionary. Rarely will an individual designer start at the beginning of the chart with a clean sheet of paper and go through the entire process.

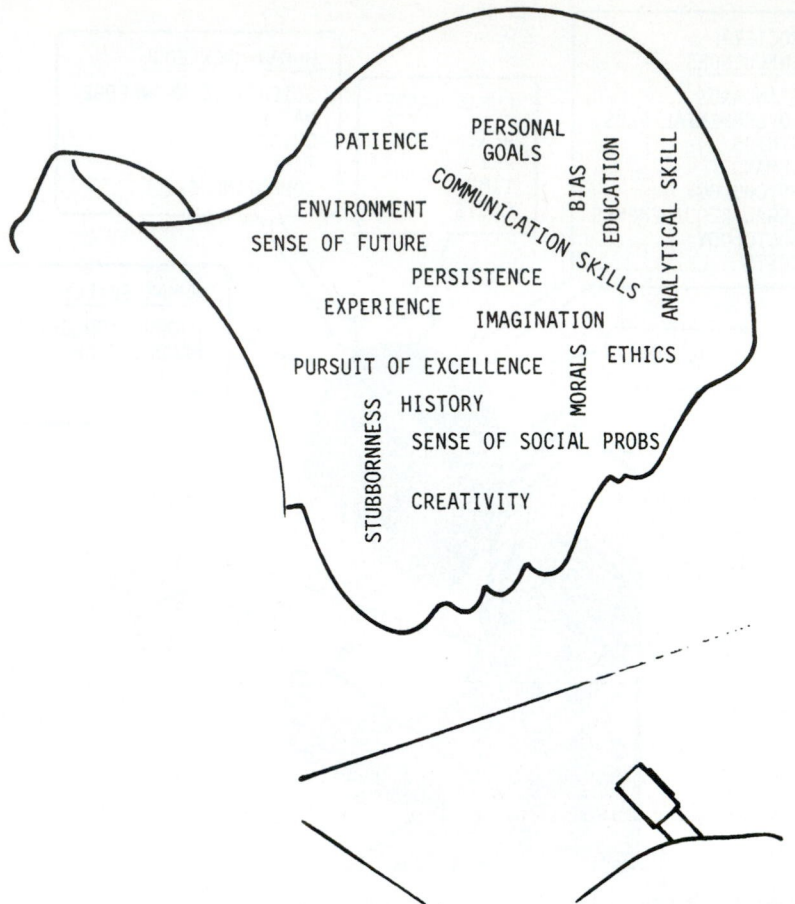

FIGURE 1.2   Internal influences on the engineering designer.

For those designers who do start at the beginning, the checklist in Table 1.1 is an example of one that may be used to organize the information required to define the design problem and aid in establishing design goals. An example list of information for a design specification based on the checklist in Table 1.1 is given in Table 1.2.

After defining the problem and setting the goals for the new design, as much search effort should be made as is feasible to gather all the information possible that applies to the design. This effort includes information on other competitive products or products of a similar nature, governmental regulations and codes, standards, field reports on failure and operation, recall, safety and accident reports, information from lawsuits, plus all the traditional technical information provided in design education (see Ref. [1.14]).

Some of these information sources have attained importance only recently. One example is governmental regulations which have been promulgated since the late 1960s and early 1970s with a major stated purpose of increasing safety both in the workplace (Occupational Safety and Health Act) and elsewhere (Consumer Prod-

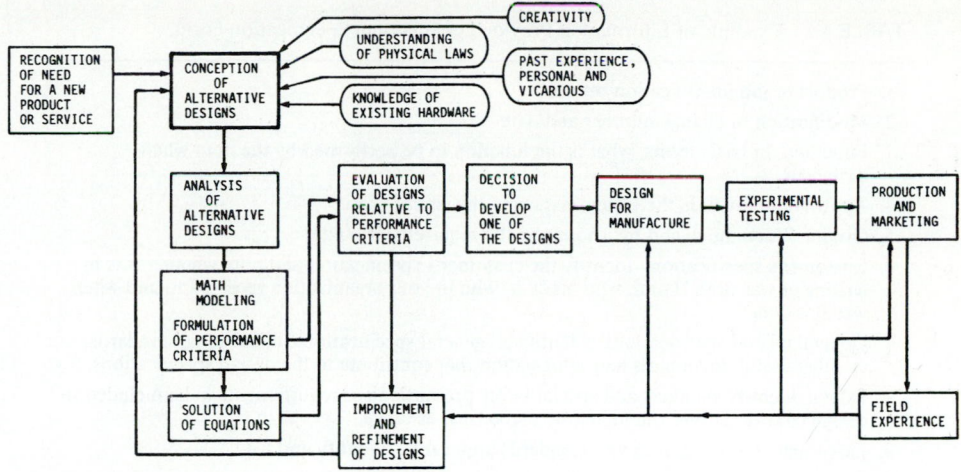

FIGURE 1.3   A flowchart for the design process. (*Adapted from Ref. [1.13]. Used by permission of Charles E. Merrill Publishing Co.*)

TABLE 1.1   Design Checklist

---

1. Function:
   A simple statement of the objective

2. Detailed functional requirements:
   Required performance stated numerically

3. Operating constraints:

Power supplies	Life
Operating procedures	Reliability
Maintenance procedures	Other operating constraints

4. Manufacturing constraints:

Manufacturing processes available	Labor available
Development facilities available	Delivery program
Permissible manufacturing cost	Number required
Other manufacturing constraints	

5. Environment:

Ambient temperature	Installation limitations
Ambient pressure	Expected operators
Climate	Effect on other parts of the parent system
Acceleration	Vibration
Contaminants	Other environmental factors

6. Other constraints:

Applicable governmental regulations	Applicable standards
Legal requirements—patents	Possible litigation

SOURCE: Adapted from Leech [1.14].

TABLE 1.2   Example of Information Provided on a Design Specification Form

---

1. Product or job identification number
2. Modification or change number and date
3. Function: In basic terms, what is the function to be performed by the item when designed?
4. Application: Include the system requiring this application.
5. Origin: When, how, and by whom was the requirement made?
6. Customer's specification: Identify the customer's specification and note whether it is in writing or was oral. If oral, who made it, who in your organization received it, and when was this done?
7. General related specifications: Identify all general specifications, definitions, standards, or other useful documents and information that contribute to the design specifications.
8. Safety: Identify standard and special safety precautions or requirements to be included in design considerations, manufacture, marketing, or usage.
9. Governmental regulations and standards applicable: Identify and list.
10. Environment: Identify and list the environmental specifications required using the items included under "Environment" in Table 1-1 as guidelines.
11. Number required and delivery schedule.
12. Desired cost or price information
13. Functional requirements:

    Life   Performance requirements with acceptable tolerance limits

    Reliability   Servicing, maintenance, or repair restrictions

    Unacceptable modes of failure   Any other functional requirements
14. Additional relevant information:

    Limitations of manufacturing facilities
    Special procedural requirements
    Any other relevant information
15. Action required: For example, preparation of proposal, preparation of detail drawings, manufacture of prototypes, or manufacture of full production quantity.

SOURCE: Adapted from Leech [1.14].

uct Safety Act). Litigation has also provided additional emphasis on including safety considerations in design. Even so, the question of how safe a product has to be is very complex and ultimately can be answered only in the courts.

Including safety considerations in the design of a product requires knowledge of the types of hazards that can occur and the application of good design principles to the product involved. One of the appropriate considerations for including safety in design is to recognize that the product will ultimately fail. If this is done, then the product can be designed in such a way that the location and mode of failure are planned and the failure and consequences can be predicted, accommodated, and controlled.

Hazards can be classified as human-caused or non-human-caused. The listings in Tables 1.3 and 1.4 are not meant to be complete or all-inclusive, but they do provide a guide for designers to hazards that they should know, appreciate, and consider in any project. To reduce the effect of these hazards in designing a product, the designer should consider the possible modes of usage; the users, operators, or bystanders; the environment of use; and the functions or requirements of expected use.

TABLE 1.3 Hazards of Human Origin

Ignorance	Smoking
Overqualification	Physical limitations
Boredom, loafing, daydreaming	Sickness
Negligence, carelessness, indifference	Exhaustion
Supervisory direction	Emotional distress
Overproduction	Disorientation
Poor judgment	Personal conflicts
Horseplay	Vandalism
Improper or insufficient training	Physical skills
Alcohol, drugs	Shortcuts

TABLE 1.4 Hazards of Nonhuman Origin

Weight	Visibility	Cold
Flammability	Pinch and crush points	Pressure and suction
Speed (high or low)	Noise	Emissions (particulates/gaseous)
Temperature	Light, strobe effect, intensity	Explosions, implosions
Toxicity (poison)	Electric shock	Vibrations
Sharp edges	Radiation	Stored energy
Rotating parts	Chemical burn	High-frequency radiowaves
Reciprocating parts	Sudden actions	Slick surfaces
Shrapnel (flying objects)	Height	Surface finish
Stability, mounting	Heat	Flames or sparks

The word *expected,* instead of *intended,* is used intentionally because society, through the courts, expects the designer and manufacturer to know and provide for *expected* usage. This will be discussed in more detail in Sec. 1.5.

Table 1.5 lists some modes of usage to include in design deliberations. Considerations for each of the modes of usage are presented in Tables 1.6 and 1.7. Naturally, not all products require consideration of all the items listed in Tables 1.3 to 1.7, and some will require even more. Further information on procedure and other aspects of a designer's tasks can be found in the references cited at the end of this chapter.

TABLE 1.5 Modes of Product Usage

Intended operation or use	Commercial and industrial use	Repair
Unintended operation or use	Assembly	Cleaning
Expected operation or use	Setup	Packaging
Misuse	Installation	Storage
Abuse	Testing/certification	Shipping/transportation
Emergency use	Maintenance/service	Starting/stopping
Changing modes of operation	Isolation	Disposal
Salvaging	Recreational use	Inspection
Repair	Servicing	Modification

TABLE 1.6 Considerations during Each Mode of Usage[†]

Life expectancy	Observation of operation	Weight and size
Duration of length of use	Materials for cleaning	Speed of operation
Complexity	Materials handling devices	Pay/compensation plan
Operator position/station	Frequency of repair	Insertion/removal of
Nonoperator position/station	Test fixtures, ancillary	workpiece
Labeling	equipment	Failure of workpiece
Misuse	Controls and human	Temperature of operation
Material used	factors	Noise of operation
Operator education/skill	Operator comfort	Emissions (particulate/
Operator mental/physical	Ratings and loadings	gaseous)
condition	Guarding and shielding	Stability
Environment or surrounding	Warnings (audible, visual)	Social restrictions
condition	Types of failure	Weather
Type of tool required	Consequences of failure	Local specific operating
Reliability	Ventilation	procedure
Waste materials	Cost	Leakage
Operating instructions	Service instructions	Light/lighting
Machine action	Power source/loss	Instructions, maintenance
Accessories/attachments	Appurtenant parts	Effects of usage/wear
Aesthetics	Government regulation	Maintenance/repair/service
		Standards

[†]There is no significance to the order in the table; various products and situations will establish the relative importance in specific cases.

TABLE 1.7 Specific Design Concepts and Philosophies

K.I.S.S.[†]	Foreign material sensing/	Deadman switches
Fail safe	elimination	Shield and guard interlocks
Design hazards out	Prevention of modification	Avoid the use of set screws
Positive lockouts	Isolation of operators from	and friction locking
Warnings	point of machine	devices
Emergency shutoffs	operation	Use self-closing lids/hatches/
Prevention of inadvertent	Controls user-friendly	closures
actuation	Provide proper safety	Consider two-handed
Prevention of unauthorized	equipment	operation for each
actuation	Provide overload/overspeed	operator
Shielding and guarding	alarms	Use load readouts when
Proper materials for	Training programs	possible
operation	High feasible factor of	Control failure mode so
Accessibility for	safety	consequences are
adjustments/service	Redundant systems	predictable
	Proper use of components	

[†]Keep it simple, stupid!

1.2 DECISIONS AND THEIR IDENTIFICATION

1.2.1 General

Decision making is a key part of the design process in which the designer tries to provide a solution to a problem faced by a customer. The customer is interested primarily in performance (including safety), time (how soon the solution will be available and how long it will last), and cost (including price, maintenance cost, and, today, litigation and insurance costs).

The designer, in order to meet the requirements of the customer, generally uses as design criteria function, safety, economy, manufacturability, and marketability. To achieve these criteria, the designer may use as a problem statement the design imperative as presented in Mischke (see Sec. 1.1 or Ref [1.2]) and then make basic product decisions of the types listed in Table 1.8. From this point on, the decisions required to establish the solution to the design problem appear to be without bound. A second level of more detailed decisions then needs to be reached. Examples are shown in Table 1.9.

Neither Table 1.8 nor Table 1.9 is represented as being complete, all-inclusive, or in any order of priority, since priority is established on a job-by-job basis.

1.2.2 Approach to Problem Solving

To make decisions effectively, a rational problem-solving approach is required. The first step in problem solving is to provide a statement defining the problem to be solved. The essential ingredients as stated and discussed in Dieter [1.15] are

- A need statement
- Goals, aims, objectives
- Constraints and allowable tradeoffs
- Definitions of terms or conditions
- Criteria for evaluating the design

TABLE 1.8 Basic Product Decisions to Be Made by the Designer[†]

Anticipated market	Expected maintenance	Controls
Component elements	Types of loadings	Materials
Fabrication methods	Target costs	Expected life
Evolutionary design or original design	Energy source(s)	Permissible stresses
		Permissible distortions

[†]No significance is to be attached to order or extent.

SOURCE: J. P. Vidosic, *Elements of Design Engineering,* The Ronald Press Company, New York, 1969.

TABLE 1.9 Second-Level Decisions to Be Made by the Designer[†]

Strength of each element	Reliability of each element	Maintenance required
Allowable distortion	Style	Noise allowable
Governing regulations	Governing standards	Governing codes
Control requirements	Surface finish	Corrosion anticipated
Friction anticipated	Lubrication required	Wear anticipated
Geometry	Tolerances	

[†]No significance is to be attached to order or extent.

All these ingredients require evaluation of safety, potential litigation, and environmental impact. Establishing each of these ingredients includes decision making from the start of the design process.

1.2.3   The Decision Maker and Decision Making

Decision makers are concerned with the consequences of their decisions for both their employers and society, as well as for their own egos and professional reputations. By themselves, these concerns may cause faulty decision making.

The decision maker may operate in one of the following ways (Janis and Mann [1.15a] as discussed by Dieter [1.15]):

- Decide to continue with current actions and ignore information about risk of losses.
- Uncritically adopt the most strongly recommended course of action.
- Evade conflict by putting off the decision, passing it off to someone else.
- Search frantically for an immediate solution.
- Search painstakingly for relevant information, digest it in an unbiased way, and evaluate it carefully before making a decision.

Unfortunately, only the last way leads to a good, effective decision, and it may be compromised by time constraints.

The basic ingredients for a good, effective decision are listed in Table 1.10, along with substitutions that may have to be made in practice. The use of these items [1.15b] is discussed at length in Dieter [1.15].

An action of some type is implied after a decision is made and may be classified as a *must* action, a *should* action, a *want* action, or an *actual* action.

A *must* action is one that has to be done and differentiates between acceptability and unacceptability. A *should* action is what ought to be done and is the expected standard of performance for meeting objectives. A *should* action is compared with an *actual* action, or what is occurring at the time the decision is being made. A *want* action does not have to be implemented but may be negotiated as reflecting desires rather than requirements (discussed in Dieter [1.15]).

The steps in [1.15b] for making a good decision are summarized by Dieter [1.15] as follows:

TABLE 1.10   Basic Decision-Making Ingredients

Ingredient	Surrogate
Fact	Information
Knowledge	Advice
Experience	Ad hoc experimentation
Analysis	Intuition
Judgment	None

SOURCE: D. Fuller, *Machine Design,* July 22, 1976, pp. 64–68.

1. Establish the objectives of the decision to be made.
2. Classify objectives by importance, identifying *musts, shoulds,* and *wants.*
3. Develop alternative actions.
4. Evaluate alternatives for meeting the objectives.
5. Choose the alternative having the most promising potential for achieving the objectives as the tentative decision.
6. Explore future consequences of tentative decision for adverse effects.
7. Control effects of final decision by taking appropriate action while monitoring both the implementation of the final decision and the consequences of the implementation.

1.2.4   Decision Theory

The following discussion is adapted from and extensively quotes Dieter [1.15], who in turn cites extensive references in the area of decision theory.

Decision theory is based on utility theory, which develops values, and probability theory, which makes use of knowledge and expectations available. A decision-making model contains the basic elements listed in Table 1.11. Decision-making models are usually classified on the basis of the state of the knowledge available, as listed in Table 1.12.

In applying decision theory, a method of determining the *utility* of a solution must be established. The *utility* of a solution is defined as being a characteristic of the pro-

TABLE 1.11   Elements of a Decision-Making Model

---

1. Alternative courses of action
2. States of nature: The environment of operation of the decision model. The designer has very little, if any, control over this element.
3. Outcome: The result of a combination of an action and a state of nature.
4. Objective: The statement of what the decision maker wishes to achieve.
5. Utility: The satisfaction of value associated with each outcome.
6. State of knowledge: Certainty associated with states of nature, usually given in terms of probabilities.

SOURCE: Adapted from Dieter [1.15].

TABLE 1.12   Classification of Decision-Making Models with Respect to State of Knowledge

---

1. Decision under certainty: Each action results in a known outcome that will occur with a probability of 1.
2. Decision under risk: Each state of nature has an assigned probability of occurrence.
3. Decision under uncertainty: Each action can result in two or more outcomes, but the probabilities for the states of nature are unknown.
4. Decision under conflict: States of nature are replaced by courses of action, determined by an opponent who is trying to maximize his or her objectives function; this is also known as *game theory.*

SOURCE: Adapted from Dieter [1.15].

posed solution that relates to a value in use or a goal of the solution that has meaning in the marketplace. Utility can be cost, price, weight, speed of performance, statistical reliability (probability of failure), factor of safety, or other like attributes. Another name for utility is *merit,* which is also discussed in Sec. 1.3.4 and is extensively presented in Ref. [1.2].

The occurrence of specific states of nature, such as those expressed as materials properties, part geometries, loadings, odors, aesthetics, or taste, may be expressed deterministically, probabilistically, or not at all. If the desired state-of-nature variable can be quantified deterministically, then the utility or merit of a given course of action (problem solution) may be determined and compared to the values of utility or merit for other solutions, allowing the decision maker to choose the better solution for each comparison and, ultimately, the best solution.

If the variables are known only probabilistically, either as a probability distribution or as a mean and variance, statistical expectation techniques as described in Haugen [1.16] or propagation of uncertainty techniques as described in Beers [1.17] have to be used to determine the statistical expectation of the value of the utility for a given course of action (solution). Decisions are then made on the basis of comparisons of expected values of utility or merit. Utility is discussed additionally in Dieter [1.15].

Decision making under risk and decision making under uncertainty are two extremes where, respectively, one does or one does not know the probabilities involved to determine the expected value of utility. Realistically, one can usually estimate the probabilities that affect the outcome, but often without much confidence.

The Bayesian theory of decision making uses the best estimate of the values of utility involved and then bases the decision on the outcome with the maximum expected utility. If probabilities are unknown or cannot be estimated, a weighting function may be established using factors developed from experience or opinion to aid in estimating the utility value for various solutions.

Decision matrices may be used to assist in making decisions where the design goals establish several measures of utility to be evaluated simultaneously for proposed solutions. An example might be a situation where low cost, small weight, and high strength are all important. Dieter [1.15] discusses creation of decision matrices, also known as *payoff matrices* or *loss tables,* and provides several examples of their use in decision making. If a utility function can be created for these cases, optimization theory (as discussed in Ref. [1.12]) may be applied through available digital computer techniques to maximize utility (or merit) functions of many variables to aid in determining the best course of action (solution).

Sometimes the utility of a given course of action cannot be quantified. One way of proceeding in this situation is to establish an arbitrary numerical scale ranging from most unacceptable to most desirable. Evaluations may then rate beauty, fragrance, odor, or whatever the utility is defined to be, on the numerical scale. The ratings may then be evaluated to assist in making the appropriate decision based on the subjective utility.

Another useful technique for exhibiting the results of a decision matrix for the case where decisions must be made in succession into the future is the *decision tree.* This technique, which appears to be an adaptation of fault-tree analysis, where utility is taken to be probability of failure, is described in an example in Dieter [1.15] and as fault-tree analysis in Scerbo and Pritchard [1.18], which also references as sources Larson [1.19], Hammer [1.20], and others. More discussion of decisions, their identification, and decision theory can be found in Wilson [1.7], Dixon [1.5], and Starr [1.10].

1.3 ADEQUACY ASSESSMENT

An *adequacy assessment* is any procedure which ensures that a design is functional, safe, reliable, competitive, manufacturable, and marketable. Usually, in the formative stages, matters of marketability, manufacturability, and competitiveness are addressed and built in, and the principal attention is focused on sustaining function, safety, and reliability. This is why quantitative concepts such as factor of safety and reliability are prominent in examining a completed design.

1.3.1 General

The designer's task is to provide a documented set of specifications for the manufacture, assembly, testing, installation, operation, repair, and use of a solution to a problem. This task may be started by considering several solution concepts, selecting one to pursue, and then generating schemes for meeting the requirements. Usually there are many iterative steps throughout such a process. At each step, decisions must be made as to which concept or detailed specification should be pursued further. This section identifies tools and other considerations necessary to assess adequacy and presents methods of rationally providing and combining information so that informed decisions can be made.

1.3.2 Criteria for Adequacy Assessment

Effective adequacy assessment requires a knowledge of all persons and organizations involved in any way with the product and an understanding of what is important to those involved. Table 1.13 lists factors to be considered and the cast of people involved in engineering adequacy assessment. The order of priority in engineering practice depends on the specific case considered.

The roles in adequacy assessment of the courts, governmental bodies, and to some extent the public as well as the criteria of governmental regulations, standards, and public expectations are addressed in some detail in Secs. 1.5 and 1.6.

TABLE 1.13 Considerations and the Cast of Characters Involved with Design Adequacy Assessment

Important considerations	Criteria	Those involved
Personal reputation	Maintainability	The designer
Keeping one's job	Serviceability	Design peers
Function	Marketability	Design supervisors
Cost	Aesthetics	Users and operators
Safety	Factor of safety	Maintenance and service personnel
Size	Manufacturability	The courts
Reliability	Standards	Governmental bodies
Factor of safety	Public expectations	The public
Government regulations		

1.3.3 Suitability-Feasibility-Acceptability Method

The suitability-feasibility-acceptability (SFA) method of evaluation (as presented in Ref. [1.2]) may be used to evaluate several proposed solutions or compare the desirability of various courses of action. The method is based on determining, in turn, the suitability, feasibility, and acceptability of the proposed solution using the following procedure and then evaluating the results:

Step 1. Develop a problem statement that is as complete as possible.

Step 2. Specify a solution as completely as possible.

Step 3. Answer the question: Is this solution suitable? In other words, with no other considerations included, does the proposed solution solve the problem?

Step 4. Answer the question: Is this solution feasible? In other words, can this solution be implemented with the personnel available, the time available, and the knowledge available without any other considerations included?

Step 5. Finally, answer the question: Is the proposed solution acceptable? In other words, are the expected results of the proposed solution worth the probable consequences to all concerned?

The results of the SFA test can only be as good as the effort and information put into the test. Done casually with inadequate information, the results will vary. Done with care and skill, it can be very effective in assessing the adequacy of proposed problem solutions.

An example of the application of the SFA test (adapted from Ref. [1.2]) is presented below:

Step 1 (Problem Statement).   Metal cans as originally designed require a special tool (can opener) to open. This was true in general, but was especially burdensome to people using beverage cans away from a kitchen or immediate source of a can opener. A method was needed to provide metal beverage cans that could be opened without a can opener or other tool.

Step 2 (Solution).   Design a can for beverages that will meet all the requirements of the original cans and, in addition, will have the top manufactured so that a ring is attached to a flap of metal that is part of the top, but is scored so that a person pulling on the ring can pull the flap out of the top of the can, thus opening the can without a tool.

Step 3.   Is this solution suitable—i.e., will it solve the stated problem? The answer is yes. For the described solution, the can may be opened generally by the user's fingers without any special tool.

Step 4.   Is this solution feasible—i.e., can it be done using available personnel, finances, facilities, time, and knowledge? The answer is yes. The state of manufacturing techniques and materials is such that the design could be produced. The additional cost appears to be reasonable. Thus this solution is feasible.

Step 5.   Is the proposed solution acceptable to all concerned? The initial decision was that the solution was acceptable to the designer, the manufacturer, the marketing organizations, and to the consumer, and so it was put into production.

However, as later events revealed, the consequences of having the ring and flap removable from the can were not generally acceptable to the public because of the consequences of the discarded flaps and rings, and so a new design, retaining the flap to the can, evolved.

1.3.4 Figure of Merit or Weighting Function Method

The *figure of merit* (FOM), also known as the *merit function* or weighting function, is applicable in problems where the important parameters can be related through a function that can be evaluated to find the "best" or "highest merit" solution to a problem. This approach differs from the SFA approach in that the SFA approach is based on more subjective factors.

The FOM lends itself well to attaining or approximating the optimal solution sought by the design imperative discussed in Sec. 1.1. Customarily, the merit function is arbitrarily written so that it is maximized in obtaining the best (highest) value of merit.

Comparing the values of the merit variables obtained for the different alternatives examined should consist only of determining which value is the largest. For situations such as the case where minimum weight or minimum cost is desired, customarily the expression for weight or cost is written either as a negative function or as a reciprocal function, thus allowing maximization techniques to be used.

Although any variable can be used as the merit variable (including an arbitrary variable which is the sum of other disparate variables), the most useful equations are written so that the function represents a characteristic of the product used as a criterion by both engineers and the marketplace. Since safety, reliability, cost, and weight are all important characteristics, useful merit variables, for example, could be the weight, cost, design factor, safety factor, reliability, or time. Equations can be either deterministic or probabilistic in nature.

Where such subjective characteristics as taste, beauty, innovation, or smell are the important characteristics, the FOM approach does not work unless some method of quantifying these characteristics is developed that will allow their mathematical representation.

Two examples will be presented to illustrate the technique involved and identify terms used in the figure-of-merit process.

Example 1.   Design and develop a package for a fragile device that will allow the packaged device to drop through a substantial distance onto concrete without the impact causing the device to fail or break. The package must be of small weight, cost, and size.

Several designs were proposed, built, and tested, and some protected the fragile device adequately. A method was then needed to determine the best of the surviving designs.

A merit function was set up which combined the three design requirements as follows:

$$M = -(A_1 w + A_2 c + A_3 d)$$

where M = merit, the sum of the three terms
    w = weight, ounces (oz)
    c = cost, cents
    d = longest dimension, inches (in)

A_1, A_2, and A_3 are factors selected to weight each of the terms consistent with the importance of the associated variable. The minus sign is used to allow the maximum value of M to be attained when the sum of the three design requirement terms is at a minimum.

The first equation relating merit (which may be a factor of safety, cost, weight, or other desired attribute) to the other variables is known as the *merit function*. It is usually expressed in the form $M = M(x_1, x_2, \ldots, x_n)$. Regional (inequality) constraints are described limits of values that each of the variables may attain in the given problem. Function (equality) constraints are relationships that exist between variables that appear in the merit function. Both types of constraints are specified as a part of the construction of the merit function.

A detailed discussion and description of the preceding method and terms can be found in Mischke [1.2]. Other discussions of this technique with somewhat different terminology may be found in Wilson [1.7] and Dixon [1.5]. A short example will be set up to illustrate the preceding terms.

Example 2. A right-circular cylindrical container is to be made from sheet steel by bending and soldering the seams. Management specifies that it wants the least expensive gallon container that can be made from a given material of a specified thickness. Specify the dimensions for the least expensive container.

Solution. If the bending and soldering are specified, then a fabrication cost per unit length of seam can be estimated. In addition, for a given material of a specific thickness, the material cost is directly proportional to the surface area. A merit function is constructed as follows:

$$M = -(\text{cost of material} + \text{cost of fabrication})$$

If h = height (in) and d = diameter (in), then

$$M = -\left[\left(\frac{2\pi d^2}{4} + \pi dh\right)k_1 + (2\pi d + h)k_2\right]$$

where k_1 = material cost (dollars/in^2) and k_2 = fabrication cost (dollars/inch of seam).

The functional constraint for this problem is the relationship between the volume of the container and the dimensions:

$$V = 1 \text{ gal} = 231 \text{ in}^3 = \frac{\pi d^2 h}{4}$$

where V = volume of container (in^3). The regional constraints are $0 < d$ and $0 < h$, which shows that we are interested only in positive values of d and h.

The next step would be to substitute the functional constraint into the merit function, which reduces the merit function to a function of one variable which may be easily maximized. A robust method such as golden section (see Mischke [1.2]) can be used for optimization.

1.3.5 Design Factor and Factor of Safety

The design factor and the factor of safety are basic tools of the engineer. Both play a role in creating figures of merit, parts and materials specifications, and performance criteria for products being designed. Both may be defined generally as the ratio of the strength or capacity to the load expected, or allowable distortion divided by existing distortion of the object or system in question. Both the design factor and the factor of safety are used to account for uncertainties resulting from manufacturing tolerances, variations in materials properties, variations in loadings, and all other unknown effects that exist when products are put into operation.

The distinction between the design factor and the factor of safety is that the first is the goal at the start of the design process and the latter is what actually exists after the design work is completed and the part or object is manufactured and put into use. The changes occur because of discreteness in sizes of available components or because of compromises that have to be made as a result of materials processing and availability, manufacturing processes, space availability, and changes in loadings and costs.

A simple example would be the design of a rigging system using wire rope to lift loads of 10 tons maximum. The designer could preliminarily specify a design factor of 5, which would be the ratio of the wire rope breaking strength to the expected load, or

$$\text{Design factor} = 5 = \frac{\text{desired breaking strength}}{\text{load}}$$

Using this criterion, a wire rope having a breaking strength of 50 tons would be selected for this application.

The engineer would then evaluate the wire rope selected for use by determining the effect of the environment; the diameters of the sheaves over which the wire rope would be running; the expected loadings, including effects of impact and fatigue; the geometry of the wire rope ends and riggings; and any other factors affecting the wire rope strength to arrive at the final strength, knowing all the application factors. The factor of safety would then be

$$\text{Factor of safety} = \frac{\text{actual breaking strength in application}}{\text{load}}$$

Mischke [1.2], Shigley and Mischke [1.21], and other machine design books discuss the design factor and the factor of safety extensively, including many more complex examples than the one presented here.

A major danger in the use of both the design factor and the factor of safety is to believe that if either is greater than 1, the product having such a factor is safe. However, Fig. 2.5 points out that the factor of safety has a statistical distribution, and that even though the mean value exceeds 1, a fraction of the devices can fail.

1.3.6 Probabilistic Techniques

Propagation-of-error techniques, as described in Chap. 3 and Beers [1.17], can be used to determine the uncertainty of the value of the factor of safety to allow the designer to better assess the adequacy of the factor of safety finally determined. The techniques of Chap. 2 and Haugen [1.16] work directly with the reliability goal.

Another method of adequacy assessment uses the load strength and geometry variables combined to form a quantity called the *stimulus parameter* and the material strength for a given design. If the mean values and the standard deviation are known for any two of the variables (i.e., load, geometry, and materials strength), the threshold value of the third variable can be estimated to provide a specified reliability. The actual value present in the design or part can then be compared to the threshold value to see if the part meets the desired reliability criteria and is then adequate for the specifications provided.

1.4 COMMUNICATION OF ENGINEERING INFORMATION

The output of an engineering department consists of specifications for a product or a process. Much of the output is in the form of drawings that convey instructions for the manufacturing of components, the assembly of components into machines, machine installations, and maintenance. Additional information is provided by parts lists and written specifications for assembly and testing of the product.

1.4.1 Drawing Identification

Drawings and machine components are normally identified by number and name, for example, Part no. 123456, Link. Each organization has its own system of numbering drawings. One system assigns numbers in sequence as drawings are prepared. In this system, the digits in the number have no significance; for example, no. 123456 would be followed by numbers 123457, 123458, etc., without regard to the nature of the drawing.

A different system of numbering detail drawings consists of digits that define the shape and nominal dimensions. This eases the task of locating an existing part drawing that may serve the purpose and thus reduces the likelihood of multiple drawings of nearly identical parts.

The generally preferred method of naming parts assigns a name that describes the nature of the part, such as piston, shaft, fender, or wheel assembly. Some organizations add descriptive words following the noun that describes the nature of its part; for example:

Bearing, roller, or bearing, ball

Piston, brake, or piston, engine

Shaft, axle, or shaft, governor

Fender, LH, or fender, RH

Wheel assembly, idler, or wheel assembly, drive

A long name that describes the first usage of a part or that ties the part to a particular model can be inappropriate if other uses are found for that part. A specific ball or roller bearing, for example, might be used for different applications and models.

1.4.2 Standard Components

Components that can be obtained according to commonly accepted standards for dimensions and strength or load capacity are known as *standard parts*. Such components can be used in many different applications, and many organizations assign part

numbers from a separate series of numbers to the components. This tends to eliminate multiple part numbers for the same component and reduces the parts inventory. Standard components include such things as antifriction bearings, bolts, nuts,
machine screws, cotter pins, rivets, and Woodruff keys.

1.4.3 Mechanical Drawings

Pictorial methods, such as perspective, isometric, and oblique projections, can be
useful for visualizing shapes of objects. These methods, however, are very rarely used
for working drawings in mechanical engineering. Orthographic projection, in which
a view is formed on a plane by projecting perpendicularly from the object to the
plane, is used almost exclusively.

In the United States, mechanical drawings are made in what is known as the
third-angle projection. An example is provided in Fig. 1.4, in which the triangular
shape can be considered to be the front view or front elevation. The top view, or
plan, appears above the front view and the side view; the side elevation, or end view,
appears alongside the front view. In this example, the view of the right-hand side is
shown; the left-hand side would be shown to the left of the front view if it were
needed.

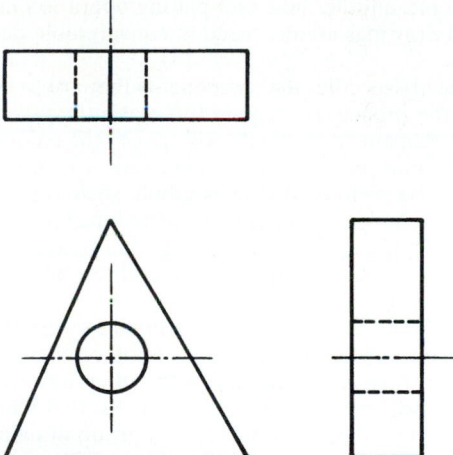

FIGURE 1.4   Arrangement of views of an object in
third-angle orthographic projection.

The first-angle projection is used in many other countries. In that arrangement,
the top view appears below the front view, and the view of the left side appears to
the right of the front view. Some organizations follow the practice of redoing drawings that are to be sent to different countries in order to eliminate the confusion that
results from an unfamiliar drawing arrangement.

Drawings, with the exception of schematics, are made to a convenient scale. The
choice of scale depends on the size and complexity of the object and fitting it on a

standard size of drawing paper. The recommended inch sizes of drawings are 8.5 × 11, 11 × 17, 17 × 22, 22 × 34, and 34 × 44. Then, sizes are multiples of the size of the commercial letterhead in general use, and folded prints will fit in letter-sized envelopes and files.

Drawings should be made to one of the standard scales in common usage. These are full, one-half, one-quarter, and one-eighth size. If a still smaller scale must be used, the mechanical engineer's or architect's rule is appropriate. These rules provide additional scales ranging from 1 in equals 1 ft to $\frac{3}{32}$ in equals 1 ft. The civil engineer's scale with decimal divisions of 20, 30, 40, 50, and 60 parts to the inch is not appropriate for mechanical drawings.

Very small parts or enlarged details of drawings are sometimes drawn larger than full size. Scales such as 2, 4, 5, 10, or 20 times normal size may be appropriate, depending on the particular situation.

Several different types of drawings are made, but in numbers produced, the detail drawing (Fig. 1.5) exceeds all other types. A *detail drawing* provides all the instructions for producing a component with a unique set of specifications. The drawing specifies the material, finished dimensions, shape, surface finish, and special processing (such as heat treatment or plating) required. Usually, each component that has a unique set of specifications is given a separate drawing. There are numbering systems, however, in which similar components are specified on the same drawing and a table specifies the dimensions that change from item to item. Sometimes the material specification consists of another part to which operations are added. For example, another hole or a plating operation might be added to an existing part. Detail drawings are discussed in considerable detail in the next portion of this section.

An *assembly drawing* specifies the components that are to be joined in a permanent assembly and the procedures required to make the assembly. An example is given in Fig. 1.6. A weldment, for example, will specify the components that are to be welded, the weld locations, and the size of weld beads. The drawing may also specify operations that are to be performed after assembly, such as machining some areas.

Another type of assembly drawing consists of an interference fit followed by subsequent machining. A bushing, for example, may be pressed into the machine bore of the upper end of an engine connecting rod, and the bushing bore may then be machined to a specified dimension.

A *group drawing* (Fig. 1.7) may resemble a layout in that it shows a number of components, in their proper relationship to one another, that are assembled to form a unit. This unit may then be assembled with other units to make a complete machine. The drawing will normally include a parts list that identifies part numbers, part names, and the required number of pieces. A group drawing might be a section through a unit that must be assembled with other equipment to make a complete machine.

A *machine outline drawing* is provided to other engineering departments or to customers who purchase that machine for installation. An example is given in Fig. 1.8. An outline may show the general shape, the location and size of holes for mounting bolts, the shaft diameter, keyseat dimensions, location of the shaft with respect to the mounting holes, and some major dimensions.

Schematic drawings, such as for electrical controls, hydraulic systems, and piping systems, show the major components in symbolic form. An example is given in Fig. 1.9. They also show the manner in which the components are connected together to route the flow of electricity or fluids. Schematic diagrams are sometimes provided for shop use, but more frequently they are used in instruction books or maintenance manuals where the functioning of the system is described.

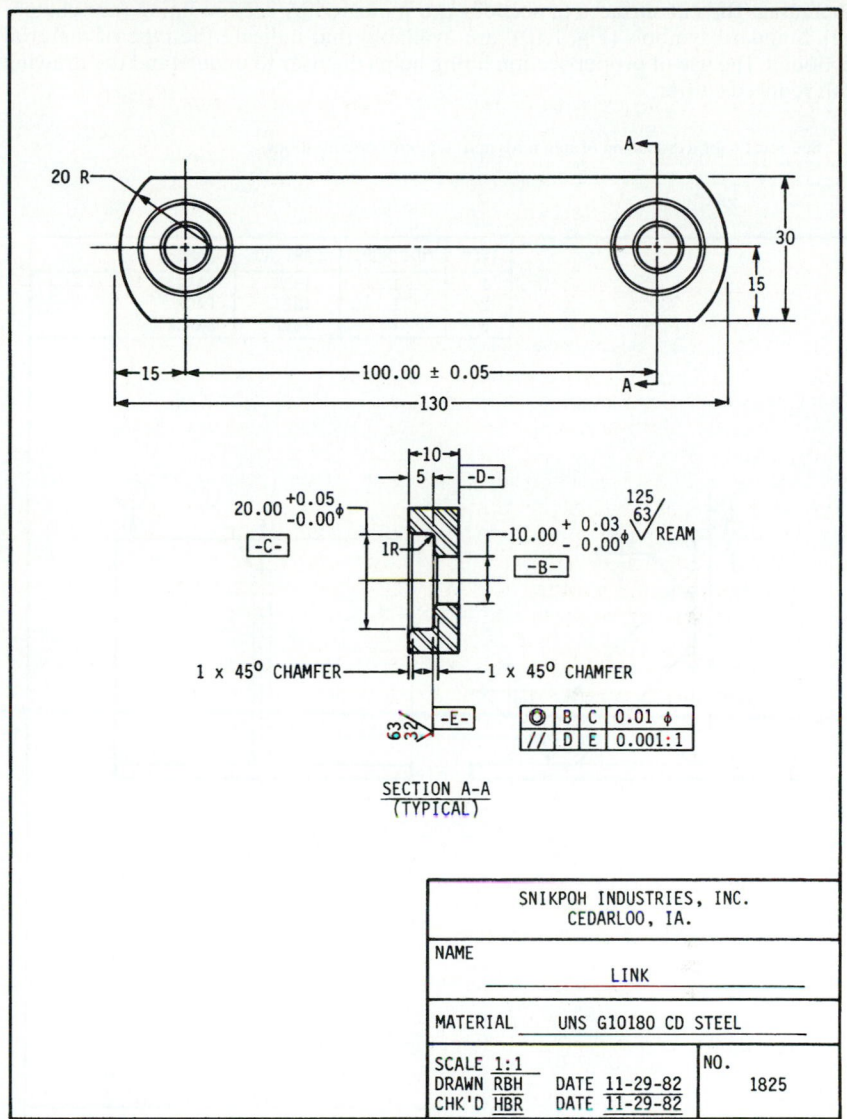

FIGURE 1.5 An example of a detail drawing.

1.4.4 Detail Drawings

A complete description of the shape of a part is provided by the views, sections, and specifications on a detail drawing. A simple part, such as a right-circular cylinder, may require only one view. A complex part, such as an engine cylinder block, may require several views and many sections for an adequate description of the geometry. The link in Fig. 1.5 is a basically simple shape with added complexity due to machining. The cut surfaces of sections are indicated by section lining (crosshatching). Standard symbols (Fig. 1.10)[†] are available that indicate the type of material sectioned. The use of proper section lining helps the user to understand the drawing with reduced clutter.

[†] See Sec. 1.6 for a discussion of standards and standards organizations.

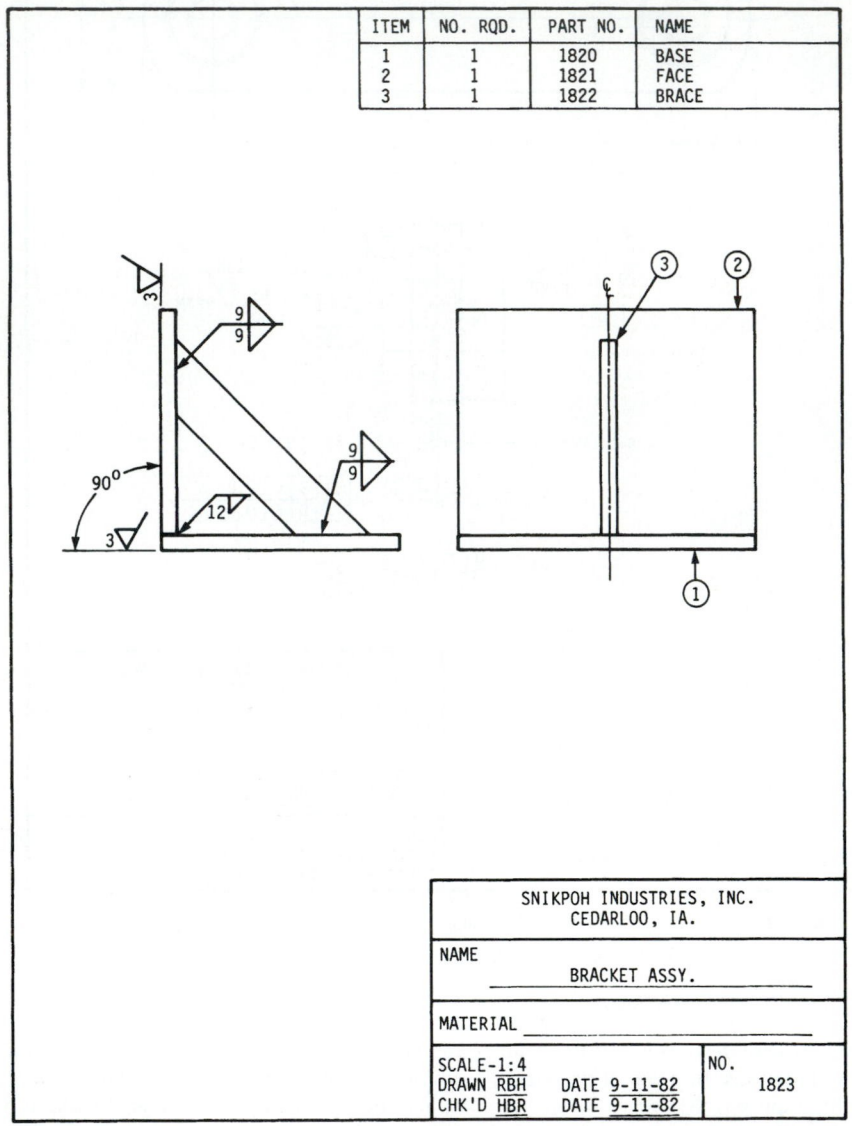

ITEM	NO. RQD.	PART NO.	NAME
1	1	1820	BASE
2	1	1821	FACE
3	1	1822	BRACE

SNIKPOH INDUSTRIES, INC.
CEDARLOO, IA.

NAME
BRACKET ASSY.

MATERIAL

SCALE–1:4		NO.
DRAWN RBH	DATE 9-11-82	1823
CHK'D HBR	DATE 9-11-82	

FIGURE 1.6   An example of an assembly drawing.

Dimensions. There are two reasons for providing dimensions: (1) to specify size and (2) to specify location. Dimensioning for sizes, in many cases, is based on the common geometric solids—cone, cylinder, prism, pyramid, and sphere. The number of dimensions required to specify these shapes varies from 1 for the sphere to 3 for the prism and frustum of a cone. Location dimensions are used to specify the positions of geometric shapes with respect to axes, surfaces, other shapes, or other refer-

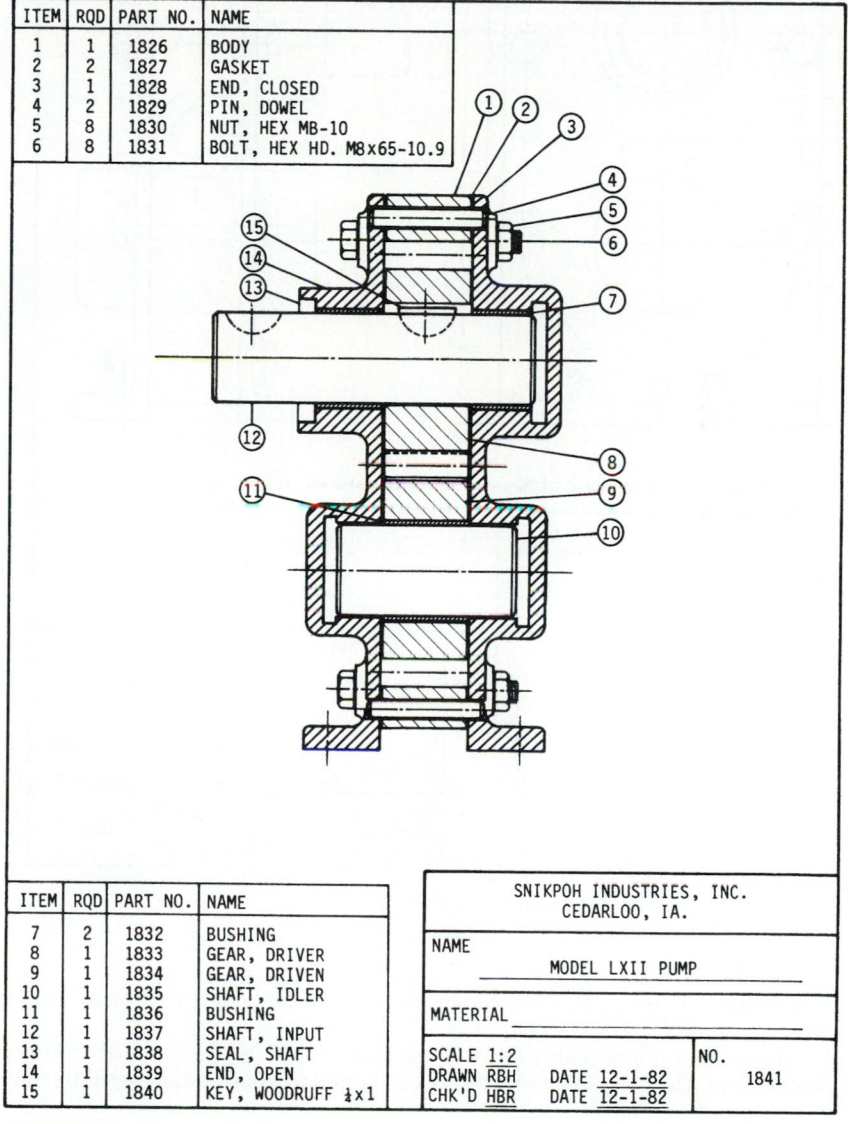

ITEM	RQD	PART NO.	NAME
1	1	1826	BODY
2	2	1827	GASKET
3	1	1828	END, CLOSED
4	2	1829	PIN, DOWEL
5	8	1830	NUT, HEX MB-10
6	8	1831	BOLT, HEX HD. M8x65-10.9

ITEM	RQD	PART NO.	NAME
7	2	1832	BUSHING
8	1	1833	GEAR, DRIVER
9	1	1834	GEAR, DRIVEN
10	1	1835	SHAFT, IDLER
11	1	1836	BUSHING
12	1	1837	SHAFT, INPUT
13	1	1838	SEAL, SHAFT
14	1	1839	END, OPEN
15	1	1840	KEY, WOODRUFF ⅛x1

SNIKPOH INDUSTRIES, INC.
CEDARLOO, IA.

NAME MODEL LXII PUMP

MATERIAL

SCALE 1:2
DRAWN RBH DATE 12-1-82
CHK'D HBR DATE 12-1-82
NO. 1841

FIGURE 1.7 An example of a group drawing.

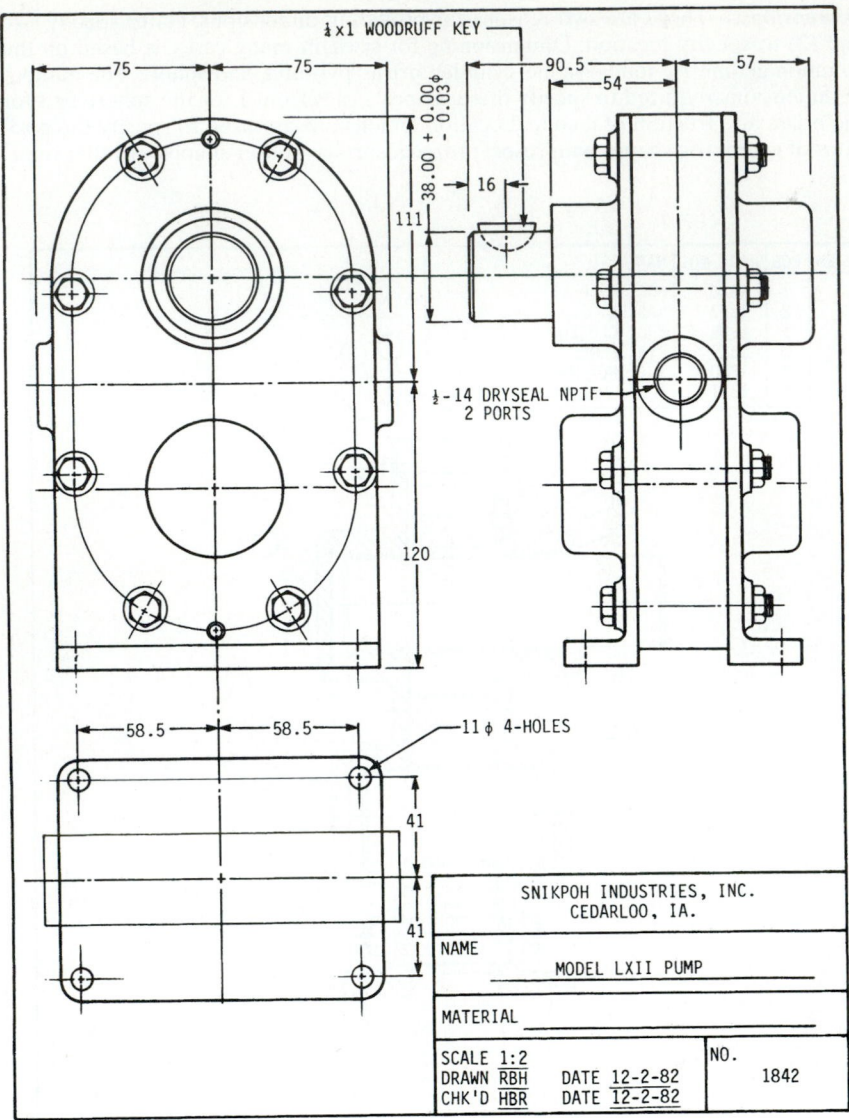

FIGURE 1.8   An example of an installation drawing.

ences. A sphere, for example, is located by its center. A cylinder is located by its axis and bases.

For many years, dimensions were stated in terms of inches and common fractions as small as ¹⁄₆₄ in. The common fractions are cumbersome when adding or subtracting dimensions, and decimal fractions are now used extensively. The decimal fractions are usually rounded to two digits following the decimal point unless a close toler-

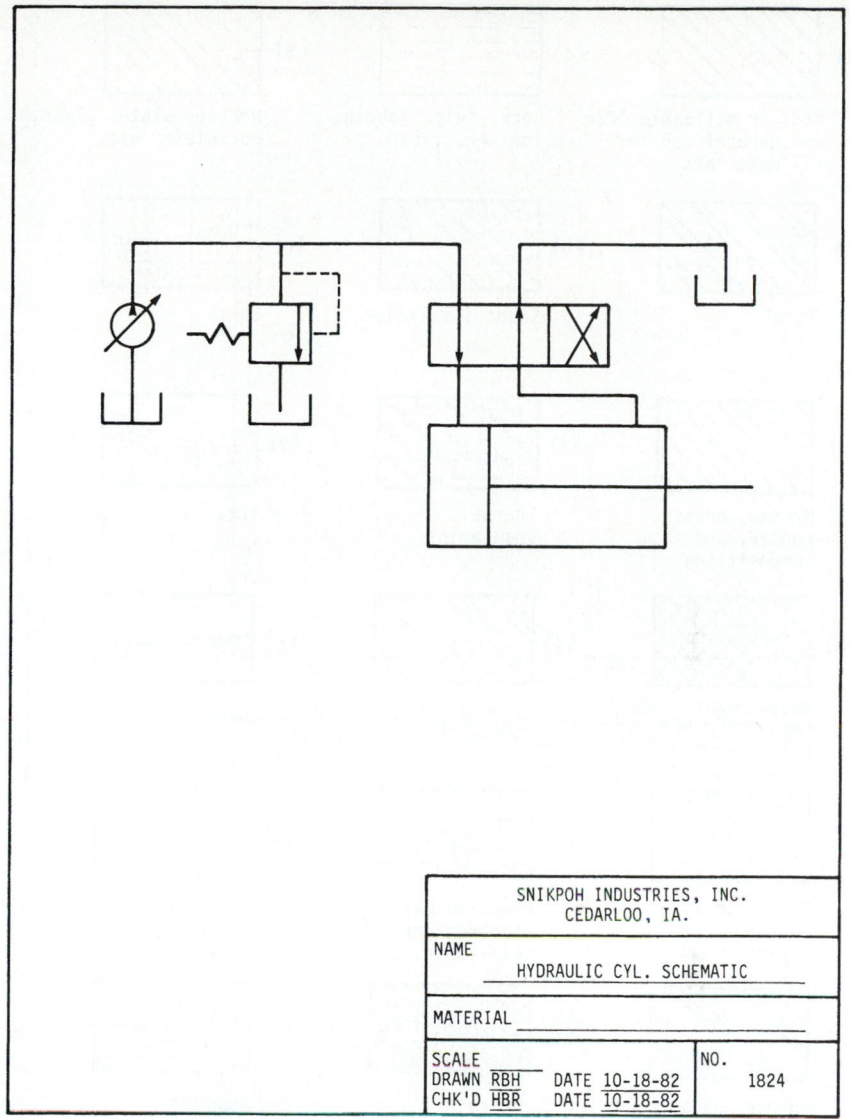

SNIKPOH INDUSTRIES, INC.
CEDARLOO, IA.

NAME
HYDRAULIC CYL. SCHEMATIC

MATERIAL

SCALE		NO.
DRAWN RBH DATE 10-18-82		1824
CHK'D HBR DATE 10-18-82		

FIGURE 1.9   A hydraulic schematic diagram.

ance is to be stated. Thus ⅜ in, which is precisely equal to 0.375 in, is normally specified by dimension as 0.38 in.

The advent of the International System of Units (SI) has led to detail drawings on
which dimensions are specified in metric units, usually millimeters (mm). Thus ½ mm
(very nearly equal to 0.020 in) is the smallest dimension ordinarily specified without
stating a tolerance. Because machine tools and measuring devices are still graduated

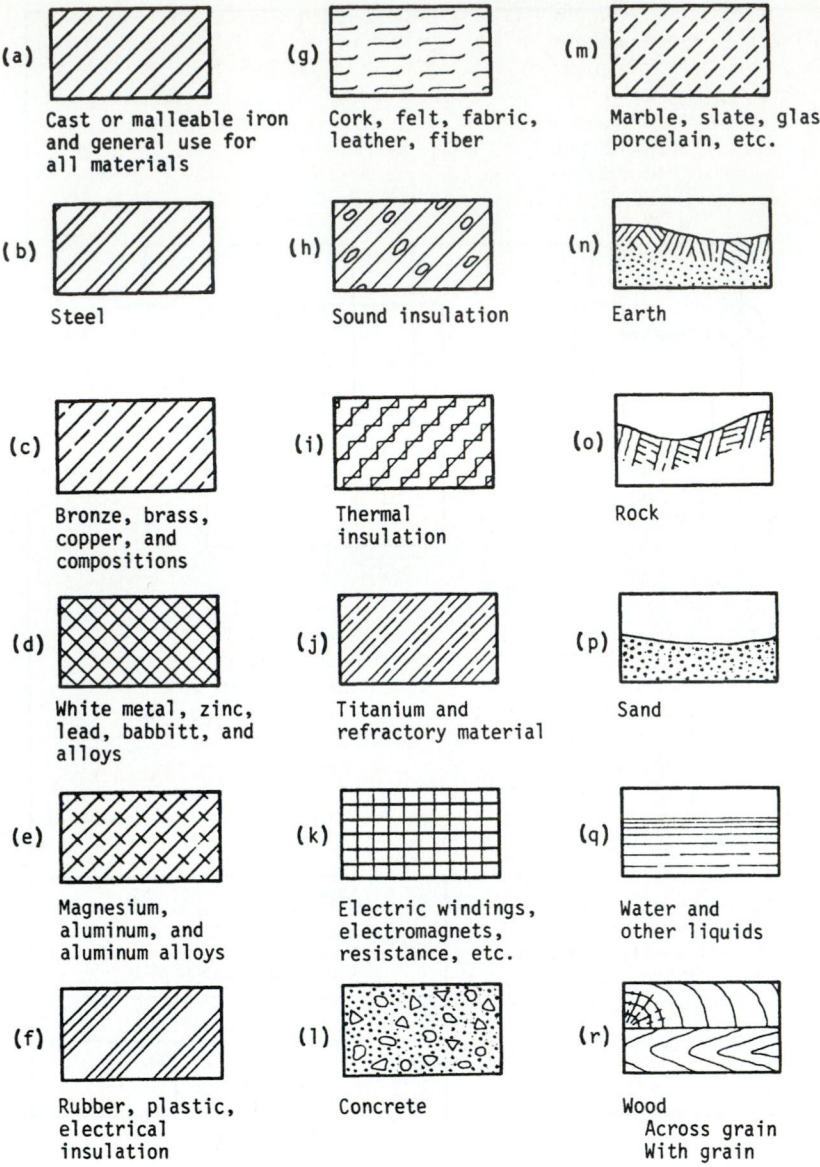

(a) Cast or malleable iron and general use for all materials

(b) Steel

(c) Bronze, brass, copper, and compositions

(d) White metal, zinc, lead, babbitt, and alloys

(e) Magnesium, aluminum, and aluminum alloys

(f) Rubber, plastic, electrical insulation

(g) Cork, felt, fabric, leather, fiber

(h) Sound insulation

(i) Thermal insulation

(j) Titanium and refractory material

(k) Electric windings, electromagnets, resistance, etc.

(l) Concrete

(m) Marble, slate, glass, porcelain, etc.

(n) Earth

(o) Rock

(p) Sand

(q) Water and other liquids

(r) Wood
   Across grain
   With grain

FIGURE 1.10   Symbols for section lining. (*ANSI standard Y14.2M-1979.*)

in inches, some organizations follow the practice of dual dimensioning. In this system, the dimensions in one system of units are followed by the dimensions in the other in parentheses. Thus a ½-in dimension might be stated as 0.50 (12.7), meaning 0.50 in or 12.7 mm.

It is poor practice to specify a shape or location more than once on a drawing. Not only can the dimensions conflict as originally stated, but the drawing may undergo subsequent changes. In making changes, the duplicate dimensions can be overlooked, and the user has the problem of determining the correct dimension.

Every dimension has either a stated or an implied tolerance associated with it. To avoid costly scrap, follow this rule: In a given direction, a surface should be located by one and only one dimension. To avoid a buildup of tolerances, it is better to locate points from a common datum than to locate each point in turn from the previous point. Standard procedures for specifying dimensions and tolerances are provided in ANSI standard Y14.5-1973.

Tolerances. Most organizations have general tolerances that apply to dimensions where an explicit tolerance is not specified on the drawing. In machined dimensions, a general tolerance might be ±0.02 in or 0.5 mm. Thus a dimension specified as 12 mm may range between 11.5 and 12.5 mm. Other general tolerances may apply to angles, drilled holes, punched holes, linear dimensions on formed metal, castings, forgings, and weld beads and fillets.

Control of dimensions is necessary for interchangeability of close-fitting parts. Consequently, tolerances are specified on critical dimensions that affect small clearances and interference fits. One method of specifying tolerances on a drawing is to state the nominal dimension followed by a permissible variation. Thus a dimension might be specified employing bilateral tolerance as 50.800 ± 0.003 mm. The limit-dimension method is to specify the maximum and minimum dimensions; for example, 50.803/50.797 mm. In this procedure, the first dimension corresponds to minimum removal of material. For a shaft, the display might be 50.803/50.797 mm and for a hole, 50.797/50.803 mm. This method of specifying dimensions and tolerances eliminates the need for each user of the drawing to perform additions and subtractions to obtain the limiting dimensions. Unilateral tolerancing has one tolerance zero, for example, $50.979 \, {}^{+0.006}_{-0.000}$ mm.

Some organizations specify center-to-center distance on a gear set unilaterally with the positive tolerance nonzero. This is done because an increase in center-to-center distance increases backlash, whereas a decrease reduces backlash. The zero backlash, or tight-meshed, condition cannot be tolerated in the operation of gears unless special precautions are taken.

Standard symbols are available (Fig. 1.11) for use in specifying tolerances on geometric forms, locations, and runout on detail drawings. Information is provided in ANSI standard Y14.5M-1982 on the proper use of these symbols.

Surface Texture. The surface characteristics depend on processing methods used to produce the surface. Surface irregularities can vary over a wide range. Sand casting and hot working of metals, for example, tend to produce highly irregular surfaces. However, the metal-removal processes of grinding, polishing, honing, and lapping can produce surfaces which are very smooth in comparison. The deviations from the nominal surface can be defined in terms of roughness, waviness, lay, and flaws. The finer irregularities of surface which result from the inherent action of the production process are called *roughness*. Roughness may be superimposed on more widely spaced variations from the nominal surface, known as *waviness*. The direction of the pattern of surface irregularities is usually established by the method of material removal and is known as *lay*. *Flaws* are unintentional variations in surface texture, such as cracks, scratches, inclusions, and blow holes. These are usually not involved in the measurement of surface texture.

Surface roughness values that can be obtained by common production methods are provided in SAE standard J449a, "Surface Texture Control." The roughness that can be tolerated depends on the function served by the surface. The roughness of a clearance hole is usually not critical, whereas a surface that moves against another, such as a piston or journal, usually needs to be smooth.

A relationship exists between permissible surface-texture variations and dimensional tolerances. Precise control of dimensions requires precise control of surface texture. Consequently, when a high degree of precision is required in a dimension, it is necessary that the variation in surface roughness and waviness also be small.

Surface texture is specified on drawings through a set of symbols (Fig. 1.12) established by ANSI standard Y14.36-1978. The basic symbol is derived from a 60° letter V which was formerly used to indicate a machined surface. Use of the symbols on a drawing is demonstrated in Fig. 1.13. It is common practice to specify a range for the surface roughness rather than a single value. In such a case, the maximum roughness is placed above the minimum value. The waviness height and width can be

SYMBOL FOR:	ANSI Y14.5	ISO
STRAIGHTNESS	—	—
FLATNESS	▱	▱
CIRCULARITY	○	○
CYLINDRICITY	⌭	⌭
PROFILE OF A LINE	⌒	⌒
PROFILE OF A SURFACE	⌓	⌓
ALL-AROUND PROFILE	⟲	NONE
ANGULARITY	∠	∠
PERPENDICULARITY	⊥	⊥
PARALLELISM	//	//
POSITION	⊕	⊕
CONCENTRICITY/COAXIALITY	◎	◎
SYMMETRY	NONE	≡
CIRCULAR RUNOUT	* ↗	↗
TOTAL RUNOUT	* ↗↗	↗↗
AT MAXIMUM MATERIAL CONDITION	Ⓜ	Ⓜ
AT LEAST MATERIAL CONDITION	Ⓛ	NONE
REGARDLESS OF FEATURE SIZE	Ⓢ	NONE
PROJECTED TOLERANCE ZONE	Ⓟ	Ⓟ
DIAMETER	∅	∅
BASIC DIMENSION	50	50
REFERENCE DIMENSION	(50)	(50)
DATUM FEATURE	-A-	*⊥ OR *⊥ Ⓐ
DATUM TARGET	∅6/A1	∅6/A1
TARGET POINT	✕	✕

* MAY BE FILLED IN

FIGURE 1.11 Symbols for geometric characteristics and tolerances on detail drawings. (*ANSI standard Y14.5M-1982.*)

specified above the horizontal line, the distance over which the roughness is measured below the horizontal line, and the direction of lay above the surface.

The use of symbols for material-removal allowance on a weldment is illustrated in Fig. 1.6, and the specifications for a range of surface finishes are given in Fig. 1.5.

Machining Information. Some parts, such as noncircular cams, gears, and involute splines, may require a table of information that is needed for machining and checking the parts. The drawing of a standard spur gear, for example, requires a list of the number of teeth, diametral pitch or module, pressure angle, pitch diameter, tooth form, circular tooth thickness, and dimensions for checking the teeth. These data are required for obtaining the proper tools, setting up for the machining, and checking the finished parts.

Joining Information. Permanent assembly of components requires instructions for joining and specification of the material for making the connection. These processes include bonding, brazing, riveting, soldering, and welding. The use of symbols to specify welds is illustrated in Fig. 1.6. Chapter 14 covers bonding, brazing, and welding, and riveting is discussed in Chap. 23.

The amount of interference in press fits and shrink fits is normally specified through the dimensions and tolerances on the mating parts. Heating or cooling of parts for ease of assembly may be specified on an assembly drawing or in assembly specifications.

Symbol	Meaning
(a) 	Basic Surface Texture Symbol. Surface may be produced by any method except when the bar or circle (Figure b or d) is specified.
(b) 	Material Removal By Machining Is Required. The horizontal bar indicates that material removal by machining is required to produce the surface and that material must be provided for that purpose.
(c) 3.5 	Material Removal Allowance. The number indicates the amount of stock to be removed by machining in millimeters (or inches). Tolerances may be added to the basic value shown or in a general note.
(d) 	Material Removal Prohibited. The circle in the vee indicates that the surface must be produced by processes such as casting, forging, hot finishing, cold finishing, die casting, powder metallurgy or injection molding without subsequent removal of material.
(e) 	Surface Texture Symbol. To be used when any surface characteristics are specified above the horizontal line or the right of the symbol. Surface may be produced by any method except when the bar or circle (Figure b and d) is specified.
(f)	LETTER HEIGHT = x

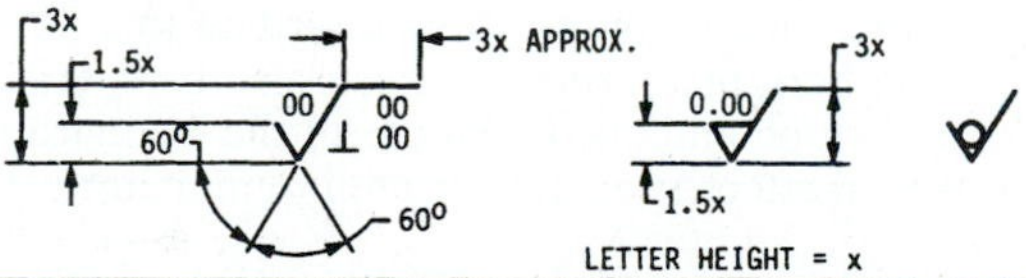

FIGURE 1.12 Surface-texture symbols and construction. (*ANSI standard Y14.36-1978.*)

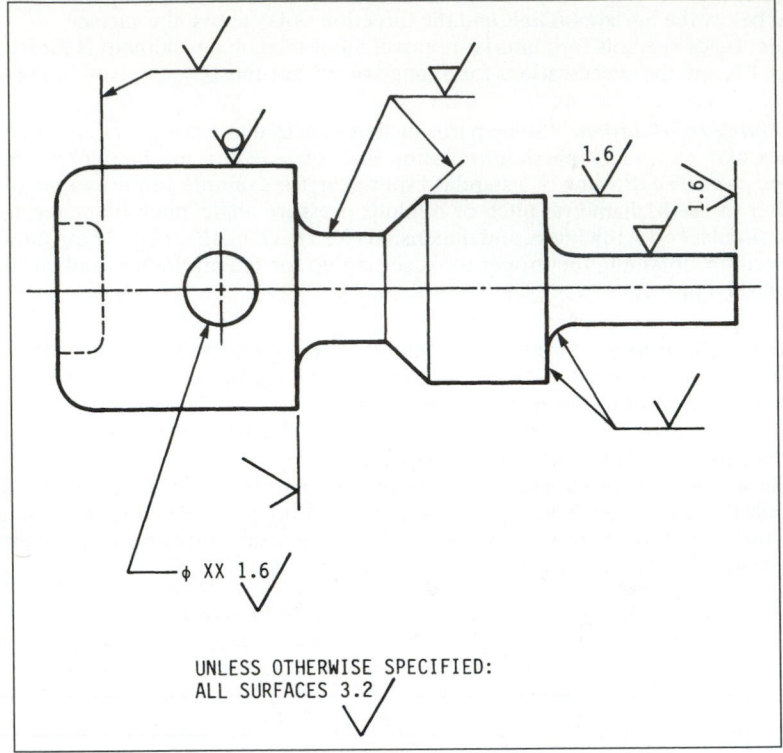

UNLESS OTHERWISE SPECIFIED:
ALL SURFACES 3.2

FIGURE 1.13   Application of surface-texture symbols. (*ANSI standard Y14.36-1978.*)

Material Specifications.   Designation of the material for a part is essential. Such ambiguous specifications as cast iron, gray iron, or mild steel should not be used. Although there may be a common understanding of the meaning of such terms within the organization, misunderstandings can arise if the drawings are sent outside the firm. The use of the term *cast iron,* for example, might be interpreted as gray iron, white iron, malleable iron, or nodular iron.

Each type of cast iron includes several grades, and so castings should be specified by both type and grade of iron. Gray iron castings can be specified according to ASTM standard A48 or SAE standard J431 AUG79, and there are similar standards for malleable iron and nodular iron. When the type and grade of cast iron have been specified, the approximate strength of the metal is known.

The composition of wrought steel bars can be specified through use of the SAE/ANSI numbering system or the newer UNS standard. Steel plate, sheet, and structural shapes are more commonly specified according to ASTM specifications. The surface condition on bars, plate, and sheet can also be specified, such as hot-rolled, cold-finished, or pickled and oiled. The use of the standard material specification and surface finish, in effect, specifies the minimum material strength and the surface condition.

Some of the larger manufacturers have their own systems of material specifications which may be very similar to the standard systems. Materials are then ordered according to the company's own specification. Such a system prevents surprises due to changes in the standard and also provides a convenient method for specifying special compositions when needed.

Heat Treatment. Processes such as annealing or normalizing may be required prior to machining and are specified on the drawings. Other treatments such as carburizing, induction hardening, or through hardening can be performed after some or all of the machining has been done and must be specified. The results desired (for example, the case depth and surface hardness after carburizing) are a better specification than processing temperatures, times, and quenching media. Especially in the case of induction hardening, it may be necessary to specify both a surface hardness and a hardness at some particular depth below the surface in order to prevent subsurface failures.

Special Processes. The use of special processes or handling, such as methods of cleaning castings, impregnation of castings to prevent leakage of fluids, degreasing of finished parts, or protection of surfaces, is frequently specified on the drawing. If the painting of internal surfaces or dipping of castings to prevent rusting is to be done, the paint color, paint type, and method of application are usually specified. Drawings of parts that are to be plated specify the plating metal and thickness of plating that is to be applied.

Weight limits may also be specified on drawings. Pistons for internal combustion engines, for example, may have provisions for metal removal to obtain the desired weight. The location of material that can be removed and the weight limits are then specified on the drawing. Engine connecting rods may have pads for weight control on each end. The maximum amount of metal that can be removed is then shown, and the weight limits at the center of each bearing journal are also specified.

Drawings of rotating parts or assemblies may have specifications for limits on static or dynamic balance. Instructions as to the location and method of metal removal or addition in order to obtain balance are then shown on the drawing.

Qualifying Tests. Drawings of parts of assemblies in which fluid leakage may be detrimental to performance may have a specification for a pressure test to evaluate leakage. A pressure vessel may have a specification for a proof test or a rotating body may have a specification for a spin test to determine that the object will meet performance requirements.

1.4.5 Release of Drawings and Specifications

A formal method of notifying other departments in the organization that drawings and specifications have been prepared is commonly used. This may be accomplished by a decision that lists parts, assemblies, and other necessary specifications for manufacture and assembly. Some organizations use a drawing release form for the same purpose. Regardless of the name by which it is known, the procedure initiates the processes in other departments to obtain tooling, purchase materials, and provide for manufacturing and assembly facilities.

Many drawings undergo changes for such purposes as to correct design or drafting errors, improve the design, or facilitate manufacturing or assembly. If the revised part is interchangeable with the previous version, the same drawing number is retained. If the part is not interchangeable, a new drawing number is assigned. Usually, the changes and the reasons for the changes are given on the decision or drawing change notice.

1.4.6 Deviations

Inevitably, situations arise in which parts do not conform to drawings. In periods of materials shortages, it may become necessary to make a materials substitution. Moreover, manufacturing errors can occur or manufacturing processes may need to be altered quickly for improvement of the part. Such temporary changes can be processed much more quickly through a deviation letter than through the decision process. A *deviation letter* specifies the part number and name, the products affected, the nature of the departure from specifications, the corrective action to be taken, and the records to be kept of the usage of deviant parts.

1.5 *LEGAL CONSIDERATIONS IN DESIGN*

Legal considerations have always been included in design to some extent, but they came to prominence in 1963 when the concept of strict liability was first enunciated in a court decision [*Greenman* v. *Yuba Power Products, Inc., 377* P. 2d 897 (1963)] and then was formally established in the Restatement of Torts (2d), Sec. 402A (1965).

In 1970, the National Commission on Product Safety issued a report which included statistics showing that the incidence of product-related injuries was very high. The report concluded that although the user, the environment, and the product were all involved, the best place to reduce the potential for injury was in the design of the products involved. This report, along with a heightened awareness of product-related problems, also contributed to the increase in product liability litigation and further delineation of the legal responsibilities of the designer and manufacturer.

The law addressing the responsibilities and duties of designers and manufacturers changes rapidly; thus details will not be presented here. Instead, the emphasis of the laws as they affect designers, manufacturers, and sellers of products will be discussed.

The law, through the various theories under which lawsuits are filed, addresses contractural representations (express warranty); implied representations of performance and operation (implied warranty); conduct of designers, manufacturers, sellers, and users (negligence); and the characteristics of the product exclusive of the conduct of all involved with the product (strict liability). Litigation affecting machines and their designers is most often filed under negligence or strict liability theories, both of which may allege the presence of a defect. Thus a major concern of designers would be to eliminate or reduce the effect of defects present in products.

A *product defect* is a characteristic of a product that makes it substandard. These characteristics, in a legal sense, lead to conditions under which a product is unreasonably dangerous or hazardous when used in certain expected or foreseeable ways.

The standards applied and the determination of whether a product (as a result of the defined characteristic) is unreasonably dangerous or hazardous is done by either a jury or a judge in court rather than by the action of the designer's peers.

The types of defects encountered may be categorized as manufacturing defects, warning defects, and design defects. *Manufacturing defects* occur when a product is not made to the designer's or manufacturer's own standards, i.e., blueprints, layouts, or specifications. Examples are holes drilled the wrong size or in the wrong place, a different material used than was specified, or welds that do not meet the designer's or manufacturer's specifications.

Warning defects occur when proper warnings are not present at hazardous locations, thus creating a defect. The warnings may be absent, insufficient in extent, unreadable, unclear, or inadequate.

Design defects occur when a product is manufactured to the designer's drawings and specifications and functions as intended by the designer and the manufacturer but is alleged to be unreasonably hazardous when used in an expected or foreseeable manner.

Since the concept of a defective design was originated in the courts, the definitions and associated tests were legal in nature rather than rooted in engineering. In an attempt to clarify the concept of a design defect, the California Supreme Court, in the case of *Barker* v. *Lull Engineering Co.,* 573 P. 2d. 443 (1978), established two tests to be applied to a product to determine if a design defect existed. If a product does not perform as safely as an ordinary user or consumer would expect when it is used in a reasonably foreseeable manner or if the benefits of a design are not greater than the risks of danger inherent in the use of the product with all things considered, then the product may be found defective.

The *consumer-expectation test* used is based on the idea that consumers expect products to operate reliably and predictably and that if the products fail, the failure will not cause harm. The risk-benefit or risk-utility analysis assumes that all factors involved in designing the product were included and evaluated in arriving at the final design chosen; thus there are no better ways of designing and manufacturing the product to accomplish its intended purposes. When the product design and manufacturing are completed, the hazards that remain have to be evaluated both on the basis of the probability that harm will occur and on all the consequences of that harm, including its seriousness and costs to all involved. Then this evaluation is balanced against the utility or benefits of the product when it is used in a foreseeable manner.

Close examination of consumer expectations and risk-benefit (or utility) considerations show that in many cases conformity to good design practices and procedures, with a heavy emphasis on safety considerations that were well known and utilized prior to the development of product liability litigation, would significantly reduce the occurrence of design defects and the resulting legal actions.

In many states, the final fault is evaluated by the jury or the judge on a comparative basis. Thus if a judgment is rendered against a manufacturer, the percentage of the fault is also established by the jury or the judge. The injured party then recovers only the same percentage of the judgment as the percentage of fault not assigned to the injured party.

The law varies from state to state on how long the injured party has after the harm is done to file the suit. This period of time is called the *statute of limitations.* If a lawsuit is not filed within the time specified by the statute of limitations, it cannot be filed at all.

Another period of time, called the *statute of repose,* is in effect in some states. This period of time starts when the product is put in service. When a product is older than the statute of repose specifies, only under certain conditions may a lawsuit be filed.

No specific lengths of time are given in this section because of the variance among states and changes occurring in the various laws involved. For such specific information as the time involved or other laws involved, either a lawyer should be consulted or an updated legal publication such as *Products Liability,* by L. R. Frumer and M. I. Friedman (Matthew Bender, N.Y.) or *American Law of Products Liability,* by R. D. Hursh and H. J. Bailey (2d ed., Lawyers Cooperative Publishing Company, Rochester, N.Y. 1976), should be consulted.

This discussion of legal considerations in design is necessarily brief and general because of the volatility of the law and the overall field. More complete discussions in the law, engineering, and all aspects of the area can be found in other publications such as Weinstein et al. [1.22], Thorpe and Middendorf [1.23], Colangelo and Thornton [1.24], Philo [1.25], Goodman [1.26], and Dieter [1.15].

1.6 STANDARDS, CODES, AND GOVERNMENTAL REGULATIONS IN DESIGN

1.6.1 Definitions and Descriptions

Design constraints, in addition to those provided by the engineer's management and sales organizations and the marketplace, now include standards, codes, and governmental regulations, both domestic and foreign.

A *standard* is defined as a criterion, rule, principle, or description considered by an authority, or by general consent or usage and acceptance, as a basis for comparison or judgment or as an approved model. The terms *standards* and *specifications* are sometimes used interchangeably; however, *standards* refer to generalized situations, whereas *specifications* refer to specialized situations. For example, a standard might refer to mechanical power transmission equipment; a specification might refer to a particular gear drive.

A *code* is a systematic collection of existing laws of a country or of rules and regulations relating to a given subject. Federal, state, or local governments may adopt engineering, design, or safety codes as part of their own laws.

Governmental regulations are the regulations developed as a result of legislation to control some area of activity. Examples are the regulations developed by the Occupational Safety and Health Administration (OSHA). These regulations, in addition to setting up various methods of operation of the areas controlled, refer to standards and codes which are then given the status and weight of laws.

Standards may be classified as mandatory or voluntary, although standards established as voluntary may be made mandatory if they become a part of a code or by themselves are referenced in governmental regulations having the effect of law.

1.6.2 Categorization by Source

Standards may be categorized by source of development as follows:

1. Governmental regulations
2. Governmental standards
3. Consensus standards
4. Technical society, trade association, and industry standards
5. Company standards
6. Standards of good engineering practice
7. Standards of consumer expectations

Governmental Regulations.   Governmental regulations function as standards and also create specific standards. Examples are OSHA regulations, CPSC regulations and standards, and the National Highway Traffic Safety Administration Motor Vehicle Safety Standards.

In addition to the regulations and standards developed by these and other governmental agencies, the regulations and standards include, by reference, other standards, such as those of the American National Standards Institute (ANSI), the Society of Automotive Engineers (SAE), and the American Society for Testing and Materials (ASTM), thus giving the referenced standards the same weight as the governmental regulations and standards. Regulations and standards developed or referenced by the government are considered as mandatory standards and have the weight of laws.

Governmental Standards. Another category of governmental standards consists of those which cover items purchased by the U.S. government and its branches. In order for an item to be considered for purchase by the U.S. government, the item must meet Air Force–Navy Aeronautical (AN or AND) standards, military standards (MS), or governmental specifications (GSA), which are standards covering all items not covered in the AN, AND, and MS standards.

Consensus Standards. Consensus standards are standards developed by a group representing all who are interested in the standard. The group is composed of representatives of the manufacturers, sellers, users, and the general or affected public. All items in the standard have to be unanimously agreed to (i.e., a consensus must be reached) before the standard is published. Since a consensus has to be reached for the standard to be accepted, many compromises have to be made. Thus consensus standards—and, for that matter, all standards developed with input from several involved parties—represent a minimum level of acceptance and are regarded generally as minimum standards. ANSI and ASTM standards generally fall into the consensus category.

Technical Societies and Trade Associations. Technical societies and trade associations develop standards which are applicable to their constituents. These standards are also known as industrial standards and are not true consensus standards unless the public or users of the products are involved in the standards formulation.

One example occurs in the agricultural equipment industry. The Farm and Industrial Equipment Institute (FIEI) is the trade association to which most of the manufacturers belong. The FIEI proposes and assists in developing standards which are published by the American Society of Agricultural Engineers or the Society of Automotive Engineers, or both. These standards include characteristics of farm crops (useful in harvesting, storing, and transporting), specifications for farm-implement mounting and operation so that farm equipment made by one manufacturer can be used with that made by another manufacturer, and safety and design specifications for items such as grain dryers, augers, and farm-implement controls.

Company Standards. Company standards are those developed by or within an individual company and include such things as specific fasteners, sizes of steel plates or shapes to be purchased, and drafting practices or design practices. Rarely are these standards used outside of a given company. These standards usually refer to or use outside standards wherever applicable.

Standards of Good Engineering Practice. The standards of good engineering practice are not as clearly defined as those previously discussed. Hammer [1.20] states that the mark of a good engineer, and inferentially, good engineering practice, is the design of a product or system to preclude failures, accidents, injuries, and damage. This increases safety and reliability when specific technical requirements do not exist or when conditions are other than ideal. Good engineering practice includes designing at least to minimum standards and generally beyond what the standards require in an effort to minimize failures and their effects, such as machine downtime, lost time, injuries, and damage. Some of the considerations in designing to good engineering practice standards are ease of operation, ease of manufacturability, accessibility for adjustments and service, ease of maintenance, ease of repair, safety, reliability, and overall economic feasibility.

Standards of Consumer and User Expectations.   Consumer and user expectations are another source of standards that are not clearly defined. In many cases, these expectation standards have been established in the marketplace and in the courts through product liability litigation.

When a consumer or user purchases or uses a product, certain expectations of performance, safety, reliability, and predictability of operation are present. For example, a person purchasing an automobile expects it to deliver the performance advertised by the manufacturer and the dealer: start reliably, stop predictably and reliably, and when in motion, speed up, slow down, and steer in a predictably reliable manner. If a brake locks when applied or the steering does not respond, the automobile has not met what would be standard consumer expectations. The failure to meet these expectations provides impetus for product liability actions, depending on the effects of not meeting the expectations. This is particularly true if personal injury, death, or property damage results. A court decision, *Barker* v. *Lull Engineering Co., Inc.,* discussed in Sec. 1.5 and accepted in many jurisdictions, established a legal criterion or standard to use in evaluating designs for meeting consumer and user expectations.

1.6.3   Categorization by Function

Functionally, all the standards discussed previously can be classified as follows:

1. Interchangeability standards
2. Performance standards
3. Construction standards
4. Safety standards
5. Test-procedure or test-method standards

There is much overlap in the functional categories. Although the standard may be listed as a safety standard, the safety may be specified in terms of machine construction or performance. For example, ANSI/ASME standard B15.1-1992 is entitled "Safety Standard for Mechanical Power Transmission Apparatus." It specifies performance requirements for the types of guarding which apply to mechanical power transmission apparatuses and shows some construction information.

Examples of interchangeability standards are SAE standard J403h, May, 1992, "Chemical Composition of SAE Carbon Steels," SAE standard J246, June 1993, "Spherical and Flanged Sleeve (Compression) Tube Fittings," and the ANSI standards in the C78 series which standardize incandescent light bulbs and screw bases. Because of these interchangeability standards, an SAE 1020 steel is the same in any part of the country, a hydraulic machine using compression fittings that were manufactured in one part of the country can be serviced or replaced with hydraulic compression tube fittings locally available in other parts of the country, and in the last case, when a bulb is blown in a lighting fixture, the fixture does not have to be taken to the store to be certain that the correct bulb is purchased.

Examples of test-procedure or test-method standards are SAE standard J406, "Methods of Determining Hardenability of Steels," ASTM standard E84-91a, "Standard Test Method for Surface Burning Characteristics of Building Materials," and ASTM standard E108-93 (reapproved 1970), "Standard Test Methods for Fire Tests of Roof Coverings." Actually, the testing standards are written to assist in achieving interchangeable or repeatable test results; thus these two categories also overlap.

1.6.4   Sources of General Information

A further discussion of the history of standards and standards-making organizations can be found in Peters [1.27]. Further information about standards in general can be found in Talbot and Stephens [1.28] and in Refs. [1.29] to [1.32], taken from Klaas [1.33].

1.6.5   Use of Standards, Codes, and Governmental Regulations in Design

In design, the development of a product or a system requires the solution of a great many repetitive problems, such as the specification of a sheet metal thickness, the selection of fasteners, the construction of welded joints, the specification of materials in noncritical areas, and other recurring problems.

Standards provide the organized solution to recurring problems. For example, an engineer does not have to design a new cap screw each time a fastener is required. All that is needed is either a company standard or an SAE standard which details the screws already designed; the engineer can quickly select one and pursue other design problems. In fact, the presence of standards allows the designer more time to create or innovate, since solutions to recurring problems of the type discussed above are provided.

Standards can also provide economy by minimizing the number of items to be carried in inventory and the number of different manufacturing operations for a given product. Henderson [1.34] cites the example of a five-sided box formed from sheet metal which had 320 different holes of nine different diameters, of which 243 were tapped. The remaining nontapped holes were for machine screws with nuts and lock washers. Sixteen different screws and rivets were required, and the labor costs required to make certain the correct fasteners were present were high.

In a design review, it was found that 304 of the 320 holes could be made the same size and that 4 different fasteners could be used rather than the original 16. Specifying a single-diameter hole for 95 percent of the cases increased production while lowering costs significantly.

Standards allow the use of technicians or drafters to do the detail work and free the designer, since company standards will generally provide analyses and sizes and finishes of raw materials either available in stock or commercially available. Other standard manuals provide tap drill sizes, bushings, standard bores and shaft sizes for bearings, and other information in this regard.

Engineers and management may perceive standards as stifling originality or creativity and being an onerous burden. In many cases, what may be meant is that the standards do not allow or recommend design practices that are detrimental in terms of pollution, safety, or some other effect on the user, consumer, or society and will require the manufacturer to spend time and money to make the proposed product meet the standards. This argument usually arises when the engineer and/or management had very little input into creation of the standard and the provisions of the standard require redesign or elimination of the product in question.

Some of these products should not have been marketed in the first place. Some standards have required conditions of performance that were beyond the state of the art of measure when insufficient or arbitrary input was used to establish the standard. However, when standards are published, there is always inertia and resistance to change or a required modification because of a standard. The other extreme of resistance is use of the standard as a design specification with very little effort made to exceed the requirements of the standard.

In general, standards are minimum requirements, particularly when proposed as consensus standards, since much compromise is required to make a standard under these conditions. The competent designer, while not always unquestioningly accepting all the standards affecting the product, uses them as a guide and as a source of information to assist in the design and to identify areas of concern.

In the case of governmental regulations and standards, the use of these and other referenced standards is required by law. The use of other consensus or industry standards as a minimum usually indicates use of the standards of good engineering practice. However, if the standard is inadequate, meeting the standard does not guarantee that the design is satisfactory. In some cases, standards-making organizations have been found liable for an inadequate standard.

The engineer should be aware that designs and applications of standards in the design process may be evaluated not by peers, but by the courts. The final evaluations will be made by nontechnical people: users, consumers, and ultimately society in general.

A standards search should be initiated in the design process either at the stage where all available information is researched or at the stage where problem-solving and solution constraints are determined. Sources for locating standards are listed at the end of this chapter. In many cases, engineering departments will be involved in developing standards that affect their product and will have a file of applicable standards.

Since standards for a specific product, such as bakery equipment, reference general standards (for example, conveyors, power transmission apparatus), the general standards should also be available in the file.

1.7 SOURCES OF STANDARDS, CODES, GOVERNMENTAL REGULATIONS, INDEXES, AND STANDARDIZATION ACTIVITIES

1.7.1 General

The information provided for sources, indexes, and activities is taken in large part from Klass [1.33] and Talbot and Stephens [1.28] and is categorized as domestic mandatory standards, domestic voluntary standards, codes and recommended practices, and foreign standards. A general source guide for regulations, codes, standards, and publications is Miller [1.35].

1.7.2 Domestic Mandatory Standards

The domestic mandatory standards are published by the U.S. government and include AN, AND, and MS series of standards. (For sources see Refs. [1.36] and [1.37].)

Reference [1.38] lists all unclassified specifications and standards adopted by the Department of Defense. This reference includes listings by title and by specification and standard numbers as well as availability, number, and date of the latest edition. A subject classification is also listed [1.39].

Reference [1.40] indexes General Services Administration (GSA) nonmilitary standards for common items used by government agencies. The listings are alphabetical by title; numerical by specification, commercial item, or standard numbers; and numerical by federal supply classification (FSC) numbers.

The executive departments and agencies of the federal government publish general and permanent rules in the *Code of Federal Regulation* (CFR) [1.41], which is published annually, and the *Federal Register* [1.42], which is published daily, providing current general and permanent rules between revisions of the CFR.

The Occupational Safety and Health Administration (OSHA), established in 1970, is responsible for producing mandatory standards for the workplace, which are available from Refs. [1.43] and [1.44] and are also published under Title 19 of the CFR [1.41].

The Consumer Product Safety Commission (CPSC), established in 1972, is responsible for producing mandatory standards for consumer products. These standards are also published in Title 16 of the CFR [1.41].

The Institute of Basic Standards of the National Institute of Standards and Technology (NIST), a part of the Department of Commerce, prepares basic standards, including those for measurement of electricity, temperature, mass, and length. These standards and other associated publications may be obtained from the Superintendent of Documents, Washington, D.C. Information on ordering these documents is in Title 15 of the CFR, parts 200–299 [1.41]. The NIST also has standards on information processing [1.45] and an *Index of State Specifications and Standards* [1.46].

1.7.3   Domestic Voluntary Standards, Codes, and Recommended Practices

Voluntary Standards.   The official coordinating organization in the United States for voluntary standards is the American National Standards Institute (ANSI) [1.47]. Other general standards organizations are the American Society for Testing and Materials (ASTM) and Underwriters Laboratories, Inc. (UL). In addition, professional societies, trade associations, and other organizations formed of people and organizations having like interests develop and promulgate voluntary standards.

The *American Society for Testing and Materials* is an international and nonprofit organization formed in 1898 to develop standards on the characteristics and performance of materials, products, systems, and services while promoting related knowledge. In addition, ASTM has become a managing organization for developing consensus standards. ASTM publishes standards and allied publications and provides a catalog and index which are continually being updated. For the latest catalogs, ASTM should be contacted directly [1.48]. Many of the ASTM standards are designated as ANSI standards also.

Underwriters Laboratories, Inc. was established in 1894 to develop standards and testing capabilities for fire resistance and electric devices. The standards were to include performance specifications and testing. A certification and testing service has evolved along with the development of safety standards for other products as well as those initially included. Many of the UL standards are also designated as ANSI standards. A listing of UL standards and other relevant information can be found in Ref. [1.49], which is available from UL.

Professional societies, trade associations, and other groups promulgate standards in their own areas of interest. Chumas [1.50] and Ref. [1.51] list the groups that fall into these categories.

Aids to finding U.S. voluntary standards are Slattery [1.52], Chumas [1.53], Parker et al. [1.54], and Hilyard et al. [1.55]. Although Slattery [1.52] is relatively old, the data base from which the reference was printed has been kept up to date and a computer printout of the up-to-date list, which provides key word access to standards, can be obtained from the National Bureau of Standards.

Standards or standards' titles and description search systems available are listed in Refs. [1.56] to [1.58]. Philo [1.25], which ostensibly is a publication for lawyers, is of particular interest in that it covers U.S. voluntary standards in chaps. 17 and 18 and international safety standards and information sources in chap. 19.

Codes.   A *code* is defined as a collection of rules or standards applying to one topic. In many cases codes become a part of federal, state, or local laws, thus becoming mandatory in application.

The National Fire Protection Association (NFPA) publishes an annual set of codes [1.59], which includes the National Electric Codes as well as NFPA standards and additional safety and design publications emphasizing fire prevention. Many of these codes and standards are also designated ANSI standards.

Other well-known codes are the *National Electrical Safety Code* [1.60], the *ASME Boiler and Pressure Vessel Code* [1.61], the *Safety Code for Elevators and Escalators* [1.62], and the *ASME Performance Test Codes* [1.63]. The *Structural Welding Code* [1.64], the *Uniform Plumbing Code* [1.65], and the *Uniform Mechanical Code* [1.66] are available and should be referred to by engineers, even though they do not appear to directly affect mechanical designers. In these and similar cases, the requirements of the codes dictate how products to be used in these areas should be designed. Another useful collection of codes was compiled by the International Labour Office and is available as *A Model Code of Safety Regulations for the Guidance of Governments and Industry* [1.67]. This discussion and listing of codes is not to be considered complete, but it does provide a listing of which mechanical designers should be aware for reference in designing products.

References for Good Engineering Practice.   There are many references that provide other standards, standard data, recommended practices, and good reference information that should be accessible to engineering designers. These and similar publications are considered standards of good engineering practice. The listing of references is not to be construed as all-encompassing, and the order listed does not indicate relative importance. It does include well-known and widely accepted and used references and data. Reference [1.20] and Refs. [1.68] to [1.78] are handbooks and compilations of reference data.

Professional Societies, Trade Associations, and Miscellaneous.   In addition to the other references presented, professional societies and trade associations publish standards in specific areas that are accepted and used by machine designers. A representative listing is found in Refs. [1.79] to [1.103].

1.7.4   Foreign Standards

Standardization activity has become worldwide in nature to facilitate international exchange of goods and services and to provide a common international framework for scientific, technologic, and economic activity. Designers of products to be sold outside the United States must include considerations of applicable international and foreign standards to effectively market their products.

The International Organization for Standardization (ISO) covers all fields except electrical and electronic engineering and is located in Geneva, Switzerland. The International Electrotechnical Commission (IEC) covers electrical and electronic engineering and is located at the same address in Geneva as the ISO. The American National Standards Institute (ANSI) is a member body of the ISO and the IEC and, as such, is the sole sales agent for foreign and international standards in the United States. Catalogs of ISO and IEC standards, as well as their standards, may be ordered from ANSI. In addition, 17 countries have standards organizations listed as correspondent members. In this case, the standards organizations are not yet the official national standards organizations for the countries in this category. The latest ISO catalog lists all the members and correspondent members.

The ISO catalog provides names, addresses, and telephone, telegraph, and telex addresses for each of the member body organizations and names and addresses for the correspondent member organizations.

There are regional standardization activities in addition to those in the countries listed in the ISO catalog. Examples are:

1. Central America Research Institute for Industry, Institute de Recherches et de Technologie, Industrielles pour d'Amerique centrale (ICAITI), Guatemala City, Guatemala. Its members are Costa Rica, El Salvador, Guatemala, Honduras, Nicaragua, and Panama.

2. European Union, which publishes *Journal Officiel des Communautés Européennes,* Rue De la Loi 200, B-1049, Bruxelles, Belgium. This journal is published daily and is the equivalent to the *U.S. Federal Register,* publishing laws, regulations, and standards.

Indexes for standards of a given country may be obtained either through ANSI or by contacting the official standards organization of the country. The most up-to-date listing of addresses is found in the ISO catalog of standards referred to previously.

Chumas [1.104] is an index by key word in context and includes addresses of standards organizations of various countries in 1974, in addition to 2700 standards titles of the ISO, IEC, the International Commission on Rules for the approval of Electrical Equipment (CEE), the International Special Committee on Radio Interference (CISPR), and the International Organization of Legal Metrology (OIML).

The *World Standards Mutual Speedy Finder* [1.105] is a six-volume set having tables of equivalent standards for the United States, the United Kingdom, West Germany, France, Japan, and the ISO in the following areas: vol. 1, Chemicals; vol. 2, Electrical and Electronics; vol. 3, Machinery; vol. 4, Materials; vol. 5, Safety, Electrical and Electronics Products; and vol. 6, Steel. The NBS Standards Information Service, library, and bibliography search referred to previously also include standards from many of the foreign countries.

REFERENCES

1.1 Edward V. Krick, *An Introduction to Engineering and Engineering Design,* John Wiley & Sons, New York, 1965.

1.2 C. R. Mischke, *Mathematical Model Building,* 2d rev. ed., Iowa State University Press, Ames, 1980.

1.3 Percy H. Hill, *The Science of Engineering Design,* Holt, Rinehart and Winston, New York, 1970.

1.4 Harold R. Buhl, *Creative Engineering Design,* Iowa State University Press, Ames, 1960.

1.5 John R. Dixon, *Design Engineering: Inventiveness, Analysis, and Decision Making,* McGraw-Hill, New York, 1966.

1.6 Thomas T. Woodson, *Introduction to Engineering Design,* McGraw-Hill, New York, 1966.

1.7   Warren E. Wilson, *Concepts of Engineering System Design,* McGraw-Hill, New York, 1965.

1.8   D. Henry Edel, Jr., *Introduction to Creative Design,* Prentice-Hall, Englewood Cliffs, N.J., 1967.

1.9   John R. M. Alger, and Carl V. Hays, *Creative Synthesis in Design,* Prentice-Hall, Englewood Cliffs, N.J., 1964.

1.10 Martin Kenneth Starr, *Production Design and Decision Theory,* Prentice-Hall, Englewood Cliffs, N.J., 1963.

1.11 Morris Asimov, *Introduction to Design,* Prentice-Hall, Englewood Cliffs, N.J., 1962.

1.12 Lee Harrisberger, *Engineersmanship. A Philosophy of Design,* Brooks/Cole, Division of Wadsworth, Inc., Belmont, Calif., 1966.

1.13 Ernest O. Doebelin, *System Dynamics: Modeling and Response,* Charles E. Merrill, New York, 1972.

1.14 D. J. Leech, *Management of Engineering Design,* John Wiley & Sons, New York, 1972.

1.15 George E. Dieter, *Engineering Design. A Materials and Processing Approach,* McGraw-Hill, New York, 1983.

1.15a T. L. Janis and L. Mann, *American Scientist,* November–December 1976, pp. 657–667.

1.15b C. H. Kepner and B. B. Tregoe, *The Rational Manager,* McGraw-Hill, New York, 1965.

1.16 E. B. Haugen, *Probabilistic Approaches to Design,* John Wiley & Sons, New York, 1968.

1.17 Yardley Beers, *Introduction to the Theory of Error,* 2d ed., Addison-Wesley, Cambridge, Mass., 1957.

1.18 F. A. Scerbo and J. J. Pritchard, *Fault Tree Analysis: A Technique for Product Safety Evaluations,* ASME paper 75-SAF-3, American Society of Mechanical Engineers, 1975.

1.19 W. F. Larson, *Fault Tree Analysis,* technical report 3822, Picatinny Arsenal, Dover, N.J., 1968.

1.20 Willie Hammer, *Handbook of System and Product Safety,* Prentice-Hall, Englewood Cliffs, N.J., 1972.

1.21 Joseph Edward Shigley and Charles R. Mischke, *Mechanical Engineering Design,* 5th ed., McGraw-Hill, New York, 1989.

1.22 Alvin S. Weinstein, Aaron D. Twerski, Henry R. Piehler, and William A. Donaher, *Products Liability and the Reasonably Safe Product,* John Wiley & Sons, New York, 1978.

1.23 James F. Thorpe and William H. Middendorf, *What Every Engineer Should Know About Product Liability,* Dekker, New York, 1979.

1.24 Vito J. Colangelo and Peter A. Thornton, *Engineering Aspects of Product Liability,* American Society for Metals, 1981.

1.25 Harry M. Philo, *Lawyers Desk Reference,* 6th ed. (2 vols.), Lawyers Cooperative Publishing Co., Rochester, 1979 (updated).

1.26 Richard M. Goodman, *Automobile Design Liability,* Lawyers Cooperative Publishing Co., 1970; cumulative supplement, 1977 (updated).

1.27 L. C. Peters, *The Use of Standards in Design,* ASME paper 82-DE-10, American Society of Mechanical Engineers, New York, 1982.

1.28 T. F. Talbot and B. J. Stephens, *Locating and Obtaining Copies of Existing Specifications and Standards,* ASME paper 82-DE-9, American Society of Mechanical Engineers, New York, 1982.

1.29 J. Brown, "Standards," in *Use of Engineering Literature,* Butterworths, Inc., Boston, 1976, chap. 7, pp. 93–114.

1.30 Rowen Gile (ed.), *Speaking of Standards,* Cahners Books, 1972.

1.31 Ellis Mount, "Specifications and Standards," in *Guide to Basic Information Sources in Engineering,* Gale Research Co., Detroit, Mich., 1965, chap. 17, pp. 133–135.

1.32 Erasmus J. Struglia, *Standards and Specifications Information Sources in Engineering*, Gale Research Co., Detroit, Mich., 1965.

1.33 Janet E. Klaas, *A Selective Guide to Standards in the Iowa State University Library*, Government Publications/Reference Department, Iowa State University Library (updated annually).

1.34 Ken L. Henderson, "Unpublished Notes on Standards," 1962; revised 1965. (Mimeographed.)

General Source Guide

1.35 David E. Miller, *Occupational Safety, Health and Fire Index* (a source guide to voluntary and obligatory regulations, codes, standards, and publications), Dekker, New York, 1976.

Sources and References for Domestic Mandatory Standards

1.36 *AN, AND and MS Series Standards*, Naval Publications and Forms Center, 5801 Tabor Avenue, Philadelphia, Pa. 19210.

1.37 *National Standards Association, AN, AND and MS Standards, Inc.,* Washington, D.C., updated, looseleaf.

1.38 U.S. Department of Defense, *Index of Specifications and Standards*, Superintendent of Documents, Washington, D.C., annual, bimonthly supplements.

1.39 U.S. Department of Defense, *Federal Supply Classification Listing of DOD Standards Documents*, Superintendent of Documents, Washington D.C., annual, bimonthly supplements.

1.40 General Services Administration Specifications and Consumer Information Distribution Section, *Index of Federal Specifications, Standards and Commercial Item Descriptions*, Superintendent of Documents, Washington D.C., annual, bimonthly supplements.

1.41 *Code of Federal Regulations*, Office of the Federal Register, Washington, D.C., annual, revised annually; Title 15, parts 200–299, *National Institute of Standards and Technology;* Title 16, parts 1000–1799, *Consumer Product Safety Commissions;* Title 29, *Department of Labor, Occupational Health and Safety Administration,* part 1910, *General Industry,* part 1915, *Ship Repairing,* part 1916, *Ship Building,* part 1917, *Ship Breaking,* part 1918, *Longshoring,* part 1926, *Construction,* part 1928, *Agriculture.*

1.42 *Federal Register*, Office of the Federal Register, Washington, D.C., daily.

1.43 Occupational Safety and Health Administration, *OSHA Safety and Health Standards*, Superintendent of Documents, U.S. Government Printing Office, Washington, D.C. 20402.

1.44 Peter Hopf, *Designers Guide to OSHA*, McGraw-Hill, New York, 1975.

1.45 U.S. National Institute of Standards and Technology, *Federal Information Processing Standards*, Washington, D.C., updated.

1.46 Linda L. Grossnickle (ed.), *An Index of State Specifications and Standards* (NIST special publication 375), National Institute of Standards and Technology, Washington, D.C., 1973 (up-to-date computer printouts of the data base for this publication may be ordered from the same source).

Sources and References for Voluntary Standards

1.47 *ANSI Catalog* and *ANSI Standards*, American National Standards Institute, 1430 Broadway, New York, N.Y. 10018.

1.48 *ASTM Publications Catalog.* American Society for Testing and Materials, 1916 Race Street, Philadelphia, Pa. 19103.

1.49 *Catalog of Standards for Safety,* Underwriters Laboratories, Inc., 207 East Ohio Street, Chicago, Ill. 60611.

1.50 Sophie J. Chumas, ed., *Directory of United States Standardization Activities* (NBS special publication 417), National Bureau of Standards, Washington, D.C., 1975.

1.51 *Encyclopedia of Associations,* Gale Research Co., Inc., Detroit, Mich., updated.

1.52 William J. Slattery, ed., *An Index of U.S. Voluntary Engineering Standards* (NBS special publication 329), with supplement 1, 1972, and supplement 2, 1975, National Bureau of Standards, Washington, D.C., 1977.

1.53 Sophie J. Chumas, ed., *Tabulation of Voluntary Standards and Certification Programs for Consumer Products* (NBS technical note 948), National Bureau of Standards, Washington, D.C., 1977.

1.54 Andrew W. Parker, Jr., Charles H. Gonnerman, and Thomas Sommer, *Voluntary Products Standards: An Index Based on Hazard Category,* National Science Foundation, Washington, D.C., 1978.

1.55 Joseph F. Hilyard, Vern L. Roberts, and James H. McElhaney, *Product Standards Index,* Product Safety News, Safety Electronics, Inc., Durham, N.C., 1976.

Standards or Standards Titles and Description Search Systems that Are Available

1.56 Information Handling Services, *Industry/International Standards Locator Index* (microfilm), Englewood, Colo., continually revised. (This index must be used in conjunction with Information Handling Services, Inc. Product/Subject Master Index.)

1.57 National Standards Association, *Standards and Specific Dialog Information Retrieval Service* (this is a computer data base), Washington, D.C., updated. (Copies of standards on paper or fiche can also be ordered.)

1.58 National Institute of Standards and Technology—Standards Information Service (NIST-SIS), Key Word Search of Computer Data Bank, Washington, D.C.

Sources and References for Codes

1.59 *National Fire Codes,* 16 volumes, annual, National Fire Protection Association, 470 Atlantic Avenue, Boston, Mass. 02210.

1.60 *National Electrical Safety Code,* annual, Institute of Electrical and Electronics Engineers, Inc., 345 East 47th St., New York, N.Y. 10017. (Also available from ANSI.)

1.61 *ASME Boiler and Pressure Vessel Code,* 11 volumes, plus Code Case Book Interpretations, updated, American Society of Mechanical Engineers, United Engineering Center, 345 East 47th Street, New York, N.Y. 10017. (Also available from ANSI.)

1.62 *Safety Code for Elevators and Escalators,* updated, American Society of Mechanical Engineers, same availability as Ref. [1.60].

1.63 *ASME Performance Test Codes,* updated, American Society of Mechanical Engineers, same availability information as Ref. [1.60].

1.64 *Structural Welding Code,* updated, American Welding Society, Miami, Fla.

1.65 *Uniform Plumbing Code,* updated, International Association of Plumbing and Mechanical Officials, 5032 Alhambra Ave., Los Angeles, Calif. 90032.

1.66 *Uniform Mechanical Code,* updated, same as Ref. [1.65].

1.67 *A Model Code of Safety Regulations for Industrial Establishments for the Guidance of Governments and Industry* (originally published by International Labour Office, Geneva, Switzerland, 1949), reprinted by Institute for Product Safety, 1410 Duke University Road, Durham, N.C. 27701.

Standards, Standard References, Standard Data, and Recommended Practices Sources and References

1.68 Theodore Baumeister (ed.), *Marks' Standard Handbook for Mechanical Engineers,* 8th ed., McGraw-Hill, New York, 1979.

1.69 Colin Carmichael (ed.), *Kent's Mechanical Engineers Handbook,* 12th ed., John Wiley & Sons, New York, 1950. (An old but still good basic reference.)

1.70 Erik Oberg, Franklin D. Jones, and Holbrook Horton, *Machinery's Handbook,* 21st ed., Industrial Press, New York, 1979.

1.71 C. B. Richey (ed.), *Agricultural Engineers Handbook,* McGraw-Hill, New York, 1961.

1.72 Harold A. Rothbart (ed.), *Mechanical Design and Systems Handbook,* McGraw-Hill, New York, 1964.

1.73 Wesley E. Woodson, *Human Factors Design Handbook,* McGraw-Hill, New York, 1981.

1.74 Henry Dreyfuss, *The Measure of Man. Human Factors in Design,* Whitney Library of Design, New York, 1967.

1.75 Albert Damon, Howard W. Staudt, and Ross A. McFarland, *The Human Body in Equipment Design,* Harvard University Press, Cambridge, Mass., 1966.

1.76 National Safety Council, *Accident Prevention Manual for Industrial Operations,* 7th ed., Chicago, Ill., 1974.

1.77 National Safety Council, *Industrial Safety, Data Sheet Series,* updated.

1.78 FMC Corporations, *Machinery Product Safety Signs and Labels,* 2d ed., Santa Clara, Calif., 1978.

1.79 Associated General Contractors of America, *Manual of Accident Prevention in Construction,* 6th ed., Washington, D.C., 1971.

References from Professional Societies, Trade Associations, and Miscellaneous

1.80 Society of Automotive Engineers, Warrendale, Pa.

    a. SAE Handbook, annual.

    b. SAE Aerospace Index and Price List of AS Standards, ARP Recommended Practices, AIR Information Reports, updated.

    c. Aerospace Material Specifications, updated.

    d. Unified Numbering System for Metals and Alloys and *Cross Index of Chemically Similar Specifications,* 2d ed., 1977.

1.81 Aerospace Industries Association, Washington, D.C.

    a. Metric NAS Standards, updated.

    b. NAS Standards, updated.

1.82 *Agricultural Engineers Yearbook,* American Society of Agricultural Engineers, St. Joseph, Mich., annual through 1983.

1.83 *Standards 1984,* American Society of Agricultural Engineers, St. Joseph, Mich., updated each year.

1.84 *NEMA Standards,* National Electrical Manufacturers Association, New York, updated.

1.85 Lois M. Ferson (ed.), *Standards and Practices for Instrumentation,* 6th ed., Instrument Society of America, Research Triangle Park, N.C., 1980.

1.86 *Engineering Materials and Process Standards,* General Motors Corporation, Warren, Mich., updated.

1.87 *ACI Manual of Concrete Practice,* American Concrete Institute, Detroit, Mich., 1982 (updated).

1.88 Robert B. Ross, *Metallic Materials Specification Handbook,* 2d ed., Chapman and Hall, London, England, 1972.

1.89 Mechanical Properties Data Center, *Structural Alloys Handbook,* Traverse City, Mich., updated.

1.90 *NACE Standards,* National Association of Corrosion Engineers, Houston, Tex., updated.

1.91 *AISC Manual of Steel Construction,* American Institute of Steel Construction, 8th ed., New York, 1980.

1.92 *Aluminum Standards and Data,* The Aluminum Society, Inc., Washington, D.C., updated.

1.93 *API Standards,* American Petroleum Institute, Dallas, Tex., updated.

1.94 American Society of Heating, Refrigerating and Air Conditioning Engineers, Inc., New York.

   a. ASHRAE Handbook and Product Directory, Systems Applications Equipment, Fundamentals, updated.

   b. ASHRAE Standards, updated.

1.95 *Standards,* Air Conditioning and Refrigeration Institute, Arlington, Va., updated.

1.96 *Fluid Power Standards,* National Fluid Power Association, Inc., Milwaukee, Wisc., updated.

1.97 *Welding Handbook,* 7th ed., American Welding Society, Miami, Fla., 1976.

1.98 *Standards,* American Nuclear Society, LaGrange Park, Ill., updated.

1.99 *Manual,* American Railway Engineering Association, Washington, D.C., updated.

1.100 John H. Callender (ed.), *Time-Saver Standards for Architectural Design Data,* 5th ed., McGraw-Hill, New York, 1974.

1.101 Hardam S. Azod (ed.), *Industrial Wastewater Management Handbook,* McGraw-Hill, New York, 1976.

1.102 *ASSE Standards,* American Society of Sanitary Engineers, Cleveland, Ohio, updated.

1.103 *Standards,* National Sanitation Foundation, Ann Arbor, Mich., updated.

Foreign Standards Indexes

1.104 Sophie J. Chumas, *Index of International Standards* (NBS special publication 390), National Bureau of Standards, Washington, D.C., 1974.

1.105 The International Technical Information Institute, *World Standards Mutual Speedy Finder,* 6 volumes, Tokyo, updated.

CHAPTER 2
STATISTICAL CONSIDERATIONS

Charles R. Mischke, Ph.D., P.E.
Professor Emeritus of Mechanical Engineering
Iowa State University
Ames, Iowa

NOMENCLATURE

A	Area, constant
a	Constant
B	Constant
b	Constant
C	Coefficient of variation
d	Diameter
F_i	ith failure, cumulative distribution function
$F(x)$	Cumulative distribution function corresponding to x
f_i	Class frequency
$f(x)$	Probability density function corresponding to x
h	Simpson's rule interval
i	failure number, index
LN	Lognormal
N	Normal
n	design factor, sample size, population
$\bar{n}$	mean of design factor distribution
P	Probability, probability of failure
R	Reliability, probability of success or survival

r Correlation coefficient

S'_{ax} Axial loading endurance limit

S'_e Rotary bending endurance limit

S_y Tensile yield strength

S'_{se} Torsional endurance limit

S_{ut} Tensile ultimate strength

x Variate, coordinate

x_i ith ordered observation

x_0 Weibull lower bound

y Companion normal distribution variable

z z variable of unit normal, $N(0, 1)$

α Constant

Γ Gamma function

Δx Histogram class interval

θ Weibull characteristic parameter

μ Population mean

$\hat{\mu}$ Unbiased estimator of population mean

σ stress

σ Standard deviation

$\hat{\sigma}$ Unbiased estimator of standard deviation

$\Phi(z)$ Cumulative distribution function of normal distribution, body of Table 2.1

ϕ Function

$\bar{\phi}$ Fatigue ratio mean

ϕ_{ax} Axial fatigue ratio variate

ϕ_b Rotary bending fatigue ratio variate

ϕ_t Torsional fatigue ratio variate

2.1 INTRODUCTION

In considering machinery, uncertainties abound. There are uncertainties as to the

- Composition of material and the effect of variations on properties
- Variation in properties from place to place within a bar of stock
- Effect of processing locally, or nearby, on properties
- Effect of thermomechanical treatment on properties
- Effect of nearby assemblies on stress conditions
- Geometry and how it varies from part to part
- Intensity and distribution in the loading
- Validity of mathematical models used to represent reality
- Intensity of stress concentrations
- Influence of time on strength and geometry
- Effect of corrosion

- Effect of wear
  :
- Length of any list of uncertainties

The algebra of real numbers produces unique single-valued answers in the evaluation of mathematical functions. It is not, by itself, well suited to the representation of behavior in the presence of variation (uncertainty). Engineering's frustrating experience with "minimum values," "minimum guaranteed values," and "safety as the absence of failure" was, in hindsight, to have been expected. Despite these not-quite-right tools, engineers accomplished credible work because any discrepancies between theory and performance were resolved by "asking nature," and nature was taken as the final arbiter. It is paradoxical that one of the great contributions to physical science, namely the search for consistency and reproducibility in nature, grew out of an idea that was only partially valid. Reproducibility in cause, effect, and extent was only approximate, but it was viewed as ideally true. Consequently, searches for invariants were "fruitful."

What is now clear is that consistencies in nature are a stability, not in magnitude, but in the pattern of variation. Evidence gathered by measurement in pursuit of uniqueness of magnitude was really a mix of systematic and random effects. It is the role of statistics to enable us to separate these and, by sensitive use of data, to illuminate the dark places.

2.2 HISTOGRAPHIC EVIDENCE

Each heat of steel is checked for chemical composition to allow its classification as, say, a 1035 steel. Tensile tests are made to measure various properties. When many heats that are classifiable as 1035 are compared by noting the frequency of observed levels of tensile ultimate strength and tensile yield strength, a histogram is obtained as depicted in Fig. 2.1a (Ref. [2.1]). For specimens taken from 1- to 9-in bars from 913 heats, observations of mean ultimate and mean yield strength vary. Simply specifying a 1035 steel is akin to letting someone else select the tensile strength randomly from a hat. When one purchases steel from a given heat, the average tensile properties are available to the buyer. The variability of tensile strength from location to location within any one bar is still present.

The loading on a floorpan of a medium-weight passenger car traveling at 20 mi/h (32 km/h) on a cobblestone road, expressed as vertical acceleration component amplitude in g's, is depicted in Fig. 2.1b. This information can be translated into load-induced stresses at critical location(s) in the floorpan. This kind of real-world variation can be expressed quantitatively so that decisions can be made to create durable products. Statistical methods permit quantitative descriptions of phenomena which exhibit consistent patterns of variability. As another example, the variability in tensile strength in bolts is shown in the histogram of the ultimate tensile strength of 539 bolts in Fig. 2.2.

The designer has decisions to make. No decisions, no product. Poor decisions, no marketable product. Historically, the following methods have been used which include varying amounts of statistical insight (Ref. [2.2]):

1. Replicate a previously successful design (Roman method).
2. Use a "minimum" strength. This is really a percentile strength often placed at the 1 percent failure level, sometimes called the ASTM minimum.
3. Use permissible (allowable) stress levels based on code or practice. For example, stresses permitted by AISC code for weld metal in fillet welds in shear are 40 percent of the tensile yield strength of the welding rod. The AISC code for structural

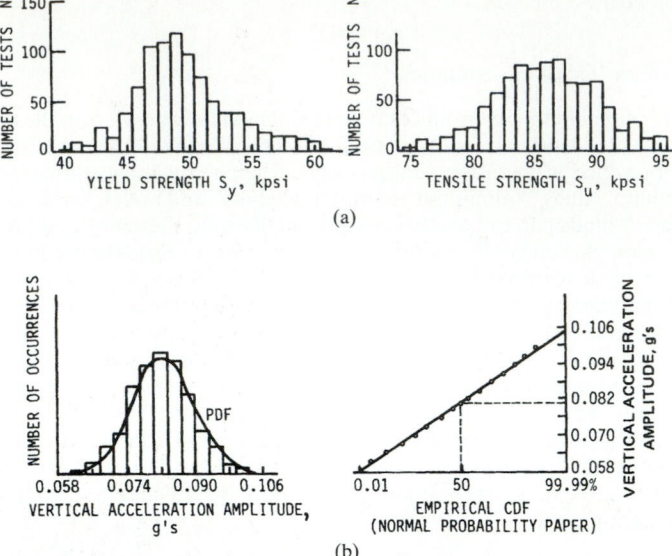

FIGURE 2.1   (*a*) Ultimate tensile strength distribution of hotrolled 1035 steel (1–9 in bars) for 913 heats, 4 mills, 21 classes, $\hat{\mu} = 86.2$ kpsi, $\hat{\sigma} = 3.92$ kpsi, and yield strength distribution for 899 heats, 22 classes, $\hat{\mu} = 49.6$ kpsi, $\hat{\sigma} = 3.81$ kpsi. (*b*) Histogram and empirical cumulative distribution function for loading of floor pan of medium weight passenger car—roadsurface, cobblestones, speed 20 mph (32 km/h).

members has an allowable stress of 90 percent of tensile yield strength in bearing. In bending, a range is offered: $0.45S_y \leq \sigma_{all} \leq 0.60S_y$.

4. Use an allowable stress based on a design factor founded on experience or the corporate design manual and the situation at hand. For example,

$$\sigma_{all} = S_y/n \tag{2.1}$$

where *n* is the design factor.

5. Assess the probability of failure by statistical methods and identify the design factor that will realize the reliability goal.

Instructive references discussing methodologies associated with methods 1 through 4 are available. Method 5 will be summarized briefly here.

In Fig. 2.3, histograms of strength and load-induced stress are shown. The stress is characterized by its mean $\overline{\sigma}$ and its upper excursion $\Delta\sigma$. The strength is characterized by its mean $\overline{S}$ and its lower excursion ΔS. The design is safe (no instances of failure will occur) if the stress margin $m = S - \sigma > 0$, or in other words, if $\overline{S} - \Delta S > \overline{\sigma} + \Delta\sigma$, since no instances of strength S are less than any instance of stress σ. Defining the design factor as $n = \overline{S}/\overline{\sigma}$, it follows that

$$n \geq \frac{1 + \Delta\sigma/\overline{\sigma}}{1 - \Delta S/\overline{S}} \tag{2.2}$$

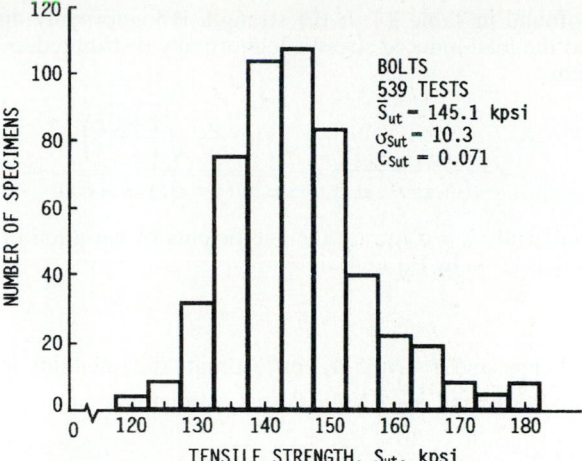

FIGURE 2.2 Histogram of bolt ultimate tensile strength based on 539 tests displaying a mean ultimate tensile strength $\overline{S}_{ut} = 145.1$ kpsi and a standard deviation of $\sigma_{S_{ut}} = 10.3$ kpsi.

As primitive as Eq. (2.2) is, it tells us that we must consider $\overline{S}$, $\overline{\sigma}$, and ΔS, $\Delta \sigma$—i.e., not just the means, but the variation as well. As the number of observations increases, Eq. (2.2) does not serve well as it stands, and so designers fit statistical distributions to histograms and estimate the risk of failure from interference of the distributions. Engineers seek to assess the chance of failure in existing designs, or to permit an acceptable risk of failure in contemplated designs.

If the strength is normally distributed, $S \sim N(\mu_S, \sigma_S)$, and the load-induced stress is normally distributed, $\sigma \sim N(\mu_\sigma, \sigma_\sigma)$, as depicted in Fig. 2.4, then the z variable of the standardized normal $N(0, 1)$ can be given by

$$z = -\frac{\mu_S - \mu_\sigma}{(\sigma_S^2 + \sigma_\sigma^2)^{1/2}} \tag{2.3}$$

and the reliability R is given by

$$R = 1 - \Phi(z) \tag{2.4}$$

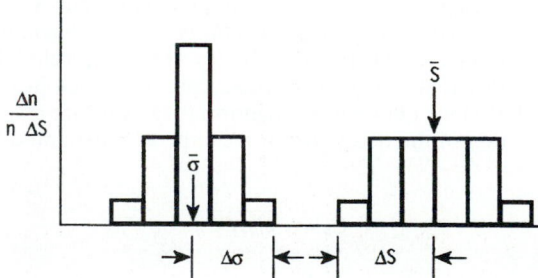

FIGURE 2.3 Histogram of a load-induced stress σ and strength S.

where $\Phi(z)$ is found in Table 2.1. If the strength is lognormally distributed, $\mathbf{S} \sim LN(\mu_S, \sigma_S)$, and the load-induced stress is lognormally distributed, $\boldsymbol{\sigma} \sim LN(\mu_\sigma, \sigma_\sigma)$, then z is given by

$$z = -\frac{\mu_{\ln S} - \mu_{\ln \sigma}}{(\sigma_{\ln S}^2 + \sigma_{\ln \sigma}^2)^{\frac{1}{2}}} = -\frac{\ln\left(\dfrac{\mu_S}{\mu_\sigma}\sqrt{\dfrac{1 + C_\sigma^2}{1 + C_S^2}}\right)}{\sqrt{\ln(1 + C_S^2)(1 + C_\sigma^2)}} \tag{2.5}$$

where $C_S = \sigma_S/\mu_S$ and $C_\sigma = \sigma_\sigma/\mu_\sigma$ are the coefficients of variation of strength and stress. Reliability is given by Eq. (2.4).

Example 1

a. If $\mathbf{S} \sim N(50, 5)$ kpsi and $\boldsymbol{\sigma} \sim N(35, 4)$ kpsi, estimate the reliability R.
b. If $\mathbf{S} \sim LN(50, 5)$ kpsi and $\boldsymbol{\sigma} \sim LN(35, 4)$ kpsi, estimate R.

   Solution

a. From Eq. (2.3),

$$z = -\frac{(50 - 35)}{\sqrt{5^2 + 4^2}} = -2.34$$

From Eq. (2.4),

$$R = 1 - \Phi(-2.34) = 1 - 0.009\,64 = 0.990$$

b. $C_S = 5/50 = 0.10$, $C_\sigma = 4/35 = 0.114$.

From Eq. (2.5),

$$z = -\frac{\ln\left(\dfrac{50}{35}\sqrt{\dfrac{1 + 0.114^2}{1 + 0.100^2}}\right)}{\sqrt{\ln(1 + 0.1^2)(1 + 0.114^2)}} = -2.37$$

and from Eq. (2.4),

$$R = 1 - \Phi(-2.37) = 1 - 0.008\,89 = 0.991$$

It is possible to design to a reliability goal. One can identify a design factor $\overline{n}$ which will correspond to the reliability goal *in the current problem*. A different problem requires a different design factor even for the same reliability goal. If the strength and stress distributions are lognormal, then the design factor $\mathbf{n} = \mathbf{S}/\boldsymbol{\sigma}$ is lognormally distributed, since quotients of lognormal variates are also lognormal. The coefficient of variation of the design factor $\mathbf{n}$ can be approximated for the quotient $\mathbf{S}/\boldsymbol{\sigma}$ as

$$C_n = \sqrt{C_S^2 + C_\sigma^2} \tag{2.6}$$

The mean and standard deviation of the companion normal to $\mathbf{n} \sim LN$ are shown in Fig. 2.5 and can be quantitatively expressed as

TABLE 2.1 Cumulative Distribution Function of Normal (Gaussian) Distribution

$$\Phi(z_\alpha) = \int_{-\infty}^{z_\alpha} \frac{1}{\sqrt{2\pi}} \exp\left(-\frac{u^2}{2}\right) du$$

$$= \begin{cases} \alpha & z_\alpha \le 0 \\ 1 - \alpha & z_\alpha > 0 \end{cases}$$

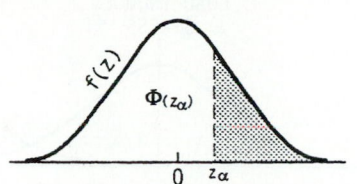

z_α	0.00	0.01	0.02	0.03	0.04	0.05	0.06	0.07	0.08	0.09
0.0	0.5000	0.4960	0.4920	0.4880	0.4840	0.4801	0.4761	0.4721	0.4681	0.4641
0.1	0.4602	0.4562	0.4522	0.4483	0.4443	0.4404	0.4364	0.4325	0.4286	0.4247
0.2	0.4207	0.4168	0.4129	0.4090	0.4052	0.4013	0.3974	0.3936	0.3897	0.3859
0.3	0.3821	0.3783	0.3745	0.3707	0.3669	0.3632	0.3594	0.3557	0.3520	0.3483
0.4	0.3446	0.3409	0.3372	0.3336	0.3300	0.3264	0.3238	0.3192	0.3156	0.3121
0.5	0.3085	0.3050	0.3015	0.2981	0.2946	0.2912	0.2877	0.2843	0.2810	0.2776
0.6	0.2743	0.2709	0.2676	0.2643	0.2611	0.2578	0.2546	0.2514	0.2483	0.2451
0.7	0.2420	0.2389	0.2358	0.2327	0.2296	0.2266	0.2236	0.2206	0.2177	0.2148
0.8	0.2119	0.2090	0.2061	0.2033	0.2005	0.1977	0.1949	0.1922	0.1894	0.1867
0.9	0.1841	0.1814	0.1788	0.1762	0.1736	0.1711	0.1685	0.1660	0.1635	0.1611
1.0	0.1587	0.1562	0.1539	0.1515	0.1492	0.1469	0.1446	0.1423	0.1401	0.1379
1.1	0.1357	0.1335	0.1314	0.1292	0.1271	0.1251	0.1230	0.1210	0.1190	0.1170
1.2	0.1151	0.1131	0.1112	0.1093	0.1075	0.1056	0.1038	0.1020	0.1003	0.0985
1.3	0.0968	0.0951	0.0934	0.0918	0.0901	0.0885	0.0869	0.0853	0.0838	0.0823
1.4	0.0808	0.0793	0.0778	0.0764	0.0749	0.0735	0.0721	0.0708	0.0694	0.0681
1.5	0.0668	0.0655	0.0643	0.0630	0.0618	0.0606	0.0594	0.0582	0.0571	0.0559
1.6	0.0548	0.0537	0.0526	0.0516	0.0505	0.0495	0.0485	0.0475	0.0465	0.0455
1.7	0.0446	0.0436	0.0427	0.0418	0.0409	0.0401	0.0392	0.0384	0.0375	0.0367
1.8	0.0359	0.0351	0.0344	0.0336	0.0329	0.0322	0.0314	0.0307	0.0301	0.0294
1.9	0.0287	0.0281	0.0274	0.0268	0.0262	0.0256	0.0250	0.0244	0.0239	0.0233
2.0	0.0228	0.0222	0.0217	0.0212	0.0207	0.0202	0.0197	0.0192	0.0188	0.0183
2.1	0.0179	0.0174	0.0170	0.0166	0.0162	0.0158	0.0154	0.0150	0.0146	0.0143
2.2	0.0139	0.0136	0.0132	0.0129	0.0125	0.0122	0.0119	0.0116	0.0113	0.0110
2.3	0.0107	0.0104	0.0102	0.00990	0.00964	0.00939	0.00914	0.00889	0.00866	0.00842
2.4	0.00820	0.00798	0.00776	0.00755	0.00734	0.00714	0.00695	0.00676	0.00657	0.00639
2.5	0.00621	0.00604	0.00587	0.00570	0.00554	0.00539	0.00523	0.00508	0.00494	0.00480
2.6	0.00466	0.00453	0.00440	0.00427	0.00415	0.00402	0.00391	0.00379	0.00368	0.00357
2.7	0.00347	0.00336	0.00326	0.00317	0.00307	0.00298	0.00289	0.00280	0.00272	0.00264
2.8	0.00256	0.00248	0.00240	0.00233	0.00226	0.00219	0.00212	0.00205	0.00199	0.00193
2.9	0.00187	0.00181	0.00175	0.00169	0.00164	0.00159	0.00154	0.00149	0.00144	0.00139

z_α	0.0	0.1	0.2	0.3	0.4	0.5	0.6	0.7	0.8	0.9
3	0.00135	0.0^3968	0.0^3687	0.0^3483	0.0^3337	0.0^3233	0.0^3159	0.0^3108	0.0^4723	0.0^4481
4	0.0^4317	0.0^4207	0.0^4133	0.0^5854	0.0^5541	0.0^5340	0.0^5211	0.0^5130	0.0^6793	0.0^6479
5	0.0^6287	0.0^6170	0.0^7996	0.0^7579	0.0^7333	0.0^7190	0.0^7107	0.0^8599	0.0^8332	0.0^8182
6	0.0^9987	0.0^9530	0.0^9282	0.0^9149	$0.0^{10}777$	$0.0^{10}402$	$0.0^{10}206$	$0.0^{10}104$	$0.0^{11}523$	$0.0^{11}260$

z_α	−1.282	−1.645	−1.960	−2.326	−2.576	−3.090	−3.291	−3.891	−4.417
$F(z_\alpha)$	0.10	0.05	0.025	0.010	0.005	0.001	0.0005	0.000 05	0.000 005
$R(z_\alpha)$	0.90	0.95	0.975	0.990	0.995	0.999	0.9995	0.999 95	0.999 995

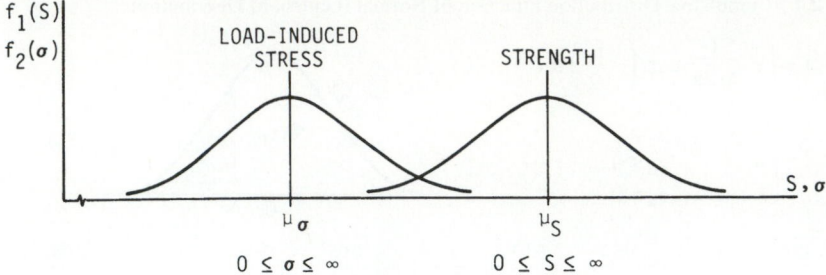

FIGURE 2.4   Probability density functions of load-induced stress and strength.

$$\mu_y = \ln \mu_n - \ln \sqrt{1 + C_n^2}$$

$$\sigma_y = \sqrt{\ln (1 + C_n^2)}$$

The z variable of $z \sim N(0, 1)$ corresponding to the abscissa origin in Fig. 2.5 is

$$z = \frac{y - \mu_y}{\sigma_y} = \frac{0 - \mu_y}{\sigma_y} = \frac{0 - (\ln \mu_n - \ln \sqrt{1 + C_n^2})}{\sqrt{\ln (1 + C_n^2)}}$$

Solving for μ_n, now denoted as $\overline{n}$, gives

$$\mu_n = \overline{n} = \exp \left[-z \sqrt{\ln (1 + C_n^2)} + \ln \sqrt{(1 + C_n^2)} \right] \qquad (2.7)$$

Equation (2.7) is useful in that it relates the mean design factor to problem variability through C_n and the reliability goal through z. Note that the design factor $\overline{n}$ is independent of the mean value of **S** or **σ**. This makes the geometric decision yet to

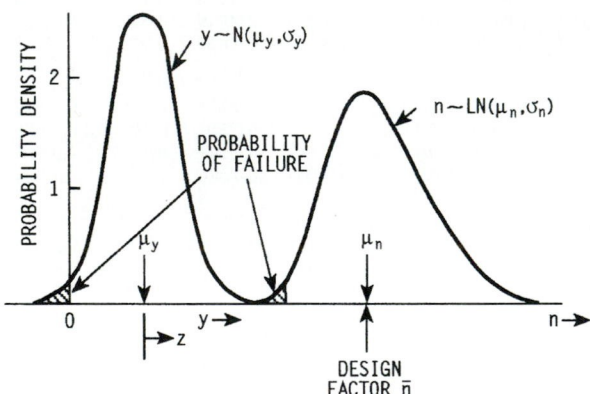

FIGURE 2.5   Lognormally-distributed design factor **n** and its companion normal **y** showing the probability of failure as two equal areas, which are easily quantified from normal probability tables.

be made independent of $\bar{n}$. If the coefficient of variation of the design factor C_n^2 is small compared to unity, then Eq. (2.7) contracts to

$$\bar{n} \doteq \exp \left[C_n(-z + C_n/2) \right] \tag{2.8}$$

Example 2. If $\mathbf{S} \sim LN(50, 5)$ kpsi and $\boldsymbol{\sigma} \sim LN(35, 4)$ kpsi, what design factor $\bar{n}$ corresponds to a reliability goal of 0.990 ($z = -2.33$)?

Solution. $C_S = 5/50 = 0.100$, $C_\sigma = 4/35 = 0.114$. From Eq. (2.6),

$$C_n = (0.100^2 + 0.114^2)^{\frac{1}{2}} = 0.152$$

From Eq. (2.7),

$$\bar{n} = \exp \left[-(-2.33) \sqrt{\ln\,(1 + 0.152^2)} + \ln \sqrt{(1 + 0.152^2)} \right]$$

$$= 1.438$$

From Eq. (2.8),

$$\bar{n} \doteq \exp \left\{ 0.152 \left[-(-2.33) + 0.152/2 \right] \right\} = 1.442$$

The role of the mean design factor $\bar{n}$ is to separate the mean strength $\bar{S}$ and the mean load-induced stress $\bar{\sigma}$ sufficiently to achieve the reliability goal. If the designer in Example 2 was addressing a shear pin that was to fail with a reliability of 0.99, then $z = +2.34$ and $\bar{n} = 0.711$. The nature of C_S is discussed in Chapters 8, 12, 13, and 37.

For normal strength–normal stress interference, the equation for the design factor $\bar{n}$ corresponding to Eq. (2.7) is

$$\bar{n} = \frac{1 \pm \sqrt{1 - (1 - z^2 C_S^2)(1 - z^2 C_\sigma^2)}}{1 - z^2 C_S^2} \tag{2.9}$$

where the algebraic sign + applies to high reliabilities ($R \geq 0.5$) and the – sign applies to low reliabilities ($R < 0.5$).

2.3 USEFUL DISTRIBUTIONS

The body of knowledge called statistics includes many classical distributions, thoroughly explored. They are useful because they came to the attention of the statistical community as a result of a pressing practical problem. A distribution is a particular pattern of variation, and statistics tells us, in simple and useful terms, the many things known about the distribution. When the variation observed in a physical phenomenon is congruent, or nearly so, to a classical distribution, one can infer all the useful things known about the classical distribution. Table 2.2 identifies seven useful distributions and expressions for the probability density function, the expected value (mean), and the variance (standard deviation squared).

TABLE 2.2 Useful Continuous Distributions

Distribution name	Parameters	Probability density function	Expected value	Variance
Uniform	$b > a$	$f(x) = \begin{cases} \dfrac{1}{b-a} & a \le x \le b \\ 0 & \text{elsewhere} \end{cases}$	$\dfrac{a+b}{2}$	$\dfrac{(b-a)^2}{12}$
Normal	$-\infty < \mu < \infty$ $\sigma > 0$	$f(x) = \dfrac{1}{\sigma\sqrt{2\pi}} \exp\left[-\dfrac{1}{2}\left(\dfrac{x-\mu}{\sigma}\right)^2\right]$ $-\infty < x < \infty$	μ	σ
Lognormal	$-\infty < \mu < \infty$ $\sigma > 0$	$f(x) = \dfrac{1}{\sigma_y \, x\sqrt{2\pi}} \exp\left[-\dfrac{1}{2}\left(\dfrac{\ln x - \mu_y}{\sigma_y}\right)^2\right]$ $\mu_y = \ln \mu_x - \ln\sqrt{1+C_x^2} \quad x \ge 0$ $\sigma_y = \sqrt{\ln\left(1 + C_x^2\right)} \quad x \ge 0$	$\exp\left[\mu_y + \dfrac{\sigma_y^2}{2}\right]$	$\exp\left[2\mu_y + 2\sigma_y^2\right] - \exp\left[2\mu_y + \sigma_y^2\right]$ or $\exp\left[2\mu_y + \sigma_y^2\right]\left(\exp\left[\sigma_y^2\right] - 1\right)$
Gamma	$\lambda > 0$ $\eta > 0$	$f(x) = \begin{cases} \dfrac{\lambda^\eta}{\Gamma(\eta)} x^{\eta-1} \exp(-\lambda x) & x \ge 0 \\ 0 & \text{elsewhere} \end{cases}$	$\dfrac{\eta}{\lambda}$	$\dfrac{\eta}{\lambda^2}$
Exponential	$\theta > 0$ $x_0 > 0$	$f(x) = \begin{cases} \dfrac{1}{\theta - x_0} \exp\left[-\left(\dfrac{x - x_0}{\theta - x_0}\right)\right] & x > x_0 \\ 0 & \text{elsewhere} \end{cases}$	θ	$(\theta - x_0)^2$
Rayleigh	$\sigma > 0$	$f(x) = \begin{cases} \dfrac{x}{\sigma^2} \exp\left(\dfrac{-x^2}{2\sigma^2}\right) & x \ge 0 \\ 0 & \text{elsewhere} \end{cases}$	$\dfrac{(\sigma^2 \pi)^{1/2}}{\sqrt{2}}$	$0.429\sigma^2$
Weibull	$x_0 > \theta$ $\theta > 0$ $b > 0$	$f(x) = \begin{cases} \dfrac{b}{\theta - x_0}\left(\dfrac{x - x_0}{\theta - x_0}\right)^{b-1} \exp\left[-\left(\dfrac{x - x_0}{\theta - x_0}\right)^b\right] & x > x_0 \\ 0 & \text{elsewhere} \end{cases}$	$x_0 + (\theta - x_0)\,\Gamma\!\left(1 + \dfrac{1}{b}\right)$	$(\theta - x_0)^2\left[\Gamma\!\left(1 + \dfrac{2}{b}\right) - \Gamma^2\!\left(1 + \dfrac{1}{b}\right)\right]$

A frequency histogram may be plotted with the ordinate $\Delta n/(n\,\Delta x)$, where Δn is the class frequency, n is the population, and Δx is the class width. This ordinate is probability density, an estimate of $f(x)$. If the data reduction gives estimates of the distributional parameters, say mean and standard deviation, then a plot of the density function superposed on the histogram will give an indication of fit. Computational techniques are available to assist in the judgment of good or bad fit. The chi-squared goodness-of-fit test is one based on the probability density function superposed on the histogram (Ref. [2.3]).

One might plot the cumulative distribution function (CDF) vs. the variate. The CDF is just the probability (the chance) of a failure at or below a specified value of the variate x. If one has data in this form, or arranges them so, then the CDF for a candidate distribution may be superposed to see if the fit is good or not. The Kolomogorov-Smirnov goodness-of-fit test is available (Ref. [2.3]). If the CDF is plotted against the variate on a coordinate system which rectifies the CDF–x locus, then the straightness of the data string is an indication of the quality of fit. Computationally, the linear regression correlation coefficient r may be used, and the corresponding r test is available (Ref. [2.3]).

Table 2.3 shows the transformations to be applied to the ordinate (variate) and abscissa (CDF, usually denoted F_i) which will rectify the data string for comparison with a suspected parent distribution.

TABLE 2.3 Transformations which Rectify CDF
Data Strings

Distribution	Transformation function to data x	Transformation to cumulative distribution function F
Uniform	x	F
Normal	x	$z(F)$
Lognormal	$\ln(x)$	$z(F)$
Weibull	$\ln(x - x_0)$	$\ln\ln[1/(1 - F)]$
Exponential	$x - x_0$	$\ln[1/(1 - F)]$

Consider a right cylindrical surface generated with an automatic screw machine turning operation. When the machine is set up to produce a diameter at the low end of the tolerance range, each successive part will be slightly larger than the last as a result of tool wear and the attendant increase in tool force due to dulling wear. If the part sequence number is n and the sequence number is n_f when the high end of the tolerance is reached, a is the initial diameter produced, and b is the final diameter produced, one can expect the following relation:

$$x = a + \frac{(b-a)n}{v_f} \tag{2.10}$$

However, suppose one measured the diameter every thousandth part and built a data set, smallest diameter to largest diameter (ordered):

n	n_1	n_2	n_3	$\cdots$
x	x_1	x_2	x_3	$\cdots$

If the data are plotted with n as abscissa and x as ordinate, one observes a rather straight data string. Consulting Table 2.2, one notes that the linearity of these untransformed coordinates indicates uniform random distribution. A word of caution: If the parts are removed and packed in roughly the order of manufacture, there is no distribution at all! Only if the parts are thoroughly mixed and we draw randomly does a distribution exist. One notes in Eq. (2.10) that the ratio n/n_f is the fraction of parts having a diameter equal to or less than a specified x, and so this ratio is the cumulative distribution function F. Substituting F in Eq. (2.10) and solving for F yields

$$F(x) = \frac{x - a}{b - a} \qquad a \le x \le b \tag{2.11}$$

From Table 2.2, take the probability density function for uniform random distribution, $f(x) = 1/(b - a)$, and integrate from a to x to obtain Eq. (2.11).

Engineers often have to identify a distribution from a small amount of data. Data transformations which rectify the data string are useful in recognizing a distribution. First, place the data in a column vector, order smallest to largest. Second, assign corresponding cumulative distribution function values F_i using median rank $(i - 0.3)/(n + 0.4)$ if seeking a median locus, or $i/(n + 1)$ if seeking a mean locus (Ref. [2-4]). Third, apply transformations from Table 2.3 and look for straightness.

Normal distributions are used for many approximations. The most likely parent of a data set is the normal distribution; however, that does not make it common. When a pair of dice is rolled, the most likely sum of the top faces is 7, which occurs in 1/6 of the outcomes, but 5/6 of the outcomes are other than 7.

Properties of materials—ultimate tensile strength, for example—can have only positive values, and so the normal cannot be the true distribution. However, a normal fit may be robust and therefore useful. The lognormal does not admit variate values which are negative, which is more in keeping with reality. Histographic data of the ultimate tensile strength of a 1020 steel with class intervals of 1 kpsi are as follows:

Class frequency f_i	2	18	23	31	83	109	138	151
Class midpoint x_i	56.5	57.5	58.5	59.5	60.5	61.5	62.5	63.5

Class frequency f_i	139	130	82	49	28	11	4	2
Class midpoint x_i	64.5	65.5	66.5	67.5	68.5	69.5	70.5	71.5

Now $\Sigma x_i f_i = 63\,625$ and $\Sigma x_i^2 f_i = 4\,054\,864$, and so $\bar{x}$ and $\hat{\sigma}$ are $\bar{x} = \Sigma x_i f_i / n = 63\,625/1000 = 63.625$ kpsi, and

$$\hat{\sigma} = \sqrt{\frac{\Sigma x_i^2 f_i - (\Sigma x_i f_i)^2/n}{n - 1}}$$

$$\sigma = \sqrt{\frac{4\,054\,864 - (63\,625)^2/1000}{(1000 - 1)}} = 2.5942 \text{ kpsi}$$

From Table 2.2, the mean and standard deviation of the companion normal to a lognormal are (Ref. [2-2])

$$\mu_y = \ln \bar{x} - \ln \sqrt{1 + C_x^2} = \ln 63.625 - \ln \sqrt{1 + 0.040\ 773^2}$$

$$= 4.1522$$

$$\sigma_y = \sqrt{\ln (1 + C_x^2)} = \sqrt{\ln (1 + 0.040\ 773^2)} = 0.0408$$

The lognormal probability density function of x is

$$g(x) = \frac{1}{x\sigma_y \sqrt{2\pi}} \exp \left[-\frac{1}{2} \left(\frac{\ln x - \mu_y}{\sigma_y} \right)^2 \right]$$

$$= \frac{1}{0.0408x \sqrt{2\pi}} \exp \left[-\frac{1}{2} \left(\frac{\ln x - 4.1522}{0.0408} \right)^2 \right]$$

A plot of the histogram and the density is shown in Fig. 2.6. A chi-squared goodness-of-fit test on a modified histogram (compacted somewhat to have 5 or more in each class) cannot reject the null hypothesis of lognormality at the 0.95 confidence level.

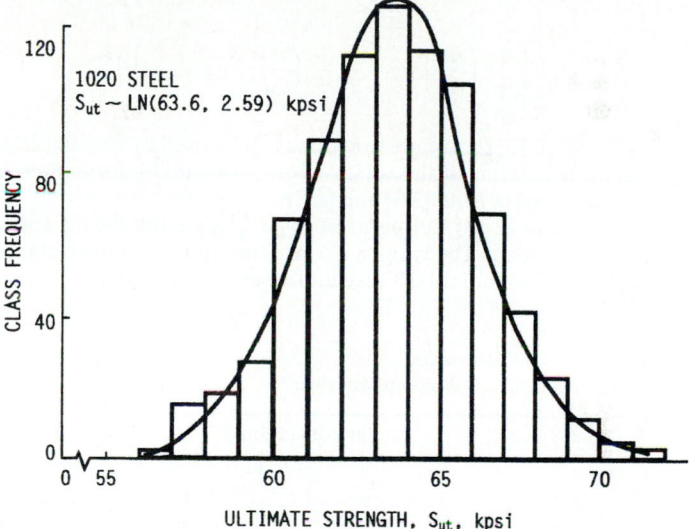

FIGURE 2.6 Histographic report of the results of 1000 ultimate tensile strength tests on a 1020 steel.

2.4 RANDOM-VARIABLE ALGEBRA

Engineering parameters which exhibit variation can be represented by random variables and characterized by distribution parameters and a distribution function. Many distributions have two parameters; the mean and standard deviation (variance) are preferred. It is common to display statistical parameters by roster between curved parentheses as (μ, σ). If the normal distribution is to be indicated, then an N is placed before the parentheses as $N(\mu, \sigma)$; this indicates a normal distribution with a mean of μ and a standard deviation of σ. Similarly, $LN(\mu, \sigma)$ is a lognormal distribution and $U(\mu, \sigma)$ is a uniform distribution. To distinguish a real-number variable w from a random variable $\mathbf{y}$, boldface is used. Thus $\mathbf{z} = \mathbf{x} + \mathbf{y}$ displays $\mathbf{z}$ as the sum of random variables $\mathbf{x}$ and $\mathbf{y}$. With knowledge of $\mathbf{x}$ and $\mathbf{y}$, of interest are the parameters and distribution of $\mathbf{z}$ (Ref. [2.6]).

For distributional information, various closure theorems and the central limit theorem of statistics are useful. The sums of normal variates are themselves normal. The quotients and products of lognormals are lognormal. Real powers of a lognormal variate are likewise lognormal. Sums of variates from any distribution tend asymptotically to approach normal. Products of variates from any distribution tend asymptotically to lognormal. In some cases a computer simulation is necessary to discover distributions resulting from an algebraic combination of variates. The mean and standard deviation of a function $\phi(x_1, x_2, \ldots, x_n)$ can be estimated by the following rapidly convergent Taylor series of expected values for unskewed (or lightly skewed) distributions (Ref. [2.7, Appendix C]):

$$\mu_\phi = \phi(x_1, x_2, \ldots, x_n)_\mu + \frac{1}{2} \sum_{i=1}^{n} \frac{\partial^2 \phi}{\partial x_i^2}\bigg|_\mu \sigma_{x_i}^2 + \cdots \tag{2.12}$$

$$\sigma_\phi = \left\{ \sum_{i=1}^{n} \left(\frac{\partial \phi}{\partial x_i}\right)_\mu^2 \sigma_{x_i}^2 + \frac{1}{2} \sum_{i=1}^{n} \left(\frac{\partial^2 \phi}{\partial x_i^2}\right)_\mu^2 \sigma_{x_i}^4 + \cdots \right\}^{1/2} \tag{2.13}$$

Equations (2.12) and (2.13) for simple functions can be used to form Table 2.4 to display the dominant first terms of the series. More expanded information, including correlation, can be found in Refs. [2.3] and [2.7].

Equations (2.12) and (2.13) can be used to propagate the means and standard deviations through functions. The various closure theorems of statistics, or computer simulation, can be used to find robust distributional information.

TABLE 2.4 Means, Standard Deviations, and Coefficients of Variation of Simple Operations with Independent (Uncorrelated) Random Variables*

Function	Mean value μ	Standard deviation σ	Coefficient of variation C
a	a	0	0
$\mathbf{x}$	μ_x	σ_x	σ_x/μ_x
$\mathbf{x} + a$	$\mu_x + a$	σ_x	σ_x/μ_x
$a\mathbf{x}$	$a\mu_x$	$a\sigma_x$	σ_x/μ_x
$\mathbf{x} + \mathbf{y}$	$\mu_x + \mu_y$	$(\sigma_x^2 + \sigma_y^2)^{1/2}$	σ_{x+y}/μ_{x+y}
$\mathbf{x} - \mathbf{y}$	$\mu_x - \mu_y$	$(\sigma_x^2 + \sigma_y^2)^{1/2}$	σ_{x-y}/μ_{x-y}
$\mathbf{xy}$	$\mu_x\mu_y$	$C_{xy}\mu_{xy}$	$(C_x^2 + C_y^2)^{1/2}$
$\mathbf{x}/\mathbf{y}$	μ_x/μ_y	$C_{x/y}\mu_{x/y}$	$(C_x^2 + C_y^2)^{1/2}$
$1/\mathbf{x}$	$1/\mu_x$	C_x/μ_x	C_x
$\mathbf{x}^2$	μ_x^2	$2C_x\mu_x^2$	$2C_x$
$\mathbf{x}^3$	μ_x^3	$3C_x\mu_x^3$	$3C_x$
$\mathbf{x}^4$	μ_x^4	$4C_x\mu_x^4$	$4C_x$

* Tabulated quantities are obtained by the partial derivative propagation method, some results of which are approximate. For a more complete listing including the first two terms of the Taylor series, see Charles R. Mischke, *Mathematical Model Building*, 2d rev. ed., Iowa State University Press, Ames, 1980, appendix C.

The first terms of Eqs. (2.12) and (2.13) are often sufficient as a first-order estimate; thus

$$\mu_\phi = \phi(\mu_{x_1}, \mu_{x_2}, \ldots, \mu_{x_n}) \tag{2.14}$$

$$\sigma_\phi = \left\{ \sum_{i=1}^{n} \left(\frac{\partial \phi}{\partial x_i} \right)_\mu^2 \sigma_{x_i}^2 \right\}^{\frac{1}{2}} \tag{2.15}$$

and if ϕ is of the form $\alpha\, x_1{}^a\, x_2{}^b\, x_3{}^c \ldots$, then C_ϕ is given by

$$C_\phi = \left(a^2 C_{x_1}^2 + b^2 C_{x_2}^2 + \cdots \right)^{\frac{1}{2}} \tag{2.16}$$

Equations (2.14), (2.15), and (2.16) are associated with the partial derivative estimation method. These equations are very important in what they suggest in general about engineering computations in stochastic situations. The estimate of the mean in a functional relationship comes from substituting mean values of the variates. This suggests that deterministic and familiar engineering computations are still useful in stochastic problems if mean values are used. Calculations such as the quotient of *minimum* strength divided by *maximum* load-induced stress are not appropriate when chance of failure is being considered.

Equation (2.15) says that the variance of ϕ is simply the sum of the weighted variances of the parameters, with the weighting factors depending on the functional relationship involved. In terms of the standard deviation, it is a weighted Pythagorean combination.

The good news is that engineering's previous deterministic experience is useful in stochastic problems provided one uses mean values. The bad news is that there is additional effort associated with propagating the variation through the same relationships and identifying the resulting distributions. The other element of bad news is that Eqs. (2.14) and (2.15) are approximations, but the corresponding good news is that they are robust approximations. In summary,

1. A random variable or function of random variables can be characterized by statistical parameters, often the mean and variance, and a distribution function, whether assumed or goodness-of-fit tested.

2. Ordinary deterministic algebra using means of variates is useful in estimating means and standard deviations of functions of variates.

3. The distribution of a function of random variables can often be determined from closure theorems.

4. Computer simulation techniques can address cases not covered (see Chap. 5).

Example 3. If 12 random selections are made from the uniform random distribution $U[0, 1]$ and the real number 6 is subtracted from the sum of the 12, what are the mean, the standard deviation, and the distribution of the result?

Solution. Note the square brackets in $U[0, 1]$. These denote parameters other than the mean and standard deviation, in this case range numbers a and b—i.e., there are no observations less than a nor more than b. The sum ϕ is defined by

$$\phi = x_1 + x_2 + \cdots + x_{12} - 6$$

From Table 2.2,

$$\mu_x = (a + b)/2 = (0 + 1)/2 = 1/2$$

$$\sigma_x^2 = (b - a)^2/12 = (1 - 0)^2/12 = 1/12$$

From Table 2.4, the mean is the sum of the means:

$$\bar{\phi} = \bar{x}_1 + \bar{x}_2 + \cdots + \bar{x}_{12} - 6 = 1/2 + 1/2 + \cdots + 1/2 - 6 = 0$$

From Table 2.4, the standard deviation of the sum of independent random variables is the square root of the sum of the variances:

$$\sigma_\phi = (1/12 + 1/12 + \cdots + 1/12 + 0)^{\frac{1}{2}} = 1$$

From the central limit theorem, the sum of random variables asymptotically approaches normality. The sum of 12 variates cannot be rejected using a null hypothesis of normality. Thus, $\phi \sim N(\mu_\phi, \sigma_\phi) = N(0, 1)$. Computing machinery manufacturers supply a machine-specific pseudo-random number generator $U[0, 1]$. The reason the program is supplied is the machine specificity involved. Such a program is the building block from which other random numbers can be generated with software.

Example 3 is the basis for a Fortran subroutine to generate pseudo-random numbers from a normal distribution N(xbar, sigmax). If RANDU is the subprogram name of the uniform random number generator in the interval $[0, 1]$, and IX and IY are seed integers, then

```
      SUBROUTINE GAUSS(IX,IY,XBAR,SIGMAX,X)
      SUM=0.
      DO 100 I=1,12
      CALL RANDU(IX,IY,U)
      SUM=SUM+U
  100 CONTINUE
      X=XBAR+(SUM-6.)*SIGMAX
      RETURN
      END
```

2.5 STOCHASTIC ENDURANCE LIMIT BY CORRELATION AND BY TEST

Designers need rational approaches to meet a variety of situations. A product can be produced in such large quantities (or be so dangerous) that elaborate testing of materials, components, and prototypes is justified. Smaller quantities can be produced and the product can be of modest value, so that less comprehensive testing of materials—perhaps only ultimate tensile strength testing—is economically justified. Or so few items can be produced that no testing of materials is done at all.

For an R. R. Moore rotating beam bending endurance test, approximately 60 specimens in a staircase test matrix method of testing are employed to find the endurance limit of a steel. Considerable time and expense is involved, using a standard specimen and a procedure that will remove the effects of surface finish, size, loading, temperature, stress concentration, et al. Since such testing is not always possible, engineers with an interest in bending, axial (push-pull), and torsional fatigue use correlations of endurance limit to mean tensile strength as a first-order estimate as follows:

$$\mathbf{S}'_e = \phi_b \bar{S}_{ut} \tag{2.17}$$

$$\mathbf{S}'_{ax} = \phi_{ax} \bar{S}_{ut} \tag{2.18}$$

$$\mathbf{S}'_{se} = \phi_t \bar{S}_{ut} \tag{2.19}$$

where ϕ_b, ϕ_{ax}, and ϕ_t are called *fatigue ratios*. Data reported by Gough are shown in Fig. 2.7. It is clear that the bending fatigue ratio ϕ_b is not constant in a class of materials and varies widely; that is to say, it is a random variable. The mean of ϕ is called the fatigue ratio, and in bending in steel it is about 0.5, which is conservative about half the time. Table 2.5 shows the mean and standard deviation of ϕ_b for classes of materials. From 133 full-scale R. R. Moore tests on steels, ϕ_b is found to be lognormally distributed.

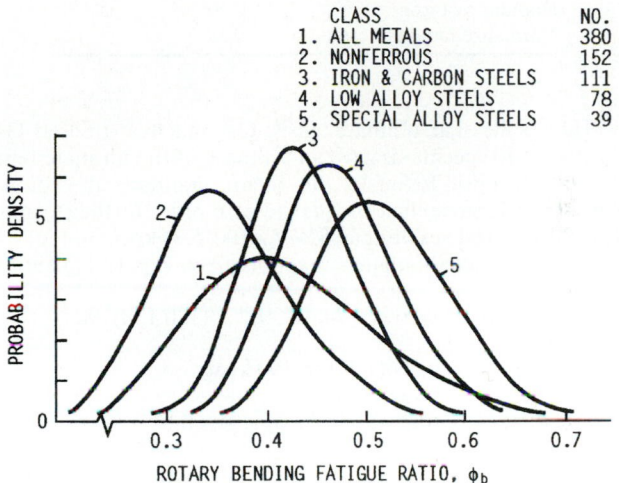

FIGURE 2.7   Probability density functions of fatigue ratio ϕ_b reported by Gough for five classifications of metals.

TABLE 2.5   Stochastic Parameters of Fatigue Ratio ϕ*

Class of metals	Number of tests	$\hat{\mu}_\phi$	$\hat{\sigma}_\phi$
All metals	380	0.44	0.10
Nonferrous	152	0.37	0.075
Irons and carbon steels	111	0.44	0.060
Low-alloy steels	78	0.475	0.063
Special-alloy steels	39	0.52	0.070

* Data from Gough reported in J. A. Pope, *Metal Fatigue,* Chapman and Hall, London, 1959 and tabulated in C. R. Mischke, "Prediction of Stochastic Endurance Strength," *Transactions of the American Society of Mechanical Engineers, Journal of Vibration, Acoustics, Stress and Reliability in Design,* vol. 109, no. 1, Jan. 1987, pp. 113–122.

$$\phi_b = 0.445 d^{-0.107}(1, 0.138) \tag{2.20}$$

When the standard specimen diameter of 0.30 in is substituted in Eq. (2.20), one obtains $\phi_{0.30} = 0.506(1, 0.138)$, which is still lognormally distributed. Multiplying the 0.506 by the mean and standard deviation, one can write $\phi_{0.30} \sim LN(0.506, 0.070)$. The coefficient of variation is 0.138. Table 2.6 shows approximate mean values of ϕ_b for several material classes.

TABLE 2.6 Typical Mean Fatigue Ratios for Several Material Classes

Material class	$\bar{\phi}_{0.30}$
Wrought steel	0.50
Cast steel	0.40
Gray cast iron	0.35
Nodular cast iron	0.40
Normalized nodular cast iron	0.33

Example 4.   The results of an ultimate tensile test on a heat-treated 4340 steel (382 Brinell) consisting of 10 specimens gave an estimate of the ultimate tensile strength of $S_{ut} \sim LN(190, 6.0)$ kpsi. Estimate the mean, standard deviation, and 99th-percentile bending endurance limit for (*a*) the case of no further testing and (*b*) an additional R. R. Moore test resulting in $S'_e \sim LN(90, 5.3)$ kpsi.

Solution.   *a.* The expected fatigue strength is, from Eqs. (2.17) and (2.20),

$$S'_e = \phi_b \bar{S}_{ut} = 0.445(0.30)^{-0.107}(1, 0.138)190$$

$$= 0.506(1, 0.138)190 \text{ kpsi}$$

The estimated mean of the endurance limit $\bar{S}'_e$ is given by

$$\bar{S}'_e = 0.506(1)(190) = 96.1 \text{ kpsi}$$

The standard deviation $\sigma_{S'_e}$ is

$$\sigma_{S'_e} = 0.506(0.138)(190) = 13.3 \text{ kpsi}$$

The coefficient of variation is $C_{S'_e} = 13.3/96.1 = 0.138$, as expected. The distribution of S'_e is lognormal because ϕ_b is lognormal. The 99th-percentile endurance limit is found from the companion normal to the endurance limit distribution as follows:

$$\mu_y = \ln \bar{S}'_e - \ln \sqrt{1 + C_{S'_e}^2} = \ln 96.1 - \ln \sqrt{1 + 0.138^2}$$

$$= 4.556$$

$$\sigma_y = \sqrt{\ln (1 + C_{S'_e}^2)} = \sqrt{\ln (1 + 0.138^2)} = 0.137$$

Now

$$_{0.99}y = \mu_y - _{0.99}z\sigma_y = 4.556 - 2.33(0.137) = 4.237$$

and $_{0.99}S'_e$ is given by

$$_{0.99}S'_e = \exp\left(_{0.99}y\right) = \exp\left(4.237\right) = 69.2 \text{ kpsi}$$

without fatigue testing from the history of the 133 steel materials ensemble embodied in ϕ_b. One can expect 99 percent of the instances of endurance limit to exceed 69.2 kpsi given that the mean tensile strength is 190 kpsi.

  b. The results of R. R. Moore testing of the 4340 gave $\mu_{S'_e} = 90$ kpsi and $\sigma_{S'_e} = 5.3$ kpsi. The coefficient of variation is 5.3/90, or 0.059. The 99th-percentile endurance limit is found from the companion normal as follows:

$$\mu_y = \ln 90 - \ln \sqrt{1 + 0.059^2} = 4.498$$

$$\sigma_y = \sqrt{\ln\left(1 + 0.059^2\right)} = 0.059$$

$$_{0.99}y = 4.498 - 2.33(0.059) = 4.361$$

$$_{0.99}S'_e = \exp\left(4.361\right) = 78.3 \text{ kpsi}$$

It is instructive to plot the density functions. The lognormal density function for part *a* is

$$g_1(S) = \frac{1}{0.137S\sqrt{2\pi}} \exp\left[-\frac{1}{2}\left(\frac{\ln S - 4.556}{0.137}\right)^2\right]$$

and that for part *b* is

$$g_2(S) = \frac{1}{0.059S\sqrt{2\pi}} \exp\left[-\frac{1}{2}\left(\frac{\ln S - 4.498}{0.059}\right)^2\right]$$

Figure 2.8 graphically depicts the two and one-half times dispersion resulting from use of the correlation rather than R. R. Moore testing. Testing is costly in money and time. It costs money to reduce dispersion, and one is never without dispersion. However, in designing to a reliability goal, dispersion in strength, loading, and geometry increases the size of parts. Using part *a* strength information results in a larger part than using part *b* information.

2.6 INTERFERENCE

In Eqs. (2.5) and (2.9), one has a way of relating geometric decisions to a reliability goal. The fundamental tactic is to separate the mean strength from the mean stress sufficiently to achieve the reliability goal through geometric decisions. The equation **n = S/σ** can be generalized. The denominator is some threatening stimulus which is resisted by some response which has a limited potential (the numerator). Defining the design factor as the quotient of the response potential divided by the stimulus is more general and useful. The stimulus might be a distortion and the response potential the deflection which compromises function. The tools discussed so far have broader application.

  Interference of normal-normal and lognormal-lognormal distributions has been presented. There is need for a general method for interference of other distribution

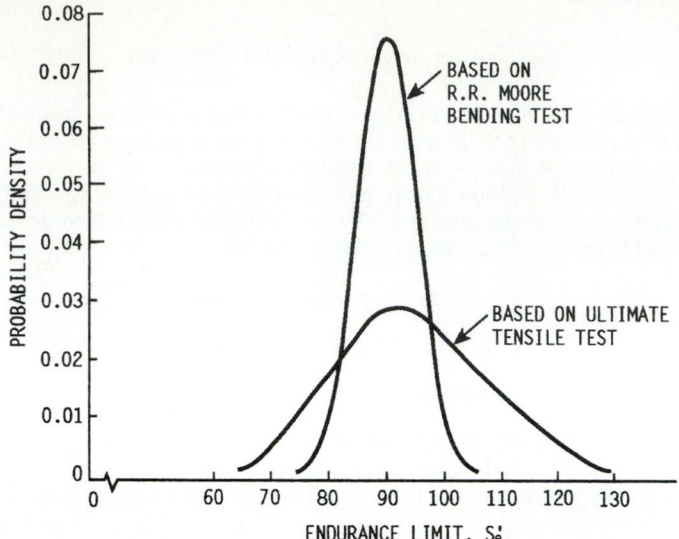

FIGURE 2.8   Probability density functions of rotary bending endurance limit based on historical knowledge of an ensemble of 133 steels, plus tensile testing on a 4340 steel, and based on R. R. Moore endurance limit testing on 4340.

combinations. In Fig. 2.9a the probability density of the response potential **S** is $f_1(S)$, and in Fig. 2.9b the density function of the stimulus **σ** is $f_2(\sigma)$. The probability that the strength exceeds a stress level x is $dP(S > x)$, which is the differential reliability dR, or

$$dR = R_1(x)\, dF_2(x) = -R_1(x)\, dR_2(x)$$

which integrates to

$$R = -\int_{x=-\infty}^{x=\infty} R_1(x)\, dR_2(x) = -\int_{R_2=1}^{R_2=0} R_1(x)\, dR_2 = \int_0^1 R_1\, dR_2 \qquad (2.21)$$

where

$$R_1(x) = \int_x^\infty f_1(S)\, dS \qquad \text{and} \qquad R_2(x) = \int_x^\infty f_2(\sigma)\, d\sigma$$

which is given geometric interpretation in Fig. 2.9c.

An alternative view is that the probability that the stress is less than the strength is expressible as $dP(\sigma < x)$, which is the differential reliability dR, or, from Fig. 2.9d and e,

$$dR = F_2(x)\, dF_1(x) = -[1 - R_2(x)]\, dR_1(x)$$

which integrates to

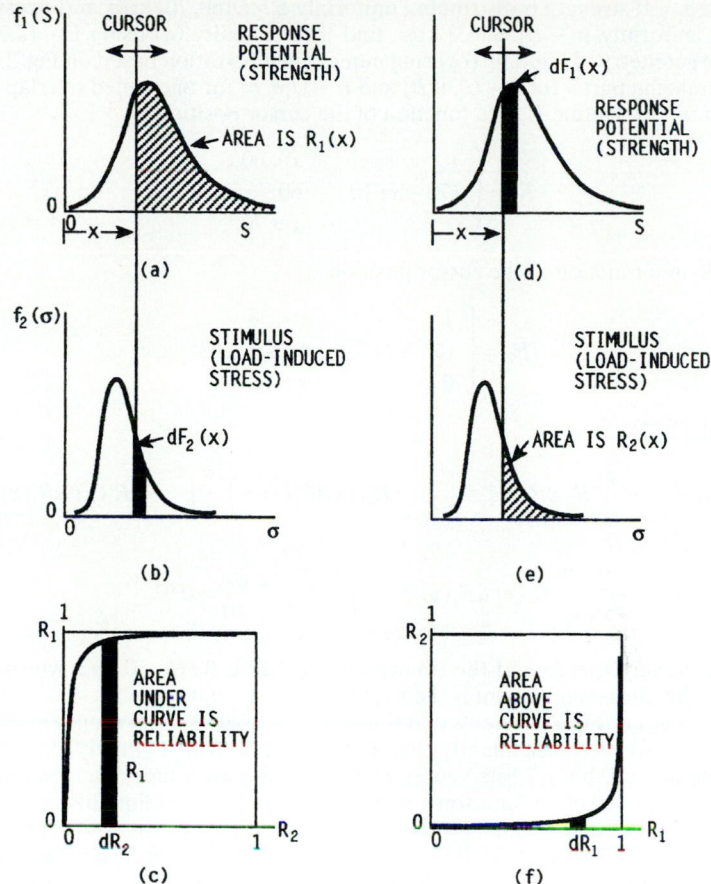

FIGURE 2.9 (a), (b), and (c) Development of the general reliability equation $\int_0^1 R_1 dR_2$ by interference; (d), (e), and (f) development of general reliability equation $1 - \int_0^1 R_2 dR_1$ by interference.

$$R = -\int_{x=-\infty}^{x=\infty} [1 - R_2(x)]\, dR_1(x) = -\int_{R_1=1}^{R_1=0} (1 - R_2)\, dR_1$$

$$= -\int_1^0 dR_1 + \int_1^0 R_2\, dR_1 = 1 - \int_0^1 R_2\, dR_1 \qquad (2.22)$$

where $R_1(x)$ and $R_2(x)$ have the definitions above. Equation (2.22) is given geometric interpretation in Fig. 2.9f. When dealing with distributions with lower bounds, such as Weibull, Eq. (2.22) is easier to integrate than Eq. (2.21).

The following example is couched in terms of geometrically simple distributions to avoid obscuring the ideas.

Example 5.   If strength is distributed uniformly, $\mathbf{S} \sim U[60, 70]$ kpsi, and stress is distributed uniformly, $\boldsymbol{\sigma} \sim U[58, 63]$ kpsi, find the reliability (*a*) using Eq. (2.22), (*b*) using the geometry of Fig. 2.9*f*, (*c*) using numerical integration based on Fig. 2.9*f*, and (*d*) generalizing part *a* for $\mathbf{S} \sim U[A, B]$ and $\boldsymbol{\sigma} \sim U[a, b]$ for one-tailed overlap.

   Solution.   *a.* Define R_1 as a function of the cursor position *x*:

$$R_1 = \begin{cases} 1 & x < 60 \\ (70 - x)/10 & 60 \le x \le 70 \\ 0 & x > 70 \end{cases}$$

Define R_2 as a function of the cursor position *x*:

$$R_2 = \begin{cases} 1 & x < 58 \\ (63 - x)/5 & 58 \le x \le 63 \\ 0 & x > 63 \end{cases}$$

From Eq. (2.22),

$$R = 1 - \int_0^1 R_2 \, dR_1 = 1 - \int_{R_1 = 0}^{R_1 = 1} R_2(x) \, dR_1(x) = 1 - \int_{x = 70}^{x = 60} R_2(x) \, dR_1(x)$$

$$= 1 - \int_{x = 63}^{x = 60} R_2(x) \, dR_1(x) = 1 - \int_{63}^{60} \frac{63 - x}{5} \frac{dx}{10} = 0.91$$

b. Geometrically, the area of the triangle in Fig. 2.10 is $0.6(1 - 0.7)/2$, which equals 0.09, and the ones complement is the reliability $R = 1 - 0.09 = 0.91$.

c. Examination of Fig. 2.9*f* shows that the largest contribution to the area under the curve is near $R_1 = 1$; consequently, the tabular method will begin with $R_1 = 1$ at the top of the table. Table 2.7 lists values of R_1 beginning with unity and decreasing in steps of 0.05 ($h = 0.05$ in Simpson's method). Column 2 contains the values of the cursor location *x* corresponding to R_1. This is obtained by solving the expression R_1 for *x*, namely $x = 70 - 10R_1$. Column 3 consists of the values of R_2 corresponding to the cursor location *x*, namely $R_2 = (63 - x)/5$. The ordinates to the curve are in the R_2 column, and values other than zero contribute to the area. At $R_1 = 0.70$, the area contributions cease. The Simpson's rule multipliers *m* are in column 4. The sum, $\Sigma m R_2$, is 5.4. The area under the curve is

$$A = (h/3) \, \Sigma \, m R_2 = (0.05/3)(5.4) = 0.09$$

and the reliability is

$$R = 1 - A = 1 - 0.09 = 0.91$$

d. The survival function R_1 is given by

$$R_1 = \begin{cases} 1 & x < A \\ (B - x)/(B - A) & A \le x \le B \\ 0 & x > B \end{cases}$$

and the survival function R_2 is given by

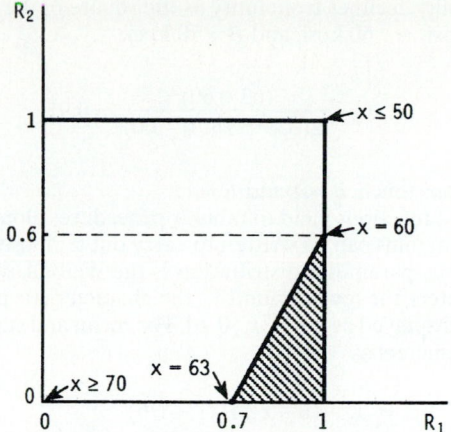

FIGURE 2.10 Assessment of reliability in Example 2.6(b) by geometric interpretation of area.

$$R_2 = \begin{cases} 1 & x < a \\ (b-x)/(b-a) & a \le x \le b \\ 0 & x > b \end{cases}$$

For one-tailed overlap, from Eq. (2.22),

$$R = 1 - \int_0^1 R_2 \, dR_1 = 1 - \int_{R_1=0}^{R_1=1} R_2(x) \, dR_1(x) = 1 - \int_{x=B}^{x=A} R_2(x) \, dR_1(x)$$

Noting that $R_2(x)$ is zero when $x < b$ allows the lower limit to be changed to b.

$$R = 1 - \int_{x=b}^{x=A} R_2(x) \, dR_1(x) = 1 - \int_b^A \frac{b-x}{b-a} \frac{dx}{B-A}$$

$$= 1 - \frac{1}{(b-a)(B-A)} \int_A^b (b-x) \, dx = 1 - \frac{(b-A)^2}{2(b-a)(B-A)}$$

TABLE 2.7 Reliability by Simpson's Rule Interference

R_1	x	R_2	Multiplier m	mR_2
1.00	60.0	0.6	1	0.6
0.95	60.5	0.5	4	2.0
0.90	61.0	0.4	2	0.8
0.85	61.5	0.3	4	1.2
0.80	62.0	0.2	2	0.4
0.75	62.5	0.1	4	0.4
0.70	63.0	0.0	1	0.0
				$\Sigma mR_2 = 5.4$

Note that the reliability declines from unity as the square of the overlap $(b - A)$. For $a = 58$ kpsi, $b = 63$ kpsi, $A = 60$ kpsi, and $B = 70$ kpsi,

$$R = 1 - \frac{(63 - 60)^2}{2(63 - 58)(70 - 60)} = 0.91$$

and when distributions touch, $b = A$ and $R = 1$.

More complicated functions yield to tabular procedures along the lines of Example 5c. Computer programs can be written to carry out tedious work.

A very useful three-parameter distribution is the Weibull, which is expressed in terms of the parameters, the lower bound x_0, the characteristic parameter θ, and the shape parameter b, displayed as $\mathbf{x} \sim W[x_0, \theta, b]$. The mean and standard deviation are found from the parameters as

$$\mu_x = x_0 + (\theta - x_0)\, \Gamma\, (1 + 1/b)$$

$$\sigma_x = (\theta - x_0)[\Gamma(1 + 2/b) - \Gamma^2(1 + 1/b)]^{\frac{1}{2}}$$

The Weibull has the advantage of being a closed-form survival function.

$$R = \exp\{-[(x - x_0)/(\theta - x_0)]^b\}$$

For interference of a Weibull strength $\mathbf{S} \sim W[x_{01}, \theta_1, b_1]$ with a Weibull stress $\boldsymbol{\sigma} \sim W[x_{02}, \theta_2, b_2]$, use a numerical evaluation of the integral in Eq. (2.22). Write the strength distribution survival equation in terms of the cursor location x as

$$R_1 = \exp\{-[(x - x_{01})/(\theta_1 - x_{01})]^{b_1}\}$$

and solve for x, which results in

$$x = x_{01} + (\theta_1 - x_{01})[\ln(1/R_1)]^{1/b_1}$$

Noting that the survival equation for the stress distribution in terms of the cursor location x is

$$R_2 = \exp\{-[(x - x_{02})/(\theta_2 - x_{02})]^{b_2}\}$$

one forms a table such as Table 2.8 to integrate the integral portion of Eq. (2.22). If $\mathbf{S} \sim W[40, 50, 3.3]$ kpsi and $\boldsymbol{\sigma} \sim W[30, 40, 2]$ kpsi, then Table 2.8 follows. The sum $\Sigma m R_2$ is 1.443 413, making the area under the $R_1 R_2$ curve by Simpson's rule

$$A = (h/3)\Sigma m R_2 = (0.1/3)(1.433\ 413) = 0.048\ 114$$

and

$$R = 1 - A = 1 - 0.048\ 114 = 0.952$$

TABLE 2.8 Weibull-Weibull Interference by Simpson's Rule,
$S \sim W[40, 50, 3.3]$ kpsi, $\sigma \sim W[30, 40, 2]$ kpsi

R_1	x	R_2	Multiplier m	mR_2
1.0	40.000 000	0.367 879	1	0.367 879
0.9	45.056 404	0.103 627	4	0.414 508
0.8	46.347 480	0.069 086	2	0.138 172
0.7	47.316 865	0.049 850	4	0.199 400
0.6	48.158 264	0.036 986	2	0.073 972
0.5	48.948 810	0.027 582	4	0.110 328
0.4	49.738 564	0.020 321	2	0.040 642
0.3	50.578 627	0.014 483	4	0.057 932
0.2	51.551 239	0.009 614	2	0.019 228
0.1	52.875 447	0.005 338	4	0.021 352
0.0		0	1	0

$$\Sigma mR_2 = 1.443\ 413$$

The means of the strength S and the stress σ are

$$\overline{S} = 40 + (50 - 40)\ \Gamma(1 + 1/3.3) = 40 + (50 - 40)(0.8970) = 48.97 \text{ kpsi}$$

$$\overline{\sigma} = 30 + (40 - 30)\ \Gamma(1 + 1/2) = 30 + (40 - 30)(0.8862) = 38.86 \text{ kpsi}$$

The design factor associated with a reliability of 0.952 is $\overline{n} = 48.57/38.86 = 1.25$. Since the distribution of the design factor as a quotient of two Weibull variates is not known, discovering the design factor corresponding to a reliability goal of (say) 0.999 becomes an iterative process, with the previous tabular integration becoming part of a root-finding process, quite tractable using a computer.

The strength distribution reflects the result of data reduction and distributional description found to be robust. Strength distributions from historical ensembles, particularly in fatigue, tend to be lognormal. Stress distributions reflect loading and geometry. Machine parts often exhibit geometries with coefficients of variation that are very small compared with that of the load. Additional useful information is to be found in the technical content of more specialized chapters and in the literature.

2.7 NUMBERS

Engineering calculations are a blend of

- Mathematical constants, such as π or e
- Toleranced dimensions
- Measurement numbers
- Mathematical functions (themselves approximate)
- Unit conversion constants
- Mechanically generated digits from calculators and computers
- Rule-of-thumb numbers

A mixture of all types of numbers can be present. It is prudent to treat all numbers as incomplete numbers, avoid serious loss of precision by using sufficient computational digits, round for brevity, and make no significant number inferences. It is well to review the kinds of numbers.

The set of all *integers* is the set of the counting numbers $1, 2, 3, \ldots$, augmented by negative integers and zero. The set of all *rational numbers m/n* is constructed from the integers (dividing by zero excepted). The set of all *real numbers* is constructed by adding limits of all bounded monotone increasing sequences (*irrational numbers*) to the set of rational numbers. Each set of numbers contains the previous set. Each point on a *number line* corresponds to a real number. The display of a real number often has the problem of economy of notation. If the true number to be expressed is $\sqrt{2} = 1.414\ 213\ 562 \ldots$, an *approximate number,* say 1.414, has a value which approximates the true number $\sqrt{2}$. It is given without qualification and is useful to someone for some purpose. A *significant number* is a number that does not differ from the true number by more than one-half in the last recorded digit. The number 1.41 is a significant number corresponding to $\sqrt{2}$ and is bounded by *range numbers* computable from the ± 0.005 implied, that is, $\sqrt{2}$ is contained in the interval $1.41 - 0.005 \leq \sqrt{2} \leq 1.41 + 0.005$ *for certain.* If the first digit to be dropped is 0 through 4, the number is simply truncated. If the first digit to be dropped is 5 through 9, the preceding digit is increased by 1. Thus 1.414 as a significant number representing $\sqrt{2}$ states that the true value of $\sqrt{2}$ lies in the interval 1.414 ± 0.0005. The numbers generated by carrying out the addition and subtraction are range numbers between which the true value of $\sqrt{2}$ is certain to lie.

A significant number is a special form of an *approximation-error number.* An approximation to the true value of $\sqrt{2}$ can be written as 1.414 ± 0.0003 with the error explicitly declared. Again, the range numbers generated by carrying out the explicitly displayed addition and subtraction bound the true value.

Another form of number is the *incomplete number.* It can be formed from the true value of the number simply by truncating after a prescribed number of digits. Nearly all who take their arithmetic seriously use incomplete numbers. Every incomplete number encountered in a calculation is treated as exact. Computational precision is assured by selecting the number of computational digits appropriately larger (double-precision rather than single-precision, for example) than the results require.

When one encounters random variables, certain properties of the distribution will be useful, such as the mean, standard deviation, etc. Range numbers as such do not exist here. The use of significant or approximation-error numbers is inappropriate and misleading. Again, incomplete numbers are used for the distribution parameters and in describing their *confidence bounds.*

Many estimation procedures used by engineers are replete with numbers. Here are some used to represent π:

3.14	(an approximate number)
3.142	(a significant number)
$3.142 \pm 0.000\ 41$	(an approximation-error number)
3.141	(an incomplete number)

The incomplete number, the significant number, and the approximate number have no tag or flag that identifies them as to type, and they are indistinguishable without qualification. In the area of machine computation and in the area of numbers resulting from measurement, significant, approximate, and approximation-error numbers are not useful. It is instructive to note that mathematical tables such as trigonometric tables are arrays of significant numbers; however, significant numbers had no role in their development, as incomplete numbers were used. Numbers that arise from measurement, such as the ultimate tensile strength of a 1020 cold-rolled steel, are properly expressed as $N(\mu, \sigma)$ or $LN(\mu, \sigma)$. Now μ is the incomplete number representing an unbiased estimate of the population mean, and σ is the incomplete number representing an unbiased estimate of the population standard deviation. The distribution is denoted by the symbol preceding the parentheses. There are no range numbers, for they contradict the entire notion of a distributed variable. Any qualification on the central tendency (the mean μ) is addressed by the standard deviation and distributional information, which give rise to confidence limits at some stated probability. These confidence limits perform the role of range numbers, in a way. All are incomplete numbers.

Since incomplete numbers are best for serious arithmetic and are used as random-variable descriptors, thoughtful engineers use them almost exclusively. They presume that any unqualified number encountered is an incomplete number. The hand-held calculator uses incomplete numbers for its arithmetic. The display option allows the user to show a *rounded* number representing the incomplete number in the register. No significance is implied, and none should be inferred. Calculations are often displayed in steps, and intermediate rounded results are recorded on paper. These results should not be reentered for the next calculation step; instead, they should be retained in calculator memory so that the full incomplete numbers are used in every calculation, and the number of correct digits in the resulting incomplete number is maximized.

Equation (2.20) particularized for the standard R. R. Moore specimen diameter gives $\phi_{0.30} \sim LN(0.506, 0.070)$. The companion normal has a mean of $\bar{y}$ and a standard deviation s_y of

$$\bar{y} = \ln 0.506 - \ln \sqrt{1 + (0.070/0.506)^2} = -0.690\ 697$$

$$s_y = \sqrt{\ln [1 + (0.070/0.506)^2]} = 0.137\ 685$$

The two-tailed confidence interval on $\bar{y}$ at the 0.99 confidence level is $y \pm 2.576(0.137\ 685)/\sqrt{133}$ or $-0.721\ 451 \le \bar{y} \le -0.659\ 943$ and $\exp(-0.721\ 451) \le \phi_{0.030} \le \exp(-0.659\ 943)$ or $0.4860 \le \phi_{0.030} \le 0.5169$. It would be misleading to consider the mean of the fatigue ratio as 0.506 to three digits or even to call it 0.5 because it is not known to three significant digits. Indeed, the digit 5 may not be correct. Dispersion in the mean is addressed by the standard deviation of 0.070. The unbiased estimator of the mean is 0.506, a rounded incomplete number, and the meaningful qualification is $0.4860 \le \phi_{0.030} \le 0.5169$ at the 0.99 confidence level.

REFERENCES

2.1 *Metals Handbook,* vol. 1, 8th ed., American Society for Testing and Materials, Philadelphia, 1961.

2.2 J. E. Shigley and C. R. Mischke, *Mechanical Engineering Design,* 5th ed., McGraw-Hill, New York, 1989.

2.3 J. B. Kennedy and A. M. Neville, *Basic Statistical Methods for Engineers and Scientists,* 3d ed., Harper and Row, Publishers, New York, 1986.

2.4 C. R. Mischke, "Fitting Weibull Strength Data and Applying It to Stochastic Mechanical Design," *Transactions of the A.S.M.E., Journal of Mechanical Design,* vol. 114, March 1992, pp. 35–41.

2.5 C. R. Mischke, "A Distribution-Independent Plotting Rule for Ordered Failures," *Transactions of the A.S.M.E., Journal of Mechanical Design,* vol. 104, no. 3, July 1982, pp. 593–597.

2.6 E. B. Haugen, *Probabilistic Mechanical Design,* Wiley-Interscience, New York, 1980.

2.7 C. R. Mischke, *Mathematical Model Building,* 2d rev. ed., Iowa State University Press, Ames, 1980.

CHAPTER 3
MEASUREMENT AND INFERENCE

Jerry Lee Hall, Ph.D., P.E.
Professor of Mechanical Engineering
Iowa State University
Ames, Iowa

3.1 THE MEASUREMENT PROBLEM

The essential purpose and basic function of all branches of engineering is design. Design begins with the recognition of a need and the conception of an idea to meet that need. One may then proceed to design equipment and processes of all varieties to meet the required needs. Testing and experimental design are now considered a necessary design step integrated into other rational procedures. Experimentation is often the only practical way of accomplishing some design tasks, and this requires measurement as a source of important and necessary information.

To measure any quantity of interest, information or energy must be transferred from the source of that quantity to a sensing device. The transfer of information can be accomplished only by the corresponding transfer of energy. Before a sensing device or transducer can detect the signal of interest, energy must be transferred to it from the signal source. Because energy is drawn from the source, the very act of measurement alters the quantity to be determined. In order to accomplish a measurement successfully, one must minimize the energy drawn from the source or the measurement will have little meaning. The converse of this notion is that without energy transfer, no measurement can be obtained.

The objective of any measurement is to obtain the most representative value $\bar{x}$ for the item measured along with a determination of its uncertainty or precision w_x. In this regard one must understand what a measurement is and how to properly select and/or design the component transducers of the measurement system. One must also understand the dynamic response characteristics of the components of the resulting measurement system in order to properly interpret the readout of the measuring system. The measurement system must be calibrated properly if one is to obtain accurate results. A measure of the repeatability or precision of the measured variable as well as the accuracy of the resulting measurement is important. Unwanted information or "noise" in the output must also be considered when using the measurement system. Until these items are considered, valid data cannot be obtained.

Valid data are defined as those data which support measurement of the most representative value of the desired quantity and its associated precision or uncertainty. When calculated quantities employ measured parameters, one must naturally ask how the precision or uncertainty is propagated to any calculated quantity. Use of appropriate propagation-of-uncertainty equations can yield a final result and its associated precision or uncertainty. Thus the generalized measurement problem requires consideration of the measuring system and its characteristics as well as the statistical analysis necessary to place confidence in the resulting measured quantity. The considerations necessary to accomplish this task are illustrated in Fig. 3.1.

First, a statement of the variables to be measured along with their probable magnitude, frequency, and other pertinent information must be formulated. Next, one brings all the knowledge of fundamentals to the measurement problem at hand. This includes the applicable electronics, engineering mechanics, thermodynamics, heat transfer, economics, etc. One must have an understanding of the variable to be measured if an effective measurement is to be accomplished. For example, if a heat flux is to be determined, one should understand the aspects of heat-energy transfer before attempting to measure entities involved with this process.

Once a complete understanding of the variable to be measured is obtained and the environment in which it is to be measured is understood, one can then consider the necessary characteristics of the components of the measurement system. This would include response, sensitivity, resolution, linearity, and precision. Consideration of these items then leads to selection of the individual instrumentation components, including at least the detector-transducer element, the signal-conditioning element, and a readout element. If the problem is a control situation, a feedback transducer would also be considered. Once the components are selected or specified, they must be coupled to form the generalized measuring system. Coupling considerations to determine the isolation characteristics of the individual transducer must also be made.

Once the components of the generalized measurement system are designed (specified), one can consider the calibration technique necessary to ensure accuracy of the measuring system.

Energy can be transferred into the measuring system by coupling means not at the input ports of the transducer. Thus all measuring systems interact with their environment, so that some unwanted signals are always present in the measuring system. Such "noise" problems must be considered and either eliminated, minimized, or reduced to an acceptable level.

If proper technique has been used to measure the variable of interest, then one has accomplished what is called a *valid measurement.* Considerations of probability and statistics then can result in determination of the precision or uncertainty of the measurement. If, in addition, calculations of dependent variables are to be made from the measured variables, one must consider how the uncertainty in the measured variables propagates to the calculated quantity. Appropriate propagation-of-uncertainty equations must be used to accomplish this task.

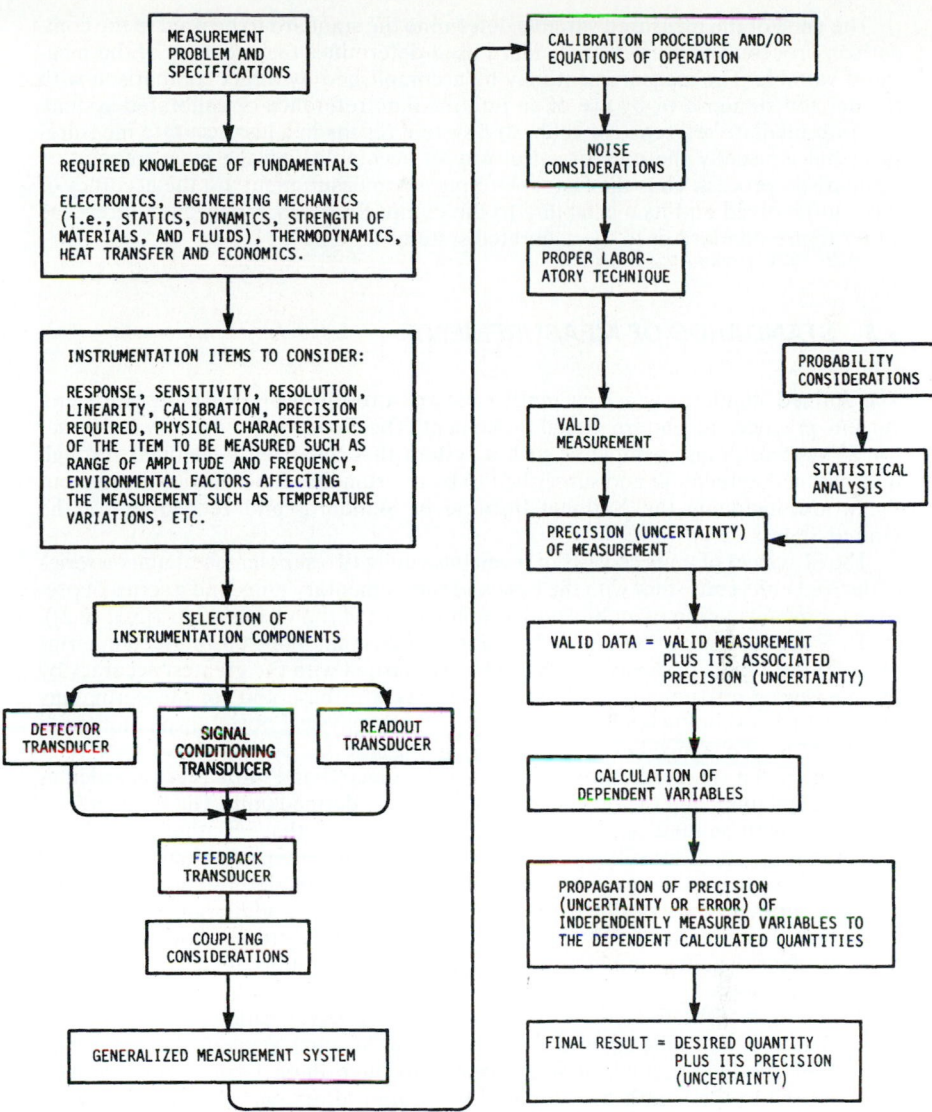

FIGURE 3.1 The generalized measurement task.

3.2 DEFINITION OF MEASUREMENT

A *measurement* is the process of comparing an unknown quantity with a predefined standard. For a measurement to be *quantitative,* the predefined standard must be accurate and reproducible. The standard must also be accepted by international agreement for it to be useful worldwide.

The units of the measured variable determine the standard to be used in the comparison process. The particular standard used determines the accuracy of the measured variable. The measurement may be accomplished by direct comparison with the defined standard or by use of an intermediate reference or calibrated system. The intermediate reference or calibrated system results in a less accurate measurement but is usually the only practical way of accomplishing the measurement or comparison process. Thus the factors limiting any measurement are the accuracy of the unit involved and its availability to the comparison process through reference either to the standard or to the calibrated system.

3.3 STANDARDS OF MEASUREMENT

The defined standards which currently exist are a result of historical development, current practice, and international agreement. The *Système International d'Unités* (or SI system) is an example of such a system that has been developed through international agreement and subscribed to by the standard laboratories throughout the world, including the National Institute of Standards and Technology of the United States.

The SI system of units consists of seven base units, two supplemental units, a series of derived units consistent with the base and supplementary units, and a series of prefixes for the formation of multiples and submultiples of the various units ([3.1], [3.2]).

The important aspect of establishing a standard is that it must be defined in terms of a physical object or device which can be established with the greatest accuracy by the measuring instruments available. The standard or base unit for measuring any physical entity should also be defined in terms of a physical object or phenomenon which can be reproduced in any laboratory in the world.

Of the seven standards, three are arbitrarily selected and thereafter regarded as fundamental units, and the others are independently defined units. The fundamental units are taken as mass, length, and time, with the idea that all other mechanical parameters can be derived from these three. These fundamental units were natural selections because in the physical world one usually weighs, determines dimensions, or times various intervals. Electrical parameters require the additional specification of current. The independently defined units are temperature, electric current, the amount of a substance, and luminous intensity. The definition of each of the seven basic units follows.

At the time of the French Revolution, the unit of *length*, called a *meter* (m), was defined as one ten-millionth of the distance from the earth's equator to the earth's pole along the longitudinal meridian passing through Paris, France. This standard was changed to the length of a standard platinum-iridium bar when it was discovered that the bar's length could be assessed more accurately (to eight significant digits) than the meridian. Today the standard meter is defined to be the length equal to 1 650 763.73 wavelengths in a vacuum of the orange-red line of krypton isotope 86.

The unit of *mass*, called a *kilogram* (kg), was originally defined as the mass of a cubic decimeter of water. The standard today is a cylinder of platinum-iridium alloy kept by the International Bureau of Weights and Measures in Paris. A duplicate with the U.S. National Bureau of Standards serves as the mass standard for the United States. This is the sole base unit still defined by an artifact.

Force is taken as a derived unit from Newton's second law. In the SI system, the unit of *force* is the *newton* (N), which is defined as that force which would give a kilogram mass an acceleration of one meter per second per second.

The unit interval of *time,* called a *second,* is defined as the duration of 9 192 631 770 cycles of the radiation associated with a specified transition of the cesium 133 atom.

The unit of *current,* called the *ampere* (A), is defined as that current flowing in two parallel conductors of infinite length spaced one meter apart and producing a force of 2×10^{-7} N per meter of length between the conductors.

The unit of *luminous intensity,* called the *candela,* is defined as the luminous intensity of one six-hundred-thousandth of a square meter of a radiating cavity at the temperature of freezing platinum (2042 K) under a pressure of 101 325 N/m^2.

The *mole* is the *amount of substance* of a system which contains as many elementary entities as there are carbon atoms in 0.012 kg of carbon 12.

Unlike the other standards, temperature is more difficult to define because it is a measure of the internal energy of a substance, which cannot be measured directly but only by relative comparison using a third body or substance which has an observable property that changes directly with temperature. The comparison is made by means of a device called a *thermometer,* whose scale is based on the *practical international temperature scale,* which is made to agree as closely as possible with the theoretical thermodynamic scale of temperature. The *thermodynamic scale of temperature* is based on the reversible Carnot heat engine and is an ideal temperature scale which does not depend on the thermometric properties of the substance or object used to measure the temperature.

The practical temperature scale currently used is based on various fixed temperature points along the scale as well as interpolation equations between the fixed temperature points. The devices to be used between the fixed temperature points are also specified between certain fixed points on the scale. See Ref. [3.3] for a more complete discussion of the fixed points used for the standards defining the practical scale of temperature.

3.4   THE MEASURING SYSTEM

A measuring system is made up of devices called *transducers.* A transducer is defined as an energy-conversion device [3.4]. A configuration of a generalized measuring system is illustrated in Fig. 3.2.

The purpose of the detector transducer in the generalized system is to sense the quantity of interest and to transform this information (energy) into a form that will be acceptable by the signal-conditioning transducer. Similarly, the purpose of the signal-conditioning transducer is to accept the signal from the detector transducer and to modify this signal in any way required so that it will be acceptable to the readout transducer. For example, the signal-conditioning transducer may be an amplifier, an integrator, a differentiator, or a filter.

The purpose of the readout transducer is to accept the signal from the signal-conditioning transducer and to present an interpretable output. This output may be in the form of an indicated reading (e.g., from the dial of a pressure gauge), or it may be in the form of a strip-chart recording, or the output signal may be passed to either a digital processor or a controller. With a control situation, the signal transmitted to the controller is compared with a desired operating point or set point. This comparison dictates whether or not the feedback signal is propagated through the feedback transducer to control the source from which the original signal was measured.

An *active transducer* transforms energy between its input and output without the aid of an auxiliary energy source. Common examples are thermocouples and piezoelectric crystals. A *passive transducer* requires an auxiliary energy source (AES) to

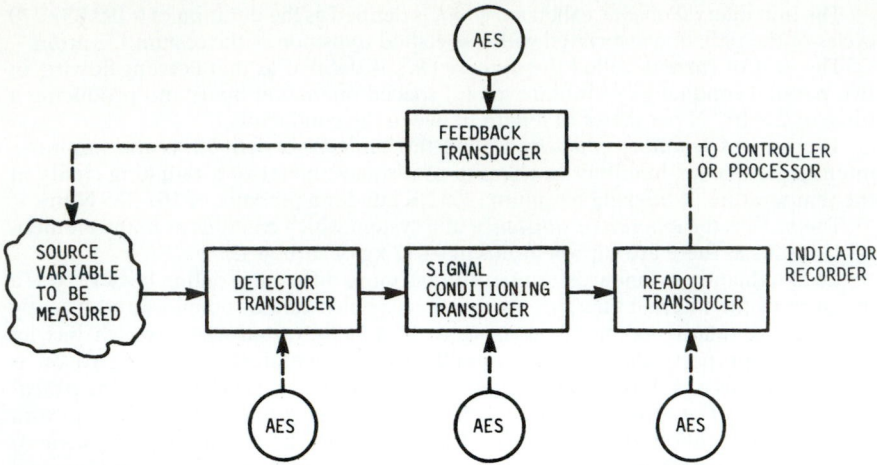

FIGURE 3.2 The generalized measurement system. AES indicates auxiliary energy source, dashed line indicates that the item may not be needed.

carry the input signal through to the output. Measuring systems using passive transducers for the detector element are sometimes called *carrier systems.* Examples of transducers requiring such an auxiliary energy source are impedance-based transducers such as strain gauges, resistance thermometers, and differential transformers. All impedance-based transducers require auxiliary energy to carry the information from the input to the output and are therefore passive transducers.

The components which make up a measuring system can be illustrated with the ordinary thermometer, as shown in Fig. 3.3. The thermometric bulb is the detector or sensing transducer. As heat energy is transferred into the thermometric bulb, the thermometric fluid (for example, mercury or alcohol) expands into the capillary tube of the thermometer. However, the small bore of the capillary tube provides a signal-conditioning transducer (in this case an amplifier) which allows the expansion of the thermometric fluid to be amplified or magnified. The readout in this case is the comparison of the length of the filament of thermometric fluid in the capillary tube with the temperature scale etched on the stem of the thermometer.

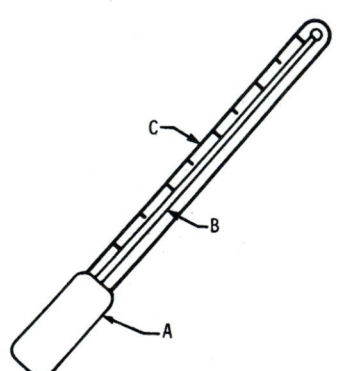

FIGURE 3.3 Components of a simple measuring system. A, detector transducer (thermometer bulb with thermometric fluid); B, signal conditioning stage (amplifier); C, readout stage (indicator).

Another example of an element of a measuring system is the Bourdon-tube pressure gauge. As pressure is applied to the Bourdon tube (a curved tube of elliptical cross section), the curved tube tends to straighten out. A mechanical linkage attached to the end of the Bourdon tube engages a gear of pinion, which in turn is attached to an indicator needle. As the Bourdon tube straightens, the mechanical linkage to the gear on the indicator needle moves, causing the gear and indicating needle to rotate, giving an indication of a change in pressure on the dial of the gauge. The magnitude of the change in pressure is indicated by a pressure scale marked on the face of the pressure gauge.

The accuracy of either the temperature measurement or the pressure measurement previously indicated depends on how accurately each measuring instrument is calibrated. The values on the readout scales of the devices can be determined by means of comparison (calibration) of the measuring device with a predefined standard or by a reference system which in turn has been calibrated in relation to the defined standard.

3.5 CALIBRATION

The *process of calibration* is comparison of the reading or output of a measuring system to the value of known inputs to the measuring system. A complete calibration of a measuring system would consist of comparing the output of the system to known input values over the complete range of operation of the measuring device. For example, the calibration of pressure gauges is often accomplished by means of a device called a *dead-weight tester* where known pressures are applied to the input of the pressure gauge and the output reading of the pressure gauge is compared to the known input over the complete operating range of the gauge.

The type of calibration signal should simulate as nearly as possible the type of input signal to be measured. A measuring system to be used for measurement of dynamic signals should be calibrated using known dynamic input signals. Static, or level, calibration signals are not proper for calibration of a dynamic measurement system because the natural dynamic characteristics of the measurement system would not be accounted for with such a calibration. A typical calibration curve for a general transducer is depicted in Fig. 3.4. It might be noted that the sensitivity of the measuring system can be obtained from the calibration curve at any level of the input signal by noting the relative change in the output signal due to the relative change in the input signal at the operating point.

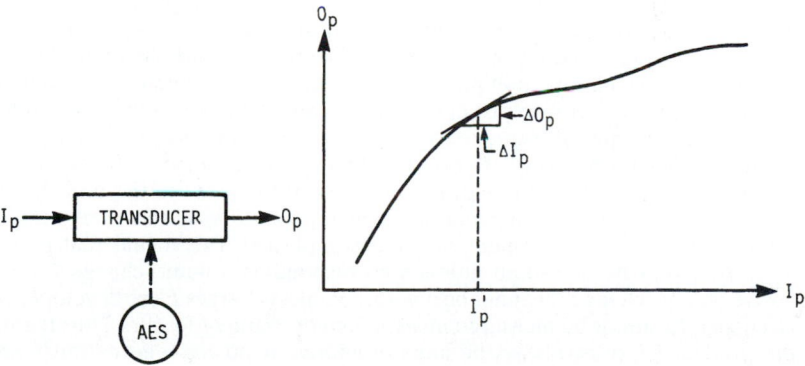

FIGURE 3.4 Typical calibration curve. Sensitivity at $I_p' = (\Delta O_p / \Delta I_p)$.

3.6 DESIGN OF THE MEASURING SYSTEM

The design of a measuring system consists of selection or specification of the transducers necessary to accomplish the detection, transmission, and indication of the desired variable to be measured. The transducers must be connected to yield an interpretable output so that either an individual has an indication or recording of the information or a controller or processor can effectively use the information at the output of the measuring system. To ensure that the measuring system will perform the measurement of the specified variable with the fidelity and accuracy required of the test, the *sensitivity, resolution, range,* and *response* of the system must be known. In order to determine these items for the measurement system, the individual transducer characteristics and the loading effect between the individual transducers in the measuring system must be known. Thus by knowing individual transducer characteristics, the system characteristics can be predicted. If the individual transducer characteristics are not known, one must resort to testing the complete measuring system in order to determine the desired characteristics.

The system characteristics depend on the mathematical order (for example, first-order, second-order, etc.) of the system as well as the nature of the input signal. If the measuring system is a first-order system, its response will be significantly different from that of a measuring system that can be characterized as a second-order system. Furthermore, the response of an individual measuring system of any order will be dependent on the type of input signal. For example, the response characteristics of either a first- or second-order system would be different for a step input signal and a sinusoidal input signal.

3.6.1 Energy Considerations

In order for a measurement of any item to be accomplished, energy must move from a source to the detector-transducer element. Correspondingly, energy must flow from the detector-transducer element to the signal-conditioning device, and energy must flow from the signal-conditioning device to the readout device in order for the measuring system to function to provide a measurement of any variable. Energy can be viewed as having intensive and extensive or primary and secondary components. One can take the primary component of energy as the quantity that one desires to detect or measure. However, the primary quantity is impossible to detect unless the secondary component of energy accompanies the primary component. Thus a force cannot be measured without an accompanying displacement, or a pressure cannot be measured without a corresponding volume change. Note that the units of the primary component of energy multiplied by the units of the secondary component of energy yield units of energy or power (an energy rate). Figure 3.5 illustrates both the active and passive types of transducers with associated components of energy at the input and output terminals of transducers. In Fig. 3.5 the primary component of energy I_p is the quantity that one desires to sense at the input to the transducer. A secondary component I_s accompanies the primary component, and energy must be transferred before a measurement can be accomplished. This means that pressure changes I_p cannot be measured unless a corresponding volume change I_s occurs. Likewise, voltage change I_p cannot be measured unless charges I_s are developed, and force change I_p cannot be measured unless a length change I_s occurs. Thus the units of the product $I_p I_s$ must always be units of energy or power (energy rate). Some important transducer characteristics can now be defined in terms of the energy

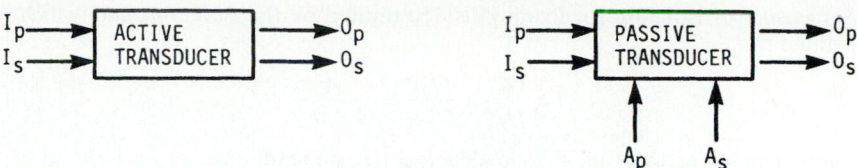

FIGURE 3.5   Energy components for active and passive transducers.

components shown in Fig. 3.5. These characteristics may have both magnitude and direction, so that generally the characteristics are complicated in mathematical nature. A more complete discussion of the following characteristics is contained in Stein [3–4].

3.6.2 Transducer Characteristics

Acceptance ratio of a transducer is defined in Eq. (3.1) as the ratio of the change in the primary component of energy at the transducer input to the change in the secondary component at the transducer input. It is similar to an input impedance for a transducer with electric energy at its input:

$$A = \frac{\Delta I_p}{\Delta I_s} \tag{3.1}$$

Emission ratio of a transducer is defined in Eq. (3.2) as the ratio of the change in the primary component of energy at the transducer output to the change in the secondary component of energy at the transducer output. This is similar to output impedance for a transducer with electric energy at its output:

$$E = \frac{\Delta O_p}{\Delta O_s} \tag{3.2}$$

Transfer ratio is defined in Eq. (3.3) as the ratio of the change in the primary component of energy at the transducer output to the change in the primary component of energy at the transducer input:

$$T = \frac{\Delta O_p}{\Delta I_p} \tag{3.3}$$

Several different types of transfer ratios may be defined which involve any output component of energy with any input component of energy. However, the main transfer ratio involves the primary component of energy at the output and the primary component of energy at the input. The main transfer ratio is similar to the *transfer function,* which is defined as that function describing the mathematical operation that the transducer performs on the input signal to yield the output signal at some operating point. The transfer ratio at a given operating point or level of input signal is also the *sensitivity* of the transducer at that operating point.

When two transducers are connected, they will interact, and energy will be transferred from the source, or first, transducer to the second transducer. When the transfer of energy from the source transducer is zero, it is said to be *isolated* or *unloaded.*

A measure of isolation (or loading) is determined by the *isolation ratio,* which is defined by

$$I = \frac{O_{p,a}}{O_{p,i}} = \frac{O_{p,L}}{O_{p,NL}} = \frac{A}{A + |E_s|} \tag{3.4}$$

where *a* means actual; *i,* ideal; *L,* loaded; and *NL,* no load.

When the *emission ratio* E_s from the source transducer is zero, the isolation ratio becomes unity and the transducers are isolated. The definition of an *infinite source* or a *pure source* is one that has an emission ratio of zero. The concept of an emission ratio approaching zero is that for a fixed value of the output primary component of energy O_p, the secondary component of energy O_s must be allowed to be as large as is required to maintain the level of O_p at a fixed value. For example, a pure voltage source of 10 V (O_p) must be capable of supplying any number (this may approach infinity) of charges (O_s) in order to maintain a voltage level of 10 V. Likewise, the pure source of force (O_p) must be capable of undergoing any displacement (O_s) required in order to maintain the force level at a fixed value.

Example 1.   The transfer ratio (measuring-system sensitivity) of the measuring system shown in Fig. 3.6 is to be determined in terms of the individual transducer transfer ratios and the isolation ratios between the transducers.
Solution

$$T = \frac{O_3}{I_1} = \frac{O_3}{O_{2,L}} \frac{O_{2,L}}{O_{2,NL}} \frac{O_{2,NL}}{O_{1,L}} \frac{O_{1,L}}{O_{1,NL}} \frac{O_{1,L}}{I_1} = T_{32} I_3 \, T_{21} I_2 \, T_1$$

$$= \text{(product of transfer ratios) (product of isolation ratios)}$$

3.6.3 Sensitivity

The *sensitivity* is defined as the change in the output signal relative to the change in the input signal at an operating point *k.* Sensitivity *S* is given by

$$S = \lim_{\Delta I_p \to 0} \left(\frac{\Delta O_p}{\Delta I_p} \right)_{I_p = k} = \left(\frac{dO_p}{dI_p} \right)_k \tag{3.5}$$

3.6.4 Resolution

The *resolution* of a measuring system is defined as the smallest change in the input signal that will yield an interpretable change in the output of the measuring system at some operating point. Resolution *R* is given by

$$R = \Delta I_{p,\min} = \frac{\Delta O_{p,\min}}{S} \tag{3.6}$$

FIGURE 3.6   Measuring-system sensitivity.

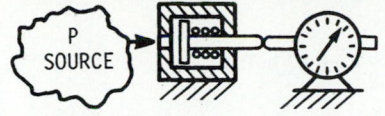

FIGURE 3.7 Pressure transducer in the form of a spring-loaded piston and a dial indicator.

It can be determined by taking the smallest change in the output signal which would be interpretable (as decided by the observer) and dividing by the sensitivity at that operating point.

Example 2. A pressure transducer is to be made from a spring-loaded piston in a cylinder and a dial indicator, as shown in Fig. 3.7. Known information concerning each element is also listed below:

Pneumatic cylinder

> Spring deflection factor = 14.28 lbf/in = K
>
> Cylinder bore = 1 in
>
> Piston stroke = ½ in

Dial indicator

> Spring deflector factor = 1.22 lbf/in = k
>
> Maximum stroke of plunger = 0.440 in
>
> Indicator dial has 100 equal divisions per 360°
>
> Each dial division represents a plunger deflection of 0.001 in

The following items are determined:

1. Block diagram of measuring system showing all components of energy (see Fig. 3.8)
2. Acceptance ratio of pneumatic cylinder:

$$A_{PC} = \frac{\Delta I_p}{\Delta I_s} = \frac{P}{V} = \frac{F/A}{AL} = \frac{K}{A^2} = \frac{14.28(16)}{\pi^2} = 23.1 \text{ psi/in}^2$$

3. Emission ratio of pneumatic cylinder:

$$E_{PC} = \frac{\Delta O_p}{\Delta O_s} = \frac{L}{F} = \frac{1}{K} = \frac{1}{14.28} = 0.070 \text{ in/lbf}$$

4. Transfer ratio of pneumatic cylinder:

$$T_{PC} = \frac{\Delta O_p}{\Delta I_p} = \frac{L}{P} = \frac{LA}{F} = \frac{A}{K} = \frac{\pi}{4(14.28)} = 0.055 \text{ in/psi}$$

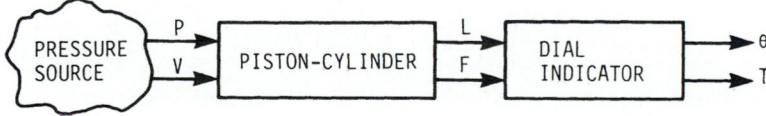

FIGURE 3.8 Pressure-transducer block diagram.

5. Acceptance ratio of dial indicator:

$$A_{DI} = \frac{\Delta I_p}{\Delta I_s} = \frac{L}{F} = \frac{1}{k} = \frac{1}{1.22} = 0.82 \text{ in/lbf}$$

6. Transfer ratio of dial indicator:

$$T_{DI} = \frac{\Delta O_p}{\Delta I_p} = \frac{\theta}{L} = (3.6° \text{ per division})/(0.001 \text{ in per division})$$
$$= 3600°/\text{in (or 1000 divisions/in)}$$

7. Isolation ratio between pneumatic cylinder and dial indicator:

$$I = \frac{A_{DI}}{A_{DI} + E_{PC}} = \frac{1/k}{1/k + 1/K} = \frac{0.82}{0.82 + 0.07} = 0.921$$

8. System sensitivity in dial divisions per psi:

$$S = \frac{\text{output}}{\text{input}} = \frac{DI \text{ output}}{DI \text{ input}} \times \frac{DI \text{ input}}{PC \text{ output}} \times \frac{PC \text{ output}}{PC \text{ input}}$$

$$= T_{DI} I T_{PC} = 0.055(0.921)(1000) = 50.7 \text{ divisions/psi}$$

9. Maximum pressure that the measuring system can sense:

$$\text{Maximum input} = \frac{\text{input}}{\text{output}} \times \text{maximum output} = \frac{1}{S}(440 \text{ dial divisions}) = 8.7 \text{ psi}$$

10. Resolution of the measuring system in psi:

$$\text{Minimum input} = \frac{\text{input}}{\text{output}} \times \text{minimum readable output} = \frac{1}{S}(1 \text{ dial division}) = 0.02 \text{ psi}$$

3.6.5 Response

When time-varying signals are to be measured, the dynamic response of the measuring system is of crucial importance. The components of the measuring system must be selected and/or designed such that they can respond to the time-varying input signals in such a manner that the input information is not lost in the measurement process. Several measures of response are important to know if one is to evaluate a measuring system's ability to detect and reproduce all the information in the input signal. Some measures of response involve time alone, whereas other measures of response are more involved. Various measures of response are defined in the following paragraphs.

Amplitude response of a transducer is defined as the ability to treat all input amplitudes uniformly [3.5]. The typical amplitude-response curve determined for either an individual transducer or a complete measuring system is depicted in Fig. 3.9.

A typical amplitude-response specification is as follows:

$$\left| \frac{O_p}{I_p} \right| = M \pm T \qquad I_{p,\min} < I_p < I_{p,\max} \tag{3.7}$$

The amplitude-response specification includes a nominal magnitude ratio M between output and input of the transducer measuring system along with an allowable tolerance T and a specification of the range of the magnitude of the primary input variable I_p over which the amplitude ratio and tolerance are valid.

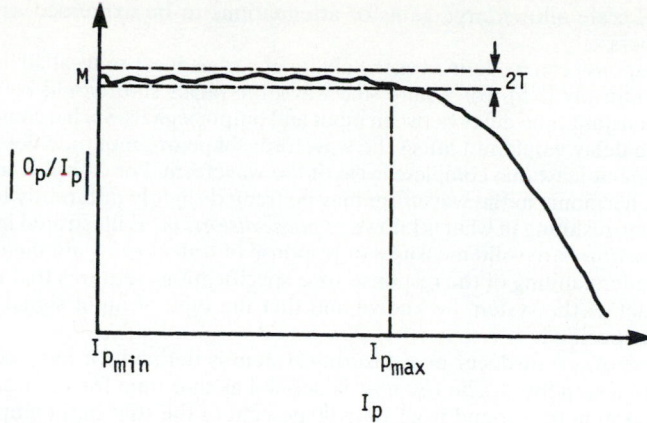

FIGURE 3.9 Typical amplitude-response characteristic.

Frequency response can be defined as the ability of a transducer to treat all input frequencies uniformly [3.5] and can be specified by a frequency-response curve such as that shown in Fig. 3.10. A typical frequency-response specification would be the nominal magnitude ratio M of output to input signals plus or minus some allowable tolerance T specified over a frequency range from the low-frequency limit f_L to the high-frequency limit f_H as follows:

$$\left|\frac{O_p}{I_p}\right| = M \pm T \qquad f_L < f < f_H \tag{3.8}$$

It is the usual practice to use the decibel (dB) rather than the actual magnitude ratio for the ordinate of the frequency-response curve. The decibel, as defined in Eq. (3.9), is used in transducers and measuring systems in specifying frequency response:

$$\text{Decibel} = 20 \log_{10} \frac{O_p}{I_p} \tag{3.9}$$

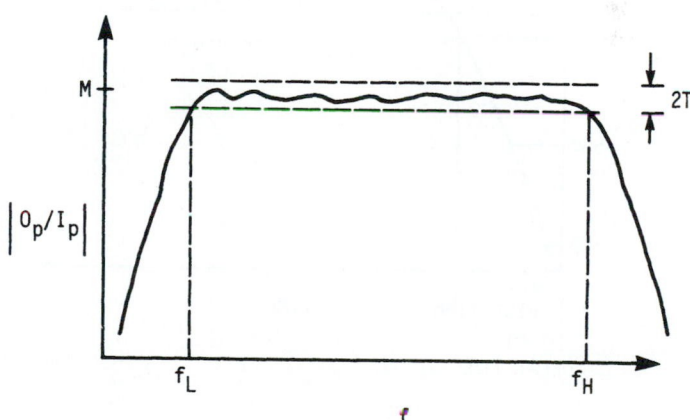

FIGURE 3.10 Typical frequency-response characteristic.

The decibel scale allows large gains or attenuations to be expressed as relatively small numbers.

Phase response can be defined as the ability of a transducer to treat all input-phase relations uniformly [3.5]. For a pure sine wave, the phase shift would be a constant angle or a constant time delay between input and output signals. Such a constant phase shift or time delay would not affect the waveform shape or amplitude determination when viewing at least one complete cycle of the waveform. For complex input waveforms, each harmonic in the waveform may be treated slightly differently in the measuring system, resulting in what is known as *phase distortion,* as illustrated in Ref. [3.5].

Response times are valid measures of response of transducers and measuring systems. An understanding of the response-time specifications requires that the mathematical order of the system be known and that the type of input signal or forcing function be specified.

Rise time of a transducer or measuring system is defined for any order system subjected to a step input. The *rise time* is defined as that time for the transducer or measuring system to respond from 10 to 90 percent of the step-input amplitude and is depicted in Fig. 3.11.

Delay time is another response time which is defined for any order system subjected to a step input. The *delay time* is defined to be that time for the transducer or measuring system to respond from 0 to 50 percent of the step-input amplitude and is depicted in Fig. 3.11.

Time constant is specifically defined for a first-order system subjected to a step input. The *time constant* τ is defined as the time for the transducer or measuring system to respond to 63.2 percent (or $1 - e^{-1}$) of the step-input amplitude. The time constant is specifically illustrated in Fig. 3.12, where the response x of the first-order system to step input x_s is known to be exponential as follows:

$$x = x_s(1 - e^{-t/\tau}) \tag{3.10}$$

When the time t is equal to the time constant τ, the first-order system has responded to 63.2 percent of the step-input amplitude. In a time span equivalent to

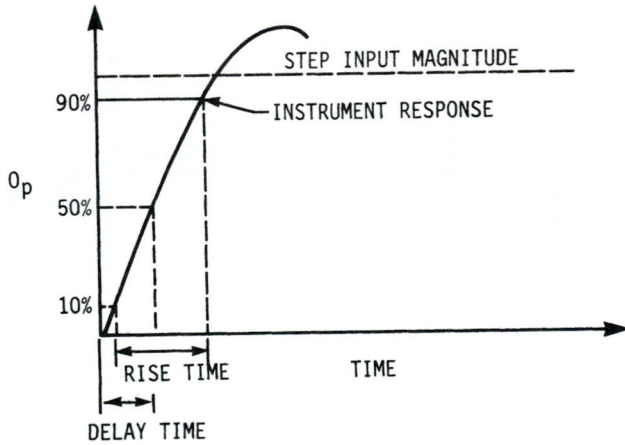

FIGURE 3.11   Rise time and delay time used as response times.

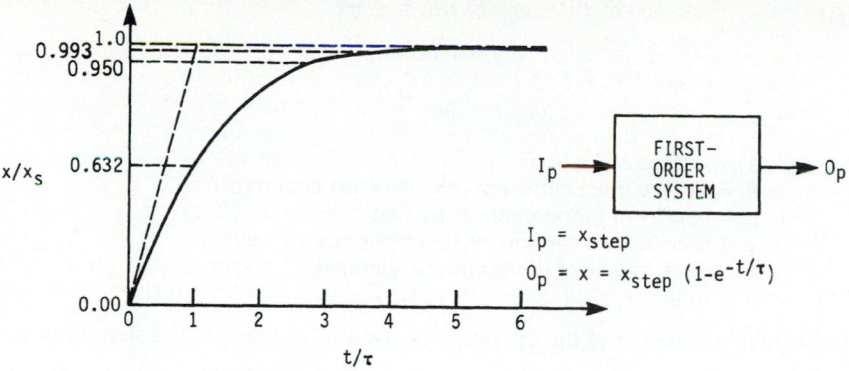

FIGURE 3.12   Response of a first-order system to a step input.

3 time constants, the system has responded to 95.0 percent of the step-input amplitude, and in a time span of 5 time constants, the system has responded to 99.3 percent of the step-input amplitude. Thus for a first-order system subjected to a step
input to yield a correct reading of the input variable, one must wait a time period of
at least 5 time constants in order for the first-order system to respond sufficiently to
give a correct indication of the measured variable.

Transducer Dynamics.   Because of the time delay or phase shift a transducer or
measuring system may have, one must be very careful to ensure that the measuring
system can respond adequately if the input signal to the measuring system is varying
with time. If the time response of the measuring system is inadequate, it may never
read the correct value. Thus if one believes the output indication of the measuring
system to be a reproduction of the actual value of the input (measured) variable
without understanding the dynamics of how the measuring system is responding to
the input signal, a crucial error can be made.

In order to understand dynamic response, one must recognize that the components of the measuring system have natural physical characteristics and that the
measuring system will tend to respond according to these natural characteristics
when perturbed by any external disturbance. In addition, the input signal supplied
to a transducer or measuring system provides a forcing function for that transducer or measuring system. The equation of operation of a transducer is a differential equation whose order is defined as the order of the system. The response of
the system is determined by solving this differential equation of operation according to the type of input signal (forcing function) supplied to the system. If the measuring system is modeled as a linear system, the differential equation of operation
will be ordinary and linear with constant coefficients. This is the type of differential equation that can be solved by well-known techniques. The nature of the solution depends on the nature of the forcing function as well as the nature of the
physical components of the system. For example, the thermometric element of the
temperature-measuring device can be modeled as shown in Fig. 3.13. For this
model,

$$q_{in} = q_{lost} + q_{stored} \ = \ \text{rate of heat energy entering control region}$$

and
$$q_{\text{in}} = hA(T_\infty - T)$$

$$q_{\text{lost}} = 0 \text{ (assumed)}$$

$$q_{\text{stored}} = \rho c v \frac{dT}{dt}$$

where A = surface area
h = surface-film coefficient of convective heat transfer
ρ = density of thermometric element
c = specific heat capacity of thermometric element
T = temperature of thermometric element
t = time

The resulting equation of the operation is given as follows for the step input $x_s = T_\infty - T_o$:

$$T - T_o = (T_\infty - T_o)(1 - e^{-t/\tau}) \tag{3.11}$$

where $\tau = \rho v c / h A$. The response $x = T - T_o$ is shown in Fig. 3.12.

Another example of a first-order system is the electric circuit composed of resistance and capacitive elements or the so-called *RC* circuit. Masses falling in viscous media also follow a similar exponential characteristic.

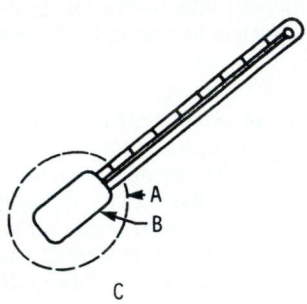

FIGURE 3.13 Thermometric element modeled as a first-order system. A, control region; B, thermometric element at temperature T; C, environment at temperature T_∞.

If the system is characterized by a second-order linear ordinary differential equation, the solution becomes more complex than that for the first-order system. The system behavior depends on the amount of friction or damping in the system. For example, the meter movement of a galvanometer or D'Arsonval movement shown in Fig. 3.14 such as exists in many electrical meters can be modeled as shown in Fig. 3.15. Applying first principles to this model yields the equation of motion

$$\Sigma T = J\ddot{\theta} = T(t) - T_s - T_f$$

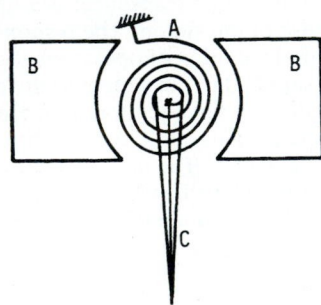

FIGURE 3.14 D'Arsonval movement. A, spring-retained armature; B, field magnets; C, indicating needle.

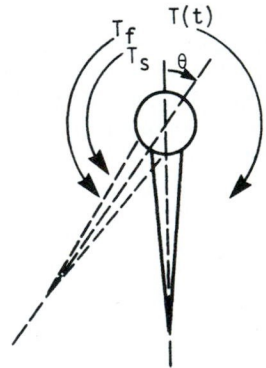

FIGURE 3.15 Torques applied to the D'Arsonval movement.

where $\quad T_s = k\theta$ for torsional damping
$\quad\quad\quad T_f = \sigma\dot\theta$ for viscous friction
$\quad\quad\quad T(t) = $ driving or forcing function

Then
$$J\ddot\theta + \sigma\dot\theta + k\theta = T(t)$$

or
$$\ddot\theta + 2\gamma\omega_n\dot\theta + \omega_n^2\theta = \frac{T(t)}{J} \tag{3.12}$$

where $\quad \omega_n = \sqrt{k/J} = $ natural undamped frequency
$\quad\quad\quad \omega_d = \omega_n\sqrt{1-\gamma^2} = $ natural damped frequency
$\quad\quad\quad \omega_p = \omega_n\sqrt{1-2\gamma^2} = $ frequency at peak of frequency response curve
$\quad\quad\quad \gamma = \sigma/\sigma_c = $ damping ratio
$\quad\quad\quad \sigma_c = \sqrt{4kJ} = $ critical value of damping
$\quad\quad\quad\quad\quad = $ lowest value of damping where no natural oscillation of system occurs

If the damping is modeled as viscous friction, the possible solutions to the equation of motion are given by Eqs. (3.13), (3.14), and (3.15) for the step input. The underdamped solution of Eq. (3.12) is shown in Fig. 3.16.

For $\sigma < 1$ (underdamped),

$$\frac{x}{x_s} = 1 - \{1 - \gamma^2\}^{-1/2}\exp(-\gamma\omega_n t)\sin(\omega_d t + \phi)$$

$$\phi = \tan^{-1}\sqrt{\frac{1-\gamma^2}{\gamma}} \tag{3.13}$$

For $\sigma = 1$ (critical damping),

$$\frac{x}{x_s} = 1 - (1 + \omega_n t)\exp(-\omega_n t) \tag{3.14}$$

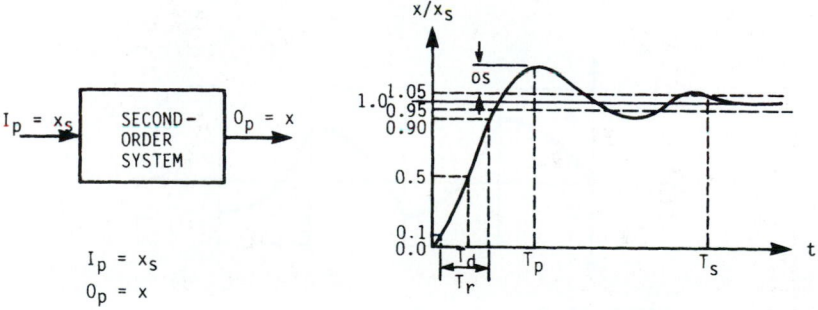

FIGURE 3.16 Response of a second-order system to a step input.

For $\sigma > 1$ (overdamped),

$$\frac{x}{x_s} = 1 - \left(\frac{\beta}{\beta - 1}\right)\left[\exp\left(\frac{-\omega_n t}{\sqrt{\beta}}\right) - \frac{1}{\beta}\exp\left(-\sqrt{\beta}\,\omega_n t\right)\right]$$

$$\beta = \frac{\gamma + \sqrt{\gamma^2 - 1}}{\gamma - \sqrt{\gamma^2 - 1}} \tag{3.15}$$

If the system is underdamped, the response of the transducer or measuring system overshoots the step-input magnitude and the corresponding oscillation occurs with a first-order decay. This type of response leads to additional response specifications which may be used by transducer manufacturers. These specifications include *overshoot OS, peak time T_p, settling time T_s, rise time T_r,* and *delay time T_d* as depicted in Fig. 3.16. If the viscous damping is at the critical value, the measuring system responds up to the step-input magnitude only after a very long period of time. If the damping is more than critical, the response of the measuring system never reaches a magnitude equivalent to the step input. Measuring-system components following a second-order behavior are normally designed and/or selected such that the damping is less than critical. With underdamping the second-order system responds with some time delay and a characteristic phase shift.

If the natural response characteristics of each measuring system are not known or understood, the output reading of the measurement system can be erroneously interpreted. Figure 3.17 illustrates the response of a first-order system to a square-wave input. Note that the system with inadequate time response never yields a valid indication of the magnitude of the step input. Figure 3.18 illustrates a first-order system with time constant adequate ($\tau \ll 1/f$) to yield a valid indication of step-input magnitude. Figure 3.19 illustrates the response of an underdamped second-order system to a square-wave input. A valid indication of the step-input magnitude is obtained after the settling time has occurred.

If the input forcing function is not a step input but a sinusoidal function instead, the corresponding differential equations of motion to the first- and second-order systems are given in Eqs. (3.16) and (3.17), respectively:

$$\dot{x} + \frac{x}{\tau} = A\cos\omega_f t \tag{3.16}$$

where A = amplitude of input signal transformed to units of the response
            variable derivative(s)
      ω_f = frequency of input signal (forcing function)
      τ = time constant

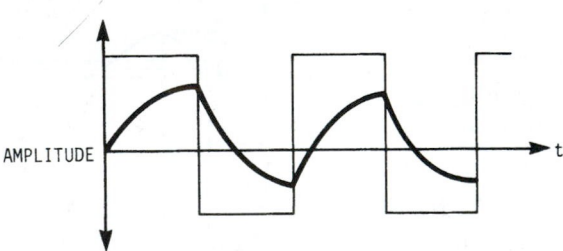

FIGURE 3.17 Response of a first-order system with inadequate response to a square-wave input ($\tau > 1/f$).

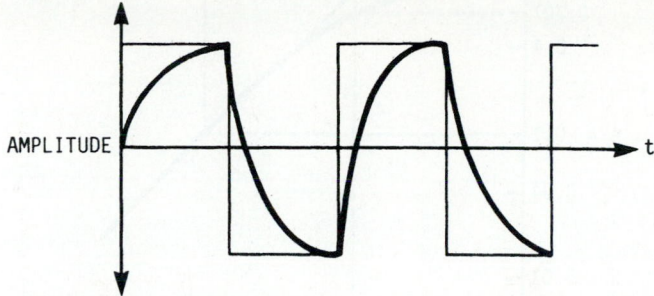

FIGURE 3.18 Response of a first-order system with barely adequate response to a square-wave input ($\tau \ll 1/f$).

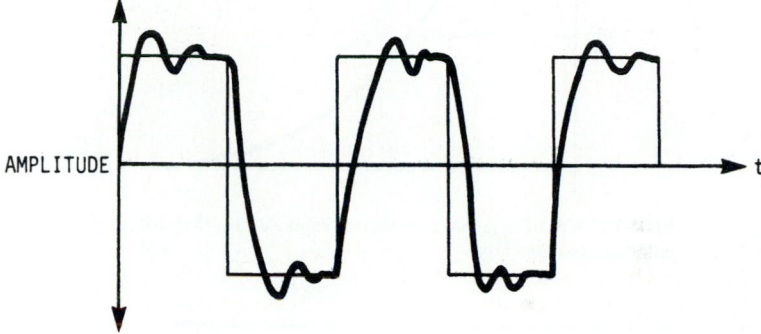

FIGURE 3.19   Response of an underdamped second-order system to a square wave.

$$\ddot{x} + 2\sigma\omega_n\dot{x} + \omega_n^2 x = A \cos \omega_f t \tag{3.17}$$

In addition, the parameters of the steady-state responses of the first- and second-order system are given by Eqs. (3.18) and (3.19), respectively, and are shown in Figs. 3.20 and 3.21. The steady-state solutions are of the form

$$x_{ss} = B \cos (\omega_f t + \phi)$$

where, for the first- and second-order systems, respectively,

$$B_1 = \frac{A}{\sqrt{(\tau\omega_f)^2 + 1}} \qquad \phi_1 = -\tan^{-1}(\tau\omega_f) \tag{3.18}$$

$$B_2 = \frac{A}{\sqrt{[1 - (\omega_f/\omega_n)^2]^2 + (2\gamma\,\omega_f/\omega_n)^2}} \qquad \phi_2 = -\tan^{-1}\frac{2\gamma\,\omega_f/\omega_n}{1 - (\omega_f/\omega_n)^2} \tag{3.19}$$

From these results it can be noted that both the first- and second-order systems, when responding to sinusoidal input functions, experience a magnitude change and a phase shift in response to the input function.

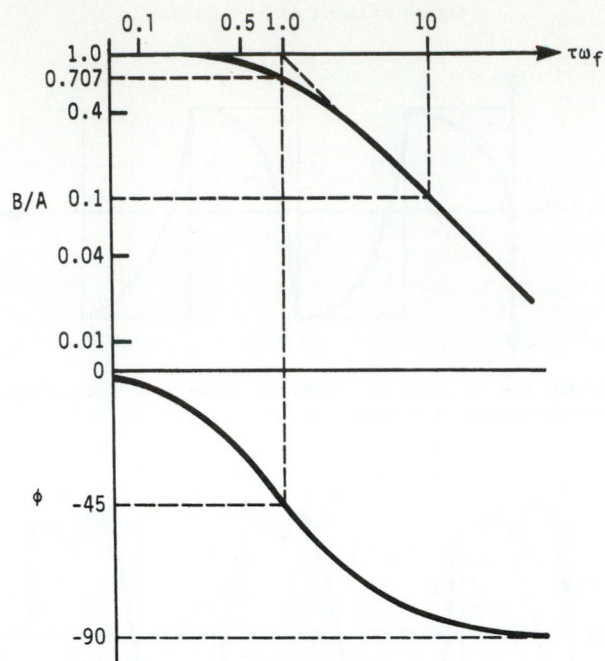

FIGURE 3.20 Frequency and phase response of a first-order system to a sinusoidal input.

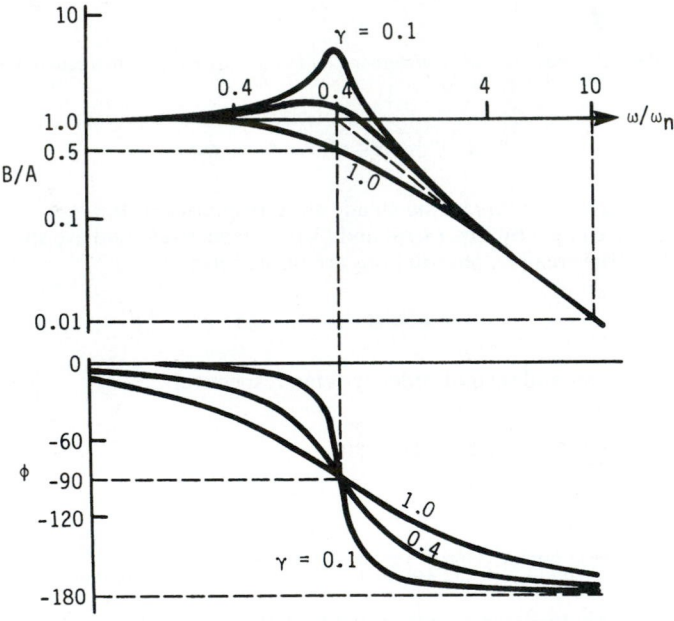

FIGURE 3.21 Frequency and phase response of a second-order system to a sinusoidal input.

Many existing transducers behave according to either a first- or second-order system. One should understand thoroughly how both first- and second-order systems respond to both the step input and sinusoidal input in order to understand how a transducer is likely to respond to such input signals. Table 3.1 is a listing of the steady-state responses of both the first- and second-order systems to a step function, ramp function, impulse function, and sinusoidal function. (See also [3.6] and [3.7].)

Understanding how a transducer might respond to a complex transient waveform can be understood by considering a sinusoidal response of the system, since any complex transient forcing function can be represented by a Fourier series equivalent [3.5]. Consideration of each separate harmonic in the input forcing function would then yield information as to how the measuring system is likely to respond.

Example 3. A thermistor-type temperature sensor is found to behave as a first-order system, and its experimentally determined time constant τ is 0.4 s. The resistance-temperature relation for the thermistor is given as

$$R = R_0 \exp\left[\beta\left(\frac{1}{T} - \frac{1}{T_0}\right)\right]$$

where β has been experimentally determined to be 4000 K. This temperature sensor is to be used to measure the temperature of a fluid by suddenly immersing the thermistor into the fluid medium.

How long one must wait to ensure that the thermometer reading will be in error by no more than *5 percent of the step change in temperature* is calculated as follows:

$$x = x_s(1 - e^{-t/\tau})$$

$$x = T - T_0 = 0.95(T_\infty - T_0)$$

$$x_s = T_\infty - T_0$$

$$\therefore 0.95 = 1 - e^{-t/0.4}$$

$$\ln 0.05 = \frac{-t}{0.4} = -2.9957$$

$$\therefore t = 1.198 \text{ s} = 1.2 \text{ s}$$

Determine the sensitivity of the thermometer at a temperature of 300 K if the resistance R is 1000 ohms (Ω) at this temperature:

$$S = \frac{dR}{dT}\bigg|_{op} = R_0 \exp\left[\beta\left(\frac{1}{T} - \frac{1}{T_0}\right)\right]\beta(-1)T^{-2}$$

$$= -\frac{R\beta}{T^2} = \frac{1000(4000)}{(300)^2}$$

$$= -44.44 \ \Omega/\text{K}$$

Determine the resolution of the thermometer if one can observe changes in resistance of 0.50 Ω on a Wheatstone bridge used as a readout device at the temperature of 300 K:

$$R = \frac{\Delta Q_{op}|_{min}}{S} = \frac{-0.50}{-44.44} = 0.0113 \text{ K}$$

TABLE 3.1 Response of First- and Second-Order Systems to Various Input Signals

First-order system	Second-order system

Equation of Motion

First-order system:

$$c\dot{x} + kx = F(t)$$

$$\tau\dot{x} + x = \frac{F(t)}{k}$$

$$\tau = \frac{c}{k}$$

Second-order system:

$$M\ddot{x} + c\dot{x} + kx = F(t)$$

$$\ddot{x} + 2\gamma\omega_n\dot{x} + \omega_n^2 x = \frac{F(t)}{M}$$

$$\text{where } \gamma = \frac{c}{c_c} \qquad c_c = \sqrt{4kM} \qquad \omega_n = \sqrt{\frac{k}{M}}$$

Step input: $F(t) = F \quad t > 0$

First-order system:

$$\frac{x}{F/k} = 1 - \exp(-t/\tau)$$

Second-order system:

(a) $\gamma < 1$: $\dfrac{x}{F/k} = 1 - \dfrac{\exp(-\gamma\omega_n t)}{\sqrt{1-\gamma^2}}\sin\left(\sqrt{1-\gamma^2}\,\omega_n t + \phi\right)$

$$\phi = \tan^{-1}\frac{\sqrt{1-\gamma^2}}{\gamma}$$

(b) $\gamma = 1$: $\dfrac{x}{F/k} = 1 - (1 + \omega_n t)\exp(-\omega_n t)$

(c) $\gamma > 1$: $\dfrac{x}{F/k} = 1 - \dfrac{\nu}{\nu - 1}\left[\dfrac{\exp\dfrac{-\omega_n t}{\sqrt{\nu}}}{\sqrt{\nu}} - \dfrac{1}{\nu}\exp\left(-\sqrt{\nu}\,\omega_n t\right)\right]$

$$\nu = \frac{\gamma + \sqrt{\gamma^2 - 1}}{\gamma - \sqrt{\gamma^2 - 1}}$$

Impulse input: $I = \int_0^t F\,dt \quad t \to 0$

$$\frac{xk\tau}{I} = \exp(-t/\tau)$$

(a) $\gamma < 1$: $\dfrac{x\sqrt{Mk}}{I} = \dfrac{\exp(-\gamma\omega_n t)}{\sqrt{1-\gamma^2}}\sin(\sqrt{1-\gamma^2}\,\omega_n t)$

(b) $\gamma = 1$: $\dfrac{x\sqrt{Mk}}{I} = \omega_n t\exp(-\omega_n t)$

(c) $\gamma > 1$: $\dfrac{x\sqrt{Mk}}{I} = \dfrac{\sqrt{\nu}}{\nu-1}\exp\left(\dfrac{-\omega_n t}{\sqrt{\nu}}\right) - \exp(\sqrt{\nu}\,\omega_n t)$

Ramp input: $F(t) = \beta t$

$$\frac{xk}{\beta\tau} = \frac{t}{\tau} - [1 - \exp(-t/\tau)]$$

(a) $\gamma < 1$: $\dfrac{x\omega_n k}{\beta} = \dfrac{1}{\sqrt{1-\gamma^2}}\exp(-\gamma\omega_n t)\sin(\sqrt{1-\gamma^2}\,\omega_n t + \phi)$
$- 2\gamma + \omega_n t$

$$\phi = \tan^{-1}\frac{\gamma\sqrt{1-\gamma^2}}{\gamma^2 - \tfrac{1}{2}}$$

(b) $\gamma = 1$: $\dfrac{x\omega_n k}{\beta} = (2 + \omega_n t)\exp(-\omega_n t) + \omega_n t - 2$

(c) $\gamma > 1$: $\dfrac{x\omega_n k}{\beta} = \dfrac{\nu\sqrt{\nu}}{\nu-1}\exp\left(\dfrac{-\omega_n t}{\sqrt{\nu}}\right) - \dfrac{1}{\nu^2}\exp(-\sqrt{\nu}\,\omega_n t)$
$- \dfrac{\nu+1}{\sqrt{\nu}} + \omega_n t$

TABLE 3.1 Response of First- and Second-Order Systems to Various Input Signals (*Continued*)

First-order system	Second-order system
	Sinusoidal input: $F(t) = F_0 \cos \Omega t$ or $F(t) = $ (real part of) $F_0 \exp(i\Omega t)$
$\dfrac{x}{(F_0/k)} = \dfrac{\cos(\Omega t + \phi)}{\sqrt{1 + (\Omega \tau)^2}}$ $-\phi = \tan^{-1} \Omega \tau$	$\dfrac{x}{F_0/k} = \dfrac{\cos(\Omega t + \phi)}{\sqrt{(1 - \beta^2)^2 + (2\gamma\beta)^2}}$ $\phi = \tan^{-1} \dfrac{-2\gamma\beta}{1 - \beta^2}$ $\beta = \dfrac{\Omega}{\omega_n}$

The expected response of the thermometer if it were subjected to step changes in temperature between 300 and 500 K in a square-wave fashion and at a frequency of 1.0 hertz (Hz) is shown in Fig. 3.22, where $x = x_s$ (0.7135). Note that the thermistor never responds sufficiently to give an accurate indication of the step-amplitude temperature. However, if the time constant of the thermistor were selected to be less than 0.1 s, the step-amplitude temperature would be indicated in 0.5 s (5 time constants).

Example 4. A strip-chart recorder (oscillograph) has been determined to behave as a second-order system with damping ratio of 0.5 and natural frequency of 60 Hz. At what frequency would the output amplitude of the recorder "peak" even with a constant-amplitude input signal? The frequency may be calculated as follows:

$$\omega_p = \omega_n \sqrt{1 - 2\gamma^2} = 60\sqrt{1 - 2(0.5)^2} = 42.4 \text{ Hz}$$

What is the maximum sine-wave frequency of input signal that would allow no more than 5 percent error in amplitude? See Fig. 3.23. The amplitude factor (AF) is calculated as follows:

$$1.05 = AF = \frac{1}{\sqrt{[1 - (\omega_f/\omega_n)^2]^2 + (2\gamma\,\omega_f/\omega_n)^2}} = \frac{1}{\sqrt{1 - z + z^2}}$$

where $z \equiv (\omega_f/\omega_n)^2$. The result is $\omega_{f_{max}} = 19.2$ Hz.

A complex waveform made up of a fundamental frequency of 10 Hz and 8 harmonics in terms of its Fourier series representation is desired to be recorded. Will the oscillograph described above suffice?

The basic equation is

$$\text{Maximum frequency} = (n + 1)(\text{fundamental}) = 90 \text{ Hz}$$

$$AF = \frac{1}{\sqrt{[(1 - (90/60)^2]^2 + (90/60)^2}} = 0.51$$

$$\psi = \tan^{-1}\frac{2(0.5)90/60}{1 - (90/60)^2} = -55.2° \qquad \text{(oscillograph will not suffice)}$$

If both the frequency and phase-response characteristics for the oscillograph are given below, show how the input signal to the oscillograph, also given below, will be changed, and give the resulting relation expected:

$$e = 10 + 5.8 \cos 5t + 3.2 \cos 10t + 1.8 \cos 20t$$

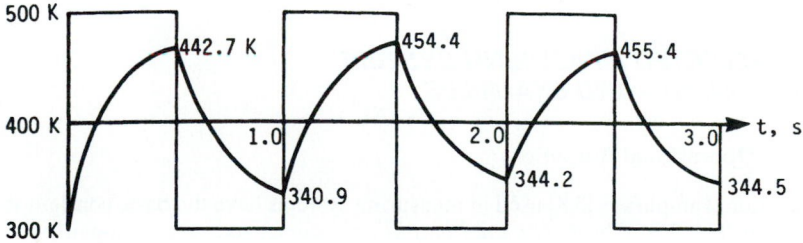

FIGURE 3.22 Thermistor temperature response of Example 3.

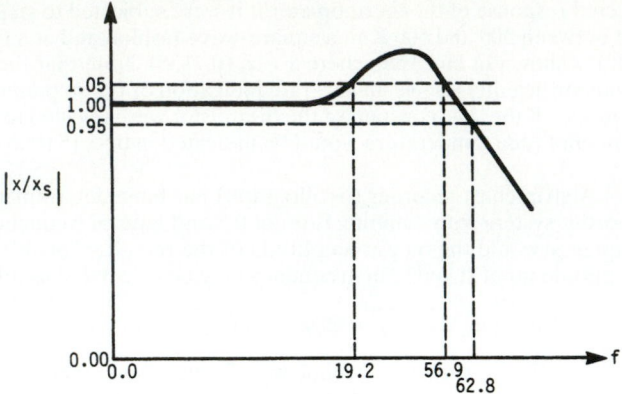

FIGURE 3.23   Frequency response of strip-chart recorder of Example 4.

Input frequency ω, rad/s	Amplitude, V		Phase angle (lag), °
	Input	Output	
0	10.0	10.0	0
5	10.0	10.0	10
10	10.0	10.2	20
15	10.0	10.6	30
20	10.0	11.0	45
25	10.0	12.2	90

It follows that

$$e_o = 10\left(\frac{10}{10}\right) + 5.8\left(\frac{10.0}{10.0}\right)\cos\left(5t - \frac{10\pi}{180}\right) + 3.2\left(\frac{10.2}{10.0}\right)\cos\left(10t - \frac{20\pi}{180}\right)$$

$$+ 1.8\left(\frac{11.0}{10.0}\right)\cos\left(20t - \frac{45\pi}{180}\right)$$

$$= 10 + 5.8\cos(5t - 0.174) + 3.26\cos(10t - 0.349) + 1.98\cos(20t - 0.785)$$

3.7   SELECTED MEASURING-SYSTEM COMPONENTS AND EXAMPLES

3.7.1   Operational Amplifiers

Operational amplifiers [3.8] used in measuring systems have the basic configuration shown in Fig. 3.24. The *operational amplifier* is composed of a high-gain voltage amplifier coupled with both input and feedback impedances. The characteristics of

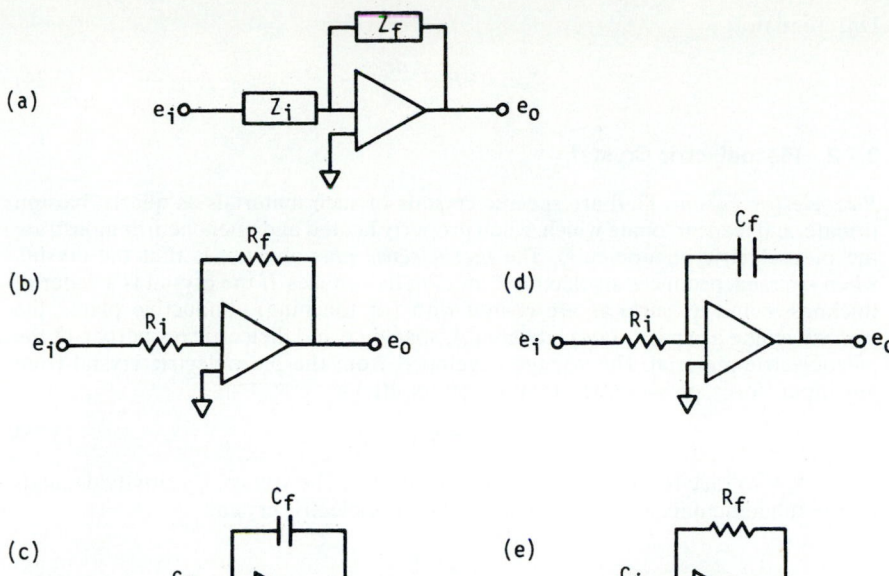

FIGURE 3.24 Operational amplifier circuit. (*a*) General; (*b*) voltage amplifier; (*c*) charge ampli-fier; (*d*) integrator; (*e*) differentiator.

the operational amplifier depend on the feedback impedance Z_f and input impedance Z_i, selected according to Eq. (3.20):

$$\frac{e_o}{e_i} = -\frac{Z_f}{Z_i} \tag{3.20}$$

The relations between input and output voltage for the specific configurations shown in Fig. 3.24 are as follows:
Voltage amplifier:

$$\frac{e_o}{e_i} = -\frac{R_f}{R_i} \tag{3.21}$$

Charge amplifier:

$$\frac{e_o}{e_i} = -\frac{C_i}{C_f} \tag{3.22}$$

Integrator:

$$e_o = -\frac{1}{R_i C_f} \int_0^t e_i \, dt + e_o(0) \tag{3.23}$$

Differentiator:

$$e_o = -R_f C_i \frac{de_i}{dt} \tag{3.24}$$

3.7.2 Piezoelectric Crystal

Piezoelectric crystals [3.9] are specific crystals of such materials as quartz, barium titinate, and lead zirconate which, when properly heated and quenched, demonstrate the piezoelectric phenomenon. The *piezoelectric phenomenon* is that the crystal, when stressed, produces an electric charge on its surfaces. If the crystal is a wafer of thickness t and its surfaces are coated with (or touching) conductive plates, the plates become a capacitor of plate area A, spacing t, and dielectric property ε of the piezoelectric material. The voltage developed from the piezoelectric crystal from any input (force, pressure, acceleration, stress, etc.) is

$$e_o = S_e x \tag{3.25}$$

where S_e = voltage sensitivity and x = input variable. The voltage sensitivity depends on the fundamental charge sensitivity of the piezoelectric crystal:

$$S_e = \frac{S_q}{C_c} \tag{3.26}$$

where $S_q = q/x$ and C_c = crystal capacitance, given by

$$C_c = \frac{KA\varepsilon}{t} \tag{3.27}$$

K is a constant which depends on the geometry and the units of the parameters in the preceding equation.

When the piezoelectric crystal is coupled via lead wires with capacitance, the voltage sensitivity and output voltage are reduced according to the relation

$$e_o = S_e x = \frac{S_q}{C_T} x \tag{3.28}$$

where C_T = total capacitance of the combination of piezoelectric crystal, lead wires, and readout device and is equal to

$$C_T = C_c + C_{lw} + C_{rd} \tag{3.29}$$

The equivalent circuits of the piezoelectric crystal are given in Fig. 3.25. The piezoelectric crystal has a dynamic response that is approximately that of an undamped second-order system. The circuit components of the piezoelectric crystal have a dynamic response that is approximately that of a first-order system. The typical frequency response of the piezoelectric transducer is that shown in Fig. 3.26 and is the combination of the crystal and circuit responses.

When the piezoelectric crystal is coupled with a voltage amplifier, the output voltage of the measuring system is dependent on lead-wire capacitance according to the relation

$$e_o = -\frac{R_f}{R_i} S_e x = -\frac{R_f}{R_i} \frac{S_q}{C_T} x \tag{3.30}$$

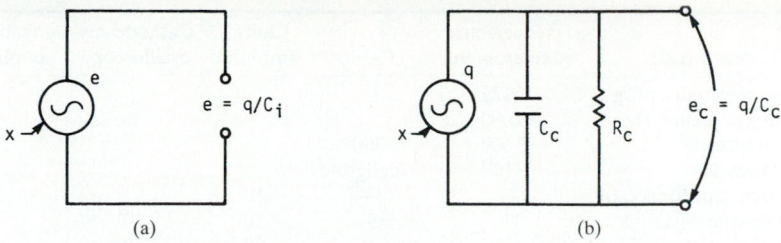

FIGURE 3.25 Equivalent circuits of a piezoelectric crystal. (*a*) Voltage generator equivalent circuit; (*b*) charge generator equivalent circuit.

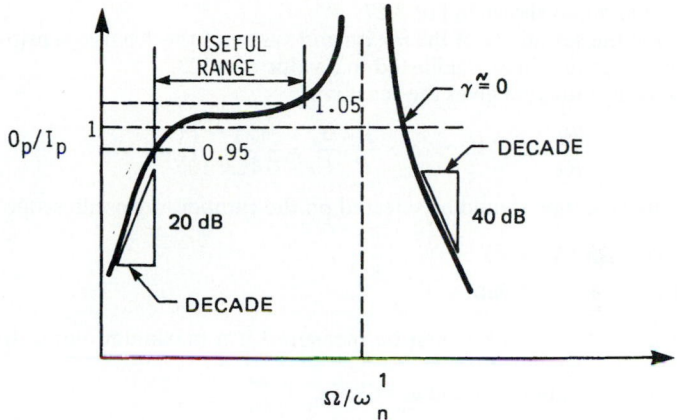

FIGURE 3.26 Composite frequency response of a piezoelectric transducer.

where $C_T = C_c + C_{lw} + C_a$ and R_f/R_i = the ratio of feedback to input resistance on the operational amplifier used for voltage amplification. Thus long lead wires or high lead-wire capacitance will significantly decrease the output voltage of the measuring system when using a voltage amplifier. The use of a charge amplifier avoids the problem of capacitance of the input lead wires, as shown by the relation

$$e_o = -\frac{C_i}{C_f} S_e x = -\frac{C_i}{C_f} \frac{S_q}{C_T} x \qquad (3.31)$$

where C_i equals C_T and is the total capacitance at the input to the charge amplifier. Thus with a charge amplifier the voltage sensitivity S_e of the system depends only on the basic crystal charge sensitivity S_q and the charge amplifier feedback capacitance C_f and not on the input capacitance.

Example 5. A piezoelectric accelerometer is to be used to measure the vibration of an automotive engine as installed in a particular test cell. The pertinent characteristics of the transducer, cable, charge amplifier, and cathode-ray oscilloscope used in the acceleration measuring system are given in the following table:

Characteristic	Piezoelectric accelerometer	Cable	Charge amplifier	Cathode-ray oscilloscope	Voltage amplifier
Charge sensitivity, pC/g	123				
Natural frequency, Hz	30 000				
Capacitance, pF	8 600	300			
Resistance, Ω	10^{12}	Negligible			
Feedback capacitance, pF	—	—	10^3		
Input resistance, Ω	—	—	10^8	10^6	
Input capacitance, pF	—	—	50	50	
$\dfrac{R_f}{R_i}$	—	—	—	—	0.123

The circuit diagram is shown in Fig. 3.27.

Determine the sensitivity of the *measuring system* if the "charge sensitivity" setting on the charge amplifier is adjusted to a value of 0.123.

The following equation gives the sensitivity:

$$\frac{R_t}{R_T} = 0.123 \qquad \frac{e_o}{x} = \frac{R_T}{R_t} \frac{S_q}{C_f} = \frac{123}{0.123} \frac{1}{10^3} = 1 \text{ V/g}$$

What sensitivity setting should be selected on the cathode-ray oscilloscope?

For 1 g/cm, use 1 V/cm.

For 0.1 g/cm, use 0.1 V/cm.

What range of acceleration can be measured if a maximum output of 10 V is available?

Range of acceleration is found as follows:

$$x_{max} = \frac{e_o|_{max}}{S} = \frac{10}{1 \text{ V/g}} = 10.0 \text{ g}$$

What is the voltage sensitivity of the accelerometer?

1. If the charge amplifier is used,

$$\frac{e_i}{x} = \frac{q/x}{C_i} = \frac{S_q}{C_i} \qquad \frac{e_o}{x} = -\frac{e_i}{x} \frac{C_i}{C_f} = -\frac{S_q}{C_f}$$

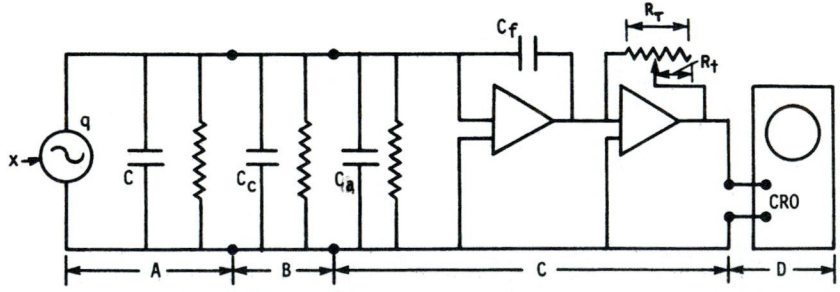

FIGURE 3.27 Circuit diagram of the vibration-measuring system. A, piezoelectric crystal; B, cable; C, charge amplifier coupled to a voltage amplifier; D, cathode-ray oscilloscope.

Thus $$\text{Voltage sensitivity} = \frac{123 \text{ pC/g}}{10^3 \text{ pF}} = 123 \text{ mV/g}$$

2. If the accelerometer is connected directly to the cathode-ray oscilloscope (no charge amplifier),

$$\frac{e_o}{e_i} = -G \qquad \frac{e_o}{x} = -G\frac{e_i}{x} = -G\frac{S_q}{C_i} = -\frac{G(123) \text{ pC/g}}{(8600 + 300 + 50) \text{ pF}}$$

Thus $$\text{Voltage sensitivity} = -13.7G \text{ mV/g}$$

If the accelerometer has zero damping, what would be the largest frequency of input vibration allowable to have no more than a 1 percent error? If the engine has eight cylinders and operates at 4000 rpm, will the measuring system work?
The computations are as follows:

$$\frac{8(4000)}{60} = 533.3 \text{ Hz} = \text{vibration frequency expected}$$

$$\text{AF}_{\text{HF}} = \frac{1}{1 - (\Omega/\omega n)^2} = 1.01 \qquad \Omega/\omega_n = 0.0995 \qquad \Omega_{\text{H}} = 2985 \text{ Hz}$$

$$\text{AF}_{\text{LF}} = \frac{1}{\sqrt{1 + (\tau\Omega)^2}} = 0.99$$

$$\tau = R_T C_T = 10^6(8650 \times 10^{-12}) = 0.00865 \text{ s}$$

$$\tau\Omega = 0.1425 \qquad \Omega_{\text{L}} = 15.83 \text{ rad/s} = 2.5 \text{ Hz}$$

The frequency response of the vibration measurement system is satisfactory and is shown in Fig. 3.28.

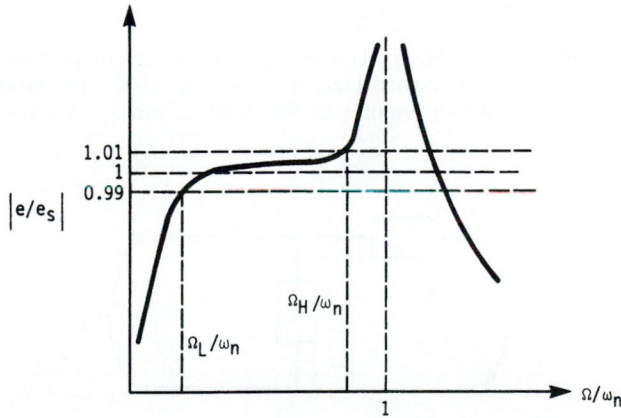

FIGURE 3.28   Frequency response of the vibration-measuring system.

3.7.3 Ballast-Type Circuit

A basic circuit used in measurement applications is the ballast-type circuit shown in Fig. 3.29. The relation between input and output voltage is given by

$$\frac{e_o}{e_i} = \frac{Z_L}{Z_B + Z_L} \tag{3.32}$$

where Z_L = load impedance and Z_B = ballast impedance.

When Z_L and Z_B are capacitance C and resistance R, respectively, the circuit is used as a low-pass filter with output voltage and phase shift given by Eqs. (3.33) and (3.34), respectively, where ω is the frequency of the input signal:

$$\left| \frac{e_o}{e_i} \right| = \sqrt{\frac{1}{1 + (RC\omega)^2}} \tag{3.33}$$

$$\phi = \tan^{-1}(RC\omega) \tag{3.34}$$

When Z_L and Z_B are resistance and capacitance, respectively, the circuit is used as a high-pass filter.

The output voltage and phase shift are then given by Eqs. (3.35) and (3.36), respectively:

$$\left| \frac{e_o}{e_i} \right| = \sqrt{\frac{(RC\omega)^2}{1 + (RC\omega)^2}} \tag{3.35}$$

$$\phi = \tan^{-1}\left(\frac{1}{RC\omega}\right) \tag{3.36}$$

An example of this type of circuit is the ac coupling circuit at the input of a cathode-ray oscilloscope. When Z_L is that of an impedance-based detector transducer such as a resistance thermometer or strain gauge, the voltage e_i is that of the auxiliary energy source and Z_B is an impedance used to limit the current flow to the detector transducer. If Joule (I^2R) heating would affect the transducer measurement, such as in resistance-thermometer or strain-gauge applications, the ability to limit current is important.

Example 6.   The circuit of Fig. 3.30a is used as a coupling circuit between a detector transducer and a readout device. Determine and sketch the amplitude and phase characteristics of the coupling circuit (see Fig. 3.30b, c, and d). Determine the load-

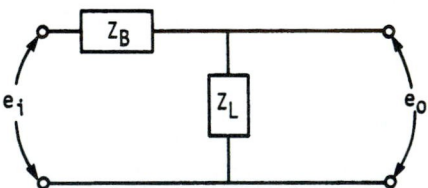

FIGURE 3.29   The ballast-type circuit.

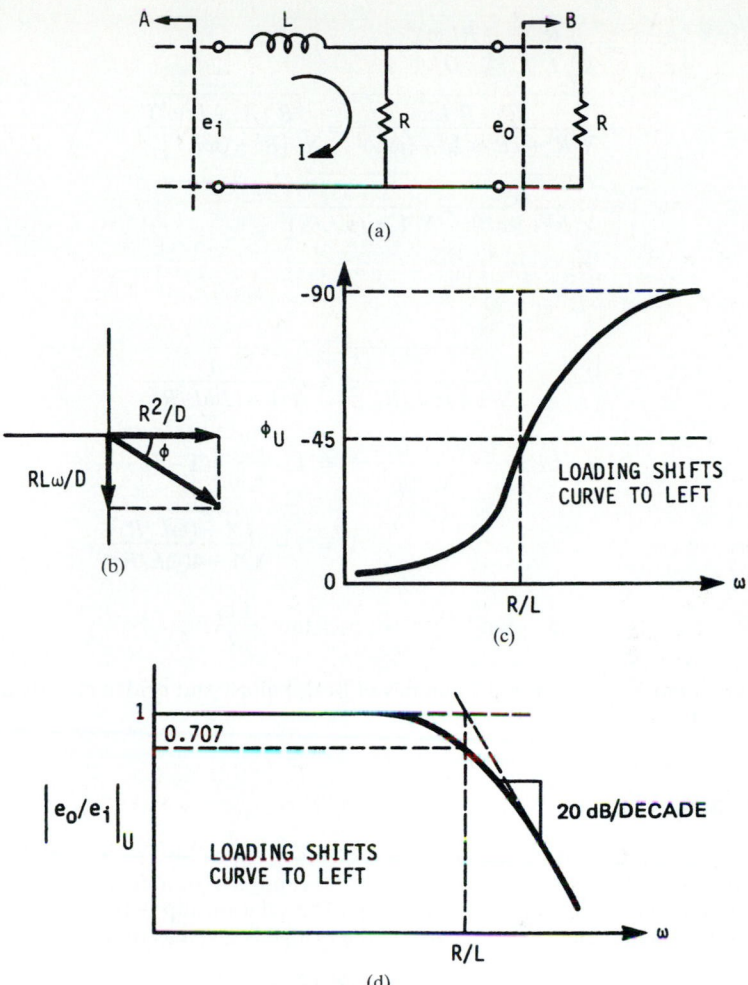

FIGURE 3.30 Coupling circuit example. A, detector transducer; B, readout. (*a*) Inductor and resistance in a ballast-type circuit; (*b*) real and complex components; (*c*) phase-shift characteristic; (*d*) frequency-response characteristic.

ing error if a readout device having an input impedance equal to R is connected to the circuit.

The equations are as follows:

$$e_o = IR$$

$$e_i = I(Z_L + R)$$

$$\left.\frac{e_o}{e_i}\right|_U = \left(\frac{R}{j\omega L + R}\right)\left(\frac{-j\omega L + R}{-j\omega L + R}\right) = \frac{R^2 - jRL\omega}{R^2 + (\omega L)^2} = \frac{R^2}{D} - \frac{jRL\omega}{D}$$

$$\left|\frac{e_o}{e_i}\right|_U = \sqrt{\left(\frac{R^2}{D}\right)^2 + \left(\frac{RL\omega}{D}\right)^2}$$

$$= \sqrt{\frac{R^4 + R^2 L^2 \omega^2}{R^4 + 2R^2\omega^2 L^2 + (\omega L)^4}} = \sqrt{\frac{R^2(R^2 + L^2\omega^2)}{[R^2 + (\omega L)^2]^2}}$$

$$= \sqrt{\frac{R^2}{R^2 + (\omega L)^2}} = \sqrt{\frac{1}{1 + (\omega L/R)^2}}$$

$$\left.\frac{e_o}{e_i}\right|_L = \frac{R_{eq}^2}{D} - \frac{jR_{eq}L\omega}{D}$$

and

$$\left|\frac{e_o}{e_i}\right|_L = \sqrt{\frac{1}{1 + (\omega L/R_{eq})^2}} = \sqrt{\frac{1}{1 + (2\omega L/R)^2}}$$

$$LE \equiv \frac{|e_o/e_i|_U - |e_o/e_i|_L}{|e_o/e_i|_U} = 1 - \frac{e_o|_L}{e_o|_U} = 1 - I$$

$$= 1 - \sqrt{\frac{1 + (\omega L/R)^2}{1 + 4(\omega L/R)^2}}$$

$$\phi_L = \tan^{-1}\frac{2\omega L}{R} \qquad \phi_U = \tan^{-1}\frac{L\omega}{R}$$

Tables 3.2 and 3.3 give several examples of both ballast and bridge circuits used in instrumentation systems.

3.7.4 Bridge Circuit

The bridge circuit used in measurement circuits is shown in Fig. 3.31. For voltage excitation, e_i, the output Δe_o corresponds to the change in output voltage due to the change in the arm impedances of the bridge. The relationship between output voltage and impedance change in one arm of the bridge is given as follows:

$$\frac{e_o + \Delta e_o}{e_i} = \frac{(Z_1 + \Delta Z_1)Z_4 - Z_2 Z_3}{(Z_1 + \Delta Z_1 + Z_2)(Z_3 + Z_4)} \tag{3.37}$$

If initially the bridge is said to be "balanced," the output voltage e_o is zero and the relationship for the impedances in the bridge is given by the balance equation

$$\frac{Z_1}{Z_2} = \frac{Z_3}{Z_4} \tag{3.38}$$

This relation is used to measure an unknown impedance connected in a bridge circuit with three other impedances which are known. The reader is referred to Prensky [3.10] and to Table 3.3 for further information in this regard.

The ability of the bridge circuit to "zero" the output at any level of input transducer impedance allows the circuit to be used for the "balance" type of measurement, which is more accurate than the "unbalance" type of measurement commonly employed when using the ballast-type circuit.

TABLE 3.2 Typical Ballast-Type Circuits Used in Instrumentation Circuits

Ballast-type circuits	Magnitude response and phase shift
(a) Low-pass RC	$\left\|\dfrac{e_o}{e_i}\right\| = \dfrac{1}{\sqrt{1 + \omega^2\tau^2}} \approx \dfrac{1}{\omega\tau}$ $\phi = -\tan^{-1} R\omega C$
(b) High-pass RC	$\left\|\dfrac{e_o}{e_i}\right\| = \dfrac{1}{\sqrt{1 + 1/\omega^2\tau^2}} \approx \omega\tau$ $\phi = \tan^{-1}\dfrac{1}{R\omega C}$
(c) Low-pass RL	$\left\|\dfrac{e_o}{e_i}\right\| = \dfrac{1}{\sqrt{1 + \omega^2\tau^2}} \approx \dfrac{1}{\omega\tau}$ $\phi = -\tan^{-1}\dfrac{\omega L}{R}$
(d) High-pass RL	$\left\|\dfrac{e_o}{e_i}\right\| = \dfrac{1}{\sqrt{1 + 1/\omega^2\tau^2}} \approx \omega\tau$ $\phi = \tan^{-1}\dfrac{R}{\omega L}$

When all impedances are initially equal, the bridge is balanced, and the impedance change from an input signal is small compared to the original impedance, the bridge output voltage is linearized to

$$\Delta e_o \approx \frac{\Delta Z}{4Z}\, e_i \tag{3.39}$$

This equation can be used to predict the output of impedance-based transducers such as variable capacitors, variable inductances, or variable resistances (such as resistance thermometers or strain gauges) used in voltage-sensitive bridge circuits.

3.7.5 Strain Gauges

The strain gauge is a resistance R (usually in the form of a grid) wire or foil that changes when strained according to the relation

$$R = \rho\,\frac{L}{A} \tag{3.40}$$

TABLE 3.3 Typical Bridge Circuits Used in Instrumentation Circuits

Bridge circuits	Balance relations

ac Wheatstone
  bridge

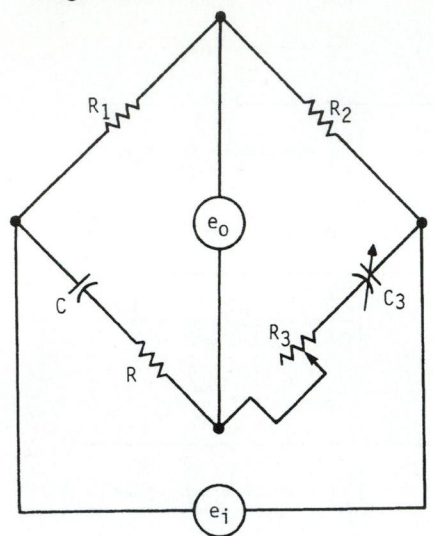

$$C = \frac{C_3 R_2}{R_1}$$

$$R = \frac{R_3 R_1}{R_2}$$

Wein bridge

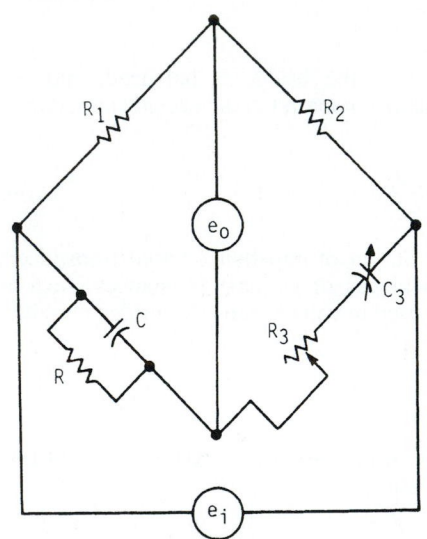

$$\frac{C}{C_3} = \frac{R_2}{R_1} - \frac{R_3}{R}$$

$$CC_3 = \frac{1}{\omega^2 R_3 R}$$

If $C_3 = C$ and $R_3 = R$,

$$f = \frac{1}{2\pi R_3 C_3}$$

TABLE 3.3 Typical Bridge Circuits Used in Instrumentation Circuits (*Continued*)

Bridge circuits	Balance relations

Resonance bridge

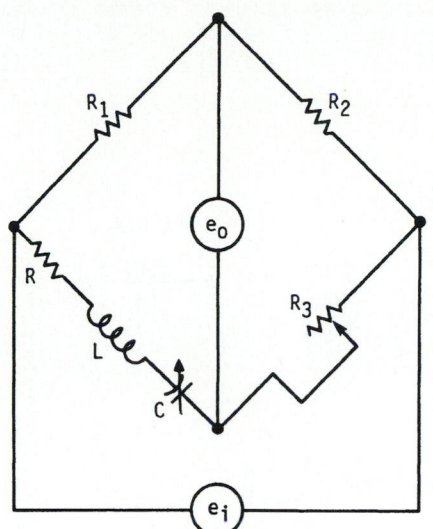

$$\omega^2 LC = 1$$

$$R = \frac{R_3 R_1}{R_2}$$

At balance

$$f = \frac{1}{2\pi \sqrt{LC}}$$

Maxwell bridge

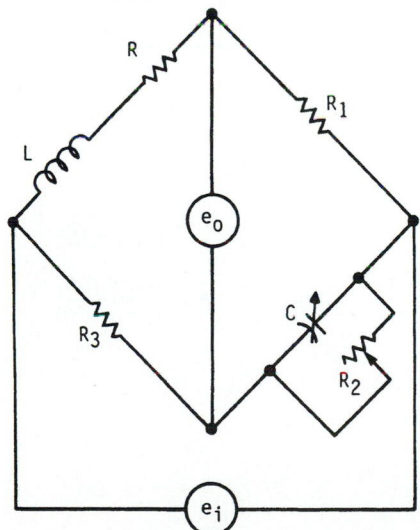

$$L = R_1 R_3 C$$

$$R = \frac{R_1 R_3}{R_2}$$

TABLE 3.3   Typical Bridge Circuits Used in Instrumentation Circuits (*Continued*)

Bridge circuits	Balance relations

Owen bridge

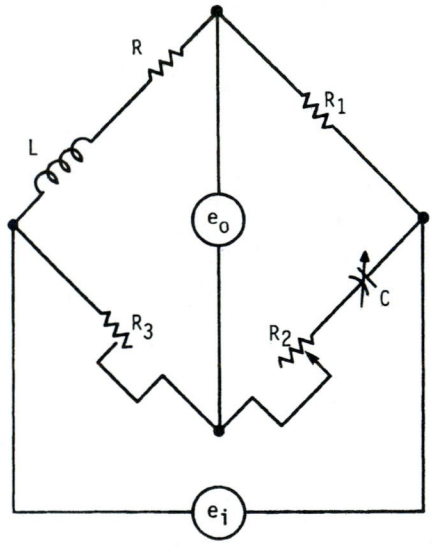

$$L = C_1 R_1 R_2$$

$$R = \frac{C_1 R_1}{C_2} - R_3$$

Hay bridge

$$L = \frac{R_1 R_3 C}{1 + \omega^2 C^2 R_2^2}$$

$$R = \frac{\omega^2 C^2 R_1 R_2 R_3}{L + \omega^2 C^2 R_2^2}$$

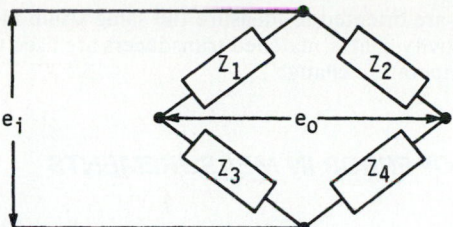

FIGURE 3.31   The bridge circuit.

where L = wire or foil length, A = wire or foil cross section, and ρ = electrical resistivity of the strain-gauge material. The strain-gauge sensitivity is the "*gauge factor*" GF given by the relation

$$\text{GF} = \frac{\Delta R/R}{\Delta L/L} = 1 + 2\mu + \frac{\Delta\rho/\rho}{\Delta L/L} \tag{3.41}$$

where μ = Poisson's ratio for the strain-gauge material.

The strain gauge is often the sensing element in force transducers (load cells), pressure transducers, and accelerometers. The use of a strain gauge is illustrated in the following example.

Example 7.   Figure 3.32 shows a rectangular cross section of a cantilever beam of width b and depth h with a bending load F applied at a distance L from where the strain is desired. A voltage-sensitive bridge circuit is used for excitation of 120-Ω, gauge factor 2.0 strain gauges.

The strain-gauge characteristics coupled with the bridge circuit and beam characteristics yield the following output voltage with respect to input load producing the strain:

$$e_o = Ke_i \left(\frac{\Delta R}{4R}\right) = \frac{Ke_i}{4}(\text{GF})\,\epsilon = \frac{K}{4}\,e_i\,(\text{GF})\,\frac{6FL}{Ebh^2} \tag{3.42}$$

K is the "bridge factor" in this equation and is a constant giving the magnification factor for using more than one active gauge in the bridge circuit. With two active gauges as shown, the bridge factor is 2 and the gauge arrangement gives complete compensation for temperature change of the beam. The temperature-induced strains are detected by the strain gauges but are effectively canceled in the bridge

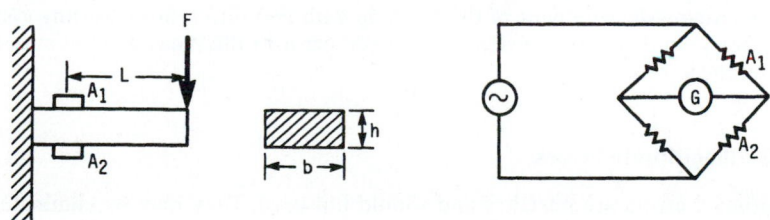

FIGURE 3.32   Cantilever beam with strain gauges.

circuit if the gauges are oriented to measure the same strain magnitude and if they have the same sensitivity; that is, matched transducers are used to cancel the "noise" signal caused by temperature change.

3.8 SOURCES OF ERROR IN MEASUREMENTS

The basic problem of every quantitative experiment is that of trying to identify the true value of the measured quantity. Philosophically, the measurement process is like viewing a deterministic event through a foggy window. Refinement of the measuring system to the ultimate should result in the measurement's being the true value. However, because errors occur in all measurements, one can never establish the true value of any quantity. Continued refinement of the methods used in any measurement will yield closer approximations to the true value, but there is always a limit beyond which refinement cannot be made. Furthermore, the fact that the measuring system draws energy from the source of the variable to be measured results in the measurement process's changing the characteristics of both the signal source and the measured variable. Thus some difference, however small, always occurs between the indicated value of the measured quantity and the original quantity to be measured.

3.8.1 Systematic Errors

Systematic errors are of consistent form. They result from conditions or procedures that cause a consistent error which is repeated every time the measurement is performed, such as faulty calibrations of the measuring system or changes in the measuring-system components due to factors such as aging.

Another form of systematic error can occur as a result of the observer. Parallax is an example of such an error. If the observer does not correctly align the indicating needle of the instrument, the reading may be consistently high or low depending on the individual observer. This type of error is difficult to detect because it is repeated with every measurement under identical conditions. However, one means of detecting systematic error is to measure something whose magnitude is accurately and independently known. For example, a check on a thermometer at one or more fixed points on the temperature scale will determine if the temperature scale on the thermometer is yielding a systematic error. Use of gauge blocks in checking micrometers and calipers is another example of checking a measuring system to eliminate systematic errors.

Calibration of the measuring system frequently, by use of accurately known input signals, can give an indication of the development of systematic errors in the measuring system. Measurement of the variable with two different measuring systems can often help detect the presence of a fixed error in the measurement system or measurement process.

3.8.2 Illegitimate Errors

Illegitimate errors are mistakes and should not exist. They may be eliminated by using care in the experimental procedure and by repetition in checking the measurement. Faulty data logging is an example of illegitimate error. This might occur by reading the wrong value from a scale, by writing down a wrong number, or by transposing the digits of a number. Another example of illegitimate error is that of using linear interpolation between scale divisions on a readout device when the scale is nonlinear. For example, some readout scales may be logarithmic, but it is quite common for observers to interpolate between these scale divisions in a linear fashion. Repeating the measurement and rechecking suspicious values can help eliminate such mistakes.

3.8.3 Random Errors

Random errors are accidental errors that occur in all measurements. They are characterized by their stochastic natures in both magnitude and time. One cannot determine their origin in the measurement process. These errors can only be estimated by statistical analysis. However, if both systematic and illegitimate errors can be eliminated, the uncertainty in the measurement due to the remaining random error can be estimated by statistical analysis of the data obtained in the experiment.

3.8.4 Loading Error

The loading error can be reduced in some measurement systems by means of a technique called *balancing*. A balance-type measurement is one where a reference signal is fed into the measurement system and a direct comparison between the reference and the measured signal is made. The reference signal is adjusted such that when its value is the same as that of the measured signal, the two signals balance one another and the output reading from the measurement system is zero. With the reference signal balancing out the measured signal, the net energy flow from the source at the balance condition is zero. Thus the balance method usually provides a more accurate measurement than the unbalance method. Examples of the balance type of measurement are the use of bridge circuits in strain-gauge measurement and the use of a voltage-balancing potentiometer with thermocouples when measuring temperatures. It should be noted that the balance type of measurement is usually difficult to achieve when dynamic or time-varying signals are being measured. The loading error is a type of fixed error in the measuring system and can be determined by appropriate calibration.

3.8.5 Noise-Measurement Systems

"Noise" in a measurement system is any output which is not generated by the input quantity to be measured ([3.4], [3.11], [3.12]). It must be remembered that all measuring systems are placed in an environment and interact in some way with that environment. Any interaction of a measuring system with the environment that is not related to the input quantity to be measured can result in an unwanted output (noise) of the measuring system. For noise to exist at a measurement-system output there must be both a source and a receiver of the noise. There must also be a coupling method between the source and the receiver of the noise.

The noise signal at the output of the measuring system can come from two general sources. One source is internally from the transducers in the measuring system, and the other source is from the environment of the transducers of the measuring system. Examples of internally generated signals are the thermal or Johnson noise

[3.4] created in a resister of an electric circuit. Another example is the shot noise [3.4] generated by tubes in electric circuits. External sources can cover a variety of possibilities, such as vibrations, electrical interference from the electromagnetic spectrum, and switching or discharge of storage elements causing transient signals to be induced in the power and signal lines of a measuring system. Any physical change in the environment can induce externally generated noise signals. These include temperature change, humidity change, pressure change, sound-level change, etc.

The noise signal can be *active* or *self-generating* in the sense that the noise is directly coupled to the measuring system without the aid of an auxiliary energy source. This might be thermoelectric or electromagnetic in nature. For example, a constantan strain gauge connected with copper lead wires could have a noise voltage generated by the thermocouple effect at the two junctions where the constantan gauge and copper lead wires are connected if the junctions are at different temperatures.

Noise effects that require the use of an auxiliary energy source to be carried into the measuring system are called *passive noise signals*. Examples of such noise signals are the temperature effects that occur with strain gauges and the strain effects that occur with resistance thermometers.

The effects of the noise on the output of the measuring system may be additive, multiplicative, or a combination of additive and multiplicative. If the noise level is an additive effect with no frequency content, the output signal due to the noise is called *zero shift*. This is a very common type of noise and can be easily detected by calibration. It is usually eliminated by a "bias" control on the measuring instrument. Noise levels that have a multiplicative effect on the system output usually affect the gain or sensitivity of the components of the measuring system. These effects can sometimes be detected by calibration.

At least four methods of handling noise are known ([3.4], [3.11], [3.12]). These include removal of the noise source, elimination of the noise by cancellation, minimization of the noise by division (filtering), and minimization of the noise by frequency-selective filtering. Removal of the noise source is not usually possible, and one must generally resort to the other techniques. However, if the noise source can be eliminated, this will be the most effective method of preventing the noise problem.

When the effects which create the noise are consistent to the extent that one can expect two matched transducers to detect identical noise signals at the same instant of time in the same environment, it is possible to arrange for cancellation of these signals by subtraction. For example, the weight of a balance-scale pan may be subtracted (or balanced out) by placing an equal weight on the other side of the balance scale. Another example is the temperature-induced noise in strain gauges illustrated in Example 7. The temperature-induced resistance change in a strain gauge can be canceled out by placing two identical strain gauges (matched transducers) in the same thermal environment and by using proper placement of the gauges in a bridge circuit to provide subtraction of the noise signals. This is called *temperature compensation*. Another example is the noise-canceling microphone, in which two sensing elements are placed opposite one another. Voice input is supplied to only one element while external noise is sensed by both elements and is effectively balanced out or canceled. This technique is often used in aircraft applications. The use of this technique for noise elimination depends on being able to have two identical sensing elements which detect and respond to the noise signal to which they are exposed in exactly the same way and at the same time. Such detector elements are called *matched transducers*.

When the effects which create the noise level are not consistent, one cannot expect that two noise sources under identical environmental conditions will emit the same noise at every instance of time. In this case, noise is minimized by division so that only a small fraction of the original noise propagates through the system. For example, contact-resistance phenomena in switching gear and slip rings cause this type of noise. Electric circuitry in the measuring system is designed so that these resistance changes will have a minimal effect on the output reading. Electromagnetic radiation can also cause this type of noise input. Appropriate shielding of lead wires and circuits is necessary and is commonly used to minimize this type of noise.

Frequency-selecting filtering can be used if the noise and the desired signal can be made to exist at different frequencies [3.4]. When this is the case, a simple filter may then be used to minimize the noise signal. If the signal and the noise exist in the same frequency range, one must resort to a technique of modulation where the signal frequency is moved to a frequency range sufficiently separated from the noise frequency that the noise frequency can be effectively filtered from the signal frequency. This technique of frequency-selective filtering can be used to minimize only active noise in a passive transducer, since the carrier wave of the auxiliary energy source is used via modulation to shift the signal frequency upward and separate it from the noise frequency. If the noise is passive noise (depends on the auxiliary source of energy) in the transducer, the signal and noise would both be modulated upward to the same frequency band, and separation could not be achieved. References [3.11] and [3.12] give specific details on how noise can be eliminated or minimized in a given measurement situation.

3.8.6 Precision and Accuracy

Accuracy is the difference between a measured variable and the true value of the measured variable. Normally, one is not able to determine accuracy. *Precision* is the difference between a measured variable and the best estimate (as obtained from the measured variable) of the true value of the measured variable. Thus precision is a measure of repeatability. It should also be noted that one may have excellent precision in a measurement but very poor accuracy. Calibration of a measuring system is essential in determining its accuracy. Precision is specified by quantities called *precision indices* (denoted by W_x) that are calculated from the random errors of a set of measurements.

When a variable is measured, it is just as important to state the precision of the measurement as it is to state the most representative value of the quantity. Thus we desire W_x to be specified for every measured variable. The confidence or probability for obtaining the range $\pm W_x$ is generally specified directly or else is implied by the particular type of precision index being used.

3.9 ANALYSIS OF DATA

The basic problem in every quantitative experiment is that of obtaining the unbiased estimate $\hat{\mu}$ of the true value μ of a quantity as well as an unbiased estimate $\hat{W}$ of the dispersion or uncertainty in the measured variable. Data sets or samples typically display the two very important characteristics of central tendency (or most representative value) and dispersion (or scatter). Other characteristics, such as skewness and flatness (or peakness), may also be of importance but are not considered in the items that follow.

If we were given a list of measured values of the same variable, the question is raised as to what value shall be taken as being nearest to the true value. In order to determine what relation the measured value has to the true value, we must be able to specify in any experiment the unbiased estimate $\hat{\mu}$ of the true value of a measurement and its uncertainty (or precision) interval W based on a given confidence level (or probability of occurrence).

3.9.1 Unbiased Sampling

An *unbiased estimator* exists if the mean of its distribution is the same as the quantity being estimated [3.13]. Thus for sample mean $\bar{x}$ to be an unbiased estimator of population mean μ, the mean of the distribution of sample means $\bar{\bar{x}}$ must be equal to the population mean.

It can be shown that the unbiased estimator $\hat{\mu}$ for the population mean μ is the sample mean $\bar{x}$. In this section the measure of dispersion is selected to be the standard deviation σ or its square σ^2, called the *variance*. Determination of the unbiased estimator of the standard deviation or variance depends on the type of sampling method used.

Figure 3.33 illustrates that different samples from a population yield slightly different estimates of population mean and variance. However, if the data from the individual samples are combined, even better estimates of population mean and variance can be achieved. The mean of the sample means $\bar{\bar{x}}$ is a better estimate of μ than any of the individual sample means $\bar{x}$. Also, the dispersion of the distribution of $\bar{x}$ values is much less than the dispersion of items within an individual sample, as indicated by the central limit theorem.

The *central limit theorem* yields the result that if one obtains random samples of size n from a large population of mean μ and variance σ^2, the distribution of sample means approaches gaussian as n becomes large with a mean μ and a variance σ^2/n. This is valid regardless of the nature of the distribution of the population from which the sample values were obtained.

A *random sample* is a sample collected such that every member of the population from which one is sampling has an equal probability of selection in every trial. This may be done with or without replacement. For a sample of size n, the probability of selection from the population for each member of the sample is $1/n$.

Unbiased estimates for determining population mean, population variance, and variance of the sample means depend on the type of sampling procedure used. Unbiased estimates of population mean μ, population variance σ^2, and variance of the mean are listed in Table 3.4 for sampling both with and without replacement. Note from the relations in Table 3.4 that sampling without replacement from an extremely large population is essentially equivalent to sampling with replacement (random sampling) since $(N-1)/N$ and $(N-n)/(N-1)$ approach unity.

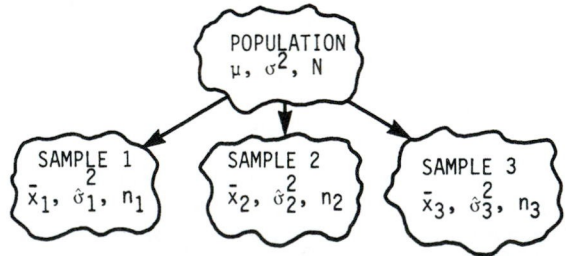

FIGURE 3.33 Sampling from a population.

TABLE 3.4 Unbiased Estimates of Mean, Variance, and Variance of the Mean[†]

Statistic	Sampling with replacement	Sampling without replacement
Mean	$\hat{\mu} = \dfrac{\Sigma x_i}{n} = \bar{x}$	$\hat{\mu} = \dfrac{\Sigma x_i}{n} = \bar{x}$
Variance	$\hat{\sigma}^2 = \dfrac{\Sigma(x_i - \bar{x})^2}{n-1} = S^2\left(\dfrac{n}{n-1}\right)$	$\hat{\sigma}^2 = S^2\left(\dfrac{n}{n-1}\right)\left(\dfrac{N-1}{N}\right)$
Variance of the mean	$\sigma_{\bar{x}}^2 = \dfrac{\hat{\sigma}^2}{n}$	$\sigma_{\bar{x}}^2 = \dfrac{\hat{\sigma}^2}{n}\left(\dfrac{N-n}{N-1}\right)$

[†]S^2 = sample variance = $\Sigma(x_i - \bar{x})^2/n$ and $n/(n-1)$ = Bessel's correction.

Example 8. The following data set is obtained by sampling from a population of 15 items:

x_i	$(x_i - \bar{x})$	$(x_i - \bar{x})^2$
65.0	−15.3	234.09
73.1	−7.2	51.84
83.0	+2.7	7.29
100.1	+19.8	392.04
$\Sigma = 321.2$	00.0	685.26

1. Determine the mean and standard deviation of the *original data*.

$$n = 4 \qquad \bar{x} = \frac{321.2}{4} = 80.3$$

$$s = \sqrt{\frac{\Sigma(x_1 - \bar{x})^2}{n}} = \sqrt{\frac{685.26}{4}} = \sqrt{171.3} = \pm 13.09$$

2. What are the estimates of *population mean* and *variance* if the sampling occurred *with* replacement?

$$\hat{\mu} = \bar{x} = 80.3$$

$$\hat{\sigma}^2 = s^2\frac{n}{n-1} = 171.3\frac{4}{3} = 228.4$$

3. What are the estimates of population mean and variance if the sampling occurred *without* replacement?

$$\hat{\mu} = \bar{x} = 80.3$$

$$\hat{\sigma}^2 = s^2\frac{n}{n-1}\frac{N-1}{N} = 171.3\frac{4}{3}\frac{14}{15} = 213.2$$

4. What is the variance associated with the distribution of $\bar{x}$ values if sampling occurred *without* replacement?

$$\hat{\sigma}_{\bar{x}}^2 = \frac{N-n}{N-1} \frac{\hat{\sigma}_x^2}{n} = \frac{(15-4)213.17}{(15-1)4} = 41.87$$

3.9.2 Uncertainty Interval

When several observations of a variable have been obtained to form a data set (multisample data), the best estimates of the most representative value (mean) and dispersion (standard deviation) are obtained from the formulas in Table 3.4. When only a single measurement exists (or when the data are taken so that they are similar to a single measurement), the standard deviation cannot be determined and the data are said to be "single-sample" data. Under these conditions, the only estimate of the true value is the single measurement, and the uncertainty interval must be estimated by the observer. Kline and McClintock [3.14] address this problem in detail. It is recommended that the precision index be estimated as the maximum error and that it correspond approximately to the 99 percent confidence level associated with multisample data.

Once the unbiased estimates of mean and variance are determined from the sample data, the *uncertainty interval* can be expressed as

$$\mu = \hat{\mu} \pm \hat{W} = \hat{\mu} \pm k(\nu, \gamma)\hat{\sigma} \qquad (3.43)$$

where $\hat{\mu}$ = the most representative value of the measured data and $\hat{W}$ = the uncertainty interval or precision index associated with the estimate of μ. The magnitude of the precision index or uncertainty interval depends on confidence level γ (or probability chosen), sample size n, and type of probability distribution governing the distribution of measured items.

The uncertainty interval $\hat{W}$ can be replaced by $k\hat{\sigma}$, where $\hat{\sigma}$ is the standard deviation (measure of dispersion) of the population as estimated from the sample and k is a constant that depends on the probability distribution, the confidence level γ, and the sample size n. For example, with a gaussian distribution, the 95 percent confidence limits are $\hat{W} = 1.96\hat{\sigma}$, where $k = 1.96$ and in this case is independent of n. For a t distribution, $k = 2.78, 2.06,$ and 1.96 for sample sizes of $5, 25,$ and ∞, respectively, at the 95 percent confidence level. The t distribution becomes the gaussian distribution as $n \rightarrow \infty$. The uncertainty interval $\hat{W}$ in Eq. (3.43) assumes a set of measured values with only random error present. Furthermore, the set of measured values is assumed to have unbounded significant digits and to have been obtained with a measuring system having infinite resolution. When *finite resolution* exists and *truncation of significant digits* occurs, the uncertainty interval will be larger than that predicted by consideration of only the random error [3.15]. The uncertainty interval can never be less than the resolution limits or truncation limits of the measured values. If $\{s_n\}$ is the theoretically possible set of measurements of unbounded resolution and $\{x_n\}$ is the actual set of measurements expressed to m significant details from a measuring system of finite resolution R, the quantity $s_i - x_i = \pm e_i$ is the resolution or truncation deficiency caused by the measurement process. The unbiased estimates of mean and variance are

$$\hat{\mu} = \frac{\Sigma s_i}{n} = \bar{s} \qquad \hat{\sigma}^2 = \frac{\Sigma(s_i - \bar{s})^2}{n-1} \qquad (3.44)$$

Noting that the experimenter has the set $\{x_n\}$ rather than $\{s_n\}$, the mean and variance become

$$\hat{\mu} = \frac{\Sigma x_i}{n} \pm \frac{\Sigma e_i}{n} = \bar{x} \pm \frac{\Sigma e_i}{n} \qquad \hat{\sigma}^2 = \frac{\Sigma(x_i - \bar{x})^2}{n-1} \tag{3.45}$$

Thus the truncation or resolution has no effect on the estimate of variance but does effect the estimate of the mean.

The truncation errors e_i are not necessarily distributed randomly and may all be of the same sign. Thus $\bar{x}$ can be biased as much as $\Sigma e_i / n = \bar{e}$ high or low from the unbiased estimate of the value of μ, so that $\hat{\mu} = \bar{x} \pm \bar{e}$.

If e_i is a random variable, such as when observing a variable with a measuring system of finite resolution, the values of e_i may be plus or minus, but their upper bound is R (the resolution of the measurement). Thus the resolution error is no larger than R, and $\mu = \bar{x} \pm R$.

If the truncation is never more than that dictated by the resolution limits R of the measuring system, the uncertainty in $\bar{x}$ as a measure of the most representative value of μ is never larger than R plus the uncertainty due to the random error. Thus $\hat{\mu} = \bar{x} \pm (\hat{W} + R)$. It should be emphasized that the uncertainty interval can never be less than the resolution bounds of the measurement no matter how small the random error might be. The resolution bounds cannot be reduced without changing the measurement system.

When x_i is observed to m significant digits, the uncertainty (except for random error) is never more than $\pm (5/10^m)$, and the bounds on s_i are equal to $x_i \pm (5/10^m)$, so that

$$x_i - \frac{5}{10^m} < s_i < x_i + \frac{5}{10^m} \tag{3.46}$$

The relation for $\hat{\mu}$ for m significant digits is then

$$\hat{\mu} = \bar{x} \pm \frac{\Sigma e_i}{n} = \bar{x} \pm \frac{\Sigma \, 5/10^m}{n} = \bar{x} \pm \frac{5}{10^m} \tag{3.47}$$

When the uncertainty due to significant digits is combined with the resolution limits and random error, the uncertainty interval on μ becomes

$$\mu = \hat{\mu} \pm \left(\hat{W} + R + \frac{5}{10^m} \right) \tag{3.48}$$

This illustrates that the number of significant digits of a measurement should be carefully chosen in relation to the resolution limits of the measuring system so that $5/10^m$ has about the same magnitude as R. Additional significant digits would imply more accuracy to the measurement than would actually exist based on the resolving ability of the measuring system.

3.9.3 Amount of Data to Take

Exactly what data to take and how much data to take are two important questions to be answered in any experiment. Assuming that the correct variables have been measured, the amount of data to obtain can be determined by using the relation

$$\mu = \hat{\mu} \pm \left(\hat{W}_x + R + \frac{5}{10^m} \right) \tag{3.49}$$

where it is presumed that several samples may exist for estimation of μ. This equation can be rewritten such that

$$\mu = \bar{\bar{x}} \pm \left[k(\nu, \alpha) \frac{\hat{\sigma}}{\sqrt{n}} + R + \frac{5}{10^m} \right] \tag{3.50}$$

If one wishes to know the value of n to achieve the difference in $\mu - \bar{x}$ within a stated percent of μ, the relation can be solved for n to yield

$$n^2 = \frac{k(\nu, \gamma)\,\hat{\sigma}}{(\text{percent}/100)\hat{\mu} - R - (5/10^m)} \tag{3.51}$$

This equation can yield valid values of n only once estimates of $\hat{\mu}, \hat{\sigma}, k, R$, and m are available. This means that the most correct value of n can be obtained only once the measurement system and data-taking procedure have been specified so that R and m are known. Furthermore, either a preliminary experiment or a portion of the actual experiment should be performed to obtain good estimates of $\hat{\mu}$ and $\hat{\sigma}$. Because k depends on the type of data distribution, the sample size n yields an iterative reduction. Thus the most valid estimates of the amount of data to take can be obtained only after the experiment has begun. This requires that estimates of the mean and the standard deviation be obtained by performing part of the experiment. However, the equation can be quite useful for prediction purposes if one wishes to estimate values of $\hat{\mu}, \hat{\sigma}, k, R$, and m. This is especially important in experiments where the cost of a single run may be relatively high.

Example 9.   The life (mileage) for a certain type of automotive tire is known to follow a gaussian (normal) distribution function. The mean and standard deviation of the mileage for these tires are estimated to be 84 000 and 2100 mi, respectively, from a sample of nine tires. Determine the 90 percent confidence limits for the means of all such tires manufactured by the company if the resolution of these measurements is 5 mi. On the basis of the sample, how much data (i.e., what is the sample size?) are required to establish the life of this type of tire to within ± 1 percent with 90 percent confidence and a resolution of 5 mi?

   Solution

$$\mu = \mu \pm (t\hat{\sigma}_{\bar{x}} + R)$$

$$\sigma_x = 2230 \text{ mi}$$

$$\hat{\sigma}_{\bar{x}} = \frac{\sigma_x}{\sqrt{n}} = \sqrt{\frac{\Sigma(x_i - \bar{x})^2}{n(n-1)}} = \pm 743 \text{ mi}$$

$$t = t(\nu, \gamma) = t(8, 0.90) = 1.860^{\dagger}$$

$$\mu = 84\,000 \pm [1.860(743) + 5]$$

$$= 84\,000 \pm 1387 \text{ mi}$$

$$\mu = \hat{\mu} \pm (t\hat{\sigma}_x + R)$$

$$\mu - \hat{\mu} = 0.01\bar{x} = t\frac{\hat{\sigma}_x}{\sqrt{n}} + R$$

$$\frac{t}{\sqrt{n}} = \frac{0.01\bar{x} - R}{\hat{\sigma}_x} = \frac{835}{2230} = 0.374$$

---

† From Ref. [3.7], Table A-8.

Use of Table A-8 in Ref. [3.7] yields the final result of $n = 122$.

3.10 CONFIDENCE LIMITS

A "confidence" limit or uncertainty interval is associated with a probability. The area under a probability-density curve between any two limits gives the value of the probability or confidence that any item sampled at random from the population will have a value between the two limits chosen. For example, the area under the gaussian (or normal) probability-density function $p(x)$ between values of x_1 and x_2 is given by

$$\int_{x_1}^{x_2} p(x)\, dx = \int_{x_1}^{x_2} \frac{1}{\sigma\sqrt{2\pi}} \exp\left[-\frac{(x-\mu)^2}{2\sigma^2}\right] dx = \gamma \qquad (3.52)$$

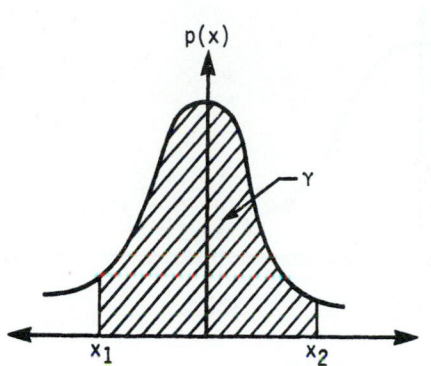

$p(x)$

FIGURE 3.34 Confidence limits on the gaussian distribution.

and represents (as shown in Fig. 3.34) the probability that any one item selected at random from the population of values will have a magnitude between x_1 and x_2. In this figure, x_1 and x_2 are called *precision indices*. The symbol W_x is used to denote the values of x to be taken as a precision index or uncertainty interval. For the gaussian distribution, a value of the precision indices of $\pm\sigma$ (one standard deviation) yields a probability or confidence of 68.3 percent. Also, $W_x = \pm 1.966$ are the 95 percent confidence limits and $W_x = \pm 2.586$ are the 99 percent confidence limits for the population of items following the gaussian distribution. The confidence limits and associated probability are illustrated in Fig. 3.34. This information is often represented by the following probabilistic statement:

$$p(x_1 < x < x_2) = \gamma = 1 - \alpha \qquad (3.53)$$

where γ = the probability of the value of x from an observation to be between the values of x_1 and x_2. The value γ is known as the *confidence level*, whereas the value α is known as the *significance level*. The meaning of the one-sided probabilistic statements

$$p(x < x_3) = \gamma \qquad \text{and} \qquad p(x > x_3) = \alpha = 1 - \gamma \qquad (3.54)$$

is illustrated in Fig. 3.35.

Example 10. Certain strain gauges are manufactured with a resistance specification of $120 \pm 0.5\ \Omega$. All gauges not meeting this specification are rejected. If all such strain gauges manufactured follow a gaussian probability function, estimate the standard deviation of the manufacturing process if 2 percent are typically rejected (refer to Fig. 3.36 and Table A-4 of Ref. [3.7]).

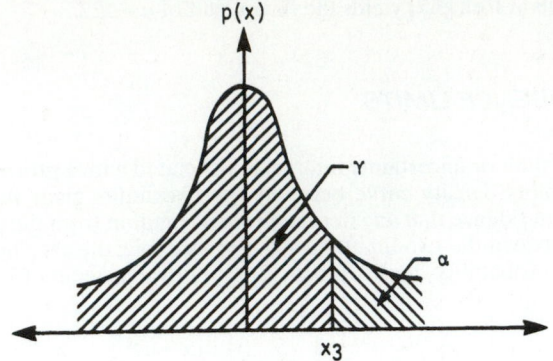

FIGURE 3.35   Single-sided probabilistic statements.

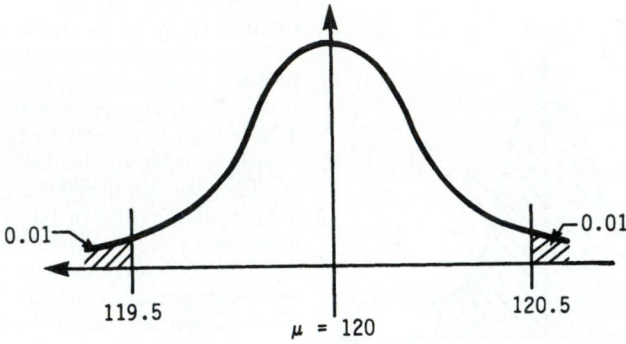

FIGURE 3.36   Gaussian distribution for Example 10.

Solution

$$p(119.5 < R < 120.5) = 0.98$$

$$0.01 = \int_{z_1}^{\infty} f(z)\, dz = 0.500 - \int_{0}^{z_1} f(z)\, dz$$

$$\int_{0}^{z_1} f(z)\, dz = 0.490 \qquad z_1 = 2.323$$

$$z = \frac{x - \mu}{\sigma} = \frac{120.5 - 120.0}{\sigma} = 2.323$$

$$\sigma = \frac{0.5}{2.323} = 0.215\ \Omega$$

3.10.1 Confidence Limits on Means

The confidence limits on the establishment of the mean μ of a population from sample data are established [3.13] by the probabilistic statement

$$P\left[-t(\nu, \gamma) < \frac{\bar{x} - \mu}{\sigma/\sqrt{n}} < t(\nu, \gamma)\right] = \gamma \tag{3.55}$$

where $t(\nu, \gamma)$ is the well-known t statistic. Rearrangement of Eq. (3.55) yields the probabilistic statement showing how to establish confidence limits on μ:

$$P\left[\left(\bar{x} - t\frac{\sigma}{\sqrt{n}}\right) < \mu < \left(\bar{x} + \frac{t\sigma}{\sqrt{n}}\right)\right] = \gamma \tag{3.56}$$

The effects of measurement resolution and significant digits can be included in the expression to yield Eq. (3.57). This equation shows that even if the random error is zero, the uncertainty on μ cannot be less than the resolution and truncation uncertainties:

$$P\left[\left(\bar{x} - \frac{t\sigma}{\sqrt{n}} - R - \frac{5}{10^m}\right) < \mu < \left(\bar{x} + \frac{t\sigma}{\sqrt{n}} + R + \frac{5}{10^m}\right)\right] = \gamma \tag{3.57}$$

Pooling or Combining Data. If one has several samples (perhaps obtained by different instrument systems) for estimating a population mean μ and variance σ^2, these data can be formally combined to obtain better estimates of these entities. If each data set comes from a normal population of mean μ and variance σ^2, the maximum-likelihood technique can be used to show how the means and variances of the samples are combined to provide the better estimates:

$$\hat{\mu}_c = \frac{\Sigma n_j \bar{x}_j / \hat{\sigma}_j^2}{\Sigma n_j / \hat{\sigma}_j^2} = \frac{n_1 \bar{x}_1 / \hat{\sigma}_1^2 + n_2 \bar{x}_2 / \hat{\sigma}_2^2 + \cdots}{n_1 / \hat{\sigma}_1^2 + n_2 / \hat{\sigma}_2^2 + \cdots} \tag{3.58}$$

$$\hat{\sigma}_c^2 = \frac{\Sigma n_j}{\Sigma n_j / \hat{\sigma}_j^2} = \frac{n_1 + n_2 + \cdots}{n_1 / \hat{\sigma}_1^2 + n_2 / \hat{\sigma}_2^2 + \cdots} \tag{3.59}$$

The reason for pooling data is to obtain more precise estimates of μ and σ^2 than any sample alone can provide. The intuitive thought, to use the more precise data and to discard the less precise data, *is not* the appropriate thing to do. Instead, the uncertainty intervals on the estimate of population mean μ and variance σ^2 are considerably reduced by pooling the data.

Example 11. The data in the sample table below represent shear strengths in pounds per square inch (psi) of an adhesive. To market this adhesive, what should its guaranteed shear strength be with a confidence level of 99.5 percent?

Sample 1	Sample 2	Sample 3	Sample 4
3566	3180	3470	3520
3630	3080	3620	3960
3800	3400	3466	4040
3880	3820	3460	3600
3840	3480	3760	3899

Solution

$$\hat{\mu}_1 = 3742.2 \qquad \hat{\mu}_2 = 3392.0 \qquad \hat{\mu}_3 = 3555.2 \qquad \hat{\mu}_4 = 3803.8$$

$$\hat{\sigma}_1^2 = 18\,880 \qquad \hat{\sigma}_2^2 = 83\,334 \qquad \hat{\sigma}_3^2 = 17\,612 \qquad \hat{\sigma}_4^2 = 70\,567$$

$$\hat{\mu}_c = 3639.7 \text{ psi} \qquad \hat{\sigma}_c^2 = 29\,431 \ (\text{psi})^2 \qquad \hat{\sigma}_c = 171.3 \text{ psi}$$

$$\mu = \hat{\mu} \pm \left(t \frac{\hat{\sigma}}{\sqrt{n}} + \frac{5}{10^m} + R \right) = 3639.7 \pm 2.921 \frac{(171.3)}{\sqrt{20}}$$

$$= 3640 \pm 108$$

where $t = t(v, \gamma) = t(16, 0.99) = 2.921$ from Ref. [3.7], Table A-8. The strength should be guaranteed to $3640 - 108 = 3532$ psi.

3.10.2 Confidence Limits on Variance

To establish the confidence limits on variance, the chi-square statistic $\chi^2 = [\Sigma(x_i - \bar{x})^2]/\sigma^2$ is used [3.13], and the probabilistic statement becomes

$$P\left[\chi_L^2 < \frac{\Sigma(x_i - \bar{x})^2}{\sigma^2} < \chi_R^2 \right] = \gamma \qquad (3.60)$$

Rearranging yields the probabilistic statement showing how to determine confidence limits on variance:

$$P\left[\frac{\Sigma(x_i - \bar{x})^2}{\chi_R^2} < \sigma^2 < \frac{\Sigma(x_i - \bar{x})^2}{\chi_L^2} \right] = \gamma \qquad (3.61)$$

Example 12. The standard deviation of the lifetimes of a sample of 200 high-pressure seals is 100 h. What are the 95 percent confidence limits on the standard deviation of all such seals? (See Ref. [3.7], Table A-7, for $v > 30$.)

Solution

$$\sqrt{2\chi^2} - \sqrt{2v - 1} \sim N(0, 1)$$

$$\therefore \ \sqrt{2\chi^2} - \sqrt{2v - 1} = z \qquad \text{or} \qquad \chi^2 = \tfrac{1}{2}(z + \sqrt{2v - 1})^2$$

$$\chi_R^2 = \chi^2(0.025, 199)$$

$$= \tfrac{1}{2}(+z_{0.025} + \sqrt{2(199) - 1})^2$$

$$= \tfrac{1}{2}(+1.96 + 19.92)^2 = 239$$

$$\chi_R = \sqrt{239} = 15.4$$

$$\chi_L^2 = \tfrac{1}{2}(+z_{0.975} + \sqrt{2(199) - 1})^2$$

$$= \tfrac{1}{2}(-1.96 + 19.92)^2 = 161$$

$$\chi_L = \sqrt{161} = 12.7$$

$$P\left(\frac{\hat{\sigma}\sqrt{n - 1}}{\chi_R} < \sigma < \frac{\hat{\sigma}\sqrt{n - 1}}{\chi_L} \right) = \gamma$$

$$\therefore \ P(91.2 < \sigma < 111.3) = 0.95$$

3.11 PROPAGATION OF ERROR OR UNCERTAINTY

In many cases the desired quantity and its uncertainty cannot be measured directly but must be calculated from the data of two or more measured variables. This is represented mathematically by

$$R = R(x_1, x_2, x_3, \ldots, x_n) \tag{3.67}$$

where the x's = measured variables and R = the dependent or calculated quantity. To determine the most representative value and uncertainty of the calculated quantity, the following equations can be used [3.15]:

$$\mu_R = R(\mu_{x_1}, \mu_{x_2}, \mu_{x_3}, \ldots, \mu_{x_n}) + \frac{1}{2}\Sigma\left(\frac{\partial^2 R}{\partial x_i^2}\right)\sigma_{x_i}^2 \tag{3.68}$$

$$\sigma_R^2 = \Sigma\left(\frac{\partial R}{\partial x_i}\right)_\mu^2 \sigma_{x_i}^2 + \frac{1}{2}\Sigma\left(\frac{\partial^2 R}{\partial x_i^2}\right)_\mu^2 \sigma_{x_i}^4 + \cdots \tag{3.69}$$

Example 13. If $r = x^n$, then μ_R and σ_R^2 become

$$\mu_R = (\mu_x)^n\left[1 + \tfrac{1}{2}(n)(n-1)\left(\frac{\sigma_x}{\mu_x}\right)^2\right] \tag{3.70}$$

$$\sigma_R^2 = n^2(\mu_x)^n\left[1 + \tfrac{1}{2}(n-1)^2\left(\frac{\sigma_x}{\mu_x}\right)^2\right] \tag{3.71}$$

The role of the coefficient of variation σ_x/μ_x should be noted. If σ_x/μ_x is small, the first terms in each of Eqs. (3.68) and (3.69) are all that need be evaluated to yield the desired results. The results for several other functions are given by Mischke [3.15].

The variances in Eqs. (3.68) and (3.69) can be replaced with confidence limits according to the equation

$$W_x = k\sigma_x \tag{3.72}$$

where k = a constant depending on the type of data distribution. The propagation-of-variance equation becomes the following propagation-of-uncertainty formula:

$$W_R^2 = \Sigma\left(\frac{\partial R}{\partial x_i}\right)_\mu^2 W_{x_i}^2 \tag{3.73}$$

Use of this equation requires that all the x_i's be independently measured, that the k values be the same for each x_i distribution, and that all W_{x_i}'s represent the same level of uncertainty or confidence.

Example 14. The strain and its uncertainty a distance L from the load F on the top surface of the cantilever beam shown in Fig. 3.32 is to be determined subject to the following measured data:

$$b = 0.500 \pm 0.001 \text{ in}$$

$$h = 0.250 \pm 0.001 \text{ in}$$

$$L = 20.00 \pm 0.01 \text{ in}$$

$$F = 10.00 \pm 0.01 \text{ lbf}$$

$$E = (30.0 \pm 0.01) \times 10^6 \text{ psi}$$

Solution.   The strain is determined by $\epsilon = 6FL/(Ebh^2) = 1280 \times 10^{-6}$ in/in, or $\epsilon = 1280$ microstrain. The uncertainty in the strain is determined by

$$W_\epsilon^2 = \left(\frac{\partial \epsilon}{\partial F}\right)_\mu^2 W_F^2 + \left(\frac{\partial \epsilon}{\partial L}\right)_\mu^2 W_L^2 + \left(\frac{\partial \epsilon}{\partial E}\right)_\mu^2 W_E^2 + \left(\frac{\partial \epsilon}{\partial b}\right)_\mu^2 W_b^2 + \left(\frac{\partial E}{\partial h}\right)_\mu^2 W_h^2 \qquad (3.74)$$

This simplifies to the following equation when all the coefficients of variation are small:

$$\left(\frac{W_\epsilon}{\overline{\epsilon}}\right)^2 = \left(\frac{W_F}{\overline{F}}\right)^2 + \left(\frac{W_L}{\overline{L}}\right)^2 + \left(\frac{W_E}{\overline{E}}\right)^2 + \left(\frac{W_b}{\overline{b}}\right)^2 + 4\left(\frac{W_h}{\overline{h}}\right)^2 \qquad (3.75)$$

Substitution of the data in Eq. (3.75) yields $W_\epsilon = \pm 11$ microstrain. When the second terms in Eqs. (3.68) and (3.69) are used, the results are $\epsilon = 1280.1$ microstrain and $W_\epsilon = \pm 11.3$ microstrain, which illustrates that when the coefficients of variation are small, the additional terms do not contribute significantly to ϵ and W_ϵ.

A simplified form of the preceding propagation-of-uncertainty equation results if the function R has the special form

$$R = (X_1^a X_2^b \cdots X_n^m)K \qquad (3.76)$$

where $K =$ any constant and the exponents a, b, and m may be positive or negative, integer or noninteger. Substitution of Eq. (3.76) into Eq. (3.73) yields

$$\left(\frac{W_R}{\overline{R}}\right)^2 = a^2\left(\frac{W_{x_1}}{\overline{x}_1}\right)^2 + b^2\left(\frac{W_{x_2}}{\overline{x}_2}\right)^2 + \cdots + m^2\left(\frac{W_{x_n}}{\overline{x}_n}\right)^2 \qquad (3.77)$$

The equation for uncertainty in the calculated quantity allows one to see the effect of the exponents a, b, . . . , m in propagation of the measured variable uncertainties to the uncertainty in the calculated quantity. For example, a squared variable x_i^2 has four times the effect on the propagation of uncertainty in the result that it would have if it had a unity exponent. It should also be noted that all terms are dimensionless in Eq. (3.77) and that the ratios $W_{x_i}/\overline{x}_i$, called *relative error*, are proportional to coefficients of variation by the factor k, since $W_x = k\sigma_x$.

If the experimental data are "single sample," the values of W_{x_i} must be estimated by the experimentor, as indicated by Kline and McClintock [3.14]. It is suggested that with single-sample data the W_{x_i} be estimated as the maximum error (which corresponds approximately to the 99 percent confidence level).

The propagation-of-uncertainty equation is extremely valuable in planning experiments. If one desires a certain precision on the calculated result R, the precision of the measured variables can be determined from this equation. Because instrument precision is directly related to its cost, one also has a method of estimating costs of a proposed measuring system by use of the propagation equation.

REFERENCES

3.1 E. A. Mechtly, *The International System of Units, Physical Constants and Conversion Factors*, 2d rev. ed., NASA sp-7012, 1973.

3.2 Anon., "Units, Standards and References," *Measurements and Control*, March–April 1970, pp. 49–80.

3.3 J. F. Swindells (ed.), *Precision Measurement and Calibration, Temperature,* vol. 2, National Bureau of Standards Special Publication 300, 1968.

3.4 P. K. Stein, *Measurement Engineering,* Stein Engineering Services, Inc., Phoenix, Ariz., 1974.

3.5 T. G. Beckwith, N. L. Buck, and R. D. Marangoni, *Mechanical Measurement,* 3d ed., Addison-Wesley, Reading, Mass., 1982.

3.6 E. Doebelin, *Measurement Systems: Application and Design,* 3d ed., McGraw-Hill, New York, 1983.

3.7 N. H. Cook and F. Rabinowicz, *Physical Measurement and Analysis,* Addison-Wesley, Reading, Mass., 1963.

3.8 H. L. Harrison and J. G. Bollinger, *Introduction to Automatic Controls,* 2d ed., Harper and Row, Publishers, New York, 1969.

3.9 Anon., *Piezoelectric Accelerometer User's Handbook,* Bulletin 4200-96, Consolidated Electrodynamics Corp., Data Instruments Division, Pasadena, Calif.

3.10 S. D. Prensky, *Electronic Instrumentation,* 2d ed., Prentice-Hall, Englewood Cliffs, N.J., 1971.

3.11 Headquarters Staff, *The Radio Amateur's Handbook,* 40th ed., American Radio Relay League, 1963.

3.12 M. R. Cereijo, "Shields and Grounds: The 'Other' Circuit Elements," *Machine Design,* Aug. 22, 1974, pp. 88–91.

3.13 J. B. Kennedy and A. M. Neville, *Basic Statistical Methods for Engineers and Scientists,* 2d ed., Harper and Row, Publishers, New York, 1976.

3.14 S. J. Kline and F. McClintock, "Describing Uncertainty in Single Sample Experiments," *Mechanical Engineering,* vol. 75, 1953, pp. 3–8.

3.15 C. R. Mischke, *Mathematical Model Building,* 2d rev. ed., Iowa State University Press, Ames, 1980.

3.16 C. Lipson and N. J. Sheth, *Statistical Design and Analysis of Engineering Experiments,* McGraw-Hill, New York, 1973.

ADDITIONAL REFERENCES

Bartee, E. M.: *Statistical Methods in Engineering Experiments,* Charles E. Merrill, Columbus, Ohio, 1966.

Cannon, R. H., Jr.: *Dynamics of Physical Systems,* McGraw-Hill, New York, 1967.

Dally, J. W., Riley, W. F., and McConnell, K. G.: *Instrumentation for Engineering Measurements,* John Wiley & Sons, New York, 1984.

Dove, R., and Adams, P.: *Experimental Stress Analysis and Morton Measurement,* Charles E. Merrill, Columbus, Ohio, 1964.

Holman, J. P.: *Experimental Methods for Engineers,* 4th ed., McGraw-Hill, New York, 1984.

Ku, H. H. (ed.): *Precision Measurement and Calibration,* vol. 1, National Bureau of Standards Special Publication 300, February 1969.

Natrella, M. G.: *Experimental Statistics,* National Bureau of Standards Handbook 91, Aug. 1, 1963.

Nilsson, J. W.: *Introduction to Circuits, Instruments and Electronics,* Harcourt, Brace and World, New York, 1968.

Snedecor, G. W., and Cochran, W. G.: *Statistical Methods,* 6th ed., Iowa State University Press, Ames, 1967.

CHAPTER 4
NUMERICAL METHODS

Ray C. Johnson, Ph.D.
Higgins Professor of Mechanical Engineering Emeritus
Worcester Polytechnic Institute
Worcester, Massachusetts

In this chapter some numerical techniques particularly useful in the field of machine design are briefly summarized. The presentations are directed toward automated calculation applications using electronic calculators and digital computers. The sequence of presentation is logically organized in accordance with the preceding table of contents, and emphasis is placed on useful equations and methods rather than on the derivation of theory.

4.1 NUMBERS

In the design and analysis of machines it is necessary to obtain quantities for various items of interest, such as dimensions, material properties, area, volume, weight, stress, and deflection. Quantities for such items are expressed by numbers accompanied by the units of measure for a meaningful perspective. Also, numbers always have an algebraic sign, which is assumed to be positive unless clearly designated as negative by a minus sign preceding the number. The various kinds of numbers are defined in Sec. 2-7, which see.

4.1.1 Real Numbers, Precision, and Rounding

Any numerical quantity is expressed by a *real number* which may be classified as an integer, a rational number, or an irrational number. For practical purposes of calculation or manufacturing, it is often necessary to approximate a real number by a specified number of digits. For some cases, significant numbers may be useful, and the following relates to the obtainable degree of precision.

Degree of Precision.   In machine design, real numbers are expressed by significant digits as related to practical considerations of accuracy in manufacturing and operation. For example, a dimension of a part may be expressed by four significant digits as 3.876 in, indicating for this number that the dimension will be controlled in manufacturing by a tolerance expressed in thousandths of an inch. As another example, the weight density of steel may be used as 0.283 lbm/in^3, indicating a level of accuracy associated with control in the manufacturing of steel stock. Both these examples illustrate numbers as basic terms in a design specification.

However, it is often necessary to analyze a design for quantities of interest using equations of various types. Generally, we wish to evaluate a dependent variable by an equation expressed in terms of independent variables. The degree of precision obtained for the dependent variable depends on the accuracy of the predominant term in the particular equation, as related to algebraic operations. In what follows, we will assume that the accuracy of the computational device is better than the number of significant figures in a determined value.

For addition and subtraction, the predominant term is the one with the least number of significant decimals. For example, suppose a dimension D in a part is determined by three machined dimensions A, B, and C using the equation $D = A + B - C$. Specifically, if the accuracy of each dimension is indicated by the significant digits in $A = 12.50$ in, $B = 1.062$ in, and $C = 12.375$ in, the predominant term is A, since it has the least number of significant decimals with only two. Thus D would be accurate to only two decimals, and we would calculate $D = A + B - C = 12.50 + 1.062 - 12.375 = 1.187$ in. We should then round this value to two decimals, giving $D = 1.19$ in as the determined value. Also, we note that D is accurate to only three significant figures, although A and B were accurate to four and C was accurate to five.

For multiplication and division, the predominant term is simply defined as the one with the least number of significant digits. For example, suppose tensile stress σ is to be calculated in a rectangular tensile bar of cross section b by h using the equation $\sigma = P/(bh)$. Specifically, if $P = 15\ 000$ lb, and as controlled by manufacturing accuracy $b = 0.375$ in and $h = 1.438$ in, the predominant term is b, since it has only three significant digits. Incidentally, we have also assumed that P is accurate to at least three significant digits. Thus we would calculate $\sigma = P/(bh) = 15\ 000/[0.375(1.438)] = 27\ 816$ psi. We should then round this value to three significant digits, giving $\sigma = 27\ 800$ psi as the determined value.

For a more rigorous approach to accuracy of dependent variables as related to error in independent variables, the theory of relative change may be applied, as explained in Sec. 4.4.

Rounding.   In the preceding examples, we note that determined values are rounded to a certain number of significant decimals or digits. For any case, the calculations are initially made to a higher level of accuracy, but rounding is made to give a more meaningful answer. Hence we will briefly summarize the rules for rounding as follows:

1. If the least significant digit is immediately followed by any digit between 5 and 9, the least significant digit is increased in magnitude by 1. (An exception to this rule is the case where the least significant digit is even and it is immediately followed by the digit 5 with all trailing zeros. In that event, the least significant digit is left unchanged.)

2. If the least significant digit is immediately followed by any digit between 0 and 4, the least significant digit is left unchanged.

For example, with three significant digits desired, 2.765 01 becomes 2.77, 2.765 becomes 2.76, −1.8743 becomes −1.87, −0.4926 becomes −0.493, and 0.003 792 8 becomes 0.003 79.

4.1.2 Complex Numbers

Complex numbers are ones that contain two independent parts, which may be represented graphically along two independent coordinate axes. The independent components are separated by introduction of the operator $j = \sqrt{-1}$. Thus we express complex number $c = a + bj$, where a and b by themselves are either integers, rational numbers, or irrational numbers. Often a is called the *real component* and bj is called the *imaginary component*. The magnitude for c is $\sqrt{a^2 + b^2}$. For example, if $c = 3.152 + 2.683j$, its magnitude is

$$|c| = \sqrt{(3.152)^2 + (2.683)^2} = 4.139$$

Algebraically, the values for a and b may be positive or negative, but the magnitude of c is always positive.

4.2 FUNCTIONS

Functions are mathematical means for expressing a definite relationship between variables. In numerical applications, generally the value of a dependent variable is determined for a set of values of the independent variables using an appropriate functional expression. Functions may be expressed in various ways, by means of tables, curves, and equations.

4.2.1 Tables

Tables are particularly useful for expressing discrete value relations in machine design. For example, a catalog may use a table to summarize the dimensions, weight, basic dynamic capacity, and limiting speed for a series of standard roller bearings. In such a case, the dimensions would be the independent variables, whereas the weight, basic dynamic capacity, and limiting speed would be the dependent variables.

For many applications of machine design, a table as it stands is sufficient for giving the numerical information needed. However, for many other applications requiring automated calculations, it may be appropriate to transform at least some of the tabular data into equations by curve-fitting techniques. For example, from the tabular data of a roller-bearing series, equations could be derived for weight, basic dynamic capacity, and limiting speed as functions of bearing dimensions. The equations would then be used as part of a total equation system in an automated design procedure.

4.2.2 Curves

Curves are particularly useful in machine design for graphically expressing continuous relations between variables over a certain range of practical interest. For the case of more than one independent variable, families of curves may be presented on a single graph. In many cases, the graph may be simplified by the use of dimensionless ratios for the independent variables. In general, curves present a valuable picture of how a dependent variable changes as a function of the independent variables.

For example, for a stepped shaft in pure torsion, the stress concentration factor K_{ts} is generally presented as a family of curves, showing how it varies with respect to the independent dimensionless variables r/d and D/d. For the stepped shaft, r is the fillet radius, d is the smaller diameter, and D is the larger diameter.

For many applications of machine design, a graph as it stands may be sufficient for giving the numerical data needed. However, for many other applications requiring automated calculations, equations valid over the range of interest may be necessary. The given graph would then be transformed to an equation by curve-fitting techniques. For example, for the stepped shaft previously mentioned, stress concentration factor K_{ts} would be expressed by an equation as a function of r, d, and D derived from the curves of the given graph. The equation would then be used as part of a total equation system in the decision-making process of an automated design procedure.

4.2.3 Equations

Equations are the most powerful means of function expression in machine design, especially when automated calculations are to be made in a decision-making procedure. Generally, equations express continuous relations between variables, where a dependent variable y is to be numerically determined from values of independent variables x_1, x_2, x_3, etc. Some commonly used types of equations in machine design are summarized next.

Linear Equations.   The general form of a linear equation is expressed as follows:

$$y = b + c_1 x_1 + c_2 x_2 + \cdots + c_n x_n \tag{4.1}$$

Constant b and coefficient $c_1, c_2, \ldots, c_n$ may be either positive or negative real numbers, and in a special case, any one of these may be zero.

For the case of one independent variable x, the linear equation $y = b + cx$ is graphically a straight line. In the case of two independent variables x_1 and x_2, the linear equation $y = b + c_1 x_1 + c_2 x_2$ is a plane on a three-dimensional coordinate system having orthogonal axes x_1, x_2, and y.

Polynomial Equations.   The general form of a polynomial equation in two variables is expressed as follows:

$$y = b + c_1 x + c_2 x^2 + \cdots + c_n x^n \tag{4.2}$$

Constant b and coefficients $c_1, c_2, \ldots, c_n$ may be either positive or negative real numbers, and in a special case, any one of these may be zero.

For the special case of $n = 1$, the equation $y = b + c_1 x$ is linear in x. For the special case of $n = 2$, the equation $y = b + c_1 x + c_2 x^2$ is known as a *quadratic equation*. For the special case of $n = 3$, the equation $y = b + c_1 x + c_2 x^2 + c_3 x^3$ is known as a *cubic equation*. In general, for $n > 3$, Eq. (4.2) is known as a *polynomial of degree n*.

Simple Exponential Equations. The general form for a type of simple exponential equation commonly used in machine design is expressed as follows:

$$y = bx_1^{c_1} x_2^{c_2} \cdots x_n^{c_n} \qquad (4.3)$$

Coefficient b and exponents $c_1, c_2, \ldots, c_n$ may be either positive or negative real numbers. However, except for the special case of any c_i being an integer, the corresponding values of x_i must be positive.

For the special case of $n = 1$ with $c_1 = 1$, the equation $y = bx$ is a simple straight line. For $n = 1$ with $c_1 = 2$, the equation $y = bx^2$ is a simple parabola. For $n = 1$ with $c_1 = 3$, the equation $y = bx^3$ is a simple cubic equation.

As a specific example of the more general case expressed by Eq. (4.3), a simple exponential equation might be as follows:

$$y = 38.69 \frac{x_1^{2.670} x_4^2}{x_2^{0.092} x_3^{1.07}}$$

For this example, $n = 4$, $b = 38.69$, $c_1 = 2.670$, $c_2 = -0.092$, $c_3 = -1.07$, and $c_4 = 2$. Also, if at a specific point we have $x_1 = 4.321$, $x_2 = 3.972$, $x_3 = 8.706$, and $x_4 = 0.0321$, the equation would give the value of $y = 0.1725$.

The general form for another type of simple exponential equation occasionally used in machine design is expressed as follows:

$$y = bc_1^{x_1} c_2^{x_2} \cdots c_n^{x_n} \qquad (4.4)$$

Coefficient b and independent variables $x_1, x_2, \ldots, x_n$ may be either positive or negative real numbers. However, except for the special case of any x_i being an integer, the corresponding values of c_i must be positive.

Transcendental Equations. The most commonly encountered types of transcendental equations are classified as being either trigonometric or logarithmic. For either case, inverse operations may be desired. In general, *transcendental equations* determine a dependent variable y from the value of an independent variable x as the argument.

The basic *trigonometric equations* are $y = \sin x$, $y = \cos x$, and $y = \tan x$. The argument x may be any real number, but it should carry angular units of radians or degrees. For electronic calculators, the units for x are generally degrees. However, for microcomputers or larger electronic computers, the units for x are generally radians.

The basic *logarithmic equation* is $y = \log x$. However, in numerical applications, care must be exercised in recognizing the base for the logarithmic system used. For natural logarithms, the Napierian base $e = 2.718\,281\,8 \ldots$ is used, and the inverse operation would be $x = e^y$. For common logarithms, the base 10 is used, and the inverse operation would be $x = 10^y$.

A special relationship of importance is recognized by taking the logarithm of both sides in the simple exponential Eq. (4.3), resulting in the following equation:

$$\log y = \log b + c_1 \log x_1 + c_2 \log x_2 + \cdots + c_n \log x_n \qquad (4.5)$$

We see that this equation is analogous to linear Eq. (4.1) by replacing $y, b, x_1, x_2, \ldots, x_n$ of Eq. (4.1) with $\log y$, $\log b$, $\log x_1$, $\log x_2$, $\ldots$, $\log x_n$, respectively. Thus the equation $y = bx^c$ will plot as a straight line on log-log graph paper, regardless of the values for constants b and c.

Combined Equations. Some basic types of equations have now been summarized, and they will be applied later in techniques of curve fitting. However, any of the more complicated equations found in machine design may be considered as special combinations of the basic equations, with the terms related by algebraic operations. Such equations might be placed in the general classification of combined equations. As a specific example of a combined equation, a *polynomial equation* is merely the sum of positive simple exponential terms, each of which has the general form of the right side of Eq. (4.3).

4.3 SERIES

A *series* is an ordered set of sequential terms generally connected by the algebraic operations of addition and subtraction. The number of terms can be either finite or infinite in scope. If the terms contain independent variables, the series is really an equation for calculating a dependent variable, such as the polynomial Eq. (4.2).

If a series is lengthy, it is often possible to approximate the series with a finite number of terms. The criterion for determining how many terms of the sequence are necessary is based on a consideration of convergence. The number of terms used must be sufficient for convergence of the determined value to an acceptable level of accuracy when compared with the entire series evaluation. This will be considered specifically in Sec. 4.4 on approximations and error.

Some commonly used series in machine design will be briefly summarized next. A more complete coverage can be found in any handbook on mathematics, and what follows is just a small sample.

4.3.1 Binomial Series

Consider the combined equation $y = (x_1 + x_2)^n$, where x_1 and x_2 are independent variables and n is an integer. The binomial series expansion of this equation is as follows:

$$y = (x_1 + x_2)^n$$

$$= x_1^n + nx_1^{n-1}x_2 + \frac{n(n-1)}{2!}x_1^{n-2}x_2^2 + \frac{n(n-1)(n-2)}{3!}x_1^{n-3}x_2^3 + \cdots \qquad (4.6)$$

In Eq. (4.6), if integer n is positive, the series consists of $n + 1$ terms. However, if integer n is negative, in general the number of terms is infinite and the series converges if $x_2^2 < x_1^2$.

4.3.2 Trigonometric Series

Some trigonometric relations will be approximated in Sec. 4.4 based on the series expansions summarized as follows:

$$y = \sin x = x - \frac{x^3}{3!} + \frac{x^5}{5!} - \frac{x^7}{7!} + \cdots \qquad (4.7)$$

$$y = \cos x = 1 - \frac{x^2}{2!} + \frac{x^4}{4!} - \frac{x^6}{6!} + \cdots \qquad (4.8)$$

In Eqs. (4.7) and (4.8), angle x must be expressed in radians.

4.3.3 Taylor's Series

If any function $y = f(x)$ is differentiable, it may be expressed by a Taylor's series expansion as follows:

$$y = f(x) = f(a) + f'(a)\frac{(x-a)}{1!} + f''(a)\frac{(x-a)^2}{2!} + f'''(a)\frac{(x-a)^3}{3!} + \cdots \qquad (4.9)$$

In Eq. (4.9), a is any feasible real number value of x, $f'(a)$ is the value of dy/dx at $x = a$, $f''(a)$ is the value of d^2y/dx^2 at $x = a$, and $f'''(a)$ is the value of d^3y/dx^3 at $x = a$. If only the first two terms in the series of Eq. (4.9) are used, we have a first-order Taylor's series expansion of $f(x)$ about a. If only the first three terms in the series of Eq. (4.9) are used, we have a second-order Taylor's series expansion of $f(x)$ about a. If $a = 0$ in Eq. (4.9), we have the special case known as a *Maclaurin's series expansion of $f(x)$.*

4.3.4 Fourier Series

Any periodic function $y = f(x) = f(x + 2\pi)$ can generally be expressed as a Fourier series expansion as follows:

$$y = f(x) = \frac{a_0}{2} + \sum_{n=1}^{\infty} [a_n \cos(nx) + b_n \sin(nx)] \qquad (4.10)$$

where

$$a_n = \frac{1}{\pi}\int_{-\pi}^{\pi} f(x) \cos(nx)\, dx \qquad \text{for } n = 0, 1, 2, 3, \ldots \qquad (4.11)$$

and

$$b_n = \frac{1}{\pi}\int_{-\pi}^{\pi} f(x) \sin(nx)\, dx \qquad \text{for } n = 1, 2, 3, \ldots \qquad (4.12)$$

Coefficients a_n and b_n of Eq. (4.10) are determined by Eqs. (4.11) and (4.12).

For the Fourier series expansion of Eq. (4.10) to be valid, the Dirichlet conditions summarized as follows must be satisfied:

1. $f(x)$ must be periodic; i.e., $f(x) = f(x + 2\pi)$, or $f(x - \pi) = f(x + \pi)$.
2. $f(x)$ must have a single, finite value for any x.
3. $f(x)$ can have only a finite number of finite discontinuities and points of maxima and minima in the interval of one period of oscillation.

Techniques of numerical integration covered later can be applied to determine the significant Fourier coefficients a_n and b_n by Eqs. (4.11) and (4.12), respectively. A corresponding finite number of terms would then be used from the Fourier series of Eq. (4.10) for approximating $y = f(x)$. Fourier series are particularly valuable when complex periodic functions expressed graphically are to be approximated by an equation for automated calculation use.

4.4 *APPROXIMATIONS AND ERROR*

In many applications of machine design and analysis, it is advantageous to simplify equations by using approximations of various types. Such approximations are often obtained by using only the significant terms of a series expansion for the function.

The approximation used must give an acceptable degree of accuracy for the dependent variable over the range of interest for the independent variables. After defining error next, we will summarize some approximations particularly useful in machine design. Some other techniques of approximation will be presented later, under curve fitting, interpolation, root finding, differentiation, and integration.

4.4.1 Error

Relative error is defined as the difference between an approximate value and the true value, divided by the true value of a variable, as in Eq. (4.13):

$$e = \frac{y_a - y_t}{y_t} \tag{4.13}$$

From this equation, error e is determined as a dimensionless decimal, y_a is an approximate value for y, and y_t is the true value for y. If y_a and y_t are expressed by equations as functions of an independent variable x, Eq. (4.13) gives an error equation as a function of x.

Also, from Eq. (4.13) we see that error e carries an algebraic sign. For positive y_t, a positive value for e means that algebraically we have the relation $y_a > y_t$, whereas for negative e we would have $y_a < y_t$. The opposite relations are true if y_t is negative. Finally, the magnitude of error is its absolute value $|e|$.

For example, for $y_a = 1.003$ in and $y_t = 1.015$ in, by Eq. (4.13) we calculate $e = (1.003 - 1.015)/1.015 = -0.0118$. This means that y_a is 1.18 percent less than its true value y_t. The magnitude of the error is $|e| = 0.0118$.

Incidentally, if error occurs at random on two or more independent variables, the accompanying error on a dependent variable may be determined statistically. This will be illustrated specifically by application of the theory of variance, as presented later under relative change.

4.4.2 Arc Sag Approximation

Consider a circular arc of radius of curvature ρ as shown in Fig. 4.1 with sag y accompanying a chordal length of $2x$. The true value for y can be calculated from the following equation ([4.5], p. 60):

$$y_t = \rho\left[1 - \sqrt{1 - \left(\frac{x}{\rho}\right)^2}\right]$$

However, from the right triangle of Fig. 4.1, we obtain the following:

$$y_t = \frac{x^2 + y_t^2}{2\rho}$$

If in this equation we drop the term y_t^2, the following approximation is derived for y (its use would obviously simplify the calculation of either sag y or radius of curvature ρ):

$$y_a = \frac{x^2}{2\rho} \tag{4.14}$$

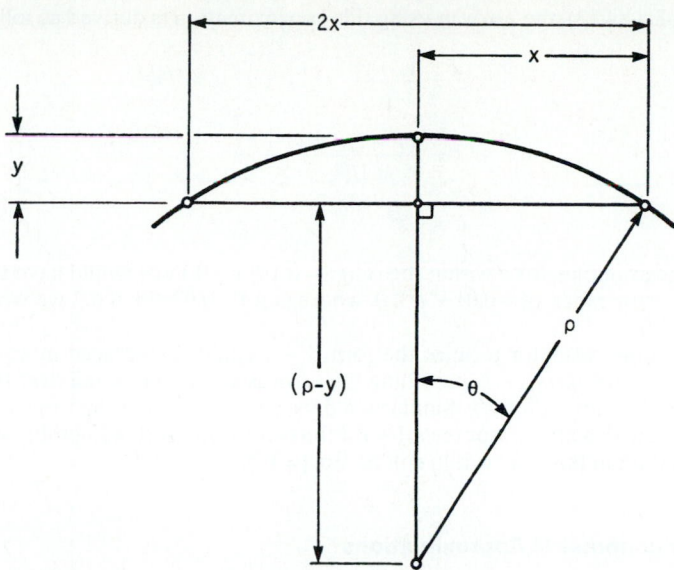

FIGURE 4.1 Circular arc of radius ρ showing sag y and chordal length $2x$.

Applying Eq. (4.13), error e in using approximate Eq. (4.14) is as follows ([4.5], p. 62):

$$e = \frac{-y_t}{2\rho} = -\sin^2 \frac{\theta}{2} \tag{4.15}$$

In Eq. (4.15), angle θ is as shown in Fig. 4.1. As specific examples, from this equation we find that y_a by Eq. (4.14) has error $e = -0.005$ for $\theta = 8.11°$, $e = -0.010$ for $\theta = 11.48°$, and $e = -0.02$ for $\theta = 16.26°$. Hence using the simple Eq. (4.14) to calculate sag would be acceptably accurate in many practical applications of machine design.

4.4.3 Approximation for $1/(1 \pm x)$

In some equations of analysis we have a term of the form $(1 + x)$ in the denominator. For purposes of simplification, as in operations of differentiation or integration, it may be desired not to have such a term in the denominator. Hence consider the true term $y_t = 1/(1 + x)$, which can be expanded into an infinite series by simple division, giving the following:

$$y_t = \frac{1}{1 + x} = 1 - x + x^2 - x^3 + \cdots$$

By dropping all but the first two terms of the series, $1/(1 + x)$ may be approximated by $1 - x$, expressed as follows:

$$\frac{1}{1 + x} \approx y_u = 1 - x \tag{4.16}$$

Applying Eq. (4.13), the error in using this approximation is derived as follows:

$$e = \frac{y_a - y_t}{y_t}$$

$$= \frac{(1 - x) - 1/(1 + x)}{1/(1 + x)}$$

$$e = -x^2 \tag{4.17}$$

As specific examples, for x within the range $-0.1 \le x \le 0.1$, we would have the corresponding error range of $-0.01 \le e \le 0$, whereas for $-0.02 \le x \le 0.2$ we would have $-0.04 \le e \le 0$.

Hence a denominator term of the form $1 + x$ could be replaced in an equation with a numerator term $1 - x$, providing the error is acceptably small over the anticipated range of variation for x. Similarly, a denominator term of the form $1 - x$ could be replaced with a numerator term $1 + x$ if the error is likewise acceptably small. The error equation in this case would still be Eq. (4.17).

4.4.4 Trigonometric Approximations

Approximations for some trigonometric functions will be summarized next, followed by the error function as derived by Eq. (4.13) in each case. For the summarized equations, angle x must be in radians. However, in the examples, ranges of angle x will be given in degrees, using the notation $x°$ in such cases.

An approximation for $\sin x$ is obtained by using only the first term in the Maclaurin's series of Eq. (4.7) as follows:

$$\sin x \approx x \tag{4.18}$$

$$e = \frac{x}{\sin x} - 1 \tag{4.19}$$

Hence for $-10° \le x° \le 10°$ we obtain positive error for e with $e \le 0.005\ 10$, whereas for $-20° \le x° \le 20°$ we have positive error $e \le 0.0206$.

A more accurate approximation for $\sin x$ is obtained by using the first two terms in the series of Eq. (4.7) as follows:

$$\sin x \approx x - \frac{x^3}{6} \tag{4.20}$$

$$e = \frac{x}{\sin x}\left(1 - \frac{x^2}{6}\right) - 1 \tag{4.21}$$

Hence for $-50° \le x° \le 50°$ we obtain negative error for e with its magnitude $|e| \le 0.005\ 41$.

An approximation for $\cos x$ is obtained by using only the first term in the Maclaurin's series of Eq. (4.8) as follows:

$$\cos x \approx 1 \tag{4.22}$$

$$e = \frac{1}{\cos x} - 1 \tag{4.23}$$

Hence for $-5° \leq x° \leq 5°$ we obtain positive error for e with $e \leq 0.003\ 82$, whereas for $-15° \leq x° \leq 15°$ we have positive error $e \leq 0.0353$.

A more accurate approximation for $\cos x$ is obtained by using the first two terms in the series of Eq. (4.8) as follows:

$$\cos x \approx 1 - \frac{x^2}{2} \tag{4.24}$$

$$e = \frac{1 - x^2/2}{\cos x} - 1 \tag{4.25}$$

Hence for $-30° \leq x° \leq 30°$ we obtain negative error for e with its magnitude $e \leq 0.003\ 58$.

An approximation for $\tan x$ is obtained by using only the first term of its Maclaurin's series expansion which follows:

$$\tan x = x + \frac{x^3}{3} + \frac{2x^5}{15} + \cdots$$

Thus the approximation and error function are as follows:

$$\tan x \approx x \tag{4.26}$$

$$e = \frac{x}{\tan x} - 1 \tag{4.27}$$

Hence for $-10° \leq x° \leq 10°$ we obtain negative error for e with its magnitude $|e| \leq 0.0102$.

A more accurate approximation for $\tan x$ is obtained by using the first two terms in its series expansion as follows:

$$\tan x \approx x + \frac{x^3}{3} \tag{4.28}$$

$$e = \frac{x}{\tan x}\left(1 + \frac{x^2}{3}\right) - 1 \tag{4.29}$$

Hence for $-30° \leq x° \leq 30°$ we obtain negative error for e with its magnitude $|e| \leq 0.0103$.

4.4.5   Taylor's Series Approximations

Consider a general differentiable function $y = f(x)$. Its first-order Taylor's series approximation about $x = a$ is obtained by using only the first two terms of the Eq. (4.9) series, resulting in the following equation:

$$y = f(x) \approx f(a) + (x - a)f'(a) \tag{4.30}$$

In Eq. (4.30), a is any feasible real number value of x, and $f'(a)$ is the value of dy/dx at $x = a$.

The accuracy of Eq. (4.30) depends on the particular function $f(x)$ and the range anticipated for x about a. For this reason, a general error function is difficult to derive and impractical to apply. The clue for best accuracy is to choose a value for a such that $(x - a)$ will be small, resulting in negligible terms beyond the second in the Eq. (4.9) series.

For example, suppose we consider $f(x) = \sin x$ and anticipate a range of $-10° \leq x° \leq 10°$ for x. A good choice for a would be $a = 0$. Equation (4.30) would then give

$$\sin x \approx \sin 0 + x \cos 0 \qquad \therefore \sin x \approx x$$

This is merely Eq. (4.18), and the error analysis for the anticipated range of x has already been made after that equation.

However, if we still consider $f(x) = \sin x$ but anticipate a range of $45° \leq x° \leq 65°$ for x, Eq. (4.18) would be highly inaccurate. Hence Eq. (4.30) will be applied, and a good choice for a would be the midpoint of the x range, with $a = 55°(\pi/180) = 0.9599$ radian. Equation (4.30) would then give the following approximation:

$$\sin x \approx \sin 0.9599 + (x - 0.9599) \cos 0.9599 \qquad \therefore \sin x \approx 0.2685 + 0.5736x$$

Hence for $x° = 45°$ we would have $y_t = \sin 45° = 0.7071$ and $y_a = 0.2685 + 0.5736(45\pi/180) = 0.7190$. For that value of x, the error by Eq. (4.13) is

$$e = \frac{0.7190 - 0.7071}{0.7071} = 0.0168$$

For $x = 55°$ we would have $y_t = \sin 55° = 0.8192$ and $y_a = 0.2685 + 0.5736(55\pi/180) = 0.8191$. For that value of x, by Eq. (4.13), the error is

$$e = \frac{0.8191 - 0.8192}{0.8192} = -0.0001$$

Finally, for $x = 65°$ we would have $y_t = \sin 65° = 0.9063$ and $y_a = 0.2685 + 0.5736(65\pi/180) = 0.9192$. For that value of x, by Eq. (4.13), the error is

$$e = \frac{0.9192 - 0.9063}{0.9063} = 0.0142$$

For any differentiable $f(x)$, a more accurate approximation can be obtained by using the first three terms of the Eq. (4.9) series, giving a second-order Taylor's series approximation about $x = a$. The technique is similar to what has been illustrated for a first-order Taylor's series approximation. An appreciably greater range of accuracy would be achieved at the expense of increased complexity for the approximation derived.

4.4.6   Fourier Series Approximation

The Fourier series of Eq. (4.10) involves an infinite number of terms, and for practical calculations, only the significant ones should be used. The clue for significance is the relative magnitude of a Fourier coefficient a_n or b_n, since the amplitudes of $\sin nx$ and $\cos nx$ in Eq. (4.10) are both unity regardless of n.

In establishing significance of a Fourier coefficient, Eqs. (4.11) and (4.12) are solved, perhaps automatically by a computer using numerical integration. The Fourier coefficients are determined for $n = 1, 2, 3, \ldots, N$, where generally a value of N equal to 10 or 12 is sufficient for the investigation. Only the coefficients of significant relative magnitude for a_n and b_n are retained. They determine the significant harmonic content of the periodic function $f(x)$, and only those coefficients are used in the Eq. (4.10) series for the approximation derived. An error analysis could then be made for the derived approximation, including perhaps a graphic presentation by a computer video display for comparative purposes.

As a final item of practical importance, a Fourier series approximation can be derived for many nonperiodic functions $f(x)$ if independent variable x is limited to a definite range corresponding to 2π. In such a case, the derived approximation is used for calculation purposes only within the confined range for x. Hence the derivation assumes hypothetical periodicity outside the confined x range. Of course, the Dirichlet conditions previously stated must be satisfied for $f(x)$ within that range.

4.4.7 Relative Change and Error Analysis

Consider a general differentiable function expressed as follows and used specifically for calculating dependent variable y in terms of independent variables $x_1, x_2, \ldots, x_n$:

$$y = f(x_1, x_2, \ldots, x_n) \tag{4.31}$$

By the theory of differentiation, we can write the following equation in terms of partial derivatives and differentials for the variables:

$$dy = \frac{\partial y}{\partial x_1} dx_1 + \frac{\partial y}{\partial x_2} dx_2 + \cdots + \frac{\partial y}{\partial x_n} dx_n \tag{4.32}$$

Small changes $\Delta x_1, \Delta x_2, \ldots, \Delta x_n$ in $x_1, x_2, \ldots, x_n$ can be substituted respectively for the differentials $dx_1, dx_2, \ldots, dx_n$ of Eq. (4.32). Thus we obtain an approximation for estimating the corresponding change in y, designated as Δy in the following equation:

$$\Delta y \approx \frac{\partial y}{\partial x_1} \Delta x_1 + \frac{\partial y}{\partial x_2} \Delta x_2 + \cdots + \frac{\partial y}{\partial x_n} \Delta x_n \tag{4.33}$$

This equation can be used to estimate the change in y corresponding to small changes or errors in $x_1, x_2, \ldots, x_n$.

As an example of application for Eq. (4.33), consider the simple exponential Eq. (4.3), since many equations in machine design are of this general form. Application of Eq. (4.33) to Eq. (4.3) results in the following simple approximation ([4.5], pp. 67–69):

$$\frac{\Delta y}{y} \approx c_1 \frac{\Delta x_1}{x_1} + c_2 \frac{\Delta x_2}{x_2} + \cdots + c_n \frac{\Delta x_n}{x_n} \tag{4.34}$$

In this equation $\Delta y/y$, $\Delta x_1/x_1$, $\Delta x_2/x_2$, $\ldots$, $\Delta x_n/x_n$ are dimensionless ratios corresponding to relative changes in the variables of Eq. (4.3).

As a specific example of application for Eq. (4.34), suppose we are given the following simple exponential equation:

$$y = \frac{5.32 x_1^{1.62} x_3^2}{x_2^{2.86}} \tag{4.35}$$

If at a point of interest we have the theoretical values $x_1 = 3.796$, $x_2 = 1.095$, and $x_3 = 2.543$, then Eq. (4.35) results in a theoretical value of $y = 230.35$. Suppose that errors exist on the theoretical values of $x_1, x_2, \ldots, x_n$, specifically given as $\Delta x_1 = $

0.005, $\Delta x_2 = 0.010$, and $\Delta x_3 = -0.020$. By Eq. (4.34) we calculate the corresponding relative change in y of Eq. (4.35) as follows:

$$\frac{\Delta y}{y} \approx 1.62 \frac{0.005}{3.796} + -2.86 \frac{0.010}{1.095} + 2 \frac{-0.020}{2.543} = -0.0397$$

Thus the given errors $\Delta x_1, \Delta x_2, \ldots, \Delta x_n$ would result in a corresponding error of $\Delta y \approx -0.0397(230.35) = -9.14$ on the theoretical value of $y = 230.35$.

In the manner illustrated by the preceding example, by application of Eq. (4.34), accuracy estimates can quickly be made for simple exponential equations of the Eq. (4.3) form. The worst possible combination of errors for $\Delta x_1, \Delta x_2, \ldots, \Delta x_n$ can be used to estimate the corresponding error Δy on the theoretical value for y. However, for cases where random errors are anticipated on the independent variables, a statistical approach is more appropriate. This will be considered next.

A Statistical Approach to Error Analysis.   Consider a general differentiable function of several variables typically expressed by Eq. (4.31). Suppose that relatively small errors are anticipated on the theoretical values of the independent variables $x_1, x_2, \ldots, x_n$, with a normal distribution of relatively small spread on any theoretical value for each variable considered as the mean. Designate the standard deviation of the normal distribution for each variable respectively by $\sigma_{x_1}, \sigma_{x_2}, \ldots, \sigma_{x_n}$. Then, for most cases, dependent variable y would approximately have a corresponding normal distribution with standard deviation σy on its theoretical value.

$$(\sigma_y)^2 \approx \left(\frac{\partial y}{\partial x_1}\right)^2 (\sigma_{x_1})^2 + \left(\frac{\partial y}{\partial x_2}\right)^2 (\sigma_{x_2})^2 + \cdots + \left(\frac{\partial y}{\partial x_n}\right)^2 (\sigma_{x_n})^2 \qquad (4.36)$$

Suppose each of the independent variables $x_1, x_2, \ldots, x_n$ has a normal distribution typically shown in Fig. 4.2 with theoretical value corresponding to the mean value $\bar{x}_i$ for variable x_i. Let Δx_i represent a tolerance band, as shown in Fig. 4.2, corresponding to, say, three standard deviations. If the tolerance band Δx_i corresponds to three standard deviations, 99.73 percent of the total population for x_i values would be within the range $x_i - \Delta x_i \le x_i + \Delta x_i$, and we would use the following relation:

$$\Delta x_i = 3\sigma_{xi} \qquad \text{for } i = 1, 2, \ldots, n \qquad (4.37)$$

Combining Eq. (4.37) with Eq. (4.36) by eliminating σ_{x_i} for $i = 1, 2, \ldots, n$, and using the corresponding relation $\Delta y = 3\sigma_y$, we obtain the following:

$$(\Delta y)^2 \approx \left(\frac{\partial y}{\partial x_1}\right)^2 (\Delta x_1)^2 + \left(\frac{\partial y}{\partial x_2}\right)^2 (\Delta x_2)^2 + \cdots + \left(\frac{\partial y}{\partial x_n}\right)^2 (\Delta x_n)^2 \qquad (4.38)$$

In this equation, all the tolerance bands $\Delta y, \Delta x_1, \Delta x_2, \ldots, \Delta x_n$ would correspond to three standard deviations and would encompass 99.73 percent of the total population for each variable.

As an example of application of Eq. (4.38), we will consider the general linear equation expressed by Eq. (4.1). Hence by calculus we obtain $\partial y/\partial x_1 = c_1$, $\partial y/\partial x_2 = c_2, \ldots, \partial y/\partial x_n = c_n$. Substituting these relations in Eq. (4.38), we obtain the following approximation for use in the case of linear Eq. (4.1):

$$(\Delta y)^2 \approx (c_1 \Delta x_1)^2 + (c_2 \Delta x_2)^2 + \cdots + (c_n \Delta x_n)^2 \qquad (4.39)$$

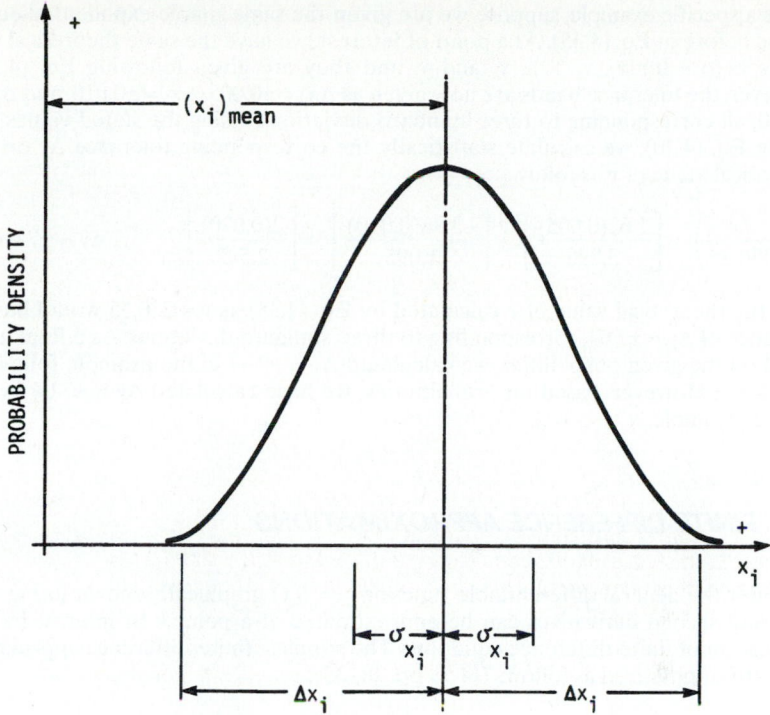

FIGURE 4.2   Typical normal distribution curve for an independent variable x_i.

As a specific example, suppose we have the following linear equation:

$$y = 2.97x_1 - 3.42x_2 + 7.81x_3$$

If tolerances of $\Delta x_1 = \pm0.005$, $\Delta x_2 = \pm0.015$, and $\Delta x_3 = \pm0.010$ exist on the theoretical values of x_1, x_2, and x_3, we calculate the corresponding tolerance Δy on the theoretical value of y statistically by Eq. (4.39) as follows:

$$(\Delta y)^2 \approx [2.97(0.005)]^2 + [-3.42(0.015)]^2 + [7.81(0.010)]^2 \qquad \therefore \Delta y \approx \pm0.0946$$

Thus the theoretical value of y calculated by the given linear equation would have a corresponding tolerance of $\Delta y \approx \pm0.0946$. All the tolerances would correspond to, say, three standard deviations.

As another example of application for Eq. (4.38), we will consider the general simple exponential equation expressed by Eq. (4.3). By application of calculus to Eq. (4.3), we obtain the expressions for $\partial y/\partial x_1, \partial y/\partial x_2, \ldots, \partial y/\partial x_n$, which are then substituted into Eq. (4.38). Dividing the left and right sides of this equation, respectively, by the left and right sides of Eq. (4.3), we obtain the following approximation for use in the case of simple exponential Eq. (4.3):

$$\left(\frac{\Delta y}{y}\right)^2 \approx \left(\frac{c_1\Delta x_1}{x_1}\right)^2 + \left(\frac{c_2\Delta x_2}{x_2}\right)^2 + \cdots + \left(\frac{c_n\Delta x_n}{x_n}\right)^2 \qquad (4.40)$$

As a specific example, suppose we are given the same simple exponential equation as before in Eq. (4.35). At a point of interest, we have the same theoretical values as before for $x_1, x_2, \ldots, x_n$ and y, and they are given following Eq. (4.35). However, the tolerance bands are now given as $\Delta x_1 = \pm0.005, \Delta x_2 = \pm0.010$, and $\Delta x_3 = \pm0.020$, all corresponding to three standard deviations. Using the stated values following Eq. (4.40), we calculate statistically the corresponding tolerance Δy on the theoretical value of y as follows:

$$\left(\frac{\Delta y}{230.35}\right)^2 \approx \left[\frac{1.62(0.005)}{3.796}\right]^2 + \left[\frac{-2.86(0.010)}{1.095}\right]^2 + \left[\frac{2(0.020)}{2.543}\right]^2 \qquad \therefore \Delta y \approx \pm 7.04$$

Thus the theoretical value of y calculated by Eq. (4.35) as $y = 230.35$ would have a tolerance of $\Delta y \approx \pm7.04$, corresponding to three standard deviations. As a final note, based on the given possibilities, we calculated $\Delta y \approx -9.47$ in the example following Eq. (4.35). However, based on probabilities, we have calculated $\Delta y \approx \pm7.04$ in the present example.

4.5 FINITE-DIFFERENCE APPROXIMATIONS

Consider the general differentiable function $y = f(x)$ graphically shown in Fig. 4.3. First and second derivatives can be approximated at a point k of interest by the application of finite-difference equations. The simplest finite-difference approximations are summarized as follows ([4.5], pp. 28–35):

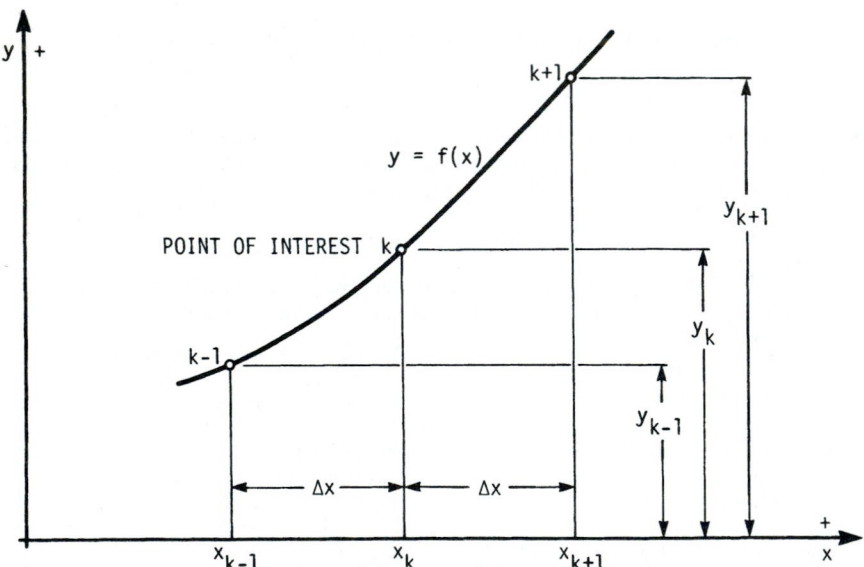

FIGURE 4.3 Graph of $y = f(x)$ showing three successive points used in finite difference equations.

$$\left(\frac{dy}{dx}\right)_k \approx \frac{y_{k+1} - y_{k-1}}{2\Delta x} \tag{4.41}$$

$$\left(\frac{d^2y}{dx^2}\right)_k \approx \frac{y_{k-1} + y_{k+1} - 2y_k}{(\Delta x)^2} \tag{4.42}$$

Equations (4.41) and (4.42) approximate the first and second derivatives of y with respect to x, respectively, at $x = x_k$. For both equations, the point of interest at x_k is surrounded by two equally spaced points, at x_{k-1} and x_{k+1}. The equal increment of spacing is Δx. The values of $y = f(x)$ at the three successive points x_{k-1}, x_k, and x_{k+1} are y_{k-1}, y_k, and y_{k+1}, respectively.

For most differentiable functions $y = f(x)$, the given finite-difference equations are reasonably accurate if the following two conditions are satisfied:

1. Spacing increment Δx, in general, should be reasonably small.

2. The values for y_{k-1}, y_k, and y_{k+1} must carry enough significant figures to give acceptable accuracy in the difference terms of Eqs. (4.41) and (4.42).

Adequate smallness of Δx can be determined by trial, that is, by successively decreasing Δx until no significant difference is determined in the calculated derivatives.

As a very simple test example, consider the function $y = \sin x$. Suppose we wish to calculate first and second derivatives at $x_k^\circ = 35°$ using Eqs. (4.41) and (4.42). We arbitrarily choose the increment $\Delta x° = 2°$, giving $x_{k-1}^\circ = 33°$ and $x_{k+1}^\circ = 37°$. Thus $y_{k-1} = \sin 33° = 0.544\ 639$, $y_k = \sin 35° = 0.573\ 576$, and $y_{k+1} = \sin 37° = 0.601\ 815$. However, for Eqs. (4.41) and (4.42), increment Δx must be expressed in radians, giving $\Delta x = 2(\pi/180) = 0.034\ 906\ 6$ radian. Hence by Eq. (4.41) we calculate

$$\left(\frac{dy}{dx}\right)_k \approx \frac{0.601\ 815 - 0.544\ 639}{2\ \Delta x}$$

$$= \frac{0.057\ 176}{2(0.034\ 906\ 6)}$$

$$= 0.818\ 99$$

Also, by Eq. (4.42), we calculate

$$\left(\frac{d^2y}{dx^2}\right)_k \approx \frac{0.544\ 639 + 0.601\ 815 - 2(0.573\ 576)}{(\Delta x)^2}$$

$$= \frac{0.000\ 698}{(0.034\ 906\ 6)^2}$$

$$= -0.573$$

To check the accuracy of the approximations, for $y = \sin x$ we know by calculus that $dy/dx = \cos x$ and $d^2y/dx^2 = -\sin x$. Therefore, the theoretically correct derivatives are calculated as $(dy/dx)_k = \cos x = \cos 35° = 0.819\ 15$ and $(d^2y/dx^2)_k = -\sin x = -\sin 35° = -0.5736$. We see that the finite-difference approximations were reasonably accurate, which could be further improved by reducing $\Delta x°$ to, say, $1°$.

Finite-difference approximations can also be used for solving differential equations. Equations (4.41) and (4.42) can be used to substitute for derivatives in such differential equations, also substituting $x = x_k$ where encountered. The range of interest for x is divided into small increments Δx. At each net point so obtained, the finite-difference–transformed differential equation is evaluated to determine the discrepancies of satisfaction, known as *residuals*. An iterative procedure is logically developed for successively relaxing the residuals by changing x values at the net points until the differential equation is approximately satisfied at each net point. Thus the solution function $y = f(x)$ is approximated at each net point by such a numerical technique. The iterative procedure of relaxation is greatly facilitated by using a digital computer.

As a final item, finite-difference Eqs. (4.41) and (4.42) may be applied to calculate partial derivatives for the case of a differentiable function of several variables. Hence, for the equation $y = f(x_1, x_2, \ldots, x_i, \ldots, x_n)$, the first and second partial derivatives may be approximated as follows:

$$\left(\frac{\partial y}{\partial x_i}\right)_k \approx \frac{(y_{k+1} - y_{k-1})_i}{2\Delta x_i} \tag{4.43}$$

$$\left(\frac{\partial^2 y}{\partial x_i^2}\right)_k \approx \frac{(y_{k-1} + y_{k+1} - 2y_k)_i}{(\Delta x_i)^2} \tag{4.44}$$

In these equations, the difference terms are subscripted by i, indicating that only x_i is incremented by Δx_i for calculating y_{k-1} and y_{k+1}, holding the other independent variables $x_1, x_2, \ldots, x_n$ constant at their k point values.

4.6 NUMERICAL INTEGRATION

Often it is necessary to evaluate a definite integral of the following form, where $y = f(x)$ is a general integrand function:

$$I = \int_{x_0}^{x_n} y \, dx \tag{4.45}$$

For the case where $y = f(x)$ is a complicated function, numerical integration will greatly facilitate obtaining the solution. If software is available for a particular computational device, the program should be directly applied. However, a commonly used numerical technique will be described next as the basis for writing a special program if necessary.

A simple and generally very accurate technique for numerical integration is based on Simpson's rule, referring to Fig. 4.4 for what follows. First, the limit range for x, between x_0 and x_n, is divided into n equal intervals by Eq. (4.46), where n must be an even number:

$$\Delta x = \frac{x_n - x_0}{n} \tag{4.46}$$

The values of y are then calculated at each of the net points so determined, giving y_0, $y_1, y_2, \ldots, y_{n-2}, y_{n-1}, y_n$. Simpson's rule, given by Eq. (4.47), is then used to approximate the definite integral I of Eq. (4.45):

$$I \approx \frac{\Delta x}{3}[(y_0 + y_n) + 4(y_1 + y_3 + \cdots + y_{n-1}) + 2(y_2 + y_4 + \cdots + y_{n-2})] \tag{4.47}$$

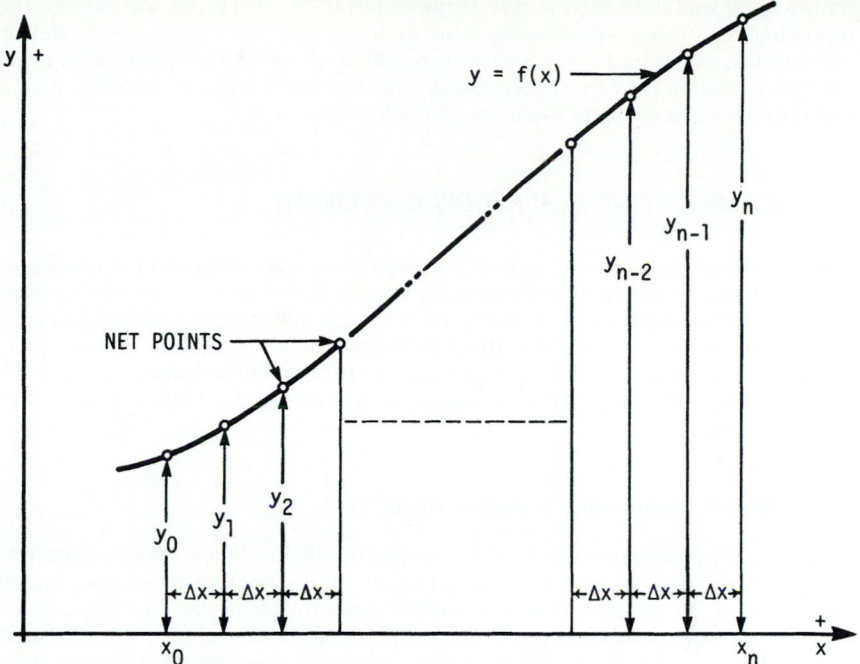

FIGURE 4.4   Graph of $y = f(x)$ divided into equal increments for numerical integration between x_0 and x_n by Simpson's rule.

With automated computation being used, probably the simplest way for determining adequacy of smallness for Δx is by trial. Hence even integer n is successively increased until the difference between successive I calculations is found to be negligible.

As a very simple test example, consider the following definite integral:

$$I = \int_{x_0}^{x_n} \sin x \, dx$$

Suppose the limits of integration are $x_0^\circ = 30^\circ$ and $x_n^\circ = 60^\circ$, giving $y_0 = \sin 30^\circ$ and $y_n = \sin 60^\circ$. For the test example, a value of $n = 20$ is arbitrarily chosen. Equation (4.46) is used to calculate Δx as follows, which must be expressed in radians for use in Eq. (4.47):

$$\Delta x = \frac{(60 - 30)(\pi/180)}{20} = 0.026\ 179\ 938\ 8$$

In degrees, the increment is $\Delta x^\circ = (60 - 30)/20 = 1.5^\circ$. The y values at the remaining net points are then calculated as $y_1 = \sin 31.5^\circ$, $y_2 = \sin 33^\circ$, . . . , $y_{n-2} = \sin 57^\circ$, and $y_{n-1} = \sin 58.5^\circ$. Simpson's rule is then applied using Eq. (4.47) to calculate the approximate value of $I = 0.366\ 025\ 404\ 7$. The described procedure, of course, is programmed for automatic calculation, and specifically the TI-59 Master Library Program ML-09 was used for the test example [4.11]. To check the accuracy of the approximation, from elementary calculus we know that $\int \sin x \, dx$ is $-\cos x$. Hence, theoretically, we obtain $I = [(-\cos 60°) - (-\cos 30°)] = 0.366\,025\,403\,8$, and we see that the approximation for I by Simpson's rule was extremely accurate. See Sec. 5.4 for Richardson's error estimate when area is not known.

4.7   CURVE FITTING FOR PRECISION POINTS

Consider the situation where we have corresponding values of x and y available for a finite number of data points. Suppose we wish to derive an equation which passes precisely through some or all of these given data sets, and these we will call *precision points*. Some techniques of curve fitting for precision points will now be presented. In each case, accuracy checks could be made for the derived equation relative to all the given data points. Validity of the equation over the range of interest could then be established.

4.7.1   Simple Exponential Equation Curve Fit

In many cases of machine design, given graphs or tabular data would plot approximately as a straight line on log-log graph paper. Stress concentration factor graphs and a table of tensile strength versus wire diameter for spring steel are good examples. In such cases, a simple exponential equation of the following form can readily be derived for passing through two precision points (it is assumed that both x and y are positive):

$$y = bx^c \tag{4.48}$$

General curve shapes which are compatible with Eq. (4.48) are summarized graphically in Fig. 2.4 of Ref. [4.5]. Taking the logarithm of both sides in Eq. (4.48) results in the following, which reveals that a straight line would be the plot on log-log graph paper:

$$\log y = c \log x + \log b \tag{4.49}$$

Suppose two precision points (x_1, y_1) and (x_2, y_2) are chosen from the given data sets. The algebraic order for x is $x_1 < x_2$. If we use these precision points in Eq. (4.49), we obtain the following:

$$\log y_1 = c \log x_1 + \log b$$

$$\log y_2 = c \log x_2 + \log b$$

Subtracting the preceding two equations gives the following relation for calculating exponent c:

$$c = \frac{\log(y_2/y_1)}{\log(x_2/x_1)} \tag{4.50}$$

Either one of the two precision points can then be used to calculate coefficient b as follows, as derived from Eq. (4.48):

$$b = \frac{y_1}{x_1^c} = \frac{y_2}{x_2^c} \tag{4.51}$$

With values of c and b so determined, the simple exponential equation is uniquely defined.

As a simple example, suppose we have available the following data for two precision points:

x	y
0.1	8.5
0.25	5.3

Equation (4.50) would then yield the following value for exponent c:

$$c = \frac{\log (5.3/8.5)}{\log (0.25/0.1)} \qquad \therefore c = -0.516$$

Equation (4.51) would then give the following value for coefficient b:

$$b = \frac{8.5}{(0.1)^{-0.516}} = 2.591$$

Therefore, the derived equation passing through the given precision points is as follows:

$$y = \frac{2.591}{x^{0.516}}$$

Accuracy checks could then be made using Eq. (4.13) for all known data points to determine the validity of the derived equation over the range of interest.

4.7.2 Polynomial Equation Curve Fit

A polynomial equation of the following form can be derived to pass through $(n + 1)$ given precision points:

$$y = b + c_1 x + c_2 x^2 + \cdots + c_n x^n \qquad (4.2)$$

The $(n + 1)$ given data sets are substituted into Eq. (4.2), giving $(n + 1)$ linear equations in terms of $b, c_1, c_2, \ldots, c_n$. These $(n + 1)$ linear equations are then solved simultaneously for the $(n + 1)$ unknowns $b, c_1, c_2, \ldots, c_n$, which uniquely defines the polynomial equation.

As a simple example, suppose we wish to derive a polynomial equation through the following four precision points. With $(n + 1) = 4$, we will obtain a polynomial equation of the third degree, since $n = 3$.

x	y
0.0	2.0
0.1	1.65
0.2	1.50
0.3	1.41

Substituting these data sets into Eq. (4.2), we obtain the following:

$$2.0 = b$$

$$\therefore 1.65 = 2.0 + c_1(0.1) + c_2(0.1)^2 + c_3(0.1)^3$$

$$1.50 = 2.0 + c_1(0.2) + c_2(0.2)^2 + c_3(0.2)^3$$

$$1.41 = 2.0 + c_1(0.3) + c_2(0.3)^2 + c_3(0.3)^3$$

Simultaneous solution of these linear equations gives $b = 2.0, c_1 = -4.97, c_2 = 17.0$, and $c_3 = -23.3$. Therefore, the derived polynomial equation passing through the four precision points is as follows:

$$y = 2.0 - 4.97x + 17.0x^2 - 23.3x^3$$

Accuracy checks could then be made using Eq. (4.13) for all known data points to determine the validity of the derived equation over the range of interest.

4.8 CURVE FITTING BY LEAST SQUARES

In many cases of machine design we wish to derive a simple equation $y = f(x)$ which approximates a large number of given data points (x_k, y_k) for $k = 1, 2, \ldots, M$, as illustrated in the following table:

x	y
x_1	y_1
x_2	y_2
$\vdots$	$\vdots$
x_k	y_k
$\vdots$	$\vdots$
x_M	y_M

The given data points are illustrated by + symbols in Fig. 4.5, which also shows the curve of the equation $y = f(x)$ to be derived. For any x_k, the difference between the given point value y_k and the corresponding equation value $f(x_k)$ is Δy_k, defined as follows:

$$\Delta y_k = y_k - f(x_k) \tag{4.52}$$

The equation $y = f(x)$ which minimizes the summation of $(\Delta y_k)^2$ terms for $k = 1$ to M of the given data set is known as the *least-squares fit*. A measure of accuracy for the derived equation is given by the dimensionless correlation coefficient r, which will have a value close to unity for the case of a "good" fit. Some simple examples will now be summarized for use in special cases of programming, although software programs are often already available for direct application [4.11].

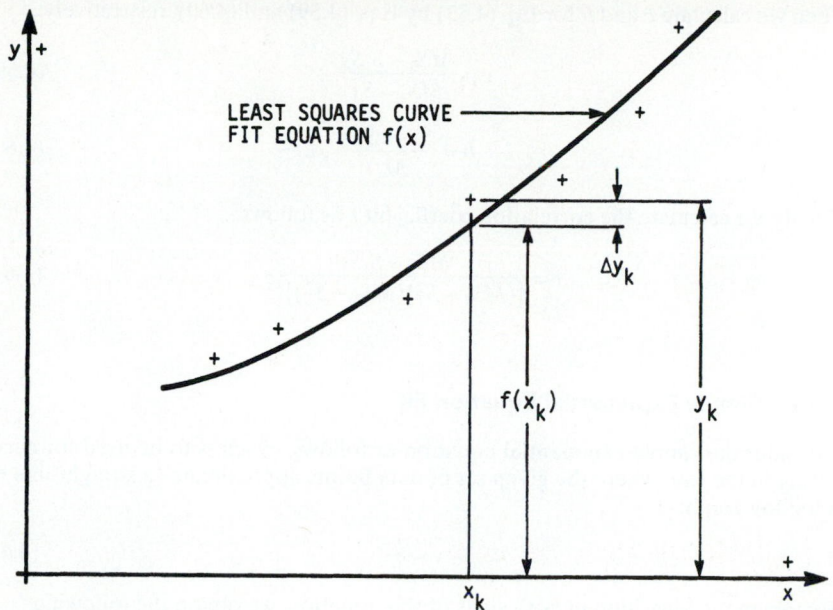

FIGURE 4.5 Least-squares curves $y = f(x)$ for given data points indicated by +.

4.8.1 Linear Equation Fit[†]

Consider the equation of a straight line as follows, which is to be used for curve fitting in the case where the given set of data points approximates a straight line on a graph:

$$y = b + cx \qquad (4.53)$$

Such an equation can be made to pass through only two precision points. However, if many data points (x_k, y_k) are given, the least-squares fit is determined as follows: First, we calculate the values of five summations as follows for S_1 through S_5. In each case, the summations are made for $k = 1$ to M, corresponding to the given data points:

$$S_1 = \Sigma x_k \qquad (4.54)$$

$$S_2 = \Sigma y_k \qquad (4.55)$$

$$S_3 = \Sigma(x_k y_k) \qquad (4.56)$$

$$S_4 = \Sigma(x_k^2) \qquad (4.57)$$

$$S_5 = \Sigma(y_k^2) \qquad (4.58)$$

[†] Ref. [4.5], pp. 55–56.

Then we calculate c and b for Eq. (4.53) by Eqs. (4.59) and (4.60), respectively:

$$c = \frac{MS_3 - S_1 S_2}{MS_4 - S_1^2} \tag{4.59}$$

$$b = \frac{S_2 - cS_1}{M} \tag{4.60}$$

Finally, we calculate the correlation coefficient r as follows:

$$r = \frac{MS_3 - S_1 S_2}{[(MS_4 - S_1^2)(MS_5 - S_2^2)]^{1/2}} \tag{4.61}$$

4.8.2 Simple Exponential Equation Fit[†]

Consider the simple exponential equation as follows, which is to be used for curve fitting in the case where the given set of data points approximates a straight line on a log-log graph:

$$y = bx^c \tag{4.48}$$

By taking the logarithm of both sides of this equation, we obtain the following:

$$\log y = \log b + c \log x \tag{4.49}$$

Hence, Eq. (4.48) would be a straight line on a log-log graph, and the least-squares fit is accomplished as follows: First, we calculate the values of three summations for S_1 through S_3 by Eqs. (4.62) to (4.64). In each case, the summations are made for $k = 1$ to M, corresponding to the given data points:

$$S_1 = \Sigma(\log x_k) \tag{4.62}$$

$$S_2 = \Sigma(\log y_k) \tag{4.63}$$

$$S_3 = \Sigma[(\log x_k)(\log y_k)] \tag{4.64}$$

Then we calculate c and b for Eq. (4.48) by Eqs. (4.65) and (4.66), respectively:

$$c = \frac{MS_3 - S_1 S_2}{2MS_1 - S_1^2} \tag{4.65}$$

$$\log b = \frac{S_2 - cS_1}{M} \tag{4.66}$$

Finally, we calculate the correlation coefficient r as follows:

$$r = \frac{MS_3 - S_1 S_2}{[(2MS_1 - S_1^2)(2MS_2 - S_2^2)]^{1/2}} \tag{4.67}$$

As a specific example, suppose we are given the following set of data points:

[†] Ref. [4.5], pp. 56–57.

k	x_k	y_k
1	0.05	1.78
2	0.10	1.65
3	0.15	1.57
4	0.20	1.50
5	0.25	1.45
6	0.30	1.41

These data fall nearly as a straight line on a log-log graph, and Eq. (4.48) should be appropriate for a least-squares fit. Hence by Eqs. (4.62) to (4.67) we calculate the following values (we use $M = 6$, corresponding to the number of given data points):

$$c = -0.1305 \qquad b = 1.2138 \qquad r = 0.9929$$

Therefore, the derived equation for the least-squares fit is as follows:

$$y = \frac{1.2138}{x^{0.1305}}$$

We note that the correlation coefficient r is close to unity, so we conclude that the derived equation is a "good" fit.

4.8.3 Polynomial Equation Fit

Polynomial Eq. (4.1) may be used for a least-squares fit, but the derivation of such an equation is appreciably more complicated than the preceding examples. If interested, the designer should consult the literature for the details of derivation ([4.2], pp. 19–21).

4.9 CURVE FITTING FOR SEVERAL VARIABLES[†]

Occasionally in machine design we wish to derive a simple equation $y = f(x_1, x_2, \ldots, x_i, \ldots, x_n)$ for the case where we have n independent variables. In such cases, the problem of curve fitting can be very difficult. However, the following simple approach is often of acceptable accuracy in practical problems.

To start, consider the case of two independent variables x_1 and x_2, and we wish to derive an equation $y = f(x_1, x_2)$ to match approximately a given set of data points. Then the function $y = f(x_1, x_2)$ represents a three-dimensional surface using the orthogonal coordinate axes x_1, x_2, and y. The simple technique requires a common precision point for the given data, designated by subscript p in what follows. First, we derive an equation $y = f_1(x_1)$ by holding x_2 constant at $(x_2)p$. Next, we derive an equation $y = f_2(x_2)$ by holding x_1 constant at $(x_1)p$. The final equation is derived using $f_1(x_1)$ and $f_2(x_2)$ satisfying the y_p, $(x_1)_p$, and $(x_2)_p$ values of the given data.

As a simple specific example, consider the problem of deriving an equation $y = f(x_1, x_2)$ for given data-point values as follows:

[†] Ref. [4.5], pp. 57–59.

For $x_2 = 4.5$		For $x_1 = 3.0$	
x_1	y	x_2	y
2.0	3.0	2.5	5.4
3.0	4.2	4.5	4.2
5.0	6.4	7.0	3.5

The common precision points in these data are $(x_1)_p = 3.0$, $(x_2)_p = 4.5$, and $y_p = 4.2$. Plots of these data for y versus x_1 and y versus x_2 fall nearly as a straight line on log-log graphs. Hence Eq. (4.3) should be appropriate for the curve fit, giving the following form for the equation to be derived:

$$y = bx_1^{c_1}x_2^{c_2}$$

Thus the first and last data points are used for both parts of the table to calculate exponents c_1 and c_2 using Eq. (4.50) as follows:

$$c_1 = \frac{\log (6.4/3.0)}{\log (5.0/2.0)} = 0.8269$$

$$c_2 = \frac{\log (3.5/5.4)}{\log (7.0/2.5)} = -0.4212$$

Coefficient b for the equation is then calculated using the common precision-point values as follows:

$$4.2 = b(3.0)^{0.8269}(4.5)^{-0.4212} \qquad \therefore\ b = 3.190$$

Therefore, the derived equation for the curve fit is as follows:

$$y = 3.190\,\frac{x_1^{0.8269}}{x_2^{0.4212}}$$

Finally, accuracy checks could be made to see if the equation is acceptable for the intended use.

The simple technique as now illustrated can be applied to curve fitting for the case of more than two independent variables. Surprisingly, often a reasonably accurate equation is derived. As a specific example in machine design, see [4.5], pp. 383–388, for the derivation of an equation for helical gears having six independent variables. Accuracy checks are also presented in that example.

4.10 INTERPOLATION

Interpolation is generally of concern when we wish to estimate the value of a variable between two known data points. Suppose the values for x and y are known at two points k and $k + 1$ in Fig. 4.6. At some intermediate point j we wish to estimate either y_j for a specified value x_j or x_j for a specified value of y_j. The problem is really an application of curve fitting by using an equation $y = f(x)$ passing through the two precision points k and $k + 1$ for obtaining the estimate. Some specific techniques will be considered next.

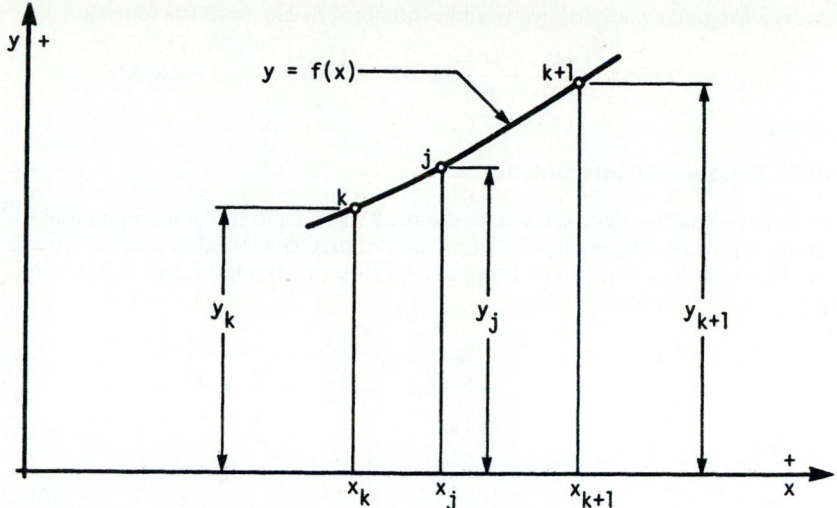

FIGURE 4.6 Curve $y = f(x)$ for interpolation at j between two data points k and $k + 1$.

4.10.1 Linear Interpolation

Consider passing a straight line between given points k and $k + 1$ in Fig. 4.6, with the assumed algebraic order for x being $x_k < x_j < x_{k+1}$. The equation for estimating y_j if x_j is specified would then be as follows:

$$y_j = y_k + \frac{y_{k+1} - y_k}{x_{k+1} - x_k} (x_j - x_k) \tag{4.68}$$

However, the equation for estimating x_j if y_j is specified would be as follows:

$$x_j = x_k + \frac{x_{k+1} - x_k}{y_{k+1} - y_k} (y_j - y_k) \tag{4.69}$$

In Eq. (4.69) it is assumed that y_k does not equal y_{k+1}, and y_j must be algebraically between y_k and y_{k+1}.

As a specific example, suppose we have the following values for x and y at two points:

Point	x	y
k	2.693	1.876
$k + 1$	2.981	2.210

For given $x_j = 2.729$, we would estimate y_j by Eq. (4.68) as follows:

$$y_j = 1.876 + \frac{2.210 - 1.876}{2.981 - 2.693} (2.729 - 2.693) \qquad \therefore \ y_j = 1.918$$

However, for given $y_j = 2.107$, we would estimate x_j by Eq. (4.69) as follows:

$$x_j = 2.693 + \frac{2.981 - 2.693}{2.210 - 1.876}(2.107 - 1.876) \qquad \therefore x_j = 2.892$$

4.10.2 Exponential Interpolation

If we believe that the given set of data points is curved and that it is compatible with a simple exponential type curve in the vicinity of precision points k and $k + 1$, we can apply Eqs. (4.50), (4.51), and (4.48) to derive Eqs. (4.70), (4.71), and (4.72), respectively, for use in the interpolation:

$$c = \frac{\log (y_{k+1}/y_k)}{\log (x_{k+1}/x_k)} \tag{4.70}$$

$$b = \frac{y_k}{x_k^c} \tag{4.71}$$

$$y_j = bx_j^c \tag{4.72}$$

It is assumed that x and y are both positive, with the algebraic order being $x_k < x_j < x_{k+1}$.

As a specific example we will use the preceding tabulated data for the points k and $k + 1$, and again we wish to estimate the value of y_j for given $x_j = 2.729$. Applying Eqs. (4.70), (4.71), and (4.72) we calculate the following values for c, b, and y_j, respectively:

$$c = \frac{\log (2.210/1.876)}{\log (2.981/2.693)} = 1.613$$

$$b = \frac{1.876}{(2.693)^{1.613}} = 0.3795$$

$$y_j = 0.3795(2.729)^{1.613} = 1.916$$

We see that this value of y_j is very close to the value of 1.918 previously calculated by linear interpolation.

4.11 ROOT FINDING

Given a function in the form $y = f(x)$, the problem is to find the values of x for which $y = 0$. For very simple functions, the roots can be found precisely. For example, for the given linear equation $y = b + cx$ we can choose any two values x_k and x_{k+1} and calculate the corresponding values y_k and y_{k+1}. Then, setting $y_j = 0$ in Eq. (4.69), we calculate the singular root x_j precisely. As another example, for the given parabolic equation $y = b + c_1 x + c_2 x^2$ we can find the two roots precisely using the quadratic equation as follows:

$$x = \frac{-c_1 \pm \sqrt{c_1^2 - 4bc_2}}{2c_2}$$

For more complicated functions $y = f(x)$, the problem of finding the roots becomes more difficult. Numerical methods may then be employed in an iterative procedure of automated calculation to approximate the values of x for which $y = 0$. For practical cases, to start, it is generally desired to determine the general characteristics of the complicated function $y = f(x)$, and this can be accomplished by execution of an exploratory search. From this initial stage, each root x_j will be bracketed in an interval $x_k \leq x_j \leq x_{k+1}$, with x and y values known at points k and $k + 1$. If only an approximate value for root x_j is needed, linear interpolation can be used by applying Eq. (4.69) directly and setting $y_j = 0$. However, if the root x_j is to be determined very accurately, an iterative numerical technique may be applied, such as interval halving or the Newton-Raphson method. The various procedures now mentioned will be outlined as follows.

4.11.1 Exploratory Search Stage

For complicated functions $y = f(x)$, it is advantageous to locate the approximate neighborhoods of the roots x_j before a more accurate determination is made for each. This is accomplished by an exploratory search stage, which calculates the y values at successive step points of the range of interest $x_{min} \leq x \leq x_{max}$. The exploratory search stage can be programmed in accordance with the flowchart in Fig. 4.7. Values of x_{min} and x_{max} are initially specified, as is the step increment Δx, which may require some trial. The increment Δx is chosen relatively large to save on computation time, but it must be small enough to identify the neighborhoods of the roots. These are recognized by an algebraic sign change in y for successive step points of the search. In this way, the roots are bracketed by known x and y values for step points which we will designate by subscripts k and $k + 1$ in what follows. First, we will consider a specific example for the exploratory search stage.

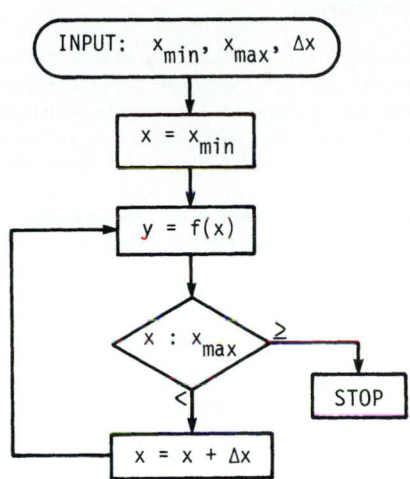

FIGURE 4.7 Exploratory search program for given equation $y = f(x)$.

Example. Consider the problem of finding the roots of the following equation:

$$y = \sqrt{x} - 3x^{1.53} + 7.656 \qquad (4.73)$$

Suppose the range of interest is $0.2 \leq x \leq 3.8$, and we choose $\Delta x = 0.2$ for the exploratory search to be made in accordance with Fig. 4.7. The results are tabulated as follows:

x	y	x	y
0.2	7.848	2.0	0.407
0.4	7.550	2.2	−0.884
0.6	7.058	2.4	−2.246
0.8	6.418	2.6	−3.674
1.0	5.656	2.8	−5.168
1.2	4.786	3.0	−6.723
1.4	3.819	3.2	−8.338
1.6	2.763	3.4	−10.011
1.8	1.624	3.6	−11.741
		3.8	−13.525

Hence by the sign change in y we recognize that there is only one root to the given equation in the range of interest, and it will be located between $x_k = 2.0$ and $x_{k+1} = 2.2$. We now have the problem of determining more accurately the value of this root x_j.

4.11.2 Approximate Roots by Linear Interpolation

In many practical applications of machine design, it is only necessary to determine approximate values for the roots of an equation. In such cases, the roots are first bracketed by an exploratory search, and linear interpolation is then made between the bracketed points k and $k + 1$ for each root using Eq. (4.69) with $y_j = 0$. As a specific illustration, consider the preceding example where we found from the exploratory search that the root of Eq. (4.73) lies between $x_k = 2.0$ and $x_{k+1} = 2.2$. The corresponding values for y are calculated accurately by Eq. (4.73), giving $y_k = 0.406\ 638$ and $y_{k+1} = -0.884\ 458$. These values are substituted in Eq. (4.69) with $y_j = 0$, giving the approximate root as $x_j = 2.063$.

4.11.3 Roots by Interval Halving

From the exploratory search for $y = f(x)$, each root is bracketed within an original interval of uncertainty $[x_k, x_{k+1}]$, as shown in Fig. 4.8. The midpoint of this interval is then determined with respect to x, and y is calculated at that point by the given function $f(x)$. Thus a new interval of uncertainty is determined based on the sign of the calculated y, as shown in the figure. Its size is one-half the original interval. The process is successively repeated until the interval of uncertainty is reduced to a size Δx which is equal to or less than a specified accuracy ε on x. The described calculation strategy is summarized in the flowchart in Fig. 4.9. In general, as the search progresses, the values of x_j and y_j are known at point A in Fig. 4.8 for an interval of uncertainty, the midpoint value x_{j+1} is determined for point C in the figure, and y_{j+1} is calculated for that point by $f(x_{j+1})$. If the product $y_j y_{j+1}$ is positive, the new interval of uncertainty is as shown in Fig. 4.8. However, if the product $y_j y_{j+1}$ is negative, the new interval of uncertainty would be within the range $[x_j, x_{j+1}]$ of the figure. For each new interval of uncertainty, its span Δx is one-half of what it was previously, and only one function evaluation is necessary for its determination, in accordance with the flowchart in Fig. 4.9.

As a specific example, consider the problem of finding the root of Eq. (4.73) with the previously tabulated results from the exploratory search now available. There-

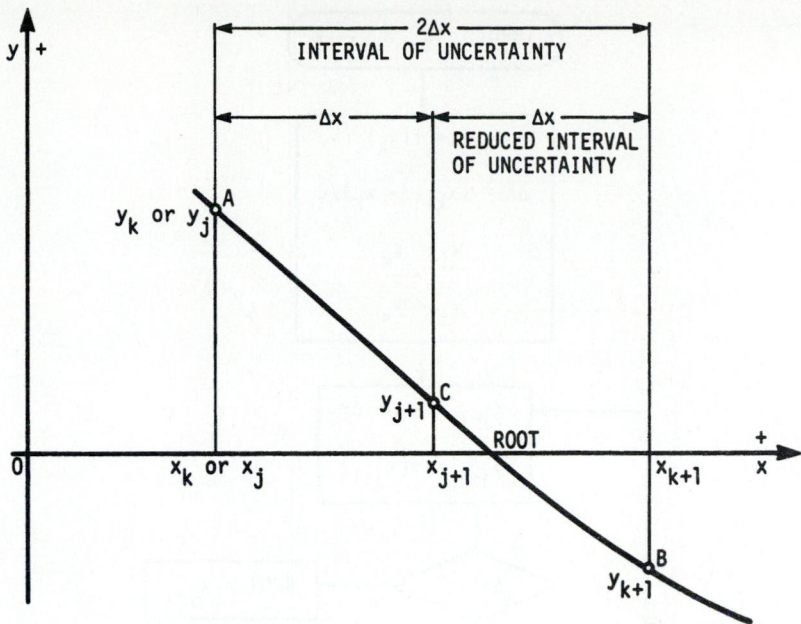

FIGURE 4.8 Reduction of interval of uncertainty by interval halving for root finding.

fore, for the input of Fig. 4.9, we would use $x_k = 2.0$ and $x_{k+1} = 2.2$, and we choose an accuracy specification of $\varepsilon = 10^{-6}$ for the root to be determined. The Fig. 4.9 calculation process was programmed on a TI-59 calculator, resulting in the following root value at the conclusion of the search (which took approximately 60 seconds for the execution time):

$$\text{Root} \approx x_{j+1} = 2.064\ 209\ 747$$

Incidentally, the corresponding value for y_{j+1} is $-7.125\ 32 \times 10^{-7}$, which we see is very close to zero, as it should be for the root.

4.11.4 Roots by the Newton-Raphson Method

The Newton-Raphson method is generally a highly efficient iterative technique for very accurately finding the roots of a given complicated function $y = f(x)$. For the method to work with some given functions it is necessary to start the search process at a point not too far from the root. The exploratory search stage will give the function characteristics necessary for choosing a good starting point.

If in the iterative search process of the Newton-Raphson method we are at some point j in Fig. 4.10, we determine an improved estimate x_{j+1} for the root by extrapolation as follows:

$$\left(\frac{dy}{dx}\right)_j = f'(x_j) = \frac{y_j}{x_j - x_{j-1}} \qquad \therefore\ x_{j+1} = x_j - \frac{y_j}{f'(x_j)} \tag{4.74}$$

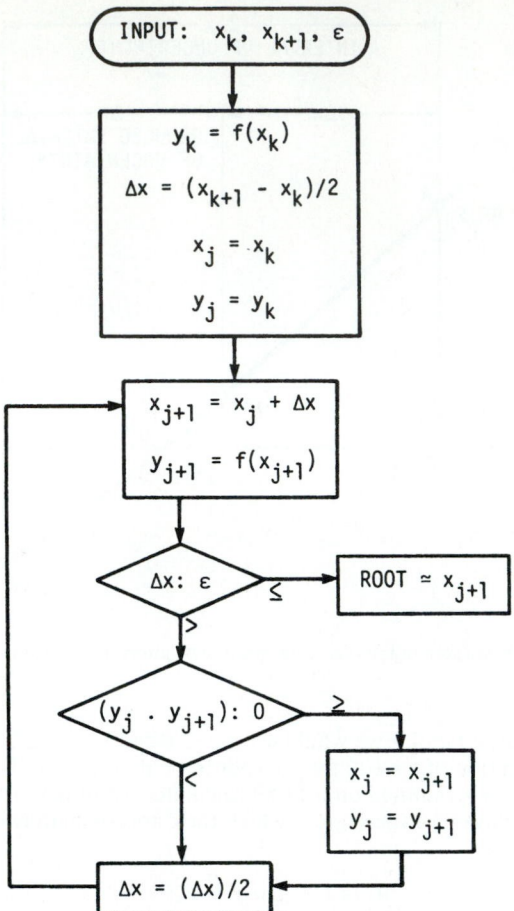

FIGURE 4.9 Interval-halving flowchart for finding root of $y = f(x)$ within bracketed interval $[x_k, x_{k+1}]$.

However, for complicated functions, the equation for $f'(x)$ is generally difficult to derive, and we circumvent this problem by resorting to a finite difference approximation instead. Thus consider an adjacent point A separated from j by a small increment δ, as shown in Fig. 4.10. Therefore, the finite difference approximation for $f'(x_j)$ is as follows:

$$f'(x_j) \approx \frac{y_A - y_j}{\delta}$$

where $y_A = f(x_j + \delta)$ and $y_j = f(x_j)$. Substituting this finite difference approximation into Eq. (4.74), we obtain Eq. (4.75) for estimating x_{j+1}:

$$x_{j+1} \approx x_j - \frac{\delta y_j}{y_A - y_j} \tag{4.75}$$

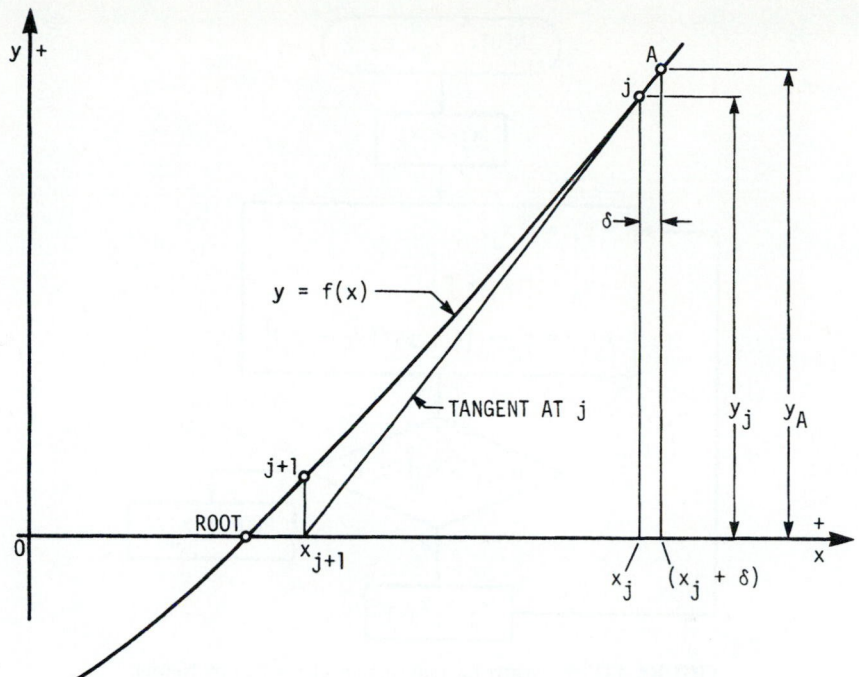

FIGURE 4.10 Newton-Raphson method of root finding.

For use of this approximation, two function evaluations are necessary for y_j and y_A as explained. The iterative calculation procedure for finding the root is summarized in the flowchart in Fig. 4.11, where start point x_k is initially specified. Also, solution accuracy ε and finite difference increment δ are generally specified as relatively small numbers compared with x_k.

As a specific example, consider the problem of finding the root of Eq. (4.73), with the previously tabulated results from the exploratory search now available. Therefore, for the input in Fig. 4.11 we could use $x_k = 2.0$ as a good start point. Also, we choose the accuracy specification for the root as $\varepsilon = 10^{-6}$ and finite difference increment as $\delta = 10^{-6}$. The Fig. 4.11 calculation process was programmed on a TI-59 calculator, resulting in the following root value at the conclusion of the search (which took approximately 15 seconds for the execution time):

$$\text{Root} \approx x_{j+1} = 2.064\ 209\ 636$$

Incidentally, the corresponding value for y_{j+1} is -1.054×10^{-9}, which we see is extremely close to zero, as it should be for the root. Finally, it should be mentioned that exactly the same root was found by the Fig. 4.11 program using other start points of $x_k = 0.2$, $x_k = 2.2$, and $x_k = 3.8$.

4.11.5 Summary of Roots Found for Eq. (4.73)

A comparison of the root findings for Eq. (4.73) from the preceding examples is given in Table 4.1. We see that linear interpolation was extremely fast, but the root

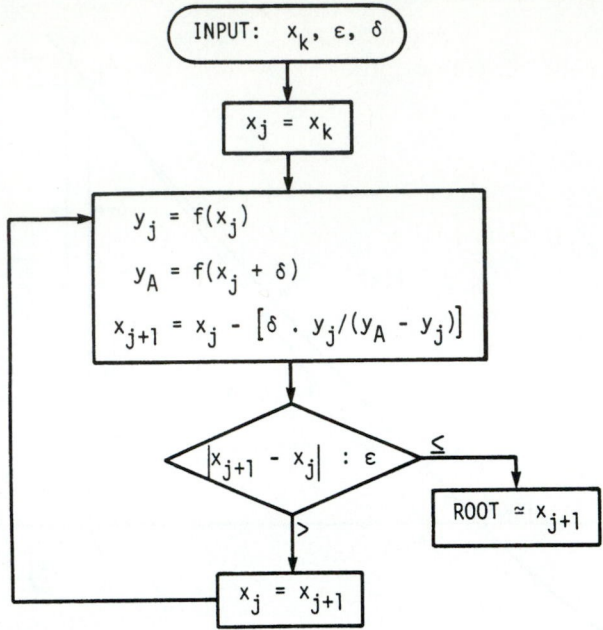

FIGURE 4.11 Flowchart for finding root of $y = f(x)$ by Newton-Raphson method.

was determined only approximately. The last two methods were extremely accurate, but more time-consuming for the solution. As indicated by the summary, the Newton-Raphson method is fast, but it does not provide range numbers as interval halving and golden section can. See Ref. [4.8], Chap. 8.

4.12 SYSTEMS OF EQUATIONS

The simultaneous solution of two or more equations can be a very difficult problem. In general, the number of unknowns cannot exceed the number of equations. We wish to find the common solution to the equation system, and some simple techniques will now be outlined.

TABLE 4.1 Root for Eq. (4.73)

Method	Root found	TI-59 execution time, s
Linear interpolation	2.063	2
Interval halving	2.064 209 747	60
Newton-Raphson	2.064 209 636	15

4.12.1 Two Equations with Two Unknowns

Case 1. Consider the problem where y is expressed by two given functions of x as follows:

$$y = f_1(x) \tag{4.76}$$

$$y = f_2(x) \tag{4.77}$$

We wish to find the common values of x and y at which the two curves cross in Fig. 4.12. To start, we could display the two curves graphically, to be sure that they do cross in the range of interest for x. Depending on the general characteristics of the functions, we could devise an iterative scheme for converging to the neighborhood of the solution point as schematically shown in Fig. 4.12. However, a more direct approach might be to equate the functions, defining $g(x)$ and giving the following single equation to work with:

$$g(x) = f_1(x) - f_2(x) = 0 \tag{4.78}$$

In this way, the problem has been simplified to one of merely finding the root of $g(x)$ expressed by Eq. (4.78).

As a specific example, suppose we are given the following two equations expressing y as a function of x:

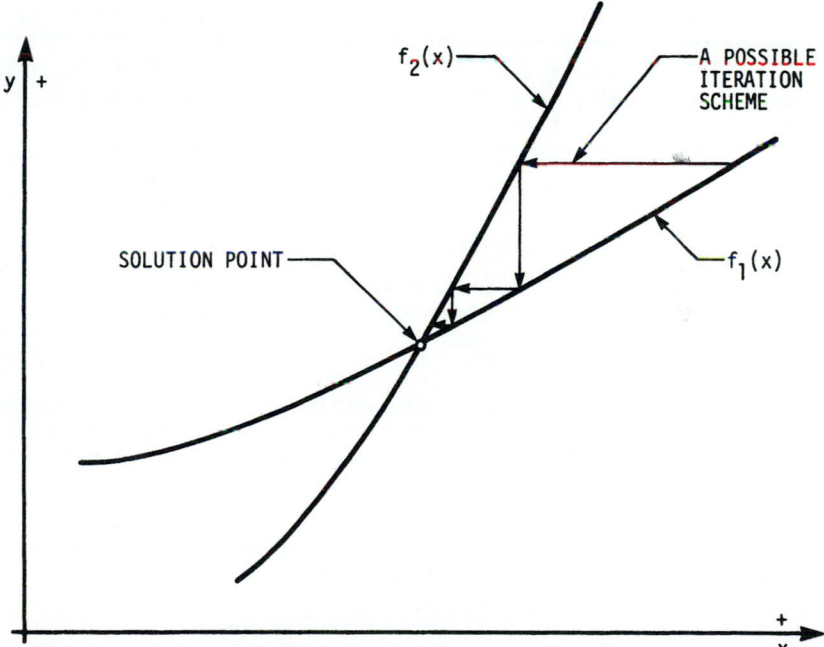

FIGURE 4.12 Simultaneous solution of two functions.

$$y = \sqrt{x} + 4.302 \tag{4.79}$$

$$y = 3x^{1.53} - 3.354 \tag{4.80}$$

By equating the two functions of x, we derive the following relation for obtaining the solution:

$$g(x) = \sqrt{x} - 3x^{1.53} + 7.656 = 0 \tag{4.81}$$

The solution value for x is merely the root of this equation, and a root-finding technique would be applied as previously explained. Since $g(x)$ of Eq. (4.81) is specifically the same as the right side of Eq. (4.73), from Table 4.1 the root is as follows:

$$x = 2.064\ 210$$

The corresponding value for y can then be calculated by either Eq. (4.79) or Eq. (4.80), giving the following specifically by Eq. (4.79):

$$y = \sqrt{2.064\ 210} + 4.302 = 5.739$$

The same value for y is obtained by Eq. (4.80).

Case 2.   Consider the problem where we have two given functions of two variables x_1 and x_2 expressed in the following form:

$$f_1(x_1, x_2) = 0 \tag{4.82}$$

$$f_2(x_1, x_2) = 0 \tag{4.83}$$

As is often feasible, combine the two given equations by eliminating either x_1 or x_2. Thereby, a single equation in one variable is obtained, such as $g(x_1) = 0$ or $h(x_2) = 0$, which is solved by a root-finding technique as previously described. The corresponding remaining value for x_1 or x_2 is then readily calculated by reversing the equation-combination procedure.

As a specific example, suppose we are given the following two functions of x_1 and x_2:

$$\sqrt{x_1} + 3x_2 + 1.053 = 0 \tag{4.84}$$

$$x_1^{1.53} + x_2 - 2.201 = 0 \tag{4.85}$$

If we multiply Eq. (4.85) by 3 and subtract what is obtained from Eq. (4.84), we eliminate x_2 and obtain the following relationship for $g(x)$:

$$g(x) = \sqrt{x_1} - 3x_1^{1.53} + 7.656 = 0 \tag{4.86}$$

A root-finding technique would then be applied to Eq. (4.86), and since $g(x)$ is specifically the same as the right side of Eq. (4.73), we may use the results from Table 4.1 in this example. Thus we have found for the solution point the following value for x_1:

$$x_1 = 2.064\ 210$$

The corresponding value for x_2 can then be calculated by either Eq. (4.84) or Eq. (4.85), giving the following specifically by Eq. (4.84):

$$x_2 = \frac{-(\sqrt{2.064\ 210} + 1.053)}{3} = -0.829\ 9$$

The same value for x_2 is obtained by Eq. (4.85).

For the situation where Eqs. (4.82) and (4.83) cannot readily be combined by eliminating either x_1 or x_2, curve-fitting techniques can generally be applied to either one of the equations, giving an explicit equation for either x_1 or x_2 expressed in terms of the other variable. This transformed equation is then substituted in the described equation-combination procedure for obtaining the solution.

4.12.2 Several Equations with Several Unknowns

The described procedures can generally be extended to solve several equations with several unknowns. For either the case 1 or case 2 type of problems previously described, the given equation system, now several in number, is reduced by equation combination to a single function of a single variable whose solution is found by a root-finding technique. The equation-combination procedure is then reversed to find the values of the other variables. If necessary, curve-fitting techniques may be employed to facilitate the equation-combination process.

4.13 OPTIMIZATION TECHNIQUES

In critical problems of design, the engineer wishes to make decisions which are as favorable as possible for the particular application at hand. In such cases, optimization of design is often worth striving for in the decision-making process. Based on the most critical aspects of the particular problem, an appropriate optimization objective must be chosen and mathematically formulated. Also, constraints of various types must be satisfied in almost all practical problems of design optimization, and these must be mathematically formulated. Hence the engineer must simultaneously address a complicated equation system of the following general form for arriving at decisions of optimal design:

$$Q^{\ddagger} = f(x_1, x_2, \ldots, x_i, \ldots, x_n) \tag{4.87}$$

subjected to

$$y_j = g_j(x_1, x_2, \ldots, x_i, \ldots, x_n) \qquad \text{for } j = 1, 2, \ldots, J \tag{4.88}$$

and

$$x_i \gtreqless c_i \qquad y_j \gtreqless c_j \tag{4.89}$$

In this compact representation of a complicated equation system, $Q^{\ddagger}$ of Eq. (4.87) is the optimization quantity to be either minimized or maximized, and $x_1, x_2, \ldots, x_i, \ldots, x_n$ are the independent variables. In Eq. (4.88), y_j represents a dependent variable, and there are J such equations in the system. Finally, the constraints of Eq. (4.89) are on the independent variables x_i and the dependent variables y_j. The general symbol $\gtreqless$ means that the required relation is one of the following, for any of the variables: >, ≥, =, ≤, or <. Specified constants c_i and c_j are the numerical limits imposed on the associated variables for an acceptable design, and any one may be zero in value.

Optimization of design is too large a subject area to describe in detail in this section. An explicit *method of optimal design* has been developed and applied to many practical examples of machine design, and it is particularly suited to mechanical elements and devices of various types (see Refs. [4.4] and [4.5]). By this technique, a calculation flowchart is derived for explicitly solving the optimization problem in numerical application. Many algorithms have also been developed based on techniques of nonlinear programming for iterative solutions to optimization problems by automated optimal design (see Refs. [4.3], [4.8], [4.9], and [4.12]).

REFERENCES

4.1 R. L. Burden, J. D. Faires, and A. C. Reynolds, *Numerical Analysis,* Prindle, Weber and Schmidt, Boston, 1978.

4.2 T. T. Furman, *Approximate Methods in Engineering Design,* Academic Press, New York, 1981.

4.3 D. M. Himmelblau, *Applied Nonlinear Programming,* McGraw-Hill, New York, 1972.

4.4 R. C. Johnson, *Mechanical Design Synthesis,* 2d ed., R. E. Krieger Co., Huntington, 1978.

4.5 R. C. Johnson, *Optimum Design of Mechanical Elements,* 2d ed., Wiley-Interscience, New York, 1980.

4.6 D. W. Kroeber, and R. L. LaForge, *Statistics and Quantitative Methods,* McGraw-Hill, New York, 1980.

4.7 H. J. Larson, *Statistics: An Introduction,* John Wiley & Sons, New York, 1975.

4.8 C. R. Mischke, *Mathematical Model Building,* Iowa State University Press, Ames, 1980.

4.9 J. N. Siddall, *Analytical Decision-Making in Engineering Design,* Prentice-Hall, Englewood Cliffs, N.J., 1972.

4.10 J. M. Smith, *Scientific Analysis on the Pocket Calculator,* John Wiley & Sons, New York, 1975.

4.11 *TI Programmable 58C/59 Master Library,* Texas Instruments, Inc., 1979.

4.12 G. V. Reklaitis, A. Ravindran, and K. M. Ragsdell, *Engineering Optimization Methods and Application,* John Wiley & Sons, New York, 1983.

CHAPTER 5
COMPUTATIONAL CONSIDERATIONS

Charles R. Mischke, Ph.D., P.E.
Professor Emeritus of Mechanical Engineering
Iowa State University
Ames, Iowa

NOMENCLATURE

a	Range number
A	Spring wire strength constant, cross-sectional area, Jacobian matrix
allow	Diametral allowance
b	Range number
B	Bushing diameter
c	Distance to outer fiber, radial clearance
C	Spring index D/d
d	Wire diameter
d_{hole}	Hole diameter
d_{rod}	Rod diameter
D	Helix diameter, journal diameter
e	Eccentricity
E	Young's modulus
fom	Figure of merit
$f(x)$	Function
F	Spring force, cumulative distribution function, function
F_1	Spring working load
F_s	Spring load at closure (soliding)

G	Shear modulus
h	Ordinate spacing in Simpson's rule
i	Subscript
k	Spring rate, successive substitution convergence parameter
ℓ	Length
ℓ_o	Free length
ℓ_s	Solid length
L	Length
L_o	Free length
L_s	Solid length
m	Spring wire strength parameter
n	Number, factor of safety
n_s	Factor of safety at soliding
$\sim N$	Normal (or gaussian) distributed
N	Number of turns
N_a	Number of active turns
N_t	Total number of turns
OD	Outside diameter of spring coil
p	Probability
P	Load, probability
Q	Spring dead coil correction
Q'	Spring dead coil correction for solid height
r	Residual, radius
R	Richardson's correction to Simpson's first rule estimate
S_{ut}	Engineering ultimate tensile strength
S_{su}	Engineering ultimate shear strength
S_y	Engineering 0.2 percent yield strength in tension
S_{sy}	Engineering 0.2 percent yield strength in shear
u	Uniform random number
$\sim U$	Uniform distributed
$\sim W$	Weibull distributed
x	Variable
y	Variable, end contraction of a spring
z	deviation in $N(0, 1)$
γ	Weight density
η	Factor of safety
θ	Weibull characteristic parameter, angle
μ	Population mean
ξ	Fractional overrun to closure, $y_s = y_1 + \xi y_1$
ρ	Link length

σ   Population standard deviation

σ   Normal stress

τ   Shear stress

τ_s Shear stress at wire surface at closure of spring

φ   angle

5.1   INTRODUCTION

Machine design is the decision-making process by which specifications for machines are created. It is from these specifications that materials are ordered and machines are manufactured. The process includes

- Inventing the concept and connectivity
- Decisions on size, material, and method of manufacture
- Secondary decisions
- Adequacy assessment
- Documentation of the design
- Construction and testing of prototype(s)
- Final design

Computer-aided engineering (CAE) means computer assistance in the major decision-making process. *Computer-aided drafting* (CAD), often confused with CAE when called *computer-aided design,* means computer assistance in creating plans and can include estimates of such geometric properties as volume, weight, centroidal coordinates, and various moments about the centroid. Three-dimensional depictions and their manipulations are often routinely available. *Computer-aided analysis* (CAA) involves use of the computer in an "if this then that" mode.

Computer-aided manufacturing (CAM) includes preparing tool passes for manufacture, including generating codes for executing complicated tool paths for numerically controlled machine tools. All kinds of auxiliary accounting associated with material and parts flow in a manufacturing line are also done by computer. The data base created during computer-aided drafting can be used by computer-aided manufacturing. This is often called *CAD/CAM.*

Some of these computer aids are commercially available and use proprietary programming. They are sometimes called "turnkey" systems. They may be used interactively by technically competent people without programming knowledge after only modest instruction. The programming detail is not important to the users. They react to displays, make decisions on the task to be accomplished, and proceed by entering appropriate system commands. Such systems are available for a number of highly repetitive tasks found in analysis, drawing, detailing, and manufacturing.

"Turnkey" systems are available from vendors to do some important work.

The machine designer's effort, however, is composed of problem-specific tasks, for many of which no commercial programming is available. The designer or his or her assistants may have to create or supervise the creation of such programs. The basis for this programming must be their understanding of the problem. This section will view computer methods of direct use to the designer in making decisions using personal or corporate resources.

It is well to keep in mind what the computer can do:

- It can remember data and programs.
- It can calculate.
- It can branch unconditionally.
- It can branch conditionally based on whether a quantity is negative, zero, or positive, or whether a quantity is true or false, or whether a quantity is larger or smaller than something else. This capability can be described as *decision making*.
- It can do a repetitive task or series of tasks a fixed number of times or an appropriate number of times based on calculations it performs. This can be called *iteration*.
- It can read and write alphabetical and numerical information.
- It can draw.
- It can pause, interact, and wait for external decisions or thoughtful input.
- It does not tire.

Humans can

- Understand the problem
- Judge what is important and unimportant
- Plan strategies and modify them as they gain experience
- Weigh intangibles
- Be skeptical, suspicious, or unconvinced
- Program computers

The designer should try to delegate to the computer those things which the computer can do well and reserve for humans those things which they do well.

5.2 AN ALGORITHMIC APPROACH TO DESIGN

A design must be functional, safe, reliable, competitive, manufacturable, and marketable. It is axiomatic that the designer must have a quantitative procedural structure in mind before computer programming is attempted. An algorithm is a step-by-step process for accomplishing a task. The designer contemplating using the computer to help in making decisions undertakes a series of tasks that include

1. Identifying the specification set
2. Identifying the decision set
3. Examining the needs to be addressed, noting the a priori decisions
4. Identifying the design variables
5. Quantifying the adequacy assessment
6. Converting the a priori decisions and design decisions into a specification set
7. Quantifying a figure of merit
8. Choosing an optimization algorithm
9. Assembling the programs

A *specification set* for a machine or component is the ensemble of drawings, text, bill of materials, and other directions that assure function, safety, reliability, competitiveness, manufacturability, and marketability no matter who builds it, assembles it, and uses it. For example, consider a helical coil compression spring for static service, such as that depicted in Fig. 5.1. The spring maker needs to know (one possible form)

- Material and its condition
- End treatment
- Coil ID or OD and tolerance
- Total turns and tolerance
- Free length and tolerance
- Wire size and tolerance

The commercial tolerances are expressible as functions of mean or median values. There are six elements in the specification set. The specification set is not couched in terms of the designer's thinking parameters or concerns, and so the designer recasts it as a decision set. The sets are equivalent, and the specification set is deducible from the decision set using ordinary deductive analytic algebraic techniques.

A *decision set* is the set of decisions which, when made, establishes the specification set. The specification set is cast the way the spring maker likes to communicate, and the decision set is expressed in such a way that the engineer can focus on function, safety, reliability, and competitiveness. In the case of the spring, a corresponding decision set is

- Material and condition
- End condition
- Function: F_1 at y_1, or F_1 at L_1
- Safety: Design factor at soliding is $n_s = 1.2$

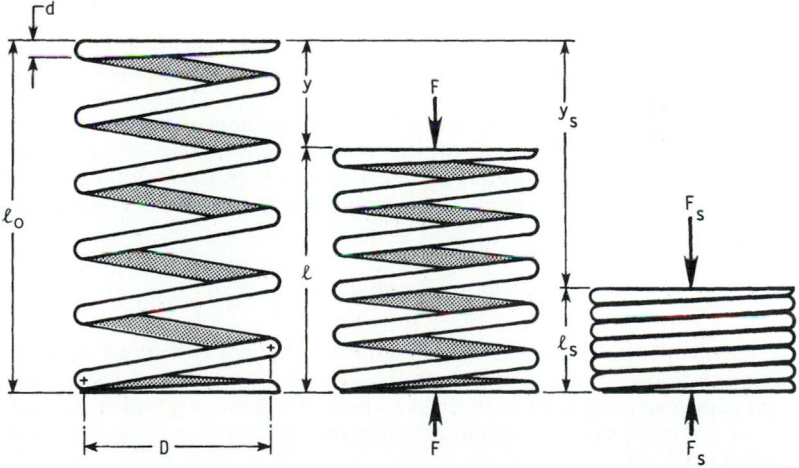

FIGURE 5.1 Nomenclature of a helical-coil compression spring with squared and ground ends.

- Robust linearity: Fractional overrun to closure $\xi = 0.15$
- Wire size: d

Note some duplication of elements in the decision set and the specification set, but also the appearance of "thinking parameters." The functional requirement of a force-geometry relationship occurs indirectly; it is important that the spring be robustly linear, and this requirement prevents the use of excess spring material while assuring no change in active turns as the coil clashes during the approach to soliding. Had the designer said that the design factor at soliding was to be greater than 1.2 (that is, $n_s > 1.2$), that would be a nondecision, and another decision would have to be added to the decision set. Any inequalities the designer is tempted to place in the decision set are moved to the adequacy assessment.

The cheapest spring is made from hard-drawn spring wire; the next stronger (and more expensive) material, 1065 OQ&T, costs 30 percent more. End treatment will almost always be squared and ground. The function is not negotiable, nor is the linearity requirement. The spring must survive closure without permanent deformation. These five decisions can be made a priori, and are consequently called *a priori decisions*. The remaining decision, that of wire size, is the decision through which the issue of competitiveness is addressed. Thus, there is one independent design variable, wire size. Note that the dimensionality of the task has been identified. Here, wire size is called the *design variable*.

An *adequacy assessment* consists of the cerebral, empirical, and related steps undertaken to determine if a specification set is satisfactory or not. In the case of the spring, the adequacy assessment can look as follows:

$4 \leq C \leq 16$ (formable and not too limber)

$3 \leq N_a \leq 15$ (sufficient turns for load precision)

$\xi \geq 0.15$ (robust linearity and little excess material)

$n_s \geq 1.2$ (spring can survive closure without permanent deformation)

$\vdots$

Additional checks can examine natural frequencies, buckling, etc., as applicable.

A *figure of merit* is a number whose magnitude is a monotonic index to the merit or desirability of a specification set (or decision set). If several satisfactory springs are discovered, the figure of merit is used to compare them and choose the best. In the case where large numbers of satisfactory specification sets are expected, an optimization strategy is needed so that the best can be identified without exhaustive examination. In the spring example, since springs sell by the pound, a useful figure of merit is

$$\text{fom} = -(\text{relative material cost}) \frac{\gamma \pi^2 d^2 N_t D}{4}$$

Often the competition is among steels and the weight density γ can be omitted.

An optimization strategy is chosen in the light of the number of design variables present (dimensionality), the number and kinds of constraints (equality, inequality, or mixed; few or many), and whether the decision variables are integer (or discrete) or continuous [5.1].

There are two related skills for the designer to master. The first skill is the ability to take a specification set and perform an adequacy assessment. The logic flow diagram for this skill is depicted in Fig. 5.2. Such a skill is analytic and deductive. The second skill is to create a specification set by surrounding the results of skill #1 with a decision set, a figure of merit, and an optimization strategy. Study Fig. 5.3 to see the interrelationships. This second skill is a synthesis procedure which is quantitative and computer-programmable. An example follows to show how simply this can be done, how a hand-held programmable calculator can be the only computational tool needed, and how some tasks can be done manually.

Example 1. A static-service helical-coil compression spring is to be made of 1085 music wire (food service application). The static load is to be 18 lbf when the spring is compressed 2.25 in. The geometric constraints are

$$0.5 \leq ID \leq 1.25 \text{ in}$$

$$0.5 \leq OD \leq 1.50 \text{ in}$$

$$0.5 \leq L_s \leq 1.25 \text{ in}$$

$$3 \leq L_o \leq 4 \text{ in}$$

Solution. The decision set with a priori decisions in place is

- Material and condition: music wire, $A = 186\ 000$ psi, $m = 0.163$, $E = 30 \times 10^6$ psi, $G = 11.5 \times 10^6$ psi
- End treatment: squared and ground, $Q = 2$, $Q' = 1$
- Function: $F_1 = 18$ lbf, $y_1 = 2.25$ in
- Soliding design factor: $n_s = 1.2$
- Fractional overrun to closure: $\xi = 0.15$
- Wire size: d

The decision variable is the wire size d. The figure of merit is cost relative to that of cold-drawn spring wire (Ref. [5.2], p. 20).

$$\text{fom} = -(\text{cost relative to CD})(\text{volume of wire used to make spring})$$

$$= -2.6 \, \frac{\pi^2 d^2 (N_a + Q)D}{4}$$

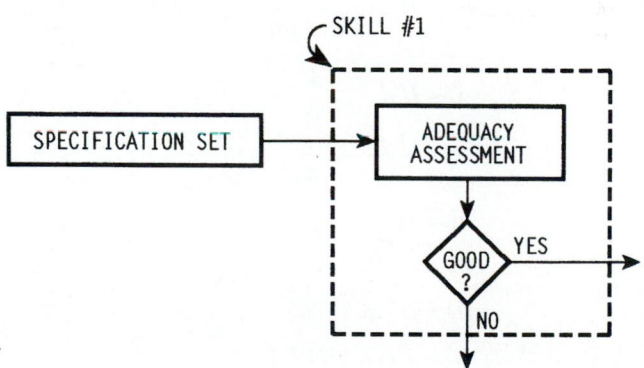

FIGURE 5.2 Designer's skill #1.

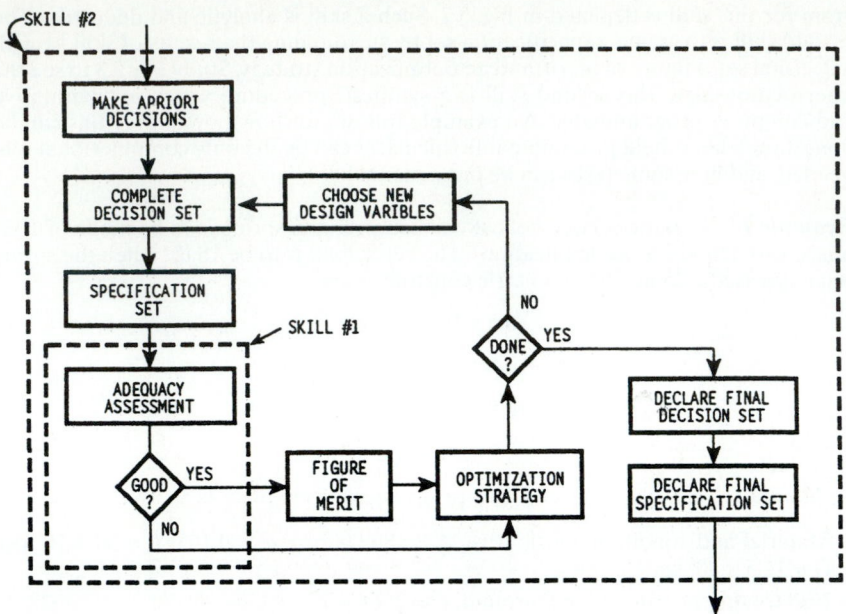

FIGURE 5.3   Designer's skill #2, which contains skill #1 imbedded.

Procedure: From the potential spring maker, get a list of available music wire sizes. Mentally choose a wire size d. The decision set is complete, so find a path to the specification set. What follows is one such path from the decision on wire size d with the three possibilities (spring works over a rod, spring is free to take on any diameter, spring works in a hole).

$$S_{sy} = 0.45A/d^m$$

Spring over a rod	Spring is free	Spring in a hole
$D = d_{rod} + d + \text{allow}$	$D = \dfrac{S_{sy}\,\pi d^3}{8(1 + \xi)F_1} - \dfrac{d}{2}$	$D = d_{hole} - d - \text{allow}$

$$C = D/d$$

$$\tau_s = \frac{(1 + 0.5/C)8(1 + \xi)F_1 D}{\pi d^3}$$

$$n_s = S_{sy}/\tau_s$$

$$\text{OD} = D + d$$

$$\text{ID} = D - d$$

$$N_a = d^4 G y_1/(8D^3 F_1)$$

$$N_t = N_a + Q$$

$$L_s = (N_a + Q')d$$

$$L_o = L_s + (1 + \xi)y_1$$

The specification set has been identified; now perform the adequacy assessment:

$$0.5 \leq ID \leq 1.25 \text{ in}$$

$$0.5 \leq OD \leq 1.50 \text{ in}$$

$$0.5 \leq L_s \leq 1.25 \text{ in}$$

$$3 \leq L_o \leq 4 \text{ in}$$

$$4 \leq C \leq 16$$

$$3 \leq N_a \leq 12$$

$$\xi \geq 0.15$$

$$n_s \geq 1.2$$

$$\vdots$$

The above computational steps can be programmed on a computer using a language such as Fortran, or on a hand-held programmable pocket calculator. The following table comes from a pocket calculator when one inputs wire size and the remaining elements of the column are presented.

d	0.031	0.041	0.063	0.067	0.071	0.075	0.080	0.085	0.090
D	0.054	0.133	0.448	0.585	0.693	0.814	0.983	1.172	1.383
C	1.740	3.244	7.742	8.729	9.766	10.85	12.28	13.79	15.37
OD	0.085	0.174	0.551	0.652	0.764	0.889	1.063	1.257	1.463
N_a	1057	215.4	24.4	18.1	13.7	10.5	7.76	5.827	4.454
L_s	32.81	8.890	1.600	1.280	1.044	0.866	0.701	0.580	0.491
L_o	35.39	11.48	4.187	3.867	3.631	3.453	3.288	3.168	3.078
fom	−0.352	−0.312	−0.328	−0.339	−0.352	−0.368	−0.394	−0.425	−0.464

Figure 5.4 shows a plot of the figure of merit vs. wire size d. Only four wire diameters result in satisfactory springs, 0.071, 0.075, 0.080, and 0.085 in, and the largest figure of merit, −0.352, of these four springs corresponds to a wire diameter of 0.071 in. Ponder the structure that identified the dimensionality of the task and guided the component computational arrangement.

5.3 ANALYSIS TASKS

In the discussion in the previous section of the adequacy-assessment task and the conversion to a specification set as illustrated by the static-service spring example, there occurred a number of routine computational chores. These were simple algebraic expressions representing mathematical models of the reality we call a spring.

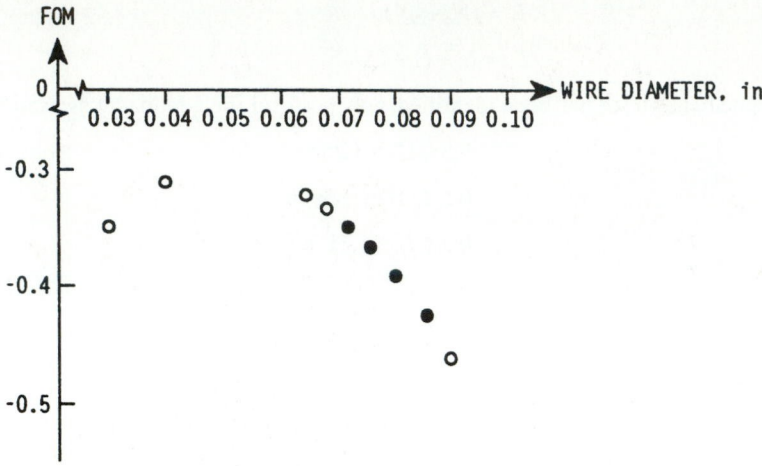

FIGURE 5.4   The figure of merit as a function of wire diameter in Example 5.1. The solid points are satisfactory springs.

The expression for spring rate is either remembered or easily found. In a more complex problem, the computational task may be more involved and harder to execute and program. However, it is of the same character. It is a calculation ritual that is known by the engineering community to be useful. It is an analysis-type "if this then that" algorithm that engineers instinctively reach for under appropriate circumstances. If this happens often, then once it is programmed, it should be available for subsequent use by anyone. Computer languages created for algebraic computational use include a feature called *subprogram capability*. The algorithm encoded is given a name and an argument list. In Fortran such a program can be a function subprogram or a subroutine subprogram. If the spring rate equation were to be coded as a Fortran subroutine with the name SPRNGK, then the coding could be

```
SUBROUTINE SPRNGK(DWIRE,DCOIL,G,EN,XK)
XK=DWIRE**4*G/8./DCOIL**3/EN
RETURN
END
```

and any program in which DWIRE, DCOIL, G, and EN have been defined can obtain XK by

```
⋮
CALL SPRNGK(DWIRE,DCOIL,G,EN,XK)
⋮
```

and XK is now defined in the calling program. This simplicity is welcome as the tasks become more complicated, such as finding the stress at the inner fiber of a curved beam of a tee cross section or locating the neutral axis of the cross section.

Such routine answers to computational chores can be added to a subroutine library to which the computer and the users have access. Usage by one designer of programs written by another person depends on documentation, error messaging, and tests. At this point and for our purposes, we will treat this as detail and retain the larger picture. A library of analysis subroutines can be created which the designer can manipulate in an executive manner simply by calling appropriate routines. Such subroutines are called *design subroutines* because through an inverse-analysis strategy they can be made to yield design decisions. Within them is the essence of the reality of the physical world. When decisions are made which completely describe a helical compression spring intended for static use, the computer can be used to examine important features. From decisions on

Material:	1085 music wire
Wire size:	0.071 in
Ends:	Ground and squared
Turns:	21.7 total
OD:	0.685 in
Length:	4.476 in

a large number of attributes can be viewed:

Ultimate strength estimate:	288 kpsi
Shearing yield strength:	125 kpsi
Spring rate:	8.01 lbf/in
Solid length:	1.47 in
Deflection to closure:	3.01 in
Working force:	7.8 lbf
Working deflection:	0.98 in
Working length:	3.5 in
Force at closure:	24.1 lbf
Spring index:	8.65
Shear stress at closure:	111 kpsi
Working shear stress:	36.1 kpsi
Static factor of safety:	3.5
Factor of safety at closure:	1.12
Critical frequency:	135 Hz
Buckling load:	9.6 lbf

Scanning these items, the designer can detect the need for a constraint to prevent buckling and observe the low factor of safety guarding against permanent set due to closure. This too can be assessed, and the computer can assist routinely. The static factor of safety at closure is given by [see Ref. [5.3], Eqs. (10.3), (10.17) and $S_y = 0.75S_{ut}$]

$$\eta_s = \frac{S_{sy}}{\tau_s} = \frac{0.577(0.75)A}{d^m} \frac{\pi d^3}{[1 + (0.5d/D)]8F_s D}$$

Substituting for force to closure

$$F_s - ky_s - \frac{d^4 G}{8D^3 N} [\ell_o - (N+1)d]$$

into the η_s equation yields

$$\eta_s = \frac{0.577(0.75)A\pi D^2 N}{[1 + (0.5d/D)]d^{1+m} G[\ell_f - (N+1)d]}$$

for ground and squared ends. Tolerances on d, D, N, and ℓ_o give rise to variation in η_s, that tabulated above being a median value. The worst-case stacking of tolerance occurs when all deviations from the midrange values are such that

$$\Delta\eta_s = \left|\frac{\partial\eta_s}{\partial d}\,\Delta d\right| + \left|\frac{\partial\eta_s}{\partial D}\,\Delta D\right| + \left|\frac{\partial\eta_s}{\partial N}\,\Delta N\right| + \left|\frac{\partial\eta_s}{\partial\ell_o}\,\Delta\ell_o\right|$$

In the case at hand with $A = 196$ kpsi (Ref. [5.3], Table 10-2), $m = 0.146$, $d = 0.071$ in, $D = 0.614$ in, $N = 19.7$, $\ell_o = 4.476$ in, and $G = 11.5 \times 10^6$ psi; $\eta_s = 1.121$:

$$\frac{\partial\eta_s}{\partial d} \doteq -11.03 \text{ in}^{-1} \qquad \frac{\partial\eta_s}{\partial D} \doteq 3.78 \text{ in}^{-1} \qquad \frac{\partial\eta_s}{\partial N} \doteq 0.083 \qquad \frac{\partial\eta_s}{\partial\ell_o} \doteq -0.372 \text{ in}^{-1}$$

where these values are obtained by taking partial derivatives in the η_s equation by numerical means. If the bilateral tolerances are

$$d = 0.071 \pm 0.001 \text{ in}$$

$$D = 0.614 \pm 0.010 \text{ in}$$

$$N = 19.7 \pm \tfrac{1}{4} \text{ turn}$$

$$\ell_o = 4.476 \pm 0.097 \text{ in}$$

we have

$$\Delta\eta_s = |-11.03(0.001)| + |3.78(0.010)| + |0.083(\tfrac{1}{4})| + |-0.372(0.097)|$$

$$= 0.011 + 0.038 + 0.021 + 0.036 = 0.106$$

The relative contribution of the various tolerances can be observed. The smallest possible value of η_s is

$$\eta_s(\min) = \eta_s - \Delta\eta_s = 1.121 - 0.106 = 1.02$$

This is the worst-case stacking of tolerances.

To make a statistical statement as to the probability of observing a value of η_s of a particular magnitude, we need an estimate of the variance of η_s.

$$\sigma_{\eta_s}^2 = \left(\frac{\partial\eta_s}{\partial d}\right)^2 \sigma_d^2 + \left(\frac{\partial\eta_s}{\partial D}\right)^2 \sigma_D^2 + \left(\frac{\partial\eta_s}{\partial N}\right)^2 \sigma_N^2 + \left(\frac{\partial\eta_s}{\partial\ell_o}\right)^2 \sigma_{\ell_o}^2$$

We can estimate the individual variances on the basis that the tolerance width represents six standard deviations as shown in Chap. 2:

$$\sigma_d = \frac{2(0.001)}{6} = 0.000\,333 \text{ in}$$

$$\sigma_D = \frac{2(0.010)}{6} = 0.003\,33 \text{ in}$$

$$\sigma_N = \frac{2(\tfrac{1}{4})}{6} = 0.0833$$

$$\sigma_{\ell_o} = \frac{2(0.097)}{6} = 0.032 \text{ in}$$

The estimate of the variance and standard deviation of η_s is

$$\sigma^2_{\eta_s} = (-11.03)^2(0.000\ 333)^2$$
$$+ (3.78)^2(0.003\ 33)^2 + (0.083)^2(0.083)^2 + (0.372)^2(0.032)^2$$
$$= 0.000\ 361$$
$$\sigma_{\eta_s} = \sqrt{0.000\ 361} = 0.019$$

For a gaussian distribution of η_s there are 3 chances in 1000 of observing a deviation from the mean of $3(0.019) = 0.057$, or about 1½ chances in 1000 of observing an instance of η_s less than $1.121 - 0.057 = 1.064$. These kinds of analysis chores are easily built into a computer-adequacy display program. This is the kind of quantitative information designers need before they commit themselves.

5.4 MATHEMATICAL TASKS

In problems which are coded for computer assistance, a number of recurring mathematical tasks are encountered which can also be discharged by the computer as they are encountered. The procedure is to identify the pertinent algorithm and then code it as appropriate to your computer. Recurring tasks can be coded as subprograms which represent convenient building blocks for use in solving larger problems.

One frequently encountered task is that of finding a root or zero place of a function of a single independent variable. An effective algorithm for this task is the successive-substitution procedure with assured convergence. The algorithm is as follows (Ref. [5.7], p. 168):

Step 1. Express the problem in the form $f(x) = 0$. Establish the largest successive difference allowable in root estimates ϵ.

Step 2. Rewrite in the form $x = F(x)$, thereby defining $F(x)$.

Step 3. Establish the convergence parameter $k = 1/[1 - F'(x)]$ or the finite-difference equivalent.

Step 4. Write the iteration equation [see Ref. [5.7], Eq. (3.25)], that is,

$$x_{i+1} = [(1-k)x + kF(x)]_i$$

and begin with root estimate x_0.

Step 5. If $|x_{i+1} - x_i| < \epsilon$, stop; otherwise go to step 4.

A simple example whose root is known is to find the root of $\ln x$:

Step 1. $f(x) = \ln x = 0$

Step 2. Solve for x by adding and subtracting x, to establish $F(x)$:

$$x - x + \ln x = 0$$
$$x = x - \ln x = F(x)$$

Step 3. Establish

$$k = \frac{1}{1 - F'(x)} = \frac{1}{1 - (1 - 1/x)} = x$$

Step 4. Write the iteration equation:

$$x_{i+1} = [(1-x)x + x(x - \ln x)]_i$$

$$= [x(1 - \ln x)]_i$$

With $x_0 = 2$, the following successive approximations are obtained:

$$2.000\ 000\ 000$$

$$0.613\ 705\ 639$$

$$0.913\ 341\ 207$$

$$0.996\ 131\ 704$$

$$0.999\ 992\ 508$$

$$1.000\ 000\ 000$$

In 5 iterations, 10 correct digits have been obtained. For a programmable hand-held calculator using reversed Polish notation, the problem-specific coding could be

```
A  STO1 lnx CHS 1 + RCL1 × R/S
```

As an example of a problem with unknown answer, consider a $2 \times \frac{1}{4}$ in tube of 1035 cold drawn steel ($S_y = 67$ kpsi) that is 48 in long and must support a column load with an eccentricity of $\frac{1}{8}$ in, as depicted in Fig. 5.5. For a design factor of 4 on the load, what allowable load is predicted by the secant column equation [Ref. [5.3], Eq. (3.54)]? The equation is

$$\frac{nP}{A} = S_y \Big/ \left[1 + \frac{ec}{r^2} \sec\left(\frac{\ell}{r} \sqrt{\frac{nP}{A} \frac{1}{4E}} \right) \right]$$

where $A = 1.374$ in^2, $r = 0.625$ in, $e = 0.125$ in, $c = 1$ in, $\ell = 48$ in, $E = 30 \times 10^6$ psi, and $S_y = 67\ 000$ psi. The secant equation is of the form $nP/A = F(nP/A)$. Choosing $\Delta = 0.001\ nP/A$, we can construct a finite-difference approximation to $F'(nP/A)$ for use in estimating the convergence parameter:

$$k = 1 \Big/ \left[1 - \frac{F(1.001nP/A) - F(nP/A)}{0.001nP/A} \right]$$

Using the iteration equation,

$$\left(\frac{nP}{A} \right)_{i+1} = \left[(1-k)\frac{nP}{A} + kF\left(\frac{nP}{A} \right) \right]_i$$

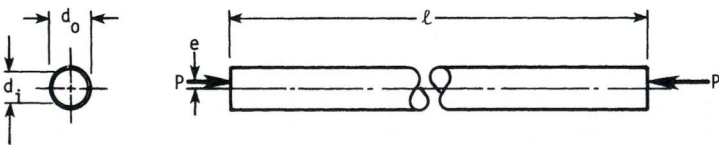

FIGURE 5.5   An eccentrically loaded hollow column.

Recalculating k every time and beginning with $(nP/A)_0 = 20\ 000$, the successive approximations are

$$20\ 000$$

$$34\ 004$$

$$32\ 548$$

$$32\ 518$$

$$32\ 518$$

It follows that

$$P = \frac{32\ 518A}{n} = \frac{32\ 518(1.374)}{4} = 11\ 170\ \text{lbf}$$

The range over which convergence is prompt is shown by using three different estimates, that is, $(nP/A)_0 = 1, 10\ 000,$ and $50\ 000$:

1	10 000	50 000
39 967	36 353	38 322
33 367	32 726	33 015
32 528	32 519	32 521
32 518	32 518	32 518

Such an effective algorithm as successive substitution deserves coding on all kinds of computers. The designer should be able to perform the algorithm manually if required.

For finding the zero place of a function of more than one variable, a somewhat different formulation is useful. If a function of x is expanded about the point x_0 in the neighborhood of the root as a Taylor series (Ref. [5.5], p. 579), we obtain

$$f(x) = f(x_0) + f'(x_0)(x - x_0) + \frac{1}{2!}f''(x_0)(x - x_0)^2 + \cdots$$

If x is the root, then $f(x) = 0$, and if the series is truncated after two terms, solution for x is a better estimate of the root than is x_0. Denoting $x - x_0$ as Δx, then

$$\Delta x = -\frac{f(x_0)}{f'(x_0)}$$

and the better estimate of the root is $x_1 = x_0 + \Delta x$. Using this pair of equations iteratively will result in finding the root. For example, if the involute of ϕ is 0.01, what is the value of ϕ? Recalling (Ref. [5.6], p. 266), that $\text{inv}\ \phi = \tan \phi - \phi$, we write

$$f(\phi) = \tan \phi - \phi - 0.01 = 0$$

$$f'(\phi) = \sec^2 \phi - 1 = \frac{1}{\cos^2 \phi} - 1$$

$$\Delta\phi = -\frac{f(\phi)}{f'(\phi)} = -\frac{\tan\phi - \phi - 0.01}{(1/\cos^2\phi) - 1}$$

$$\phi_{i+1} = \phi_i + \Delta\phi_i$$

For an initial estimate of $\phi_0 = 0.35$, we obtain

$$0.350\ 000\ 000$$
$$0.312\ 261\ 514$$
$$0.306\ 874\ 838$$
$$0.306\ 772\ 584$$
$$0.306\ 772\ 547$$
$$0.306\ 772\ 547$$

observing convergence to be rapid. For details, see any textbook on numerical methods, such as Carnahan et al. (Ref. [5.4], p. 171).

For the problem of two functions of two independent variables, namely, $f_1(x, y)$ and $f_2(x, y)$, the Taylor-series expansions are (Ref. [5.5], p. 580):

$$f_1(x, y) = f_1(x_0, y_0) + \frac{\partial f_1(x_0, y_0)}{\partial x}(x - x_0) + \frac{\partial f_1(x_0, y_0)}{\partial y}(y - y_0) + \cdots$$

$$f_2(x, y) = f_2(x_0, y_0) + \frac{\partial f_2(x_0, y_0)}{\partial x}(x - x_0) + \frac{\partial f_2(x_0, y_0)}{\partial y}(y - y_0) + \cdots$$

If x and y represent the roots of $f_1(x, y) = 0$ and $f_2(x, y) = 0$, and identifying $(x - x_0) = \Delta x$ and $(y - y_0) = \Delta y$, then the preceding equations can be written as

$$\frac{\partial f_1}{\partial x}\Delta x + \frac{\partial f_1}{\partial y}\Delta y = -f_1 = r_1$$

$$\frac{\partial f_2}{\partial x}\Delta x + \frac{\partial f_2}{\partial y}\Delta y = -f_2 = r_2$$

where r_1 and r_2 are called *residuals*. Solving the preceding equations simultaneously for Δx and Δy, we obtain

$$\Delta x = \frac{1}{A}\left(r_1\frac{\partial f_2}{\partial y} - r_2\frac{\partial f_1}{\partial y}\right)$$

$$\Delta y = \frac{1}{A}\left(r_2\frac{\partial f_1}{\partial x} - r_1\frac{\partial f_2}{\partial x}\right)$$

where

$$A = \frac{\partial f_1}{\partial x}\frac{\partial f_2}{\partial y} - \frac{\partial f_2}{\partial x}\frac{\partial f_1}{\partial y}$$

Better estimates of x and y than x_0 and y_0 are

$$x_{i+1} = x_i + \Delta x_i \qquad y_{i+1} = y_i + \Delta y_i$$

The solution algorithm is

Step 1. Decide the value of ϵ that $|\Delta x|$ and $|\Delta y|$ must not exceed. Then write equations in the form

$$f_1(x, y) = 0 \qquad f_2(x, y) = 0$$

Step 2. Calculate the residuals using starting estimates $x = x_0$ and $y = y_0$ for first evaluation:

$$r_1 = -f_1(x, y) \qquad r_2 = -f_2(x, y)$$

Step 3. Evaluate the Jacobian:

$$A = \frac{\partial f_1}{\partial x} \frac{\partial f_2}{\partial y} - \frac{\partial f_2}{\partial x} \frac{\partial f_1}{\partial y}$$

Step 4.

$$\Delta x = \frac{1}{A} \left(r_1 \frac{\partial f_2}{\partial y} - r_2 \frac{\partial f_1}{\partial y} \right)$$

$$\Delta y = \frac{1}{A} \left(r_2 \frac{\partial f_1}{\partial x} - r_1 \frac{\partial f_2}{\partial x} \right)$$

Step 5. Estimate

$$x \leftarrow x + \Delta x \qquad y \leftarrow y + \Delta y$$

Step 6. If $|\Delta x| < \epsilon$ and $|\Delta y| < \epsilon$, stop; otherwise go to step 2 with the new estimates of x and y.

As an example of the use of the Newton-Raphson method, consider a position analysis of the four-bar linkage depicted in Fig. 5.6, wherein $\rho_1 = 2$ in, $\rho_2 = 1$ in, $\rho_3 = 2.5$ in, and $\rho_4 = 3$ in. For a crank angle of $\theta_2 = 90°$, what are the abscissa angles θ_3 and θ_4? The vector equation

$$\boldsymbol{\rho}_2 + \boldsymbol{\rho}_3 = \boldsymbol{\rho}_1 + \boldsymbol{\rho}_4$$

gives rise to the pair of scalar equations

$$\rho_3 \cos \theta_3 = \rho_4 \cos \theta_4 - \rho_2 \cos \theta_2 + \rho_1$$

$$\rho_3 \sin \theta_3 = \rho_4 \sin \theta_4 - \rho_2 \sin \theta_2$$

Rewrite the equations in the form $f_1(\theta_3, \theta_4) = 0$ and $f_2(\theta_3, \theta_4) = 0$:

$$f_1(\theta_3, \theta_4) = \rho_4 \cos \theta_4 - \rho_2 \cos \theta_2 + \rho_1 - \rho_3 \cos \theta_3$$

$$f_2(\theta_3, \theta_4) = \rho_4 \sin \theta_4 - \rho_2 \sin \theta_2 - \rho_3 \sin \theta_3$$

The residuals are

$$r_1 = -f_1 = -\rho_4 \cos \theta_4 + \rho_2 \cos \theta_2 - \rho_1 + \rho_3 \cos \theta_3$$

$$r_2 = -f_2 = -\rho_4 \sin \theta_4 + \rho_2 \sin \theta_2 + \rho_3 \sin \theta_3$$

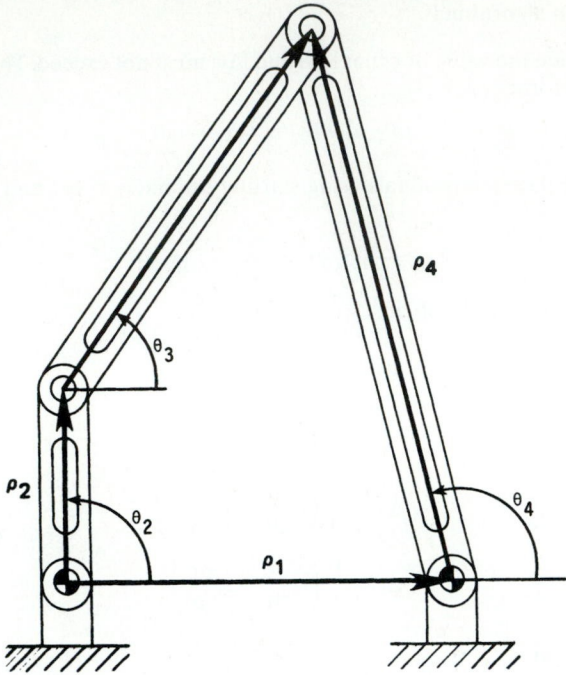

FIGURE 5.6 A vector model of a four-bar linkage with link 1 grounded, link 2 as a crank, link 3 as a coupler, and link 4 as a follower.

The value of A is determined:

$$\frac{\partial f_1}{\partial \theta_3} = \rho_3 \sin \theta_3 \qquad \frac{\partial f_1}{\partial \theta_4} = -\rho_4 \sin \theta_4$$

$$\frac{\partial f_2}{\partial \theta_3} = -\rho_3 \cos \theta_3 \qquad \frac{\partial f_2}{\partial \theta_4} = \rho_4 \cos \theta_4$$

$$A = (\rho_3 \sin \theta_3)(\rho_4 \cos \theta_4) - (-\rho_3 \cos \theta_3)(-\rho_4 \sin \theta_4)$$

$$= \rho_3 \rho_4 \sin (\theta_3 - \theta_4)$$

Find $\Delta\theta_3$ and $\Delta\theta_4$:

$$\Delta\theta_3 = \frac{1}{A} (r_1\rho_4 \cos \theta_4 + r_2\rho_4 \sin \theta_4)$$

$$= \frac{\rho_4}{A} (r_1 \cos \theta_4 + r_2 \sin \theta_4)$$

$$\Delta\theta_4 = \frac{1}{A} [r_2\rho_3 \sin \theta_3 - r_1(-\rho_3 \cos \theta_3)]$$

$$= \frac{\rho_3}{A} (r_2 \sin \theta_3 + r_1 \cos \theta_3)$$

Improve the estimate of θ_3 and θ_4:

$$\theta_3 \leftarrow \theta_3 + \Delta\theta_3 \qquad \theta_4 \leftarrow \theta_4 + \Delta\theta_4$$

For initial values $(\theta_3)_0 = 0.8$ and $(\theta_4)_0 = 1.6$, we obtain in four iterations

θ_3	θ_4
0.800 000	1.600 000
0.911 666	1.723 684
0.904 495	1.722 968
0.904 519	1.722 977
0.904 519	1.722 977

converging on $\theta_3 = 0.904\ 519$ rad or $51.83°$ and $\theta_4 = 1.722\ 977$ rad or $98.72°$. For initial values $(\theta_3)_0 = 5$ and $(\theta_4)_0 = 4$, we obtain in five iterations

θ_3	θ_4
5.000 000	4.000 000
4.477 696	3.743 041
4.440 784	3.626 671
4.451 343	3.632 940
4.451 371	3.632 893
4.451 371	3.632 893

identifying θ_3 as $255.0°$ and θ_4 as $208.1°$, which represents the configuration where the coupler crosses the grounded link.

Such a solution algorithm for simultaneous equations can be generalized to n equations. The previous kinematic problem can be coded for a hand-held calculator in approximately a hundred steps.

Another recurring task is that of integration. A powerful numerical tool is Simpson's first rule:

$$\int_{x_0}^{x_2} f(x)\ dx = \frac{h}{3}\left[f(x_0) + 4f(x_1) + f(x_2)\right] - \frac{h^5}{90} f''''(\xi) \qquad x_0 \le \xi \le x_2$$

The error term is exact for some (generally unavailable) value of ξ. If the number of repetitions of this rule made in an interval a, b is n, then the number of panels is $N = 2n$. Richardson (Ref. [5.4], p. 78) showed that if an integration is performed in interval a, b with N_2 panels and then repeated with $N_1 = N_2/2$ panels (using every other ordinate), then the value of the integral is given by

$$I = I_{N_2} + \frac{I_{N_2} - I_{N_1}}{15} = I_{N_2} + R_{N_2}$$

where the last term is called *Richardson's error estimate*. The number of panels N_2 must be divisible by 4. The approximate relation between the number of panels and the error is

$$N_j \doteq N_i \left|\frac{E_i}{E_j}\right|^{1/4} = N_i \left|\frac{I_{N_2} - I_{N_1}}{15E_j}\right|^{1/4} = N_i \left|\frac{R_i}{E_j}\right|^{1/4}$$

Solved Problem. Estimate the value of $\int_0^\pi \sin x \, dx$ to five significant digits to the right of the decimal point.

Step 1. Perform the integration with 2 panels, obtaining $I_2 = 2.0944$.

Step 2. Perform the integration with 4 panels, obtaining $I_4 = 2.00456$.

Step 3. Estimate the error in I_4 as

$$R_i = \frac{I_4 - I_2}{15} = \frac{2.004\,56 - 2.0944}{15} = -0.005\,989$$

Step 4. Estimate the number of panels necessary to attain requisite accuracy:

$$N_j \doteq N_i \left| \frac{R_i}{E_j} \right|^{1/4} = 4 \left| \frac{0.005\,989}{0.5 \times 10^{-5}} \right|^{1/4}$$

$$= 23.5$$

say, 24 panels.

Step 5. Integrate using 24 panels, obtaining $I_{24} = 2.000\,003\,269$.

This result is high, as indicated by the sign of Richardson's correction and by an estimated amount of $-0.000\,003\,290$. Note that the objective has been achieved. An improved estimate of the value of the integral might be $I_{24} + R_{24} = 1.999\,999\,979$, which rounded to five significant digits to the right of the decimal point is still 2.000 00.

Solved Problem. An electric motor has a torque-rpm characteristic of $36(1 - n/1800)$ ft·lbf and a moment of inertia of 1 slug·ft². Estimate within a tenth of a second the time to come up to a speed of 1600 rpm from rest in the absence of load.

The expression for the time estimate is in the form of an integral:

$$t = \int \frac{I}{T} \, d\omega = \frac{2\pi}{60} \int_0^{1600} \frac{dn}{36(1 - n/1800)}$$

Integrating with four panels using Simpson's rule, the Richardson correction is $R_4 = -0.094\,561\,754$. The estimated number of panels to assess the starting time to within a tenth of a second ($E = 0.05$ s) is

$$N_j \doteq N_i \left| \frac{R_i}{E_j} \right|^{1/4} = 4 \left| \frac{0.094\,561\,754}{0.05} \right|^{1/4} = 4.69$$

This is rounded to the next larger integer divisible by 4. Using eight panels:

$$I = I_8 + R_8 = 11.571\,809\,98 - 0.023\,455\,747$$

$$= 11.548\,354\,23$$

The result shows the objective achieved with a result of 11.5 s. Simpson's first rule should be coded and available to any user of a computer.

In computer-aided engineering, a number of routine mathematical tasks are encountered, and these should be available to the programmer in an executive fashion, discharged, if possible, by a one-line call statement (in Fortran).

5.5 STATISTICAL TASKS

There are innumerable statistical tasks to be performed incidental to engineering calculations; for example:

- Descriptive statistics such as means, medians, variances, and ranks have to be developed from data.
- Probabilities of observations from binomial, hypergeometric, Poisson, normal, lognormal, exponential, and Weibullian distributions need to be found.
- Inferential statistics must be developed for distributional parameters such as means, variances, and proportions.
- Data need to be fitted to distributional curves using least-squares lines, polynomials, or distributional functions.
- Goodness-of-fit tests for conformity must be made.

There exist programs for large computational machines which can be imitated or approximated on smaller machines. An important thing to be remembered concerning statistical computations conducted with paper and pencil as compared to those conducted with a computer is that computer programs are executed out of sight and supervision of a human and there is no experienced eye monitoring intermediate results and exhibiting a healthy skepticism when the occasion warrants.

It is so easy to calculate a correlation coefficient, be impressed with its nearness to unity, have it indicate statistical significance of fit, and be wrong without a warning signal. The correlation coefficient has meaning *only* if the data fall randomly about the regression curve. When using the computer, it is important to inspect a graphic presentation or to conduct a run test prior to testing for significance of fit. This test detects randomness or the lack of it in the case of dichotomous events (heads or tails, larger or smaller than the mean, above or below a regression curve). We can observe successes (above), failures (below), and runs (sequences of successes or failures). If n_1 is the number of successes, n_2 is the number of failures, and $n_1 \geq 10$ and $n_2 \geq 10$, then the sampling distribution of the number of runs n is approximately gaussian, with (Ref. [5.17], p. 414)

$$\mu_n = \frac{2n_1 n_2}{n_1 + n_2} + 1$$

$$\sigma_n = \left[\frac{2n_1 n_2 (2n_1 n_2 - n_1 - n_2)}{(n_1 + n_2)^2 (n_1 + n_2 - 1)} \right]^{1/2}$$

The null hypothesis that the sample is random can be based on the statistic

$$z = \begin{cases} \dfrac{n + \frac{1}{2} - \mu_n}{\sigma_n} & n > \mu_n \\[3mm] \dfrac{n - \frac{1}{2} - \mu_n}{\sigma_n} & n < \mu_n \end{cases}$$

where the $\frac{1}{2}$ improves the gaussian continuous fit to the discrete PDF of n.

If a regression line were determined from data, using a for above the line and b for below the line, then as we move along the abscissa we would observe

$$\underline{aa}\ \underline{b}\ \underline{a}\ \underline{b}\ \underline{a}\ \underline{b}\ \underline{a}\ \underline{b}\ \underline{aa}\ \underline{b}\ \underline{a}\ \underline{b}\ \underline{aa}\ \underline{bb}\ \underline{aa}\ \underline{b}\ \underline{aa}\ \underline{bb}\ \underline{a}\ \underline{b}\ \underline{a}\ \underline{bb}\ \underline{a}\ \underline{b}\ \underline{a}\ \underline{b}\ \underline{a}$$

detecting $n = 27$ runs, $n_1 = 19$ above, $n_2 = 16$ below. The mean number of runs expected μ_n and the standard deviation expected σ_n are

$$\mu_n = \frac{2(19)16}{19+16} + 1 = 18.37$$

$$\sigma_n = \left[\frac{2(19)16[2(19)16 - 19 - 16]}{(19+16)^2(19+16-1)} \right]^{1/2} = 2.89$$

Since the number of runs, that is, 27, is greater than the mean of 18.37,

$$z = \frac{27 + \frac{1}{2} - 18.37}{2.89} = 3.16$$

If the null hypothesis is H_0: runs random, then under H_0, $z = 3.16$, and z (tabulated two-tailed) = 1.96, and we can reject H_0 at 0.95 confidence level and embrace the alternative that the runs are not random. Similarly, if a straight line is fitted to parabolic data, the number of runs might be 3 or 4 or 5. If it is as much as 9, then

$$z = \frac{9 - \frac{1}{2} - 18.37}{2.89} = -3.42$$

and we can still reject randomness at the 0.95 confidence level. Inasmuch as no one is looking, it is important that we build in sentinels such as this.

If the differences between y and $\hat{y}$ are ranked in the order of their corresponding abscissas and placed in the column vector DY, then the number of runs can be detected with the following Fortran coding:

```
      NUMBER=1
      DO 100 I=2,N
      A=DY(I)/ABS(DY(I))
      B=DY(I-1)/ABS(DY(I-1))
      IF(A/B.LT.0.)NUMBER=NUMBER+1
  100 CONTINUE
```

where the integer NUMBER has a magnitude equal to the number of runs.

5.6 OPTIMIZATION TASKS

The structure of the design-decision problem and that of the optimization problem are similar. Many ideas and techniques of the latter are applicable to the former. The optimization problem can be posed as

$$\text{Maximize (or minimize) } M(x_1, x_2, \ldots, x_n)$$

subject to

$$g_1(x_1, x_2, \ldots, x_n) = 0$$

$$g_2(x_1, x_2, \ldots, x_n) = 0$$

$$\ldots\ldots\ldots\ldots\ldots$$

$$g_m(x_1, x_2, \ldots, x_n) = 0$$

and

$$z_1 \le f_1(x_1, x_2, \ldots, x_n) \le Z_1$$
$$z_2 \le f_2(x_1, x_2, \ldots, x_n) \le Z_2$$

$$\cdots\cdots\cdots\cdots\cdots$$

$$z_\lambda \le f_\lambda(x_1, x_2, \ldots, x_n) \le Z_\lambda$$

The functions $g_i\{x_n\} = 0$ are called *equality* or *functional constraints*. The functions $z_i \le f_i\{x_n\} \le Z_i$ are called *inequality* or *regional constraints*. The set $\{x_n\}$ is called the *decision set*. In terms of the ideas in Sec. 5.2, the *specification set* consists of

$$P_1, P_2, \ldots, P_k$$

which for a helical-coil compression spring can consist of (1) material and its condition, that is, A, m, G, E, (2) wire size d, (3) end treatment, that is, Q, Q', (4) total number of turns T, (5) coil outside diameter OD, and (6) free length ℓ_o. The adequacy assessment can be performed by a Fortran subroutine:

```
ADEQ(A,m,G,E,d,Q,Q',T,OD,d,l_o,J)
```

where J is returned as $\ne 0$ if inadequate and as 0 if adequate. If the a priori decisions are (1) material and condition, (2) end treatment, (3) total turns, (4) coil outside diameter, and (5) free length, and the wire diameter is chosen as the sole decision variable, then the tasks are

1. Choose d. (This completes the decision set.)
2. Call CONVERT. (Change the decision set into the equivalent specification set.)
3. Call ADEQ. (Establish the adequacy of the decision set.)
4. Call FOM. (Evaluate the figure of merit if the decision set is adequate.)

The choice of d is provided either manually (interactively) or by an appropriate optimization algorithm which makes successive choices of d which have superior merit. The program CONVERT might be problem-specific and need to be created for each type of design problem. The program ADEQ is durable and once programmed can be used. The program FOM is durable as long as the merit criterion (say spring cost) is unchanged.

Figure 5.7 shows the interrelationships of programs FOM, CONVERT, ADEQ, OPT, and the executive program for the helical-spring example.

Optimization programs have to be chosen with care because highly constrained problems can defeat classical strategies. The issue is further complicated by the mixture of discrete and continuous variables. In a spring design, the wire size, end treatment, and material parameters are discrete, whereas the other variables are usually continuous. The user is solving a problem to which the answer is not known and cannot be sure of having attained the global extreme of the figure-of-merit function. Since multidimensional optimization strategies involve some gradient sensitivity, it is judicious to ensure that discrete variables in a problem are among the a priori decisions and the decision set and not buried within. Sometimes an exhaustive search over discrete variables, although computationally inelegant, will attain a global maximum with the least expenditure of computer plus engineering costs. For example, in a spring design with the wire size d as the sole decision variable, marching through the discrete preferred (or available) wire diameters will solve the problem efficiently:

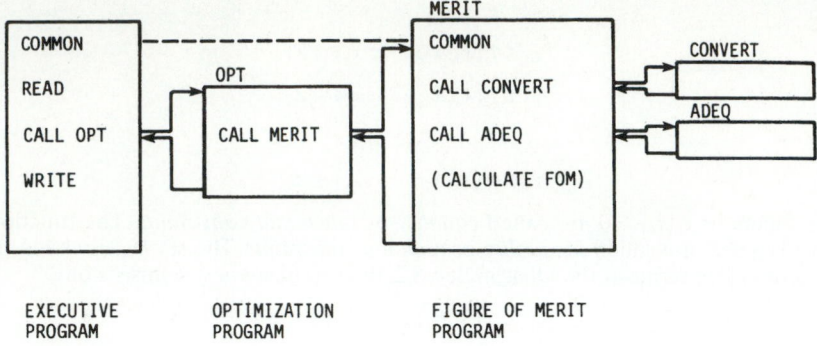

EXECUTIVE   OPTIMIZATION   FIGURE OF MERIT
PROGRAM   PROGRAM   PROGRAM

FIGURE 5.7   Organization and subordination of programs for the helical-coil compression spring example.

Washburn and Moen gage: No. 40, 39, 38, 37, . . .
Decimal inch preferred: 0.004, 0.005, 0.006, 0.008, . . .
Decimal millimeter preferred: 0.1, 0.12, 0.16, 0.20, 0.25, . . .

Suppose two decision variables remain after four a priori decisions, and these are wire diameter d and free length ℓ_o (d is discrete and ℓ_o is continuous). The steps might be

1. Enter d and bounds on ℓ_o.
2. For an available wire size d, use a do-loop to show 11 springs of different lengths, displaying d, ℓ_o, FOM, and NG.
3. If there exists a feasible range, enter the ℓ_o bounds which define the range.
4. Use a golden-section search strategy (Ref. [5.8]) on ℓ_o to find the maximum figure of merit.
5. Display the specification set and figure of merit.
6. Repeat for all possible wire sizes.
7. Select best of field as global optimum.

This procedure is best made interactive and can be presented to a user without requiring a knowledge of programming.

For the static-service helical-coil compression spring using one decision variable d and two decision variables d and ℓ_o, the specification sets are

	Case 1. One decision variable d	Case 2. Two decision variables d and ℓ_o
Material and condition	Cold-drawn spring wire 1066	Cold-drawn 1066
Wire size	W&M no. 11 (0.1205 in)	W&M no. 11 (0.1205 in)
End condition	Squared and ground	Squared and ground
Total turns	14.4	13.4
OD	1.091 in	1.091 in
Free length	3.25 in	3.19 in
(Wire volume)	(0.502 in³)	(0.467 in³)

Note that both springs are optimal for the conditions, but case 2 is a superior spring in that less material, and therefore less cost, is involved.

Reference [5.1] is a comprehensive introduction to the optimization problem with many examples.

5.7 SIMULATION

Statistics is the science of empiricism. Armed with rationales from probability theory, statistics developed methods for gathering, analyzing, and summarizing data and formulating inferences to learn of systematic relationships, together with an estimate of the chance of being incorrect. Drawing balls from an urn to learn about its contents is a statistical experiment, the balls simulating a universe (a distribution) and those withdrawn constituting a sample. To simulate is to mimic some or all of the behavior of one system with another, with equipment, or with a computer using random numbers.

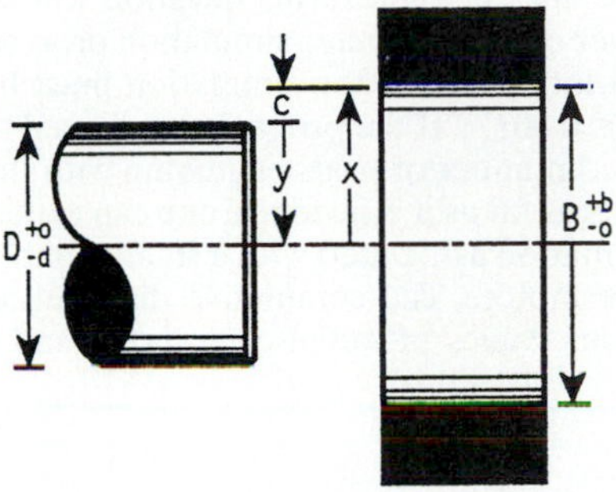

Random numbers from the uniform distribution $U[0, 1]$ can be selected using a machine-specific subprogram supplied by the computer manufacturer. These numbers can be transformed into another distribution of interest using software. Through selection of random numbers and calculations performed with them, data can be gathered and answers to useful questions obtained. Consider the unilateral tolerances on a journal and a bushing as depicted in Fig. 5.8. If the journal is generated in a turning or grinding operation, each successive part created on automatic machinery is slightly larger as a result of tool wear and dulling, which increases tool forces and workpiece deflection. The distribution of journal diameter produced between setups and thoroughly mixed is uniform. The reamed bushing bores are uniform for similar reasons. The radii **x** and **y** and the radial clearance **c** are random variables related through the equation $\mathbf{x} - \mathbf{c} - \mathbf{y} = 0$ or

FIGURE 5.8 Dimensions and unilateral tolerances on a journal and bushing. In a specific bearing assembly the bushing radius x, the journal radius y, and the radial clearance c are signed variates related as shown.

$$\mathbf{c} = \mathbf{x} - \mathbf{y}$$

What is the distribution of the radial clearance **c**? What is its cumulative distribution function $F(c)$? It takes more than elementary statistics to go straight to the answers. A simulation can be run to obtain robust answers without knowing or using the statistical knowledge.

Example 2. The bearing formed by the journal and bushing of Fig. 5.8 has $D = 2.002$ in, $d = 0.002$ in, $B = 2.004$ in, and $b = 0.003$ in. Estimate the probability that the random assembly radial clearance c is less than or equal to 0.0015 in.

Solution.   Given that your computer has a uniform random number generator $u \sim U[0, 1]$ subroutine named RANDU, set up a do-loop which calls RANDU twice, returning u_1 and u_2, and create instances of

$$\vdots$$

$$x = (B + bu_1)/2$$

$$y = (D - d + du_2)/2$$

$$c = x - y$$

$$\vdots$$

Note whether c is less than or equal to 0.0015 in. If the do-loop is executed n times, then the probability of encountering radial clearances less than or equal to 0.0015 is $p = n_1/n$, where n_1 is the number of instances of $c \leq 0.0015$ in. In one million trials, $p = 0.083\,395$. This was accomplished without any special statistical knowledge.

How close is the simulation answer to the correct value? This question will be answered later. Example 2 illustrates the power of the computer simulation process. The price of not knowing the closed-form solution is that the simulation must be repeated to answer another question, say $p(c \leq 0.0017)$. If the program is adjusted to run the simulation to an interactively specified number of trials beginning with the same seed(s) (the same list of random numbers as far as it is needed), one can gather data for the plot of Fig. 5.9. Note the poor estimation associated with a small number of trials, and the eventual approach to an asymptote. The cumulative distribution function can be well approximated for various values of radial clearance c and a polynomial fitted to the data.

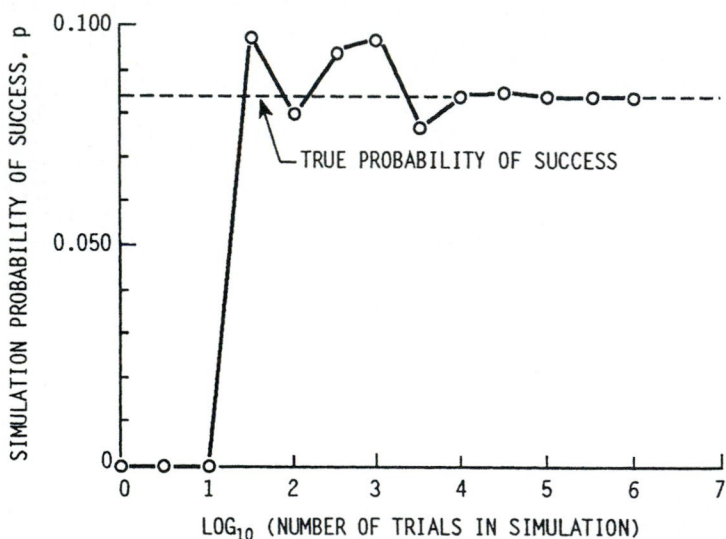

FIGURE 5.9   Simulation convergence in Example 2.

Example 3. For the assembly of Example 2, find an expression for the cumulative distribution function $F(c)$ in the range $0.001 \leq c \leq 0.002$.

Solution. For simulations of one million trials, the following data were obtained:

c, in	$F(c)$
0.0010	0
0.0012	0.013 308
0.0014	0.053 201
0.0016	0.120 140
0.0018	0.213 820
0.0020	0.333 989

A least-squares quadratic fit of the form $F(c) = a_0 + a_1 c + a_2 c^2$ gives

$$F(c) = 0.335\,084 - 669.672c + 334577c^2$$

Using this equation to predict the answer to Example 2 yields

$$F(0.0015) = 0.335\,084 - 669.672(0.0015) + 334\,577(0.0015)^2$$

$$= 0.083\,374$$

It is useful to have subroutines to generate random numbers from other useful distributions using RANDU. For a uniform distribution in the interval a, b,

```
SUBROUTINE UNIF1(IX,IY,A,B,U)
CALL RANDU(IX,IY,R)
U=A+(B-A)*R
RETURN
END
```

For a mean of $\bar{x}$, coded XBAR, and a standard deviation of s_x, coded SX, a uniform random number U is returned by

```
SUBROUTINE UNIF2(IX,IY,XBAR,SX,U)
CALL RANDU(IX,IY,R)
A=XBAR-SQRT(3.)*SX
B=XBAR+SQRT(3.)*SX
U=A+(B-A)*R
RETURN
END
```

For a mean of $\bar{x}$, coded XBAR, and a standard deviation s_x, coded SX, a normally distributed random number G is returned by

```
SUBROUTINE GAUSS(IX,IY,XBAR,SX,G)
SUM=0.
DO 100 I=1,12
CALL RANDU(IX,IY,U)
SUM=SUM+U
100 CONTINUE
G=XBAR+(SUM-6.)*SX
RETURN
END
```

For a mean of $\bar{x}$, coded XBAR, and a standard deviation of s_x, coded SX, a lognormally distributed random number XLOG is returned by

```
SUBROUTINE LOGNOR(IX,IY,XBAR,SX,XLOG)
CX=SX/XBAR
YBAR=ALOG(XBAR)-ALOG(SQRT(1.+CX**2))
SY=SQRT(ALOG(1.+CX**2))
CALL GAUSS(IX,IY,YBAR,SY,G)
XLOG=EXP(G)
RETURN
END
```

For Weibull parameters x_0, θ, and b, a Weibull-distributed random number W is returned by

```
SUBROUTINE WEIBUL(IX,IY,XO,THETA,B,W)
CALL RANDU(IX,IY,U)
W=XO+(THETA-XO)*(ALOG(1./U))**(1./B)
RETURN
END
```

For distributions with survival equations that can be explicitly solved for R or F, solve for variate x and substitute random numbers $u \sim U[0, 1]$ for either R or F, and x will be random in that distribution.

Example 4.   Interfere a Weibull-distributed strength $S \sim W[40, 50, 3.3]$ kpsi with a Weibull-distributed stress $\sigma \sim W[30, 40, 2]$ kpsi by simulation.

Solution.   The body of a Fortran program using the Weibull random number generator WEIBUL is

```
        ⋮
     NSUCC=0
     DO 100 I=1,N
     CALL WEIBUL(IX,IY,XO1,THETA1,B1,W1)
     CALL WEIBUL(IX,IY,XO2,THETA2,B2,W2)
     IF(W1.GE.W2)NSUCC=NSUCC+1
100  CONTINUE
     R=FLOAT(NSUCC)/FLOAT(N)
        ⋮
```

For one million trials, the reliability estimate was 0.956 815.

The question of the accuracy of the reliability estimate is addressed as follows. An interference reliability simulation is really the construction of a column vector $\{x\}$ composed of zeros and ones. The sum of the elements in $\{x\}$ is np, where n is the number of entries (trials) and p is the probability of success. The mean of the elements in $\{x\}$ is

$$\bar{x} = \frac{\Sigma x}{n} = \frac{np}{n} = p \tag{5.1}$$

so $\bar{x}$ is an estimator of the probability of success p. The column vector of the squares of the elements in $\{x\}$ is identical to the elements in $\{x\}$. The sum of the squares of the elements in $\{x\}$ is also np. The standard deviation of x is

$$\sigma_x = \sqrt{\frac{\Sigma x^2 - (\Sigma x)^2/n}{n-1}} \doteq \sqrt{\frac{np - n^2 p^2/n}{n}} = \sqrt{p(1-p)} \tag{5.2}$$

The standard deviation of the mean $\bar{x}$ is

$$\sigma_{\bar{x}} = \frac{\sigma_x}{\sqrt{n}} = \frac{\sqrt{p(1-p)}}{\sqrt{n}} = \sqrt{\frac{p(1-p)}{\sqrt{n}}} \tag{5.3}$$

In a simulation involving n_1 trials, the bilateral tolerance at the α confidence level (two-tailed) $z_\alpha \sigma_{\bar{x}}$ is denoted as error e_1:

$$e_1 = z_\alpha(\sigma_{\bar{x}})_1 = z_\alpha \sqrt{\frac{p_1(1-p_1)}{n_1}} \tag{5.4}$$

The number of trials n_2 necessary to attain an error $e_2 = \frac{1}{2}(10^{-m})$ associated with m significant digits to the right of the decimal point is

$$e_2 = \frac{1}{2}(10^{-m}) = z_\alpha \sqrt{\frac{p_2(1-p_2)}{n_2}} \tag{5.5}$$

Arguing that $p_1(1-p_1) \doteq p_2(1-p_2)$ allows Eqs. (5.4) and (5.5) to be combined as

$$e_1\sqrt{n_1} = \frac{1}{2}(10^{-m})\sqrt{n_2}$$

or

$$n_2 = n_1\left[\frac{e_1}{\frac{1}{2}(10^{-m})}\right]^2 = 4n_1(10^{2m})z_\alpha^2(\sigma_{\bar{x}})_1^2 \tag{5.6}$$

Using Eq. (5.3),

$$n_2 = 4(10^{2m})z_\alpha^2 \hat{p}(1-\hat{p}) \tag{5.7}$$

Solving Eq. (5.7) for m gives

$$m = \frac{1}{2}\log\frac{n_2}{4z_\alpha^2\hat{p}(1-\hat{p})} \tag{5.8}$$

In Example 4, $p = 0.956\ 815$ in 10^6 trials. The digits $0.956\ 815$ were in the computer arithmetic register. Some of the left-hand digits are correct. Using Eq. (5.8),

$$m = \frac{1}{2}\log\frac{10^6}{4(1.96)^2(0.956\ 815)(1-0.956\ 815)} = 3.1$$

At the 0.95 confidence level, the left two digits are correct; the third may be correct or rounded.

Alternatively, one may use Eq. (5.5) to find the bilateral tolerance:

$$e_2 = 1.96\sqrt{\frac{0.956\ 815(1-0.956\ 815)}{10^6}} = 0.000\ 398$$

and display $p = 0.956\ 815 \pm 0.000\ 398$ at the 0.95 confidence level.

There is an advantage to writing a computer code to start a string of random numbers from a seed or seeds, then proceeding with the simulation in steps.

- Using a few thousand trials, find n_1 and p_1, and from Eq. (5.4) estimate e_1.
- Use Eq. (5.6) to estimate n_2. The number of additional trials needed to reach the accuracy goal is $n_2 - n_1$. Conduct the additional trials.

- Using n_2 and p_2, find m using Eq. (5.8) or find e_2 from Eq. (5.5). Use Eq. (5.7) and n_2 results as a check.

The net effect of this plan is that no more trials than necessary are performed. In small problems this is a minor consideration, but if a single trial has consequential cost, economy of effort is important.

A Fortran program for solving Example 4 using incremental trials, rather than ever-larger simulations until the goal is reached, follows.

```
c Program wsim.f
c Simulation program for Weibull-Weibull interference C. Mischke Nov 93
    1 print*,'Simulation program Weibull-Weibull interference,Mischke'
      print*,' '
      print*,'Enter RANDU seeds ix, iy (odd, five digits or more)'
      read*,ix,iy
c Initialize counters so simulation proceeds in steps (economically)
c under the control of the user.
      sum1=0.
      sum2=0.
      nused=0
c Enter distribution parameters
      print*,'Enter strength Weibull parameters x0, theta, b'
      read*,x01,theta1,b1
      print*,'Enter stress Weibull parameters x0,theta,b'
      read*,x02,theta2,b2
      print*,'Enter z-variable and corresponding confidence level'
      print*,'for bilateral tolerance on result'
      read*,z,alpha
c Conduct simulation
    2 print*,'Enter number of trials in similation n'
      read*,nmore
      print*,'Enter bilateral tolerance allowable on simulation result'
      read*,error
      do 100 i=1,nmore
      call weibul(ix,iy,x01,theta1,b1,w1)
      call weibul(ix,iy,x02,theta2,b2,w2)
      if(w1.ge.w2)x=1
      if(w1.lt.w2)x=0.
      sum1=sum1+x
      sum2=sum2+x*x
  100 continue
      nused=nmore+nused
      p=sum1/float(nused)
      sigmap=sqrt((sum2-sum1**2/float(nused))/float(nused-1))
      sigmamu=sigmap/sqrt(float(nused))
      tol=z*sigmamu
      print*,'For',nused,' trials probability of success p is',p
      print*,'sigmap=',sigmap,' sigmapbar=',sigmamu,' bilat. tol.=',tol
      print*,'Largest allowable bilateral error is',error
      xn2= float(nused)*tol**2/error**2
      if(nused.gt.xn2)go to 10
      print*,'Total trials needed =',xn2,' or ',xn2-float(nused),' more'
      go to 11
   10 print*,'Simulation complete in',nused,' trials'
      correct=alog10(1./2./tol)
      print*,'Correct digits to right of decimal in p is',correct
      print*,'at',alpha,' confidence level.'
      ncorrect=correct
      np=p*10.**ncorrect
      p=float(np)/10.**ncorrect
      print*,'Correct digits are',p,' at',alpha,' confidence level'
   11 print*,' '
      print*,'For a new problem,        enter 1'
      print*,'To continue simulation,   enter 2'
      print*,'To quit                   enter 3'
      read*,index
      go to(1,2,3),index
    3 call exit
      end
```

REFERENCES

5.1 G. V. Reklaitis, A. Ravindran, and K. M. Ragsdell, *Engineering Optimization, Methods and Applications,* John Wiley & Sons, New York, 1983.

5.2 *Design Handbook,* Associated Spring-Barnes Group, Bristol, Conn., 1981.

5.3 J. E. Shigley and C. R. Mischke, *Mechanical Engineering Design,* 5th ed., McGraw-Hill, New York, 1989.

5.4 B. Carnahan, H. A. Luther, and J. O. Wilkes, *Applied Numerical Methods,* John Wiley & Sons, New York, 1969, p. 168.

5.5 R. C. Weast and S. M. Shelby, eds., *Handbook of Tables for Mathematics,* 3d ed., The Chemical Rubber Company, 1967, p. 579.

5.6 J. E. Shigley and J. J. Uicker, *Theory of Machines and Mechanisms,* McGraw-Hill, New York, 1980, p. 266.

5.7 W. H. Beyer, *Handbook of Tables for Probability and Statistics,* 2d ed., The Chemical Rubber Company, 1968, p. 414.

5.8 C. R. Mischke, *Mathematical Model Building,* 2d rev. ed., Iowa State University Press, Ames, 1980, chap. 3.

References

1.
2.
3.
4.
5.
6.
7.

CHAPTER 6
WEAR

Kenneth C. Ludema
Professor of Mechanical Engineering
Department of Mechanical Engineering and Applied Mechanics
The University of Michigan
Ann Arbor, Michigan

There is no shorthand method of designing machinery for a specified wear life. Thus a step-by-step method is given for designers to follow. The method begins with an examination of worn parts of the type to be improved. The next step is an estimate of stresses, temperatures, and likely conditions of operation of the redesigned machinery. Material testing for wear resistance is discussed, and finally, a procedure is given for selecting materials for wear resistance.

6.1 GENERAL PRINCIPLES IN DESIGN FOR WEAR RESISTANCE

The wear life of mechanical components is affected by nearly as many variables as human life. Wearing surfaces are composed of substrate material, oxide, absorbed gas, and dirt. They respond to their environment, method of manufacture, and conditions of operation. They suffer acute and/or progressive degeneration, and they can often be partially rehabilitated by either a change in operating conditions or some intrusive action.

The range of wearing components and devices is endless, including animal teeth and joints, cams, piston rings, tires, roads, brakes, dirt seals, liquid seals, gas seals, belts, floors, shoes, fabrics, electrical contacts, disks and tapes, tape heads, printer heads, tractor tracks, cannon barrels, rolling mills, dies, sheet products, forgings, ore crushers, conveyors, nuclear machinery, home appliances, sleeve bearings, rolling-element bearings, door hinges, zippers, drills, saws, razor blades, pump impellers, valve seats, pipe bends, stirring paddles, plastic molding screws and dies, and erasers. There is not a single universal approach to designing all these components for an acceptable wear life, but there are some rational design steps for some. There are no equations, handbooks, or material lists of broad use, but there are guidelines for some cases. Several will be given in this section.

6.1.1 Types, Appearances, and Mechanisms of Wear

Wear is a loss or redistribution of surface material from its intended location by definition of the ASTM. Using this definition, we could develop a simple explanation for wear as occurring either by chemical reaction (that is, corrosion), by melting, or by mechanical straining. Thus to resist wear, a material should be selected to resist the preceding individual causes of wear or else the environment should be changed to reduce surface stress, temperature, or corrosiveness.

The preceding three natural processes are too broad to be useful for material selection in light of the known properties of materials. A more detailed list of material properties appropriate to the topic of wear is given in Table 6.1.

The preceding methods of material removal are usually not classified among the "mechanisms" of wear. Usually a *mechanism* is defined as a fundamental cause. Thus a fundamental argument might be that wear would not occur if there were no contact. If this were so, then mere contact could be called a mechanism of wear. However, if we define a *mechanism* as that which is capable of explanation by the laws of physics, chemistry, and derivative sciences, then mere contact becomes a statement of the condition in which surfaces exist and not a mechanism. But if stresses, lattice order, hydrogen-ion concentration, fugacity, or index of refraction were known, *and if* the effect of these variables on the wear rate were known, then a mechanism of wear has been given. Most terms used to describe wear therefore do not suggest a mechanism. Rather, most terms describe the condition under which wearing occurs or they describe the appearance of a worn surface. Terms of the former type include dry wear, metal-to-metal wear, hot wear, frictional wear, mechanical wear, and impact wear. Closer observation may elicit descriptions such as erosion, smooth

TABLE 6.1   Material Properties Involved in Wear

Chemical action
1. Chemical dissolution
2. Oxidation (corrosion, etc.)

Mechanical straining
3. Brittle fracture (as in spalling; see below)

4. Ductile deformation:
   a. To less than fracture strain (as in indentation)
   b. To fracture (as in cutting, galling, transfer, etc.)

5. High-cycle fatigue (as occurs in rolling contacts)
6. Low-cycle fatigue (as in scuffing, dry wear, etc.)
7. Melting

SOURCE: From Ludema [6.2].

wear, polishing wear, cavitation, corrosive wear, false brinelling, friction oxidation, chafing fatigue, fretting, and chemical wear. Still closer observation may reveal spalling, fatigue wear, pitting corrosion, delamination, cutting wear, deformation wear, gouging wear, galling, milling wear, plowing wear, scratching, scouring, and abrasion. The latter is often subdivided into two-body or three-body abrasion and low-stress or high-stress abrasion. Finally, some of the terms that come from the literature on "lubricated" wear include scuffing, scoring, and seizure. Most of these terms have specific meanings in particular products and in particular industries, but few find wide use.

Valiant attempts are continuously being made to define wear terms in the professional societies, but progress is slow. Researchers have attempted to classify most of the terms as either abrasive or adhesive mechanisms primarily, with a few terms classified as a fatigue mechanism. It is interesting that adhesiveness or abrasiveness is not often proven in real problems. Rather, a given wear process is simply modeled as abrasive *or* adhesive and often considered as exclusively so. Some authors attempt to escape such categories by separating wear into the mild and severe categories, which introduces value judgments on wear rates not inherently found in the other terms. Mechanisms of wear will be discussed at greater length below.

6.1.2 Design Philosophy

Most wearing surfaces are redesigned rather than designed for the first time. Thus designers will usually have access to people who have experience with previous products. Designing a product for the first time requires very mature skills, not only in materials and manufacturing methods, but also in design philosophy for a particular product.

The philosophy by which wear resistance or wear life of a product is chosen may differ strongly within and between various segments of industry. Such considerations as acceptable modes of failure, product repair, controllability of environment, product cost, nature of product users, and the interaction between these factors receive different treatment for different products. For example, since automobile tires are easier to change than is an engine crankshaft, the wear life of tires is not a factor in discussions of vehicle life. The opposite philosophy must apply to drilling bits used in the oil-well industry. The cone teeth and the bearing upon which the cone rotates must be designed for equal life, since both are equally inaccessible while wearing.

In some products or machines, function is far more important than manufacturing costs. One example is the sliding elements in nuclear reactors. The temperature environment of the nuclear reactor is moderate, lubricants are not permitted, and the result of wear is exceedingly detrimental to the function of the system. Thus expensive metal-ceramic coatings are frequently used. This is an example of a highly specified combination of materials and wearing conditions. Perhaps a more complex example is that of artificial teeth. The surrounding system is very adaptable, a high cost is relatively acceptable, but durability may be strongly influenced by body chemistry and choice of food, all beyond the range of influence by the designers.

Thus there is no general rule whereby designers can quickly proceed to select a wear-resisting material for a product. One often heard but misleading simple method of reducing wear is to increase the hardness of the material. There are, unfortunately, too many exceptions to this rule to have high confidence in it except for some narrowly defined wearing systems. One obvious exception is the case of bronzes, which are more successful as a gear material against a hardened-steel pinion than is a hardened-steel gear. The reason usually given for the success of bronze is that dirt particles are readily embedded into the bronze and therefore do not cut or wear the steel away, but this is more of an intuitive argument than fact. Another exception to the hardness rule is the cam in automotive engines. They are hardened in the range of 50 Rockwell C instead of to the maximum available, which may be as high as 67 R_C. A final example is that of buckets and chutes for handling some ores. Rubber is sometimes found to be superior to very hard white cast iron in these applications.

We see in the preceding examples the possibility of special circumstances requiring special materials. The rubber offers resilience, and the cam material resists fatigue failure if it is not fully hardened. It is often argued that special circumstances are rare or can be dealt with on a case-by-case basis. This attitude seems to imply that most wearing systems are "standard," thus giving impetus to specifying a basic wear resistance of a material as one of its intrinsic properties. Little real progress has been made in this effort, and very little is likely to be made in the near future. Wear resistance is achieved by a balance of several very separate properties, not all of them intrinsic, that are different for each machine component or wear surface. Selecting material for wear resistance is therefore a complex task, and guidelines are needed in design. Such guidelines will be more useful as our technology becomes more complex, but some guidelines are given in the next section.

6.2 STEPS IN DESIGN FOR WEAR LIFE WITHOUT SELECTING MATERIALS

6.2.1 The Search for Standard Components

Designers make most of the decisions concerning material selection. Fortunately, for many cases and for most designers, the crucial components in a machine in which wear may limit useful machine life are available as separate packages with fairly well specified performance capabilities. Examples are gear boxes, clutches, and bearings. Most such components have been well tested in the marketplace, having been designed and developed by very experienced designers. For component designers, very general rules for selecting materials are of little value. They must build devices with a predicted wear life of ±10 percent accuracy or better. They know the range of capability of lubricants, they know the reasonable range of temperature in which their products will survive, and they know how to classify shock loads and other real operating conditions. Their specific expertise is not available to the general designer except in the form of the shapes and dimensions of hardware, the materials selected, and the recommended practices for use of their product. Some of these selections are based on tradition, and some are based on reasoning, strongly tempered by experience. The makers of specialized components usually also have the facilities to test new designs and materials extensively before risking their product in real use. General designers, however, must often proceed without extensive testing.

General designers must then decide whether to avail themselves of standard specialized components or to risk designing every part. Sometimes the choice is based on economics, and sometimes desired standard components are not available. In such cases, components as well as other machine parts must be designed in-house.

6.2.2 In-House Design

If a designer is required to design for wear resistance, it is logical to follow the methods used in parallel activities, such as in determining the strength and vibration characteristics of new machinery. This is often done by interpolating within or extrapolating beyond experience, if any, using

1. Company practice for similar items
2. Vendors of materials, lubricants, and components
3. Handbooks

Company Practice. If good information is available on similar items, a prediction of the wear life of a new product can be made with ± 20 percent accuracy unless the operating conditions of the new design are much beyond standard experience. Simple scaling of sizes and loads is often successful, but usually this technique fails after a few iterations. Careless comparison of a new design with "similar" existing items can produce very large errors for reasons discussed below.

When a new product must be designed that involves loads, stresses, or speeds beyond those previously experienced, it is often helpful to examine the worn surface of a well-used previous model in detail. It is also helpful to examine unsuccessful prototypes or wear-test specimens, as will be discussed below. An assessment should be made of the modes or mechanisms of wear of each part of the product. For this purpose, it is also useful to examine old lubricants, the contents of the lubricant sump, and other accumulations of residue.

Vendors of Materials. Where a new product requires bearings or materials of higher capacity than now in use, it is frequently helpful to contact vendors of such products. When a vendor simply suggests an existing item or material, the wear life of a new product may not be predictable to an accuracy of better than ± 50 percent of the desired life. This accuracy is worse than the ± 20 percent accuracy given earlier, especially where there is inadequate communication between the designer and the vendor. Accuracy may be improved where an interested vendor carefully assesses the needs of a design, supplies a sample for testing, and follows the design activity to the end.

Contact with vendors, incidentally, often has a general beneficial effect. It encourages designers to revise their thinking beyond the logical projection of their own experience. Most designers need a steady flow of information from vendors to remain informed on both the new products and the changing capability of products.

Handbooks. There are very few handbooks on selecting materials for wear resistance. Materials and design handbooks usually provide lists of materials, some of which are highlighted as having been successfully used in wearing parts of various products. They usually provide little information on the rates of wear of products, the mode of wear failure, the limits on operating conditions, or the method by which the wear-resisting parts should be manufactured or run in (if necessary).

Some sources will give wear coefficients, which are purported to be figures of merit, or rank placing of materials for wear resistance. A major limitation of wear coefficients of materials as given in most literature is that there is seldom adequate information given on how the data were obtained. Usually this information is taken from standard laboratory bench tests, few of which simulate real systems. The final result of the use of handbook data is a design which will probably not perform to an accuracy of better than ±95 percent.

6.3   WEAR EQUATIONS

There is a great need for wear equations. Ideally, a wear equation would provide a numerical value for material loss or transfer for a wide range of materials and operating conditions of the wearing parts.

Useful equations derived from fundamental principles are not yet available. Some empirical equations are available for very special conditions. The strictly empirical equations usually contain very few variables and are of the form

$$VT^n = f^a d^b K \tag{6.1}$$

which applies to metal cutting, and in which V = cutting speed, T = tool life, f = feed rate, and d = depth of cut. Experiments are done, measuring T over a range of f while holding V and d fixed at some arbitrary values, from which a can be obtained. The experiments are repeated over ranges of d and V to obtain b and K. It is generally assumed that the results will not depend on the selection of the variables to hold constant, which therefore assumes that there is neither any limit to the range of valid variables nor any interdependence between variables, which ultimately means that there is no change of wearing mechanisms over any chosen range of the variables. Wear equations built by strictly empirical methods are therefore seen to be limited to the case under present study; they have limited ability to predict conditions beyond those of the tests from which they were derived, and they have little applicability to other sliding systems.

A common method of building equations from fundamental principles is to assume that wearing will take place in direct proportion to the real (microscopic) contact area. These equations omit such important considerations as the presence of oxides and adsorbed gases on surfaces, and few of them recognize the role of repeated contact on sliding surfaces, which may lead to fatigue modes of material loss (wear).

In a recent study [6.1], over 180 wear equations were analyzed as to content and form. Though the authors collectively cited over 100 variables to use in these equations, few authors cited more than 5. The fact, then, that quantities such as hardness are found in the numerator of some equations and in the denominator of others leads to some confusion. Overall, no way was found to harmonize any selected group of equations, nor was there any way to determine which material properties are important to the wearing properties.

The parameters that may be included in the equation are of three types, as listed in Table 6.2. It may be readily seen from Table 6.2 that many of the parameters are difficult to quantify, and yet these (and perhaps several more) are known to affect the wear rate. Further complexity is added in cases where wear mechanisms, and therefore wear rates, change with time of sliding.

This state of affairs seems incomprehensible to designers who are steeped in mathematical methods that promise precise results. To use a specific example: For calculating the deflections of beams, simple equations are available that require only one material property, namely, Young's modulus. All other quantities in these equations are very specific; that is, they are measured in dimensions which not only seem available in four or five significant figures, but have compatible units.

Wear is far more complex, involving up to seven basic mechanisms that are operative in different balances or ratios under various conditions. Moreover, many of the mechanisms produce wear rates that are not linear in the simple parameters, such as applied load, sliding speed, surface finish, etc. Thus, in summary, there are at this time

TABLE 6.2   Parameters Often Seen in Wear Equations

---

a. Operational parameters

---

1. Surface topography
2. Contact geometry
3. Applied load
4. Slide/role speed
5. Coefficient of friction
6. Etc.

---

b. Material parameters

---

1. Hardness, cold and hot
2. Ductility
3. Fracture toughness
4. Strength
5. Work hardenability
6. Elastic moduli
7. Material morphology
8. Type and thickness of surface film
9. Thermal properties
10. Etc.

---

c. Environmental parameters

---

1. Type and amount of lubricant
2. Type and amount of dirt and debris
3. Rigidity of supporting structure
4. Ambient temperature
5. Multiple pass of continuous contact
6. Continuous, stop-start, reciprocating
7. Clearance, alignment, and fit
8. Matched or dissimilar material pair
9. Etc.

SOURCE: From Ludema [6.2].

no complete first principles or models available to use in selecting materials for wear resistance. However, there are good procedures to follow in selecting materials for wear resistance.

6.4   STEPS IN SELECTING MATERIALS FOR WEAR RESISTANCE

When designing for wear resistance, it is necessary to ascertain that wear will proceed by the same mechanism throughout the substantial portion of the life of the product. Only then is some reasonable prediction of life possible.

Certain considerations are vital in selecting materials, and these may be more important than selecting a material for the best wear resistance. These considerations are

1. The restriction on material use
2. Whether the sliding surface can withstand the expected static load
3. Whether the materials can withstand the sliding severity
4. Whether a break-in procedure is necessary or prohibited
5. The acceptable modes of wear failure or surface damage
6. The possibility of testing candidate materials in bench tests or in prototype machines

These considerations are discussed in detail in the next several pages.

6.4.1   Restrictions on Material Use

The first step in selecting materials for wear resistance is to determine whether there are any restrictions on material use. In some industries it is necessary for economic and other purposes to use, for example, a gray cast iron, or a material that is compatible with the human body, or a material with no cobalt in it such as is required in a nuclear reactor, or a material with high friction, or a selected surface treatment applied to a low-cost substrate. Furthermore, there may be a limitation on the surface finish available or the skill of the personnel to manufacture or assemble the product. Finally, there may be considerations of delivery or storage of the item before use, leading to corrosion, or false brinelling, or several other events that may befall a wear surface.

6.4.2   Static Load

The second step is to determine whether the sliding surface can withstand the expected static load without indentation or excessive distortion. Generally, this would involve a simple stress analysis.

6.4.3   Sliding Severity

The materials used must be able to withstand the severity of sliding. Factors involved in determining sliding severity include the contact pressure or stress, the temperature due to ambient heating and frictional temperature rise, the sliding speed, misalignment, duty cycle, and type of maintenance the designed item will receive. These factors are explained as follows:

Contact Stress.   Industrial standards for allowable contact pressure vary considerably. Some specifications in the gear and sleeve bearing industries limit the average contact pressures for bronzes to about 1.7 MPa, which is about 1 to 4 percent of the yield strength of bronze. Likewise, in pump parts and valves made of tool steel, the contact pressures are limited to about 140 MPa, which is about 4 to 6 percent of the yield strength of the hardest state of tool steel.

However, one example of high contact pressure is the sleeve bearings in the landing gear of modern commercial aircraft. These materials again are bronzes and have yield strengths up to 760 MPa. The design bearing stress is 415 MPa but with expectations of peak stressing up to 620 MPa. Another example is the use of tool steel in lubricated sheet-metal drawing. Dies may be expected to be used for 500 000 parts with contact pressures of about 860 MPa, which is half the yield strength.

Temperature. The life of some sliding systems is strongly influenced by temperature. Handbooks often specify a material for "wear" conditions without stating a range of temperature within which the wear-resistance behavior is satisfactory. The influence of temperature may be its effect on the mechanical properties of the sliding parts. High temperatures soften most materials and low temperatures embrittle some. High temperature will produce degradation of most lubricants, but low temperature will solidify a liquid lubricant.

Ambient temperature is often easy to measure, but the temperature rise due to sliding may have a larger influence. For a quick view of the factors that influence temperature rise ΔT of asperities on rubbing surfaces, we may reproduce one simple equation:

$$\Delta T = \frac{fWV}{2a(k_1 + k_2)J} \tag{6.2}$$

where f = coefficient of friction, W = applied load, V = sliding speed, and k_1 and k_2 = thermal conductivities of the sliding materials. The quantity a is related to junction size, that is, the few, widely scattered points of contact between sliding parts.

From Eq. (6.2) it may seem that thermal conductivity of the materials could be influential in controlling temperature rise in some cases, but a more important factor is f, the coefficient of friction. If a temperature-sensitive wear mechanism is operative in a particular case, then high friction may contribute to a high wear rate, if not cause it. There is at least a quantitative connection between wear rate and the coefficient of friction when one compares dry sliding with adequately lubricated sliding, but there is no formal way to connect the coefficient of friction with the temperature rise.

Sliding Speed. Both the sliding speed and the *PV* limits are involved in determining the sliding severity. Maximum allowable loads and sliding speeds for materials are often specified in catalogs in the form of *PV* limits. In the *PV* product, *P* is the calculated average contact pressure (in psi) and *V* is the sliding speed (in ft/min). Plastics to be used in sleeve bearings and bronze bushings are the most common material to have *PV* limits assigned to them. A common range of *PV* limits for plastics is from 500 to 10 000, and these data are usually taken from simple laboratory test devices. The quantity *P* is calculated from *W/A*, where *W* = applied load and *A* = projected load-carrying area between sliding members. Thus *PV* could be written as *WV/A*. Returning to Eq. (6.2) for the temperature rise, it may be seen that the product *WV* influences ΔT directly, and it would seem that a *PV* limit might essentially be a limit on surface-temperature rise. This is approximately true, but not useful. That is, wear resistance of materials cannot be related in a simple way to the melting point or softening temperature of materials. The wide ranges of f, k, and other properties of materials prevent formulating a general rule on the relationship between *PV* limits and melting temperature. Indeed, a *PV* limit indicates nothing about the actual rate of wear of materials; it indicates only that above a given *PV* limit a very severe form of wear may occur. However, the *PV* limit for one material has meaning relative to that of other materials, at least in test machinery.

Misalignment.   The difficulty with misalignment is that it is an undefined condition other than that for which contact pressure between two surfaces is usually calculated. Where some misalignment may exist, it is best to use materials that can adjust or accommodate themselves, that is, break in properly.

Misalignment arises from manufacturing errors or from a deflection of the system-producing loading at one edge of the bearing, or it may arise from thermal distortion of the system, etc. Thus a designer must consider designing a system such that a load acts at the expected location in a bearing under all conditions. This may involve designing a flexible bearing mount, or several bearings along the length of a shaft, or a distribution of the applied loading, etc.

Designers must also consider the method of assembly of a device. A perfectly manufactured set of parts can be inappropriately or improperly assembled, producing misalignment or distortion. A simple tapping of a ball bearing with a hammer to seat the race may constitute more severe service than occurs in the lifetime of the machine and often results in early failure.

Misalignment may result from wear. If abrasive species can enter a bearing, the fastest wear will occur at the point of entry of the dirt. In that region, the bearing will wear away and transfer the load to other locations. A successful design must account for such events.

Duty Cycle.   Important factors in selecting materials for wear resistance are the extent of shock loading of sliding systems, stop-start operations, oscillatory operation, etc. It is often useful to determine also what materials surround the sliding system, such as chemical or abrasive particles.

Maintenance.   A major consideration that may be classified under sliding severity is maintenance. Whereas most phosphor bronze bushings are allowed a contact stress of about 1.4 to 7 MPa, aircraft wheel bushings made of beryllium bronze are allowed a maximum stress of 620 MPa, as mentioned before. The beryllium bronze has a strength only twice that of the phosphor bronze, but the difference between industrial and aircraft use includes different treatment of bearings in maintenance. Industrial goals are to place an object into service and virtually ignore it or provide infrequently scheduled maintenance. Aircraft maintenance, however, is more rigorous, and each operating part is under regular scrutiny by the flight crew and ground crew. There is scheduled maintenance, but there is also careful continuous observation of the part and supplies. Thus it is easier for an error to be made in selection of the lubricant in industry than with aircraft, for example. Second, the aircraft wheel bearing operates in a much more standard or narrowly defined environment. Industrial machinery must operate in the dirtiest and hottest of places and with the poorest care. These must be considered as severity conditions by the designer.

6.4.4 Break-In Procedure

Another vital consideration in the selection of materials is to determine whether or not a break-in procedure is necessary or prohibited. It cannot be assumed that the sliding surfaces made to a dimensional accuracy and specified surface finish are ready for service. Sliding alters surfaces. Frequently, sliding under controlled light loads can prepare a surface for a long life of high loading, whereas immediate operation at moderate loads may cause early failure.

It is useful here to distinguish between two surface-altering strategies. The first we refer to as *break-in,* where a system is immediately loaded or operated to its design load. The incidence of failure of a population of such parts decreases with time of operation as the sliding surfaces change, and frequently the ability of the system to accommodate an overload or inadequate lubricant increases in the same time. The surfaces have changed in some way during running, and this is *break-in.* *Run-in,* however, is the deliberate and planned action that is necessary to prepare surfaces for normal service.

The wear that occurs during run-in or break-in can be considered a final modification to the machine surface. This leads to the possibility that a more careful specification of manufacturing practice may obviate the need for run-in or break-in. This has been the case with the automobile engine in particular, although part of a successful part surface-finish specification often includes the exact technique for making the surface. Only 30 years ago it was necessary to start and run an engine carefully for the first few thousand miles to ensure a reasonable engine life. If run-in were necessary today, one would not see an engine survive the short trip from the assembly plant to the haul-away trucks.

It is difficult to determine whether or not some of the present conservative industrial design practices result from the impracticality of effecting a run-in of some products. For example, a gear box on a production machine is expected to function immediately without run-in. If it were run in, its capacity might be greatly increased. But it is also well known that for each expected severity of operation of a device, a different run-in procedure is necessary. Thus a machine that has been operating at one level of severity may be no more prepared for a different state of severity than if it had never been run. A *safe* procedure, therefore, is to operate a device below the severity level at which run-in is necessary, but the device could actually be overdesigned simply to avoid run-in.

6.4.5 Modes of Wear Failure

The fifth consideration is to determine acceptable modes of wear failure or surface damage of machinery. To specify a wear life in terms of a rate of loss of material is not sufficient. For example, when an automotive engine seizes up, there is virtually no loss of material, only a rearrangement such that function is severely compromised. Thus in an engine, as well as on other precision slideways of machines, surface rearrangement or change in surface finish is less acceptable than attrition or loss of material from the system. Again, in metal-working dies, loss of material from the system is less catastrophic than is scratching of the product.

In truck brakes, some abrasiveness of brake linings is desirable, even though it wears brake drums away. This wear removes microcracks and avoids complete thermal fatigue cracking. However, in cutting tools, ore-crushing equipment, and amalgam filling in teeth, surface rearrangement is of little consequence, but material loss is to be avoided.

A final example of designing for an acceptable wear failure is a sleeve bearing in engines. Normally it should be designed against surface fatigue. However, in some applications corrosive conditions may seriously accelerate fatigue failure. This may require the selection of a material that is less resistant to dry fatigue than is the best bearing material, and this applies especially to two-layer bearing materials. In all these examples a study of acceptable modes of wear may result in a different selection of material than if the goal were simply to minimize wear.

6.4.6 Testing Materials

Finally, it is necessary to consider the possibility of testing candidate materials in bench tests or in prototypes. After some study of worn parts from a device or machine that most nearly approximates the new or improved product, one of several conclusions could be reached:

1. The same design and materials in the wearing parts of the example device will perform adequately in the redesign, in terms of function, cost, and all other attributes.
2. A slight change in size, lubrication, or cooling of the example parts will be adequate for the design.
3. A significant change in size, lubrication, or cooling of the example parts will be necessary for the redesign.
4. A different material will be needed in the redesign.

The action to be taken after reaching one of the preceding conclusions will vary. The first conclusion can reasonably be followed by production of a few copies of the redesign. These should be tested and minor adjustments made to ensure adequate product life. The second conclusion should be followed by cautious action, and the third conclusion should invoke the building and exhaustive testing of a prototype of the redesign. The fourth conclusion may require tests in bench-test devices in conjunction with prototypes.

It is usually costly and fruitless to purchase bench-test machinery and launch into testing of materials or lubricants without experience and preparation. It is doubly futile for the novice to run accelerated wear tests with either bench tests, prototypes, or production parts.

Experience shows time after time that simple wear tests complicate the prediction of product life. The problem is correlation. For example, automotive company engineers have repeatedly found that engines on dynamometers must be run in a completely unpredictable manner to achieve the same type of wear as seen in engines of cars in suburban service. Engines turned by electric motors, though heated, wear very differently from fired engines. Separate components such as a valve train can be made to wear in a separate test rig nearly the same way as in a fired engine, with some effort, but cam materials rubbing against valve-lifter materials in a bench test inevitably produce very different results from those in a valve-train test rig.

Most machines and products are simpler than engines, but the principles of wear testing are the same; namely, the wear mechanisms must be very similar in each of the production designs, the prototype test, the subcomponent test, and the bench test. The wear rate of each test in the hierarchy should be similar, the worn surfaces must be nearly identical, and the transferred and loose wear debris should contain the same range of particle sizes, shapes, and composition. Thus it is seen that the prototype, subcomponent, and bench tests must be designed to correlate with the wear results of the final product. This requires considerable experience and confidence where the final product is not yet available. This is the reason for studying the worn parts of a product nearest to the redesign and a good reason for retaining resident wear expertise in every engineering group.

A clear indication of the problem with bench tests may be seen in some results with three test devices. These are:

1. Pin-V test in which a ¼-in-diameter pin of AISI 3135 steel rotates at 200 rpm with four-line contact provided by two V blocks made of AISI 1137 steel.
2. Block-on-ring test where a rectangular block of a chosen steel slides on the outer (OD) surface of a ring of hard case-carburized AISI 4620 steel.
3. The four-ball test where a ball rotates in contact with three stationary balls, all of hardened AISI 52100 steel.

The four-ball test and the ring-on-block test were run over a range of applied loads and speeds. The pin-V test was run over a range of loads only. All tests were run continuously, that is, not in an oscillating or stop-start sequence mode. All tests were run with several lubricants.

Results from the ring-block test were not sufficiently reproducible or consistent for reliable analysis. Results from the other two tests were adequate for the formulation of a wear equation from each, as follows:

Pin-V test: Wear rate $\propto$ (load)^2

Four-ball test: Wear rate $\propto$ (load)$^{4.75}$ $\times$ (speed)$^{2.5}$

These results may be compared with linear laws of wear discussed frequently in the literature, which would be of the form

Linear law: Wear rate $\propto$ (load)$^{1.0}$ $\times$ (speed)$^{1.0}$

There are several points about the usefulness of published wear data to be derived from these results:

1. Practical wear rates are probably not linear in any parameter or variable of operation.
2. If three standard wear tests respond differently to changes in load and speed, then a practical part will probably respond differently again. Furthermore, an accelerated test with a standard wear tester will likely be misleading, since the influence of doubling load or speed would most likely be different between the test device and the product. In fact, the effect of variation in load and speed produces very irregular results with the block-on-ring test machine, which renders extrapolated values of tests useless.
3. It is not known whether the different results from the three wear testers are due to the use of different specimen materials or different specimen shapes or both. Thus rank ordering of materials from wear tests is likely to change among test devices and different testing conditions.

The point of the preceding discussion is that wear testing of materials and components must be done, but it must be done carefully. Testing should be done by experienced engineers and with a strong insistence upon close analysis of worn surfaces and wear debris. It would be useful to be able to compare one's observations with a large and comprehensive atlas of photographs of surfaces in various stages of wear, but none is available. Photographs are scattered through published papers and handbooks and are of some value only when properly described and understood.

Engineers must therefore solve most wear problems themselves by analysis of worn surfaces and wear debris from existing machinery and wear testers. Guidelines for selecting wear-resisting materials and for indirectly selecting lubricants are given in the next section using the methods of product analysis.

6.5 MATERIAL-SELECTION PROCEDURE

The previous sections have established the important point that selecting materials for wear resistance requires a study of the details of wear in order to determine which of the several conventional properties of material can best resist a particular mode of wear. The best way to proceed, therefore, is to examine the most appropriate wear system (including the solids, the lubricant, and all the wear debris), such as an old product being redesigned or a wear tester. The best tools to use are microscopes, with some photography. The most useful microscope is a stereozoom type with a magnification range of 1× to 7×, with 5× or 10× eyepieces and a 100-W movable external light source. Stereo viewing gives a perspective on the shapes of surface features, such as grooves, folds, flakes, etc. The next most useful tool is the scanning (reflecting) electron microscope (SEM). The novice should use this instrument in conjunction with optical microscopes because the SEM and optical devices produce very different pictures of surfaces. Frequently the most useful SEM observations are those done at low magnification, between 20× and 200×, although it is fun to "see" surfaces at 20 000×. The virtue of the SEM is that very rough surfaces can be seen without the high regions and low regions being out of focus, as occurs in optical microscopy. The major problem is that the SEM usually accepts only small specimens [for example, ½ in (12.5 mm) thick by 2 in (50 mm) in diameter], and the surfaces must be clean because the SEM requires a vacuum (about 10^{-5} torr).

For a more detailed analysis of surface chemistry and substrate composition, an SEM with an x-ray dispersion (EDAX, etc.) attachment can be used. The operation of these instruments requires some combination of skill and expensive automation. Optical metallurgical microscopes may be useful as well, but usually not in the conventional bright-field, reflected-light mode. These microscopes often have several special objectives for phase contrast, polarized light, interference, and dark field, all requiring skill in use and in the interpretation of results.

Sometimes it is useful to obtain a topographic profile of worn surfaces. This can be done with the stylus tracer used in conventional surface-finish measurement, but it should be connected to a strip-chart recorder. It is the surface shape rather than a numerical value of surface finish that is most useful. Traces should be made in several places and in several directions during the progression of wearing, if possible. A major precaution to observe in analysis of the strip-chart data is that the representation of the height of surface shapes is often magnified from 10 to 1000 times greater than is the "horizontal" dimension. This leads to the sketching of surfaces as very jagged peaks upon which no one would care to sit. Actually, solid surfaces are more like the surfaces of a body of water in a 10-mi/h breeze.

Having examined a wear system, the designer can proceed through Table 6.3 and make a first attempt to select a material for wear resistance.

TABLE 6.3   Guide for Determining the Material Properties that Resist Wear

How to use the table:

1. Observe the nature of wear in existing equipment or of similar materials from appropriate wear-testing machines.
2. Check the lists in Section A for an applicable general description of worn surfaces or type of service and note the code that follows the selected term.
3. Proceed to Section B which lists 6 terms† that describe three scales of superimposed surface changes. Verify that the code listing is an adequate description of the worn surface. (It is possible to use Section B without reference to Section A.) From Section B, find the major term (capitalized).
4. In Section C, find the detailed description of the capitalized term from Section B and note which material-loss mechanism is applicable and confirm from the nature or description of wear debris.
5. Find the material-loss mechanism in Section D, note the material characteristics and microstructure that should influence wear resistance of material, and note the precautions in material selection to prevent failure.
6. Select materials in conjunction with materials specialists.

Section A   Description of worn surfaces and type of service with code for use in Section B

General surface appearance‡	Some types of service‡
Stained: f	Surface corrosion ⎤ ⎧ in solid machinery: $a1 + c$
	or ⎱ ⎰
Polished or smooth wear: $a1 + c$ $+ e$ or $a2 + c + e$	Erosion/corrosion ⎦ ⎩ in fluids: $a2 + d2$
Scratched (short grooves): $b3 + c$ $+ e$	Abrasive wear (multiple scratches): $b3 + c$
Gouged: $b3 + d1$	
Scuffed: $a1 +$ initiated and periodically perpetuated by $d3 + e$	Gouging: $b1 + d1 + e$
Galled: $b1 + d3 + 3$ (usually very rough)	Dry wear or unlubricated sliding: $b1 + d3 + e$ or $a1$ $+ c + e$
Grooved (smooth or rough): $a1 +$ periodically advanced by $d1 + $ e	Metal-to-metal wear or adhesive wear: $b1 + d3 + e$
Hazy: $b2$	Erosion at high angle: $b2 + d4$
Exfoliated or delaminated: $d4 + e$	Erosion at low angle: $b3 + d1$ or $d2$
Pitted: $b2$ and/or $d5$	
Spalled: $d4$	
Melted: $a3$	
Fretted: $a1 + d5 + f$	Fretting: $a1 + d5 + f$

†Surface geometries can usually be described in three scales, namely, macro-, micro-, and submicro. The first two scales can describe roughness; the third describes reflectivity. The worn surfaces in Section A may be described in terms of the three scales; e.g., polished surfaces are usually microsmooth (a), macrosmooth (c), and shiny (e). The numbers following the code letters explain how the suggested scale of surface geometry wa achieved, i.e., by abrasion which left a very thin film on the surface. Thus the code, polished wear—$a1$ $+ c + e$, etc. Where a scale of geometry is not given, that scale may not be of consequence in the description of the worn surface.

‡Rigorous connection cannot always be made between the terms in the two columns in Section A because of the wide diversity of use and meaning of terms.

TABLE 6.3 Guide for Determining the Material Properties that Resist Wear *(Continued)*

Section B Code listing

a. Microsmooth, caused by

    1. Progressive loss and reformation of surface films by fine *abrasion* and/or by tractive stresses imposed by *adhesive* or viscous interaction, or by

    2. Very fine *abrasion,* with loss of substrate in addition to loss of surface film, if any, or

    3. From *melting.*

c. Macro-smooth, caused by, abrasive particles held on or between solid, smooth backing

e. Shiny, due to very thin (<25nm?) surface films of oxide, hydroxide, sulfide, chloride, or other species

b. Microrough, caused by

    1. Tractive stresses resulting from *adhesion,* or by

    2. Micropitting by *fatigue,* or by

    3. *Abrasion* by medium-coarse particles

d. Macrorough, caused by

    1. *Abrasion* with coarse particles, including carbide and other hard inclusions in the sliding materials that are removed by sliding action as the wear of matrix progresses, or by

    2. *Abrasion* by fine particles in turbulent fluid, producing scallops, waves, etc., or by

    3. Severe *adhesion* in early stages of damage, or by

    4. Local *fatigue* failure resulting in pits or depressions due to repeated rolling-contact stress, repeated thermal gradients, high-friction sliding, or impact by hard particles as in erosion, or in

    5. Advanced stages of microroughening, where little unaffected surface remains between pits.

f. Dull or matte, due to films of perhaps greater than 25-nm thickness (resulting from aggressive environments, including high temperatures), i.e., due to *corrosion*

Section C Material-loss mechanisms and nature of debris

Material-loss mechanisms (italic)	Nature of debris
Corrosion (of surface): Chemical combination of material surface atoms with passing or deposited active species to form a new compound, i.e., oxide, etc.	Newly formed chemical compound, usually agglomerated and sometimes mixed with fragments of the original surface material
Abrasion: Involves particles (or acute angular shapes but mostly obtuse) that produce wear debris, some of which forms ahead of the abrasive particle, which mechanism is called *cutting,* but most of which is material that has been plowed aside repeatedly by passing particles, and breaks off by *low-cycle fatigue.*	Long, often curly chips or strings

Section C   Material-loss mechanisms and nature of debris *(Continued)*

Material-loss mechanisms (italic)	Nature of debris
Adhesion: A strong bond that develops between two surfaces (either between coatings and/or substrate materials) that, with relative motion, produces tractive stress that may be sufficient to deform materials to fracture. The mode of fracture will depend on the property of the material, involving various amounts of energy loss or ductility to fracture, that is,	
Low energy and ductility → *brittle fracture*	Solid particles, often with cleavage surfaces
High energy and ductility → *ductile fracture*	Severely deformed solids, sometimes with oxide clumps mixed in
Fatigue: Due to cyclic strains, usually at stress levels below the yield strength of the material, also called *high-cycle fatigue*	Solid particles, often with cleavage surfaces and ripple pattern
Melting: From very high-speed sliding	Spheres, solid or hollow, and "splat" particles

Section D   Material-selection characteristics

Material-loss mechanisms	Appropriate material characteristics to resist wear	Precautions to be observed when selecting a material†
Corrosion	Reduce corrosiveness of surrounding region; increase corrosion resistance of material by alloy addition or by selection of soft, homogeneous material	Total avoidance of new surface species can result in high adhesion of contacting surfaces; soft materials tend to promote galling and seizure.
Cutting	Use material of high hardness, with very hard particles or inclusions, such as carbides, nitrides, etc., and/or overlaid or coated with materials that are hard or contain very hard particles	All methods of increasing cutting resistance cause brittleness or lower fatigue resistance.
Ductile fracture	High strength achieved by any method other than cold working or heat treatments that produce internal cracks and large, poorly bonded intermetallic compounds	
Brittle fracture	Minimize tensile residual stress for cold temperature; ensure low-temperature brittle transition; temper all martensites; use deoxidized metal; avoid carbides such as in pearlite, etc.; effect good bond between fillers and matrix to deflect cracks	In essence, soft materials will not fail through brittleness and will not resist cutting.
Low-cycle fatigue	Use homogeneous and high-strength materials that do not strain-soften; avoid overaged materials and two-phase systems with poor adhesion between filler and matrix	

TABLE 6.3   Guide for Determining the Material Properties that Resist Wear *(Continued)*

Section D   Material-selection characteristics

Material-loss mechanisms	Appropriate material characteristics to resist wear	Precautions to be observed when selecting a material†
High-cycle fatigue	For steel and titanium, use stresses less than half the tensile strength (however achieved); for other materials to be load-cycled less than 10^8 times, allow stresses less than one-fourth the tensile strength (however achieved); avoid retained austenite; use spherical pearlite rather than plate structure; avoid poorly bonded second phases; avoid decarburization of surfaces; avoid platings with cracks; avoid tensile residual stresses or form-compressive residual stresses by carburizing or nitriding	Calculation of stress should include the influence of tractive stress.
Melting	Use material of high melting point and/or high thermal conductivity	

†Materials of high hardness or strength usually have decreased corrosion resistance, and all materials with multiple and carefully specified properties and structures are expensive.

SOURCE: From Ludema [6.2].

REFERENCES

6.1 H. C. Meng and K. C. Ludema, "Wear Models and Predictive Equations: Their Form and Content," *Wear,* vol. 181–183, pp. 443–457, 1995.

6.2 K. C. Ludema, "Selecting Materials for Wear Resistance," Conference on Wear of Materials, San Francisco, 1981, ASME, New York.

BIBLIOGRAPHY

The previous sections are composite views of many authors, so that a reference list would be very long. Interested readers could consult the many journals containing papers on wear, but that literature is potentially very confusing. The most useful journals are

Wear, Elsevier, Lausanne, starting in 1957.

Tribology International, Butterworth Scientific, Ltd., London, starting in 1968.

Tribology Transactions of the Society of Tribologists and Lubrication Engineers (formerly *Transactions* of the American Society of Lubrication Engineers).

Journal of Tribology, also identified as *Transactions F,* American Society of Mechanical Engineers.

CHAPTER 7
SOLID MATERIALS

Joseph Datsko
Professor Emeritus of Mechanical Engineering
The University of Michigan
Ann Arbor, Michigan

This chapter summarizes the structure of solids, including atomic bonding forces, atomic structures, crystal imperfections, slip, and mechanical strength. The section on mechanical properties and tests discusses all the hardness tests and includes a detailed explanation of the tensile test and tensile properties. The section on strength, stress, and strain relations includes many new relationships that have been developed during the past two decades and are not found in other handbooks. The mechanical property data presented in this section are in a new format that is well suited for use in computer-aided-engineering (CAE) applications.

7.1 STRUCTURE OF SOLIDS

A study of the mechanical properties of materials must begin with an understanding of the structure of solid materials. In this context, *structure* refers to the atomistic and crystalline patterns of which the solid material is composed. The definitions of the mechanical properties given in the following sections are on the basis of the crystalline structure of material. For example, *strength* (and *hardness*) is defined as the ability of the material to resist slip along its crystallographic planes. Thus, in order to increase the strength of a material, something must be done to it which will make slip more difficult to initiate. The following sections will explain the manner in which the various thermal and mechanical processes affect the structure of a material, which in turn determines the mechanical properties. The next section presents a brief review of atomic structure.

7.2 ATOMIC BONDING FORCES

The smallest particles that must be considered in the preceding context are atoms. The manner in which atoms are arranged in a solid material determines the material's crystal structure. The crystal structure and the type of interatomic bonding forces determine the strength and ductility of the material.

The simple model of an atom is a dense *nucleus,* consisting of *protons* and *neutrons,* surrounded by discrete numbers of planetary *electrons* orbiting in shells at specific distances from the nucleus. Each proton has a positive electric charge of unity (1+). The number of protons in the nucleus determines the nuclear charge of the atom and is called the *atomic number.* The neutrons have no charge, but they do have mass. The *atomic weight* of an atom is the sum of the number of protons and neutrons. The electrons have negligible mass and a negative charge of unity (1−). The number of electrons in a given type of atom is also equal to the atomic number of that element. The maximum number of electrons in any shell is $2n^2$, where n is the *quantum number* of the shell. Thus the maximum number of electrons that can be present in the first (innermost) shell is 2, and 8 is the maximum in the second shell. However, no more than 8 electrons are ever present in the outermost shell of an atom. The *valence* of an element is either the number of electrons in its outermost shell or the number of electrons necessary to fill that shell, whichever number is lower.

The interatomic bonding forces are determined by the valence, or outer-shell, electrons. There are four types of atomic bonding forces that hold the atoms of a solid material in their relatively fixed positions. The three strongest (*ionic, covalent,* and *metallic*) types of bond are referred to as *primary;* the fourth (*molecular*) is referred to as a *secondary* type of bond.

7.2.1 Ionic Bonds

From the preceding brief description of atomic structure, it is evident that the uncombined atom is electrically neutral—the number of protons (+ charges) in the nucleus exactly equals the number of electrons (− charges). When atoms combine, only the valence electrons are involved and not the nuclei. When a metal combines with a nonmetal, each metal atom "loses" its valence electrons and thus acquires a positive charge that is equal to the number of electrons so lost. Likewise each nonmetallic atom "gains" a number of electrons equal to its valence and acquires an equal negative charge. While in this state, the positively charged metallic atom and the negatively charged nonmetallic atom are called *ions.*

Like-charged particles repel each other and oppositely charged particles attract each other with an electric force called the *Coulomb force.* When a material is maintained in the solid state by the mutual attraction of positively and negatively charged ions, the interatomic bonding force is called *ionic.*

The Coulomb forces attracting oppositely charged ions are very large. Therefore, ionic-bonded solids exhibit very high strength and relatively low melting temperatures. However, they exhibit very low ductility under normal conditions because the interatomic bonds must be broken in order for the atoms to slide past each other. This is one of the most important distinctions between ionic (or covalent) bonding and metallic bonding and is discussed later.

7.2.2 Covalent Bonds

Covalent bonds are those in which the atoms reach a stable configuration (filled outer shell) by *sharing* valence electrons. Unlike ionic bonds, which are nondirectional, covalent bonds act between specific pairs of atoms and thus form molecules. Covalent bonds are most prevalent in gas molecules. Covalent bonding also results in the formation of very large molecules which are present as solids rather than as liquids and gases. Diamond, silicon, and silicon carbide are examples of such covalent-bonded solids. They are characterized by high strength and melting temperature and low ductility. The atoms in the diamond structure are arranged on two interpenetrating face-centered cubic lattices. The entire crystal is composed of only one molecule, and in order to fracture the crystal, the strong covalent interatomic bonds must be broken.

7.2.3 Metallic Bonds

Of the three primary bonding forces, the metallic bond is by far the most important for an understanding of the mechanical properties of the materials with which the practicing engineer is concerned. The *metallic bond* is a special type of covalent bond wherein the positively charged nuclei of the metal atoms are attracted by electrostatic forces to the valence electrons that surround them. Unlike the common covalent bond, which is directional, i.e., between a pair of atoms, the metallic bond is nondirectional, and each nucleus attracts as many valence electrons as possible. This leads to a dense packing of the atoms, and thus the most common crystal structures of the metals are the close-packed ones: face- and body-centered cubic and hexagonal close-packed structures.

The reason that metal atoms have their own unique type of bonding force is the looseness with which their valence electrons are held in the outer shell. This is evident from the fact that the ionization potential of metal atoms is one-half to two-thirds that of nonmetal atoms. The mean radius of the valence electrons in a free (isolated) metal atom is larger than the interatomic distance of that metal in the solid crystalline state. This means that the valence electrons are closer to a nucleus in the solid metal than they are in a free atom, and thus their potential energy is lower in the solid.

Since the valence electrons are not localized between a pair of positive ions, they are free to move through the solid. Thus the structure of the solid metal is a close-packed arrangement of positive ion "cores" (the nucleus plus the nonvalence electrons) that is permeated by an electron "gas" or "cloud." This ability of the valence electrons to move freely through the solid explains the high thermal and electrical conductivities of metals. Also, the fact that the valence electrons are nondirectional (not shared by only two atoms) explains the relatively low strength and high ductility of elemental metals, since the positive ions can move relative to one another without breaking any primary bonds. This mechanism is referred to as *slip* and is discussed in more detail in a following section on crystal structures.

7.2.4   Molecular or van der Waals Bonds

In addition to the three strong primary bonds discussed above, there are also several much weaker (and therefore called *secondary*) bonds which provide the interatomic attractive forces that hold some types of atoms together in a solid material. These forces are referred to as either *secondary* bonds, *molecular* bonds, or *van der Waals* bonds. These bonds are due to residual electrostatic fields between neutral molecules whose charge distribution is not uniform.

Covalently bonded atoms frequently form molecules that behave as electric or magnetic *dipoles*. Although the molecule itself is electrically neutral, there is an electrical imbalance within the molecule. That is, the center of the positive charge and the center of the negative charge do not coincide, and it is this dipole that creates molecular bonding.

7.3   ATOMIC STRUCTURES

Whereas the electrical properties of a material depend on the internal structure of the atoms, the mechanical properties depend on the types of structures that groups of atoms form. In this context, *atomic structures* refer to the structures that are built by particular arrangements of atoms, not to the internal structure of individual atoms. All solid materials can be classified on the basis of atomic structure into three groups: amorphous, molecular, or crystalline (in order of increasing importance to mechanical properties). Knowledge of the atomic structure of solids makes it possible to understand why a given material has its unique properties and thus to be able to specify the type of material and the condition it should be in to achieve optimum mechanical properties.

7.3.1   Amorphous Solids

Amorphous materials are those whose structure has no repetitive arrangement of the atoms of which it is comprised. In a sense, they have no "structure." Although gases and liquids are amorphous materials, the only important amorphous solids are the glasses, and they are frequently considered simply as supercooled liquids.

Glass behaves as a typical liquid at high temperatures. The atoms are very mobile and do not vibrate in a fixed location in space. A given mass of hot glass, like any liquid, takes the shape of the container in which it is placed.

As a hot glass cools, its atoms vibrate at lower amplitudes and come closer together, resulting in an overall thermal contraction or decrease in specific volume. This decrease in specific volume of a liquid as temperature decreases is approximately linear and occurs with all liquids, including liquid metals. This is illustrated in Fig. 7.1.

When any unalloyed liquid metal (a pure metallic element) or chemical compound is cooled to its freezing (or melting) temperature T_m, the atoms come much closer together and become relatively immobile with respect to one another. They form a crystalline structure with very efficient packing, and thus there is a very marked decrease in specific volume at this temperature, as shown in Fig. 7.1. When an alloyed liquid metal freezes to form a solid solution, the transition from liquid to solid takes place in the range of temperatures between the liquidus and the solidus. Further cooling of both solid metals results in a further decrease in specific volume, also linear but of lower slope than in the liquid state.

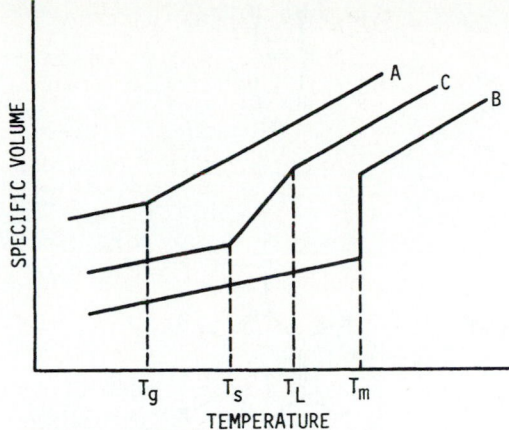

FIGURE 7.1 Specific volume versus temperature. (A) Glass with a transition temperature T_g; (B) a crystal that melts at a fixed temperature T_m, such as a pure element or a compound; (C) a crystal that melts over a range of temperature, such as a solid-solution alloy with T_L the liquidus temperature and T_s the solidus temperature.

When hot liquid glass is cooled to some temperature T_g, called the *glass transition temperature,* there is an abrupt change in the slope of the specific volume versus temperature curve. Unlike crystalline solids, the glass shows no marked decrease in specific volume at this temperature. Below T_g, glass behaves as a typical solid.

7.3.2 Molecular Solids

A *molecule* is a group of atoms that are held together by strong ionic or covalent bonds. A *molecular solid* is a structure made up of molecules that are attracted to each other by weak van der Waals forces. The two most common types of molecular solids are silicates and polymers. The silicates have ionic intramolecular bonds, and the polymers have covalent ones. Since it is the latter materials that are more important in terms of mechanical properties, they will be discussed in more detail.

Polymers are organic compounds of carbon, hydrogen, and oxygen to which other elements such as chlorine or fluorine may be added. They cover a wide range of structural arrangements, with resulting variations in properties. Large molecules are constructed from a repeating pattern of small structural units. The hydrocarbons have repeating structural units of carbon and hydrogen atoms.

Figure 7.2 shows some of the more common monomers or unsaturated molecules that are used in the building of macromolecules. The simplest monomer is ethylene (C_2H_4); it is shown in Fig. 7.2a. It is the base of the group of hydrocarbons called *olefins.* The olefins have the chemical formula C_nH_{2n}. The benzene molecule, shown in Fig. 7.2d, is another important building unit. Because of the shape of the molecule, it is described as a ring molecule or compound. The benzene group is also called the *aromatic* hydrocarbons.

Figure 7.3 illustrates the addition polymerization of the ethylene monomer. The double bonds of ethylene are broken in the presence of a catalyst such as boron tri-

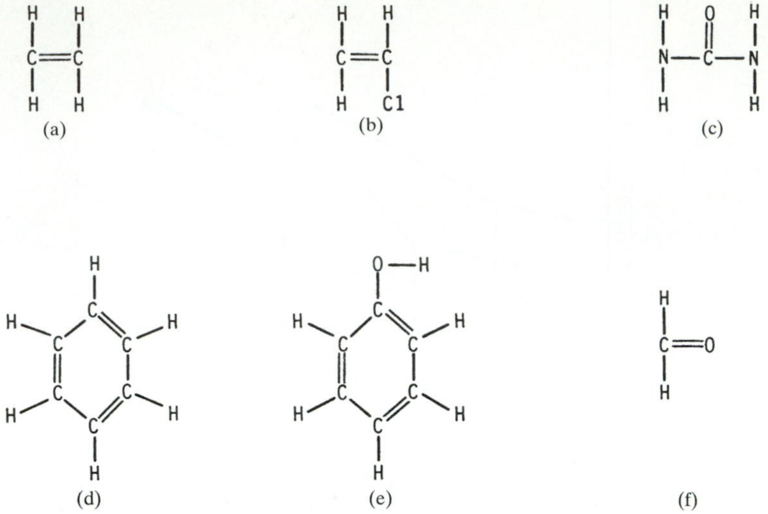

FIGURE 7.2 Monomers: Small unsaturated (double-bonded) molecules that are building units for large polymer molecules. (*a*) Ethylene; (*b*) vinyl chloride; (*c*) urea; (*d*) benzene; (*e*) phenol; (*f*) formaldehyde.

fluoride. The vinyl chloride monomer, as shown in Fig. 7.2*b*, is similar to ethylene except that one of the hydrogen atoms is replaced with a chlorine atom. The polymerization of this monomer results in polyvinyl chloride. These macromolecules resemble, more or less, smooth strings or chains, as can be seen from their structural arrangement.

Some macromolecules resemble rough chains—that is, chains with many short side arms branching from them. Polystyrene, which is a very important industrial polymer, is of this type. The styrene monomer is made from the benzene ring (C_6H_6) with one of the hydrogen atoms replaced with a $CH=CH_2$ molecule, as shown in Fig. 7.4*a*. Polymerization then occurs by breaking the double bond in the $CH=CH_2$ group with the help of a peroxide catalyst and joining two of them together, as shown in Fig. 7.4*b*.

The polymers just described are *thermoplastic;* they melt or soften when they are heated. This is due to the fact that the individual macromolecules are stable and the linkages to other macromolecules are loose (since they are attracted to each

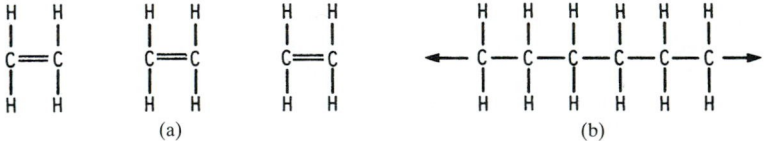

(a)   (b)

FIGURE 7.3 Addition polymerization. (*a*) Three individual monomers of ethylene; (*b*) a portion of a polyethylene molecule formed when each double bond of the monomers is broken by a catalyst to form two single bonds and join the individual molecules together.

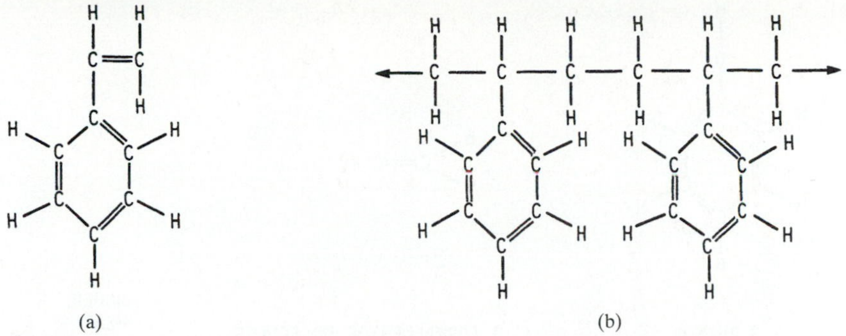

FIGURE 7.4 (*a*) Styrene structure; (*b*) polystyrene structure. The polymerization takes place in the presence of a peroxide catalyst.

other by weak van der Waals forces). Some polymers are *thermosetting;* they do not soften when they are heated, but retain their "set" or shape until charred. This is due to the fact that the individual macromolecules unite with each other and form many cross-linkages. Bakelite (phenol formaldehyde) is such a polymer. Figure 7.5 shows how each formaldehyde monomer joins two phenol monomers together, under suitable heat and pressure, to form a macromolecule. This is a condensation type of polymerization because one water molecule is formed from the oxygen atom of each formaldehyde molecule and a hydrogen atom from each of the two phenol molecules.

7.3.3 Mechanical Properties of Molecular Structures

The mechanical properties of polymers are determined by the types of forces acting between the molecules. The polymers are amorphous with random chain orientations while in the liquid state. This structure can be retained when the polymer is cooled rapidly to the solid state. In this condition, the polymer is quite *isotropic.* However, with slow cooling or plastic deformation, such as stretching or extruding, the molecules can become aligned. That is, the long axes of the chains of all the molecules tend to be parallel. A material in this condition is said to be "oriented" or "crystalline," the degree of orientation being a measure of the crystallinity. When the molecular chains of a polymer have this type of directionality, the mechanical properties are also directional and the polymer is *anisotropic*. The strength of an aligned polymeric material is stronger along the axis of the chains and much lower in the perpendicular directions. This is due to the fact that only weak van der Waals forces hold the individual, aligned macromolecules together, whereas the atoms along the axes of the chains are held together by strong and covalent bonds. The intermolecular strength of linear polymers can be increased by the addition of polar (dipole) groups along the length of the chain. The most frequently used polar groups are chlorine, fluorine, hydroxyl, and carboxyl.

The thermosetting (cross-linked) types of polymers have all the macromolecules connected together in three directions with strong covalent bonds. Consequently, these polymers are stronger than thermoplastic ones, and they are also more isotropic.

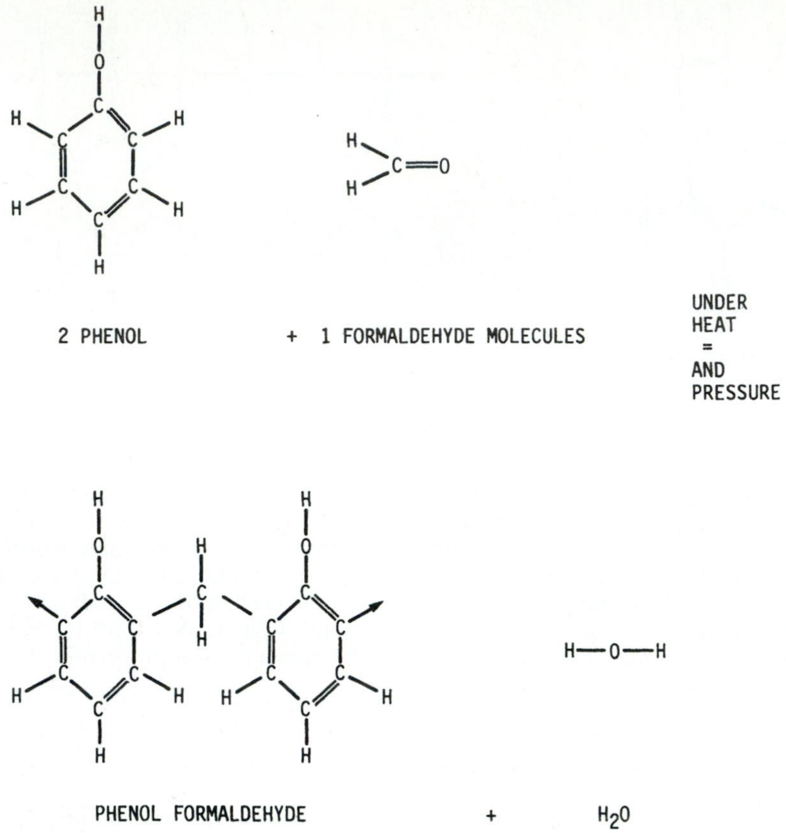

2 PHENOL   +   1 FORMALDEHYDE MOLECULES

UNDER
HEAT
=
AND
PRESSURE

PHENOL FORMALDEHYDE   + H_2O

FIGURE 7.5   Condensation polymerization of phenol and formaldehyde into bakelite.

7.3.4 Crystalline Solids

Crystalline solids are by far the most frequently used ones on the basis of mechanical properties or load-carrying capacity. Moreover, of all the crystalline solids, metals are the most important. A *crystal* (or crystalline solid) is an orderly array of atoms having a repeating linear pattern in three dimensions. The atoms are represented as spheres of radius *r*. A *space lattice* is the three-dimensional network of straight lines that connects the centers of the atoms along three axes. The intersections of the lines are *lattice points,* and they designate the locations of the atoms. Although the atoms vibrate about their centers, they occupy the fixed positions of the lattice points. Figure 7.6 is a sketch of a space lattice, with the circles representing the centers of the atoms. A space lattice has two important characteristics: (1) the space-lattice network divides space into equal-sized prisms whose faces contact one another in such a way that no void spaces are present, and (2) every lattice point of a space lattice has identical surroundings.

The individual prisms that make up a space lattice are called *unit cells.* Thus a unit cell is the smallest group of atoms which, when repeated in all three directions, make up the space lattice, as illustrated by the dark-lined parallelepiped in Fig. 7.6.

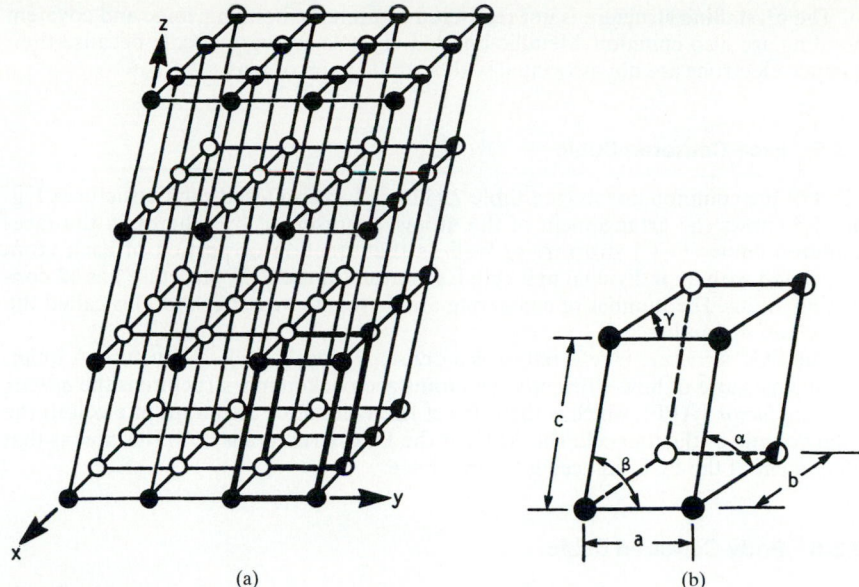

FIGURE 7.6 A space lattice. (*a*) A unit cell is marked by the heavy lines. Black circles are on the front face; horizontal shading on the top face; vertical shading on the right side face; hidden circles are white. (*b*) An isolated unit cell showing dimensions *a*, *b*, and *c* and angles α, β, and γ.

Only 14 different space lattices and 7 different systems of axes are possible. Most of the metals belong to three of the space-lattice types: face-centered cubic, body-centered cubic, and hexagonal close-packed. They are listed in Table 7.1, along with four metals that have a rhombohedral and two that have orthorhombic structures.

TABLE 7.1 Lattice Structure of Metal Crystals

Face-centered cubic	Body-centered cubic	Hexagonal close-packed	Rhombohedral	Orthorhombic
Ag	Cb	Be	As	Ga
Al	α-Cr	Cd	Bi	U
Au	Cs	α-Co	Hg	
Ce	α-Fe	β-Cr	Sb	
β-Co	δ-Fe	Hf		
Cu	K	Mg		
γ-Fe	Li	Os		
Ir	Mo	Ru		
Ni	Na	Se		
Pb	Ta	Te		
Pd	V	Ti		
Pt	W	Tl		
Rh		Y		
Sc		Zn		
Th		Zr		
β-Tl				

The crystalline structure is not restricted to metallic bonding; ionic and covalent bonding are also common. Metallic-bonded crystals are very ductile because their valence electrons are not associated with specific pairs of ions.

7.3.5 Face-Centered Cubic

Most of the common metals (see Table 7.1) have face-centered cubic structures. Figure 7.7 shows the arrangement of the atoms, represented by spheres, in the face-centered cubic (FCC) structure as well as that fraction or portion of each atom associated with an individual unit cell. Each atom in the FCC structure has 12 contacting atoms. The number of contacting atoms (or nearest neighbors) is called the *coordination number.*

The FCC structure is referred to as a dense or closely packed structure. A quantitative measure of how efficiently the atoms are packed in a structure is the *atomic packing factor* (APF), which is the ratio of the volume of the atoms in a cell to the total volume of the unit cell. The APF for the FCC structure is 0.74. This means that 26 percent of the FCC unit cell is "void" space.

7.3.6 Body-Centered Cubic

Many of the stronger metals (Cr, Fe, Mo, W) have body-centered cubic (BCC) lattice structures, whereas the softer, more ductile metals (Ag, Al, Au, Cu, Ni) have the FCC structure (see Table 7.1). Figure 7.8 shows the arrangement of atoms in the BCC structure. There are two atoms per unit cell: one in the center (body center) and ⅛ in each of the eight corners. As can be seen in Fig. 7.8, each atom is contacted by eight other atoms, and so its coordination number is 8. The atomic packing factor for the BCC structure is 0.68, which is a little lower than that for the FCC structure.

The *Miller indices* are used to designate specific crystallographic planes with respect to the axes of the unit cell. They do not fix the position in terms of distance from the origin; thus, parallel planes have the same designation. The Miller indices are determined from the three intercepts that the plane makes with the three axes of the crystal. Actually it is the reciprocal of the distances between the intercepts with

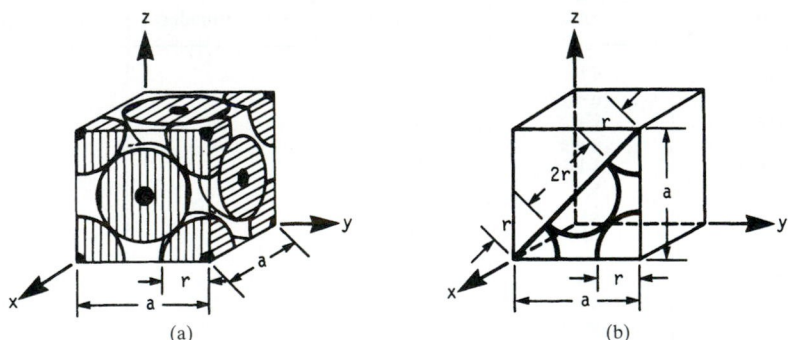

(a)   (b)

FIGURE 7.7   Unit cell of face-centered cubic structure. (*a*) The unit cell has 8 corners with ⅛ atom at each plus 6 faces with ½ atom, for a total of 4 atoms per unit cell; (*b*) one half of the front face showing the relationship between the lattice parameter *a* and the atomic radius *r*.

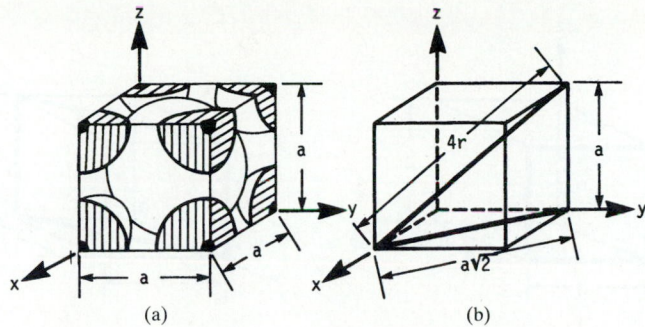

FIGURE 7.8 Unit cell of body-centered cubic structure. (*a*) The unit cell has ⅛ atom at each of 8 corners and 1 atom at the geometric center of the cell, for a total of 2 atoms; (*b*) the relationship of the lattice parameter *a* and atomic radius *r*.

the axis and the origin measured in terms of multiples or fractions of the unit cell lengths *a, b,* and *c* used in the determination. The final steps in specifying the Miller indices are to reduce the three reciprocals to the lowest integers having the same ratio and then to enclose them in parentheses. As is true with direction indices, the sequence of integers relates to the distances along the *x, y,* and *z* axes, respectively. The following examples should make this procedure clear.

Figure 7.9*a* identifies the front face of the crystal with the Miller indices (100). This notation is arrived at as follows: The front face intercepts the *x* axis at one *a* distance, and it does not intercept the *y* and *z* axes (or it intercepts at zero *b* and zero *c* units). If the side lengths are dropped, the intercepts are 1, 0, and 0. The reciprocals of these are also 1, 0, and 0. Since these are already the smallest integers, the Miller indices are specified by enclosing them in parentheses: (100). The commas are not included because they are simply part of the sentence structure.

Figure 7.9*b* shows the (110) plane that is parallel to the *z* axis and is a face diagonal on the top and bottom faces of the unit cell. This plane intercepts the *x* axis at one *a* distance, the *y* axis at one *b* distance, and the *z* axis at zero *c* distance. The intercepts are 1, 1, and 0, and so are the reciprocals. Since these are the smallest integers, the Miller indices are specified as (110).

Figure 7.9*d* shows the crystallographic plane that intercepts the *x* axis at ½*a*, the *y* axis at one *b*, and the *z* axis at one *c*. The reciprocals are therefore 2, 1, and 1, and so this plane is identified as the (211) plane.

Parentheses are used, as in the preceding examples, to specify a single plane or a family of parallel planes. Thus (100) represents all the planes that are parallel to the *yz* axes and intercept the lattice structure at one *a,* two *a,* three *a,* etc. distances. Wavy brackets, or braces, are used to designate all planes in a crystal that are equivalent. For example, the six face planes of a unit cell such as that in Fig. 7.9*a* are (100), (010), (001), ($\bar{1}$00), ($0\bar{1}0$), and ($00\bar{1}$). The notation {100} includes all these six planes.

7.4 CRYSTAL IMPERFECTIONS

The previous discussions on crystal structure assumed a perfect lattice; that is, an atom occupied each and every lattice point, and the distances between equivalent

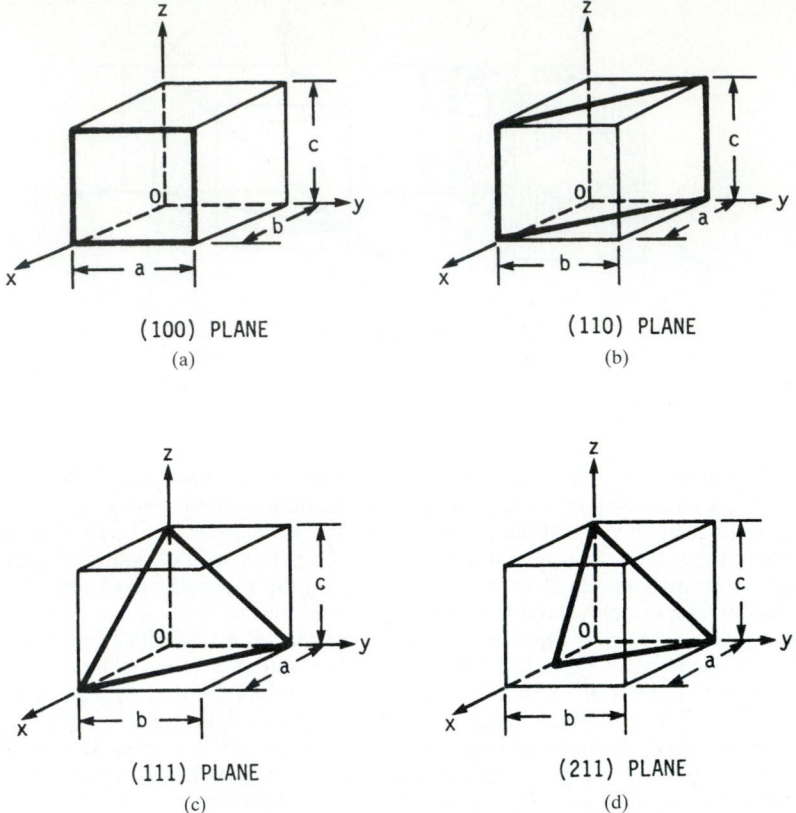

FIGURE 7.9   Miller indices for some crystallographic planes.

lattice points were all exactly the same. In the early 1900s it was found that real crystals did not have perfect properties. Notable among these properties was a mechanical strength much lower than crystals should have. As early as 1928 Prandtl suggested that slip (plastic deformation) in a crystal and the strength of a crystal are related to the presence of linear imperfections within the crystal. This type of imperfection is now called a *dislocation*. At the present time, the terms *imperfection* and *defect* refer to a deviation from a perfectly ordered lattice structure.

Lattice imperfections are classified into three types: *point defects,* where the imperfection is localized about a single lattice point and involves only a few atoms; *line defects,* where the imperfection lies along a line of finite length involving a row (line) or many atoms; and *planar defects* or *boundaries,* where the imperfections involve entire planes or atoms such as the interface between adjacent crystals.

7.4.1   Point Defects

Point defects are caused by (1) the absence of an atom from a lattice point, (2) the presence of an extra atom (usually a small foreign one) in the "void" spaces of the lattice, (3) the presence of a foreign atom at one of the lattice sites, or (4) atoms that are displaced from their normal positions in the array. Figure 7.10 illustrates these defects.

The first type of point defect, the absence of an atom from a lattice point, is called a *vacancy*. Figure 7.10a shows a vacancy on the (100) planes of an FCC lattice. The sketch was not made to be a true representation in order to make the defect more apparent. In reality, the atoms that are near neighbors to the vacant site would be displaced from their normal positions in the array toward centers closer to the vacancy. Thus the lattice lines joining the centers of the atoms are not straight in the vicinity of the vacancy. In three dimensions this means that the crystallographic planes are warped inward near a vacancy. Individual vacancies can cluster together to form larger voids. Vacancies have no effect on the metallurgical control of the mechanical properties discussed in later sections. However, they do affect properties such as conductivity and diffusivity.

The second type of point defect, the presence of an extra atom at the interstices of the lattice, is known as an *interstitial defect*. This type of defect in the lattice structure is the basis for the strengthening mechanism known as *interstitial alloying*, where the solute atom fits in the interstices of the solvent lattice, and it accounts for the high strength in fully hardened (heat-treated) steel. Commercially pure iron (ferrite with a BCC structure) has a yield strength of 70 to 140 MPa. However, with 0.8 percent carbon dissolved interstitially in the iron and stress relieved after heat treating, it has a yield strength of 2400 to 2800 MPa. No other metal has such high room-temperature strength, and no other strengthening mechanism has a greater effect than the interstitial alloying of carbon in iron. The details of these strengthening mechanisms are discussed later.

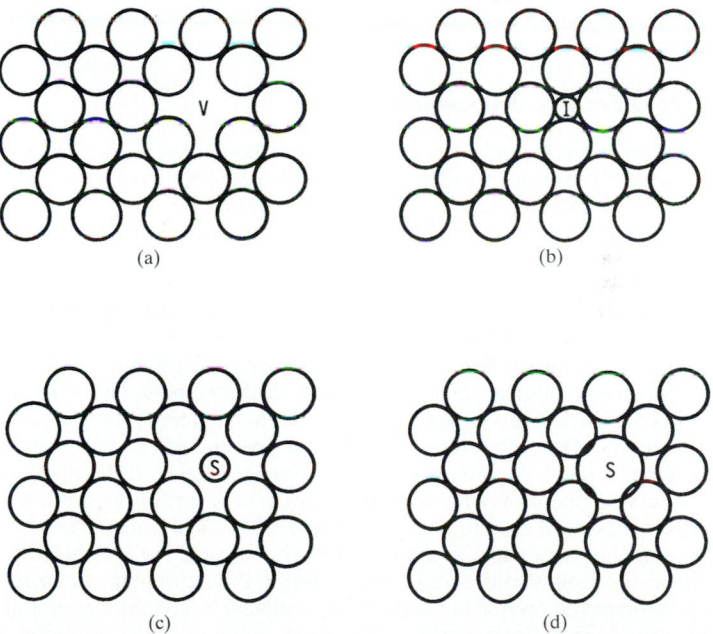

FIGURE 7.10　Some common point defects. (*a*) Vacancy; (*b*) interstitial atom I; (*c*) substitution of a smaller atom S; (*d*) substitution of a larger atom S.

Figure 7.10*b* shows an interstitial atom I in the solvent matrix. However, it does not lie in the plane of the solvent lattice but lies either above or below the sketched plane. Also, the foreign atom is always larger than the "void" space it occupies, and so it necessarily forces the surrounding solvent atoms out of their normal array. Therefore, the crystallographic planes are warped outward in the vicinity of an interstitial defect.

The third type of point defect, the presence of a foreign atom at one of the lattice points, is referred to as a *substitutional defect.* When an alloy is made by adding solute atoms that replace (substitute for) solvent atoms in the lattice structure, it is called a *substitutional alloy.* This type is the most common one in the metal system. Figure 7.10*c* and *d* shows the substitution of a smaller and a larger atom S at one of the lattice points. Unlike the interstitial atom, the substitutional one is in the plane of the solvent matrix. The crystallographic planes are also warped in the vicinity of the substitutional defect, inward for the smaller atom and outward for the larger atom. The distortion of the crystallographic planes is very important to an understanding of control of the strength of materials, which is presented later.

The fourth type of point defect, atoms that are displaced from their normal position, occurs in several forms. The atoms in the "contacting" planes of two adjoining crystals are not in their normal positions as a result of the crystals having solidified from the liquid without being in perfect registry with each other. This is considered to be a *grain boundary defect,* which has a significant effect on the strength of all polycrystalline materials.

Two additional types of atom displacement defects occur in ionic crystals that are not present in metallic-bonded crystals. A vacancy in an ionic crystal that is associated with a displaced pair, one cation and one anion, is called a *Schottky defect.* A *Frenkel defect* occurs when a small cation moves from a lattice point, leaving a vacancy, and occupies an interstitial site.

7.4.2 Line Defects or Dislocations

Examinations of crystals under the electron microscope have shown interruptions in the periodicity of the lattice structure in certain directions. In a two-dimensional representation these interruptions appear as lines; hence the name *line defects.* It is believed that a perfect crystal of a metal such as pure iron should have a strength of 1 or 2 million pounds force per square inch, whereas in reality such perfect crystals have a yield strength of only a few thousand. The reason given for the three orders of magnitude difference between the postulated and actual strength of metal crystals is the presence of these line defects.

The two most common line defects are edge dislocation and screw dislocation. An *edge dislocation* is the line defect that results from the presence of an extra plane of atoms in one portion of a crystal compared to the adjacent part. Actually, it is the edge of this extra plane of atoms and runs from one end of the crystal to the other. When looking at the crystalline plane that is perpendicular to the dislocation line, the imperfection appears as an extra row of atoms in a part of the crystal.

An edge dislocation is customarily represented by a symbol in which the vertical leg designates the extra plane of atoms. When the vertical leg is above the horizontal leg, the dislocation is considered positive. When the extra plane of atoms is in the bottom portion of the crystal, the vertical leg is placed below the horizontal one and the dislocation is said to be negative. The part of the crystal containing the extra plane of atoms is in compression, whereas that portion on the other side of the dislocation line is in tension. Since one dislocation line runs completely across a crystal, it deforms the lattice structure to a greater extent than does one point defect.

A *screw dislocation* is a crystal defect in which the lattice points lie on a spiral or helical surface that revolves around a center line that is called the *dislocation line.* A screw dislocation terminates at a crystal surface. Shear stresses are set up in the lattice surrounding a screw dislocation as a result of the distortion in atomic array that the defect causes.

The *Burgers vector* is the distance, measured in multiples of the lattice parameter, that is needed to close a straight-sided loop around a dislocation when going the same number of lattice distances in all four directions. It is the term used to define the size of a dislocation and is designated by the letter *b.* A characteristic of an edge dislocation is that it lies perpendicular to its Burgers vector, whereas a screw dislocation lies parallel to its Burgers vector.

7.4.3 Planar Defects

There are several types of planar (or surface) defects that occur from a change in the orientation of crystallographic planes across a surface boundary. The most important planar defect is the *grain boundary,* which is the imperfect plane surface that separates two crystals of different orientation in a polycrystalline solid. Grain boundaries originate when the last few remaining atoms of a liquid freeze onto the meeting faces of two adjacent crystals that have grown from the melt or, similarly, when two adjacent crystals that grow by recrystallization meet each other.

The material in the grain boundary is at a higher energy level than the material near the center of the grain because of the increased elastic strain energy of the atoms that are forced from their normal (lowest-energy) sites in a perfect lattice. This higher energy level and lattice distortion cause the grain boundary material to be stronger, have a higher diffusion rate, and serve as a more favorable site for the nucleation of second phases than the interior materials.

Another important planar defect is the *twin boundary,* which is the plane that separates two portions of a single crystal having slightly different orientations. The two twins are mirror images of each other. The distortion of the twinned lattice is low in comparison to that at a grain boundary. Twins which form in most FCC metal crystals, especially the copper- and nickel-base alloys, during freezing from the melt or recrystallization are called *annealing twins.* Twins which form in some metals during cold work (plastic deformation) are called *mechanical twins.*

A third planar defect is the *low-angle grain boundary* or *low-angle tilt boundary,* where the angular misalignment of the two grains is very small, on the order of a few degrees. In a sense it is a very undistorted grain boundary. The angular mismatch of the crystal planes is due to a row of dislocations piled above each other.

A *stacking fault* is a planar defect that occurs when one crystalline plane is stacked out of its normal sequence in the lattice array. The lattice on both sides of the defect is normal. For example, the normal FCC stacking of planes may be interrupted by one layer of a hexagonal close-packed (HCP) plane, since both are close-packed structures with atomic packing factors of 0.74. Such stacking faults can occur during the formation of a crystal or by plastic deformation.

7.5   SLIP IN CRYSTALLINE SOLIDS

Slip can be defined as the massive sliding movement of one large body of atoms with respect to the remaining body of atoms of the crystal along crystallographic planes. Slip can also be considered as an avalanche of dislocations along one plane that pile up at grain boundaries or inclusions. The planes along which slip occurs are called *slip planes*. Slip occurs only with relatively high stresses, greater than the yield strength, and it causes plastic deformation.

When a crystalline solid or a single crystal is subjected to low loads, the atoms move slightly from their normal lattice sites and return to their proper positions when the load is removed. The displacements of the individual atoms are very small during elastic deformation. They are submicroscopic, a fraction of an atomic distance. Although there are some dislocation movements, they are few in number, involve very short distances, and are reversible.

Slip, however, is microscopic in size and causes plastic (permanent) deformation that is macroscopic. Figure 7.11 contains several two-dimensional lattice arrays which, in a simplified manner, illustrate the mechanism by means of which slip takes place. A typical perfect cubic lattice is shown in Fig. 7.11*a*, which is a small part of a single crystal. If sufficiently large shear stresses τ are placed on the crystal, all the atoms above the labeled slip plane move to the right simultaneously with respect to the atoms below the slip plane, as shown in Fig. 7.11*b*. The lattice is still a perfect cubic structure; only the outline or exterior shape of the single crystal has changed. It is believed, on the basis of the theories of elasticity, that the shear stress must be equal to the value of $G/2\pi$, where G is the shear modulus of elasticity. Young's modulus of elasticity E, Poisson's ratio ν, and G are related to one another by the equation

$$G = \frac{E}{2(1 + \nu)} \tag{7.1}$$

For iron, $E = 30$ Mpsi and $\nu = 0.30$, and so $G = 11.5$ Mpsi. Therefore, the so-called theoretical shear strength for slip to occur in iron is

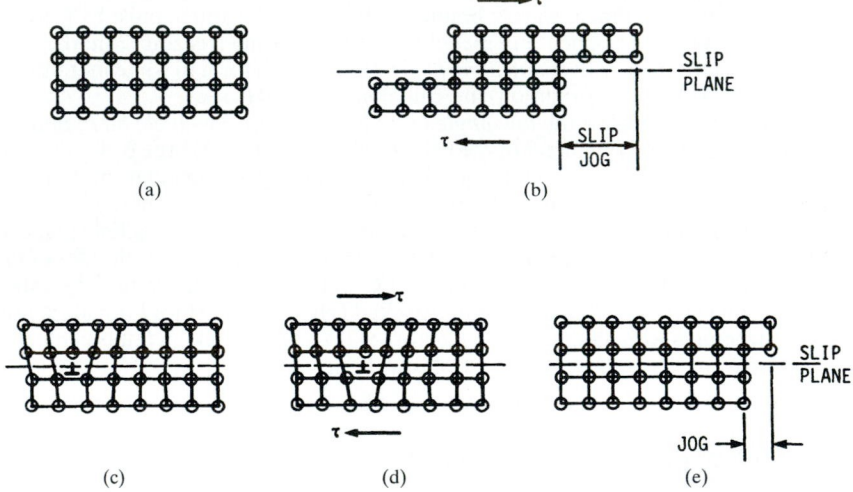

FIGURE 7.11   Two-dimensional sketch of the slip mechanism. (*a*) A perfect crystal; (*b*) idealized slip in a perfect crystal; (*c*) part of a crystal with one edge dislocation; (*d*) movement of dislocation subject to shear stress; (*e*) jog produced in the crystal face by dislocation motion.

$$\tau = \frac{G}{2\pi} = \frac{11.5}{2\pi} = 1.83 \text{ Mpsi}$$

However, slip occurs in iron crystals with shear stresses of only 4 to 5 kpsi, which is more than two orders of magnitude smaller. The "theoretical" shear strength of the other pure metals is also 400 to 500 times larger than the actual shear strength. The commonly accepted explanation of why the actual shear stress is so much lower than the theoretical value is that slip does not occur by the simultaneous movement of all the atoms along the slip plane; rather, it occurs by the movement of individual rows (the dislocation row or plane) of atoms. Thus it is the movement of dislocations along the slip plane to the grain boundary that causes the actual shear stress for plastic deformation to be so low. Figure 7.11c, d, and e illustrates the movement of a dislocation that results in slip.

In real crystals of metals, slip terminates at the grain boundaries or the free-surface faces and causes substantial jogs or steps, much larger than shown in Fig. 7.11. Experimental study of the spacings of the slip planes and the sizes of the jog have been made on some of the common metals. The spacing of the parallel planes along which slip occurs varies randomly, with an average distance between slip planes of about 2000 atom diameters. The length of the step or jog at the surface of the grain is approximately 200 to 700 atom diameters.

The atomic displacements associated with slip, unlike those of the initial movements of dislocations, are irreversible in that the slip jog remains when the shear stresses are removed. That is, slip causes a permanent change in shape, or *plastic deformation,* as it is called.

The evidence of slip is seen on metallurgically prepared samples as slip lines when examined under a microscope. The slip lines are the intersection of the crystallographic planes along which slip occurred with the etched surface of the specimen. Slip results in a narrow band on either side of the slip plane within which the lattice structure is severely distorted. These slip lines do not appear on the face of a specimen that is metallurgically polished after slip occurs; they appear only after etching with a suitable chemical reagent that dissolves the metal being studied. The slip lines become visible for the same reason that grain boundaries are visible after etching: The internal energy of the material within the distorted area is considerably higher than that of the material within the rest of the crystal. The metal in the higher energy level dissolves into the reagent much more rapidly than the rest of the crystal, leaving a narrow groove where the severely distorted band intersects the surface. Slip lines can also be seen on specimens that are polished prior to being plastically deformed and that have not been etched.

7.6　MECHANICAL STRENGTH

Although the specific mechanical properties of real materials are discussed in detail in the material that follows, it is very appropriate at this time to relate the concepts of the strengthening mechanisms to the previously described crystalline structures. Mechanical properties can best be studied on the basis of three precepts which encompass all the strengthening mechanisms. These three principles are stated here because they involve the distortion of the lattice structure that has just been discussed.

7.6.1 Principles of Mechanical Strength

Strength can be defined as a material's ability to resist slip. Thus it follows that the first principle of mechanical strength is this: *A material is strengthened when slip is made more difficult to initiate.* Therefore, to make a material stronger, it must be given a treatment that retards the avalanche of dislocations or, in other words, "pegs" the slip planes.

The second principle of mechanical strength is this: *Slip is retarded by inducing mechanical strains, or distortions, in the lattice structure of the material.* These distortions were discussed previously as lattice imperfections or defects. Thus it is a paradox that the source of strength in real polycrystalline materials is "crystal imperfections" or "crystal defects."

The third principle of mechanical strength is this: *There are four methods to induce mechanical strains or lattice distortions in a material, namely, decreasing the grain size, low-temperature plastic deformation (cold work), single-phase alloying, and multiple-phase alloying.*

7.6.2 Grain Size

Local distortion of the lattice structure at the grain boundaries induces substantial strain energy in those regions. This distortion impedes slip, or causes the dislocations to pile up, and consequently, the grain-boundary material is stronger than the material at the central portions of the crystal. This is true for most metals at room temperature. However, as additional energy is added to a polycrystalline material by raising the temperature, the grain-boundary material softens (and also melts) sooner or at a lower temperature than the bulk of the grain. At some temperature, called the *equicohesive temperature,* the strengths at these two regions are equal. Above the equicohesive temperature, the grain-boundary material is the weaker of the two. This explains why materials that are used at elevated temperatures have higher creep strengths when their grains are coarse rather than fine.

The surface-area-to-volume ratio of a sphere is inversely proportional to its diameter. Therefore, as the diameter of a sphere decreases, its ratio of surface area to volume increases. This means that for a given weight or volume of a polycrystalline solid, the total grain-boundary surface increases as the grain size decreases. Since the grain-boundary material is stronger than the interior material, the strength also varies inversely with the grain size. Also, since the surface area of a sphere is proportional to the square of its diameter, it can be assumed as a first approximation that the yield strength is proportional to the reciprocal of the square of the grain diameter.

Figure 7.12*a* shows how the 0.2 percent offset yield strength of 70Cu-30Zn brass varies with grain size. In this case, the yield strength increases by a factor of 4 with a grain diameter ratio of 24. The strengths of some materials, such as aluminum or steel, are not so greatly affected by grain size alone.

7.6.3 Cold Work

Cold work is a more significant strengthening mechanism than decreasing the grain size for most metals. When a crystalline material is plastically deformed, there is an avalanche of dislocations (called *slip*) that terminates at the grain boundaries. It is a mass movement of a body of atoms along a crystallographic plane. This movement in a polycrystalline material distorts both the grain boundaries and the crystalline planes in the grain so that slip occurs in the adjacent grains as well. Actually, a por-

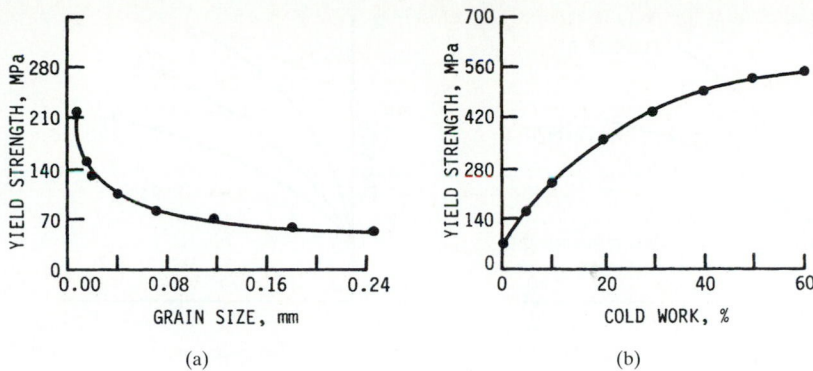

FIGURE 7.12   Yield strength versus grain size (*a*) and percent cold work (*b*) for 70Cu-30Zn brass.

tion of one grain intrudes into the space that was previously occupied by another grain, with a resulting distortion of the lattice in both grains.

Figure 7.12*b* illustrates the effect of cold work on the yield strength of 70Cu-30Zn brass. With only 10 percent cold work, the yield strength is raised by a factor of 3.5, and 60 percent cold work increases the strength nearly 8 times. In general, 10 percent cold work more than doubles the yield strength of most metals.

7.6.4   Single-Phase Alloying

Alloying (single- and multiple-phase) is the most important of the methods available to control or manipulate the mechanical properties of materials. The greatest increase in strength known today occurs when iron having a yield strength of 10 to 20 kpsi (70 to 140 MPa) is alloyed with less than 1.0 percent carbon to form a single phase (martensite) that has a yield strength of nearly 435 kpsi (3000 MPa).

The lattice is distorted and dislocation movement is impeded when foreign (solute) atoms are added to the lattice structure of a pure material. Figure 7.10*b* through *d* illustrates this condition, which was discussed previously under the heading "Point Defects." Vacancy defects, as shown in Fig. 7.10*a*, are not a practical strengthening method. The reason single-phase alloying has such a great effect on strength is that the entire lattice network is distorted, and uniformly, whereas in the other mechanism, there are regions in each crystal that are severely distorted and other regions that are hardly distorted at all.

Figure 7.13*a* shows the effect on the strength of the material of adding a foreign element B or C to the lattice structure of element A. From this figure it is clear that not all elements have the same strengthening effect. In general, the further the ratio of diameters of solute to solvent atoms is from unity, the greater will be the strengthening effect. However, the further this ratio is from unity, as previously explained, the less soluble the two atoms are in each other's lattice.

7.6.5   Multiple-Phase Alloying

This mechanism is sometimes referred to as *fine-particle strengthening*. In a sense, multiple-phase alloying is a combination of single-phase alloying and grain-boundary

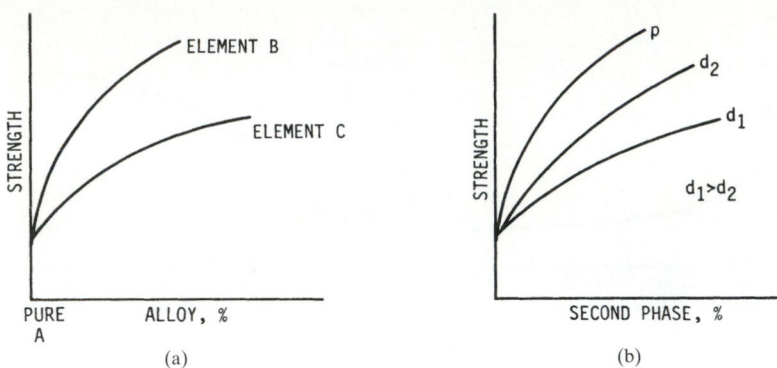

FIGURE 7.13   The effect of alloying on strength. (*a*) Single-phase alloying; atomic diameter ratio $B/A > C/A$; (*b*) multiple-phase alloying; *p* is a nonspherical-shaped particle; *d* is the spherical particle diameter.

strengthening. That is, some of the added element goes into solution in the solvent lattice and thus has a strengthening effect; the remainder of the added element forms a second phase (either another solid solution or a compound) that is present as small grains or crystals.

Multiple-phase alloys can be made in three different ways. One method is by *annealing*. In this case, the alloy is heated to a one-phase region where the second element is completely soluble in the first. On slow cooling, the second phase precipitates as a massive network in the grain-boundary regions of the solvent matrix. This is the least beneficial form of alloying. The second method is similar except that the alloy is rapidly cooled from the high one-phase region so that a supersaturated solid phase occurs at room temperature. This material is then reheated to a relatively low temperature so that the second phase precipitates throughout the entire crystal as extremely fine particles rather than concentrating at the grain boundaries. This is the common *precipitation-hardening procedure*. The third method is to add a compound, in the form of small particles, that is insoluble in the parent material. Thus the two phases must be mixed in powder form and then sintered. This method is called *dispersion hardening*. At the present time there are only about a half dozen dispersion-hardenable alloys in commercial use. The most notable ones are Al_2O_3 particles in aluminum (called SAP, for sintered aluminum power) and ThO in nickel.

7.7 MECHANICAL PROPERTIES AND TESTS

Most mechanical properties are structure-sensitive; that is, they are affected by changes in either the lattice structure or the microstructure. However, modulus of elasticity is one property that is structure-insensitive. For example, ductility and toughness of any material (regardless of whether it is a pure element such as copper, a simple alloy such as AISI 1080 steel, or a complex alloy such as a cobalt-base superalloy) vary with grain size, amount of cold work if any, or the microstructure if heat-treated. The modulus of elasticity of any material is the same regardless of grain size, amount of cold work, or microstructure.

Mechanical properties are discussed individually in the sections that follow. Several new quantitative relationships for the properties are presented here which make it possible to understand the mechanical properties to a depth that is not possible by means of the conventional tabular listings, where the properties of each material are listed separately.

7.8 HARDNESS

Hardness is used more frequently than any other of the mechanical properties by the design engineer to specify the final condition of a structural part. This is due in part to the fact that hardness tests are the least expensive in time and money to conduct. The test can be performed on a finished part without the need to machine a special test specimen. In other words, a hardness test may be a nondestructive test in that it can be performed on the actual part without affecting its service function.

Hardness is frequently defined as a measure of the ability of a material to resist plastic deformation or penetration by an indenter having a spherical or conical end. At the present time, hardness is more a technological property of a material than it is a scientific or engineering property. In a sense, hardness tests are practical shop tests rather than basic scientific tests. All the hardness scales in use today give relative values rather than absolute ones. Even though some hardness scales, such as the Brinell, have units of stress (kg/mm^2) associated with them, they are not absolute scales because a given piece of material (such as a 2-in cube of brass) will have significantly different Brinell hardness numbers depending on whether a 500-kg or a 3000-kg load is applied to the indenter.

7.8.1 Rockwell Hardness

The *Rockwell hardnesses* are hardness numbers obtained by an indentation type of test based on the depth of the indentation due to an increment of load. The Rockwell scales are by far the most frequently used hardness scales in industry even though they are completely relative. The reasons for their large acceptance are the simplicity of the testing apparatus, the short time necessary to obtain a reading, and the ease with which reproducible readings can be obtained, the last of these being due in part to the fact that the testing machine has a "direct-reading" dial; that is, a needle points directly to the actual hardness value without the need for referring to a conversion table or chart, as is true with the Brinell, Vickers, or Knoop hardnesses. Table 7.2 lists the most common Rockwell hardness scales.

TABLE 7.2 Rockwell Hardness Scales

	Scale								
	A	B	C	D	E	F	G	H	K
Indenter	1	2	1	1	3	2	2	3	3
Load, kg	60	100	150	100	100	60	150	60	150

Indenter 1 is a diamond cone having an included angle of 120° and a spherical end radius of 0.008 in. Indenters 2 and 3 are ¹⁄₁₆-in-diameter and ⅛-in-diameter balls, respectively. In addition to the preceding scales, there are several others for testing very soft bearing materials, such as babbit, that use ¼-in-diameter and ½-in-diameter balls. Also, there are several "superficial" scales that use a special diamond cone with loads less than 50 kg to test the hardness of surface-hardened layers.

The particular materials that each scale is used on are as follows: the A scale on the extremely hard materials, such as carbides or thin case-hardened layers on steel; the B scale on soft steels, copper and aluminum alloys, and soft-case irons; the C scale on medium and hard steels, hard-case irons, and all hard nonferrous alloys; the E and F scales on soft copper and aluminum alloys. The remaining scales are used on even softer alloys.

Several precautions must be observed in the proper use of the Rockwell scales. The ball indenter should not be used on any material having a hardness greater than $50\ R_C$; otherwise the steel ball will be plastically deformed or flattened and thus give erroneous readings. Readings taken on the sides of cylinders or spheres should be corrected for the curvature of the surface. Readings on the C scale of less than 20 should not be recorded or specified because they are unreliable and subject to much variation.

The hardness numbers for all the Rockwell scales are an inverse measure of the depth of the indentation. Each division on the dial gauge of the Rockwell machine corresponds to an 80×10^6 in depth of penetration. The penetration with the C scale varies between 0.0005 in for hard steel and 0.0015 in for very soft steel when only the minor load is applied. The total depth of penetration with both the major and minor loads applied varies from 0.003 in for the hardest steel to 0.008 in for soft steel (20 R_C). Since these indentations are relatively shallow, the Rockwell C hardness test is considered a nondestructive test and it can be used on fairly thin parts.

Although negative hardness readings can be obtained on the Rockwell scales (akin to negative Fahrenheit temperature readings), they are usually not recorded as such, but rather a different scale is used that gives readings greater than zero. The only exception to this is when one wants to show a continuous trend in the change in hardness of a material due to some treatment. A good example of this is the case of the effect of cold work on the hardness of a fully annealed brass. Here the annealed hardness may be $-20\ R_B$ and increase to $95\ R_B$ with severe cold work.

7.8.2 Brinell Hardness

The *Brinell hardness* H_B is the hardness number obtained by dividing the load that is applied to a spherical indenter by the surface area of the spherical indentation produced; it has units of kilograms per square millimeter. Most readings are taken with a 10-mm ball of either hardened steel or tungsten carbide. The loads that are applied vary from 500 kg for soft materials to 3000 kg for hard materials. The steel ball should not be used on materials having a hardness greater than about 525 H_B (52 R_C) because of the possibility of putting a flat spot on the ball and making it inaccurate for further use.

The Brinell hardness machine is as simple as, though more massive than, the Rockwell hardness machine, but the standard model is not direct-reading and takes a longer time to obtain a reading than the Rockwell machine. In addition, the indentation is much larger than that produced by the Rockwell machine, and the machine cannot be used on hard steel. The method of operation, however, is simple. The prescribed load is applied to the 10-mm-diameter ball for approximately 10 s. The part is then withdrawn from the machine and the operator measures the diameter of the indentation by means of a millimeter scale etched on the eyepiece of a special Brinell microscope. The Brinell hardness number is then obtained from the equation

$$H_B = \frac{L}{(\pi D/2)[D - (D^2 - d^2)^{1/2}]}$$ (7.2)

where L = load, kg
  D = diameter of indenter, mm
  d = diameter of indentation, mm

The denominator in this equation is the spherical area of the indentation.

The Brinell hardness test has proved to be very successful, partly due to the fact that for some materials it can be directly correlated to the tensile strength. For example, the tensile strengths of all the steels, if stress-relieved, are very close to being 0.5 times the Brinell hardness number when expressed in kilopounds per square inch (kpsi). This is true for both annealed and heat-treated steel. Even though the Brinell hardness test is a technological one, it can be used with considerable success in engineering research on the mechanical properties of materials and is a much better test for this purpose than the Rockwell test.

The Brinell hardness number of a given material increases as the applied load is increased, the increase being somewhat proportional to the strain-hardening rate of the material. This is due to the fact that the material beneath the indentation is plastically deformed, and the greater the penetration, the greater is the amount of cold work, with a resulting high hardness. For example, the cobalt base alloy HS-25 has a hardness of 150 H_B with a 500-kg load and a hardness of 201 H_B with an applied load of 3000 kg.

7.8.3 Meyer Hardness

The *Meyer hardness* H_M is the hardness number obtained by dividing the load applied to a spherical indenter by the projected area of the indentation. The Meyer hardness test itself is identical to the Brinell test and is usually performed on a Brinell hardness-testing machine. The difference between these two hardness scales is simply the area that is divided into the applied load—the projected area being used for the Meyer hardness and the spherical surface area for the Brinell hardness. Both are based on the diameter of the indentation. The units of the Meyer hardness are also kilograms per square millimeter, and hardness is calculated from the equation

$$H_M = \frac{4L}{\pi d^2}$$ (7.3)

Because the Meyer hardness is determined from the projected area rather than the contact area, it is a more valid concept of stress and therefore is considered a more basic or scientific hardness scale. Although this is true, it has been used very little since it was first proposed in 1908, and then only in research studies. Its lack of acceptance is probably due to the fact that it does not directly relate to the tensile strength the way the Brinell hardness does.

Meyer is much better known for the original strain-hardening equation that bears his name than he is for the hardness scale that bears his name. The strain-hardening equation for a given diameter of ball is

$$L = Ad^p \tag{7.4}$$

where L = load on spherical indenter
$\quad d$ = diameter of indentation
$\quad p$ = Meyer strain-hardening exponent

The values of the strain-hardening exponent for a variety of materials are available in many handbooks. They vary from a minimum value of 2.0 for low-work-hardening materials, such as the PH stainless steels and all cold-rolled metals, to a maximum of about 2.6 for dead soft brass. The value of p is about 2.25 for both annealed pure aluminum and annealed 1020 steel.

Experimental data for some metals show that the exponent p in Eq. (7.4) is related to the strain-strengthening exponent m in the tensile stress-strain equation $\sigma = \sigma_0 \varepsilon^m$, which is to be presented later. The relation is

$$p - 2 = m \tag{7.5}$$

In the case of 70-30 brass, which had an experimentally determined value of $p = 2.53$, a separately run tensile test gave a value of $m = 0.53$. However, such good agreement does not always occur, partly because of the difficulty of accurately measuring the diameter d. Nevertheless, this approximate relationship between the strain-hardening and the strain-strengthening exponents can be very useful in the practical evaluation of the mechanical properties of a material.

7.8.4   Vickers or Diamond-Pyramid Hardness

The *diamond-pyramid hardness* H_p, or the *Vickers hardness* H_V, as it is frequently called, is the hardness number obtained by dividing the load applied to a square-based pyramid indenter by the surface area of the indentation. It is similar to the Brinell hardness test except for the indenter used. The indenter is made of industrial diamond, and the area of the two pairs of opposite faces is accurately ground to an included angle of 136°. The load applied varies from as low as 100 g for microhardness readings to as high as 120 kg for the standard macrohardness readings. The indentation at the surface of the workpiece is square-shaped. The diamond pyramid hardness number is determined by measuring the length of the two diagonals of the indentation and using the average value in the equation

$$H_p = \frac{2L \sin{(\alpha/2)}}{d^2} = \frac{1.8544L}{d^2} \tag{7.6}$$

where L = applied load, kg
$\quad d$ = diagonal of the indentation, mm
$\quad \alpha$ = face angle of the pyramid, 136°

The main advantage of a cone or pyramid indenter is that it produces indentations that are geometrically similar regardless of depth. In order to be geometrically similar, the angle subtended by the indentation must be constant regardless of the depth of the indentation. This is not true of a ball indenter. It is believed that if geometrically similar deformations are produced, the material being tested is stressed to the same amount regardless of the depth of the penetration. On this basis, it would be expected that conical or pyramidal indenters would give the same hardness number regardless of the load applied. Experimental data show that the pyramid hardness number is independent of the load if loads greater than 3 kg are applied. However, for loads less than 3 kg, the hardness is affected by the load, depending on the strain-hardening exponent of the material being tested.

7.8.5 Knoop Hardness

The *Knoop hardness* H_K is the hardness number obtained by dividing the load applied to a special rhombic-based pyramid indenter by the projected area of the indentation. The indenter is made of industrial diamond, and the four pyramid faces are ground so that one of the angles between the intersections of the four faces is 172.5° and the other angle is 130°. A pyramid of this shape makes an indentation that has the projected shape of a parallelogram having a long diagonal that is 7 times as large as the short diagonal and 30 times as large as the maximum depth of the indentation.

The greatest application of Knoop hardness is in the microhardness area. As such, the indenter is mounted on an axis parallel to the barrel of a microscope having magnifications of 100× to 500×. A metallurgically polished flat specimen is used. The place at which the hardness is to be determined is located and positioned under the hairlines of the microscope eyepiece. The specimen is then positioned under the indenter and the load is applied for 10 to 20 s. The specimen is then located under the microscope again and the length of the long diagonal is measured. The Knoop hardness number is then determined by means of the equation

$$H_K = \frac{L}{0.070\ 28d^2} \tag{7.7}$$

where L = applied load, kg
d = length of long diagonal, mm

The indenter constant 0.070 28 corresponds to the standard angles mentioned above.

7.8.6 Scleroscope Hardness

The *scleroscope hardness* is the hardness number obtained from the height to which a special indenter bounces. The indenter has a rounded end and falls freely a distance of 10 in in a glass tube. The rebound height is measured by visually observing the maximum height the indenter reaches. The measuring scale is divided into 140 equal divisions and numbered beginning with zero. The scale was selected so that the rebound height from a fully hardened high-carbon steel gives a maximum reading of 100.

All the previously described hardness scales are called *static hardnesses* because the load is slowly applied and maintained for several seconds. The scleroscope hardness, however, is a *dynamic hardness*. As such, it is greatly influenced by the elastic modulus of the material being tested.

7.9   THE TENSILE TEST

The tensile test is conducted on a machine that can apply uniaxial tensile or compressive loads to the test specimen, and the machine also has provisions for accurately registering the value of the load and the amount of deformation that occurs to the specimen. The tensile specimen may be a round cylinder or a flat strip with a reduced cross section, called the *gauge section,* at its midlength to ensure that the fracture does not occur at the holding grips. The minimum length of the reduced section for a standard specimen is four times its diameter. The most commonly used specimen has a 0.505-in-diameter gauge section (0.2 in^2 cross-sectional area) that is 2¼ in long to accommodate a 2-in-long gauge section. The overall length of the specimen is 5½ in, with a 1-in length of size ¾-10NC screw threads on each end. The ASTM specifications list several other standard sizes, including flat specimens.

In addition to the tensile properties of strength, rigidity, and ductility, the tensile test also gives information regarding the stress-strain behavior of the material. It is very important to distinguish between *strength* and *stress* as they relate to material properties and mechanical design, but it is also somewhat awkward, since they have the same units and many books use the same symbol for both.

Strength is a property of a material—it is a measure of the ability of a material to withstand stress or it is the load-carrying capacity of a material. The numerical value of strength is determined by dividing the appropriate load (yield, maximum, fracture, shear, cyclic, creep, etc.) by the original cross-sectional area of the specimen and is designated as *S.* Thus

$$S = \frac{L}{A_0} \tag{7.8}$$

The subscripts *y, u, f,* and *s* are appended to *S* to denote yield, ultimate, fracture, and shear strength, respectively. Although the strength values obtained from a tensile test have the units of stress [psi (Pa) or equivalent], they are not really values of stress.

Stress is a condition of a material due to an applied load. If there are no loads on a part, then there are no stresses in it. (Residual stresses may be considered as being caused by unseen loads.) The numerical value of the stress is determined by dividing the actual load or force on the part by the actual cross section that is supporting the load. Normal stresses are almost universally designated by the symbol σ, and the stresses due to tensile loads are determined from the expression

$$\sigma = \frac{L}{A_i} \tag{7.9}$$

where A_i = instantaneous cross-sectional area corresponding to that particular load. The units of stress are pounds per square inch (pascals) or an equivalent.

During a tensile test, the stress varies from zero at the very beginning to a maximum value that is equal to the true fracture stress, with an infinite number of stresses in between. However, the tensile test gives only three values of strength: yield, ultimate, and fracture. An appreciation of the real differences between strength and stress will be achieved after reading the material that follows on the use of tensile-test data.

7.9.1 Engineering Stress-Strain

Traditionally, the tensile test has been used to determine the so-called engineering stress-strain data that are needed to plot the engineering stress-strain curve for a given material. However, since engineering stress is not really a stress but is a measure of the strength of a material, it is more appropriate to call such data either *strength–nominal strain* or *nominal stress–strain data*. Table 7.3 illustrates the data that are normally collected during a tensile test, and Fig. 7.14 shows the condition of a standard tensile specimen at the time the specific data in the table are recorded. The load-versus-gauge-length data, or an elastic stress-strain curve drawn by the machine, are needed to determine Young's modulus of elasticity of the material as well as the proportional limit. They are also needed to determine the yield strength if the offset method is used. All the definitions associated with engineering stress-strain, or, more appropriately, with the strength–nominal strain properties, are presented in the section which follows and are discussed in conjunction with the experimental data for commercially pure titanium listed in Table 7.3 and Fig. 7.14.

The elastic and elastic-plastic data listed in Table 7.3 are plotted in Fig. 7.15 with an expanded strain axis, which is necessary for the determination of the yield strength. The nominal (approximate) stress or the strength S which is calculated by means of Eq. (7.8) is plotted as the ordinate.

The abscissa of the engineering stress-strain plot is the *nominal strain,* which is defined as the unit elongation obtained when the change in length is divided by the original length and has the units of inch per inch and is designated as n. Thus, for tension,

$$n = \frac{\Delta \ell}{\ell} = \frac{\ell_f - \ell_0}{\ell_0} \tag{7.10}$$

where ℓ = gauge length and the subscripts 0 and f designate the original and final state, respectively. This equation is valid for deformation strains that do not exceed the strain at the maximum load of a tensile specimen.

It is customary to plot the data obtained from a tensile test as a stress-strain curve such as that illustrated in Fig. 7.16, but without including the word *nominal*. The reader then considers such a curve as an actual stress-strain curve, which it obviously is not. The curve plotted in Fig. 7.16 is in reality a load-deformation curve. If the ordinate axis were labeled load (lb) rather than stress (psi), the distinction between

TABLE 7.3 Tensile Test Data

Material: A40 titanium; condition: annealed; specimen size: 0.505-in diameter by 2-in gauge length; $A_0 = 0.200$ in^2

Yield load	9 040 lb	Yield strength	45.2 kpsi
Maximum load	14 950 lb	Tensile strength	74.75 kpsi
Fracture load	11 500 lb	Fracture strength	57.5 kpsi
Final length	2.480 in	Elongation	24%
Final diameter	0.352 in	Reduction of area	51.15%

Load, lb	Gauge length, in	Load, lb	Gauge length, in
1 000	2.0006	6 000	2.0044
2 000	2.0012	7 000	2.0057
3 000	2.0018	8 000	2.0070
4 000	2.0024	9 000	2.0094
5 000	2.0035	10 000	2.0140

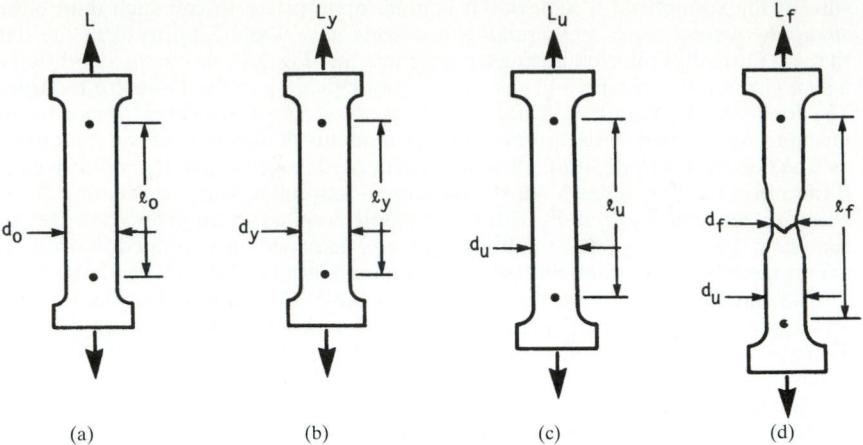

FIGURE 7.14 A standard tensile specimen of A40 titanium at various stages of loading. (*a*) Unloaded, $L = 0$ lb, $d_0 = 0.505$ in, $\ell_0 = 2.000$ in, $A_0 = 0.200$ in^2; (*b*) yield load $L_y = 9040$ lb, $d_y = 0.504$ in, $\ell_y = 2.009$ in, $A_y = 0.1995$ in^2; (*c*) maximum load $L_u = 14\,950$ lb, $d_u = 0.470$ in, $\ell_u = 2.310$ in, $A_u = 0.173$ in^2; (*d*) fracture load $L_f = 11\,500$ lb, $d_f = 0.352$ in, $\ell_f = 2.480$ in, $A_f = 0.097$ in^2, $d_u = 0.470$ in.

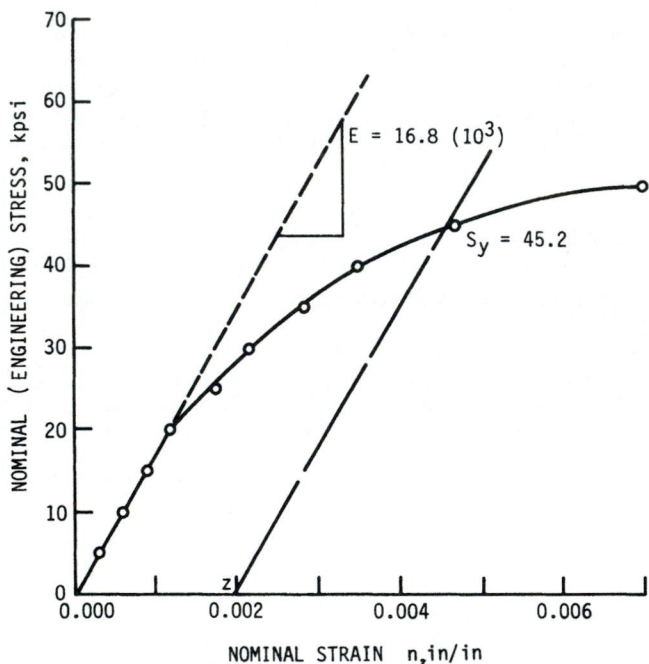

FIGURE 7.15 The elastic-plastic portion of the engineering stress-strain curve for annealed A40 titanium.

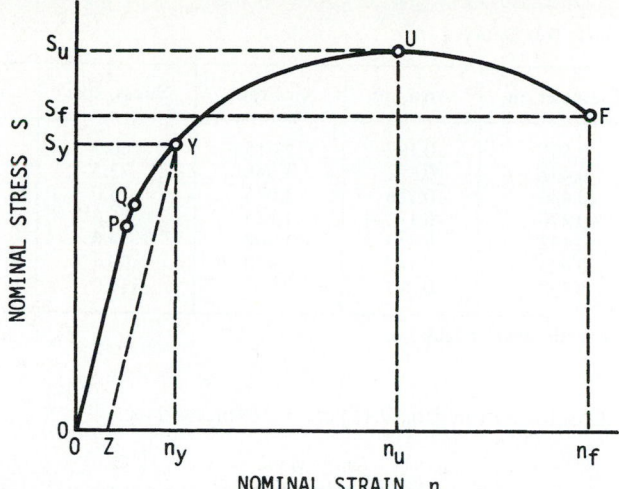

FIGURE 7.16   The engineering stress-strain curve. P = proportional limit, Q = elastic limit, Y = yield load, U = ultimate (maximum) load, and F = fracture load.

strength and stress would be easier to make. Although the fracture load is lower than the ultimate load, the stress in the material just prior to fracture is much greater than the stress at the time the ultimate load is on the specimen.

7.9.2   True Stress-Strain

The tensile test is also used to obtain true stress-strain or true stress–natural strain data to define the plastic stress-strain characteristics of a material. In this case it is necessary to record simultaneously the cross-sectional area of the specimen and the load on it. For round sections it is sufficient to measure the diameter for each load recorded. The load-deformation data in the plastic region of the tensile test of an annealed titanium are listed in Table 7.4. These data are a continuation of the tensile test in which the elastic data are given in Table 7.3.

The load-diameter data in Table 7.4 are recorded during the test and the remainder of the table is completed afterwards. The values of stress are calculated by means of Eq. (7.9). The strain in this case is the *natural strain* or *logarithmic strain*, which is the sum of all the infinitesimal nominal strains, that is,

$$\varepsilon = \frac{\Delta \ell_1}{\ell_0} + \frac{\Delta \ell_2}{\ell_0 + \Delta \ell_1} + \frac{\Delta \ell_3}{\ell_0 + \Delta \ell_1 + \Delta \ell_2} + \cdots$$

$$= \ln \frac{\ell_f}{\ell_0} \tag{7.11}$$

The volume of material remains constant during plastic deformation. That is,

$$V_0 = V_f \qquad \text{or} \qquad A_0 \ell_0 = A_f \ell_f$$

TABLE 7.4 Tensile Test Data†

Load, lb	Diameter, in	Area, in^2	Area ratio	Stress, kpsi	Strain, in/in
12 000	0.501	0.197	1.015	60.9	0.0149
14 000	0.493	0.191	1.048	73.5	0.0473
14 500	0.486	0.186	1.075	78.0	0.0724
14 950	0.470	0.173	1.155	86.5	0.144
14 500	0.442	0.153	1.308	94.8	0.268
14 000	0.425	0.142	1.410	99.4	0.344
11 500	0.352	0.097	2.06	119.0	0.729

†This table is a continuation of Table 7-3.

Thus, for tensile deformation, Eq. (7.11) can be expressed as

$$\varepsilon = \ln \frac{A_0}{A_f} \tag{7.12}$$

Quite frequently, in calculating the strength or the ductility of a cold-worked material, it is necessary to determine the value of the strain ε that is equivalent to the amount of the cold work. The *amount of cold work* is defined as the percent reduction of cross-sectional area (or simply the percent reduction of area) that is given the material by a plastic-deformation process. It is designated by the symbol W and is determined from the expression

$$W = \frac{A_0 - A_f}{A_0} (100) \tag{7.13}$$

where the subscripts 0 and f refer to the original and the final area, respectively. By solving for the A_0/A_f ratio and substituting into Eq. (7.12), the appropriate relationship between strain and cold work is found to be

$$\varepsilon_W = \ln \frac{100}{100 - W} \tag{7.14}$$

The stress-strain data of Table 7.4 are plotted in Fig. 7.17 on cartesian coordinates. The most significant difference between the shape of this stress-strain curve and that of the load-deformation curve in Fig. 7.16 is the fact that the stress continues to rise until fracture occurs and does not reach a maximum value as the load-deformation curve does. As can be seen in Table 7.4 and Fig. 7.17, the stress at the time of the maximum load is 86 kpsi, and it increases to 119 kpsi at the instant that fracture occurs. A smooth curve can be drawn through the experimental data, but it is not a straight line, and consequently many experimental points are necessary to accurately determine the shape and position of the curve.

The stress-strain data obtained from the tensile test of the annealed A40 titanium listed in Tables 7.3 and 7.4 are plotted on logarithmic coordinates in Fig. 7.18. The elastic portion of the stress-strain curve is also a straight line on logarithmic coordinates as it is on cartesian coordinates. When plotted on cartesian coordinates, the slope of the elastic modulus is different for the different materials. However, when

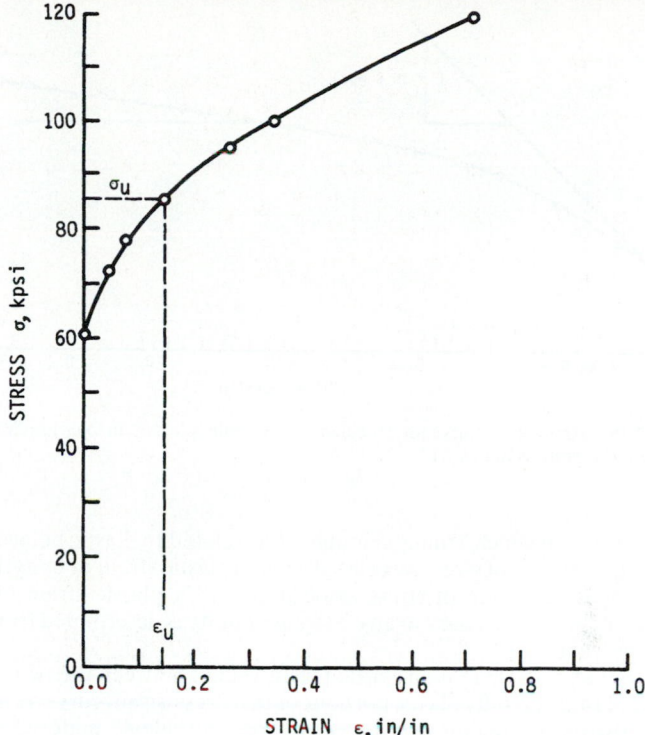

FIGURE 7.17 Stress-strain curve for annealed A40 titanium. The strain is the natural or logarithmic strain and the data of Tables 7.3 and 7.4 are plotted on cartesian coordinates.

plotted on logarithmic coordinates, the slope of the elastic modulus is 1 (unity) for all materials—it is only the height, or position, of the line that is different for different materials. In other words, the elastic moduli for all the materials are parallel lines making an angle of 45° with the ordinate axis.

The experimental points in Fig. 7.18 for strains greater than 0.01 (1 percent plastic deformation) also fall on a straight line having a slope of 0.14. The slope of the stress-strain curve in logarithmic coordinates is called the *strain-strengthening exponent* because it indicates the increase in strength that results from plastic strain. It is sometimes referred to as the *strain-hardening exponent,* which is somewhat misleading because the real strain-hardening exponent is the Meyer exponent p, discussed previously under the subject of strain hardening. The strain-strengthening exponent is represented by the symbol m.

The equation for the plastic stress-strain line is

$$\sigma = \sigma_0 \varepsilon^m \tag{7.15}$$

and is known as the *strain-strengthening equation* because it is directly related to the yield strength. The proportionality constant σ_0 is called the *strength coefficient.* The strength coefficient σ_0 is related to the plastic behavior of a material in exactly

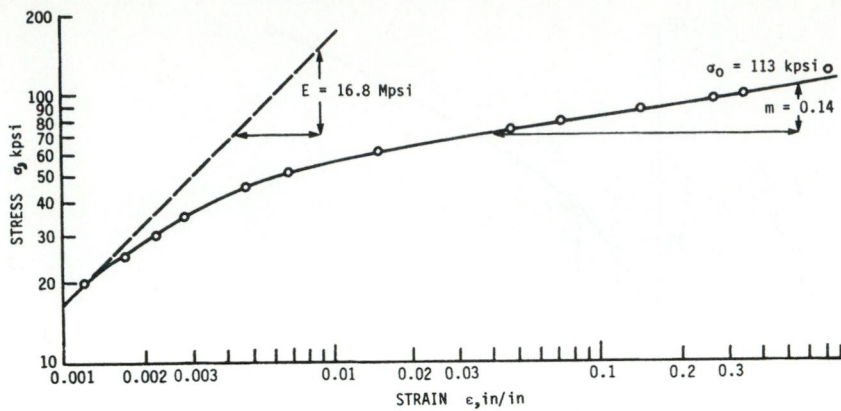

FIGURE 7.18   Stress-strain curve for annealed A40 titanium plotted on logarithmic coordinates. The data are the same as in Fig. 7.17.

the same manner in which Young's modulus E is related to elastic behavior. Young's modulus E is the value of stress associated with an elastic strain of unity; the strength coefficient σ_0 is the value of stress associated with a plastic strain of unity. The amount of cold work necessary to give a strain of unity is determined from Eq. (7.14) to be 63.3 percent.

For most materials there is an elastic-plastic region between the two straight lines of the fully elastic and fully plastic portions of the stress-strain curve. A material that has no elastic-plastic region may be considered an "ideal" material because the study and analysis of its tensile properties are simpler. Such a material has a complete stress-strain relationship that can be characterized by two intersecting straight lines, one for the elastic region and one for the plastic region. Such a material would have a stress-strain curve similar to the one labeled I in Fig. 7.19. A few real materials have a stress-strain curve that approximates the "ideal" curve. However, most engineering materials have a stress-strain curve that resembles curve O in Fig. 7.19. These materials appear to "overyield"; that is, they have a higher yield strength than the "ideal" value, followed by a region of low or no strain strengthening before the fully plastic region begins. Among the materials that have this type of curve are steel, stainless steel, copper, brass alloys, nickel alloys, and cobalt alloys.

Only a few materials have a stress-strain curve similar to that labeled U in Fig. 7.19. The characteristic feature of this type of material is that it appears to "underyield"; that is, it has a yield strength that is lower than the "ideal" value. Some of the fully annealed aluminum alloys have this type of curve.

7.10   TENSILE PROPERTIES

Tensile properties are those mechanical properties obtained from the tension test; they are used as the basis of mechanical design of structural components more frequently than any other of the mechanical properties. More tensile data are available for materials than any other type of material property data. Frequently the design engineer must base his or her calculations on the tensile properties even under

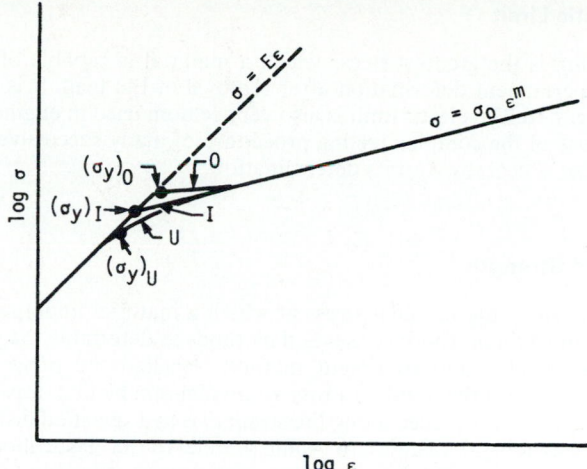

FIGURE 7.19 Schematic representation of three types of stress-strain curves. *I* is an "ideal" curve, and *O* and *U* are two types of real curve.

cyclic, shear, or impact loading simply because the more appropriate mechanical property data are not available for the material he or she may be considering for a specific part. All the tensile properties are defined in this section and are briefly discussed on the basis of the tensile test described in the preceding section.

7.10.1 Modulus of Elasticity

The *modulus of elasticity,* or *Young's modulus,* is the ratio of stress to the corresponding strain during elastic deformation. It is the slope of the straight-line (elastic) portion of the stress-strain curve when drawn on cartesian coordinates. It is also known, as indicated previously, as Young's modulus, or the proportionality constant in Hooke's law, and is commonly designated as E with units of pounds per square inch (pascals) or the equivalent. The modulus of elasticity of the titanium alloy whose tensile data are reported in Table 7.3 is shown in Fig. 7.15, where the first four experimental data points fall on a straight line having a slope of 16.8 Mpsi.

7.10.2 Proportional Limit

The *proportional limit* is the greatest stress which a material is capable of developing without any deviation from a linear proportionality of stress to strain. It is the point where a straight line drawn through the experimental data points in the elastic region first departs from the actual stress-strain curve. Point P in Fig. 7.16 is the proportional limit (20 kpsi) for this titanium alloy. The proportional limit is very seldom used in engineering specifications because it depends so much on the sensitivity and accuracy of the testing equipment and the person plotting the data.

7.10.3   Elastic Limit

The *elastic limit* is the greatest stress which a material is capable of withstanding without any permanent deformation after removal of the load. It is designated as point Q in Fig. 7.16. The elastic limit is also very seldom used in engineering specifications because of the complex testing procedure of many successive loadings and unloadings that is necessary for its determination.

7.10.4   Yield Strength

The *yield strength* is the nominal stress at which a material undergoes a specified permanent deformation. There are several methods to determine the yield strength, but the most reliable and consistent method is called the *offset method*. This approach requires that the nominal stress-strain diagram be first drawn on cartesian coordinates. A point z is placed along the strain axis at a specified distance from the origin, as shown in Figs. 7.15 and 7.16. A line parallel to the elastic modulus is drawn from Z until it intersects the nominal stress-strain curve. The value of stress corresponding to this intersection is called the *yield strength* by the offset method. The distance $0Z$ is called the *offset* and is expressed as percent. The most common offset is 0.2 percent, which corresponds to a nominal strain of 0.002 in/in. This is the value of offset used in Fig. 7.15 to determine the yield strength of the A40 titanium. An offset of 0.01 percent is sometimes used, and the corresponding nominal stress is called the *proof strength*, which is a value very close to the proportional limit. For some nonferrous materials an offset of 0.5 percent is used to determine the yield strength.

Inasmuch as all methods of determining the yield strength give somewhat different values for the same material, it is important to specify what method, or what offset, was used in conducting the test.

7.10.5   Tensile Strength

The *tensile strength* is the value of nominal stress obtained when the maximum (or ultimate) load that the tensile specimen supports is divided by the original cross-sectional area of the specimen. It is shown as S_u in Fig. 7.16 and is sometimes called the *ultimate strength*. The tensile strength is a commonly used property in engineering calculations even though the yield strength is a measure of when plastic deformation begins for a given material. The real significance of the tensile strength as a material property is that it indicates what maximum load a given part can carry in uniaxial tension without breaking. It determines the absolute maximum limit of load that a part can support.

7.10.6   Fracture Strength

The *fracture strength,* or *breaking strength,* is the value of nominal stress obtained when the load carried by a tensile specimen at the time of fracture is divided by its original cross-sectional area. The breaking strength is not used as a material property in mechanical design.

7.10.7 Reduction of Area

The *reduction of area* is the maximum change in area of a tensile specimen divided by the original area and is usually expressed as a percent. It is designated as A_r and is calculated as follows:

$$A_r = \frac{A_0 - A_f}{A_0} \, (100) \tag{7.16}$$

where the subscripts 0 and f refer to the original area and area after fracture, respectively. The percent reduction of area and the strain at ultimate load ε_u are the best measure of the ductility of a material.

7.10.8 Fracture Strain

The *fracture strain* is the true strain at fracture of the tensile specimen. It is represented by the symbol ε_f and is calculated from the definition of strain as given in Eq. (7.12). If the percent reduction of area A_r is known for a material, the fracture strain can be calculated from the expression

$$\varepsilon_f = \ln \frac{100}{100 - A_r} \tag{7.17}$$

7.10.9 Percentage Elongation

The *percentage elongation* is a crude measure of the ductility of a material and is obtained when the change in gauge length of a fractured tensile specimen is divided by the original gauge length and expressed as percent. Because of the ductility relationship, we express it here as

$$D_e = \frac{\ell_f - \ell_0}{\ell_0} \, (100) \tag{7.18}$$

Since most materials exhibit nonuniform deformation before fracture occurs on a tensile test, the percentage elongation is some kind of an average value and as such cannot be used in meaningful engineering calculations.

The percentage elongation is not really a material propety, but rather it is a combination of a material property and a test condition. A true material property is not significantly affected by the size of the specimen. Thus a ¼-in-diameter and a ½-in-diameter tensile specimen of the same material give the same values for yield strength, tensile strength, reduction of area or fracture strain, modulus of elasticity, strain-strengthening exponent, and strength coefficient, but a 1-in gauge-length specimen and a 2-in gauge-length specimen of the same material do not give the same percentage elongation. In fact, the percentage elongation for a 1-in gauge-length specimen may actually be 100 percent greater than that for the 2-in gauge-length specimen even when they are of the same diameter.

7.11 STRENGTH, STRESS, AND STRAIN RELATIONS

The following relationships between strength, stress, and strain are very helpful to a complete understanding of tensile properties and also to an understanding of their use in specifying the optimum material for a structural part. These relationships also help in solving manufacturing problems where difficulty is encountered in the fabrication of a given part because they enable one to have a better concept of what can be expected of a material during a manufacturing process. A further advantage of these relations is that they enable an engineer to more readily determine the mechanical properties of a fabricated part on the basis of the original properties of the material and the mechanisms involved with the particular process used.

7.11.1 Natural and Nominal Strain

The relationship between these two strains is determined from their definitions. The expression for the natural strain is $\varepsilon = \ln (\ell_f/\ell_0)$. The expression for the nominal strain can be rewritten as $\ell_f/\ell_0 = n + 1$. When the latter is substituted into the former, the relationship between the two strains can be expressed in the two forms

$$\varepsilon = \ln (n + 1) \qquad \exp (\varepsilon) = n + 1 \tag{7.19}$$

7.11.2 True and Nominal Stress

The definition of true stress is $\sigma = L/A_i$. From constancy of volume it is found that $A_i = A_0(\ell_0/\ell_i)$, so that

$$\sigma = \frac{L}{A_0} \left(\frac{\ell_i}{\ell_0} \right)$$

which is the same as

$$\sigma = \begin{cases} S(n + 1) \\ S \exp (\varepsilon) \end{cases} \tag{7.20}$$

7.11.3 Strain-Strengthening Exponent and Maximum-Load Strain

One of the more useful of the strength-stress-strain relationships is the one between the strain-strengthening exponent and the strain at maximum load. It is also the simplest, since the two are numerically equal, that is, $m = \varepsilon_u$. This relation is derived on the basis of the load-deformation curve shown in Fig. 7.20. The load at any point along this curve is equal to the product of the true stress on the specimen and the corresponding area. Thus

$$L = \sigma A$$

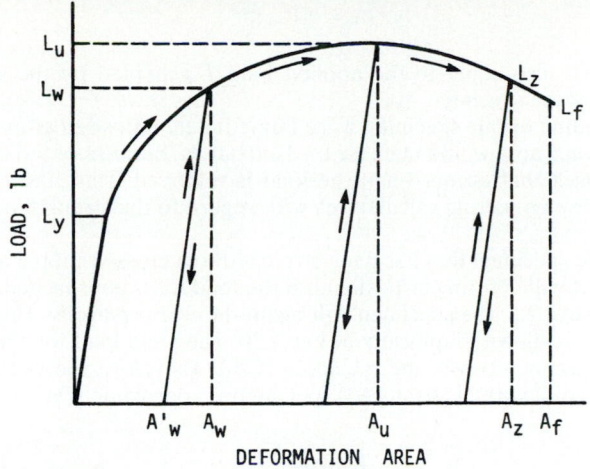

FIGURE 7.20   A typical load-deformation curve showing unloading and reloading cycles.

Now, since

$$\sigma = \sigma_0 \varepsilon^m$$

and

$$\varepsilon = \ln \frac{A_0}{A} \qquad \text{or} \qquad A = \frac{A_0}{\exp(\varepsilon)}$$

the load-strain relationship can be written as

$$L = \sigma_0 A_0 \varepsilon^m \exp(-\varepsilon)$$

The load-deformation curve shown in Fig. 7.20 has a maximum, or zero-slope, point on it. Differentiating the last equation and equating the result to zero gives the simple expression $\varepsilon = m$. Since this is the strain at the ultimate load, the expression can be written as

$$\varepsilon_u = m \tag{7.21}$$

7.11.4   Yield Strength and Percent Cold Work

The stress-strain characteristics of a material obtained from a tensile test are shown in Fig. 7.18. In the region of plastic deformation, the relationship between stress and strain for most materials can be approximated by the equation $\sigma = \sigma_0 \varepsilon^m$. When a load is applied to a tensile specimen that causes a given amount of cold work W (which is a plastic strain of ε_W), the stress on the specimen at the time is σ_W and is defined as

$$\sigma_w = \sigma_0 (\varepsilon_w)^m \tag{7.22}$$

Of course, σ_w is also equal to the applied load L_w divided by the actual cross-sectional area of the specimen A_w.

If the preceding tensile specimen were immediately unloaded after reading L_w, the cross-sectional area would increase to A'_w from A_w because of the elastic recovery or springback that occurs when the load is removed. This elastic recovery is insignificant for engineering calculations with regard to the strength or stresses on a part.

If the tensile specimen that has been stretched to a cross-sectional area of A'_w is now reloaded, it will deform elastically until the load L_w is approached. As the load is increased above L_w, the specimen will again deform plastically. This unloading-reloading cycle is shown graphically in Fig. 7.20. The yield load for this previously cold-worked specimen before the reloading is A'_w. Therefore, the yield strength of the previously cold-worked (stretched) specimen is approximately

$$(S_y)_w = \frac{L_w}{A'_w}$$

But since $A'_w = A_w$, then

$$(S_y)_w = \frac{L_w}{A_w}$$

By comparing the preceding equations, it is apparent that

$$(S_y)_w \cong \sigma_w$$

And by substituting this last relationship into Eq. (7.22), we get

$$(S_y)_w = \sigma_0 (\varepsilon_w)^m \tag{7.23}$$

Thus it is apparent that *the plastic portion of the $\sigma - \varepsilon$ curve is approximately the locus of yield strengths for a material as a function of the amount of cold work.* This relationship is valid only for the axial tensile yield strength after tensile deformation or for the axial compressive yield strength after axial deformation.

7.11.5 Tensile Strength and Cold Work

It is believed by materials and mechanical-design engineers that the only relationships between the tensile strength of a cold-worked material and the amount of cold work given it are the experimentally determined tables and graphs that are provided by the material manufacturers and that the results are different for each family of materials. However, on the basis of the concepts of the tensile test presented here, two relations are derived in Ref. [7.1] between tensile strength and percent cold work that are valid when the prior cold work is tensile. These relations are derived on the basis of the load-deformation characteristics of a material as represented in Fig. 7.20. This model is valid for all metals that do not strain age.

Here we designate the tensile strength of a cold-worked material as $(S_u)_w$, and we are interested in obtaining the relationship to the percent cold work W. For any specimen that is given a tensile deformation such that A_W is equal to or less than A_u, we have, by definition, that

$$(S_u)_W = \frac{L_u}{A_W'}$$

And also, by definition,

$$L_u = A_0(S_u)_0$$

where $(S_u)_0$ = tensile strength of the original non-cold-worked specimen and A_0 = its original area.

The percent cold work associated with the deformation of the specimen from A_0 to A_W' is

$$W = \frac{A_0 - A_W'}{A_0}(100) \qquad \text{or} \qquad w = \frac{A_0 - A_W'}{A_0}$$

where $w = W/100$. Thus

$$A_W' = A_0(1 - w)$$

By substitution into the first equation,

$$(S_u)_W = \frac{A_0(S_u)_0}{A_0(1 - w)} = \frac{(S_u)_0}{1 - w} \qquad (7.24)$$

Of course, this expression can also be expressed in the form

$$(S_u)_W = (S_u)_0 \exp(\varepsilon) \qquad (7.25)$$

Thus *the tensile strength of a material that is prestrained in tension to a strain less than its ultimate load strain is equal to its original tensile strength divided by one minus the fraction of cold work.* This relationship is valid for deformations less than the deformation associated with the ultimate load. That is, for

$$A_W \leq A_u \qquad \text{or} \qquad \varepsilon_W \leq \varepsilon_u$$

Another relationship can be derived for the tensile strength of a material that has been previously cold-worked in tension by an amount greater than the deformation associated with the ultimate load. This analysis is again made on the basis of Fig. 7.20. Consider another standard tensile specimen of 1020 steel that is loaded beyond L_u (12 000 lb) to some load L_z, say, 10 000 lb. If dead weights were placed on the end of the specimen, it would break catastrophically when the 12 000-lb load was applied. But if the load had been applied by means of a mechanical screw or a hydraulic pump, then the load would drop off slowly as the specimen is stretched. For this particular example the load is considered to be removed instantly when it drops to L_z or 10 000 lb. The unloaded specimen is not broken, although it may have a "necked" region, and it has a minimum cross-sectional area $A_z = 0.100$ in^2 and a diameter of 0.358 in. Now when this same specimen is again loaded in tension, it deforms elastically until the load reaches L_z (10 000 lb) and then it deforms plastically. But L_z is also the maximum value of load that this specimen reaches on reloading. It never again will support a load of $L_u = 12\ 000$ lb. On this basis, the yield strength of this specimen is

$$(S_y)_W = \frac{L_z}{A_z'} = \frac{10\,000}{0.101} = 99\ 200 \text{ psi}$$

And the tensile strength of this previously deformed specimen is

$$(S_u)_W = \frac{L_z}{A_z} = \frac{10\,000}{0.101} = 99\ 200 \text{ psi}$$

7.11.6   Ratio of Tensile Strength to Brinell Hardness

It is commonly known by mechanical-design engineers that the tensile strength of a steel can be estimated by multiplying its Brinell hardness number by 500. As stated earlier, this fact led to the wide acceptance of the Brinell hardness scale. However, this ratio is not 500 for all materials—it varies from as low as 450 to as high as 1000 for the commonly used metals. The ratio of the tensile strength of a material to its Brinell hardness number is identified by the symbol K_B, and it is a function of both the load used to determine the hardness and the strain-strengthening exponent of the material.

Since the Brinell hardness number of a given material is not a constant but varies in proportion to the applied load, it then follows that the proportionality coefficient K_B is not a constant for a given material, but it too varies in proportion to the load used in determining the hardness. For example, a 50 percent cobalt alloy (L605 or HS25) has a Brinell hardness number of 201 when tested with a 3000-kg load and a hardness of only 150 when tested with a 500-kg load. Since the tensile strength is about 145 000 psi for this annealed alloy, the value for K_B is about 970 for the low load and about 730 for the high load.

Since the material is subjected to considerable plastic deformation when both the tensile strength and the Brinell hardness are measured, these two values are influenced by the strain-strengthening exponent m for the material. Therefore, K_B must also be a function of m.

Figure 7.21 is a plot of experimental data obtained by this author over a number of years that shows the relationships between the ratio K_B and the two variables strain-strengthening exponent m and diameter of the indentation, which is a function of the applied load. From these curves it is apparent that K_B varies directly with m and inversely with the load or diameter of the indentation d. The following examples will illustrate the applicability of these curves.

A test was conducted on a heat of alpha brass to see how accurately the tensile strength of a material could be predicted from a hardness test when the strain-strengthening exponent of the material is not known. Loads varying from 200 to 2000 kg were applied to a 10-mm ball, with the following results:

Load, kg	200	500	1000	1500	2000
Diameter, mm	2.53	3.65	4.82	5.68	6.30

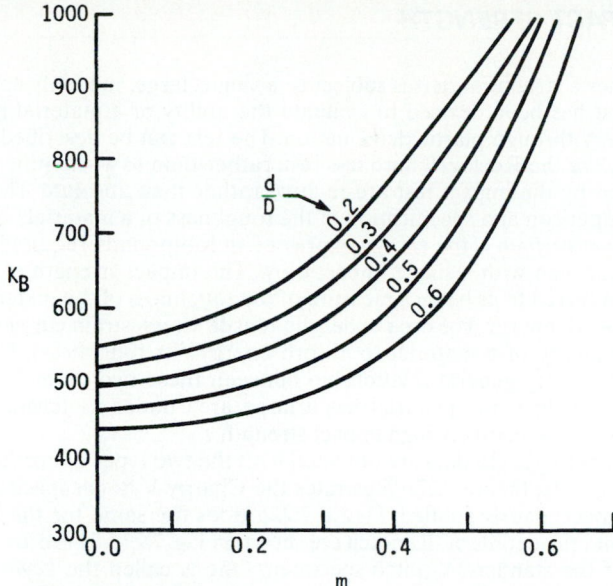

FIGURE 7.21 Relationships between the S_u/H_B ratio (K_B) and the strain-strengthening exponent m. D = diameter of the ball, and d = diameter of the indentation. Data are based on experimental results obtained by the author.

When plotted on log-log paper, these data fall on a straight line having a slope of 2.53, which is the Meyer strain-hardening exponent n. The equation for this straight line is

$$L = 18.8d^{2.53}$$

Since, for some metals, $m = n - 2$, the value of m is 0.53.

For ease in interpreting Fig. 7.21, the load corresponding to an indentation of 3 mm is calculated from Eq. (7.2) as 43. K_B can now be determined from Fig. 7.21 as 890. Thus the tensile strength is $S_u = K_B H_B = 890(43) = 38\,300$ psi. In a similar fashion, the load for a 5-mm diameter is 110 kg, and the corresponding Brinell hardness number is 53. From Fig. 7.21, the value of K_B is found to be 780, and the tensile strength is estimated as $S_u = K_B H_B = 780(53) = 41\,300$ psi. The average value of these two calculated tensile strengths is 39 800 psi. The experimentally determined value of the tensile strength for this brass was 40 500 psi, which is just 2 percent lower than the predicted value.

As another example, consider the estimation of tensile strength for a material when its typical strain-strengthening exponent is known. Annealed 3003 aluminum has an average m value of 0.28. What is the tensile strength of a heat that has a Brinell hardness number of 28 when measured with a 500-kg load? The diameter of the indentation for this hardness number is 4.65. Then from Fig. 7.21 the value of K_B is determined as 535. The tensile strength can then be calculated as $S_u = K_B H_B = 535(28) = 15\,000$ psi.

7.12 IMPACT STRENGTH

In some cases a structural part is subject to a single, large, suddenly applied load. A standard test has been devised to evaluate the ability of a material to absorb the impact energy through plastic deformation. The test can be described as a technological one, like the Rockwell hardness test, rather than as a scientific one. The values obtained by the impact test are relative rather than absolute. They serve as a basis of comparison and specification of the toughness of a material.

The *impact strength* is the energy, expressed in footpounds, required to fracture a standard specimen with a single-impact blow. The impact strength of a material is frequently referred to as being a measure of the toughness of the material, that is, its ability to absorb energy. The area under the tensile stress–strain curve is also a measure of the ability of a material to absorb energy (its toughness). Unfortunately, there is only a very general relationship between these two different measures of toughness; namely, if the material has a large area under its tensile stress–strain curve, it also has a relatively high impact strength.

Most impact-strength data are obtained with the two types of notched specimens shown in Fig. 7.22. Figure 7.22*a* illustrates the Charpy V-notch specimen as well as how the impact load is applied. Figure 7.22*b* does the same for the Izod V-notch specimen, and the details of the notch are shown in Fig. 7.22*c*. There are several modifications of the standard V-notch specimen. One is called the *keyhole notch* and another the *U-notch*. Both have a 1-mm radius at the bottom rather than the 0.25-mm radius of the V-notch. There is no correlation between the various types of notch-bar impact-strength values. However, the Charpy V-notch impact-strength value is considerably greater than the Izod V-notch value, particularly in the high toughness range.

The impact-testing machine consists of a special base mounted on the floor to support the specimen and a striking hammer that swings through an arc of about 32-in radius, much like a pendulum. When the hammer is "cocked" (raised to a locked elevation), it has a potential energy that varies between 25 and 250 ft · lb, depending on the mass of the hammer and the height to which it is raised. When the hammer is released and allowed to strike the specimen, a dial registers the energy that was absorbed by the specimen. The standards specify that the striking velocity must be in the range of 10 to 20 ft/s because velocities outside this range have an effect on the impact strength.

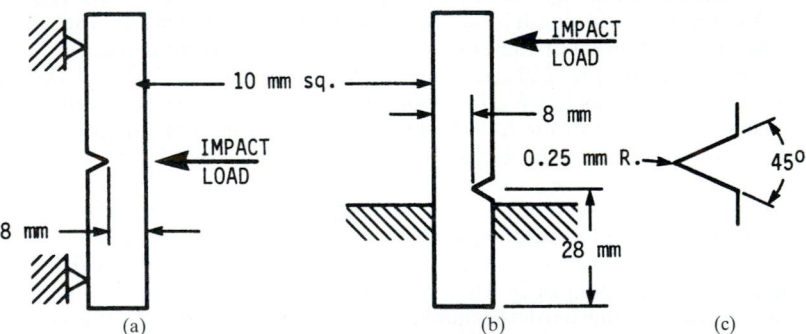

FIGURE 7.22 Impact tests and specimens. (*a*) Charpy $L = 55$ mm; (*b*) Izod $L = 75$ mm; (*c*) details of the notch.

The impact strengths of some materials, particularly steel, vary significantly with the testing temperature. Figure 7.23 shows this variation for a normalized AISI 1030 steel. At the low testing temperature the fracture is of the cleavage type, which has a bright, faceted appearance. At the higher temperatures the fractures are of the shear type, which has a fibrous appearance. The *transition temperature* is that temperature that results in 50 percent cleavage fracture and 50 percent shear fracture, or it may be defined as the temperature at which the impact strength shows a marked drop. The *nil-ductility temperature* is the highest temperature at which the impact strength starts to increase above its minimum value. These two temperatures are shown in Fig. 7.23.

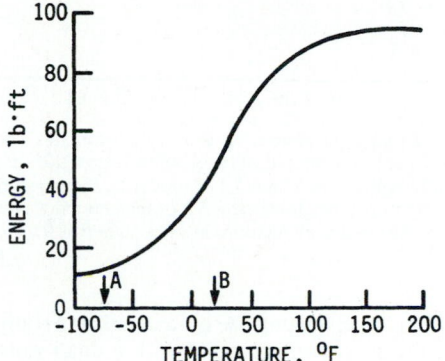

FIGURE 7.23   Charpy V-notch impact strength of 1030 steel versus temperature. A = nil-ductility temperature; B = transition temperature.

7.13   CREEP STRENGTH

A part may fail with a load that induced stresses in it that lie between the yield strength and the tensile strength of the material even if the load is steady and constant rather than alternating and repeating as in a fatigue failure. This type of constant loading causes the part to elongate or creep. The failure point may be when the part stretches to some specified length, or it may be when the part completely fractures.

The *creep strength* of a material is the value of nominal stress that will result in a specified amount of elongation at a specific temperature in a given length of time. It is also defined as the value of nominal stress that induces a specified creep rate at a specific temperature. The creep strength is sometimes called the *creep limit*. The *creep rate* is the slope of the strain-time creep curve in the steady-creep region, referred to as a *stage 2 creep*. It is illustrated in Fig. 7.24.

Most creep failures occur in parts that are exposed to high temperatures rather than room temperature. The stress necessary to cause creep at room temperature is considerably higher than the yield strength of a material. In fact, it is just slightly less than the tensile strength of a material. The stress necessary to induce creep at a temperature that is higher than the recrystallization temperature of a material, however, is very low.

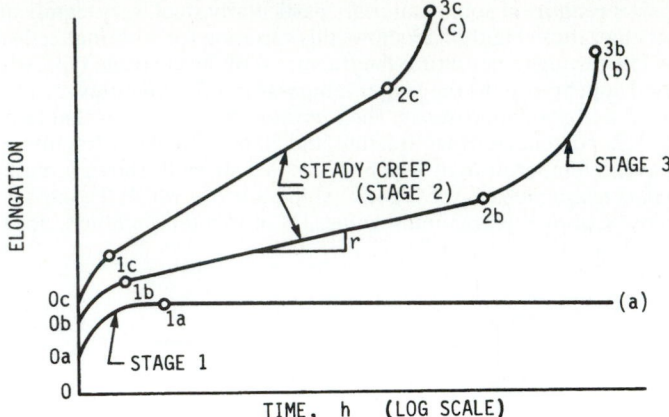

FIGURE 7.24   Creep data plotted on semilog coordinates. (*a*) Low stress (slightly above S_y) or low temperature (well below recrystallization); (*b*) moderate stress (midway between S_y and S_u) or moderate temperature (at recrystallization); (*c*) high stress (slightly below S_u) or high temperature (well above recrystallization). The elastic elongations are designated as 0*a*, 0*b*, and 0*c*.

The specimens used for creep testing are quite similar to round tensile specimens. During the creep test the specimen is loaded with a dead weight that induces the required nominal stress applied throughout the entire test. The specimen is enclosed in a small round tube-type furnace to maintain a constant temperature throughout the test, and the gauge length is measured after various time intervals. Thus the three variables that affect the creep rate of the specimen are (1) nominal stress, (2) temperature, and (3) time.

Figure 7.24 illustrates the most common method of presenting creep-test data. Three different curves are shown. Curve (*a*) is typical of a creep test conducted at a temperature well below the recrystallization temperature of the material (room temperature for steel) and at a fairly high stress level, slightly above the yield strength. Curve (*a*) is also typical of a creep test conducted at a temperature near the recrystallization temperature of a material but at a low stress level. Curve (*c*) is typical of either a high stress level, such as one slightly below S_u, at a low temperature, or else a low stress level at a temperature significantly higher than the recrystallization temperature of the material. Curve (*b*) illustrates the creep rate at some intermediate combination of stress and temperature.

A creep curve consists of four separate parts, as illustrated with curve (*b*) in Fig. 7.24. These are explained as follows:

1. An initial elastic extension from the origin 0 to point 0*b*.

2. A region of primary creep, frequently referred to as *stage 1 creep*. The extension occurs at a decreasing rate in this portion of the creep curve.

3. A region of secondary creep, frequently called *stage 2 creep*. The extension occurs at a constant rate in this region. Most creep design is based on this portion of the creep curve, since the creep rate is constant and the total extension for a given number of hours of service can be easily calculated.

4. A region of tertiary creep or *stage 3 creep*. The extension occurs at an increasing rate in this region until the material fractures.

Another practical way of presenting creep data is illustrated in Fig. 7.25, which is a log-log plot of nominal stress versus the second-stage creep rate expressed as percent per hour with the temperature as a parameter. Figure 7.26 illustrates still another type of plot that is used to present creep data where both the stress and temperature are drawn on cartesian coordinates.

The mechanism of creep is very complex inasmuch as it involves the movements of vacancies and dislocations, strain hardening, and recrystallization, as well as grain-boundary movements. At low temperatures, creep is restricted by the pile-up of dislocations at the grain boundaries and the resulting strain hardening. But at higher temperatures, the dislocations can climb out of the original slip plane and thus permit further creep. In addition, recrystallization, with its resulting lower strength, permits creep to occur readily at high temperatures.

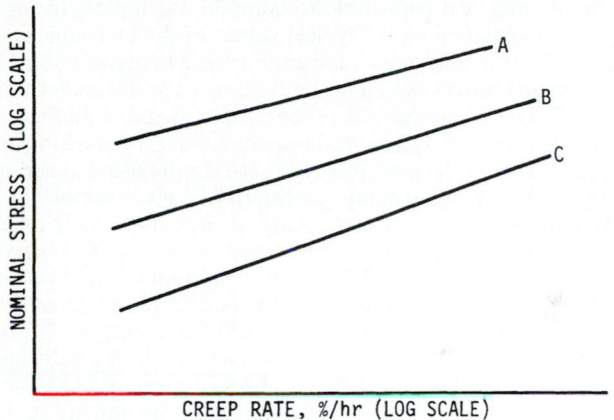

FIGURE 7.25   Second-stage creep rate versus nominal stress. *A, B,* and *C* are for low, medium, and high temperatures, respectively.

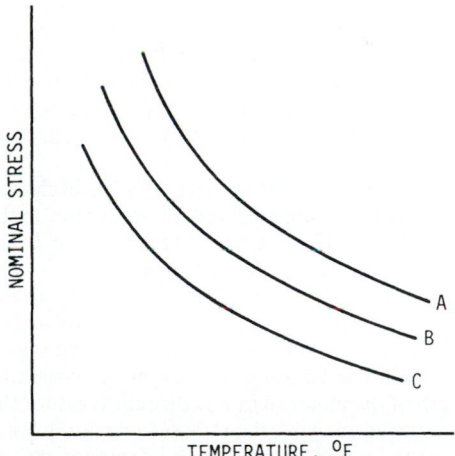

FIGURE 7.26   Second-stage creep rate versus temperature and nominal stress. *A,* 1%/h creep rate; *B,* 0.1%/h creep rate; *C,* 0.001%/h creep rate.

As explained in an earlier section, the grain-boundary material is stronger than the material at the interior portions at low temperatures, but the opposite is true at high temperatures. The temperature where these two portions of the grains are equal is called the *equicohesive temperature*. Consequently, parts that are exposed to high temperatures have lower creep rates if they are originally heat-treated to form coarse grains.

7.14 MECHANICAL-PROPERTY DATA

The number of different combinations of thermal and mechanical treatments for each family of materials plus the large number of individual material compositions within one family makes it impossible to compile a complete listing in one handbook. For more information on the typical values of the mechanical properties of a specific material, you should consult the engineering manuals published by the various material manufacturers as well as the references at the end of this chapter. The total number of pages of mechanical-property data listed in all the sources cited at the end of this chapter runs into the thousands, making it impossible to include it all in one handbook. The mechanical properties are simply listed as the experimentally obtained values for each of the many conditions of a given metal.

This section includes some select mechanical-property data in tabular form for a variety of materials in Table 7.5. Both the format of the data and the actual mechanical properties listed are different from the traditional handbook presentations. There are several advantages to presenting the data in this manner. One of the main advantages is that it requires much less space. For example, as shown in Table 7.5, only five numbers have to be recorded for each metal in its original non-cold-worked condition. These numbers refer to the five primary mechanical properties. From these five original values it is very easy to calculate the new strengths of the metal after any specific amount of cold work or to construct tables or graphs of strength versus percent cold work. This is in sharp contrast to the large amount of space required to store the tables or graphs in any handbook or manual of properties.

A second advantage of presenting the data in this manner is that it is possible to make use of the rules and relationships included in Chap. 8 to calculate both the compressive and tensile properties in all directions in a cold-worked part. This is extremely important because for some materials (those having a high strain-strengthening exponent m) it is possible to have a compressive yield strength for a given amount of cold work that is one-half the tensile yield strength that may be tabulated in a materials handbook.

Table 7.6 includes some of the properties of very-high-strength steels. To illustrate the versatility of documenting mechanical properties in the format of Table 7.5, consider the annealed 303 stainless steel listed in the table. In the annealed condition it has the following listed properties: $S_y = 35$ kpsi, $S_u = 87.3$ kpsi, strength coefficient $\sigma_0 = 205$ kpsi, strain-strengthening exponent $m = 0.51$, and fracture strain $\varepsilon_f = 1.16$. In an upsetting operation, a 2-in-diameter bar is upset to a diameter of $2\frac{1}{2}$ in for a length of $1\frac{1}{2}$ in prior to having splines machined on the end. Since the splines are cantilevered beams subject to bending stresses in the circumferential or transverse direction, the strength of the material in this direction rather than in the axial or longitudinal direction is required. Also, the stresses are compressive on one side of the splines and tensile on the other. Therefore, the designer should know both the tensile and compressive strengths in the transverse direction. The following sample calculations will demonstrate how this can be done.

TABLE 7.5 Tensile Properties of Some Metals[a]

Material	Condition	Strength			Strain-strengthening exponent m	Fracture strain ε_f
		Yield S_y, kpsi	Ultimate S_u, kpsi	Coefficient σ_0, kpsi		
Carbon and alloy steels						
1002	1500°F @ 1 h, A 0.032 in	22.0	39.5	76.0	0.29	1.25
1002[a]	1800°F @ 1 h, A	19.0	42.0	78.0	0.27	1.25
1008 DQ	As rec'd 0.024 in	25.0	39.0	70.0	0.24	1.20
1008 DQ	As above—trans	27.0	43.0	70.0	0.24	1.10
1008 DQ	1600°F @ 1 h A	26.5	40.0			
1010	0.024-in CD strip	33.2	47.5	84.0	0.23	1.20
1010	As above—trans	36.8	48.5	88.0	0.26	1.00
1010	1600°F @ 1 h A	28.6	44.2	82.0	0.23	1.20
1010	As above—trans	29.1	43.8	82.0	0.23	1.20
1018	A	32.0	49.5	90.0	0.25	1.05
1020	HR	42.0	66.2	115.0	0.22	0.90
1045	HR	60.0	92.5	140.0	0.14	0.58
1144	A	52.0	93.7	144.0	0.14	0.49
1144[b]	A	50.0	93.7	144.0	0.14	0.05
1212	HR	28.0	61.5	110.0	0.24	0.85
4340	HR	132.0	151.0	210.0	0.09	0.45
52100	Spher. A	80.0	101.0	165.0	0.18	0.58
52100	1500°F A	131.0	167.0	210.0	0.07	0.40
Stainless steels						
18-8	1600°F @ 1 h A	37.0	89.5	210.0	0.51	1.08
18-8	1800°F @ 1 h A	37.5	96.5	230.0	0.53	1.38
302	1800°F @ 1 h A	34.0	92.4	210.0	0.48	1.20
303	A	35.0	87.3	205.0	0.51	1.16
304	A	40.0	82.4	185.0	0.45	1.67
202	1900°F @ 1 h A	55.0	105.0	195.0	0.30	1.00
17-4 PH	1100°F aged	240.0	246.0	260.0	0.01	0.65
17-4 PH	A	135.0	142.0	173.0	0.05	1.20
17-7 PH	1050°F aged	155.0	185.0	225.0	0.05	0.90
17-7 PH	900°F aged	245.0	255.0	300.0	0.04	0.50
440 C	Solution H T	63.5	107.0	153.0	0.11	0.36
440 C	A 1600°F–50°F/h	67.6	117.0	180.0	0.14	0.12
Aluminum alloys						
1100	900°F @ 1 h A	4.5	12.1	22.0	0.25	2.30
3003	800°F @ 1 h A	6.0	15.0	29.0	0.30	1.50
2024[c]	T-351	52.0	68.8	115.0	0.20	0.37
2024	T-4	43.0	64.8	100.0	0.15	0.18
7075	800°F A	14.3	33.9	61.0	0.22	0.53
7075	T-6	78.6	86.0	128.0	0.13	0.18
2011	800°F @ 1 h A	7.0	25.2	41	0.18	0.35
2011	T-6	24.5	47.0	90	0.28	0.10

TABLE 7.5   Tensile Properties of Some Metals[a] (*Continued*)

Material	Condition	Strength			Strain-strengthening exponent m	Fracture strain ε_f
		Yield S_y, kpsi	Ultimate S_u, kpsi	Coefficient σ_0, kpsi		
Magnesium alloys						
HK31 XA	800°F @ 1 h A	19.0	25.5	49.5	0.22	0.33
HK31 XA	H-24	31.0	36.2	48.0	0.08	0.20
Copper alloys						
ETP Cu	100°F @ 1 h A	4.7	31.0	78.0	0.55	1.19
ETP Cu	1250°F @ 1 h A	4.6	30.6	72.0	0.50	1.21
ETP Cu	1500°F @ 1 h A	4.2	30.0	68.0	0.48	1.26
OFHC Cu	1250°F @ 1 h A	5.3	33.1	67.0	0.35	1.00
90-10 brass	As rec'd[c]	12.8	38.0	85.0	0.43	
90-10 brass	1200°F @ 1 h A	8.4	36.4	83.0	0.46	
90-10 brass	As above, 10% CW, 1200°F A	6.9	35.0	87.0	1.83	
80-20 brass	1200°F @ 1 h A	7.2	35.8	84.0	0.48	
80-20 brass	As above, 10% CW, 1200°F A	6.4	34.6	85.0	0.51	1.83
70-30 brass	1200°F @ 1 h A	12.1	44.8	112.0	0.59	
70-30 brass	As above, 10% CW, 1200°F A	10.7	43.4	107.0	0.59	1.62
70-30 brass[b]	1000°F @ 1 h A	11.5	45.4	110.0	0.56	1.50
70-30 brass[b]	1200°F @ 1 h A	10.5	44.0	105.0	0.52	1.55
70-30 brass[b]	1400°F @ 1 h A	8.8	42.3	105.0	0.60	1.60
70-30 leaded brass	1250°F @ 1 h A	11.0	45.0	105.0	0.50	1.10
Naval brass[d]	1350°F @ $\frac{1}{2}$ h A	17.0	54.5	125.0	0.48	1.00
Naval brass[d]	1350°F @ $\frac{1}{2}$ h WQ	27.0	66.2	135.0	0.37	0.50
Naval brass[d]	850°F @ $\frac{1}{2}$ h A	17.5	56.0	125.0	0.48	0.90
Naval brass[d]	850°F @ $\frac{1}{2}$ h WQ	31.5	64.5	135.0	0.37	0.80
Naval brass[d]	1500°F @ 3 h A	11.0	48.0			0.74
Nickel alloys						
Ni 200	1700°F @ $\frac{1}{4}$ h WQ	16.2	72.1	150.0	0.375	1.805
99.44% Ni	CD, A	20.5	73.7	160.0	0.40	1.47
Monel 400	1700°F @ $\frac{1}{4}$ h WQ	26.5	77.7	157.0	0.337	1.184
Monel K500	1700°F @ $\frac{1}{4}$ h WQ	34.4	92.6	182.0	0.32	1.305
Inconel 600	1700°F @ $\frac{1}{4}$ h WQ	46.6	102.5	201.0	0.315	1.14
Inconel 625	1700°F @ $\frac{1}{2}$ h WQ	77.1	139.7	297.0	0.395	0.75
Inconel 718	1750°F @ 20 min AC	43.6	99.4	205.0	0.363	1.337
Inconel X750	2050°F @ 45 min WQ	36.4	106.4	230.0	0.415	1.27
Incoloy 800	2050°F @ 2 h AC	22.2	77.1	169.0	0.420	1.262
Incoloy 825	1700°F @ 20 min WQ	66.7	138.0	283.0	0.353	0.715
Ni, 2% Be	1800°F sol. T WQ	41.0	104.0	222.0	0.39	1.00

TABLE 7.5 Tensile Properties of Some Metals[a] (*Continued*)

Material	Condition	Yield S_y, kpsi	Ultimate S_u, kpsi	Coefficient σ_0, kpsi	Strain-strengthening exponent m	Fracture strain ε_f
				Strength		
Nickel alloys (Continued)						
Ni, 2% Be	As above + 1070°F @ 2 h aged	140.0	195.0	300.0	0.15	0.18
Ni, 15.8% Cr, 7.2% Fe	A	36.0	90.0	203.0	0.45	0.92
Special alloys						
Cobalt alloy[f]	2250°F solution HT	65.0	129.0	300.0	0.50	0.51
Cobalt alloy[f]	As above—trans[c]	65.0	129.0	300.0	0.50	0.40
Cobalt alloy[e,g]	As rec'd, annealed	62.8	119.5	283.0	0.52	0.75
Cobalt alloy[e,g]	Machined, 2250°F sol. HT	48.0	112.5	283.0	0.62	0.70
Cobalt alloy[e,g]	2250°F sol. HT, 925°F aged	48.0	107.5	270.0	0.63	1.00
Molybdenum	Extr'd A	49.5	70.7	106.0	0.12	0.38
Vanadium	A	45.0	63.0	97.0	0.17	1.10

[a]All values are for longitudinal specimens except as noted. These are values obtained from only one or two heats. The values will vary from heat to heat because of the differences in composition and annealing temperatures. The fracture strain may vary as much as 100 percent.
[b]$\frac{3}{4}$-in-diameter bar.
[c]Tensile specimen machined from a 4-in-diameter bar transverse to rolling direction.
[d]Specimens cut from $\frac{1}{2}$-in hot-rolled plate.
[e]$\frac{1}{2}$-in-diameter bar.
[f]HS 25 or L 605 alloy; 50 Co, 20 Cr, 15 W, 10 Ni, 3 Fe.
[g]Eligiloy; 50 Co, 20 Cr, 15 Ni, 7 Mo, 15 Fe.
SOURCE: From Datsko [7-1].

The strain associated with upsetting a 2-in-diameter bar to a 2½-in-diameter can be calculated by means of Eq. (7.12). Thus

$$\varepsilon = -\ln\left(\frac{2}{2.5}\right)^2 = 0.45$$

The negative sign in front of the function is needed because Eq. (7.12) is for tensile deformation, whereas in this problem the deformation is axial compression. The equivalent amount of cold work can be calculated from Eq. (7.14) as 36.2 percent.

The axial compressive yield strength can be approximated by means of Eq. (7.23). Thus

$$(S_y)_c = \sigma_0(\varepsilon_w)^m = 205(0.45)^{0.51} = 136 \text{ kpsi}$$

If one were to interpolate in a table of yield strength versus cold work in a handbook, this value of 136 kpsi would be approximately the value that would be

TABLE 7.6 Properties of Some High-Strength Steels

AISI number	Processing[a]	Brinell hardness H_B	Modulus of elasticity E, Mpsi	Yield strength[b] S_y, kpsi	Ultimate strength S_u, kpsi	Reduction in area, %	True fracture strength[c] σ_F, kpsi	True fracture ductility[c] ϵ_F	Strain-strengthening exponent m
1045	Q & T 80°F	705	29	265T[d] 300C[d]	300	2	310T 420C	0.02	0.186
1045	Q & T 360°F	595	30	270	325	41	430/ 395	0.52	0.071
1045	Q & T 500°F	500	30	245	265	51	370/ 330	0.71	0.047
1045	Q & T 600°F	450	30	220	230	55	345/ 305	0.81	0.041
1045	Q & T 720°F	390	30	185	195	59	315/ 270	0.89	0.044
4142	Q & T 80°F	670	29	235T 275C	355	6	375	0.06	0.136
4142	Q & T 400°F	560	30	245	325	27	405/ 385	0.31	0.091
4142	Q & T 600°F	475	30	250	280	35	340/ 315	0.43	0.048
4142	Q & T 700°F	450	30	230	255	42	320/ 290	0.54	0.043
4142	Q & T 840°F	380	30	200	205	48	295/ 265	0.66	0.051
4142[e]	Q & D 550°F	475	29	275T 225C	295	20	310/ 300	0.22	0.101T 0.060C
4142	Q & D 650°F	450	29	270T 205C	280	37	330/ 305	0.46	0.016T 0.070C
4142	Q & D 800°F	400	29	210T[f] 175C	225	47	305/ 275	0.63	0.032T 0.085C

[a]AISI 1045: Cold drawn to $\frac{7}{8}$-in rounds from hot-rolled rod. Austenized 1500°F (oxidizing atmosphere) 20 min, water quenched at 70°F. AISI 4142: Cold-drawn to $\frac{7}{8}$-in rounds from annealed rod. Austenized at 1500°F (neutral atmosphere), quenched in agitated oil at 180°F. AISI 4142 Def: Austenized at 1500°F, oil quenched. Reheated in molten lead, drawn 14 percent through die at reheating temperature to $\frac{5}{8}$-in rods.

[b]0.2 percent offset method.

[c]Bridgman's correction for necking.

[d]T, tension; C, compression.

[e]Deformed 14 percent.

[f]Proportional limit in tension.

SOURCE: Data from R. W. Landgraf, *Cyclic Deformation and Fatigue Behavior of Hardened Steels*, Report no. 320, Dept. of Theoretical and Applied Mechanics, University of Illinois, Urbana, 1968.

obtained for 36 percent cold work. And the handbook would not indicate whether it was a compressive or tensile yield strength, nor in what direction it was applied.

However, for this problem, the designer really needs both the tensile and compressive yield strengths in the transverse, i.e., circumferential, direction. These values can be closely approximated by means of Table 8.1. The tensile yield strength in the transverse direction is designated by the code $(S_y)_{tTt}$, which is in group 2 of Table 8.1. Since the bar was given only one cycle of deformation (a single upset), ε_{qus} is 0.45 and the tensile yield strength is calculated to be 123 kpsi. The compressive yield strength in the transverse direction is designated by the code $(S_y)_{cTt}$, which is in group 4 of Table 8.1. The compressive yield strength is then calculated to be $0.95(S_y)_{tTt} = 0.95(123) = 117$ kpsi; this is 14 percent lower than the 136 kpsi that would normally be listed in a materials handbook.

In some design situations, the actual value of the yield strength in a given part for a specific amount of cold work may be 50 percent less than the value that would be listed in the materials handbook. In order to have a reliable design, the designer must be able to determine the strength of the material in a part in the direction and sense of the induced stresses. The information in this chapter and in Chap. 8 makes it possible for the design engineer to make a reasonable prediction of the mechanical properties of a fabricated part. However, it must be recognized that the original non-cold-worked properties of a given metal vary from heat to heat, and that the calculations are valid only for a part having the original properties that are used in the calculations.

7.15 NUMBERING SYSTEMS[†]

7.15.1 AISI and SAE Designation of Steel

Carbon and alloy steels are specified by a code consisting of a four-digit (sometimes five) number, as illustrated below with the substitution of the letters X, Y, and Z for the numbers. A steel specification of XYZZ (or XYZZZ) has the following meaning:

X indicates the type of alloy or alloys present.

Y indicates the total percent of the alloys present.

ZZ (or ZZZ) indicates the "points" of carbon in the steel (points of carbon equals the percent carbon times 100). For example, if ZZ is 40, then the steel has 0.40 percent carbon (C). If ZZZ is 120, then the steel has 1.20 percent carbon.

Table 7.7 identifies the number X corresponding to the alloy or alloys present. In addition, the following two special classes are included. A resulfurized free-machining steel is identified as 11ZZ and 12ZZ. These steels have a high sulfur content, which combines with the manganese to form the compound manganese sulfide. It is the presence of this compound that makes the steel more machinable. The 13ZZ and 15ZZ groups are plain carbon steels that have high and moderate amounts of manganese, respectively.

[†] This section presents the numbering systems now in general use in order to correspond with those used in other sections of this handbook. See Ref. [7.2] for details of the unified numbering system (UNS).

TABLE 7.7 Alloy Designations for Steels

Number X	Alloying elements
1	None (plain carbon)
2	Nickel
3	Nickel-chromium
4	Molybdenum-nickel-chromium
5	Chromium
6	Chromium-vanadium
8	Nickel-chromium-molybdenum
9	Silicon-manganese

Some examples:

2130 is a steel with 1 percent nickel and 0.3 percent carbon.

4340 is a steel with a total of 3 percent Mo, Ni, and Cr and 0.4 percent C.

52100 is a steel with 2 percent Cr and 1 percent C.

7.15.2 Designation System for Aluminum Alloys

Wrought aluminum alloys are specified by a code consisting of four-digit numbers such as 1100, 2024, or 7075. To explain this code, the letters XYZZ are substituted for the four digits. The types of alloys present in the aluminum are identified by the letter X from Table 7.8.

The second digit in the code, Y, indicates alloy modifications. When Y is zero, it indicates the original alloy, or in the 1YZZ series it indicates that the alloy is made to the standard impurity limits. When Y is any digit from 1 to 9, it indicates that a modification has been made to the original alloy and then designates which of the sequential changes were made. For example, 7075 refers to the original zinc alloy, whereas 7175 and 7475 refer to the first and fourth modifications made to it.

The third and fourth digits (ZZ in the code) have no numerical significance but simply relate to the chemical composition of the alloys.

TABLE 7.8 Alloy Designations
for Wrought Aluminum Alloys

Number X	Alloying elements
1	None (99.00% Al min.)
2	Copper
3	Manganese
4	Silicon
5	Magnesium
6	Magnesium-silicon
7	Zinc

Temper Designation.   The temper designation for aluminum alloys consists of a suffix which is a letter that may be followed by several digits. The suffix is separated from the alloy designation by a hyphen or dash. For example, 7075-T4 identifies both the alloy composition and its temper. The T in this suffix identifies the tempering treatment as a heat-treating process. Table 7.9 shows the letters used to identify the type of process used in the tempering treatment.

In addition to the T temper designations, other two- or three-digit numbers have been assigned to some specific treatments to certain special alloys or types of products.

7.15.3   Designation System for Copper Alloys

The designation system for copper alloys is not based on a coded system as those for steel and aluminum alloys are. It is simply a means of defining the chemical composition of the specific alloys. Table 7.10 identifies the principal alloying elements for the common classes of copper alloys.

Temper Designation.   The temper designation for copper alloys refers to the amount of cold work given to the metal. Table 7.11 defines the amount of cold work associated with each temper designation.

TABLE 7.9   Tempering Processes and Designations for Aluminum Alloys

Designation	Process
F	As fabricated
O	Annealed
H	Strain hardened; the H is followed by two or more digits to indicate the amount of strain hardening
H1	Strain hardened only
H2	Strain hardened and partially annealed
H3	Strain hardened and stabilized
W	Solution heat treated
T	Heat treated; the T is always followed by one or more digits to specify the particular process used
T1	Cooled from a high-temperature forming process and naturally aged
T2	Cooled from a high-temperature forming process, cold worked, and naturally aged
T3	Solution heat treated, cold worked, and naturally aged
T4	Solution heat treated and naturally aged
T5	Cooled from a high-temperature forming process and artificially aged
T6	Solution heat treated and artifically aged
T7	Solution heat treated
T8	Solution heat treated, cold worked, and artifically aged
T9	Solution heat treated, artifically aged, and cold worked
T10	Cooled from a high-temperature forming process, cold worked, and artifically aged

TABLE 7.10 Designation of Copper Alloys

UNS Number	Alloy	Class or name
C10000–C13000	None	Commercially pure (ETP or OFHC)
C21000	Zn	Gilding brass (95% Cu)
C22000	Zn	Commercial bronze (90% Cu)
C23000	Zn	Red brass (85% Cu)
C24000	Zn	Low brass (80% Cu)
C26000	Zn	Cartridge brass (70% Cu)
C28000	Zn	Muntz metal (60% Cu)
C50000	Sn	Phosphor bronze
C60600–C64200	Al	Aluminum bronze
C64700–C66100	Li	Silicon bronze
C70000	Ni	Copper-nickel

TABLE 7.11 Temper Designation of Copper Alloys

Temper	Percent cold work	
	Rolled sheet	Drawn wire
¼ hard	10.9	20.7
½ hard	20.7	37.1
¾ hard	29.4	50.1
Hard	37.1	60.5
Extra hard	50.1	75.1
Spring	60.5	84.4
Extra spring	68.6	90.2
Special spring	75.1	93.8

7.15.4 Designation System for Magnesium Alloys

The designation system for magnesium alloys consists of four parts that include a combination of letters and digits. A typical example is AZ31B-H24.

The first part of the designation consists of two letters representing the two main alloying elements in order of decreasing amounts. The 10 principal alloying elements are given the following letters: A, aluminum; E, rare earth; H, thorium; K, zirconium; L, lithium; M, manganese; Q, silver; S, silicon; T, tin; and Z, zinc. Thus, in the preceding alloy, the main element is aluminum and the second one is zinc.

The second part consists of two digits corresponding to rounded-off percentages of the two main alloying elements. In the preceding example the alloy contains 3 percent aluminum and 1 percent zinc.

The third part of the designation consists of a letter that indicates the chronologic order of when that particular composition became a standard one. In the preceding example the letter B indicates that this particular alloy is the second one, having 3 percent aluminum and 1 percent zinc, that became an industry standard.

The fourth part consists of a letter preceded by a hyphen and followed by a number. It indicates the specific condition or temper that the alloy is in. Table 7.12 specifies the symbols that are used for each temper.

TABLE 7.12 Temper Designation of Magnesium Alloys

Designation	Process
F	As fabricated
O	Annealed
H10, H11	Slightly strain hardened
H23, H24, H26	Strain hardened and partially annealed
T4	Solution heat treated
T5	Solution heat treated and artificially aged
T8	Solution heated treated, cold worked, and artificially aged

REFERENCES

7.1 J. Datsko, *Materials in Design and Manufacture,* J. Datsko Consultants, Ann Arbor, Mich., 1978.

7.2 *Metals and Alloys in the Unified Numbering System,* 3d ed., Society of Automotive Engineers, Inc. (SAE), 1983.

7.3 *Metals Handbook Desk Edition,* American Society for Metals (ASM), Metals Park, Ohio, 1984.

7.4 R. M. Brick, A. W. Pense, and R. B. Gordon, *Structure and Properties of Engineering Materials,* 4th ed., McGraw-Hill, New York, 1977.

7.5 M. M. Schwartz, *Composite Materials Handbook,* McGraw-Hill, New York, 1984.

7.6 L. H. Van Vlack, *Elements of Materials Science,* 4th ed., Addison-Wesley, Reading, Mass., 1980.

CHAPTER 8

THE STRENGTH OF COLD-WORKED AND HEAT-TREATED STEELS

Charles R. Mischke, Ph.D., P.E.
Professor Emeritus of Mechanical Engineering
Iowa State University
Ames, Iowa

GLOSSARY

AR	Fractional area reduction
A	Area
B	Critical hardness for carbon content and tempering temperature, Rockwell C scale
d	Diameter
D	Tempering decrement, Rockwell C scale; carbon ideal diameter, in
D_I	Ideal critical diameter, in
DH	Distant hardness, Rockwell C scale
EJD	Equivalent Jominy distance, sixteenths of inch
f	Tempering factor for carbon content and tempering temperature
F	Load, temperature, degrees Fahrenheit
H	Quench severity, in^{-1}
IH	Initial hardness, Rockwell C scale

m	Strain-strengthening exponent
n	Design factor
r	Radius
R_{max}	Maximum hardness attainable, Rockwell C scale
R_Q	As-quenched Jominy test hardness, Rockwell C scale
R_T	Tempered hardness, Rockwell C scale
S_e'	Engineering endurance limit
S_u	Engineering ultimate strength in tension
S_y	Engineering yield strength, 0.2 percent offset
t	Time
ε	True strain
η	Factor of safety
$\overline{\sigma}_0$	Strain-strengthening coefficient
σ	Normal stress
ΣA	Sum of alloy increments, Rockwell C scale
τ_o	Octahedral shear stress
τ	Shearing stress

Subscripts

a	Axial
B	Long traverse
c	Compression
C	Circumferential
D	Short traverse
e	Endurance
f	Fracture
L	Longitudinal
R	Radial
s	Shear
t	Tension
u	Ultimate
y	Yield
0	No prior strain

8.1 INTRODUCTION

The mechanical designer needs to know the yield strength of a material so that a suitable margin against permanent distortion can be provided. The yield strength provided by a standardized tensile test is often not helpful because the manufacturing process has altered this property. Hot or cold forming and heat treatment (quenching and tempering) change the yield strength. The designer needs to know the yield strength of the material at the critical location in the geometry and at condition of use.

The designer also needs knowledge of the ultimate strength, principally as an estimator of fatigue strength, so that a suitable margin against fracture or fatigue can be provided. Hot and cold forming and various thermomechanical treatments during manufacture have altered these properties too. These changes vary within the part and can be directional. Again, the designer needs strength information for the material at the critical location in the geometry and at condition of use.

This chapter addresses the effect of plastic strain or a sequence of plastic strains on changes in yield and ultimate strengths (and associated endurance limits) and gives quantitative methods for the estimation of these properties. It also examines the changes in ultimate strength in heat-treated plain carbon and low-alloy steels.

8.2   STRENGTH OF PLASTICALLY DEFORMED MATERIALS

Methods for strength estimation include the conventional uniaxial tension test, which routinely measures true and engineering yield and ultimate strengths, percentage elongation and reduction in area, true ultimate and fracture strains, strain-strengthening exponent, strain-strengthening coefficient, and Young's modulus. These results are for the material in specimen form. Machine parts are of different shape, size, texture, material treatment, and manufacturing history and resist loading differently. Hardness tests can be made on a prototype part, and from correlations of strength with hardness and indentor size ([8.1], p. 5–35) and surface, ultimate strength can be assessed. Such information can be found in corporate manuals and catalogs or scattered in the literature. Often these are not helpful.

In the case of a single plastic deformation in the manufacturing process, one can use the true stress-strain curve of the material in the condition prior to straining provided the plastic strain can be determined. The results are good. For a sequence of successive strains, an empirical method is available which approximates what happens but is sometimes at variance with test results.

Cold work or *strain strengthening* is a common result of a cold-forming process. The process changes the properties, and such changes must be incorporated into the application of a theory of failure. The important strength is that of the part in the critical location in the geometry and at condition of use.

8.2.1   Datsko's Notation

In any discussion of strength it is necessary to identify

1. The kind of strength: ultimate, u; yield, y; fracture, f; endurance, e.
2. The sense of the strength: tensile, t; compressive, c; shear, s.
3. The direction or orientation of the strength: longitudinal, L; long transverse, B; short transverse, D; axial, a; radial, R; circumferential, C.
4. The sense of the most recent prior strain in the axial direction of the envisioned test specimen: tension, t; compression, c. If there is no prior strain, the subscript 0 is used.

8.2.2 Datsko's Rules

Datsko [8.1] suggests a notation $(S_1)_{234}$, where the subscripts correspond to 1, 2, 3, and 4 above. In Fig. 8.1 an axially deformed round and a rolled plate are depicted. A strength $(S_u)_{tLc}$ would be read as the engineering ultimate strength S_u, in tension $(S_u)_t$, in the longitudinal direction $(S_u)_{tL}$, after a last prior strain in the specimen direction that was compressive $(S_u)_{tLc}$. Datsko [8.1] has articulated rules for strain strengthening that are in approximate agreement with data he has collected. Briefly,

>   *Rule 1.* Strain strengthening is a bulk mechanism, exhibiting changes in strength in directions free of strain.
>
>   *Rule 2.* The maximum strain that can be imposed lies between the true strain at ultimate load ε_u and the true fracture strain ε_f. In upsetting procedures devoid of flexure, the limit is ε_f, as determined in the tension test.
>
>   *Rule 3.* The significant strain in a deformation cycle is the largest absolute strain, denoted ε_w. In a round $\varepsilon_w = \max(|\varepsilon_r|, |\varepsilon_\theta|, |\varepsilon_x|)$. The largest absolute strain ε_w is used in calculating the equivalent plastic strain ε_q, which is defined for two categories of strength, ultimate and yield, and in four groups of strength in Table 8.1.
>
>   *Rule 4.* In the case of several strains applied sequentially (say, cold rolling then upsetting), in determining ε_{qu}, the significant strains in each cycle ε_{wl} are added in decreasing order of magnitude rather than in chronological order.
>
>   *Rule 5.* If the plastic strain is imposed below the material's recrystallization temperature, the ultimate tensile strength is given by

$$S_u = (S_u)_o \exp \varepsilon_{qu} \qquad \varepsilon_{qu} < m$$

$$= \overline{\sigma}_0 (\varepsilon_{qu})^m \qquad \varepsilon_{qu} > m$$

>   *Rule 6.* The yield strength of a material whose recrystallization temperature was not exceeded is given by

$$S_y = \overline{\sigma}_0 (\varepsilon_{qy})^m$$

Table 8.1 summarizes the strength relations for plastically deformed metals.

8.3 ESTIMATING ULTIMATE STRENGTH AFTER PLASTIC STRAINS

This topic is best illuminated by example, applying ideas expressed in Secs. 8.2.1 and 8.2.2.

Example 1.   A 1045HR bar has the following properties from tension tests:

$$S_y = 60 \text{ kpsi} \qquad S_u = 92.5 \text{ kpsi}$$

$$\text{AR} = 0.44 \qquad m = 0.14$$

The material is to be used to form an integral pinion on a shaft by cold working from 2¼ in to 2 in diameter and then upsetting to 2½ in to form a pinion blank, as depicted in Fig. 8.2. Find, using Datsko's rules, an estimate of the ultimate strength in a direction resisting tooth bending at the root of the gear tooth to be cut in the blank.

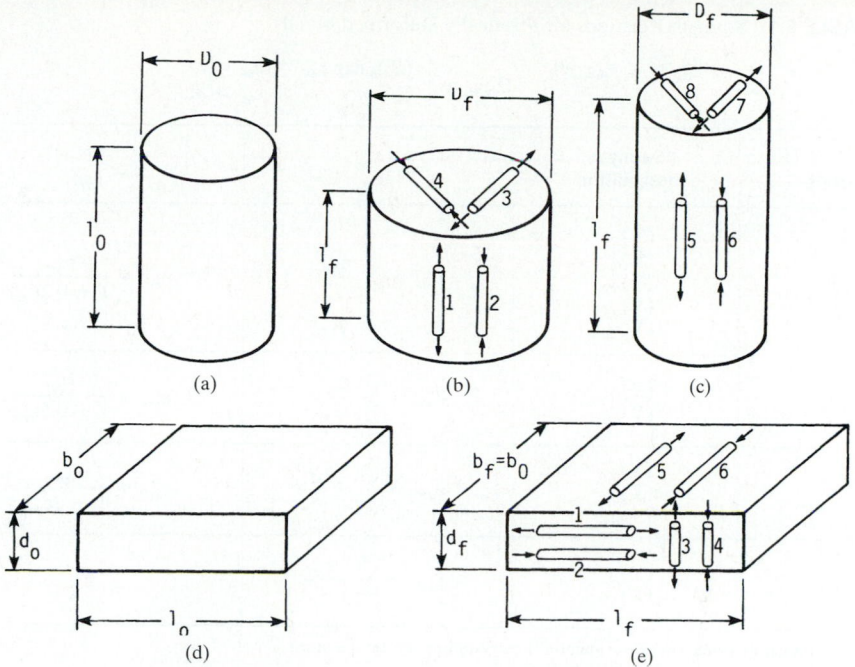

FIGURE 8.1 Sense of strengths in bar and plate. *(Adapted from [8.1], p. 7-7 with permission.)*
(*a*) Original bar before axial deformation.

	Specimen	Sense of strength	Direction in the bar	Prior strain	Designation
(*b*)	1	t	L	c	$(S)_{tLc}$
	2	c	L	c	$(S)_{cLc}$
	3	t	T	t	$(S)_{tTt}$
	4	c	T	t	$(S)_{cTt}$
(*c*)	5	t	L	t	$(S)_{tLt}$
	6	c	L	t	$(S)_{cLt}$
	7	t	T	c	$(S)_{tTc}$
	8	c	T	c	$(S)_{cTc}$

(*d*) Plate prior to rolling.

	Specimen	Sense of strength	Direction in the bar	Prior strain	Designation
(*e*)	1	t	L	t	$(S)_{tLt}$
	2	c	L	t	$(S)_{cLt}$
	3	t	D	c	$(S)_{tDc}$
	4	c	D	c	$(S)_{cDc}$
	5	t	B	0	$(S)_{tB0}$
	6	c	B	0	$(S)_{cB0}$

TABLE 8.1 Strength Relations for Plastically Deformed Metals[†]

$$(s_y)_w = \bar{\sigma}_0(\varepsilon_{qy})^m \qquad (S_u)_w = \begin{cases} (S_u)_0 \exp \varepsilon_{qu} & \varepsilon_{qu} < m \\ \bar{\sigma}_w & \varepsilon_{qu} > m \end{cases}$$

Group	Strength designation	ε_{qu}	ε_{qy}
1	$(S)_{cLc}$ $(S)_{tLt}$ $(S)_{tB0}$ $(S)_{cB0}$ $(S)_{cDc}$	$\varepsilon_{qus} = \sum_{i=1}^{n} \dfrac{\varepsilon_{wi}}{i}$	$\varepsilon_{qys} = \dfrac{\varepsilon_{qus}}{1 + 0.2\varepsilon_{qus}}$
2	$(S)_{tTt}$ $(S)_{cTc}$	$\varepsilon_{qus} = \sum_{i=1}^{n} \dfrac{\varepsilon_{wi}}{i}$	$\varepsilon_{qys} = \dfrac{\varepsilon_{qus}}{1 + 0.5\varepsilon_{qus}}$
3	$(S)_{cLt}$ $(S)_{tLc}$ $(S)_{tDc}$	$\varepsilon_{qu0} = \sum_{i=1}^{n} \dfrac{\varepsilon_{wi}}{i+1}$	$\varepsilon_{qy0} = \dfrac{\varepsilon_{qu0}}{1 + 2\varepsilon_{qu0}}$
4	$(S)_{tTc}$ $(S)_{cTt}$	$\varepsilon_{qu0} = \sum_{i=1}^{n} \dfrac{\varepsilon_{wi}}{i+1}$	‡

† Plastic deformation below material's recrystallization temperature.
‡ $(S_y)_{tTc} = (S_y)_{cTt} = 0.95(S_y)_{tTt}$ or $0.95(S_y)_{cTc}$
    ε_{qus} = equivalent strain when prestrain sense is same as sense of strength
    ε_{qu0} = equivalent strain when prestrain sense is opposite to sense of strength
SOURCE: From Datsko [8.1] and Hertzberg [8.2].

The strain-strengthening coefficient $\bar{\sigma}_0$ is, after [8.3],

$$\bar{\sigma}_0 = S_u \exp(m)m^{-m} = 92.5 \exp(0.14)0.14^{-0.14} = 140.1 \text{ kpsi}$$

The fracture strain (true) of the hot-rolled material from the tension test is

$$\varepsilon_f = \ln \frac{1}{1 - \text{AR}} = \ln \frac{1}{1 - 0.44} = 0.58$$

which represents limiting strain in deformation free of bending (rule 2). In the first step (cold rolling), the largest strain is axial, and it has a magnitude of (rule 3)

$$\varepsilon_1 = \left| \ln \left(\frac{D_0}{D_1} \right)^2 \right| = \left| \ln \left(\frac{2.25}{2} \right)^2 \right| = 0.236$$

In the second step (upsetting), the largest strain is axial, and it has a magnitude (rule 3) of

$$\varepsilon_2 = \left| \ln \left(\frac{D_1}{D_2} \right)^2 \right| = \left| \ln \left(\frac{2}{2.5} \right)^2 \right| = |-0.446| = 0.446$$

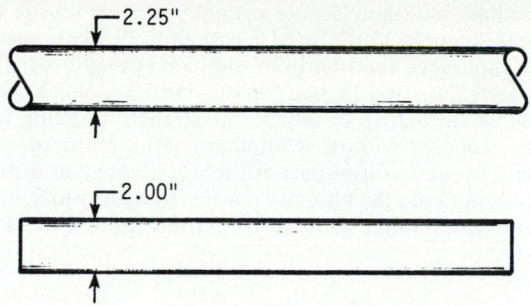

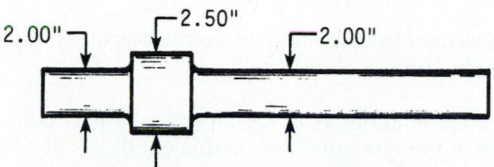

FIGURE 8.2   Cold working bar stock in two steps to form integral pinion blank on spindle.

The significant strains ε_{w1} and ε_{w2} are (rule 4) $\varepsilon_{w1} = 0.446$ and $\varepsilon_{w2} = 0.236$. Strengths will be carried with four computational digits until numerical work is done. For group 1 strengths,

$$\varepsilon_{qu} = \sum \frac{\varepsilon_{wi}}{i} = \frac{0.446}{1} + \frac{0.236}{2} = 0.564$$

$$S_u = \overline{\sigma}_0(\varepsilon_{qu})^m = 140.1(0.564)^{0.14} = 129.3 \text{ kpsi}$$

According to rule 5, $\varepsilon_{qu} > m$.

For group 2 strengths,

$$\varepsilon_{qu} = \sum \frac{\varepsilon_{wi}}{i} = \frac{0.446}{1} + \frac{0.236}{2} = 0.564$$

$$S_u = \overline{\sigma}_0(\varepsilon_{qu})^m = 140.1(0.564)^{0.14} = 129.3 \text{ kpsi}$$

For group 3 strengths,

$$\varepsilon_{qu} = \sum \frac{\varepsilon_{wi}}{1+i} = \frac{0.446}{2} + \frac{0.236}{3} = 0.302$$

$$S_u = \overline{\sigma}_0(\varepsilon_{qu})^m = 140.1(0.302)^{0.14} = 118.5 \text{ kpsi}$$

For group 4 strengths,

$$\varepsilon_{qu} = \sum \frac{\varepsilon_{wi}}{1+i} = \frac{0.446}{2} + \frac{0.236}{3} = 0.302$$

$$S_u = \overline{\sigma}_0(\varepsilon_{qu})^m = 140.1(0.302)^{0.14} = 118.5 \text{ kpsi}$$

The endurance limit and the ultimate strength resisting tensile bending stresses are $(S'_e)_{tTt}$ and $(S_u)_{tTt}$, namely, 129.3/2 = 64.7 kpsi and 129.3 kpsi, respectively (group 2 strengths). The endurance limit and the ultimate strength resisting compressive bending stresses are $(S'_e)_{cTt}$ and $(S_u)_{cTt}$, namely, 118.5/2 = 59.3 kpsi and 118.5 kpsi, respectively (group 4 strengths). In fatigue the strength resisting tensile stresses is the significant one, namely, 64.7 kpsi. A summary of this information concerning the four group ultimate strengths forms part of Table 8.2. Note that these two successive plastic strains have improved the ultimate tensile strength (which has become directional). The pertinent endurance limit has risen from 92.5/2 = 46.3 kpsi to 59.3 kpsi.

8.4 ESTIMATING YIELD STRENGTH AFTER PLASTIC STRAINS

This topic is best presented by extending the conditions of Example 1 to include the estimation of yield strengths.

Example 2. The same material as in Example 1 is doubly cold-worked as previously described. The strain-strengthening coefficient $\bar{\sigma}_0$ is still 140.1 kpsi, true fracture strain ε_f is 0.58, and $\varepsilon_1 = 0.236$, $\varepsilon_2 = 0.446$, $\varepsilon_{w1} = 0.446$, and $\varepsilon_{w2} = 0.236$ as before. For group 1 strengths,

$$\varepsilon_{qy} = \frac{\varepsilon_{qu}}{1 + 0.2\varepsilon_{qu}} = \frac{0.564}{1 + 0.2(0.564)} = 0.507$$

$$S_y = \bar{\sigma}_0(\varepsilon_{qy})^m = 140.1(0.507)^{0.14} = 127.4 \text{ kpsi} \qquad \text{(rule 6)}$$

For group 2 strengths,

$$\varepsilon_{qy} = \frac{\varepsilon_{qu}}{1 + 0.5\varepsilon_{qu}} = \frac{0.564}{1 + 0.5(0.564)} = 0.440$$

$$S_y = \bar{\sigma}_0(\varepsilon_{qy})^m = 140.1(0.440)^{0.14} = 124.9 \text{ kpsi}$$

For group 3 strengths,

$$\varepsilon_{qy} = \frac{\varepsilon_{qu}}{1 + 2\varepsilon_{qu}} = \frac{0.302}{1 + 2(0.302)} = 0.188$$

$$S_y = \bar{\sigma}_0(\varepsilon_{qy})^m = 140.1(0.188)^{0.14} = 110.9 \text{ kpsi}$$

TABLE 8.2 Summary of Ultimate and Yield Strengths for Groups 1 to 4 for Upset Pinion Blank

Group	ε_{qu}	S_u, kpsi	ε_{qy}	S_y, kpsi
1	0.564	129.3	0.507	127.4
2	0.564	129.3	0.440	124.9
3	0.302	118.5	0.188	110.9
4	0.302	118.5	. . .	118.7

Group 4 yield strengths are 0.95 of group 2:

$$S_y = 0.95(S_y)_2 = 0.95(124.9) = 118.7 \text{ kpsi}$$

Table 8.2 summarizes the four group strengths.

The yield strength resisting tensile bending stresses is $(S_y)_{tTt}$, a group 2 strength equaling 124.9 kpsi. The yield strength resisting compressive bending stresses is $(S_y)_{cTt}$, a group 4 strength equaling 118.7 kpsi. Yielding will commence at the weaker of the two strengths. If the bending stress level is 60 kpsi, the factor of safety against yielding is

$$\eta_y = \frac{(S_y)_{cTt}}{\sigma} = \frac{118.7}{60} = 1.98$$

If the estimate were to be based on the original material,

$$\eta_y = \frac{(S_y)_0}{\sigma} = \frac{60}{60} = 1$$

Datsko reports that predictions of properties after up to five plastic strains are reasonably accurate. For a longer sequence of different strains, Datsko's rules are approximate. They give the sense (improved or impaired) of the strength change and a prediction of variable accuracy. This is the only method of estimation we have, and if it is used cautiously, it has usefulness in preliminary design and should be checked by tests later in the design process.

8.5 ESTIMATING ULTIMATE STRENGTH OF HEAT-TREATED PLAIN CARBON STEELS

For a plain carbon steel the prediction of heat-treated properties requires that Jominy tests be carried out on the material. The addition method of Crafts and Lamont [8.4] can be used to estimate tempered-part strengths. Although the method was devised over 30 years ago, it is still the best approximation available, in either graphic or tabular form. The method uses the Jominy test, the ladle analysis, and the tempering time and temperature.

A 1040 steel has a ladle analysis as shown in Table 8.3 and a Jominy test as shown in Table 8.4. The symbol R_Q is the Jominy-test Rockwell C-scale hardness. The Jominy distance numbers are sixteenths of an inch from the end of the standard Jominy specimen. The tempered hardness after 2 hours (at 1000°F, for example) may be predicted from

$$R_T = (R_Q - D - B)f + B + \Sigma A \qquad R_T < R_Q - D \qquad (8.1)$$

$$R_T = R_Q - D \qquad\qquad\qquad\quad R_T > R_Q - D \qquad (8.2)$$

TABLE 8.3 Ladle Analysis of a 1040 Steel

Element	C	Mn	P	S	Si
Percent	0.39	0.71	0.019	0.036	0.15

TABLE 8.4　Jominy Test of a 1040 Steel

Station	1	2	3	4	5	6	7	8	9	10	11	12	13	14	15	16	20	24	28	32
R_Q	55	49	29	25	25	24	23	22	21	20	19	18	17	17	16	16	14	12	11	9

where　R_T = tempered hardness, Rockwell C scale
　　　　R_Q = as-quenched hardness, Rockwell C scale
　　　　D = tempering decrement, Rockwell C scale
　　　　B = critical hardness for carbon content and tempering temperature,
　　　　　　Rockwell C scale
　　　　f = tempering factor of carbon content and tempering temperature
　　　　ΣA = sum of alloy increments, Rockwell C scale

From the appropriate figures for tempering for 2 hours at 1000°F, we have

$$D = 5.4 \quad \text{(Fig. 8.3)} \qquad A_{\text{Mn}} = 1.9 \quad \text{(Fig. 8.6)}$$

$$B = 10 \quad \text{(Fig. 8.4)} \qquad A_{\text{Si}} = 0.7 \quad \text{(Fig. 8.7)}$$

$$f = 0.34 \quad \text{(Fig. 8.5)} \qquad \overline{\Sigma A = 2.6}$$

The transition from Eq. (8.1) to Eq. (8.2) occurs at a Rockwell hardness determined by equating these two expressions:

$$(R_Q - 5.4 - 10)0.34 + 10 + 2.6 = R_Q - 5.4$$

from which $R_Q = 19.3$, Rockwell C scale. The softening at each station and corresponding ultimate tensile strength can be found using Eq. (8.1) or Eq. (8.2) as appropriate and converting R_T to Brinell hardness and then to tensile strength or converting directly from R_T to tensile strength. Table 8.5 displays the sequence of steps in estimating the softening due to tempering at each Jominy distance of interest.

　　A shaft made from this material, quenched in oil ($H = 0.35$)[†] and tempered for 2 hours at 1000°F would have surface properties that are a function of the shaft's diameter. Figures 8.8 through 8.11 express graphically and Tables 8.6 through 8.9 express numerically the equivalent Jominy distance for the surface and interior of rounds for various severities of quench. A 1-in-diameter round has a rate of cooling at the surface that is the same as at Jominy distance 5.1 (see Table 8.6). This means an as-quenched hardness of about 15.9 and a surface ultimate strength of about 105.7 kpsi. Similar determinations for other diameters in the range 0.1 to 4 in leads to the display that is Table 8.10. A table such as this is valuable to the designer and can be routinely produced by computer [8.5]. A plot of the surface ultimate strength versus diameter from this table provides the 1000°F contour shown in Fig. 8.12. An estimate of 0.2 percent yield strength at the surface can be made (after Ref. [8.4], p. 191):

$$S_y = [0.92 - 0.006(R_{\text{max}} - R_Q)]S_u \qquad (8.3)$$

[†] The quench severity H is the ratio of the film coefficient of convective heat transfer h [Btu/(h·in²·°F)] to the thermal conductivity of the metal k [Btu/(h·in·°F)], making the units of H in $^{-1}$.

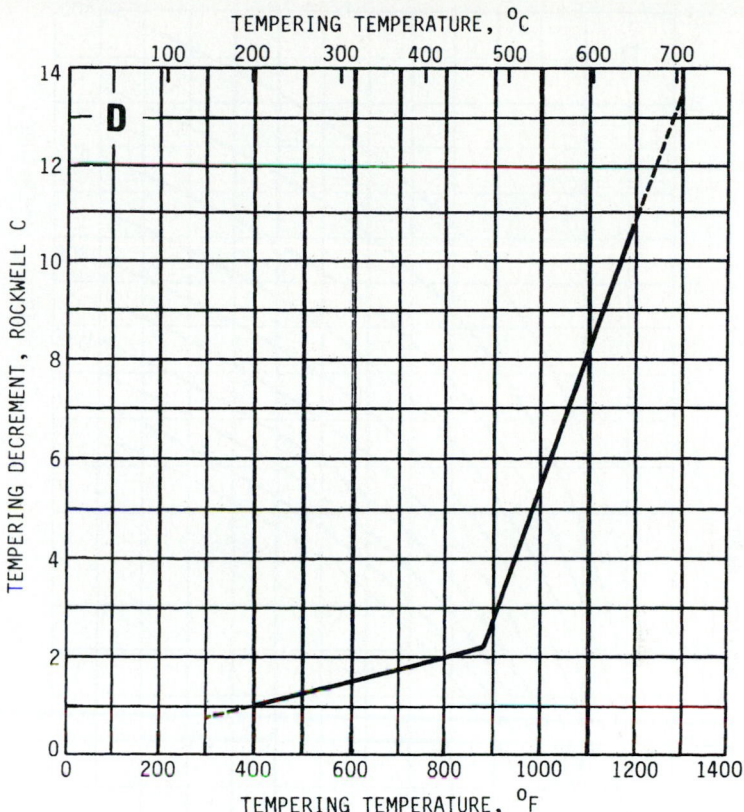

FIGURE 8.3 Hardness decrement D caused by tempering for "unhardened" steel. *(From [8.4] with permission of Pitman Publishing Ltd., London.)*

where R_{max} = maximum Rockwell C-scale hardness attainable for this steel, $32 + 60(\%C)$, and R_Q = as-quenched hardness. An estimate of yield strength at the surface of a 1-in round of this material is as follows (equivalent Jominy distance is 5.1):

$$S_y = [0.92 - 0.006(55 - 25)]105.7 = 78.2 \text{ kpsi}$$

Different properties exist at different radii. For example, at the center of a 1-in round the properties are the same as at Jominy distance 6.6, namely, a predicted ultimate strength of 104.5 kpsi and a yield strength of 76.3 kpsi, which are not very different from surface conditions. This is not always the case.

8.6 ESTIMATING ULTIMATE STRENGTH OF HEAT-TREATED LOW-ALLOY STEELS

For heat-treated low-alloy steels, the addition method of Crafts and Lamont changes only in that additional constituents are present in the ΣA term if a Jominy test is

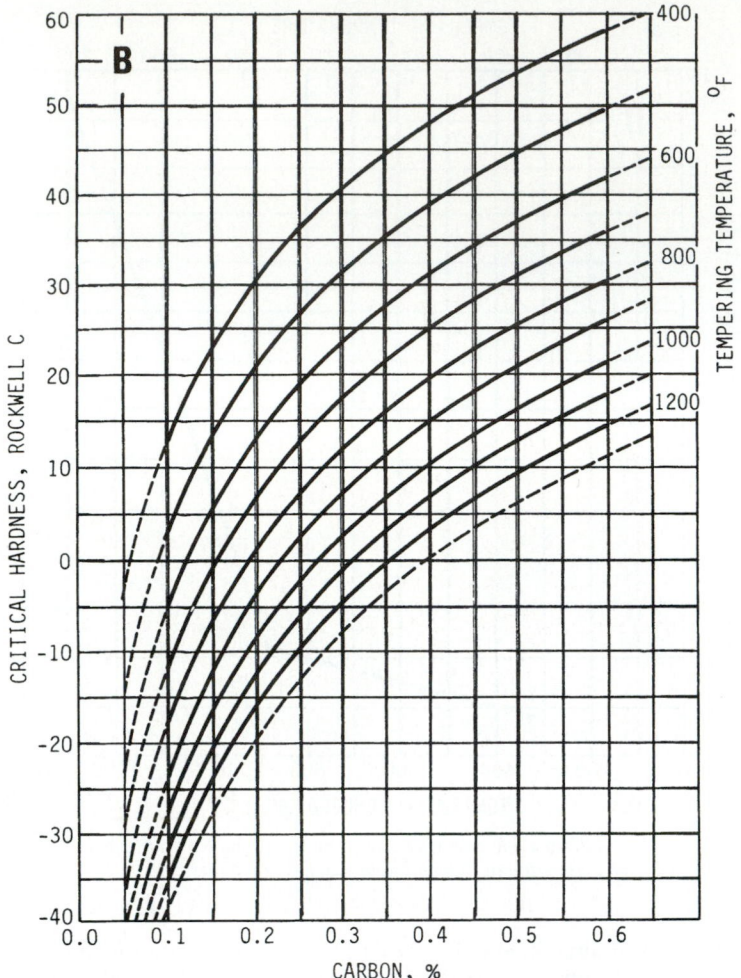

FIGURE 8.4 Critical hardness B for alloy-free steel as affected by carbon content and tempering temperature. *(From [8.4] with permission of Pitman Publishing Ltd., London.)*

available. However, for heat-treated low-alloy steels, the Jominy test may be replaced by an estimate based on the multiplication method of Grossmann and Fields coupled with knowledge of grain size and ladle analysis. Again, although the method was devised over 30 years ago, it is still the best approach available, in either graphic or tabular form. The multiplying factors for sulfur and phosphorus in this method are close to unity in the trace amounts of these two elements. The basic equation is

$$
\begin{matrix}
\text{Ideal} \\
\text{critical} \\
\text{diameter } D_I
\end{matrix}
=
\left(
\begin{matrix}
\text{carbon} \\
\text{ideal} \\
\text{diameter } D
\end{matrix}
\right)
\left(
\begin{matrix}
\text{Mn} \\
\text{multiplying} \\
\text{factor}
\end{matrix}
\right)
\left(
\begin{matrix}
\text{Cr} \\
\text{multiplying} \\
\text{factor}
\end{matrix}
\right)
\cdots
$$

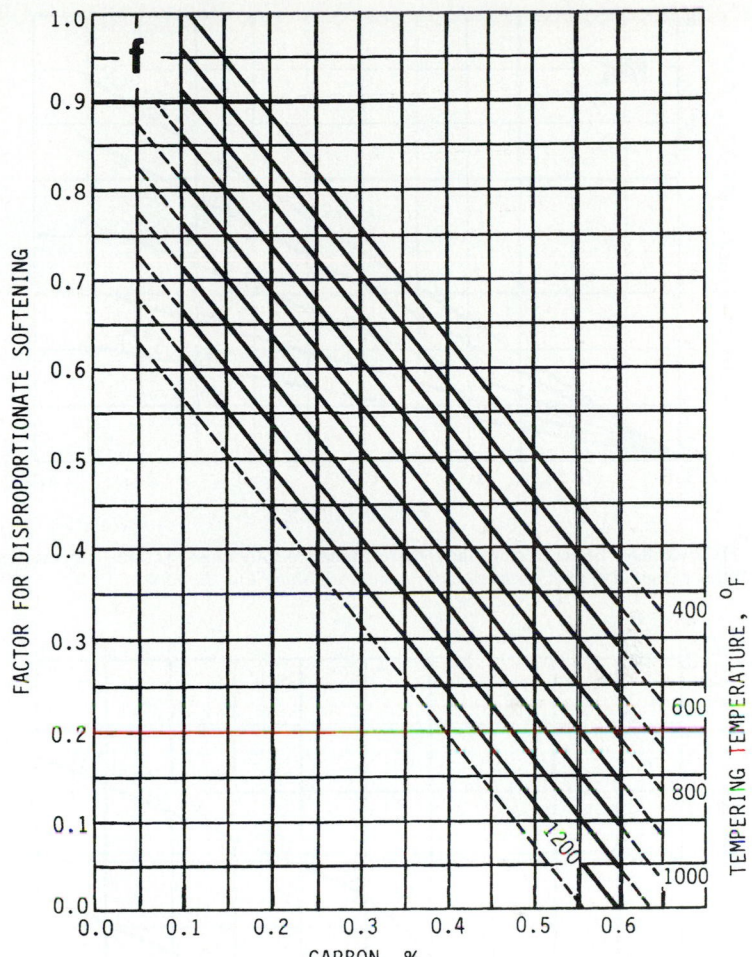

FIGURE 8.5 Factor f for disproportionate softening in "hardened" steel as affected by carbon content and tempering temperature. *(From [8.4] with permission of Pitman Publishing Ltd., London.)*

The multiplying factors for the elements Mn, Si, Cr, Ni, Mo, and Cu are presented in Fig. 8.13. The carbon ideal diameter D is available from Fig. 8.14 as a function of percent carbon and grain size of the steel.

Example 3.   Determine the surface properties of an 8640 steel with average grain size 8 that was oil-quenched ($H = 0.35$) and tempered 2 hours at 1000°F. The ladle analysis and the multiplying factors are shown in Table 8.11. The multiplying factors are determined from Figs. 8.13 and 8.14. If boron were present, the multiplying factor would be

$$B = 17.23(\text{percent boron})^{-0.268}$$

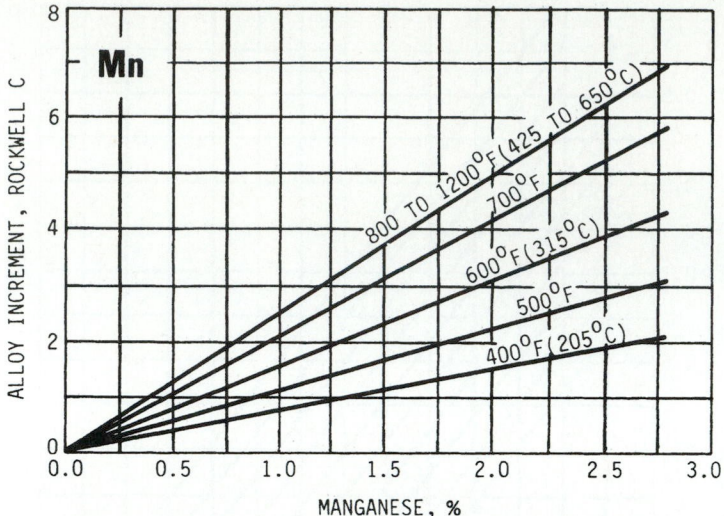

FIGURE 8.6 Effect of manganese on resistance to softening at various temperatures. *(From [8.4] with permission of Pitman Publishing Ltd., London.)*

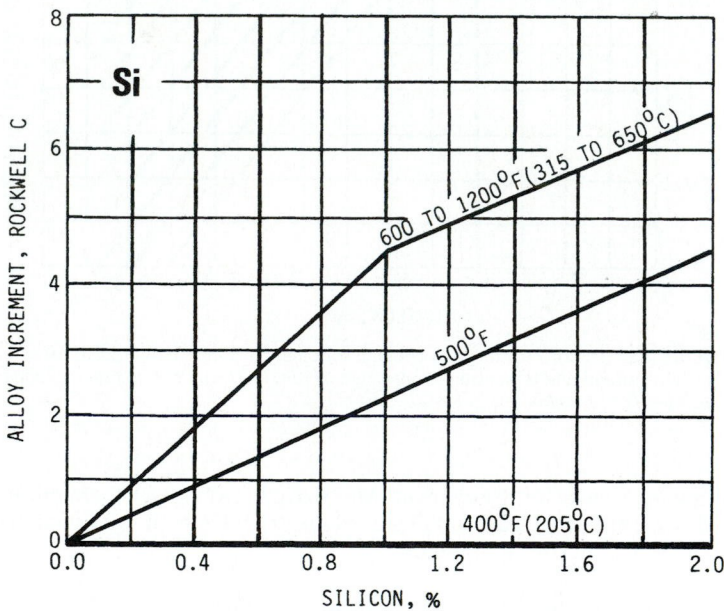

FIGURE 8.7 Effect of silicon on resistance to softening at various tempering temperatures. *(From [8.4] with permission of Pitman Publishing Ltd., London.)*

TABLE 8.5 Softening of 1040 Round Due to Tempering at 1000°F for 2 Hours

Jominy distance	R_Q	R_T	H_B	S_u, kpsi
1	55	26.1	258.6	129.3
2	49	24.0	247.0	123.5
3	29	17.2	216.2	108.1
4	25	15.9	211.6	105.8
5	25	15.9	211.6	105.8
6	24	15.3	209.8	104.9
7	23	15.2	208.4	104.2
8	22	14.8	206.6	103.3
9	21	14.5	205.3	102.6
10	20	14.2	203.9	102.0
				←Transition
11	19	13.6	201.2	100.6
12	18	12.6	196.7	98.4

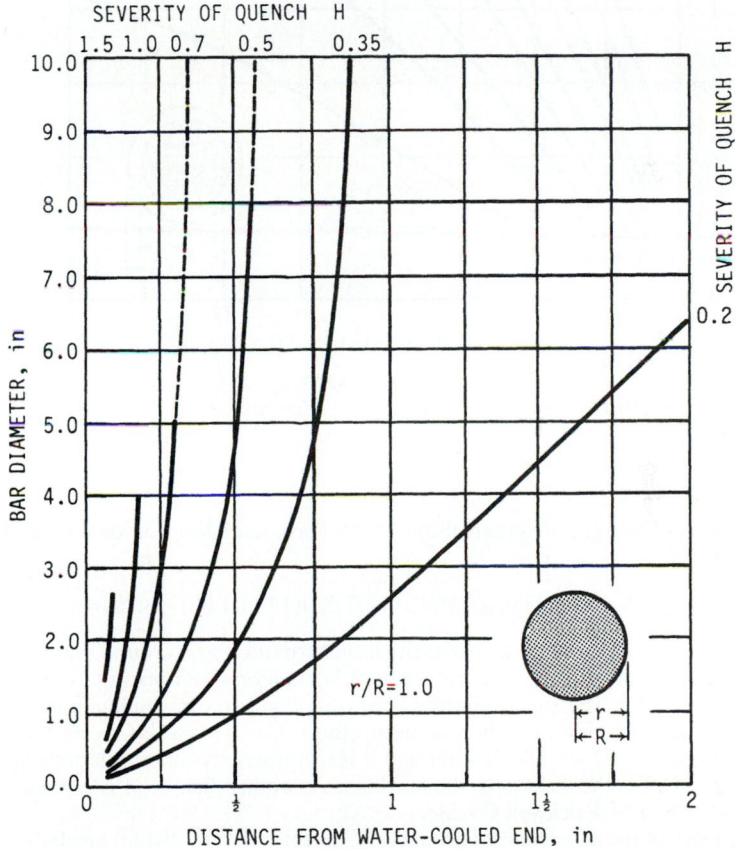

FIGURE 8.8 Location on end-quenched Jominy hardenability specimen corresponding to the surface of round bars. *(From [8.4] with permission of Pitman Publishing Ltd., London.)*

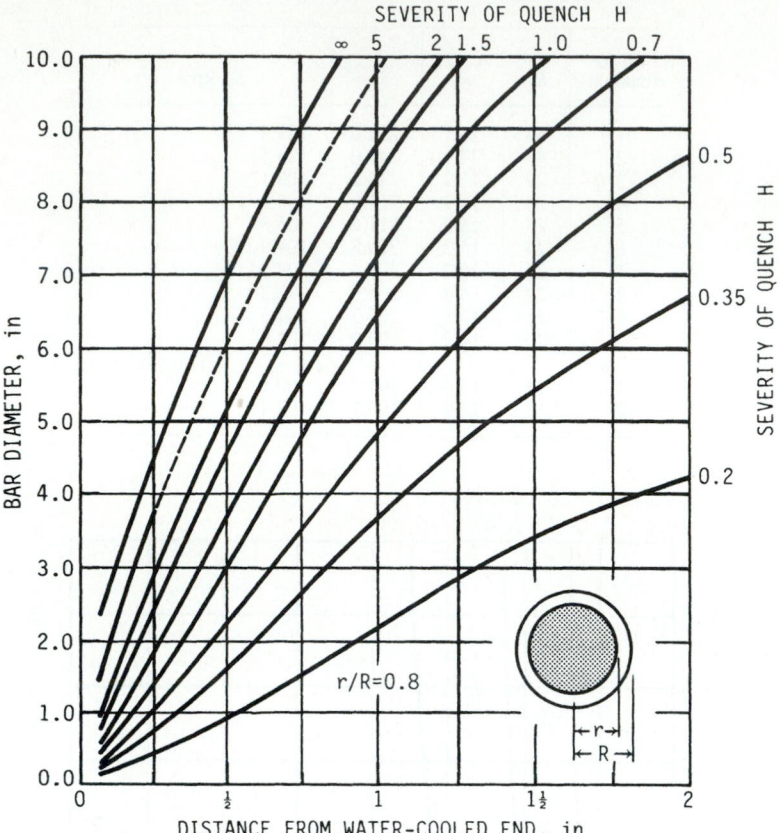

FIGURE 8.9 Location on end-quenched Jominy hardenability specimen corresponding to 80 percent from center of round. *(From [8.4] with permission of Pitman Publishing Ltd., London.)*

where percent boron is less than about 0.002. The calculation for ideal critical diameter D_I is

$$D_I = 0.197(3.98)(1.18)(2.08)(1.20)(1.60)(1.00) = 3.70 \text{ in}$$

The meaning of D_I is that it describes the largest diameter of a round that has at least 50 percent martensite structure everywhere in the cross section and exactly 50 percent at the center. The surface hardness of quenched steels is independent of alloy content and a function of carbon content alone. The Rockwell C-scale hardness is approximated by $32 + 60(\%C)$, although it is not a strictly linear relationship ([8.4], p. 88; Fig. 8.4). For the 8640 steel, the hardness at Jominy distance 1 is estimated to be $32 + 60(0.40)$ or 56 Rockwell C scale.

The ratio of initial hardness (distance 1), denoted IH, to distant hardness (at any other Jominy distance), denoted DH, is available as a function of the ideal critical

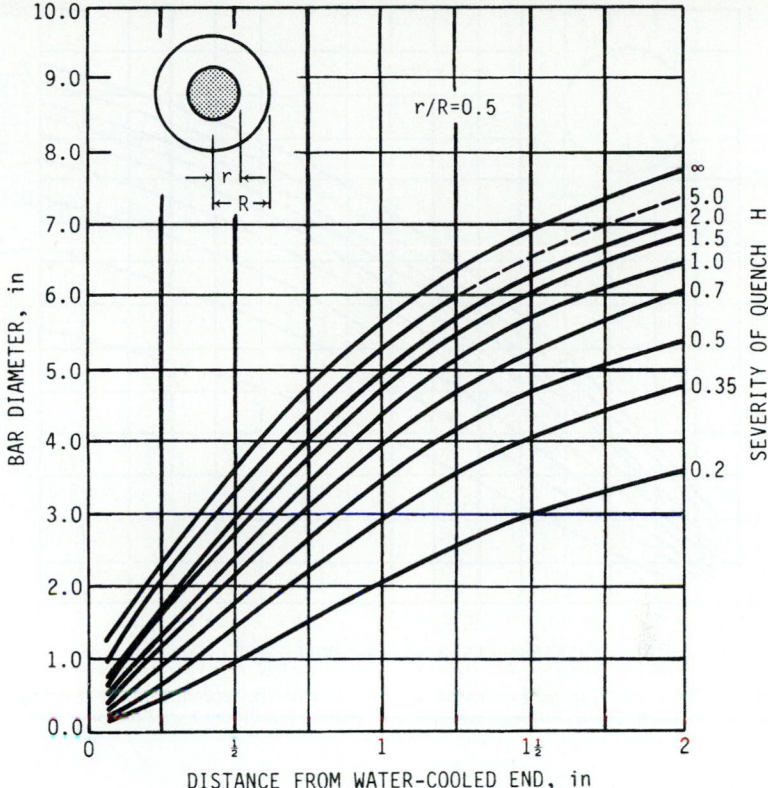

FIGURE 8.10 Location on end-quenched Jominy hardenability specimen correspond-ing to 50 percent from the center of round bars. *(From [8.4] with permission of Pitman Publishing Ltd., London.)*

diameter and the Jominy distance (Fig. 8.15). For the 8640 steel the Jominy hard-nesses are estimated as displayed in Table 8.12. The Rockwell C-scale hardness is plotted against Jominy distance in Fig. 8.16, upper contour. The softening due to 2 hours of tempering at 1000°F can be estimated as before using the addition method of Crafts and Lamont. The ΣA term is evaluated as follows:

$$D = 5.31 \quad \text{(Fig. 8.3)} \qquad A_{Mn} = 2.25 \quad \text{(Fig. 8.6)}$$

$$B = 9.90 \quad \text{(Fig. 8.4)} \qquad A_{Si} = 1.13 \quad \text{(Fig. 8.7)}$$

$$f = 0.34 \quad \text{(Fig. 8.5)} \qquad A_{Cr} = 2.59 \quad \text{(Fig. 8.17)}$$

$$A_{Ni} = 0.11 \quad \text{(Fig. 8.18)}$$

$$A_{Mo} = 3.60 \quad \text{(Fig. 8.19)}$$

$$\overline{\Sigma A = 9.67}$$

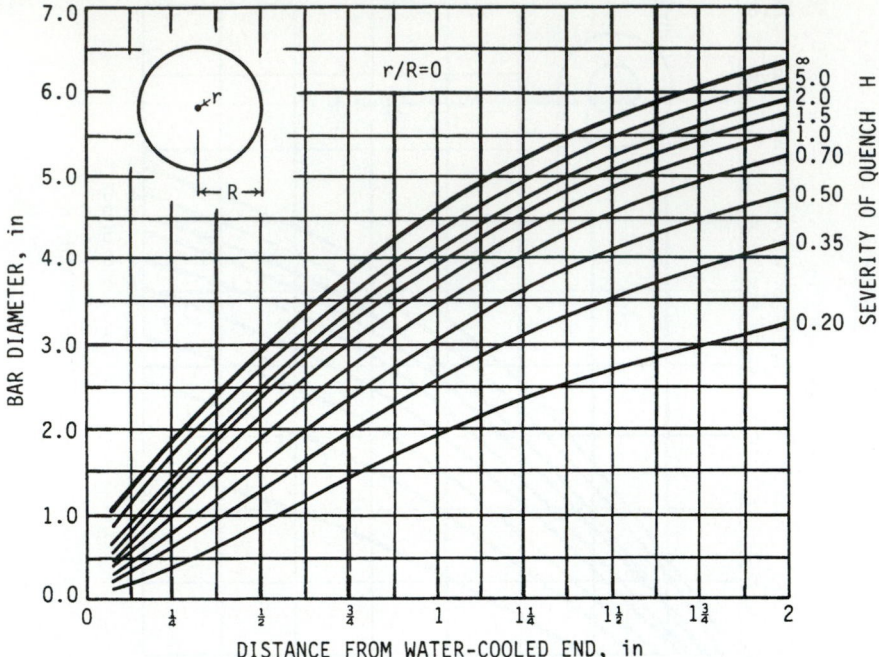

FIGURE 8.11   Location on end-quenched Jominy hardenability specimen corresponding to the center of round bars. *(From [8.4] with permission of Pitman Publishing Ltd., London.)*

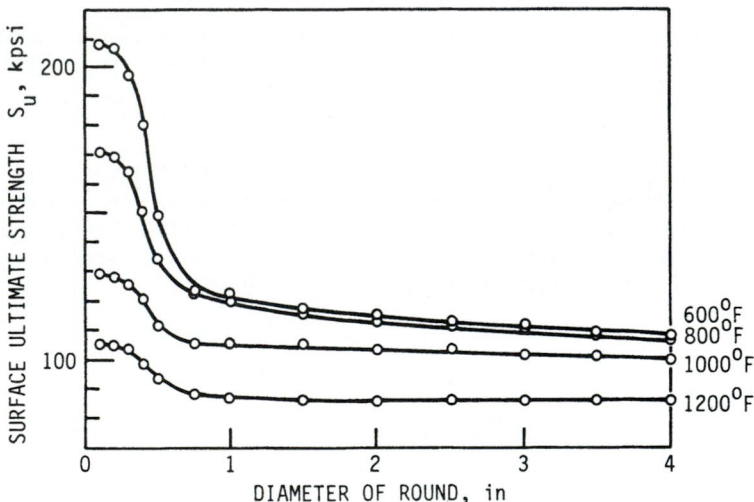

FIGURE 8.12   Variation of surface ultimate strength with diameter for a 1040 steel oil-quenched ($H = 0.35$) from 1575°F and tempered 2 hours at 1000°F.

TABLE 8.6 Equivalent Jominy Distances for Quenched Rounds at $r/R = 1$

Diameter, in	Severity of quench H, in^{-1}						
	0.20	0.30	0.35	0.40	0.50	0.60	0.70
0.1	1.0	1.0	1.0	1.0	1.0	1.0	1.0
0.2	1.8	1.3	1.1	1.0	1.0	1.0	1.0
0.3	2.7	1.9	1.6	1.4	1.2	1.0	1.0
0.4	3.6	2.5	2.2	1.9	1.5	1.2	1.0
0.5	4.5	3.2	2.7	2.3	1.9	1.5	1.2
0.75	6.7	4.8	4.0	3.5	2.9	2.2	1.7
1.0	8.3	6.0	5.1	4.4	3.5	2.7	2.2
1.5	10.7	8.0	6.9	6.0	4.6	3.5	2.8
2.0	13.2	9.6	8.2	7.1	5.4	4.2	3.3
2.5	15.4	11.0	9.2	7.8	6.1	4.6	3.7
3.0	17.6	12.1	10.0	8.4	6.6	5.0	4.0
3.5	19.8	13.1	10.7	8.9	7.0	5.4	4.3
4.0	22.1	14.2	11.4	9.4	7.6	5.7	4.5

The tempered hardness equations become either

$$R_T = (R_Q - 5.31 - 9.90)0.34 + 9.90 + 9.67$$

$$= 0.34R_Q + 14.4$$

or

$$R_T = R_Q - 5.31$$

TABLE 8.7 Equivalent Jominy Distances for Quenched Rounds at $r/R = 0.8$

Diameter, in	Severity of quench H, in^{-1}						
	0.20	0.30	0.35	0.40	0.50	0.60	0.70
0.1	1.0	1.0	1.0	1.0	1.0	1.0	1.0
0.2	1.8	1.3	1.1	1.0	1.0	1.0	1.0
0.3	2.7	1.9	1.7	1.5	1.3	1.1	1.0
0.4	3.6	2.6	2.2	2.0	1.7	1.4	1.2
0.5	4.5	3.2	2.8	2.5	2.1	1.8	1.5
0.75	6.7	4.9	4.2	3.7	3.2	2.6	2.2
1.0	8.3	6.2	5.4	4.8	4.0	3.4	3.0
1.5	11.5	8.7	7.6	6.7	5.6	4.8	4.4
2.0	14.6	10.9	9.6	8.5	7.3	6.3	5.7
2.5	17.7	13.1	11.4	10.2	8.9	7.7	7.0
3.0	21.0	15.4	13.4	11.9	10.4	9.0	8.1
3.5	24.9	18.0	15.5	13.7	12.0	10.3	9.3
4.0	29.4	21.1	18.0	15.9	13.4	11.5	10.3

TABLE 8.8   Equivalent Jominy Distances for Quenched Rounds at $r/R = 0.5$

Diameter, in	Severity of quench H, in^{-1}						
	0.20	0.30	0.35	0.40	0.50	0.60	0.70
0.1	1.0	1.0	1.0	1.0	1.0	1.0	1.0
0.2	1.8	1.3	1.2	1.1	1.0	1.0	1.0
0.3	2.7	2.0	1.8	1.6	1.4	1.2	1.0
0.4	3.6	2.7	2.4	2.1	1.8	1.6	1.4
0.5	4.5	3.4	3.0	2.6	2.3	2.0	1.1
0.75	6.7	5.0	4.4	4.0	3.4	2.9	2.6
1.0	8.3	6.4	5.7	5.2	4.5	4.0	3.5
1.5	11.9	9.2	8.3	7.5	6.7	5.9	5.4
2.0	15.4	12.0	10.8	9.8	8.9	8.0	7.3
2.5	19.3	15.0	13.4	12.2	11.1	10.1	9.3
3.0	24.2	18.4	16.3	14.8	13.6	12.3	11.5
3.5	30.3	22.4	19.6	17.7	16.2	14.7	13.8
4.0	32.0	25.9	23.5	21.6	19.1	17.3	16.4

The transition hardness obtained by equating the preceding pair of equations is $R_Q = 29.9$. The Jominy curve may be corrected for tempering. Table 8.13 shows the tempered hardness and ultimate strength corresponding to the Jominy distances. The column R_T is plotted against Jominy distance as the lower curve in Fig. 8.16. The surface ultimate strength can be estimated for diameters 0.5, 1, 2, 3, and 4 in. At a diameter of 2 in, the equivalent Jominy distance is 8.2 from Table 8.6. The surface ultimate strength as a function of diameter of round is displayed in Table 8.14. The ultimate tensile strength is found by interpolation in the prior display, entering with equivalent Jominy distance. The tensile ultimate strength at the surface versus diam-

TABLE 8.9   Equivalent Jominy Distances for Quenched Rounds at $r/R = 0$

Diameter, in	Severity of quench H, in^{-1}						
	0.20	0.30	0.35	0.40	0.50	0.60	0.70
0.1	1.0	1.0	1.0	1.0	1.0	1.0	1.0
0.2	1.8	1.5	1.4	1.2	1.0	1.0	1.0
0.3	2.7	2.2	2.0	1.9	1.5	1.3	1.2
0.4	3.6	3.0	2.7	2.5	2.0	1.8	1.6
0.5	4.5	3.7	3.4	3.1	2.6	2.2	2.0
0.75	6.7	5.6	5.1	4.6	3.8	3.3	3.0
1.0	8.3	7.1	6.6	6.1	5.1	4.5	4.1
1.5	12.4	10.3	9.5	8.7	7.7	6.9	6.4
2.0	16.7	13.5	12.3	11.4	10.2	9.2	8.6
2.5	21.8	17.2	15.5	14.2	12.9	11.7	11.0
3.0	28.1	21.6	19.3	17.5	15.8	14.4	13.6
3.5	32.0	26.2	23.9	21.9	19.2	17.3	16.5
4.0	32.0	30.9	29.7	27.9	23.0	20.6	19.9

TABLE 8.10   Surface Ultimate Strength of a 1040 Steel
Heat-Treated Round as a Function of Diameter[†]

Diameter, in	Equivalent Jominy distance, $\frac{1}{16}$ in	Surface ultimate strength S_u, kpsi
0.1	1.0	129.3
0.2	1.1	128.7
0.3	1.6	125.8
0.4	2.2	120.4
0.5	2.7	112.7
1.0	5.1	105.7
1.5	6.9	104.3
2.0	8.2	103.2
3.0	10.0	102.0
4.0	11.4	99.7

† Round quenched from 1575°F in still oil ($H = 0.35$) tempered for
2 hours at 1000°F. Predictions by the addition method of Crafts and
Lamont.

eter of round is plotted in Fig. 8.20. Note the greater hardening ability of the 8640
compared to the 1040 steel of the previous section. Local interior properties are
available using Figs. 8.9, 8.10, and 8.11. An estimate of the variation of properties
across the section of a round 4 in in diameter will be made. The equivalent Jominy
distances are 11.2 at $r = 2$ in, 18.0 at $r = 1.6$ in, 23.5 at $r = 1$ in, and 29.7 at $r = 0$. Thus
Table 8.15 may be formed. The values of S_u are obtained by interpolation; the values
of S_y are estimated using Eq. (8.3). A plot is shown in Fig. 8.21.

A common source for properties of steels is *Modern Steels and Their Properties*
[8.5]. It is well to note that hardness was taken in this reference at the surface of a

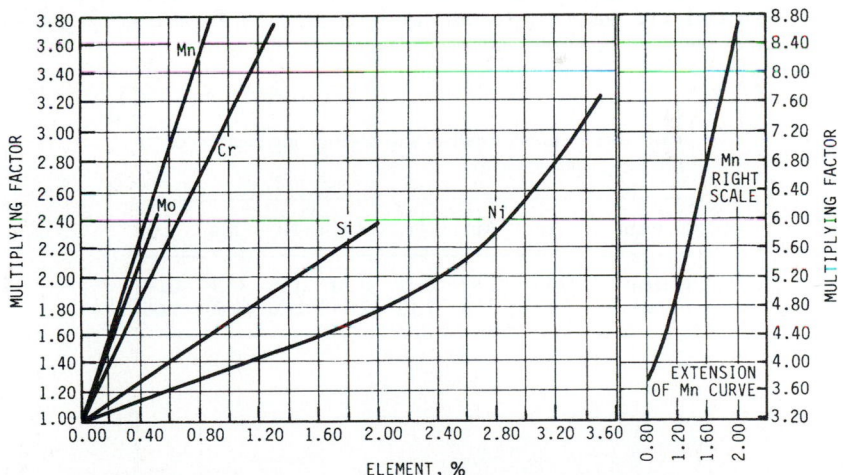

FIGURE 8.13   Multiplying factors for five common alloying elements (for trace copper, use
nickel curve). *(From [8.4] with permission of Pitman Publishing Ltd., London).*

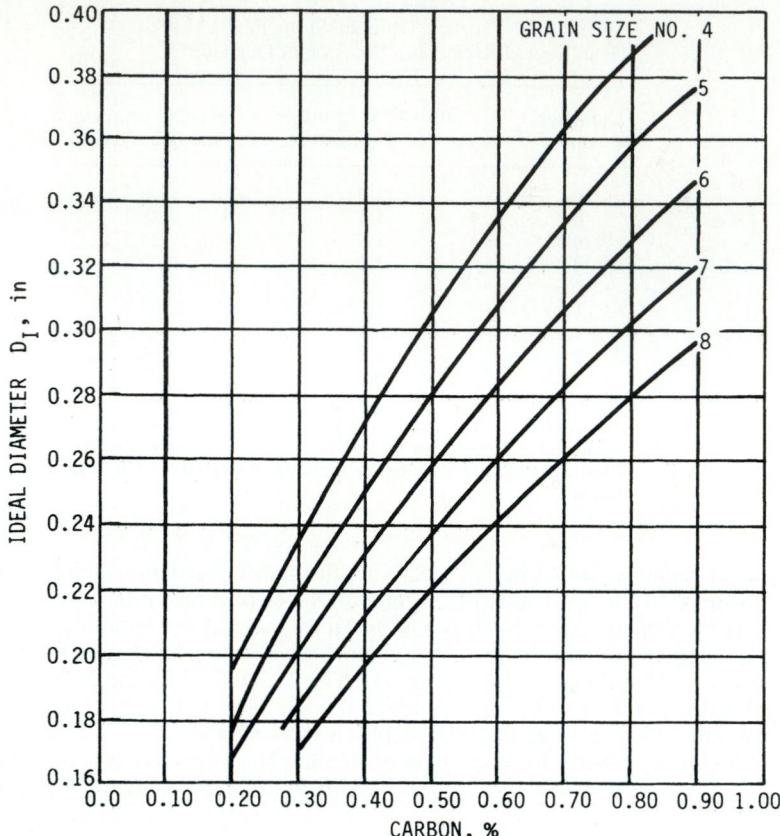

FIGURE 8.14 Relationship between ideal diameter D_I, carbon content, and grain size. *(From [8.4] with permission of Pitman Publishing Ltd., London.)*

TABLE 8.11 Ladle Analysis and Multiplying Factors for 8640 Steel, Grain Size 8

Element	C	Mn	Si	Cr	Ni	Mo	Cu
Percent	0.40	0.90	0.25	0.50	0.55	0.20	0.00
Factor	0.197	3.98	1.18	2.08	1.20	1.60	1.00

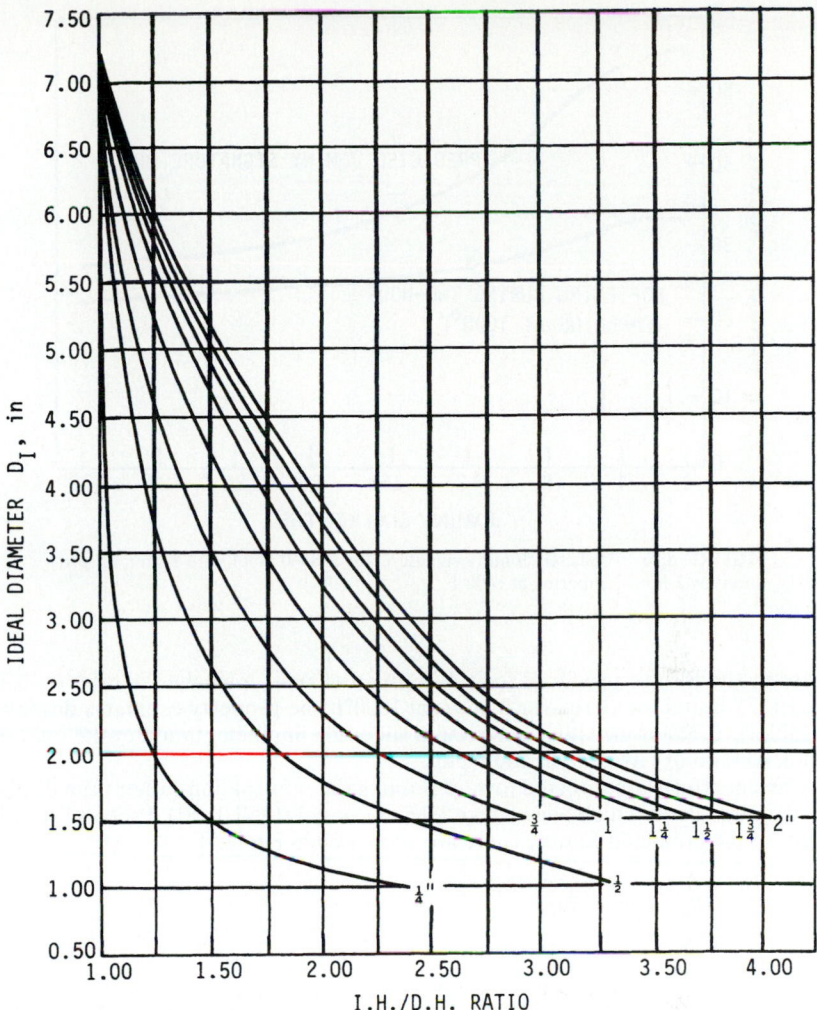

FIGURE 8.15 Relation between ideal critical diameter and the ratio of initial hardness IH to distant hardness DH. *(From [8.4] with permission of Pitman Publishing Ltd., London.)*

1-in-diameter quenched and tempered bar, and that the tensile specimen was taken from the center of that bar for plain carbon steels. Alloy-steel quenched and tempered bars were 0.532 in in diameter machined to a standard 0.505-in-diameter specimen. From the traverse of strengths in the previous array, it is clear that central and surface properties differ. In addition, the designer needs to know the properties of the critical location in the geometry and at condition of use. Methods of estimation such as the Crafts and Lamont addition method and the Grossmann and Fields multiplication method are useful prior to or in the absence of tests on the machine part.

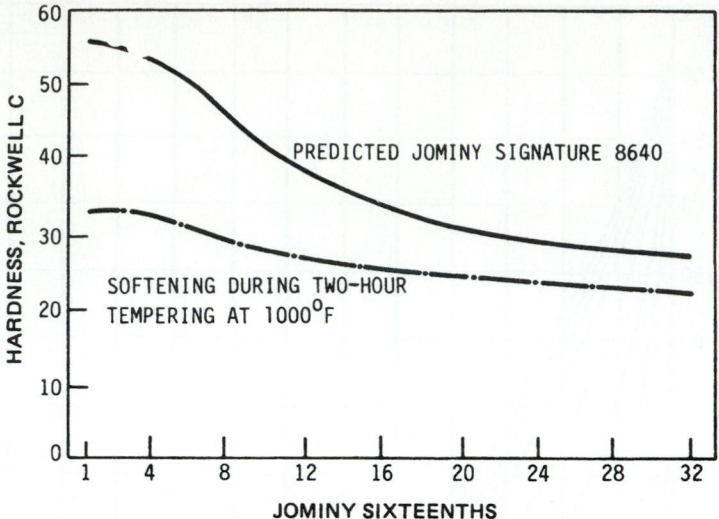

FIGURE 8.16 Predicted Jominy signature for a 8640 steel with softening produced by 2-hour tempering at 1000°F.

These methods have produced for a 4-in round of 8640, quenched in oil ($H = 0.35$) from 1575°F, and tempered for 2 hours at 1000°F, the property estimates displayed as Table 8.16. Reference [8.6] is a circular slide rule implementation of the multiplication method of Grossmann and Fields.

Current efforts are directed toward refining the information rather than displacing the ideas upon which Secs. 8.5 and 8.6 are based ([8.7], [8.8]). Probabilistic elements of the predicted Jominy curve are addressed in Ho [8.9].

TABLE 8.12 Prediction of Jominy Curve for 8640 Steel by Multiplication Method of Grossmann and Fields

Jominy distance	$\dfrac{IH}{DH}$	$R_Q = \dfrac{IH}{(IH/DH)}$
1	1.00	56.0
4	1.03	54.3
8	1.24	45.0
12	1.46	38.4
16	1.67	33.6
20	1.82	30.7
24	1.92	29.2
28	2.00	28.0
32	2.04	24.7

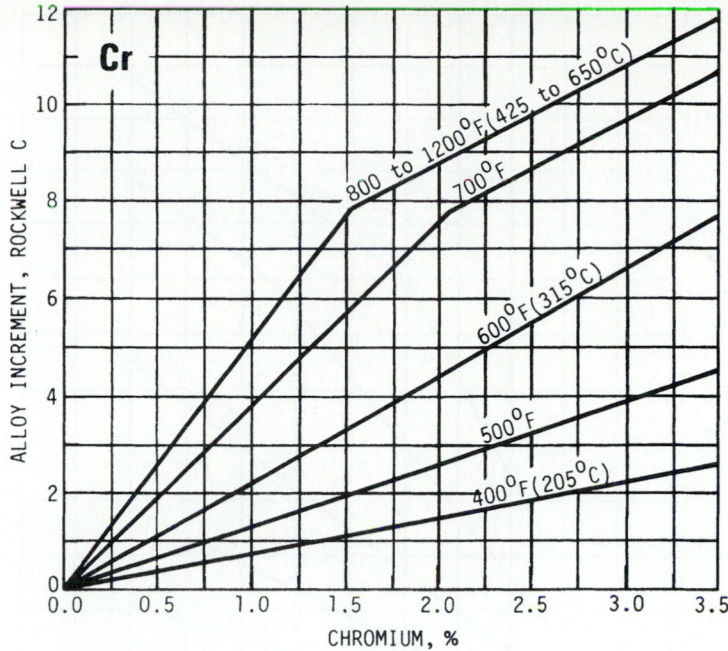

FIGURE 8.17 Effect of chromium on resistance to softening at various tempering temperatures. *(From [8.4] with permission of Pitman Publishing Ltd., London.)*

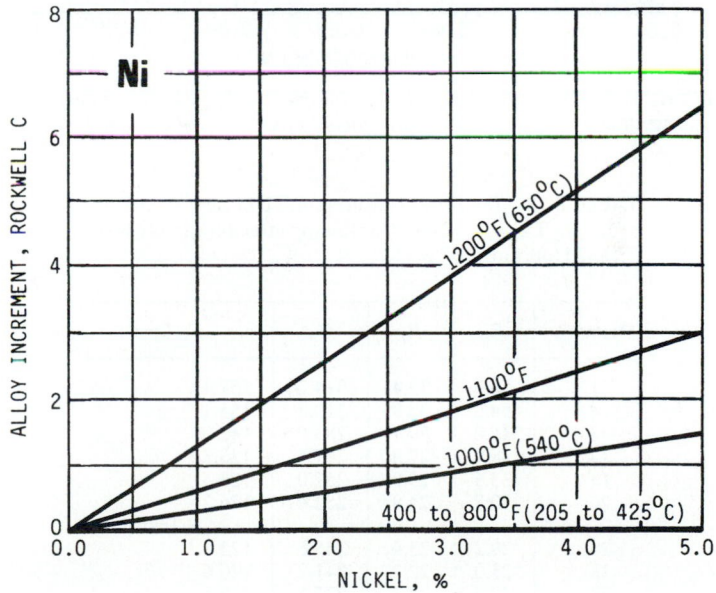

FIGURE 8.18 Effect of nickel on resistance to softening at various tempering temperatures. *(From [8.4] with permission of Pitman Publishing Ltd., London.)*

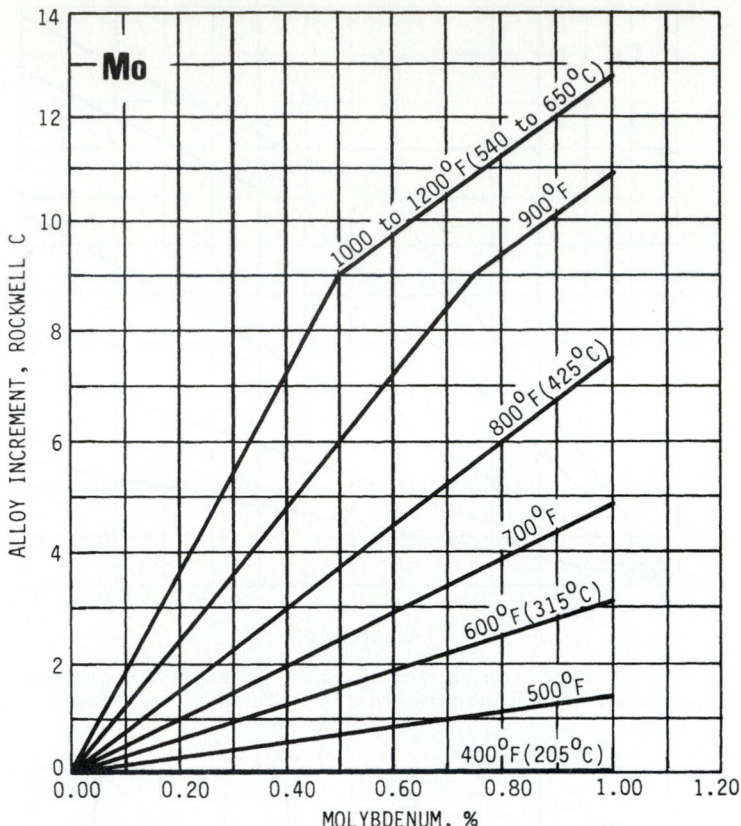

FIGURE 8.19   Effect of molybdenum on resistance to softening at various tempering temperatures. *(From [8.4] with permission of Pitman Publishing Ltd., London.)*

TABLE 8.13   Tempered Hardness and Ultimate Strength at Jominy Distances Due to Softening after Tempering 8640 Steel 2 Hours at 1000°F

Distance	R_Q	R_T	H_B	S_u, kpsi
1	56.0	33.4	314.2	157.1
4	54.3	32.9	310.2	155.1
8	45.0	29.7	283.9	142.0
12	38.4	27.5	267.5	133.8
16	33.6	25.8	257.0	128.5
20	30.7	24.8	252.4	126.2
				←Transition
24	29.2	23.9	246.6	123.3
28	28.0	22.7	241.2	120.6
32	27.4	22.1	237.6	118.8

TABLE 8.14   Surface Ultimate Strength of 8640 Steel Tempered for 2 Hours at 1000°F as a Function of Diameter of Round

Diameter, in	Equivalent Jominy distance, $\frac{1}{16}$ in	S_u, kpsi
0.5	2.7	156.0
1	5.1	151.5
2	8.2	141.6
3	10.0	137.3
4	11.4	135.0

TABLE 8.15   Ultimate and Yield Strength Traverse of a 4-in-Diameter Round of 8640 Steel Tempered 2 Hours at 1000°F

Location r, in	Equivalent Jominy distance, $\frac{1}{16}$ in	S_u, kpsi	S_y, kpsi
2	11.4	135.0	110.8
1.6	18.0	127.4	99.0
1	23.5	123.7	94.1
0	29.7	119.8	89.9

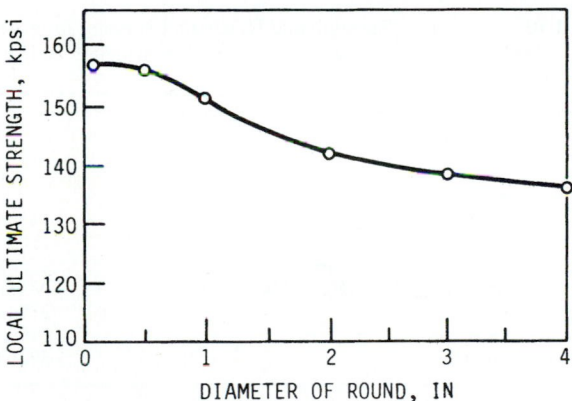

FIGURE 8.20   Variation on surface ultimate strength for 8640 steel oil-quenched ($H = 0.35$) from 1575°F and tempered for 2 hours at 1000°F as a function of diameter of round.

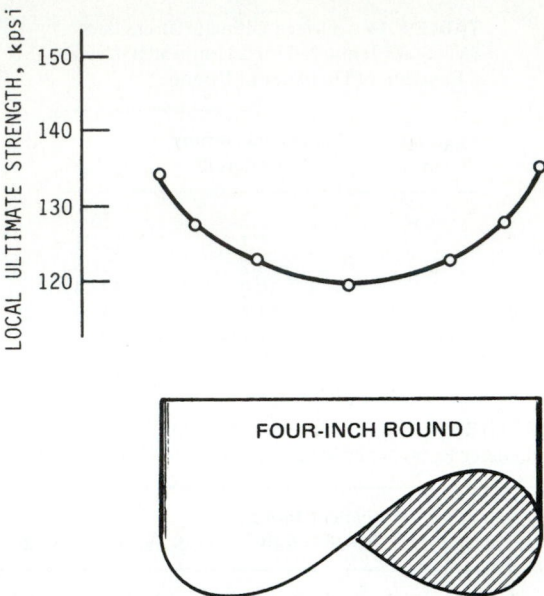

FIGURE 8.21   Variation in surface ultimate strength across a section of a 4-in round of 8640 steel oil-quenched ($H = 0.35$) from 1575°F and tempered for 2 hours at 1000°F as a function of radial position.

TABLE 8.16   Summary of Strength and Hardness Estimates for a 4-in Round of 8640 Steel Quenched in Oil ($H = 0.35$) from 1575°F and Tempered 2 Hours at 1000°F

Property	Estimate
Surface hardness	270 Brinell
Surface ultimate strength	135 kpsi
Surface yield strength	110.8 kpsi
Surface R. R. Moore endurance limit	67.5 kpsi
Contact endurance strength ($0.4H_B - 10$)	98 kpsi†
Central hardness	239.6 Brinell
Central ultimate strength	119.8 kpsi
Central yield strength	89.9 kpsi

† 10^8 cycles.

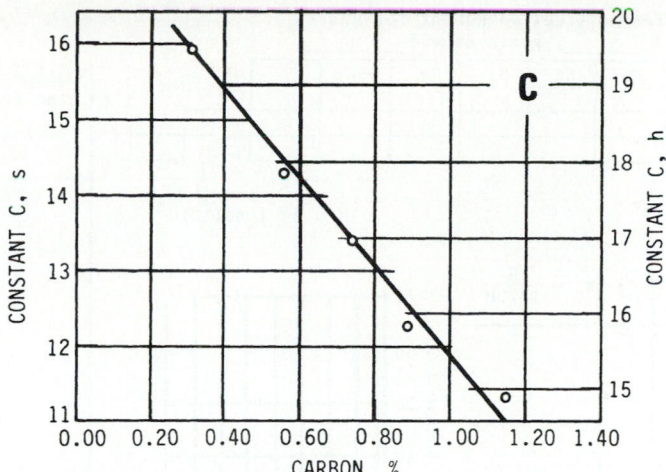

FIGURE 8.22 Variation with carbon content of constant C in time-temperature tradeoff equation for tempered, fully quenched plain carbon steels. *(From [8.4] with permission of Pitman Publishing Ltd., London.)*

8.7 TEMPERING TIME AND TEMPERATURE TRADEOFF RELATION

The tempering-temperature/time tradeoff equation is

$$(459 + F_1)(C + \log_{10} t_1) = (459 + F_2)(C + \log_{10} t_2) \tag{8.4}$$

where C is a function of carbon content determinable from Fig. 8.22. For 8640 steel, the value of C is 18.85 when the time is measured in hours. For a tempering temperature of 975°F, the tempering time is

$$(459 + 1000)(18.85 + \log_{10} 2) = (459 + 975)(18.85 + \log_{10} t_2)$$

from which $t_2 = 4.3$ h.

Since steel is bought in quantities for manufacturing purposes and the heat from which it came is identified as well as the ladle analysis, once such an estimation of properties procedure is carried out, the results are applicable for as long as the material is used. It is useful to employ a worksheet and display the results. Such a sheet is depicted in Fig. 8.23.

8.8 COMPUTER PROGRAMS

It is possible to program the digital computer to give mean values of the ultimate strength predictions [8.10]. An example of a computer-generated worksheet is displayed as Table 8.17. For Jominy distances 1 to 32, the Rockwell C-scale hardness is displayed as an ultimate strength as a function of the 2-hour tempering temperature in both IPS and SI units. The time-temperature tradeoff equation is displayed. Quench severities ($H = 0.35$ and $H = 0.50$) generate equivalent Jominy distances as

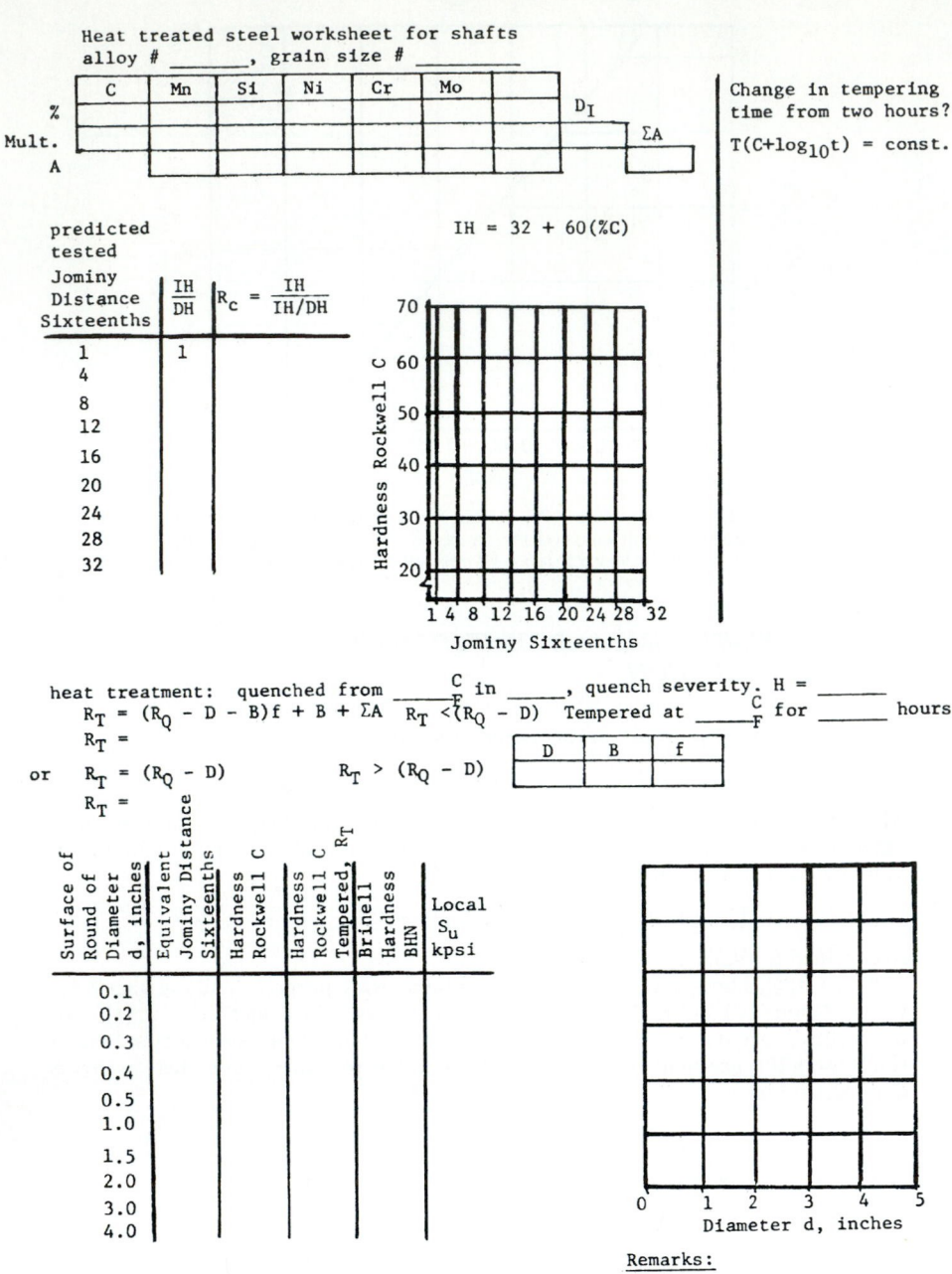

FIGURE 8.23   Heat-treated-steel worksheet for shafts.

a function of diameter and position within the round. These are converted to ultimate strengths for a 2-hour 1000°F tempering temperature. For this 1340 steel we can expect for a still-oil quench ($H = 0.35$) that is tempered 2 hours at 1000°F a surface ultimate strength on a 3-in round of 121.4 kpsi. The standard deviation is largest at Jominy Station 8 and is about two points on the Rockwell C scale [8.9].

TABLE 8.17   Computer-Generated Worksheet, CADET Program HTTREAT

```
HEAT TREATMENT WORKSHEET: 1340 STEEL, GRAINSIZE 7.0

CONSTITUENT          C     MN    SI    CR    NI    MO    CU
PER CENT          0.400 1.770 0.250 0.120 0.100 0.010 0.000  DI
MULTIPLYING FACTOR 0.213 7.933 1.175 1.259 1.037 1.030 1.000 2.675

HARDENABILITY DATA FOR TWO HOUR TEMPERING EXPRESSED AS ULTIMATE STRENGTH

JOM  JOM   400F  600F  800F  1000F 1200F JOM  200C  300C  400C  500C  600C
STA  RC    KPSI  KPSI  KPSI  KPSI  KPSI  RC   MPa   MPa   MPa   MPa   MPa

 1   56.0  265.6 224.1 189.1 144.2 112.5 56.0 1831. 1588. 1365. 1104.  859.
 4   50.6  238.1 207.7 179.7 137.3 110.2 50.6 1641. 1468. 1282. 1043.  834.
 8   37.8  171.2 169.2 154.1 123.6 104.4 37.8 1180. 1168. 1104.  918.  769.
12   31.5  145.0 143.0 141.2 118.1 100.9 31.5 1000.  988.  977.  872.  747.
16   27.2  129.6 128.2 127.0 114.0  98.5 27.2  893.  885.  878.  843.  732.
20   24.5  122.6 121.6 120.1 111.6  97.0 24.5  845.  839.  830.  797   720.
24   23.4  119.7 118.2 116.7 109.7  96.5 23.4  825.  816.  807.  778.  715.
28   22.5  117.1 115.6 114.4 108.2  96.2 22.5  808.  799.  791.  768.  703.
32   21.9  115.2 114.0 112.8 107.0  95.6 21.9  794.  787.  779.  760.  693.


TIME-TEMPERATURE TRADE-OFF FOR TEMPERING

(FAHR1+459)(18.85+LOG10(HOUR1))=(FAHR2+459)(18.85+LOG10(HOUR2))

EQUIVALENT JOMINY DISTANCES WITH OR WITHOUT TEMPERING
QUENCH SEVERITY H=0.35                   QUENCH SEVERITY =0.50

  DIA  0.0R 0.2R 0.4R 0.6R 0.8R 1.0R     0.0R 0.2R 0.4R 0.6R 0.R 1.0R

0.75  5.1  4.7  4.5  4.4  4.2  4.0      3.8  3.6  3.5  3.4  3.2  2.9
1.00  6.6  6.1  5.9  5.6  5.4  5.1      5.1  4.9  4.6  4.4  4.0  3.5
1.50  9.5  8.9  8.6  8.1  7.6  6.9      7.7  7.3  6.9  6.4  5.6  4.6
2.00 12.3 11.8 11.3 10.5  9.6  8.2     10.2  9.7  9.3  8.5  7.3  5.4
2.50 15.5 15.0 14.2 13.0 11.4  9.2     12.9 12.2 11.7 10.7  8.9  6.1
3.00 19.3 18.5 17.4 15.6 13.4 10.0     15.8 15.0 14.3 12.8 10.4  6.6
3.50 23.9 22.6 21.1 18.5 15.5 10.7     19.2 18.2 17.2 15.0 12.0  7.0
4.00 29.7 27.4 25.4 21.8 18.0 11.4     23.0 22.0 20.2 17.2 13.4  7.6

LOCAL ULTIMATE STRENGTH (KPSI) IN 1000F TEMPERED ROUNDS
QUENCH SEVERITY H=0.35                   QUENCH SEVERITY H=0.50
D= 5.31    B=  9.90   f=0.34   SUM A= 6.38
DIA   0.0R  0.2R  0.4R  0.6R   0.8R  1.0R    0.0R  0.2R  0.4R  0.6R  0.8R  1.0R

0.75 132.8 134.4 135.0 135.6 136.4 137.1    137.7 138.3 138.5 138.8 139.3 140.0
1.00 127.9 129.4 130.0 130.9 131.6 132.7    132.7 133.6 134.6 135.6 137.3 138.4
1.50 122.0 122.6 122.9 123.5 124.9 127.1    124.8 125.8 127.0 128.4 130.9 134.7
2.00 117.8 118.4 119.2 120.6 121.9 123.3    121.1 121.7 122.2 123.0 126.0 131.6
2.50 114.5 115.0 115.7 117.1 119.1 122.3    117.1 117.9 118.6 120.3 122.6 129.5
3.00 112.0 112.4 113.0 114.4 116.6 121.4    114.2 114.9 115.5 117.2 120.7 128.0
3.50 109.7 110.4 111.1 112.4 114.4 120.3    112.1 112.6 113.1 114.9 118.2 126.7
4.00 107.7 108.4 109.1 110.8 112.6 119.1    110.2 110.6 111.5 113.1 116.5 125.1

WATER-QUENCHING OF THIS STEEL NOT RECOMMENDED
WITHOUT CAREFUL EXPLORATION FOR QUENCH CRACKING
```

A simple interactive Fortran program can take the detailed drudgery from the cold-worked strength estimation procedure based on Datsko's method, Secs. 8.1 through 8.4. It is based on Datsko's 35 years of research on aluminum, copper, brass, carbon steel, stainless steel, and high-strength steels. The average error is about 5 percent when the strengths were increased 400 to 600 percent. The Fortran source program follows.

```
C EFFECT OF COLDWORK ON STRENGTH ALA DATSKO.  C. MISCHKE JAN 1987
      CHARACTER*10,FILNAM
      CHARACTER*9,DAY
      DIMENSION equ(4),eqy(4),ew(20),e(100),Su(4),Sy(4),iabc(60)
      real m
      CALL DATE(DAY)
      icount=1
      print*,'CADET Program COLDWORK based on method of J. Datsko'
      print*,'Iowa State University, Mechanical Engineering, C.Mischke'
    1 iflag=0
      write(*,600)
  600 format('$Do you want the output to go to a file (y/n)? ')
      READ(*,900),FANS
  900 FORMAT(A1)
      if(fans.eq.'y'.or.fans.eq.'Y')iflag=1
      if(fans.eq.'n'.or.fans.eq.'N')iflag=2
      if(iflag.eq.0) go to 1
      if(iflag.eq.1) then
      write(*,601)
  601 format('$Enter the file name   xxxxxx.DAT:     ')
      read(*,602),filnam
  602 format(a10)
      else
      filnam='sys$output'
      end if
      print*,'Enter material description, up to 60 characters'
      read(5,10)iabc
   10 format(60a1)
      print*,'Enter engineering yield strength Sy in kpsi and offset'
      read*,Syo,offset
      print*,'Enter engineering ultimate strength Su in kpsi'
      read*,Suo
      print*,'Enter fractional reduction in area from tensile test'
      read*,AR
      epsif=alog(1./(1.-AR))
      print*,'Strain-strengthening exponent m '
      print*,'If it is known to you,        enter 1'
      print*,'For computer to estimate it, enter 2'
      read*,index
      If(index.eq.1)Print*,'Enter exponent m'
      if(index.eq.1)read*,m
      if(index.eq.1)go to 11
      m=0.1
      do 12 i=i,10
      m=alog(suo/Syo)/log(m/(offset*2.718))
   12 continue
   11 sigmao=Suo*exp(m)*m**(-m)
      print*,'If bending is present in plastic strain cycle, enter 1'
      print*,'otherwise enter 0'
      read*,k
      print*,'Enter the number of successive plastic strains in cycle'
      read*,j
      do 13 i=1,j
      print*,'Enter largest strain in step(',i,') of cycle'
      read*,ew(i)
   13 continue
      if(iflag.eq.1)open(unit=6,file=filnam,status='new',err=987)
      if(iflag.eq.1)write(6,899)DAY
      write(*,899),DAY
  899 format(' CADET Program COLDWORK, Method of Datsko',9X,'DATE:',A9)
      if(iflag.eq.1)write(6,50)(ew(i),i=1,j)
      write(*,50),(ew(i),i=1,j)
   50 format(' Strain sequence is ',10f7.3)
      do 14 i=1,j
      ew(i)=abs(ew(i))
      ew(i)=-ew(i)
```

```
   14 continue
      call me0200(j,ew)
      do 15 i=1,j
      e(i)=abs(ew(i))
   15 continue
c Group 1 strengths
      equ(1)=0.
      do 16 i=1,j
      equ(1)=equ(1)+e(i)/float(i)
   16 continue
      if(equ(1).gt.epsif)print*,'Strain cycle too severe, rupture'
      if(equ(1).gt.epsif)close(unit=6)
      if(equ(1).gt.epsif)call exit
      eqy(1)=equ(1)/(1.+0.2*equ(1))
      Sy(1)=sigmao*eqy(1)**m
      if(equ(1).le.m)Su(1)=Suo*exp(equ(1))
      if(equ(1).gt.m)Su(1)=sigmao*equ(1)**m
      if(k.eq.1.and.equ(1).gt.m)write(*,901)
      if(k.eq.1.and.equ(1).gt.m.and.iflag.eq.1)write(6,901)
  901 format(' PRESENCE OF BENDING REQUIRES CRACK TESTS!')
      WRITE(*,902)
      IF(IFLAG.EQ.1)WRITE(6,902)
  902 format(' Group equ     Su      eqy      Sy')
      if(iflag.eq.1)write(6,17)icount,equ(1),Su(1),eqy(1),Sy(1)
      write(*,17),icount,equ(1),Su(1),eqy(1),Sy(1)
   17 format(2x,I2,2x,F5.3,2x,F5.1,2x,F5.3,2x,F5.1)
      icount=icount+1
c Group 2 strengths
      equ(2)=equ(1)
      eqy(2)=equ(2)/(1.+0.5*equ(2))
      Sy(2)=sigmao*eqy(2)**m
      Su(2)=Su(1)
      if(iflag.eq.1)write(6,17)icount,equ(2),Su(2),eqy(2),Sy(2)
      write(*,17),icount,equ(2),Su(2),eqy(2),Sy(2)
      icount=icount+1
c Group 3 strengths
      equ(3)=0.
      do 18 i=1,j
      equ(3)=equ(3)+e(i)/float(1+i)
   18 continue
      eqy(3)=equ(3)/(1.+2.*equ(3))
      Sy(3)=sigmao*eqy(3)**m
      if(equ(3).le.m)Su(3)=Suo*exp(equ(3))
      if(equ(3).gt.m)Su(3)=sigmao*equ(3)**m
      if(iflag.eq.1)write(6,17)icount,equ(3),Su(3),eqy(3),Sy(3)
      write(*,17),icount,equ(3),Su(3),eqy(3),Sy(3)
      icount=icount+1
c Group 4 strengths
      equ(4)=equ(3)
      SytTc=0.95*Sy(2)
      Su(4)=Su(3)
      if(iflag.eq.1)write(6,19)icount,equ(4),Su(4),SytTc
      write(*,19),icount,equ(4),Su(4),SytTc
   19 format(2x,I2,2x,F5.3,2x,F5.1,9x,F5.1)
      if(iflag.eq.1)write(6,904)
      write(*,904)
  904 format(/,
     1' Strengths are grouped as follows:',/,
     2' Group 1      Group 2     Group 3     Group 4',/,
     3' (S )cLc      (S )tTt     (S )cLt     (S )tTc',/,
     4' (S )tLt      (S )cTc     (S )tLc     (S )cTt',/,
     4' (S )tBo                  (S )tDc',/,
     5' (S )cBo',/,
     6' (S )cDc',/)
      if(iflag.eq.1)write(6,25)iabc
      write(*,25),iabc
   25 format(1x,60a1)
      if(iflag.eq.1)write(6,905)Syo,offset,Suo
      write(*,905),Syo,offset,Suo
  905 format(' Syo=',F6.1,' kpsi, offset=',f6.3,', Suo=',f6.1,' kpsi')
      if(iflag.eq.1)write(6,916)AR
      write(*,916),AR
  916 format(' Fractional area reduction =', F5.2)
      if(iflag.eq.1)write(6,906)epsif
      write(*,906),epsif
  906 format(' True strain at fracture is epsilon =',F6.3)
```

```
      if(iflag.eq.1)write(6,907)sigmao
      write(*,907),sigmao
907   format(' Strain-strengthening coeff. is sigmao =',f6.1,' kpsi')
      if(index.eq.1.and.iflag.eq.1)write(6,908)m
      if(index.eq.1)write(*,908),m
908   format(' Strain-strengthening exponent is m =',f5.2)
      if(index.eq.2.and.iflag.eq.1)write(6,909)m
      if(index.eq.2)write(*,909),m
909   format(' IDEAL BEHAVIOR exponent m =',f5.2,' CAREFUL!')
      close(unit=6)
      if(iflag.eq.1)print*,'Output in directory under name ',filnam
      call exit
987   print*,'SYSTEM ERROR WRITING TO FILE',FILNAM
      call exit
      end
```

REFERENCES

8.1 J. Datsko, *Materials in Design and Manufacturing,* published by the author, Ann Arbor, Mich., 1977.

8.2 M. P. Borden, "Multidimensional Tensile Properties of Materials Subjected to Large Cyclic Strains," Ph.D. thesis, University of Michigan, Ann Arbor, 1975.

8.3 R. W. Hertzberg, *Deformation and Fracture Mechanics of Engineering Materials,* John Wiley & Sons, New York, 1976.

8.4 W. Crafts and J. L. Lamont, *Hardenability and Steel Selection,* Pitman & Sons, London, 1949.

8.5 *Modern Steels and Their Properties,* Bethlehem Steel Corporation, 1972.

8.6 *Bethlehem Alloy Steel Hardenability Calculator, Calc 96,* Bethlehem Steel Corporation, 1966.

8.7 D. V. Doane and J. J. Kirkaldy (eds.), *Hardenability Concepts with Applications to Steel,* American Institute of Mining, Metallurgical and Petroleum Engineers, Warrendale, Pa., 1978.

8.8 C. A. Siebert, D. V. Doane, and D. H. Breen, *The Hardenability of Steels,* American Society for Metals, Metals Park, Ohio, 1977.

8.9 T. K. Ho, "Probabilistic Prediction of the Jominy Curve of Low Alloy Steels from Composition and Grain Size," Ph.D. thesis, Iowa State University, Ames, 1978.

8.10 *Iowa CADET Documentation Manual,* Iowa State University Bookstore, Ames, revised annually.

RECOMMENDED READING

P. Harvey (ed.), *Engineering Properties of Steel,* American Society for Metals, Metals Park, Ohio, 1982.

C. S. Brady and H. R. Caluser, *Metals Handbook,* 11th ed., McGraw-Hill, New York, 1977.

C. R. Brooks, *Heat Treatment, Structure and Properties of Non Ferrous Alloys,* American Society for Metals, Metals Park, Ohio, 1982.

G. Krauss, *Principles of Heat Treatment of Steel,* American Society for Metals, Metals Park, Ohio, 1980.

Metals Handbook, 9th ed., Vol. 1: *Properties and Selection: Irons and Steels;* Vol. 4: *Heat Treating,* American Society for Metals, Metals Park, Ohio, 1981.

E. A. Brandes (ed.), *Smithells Metal Reference Book,* 6th ed., Butterworths, London, 1983.

CHAPTER 9
USABILITY

Karl H. E. Kroemer, Ph.D.
Professor of Industrial and Systems Engineering
Virginia Tech
Blacksburg, Virginia

Not only must tools, equipment, and machines function, but in many cases their effectiveness depends on how well a human can use and operate them. A pair of pliers is useless unless it is held in the human hand; a lathe (if not run automatically) needs an operator to observe the cutting edge, to operate controls, and to feed and unload; maintenance and repair of equipment must be facilitated by proper design.

Of course, fitting tools and work to human capabilities and limitations has always been done, but this was formally established as "work physiology" and "industrial psychology" early in the twentieth century. During the Second World War, "human engineering" was systematically applied to weapon systems, and since then it has been increasingly applied to technical products and human-machine systems. *Ergonomics,* the current generally used term, is rooted in safety and ease of use; its desired outcome is the optimization of work, especially of the interface between the human and the technical product.

Designing for human use is the field of ergonomics, or human (factors) engineering. The term *ergonomics* was coined in 1950 from two Greek words: *ergon* for human work and *nomos* for rules. In the United States, the Human Factors and Ergonomics Society is the professional organization; the worldwide umbrella organization is the International Ergonomics Association, with nearly three dozen national member societies. Courses in ergonomics or human engineering are taught in more than fifty engineering departments (mostly industrial engineering) and psychology departments (engineering psychology) in North American universities.

Books provide encompassing information about ergonomics and its engineering applications; in English, for example, there are publications by Boff, Kaufman, and Thomas [9.1]; Cushman and Rosenberg [9.2]; Eastman Kodak Company [9.3]; Fraser [9.4]; Grandjean [9.5]; Helander [9.6]; Kroemer, Kroemer, and Kroemer-Elbert [9.7],

[9.8]; Proctor and Van Zandt [9.9]; Pulat [9.10]; Salvendy [9.11]; Sanders and McCormick [9.12]; Weimer [9.13]; Wilson and Corlett [9.14]; and Woodson, Tillman, and Tillman [9.15]. Furthermore, standards offer practical information, in particular U.S. Military Standards 759 and 1472, as well as more specific issues by the U.S. Air Force, Army, and Navy, and NASA Standard 3000. The American Society of Safety Engineers (ASSE), the Society of Automotive Engineers (SAE), and the American Society of Heating, Refrigerating, and Air Conditioning Engineers (ASHRAE) as well as the American National Standards Institute (ANSI) and the Occupational Safety and Health Agency (OSHA) issue ergonomic standards on specific topics. (Addresses are given in the References section.)

9.1 DESIGNING FOR HUMAN BODY SIZE

"Fitting" a hand tool, a machine, or a complex technical system to the operator is very important: Pliers are hard to use if the handles hurt the hand; a caulking gun that has handles so far apart that persons with small hands cannot grasp it is unusable for many; gloves that don't fit won't be used. Tools, machines, and systems can be designed to fit the body, whereas genetic engineering of the body to fit ill-designed equipment is not practical. The axiom is, "Fit tool and task to the human."

Four steps assure that the product or system fit the operator (see Ref. [9.8] for more details):

Step 1. Select those body dimensions that directly relate to equipment dimensions. For example, hand size should be related to handle size; shoulder and hip breadth to an opening through which a repair person must enter; head length and breadth to helmet size; eye height to the height of an object that must be seen, such as a computer display; knee height and hip breadth to the leg room needed by a seated operator.

Step 2. For each of these pairings, decide whether the design must fit only one given body dimension or a range of body dimensions. For example, an opening must be large enough to allow the person with the largest shoulder and hip breadths to pass through, even when wearing bulky clothing and equipment; pliers can come in different sizes to fit either small or large hands; the height of a seat should be adjustable to accommodate persons ranging from short to tall, with different lower leg lengths.

Step 3. Combine all selected design values in a careful drawing, computer model, or mock-up to ascertain that they are compatible. For example, the leg-room clearance height needed for a seated person with long lower legs might be very close to the height of the working object, which is related to elbow height.

Step 4. Determine whether one design will fit all users; if not, several sizes or adjustability are needed. For example, a large opening will allow all users to pass through; work clothes must come in different sizes; pilot seats are adjustable to fit female and male, small and big air crew members.

9.1.1 Available Anthropometric Information

Human body dimensions are measured by anthropometrists. Unfortunately, large surveys of national populations have been performed almost exclusively on soldiers;

very few large civilian groups have been measured in recent years. Thus, the available information is usually derived from soldier anthropometry, and these data are then applied to the adult population in general.

Table 9.1 contains body dimensions of U.S. adults. These numbers have been extracted from recent compilations by Gordon et al. [9.16] and Greiner [9.17], who reported a large number of U.S. Army body dimensions. Some information on the body dimensions of elderly persons, of children, and of pregnant women is available as well—see, for example, tables published recently by Kroemer, Kroemer, and Kroemer-Elbert [9.8] and Roebuck [9.18].

Fortunately, measurements of human body dimensions usually fall into "normal" (Gaussian) distributions which can be described statistically in terms of average (mean) and standard deviation, provided that a sufficient number of people is included in the survey. Hence, one can apply regular parametric statistics.

9.1.2   Use of Percentiles

Percentile values can be determined from anthropometric data. The 50th percentile coincides, in a normal distribution, with the average. Average values for important body dimensions are given in Table 9.1 (in the column labeled 50th percentile), together with the standard deviation. If one multiplies the standard deviation S by the factor k presented in Table 9.2, one can determine percentile values below or above which lie known subsamples. For example, below the 2d percentile are 2 percent of all data and the remaining 98 percent are above; conversely, 98 percent of all data lie below the 98th percentile and 2 percent of all data are above. To determine the 2d percentile, or the 98th percentile, one multiplies the standard deviation of the anthropometric dimensions by the factor 2.06 (as shown in Table 9.2). For the 2d percentile, the product is deducted from the average; it is added to the average in order to determine the 98th percentile. In the range between the 2d and 98th percentiles, 96 percent of all data are contained.

Percentiles serve the designer/engineer in several ways [9.8]. First, they help to establish the portion of a user population that will be able to make (or excluded from making) proper use of a specific piece of equipment. Second, knowledge of percentile values can be used to select subjects for fit tests. Third, any design value or a body dimension can be exactly located on the range for that specific dimension.

9.1.3   Models of Operator Size

Some body dimensions are highly correlated, such as eye height and stature. Other dimensions are practically unrelated, such as stature and hip breadth. In the case of high correlations, one can use one dimension to predict another: If eye height is unknown but stature has been measured, one can predict eye height from stature with high accuracy. However, some height dimensions and almost all width and depth dimensions are practically unrelated to stature; thus, one cannot assume, with sufficient certainty, that a short person must have narrow hips or small wrists, or be of light weight.

Therefore, one must be careful when estimating body dimensions from others. If needed body dimensions are unknown, one has to take specific body size measurements of the equipment operators and product users; it may be necessary to use the expertise of ergonomists or anthropometrists. A common mistake is using "the average person," a phantom who is assumed to possess average dimensions throughout.

TABLE 9.1 Selected Anthrometric Data of the U.S. Adult Population, Females/Males

All values in cm, except weight in kg.

Dimensions	5th	50th	95th	Standard deviation
		Percentile		
Heights, standing				
Stature ("height")	152.8/164.7	162.94/175.58	173.7/186.6	6.36/6.68
Eye	141.5/152.8	151.61/163.39	162.1/174.3	6.25/6.57
Shoulder (acromion)	124.1/134.2	133.36/144.25	143.2/154.6	5.79/6.20
Elbow	92.6/99.5	99.79/107.25	107.4/115.3	4.48/4.81
Wrist	72.8/77.8	79.03/84.65	85.5/91.5	3.86/4.15
Crotch	70.0/76.4	77.14/83.72	84.6/91.6	4.41/4.62
Overhead fingertip reach (on toes)	200.6/216.7	215.34/132.80	231.3/249.4	9.50/9.99
Heights, sitting				
Sitting height	79.5/85.5	85.20/91.39	91.0/97.2	3.49/3.56
Eye	68.5/73.5	73.87/79.02	79.4/84.8	3.32/3.42
Shoulder (acromion)	50.9/54.9	55.55/59.78	60.4/64.6	2.86/2.96
Elbow rest	17.6/18.4	22.05/23.06	27.1/27.4	2.68/2.72
Knee	47.4/51.4	51.54/55.88	56.0/60.6	2.63/2.79
Popliteal	35.1/39.5	38.94/43.41	42.9/47.6	2.37/2.49
Thigh clearance	14.0/14.9	15.89/16.82	18.0/19.0	1.21/1.26
Depths				
Chest	20.9/21.0	23.94/24.32	27.8/28.0	2.11/2.15
Elbow–fingertip	40.6/44.8	44.35/48.40	48.3/52.5	2.36/2.33
Buttock–knee sitting	54.2/56.9	58.89/61.64	64.0/66.7	2.96/2.99
Buttock–popliteal sitting	44.0/45.8	48.17/50.04	52.8/54.6	2.66/2.66
Thumbtip reach	67.7/73.9	73.46/80.08	79.7/86.7	3.64/3.92
Breadths				
Forearm–forearm	41.5/47.7	46.85/54.61	52.8/62.1	3.47/4.36
Hip, sitting	34.3/32.9	38.45/36.68	43.2/41.2	2.72/2.52
Head dimensions				
Length	17.6/18.5	18.72/19.71	19.8/20.9	0.64/0.71
Breadth	13.7/14.3	14.44/15.17	15.3/16.1	0.49/0.54
Circumference	52.3/54.3	54.62/56.77	57.1/59.4	1.46/1.54
Interpupillary breadth	5.7/5.9	6.23/6.47	6.9/7.1	0.36/0.37
Hand dimensions				
Wrist circumference	14.1/16.2	15.14/17.43	16.3/18.8	0.69/0.82
Length, stylion to tip 3	16.5/17.8	18.07/19.41	19.8/21.1	0.98/0.99
Breadth, metacarpal	7.4/8.4	7.95/9.04	8.6/9.8	0.38/0.42
Circumference, metacarpal	17.3/19.8	18.65/21.39	20.1/23.1	0.86/0.98
Digit 1: breadth, distal joint	1.9/2.2	2.06/2.40	2.3/2.6	0.13/0.13
Length	5.6/6.2	6.35/6.97	7.2/7.8	0.48/0.48
Digit 2: breadth, distal joint	1.5/1.8	1.73/2.01	1.9/2.3	0.12/0.15
Length	6.2/6.7	6.96/7.53	7.7/8.4	0.46/0.49
Digit 3: breadth, distal joint	1.5/1.7	1.71/1.98	1.9/2.2	0.11/0.14
Length	6.9/7.5	7.72/8.38	8.6/9.3	0.51/0.54
Digit 4: breadth, distal joint	1.4/1.6	1.58/1.85	1.8/2.1	0.11/0.14
Length	6.4/7.1	7.22/7.92	8.1/8.8	0.50/0.52
Digit 5: breadth, distal joint	1.3/1.5	1.47/1.74	1.7/2.0	0.11/0.13
Length	5.1/5.7	5.83/6.47	6.6/7.3	0.46/0.49
Foot dimensions				
Length	22.4/24.9	24.44/26.97	26.5/29.2	1.22/1.31
Breadth	8.2/9.2	8.97/10.06	9.8/11.0	0.49/0.53
Lateral malleolus height	5.2/5.8	6.06/6.71	7.0/7.6	0.53/0.55
Weight (kg), U.S. Army	49.6/61.6	62.01/78.49	77.0/98.1	8.35/11.10
Weight (kg), civilians[†]	39/58[†]	62.0/78.5[†]	85/99[†]	13.8/12.6[†]

[†] Estimated (from Kroemer, 1981).

Note that all values (except for civilians' weight) are based on measured, not estimated, data that may be slightly different from values calculated from average plus or minus 1.65 standard deviation.

Source: Adapted from [9.15] and [9.16].

9.4

TABLE 9.2 Calculation of Percentiles Using the Average and Multiples of the Standard Deviation

Percentile p associated with			
$x_i = \bar{x} - kS$ (below mean)	$x_j = \bar{x} + kS$ (above mean)	Central percent included in the range x_i to x_j	k
0.5	99.5	99	2.576
1	99	98	2.326
2	98	96	2.06
2.5	97.5	95	1.96
3	97	94	1.88
5	95	90	1.65
10	90	80	1.28
15	85	70	1.04
16.5	83.5	67	1.00
20	80	60	0.84
25	75	50	0.67
37.5	62.5	25	0.32
50	50	0	0

(People who are all 5th, or nth, percentile are figments of the imagination as well.) As discussed above, it is necessary to consider ranges of body dimensions, and to ascertain whether correlations exist between sets of body dimensions. For example, there is only a very small statistical correlation (about 0.4) between body height and body weight, contradicting the popular image of ideal height/weight ratios. Several such misleading body-proportion models have been used in the past, including design templates with fixed body proportions or CAD/CAM programs that utilize single-percentile constructs of the human body.

Human bodies come in a variety of sizes and proportions. Information about these is available (see especially Refs. [9.8], [9.16], [9.17], and [9.18]), and this can and must be used by the engineer to assure that the design fits the user.

9.2 DESIGNING FOR HUMAN BODY POSTURE

People seldom do work when lying supine or prone, but such postures do occur—for example, in repair jobs, or in low-seam underground mining. In some fighter airplanes and tanks, or in low-seam mining equipment, pilots or drivers are semireclining. There are also transient or temporary work postures such as kneeling on one or both knees, squatting, or stooping, often in confined spaces such as the cargo holds of aircraft; these postures as well as reaching, bending, and twisting the body should be avoided even in short-term activities to avert fatigue or injury. Proper equipment design is the task of the design engineer; proper equipment use is the responsibility of the manager.

By itself, lying is the least strenuous posture in terms of physical effort as measured by oxygen consumption or heart rate. Yet it is not well suited for performing physical work with the arms and hands because they must be elevated for most activities. Standing is much more energy-consuming, but it allows free use of the arms and hands, and, if one walks around, much space can be covered. Walking facilitates dynamic use of the body and is suitable for the development of fairly large energies and impact forces.

Sitting is, in most respects, between these two postures. Body weight is partially supported by a seat; energy consumption and circulatory strain are higher than when lying, but lower than when standing. Arms and hands can be used freely, although the work space they can cover is more limited than when walking. The energy that can be developed is smaller than when standing, but because of the stability of the trunk when it is supported on the seat, performing finely controlled manipulations is easier. Operation of pedals and controls with the feet is easy in the sitting posture: The feet are fairly mobile, since they are little needed to stabilize the posture and support the body weight.

Sitting and standing are usually thought to involve a more or less "upright" or "erect" trunk. The model of all major body joints at 0, 90, or 180 degrees is used for standardization of body measurements, but it is neither commonly employed, nor even proven to be healthy. Thus, the convenient model of the "0–90–180 posture" at work is just another phantom, like the "average person." In fact, deviations are common, subjectively preferred, and desirable in terms of variations in posture; moving about breaks maintained static muscle efforts and provides physiological stimuli and exercise.

9.2.1 Designing for the Standing Operator

Standing is used as a working posture if sitting is not suitable, either because the operator has to cover a fairly large work area or because very large forces must be exerted with the hands, particularly if these conditions prevail only for a limited period of time. Forcing a person to stand simply because the work object is customarily put high above the floor is usually not a sufficient justification; for example, in automobile assembly, car bodies can be turned or tilted, and parts redesigned, so that the worker does not have to stand and bend in order to reach the work object. Some work stations are designed for standing operators because of a need to exert large forces over large spaces, make strong exertions with visual control, or work with large objects are shown in Fig. 9.1.

People should never be forced to stand still at a work station just because the equipment was originally badly designed or badly placed, as is unfortunately too often the case with drill presses used in continuous work. Also, many other machine tools, such as lathes, have been so constructed that the operator must stand and lean forward to observe the cutting action, and at the same time extend the arms to reach the controls on the machine.

The height of the work station depends largely on the activities to be performed with the hands and the size of the object. In fitting the work station to the operator, the main reference point is the operator's individual elbow height, as further discussed below. The support surface (for example, workbench or table) is determined by the working height of the hands and the size of the object on which the person works.

Sufficient room for the operator's feet must be provided, including toe and knee space to allow him or her to move up close to the work area. Of course, the floor should be flat and free of obstacles; use of platforms to stand on should be avoided, if possible, because the operator may stumble over the edge. While movements of the body associated with dynamic work are, basically, a desirable physiological feature, they should not involve excessive bends and reaches, and especially should not include twisting motions of the trunk; these can cause overexertions and injury, often to the low back [9.8].

FIGURE 9.1 Work stations designed for standing operators. (With permission from K. H. E. Kroemer, H. B. Kroemer, and K. E. Kroemer-Elbert, (1994), *Ergonomics: How to Design for Ease and Efficiency*. All rights retained by the publisher, Prentice Hall, Englewood Cliffs, NJ.)

9.2.2 Designing for the Sitting Operator

Sitting is a much less stressful posture than standing. It allows better-controlled hand movements, but permits coverage of only a smaller area and exertion of smaller forces with the hands. A sitting person can easily operate controls with the feet and do so, if suitably seated, with much force (see below). When designing a work station for a seated operator, one must particularly consider the free space required for the legs. If this space is severely limited, very uncomfortable and fatiguing body postures result, as shown in Fig. 9.2.

The height of the working area for the hands is mostly determined by elbow height. However, many activities require close visual observation; thus eye height

FIGURE 9.2 Missing leg room makes for an awkward sitting posture.

co-determines the proper height of the manipulation area, depending on the operator's preferred visual distance and direction of gaze. The design principles for accommodating a seated person are discussed in more detail later in this chapter. In some work stations, sit-stand transitions are suitable, as shown in Fig. 9.3.

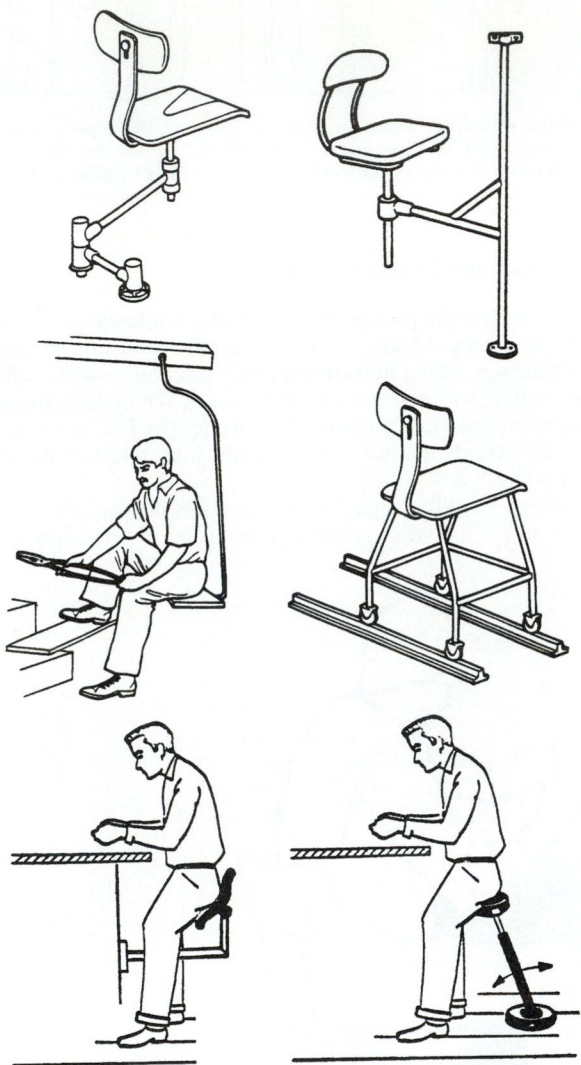

FIGURE 9.3 Stools and body props for sit–stand transitions. (With permission from K. H. E. Kroemer, H. B. Kroemer, and K. E. Kroemer-Elbert, (1994), *Ergonomics: How to Design for Ease and Efficiency*. All rights retained by the publisher, Prentice Hall, Englewood Cliffs, NJ.)

9.3 DESIGNING FOR REACH AND MOBILITY

Reach is the ability to extend hands and arms, or feet and legs, to touch and operate a control. Objects at the periphery of one's reach can just barely be pushed, pulled, turned, but more complex operations can be performed within the reach envelope.

The utmost reach envelope depends on the location of the body joint about which the limb moves; usually, this is the shoulder for hand reaches and the hip for foot reaches. The radius is the length of arm or leg. The contours of reach envelopes are nearly spherical in front and to the sides, and above and below the joint; but to the rear of the body, these envelopes become much reduced, as shown in Figs. 9.4 and 9.5.

The most preferred working areas are sections of the reach envelope in front of the body and close to the body, as shown in Fig. 9.6. For the hands, the preferred areas are directly in front of the chest at about elbow height, with the arm more or less bent. In these areas, motions can be performed most quickly, with best accuracy, and with least effort. (These areas are also suitable for exertion of moderate to large

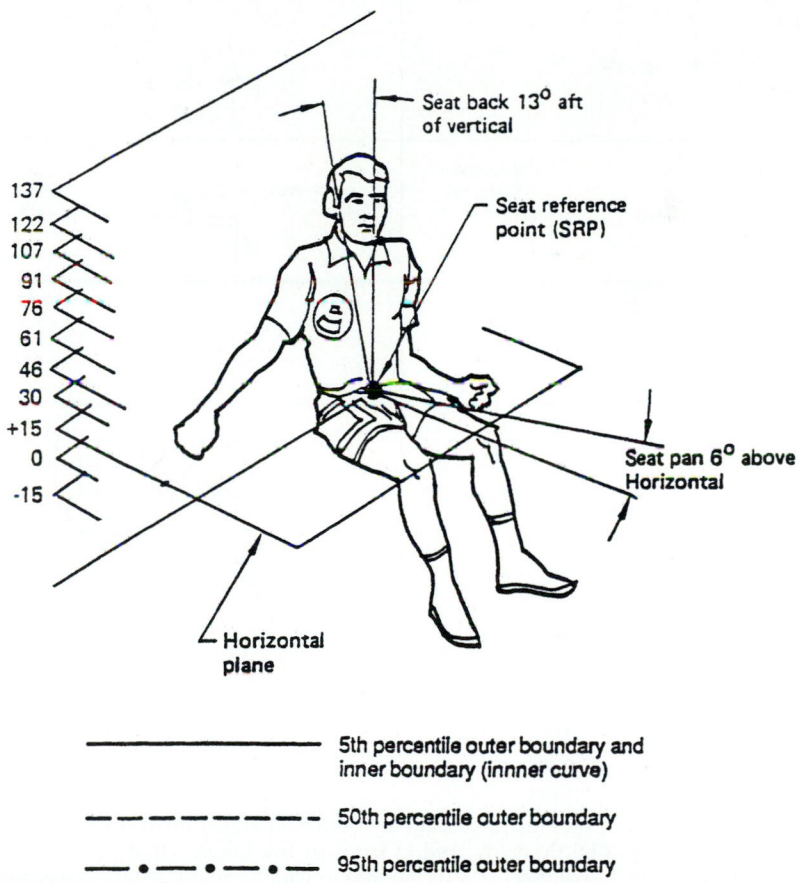

FIGURE 9.4 Reference planes for reaches. (Adapted from NASA STD 3000.)

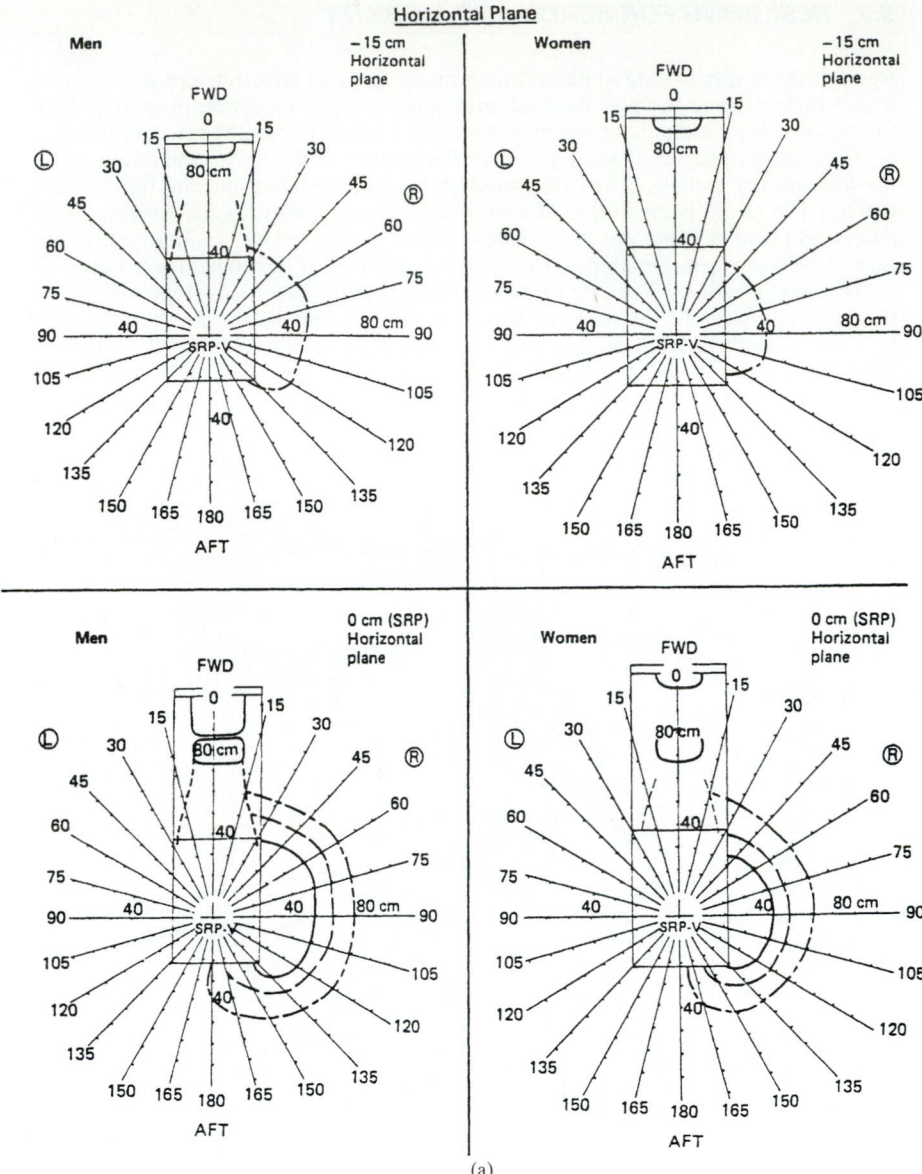

FIGURE 9.5 Examples of reach envelopes of seated operators. (Adapted from NASA STD 3000.)

hand forces, as discussed in the next section.) For the feet, the most suitable area for a seated operator is slightly below and in front of the knees—that is, with a knee angle of about 90 to 120 degrees. This is an area in which relatively fast and accurate foot motions can be made. (Foot forces in this posture are only small to moderate, however; see below.)

Horizontal Plane

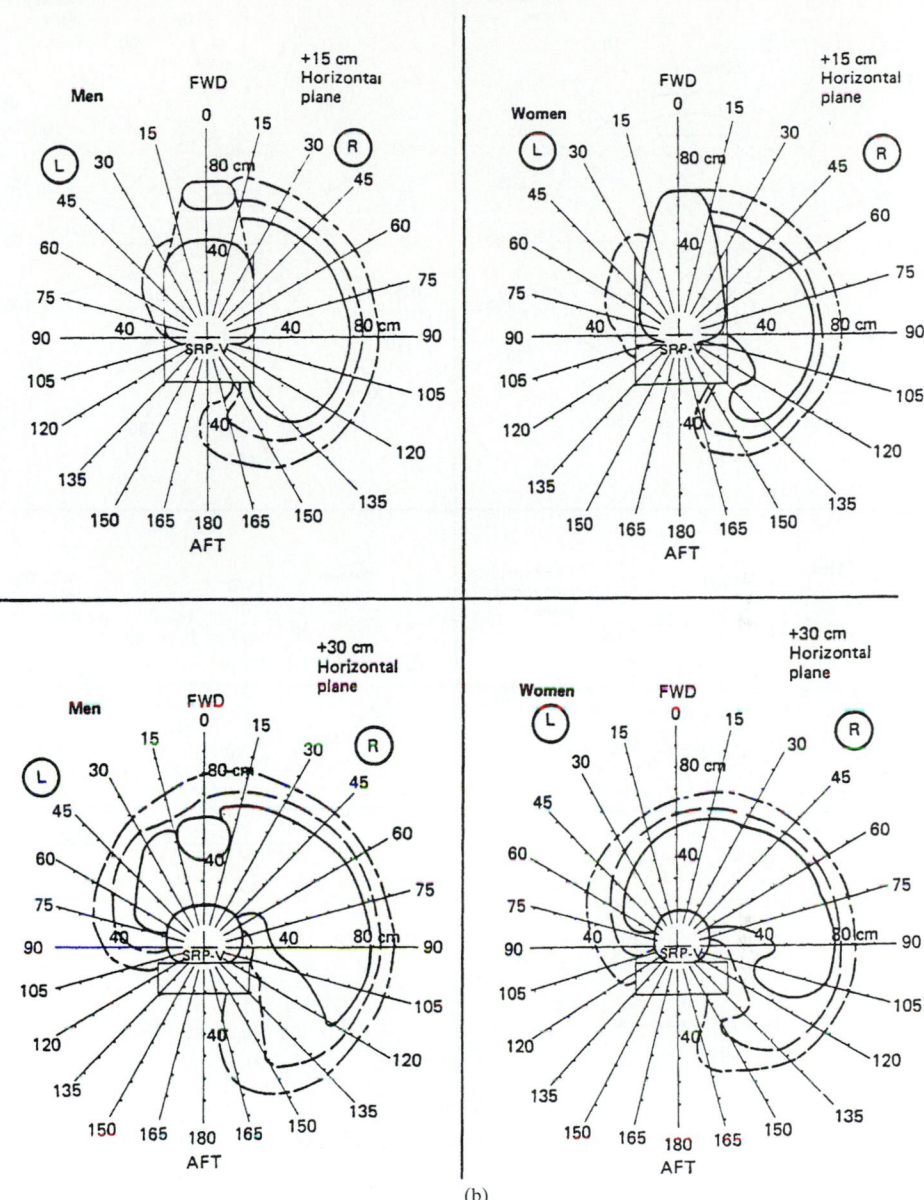

FIGURE 9.5 (*Continued*)   Examples of reach envelopes of seated operators. (Adapted from NASA STD 3000.)

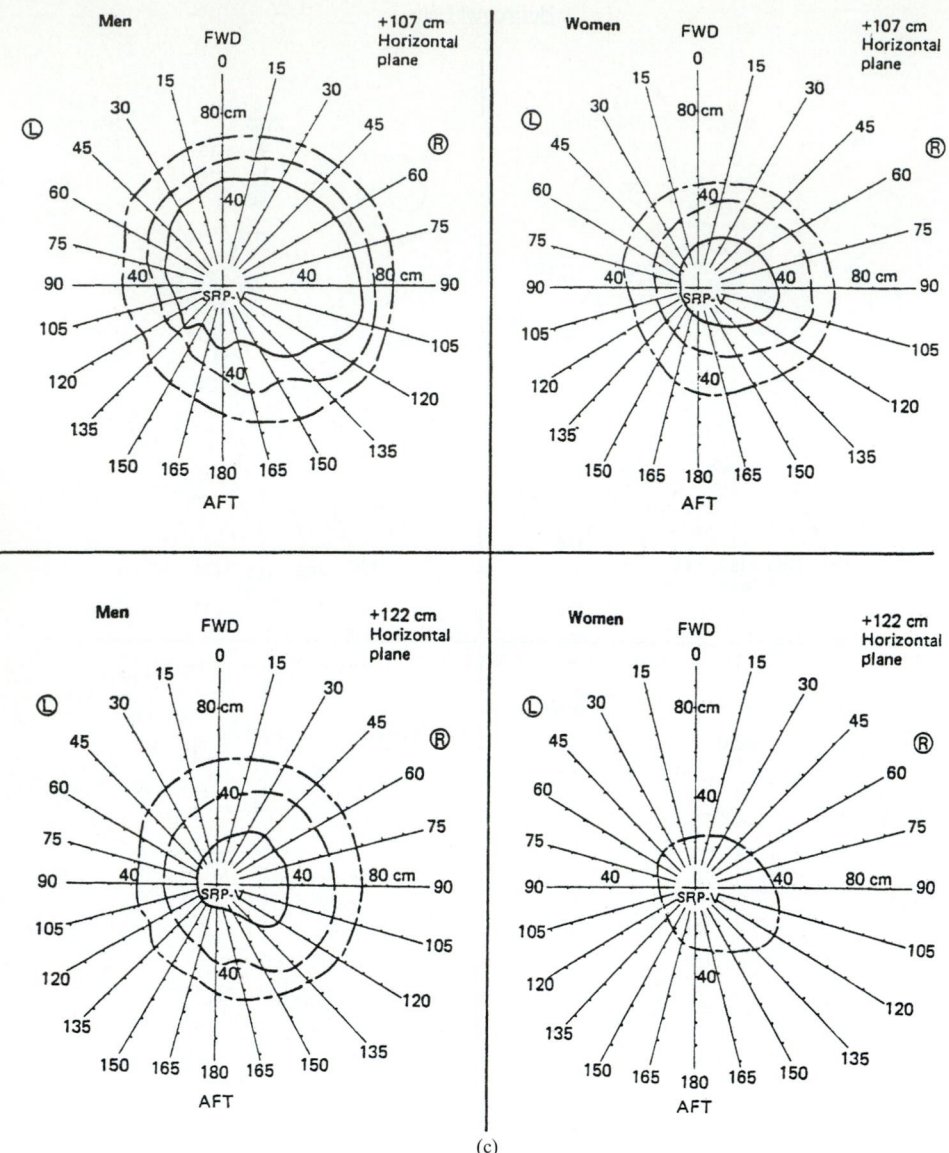

(c)

FIGURE 9.5 (*Continued*)   Examples of reach envelopes of seated operators. (Adapted from NASA STD 3000.)

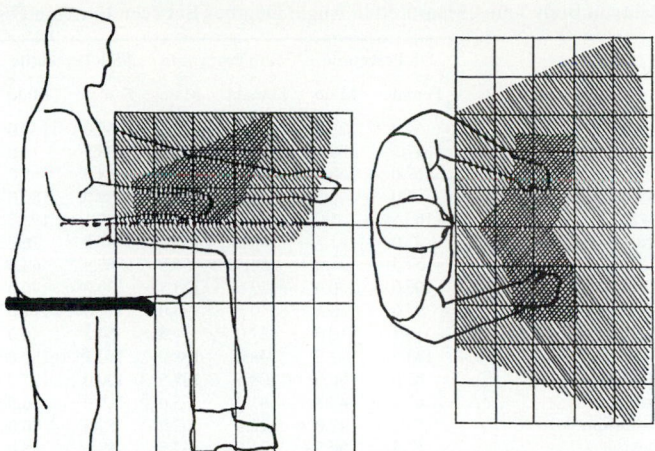

FIGURE 9.6   The normal and preferred (cross-hatched) work space for the hands. (With permission from K. H. E. Kroemer, H. B. Kroemer, and K. E. Kroemer-Elbert, (1994), *Ergonomics: How to Design for Ease and Efficiency*. All rights retained by the publisher, Prentice Hall, Englewood Cliffs, NJ.)

Mobility (often also called flexibility) refers to the range of motion that can be achieved about a body articulation. The boundaries are measured by angles from a known reference position (often, but not always, the so-called neutral position) as the difference between the smallest and largest angular excursions by a body segment about its body-next (proximal) articulation. Extremes of such displacements in body joints are described and listed in Table 9.3.

Specific zones of preference, convenience, or expediency need to be defined for each given condition and task; they will normally fall in the crosshatched areas of Figs. 9.6 and 9.7. For unusual or seldom done tasks, controls and tools can be located away from those normally preferred zones. In fact, in some cases one purposely locates objects outside those zones, beyond restrictive guards, walls, or other barriers, so that a "safe distance" between a danger point and the body is achieved, as shown in Fig. 9.8.

9.4   DESIGNING FOR HUMAN FORCE AND POWER

In energy terms, the human is very inefficient at doing heavy physical work; in most of our daily tasks, our energy efficiency is only about 5 percent. The human body also is not built for large force exertions, but rather for the exertion of fast, exact, well-controlled movements. Nevertheless, there are occasions at work on which the human must generate large torques or forces; however, these should be required only occasionally and for short periods of time. Biomechanically and psychologically, the human body is better able to perform rhythmic dynamic work, such as walking, pedaling, or turning a hand crank or lever, than to perform continual efforts with little or no movement. Static efforts (called *isometric* in physiological terminology) quickly lead to fatigue; for example, a human can maintain a maximal muscle

TABLE 9.3   Mobility in Body Joints, Measured in Angle Degrees Between Extreme Positions

Joint	Movement	5th Percentile		50th Percentile		95th Percentile		Difference[†]	
		Female	Male	Female	Male	Female	Male	Female	Male
Neck	Ventral flexion	34.0	25.0	51.5	43.0	69.0	60.0	+8.5	
	Dorsal flexion	47.5	38.0	70.5	56.5	93.5	74.0	+14.0	
	Right rotation	67.0	56.0	81.0	74.0	95.0	85.0	+7.0	
	Left rotation	64.0	67.5	77.0	77.0	90.0	85.0	NS	
Shoulder	Flexion	169.5	161.0	184.5	178.0	199.5	193.5	+6.5	
	Extension	47.0	41.5	66.0	57.5	85.0	76.0	+8.5	
	Adduction	37.5	36.0	52.5	50.5	67.5	63.0	NS	
	Abduction	106.0	106.0	122.5	123.5	139.0	140.0	NS	
	Medial rotation	94.0	68.5	110.5	95.0	127.0	114.0	+15.5	
	Lateral rotation	19.5	16.0	37.0	31.5	54.5	46.0	+5.5	
Elbow-forearm	Flexion	135.5	122.5	148.0	138.0	160.5	150.0	+10.0	
	Supination	87.0	86.0	108.5	107.5	130.0	135.0	NS	
	Pronation	63.0	42.5	81.0	65.0	99.0	86.5	+16.0	
Wrist	Extension	56.5	47.0	72.0	62.0	87.5	76.0	+10.0	
	Flexion	53.5	50.5	71.5	67.5	89.5	85.0	+4.0	
	Adduction	16.5	14.0	26.5	22.0	36.5	30.0	+4.5	
	Abduction	19.0	22.0	28.0	30.5	37.0	40.0	−2.5	
Hip	Flexion	103.0	95.0	125.0	109.5	147.0	130.0	+15.5	
	Adduction	27.0	15.5	38.5	26.0	50.0	39.0	+12.5	
	Abduction	47.0	38.0	66.0	59.0	85.0	81.0	+7.0	
	Medial rotation (prone)	30.5	30.0	44.5	46.0	58.5	62.5	NS	
	Lateral rotation (prone)	29.0	21.5	45.5	33.0	62.0	46.0	+12.5	
	Medial rotation (sitting)	20.5	18.0	32.0	28.0	43.5	43.0	+4.0	
	Lateral rotation (sitting)	20.5	18.0	33.0	26.5	45.5	37.0	+6.5	
Knee	Flexion (standing)	99.5	87.0	113.5	103.5	127.5	122.0	+10.0	
	Flexion (prone)	116.0	99.5	130.0	117.0	144.0	130.0	+13.0	
	Medial rotation	18.5	14.5	31.5	23.0	44.5	35.0	+8.5	
	Lateral rotation	28.5	21.0	43.5	33.5	58.5	48.0	+10.0	
Ankle	Flexion	13.0	18.0	23.0	29.0	33.0	34.0	−6.0	
	Extension	30.5	21.0	41.0	35.5	51.5	51.5	+5.5	
	Adduction	13.0	15.0	23.5	25.0	34.0	38.0	NS	
	Abduction	11.5	11.0	24.0	19.0	36.5	30.0	+5.0	

[†] Listed are only differences at the 50th percentile, and if significant ($\alpha < 0.5$).
Source: With permission from Ref. [9.7]. All rights reserved by the publisher, Van Nostrand Reinhold.

exertion for only a few seconds, and even half of the maximally possible contraction can be endured for only about a minute. This explains why it is so difficult to work with the hands overhead or to keep one's back bent.

Unfortunately, most of the existing information on human strength comes from measurements made under static (isometric) conditions, mostly because dynamic conditions are difficult to control experimentally. However, an increasing amount of information on dynamic exertions of force, energy, and power is becoming available; one needs to check the ergonomic literature for emerging information.

9.4.1   Foot Strength

Maximal static body forces are exerted with the foot by an operator who is sitting on a chair with a solid backrest [9.8], [9.12]. The backrest provides resistance to the force exerted with the foot, especially when the operator is pushing forward at about seat height, with the leg almost fully extended. Forces directed more downward are

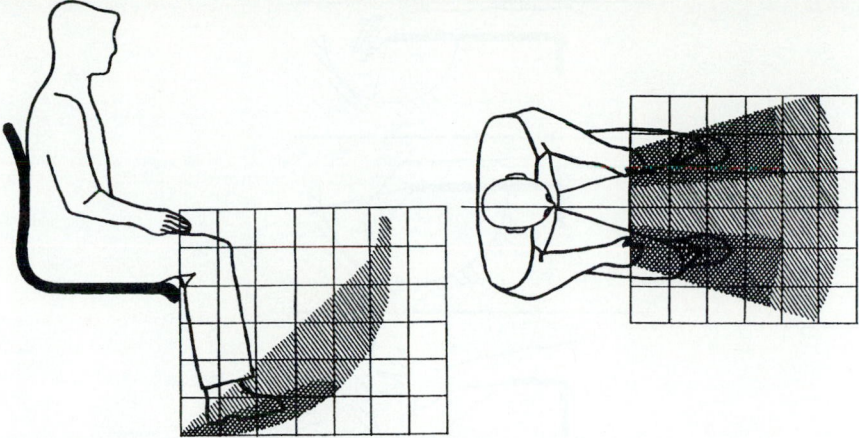

FIGURE 9.7 The normal and preferred (cross-hatched) work space for the feet. (With permission from K. H. E. Kroemer, H. B. Kroemer, and K. E. Kroemer-Elbert, (1994), *Ergonomics: How to Design for Ease and Efficiency*. All rights retained by the publisher, Prentice Hall, Englewood Cliffs, NJ.)

weaker and are limited by the counterforce provided by the body mass (according to Newton's third law, "action equals reaction") of the operator. Figure 9.9 illustrates these conditions.

Except for brief and occasional exertions, one should not expect a standing person to exert force with the foot because the body must be balanced on the other foot alone while doing so.

9.4.2 Hand Strength

Force exertion with the hands is weaker than that with the feet, but the effort is, as a rule, better controlled and more exact. (Yet, consider the driver of an automobile, who customarily controls the speed of the vehicle via pedals.) The strength of arm exertion depends on the direction of force and the location of the handle with respect to the body [9.8], [9.13]. Figure 9.10 provides detailed information; in some cases the force is largest with the arm extended, in other cases with the arm bent. (Note that the numbers given in Fig. 9.10 are 5th-percentile values.) In terms of the preferred working area for the hands and feet discussed earlier, it is obvious that in many cases the largest exertions of strength are at or near the periphery of the reach envelope.

9.4.3 Whole-Body Strength

The weakest link in the chain of force vectors transmitted from the point of application (e.g., the hand) through the body to the point of reactive resistance (e.g., the foot) determines the possible output of the human body. Figure 9.11 presents static forces that have been exerted horizontally by males. These figures show, again, the dependence of strength on location of the object, on body posture, on the kind of reaction force provided (Newton's third law), and, of course, on individual strength.

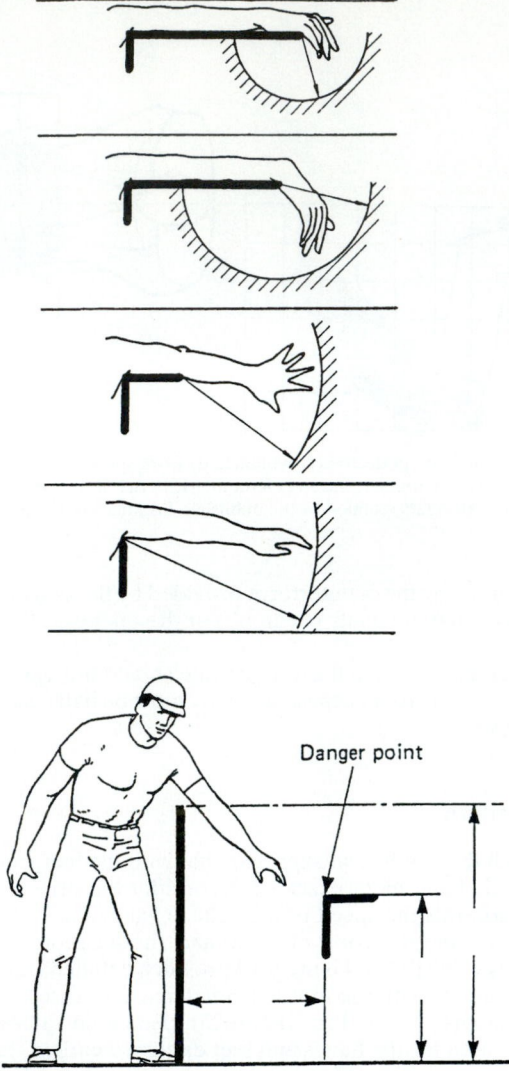

FIGURE 9.8 Examples of "safe distances". (With permission from K. H. E. Kroemer, H. B. Kroemer, and K. E. Kroemer-Elbert, (1994), *Ergonomics: How to Design for Ease and Efficiency*. All rights retained by the publisher, Prentice Hall, Englewood Cliffs, NJ.)

More information about human strengths, such as that of the fingers and thumb of the hand or of the hand, and of the forces and torques involved in the operation of specific controls is available in the ergonomic literature. This has been compiled most recently by Kroemer, Kroemer, and Kroemer-Elbert [9.8] and by Weimer [9.13]; military standards also provide such information.

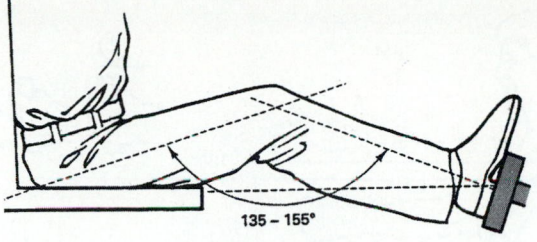

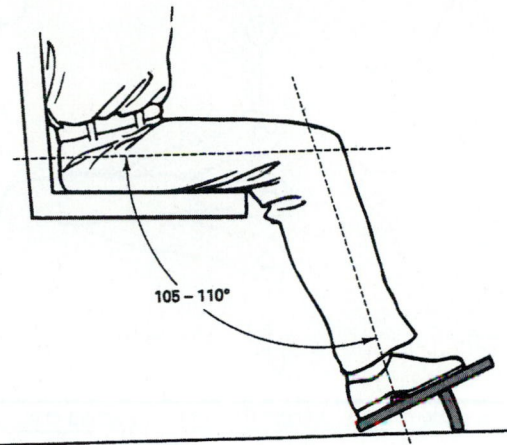

FIGURE 9.9 A sitting operator can exert the largest foot forces in the forward direction, with the leg almost extended and the seat back providing resistance. A downward directed force with a bent knee is weaker, but more comfortable. (Adapted form VanCott and Kinkade, 1972.)

9.5 DESIGNING FOR FAST AND ACCURATE CONTROL ACTIVATION

Different controls are appropriate for hand or foot operation, and for transmission of different amounts of energies and forces. Selection criteria depend on the purposes of control operation, such as

- Activation or shutting down of equipment, as with an on-off control
- Choosing among "discrete settings," as in ratcheting a control knob or putting the shift selector into one of several positions
- Making "quantitative settings," such as selecting a temperature on a thermostat (however, this may be a special case of a discrete setting)
- Applying "continuous control," as in turning a steering wheel in a vehicle
- Entering data, as on a computer keyboard

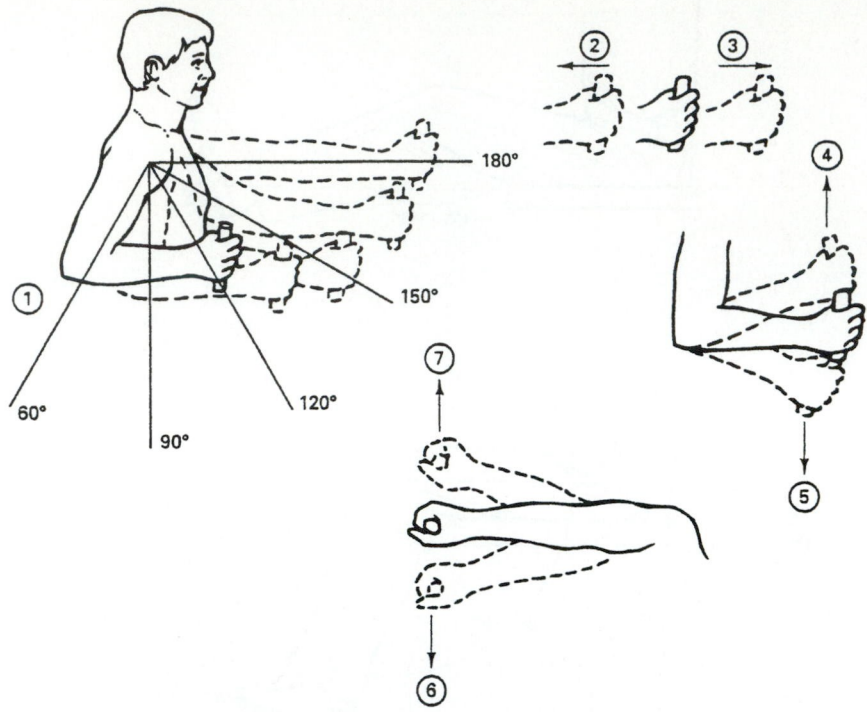

Fifth-percentile arm strength (N) exerted by sitting men													
(1)	(2)		(3)		(4)		(5)		(6)		(7)	
elbow flexion (deg)	Pull		Push		Up		Down		In		Out	
	Left	Right	L	R	L	R	L	R	L	R	L	R
180	222	231	187	222	40	62	58	76	58	89	36	62
150	187	249	133	187	67	80	80	89	67	89	36	67
120	151	187	116	160	76	107	93	116	89	98	45	67
90	142	165	98	160	76	89	93	116	71	80	45	71
60	116	107	96	151	67	89	80	89	76	89	53	71

FIGURE 9.10 Hand forces exerted with the arm in different positions. (Adapted form MIL-HBK 759.)

Most ergonomic recommendations for selection and arrangements of controls (and displays) have been derived from existing devices and western stereotypes (action-effect expectations)—see Refs. [9.2], [9.8], [9.11], [9.12], [9.13], and [9.14]. Thus, for new types of controls and tasks, tradition may not apply.

9.5.1 Control Selection

In general, controls shall be selected for their functional usefulness. The major rules are:

	Force-plate[1] height	Distance[2]	Force, N	
			Mean	SD
	50	80	664	177
	50	100	772	216
	50	120	780	165
	70	80	716	162
	70	100	731	233
	70	120	820	138
	90	80	625	147
	90	100	678	195
	90	120	863	141
	Percent of shoulder height		Both hands	
	60	70	761	172
	60	80	854	177
	60	90	792	141
	70	60	580	110
	70	70	698	124
	70	80	729	140
	80	60	521	130
	80	70	620	129
	80	80	636	133
	Percent of shoulder height			
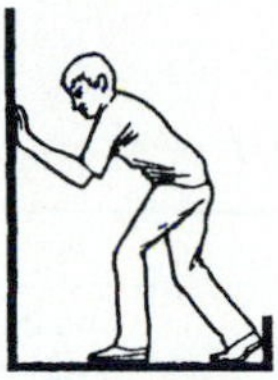	70	70	623	147
	70	80	688	154
	70	90	586	132
	80	70	545	127
	80	80	543	123
	80	90	533	81
	90	70	433	95
	90	80	448	93
	90	90	485	80
	Percent of shoulder height		Both hands	
			Both hands	
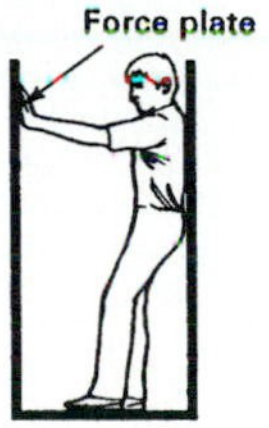 Force plate	100 percent of shoulder height	50	581	143
		60	667	160
		70	981	271
		80	1285	398
		90	980	302
		100	646	254
			Preferred hand	
		50	262	67
		60	298	71
		70	360	98
		80	520	142
		90	494	169
		100	427	173
		Percent of thumb-tip reach*		
	100 percent of shoulder height	50	367	136
		60	346	125
		70	519	164
		80	707	190
		90	325	132
		Percent of span**		

[1]Height of the center of the force plate – 20 cm high by 25 cm long – upon which force is applied.
[2]Horizontal distance between the vertical surface of the force plate and the opposing vertical surface (wall or footrest, respectively) against which the subjects brace themselves.

*Thumb-tip reach – distance from backrest to tip of subject's thumb as arm and hand are extended forward.
**Span – the maximal distance between a person's fingertips when arms and hands are extended to each side.

FIGURE 9.11 Whole-body strength exerted in various postures. (Adapted from NASA STD 3000, using data measured by the author.)

- The control type shall be compatible with common expectations. Thus, one would use a pushbutton or a toggle switch to turn on a light, not a rotary knob.
- The size of the control and its motion characteristics shall be compatible with past experience and practice. Thus, one would expect to have a fairly large steering wheel for two-handed operation in an automobile, not a small rotary or linear control.
- The direction of operation shall be compatible with stereotypical or common expectations. This means that an on control is pushed or pulled, not turned to the left.
- The control shall be "safe" against inadvertent, false, or excessive operation.

Accordingly, the information contained in Tables 9.4, 9.5, and 9.6 is helpful in selecting proper controls.

More detailed information, such as about the proper control size, spacing, location on control panels, and association with displays, is contained in the ergonomic literature, especially in standards and in design compilations by Cushman and Rosenberg [9.2], Kroemer, Kroemer, and Kroemer-Elbert [9.8], and Woodson, Tillman, and Tillman [9.15].

TABLE 9.4 Control Movements and Expected Effects

Function	Up	Right	Forward	Clockwise	Press,[†] Squeeze	Down	Left	Rearward	Back	Counter-clockwise	Pull[†]	Push[‡]
On	1[§]	1	1	1	2	1[§]	—	—	—	—	1	—
Off	—	—	—	—	—	1	2	2	—	1	—	2
Right	—	1	—	2	—	—	—	—	—	—	—	—
Left	—	—	—	—	—	—	1	—	2	—	—	—
Raise	1	—	—	—	—	—	—	2	—	—	—	—
Lower	—	—	2	—	—	1	—	—	—	—	—	—
Retract	2	—	—	—	—	—	—	1	—	—	2	—
Extend	—	—	1	—	—	2	—	—	—	—	—	2
Increase	2	2	1	2	—	—	—	—	—	—	—	—
Decrease	—	—	—	—	—	2	2	1	—	2	—	—
Open Valve	—	—	—	—	—	—	—	—	—	1	—	—
Close Valve	—	—	—	1	—	—	—	—	—	—	—	—

1 = most preferred; 2 = less preferred.
[†] With trigger-type control.
[‡] With push-pull switch.
[§] Up in United States, down in Europe.
Source: Modified from K. H. E. Kroemer, "Ergonomics," chap. 13 in B. A. Plog (ed.), *Fundamentals of Industrial Hygiene*, 3d ed., 1988, pp. 521–539, with permission by the publisher, National Safety Council, Chicago.

9.5.2 Avoiding Inadvertent Operation

There are design procedures to guard against inadvertent control activation [9.2]:

- Locate and orient the control in such a way that the operator is unlikely to strike it or move it accidentally in the normal sequence of operations.
- Recess or shield the control or surround it by physical barriers.
- Cover or guard the control by providing a cover, a pin, a lock, or some other device that must be removed or broken before the control can be operated.

TABLE 9.5 Control-Effect Relations of Common Hand Controls

Effect	Keylock	Toggle switch	Push-button	Bar knob	Round knob	Thumbwheel Discrete	Thumbwheel Continuous	Crank	Rocker switch	Lever	Joystick or ball	Legend switch	Slide†
Select ON/OFF	1	1	1	3	—	—	—	—	1	—	—	1	1
Select ON/STANDBY/OFF	—	2	1	1	—	—	—	—	—	1	—	1	1
Select OFF/MODE 1/MODE 2	—	3	2	1	—	—	—	—	—	1	—	1	1
Select one function of several related functions	—	2	1	—	—	—	—	—	2	—	—	—	3
Select one of three or more discrete alternatives	—	—	—	1	—	—	—	—	—	—	—	—	1
Select operating condition	—	1	1	2	—	—	—	—	1	1	—	1	2
Engage or disengage	—	—	—	—	—	—	—	—	—	1	—	—	—
Select one of mutually exclusive functions	—	—	1	—	1	—	—	—	—	—	—	1	—
Set value on scale	—	—	—	—	1	—	2	3	—	3	3	—	1
Select value in discrete steps	—	—	1	1	—	1	—	—	—	—	—	—	1

1 = most preferred; 3 = least preferred.
† Estimated, no experiments known.
Source: Modified from K. H. E. Kroemer, "Ergonometrics," chap. 13 in B. A. Plog (ed.), *Fundamentals of Industrial Hygiene*, 3d ed., 1988, pp. 521–539, with permission of the publisher, National Safety Council, Chicago.

TABLE 9.6 Guide for Selection of Controls

Small operating force	
2 discrete positions	Keylock, hand-operated
	Toggle switch, hand-operated
	Pushbutton, hand-operated
	Rocker switch, hand-operated
	Legend switch, hand-operated
	Bar knob, hand-operated
	Slide, hand-operated
	Push-pull switch, hand-operated
3 discrete positions	Toggle switch, hand-operated
	Bar knob, hand-operated
	Legend switch, hand-operated
	Slide, hand-operated
4 to 24 discrete positions, or continuous operation	Bar knob, hand-operated
	Round knob, hand-operated
	Joystick, hand-operated
	Continuous thumbwheel, hand-operated
	Crank, hand-operated
	Lever, hand-operated
	Slide, hand-operated
	Track ball, hand-operated
	Mouse, hand-operated
	Light pen, hand-operated
Continuous slewing, fine adjustments	Crank, hand-operated
	Round knob, hand-operated
	Track ball, hand-operated
Large operating force	
2 discrete positions	Pushbutton, foot-operated
	Pushbutton, hand-operated
	Detent lever, hand-operated
3 to 24 discrete positions	Detent lever, hand-operated
	Bar knob, hand-operated
Continuous operation	Hand wheel, hand-operated
	Lever, hand-operated
	Joystick, hand-operated
	Crank, hand-operated
	Pedal, foot-operated

Source: Modified from K. H. E. Kroemer, "Ergonomics," chap. 13 in *Fundamentals of Industrial Hygiene,* 3d ed., 1988, pp. 521–539, with permission of the publisher, National Safety Council, Chicago.

- Provide extra resistance (viscous or coulomb friction, by spring loading or inertia) so that an unusual effort is required for activation.
- Provide a delaying means so that the control must pass through a critical position with an unusual movement, such as in manual gear shifting in automobiles.
- Provide interlocking between controls so that prior operation of a related control is required before the critical control can be activated.

Some of these measures may be combined. Note, however, that such design features usually slow the operation, which may be detrimental in an emergency.

9.6 DESIGNING LABELS AND WARNINGS

Ideally, it should not be necessary to explain the use of a control or piece of equipment or to warn about possible hazards. However, in many cases an ideal design solution is impossible, and it is therefore necessary to use labels, symbols, and warnings so that one may locate, identify, activate and correctly manipulate the equipment [9.8].

9.6.1 Labels

Labeling must be done in such a way that the information is accurately provided and rapidly understood. The guidelines in Table 9.7 apply.

The font (typeface) should be simple, bold, and vertical, as is the case with Futura, Helvetica, Namel, Tempo, and Vega. Most electronically generated fonts are, unfortunately, inferior to printed fonts, but special effort and attention help to make these as legible as possible.

- The recommended *height of characters* depends on the viewing distance; for example,

TABLE 9.7   Guidelines for Labeling

- *Orientation.* A label and the information printed on it shall be oriented horizontally so that it can be read quickly and easily. (Note that this applies if the operator is used to reading horizontally, as in western countries.)
- *Location.* A label shall be placed on or very near the item that it identifies.
- *Standardization.* Placement of all labels shall be consistent throughout the equipment and system.
- *Equipment Functions.* A label shall primarily describe the function ("what does it do?") of the labeled item.
- *Abbreviations.* Common abbreviations may be used. If a new abbreviation is necessary, its meaning shall be obvious to the reader. The same abbreviation shall be used for all tenses and for the singular and plural forms of a word. Capital letters shall be used, with periods normally omitted.
- *Brevity.* The label inscription shall be as concise as possible without distorting the intended meaning or information. The texts shall be unambiguous, with redundancy minimized.
- *Familiarity.* Words that are familiar to the operator shall be chosen, if possible.
- *Visibility* and *Legibility.* The operator shall be able to read easily and accurately at the anticipated actual reading distances, at the anticipated worst illumination level, and within the anticipated vibration and motion environment. Important are the contrast between the lettering and its background; the height, width, stroke width, spacing, and style of letters; and the specular reflection of the background, cover, or other components.
- *Font* and *Size.* Typography determines the legibility of written information; it refers to style, font, arrangement, and appearance.

Source: Adapted with permission from K. H. E. Kroemer, H. B. Kroemer, and K. E. Kroemer-Elbert, *Ergonomics: How to Design for Ease and Efficiency,* 1994 (Ref. [9.8]). All rights retained by the publisher, Prentice-Hall, Englewood Cliffs, N.J.

Viewing distance 35 cm, suggested height 22 mm
Viewing distance 70 cm, suggested height 50 mm
Viewing distance 1 m, suggested height 70 mm
Viewing distance 1.5 m, suggested height at least 1 cm

- The *ratio of stroke width to character height* should be between 1:8 and 1:6 for black letters on a white background, and between 1:10 and 1:8 for white letters on a black background.
- The *ratio of character width to character height* should be about 3:5.
- The *space between letters* should be at least one stroke width.
- The *space between words* should be at least one character width.
- For continuous text, mix upper- and lowercase letters. For labels, use uppercase letters only.
- Text should be horizontal.

9.6.2 Warnings

All devices should be safe to use, but realistically this cannot always be achieved through design. If it cannot be, one must warn users of dangers associated with the product and provide instructions for safe use to prevent injury or damage.

It is preferable to have an active warning, consisting of a sensor that notices inappropriate use and an alerting device that warns the human of an impending danger. Yet, in most cases, the warnings are passive, often simply a label attached to the product and instructions written in the user manual. Such passive warnings rely completely on the human to recognize an existing or potential dangerous situation, to remember the warning, and to behave prudently, and so passive warnings are often ineffectual. Thus, passive warnings must be carefully designed, in conformance with the most recent government laws and regulations, national and international standards, and the best applicable human engineering information.

Warning labels and placards usually contain text and/or graphics. Pictures, pictograms, and icons can communicate information to persons from various cultures who speak different languages; however, users may have rather different perceptions depending on their ages, experiences, and ethnic and educational backgrounds. Therefore, designing a "safe" product is much preferable to applying warnings to an inferior product.

9.7   DESIGNING FOR VISION

For an erect operator, the preferred direction of the line of sight to distant targets is about horizontal. However, if one must focus on a close target, the preferred line of sight is distinctly lower. The best way of describing the angle of gaze is against a reference line that moves with the skull, because the head may be held at different pitch angles. An easy way to establish such a reference is to run a straight line through the earhole and the juncture of the upper and lower eyelids, in the side view from the right. This *ear-eye line* is shown in Fig. 9.12. An approximately upright head has an angle (P in Fig. 9.12) between the ear-eye line and the horizon of about 15 degrees. The angle of the line of sight is easily described in relation to the ear-eye line. For close targets, such as at reading distance, most people prefer the line-of-

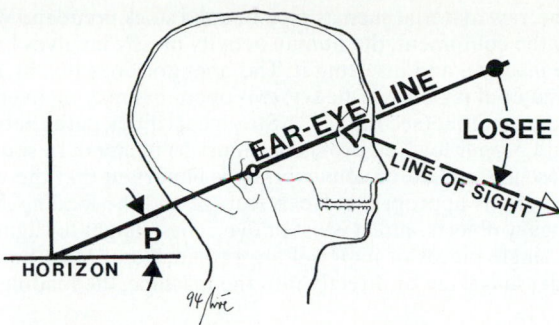

FIGURE 9.12 The Ear–Eye Line runs, in the side view, through the ear hole and the junction of the eyelids. The head is held "erect" (or "upright") when the angle between the ear–eye line and the horizon is approximately 15°. Distant targets can be easily viewed when "straight ahead," i.e., with a LOSEE angle of about 15° below the ear–eye line. Close targets, however, are best located distinctly below, that is at an average LOSEE of 45°, plus or minus 20° to accommodate individual preferences.

sight angle below the ear-eye line (LOSEE in Fig. 9.12) to be between 25 and 65 degrees (the average is 45). Note that this differs from some erroneous statements in the older literature, which indicate a flatter LOSEE angle. Optometrists always knew better and habitually ground the reading section of bifocal lenses into their lower part. They knew: "the closer, the lower."

Especially in the design of computer work stations, but also in many other circumstances, one must consider all equipment components as parts of an interactive system in which the visual targets (for example, the display), the manipulation areas (for example, controls or the keyboard), and the body support (the chair) interact with one another in determining the postures and habits of the worker. Of course, the work task and the environment (illumination, sound, climate) also strongly affect the worker; if just one element is badly placed, the resulting system condition may be unacceptable. Recommendations for the design of ergonomically acceptable computer work stations are provided in recent publications—see especially Refs. [9.6], [9.8], and [9.19].

9.8 DESIGNING FOR MATERIAL HANDLING

Handling objects is a common task; examples are loading and unloading machines, manipulating objects in assembly and inspection, or using hand tools. Design for such tasks will be discussed in this section; the more general topic of manual material handling (lifting, lowering, carrying, etc.) is treated in other publications, especially by Ayoub and Mital [9.20] and in special 1992 issues of the journal *Ergonomics* [9.21]. The 1981 and 1991 NIOSH recommendations for safe lifting [9.22] and the 1991 guidelines by Snook and Ciriello [9.23] were compared by Kroemer, Kroemer, and Kroemer-Elbert [9.8].

In focusing on material handling at the workplace, especially the design of machines for proper manual feeding and unloading, one can distinguish between two cases. In one, raw material such as pipe or solid stock is cut and worked (milled, turned, etc.) by the equipment; the human activity mostly involves bringing the raw material to the machine and inserting it. The operator does this by hand and body motion; if the material is bulky or heavy, two operators may be needed to manipulate and move the material (see Fig. 9.13). Storage facilities, paths between there and the machine, and proper loading provisions must be planned. In some cases, it may be possible to use jigs, rollers, or conveyors. It is important that the work height for the material handler is appropriate: Keep storage and the loading station at about hip height for heavy objects, and at waist or even chest height for light material. If, in the other case, single pieces of material are provided and must be loaded, one by one, into a feeder magazine or directly into the machine, the loading fixtures of the

FIGURE 9.13   Bulky and heavy material should be so stored that it can be moved easily by two operators without bending and twisting. (Adapted, with permission, from [9.22].)

equipment itself must be so designed that the operator does not have to bend or twist the body. The manipulation area should be close to the body, in front of the trunk, as discussed before.

The location of containers for raw materials or for machined items is of great concern if overexertion injuries of the material handler are to be avoided. Containers should not be put on the floor if this requires the worker to bend or "dive" into them in order to retrieve or deposit material (see Fig. 9.14). Instead, the bins, trays,

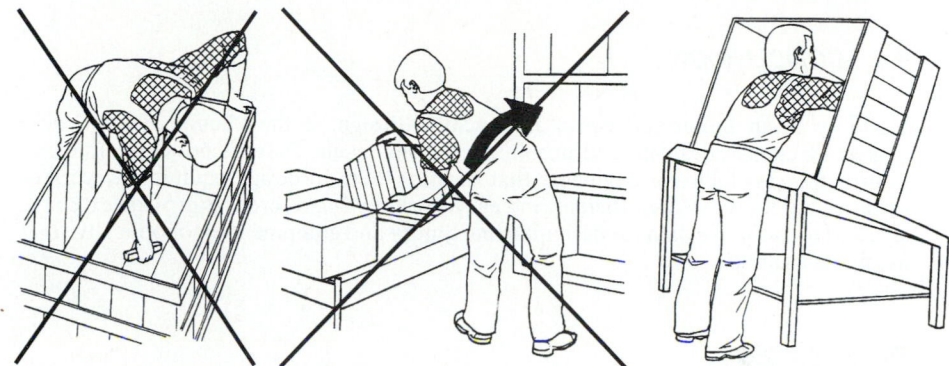

FIGURE 9.14   Avoid twisting and bending the trunk by storing containers at proper height and angle. (With permission from K. H. E. Kroemer, H. B. Kroemer, and K. E. Kroemer-Elbert, (1994), *Ergonomics: How to Design for Ease and Efficiency*. All rights retained by the publisher, Prentice Hall, Englewood Cliffs, NJ.)

or containers should be placed in such a way (raised, tilted) and be of such a size that the operator can work with an upright posture, grasping and moving the objects in front of the trunk and close to the body. The axiom is, "Keep the body upright; avoid body bending and twisting."

Methods commonly used in industrial engineering are suitable for the evaluation of planned or existing material handling. Employ flowcharts and flow diagrams as well as motion analyses and other work study techniques familiar to the industrial engineer (see Ref. [9.8]). These techniques allow a motion-by-motion determination of the activities involved, and of their ease or difficulty for the operator, which in turn facilitates determination of the best work method and equipment design.

Various types of equipment and machines provide assistance to the material handler. There are lift tables, hoists, cranes, turntables, dollies, walkies, trolleys, trucks, stackers, and even forklift trucks (Refs. [9.8], [9.24]). Not only should such equipment be able to move the material, but it should also fit the human operator, and must be easy and safe to operate. In the past, some material movement machinery, such as cranes, conveyors, and hand and power trucks, has been produced that showed an alarming lack of consideration for human factors and safety principles. The worst example was a forklift which obstructed the operator's view and also transmitted shocks and impacts to the operator, who was forced into contorted body postures. The following excerpt from a letter published in the *Human Factors Society Bulletin* of January 1984 on page 7 described this condition:

I was shocked, dismayed, and perturbed. Recently, I attended a regional industrial exhibition that had an emphasis on materials-handling equipment. I intentionally went around looking for bad or lacking human engineering. I found plenty . . . inadequate labels; wrong-size controls; lack of shape, position, color coding; controls that could be inadvertently actuated; absence of guard rails; unintelligible instructions; slippery surfaces; impossible reach requirements; sharp edges; unguarded pinch-points; extreme strength requirements; lack of guards; spaces needed for maintenance too small for the human hand; poorly located emergency switches; and so on, and so on! . . . Spacecraft and supersonic aircraft and missile monitoring equipment need human engineering; so, too, do hydraulic hoists and forklift trucks and conveyor systems and ladders.

9.9   CONCLUSION

Usability is an important aspect in machine design; if the "human factor" is neglected, clumsy, difficult, and unsafe conditions result. There is now ergonomic information available for engineers that enables them to design equipment, work stations, and tools for safe, efficient, and easy working. The references provide extensive and quantitative data for designing machinery and equipment ergonomically, to assure its usability.

REFERENCES

9.1 K. R. Boff, L. Kaufman, and J. P. Thomas (eds.), *Handbook of Perception and Human Performance,* John Wiley & Sons, New York, 1986.

9.2 W. H. Cushman and D. J. Rosenberg, *Human Factors in Product Design,* Elsevier, Amsterdam, 1991.

9.3 Eastman Kodak Company, *Ergonomic Design for People at Work,* 2 vols. Van Nostrand Reinhold, New York, 1983, 1986.

9.4 T. M. Fraser, *The Worker at Work,* Taylor & Francis, London, 1989.

9.5 E. Grandjean, *Fitting the Task to the Man,* 4th ed., Taylor & Francis, London, 1988.

9.6 M. G. Helander, *Handbook of Human-Computer Interaction,* Elsevier, Amsterdam, 1991.

9.7 K. H. E. Kroemer, H. J. Kroemer, and K. E. Kroemer-Elbert, *Engineering Physiology,* 2d ed., Van Nostrand Reinhold, New York, 1990.

9.8 K. H. E. Kroemer, H. B. Kroemer, and K. E. Kroemer-Elbert, *Ergonomics: How to Design for Ease and Efficiency,* Prentice-Hall, Englewood Cliffs, N.J., 1994.

9.9 R. W. Proctor and T. Van Zandt, *Human Factors in Simple and Complex Systems,* Allyn & Bacon, Boston, 1994.

9.10 B. M. Pulat, *Fundamentals of Industrial Ergonomics,* Prentice-Hall, Englewood Cliffs, N.J., 1992.

9.11 G. Salvendy (ed.), *Handbook of Human Factors,* John Wiley & Sons, New York, 1987.

9.12 M. S. Sanders and E. J. McCormick, *Human Factors in Engineering and Design,* 7th ed., McGraw-Hill, New York, 1993.

9.13 J. Weimer, *Handbook of Ergonomic and Human Factors Tables,* Prentice-Hall, Englewood Cliffs, N.J., 1993.

9.14 J. R. Wilson and E. N. Corlett, *Evaluation of Human Work,* Taylor & Francis, London, 1990.

9.15 W. E. Woodson, B. Tillman, and P. Tillman, *Human Factors Design Handbook,* 2d ed., McGraw-Hill, New York, 1991.

9.16 C. C. Gordon, T. Churchill, C. E. Clauser, B. Bradtmiller, J. T. McConville, I. Tebbetts, and R. A. Walker, *1988 Anthropometric Survey of U.S. Army Personnel: Summary Statistics Interim Report,* Technical Report NATICK/TR-89-027, U.S. Army Natick Research, Development and Engineering Center, Natick, Mass., 1989.

9.17 T. M. Greiner, *Hand Anthropometry of U.S. Army Personnel,* Technical Report TR-92/011, U.S. Army Natick Research, Development and Engineering Center, Natick, Mass., 1991.

9.18 J. A. Roebuck, *Anthropometric Methods,* Human Factors and Ergonomics Society, Santa Monica, Calif., 1995.

9.19 R. Lueder and K. Noro (eds.), *Hard Facts about Soft Machines. The Ergonomics of Seating,* Taylor & Francis, London, 1994.

9.20 M. M. Ayoub and A. Mital, *Manual Materials Handling,* Taylor & Francis, London, 1989.

9.21 *Ergonomics,* vol. 35, nos. 7 and 8, 1992.

9.22 T. R. Waters, V. Putz-Anderson, A. Gary, and L. J. Fine, "Revised NIOSH Equation for the Design and Evaluation of Manual Lifting Tasks," *Ergonomics,* vol. 36, no. 7, 1993, pp. 749–776.

9.23 S. H. Snook and V. M. Ciriello, "The Design of Manual Handling Tasks: Revised Tables of Maximum Acceptable Weights and Forces," *Ergonomics,* vol. 34, no. 9, 1991, 1197–1213.

9.24 ILO (ed.), *Maximum Weights in Load Lifting and Carrying,* Occupational Safety and Health Series, No. 59, International Labour Office, Geneva, Switzerland, 1988.

ADDRESSES

American National Standards Institute (ANSI),
11 West 42nd St., 13th Floor, New York, NY 10036.

American Society of Heating, Refrigerating, and Air Conditioning
Engineers (ASHRAE), 1791 Tullie Circle, Atlanta, GA 30329.

American Society of Safety Engineers, (ASSE),
1800 East Oakton St., Des Plaines, IL 60018-2187.

National Aeronautics and Space Administration (NASA),
SP 34-MSIS, LBJ Space Center, Houston, TX 77058.

Occupational Safety and Health Agency (OSHA),
200 Constitution Ave., NW, N3651, Washington, DC 20210.

Society of Automotive Engineers (SAE),
400 Commonwealth Dr., Warrendale, PA 15096-0001.

U.S. Military Standards: National Technical Information Service,
5285 Port Royal Rd., Springfield, VA 22161.

CHAPTER 10
SAFETY

Charles O. Smith, Sc.D., P.E.
Professor Emeritus of Mechanical Engineering
Consultant, Terre Haute, Indiana

10.1 WHY SAFETY?

The ASME Code of Ethics says: "Engineers shall hold paramount the safety, health and welfare of the public in the performance of their professional duties." This consideration is not new. Tacitus [10.1], about the first century A.D., said: "The desire for safety lies over and against every great and noble enterprise." Even some 2000 years earlier, the first known written law code [10.2], while not specifically mentioning safety, clearly implied a necessity for a builder to consider safety.

The National Safety Council [10.3] says:

> Each year, accidental deaths and injuries cost our society in excess of $399 billion—in the United States alone. This figure includes lost wages, medical outlays, property damage and other expenses. The cost in human misery is incalculable. Accidents are the fifth leading cause of death. The Council believes that accidents are not just random occurrences, but instead result mostly from poor planning or adverse conditions of the environments in which people live, work, drive and play. In our view, "accidents" nearly always are preventable—as are many illnesses.

If for no other reason, one should emphasize safety as a matter of enlightened self-interest.

Those who design machines and who have an interest in productivity and cost control serve their "customers" well if risks are at a minimum, as interruptions called accidents will also be at a minimum.

10.2 WHAT IS SAFETY?

One dictionary [10.4] definition is: "The quality or condition of being safe; freedom from danger, injury or damage." Most other dictionary definitions are similar. Hammer [10.5] says: "Safety is frequently defined as 'freedom from hazards.' However, it is practically impossible to completely eliminate all hazards. Safety is therefore a matter of relative protection from exposure to hazards: the antonym of danger." Lowrance [10.6] says: "A thing is safe if its risks are judged to be acceptable." This definition contrasts sharply with the Webster definition (which indicates "zero" risk) and, like Hammer's, implies that nothing is absolutely free of risk. Safety is a relative attribute that changes from time to time and is often judged differently in different contexts. For example, a power saw, a lawnmower, or similar powered equipment that may be "safe" for an adult user may not be "safe" in the hands of a child.

Lowrance's definition [10.6] emphasizes the relativistic and judgmental nature of the concept of safety. It further implies that two very different activities are required in determining how safe a thing is: *measuring risk,* an objective but probabilistic effort, and *judging the acceptability* of that risk, a matter of personal and/or societal value judgment. In addition, the level of acceptable risk involves moral, technical, economic, political, and legal issues.

Technical people are capable of measuring risks, and are generally qualified to do so. The decision as to whether the general public, with all its individual variations of need, desire, taste, tolerance, and adventurousness, might be (or should be) willing to assume the estimated risks is a value judgment that technical people are no better qualified (and perhaps less qualified) to make than anyone else.

10.3 HAZARD, RISK, AND DANGER

There is substantial confusion about the meaning of words such as *hazard, risk,* and *danger.* Webster [10.4] defines *danger* as "liability to injury, pain, damage or loss; hazard; peril; risk." Webster [10.4] makes some distinction by further saying, "Hazard arises from something fortuitous or beyond our control. Risk is doubtful or uncertain danger, often incurred voluntarily."

One can also consider a hazard to be (1) any aspect of technology or activity that produces risk or (2) the potential for harm or damage to people, property, or the environment, including (3) the characteristics of things and the actions (or inactions) of individuals. One can also consider risk to be a measure of the probability and severity of adverse effects.

With all the products liability litigation in the United States, a clear distinction among these three words for legal purposes has developed. In this context, a *hazard* is a condition or changing set of circumstances which presents an injury potential, such as a railroad crossing at grade, a toxic chemical, a sharp knife, or the jaws of a power press. *Risk* is the probability of injury and is affected by proximity, exposure, noise, light, experience, attention arresters, intelligence of an involved individual, etc. Risk (probability of exposure) is obviously much higher with a consumer product than with an industrial product to be used by trained workers in a shop environment. *Danger* is the unreasonable or unacceptable combination of hazard and risk. The U.S. courts generally hold as unreasonable and unacceptable any risk which can be eliminated by reasonable accident prevention methods. A high risk of injury could be considered reasonable and acceptable *if* the injury is minimal and the risk is recognized by the individual concerned. (Lowrance's use of *risk* seems close to the legal definition of *danger.*)

As might be expected, there is extensive and ongoing debate over the meaning of "reasonable" or "unreasonable." The American Law Institute [10.7] says *unreasonably dangerous* means that

The article sold must be dangerous to an extent beyond that which would be contemplated by the ordinary consumer who purchases it, with the ordinary knowledge common to the community as to its characteristics. Good whiskey is not unreasonably dangerous merely because it will make some people drunk, and is especially dangerous to alcoholics; but bad whiskey, containing a dangerous amount of fusel oil, is unreasonably dangerous.

The American Law Institute further says:

There are some products which, in the present state of human knowledge, are quite incapable of being made safe for their intended and ordinary use. . . . Such a product, properly prepared, and accompanied by proper directions and warnings, is not defective, nor is it unreasonably dangerous.

The American Law Institute [10.7] says that a product is in a defective condition if "it leaves the seller's hands, in a condition not contemplated by the ultimate user, which will be unreasonably dangerous to him." Peters [10.8] indicates that a California Supreme Court decision, *Barker* v. *Lull* [10.9], established a good assessment of "defective condition." This provides three definitions (or criteria) for manufacturing defects and two for design defects.

Defective Conditions

Manufacturing defects
1. Nonconformance with specifications
2. Nonsatisfaction of user requirements
3. Deviation from the norm

Design defects
1. Less safe than expected by ordinary consumer
2. Excessive preventable danger

Manufacturing Defects. A failure to conform with stated specifications is an obvious manufacturing defect; this is not a new criterion. The aspect of user satisfaction may not be as well known, but in the legal context it has long been recognized that a manufacturing defect exists when there is such a departure from some quality characteristic that the product or service does not satisfy user requirements. Under the third criterion (deviation from the norm), added by Barker, a manufacturing defect occurs (1) when a product leaves the assembly line in a substandard condition, (2) when the product differs from the manufacturer's intended result, or (3) when the product differs from other ostensibly identical units of the same product.

Design Defects. A product may be considered to have a design defect if it fails to perform as safely as an ordinary consumer would expect. This failure to perform safely is interpreted in the context of intended use (or uses) in a reasonably foreseeable manner, where *foreseeable* has the same meaning as *predicted* in failure-modes-and-effects, fault-tree, or hazard analyses. It appears that many "ordinary" consumers would have no concept of how safe a product should, or could, be without the expectations created by statements in sales material, inferences from mass media, general assumptions regarding modern technology, and faith in corporate enterprise.

A design defect also exists if there is excessive preventable danger. The real question is whether the danger outweighs the benefits; this can be answered by a risk-benefit analysis which should include at least five factors: (1) gravity of the danger posed by the design (i.e., severity of the consequences in the event of injury or failure), (2) probability (including frequency of and exposure to the failure mode) that such a danger will occur, (3) technical feasibility of a safer alternative design, including possible remedies or corrective action, (4) economic feasibility of these possible alternatives, and (5) possible adverse consequences to the product and consumer which would result from alternative designs. Additional relevant factors may be included, but design adequacy is evaluated in terms of a balance between benefits from the product and the probability of danger. For example, an airplane propeller and a fan both move air. The fan is guarded or shielded, whereas the propeller is not. Quantification is not required but may be desirable.

10.4 DESIGNER'S OBLIGATION

The designer or manufacturer of any product—consumer product, industrial machinery, tool, system, etc.—has a major obligation to make this product safe, that is, to reduce the risks associated with the product to an acceptable level. In this context, *safe* means a product with an irreducible minimum of danger (as defined in the legal sense); that is, the product is safe with regard not only to its intended use (or uses) but also to all unintended but foreseeable uses. For example, consider the common flat-tang screwdriver. Its intended use is well known. Can anyone say that he or she has never used such a screwdriver for any other purpose? It must be designed and manufactured to be safe in all these uses. It can be done.

There are three aspects, or stages, in designing for safety.

1. Make the product safe; that is, design all hazards out of the product.
2. If it is impossible to design out all hazards, provide guards which eliminate the danger.
3. If it is impossible to provide proper and complete guarding, provide appropriate directions and warnings.

10.4.1 Make It Safe

In designing any product, the designer is concerned with many aspects, such as function, safety, reliability, producibility, maintainability, environmental impact, quality, unit cost, etc. With regard to safety, consideration of hazards and their elimination must start with the first concept of the design of the product. This consideration must be carried through the entire life cycle. As Hunter [10.10] says,

> This must include hazards which occur during the process of making the product, the hazards which occur during the expected use of the product, the hazards which occur during foreseeable misuse and abuse of the product, hazards occurring during the servicing of the product, and the hazards connected with the disposal of the product after it has worn out.

Since each design is different, the designer needs to give full consideration to safety aspects of the product, even if it is a modification of an existing product. There is no fixed, universal set of rules which tells the designer how to proceed. There are, however, some general considerations and guidelines.

Hazard Recognition. Hazard recognition needs to start at the earliest possible stage in a design. Hazard recognition requires much background and experience in accident causation. There is extremely little academic training available, although the National Safety Council (NSC) and many other organizations publish information on this topic. Any threat to personal safety should be regarded as a hazard and treated as such. These threats come from several sources.

Kinematic/Mechanical Hazards. Any location where moving components come together, with resulting possible pinching, cutting, or crushing, is in this class. Examples are belts and pulleys, sets of gears, mating rollers, shearing operations, and stamping operations with closing forming dies. The author can remember working in a machine shop where individual machines (lathes, grinders, shapers, planers, etc.) were driven by belts and pulleys supplied by power from a large prime mover. Such shops had (1) a great number of nip-point hazards where belts ran onto pulleys and (2) a possible flying object hazard if a belt came apart or slipped off the pulley. Development of low-cost, reliable electric motors which could be used to drive individual machines removed the belt-pulley hazards but introduced a new electrical hazard.

Electrical Hazards. Shock hazard, possibly causing an undesirable involuntary motion, and electrocution hazard, causing loss of consciousness or death, are the principal electrical hazards for people. Electrical faults ("short circuits") are the major hazard to property. Massive arcing, cascading sparks, and molten metal often start fires in any nearby combustible material. Any person in the vicinity of a large electrical fault could be severely injured, even though the danger of electric shock has been reduced by ground fault devices.

Energy Hazards. Any stored energy is a potential energy hazard if the energy is suddenly released in an unexpected manner. Compressed or stretched springs, compressed gas containers, counterbalancing weights, electrical capacitors, etc., are all possible sources of energy hazards. Energy hazards are of major importance during servicing of equipment. A designer must develop methods and procedures for placing the product in a "zero-energy state" while it is being serviced.

Flywheels, fan blades, loom shuttles, conveyor components, and, in general, any parts with substantial mass which move with significant velocity are kinematic energy hazards which can damage any objects (including humans) which interfere with their motion.

Human Factors/Ergonomic Hazards. All consumer products and most industrial and commercial equipment is intended to be used by humans. Ergonomics, defined as the art and science of designing work and products to fit the worker and product user, is a top-priority consideration in the design process.

The human is a wonderful creation, capable, in many ways, of exceeding the machine's capability. The human can adjust to unusual situations; the machine cannot. The human can decide to go over, under, or around an obstacle, and do it; the machine cannot. In an emergency situation, the human can exceed normal performance to a degree that would cause a machine to fail (blow a fuse, pop a gasket, etc.). Unfortunately, the human can also make mistakes which lead to accidents.

Human beings exhibit a multitude of variations: height, weight, physical strength, visual acuity, hearing, computational capability, intelligence, education, etc. Designers must consider all these variables, and their ranges, as they recognize that their product will ultimately be used by humans.

The designer certainly must consider the hazards in the design when it is used or operated in the intended manner. The designer must also recognize that the product may be used in other, unintended but foreseeable, ways. As noted above, a hazard is any aspect of technology or activity that produces risk. The designer must provide protection against the hazards in all uses which can be foreseen by the designer. Unfortunately, a most diligent and careful search for foreseeable uses may still leave a mode of use undiscovered. In litigation, a key question is often whether the specific use was foreseeable by a reasonably diligent designer.

When humans are involved, there will be errors and mistakes. Some errors are extremely difficult, if not impossible, to anticipate. In many situations, people will abuse equipment. This is commonly a result of poor operating practices or lack of maintenance. In other situations, the user may take deliberate action to fit two components together in a manner which is not intended, e.g., to make and install thread adapters on pressurized gas containers. There is no question that the designer cannot anticipate all these possibilities and provide protection. Nevertheless, the designer is not relieved of a substantial effort to anticipate such actions and to try to thwart them.

Environmental Hazards. Internal environmental hazards are things which can damage the product as a result of changes in the surrounding environment. For example, in a water-cooled engine, the water can freeze and rupture the cylinder block if the temperature goes below the freezing point. This freezing problem can be alleviated by using freeze plugs which are forced out of an engine block if the water freezes, adding antifreeze to the cooling water, or using an electrical heating coil in place of the oil drain plug (standard winter equipment in cities like Fairbanks, Alaska).

External environmental hazards are adverse effects the product may have on the surrounding environment. These include such items as noise; vibrations, such as those from forging and stamping operations; exhaust products from internal combustion engines; various chemicals such as chlorinated fluorocarbons (Freon); polychlorinated biphenyls (PCBs); electronic switching devices which radiate electromagnetic disturbances; hot surfaces which can burn a human or cause thermal pollution; etc.

Hazard Analysis. Hazards are more easily recognized by conducting a complete hazard analysis, which is the investigation and evaluation of

1. The interrelationships of primary, initiating, and contributory hazards which may be present
2. The circumstances, conditions, equipment, personnel, and other factors involved in the safety of a product or the safety of a system and its operation
3. The means of avoiding or eliminating any specific hazard by use of suitable design, procedures, processes, or material
4. The controls that may be required to avoid or eliminate possible hazards and the best methods for incorporating these controls into the product or system
5. The possible damaging effects resulting from lack, or loss, of control of any hazard that cannot be avoided or eliminated
6. The safeguards for preventing injury or damage if control of the hazard is lost

Various approaches to hazard analyses are found in many places. Hammer [10.11], [10.12], [10.13], Roland and Moriarty [10.14], and Stephenson [10.15] present typical approaches. Additional techniques are discussed below.

For those concerned with consumer products, the Consumer Product Safety Commission (CPSC) publishes much of the results of its accident data collections and analyses in the form of Hazard Analyses, Special Studies, and Data Summaries.

These identify hazards and report accident patterns by types of products. Information is available from the National Injury Information Clearinghouse, CPSC, 5401 Westbard Avenue, Washington, DC 20207.

Consumer products, as the term implies, are those products used by the ultimate consumer, usually a member of the general public. Service life, in most instances, is relatively short, although some items such as household refrigerators and clothes washers and dryers may operate for many years. In contrast to consumer products, industrial and commercial products are intended to provide revenue for their owners and normally have a relatively long service life. This long life is an advantage from the economic viewpoint. From the safety aspect, however, it tends to perpetuate safety design problems for years after safer designs have been developed and distributed in the marketplace. Because of this long life, extra care is required in designing for safety.

Failure Modes and Effects Analysis (FMEA).   Failure modes and effects analyses are performed at the individual component level very early in the design phase to find all possible ways in which equipment can fail and to determine the effect of such failures on the system, that is, what the user will experience. FMEA is an inductive process which asks: What if? An FMEA is used to assure that (1) all component failure modes and their effects have been considered and either eliminated or controlled; (2) information for design reviews, maintainability analysis, and quantitative reliability analysis is generated; (3) data for maintenance and operational manuals are provided; and (4) inputs to hazard analyses are available.

Failure Modes and Criticality Analysis (FMECA).   In any product, some components or assemblies are especially critical to the product's function and the safety of operators. These should be given special attention, with more detailed analysis than others. Which components are critical can be established through experience or as a result of analysis. Criticality is rated in more than one way and for more than one purpose. For example, the Society of Automotive Engineers (SAE) has an Aerospace Recommended Practice (ARP 926). The method described in ARP 926 establishes four categories of criticality (as a function of the seriousness of the consequences of failure) and is essentially an extension of FMEA which is designated failure modes, effects, and criticality analysis (FMECA).

Fault-Tree Analysis (FTA).   Fault-tree analysis is substantially different from FMEA in that it is deductive rather than inductive. FTA starts with what the user experiences and traces back through the system to determine possible alternative causes. The focus is on the product, system, or subsystem as a complete entity. FTA can provide an objective basis for (1) analyzing system design, (2) performing trade-off studies, (3) analyzing common-cause failures, (4) demonstrating compliance with safety requirements, and (5) justifying system changes and additions.

Fault Hazard Analysis (FHA).   FMEA considers only malfunctions. FHA has been developed to assess the other categories of hazards. FHA was developed at about the same time as FTA, but it does not use the same logic principles as FTA or have the quantitative aspects of FMEA. It was first used by analysts with no knowledge of FTA and by those desiring a tabulated output, which FTA does not provide. FHA is qualitative. It is used mainly as a detailed extension of a preliminary hazard analysis.

Operating Hazards Analysis (OHA).   FMEA, FMECA, FTA, and FHA are primarily concerned with problems with hardware. OHA, on the other hand, intensively studies the actions of operators involved in activities such as operating a product, testing, maintaining, repairing, transporting, handling, etc. Emphasis is primarily on personnel performing tasks, with equipment a secondary consideration. The end result is usually recommendations for design or operational changes to eliminate hazards or better control them. OHAs should be started early enough to allow time for consideration and incorporation of changes prior to release of a product for production.

Design Review.   Design review is an effort, through group examination and discussion, to ensure that a product (and its components) will meet all requirements. In a design of any complexity, there is a necessity for a minimum of three reviews: conceptual, interim, and final. Conceptual design reviews have a major impact on the design, with interim and final reviews having relatively less effect as the design becomes more fixed and less time is available for major design changes. *It is much easier and much less expensive to design safety in at the beginning than to include it retroactively.*

A more sophisticated product may require several design reviews during the design process. These might be conceptual, definition, preliminary (review of initial design details), critical (or interim review, or perhaps several reviews in sequence—review details of progress, safety analyses, progress in hazard elimination, etc.), prototype (review of design before building a prototype), prototype function, and preproduction (final review—the last complete review before release of the design to production).

These periodic design reviews should (1) review the progress of the design, (2) monitor design and development, (3) assure that all requirements are met, and (4) provide feedback of information to all concerned.

A design review is conducted by an ad hoc design review board composed of mechanical designers, electrical designers, reliability engineers, safety engineers, packaging engineers, various other design engineers as appropriate, a management representative, a sales representative, an insurance consultant, an attorney specializing in products liability, outside "experts" (be sure they are truly expert!), etc. Members of the design review board should not be direct participants in day-to-day design and development of the product under review, but should have technical capability at least equal to that of the actual design team. Vendor participation is highly desirable, especially in conceptual and final design reviews. Design review checklists should be prepared well in advance of actual board meetings. These checklists should be thoroughly detailed, covering all aspects of the design and expected performance. They should include all phases of production and distribution as well as design. Checklists should be specific, detailed, and not used for any other product. New checklists should be developed for each new product. It is good practice for a designer or manufacturer to have some sort of permanent review process in addition to the ad hoc board for each individual product. This permanent group should evaluate all new products, reevaluate old products, and keep current with trends, standards, and safety devices.

If properly conducted, a design review can contribute substantially to avoiding serious problems by getting the job done right the first time. Formal design review processes are effective barriers to "quick and dirty" designs based on intuition (or "educated guesses") without adequate analyses.

Standards.   Once a design problem is formulated and the intended function is clear, the designer should collect, review, and analyze all pertinent information relative to standards, codes, regulations, industry practice, etc. From this study, the designer can usually get assistance in hazards analysis and formulate the design constraints resulting from the known requirements. One must be clear on which requirements are voluntary and which are mandatory. Standards published by the American National Standards Institute (ANSI) are considered voluntary, consensus standards. A voluntary standard need not necessarily be followed in designing and manufacturing a product, although it is strongly recommended that such standards be followed, or exceeded, in the design. However, if a municipality, state, or federal agency includes a given standard in its requirements, then that standard becomes mandatory, with the force of law. For example, ANSI Standard A17.1, *Safety Code for Elevators, Dumbwaiters, Escalators, and Moving Walks,* is a voluntary standard. If a city incorporates that standard in its building code, then the standard is mandatory and must be followed in constructing a building in that city.

Standards are published by many different organizations. Some of the better known are the American National Standards Institute (ANSI), 11 West 42nd St., New York, NY 10036; American Society for Testing and Materials (ASTM), 1919 Race St., Philadelphia, PA 19103; Underwriters Laboratories, Inc. (UL), 333 Pfingsten Road, Northbrook, IL 60062; and National Fire Protection Association (NFPA), 1 Batterymarch Park, Quincy, MA 02269. The federal government has many agencies which establish and publish a large number of standards and regulations. Proposed regulations are published in the *Federal Register,* with the public invited to comment. After the comment period is over and all hearings have been held, the final version is published in the *Federal Register* with a date when the regulation becomes effective. All approved and published federal regulations are collected in the *Code of Federal Regulations (CFR).* There are 50 CFR titles covering all areas of the federal government. All published regulations are reviewed and revised annually. The *Index of Federal Specifications, Standards, and Commercial Item Descriptions,* issued annually in April by the General Services Administration, is available from the Superintendent of Documents, U.S. Government Printing Office, Washington, DC 20402.

More than 35,000 documents have been generated by nearly 350 standards-writing organizations in the United States. There is a two-volume *Index and Directory of U.S. Industry Standards.* Volume 1 contains the subject index and lists all applicable standards from all sources for any selected subject. Volume 2 contains a listing of all standards-publishing organizations in alphabetical order of their acronyms. The index is published by Information Handling Services of Englewood, Colorado. It is available from Global Engineering Documents, which has offices at 2805 McGaw Ave., Irvine, CA 92714 and 4351 Garden City Drive, Landover, MD 20785. Global can also supply copies of any desired document for a fee.

The Department of Defense (DoD) has a large number of military handbooks, military standards, and military specifications which can be applied to civilian and commercial products as well as to military needs. (These require that all the desirable features be designed into the product from the start of the design effort rather than being added at the end after testing and evaluations have shown deficiencies. This design approach is totally applicable to nonmilitary products.) These DoD documents are available from the Naval Publications and Forms Center, 5801 Tabor Ave., Philadelphia, PA 19210.

Occupational Safety and Health Administration (OSHA). The federal Occupational Safety and Health Act establishing the Occupational Safety and Health Administration (OSHA) was passed in 1970. One of its goals was "to assure so far as possible every working man and woman in the nation safe and healthful working conditions." OSHA regulations have the force of law, which means that employers must provide a workplace with no recognized hazards. Thus employers cannot legally operate equipment which exposes workers to unprotected hazards. Consequently, designers must design hazards out of their products before these products reach the market. The regulations are published in title 29 of the CFR. Section 1910 applies to general industry. As the act went into effect, the administrators were allowed to draw on the large number of existing safety standards and adopt them as they saw fit over a period of two years. Many of these standards were adopted by reference when the act became effective in 1971. Today, many of these standards are obsolete but, unfortunately, are still being used as the basis for OSHA regulations. In addition, there are many products which did not exist in 1971, and new standards have been developed for such products. For example, OSHA standards for mechanical power presses are based on the 1971 edition of ANSI B11.1. Since that time, the B11 Committee of ANSI has published at least 18 standards relating to the larger field of machine tools. Designers should not rely on OSHA regulations alone, but should determine the availability and applicability of the latest published standards. OSHA regulations obviously must be used with caution. Even though many are obsolete, they still have the force of law. OSHA regulations can be obtained from the U.S. Government Printing Office.

Maintenance.   Maintenance safety problems can be separated into those that occur during maintenance, from lack of maintenance, or from improper maintenance. Improper maintenance, for example, might be a situation in which electrical connections on a metal case were not installed correctly, thus producing a hazardous condition where none had existed previously. There seems to be little the designer can do to prevent a lack of maintenance. Much improper maintenance can be avoided by designing products in such a way that it is extremely difficult to reassemble them incorrectly.

There is no question that equipment of all kinds does require periodic adjustment or replacement of parts. There is much evidence that designers have too often failed to consider the hazards to which maintenance personnel will be exposed, even in routine maintenance. During maintenance, safety devices must often be disconnected and/or protective guards removed to permit the necessary access. In this context, maintenance personnel may need to put parts of their bodies in hazardous locations which were protected by the necessarily inoperative safety devices. It is the responsibility of the designer to provide protection in this situation.

Lockouts, Lockins, and Interlocks.   Many injuries and fatalities have occurred when a worker unwittingly started equipment while a maintenance worker was in the equipment. It is necessary to make it impossible for machinery undergoing maintenance to be started by anyone other than the maintenance worker. CFR 1910.147(c)(2)(iii) [OSHA] requires the designer to provide lockout protection.

A lockout prevents an event from happening or prevents an individual, object, force, or other factor from entering a dangerous zone. A lockin maintains an event or prohibits an individual, object, force, or other factor from leaving a safe zone. Locking a switch on a live circuit to prevent the current being shut off is a lockin; a similar lock on a switch on an open circuit to prevent it being energized is a lockout. Both lockouts and lockins can be accomplished by giving each individual worker a personal padlock and key (any duplicate key would be in a central office in a locked cabinet). This procedure can mean placing multiple locks on a lockout panel.

Interlocks are provided to ensure that an event does not occur inadvertently or where a sequence of operations is important or necessary and a wrong sequence could cause a mishap. The most common interlock is an electrical switch which must be in the closed position for power to be supplied to the equipment. If a guard, cover, or similar device is opened or left open, the machine will not operate. Smith [10.16] comments on two accidents, one involving a screw auger for mixing core sand in a foundry, the other involving a large batch mixer. In both cases, maintenance workers suffered permanent disabling injuries when another worker switched on the equipment. In both cases, a lockout or an interlock which would function when the cover was lifted would have prevented the injuries. Although interlocks are usually very effective, they can be rather easily bypassed by using some means to keep the switch closed.

Zero Energy. Many products require storage of energy for operation. For example, energy is stored in any spring which is changed during assembly from its free, unstressed dimensions. This energy storage also exists in cables, cords, and chains which are loaded in tension. Other sources of stored energy are compressed gases, energized electronic power sources, lifted counterweights, etc. The zero-energy concept requires the designer to provide protection for any operator or maintainer of equipment against the consequences of the unanticipated release of stored energy; that is, there must be a means of neutralizing these energy sources in an emergency situation or during maintenance work.

Fail-Safe Designs. Product failures produce a significant fraction of accidents. Fail-safe design seeks to ensure that a failure (1) will not affect the product or (2) will change it to a state in which no injury or damage will occur.

1. Fail-passive designs reduce the system to its lowest energy level. The product will not operate until corrective action is taken, but the failure-initiating hazard will cause no further damage. Circuit breakers are a good example of fail-passive devices.

2. Fail-active designs maintain an energized condition that keeps the system in a safe mode of operation until corrective action can be taken or the system is replaced by an alternative system. Redundancy using standby equipment is an example of a fail-active system.

3. Fail-operational designs allow safe continuation of function until corrective action can be taken. Fail-operational is obviously preferred, if possible. The ASME requires fail-operational feedwater valves for boilers. Water must first flow under, rather than over, the valve disk. If the disk is detached from the valve stem, water will continue to flow and allow the boiler to function normally. Designs should be made fail-safe to the greatest degree possible.

General Principles. Hunter [10.10] gives the following statements as general principles or guidelines for designing safe products:

1. Recognize and identify actual or potential hazards, then design them out of the product.

2. Thoroughly test and evaluate prototypes of the product to reveal any hazard missed in the preliminary design stages.

3. Make certain that the product will actually perform its intended function in an acceptable manner so that the user will not be tempted to modify it or need to improvise possibly unsafe methods for using it.

4. If field experience reveals a safety problem, determine its real cause, develop a corrective action to eliminate the hazard, and follow up to make certain that the corrective action is successful.

5. Design equipment so that it is easier to use safely than unsafely.

6. Realize that most product safety problems arise from improper product use rather than product defects.

Safety Checklists. Hammer [10.12], [10.13] and the National Safety Council [10.17] give lists of basic safety requirements for use in developing safe designs. For example, at the top of his list, Hammer [10.12], [10.13] says: "Sharp corners, projections, edges, and rough surfaces which can cause cuts, scratches, or puncture wounds will be eliminated unless required for a specific function." There are 21 more items in the list.

Acceptable Conditions.   Hammer [10.12], [10.13] notes that safety engineers (perhaps no one else?) generally consider the following conditions acceptable and indicative of good design:

1. Any design which requires at least (a) two *independent* malfunctions, or (b) two *independent* errors, or (c) a malfunction and an error which are *independent* to cause an accident

2. Any design which positively prevents an error in assembly, installation, connection, or operation that analysis shows would be safety-critical

3. Any design which positively prevents a malfunction of one component (or assembly) from causing other failures which could cause injury or damage (fail-safe)

4. Any design which limits and controls the operation, interaction, or sequencing of components (or subassemblies) when an error or malfunction could cause an accident—for example, when activating switch B before activating switch A could cause damage (interlock)

5. Any design which will safely withstand a release of greater energy than expected, or normally required

6. Any design that positively controls buildup of energy to a level which could potentially cause damage (for example, use of a shear pin to protect a shaft)

10.4.2   Guarding

As indicated above, if it is impossible to design out all hazards, it is necessary to provide guards. The basic legal requirements are set forth in CFR 1910.212, *General Requirements for All Machines* (OSHA), which says:

> (a) Machine guarding (1) Types of guarding. One or more methods of machine guarding shall be provided to protect the operator and other employees in the machine area from hazards such as those created by point of operation, ingoing nip points, rotating parts, flying chips and sparks. Examples of guarding methods are barrier guards, two-hand tripping devices, electronic safety devices, etc.
>
>   (2) General requirements for machine guards. Guards shall be affixed to the machine where possible and secured elsewhere if for any reason attachment to the machine is not possible. The guard shall be such that it does not offer an accident hazard in itself.

One should note the key word *all* in the heading. Further, the use of *shall* makes the requirement for guards mandatory.

Most of the dangerous hazards from moving parts of machines occur in three areas:

1. *Point of operation.* This is where the machine works on the workpiece to shape, cut, etc.

2. *Power train.* This is the set of moving parts which delivers power to the point of operation. These parts include shafts, gears, chains, pulleys, cams, etc.

3. *Auxiliary components.* These are such items as feeding mechanisms and other components which move when the machine is in operation.

All of these have obvious nip points. Less obvious nip points are between an augur screw conveyor and the trough, between a tool rest and a grinding wheel or part being turned on a lathe, between the spokes of a handwheel and the guide or support behind it, and between a translating component and a fixed component close to it, (that is, a shear of any kind) (see Smith [10.18]). In general, a nip point occurs when two components are in close proximity with relative motion which reduces the separation between them. There are other hazards, such as potential pressure vessel explosions and bursting flywheels, but one can take the position that these kinds of hazards should be eliminated in the original design.

The general requirement for a guard is that the point of hazard be substantially enclosed, screened, barricaded, or otherwise protected so that persons, whether workers or bystanders, cannot inadvertently come in contact with the hazard.

Mechanical Guards.   Mechanical guards, the most common type, can be fixed, adjustable, or interlocked. Grimaldi and Simonds [10.19] give the basic requirements for a mechanical guard as:

1. It must be sturdy to prevent damage to the guard from external sources or interference with the operation of the machine. Either of these possibilities would probably result in the operator removing the guard and not arranging to have it repaired and replaced.

2. It must permit required maintenance operations without necessitating excessive labor for dismantling and reassembling the guard, or else there will be a tendency to omit its installation.

3. It must be properly mounted. The mounting must be rigid to prevent objectionable rattles or interference with working parts. The mountings should be strong enough so that they will not fail under use.

4. It should be designed so that there are no detachable parts, which if removed and not replaced would reduce its guarding effectiveness.

5. It should be easy to inspect, and a periodic checkup program, as a part of the maintenance procedure for shop equipment, should be established in order to continue its effectiveness.

Fixed guards should be used wherever possible, since they provide permanent protection against hazardous machinery components. Adjustable guards are used when the mode of operation of the machine is expected to change and adjustment will be necessary to accommodate a new set of dimensions. Once adjusted, the guard should function as a fixed guard. Interlocked guards prevent operation of the machine until the guards have moved into positions which keep the worker out of the hazardous zone. It is essential that the guard put the machine in a safe mode if the guard should fail for any reason (fail-safe).

Pullbacks are bands strapped around the operator's wrists with cords or cables running from the bands to a pulling mechanism synchronized with the down stroke of a power press. If the operator does not remove his or her hands from the hazard area, they are automatically pulled away. This pullback occurs even if the press recycles on its own. Pullbacks are not complete protection, however; the author knows of at least one situation in which injury to the worker resulted from a recycle. Pullbacks require adjustment to each operator, frequent inspection, and diligent maintenance. They are often objectionable to the worker, who feels tied to the machine.

Barrier gates are simple mechanical devices which are opened and shut by machine motion during the operating cycle. This allows the operator to approach the point of operation, e.g., to feed work stock, but protects against any part of the body being in the hazard zone when the machine is activated. In most cases, there is an interlock that shuts off the power when the gate is open and prevents opening the gate when the machine is in motion.

Electromechanical Devices.   Presence-sensing devices commonly use (1) a light beam and a photoelectric cell ("electric eye") to stop the machine if the light beam is interrupted or (2) a radio-frequency electromagnetic field which is disturbed by the capacitance effect of the intruding body.

Distance/Separation Guarding.   A very logical and effective way of guarding is by separation or distance. The question of location must be considered by the designer. For example, tables of distances and the corresponding openings permitted are given in CFR 1910.217(c)(2)(vi) (OSHA) and in ANSI Standard B11.1. As a sample, if the distance from the point of operation is 1.50 to 2.50 in, the maximum width of opening is 0.375 in; if the distance is 5.50 to 6.50 in, the maximum width of opening is 0.75 in. The dimensions in the tables have been chosen to prevent the fingers of the average-size operator from reaching the point of operation.

Input/Output Systems.   Systems for feeding stock and ejecting workpieces can provide more safety if semiautomatic or fully automatic systems are used. Perhaps the most desirable is a robotic system for mechanical feeding of stock and retrieval of parts. Although more expensive, robots can work where there is a high noise level, can work at a higher temperature than is tolerable for most humans, and can perform repetitive monotonous tasks indefinitely. One hazard is that the robot may strike a bystander. This hazard, however, can be avoided by barriers or presence sensors.

Auxiliary Equipment.   Auxiliary equipment is generally used in connection with other protective devices to give an additional measure of safety. For example, it is very difficult to provide complete point-of-operation guarding for a band saw, since the saw blade must be exposed in order to accomplish the desired function of cutting material. When small or narrow pieces are being cut, the operator's fingers can get too close unless a push stick or push block is used. The block allows control over the workpiece to get the desired result but keeps the operator's fingers away from the hazard zone. A great variety of pliers, tongs, tweezers, magnetic lifters, suction cup lifters, etc., are available for use as auxiliary equipment. Such auxiliary equipment may need to be adjusted for use in different applications.

Controls.   Operating controls can be designed to ensure that the operator is out of the hazard zone, such as the point of operation. If only one pushbutton is provided, the operator's other hand could be in the hazard zone. To prevent this, two buttons are provided, far enough apart to require use of both hands and arranged in series so that both must be pushed to activate the machine. If the stroke time is long enough for the operator to push the buttons and still get a hand into the hazard zone, a requirement that both buttons be held down until the stroke is essentially completed can be incorporated. There is a temptation for workers to tie down one of the buttons, which obviously defeats the two-button safety feature. To circumvent this, both buttons must be pressed within a short time period. If the allowable delay is exceeded, the machine will not operate. While most machines should have a two-button control system, there are situations, such as control of an overhead crane, in which a single set of on-jog-off buttons is acceptable because the operator is physically distant from the hazard zone.

Another aspect of control buttons is that the start, or operate, button (or buttons) should be recessed to reduce the possibility of inadvertent operation. Start buttons are also usually green in color. A stop button should have a large, mushroom-shaped head which is not recessed. This stop button should be easily reachable from the normal operating position for use in case of an emergency. The usual color for stop buttons is red.

In cases where a machine runs continuously, while the operator is exposed to hazards in any manner, use of a control which can immediately trip the switch—that is, stop the machine—is necessary. The stop button, noted above, is one possibility. In other cases, a trip wire is placed where a worker can easily reach it from any location of the work station. Pulling on this wire will stop the operation. In one situation (Smith [10.18]), there was a trip wire, but it was not close enough to be effective when a worker had a hand caught in a shear nip point. In other situations, a force- or pressure-sensitive bar has been used. When the bar is pushed (for example, if the operator stumbles, loses balance, or is pulled into the machine), the machine will be deactivated. The location of the bar is critical. It must be located where it will be effective in an emergency but will not be inadvertently activated by the material being processed. Presence-sensing devices, "electric eyes," IR beams, etc., can also be used to deactivate equipment. Machines which continue to run after power is cut off require a brake for quick stopping.

Data Sources. As noted above, OSHA regulations and ANSI standards are available that can provide much information on guarding. Pertinent data can be found in many other publications, such as Hunter [10.10] and Grimaldi and Simonds [10.19]. Information is also available from the National Safety Council [10.17], [10.20], [10.21]. It might also be noted that the National Safety Council has videos available for employee training.

10.4.3 Warnings

As noted above, in those situations in which it is not possible to provide complete and effective guarding, or in those situations where such guarding would severely impair the intended function of the product, it is necessary to provide appropriate directions and warnings.

It is obvious that eliminating all the potential hazards in a design and/or providing effective guarding is not a simple task. In some cases, it is impossible. Developing a proper, effective warning is generally considered even more difficult. In large measure, this is because there is hardly consensus, let alone anything approaching unanimity, on what is a truly adequate and acceptable warning for a given situation. Nonetheless, a full-scale effort must be made.

Directions are instructions intended to ensure *effective* use of a product. *Warnings,* in contrast, are intended to ensure *safe* use, that is, to inform of hazards and of improper use, and to instruct how to guard against these, if possible. The distinction is clear in concept, but it is not always possible to tell whether a given statement is a direction or a warning. Lehto and Miller [10.22] say:

> Perhaps the best way to initially distinguish between warnings and other forms of safety-related information is to state that warnings are specific stimuli which alert a user to the presence of a hazard, thereby triggering the processing of additional information regarding the nature, probability, and magnitude of the hazard. This additional information may be within the user's memory or may be provided by other sources external to the user. Much of the current controversy regarding warnings is actually related to the need for this additional information.

There are three criteria which must be met for a warning to be fully effective:

1. The message must be received.
2. The message must be understood.
3. The endangered person must act in accordance with the message.

A warning is not effective unless it changes the potential behavior of the endangered individual.

Types of Warnings. Injury or damage can often be avoided by a focus on the existence of a hazard and the need for careful action. Every method for calling attention to a hazard requires communication; each of the human senses, singly or sometimes in concert, has been used for this purpose.

Visual Warnings. It is widely recognized that most information on hazards, perhaps as much as 80 percent, is visually transmitted to personnel. There are more variations of visual methods than of methods involving the other senses. A hazardous area is often more brightly illuminated than other areas in order to focus attention on it. A piece of equipment can be painted in alternating stripes or in a bright, distinctive color; for example, fire trucks are now being painted greenish-yellow rather than red for better visibility. Signal lights are often used—for example, on emergency vehicles and at railroad crossings at grade. Flags and streamers can be used. Signs are common, eg, highway signs.

Auditory Warnings. Auditory warnings may have a shorter range of effectiveness than visual warnings, but their effectiveness may be greater in that short range. Auditory warnings are often coupled with visual warnings, as on emergency vehicles. Typical devices are sirens, bells, or buzzers; an example is the intermittent sound of a horn on heavy equipment which is backing up.

Olfactory Warnings. Odorants can be used in some limited, although effective, ways, such as the addition of small amounts of a gaseous odorant to natural gas to warn of leaks.

Tactile Warnings. Vibration is the major tactile means of warning; an example is rumble strips on highways. Vibration in machinery may mean the beginning of serious wear or lubrication failure. Temperature sensing, or at least an indication of significant temperature change, can also be included in this category.

Tastable Warnings. These may have little use in machine applications, but they have been used in various ways to provide warnings concerning foods and medicines.

Written Warnings—Labels. Much confusion exists, especially within the legal system, concerning the meaning of *warning* when applied to products and their uses. The major reason may be that warnings are usually considered to be synonymous with the explicit "warning labels" which are sometimes placed on products. One consequence is that sources of information which do not explicitly (in words) describe the hazard, specify its intensity, provide instructive countermeasures, and strongly advocate adherence may not be considered adequate warnings. Another reason for the confusion is that society seems to expect warnings to perform multiple functions.

Warnings should supplement the safety-related design features of a product by indicating how to avoid injury or damage from the hazards which could not be (1) feasibly designed out of the product, (2) designed out without seriously compromising its utility, or (3) protected against by providing adequate guards. In theory, providing such information will reduce danger by altering people's behavior while using a product or by causing people to avoid using a product. From the litigation viewpoint, warnings often perform functions that have little to do with either safety or transfer of safety-related information. A manufacturer may view warnings as a defense against litigation. One consequence is extensive use of warning labels. Such use often means products with warning labels which yield no increase in safety. Even more unfortunately, some manufacturers may use warnings instead of careful design, which is absolutely unacceptable.

As indicated above, for a warning to be effective, the endangered person must receive the message, understand it, and act in accordance with it. The designer and manufacturer obviously have no control over the action, but they do have substantial control over sending the message and making it understandable. Failure on the part of the endangered person to do any one of the above results in failure of the communication process and the warning being ineffective. Consider, for example, a situation in which (1) 40 percent of the users read the warning, (2) 50 percent of those readers truly comprehend and understand the warning, (3) 40 percent of those act properly in accordance with the warning, and (4) the action is sufficient to avoid injury 90 percent of the time. On the basis of these numbers, the probability of the warning being completely effective is 7 percent. Whatever numbers one may use, the probability of a warning being effective is relatively low. This probability is certainly no higher than the percentage of users who read the warning. There is general agreement that many people who see a warning label do not read it. Many do not even see the label. This obviously can be discouraging to someone trying to develop a proper, effective, warning label. Nonetheless, a major effort must be made.

Every warning, including labels, has an *alerting function*. The warning label must be prominently located, that is, in a position such that the user has great difficulty avoiding seeing it. The warning label must be distinctive; that is, it must be sufficiently different from other labels that there is no question of its identity. Shape has an influence; shapes with rounded or curved boundaries are not as effective in attracting attention as shapes with sharp corners. Rectangles seem to be more effective than squares or triangles. Labels with five or more sides are rarely used on industrial or consumer products.

Three signal words (in relatively large letters) and color combinations are normally used to attract attention.

1. *DANGER.* The hazard can immediately cause (1) death or (2) severe injury upon contact with the source of hazard. Letters should be white on a red background.

2. *WARNING.* (1) The hazard can immediately cause moderate injury, or (2) death or severe injury may eventually result from contact with the source of hazard. Letters should be black on an orange background.

3. *CAUTION.* (1) The hazard can immediately cause minor injury, or (2) moderate injury may eventually result from contact with the source of hazard. Letters should be black on a yellow background.

Every warning, especially a label, has a *message*. This message must be clear, simple (unambiguous), succinct, and convincing. Short words are preferred, and there should be as few words as possible. Long sentences with technical terms should be avoided. There are indications in the literature that directions and warnings should be written at sixth-grade level. The use of indices such as the Flesch Reading Ease Formula, Gunning's Fog Index, or McElroy's Fog Count (Klare [10.24]) can be helpful in this respect. For products which will be used only within a country or region in which there is one common language, the choice of language is obvious. For products which will be used in regions with different languages, warning labels must be in those languages. Those who write labels in languages different from that of the manufacturer must be knowledgeable about the linguistic characteristics of those regions.

A partial solution to the problem of the need for multiple languages is the increasing use of pictographs. A pictograph communicates an idea or concept in one symbol which is universally recognized. For example, there is general recognition that a 45° red diagonal line (from upper left to lower right) through an annulus forbids whatever is displayed within the annulus (for example, a lighted cigarette within the annulus indicates that smoking is not permitted). General guidelines for pictographs are

1. Use a simple design for the symbol.
2. Use only one idea per pictograph.
3. Use only correct colors and shapes.
4. Locate the symbol as close as possible to any related words.

Words on the label must be legible by the average person, some of whom may have uncorrected visual impairments. ANSI Standard Z535.4 [10.25] gives requirements for wording and colors to be used. These differ from those in the standards issued by OSHA and the CPSC. ANSI Z535.4 also specifies letter size. Signal words must be at least 3 mm high (9-point type), and the text must be at least 1.5 mm high (5 points minimum). (*Point* is a measure of type size equal to 0.013837 inch; there are essentially 72 points per inch.) This is a consensus standard and represents the minimum acceptable to those involved in developing the document. There are many who believe that lettering should be larger. Bailey [10.26], for example, notes that "type size in books and magazines usually ranges from 7 to 14 points with the majority being about 10 to 11 points. Probably the optimum range is from 9 to 11 points— sizes smaller or larger can slow reading speed."

A warning label should be *permanent*. It should not fade or fall off before the end of the product's service life. Most labels are decalcomanias. Fortunately, they are available with a base of tough, wear-resistant material and good adhesive backing. Some products have warnings on stamped or embossed plates that are permanently secured to the product. Operator's manuals and/or maintenance manuals commonly accompany the product when it is shipped from the manufacturer but do not always find their way to the product in its operational situation. Providing a tough, dirt- and lubrication-resistant envelope which contains the manual and is "permanently" attached to the product (such as a power press or similar machinery) by a short chain can be useful for the worker.

CFR 1910.145 (OSHA) specifies requirements for accident prevention signs. By reference, two ANSI standards, Z35.1, *Specifications for Accident Prevention Signs,* and Z53.1, *Safety Color Code for Marking Physical Hazards,* were incorporated. Designers should consult these as soon as a decision is made to incorporate warnings. It should be noted, however, that in 1979, ANSI Z53 Committee on Safety Colors was combined with ANSI Z35 Committee on Safety Signs to form ANSI Z535 Committee on Safety Signs and Colors. Five subcommittees were formed to update the Z35 and Z53 standards and write two new standards. These are listed in References [10.25], [10.27], [10.28], [10.29], and [10.30]. One might note that the Society of Automotive Engineers (SAE) has a recommended practice, J115 [10.31], relating to safety signs. This is generally consistent with the ANSI 535 series, but there are some differences. (This situation is an example of old standards still having the force of law in OSHA standards, even though these old standards have been replaced by much more recent standards.)

Figure 10.1 shows a label (full size) which was used on a fiberglass ladder about 20 ft (6 m) long. It is suggested that this label be critiqued in light of the above comments before reading further. How good is it? How effective is it? Assuming that

INSPECTION

1. Inspect upon receipt and before use.
2. Never climb a damaged ladder. Return for repair or discard.
3. Check all working parts, rivets, bolts, rope and cable for good working order.
4. Never use ladder with missing parts.
5. Discard if exposed to fire or chemicals.

SELECTION

1. Use 300 lb., and 200 lb. Duty-Rated Ladder for maintenance and heavy-duty work. Never use ladder jacks on 200 lb. or 225 lb. Duty-Rated Ladders.
2. Use ladder with correct duty rating to support combined weight of the user and material. Ladders are available with duty ratings of 200, 225, 250, 300 lb.

SET-UP AND USE

1. Set up ladder at 75½° by placing bottom ¼ of length being used out from vertical resting point.
2. Set ladder on firm level ground. Never lean sideways and never use on ice or snow.
3. Use proper size ladder. Never use temporary supports to increase length or to adjust for uneven surfaces.
4. Keep rungs free from wet paint, mud, snow, grease, or other slippery material.
5. Extend only from ground. Never extend from top or by bouncing.
6. Never walk or jog ladder while on it.
7. Securely engage ladder locks before climbing.
8. Erect ladder with fly (upper) section above and resting on base (lower) section.
9. Each section of a multi-section ladder shall overlap the adjacent section by 3 ft. up to and including 36 ft.; by 4 ft. over 36 ft., up to and including 48 ft.; by 5 ft. over 48 ft., up to and including 60 ft.
10. Always have the four ends of the ladder rails firmly supported.
11. Always tie top and base to building.
12. Project ladder minimum of 3 feet above roof edge.
13. Tie down ladder before stepping onto roof.
14. Never over-reach. Move ladder instead. Keep belt buckle inside ladder side rails.
15. Never use in high winds.
16. Never overload. Ladder designed to support one person when properly used.
17. Never use as a horizontal platform, plank or material hoist.
18. Never use on a scaffold.
19. Never fasten different ladders together to increase length.
20. Never apply a side load to ladder to push or pull anything while on ladder.
21. Never drop or apply impact load to ladder.
22. Never sit on end of ladder rails.
23. When reassembling, properly engage all guide brackets and lock prior to use.
24. Never use in front of unlocked doors.
25. Fly section must have safety shoes if used as a single ladder.
26. Hooks may be attached at or near top for added security.
27. To support the top of a ladder at a window opening, a stabilizer should be attached to span the window.
28. Never use ladder when you are in poor health.
29. Never use if taking drugs or alcoholic beverages.
30. Recommend never using if over 65 years of age.

CLIMBING INSTRUCTIONS

1. Never climb onto ladder from the side or from one ladder to another.
2. Face ladder when ascending or descending. Maintain a firm grip and stand on middle of rung.
3. Never stand above 3rd rung from top.
4. Never climb above support point.

STORAGE

1. Support ladder on racks when stored.
2. Never store material on ladder.
3. Properly support ladder in transit.

FIGURE 10.1 A black-and-white reproduction of a decalcomania label to be placed on the inside of a side rail of a fiberglass ladder. The heading was yellow lettering on a black background. The text lettering was black on a yellow background. The reproduction is 100% of original size. See page 10.18 for discussion.

users do indeed see the label, how many will read it, especially with that length and type size? Of those who do read it, how many will really comprehend what the manufacturer is trying to say? This label was not well thought out, either in content or in phrasing, which is ambiguous or without clear meaning in several statements. The label does not provide clear instruction on use or explicitly warn of the consequences of hazards. It appears to use direction and warning statements without distinguishing between them. This label is clearly inferior and essentially ineffective. The inference (Smith and Talbot [10.32]) is that the manufacturer was trying to cover all possibilities to provide "protection" against product liability suits.

A warning that helps prevent an injury may not make great advertising copy, but it should be considered a necessity. One might note that warnings are not new. When Samuel Jones began manufacturing "Lucifer" matches (smelling of "hellfire and brimstone") in 1829, he printed the following warning on the boxes: "If possible, avoid inhaling the gas that escapes from the combustion of the black composition. Persons whose lungs are delicate should by no means use Lucifers." In terms of the above discussion, this is a relatively good statement.

Sources.   There has been much written with regard to warnings in both the technical and the legal literature. The best (technical) source currently available for understanding the nature of warnings and the difficulty in writing them is Lehto and Miller [10.22], [10.23].

10.5   HUMAN FACTORS/ERGONOMICS

Human beings interact with all products in designing, manufacturing, operating, and maintaining them. Human beings constitute the most complex subsystem in any system because of their abilities and limitations. In addition, the number and variety of actions that people, either as individuals or as a group, can take in any situation generates a high probability that any deficiency in the system will be linked to, and affected by, personal factors that can generate an accident. In other words, the most erratic, and the least controllable, parameter in any system is the human being.

In the design and development of a new product or system, the majority of the most critical decisions to be made are related to human performance. Informed decisions require the designer to have a good understanding of human engineering, human factors, and ergonomics. These three terms are often used interchangeably, but there are differences. Perhaps the broadest in scope is human engineering, which is a technical discipline primarily concerned with the interdependencies and interactions of humans and machines. Problems are highly likely when the two come in contact. Human engineering attempts to minimize these problems and obtain maximum effectiveness in any human-machine operation by integrating the best capabilities of both.

The designer must avoid any design which expects, or requires, individual operators to (1) exceed their available physical strength, (2) perform too many functions simultaneously, (3) perceive (or detect) and process more information than is possible, (4) perform meticulous tasks under difficult environmental conditions, (5) work at peak performance (or capability) for long periods, (6) work with tools in cramped spaces, etc. Insofar as possible, *the designer should adapt the machine to the human.*

The designer may think in terms of the "typical" or "average" human. This view is much too simple. People come in assorted sizes, shapes, capabilities, and varieties. Even when it may be appropriate to design for the "average," the designer must remember there is a range of differences from that average.

Some products are designed for limited groups, such as infants, children, teen-agers, the elderly, or the infirm. In such cases, the characteristics of the specific group must be emphasized. When designing for the "public," the designer needs to provide for the characteristics of the entire range of people, from babes in arms to nonagenarians. For example, doors, ramps, escalators, entries, etc., must be appropriate for a baby in a perambulator, a healthy and active man or woman, and a handicapped or elderly person with a walker or in a wheelchair. The task is not easy.

How does one proceed? The designer must be well informed on anthropometrics (physical characteristics), how people tend to behave or perform, and how to combine such data to achieve a suitable, effective, and safe design. A wealth of literature is available.

Hunter [10.10] includes enough anthropometric data to give insight into the kind of data to expect. He also provides much information on sources of information. He comments on Department of Defense documents which provide substantial and significant information. The objectives of these various documents can be applied with equal validity to both civilian and military products.

The aspect of human behavior is largely a question of psychology, a topic about which most engineers know little. Little information which is directed toward engineers seems to be readily available. One possible source is Grandjean [10.33].

There are many publications which provide varying degrees of insight and help in applying human factors information to design. Two which are particularly useful are Woodson [10.34] and Salvendy [10.35].

One of the many objectives of the designer is to minimize the probability of "human error," where human error is any personnel action (1) that is inconsistent with established behavioral patterns considered to be normal or (2) that differs from prescribed procedures. Predictable errors are those which experience shows will occur and reoccur under similar circumstances. The designer must minimize the possibility of such errors.

It is recognized that people have a strong tendency to follow procedures which require a minimum of physical and mental effort, discomfort, and/or time. Any task which conflicts with this tendency is highly likely to be modified or ignored by the person who is expected to execute it.

One of many important considerations in design is to follow common stereotypical expectations as much as possible. Consider a few examples:

1. Clockwise rotation of a rotary control (knob) is expected to increase the output.

2. Movement of a lever forward, upward, or to the right is expected to increase the output.

3. On a vertically numbered scale, the higher-value numbers are expected to be at the top.

4. On vehicles, depressing the accelerator is expected to increase speed, and depressing the brake is expected to decrease speed. One expects the right foot to be used to apply force to the accelerator, then moved to the brake pedal.

Smith [10.36] tells of a forklift truck which violated this fourth item: The left foot depressed a pedal which increased speed but applied a brake when the foot was lifted.

Sources.   Hunter [10.10] cites SAE Recommended Practice J833, *Human Physical Dimensions,* and other SAE documents. NASA has a three-volume *Anthropometrics Source Book* (Volume 1 has data for the designer, Volume 2 is a handbook of anthropometric data, and Volume 3 is an annotated bibliography) available from the NASA Scientific and Technical Information Office, Yellow Springs, OH 45387. The Department of Defense has a basic handbook, *Human Engineering Procedures Guide,* DOD-HDBK-763. One of the basic military specifications is *Human Engineering Design Criteria,* MIL-H-1472. DoD documents normally refer to additional references; MIL-H-1472, for example, refers to 54 other documents. All DoD and MIL documents can be obtained from the Standardization Documents Order Desk, 700 Robbins Ave., Philadelphia, PA 19111. A limited set of references is given following the references cited in this chapter.

10.6 SUMMARY

The designer or manufacturer has a moral, ethical, and legal obligation to provide safe products. If that is not enough motivation, there is a matter of enlightened self-interest. There are three aspects to this obligation: (1) The product must be made safe. (2) If it is not possible to design out all hazards, guarding must be provided. (3) If complete and proper guarding cannot be provided, appropriate directions and warnings must be provided. It is absolutely unacceptable to use a warning in a situation where safe design or proper guarding is possible. It is not an easy task to write a proper and effective warning, since no warning is effective unless it changes the potential behavior of the endangered individual.

The most difficult variable in product design is the human in the human-machine system. Perhaps the designer needs to keep Murphy's law in mind: If anything can go wrong, it will. If that is not enough, there is O'Toole's law: Murphy was an optimist. Developing a truly safe product is not an easy task, but it can be done.

REFERENCES

10.1 Tacitus, Publius Cornelius, *Annals,* Vol. 15.

10.2 *The Code of Hammurabi,* University of Chicago Press, 1904.

10.3 *Information Bulletin 000080021,* National Safety Council, Itasca, Ill., 1994.

10.4 *Webster's New Twentieth Century Dictionary,* Unabridged, 2d ed., Simon and Schuster, New York, 1979.

10.5 Willie Hammer, *Occupational Safety Management and Engineering,* Prentice-Hall, Englewood Cliffs, N.J., 1976.

10.6 W. W. Lowrance, *Of Acceptable Risk,* William Kaufman, Los Altos, Calif., 1976.

10.7 *American Law Institute, Restatement of the Law, Second, Torts, 2d, Vol. 2,* American Law Institute Publishers, St. Paul, Minn., 1965.

10.8 G. A. Peters, "New Product Safety Legal Requirements," *Hazard Prevention,* September–October 1978, pp. 21–23.

10.9 *Barker* v. *Lull Engineering Co.,* 20C. 3d 413.

10.10 Thomas A. Hunter, *Engineering Design for Safety,* McGraw-Hill, New York, 1992. Provides good guidance and supplies many information sources.

10.11 Willie Hammer, *Handbook of System and Product Safety,* Prentice-Hall, Englewood Cliffs, N.J., 1972.

10.12 Willie Hammer, *Product Safety Management and Engineering,* Prentice-Hall, Englewood Cliffs, N.J., 1980.

10.13 Willie Hammer, *Product Safety Management and Engineering,* 2d ed., ASSE, Des Plaines, Ill., 1993.

10.14 Harold E. Roland and Brian Moriarty, *System Safety Engineering and Management,* 2d ed., Wiley, New York, 1990.

10.15 Joe Stephenson, *Systems Safety 2000,* Van Nostrand Reinhold, New York, 1991.

10.16 C. O. Smith, *Problems in Machine Guarding,* ASME Paper No. 87-WA/DE-6.

10.17 *Accident Prevention Manual for Business and Industry,* 10th ed., National Safety Council, Itasca, Ill., 1992.

Volume 1 includes chapters on government regulations and standards, ergonomics, personal protective equipment, industrial sanitation, and more. There are completely new chapters on environmental management and employee assistance programs.

Volume 2 focuses on one of the most vital safety and health issues: engineering safety into the design, construction, and maintenance of industrial facilities. Topics include equipment safeguarding, materials handling and storage, hoists and cranes, and powered industrial trucks. There is a completely new chapter on automated processes and a new safety and health glossary.

Volume 3 is a study guide for Volumes 1 and 2.

10.18 C. O. Smith, *System Unsafety in a Transfer Machine,* Proceedings, System Safety Society, 4th International Conference, San Francisco, July 9–13, 1979.

10.19 John V. Grimaldi and Rollin H. Simonds, *Safety Management,* 5th ed., Irwin, Homewood, Ill., 1989.

10.20 *Safeguarding Concepts Illustrated,* 6th ed., National Safety Council, Itasca, Ill. This comprehensive handbook discusses conventional and high-tech safeguarding techniques, with over 300 photographs and line illustrations.

10.21 *Power Press Safety Manual,* 4th ed., National Safety Council, Itasca, Ill. Safeguard power press operations with the information contained in this fully illustrated manual. It includes basic press construction, employee training, noise abatement, ergonomics, point-of-operation safeguards, and power press operations.

10.22 M. R. Lehto and J. M. Miller, *Warnings: Volume I, Fundamentals, Design, and Evaluation Methodologies,* Fuller Technical Publications, Ann Arbor, Mich., 1986.

10.23 M. R. Lehto and J. M. Miller, *Warnings: Volume II,* An Annotated Bibliography, Fuller Technical Publications, Ann Arbor, Mich., 1986.

10.24 George R. Klare, *The Measurement of Readability,* Iowa State University Press, Ames, 1963. This contains several indices of readability in addition to the three cited in the text.

10.25 ANSI Z535.4, *American National Standard for Product Safety Signs and Labels,* American National Standards Institute, New York, 1991.

10.26 R. W. Bailey, *Human Performance Engineering: A Guide for System Designers,* Prentice-Hall, Englewood Cliffs, N.J., 1982.

10.27 ANSI Z535.1, *American National Standard Safety Color Code,* American National Standards Institute, New York, 1991 (updates Z53.1-1979).

10.28 ANSI Z535.2, *American National Standard for Environmental and Facility Safety Signs,* American National Standards Institute, New York, 1991 (updates Z35.1-1972).

10.29 ANSI Z535.3, *Criteria for Safety Symbols,* American National Standard Institute, New York, 1991.

10.30 ANSI Z535.5, *Specifications for Accident Prevention Tags,* American National Standards Institute, New York, 1991 (updates Z35.2-1976).

10.31 SAE J115, *Safety Signs,* SAE Recommended Practice, Society of Automotive Engineers, Warrendale, Pa. Approved by Human Factors Technical Committee, January 1987.

10.32 C. O. Smith and T. F. Talbot, *Product Design and Warnings,* ASME Paper No. 91-WA/DE-7.

10.33 Etienne, Grandjean, *Fitting the Task to the Man,* 4th ed., Taylor and Francis, New York, 1988.

10.34 Wesley E. Woodson, *Human Factors Design Handbook,* McGraw-Hill, New York, 1981.

10.35 Gavriel Salvendy, (ed.), *Handbook of Human Factors,* Wiley-Interscience, New York, 1987.

10.36 C. O. Smith, *Two Industrial Products—Defective Design?,* ASME Paper No. 93-WA/DE-11.

RECOMMENDED READING

Human Engineering

P. Tillman and B. Tillman, *Human Factors Essentials,* McGraw-Hill, New York, 1991.

M. S. Sanders and E. J. McCormick, *Human Factors in Engineering Design,* McGraw-Hill, New York, 1987.

Eastman Kodak Co., E. M. Eggleton (Ed.), *Ergonomic Design for People at Work,* 2 vols., Van Nostrand Reinhold, New York, 1983, 1986.

C. D. Wickens, *Engineering Psychology and Human Performance,* 2d ed., Harper-Collins, New York, 1992.

B. H. Kantowicz and R. D. Sorkin, *Human Factors: Understanding People-System Relationships,* John Wiley & Sons, New York, 1983.

J. H. Burgess, *Designing for Humans: The Human Factor in Engineering,* Petrocelli Books, Princeton, N.J., 1986.

System Safety

Safety, Health and Environmental Resources Catalog, National Safety Council, Itasca, Ill., current annual copy.

Publications of the Institute for Product Safety, P.O. Box 1931, Durham, NC 27702.

Fred A. Manuele, *On the Practice of Safety,* Van Nostrand Reinhold, New York, 1993.

William G. Johnson, *MORT Safety Assurance Systems,* Marcel Dekker, New York, 1980.

Roger L. Brauer, *Safety and Health for Engineers,* Van Nostrand Reinhold, New York, 1990.

Willie Hammer, *Occupational Safety Management and Engineering,* Prentice-Hall, Englewood Cliffs, N.J., 1989.

R. A. Wadden and P. A. Scheff, *Engineering Design for the Control of Workplace Hazards,* McGraw-Hill, New York, 1987.

CHAPTER 11
MINIMIZING ENGINEERING EFFORT

Charles R. Mischke, Ph.D., P.E.
Professor Emeritus of Mechanical Engineering
Iowa State University
Ames, Iowa

NOMENCLATURE

a Distance, range number, bilaterial tolerance

b Width, range number

C Constant

D Helix diameter

dim   Dimensional operator

E Young's modulus

E_n Error using n applications of Simpson's rule

e_i The ith exponent

f Function

$f^{(i)}$ The ith derivative of function f

F Fundamental dimension of force, fractional reduction of interval of uncertainty

g Function

h Function, ordinate spacing

i Index

I Second area moment, value of integral

I_i Approximate value of integral using i applications of Simpson's rule

k Spring rate

K_{ij} Exponent of fundamental dimension in row i, of parameter j in dimensional matrix

L Fundamental dimension of length

ℓ Span, left

ln   Natural logarithm

m Mass, subscript of model

n Number

N Number of experiments to establish a robust functional relationship among n parameters, number of function evaluations

N' Number of experiments to establish a robust functional relationship among dimensionless parameters

N_a Number of active turns in a spring

N_π Number of pi terms in a complete set

p Number of points necessary to establish a robust functional relationship between two parameters

P Load

Q Fundamental dimension of charge

r Rank of dimensional matrix, right

s Scale factor, the ratio of model over prototype dimension

T Fundamental dimension of time

x Location parameter

x^* Abscissa of extreme of a function

x_ℓ, x_r Range numbers on left and right, respectively

y Transverse beam deflection

Δ Tolerable error

θ Fundamental dimension of temperature

π_i The ith pi term

ξ Location in Simpson's rule application interval where error term is exact

11.1   INTRODUCTION

The old carpenter's admonition "Measure twice, cut once" reminds us that sound preparatory effort avoids later grief in terms of redoing or scrapping prior work effort. In technical undertakings, engineering effort is required long before work starts. Not only must it be done correctly, but since it is an overhead cost, it is important that it be accomplished in a cost-efficient manner without compromising the quality of the result. In order to accomplish this routinely, engineers have developed and adopted strategies, manners of approach that are routinely mindful of effective use of engineering resources.

One such strategy is the mathematical model. It gives us quantitative insight into domains that are new to us. It is unfortunate that the name *mathematical model* is commonly applied to this tool, for mathematics does not intrinsically contain the reality. It has to be carefully built in if the model is to satisfactorily describe nature.

Attention focus for thinking and communicative processes is rooted in and well served by concepts of system, boundary, and surroundings or control region, control surface, and surroundings. There are also notions of cause, effect, and extent as systems interact with their surroundings. We recognize heat and work effects, tractive effects, charge effects, chemical effects, and ballistic effects related to nuclear phenomena. It is in these effects (and their quantitative expression) that reality is modeled. It is when these effects are combined with notions of accountability, or balances, and first principles that reality can be incorporated into mathematical models ([11.1], Chaps. 6, 7).

Deterministic, deductive mathematical models are usually created using the following steps ([11.1], p. 228):

1. *Isolate* a finite or infinitesimal system or control region.

2. *Identify* the significant influences of the surroundings, or changes within the isolated system or control region.

3. *Qualify* significant influences or changes with mathematical models of effects.

4. *Relate* influences to system or control-region behavior by using first principles.

5. *Limit,* if necessary, as Δx, Δy, Δz, Δt, etc., approach zero.

6. *Solve* the resulting equation(s) for variable(s) of interest. Assumptions or judgments may be required to make a solution possible.

7. *Check* your work (see Sec. 11.6).

Engineers recognize that variability is omnipresent in nature and that measured quantities are knowable only in terms of estimates of means and variances, distributional forms, and confidence limits. This variability or uncertainty must be considered when judging the worth of the model results.

11.2 REDUCING THE NUMBER OF EXPERIMENTS

In describing the functional relationship between variables x_1 and x_2, it takes a number of experiments (points) to establish a satisfactory approximation to the functional relationship. Consider that number of experiments to be p. At this point we are concerned not with the method of establishing the working approximation (least-square curve fits, for example) but with the amount of effort associated with gathering the data points used to establish that relationship. If the level of effort in time and expense is proportional to the number of points p, we use the magnitude of p as our index to cost. The relationship between x_1 and x_2 can be displayed as a data string on a sheet of graph paper ([11.1], pp. 139–160).

How many experiments are necessary to describe a phenomenon involving n parameters $x_1, x_2, \ldots, x_n$? During the experiments necessary to relate x_1 to x_2, all other parameters were held constant. The role of x_3 is then introduced by performing p experiments at level $(x_3)_1, (x_3)_2, \ldots, (x_3)_p$. This places p contours on the $x_1 x_2$ graph. Up to this point there have been p^2 experiments. The introduction of the third parameter increased the level of effort *exponentially*. Similarly, the fourth parameter requires p pages of p curves of p points each. The total number of experiments N necessary for n parameters using p points for each curve is, therefore,

$$N - p^{n-1} \tag{11.1}$$

If $p = 6$ and $n = 5$, then $N = 6^{5-1} = 1296$ experiments. If the cost of experimental determination is \$100 or \$1000 per point, then quantitative understanding is prohibitively expensive. Is there any alternative to this investment of time and effort?

We are indebted to Buckingham, who suggested clustering parameters in dimensionless groups. Instead of finding the relationship among

$$f(x_1, x_2, \ldots, x_n) = 0$$

Buckingham suggested finding the relationship among

$$g(\pi_1, \pi_2, \ldots, \pi_{n-r}) = 0$$

where r is the rank of the matrix of dimensions. The level of effort N' is now given by, after Eq. (11.1),

$$N' = p^{n-r-1} \tag{11.2}$$

The ratio N'/N is, using Eqs. (11.1) and (11.2),

$$\frac{N'}{N} = \frac{p^{n-r-1}}{p^{n-1}} = \frac{1}{p^r} \tag{11.3}$$

If the rank of the matrix of dimensions is 2 and 10 points are necessary, then

$$\frac{N'}{N} = \frac{1}{10^2} = \frac{1}{100}$$

and the level of effort has been reduced by a factor of 100.

Pi terms are multiplicative clusters of parameters, formed by exploiting the rule of dimensional homogeneity. The set of *fundamental dimensions* consists of the irreducible set of force F, length L, time T, temperature θ, and charge Q. Mass can be used instead of force. A velocity V has the dimensions of length/time, or L/T, and such quantities are called *secondary* or *derived* quantities. We can say that the dimensions of V, dim(V), are L/T or $L^1 T^{-1}$, or, more completely,

$$\dim(V) = F^0 L^1 T^{-1} \theta^0 Q^0 \tag{11.4}$$

Care has to be taken to establish a *complete set* of dimensionless clusters, or pi terms. A complete set means that the pi-term set is the exact counterpart of the parameter set. The first step is to construct a *matrix of dimensions* for the parameter set. If the parameters are $x_1, x_2, \ldots, x_n$ and the fundamental dimensions involved are force F and length L, then the matrix of dimensions is displayed as

	x_1	x_2	$\cdots$	x_n
F	K_{11}	K_{12}	$\cdots$	K_{1n}
L	K_{21}	K_{22}	$\cdots$	K_{2n}

For example, for a helical compression spring, the spring rate k is affected by the number of active turns N_a, wire diameter d, torsional modulus G, and helix diameter D. The dimensions are

$$\dim(k) = F^1 L^{-1} \qquad\qquad \dim(G) = F^1 L^{-2}$$

$$\dim(N_a) = F^0 L^0 \qquad\qquad \dim(D) = F^0 L^1$$

$$\dim(d) = F^0 L^1$$

The matrix of dimensions for the spring consists of the display of the exponents of the fundamental dimensions in each of the parameters:

	k	N_a	d	G	D
F	1	0	0	1	0
L	−1	0	1	−2	1

The *rank* of this matrix is the order of the largest nonzero determinant that can be found in the matrix. Since the right-hand determinant

$$\begin{vmatrix} 1 & 0 \\ -2 & 1 \end{vmatrix} = 1 - 0 = 1 \neq 0$$

is nonzero, the rank r is 2. There may be several of these depending on the sequencing of parameters across the top. It is important for completeness that a nonzero determinant be placed on the right in the matrix of dimensions. The number of multiplicative dimensionless clusters or pi terms N_π is given by

$$N_\pi = n - r \tag{11.5}$$

A pi term is formed by writing

$$\pi_i = k^{e_1} N_a^{e_2} d^{e_3} G^{e_4} D^{e_5} \tag{11.6}$$

The dimensional operator is applied as follows:

$$\dim(\pi_i) = \dim(k^{e_1})\dim(N_a^{e_2})\dim(d^{e_3})\dim(G^{e_4})\dim(D^{e_5})$$

$$= (F^1 L^{-1})^{e_1}(F^0 L^0)^{e_2}(F^0 L^1)^{e_3}(F^1 L^{-2})^{e_4}(F^0 L^1)^{e_5}$$

For the force dimension,

$$F^0 = F^{e_1} F^0 F^0 F^{e_4} F^0$$

The exponent of F must be the same on both sides:

$$0 = (1)e_1 + (0)e_2 + (0)e_3 + (1)e_4 + (0)e_5$$

Note that the coefficients of the exponential equation agree with the first row of the dimensional matrix. In other words, the exponential equation associated with any fundamental dimension can be written *by inspection* from the matrix of dimensions. The two exponential equations are

$$e_1 + e_4 = 0 \quad \text{(for force dimension)} \tag{11.7}$$

$$-e_1 + e_3 - 2e_4 + e_5 = 0 \quad \text{(for length dimension)} \tag{11.8}$$

There are two exponential equations (r is 2) and five exponents (n is 5), and so three exponents are mathematically arbitrary. We will choose them so that the first three parameters k, N_a, and d each appear in only one pi term. Such parameters are used to control their pi terms independently, if necessary.

It is useful to display a *matrix of solutions*. There are $n - r = 5 - 2 = 3$ pi terms.

	(k)	(N_a)	(d)	(G)	(D)
	e_1	e_2	e_3	e_4	e_5
π_1	1	0	0		
π_2	0	1	0		
π_3	0	0	1		

Solving Eqs. (11.7) and (11.8) to complete the matrix of solutions is done as follows:

$$e_4 = -e_1$$

$$e_5 = 2e_4 + e_1 - e_3$$

For $e_1 = 1, e_2 = 0, e_3 = 0$,   For $e_1 = 0, e_2 = 1, e_3 = 0$,   For $e_1 = 0, e_2 = 0, e_3 = 1$,

$e_4 = -1$ $e_4 = 0$ $e_4 = 0$

$e_5 = -2 + 1 = -1$ $e_5 = 0$ $e_5 = -1$

The completed matrix of solutions is

	(k)	(N_a)	(d)	(G)	(D)	
	e_1	e_2	e_3	e_4	e_5	
π_1	1	0	0	-1	-1	$\Rightarrow \pi_1 = k^1 G^{-1} D^{-1}$
π_2	0	1	0	0	0	$\Rightarrow \pi_2 = N_a^1$
π_3	0	0	1	0	-1	$\Rightarrow \pi_3 = d^1 D^{-1}$

and the pi terms can be displayed as

$$\pi_1 = \frac{k}{GD} \qquad \pi_2 = N_a \qquad \pi_3 = \frac{d}{D}$$

Recall that if $p = 10$, then the number of experiments from Eq. (11.1) is $N = p^{n-1} = 10^{5-1} = 10{,}000$. By using Buckingham's multiplicative dimensionless clusters, Eq. (11.2) gives $N' = 10^{5-2-1} = 100$.

Can we reduce the hundred experiments even more? If we can introduce information we already know, we can. Two identical springs in series (end to end) have twice the turns and half the spring rate; in other words, $\pi_1 \pi_2 = C_1(\pi_3)$. The problem reduces to finding

$$\pi_1 \pi_2 = h(\pi_3)$$

Now there are only 10 experiments to be performed. As an aid to partitioning our thinking so that we can deal with one thing at a time, we can use the method of derivatives. Since there are three pi terms in the spring problem, we seek the function

$$\pi_1 = h_1(\pi_2, \pi_3)$$

It follows then that

$$d\pi_1 = \frac{\partial \pi_1}{\partial \pi_2} d\pi_2 + \frac{\partial \pi_1}{\partial \pi_3} d\pi_3 \qquad (11.9)$$

In noting the inverse proportionality between π_1 and π_2 from before, we write

$$\pi_1 = \frac{C_1}{\pi_2} \qquad \frac{\partial \pi_1}{\partial \pi_2} = -\frac{C_1}{\pi_2^2} = -\frac{\pi_1 \pi_2}{\pi_2^2} = -\frac{\pi_1}{\pi_2} \qquad \text{(from prior experience)}$$

When we conduct the p experiments and find $\pi_1/\pi_3^4 = C_2(\pi_2)$ at constant π_2, we have

$$\pi_1 = C_2\, \pi_3^4 \qquad \frac{\partial \pi_1}{\partial \pi_3} = 4C_2\pi_3^3 = 4\,\frac{\pi_1}{\pi_3^4}\,\pi_3^3 = 4\,\frac{\pi_1}{\pi_3} \qquad \text{(from test)}$$

Thus Eq. (11.9) becomes

$$d\pi_1 = -\frac{\pi_1\, d\pi_2}{\pi_2} + 4\,\frac{\pi_1}{\pi_3}\, d\pi_3$$

Dividing through by π_1 renders the equation exact and integrable term by term:

$$\frac{d\pi_1}{\pi_1} = -\frac{d\pi_2}{\pi_2} + 4\,\frac{d\pi_3}{\pi_3}$$

$$\ln \pi_1 = -\ln \pi_2 + \ln \pi_3^4 + \ln C$$

or

$$\pi_1 = C\,\frac{\pi_3^4}{\pi_2} \qquad (11.10)$$

The constant C can be found from the p experiments. Equation (11.10) can be written as

$$k = \frac{d^4 G}{8D^3 N_a}$$

Do not underestimate the power of Buckingham's suggestion *and* the incorporation of a priori knowledge with test results to enormously reduce the effort.

11.3 SIMILITUDE

The first similitude equation of which we have a record dates to the fourth century B.C., when it was recorded by Philon of Byzantium for the ballista [11.2]. It related what we now call the mass of the projectile to be thrown to the diameter of the torsional springs used as

$$\frac{d_1}{d_2} = \left(\frac{m_1}{m_2}\right)^{1/3}$$

(11.11)†

Ever since, engineers have embroidered on this idea with useful results. In the context of Sec. 11.2, this is a relationship between two pi terms. The idea that will be useful to us can be related to the helical spring example of Sec. 11.2. With a spring in hand, one can quantitatively express $\pi_1 = k/GD$. However, knowing that π_1 is 0.5×10^{-5} will not identify the spring parameters. What constructing the pi term has done is map all springs with $\pi_1 = 0.5 \times 10^{-5}$ onto a single coordinate. This suggests that one can *model* one spring with $\pi_1 = 0.5 \times 10^{-5}$ with another that also has $\pi_1 = 0.5 \times 10^{-5}$, but is of differing material, spring rate, and helix diameter. This can be useful in adjusting to size and capacity constraints on test instrumentation.

For a timber beam of cross section b wide and d deep, with a concentrated load P located a distance a from the left support, and a span of ℓ, the transverse deflection y at a distance x from the left support is described by

$$f(y, a, b, d, x, P, E, \ell) = 0$$

or equally as well by Buckingham's pi terms as

$$g\left(\frac{y}{\ell}, \frac{a}{\ell}, \frac{b}{\ell}, \frac{d}{\ell}, \frac{x}{\ell}, \frac{P}{E\ell^2}\right) = 0$$

Suppose we wish to model the timber beam in a different size and material. The function g in model terms is written

$$g\left(\frac{y_m}{\ell_m}, \frac{a_m}{\ell_m}, \frac{b_m}{\ell_m}, \frac{d_m}{\ell_m}, \frac{x_m}{\ell_m}, \frac{P_m}{E_m\ell_m^2}\right) = 0$$

In order for this to be a model, corresponding pi terms must be identical. Since $y_m/\ell_m = y/\ell$, it follows that

$$y_m = \frac{\ell_m}{\ell} y = sy$$

where s is the *scale factor*, $s = \ell_m/\ell$. The other linear dimensions are

$$a_m = sa \qquad b_m = sb \qquad d_m = sd \qquad x_m = sx$$

The sixth pi terms are equated, from which

$$P_m = \frac{PE_m\ell_m^2}{E\ell^2} = s^2 \frac{E_m}{E} P$$

The load P_m is the mandatory load on the model corresponding to P. The location at which to measure the transverse deflection is $x_m = sx$. If a steel model is 1/10 size and the prototype load is 4800 lbf, the model load P_m is

---

† In a book addressing machine design, shouldn't this be Eq. (1.1)?

$$P_m = 0.1^2 \frac{30 \times 10^6}{1.5 \times 10^6} 4800 = 960 \text{ lbf}$$

and the prototype deflection is $y = y_m/s$.

11.4 OPTIMALITY

The subject of optimality is extensive [11.3], [11.4]. Our purpose here is to examine the efficiency of an optimization process itself, for any internal wasted effort in a computer-coded algorithm is incessantly repeated. A unimodal function is one that monotonically increases, monotonically decreases, or monotonically increases then decreases. If the original interval containing a maximum has the range numbers x_ℓ, x_r and there are n ordinates equally spaced within the interval (but no ordinates at x_ℓ or x_r), then the ordinate spacing is

$$h = \frac{x_r - x_\ell}{n + 1}$$

By examining the ordinates, the final interval of uncertainty is reduced to $2h$, and the fractional reduction in the interval of uncertainty is

$$F = \frac{2h}{x_r - x_\ell} = \frac{2(x_r - x_\ell)/(n + 1)}{x_r - x_\ell} = \frac{2}{n + 1}$$

Solving for n gives, for fractional reduction F and bilateral tolerance, $x^* \pm a$ locations of the extreme, respectively:

$$n = \left[\frac{2}{F} - 1 \right]_+ = \left[\frac{x_r - x_\ell}{a - 1} \right]_+ \tag{11.12}$$

When n is not an integer, it is rounded up. For $F = 0.001$,

$$n = \left[\frac{2}{0.001} - 1 \right]_+ = [2000 - 1]_+ = 1999$$

Thus, 1999 function evaluations are required. See Ref. [11.1], pp. 278–290.

Instead of expending all ordinates simultaneously, one can spend a few, reduce the interval somewhat, and keep repeating the process. For equally spaced ordinates, the optimal procedure ([11.3], p. 282) is spending n as $3 + 2 + 2 + \cdots$. This is called *interval halving*. The total number of function evaluations N spent this way is

$$N = \left[1 + \frac{2 \ln 1/F}{\ln 2} \right]_{\text{odd} +} = \left[2.88 \ln \frac{x_r - x_\ell}{a} - 1 \right]_{\text{odd} +} \tag{11.13}$$

For $F = 0.001$,

$$N = \left[1 + \frac{2 \ln 1/1000}{\ln 2} \right]_{\text{odd} +} = [20.93]_{\text{odd} +} = 21$$

This is a remarkable reduction in effort. One can do better by relaxing the equal spacing stipulation and spending ([11.1], pp. 284–289) ordinates $2 + 1 + 1 + \cdots$. Under these circumstances, for fractional and bilateral tolerance reductions, respectively,

$$N = \left[1 + \frac{\ln F}{\ln 0.618\,033\,989} \right]_+ = \left[2.08 \ln \frac{x_r - x_\ell}{a} \right]_+ \qquad (11.14)$$

and the method is called *golden section*. For $F = 0.001$,

$$N = \left[1 + \frac{\ln 0.001}{\ln 0.618\,033\,989} \right]_+ = [15.35]_+ = 16$$

which is approximately three-fourths as many function evaluations as were required for interval halving. Can one do better? The answer is a qualified yes. A Fibonacci search will reduce effort by about one function evaluation at the $F = 0.001$ level, but it is not amenable to predicting the number of function evaluations in advance.

While interval halving may be easier to apply manually, golden section should be coded for the computer. Golden section is used for real root finding of $f(x)$ by maximizing $-|f(x)|$. The root is at the zero-ordinate cusp. Figure 11.1 is the documentation sheet for a golden section subroutine named GOLD. This subroutine has served thousands of users over several decades at Iowa State University and elsewhere. The Fortran coding follows.

```
      SUBROUTINE GOLD(K,XA,XB,F,MERIT1,YBIG,XBIG,XL1,XR1,N)
C     IOWA CADET, IOWA STATE UNIVERSITY, C. MISCHKE
      XL=XA
      XR=XB
      Q=10.E-07
      IF(F.LT.-Q) GO TO 41
      IF(F.GT.Q) GO TO 42
      IF(F.GT.-Q.AND.F.LT.Q) GO TO 43
   41 ICODE=-1
      GO TO 100
   42 ICODE=1
      GO TO 100
   43 ICODE=0
      F=ICODE
      GO TO 100
  111 IF(K)32,31,32
   32 WRITE(6,33)
   33 FORMAT(' CONVERGENCE MONITOR IOWA CADET SUBROUTINE GOLD',/,
     1' VERSION 11/76 C. MISCHKE',/,/,
     2'     N          Y1             Y2              X1              X2'
     3,/,/)
   31 N=0
      XLEFT=XL
      XRIGHT=XR
   13 SPAN=XR-XL
      DELTA=ABS(SPAN)
   14 X1=XL+0.381966*DELTA
      X2=XL+0.618034*DELTA
      CALL MERIT1(X1,Y1)
      CALL MERIT1(X2,Y2)
      N=N+2
    3 IF(K)34,9,34
   34 WRITE(6,35)N,Y1,Y2,X1,X2
   35 FORMAT(I5,4(1X,G15.7))
    9 IF(ICODE)50,50,51
   50 IF(0.381966*DELTA-ABS(F))4,4,8
   51 IF(0.618034*(XR-XL)-F*SPAN)4,4,8
    8 DELTA=0.618034*DELTA
      IF(Y1-Y2)1,10,2
    1 XL=X1
      X1=X2
      Y1=Y2
      X2=XL+0.618034*DELTA
      CALL MERIT1(X2,Y2)
      N=N+1
      GO TO 3
```

```
  2 XR=X2
    Y2=Y1
    X2=X1
    X1=XL+0.381966*DELTA
    CALL MERIT1(X1,Y1)
    N=N+1
    GO TO 3
  4 IF(Y2-Y1)5,5,6
  5 YBIG=Y1
    XBIG=X1
    XL1=XL
    XR1=X2
    GO TO 39
  6 YBIG=Y2
    XBIG=X2
    XL1=X1
    XR1=XR
    GO TO 39
 10 XL=X1
    XR=X2
    DELTA=XR-XL
    GO TO 14
 39 IF(K)40,40,37
 37 IF(ICODE)60,60,61
 60 A=-F
    WRITE(6,138)A
138 FORMAT(/,/,
   1' ACCEPTABLE BILATERAL TOLERANCE ON XSTAR ..............',G15.7)
    GO TO 140
 61 WRITE(6,139)F
139  FORMAT(/,/,
   1' FRACTIONAL REDUCTION IN INTERVAL OF UNCERTAINTY ......',G15.7)
140 WRITE(6,38)XLEFT,XRIGHT,YBIG,XBIG,XL1,XR1,N
 38 FORMAT(/,
   1' LEFTHAND ABSCISSA OF INTERVAL OF UNCERTAINTY .........',G15.7,/,
   2' RIGHTHAND ABSCISSA OF INTERVAL OF UNCERTAINTY ........',G15.7,/,
   3' EXTREME ORDINATE DISCOVERED DURING SEARCH ............',G15.7,/,
   4' ABSCISSA OF EXTREME ORDINATE .........................',G15.7,/,
   5' NEW LEFTHAND ABSCISSA OF INTERVAL OF UNCERTAINTY .....',G15.7,/,
   6' NEW RIGHTHAND ABSCISSA OF INTERVAL OF UNCERTAINTY ....',G15.7,/,
   7' NUMBER OF FUNCTION EVALUATIONS EXPENDED DURING SEARCH ',I11,/,/)
 40 XL=XLEFT
    XR=XRIGHT
112 RETURN
100 IERROR=0
    IF(K)102,101,101
101  IF(K-1)104,104,102
102 WRITE(6,103)K
103 FORMAT(' *****ERROR MESSAGE SUBROUTINE GOLD*****',/,
   1'       I1,',',I15,' IS NOT 0 OR 1')
    IERROR=IERROR+1
104 IF(XR-XL)105,105,1070
105 WRITE(6,106)XL,XR
106 FORMAT(' *****ERROR MESSAGE SUBROUTINE GOLD*****',/,
   1'       A2,',',G15.7,' CANNOT BE .GE. A3,',G15.7)
    IERROR=IERROR+1
1070 IF(ICODE.NE.0) GO TO 107
120 WRITE(6,121)F
121 FORMAT(' *****ERROR MESSAGE SUBROUTINE GOLD*****',/,
   1'        VALUE OF A4,',G15.7,' CANNOT BE ZERO')
    IERROR=IERROR+1
107 IF(ICODE)113,115,1107
1107 IF(F.GT.0..AND.F.LT.1.) GO TO 115
    WRITE(6,110)F
110 FORMAT(' *****ERROR MESSAGE SUBROUTINE GOLD*****',/,
   1'        A4,'G15.7,' DOES NOT LIE BETWEEN 0. AND 1.')
    IERROR=IERROR+1
    GO TO 115
113 IF(ABS(F).LT.ABS(XR-XL)/2.) GO TO 115
    WRITE(6,114)F
114 FORMAT(' *****ERROR MESSAGE SUBROUTINE GOLD*****',/,
   1'        ABSOLUTE VALUE OF A4,',G15.7, ' .GE.(A2 MINUS A3)/2')
    IERROR=IERROR+1
115 IF(IERROR)111,111,112
    END
```

One-Dimensional Golden Section Search
GOLD(I1, A2, A3, A4, A5, B1, B2, B3, B4, J5)   Mischke

This subroutine will search over a one-dimensional unimodal function and report the largest ordinate found, its abscissa, final abscissas bound in the interval of uncertainty, and the number of function evaluations expended during the search.

The subroutine requires the specification of the present interval of uncertainty, the fractional reduction required in the interval of uncertainty (or bilateral tolerance on abscissa of the extreme), and whether or not a convergence monitor printout is desired. The necessary number of function evaluations may be predicted from

$$N = \left[1 + 2.08 \ln \frac{1}{F} \right]_+$$

when the fractional reduction in interval of uncertainty F is given, or

$$N = \left[2.08 \ln \frac{x_r - x_\ell}{a} \right]_+$$

when the bilateral tolerance a on abscissa of extreme is given.

See *Introduction to Computer-Aided Design*, C. Mischke, Prentice-Hall, 1968, p. 64, or *Mathematical Model Building*, 2nd rev. ed., C. Mischke, Iowa State University Press, 1980, pp. 282–290.

CALLING PROGRAM REQUIREMENTS

Provide a subroutine A5(X,Y) which returns the ordinate Y when the abscissa X is tendered.

Provide the equivalent of the following statement: EXTERNAL A5

CALL LIST ARGUMENTS:

I1 = 0,   convergence monitor will not print
   = 1   convergence monitor will print
A2 = x_ℓ original left-hand abscissa of interval of uncertainty
A3 = x_r original right hand abscissa of interval of uncertainty
A4 =   fractional reduction in interval of uncertainty desired, F, entered positive or
            bilateral tolerance on abscissa at extreme, a, entered negative
A5 =   name of the one-dimensional unimodal function SUBROUTINE

B1 = y^*,   extreme ordinate discovered during search (maximum)
B2 = x^*,   abscissa of extreme ordinate
B3 = x_1,   final left-hand abscissa of interval of uncertainty
B4 = x_2,   final right-hand abscissa of interval of uncertainty
J5 = N,   number of function evaluations expended during search

PREEMPTED NAMES:

   None

SIZE:

   4264 bytes WATFIV compiler.

FIGURE 11.1   The documentation page of subroutine GOLD.

11.5 QUADRATURE

Another numerical chore is integration. Fortunately, Simpson's first rule is simple, robust, and surprisingly accurate. It is applied two panels at a time with equally spaced ordinates. A parabola is passed through the three ordinate points ([11.5], p. 79). If h is the ordinate spacing, then for the three abscissas $x_0, x_1,$ and x_2,

$$\int_{x_0}^{x_2} f(x)\, dx = \frac{h}{3}[f(x_0) + 4f(x_1) + f(x_2)] - \frac{h^5}{90} f^{(4)}(\xi) \tag{11.15}$$

where ξ is in the interval (x_0, x_2). The right-hand term is Richardson's error term, which is exact for some ξ which is generally unknown a priori. When this two-panel, three-ordinate operation is repeated a number of times in the interval a, b, then

$$\int_a^b f(x)\, dx = \frac{h}{3}\left[f(x_0) + 4f(x_1) + 2f(x_2) + \cdots \right.$$

$$\left. + 4f(x_{2n-1}) + f(x_{2n})\right] - \frac{h^5}{90}\sum_{i=1}^{n} f^{(4)}(\xi_i) \tag{11.16}$$

where n is the number of applications of the two-panel ritual, $2n$ is the number of panels, and $2n + 1$ is the number of function evaluations. There is great merit in making the number of panels an even number divisible by 4. The ordinate spacing h is given by $h = (b - a)/(2n)$, and so the error term becomes, removing the summation sign,

$$E_n \doteq \frac{(b-a)^5}{2880n^4} f^{(4)}(\xi) \tag{11.17}$$

By evaluating the integral using n_2 applications, then again with n_1 applications, from Eq. (11.17),

$$\frac{E_{n_2}}{E_{n_1}} = \left(\frac{n_2}{n_1}\right)^4 \tag{11.18}$$

from which

$$n_2 = n_1 \left|\frac{E_{n_1}}{E_{n_2}}\right|^{1/4} = n_1 \left|\frac{E_{n_1}}{\Delta}\right|^{1/4} \tag{11.19}$$

where Δ is the tolerable error. If n_1 is the number of applications of the rule in the interval a, b and n_2 is the number of applications in another evaluation in the same interval a, b, then the value of the integral I is

$$I \doteq I_{n_1} + E_{n_1} \doteq I_{n_2} + E_{n_2}$$

If n_2 is one-half of n_1, then combining Eq. (11.18) with the above equation results in

$$I \doteq I_{n_1} + \frac{I_{n_1} - I_{n_2}}{15} \tag{11.20}$$

Example 1. Evaluate the integral $\int_1^2 dx/x$.

(a) Using two applications of Simpson's rule, estimate the error in $I_{n=2}$. (b) For an error of the magnitude 0.000 01, estimate the number of applications necessary, and (c) integrate and examine the error.

Solution. (a) Using two applications of Simpson's rule,

x	$1/x$	Mult.	
1.00	1.000 000 000	1	1.000 000 000
1.25	0.800 000 000	4	3.200 000 000
1.50	0.666 666 667	2	1.333 333 333
1.75	0.571 428 571	4	2.285 714 286
2.00	0.500 000 000	1	0.500 000 000
		$\Sigma =$	8.319 047 619

From the first part of Eq. (11.16),

$$I_{n=2} = \frac{0.25}{3}(8.319\ 047\ 619) = 0.693\ 253\ 968$$

For one application of Simpson's rule,

x	$1/x$	Mult.	
1.00	1.000 000 000	1	1.000 000 000
1.50	0.666 666 667	4	2.666 666 667
2.00	0.500 000 000	1	0.500 000 000
		$\Sigma =$	4.166 666 667

$$I_{n=1} = \frac{0.5}{3}(4.166\ 666\ 667) = 0.694\ 444\ 445$$

From the second part of Eq. (11.20),

$$E_{n=2} = \frac{I_{n=2} - I_{n=1}}{15} = \frac{0.693\ 253\ 968 - 0.694\ 444\ 445}{15}$$

$$= -0.000\ 079\ 565$$

(b) From Eq. (11.19),

$$n_2 = n_1 \left| \frac{E_{n_1}}{E_{n_2}} \right|^{1/4} = 2 \left| \frac{-0.000\ 079\ 365}{0.000\ 01} \right|^{1/4} = 3.36 => 4$$

(c) Using four applications,

x	$1/x$	Mult.	
1.000	1.000 000 000	1	1.000 000 000
1.125	0.888 888 889	4	3.555 555 556
1.250	0.800 000 000	2	1.600 000 000
1.375	0.727 272 727	4	2.909 090 903
1.500	0.666 666 667	2	1.333 333 333
1.625	0.615 384 615	4	2.461 538 462
1.750	0.571 428 571	2	1.142 857 143
1.875	0.533 333 333	4	2.133 333 333
2.000	0.500 000 000	1	0.500 000 000
		$\Sigma =$	16.635 708 73

$$I_{n\,=\,4} = \frac{0.125}{3}\,(16.635\,708\,73) = 0.693\,154\,530$$

The true value of the integral is $\ln 2 = 0.693\,147\,181$. The value of $I_{n\,=\,4}$ differs by 1 in the fifth decimal place. Furthermore, we can improve $I_{n\,=\,4}$ using Eq. (11.20):

$$I \doteq I_{n\,=\,4} + E_{n\,=\,4} \doteq I_{n\,=\,4} + \frac{I_{n\,=\,4} - I_{n\,=\,2}}{15}$$

$$= 0.693\,154\,530 - 0.000\,006\,629 = 0.693\,147\,901$$

and we have six correct digits, a bonus since we estimated $E_{n\,=\,4}$ along the way.

The use of Simpson's rule to evaluate an integral is both controllably accurate and relatively simple.

11.6   CHECKING

It is useful to check intermediate results on an as-you-go basis as well as upon completion. If there are one hundred subtasks involved in completing an engineering task, ponder this: If your average reliability in performing each subtask correctly is 0.99, then the probability of performing them in sequence correctly is 0.99^{100}, or 0.37. How does one improve such a performance? One way is by checking. If the hundred steps are concatenated to reach the result, where in the chain of events is a mistake most wasteful in effort because work has to be repeated? Early! When do most people think of checking? At the end. Checking steps as they are done, checking groups of steps, and checking the final result is an appropriate and wise course of action.

11.6.1   Limiting-Case Check

This method of checking a derived equation allows the parameters, in turn, to range from the point of vanishing to increasing without upper bound. Do the results still make sense? Do the results contract to a previously known correct result? Making sense is a necessary but not sufficient condition.

11.6.2   Dimensional Check

Equations should be dimensionally homogeneous. Apply the dimensional operator dim( ) to every term in an equation, substituting the fundamental dimensions term by term. Remember that the result of applying the operator to a dimensionless term is unity. Dimensional homogeneity is a necessary but not sufficient condition.

11.6.3   Experience

Our lifetime experience with similar things will suggest "expected relationships": symmetries of certain forms, indirect and direct proportionalities, and nonlinearities of a particular order, such as proportionality to the cube of some parameter. All these little tidbits of reality from prior contexts can be examined for applicability to the case at hand. Congruence with experience is a necessary but not sufficient condition.

11.6.4 Robustness of Assumptions

Deductions from first principles and cause-effect-extent mathematical models depend on assumptions such as "friction is negligible" or "radiation is secondary." What we are really saying is that the result will be useful to our purpose—that is, *robust*—even if the influences of friction or radiation are ignored. The mathematical meaning of the word *assumption* does not fully apply here, nor does it serve us well. In reality, we have made a *decision* based on an experiential value judgment that acting on the result is prudent and resources are risked at a very small, acceptable level. Engineers should treat all such "assumptions" as the decisions they really are. It is useful to ask

- Was it necessary to make this decision (assumption)?
- Has embracing it hidden an important influence of the surroundings on matter in the system or control region?
- Did I qualify my result with an explicit statement of this decision (assumption)?
- Is this decision (assumption) defendable at all values of the parameters that will be encountered?
- Did this decision (assumption) make the model sufficiently incongruent with nature to lose robustness?

Thoughtful responses to questions such as these can help uncover sketchy work. Engineers are responsible for all the decisions they make, whether by commission or by omission. It is prudent to list (call out) all such decisions (assumptions) in the design notebook, and the responses to queries such as those above in the check steps, so that the original engineer at a later date, or another engineer at any time, can understand, appreciate, and possibly challenge them.

11.6.5 Experiment

Results can be verified by experiment. In order to check a spring rate formulation, such as

$$k = \frac{d^4 G}{8 D^3 N_a}$$

we express it in dimensionless form, Eq. (11.10):

$$\frac{k}{GD} = \frac{1}{8 N_a} \left(\frac{d}{D}\right)^4 = \pi_1 = \frac{\pi_3^{\,4}}{8\pi_2}$$

and check the two nuggets of reality, $\pi_1 \pi_2 = C_1$ and $\pi_1/\pi_3^4 = C_2$, which were the bases for the evaluation of the partial derivatives of Eq. (11.9). The first lends itself to a linear plot of the form $\pi_1 = C_1/\pi_2$. Ideally this should lead to a straight data string on a plot of π_1 versus $1/\pi_2$ which lines up with the origin. If one has several springs which differ in turns count only, placing known weights on the spring and measuring the deflection with a dial indicator can supply some data points.

If a least-squares fit of the form $\pi_1 = a + b/\pi_2$ misses the origin, has the origin really been missed? There are statistical methods for saying that with the data you have, the origin has been hit (statistically) or missed (statistically) and quantifying the level of risk in believing either. For example, the number of dead turns N_d comes from the equation $N_t = N_d + N_a$. The number of dead turns N_d may not be precisely 2, depending on how the squared and ground end turns are actually formed. The determination of N_a appears to be a counting procedure; that is, $N_a = N_t - N_d = N_t - 2$ implies great precision, except that the number 2 can be suspect.

The second experimental check on $\pi_1 = C_2 \pi_3^4$ can be done on a log-log plot.

The final form constant C in Eq. (11.10) can be found from the experimental data, and again, statistical methods will develop the chances that C is 1/8.

Notice that the economy of effort of Sec. 11.2 is used to make an experimental check one of least cost. An equation that is an old familiar friend, when applied outside its domain of validity, can't play the game well, or at all. The experimental check can detect this.

11.6.6 Alternative Method of Derivation

If one is truly at "the cutting edge," one rejoices at achieving a result, and urging a second approach, say an energy method, may not be helpful. Nevertheless, we are rarely at the edge, and an alternative approach is a possible and useful method of check. For example, an analog equation can be found.

11.6.7 Have a Colleague Check Your Work

Often, a colleague was educated at a different school by a different faculty using differing emphases and methodologies. Some of these may not be familiar to you, or of first choice. Having a colleague check brings not only a fresh viewpoint but a different ensemble of experience to the problem. While a challenge to some of your decisions may result, it should be welcomed in the pursuit of soundness and completeness of analysis and documentation.

11.6.8 The Insufficiency of Checking Methods

Methods of checking are directed toward verification of matters of mathematical necessity but not sufficiency. Additionally, the limiting-case check, dimensional check, experience check, and assumptions check will not uncover an error such as in the 8 of the spring example. The experimental check *can,* the alternative method check *may,* and the colleague check *might* detect it.

Methods of checking are ways of detecting troubles. In themselves they do not rectify troubles. Being unable to assure infallibility, engineers check, check, and check again.

11.6.9 Checking the Problem-Solving Strategy

Failure to achieve a solution or ineffective progress in the pursuit of a solution can be traceable to problem-solving methodology. Previously identified checks were focused on technical matters, usually mathematical modeling. Problem-solving strategies are more global, more qualitative, and less tangible. The following ideas and questions can be useful in encouraging a healthy skepticism.

There are three clearly identifiable steps in problem solving: (1) defining the problem, (2) planning its treatment, and (3) executing the plan. There are two more steps which are not sequential, but are woven into defining, planning, and executing as necessary. They are also the final two steps following the completion of the execution step. These are (4) checking and (5) learning and generalizing. In more detail, the steps consist of asking the following questions:

Defining the Problem

- What is the real problem or issue?
- What questions are to be answered?
- What are the pertinent facts?
- If several problems are present, which should be addressed first?

Planning Its Treatment

- How can I solve the problem?
- What fundamental principles apply?
- What general truths will help toward a solution?
- What is my plan to move from what is known to what I want?
- Is my plan sufficiently complete for execution? Has any other work been done on this problem?

Executing the Plan

- What is the result of my plan?
- How do I get a useful result from the principle applied?
- Where am I with respect to my plan?

Checking

- Is my work correct in every detail?
- Are assumptions (decisions) reasonable?
- Have I considered *all* important factors?
- Do the results make good sense?
- Have I applied all methods of checking?

Learning and Generalizing

- What have I found out?
- What does the result tell me about the answer to the original problem?
- What does the result mean, and what is its interpretation in common terms?

- How may my results have been affected by my assumptions (decisions)?
- Is the result good enough to act upon, or must the solution be refined?

In moments of doubt as to what to do next, ask

- What do I really want to know?
- What am I doing now?
- Why?
- Will it help?

All this is a demonstration of the sagacity of the adage, "There are no right answers, only right questions."

11.6.10   Checking Cause-Effect-Extent Models

Engineers tend to be self-sufficient in mastering cause-effect-extent models of system/surroundings interactions. This in turn leads them to rarely check to see if some relevant caveat has been ignored. For example, changes in system *internal* energy are those that occur in the absence of gravitation, motion, charge, magnetism, and capillarity. If internal energy is a consideration, one should check to be sure that these things are absent, inconsequential in magnitude, or accounted for in some other way. Since this kind of information is scattered in many books on various subjects, it can be helpful to consult Ref. [11.6], which treats more than a hundred effects by providing descriptions, illustrations, magnitude relations, and references.

11.6.11   Checking Personal Competence

There are times when every engineer is "in over his or her head" and outside his or her personal knowledge and experience base. This happens occasionally because no one can predict where a solution will lead. Engineers do not like to talk about this. The best remedy is knowing when to seek help, or the resources to acquire that help.

11.6.12   The Final Adequacy Assessment as a Final Check

An adequacy assessment (Sec. 5.2) consists of those cerebral and empirical steps that convince the designer that the specification set represents a robust design. The recommended step before "turning work in" is a final adequacy assessment. It should begin not with what is in your head, but with information taken directly from the report and drawings that will leave your desk. Engineers think in terms of significant attributes (a midrange length, a smallest diameter, etc.). These should not be remembered, but reconstructed from your specifications. If you have been thinking in terms of, say, active spring turns, and you have entered that in the spring maker's form blank for total turns, you will set in motion mass production of springs that are not what you had in mind. By starting the final adequacy assessment check with what leaves your desk, you can catch these kinds of errors.

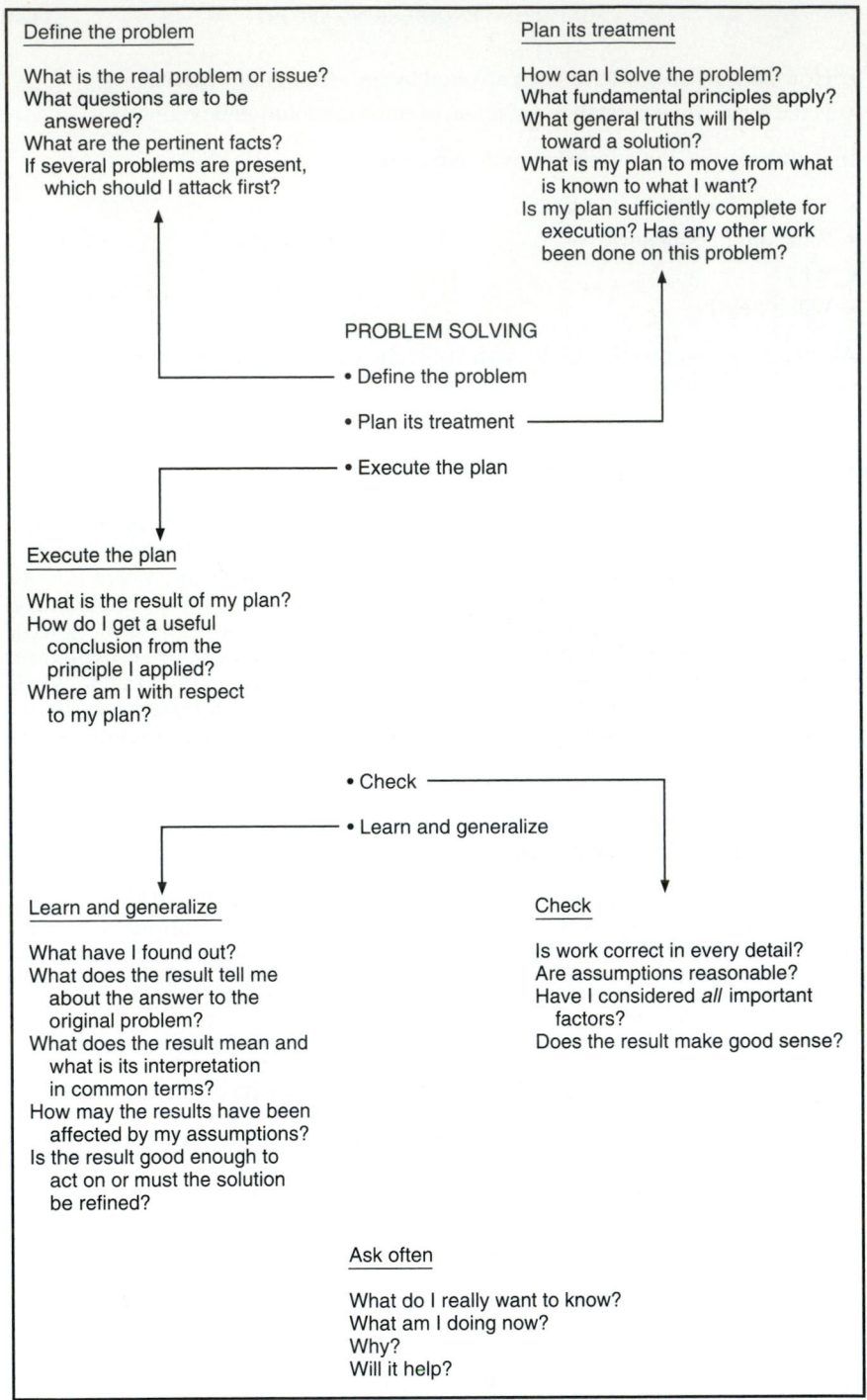

Define the problem

What is the real problem or issue?
What questions are to be
    answered?
What are the pertinent facts?
If several problems are present,
    which should I attack first?

Plan its treatment

How can I solve the problem?
What fundamental principles apply?
What general truths will help
    toward a solution?
What is my plan to move from what
    is known to what I want?
Is my plan sufficiently complete for
    execution? Has any other work
    been done on this problem?

PROBLEM SOLVING

• Define the problem

• Plan its treatment

• Execute the plan

Execute the plan

What is the result of my plan?
How do I get a useful
    conclusion from the
    principle I applied?
Where am I with respect
    to my plan?

• Check

• Learn and generalize

Learn and generalize

What have I found out?
What does the result tell me
    about the answer to the
    original problem?
What does the result mean and
    what is its interpretation
    in common terms?
How may the results have been
    affected by my assumptions?
Is the result good enough to
    act on or must the solution
    be refined?

Check

Is work correct in every detail?
Are assumptions reasonable?
Have I considered *all* important
    factors?
Does the result make good sense?

Ask often

What do I really want to know?
What am I doing now?
Why?
Will it help?

FIGURE 11.2 Some pertinent questions to ask oneself while solving problems. The check, learn and generalize steps are to be woven into the first three, and are the last ones.

REFERENCES

11.1 C. R. Mischke, *Mathematical Model Building,* 2d rev. ed., Iowa State University Press, Ames, 1980.

11.2 A. D. Dimarogonas, "Origins of Engineering Design," in *Design Engineering,* vol. 62, *Vibrations in Mechanical Systems and the History of Engineering,* American Society of Mechanical Engineers Design Conference Plenary Session presentation, Albuquerque, September 1993.

11.3 C. R. Mischke, *An Introduction to Computer-Aided Design,* Prentice-Hall, Englewood Cliffs, N.J., 1968.

11.4 G. V. Reklaitis, A. Ravindran, and K. M. Ragsdell, *Engineering Optimization,* Wiley-Interscience, New York, 1983.

11.5 B. Carnahan, H. A. Luther, and J. O. Wilkes, *Applied Numerical Methods,* John Wiley & Sons, New York, 1969.

11.6 C. F. Hix, Jr., and R. P. Alley, *Physical Laws and Effects,* John Wiley & Sons, New York, 1958.

CHAPTER 12
STRENGTH UNDER STATIC CIRCUMSTANCES

Charles R. Mischke, Ph.D., P.E.
Professor Emeritus of Mechanical Engineering
Iowa State University
Ames, Iowa

Joseph E. Shigley
Professor Emeritus
The University of Michigan
Ann Arbor, Michigan

GLOSSARY OF SYMBOLS

a	Crack semilength
A	Area
D, d	Diameter
F	Force or load
I	Second moment of area
J	Second polar moment of area
K	Stress-intensity factor
K'	Stress-concentration factor for static loading
K_c	Critical-stress-intensity factor
K_t	Normal-stress-concentration factor
K_{ts}	Shear-stress-concentration factor
M	Moment
n	Design factor
q_s	Sensitivity index

r Radius or ratio

S_{sy} Yield strength in shear

S_{uc} Ultimate compressive strength

S_{ut} Ultimate tensile strength

S_y Yield strength

η Factor of safety

σ Normal stress

σ' von Mises stress

τ Shear stress

τ_o Octahedral shear stress or nominal shear stress

12.1 PERMISSIBLE STRESSES AND STRAINS

The discovery of the relationship between stress and strain during elastic and plastic deformation allows interpretation either as a stress problem or as a strain problem. Imposed conditions on machine elements are more often loads than deformations, and so the usual focus is on stress rather than strain. Consequently, when durability under static conditions is addressed, attention to permissible stress is more common than attention to permissible strain.

Permissible stress levels are established by

• Experience with successful machine elements
• Laboratory simulations of field conditions
• Corporate experience manifested as a design-manual edict
• Codes, standards, and state of the art

During the design process, permissible stress levels are established by dividing the significant strength by a *design factor n*. The design factor represents the original intent or goal. As decisions involving discrete sizes are made, the stress levels depart from those intended. The quotient, obtained by dividing the significant strength by the load-induced stress at the critical location, is the *factor of safety* η, which is unique to the completed design. The design factor represents the goal and the factor of safety represents attainment. The adequacy assessment of a design includes examination of the factor of safety. Finding a permissible stress level which will provide satisfactory service is not difficult. Competition forces a search for the highest stress level which still permits satisfactory service. This is more difficult.

Permissible stress level is a function of material strength, which is assessible only by test. Testing is costly. Where there is neither time nor money available or testing the part is impossible, investigators have proposed theories of failure for guidance of designers. Use of a theory of failure involves (1) identifying the significant stress at the critical location and (2) comparing that stress condition with the strength of the part at that location in the condition and geometry of use. Standardized tests, such as the simple tension test, Jominy test, and others, provide some of the necessary information. For example, initiation of general yielding in a ductile part is predicted on the basis of yield strength exhibited in the simple tension test and modified by the manufacturing process. Rupture of brittle parts is predicted on the basis of ultimate strength (see Chap. 8).

Estimates of permissible stress level for long and satisfactory performance of function as listed above are based on design factors reflecting these experiences and are modified by the following:

1. Uncertainty as to material properties within a part, within a bar of steel stock, and within a heat of steel or whatever material is being considered for the design. Properties used by a designer may come not from an actual test, but from historical experience, since parts are sometimes designed before the material from which they will be made has even been produced.

2. Uncertainty owing to the discrepancy between the designed part and the necessarily small size of the test specimen. The influence of size on strength is such that smaller parts *tend* to exhibit larger strengths.

3. Uncertainty concerning the actual effects of the manufacturing process on the local material properties at the critical locations in the part. Processes such as upsetting, cold or hot forming, heat treatment, and surface treatment change strengths and other properties.

4. Uncertainties as to the true effect of peripheral assembly operations on strengths and other properties. Nearby weldments, mechanical fasteners, shrink fits, etc., all have influences that are difficult to predict with any precision.

5. Uncertainty as to the effect of elapsed time on properties. Aging in steels, aluminums, and other alloys occurs, and some strengthening mechanisms are time-dependent. Corrosion is another time-dependent enemy of integrity.

6. Uncertainty as to the actual operating environment.

7. Uncertainty as to the validity and precision of the mathematical models employed in reaching decisions on the geometric specifications of a part.

8. Uncertainty as to the intensity and dispersion of loads that may or will be imposed on a machine member and as to the understanding of the effect of impact.

9. Uncertainty as to the stress concentrations actually present in a manufactured part picked at random for assembly and use. Changes in tool radius due to wear, regrinding, or replacement can have a significant influence on the stress levels actually attained in parts in service.

10. Company design policies or the dictates of codes.

11. Uncertainty as to the completeness of a list of uncertainties.

Although specific recommendations that suggest design factors qualified by usage are to be found in many places, such factors depend on the stochastic nature of properties, loading, geometry, the form of functional relationships between them, and the reliability goal.

12.2 THEORY OF STATIC FAILURE

For ductile materials, the best estimation method for predicting the onset of yielding, for materials exhibiting equal strengths in tension and compression, is the octahedral shear theory (distortion energy or Hencky-von Mises). The *octahedral shear stress* is

$$\tau_o = \tfrac{1}{3}[(\sigma_1 - \sigma_2)^2 + (\sigma_2 - \sigma_3)^2 + (\sigma_3 - \sigma_1)^2]^{1/2}$$

where $\sigma_1, \sigma_2,$ and σ_3 are ordered principal stresses (see Chap. 49). In terms of orthogonal stress components in any other directions, the octahedral shear stress is

$$\tau_o = \tfrac{1}{3}\left[(\sigma_x - \sigma_y)^2 + (\sigma_y - \sigma_z)^2 + (\sigma_z - \sigma_x)^2 + 6(\tau_{xy}^2 + \tau_{yz}^2 + \tau_{zx}^2)\right]^{1/2}$$

The limiting value of the octahedral shear stress is that which occurs during uniaxial tension at the onset of yield. This limiting value is

$$\tau_o = \frac{\sqrt{2}S_y}{3}$$

By expressing this in terms of the principal stresses and a design factor, we have

$$\frac{S_y}{n} = \frac{3}{\sqrt{2}}\,[\tau_o]_{\text{lim}} = \frac{1}{\sqrt{2}}[(\sigma_1 - \sigma_2)^2 + (\sigma_2 - \sigma_3)^2 + (\sigma_3 - \sigma_1)^2]^{1/2} = \sigma' \qquad (12.1)$$

The term σ' is called the *von Mises stress*. It is the uniaxial tensile stress that induces the same octahedral shear (or distortion energy) in the uniaxial tension test specimen as does the triaxial stress state in the actual part.

For plane stress, one principal stress is zero. If the larger nonzero principal stress is σ_A and the smaller σ_B, then

$$\sigma' = (\sigma_A^2 + \sigma_B^2 - \sigma_A\sigma_B)^{1/2} = \frac{S_y}{n} \qquad (12.2)$$

By substituting the relation

$$\sigma_{A,B} = \frac{\sigma_x - \sigma_y}{2} \pm \sqrt{\left(\frac{\sigma_x - \sigma_y}{2}\right)^2 + \tau_{xy}^2}$$

we get a more convenient form:

$$\sigma' = (\sigma_x^2 + \sigma_y^2 - \sigma_x\sigma_y + 3\tau_{xy}^2)^{1/2} = \frac{S_y}{n} \qquad (12.3)$$

Example 1. A thin-walled pressure cylinder has a tangential stress of σ and a longitudinal stress of $\sigma/2$. What is the permissible tangential stress for a design factor of n?

Solution

$$\sigma' = (\sigma_A^2 + \sigma_B^2 - \sigma_A\sigma_B)^{1/2}$$

$$= \left[\sigma^2 + \left(\frac{\sigma}{2}\right)^2 - \sigma\left(\frac{\sigma}{2}\right)\right]^{1/2} = \frac{S_y}{n}$$

From which

$$\sigma = \frac{2}{\sqrt{3}}\frac{S_y}{n}$$

Note especially that this result is larger than the uniaxial yield strength divided by the design factor.

Example 2. Estimate the shearing yield strength from the tensile yield strength.
  Solution. Set $\sigma_A = \tau$, $\sigma_B = -\tau$, and at yield, $\tau = S_{sy}$, so

$$\sigma' = (\sigma_A^2 + \sigma_B^2 - \sigma_A\sigma_B)^{1/2}$$

$$= [S_{sy}^2 + (-S_{sy})^2 - S_{sy}(-S_{sy})]^{1/2} = S_y$$

Solving gives

$$S_{sy} = \frac{S_y}{\sqrt{3}} = 0.577S_y$$

12.2.1 Brittle Materials

To define the criterion of failure for brittle materials as *rupture*, we require that the fractional reduction in area be less than 0.05; this corresponds to a true strain at fracture of about 0.05. Brittle materials commonly exhibit an ultimate compressive strength significantly larger than their ultimate tensile strength. And unlike with ductile materials, the ultimate torsional strength is approximately equal to the ultimate tensile strength. If σ_A and σ_B are ordered-plane principal stresses, then there are five points on the rupture locus in the $\sigma_A\sigma_B$ plane that can be immediately identified (Fig. 12.1). These are

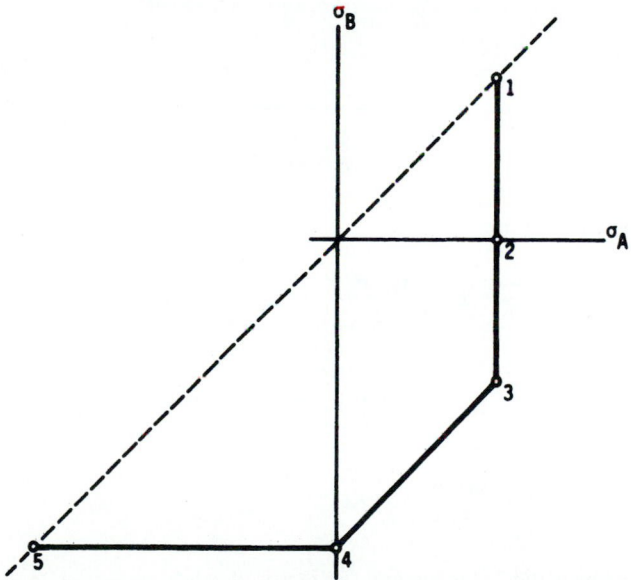

FIGURE 12.1 $\sigma_A\sigma_B$ plane with straight-line Coulomb-Mohr strength locus.

Locus 1–2: $\sigma_A = S_{ut}, \sigma_A > \sigma_B > 0$
Point 3: $\sigma_A = S_{ut} = S_{su}, \sigma_B = -S_{ut} = -S_{su}$
Point 4: $\sigma_A = 0, \sigma_B = S_{uc}$
Locus 4–5: $\sigma_B = S_{uc}, \sigma_A < 0$

Connecting points 2, 3, and 4 with straight-line segments defines the *modified Mohr theory of failure*. This theory evolved from the *maximum normal stress theory* and the *Coulomb-Mohr internal friction theory*. We can state this in algebraic terms by defining $r = \sigma_B/\sigma_A$. The result is

$$\sigma_A = \frac{S_{ut}}{n} \qquad \text{when } \sigma_A > 0, \sigma_B > -\sigma_A$$

$$\sigma_A = \frac{S_{uc}S_{ut}/n}{(1+r)S_{ut}+S_{uc}} \qquad \text{when } \sigma_A > 0, \sigma_B < 0, r < -1 \qquad (12.4)$$

$$\sigma_B = \frac{S_{uc}}{n} \qquad \text{when } \sigma_A < 0$$

Figure 12.2 shows some experimental points from tests on gray cast iron.

Example 3. A ¼-in-diameter ASTM No. 40 cast iron pin with $S_{ut} = 40$ kpsi and $S_{uc} = -125$ kpsi is subjected to an axial compressive load of 800 lb and a torsional moment of 100 lb · in. Estimate the factor of safety.
   Solution.   The axial stress is

$$\sigma_x = \frac{F}{A} = \frac{-800}{\pi(0.25)^2/4} = -16.3 \text{ kpsi}$$

The surface shear stress is

$$\tau_{xy} = \frac{16T}{\pi d^3} = \frac{16(100)}{\pi(0.25)^3} = 32.6 \text{ kpsi}$$

The principal stresses are

$$\sigma_{A,B} = \frac{\sigma_x + \sigma_y}{2} \pm \sqrt{\left(\frac{\sigma_x - \sigma_y}{2}\right)^2 + \tau_{xy}^2}$$

$$= \frac{-16.3}{2} \pm \sqrt{\left(\frac{-16.3}{2}\right)^2 + (32.6)^2} = 25.45, -41.25 \text{ kpsi}$$

$$r = \frac{\sigma_B}{\sigma_A} = \frac{-41.25}{25.45} = -1.64$$

The rupture line is the 3–4 locus, and the factor of safety is

$$\eta = \frac{S_{uc}S_{ut}}{(1+r)S_{ut}+S_{uc}} \frac{1}{\sigma_A}$$

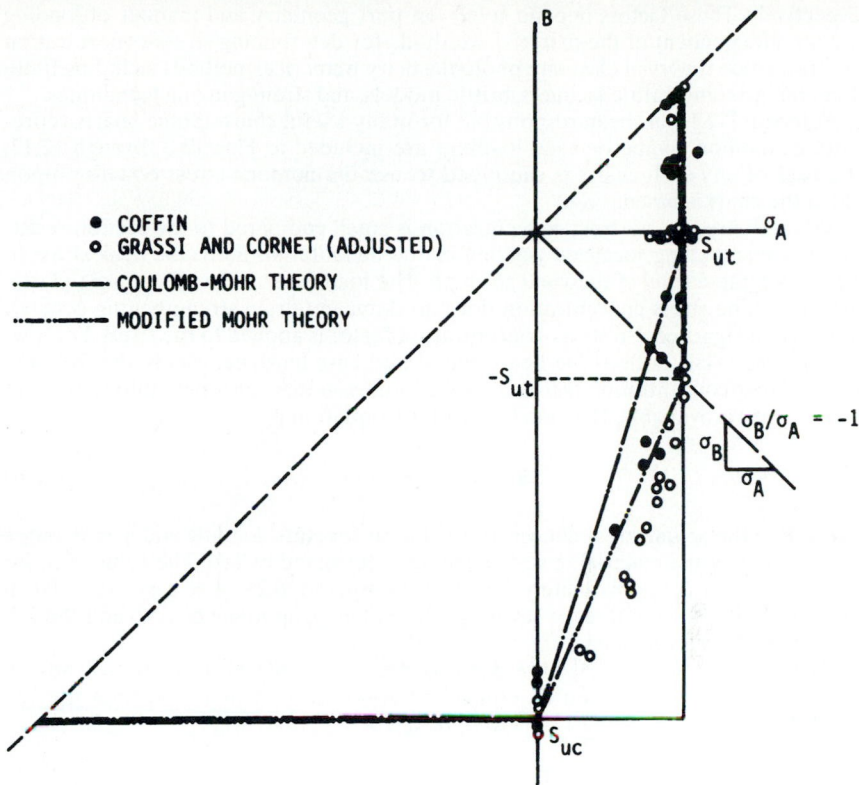

FIGURE 12.2   Experimental data from tests of gray cast iron subjected to biaxial stresses. The data were adjusted to correspond to $S_{ut} = 32$ kpsi and $S_{uc} = 105$ kpsi. Superposed on the plot are graphs of the maximum-normal-stress theory, the Coulomb-Mohr theory, and the modified Mohr theory. (*Adapted from J. E. Shigley and L. D. Mitchell,* Mechanical Engineering Design, *4th ed., McGraw-Hill, 1983, with permission.*)

$$= \frac{(-125)(40)}{[(1 - 1.64)(40) - 125](25.45)} = 1.30$$

12.3   STRESS CONCENTRATION

Geometric discontinuities increase the stress level beyond the nominal stresses, and the elementary stress equations are inadequate estimators. The geometric discontinuity is sometimes called a *stress raiser,* and the domains of departure from the elementary equation are called the *regions of stress concentration.* The multiplier applied to the nominal stress to estimate the peak stress is called the *stress-concentration factor,* denoted by K_t or K_{ts}, and is defined as

$$K_t = \frac{\sigma_{max}}{\sigma_o} \qquad K_{ts} = \frac{\tau_{max}}{\tau_o} \qquad (12.5)$$

respectively. These factors depend solely on part geometry and manner of loading and are independent of the material. Methods for determining stress-concentration factors include theory of elasticity, photoelasticity, numerical methods including finite elements, gridding, brittle lacquers, brittle models, and strain-gauging techniques.

Peterson [12.1] has been responsible for many useful charts. Some charts representing common geometries and loadings are included as Figs. 12.3 through 12.17. The user of any such charts is cautioned to use the nominal stress equation upon which the chart is based.

When the region of stress concentration is small compared to the section resisting the static loading, localized yielding in ductile materials limits the peak stress to the approximate level of the yield strength. The load is carried without gross plastic distortion. The stress concentration does no damage (strain strengthening occurs), and it can be ignored. No stress-concentration factor is applied to the stress. For low-ductility materials, such as the heat-treated and case-hardened steels, the full geometric stress-concentration factor is applied unless notch-sensitivity information to the contrary is available. This notch-sensitivity equation is

$$K' = 1 + q_s(K_t - 1) \tag{12.6}$$

where K' = the actual stress-concentration factor for static loading and q_s = an index of sensitivity of the material in static loading determined by test. The value of q_s for hardened steels is approximately 0.15 (if untempered, 0.25). For cast irons, which have internal discontinuities as severe as the notch, q_s approaches zero and the full value of K_t is rarely applied.

Kurajian and West [12.3] have derived stress-concentration factors for hollow stepped shafts. They develop an equivalent solid stepped shaft and then use the usual charts (Figs. 12.10 and 12.11) to find K_t. The formulas are

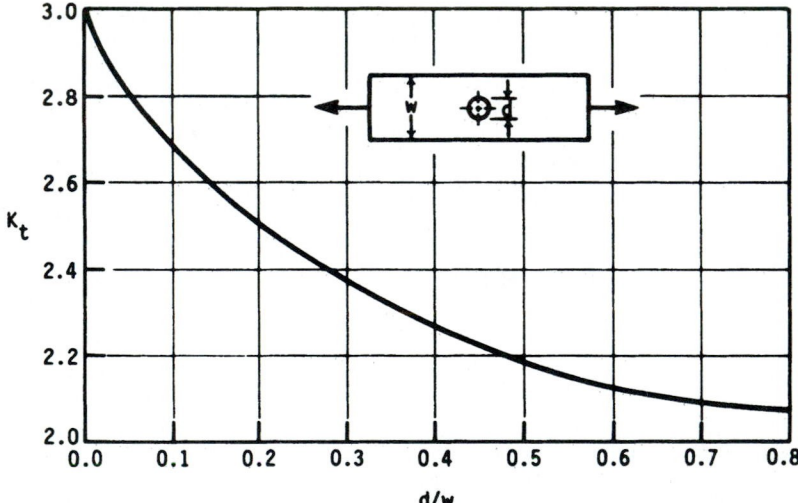

FIGURE 12.3   Bar in tension or simple compression with a transverse hole. $\sigma_o = F/A$, where $A = (w - d)t$, and t = thickness. *(From Peterson [12.2].)*

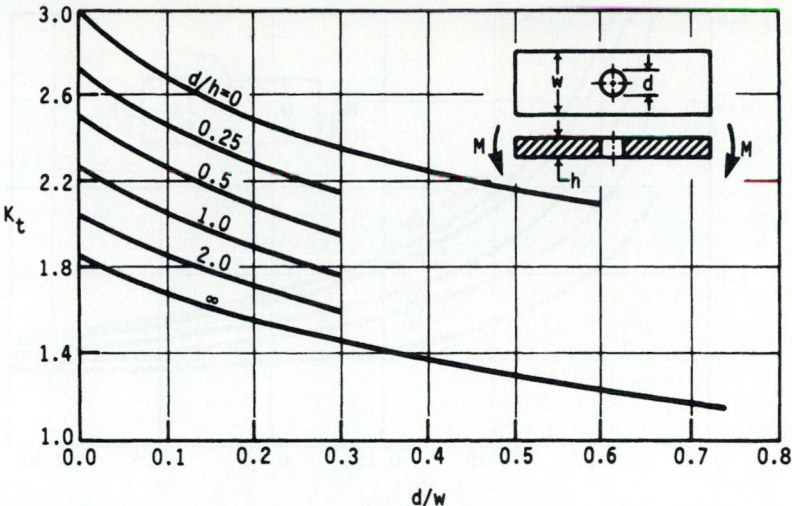

FIGURE 12.4 Rectangular bar with a transverse hole in bending. $\sigma_o = Mc/I$, where $I = (w - d)h^3/12$. *(From Peterson [12.2].)*

$$D = \left(\frac{D_o^4 - d_i^4}{D_o}\right)^{1/3} \quad d = \left(\frac{d_o^4 - d_i^4}{d_o}\right)^{1/3} \tag{12.7}$$

where D, d = diameters of solid stepped shaft (Fig. 12.10)

$\quad\quad D_o, d_o$ = diameters of hollow stepped shaft

$\quad\quad d_i$ = hole diameter

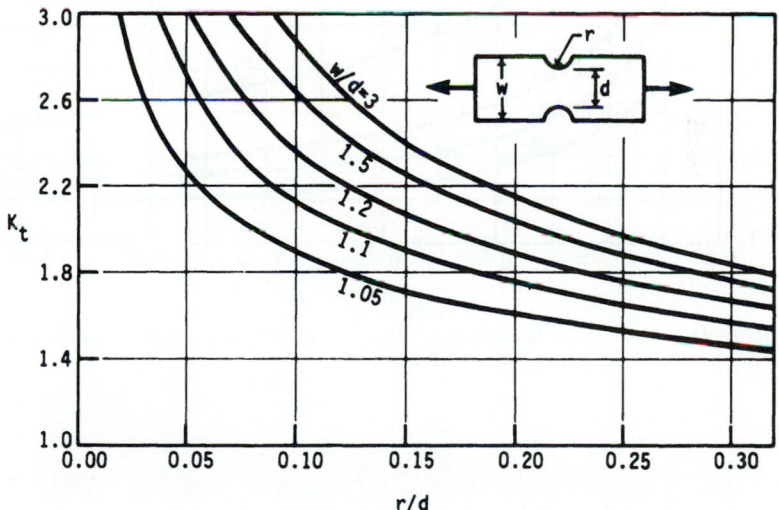

FIGURE 12.5 Notched rectangular bar in tension or simple compression. $\sigma_o = F/A$, where $A = td$ and t = thickness. *(From Peterson [12.2].)*

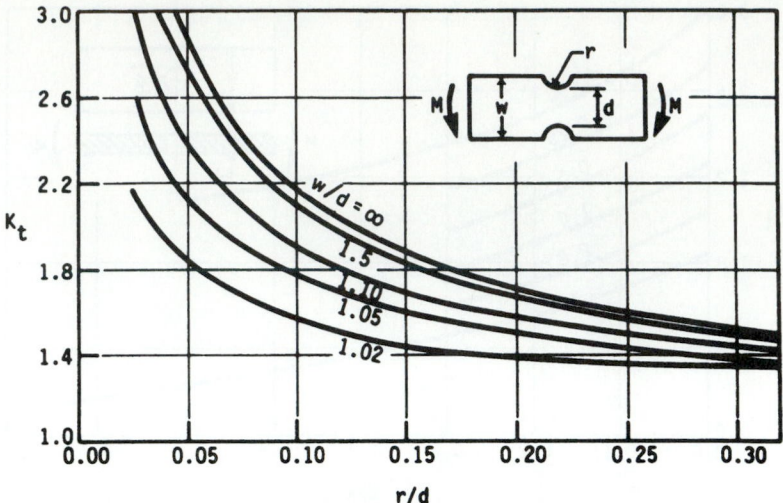

FIGURE 12.6 Notched rectangular bar in bending. $\sigma_o = Mc/I$, where $c = d/2$, $I = td^3/12$, and t = thickness. *(From Peterson [12.2].)*

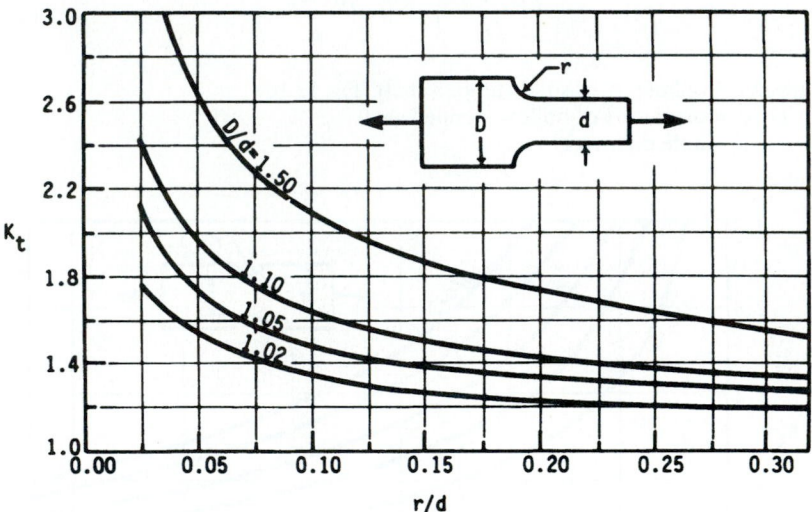

FIGURE 12.7 Rectangular filleted bar in tension or simple compression. $\sigma_o = F/A$, where $A = td$ and t = thickness. *(From Peterson [12.2].)*

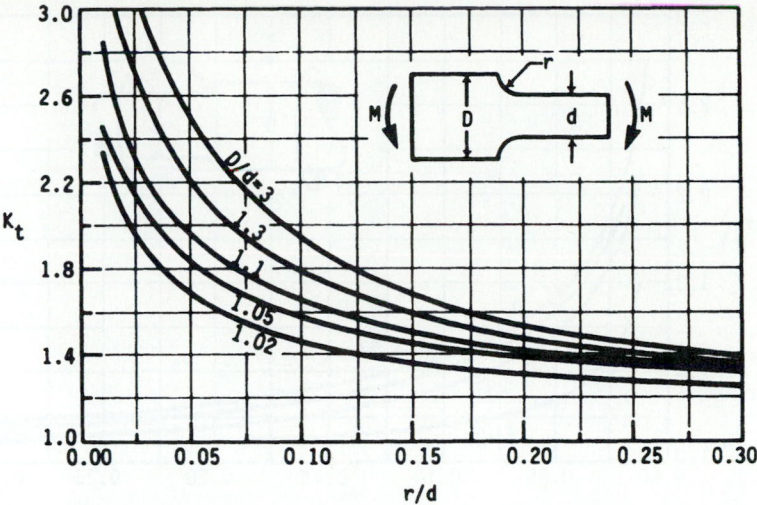

FIGURE 12.8   Rectangular filleted bar in bending. $\sigma_o = Mc/I$, where $c = d/2$, $I = td^3/12$, and t = thickness. *(From Peterson [12.2].)*

The fillet radius is unchanged. No change is necessary for axial loading because of the uniform stress distribution.

12.4   FRACTURE MECHANICS

The use of stress-concentration factors is really of little use when brittle materials are used or when a very small crack or flaw exists in the material. Ductile materials

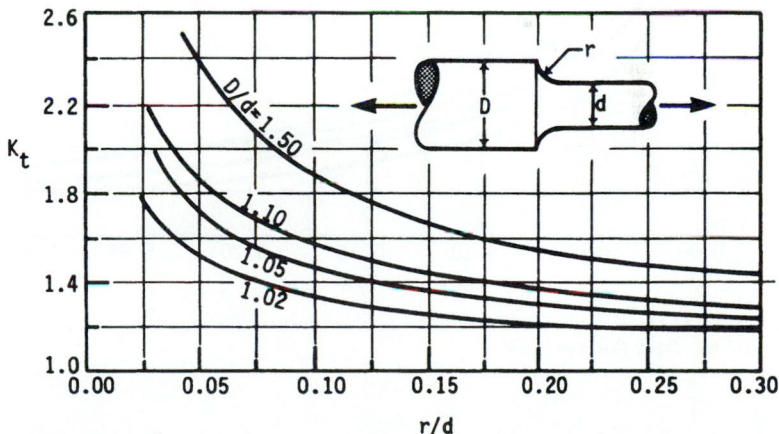

FIGURE 12.9   Round shaft with shoulder fillet in tension. $\sigma_o = F/A$, where $A = \pi d^2/4$. *(From Peterson [12.2].)*

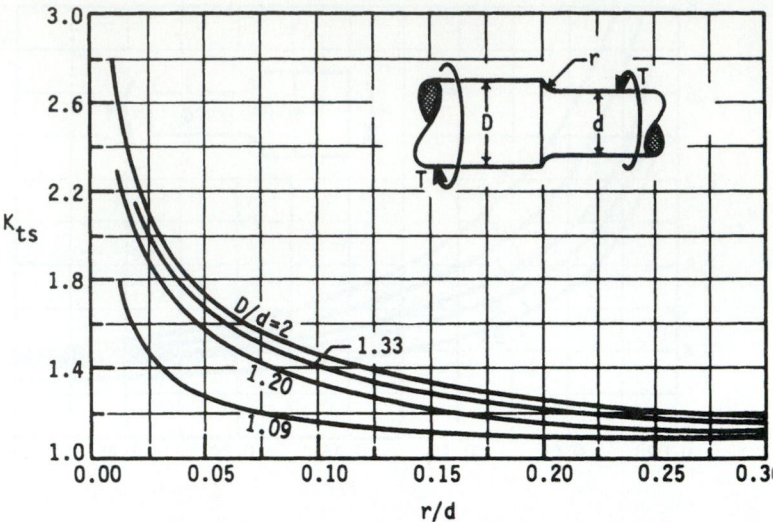

FIGURE 12.10   Round shaft with shoulder fillet in torsion. $\tau_o = Tc/J$, where $c = d/2$ and $J = \pi d^4/32$. *(From Peterson [12.2].)*

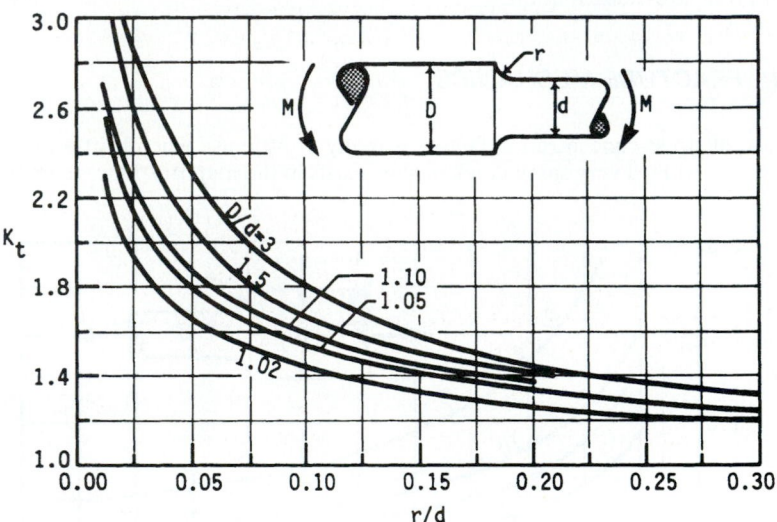

FIGURE 12.11   Round shaft with shoulder fillet in bending. $\sigma_o = Mc/I$, where $c = d/2$ and $I = \pi d^4/64$. *(From Peterson [12.2].)*

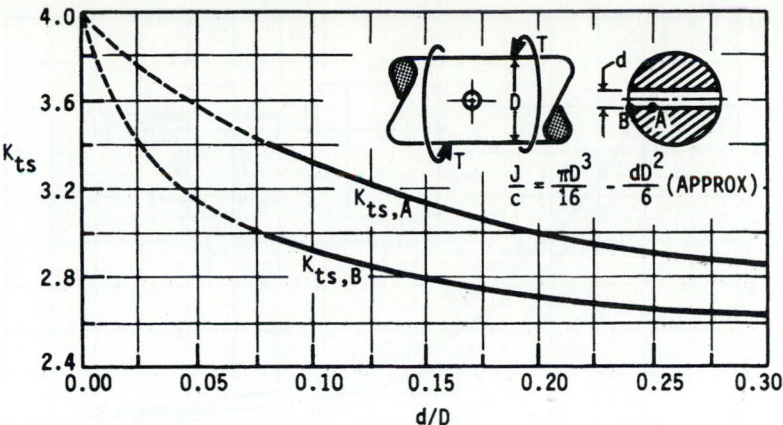

FIGURE 12.12   Round shaft in torsion with transverse hole. *(From Peterson [12.2].)*

also may fail in a brittle manner, possibly because of low temperature or other causes. So another method of analysis is necessary for all materials that cannot yield and relieve the stress concentration at a notch, defect, or crack.

Fracture mechanics can be used to determine the average stress in a part that will cause a crack to grow; energy methods of analysis are used (see Ref. [12.4]).

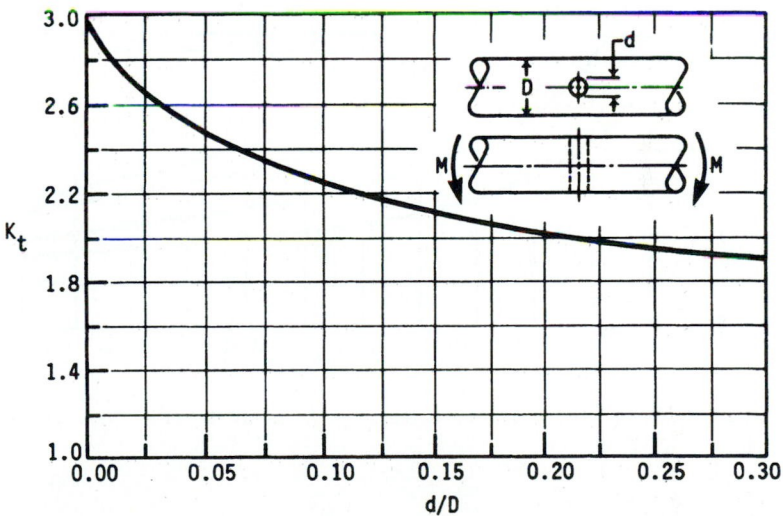

FIGURE 12.13   Round shaft in bending with a transverse hole. $\sigma_o = M/[(\pi D^3/32) - (dD^2/6)]$, approximately. *(From Peterson [12.2].)*

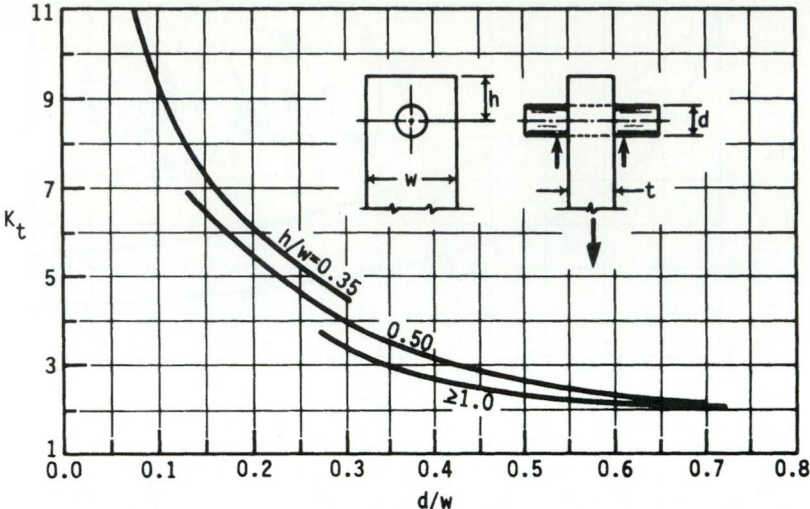

FIGURE 12.14   Plate loaded in tension by a pin through a hole. $\sigma_o = F/A$, where $A = (w - d)t$. When clearance exists, increase K_t by 35 to 50 percent. (*From M. M. Frocht and H. N. Hill, "Stress Concentration Factors around a Central Circular Hole in a Plate Loaded through a Pin in Hole," Journal of Applied Mechanics, vol. 7, no. 1, March 1940, p. A-5, with permission.*)

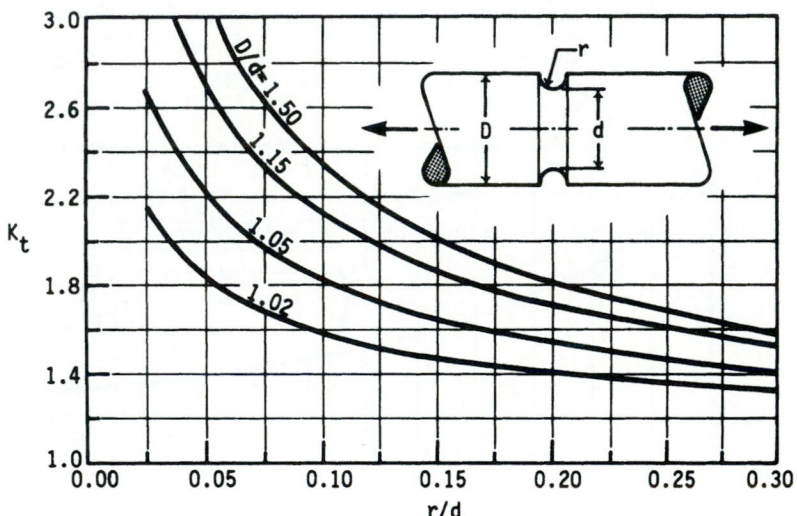

FIGURE 12.15   Grooved round bar in tension. $\sigma_o = F/A$, where $A = \pi d^2/4$. (*From Peterson [12.2].*)

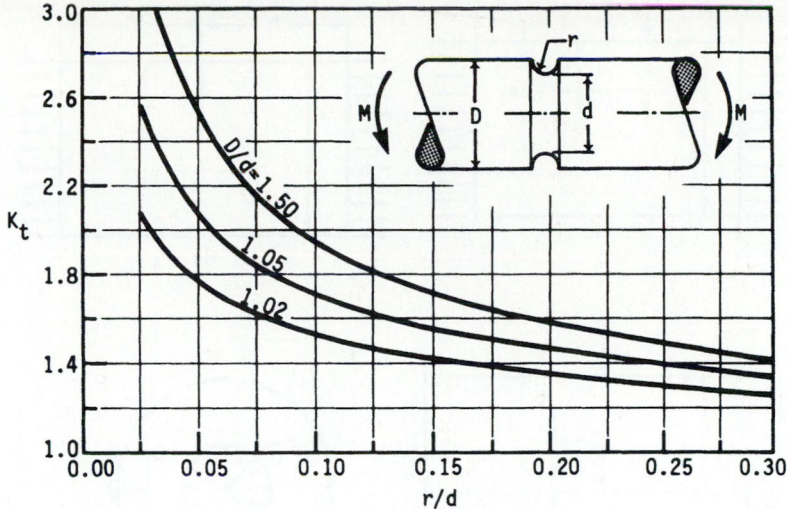

FIGURE 12.16 Grooved round bar in bending. $\sigma_o = Mc/I$, where $c = d/2$ and $I = \pi d^4/64$. *(From Peterson [12.2].)*

12.4.1 Stress Intensities

In Fig. 12.18a, suppose the length of the tensile specimen is large compared to the width $2b$. Also, let the crack, of length $2a$, be centrally located. Then a stress-intensity factor K can be defined by the relation

$$K_0 = \sigma(\pi a)^{1/2} \tag{12.8}$$

where σ = average tensile stress. The units of K_0 are kpsi $\cdot$ in$^{1/2}$ or, in SI, MPa $\cdot$ m$^{1/2}$.

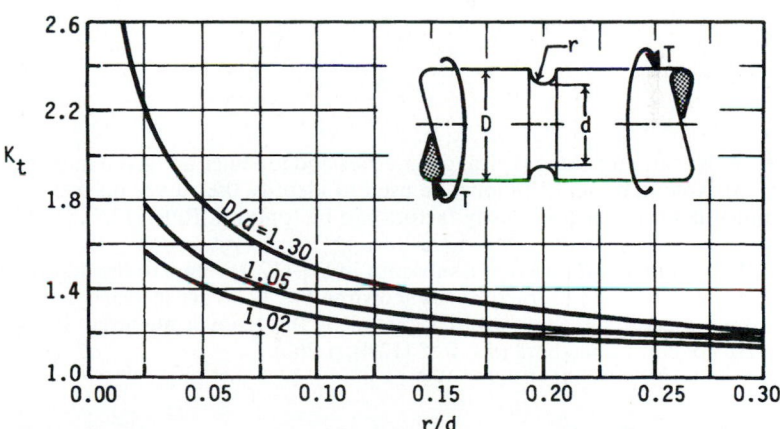

FIGURE 12.17 Grooved round bar in torsion. $\tau_o = Tc/J$, where $c = d/2$ and $J = \pi d^4/32$. *(From Peterson [12.2].)*

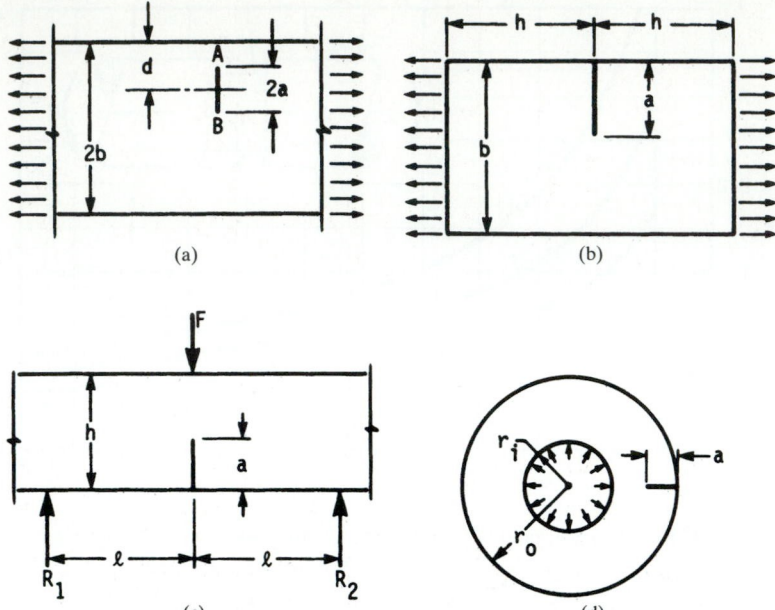

FIGURE 12.18   Typical crack occurrences. (*a*) Bar in tension with interior crack; (*b*) bar in tension with edge crack; (*c*) flexural member of rectangular cross section with edge crack; (*d*) pressurized cylinder with radial edge crack parallel to cylinder axis.

Since the actual value of K for other geometries depends on the loading too, it is convenient to write Eq. (12.8) in the form

$$K_I = C\sigma(\pi a)^{1/2} \tag{12.9}$$

where

$$C = \frac{K_I}{K_0} \tag{12.10}$$

Values of this ratio for some typical geometries and loadings are given in Figs. 12.19 and 12.20. Note that Fig. 12.18 must be used to identify the curves on these charts. Additional data on stress-intensity factors can be found in Refs. [12.5], [12.6], and [12.7].

The Roman numeral I used as a subscript in Eq. (12.9) refers to the deformation mode. Two other modes of fracture not shown in Fig. 12.18 are in-plane and out-of-plane shear modes, and these are designated by the Roman numerals II and III. These are not considered here (see Ref. [12.4], p. 262).

12.4.2 Fracture Toughness

When the stress σ of Eq. (12.9) reaches a certain critical value, crack growth begins, and the equation then gives the *critical-stress-intensity factor* K_{Ic}. This is also called the

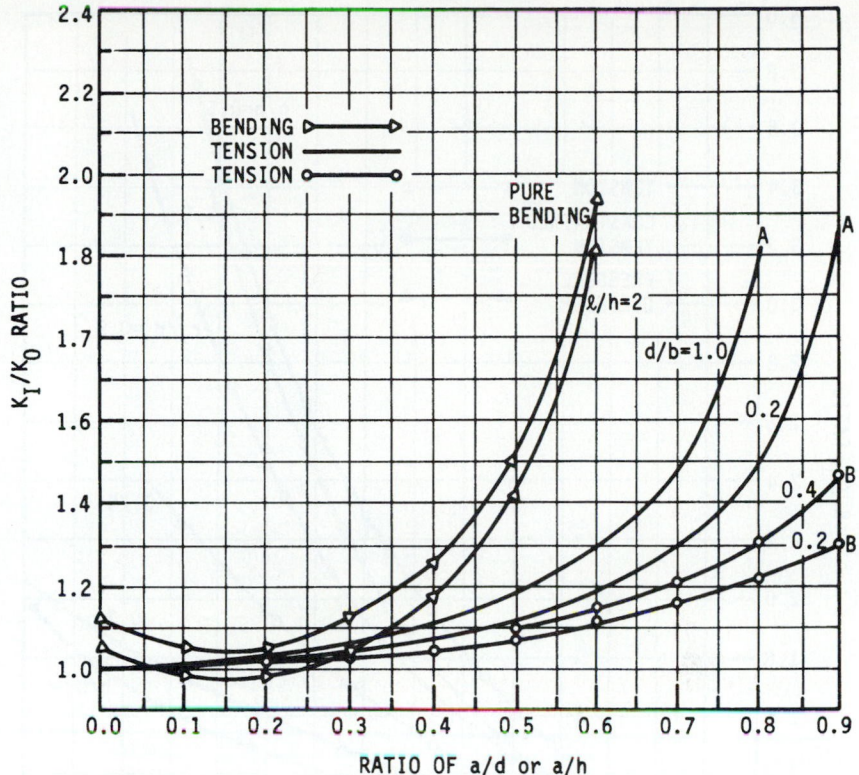

FIGURE 12.19 Stress-intensity charts for cracks shown in Fig. 12.18a and c. Letters A and B identify the ends of the crack shown in Fig. 12.18a. Values of $\ell/h > 2$ will produce curves closer to the curve for pure bending.

fracture toughness. Since it is analogous to strength, we can define design factor as

$$n = \frac{K_c}{K} \tag{12.11}$$

Some typical values of K_c are given in Table 12.1. For other materials, see Ref. [12.8].

12.5 NONFERROUS METALS

Designing for static loads with aluminum alloys is not much different from designing for the steels. Aluminum alloys, both cast and wrought, have strengths in tension and compression that are about equal. The yield strengths in shear vary from about 55 to 65 percent of the tensile yield strengths, and so the octahedral shear theory of failure is valid.

The corrosion resistance (see Chap. 44), workability, and weldability obtainable from some of the alloys make this a very versatile material for design. And the extrusion capability means that a very large number of wrought shapes are available.

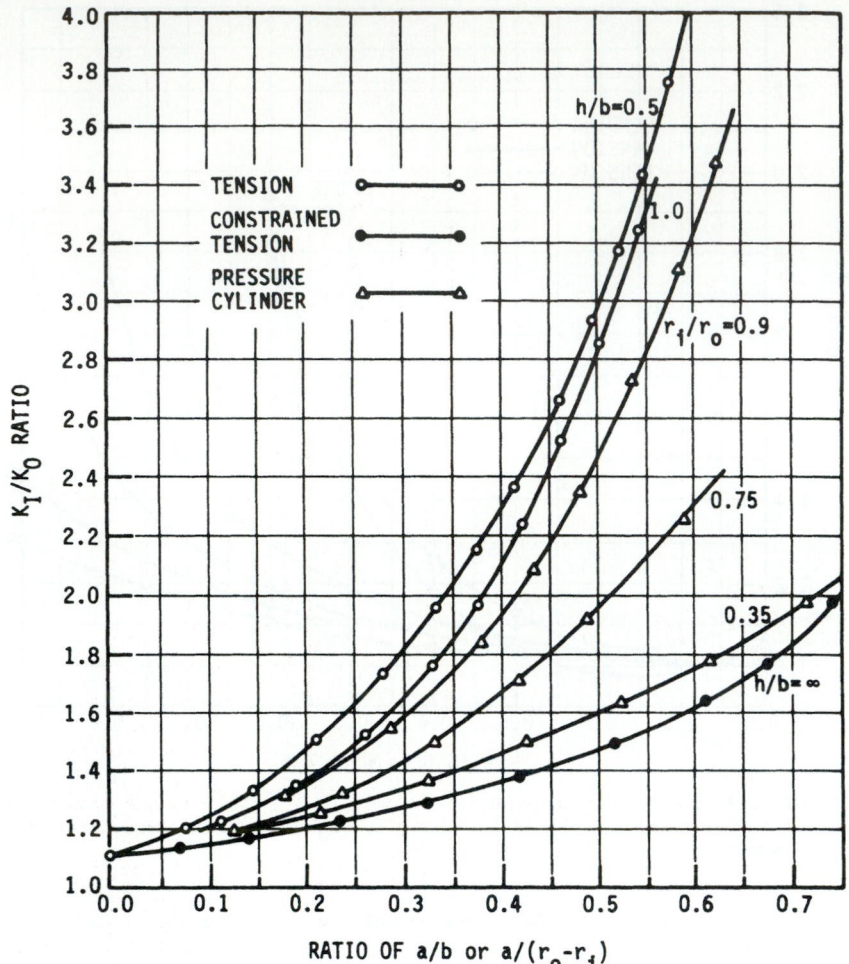

FIGURE 12.20   Stress-intensity chart for cracks shown in Figs. 12.18*b* and *d*. The curve $h/b = \infty$ has bending constraints acting on the member.

However, these alloys do have a temperature problem, as shown by the curves of strength versus temperature in Fig. 12.21. Other aluminum alloys will exhibit a similar characteristic.

Alloying elements used with copper as the base element include zinc, lead, tin, aluminum, silicon, manganese, phosphorus, and beryllium. Hundreds of variations in the percentages used are possible, and consequently, the various copper alloys may have widely differing properties. The primary consideration in selecting a copper alloy may be the machinability, ductility, hardness, temperature properties, or corrosion resistance. Strength is seldom the primary consideration. Because of these variations in properties, it is probably a good idea to consult the manufacturer concerning new applications until a backlog of experience can be obtained.

TABLE 12.1   Values of the Fracture Toughness K_{Ic} for a Few Engineering Materials

Material	Designation	UNS no.	Yield strength		Fracture toughness	
			MPa	kpsi	MPa·m$^{1/2}$	kpsi·in$^{1/2}$
Aluminum	2024-T851	A92094-T851	455	66	26	24
	7075-T651	A97075-T651	495	72	24	22
Titanium	Ti-6AL-4V	R56401	910	132	115	105
	Ti-6AL-4V	R56401	1035	150	55	50
Steel	AISI 4340	G43400	860	125	99	90
	AISI 4340	G43400	1515	220	60	55
	AISI 52100	G52986	2070	300	14	13

SOURCE: Professor David K. Felbeck, The University of Michigan, by personal communication.

Magnesium alloys have a weight about two-thirds that of aluminum and one-fourth that of steel. Magnesium alloys are not very strong and are characterized by having a compressive strength that is somewhat less than the tensile strength. They are also so sensitive to temperature that they are weakened even by contact with boiling water.

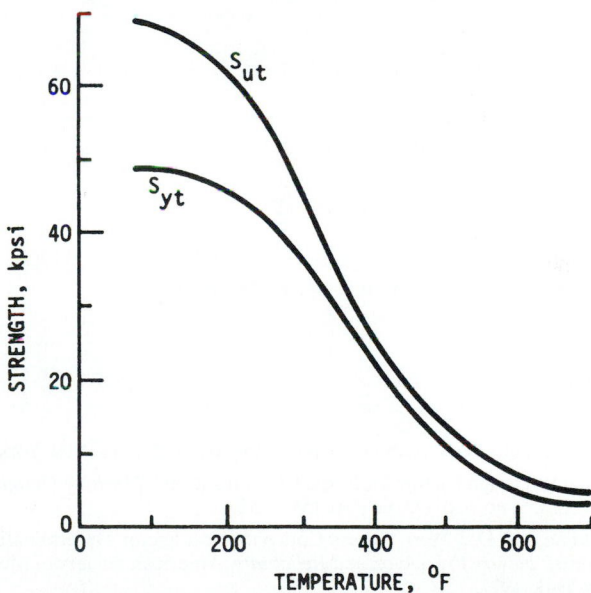

FIGURE 12.21   Effect of temperature on the yield strength and tensile strength of aluminum alloy A92024-T4. (*ALCOA.*)

12.6 STOCHASTIC CONSIDERATIONS

It is helpful to understanding to recall the nature of deterministic theories of failure. All the stresses at a point are consequences of geometry and loading. In linear systems, all the stresses at a point are proportional to load. If there is variability in load, there is variability in the stresses at a point. When variability in geometry can be ignored compared to variability in loading, the distribution type and the coefficients of variation agree in both stress and load. In a deterministic theory of failure, such as distortion energy, the von Mises σ' is set equal to the loss-of-function stress called yield strength S_y, or to S_y/n. One cannot do this directly with stochastic variables. The distribution type of the von Mises stress is that of the load, and is independent of the distribution type of the yield strength. An equation such as Eq. (12.3) can be constructed only for the approximate relationship of the means (strictly, the medians).

In examining histograms obtained in yield strength testing, normal, lognormal, and Weibull fits can be successful. The coefficient of variation averages 0.064 for yield strengths of 31 steels that differ in composition or thermomechanical treatment, and the range is 0.02 to 0.10. More data can be found in Ref. [12.9]. The dispersion is wider than that of the corresponding ultimate tensile strength, which averages 0.047 and ranges from 0.017 to 0.077; this may be due in part to testing methodology differences.

Since one often encounters lognormal loading and yield strength, Eq. (2.6), with $S = S_y$, the yield strength, and $\sigma = \sigma'$, the von Mises stress, is usefully written as

$$C_n = \sqrt{C_{S_y}^2 + C_{\sigma'}^2} = \sqrt{C_{S_y}^2 + C_P^2}$$

where **P** is the load causing σ'. It follows that

$$\bar{n} = \exp\left[-z\sqrt{\ln(1 + C_n^2)} + \ln\sqrt{1 + C_n^2}\right] \doteq \exp\left[C_n(-z + C_n/2)\right]$$

and Eq. (12.3) can now be expressed as

$$\bar{\sigma}' = \frac{\bar{S_y}}{\bar{n}}$$

If reliability is sought, Eqs. (2.5) and (2.4) can be used.

REFERENCES

12.1 R. E. Peterson, *Stress Concentration Factors,* John Wiley & Sons, New York, 1974.

12.2 R. E. Peterson, "Design Factors for Stress Concentration," *Machine Design,* vol. 23, no. 2, p. 161; no. 5, p. 159; no. 6, p. 173; no. 7, p. 155; 1951.

12.3 G. M. Kurajian and D. J. West, "Stress Concentration Factor Determination in Stepped Hollow Shafts," *Failure Prevention and Reliability,* American Society of Mechanical Engineers, New York, 1977.

12.4 R. W. Hertzberg, *Deformation and Fracture Mechanics of Engineering Materials,* John Wiley & Sons, New York, 1976.

12.5 H. Tada, P. C. Paris, and G. R. Irwin, *The Stress Analysis of Cracks Handbook,* Del Research, Hellertown, Pa., 1973.

12.6 G. C. M. Sih, *Handbook of Stress Intensity Factors,* Lehigh University, Bethlehem, Pa., 1973.

12.7 D. P. Rooke and D. J. Cartwright, *Compendium of Stress Intensity Factors,* Hillingdon Press, Uxbridge, England, 1976.

12.8 *Damage Tolerant Handbook,* Metals and Ceramics Information Center, Battelle, Columbus, Ohio, 1975.

12.9 C. R. Mischke, "Some Property Data and Corresponding Weibull Parameters for Stochastic Mechanical Design, *Transactions of A.S.M.E., Journal of Mechanical Design,* vol. 114, March 1992, pp. 29–34.

CHAPTER 13
STRENGTH UNDER DYNAMIC CONDITIONS

Charles R. Mischke, Ph.D., P.E.
Professor Emeritus of Mechanical Engineering
Iowa State University
Ames, Iowa

NOMENCLATURE

a	Distance, exponent, constant
A	Area, addition factor, $\Sigma i N_i$
b	Distance, width, exponent
B	$\Sigma i^2 N_i$
bhn	Brinell hardness, roller or pinion
BHN	Brinell hardness, cam or gear
c	Exponent
C	Coefficient of variation
C_p	Materials constant in rolling contact
d	Difference in stress level, diameter
d_e	Equivalent diameter
D	Damage per cycle or block of cycles
D_I	Ideal critical diameter
E	Young's modulus
f	Fraction of mean ultimate tensile strength
f_i	Fraction of life measure

F	Force
$\mathscr{F}$	Significant force in contact fatigue
h	Depth
H_B	Brinell hardness
I	Second area moment
$\mathbf{k}_a, k_a$	Marin surface condition modification factor
k_b	Marin size modification factor
$\mathbf{k}_c, k_c$	Marin load modification factor
$\mathbf{k}_d, k_d$	Marin temperature modification factor
$\mathbf{k}_e, k_e$	Marin miscellaneous-effects modification factor
K	Load-life constant
$\mathbf{K}_f$	Fatigue stress concentration factor
$\mathbf{K}_t$	Geometric (theoretical) stress concentration factor
ℓ	Length
log	Base 10 logarithm
ln	Natural logarithm
L	Life measure
LN	Lognormal
m	Strain-strengthening exponent, revolutions ratio
M	Bending moment
$\mathbf{n}, n$	Design factor
N	Cycles
N_f	Cycles to failure
$N(\mu, \sigma)$	Normal distribution with mean μ and standard deviation σ
p	Pressure
P	Axial load
q	Notch sensitivity
r	Notch radius, slope of load line
r_i	Average peak-to-valley distance
R	Reliability
R_a	Average deviation from the mean
R_{rms}	Root-mean-squared deviation from the mean
RA	Fraction reduction in area
R_Q	As-quenched hardness, Rockwell C scale
R_T	Tempered hardness, Rockwell C scale
S	Strength
S'_{ax}	Axial endurance limit
S'_e	Rotating-beam endurance limit
S_f	Fatigue strength
S'_{se}	Torsional endurance limit

S_u, S_{ut}	Ultimate tensile strength
S_y	Yield strength
t_f	Temperature, °F
T	Torque
w	Width
x	Variable, coordinate
y	Variable, coordinate
z	Variable, coordinate, variable of $N(0, z)$
α	Prot loading rate, psi/cycle
β	Rectangular beam width
Δ	Approach of center of roller
ε	True strain
ε_f	True strain at fracture
η	Factor of safety
θ	Angle, misalignment angle
λ	Lognormally distributed
μ	Mean
ν	Poisson's ratio
ξ	Normally distributed
σ	Normal stress
σ_a	Normal stress amplitude component
σ'_f	Fatigue strength coefficient
σ_m	Steady normal stress component
σ_{max}	Largest normal stress
σ_{min}	Smallest normal stress
σ_0	Nominal normal stress
$\bar{\sigma}_0$	Strain-strengthening coefficient
σ	Standard deviation
τ	Shear stress
ϕ	Pressure angle
ϕ	Fatigue ratio: ϕ_b, beading; ϕ_{ax}, axial; ϕ_t, torsion; $\phi_{0.30}$, bending with 0.30-in-diameter rotating specimen
$\Phi(z)$	Cumulative distribution function of the standardized normal

13.1   TESTING METHODS AND PRESENTATION OF RESULTS

The designer has need of knowledge concerning endurance limit (if one exists) and endurance strengths for materials specified or contemplated. These can be estimated from the following:

- Tabulated material properties (experience of others)
- Personal or corporate R. R. Moore endurance testing
- Uniaxial tension testing and various correlations
- For plain carbon steels, if heat treating is involved, Jominy test and estimation of tempering effects by the method of Crafts and Lamont
- For low-alloy steels, if heat treating is involved, prediction of the Jominy curve by the method of Grossmann and Fields and estimation of tempering effects by the method of Crafts and Lamont
- If less than infinite life is required, estimation from correlations
- If cold work or plastic strain is an integral part of the manufacturing process, using the method of Datsko

The representation of data gathered in support of fatigue-strength estimation is best made probabilistically, since inferences are being made from the testing of necessarily small samples. There is a long history of presentation of these quantities as deterministic, which necessitated generous design factors. The plotting of cycles to failure as abscissa and corresponding stress level as ordinate is the common SN curve. When the presentation is made on logarithmic scales, some piecewise rectification may be present, which forms the basis of useful curve fits. Some ferrous materials exhibit a pronounced knee in the curve and then very little dependency of strength with life. Deterministic researchers declared the existence of a zero-slope portion of the curve and coined the name *endurance limit* for this apparent asymptote. Probabilistic methods are not that dogmatic and allow only such statements as, "A null hypothesis of zero slope cannot be rejected at the 0.95 confidence level."

Based on many tests over the years, the general form of a steel SN curve is taken to be approximately linear on log-log coordinates in the range 10^3 to 10^6 cycles and nearly invariant beyond 10^7 cycles. With these useful *approximations* and knowledge that cycles-to-failure distributions at constant stress level are lognormal (cannot be rejected) and that stress-to-failure distributions at constant life are likewise lognormal, specialized methods can be used to find some needed attribute of the SN picture. The cost and time penalties associated with developing the complete picture motivate the experimentor to seek only what is needed.

13.1.1 Sparse Survey

On the order of a dozen specimens are run to failure in an R. R. Moore apparatus at stress levels giving lives of about 10^3 to 10^7 cycles. The points are plotted on log-log paper, and in the interval $10^3 < N < 10^7$ cycles, a "best" straight line is drawn. Those specimens which have not failed by 10^8 or 5×10^8 cycles are used as evidence of the existence of an endurance limit. All that this method produces is estimates of two median lines, one of the form

$$S_f' = CN^b \qquad 10^3 < N < 10^6 \tag{13.1}$$

and the other of the form

$$S_f' = S_e' \qquad N > 10^6 \tag{13.2}$$

This procedure "roughs in" the SN curve as a gross estimate. No standard deviation information is generated, and so no reliability contours may be created.

13.1.2 Constant-Stress-Level Testing

If high-cycle fatigue strength in the range of 10^3 to 10^6 cycles is required and reliability (probability of survival) contours are required, then constant-stress-level testing is useful. A dozen or more specimens are tested at each of several stress levels. These results are plotted on lognormal probability paper to "confirm" by inspection the lognormal distribution, or a statistical goodness-of-fit test (Smirnov-Kolomogorov, chi-squared) is conducted to see if lognormal distribution can be rejected. If not, then reliability contours are established using lognormal statistics. Nothing is learned about endurance limit. Sixty to 100 specimens usually have been expended.

13.1.3 Probit Method

If statistical information (mean, standard deviation, distribution) concerning the endurance limit is needed, the probit method is useful. Given a priori knowledge that a "knee" exists, stress levels are selected that at the highest level produce one or two runouts and at the lowest level produce one or two failures. This places the testing at the "knee" of the curve and within a couple of standard deviations on either side of the endurance limit. The method requires exploratory testing to estimate the stress levels that will accomplish this. The results of the testing are interpreted as a lognormal distribution of stress either by plotting on probability paper or by using a goodness-of-fit statistical reduction to "confirm" the distribution. If it is confirmed, the mean endurance limit, its variance, and reliability contours can be expressed. The existence of an endurance limit has been assumed, not proven. With specimens declared runouts if they survive to 10^7 cycles, one can be fooled by the "knee" of a nonferrous material which exhibits no endurance limit.

13.1.4 Coaxing

It is intuitively appealing to think that more information is given by a failed specimen than by a censored specimen. In the preceding methods, many of the specimens were unfailed (commonly called *runouts*). Postulating the existence of an endurance limit and no damage occurring for cycles endured at stress levels less than the endurance limit, a method exists that raises the stress level of unfailed (by, say, 10^7 cycles) specimens to the next higher stress level and tests to failure starting the cycle count again. Since every specimen fails, the specimen set is smaller. The results are interpreted as a normal stress distribution. The method's assumption that a runout specimen is neither damaged nor strengthened complicates the results, since there is evidence that the endurance limit can be enhanced by such coaxing [13.1].

13.1.5 Prot Method[†]

This method involves steadily increasing the stress level with every cycle. Its advantage is reduction in number of specimens; its disadvantage is the introduction of (1) coaxing, (2) an empirical equation, that is,

$$S_\alpha = S_e' + K\alpha^n \tag{13.3}$$

[†] See Ref. [13.2].

where S_α = Prot failure stress at loading rate, α psi/cycle
S'_e = material endurance limit
K, n = material constants
α = loading rate, psi/cycle

and (3) an extrapolation procedure. More detail is available in Collins [13.3].

13.1.6   Up-Down Method[†]

The up-down method of testing is a common scheme for reducing R. R. Moore data to an estimate of the endurance limit. It is adaptable to seeking endurance strength at any arbitrary number of cycles. Figure 13.1 shows the data from 54 specimens

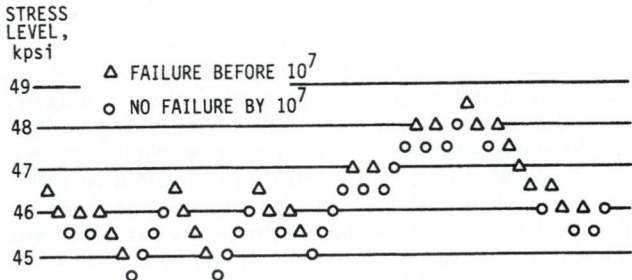

FIGURE 13.1   An up-down fatigue test conducted on 54 specimens. *(From Ransom [13.5], with permission.)*

gathered for determining the endurance strength at 10^7 cycles. The step size was 0.5 kpsi. The first specimen at a stress level of 46.5 kpsi failed before reaching 10^7 cycles, and so the next lower stress level of 46.0 kpsi was used on the subsequent specimen. It also failed before 10^7 cycles. The third specimen, at 45.5 kpsi, survived 10^7 cycles, and so the stress level was increased. The data-reduction procedure eliminates specimens until the first runout-fail pair is encountered. We eliminate the first specimen and add as an observation the next (no. 55) specimen, $\sigma = 46.5$ kpsi. The second step is to identify the least-frequent event—failures or runouts. Since there are 27 failures and 27 runouts, we arbitrarily choose failures and tabulate N_i, iN_i, and i^2N_i as shown in Table 13.1. We define $A = \Sigma iN_i$ and $B = \Sigma i^2N_i$. The estimate of the mean of the 10^7-cycle strength is

$$\hat{\mu} = S_0 + d\left(\frac{A}{\Sigma N_i} \pm \frac{1}{2}\right) \tag{13.4}$$

where S_0 = the lowest stress level on which the less frequent event occurs, d = the stress-level increment or step, and N_i = the number of less frequent events at stress level σ_i. Use $+\frac{1}{2}$ if the less frequent event is runout and $-\frac{1}{2}$ if it is failure. The estimate of the mean 10^7-cycle strength is

---

[†] See Refs. [13.4] and [13.5].

TABLE 13.1 Extension of Up-Down Fatigue Data

Stress level, kpsi	Coded level	Class failures		
		N_i	iN_i	i^2N_i
48.5	7	1	7	49
48.0	6	4	24	144
47.5	5	1	5	25
47.0	4	3	12	48
46.5	3	5	15	45
46.0	2	8	16	32
45.5	1	3	3	3
45.0	0	2	0	0
	Σ	27	82	346

$$\hat{\mu} = 45.0 + 0.5 \left(\frac{82}{27} - \frac{1}{2} \right) = 46.27 \text{ kpsi}$$

The standard deviation is

$$\hat{\sigma} = 1.620d \left[\frac{B\Sigma N_i - A^2}{(\Sigma N_i)^2} + 0.029 \right] \tag{13.5}$$

as long as $(B\Sigma N_i - A^2)/(\Sigma N_i)^2 \geq 0.3$. Substituting test data into Eq. (13.5) gives

$$\hat{\sigma} = 1.620(0.5) \left[\frac{342(27) - 82^2}{27^2} + 0.029 \right] = 2.93 \text{ kpsi}$$

The result of the up-down test can be expressed as $(S_f')_{10^7}(\hat{\mu}, \hat{\sigma})$ or $(S_f')_{10^7}(46.27, 2.93)$. Consult Refs. [13.3] and [13.4] for modification of the usual t-statistic method of placing a confidence interval on μ and Ref. [13.4] for placing a confidence interval on σ. A point estimate of the coefficient of variation is $\sigma/\mu = 2.93/46.27$, or 0.063. Coefficients of variation larger than 0.1 have been observed in steels. One must examine the sources of tables that display a single value for an endurance strength to discover whether the mean or the smallest value in a sample is being reported. This can also reveal the coefficient of variation. This is still not enough information upon which a designer can act.

13.2 SN DIAGRAM FOR SINUSOIDAL AND RANDOM LOADING

The usual presentation of R. R. Moore testing results is on a plot of S_f' (or S_f'/S_u) versus N, commonly on log-log coordinates because segments of the locus appear to be rectified. Figure 13.2 is a common example. Because of the dispersion in results, sometimes a $\pm 3\sigma$ band is placed about the median locus or (preferably) the data points are shown as in Fig. 13.3. In any presentation of this sort, the only things that might be true are the observations. All lines or loci are curve fits of convenience, there being no theory to suggest a rational form. What will endure is the data and not

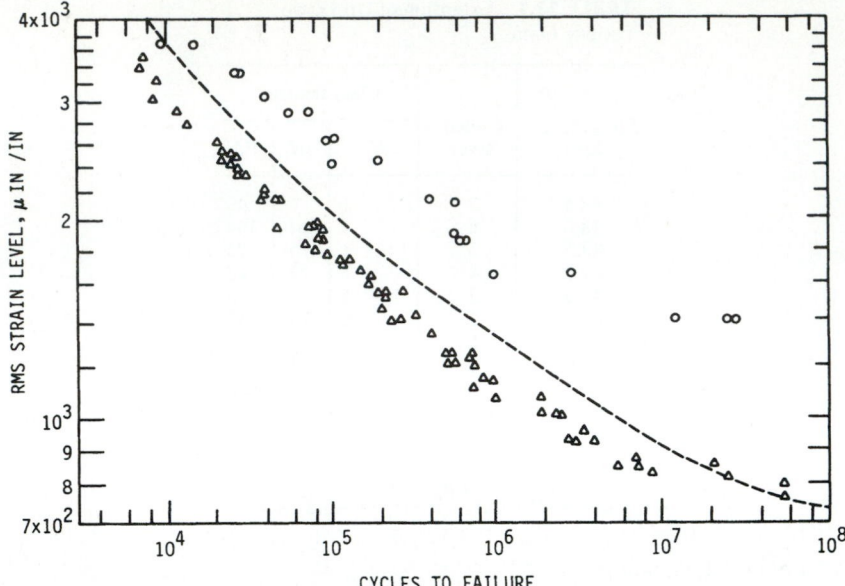

FIGURE 13.2 Fatigue data on 2024-T3 aluminum alloy for narrow-band random loading, △, and for constant-amplitude loading, ○. *(Adapted with permission from Haugen [13.14], p. 339.)*

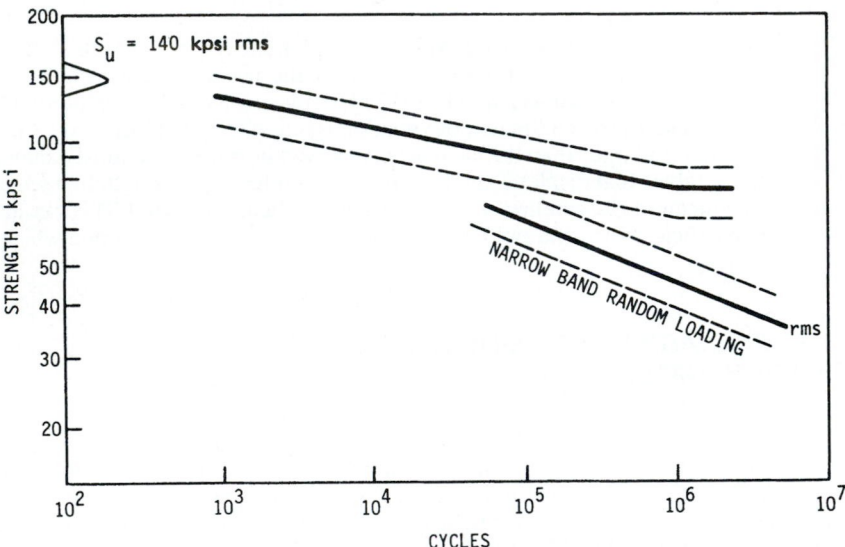

FIGURE 13.3 Statistical SN diagram for constant-amplitude and narrow-band random loading for a low-alloy steel. Note the absence of a "knee" in the random loading.

the loci. Unfortunately, too much reported work is presented without data; hence early effort is largely lost as we learn more.

The R. R. Moore test is a sinusoidal completely reversed flexural loading, which is typical of much rotating machinery, but not of other forms of fatigue. Narrow-band random loading (zero mean) exhibits a lower strength than constant-amplitude sine-wave loading. Figure 13.3 is an example of a distributional presentation, and Fig. 13.2 shows the difference between sinusoidal and random-loading strength.

13.3 FATIGUE-STRENGTH MODIFICATION FACTORS

The results of endurance testing are often made available to the designer in a concise form by metals suppliers and company materials sections. Plotting coordinates are chosen so that it is just as easy to enter the plot with maximum-stress, minimum-stress information as steady and alternating stresses. The failure contours are indexed from about 10^3 cycles up to about 10^9 cycles. Figures 13.4, 13.5, and 13.6 are

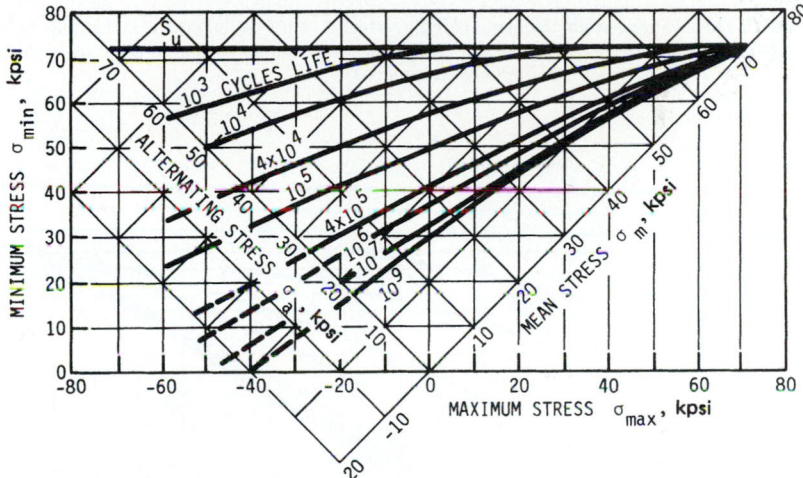

FIGURE 13.4　Fatigue-strength diagram for 2024-T3, 2024-T4, and 2014-T6 aluminum alloys, axial loading. Average of test data for polished specimens (unclad) from rolled and drawn sheet and bar. Static properties for 2024: $S_u = 72$ kpsi, $S_y = 52$ kpsi; for 2014: $S_u = 72$ kpsi, $S_y = 63$ kpsi. *(Grumman Aerospace Corp.)*

examples. The usual testing basis is bending fatigue, zero mean, constant amplitude. Figure 13.6 represents axial fatigue. The problem for the designer is how to adjust this information to account for all the discrepancies between the R. R. Moore specimen and the contemplated machine part. The Marin approach [13.6] is to introduce multiplicative modification factors to adjust the endurance limit, in the deterministic form

$$S_e = k_a k_b k_c k_d k_e S_e'$$
(13.6)

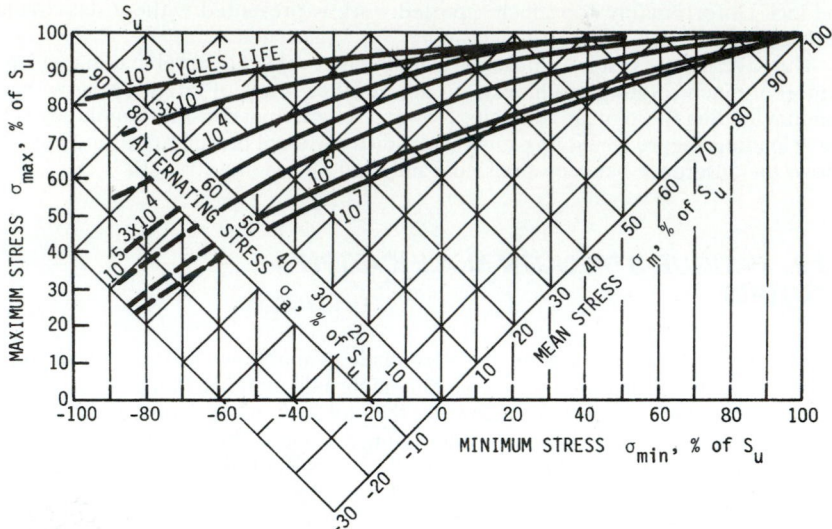

FIGURE 13.5 Fatigue-strength diagram for alloy steel, S_u = 125 to 180 kpsi, axial loading. Average of test data for polished specimens of AISI 4340 steel (also applicable to other alloy steels, such as AISI 2330, 4130, 8630). *(Grumman Aerospace Corp.)*

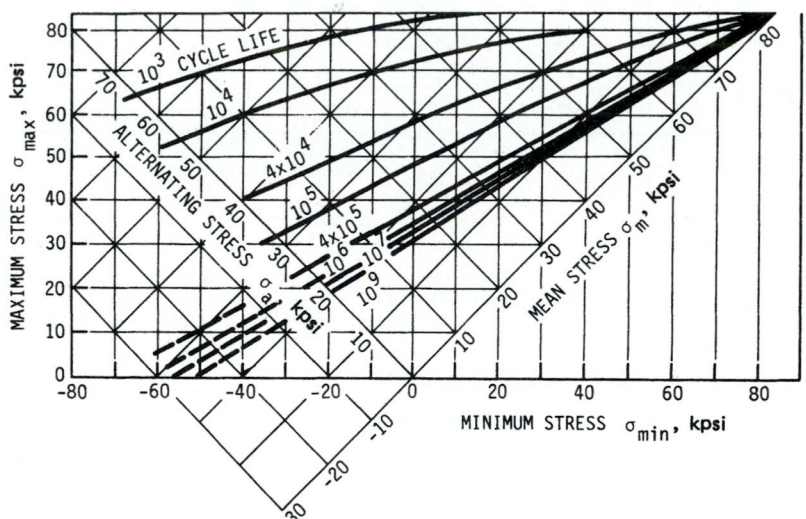

FIGURE 13.6 Fatigue-strength diagram for 7075-T6 aluminum alloy, axial loading. Average of test data for polished specimens (unclad) from rolled and drawn sheet and bar. Static properties: S_u = 82 kpsi, S_y = 75 kpsi. *(Grumman Aerospace Corp.)*

where k_a = surface condition modification factor
k_b = size modification factor
k_c = loading modification factor
k_d = temperature modification factor
k_e = miscellaneous-effects modification factor
S_e' = endurance limit of rotating-beam specimen
S_e = endurance limit at critical location of part in the geometry and condition of use

The stochastic Marin equation is expressed as

$$\mathbf{S}_e = \mathbf{k}_a k_b \mathbf{k}_c \mathbf{k}_d \mathbf{k}_e S_e' \tag{13.7}$$

where $\mathbf{k}_a \sim LN(\mu_{ka}, \sigma_{ka})$, k_b is deterministic, $\mathbf{k}_c \sim LN(\mu_{kc}, \sigma_{kc})$, $\mathbf{k}_d \sim LN(\mu_{kd}, \sigma_{kd})$, $\mathbf{k}_e$ distribution depends on the effect considered, $\mathbf{S}_e' \sim LN(\mu_{S_e'}, \sigma_{S_e'})$ and $\mathbf{S}_e \sim LN(\mu_{Se}, \sigma_{Se})$ by the central limit theorem of statistics. Where endurance tests are not available, estimates of R. R. Moore endurance limits are made by correlation of the mean ultimate tensile strength to the endurance limit through the fatigue ratio $\boldsymbol{\phi}$:

$$\mathbf{S}_e' = \boldsymbol{\phi} \bar{S}_{ut} \tag{13.8}$$

The fatigue ratios of Gough are shown in Fig. 13.7. He reports the coefficients of variation as for all metals, 0.23; for nonferrous metals, 0.20; for iron and carbon steels, 0.14, for low-alloy steels, 0.13, and for special-alloy steels, 0.13. Since the materials involved were of metallurgical interest, there are members in his ensembles other than engineering materials. Nevertheless, it is clear that knowledge of the mean of $\boldsymbol{\phi}$ is not particularly useful without the coefficient of variation. Coefficients of variation of engineering steel endurance limits may range from 0.03 to 0.10 individually, and the coefficient of variation of the fatigue ratio of the ensemble about 0.15. For more detail, see Sec. 13.3.7.

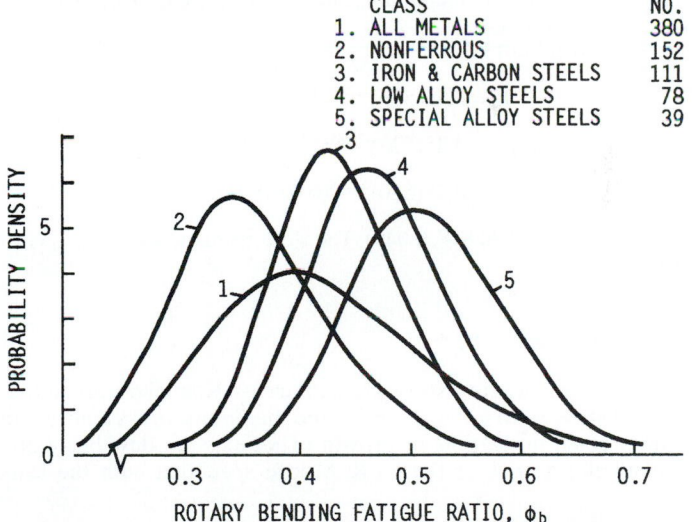

CLASS	NO.
1. ALL METALS	380
2. NONFERROUS	152
3. IRON & CARBON STEELS	111
4. LOW ALLOY STEELS	78
5. SPECIAL ALLOY STEELS	39

FIGURE 13.7   The lognormal probability density function of the fatigue ratio ϕ_b of Gough.

13.3.1 Marin Surface Factor k_a

The Marin surface condition modification factor for steels may be expressed in the form

$$\mathbf{k}_a = a\bar{S}_{ut}^b(1, C) \tag{13.9}$$

The mean and standard deviation are

$$\mu_{ka} = a\bar{S}_{ut}^b \qquad \sigma_{k_a} = C\mu_{k_a} \tag{13.10}$$

Table 13.2 gives values of a, b, and C for various surface conditions. See also Fig. 13.8.

TABLE 13.2   Parameters of Marin Surface Condition Factor

$$\mathbf{k}_a = a\bar{S}_{ut}^b(1, C)$$

Surface finish	a		b	Coefficient of variation C
	kpsi	MPa		
Ground[†]	1.34	1.58	−0.086	0.120
Machined, Cold-rolled	2.67	4.45	−0.265	0.058
Hot-rolled	14.5	58.1	−0.719	0.110
As-forged	39.8	271	−0.995	0.146

[†] Because of scatter in ground surface data, an alternative function is the stochastic constant 0.878(1, 0.120).

Source: Data from C. G. Noll and C. Lipson, "Allowable Working Stresses," *Society of Experimental Stress Analysis,* vol. 3, no. 2, 1946, p. 49, reduced from their graphed data points.

Example 1.   A steel has a mean ultimate tensile strength of 520 MPa and a machined surface. Estimate k_a.
   Solution: From Table 13.2,

$$\mathbf{k}_a = 4.45(520)^{-0.265}(1, 0.058)$$

$$\mu_{k_a} = 4.45(520)^{-0.265}(1) = 0.848$$

$$\sigma_{k_a} = 4.45(520)^{-0.265}(0.058) = 0.049$$

The distribution is $\mathbf{k}_a \sim LN(0.848, 0.049)$. The deterministic value of k_a is simply the mean, 0.848.

13.3.2 Marin Size Factor k_b

In bending and torsion, where a stress gradient exists, Kuguel observed that the volume of material stressed to 0.95 or more of the maximum stress controls the risk of encountering a crack nucleation, or growth of an existing flaw becoming critical. The equivalent diameter d_e of the R. R. Moore specimen with the same failure risk is

$$d_e = \sqrt{\frac{A_{0.95}}{0.076\,576}} \tag{13.11}$$

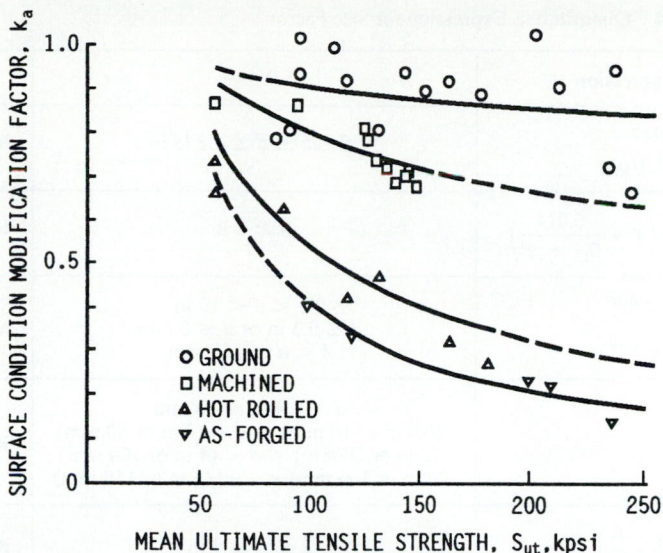

FIGURE 13.8 Marin endurance limit fatigue modification factor k_a for various surface conditions of steels. See also Table 13.2.

where $A_{0.95}$ is the cross-sectional area exposed to 95 percent or more of the maximum stress. For a round in rotating bending or torsion, $A_{0.95} = 0.075\ 575d^2$. For a round in nonrotating bending, $A_{0.95} = 0.010\ 462d^2$. For a rectangular section $b \times h$ in bending, $A_{0.95} = 0.05bh$. See [13.6], p. 284 for channels and I-beams in bending. Table 13.3 gives useful relations. In bending and torsion,

$$k_b = \begin{cases} (d_e/0.30)^{-0.107} = 0.879d_e^{-0.107} & d_e \text{ in inches} & (13.12) \\ (d_e/7.62)^{-0.107} = 1.24d_e^{-0.107} & d_e \text{ in mm} & (13.13) \end{cases}$$

For axial loading, $k_b = 1$. Table 13.4 gives various expressions for k_b. The Marin size factor is scalar (deterministic). At less than standard specimen diameter (0.30 in), many engineers set $k_b = 1$.

13.3.3 Marin Loading Factor k_c

The Marin loading factor k_c can be expressed as

$$k_c = \alpha \bar{S}_{ut}^\beta(1, C) \qquad (13.14)$$

TABLE 13.3 Equivalent Diameters for Size Factor

Section	Equivalent diameter d_e
Round, rotary bending, torsion	d
Round, nonrotating bending	$0.37d$
Rectangle, nonrotating bending	$0.808bh$

TABLE 13.4   Quantitative Expression for Size Factor

Expression	Range	Proposer
$k_b = \dfrac{0.947}{1 - 0.016/d}$	$0.125 \leq d \leq 1.875$ in	Moore
$k_b = 0.931 \left(1 + \dfrac{0.014}{0.1 + d^2}\right)$	$d \geq 2$ in	Heywood
$k_b = \begin{cases} 0.869d^{-0.097} \\ 1 \\ 1.189d^{-0.97} \end{cases}$	$0.3 < d < 10$ in $d \leq 0.3$ in or $d \leq 8$ mm $8 < d \leq 250$ mm	Shigley and Mitchell
$k_b = \begin{cases} 1 \\ 0.9 \\ 0.8 \\ 0.7 \end{cases}$	$d < 0.4$ in or 10 mm (0.4 in or 10 mm) $< d <$ (2 in or 50 mm) (2 in or 50 mm) $< d <$ (4 in or 100 mm) (4 in or 100 mm) $< d <$ (5 in or 150 mm)	Juvinall
$k_b = 1 - \dfrac{d - 0.3}{15}$	$2 \leq d \leq 9$ in	Roark

Table 13.5 gives values for α, β, and C. The distribution of $\mathbf{k}_c$ is lognormal. For axial loading of steel based on Table 13.5, Table 13.6 was prepared. Juvinall [13.12] reports that for steel, $0.75 < \bar{k}_c < 1.0$, and suggests using $k_c = 0.90$ for accurate control of loading eccentricity and $0.60 < k_c < 0.85$ otherwise. The problem of load eccentricity plagues testing as well as the designer. Axial loading in fatigue requires caution. As shown in Fig. 13.9 $\bar{k}_c$ is involved in finite life fatigue also.

For torsion, Table 13.7 summarizes experimental experience. In metals described by distortion-energy failure theory, the average value of k_c would be 0.577. The distribution of $\mathbf{k}_c$ is lognormal.

13.3.4 Marin Temperature Factor k_d

Steels exhibit an increase in endurance limit when temperatures depart from ambient. For temperatures up to about 600°F, such an increase is often seen, but above 600°F, rapid deterioration occurs. See Fig. 13.8. If specific material endurance

TABLE 13.5   Parameters of Marin Loading Factor

$$\mathbf{k}_c = \alpha \bar{S}_{ut}^{\beta}(1, C)$$

Mode of loading	α		β	Coefficient of variation C
	kpsi	MPa		
Bending	1	1	0	0
Axial	1.23	1.43	−0.078	0.126
Torsion	0.328	0.258	0.125	0.125

TABLE 13.6 Marin Loading Factor for Axial Loading

S_{ut}, kpsi	$\bar{k}_c$[†]
50	0.907
100	0.859
150	0.822
200	0.814

[†] Average entry is 0.853.

TABLE 13.7 Torsional Loading, k_c

Material	Range k_c	$(k_c)_{avg}$
Wrought steels	0.52–0.69	0.60
Wrought Al	0.43–0.74	0.55
Wrought Cu and alloys	0.41–0.67	0.56
Wrought Mg and alloys	0.49–0.61	0.54
Titaniums	0.37–0.57	0.48
Cast irons	0.79–1.01	0.90
Cast Al, Mg, and alloys	0.71–0.91	0.85

limit–temperature testing is not available, then an ensemble of 21 carbon and alloy steels gives, for t_f in °F,

$$\mathbf{k}_d = [0.975 + (0.432 \times 10^{-3})t_f - (0.115 \times 10^{-5})t_f^2$$
$$+ (0.104 \times 10^{-8})t_f^3 - (0.595 \times 10^{-12})t_f^4](1, 0.11)$$
$$70 < t_f < 600°F \tag{13.15}$$

which equation may be useful. The distribution of k_d is lognormal. See Fig. 13.10 for some specific materials.

13.3.5 Stress Concentration and Notch Sensitivity

The modified Neuber equation (after Heywood) is

$$\mathbf{K}_f = \frac{K_t(1, C_{K_f})}{1 + \dfrac{2}{\sqrt{r}} \dfrac{K_t - 1}{K_t} \sqrt{a}} \tag{13.16}$$

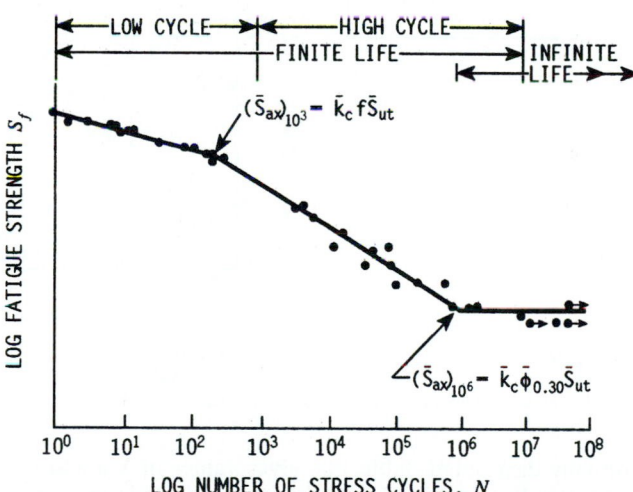

FIGURE 13.9 An S-N diagram plotted from the results of completely reversed axial fatigue tests on a normalized 4130 steel, $S_{ut} = 116$ kpsi (data from NACA *Tech. Note* 3866, Dec. 1966).

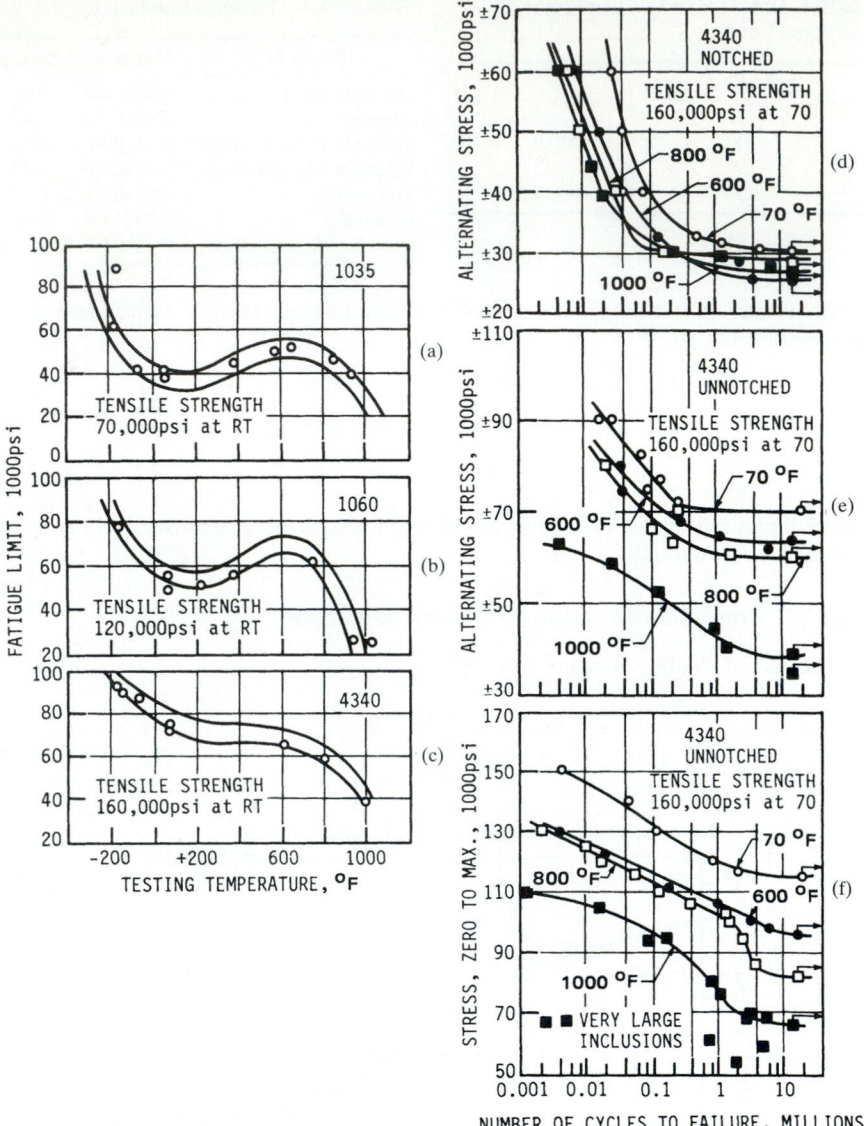

FIGURE 13.10   Effect of temperature on the fatigue limits of three steels: (*a*) 1035; (*b*) 1060; (*c*) 4340; (*d*) 4340, completely reversed loading $K = 3$ to 3.5; (*e*) 4340, completely reversed loading, unnotched; (*f*) 4340, repeatedly applied tension. (*American Society for Metals.*)

and is lognormally distributed. Table 13.8 gives values of $\sqrt{a}$ and C_K. The stress-concentration factor $\mathbf{K}_f$ may be applied to the nominal stress $\boldsymbol{\sigma}_0$ as $\mathbf{K}_f\boldsymbol{\sigma}_0$ as an augmentation of stress (preferred) or as a strength reduction factor $\mathbf{k}_e = 1/\mathbf{K}_f$ (sometimes convenient). Both $\mathbf{K}_f$ and $\mathbf{k}_e$ are lognormal with the same coefficient of variation. For stochastic methods, Eq. (13.16) is preferred, as it has a much larger statistical base than notch sensitivity.

TABLE 13.8 Heywood's Parameters, Eq. (13.16), Stress Concentration

Feature	$\sqrt{a}$				C_{Kf}	
	Using $\overline{S}_{ut}$, kpsi	Using $\overline{S}'_e$, kpsi	Using $\overline{S}_{ut}$, MPa	Using $\overline{S}'_e$, MPa	Using $\overline{S}_{ut}$	Using $\overline{S}'_e$
Transverse hole	$5/\overline{S}_{ut}$	$2.5/\overline{S}'_e$	$174/\overline{S}_{ut}$	$87/\overline{S}'_e$	0.10	0.11
Shoulder	$4/\overline{S}_{ut}$	$2/\overline{S}'_e$	$139/\overline{S}_{ut}$	$69.5/\overline{S}'_e$	0.11	0.08
Groove	$3/\overline{S}_{ut}$	$1.5/\overline{S}'_e$	$104/\overline{S}_{ut}$	$52/\overline{S}'_e$	0.15	0.13

The finite life stress-concentration factor for steel for N cycles is obtained from the notch sensitivities $(q)_{10^3}$ and $(q)_{10^6}$. For 10^3 cycles,

$$(q)_{10^3} = \frac{(K_f)_{10^3} - 1}{K_t - 1}$$

$$= -0.18 + (0.43 \times 10^{-2})\overline{S}_{ut} - (0.45 \times 10^{-5})\overline{S}_{ut}^2 \tag{13.17}$$

where $\overline{S}_{ut} < 330$ kpsi. For 10^6 cycles,

$$(q)_{10^6} = \frac{(K_f)_{10^6} - 1}{K_t - 1} \tag{13.18}$$

and for N cycles,

$$(K_f)_N = (K_f)_{10^3} \left[\frac{(K_f)_{10^6}}{(K_f)_{10^3}} \right]^{(1/3 \log N - 1)} \tag{13.19}$$

There is some evidence that $\overline{K}_f$ does not exceed 4 for Q&T steels and 3 for annealed steels [13.11]. Figure 13.11 shows scatter bands, and Fig. 13.12 relates notch radius to notch sensitivity.

13.3.6 Miscellaneous-Effects Modification Factor k_e

There are other effects in addition to surface texture, size loading, and temperature that influence fatigue strength. These other effects are grouped together because their influences are not always present, and are not well understood quantitatively in any comprehensive way. They are largely detrimental ($\overline{k}_e < 1$), and consequently cannot be ignored. For each effect present, the designer must make an estimate of the magnitude and probable uncertainty of $\mathbf{k}_e$. Such effects include

- Introduction of complete stress fields due to press or shrink fits, hydraulic pressure, and the like
- Unintentional residual stresses that result from grinding or heat treatment and intentional residual stresses that result from shot peening or rolling of fillets
- Unintended coatings, usually corrosion products, and intentional coatings, such as plating, paint, and chemical sheaths
- Case hardening for wear resistance by processes such as carborization, nitriding, tuftriding, and flame and induction hardening
- Decarborizing of surface material during processing

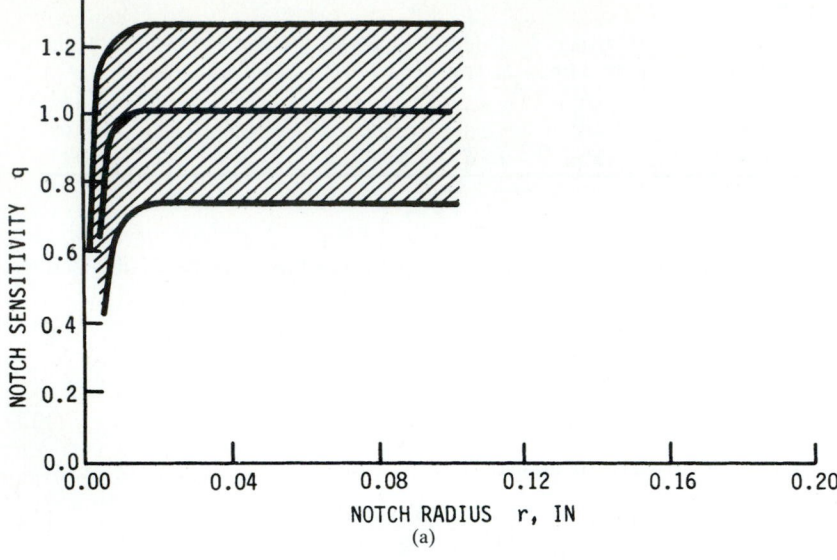

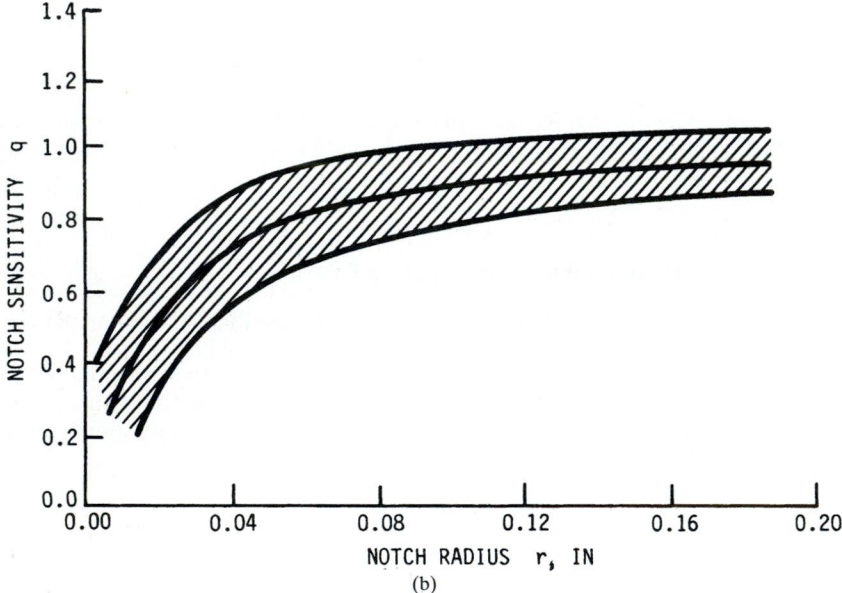

FIGURE 13.11 Scatterbands of notch sensitivity q as a function of notch radius and heat treatment for undifferentiated steels. (*a*) Quenched and tempered; (*b*) normalized or annealed. (*Adapted from Sines and Waisman [13.16], with permission of McGraw-Hill, Inc.*)

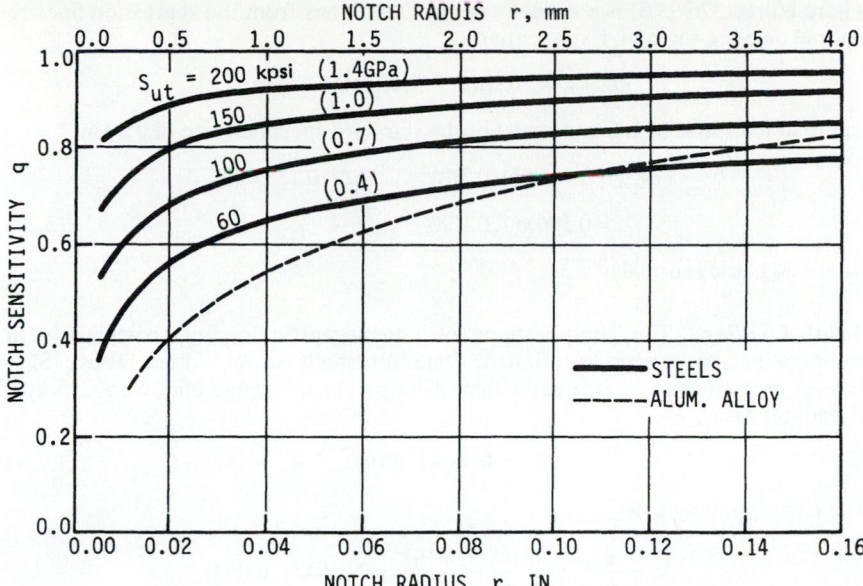

FIGURE 13.12   Notch-sensitivity chart for steels and UNS A92024T wrought aluminum alloys subjected to reversed bending or reversed axial loads. For larger notch radii, use values of q corresponding to $r = 0.16$ in (4 mm). *(From Sines and Waisman [13.16], with permission of McGraw-Hill, Inc.)*

When these effects are present, tests are needed in order to assess the extent of such influences and to form a rational basis for assignment of a fatigue modification factor k_e.

13.3.7   Correlation Method

The scalar fatigue ratio $\overline{\phi}$ is defined as the mean endurance limit divided by the mean ultimate tensile strength. The stochastic fatigue ratio ϕ is defined as $S_e'/\overline{S}_{ut}$. Engineers can estimate endurance limit by multiplying a random variable ϕ by the mean ultimate tensile strength. Since designers have an interest in bending, axial (push-pull), and torsional fatigue, the appropriate endurance limit is correlated to the mean ultimate tensile strength as follows:

$$S_e' = \phi_b \overline{S}_{ut} \tag{13.20}$$

$$S_{ax}' = \phi_{ax} \overline{S}_{ut} \tag{13.21}$$

$$S_{se}' = \phi_t \overline{S}_{ut} \tag{13.22}$$

Rotating Bending.   Data for this mode of loading exist for various specimen diameters, and so the size effect is mixed in. With data in the form of S_{ut}, S_e', and specimen diameter d from 133 full-scale R. R. Moore tests, plotting $\ln \phi$ versus $\ln d$ leads to the regression equation

$$\ln \phi_b = -0.819\ 005\ 7 + \ln d^{-0.106\ 951} + \xi(0, 0.137\ 393\ 3)$$

where $\xi(0, 0.137\,393\,3)$ is a normal variate. Deviations from the regression line are normal on a log-log plot. Exponentiating,

$$\boldsymbol{\phi}_b = 0.445\,031 d^{-0.106\,951}\,\boldsymbol{\lambda}(1, 0.138\,050)$$

where $\boldsymbol{\lambda}(1, 0.138\,050)$ is lognormal. For the standard specimen size of 0.30 in,

$$\boldsymbol{\phi}_{0.30} = 0.445\,031 (0.30)^{-0.106\,951}\,\boldsymbol{\lambda}(1, 0.138\,050)$$

$$= 0.506 \boldsymbol{\lambda}(1, 0.138) \tag{13.23}$$

where $\boldsymbol{\phi}_{0.30}$ is lognormal.

Axial Loading. The complications of nonconcentric loading, particularly in compression, have been mentioned. Data for steels show a knee above $\bar{S}'_{ax} = 106.7(1, 0.089)$ when $\bar{S}_{ut}$ is greater than 220 kpsi. In the range $60 \le \bar{S}_{ut} \le 213$ kpsi (Landgraf data),

$$\mathbf{S}'_{ax} = 0.465(1, 0.19)\bar{S}_{ut} \tag{13.24}$$

The corresponding $\mathbf{k}_c$ is

$$\mathbf{k}_c = \frac{\boldsymbol{\phi}_{ax}}{\boldsymbol{\phi}_{0.30}} = \frac{0.465(1, 0.19)}{0.506(1, 0.138)} = 0.919\boldsymbol{\lambda}(1, 0.052)$$

The random variable $\boldsymbol{\lambda}(1, 0.052)$ is lognormal. A larger data base combining data from Landgraf and Grover, et al., gives

$$\boldsymbol{\phi}_{ax} = 0.623\bar{S}_{ut}^{-0.0778}\boldsymbol{\lambda}(1, 0.264) \tag{13.25}$$

where $\boldsymbol{\phi}_{ax}$ is lognormal. Note the mild influence of tensile strength. Now $\mathbf{k}_c$ is

$$\mathbf{k}_c = \frac{\boldsymbol{\phi}_{ax}}{\boldsymbol{\phi}_{0.30}} = \frac{0.623 S_{ut}^{-0.0778}\boldsymbol{\lambda}(1, 0.264)}{0.506\boldsymbol{\lambda}(1, 0.138)}$$

$$= 1.23 S_{ut}^{-0.0778}\boldsymbol{\lambda}(1, 0.126) \tag{13.26}$$

where $\mathbf{k}_c$ is lognormally distributed. This value of $\mathbf{k}_c$ is reported in Table 13.5 because of the larger data base.

Torsional Loading. From distortion energy theory, a Marin torsional loading modification factor is to be expected. From steel data from Grover et al.,

$$\boldsymbol{\phi}_t = \frac{\mathbf{S}'_{se}}{\mathbf{S}_{ut}} = 0.294\boldsymbol{\lambda}(1, 0.270)$$

where $\boldsymbol{\phi}_t$ is lognormal. The loading factor is

$$\mathbf{k}_c = \frac{\boldsymbol{\phi}_t}{\boldsymbol{\phi}_{0.30}} = \frac{0.294\boldsymbol{\lambda}(1, 0.270)}{0.506\boldsymbol{\lambda}(1, 0.138)} = 0.581\boldsymbol{\lambda}(1, 0.132)$$

where $\mathbf{k}_c$ is lognormal. Reducing data to reveal any material strength influence gives

$$\boldsymbol{\phi}_t = 0.166\bar{S}_{ut}^{0.125}\boldsymbol{\lambda}(1, 0.263) \tag{13.27}$$

where $\lambda(1, 0.263)$ is lognormal. It follows that

$$\mathbf{k}_c = \frac{\phi_t}{\phi_{0.30}} = \frac{0.166\overline{S}_{ut}^{0.125}\lambda(1, 0.263)}{0.506\lambda(1, 0.138)} = 0.328\overline{S}_{ut}^{0.125}\lambda(1, 0.125) \qquad (13.28)$$

where $\mathbf{k}_c$ is lognormal. This is also reported in Table 13.5. Tables 13.9 and 13.10 are useful summaries of fatigue equation details.

TABLE 13.9 Summary of Fatigue Equation $\mathbf{S}_e = \mathbf{k}_a\mathbf{k}_b\mathbf{k}_c\mathbf{k}_d\mathbf{k}_e\phi_{0.30}\overline{S}_{ut}$, Customary Engineering Units[†]

Quantity	Relation[‡]	Equation or table no.
Ultimate strength	$\overline{S}_{ut} = 0.5\overline{H}_B$	
Fatigue ratio	$\phi_{0.30} = 0.506\lambda(1, 0.138)$	Eq. (13.23)
Endurance limit		
Bending	$\mathbf{S}'_e = \phi_{0.30}\overline{S}_{ut} = 0.506\overline{S}_{ut}\lambda(1, 0.138)$	Eq. (13.8)
Axial	$(\mathbf{S}'_e)_{ax} = 0.623\overline{S}_{ut}^{-0.0778}\lambda(1, 0.264)$	Eq. (13.25)
Torsion	$\mathbf{S}'_{se} = 0.166\overline{S}_{ut}^{0.125}\lambda(1, 0.263)$	Eq. (13.27)
Surface factor		
Ground	$\mathbf{k}_a = 1.34\overline{S}_{ut}^{-0.086}\lambda(1, 0.120)$	Table 13.2
Machined	$\mathbf{k}_a = 2.67\overline{S}_{ut}^{-0.265}\lambda(1, 0.058)$	Table 13.2
Hot-rolled	$\mathbf{k}_a = 14.5\overline{S}_{ut}^{-0.719}\lambda(1, 0.110)$	Table 13.2
As-forged	$\mathbf{k}_a = 39.8\overline{S}_{ut}^{-0.995}\lambda(1, 0.146)$	Table 13.2
Size factor	$k_b = (d_e/0.30)^{-0.107} = 0.879d_e^{-0.107}$	Eq. (13.12)
Loading factor		
Bending	$\mathbf{k}_c = \lambda(1, 0)$	Table 13.5
Axial	$\mathbf{k}_c = 1.23\overline{S}_{ut}^{-0.078}\lambda(1, 0.126)$	Table 13.5
Torsion	$\mathbf{k}_c = 0.328\overline{S}_{ut}^{0.125}\lambda(1, 0.125)$	Table 13.5
Temperature factor	(as appropriate)	Eq. (13.15)
Miscellaneous factor	(as appropriate)	
Stress concentration	$\mathbf{K}_f = \dfrac{K_t\lambda(1, C_{K_f})}{1 + \dfrac{2}{\sqrt{r}}\dfrac{K_t - 1}{K_t}\sqrt{a}}$	Eq. (13.16)
Transverse hole	$a = 5/\overline{S}_{ut},\ C_{K_f} = 0.10$	Table 13.8
Shoulder	$a = 4/\overline{S}_{ut},\ C_{K_f} = 0.11$	Table 13.8
Groove	$a = 3/\overline{S}_{ut},\ C_{K_f} = 0.15$	Table 13.8

[†] Units: strengths, kpsi; diameter or radius, in; $\sqrt{a}$, in$^{1/2}$.
[‡] Deterministic values are simply the means, obtained by substituting unity for the lognormal variate λ.

Cast iron in torsion, when behavior is described by the maximum principal stress theory, has $k_c = 1$, and when behavior is described by the maximum principal strain theory, exhibits

$$k_c = \frac{1}{1 + \nu} = \frac{1}{1 + 0.211} = 0.826 \qquad (13.29)$$

TABLE 13.10 Summary of Fatigue Equation $\mathbf{S}_e = \mathbf{k}_a\mathbf{k}_b\mathbf{k}_c\mathbf{k}_d\mathbf{k}_e\phi_{0.30}\overline{S}_{ut}$, SI Units[†]

Quantity	Relation[‡]	Equation or table no.
Ultimate strength	$\overline{S}_{ut} = 3.45\overline{H}_B$	
Fatigue ratio	$\phi_{0.30} = 0.506\lambda(1, 0.138)$	Eq. (13.23)
Endurance limit		
Bending	$\mathbf{S}'_e = \phi_{0.30}\overline{S}_{ut} = 0.506\overline{S}_{ut}\lambda(1, 0.138)$	Eq. (13.8)
Axial	$(\mathbf{S}'_e)_{ax} = 0.724\overline{S}_{ut}^{-0.0778}\lambda(1, 0.264)$	Eq. (13.25)
Torsion	$\mathbf{S}'_{se} = 0.130\overline{S}_{ut}^{0.125}\lambda(1, 0.263)$	Eq. (13.27)
Surface factor		
Ground	$\mathbf{k}_a = 1.58\overline{S}_{ut}^{-0.086}\lambda(1, 0.120)$	Table 13.2
Machined	$\mathbf{k}_a = 4.45\overline{S}_{ut}^{-0.265}\lambda(1, 0.058)$	Table 13.2
Hot-rolled	$\mathbf{k}_a = 58.1\overline{S}_{ut}^{-0.719}\lambda(1, 0.110)$	Table 13.2
As-forged	$\mathbf{k}_a = 271\overline{S}_{ut}^{-0.995}\lambda(1, 0.146)$	Table 13.2
Surface factor	$k_b = (d_e/7.62)^{-0.107} = 1.24d_e^{-0.107}$	Eq. (13.13)
Loading factor		
Bending	$\mathbf{k}_c = \lambda(1, 0)$	Table 13.5
Axial	$\mathbf{k}_c = 1.43\overline{S}_{ut}^{-0.078}\lambda(1, 0.126)$	Table 13.5
Torsion	$\mathbf{k}_c = 0.258\overline{S}_{ut}^{0.125}\lambda(1, 0.125)$	Table 13.5
Temperature factor $\mathbf{k}_d$	(as appropriate)	
Miscellaneous factor $\mathbf{k}_e$	(as appropriate)	
Stress concentration	$\mathbf{K}_f = \dfrac{K_t\lambda(1, C_{Kf})}{1 + \dfrac{2}{\sqrt{r}}\dfrac{K_t-1}{K_t}\sqrt{a}}$	Eq. (13.16)
Transverse hole	$\sqrt{a} = 174/\overline{S}_{ut}, C_{Kf} = 0.10$	Table 13.8
Shoulder	$\sqrt{a} = 139/\overline{S}_{ut}, C_{Kf} = 0.11$	Table 13.8
Groove	$\sqrt{a} = 104/\overline{S}_{ut}, C_{Kf} = 0.15$	Table 13.8

[†] Units: strengths, MPa; diameter or radius, mm; $\sqrt{a}$, mm$^{1/2}$.

[‡] Deterministic values are simply means, obtained by substituting unity for the lognormal variate λ.

Many tests show the behavior of cast iron falling between these two theoretical models. Also,

$$\nu = 0.225 - 0.003\ 04\overline{S}_{ut}$$

$$S_{ut} = 5.76 + 0.179H_B \pm 5 \text{ kpsi} \qquad \text{(for gray cast iron)} \tag{13.30}$$

Example 2. Cycles to failure of 70 000 in rotary bending at 550°F is intended for a round machined steel part, 1 in in diameter, in the presence of a notch ($K_t = 2.1$) with a 0.1-in radius. The tensile strength is $\overline{S}_{ut} = 100$ kpsi. The nominal stress is $\boldsymbol{\sigma}_0 \sim LN(12, 1.2)$ kpsi. Estimate the reliability in attaining the cycles-to-failure goal.

Solution. Estimate fatigue strength $\mathbf{S}_f$.

From Eqs. (13.20) and (13.23),

$$\mathbf{S}'_e = \phi_{0.30}\overline{S}_{ut} = 0.506(1, 0.138)100$$

$$\overline{S}'_e = 0.506(1)100 = 50.6 \text{ kpsi}$$

$$C_{S'_e} = 0.138$$

From Eq. (13.12),

$$\bar{k}_a = 2.67(100)^{-0.265} = 0.788$$

$$C_{k_a} = 0.058$$

From Eq. (13.9), $k_b = 0.879(1)^{-0.107} = 0.879$

From Eq. (13.14),

$$\mathbf{k}_c = (1)100^0(1,0)$$

$$\bar{k}_c = 1 \qquad C_{k_c} = 0$$

From Eq. (13.15),

$$\mathbf{k}_d = [0.975 + (0.432 \times 10^{-3})550 - (0.115 \times 10^{-5})550^2$$

$$+ (0.104 \times 10^{-8})550^3 - (0.595 \times 10^{-12})550^4](1, 0.11)$$

$$\bar{k}_d = 0.983 \qquad C_{k_d} = 0.11$$

$$\mathbf{k}_e = (1, 0)$$

From Eq. (13.7),

$$\mathbf{S}_f = [0.788(1, 0.058)][0.879(1, 0)][0.983(1, 0.11)](1, 0)[50.6(1, 0.138)]$$

$$\bar{S}_f = [0.788(0.879)][0.983(50.6)] = 34.45 \text{ kpsi}$$

$$C_{S_f} = (0.058^2 + 0.11^2 + 0.138^2)^{1/2} = 0.186$$

$$\mathbf{S}_f \sim LN[34.45(1, 0.186)]$$

Stress:

From Eq. (13.17),

$$(q)_{10^3} = -0.18 + (0.43 \times 10^{-2})100 - (0.45 \times 10^{-5})100^2 = 0.21$$

$$(K_f)_{10^3} = 1 + (q)_{10^3}(K_t - 1) = 1 + 0.21(2.1 - 1) = 1.231$$

From Fig. 13.12,

$$(q)_{10^6} = 0.82$$

$$(K_f)_{10^6} = 1 + 0.82(2.1 - 1) = 1.902$$

From Eq. (13.19), $(K_f)_{70\,000} = 1.231 \left[\dfrac{1.902}{1.231} \right]^{(1/3 \log 70\,000 - 1)} = 1.609$

From Table 13.8, $C_{K_f} = 0.15$. Thus $\mathbf{K}_f = 1.609(1, 0.15)$.

$$\boldsymbol{\sigma} = \mathbf{K}_f \boldsymbol{\sigma}_0 \qquad C_\sigma = 1.2/12 = 0.10$$

$$\boldsymbol{\sigma} = 1.609(1, 0.15)12(1, 0.10)$$

$$\bar{\sigma} = 1.609(12) = 19.3 \text{ kpsi}$$

$$C_\sigma = (0.15^2 + 0.10^2)^{1/2} = 0.180$$

Reliability:

From Eq. (2.5), $z = -\dfrac{\left(\dfrac{34.45}{19.3}\sqrt{\dfrac{1 + 0.180^2}{1 + 0.186^2}}\right)}{\sqrt{\ln{(1 + 0.186^2)(1 + 0.180^2)}}} = -2.25$

From Eq. (2.4), $R = 1 - \Phi(-2.25) = 1 - 0.0122 = 0.988$

A deterministic analysis would produce all the mean values above and a factor of safety η of

$$\eta = \frac{\overline{S}_f}{\overline{\sigma}} = \frac{34.45}{19.3} = 1.78$$

with no feel for the chance of survival to 70 000 cycles.

13.4 FLUCTUATING STRESS

Variable loading is often characterized by an amplitude component σ_a as ordinate and a steady component σ_m as abscissa. Defined in terms of maximum stress σ_{max} and minimum stress σ_{min} the coordinates are as follows:

$$\sigma_a = \frac{1}{2}\,|\sigma_{max} - \sigma_{min}| \qquad \sigma_m = \frac{1}{2}(\sigma_{max} + \sigma_{min}) \tag{13.31}$$

The designer's fatigue diagram is depicted in Fig. 13.13.

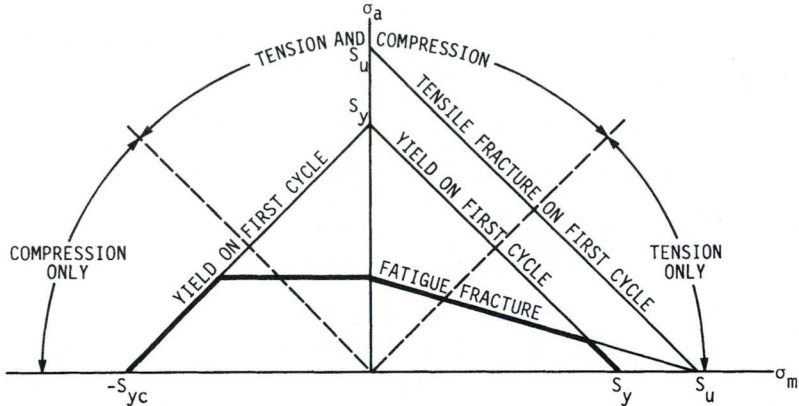

FIGURE 13.13   Designer's fatigue diagram using a Goodman failure locus for push-pull (axial) fatigue.

If one plots combinations σ_a and σ_m that partition part survival from part failure, a failure locus is defined. Properties such as ultimate tensile strength $\overline{S}_{ut}$, yield strength in tension $\overline{S}_y$, yield strength in compression $\overline{S}_{yc}$, endurance strength $\overline{S}_e$, or fatigue strength $\overline{S}_f$ appear as axis intercepts. In the range in which σ_m is negative and fatigue fracture is a threat, the failure locus is taken as a horizontal line unless specific experimental data allow another locus. In the range in which σ_m is positive, the preferred locus is Gerber or ASME-Elliptic. The Goodman and Soderberg loci have historical importance and algebraic simplicity, but these loci do not fall centrally among the data, and they are not dependably conservative.

In general, the factor of safety **n** is the quotient **S/σ**, Fig. 13.14. The coefficient of variation of **n**, C_n, is well estimated by

$$C_n = (C_S^2 + C_\sigma^2)^{1/2} \tag{13.32}$$

where C_S is the coefficient of variation of the significant strength and C_σ is the coefficient of variation of the load-induced stress. For a radial load line from the origin on the designer's fatigue diagram, either $\bar{S}_a/\bar{\sigma}_a$ or $\bar{S}_m/\bar{\sigma}_m$ will give $\bar{n}$. For a load line not containing the origin (as in bolted joints and extension springs), only $\bar{S}_a/\bar{\sigma}_a$ gives $\bar{n}$.

The nature of C_S depends on material properties and on the fatigue locus used. For the Gerber locus,

$$\frac{\sigma_a}{S_e} + \left(\frac{\sigma_m}{S_{ut}}\right)^2 = 1 \tag{13.33}$$

For a radial load line,

$$\bar{S} = \bar{S}_a = \frac{r^2 \bar{S}_{ut}^2}{2\bar{S}_e}\left[-1 + \sqrt{1 + \left(\frac{2\bar{S}_e}{r\bar{S}_{ut}}\right)^2}\right] \tag{13.34}$$

$$C_S = C_{S_a} = \frac{(1 + C_{S_{ut}})^2}{1 + C_{S_e}}\left\{\frac{-1 + \left[1 + \dfrac{4\bar{S}_e^2}{r^2\bar{S}_{ut}^2}\dfrac{(1 + C_{S_e})^2}{(1 + C_{S_{ut}})^2}\right]^{1/2}}{-1 + \left[1 + \dfrac{4\bar{S}_e^2}{r^2\bar{S}_{ut}^2}\right]^{1/2}} \right\} - 1 \tag{13.35}$$

and $\sigma_{S_a} = C_{S_a}\bar{S}_a$. Although Eq. (13.35) is algebraically complicated, it is only a weighting of C_{S_e}, $C_{S_{ut}}$, and load line slope r. For the ASME-elliptic fatigue locus,

$$\left(\frac{\sigma_a}{S_e}\right)^2 + \left(\frac{\sigma_m}{S_y}\right)^2 = 1 \tag{13.36}$$

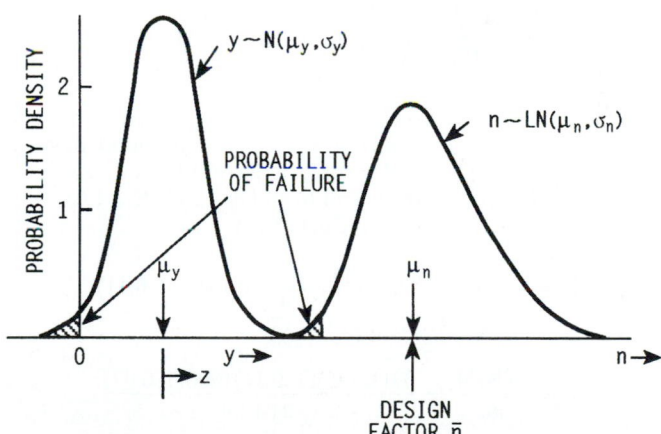

FIGURE 13.14 The probability density function of a lognormally distributed design factor **n**, and its companion normal **y** = ln **n**.

For a radial load line,

$$\overline{S} = \overline{S}_a = \frac{r\overline{S}_y\overline{S}_e}{\sqrt{r^2\overline{S}_y^2 + \overline{S}_e^2}} \qquad (13.37)$$

$$C_S = C_{S_a} = (1 + C_{S_y})(1 + C_{S_e}) \sqrt{\frac{r^2\overline{S}_y^2 + \overline{S}_e^2}{r^2\overline{S}_y^2(1 + C_{S_y})^2 + \overline{S}_e^2(1 + C_{S_e})^2}} - 1 \qquad (13.38)$$

and $\sigma_{S_a} = C_{S_a}\overline{S}_a$. Brittle materials subject to fluctuating stresses which follow the Smith-Dolan fatigue locus in the first quadrant of the designer's fatigue diagram have

$$\frac{\sigma_a}{\overline{S}_e} = \frac{1 - \sigma_m/\overline{S}_{ut}}{1 + \sigma_m/\overline{S}_{ut}} \qquad (13.39)$$

For a radial load line,

$$\overline{S} = \overline{S}_a = \frac{r\overline{S}_{ut} + \overline{S}_e}{2} \left[-1 + \sqrt{1 + \frac{4r\overline{S}_{ut}\overline{S}_e}{(r\overline{S}_{ut} + \overline{S}_e)^2}} \right] \qquad (13.40)$$

$$C_S = C_{S_a} = \frac{r\overline{S}_{ut}(1 + C_{S_{ut}}) + \overline{S}_e(1 + C_{S_e})}{2\overline{S}_e}$$

$$\times \left[-1 + \sqrt{1 + \frac{4r\overline{S}_e\overline{S}_{ut}(1 + C_{S_e})(1 + C_{S_{ut}})}{[r\overline{S}_{ut}(1 + C_{S_{ut}}) + \overline{S}_e(1 + C_{S_e})]^2}} \right] - 1 \qquad (13.41)$$

and $\sigma_{S_a} = C_{S_a}\overline{S}_a$.

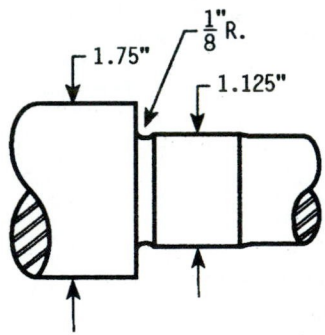

FIGURE 13.15

Example 3. Estimate the reliability at the critical location on a shaft at the fillet of Fig. 13.15. Imposed is a steady lognormal torque $\mathbf{T}_m$ of $2000(1, 0.07)$ in · lbf and a lognormal bending moment $\mathbf{M}_a$ of $1800(1, 0.07)$ in · lbf. The shaft is machined from 1035 HR with an ultimate tensile strength of $\mathbf{S}_{ut} \sim LN(86.2, 3.92)$ kpsi and a yield strength of $\mathbf{S}_y \sim LN(49.6, 3.81)$ kpsi. Use an ASME-elliptic fatigue failure locus for this adequacy assessment.

Solution. *Load-induced stress.* The stress concentration factor is found from noting that $r/d = 0.125/1.125 = 0.11$ and $D/d = 1.75/1.125 = 1.56$, and from Fig. 12.11, $K_t = 1.65$. From Eq. (13.16),

$$K_f = \frac{1.65(1, 0.11)}{1 + \dfrac{2}{\sqrt{0.125}} \dfrac{1.65 - 1}{1.65} \dfrac{4}{86.2}} = 1.495(1, 0.11)$$

$$\sigma_a = \frac{32K_f\mathbf{M}_a}{\pi d^3} = \frac{32(1.495)(1, 0.11)1800(1, 0.07)}{\pi 1.125^3}$$

$$\overline{\sigma}_a = \frac{32(1.495)1800}{\pi 1.125^3} = 19.251 \text{ kpsi}$$

$$C_{S_a} = (0.11^2 + 0.07^2)^{1/2} = 0.130$$

$$\sigma_{\sigma_a} = C_{\sigma_a}\overline{\sigma}_a = 0.130(19.251) = 2.503 \text{ kpsi}$$

The load-induced stress is $\boldsymbol{\sigma}_a \sim LN(19.25, 2.50)$ kpsi.
   Strength: Note that $C_{S_y} = 3.81/49.6 = 0.0768$.

From Eq. (13.9), $\mathbf{k}_a = 2.67(1, 0.058)86.2^{-0.265} = 0.820(1, 0.058)$

From Eq. (13.12), $k_b = (1.125/0.30)^{-0.107} = 0.868$

From Eq. (13.13),

$$\mathbf{k}_c = (1, 0)$$

$$\mathbf{k}_d = \mathbf{k}_e = (1, 0)$$

From Eq. (13.7),

$$\mathbf{S}_e = [0.820(1, 0.058)](0.868)[0.506(1, 0.138)]86.2$$

$$\overline{S}_e = [0.820(0.868)][0.506(86.2)] = 31.0 \text{ kpsi}$$

$$C_{S_e} = (0.058^2 + 0.138^2)^{1/2} = 0.150$$

$$\sigma_{S_e} = 0.150(31.0) = 4.65 \text{ kpsi}$$

The endurance strength $\mathbf{S}_e \sim LN(31.0, 4.64)$ kpsi. The slope of the load line r is

$$r = \frac{2\overline{K}_f\overline{M}_a}{\sqrt{3}\,\overline{T}_m} = \frac{2(1.495)(1800)}{\sqrt{3}(2000)} = 1.554$$

The ASME-elliptic locus is given by Eq. (13.36), from which, from Eq. (13.37),

$$\overline{S}_a = \frac{1.554(49.6)31.0}{\sqrt{1.554^2 49.6^2 + 31.0^2}} = 28.76 \text{ kpsi}$$

From Eq. (13.38),

$$C_{S_a} = (1 + 0.0768)(1 + 0.050)$$

$$\times \sqrt{\frac{1.554^2 49.6^2 + 31.0^2}{1.554^2 49.6^2 (1 + 0.0768)^2 + 31.0^2 (1 + 0.150)^2} - 1}$$

$$= 0.1389$$

$$\sigma_{S_a} = 0.1389(28.76) = 4.00 \text{ kpsi}$$

The strength amplitude is $\mathbf{S}_a \sim LN(28.76, 4.00)$ kpsi.
   Reliability estimate: For the interference of $\mathbf{S}_a$ with $\boldsymbol{\sigma}_a$, from Eq. (2.5),

$$z = -\frac{\ln\left(\dfrac{28.76}{19.25}\sqrt{\dfrac{1 + 0.130^2}{1 + 0.1389^2}}\right)}{\sqrt{\ln\,(1 + 0.1389^2)(1 + 0.130^2)}} = -2.11$$

From Eq. (2.4), $R = 1 - \Phi(-2.11) = 1 - 0.0174 = 0.9826$

Example 4. Invert Example 3 into a design task to make the decision for the diameter d if the reliability goal is 0.9826.

Solution. From calculations already made in Example 3, the coefficient of variation C_n of the design factor n is given by [from Eq. (13.32)]

$$C_n = (0.1389^2 + 0.130^2)^{1/2} = 0.1902$$

From Eq. (2.7),

$$\bar{n} = \exp[-(-2.11)\,\sqrt{\ln(1 + 0.1902^2)} + \ln\sqrt{1 + 0.1902^2}]$$

$$= 1.46$$

The mean load-induced stress is given by

$$\bar{\sigma}_a = \frac{32\overline{K}_f\overline{M}_a}{\pi d^3}$$

Solving for the diameter d, substituting $\overline{S}_a/\bar{n}$ for $\bar{\sigma}_a$, gives

$$d = \left(\frac{32\overline{K}_f\overline{M}_a}{\pi \overline{S}_a/\bar{n}}\right)^{1/3} = \left[\frac{32(1.495)(1800)}{\pi 28\,760/1.46}\right]^{1/3} = 1.12 \text{ in}$$

Clearly the design decision for the diameter d cannot proceed as directly as shown using parts of Example 3. The solution would have to be an iterative process.

1. Choose a material and condition, say 1035HR.
2. Choose a trial diameter d.
3. Develop $\sigma_a \sim LN(\bar{\sigma}_a, \sigma_{\sigma_a})$.
4. Develop $\mathbf{S}_e \sim LN(\overline{S}_e, \sigma_{S_e})$.
5. Develop $\mathbf{S}_a \sim LN(\overline{S}_a, \sigma_{S_a})$.
6. Find C_n and $\bar{n}$.
7. Find d'.
8. Compare d' with d, and with a root-finding algorithm reduce $d' - d$ to zero (or closely).

Alternatively, one can perform steps 1 to 5 above, then

6'. Find z and reliability R_d.
7'. Compare R_d with R_{goal}, and reduce $R_d - R_{goal}$ to zero, omitting any consideration of the design factor. After the decision for d is made, as part of the adequacy assessment, estimate the factor of safety.

When endurance strengths of steels are plotted against cycles to failure, a logarithmic transformation of each coordinate results in a linear data string in the interval $10^3 \le N \le 10^6$ cycles (see Fig. 13.9). At 10^6 cycles, $\overline{S}_f = \overline{S}_e = \phi_{0.30}\overline{S}_{ut}$. At 10^3 cycles, $(\overline{S}_f)_{10^3} = f\overline{S}_{ut}$, where f is a fraction. Arguments in [13.6], p. 279, give

$$f = \frac{S_e}{S_{ut}}\left(\frac{\sigma'_f}{S_e}\right)^{1/2.1} = \frac{S_e}{S_{ut}}\left(\frac{\bar{\sigma}_0\varepsilon_f^m}{S_e}\right)^{1/2.1} \tag{13.42}$$

The fraction of ultimate tensile strength realized at 10^3 cycles is a function of the fatigue ratio $\phi_{0.30}$ and the ultimate tensile strength. The fraction f is in the range

$0.8 < f < 1$ for steels. One can estimate f from Eq. (13.42). The constants a and b of $S_f = aN^b$ can then be found from

$$a = \frac{f^2 \overline{S}_{ut}^2}{\overline{S}_e} \tag{13.43}$$

$$b = -\frac{1}{3} \log \frac{f \overline{S}_{ut}}{\overline{S}_e} \tag{13.44}$$

For axial fatigue, the $(\overline{S}_{ax})_{10^3}$ ordinate in Fig. 13.9 is $\overline{k}_c f \overline{S}_{ut}$ and the $(\overline{S}_{ax})_{10^6}$ ordinate is $\overline{k}_c \phi_{0.30} \overline{S}_{ut}$, where $\overline{k}_c$ is taken from the Table 13.5 equation or determined by interpolation in Table 13.6. For torsional fatigue, the $(\overline{S}_{sf})_{10^3}$ ordinate is $\overline{k}_c f \overline{S}_{ut}$ and the $(\overline{S}_{sf})_{10^6}$ ordinate is $\overline{k}_c \phi_{0.30} \overline{S}_{ut}$, where $\overline{k}_c$ is taken from Table 13.7 or, for materials following the distortion-energy theory of failure closely, $\overline{k}_c = 0.577$.

13.5 COMPLICATED STRESS-VARIATION PATTERNS

Many loading patterns depart from the sinusoidal character of smoothly rotating machinery and the convenient equation pair displayed as Eq. (13.31). In complicated patterns, characterization in the form of maximum force F_{max} (or maximum stress σ_{max}) and minimum force F_{min} (or minimum stress σ_{min}) is more useful. In fact, max-min (max/min/same max or min/max/same min) will avoid losing a damaging cycle.

Consider a full cycle with stresses varying 60, 80, 40, 60 kpsi and another cycle with stresses −40, −60, −20, −40 kpsi, as depicted in Fig. 13.16a. These two cycles cannot be imposed on a part by themselves, but in order for this to be a repetitive block, it is necessary to acknowledge the loading shown by the dashed lines. This adds a hidden cycle that is often ignored. To ensure not losing the hidden cycle, begin the analysis block with the largest (or smallest) stress, adding any preceding history to the right end of the analysis block as shown in Fig. 13.16b. One now searches for cycles using max-min characterizations. Taking the "hidden" cycle first so that it is not lost, one moves along the trace as shown by the dashed line in Fig. 13.16b, iden-

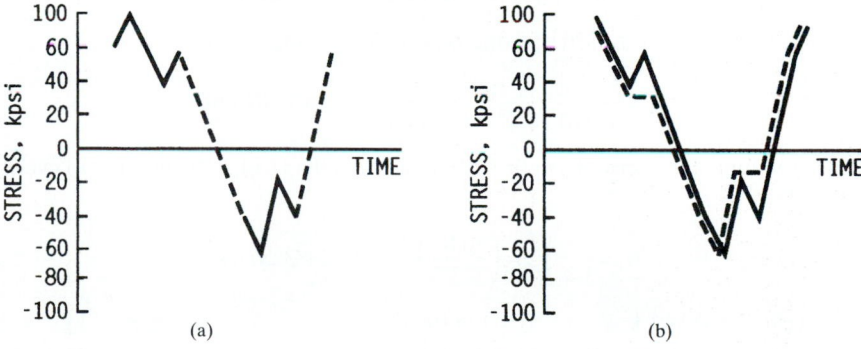

(a)   (b)

FIGURE 13.16

tifying a cycle with a maximum of 80 kpsi and a minimum of –60 kpsi. Looking at the remaining cycles, one notes a cycle with a maximum of 60 and a minimum of 40 kpsi. There is another with a maximum of –20 and a minimum of –40 kpsi. Since failure theories are expressed in terms of σ_a and σ_m components, one uses Eqs. (13.31) and constructs the table below.

Cycle	σ_{max}	σ_{min}	σ_a	σ_m
1	80	–60	70	10
2	60	40	10	50
3	–20	–40	10	–30

Note that the most damaging cycle, cycle 1, has been identified because care was taken not to lose it. The method used is a variation of the *rainflow counting* technique. One is now ready to apply Miner's rule. Note that if the original cycles were doubled in the block, there would be five cycles, with the additional two duplicating cycles 2 and 3 above.

Example 5.   The loading of Fig. 13.16a is imposed on a part made of 1045 HR steel. Properties at the critical location are $\bar{S}_{ut} = 92.5$ kpsi, $\bar{S}'_e = 46$ kpsi, strain-strengthening coefficient $\bar{\sigma}_0 = 140$ kpsi, true strain at fracture $\varepsilon_f = 0.58$, and strain-strengthening exponent $m = 0.14$. Estimate how many repetitions of the loading block may be applied if a Gerber fatigue locus is used.
    Solution.   From Eq. (13.42),

$$f = \frac{S_e}{S_{ut}}\left(\frac{\sigma'_f}{S_e}\right)^{1/2.1} = \frac{S_e}{S_{ut}}\left(\frac{\bar{\sigma}_0 \varepsilon_f^m}{S_e}\right)^{1/2.1}$$

$$= \frac{46}{92.5}\left(\frac{140(0.58)^{0.14}}{46}\right)^{1/2.1} = 0.815$$

From Eq. (13.43), $a = \dfrac{0.815^2 92.5^2}{46} = 123.5$ kpsi

From Eq. (13.44),

$$b = -\frac{1}{3}\log\frac{0.815(92.5)}{46} = -0.0715$$

$$(S_f)_{10^3} = fS_{ut} = 0.815(92.5) = 75.39 \text{ kpsi}$$

For cycle 1, which has $\sigma_a = 70$ kpsi and $\sigma_m = 10$ kpsi, using Eq. (13.33),

$$S_f = \frac{\sigma_a}{1 - (\sigma_m/S_{ut})^2} = \frac{70}{1 - (10/92.5)^2} = 70.83 \text{ kpsi}$$

Since required endurance strength 70.83 kpsi is less than $(S_f)_{10^3}$, the number of cycles exceeds 10^3.

$$N_1 = \left(\frac{S_f}{a}\right)^{1/b} = \left(\frac{70.83}{123.5}\right)^{-1/0.0715} = 2383 \text{ cycles}$$

For cycle 2, $S_f = 14.1$ kpsi and $N_2 = 1.5(10^{13}) \doteq \infty$. For cycle 3, $S_f = \sigma_a = 10$ kpsi and $N_3 = 1.8(10^{15}) \doteq \infty$. Extend the previous table:

Cycle	S_f	N
1	70.8	2383
2	14.1	∞
3	10.0	∞

The damage per block application according to Miner's rule is

$$D = \Sigma\,(1/N_i) = 1/2383 + 1/\infty + 1/\infty = 1/2383$$

The number of repetitions of the block is $1/D = 1/(1/2383) = 2383$. For the original two cycles, the damage per block application is $1/\infty + 1/\infty = 0$ and the number of repetitions is infinite. Note the risk of an analysis conclusion associated with not drawing how the cycles connect.

13.6 STRENGTH AT CRITICAL LOCATIONS

The critical locations of strength-limited designs can be identified as regions in which load-induced stresses peak as a result of distribution of bending moment and/or changes in geometry. Since the strength at the critical location in the geometry and at condition of use is required, it is often necessary to reflect the manufacturing process in this estimation. For heat-treatable steels under static loading, an estimate of yield or proof strength is required, and under fatigue loading, an estimate of the ultimate strength of the endurance limit is needed for an adequacy assessment. For the design process, strength as a function of intensity of treatment is required. In Chap. 8, the quantitative estimation methods of Crafts and Lamont and of Grossmann and Field for heat-treatable steels are useful. For cold work and cold heading, the methods of Datsko and Borden give useful estimates.

Consider an eyebolt of cold-formed 1045 steel hot-rolled quarter-inch-diameter rod in the geometry of Fig. 13.17. The properties of the hot-rolled bar are

$$S_y = 60 \text{ kpsi} \qquad m = 0.14 \qquad \overline{\sigma}_o = 140 \text{ kpsi}$$

$$S_u = 92.5 \text{ kpsi} \qquad \varepsilon_f = 0.58$$

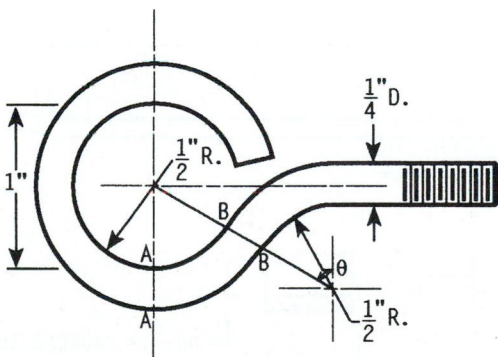

FIGURE 13.17 Geometry of a cold-formed eyebolt.

At section AA on the inside surface, the true strain is estimated as

$$\varepsilon_i = \left| -\frac{1}{2}\ln\left(1 + \frac{2d}{D}\right)\right| = \left| -\frac{1}{2}\ln\left[1 + \frac{2(0.25)}{1}\right]\right|$$

$$= |-0.203| = 0.203$$

The yield strength of the surface material at this location is estimated as, from Table 8.1

$$\varepsilon_{quo} = \frac{0.203}{1+1} = 0.1015 \qquad \varepsilon_{qyo} = \frac{0.1015}{1 + 2(0.1015)} = 0.0844$$

$$(S_y)_{tLc} = \overline{\sigma}_o\,(\varepsilon_{qyo})^m = 140(0.0844)^{0.14} = 99 \text{ kpsi}$$

The ultimate strength at this location is estimated as, from Table 8.1,

$$(S_u)_{tLc} = (S_u)_o \exp(\varepsilon_{quo}) = 92.5 \exp(0.1015) = 102.4 \text{ kpsi}$$

Both the yield strength and the ultimate strength have increased. They are nominally equal because the true strain of 0.203 exceeds the true strain of 0.14 which occurs at ultimate load. The yield strength has increased by 65 percent and the ultimate strength has increased by 11 percent at this location. The strength at the inside and outside surface at section BB has not changed appreciably. The changes at the sections above BB are improvements in accord with the local geometry. For dynamic strength, the endurance limits have changed in proportion to the changes in ultimate strength. At section AA the R. R. Moore endurance limit is estimated to be

$$S'_e = \frac{S_u}{2} = \frac{102.4}{2} = 51.2 \text{ kpsi}$$

an improvement of 11 percent. Since the strengths vary with position and stresses vary with position also, a check is in order to see if section AA or section BB is critical in a tensile loading of the eyebolt.

The increase in yield strength and endurance limit due to cold work, while present, may not be helpful. Consider the strip spring formed from bar stock to the geometry of Fig. 13.18. Just to the right of the radius the original properties prevail, and the bending moment is only slightly less than to the left of section AA. In this case, the increased strength at the critical location is not really exploitable.

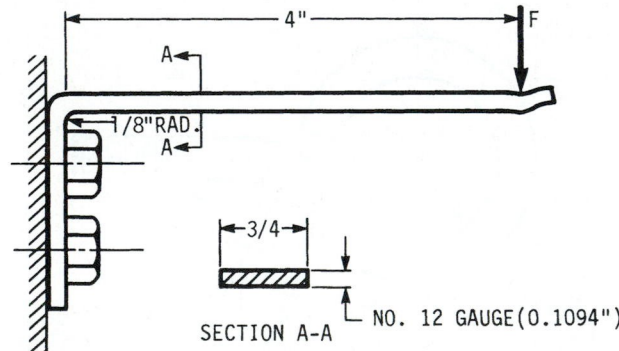

FIGURE 13.18 A latching spring cold formed from ¾-in-wide No. 12 gauge strip.

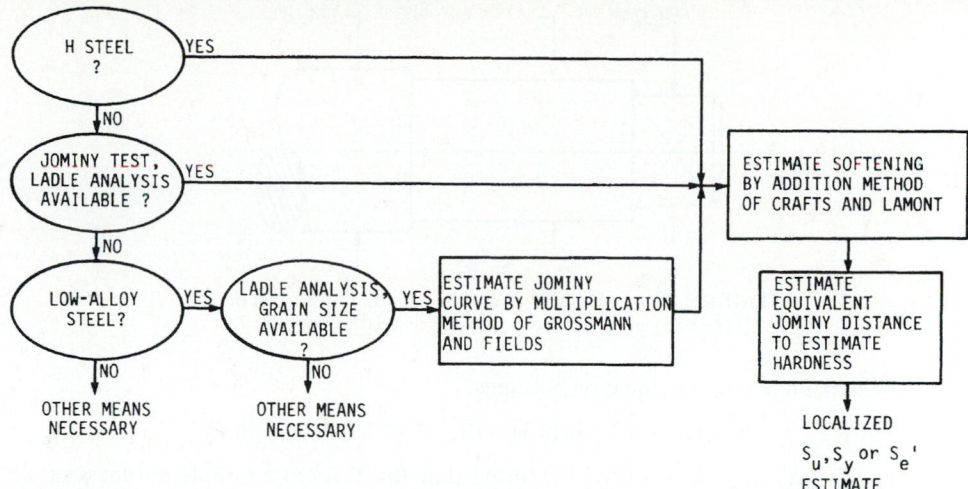

FIGURE 13.19 Logic flowchart for estimation of localized ultimate strength or endurance limit for heat-treated steels.

For parts that are heat-treated by quenching and tempering, the methods and procedures are given in Fig. 13.19 (see Chap. 8). If a shaft has been designed, an adequacy assessment is required. An estimate of the strength at a location where the shaft steps in Fig. 13.20 from 1 to 1.125 in is necessary. The specifications include the material to be 4140 steel quenched in still oil with mild part agitation and tempered for 2 hours at 1000°F. The material earmarked for manufacture has a ladle analysis of

	C	Mn	P	S	Si	Ni	Cr	Mo
Percent	0.40	0.83	0.012	0.009	0.26	0.11	0.94	0.21
Multiplier	0.207	3.87	—	—	1.18	1.04	3.04	1.65

The experience is that a grain size of 7½ can be maintained with this material and heat treatment. The multipliers are determined by the methods of Chap. 8. The ideal critical diameter is estimated as

$$D_I = 0.207(3.87)(1.18)(1.04)(3.04)(1.65) = 4.93 \text{ in}$$

The factors are $D = 5.3$, $B = 10$, and $f = 0.34$. The addition factors are

$$A_{Mn} = 2.1$$

$$A_{Si} = 1.1$$

$$A_{Ni} = 0.03$$

$$A_{Cr} = 4.9$$

$$A_{Mo} = 3.78$$

$$\Sigma A = 11.91$$

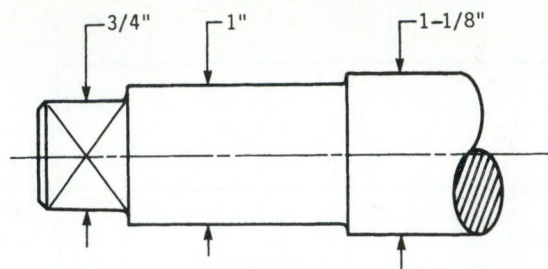

FIGURE 13.20 A portion of a 4140 steel shaft quenched in still oil ($H = 0.35$) and tempered for 2 hours at 1000°F, grain size 7.5.

The tempered hardness equation becomes

$$R_T = (R_Q - 5.3 - 10)0.34 + 10 + 11.91 = 0.34R_Q + 16.71$$

The Jominy curve is predicted by noting that the Rockwell C-scale hardness at Jominy station 1 is

$$(R_Q)_1 = 32 + 60(\%C) = 32 + 60(0.40) = 56.0$$

and a table is prepared as depicted in Table 13.11 for 1000°F tempering temperature. The variation of surface strength with size of round is prepared using equivalent Jominy distances as depicted in Table 13.12. Table 13.13 shows an ultimate-strength traverse of a 1⅛-in round. There is only a mild strength profile in the traverse. The significant strength for bending and torsion is at the surface. For the 1⅛-in round, the surface ultimate strength is estimated by interpolation to be 164.3 kpsi. The R. R. Moore endurance limit at this location is estimated to be 164.3/2, or 82 kpsi. Steels in large sections or with less alloying ingredients (smaller ideal critical diameters) exhibit larger transverse strength changes. For sections in tension, significant strength is at the center. When testing, machined specimens from the center of a round say little about the strength at the surface. Heat treating a specimen proves little about strengths in the actual part. Some variations in strength attributed to size result from differences in cooling rates.

When the number of cycles is less than 10^7, the endurance strength must be estimated. Reference [13.8] gives a useful curve fit for steels:

$$S_f' = \begin{cases} S_u m^{-m} \exp(m) \varepsilon_f^m N_f^{cm} \exp(-\varepsilon_f N_f^c) & \varepsilon_f N_f^c \leq m \\ S_u & \varepsilon_f N_f^c > m \end{cases} \tag{13.45}$$

TABLE 13.11 Surface Ultimate Strength as a Function of Jominy Distance for 4140 Steel Oil Quenched ($H = 0.35$) and Tempered 2 Hours at 1000°F, Grain Size 7½

Jominy distance, $\frac{1}{16}$ in	IH/DH	Predicted Jominy R_Q, Rockwell C	Tempered hardness R_T, Rockwell C	Surface ultimate strength S_u, kpsi
1	1	56.0	44.1	206.6
4	1	56.0	44.1	206.6
8	1.09	51.4	42.0	196.0
12	1.18	47.5	40.3	187.5

TABLE 13.12   Variation of Surface Strength with Diameter of 4140 Steel Round Quenched in Still Oil ($H = 0.35$) and Tempered for 2 Hours at 1000°F, Grain Size 7½

Diameter, in	Equivalent Jominy distance, $\frac{1}{16}$ in	Surface ultimate strength, kpsi
0.1	1.0	167
0.5	2.7	167
1	5.1	165
2	8.2	159.1
3	10.0	156.1
4	11.4	153.2

where m = strain-strengthening exponent
ε_f = true strain at fracture
c = an exponent commonly in the neighborhood of $-\frac{1}{2}$
N_f = the number of cycles to failure
S_u = ultimate tensile strength

Example 6.   Estimate the finite-life engineering fatigue strength of an annealed 4340 steel with the following properties:

$$S_u = 103 \text{ kpsi} \qquad S_y = 65.6 \text{ kpsi at 0.2 percent offset}$$

$$RA = 0.56$$

Solution.   The endurance limit is estimated as $S_u/2 = 103/2 = 51.5$ kpsi at 10^7 cycles. Because no strain-hardening information is supplied, it is necessary to estimate m from

$$\frac{S_u}{S_y} = \left[\frac{m}{(\text{offset}) \exp 1} \right]^m$$

or

$$\frac{103}{65.6} - \left[\frac{m}{0.002(2.718)} \right]^m = 0$$

from which $m = 0.14$. The true strain at fracture can be assessed from the reduction in area:

$$\varepsilon_f = \ln \frac{1}{1 - RA} = \ln \frac{1}{1 - 0.56} = 0.821$$

TABLE 13.13   Variation of Local Strength in a 1.125-in Round of 4140 Steel Quenched in Still Oil ($H = 0.35$) and Tempered for 2 Hours at 1000°F, Grain Size 7½

Radial position	Equivalent Jominy distance, $\frac{1}{16}$ in	Local ultimate strength, kpsi
(0)r	7.33	161.0
0.5r	6.35	162.8
0.8r	5.95	163.5
r	5.55	164.2

The true stress coefficient of the strain-strengthening equation $\overline{\sigma} = \overline{\sigma}_o \varepsilon^m$ is

$$\overline{\sigma}_o = S_u m^{-m} \exp m = 103(0.14)^{-0.14} \exp 0.14$$

$$= 156.0 \text{ kpsi}$$

The constructive strain ε_1 is a root of the equation

$$\frac{\overline{\sigma}_o}{S'_e} \varepsilon_1^m \exp(-\varepsilon_1) - 1 = 0$$

or alternatively,

$$\frac{S_u}{S'_e} m^{-m} \exp(m)\varepsilon_1^m \exp(-\varepsilon_1) - 1 = 0$$

When ε_1 is small, the term $\exp -\varepsilon_1$ approaches 1, and ε_1 can be found explicitly from

$$\varepsilon_1 = \left(\frac{S'_e}{\overline{\sigma}_o}\right)^{1/m} \qquad \text{or} \qquad \varepsilon_1 = \frac{m}{2.718}\left(\frac{S'_e}{S_u}\right)^{1/m}$$

From the first,

$$\varepsilon_1 = \left(\frac{51.5}{156}\right)^{1/0.14} = 0.000365$$

This value of the constructive strain allows estimation of the exponent c from $\varepsilon_1 = \varepsilon_f N_f^c$:

$$c = \frac{\log \varepsilon_1/\varepsilon_f}{\log N_e} = \frac{\log(0.000\ 365/0.821)}{\log 10^7} = -0.4788$$

Now Eq. (13.45) can be written as

$$S'_f = 103(0.14)^{-0.14} \exp(0.14)0.821^{0.14} N_f^{-0.4778(0.14)} \exp(0.821 N_f^{-0.4788})$$

which simplifies to

$$S'_f = 151.8 N_f^{-0.067} \exp(0.821 N_f^{-0.4788})$$

Table 13.14 can be constructed. See Ref. [13.8] for notch-sensitivity corrections for low cycle strengths.

13.7 COMBINED LOADING

Simple loading is regarded as an influence that results in tension, compression, shear, bending, or torsion, and the stress fields that result are regarded as simple. *Combined loading* is the application of two or more of these simple loading schemes. The stresses that result from both simple and combined loading are three-dimensional. Applying the adjective *combined* to stresses is inappropriate. The nature of both yielding and fatigue for ductile materials is best explained by distortion-energy (octahedral shear, Henckey-von Mises) theory. For variable loading, the stress state is plotted on a modified Goodman diagram that has tensile mean stresses as abscissa and

TABLE 13.14   Fatigue Strength Ratio S_f'/S_u as a Function of Cycles to Failure for Annealed 4340 Steel[†]

Number of cycles-to-failure N_f	Constructive true strain ε_1	Endurance strength S_f', kpsi	Ratio S_f'/S_u
10^0	0.821	103‡	1‡
10^1	0.273	103‡	1‡
10^2	0.091	101.8	0.99
10^3	0.030	92.7	0.90
10^4	0.010	81.0	0.79
10^5	0.0033	69.9	0.68
10^6	0.0011	60.1	0.58
10^7	0.000 37	51.5	0.50

[†]$S_u = 103$ kpsi, $S_y = 65.6$ kpsi (0.002 offset), reduction in area 56 percent.
‡Since $\varepsilon_1 > m$, $S_f' = S_u$ and $S_f'/S_u = 1$.

tensile stress amplitude as ordinate. The stress amplitude is that present in a uniform tension that induces the same distortion-energy amplitude (octahedral shear amplitude) as is present in the critical location of the machine part. The steady stress is that stress present in a uniform tension that induces the same steady distortion energy (steady octahedral shear) as is present in the critical location of the machine part. The plotting process involves conversion of the actual stress state to the equivalent uniform tension circumstances.

The von Mises axial tensile stress that has the same distortion energy as a general three-dimensional stress field is, in terms of the ordered principal stresses σ_1, σ_2, and σ_3,

$$\sigma_v = \left[\frac{(\sigma_1 - \sigma_2)^2 + (\sigma_2 - \sigma_3)^2 + (\sigma_3 - \sigma_1)^2}{2} \right]^{1/2} \tag{13.46}$$

If one of the principal stresses is zero and the other two are σ_A and σ_B, then

$$\sigma_v = (\sigma_A^2 + \sigma_B^2 - \sigma_A \sigma_B)^{1/2} \tag{13.47}$$

If the axes xy are not principal, then

$$\sigma_v = (\sigma_x^2 + \sigma_y^2 - \sigma_x \sigma_y + 3\tau_{xy}^2)^{1/2} \tag{13.48}$$

If the concern is yielding, then yielding begins when the von Mises stress equals the tensile value of S_y. If the concern is fatigue, then failure occurs when the von Mises steady stress and amplitude equal the simple steady tension and amplitude that result in failure. If the Eq. (13.47) is equated to a critical value σ_{cr}, then

$$\sigma_A^2 + \sigma_B^2 - \sigma_A \sigma_B = \sigma_{cr}^2$$

Treating the preceding equation as a quadratic in σ_A, we have

$$\sigma_A = \frac{1}{2}\sigma_B \pm \frac{1}{2}\sqrt{(2\sigma_{cr})^2 - 3\sigma_B^2} \tag{13.49}$$

On a plot in the $\sigma_A \sigma_B$ plane, the critical-stress magnitude can be observed at six places, three tensile and three compressive. The locus is an ellipse with the major axis

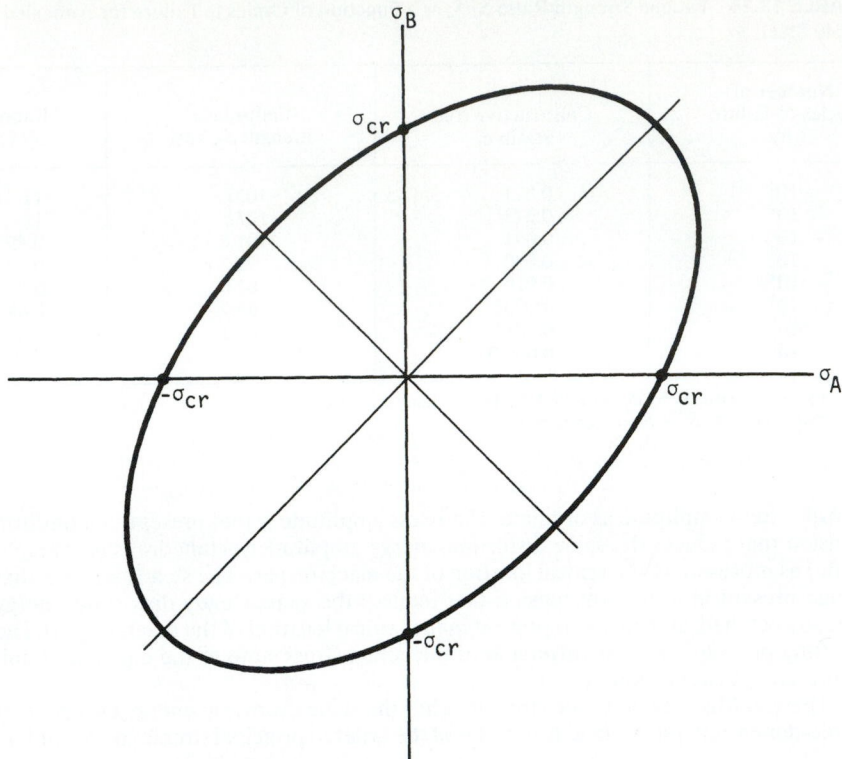

FIGURE 13.21   The distortion-energy critical-stress ellipse. For any point σ_A, σ_B on the ellipse, the uniaxial tension with the same distortion energy is the positive abscissa intercept σ_{cr}, 0.

having a unity slope, as depicted in Fig. 13.21. The octahedral stress (von Mises stress) tensile equivalent is the σ_A-axis intercept of the ellipse. For the Goodman diagram, the transformation is to move the point representing the stress condition σ_A, σ_B to the abscissa, while staying on the ellipse. This is done by Eq. (13.47). For three-dimensional stress, the surface is an ellipsoid in the $\sigma_1\sigma_2\sigma_3$ space, and the transformation is accomplished by Eq. (13.46). Figure 13.22 shows the conversions of the steady-stress condition and the stress-amplitude condition to the respective simple-tension equivalents for the purposes of plotting a point representing the equivalent stress state on the modified Goodman diagram. In Fig. 13.23 an element on a shaft sees a steady torque and fully reversed bending stresses. For the steady-stress element,

$$\tau_{xym} = \frac{16T}{\pi d^3} \qquad \sigma_{Am} = \frac{16T}{\pi d^3} \qquad \sigma_{Bm} = -\frac{16T}{\pi d^3}$$

and the corresponding von Mises stress is

$$\sigma_{vm} = (\sigma_{Am}^2 + \sigma_{Bm}^2 - \sigma_{Am}\sigma_{Bm})^{1/2} = \frac{\sqrt{3}\,16T}{\pi d^3}$$

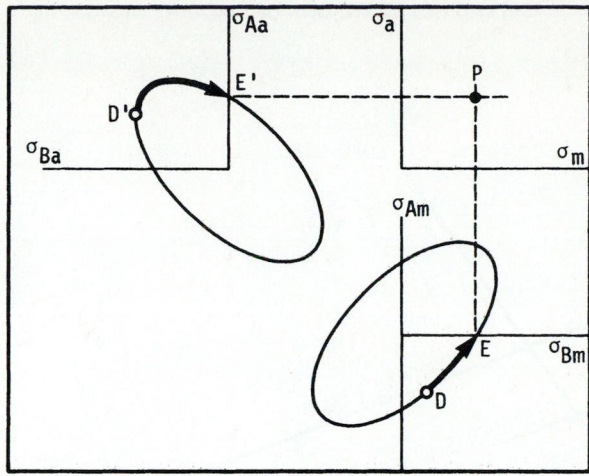

FIGURE 13.22 The principal stresses due to the steady stresses σ_{Am} and σ_{Bm} appear on the distortion-energy ellipse as point D. The transform to equivalent distortion energy in tension is point E, which becomes the abscissa of point P. The principal stresses due to stress amplitude σ_{Aa} and σ_{Ba} appear as point D'; the transform is E', which becomes the ordinate of point P.

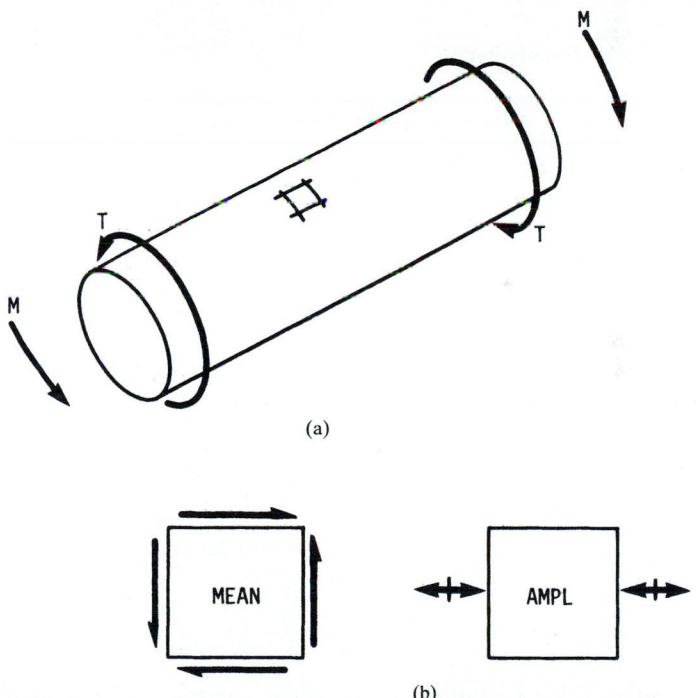

FIGURE 13.23 (*a*) A shaft subjected to a steady torque T and completely reversed flexure due to bending moment M; (*b*) the mean-stress element and the stress-amplitude element.

13.39

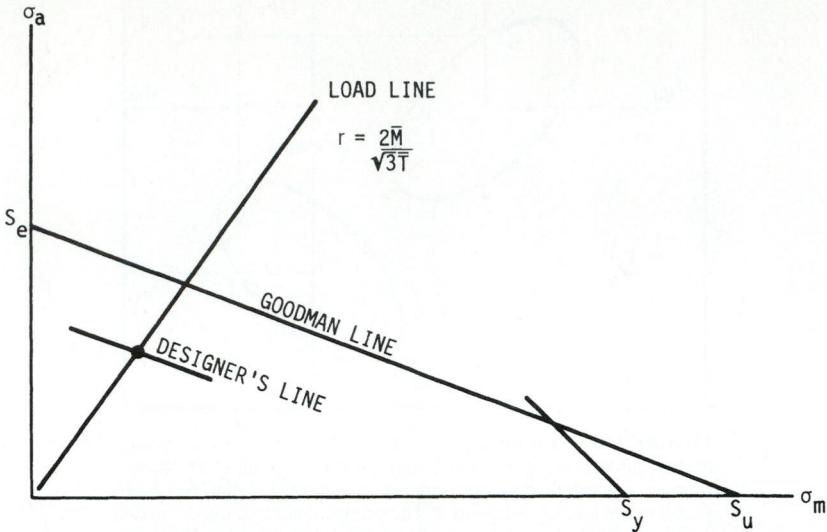

FIGURE 13.24   Designer's fatigue diagram for geared shaft showing load line of slope $r = 2M/(\sqrt{3}T)$, the operating point P, using the Goodman failure locus, and the designer's line reflecting a design factor of n.

For the amplitude-stress element,

$$\sigma_{x,\max} = \frac{32M}{\pi d^3} \qquad \sigma_{x,\min} = -\frac{32M}{\pi d^3}$$

$$\sigma_{Aa} = \frac{32M}{\pi d^3} \qquad \sigma_{Ba} = 0$$

and the corresponding von Mises stress is

$$\sigma_{va} = (\sigma_{Aa}^2 + \sigma_{Ba}^2 - \sigma_{Aa}\sigma_{Ba})^{1/2} = \frac{32M}{\pi d^3}$$

If this is an element of a geared shaft, then M and T are proportional and the locus of possible points is a radial line from the origin with a slope of

$$r = \frac{\sigma_{va}}{\sigma_{vm}} = \frac{32M}{\pi d^3} \frac{\pi d^3}{\sqrt{3}\,16T} = \frac{2M}{\sqrt{3}T} \tag{13.50}$$

This is called the *load line*. If data on failures have been collected and converted to von Mises components and a Goodman line is an adequate representation of the failure locus, then for the designer's line in Fig. 13.24,

$$\frac{\sigma_{va}}{S_e} + \frac{\sigma_{vm}}{S_u} = \frac{1}{n} \tag{13.51}$$

where n = design factor. Substituting for σ_{va} and σ_{vm} and solving for d, we obtain

$$d = \left[\frac{32n}{\pi}\left(\frac{M}{S_e} + \frac{\sqrt{3}T}{2S_u}\right)\right]^{1/3} \tag{13.52}$$

Data points representing failure are plotted embracing a significant stress commitment, and the plotted load line represents the same belief. It is appropriate that equations such as Eq. (13.52) be labeled with two adjectives: (1) *significant stress* and (2) *failure locus*. For example, Eq. (13.52) could be called a distortion-energy Goodman equation.

For the case where moments, torques, and thrusts contribute both steady and alternating components of stress, then the distortion-energy Goodman equation for the critical location is

$$\frac{16}{\pi d^3 S_e}\left[\left(2M_a + \frac{P_a d}{4}\right)^2 + 3T_a^2\right]^{1/2} + \frac{16}{\pi d^3 S_u}\left[\left(2M_m + \frac{P_m d}{4}\right)^2 + 3T_m^2\right]^{1/2} - \frac{1}{n} = 0 \quad (13.53)$$

where n = design factor
    M_a = component of bending moment causing flexural stress amplitude
    M_m = component of bending moment causing flexural stress, steady
    T_a = component of torque causing shear-stress amplitude
    T_m = component of torque causing shear stress, steady
    P_a = component of axial thrust causing tensile-stress amplitude
    P_m = component of axial thrust causing tensile stress, steady
    S_e = local fatigue strength of shaft material
    S_u = ultimate local tensile strength
    d = local shaft diameter

Since the equation cannot be solved for d explicitly, numerical methods (see Chap. 4) are used.

13.8 SURFACE FATIGUE

When cylinders are in line contact, sustained by a force F, a flattened rectangular zone exists in which the pressure distribution is elliptical. The half width of the contact zone b is

$$b = \sqrt{\frac{2F}{\pi \ell}\frac{[1 - v_1^2]/E_1 + [1 - v_2^2]/E_2}{(1/d_1) + (1/d_2)}} \quad (13.54)$$

The largest stress in magnitude is compressive and exists on the z axis. As a pressure, its magnitude is

$$p_{max} = \frac{2F}{\pi b \ell} \quad (13.55)$$

Along the z axis, the orthogonal stresses are [13.6]

$$\sigma_x = -2p_{max}\left[\sqrt{1 + \left(\frac{z}{b}\right)^2} - \frac{z}{b}\right] \quad (13.56)$$

$$\sigma_y = -p_{max}\left\{\left[2 - \frac{1}{1 + (z/b)^2}\right]\sqrt{1 + \left(\frac{z}{b}\right)^2} - 2\frac{z}{b}\right\} \quad (13.57)$$

$$\sigma_z = \frac{-p_{max}}{\sqrt{1 + (z/b)^2}} \quad (13.58)$$

These equations can be useful with rolling contact, such as occurs in cams, roller bearings, and gear teeth. The approach of the center of the rollers is

$$\Delta = \frac{2F}{\pi \ell} \left(\frac{1 - v_1^2}{E_1} + \frac{1 - v_2^2}{E_2} \right) \left(\ln \frac{d_1}{b} + \ln \frac{d_2}{b} + \frac{2}{3} \right) \tag{13.59}$$

The largest principal stress is compressive and located at the center of the rectangular flat and is $-p_{max}$ in magnitude. The largest shear stress is approximately $0.30p_{max}$ and is located at about $0.78b$ below the surface. The maximum compressive stress is repeatedly applied in rolling cylinders. At a position of $z = 0.4b$, $y = 0.915b$ the shear stress has a magnitude of $0.242p_{max}$ but is completely reversed in rolling cylinders.

The loss of surface integrity due to repeated application of hertzian contact stresses is called *surface fatigue*. The phenomenon is marked by the loss of material from the surface, leaving pits or voids. The onset of fatigue is often defined as the appearance of craters larger than a specified breadth. If Eq. (13.54) is substituted into Eq. (13.55) and the magnitude of p_{max} associated with the first tangible evidence of fatigue at a specified number of cycles is called the *surface endurance strength* S_{fe}, then

$$\frac{F}{\ell} \left(\frac{2}{d_1} + \frac{2}{d_2} \right) = \pi S_{fe}^2 \left(\frac{1 - v_1^2}{E_1} + \frac{1 - v_2^2}{E_2} \right) = K \tag{13.60}$$

The left-hand side of the equation consists of parameters under the designer's control. The right-hand side consists of materials properties. The factor K is called *Buckingham's load-stress factor* and is associated with a number of cycles. In gear studies a similar K factor is used and is related to K through

$$K_g = \frac{K}{4} \sin \phi \tag{13.61}$$

where ϕ = gear-tooth pressure angle. Note that p_{max} is proportional to other stresses present, and it is conventional to describe surface fatigue in terms of the strength S_{fe}. Reference [13.1] gives K_g information for pressure angles of $\phi = 14\frac{1}{2}$ degrees and $\phi = 20$ degrees, as well as S_{fe} for various materials and numbers of cycles. The implication in the literature that above various cycles K_g does not change is unsupported. Log-log plots of K or K_g versus cycles to failure produce "parallel" lines with some consistency in slope for classes of material. AGMA standard 218.01 (Dec. 1982) suggests allowable contact-stress numbers for steel as high as

$$(\sigma)_{10^7} = 0.364 H_B + 27 \text{ kpsi} \tag{13.62}$$

for 10^7 cycles, and for the curve fit for other than 10^7 cycles,

$$\sigma_N = C_L \sigma_{10^7} = 2.46 N^{-0.056} (0.364 H_B + 27) \text{ kpsi} \tag{13.63}$$

For applications where little or no pitting is permissible, the slope of -0.056 persists to 10^{10} (end of presentation). Another curve fit to S_{fe} data for steel at 10^8 cycles is

$$(S_{fe})_{10^8} = 0.4 H_B - 10 \text{ kpsi} \tag{13.64}$$

When a gear and pinion are in mesh, the number of cycles to which a point on the tooth surface is exposed is different. Consequently, the material strengths can be tuned to have both pinion and gear show tangible signs of wear simultaneously. For steel gears, using Eq. (13.64), the appropriate gear hardness BHN for a stipulated pinion hardness bhn is

$$\text{BHN} = m_G^{b/2}(\text{bhn} - 25) + 25 \tag{13.65}$$

where m_G = gear ratio, i.e., teeth on the gear per tooth on the pinion. This matching can be done by controlling surface hardness. Strength matching for bending resistance is accomplished by control of core properties.

When needle bearings are specified, the needle assembly is a vendor's product, but the roller track is supplied by the user and is a design concern. The equivalent load F_{eq} accumulating the same damage as a variable radial load F is

$$F_{eq} = \left(\frac{1}{2\pi} \int_0^{2\pi} F^a \, d\theta \right)^{1/a}$$

If the entire assembly is vendor-supplied as a needle-bearing cam follower, then the average load is dictated by the cam-follower force F_{23}. The following makes m turns per cam revolution. The follower's first turn has an average load to the a power of

$$(F_{23})_1^a = \frac{1}{2\pi/m} \int_0^{2\pi/m} F_{23}^a \, d\theta$$

where $d\theta$ = cam angular displacement. The subsequent averages are

$$(F_{23})_2^a = \frac{1}{2\pi/m} \int_{2\pi/m}^{4\pi/m} F_{23}^a \, d\theta$$

$$\vdots$$

The global average to the a power is

$$F_{global}^a = \frac{1}{m} \sum_{i=1}^{m} (F_{23})_i^a = \frac{1}{2\pi} \int_0^{2\pi} F_{23}^a \, d\theta$$

Consequently, the roller average radial load is identical to the cam's, but the follower makes m times as many turns.

The follower contact surface between the cam and follower has an endurance strength cycles-to-failure relation of the form of $S^{-1/b}N = K$, and so the average hertzian stress can be written

$$\bar{\sigma}_H = \left(\frac{1}{2\pi} \int_0^{2\pi} \sigma_H^{-1/b} \, d\theta \right)^{-b}$$

$$= \frac{C_p}{\sqrt{w}} \left\{ \frac{1}{2\pi} \int_0^{2\pi} [F_{23}(K_c + K_r)]^{-1/2b} \, d\theta \right\}^{-b} \tag{13.66}$$

where θ = cam rotation angle, b = slope of the rectified SN locus, w = width of the roller or cam (whichever is less), C_p = a joint materials constant

$$C_p = \sqrt{\frac{1}{\pi \left(\dfrac{1 - v_1^2}{E_1} + \dfrac{1 - v_2^2}{E_2} \right)}} \tag{13.67}$$

and the parameters K_c and K_r = the curvatures of the cam and roller surfaces, respectively.

One surface location on the cam sees the most intense hertzian stress every revolution. That spot has a hertzian stress of

$$\sigma_H = \frac{C_p}{\sqrt{w}} [F_{23}(K_c + K_r)]_{max}^{1/2} \tag{13.68}$$

Relative strengths can be assessed by noting that the cam requires a strength of $(S_{fe})_{mN}$ at the critical location in order to survive N cycles. The roller sees the average stress everywhere on its periphery mN times, and its strength requirement is $(S_{fe})_{mN}$. The strength ratio is

$$\frac{[(S_{fe})_N]_{\text{cam}}}{[(S_{fe})_{mN}]_{\text{roller}}} = \frac{[F_{23}(K_c + K_r)]_{\text{max}}^{1/2}}{\left\{ \dfrac{1}{2\pi} \displaystyle\int_0^{2\pi} [F_{23}(K_c + K_r)]^{-1/2b} \, d\theta \right\}^{-b}}$$

$$= \frac{\sqrt{\mathscr{F}}_{\text{max}}}{(\sqrt{\mathscr{F}})_{\text{avg}}}$$

If cam and roller are steel,

$$(S_{fe})_{mN} = m^b (S_{fe})_N$$

enabling us to place endurance strengths on the same life basis, namely, N, which is convenient when consulting tables of Buckingham load-strength data giving K or its equivalent. Thus,

$$\frac{[(S_{fe})_N]_{\text{cam}}}{[(S_{fe})_N]_{\text{roller}}} = \frac{m^b \sqrt{\mathscr{F}}_{\text{max}}}{(\sqrt{\mathscr{F}})_{\text{avg}}}$$

For steel, a 10^8 cycle expression, that is, $(S_{fe}) = 0.4H_B - 10$ kpsi, can be used. Using bhn for roller Brinell hardness and BHN for cam Brinell hardness, we can write

$$\text{BHN} = \frac{m^b \sqrt{\mathscr{F}}_{\text{max}}}{(\sqrt{\mathscr{F}})_{\text{avg}}} \, (\text{bhn} - 25) + 25 \qquad (13.69)$$

This form is convenient because the roller follower is often a vendor-supplied item and the cam is manufactured elsewhere. Since $\sqrt{\mathscr{F}}_{\text{max}}$ is larger than $(\sqrt{\mathscr{F}})_{\text{avg}}$, this alone tends to require that the cam be harder, but since the roller endures more turns, the roller should be harder, since $m > 1$ and $b < 0$. Matching (tuning) the respective hardnesses so that the cam and roller will wear out together is often a design goal.

Design factor can be introduced by reducing strength rather than increasing load (not equivalent when stress is not directly proportional to load). Since the loads are often more accurately known than strengths in these applications, design factors are applied to strength. The relative hardnesses are unaffected by design factor, but the necessary widths are

$$w = \left(\frac{C_p}{S_{fe}/n} \right)^2_{\text{cam}} \mathscr{F}_{\text{max}} \qquad w = \left(\frac{C_p}{S_{fe}/n} \right)^2_{\text{roller}} (\sqrt{\mathscr{F}})_{\text{avg}} m^{-2b} \qquad (13.70)$$

Either equation may be used. The width decision controls the median cycles to failure.

REFERENCES

13.1 R. C. Juvinall, *Stress Strain and Strength*, McGraw-Hill, New York, 1967, p. 218.

13.2 E. M. Prot, "Fatigue Testing under Progressive Loading: A New Technique for Testing Materials," E. J. Ward (trans.), Wright Air Development Center Tech. Rep., TR52-148, September 1952.

13.3 J. A. Collins, *Failure of Materials in Mechanical Design,* Wiley-Interscience, New York, 1981, chap. 10.

13.4 W. J. Dixon and F. J. Massey, Jr., *Introduction to Statistical Analysis,* 3d ed., McGraw-Hill, New York, 1969, p. 380.

13.5 J. T. Ransom, *Statistical Aspects of Fatigue,* Special Technical Publication No. 121, American Society for Testing Materials, Philadelphia, Pa., 1952, p. 61.

13.6 J. E. Shigley and C. R. Mischke, *Mechanical Engineering Design,* 5th ed., McGraw-Hill, New York, 1989.

13.7 C. R. Mischke, "A Probabilistic Model of Size Effect in Fatigue Strength of Rounds in Bending and Torsion," *Transactions of ASME, Journal of Machine Design,* vol. 102, no. 1, January 1980, pp. 32–37.

13.8 C. R. Mischke, "A Rationale for Mechanical Design to a Reliability Specification," *Proceedings of the Design Engineering Technical Conference of ASME,* American Society of Mechanical Engineers, New York, 1974, pp. 221–248.

13.9 L. Sors, *Fatigue Design of Machine Components,* Part 1, Pergamon Press, Oxford, 1971, pp. 9–13.

13.10 H. J. Grover, S. A. Gordon, and L. R. Jackson, *Fatigue of Metals and Structures,* Bureau of Naval Weapons Document NAVWEPS 00-25-534, Washington, D.C., 1960, pp. 282–314.

13.11 C. Lipson and R. C. Juvinall, *Handbook of Stress and Strength,* Macmillan, New York, 1963.

13.12 R. C. Juvinall, *Fundamentals of Machine Component Design,* John Wiley & Sons, New York, 1983.

13.13 A. D. Deutschman, W. J. Michels, and C. E. Wilson, *Machine Design,* MacMillan, New York, 1975.

13.14 E. B. Haugen, *Probabilistic Mechanical Design,* John Wiley & Sons, New York, 1980.

13.15 C. R. Mischke, *Mathematical Model Building,* 2d rev. ed., Iowa State University Press, Ames, 1980.

13.16 G. Sines and J. L. Waisman (eds.), *Metal Fatigue,* McGraw-Hill, New York, 1959.

13.17 H. O. Fuchs and R. I. Stephens, *Metal Fatigue in Engineering,* John Wiley & Sons, New York, 1980.

RECOMMENDED READING

Proceedings of the Society of Automotive Engineers Fatigue Conference, P109, Warrendale, Pa. April 1982.

CHAPTER 14
FUNDAMENTALS OF ARC WELDING

Richard S. Sabo
Manager, Educational Services
The Lincoln Electric Company
Cleveland, Ohio

Omer W. Blodgett
Design Consultant
The Lincoln Electric Company
Cleveland, Ohio

14.1 DEFINITIONS AND TERMINOLOGY

Arc welding is one of several fusion processes for joining metals. By the application of intense heat, metal at the joint between two parts is melted and caused to intermix—directly or, more commonly, with an intermediate molten filler metal. Upon cooling and solidification, a metallurgical bond results. Since the joining is by intermixture of the substance of one part with the substance of the other part, with or without an intermediate of like substance, the final weldment has the potential for exhibiting at the joint the same strength properties as the metal of the parts. This is in sharp contrast to nonfusion processes of joining—such as soldering, brazing, or adhesive bonding—in which the mechanical and physical properties of the base materials cannot be duplicated at the joint.

In arc welding, the intense heat needed to melt metal is produced by an electric arc. The arc is formed between the work to be welded and an electrode that is manually or mechanically moved along the joint (or the work may be moved under a stationary electrode). The electrode may be a carbon or tungsten rod, the sole purpose of which is to carry the current and sustain the electric arc between its tip and the workpiece. Or it may be a specially prepared rod or wire that not only conducts the current and sustains the arc, but also melts and supplies filler metal to the joint. If the electrode is a carbon or tungsten rod and the joint requires added metal for fill, that metal is supplied by a separately applied filler-metal rod or wire. Most welding in the manufacture of steel products where filler metal is required, however, is accomplished with the second type of electrode—the type that supplies filler metal as well as providing the conductor for carrying electric current.

14.2 BASIC WELDING CIRCUIT

The basic arc-welding circuit is illustrated in Fig. 14.1. An ac or dc power source fitted with whatever controls may be needed is connected by a ground-work cable to the workpiece and by a "hot" cable to an electrode holder of some type, which makes electrical contact with the welding electrode. When the circuit is energized and the electrode tip is touched to the grounded workpiece and then withdrawn and held close to the spot of contact, an arc is created across the gap. The arc produces a temperature of about 6500°F at the tip of the electrode, a temperature more than adequate for melting most metals. The heat produced melts the base metal in the vicinity of the arc and any filler metal supplied by the electrode or by a separately introduced rod or wire. A common pool of molten metal is produced, called a *crater.* This crater solidifies behind the electrode as it is moved along the joint being welded. The result is a fusion bond and the metallurgical unification of the workpieces.

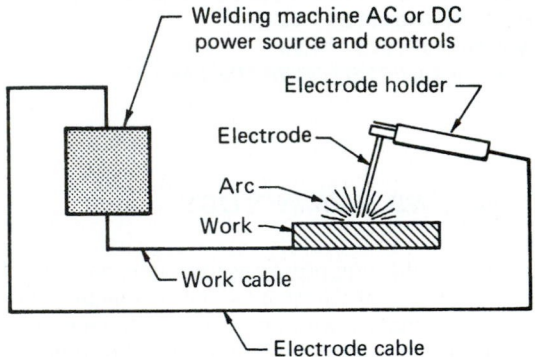

FIGURE 14.1 The basic arc-welding circuit. *(The Lincoln Electric Company.)*

14.3 ARC SHIELDING

Using the heat of an electric arc to join metals, however, requires more than the moving of the electrode with respect to the weld joint. Metals at high temperatures are chemically reactive with the main constituents of air—oxygen and nitrogen. Should the metal in the molten pool come in contact with air, oxides and nitrides would be formed, which upon solidification of the molten pool would destroy the strength properties of the weld joint. For this reason, the various arc-welding processes provide some means for covering the arc and the molten pool with a protective shield of gas, vapor, or slag. This is referred to as *arc shielding,* and such shielding may be accomplished by various techniques, such as the use of a vapor-generating covering on filler-metal-type electrodes, the covering of the arc and molten pool with a separately applied inert gas or a granular flux, or the use of materials within the cores of tubular electrodes that generate shielding vapors.

Whatever the shielding method, the intent is to provide a blanket of gas, vapor, or slag that prevents or minimizes contact of the molten metal with air. The shielding method also affects the stability and other characteristics of the arc. When the shielding is produced by an electrode covering, by electrode core substances, or by separately applied granular flux, a fluxing or metal-improving function is usually also provided. Thus the core materials in a flux-core electrode may perform a deoxidizing function as well as a shielding function, and in submerged-arc welding, the granular flux applied to the joint ahead of the arc may add alloying elements to the molten pool as well as shielding it and the arc.

Figure 14.2 illustrates the shielding of the welding arc and molten pool with a covered "stick" electrode—the type of electrode used in most manual arc welding. The extruded covering on the filler metal rod, under the heat of the arc, generates a gaseous shield that prevents air from coming in contact with the molten metal. It also supplies ingredients that react with deleterious substances on the metals, such as oxides and salts, and ties these substances up chemically in a slag that, being lighter than the weld metal, rises to the top of the pool and crusts over the newly solidified metal. This slag, even after soldification, has a protective function: It minimizes contact of the very hot solidified metal with air until the temperature lowers to a point where reaction of the metal with air is lessened.

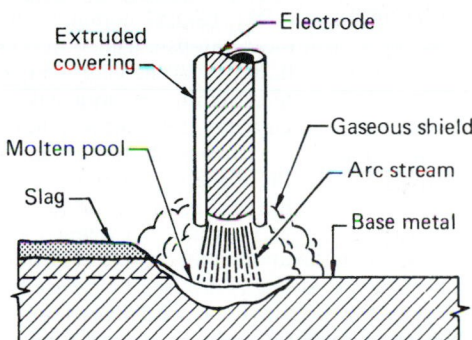

FIGURE 14.2 How the arc and molten pool are shielded by a gaseous blanket developed by the vaporization and chemical breakdown of the extruded covering on the electrode in stick-electrode welding. Fluxing material in the electrode covering reacts with unwanted substances in the molten pool, tying them up chemically and forming a slag that crusts over the hot solidified metal. The slag, in turn, protects the hot metal from reaction with the air while it is cooling. *(The Lincoln Electric Company.)*

While the main function of the arc is to supply heat, it has other functions that are important to the success of arc-welding processes. It can be adjusted or controlled to transfer molten metal from the electrode to the work, to remove surface films, and to bring about complex gas-slag-metal reactions and various metallurgical changes.

14.4   NATURE OF THE ARC

An arc is an electric current flowing between two electrodes through an ionized column of gas called a *plasma*. The space between the two electrodes—or, in arc welding, the space between the electrode and the work—can be divided into three areas of heat generation: the *cathode*, the *anode*, and the arc *plasma*.

The welding arc is characterized as a high-current, low-voltage arc that requires a high concentration of electrons to carry the current. Negative electrons are emitted from the cathode and flow—along with the negative ions of the plasma—to the positive anode, as shown in Fig. 14.3. Positive ions flow in the reverse direction. A *negative ion* is an atom that has picked up one or more electrons beyond the number needed to balance the positive charge on its nucleus—thus the negative charge. A *positive ion* is an atom that has lost one or more electrons—thus the positive charge. However, just as in a solid conductor, the principal flow of current in the arc is by electron travel.

Heat is generated in the cathode area mostly by the positive ions striking the surface of the cathode. Heat at the anode is generated mostly by electrons. These have been accelerated as they pass through the plasma by the arc voltage, and they give up their energy as heat when striking the anode.

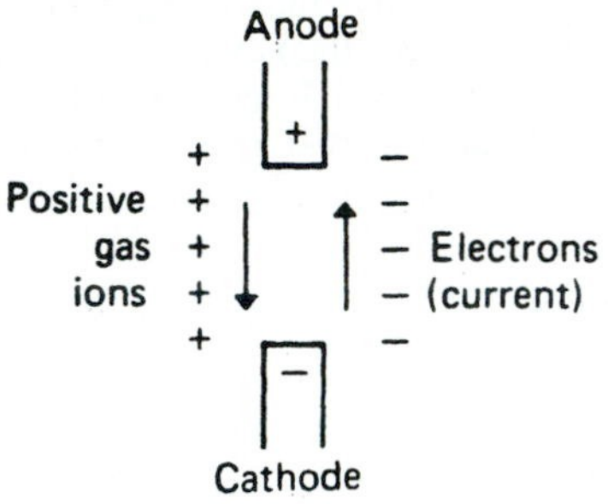

The plasma, or arc column, is a mixture of neutral and excited gas atoms. In the central column of the plasma, electrons, atoms, and ions are in accelerated motion and are constantly colliding. The hottest part of the plasma is the central column, where the motion is most intense. The outer portion or the arc flame is somewhat cooler and consists of recombining gas molecules that were disassociated in the central column.

The distribution of heat or voltage drop in the three heat zones can be changed. Changing the arc length has the greatest effect on the arc plasma. Changing the shielding gas can change the heat balance between the anode and cathode. The addition of potassium salts to the plasma reduces the arc voltage because of increased ionization.

FIGURE 14.3 Characteristics of the arc. *(The Lincoln Electric Company.)*

In welding, not only does the arc provide the heat needed to melt the electrode and the base metal, but under certain conditions it must also supply the means to transport the molten metal from the tip of the electrode to the work. Several mechanisms for metal transfer exist. In one, the molten drop of metal touches the molten metal in the crater, and transfer is by surface tension. In another, the drop is ejected from the molten metal at the electrode tip by an electric pinch. It is ejected at high speed and retains this speed unless slowed by gravitational forces. It may be accelerated by the plasma, as in the case of a pinched-plasma arc. These forces are the ones that transfer the molten metal in overhead welding. In flat welding, gravity is also a significant force in metal transfer.

If the electrode is consumable, the tip melts under the heat of the arc, and molten droplets are detached and transported to the work through the arc column. Any arc-welding system in which the electrode is melted off to become part of the weld is described as *metal arc*. If the electrode is refractory—carbon or tungsten—there are no molten droplets to be forced across the gap and onto the work. Filler metal is melted into the joint from a separate rod or wire.

More of the heat developed by the arc ends up in the weld pool with consumable electrodes than with nonconsumable electrodes, with the result that higher thermal efficiencies and narrower heat-affected zones are obtained. Typical thermal efficiencies for metal-arc welding are in the 75 to 80 percent range; for welding with nonconsumable electrodes, efficiencies are 50 to 60 percent.

Since there must be an ionized path to conduct electricity across a gap, the mere switching on of the welding current with a cold electrode poised over the work will not start the arc. The arc must first be *ignited*. This is accomplished either by supplying an initial voltage high enough to cause a discharge or by touching the electrode to the work and then withdrawing it as the contact area becomes heated. High-frequency spark discharges are frequently used for igniting gas-shielded arcs, but the most common method of striking an arc is the touch-and-withdraw method.

Arc welding may be done with either alternating or direct current and with the electrode either positive or negative. The choice of current and polarity depends on the process, the type of electrode, the arc atmosphere, and the metal being welded. Whatever the current, it must be controlled to satisfy the variables—amperage and voltage—which are specified by the welding procedures.

14.5 OVERCOMING CURRENT LIMITATIONS

The objective in commercial welding is to get the job done as fast as possible so as to lessen the time costs of skilled workers. One way to speed the welding process is to raise the current—use a higher amperage—since the faster electrical energy can be induced in the weld joint, the faster will be the welding rate.

With manual stick-electrode welding, however, there is a practical limit to the current. The covered electrodes are from 9 to 18 in long, and if the current is raised too high, electrical resistance heating within the unused length of electrode will become so great that the covering overheats and "breaks down"—the covering ingredients react with each other or oxidize and do not function properly at the arc. Also, the hot core wire increases the melt-off rate and the arc characteristics change. The mechanics of stick-electrode welding are such that electric contact with the electrode cannot be made immediately above the arc—a technique that would circumvent much of the resistance heating.

Not until semiautomatic guns and automatic welding heads (which are fed by continuous electrode wires) were developed was there a way of solving the resistance-heating problem and thus making feasible the use of high currents to speed the welding process. In such guns and heads, electric contact with the electrode is made close to the arc. The length between the tip of the electrode and the point of electric contact is then inadequate for enough resistance heating to take place to overheat the electrode in advance of the arc, even with currents two or three times those usable with stick-electrode welding.

This solving of the point-of-contact problem and circumventing of the effects of resistance heating in the electrode constituted a breakthrough that substantially lowered welding costs and increased the use of arc welding in industrial metals joining. In fact, through the ingenuity of welding equipment manufacturers, the resistance-heating effect has been put to work constructively in a technique known as long-stickout welding. Here, the length of electrode between the point of electric contact in the welding gun or head and the arc is adjusted so that resistance heating almost—but not quite—overheats the protruding electrode. Thus when a point on the electrode reaches the arc, the metal at that point is about ready to melt and less arc heat is required to melt it. Because of this, still higher welding speeds are possible.

14.6 COMMERCIAL ARC-WELDING PROCESSES

14.6.1 Shielded Metal-Arc Welding

The *shielded metal-arc process*—commonly called *stick-electrode welding* or *manual welding*—is the most widely used of the various arc-welding processes. It is characterized by application versatility and flexibility and relative simplicity in equipment. It is the process used by the small welding shop, by the home mechanic, and by the farmer for repair of equipment; it is also a process having extensive application in industrial fabrication, structural steel erection, weldment manufacture, and other commercial metals joining. Arc welding, to persons only casually acquainted with welding, usually means shielded metal-arc welding.

With this process, an electric arc is struck between the electrically grounded work and a 9- to 18-in length of covered metal rod—the electrode. The electrode is clamped in an electrode holder, which is joined by a cable to the power source. The welder grips the insulated handle of the electrode holder and maneuvers the tip of the electrode with respect to the weld joint. When the welder touches the tip of the electrode against the work and then withdraws it to establish the arc, the welding circuit is completed. The heat of the arc melts base metal in the immediate area, the electrode's metal core, and any metal particles that may be in the electrode's covering. It also melts, vaporizes, or breaks down chemically nonmetallic substances incorporated in the covering for arc-shielding, metal-protection, or metal-conditioning purposes. The mixing of molten base metal and filler metal from the electrode provides the coalescence required to effect joining (see Fig. 14.2).

As welding progresses, the covered rod becomes shorter and shorter. Finally, the welding must be stopped to remove the stub and replace it with a new electrode. This periodic changing of electrodes is one of the major disadvantages of the process in production welding. It decreases the *operating factor,* or the percent of the welder's time spent in the actual laying of weld beads.

Another disadvantage of shielded metal-arc welding is the limitation placed on the current that can be used. High amperages, such as those used with semiautomatic guns or automatic welding heads, are impractical because of the long (and varying) length of electrode between the arc and the point of electric contact in the jaws of the electrode holder. The welding current is limited by the resistance heating of the electrode. The electrode temperature must not exceed the *breakdown temperature* of the covering. If the temperature is too high, the covering chemicals react with each other or with air and therefore do not function properly at the arc.

The versatility of the process—plus the simplicity of equipment—is viewed by many users whose work would permit some degree of mechanized welding as overriding its inherent disadvantages. This point of view was formerly well taken, but now that semiautomatic self-shielded flux-cored arc welding has been developed to a similar (or even superior) degree of versatility and flexibility, there is less justification for adhering to stick-electrode welding in steel fabrication and erection wherever substantial amounts of weld metals must be placed.

14.6.2 Self-Shielded Flux-Cored Welding

The self-shielded flux-cored arc-welding process is an outgrowth of shielded metal-arc welding. The versatility and maneuverability of stick electrodes in manual welding stimulated efforts to mechanize the shielded metal-arc process. The thought was that if some way could be found to put an electrode with self-shielding characteristics in coil form and to feed it mechanically to the arc, welding time lost in changing electrodes and the material lost as electrode stubs would be eliminated. The result of these efforts was the development of the semiautomatic and full-automatic processes for welding with continuous flux-cored tubular electrode "wires." Such fabricated wires (Fig. 14.4) contain in their cores the ingredients for fluxing and deoxidizing molten metal and for generating shielding gases and vapors and slag coverings.

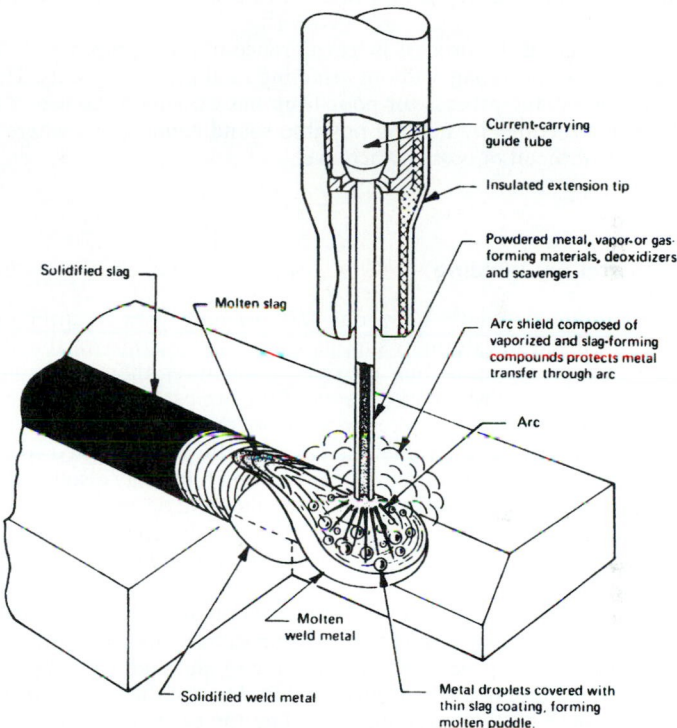

Current-carrying guide tube

Insulated extension tip

Powdered metal, vapor- or gas-forming materials, deoxidizers and scavengers

Arc shield composed of vaporized and slag-forming compounds protects metal transfer through arc

Arc

Solidified slag

Molten slag

Molten weld metal

Solidified weld metal

Metal droplets covered with thin slag coating, forming molten puddle.

FIGURE 14.4 Principles of the self-shielded flux-cored arc-welding process. The electrode may be viewed as an *inside-out* construction of the stick electrode used in shielded metal-arc welding. Putting the shield-generating materials inside the electrode allows the coiling of long, continuous lengths of electrode and gives an outside conductive sheath for carrying the welding current from a point close to the arc. *(The Lincoln Electric Company.)*

In essence, semiautomatic welding with flux-cored electrodes is manual shielded metal-arc welding with an electrode many feet long instead of just a few inches long.

By pressing the trigger that completes the welding circuit, the operator activates the mechanism that feeds the electrode to the arc. The operator uses a gun instead of an electrode holder, but it is similarly light in weight and easy to maneuver. The only other major difference is that the weld metal of the electrode surrounds the shielding and fluxing chemicals rather than being surrounded by them.

Full-automatic welding with self-shielded flux-cored electrodes goes one step further in mechanization—the removal of direct manual manipulation in the utilization of the open-arc process.

One of the advantages of the self-shielded flux-cored arc-welding process is the high deposition rates that are made possible with the hand-held semiautomatic gun. Higher deposition rates, plus automatic electrode feed and elimination of lost time for changing electrodes, have resulted in substantial production economies wherever the semiautomatic process has been used to replace stick-electrode welding. Decreases in welding costs as great as 50 percent have been common, and in some production welding, deposition rates have been increased as much as 400 percent.

Another advantage of the process is its tolerance of poor fitup, which in shops often reduces rework and repair without affecting final product quality. The tolerance of the semiautomatic process for poor fitup has expanded the use of tubular steel members in structures by making possible sound connections where perfect fitup would be too difficult or costly to achieve.

14.6.3   Gas Metal-Arc Welding

Gas metal-arc welding, popularly known as *MIG welding,* uses a continuous electrode for filler metal and an externally supplied gas or gas mixture for shielding. The shielding gas—helium, argon, carbon dioxide, or mixtures thereof—protects the molten metal from reacting with constituents of the atmosphere. Although the gas shield is effective in shielding the molten metal from the air, deoxidizers are usually added as alloys in the electrode. Sometimes light coatings are applied to the electrode for arc stabilizing or other purposes. Lubricating films may also be applied to increase the electrode feeding efficiency in semiautomatic welding equipment. Reactive gases may be included in the gas mixture for arc-conditioning functions. Figure 14.5 illustrates the method by which shielding gas and continuous electrode are supplied to the welding arc.

MIG welding may be used with all the major commercial metals, including carbon, alloy, and stainless steels and aluminum, magnesium, copper, iron, titanium, and zirconium. It is a preferred process for the welding of aluminum, magnesium, copper, and many of the alloys of these reactive metals. Most of the irons and steels can be satisfactorily joined by MIG welding, including the carbon-free irons, the low-carbon and low-alloy steels, the high-strength quenched and tempered steels, the chromium irons and steels, the high-nickel steels, and some of the so-called super-alloy steels. With these various materials, the welding techniques and procedures may vary widely. Thus carbon dioxide or argon-oxygen mixtures are suitable for arc shielding when welding the low-carbon and low-alloy steels, whereas pure inert gas may be essential when welding highly alloyed steels. Copper and many of its alloys and the stainless steels are successfully welded by this process.

Welding is either semiautomatic, using a hand-held gun to which electrode is fed automatically, or done with fully-automatic equipment. The welding guns or heads are similar to those used with gas-shielded flux-cored welding.

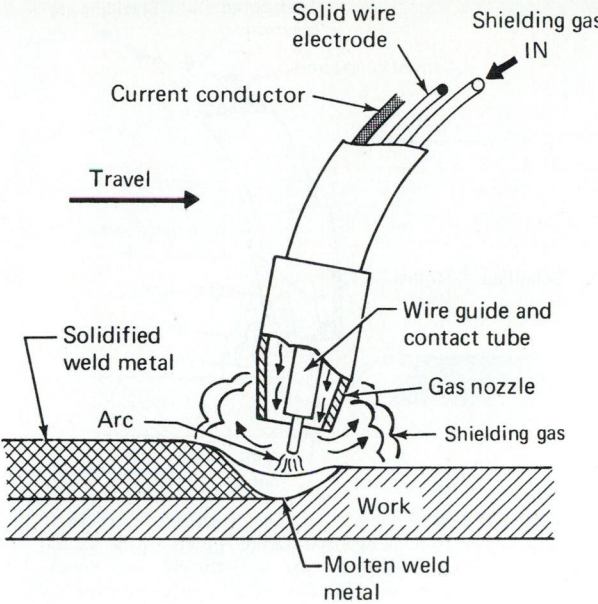

FIGURE 14.5 Principle of the gas metal-arc process. Continuous solid-wire electrode is fed to the gas-shielded arc. *(The Lincoln Electric Company.)*

14.6.4 The Gas-Shielded Flux-Cored Process

The *gas-shielded flux-cored process* may be looked on as a hybrid between self-shielded flux-cored arc welding and gas metal-arc welding. Tubular electrode wire is used (Fig. 14.6), as in the self-shielded process, but the ingredients in its core are for fluxing, deoxidizing, scavenging, and sometimes alloying additions rather than for these functions plus the generation of protective vapors. In this respect, the process has similarities to the self-shielded flux-cored electrode process, and the tubular electrodes used are classified by the American Welding Society (AWS) along with electrodes used in the self-shielded process. However, the process is similar to gas metal-arc welding in that a gas is separately applied to act as arc shield.

The gas-shielded flux-cored process is used for welding mild and low-alloy steels. It gives high deposition rates, high deposition efficiencies, and high operating factors. Radiographic-quality welds are easily produced, and the weld metal with mild and low-alloy steels has good ductility and toughness. The process is adaptable to a wide variety of joints and has the capability for all-position welding.

14.6.5 Gas Tungsten-Arc Welding

The AWS definition of *gas tungsten-arc* (TIG) welding is "an arc-welding process wherein coalescence is produced by heating with an arc between a tungsten electrode and the work." A filler metal may or may not be used. Shielding is obtained with a gas or a gas mixture.

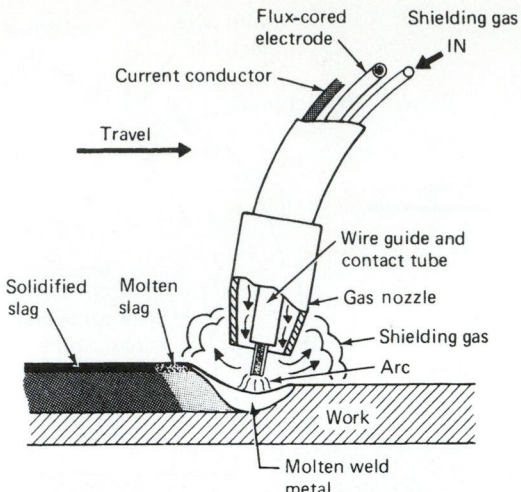

FIGURE 14.6 Principles of the gas-shielded flux-cored process. Gas from an external source is used for the shielding; the core ingredients are for fluxing and metal-conditioning purposes. *(The Lincoln Electric Company.)*

Essentially, the nonconsumable tungsten electrode is a *torch*—a heating device. Under the protective gas shield, metals to be joined may be heated above their melting points so that material from one part coalesces with material from the other part. Upon solidification of the molten area, unification occurs. Pressure may be used when the edges to be joined are approaching the molten state to assist coalescence. Welding in this manner requires no filler metal.

If the work is too heavy for the mere fusing of abutting edges, and if groove joints or reinforcements such as fillets are required, filler metal must be added. This is supplied by a filler rod that is manually or mechanically fed into the weld puddle. Both the tip of the nonconsumable tungsten electrode and the tip of the filler rod are kept under the protective gas shield as welding progresses.

Figure 14.7 illustrates the TIG torch. In automatic welding, filler wire is fed mechanically through a guide into the weld puddle. When running heavy joints manually, a variation in the mode of feeding is to lay or press the filler rod in or along the joint and melt it along with the joint edges. All the standard types of joints can be welded with the TIG process and filler metal.

Materials weldable by the TIG process are most grades of carbon, alloy, and stainless steels; aluminum and most of its alloys; magnesium and most of its alloys; copper and various brasses and bronzes; high-temperature alloys of various types; numerous hard-surfacing alloys; and such metals as titanium, zirconium, gold, and silver. The process is especially adapted for welding thin materials where the requirements for quality and finish are exacting. It is one of the few processes that is satisfactory for welding such tiny and thin-walled objects as transistor cases, instrument diaphragms, and delicate expansion bellows.

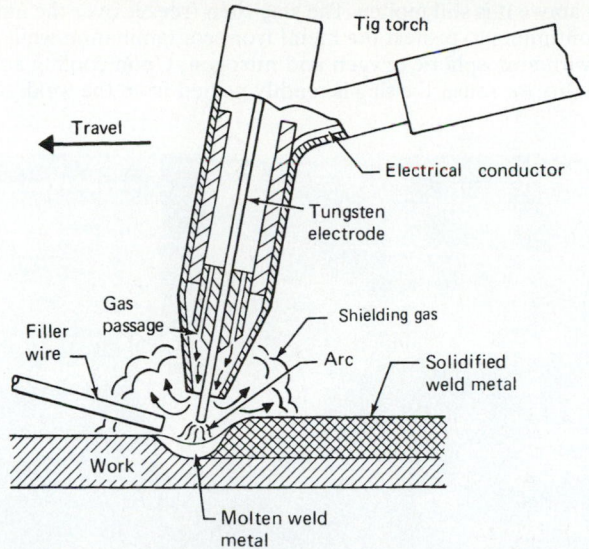

FIGURE 14.7 Principles of the gas tungsten-arc process. If filler metal is required, it is fed into the pool from a separate filler rod. *(The Lincoln Electric Company.)*

14.6.6 Submerged-Arc Welding

Submerged-arc welding differs from other arc-welding processes in that a blanket of fusible granular material—commonly called *flux*—is used for shielding the arc and the molten metal. The arc is struck between the workpiece and a bare wire electrode, the tip of which is submerged in the flux. Since the arc is completely covered by the flux, it is not visible, and the weld is run without the flash, spatter, and sparks that characterize the open-arc process. The nature of the flux is such that very little smoke or visible fumes are developed.

The process is either semiautomatic or fully automatic, and the electrode is fed mechanically to the welding gun, head, or heads. In semiautomatic welding, the welder moves the gun, usually equipped with a flux-feeding device, along the joint. Flux feed may be by gravity flow through a nozzle concentric with the electrode from a small hopper atop the gun, or it may be through a concentric nozzle tube connected to an air-pressurized flux tank. Flux may also be applied in advance of the welding operation or ahead of the arc from a hopper run along the joint. In fully automatic submerged-arc welding, flux is fed continuously to the joint ahead of or concentric with the arc, and fully automatic installations are commonly equipped with vacuum systems to pick up the unfused flux left by the welding head or heads for cleaning and reuse.

During welding, the heat of the arc melts some of the flux along with the tip of the electrode, as illustrated in Fig. 14.8. The tip of the electrode and the welding zone are always surrounded and shielded by molten flux, surmounted by a layer of unfused flux. The electrode is held a short distance above the workpiece. As the electrode progresses along the joint, the lighter molten flux rises above the molten metal in the form of a slag. The weld metal, having a higher melting (freezing) point, solidifies while the slag above it is still molten. The slag then freezes over the newly solidified weld metal, continuing to protect the metal from contamination while it is very hot and reactive with atmospheric oxygen and nitrogen. Upon cooling and removal of any unmelted flux for reuse, the slag is readily peeled from the weld.

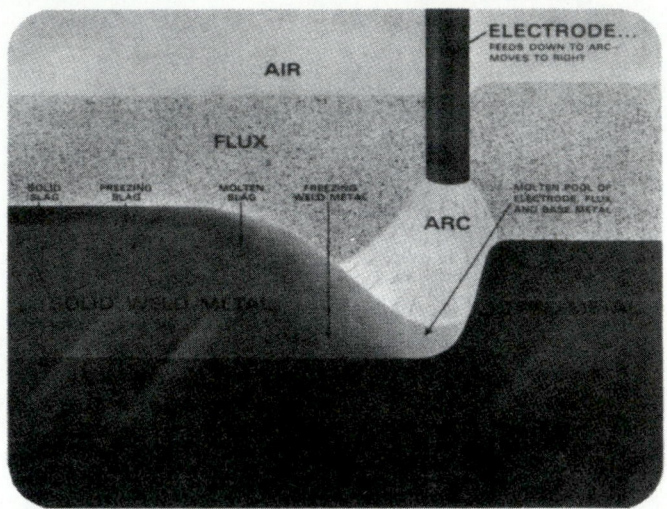

FIGURE 14.8   The mechanics of the submerged-arc process. The arc and the molten weld metal are buried in the layer of flux, which protects the weld metal from contamination and concentrates the heat into the joint. The molten flux arises through the pool, deoxidizing and cleansing the molten metal, and forms a protective slag over the newly deposited weld. *(The Lincoln Electric Company.)*

There are two general types of submerged-arc fluxes: bonded and fused. In *bonded* fluxes, the finely ground chemicals are mixed, treated with a bonding agent, and manufactured into a granular aggregate. The deoxidizers are incorporated in the flux. *Fused* fluxes are a form of glass resulting from fusing the various chemicals and then grinding the glass to a granular form. Fluxes are available that add alloying elements to the weld metal, enabling alloy weld metal to be made with mild-steel electrodes.

High currents can be used in submerged-arc welding, and extremely high heat can be developed. Because the current is applied to the electrode a short distance above its tip, relatively high amperages can be used on small-diameter electrodes. This results in extremely high current densities on relatively small cross sections of electrode. Currents as high as 600 A can be carried on electrodes as small as 5⁄64 in, giving a density of the order of 100 000 A/in^2—6 to 10 times that carried on stick electrodes.

Because of the high current density, the melt-off rate is much higher for a given electrode diameter than with stick-electrode welding. The melt-off rate is affected by the electrode material, the flux, the type of current, the polarity, and the length of wire beyond the point of electric contact in the gun or head.

The insulating blanket of flux above the arc prevents rapid escape of heat and concentrates it in the welding zone. Not only are the electrode and base metal melted rapidly, but the fusion is deep into the base metal. The deep penetration allows the use of small welding grooves, thus minimizing the amount of filler metal per foot of joint and permitting fast welding speeds. Fast welding, in turn, minimizes the total heat input into the assembly and thus tends to prevent problems of heat distortion. Even relatively thick joints can be welded in one pass by the submerged-arc process.

Welds made under the protective layer of flux have good ductility and impact resistance and uniformity in bead appearance. Mechanical properties at least equal to those of the base metal are consistently obtained. In single-pass welds, the amount of fused base material is large compared to the amount of filler metal used. Thus in such welds the base metal may greatly influence the chemical and mechanical properties of the weld. For this reason, it is sometimes unnecessary to use electrodes of the same composition as the base metal for welding many of the low-alloy steels.

With proper selection of equipment, submerged-arc welding is widely applicable to the welding requirements of industry. It can be used with all types of joints and permits welding a full range of carbon and low-alloy steels, from 16-gauge (1.5-mm) sheet to the thickest plate. It is also applicable to some high-alloy, heat-treated, and stainless steels and is a favored process for rebuilding and hard surfacing. Any degree of mechanization can be used—from the hand-held semiautomatic gun to boom- or track-carried and fixture-held multiple welding heads.

The high quality of submerged-arc welds, the high deposition rates, the deep penetration, the adaptability of the process to full mechanization, and the comfort characteristics (no glare, sparks, spatter, smoke, or excessive heat radiation) make it a preferred process in steel fabrication. It is used extensively in ship and barge building, in railroad car building, in pipe manufacture, and in fabricating structural beams, girders, and columns where long welds are required. Automatic submerged-arc installations are also key features of the welding areas of plants turning out mass-produced assemblies joined with repetitive short welds.

The high deposition rates attained with submerged-arc welding are chiefly responsible for the economies achieved with the process. The cost reductions from changing from the manual shielded metal-arc process to the submerged-arc process are frequently dramatic. Thus a hand-held submerged-arc gun with mechanized travel may reduce welding costs more than 50 percent; with fully automatic multiarc equipment, it is not unusual for the costs to be but 10 percent of those attained with stick-electrode welding.

14.6.7 Other "Arc-Welding" Processes

Various adaptations of the arc-welding processes described have been made to meet specialized joining needs. In addition, there are processes using electrical energy to join metals that do not fall under the category of arc welding—including electrical resistance welding and ultrasonic, electron beam, and electrodeposition welding.

Electroslag welding is an adaptation of the submerged-arc process for joining thick materials in a vertical position. Figure 14.9 is a diagrammatic sketch of the electroslag process. It will be noted that whereas some of the principles of submerged-arc welding apply, in other respects the process resembles a casting operation.

In Fig. 14.9, a square butt joint in heavy plate is illustrated, but the electroslag process—with modifications in equipment and technique—is also applicable to T joints, corner joints, girth seams in heavy-wall cylinders, and other joints. The process is suited best for materials at least 1 in in thickness and can be used with multiple electrodes on materials up to 10 in thick without excessive difficulties.

As illustrated by the open square butt joint, the assembly is positioned for the vertical deposition of weld metal. A starting pad at the bottom of the joint prevents

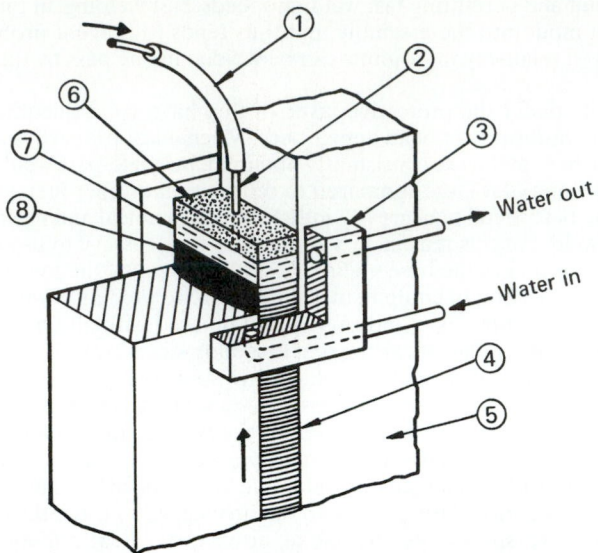

FIGURE 14.9 Schematic sketch of electroslag welding: (1) electrode guide tube, (2) electrode, (3) water-cooled copper shoes, (4) finished weld, (5) base metal, (6) molten slag, (7) molten weld metal, and (8) solidified weld metal. *(The Lincoln Electric Company.)*

the fall-out of the initially deposited weld metal and, since it is penetrated, ensures a full weld at this point. Welding is started at the bottom and progresses upward. Water-cooled dams, which may be looked on as molds, are placed on each side of the joint. These dams are moved upward as the weld-metal deposition progresses. The joint is filled in one *pass*—a single upward progression—of one or more consumable electrodes. The electrode or electrodes may be oscillated across the joint if the width of the joint makes this desirable.

At the start of the operation, a layer of flux is placed in the bottom of the joint and an arc is struck between the electrode (or electrodes) and the work. The arc melts the slag, forming a molten layer, which subsequently acts as an electrolytic heating medium. The arc is then quenched or shorted-out by this molten conductive layer. Heat for melting the electrode and the base metal subsequently results from the electrical resistance heating of the electrode section extending from the contact tube and from the resistance heating within the molten slag layer. As the electrode (or electrodes) is consumed, the welding head (or heads) and the cooling dams move upward.

In conventional practice, the weld deposit usually contains about one-third melted base metal and two-thirds electrode metal—which means that the base metal substantially contributes to the chemical composition of the weld metal. Flux consumption is low, since the molten flux and the unmelted flux above it "ride" above the progressing weld.

The flux used has a degree of electrical conductivity and low viscosity in the molten condition and a high vaporization temperature. The consumable electrodes may be either solid wire or tubular wire filled with metal powders. Alloying elements may be incorporated into the weld by each of these electrodes.

Weld quality with the electroslag process is generally excellent, because of the protective action of the heavy slag layer. Sometimes, however, the copper dams are provided with orifices just above the slag layer through which a protective gas—argon or carbon dioxide—is introduced to flush out the air above the weld and thus give additional assurance against oxidation. Such provisions are sometimes considered worthwhile when welding highly alloyed steels or steels that contain easily oxidized elements.

Electrogas welding is very similar to electroslag welding in that the equipment is similar and the joint is in the vertical position. As the name implies, the shielding is by carbon dioxide or an inert gas. A thin layer of slag, supplied by the flux-cored electrode, covers the molten metal, and the heat is supplied by an arc rather than by resistance heating, as in the electroslag process.

A disadvantage of the process is that it requires an external source of shielding gas. However, one advantage is that if the welding is stopped, the electrogas process can be started again with less difficulty than the electroslag process.

Stud arc welding is a variation of the shielded metal-arc process that is widely used for attaching studs, screws, pins, and similar fasteners to a large workpiece. The *stud* (or small part) itself—often plus a ceramic ferrule at its tip—is the arc-welding electrode during the brief period of time required for studding.

In operation, the stud is held in a portable pistol-shaped tool called a *stud gun* and positioned by the operator over the spot where it is to be weld-attached. At a press of the trigger, current flows through the stud, which is lifted slightly, creating an arc. After a very short arcing period, the stud is then plunged down into the molten pool created on the base plate, the gun is withdrawn from it, and the ceramic ferrule—if one has been used—is removed. The timing is controlled automatically, and the stud is welded onto the workpiece in less than a second. The fundamentals of the process are illustrated in Fig. 14.10.

Studs are of many shapes. All may be weld-attached with portable equipment. The stud may be used with a ceramic arc-shielding ferrule, as shown in Fig. 14.10, which prevents air infiltration and also acts as a dam to retain the molten metal, or it may have a granular flux, flux coating, or solid flux affixed to the welding end, as illustrated in Fig. 14.11. The flux may include any of the agents found in a regular electrode covering; most important to stud welding is a deoxidizer to guard against porosity.

Plasma-arc (or plasma-torch) welding is one of the newer welding processes which is used industrially, frequently as a substitute for the gas tungsten-arc process. In some applications, it offers greater welding speeds, better weld quality, and less sensitivity to process variables than the conventional processes it replaces. With the plasma torch, temperatures as high as $60\,000°F$ are developed, and theoretically, temperatures as high as $200\,000°F$ are possible.

The heat in plasma-arc welding originates in an arc, but this arc is not diffused as is an ordinary welding arc. Instead, it is constricted by being forced through a relatively small orifice. The *orifice,* or plasma gas, may be supplemented by an auxiliary source of shielding gas.

Orifice gas refers to the gas that is directed into the torch to surround the electrode. It becomes ionized in the arc to form the plasma and emerges from the orifice in the torch nozzle as a plasma jet. If a shielding gas is used, it is directed onto the workpiece from an outer shielding ring.

The workpiece may or may not be part of the electric circuit. In the *transferred-arc system,* the workpiece is a part of the circuit, as in other arc-welding processes. The arc transfers from the electrode through the orifice to the work. In the *non-transferred system,* the constricting nozzle surrounding the electrode acts as an elec-

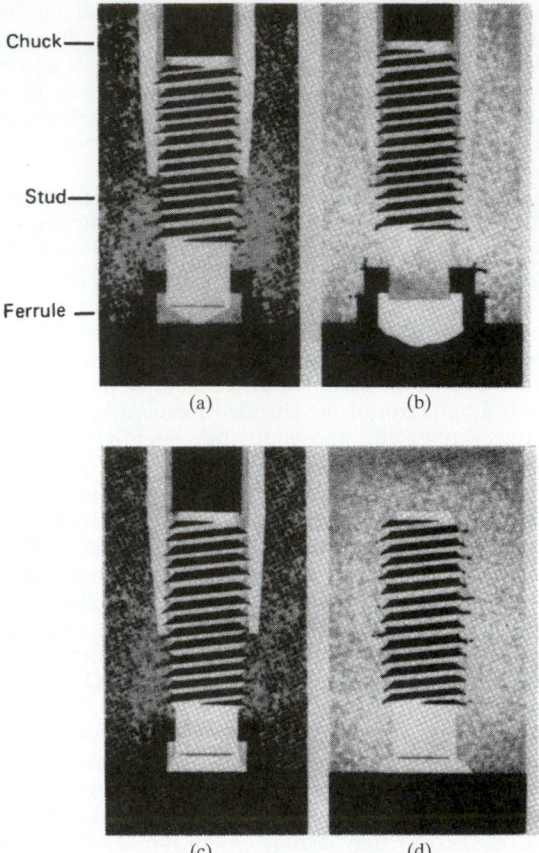

Chuck

Stud

Ferrule

(a)   (b)   (c)   (d)

FIGURE 14.10 Principles of stud welding, using a ceramic ferrule to shield the pool. (*a*) The stud with ceramic ferrule is grasped by the chuck of the gun and positioned for welding. (*b*) The trigger is pressed, the stud is lifted, and the arc is created. (*c*) With the brief arcing period completed, the stud is plunged into the molten pool on the base plate. (*d*) The gun is withdrawn from the welded stud and the ferrule is removed. (*The Lincoln Electric Company.*)

tric terminal, and the arc is struck between it and the electrode tip; the plasma gas then carries the heat to the workpiece. Figure 14.12 illustrates transferred and nontransferred arcs.

The advantages gained by using a constricted-arc process rather than the gas tungsten-arc process include greater energy concentration, improved arc stability, higher welding speeds, and lower width-to-depth ratio for a given penetration. *Keyhole welding*—or penetrating completely through the workpiece—is possible.

The *atomic-hydrogen process of arc welding* may be regarded as a forerunner of gas-shielded and plasma-torch arc welding. Although largely displaced by other pro-

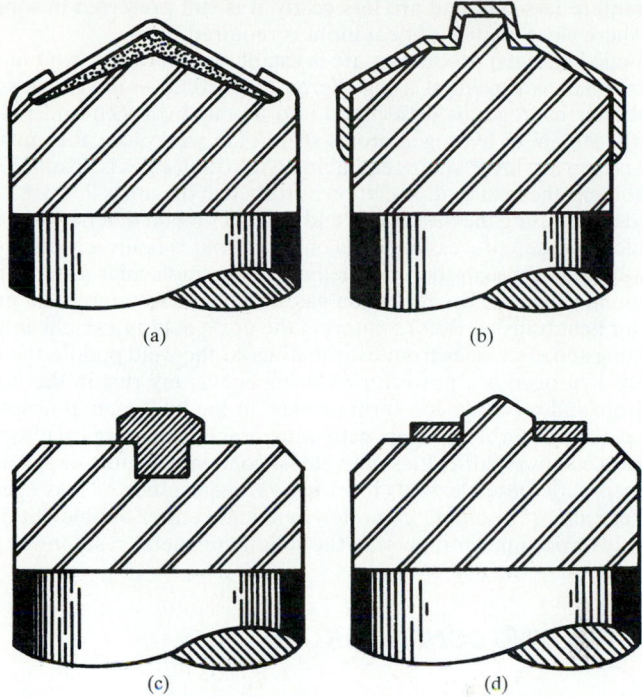

(a)   (b)

(c)   (d)

FIGURE 14.11 Three methods of containing flux on the end of a welding stud: (*a*) granular flux; (*b*) flux coating; (*c*) and (*d*) solid flux. *(The Lincoln Electric Company.)*

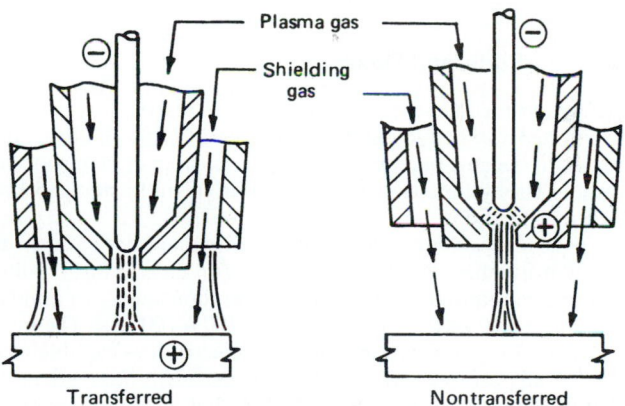

Plasma gas

Shielding gas

Transferred   Nontransferred

FIGURE 14.12 Transferred and nontransferred arcs. *(The Lincoln Electric Company.)*

cesses that require less skill and are less costly, it is still preferred in some manual operations where close control of heat input is required.

In the atomic-hydrogen process, an arc is established between two tungsten electrodes in a stream of hydrogen gas using alternating current. As the gas passes through the arc, molecular hydrogen is dissociated into atomic hydrogen under the intense heat. When the stream of hydrogen atoms strikes the workpiece, the environmental temperature is then at a level where recombining into molecules is possible. As a result of the recombining, the heat of dissociation absorbed in the arc is liberated, supplying the heat needed for fusing the base metal and any filler metal that may be introduced.

The atomic-hydrogen process depends on an arc, but is really a heating torch. The arc supplies the heat through the intermediate of the molecular-dissociation, atom-recombination mechanism. The hydrogen gas, however, does more than provide the mechanism for heat transfer. Before entering the arc, it acts as a shield and a coolant to keep the tungsten electrodes from overheating. At the weld puddle, the gas acts as a shield. Since hydrogen is a powerful reducing agent, any rust in the weld area is reduced to iron, and no oxide can form or exist in the hydrogen atmosphere. Weld metal, however, can absorb hydrogen, with unfavorable metallurgical effects. For this reason, the process gives difficulties with steels containing sulfur or selenium, since hydrogen reacts with these elements to form hydrogen sulfide or hydrogen selenide gases. These are almost insoluble in molten metal and either bubble out of the weld pool vigorously or become entrapped in the solidifying metal, resulting in porosity.

14.7 ARC-WELDING CONSUMABLES

Arc-welding consumables are the materials used up during welding, such as electrodes, filler rods, fluxes, and externally applied shielding gases. With the exception of the gases, all the commonly used consumables are covered by AWS specifications.

Twenty specifications in the AWS A5.x series prescribed the requirements for welding electrodes, rods, and fluxes.

14.7.1 Electrodes, Rods, and Fluxes

The first specification for mild-steel-covered electrodes, A5.1, was written in 1940. As the welding industry expanded and the number of types of electrodes for welding steel increased, it became necessary to devise a system of electrode classification to avoid confusion. The system used applies to both the mild-steel A5.1 and the low-alloy steel A5.5 specifications.

Classifications of *mild and low-alloy steel electrodes* are based on an E prefix and a four- or five-digit number. The first two digits (or three, in a five-digit number) indicate the minimum required tensile strength in thousands of pounds per square inch. For example, 60 = 60 kpsi, 70 = 70 kpsi, and 100 = 100 kpsi. The next to the last digit indicates the welding position in which the electrode is capable of making satisfactory welds: 1 = all positions—flat, horizontal, vertical, and overhead; 2 = flat and horizontal fillet welding (see Table 14.1). The last digit indicates the type of current to be used and the type of covering on the electrode (see Table 14.2).

Originally a color identification system was developed by the National Electrical Manufacturers Association (NEMA) in conjunction with the AWS to identify the electrode's classification. This was a system of color markings applied in a specific relationship on the electrode, as in Fig. 14.13a. The colors and their significance are

TABLE 14.1   AWS A5.1-69 and A5.5-69 Designations for Manual Electrodes

---

a. The prefix *E* designates arc-welding electrode.

b. The first two digits of four-digit numbers and the first three digits of five-digit numbers indicate minimum tensile strength:

    E 60XX   60 000 psi minimum tensile strength
    E 70XX   70 000 psi minimum tensile strength
    E110XX   110 000 psi minimum tensile strength

c. The next-to-last digit indicates position:

    EXX1X   All positions
    EXX2X   Flat position and horizontal fillets

d. The suffix (for example, EXXXX- *A1*) indicates the approximate alloy in the weld deposit:

-A1	0.5% Mo
-B1	0.5% Cr, 0.5% Mo
-B2	1.25% Cr, 0.5% Mo
-B3	2.25% Cr, 1% Mo
-B4	2% Cr, 0.5% Mo
-B5	0.5% Cr, 1% Mo
-C1	2.5% Ni
-C2	3.25% Ni
-C3	1% Ni, 0.35% Mo, 0.15% Cr
-D1 and D2	0.25 to 0.45% Mo, 1.75% Mn
-G	0.5% min Ni, 0.3% min Cr, 0.2% min Mo, 0.1% min V, 1% min Mn (only one element required)

---

listed in Tables 14.3 and 14.4. The NEMA specification also included the choice of imprinting the classification number on the electrode, as in Fig. 14.13*b*.

Starting in 1964, new and revised AWS specifications for covered electrodes required that the classification number be imprinted on the covering, as in Fig. 14.13*b*. However, some electrodes can be manufactured faster than the imprinting equipment can mark them, and some sizes are too small to be legibly marked with an imprint. Although AWS specifies an imprint, the color code is accepted on electrodes if imprinting is not practical.

Bare mild-steel electrodes (electrode wires) for submerged-arc welding are classified on the basis of chemical composition, as shown in Table 14.5. In this classifying system, the letter *E* indicates an electrode as in the other classifying systems, but

TABLE 14.2   AWS A5.1-69 Electrode Designations for Covered Arc-Welding Electrodes

Designation	Current	Covering type
EXX10	dc+ only	Organic
EXX11	ac or dc+	Organic
EXX12	ac or dc−	Rutile
EXX13	ac or dc±	Rutile
EXX14	ac or dc±	Rutile, iron-powder (approx. 30%)
EXX15	dc+ only	Low-hydrogen
EXX16	ac or dc+	Low-hydrogen
EXX18	ac or dc+	Low-hydrogen, iron-powder (approx. 25%)
EXX20	ac or dc±	High iron-oxide
EXX24	ac or dc±	Rutile, iron-powder (approx. 50%)
EXX27	ac or dc±	Mineral, iron-powder (approx. 50%)
EXX28	ac or dc+	Low-hydrogen, iron-powder (approx. 50%)

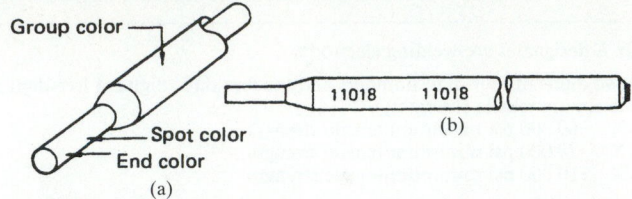

FIGURE 14.13 (*a*) National Electrical Manufacturers Association color-code method to identify an electrode's classification. (*b*) American Welding Society imprint method. *(The Lincoln Electric Company.)*

here the similarity stops. The next letter, *L, M,* or *H,* indicates low, medium, or high manganese, respectively. The following number or numbers indicate the approximate carbon content in hundredths of one percent. If there is a suffix *K,* this indicates a silicon-killed steel.

Fluxes for submerged-arc welding are classified on the basis of the mechanical properties of the weld deposit made with a particular electrode. The classification designation given to a flux consists of a prefix *F* (indicating a flux) followed by a two-digit number representative of the tensile-strength and impact requirements for test welds made in accordance with the specification. This is then followed by a

TABLE 14.3   Color Identification for Covered Mild-Steel and Low-Alloy Steel Electrodes

Spot color	End color			
	No color	Blue	Black	Orange
	Group color—No color			
	XX10, XX11, XX14, XX24, XX27, XX28, and all 60 XX			
No color	E6010	E7010G	. . . .	EST
White	E6012	E7010-Ai	. . . .	EC1
Brown	E6013	. . . .	E7014	
Green	E6020			
Blue	E6011	E7011G		
Yellow	. . . .	E7011-A1	E7024	
Black	. . . .	. . . .	E7028	
Silver	E6027			
	Group color—Silver			
	All XX13 and XX20 except E6013 and E6020			
Brown				
White				
Green	. . . .	E7020G		
Yellow	. . . .	E7020-A1		

TABLE 14.4 Color Identification for Covered Low-Hydrogen Low-Alloy Electrodes

Group color—Green
XX15, XX16, and XX18, except E6015 and E6016

Spot color	End color									
	No color	Blue	Black	White	Gray	Brown	Violet	Green	Red	Orange
Red	E7015G	E7015	...	...	E8015G	E9015G	...	E10015G	...	E12015G
White	...	E7015-A1	E90150-B3L	...	...	E9015-D1	...	...	...	
Brown	...	...	E8015-B2L	...	...	E9015-B3	...			
Green	...	...	E8015-B4L	...	...	E8015-B4	...			
Bronze	E7016G	E7016	E7018	E8016-C3	...	E9016G	...	E10016G	...	E12016G
Orange	...	E7016-A1	E7018-A1	E8016G	...	E9016-D1	...	E10015-D2	E11016G	
Yellow	...	...	E8018-C3	E8016-B1	E8018-B1	...	E9018-B3	E10018G	E11018G	E12018G
Black	E7018G	...	E8018G	E8016-C1	E8018-C1	E9016-B3	E9018G	E10018-D2		
Blue	...	...	...	E8016-C2	E8018-C2	E8016-B4	E9018-D1	E10016-D2		
Violet	...	...	E8018-B4	E8016-B2	E8018-B2	...				
Gray	...	...	Mil-12018							
Silver	...	...								

TABLE 14.5 AWS A5.17-69 Chemical-Composition Requirements for Submerged-Arc Electrodes

AWS classification	Chemical composition, percent						
	Carbon	Manganese	Silicon	Sulfur	Phosphorus	Copper†	Total other elements
Low manganese classes:							
EL8	0.10	0.30–0.55	0.05	0.035	0.03	0.15	0.50
EL8K	0.10	0.30–0.55	0.10–0.20	0.035	0.03	0.15	0.50
EL12	0.07–0.15	0.35–0.60	0.05	0.035	0.03	0.15	0.50
Medium manganese classes:							
EM5K‡	0.06	0.90–1.40	0.40–0.70	0.035	0.03	0.15	0.50
EM12	0.07–0.15	0.85–1.25	0.05	0.035	0.03	0.15	0.50
EM12K	0.07–0.15	0.85–1.25	0.15–0.35	0.035	0.03	0.15	0.50
EM13K	0.07–0.19	0.90–1.40	0.45–0.70	0.035	0.03	0.15	0.50
EM15K	0.12–0.20	0.85–1.25	0.15–0.35	0.035	0.03	0.15	0.50
High manganese class:							
EH14	0.10–0.18	1.75–2.25	0.05	0.035	0.03	0.15	0.50

†The copper limit is independent of any copper or other suitable coating which may be applied to the electrode.
‡This electrode contains 0.05 to 0.15 percent titanium, 0.02 to 0.12 percent zirconium, and 0.05 to 0.15 percent aluminum, which is exclusive of the "Total other elements" requirement.

Note: Analysis shall be made for the elements for which specific values are shown in this table. If, however, the presence of other elements is indicated in the course of routine analysis, further analysis shall be made to determine that the total of these other elements is not present in excess of the limits specified for "Total other elements" in the last column of the table. Single values shown are maximum percentages.

set of letters and numbers corresponding to the classification of the electrode used with the flux.

Gas-shielded flux-cored electrodes are available for welding the low-alloy high-tensile steels. *Self-shielded flux-cored electrodes* are available for all-position welding, as in building construction. Fabricators using or anticipating using the flux-cored arc-welding processes should keep in touch with the electrode manufacturers for new or improved electrodes not included in present specifications.

Mild-steel electrodes for gas metal-arc welding of mild and low-alloy steels are classified on the basis of their chemical compositions and the as-welded mechanical properties of the weld metal. Tables 14.6 and 14.7 are illustrative.

AWS specifications for electrodes also cover those used for welding the stainless steels, aluminum and aluminum alloys, and copper and copper alloys, as well as for weld surfacing.

Shielding gases are consumables used with the MIG and TIG welding processes. The AWS does not write specifications for gases. There are federal specifications, but the welding industry usually relies on *welding grade* to describe the required purity.

The primary purpose of a shielding gas is to protect the molten weld metal from contamination by the oxygen and nitrogen in air. The factors, in addition to cost, that affect the suitability of a gas include the influence of the gas on the arcing and metal-transfer characteristics during welding, weld penetration, width of fusion and surface shape, welding speed, and the tendency to undercut. Among the inert gases—helium, argon, neon, krypton, and xenon—the only ones plentiful enough for practical use in welding are helium and argon. These gases provide satisfactory shielding for the more reactive metals, such as aluminum, magnesium, beryllium, columbium, tantalum, titanium, and zirconium.

Although *pure* inert gases protect metal at any temperature from reaction with constituents of the air, they are not suitable for all welding applications. Controlled quantities of reactive gases mixed with inert gases improve the arc action and metal-transfer characteristics when welding steels, but such mixtures are not used for reactive metals.

Oxygen, nitrogen, and carbon dioxide are reactive gases. With the exception of carbon dioxide, these gases are not generally used alone for arc shielding. Carbon dioxide can be used alone or mixed with an inert gas for welding many carbon and low-alloy steels. Oxygen is used in small quantities with one of the inert gases—usually argon. Nitrogen is occasionally used alone, but it is usually mixed with argon as a shielding gas to weld copper. The most extensive use of nitrogen is in Europe, where helium is relatively unavailable.

14.8 DESIGN OF WELDED JOINTS

While designers need some basic knowledge of welding processes, equipment, materials, and techniques, their main interest is in how to transfer forces through welded joints most effectively and efficiently. Proper joint design is the key to good weld design.

The loads in a welded-steel design are transferred from one member to another through welds placed in weld joints. Both the type of joint and the type of weld are specified by the designer.

Figure 14.14 shows the joint and weld types. Specifying a joint does not by itself describe the type of weld to be used. Thus 10 types of welds are shown for making a

TABLE 14.6 AWS A5.18-69 Mechanical Property Requirements for Gas Metal-Arc Welding Weld Metal[†]

Electrode group	AWS classification	Shielding gas[a]	Current and polarity[b]	Tensile strength[d] min., kpsi	Yield strength[e] min., kpsi	Elongation in 2 in[d] min., %
A. Mild steel	E70S-1	AO	dc, reverse	72	60	22
	E70S-2 E70S-3	AO and CO[c]2				
	E70S-4 E70S-5 E70S-6	CO2				
	E70s-g	Not specified				
B. Low-alloy steel	E70S-1b E70S-GB	CO2 Not specified	dc, reverse Not specified	72 72	60 60	17 22
C. Emissive	E70U-1	AO and A[c]	dc, straight	72	60	22

[†]As-welded mechanical properties determined from an all-weld-metal tension-test specimen.
[a]Shielding gases are AO, argon plus 1 to 5 percent oxygen; CO2, carbon dioxide; A, argon.
[b]*Reverse polarity* means electrode is positive; *straight polarity* means electrode is negative.
[c]Where two gases are listed as interchangeable (that is, AO and CO2 and AO and A) for classification of a specific electrode, the classification may be conducted using either gas.
[d]For each increase of one percentage point in elongation over the minimum, the yield strength or tensile strength, or both, may decrease 1 kpsi to a minimum of 70 kpsi for the tensile strength and 58 kpsi for the yield strength, except for group C electrodes.
[e]0.2 percent offset value.

TABLE 14.7 AWS A5.18-69 Chemical-Composition Requirements for Gas Metal-Arc Welding Electrode

AWS classification	Chemical composition, percent											
	Carbon	Man-ganese	Silicon	Phos-phorus	Sulfur	Nickel†	Chro-mium†	Molyb-denum†	Vana-dium†	Tita-nium	Zirco-nium	Alumi-num
Group A: Mild-steel electroces												
E70S-1	0.07–0.19	0.90–1.40	0.30–0.50	0.025	0.035							
E70S-2	0.06	0.90–1.40	0.40–0.70	0.025	0.035					0.05–0.15	0.02–0.12	0.05–0.15
E70S-3	0.06–0.15	0.90–1.40	0.45–0.70	0.025	0.035							
E70S-4	0.07–0.15	0.90–1.40	0.65–0.85	0.025	0.035							
E70S-5	0.07–0.19	0.90–1.40	0.30–0.60	0.025	0.035							0.50–0.90
E70S-6	0.07–0.15	1.40–1.85	0.80–1.15	0.025	0.035							
E70S-G	No chemical requirements‡											
Group B: Low-alloy steel electrodes												
E70S-1B	0.07–0.12	1.60–2.10	0.50–0.80	0.025	0.035	0.15		0.40–0.60				
E70S-GB	No chemical requirements											
Group C: Emissive electrode												
E70U-1	0.07–0.15	0.80–1.40	0.15–0.35	0.025	0.035							

†For groups A and C these elements may be present but are not intentionally added.
‡For this classification there are no chemical requirements for the elements listed with the exception that there shall be no intentional addition of Ni, Cr, Mo or V.
Note: Single values shown are maximums.

14.25

Type of welds

Type of joints

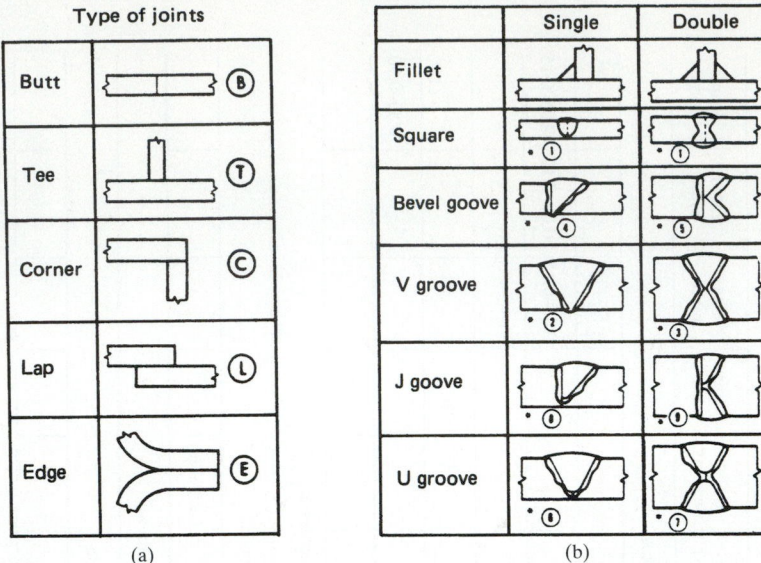

FIGURE 14.14   (*a*) Joint design; (*b*) weld grooves. *(The Lincoln Electric Company.)*

butt joint. Although all but two welds are illustrated with butt joints here, some may be used with other types of joints. Thus a single-bevel weld may also be used in a T or corner joint (Fig. 14.15), and a single-V weld may be used in a corner, T, or butt joint.

14.8.1   Fillet-Welded Joints

The fillet weld, requiring no groove preparation, is one of the most commonly used welds. Corner welds are also widely used in machine design. Various corner arrangements are illustrated in Fig. 14.16. The corner-to-corner joint, as in Fig. 14.16*a*, is difficult to assemble because neither plate can be supported by the other. A small electrode with low welding current must be used so that the first welding pass does not burn through. The joint requires a large amount of metal. The corner joint shown in Fig. 14.16*b* is easy to assemble, does not easily burn through, and requires just half

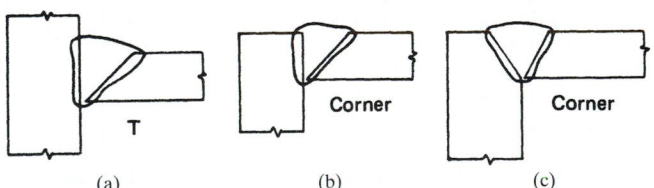

FIGURE 14.15   (*a*) Single-bevel weld used in T joint and (*b*) corner joint; (*c*) single-V weld in corner joint. *(The Lincoln Electric Company.)*

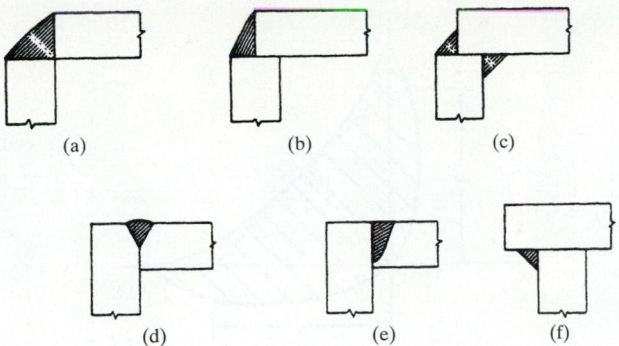

FIGURE 14.16 Various corner joints. *(The Lincoln Electric Company.)*

the amount of the weld metal as the joint in Fig. 14.16*a*. However, by using half the weld size but placing two welds, one outside and the other inside, as in Fig. 14.16*c*, it is possible to obtain the same total throat as with the first weld, but only half the weld metal need be used.

With thick plates, a partial-penetration groove joint, as in Fig. 14.16*d*, is often used. This requires beveling. For a deeper joint, a J preparation, as in Fig. 14.16*e*, may be used in preference to a bevel. The fillet weld in Fig. 14.16*f* is out of sight and makes a neat and economical corner.

The size of the weld should always be designed with reference to the size of the thinner member. The joint cannot be made any stronger by using the thicker member for the weld size, and much more weld metal will be required, as illustrated in Fig. 14.17.

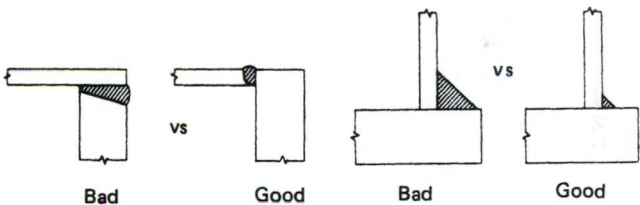

FIGURE 14.17 Size of weld should be determined with reference to thinner member. *(The Lincoln Electric Company.)*

In the United States, a fillet weld is measured by the leg size of the largest right triangle that may be inscribed within the cross-sectional area (Fig. 14.18). The throat, a better index to strength, is the shortest distance between the root of the joint and the face of the diagrammatical weld. As Fig. 14.18 shows, the leg size used may be shorter than the actual leg of the weld. With convex fillets, the actual throat may be longer than the throat of the inscribed triangle.

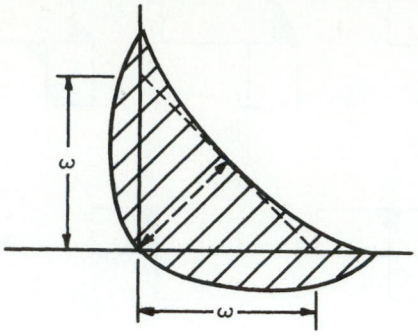

FIGURE 14.18   Leg size ω of a fillet weld. *(The Lincoln Electric Company.)*

14.8.2   Groove and Fillet Combinations

A combination of a partial-penetration groove weld and a fillet weld (Fig. 14.19) is used for many joints. The AWS prequalified single-bevel groove T joint is reinforced with a fillet weld.

The designer is frequently faced with the question of whether to use fillet or groove welds (Fig. 14.20). Here cost becomes a major consideration. The fillet welds in Fig. 14.20*a* are easy to apply and require no special plate preparation. They can be made using large-diameter electrodes with high welding currents, and as a consequence, the deposition rate is high. The cost of the welds increases as the square of the leg size.

FIGURE 14.19   Combined groove- and fillet-welded joints. *(The Lincoln Electric Company.)*

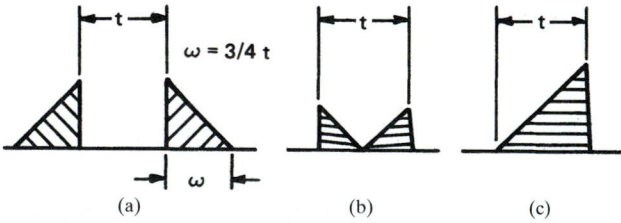

FIGURE 14.20   Comparison of fillet welds and groove welds. *(The Lincoln Electric Company.)*

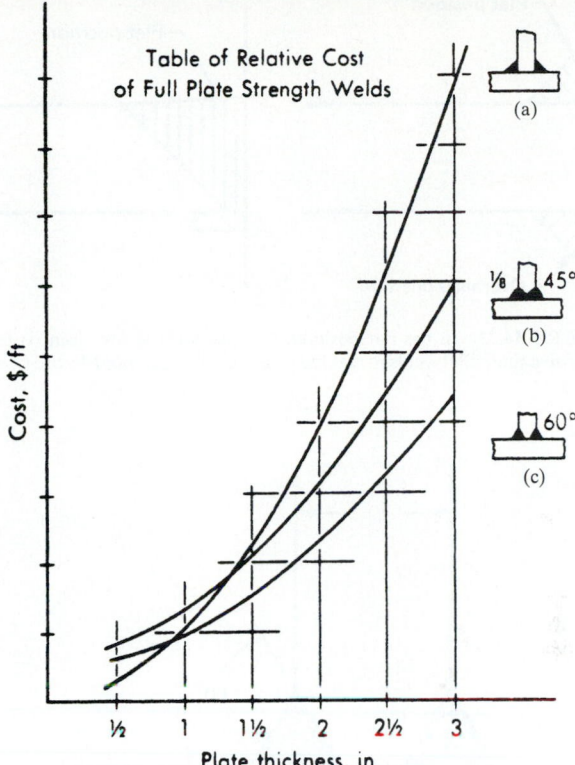

Table of Relative Cost
of Full Plate Strength Welds

FIGURE 14.21   Relative cost of welds having the full strength of the plate. *(The Lincoln Electric Company.)*

In comparison, the double-bevel groove weld in Fig. 14.20*b* has about one-half the weld area of the fillet welds. However, it requires extra preparation and the use of smaller-diameter electrodes with lower welding currents to place the initial pass without burning through. As plate thickness increases, this initial low-deposition region becomes a less important factor and the higher cost factor decreases in significance. The construction of a curve based on the best possible determination of the actual cost of welding, cutting, and assembling, such as that illustrated in Fig. 14.21, is a possible technique for deciding at what point in plate thickness the double-bevel groove weld becomes less costly. The point of intersection of the fillet-weld curve with the groove-weld curve is the point of interest. The accuracy of this device is dependent on the accuracy of the cost data used in constructing the curves.

Referring to Fig. 14.20*c*, it will be noted that the single-bevel groove weld requires about the same amount of weld metal as the fillet welds deposited in Fig. 14.20*a*. Thus there is no apparent economic advantage. There are some disadvantages, though. The single-bevel joint requires bevel preparation and initially a lower deposition rate at the root of the joint. From a design standpoint, however, it offers a direct transfer of force through the joint, which means that it is probably better under fatigue loading. Although the illustrated full-strength fillet weld, having leg

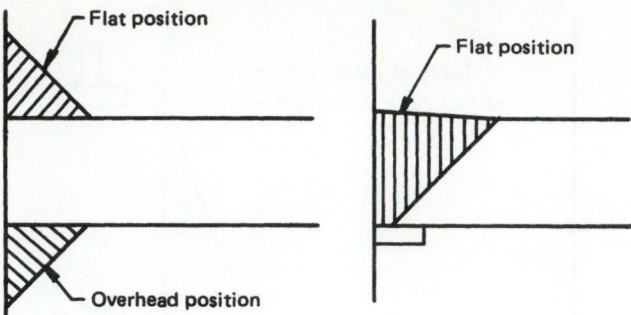

FIGURE 14.22   In the flat position, a single-bevel groove joint is less expensive than fillet welds in making a T joint. *(The Lincoln Electric Company.)*

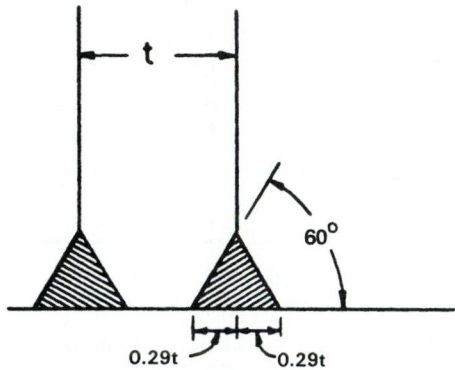

FIGURE 14.23   Partial-penetration double-bevel groove joint. *(The Lincoln Electric Company.)*

sizes equal to three-quarters the plate thickness, would be sufficient, some codes have lower allowable limits for welds, and many require a leg size equal to the plate thickness. In this case, the cost of the fillet-welded joint may exceed the cost of the single-bevel groove-welded joint in thicker plates. Also, if the joint is so positioned that the weld can be made in the flat position, a single-bevel groove weld would be less expensive than fillet welds. As can be seen in Fig. 14.22, one of the fillets would have to be made in the overhead position—a costly operation.

The partial-penetration double-bevel groove joint shown in Fig. 14.23 has been suggested as a full-strength weld. The plate is beveled to 60 degrees on both sides to give a penetration of at least 29 percent of the thickness of the plate ($0.29t$). After the groove is filled, it is reinforced with a fillet weld of equal cross-sectional area and shape. This partial-penetration double-bevel groove joint uses 57.8 percent of the weld metal used by the full-strength fillet weld. It requires joint preparation, but the 60-degree angle allows the use of large electrodes and high welding current.

Full-strength welds are not always required in the design, and economies can often be achieved by using partial-strength welds where these are applicable and

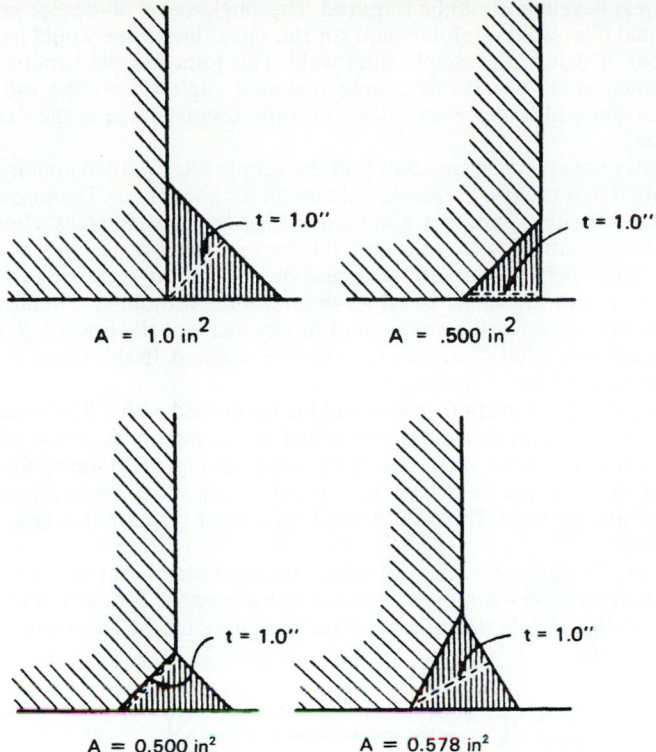

t = 1.0″

A = 1.0 in²

t = 1.0″

A = .500 in²

t = 1.0″

A = 0.500 in²

t = 1.0″

A = 0.578 in²

FIGURE 14.24 Comparison of weld joints having equal throats. *(The Lincoln Electric Company.)*

permissible. Referring to Fig. 14.24, it can be seen that on the basis of an unreinforced 1-in throat, a 45-degree partial-penetration single-bevel groove weld requires just one-half the weld area needed for a fillet weld. Such a weld may not be as economical as the same-strength fillet weld, however, because of the cost of edge preparation and the need to use a smaller electrode and lower current on the initial pass.

If the single-bevel groove joint were reinforced with an equal-leg fillet weld, the cross-sectional area for the same throat size would still be one-half the area of the

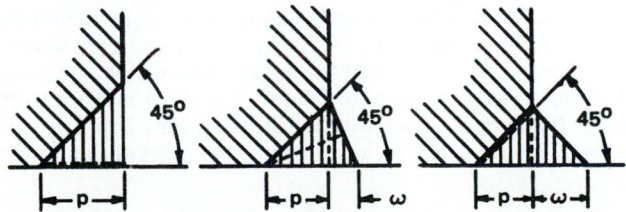

45°

45°

45°

p

p   ω

p   ω

FIGURE 14.25 Comparison of weld joints with and without reinforcing fillet welds. *(The Lincoln Electric Company.)*

fillet, and less beveling would be required. The single-bevel 60-degree groove joint with an equal fillet-weld reinforcement for the same throat size would have an area 57.8 percent of that of the simple fillet weld. This joint has the benefit of smaller cross-sectional area—yet the 60-degree included angle allows the use of higher welding current and larger electrodes. The only disadvantage is the extra cost of preparation.

From this discussion it is apparent that the simple fillet-welded joint is the easiest to make, but it may require excessive weld metal for larger sizes. The single-bevel 45-degree included-angle joint is a good choice for larger weld sizes. However, one would miss opportunities by selecting the two extreme conditions of these two joints. The joints between these two should be considered. Referring to Fig. 14.25, one may start with the single-bevel 45-degree joint without the reinforcing fillet weld, gradually add a reinforcement, and finally increase the lower leg of the fillet reinforcement until a full 45-degree fillet weld is reached. In this figure, p = depth of preparation and ω = leg of reinforcing fillet.

When a partial-penetration groove weld is reinforced with a fillet weld, the minimum throat is used for design purposes, just as the minimum throat of a fillet or partial-penetration groove weld is used. However, as Fig. 14.26 shows, the allowable load for this combination weld is not the sum of the allowable limits for each portion of the combination weld. This would result in a total throat much larger than the actual throat.

Figure 14.27a shows the effect of using the incorrect throat in determining the allowable unit force on a combination weld. The allowable[†] for each weld was added separately. In Fig. 14.27b, weld size is correctly figured on the minimum throat.

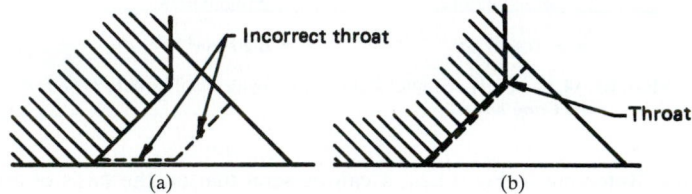

FIGURE 14.26   Determining minimum throat. (*a*) Incorrect result; (*b*) correct result. (*The Lincoln Electric Company.*)

14.8.3 Sizing of Fillets

Table 14.8 gives the sizing of fillet welds for rigidity at various strengths and plate thicknesses, where the strength of the weld metal matches the plate.

In machine design work, where the primary design requirement is rigidity, members are often made with extra-heavy sections, so that movement under load will be within very close tolerances. Because of the heavy construction, stresses are very low. Often the allowable stress in tension for mild steel is given as 20 kpsi, yet the welded machine base or frame may have a working stress of only 2 to 4 kpsi. The question arises as to how to determine the weld sizes for these types of rigidity designs.

[†] The term *allowable* is often used in the welding industry to indicate allowable load, allowable stress, or unit allowable load—EDS.

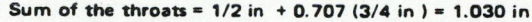

Sum of the throats = 1/2 in + 0.707 (3/4 in) = 1.030 in

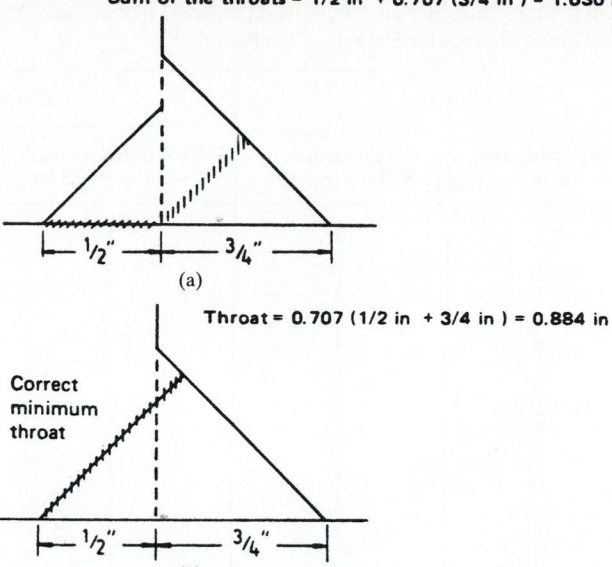

(a)

Throat = 0.707 (1/2 in + 3/4 in) = 0.884 in

Correct
minimum
throat

(b)

FIGURE 14.27 Examples showing the effect of correct and incorrect throat dimension in determining the allowable load on a combination weld. (*a*) The weld allowable load would be incorrectly figured by adding each weld throat separately; (*b*) weld allowable load is correctly figured using the minimum throat. *(The Lincoln Electric Company.)*

It is not very practical to first calculate the stresses resulting in a weldment when the unit is loaded within a predetermined dimensional tolerance and then use these stresses to determine the forces that must be transferred through the connecting welds. A very practical method, however, is to design the weld for the thinner plate, making it sufficient to carry one-third to one-half the carrying capacity of the plate. This means that if the plate were stressed to one-third to one-half its usual value, the weld would be sufficient. Most rigidity designs are stressed much below these values; however, any reduction in weld size below one-third the full-strength value would give a weld too small an appearance for general acceptance.

14.8.4 Groove Joints

Figure 14.28*a* indicates that the *root opening R* is the separation between the members to be joined. A root opening is used for electrode accessibility to the base or root of the joint. The smaller the angle of the bevel, the larger the root opening must be to get good fusion at the root. If the root opening is too small, root fusion is more difficult to obtain, and smaller electrodes must be used, thus slowing down the welding process. If the root opening is too large, weld quality does not suffer, but more weld metal is required; this increases welding cost and will tend to increase distortion.

TABLE 14.8 Rule-of-Thumb Fillet-Weld Sizes for Use in Cases Where the Strength of the Weld Metal Matches the Strength of the Plate

Plate thickness t, in	Strength design, full-strength weld, $\omega = 0.75t$	Rigidity design	
		50% of full-strength weld, $\omega = 0.375t$	33% of full-strength weld, $\omega = 0.25t$
$< \frac{1}{4}$	$\frac{1}{8}$	$\frac{1}{8}t$†	$\frac{1}{8}t$†
$\frac{1}{4}$	$\frac{3}{16}$	$\frac{3}{16}t$†	$\frac{3}{16}t$†
$\frac{5}{16}$	$\frac{1}{4}$	$\frac{3}{16}t$†	$\frac{3}{16}t$†
$\frac{3}{8}$	$\frac{5}{16}$	$\frac{3}{16}t$†	$\frac{3}{16}t$†
$\frac{7}{16}$	$\frac{3}{8}$	$\frac{3}{16}$	$\frac{3}{16}t$†
$\frac{1}{2}$	$\frac{3}{8}$	$\frac{3}{16}$	$\frac{3}{16}t$†
$\frac{9}{16}$	$\frac{7}{16}$	$\frac{1}{4}$	$\frac{1}{4}t$†
$\frac{5}{8}$	$\frac{1}{2}$	$\frac{1}{4}$	$\frac{1}{4}t$†
$\frac{3}{4}$	$\frac{9}{16}$	$\frac{5}{16}$	$\frac{1}{4}t$†
$\frac{7}{8}$	$\frac{5}{8}$	$\frac{3}{8}$	$\frac{3}{16}t$†
1	$\frac{3}{4}$	$\frac{3}{8}$	$\frac{3}{16}t$†
$1\frac{1}{8}$	$\frac{7}{8}$	$\frac{7}{16}$	$\frac{5}{16}$
$1\frac{1}{4}$	1	$\frac{1}{2}$	$\frac{5}{16}$
$1\frac{3}{8}$	1	$\frac{1}{2}$	$\frac{3}{8}$
$1\frac{1}{2}$	$1\frac{1}{8}$	$\frac{9}{16}$	$\frac{3}{8}$
$1\frac{5}{8}$	$1\frac{1}{4}$	$\frac{5}{8}$	$\frac{7}{16}$
$1\frac{3}{4}$	$1\frac{3}{8}$	$\frac{3}{4}$	$\frac{7}{16}$
2	$1\frac{1}{2}$	$\frac{3}{4}$	$\frac{1}{2}$
$2\frac{1}{8}$	$1\frac{5}{8}$	$\frac{7}{8}$	$\frac{9}{16}$
$2\frac{1}{4}$	$1\frac{3}{4}$	$\frac{7}{8}$	$\frac{9}{16}$
$2\frac{3}{8}$	$1\frac{3}{4}$	1	$\frac{5}{8}$
$2\frac{1}{2}$	$1\frac{7}{8}$	1	$\frac{5}{8}$
$2\frac{5}{8}$	2	1	$\frac{3}{4}$
$2\frac{3}{4}$	2	1	$\frac{3}{4}$
3	$2\frac{1}{4}$	$1\frac{1}{8}$	$\frac{3}{4}$

†These values have been adjusted to comply with AWS recommended minimums.
SOURCE: The Lincoln Electric Company, Cleveland, Ohio.

Figure 14.28b indicates how the root opening must be increased as the included angle of the bevel is decreased. Backup strips are used on larger root openings. All three preparations are acceptable; all are conducive to good welding procedure and good weld quality. Selection, therefore, is usually based on cost.

Root openings and joint preparation will directly affect weld cost (mass of weld metal required), and the choice should be made with this in mind. Joint preparation involves the work required on plate edges prior to welding and includes beveling and providing a root face.

Using a double-groove joint in preference to a single-groove joint (Fig. 14.29) cuts in half the amount of welding. This reduces distortion and makes possible alternating the weld passes on each side of the joint, again reducing distortion.

In Fig. 14.30a, if the bevel or gap is too small, the weld will bridge the gap, leaving slag at the root. Excessive back-gouging is then required. Figure 14.30b shows how

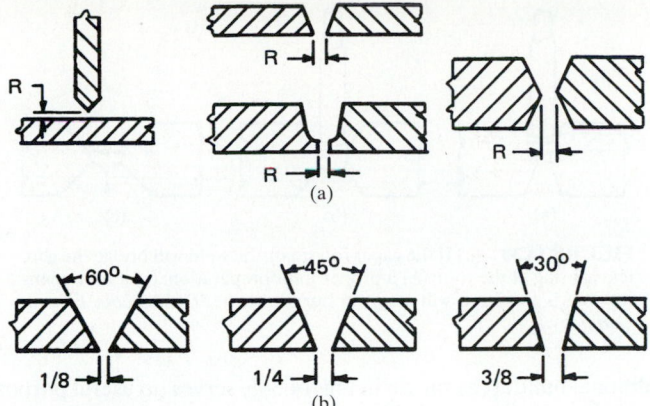

FIGURE 14.28 (*a*) Root opening is designated as *R*; (*b*) size of root opening depends on bevel angle. (*The Lincoln Electric Company.*)

proper joint preparation and procedure will produce good root fusion and will minimize back-gouging. In Fig. 14.30*c*, a large root opening will result in burnthrough. Spacer strip may be used, in which case the joint must be back-gouged.

Backup strips are commonly used when all welding must be done from one side or when the root opening is excessive. Backup strips, shown in Fig. 14.31*a* through *c*, are generally left in place and become an integral part of the joint. Spacer strips may be used, especially in the case of double-V joints, to prevent burnthrough. The spacer in Fig. 14.31*d* used to prevent burnthrough will be gouged out before welding the second side.

14.8.5 Backup Strips

Backup strip material should conform to the base metal. Feather edges of the plate are recommended when using a backup strip.

Short, intermittent tack welds should be used to hold the backup strip in place, and these should preferably be staggered to reduce any initial restraint on the joint. They should not be directly opposite one another (Fig. 14.32).

The backup strip should be in intimate contact with both plate edges to avoid trapped slag at the root, as shown in Fig. 14.33. On a butt joint, a nominal weld reinforcement (approximately ¹⁄₁₆ in above flush) is all that is necessary, as shown in Fig.

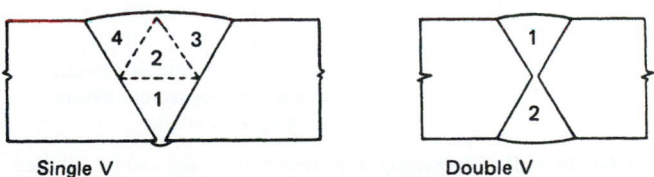

FIGURE 14.29 Using a double-groove joint in place of a single-groove joint reduces the amount of welding. (*The Lincoln Electric Company.*)

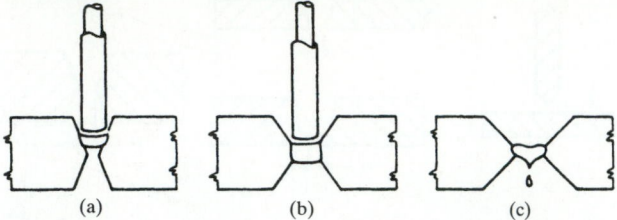

(a)   (b)   (c)

FIGURE 14.30   (*a*) If the gap is too small, the weld will bridge the gap, leaving slag at the root; (*b*) a proper joint preparation; (*c*) a root opening that is too large will result in burnthrough. *(The Lincoln Electric Company.)*

14.34*a*. Additional buildup, as shown in Fig. 14.34*b*, serves no useful purpose and will increase the weld cost. Care should be taken to keep both the width and the height of the reinforcement to a minimum.

14.8.6 Edge Preparation

The main purpose of a root face (Fig. 14.35*a*) is to provide an additional thickness of metal, as opposed to a feather edge, in order to minimize any burnthrough tendency. A feather-edge preparation is more prone to burnthrough than a joint with a root face, especially if the gap gets a little too large (Fig. 14.35*b*).

A root face is not as easily obtained as a feather edge. A feather edge is generally a matter of one cut with a torch, whereas a root face will usually require two cuts or possibly a torch cut plus machining.

A root face usually requires back-gouging if a 100 percent weld is required. A root face is not recommended when welding into a backup strip, since a gas pocket would be formed.

Plate edges are beveled to permit accessibility to all parts of the joint and to ensure good fusion throughout the entire weld cross section. Accessibility can be gained by compromising between maximum bevel and minimum root opening (Fig. 14.36).

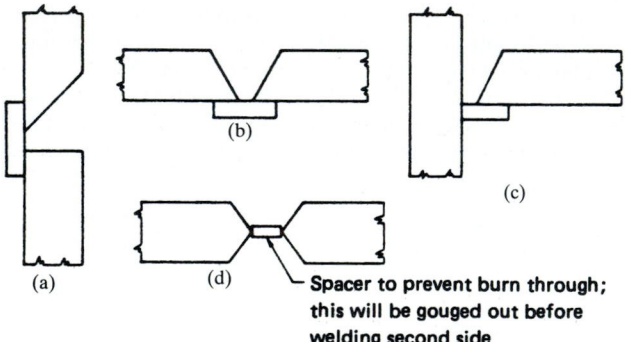

(b)

(c)

(a)   (d)   Spacer to prevent burn through; this will be gouged out before welding second side

FIGURE 14.31   The backup strips shown in (*a*), (*b*), and (*c*) are used when all welding is done from one side or when the root opening is excessive; a spacer to prevent burnthrough as shown in (*d*) will be gouged out before welding the second side. *(The Lincoln Electric Company.)*

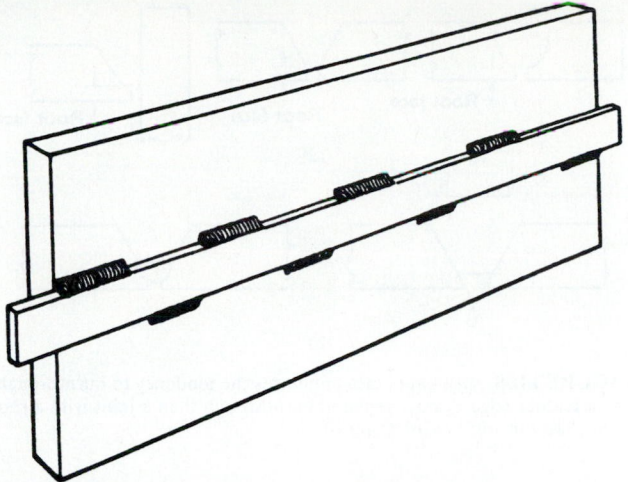

FIGURE 14.32 Short, intermittent tack welds should be used to hold the backup strip in place. *(The Lincoln Electric Company.)*

Degree of bevel may be dictated by the importance of maintaining proper electrode angle in confined quarters (Fig. 14.37). For the joint illustrated, the minimum recommended bevel is 45 degrees.

J and U preparations are excellent to work with, but economically they may have little to offer because preparation requires machining as opposed to simple torch cutting. Also, a J or U groove requires a root face (Fig. 14.38) and thus back-gouging.

To consistently obtain complete fusion when welding a plate, back-gouging is required on virtually all joints except bevel joints with a feather edge. This may be done by any convenient means: grinding, chipping, or gouging. The latter method is generally the most economical and leaves an ideal contour for subsequent beads.

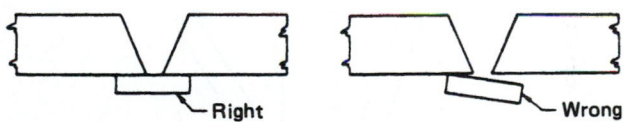

FIGURE 14.33 The backup strip should be in intimate contact with both edges of the plate. *(The Lincoln Electric Company.)*

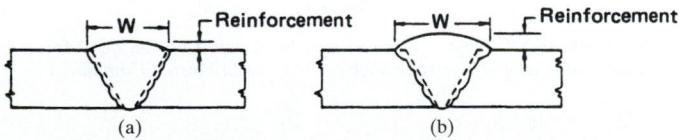

FIGURE 14.34 *(a)* A minimum reinforcement on a butt joint is preferred; *(b)* too much reinforcement. *(The Lincoln Electric Company.)*

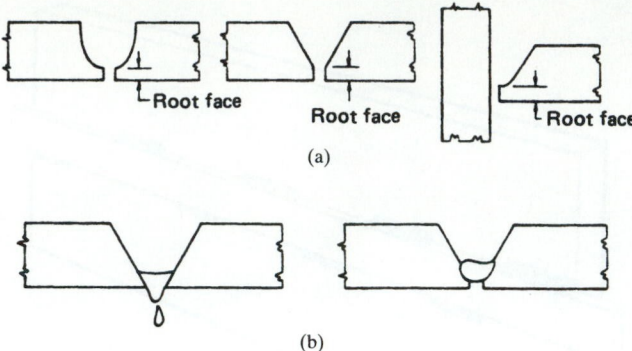

FIGURE 14.35   (*a*) A root face minimizes the tendency to burnthrough; (*b*) a feather edge is more prone to burnthrough than a joint with a root face. (*The Lincoln Electric Company.*)

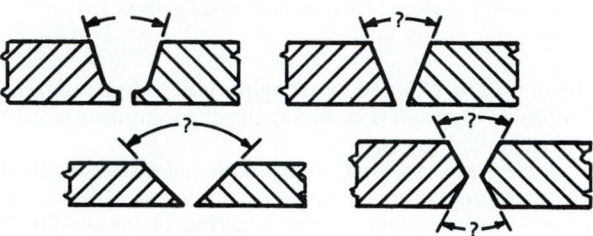

FIGURE 14.36   Accessibility is gained by compromising between bevel and root opening. (*The Lincoln Electric Company.*)

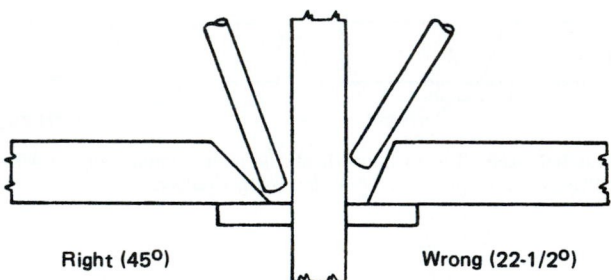

FIGURE 14.37   Degree of bevel may be dictated by the need for maintaining proper electrode angle. (*The Lincoln Electric Company.*)

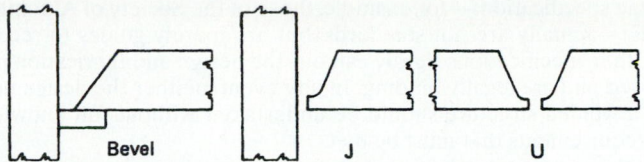

FIGURE 14.38 A bevel preparation with a backup strip may be more economical than a J or U groove. *(The Lincoln Electric Company.)*

Without back-gouging, penetration is incomplete (Fig. 14.39*a*). Proper back-chipping should be deep enough to expose sound weld metal, and the contour should permit the electrode complete accessibility (Fig. 14.39*b*).

14.9 CODES AND SPECIFICATIONS FOR WELDS

Welds are designed and executed in accordance with codes, standards, and specifications intended to enhance the integrity of the product and its safe performance in use. Codes and specifications are generally written by industrial groups, trade or professional organizations, or government bureaus, and each code or specification deals with applications pertaining specifically to the interest of the authoring body. Large manufacturing organizations may prepare their own specifications to meet their specific needs.

Among the major national organizations that write codes that involve arc welding are the American Welding Society (AWS), the American Institute of Steel Construction (AISC), the American Society for Testing Materials (ASTM), the American Society of Mechanical Engineers (ASME), and the American Petroleum Institute (API).

Among government agencies, the Interstate Commerce Commission (ICC) has rules for the fabrication of over-the-road vehicles and for containers used in interstate commerce. The various branches of the military services also prepare specifi-

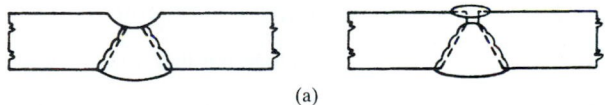

(a)

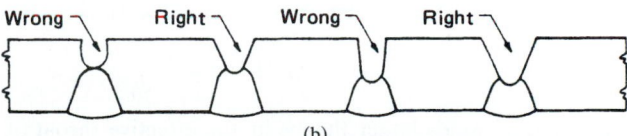

(b)

FIGURE 14.39 (*a*) Without back-gouging, penetration is incomplete; (*b*) proper back-gouging should be deep enough to expose sound weld metal. *(The Lincoln Electric Company.)*

cations. Some specifications—for example, those of the Society of Automotive Engineers (SAE)—actually are not standards, but are merely guides to recommended practices. Other specifications rigidly call out the design and fabrication procedures to be followed and are legally binding. In any event, neither the design nor the fabrication of a welded structure should be undertaken without full knowledge of all codes and requirements that must be met.

Meeting the requirements of a code does not protect anyone against liability concerning the performance of the welds or structure. Nor, in general, does any code-writing body approve, endorse, guarantee, or in any way attest to the correctness of the procedures, designs, or materials selected for code application.

The strength values permitted by governing codes are called *allowables*. Thus there are specified allowables for shear stress and unit force on various sizes of fillet welds, and there are fatigue allowables for various welds in reference to the geometry of the joint. Most weldments used in machinery are made in accordance with AWS and AISC specifications, with ASME and API rules applicable where pressure vessels and piping are involved.

14.9.1 Allowable Shear and Unit Forces

The basic formula for allowable shear stress τ for weld metal in a fillet or partial-penetration bevel-groove weld has been established by the AWS and AISC as

$$\tau = 0.30S_t \tag{14.1}$$

where S_t = minimum tensile strength. Table 14.9 shows the values for various weld-metal strength levels obtained by this formula and the more common fillet-weld sizes. These values are for equal-leg fillet welds where the effective throat t_e = 0.707ω, where ω is the leg size. With Table 14.9 one can calculate the allowable unit force f per linear inch for a weld size made with a particular electrode type. For example, calculating the allowable unit force f per inch for a ½-in fillet weld made with an E70 electrode gives

$$f = 0.707\omega\tau = 0.707\omega\tau(0.30S_t)$$

$$= 0.707(\tfrac{1}{2})(0.30)(70)(10)^3$$

$$= 7420 \text{ lb per linear inch}$$

An AISC provision gives limited credit for penetration beyond the root of a fillet weld made with the submerged-arc process. Since penetration increases the effective throat thickness of the weld, as shown in Fig. 14.40, the provision permits an increase in this value when calculating weld strength. For fillet welds ⅜ in and smaller, the effective throat t_e is now equal to the leg size of the weld ω. Thus,

$$t_e = \omega \qquad \omega \leq \tfrac{3}{8} \text{ in} \tag{14.2}$$

For submerged-arc fillet welds larger than ⅜ in, the effective throat of the weld is obtained by adding 0.11 to 0.707ω. Thus,

$$t_e = 0.707\omega + 0.11 \qquad \omega > \tfrac{3}{8} \text{ in} \tag{14.3}$$

TABLE 14.9 Allowable Unit Load for Various Sizes of Fillet Welds

Tensile strength of weld metal, kpsi						
$S_t =$	60	70	80	90	100	110
Allowable shear stress on throat of fillet weld or partial-penetration groove weld, kpsi						
$\tau =$	18.0	21.0	24.0	27.0	30.0	33.0
Allowable unit force on fillet weld kip/linear in						
$f =$	12.73	14.85	16.97	19.09	21.21	23.33
Leg size ω, in	Allowable unit force for various sizes of fillet welds, kip/linear in					
1	i2.73	14.85	16.97	19.09	21.21	23.33
$\frac{7}{8}$	11.14	12.99	14.85	16.70	18.57	20.41
$\frac{3}{4}$	9.55	11.14	12.73	14.32	15.92	17.50
$\frac{5}{8}$	7.96	9.28	10.61	11.93	13.27	14.58
$\frac{1}{2}$	6.37	7.42	8.48	9.54	10.61	11.67
$\frac{7}{16}$	5.57	6.50	7.42	8.35	9.28	10.21
$\frac{3}{8}$	4.77	5.57	6.36	7.16	7.95	8.75
$\frac{5}{16}$	3.98	4.64	5.30	5.97	6.63	7.29
$\frac{1}{4}$	3.18	3.71	4.24	4.77	5.30	5.83
$\frac{3}{16}$	2.39	2.78	3.18	3.58	3.98	4.38
$\frac{1}{8}$	1.59	1.86	2.12	2.39	2.65	2.92
$\frac{1}{16}$	0.795	0.930	1.06	1.19	1.33	1.46

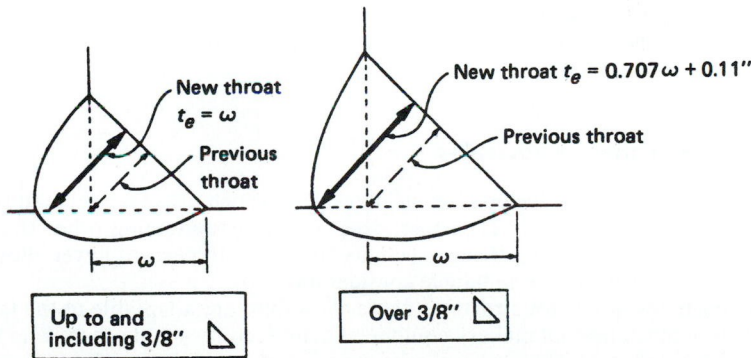

FIGURE 14.40 The AISC gives credit for penetration beyond the root of fillets made with the submerged-arc process. (*The Lincoln Electric Company.*)

where t_e is in inches. Note that allowance for penetration applies only to fillet welds made by the submerged-arc welding process. Electrode polarity will provide this penetration.

14.9.2   Minimum Fillet-Weld Size

The minimum sizes of fillet welds for specific material thicknesses are shown in Table 14.10. In the AISC Specifications and the AWS Structural Welding Code, this table has been expanded to include material less than ¼ in thick and ⅛-in fillets. Where materials of different thicknesses are being joined, the minimum fillet weld size is governed by the thicker material, but this size does not have to exceed the thickness of the thinner material unless required by the calculated stress.

TABLE 14.10   Minimum Fillet-Weld Size ω in Inches

Material thickness of thicker part joined	Minimum fillet size
To ¼ inclusive	⅛
Over ¼ to ½	3/16
Over ½ to ¾	¼
Over ¾ to 1½	5/16
Over 1½ to 2¼	⅜
Over 2¼ to 6	½
Over 6	⅝

SOURCE: AISC Specifications, Sec. 1.17.5.

14.9.3   Allowables for Weld Metal—A Handy Reference

Table 14.11 summarizes the AWS Structural Welding Code and AISC allowables for weld metal. It is intended to provide a ready reference for picking the proper strength levels for the various types of steels. Once this selection has been made, the allowables can be quickly found for the various types of welds that may be required for the specific assembly.

14.9.4   AISC Fatigue Allowables

The AISC Specifications include fatigue allowables, which also are accepted by the AWS Building Code, Sec. 8. Therefore, designers have something other than the AWS Building Code, Sec. 10, Bridges, with its automatic 10 percent lower allowable design stress, on which to base fatigue considerations.

Although developed for structures, these allowables are adaptable to the fatigue problems of machine-tool makers, equipment manufacturers, and others who fabricate with welded steel. They cover a wide range of welded joints and members and not only provide values for various types of welds, but also take into consideration the strength of members attached by welds.

The conventional method of handling fatigue is based on a maximum fatigue stress. The AISC-suggested method is based on the range of stress. Either may be used in design; they will give comparable values. The AISC method is generally quicker.

Under the new approach, the allowables for members are designed M and for welds W. A tensile load is T, a compressive load C, a reversal R, and shear S. In the chart used for determining values for allowable range of stress (Fig. 14.41), there are four groups representing life. These are

1. 20 000 to 100 000 cycles
2. Over 100 000 to 500 000 cycles
3. Over 500 000 to 2 000 000 cycles
4. Over 2 000 000 cycles

And there are eight different categories representing type of joint and detail of member. The chart provides the allowable range in stress σ_{sr} or τ_{sr}, which value may be used in the conventional fatigue formulas. These formulas are

$$\sigma_{max} = \frac{\sigma_{sr}}{1 - K} \quad \text{or} \quad \sigma_{max} = \frac{\tau_{sr}}{1 - K} \tag{14.4}$$

where

$$K = \frac{\text{min. stress}}{\text{max. stress}} = \frac{\text{min. force}}{\text{max. force}}$$

$$= \frac{\text{min. moment}}{\text{max. moment}} = \frac{\text{min. shear}}{\text{max. shear}} \tag{14.5}$$

Of course, the maximum allowable fatigue value used should not exceed the allowable for steady loading.

An alternative use of the allowable range of stress—taken from the table—is to divide it into the range of applied load. This will provide the required property of the section—area or section modulus. The section, as determined, must additionally be large enough to support the total load (dead and live load) at steady allowable stresses.

Reference to the chart of joint types and conditions and the table of allowable range of stress for the different categories (Fig. 14.41) will help make clear their use. Such reference also points up some of the new ideas introduced.

One new concept is that the fatigue allowable of a member, for example, a welded plate girder as shown by (2) in the chart (Fig. 14.41), is now determined by the allowable of the plate when connected by the fillet welds parallel to the direction of the applied stress. M and W are equal, and the applicable category is B, rather than the allowable of plate without welds, category A.

If stiffeners are used on the girder, as in (4), the fatigue allowable of the web or flange is determined by the allowable in the member at the termination of the weld or adjacent to the weld, category C or D, depending on the shear value in the web.

The fatigue allowable of a flange plate at the termination of a cover plate, either square or tapered end, is represented by (5). The applicable category is E. The same category also applies to a plate or cover plate adjacent to the termination of an intermittent fillet weld, as in (6) and (39).

Groove welds in butt joints of plate loaded transversely to the weld are shown in (8) to (14). In (15), the groove weld is parallel to the load. In (10), (13), (14), (15), and

TABLE 14.11 Permissible Stress of Weld†

Type of Weld Stress	Permissible Stress	Required Strength Level (1)(2)
COMPLETE PENETRATION GROOVE WELDS		
Tension normal to the effective throat.	Same as base metal.	Matching weld metal must be used. See Table below.
Compression normal to the effective throat.	Same as base metal.	Weld metal with a strength level equal to or one classification (10 ksi) less than matching weld metal may be used.
Tension or compression parallel to the axis of the weld.	Same as base metal.	Weld metal with a strength level equal to or less than matching weld metal may be used.
Shear on the effective throat.	.30 x Nominal Tensile strength of weld metal (ksi) except stress on base metal shall not exceed .40 x yield stress of base metal.	
PARTIAL PENETRATION GROOVE WELDS		
Compression normal to effective throat.	Designed not to bear — .50 x Nominal Tensile strength of weld metal (ksi) except stress on base metal shall not exceed .60 x yield stress of base metal. Designed to bear. Same as base metal.	Weld metal with a strength level equal to or less than matching weld metal may be used.
Tension or compression parallel to axis of the weld. (3)	Same as base metal.	
Shear parallel to axis of weld.	.30 x Nominal Tensile strength of weld metal (ksi) except stress on base metal shall not exceed .40 x yield stress of base metal.	
Tension normal to effective throat. (4)	.30 x Nominal Tensile strength of weld metal (ksi) except stress on base metal shall not exceed .60 x yield stress of base metal.	

14.44

FILLET WELDS (3)

Stress on effective throat, regardless of direction of application of load.	.30 x Nominal Tensile strength of weld metal (ksi) except stress on base metal shall not exceed .40 x yield stress of base metal.
Tension or compression parallel to axis of weld.	Same as base metal.

PLUG AND SLOT WELDS

Shear parallel to faying surfaces.	.30 x Nominal Tensile strength of weld metal (ksi) except stress on base metal shall not exceed .40 x yield stress of base metal.

Weld metal with a strength level equal to or less than matching weld metal may be used.

Weld metal with a strength level equal to or less than matching weld metal may be used.

(1) For matching weld metal, see AISC Table 1.17.2 or AWS Table 4.1.1 or table below.
(2) Weld metal, one strength level (10 KSI) stronger than matching weld metal may be used when using alloy weld metal on A242 or A588 steel to match corrosion resistance or coloring characteristics (Note 3 of Table 4.1.4 or AWS D1.1).
(3) Fillet welds and partial penetration groove welds joining the component elements of built up members (ex. flange to web welds) may be designed without regard to the axial tensile or compressive stress applied to them.
(4) Cannot be used in tension normal to their axis under tensile loading (AWS 2.5). AWS Bridge prohibits their use on any butt joint (9.12.1.1), or any splice in a tension or compression member (9.17), or splice in beams or girders (9.21), however, are allowed on corner joints parallel to axial force of components of built up members (9.12.1.2 (2). Cannot be used in girder splices (AISC 1.10.8).

MATCHING WELD METAL AND BASE METAL

Weld Metal	60 or 70	70	80	100	110
Type of Steel	A36; A53, Gr. B; A106, Gr. B; A131, Gr. A, B, C, CS, D, E; A139, Gr. B; A381, Gr. Y35; A500, Gr. A, B; A501; A516, Gr. 55, 60; A524, Gr. I, II; A529; A570, Gr. D, E; A573, Gr. 65; A709, Gr. 36; API 5L, Gr. B; API 5LX, Gr. 42; ABS, Gr. A, B, D, CS, DS, E	A131, Gr. AH32, DH32, EH32, AH36, DH36, EH36; A242; A441; A516, Gr. 65; 70; A537, Class 17; A572, Gr. 42, 45, 50, 55; A588 (4 in and under); A595, Gr. A, B, C; A606; A607, Gr. 45, 50, 55; A618; A633, Gr. A, B, C, D (2-1/2 in and under); A709, Gr. 50, 50W; API 2H; ABS Gr. AH32, DH32, EH32, AH36, DH36, EH36.	A572, Gr. 60, 65; A537, Class 2; A63, Gr. E	A514 [over 2-1/2 in (63 mm)]; A709, Gr. 100, 100W [2-1/2 to 4 in (63 to 102 mm)]	A514 [2-1/2 in (63 mm) and under]; A517; A709, Gr. 100, 100W [2-1/2 in (63 mm) and under]

†This table summarizes the AISC Specifications and the AWS Structural Welding Code ("Specification for the Design, Fabrication and Erection of Structural Steel for Buildings," American Institute of Steel Construction; AWS D.1-82, American Welding Society).

SOURCE: The James F. Lincoln Arc Welding Foundation, Cleveland, Ohio.

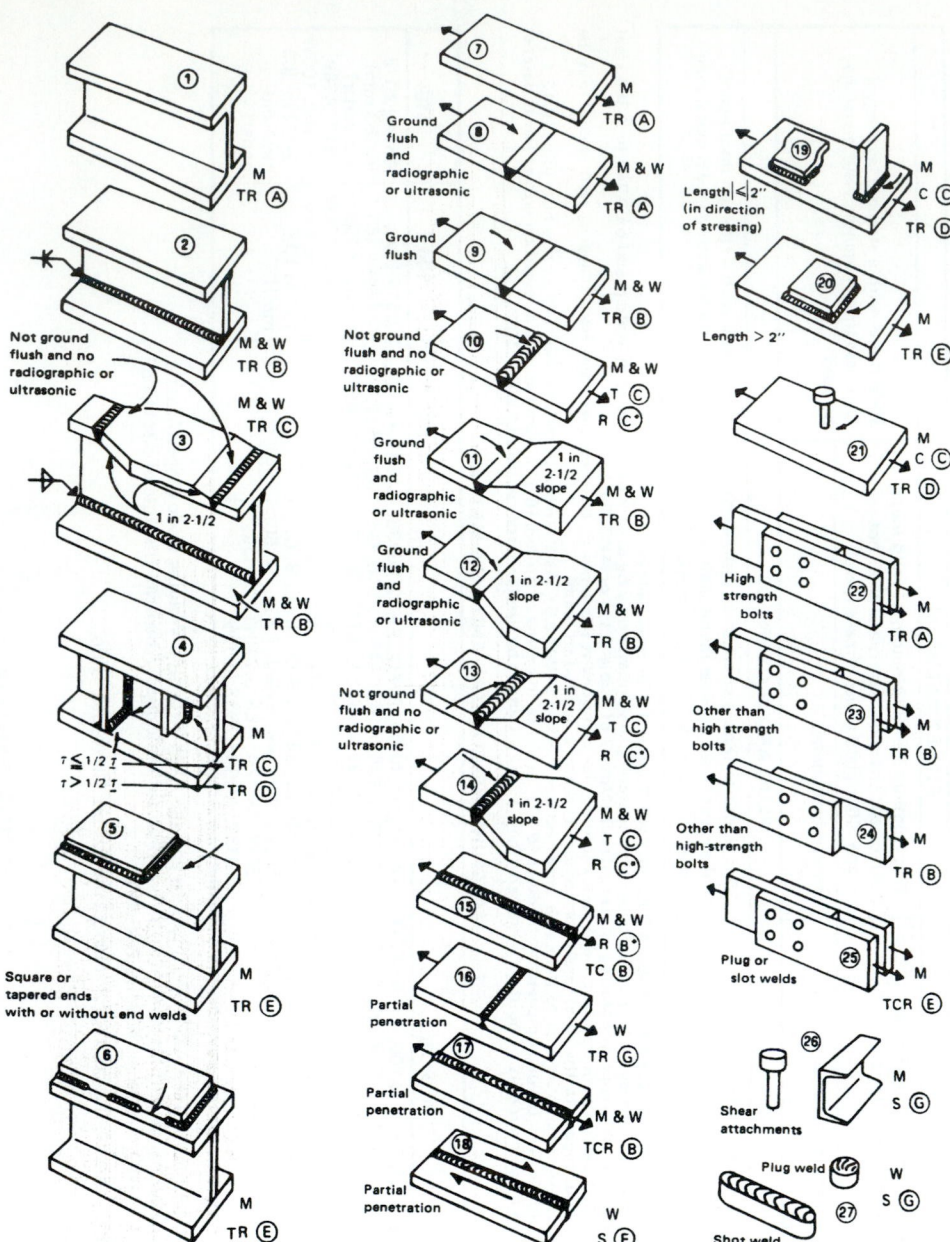

FIGURE 14.41   The AISC allowable range of stress σ_{sr} or τ_{sr}. *(The Lincoln Electric Company.)*

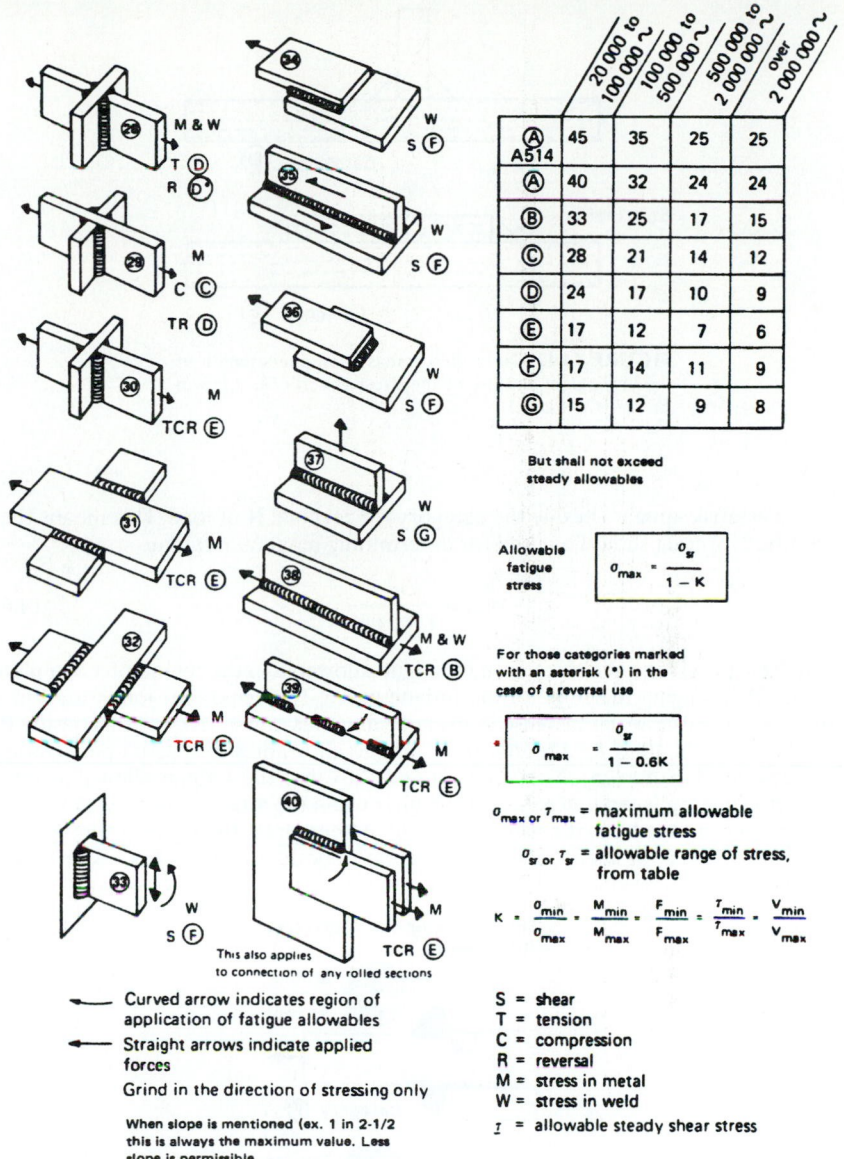

		20 000 to 100 000 ~	100 000 to 500 000 ~	500 000 to 2 000 000 ~	over 2 000 000 ~
(A) A514		45	35	25	25
(A)		40	32	24	24
(B)		33	25	17	15
(C)		28	21	14	12
(D)		24	17	10	9
(E)		17	12	7	6
(F)		17	14	11	9
(G)		15	12	9	8

But shall not exceed
steady allowables

Allowable fatigue stress

$$\sigma_{max} = \frac{\sigma_{sr}}{1 - K}$$

For those categories marked with an asterisk (*) in the case of a reversal use

$$\sigma_{max} = \frac{\sigma_{sr}}{1 - 0.6K}$$

σ_{max} or τ_{max} = maximum allowable fatigue stress

σ_{sr} or τ_{sr} = allowable range of stress, from table

$$K = \frac{\sigma_{min}}{\sigma_{max}} = \frac{M_{min}}{M_{max}} = \frac{F_{min}}{F_{max}} = \frac{\tau_{min}}{\tau_{max}} = \frac{V_{min}}{V_{max}}$$

Curved arrow indicates region of application of fatigue allowables

Straight arrows indicate applied forces

Grind in the direction of stressing only

When slope is mentioned (ex. 1 in 2-1/2) this is always the maximum value. Less slope is permissible.

S = shear
T = tension
C = compression
R = reversal
M = stress in metal
W = stress in weld
τ = allowable steady shear stress

This also applies to connection of any rolled sections

FIGURE 14.41 (*Continued*) The AISC allowable range of stress σ_{sr} or τ_{sr}. (*The Lincoln Electric Company.*)

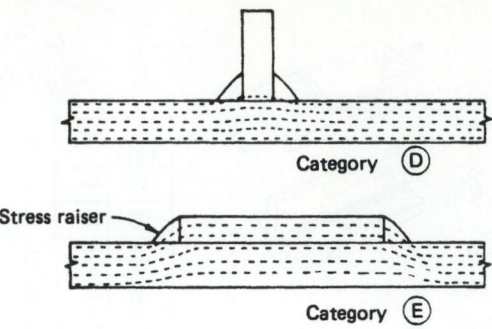

FIGURE 14.42 Note the decreased fatigue strength of the lower joint because of the stress raiser. *(The Lincoln Electric Company.)*

(28), an asterisk appears beside the category for reversal R of load. This means that a modified formula should be used for determining maximum fatigue stress:

$$\sigma_{max} = \frac{\sigma_{sr}}{1 - 0.6K} \tag{14.6}$$

Using $0.6K$ provides a slight increase in fatigue allowable in the region of a complete reversal by changing the slope of the fatigue curve. The same butt joints used in a girder (3) do not show this increase in strength, and thus no asterisk appears beside R.

This approach gives, for the first time, fatigue allowables for partial-penetration groove welds, (16) to (18). Note by (19) and (20) that the fatigue allowable for a member with a transverse attachment is higher when the attachment is less than 2 in long, measured parallel to the axis of the load. Although there may be a similar geometric notch effect or abrupt change in section in both, it is the stress raiser that is

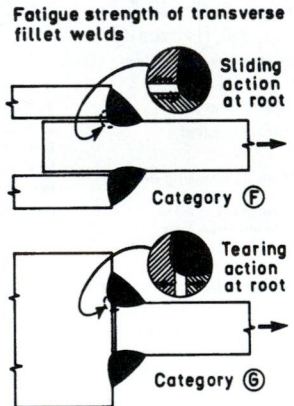

FIGURE 14.43 There is a greater tearing action at the root in category G, warranting a lower fatigue allowable. *(The Lincoln Electric Company.)*

important. The transverse bar in (19) is so short as far as the axis of the member and load are concerned that very little of the force is able to swing up and into the bar and then back down again. Consequently, the stress raiser is not severe. The longer bar attachment in (20), however, is sufficiently long to provide a path for the force through it and the connecting welds. Because of this force transfer through the welds, there will be a higher stress raiser and, as a result, a reduction of the fatigue strength of the member. The difference is illustrated in Fig. 14.42.

Item (30) of the chart, which falls into category E, should not be confused with (37), category G. Both depict transverse fillet welds, but (30) provides a fatigue allowable for the member adjacent to the fillet weld, whereas (37) provides a fatigue shear allowable for the throat of the fillet weld.

Knowing that the steady strength of a transverse fillet is about a third stronger than that of a parallel fillet, one might question why the fatigue allowable for a parallel fillet, (34) and (35), category F, is the same as that for a transverse fillet (36) and higher than that for a transverse fillet (37), category G. The fatigue strength of the transverse fillet (36) is actually higher than that of a parallel fillet (34), but they both fall into the range covered by category F. However, there is a difference in the two transverse fillet welds in (36) and (37). In (36) there may be a slight stress raiser because of the pinching together of forces as they pass through the weld. But in (37) there is a greater tearing action at the root of the weld, thus producing a lower fatigue strength and warranting a lower fatigue allowable. This is illustrated by Fig. 14.43.

CHAPTER 15
INSTABILITIES IN BEAMS AND COLUMNS

Harry Herman
Professor of Mechanical Engineering
New Jersey Institute of Technology
Newark, New Jersey

NOTATION

A	Area of cross section
$B(n)$	Arbitrary constants
$c(n)$	Coefficients in series
$c(y), c(z)$	Distance from y and z axis, respectively, to outermost compressive fiber
e	Eccentricity of axial load P
E	Modulus of elasticity of material
$E(t)$	Tangent modulus for buckling outside of elastic range
$F(x)$	A function of x
G	Shear modulus of material
h	Height of cross section
H	Horizontal (transverse) force on column
I	Moment of inertia of cross section
$I(y), I(z)$	Moment of inertia with respect to y and z axis, respectively
J	Torsion constant; polar moment of inertia
k^2	P/EI
K	Effective-length coefficient
$K(0)$	Spring constant for constraining spring at origin

$K(T, 0), K(T, L)$	Torsional spring constants at $x = 0, L$, respectively
l	Developed length of cross section
L	Length of column or beam
L_{eff}	Effective length of column
M, M'	Bending moments
$M(0), M(L), M_{\text{mid}}$	Bending moments at $x = 0, L$, and midpoint, respectively
$M(0)_{\text{cr}}$	Critical moment for buckling of beam
M_{tr}	Moment due to transverse load
$M(y), M(z)$	Moment about y and z axis, respectively
n	Integer; running index
P	Axial load on column
P_{cr}	Critical axial load for buckling of column
r	Radius of gyration
R	Radius of cross section
s	Running coordinate, measured from one end
t	Thickness of cross section
T	Torque about x axis
x	Axial coordinate of column or beam
y, z	Transverse coordinates and deflections
Y	Initial deflection (crookedness) of column
Y_{tr}	Deflection of beam-column due to transverse load
η	Factor of safety
σ	Stress
ϕ	Angle of twist

As the terms *beam* and *column* imply, this chapter deals with members whose cross-sectional dimensions are small in comparison with their lengths. Particularly, we are concerned with the stability of beams and columns whose axes in the undeformed state are substantially straight. Classically, instability is associated with a state in which the deformation of an idealized, perfectly straight member can become arbitrarily large. However, some of the criteria for stable design which we will develop will take into account the influences of imperfections such as the eccentricity of the axial load and the crookedness of the centroidal axis of the column. The magnitudes of these imperfections are generally not known, but they can be estimated from manufacturing tolerances. For axially loaded columns, the onset of instability is related to the moment of inertia of the column cross section about its minor principal axis. For beams, stability design requires, in addition to the moment of inertia, the consideration of the torsional stiffness.

15.1 EULER'S FORMULA

We will begin with the familiar Euler column-buckling problem. The column is idealized as shown in Fig. 15.1. The top and bottom ends are pinned; that is, the moments at the ends are zero. The bottom pin is fixed against translation; the top pin is free to move in the vertical direction only; the force P acts along the x axis, which coincides with the centroidal axis in the undeformed state. It is important to keep in mind that the analysis which follows applies only to columns with cross sections and loads that are symmetrical about the xy plane in Fig. 15.1 and satisfy the usual assumptions of linear beam theory. It is particularly important in this connection to keep in mind that this analysis is valid only when the deformation is such that the square of the slope of the tangent at any point on the deflection curve is negligibly small compared to unity (fortunately, this is generally true in design applications). In such a case, the familiar differential equation for the bending of a beam is applicable. Thus,

$$EI \frac{d^2y}{dx^2} = M \tag{15.1}$$

For the column in Fig. 15.1,

$$M = -Py \tag{15.2}$$

We take E and I as constant, and let

$$\frac{P}{EI} = k^2 \tag{15.3}$$

Then we get, from Eqs. (15.1), (15.2), and (15.3),

$$\frac{d^2y}{dx^2} + k^2y = 0 \tag{15.4}$$

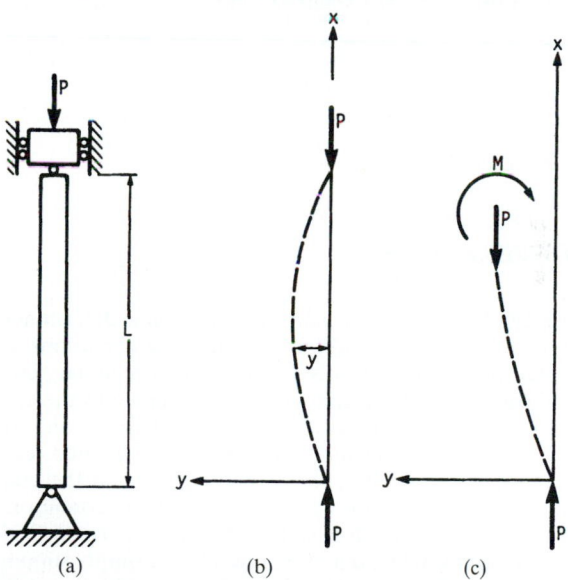

FIGURE 15.1 Deflection of a simply supported column. (*a*) Ideal simply supported column; (*b*) column-deflection curve; (*c*) free-body diagram of deflected segment.

The boundary conditions at $x = 0$ and $x = L$ are

$$y(0) = y(L) = 0 \tag{15.5}$$

In order that Eqs. (15.4) and (15.5) should have a solution $y(x)$ that need not be equal to zero for all values of x, k must take one of the values in Eq. (15.6):

$$k(n) = \frac{n\pi}{L} \qquad n = 1, 2, 3, \dots \tag{15.6}$$

which means that the axial load P must take one of the values in Eq. (15.7):

$$P(n) = \frac{n^2 \pi^2 EI}{L^2} \qquad n = 1, 2, 3, \dots \tag{15.7}$$

For each value of n, the corresponding nonzero solution for y is

$$y(n) = B(n) \sin\left(\frac{n\pi x}{L}\right) \qquad n = 1, 2, 3, \dots \tag{15.8}$$

where $B(n)$ is an arbitrary constant.

In words, the preceding results state the following: Suppose that we have a perfectly straight prismatic column with constant properties over its entire length. If the column is subjected to a perfectly axial load, there is a set of load values, together with a set of sine-shaped deformation curves for the column axis, such that the applied moment due to the axial load and the resisting internal moment are in equilibrium everywhere along the column, no matter what the amplitude of the sine curve may be. From Eq. (15.7), the smallest load at which such deformation occurs, called the *critical load,* is

$$P_{cr} = \frac{\pi^2 EI}{L^2} \tag{15.9}$$

This is the familiar Euler formula.

15.2   EFFECTIVE LENGTH

Note that the sinusoidal shape of the solution function is determined by the differential equation and does not depend on the boundary conditions. If we can find a segment of a sinusoidal curve that satisfies our chosen boundary conditions and, in turn, we can find some segment of that curve which matches the curve in Fig. 15.1, we can establish a correlation between the two cases. This notion is the basis for the "effective-length" concept. Recall that Eq. (15.9) was obtained for a column with both ends simply supported (that is, the moment is zero at the ends). Figure 15.2 illustrates columns of length L with various idealized end conditions. In each case, there is a multiple of L, KL, which is called the *effective length of the column* L_{eff}, that has a shape which is similar to and behaves like a simply supported column of that length. To determine the critical loads for columns whose end supports may be idealized as shown in Fig. 15.2, we can make use of Eq. (15.9) if we replace L by KL, with the appropriate value of K taken from Fig. 15.2. Particular care has to be taken to distinguish between the case in Fig. 15.2c, where both ends of the column are secured against rotation and transverse translation, and the case in Fig. 15.2e, where the ends do not rotate, but relative transverse movement of one end of the column with respect to the other end is possible. The effective length in the first case is half that in the second case, so that the critical load in the first case is four times that in the second case. A major difficulty with using the results in Fig. 15.2 is that in real problems a column end is seldom perfectly fixed or perfectly free (even approximately) with regard to translation or rotation. In addition, we must remember that the critical load is inversely proportional to the square of the effective length. Thus a change of 10 percent in L_{eff} will result in a change of about 20 percent in the critical load, so that a fair approximation of the effective length produces an unsatisfactory approximation of the critical load. We will now develop more general results that will allow us to take into account the elasticity of the structure surrounding the column.

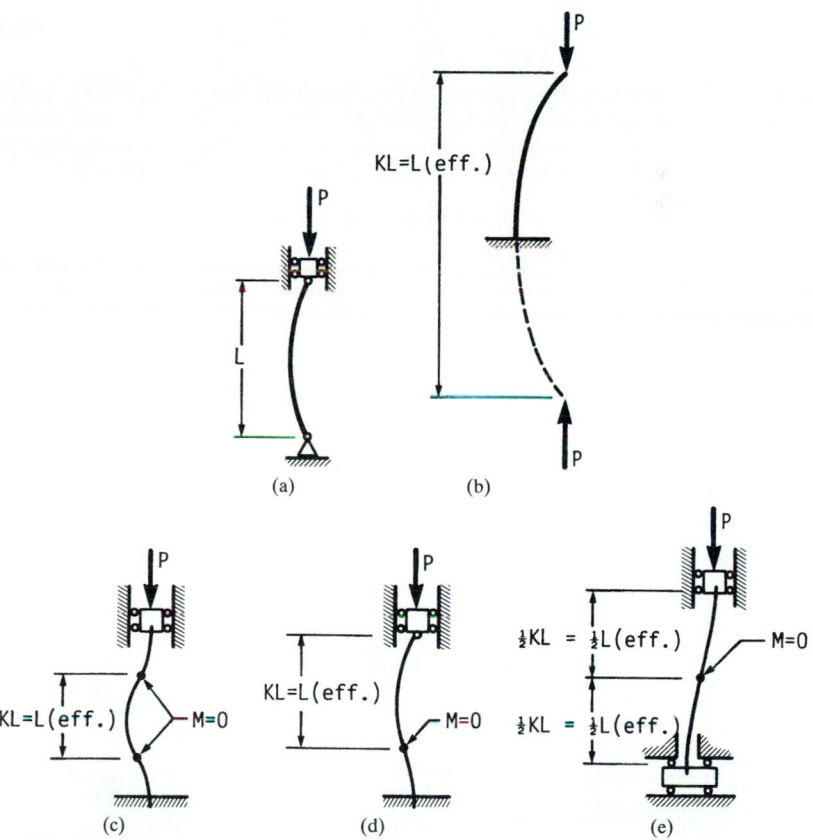

FIGURE 15.2 Effective column lengths for different types of support. (a) Simply supported, $K = 1$; (b) fixed-free, $K = 2$; (c) fixed-fixed, $K = \frac{1}{2}$; (d) fixed-pinned, $K = 0.707$; (e) ends nonrotating, but have transverse translation.

15.3   GENERALIZATION OF THE PROBLEM

We will begin with a generalization of the case in Fig. 15.2e. In Fig. 15.3, the lower end is no longer free to translate, but instead is elastically constrained. The differential equation is

$$EI \frac{d^2y}{dx^2} = M = M(0) + P[y(0) - y] + Hx \qquad (15.10)$$

Here H, the horizontal force at the origin, may be expressed in terms of the deflection at the origin $y(0)$ and the constant of the constraining spring $K(0)$:

$$H = -K(0)y(0) \qquad (15.11)$$

$M(0)$ is the moment which prevents rotation of the beam at the origin. The moment which prevents rotation of the beam at the end $x = L$ is $M(L)$. The boundary conditions are

$$y(L) = 0 \qquad \frac{dy(L)}{dx} = 0 \qquad \frac{dy(0)}{dx} = 0 \qquad (15.12)$$

The rest of the symbols are the same as before. We define k as in Eq. (15.3). As in the case of the simply supported column, Eqs. (15.10), (15.11), and (15.12) have solutions in which $y(x)$ need not be zero for all values of x, but again these solutions occur only for certain values of kL. Here these values of kL must satisfy Eq. (15.13):

$$[2(1 - \cos kL) - kL \sin kL]L^3 K(0) + EI(kL)^3 \sin kL = 0 \qquad (15.13)$$

The physical interpretation is the same as in the simply supported case. If we denote the lowest value of kL that satisfies Eq. (15.13) by $(kL)_{cr}$, then the column buckling load is given by

$$P_{cr} = \frac{EI(kL)_{cr}^2}{L^2} \qquad (15.14)$$

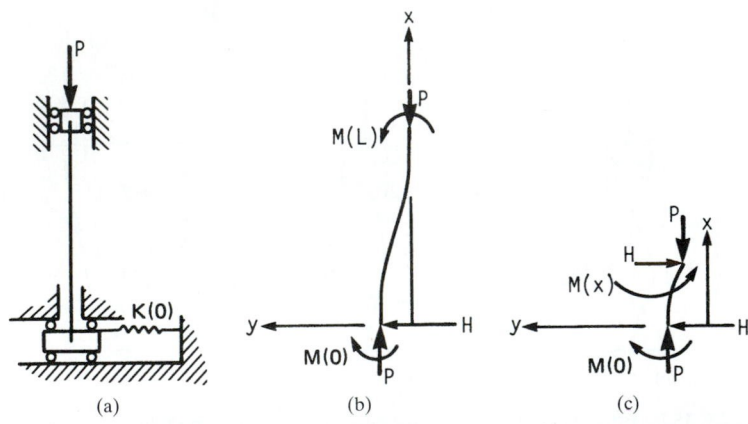

FIGURE 15.3 Column with ends fixed against rotation and an elastic end constraint against transverse deflection. (*a*) Undeflected column; (*b*) deflection curve; (*c*) free-body diagram of deflected segment.

Since the column under consideration here has greater resistance to buckling than the case in Fig. 15.2e, the $(kL)_{cr}$ here will be greater than π. We can therefore evaluate Eq. (15.13) beginning with $kL = \pi$ and increasing it slowly until the value of the left side of Eq. (15.13) changes sign. Since $(kL)_{cr}$ lies between the values of kL for which the left side of Eq. (15.13) has opposite signs, we now have bounds on $(kL)_{cr}$. To obtain improved bounds, we take the average of the two bounding values, which we will designate by $(kL)_{av}$. If the value of the left side of Eq. (15.13) obtained by using $(kL)_{av}$ is positive (negative), then $(kL)_{cr}$ lies between $(kL)_{av}$ and that value of kL for which the left side of Eq. (15.13) is negative (positive). This process is continued, using the successive values of $(kL)_{av}$ to obtain improved bounds on $(kL)_{cr}$, until the desired accuracy is obtained.

The last two equations in Eq. (15.12) imply perfect rigidity of the surrounding structure with respect to rotation. A more general result may be obtained by taking into account the elasticity of the surrounding structure in this respect. Suppose that the equivalent torsional spring constants for the surrounding structure are $K(T, 0)$ and $K(T, L)$ at $x = 0$ and $x = L$, respectively. Then Eq. (15.12) is replaced by

$$y(L) = 0$$

$$M(0) = K(T, 0)\,\frac{dy(0)}{dx} \qquad (15.15)$$

$$M(L) = -K(T, L)\,\frac{dy(L)}{dx}$$

Proceeding as before, with Eq. (15.15) replacing Eq. (15.12), we obtain the following equation for kL:

$$\left\{\left[\frac{L^3}{EI(kL)^3}\right][K(T, 0) + K(T, L)]K(0) - \left[\frac{L^4 K(0)K(T, 0)K(T, L)}{(EI)^2(kL)^3}\right]\right.$$
$$+ \left[\frac{K(0)L^2}{(kL)}\right] + \left[\frac{LK(T, 0)K(T, L)}{EI(kL)}\right] - \left[\frac{EI(kL)}{L}\right]\right\}\sin kL$$
$$+ \left\{K(T, 0) + K(T, L) - \left[\frac{L^3}{EI(kL)^2}\right][K(T, 0) + K(T, L)]K(0)\right\}\cos kL$$
$$+ 2\left[\frac{L^4 K(0)K(T, 0) + K(T, L)}{(EI)^2(kL)^4}\right](1 - \cos kL) = 0 \quad (15.16)$$

The lowest value of kL satisfying Eq. (15.16) is the $(kL)_{cr}$ to be substituted in Eq. (15.14) in order to obtain the critical load. Here there is no apparent good guess with which to begin computations. Considering the current accessibility of computers, a convenient approach would be to obtain a plot of the left side of Eq. (15.16) for $0 \le kL < \pi$, and if there is no change in sign, extend the plot up to $kL = 2\pi$, which is the solution for the column with a perfectly rigid surrounding structure (Fig. 15.2c). However, see also Chap. 4.

15.4 *MODIFIED BUCKLING FORMULAS*

The critical-load formulas developed above provide satisfactory values of the allowable load for very slender columns for which buckling, as manifested by unacceptably large deformation, will occur within the elastic range of the material. For more massive columns, the deformation enters the plastic region (where strain increases more rapidly with stress) prior to the onset of buckling. To take into account this change in the stress-strain relationship, we modify the Euler formula. We define the *tangent modulus* $E(t)$ as the slope of the tangent to the stress-strain curve at a given strain. Then the modified formulas for the critical load are obtained by substituting $E(t)$ for E in Eq. (15.9) and Eq. (15.13) plus Eq. (15.14) or Eq. (15.16) plus Eq. (15.14). This will produce a more accurate prediction of the buckling load. However, this may not be the most desirable design approach. In general, a design which will produce plastic deformation under the operating load is undesirable. Hence, for a column which will undergo plastic deformation prior to buckling, the preferred design-limiting criterion is the onset of plastic deformation, not the buckling.

15.5 STRESS-LIMITING CRITERION

We will now develop a design criterion which will enable us to use the yield strength as the upper bound for acceptable design regardless of whether the stress at the onset of yielding precedes or follows buckling. Here we follow Ref. [15.1]. This approach has the advantage of providing a single bounding criterion that holds irrespective of the mode of failure. We begin by noting that, in general, real columns will have some imperfection, such as crookedness of the centroidal axis or eccentricity of the axial load. Figure 15.4 shows the difference between the behavior of an ideal, perfectly straight column subjected to an axial load, in which case we obtain a distinct critical point, and the behavior of a column with some imperfection.

It is clear from Fig. 15.4 that the load-deflection curve for an imperfect column has no distinct critical point. Instead, it has two distinct regions. For small axial loads, the deflection increases slowly with load. When the load is approaching the critical value obtained for a perfect column, a small increment in load produces a large change in deflection. These two regions are joined by a "knee." Thus the advent of buckling in a real column corresponds to the entry of the column into the second, above-the-knee, load-deflection region. A massive column will reach the stress at the yield point prior to buckling, so that the yield strength will be the limiting criterion for the maximum allowable load. A slender column will enter the above-the-

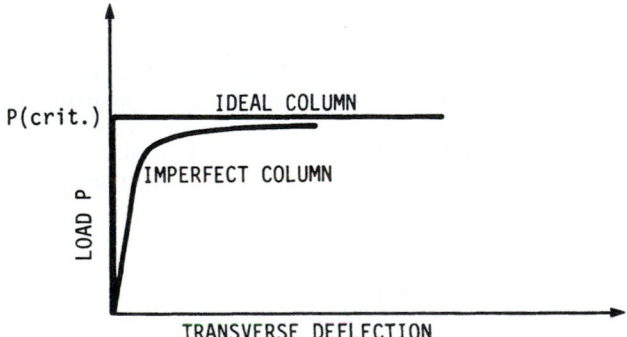

FIGURE 15.4   Typical load-deflection curves for ideal and real columns.

knee region prior to reaching the stress at the yield point, but once in the above-the-knee region, it requires only a small increment in load to produce a sufficiently large increase in deflection to reach the yield point. Thus the corresponding yield load may be used as an adequate approximation of the buckling load for a slender column as well. Hence the yield strength provides an adequate design bound for both massive and slender columns. It is also important to note that, in general, columns found in applications are sufficiently massive that the linear theory developed here is valid within the range of deflection that is of interest.

Application of Eq. (15.1) to a simply supported imperfect column with constant properties over its length yields a modification of Eq. (15.4). Thus,

$$\frac{d^2y}{dx^2} + k^2y = k^2(e - Y) \tag{15.17}$$

where e = eccentricity of the axial load P (taken as positive in the positive y direction) and Y = initial deflection (crookedness) of the unloaded column. The x axis is taken through the end points of the centroidal axis, so that Eq. (15.5) still holds and Y is zero at the end points. Note that the functions in the right side of Eq. (15.8) form a basis for a trigonometric (Fourier) series, so that any function of interest may be expressed in terms of such a series. Thus we can write

$$Y = \sum_{n=1}^{\infty} c(n) \sin \frac{n\pi x}{L} \tag{15.18}$$

where

$$c(n) = \frac{2}{L} \int_0^L Y(x) \sin \frac{n\pi x}{L} \, dx \tag{15.19}$$

The solution for the deflection y in Eq. (15.17) is given by

$$y = P \sum_{n=1}^{\infty} \left[c(n) - \frac{4e}{n\pi} \right] \frac{\sin(n\pi x/L)}{[EI(n\pi/L)^2 - P]} \tag{15.20}$$

The maximum deflection y_{max} of a simply supported column will usually (except for cases with a pronounced and asymmetrical initial deformation or antisymmetrical load eccentricity) occur at the column midpoint. A good approximation (probably within 10 percent) of y_{max} in the above-the-knee region that may be used in deflection-limited column design is given by the coefficient of the first term in Eq. (15.20):

$$y_{max} = \frac{P[c(1) - (4e/\pi)]}{EI(\pi/L)^2 - P} \tag{15.21}$$

The maximum bending moment will also usually occur at the column midpoint and at incipient yielding is closely approximated by

$$M_{max} = P \left\{ e - Y_{mid} + \left[\frac{4e}{\pi} - c(1) \right] \frac{P}{[EI(\pi/L)^2 - P]} \right\} \tag{15.22}$$

The immediately preceding analysis deals with the bending moment about the z axis (normal to the paper). Clearly, a similar analysis can be made with regard to bending about the y axis (Fig. 15.1). Unlike the analysis of the perfect column, where it is merely a matter of finding the buckling load about the weaker axis, in the present approach the effects about the two axes interact in a manner familiar from analysis of an eccentrically loaded short strut. We now use the familiar expression for combining direct axial stresses and bending stresses about two perpendicular axes. Since there is no ambiguity, we will suppress the negative sign associated with compressive stress:

$$\sigma = \frac{P}{A} + \frac{M(z)c(y)}{I(z)} + \frac{M(y)c(z)}{I(y)} \tag{15.23}$$

where $c(y)$ and $c(z)$ = perpendicular distances from the z axis and y axis, respectively (these axes meet the x axis at the cross-section centroid at the origin), to the outermost fiber in compression; A = cross-sectional area of the column; and σ = total compressive stress in the fiber which is farthest removed from both the y and z axes. For an elastic design limited by yield strength, σ is replaced by the yield strength; $M(z)$ in the right side of Eq. (15.23) is the magnitude of the right side of Eq. (15.22); and $M(y)$ is an expression similar to Eq. (15.22) in which the roles of the y and z axes interchange.

Usually, in elastic design, the yield strength is divided by a chosen factor of safety η to get a permissible or allowable stress. In problems in which the stress increases linearly with the load, dividing the yield stress by the factor of safety is equivalent to multiplying the load by the factor of safety. However, in the problem at hand, it is clear from the preceding development that the stress is not a linear function of the axial load and that we are interested in the behavior of the column as it enters the above-the-knee region in Fig. 15.4. Here it is necessary to multiply the applied axial load by the desired factor of safety. The same procedure applies in introducing a factor of safety in the critical-load formulas previously derived.

Example 1. We will examine the design of a nominally straight column supporting a nominally concentric load. In such a case, a circular column cross section is the most reasonable choice, since there is no preferred direction. For this case, Eq. (15.23) reduces to

$$\sigma = \frac{P}{A} + \frac{Mc}{I} \tag{1}$$

For simplicity, we will suppose that the principal imperfection is due to the eccentric location of the load and that the column crookedness effect need not be taken into account, so that Eq. (15.22) reduces to

$$M_{\max} = P\left\{ e + \left(\frac{4e}{\pi}\right) \frac{P}{[EI(\pi/L)^2 - P]} \right\} \tag{2}$$

Note that for a circular cross section of radius R, the area and moment of inertia are, respectively,

$$A = \pi R^2 \quad \text{and} \quad I = \frac{\pi R^4}{4} = \frac{A^2}{4\pi} \tag{3}$$

We will express the eccentricity of the load as a fraction of the cross-section radius. Thus,

$$e = \varepsilon R \tag{4}$$

Then we have, from Eqs. (1) through (4),

$$\sigma_{\text{allow}} = \frac{P}{A} + \frac{4P\varepsilon}{A}\left\{ 1 + \left(\frac{4}{\pi}\right)\frac{P}{[(EA^2)/(4\pi)](\pi/L)^2 - P}\right\} \tag{5}$$

Usually P and L are given, σ and E are the properties of chosen material, and ε is determined from the clearances, tolerances, and kinematics involved, so that Eq. (5) is reduced to a cubic in A.

At the moment, however, we are interested in comparing the allowable nominal column stress P/A with the allowable stress of the material σ_{allow} for columns of different lengths. Keeping in mind that the radius of gyration r of a circular cross section of geometric radius R is $R/2$, we will define

$$\frac{R}{2} = r$$

$$\frac{\sigma_{\text{allow}}}{P/A} = p \tag{6}$$

$$\frac{E}{\sigma_{\text{allow}}} = q$$

Then Eq. (5) may be written as

$$p = 1 + 4\varepsilon + \frac{16\varepsilon}{\pi[\pi^2 pq(r/L)^2 - 1]} \tag{7}$$

The first term on the right side of Eq. (7) is due to direct compressive stress; the second term is due to the bending moment produced by the load eccentricity; the third term is due to the bending moment arising from the column deflection. When ε is small, p will be close to unity unless the denominator in the third term on the right side of Eq. (7) becomes small—that is, the moment due to the column deflection becomes large. The ratio L/r, whose reciprocal appears in the denominator of the third term, is called the *slenderness ratio*. Equation (7) may be rewritten as a quadratic in p. Thus,

$$\pi^2 q\left(\frac{r}{L}\right)^2 p^2 - \left[(1 + 4\varepsilon)\pi^2 q\left(\frac{r}{L}\right)^2 + 1\right]p + (1 + 4\varepsilon) - \frac{16\varepsilon}{\pi} = 0 \tag{8}$$

We will take for q the representative value of 1000 and tabulate $1/p$ for a number of values of L/r and ε. To compare the value of $1/p$ obtained from Eq. (8) with the corresponding result from Euler's formula, we will designate the corresponding result obtained by Euler's formula as $1/p_{\text{cr}}$ and recast Eq. (15.9) as

$$\frac{1}{p_{\text{cr}}} = \pi^2 q\left(\frac{r}{L}\right)^2 \tag{9}$$

To interpret the results in Table 15.1, note that the quantities in the second and third columns of the table are proportional to the allowable loads calculated from the respective equations. As expected, the Euler formula is completely inapplicable when L/r is 50. Also, as expected, the allowable load decreases as the eccentricity increases. However, the effect of eccentricity on the allowable load decreases as the slenderness ratio L/r increases. Hence when L/r is 250, the Euler buckling load, which is the limiting case for which the eccentricity is zero, is only about 2 percent higher than when the eccentricity is 2 percent.

TABLE 15.1   Influence of Eccentricity and
Slenderness Ratio on Allowable Load

ε	L/r	p^{-1}	p_{cr}^{-1}
0.02	50	0.901	3.95
	150	0.408	0.439
	250	0.155	0.158
0.05	50	0.791	3.95
	150	0.374	0.439
	250	0.151	0.158
0.10	50	0.665	3.95
	150	0.333	0.439
	250	0.145	0.158

15.6   BEAM-COLUMN ANALYSIS

A member that is subjected to both a transverse load and an axial load is frequently
called a *beam-column*. To apply the immediately preceding stress-limiting criterion
to a beam-column, we first determine the moment distribution, say, M_{tr}, and the corresponding deflection, say, Y_{tr}, resulting from the transverse load acting alone. Suppose that the transverse load is symmetrical about the column midpoint, and let
$Y_{tr,mid}$ and $M_{tr,mid}$ be the values of Y_{tr} and M_{tr} at the column midpoint. Then the only
modifications necessary in the preceding development are to replace Y by $Y + Y_{tr}$ in
Eqs. (15.17) and (15.20), and to replace Y_{mid} by $Y_{mid} + Y_{tr,mid}$ and add $M_{tr,mid}$ on the
right side of Eq. (15.23). If the transverse load is not symmetrical, then it is necessary
to determine the maximum moment by using an approach which will now be developed.

Note that, at any point, x, the moment about the z axis is

$$M(z) = P(e - y - Y) + M(z)_{tr} \qquad (15.24)$$

where Y includes the deflection due to $M(z)_{tr}$. $M(y)$ has the same form as Eq.
(15.24), but with the roles of y and z interchanged. The maximum stress for any given
value of x is given by Eq. (15.23). We seek to apply this equation at that value of x
which yields the maximum value of σ. A method that is reasonably efficient in locating a minimum or maximum to any desired accuracy is the golden-section search.
However, this method is limited to finding the minimum (maximum) of a unimodal
function, that is, a function which has only one minimum (maximum) in the interval
in which the search is conducted. We therefore have to conduct some exploratory
calculation to find the stress at, say, a dozen points on the beam-column in order to
locate the unimodal interval of interest within which to apply the golden-section
search. The actual number of exploratory calculations will depend on the individual
case. For example, in a simply supported case with a unimodal transverse moment,
there is clearly only one maximum. But, in general, we must check enough points to
be sure that a potential maximum is not overlooked.

The golden-section search procedure is as follows: Suppose that we seek the minimum value of $F(x)$ in Fig. 15.5 within the interval D (note that if we sought a maximum in Fig. 15.5, we would have to conduct two searches). We locate two points $x(1)$
and $x(2)$. The first is $0.382D$ from the left end of the interval; the second is $0.382D$

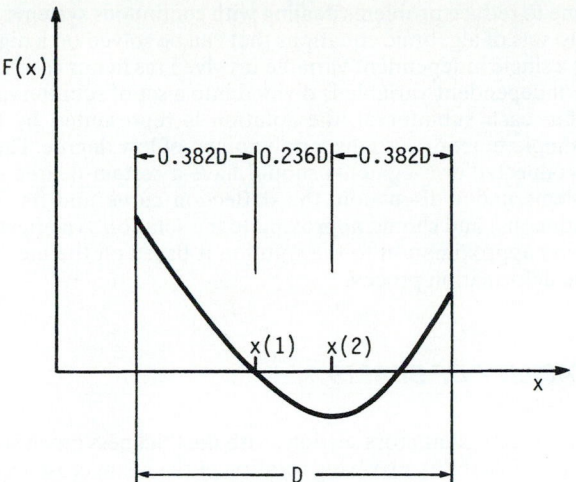

FIGURE 15.5   Finding min $F(x)$ in the interval D.

from the right end. We evaluate $F(x)$ at the chosen points. If the value of $F(x)$ at $x(1)$ is algebraically smaller than at $x(2)$, then we eliminate the subinterval between $x(2)$ and the right end of the search interval. In the oppositive event, we eliminate the subinterval between $x(1)$ and the left end. The distance of $0.236D$ between $x(1)$ and $x(2)$ equals 0.382 of the new search interval, which is $0.618D$ overall. Thus we have already located one point at 0.382 of the new search interval, and we have determined the value of $F(x)$ at that point. We need only locate a point that is 0.382 from the opposite end of the new search interval, evaluate $F(x)$ at that point, and go through the elimination procedure, as before, to further reduce the search interval. The process is repeated until the entire remaining interval is small and the value of $F(x)$ at the two points at which it is evaluated is the same within the desired accuracy. The preceding method may run into difficulty if the function $F(x)$ is flat, that is, if $F(x)$ does not vary over a substantial part of the search interval or if $F(x)$ happens to have the same value at the two points that are compared with each other. The first case is not likely to occur in the types of problems under consideration here. In the second case, we calculate $F(x)$ at another point, near one of the two points of comparison, and use the newly chosen point in place of the nearby original point. Reference [15.2] derives the golden-section search strategy and provides equations (p. 289) for predicting the number of function evaluations required to attain a specified fractional reduction of search interval or an absolute final search interval size.

15.7 APPROXIMATE METHOD

The reason why we devoted so much attention to uniform prismatic column problems is that their solution is analytically simple, so that we could obtain the results directly. In most other cases, we have to be satisfied with approximate formulations for computer (or programmable calculator) calculation of the solution. The finite-element method is the approximation method most widely used in engineering at the present time to reduce problems dealing with continuous systems, such as beams and columns, to sets of algebraic equations that can be solved on a digital computer. When there is a single independent variable involved (as in our case), the interval of interest of the independent variable is divided into a set of subintervals called *finite elements*. Within each subinterval, the solution is represented by an arc that is defined by a simple function, usually a polynomial of low degree. The curve resulting from the connected arc segments should have a certain degree of smoothness (for the problems under discussion, the deflection curve and its first derivative should be continuous) and should approximate the solution. An effective method of obtaining a good approximation to the solution is based on the mechanical energy involved in the deformation process.

15.8   INSTABILITY OF BEAMS

Beams that have rectangular cross sections with the thickness much smaller than the depth are prone to instability involving rotation of the beam cross section about the beam axis. This tendency to instability arises because such beams have low resistance to torsion about their axes. In preparation for the analysis of this problem, recall that for a circular cylindrical member of length L and radius R, subjected to an axial torque T, the angle of twist ϕ is given by

$$\phi = \frac{TL}{JG} \tag{15.25}$$

where G = shear modulus and J = polar moment of inertia of the cross section. For noncircular cross sections, the form of the right side of Eq. (15.25) does not change; the only change is in the expression for J, the torsion constant of the cross section. If the thickness of the cross section, to be denoted by t, is small and does not vary much, then J is given by

$$3J = \int_0^l t^3 \, ds \tag{15.26}$$

where s = running coordinate measured from one end of the cross section and l = total developed length of the cross section. Thus, for a rectangular cross section of depth h,

$$J = \frac{ht^3}{3} \tag{15.27}$$

Clearly, J decreases rapidly as t decreases. To study the effect of this circumstance on beam stability, we will examine the deformation of a beam with rectangular cross section subjected to end moments $M(0)$, taking into account rotation of the cross section about the beam axis. The angle of rotation ϕ is assumed to be small, so that $\sin \phi$ may be replaced by ϕ and $\cos \phi$ by unity.

The equations of static equilibrium may be written from Fig. 15.6, where the moments are shown as vectors, using the right-hand rule:

$$-M'(z) + M(0) = 0$$

$$M'(y) + M(0)\phi = 0 \tag{15.28}$$

$$\frac{dT}{dx} + \frac{dM'(z)}{dx} = 0$$

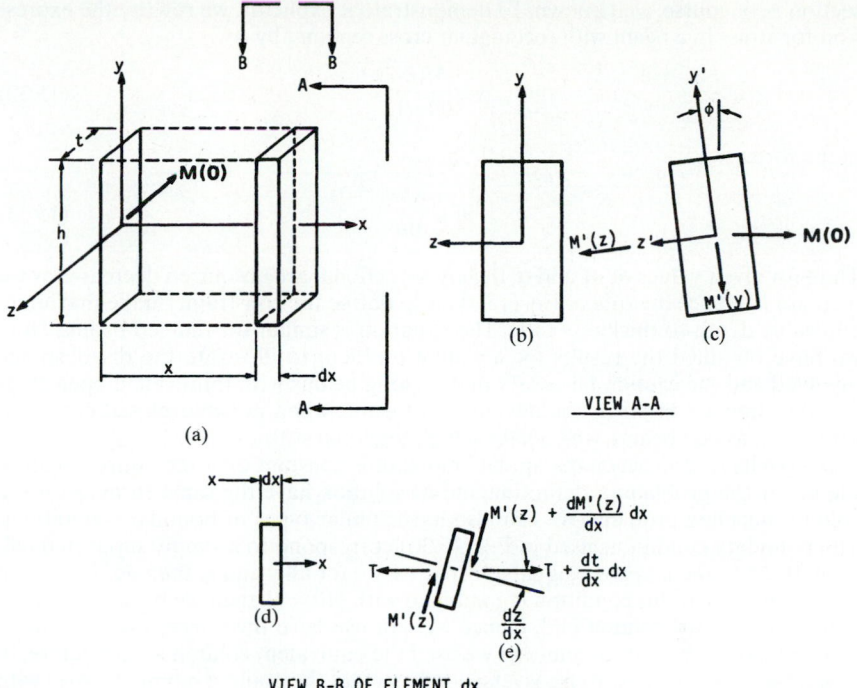

FIGURE 15.6 Instability of beams. (*a*) Segment of a beam with applied moment $M(0)$ at the ends; (*b*) cross section of undeformed beam; (*c*) cross section after onset of beam instability; (*d*) top view of undeformed differential beam element; (*e*) the element after onset of instability.

These, combined with Eq. (15.25) and the standard moment-curvature relation as given by Eq. (15.1), lead to the defining differential equation (15.29) for the angle ϕ:

$$GJ \frac{d^2\phi}{dx^2} + M^2(0)\phi \left[\frac{1}{EI(y)} - \frac{1}{EI(z)} \right] = 0 \qquad (15.29)$$

Suppose that the end faces of the beam are fixed against rotation about the x axis. Then the boundary conditions are

$$\phi(0) = \phi(L) = 0 \qquad (15.30)$$

Noting the similarity between Eq. (15.4) with its boundary conditions [Eq. (15.5)] and Eq. (15.29) with its boundary conditions [Eq. (15.30)], the similarity of the solutions is clear. Thus we obtain the expression for $M(0)_{cr}$:

$$[M(0)_{cr}]^2 = \frac{\pi^2 E}{L^2} GJ \left[\frac{I(z)I(y)}{I(z) - I(y)} \right] \qquad (15.31)$$

It may be seen from Eq. (15.31) that if the torsion constant J is small, the critical moment is small. In addition, it may be seen from the bracketed term in Eq. (15.31) that as the cross section approaches a square shape, the denominator becomes small, so that the critical moment becomes very large. The disadvantage of a square cross section is, of course, well known. To demonstrate it explicitly, we rewrite the expression for stress in a beam with rectangular cross section t by h:

$$\sigma = \frac{M(h/2)}{th^3/12} \tag{15.32}$$

in the form

$$th = \frac{6M}{\sigma h} \tag{15.33}$$

Thus, for given values of M and σ, the cross-sectional area required decreases as we increase h. Hence the role of Eq. (15.31) is to define the constraint on the maximum allowable depth-to-thickness ratio. The situation is similar for flanged beams. Here we have obtained the results for a simple problem to illustrate the disadvantage involved and the caution necessary in designing beams with thin-walled open cross sections. Implicit in this is the advantage of using, when possible, closed cross sections, such as box beams, which have a high torsional stiffness.

As we have seen, when the applied moment is constant over the entire length of the beam, the problem of definition and its solution have the same form as for the column-buckling problem. We can also have similar types of boundary conditions. The boundary conditions used in Eq. (15.30) correspond to a simply supported column. If dy/dx and dz/dx are equal to zero at $x = 0$ for all y and z, then $d\phi/dx$ is equal to zero at $x = 0$. If this condition is combined with $\phi(0) = 0$, then we have the equivalent of a clamped column end. Hence we can use here the concept of equivalent beam length in the same manner as we used the equivalent column length before. In case a beam is subjected to transverse loads, so that the applied moment varies with x, the problem is more complex. For the proportioning of flanged beams, Ref. [15.3] should be used as a guide. This reference deals with structural applications, so that the size range of interest dealt with is different from the size range of interest in machine design. But the underlying principles of beam stability are the same, and the proportioning of the members should be similar.

Example 2. We will examine the design of a beam of length L and rectangular cross section t by h. The beam is subjected to an applied moment M (we will not use any modifying symbols here, since there is no ambiguity), which is constant over the length of the beam. As noted previously, the required cross-sectional area th will decrease as h is increased. We take the allowable stress in the material to be σ. The calculated stress in the beam is not to exceed this value. Thus,

$$\sigma \geq \frac{M(h/2)}{th^3/12} \qquad \sigma \geq \frac{6M}{th^2} \qquad th \geq \frac{6M}{h\sigma} \tag{1}$$

We want the cross-sectional area th as small as possible. Hence h should be as large as possible. We can, therefore, replace the inequality in Eq. (1) by the equality

$$th = \frac{6M}{h\sigma} \tag{2}$$

The maximum value of h that we can use is subject to a constraint based on Eq. (15.31). We will use a factor of safety η in this connection. Thus,

$$(\eta M)^2 \leq \frac{\pi^2 E}{L^2} GJ \left[\frac{I(z)I(y)}{I(z) - I(y)} \right] \tag{3}$$

Here

$$I(z) = \frac{th^3}{12} \qquad I(y) = \frac{ht^3}{12} \qquad J = \frac{ht^3}{3} \tag{4}$$

Using Eq. (4), we may write Eq. (3) as

$$(\eta M)^2 \le \frac{\pi^2 EG}{(6L)^2} \left(\frac{t}{h}\right)^2 \left[\frac{1}{1 - (t/h)^2}\right](th)^4 \tag{5}$$

or, from Eq. (2),

$$(\eta M)^2 \le \frac{\pi^2 EG}{(6L)^2} \left(\frac{6M}{h^3\sigma}\right)^2 \left[\frac{1}{1 - [(6M/(h^3\sigma)]^2}\right]\left(\frac{6M}{h\sigma}\right)^4 \tag{6}$$

Since we seek to minimize th and maximize h, it may be seen from Eqs. (5) and (6) that the inequality sign may be replaced by the equality sign in those two equations. In Eq. (6), h is the only unspecified quantity. Further, since the square of t/h may be expected to be small compared to unity, we can obtain substantially simpler approximations of reasonable accuracy. As a first step, we have

$$\frac{1}{1 - (t/h)^2} = 1 + \left(\frac{t}{h}\right)^2 + \left(\frac{t}{h}\right)^4 + \cdots \tag{7}$$

If we retain only the first two terms in the right side of Eq. (7), we have

$$\eta M = \frac{\pi(EG)^{1/2}}{6L} \left[1 + \left(\frac{6M}{h^3\sigma}\right)^2\right]\left(\frac{6M}{h^3\sigma}\right)\left(\frac{6M}{h\sigma}\right)^2 \tag{8}$$

If we also neglect the square of t/h in comparison with unity, we obtain

$$h = \left[\frac{\pi(6M)^2(EG)^{1/2}}{\eta L\sigma^3}\right]^{1/5} \tag{9}$$

as a reasonable first approximation. Thus if we take the factor of safety η as 1.5, we have, for a steel member with $E = 30$ Mpsi, $G = 12$ Mpsi, and $\sigma = 30$ kpsi,

$$h = \left[\frac{\pi(36)[(30 \times 10^6)(12 \times 10^6)]^{1/2}}{(1.5)(30\,000)} \frac{M^2}{L}\right]^{1/5} = 8.62\left(\frac{M^2}{L}\right)^{1/5} \tag{10}$$

This is a reasonable approximation to the optimal height of the beam cross section. It may also be used as a starting point for an iterative solution to the exact expression, Eq. (6). For the purpose of iteration, we rewrite Eq. (6) as

$$h = \left\{\frac{\pi(6M)^2}{\eta L\sigma^3}\left[\frac{EG}{1 - [(6M)/(h^3\sigma)]^2}\right]^{1/2}\right\}^{1/5} \tag{11}$$

The value of h obtained from Eq. (9) is substituted into the right side of Eq. (11). The resultant value of h thus obtained is then resubstituted into the right side of Eq. (11); the iterative process is continued until the computed value of h coincides with the value substituted into the right side to the desired degree of accuracy. Having determined h, we can determine t from Eq. (2).

REFERENCES

15.1 H. Herman, "On the Analysis of Uniform Prismatic Columns," *Transactions of the ASME, Journal of Mechanical Design,* vol. 103, 1981, pp. 274–276.

15.2 C. R. Mischke, *Mathematical Model Building,* 2d rev. ed., Iowa State University Press, Ames, 1980.

15.3 *Manual of Steel Construction,* 8th ed., American Institute of Steel Construction, New York.

CHAPTER 16
CURVED BEAMS AND RINGS

Joseph E. Shigley
Professor Emeritus
The University of Michigan
Ann Arbor, Michigan

NOTATION

A	Area, or a constant
B	Constant
C	Constant
E	Modulus of elasticity
e	Eccentricity
F	Force
G	Modulus of rigidity
I	Second moment of area (Table 48.1)
K	Shape constant (Table 49.1), or second polar moment of area
M	Bending moment
P	Reduced load
Q	Fictitious force
R	Force reaction
r	Ring radius
$\bar{r}$	Centroidal ring radius
T	Torsional moment
U	Strain energy
V	Shear force
W	Resultant of a distributed load
w	Unit distributed load
X	Constant

Y	Constant
y	Deflection
Z	Constant
γ	Load angle
ϕ	Span angle, or slope
σ	Normal stress
θ	Angular coordinate or displacement

Methods of computing the stresses in curved beams for a variety of cross sections are included in this chapter. Rings and ring segments loaded normal to the plane of the ring are analyzed for a variety of loads and span angles, and formulas are given for bending moment, torsional moment, and deflection.

16.1 BENDING IN THE PLANE OF CURVATURE

The distribution of stress in a curved member subjected to a bending moment in the plane of curvature is hyperbolic ([16.1], [16.2]) and is given by the equation

$$\sigma = \frac{My}{Ae(r - e - y)} \tag{16.1}$$

where $r =$ radius to centroidal axis
$y =$ distance from neutral axis
$e =$ shift in neutral axis due to curvature (as noted in Table 16.1)

The moment M is computed about the *centroidal axis,* not the neutral axis. The maximum stresses, which occur on the extreme fibers, may be computed using the formulas of Table 16.1.

In most cases, the bending moment is due to forces acting to one side of the section. In such cases, be sure to add the resulting axial stress to the maximum stresses obtained using Table 16.1.

16.2 CASTIGLIANO'S THEOREM

A complex structure loaded by any combination of forces, moments, and torques can be analyzed for deflections by using the elastic energy stored in the various components of the structure [16.1]. The method consists of finding the total strain energy stored in the system by all the various loads. Then the displacement corresponding to a particular force is obtained by taking the partial derivative of the total energy with respect to that force. This procedure is called *Castigliano's theorem.* General expressions may be written as

$$y_i = \frac{\partial U}{\partial F_i} \qquad \theta_i = \frac{\partial U}{\partial T_i} \qquad \phi_i = \frac{\partial U}{\partial M_i} \tag{16.2}$$

where $U =$ strain energy stored in structure
$y_i =$ displacement of point of application of force F_i in the direction of F_i

$\theta_i =$ angular displacement at T_i

$\phi_i =$ slope or angular displacement at moment M_i

If a displacement is desired at a point on the structure where no force or moment exists, then a fictitious force or moment is placed there. When the expression for the corresponding displacement is developed, the fictitious force or moment is equated to zero, and the remaining terms give the deflection at the point where the fictitious load had been placed.

Castigliano's method can also be used to find the reactions in indeterminate structures. The procedure is simply to substitute the unknown reaction in Eq. (16.2) and use zero for the corresponding deflection. The resulting expression then yields the value of the unknown reaction.

It is important to remember that the displacement-force relation must be linear. Otherwise, the theorem is not valid.

Table 16.2 summarizes strain-energy relations.

16.3 RING SEGMENTS WITH ONE SUPPORT

Figure 16.1 shows a cantilevered ring segment fixed at C. The force F causes bending, torsion, and direct shear. The moments and torques at the fixed end C and at any section B are shown in Table 16.3. The shear at C is $R_C = F$. Stresses in the ring can be computed using the formulas of Chap. 49.

To obtain the deflection at end A, we use Castigliano's theorem. Neglecting direct shear and noting from Fig. 16.1b that $l = r\, d\theta$, we determine the strain energy from Table 16.2 to be

$$U = \int_0^\phi \frac{M^2 r\, d\theta}{2EI} + \int_0^\phi \frac{T^2 r\, d\theta}{2GK} \tag{16.3}$$

Then the deflection y at A and in the direction of F is computed from

$$y = \frac{\partial U}{\partial F} = \frac{r}{EI} \int_0^\phi M \frac{\partial M}{\partial F}\, d\theta + \frac{r}{GK} \int_0^\phi T \frac{\partial T}{\partial F}\, d\theta \tag{16.4}$$

The terms for this relation are shown in Table 16.3. It is convenient to arrange the solution in the form

$$y = \frac{Fr^3}{2} \left(\frac{A}{EI} + \frac{B}{GK} \right) \tag{16.5}$$

where the coefficients A and B are related only to the span angle. These are listed in Table 16.3.

Figure 16.2a shows another cantilevered ring segment, loaded now by a distributed load. The resultant load is $W = wr\phi$; a shear reaction $R = W$ acts upward at the fixed end C, in addition to the moment and torque reactions shown in Table 16.3.

A force $W = wr\theta$ acts at the centroid of segment AB in Fig. 16.2b. The centroidal radius is

$$\bar{r} = \frac{2r \sin(\theta/2)}{\theta} \tag{16.6}$$

TABLE 16.1 Eccentricities and Stress Factors for Curved Beams[†]

1. Rectangle

$$e = r - \frac{h}{\ln\left(\dfrac{r+c}{r-c}\right)} \qquad K_i = \frac{c(c-e)}{3e(r-c)} \qquad K_o = \frac{c(c+e)}{3e(r+c)}$$

2. Solid round

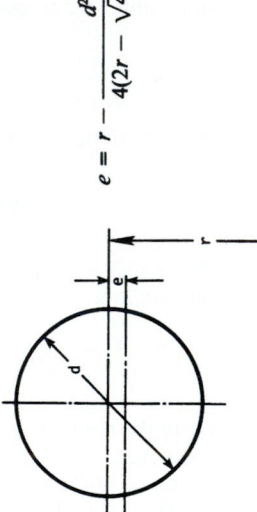

$$e = r - \frac{d^2}{4(2r - \sqrt{4r^2 - d^2})} \qquad K_i = \frac{d(d-2e)}{8e(2r-d)} \qquad K_o = \frac{d(d+2e)}{8e(2r+d)}$$

16.4

3. Hollow round

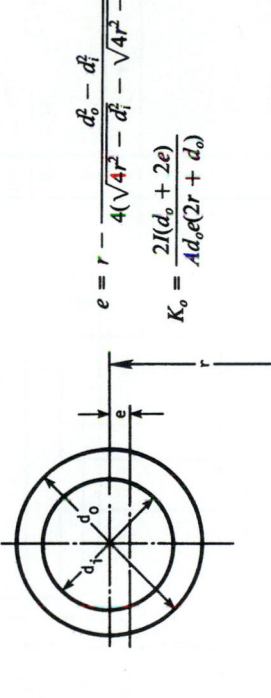

$$e = r - \frac{d_o^2 - d_i^2}{4\left(\sqrt{4r^2 - d_i^2} - \sqrt{4r^2 - d_o^2}\right)} \qquad K_i = \frac{2I(d_o - 2e)}{Ad_o e(2r - d_o)}$$

$$K_o = \frac{2I(d_o + 2e)}{Ad_o e(2r + d_o)}$$

4. Hollow rectangle

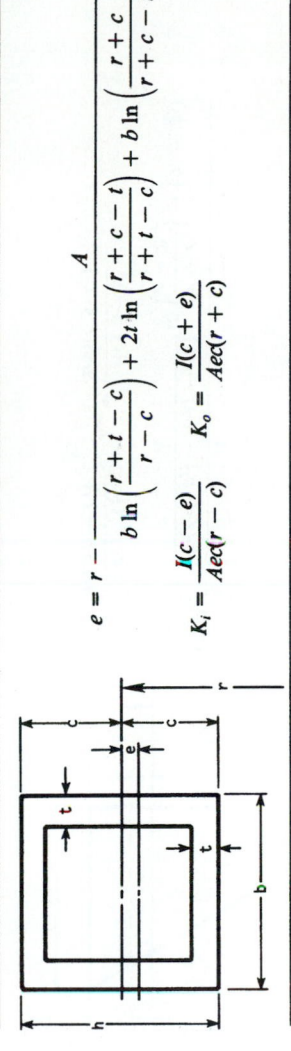

$$e = r - \frac{A}{b\ln\left(\dfrac{r+t-c}{r-c}\right) + 2t\ln\left(\dfrac{r+c-t}{r+t-c}\right) + b\ln\left(\dfrac{r+c}{r+c-t}\right)}$$

$$K_i = \frac{I(c - e)}{Aec(r - c)} \qquad K_o = \frac{I(c + e)}{Aec(r + c)}$$

16.5

5. Trapezoid

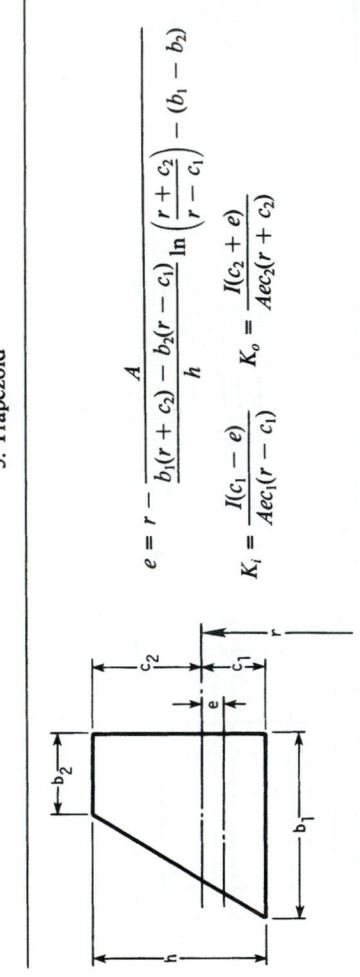

$$e = r - \frac{A}{\dfrac{b_1(r + c_2) - b_2(r - c_1)}{h}\ln\left(\dfrac{r + c_2}{r - c_1}\right) - (b_1 - b_2)}$$

$$K_i = \frac{I(c_1 - e)}{Aec_1(r - c_1)} \qquad K_o = \frac{I(c_2 + e)}{Aec_2(r + c_2)}$$

6. T Section

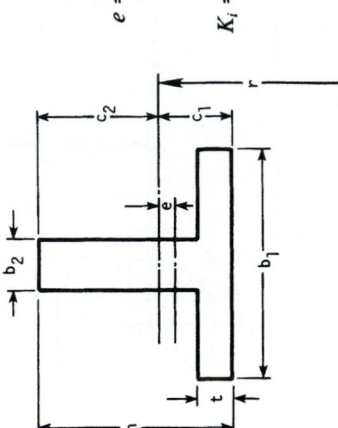

$$e = r - \frac{A}{b_1\ln\left(\dfrac{r + t - c_1}{r - c_1}\right) + b_2\ln\left(\dfrac{r + c_2}{r + t - c_1}\right)}$$

$$K_i = \frac{I(c_1 - e)}{Aec_1(r - c_1)} \qquad K_o = \frac{I(c_2 + e)}{Aec_2(r + c_2)}$$

7. U Section

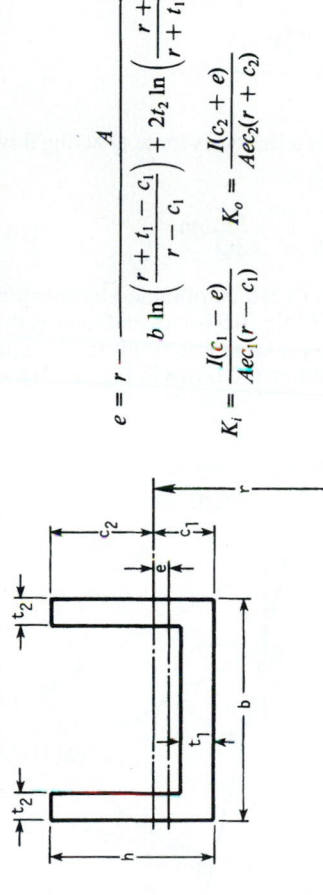

$$e = r - \cfrac{A}{b \ln\left(\cfrac{r + t_1 - c_1}{r - c_1}\right) + 2t_2 \ln\left(\cfrac{r + c_2}{r + t_1 - c_1}\right)}$$

$$K_i = \frac{I(c_1 - e)}{Aec_1(r - c_1)} \qquad K_o = \frac{I(c_2 + e)}{Aec_2(r + c_2)}$$

†Notation: r = radius of curvature to centroidal axis of section; A = area; I = second moment of area; e = distance from centroidal axis to neutral axis; $\sigma_i = K_i\sigma$ and $\sigma_o = K_o\sigma$ where σ_i and σ_o are the normal stresses on the fibers having the smallest and largest radii of curvature, respectively, and σ are the corresponding stresses computed on the same fibers of a straight beam. (Formulas for A and I can be found in Table 48.1.)

16.7

TABLE 16.2 Strain Energy Formulas

Loading	Formula
1. Axial force F	$U = \dfrac{F^2 l}{2AE}$
2. Shear force F	$U = \dfrac{F^2 l}{2AG}$
3. Bending moment M	$U = \displaystyle\int \dfrac{M^2\, dx}{2EI}$
4. Torsional moment T	$U = \dfrac{T^2 l}{2GK}$

To determine the deflection of end A, we employ a fictitious force Q acting down at end A. Then the deflection is

$$y = \frac{\partial U}{\partial Q} = \frac{r}{EI} \int_0^\phi M \frac{\partial M}{\partial Q} \, d\theta + \frac{r}{GK} \int_0^\phi T \frac{\partial T}{\partial Q} \, d\theta \tag{16.7}$$

The components of the moment and torque due to Q can be obtained by substituting Q for F in the moment and torque equations in Table 16.3 for an end load F; then the total of the moments and torques is obtained by adding this result to the equations for M and T due only to the distributed load. When the terms in Eq. (16.7) have

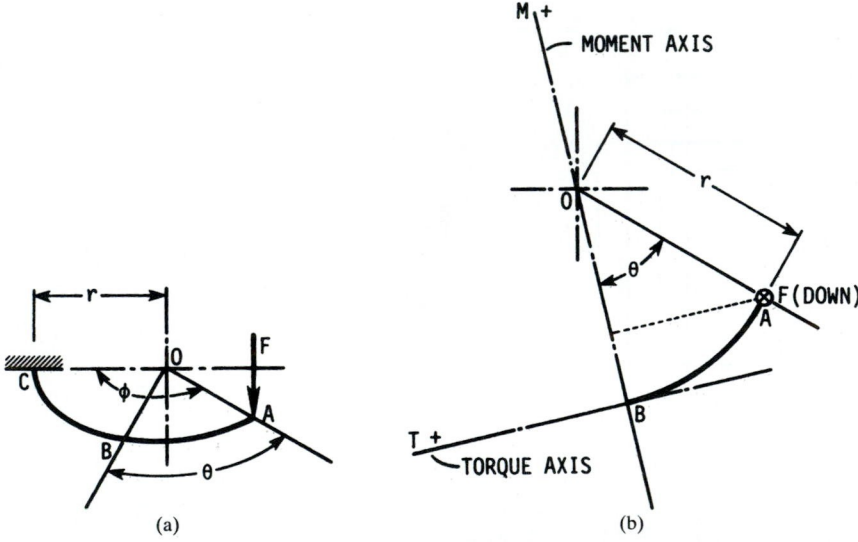

FIGURE 16.1 (*a*) Ring segment of span angle ϕ loaded by force F normal to the plane of the ring. (*b*) View of portion of ring AB showing positive directions of the moment and torque for section at B.

TABLE 16.3 Formulas for Ring Segments with One Support

Loading	Term	Formula
End load F	Moment Torque	$M = Fr \sin \theta \qquad M_C = Fr \sin \phi$ $T = Fr(1 - \cos \theta) \qquad T_C = Fr(1 - \cos \phi)$
	Derivatives	$\dfrac{\partial M}{\partial F} = r \sin \theta \qquad \dfrac{\partial T}{\partial F} = r(1 - \cos \theta)$
	Deflection coefficients	$A = \phi - \sin \phi \cos \phi$ $B = 3\phi - 4 \sin \phi + \sin \phi \cos \phi$
Distributed load w; fictitious load Q	Moment Torque	$M = wr^2(1 - \cos \theta) \qquad M_C = wr^2(1 - \cos \phi)$ $T = wr^2(\theta - \sin \theta) \qquad T_C = wr^2(\phi - \sin \phi)$
	Derivatives	$\dfrac{\partial M}{\partial Q} = r \sin \theta \qquad \dfrac{\partial T}{\partial Q} = r(1 - \cos \theta)$
	Deflection coefficients	$A = 2 - 2 \cos \phi - \sin^2 \phi$ $B = \phi^2 - 2\phi \sin \phi + \sin^2 \phi$

been formed, the force Q can be placed equal to zero prior to integration. The deflection equation can then be expressed as

$$y = \frac{wr^4}{2}\left(\frac{A}{EI} + \frac{B}{GK}\right) \tag{16.8}$$

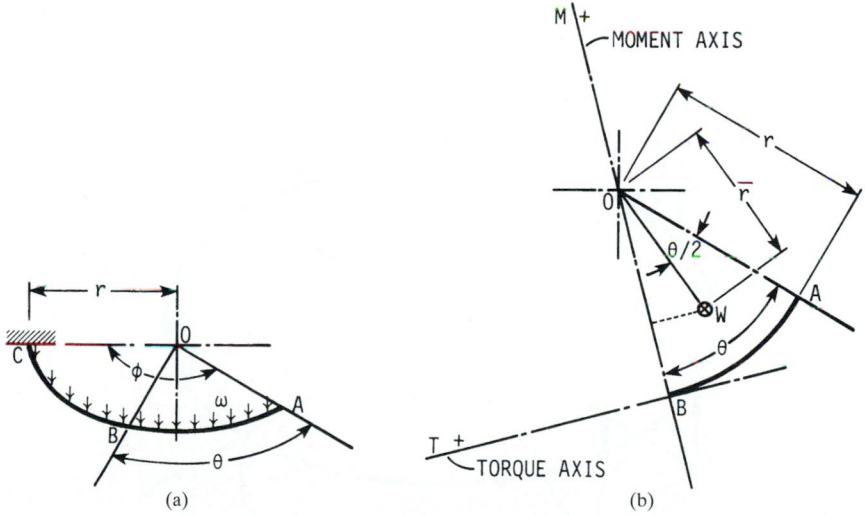

FIGURE 16.3 (a) Ring segment of span angle ϕ loaded by a uniformly distributed load w acting normal to the plane of the ring segment; (b) view of portion of ring AB; force W is the resultant of the distributed load w acting on portion AB of ring, and it acts at the centroid.

16.4 RINGS WITH SIMPLE SUPPORTS

Consider a ring loaded by any set of forces F and supported by reactions R, all normal to the ring plane, such that the force system is statically determinate. The system shown in Fig. 16.3, consisting of five forces and three reactions, is statically determinate and is such a system. By choosing an origin at any point A on the ring, all forces and reactions can be located by the angles ϕ measured counterclockwise from A. By treating the reactions as negative forces, Den Hartog [16.3], pp. 319–323, describes a simple method of determining the shear force, the bending moment, and the torsional moment at any point on the ring. The method is called *Biezeno's theorem*.

A term called the *reduced load P* is defined for this method. The reduced load is obtained by multiplying the actual load, plus or minus, by the fraction of the circle corresponding to its location from A. Thus for a force F_i, the reduced load is

$$P_i = \frac{\phi_i}{360°} F_i \tag{16.9}$$

Then Biezeno's theorem states that the shear force V_A, the moment M_A, and the torque T_A at section A, all statically indeterminate, are found from the set of equations

$$V_A = \sum_n P_i$$

$$M_A = \sum_n P_i r \sin \phi_i \tag{16.10}$$

$$T_A = \sum_n P_i r(1 - \cos \phi_i)$$

where n = number of forces and reactions together. The proof uses Castigliano's theorem and may be found in Ref. [16.3].

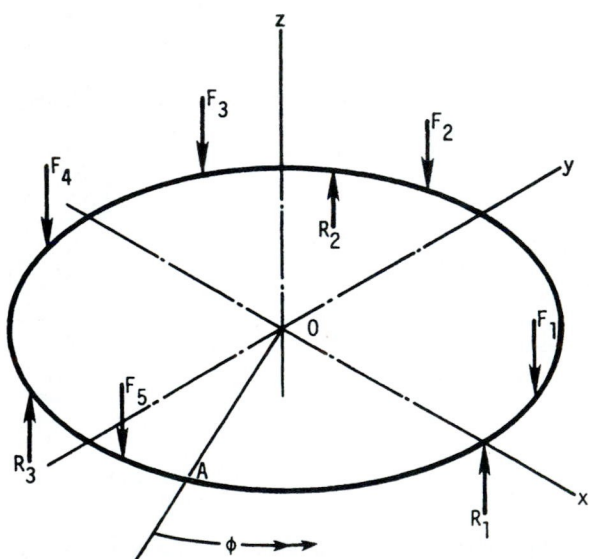

FIGURE 16.3   Ring loaded by a series of concentrated forces.

Example 1.   Find the shear force, bending moment, and torsional moment at the location of R_3 for the ring shown in Fig. 16.4.

  Solution.   Using the principles of statics, we first find the reactions to be

$$R_1 = R_2 = R_3 = \frac{2}{3} F$$

Choosing point A at R_3, the reduced loads are

$$P_0 = -\frac{0°}{360°} R_3 = 0 \qquad P_1 = \frac{30}{360} F = 0.0833F$$

$$P_2 = -\frac{120}{360} R_1 = -\frac{120}{360} \frac{2}{3} F = -0.2222F$$

$$P_3 = \frac{210}{360} F = 0.5833F$$

$$P_4 = -\frac{240}{360} R_2 = -\frac{240}{360} \frac{2}{3} F = -0.4444F$$

Then, using Eq. (16.10), we find $V_A = 0$. Next,

$$M_A = \sum_5 P_i r \sin \phi_i$$

$$= Fr\,(0 + 0.0833 \sin 30° - 0.2222 \sin 120° + 0.5833 \sin 210°$$

$$- 0.4444 \sin 240°)$$

$$= -0.0576Fr$$

In a similar manner, we find $T_A = 0.997Fr.$

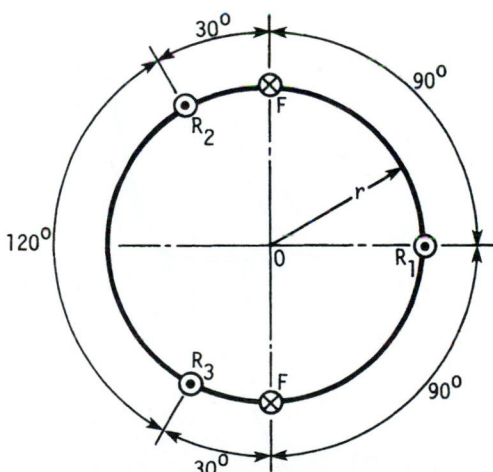

FIGURE 16.4   Ring loaded by the two forces F and supported by reactions R_1, R_2, and R_3. The crosses indicate that the forces act downward; the heavy dots at the reactions R indicate an upward direction.

The task of finding the deflection at any point on a ring with a loading like that of Fig. 16.3 is indeed difficult. The problem can be set up using Eq. (16.2), but the resulting integrals will be lengthy. The chances of making an error in signs or in terms during any of the simplification processes are very great. If a computer or even a programmable calculator is available, the integration can be performed using a numerical procedure such as Simpson's rule (see Chap. 4). Most of the user's manuals for programmable calculators contain such programs in the master library. When this approach is taken, the two terms behind each integral should not be multiplied out or simplified; reserve these tasks for the computer.

16.4.1   A Ring with Symmetrical Loads

A ring having three equally spaced loads, all equal in magnitude, with three equally spaced supports located midway between each pair of loads, has reactions at each support of $R = F/2$, $M = 0.289Fr$, and $T = 0$ by Biezeno's theorem. To find the moment and torque at any location θ from a reaction, we construct the diagram shown in Fig. 16.5. Then the moment and torque at A are

$$M = M_1 \cos \theta - R_1 r \sin \theta$$
$$= Fr \, (0.289 \cos \theta - 0.5 \sin \theta) \qquad (16.11)$$

$$T = M_1 \sin \theta - R_1 r \, (1 - \cos \theta)$$
$$= Fr \, (0.289 \sin \theta - 0.5 + 0.5 \cos \theta) \qquad (16.12)$$

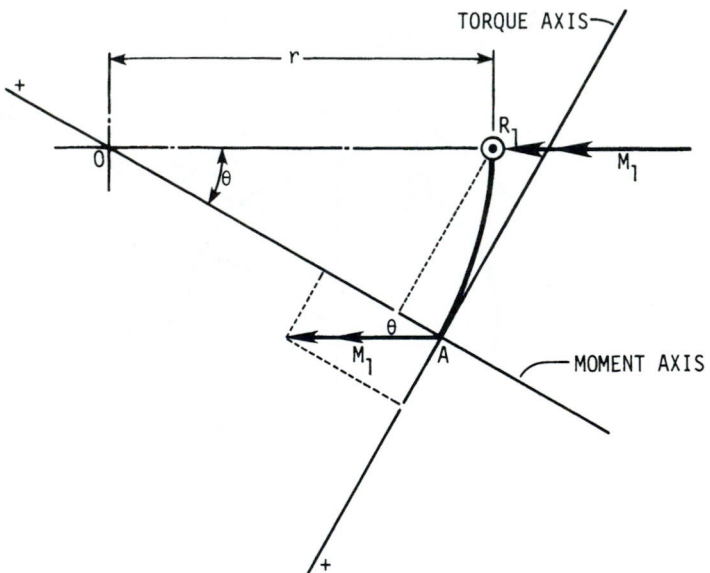

FIGURE 16.5   The positive directions of the moment and torque axes are arbitrary. Note that $R_1 = F/2$ and $M_1 = 0.289Fr$.

Neglecting direct shear, the strain energy stored in the ring between any two supports is, from Table 16.2,

$$U = 2 \int_0^{\pi/3} \frac{M^2 r\, d\theta}{2EI} + 2 \int_0^{\pi/3} \frac{T^2 r\, d\theta}{2GK} \tag{16.13}$$

Castigliano's theorem states that the deflection at the load F is

$$y = \frac{\partial U}{\partial F} = \frac{2r}{EI} \int_0^{\pi/3} M \frac{\partial M}{\partial F}\, d\theta + \frac{2r}{GK} \int_0^{\pi/3} T \frac{\partial T}{\partial F}\, d\theta \tag{16.14}$$

From Eqs. (16.11) and (16.12), we find

$$\frac{\partial M}{\partial F} = r(0.289 \cos \theta - 0.5 \sin \theta)$$

$$\frac{\partial T}{\partial F} = r(0.289 \sin \theta - 0.5 + 0.5 \cos \theta)$$

When these are substituted into Eq. (16.14), we get

$$y = \frac{Fr^3}{2} \left(\frac{A}{EI} + \frac{B}{GK} \right) \tag{16.15}$$

which is the same as Eq. (16.5). The constants are

$$A = 4 \int_0^{\pi/3} (0.289 \cos \theta - 0.5 \sin \theta)^2\, d\theta$$
$$\tag{16.16}$$
$$B = 4 \int_0^{\pi/3} (0.289 \sin \theta - 0.5 + 0.5 \cos \theta)^2\, d\theta$$

These equations can be integrated directly or by a computer using Simpson's rule. If your integration is rusty, use the computer. The results are $A = 0.1208$ and $B = 0.0134$.

16.4.2 Distributed Loading

The ring segment in Fig. 16.6 is subjected to a distributed load w per unit circumference and is supported by the vertical reactions R_1 and R_2 and the moment reactions M_1 and M_2. The zero-torque reactions mean that the ring is free to turn at A and B. The resultant of the distributed load is $W = wr\phi$; it acts at the centroid:

$$\bar{r} = \frac{2r \sin (\phi/2)}{\phi} \tag{16.17}$$

By symmetry, the force reactions are $R_1 = R_2 = W/2 = wr\phi/2$. Summing moments about an axis through BO gives

$$\Sigma M(BO) = -M_2 + W\bar{r} \sin \frac{\phi}{2} - M_1 \cos (\pi - \phi) - \frac{wr^2\phi}{2} \sin \phi = 0$$

Since M_1 and M_2 are equal, this equation can be solved to give

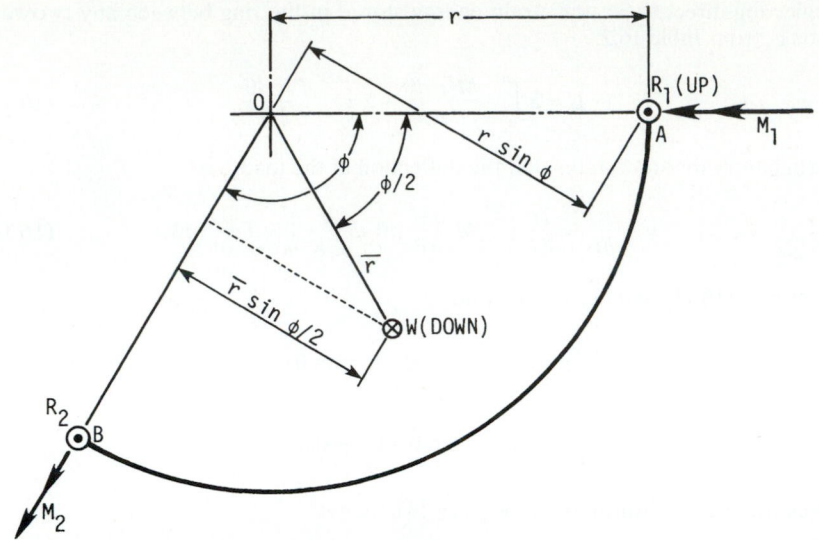

FIGURE 16.6   Section of ring of span angle ϕ with distributed load.

$$M_1 = wr^2 \left[\frac{1 - \cos \phi - (\phi/2) \sin \phi}{1 - \cos \phi} \right] \tag{16.18}$$

Example 2.   A ring has a uniformly distributed load and is supported by three equally spaced reactions. Find the deflection midway between supports.

   Solution.   If we place a load Q midway between supports and compute the strain energy using half the span, Eq. (16.7) becomes

$$y = \frac{\partial U}{\partial Q} = \frac{2r}{EI} \int_0^{\phi/2} M \frac{\partial M}{\partial Q} \, d\theta + \frac{2r}{GK} \int_0^{\phi/2} T \frac{\partial T}{\partial Q} \, d\theta \tag{16.19}$$

Using Eq. (16.18) with $\phi = 2\pi/3$ gives the moment at a support due only to w to be $M_1 = 0.395 \, wr^2$. Then, using a procedure quite similar to that used to write Eqs. (16.11) and (16.12), we find the moment and torque due only to the distributed load at any section θ to be

$$M_w = wr^2 \left(1 - 0.605 \cos \theta - \frac{\pi}{3} \sin \theta \right)$$

$$T_w = wr^2 \left(\theta - 0.605 \sin \theta - \frac{\pi}{3} + \frac{\pi}{3} \cos \theta \right) \tag{16.20}$$

In a similar manner, the force Q results in additional components of

$$M_Q = \frac{Qr}{2} (0.866 \cos \theta - \sin \theta)$$

$$T_Q = \frac{Qr}{2} (0.866 \sin \theta - 1 + \cos \theta) \tag{16.21}$$

Then
$$\frac{\partial M_Q}{\partial Q} = \frac{r}{2}(0.866 \cos \theta - \sin \theta)$$

$$\frac{\partial T_Q}{\partial Q} = \frac{r}{2}(0.866 \sin \theta - 1 + \cos \theta)$$

And so, placing the fictitious force Q equal to zero, Eq. (16.19) becomes

$$y = \frac{wr^4}{EI}\int_0^{\pi/3}\left(1 - 0.605 \cos \theta - \frac{\pi}{3}\sin \theta\right)(0.866 \cos \theta - \sin \theta)\, d\theta$$

$$+ \frac{wr^4}{GK}\int_0^{\pi/3}\left(\theta - 0.605 \sin \theta - \frac{\pi}{3} + \frac{\pi}{3}\cos \theta\right)(0.866 \sin \theta - 1 + \cos \theta)\, d\theta \qquad (16.22)$$

When this expression is integrated, we find

$$y = \frac{wr^4}{2}\left(\frac{0.141}{EI} + \frac{0.029}{GK}\right) \qquad (16.23)$$

16.5 RING SEGMENTS WITH FIXED ENDS

A ring segment with fixed ends has a moment reaction M_1, a torque reaction T_1, and a shear reaction R_1, as shown in Fig. 16.7a. The system is indeterminate, and so all three relations of Eq. (16.2) must be used to determine them, using zero for each corresponding displacement.

16.5.1 Segment with Concentrated Load

The moment and torque at any position θ are found from Fig. 16.7b as

$$M = T_1 \sin \theta + M_1 \cos \theta - R_1 r \sin \theta + Fr \sin (\theta - \gamma)$$

$$T = -T_1 \cos \theta + M_1 \sin \theta - R_1 r(1 - \cos \theta) + Fr[1 - \cos (\theta - \gamma)]$$

These can be simplified; the result is

$$M = T_1 \sin \theta + M_1 \cos \theta - R_1 r \sin \theta + Fr \cos \gamma \sin \theta - Fr \sin \gamma \cos \theta \qquad (16.24)$$

$$T = -T_1 \cos \theta + M_1 \sin \theta - R_1 r(1 - \cos \theta)$$

$$-Fr \cos \gamma \cos \theta - Fr \sin \gamma \sin \theta + Fr \qquad (16.25)$$

Using Eq. (16.3) and the third relation of Eq. (16.2) gives

$$\frac{\partial U}{\partial M_1} = \frac{r}{EI}\int_0^\phi M \frac{\partial M}{\partial M_1}\, d\theta + \frac{r}{GK}\int_0^\phi T \frac{\partial T}{\partial M_1}\, d\theta = 0 \qquad (16.26)$$

Note that
$$\frac{\partial M}{\partial M_1} = \cos \theta$$

$$\frac{\partial T}{\partial M_1} = \sin \theta$$

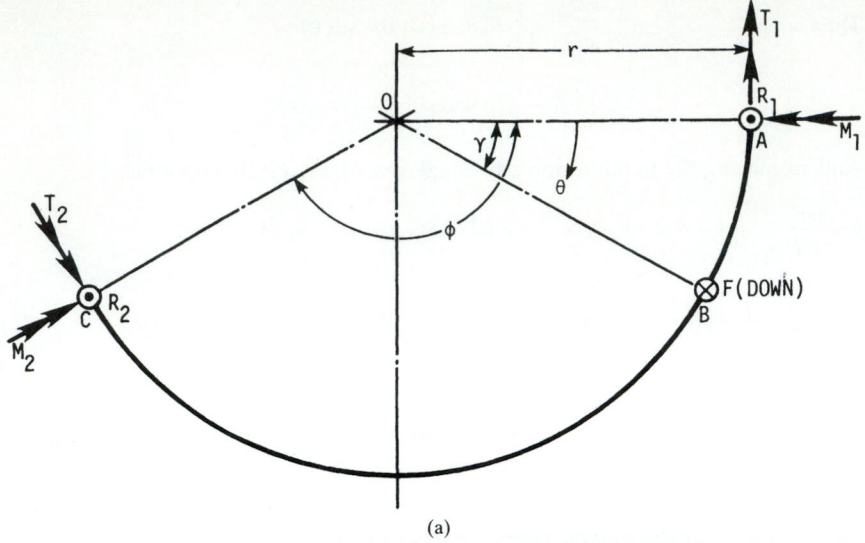

(a)

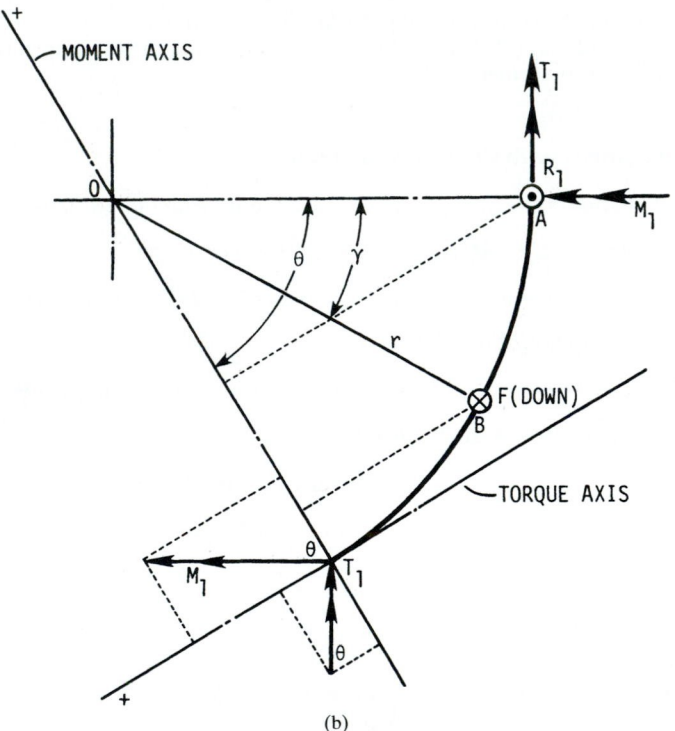

(b)

FIGURE 16.7   (*a*) Ring segment of span angle ϕ loaded by force *F*. (*b*) Portion of ring used to compute moment and torque at position θ.

Now multiply Eq. (16.26) by EI and divide by r; then substitute. The result can be written in the form

$$\int_0^\phi (T_1 \sin \theta + M_1 \cos \theta - R_1 r \sin \theta) \cos \theta \, d\theta$$

$$+ Fr \int_\gamma^\phi (\cos \gamma \sin \theta - \sin \gamma \cos \theta) \cos \theta \, d\theta$$

$$+ \frac{EI}{GK} \left\{ \int_0^\phi [-T_1 \cos \theta + M_1 \sin \theta - R_1 r(1 - \cos \theta)] \sin \theta \, d\theta \right.$$

$$\left. - Fr \int_\gamma^\phi (\cos \gamma \cos \theta + \sin \gamma \sin \theta - 1) \sin \theta \, d\theta \right\} = 0 \qquad (16.27)$$

Similar equations can be written using the other two relations in Eq. (16.2). When these three relations are integrated, the results can be expressed in the form

$$\begin{bmatrix} a_{11} & a_{12} & a_{13} \\ a_{21} & a_{22} & a_{23} \\ a_{31} & a_{32} & a_{33} \end{bmatrix} \begin{bmatrix} T_1/Fr \\ M_1/Fr \\ R_1/F \end{bmatrix} = \begin{bmatrix} b_1 \\ b_2 \\ b_3 \end{bmatrix} \qquad (16.28)$$

where

$$a_{11} = \sin^2 \phi - \frac{EI}{GK} \sin^2 \phi \qquad (16.29)$$

$$a_{21} = (\phi - \sin \phi \cos \phi) + \frac{EI}{GK} (\phi + \sin \phi \cos \phi) \qquad (16.30)$$

$$a_{31} = (\phi - \sin \phi \cos \phi) + \frac{EI}{GK} (\phi + \sin \phi \cos \phi - 2 \sin \phi) \qquad (16.31)$$

$$a_{12} = (\phi + \sin \phi \cos \phi) + \frac{EI}{GK} (\phi - \sin \phi \cos \phi) \qquad (16.32)$$

$$a_{22} = a_{11} \qquad (16.33)$$

$$a_{32} = \sin^2 \phi + \frac{EI}{GK} [2(1 - \cos \phi) - \sin^2 \phi] \qquad (16.34)$$

$$a_{13} = -a_{32} \qquad (16.35)$$

$$a_{23} = -a_{31} \qquad (16.36)$$

$$a_{33} = -(\phi - \sin \phi \cos \phi) - \frac{EI}{GK} (3\phi - 4 \sin \phi + \sin \phi \cos \phi) \qquad (16.37)$$

$$b_1 = \sin \gamma \sin \phi \cos \phi - \cos \gamma \sin^2 \phi + (\phi - \gamma) \sin \gamma + \frac{EI}{GK} [\cos \gamma \sin^2 \phi$$

$$- \sin \gamma \sin \phi \cos \phi + (\phi - \gamma)\sin \gamma + 2 \cos \phi - 2 \cos \gamma] \qquad (16.38)$$

$$b_2 = (\gamma - \phi)\cos\gamma - \sin\gamma + \cos\gamma \sin\phi \cos\phi + \sin\gamma \sin^2\phi$$

$$+ \frac{EI}{GK}[(\gamma - \phi)\cos\gamma - \sin\gamma + 2\sin\phi - \cos\gamma \sin\phi \cos\phi - \sin\gamma \sin^2\phi] \qquad (16.39)$$

$$b_3 = (\gamma - \phi)\cos\gamma - \sin\gamma + \cos\gamma \sin\phi \cos\phi + \sin\gamma \sin^2\phi$$

$$+ \frac{EI}{GK}[(\gamma - \phi)\cos\gamma - \sin\gamma - \cos\gamma \sin\phi \cos\phi - \sin\gamma \sin^2\phi$$

$$+ 2(\sin\phi - \phi + \gamma + \cos\gamma \sin\phi - \sin\gamma \cos\phi)] \qquad (16.40)$$

For tabulation purposes, we indicate these relations in the form

$$a_{ij} = X_{ij} + \frac{EI}{GK}Y_{ij} \qquad b_k = X_k + \frac{EI}{GK}Y_k \qquad (16.41)$$

Programs for solving equations such as Eq. (16.28) are widely available and easy to use. Tables 16.4 and 16.5 list the values of the coefficients for a variety of span and load angles.

TABLE 16.4 Coefficients a_{ij} for Various Span Angles

Coefficients		Span angle ϕ						
		$3\pi/2$	π	$3\pi/4$	$2\pi/3$	$\pi/2$	$\pi/3$	$\pi/4$
a_{11}	X_{11}	1	0	0.5	0.75	1	0.75	0.5
	Y_{11}	-1	0	-0.5	-0.75	-1	-0.75	-0.5
a_{21}	X_{21}	4.7124	π	2.8562	2.5274	1.5708	0.6142	0.2854
	Y_{21}	4.7124	π	1.8562	1.6614	1.5708	1.4802	1.2854
a_{31}	X_{31}	4.7124	π	2.8562	2.5274	1.5708	0.6142	0.2854
	Y_{31}	6.7124	π	0.4420	-0.0707	-0.4292	-0.2518	-0.1288
a_{12}	X_{12}	4.7124	π	1.8562	1.6614	1.5708	1.4802	1.2854
	Y_{12}	4.7124	π	2.8562	2.5274	1.5708	0.6142	0.2854
a_{22}	X_{22}	1	0	0.5	0.75	1	0.75	0.5
	Y_{22}	-1	0	-0.5	-0.75	-1	-0.75	-0.5
a_{32}	X_{32}	1	0	0.5	0.75	1	0.75	0.5
	Y_{32}	1	4	2.9142	2.25	1	0.25	0.0858
a_{13}	X_{13}	-1	0	-0.5	-0.75	-1	-0.75	-0.5
	Y_{13}	-1	-4	-2.9142	-2.25	-1	-0.25	-0.0858
a_{23}	X_{23}	-4.7124	$-\pi$	-2.8562	-2.5274	-1.5708	-0.6142	-0.2854
	Y_{23}	-6.7124	$-\pi$	-0.4420	0.0707	0.4292	0.2518	0.1288
a_{33}	X_{33}	-4.7124	$-\pi$	-2.8562	-2.5274	-1.5708	-0.6142	-0.2854
	Y_{33}	-18.1372	-3π	-3.7402	-2.3861	-0.7124	-0.1105	-0.0277

TABLE 16.5 Coefficients b_k for Various Span Angles ϕ and Load Angles γ in Terms of ϕ

Coefficients, load angles γ		Span angle ϕ						
		$3\pi/2$	π	$3\pi/4$	$2\pi/3$	$\pi/2$	$\pi/3$	$\pi/4$
$\dfrac{\phi}{4}$ b_1	X_1	2.8826	1.6661	0.2883	−0.0806	−0.4730	−0.4091	−0.2780
	Y_1	2.8826	−1.7481	−1.4019	1.0806	−0.4730	−0.1162	−0.0396
b_2	X_2	−1.3525	−2.3732	−2.1628	−1.8603	−1.0884	−0.4051	−0.1849
	Y_2	−5.2003	−2.3732	−0.4727	−0.1283	0.1462	0.1022	0.0535
b_3	X_3	−1.3525	−2.3732	−2.1628	−1.8603	−1.0884	−0.4051	−0.1849
	Y_3	−13.0342	−5.6714	−2.0455	−1.2699	−0.3622	−0.0544	−0.0135
$\dfrac{\phi}{3}$ b_1	X_1	3.1416	1.8138	0.4036	0.0446	−0.3424	−0.3179	−0.2180
	Y_1	3.1416	−1.1862	−1.0106	−0.7817	−0.3424	−0.0839	−0.0286
b_2	X_2	0	−1.9132	−1.8178	−1.5620	−0.9069	−0.3346	−0.1522
	Y_2	−4	−1.9132	−0.4036	−0.1307	0.0931	0.0706	0.0373
b_3	X_3	0	−1.9132	−1.8178	−1.5620	−0.9069	−0.3346	−0.1522
	Y_3	−10.2832	−4.3700	−1.5452	−0.9536	−0.2692	−0.0401	−0.0099
$\dfrac{\phi}{2}$ b_1	X_1	2.3732	1.5708	0.4351	0.1569	−0.1517	−0.1712	−0.1203
	Y_1	2.3732	−0.4292	−0.4379	−0.3431	−0.1517	−0.0372	−0.0127
b_2	X_2	1.6661	−1	−1.1041	−0.9566	−0.5554	−0.2034	−0.0922
	Y_2	−1.7481	−1	−0.2311	−0.0906	0.0304	0.0286	0.0154
b_3	X_3	1.6661	−1	−1.1041	−0.9566	−0.5554	−0.2034	−0.0922
	Y_3	−5.0463	−2.1416	−0.7395	−0.4529	−0.1262	−0.0186	−0.0046

16.5.2 Deflection Due to Concentrated Load

The deflection of a ring segment at a concentrated load can be obtained using the first relation of Eq. (16.2). The complete analytical solution is quite lengthy, and so a result is shown here that can be solved using computer solutions of Simpson's approximation. First, define the three solutions to Eq. (16.28) as

$$T_1 = C_1 Fr \qquad M_1 = C_2 Fr \qquad R_1 = C_3 F \tag{16.42}$$

Then Eq. (16.2) will have four integrals, which are

$$A_F = \int_0^\phi [(C_1 - C_3)\sin\theta + C_2\cos\theta]^2 \, d\theta \tag{16.43}$$

$$B_F = \int_0^\phi (\cos\gamma\sin\theta - \sin\gamma\cos\theta)^2 \, d\theta \tag{16.44}$$

$$C_F = \int_0^\phi [(C_3 - C_1)\cos\theta + C_2\sin\theta - C_3]^2 \, d\theta \tag{16.45}$$

$$D_F = \int_0^\phi [1 - (\cos \gamma \cos \theta + \sin \gamma \sin \theta)]^2 \, d\theta \qquad (16.46)$$

The results of these four integrations should be substituted into

$$y = \frac{Fr^3}{EI} \left[A_F + B_F + \frac{EI}{GK} (C_F + D_F) \right] \qquad (16.47)$$

to obtain the deflection due to F and at the location of the force F.

It is worth noting that the point of maximum deflection will never be far from the middle of the ring, even though the force F may be exerted near one end. This means that Eq. (16.47) will not give the maximum deflection unless $\gamma = \phi/2$.

16.5.3 Segment with Distributed Load

The resultant load acting at the centroid B' in Fig. 16.8 is $W = wr\phi$, and the radius $\bar{r}$ is given by Eq. (16.6), with ϕ substituted for θ. Thus the shear reaction at the fixed end A is $R_1 = wr\phi/2$. M_1 and T_1, at the fixed ends, can be determined using Castigliano's method.

We use Fig. 16.9 to write equations for moment and torque for any section, such as the one at D. When Eq. (16.6) for $\bar{r}$ is used, the results are found to be

$$M = T_1 \sin \theta + M_1 \cos \theta - \frac{wr^2\phi}{2} \sin \theta + wr^2(1 - \cos \theta) \qquad (16.48)$$

$$T = -T_1 \cos \theta + M_1 \sin \theta - \frac{wr^2\phi}{2} (1 - \cos \theta) + wr^2(\theta - \sin \theta) \qquad (16.49)$$

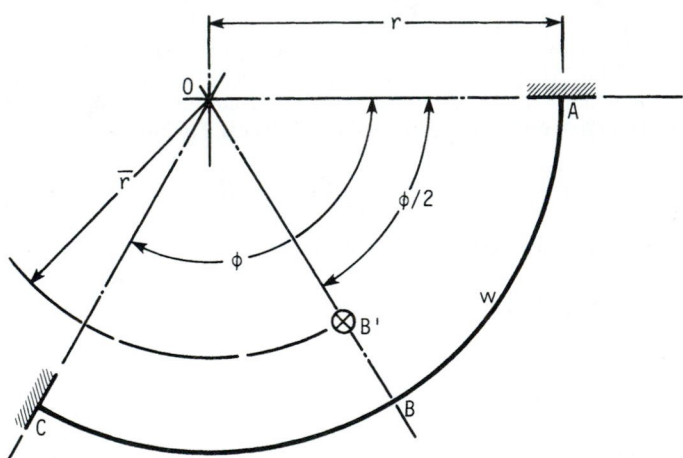

FIGURE 16.8 Ring segment of span angle ϕ subjected to a uniformly distributed load w per unit circumference acting downward. Point B' is the centroid of the load. The ends are fixed to resist bending moment and torsional moment.

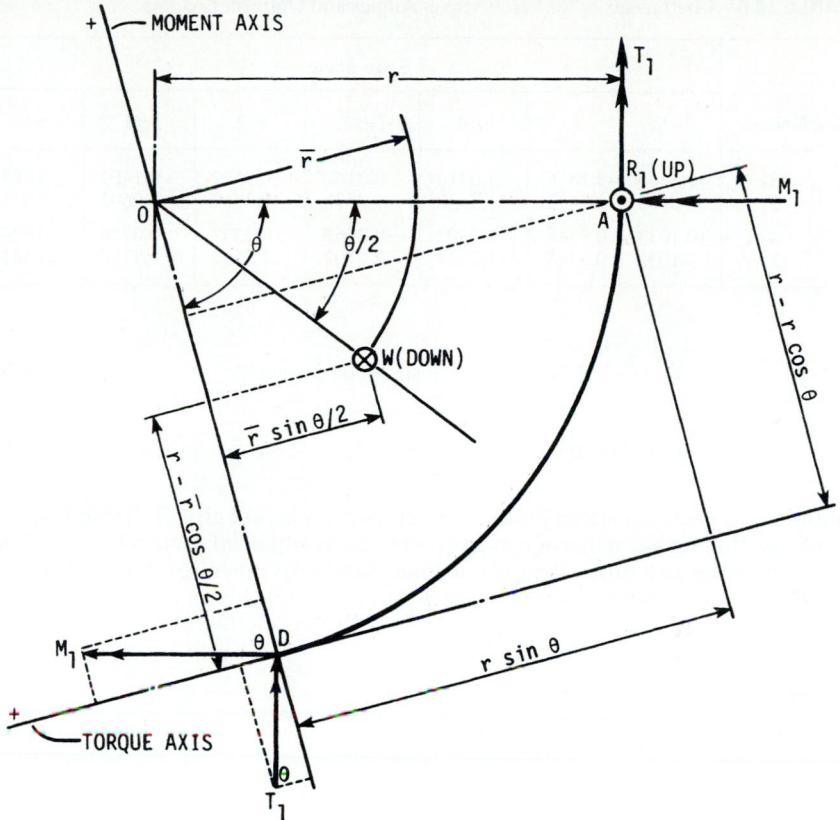

FIGURE 16.9 A portion of the ring has been isolated here to determine the moment and torque at any section D at angle θ from the fixed end at A.

These equations are now employed in the same manner as in Sec. 16.5.1 to obtain

$$\begin{bmatrix} a_{11} & a_{12} \\ a_{21} & a_{22} \end{bmatrix} \begin{bmatrix} T_1/wr^2 \\ M_1/wr^2 \end{bmatrix} = \begin{bmatrix} b_1 \\ b_2 \end{bmatrix} \tag{16.50}$$

It turns out that the a_{ij} terms in the array are identical with the same coefficients in Eq. (16.28); they are given by Eqs. (16.29), (16.30), (16.32), and (16.33), respectively. The coefficients b_k are

$$b_k = X_k + \frac{EI}{GK} Y_k \tag{16.51}$$

where

$$X_1 = \frac{\phi}{2} \sin^2 \phi + \sin \phi \cos \phi + \phi - 2 \sin \phi \tag{16.52}$$

$$Y_1 = \psi - 2 \sin \psi - \frac{\phi}{2} \sin^2 \psi - \sin \psi \cos \phi + \phi(1 + \cos \phi) \tag{16.53}$$

TABLE 16.6   Coefficients b_k for Various Span Angles and Uniform Loading

Coefficients		Span angle ϕ						
		$3\pi/2$	π	$3\pi/4$	$2\pi/3$	$\pi/2$	$\pi/3$	$\pi/4$
b_1	X_1	9.0686	3.1416	1.0310	0.7147	0.3562	0.1409	0.0675
	Y_1	9.0686	3.1416	1.5430	1.0572	0.3562	0.0602	0.0156
b_2	X_2	10.1033	0.9348	0.4507	0.3967	0.2337	0.0716	0.0263
	Y_2	8.1033	0.9348	−1.7274	−2.0102	−1.7663	−0.9750	−0.5810

$$X_2 = \frac{\phi^2}{2} - 2(1 - \cos\phi) - \frac{\phi}{2}\sin\phi\cos\phi + \sin^2\phi \tag{16.54}$$

$$Y_2 = \frac{\phi^2}{2} - 2(1 - \cos\phi) + \frac{\phi}{2}\sin\phi\cos\phi - \sin^2\phi + \phi\sin\phi \tag{16.55}$$

Solutions to these equations for a variety of span angles are given in Table 16.6.

A solution for the deflection at any point can be obtained using a fictitious load Q at any point and proceeding in a manner similar to other developments in this chapter. It is, however, a very lengthy analysis.

REFERENCES

16.1 Raymond J. Roark and Warren C. Young, *Formulas for Stress and Strain,* 6th ed., McGraw-Hill, New York, 1984.

16.2 Joseph E. Shigley and Charles R. Mischke, *Mechanical Engineering Design,* 5th ed., McGraw-Hill, New York, 1989.

16.3 J. P. Den Hartog, *Advanced Strength of Materials,* McGraw-Hill, New York, 1952.

CHAPTER 17
SEALS

R. Bruce Hopkins, Ph.D., P.E.

The Hopkins Engineering Co., P.C.
Cedar Falls, Iowa

17.1 ELASTOMERIC SEAL RINGS

Seal rings of the O-ring type are used as both static and dynamic seals. Static seals serve the same purpose as gaskets; that is, they provide a seal between two members that are not intended to undergo relative motion. Dynamic seals, however, are used where rotating or reciprocating motion is intended to occur.

O-rings are molded to the size of the elastomeric material with a circular cross section, as shown in Fig. 17.1a. The *size* is designated by the cross-sectional diameter w and the nominal inside diameter (ID). The standard sizes specified in SAE J120a are summarized in Table 17.1. The first size number in a group is associated with the minimum inside diameter, and the last size number is associated with the maximum inside diameter. Some manufacturers provide additional sizes that extend the range of inside diameters for a particular cross-section size. The nominal inside diameters were selected to provide dynamic seals in cylinder bores dimensioned in inches and common fractions of inches. Either SAE J120a or manufacturers' recommendations should be consulted to obtain the recommended compression of the cross section. The compression is different for static and dynamic applications.

The rectangular-section ring in Fig. 17.1b is manufactured by cutting lengths from a tube of molded material. The standard sizes listed in SAE J120a are summarized in Table 17.2. The first size number in a group is associated with the minimum inside diameter, and the last size number is associated with the maximum inside diameter. Rectangular-section rings are suitable for static applications with pressures up to 1500 psi [10.3 newtons per square millimeter (N/mm^2)].

The standard shape of groove for sealing rings is shown in Fig. 17.2. The actual groove dimensions depend on the type and size of the seal ring cross section and the nature of the application. Recommended groove dimensions are provided in SAE J120a and in the manufacturers' literature. Because elastomeric materials are almost incompressible, it is necessary to provide sufficient volume for the seal ring in the groove. The recommended groove dimensions do so.

For static seals, a finish on surfaces contacted by the seal ring that is rougher than 32 μin (0.8 μm) may lead to leakage. Because rough finishes accelerate seal wear in dynamic seals, a surface finish of 5 to 16 μin (0.13 to 0.4 μm) is preferred. Friction is

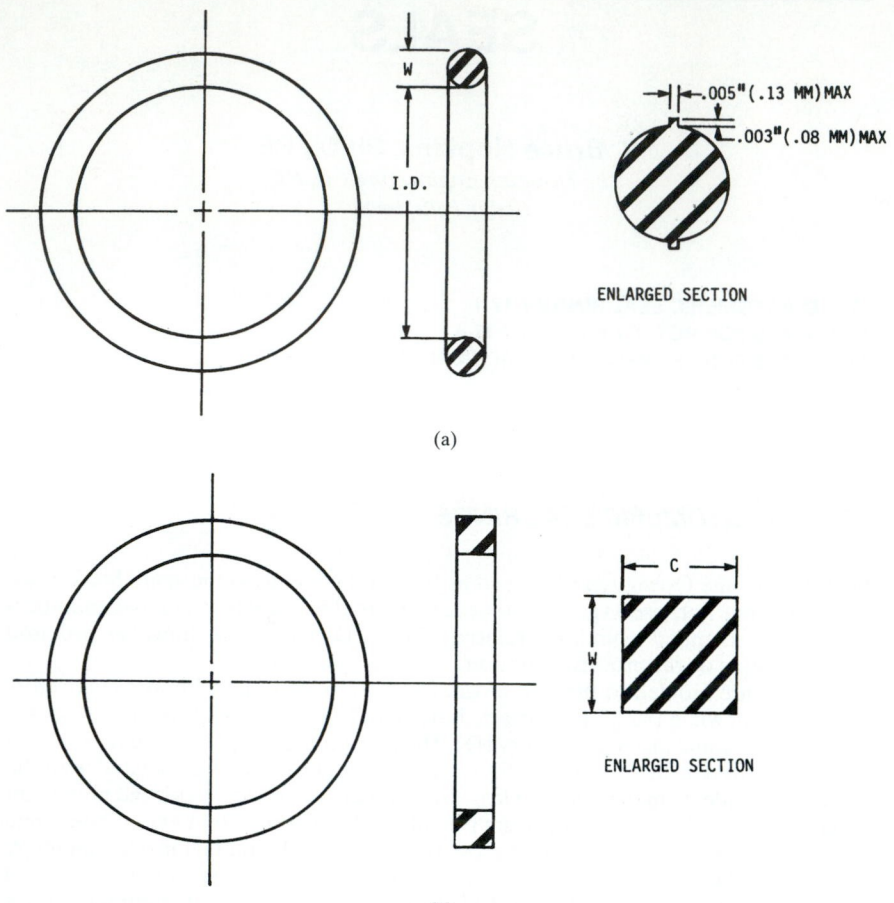

FIGURE 17.1   Seal rings. (*a*) O-ring; (*b*) rectangular-section ring.

TABLE 17.1   Standard Sizes of O-Rings

Size no.	*w*, in	Actual ID, in	
		Minimum	Maximum
006 to 045	0.070 ± 0.003	0.114 ± 0.005	3.989 ± 0.015
110 to 163	0.103 ± 0.003	0.362 ± 0.005	5.987 ± 0.023
210 to 281	0.139 ± 0.004	0.734 ± 0.006	14.984 ± 0.060
325 to 349	0.210 ± 0.005	1.475 ± 0.010	4.475 ± 0.010
425 to 460	0.275 ± 0.006	4.475 ± 0.015	15.475 ± 0.015

TABLE 17.2 Standard Sizes of Rectangular-Section Rings

Size no.	w, in	c, in	Actual ID, in	
			Minimum	Maximum
R006 to R045	0.066 ± 0.004	0.066 ± 0.003	0.114 ± 0.005	3.989 ± 0.015
R110 to R163	0.099 ± 0.004	0.099 ± 0.003	0.362 ± 0.005	5.987 ± 0.023
R210 to R281	0.134 ± 0.004	0.134 ± 0.004	0.734 ± 0.006	14.984 ± 0.060
R325 to R349	0.203 ± 0.005	0.203 ± 0.005	1.475 ± 0.010	4.475 ± 0.015
R425 to R460	0.265 ± 0.005	0.265 ± 0.005	4.475 ± 0.015	15.475 ± 0.030

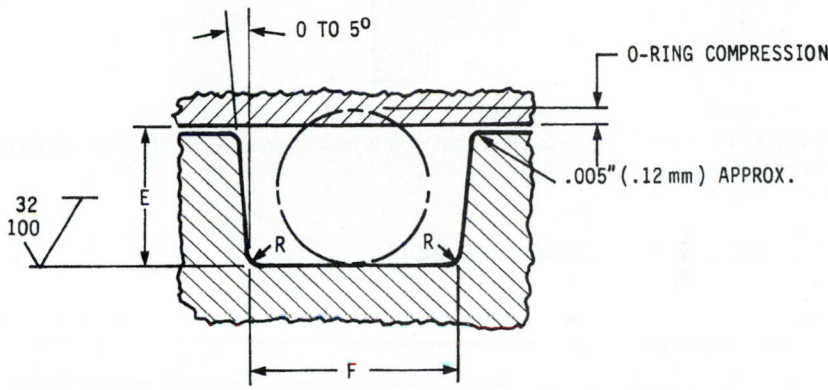

FIGURE 17.2 Shape of groove for seal rings.

reduced with the smoother finish, but surfaces smoother than 5 μin (0.13 μm) may not be satisfactory for reciprocating motion.

A static seal ring application in which the joint is subject to internal pressure only is shown in Fig. 17.3a. The groove design in Fig. 17.3b is for a joint subject to external pressure or internal vacuum only. It is generally advisable in these applications to use as large a seal ring cross section as possible because the tolerance on the groove depth is greater with larger cross sections. This requires less precise machining and tends to reduce manufacturing costs.

O-rings are also used as static seals for hydraulic tube fittings that are screwed into tapped holes. Recommended machining dimensions are provided in SAE J514 (June 1993).

Elastomeric sealing rings are most commonly made of nitrile (Buna N) compounds. These compounds are low in cost and are compatible with alcohol, gasoline, hydraulic fluids, lubricating oils, and water. They also are suitable for temperatures ranging from −67 to 257°F (−55 to 125°C). For resistance to higher temperatures or compatibility with other fluids, other compounds are employed. Among these compounds are butyl, ethylene propylene, neoprene, fluorocarbon, silicone, and polyurethane.

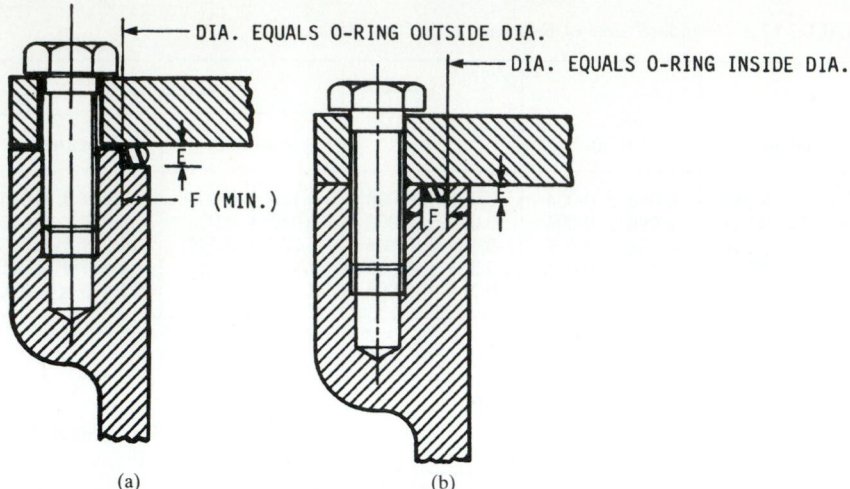

FIGURE 17.3   Static O-ring seals. (*a*) Joint subject to internal pressure only; (*b*) joint subject to external pressure.

17.2   SEALS FOR ROTARY MOTION

Seals are required on rotating shafts to retain working fluids, to retain lubricants, and to exclude dirt. The selection of a seal type depends on fluid pressure, shaft speed, and whether any leakage can be permitted. There are many variations of the basic seal types that are available from various manufacturers.

17.2.1   O-Rings

Attempts to use O-rings as seals for rotating shafts have not always been successful because the elastomers shrink when heated. If an O-ring is under tension, friction between the ring and the shaft generates heat that makes the ring shrink. Contraction of the ring creates additional heat, and failure occurs rapidly.

O-rings have been used successfully on rotating shafts when they are installed under compression by using a smaller-than-normal groove diameter in the housing. Satisfactory life can then be obtained at shaft speeds up to 750 feet per minute (ft/min) [3.8 meters per second (m/s)] and sealed pressures up to 200 psi (1.38 N/mm²). Recommended O-ring cross sections are 0.139 in (3.53 mm) for speeds up to 400 ft/min (2.0 m/s), 0.103 in (2.62 mm) for speeds from 400 to 600 ft/min (2.0 to 3.0 m/s), and 0.070 in (1.78 mm) for speeds exceeding 600 ft/min (3.0 m/s) [17.1].

17.2.2   Radial Lip Seals

A section through a radial lip seal is shown in Fig. 17.4. This type is used primarily for retention of lubricants and exclusion of dirt. It is suited for conditions of low lubricant pressure, moderate shaft speeds, less-than-severe environmental conditions, and situations where slight leakage may be permitted. Radial lip seals are compact, effective, inexpensive, and easily installed.

The outer case is held in the bearing housing by an interference fit. The garter spring provides a uniform radial force to maintain contact between the elastomeric sealing lips and the shaft. Lubricant leakage is reduced when hydrodynamic sealing lips are used. Such lips have very shallow grooves molded into the primary sealing lip to pump lubricant out of the contact area. Hydrodynamic sealing lips are manufactured for rotation in one direction only or for rotation in either direction.

Sealing lips are most commonly made of nitrile (Buna N) rubber compounds because of their compatibility with greases, lubricating oils, and hydraulic fluids. The nitrile compounds have poor to fair compatibility with extreme-pressure (EP) additives used in some gear lubricants. A polyacrylate or fluoroelastomer compound is a better choice with EP lubricants.

Radial lip seal terminology is presented in SAE J111 (Jun. 88), and recommendations for applications are made in SAE J946 (Oct. 91). One of the purposes of the secondary sealing lip shown in Fig. 17.4 is to exclude dust. That lip, however, leads to higher seal temperatures because of the additional friction, and the higher temperatures lead to earlier seal failure. Dual lip seals are not recommended for shaft speeds exceeding 150 ft/min (0.76 m/s) [17.2].

A minimum hardness of Rockwell C 30 is recommended for the portions of shafts that contact the sealing lips in order to prevent scoring of the shaft. If the shaft may be damaged in handling, a minimum hardness of Rockwell C 45 will provide protection against damage. A hard surface can be provided for soft shafts by use of a hardened wear sleeve of thin steel that is held in place by an interference fit.

Radial lip seals function best with carbon-, alloy-, or stainless-steel shafts or nickel-plated surfaces. Use with aluminum alloys, brass, bronze, magnesium, zinc, or similar metals is not recommended. Shaft surface texture should be in the range of 10 to 20 µin (0.25 to 0.50 µm). This condition can best be met by plunge grinding.

This type of seal is limited to sealing pressures of 3 psig (0.02 N/mm² gauge) at shaft speeds exceeding 2000 ft/min (10.2 m/s) and 7 psig (0.05 N/mm² gauge) at speeds up to 1000 ft/min (5.1 m/s). When pressures exceed these limits, a mechanical face seal is preferable.

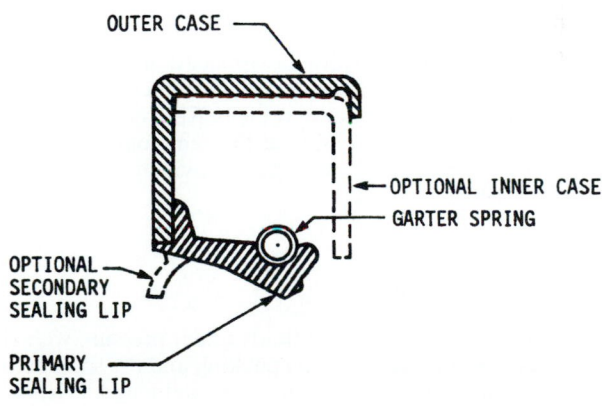

FIGURE 17.4 Cross section of radial lip seal.

17.2.3   Face Seals

Whereas a radial lip seal contacts the shaft circumference, a face seal acts against a surface perpendicular to the shaft axis. The seal may be mounted in a housing and seal against a shoulder or collar on the shaft. The seal also may be mounted on the shaft and seal against a surface on the housing. Some elastomeric face-sealing elements are loaded by mechanical springs, whereas others are not.

Elastomeric face seals (Fig. 17.5a) are used to retain lubricants. For high-speed applications, the lubricant should be on the side where centrifugal force throws the lubricant into the sealing area. These seals have the disadvantage of requiring rather precise location with respect to the sealing surface in order to provide the proper force on the seal. The sealing surface must be flat and smooth, with a surface finish of 10 to 20 μin (0.25 to 0.50 μm). Less rigid control of surface flatness is required for low speeds.

Mechanical face seals are used in situations where a radial lip seal or elastomeric face seal will not be satisfactory. Abrasive conditions, such as those encountered by earth-moving machinery and mining machinery, may dictate the use of a mechanical face seal. A mechanical seal is frequently used in the automotive coolant pump (Fig. 17.5b), in which the pressure is relatively low. Here, the stationary spring-loaded seal ring contacts a flat surface on the rotor.

If the seal in Fig. 17.5b were used with high-pressure fluid, a high axial sealing force would result. Friction between the rotor and stator could possibly cause overheating of the seal elements. Consequently, a balanced mechanical face seal, such as in Fig. 17.5c, is used for higher pressures. These seals are proportioned so that much of the force due to pressure is balanced. This leaves a small net force to provide contact between rotor and stator.

Another type of face seal, Fig. 17.6, is used to seal lubricants in rotating elements of machines that operate in environments where water, mud, or dust must be excluded. The seal consists of two hardened steel rings with lapped sealing surfaces that are forced together by surrounding elastomeric rings. The seal surfaces are tapered so that the point of contact moves inward as wear occurs. The seal rings do not contact the shaft that they surround; instead, one seal ring is driven by the rotating component through the elastomeric ring.

17.2.4   Metal Sealing Rings

Cast-iron sealing rings are used in hydraulic applications where oil must be introduced through a rotating shaft (Fig. 17.7). A typical application is to operate a clutch in an automatic transmission for a motor vehicle. Ring cross-sectional dimensions are similar to those for engine piston rings of the same outside diameter. Information on designing to accommodate these rings is provided in SAE J281 (Sept. 1980) and SAE J1236 (Apr. 1980).

17.2.5   Compression Packings

The stuffing box (Fig. 17.8) is used to seal fluids under pressure with either rotating or reciprocating shafts. Sealing between the packing and the shaft occurs as a result of axial movement of the gland when the nuts are tightened. Friction between the packing and the shaft causes wear, and so periodic tightening of the nuts is required.

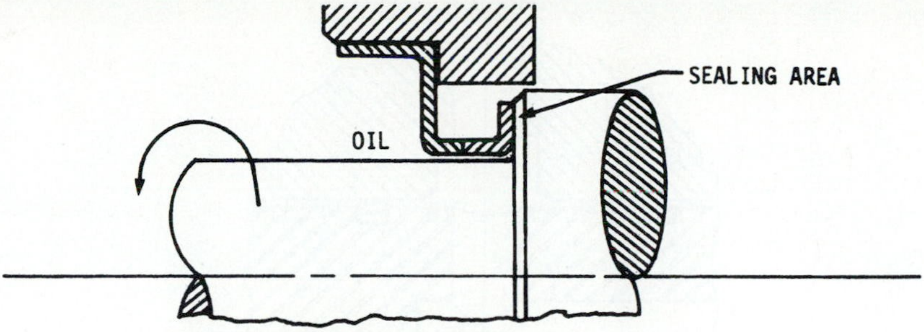

SEALING AREA

OIL

(a)

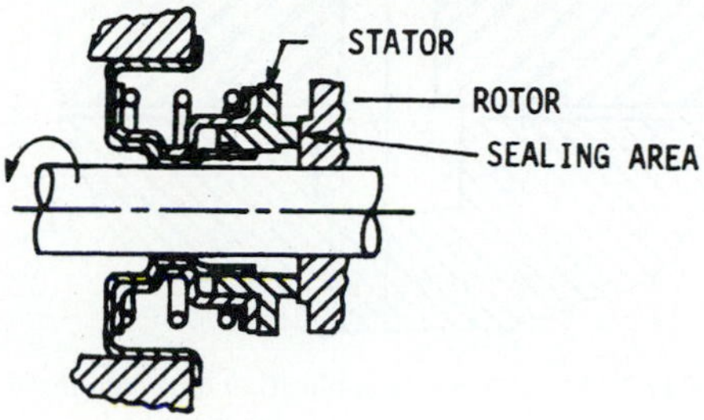

STATOR
ROTOR
SEALING AREA

(b)

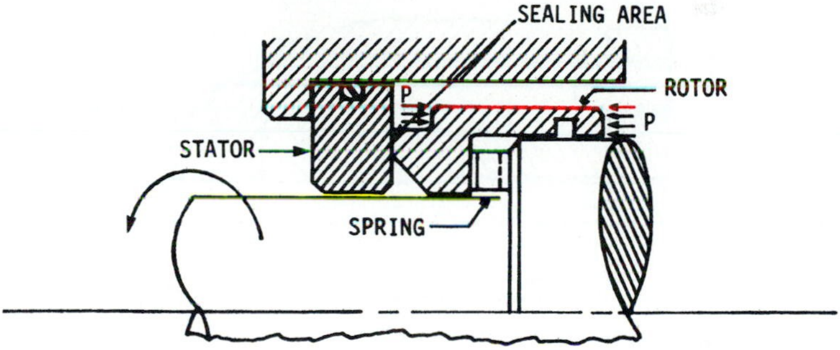

SEALING AREA

ROTOR

P

P

STATOR

SPRING

(c)

FIGURE 17.5 Face seals. (*a*) Housing-mounted elastomeric seal; (*b*) mechanical seal for engine coolant pump; (*c*) balanced mechanical seal.

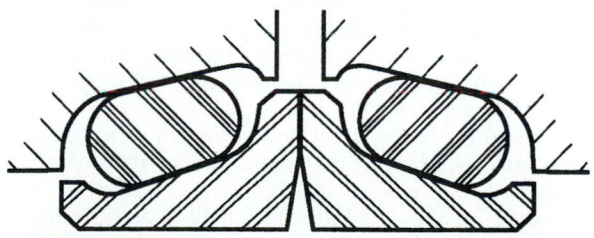

FIGURE 17.6 Cross section of a face seal for severe operating environments.

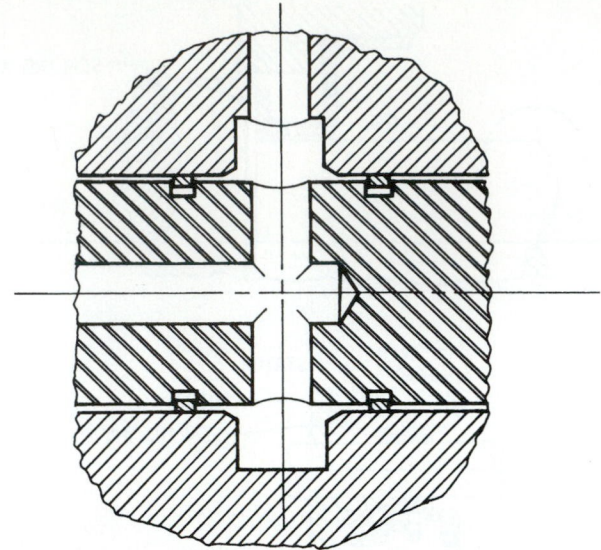

FIGURE 17.7   Metal seal ring application on a rotating shaft.

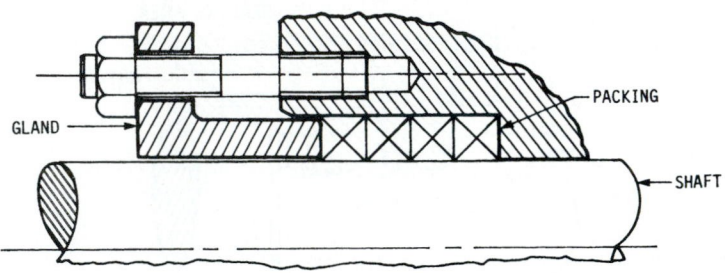

FIGURE 17.8   Stuffing box for a rotating shaft.

Packing material is usually obtained in straight lengths of square or rectangular cross section. Pieces are cut off and formed into rings that fit the stuffing box. The choice of packing material depends on the fluid to be sealed. Available packing materials include artificial fibers, asbestos, cotton, graphite, jute, leather, and metals. The metal packings are used for temperature conditions where the other materials are inadequate. The metal packings are formed from foil which is compressed into the proper packing shape.

In the design of stuffing boxes, small clearances are provided between the shaft and surrounding parts. The small clearances minimize extrusion of the packing into the clearance spaces.

Valve stems undergo a helical motion rather than a rotary motion when the valve is opened or closed. Investigations into the prevention of valve leakage and wear of valve stems resulted in a procedure for establishing packing dimensions for valve stems [17.3].

17.2.6 Noncontacting Seals

Frictional losses occur with sealing methods that utilize physical contact between a rotating and a stationary part. With high rubbing velocities, friction losses may be a significant factor. Those losses can be eliminated by using a seal that does not require physical contact. A noncontacting seal, however, cannot prevent leakage completely, although it does reduce it to a tolerable level.

One method of achieving a low leakage rate is to provide a very small clearance between the shaft and the surrounding housing or bushing. The longer the low-clearance passage, the greater the reduction in leakage.

A type of noncontacting seal called the *labyrinth seal* (Fig. 17.9) is used on such machines as large blowers and steam turbines. It can be used to retain lubricant in the bearings or to seal the working fluid in the machine. Effectiveness of the seal depends on small clearances between the seal and the shaft. In sealing the working fluid, the small clearances create a series of pressure drops between the working fluid and the atmosphere.

Labyrinth seals are usually made from a relatively soft metal such as aluminum or bronze so that the shaft is not damaged if contact between shaft and seal occurs. The simplest type is shown in Fig. 17.9, but other types are also used.

17.3 SEALS FOR RECIPROCATING MOTION

Some of the sealing methods used for rotary motion are also satisfactory for reciprocating motion. O-rings, compression packings, metal sealing rings, and some additional types are used to seal reciprocating rods, shafts, and pistons.

17.3.1 O-Rings

The O-ring is used extensively because of the low installed cost and effectiveness as a seal. It is well adapted to sealing reciprocating motion as well as for use as a static seal. Figure 17.10 shows applications on a piston and piston rod as well as a static seal application. Many such applications are for hydraulic cylinders in which the hydraulic oil acts as the lubricant for the O-rings.

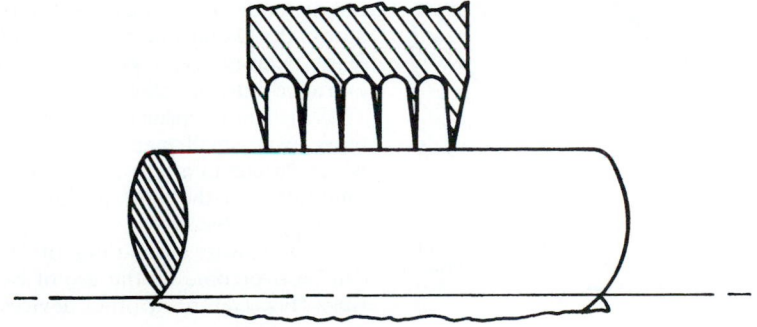

FIGURE 17.9 Labyrinth type of noncontacting seal.

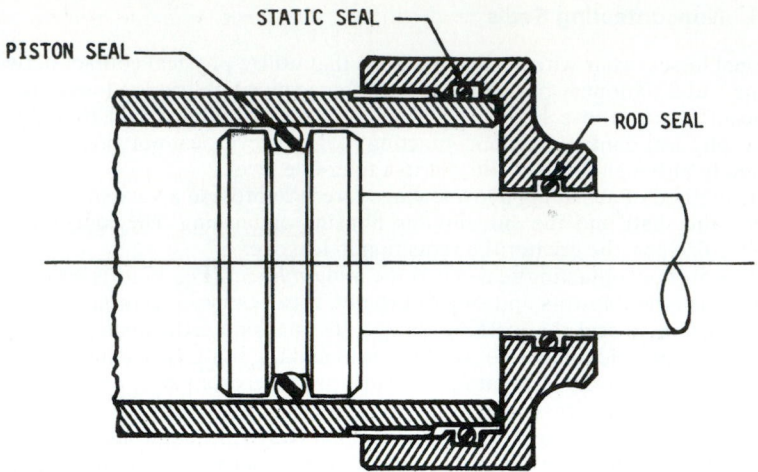

FIGURE 17.10   Applications of O-rings as a static seal and as seals for reciprocating motion.

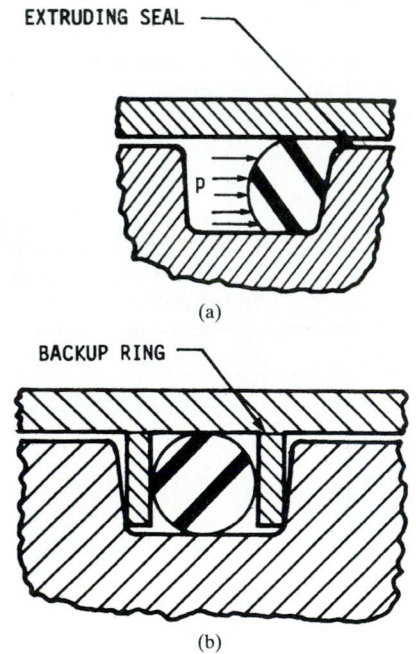

FIGURE 17.11 (*a*) Extrusion of O-ring into clearance space due to pressure. (*b*) Use of backup ring to prevent extrusion.

Rectangular-section rings are not suited for reciprocating motion and are used only in static applications. The shape of the groove for circular-section O-rings for sealing reciprocating motion is the same as that for static applications (Fig. 17.2). The recommended groove depth E, however, is slightly different for reciprocating motion. Recommended dimensions are available in SAE J120a and in the manufacturers' literature.

An O-ring must seal the clearance space between the reciprocating and stationary parts, for example, between the piston and the cylinder in Fig. 17.10. The amount of clearance that can be permitted depends on the pressure differential across the O-ring and the ring hardness. If the clearance is too great, the O-ring is extruded into the clearance space (Fig. 17.11*a*). The reciprocating motion then tears away small pieces of the O-ring, which results in a leaking seal and contamination of the working fluid by the O-ring particles.

The pressure limitations of O-rings can be overcome by the use of backup rings (Fig. 17.11*b*) or other devices that prevent O-ring extrusion. Backup rings are made from leather, plastics, or metal. The metal rings are split like a piston ring for radial compression during assembly into the cylinder.

Recommendations on the combination of clearance and pressure for which backup rings are required vary to some extent among the various O-ring suppliers. Figure 17.12 provides one such recommendation [17.4]. In this figure, if the combination of fluid pressure and maximum gap falls to the right of a hardness curve, backup rings are required. If a piston or rod can be forced to one side of the bore, the maximum gap is the difference between the two diameters. If, however, the radial position of the piston or rod is restrained, as by bearings, the radial clearance is the maximum gap.

Both the compression of the O-ring cross section to effect a seal and fluid pressure acting on the seal cause friction forces that oppose reciprocating motion. Information for estimating friction factors is provided in Fig. 17.13 [17.4]. The friction factor due to compression of the cross section can be obtained from Fig. 17.13a. The seal compression is expressed as a percentage of the O-ring cross section. The friction factor is multiplied by the circumference of the surface where relative motion occurs to obtain the friction force. For a piston, the circumference of the cylinder bore is used; for a piston rod, the rod circumference applies.

The friction factor for pressure differential across the O-ring is obtained from Fig. 17.13b. That factor is multiplied by the projected area of the O-ring to obtain the friction force. For an O-ring in a piston, the projected area is the product of the diameter of the ring cross section and the circumference of the cylinder bore. The total estimated friction force is the sum of the friction forces due to compression and fluid pressure.

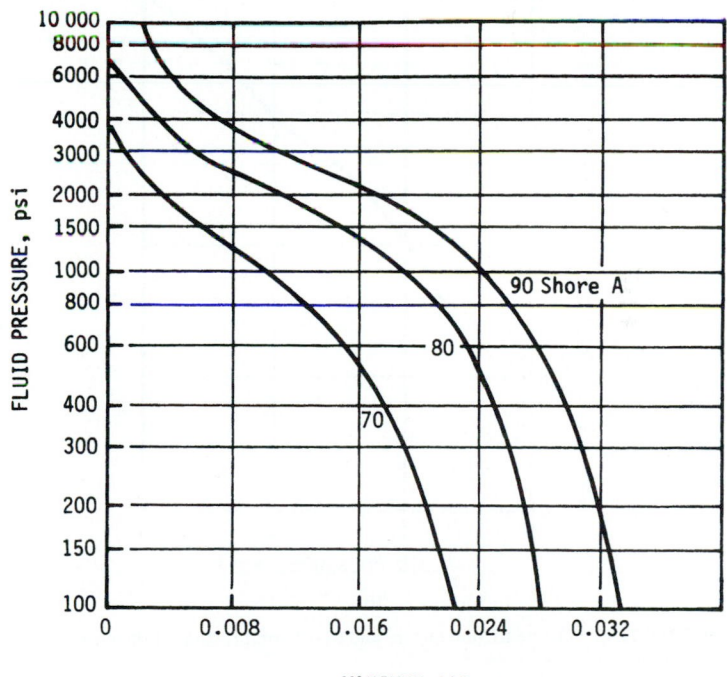

FIGURE 17.12 Extrusion limits for O-rings. *(From Ref. [17.4].)*

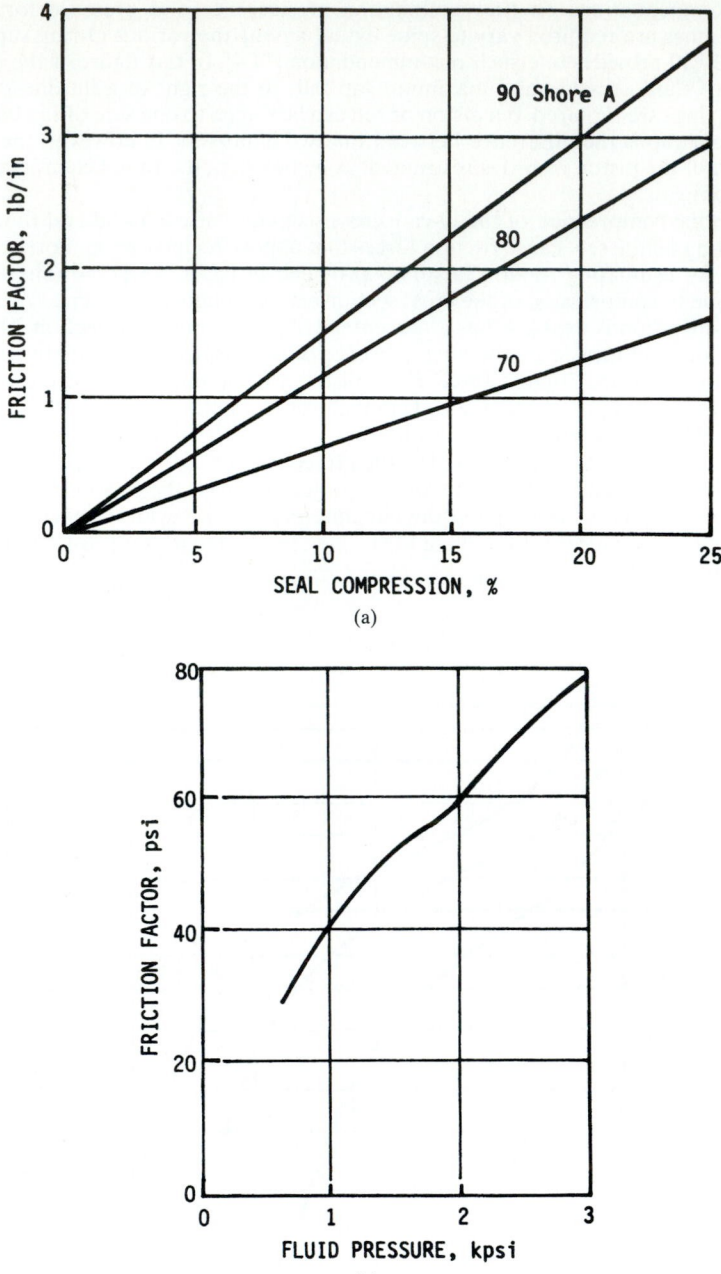

FIGURE 17.13 O-ring friction factors due to (*a*) compression of the cross section and (*b*) fluid pressure. *(From Ref. [17.4].)*

The finish of rubbing surfaces should be 8 to 16 μin (0.2 to 0.4 μm) for O-rings with a hardness of 70 Shore A. For rougher surfaces, a higher O-ring hardness should be used to ensure reasonable life.

O-rings may be damaged if they are forced over sharp corners during assembly. The addition of chamfers to corners is an inexpensive method of reducing damage. Serious damage to O-rings occurs if they are forced to pass over a hole in a cylinder wall while under pressure. If this occurs, the O-ring expands into the hole and must later be forced back into the groove. This action tends to shear pieces off the ring and thus destroy its ability to seal.

17.3.2   Lip Packings

Cup packings, U-seals, V-ring packings, and other forms of lip packings are used primarily to seal reciprocating motion. The packing material is usually leather, solid rubber, or fabric-reinforced rubber, although other compounds are available for difficult applications. An advantage of leather packings is a low coefficient of friction, on the order of 0.006 to 0.008 depending on the tanning process. Low friction increases the life of a packing because less heat is generated.

Cup packings (Fig. 17.14) were one of the first types of piston seals for hydraulic and pneumatic applications. The fluid pressure expands the cup outward against the cylinder wall and thus seals the piston in the cylinder. This action requires that a double-acting cylinder have two packings in order to seal the pressure in both directions of operation. The inner portion of the piston in Fig. 17.14 is a boss to prevent excessive tightening of the washer against the cup. If the cup is crushed against the piston, good sealing will not be obtained.

Figure 17.15 shows elastomeric U-seals on a double-acting piston. This type of seal is also used on piston rods. They have approximately the same pressure limitations as O-rings, and backup rings are required for higher pressures. When a U-seal is made of leather, a filler is required between the lips to prevent collapse of the seal.

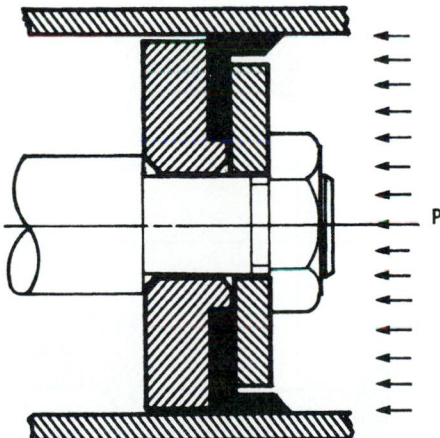

FIGURE 17.14 Cup packing for single-acting cylinder.

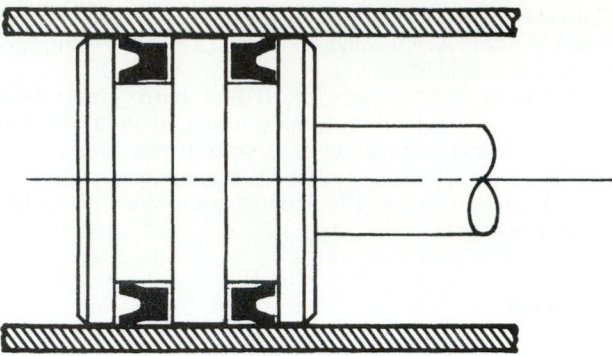

FIGURE 17.15   U-seals for a double-acting piston.

Rod scrapers are used on piston rods of hydraulic cylinders that are exposed to harsh environments. The purpose is to exclude mud, dust, and ice from the cylinder. A typical rod scraper is composed of a polyurethane element bonded to a metal shell which is pressed into the end cap of the cylinder. Molybdenum disulfide is sometimes added to the polyurethane to reduce friction. A rod scraper added to the end cap of a cylinder is shown in Fig. 17.16. The sealing lip is pointed outward to remove foreign material when the piston rod is retracted.

The chief use of the V-ring packing (Fig. 17.17) is for sealing piston rods or reciprocating shafts, although it can also be used to seal pistons. The ability to seal fluids under pressure depends on the type of packing material and number of packings used. The V-ring packing is considered superior to other lip types for sealing high pressures, especially above 50 000 psi (345 N/mm^2).

The V shape of the packing is obtained through the adapters that support the rings. Fluid pressure then expands the packing against the shaft and housing to seal

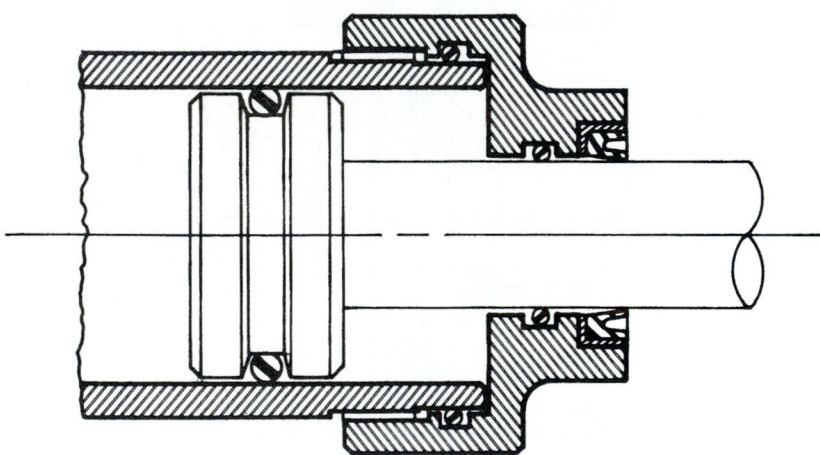

FIGURE 17.16   Rod scraper on piston rod.

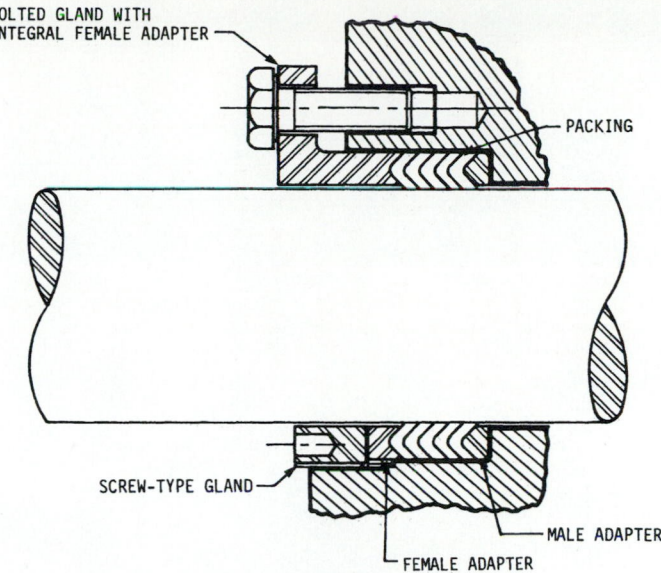

FIGURE 17.17   V-ring packing for a reciprocating shaft.

the fluid. A continuous packing provides better sealing than a series of split rings, although the latter are easier to install and remove.

17.3.3 Piston Rings

Piston rings for automotive engine applications are made of gray cast iron. The rings are used to seal gases in the cylinder and to restrict oil to the crankcase.

Piston rings must be split for assembly over the piston. This requires shaping the ring so that it will provide a uniform radial force against the cylinder wall. Piston ring manufacturers have developed methods of attaining this objective.

REFERENCES

17.1 Leonard J. Martini, "Sealing Rotary Shafts with O-Rings," *Machine Design,* May 26, 1977, pp. 97–99.

17.2 Bert Robins, "Radial Lip Seals—Are Two Too Many?" *Power Transmission Design,* October 1982, pp. 73, 74.

17.3 L. I. Ezekoye and J. A. George, "Valve Packings that Don't Leak," *Machine Design,* Jan. 20, 1977, pp. 142, 143.

17.4 Wes J. Ratelle, "Seal Selection: Beyond Standard Practice," *Machine Design,* Jan. 20, 1977, pp. 133–137.

CHAPTER 18
FLYWHEELS

Daniel M. Curtis, Ph.D.
Senior Mechanical Engineer
NKF Engineering, Inc.
Reston, Virginia

LIST OF SYMBOLS

a	Constant, $\text{lb}\cdot\text{s}\cdot\text{ft/rad}$ ($\text{J}\cdot\text{s/rad}$)
A	Cross-sectional area of rim, in^2 (m^2)
A_s	Cross-sectional area of spoke, in^2 (m^2)
A_j, B_j, C_j	Difference coefficients
b	Constant, $\text{lb}\cdot\text{ft}$ (J)
C_s	Coefficient of speed fluctuation
C_u	Coefficient of energy fluctuation
D_j	Difference coefficient, lb (kN)
f_i	Intermediate variable
F	Stress function ($= rt\sigma_r$), lb (kN)
F_s	Geometric shape factor
g	Acceleration of gravity, 32.2 ft/s^2 (9.80 m/s^2)
I	Second moment of area, in^4 (m^4)
j	Index
J	Polar-mass moment of inertia, $\text{lb}\cdot\text{s}^2\cdot\text{ft}$ ($\text{N}\cdot\text{s}^2\cdot\text{m}$)
K	Coefficient, $33\,000 \text{ lb}\cdot\text{ft}\cdot\text{rpm/hp}$ $[2\pi \text{ J}\cdot\text{rad/(W}\cdot\text{s)}]$
n	Engine speed, rpm (rad/s)
N_c	Number of cylinders
N_s	Number of spokes
P	Power, hp (W)
r	Radial distance, in (m)

Δr	Radial-distance increment, in (m)
r_h	Hub radius, in (m)
r_a	Average radius of rim, in (m)
r_i	Inner radius of rim, in (m)
r_o	Outer radius of rim, in (m)
R	Specific energy, in (m)
S_y	Yield strength, psi (MPa)
t	Time, s
Δt	Time increment, s
T	Torque, lb·ft (J)
T_a	Angle-dependent torque, lb·ft (J)
T_s	Speed-dependent torque, lb·ft (J)
T_j	Torque at end of interval j, lb·ft (J)
U	Difference between the flywheel energy at maximum speed and at minimum speed, lb·ft (J)
V	Rim velocity, ft/s (m/s)
W	Weight, lb (kN)
z	Thickness, in (m)
z'	Radial derivative of flywheel thickness $(= dz/dr)$
z_0	Thickness at center, in (m)
Z_r	Section modulus of rim, in^3 (m^3)
Z_s	Section modulus of spoke, in^3 (m^3)
2α	Angle between adjacent spokes (see Fig. 18.6), rad
β	Angle, rad
θ	Angular position, rad
θ_{max}	Maximum angular deviation from constant-speed position, rad
θ_0	Angular position at start of machine cycle, rad
$\Delta\theta$	Angular increment, rad
ν	Poisson's ratio
ξ	Time, s
ρ	Weight density, lb/in^3 (kN/m^3)
σ	Stress, psi (MPa)
σ_0	Stress constant, psi (MPa)
σ_r	Radial stress, psi (MPa)
σ_t	Tangential stress, psi (MPa)
ω	Rotational speed, rad/s
$\omega_{max}, \omega_{min}$	Maximum and minimum speed, rad/s
ω_{avg}	Average speed $[= 0.5(\omega_{max} + \omega_{min})]$, rad/s
ω_0	Speed at start of machine cycle, rad/s
$\Delta\omega$	Maximum deviation of speed from average value, rad/s
ω_j	Speed at end of interval j, rad/s

The energy-storage capacity of a flywheel is determined from its polar moment of inertia J and its maximum safe running speed. The necessary inertia depends on the cyclic torque variation and the allowable speed variation or, in the case of energy-storage flywheels, the maximum energy requirements. The safe running speed depends on the geometry and material properties of the flywheel.

18.1 FLYWHEEL USAGE

Flywheels store energy. Indeed, flywheels are used as energy reservoirs, and this use will be discussed in Sec. 18.4. Their principal use in machine design, however, is to smooth the variations in shaft speed that are caused by loads or power sources that vary in a cyclic fashion. By using its stored kinetic energy $0.5J\omega^2$ to absorb the variations in torque during a machine cycle, a flywheel smooths the fluctuating speed of a machine and reduces undesirable transient loads. The effect of a flywheel is therefore fundamentally different from that of a regulator: A flywheel limits the speed variation over one cycle and has minimal effect on the average speed; a regulator uses negative feedback to maintain a selected average speed with only secondary effects on the speed during a cycle.

The flywheel has other features which have to be considered in design. Its size, speed, and windage effect can all be used to advantage in providing a secondary function as part of a clutch, gear, belt pulley, cooling fan, pump, gyroscope, or torsional damper.

18.2 SIZING THE FLYWHEEL

18.2.1 Coefficient of Speed Variation

A certain amount of fluctuation in shaft speed will not cause harmful torques or reduce the usefulness of a machine. The *coefficient of speed fluctuation* C_s is defined as

$$C_s = \frac{\omega_{max} - \omega_{min}}{\omega_{avg}} \tag{18.1}$$

where ω = rotational speed at the flywheel and ω_{avg} = average of ω_{max} and ω_{min}. Ranges for C_s for several categories of speed variation are given in Table 18.1. Assume that the system is stiff (the speeds of all shafts are proportional), that the external torque input or load is constant, and that ω_{avg} is close to the constant speed at which the energy from the average torque balances the external energy (this is usually a good assumption for values of C_s up to about 0.2). The energy equation $U = 0.5J(\omega_{max}^2 - \omega_{min}^2)$ and the definition of C_s combine to give the equation for the required mass moment of inertia [18.11]:

$$J = \frac{U}{\omega_{avg}^2 C_s} \tag{18.2}$$

This inertia includes the flywheel inertia and the inertia of all rotating parts, referred to the flywheel speed by multiplying by the square of the ratio of the shaft speeds (see Chap. 38).

TABLE 18.1 Suggested Values for the
Coefficient of Speed Fluctuation C_s

Required speed uniformity	C_s
Very uniform	≤ 0.003
Moderately uniform	0.003–0.012
Some variation acceptable	0.012–0.05
Moderate variation	0.05–0.2
Large variation acceptable	≥ 0.2

Example 1.   During each punching cycle, the cranking shaft for a punching operation does 270 J of work while rotating 30 degrees, as shown in Fig. 18.1. No work is done during the remaining 330 degrees. What size flywheel is necessary if the speed at the location of the flywheel is 20 rad/s and the inertia of the other rotating parts referred to the flywheel is 0.51 N·s²·m?

The average work required is $270/(2\pi) = 43.0$ J/rad. The motor will supply this constant torque throughout the cycle. Referring to Fig. 18.1, the flywheel will give up some of its stored energy during the 30 degrees of actual punching. This is the shaded area above the average-torque line; the motor will supply the additional 43 J/rad. During the remaining 330 degrees, the motor will resupply the flywheel, as shown by the shaded area below the average-torque line. The flywheel speed reaches its maximum and minimum where the loading torque crosses the average-torque line.

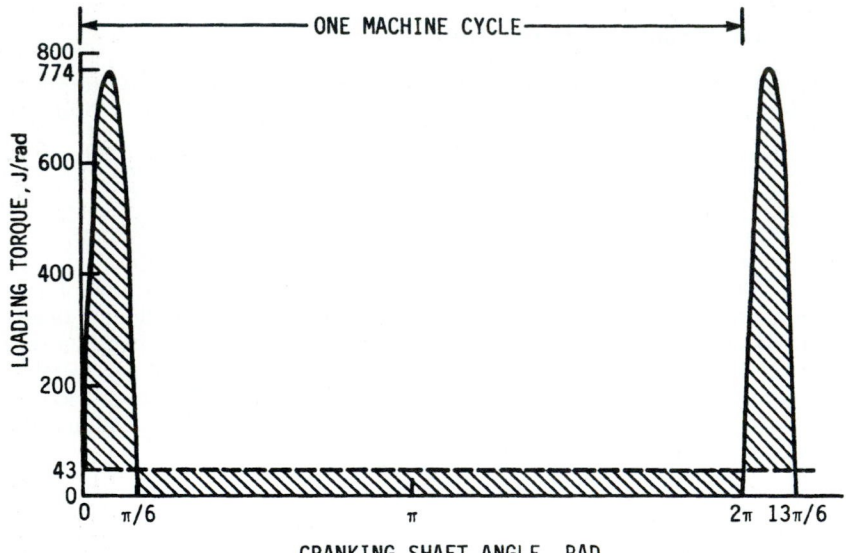

FIGURE 18.1 Torque-angle curve for punching operation in Example 1. The dashed line indicates the average torque of 43 J/rad, and the shaded areas are each equal to the maximum energy variation of 247 J.

Since the torque-angle curve is steep at both ends, the area above the average torque line is the total area minus the small rectangle below the average. The energy variation is then

$$U = 270 - 43.0 \left(\frac{\pi}{6} \right) = 247 \text{ J} \tag{18.3}$$

The relative speed between the cranking shaft and the flywheel is unimportant since the energy (torque times angle) is the same at either speed. Letting $C_s = 0.10$ for a moderate speed variation, Eq. (18.2) gives the necessary flywheel inertia:

$$J = \frac{247}{20^2(0.10)} - 0.51 = 5.67 \text{ N} \cdot \text{s}^2 \cdot \text{m} \tag{18.4}$$

For a steel-rim-type flywheel, assuming that 10 percent of the inertia is provided by the hub and spokes, the flywheel rim with a 0.5-m average rim diameter will weigh approximately

$$W = \frac{Jg}{r_a^2} = \frac{0.9(5.67)(9.80)}{0.25^2(1000)} = 0.800 \text{ kN} \tag{18.5}$$

Using a density of 76.5 kN/m^3, the necessary cross-sectional area is then given by

$$A = \frac{Jg}{2\pi\rho r_a^3} = \frac{0.9(5.67)(9.80)}{2\pi(76.5)(0.25)^3(1000)} = 0.006\ 66 \text{ m}^2 \tag{18.6}$$

Assuming that the speed is at its average during the peak torque of 774 J, the peak power required without any flywheel effect would be

$$P = T\omega = 774(20)(0.001) = 15.5 \text{ kW} \tag{18.7}$$

Without a flywheel, the design limitation on the speed fluctuation would have to be met using a nonuniform input torque. With the flywheel, the required power is determined from the average torque of 43.0 J:

$$P = T\omega = 43.0(20)(0.001) = 0.9 \text{ kW} \tag{18.8}$$

This shows that in addition to smoothing the machine operation, the flywheel actually reduces the size of the motor required.

18.2.2   Integration of the Torque-Angle Relation

If the torque-angle curve for a machine cycle is available from experimental data or a dynamic analysis, U is determined from the areas between the curve and the average-torque line. If the external torque input or load is not constant, it can be combined with the torque-angle curve for the machine. If the loading torque and the driving torque are not synchronized or have an unknown phase difference, a worst-case combination should be used. The areas under the curve can be determined using a planimeter or by graphic or numerical integration as shown in Chap. 4. (See also Example 4 or consult the user handbook for your programmable calculator or computer.) Unless C_s is accurately known and the curve is from a worst case or is highly repeatable, precision in integrating is not warranted.

Example 2.   An engine has the torque-angle curve given in Fig. 18.2. If the average speed at the flywheel is 2000 rpm and the output speed is allowed to vary by ±2.5 percent, how large a flywheel is necessary if the loading torque is assumed constant? The inertia of the other rotating parts, referred to the flywheel, is 0.11 lb·s²·ft.

The net area under the curve, using a planimeter, is 1156 lb·ft. One machine cycle for the four-stroke engine consists of two crankshaft cycles. The average torque is therefore $1156/(4\pi) = 92.0$ lb·ft. This average torque is shown as the dashed line in Fig. 18.2. The maximum and minimum velocities will occur at the points where the curve crosses this line. Each area between crossover points is measured and tabulated (see Table 18.2). The relative maxima and minima of the speed occur at the crossover points; therefore, the largest energy difference between any two crossover points will determine U. Since these two points will not necessarily be adjacent to each other, a running sum of the individual areas $A + I$ through H is formed, starting at an arbitrary crossover point. The largest energy difference is then the maximum sum minus the minimum sum; in this case, $U = 1106 - (-95) = 1201$ lb·ft. With $\omega = 2\pi(2000)/60 = 209.4$ rad/s and $C_s = 2(0.025) = 0.05$, Eq. (18.2) gives

$$J = \frac{1201}{209.4^2(0.05)} - 0.11 = 0.438 \text{ lb·s}^2\text{·ft} \tag{18.9}$$

Note that if the engine were operated at a slower speed, Eq. (18.2) indicates that a larger flywheel would be necessary even if the torque-angle curve did not change.

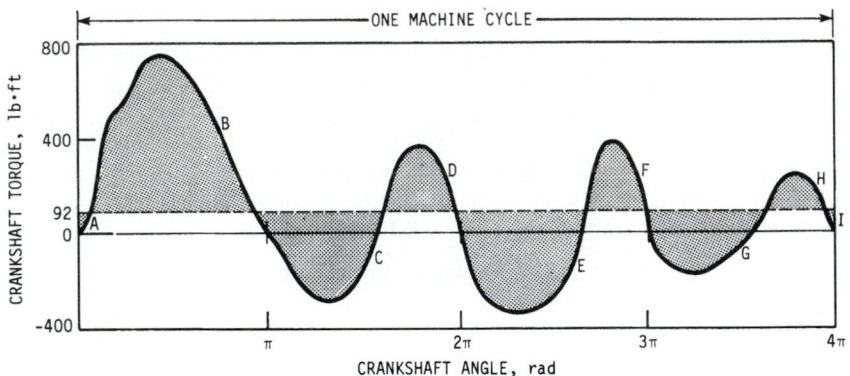

FIGURE 18.2   Torque-angle curve for the engine in Example 2. The dashed line indicates the average torque of 92.0 lb·ft, and the shaded areas are the energy variations from the average speed. Sections A and I correspond to the values given in Table 18.2.

TABLE 18.2   Area Sums for Example 2

Section	$A + I$	B	C	D	E	F	G	H
Area, lb·ft	−18	1124	−550	242	−710	208	−391	95
Sum, lb·ft	−18	1106	556	798	88	296	−95	0
Extreme values, lb·ft		1106					−95	

18.2.3 Coefficient of Energy Variation

The torque-angle relationship for an engine depends on the fuel, gas pressures, reciprocating masses, speed, and engine geometry [18.2]. The large variation that is possible between different engine designs shows that dynamic measurement or kinematic analysis is necessary to determine the torque fluctuation. It is often necessary, however, to come up with a rough estimate for preliminary design purposes or for checking the reasonableness of calculated values. For these purposes, the energy variation for an internal-combustion engine can be estimated by

$$U = C_u \frac{KP}{\omega} \qquad (18.10)$$

where $K = 33\,000$ lb·ft·rpm/hp [2π J·rad/(W·s)]. The coefficient of energy variation C_u can be approximated for a two-stroke engine with from 1 to 8 cylinders using the equation

$$C_u = \frac{7.46}{(N_c + 1)^3} \qquad (18.11)$$

and for a four-stroke engine with from 1 to 16 cylinders using the two-branched equation

$$C_u = \frac{0.8}{|N_c - 1.4|^{1.3}} - 0.015 \qquad (18.12)$$

Example 3. A 150-hp four-cylinder, four-stroke engine has a flywheel speed of 1000 rpm. Estimate the flywheel necessary for a 2 percent speed variation with a uniform load at an engine speed of 3000 rpm, neglecting the flywheel effect of the other rotating parts.
Using Eq. (18.12),

$$C_u = \frac{0.8}{|4 - 1.4|^{1.3}} - 0.015 = 0.22 \qquad (18.13)$$

Then from Eq. (18.10),

$$U = 0.22 \frac{33\,000(150)}{3000} = 363 \text{ lb·ft} \qquad (18.14)$$

so that from Eq. (18.2), with $\omega = 2\pi(1000)/60 = 105$ rad/s,

$$J = \frac{363}{105^2(0.02)} = 1.6 \text{ lb·s}^2\text{·ft} \qquad (18.15)$$

18.2.4 Angular Fluctuation

Certain machines, such as electric generators and magnetic digital storage systems, must maintain their angular position within a close tolerance of the constant-speed position. If the torque is known as a function of time, it can be integrated to determine the angular velocity, and then the angular velocity can be integrated to give the angular position:

$$\omega(t) = \int_0^t \frac{T(\xi)}{J} \, d\xi + \omega_0 \tag{18.16}$$

$$\theta(t) = \int_0^t \omega(\xi) \, d\xi + \omega_0 t + \theta_0 \tag{18.17}$$

where the $\omega_0 t + \theta_0$ term represents the constant-speed position.

In the more usual instance, the torque is known only as a function of angle. For small values of C_s, however, the torque-time curve is indistinguishable from the torque-angle curve with the angle coordinate divided by ω_{avg}.

Example 4.   A generator with the input torque given in Fig. 18.3a must maintain an angular position within ±0.25 degrees of the uniform 200-rpm position. Assuming a uniform load, what flywheel inertia is necessary?

For illustration purposes, the machine cycle will be divided into 10 intervals of $\Delta t = 0.03$ s each, as shown in Fig. 18.3a. For an accurate solution, the problem would be programmed with perhaps 20 intervals.

The torque at each step is tabulated (column 3 in Table 18.3), and then the average torque in each interval is placed in column 4. This value, if multiplied by Δt, would be the area below the curve using the trapezoid rule. Adding these average torques (column 5) and dividing by 10 intervals gives the average torque for the curve, 902 lb·ft (column 6), shown as the dashed line in Fig. 18.3a. Subtracting this average, the constant loading torque, from column 4 gives column 7, the average excess of supplied torque in each interval. The running sum of these values (column 8) performs the integration, to give $J\omega/\Delta t$ (see Fig. 18.3b). The relative speed at the end of each interval is therefore the value in column 8 times $\Delta t/J$.

The procedure is repeated for the second integration, giving columns 9 through 13. Column 13 is then $J\theta/(\Delta t)^2$ (Fig. 18.3c), so that the relative angular position is the value in column 13 times $\Delta t^2/J$. The maximum range in column 13 is $6915 - (-7725) = 14\ 640$ lb·ft. The maximum angular deviation from the mean position is calculated from half the maximum range, so that

$$\theta_{max} = \frac{(\Delta t)^2(14\ 640)}{J(2)} \tag{18.18}$$

For $\theta_{max} = 0.25$ degrees $= 0.004\ 36$ rad deviation, this gives

$$J = \frac{0.03^2(14\ 640)}{0.004\ 36(2)} = 1511\ \text{lb·s}^2\text{·ft} \tag{18.19}$$

The speed variation is determined as a by-product of the process. The maximum range in column 8 is $9878 - 0 = 9878$ lb·ft. The maximum speed variation is then

$$\omega_{max} - \omega_{min} = \frac{\Delta t(9878)}{J} = \frac{0.03(9878)}{1511} = 0.196\ \text{rad/s} \tag{18.20}$$

For $\omega_{avg} = 2\pi(200)/60 = 20.94$ rad/s, the coefficient of speed fluctuation is then, from Eq. (18.1),

$$C_s = \frac{0.196}{20.94} = 0.009\ 36 \tag{18.21}$$

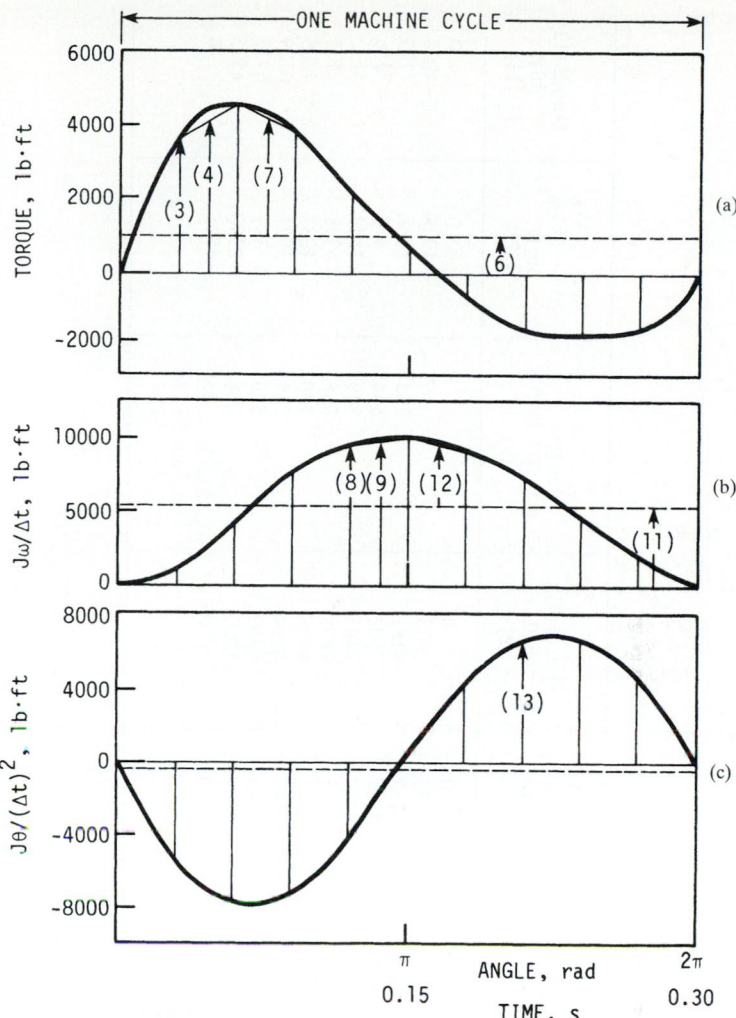

FIGURE 18.3 (*a*) Torque-angle or torque-time curve for Example 4. Dashed lines indicate average values, and the numbers in parentheses represent typical values found in Table 18.3. (*b*) Calculated rotational speed. (*c*) Calculated rotation angle.

18.2.5 Speed-Dependent Torques

The input and loading torques are in reality a function of angle, time, speed, acceleration, and other factors. In most cases, the assumption that they are functions of angle only is a good one. In some applications, however, different assumptions are necessary.

Figure 18.4 shows the torque-speed relationship for an induction motor. The curve can be approximated by a straight line $T_r = a\omega + b$ in the recommended oper-

TABLE 18.3 Numerical Integration for Example 4

j (1)†	t, s (2)	T, lb·ft (3)	Area, $[(3)_j + (3)_{j-1}]/2$, lb·ft (4)	Area − Avg. $[(4) − (6)]$, lb·ft (7)	Sum Σ (7), lb·ft (8)	Area $[(8)_j + (8)_{j-1}]/2$, lb·ft (9)	Area − Avg. $[(9) − (11)]$, lb·ft (12)	Sum Σ (12), lb·ft (13)
0	0	0		0	0		0	0
1	0.03	3670	1835‡	933	933	467	−4908	−4908
2	0.06	4628	4149	3247	4181	2557	−2818	−7725
3	0.09	3830	4229	3327	7508	5845	470	−7256
4	0.12	1931	2881	1979	9487	8498	3123	−4132
5	0.15	654	1293	391	9878	9683	4308	175
6	0.18	−734	−40	−942	8936	9407	4033	4208
7	0.21	−1628	−1181	−2083	6854	7895	2521	6729
8	0.24	−1739	−1684	−2585	4269	5561	187	6915
9	0.27	−1596	−1668	−2569	1700	2984	−2390	4525
10	0.30	0	−798	−1700	0	850	−4525	0
			(5) Sum 9016			(10) Sum 53 746		
			(6) Avg. 902			(11) Avg. 5375		

†See Example 4 and Fig. 18-3 for a description of each column.
‡The table entries were calculated to a higher precision and rounded.

ating range [18.7]. If the torque-angle curve is known for the loading torque T_a, the speed equation is

$$\omega^2(\theta) - \omega_0^2 = \frac{2}{J} \int_0^\theta [a\omega(\beta) + b - T_a(\beta)] \, d\beta \qquad (18.22)$$

This equation is implicit in ω, and the initial value ω_0 is unknown. Dividing the angle for one cycle into intervals $\Delta\theta$ long and letting ω_j denote the speed at the end of j intervals, the trapezoidal rule gives

$$\omega_j^2 - \omega_0^2 = \frac{2}{J} \left(\frac{a(\omega_0 + 2\omega_1 + 2\omega_2 + \cdots + 2\omega_{j-1} + \omega_j)}{2} \right.$$

$$\left. + jb - \frac{T_0 + 2T_1 + 2T_2 + \cdots + 2T_{j-1} + T_j}{2} \right) \Delta\theta \qquad (18.23)$$

where T_0 through T_j are the values of T_a at the end of each interval. The unknowns in this equation are ω_j and ω_0. If a value for ω_0 is assumed, the quadratic equation gives

$$\omega_j = A_j + \sqrt{A_j^2 + B_j} \qquad (18.24)$$

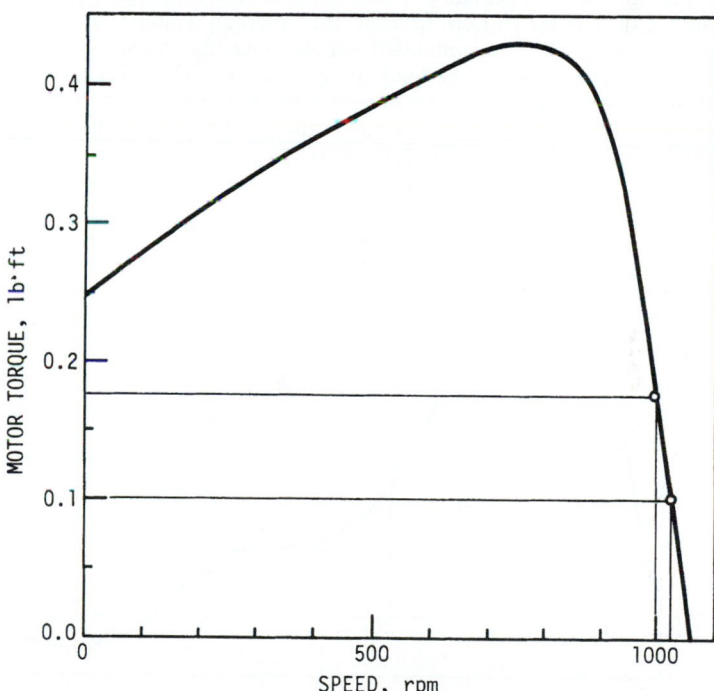

FIGURE 18.4   Torque-speed curve for the induction motor of Example 5. The curve can be approximated by a straight line near the recommended full-torque operating speed of 1000 rpm.

where

$$A_j = \frac{a\,\Delta\theta}{2J} \tag{18.25}$$

$$B_j = \omega_0^2 + \left[\frac{a}{2}\,(\omega_0 + 2\omega_1 + \cdots + 2\omega_{j-1}) + C_j \right] \frac{2\,\Delta\theta}{J} \tag{18.26}$$

and

$$C_j = jb - \frac{T_0 + 2T_1 + 2T_2 + \cdots + 2T_{j-1} + T_j}{2} \tag{18.27}$$

Equation (18.24) is solved successively for each j. The calculated value of ω at the end of the final interval will not in general equal the value that was originally assumed for ω_0. If the calculation is rerun with this final speed as the assumed value for ω_0, the numbers will be closer. After a few calculation cycles, the answers will converge.

Example 5.　　An induction motor provides 0.175 lb·ft of torque at the recommended full-load operating speed of 1000 rpm. The linear portion of the torque-speed curve also goes through the point of 0.102 lb·ft torque at 1025 rpm. Using the torque-angle curve for the load given in Fig. 18.5, find C_s if the load has the same 200-rpm speed as the flywheel and $J = 0.085$ lb·s²·ft referred to the flywheel.

The motor has been chosen so that the average loading torque balances the motor torque at the recommended speed; since the speed of the motor is $1000/200 = 5$ times the speed of the load, the average load should be approximately $T = 5(0.175) = 0.875$ lb·ft.

Referring the motor torques and speeds to the flywheel speed, the two points on the torque-speed curve become $0.175(5) = 0.875$ lb·ft at $1000/5 = 200$ rpm and $0.102(5) = 0.510$ lb·ft at $1025/5 = 205$ rpm. With 200 rpm $= 2\pi(200)/60 = 20.944$ rad/s

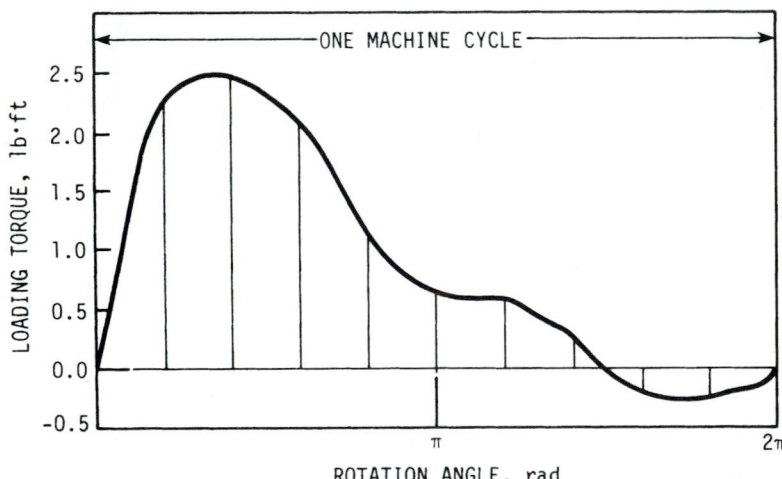

FIGURE 18.5　Torque-angle curve for the load in Example 5.

and 205 rpm $= 2\pi(205)/60 = 21.468$ rad/s, the equation for the linear part of the torque-speed curve for the motor, referred to the flywheel, is given by the two-point formula

$$\frac{T_s - 0.875}{\omega - 20.944} = \frac{0.510 - 0.875}{21.468 - 20.944} \tag{18.28}$$

or

$$T_s = -0.697\omega + 15.46 \tag{18.29}$$

Therefore, $a = -0.697$ lb·s·ft/rad and $b = 15.46$ lb·ft.

For illustration purposes, the machine cycle will be divided into 10 intervals of $\pi/5$ rad each (see Fig. 18.5). For an accurate solution, the problem would be programmed on a computer with perhaps 50 intervals. The loading torque at the end of each interval is first tabulated as shown in column 3 in Table 18.4. The average torque in each interval, $0.5(T_{j-1} + T_j)$, is placed in column 4, and the cumulative sum of these numbers is placed in column 5. The values in column 5 thus represent the T summation in Eq. (18.27). In column 6 these values are subtracted from $jb = j(15.46)$ to form C_j. The values in columns 1 through 6 will not be changed in the iteration process.

For the first iteration, a value of $\omega_0 = 20.944$ rad/s is chosen as an initial guess of the speed at the start of the cycle. For $j = 1$, $\omega_{j-1} = \omega_0 = 20.944$ rad/s is placed in column 7. As column 7 is filled, column 8 will be formed as the summation $\omega_0 + 2\omega_1 + 2\omega_2 + \cdots + 2\omega_{j-1}$ required in Eq. (18.26). B_j is then evaluated from Eq. (18.26) using the values of a, C_j, ω_0, and $2\,\Delta\theta/J = 2(\pi/5)/0.085 = 14.784$ rad/(lb·s²·ft) and placed in column 9. With $A_j = -0.697(\pi/5)(0.5)/0.085 = -2.576$ rad/s, Eq. (18.24) can be evaluated to give ω_j. For $j = 1$, $\omega_1 = 20.87$ rad/s is thus placed in column 10. This value is then brought over to column 7 for $j = 2$ and the process is repeated until column 10 is filled. The last value in column 10, 21.59 rad/s, is the speed at the end of the cycle using the initial guess for ω_0. Then 21.59 rad/s is used as the ω_0 value for the second iteration, producing columns 11 through 14. This process is repeated until the final speed for the cycle equals the initial speed that was assumed. In this example this takes three iterations, giving the speed variation given in column 18.

The maximum and minimum speeds, from column 18, are 21.64 rad/s and 20.32 rad/s. From Eq. (18.1), C_s is then this range divided by the average:

$$C_s = \frac{\omega_{max} - \omega_{min}}{0.5(\omega_{max} + \omega_{min})} = \frac{21.64 - 20.32}{0.5(21.64 + 20.32)} = 0.063 \tag{18.30}$$

18.3 STRESS

18.3.1 Rim-Type: No Bending

If interaction with the spokes is ignored, the stress in a thin-rim flywheel is the uniform hoop stress:

$$\sigma = \frac{\rho r_a^2 \omega^2}{g} = \frac{\rho V^2}{g} \tag{18.31}$$

Since the stress is a function of velocity, the strength of flywheel materials can be given in terms of maximum rim velocity (see Sec. 18.5.1).

TABLE 18.4 Numerical Tabulation for Example 5

(1)‡ j	(2) θ, rad	(3) T	(4) Avg. lb·ft	(5) Sum lb·ft	(6) C_j	(7) ω_{j-1} rad/s	(8) Sum† rad/s	(9) B_p (rad/s)²	(10) ω_p rad/s	(11) ω_{j-1} rad/s	(12) Sum† rad/s	(13) B_p (rad/s)²	(14) ω_p rad/s	(15) ω_{j-1} rad/s	(16) Sum† rad/s	(17) B_p (rad/s)²	(18) ω_p rad/s
0	0	0.00	0.000	0.000	....	....	....	....	20.94	....	....	....	21.59	....	....	....	21.64
1	π/5	2.22	1.110§	1.110	14.35	20.94	20.94	542.89	20.87	21.59	21.59	566.90	21.37	21.64	21.64	568.89	21.41
2	2π/5	2.43	2.325	3.435	27.49	20.87	62.68	522.07	20.42	21.37	64.33	540.86	20.82	21.41	64.47	542.42	20.86
3	3π/5	2.06	2.245	5.680	40.70	20.42	103.51	507.05	20.09	20.82	105.98	521.67	20.41	20.86	106.18	522.88	20.44
4	4π/5	1.08	1.570	7.250	54.59	20.09	143.69	505.40	20.05	20.41	146.79	516.71	20.30	20.44	147.05	517.66	20.32
5	π	0.62	0.850	8.100	69.20	20.05	183.79	514.77	20.26	20.30	187.40	523.52	20.45	20.32	187.69	524.25	20.46
6	6π/5	0.55	0.585	8.685	84.08	20.26	224.31	525.93	20.50	20.45	228.29	532.72	20.65	20.46	228.62	533.29	20.66
7	7π/5	0.23	0.390	9.075	99.15	20.50	265.31	537.47	20.75	20.65	269.59	542.75	20.86	20.66	269.94	543.19	20.87
8	8π/5	−0.20	0.015	9.090	114.59	20.75	306.81	551.99	21.06	20.86	311.31	556.11	21.15	20.87	311.69	556.45	21.15
9	9π/5	−0.26	−0.230	8.860	130.28	21.06	348.93	566.95	21.37	21.15	353.61	570.17	21.44	21.15	353.99	570.44	21.45
10	2π	0.00	−0.130	8.730	145.87	21.37	391.68	577.19	21.59	21.44	396.49	579.72	21.64	21.45	396.89	579.93	21.64

$\omega_0 = 21.59$ $\omega_0 = 21.64$ $\omega_0 = 21.64$

†$\omega_1 + 2\omega_2 + 2\omega_3 + \cdots + 2\omega_{j-1}$
‡See Example 5 for a description of each column.
§The table entries were calculated to a higher precision and rounded.

Example 6. An alloy of density 26.6 kN/m³ has an allowable stress of 70 MPa. For a flywheel speed of 200 rad/s, what is the largest possible average radius, overlooking the additional stress caused by the spokes?

From Eq. (18.31),

$$V = \sqrt{\frac{g\sigma}{\rho}} = \sqrt{\frac{9.80(70)(1000)}{26.6}}$$

$$= 160.6 \text{ m/s} \tag{18.32}$$

and

$$r_a = \frac{V}{\omega} = \frac{160.6}{200} = 0.803 \text{ m} \tag{18.33}$$

18.3.2 Rim-Type: With Bending

If the effects of the spoke are taken into account, Timoshenko showed that the tensile stress in the thin rim (Fig. 18.6), including bending but neglecting the effect of the curvature on the bending stress and assuming spokes of constant cross-sectional area, can be calculated from beam theory using Castigliano's theorem [18.10], giving

$$\sigma = \frac{\rho r_a^2 \omega^2}{g} \left[1 - \frac{f_4}{3} + \frac{A r_a}{3 Z_r} \left(f_4 - \frac{1}{f_3 \alpha} \right) \right] \tag{18.34}$$

where

$$f_1 = \frac{1}{2 \sin^2 \alpha} \left(\frac{\sin 2\alpha}{4} + \frac{\alpha}{2} \right) \qquad f_2 = f_1 - \frac{1}{2\alpha} \tag{18.35}$$

$$f_3 = \frac{A r_a^2 f_2}{I} + f_1 + \frac{A}{A_s} \qquad f_4 = \frac{\cos \beta}{f_3 \sin \alpha} \tag{18.36}$$

and the section modulus of the rim Z_r is positive for the outer face of the rim and negative for the inner face. The stress in the spokes is given by

$$\sigma = \frac{\rho r_a^2 \omega^2}{6g} \left(3 + \frac{4A}{f_3 A_s} - \frac{3 r^2}{r_a^2} \right) \tag{18.37}$$

In addition, there will be bending stresses in the spokes as power is exchanged with the flywheel. This torque might be determined by the maximum shaft torque, by the minimum braking time from full speed, or, if the flywheel also serves as a pulley, by the difference in belt force between the power and slack sides.

Depending on how thick the rim is, the spoke will behave as something between an end-loaded cantilever and a guided cantilever. For a thin rim, the maximum bending stress on the spokes will occur at the hub, assuming that they do not taper more than the usual 10 to 25 percent. This bending stress is

$$\sigma = \frac{T(r_a - r_h)}{Z_s N_s r_a} \tag{18.38}$$

If the flywheel serves as a pulley, the torque is not equally distributed to the spokes; the value for N_s should be halved in this case. For a flywheel with a large rim

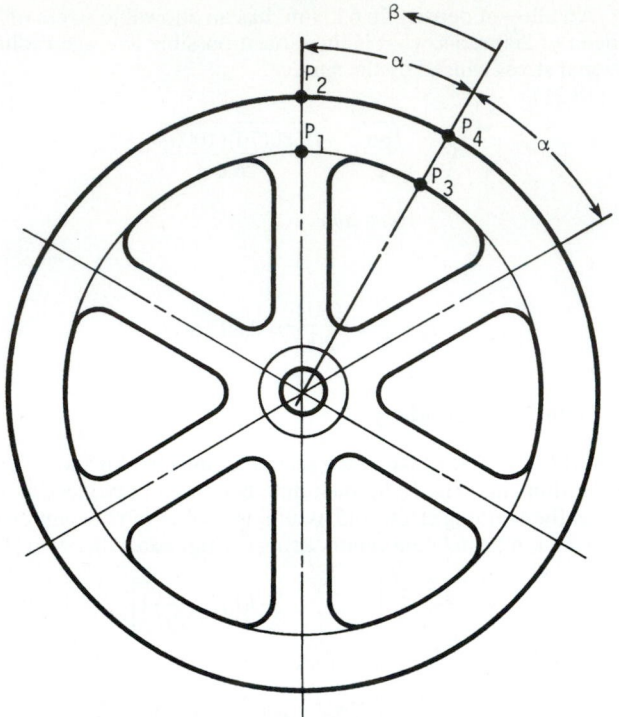

FIGURE 18.6   Rim-type flywheel. The angular position β is measured from the center line bisecting the adjacent spoke locations. Points P_1 through P_4 correspond to the values given in Table 18.5.

section in comparison to the spokes, the spokes act as guided cantilevers and have a maximum bending stress that is half the value given by Eq. (18.38).

Example 7.   A cast-iron flywheel with a rectangular rim 5 in thick and 11 in wide has an average radius of 48 in. If the hub has a 5.2-in radius and each of the 8 spokes has an area of 11 in², what are the maximum stresses at a speed of 400 rpm? The maximum torque possible is 4000 lb·ft at full speed.

The area $A = (5)(11) = 55$ in², $\omega = 2\pi(400)/60 = 41.9$ rad/s, $I = 11(5)^3/12 = 114.6$ in⁴, and $Z_r = 11(5)^2/6 = 45.8$ in³. The angle α is 22.5 degrees, or $\pi/8$, and the maximum rim stress occurs at the inner surface of the rim with $\beta = \pi/8$. From Eqs. (18.34) through (18.36), using a density of 0.283 lb/in³,

$$f_1 = \frac{1}{2\sin^2(\pi/8)}\left[\frac{\sin(\pi/4)}{4} + \frac{\pi/8}{2}\right] = 1.274 \tag{18.39}$$

$$f_2 = 1.274 - \frac{1}{\pi/4} = 7.600 \times 10^{-4} \tag{18.40}$$

$$f_3 = \frac{55(48)^2 7.600 \times 10^{-4}}{114.6} + 1.274 + \frac{55}{11} = 7.11 \tag{18.41}$$

$$f_4 = \frac{\cos (\pi/8)}{7.11 \sin (\pi/8)} = 0.340 \tag{18.42}$$

$$\sigma = \frac{0.283(48)^2(41.9)^2}{32.2(12)} \left[1 - \frac{0.340}{3} + \frac{55(48)}{3(-45.8)} \left(0.340 - \frac{1}{7.11\pi/8} \right) \right] = 3660 \text{ psi} \tag{18.43}$$

Table 18.5 shows the values for points P_1, P_2, P_3, and P_4, giving additional digits for checking computer programs. Although the stress formulation was based on a spoke of length r_a, the maximum tension stress in the spoke is found for the hub radius, 5.2 in. From Eq. (18.37),

$$\sigma = \frac{0.283(48)^2(41.9)^2}{6(32.2)(12)} \left[3 + \frac{4(55)}{7.11(11)} - \frac{3(5.2)^2}{(48)^2} \right] = 2850 \text{ psi} \tag{18.44}$$

If the spokes are of standard design, an elliptical cross section twice as long in the direction of motion as in the axial direction, the dimensions for an 11-in^2 spoke would be $\sqrt{2(11)/\pi} = 2.65$ in by 5.29 in. The section modulus is then $Z_s = \pi(2.65)(5.29)^2/32 = 7.28$ in^3, and the bending stress in the spokes, from Eq. (18.38), would be

$$\sigma = \frac{4000(48 - 5.2)}{7.28(8)(48)} = 61 \text{ psi} \tag{18.45}$$

The combined stress for the spoke is then $\sigma = 2850 + 61 = 2910$ psi.

18.3.3   Thin Disk

If a thin disk has a large radius in comparison to its thickness, the stress can be assumed constant across the thickness. Defining the stress function F as

$$F = rz\sigma_r \tag{18.46}$$

the stress equation for a thin disk of variable thickness $z(r)$ is

$$r^2 \frac{d^2F}{dr^2} + r \frac{dF}{dr} - F + (3 + v) \frac{\rho\omega^2 r^3 z}{g} - \frac{r}{z} \frac{dz}{dr} \left(r \frac{dF}{dr} - vF \right) = 0 \tag{18.47}$$

and

$$\sigma_t = \frac{1}{z} \frac{dF}{dr} + \rho\omega^2 r^2 \tag{18.48}$$

TABLE 18.5   Rim Stress for Example 7

Term	P_1	P_2	P_3	P_4
Z_r, in^3	$-45.833\ 3$	$45.844\ 3$	$-45.833\ 3$	$45.833\ 3$
β	$\pi/8$	$\pi/8$	0	0
f_1	1.273 93	1.273 93	1.273 93	1.273 93
f_2 ($\times 10^{-4}$)	6.932 02	6.932 02	6.932 02	6.932 02
f_3	7.040 56	7.040 56	7.040 56	7.040 56
f_4	0.342 901	0.342 901	0.371 153	0.371 153
σ_r, psi	3690.32	1554.43	2056.37	3132.62

The boundary conditions for the center or inner radius and the outer radius are both $F = 0$. If there is a central bore, a finite value for r_i is used in the formulation, and for an infinitesimal hole, the limit is taken later. At the center of the solid disk, $\sigma_r = \sigma_t$. The inertia for a disk is given by

$$J = \int_{r_i}^{r_o} \frac{2\pi\rho r^3 z}{g}\, dr \qquad (18.49)$$

Equations (18.46) through (18.48) can be solved explicitly for the uniform and hyperbolic profiles [18.4]. For the general case, using a difference approximation for dz/dr and F leads to a tractable banded set of linear simultaneous equations. Divide the radius into equal intervals of length Δr. Determine the z derivative for each point using

$$z_j' = \frac{z_{j+1} - z_{j-1}}{2\,\Delta r} \qquad (18.50)$$

except at the two boundaries, where, for the same accuracy,

$$z_j' = \frac{2z_{j+3} - 9z_{j+2} + 18z_{j+1} - 11z_j}{6\,\Delta r} \qquad (18.51)$$

at the center or inner radius and

$$z_j' = \frac{11z_j - 18z_{j-1} + 9z_{j-2} - 2z_{j-3}}{6\,\Delta r} \qquad (18.52)$$

at the outer radius. The stress equation then becomes the set of equations

$$A_j F_{j-1} + B_j F_j + C_j F_{j+1} = D_j \qquad (18.53)$$

where

$$A_j = \frac{r_j}{\Delta r}\left(\frac{r_j}{\Delta r} - \frac{1}{2} + \frac{r_j}{2}\frac{z_j'}{z_j}\right) \qquad (18.54)$$

$$B_j = -2\left(\frac{r_j}{\Delta r}\right)^2 + v r_j\left(\frac{z_j'}{z_j}\right) - 1 \qquad (18.55)$$

$$C_j = \frac{r_j}{\Delta r}\left(\frac{r_j}{\Delta r} + \frac{1}{2} - \frac{r_j}{2}\frac{z_j'}{z_j}\right) \qquad (18.56)$$

$$D_j = -(3 + v)\frac{r_j^3 z_j \rho \omega^2}{g} \qquad (18.57)$$

At the two boundary points the equation is identically satisfied, leaving $(r_o - r_i)/\Delta r - 1$ equations. For the inner equation, $A_j = 0$, and for the outer equation, $C_j = 0$. The simultaneous equations are solved for F_j, and the stresses are found from Eqs. (18.46) and (18.48). See the user's guide for your computer for a convenient method of solving linear simultaneous equations.

18.3.4 Disk of Constant Thickness

The solutions for a disk of uniform thickness are simplified if a constant is defined as

$$\sigma_0 = \frac{\rho r_o^2 \omega^2 (3 + v)}{8g} \qquad (18.58)$$

For a constant-thickness disk without a central hole, the radial and tangential stresses are then given by

$$\sigma_r = \sigma_0\left(1 - \frac{r^2}{r_o^2}\right) \tag{18.59}$$

$$\sigma_t = \sigma_0\left(1 - \frac{1+3v}{3+v}\frac{r^2}{r_o^2}\right) \tag{18.60}$$

and the maximum stress, at $r = 0$, is

$$\sigma_{r,\max} = \sigma_{t,\max} = \sigma_0 \tag{18.61}$$

For a disk with a central hole, the stresses are given by

$$\sigma_r = \sigma_0\left(1 - \frac{r^2}{r_o^2} + \frac{r_i^2}{r_o^2} - \frac{r_i^2}{r^2}\right) \tag{18.62}$$

$$\sigma_t = \sigma_0\left(1 - \frac{1+3v}{3+v}\frac{r^2}{r_o^2} + \frac{r_i^2}{r_o^2} + \frac{r_i^2}{r^2}\right) \tag{18.63}$$

The maximum radial stress, at $r = \sqrt{r_o r_i}$, is

$$\sigma_{r,\max} = \sigma_0\left(1 - \frac{r_i}{r_o}\right)^2 \tag{18.64}$$

and the maximum tangential stress, at $r = r_i$, is

$$\sigma_{t,\max} = \sigma_0\left(2 + \frac{2-2v}{3+v}\frac{r_i^2}{r_o^2}\right) \tag{18.65}$$

The inertia is

$$J = \frac{\pi \rho z}{2g}(r_o^4 - r_i^4) \tag{18.66}$$

Example 8. A steel disk of uniform thickness and outer radius 0.6 m rotates at 30 rad/s. Find the maximum stress in the disk if it has an integral shaft, neglecting stress rises due to the geometry change at the shaft. What is the maximum disk stress if the disk is bored for a shaft of 0.025-m radius?

The stress is independent of thickness. From Eqs. (18.58) and (18.61) with $\rho = 76.5$ kN/m^3 and $v = 0.3$,

$$\sigma_{r,\max} = \sigma_{t,\max} = \sigma_0$$

$$= \frac{76.5(0.6)^2(30)^2(3+0.3)}{8(9.80)(1000)} = 1.043 \text{ MPa} \tag{18.67}$$

For the bored disk, the maximum radial stress occurs at $r = \sqrt{0.6(0.025)} = 0.122$ m. From Eq. (18.64),

$$\sigma_{r,\max} = 1.043\left(1 - \frac{0.025}{0.6}\right)^2 = 0.958 \text{ MPa} \tag{18.68}$$

The maximum tangential stress occurs at the hub. From Eq. (18.65),

$$\sigma_{t,\,max} = 1.043 \left[2 + \frac{2 - 2(0.3)}{3 + 0.3} \frac{(0.025)^2}{(0.6)^2} \right]$$

$$= 2.09 \text{ MPa} \tag{18.69}$$

18.4   FLYWHEELS FOR ENERGY STORAGE

The flywheel can be used as an energy reservoir, with energy being supplied at a slow constant rate or when it is available and being withdrawn when desired. A flywheel might, for example, be used to give good acceleration to an automobile that is underpowered by present standards. Regenerative breaking, power storage for peak-demand periods, and mechanical replacements for battery banks are all potential uses for the flywheel. The high charging and discharging rates of a flywheel system give it an advantage over other portable sources of power, such as batteries.

Although the concepts developed in the previous sections are still true for energy-storage flywheels, the purpose is now to store as much kinetic energy, $0.5J\omega^2$, as possible. In most applications, the flywheel speed does not vary over 50 percent, so that only about 75 percent of this total energy is actually recoverable. The design of the ordinary flywheel is usually dictated by the allowable diameter, governed by the machine size, and the maximum speed, governed by the practicalities of a speed-increasing drive and higher bearing speeds. These constraints can result in a low peripheral speed, causing the economics to favor a rim-type flywheel design. The economics change with the energy-storage flywheel, since (1) larger values of total stored energy are usually involved, requiring heavier flywheels or more energy per unit weight of flywheel, (2) the weight of a heavy flywheel and the correspondingly heavy bearings and other components may be unacceptable, especially in mobile applications, and (3) the design constraints imposed in a machine where the flywheel limits the speed variation can be relaxed when the flywheel is the main component, encouraging optimization. Depending on the application, the energy per dollar, energy per weight, or energy per swept volume is usually maximized [18.1].

18.4.1   Isotropic and Anisotropic Designs

The stress equations for the thin disk given in Sec. 18.3.3 can be solved with $\omega_r = \omega_t$ to give the shape for a fully stressed thin isotropic disk with no central bore:

$$z = z_0 \exp\left(-\frac{\rho\omega^2 r^2}{2gS_y}\right) \tag{18.70}$$

where S_y = allowable strength, for example, the yield strength of the material. Define the energy stored per unit weight as

$$R = F_s \frac{S_y}{\rho} \tag{18.71}$$

where F_s is a dimensionless factor that depends only on the shape of the flywheel. Using Eqs. (18.46) through (18.49), it turns out that the efficiency or geometric shape factor $F_s = 1.0$ for the fully stressed profile. As Eq. (18.70) indicates, this profile has an infinite outer radius; truncating the profile reduces F_s substantially. Modifying the profile can improve the efficiency, increasing F_s to 0.97 (see Table 18.6), but that is the limit for a homogeneous isotropic design. To avoid this basic limitation and to take advantage of the higher strength-to-weight and stiffness-to-weight ratio of composite materials, recent efforts have concentrated on anisotropic designs ([18.1], [18.8]). Figure 18.7 describes a few of the many designs that are being developed.

18.4.2 Special Considerations

Energy-storage flywheels have special problems which are related to their high speed, flexibility, and anisotropy. The increased operational speed requires high-speed power transmission and bearings as well as special attention to drag forces and the critical vibration speeds of the torsional system. A continuously variable transmission and operation in a vacuum are usually required. Forward whirling is a potentially serious problem that may limit the operating range or require external damping [18.3]. The flexibility can cause flywheel vibration modes with frequencies in the operating range and can also cause significant imbalance to occur as the flywheel deforms with speed. Anisotropy can cause failure in the weak direction (matrix failure or delamination) before the full strength in the strong direction is utilized.

18.5 STRENGTH AND SAFETY

18.5.1 Materials

Neglecting gravity and other secondary loads, the stresses in a flywheel will be proportional to $V^2 = (r_o\omega)^2$. Two flywheels of the same design but of different size will therefore have the same rim stress when their rim velocities are equal. Also, since the stress is proportional to V^2, a 10 percent increase in rotational speed will cause a 21 percent increase in stress. Although the strengths of flywheel materials are sometimes given in terms of their maximum rim velocities, these strengths include generous factors of safety to account for all the possible variations in the material properties, design details, and methods of manufacture. A rational approach would

TABLE 18.6   Shape Factor for Several Isotropic Flywheel Shapes

Shape	F_s
Fully stressed, infinite radius	1.00
Optimum, finite radius	0.97
Exponential or truncated conical (approximate)	0.8
Uniform thickness	0.61
Thin-rim type	0.50
Bar	0.33
Uniform thickness, central bore	0.31

SOURCE: From Gilbert et al. [18.5], by permission.

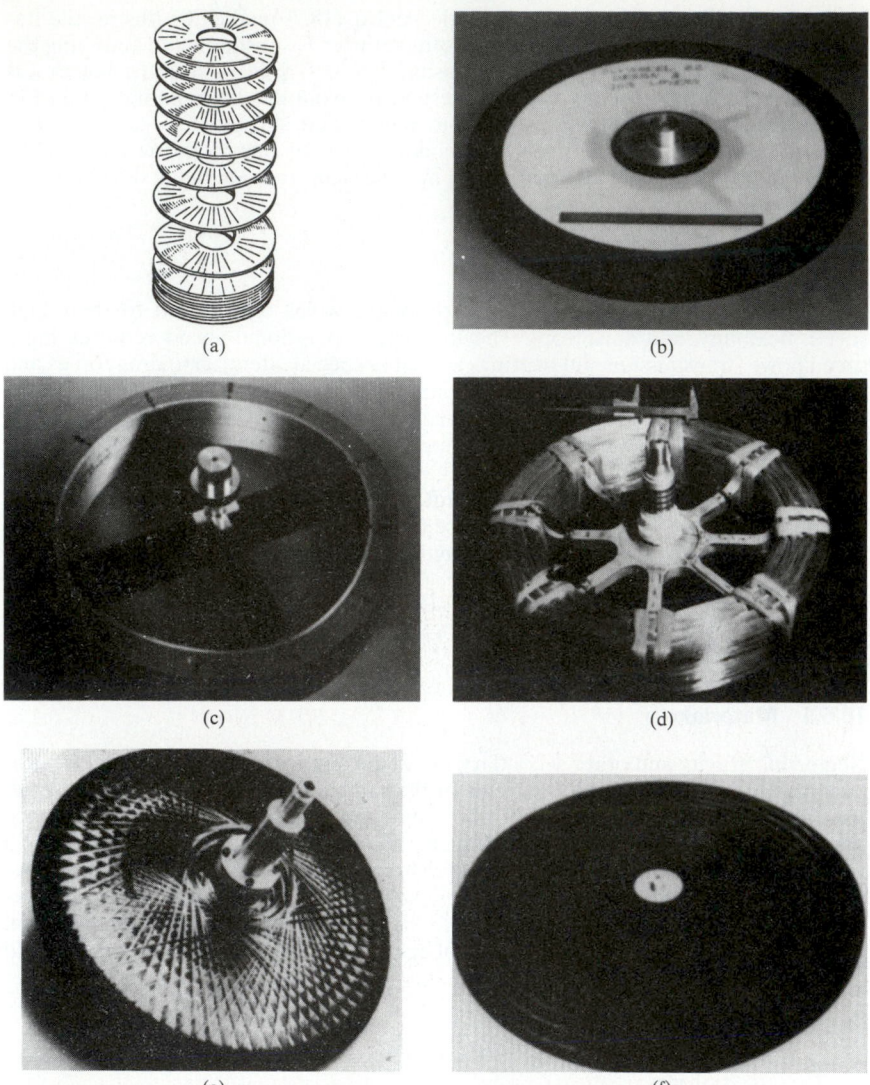

(a)

(b)

(c)

(d)

(e)

(f)

FIGURE 18.7 Energy-storage flywheel designs. (*a*) Flywheel of helically woven fabric with variable properties in the radial and circumferential directions (*exploded view*) (*Avco Systems Division*); (*b*) filament-wound graphite/epoxy rim with laminated S2-glass/epoxy disk (*General Electric Company*); (*c*) subcircular multilayer rim of S-glass and Kevlar, with graphite/epoxy spokes (*Garrett AiResearch Corporation*); (*d*) bare-filament Kevlar rim with aluminum hub (*Istituto della Motorizzazione, Torino, Italy, built by Industrie Pirelli S.p.A. and sponsored by the Italian National Research Council*); (*e*) graphite/epoxy and steel-filament/epoxy rim with a woven graphite/epoxy overwrap (*MAN Advanced Technology, Munich, Germany*); (*f*) variable-thickness graphite/epoxy laminated disk (*Lawrence Livermore National Laboratory*).

FIGURE 18.7 *(Continued)* Energy-storage flywheel designs. (*g*) bare-filament high-strength steel-wire rim, as used in steel-belted radial tires, with fiberglass hub (*The Johns Hopkins University Applied Physics Laboratory*); (*h*) multilayer rim of S-glass and Kevlar, with catenary tension-balanced spokes (*William M. Brobeck and Associates*); (*i*) graphite/epoxy-wound rim with overwrap over an aluminum liner with two contoured-aluminum hubs (*Rockwell International/Rocketdyne Division*); (*j*) vinyl-coated S-glass rim with Russian birch hub (*The Johns Hopkins University Applied Physics Laboratory*); (*k*) Metglas (amorphous steel) ribbon rim with aluminum spokes (*The Johns Hopkins University Applied Physics Laboratory*); (*l*) contoured graphite/epoxy filament-wound disk (*Hercules Aerospace*). (Kevlar is a registered trademark of E. I. du Pont de Nemours and Company, Inc.; Metglas is a registered trademark of The Allied Corporation.) (*Photographs assembled with the assistance of W. Wilkinson, The Johns Hopkins University Applied Physics Laboratory, and S. Kulkarni, The Lawrence Livermore National Laboratory; by permission.*)

be to assign risk factors to each unknown in the specific flywheel design and multiply the resulting factors of safety or, if sufficient data are available, to use a probabilistic or statistical approach [18.6] (see also Chap. 2).

Cast iron has often been chosen as a flywheel material on the basis of cost per pound. However, this criterion is valid only if the design constraints on radius and speed dictate a low rim stress. Otherwise the higher strength-to-weight ratio of other materials may make them less expensive. With any cast flywheel there is the possibility of brittleness, of blowholes, and of other casting flaws; shrinkage stresses in the casting must also be controlled. To reduce shrinkage stresses, one-piece flywheels larger than a few feet in diameter are often cast with a split hub, i.e., with each separate arc of the hub allowed to move with the attached spokes. After cooling, the hub is then bolted together using spacer plates.

The rim is sometimes cast separately from the spokes, and special care must be taken to design an efficient joint between the two. On flywheels larger than 10 ft in diameter, the rim is sometimes fabricated in two or more sections, usually for transportation reasons. The joint efficiency in these cases can be as low as 30 percent, and so extreme care should be taken in designing the joint. The more efficient designs rely on bow tie or ring shrink-fit connectors placed at the spoke locations. A more advanced study should be made unless the calculated stresses are markedly low; a photoelastic or finite-element model [18.9] might be constructed, for example.

Cast steel is stronger than cast iron and is widely used. Flame-cut steel-plate flywheels are relatively inexpensive to manufacture, and their simplicity may make up for their inefficient shape. Flywheels welded from steel plate are inherently stronger than cast flywheels as long as proper care is taken to control flaws and residual weld stresses.

18.5.2 Safety

If the designer can reduce the uncertainty in any of the design unknowns, the necessary factor of safety will be less and a lighter flywheel will result. Design unknowns include (1) material properties, (2) analysis error, (3) loading, and (4) installation. The material properties of the finished flywheel may vary due to welds, porous casting, or lamination flaws. The analysis may have been an elementary beam calculation or a detailed finite-element/fatigue/fracture mechanics study. The input and output loads and speeds may be estimated, calculated, or measured values. The skill or care used in installation may be unknown or well controlled. Improvement in any of these areas will give a better design.

Another way to reduce the necessary factor of safety is through testing. Both destructive and nondestructive tests of the material are excellent tools in answering questions about material properties and fabrication technique. The spin test, of course, is the primary proof of the material, fabrication, and design. Although it will not ordinarily answer questions of fatigue or cycled loads, a destructive spin test gives the best indication of the true safety margin. Long-term cyclic life tests are very helpful if a realistic load cycle can be devised.

A third way to reduce the necessary factor of safety is to reduce the physical and financial harm that would be caused by a flywheel failure. No matter what material is used, any flywheel will break if it is rotated fast enough. If this failure can be controlled in some way, the design speed of the flywheel can be closer to the failure speed without undue risk. Standard steel rim-type and disk flywheels usually burst into three to six pieces at failure; each piece travels with quite a bit of kinetic energy. One method to control failure is therefore to force a breakup into smaller pieces;

many composite flywheels shred, for example. Another method used with the energy-storage flywheel relies on centrifugal force to separate the outer circular sections of the flywheel from the spokes or inner hub section before the material begins to fail. If a benign failure mode cannot be introduced into the design, a heavier containment enclosure could serve the same purpose, lessening the consequence of a flywheel failure.

REFERENCES

18.1 *1980 Flywheel Technology Symposium,* October 1980, Scottsdale, Arizona, University of California Lawrence Livermore National Laboratory, Livermore, Calif., NTIS CONF-801022.

18.2 Rowland S. Benson and N. D. Whitehouse, *Internal Combustion Engines,* Pergamon, New York, 1979.

18.3 C. W. Bert and G. Ramunujam, "Design Guide for Composite-Material Flywheels: Rotor Dynamic Considerations, Part I," University of Oklahoma, Norman, Okla., September 1981, UCRL-15420.

18.4 J. P. Den Hartog, *Advanced Strength of Materials,* McGraw-Hill, New York, 1952.

18.5 R. R. Gilbert, et al., "Flywheel Feasibility Study and Demonstration," Lockheed Missiles and Space Company, Sunnyvale, Calif., April 1971, NTIS PB-200143.

18.6 Edward B. Haugen, *Probabilistic Approaches to Design,* John Wiley & Sons, New York, 1968.

18.7 Charles R. Mischke, *Elements of Mechanical Analysis,* Addison-Wesley, Reading, Mass., 1963.

18.8 R. P. Nimmer, K. Torossian, and W. W. Wilkening, "Laminated Composite Disk Flywheel Development," General Electric Company, Schenectady, N.Y., UCRL-15383.

18.9 M. Saraph, A. Midha, and J. C. Wambold, "Automated Stress Analysis of Mechanical Sheaves and Pulleys," *Computers in Mechanical Engineering,* pp. 34–42, October 1982.

18.10 Joseph E. Shigley and Charles R. Mischke, *Mechanical Engineering Design,* 5th ed., McGraw-Hill, New York, 1989.

18.11 Joseph Edward Shigley and John Joseph Uicker, Jr., *Theory of Machines and Mechanisms,* 2d ed. McGraw-Hill, New York, 1995.

CHAPTER 19
LIMITS AND FITS

Joseph E. Shigley
Professor Emeritus
The University of Michigan
Ann Arbor, Michigan

Charles R. Mischke, Ph.D., P.E.
Professor Emeritus of Mechanical Engineering
Iowa State University
Ames, Iowa

NOMENCLATURE

a	Radius
B	Smallest bore diameter
b	Radius
c	Radius radial clearance
D	Diameter, mean of size range, largest journal diameter
E	Young's modulus
e	Bilateral tolerance expressing error
L	Upper or lower limit
p	Probability
p_f	Probability of failure
t	Bilateral tolerance of dimension
w	Left-tending vector representing gap
x	Right-tending dimensional vector magnitude
y	Left-tending dimensional vector magnitude
δ	Radial interference
υ	Poisson's ratio
σ	Normal stress
σ	Standard deviation

19.1   INTRODUCTION

Standards of limits and fits for mating parts have been approved for general use in the United States for use with U.S. customary units [19.1] and for use with SI units [19.2]. The tables included in these standards are so lengthy that formulas are presented here instead of the tables to save space. As a result of rounding and other variations, the formulas are only close approximations. The nomenclature and symbols used in the two standards differ from each other, and so it is necessary to present the details of each standard separately.

19.2   METRIC STANDARDS

19.2.1   Definitions

Terms used are illustrated in Fig. 19.1 and are defined as follows:

1. *Basic size* is the size to which limits or deviations are assigned and is the same for both members of a fit. It is measured in millimeters.
2. *Deviation* is the algebraic difference between a size and the corresponding basic size.
3. *Upper deviation* is the algebraic difference between the maximum limit and the corresponding basic size.
4. *Lower deviation* is the algebraic difference between the minimum limit and the corresponding basic size.
5. *Fundamental deviation* is either the upper or the lower deviation, depending on which is closest to the basic size.
6. *Tolerance* is the difference between the maximum and minimum size limits of a part.
7. *International tolerance grade* (IT) is a group of tolerances which have the same relative level of accuracy but which vary depending on the basic size.
8. *Hole basis* represents a system of fits corresponding to a basic hole size.
9. *Shaft basis* represents a system of fits corresponding to a basic shaft size.

19.2.2   International Tolerance Grades

The *variation in part size,* also called the *magnitude of the tolerance zone,* is expressed in grade or IT numbers. Seven grade numbers are used for high-precision parts; these are

$$IT01, IT0, IT1, IT2, IT3, IT4, IT5$$

The most commonly used grade numbers are IT6 through IT16, and these are based on the Renard R5 geometric series of numbers (see Sec. 48.3). For these, the basic equation is

$$i = \frac{1}{1000}(0.45D^{1/3} + 0.001D) \tag{19.1}$$

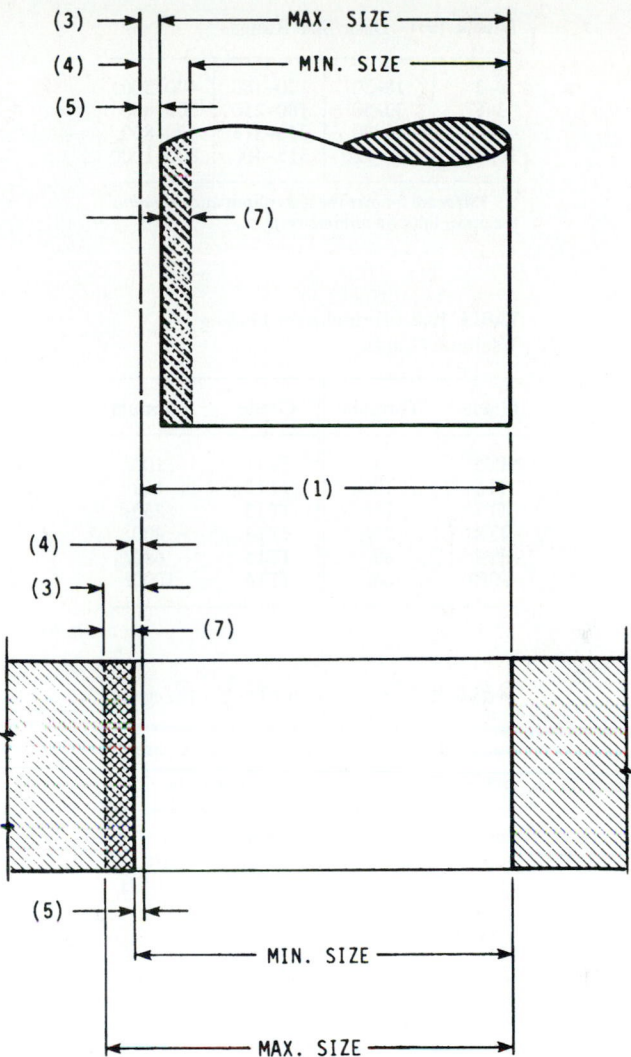

FIGURE 19.1 Definitions applied to a cylindrical fit. The numbers in parentheses are the definitions in Sec. 19.2.1.

where D is the geometric mean of the size range under consideration and is obtained from the formula

$$D = \sqrt{D_{max}D_{min}} \qquad (19.2)$$

The ranges of basic sizes up to 1000 mm for use in this equation are shown in Table 19.1. For the first range, use $D_{min} = 1$ mm in Eq. (19.2).

With D determined, tolerance grades IT5 through IT16 are found using Eq. (19.1) and Table 19.2. The grades IT01 to IT4 are computed using Table 19.3.

TABLE 19.1 Basic Size Ranges[†]

0–3	18–30	120–180	400–500
3–6	30–50	180–250	500–630
6–10	50–80	250–315	630–800
10–18	80–120	315–400	800–1000

[†]Sizes are for *over* the lower limit and *including* the upper limit (in millimeters).

TABLE 19.2 Formulas for Finding Tolerance Grades

Grade	Formula	Grade	Formula
IT5	$7i$	IT11	$100i$
IT6	$10i$	IT12	$160i$
IT7	$16i$	IT13	$250i$
IT8	$25i$	IT14	$400i$
IT9	$40i$	IT15	$640i$
IT10	$64i$	IT16	$1000i$

TABLE 19.3 Formulas for Higher-Precision Tolerance Grades

Grade	Formula
IT01	$(0.008D + 0.3)/1000$
IT0	$(0.012D + 0.5)/1000$
IT1	$(0.02D + 0.8)/1000$
IT2	$(IT1)[7i/(IT1)]^{1/4}$
IT3	$(IT2)^2$
IT4	$(IT2)^3$

19.2.3 Deviations

Fundamental deviations are expressed by *tolerance position letters* using capital letters for internal dimensions (holes) and lowercase letters for external dimensions (shafts). As shown by item 5 in Fig. 19.1, the fundamental deviation is used to position the tolerance zone relative to the basic size (item 1).

Figure 19.2 shows how the letters are combined with the tolerance grades to establish a fit. If the basic size for Fig. 19.2 is 25 mm, then the hole dimensions are defined by the ISO symbol

$$25D9$$

where the letter D establishes the fundamental deviation for the holes, and the number 9 defines the tolerance grade for the hole.

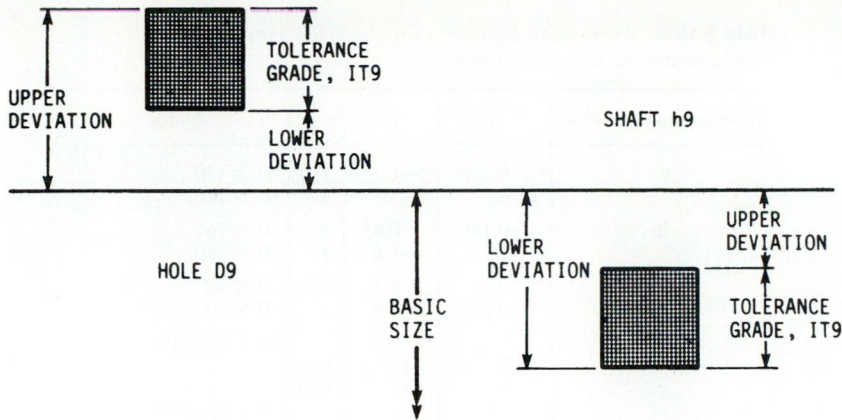

FIGURE 19.2 Illustration of a shaft-basis free-running fit. In this example the upper deviation for the shaft is actually zero, but it is shown as nonzero for illustrative purposes.

Similarly, the shaft dimensions are defined by the symbol

$$25h9$$

The formula for the fundamental deviation for shafts is

$$\text{Fundamental deviation} = \alpha + \frac{\beta D^{\gamma}}{1000} \qquad (19.3)$$

where D is defined by Eq. (19.2), and the three coefficients are obtained from Table 19.4.

Shaft Deviations. For shafts designated a through h, the upper deviation is equal to the fundamental deviation. Subtract the IT grade from the fundamental deviation to get the lower deviation. Remember, the deviations are defined as algebraic, so be careful with signs.

Shafts designated j through zc have the lower deviation equal to the fundamental deviation. For these, the upper deviation is the sum of the IT grade and the fundamental deviation.

Hole Deviations. Holes designated A through H have a lower deviation equal to the negative of the upper deviation for shafts. Holes designated as J through ZC have an upper deviation equal to the negative of the lower deviation for shafts.

An exception to the rule occurs for a hole designated as N having an IT grade from 9 to 16 inclusive and a size over 3 mm. For these, the fundamental deviation is zero.

A second exception occurs for holes J, K, M, and N up to grade IT8 inclusive and holes P through ZC up to grade 7 inclusive for sizes over 3 mm. For these, the upper deviation of the hole is equal to the negative of the lower deviation of the shaft plus the change in tolerance of that grade and the next finer grade. In equation form, this can be written

Upper deviation (hole)

$$= -\text{lower deviation (shaft)} + \text{IT (shaft)} - \text{IT (next finer shaft)} \qquad (19.4)$$

TABLE 19.4 Coefficients for Use in Eq. (19.3) to Compute the Fundamental Deviations for Shafts[†]

Fundamental deviation	α	β	γ	Notes
a	−0.265	−1.3	1	$D \leq 120$
	0	−3.5	1	$D > 120$
b	−0.140	−0.85	1	$D \leq 160$
	0	−1.8	1	$D > 160$
c	0	−5.2	0.2	$D \leq 40$
	−0.095	−0.8	1	$D > 40$
cd				$cd = (c \cdot d)^{1/2}$
d	0	−16	0.44	
e	0	−11	0.41	
ef				$ef = (e \cdot f)^{1/2}$
f	0	−5.5	0.41	
fg				$fg = (f \cdot g)^{1/2}$
g	0	−2.5	0.34	
h	0	0	0	
j				No formula
js				$js = IT/2$
k	0	0.6	0.33	IT4 to IT7, $D \leq 500$
	0	0	0	IT8 to IT16, $D > 500$
m	IT7/1000	−IT6	0	$D \leq 500$
	0.013	0.024	1	$D > 500$
n	0	5	0.34	$D \leq 500$
	0.021	0.04	1	$D > 500$
p	IT7	2	0	$D \leq 500$
	0.038	0.072	D	$D > 500$
r				$r = (p \cdot s)^{1/2}$
s	IT8	2	0	$D \leq 50$
	IT7	0.4	1	$D > 50$
t	IT7	0.63	1	
u	IT7	1	1	
v	IT7	1.25	1	
x	IT7	1.6	1	
y	IT7	2	1	
z	IT7	2.5	1	
za	IT8	3.15	1	
zb	IT9	4	1	
zc	IT10	5	1	

[†]These coefficients will give results that may not conform exactly to the fundamental deviations tabulated in the standards. Use the standards if exact conformance is required.
SOURCE: From Ref. [19.2].

TABLE 19.5 Preferred Fits

Type	Hole basis	Shaft basis†	Name and application
Clearance	H11/c11	C11/h11	*Loose-running fit* for wide commercial tolerances or allowances on external members
	H9/d9	D9/h9	*Free-running fit* not for use where accuracy is essential, but good for large temperature variations, high running speeds, or heavy journal pressures
	H8/f7	F8/h7	*Close-running fit* for running on accurate machines and for accurate location at moderate speeds and journal pressures
	H7/g6	G7/h6	*Sliding fit* not intended to run freely, but to move and turn freely and locate accurately
	H7/h6	H7/h6	*Locational-clearance fit* provides snug fit for locating stationary parts, but can be freely assembled and disassembled
Transition	H7/k6	K7/h6	*Locational-transition fit* for accurate location, a compromise between clearance and interference
	H7/n6	N7/h6	*Locational-transition fit* for more accurate location where greater interference is permissible
Interference	H7/p6	P7/h6	*Locational-interference fit* for parts requiring rigidity and alignment with prime accuracy of location but without special bore pressure requirements
	H7/s6	S7/h6	*Medium-drive fit* for ordinary steel parts or shrink fits on light sections, the tightest fit usable with cast iron
	H7/u6	U7/h6	*Force fit* suitable for parts which can be highly stressed or for shrink fits where the heavy pressing forces required are impracticable

†The transition and interference shaft-basis fits shown do not convert to exactly the same hole-basis fit conditions for basic sizes from 0 to 3 mm. Interference fit P7/h6 converts to a transition fit H7/p6 in the size range 0 to 3 mm.
SOURCE: From Ref. [19.2].

19.2.4 Preferred Fits

Table 19.5 lists the preferred fits for most common applications. Either first or second choices from Table 19.3 should be used for the basic sizes.

Example 1.   Using the shaft-basis system, find the limits for both members using a basic size of 25 mm and a free-running ft.
    Solution.   From Table 19.5, we find the fit symbol as D9/h9, the same as Fig. 19.2. Table 19.1 gives $D_{min} = 18$ and $D_{max} = 30$ for a basic size of 25. Using Eq. (19.2), we find

$$D = \sqrt{D_{max}D_{min}} = \sqrt{30(18)} = 23.2 \text{ mm}$$

Then, from Eq. (19.1) and Table 19.2,

$$40i = \frac{40}{1000}(0.45D^{1/3} + 0.001D)$$

$$= \frac{40}{1000}[0.45(23.2)^{1/3} + 0.001(23.2)] = 0.052 \text{ mm}$$

This is the IT9 tolerance grade for the size range 18 to 30 mm.
    We proceed next to find the limits on the 25D9 hole. From Table 19.4, for a d shaft, we find $\alpha = 0$, $\beta = -16$, and $\gamma = 0.44$. Therefore, using Eq. (19.3), we find the fundamental deviation for a d shaft to be

$$\text{Fundamental deviation} = \alpha + \frac{\beta D^{\gamma}}{1000} = 0 + \frac{-16(23.2)^{0.44}}{1000}$$

$$= -0.064 \text{ mm}$$

But this is also the upper deviation for a d shaft. Therefore, for a D hole, we have

$$\text{Lower deviation (hole)} = -\text{upper deviation (shaft)}$$

$$= -(-0.064) = 0.064 \text{ mm}$$

The upper deviation for the hole is the sum of the lower deviation and the IT grade. Thus

$$\text{Upper deviation (hole)} = 0.064 + 0.052 = 0.116 \text{ mm}$$

The two limits of the hole dimensions are therefore

$$\text{Upper limit} = 25 + 0.116 = 25.116 \text{ mm}$$

$$\text{Lower limit} = 25 + 0.064 = 25.064 \text{ mm}$$

For the h shaft, we find from Table 19.4 that $\alpha = \beta = \gamma = 0$. Therefore, the fundamental deviation, which is the same as the upper deviation, is zero. The lower deviation equals the upper deviation minus the tolerance grade, or

$$\text{Lower deviation (shaft)} = 0 - 0.052 = -0.052 \text{ mm}$$

Therefore, the shaft limits are

$$\text{Upper limit} = 25 + 0 = 25.000 \text{ mm}$$

$$\text{Lower limit} = 25 - 0.052 = 24.948 \text{ mm}$$

19.3 U.S. STANDARD—INCH UNITS

The fits described in this section are all on a *unilateral hole basis*. The kind of fit obtained for any one class will be similar throughout the range of sizes. Table 19.6 describes the various fit designations. Three classes, RC9, LC10, and LC11, are described in the standards [19.1] but are not included here. These standards include recommendations for fits up to a basic size of 200 in. However, the tables included here are valid only for sizes up to 19.69 in; this is in accordance with the American-British-Canadian (ABC) recommendations.

The coefficients listed in Table 19.7 are to be used in the equation

$$L = CD^{1/3} \tag{19.5}$$

where L is the limit in thousandths of an inch corresponding to the coefficient C and the basic size D in inches. The resulting four values of L are then summed algebraically to the basic hole size to obtain the four limiting dimensions.

It is emphasized again that the limits obtained by the use of these equations and tables are only close approximations to the standards.

19.4 INTERFERENCE-FIT STRESSES

The assembly of two cylindrical parts by press-fitting or shrinking one member onto another creates a contact pressure between the two members. The stresses resulting from the interference fit can be computed when the contact pressure is known. This pressure may be obtained from Eq. (2.67) of Ref. [19.3]. The result is

$$p = \frac{\delta}{bA} \tag{19.6}$$

where δ = radial interference and A is given by

$$A = \frac{1}{E_i}\left(\frac{b^2 + a^2}{b^2 - a^2} - v_i\right) + \frac{1}{E_o}\left(\frac{c^2 + b^2}{c^2 - b^2} + v_o\right) \tag{19.7}$$

The dimensions a, b, and c are the radii of the members, as shown in Fig. 19.3. The terms E_i and E_o are the elastic moduli for the inner and outer cylinders, respectively. If the inner cylinder is solid, then $a = 0$ and Eq. (19.7) becomes

$$A = \frac{1}{E_i}(1 - v_i) + \frac{1}{E_o}\left(\frac{c^2 + b^2}{c^2 - b^2} + v_o\right) \tag{19.8}$$

Sometimes the mating parts have identical moduli. In this case, Eq. (19.6) becomes

$$p = \frac{E\delta}{b}\left[\frac{(c^2 - b^2)(b^2 - a^2)}{2b^2(c^2 - a^2)}\right] \tag{19.9}$$

This equation simplifies still more if the inner cylinder is solid. We then have

$$p = \frac{E\delta}{2bc^2}(c^2 - b^2) \tag{19.10}$$

TABLE 19.6 Standard Fits

Designation	Name and application
RC1	*Close sliding fits* are intended for the accurate location of parts which must be assembled without perceptible play.
RC2	*Sliding fits* are intended for accurate location, but with greater maximum clearance than an RC1 fit.
RC3	*Precision running fits* are about the loosest fits which can be expected to run freely and are intended for precision work at slow speeds and light journal pressures but are not suitable where appreciable temperature differences are likely.
RC4	*Close-running fits* are intended chiefly for running fits on accurate machinery with moderate surface speeds and journal pressure, where accurate location and minimum play are desired.
RC5	*Medium-running fits* are intended for higher running speeds or heavy journal pressures, or both.
RC6	*Medium-running fits* are intended for applications where more play than RC5 is required.
RC7	*Free-running fits* are intended for use where accuracy is not essential or where large temperature variations are likely, or both.
RC8	*Loose-running fits* are intended for use where wide commercial tolerances may be necessary, together with an allowance, on the hole.
LC1 to LC9	*Locational-clearance fits* are intended for parts which are normally stationary, but which can be freely assembled or disassembled. Snug fits are for parts requiring accuracy of location. Medium fits are for parts such as ball, race, and housings. The looser-fastener fits are needed where freedom of assembly is of first importance.
LT1 to LT6	*Locational-transitional fits* are a compromise between clearance and interference fits for application where accuracy of location is important but either a small amount of clearance or interference is permissible.
LN1 to LN3	*Locational-interference fits* are used where accuracy of location is of prime importance and for parts requiring rigidity and alignment with no special requirements for bore pressure. These fits are not intended for parts that must transmit frictional loads to one another.
FN1	*Light-drive fits* are those requiring light assembly pressures and produce more or less permanent assemblies. They are suitable for thin sections or long fits or in cast-iron external members.
FN2	*Medium-drive fits* are suitable for ordinary steel parts or for shrink fits on light sections. They are about the tightest fits that can be used with high-grade cast-iron external members.
FN3	*Heavy-drive fits* are suitable for heavier steel parts or for shrink fits in medium sections.
FN4 and FN5	*Force fits* are suitable for parts which can be highly stressed or for shrink fits where the heavy pressing forces required are impractical.

The maximum stresses occur at the contact surface. Here the stresses are biaxial, if the longitudinal direction is neglected, and for the outer member are given in Ref. [19.3] as

$$\sigma_{ot} = p\,\frac{c^2 + b^2}{c^2 - b^2} \qquad \sigma_{or} = -p \tag{19.11}$$

TABLE 19.7   Coefficients C for Use in Eq. (19.5)

Class of fit	Hole limits		Shaft limits	
	Lower	Upper	Lower	Upper
RC1	0	+0.392	−0.588	−0.308
RC2	0	+0.571	−0.700	−0.308
RC3	0	+0.907	−1.542	−0.971
RC4	0	+1.413	−1.879	−0.971
RC5	0	+1.413	−2.840	−1.932
RC6	0	+2.278	−3.345	−1.932
RC7	0	+2.278	−4.631	−3.218
RC8	0	+3.570	−7.531	−5.253
LC1	0	+0.571	−0.392	0
LC2	0	+0.907	−0.571	0
LC3	0	+1.413	−0.907	0
LC4	0	+3.570	−2.278	0
LC5	0	+0.907	−0.879	−0.308
LC6	0	+2.278	−2.384	−0.971
LC7	0	+3.570	−4.211	−1.933
LC8	0	+3.570	−5.496	−3.218
LC9	0	+5.697	−8.823	−5.253
LT1	0	+0.907	−0.281	+0.290
LT2	0	+1.413	−0.442	+0.465
LT3†	0	+0.907	+0.083	+0.654
LT4†	0	+1.413	+0.083	+0.990
LT5	0	+0.907	+0.656	+1.227
LT6	0	+0.907	+0.656	+1.563
LN1	0	+0.571	+0.656	+1.048
LN2	0	+0.907	+0.994	+1.565
LN3	0	+0.907	+1.582	+2.153
FN1	0	+0.571	+1.660	+2.052
FN2	0	+0.907	+2.717	+3.288
FN3‡	0	+0.907	+3.739	+4.310
FN4	0	+0.907	+5.440	+6.011
FN5	0	+1.413	+7.701	+8.608

†Not for sizes under 0.24 in.
‡Not for sizes under 0.95 in.

where t and r designate the tangential and radial directions, respectively.
For the inner member, the stresses at the contact surface are

$$\sigma_{it} = -p\,\frac{b^2 + a^2}{b^2 - a^2} \qquad \sigma_{ir} = -p \tag{19.12}$$

A stress-concentration factor may be needed for certain situations. A hub press-fitted to a shaft, for example, would be likely to have an increased pressure at the ends. So if either a brittle fracture or a fatigue failure is a possibility, then for such cases a stress-concentration factor in the range from 1.5 to 2 should be used.

Example 2.   A 1½-in solid-steel shaft is fitted to a steel forging having an outside diameter of 2½ in using a class FN3 fit. Determine the worst-condition stresses for each member.

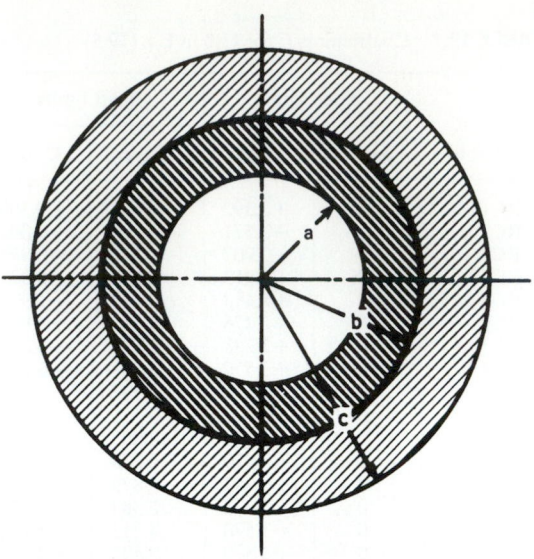

FIGURE 19.3 A press-fitted assembly. Inner member has hole of radius a. Contact surface has radius b. Outer member has outside radius c.

Solution.   The worst condition would occur when the hole is minimum and the shaft is maximum. From Table 19.7, we find $C = 0$ and $C = +4.310$ for the lower limit of the hole and upper limit of the shaft, respectively. Using Eq. (19.5), we find

$$L = CD^{1/3} = \frac{+4.310(1.5)^{1/3}}{1000} = 0.0049 \text{ in}$$

Therefore, the maximum shaft has a diameter $d_i = 1.5 + 0.0049 = 1.5049$ in. Similarly, the minimum hole is $d_o = 1.5000$ in. The radial interference is $\delta = 0.5(0.0049) = 0.00245$ in. For use in Eq. (19.10), we observe that $b = 0.75$ in and $c = 1.25$ in based on the nominal dimensions. Using $E = 30$ Mpsi, we find the contact pressure to be

$$p = \frac{E\delta}{bc^2}(c^2 - b^2)$$

$$= \frac{30(10)^6(0.00245)}{0.75(1.25)^2}[(1.25)^2 - (0.75)^2]$$

$$= 62.7 \text{ kpsi}$$

Using Eq. (19.11) to get the stresses in the outer member gives

$$\sigma_{ot} = p\,\frac{c^2 + b^2}{c^2 - b^2} = 62.7\left[\frac{(1.25)^2 + (0.75)^2}{(1.25)^2 - (0.75)^2}\right] = 133.2 \text{ kpsi}$$

$$\sigma_{or} = -p = -62.7 \text{ kpsi}$$

For the inner member, the worst stress is given by

$$\sigma_{it} = \frac{-pb^2}{b^2 - a^2} = -p = -62.7 \text{ kpsi}$$

and the result is

$$\sigma_{ir} = \sigma_{it} = -62.7 \text{ kpsi}$$

19.5   ABSOLUTE TOLERANCES[†]

When an aggregate of several parts is assembled, the gap, grip, or interference is related to dimensions and tolerances of the individual parts. Consider an array of parallel vectors as depicted in Fig. 19.4, the x's directed to the right and the y's directed to the left. They may be treated as scalars and represented algebraically. Let t_i be the bilateral tolerance on $\overline{x}_i$ and t_j be the bilateral tolerance on $\overline{y}_j$, all being positive numbers. The gap remaining short of closure is called w and may be viewed as the slack variable permitting summation to zero. Thus,

$$(x_1 + x_3 + \cdots) - (y_2 + y_4 + \cdots) - w = 0$$

or

$$w = \Sigma x_i - \Sigma y_j \tag{19.13}$$

The largest gap w exists when the right-tending vectors are the largest possible and the left-tending vectors are the smallest possible. Expressing Eq. (19.13) in terms of the greatest deviations from the means gives

$$w_{\max} = \Sigma(\overline{x}_i + t_i) - \Sigma(\overline{y}_j - t_j) = \Sigma\overline{x}_i - \Sigma\overline{y}_j + \sum_{\text{all}} t \tag{19.14}$$

[†] See Ref. [19.4].

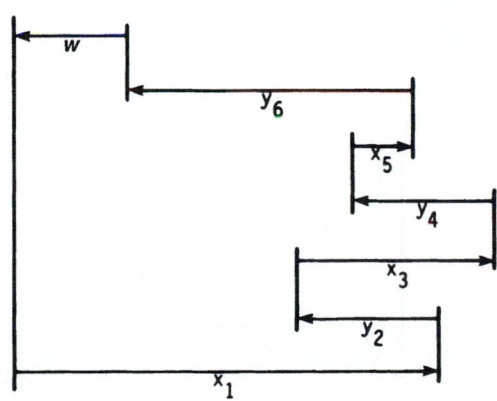

FIGURE 19.4   An array of parallel vectors.

Similarly, for the smallest gap,

$$w_{min} = \Sigma(\bar{x}_i - t_i) - \Sigma(\bar{y}_j + t_j) = \Sigma\bar{x}_i - \Sigma\bar{y}_j - \sum_{all} t \qquad (19.15)$$

The mean of w is

$$\bar{w} = \frac{1}{2}(z_{max} + z_{min}) = \frac{1}{2}[(\Sigma\bar{x}_i - \Sigma\bar{y}_j + \Sigma t) + (\Sigma\bar{x}_i - \Sigma\bar{y}_j - \Sigma t)] \qquad (19.16)$$

$$\bar{w} = \Sigma\bar{x}_i - \Sigma\bar{y}_j$$

The bilateral tolerance of w is

$$t_w = \frac{1}{2}(w_{max} - w_{min}) = \frac{1}{2}[(\Sigma\bar{x}_i - \Sigma\bar{y}_j + \Sigma t) - (\Sigma\bar{x}_i - \Sigma\bar{y}_j - \Sigma t)] \qquad (19.17)$$

$$t_w = \sum_{all} t$$

Equation (19.15) gives rise to expressions such as "the stacking of tolerances" in describing the conditions at the gap. All the bilateral tolerances of the constituent x's and y's add to the tolerance of the gap. If the gap is an interference, then w is a right-tending vector (negative). For all instances to be interference fits, both w_{max} and w_{min} have to be negative.

Example 3. In the pin-washer-sleeve–snap-ring assembly depicted in Fig. 19.5, identify the mean gap $\bar{w}$, gap tolerance t_w, maximum gap w_{max}, and minimum gap w_{min} if $x_1 = 1.385 \pm 0.005$, $y_2 = 0.125 \pm 0.001$, $y_3 = 1.000 \pm 0.002$, and $y_4 = 0.250 \pm 0.001$ in.

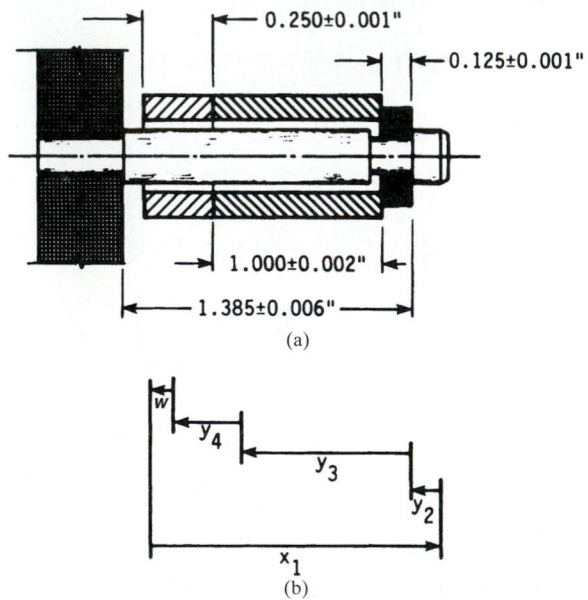

FIGURE 19.5 (*a*) A pin-washer–snap-ring assembly and associated gap; (*b*) parallel vectors describing gap.

Solution. From Eq. (19.16),

$$\overline{w} = \Sigma x_i - \Sigma y_j = 1.385 - 0.125 - 1.000 - 0.250 = 0.010 \text{ in}$$

From Eq. (19.17), $t_w = \displaystyle\sum_{\text{all}} t = 0.005 + 0.001 + 0.002 + 0.001 = 0.009 \text{ in}$

From Eq. (19.14), $w_{\max} = \overline{w} + t_w = 0.010 + 0.009 = 0.019 \text{ in}$

From Eq. (19.15), $w_{\min} = \overline{w} - t_w = 0.010 - 0.009 = 0.001 \text{ in}$

All instances of the gap w are positive, and therefore noninterfering.

Example 4. In Example 3, the washer, sleeve, and snap ring are vendor-supplied parts, and the pin is machined in-house. To assure a noninterfering assembly, what should the pin tolerance t_1 be?

Solution. From Eq. (19.16),

$$\overline{w} = \Sigma x_i - \Sigma y_j = 1.385 - 0.125 - 1.000 - 0.250 = 0.010 \text{ in}$$

From Eq. (19.17), $\quad t_w = \displaystyle\sum_{\text{all}} t = t_1 + 0.001 + 0.002 + 0.001 = t_1 + 0.004$

$$t_1 = t_w - 0.004$$

As long as $t_w \le \overline{w}$—that is, $t_w \le 0.010$ in—there will be a gap.

$$t_w = t_1 + 0.004 \le 0.010$$

$$t_1 \le 0.006 \text{ in}$$

If t_1 cannot be economically maintained at 0.006 or less, but may be 0.007 in or more, then there will be instances of interference, unless

1. Vendors can reduce the tolerance on the washer, spacer, and snap ring.
2. Inspection and selective assembly is acceptable.
3. Some interference, when detected, is solved by selective assembly for some parts, or scrapping.

Important to alternatives 2 and 3 is a prediction of the chance of encountering an interference fit. See Sec. 19.6.

Example 5. Figure 19.6 shows a journal-bushing assembly with unilateral tolerances. What is the description of the radial clearances resulting from these specifications?

Solution. From Eq. (19.14),

$$\overline{w} = \overline{c} = \Sigma x_i - \Sigma y_j = \left(\frac{B}{2} + \frac{b}{4} \right) - \left(\frac{D}{2} + \frac{d}{4} \right)$$

$$= \frac{B - D}{2} + \frac{b + d}{4} \tag{19.18}$$

From Eq. (19.17), $\qquad t_w = \displaystyle\sum_{\text{all}} t = \frac{b}{4} + \frac{d}{4} \tag{19.19}$

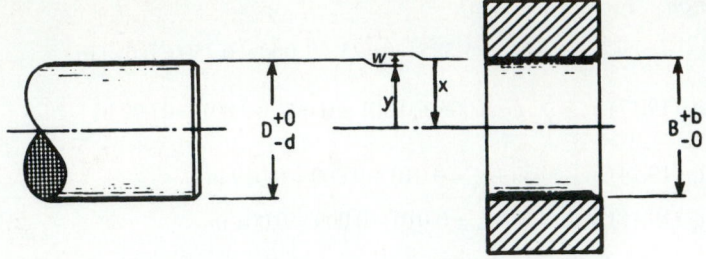

FIGURE 19.6   A journal-bushing assembly with unilateral tolerances.

From Eq. (19.14), $w_{max} = c_{max} = \overline{w} + t_w = \left(\dfrac{B-D}{2} + \dfrac{b+d}{4} \right)$

$$+ \frac{b+d}{4} = \frac{B-D}{2} + \frac{b+d}{2} \qquad (19.20)$$

From Eq. (19.15), $w_{min} = c_{min} = \overline{w} - t_w = \left(\dfrac{B-D}{2} + \dfrac{b+d}{4} \right)$

$$- \frac{b+d}{4} = \frac{B-D}{2} \qquad (19.21)$$

Table 19.8 is an absolute tolerance worksheet, a convenient nonalgebraic form suitable to the manufacturing floor.

19.6   STATISTICAL TOLERANCES

Examples 3, 4, and 5 describe situations where no scrap would be produced. This is a feature of absolute tolerances. However, gap dimensions near the gap limits would require observations near the bounds of all four intervals simultaneously, with all observations on the same side of the means. This is a rare event, leading to the con-

TABLE 19.8   Absolute Tolerance Worksheet

i	t	x_i	y_i
1	0.006	1.385	
2	0.001		0.125
3	0.002		1.000
4	0.001		0.250
	$\Sigma = 0.010$	1.385	1.375
		-1.375	
		$\overline{w} = 0.010$	

sideration of statistical tolerancing schemes. The operative equations are based on the fundamental equation (19.13)

$$\mathbf{w} = \Sigma \mathbf{x}_i - \Sigma \mathbf{y}_j \qquad (19.22)$$

where all the elements are random variables. The mean gap $\overline{w}$ is

$$\overline{w} = \Sigma \overline{x}_i - \Sigma \overline{y}_j$$

which is the same as Eq. (19.16). The variance of the algebraic sum of uncorrelated random variables is the *sum* of all constituent variances, or

$$\sigma_w^2 = \Sigma \sigma_{x_i}^2 + \Sigma \sigma_{y_j}^2 = \sum_{\text{all}} \sigma^2$$

It follows that

$$\sigma_w = \sqrt{\Sigma_{\text{all}}\, \sigma^2} \qquad (19.23)$$

A common formulation,

$$t_w = \sqrt{\sum_{\text{all}} t^2}$$

is less general, as it can be derived from Eq. (19.23) *only* if all t's are the same consistent multiple of σ, which occurs in an additive situation when the individual parts distributions are already normal. Since this is a rare occurrence, the above equation (unnumbered) is to be used with caution.

The distribution of the gap $\mathbf{w}$ depends on the distributions of the individual $\mathbf{x}$'s and $\mathbf{y}$'s. The common presumption of normality is often not borne out in reality, and errors due to this unjustified presumption are counterproductive. Geometric dimensions produced by automatic tooling (turning, grinding, reaming, broaching) often follow Eqs. (2.10) and (2.11), which lead to uniform (not normal) distributions. The best procedure is to monitor the distribution of in-house manufactures and to inspect vendor-supplied geometries. The second unjustified presumption is that the sum of several uniform random variables tends toward normality. Tend it does, but one is usually far from the critical number necessary to realize the presumption. The procedure that does not rely on normality presumptions is computer simulation.

Example 6. In Example 4, let the lengths $\mathbf{y}_2$, $\mathbf{y}_3$, and $\mathbf{y}_4$ be vendor-supplied with uniform distributions. Relate the pin tolerance t_1 to the probability of interference p_f.

Solution. The standard deviations are (range of uniform distribution is $\pm \sqrt{3}\,\sigma$)

$$\sigma_1 = \frac{t_1}{\sqrt{3}} \qquad \sigma_2 = \frac{0.001}{\sqrt{3}} \qquad \sigma_3 = \frac{0.002}{\sqrt{3}} \qquad \sigma_4 = \frac{0.001}{\sqrt{3}}$$

By computer simulation, relate in a table the tolerance t_1 to probability of interference p_f using 10^6 trials and a confidence level of 0.95:

t_1	p_f	e, Eq. (5.5)	Normal p_f
0.006	0	0	0.003 763
0.0065	0.000 053	0.000 014	0.006 324
0.007	0.000 257	0.000 031	0.009 759
0.008	0.002 749	0.000 103	0.019 211
0.009	0.011 352	0.000 208	0.031 659
0.010	0.028 680	0.000 327	0.046 254
0.011	0.054 429	0.000 495	0.062 154

There is a great variation in the relative magnitude of p_f with a change of tolerance t_1. A least-squares polynomial is difficult to fit well, and so interpolation is recommended, if needed. Note how poor the normal presumption prediction is for small probabilities. The use of Eq. (5.5) for the bilateral error bounds on p_f gives an index to simulation accuracy. The error in $p_f = 0.011\ 352$, corresponding to $t_1 = 0.009$ in, by Eq. (5.5) is

$$e = 1.96 \sqrt{\frac{0.011\ 352(1 - 0.011\ 352)}{10^6}} = 0.000\ 208$$

and so $p_f = 0.011\ 352 \pm 0.000\ 208$ at 0.95 confidence level.

If setting $t_1 = 0.007$ in is contemplated, then at the 0.95 confidence level, one can state that there is a (one-tailed) probability of the gap w being less than zero (interference) of $p = 0.000\ 257 \pm 0.000\ 031$.

REFERENCES

19.1 "Preferred Limits and Fits for Cylindrical Parts," ANSI B4.1-1967 (R1979).[†]

19.2 "Preferred Metric Limits and Fits," ANSI B4.2-1978.

19.3 Joseph E. Shigley and Charles R. Mischke, *Mechanical Engineering Design*, 5th ed., McGraw-Hill, New York, 1989.

19.4 M. F. Spotts, *Dimensioning and Tolerancing for Quality Production*, Prentice-Hall, Englewood Cliffs, N.J., 1983. (Excellent bibliography on standards and handbooks, dimensioning and tolerancing, quality control, gauging and shop practice, probability and statistics.)

19.5 C. R. Mischke, *Mathematical Model Building*, 2d rev. ed., Iowa State University Press, Ames, 1980.

[†] The symbol R indicates that the standard has been reaffirmed as up-to-date.

CHAPTER 20
POWER SCREWS

Rudolph J. Eggert, Ph.D., P.E.

Associate Professor of Mechanical Engineering
University of Idaho
Boise, Idaho

LIST OF SYMBOLS

A	Area
$A(t)$	Screw translation acceleration
C	End condition constant
d	Major diameter
d_c	Collar diameter
d_m	Mean diameter
d_r	Root or minor diameter
E	Modulus of elasticity
F	Load force
F_c	Critical load force
G	Shear modulus
h	Height of engaged threads
I	Second moment of area
J	Polar second moment of area
k	Radius of gyration
L	Thread lead
L_c	Column length
n	Angular speed, r/min
n_s	Number of thread starts

N_e	Number of engaged threads
P_i	Basic load rating
p	Thread pitch
S_y	Yield strength
T_c	Collar friction torque
T_i	Basic static thrust capacity
T_R	Raising torque
T_L	Lowering torque
t	Time
$V(t)$	Screw translation speed
w	Thread width at root
W_i	Input work
W_o	Output work
α	Flank angle
α_n	Normalized flank angle
β	Thread geometry parameter
Δx	Screw translation
$\Delta\theta$	Screw rotation
η	Efficiency
λ	Lead angle
μ_t	Coefficient of thread friction
μ_c	Coefficient of collar friction
σ	Normal stress
σ'	von Mises stress
τ	Shear stress
Ψ	Helix angle

20.1 INTRODUCTION

Power screws convert the input rotation of an applied torque to the output translation of an axial force. They find use in machines such as universal tensile testing machines, machine tools, automotive jacks, vises, aircraft flap extenders, trench braces, linear actuators, adjustable floor posts, micrometers, and C-clamps. The mechanical advantage inherent in the screw is exploited to produce large axial forces in response to small torques. Typical design considerations, discussed in the following sections, include kinematics, mechanics, buckling and deflection, and stresses.

Two principal categories of power screws are machine screws and recirculating-ball screws. An example of a machine screw is shown in Fig. 20.1. The screw threads are typically formed by thread rolling, which results in high surface hardness, high strength, and superior surface finish. Since high thread friction can cause self-locking when the applied torque is removed, protective brakes or stops to hold the load are usually not required.

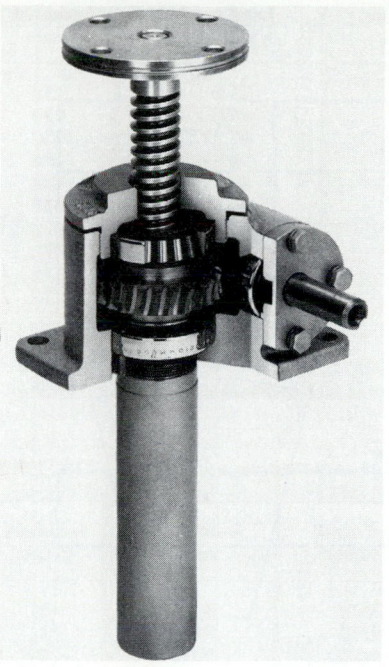

Three thread forms that are often used are the *Acme thread,* the *square thread,* and the *buttress thread.* As shown in Fig. 20.2, the Acme thread and the square thread exhibit symmetric leading and trailing flank angles, and consequently equal strength in raising and lowering. The Acme thread is inherently stronger than the square thread because of the larger thread width at the root or minor diameter. The general-purpose Acme thread has a 14½-degree flank angle and is manufactured in a number of standard diameter sizes and thread spacings, given in Table 20.1. The buttress thread is proportionately wider at the root than the Acme thread and is typically loaded on the 7-degree flank rather than the 45-degree flank. See Refs. [20.1], [20.2], [20.3], and [20.4] for complete details of each thread form.

Ball screws recirculate ball bearings between the screw rod and the nut, as shown in Fig. 20.3. The resulting rolling friction is significantly less than the sliding friction of the machine screw type. Therefore less input torque and power are needed. However, motor brakes or screw stops are usually required to prevent ball screws from *self-lowering* or *overhauling.*

FIGURE 20.1 Power screw assembly using rolled thread load screw driven by worm shaft and gear nut. *(Simplex Uni-Lift catalog UC-101, Templeton, Kenly & Co., Inc., Broadview, Ill., with permission.)*

20.2 KINEMATICS

The primary function or design requirement of a power screw is to move an axial load F through a *specified* linear distance, called the *travel.* As a single-degree-of-freedom mechanism, screw travel is constrained between the fully *extended* position $x_{\max}$ and the closed or *retracted* position $x_{\min}$. The output *range of motion,* therefore, is $x_{\max} - x_{\min}$. As the input torque T is applied through an angle of rotation $\Delta\theta$, the screw travels Δx in proportion to the screw lead L or total number of screw turns N_t as follows:

$$\Delta x = L \frac{\Delta\theta}{2\pi} = LN_t \qquad (20.1)$$

In addition to range of motion specifications, other kinematic requirements may be prescribed, such as velocity or acceleration. The linear screw speed V, in/min, is obtained for a constant angular speed of n, r/min, as

$$V = nL \qquad (20.2)$$

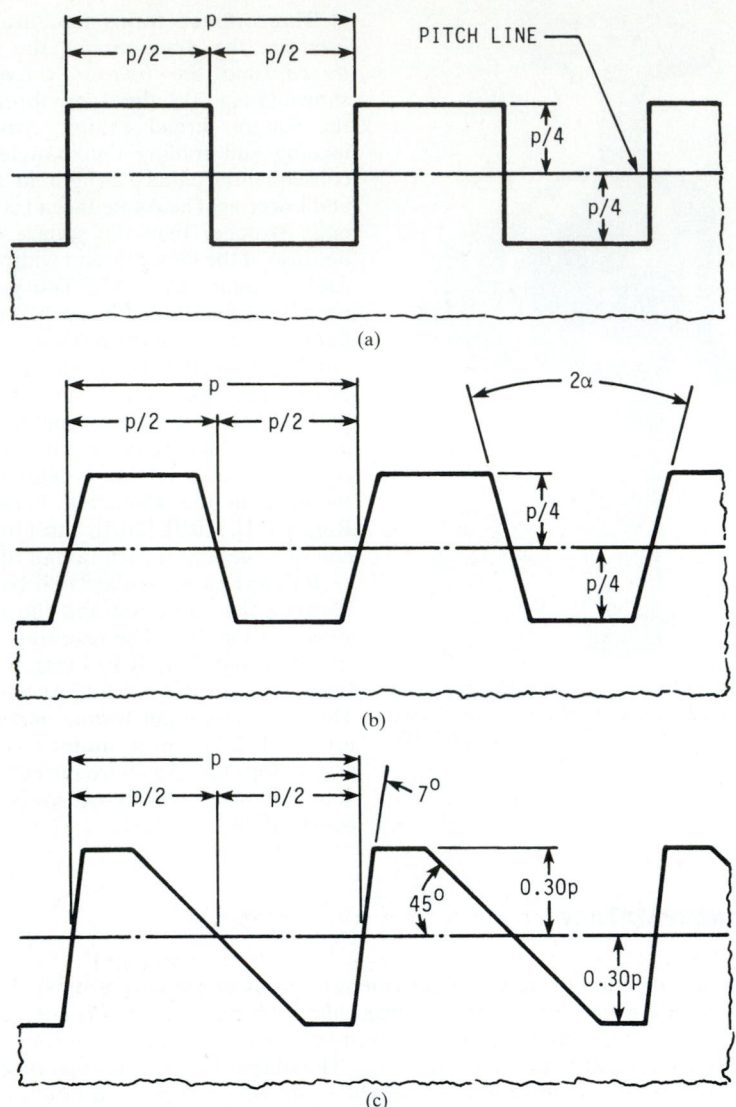

FIGURE 20.2 Basic thread forms. (*a*) Square; (*b*) general-purpose Acme; (*c*) buttress. The stub Acme thread height is 0.3*p*.

TABLE 20.1 Standard Thread Sizes for Acme Thread Form[†]

Size D, in	Threads per inch n
¼	**16**
⁵⁄₁₆	16, **14**
⅜	16, 14, **12, 10**
⁷⁄₁₆	16, 14, **12, 10**
½	16, 14, 12, **10**, 8
⅝	16, 14, 12, 10, **8**
¾	16, 14, 12, 10, 8, **6**
⅞	14, 12, 10, 8, **6**, 5
1	14, 12, 10, 8, 6, **5**
1⅛	12, 10, 8, 6, **5**, 4
1¼	12, 10, 8, 6, **5**, 4
1⅜	10, 8, 6, 5, **4**
1½	10, 8, 6, 5, **4**, 3
1¾	10, 8, 6, 4, **4**, 3, 2½
2	8, 6, 5, **4**, 3, 2½, 2
2¼	6, 5, 4, **3**, 2½, 2
2½	5, 4, **3**, 2½, 2
2¾	4, **3**, 2½, 2
3	4, 3, 2½, **2**, 1½, 1⅓
3½	4, 3, 2½, **2**, 1½, 1⅓, 1
4	4, 3, 2½, **2**, 1½, 1⅓, 1
4½	3, 2½, **2**, 1½, 1⅓, 1
5	3, 2½, **2**, 1½, 1⅓, 1

[†] The preferred size is shown in boldface.

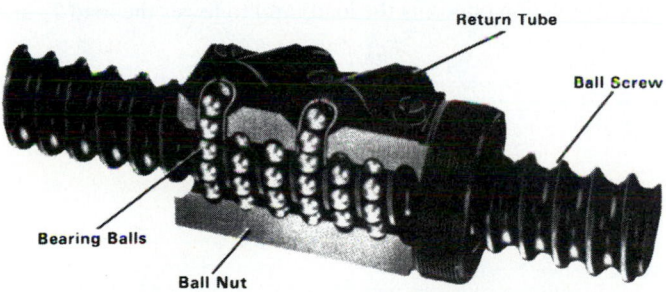

FIGURE 20.3 Ball screw assembly. *(Saginaw Steering Gear Division, General Motors Corporation.)*

The input speed may vary with respect to time t, resulting in a proportional change in output speed according to

$$V(t) = \dot{x}(t) = \frac{L}{2\pi}\,\dot{\theta}(t) \tag{20.3}$$

Similarly, the linear and angular accelerations of the load screw are related as follows:

$$A(t) = \ddot{x}(t) = \frac{L}{2\pi}\,\ddot{\theta}(t) \tag{20.4}$$

Inertia forces and *torques* are often neglected for screw systems which have small accelerations or masses. If the screw accelerates a large mass, however, or if a nominal mass is accelerated quickly, then inertia forces and torques should be analyzed. The total required input torque is obtained by *superposing* the static equilibrium torque, the torque required to accelerate the load, and the inertia torque of the screw rod itself. The inertia torque of the screw is sometimes significant for high-speed linear actuators. And lastly, impacts resulting from jerks can be analyzed using strain-energy methods or finite-element methods.

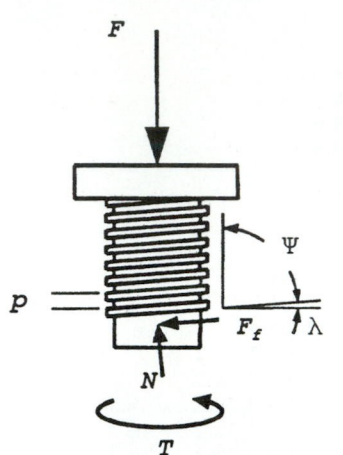

FIGURE 20.4 Free-body diagram of load screw.

20.3 MECHANICS

Under static equilibrium conditions, the screw rotates at a constant speed in response to the input torque T shown in the free-body diagram of Fig. 20.4. In addition, the load force F, normal force N, and sliding friction force F_f act on the screw. The friction force opposes relative motion. Therefore, the direction of the friction force F_f will reverse when the screw translates in the direction of the load rather than against it. The torques required to raise the load T_R (i.e., move the screw in the direction opposing the load) and to lower the load T_L are

$$T_R = \frac{Fd_m}{2} \left(\frac{\pi \mu_t d_m + L\beta}{\pi d_m \beta - \mu_t L} \right) \tag{20.5}$$

$$T_L = \frac{Fd_m}{2} \left(\frac{\pi \mu_t d_m - L\beta}{\pi d_m \beta + \mu_t L} \right) \tag{20.6}$$

where
$$d_m = d - p/2$$
$$L = p n_s$$
$$\tan \lambda = \frac{L}{\pi d_m}$$
$$\tan \alpha_n = \tan \alpha \cos \lambda$$
$$\beta = \cos \alpha_n \quad (\beta = 1 \text{ for square threads})$$

The thread geometry parameter β includes the effect of the flank angle α as it is projected normal to the thread and as a function of the lead angle. For general-purpose single-start Acme threads, α is 14.5 degrees and β is approximately 0.968, varying less than 1 percent for diameters ranging from ¼ in to 5 in and thread spacing ranging from 2 to 16 threads per inch. For square threads, $\beta = 1$.

In many applications, the load slides relative to a collar, thereby requiring an additional input torque T_c:

$$T_c = \frac{F \mu_c d_c}{2} \tag{20.7}$$

Ball and tapered-roller thrust bearings can be used to reduce the collar torque.

The *starting torque* is obtained by substituting the *static* coefficients of friction into the above equations. Since the sliding coefficient of friction is roughly 25 percent less than the static coefficient, the *running torque* is somewhat less than the starting torque. For precise values of friction coefficients, specific data should be obtained from the published technical literature and verified by experiment.

Power screws can be *self-locking* when the coefficient of friction is high or the lead is small, so that $\pi\mu_t d_m > L$ or, equivalently, $\mu_t > \tan\lambda$. When this condition is not met, the screw will *self-lower* or *overhaul* unless an opposing torque is applied.

A measure of screw efficiency η can be formulated to compare the work output W_o with the work input W_i:

$$\eta = \frac{W_o}{W_i} = \frac{F\,\Delta x}{T\,\Delta\theta} \tag{20.8}$$

where T is the total screw and collar torque. Similarly, for one revolution or 2π radians and screw translation L,

$$\eta = \frac{FL}{2\pi T} \tag{20.9}$$

Screw manufacturers often list output travel speed V, in in/min, as a function of required motor torque T in lbf·in, operating at n r/min, to lift the rated capacity F, in lbf. The actual efficiency for these data is therefore

$$\eta = \frac{FV}{2\pi n T} \tag{20.10}$$

Efficiency of a square-threaded power screw with respect to lead angle λ, as shown in Fig. 20.5, is obtained from

$$\eta = \frac{1 - \mu\tan\lambda}{1 + \mu\cot\lambda} \tag{20.11}$$

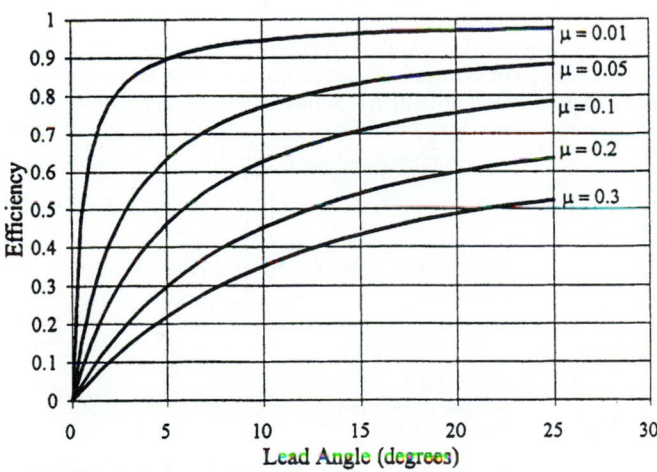

FIGURE 20.5   Screw efficiency η versus thread lead angle λ.

Note the importance of proper lubrication. For example, for $\lambda = 10$ degrees and $\mu = 0.05$, η is over 75 percent. However, as the lubricant becomes contaminated with dirt and dust or chemically breaks down over time, the friction coefficient can increase to $\mu = 0.30$, resulting in an efficiency $\eta = 35$ percent, thereby doubling the torque, horsepower, and electricity requirements.

20.4 BUCKLING AND DEFLECTION

Power screws subjected to compressive loads may buckle. The Euler formula can be used to estimate the critical load F_c at which buckling will occur for relatively long screws of column length L_c and second moment of area $I = \pi d_r^4/64$ as

$$F_c = \frac{C\pi^2 E}{L_c^2}\left(\frac{\pi d_r^4}{64}\right) \tag{20.12}$$

where C is the theoretical end-condition constant for various cases given in Table 20.2. Note that the critical buckling load F_c should be reduced by an appropriate load factor of safety as conditions warrant. See Chap. 15 for an illustration of various end conditions and effective length factor K, which is directly related to the end-condition constant by $C = 1/K^2$.

A column of length L_c and radius of gyration k is considered long when its slenderness ratio L_c/k is larger than the critical slenderness ratio:

$$\frac{L_c}{k} > \left(\frac{L_c}{k}\right)_{\text{critical}} \tag{20.13}$$

$$\frac{L_c}{k} > \left(\frac{2\pi^2 CE}{S_y}\right)^{1/2} \tag{20.14}$$

The radius of gyration k, cross-sectional area A, and second moment of area I are related by $I = Ak^2$, simplifying the above expression to

$$\frac{L_c}{d_r} > \frac{1}{4}\left(\frac{2\pi^2 CE}{S_y}\right)^{1/2} \tag{20.15}$$

For a steel screw whose yield strength is 60 000 psi and whose end-condition constant is 1.0, the critical slenderness ratio is about 100, and L_c/d_r is about 25. For steels whose slenderness ratio is less than critical, the Johnson parabolic relation can be used:

$$\frac{F_c}{A} = S_y - \frac{1}{CE}\left(\frac{S_y L_c}{2\pi k}\right)^2 \tag{20.16}$$

TABLE 20.2 Buckling End-Condition Constants

End condition	C
Fixed-free	¼
Rounded-rounded	1
Fixed-rounded	2
Fixed-fixed	4

which can be solved for a circular cross section of minor diameter d_r as

$$d_r = \sqrt{\frac{F_c}{\pi S_y} + \frac{S_y L_c^2}{\pi^2 C E}}$$ (20.17)

The load should be externally guided for long travels to prevent eccentric loading. Axial compression or extension δ can be approximated by

$$\delta = \frac{F L_c}{A E} = \frac{4 F L_c}{\pi d_r^2 E}$$ (20.18)

And similarly, angle of twist ϕ, in radians, can be approximated by

$$\phi = \frac{T L_c}{J G} = \frac{32 T L_c}{\pi d_r^4 G}$$ (20.19)

20.5 STRESSES

Using St. Venants' principle, the nominal *shear* and *normal stresses* for cross sections of the screw rod away from the immediate vicinity of the load application may be approximated by

$$\tau = \frac{T r}{J} = \frac{16 T}{\pi d_r^4}$$ (20.20)

$$\sigma_x = \frac{F}{A} = \frac{4 F}{\pi d_r^2}$$ (20.21)

Failure due to yielding can be estimated by the ratio of S_y to an equivalent, von Mises stress σ' obtained from

$$\sigma' = \sqrt{\left(\frac{4F}{\pi d_r^2}\right)^2 + 3\left(\frac{16T}{\pi d_r^3}\right)^2} = \frac{4}{\pi} \sqrt{\left(\frac{F}{d_r^2}\right)^2 + 48\left(\frac{T}{d_r^3}\right)^2}$$ (20.22)

The nominal *bearing* stress σ_b on a nut or screw depends on the number of engaged threads $N_e = h/p$ of pitch p and engaged thickness h and is obtained from

$$\sigma_b = \frac{F}{A_{\text{projected}}} = \frac{4F}{\pi (d^2 - d_r^2)} \left(\frac{p}{h}\right)$$ (20.23)

Threads may also shear or strip off the screw or nut because of the load force, which is approximately parabolically distributed over the cylindrical surface area A_{cyl}. The area depends on the width w of the thread at the root and the number of engaged threads N_e according to $A_{\text{cyl}} = \pi d w N_e$. The maximum shear stress is estimated by

$$\tau = \frac{3}{2} \frac{F}{A_{\text{cyl}}}$$ (20.24)

For square threads such that $w = p/2$, the maximum shear stress for the nut thread is

$$\tau = \frac{3F}{\pi d h}$$ (20.25)

To obtain the shear stress for the screw thread, substitute d_r for d. Since d_r is slightly less than d, the stripping shear stress for the screw is somewhat larger.

Note that the load flows from the point of load application through the thread geometry to the screw rod. Because of the nonlinear strains induced in the threads at the point of load application, each thread carries a *disproportionate* share of the load. A detailed analytical approach such as finite-element methods, backed up by experiments, is recommended for more accurate estimates of the above stresses and of other stresses, such as a thread bending stress and hoop stress induced in the nut.

20.6 BALL SCREWS

The design of ball screw assemblies is similar to that of machine screw systems. Kinematic considerations such as screw or nut travel, velocity, and acceleration can be estimated following Sec. 20.2. Similarly input torque, power, and efficiency can be approximated using formulas from Sec. 20.3. Critical buckling loads can be estimated using Eq. (20.12) or (20.16). Also, nominal shear and normal stresses of the ball screw shaft (or rod) can be estimated using Eqs. (20.20) and (20.21).

Design for strength, however, is typically completed using a catalog selection procedure rather than analytical stress-versus-strength analysis. Ball screw manufacturers usually list static and dynamic load capacities for a variety of screw shaft (rod) diameters, ball diameters, and screw leads; an example is shown in Table 20.3. The static capacity for *basic static thrust capacity* T_i, lbf, is the load which will produce a ball track deformation of 0.0001 times the ball diameter. The dynamic capacity or *basic load rating* P_i, lbf, is the constant axial load that a group of ball screw assemblies can endure for a rated life of one million inches of screw travel. The *rated life* is the length of travel that 90 percent of a group of assemblies will complete or exceed before any signs of fatigue failure appear. The catalog ratings, developed from laboratory test results, therefore involve the effects of hertzian contact stresses, manufacturing processes, and surface fatigue failure.

The catalog selection process requires choosing the appropriate combination of screw diameter, ball diameter, and lead, so that the axial load F will be sufficiently less than the basic static thrust capacity or the basic load rating for the rated axial travel life. For a different operating travel life of X inches, the modified basic load rating P_{iX}, lbf, is obtained from

$$P_{iX} = P_i \left(\frac{10^6}{X} \right)^{1/3}$$

(20.26)

An equivalent load rating P can be obtained for applications involving loads P_1, P_2, P_3, ..., P_n that occur for C_1, C_2, C_3, ..., C_n percent of the life, respectively:

$$P = \sqrt[3]{\frac{C_1 P_1^3 + C_2 P_2^3 + \cdots + C_n P_n^3}{100}}$$

(20.27)

For the custom design of a ball screw assembly, see Ref. [20.5], which provides a number of useful relations.

TABLE 20.3 Sizes and Capacities of Ball Screws[†]

Major diameter, in	Lead, in, **mm**	Ball diameter, in	Dynamic capacity, lb	Static capacity, lb
0.750	0.200	0.125	1 242	4 595
	0.250	0.125	1 242	4 495
0.875	0.200	0.125	1 336	5 234
	0.250	0.125	1 336	5 234
1.000	0.200	0.125	1 418	5 973
	0.200[†]	0.156	1 909	7 469
	0.250	0.125	1 418	5 973
	0.250	0.156	1 909	7 469
	0.250	0.187	—	—
	0.400	0.125	1 418	5 973
	0.400	0.187	—	—
1.250	0.200	0.125	1 904	9 936
	0.200[†]	0.156	2 583	12 420
	0.250	0.125	1 904	9 936
	0.250	0.156	2 583	12 420
	0.250	0.187	3 304	15 886
1.500	0.200	0.125	2 046	11 908
	0.200[†]	0.156	2 786	14 881
	0.250	0.156	2 786	14 881
	0.250	0.187	3 583	18 748
	0.500	0.156	2 786	14 881
	0.500	0.250	5 290	24 762
1.500	**5**[†]	0.125	2 046	11 908
	5	0.156	2 787	14 881
	10	0.156	2 786	14 881
	10	0.250	5 290	24 762
	10	0.312	7 050	29 324
1.750	0.200	0.125	2 179	13 879
	0.200[†]	0.156	2 968	17 341
	0.250	0.156	2 968	17 341
	0.250	0.187	3 829	20 822
	0.500	0.187	3 829	20 882
	0.500	0.250	5 664	27 917
	0.500	0.312	7 633	33 232
2.000	0.200	0.125	2 311	15 851
	0.200[†]	0.156	3 169	19 801
	0.250	0.156	3 169	19 801
	0.250	0.187	4 033	23 172
	0.400	0.250	6 043	31 850
	0.500	0.312	8 135	39 854
	5	0.125	2 311	15 851
	5[†]	0.156	3 169	19 801
	6	0.156	3 169	19 801
	6	0.187	4 033	23 172
	10	0.250	6 043	31 850
	10	0.312	8 135	39 854

TABLE 20.3   Sizes and Capacities of Ball Screws[†] *(Continued)*

Major diameter, in	Lead, in, **mm**	Ball diameter, in	Dynamic capacity, lb	Static capacity, lb
2.250	0.250	0.156	3 306	22 262
	0.250	0.187	4 266	26 684
	0.500	0.312	8 593	44 780
	0.500	0.375	10 862	53 660
2.500	0.200	0.125	2 511	19 794
	0.200	0.156	3 134	24 436
	0.250	0.187	4 410	29 671
	0.400	0.250	6 633	39 746
	0.500	0.312	9 015	49 701
	0.500	0.375	10 367	59 308
	5	0.125	2 511	19 794
	5[†]	0.156	3 134	24 436
	10	0.250	6 633	39 746
	10	0.312	9 015	49 701
3.000	0.250	0.187	4 810	35 570
	0.400	0.250	7 125	47 632
	0.500	0.375	12 560	71 685
	0.660	0.375	12 560	71 685
	10	0.250	7 125	47 632
	10	0.312	9 744	58 648
3.500	0.500	0.312	10 360	69 287
	0.500	0.375	13 377	83 514
	1.000	0.500	19 812	111 510
	1.000	0.625	26 752	139 585
4.000	0.500	0.375	14 088	95 343
	1.000	0.500	21 066	127 282

[†] These values are not recommended; consult manufacturer.
Source: 20th Century Machine Company, Sterling Heights, Mich., by permission.

20.7   *OTHER CONSIDERATIONS*

A number of other important design factors should also be considered. Principal among these is lubrication. Greases using lithium thickeners with antioxidants and EP additives are effective in providing acceptable coefficients of sliding friction and corrosion protection. For operating environments which expose the screw threads to dust, dirt, or water, a protective boot, made of a compatible material, is recommended. Maintenance procedures should ensure that the screw threads are free of contaminants and have a protective film of grease. Operation at ambient temperatures in excess of 200°F requires special lubricants and boot materials as recommended by the manufacturer.

Screw and nut threads will wear with use, especially in heavy-duty-cycle applications, increasing the backlash from the as-manufactured allowance. Use of adjustable split nuts and routine inspection of thread thickness is recommended.

Power screws employing electric motors are often supplied with integral limit switches to control extension and retraction. To prevent ejection of the screw in case of a limit switch failure, a stop nut can be added. In addition, a torque-limiting clutch can be integrated at the motor to prevent equipment damage.

REFERENCES

20.1 ANSI B1.7M-1984 (R1992), "Screw Threads, Nomenclature, Definitions, and Letter Symbols," American Society of Mechanical Engineers, New York, 1992.

20.2 ANSI B1.5-1977, "Acme Screw Threads," American Society of Mechanical Engineers, New York, 1977.

20.3 ANSI B1.8-1977, "Stub Acme Screw Threads," American Society of Mechanical Engineers, New York, 1977.

20.4 ANSI B1.9-1973 (R1979), "Buttress Screw Threads," American Society of Mechanical Engineers, New York, 1973.

20.5 ANSI B5.48-1977 (R1988), "Ball Screws," American Society of Mechanical Engineers, New York, 1977.

CHAPTER 21
THREADED FASTENERS

Joseph E. Shigley
Professor Emeritus
The University of Michigan
Ann Arbor, Michigan

This chapter is intended to cover the description, uses, materials, and sizes of threaded fasteners. The amount of data available concerning this subject is extremely large, so the intent here is to provide the information necessary for the usual machine-design task of selecting such fasteners. The data contained in this chapter have been compiled in part from the standards listed in Ref. [21.1].

21.1 SCREW THREADS

Standard screw threads consist of the *Unified inch series* and the *metric series*. Two profiles have been standardized in the metric series; these are called the M and MJ profiles. Figure 21.1 shows that both the Unified and metric M threads utilize the same profile.

The metric MJ profile has a rounded fillet at the root of the external thread and a larger minor diameter of both the internal and external threads. This profile is used for applications requiring a high fatigue strength and is also employed in aerospace applications.

The Unified-series profile, shown in Fig. 21.1, is designated as UN. Another unified profile, designated as UNR, has a rounded root on the external thread.

Unified thread standards are based on the nominal size (major diameter) and the number of threads per inch. The three standards *coarse* (UNC), *fine* (UNF), and *extra fine* (UNEF) are listed in Table 21.1 and are called the *standard series*. Typical specifications would be written

$$\tfrac{1}{4}\text{–}20 \text{ UNC} \qquad \text{or} \qquad \tfrac{1}{4}\text{–}20 \text{ UNRC}$$

Both these designations specify a nominal size of ¼ in and 20 threads per inch.

A *constant-pitch* unified series consisting of 4, 6, 8, 12, 16, 20, 28, and 32 threads per inch has also been standardized. These are used mostly for sizes over 1 in, and 8 UN, 12 UN, and 16 UN are the preferred pitches.

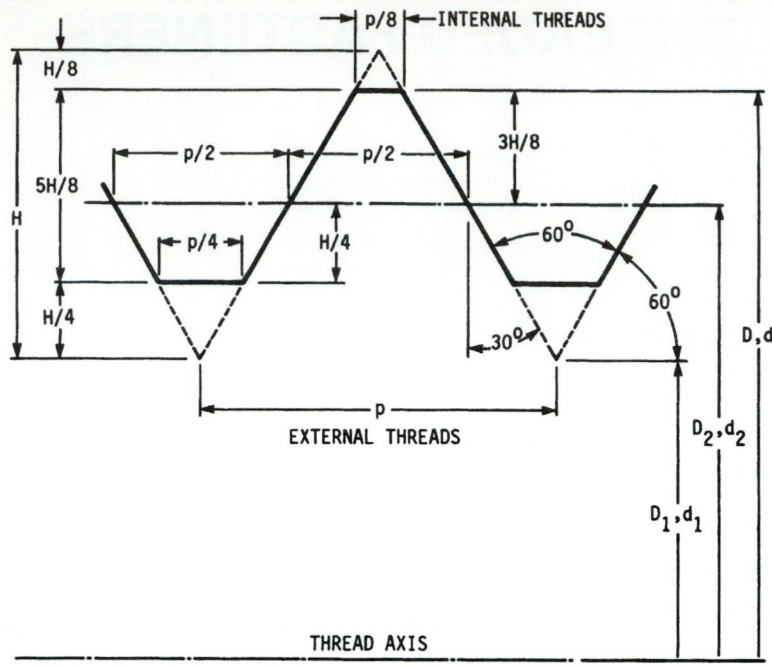

FIGURE 21.1 Basic thread profile for unified (UN) and metric (M) threads (ISO 68). $D(d)$ = basic major diameter of internal (external) thread; $D_1(d_1)$ = basic minor diameter of internal (external) thread; $D_2(d_2)$ = basic pitch diameter of internal (external) thread; p = pitch; $H = 0.5\sqrt{3}p$.

As shown in Table 21.2, the metric series consists of a coarse thread and, often, several fine threads. These are specified by giving the size or major diameter and the pitch (see Fig. 21.1). Typical specifications would be written

$$\text{M } 70 \times 1.5 \qquad \text{or} \qquad \text{MJ } 70 \times 1.5$$

which specifies a major diameter of 70 mm and a pitch of 1.5 mm.

Unified threads may be further designated as UN A for external threads and UN B for internal threads. The tolerance classes are 1A, 2A, and 3A for external threads and 1B, 2B, and 3B for internal threads. Class 2 is for general use, class 3 is a tight fit used where great accuracy is required, and class 1 is a loose fit which permits very easy assembly and allows the possibility of nicks on the threads.

Metric threads utilize the international tolerance grades (see Chap. 19).

21.1.1 Choosing the Pitch

The Unified coarse-thread series (UNC or UNRC) and the metric coarse-thread series (M or MJ) provide the most resistance to internal thread stripping. Consequently, coarse threads should be used for materials such as brass, cast iron, aluminum, and other lower-strength materials. However, the coarse-thread series are

TABLE 21.1 Standard Series of UN and UNR Screw Threads[†]

Nominal size	Basic major diameter	Threads per inch		
		Coarse, UNC	Fine, UNF	Extra-fine, UNEF
0	0.0600		80	
1	0.0730	64	72	
2	0.0860	56	64	
3	0.0990	48	56	
4	0.1120	40	48	
5	0.1250	40	44	
6	0.1380	32	40	
8	0.1640	32	36	
10	0.1900	24	32	
12	0.2160	24	28	32
$\frac{1}{4}$	0.2500	20	28	32
$\frac{5}{16}$	0.3125	18	24	32
$\frac{3}{8}$	0.3750	16	24	32
$\frac{7}{16}$	0.4375	14	20	28
$\frac{1}{2}$	0.500	13	20	28
$\frac{9}{16}$	0.5625	12	18	24
$\frac{5}{8}$	0.6250	11	18	24
$\frac{3}{4}$	0.7500	10	16	20
$\frac{7}{8}$	0.8750	9	14	20
1	1.0000	8	12	20
$1\frac{1}{8}$	1.1250	7	12	18
$1\frac{1}{4}$	1.2500	7	12	18
$1\frac{3}{8}$	1.3750	6	12	18
$1\frac{1}{2}$	1.5000	6	12	18

[†]All dimensions in inches.

TABLE 21.2 Standard Diameter-Pitch Combinations for Metric M Screw Threads[†]

Basic major diameter			Pitch	
Preferred	First option	Second option	Coarse	Fine
1.6	. . .	. . .	0.35	
2	. . .	. . .	0.4	
2.5	. . .	. . .	0.45	
3	. . .	. . .	0.5	
	3.5	. . .	0.6	
4	. . .	. . .	0.7	
5	. . .	. . .	0.8	
6	. . .	. . .	1	

TABLE 21.2   Standard Diameter-Pitch Combinations for Metric M Screw Threads[†]
(*Continued*)

Basic major diameter			Pitch	
Preferred	First option	Second option	Coarse	Fine
8	...	...	1.25	1
10	...	...	1.5	1.25 or 0.75
12	...	...	1.75	1.25 or 1
	14	...	2	1.5 or 1.25‡
	15	...	...	1
16	...	...	2	1.5
	...	17	...	1
	18	...	...	1.5
20	...	...	2.5	1.5 or 1
	22	...	2.5§	1.5
24	...	...	3	2
	...	25	...	1.5
	27	...	3¶	2
30	...	...	3.5	2 or 0.5
	33	...	...	2
	...	35¶	...	1.5
36	...	...	4	2
	39	...	...	2
	...	40	...	1.5
42	...	...	4.5	2
	45	...	...	1.5
48	...	...	5	2
	...	50	...	1.5
	...	55	...	1.5
56	...	...	5.5	2
	60	...	...	1.5
64	...	...	6	2
	...	65	...	1.5
	...	70	...	1.5
72	...	...	6	2
	...	75	...	1.5
80	...	...	6	1.5
	85	...	...	2
90	...	...	6	2
	95	...	...	2
100	...	...	6	2
	105	...	...	2
110	...	...	...	2
	120	...	...	2
	130	...	...	2
140	...	...	...	2
	150	...	...	2
160	...	...	...	3
	170	...	...	3
180	...	...	...	3
	190	...	...	3
200	...	...	...	3

[†]All dimensions in millimeters.
[‡]Only for engine spark plugs.
[§]Only for high-strength structural steel bolts.
[¶]Only for nuts for bearings.

also widely used with other materials because mass-produced fasteners are usually made with coarse threads and hence are the most economical. The coarse-thread series should also be used whenever fast assembly is needed or when dropping or handling the fasteners may damage the threads by causing nicks or dents.

The Unified fine-thread series (UNF or UNRF) and metric fine-thread series (M and MJ) find their greatest use where a high fastener strength is required and where vibration may be a problem. The shallow depth of thread, and hence larger minor diameter, increases the strength of the external member. It also permits a smaller wall thickness for the internal member.

Extra-fine-series screw threads are useful for thin nuts, on thin-wall tubing, and where parts may require a very fine adjustment.

21.1.2 Pipe Threads

The profile of pipe threads is similar to the UN profile except that there is a taper of 1 on 16 based on the outside diameter. The last few threads will be imperfect because of the taper and the chamfer on the thread-cutting die. Table 21.3 gives the basic dimensions of Unified-inch-series standard pipe threads.

21.2 BOLTS

The symbols used to indicate the dimensions of square and hex bolt heads are shown in Fig. 21.2. See Table 21.4 for head dimensions. The washer or bearing face shown in Fig. 21.2b is standard for the *heavy structural hex bolt* (Table 21.4) and for the *finished hex bolt*. A finished hex bolt is identical to a hex cap screw (see Table 21.13). The basic thread length for bolts is

$$L_T = \begin{cases} 2D + 0.25 & L \le 6 \\ 2D + 0.50 & L > 6 \end{cases} \tag{21.1}$$

TABLE 21.3 Basic Dimensions of Standard Pipe Threads[†]

Nominal pipe size	Outside diameter	Threads per inch	Thread length on OD (approx.)
$\frac{1}{16}$	0.3125	27	0.261
$\frac{1}{8}$	0.405	27	0.264
$\frac{1}{4}$	0.540	18	0.402
$\frac{3}{8}$	0.675	18	0.408
$\frac{1}{2}$	0.840	14	0.534
$\frac{3}{4}$	1.050	14	0.546
1	1.315	$11\frac{1}{2}$	0.683
$1\frac{1}{4}$	1.660	$11\frac{1}{2}$	0.707
$1\frac{1}{2}$	1.900	$11\frac{1}{2}$	0.723
2	2.375	$11\frac{1}{2}$	0.757
$2\frac{1}{2}$	2.875	8	1.138
Over $2\frac{1}{2}$	. . . . .	8	

†All dimensions in inches.

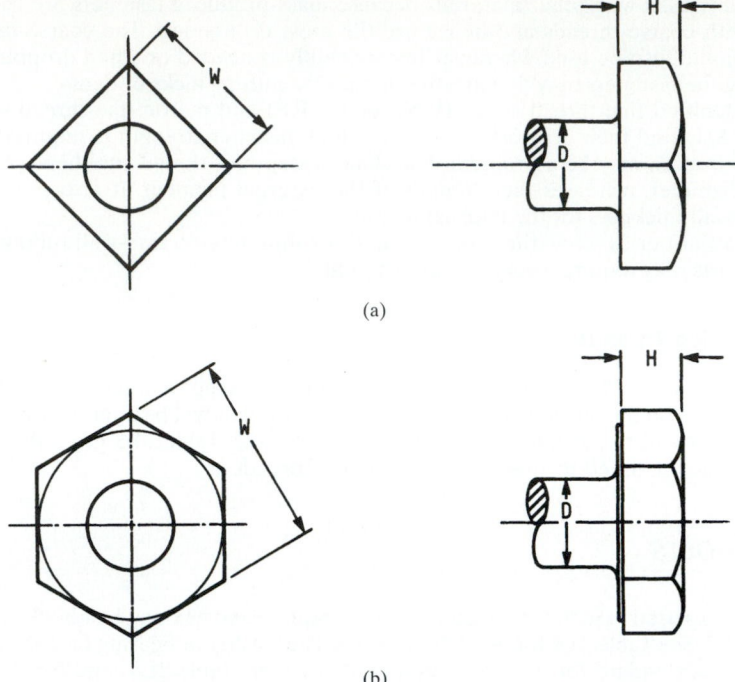

FIGURE 21.2   Bolt heads. (*a*) Square; (*b*) hex; the washer or bearing face is used only on heavy hex structural bolts.

where L = bolt length, measured under the head, and L_T = thread length, in inches.

Head dimensions for metric hex bolts are listed in Table 21.5. The heavy hex structural bolt is the only one of these with a bearing face. The thread length is

$$L_T = \begin{cases} 2D + 6 & L \leq 125 \quad D \leq 48 \\ 2D + 12 & 125 \leq L \leq 200 \\ 2D + 25 & L > 200 \end{cases} \tag{21.2}$$

Here L and L_T are in millimeters.

Standards for bolt materials and the corresponding head markings are listed in Tables 21.6, 21.7, and 21.8. The property class number in Table 21.8 is a code derived from the tensile strength S_{ut} and the yield strength S_y. If we designate the class number by the symbol $X.Y$, then $X = S_{ut}/100$ and $Y = S_y/S_{ut}$. Bolts in metric sizes are normally manufactured to SAE and ASTM specifications too. For fillet dimensions, see Tables 21.9 and 21.10.

Typical heads for *round-head* or *carriage bolts* are shown in Fig. 21.3, and head dimensions are given in Tables 21.11 and 21.12. Other standard bolts are step bolts, which have a square neck with a larger-diameter head, and several countersunk-head bolts with and without square necks. The bolts listed in Table 21.12 are the only round-head metric bolts that are standardized at this writing. Round-head bolts are made to the same material specifications as hex bolts and use the same head markings.

TABLE 21.4 Dimensions of Square- and Hex-Head Bolts (Inch Series)

Nominal size	Head type							
	Square		Regular hex†		Heavy hex‡		Structural hex	
	W	H	W	H	W	H	W	H
$\frac{1}{4}$	$\frac{3}{8}$	$\frac{11}{64}$	$\frac{7}{16}$	$\frac{11}{64}$				
$\frac{5}{16}$	$\frac{1}{2}$	$\frac{13}{64}$	$\frac{1}{2}$	$\frac{7}{32}$				
$\frac{3}{8}$	$\frac{9}{16}$	$\frac{1}{4}$	$\frac{9}{16}$	$\frac{1}{4}$				
$\frac{7}{16}$	$\frac{5}{8}$	$\frac{19}{64}$	$\frac{5}{8}$	$\frac{19}{64}$				
$\frac{1}{2}$	$\frac{3}{4}$	$\frac{21}{64}$	$\frac{3}{4}$	$\frac{11}{32}$	$\frac{7}{8}$	$\frac{7}{32}$	$\frac{7}{8}$	$\frac{5}{16}$
$\frac{5}{8}$	$\frac{15}{16}$	$\frac{27}{64}$	$\frac{15}{16}$	$\frac{27}{64}$	$1\frac{1}{16}$	$\frac{27}{64}$	$1\frac{1}{16}$	$\frac{25}{64}$
$\frac{3}{4}$	$1\frac{1}{8}$	$\frac{1}{2}$	$1\frac{1}{8}$	$\frac{1}{2}$	$1\frac{1}{4}$	$\frac{1}{2}$	$1\frac{1}{4}$	$\frac{15}{32}$
$\frac{7}{8}$	$1\frac{5}{16}$	$\frac{19}{32}$	$1\frac{5}{16}$	$\frac{37}{64}$	$1\frac{7}{16}$	$\frac{37}{64}$	$1\frac{7}{16}$	$\frac{35}{64}$
1	$1\frac{1}{2}$	$\frac{21}{32}$	$1\frac{1}{2}$	$\frac{43}{64}$	$1\frac{5}{8}$	$\frac{43}{64}$	$1\frac{5}{8}$	$\frac{39}{64}$
$1\frac{1}{8}$	$1\frac{11}{16}$	$\frac{3}{4}$	$1\frac{11}{16}$	$\frac{3}{4}$	$1\frac{13}{16}$	$\frac{3}{4}$	$1\frac{13}{16}$	$\frac{11}{16}$
$1\frac{1}{4}$	$1\frac{7}{8}$	$\frac{27}{32}$	$1\frac{7}{8}$	$\frac{27}{32}$	2	$\frac{27}{32}$	2	$\frac{25}{32}$
$1\frac{3}{8}$	$2\frac{1}{16}$	$\frac{29}{32}$	$2\frac{1}{16}$	$\frac{29}{32}$	$2\frac{1}{16}$	$\frac{29}{32}$	$2\frac{1}{16}$	$\frac{27}{32}$
$1\frac{1}{2}$	$2\frac{1}{4}$	1	$2\frac{1}{4}$	1	$2\frac{3}{8}$	1	$2\frac{3}{8}$	$\frac{13}{16}$

†Also available in standard sizes up to 4 in.
‡Also available in standard sizes up to 3 in.

TABLE 21.5 Dimensions of Metric Hex Bolts (Metric Series)†

Nominal diameter	Thread pitch	Type of bolt					
		Regular‡		Heavy		Structural	
		W	H	W	H	W	H
M5	0.8	8	3.58				
M6	1	10	4.38				
M8	1.25	13	5.68				
M10	1.5	16	6.85				
M12	1.75	18	7.95	21	7.95		
M14	2	21	9.25	24	9.25		
M16	2	24	10.75	27	10.75	27	10.75
M20	2.5	30	13.40	34	13.40	34	13.40
M22	2.5	. .	. . . .	. .	. . . .	36	14.90
M24	3	36	15.90	41	15.90	41	15.90
M27	3	. .	. . . .	. .	. . . .	46	17.90
M30	3.5	46	19.75	50	19.75	50	19.75
M36	4	55	23.55	60	23.55	60	23.55

†Head dimensions are maximum. All dimensions in millimeters.
‡Also available in standard sizes to 100 mm.

TABLE 21.6 SAE Grade Markings for Steel Bolts

SAE grade no.	Size range incl.	Proof strength,[†] kpsi	Tensile strength,[†] kpsi	Material	Head marking
1	$\frac{1}{4}$–$1\frac{1}{2}$			Low- or medium-carbon steel	
2	$\frac{1}{4}$–$\frac{3}{4}$	55	74		
	$\frac{7}{8}$–$1\frac{1}{2}$	33	60		
5	$\frac{1}{4}$–1	85	120	Medium-carbon steel, Q & T	
	$1\frac{1}{8}$–$1\frac{1}{2}$	74	105		
5.2	$\frac{1}{4}$–1	85	120	Low-carbon martensite steel, Q & T	
7	$\frac{1}{4}$–$1\frac{1}{2}$	105	133	Medium-carbon alloy steel, Q & T‡	
8	$\frac{1}{4}$–$1\frac{1}{2}$	120	150	Medium-carbon alloy steel, Q & T	
8.2	$\frac{1}{4}$–1	120	150	Low-carbon martensite steel, Q & T	

†Minimum values.
‡Roll threaded after heat treatment.
SOURCES: See "Helpful Hints," by Russell, Burdsall & Ward Corp., Mentor, Ohio 44060; and Chap. 23.

TABLE 21.7 ASTM Grade Markings for Steel Bolts

ASTM designation	Size range incl.	Proof strength,† kpsi	Tensile strength,† kpsi	Material	Head marking
A307	$\frac{1}{4}$ to 4			Low-carbon steel	
A325 type 1	$\frac{1}{2}$ to 1	85	120	Medium-carbon steel, Q & T	
	$1\frac{1}{8}$ to $1\frac{1}{2}$	74	105		A325
A325 type 2	$\frac{1}{2}$ to 1	85	120	Low-carbon martensite steel, Q & T	
	$1\frac{1}{8}$ to $1\frac{1}{2}$	74	105		A325
A325 type 3	$\frac{1}{2}$ to 1	85	120	Weathering steel, Q & T	
	$1\frac{1}{8}$ to $1\frac{1}{2}$	74	105		A325
A354 grade BC				Alloy steel, Q & T	BC
A354 grade BD	$\frac{1}{4}$ to 4	120	150	Alloy steel, Q & T	
A449	$\frac{1}{4}$ to 1	85	120	Medium-carbon steel, Q & T	
	$1\frac{1}{8}$ to $1\frac{1}{2}$	74	105		
	$1\frac{3}{4}$ to 3	55	90		
A490 type 1	$\frac{1}{2}$ to $1\frac{1}{2}$	120	150	Alloy steel, Q & T	A490
A490 type 3				Weathering steel, Q & T	A490

†Minimum values.
SOURCES: See "Helpful Hints," by Russell, Burdsall & Ward Corp., Mentor, Ohio 44060; and Chap. 23.

TABLE 21.8 Metric Mechanical-Property Classes for Steel Bolts, Screws, and Studs

Property class	Size range incl.	Proof strength, MPa	Tensile strength, MPa	Material	Head marking
4.6	M5–M36	225	400	Low- or medium-carbon steel	4.6
4.8	M1.6–M16	310	420	Low- or medium-carbon steel	4.8
5.8	M5–M24	380	520	Low- or medium-carbon steel	5.8
8.8	M16–M36	600	830	Medium-carbon steel, Q & T	8.8
9.8	M1.6–M16	650	900	Medium-carbon steel, Q & T	9.8
10.9	M5–M36	830	1040	Low-carbon martensite steel, Q & T	10.9
12.9	M1.6–M36	970	1220	Alloy steel, Q & T	12.9

SOURCES: "Helpful Hints," by Russell, Burdsall & Ward Corp., Mentor, Ohio 44060; see also Chap. 23 and SAE standard J1199, and ASTM standard F568.

TABLE 21.9 Under-the-Head Fillet Radii for Hex Bolts (Inch Series)

	Regular and heavy		Heavy structural	
Size	Maximum	Minimum	Maximum	Minimum
$\frac{1}{4}-\frac{1}{2}$	0.03	0.01	0.031	0.009
$\frac{5}{8}-\frac{7}{8}$	0.06	0.02	0.062	0.021
$1-1\frac{1}{2}$	0.09	0.03	0.093	0.062

TABLE 21.10 Under-the-Head Fillet Radii for Hex Bolts (Metric Series)[†]

Size	Regular and heavy (min.)	Heavy structural (min.)
M5	0.2	
M6	0.3	
M8–M10	0.4	
M12–M16	0.6	0.6
M20–M22	0.8	0.8
M24	0.8	1.0
M27	. .	1.2
M30	1.0	1.2
M36	1.0	1.5

†All dimensions in millimeters.

21.3 SCREWS

21.3.1 Hexagon Head

Hex screw heads resemble those shown in Fig. 21.2b, and they all have a washer or bearing face. Basic dimensions of the *cap screw* and the *heavy screw* in the inch series are given in Table 21.13. Three metric series are standardized. These are the *cap screw*, the *formed screw*, which has an indentation in the head, and the *heavy screw*; see Table 21.14 for basic dimensions of these. Hex screws are made to the same material specifications as bolts and utilize the same head markings (see Tables 21.6, 21.7, and 21.8). Use Eq. (21.1) or (21.2) to determine the basic length of thread.

21.3.2 Sockets and Keys

Figure 21.4 illustrates the standard hex and spline socket, and the products in Fig. 21.5 illustrate the variety. Socket screws are driven with a socket key, as in Fig. 21.4c, or with a length of hex or spline stock, called a *bit*. The bit is used for driving by inserting it into a standard socket wrench or power driver. Dimensions of standard keys are given in Tables 21.15, 21.16, and 21.17. Metric splines have not as yet been standardized.

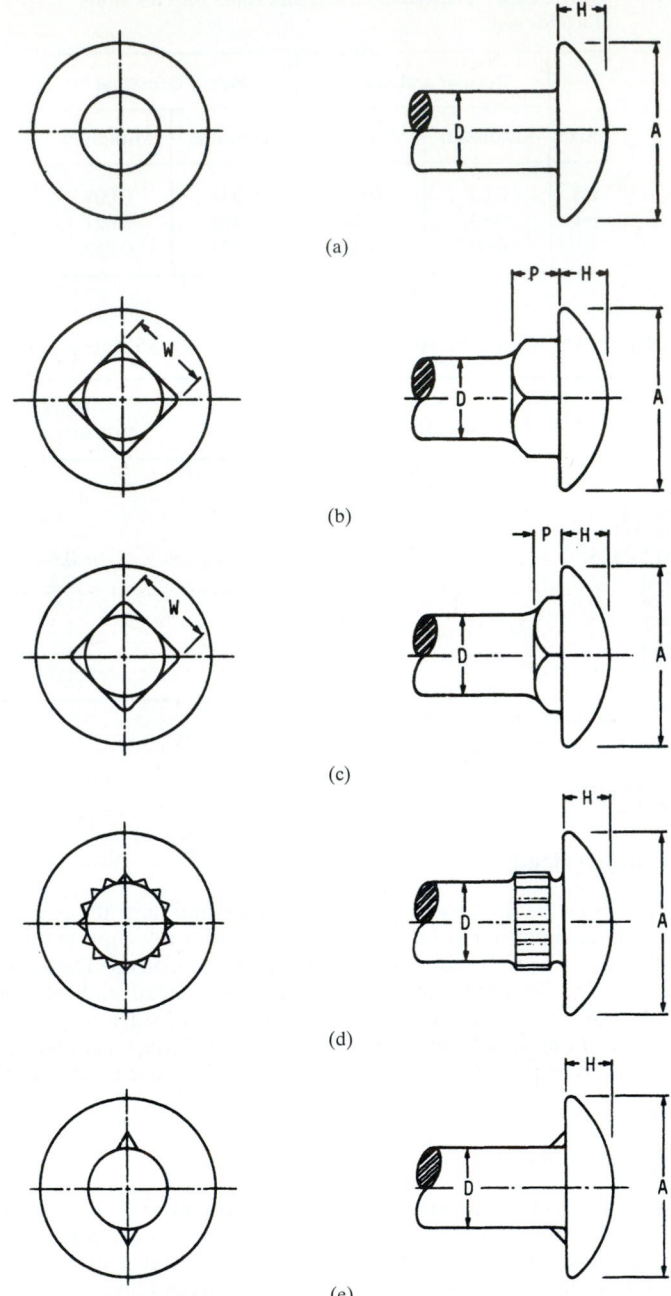

FIGURE 21.3 Some types of round-head bolts. (*a*) Plain; (*b*) regular square-neck; (*c*) short square-neck; (*d*) rib-neck; (*e*) fin-neck.

TABLE 21.11   Some Basic Dimensions of Round-Head Bolts (Inch Series)[†]

Nominal size	Max. head diameter A	Max. head height H	Max. square width W[‡]
No. 10	0.469	0.114	0.199
$\frac{1}{4}$	0.594	0.145	0.260
$\frac{5}{16}$	0.719	0.176	0.324
$\frac{3}{8}$	0.844	0.208	0.388
$\frac{7}{16}$	0.969	0.239	0.452
$\frac{1}{2}$	1.094	0.270	0.515
$\frac{5}{8}$	1.344	0.344	0.642
$\frac{3}{4}$	1.594	0.406	0.768
$\frac{7}{8}$	1.844	0.469	0.895
1	2.094	0.531	1.022

†Short square-neck and rib-neck bolts are standardized only to $\frac{3}{4}$ in; fin-neck bolts are standard only to $\frac{1}{2}$ in.
‡Not applicable to plain, rib-neck, or fin-neck bolts.

TABLE 21.12   Some Basic Dimensions of Round-Head Short Square-Neck Metric Bolts

Nominal size	Thread pitch	Max. head diameter A	Max. head height H	Max. square width W
M6	1	14.2	3.6	6.48
M8	1.25	18.0	4.8	8.58
M10	1.5	22.3	5.8	10.58
M12	1.75	26.6	6.8	12.70
M14	2	30.5	7.9	14.70
M16	2	35.0	8.9	16.70
M20	2.5	43.0	10.9	20.84

21.3.3 Socket-Head Cap Screws

Figure 21.6a illustrates a *socket-head cap screw,* and Fig. 21.6c shows a *flat counter-sunk-head cap screw. Socket button-head cap screws* resemble Fig. 21.8a, but have a hex or spline driving socket instead of the slot. Head dimensions for these, in inch and metric sizes, are given in Tables 21.18 to 21.23 inclusive.

Thread-length formulas are

$$L_T = 2D + 0.50 \text{ in} \qquad L_T = 2D + 12 \text{ mm} \qquad (21.3)$$

Shorter cap screws are threaded full length.

Alloy-steel cap screws, in both inch and metric sizes, should contain an alloying element, such as chromium, nickel, molybdenum, or vanadium, in such quantities as to ensure that a hardness range of 36 to 45 R_C is achieved.

TABLE 21.13 Basic Dimensions of Hex Cap Screws (Finished Hex Bolts) and Heavy Hex Screws (Inch Series)

Nominal size	Fillet radii		Fastener type			
			Hex cap screw		Heavy hex screw	
	Maximum	Minimum	W	H	W	H
$\frac{1}{4}$	0.025	0.015	$\frac{7}{16}$	$\frac{5}{32}$		
$\frac{5}{16}$	0.025	0.015	$\frac{1}{2}$	$\frac{13}{64}$		
$\frac{3}{8}$	0.025	0.015	$\frac{9}{16}$	$\frac{15}{64}$		
$\frac{7}{16}$	0.025	0.015	$\frac{5}{8}$	$\frac{9}{32}$		
$\frac{1}{2}$	0.025	0.015	$\frac{3}{4}$	$\frac{5}{16}$	$\frac{7}{8}$	$\frac{5}{16}$
$\frac{9}{16}$	0.045	0.020	$\frac{13}{16}$	$\frac{23}{64}$		
$\frac{5}{8}$	0.045	0.020	$\frac{15}{16}$	$\frac{25}{64}$	$1\frac{1}{16}$	$\frac{25}{64}$
$\frac{3}{4}$	0.045	0.020	$1\frac{1}{8}$	$\frac{15}{32}$	$1\frac{1}{4}$	$\frac{15}{32}$
$\frac{7}{8}$	0.065	0.040	$1\frac{5}{16}$	$\frac{35}{64}$	$1\frac{7}{16}$	$\frac{35}{64}$
1	0.095	0.060	$1\frac{1}{2}$	$\frac{39}{64}$	$1\frac{5}{8}$	$\frac{39}{64}$
$1\frac{1}{8}$	0.095	0.060	$1\frac{11}{16}$	$\frac{11}{16}$	$1\frac{13}{16}$	$\frac{11}{16}$
$1\frac{1}{4}$	0.095	0.060	$1\frac{7}{8}$	$\frac{25}{32}$	2	$\frac{25}{32}$
$1\frac{3}{8}$	0.095	0.060	$2\frac{1}{16}$	$\frac{27}{32}$	$2\frac{3}{16}$	$\frac{27}{32}$
$1\frac{1}{2}$	0.095	0.060	$2\frac{1}{4}$	$1\frac{15}{16}$	$2\frac{3}{8}$	$\frac{15}{16}$

TABLE 21.14 Basic Dimensions of Metric Hex Screws[†]

Nominal diameter	Thread pitch	Type of screw			Height H	Fillet radius§
		Cap‡ W	Formed‡ W	Heavy‡ W		
M5	0.8	8	8	..	3.65	0.2
M6	1	10	10	..	4.15	0.3
M8	1.25	13	13	..	5.50	0.4
M10	1.5	16	16	..	6.63	0.4
M12	1.75	18	18	21	7.76	0.6
M14	2	21	21	24	9.09	0.6
M16	2	24	24	27	10.32	0.6
M20	2.5	30	30	34	12.88	0.8
M24	3	36	36	41	15.44	0.8
M30	3.5	46	..	50	19.48	1.0
M36	4	55	..	60	23.38	1.0

†All dimensions in millimeters.
‡Maximum.
§Minimum.

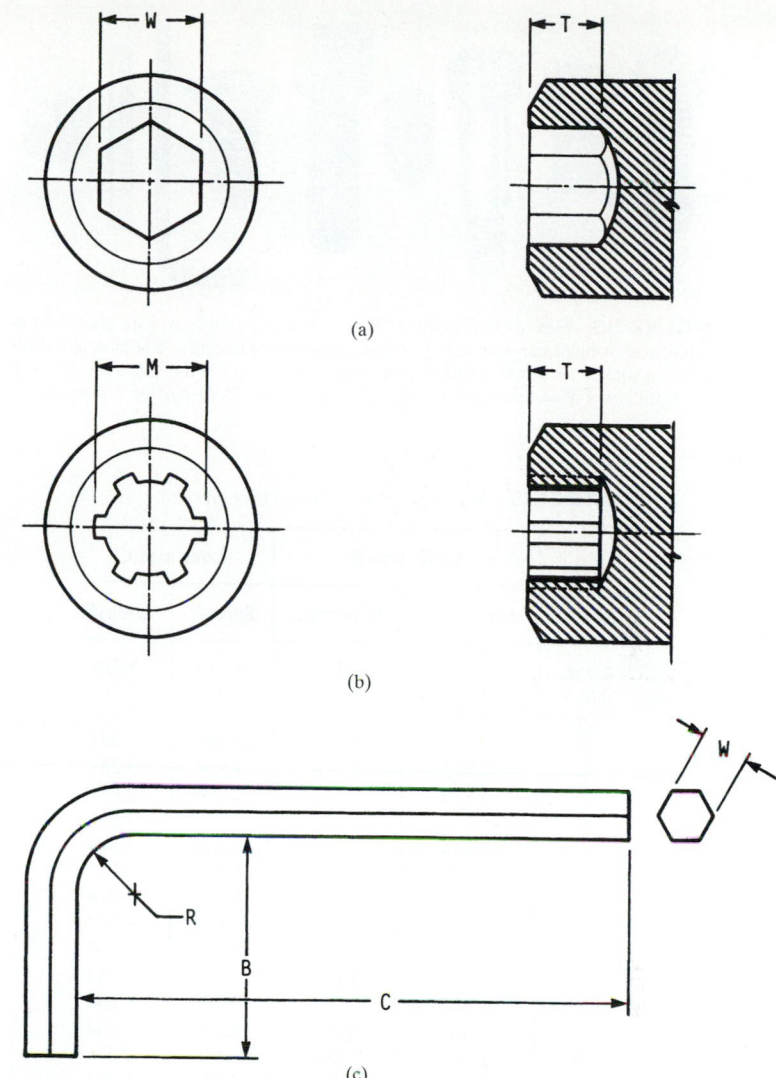

FIGURE 21.4 Standard socket shapes. (*a*) Forged hex socket; (*b*) forged spline; (*c*) hex-socket key.

21.3.4 Shoulder Screws

The nominal size D_S shown for the shoulder screw in Fig. 21.6*b* is related to the maximum and minimum shoulder diameters by the relation

$$D_S(\max) = D_S - 0.002 \qquad D_S(\min) = D_s - 0.004 \qquad (21.4)$$

Sizes, in the inch series, are tabulated in Table 21.24.

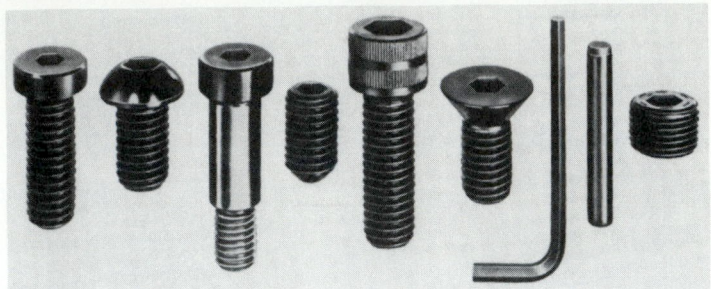

FIGURE 21.5 Hex-socket fasteners. From left to right, the parts are identified as a low socket-head cap screw, a button-head socket cap screw, a socket shoulder screw, a socket set screw, a socket-head cap screw, a socket flat-head cap screw, a hexagon key, a dowel pin, and a socket pressure plug. *(Holo-Krome Company.)*

TABLE 21.15 Hex-Socket Key Sizes (Inch Series)

Nominal width W	Short arm B		Long arm C	
	Maximum	Minimum	Shorts†	Longs†
0.028	0.312	0.125	1.312	2.688
0.035	0.438	0.250	1.312	2.766
0.050	0.625	0.438	1.750	2.938
$\frac{1}{16}$	0.656	0.469	1.844	3.094
$\frac{5}{64}$	0.703	0.516	1.969	3.281
$\frac{3}{32}$	0.750	0.562	2.094	3.469
$\frac{7}{64}$	0.797	0.609	2.219	3.656
$\frac{1}{8}$	0.844	0.656	2.344	3.844
$\frac{9}{64}$	0.891	0.703	2.469	4.031
$\frac{5}{32}$	0.938	0.750	2.954	4.219
$\frac{3}{16}$	1.031	0.844	2.844	4.594
$\frac{7}{32}$	1.125	0.938	3.094	4.969
$\frac{1}{4}$	1.219	1.031	3.344	5.344
$\frac{5}{16}$	1.344	1.156	3.844	6.094
$\frac{3}{8}$	1.469	1.281	4.344	6.844
$\frac{7}{16}$	1.594	1.406	4.844	7.594
$\frac{1}{2}$	1.719	1.531	5.344	8.344
$\frac{9}{16}$	1.844	1.656	5.844	9.094
$\frac{5}{8}$	1.969	1.781	6.344	9.844
$\frac{3}{4}$	2.219	2.031	7.344	11.344
$\frac{7}{8}$	2.469	2.281	8.344	12.844
1	2.719	2.531	9.344	14.344

†Maximum.

TABLE 21.16 Spline-Socket Key Sizes (Inch Series)

Nominal size M	Short arm B		Long arm C	
	Maximum	Minimum	Shorts†	Longs†
0.033‡	0.312	0.125	1.312	
0.048	0.438	0.250	1.312	
0.060	0.625	0.438	1.750	
0.072	0.656	0.469	1.844	
0.096	0.703	0.516	1.969	
0.111	0.750	0.562	2.094	
0.133	0.797	0.609	2.219	3.656
0.145	0.844	0.656	2.344	3.844
0.168	0.891	0.703	2.469	4.031
0.183	0.938	0.750	2.594	4.219
0.216	1.031	0.844	2.844	4.594
0.251	1.125	0.938	3.094	4.969
0.291	1.219	1.031	3.344	5.344
0.372	1.344	1.156	3.844	6.094
0.454	1.469	1.281	4.344	6.844
0.595	1.719	1.531	5.344	8.344
0.620	1.844	1.656	5.844	9.094
0.698	1.844	1.656	5.844	
0.790	1.969	1.781	6.344	

†Maximum.
‡This size has only four splines.

TABLE 21.17 Basic Maximum Dimensions of Metric Hex Keys†

Nominal size W	Short arm B	Long arm C	
		Shorts	Longs
0.7	5.5	34	62
0.9	9	34	62
1.3	13.5	44	84
1.5	14	45	90
2	16	50	100
2.5	18	56	112
3	20	63	126
4	25	70	142
5	28	80	160
6	32	90	180
8	36	100	200
10	40	112	224
12	45	125	250
14	56	140	280
17	63	160	320
19	70	180	360
22	80	200	400
24	90	224	448
27	100	250	500

†All dimensions in millimeters.

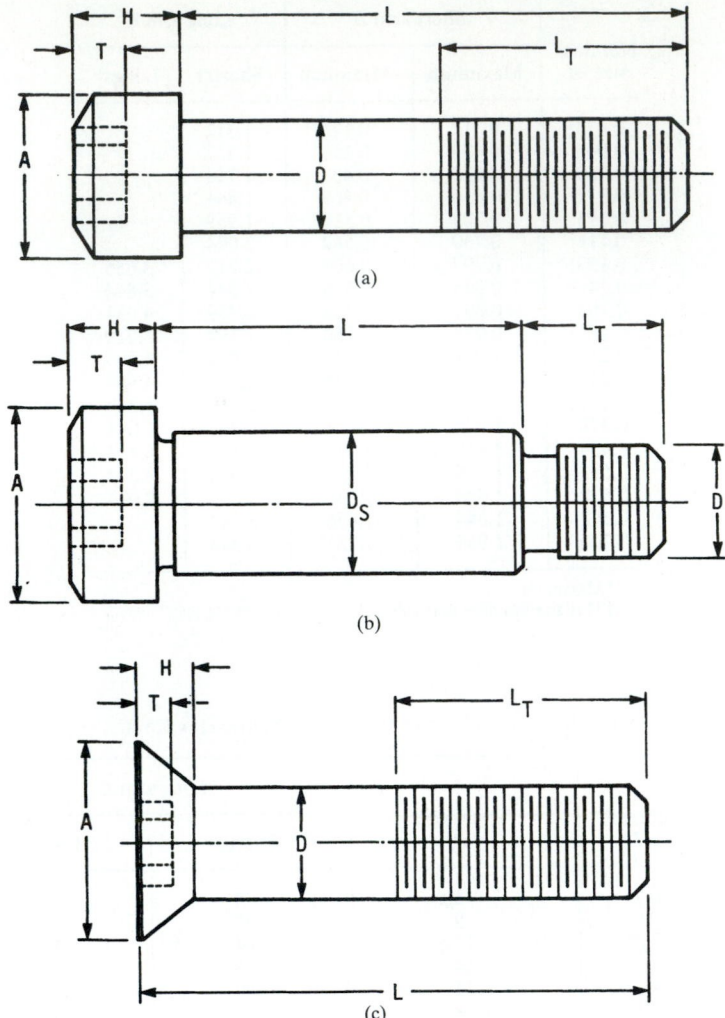

FIGURE 21.6   Socket screws. (*a*) Cap screw; (*b*) shoulder screw; (*c*) flat-head screw.

The maximum and minimum shoulder diameters for metric sizes are

$$D_S(\max) = D_S - \begin{cases} 0.013 & D_S \le 10 \\ 0.016 & 10 < D_S \le 20 \\ 0.020 & D_S > 20 \end{cases}$$

$$D_S(\min) = D_S - \begin{cases} 0.049 & D_S \le 10 \\ 0.059 & 10 < D_S \le 20 \\ 0.072 & D_S > 20 \end{cases} \tag{21.5}$$

THREADED FASTENERS**

TABLE 21.18 Basic Dimensions of Socket-Head Cap Screws (Inch Series)

Nominal size D	Max. head diameter A	Max. head height H	Hex size W	Spline size M	Socket depth† T
0	0.096	0.060	0.050	0.060	0.025
1	0.118	0.073	$\frac{1}{16}$	0.072	0.031
2	0.140	0.086	$\frac{5}{64}$	0.096	0.038
3	0.161	0.099	$\frac{5}{64}$	0.096	0.044
4	0.183	0.112	$\frac{3}{32}$	0.111	0.051
5	0.205	0.125	$\frac{3}{32}$	0.111	0.057
6	0.226	0.138	$\frac{7}{64}$	0.133	0.064
8	0.270	0.164	$\frac{9}{64}$	0.168	0.077
10	0.312	0.190	$\frac{5}{32}$	0.183	0.090
$\frac{1}{4}$	0.375	0.250	$\frac{3}{16}$	0.216	0.120
$\frac{5}{16}$	0.469	0.312	$\frac{1}{4}$	0.291	0.151
$\frac{3}{8}$	0.562	0.375	$\frac{5}{16}$	0.372	0.182
$\frac{7}{16}$	0.656	0.438	$\frac{3}{8}$	0.454	0.213
$\frac{1}{2}$	0.750	0.500	$\frac{3}{8}$	0.454	0.245
$\frac{5}{8}$	0.938	0.625	$\frac{1}{2}$	0.595	0.307
$\frac{3}{4}$	1.125	0.750	$\frac{5}{8}$	0.620	0.370
$\frac{7}{8}$	1.312	0.875	$\frac{3}{4}$	0.698	0.432
1	1.500	1.000	$\frac{3}{4}$	0.790	0.495
$1\frac{1}{8}$	1.688	1.125	$\frac{7}{8}$	....	0.557
$1\frac{1}{4}$	1.875	1.250	$\frac{7}{8}$	....	0.620
$1\frac{3}{8}$	2.062	1.365	1	....	0.682
$1\frac{1}{2}$	2.250	1.500	1	....	0.745

†Minimum.

where D_S is, of course, in millimeters.

See Tables 21.24 and 21.25 for basic dimensions of shoulder screws. These are made of the same material and of the same hardness as specified for cap screws.

21.3.5 Set Screws

Socket set screws (Fig. 21.7) are available in both inch and metric sizes with either hex or spline sockets for the inch series and hex sockets for the metric series. The cone point in Fig. 21.7*b* comes in seven different variations.

Square-head set screws (not shown) have a width across flats equal to the nominal size of the screw. The head height is three-quarters of the nominal size. These have a reduced-diameter neck just below the head.

21.3.6 Slotted-Head Cap Screws

The three standard head styles of the inch-series slotted-head cap screws are shown in Fig. 21.8, and the basic head dimensions are given in Table 21.26. The slot width is

TABLE 21.19   Basic Dimensions of Socket-Head Cap Screws (Metric Series)[†]

Nominal size D	Max. head diameter A	Max. head height H	Hex size W	Spline size M	Socket depth[‡] T
M1.6	3.00	1.60	1.5	1.829	0.80
M2	3.80	2.00	1.5	1.829	1.00
M2.5	4.50	2.50	2.0	2.438	1.25
M3	5.50	3.00	2.5	2.819	1.50
M4	7.00	4.00	3.0	3.378	2.00
M5	8.50	5.00	4.0	4.648	2.50
M6	10.00	6.00	5.0	5.486	3.00
M8	13.00	8.00	6.0	7.391	4.00
M10	16.00	10.00	8.0	....	5.00
M12	18.00	12.00	10.0	....	6.00
M16	24.00	16.00	14.0	....	8.00
M20	30.00	20.00	17.0	....	10.00
M24	36.00	24.00	19.0	....	12.00
M30	45.00	30.00	22.0	....	15.00
M36	54.00	36.00	27.0	....	18.00

†All dimensions in millimeters.
‡Minimum.

TABLE 21.20   Basic Dimensions of Socket Flat-Head Cap Screws (Inch Series)

Nominal size D	Max. head diameter A	Max. head height H	Hex size W	Spline size M	Socket depth[†] T
0	0.138	0.044	0.035	0.048	0.025
1	0.168	0.054	0.050	0.060	0.031
2	0.197	0.064	0.050	0.060	0.038
3	0.226	0.073	$\frac{1}{16}$	0.072	0.044
4	0.255	0.083	$\frac{1}{16}$	0.072	0.055
5	0.281	0.090	$\frac{5}{64}$	0.096	0.061
6	0.307	0.097	$\frac{5}{64}$	0.096	0.066
8	0.359	0.112	$\frac{3}{32}$	0.111	0.076
10	0.411	0.127	$\frac{1}{8}$	0.145	0.087
$\frac{1}{4}$	0.531	0.161	$\frac{5}{32}$	0.183	0.111
$\frac{5}{16}$	0.656	0.198	$\frac{3}{16}$	0.216	0.135
$\frac{3}{8}$	0.781	0.234	$\frac{7}{32}$	0.251	0.159
$\frac{7}{16}$	0.844	0.234	$\frac{1}{4}$	0.291	0.159
$\frac{1}{2}$	0.938	0.251	$\frac{5}{16}$	0.372	0.172
$\frac{5}{8}$	1.188	0.324	$\frac{3}{8}$	0.454	0.220
$\frac{3}{4}$	1.438	0.396	$\frac{1}{2}$	0.454	0.220
$\frac{7}{8}$	1.688	0.468	$\frac{9}{16}$	....	0.248
1	1.938	0.540	$\frac{5}{8}$	....	0.297
$1\frac{1}{8}$	2.188	0.611	$\frac{3}{4}$	....	0.325
$1\frac{1}{4}$	2.438	0.683	$\frac{7}{8}$	....	0.358
$1\frac{3}{8}$	2.688	0.755	$\frac{7}{8}$	....	0.402
$1\frac{1}{2}$	2.938	0.827	1	....	0.435

†Minimum.

TABLE 21.21 Basic Dimensions of Socket Flat-Head Cap Screws (Metric Series)[†]

Nominal size D	Max. head diameter A	Max. head height H	Hex size W	Socket depth‡ T
M3	6.72	1.85	2	18
M4	8.96	2.69	2.5	20
M5	11.20	3.18	3	22
M6	13.44	3.58	4	24
M8	17.92	4.42	5	28
M10	22.40	6.01	6	32
M12	26.88	6.85	8	36
M16	33.60	8.10	10	44
M20	40.32	8.70	12	52
M24	40.42	16.05	14	60

†All dimensions in millimeters.
‡Minimum.
SOURCE: Unbrako, Division of SPS, Jenkintown, Pa. 19046

TABLE 21.22 Basic Dimensions of Socket Button-Head Cap Screws (Inch Series)

Nominal size D	Max. head diameter A	Max. head height H	Hex size W	Spline size M	Socket depth† T
0	0.114	0.032	0.035	0.048	0.020
1	0.139	0.039	0.050	0.060	0.028
2	0.164	0.046	0.050	0.060	0.028
3	0.188	0.052	$\frac{1}{16}$	0.072	0.035
4	0.213	0.059	$\frac{1}{16}$	0.072	0.035
5	0.238	0.066	$\frac{5}{64}$	0.096	0.044
6	0.262	0.073	$\frac{5}{64}$	0.096	0.044
8	0.312	0.087	$\frac{3}{32}$	0.111	0.052
10	0.361	0.101	$\frac{1}{8}$	0.145	0.070
$\frac{1}{4}$	0.437	0.132	$\frac{5}{32}$	0.183	0.087
$\frac{5}{16}$	0.547	0.166	$\frac{3}{16}$	0.216	0.105
$\frac{3}{8}$	0.656	0.199	$\frac{7}{32}$	0.251	0.122
$\frac{1}{2}$	0.875	0.265	$\frac{5}{16}$	0.372	0.175
$\frac{5}{8}$	1.000	0.331	$\frac{3}{8}$	0.454	0.210

†Minimum.

TABLE 21.23 Basic Dimensions of Socket Button-Head Cap Screws (Metric Series)[†]

Nominal size D	Max. head diameter A	Max. head height H	Hex size W	Socket depth‡ T
M3	5.70	1.65	2	1.04
M4	7.60	2.20	2.5	1.30
M5	9.50	2.75	3	1.56
M6	10.50	3.30	4	2.08
M8	14.00	4.40	5	2.60
M10	17.50	5.50	6	3.12
M12	21.00	6.60	8	4.16
M16	28.00	8.80	10	5.20

†All dimensions in millimeters.
‡Minimum.

21.21

TABLE 21.24 Basic Dimensions of Socket Shoulder Screws (Inch Series)

Shoulder diameter D_S	Max. head diameter A	Max. head height H	Hex size W	Socket depth† T	Thread size D	Thread length L_T
1/4	0.375	0.188	1/8	0.094	10	0.375
5/16	0.438	0.219	5/32	0.117	1/4	0.438
3/8	0.562	0.250	3/16	0.141	5/16	0.500
1/2	0.750	0.312	1/4	0.188	3/8	0.625
5/8	0.875	0.375	5/16	0.234	1/2	0.750
3/4	1.000	0.500	3/8	0.281	5/8	0.875
1	1.312	0.625	1/2	0.375	3/4	1.000
1¼	1.750	0.750	5/8	0.469	7/8	1.125
1½	2.125	1.000	7/8	0.656	1⅛	1.500
1¾	2.375	1.125	1	0.750	1¼	1.750
2	2.750	1.250	1¼	0.937	1½	2.000

†Minimum.

TABLE 21.25 Basic Dimensions of Socket-Head Shoulder Screws (Metric Series)†

Shoulder diameter D_S	Max. head diameter A	Max. head height H	Hex size W	Socket depth‡ T	Thread size D	Thread length L_T
6.5	10.00	4.50	3	2.4	M5	9.75
8.0	13.00	5.50	4	3.3	M6	11.25
10.0	16.00	7.00	5	4.2	M8	13.25
13.0	18.00	9.00	6	4.9	M10	16.40
16.0	24.00	11.00	8	6.6	M12	18.40
20.0	30.00	14.00	10	8.8	M16	22.40
25.0	36.00	16.00	12	10.0	M20	27.40

†All dimensions in millimeters.
‡Minimum.

$$J = \begin{cases} 0.160D + 0.024 & D \leq 1 \\ 0.160D & 1 < D \leq 1\frac{1}{2} \end{cases} \qquad (21.6)$$

The slot depth varies, depending on the head type and the nominal size.

Slotted-head cap screws are normally made from carbon steel conforming to ASTM A307 properties (see Table 21.7). However, they can also be obtained in grade ASTM A449 material and properties.

21.3.7 Machine Screws

We can keep track of the many types of machine screws by classifying them as follows:

1. Flat countersunk head (80 degrees)
   a. Regular or undercut
   b. Slotted or cross-recessed

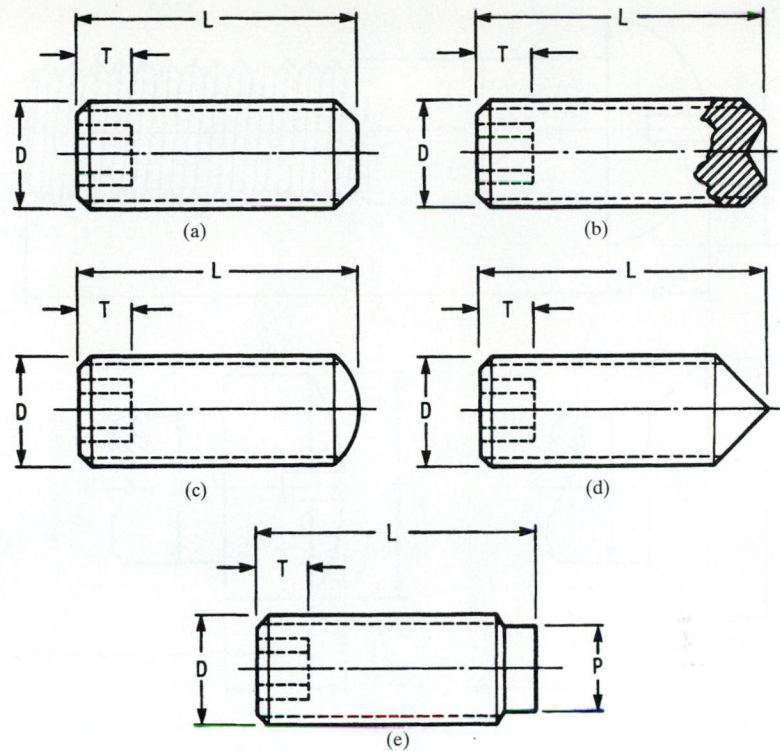

FIGURE 21.7 Socket set screws. (*a*) Flat point; (*b*) cup point; (*c*) oval point; (*d*) cone point; (*e*) half-dog point.

2. 100-degree flat countersunk head
  a. Regular or close tolerance
  b. Slotted or cross-recessed
3. Oval countersunk head
  a. Regular or undercut
  b. Slotted or cross-recessed
4. Flat countersunk trim head
  a. Regular or short
  b. Cross-recessed
5. Oval countersunk trim head
  a. Regular or short
  b. Cross-recessed
6. Pan head
  a. Slotted or cross-recessed
7. Fillister head
  a. Slotted and cross-drilled
  b. Slotted or cross-recessed
8. Truss head
  a. Slotted or cross recessed

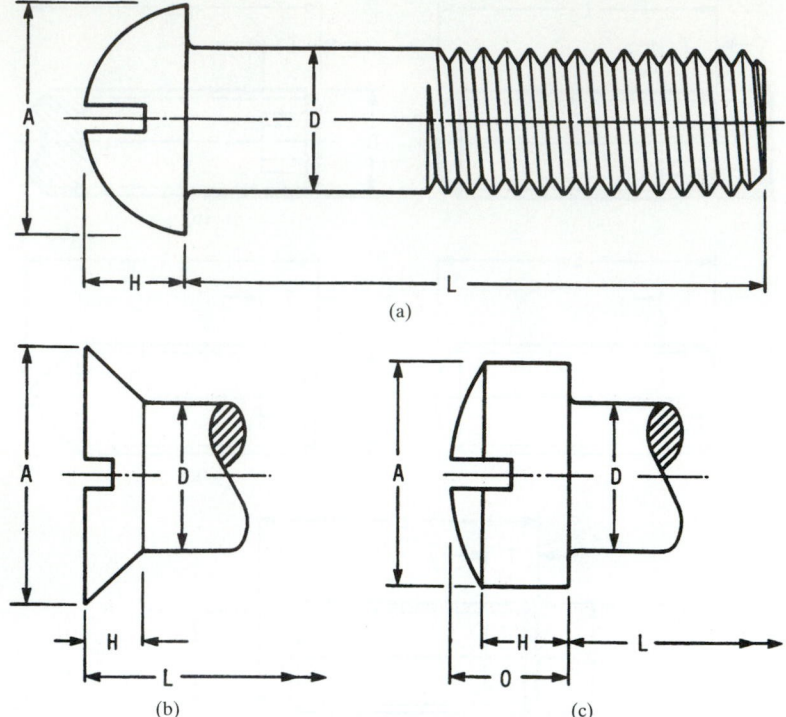

FIGURE 21.8 Slotted-head cap screws. (*a*) Round head; (*b*) flat countersunk head; (*c*) fillister head.

TABLE 21.26 Basic Head Dimensions of Slotted-Head Cap Screws (Inch Series)

Nominal size	Flat head†		Round head†		Fillister head†		
	A	*H*	*A*	*H*	*A*	*H*	*O*
$\frac{1}{4}$	0.500	0.140	0.437	0.191	0.375	0.172	0.216
$\frac{5}{16}$	0.625	0.177	0.562	0.245	0.437	0.203	0.253
$\frac{3}{8}$	0.750	0.210	0.625	0.273	0.562	0.250	0.314
$\frac{7}{16}$	0.812	0.210	0.750	0.328	0.625	0.297	0.368
$\frac{1}{2}$	0.875	0.210	0.812	0.354	0.750	0.328	0.413
$\frac{9}{16}$	1.000	0.244	0.937	0.409	0.812	0.375	0.467
$\frac{5}{8}$	1.125	0.281	1.000	0.437	0.875	0.422	0.521
$\frac{3}{4}$	1.375	0.352	1.250	0.546	1.000	0.500	0.612
$\frac{7}{8}$	1.625	0.423	....	....	1.125	0.594	0.720
1	1.875	0.494	....	....	1.312	0.656	0.803
$1\frac{1}{8}$	2.062	0.529					
$1\frac{1}{4}$	2.312	0.600					
$1\frac{3}{8}$	2.562	0.665					
$1\frac{1}{2}$	2.812	0.742					

†Maximum.

9. Binding head
   a. Slotted or cross-recessed
10. Hex head
   a. Indented
   b. Slotted
   c. Indented and slotted
11. Hex washer head
   a. Indented
   b. Indented and slotted

The *round-head* machine screw is obsolete. Use the *pan-head* screw, instead; it has more driving power. Most of the head types outlined here are illustrated in Figs. 21.9 to 21.11, some slotted and some with cross-recesses.

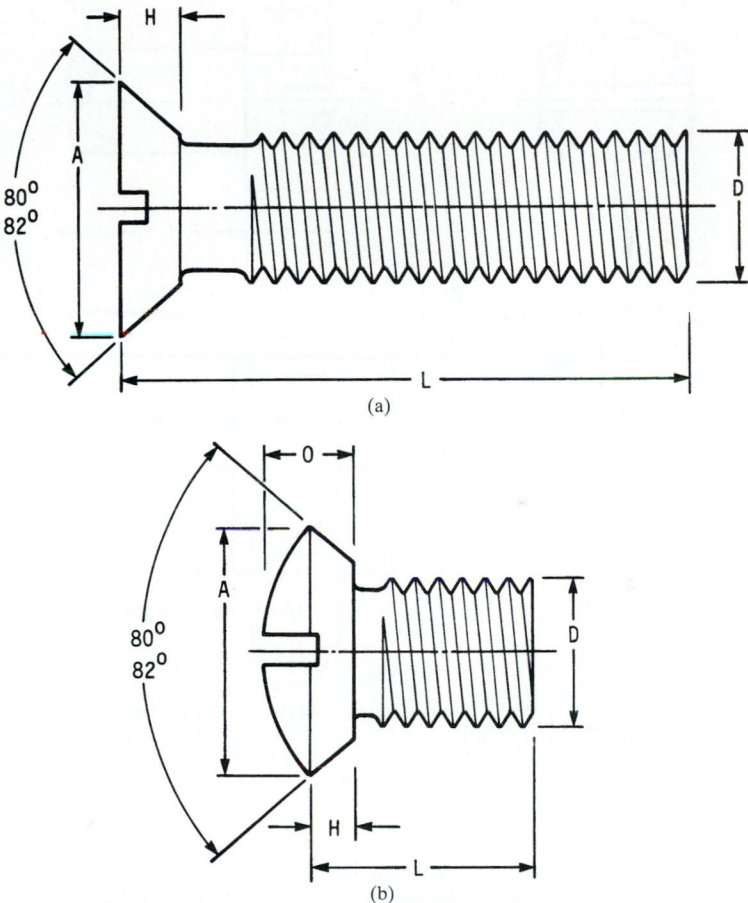

FIGURE 21.9 (*a*) Slotted flat countersunk-head machine screw; also available with 100-degree head; (*b*) short or undercut slotted oval countersunk-head machine screw. Note the difference between the body of a machine screw and that of a cap screw. Compare this figure with Fig. 21.8*a*.

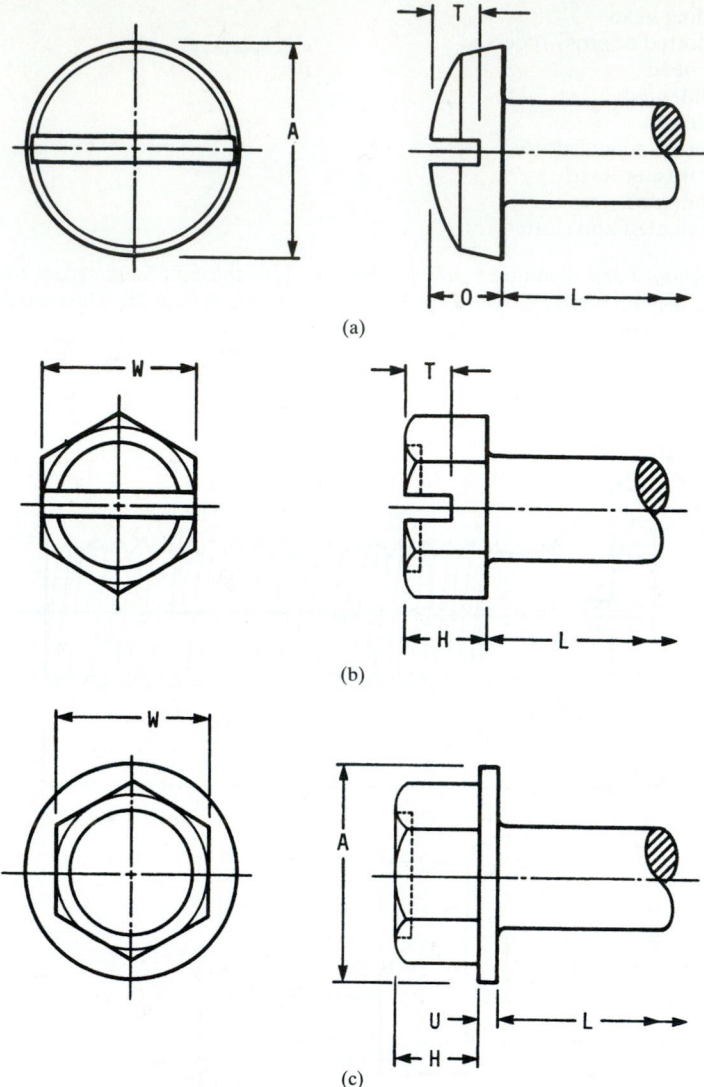

FIGURE 21.10   (*a*) Slotted binding-head machine screw; the edge angle is 5 degrees; (*b*) slotted and indented hex-head machine screw; (*c*) hex washer-head machine screw. Both hex screws in this figure may be obtained with or without slots and/or indentations.

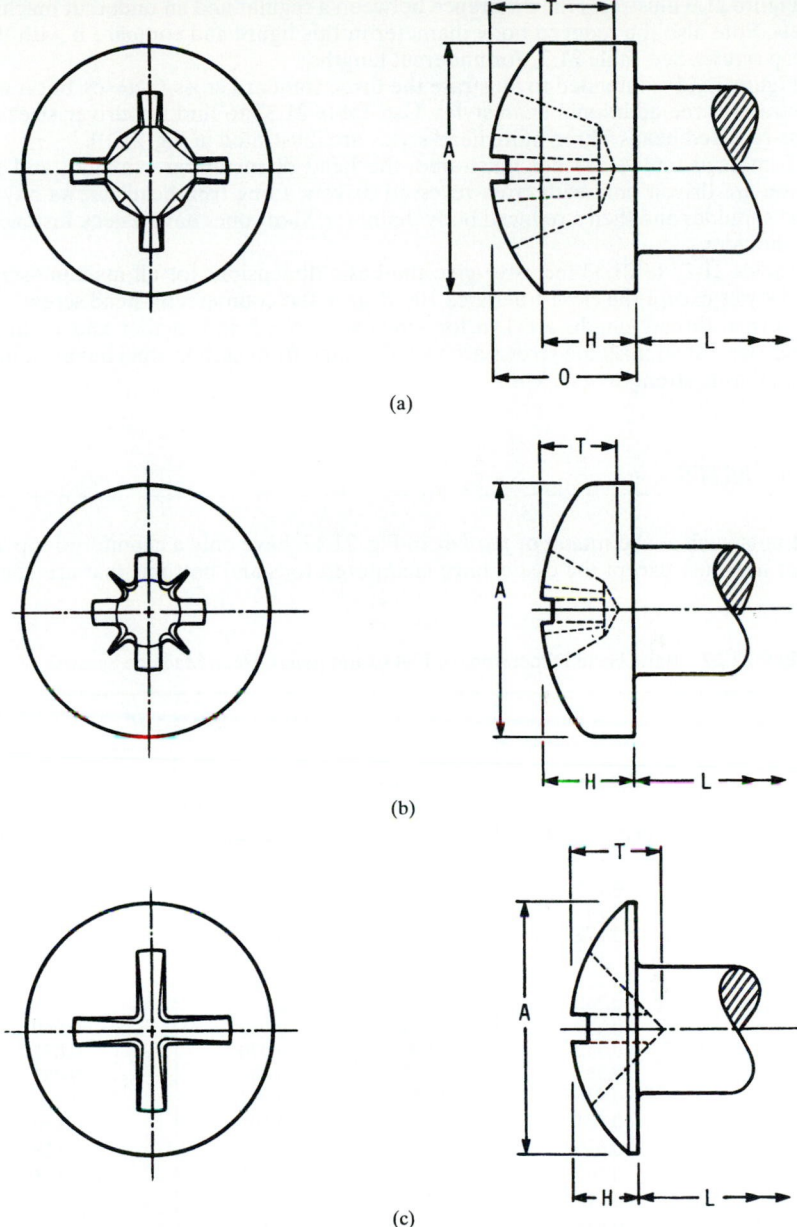

(a)

(b)

(c)

FIGURE 21.11 The three standard recesses. (*a*) Fillister-head machine screw with Type 1 cross-recess. This recess has a large center opening, tapered wings, and a blunt bottom. (*b*) Pan-head machine screw with Type 1A cross-recess. This recess has a large center opening, wide straight wings, and a blunt bottom. (*c*) Truss-head machine screw with Type 2 cross-recess. This recess has two intersecting slots with parallel sides. The sides converge to a slightly truncated apex at the bottom.

Figure 21.9 illustrates the difference between a regular and an undercut machine screw. Note also the reduced body diameter in this figure and compare it with that of cap screws. See Table 21.32 for undercut lengths.

Figure 21.11 is intended to illustrate the three standard cross-recesses, but it also illustrates three additional head styles. Use Table 21.32 to find the driver sizes for cross-recessed heads. Three more head styles are illustrated in Fig. 21.10.

Trim-head screws are not illustrated; the head diameters are smaller, and the screws are driven only with cross-recessed drivers. Long trim-head screws have a short shoulder and then a reduced body diameter. Short ones have a neck instead of the shoulder.

Tables 21.27 to 21.33 inclusive give the basic dimensions for all machine-screw head styles except the close-tolerance 100-degree flat countersunk-head screw.

Normal thread lengths are 1 in for screw sizes No. 5 and smaller and 1½ in for larger sizes. Steel machine screws are usually made from carbon steel having a minimum tensile strength of 60 kpsi.

21.4   NUTS

Flat nuts, such as the *square* or *flat hex* in Fig. 21.12, have only a chamfered top. All other hex nuts, except the castle, have chamfered tops and bottoms that are cham-

TABLE 21.27   Basic Head Dimensions of Flat Countersunk-Head Machine Screws[†]

| Nominal size D | Head diameter[‡] A | Head thickness[‡] H | | |
| | | Regular, 80 degrees | | |
		Long	Undercut	100-degree head
0	0.119	0.035	0.025	0.026
1	0.146	0.043	0.031	0.031
2	0.172	0.051	0.036	0.037
3	0.199	0.059	0.042	0.043
4	0.225	0.067	0.047	0.049
5	0.252	0.075	0.053	
6	0.279	0.083	0.059	0.060
8	0.332	0.100	0.070	0.072
10	0.385	0.116	0.081	0.083
12	0.438	0.132	0.092	
$\frac{1}{4}$	0.507	0.153	0.107	0.110
$\frac{5}{16}$	0.635	0.191	0.134	0.138
$\frac{3}{8}$	0.762	0.230	0.161	0.165
$\frac{7}{16}$	0.812	0.223	0.156	
$\frac{1}{2}$	0.875	0.223	0.156	
$\frac{9}{16}$	1.000	0.260		
$\frac{5}{8}$	1.125	0.298		
$\frac{3}{4}$	1.375	0.372		

†All dimensions in inches.
‡Maximum.

TABLE 21.28 Total Head Heights O for Oval Countersunk-Head Machine Screws[†]

Nominal size D	Total head height ‡ O		Nominal size D	Total head height‡ O	
	Long	Undercut		Long	Undercut
0	0.056	0.046	12	0.200	0.161
1	0.068	0.056	$\frac{1}{4}$	0.232	0.186
2	0.080	0.065	$\frac{5}{16}$	0.290	0.232
3	0.092	0.075	$\frac{3}{8}$	0.347	0.278
4	0.104	0.084	$\frac{7}{16}$	0.345	0.279
5	0.116	0.094	$\frac{1}{2}$	0.354	0.288
6	0.128	0.104	$\frac{9}{16}$	0.410	
8	0.152	0.123	$\frac{5}{8}$	0.467	
10	0.176	0.142	$\frac{3}{4}$	0.578	

†Head dimensions A and H are the same as for regular 80-degree flat-head machine screws (Table 21-27). All dimensions are in inches.
‡Maximum.

TABLE 21.29 Basic Head Dimensions of Flat and Oval Countersunk Trim-Head Machine Screws (Available Only with Cross-Recessed Heads)[†]

Nominal size D	Shoulder diameter‡§	Head diameter‡ A	Flat head height H	Oval head height¶ O
4	0.112	0.199	0.052	0.086
5	0.125	0.225	0.060	0.099
6	0.138	0.225	0.052	0.091
6	0.138	0.252	0.068	0.112
8	0.164	0.252	0.052	0.096
8	0.164	0.279	0.069	0.117
10	0.190	0.332	0.085	0.141
12	0.216	0.332	0.069	0.125
12	0.216	0.385	0.101	0.166
$\frac{1}{4}$	0.250	0.385	0.080	0.146
$\frac{1}{4}$	0.250	0.438	0.112	0.187
$\frac{5}{16}$	0.312	0.438	0.075	0.150
$\frac{5}{16}$	0.312	0.507	0.116	0.202
$\frac{3}{8}$	0.375	0.635	0.155	0.265

†All dimensions are in inches.
‡Maximum.
§Screws having nominal lengths over $1\frac{1}{8}$ in for sizes No. 5 and smaller and over 2 in for sizes No. 6 and larger have a shoulder of this diameter and about $1\frac{1}{16}$ in long beneath the head.
¶This is the total height. The side height H is the same as for flat-head trim screws.

TABLE 21.30 Basic Head Dimensions of Pan- and Truss-Head Machine Screws[†]

Nominal size D	Pan head		Truss head	
	Diameter A	Height H	Diameter A	Height H
0	0.116	0.044	0.131	0.037
1	0.142	0.053	0.164	0.045
2	0.167	0.062	0.194	0.053
3	0.193	0.071	0.226	0.061
4	0.219	0.080	0.257	0.069
5	0.245	0.089	0.289	0.078
6	0.270	0.097	0.321	0.086
8	0.322	0.115	0.384	0.102
10	0.373	0.133	0.448	0.118
12	0.425	0.151	0.511	0.134
$\frac{1}{4}$	0.492	0.175	0.573	0.150
$\frac{5}{16}$	0.615	0.218	0.698	0.183
$\frac{3}{8}$	0.740	0.261	0.823	0.215
$\frac{7}{16}$	0.863	0.305	0.948	0.248
$\frac{1}{2}$	0.987	0.348	1.073	0.280
$\frac{9}{16}$	1.041	0.391	1.198	0.312
$\frac{5}{8}$	1.172	0.434	1.323	0.345
$\frac{3}{4}$	1.435	0.521	1.573	0.410

†All values are maximum; all dimensions are in inches.

TABLE 21.31 Basic Head Dimensions of Binding- and Fillister-Head Machine Screws[†]

Nominal size D	Fillister head			Binding head	
	Diameter A	Height H	Height O	Diameter A	Height O
0	0.096	0.043	0.055	0.126	0.032
1	0.118	0.053	0.066	0.153	0.041
2	0.140	0.062	0.083	0.181	0.050
3	0.161	0.070	0.095	0.208	0.059
4	0.183	0.079	0.107	0.235	0.068
5	0.205	0.088	0.120	0.263	0.078
6	0.226	0.096	0.132	0.290	0.087
8	0.270	0.113	0.156	0.344	0.105
10	0.313	0.130	0.180	0.399	0.123
12	0.357	0.148	0.205	0.454	0.141
$\frac{1}{4}$	0.414	0.170	0.237	0.525	0.165
$\frac{5}{16}$	0.518	0.211	0.295	0.656	0.209
$\frac{3}{8}$	0.622	0.253	0.355	0.788	0.253
$\frac{7}{16}$	0.625	0.265	0.368		
$\frac{1}{2}$	0.750	0.297	0.412		
$\frac{9}{16}$	0.812	0.336	0.466		
$\frac{5}{8}$	0.875	0.375	0.521		
$\frac{3}{4}$	1.000	0.441	0.612		

†All values are maximum; all dimensions are in inches.

TABLE 21.32   Undercut Lengths for Flat and Oval Countersunk-Head Machine Screws and Driver Sizes for Type I and IA Cross-Recesses[†]

Nominal size	Undercut lengths (or less)	Driver size	Nominal size	Undercut lengths (or less)	Driver size
0	$\frac{1}{8}$	0	10	$\frac{5}{16}$	2
1	$\frac{1}{8}$	0	12	$\frac{3}{8}$	3
2	$\frac{1}{8}$	1	$\frac{1}{4}$	$\frac{7}{16}$	3
3	$\frac{1}{8}$	1	$\frac{5}{16}$	$\frac{1}{2}$	4
4	$\frac{3}{16}$	1	$\frac{3}{8}$	$\frac{9}{16}$	4
5	$\frac{3}{16}$	2	$\frac{7}{16}$	$\frac{5}{8}$	4
6	$\frac{3}{16}$	2	$\frac{1}{2}$	$\frac{3}{4}$	4
8	$\frac{1}{4}$	2	$\frac{9}{16}$	. . . .	4

†Type II drivers have the same point size for all screw sizes.

TABLE 21.33   Basic Head Dimensions for Regular-Hex- and Washer-Hex-Head Machine Screws[†]

Nominal size D	Head height H	Width across faults W		Washer face	
		Regular hex	Washer hex	Diameter A	Thickness U
1	0.044	0.120			
2	0.050	0.120	0.125	0.166	0.016
3	0.055	0.181	0.125	0.177	0.016
4	0.060	0.181	0.188	0.243	0.019
5	0.070	0.181	0.188	0.260	0.025
6	0.093	0.244	0.250	0.328	0.025
8	0.110	0.244	0.250	0.348	0.031
10	0.120	0.305	0.312	0.414	0.031
12	0.155	0.305	0.312	0.432	0.039
$\frac{1}{4}$	0.190	0.367	0.375	0.520	0.050
$\frac{5}{16}$	0.230	0.489	0.500	0.676	0.055
$\frac{3}{8}$	0.295	0.551	0.562	0.780	0.063

†All values are maximum; all dimensions are in inches.

fered or washer-faced, as in Fig. 21.12c and d. *Castle nuts* (Fig. 21.12f) are made in both styles, washer-faced or with chamfered bottoms.

Most styles can also be classified as regular, thick, or heavy. A thick nut has the same width W across flats as a regular nut, but the height H is greater. A *jam nut* is a thin hex nut. A *heavy nut* is larger in both dimensions, W and H, than the regular style.

It is convenient to outline these varieties of head styles with their accepted names and indicate in which table the dimensions are to be found:

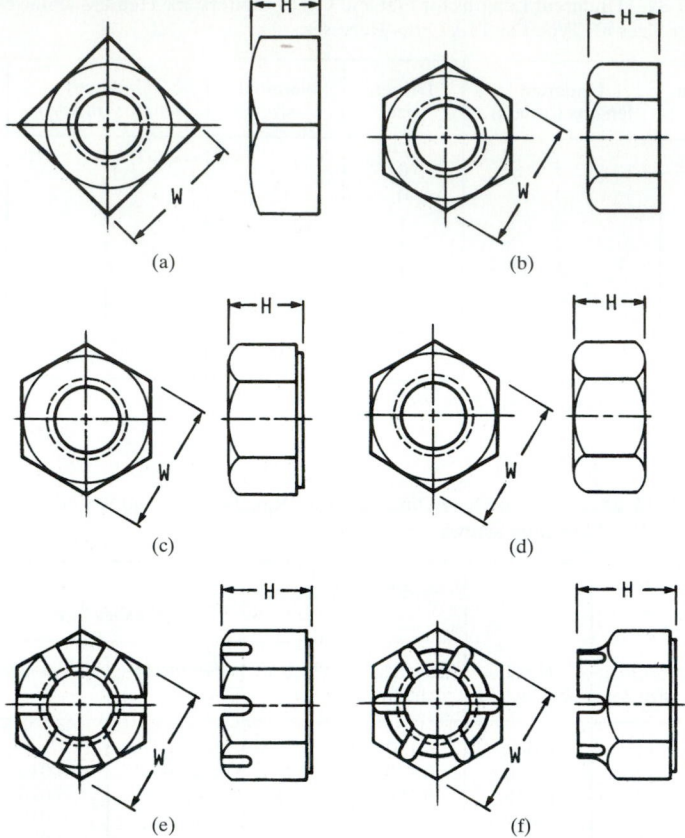

FIGURE 21.12   Types of nuts. (*a*) Square; (*b*) hex flat; (*c*) hex with washer face; (*d*) double-chamfered hex; (*e*) slotted hex; (*f*) hex castle.

1. Flat nuts
    a. Square (Table 21.34)
    b. Heavy square (Table 21.34)
    c. Hex (Table 21.35)
    d. Hex jam (Table 21.35)
    e. Heavy hex (Table 21.39)
    f. Heavy hex jam (Table 21.39)
2. Nuts with washer-faced or chamfered bottoms
    a. Hex (Table 21.35)
        (1) Metric style 1 (Table 21.36)
    b. Hex jam (Table 21.35)
        (1) Metric (Table 21.36)
    c. Hex slotted (Table 21.35)
        (1) Metric (Table 21.36)

TABLE 21.34 Basic Dimensions of Square Nuts (Inch Series)

Nominal size	Regular W	Regular H	Heavy W	Heavy H
$\frac{1}{4}$	$\frac{7}{16}$	$\frac{7}{32}$	$\frac{1}{2}$	$\frac{1}{4}$
$\frac{5}{16}$	$\frac{9}{16}$	$\frac{17}{64}$	$\frac{9}{16}$	$\frac{5}{16}$
$\frac{3}{8}$	$\frac{5}{8}$	$\frac{21}{64}$	$\frac{11}{16}$	$\frac{3}{8}$
$\frac{7}{16}$	$\frac{3}{4}$	$\frac{3}{8}$	$\frac{3}{4}$	$\frac{7}{16}$
$\frac{1}{2}$	$\frac{13}{16}$	$\frac{7}{16}$	$\frac{7}{8}$	$\frac{1}{2}$
$\frac{5}{8}$	1	$\frac{35}{64}$	$1\frac{1}{16}$	$\frac{5}{8}$
$\frac{3}{4}$	$1\frac{1}{8}$	$\frac{21}{32}$	$1\frac{1}{4}$	$\frac{3}{4}$
$\frac{7}{8}$	$1\frac{5}{16}$	$\frac{49}{64}$	$1\frac{7}{16}$	$\frac{7}{8}$
1	$1\frac{1}{2}$	$\frac{7}{8}$	$1\frac{5}{8}$	1
$1\frac{1}{8}$	$1\frac{11}{16}$	1	$1\frac{13}{16}$	$1\frac{1}{8}$
$1\frac{1}{4}$	$1\frac{7}{8}$	$1\frac{3}{32}$	2	$1\frac{1}{4}$
$1\frac{3}{8}$	$2\frac{1}{16}$	$1\frac{13}{64}$	$2\frac{3}{16}$	$1\frac{3}{8}$
$1\frac{1}{2}$	$2\frac{1}{4}$	$1\frac{5}{16}$	$2\frac{3}{8}$	$1\frac{1}{2}$

TABLE 21.35 Basic Dimensions of Hex, Hex Jam, Hex Flat, Hex Flat Jam, Hex Slotted, Hex Thick, and Hex Thick Slotted Nuts (Inch Series)

Nominal size D	Width W	Height H — Hex and hex slotted	Hex jam	Hex flat	Hex flat jam	Hex thick and thick slotted
$\frac{1}{4}$	$\frac{7}{16}$	$\frac{7}{32}$	$\frac{5}{32}$	...	...	$\frac{9}{32}$
$\frac{5}{16}$	$\frac{1}{2}$	$\frac{17}{64}$	$\frac{3}{16}$	...	...	$\frac{21}{64}$
$\frac{3}{8}$	$\frac{9}{16}$	$\frac{21}{64}$	$\frac{7}{32}$	...	...	$\frac{13}{32}$
$\frac{7}{16}$	$\frac{11}{16}$	$\frac{3}{8}$	$\frac{1}{4}$	...	...	$\frac{29}{64}$
$\frac{1}{2}$	$\frac{3}{4}$	$\frac{7}{16}$	$\frac{5}{16}$	...	...	$\frac{9}{16}$
$\frac{9}{16}$	$\frac{7}{8}$	$\frac{31}{64}$	$\frac{5}{16}$	...	...	$\frac{39}{64}$
$\frac{5}{8}$	$\frac{15}{16}$	$\frac{35}{64}$	$\frac{3}{8}$	...	...	$\frac{23}{32}$
$\frac{3}{4}$	$1\frac{1}{8}$	$\frac{41}{64}$	$\frac{27}{64}$	...	...	$\frac{13}{16}$
$\frac{7}{8}$	$1\frac{5}{16}$	$\frac{3}{4}$	$\frac{31}{64}$	...	...	$\frac{29}{32}$
1	$1\frac{1}{2}$	$\frac{55}{64}$	$\frac{35}{64}$	...	...	1
$1\frac{1}{8}$	$1\frac{11}{16}$	$\frac{31}{32}$	$\frac{39}{64}$	1	$\frac{5}{8}$	$1\frac{5}{32}$
$1\frac{1}{4}$	$1\frac{7}{8}$	$1\frac{1}{16}$	$\frac{23}{32}$	$1\frac{3}{32}$	$\frac{3}{4}$	$1\frac{1}{4}$
$1\frac{3}{8}$	$2\frac{1}{16}$	$1\frac{11}{64}$	$\frac{25}{32}$	$1\frac{13}{64}$	$\frac{13}{16}$	$1\frac{3}{8}$
$1\frac{1}{2}$	$2\frac{1}{4}$	$1\frac{9}{32}$	$\frac{27}{32}$	$1\frac{5}{16}$	$\frac{7}{8}$	$1\frac{1}{2}$

TABLE 21.36 Basic Dimensions of Metric Hex Nuts[†]

Nominal size D	Width across flats‡ W	Height H‡		
		Style 1	Style 2 or slotted	Jam
M1.6	3.2	1.3		
M2	4	1.6		
M2.5	5	2.0		
M3	5.5	2.4	2.9§	
M3.5	6	2.8	3.3§	
M4	7	3.2	3.8§	
M5	8	4.7	5.1	2.7
M6	10	5.2	5.7	3.2
M8	13	6.8	7.5	4.0
M10	16	8.4	9.3	5.0
M12	18	10.8	12.0	6.0
M14	21	12.8	14.1	7.0
M16	24	14.8	16.4	8.0
M20	30	18.0	20.3	10.0
M24	36	21.5	23.9	12.0
M30	46	25.6	28.6	15.0
M36	55	31.0	34.7	18.0

†All dimensions are in millimeters.
‡Maximum.
§Not standard in slotted style.

TABLE 21.37 Basic Dimensions of Metric Heavy Hex Nuts[†]

Nominal size D	Width across flats‡ W	Height‡ H	Nominal size D	Width across flats‡ W	Height‡ H
M12	21	12,3	M24	41	24.2
M14	24	14.3	M27	46	27.6
M16	27	17.1	M30	50	30.7
M20	34	20.7	M36	60	36.6
M22	36	23.6	M42	70	42.0

†All dimensions are in millimeters.
‡Maximum.

    d. Hex thick (Table 21.35)
      (1) Metric style 2 (Table 21.36)
    e. Hex thick slotted (Table 21.35)
    f. Heavy hex (Table 21.38)
      (1) Metric (Table 21.37)
    g. Heavy hex jam (Table 21.38)
    h. Hex castle (Table 21.38)
3. Machine screw nuts
    a. Hex (Table 21.40)
    b. Square (Table 21.40)

TABLE 21.38   Basic Dimensions of Heavy Hex, Heavy Hex Jam, Heavy Hex Slotted, and Heavy Hex Castle Nuts (Inch Series)

Nominal size D	Heavy hex Width W	Heavy hex Height H Plain or slotted	Heavy hex Height H Jam	Hex castle Width W	Hex castle Height H
$\frac{1}{4}$	$\frac{1}{2}$	$\frac{15}{64}$	$\frac{11}{64}$	$\frac{7}{16}$	$\frac{9}{32}$
$\frac{5}{16}$	$\frac{9}{16}$	$\frac{19}{64}$	$\frac{13}{64}$	$\frac{1}{2}$	$\frac{21}{64}$
$\frac{3}{8}$	$\frac{11}{16}$	$\frac{23}{64}$	$\frac{15}{64}$	$\frac{9}{16}$	$\frac{13}{32}$
$\frac{7}{16}$	$\frac{3}{4}$	$\frac{27}{64}$	$\frac{17}{64}$	$\frac{11}{16}$	$\frac{29}{64}$
$\frac{1}{2}$	$\frac{7}{8}$	$\frac{31}{64}$	$\frac{19}{64}$	$\frac{3}{4}$	$\frac{9}{16}$
$\frac{9}{16}$	$\frac{15}{16}$	$\frac{35}{64}$	$\frac{21}{64}$	$\frac{7}{8}$	$\frac{39}{64}$
$\frac{5}{8}$	$1\frac{1}{16}$	$\frac{39}{64}$	$\frac{23}{64}$	$\frac{13}{16}$	$\frac{23}{32}$
$\frac{3}{4}$	$1\frac{1}{4}$	$\frac{47}{64}$	$\frac{27}{64}$	$1\frac{1}{8}$	$\frac{13}{16}$
$\frac{7}{8}$	$1\frac{7}{16}$	$\frac{55}{64}$	$\frac{31}{64}$	$1\frac{5}{16}$	$\frac{29}{32}$
1	$1\frac{5}{8}$	$\frac{63}{64}$	$\frac{33}{64}$	$1\frac{1}{2}$	1
$1\frac{1}{8}$	$1\frac{13}{16}$	$1\frac{7}{64}$	$\frac{39}{64}$	$1\frac{11}{16}$	$1\frac{5}{32}$
$1\frac{1}{4}$	2	$1\frac{7}{32}$	$\frac{23}{32}$	$1\frac{7}{8}$	$1\frac{1}{4}$
$1\frac{3}{8}$	$2\frac{3}{16}$	$1\frac{11}{32}$	$\frac{25}{32}$	$2\frac{1}{16}$	$1\frac{3}{8}$
$1\frac{1}{2}$	$2\frac{3}{8}$	$1\frac{15}{32}$	$\frac{27}{32}$	$2\frac{1}{4}$	$1\frac{1}{2}$

TABLE 21.39   Basic Dimensions of Heavy Hex Flat and Heavy Hex Flat Jam Nuts (Inch Series)

Nominal size D	Width across flats W	Height H Regular	Height H Jam
$1\frac{1}{8}$	$1\frac{13}{16}$	$1\frac{1}{8}$	$\frac{5}{8}$
$1\frac{1}{4}$	2	$1\frac{1}{4}$	$\frac{3}{4}$
$1\frac{3}{8}$	$2\frac{13}{16}$	$1\frac{3}{8}$	$\frac{13}{16}$
$1\frac{1}{2}$	$2\frac{3}{8}$	$1\frac{1}{2}$	$\frac{7}{8}$
$1\frac{3}{4}$	$2\frac{3}{4}$	$1\frac{3}{4}$	1
2	$3\frac{1}{8}$	2	$1\frac{1}{8}$

Carbon steel nuts usually are made to conform to ASTM A563 Grade A specifications or to SAE Grade 2.

21.5   TAPPING SCREWS

Self-tapping screws are available in all head styles and in sizes up to and including ⅜ in. They are hardened sufficiently to form their own mating threads when driven. In

TABLE 21.40 Basic Dimensions of Machine-Screw Nuts[†]

Nominal size D	Width W	Height‡ H	Nominal size D	Width W	Height‡ H
0	$\frac{5}{32}$	0.050	8	$\frac{11}{32}$	0.130
1	$\frac{5}{32}$	0.050	10	$\frac{3}{8}$	0.130
2	$\frac{3}{16}$	0.066	12	$\frac{7}{16}$	0.161
3	$\frac{3}{16}$	0.066	$\frac{1}{4}$	$\frac{7}{16}$	0.193
4	$\frac{1}{4}$	0.098	$\frac{5}{16}$	$\frac{9}{16}$	0.225
5	$\frac{5}{16}$	0.114	$\frac{3}{8}$	$\frac{5}{8}$	0.257
6	$\frac{5}{16}$	0.114			

†All dimensions are in inches; dimensions apply to both hex and square nuts.
‡Maximum.

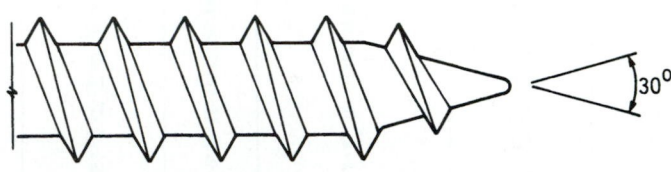

TYPE A

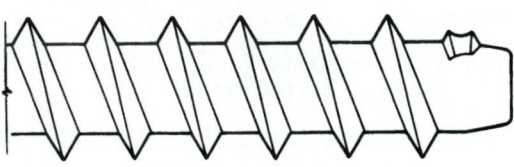

TYPE B

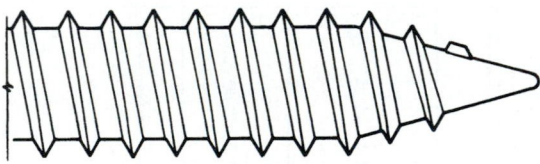

TYPE AB

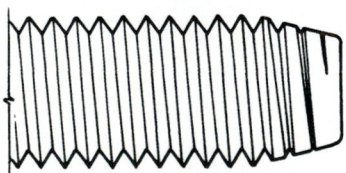

TYPE C

FIGURE 21.13 Thread-forming screws.

fact, at least one manufacturer has developed a screw which both drills and taps in a single operation.

Thread-forming screws are used when sufficient joint stresses can be developed to guard against loosening. The Type A screw (Fig. 21.13) has a gimlet point and is used to join sheet metal. It is sometimes used instead of a wood screw.

The Type B screw has a blunt point and, sometimes, a finer pitch than the Type A. With the finer pitch, it has a greater range of applications and is used in heavier sheet metal and nonferrous castings.

The Type AB screw is similar to Type A but has a finer pitch, which permits it to be used in more brittle materials, such as plastics or zinc die castings. It is also used as a wood screw.

The Type C screw has machine-screw threads, a blunt point, and tapered threads at the start, as shown in Fig. 21.13. This screw is useful for thick sections but may require large driving torques.

Thread-cutting screws (shown in Fig. 21.14) are used instead of thread-forming screws to lessen the driving torque and the internal stresses. Types D, G, F, and T all have machine-screw threads and an end taper with a blunt point. They are used for various die castings, cast iron, brass, plastics, and sheet steel.

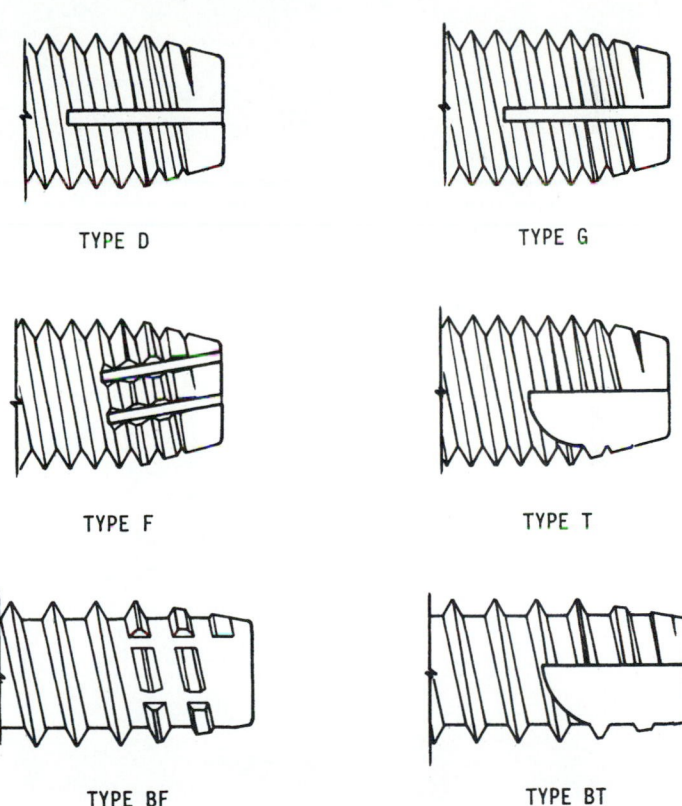

TYPE D

TYPE G

TYPE F

TYPE T

TYPE BF

TYPE BT

FIGURE 21.14   Thread-cutting screws.

Types BF and BT screws (shown in Fig. 21.14) have additional cutting edges and a greater chip-storage capacity. They are useful for various plastics and compositions.

Thread-rolling screws (not shown) have a unique point and body shape to make starting easier and to lessen the driving torque. The thread rolling cold-works the material, thus contributing to the strength of the joint.

REFERENCE

21.1 "Codes and Standards; Fasteners," in *ASME Publications Catalog,* American Society of Mechanical Engineers, New York, 1985.

CHAPTER 22
UNTHREADED FASTENERS

Joseph E. Shigley
Professor Emeritus
The University of Michigan
Ann Arbor, Michigan

22.1 RIVETS

A *rivet* is a fastener that has a head and a shank and is made of a deformable material. It is used to join several parts by placing the shank into holes through the several parts and creating another head by upsetting or deforming the projecting shank.

During World War II, Rosie the Riveter was a popular cartoon character in the United States. No better image can illustrate the advantages of riveted joints. These are

1. Low cost
2. Fast automatic or repetitive assembly
3. Permanent joints
4. Usable for joints of unlike materials such as metals and plastics
5. Wide range of rivet shapes and materials
6. Large selection of riveting methods, tools, and machines

Riveted joints, however, are not as strong under tension loading as are bolted joints (see Chap. 23), and the joints may loosen under the action of vibratory tensile or shear forces acting on the members of the joint. Unlike with welded joints, special sealing methods must be used when riveted joints are to resist the leakage of gas or fluids.

22.1.1 Head Shapes

A group of typical rivet-head styles is shown in Figs. 22.1 and 22.2. Note that the button head, the oval head, and the truss head are similar. Of the three, the oval head has an intermediate thickness.

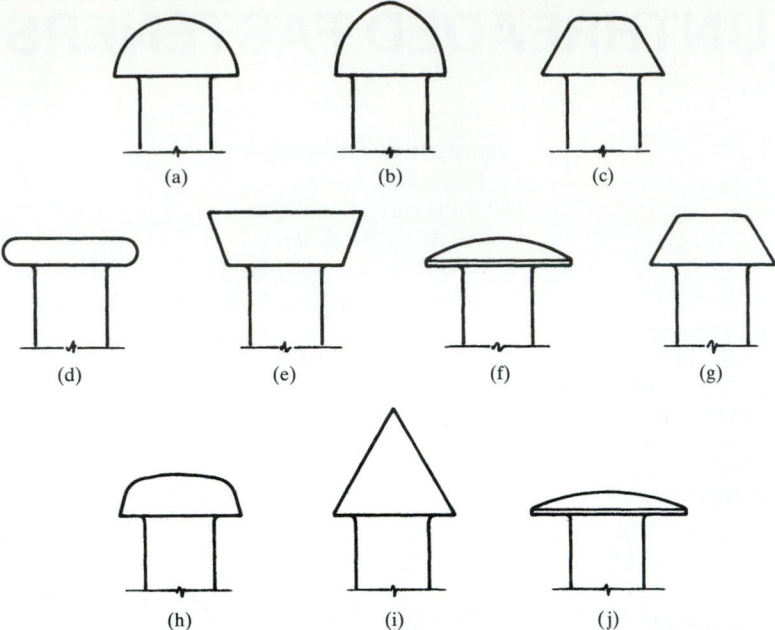

FIGURE 22.1 Standard rivet heads with flat bearing surfaces. (*a*) Button or round head; (*b*) high button or acorn head; (*c*) cone head; (*d*) flat head; (*e*) machine head; (*f*) oval head; (*g*) large pan head; (*h*) small pan head; (*i*) steeple head; (*j*) truss head, thinner than oval head.

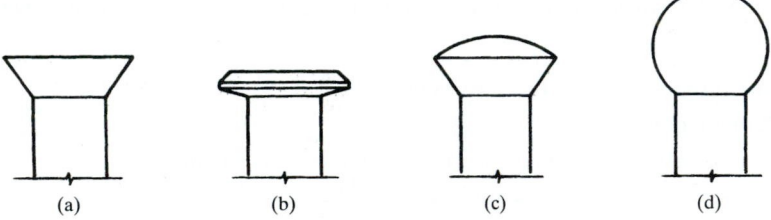

FIGURE 22.2 Various rivet heads. (*a*) Countersunk head; (*b*) countersunk head with chamfered top; (*c*) countersunk head with round top; (*d*) globe head.

A large rivet is one that has a shank diameter of ½ in or more; such rivets are mostly hot-driven. Head styles for these are button, high button, cone, countersunk, and pan. Smaller rivets are usually cold-driven. The countersunk head with chamfered flat top and the countersunk head with round top are normally used only on large rivets.

22.1.2 Rivet Types

The standard structural or machine rivet has a cylindrical shank and is either hot- or cold-driven.

A *boiler rivet* is simply a large rivet with a cone head.

A *cooper's rivet,* used for barrel-hoop joints, is a solid rivet with a head like that in Fig. 22.2*b* which has a shank end that is chamfered.

A *shoulder rivet* has a shoulder under the head.

A *tank rivet,* used for sheet-metal work, is a solid rivet with a button, counter-sunk, flat, or truss head.

A *tinner's rivet,* used for sheet-metal work, is a small solid rivet with a large flat head (Fig. 22.1*d*).

A *belt rivet,* shown in Fig. 22.3*a,* has a *riveting burr* and is used for leather or fabric joints.

A *compression* or *cutlery rivet,* shown in Fig. 22.3*b,* consists of a tubular rivet and a solid rivet. The hole and shank are sized to produce a drive fit when the joint is assembled.

A *split* or *bifurcated rivet,* shown in Fig. 22.3*c,* is a small rivet with an oval or countersunk head. The prongs cut their own holes when driven through softer metals or fibrous materials such as wood.

A *swell-neck rivet,* shown in Fig. 22.3*d,* is a large rivet which is used when a tight fit with the hole is desired.

A *tubular rivet,* shown in Fig. 22.3*e,* is a small rivet with a hole in the shank end. The rivet is cold-driven with a punchlike tool that expands or curls the shank end. *Semitubular rivets* are classified as those having hole depths less than 112 percent of the shank diameter.

A *blind rivet* is intended for use where only one side of the joint is within reach. The blind side is the side that is not accessible. However, blind rivets are also used where both sides of the joint can be accessed because of the simplicity of the assembly, the appearance of the completed joint, and the portability of the riveting tools. The rivets shown in Figs. 22.4 to 22.8 are typical of the varieties available.

22.1.3 Sizes and Materials

Large rivets are standardized in sizes from ½ to 1¾ in in ⅛-in increments. The nominal head dimensions may be calculated using the formulas in Table 22.1. The tolerances are found in Ref. [22.2]. The materials available are specified according to the following ASTM Specifications:

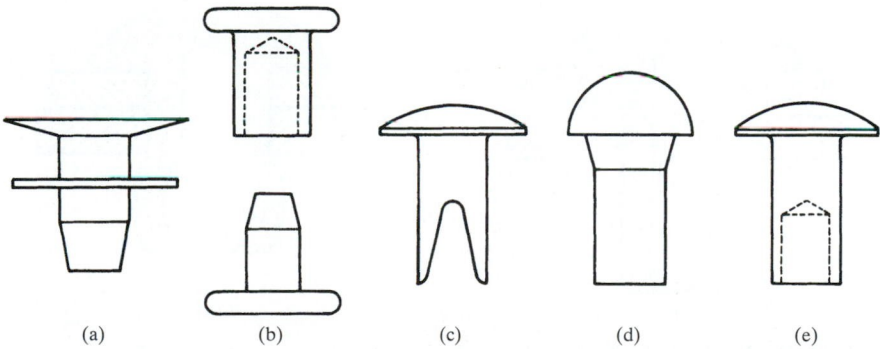

(a)   (b)   (c)   (d)   (e)

FIGURE 22.3 (*a*) Belt rivet; (*b*) compression rivet; (*c*) split rivet; (*d*) swell-neck rivet; (*e*) tubular rivet

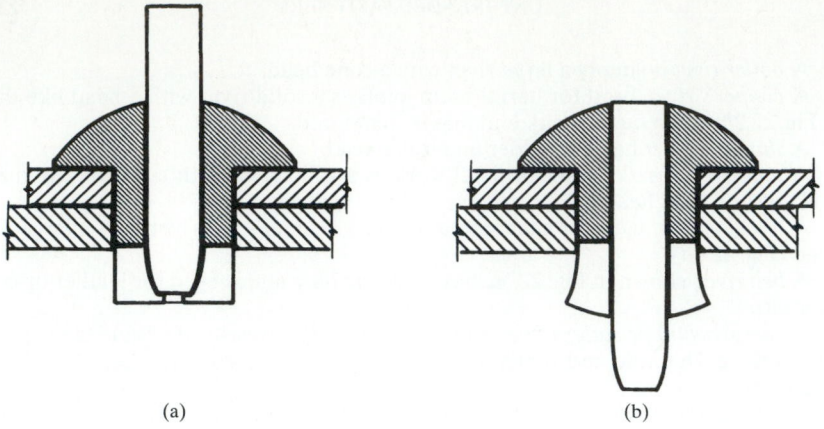

FIGURE 22.4 Drive-pin type of blind rivet. (*a*) Rivet assembled into parts; (*b*) ears at end of rivet expand outward when pin is driven.

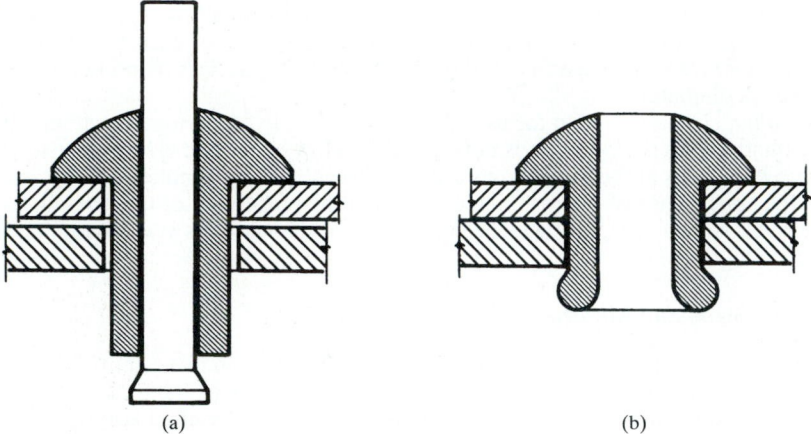

FIGURE 22.5 Pull-through-type blind riveting. (*a*) Before riveting; (*b*) after riveting.

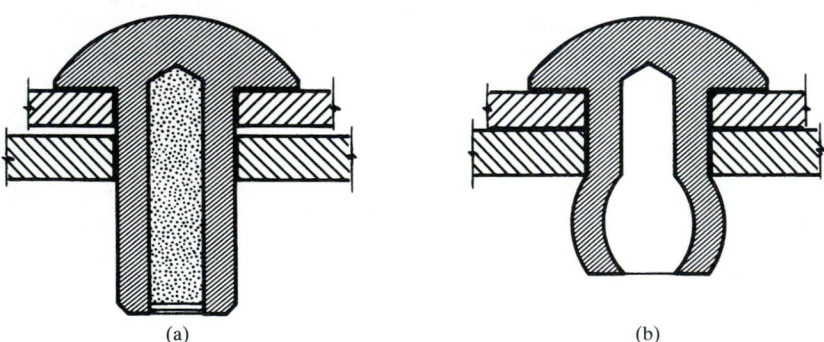

FIGURE 22.6 Explosive blind rivet. (*a*) Before explosion; (*b*) after; notice that the explosion clamps the joint.

22.4

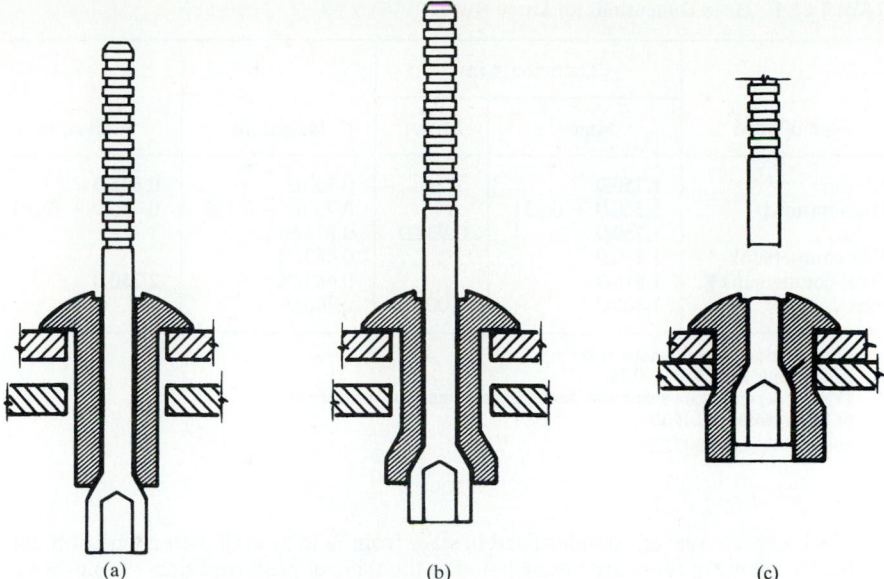

FIGURE 22.7   Self-plugging blind rivet. (*a*) Rivet inserted into prepared hole with power tool; (*b*) axial pull with power tool fills holes completely and clamps work pieces together; (*c*) stem separates flush with head and remaining section is locked in place. (*Avdel Corporation.*)

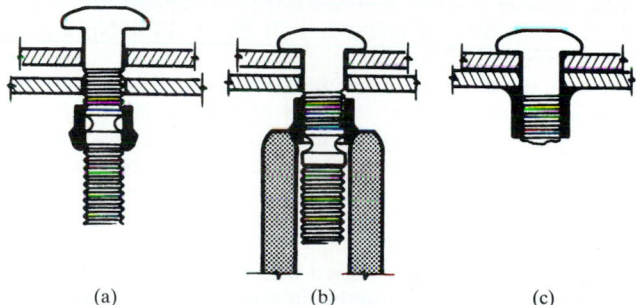

FIGURE 22.8   Lock-bolt or collar-type blind rivet. (*a*) Pin inserted through holes and collar placed over the pin tail; (*b*) nose tool pulls on the pin and reacts against the collar, clamping the work tightly; (*c*) installation finished by swaging the collar into the annular locking grooves and separating the pin at the breaker groove. (*Avdel Corporation.*)

A31 Boiler rivet steel.

A131 Rivet steel for ships.

A152 Wrought-iron rivets.

A502 Grade 1 carbon structural steel for general purposes. Grade 2 carbon-manganese steel for use with high-strength carbon and low alloy steels.

TABLE 22.1   Head Dimensions for Large Rivets

| Type of head | Diameter,† in | | Height, in | Radius, in |
	Major	Minor		
Button	1.750D	......	0.750D	0.885D
High button‡	1.500D + 0.031	......	0.750D + 0.125	0.750D + 0.281
Cone	1.750D	0.938D	0.875D	
Flat countersunk	1.810D	......	0.483D§	
Oval countersunk¶	1.810D	......	0.483D§	2.250D
Pan	1.600D	1.000D	0.700D	

†The nominal rivet diameter is D.
‡Side radius is $0.750D - 0.281$.
§Varies, depending on shank and head diameters and the included angle.
¶Crown radius is $0.190D$.
SOURCE: From Ref. [22.2].

Small solid rivets are standardized in sizes from $\frac{1}{16}$ to $\frac{7}{16}$ in in increments of $\frac{1}{32}$ in. Note that some of these are not included in the table of preferred sizes (Table 48.4). Table 22.2 is a tabulation of standard head styles available and formulas for head dimensions. ASTM standard A31 Grade A or the SAE standard J430 Grade 0 are used for small steel rivets. But other materials, such as stainless steel, brass, or aluminum may also be specified.

Tinner's and cooper's rivets are sized according to the weight of 1000 rivets. A 5-lb rivet has a shank diameter of about $\frac{3}{16}$ in. See Ref. [22.1] for sizes and head dimensions.

Belt rivets are standardized in gauge sizes from No. 14 to No. 4 using the Stubs iron-wire gauge (Table 48.17).

Tubular rivets are standardized in decimals of an inch; sizes corresponding to various head styles are listed in Tables 22.3 and 22.4. These are used with rivet caps, which are available in several styles and diameters for each rivet size. These rivets are manufactured from ductile wire using a cold-heading process. Thus any ductile material, such as steel, brass, copper, aluminum, etc., can be used. For standard tolerances, see Ref. [22.3].

Split rivet sizes are shown in Table 22.5. Split rivets are available in the same materials as tubular rivets and may be used with rivet caps too.

Some types of *blind rivets* are available in sizes from $\frac{3}{32}$ to $\frac{3}{8}$ in in diameter. The usual materials are carbon steel, stainless steel, brass, and aluminum. A variety of

TABLE 22.2   Head Dimensions for Small Solid Rivets

Head type	Diameter,† in	Height, in	Radius, in
Flat	2.000D	0.330D	
Flat countersunk	1.850D	0.425D	
Button	1.750D	0.750D	0.885D
Pan	1.720D	0.570D	3.430D‡
Truss	2.300D	0.330D	2.512D

TABLE 22.3 Sizes of Standard Semitubular Rivets[†]

Nominal size	Oval head		Truss head		Flat countersunk[‡]		Hole diameter[§]	Length increment
	Diameter	Thickness	Diameter	Thickness	Diameter	Thickness		
0.061	0.114	0.019	0.130	0.019	...	...	0.046	0.016
0.089	0.152	0.026	0.192	0.026	0.223	0.039	0.068	0.016
0.099	0.192	0.032	...	...	...	...	0.076	0.016
0.123	0.223	0.038	0.286	0.038	0.271	0.043	0.095	0.016
0.146	0.239	0.045	0.318	0.045	0.337	0.056	0.112	0.031
0.188	0.318	0.065	0.381	0.065	0.404	0.063	0.145	0.031
0.217	0.444	0.075	...	...	0.472	0.075	0.166	0.062
0.252	0.507	0.085	...	...	0.540	0.084	0.191	0.062
0.310	0.570	0.100	...	...	...	...	0.235	0.062

[†]Dimensions in inches; all values are maximums.
[‡]120-degree included angle; also available in 150-degree angle with chamfered top for friction materials.
[§]For Type T tapered hole; diameter is at end of rivet; also available as Type S straight hole.
SOURCE: From Ref. [22.3].

22.7

TABLE 22.4   Sizes of Standard Full Tubular Rivets[†]

| Head shape | Nominal size | Head | | Hole diameter |
		Diameter	Thickness	
Oval	0.146	0.239	0.045	0.107
Truss	0.146	0.318	0.045	0.107
	0.188	0.381	0.065	0.141
Flat countersunk	0.146	0.317	0.050	0.107
	0.188	0.364	0.060	0.141

[†]Dimensions in inches; all values are maximum; maximum hole depth is to head.
[‡]Chamfered.
SOURCE: From Ref. [22.3].

TABLE 22.5   Sizes of Standard Split Rivets[†]

| Nominal size | Oval head | | Flat countersunk head | |
	Diameter	Thickness	Diameter	Thickness
0.092	0.152	0.026		
0.125	0.223	0.035	0.223	0.036
0.152	0.318	0.045	0.317	0.053
0.152	. . . .	. . . .	0.380[‡]	0.062[‡]
0.190	0.349	0.055	0.443	0.061

[†]Dimensions in inches; all values are maximum.
[‡]Designates a *large flat countersunk head rivet.*
SOURCE: From Ref. [22.3].

head styles are available, but many of these are modifications of the countersunk head, the truss head, and the pan head. Head dimensions, lengths, and grips may be found in the manufacturer's catalogs.

22.2   PINS

When a joint is to be assembled in which the principal loading is shear, then the use of a pin should be considered because it may be the most cost-effective method. While a special pin can be designed and manufactured for any situation, the use of a standard pin will be cheaper.

   Taper pins (Fig. 22.9a) are sized according to the diameter at the large end, as shown in Table 22.6. The diameter at the small end can be calculated from the equation

$$d = D - 0.208L$$

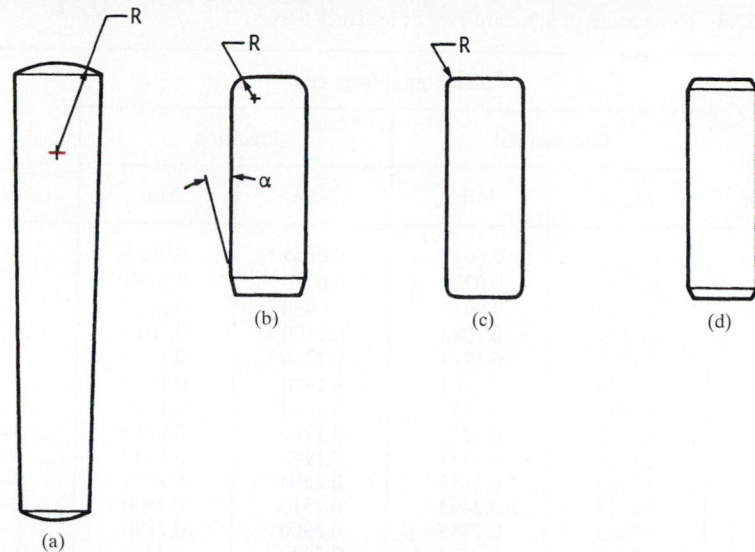

FIGURE 22.9   (*a*) Taper pin has crowned ends and a taper of 0.250 in/ft based on the diameter. (*b*) Hardened and ground machine dowel pin; the range of α is 4 to 16 degrees. (*c*) Hardened and ground production pin; corner radius is in the range 0.01 to 0.02 in. (*d*) Ground unhardened dowel pin or straight pin, both ends chamfered. Straight pins are also made with the corners broken.

where d = diameter at small end, in
D = diameter at large end, in
L = length, in

The constant in this equation is based on the taper. Taper pins can be assembled into drilled and taper-reamed holes or into holes which have been drilled by section. For the latter method, the first drill would be the smallest and would be drilled through. The next several drills would be successively larger and be drilled only part way (see Ref. [22.5]).

Dowel pins (Fig. 22.9*b*, *c*, and *d*) are listed in Tables 22.7 to 22.9 by dimensions and shear loads. They are case-hardened to a minimum case depth of 0.01 in and should have a single shear strength of 102 kpsi minimum. After hardening, the ductility should be such that they can be press-fitted into holes 0.0005 in smaller without cracking. See Chap. 19 for press fits.

Drive pins and studs are illustrated in Fig. 22.10 and tabulated in Tables 22.10 and 22.11. There are a large number of variations of these grooved drive pins. See Ref. [22.5] and manufacturers' catalogs. The standard grooved drive pin, as in Fig. 22.10*a* and *b*, has three equally spaced grooves. These pins are made from cold-drawn carbon-steel wire or rod, and the grooves are pressed or rolled into the stock. This expands the pin diameter and creates a force fit when assembled.

Spring pins are available in two forms. Figure 22.11*a* shows the slotted type of tubular spring pin. Another type, not shown, is a tubular pin made as a spiral by wrapping about 2¼ turns of sheet steel on a mandrel. This is called a *coiled spring pin*. Sizes and loads are listed in Tables 22.12 to 22.14.

TABLE 22.6   Dimensions of Standard Taper Pins (Inch Series)

Size no.	Diameter at large end				Lengths†
	Commercial		Precision		
	Max.	Min.	Max.	Min.	
7/0	0.0638	0.0618	0.0635	0.0625	$\frac{1}{4}$–1
6/0	0.0793	0.0773	0.0790	0.0780	$\frac{1}{4}$–1$\frac{1}{2}$
5/0	0.0953	0.0933	0.0950	0.0940	$\frac{1}{4}$–1$\frac{1}{2}$
4/0	0.1103	0.1083	0.1100	0.1090	$\frac{1}{4}$–2
3/0	0.1263	0.1243	0.1260	0.1250	$\frac{1}{4}$–2
2/0	0.1423	0.1403	0.1420	0.1410	$\frac{1}{4}$–2$\frac{1}{2}$
0	0.1573	0.1553	0.1570	0.1560	$\frac{1}{4}$–3
1	0.1733	0.1713	0.1730	0.1720	$\frac{3}{8}$–3
2	0.1943	0.1923	0.1940	0.1930	$\frac{3}{8}$–3
3	0.2203	0.2183	0.2200	0.2190	$\frac{1}{2}$–4
4	0.2513	0.2493	0.2510	0.2500	$\frac{1}{2}$–4
5	0.2903	0.2883	0.2900	0.2890	1–6
6	0.3423	0.3403	0.3420	0.3410	1$\frac{1}{4}$–6
7	0.4103	0.4083	0.4100	0.4090	1$\frac{1}{4}$–8
8	0.4933	0.4913	0.4930	0.4920	1$\frac{1}{4}$–8
9	0.5923	0.5903	0.5920	0.5910	1$\frac{1}{4}$–8
10	0.7073	0.7053	0.7070	0.7060	1$\frac{1}{4}$–8
11	0.8613	0.8593	. . . . .	. . . . .	2–8
12	1.0333	1.0313	. . . . .	. . . . .	2–9
13	1.2423	1.2403	. . . . .	. . . . .	3–11
14	1.5223	1.5203	. . . . .	. . . . .	3–13

†In preferred sizes but not in $\frac{1}{16}$-in increments; see Table 48.4 for list of preferred sizes in fractions of inches.

SOURCE: From Ref. [22.5].

Slotted tubular pins can be used inside one another to form a double pin, thus increasing the strength and fatigue resistance. When this is done, be sure the slots are not on the same radial line when assembled.

Clevis pins, shown in Fig. 22.11*b,* have standard sizes listed in Table 22.15. They are made of low-carbon steel and are available soft or case-hardened.

Cotter pins are listed in Table 22.16. These are available in the square-cut type, as in Fig. 22.11*c,* or as a hammer-lock type, in which the extended end is bent over the short end.

22.3   EYELETS AND GROMMETS

For some applications, eyelets are a trouble-free and economical fastener. They can be assembled very rapidly using special eyeling and grommeting machines, which punch the holes, if necessary, and then set the eyelets. The eyelets are fed automatically from a hopper to the work point.

TABLE 22.7 Dimensions of Hardened Ground Machine Dowel Pins (Inch Series) (Fig. 22.9*b*)

Nominal size	Diameter				Shear load,† kip	Length‡
	Standard series		Oversize series			
	Max.	Min.	Max.	Min.		
$\frac{1}{16}$	0.0628	0.0626	0.0636	0.0634	0.80	$\frac{1}{16}-\frac{3}{4}$
$\frac{3}{32}$	0.0941	0.0939	0.0949	0.0947	1.80	$\frac{5}{16}-1$
$\frac{1}{8}$	0.1253	0.1251	0.1261	0.1259	3.20	$\frac{3}{8}-2$
$\frac{3}{16}$	0.1878	0.1876	0.1886	0.1884	7.20	$\frac{1}{2}-2$
$\frac{1}{4}$	0.2503	0.2501	0.2511	0.2509	12.8	$\frac{1}{2}-2\frac{1}{2}$
$\frac{5}{16}$	0.3128	0.3126	0.3136	0.3134	20.0	$\frac{1}{2}-2\frac{1}{2}$
$\frac{3}{8}$	0.3753	0.3751	0.3761	0.3759	28.7	$\frac{1}{2}-3$
$\frac{7}{16}$	0.4378	0.4376	0.4386	0.4384	39.1	$\frac{7}{8}-3$
$\frac{1}{2}$	0.5003	0.5001	0.5011	0.5009	51.0	$\frac{3}{4}-4$
$\frac{5}{8}$	0.6253	0.6251	0.6261	0.6259	79.8	$1\frac{1}{4}-5$
$\frac{3}{4}$	0.7503	0.7501	0.7511	0.7509	114.0	$1\frac{1}{2}-6$
$\frac{7}{8}$	0.8753	0.8751	0.8761	0.8759	156.0	$2-6$
1	1.0003	1.0001	1.0011	1.0009	204.0	$2-6$

†Minimum double shear load for carbon or alloy steel, manufacturer's responsibility to achieve.
‡Use Table 48.4 for preferred sizes in range given.
SOURCE: From Ref. [22.5].

TABLE 22.8 Dimensions of Hardened Ground Production Dowel Pins (Inch Series) (Fig. 22.9*c*)

Nominal size	Diameter		Load,† kip	Length‡
	Max.	Min.		
$\frac{1}{16}$	0.0628	0.0626	0.79	$\frac{3}{16}-1$
$\frac{3}{32}$	0.0940	0.0938	1.40	$\frac{3}{16}-2$
$\frac{7}{64}$	0.1096	0.1094	1.90	$\frac{3}{16}-2$
$\frac{1}{8}$	0.1253	0.1251	2.60	$\frac{3}{16}-2$
$\frac{5}{32}$	0.1565	0.1563	4.10	$\frac{3}{16}-2$
$\frac{3}{16}$	0.1878	0.1876	5.90	$\frac{3}{16}-2$
$\frac{7}{32}$	0.2190	0.2188	7.60	$\frac{1}{4}-2$
$\frac{1}{4}$	0.2503	0.2501	10.0	$\frac{1}{4}-2\frac{1}{2}$
$\frac{5}{16}$	0.3128	0.3126	16.0	$\frac{5}{16}-2\frac{1}{2}$
$\frac{3}{8}$	0.3753	0.3751	23.0	$\frac{3}{8}-3$

†Minimum double shear load for carbon steel, manufacturer's responsibility to achieve.
‡See Table 48.4 for preferred sizes in range given.
SOURCE: From Ref. [22.5].

TABLE 22.9 Dimensions of Unhardened Dowel Pins and Straight Pins (Inch Series) (Fig. 22.9d)

Nominal size	Unhardened dowel pins					Straight pins	
	Diameter		Load,† kip			Diameter	
	Max.	Min.	Steel	Brass	Length‡	Max.	Min.
$\frac{1}{16}$	0.0600	0.0595	0.35	0.22	$\frac{1}{4}$–1	0.0625	0.0605
$\frac{3}{32}$	0.0912	0.0907	0.82	0.51	$\frac{1}{4}$–1$\frac{1}{2}$	0.0937	0.0917
$\frac{1}{8}$	0.1223	0.1218	1.49	0.93	$\frac{1}{4}$–2	0.1250	0.1230
$\frac{5}{32}$	0.1535	0.1530	2.35	1.47	$\frac{1}{4}$–2	0.1562	0.1542
$\frac{3}{16}$	0.1847	0.1842	3.41	2.13	$\frac{1}{4}$–2	0.1875	0.1855
$\frac{7}{32}$	0.2159	0.2154	4.66	2.91	$\frac{1}{4}$–2	0.2187	0.2167
$\frac{1}{4}$	0.2470	0.2465	6.12	3.81	$\frac{1}{4}$–2$\frac{1}{2}$	0.2500	0.2480
$\frac{5}{16}$	0.3094	0.3089	9.59	5.99	$\frac{5}{16}$–2$\frac{1}{2}$	0.3125	0.3105
$\frac{3}{8}$	0.3717	0.3712	13.85	8.65	$\frac{3}{8}$–2$\frac{1}{2}$	0.3750	0.3730
$\frac{7}{16}$	0.4341	0.4336	18.90	11.81	$\frac{7}{16}$–2$\frac{1}{2}$	0.4375	0.4355
$\frac{1}{2}$	0.4964	0.4959	24.72	15.45	$\frac{1}{2}$–3	0.5000	0.4980
$\frac{5}{8}$	0.6211	0.6206	38.71	24.19	$\frac{5}{8}$–4	0.6250	0.6230
$\frac{3}{4}$	0.7548	0.7453	55.84	34.90	$\frac{3}{4}$–4	0.7500	0.7480
$\frac{7}{8}$	0.8705	0.8700	76.09	47.55	$\frac{7}{8}$–4	0.8750	0.8730
1	0.9952	0.9947	99.46	62.16	1–4	1.0000	0.9980

†Minimum double shear load, manufacturer's responsibility to achieve.
‡See Table 48.4 for preferred sizes in range given.
SOURCE: From Ref. [22.5].

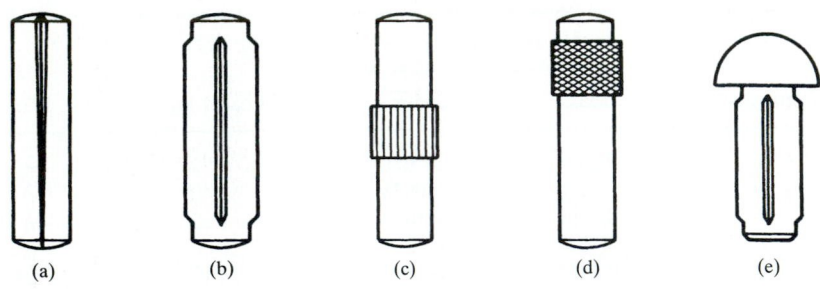

(a)   (b)   (c)   (d)   (e)

FIGURE 22.10 An assortment of drive pins. (a) Standard drive pin has three equally spaced grooves; (b) standard grooved drive pin with relief at each end; (c) (d) annular grooved and knurled drive pins; these may be obtained in a variety of configurations (*DRIV-LOK, Inc.*); (e) standard round head grooved stud.

TABLE 22.10 Dimensions of Grooved Drive Pins (Inch Series) (Fig. 22.10a, b)[†]

Basic size	Diameter Max.	Diameter Min.	Expanded diameter[‡]	Length[§]
1/32	0.0312	0.0302	0.035	1/8–1/2
3/64	0.0469	0.0459	0.051	1/8–5/8
1/16	0.0625	0.0615	0.067	1/8–1
5/64	0.0781	0.0771	0.083	1/4–1
3/32	0.0938	0.0928	0.100	1/4–1 1/4
7/64	0.1094	0.1074	0.115	1/4–1 1/4
1/8	0.1250	0.1230	0.132	1/4–1 1/2
5/32	0.1563	0.1543	0.163	3/8–2
3/16	0.1875	0.1855	0.196	3/8–2 1/4
7/32	0.2188	0.2168	0.227	1/2–2 3/4
1/4	0.2500	0.2480	0.260	1/4–3
5/16	0.3125	0.3105	0.326	5/8–3 1/4
3/8	0.3750	0.3730	0.390	3/4–4 1/4
7/16	0.4375	0.4355	0.454	7/8–4 1/2
1/2	0.5000	0.4980	0.520	1–4 1/2

[†]Reference [22.5] lists a total of seven different types of grooved drive pins.
[‡]Minimum; varies a few thousandths with length; ±0.002 in; not for Monel or stainless steel pins.
[§]In 1/8-in increments only to 1 in.
SOURCE: From Ref [22.5].

TABLE 22.11 Dimensions of Round-Head Grooved Drive Studs (Inch Series) (Fig. 22.10e)

Size no.	Basic diameter	Head diameter max.	Head thickness max.	Expanded diameter[†]	Length
0	0.067	0.130	0.050	0.074	1/8–1/4
2	0.086	0.162	0.070	0.095	1/8–1/4
4	0.104	0.211	0.086	0.113	3/16–5/16
6	0.120	0.260	0.103	0.130	1/4–3/8
7	0.136	0.309	0.119	0.144	5/16–1/2
8	0.144	0.309	0.119	0.153	3/8–5/8
10	0.161	0.359	0.136	0.171	3/8–5/8
12	0.196	0.408	0.152	0.204	1/2–3/4
14	0.221	0.457	0.169	0.232	1/2–3/4
16	0.250	0.472	0.174	0.263	1/2 only

[†]Minimum; ±0.002 in.
SOURCE: From Ref. [22.5].

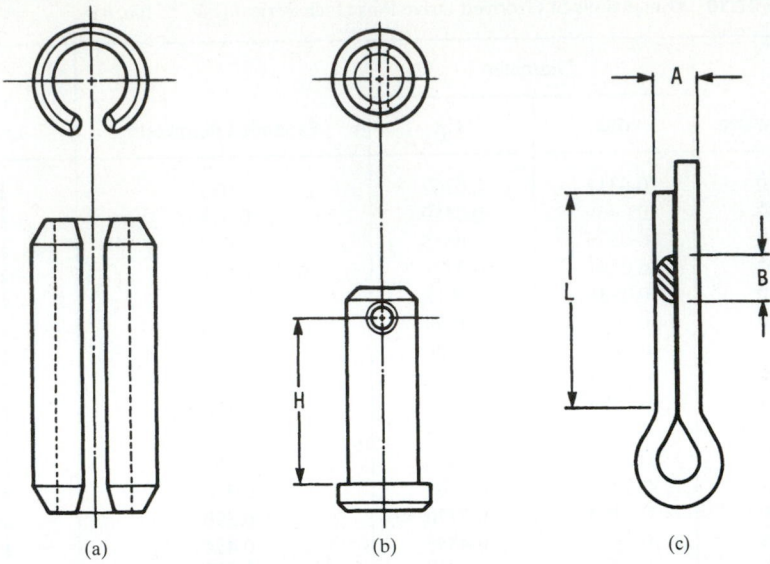

FIGURE 22.11   (*a*) Slotted spring pin; (*b*) clevis pin; (*c*) cotter pin.

TABLE 22.12   Dimensions and Safe Loads for Slotted Spring Pins (Inch Series) (Fig. 22.11*a*)

Size	Diameter		Hole size		Shear load,† kip		
	Max.	Min.	Max.	Min.	AISI 1070, AISI 1095, AISI 420	AISI 302	Beryllium copper
$\frac{1}{16}$	0.069	0.066	0.065	0.062	0.425	0.350	0.270
$\frac{5}{64}$	0.086	0.083	0.081	0.078	0.650	0.550	0.400
$\frac{3}{32}$	0.103	0.099	0.097	0.094	1.000	0.800	0.660
$\frac{1}{8}$	0.135	0.131	0.129	0.125	2.100	1.500	1.200
$\frac{9}{64}$	0.149	0.145	0.144	0.140	2.200	1.600	1.400
$\frac{5}{32}$	0.167	0.162	0.160	0.156	3.000	2.000	1.800
$\frac{3}{16}$	0.199	0.194	0.192	0.187	4.400	2.800	2.600
$\frac{7}{32}$	0.232	0.226	0.224	0.219	5.700	3.550	3.700
$\frac{1}{4}$	0.264	0.258	0.256	0.250	7.700	4.600	4.500
$\frac{5}{16}$	0.328	0.321	0.318	0.312	11.500	7.095	6.800
$\frac{3}{8}$	0.392	0.385	0.382	0.375	17.600	10.000	10.100
$\frac{7}{16}$	0.456	0.448	0.445	0.437	20.000	12.000	12.200
$\frac{1}{2}$	0.521	0.513	0.510	0.500	25.800	15.500	16.800
$\frac{5}{8}$	0.650	0.640	0.636	0.625	46.000‡	18.800	
$\frac{3}{4}$	0.780	0.769	0.764	0.750	66.000‡	23.200	

†Minimum double shear load, manufacturer's responsibility to achieve.
‡Sizes $\frac{5}{8}$ in and larger are produced only in AISI 6150H.
SOURCE: From Ref. [22.5].

TABLE 22.13 Dimensions and Safe Loads for Coiled Spring Pins (Inch Series)

Size	Light duty Diameter Max.	Light duty Diameter Min.	Light duty Safe load† kip Mat. A‡	Light duty Safe load† kip Mat. B§	Standard duty Diameter Max.	Standard duty Diameter Min.	Standard duty Safe load† kip Mat. A‡	Standard duty Safe load† kip Mat. B§	Heavy duty Diameter Max.	Heavy duty Diameter Min.	Heavy duty Safe load† kip Mat. A‡	Heavy duty Safe load† kip Mat. B§	Hole size Max.	Hole size Min.
1/32	…	…	…	…	0.035	0.033	0.075	0.060	…	…	…	…	0.032	0.031
3/64	…	…	…	…	0.052	0.049	0.170	0.140	…	…	…	…	0.048	0.046
1/16	0.073	0.067	…	0.135	0.072	0.067	0.300	0.250	0.070	0.066	0.450	0.350	0.065	0.061
5/64	0.089	0.083	…	0.225	0.088	0.083	0.475	0.400	0.086	0.082	0.700	0.550	0.081	0.077
3/32	0.106	0.099	0.375	0.300	0.105	0.099	0.700	0.550	0.103	0.099	1.000	0.800	0.097	0.093
7/64	0.121	0.114	0.525	0.425	0.120	0.114	0.950	0.750	0.118	0.113	1.400	1.250	0.112	0.108
1/8	0.139	0.131	0.675	0.550	0.138	0.131	1.250	1.000	0.136	0.130	2.100	1.700	0.129	0.124
5/32	0.172	0.163	1.100	0.875	0.171	0.163	1.925	1.550	0.168	0.161	3.000	2.400	0.160	0.155
3/16	0.207	0.196	1.500	1.200	0.205	0.196	2.800	2.250	0.202	0.194	4.400	3.500	0.192	0.185
7/32	0.240	0.228	2.100	1.700	0.238	0.228	3.800	3.000	0.235	0.226	5.700	4.600	0.224	0.217
1/4	0.273	0.260	2.700	2.200	0.271	0.260	5.000	4.000	0.268	0.258	7.700	6.200	0.256	0.247
5/16	0.339	0.324	4.440	3.500	0.337	0.324	7.700	6.200	0.334	0.322	11.500	9.200	0.319	0.308
3/8	0.405	0.388	6.000	5.000	0.403	0.388	11.200	9.000	0.400	0.386	17.600	14.000	0.383	0.370
7/16	0.471	0.452	8.400	6.700	0.469	0.452	15.200	13.000	0.466	0.450	22.500	18.000	0.446	0.431
1/2	0.537	0.516	11.000	8.800	0.535	0.516	20.000	16.000	0.532	0.514	30.000	24.000	0.510	0.493
5/8	…	…	…	…	0.661	0.642	31.000	25.000	0.658	0.640	46.000	37.000	0.635	0.618
3/4	…	…	…	…	0.787	0.768	45.000	36.000	0.784	0.766	66.000	53.000	0.760	0.743

†Minimum double shear load, manufacturer's responsibility to achieve.
‡Material A is AISI 1070, AISI 1095, or AISI 420; sizes ½in and ⅝in are available only in AISI 420; sizes ¾in and larger are available only in AISI 6150 steel.
§Material B is AISI 302.
SOURCE: From Ref. [22.5].

TABLE 22.14   Standard Lengths of Coiled and Slotted Spring Pins (Inch Series)[†]

Size	Length	Size	Length	Size	Length
$\frac{1}{32}$	$\frac{1}{4}$–$\frac{5}{8}$	$\frac{1}{8}$	$\frac{5}{16}$–2	$\frac{5}{16}$	$\frac{3}{4}$–4
$\frac{3}{64}$	$\frac{1}{4}$–$\frac{5}{8}$	$\frac{9}{64}$	$\frac{3}{8}$–2	$\frac{3}{8}$	$\frac{3}{4}$–4
$\frac{1}{16}$	$\frac{3}{16}$–1	$\frac{5}{32}$	$\frac{7}{16}$–$2\frac{1}{2}$	$\frac{7}{16}$	1–4
$\frac{5}{64}$	$\frac{3}{16}$–$1\frac{1}{2}$	$\frac{3}{16}$	$\frac{1}{2}$–$2\frac{1}{2}$	$\frac{1}{2}$	$1\frac{1}{4}$–4
$\frac{3}{32}$	$\frac{3}{16}$–$1\frac{1}{2}$	$\frac{7}{32}$	$\frac{1}{2}$–3	$\frac{5}{8}$	2–6
$\frac{7}{64}$	$\frac{1}{4}$–$1\frac{3}{4}$	$\frac{1}{4}$	$\frac{1}{2}$–$3\frac{1}{2}$		

[†]See Table 48.4 for list of preferred lengths.
SOURCE: From Ref. [22.5].

TABLE 22.15   Dimensions of Clevis Pins (Inch Series)

Size	Diameter		Maximum head		Hole Minimum	Distance H[†]		Length	Cotter pin size
	Max.	Min.	Diameter	Thickness		Max.	Min.		
$\frac{3}{16}$	0.186	0.181	0.32	0.07	0.073	0.504	0.484	0.58	$\frac{1}{16}$
$\frac{1}{4}$	0.248	0.243	0.38	0.10	0.073	0.692	0.672	0.77	$\frac{1}{16}$
$\frac{5}{16}$	0.311	0.306	0.44	0.10	0.104	0.832	0.812	0.94	$\frac{3}{32}$
$\frac{3}{8}$	0.373	0.368	0.51	0.13	0.104	0.958	0.938	1.06	$\frac{3}{32}$
$\frac{7}{16}$	0.436	0.431	0.57	0.16	0.104	1.082	1.062	1.19	$\frac{3}{32}$
$\frac{1}{2}$	0.496	0.491	0.63	0.16	0.136	1.223	1.203	1.36	$\frac{1}{8}$
$\frac{5}{8}$	0.621	0.616	0.82	0.21	0.136	1.473	1.453	1.61	$\frac{1}{8}$
$\frac{3}{4}$	0.746	0.741	0.94	0.26	0.167	1.739	1.719	1.91	$\frac{3}{32}$
$\frac{7}{8}$	0.871	0.866	1.04	0.32	0.167	1.989	1.969	2.16	$\frac{3}{32}$
1	0.996	0.991	1.19	0.35	0.167	2.239	2.219	2.41	$\frac{3}{32}$

[†]To hole center; see Fig. 22.11b.
SOURCE: From Ref. [22.4].

Figure 22.12 illustrates some of the more common eyelets and grommets. These are available in many other styles and in thousands of sizes. The usual materials are brass, copper, zinc, aluminum, steel, and nickel silver. Various finishing operations such as plating, anodizing, or lacquering can also be employed.

22.4   RETAINING RINGS

Shoulders are used on shafts and on the interior of bored parts to accurately position or retain assembled parts to prevent axial motion or play. It is often advantageous to use retaining rings as a substitute for these machined shoulders. Such rings can be used to axially position parts on shafts and in housing bores and often save a great deal in machining costs.

TABLE 22.16 Dimensions of Cotter Pins (Inch Series) (Fig. 22.11c)

	Shank diameter A		Wire width B		Hole size
Size	Max.	Min.	Max.	Min.	
1/32	0.032	0.028	0.032	0.022	0.047
3/64	0.048	0.044	0.048	0.035	0.062
1/16	0.060	0.056	0.060	0.044	0.078
5/64	0.076	0.072	0.076	0.057	0.094
3/32	0.090	0.086	0.090	0.069	0.109
7/64	0.104	0.100	0.104	0.080	0.125
1/8	0.120	0.116	0.120	0.093	0.141
9/64	0.134	0.130	0.134	0.104	0.156
5/32	0.150	0.146	0.150	0.116	0.172
3/16	0.176	0.172	0.176	0.137	0.203
7/32	0.207	0.202	0.207	0.161	0.234
1/4	0.225	0.220	0.225	0.176	0.266
5/16	0.280	0.275	0.280	0.220	0.312
3/8	0.335	0.329	0.335	0.263	0.375
7/16	0.406	0.400	0.406	0.320	0.438
1/2	0.473	0.467	0.473	0.373	0.500
5/8	0.598	0.590	0.598	0.472	0.625
3/4	0.723	0.715	0.723	0.572	0.750

SOURCE: From Ref. [22.4].

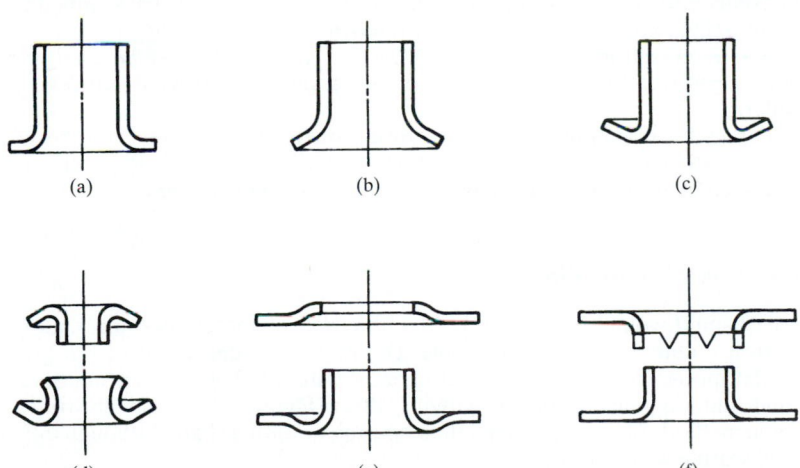

FIGURE 22.12 (a) Flat-flange eyelet; (b) funnel-flange eyelet; (c) rolled-flange eyelet; (d) telescoping eyelet with neck washer; (e) plain grommet; (f) toothed grommet.

Retaining rings may be as simple as a hardened spring wire bent into a C or U shape and fitted into a groove on a shaft or a housing. *Spiral-wound* and *stamped retaining rings* have been standardized (Refs. [22.7], [22.8], and [22.9]), and they are available in many shapes and sizes from various manufacturers.

22.4.1 Stamped Retaining Rings

Figure 22.13 shows a large variety of retaining rings. These are designated using the catalog numbers of a manufacturer, but can be changed to military standard numbers using Table 22.17.

The E rings shown in Fig. 22.13*a, b,* and *c* are intended to provide wide shoulders on small-diameter shafts. They are assembled by snapping them on in a radial direction. They are very satisfactory substitutes for cotter pins or the more expensive shaft shoulders or collars secured by set screws. Figure 22.14 shows typical mounting details for the rings in Fig. 22.13*a* and *b*. The ring in Fig. 22.13*c* is similar but is reinforced with tapered web sections for greater resistance to vibration and shock loads.

The C ring in Fig. 22.13*d* is also assembled radially, as will be shown in Fig. 22.17*a*. This ring is useful when axial access to the groove is difficult and for applications in which only a small shoulder is desired.

The internal rings in Fig. 22.13*e* and *f* are shown assembled in Fig. 22.15*a* and *b*. These are applied axially into grooved housings using specially designed pliers.

The external rings shown in Fig. 22.13*g* and *h* are shown assembled in Fig. 22.16. They are also assembled axially using pliers. Note how the bowed or dished ring in Fig. 22.16*b* can be used to take up end play or allow for temperature-induced dimensional changes.

The self-locking rings in Fig. 22.13*k* and *l* do not require grooves. They provide shoulders in soft materials, such as low-carbon steels or plastics, merely by pushing them axially into position. When a reverse force is applied, the prongs embed themselves into the mating material and resist removal.

The external self-locking ring in Figs. 22.13*m* and 22.17*b* may be used with or without a groove. This ring resists moderate thrust and provides an adjustable shoulder.

Materials for retaining rings are the spring steels, stainless steel, and beryllium copper. For dimensions and loads, see Refs. [22.7], [22.8], and [22.9] and the manufacturers' catalogs. They are available in both inch and metric sizes.

22.4.2 Spiral Wound Rings

Standard spiral-wound rings (Ref. [22.7]) have approximately two turns, although three-turn retaining rings are available. The rings are edge-wound from pretempered flat spring wire. The crimp or offset of the wire (see Fig. 22.18) produces a better seat, but rings are available without offset. Figure 22.18 also illustrates the machine methods of seating a ring into a housing or onto a shaft. Although difficult, manual seating is also possible.

Spiral-wound rings are sized by the inside diameter when they are to be used on a shaft and by the outside diameter when they are to be used in a housing. For sizes and thrust loads, see the manufacturers' catalogs. Usual materials are the plain carbon spring steels, stainless steel, nickel alloys, and beryllium copper.

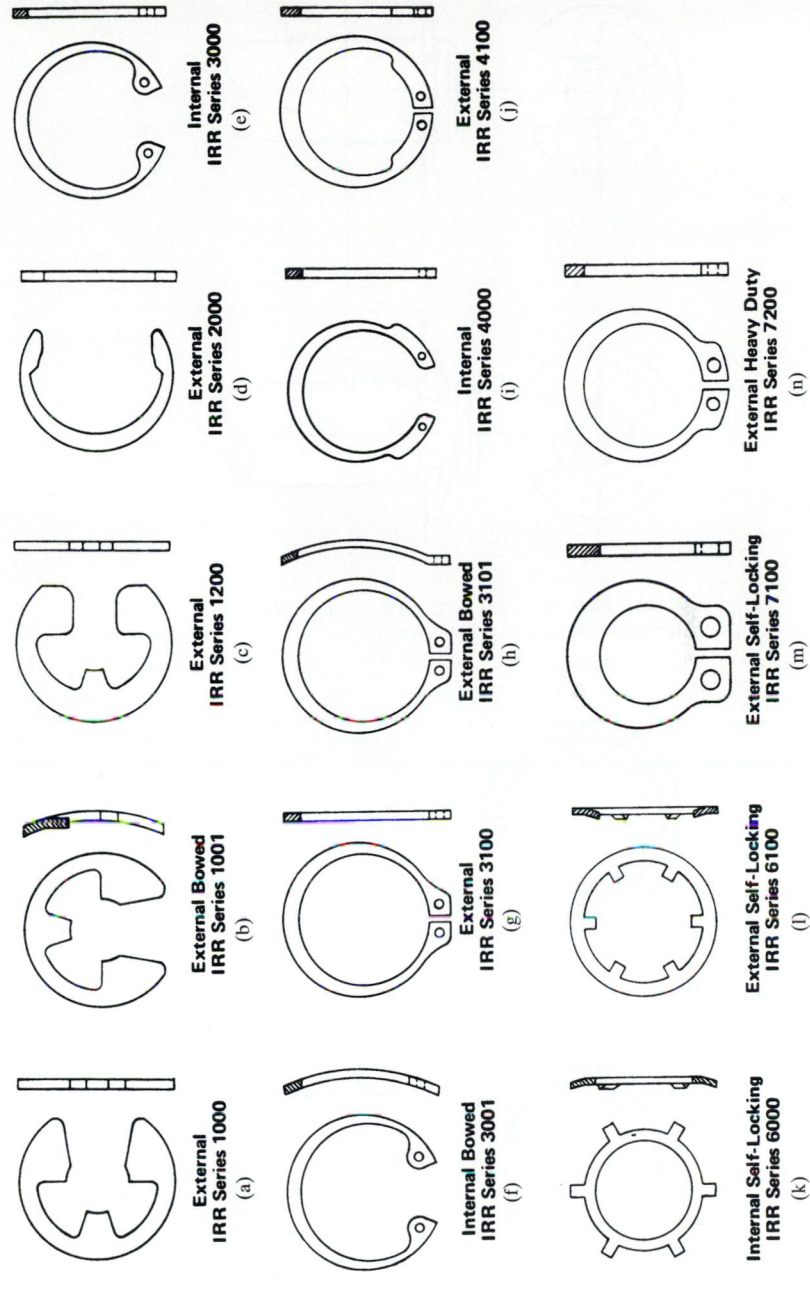

FIGURE 22.13 Retaining rings. The IRR numbers are catalog numbers. See Table 22.17 for conversion to military standard numbers. (*Industrial Retaining Ring Company.*)

External
IRR Series 1000
(a)

External Bowed
IRR Series 1001
(b)

External
IRR Series 1200
(c)

External
IRR Series 2000
(d)

Internal
IRR Series 3000
(e)

Internal Bowed
IRR Series 3001
(f)

External
IRR Series 3100
(g)

External Bowed
IRR Series 3101
(h)

Internal
IRR Series 4000
(i)

External
IRR Series 4100
(j)

Internal Self-Locking
IRR Series 6000
(k)

External Self-Locking
IRR Series 6100
(l)

External Self-Locking
IRR Series 7100
(m)

External Heavy Duty
IRR Series 7200
(n)

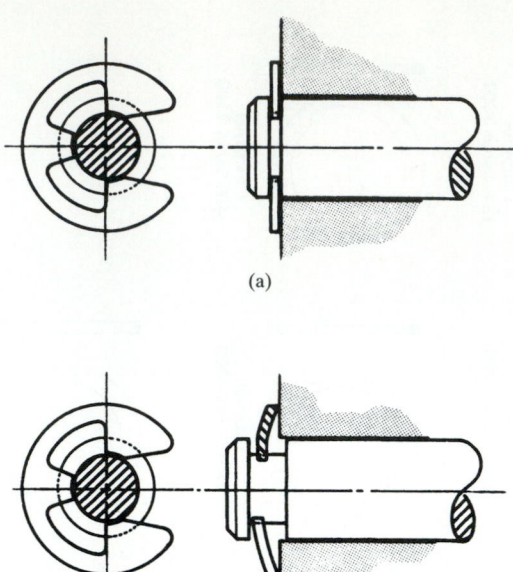

(a)

(b)

FIGURE 22.14 Open-type E rings. (*a*) Flat; (*b*) bowed.
(Industrial Retaining Ring Company.)

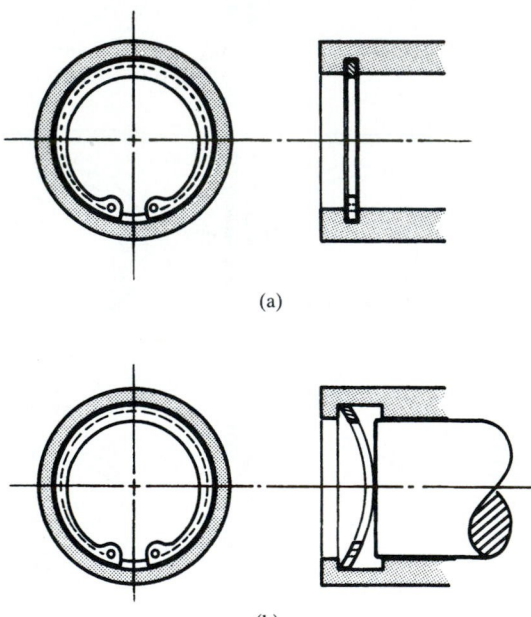

(a)

(b)

FIGURE 22.15 Internal rings. (*a*) Flat type (see Fig. 22.13*e*
for shape before assembly); (*b*) bowed type (see Fig. 22.13*f* for
shape before assembly).

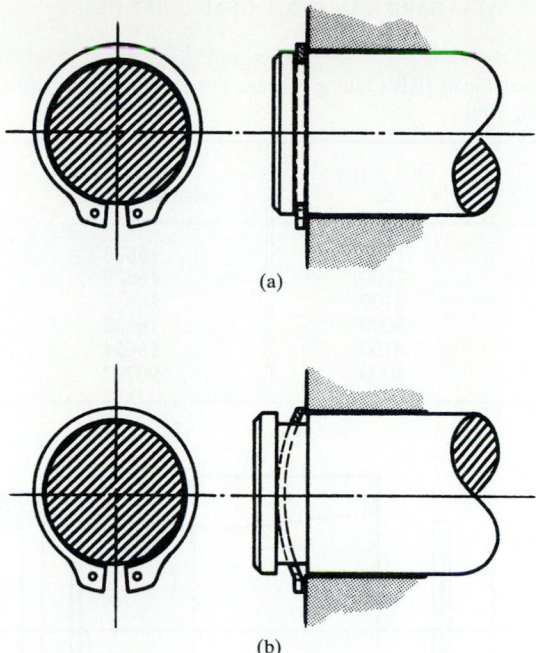

(a)

(b)

FIGURE 22.16 External rings. (*a*) Flat; (*b*) bowed. *(Industrial Retaining Ring Company.)*

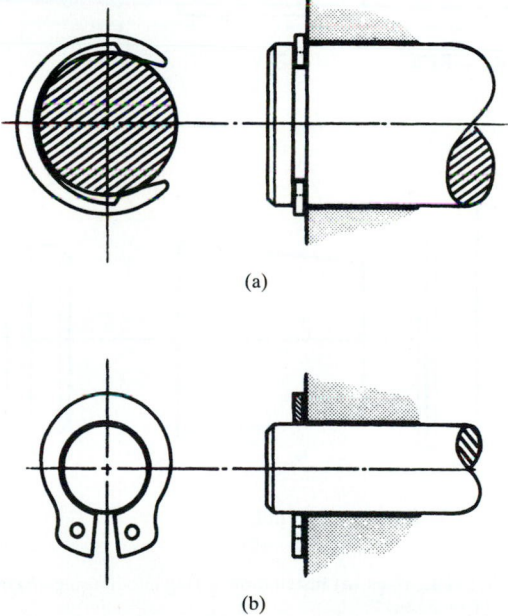

(a)

(b)

FIGURE 22.17 (*a*) External C-ring; (*b*) self-locking external ring. *(Industrial Retaining Ring Company.)*

TABLE 22.17 Conversion of IRR Catalog Numbers to Corresponding Military Standard Numbers of Retaining Rings

Government standard MS no.	IRR series no.	Government standard MS no.	IRR series no.
3215	1200	16628	3101
3217	7200	16629	3001
16624	3100	16632	2000
16625	3000	16633	1000
16626	4100	16634	1001
16627	4000	90707	7100

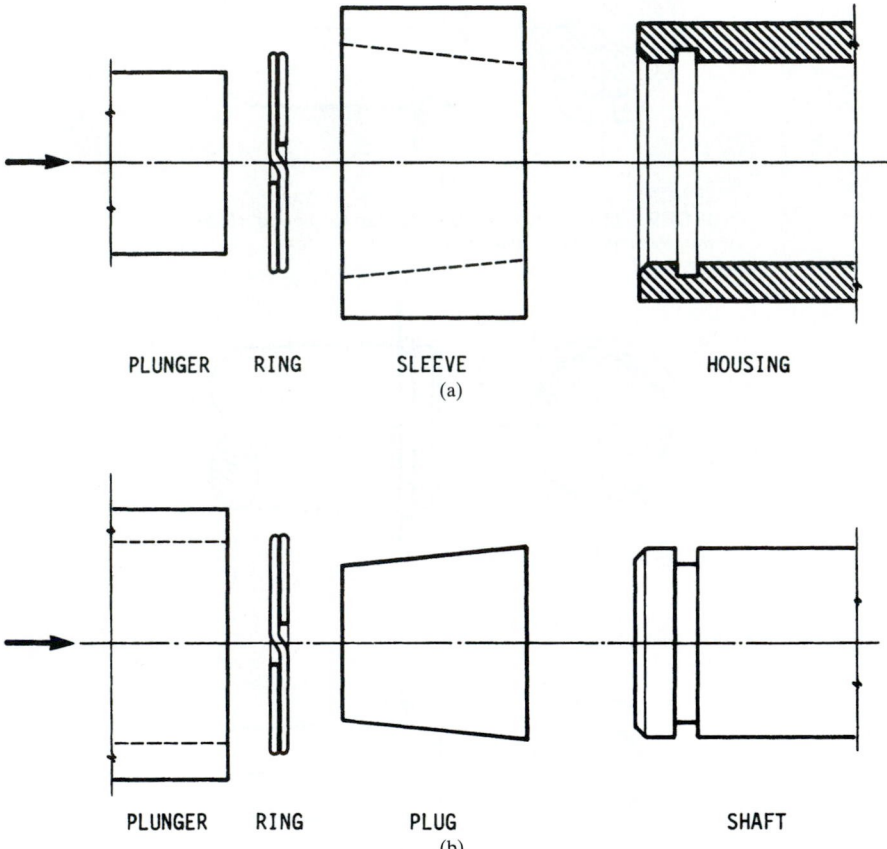

FIGURE 22.18 Spiral retaining rings. (*a*) Installation of ring into housing; (*b*) installation of ring onto shaft. *(Smalley Steel Ring Company.)*

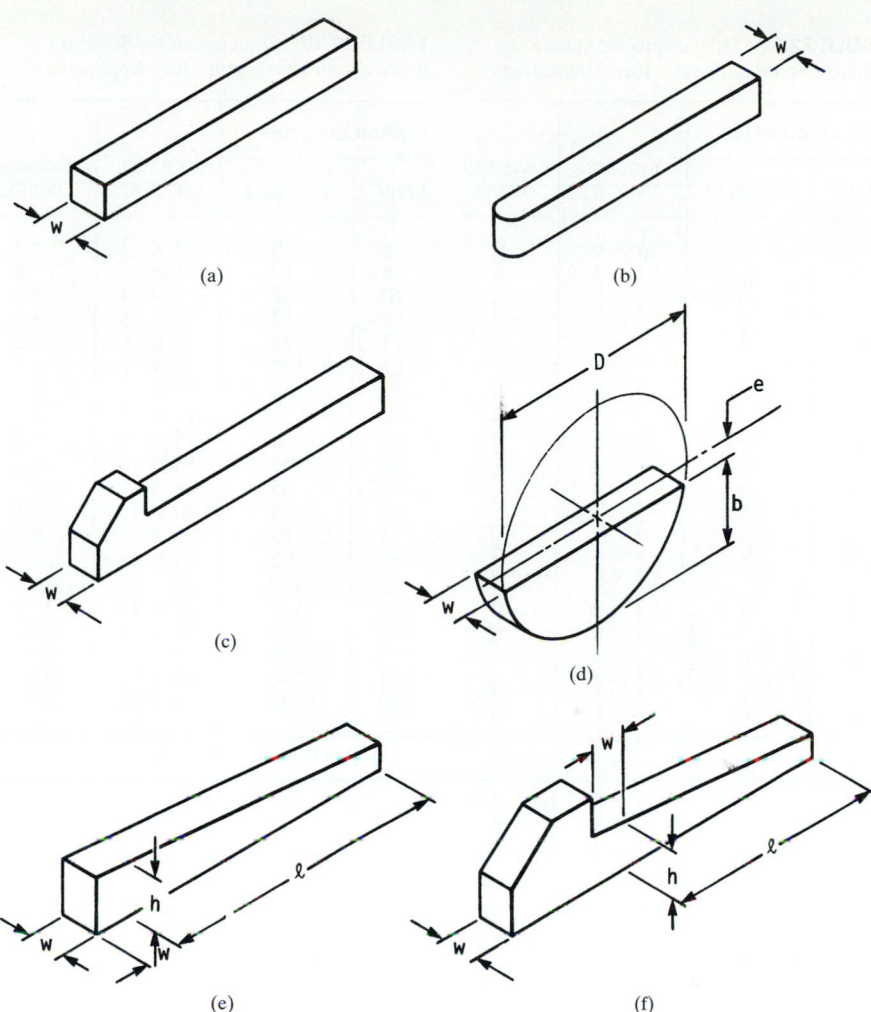

FIGURE 22.19 (*a*) Square or rectangular key. (*b*) Square or rectangular key with one end rounded; also available with both ends rounded. (*c*) Square or rectangular key with gib head. (*d*) Woodruff key; also available with flattened bottom. (*e*) Tapered rectangular key; ℓ = hub length, h = height; taper is ⅛ in for 12 in or 1 for 100 for metric sizes. (*f*) Tapered gib-head key; dimensions and taper same as in (*e*).

TABLE 22.18 Dimensions for Standard Square- and Rectangular-Key Applications[†]

Shaft diameter		Key size, $w \times h$	Keyway depth
Over	To (incl.)		
$\frac{5}{16}$	$\frac{7}{16}$	$\frac{3}{32} \times \frac{3}{32}$	$\frac{3}{64}$
$\frac{7}{16}$	$\frac{9}{16}$	$\frac{1}{8} \times \frac{3}{32}$	$\frac{3}{64}$
		$\frac{1}{8} \times \frac{1}{8}$	$\frac{1}{16}$
$\frac{9}{16}$	$\frac{7}{8}$	$\frac{3}{16} \times \frac{1}{8}$	$\frac{1}{16}$
		$\frac{3}{16} \times \frac{3}{16}$	$\frac{3}{32}$
$\frac{7}{8}$	$1\frac{1}{4}$	$\frac{1}{4} \times \frac{3}{16}$	$\frac{3}{32}$
		$\frac{1}{4} \times \frac{1}{4}$	$\frac{1}{8}$
$1\frac{1}{4}$	$1\frac{3}{8}$	$\frac{5}{16} \times \frac{1}{4}$	$\frac{1}{8}$
		$\frac{5}{16} \times \frac{5}{16}$	$\frac{5}{32}$
$1\frac{3}{8}$	$1\frac{3}{4}$	$\frac{3}{8} \times \frac{1}{4}$	$\frac{1}{8}$
		$\frac{3}{8} \times \frac{3}{8}$	$\frac{3}{16}$
$1\frac{3}{4}$	$2\frac{1}{4}$	$\frac{1}{2} \times \frac{3}{8}$	$\frac{3}{16}$
		$\frac{1}{2} \times \frac{1}{2}$	$\frac{1}{4}$
$2\frac{1}{4}$	$2\frac{3}{4}$	$\frac{5}{8} \times \frac{7}{16}$	$\frac{7}{32}$
		$\frac{5}{8} \times \frac{5}{8}$	$\frac{5}{16}$
$2\frac{3}{4}$	$3\frac{1}{4}$	$\frac{3}{4} \times \frac{1}{2}$	$\frac{1}{4}$
		$\frac{3}{4} \times \frac{3}{4}$	$\frac{3}{8}$
$3\frac{1}{4}$	$3\frac{3}{4}$	$\frac{7}{8} \times \frac{5}{8}$	$\frac{5}{16}$
		$\frac{7}{8} \times \frac{7}{8}$	$\frac{7}{16}$
$3\frac{3}{4}$	$4\frac{1}{2}$	$1 \times \frac{3}{4}$	$\frac{3}{8}$
		1×1	$\frac{1}{2}$
$4\frac{1}{2}$	$5\frac{1}{2}$	$1\frac{1}{4} \times \frac{7}{8}$	$\frac{7}{16}$
		$1\frac{1}{4} \times 1\frac{1}{4}$	$\frac{5}{8}$
$5\frac{1}{2}$	$6\frac{1}{2}$	$1\frac{1}{2} \times 1$	$\frac{1}{2}$
		$1\frac{1}{2} \times 1\frac{1}{2}$	$\frac{3}{4}$

[†]Dimensions in inches
SOURCE: From Ref. [22.10].

TABLE 22.19 Dimensions for Standard Square- and Rectangular-Key Applications[†]

Shaft diameter		Key size, $w \times h$	Keyway depth
Over	To (incl.)		
6	8	2 × 2	1.2
8	10	3 × 3	1.8
10	12	4 × 4	2.5
12	17	5 × 5	3
17	22	6 × 6	3.5
22	30	8 × 7	4
30	38	10 × 8	5
38	44	12 × 8	5
44	50	14 × 9	5.5
50	58	16 × 10	6
58	65	18 × 11	7
65	75	20 × 12	7.5
75	85	22 × 14	9
85	95	25 × 14	9
95	110	28 × 16	10
110	130	32 × 18	11
130	150	36 × 20	12
150	170	40 × 22	13
170	200	45 × 25	15
200	230	50 × 28	17

[†]Dimensions in millimeters.

A *wave spring* is a one-turn edge-wound spring washer also made from flat spring wire. A thrust load tends to flatten the spring, and hence such springs can be used to take up end play or to allow for expansion. Several of these can be used together, either crest-to-crest or nested, depending on the requirements for thrust loads or axial motion.

22.5 KEYS

All standard plain, tapered, and Woodruff keys are illustrated in Fig. 22.19. These are usually made with edges broken, but they may be chamfered if fillets are used in the

TABLE 22.20 Dimensions for Woodruff-Key Applications (Fig. 22.19d)[†]

Key size, $w \times D$	Height‡ b	Offset e	Keyseat depth Shaft	Keyseat depth Hub
$\frac{1}{16} \times \frac{1}{4}$	0.109	$\frac{1}{64}$	0.0728	0.0372
$\times \frac{5}{16}$	0.140	$\frac{1}{64}$	0.1038	0.0372
$\times \frac{3}{8}$	0.172	$\frac{1}{64}$	0.1358	0.0372
$\frac{3}{32} \times \frac{5}{16}$	0.140	$\frac{1}{64}$	0.0882	0.0529
$\times \frac{3}{8}$	0.172	$\frac{1}{64}$	0.1202	0.0529
$\times \frac{1}{2}$	0.203	$\frac{3}{64}$	0.1511	0.0529
$\times \frac{5}{8}$	0.250	$\frac{1}{16}$	0.1981	0.0529
$\frac{1}{8} \times \frac{3}{8}$	0.172	$\frac{1}{64}$	0.1045	0.0685
$\times \frac{1}{2}$	0.203	$\frac{3}{64}$	0.1355	0.0685
$\times \frac{5}{8}$	0.250	$\frac{1}{16}$	0.1825	0.0685
$\times \frac{3}{4}$	0.313	$\frac{1}{16}$	0.2455	0.0685
$\frac{5}{32} \times \frac{5}{8}$	0.250	$\frac{1}{16}$	0.1669	0.0841
$\times \frac{3}{4}$	0.313	$\frac{1}{16}$	0.2299	0.0841
$\times \frac{7}{8}$	0.375	$\frac{1}{16}$	0.2919	0.0841
$\frac{3}{16} \times \frac{5}{8}$	0.250	$\frac{1}{16}$	0.1513	0.0997
$\times \frac{3}{4}$	0.313	$\frac{1}{16}$	0.2143	0.0997
$\times \frac{7}{8}$	0.375	$\frac{1}{16}$	0.2763	0.0997
$\times 1$	0.438	$\frac{1}{16}$	0.3393	0.0997
$\times 1\frac{1}{8}$	0.484	$\frac{5}{64}$	0.3853	0.0997
$\times 1\frac{1}{4}$	0.547	$\frac{5}{64}$	0.4483	0.0997
$\times 2\frac{1}{8}$	0.406	$\frac{21}{32}$	0.3073	0.0997
$\frac{7}{32} \times \frac{7}{8}$	0.375	$\frac{1}{16}$	0.2607	0.1153
$\times 1$	0.438	$\frac{1}{16}$	0.3237	0.1153
$\times 1\frac{1}{8}$	0.484	$\frac{5}{64}$	0.3697	0.1153
$\times 1\frac{1}{4}$	0.547	$\frac{5}{64}$	0.4327	0.1153
$\frac{1}{4} \times \frac{3}{4}$	0.313	$\frac{1}{16}$	0.1830	0.1310
$\times \frac{7}{8}$	0.375	$\frac{1}{16}$	0.2450	0.1310
$\times 1$	0.438	$\frac{1}{16}$	0.3080	0.1310
$\times 1\frac{1}{8}$	0.484	$\frac{5}{64}$	0.3540	0.1310
$\times 1\frac{1}{4}$	0.547	$\frac{5}{64}$	0.4170	0.1310
$\times 1\frac{3}{8}$	0.594	$\frac{3}{32}$	0.4640	0.1310
$\times 1\frac{1}{2}$	0.641	$\frac{7}{64}$	0.5110	0.1310
$\times 2\frac{1}{8}$	0.531	$\frac{17}{32}$	0.4010	0.1310
$\times 2\frac{3}{4}$	0.750	$\frac{5}{8}$	0.4640	0.1310
$\frac{5}{16} \times 1$	0.438	$\frac{1}{16}$	0.2768	0.1622
$\times 1\frac{1}{8}$	0.484	$\frac{5}{64}$	0.3228	0.1622
$\times 1\frac{1}{4}$	0.547	$\frac{5}{64}$	0.3858	0.1622
$\times 1\frac{3}{8}$	0.594	$\frac{3}{32}$	0.4328	0.1622
$\times 1\frac{1}{2}$	0.641	$\frac{7}{64}$	0.4798	0.1622
$\times 2\frac{1}{8}$	0.531	$\frac{17}{32}$	0.3698	0.1622
$\times 2\frac{3}{4}$	0.750	$\frac{5}{8}$	0.5888	0.1622
$\frac{3}{8} \times 1$	0.438	$\frac{1}{16}$	0.2455	0.1935
$\times 1\frac{1}{4}$	0.547	$\frac{5}{64}$	0.3545	0.1935
$\times 1\frac{3}{8}$	0.594	$\frac{3}{32}$	0.4015	0.1935
$\frac{3}{8} \times 1\frac{1}{2}$	0.641	$\frac{7}{64}$	0.4485	0.1935

TABLE 22.20 Dimensions for Woodruff-Key Applications (Fig. 22.19d)[†]
(Continued)

Key size, $w \times D$	Height‡ b	Offset e	Keyseat depth Shaft	Keyseat depth Hub
× 2⅛	0.531	17/32	0.3385	0.1935
× 2¾	0.750	⅝	0.5575	0.1935
× 3½	0.938	13/16	0.7455	0.1935
⁷⁄₁₆× 2¾	0.750	⅝	0.5263	0.2247
× 3½	0.938	13/16	0.7143	0.2247
½ × 2¾	0.750	⅝	0.4950	0.2560
× 3½	0.938	13/16	0.6830	0.2560
⁹⁄₁₆× 3½	0.938	13/16	0.6518	0.2872
⅝ × 3½	0.938	13/16	0.6205	0.3185
¹¹⁄₁₆× 3½	0.938	13/16	0.5893	0.3497
¾ × 3½	0.938	13/16	0.5580	0.3810

†All dimensions in inches. If catalog or key numbers are given, the last two digits correspond to the nominal diameter D in eighths of an inch. The preceding digits give the nominal width w in thirty-seconds of an inch. Thus key no. 1208 is a size ⅜× 1.

‡This is the maximum height for a full-radius key; this dimension will be slightly less for a flat-bottom key.

SOURCE: From Ref. [22.11].

keyseats. Standard sizes and keyseat dimensions needed for design are given in Tables 22.18 to 22.20.

22.6 WASHERS

Plain washers, shown in Fig. 22.20a, are flat and circular and are used on bolts and screws. They are applied under the nut, under the head, or both. Plain washers can also be made square or triangular and are sometimes beveled for use on an inclined surface.

Cylindrically curved or bent washers, shown in Fig. 22.20b, are useful in certain applications as a means of obtaining additional bolt tension in the joint.

Conical or Belleville washers, shown in Fig. 22.20c and d, are springs and are useful for heavy loads with small deflections and where a nonlinear force-deflection relation is desired. See Chap. 24 for more details.

Spring washers, shown in Fig. 22.20e and f, are hardened circular washers that are split and then bent out of a flat plane. They are sometimes called *lock washers,* although their principal purpose is to take up for relaxing bolt tension or looseness in the joint.

Wood-grip washers, shown in Fig. 22.21a, are useful on soft materials, such as wood. When the joint is tightened, the bent-over end penetrates and grips the mating material.

Horseshoe or C-washers are useful where it is desirable to remove the washer without unbolting the connection (see Fig. 22.21b).

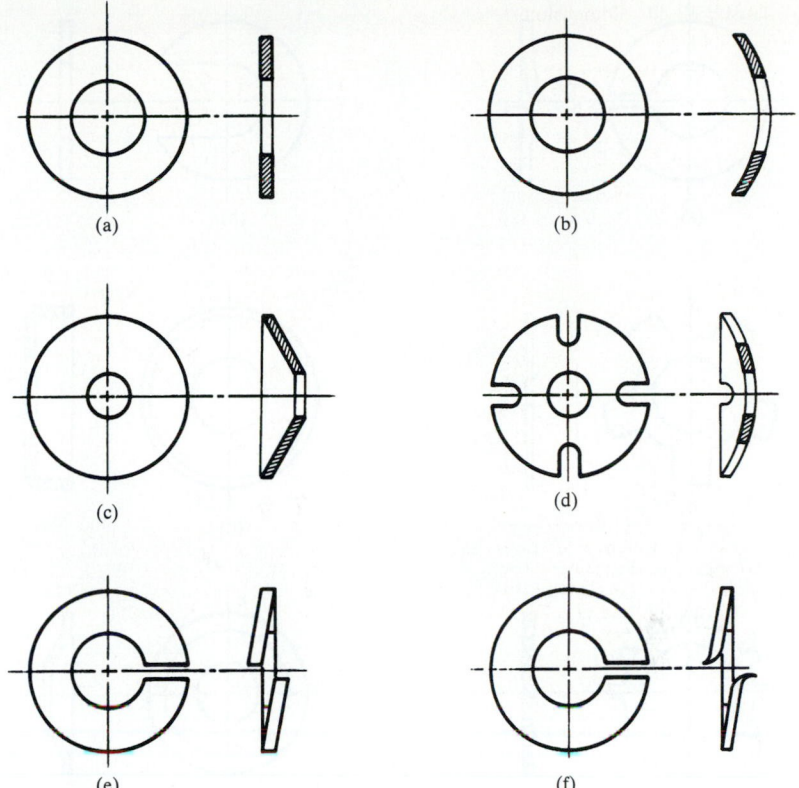

FIGURE 22.20 Washers. (*a*) Plain; (*b*) cylindrically curved; (*c*) conical or Belleville; (*d*) slotted; (*e*) spring; (*f*) spring-locking.

Lockplate or eared washers, shown in Fig. 22.21*c,* are used for locking purposes by bending some of the ears *up* against the flats of the nut or bolt head and others *down* over the edges of the joint members so as to prevent rotation of the nut or bolt head.

Cup washers, shown in Fig. 22.21*d,* are also available with a flange. When the depth of the cup is shallow, they are also called *back-up washers.*

Toothed lock washers, shown in Fig. 22.21*e,* have the teeth or prongs twisted so as to bite or penetrate the nut face as well as the adjoining part. These are hardened and made either with internal teeth or as internal-external toothed washers.

Countersunk washers, shown in Fig. 22.21*f,* serve the same purpose as plain washers when used with oval-head or countersunk-head screws.

Finish washers, shown in Fig. 22.22, are used under oval-head and flat-head screws to provide a more finished appearance and to increase the bearing surface between the fastener and the joint material.

Tables of washer sizes are not included here because of the large amount of space that would be required. Some manufacturers have as many as 60 000 stock dies, and so almost any size needed can be obtained. Washer materials include almost all the metals and many nonmetals as well.

(a)

(b)

(c)

(d)

(e)

(f)

FIGURE 22.21   Washer. (*a*) Wood-grip; (*b*) C or horseshoe; (*c*) lockplate; (*d*) cup; (*e*) external-tooth locking; (*f*) countersunk.

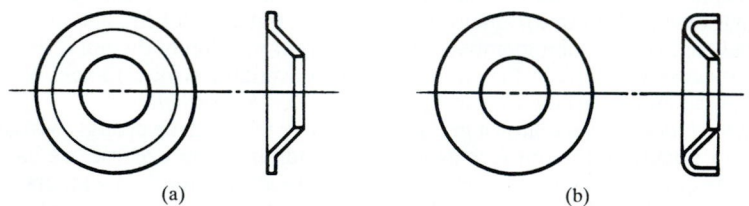

(a)

(b)

FIGURE 22.22   Finish washers. (*a*) Flush; (*b*) raised.

REFERENCES†

22.1 ANSI B18.1.1-1972 (R1977), "Small Solid Rivets."

22.2 ANSI B18.1.2-1972 (R1977), "Large Rivets."

22.3 ANSI B18.7-1972 (R1980), "General Purpose Semi-Tubular Rivets, Full Tubular Rivets, Split Rivets, and Rivet Caps."

22.4 ANSI B18.8.1-1972 (R1977), "Clevis Pins and Cotter Pins."

22.5 ANSI B18.8.2-1978, "Taper Pins, Dowel Pins, Straight Pins, Grooved Pins, and Spring Pins (Inch Series)."

22.6 ASA B18.12-1962 (R1981), "Glossary of Terms for Mechanical Fasteners."

22.7 ANSI B27.6-1972 (R1977), "General Purpose Uniform Cross Section Spiral Retaining Rings."

22.8 ANSI B27.7-1977, "General Purpose Tapered and Reduced Cross Section Retaining Rings (Metric)."

22.9 ANSI B27.8M-1978, "General Purpose Metric Tapered and Reduced Cross Section Retaining Rings."

22.10 ANSI B17.7-1967 (R1973), "Keys and Keyseats."

22.11 ANSI B17.2-1967 (R1978), "Woodruff Keys and Keyseats."

---

† References [22.1] to [22.5] and [22.7] to [22.11] are published by American Society of Mechanical Engineers; Ref. [22.6] is published by American Standards Association

CHAPTER 23
BOLTED AND RIVETED JOINTS

John H. Bickford
Vice President, Manager of the Power-Dyne Division, Retired
Raymond Engineering Inc.
Middletown, Connecticut

SYMBOLS AND UNITS

A	Cross-sectional area, in^2 (mm^2)
A_B	Cross-sectional area of the body of a bolt, in^2 (mm^2)
A_r	Cross-sectional area of the body of the rivet, in^2 (mm^2)
A_S	Cross-sectional area of the tensile stress area of the threaded portion of a bolt, in^2 (mm^2)
A_1, A_2, A_3, etc.	Cross-sectional areas of individual fasteners, in^2 (mm^2)
b	Number of shear planes which pass through the fastener; and/or the number of slip surfaces in a shear joint
d	Nominal diameter of the bolt, in (mm)
E	Modulus of elasticity, psi (MPa)
F	Force, lb (kN)
F_B	Tension in a bolt, lb (kN)
F_b	Primary shear force on a bolt, lb (kN)
$F_B(max)$	Maximum anticipated tension in the bolt, lb (kN)
F_{BY}	Tension in a bolt at yield, lb (kN)
F_C	Clamping force on the joint, lb (kN)
$F_C(min)$	Minimum acceptable clamping force on a joint, lb (kN)
$F_J(min)$	Minimum anticipated clamping force on the joint, lb (kN)

F_n	Reaction moment force seen by the nth bolt in an eccentrically loaded shear joint, lb (kN)
F_{PA}	Average preload in a group of bolts, lb (kN)
$F_P(\max)$	Maximum anticipated initial preload in a bolt, lb (kN)
$F_P(\min)$	Minimum anticipated initial preload in a bolt, lb (kN)
F_{PT}	Target preload, lb (kN)
F_{TR}	Maximum external transverse load on the joint, per bolt, lb (kN)
F_r	External shear load on the rivet, lb (kN)
$F_T(\max)$	Maximum acceptable tension in a bolt, lb (kN)
F_X	External tension load on a joint, lb (kN)
$F_1, F_2, F_3,$ etc.	Secondary shear or reaction moment forces seen by individual bolts in an eccentric joint, lb (kN)
H	Distance between the centerline of the bolt holes nearest to the edge of a joint or splice plate and that edge, in (mm)
k_B	Stiffness of a bolt or rivet, lb/in (kN/mm)
k_G	Stiffness of a gasket, lb/in (kN/mm)
k_J	Stiffness of the joint members, lb/in (kN/mm)
k_T	Stiffness of gasketed joint, lb/in (kN/mm)
K	Nut factor
l_G	Grip length of the fasteners, in (mm)
L	Distance between the bolt and the nearest edge of the connected part, or to the nearest edge of the next bolt hole, measured in the direction of the force on the joint in (mm)
L_B	Effective length of the body of a bolt (the length of body in the grip plus one-half the thickness of the head, for example), in (mm)
L_S	Effective length of the threaded portion of a bolt [the length of the threads within the grip plus one-half the thickness of the nut(s), for example], in (mm)
m	Number of fasteners in the joint
M	Moment exerted on a shear joint by an external force, lb · in (N · m)
n	Number of threads per inch
N	Number of cycles achieved in fatigue life test
P	Pitch of the threads, in (mm)
P_S	Scatter in preload anticipated from bolting tool used for assembly (expressed as a decimal)
P_Z	Percentage loss (expressed as a decimal) in initial preload as a result of short-term relaxation and/or elastic interactions
r	Radial distance from the centroid of a group of fasteners to a fastener, in (mm)
r_n	Radial distance to the nth fastener, in (mm)

r_s	Bolt slenderness ratio (l_G/d)
r_1, r_2, r_3, etc.	Radial distance of individual fasteners, in (mm)
R_{JB}	Stiffness ratio (k_J/k_B)
R_S	Slip resistance of a friction-type joint, lb (kN)
S	Ratio of the ultimate shear strength of the bolt material to its ultimate tensile strength
S_u	Minimum ultimate tensile strength, psi (MPa)
S_{YB}	Yield strength of the bolt, psi (MPa)
t	Thickness of a joint or a splice plate, in (mm)
t_J	Total thickness of a joint, in (mm)
T	Torque, lb $\cdot$ in (N $\cdot$ m)
W	Width of a joint plate, in (mm)
x	Coordinate distance, in (mm)
$\bar{x}$	Coordinate distance to the centroid of a bolt group, in (mm)
x_1, x_2, x_3, etc.	x coordinates for individual fasteners, in (mm)
y	Coordinate distance, in (mm)
$\bar{y}$	Coordinate distance to the centroid of a bolt group, in (mm)
y_1, y_2, y_3, etc.	y coordinates for individual fasteners, in (mm)
Δ	Incremental change or variation
λ	Ratio of shear stress in a bolt to the ultimate tensile strength
μ_S	Slip coefficient of a friction-type joint
σ	Stress, psi (MPa)
σ_B	Bearing stress, psi (MPa)
$\sigma(\text{max})$	Maximum tensile stress imposed during fatigue tests, psi (MPa)
σ_T	Allowable tensile stress, psi (MPa)
$\sigma_T(\text{max})$	Maximum acceptable tensile stress in a bolt, psi (MPa)
σ^2	Statistical variance (standard deviation squared)
σ_O^2	Statistical variance of the tension errors created by operator variables
σ_T^2	Statistical variance of the tension errors created by tool variables
τ	Shear stress, psi (MPa)
τ_A	Allowable shear stress, psi (MPa)
τ_B	Shear stress in a bolt, psi (MPa)
ϕ	Ratio of tensile stress in a bolt to the ultimate tensile strength

Joints are an extremely important part of any structure. Whether held together by bolts or rivets or weldments or adhesives or something else, joints make complex structures and machines possible. Bolted joints, at least, also make disassembly and reassembly possible. And many joints are critical elements of the structure, the thing most likely to fail. Because of this, it is important for the designer to understand joints. In this chapter we will deal specifically with bolted and riveted joints, starting with a discussion of joints loaded in shear (with the applied loads at right angles to the axes of the fasteners) and continuing with tension joints in which the loads are applied more or less parallel to the axes of fasteners. As we shall see, the design procedures for shear joints and tension joints are quite different.

23.1 SHEAR LOADING OF JOINTS

Now let us look at joints loaded in shear. I am much indebted, for the discussion of shear joints, to Shigley, Fisher, Higdon, and their coauthors ([23.1], [23.2], [23.3]).

23.1.1 Types of Shear Joints

Shear joints are found almost exclusively in structural steel work. Such joints can be assembled with either rivets or bolts. Rivets used to be the only choice, but since the early 1950s, bolts have steadily gained in popularity.

Two basic types of joint are used, *lap* and *butt,* each of which is illustrated in Fig. 23.1. These are further defined as being either (1) friction-type joints, where the fasteners create a significant clamping force on the joint and the resulting friction between joint members prevents joint slip, or (2) bearing-type joints, where the fasteners, in effect, act as points to prevent slip.

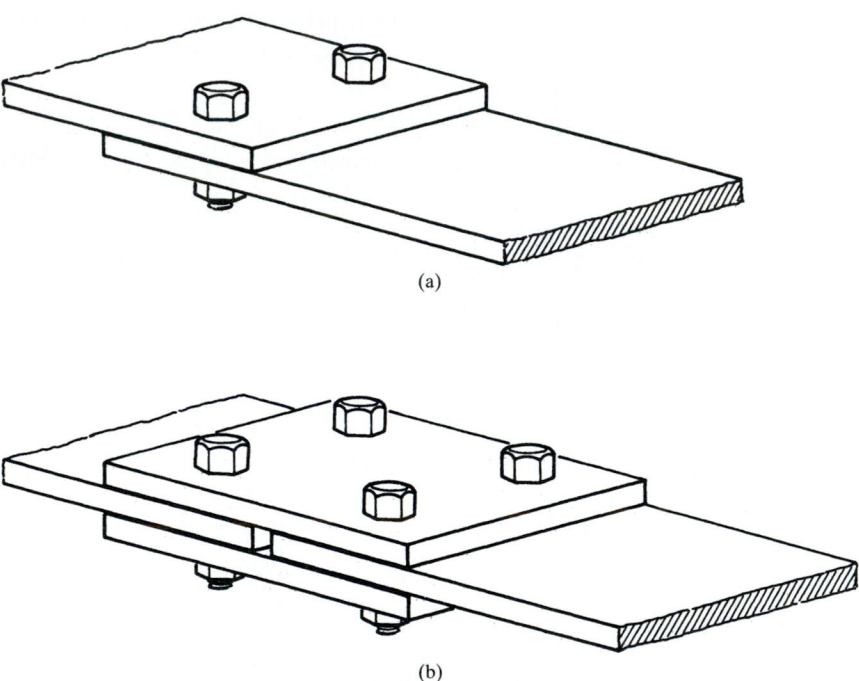

(a)

(b)

FIGURE 23.1   Joints loaded in shear. (*a*) Lap joint; (*b*) butt joint.

Only bolts can be used in friction-type joints, because only bolts can be counted on to develop the high clamping forces required to produce the necessary frictional resistance to slip. Rivets or bolts can be used in bearing-type joints.

23.1.2 Allowable-Stress Design Procedure

In the *allowable-stress design procedure,* all fasteners in the joint are assumed to see an equal share of the applied loads. Empirical means have been used to determine the maximum working stresses which can be allowed in the fasteners and joint members under these assumptions. A typical allowable shear stress might be 20 percent of the ultimate shear strength of the material. A factor of safety (in this case 5:1) has been incorporated into the selection of allowable stress.

We should note in passing that the fasteners in a shear joint do not, in fact, all see equal loads, especially if the joint is a long one containing many rows of fasteners. But the equal-load assumption greatly simplifies the joint-design procedure, and if the assumption is used in conjunction with the allowable stresses (with their built-in factors of safety) derived under the same assumption, it is a perfectly safe procedure.

Bearing-type Joints.   To design a successful bearing-type joint, the designer must size the parts so that the fasteners will not shear, the joint plates will not fail in tension nor be deformed by bearing stresses, and the fasteners will not tear loose from the plates. None of these things will happen if the allowable stresses are not exceeded in the fasteners or in the joint plates. Table 23.1 lists typical allowable stresses specified for various rivet, bolt, and joint materials. This table is for reference only. It is always best to refer to current engineering specifications when selecting an allowable stress for a particular application.

Here is how the designer determines whether or not the stresses in the proposed joint are within these limits.

Stresses within the Fasteners.   The shear stress within a rivet is

$$\tau = \frac{F}{bmA_r} \tag{23.1}$$

The shear stress within each bolt in the joint will be

$$\tau = \frac{F}{A_T} \tag{23.2}$$

A bolt can have different cross-sectional areas. If the plane passes through the unthreaded body of the bolt, the area is simply

$$A_B = \frac{\pi d^2}{4} \tag{23.3}$$

If the shear plane passes through the threaded portion of the bolt, the cross-sectional area is considered to be the tensile-stress area of the threads and can be found for Unified [23.4] or metric [23.5] threads from

TABLE 23.1 Allowable Stresses

Material	Source	Comments	Tension kpsi (MPa)	Shear kpsi (MPa)	Bearing† kpsi (MPa)
ASTM A325 bolts	1	Used in bearing-type joints with slotted or standard holes, and			†
		some threads in shear planes	....	21.0 (145)	
		no threads in shear planes	....	30.0 (207)	
ASTM A325 bolts	1	Used in friction-type joints with standard holes and surfaces of			†
		clean mill scale	....	17.5 (52)	
		blast-cleaned carbon or low-alloy steel	....	27.5 (190)	
		blast-cleaned inorganic zinc rich paint		29.5 (203)	
ASTM A490 bolts	1	Bearing-type joints with slotted or standard holes, and			†
		some threads in shear planes	....	28.0 (193)	
		no threads in shear planes	....	40.0 (276)	
ASTM A490 bolts	1	Friction-type joints with standard holes and surfaces of			†
		clean mill scale	....	22.0 (152)	
		blast-cleaned carbon or alloy steel	....	34.5 (238)	
		blast-cleaned inorganic zinc-rich paint		37.0 (255)	
ASTM SA193 Grade B7 at an operating temperature of	2	Used for bolts‡			
−20°F	...	....	18.8–25 (130–172)		
+650°F	...	....	18.8–25.0 (130–172)		
+850°F	...	....	16.3–17.0 (112–117)		
+1000°F	...	....	4.5 (31)		

TABLE 23.1   Allowable Stresses (*Continued*)

Material	Source	Comments	Allowable stress		
			Tension kpsi (MPa)	Shear kpsi (MPa)	Bearing† kpsi (MPa)
ASTM SA31 rivets	3	Used in SA515 plate	....	9 (62)	18 (124)
ASTM A502-1 rivets	3	Used in A36 plate		13 (93)	401 (276)
ASTM A36 joint material	4		22 (152)	14.5 (100)	48.6 (335)
58-kpsi ultimate tensile steel: joint material	5	Joint length 25 in (with A325 bolts) Joint length 80 in (with A325 bolts)	....	23.2 (160) 29 (200)	
100-kpsi ultimate tensile strength steel: joint material	6	Joint length 20 in (with A490 bolts) Joint length 90 in (with A490 bolts)	....	50 (345) 40 (276)	
ASTM A440 joint material	7	Based on a safety factor of 2.48:1 (S_u/σ_T)	25.4–28.2 (175–194)		
ASTM A514 joint material	7	Based on a safety factor of 2:00:1 (S_u/σ_T)	50–65 (345–448)		
ASTM A515 joint material	3	Stress in net section	14 (95)		

†The allowable bearing stress for either A325 or A490 bolts is either $LS_u/2d$ or $1.5S_u$, whichever is least.
‡The stress allowed depends on the diameter of the bolts. The material cannot be through-hardened, so larger sizes will support less stress.

SOURCES:
1. "Structural Joints Using ASTM A325 or A490 Bolts." AISC specification, April 14, 1980, pp. 4–5.
2. "ASME Boiler and Pressure Vessel Code," Sec. VIII, Div. I, American Society of Mechanical Engineers, New York, 1977. Table UCS-23, pp. 208–209.
3. Archie Higdon, Edward H. Ohlsen, William B. Stiles, John A. Weese, and William F. Riley, *Mechanics of Materials,* 3d ed., John Wiley and Sons, New York, 1978, p. 632.
4. John W. Fisher, "Design Examples for High Strength Bolting," *High Strength Bolting for Structural Joints,* Bethlehem Steel Co., Bethlehem, Pennsylvania, 1970, p. 52.
5. John W. Fisher and John H. A. Struik, *Guide to Design Criteria for Bolted and Riveted Joints,* John Wiley and Sons, New York, 1974, p. 124.
6. Ibid., p. 127.
7. Ibid., p. 123.

Unified:
$$A_S = \frac{\pi}{4}\left(d - \frac{0.9743}{n}\right)^2$$

(23.4)

Metric:
$$A_S = \frac{\pi}{4}(d - 0.9382P)^2$$

Here is an example based on Fig. 23.2. The bolts are ASTM A325 steel, $m = 5$ bolts, $F = 38\,250$ lb (170.1 kN), $d = \frac{3}{4}$ in (19.1 mm), $b = 2$ (one through the body of each bolt, one through the threads), and $n = 12$ threads per inch (2.12 mm per thread).

The total cross-sectional area through the bodies of all five bolts and then through the threads is

$$5A_B = \frac{5\pi(0.75)^2}{4} = 2.209 \text{ in}^2 \text{ (1425 mm}^2\text{)}$$

$$5A_S = \frac{5\pi}{4}\left(0.75 - \frac{0.9743}{12}\right)^2 = 1.757 \text{ in}^2 \text{ (1133 mm}^2\text{)}$$

The shear stress in each bolt will be

$$\tau = \frac{F}{A_T} = \frac{38\,250}{2.209 + 1.757} = 9646 \text{ psi (66.5 MPa)}$$

which is well within the shear stress allowed for A325 steel bolts (Table 23.1).

Tensile Stress in the Plate. To compute the tensile stress in the plates (we will assume that these are made of A36 steel), we first compute the cross-sectional area of a row containing the most bolts. With reference to Figs. 23.2 and 23.3, that area will be

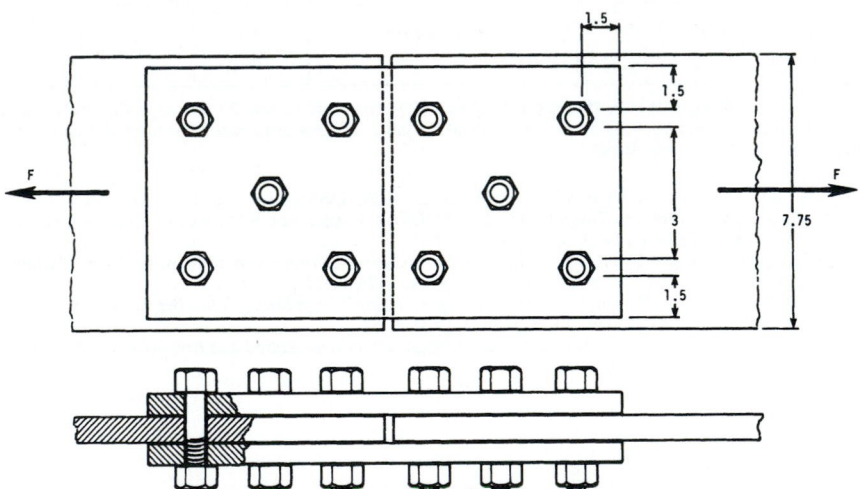

FIGURE 23.2 Shear joint example. The joint and splice plates here are each $\frac{3}{4}$ in (19.1 mm) thick. Dimensions given are in inches. To convert to millimeters, multiply by 25.4.

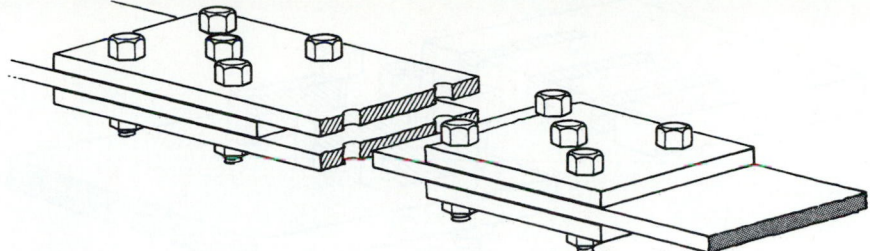

FIGURE 23.3 Tensile failure of the splice plates. Tensile failure in the plates occurs in the cross section intersecting the most bolt holes.

$$A = 0.75(1.5) + 0.75(3) + 0.75(1.5) = 4.5 \text{ in}^2 \ (2903 \text{ mm}^2)$$

The stress in two such cross sections (there are two splice plates) will be

$$\sigma = \frac{F}{A} = \frac{38\,250}{(4.5)2} = 4250 \text{ psi } (29.3 \text{ MPa})$$

These plates will not fail; the stress level in them is well within the allowable tensile-stress value of 21.6 kpsi for A36 steel. In some joints we would want to check other sections as well, perhaps a section in the splice plate.

Bearing Stresses on the Plates. If the fasteners exert too great a load on the plates, the latter can be deformed; bolt holes will elongate, for example. To check this possibility, the designer computes the following (see Fig. 23.4):

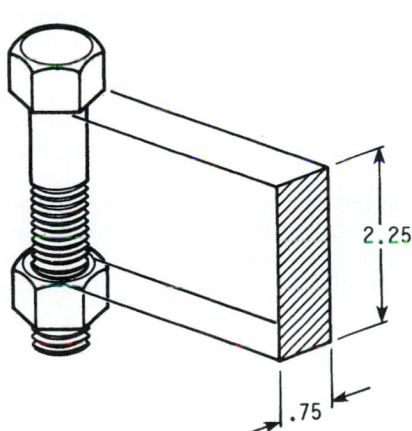

$$\sigma_B = \frac{F}{mdl_G}$$

For our example, $l_G = 2.25$ in (57.2 mm), $m = 5$, and $d = 0.75$ in (19.1 mm). Then

$$\sigma_B = \frac{38\,250}{5(0.75)(2.25)} = 4533 \text{ psi } (31.3 \text{ MPa})$$

Note that the allowable bearing stresses listed in Table 23.1 are greater than the allowable shear stresses for the same plate material.

Tearout Stress. Finally, the designer should determine whether or not the fasteners will tear out of a joint plate, as in the lap joint shown in Fig. 23.5. In the example shown there are six shear areas. The shear stress in the tearout sections will be

FIGURE 23.4 The bearing area of a bolt. The dimensions given are those used in the example in the text for the joint shown in Fig. 23.2. Dimensions are in inches. Multiply by 25.4 to convert to millimeters.

$$\tau = \frac{100\,000}{6(0.75)(2)} = 11\,111 \text{ psi } (76.6 \text{ MPa})$$

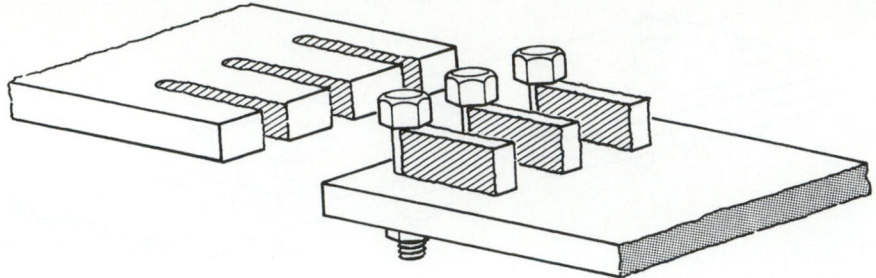

FIGURE 23.5 Tearout. The pieces torn from the margin of the plate can be wedge-shaped as well as rectilinear, as shown here.

where F = 100 kip (445 kN)
$\quad H$ = 2 in (50.8 mm)
$\quad t$ = ¾ in (19.1 mm)

Friction-type Joints. Now let us design a friction-type joint using the same dimensions, materials, and bolt pattern as in Fig. 23.1, but this time preloading the bolts high enough so that the friction forces between joint members (between the so-called faying surfaces) become high enough to prevent slip under the design load.

Computing Slip Resistance. To compute the slip resistance of the joint under a shear load, we use the following expression (from Ref. [23.6], p. 72):

$$R_S = \mu_S F_{PA} bm \qquad (23.5)$$

Typical slip coefficients are tabulated in Table 23.2. Note that engineering specifications published by the AISC and others carefully define and limit the joint surface conditions that are permitted for structural steel work involving friction-type joints. The designer cannot arbitrarily paint such surfaces, for example; if they are painted, they must be painted with an approved material. In most cases they are not painted. Nor can such surfaces be polished or lubricated, since these treatments would alter the slip coefficient. A few of the surface conditions permitted under current specifications are listed in Table 23.2. Further conditions and coating materials are under investigation.

To continue our example, let us assume that the joint surfaces will be grit blasted before use, resulting in an anticipated slip coefficient of 0.493. Now we must estimate the average preload in the bolts. Let us assume that we have created an average preload of 17 kip in each of the five bolts in our joint. We can now compute the slip resistance as

$$R_S = \mu_S F_{PA} bm = 0.493 \ (17 \ 000)(2)(5)$$

$$= 83 \ 810 \ \text{lb} \ (373 \ \text{kN})$$

Comparing Slip Resistance to Strength in Bearing. The ultimate strength of a friction-type joint is considered to be the lower of its slip resistance or bearing strength. To compute the bearing strength, we use the same equations we used earlier. This time, however, we enter the allowable shear stress for each material and

TABLE 23.2 Slip Coefficients

Surfaces	Source	Typical slip coefficient μ_S
Free of paint or other applied finish, oil, dirt, loose rust or scale, burrs, or defects. Tight mill scale permitted	1	0.45
Clean mill scale	2	0.35
Hot dip galvanized	2	0.16
Hot dip galvanized, wire brushed	2	0.3–0.4
Grit blasted	3	0.331–0.527
Sand blasted	3	0.47
Metallized zinc sprayed (hot) onto grit blasted surface	4	0.422

SOURCES:
1. Specification BS 4604: Part 1: 1970, British Standards Institution, London, 1970.
2. *High Strength Bolting for Structural Joints,* Bethlehem Steel Co., Bethlehem, Pennsylvania, 1970, p. 14.
3. John W. Fisher and John H. A. Struik, *Guide to Design Criteria for Bolted and Riveted Joints,* John Wiley and Sons, New York, 1974, p. 78.
4. Ibid., p. 200.

then compute the force which would produce that stress. These forces are computed separately for the fasteners, the net section of the plates, the fasteners bearing against the plates, and tearout. The least of these forces is then compared to the slip resistance to determine the ultimate design strength of the joint. If you do this for our example, you will find that the shear strength of the bolts determines the ultimate strength of this joint.

23.2 ECCENTRIC LOADS ON SHEAR JOINTS

23.2.1 Definition of an Eccentric Load

If the resultant of the external load on a joint passes through the centroid of the bolt pattern, such a joint is called an *axial shear joint.* Under these conditions, all the fasteners in the joint can be assumed to see an equal shear load.

If the resultant of the applied load passes through some point other than the centroid of the bolt group, as in Fig. 23.6, there will be a net moment on the bolt pattern. Each of the bolts will help the joint resist this moment. A joint loaded this way is said to be under an *eccentric shear load.*

23.2.2 Determine the Centroid of the Bolt Group

To locate the centroid of the bolt group, we arbitrarily position *xy* reference axes near the joint, as shown in Fig. 23.7. We then use the following equations to locate the centroid within the group (Ref. [23.1], p. 360):

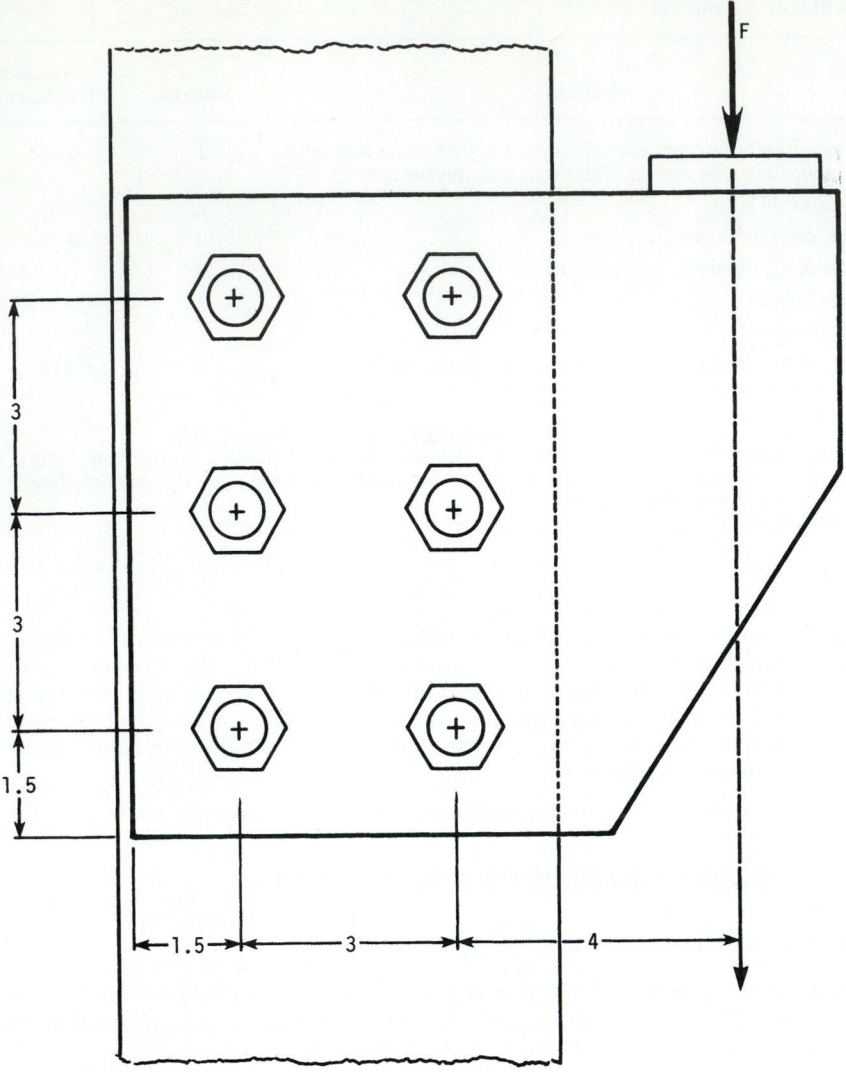

FIGURE 23.6 Eccentrically loaded shear joint. For the example used in the text, it is assumed that the bolts are ¾—12 × 3, ASTM A325; the plates are made of A36 steel; the eccentric applied load F is 38.25 kip (170 kN).

$$\bar{x} = \frac{A_1 x_1 + A_2 x_2 + \cdots + A_6 x_6}{A_1 + A_2 + \cdots + A_6}$$

$$\bar{y} = \frac{A_1 y_1 + A_2 y_2 + \cdots + A_6 y_6}{A_1 + A_2 + \cdots + A_6}$$

(23.6)

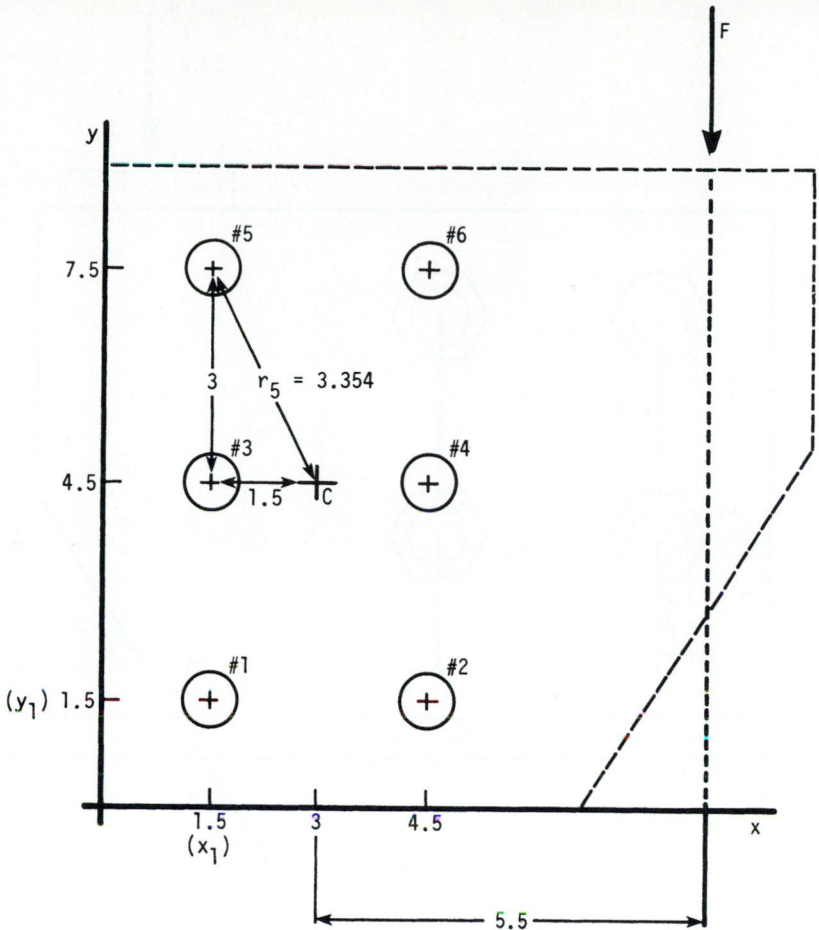

FIGURE 23.7   The centroid of a bolt pattern. To determine the centroid of a bolt pattern, one arbitrarily positions coordinate axes near the pattern and then uses the procedure given in the text. I have used the edges of the splice plate for the x and y axes in this case. Multiply the dimensions shown (which are in inches) by 25.4 to convert them to millimeters.

For the joint shown in Fig. 23.6 we see, assuming that $A_1 = A_2 =$ etc. $= 0.442$ in^2 (285 mm^2),

$$\bar{x} = \frac{0.442(1.5 + 4.5 + 1.5 + 4.5 + 1.5 + 4.5)}{6(0.442)} = 3 \text{ in (76.2 mm)}$$

Similarly, we find that $\bar{y} = 4.5$ in (114.3 mm)

23.2.3   Determining the Stresses in the Bolts

Primary Shear Force.   We compute the primary shear forces on the fasteners as simply (see Fig. 23.8)

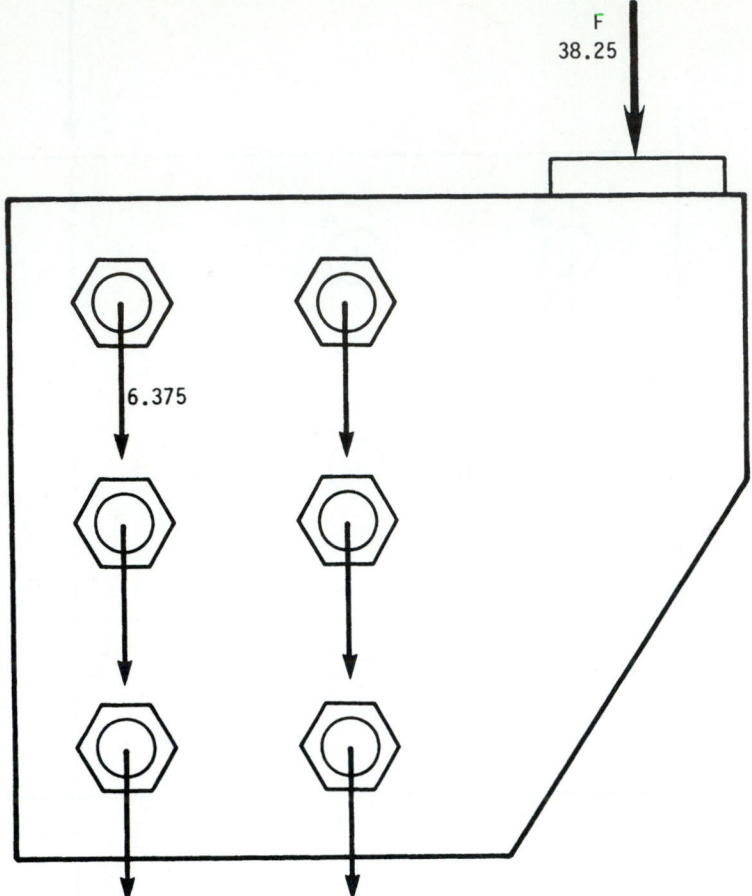

FIGURE 23.8   Primary shear forces on the bolts. The primary forces on the bolts are equal and are parallel. Forces shown are in kilopounds; multiply by 4.448 to convert to kilonewtons.

$$F_b = \frac{F}{m} = \frac{38\ 250}{6} = 6375 \text{ lb (28.4 kN)}$$

Secondary Shear Forces.   We next determine the reaction moment forces in each fastener using the two equations (Ref. [23.1], p. 362):

$$M = F_1 r_1 + F_2 r_2 + \cdots + F_6 r_6 \tag{23.7}$$

$$\frac{F_1}{r_1} = \frac{F_2}{r_2} = \frac{F_3}{r_3} = \cdots = \frac{F_6}{r_6} \tag{23.8}$$

Combining these equations, we determine that the reaction force seen on a given bolt is

$$F_n = \frac{Mr_n}{r_1^2 + r_2^2 + \cdots + r_6^2} \qquad (23.9)$$

Let us continue our example. As we can see from Fig. 23.7, we have an external force of 38 250 lb (170.1 kN) acting at a distance from the centroid of 5.5 in (140 mm). The input moment, then, is 210 kip · in (23.8 N · m). The radial distance from the centroid to bolt 5 (one of the four bolts which are most distant from the centroid) is 3.354 in (85.2 mm). The reaction force seen by each of these bolts is (see Fig. 23.9)

$$F_5 = \frac{210\,375(3.354)}{4(3.354)^2 + 2(1.5)^2} = 14\,255 \text{ lb (63.4 kN)}$$

Combining Primary and Secondary Shear Forces. The primary and secondary shear forces on bolt 5 are shown in Fig. 23.9. Combining these two forces by vectorial means, we see that the total force F_{R5} on this bolt is 12 750 lb (56.7 kN).

Let us assume that there are two slip planes here—that one of them passes through the body of the bolt and the other passes through the threads as in the ear-

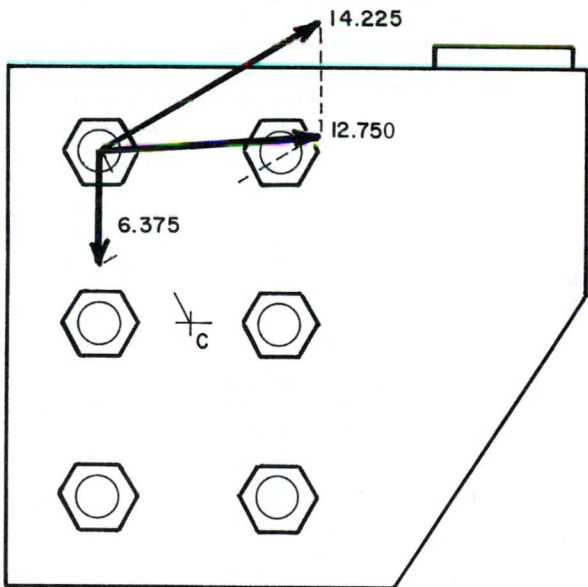

FIGURE 23.9 Combining primary and shear forces. I have selected one of the four *most distant* bolts to calculate the secondary shear force, 14.255 kip (63.4 kN), which has a line of action at right angles to the radial line connecting the bolt to the centroid. The resultant of primary and secondary forces on this bolt is 12.750 kip (56.7 kN).

lier example illustrated in Fig. 23.2. The shear area of bolt 5, therefore, is (see Sec. 23.1.2 for the equations) 0.793 in² (511 mm²).

We can now compute the shear stress within this bolt:

$$\tau = \frac{F_{R5}}{A_5} = \frac{12\,750}{0.793} = 16\,078 \text{ psi}$$

This is less than the maximum shear stress allowed for A325 steel bolts (see Table 23.1), and so the design is acceptable.

It is informative to compare these results with those obtained in Sec. 23.1.3, where we analyzed a joint having similar dimensions, the same input load, and one less bolt. The axial load in the earlier case created a shear stress of only 9646 psi (66.5 MPa) in each bolt. When the same load is applied eccentrically, passing 5.5 in from the centroid, it creates 16 078/9646 times as much stress in the most distant bolts, even though there are more bolts this time to take the load. Be warned!

23.3 TENSION-LOADED JOINTS: PRELOADING OF BOLTS

In the joints discussed so far, the bolts or rivets were loaded in shear. Such joints are usually encountered in structural steel work. Most other bolted joints in this world are loaded primarily in tension—with the applied loads more or less parallel to the axis of the bolts.

The analysis of tension joints usually centers on an analysis of the tension in the fasteners: first with the initial or *pre*load in the fasteners when they are initially tightened, and then with the working loads that exist in the fasteners and in the joint members when external forces are applied to the joint as the product or structure is put into use. These working loads consist of the preload plus or minus some portion of the external load seen by the joint in use.

Because clamping force is essential when a joint has to resist tension loads, rivets are rarely used. The following discussion, therefore, will focus on bolted joints. The analytical procedure described, however, could be used with riveted joints if the designer is able to estimate the initial preload in the rivets.

23.3.1 Preliminary Design and Calculations

Estimate External Loads.   The first step in the design procedure is to estimate the external loads which will be seen by each bolted joint. Such loads can be static, dynamic, or impact in nature. They can be created by weights such as snow, water, or other parts of the structure. They can be created by inertial forces, by shock or vibration, by changes in temperature, by fluid pressure, or by prime movers.

Fastener Stiffness.   The next step is to compute the stiffness or spring rate of the fasteners. Using the following equation,

$$k_B = \frac{A_S A_B E}{L_S A_B + L_B A_S} \tag{23.10}$$

Example. With reference to Fig. 23.10, $A_S = 0.232$ in^2 (150 mm^2), $L_B = 2.711$ in (68.9 mm), $A_B = 0.307$ in^2 (198 mm^2), $E = 30 (10)^6$ psi (207 GPa), and $L_S = 1.024$ in (26 mm). Thus

$$k_B = \frac{0.232(0.307)(30 \times 10^6)}{1.024(0.307) + 2.711(0.232)} = 2.265 \times 10^6 \text{ lb/in (0.396 N/mm)}$$

Stiffness of a Nongasketed Joint. The only accurate way to determine joint stiffness at present is by experiment. Apply an external tension load to a fastener in an actual joint. Using strain gauges or ultrasonics, determine the effect which this external load has on the tension in the bolt. Knowing the stiffness of the bolt (which must be determined first), use joint-diagram techniques (which will be discussed soon) to estimate the stiffness of the joint.

Although it is not possible for me to give you theoretical equations, I can suggest a way in which you can make a rough estimate of joint stiffness. This procedure is based on experimental results published by Motosh [23.7], Junker [23.8], and Osgood [23.9], and can be used only if the joint members and bolts are made of steel with a modulus of approximately 30×10^6 psi (207 GPa).

First compute the slenderness ratio for the bolt (l_G/d). If this ratio is greater than 1/1, you next compute a stiffness ratio R_{JB} using the empirical equation

$$R_{JB} = 1 + \frac{3(l_G)}{7d} \tag{23.11}$$

The final step is to compute that portion of the stiffness of the joint which is loaded by a single bolt from

$$k_J = R_{JB}k_B$$

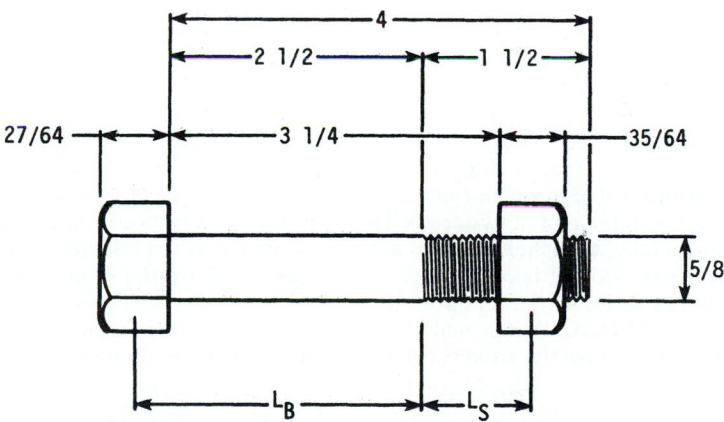

FIGURE 23.10 Computing the stiffness of a bolt. The dimensions given are those used in the example in the text. This is a ⅜—12 × 4, SAE J429 Grade 8 hexagon-head bolt with a 3.25-in (82.6-mm) grip. Other dimensions shown are in inches. Multiply them by 25.4 to convert to millimeters.

If the slenderness ratio l_G/d falls between 0.4 and 1.0, it is reasonable to assume a stiffness ratio R_{JB} of 1.0. When the slenderness ratio l_G/d falls below 0.4, the stiffness of the joint increases dramatically. At a slenderness ratio of 0.2, for example, R_{JB} is 4.0 and climbing rapidly (Ref. [23.6], pp. 199–206).

Example.   For the bolt shown in Fig. 23.10 used in a 3.25-in (82.6-mm) thick joint,

$$R_{JB} = \frac{3(3.25)}{7(0.625)} = 3.23$$

Since we computed the bolt stiffness earlier as 2.265×10^6 lb/in (396 kN/mm), the joint stiffness will be

$$k_J = 3.23(2.265 \times 10^6) = 7.316 \times 10^6 \text{ lb/in (1280 kN/mm)}.$$

Stiffness of Gasketed Joints.   The procedure just defined allows you to determine the approximate stiffness of a nongasketed joint. If a gasket is involved, you should use the relationship

$$\frac{1}{k_T} = \frac{1}{k_J} + \frac{1}{k_G} \tag{23.12}$$

You may have to determine the compressive stiffness of the gasket by making an experiment or by contacting the gasket manufacturer, since very little information has been published on this subject (but see Chap. 26). A few typical values for pressure-vessel gasket materials are given in Table 23.3, but these values should be used for other gaskets with caution.

Note that the stiffness of a gasket, like the stiffness of everything else, depends on its cross-sectional area. The values given in Table 23.3 are in terms of pressure or stress on the gasket versus deflection, not in terms of force versus deflection. Before you can combine gasket stiffness with joint stiffness, therefore, you must determine how large an area of the gasket is loaded by a single bolt (total gasket area divided by the number of bolts). This per-bolt area is multiplied by stress to determine the stiffness in terms of force per unit deflection. For example, the compressed asbestos gasket listed in Table 23.3 has a total surface area of 11.2 in^2 (7219 mm^2). If it is clamped by eight bolts, the per-bolt area is 1.4 in^2 (903 mm^2). The stiffness is listed in Table 23.3 as 6.67×10^2 ksi/in (181 MPa/mm). In force terms, per bolt, this becomes 6.67×10^5 (1.4) $= 9.338 \times 10^5$ lb/in (1.634×10^2 kN/mm).

The stiffness values given in Table 23.3 are for gaskets in use, after initial preloading. Gaskets exhibit a lot of hysteresis. Their stiffness during initial compression is a lot less (generally) than their stiffness as they are unloaded and reloaded. As long as the usage cycles do not take the stress on the gasket above the original assembly stress, their behavior will be repetitive and elastic, with only a little hysteresis, as suggested by Fig. 23.11. And when analyzing the effect of loads on joint behavior, we are interested only in how the gaskets act as they are used, not in their behavior during assembly.

23.3.2   Selecting the Target Preload

Our joint will perform as intended only if it is properly clamped together by the fasteners. We must, therefore, select the preload values very carefully.

TABLE 23.3 Gasket Stiffness

Source	Gasket	Dimensions, in (mm)				Stiffness kpsi/in (MPa/mm)
		ID	OD	w	t	
1	Spiral-wound, asbestos-filled (300-lb class)	5 (127)	5.75 (146)	0.375 (9.52)	0.175 (4.45)	4.71×10^2 (127.6)
1	Spiral-wound asbestos-filled (600-lb class)	4.75 (121)	5.75 (146)	0.5 (12.7)	0.175 (4.45)	6.95×10^2 (188.3)
1	Compressed asbestos	4 (102)	5.5 (140)	0.75 (19)	0.062 (1.59)	6.67×10^2 (180.7)
2	Flat stainless steel double-jacketed asbestos-filled	6.5 (191)	7.5 (216)	0.5 (12.7)	0.125 (3)	43.3×10^2 (1176)
2	Solid oval ring-style 950 soft iron	5.438 (138)	6.314 (160)	0.469 (9.7)	0.688 (14.3)	27.5×10^2 (747)

SOURCE:
1. H. D. Raut, André Bazergui, and Luc Marchand, "Gasket Leakage Behavior Trends: Results of 1977–79 PVRC Exploratory Gasket Test Program," *Welding Research Council Bulletin* no. 271, WRC, New York, October 1981, Figs. 16 and 18.
2. André Bazergui and Luc Marchand, "PVRC Milestone Gasket Tests—First Results," report submitted to the Special Commission on Bolted Flanged Connections of the Pressure Vessel Research Committee of the Welding Research Council, September 1982, Figs. 12 and 13.

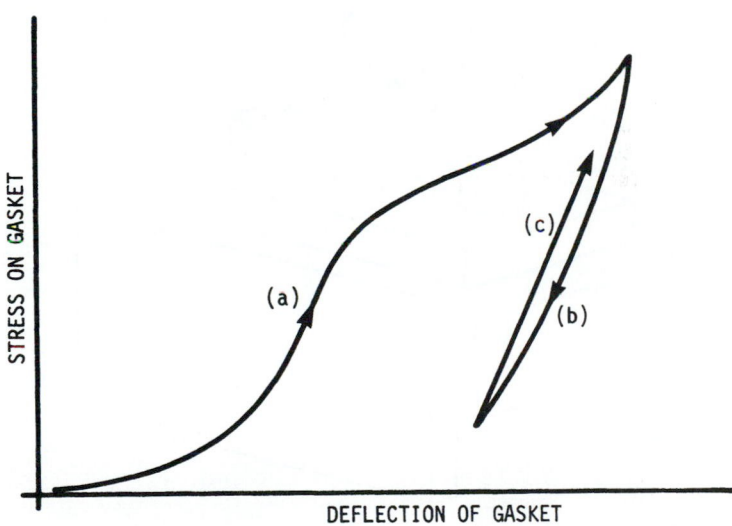

FIGURE 23.11 Typical stress versus deflection characteristics of a spiral-wound, asbestos-filled gasket during (a) initial loading, (b) unloading, and (c) reloading.

Acceptable Upper Limit for the Tension in the Bolts.   In general, we always want
the greatest preload in the bolts which the parts (bolts, joint members, and gasket)
can stand. To determine the maximum acceptable tension in the bolt, therefore, we
start by determining the yield load of each part involved in terms of bolt tension. The
force that will cause the bolt material to yield is

$$F_{BY} = S_{YB} A_S \tag{23.13}$$

Let us begin an example using the joint shown in Fig. 23.12. We will use the bolt
illustrated in Fig. 23.10. Let us make the joint members of ASTM A441 steel. The
yield strength of our J429 Grade 8 bolts is 81 kpsi (558 MPa), worst case. For the
bolts, with $A_S = 0.232$ in^2 (150 mm^2),

$$F_{BY} = 81 \times 10^3 (0.232) = 18.8 \times 10^3 \text{ lb (83.6 kN)}$$

For the joint, we determine the yield load of that portion of the joint which lies
under the head of the bolt or under the washer (using the distance across flats of the
head or nut to compute the bearing area). If our joint material is ASTM A441 steel

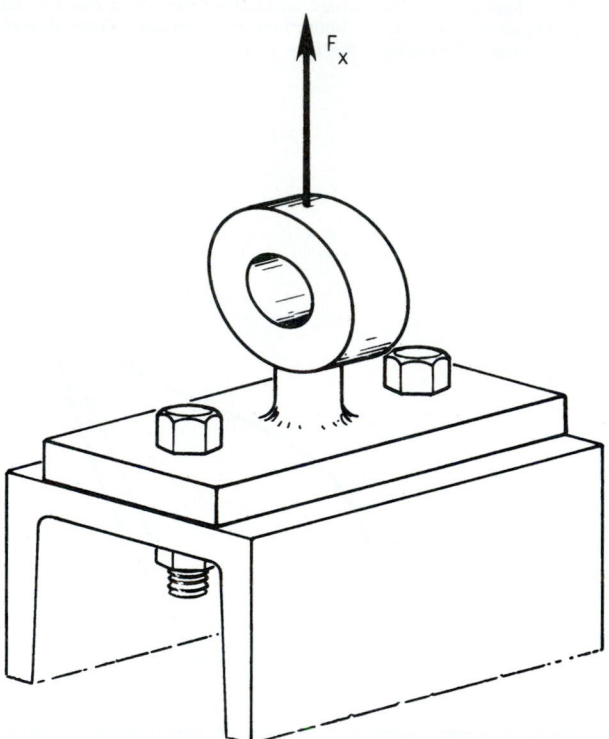

FIGURE 23.12   Joint loaded in tension. This is the joint analyzed in
the text. The bolts used here are those shown in Fig. 23.10.

with a yield strength of 40 kpsi (276 MPa), we would find that a *bolt* force of 27.6 kip (122 kN) would yield the joint. We take the lesser of the joint or bolt yield loads, 18.8×10^3 lb (83.6 kN) as the yield load of the system.

Next, since we are planning to tighten these fasteners by applying torque to the nuts, we subtract 10 percent from the yield strength of the fastener to allow for the torsional stresses which will be developed in the fastener as it is tightened. If we were planning to use a high-pressure lubricant on the threads, we might subtract only 2 to 4 percent for torsion. Since no lube will be used in our example, we subtract 10 percent, making the upper limit 16.9×10^3 lb (75.2 kN).

This will remain our upper limit unless the fasteners will also be subjected to shear stress; or unless code limits, stress corrosion problems, or the desire for a safety factor forces us to reduce it further. We will assume, in our example, that they do not.

Before continuing, however, let us see what we would have to do if the bolts *did* see a combined tension and shear load. This might happen, for example, in a bearing-type joint in which we planned to preload (tension) the bolts to a significant percentage of yield. There are other types of joint, of course, which are subjected to both tensile and shear loads in use. Any shear load on the bolt will "use up" part of the strength of the bolt, leaving less capacity for tensile loads (Ref. [23.6], p. 226).

Under these conditions, the maximum acceptable tensile stress in a bolt can be determined using any of the static failure theories of Chap. 12. Here, we select the equation

$$\left(\frac{\lambda}{S}\right)^2 + \phi^2 = 1 \qquad (23.14)$$

where S = the ratio of the shear strength to the tensile strength (typically 0.6) for bolt steels. Equation (23.14) is a form of the maximum-shear-stress theory.

Acceptable Lower Limit for the Clamping Force on the Joint. When we are computing the maximum acceptable forces, we focus on the joint, because its behavior can be seriously affected if the interface forces become too small. The joint, for example, might leak, it might vibrate loose, or it might have a short fatigue life. To determine the lower acceptable limit, we must consider each potential failure mode separately, estimate the minimum preload required to control that particular problem, and then select the *highest* of these several minimum requirements to establish the minimum for the system.

This is one of the more difficult steps of our procedure. In fact, we may be able to determine the acceptable minimum preload only by making fatigue or vibration tests or the like. (We will consider fatigue problems at length in Sec. 23.5.) There are some rules of thumb, however, which we can apply.

If our joint is a friction-type shear joint, or if it will be subjected to transverse vibration, we want a minimum preload which will prevent joint slip under the maximum anticipated shear load. This load is

$$F_C(\min) \geq \frac{F_X}{S} \qquad (23.15)$$

If we are dealing with a foundation bolt or something where it is only necessary to avoid separation of the joint members, the minimum acceptable preload can be zero. If we are dealing with a gasketed joint, we will have to worry about the minimum acceptable gasket pressure required to keep that gasket from leaking.

Example.   Let us assume for our ongoing example that we are concerned only about separation of the joint members. Minimum acceptable clamping force, therefore, is zero.

Select an Initial Preload Target.   We now, rather arbitrarily, select an initial preload target that is somewhere between the acceptable minimum and acceptable maximum bolt tensions which we computed earlier. Let us try 60 percent of the acceptable maximum of 16.9 kip (75.2 kN) or, in our example, 10.1×10^3 lb (45.1 kN).

23.3.3   Estimating Actual Upper Limit on Bolt Tension

We must now determine whether or not the tension we will actually develop in any bolt will exceed the maximum acceptable tension, given our preliminary target preload and a consideration of the tools, lubricants, and procedures we are planning to use during assembly. We must also consider the effects of the external tension loads which will be placed on this joint after assembly.

Tool Errors.   We can select many different types of assembly tools. Each choice carries with it certain accuracy implications; some tools can produce preload in the fasteners with far greater precision than can other tools. Table 23.4 lists some of the many possibilities. We will assume for our example that we are going to use a manual torque wrench and must face a potential scatter in preload for a given torque of ±30 percent.

Operator Problems.   Even if we used perfect tools, we would see some scatter in the resulting preload because of operator problems. Are the operators skilled, properly trained, or tired? Do they care about their work? Are the bolts readily accessible?

Let us assume for purposes of our example that the operators will contribute an estimated ±10 percent additional scatter in preload. We do not just add this 10 percent to the 30 percent we assigned to the torque wrench when we assess the combined impact of tools and operators. We use the statistician's method for combining the variances of two variables, as follows:

$$\sigma^2 = \sigma_T^2 + \sigma_O^2 \tag{23.16}$$

In our example this suggests that the combined variance will be

$$\sigma^2 = 30^2 + 10^2 = 1000$$

giving us a ±3 sigma "deviation" (square root of the variance) of ±31.6 percent.

We have selected a target preload of 10.1×10^3 lb. Consideration of tool and operator scatter gives us

$$F_P(\text{max}) = F_{PT} + 0.316 F_{PT} = 13.3 \times 10^3 \text{ lb } (59.2 \text{ kN})$$

$$F_P(\text{max}) = F_{PT} - 0.316 F_{PT} = 6.91 \times 10^3 \text{ lb } (30.7 \text{ kN})$$

Effects of External Tension Load.   Now let us see what happens when an external tension load is placed on the preloaded joint. Although it is difficult to do this in

TABLE 23.4 Tool Accuracy

Control parameter and type of tool	Source	Reported scatter in preload, %
Torque control with		
1. Power wrench	1	±23 to 28
2. Hand wrench	2	±21 to 81
3. Hand wrench	3	−20. 4 to 99
4. Hand wrench plus torque multiplier	4	±70 to150
5. Dial or click wrench	4	±60 to 80
6. Wrench with electronic readout	4	±40 to 60
7. Hand wrench	2	±12
8. Air-powered tool with torque feedback	5	±20
9. Hand wrench in laboratory conditions	6	±30
10. Air tool with one shot clutch	7	±30
11. Stall torque air tool	7	±35
12. Hand wrench	7	±30
Torque-turn control with		
13. Yield control system (computer-controlled air tool)	8	±8
14. Turn-of-the-nut procedure used in structural steel work	9	±15
15. Logarithmic rate method (LRM†) controlled air tool	11	±2.2 to 2.6
16. Turn-of-the nut	6	±15
Miscellaneous methods		
17. Strain-gauged load washers	10	±15
18. Strain-gauged bolts	10	±1
19. Swaged lockbolts	10	±5
20. Bolt heaters	12	±15
21. Air-powered impact wrench	5	±50
22. Manual slug wrench	3	−48 to + 50
23. Air-powered impact wrench	4	−300 to +150
24. Hydraulic tensioners controlled by "large vernier gauge readout"	4	±20
25. Torque-time control on air tool	5	±11
26. Operator feel	6	±35
27. Fastener elongation	6	±3 to 5
28. Ultrasonic control of preload		±1 to 10

†Trademark of Rockwell International

SOURCES:
1. Robert J. Finkelston and P. W. Wallace, "Advances in High Performance Mechanical Fastening," SAE Paper No. 800451, 1980, p. 6.
2. Results of tests conducted privately.
3. Roly Laird, "The Nuts and Bolts of Preventing Failure Where Pressure Is Hot," *Engineer* (GB), vol. 245, no. 6330, July 21, 1977, p. 37.
4. Steven F. Aaronson, "Analyzing Critical Joints," *Machine Design*, January 21, 1982, p. 95.
5. *Investigation of Threaded Fastener Structural Integrity*, Southwest Research Institute, San Antonio, Texas, October 1977, p. 3.
6. *Fastener Standards*, 5th ed., Industrial Fastener Institute, Cleveland, Ohio, 1970, p. N-12.
7. Edwin Rodkey, "Making Fastened Joints Reliable—Way to Keep 'em Tight," *Assembly Engineering*, March 1977, pp. 24–27.
8. Robert J. Finkelston, "Optimized Bolt Tightening Takes to the Field," SPS Laboratories, Jenkintown, PA., March 1981.
9. E. Donald, "Fatigue-Indicating Fasteners; A New Dimension in Quality Control," *Fastener Technology*, March 1979.
10. Larry D. Mercer, "How Swaged Lockbolts Optimize Fastener Preload," National Design Engineering Conference, Chicago, Ill., 1982.
11. S. Eshghy, "Tension by Ultrasonic Stretch," privately published, June 1982.
12. Carl Osgood, "How Elasticity Influences Bolted Joint Design," *Machine Design*, March 1972, p. 106.

practice, we usually assume that a tension load is applied between the head of the fastener and the nut or tapped hole at the other end. Such a load would have a *worst-case effect* on the tension in the fasteners and on the clamping force on joint members, so this assumption is a safe and conservative one.

Any such tension load, no matter how small, will add to the tension in the bolts, increasing the length of the bolts slightly and thereby reducing the clamping force on the joint interface.

Not all the external load applied to the bolt is seen by the bolt, however. Part of the external load merely replaces part of the outward force which the joint initially exerted on the bolts that were clamping it. This can be illustrated by what engineers call a *joint diagram,* such as that shown in Fig. 23.13. Note that the external load applied to the bolt is equal to the sum of the changes which occur in the bolt and joint. It is equal, in other words, to the increase in tension in the fasteners plus the decrease in compressive load in the joint. We say that one part of the external load has been "absorbed" by the bolts; the rest has been absorbed by the joint (Ref. [23.6], pp. 199*ff*).

The relative stiffness or spring rate of bolt and joint determines how much of the load each will absorb. In Fig. 23.13 the joint stiffness is 3.23 times that of the bolt, as determined by our previous calculation of the stiffness ratio R_{JB} for the bolt in Fig. 23.10. The joint, therefore, will absorb approximately seven-tenths of any applied external tension load.

We should note in passing that the effects of an external *compressive* load can also be illustrated by a joint diagram, such as that shown in Fig. 23.14. This time the tension in the bolt is reduced and the compression in the joint is increased simultaneously by the single external load. The portions absorbed by fasteners and joint are again proportional to their relative stiffness.

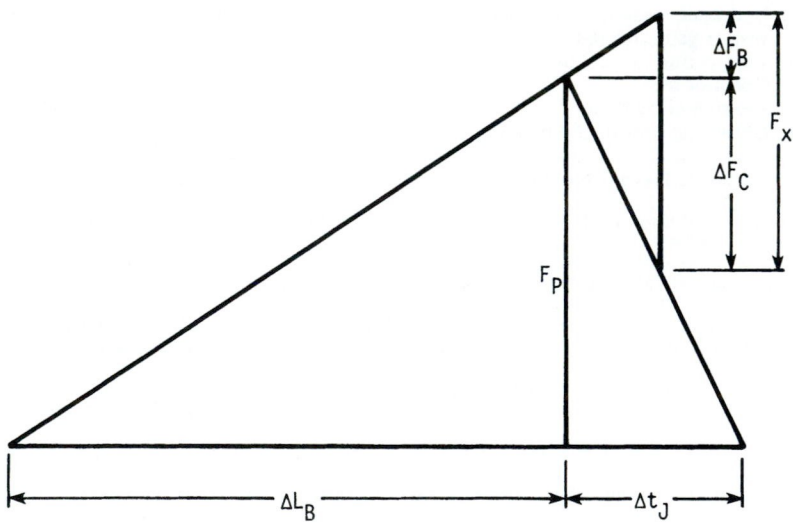

FIGURE 23.13   Joint diagram for a joint loaded in tension. A joint diagram consists of two force-elongation diagrams, one for the bolt and one for the joint material loaded by that bolt put front-to-front (Ref. [23.6], pp. 199–206). It illustrates the combined elastic behavior of bolt and joint.

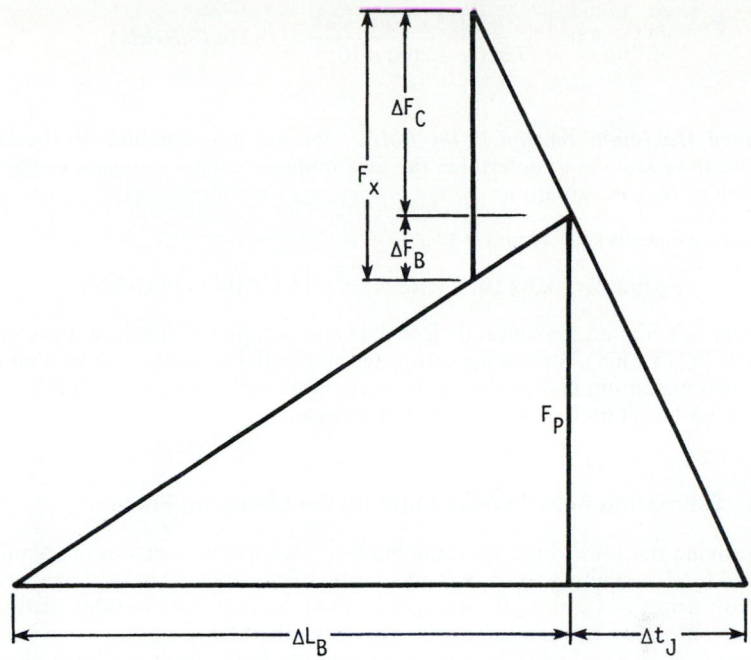

FIGURE 23.14   Joint diagram for a joint loaded in compression.

Note that *any* external compression or tension load will alter both the tension in the fasteners and the compression in the joint. This is contrary to the widely held belief that there will be no such change in either member until and unless the external tensile load exceeds the preload in magnitude.

If the external tensile load exceeds the initial preload, then all clamping force will have been removed from the joint and the bolts will see, in its entirety, any additional tension load placed on that joint.

One nice feature of the joint diagram is that it allows us to derive expressions which define joint behavior, such as the following:

$$k_B = \frac{F_P}{\Delta L_B} \qquad k_J = \frac{F_P}{\Delta t_j} \tag{23.17}$$

$$\Delta F_B = F_x \left(\frac{k_B}{k_J + k_B} \right) \tag{23.18}$$

$$\Delta F_C = F_x \left(1 - \frac{k_B}{k_J + k_B} \right) \tag{23.19}$$

Let us continue the example we started in Fig. 23.12 by assuming an external tensile load of 10 kip (44.5 kN) per bolt has been placed on the system and computing the estimated effect of the external load on the bolt tension:

$$F_B = 5000 \left[\frac{2.265 \times 10^6}{(7.316 + 2.265) \times 10^6} \right] = 1182 \text{ lb } (5.26 \text{ kN})$$

Estimated Maximum Tension in the Bolts.   We can now combine all the effects which we have studied to determine the maximum anticipated tension in the bolts under the worst-case situation:

$$F_B(\text{max}) = F_P(\text{max}) + \Delta F_B$$

$$F_B(\text{max}) = 13.3 \times 10^3 + 1.182 \times 10^3 = 14.5 \times 10^3 \text{ lb } (64.5 \text{ kN})$$

So our anticipated maximum is less than the acceptable maximum preload of 16.9×10^3 lb (75.2 kN). We can therefore continue with our analysis. Note that if the anticipated maximum had exceeded the acceptable maximum, we would have had to lower our target preload somewhat and try again.

23.3.4   Estimating Actual Lower Limit on the Clamping Force

To determine the lower limit on clamping force, we follow a procedure similar to that used for determining the maximum tension to be expected in the fasteners, this time subtracting, not adding, from target preload the tool and operator scatter and that portion of the external load which reduces the clamping force on the joint members (ΔF_C in Fig. 23.13). When considering the lower limit, we must also consider one other effect: the short-term relaxation of the joint following or during initial tightening.

Relaxation Effects.
  Embedment Relaxation.   When joint and fastener are first assembled, especially if we are dealing with new parts, they contact each other only on the microscopically rough high spots of their contact surfaces. These high spots will be loaded past the yield point and will creep and flow when first placed under load until enough additional surface area has been brought into play to stabilize the process. Other plastic flow can occur in thread roots, in the head-to-body fillet, and even in some whole threads, causing further relaxation. All these effects are lumped together under the term *embedment relaxation.* Typically they will lead to a loss of 2 to 10 percent of initial preload in the first few seconds or minutes after initial tightening. Let us assume a 5 percent loss of our ongoing example.
  Elastic Interactions in the Joint.   Achieving perfect initial preload is not our only problem when we tighten a multibolt joint. When we tighten the first fasteners, we stretch them and partially compress the joint. When we subsequently tighten other fasteners in the same joint, we further compress the joint. Because this act allows previously tightened fasteners to contract a little, it reduces the tension in those fasteners even if we achieved perfect *initial* preload in each when we first tightened it.
  The effect a fastener has on a previously tightened fastener is illustrated in Fig. 23.14. The new fastener applies a compressive load to the joint which had been previously preloaded by the first bolt.

Estimating Minimum Clamping Force.   Elastic interactions can be a special problem in large-diameter and/or gasketed joints. Since our example is neither, and involves only two bolts, we will assume that these interactions cost us, worst case, only 25 percent of initial preload in that bolt tightened first. Worst-case minimum clamping force in the vicinity of that bolt, therefore, becomes

$$F_C(\text{min}) = F_P(\text{min}) - F_X\left(1 - \frac{k_B}{k_J + k_B}\right) - F_P(\text{min})P_Z$$

$$F_C(\text{min}) = 6.91 \times 10^3 - 5000\left[1 - \frac{2.265 \times 10^6}{(7.316 + 2.265) \times 10^6}\right] \qquad (23.20)$$

$$- 6.91 \times 10^3(0.05 + 0.25) = 1019 \text{ lb } (4.53 \text{ kN})$$

So our computed anticipated minimum is greater than the zero minimum acceptable clamping force determined earlier. If it had not been acceptable, we would, of course, have had to readjust our target preload and/or select more accurate tools and/or revise our design.

23.3.5   Achieving a Desired Preload

In most cases we will try to achieve our target preload by using a wrench of some sort to apply torque to the nut or to the head of the bolt, accepting the resulting scatter in preload which this implies. We will discuss the torque-preload relationship at length in the next section. First, however, let us take a brief look at some of our other options for tightening bolts.

Hydraulic Tensioners and Bolt Heaters.   On large bolts we do not have to use a wrench. We can use a hydraulic tensioner which exerts a pure tension on the bolt, grabbing a few threads which stick out past the nut. Once the tool has stretched the bolt by the desired amount, the nut is run down to retain the tension. Then the tool lets go. At first glance this sounds like a perfect answer to some of the torquing and friction uncertainties we shall consider in Sec. 23.4, but there are other problems. The amount of preload retained by the fastener's nut is never the same as the preload introduced by the tool, because the nut must embed itself in the joint to pick up the loads originally supported by the much larger feet of the tensioner. This *elastic recovery loss* can equal 10 to 80 percent of the initial tension, depending on whether the fastener is relatively long (smaller loss) or short, and depending on how much torque was applied to the nut when it was run down (more torque, less loss).

Hydraulic tensioners, however, are superb tools when it comes to preloading large fasteners. They can be gang-driven from a single hydraulic pump and so can tighten several fasteners simultaneously with the same initial tension in each. This can be a very important feature when you are tightening large joints, especially if they are gasketed.

Tensioners also eliminate the galling problems often encountered when we attempt to torque large fasteners (3 in in diameter or more). The male and female threads are not turned relative to each other under heavy contact pressure with the tensioner. So tensioners have a place, but they do not provide perfect control of bolt preload.

Another nonwrench sometimes used to preload large fasteners is a bolt heater. This is inserted in an axial hole running down the center of the bolt. The bolt gets longer as it gets hot. When it is hot, the nut is run down against the joint to retain the increase in length produced by the heat. Since this is a crude way to preload bolts, the process must be controlled by other means. Dial gauges or micrometers are usually used to measure the net change in length of the bolts after they have cooled. If they have been stretched too much or too little, the bolts must be reheated and the nuts run down again. The process takes skill but is widely used on large fasteners (again 3 in or so and larger).

Microprocessor-Controlled Torque-Turn Tools.   Hydraulic tensioners and bolt heaters make it possible to tighten fasteners without suffering the uncertainties of the torque-preload relationship, but they can be used only on large-diameter fasteners. For smaller ones, we need something else. One relatively new approach—microprocessor-controlled tools—measures both applied torque and the turn of the nut to monitor and/or control fastener preload. Most of the presently available systems are designed for automatic or semiautomatic assembly in mass-production operations (automotive, primarily), but there are manual versions of some of them. They can control preload to ±2 to 5 percent if the joints are relatively soft (preload builds up smoothly as the bolts are tightened) and reasonable control is maintained over fastener dimensions and lubricity.

Some of these systems are designed to tighten every fastener to the yield point. This provides good preload accuracy (control is based on the act of yield rather than the torque-friction-preload relationship). But not all joints can be tightened safely to the yield point of the fastener. Although there is considerable debate on this point, many designers feel that yield-point tightening can lead to fatigue failure or rupture unless future external loads can be predicted and controlled.

Turn-of-the-Nut Control.   There is one place where tightening to or past the yield point is the norm; structural steel joints have been tightened this way for half a century using a carefully designed process called *turn-of-the-nut*. The fastener is first tightened to 60 to 80 percent of yield by the application of torque (usually with an air-powered impact wrench). The location of one corner of the nut is then noted, and a wrench is used to give the nut a specified half turn or so (depending on the size of the fastener and whether or not it is being used on a flat or tapered joint member). This amount of turn always takes the bolt beyond yield. Since external loads can be predicted, however, and are generally static rather than dynamic, the process is a safe and effective way to control preload.

Ultrasonic Control of Preload.   Ultrasonic instruments are sometimes used instead of torque and/or turn-of-the-nut to control preload (Ref. [23.6], p. 157). This technology has been used in a few aerospace applications for nearly a decade and is just starting to emerge in the commercial marketplace.

Presently available instruments send bursts of sound through the fastener and measure the time it takes for these wavefronts to travel through the fastener, echo off the far end, and return to the transducer. As the fastener is tightened, the time required for this round trip increases because the fastener gets longer, and so the path length is increased. Also, the velocity of the sound waves decreases as the stress level increases.

Microprocessors in the instruments sort out the change-in-length effect from the velocity effect and display either the change in length of the fastener or the average stress level in the tensile-stress area of the threads. Either of these quantities can be used to estimate fastener preload with better accuracy than is possible with torque or torque-turn controls.

One advantage of ultrasonics is that it can also be used in some cases to measure residual or working loads in the fasteners, as well as initial loads. You can use it to detect the effects of elastic interactions, for example, and therefore to compensate for such interactions. You can measure residual loads days or even years after initial tightening, which is never possible with torque and/or turn means.

Ultrasonics can be used with any sort of wrench, as well as with tensioners or heaters, to tighten fasteners.

23.4 BOLT TORQUE REQUIREMENTS

23.4.1 The Problem

Although torque is the most common way to tighten a fastener, it is not a very good way, usually, to control the preload developed within the fastener. As we saw in Table 23.4, we must expect to see a scatter of ±30 percent or worse in the preload we achieve if we are using torque tools to tighten the fasteners. This scatter is acceptable in most applications, however. We compensate for it by overdesign, using larger bolts than might otherwise be necessary, for example.

Many factors affect this scatter in preload. These include such things as the finish on nuts, bolts, and joint members; the age, temperature, quantity, condition, and type of the lubricants used, if any; the speed with which the fasteners are tightened; the fit between male and female threads; the size of the holes and their perpendicularity with respect to joint surfaces; and the hardness of all parts. There is no way in which we can control or predict all the variables in a given situation, and so we must always expect and accept a considerable scatter in preload results when we use torque to control the tightening operation.

23.4.2 Selecting the Correct Torque

Having said all this, we must still select an appropriate torque to produce, or attempt to produce, the target preload we have established for our design. Our best bet is to use the so-called short-form torque equation to make an estimate. This equation is

$$T = KdF_{PT} \qquad (23.21)$$

The nut factor K is an experimental constant, a *bugger factor*, if you will, which defines the relationship which exists between applied torque and achieved preload in a given situation. The only way to determine what K should be in your application is to make some actual experiments in which you measure both torque *and* preload and compute the mean K and the scatter in K. If accuracy is not a big concern or you are merely trying to select the proper size of wrench or determine the approximate preloads you will achieve, then it is safe to use a nut factor listed in Table 23.5.

23.5 FATIGUE LOADING OF BOLTED AND RIVETED JOINTS

When a bolt or joint member suddenly and unexpectedly breaks, it has probably failed because of fatigue. This is certainly one of the most common modes of failure for bolted joints. The designer, therefore, should learn how to cope with it.

TABLE 23.5   Nut Factors

Lubricant or coating on the fastener	Source	Nut factor	
		Reported mean	Reported range
1. Cadmium plate	1	0.194–0.246	0.153–0.328
2. Zinc plate	5	0.332	0.262–0.398
3. Black oxide	1	0.163–0.194	0.109–0.279
4. Baked on PTFE	1	0.092–0.112	0.064–0.142
5. Molydisulfide paste	2	0.155	0.14–0.17
6. Machine oil	2	0.21	0.20–0.225
7. Carnaba wax (5% emulsion)	2	0.148	0.12–0.165
8. 60 Spindle oil	2	0.22	0.21–0.23
9. As received steel fasteners	3	0.20	0.158–0.267
10. Molydisulfide grease	3	0.137	0.10–0.16
11. Phosphate and oil	3	0.19	0.15–0.23
12. Copper-based anti seize compound	3	0.132	0.08–0.23
13. As received steel fasteners	4	0.20	0.161–0.267
14. Plated fasteners	4	0.15	
15. Grease, oil, or wax	4	0.12	

SOURCES:
1. Values given are typical results from a very large and unpublished set of experiments on ASTM A193 B7 studs treated with various coatings. The tests were made in 1979–1980. Mean values for K varied with the diameter of the studs tested and the torques applied in various test series.
2. Kazuo Maruyama, Makoto Masuda, and Nobutoshi Ohashi, "Study of Tightening Control Methods for High Strength Bolts," *Bulletin of the Research Laboratory Precision Machine Selection,* Tokyo Institute of Technology, N46, September 1980, pp. 27–32.
3. John H. Bickford, *An Introduction to the Design and Behavior of Bolted Joints,* Marcel Dekker, Inc., New York, 1981, p. 429.
4. *Fastener Standards,* 5th ed., Industrial Fastener Institute, Cleveland, Ohio, 1970, p. N-16.
5. Edwin Rodkey, "Making Fastened Joints Reliable—Ways to Keep 'em Tight," *Assembly Engineering,* March 1977, p. 24.

23.5.1   Spotting a Fatigue Problem

It is usually easy to diagnose a fatigue failure. Here are the clues:

1. *Cyclic Loads*   Fatigue failures always occur under cyclic tension loads.

2. *No Advance Warning*   Fatigue failure is always sudden and almost always unexpected. The parts do not neck-down or wear out before they fail.

3. *Appearance of the Break Surface*   If you examine the surface of a part which has failed in fatigue, you will usually find that a section of the surface is smooth, sometimes almost polished. Another portion of the surface, surrounding the first, may be a little rougher but is still basically smooth. The remainder of the surface will be very rough indeed.

4. *Typical Failure Points*   The parts tend to fail at points of high stress concentration. Figures 23.15 and 23.16 show the most common failure points.

23.5.2   Estimating Fatigue Life

Many factors affect the fatigue life of any machine part, including fasteners. Such things as shape, heat treatment, surface finish, the mean load stress, the magnitude of load excursions, and the material all play a role. If you know the basic strength of the part, however, you can use the methods of Chap. 13 to estimate fatigue strength or endurance limit. Table 23.6 gives you the strength information you will need to do this for fasteners. You will find other information pertinent to the fatigue of joint members in Chaps. 7 and 13.

The term *proof strength* in Table 23.6 deserves explanation. It is common to test the strength of fasteners by applying tension loads to them. The *proof load* of a given fastener is the highest tensile force which can be applied to it without causing a permanent set to the fastener. The *proof strength* can then be determined by dividing the proof load by the tensile-stress area of the threads [Eq. (23.4)].

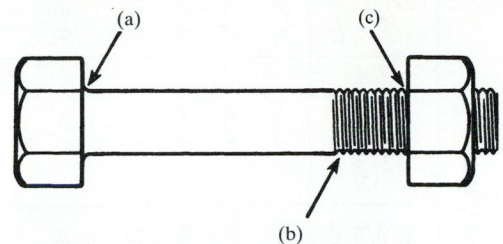

FIGURE 23.15   Typical failure points of a bolt. (*a*) Failure at head fillet; (*b*) failure at thread runout; (*c*) failure at first thread to engage the nut.

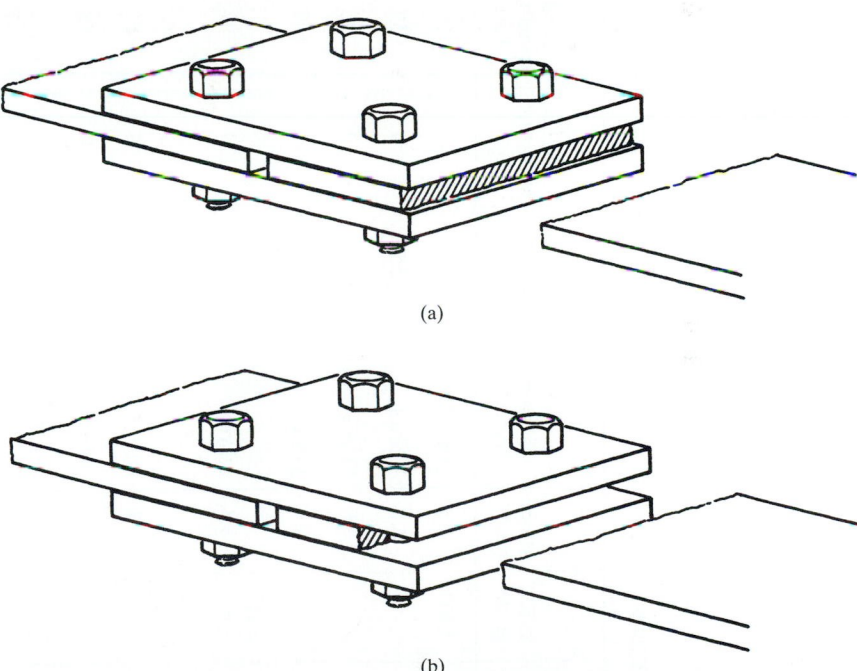

FIGURE 23.16   Typical fatigue failures in joints loaded in shear. (*a*) Failure occurs in the gross cross section, near the place where the splice plates and joint plates meet, in a friction-type joint; (*b*) failure occurs in a net cross section, through a line of bolts, in a bearing-type joint.

TABLE 23.6 Specifications and Identification Markings for Bolts, Screws, Studs, Sems,[a] and U Bolts[b]

SAE grade	ASTM grade	Metric grade[c]	Nominal diameter in	Proof strength,[d] kpsi	Tensile strength,[d] kpsi	Yield strength,[d,e] kpsi	Core hardness Rockwell min/max	Grade identification marking	Products[f]	Material
1	A307	4.6	$\frac{1}{4}$ thru $1\frac{1}{2}$	33	60	36	B70/B100	None	B, Sc, St	Low- or medium-carbon steel
2	...	5.8	$\frac{1}{4}$ thru $\frac{3}{4}$	55	74	57	B80/B100	None	B, Sc, St	Low- or medium-carbon steel
		4.6	Over $\frac{3}{4}$ thru $1\frac{1}{2}$	33	60	36	B70/B100	None	B, Sc, St	Low- or medium-carbon steel
4	...	8.9	$\frac{1}{4}$ thru $1\frac{1}{2}$	65[g]	115	100	C22/C32	None	St	Medium-carbon, cold-drawn steel
5	A449 or A325 Type 1	8.8	$\frac{1}{4}$ thru 1	85	120	92	C25/C34	⌄	B, Sc, St	Medium-carbon steel, Q&T
		7.8	Over 1 thru $1\frac{1}{2}$	74	105	81	C19/C30	⌄	B, Sc, St	Medium-carbon steel, Q&T
		8.6	Over $1\frac{1}{2}$ to 3	55	90	58	...	⌄	B, Sc, St	Medium-carbon steel, Q&T
5.1	...	8.8	No. 6 thru $\frac{3}{8}$	85	120	...	C25/C40	⊥	Se	Low- or medium-carbon, Q&T
		8.8	No. 6 thru $\frac{1}{2}$	85	120	...	C25/C40	⊥	B, Sc, St	Low- or medium-carbon, Q&T

SAE grade	ASTM spec.	Metric grade	Size range, in	Min. proof strength	Min. tensile strength S_{ut}	Min. yield strength S_y	Core hardness Rockwell C	Head marking	Type	Material
5.2	A325 Type 2	8.8	¼ thru 1	85	120	92	C26/C36	(marking)	B, Sc	Low-carbon martensite steel, fully killed, fine-grained, Q&T
7[h]	. . .	10.9	¼ thru 1½	105	133	115	C28/C34	(marking)	B, Sc	Medium-carbon alloy steel, Q&T
8	A354 Grade BD	10.9	¼ thru 1½	120	150	130	C33/C39	(marking)	B, Sc, St	Medium-carbon alloy steel, Q&T
8.1	. . .	10.9	¼ thru 1½	120	150	130	C31/C38	None	St	Elevated temperature drawn steel-medium carbon alloy or G15410
8.2	. . .	10.9	¼ thru 1	120	150	130	C35/C42	(marking)	B, Sc	Low-carbon martensite steel, fully killed, fine-grained, Q&T
. . .	A574	12.9	0 thru ½	140	180	160	C39/C45	12.9	SHCS	Alloy steel, Q&T
. . .	A574	12.9	⅝ thru 1½	135	170	160	C37/C45	12.9	SHCS	Alloy steel, Q&T

[a] Sems = screw and washer assemblies.

[b] Compiled from ANSI/SAE J429j; ANSI B18.3.1-1978; and ASTM A307, A325, A354, A449, and A574.

[c] Metric grade is $xx.x$, where xx is approximately $0.01S_{ut}$ in MPa and $.x$ is the ratio of the minimum S_y to S_{ut}.

[d] Multiply the strengths in kilopounds per square inch by 6.89 to get strength in megapascals.

[e] Yield strength is stress at which a permanent set of 0.2% of gauge length occurs.

[f] B = bolt, Sc = screws, St = studs, Sc = Sems, and SHCS = socket head cap screws.

[g] Entry appears to be in error but conforms to the standard ANSI/SAE J429j.

[h] Grade 7 bolts and screws are roll threaded after heat treatment.

SOURCE: From Joseph E. Shigley and Charles R. Mischke, *Mechanical Engineering Design*, 5th ed., McGraw-Hill, 1989; reproduced by permission of the authors and the publisher.

23.5.3 Reducing Fatigue Problems

There are a lot of things you can do to minimize fatigue problems.

Material and Part Selection and Care. Materials with higher tensile strengths tend to have better fatigue lives than those with lower tensile strengths, at least up to an ultimate tensile strength of 200 kpsi (1379 MPa) or so. It also helps to select a material having low-notch sensitivity.

Avoid decarburization of the parts. Decarburization can weaken part surfaces and make it much easier for initial cracks to form.

Make sure that nut faces and the undersurface of the bolt head are perpendicular to the axis of the bolt threads and that the holes are perpendicular to the surfaces of the joints [23.10]. Two degrees of angularity can reduce fatigue life to only 25 percent of normal.

Lubricate the threads [23.10]. If nothing else, this can reduce corrosion problems, and corrosion is a main source of initial cracks.

If using fasteners with a tensile strength above 150 kpsi (1034 MPa), do *not* use lubricants containing sulfides, since these can contribute to stress-corrosion cracking, which will accelerate fatigue failure [23.11].

Grit blast the surfaces of joints loaded in shear before assembling because anything which increases the slip resistance improves fatigue life ([23.2], p. 120).

Prevent Crack Initiation. Polish, but do not hard coat, bolt surfaces, or shot peen the surfaces, or roll bolt threads after heat treatment. Do anything and everything possible to avoid corrosion of bolts or joint members (see Chap. 44).

Reduce Load Excursions. Even if the magnitude of external loads imposed on a joint are beyond the designer's control, there are many things which he or she can do to reduce the variations in load seen by a given joint. And these variations, or *load excursions,* are a key issue. We always want to keep the ratio between minimum load and maximum load seen by the parts as close to unity as possible.

Some say the minimum bolt tension should always be more than half the maximum bolt tension. Others recommend a preload that is at least two to three times the magnitude of the worst-case external load to be applied to the joint. Because of the large number of variables involved, such rules will apply only to certain applications. Nevertheless, they give you an idea of the importance of minimizing load excursions.

It helps to increase the ratio between the stiffness of the joint and the stiffness of the bolt (k_J/k_B) so that the joint will absorb a larger percentage of the applied load excursions. There are many ways to do this. For examples, see Fig. 23.17a and b. Reducing the body of the bolt to nine-tenths of the nominal diameter is sometimes recommended.

It helps to compensate for initial preload loss and relaxation effects by retightening the bolts after they have relaxed. By the same token, try to avoid vibration loss of preload by providing damping and/or by periodic retightening of the nuts and/or by using special vibration-resistant fasteners.

Reduce Stresses in Parts. Make sure there are at least three threads above and below the faces of the nut (Fig. 23.17c). Do not let the thread run-out point coincide with the shear plane of the joint (Fig. 23.17d). Roll the threads instead of cutting

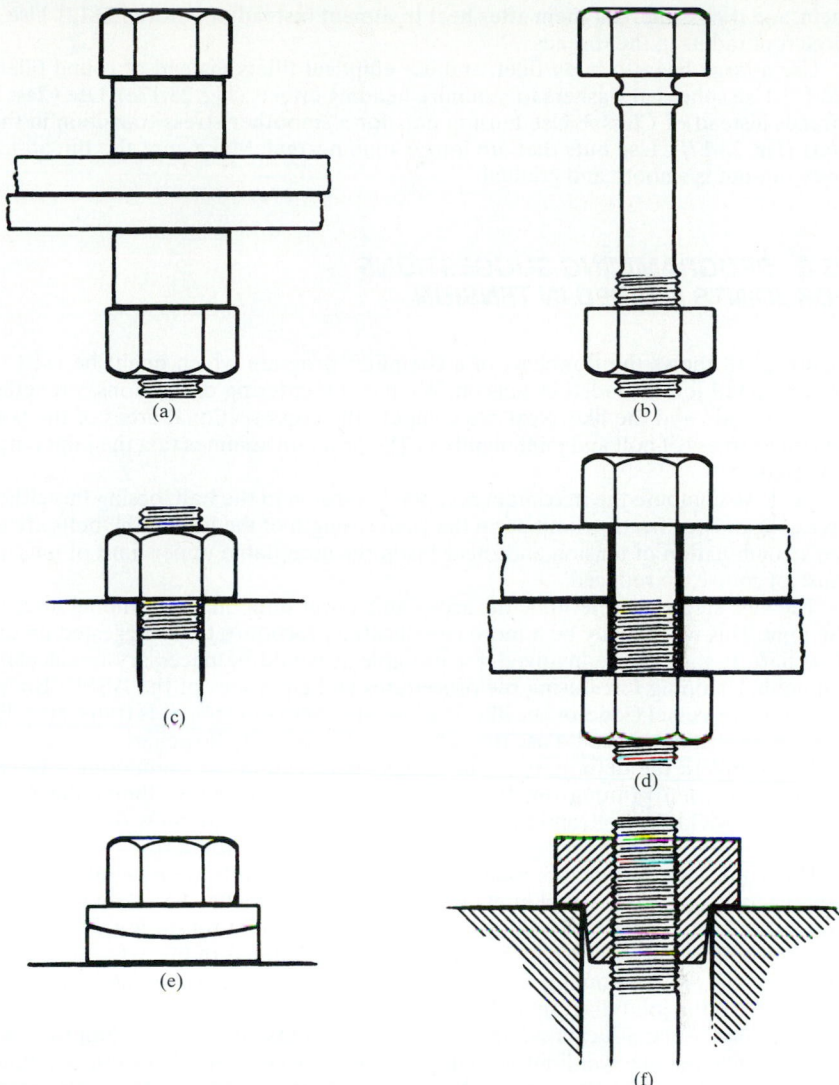

FIGURE 23.17 Ways to improve the fatigue life of bolts. (*a*) Use collars to increase the length-to-diameter ratio of the bolts; (*b*) turn down the body of a bolt to reduce its stiffness; (*c*) make sure that there are at least three threads above and below a nut to reduce thread stress concentrations; (*d*) it also helps to avoid the situation, shown here, where thread run-out coincides with a shear plane in the joint, or (*e*) to use spherical washers to help a bolt adjust to bending loads, or (*f*) to use tension nuts to reduce thread stress levels. All figures shown are improvements except (*d*).

them, and if possible, roll them after heat treatment instead of before [23.12]. Use a large root radius in the threads.

Use a large head-to-body fillet, and use elliptical fillets instead of round fillets [23.13]. Use spherical washers to minimize bending effects (Fig. 23.17e). Use Class 2 threads instead of Class 3. Use tension nuts for a smoother stress transition in the bolts (Fig. 23.17f). Use nuts that are longer than normal. Make sure the thread-to-body run-out is smooth and gradual.

23.6 PROGRAMMING SUGGESTIONS FOR JOINTS LOADED IN TENSION

Figure 23.18 shows the flowchart of a computer program which might be used to design bolted joints loaded in tension. We start by entering dimensions, strengths, external loads, and the like. Next, we compute the cross-sectional areas of the bolt and the stiffness of bolt and joint members. The program assumes that the joint is not gasketed.

Next, we compute the maximum acceptable tension in the bolt, basing this either on a code or specification limit or on the yield strength of the bolt. If the bolts are to see a combination of tension and shear loads, the acceptable upper limit of tension must, of course, be reduced.

The next step is to determine the acceptable lower limit on the clamping force in the joint. This will usually be a more complicated procedure than suggested by the flowchart. If gaskets are involved, for example, it would be necessary to calculate minimum clamping force using the procedures and equations of the ASME Boiler and Pressure Vessel Code, or the like. If all we are concerned about is transverse slip or total separation, we could use the equations shown in the flowchart.

We complete the definition of our design specifications by computing a target preload and then printing out the upper and lower acceptable limits, the force required to yield the bolt, and the target preload. It is useful to know these things if we need to revise the target preload at a later point in the program.

Having determined the acceptable upper and lower limits, we now take a series of steps to estimate the actual limits we will achieve in practice based on our target preload and estimates of such things as tool scatter and joint relaxation. During this part of the procedure we also introduce the estimated effects of the external tension loads on the joint, assuming linear joint behavior. The equations used here are derived from the joint diagram in Fig. 23.13.

We compute the anticipated upper limit on bolt tension first, and then we compute the anticipated lower limit on clamping force. We compare them, one at a time, to the acceptable limits. We recycle, choosing a new target preload, if the anticipated limits fall outside of the acceptable limits. In some cases we will not be able to satisfy our specifications merely by modifying the target preload; we may have to choose new joint dimensions to enlarge the range between upper and lower limits or choose more accurate tools to reduce the range between the upper and lower limits anticipated in practice.

When our conditions are satisfied, we complete the program by computing the torque required to aim for the target preload. Then we print out the final values of the parameters computed.

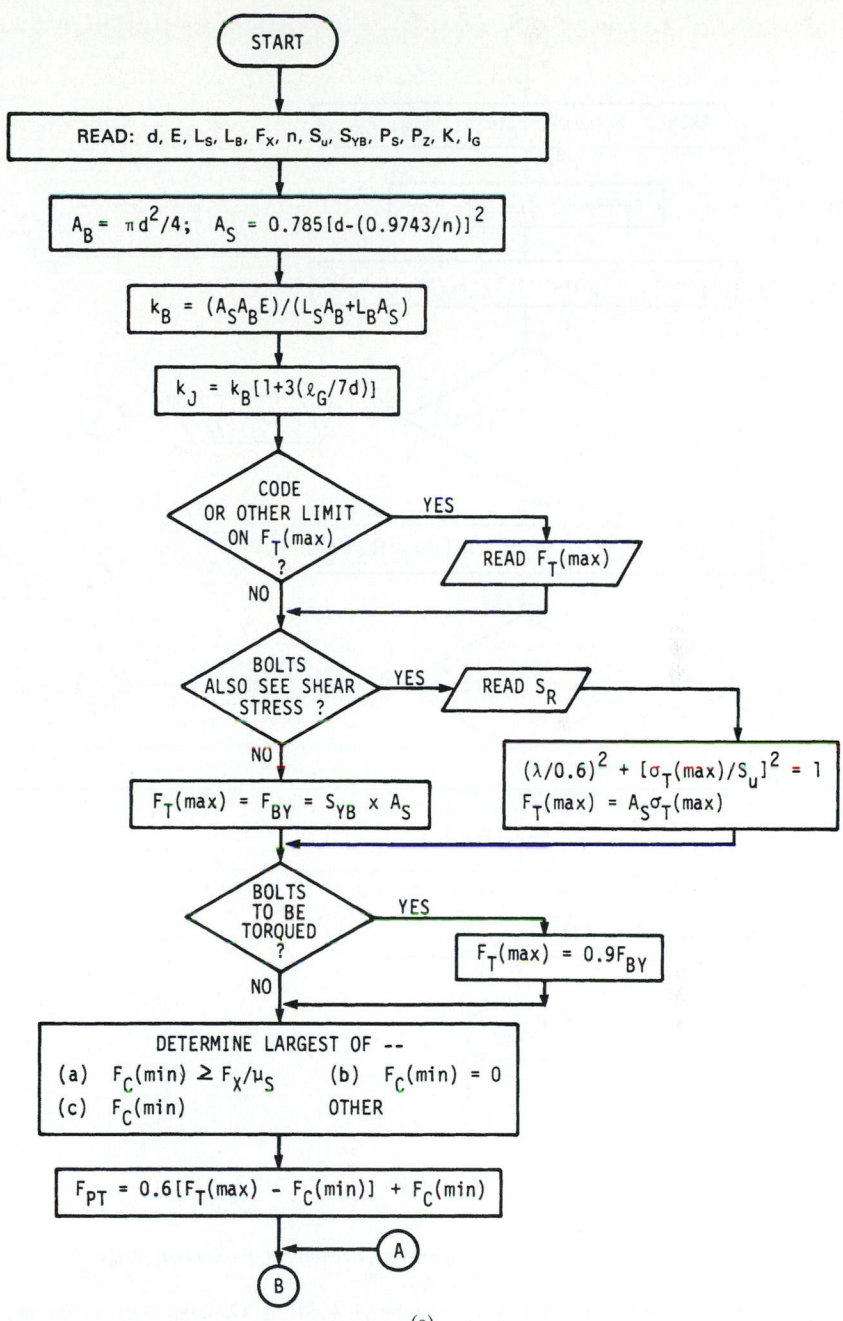

FIGURE 23.18 Flowchart describing a computer program which could be used to design non-gasketed joints loaded in tension. Chart continues on next page.

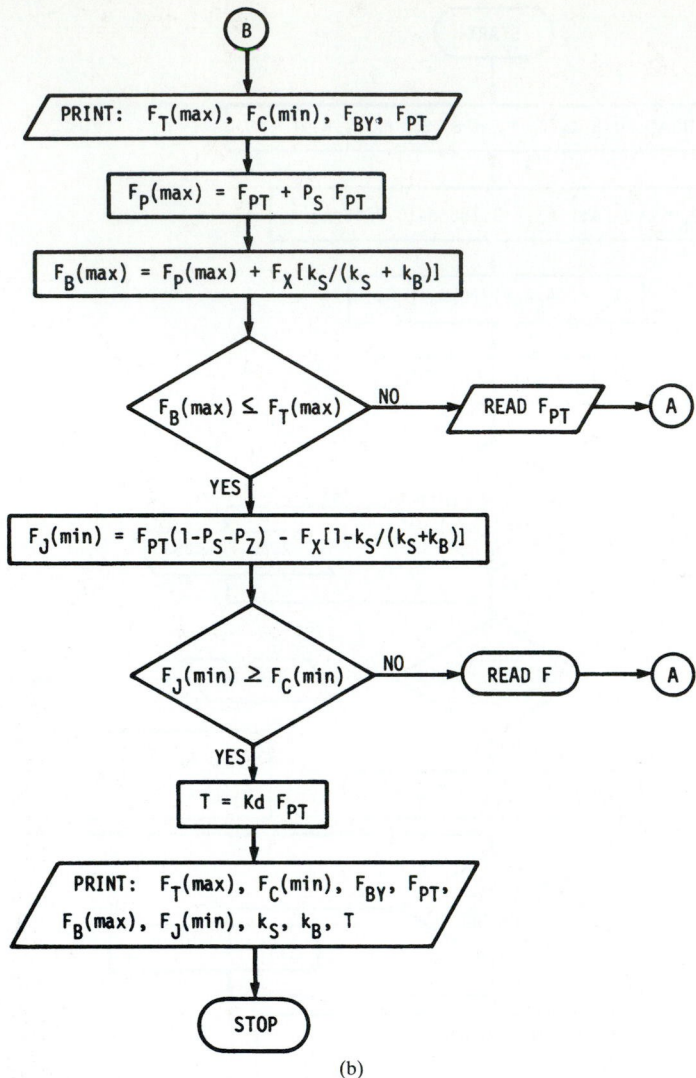

(b)

FIGURE 23.18   *(Continued)*

REFERENCES

23.1 Joseph E. Shigley and Charles R. Mischke, *Mechanical Engineering Design,* 5th ed., McGraw-Hill Book Company, New York, 1989.

23.2 Goeffrey L. Kulak, John W. Fisher, and John H. A. Struik, *Guide to Design Criteria for Bolted and Riveted Joints,* 2d ed., John Wiley & Sons, New York, 1987.

23.3 Archie Higdon, Edward H. Ohlsen, William B. Stiles, John A. Weese, and William F. Riley, *Mechanics of Materials,* 3d ed., John Wiley & Sons, New York, 1978.

23.4 ANSI Specification B1.1-1974, "Screw Threads," American Society of Mechanical Engineers, New York, 1974, p. 80.

23.5 Specifications BS 3643: Part 2: 1981, "ISO Metric Screw Threads," British Standards Institute, London, 1981, p. 10.

23.6 John H. Bickford, *An Introduction to the Design and Behavior of Bolted Joints,* 2d ed., Marcel Dekker, New York, 1990, p. 106.

23.7 Nabil Motosh, "Determination of Joint Stiffness in Bolted Connections," *Trans. ASME,* August 1976, p. 859.

23.8 G. H. Junker, "Principle of the Calculation of High Duty Bolted Joints: Interpretation of VDI Directive 2230," Unbrako Technical Thesis, SPS, Jenkintown, Pa., 1974, p. 8.

23.9 Carl C. Osgood, *Fatigue Design,* Wiley-Interscience, New York, 1970, p. 196.

23.10 Joseph Viglione, "Nut Design Factors for Long Life," *Machine Design,* August 1965, p. 138.

23.11 "Degradation of Threaded Fasteners in the Reactor Coolant Pressure Boundary of PWR Plants," U.S. Nuclear Regulatory Commission, Washington, D.C., June 1982, p. 2.

23.12 Edwin Rodkey, "Making Fastened Joints Reliable—Ways to Keep 'em Tight," *Assembly Engineering,* March 1977, pp. 24–27.

23.13 "Thread Forms and Torque Systems Boost Reliability of Bolted Joints," *Product Engineering,* December 1977, p. 38.

CHAPTER 24
SPRINGS

Robert E. Joerres
Applications Engineering Manager
Associated Spring, Barnes Group, Inc.
Bristol, Connecticut

GENERAL NOMENCLATURE[†]

A	Area, mm^2 (in^2)
b	Width, mm (in)
C	Spring index, D/d
d	Wire diameter, mm (in)
D	Mean diameter (OD minus wire diameter), mm (in)
E	Modulus of elasticity in tension or Young's modulus, MPa (psi)
f	Deflection, mm (in)
g	Gravitational constant, 9.807 m/s^2 (386.4 in/s^2)
G	Shear modulus or modulus of rigidity, MPa (psi)
I	Moment of inertia, mm^4 (in^4)
ID	Inside diameter, mm (in)
k	Spring rate, N/mm (lb/in) or $N \cdot mm/r$ ($lb \cdot in/r$)

[†] The symbols presented here are used extensively in the spring industry. They may differ from those used elsewhere in this Handbook.

K	Design constant
K_w	Stress correction factor for helical springs
L	Length, mm (in)
L_f	Free length, mm (in)
L_s	Length at solid, mm (in)
M	Moment or torque, N·mm (lb·in)
n	Frequency, Hz
N_a	Number of active coils or waves
N_t	Total number of coils
OD	Outside diameter, mm (in)
P	Load, N (lbf)
r	Radius, mm (in)
S	Stress, MPa (psi)
TS	Tensile strength, MPa (psi)
t	Thickness, mm (in)
YS	Yield strength, MPa (psi)
ρ	Density, g/cm^3 (lb/in^3)
θ	Angular deflection, expressed in number of revolutions
μ	Poisson's ratio

24.1 INTRODUCTION

Spring designing is a complex process. It is an interactive process which may require several iterations before the best design is achieved. Many simplifying assumptions have been made in the design equations, and yet they have proved reliable over the years. When more unusual or complex designs are required, designers should rely on the experience of a spring manufacturer.

The information in this chapter is offered for its theoretical value and should be used accordingly.

24.2 GLOSSARY OF SPRING TERMINOLOGY

active coils: those coils which are free to deflect under load.

baking: heating of electroplated springs to relieve hydrogen embrittlement.

buckling: bowing or lateral displacement of a compression spring; this effect is related to slenderness ratio L/D.

closed and ground ends: same as *closed ends,* except that the first and last coils are ground to provide a flat bearing surface.

closed ends: compression spring ends with coil pitch angle reduced so that they are square with the spring axis and touch the adjacent coils.

close-wound: wound so that adjacent coils are touching.

deflection: motion imparted to a spring by application or removal of an external load.

elastic limit: maximum stress to which a material may be subjected without permanent set.

endurance limit: maximum stress, at a given stress ratio, at which material will operate in a given environment for a stated number of cycles without failure.

free angle: angular relationship between arms of a helical torsion spring which is not under load.

free length: overall length of a spring which is not under load.

gradient: see *rate*.

heat setting: a process to prerelax a spring in order to improve stress-relaxation resistance in service.

helical springs: springs made of bar stock or wire coiled into a helical form; this category includes compression, extension, and torsion springs.

hooks: open loops or ends of extension springs.

hysteresis: mechanical energy loss occurring during loading and unloading of a spring within the elastic range. It is illustrated by the area between load-deflection curves.

initial tension: a force that tends to keep coils of a close-wound extension spring closed and which must be overcome before the coils start to open.

loops: formed ends with minimal gaps at the ends of extension springs.

mean diameter: in a helical spring, the outside diameter minus one wire diameter.

modulus in shear or torsion (modulus of rigidity G): coefficient of stiffness used for compression and extension springs.

modulus in tension or bending (Young's modulus E): coefficient of stiffness used for torsion or flat springs.

moment: a product of the distance from the spring axis to the point of load application and the force component normal to the distance line.

natural frequency: lowest inherent rate of free vibration of a spring vibrating between its own ends.

pitch: distance from center to center of wire in adjacent coils in an open-wound spring.

plain ends: end coils of a helical spring having a constant pitch and with the ends not squared.

plain ends, ground: same as *plain ends*, except that wire ends are ground square with the axis.

rate: spring gradient, or change in load per unit of deflection.

residual stress: stress mechanically induced by such means as set removal, shot peening, cold working, or forming; it may be beneficial or not, depending on the spring application.

set: permanent change of length, height, or position after a spring is stressed beyond material's elastic limit.

set point: stress at which some arbitrarily chosen amount of set (usually 2 percent) occurs; set percentage is the set divided by the deflection which produced it.

set removal: an operation which causes a permanent loss of length or height because of spring deflection.

solid height: length of a compression spring when deflected under load sufficient to bring all adjacent coils into contact.

spiral springs: springs formed from flat strip or wire wound in the form of a spiral, loaded by torque about an axis normal to the plane of the spiral.

spring index: ratio of mean diameter to wire diameter.

squared and ground ends: see *closed and ground ends.*

squared ends: see *closed ends.*

squareness: angular deviation between the axis of a compression spring in a free state and a line normal to the end planes.

stress range: difference in operating stresses at minimum and maximum loads.

stress ratio: minimum stress divided by maximum stress.

stress relief: a low-temperature heat treatment given springs to relieve residual stresses produced by prior cold forming.

torque: see *moment.*

total number of coils: the sum of the number of active and inactive coils in a spring body.

24.3 SELECTION OF SPRING MATERIALS

24.3.1 Chemical and Physical Characteristics

Springs are resilient structures designed to undergo large deflections within their elastic range. It follows that the materials used in springs must have an extensive elastic range.

Some materials are well known as spring materials. Although they are not specifically designed alloys, they do have the elastic range required. In steels, the medium- and high-carbon grades are suitable for springs. Beryllium copper and phosphor bronze are used when a copper-base alloy is required. The high-nickel alloys are used when high strength must be maintained in an elevated-temperature environment.

The selection of material is always a cost-benefit decision. Some factors to be considered are costs, availability, formability, fatigue strength, corrosion resistance, stress relaxation, and electric conductivity. The right selection is usually a compromise among these factors. Table 24.1 lists some of the more commonly used metal alloys and includes data which are useful in material selection.

Surface quality has a major influence on fatigue strength. This surface quality is a function of the control of the material manufacturing process. Materials with high surface integrity cost more than commercial grades but must be used for fatigue applications, particularly in the high cycle region.

24.3.2 Heat Treatment of Springs

Heat treatment is a term used in the spring industry to describe both low- and high-temperature heat treatments. Low-temperature heat treatment, from 350 to 950°F (175 to 510°C), is applied to springs after forming to reduce unfavorable residual stresses and to stabilize parts dimensionally.

When steel materials are worked in the spring manufacturing process, the yield point is lowered by the unfavorable residual stresses. A low-temperature heat treatment restores the yield point. Most heat treatment is done in air, and the minor oxide that is formed does not impair the performance of the springs.

When hardened high-carbon-steel parts are electroplated, a phenomenon known as *hydrogen embrittlement* occurs, in which hydrogen atoms diffuse into the metallic lattice, causing previously sound material to crack under sustained stress. Low-temperature baking in the range of 375 to 450°F (190 to 230°C) for times ranging from 0.5 to 3 h, depending on the type of plating and the degree of embrittlement, will reduce the concentration of hydrogen to acceptable levels.

High-temperature heat treatments are used to strengthen annealed material after spring forming. High-carbon steels are austenitized at 1480 to 1652°F (760 to 900°C), quenched to form martensite, and then tempered to final hardness. Some nickel-base alloys are strengthened by high-temperature aging. Oxidation will occur at these temperatures, and it is advisable to use a protective atmosphere in the furnace.

Heat treatments for many common materials are listed in Table 24.2. Unless otherwise noted, 20 to 30 min at the specified temperature is sufficient. Thin, flimsy cross-sectional springs can be distorted by the heat-treatment operation. Pretempered materials are available for use in such cases.

24.3.3 Relaxation

The primary concern in elevated-temperature applications is stress relaxation. *Stress relaxation* is the loss of load or spring length that occurs when a spring is held at load or cycled under load. Heat affects modulus and tensile strength. In addition to the factors of stress, time, and temperature which affect relaxation, other controllable factors are

1. Alloy type—the highly alloyed materials are generally more temperature-resistant.
2. Residual stresses—such stresses remaining from forming operations are detrimental to relaxation resistance. Use the highest practical stress-relief temperature.
3. Heat setting—procedures employed to expose springs under some load to stress and heat to prepare them for a subsequent exposure. The effect is to remove the first stage of relaxation.

24.3.4 Corrosion

The specific effect of a corrosive environment on spring performance is difficult to predict. In general, if the environment causes damage to the spring surface, the life and the load-carrying ability of the spring will be reduced.

The most common methods of combating corrosion are to use materials that are resistant or inert to the particular corrosive environment or to use coatings that slow

TABLE 24.1 Typical Properties of Common Spring Materials

Common Name	Young's Modulus E (1) MPa 10^3	psi 10^6	Modulus of Rigidity G (1) MPa 10^3	psi 10^6	Density (1) g/cm^3	(lb/in^3)	Electrical Conductivity (1) %IACS	Sizes Normally Available (2) Min. mm (in.)	Max. mm (in.)	Typical Surface Quality (3)	Maximum Service Temperature (4) °C	°F
Carbon Steel Wires:												
Music (5)	207	(30)	79.3	(11.5)	7.86	(0.284)	7	0.10 (0.004)	6.35 (0.250)	a	120	250
Hard Drawn (5)	207	(30)	79.3	(11.5)	7.86	(0.284)	7	0.13 (0.005)	16 (0.625)	c	150	250
Oil Tempered	207	(30)	79.3	(11.5)	7.86	(0.284)	7	0.50 (0.020)	16 (0.625)	c	150	300
Valve Spring	207	(30)	79.3	(11.5)	7.86	(0.284)	7	1.3 (0.050)	6.35 (0.250)	a	150	300
Alloy Steel Wires:												
Chrome Vanadium	207	(30)	79.3	(11.5)	7.86	(0.284)	7	0.50 (0.020)	11 (0.435)	a,b	220	425
Chrome Silicon	207	(30)	79.3	(11.5)	7.86	(0.284)	5	0.50 (0.020)	9.5 (0.375)	a,b	245	475
Stainless Steel Wires:												
Austenitic Type 302	193	(28)	69.0	(10.)	7.92	(0.286)	2	0.13 (0.005)	9.5 (0.375)	b	260	500
Precipitation Hardening 17-7 PH	203	(29.5)	75.8	(11)	7.81	(0.282)	2	0.08 (0.002)	12.5 (0.500)	b	315	600
NiCr A286	200	(29)	71.7	(10.4)	8.03	(0.290)	2	0.40 (0.016)	5 (0.200)	b	510	950
Copper Base Alloy Wires:												
Phosphor Bronze (A)	103	(15)	43.4	(6.3)	8.86	(0.320)	15	0.10 (0.004)	12.5 (0.500)	b	95	200
Silicon Bronze (A)	103	(15)	38.6	(5.6)	8.53	(0.308)	7	0.10 (0.004)	12.5 (0.500)	b	95	200
Silicon Bronze (B)	117	(17)	44.1	(6.4)	8.75	(0.316)	12	0.10 (0.004)	12.5 (0.500)	b	95	200
Beryllium Copper	128	(18.5)	48.3	(7.0)	8.26	(0.298)	21	0.08 (0.003)	12.5 (0.500)	b	205	400
Spring Brass, CA260	110	(16)	42.0	(6.0)	8.53	(0.308)	17	0.10 (0.004)	12.5 (0.500)	b	95	200
Nickel Base Alloys:												
Inconel® Alloy 600	214	(31)	75.8	(11)	8.43	(0.304)	1.5	0.10 (0.004)	12.5 (0.500)	b	320	700
Inconel® Alloy X750	214	(31)	79.3	(11.5)	8.25	(0.298)	1	0.10 (0.004)	12.5 (0.500)	b	595	1100
Ni-Span-C®	186	(27)	62.9	(9.7)	8.14	(0.294)	1.6	0.10 (0.004)	12.5 (0.500)	b	95	200
Monel® Alloy 400	179	(26)	66.2	(9.6)	8.83	(0.319)	3.5	0.05 (0.002)	9.5 (0.375)	b	230	450
Monel Alloy K500	179	(26)	66.2	(9.6)	8.46	(0.306)	3	0.05 (0.002)	9.5 (0.375)	b	260	500

Carbon Steel Strip:											
AISI 1050	207 (30)	79.3 (11.5)	7.86 (0.284)	7	0.25 (0.010)	3 (0.125)	b	95	200
1065	207 (30)	79.3 (11.5)	7.86 (0.284)	7	0.08 (0.003)	3 (0.125)	b	95	200
1074, 1075	207 (30)	79.3 (11.5)	7.86 (0.284)	7	0.08 (0.003)	3 (0.125)	b	120	250
1095	207 (30)	79.3 (11.5)	7.86 (0.284)	7	0.08 (0.003)	3 (0.125)	b	120	250
Stainless Steel Strip:									
Austenitic Types 301, 302	193 (28)	69.0 (10)	7.92 (0.286)	2	0.08 (0.003)	1.5 (0.063)	b	315	600
Precipitation Hardening 17-7 PH	203 (29.5)	75.8 (11)	7.81 (0.282)	2	0.08 (0.003)	3 (0.125)	b	370	700
Copper Base Alloy Strip:									
Phosphor Bronze (A)	103 (15)	43 (6.3)	8.86 (0.320)	15	0.08 (0.003)	5 (0.188)	b	95	200
Beryllium Copper	128 (18.5)	48 (7.0)	8.26 (0.298)	21	0.08 (0.003)	9.5 (0.375)	b	205	400

(1) Elastic moduli, density and electrical conductivity can vary with cold work, heat treatment and operating stress. These variations are usually minor but should be considered if one or more of these properties is critical.

(2) Sizes normally available are diameters for wire; thicknesses for strip.

(3) Typical surface quality ratings. (For most materials, special processes can be specified to upgrade typical values.)

a. Maximum defect depth: 0 to 0.5% of d or t.

b. Maximum defect depth: 1.0% of d or t.

c. Defect depth: less than 3.5% of d or t.

(4) Maximum service temperatures are guidelines and may vary due to operating stress and allowable relaxation.

(5) Music and hard drawn are commercial terms for patented and cold-drawn carbon steel spring wire.

INCONEL, MONEL and NI-SPAN-C are registered trademarks of International Nickel Company, Inc.

SOURCE: Associated Spring, Barnes Group Inc.

TABLE 24.2   Typical Heat Treatments for Springs after Forming

Materials	Heat Treatment °C	°F
Patented and Cold-Drawn Steel Wire	190–230	375–450
Tempered Steel Wire:		
Carbon	260–400	500–750
Alloy	315–425	600–800
Austenitic Stainless Steel Wire	230–510	450–950
Precipitation Hardening Stainless Wire (17–7 PH):		
Condition C	480/1 hour	900/1 hour
Condition A to TH 1050	760/1 hour cool to 15°C followed by 565/1 hour	1400/1 hour, cool to 60°F followed by 1050/1 hour
Monel:		
Alloy 400	300–315	575–600
Alloy K500, Spring Temper	525/4 hours	980/4 hours
Inconel:		
Alloy 600	400–510	750–950
Alloy X-750:		
#1 Temper	730/16 hours	1350/16 hours
Spring Temper	650/4 hours	1200/4 hours
Copper Base, Cold Worked (Brass, Phosphor Bronze, etc.)	175–205	350–400
Beryllium Copper:		
Pretempered (Mill Hardened)	205	400
Solution Annealed,		
Temper Rolled or Drawn	315/2-3 hours	600/2-3 hours
Annealed Steels:		
Carbon (AISI 1050 to 1095)	800–830*	1475–1525*
Alloy (AISI 5160H 6150, 9254)	830–885*	1525–1625*

*Time depends on heating equipment and section size. Parts are austenitized then quenched and tempered to the desired hardness.
SOURCE: Associated Spring, Barnes Group Inc.

down the rate of corrosion attack on the base metal. The latter approach is most often the most cost-effective method.

Spring Wire.   The tensile strength of spring wire varies inversely with the wire diameter (Fig. 24.1).

Common spring wires with the highest strengths are ASTM A228 (music wire) and ASTM A401 (oil-tempered chrome silicon). Wires having slightly lower tensile strength and with surface quality suitable for fatigue applications are ASTM A313 type 302 (stainless steel), ASTM A230 (oil-tempered carbon valve-spring-quality steel), and ASTM A232 (oil-tempered chrome vanadium). For most static applica-

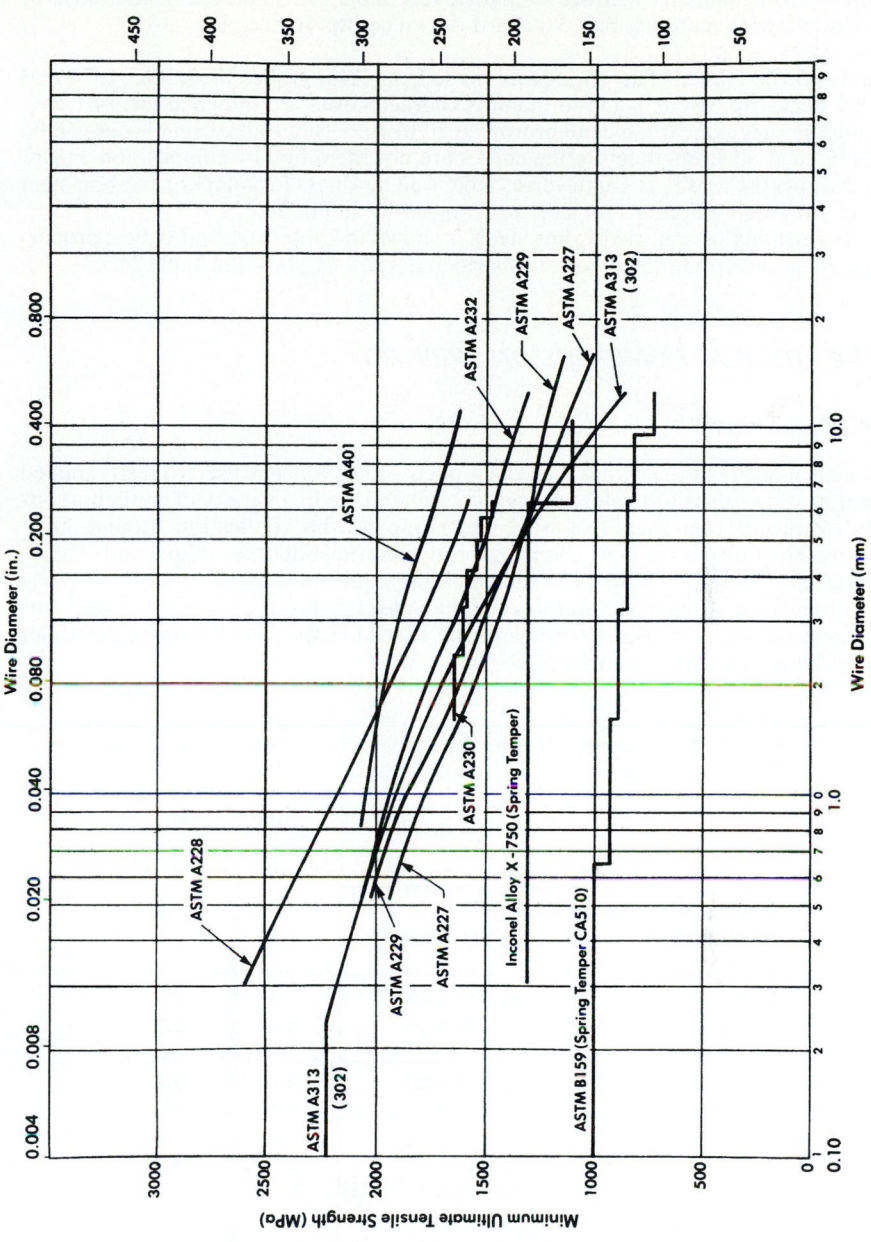

FIGURE 24.1 Minimum tensile strengths of spring wire. (*Associated Spring, Barnes Group Inc.*)

24.9

tions ASTM A227 (hard-drawn carbon steel) and ASTM A229 (oil-tempered carbon steel) are available at lower strength levels. Table 24.3 ranks the relative costs of common spring materials based on hard-drawn carbon steel as 1.0.

Spring Strip.   Most "flat" springs are made from AISI grades 1050, 1065, 1074, and 1095 steel strip. Strength and formability characteristics are shown in Fig. 24.2, covering the range of carbon content from 1050 to 1095. Since all carbon levels can be obtained at all strength levels, the curves are not identified by composition. Figure 24.3 shows the tensile strength versus Rockwell hardness for tempered carbon-steel strip. Edge configurations for steel strip are shown in Fig. 24.4.

Formability of annealed spring steels is shown in Table 24.4, and typical properties of various spring-tempered alloy strip materials are shown in Table 24.5.

24.4   HELICAL COMPRESSION SPRINGS

24.4.1   General

A helical compression spring is an open-pitch spring which is used to resist applied compression forces or to store energy. It can be made in a variety of configurations and from different shapes of wire, depending on the application. Round, high-carbon-steel wire is the most common spring material, but other shapes and compositions may be required by space and environmental conditions.

Usually the spring has a uniform coil diameter for its entire length. Conical, barrel, and hourglass shapes are a few of the special shapes used to meet particular load-deflection requirements.

TABLE 24.3   Ranking of Relative Costs of Common Spring Wires

Wire	Specification	Relative Cost of 2 mm (0.079″) Dia. Mill Quantities	Relative Cost of 2 mm (0.079″) Dia. Ware-House Lots
Patented and Cold Drawn	ASTM A227	1.0	1.0
Oil Tempered	ASTM A229	1.3	1.3
Music	ASTM A228	2.6	1.4
Carbon Valve Spring	ASTM A230	3.1	1.9
Chrome Silicon Valve	ASTM A401	4.0	3.9
Stainless Steel (Type 302)	ASTM A313 (302)	7.6	4.7
Phosphor Bronze	ASTM	8.0	6.7
Stainless Steel (Type 631) (17–7 PH)	ASTM A 313 (631)	11	8.7
Beryllium Copper	ASTM B197	27	17
Inconel Alloy X–750		44	31

SOURCE: Associated Spring, Barnes Group Inc.

Helical compression springs are stressed in the torsional mode. The stresses, in the elastic range, are not uniform about the wire's cross section. The stress is greatest at the surface of the wire and, in particular, at the inside diameter (ID) of the spring.

In some circumstances, residual bending stresses are present as well. In such cases, the bending stresses become negligible after set is removed (or the elastic limit is exceeded) and the stresses are redistributed more uniformly about the cross section.

24.4.2 Compression Spring Terminology

The definitions that follow are for terms which have evolved and are commonly used in the spring industry. Figure 24.5 shows the relationships among the characteristics.

Wire Diameter d. Round wire is the most economical form. Rectangular wire is used in situations where space is limited, usually to reduce solid height.

Coil Diameter. The outside diameter (OD) is specified when a spring operates in a cavity. The inside diameter is specified when the spring is to operate over a rod. The mean diameter D is either OD minus the wire size or ID plus the wire size.

The coil diameter increases when a spring is compressed. The increase, though small, must be considered whenever clearances could be a problem. The diameter increase is a function of the spring pitch and follows the equation

$$\text{OD}_{\text{at solid}} = \sqrt{D^2 + \frac{p^2 - d^2}{\pi^2}} + d \tag{24.1}$$

where p = pitch and d = wire size.

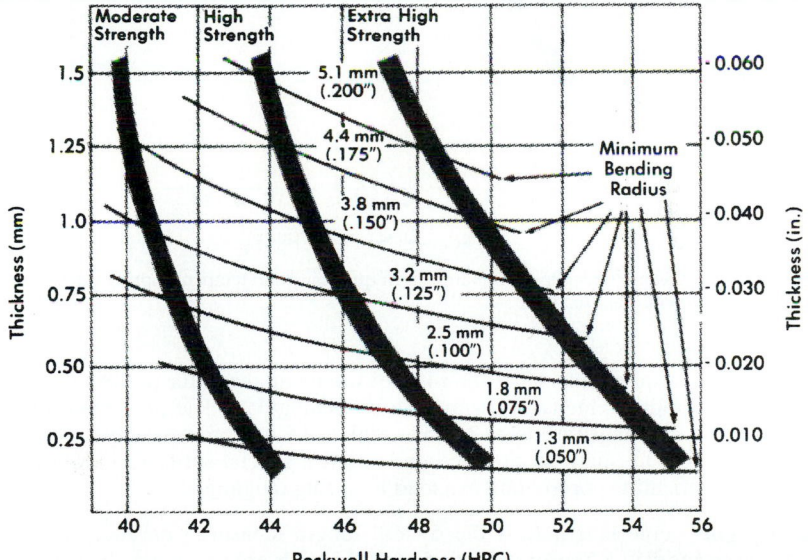

FIGURE 24.2 Minimum transverse bending radii for various tempers and thicknesses of tempered spring steel. (*Associated Spring, Barnes Group Inc.*)

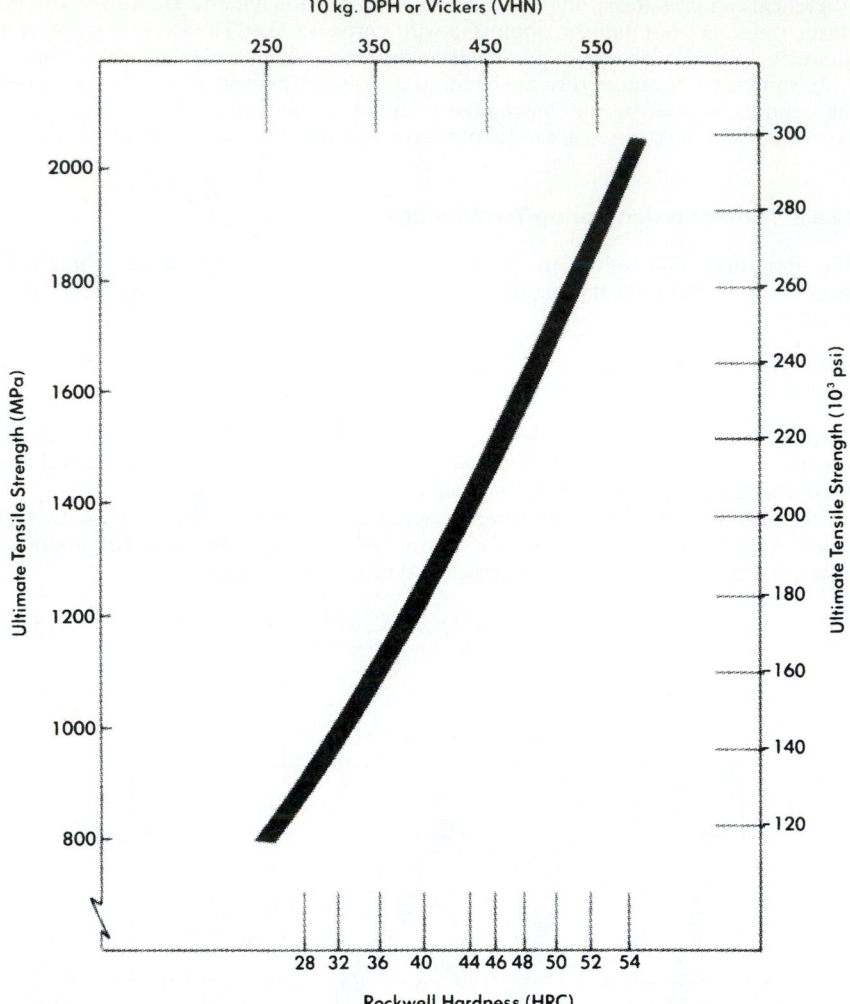

FIGURE 24.3   Tensile strength versus hardness of quenched and tempered spring steel. *(Associated Spring, Barnes Group Inc.)*

Spring Index.   Spring index C is the ratio of the mean diameter to the wire diameter (or to the radial dimension if the wire is rectangular). The preferred range of index is 5 to 9, but ranges as low as 3 and as high as 15 are commercially feasible. The very low indices are hard to produce and require special setup techniques. High indices are difficult to control and can lead to spring tangling.

Free Length.   Free length L_f is the overall length measured parallel to the axis when the spring is in a free, or unloaded, state. If loads are not given, the free length should be specified. If they are given, then free length should be a reference dimension which can be varied to meet the load requirements.

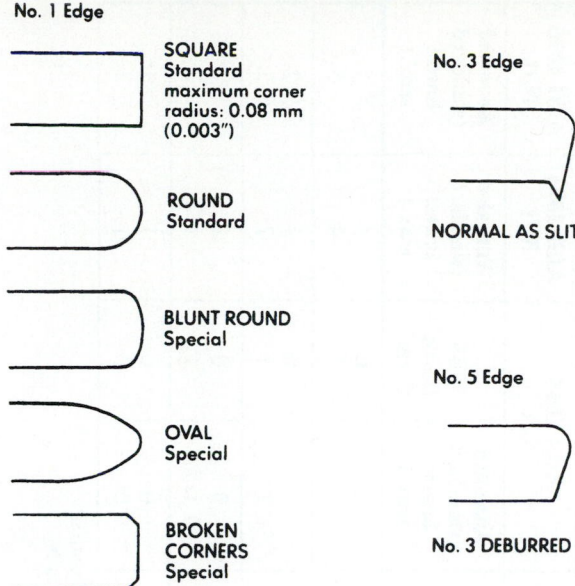

FIGURE 24.4 Edges available on steel strip. *(Associated Spring, Barnes Group Inc.)*

Types of Ends. Four basic types of ends are used: closed (squared) ends, closed (squared) ends ground, plain ends, and plain ends ground. Figure 24.6 illustrates the various end conditions. Closed and ground springs are normally supplied with a ground bearing surface of 270 to 330°.

Number of Coils. The number of coils is defined by either the total number of coils N_t or the number of active coils N_a. The difference between N_t and N_a equals the number of inactive coils, which are those end coils that do not deflect during service.

Solid Height. The solid height L_s is the length of the spring when it is loaded with enough force to close all the coils. For ground springs, $L_s = N_t d$. For unground springs, $L_s = (N_t + 1)d$.

Direction of the Helix. Springs can be made with the helix direction either right or left hand. Figure 24.7 illustrates how to define the direction. Springs that are nested one inside the other should have opposite helix directions. If a spring is to be assembled onto a screw thread, the direction of the helix must be opposite to that of the thread.

Spring Rate. Spring rate k is the change in load per unit deflection. It is expressed as

$$k = \frac{P}{f} = \frac{Gd^4}{8D^3 N_a} \tag{24.2}$$

where G = shear modulus.

TABLE 24.4 Formability of Annealed Spring Steels

Thickness (t) mm (in.)	Direction of Bend	AISI 1050 N_t/t		AISI 1065 N_t/t		AISI 1074 N_t/t	AISI 1095 N_t/t
		Annealed (standard lowest max.)	WBS* Barco-Form®	Annealed (standard lowest max.)	WBS Barco-Form	Annealed (standard lowest max.)	Annealed (standard lowest max.)
1.9 mm (0.076)-over	⊥	2	0	2	0	2	3
	∥	4	3	4	3	4	5
0.9–1.89 mm (0.036–0.075")	⊥	1	0	1	0	1	2
	∥	2	1	2	1	2	3
0.37–0.89 mm (0.015–0.035")	⊥	0	0	0	0	1 1/2	1
	∥	1	0	1 1/2	1		2
0.2–0.36 mm (0.008–0.014")	⊥	0	0	0	0	1	1
	∥	0	0	0	0	1	1

Formability is determined by slowly bending a sample over 180° until its ends are parallel. The measured distance between the ends is N_t.
For example, if $N_t = 4$ and t = 2, then $N_t/t = 2$
*Wallace Barnes Steel.

SOURCE: Associated Spring, Barnes Group Inc.

TAB_E 24.5 Typical Properties of Spring-Tempered Alloy Strip

Material	Tensile Strength MPa (10^3 psi)	Rockwell Hardness	Elongation(1) Percent	Bend Factor (1) ($2r/t$ trans. bends)	Modulus of Elasticity 10^4 MPa (10^6 psi)	Poisson's Ratio
Steel, spring temper	1700 (246)	C50	2	5	20.7 (30)	0.30
Stainless 301	1300 (189)	C40	8	3	19.3 (28)	0.31
Stainless 302	1300 (189)	C40	5	4	19.3 (28)	0.31
Monel 400	690 (100)	B95	2	5	17.9 (26)	0.32
Monel K500	1200 (174)	C34	40	5	17.9 (26)	0.29
Inconel 600	1040 (151)	C30	2	2	21.4 (31)	0.29
Inconel X-750	1050 (152)	C35	20	3	21.4 (31)	0.29
Copper-Beryllium	1300 (189)	C40	2	5	12.8 (18.5)	0.33
Ni- Span-C	1400 (203)	C42	6	2	18.6 (27)	—
Brass CA 260	620 (90)	B90	3	3	11 (16)	0.33
Phosphor Bronze	690 (100)	B90	3	2.5	10.3 (15)	0.20
17–7 PH RH950	1450 (210)	C44	6	flat	20.3 (29.5)	0.34
17–7 PH Condition C	1650 (239)	C46	1	2.5	20.3 (29.5)	0.34

(1) Before heat treatment.

SOURCE: Associated Spring, Barnes Group Inc.

24.15

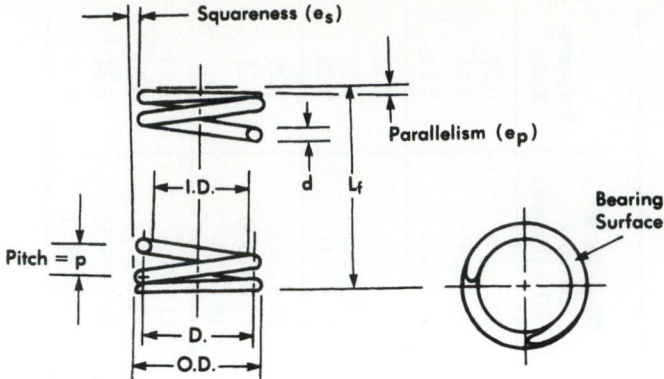

FIGURE 24.5 Dimensional terminology for helical compression springs. *(Associated Spring, Barnes Group Inc.)*

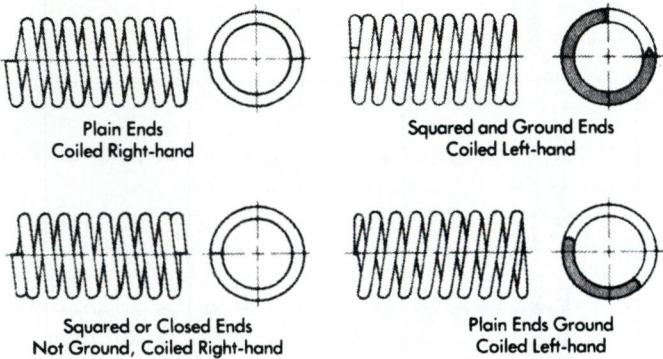

FIGURE 24.6 Types of ends for helical compression springs. *(Associated Spring, Barnes Group Inc.)*

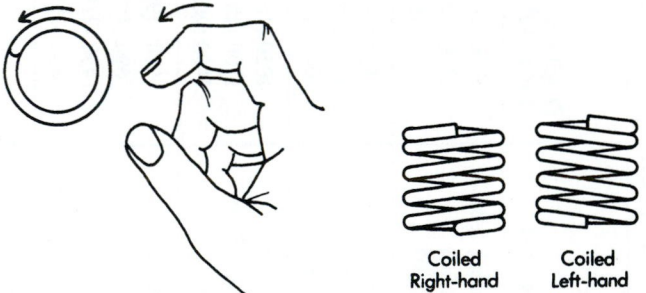

FIGURE 24.7 Direction of coiling of helical compression springs. *(Associated Spring, Barnes Group Inc.)*

The rate equation is accurate for a deflection range between 15 and 85 percent of the maximum available deflection. When compression springs are loaded in parallel, the combined rate of all the springs is the sum of the individual rates. When the springs are loaded in series, the combined rate is

$$k = \frac{1}{1/k_1 + 1/k_2 + 1/k_3 + \cdots + 1/k_n} \tag{24.3}$$

This relationship can be used to design a spring with variable diameters. The design method is to divide the spring into many small increments and calculate the rate for each increment. The rate for the whole spring is calculated as in Eq. (24.3).

Stress.   Torsional stress S is expressed as

$$S = \frac{8K_w PD}{\pi d^3} \tag{24.4}$$

Under elastic conditions, torsional stress is not uniform around the wire's cross section because of the coil curvature and direct shear loading.

The highest stress occurs at the surface in the inside diameter of the spring, and it is computed by using the stress factor K_w. In most cases, the correction factor is expressed as

$$K_{w_1} = \frac{4C - 1}{4C - 4} + \frac{0.615}{C} \tag{24.5}$$

The stress-concentration factor K_{w_1} becomes K_{w_2} after a spring has been set out because stresses become more uniformly distributed after subjecting the cross section to plastic flow during set-out:

$$K_{w_2} = 1 + \frac{0.5}{C} \tag{24.6}$$

The appropriate stress correction factor is discussed in Sec. 24.4.3.

Loads.   If deflection is known, the load is found by multiplying deflection by the spring rate. When the stress is either known or assumed, loads can be obtained from the stress equation.

Loads should be specified at a test height so that the spring manufacturer can control variations by adjustments of the free length. The load-deflection curve is not usually linear at the start of deflection from free position or when the load is very close to solid height. It is advisable to specify loads at test heights between 15 and 85 percent of the load-deflection range.

Loads can be conveniently classified as static, cyclic, and dynamic. In static loading, the spring will operate between specified loads only a few times. In other instances, the spring may remain under load for a long time. In cyclic applications, the spring may typically be required to cycle between load points from 10^4 to more than 10^9 times. During dynamic loading, the rate of load application is high and causes a surge wave in the spring which usually induces stresses higher than calculated from the standard stress equation.

Buckling.   Compression springs with a free length more than 4 times the mean coil diameter may buckle when compressed. Guiding the spring, either in a tube or over a rod, can minimize the buckling but can result in additional friction which will affect loads, especially when the L_f/D ratio is high.

Buckling conditions are shown in Figs. 24.8 and 24.9 for springs loaded axially and with squared and ground ends. Buckling occurs at points above and to the right of the curves. Curve A is for the springs with one end on a fixed, flat surface and the other end free to tip. Curve B is for springs with both ends on fixed, flat surfaces. The tendency to buckle is clearly less for curve B springs.

24.4.3   Choice of Operating Stress

The choice of operating stress depends on whether the application is static or cyclic. For static applications, yield strength or stress-relaxation resistance of the material limits the load-carrying ability of the springs. The required cycles are few, if any, and the velocity of the end coils is so low as to preclude surging or impact conditions.

The maximum allowable torsional stresses for static applications are shown in Table 24.6 as percentages of tensile strengths for common spring materials. To cal-culate the stress before set removal, use the K_{w_1} correction factor. If the calculated stress is greater than the indicated percentage of the tensile strength, then the spring will take a permanent set when deflected to solid. The amount of set is a function of the amount by which the calculated stress exceeds the tabular percentage.

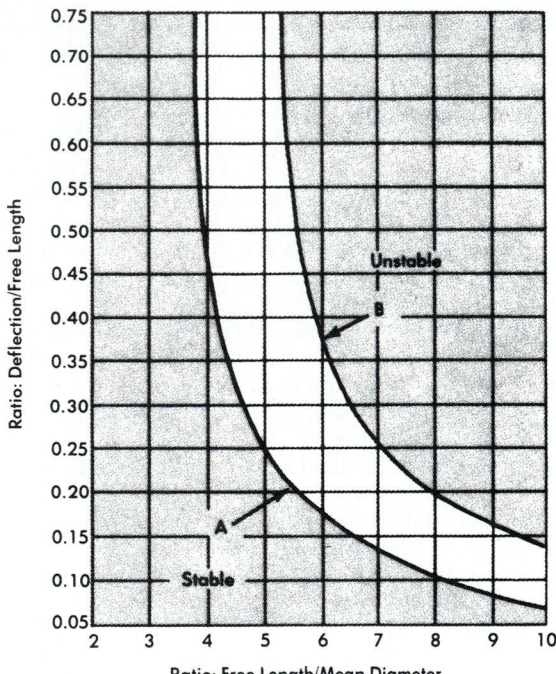

FIGURE 24.8 Critical buckling curves. *(Associated Spring, Barnes Group Inc.)*

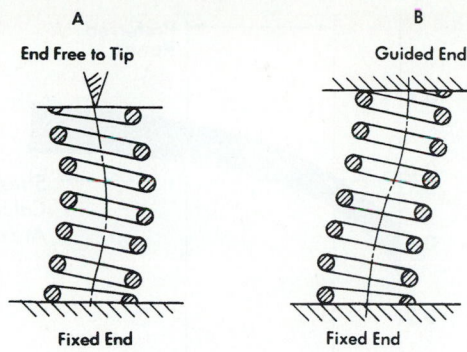

FIGURE 24.9   End conditions used to determine critical buckling. *(Associated Spring, Barnes Group Inc.)*

It is common practice, in static applications, to increase the load-carrying capability of a spring by making it longer than the desired free length and then compressing it to solid. The spring *sets* to its final desired length. This procedure is called *removing set*. It induces favorable residual stresses which allow for significantly higher stresses than in springs not having the set removed. The loss of the length should be at least 10 percent to be effective (see Fig. 24.10).

Note that set removal causes stresses to be more uniformly distributed about the cross section. Therefore, stress after set removal is calculated by using the K_{w_2} correction factor. If the stress calculated by using the K_{w_2} correction factor exceeds the percentage of tensile strength shown in Table 24.6, the spring cannot be made. It is then necessary either to lower the design stress or to select a higher-strength material.

For cyclic applications, the load-carrying ability of the spring is limited by the fatigue strength of the material. To select the optimum stress level, spring costs must be balanced against reliability. The designer should know the operating environ-

TABLE 24.6   Maximum Allowable Torsional Stresses for Helical Compression Springs in Static Applications

Materials	Maximum % of Tensile Strength	
	Before Set Removed (K_{W1})	After Set Removed (K_{W2})
Patented and cold drawn carbon steel	45%	
Hardened and tempered carbon and low alloy steel	50%	65–75%
Austenitic stainless steels	35%	
Nonferrous alloys	35%	

SOURCE: Associated Spring, Barnes Group Inc.

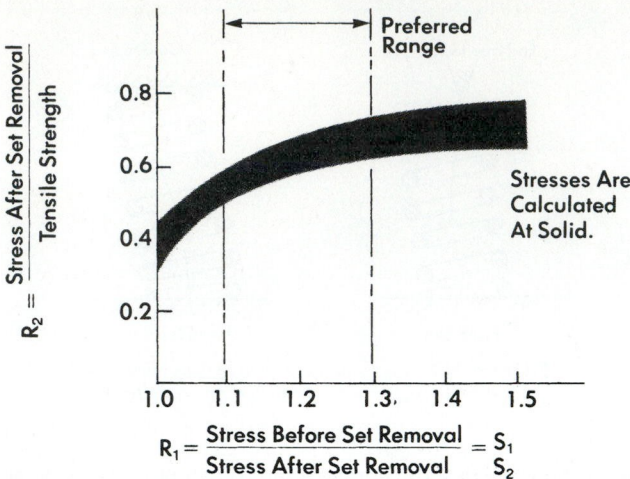

FIGURE 24.10 Spring load-carrying ability versus amount of set removed. (*Associated Spring, Barnes Group Inc.*)

ment, desired life, stress range, frequency of operation, speed of operation, and permissible levels of stress relaxation in order to make a cost-reliability decision.

Fatigue life can be severely reduced by pits, seams, or tool marks on the wire surface where stress is at a maximum. Shot peening improves fatigue life, in part, by minimizing the harmful effects of surface defects. It does not remove them. Additionally, shot peening imparts favorable compression stresses to the surface of the spring wire.

Maximum allowable stresses for fatigue applications should be calculated by using the K_{w_1} stress correction factor. Table 24.7 shows the estimated fatigue life for common spring materials. Note the significant increase in fatigue strength from shot peening.

TABLE 24.7 Maximum Allowable Torsional Stress for Round-Wire Helical Compression Springs in Cyclic Applications

Fatigue Life (cycles)	Percent of Tensile Strength			
	ASTM A228, Austenitic Stainless Steel and Nonferrous		ASTM A230 and A232	
	Not Shot-Peened	Shot-Peened	Not Shot-Peened	Shot-Peened
10^5	36	42	42	49
10^6	33	39	40	47
10^7	30	36	38	46

This information is based on the following conditions: no surging, room temperature and noncorrosive environment.

Stress ratio in fatigue $= \dfrac{S \text{ minimum}}{S \text{ maximum}} = 0$

SOURCE: Associated Spring, Barnes Group Inc.

The fatigue life estimates in Table 24.7 are guideline values which should be used only where specific data are unavailable. The values are conservative, and most springs designed using them will exceed the anticipated lives.

24.4.4 Dynamic Loading under Impact

When a spring is loaded or unloaded, a surge wave is established which transmits torsional stress from the point of load along the spring's length to the point of restraint. The surge wave will travel at a velocity approximately one-tenth that of a normal, torsional-stress wave. The velocity of the torsional-stress wave V_T, in meters per second (m/s) [inches per second (in/s)], is given by

$$V_T = 10.1 \sqrt{\frac{Gg}{\rho}} \quad \text{m/s} \quad \text{or} \quad V_T = \sqrt{\frac{Gg}{\rho}} \quad \text{in/s} \qquad (24.7)$$

The velocity of the surge wave V_s varies with material and design but is usually in the range of 50 to 500 m/s. The surge wave limits the rate at which a spring can absorb or release energy by limiting the impact velocity V. *Impact velocity* is defined as the spring velocity parallel to the spring axis and is a function of stress and material as shown:

$$V \simeq 10.1S \sqrt{\frac{g}{2\rho G}} \quad \text{m/s} \quad \text{or} \quad V \simeq S \sqrt{\frac{g}{2\rho G}} \quad \text{in/s} \qquad (24.8)$$

For steel, this reduces to

$$V = \frac{S}{35.5} \quad \text{m/s} \quad \text{or} \quad V = \frac{S}{131} \quad \text{in/s} \qquad (24.9)$$

If a spring is compressed to a given stress level and released instantaneously, the maximum spring velocity is the stress divided by 35.5. Similarly, if the spring is loaded at known velocity, the instantaneous stress can be calculated. At very high load velocities, the instantaneous stress will exceed the stress calculated by the conventional equation. This will limit design performance. Since the surge wave travels the length of the spring, springs loaded at high velocity often are subject to resonance.

24.4.5 Dynamic Loading—Resonance

A spring experiences resonance when the frequency of cyclic loading is near the natural frequency or a multiple of it. Resonance can cause an individual coil to deflect to stress levels above those predicted by static stress analysis. Resonance can also cause the spring to bounce, resulting in loads lower than calculated. To avoid these effects, the natural frequency should be a minimum of 13 times the operating frequency.

For a compression spring with both ends fixed and no damper, the natural frequency in International System (SI) units is

$$n = \frac{1.12(10^3)d}{D^2 N_a} \sqrt{\frac{Gg}{\rho}} \qquad (24.10)$$

For steel, this equation becomes

$$n = \frac{3.5(10^5)d}{D^2 N_a} \qquad (24.11)$$

where n = frequency in hertz (Hz). The corresponding equation in U.S. Customary System (USCS) units is

$$n = \frac{d}{9D^2 N_a} \sqrt{\frac{Gg}{\rho}} \qquad (24.12)$$

and for steel we have

$$n = \frac{14(10^3)d}{D^2 N_a} \qquad (24.13)$$

If the spring cannot be designed to have a natural frequency more than 13 times the operating frequency, energy dampers may be employed. They are generally friction devices which rub against the coils. Often, variable-pitch springs are used to minimize resonance effects.

24.4.6 Rectangular-Wire Springs

In applications where high loads and relatively low stresses are required but solid height is also restricted, rectangular wire can be used to increase the material volume while maintaining the maximum solid-height limitation.

Springs made of rectangular wire with the long side of the wire cross section perpendicular to the axis of the coils can store more energy in a smaller space than an equivalent, round-wire spring.

When rectangular wire is coiled, it changes from a rectangular to a keystone shape, as shown in Fig. 24.11. Similarly, if the wire is made to the keystone shape, it will become rectangular after coiling. The cross-sectional distortion can be approximated by

$$t_1 = t \frac{C + 0.5}{C} \qquad (24.14)$$

where t_1 = wider end of keystone section and t = original, smaller dimension of rectangle.

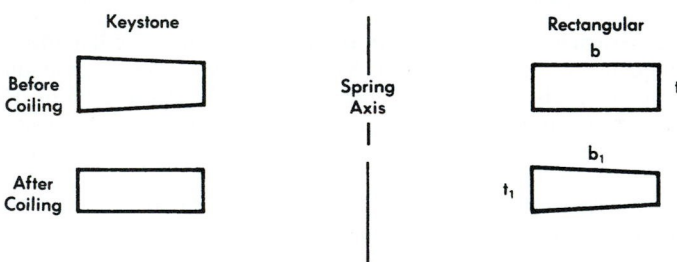

FIGURE 24.11 Wire cross section before and after coiling. *(Associated Spring, Barnes Group Inc.)*

The spring rate for a rectangular-wire spring is

$$k = \frac{P}{f} = \frac{K_2 G b t^3}{N_a D^3} \tag{24.15}$$

Since the wire is loaded in torsion, it makes no difference whether the wire is wound on the flat or on edge. See Fig. 24.12.
    Stress is calculated by

$$S = \frac{K_E P D}{K_1 b t^2} \quad \text{or} \quad S = \frac{K_F P D}{K_1 b t^2} \tag{24.16}$$

Values for K_1 and K_2 are found in Fig. 24.13, and those for K_E and K_F are found in Figs. 24.15 and 24.14, respectively.
    When a round wire cannot be used because the solid height exceeds the specification, the approximate equivalent rectangular dimensions are found from

$$t = \frac{2d}{1 + b/t} \tag{24.17}$$

where d = round-wire diameter.

24.4.7 Variable-Diameter Springs

Conical, hourglass, and barrel-shaped springs, shown in Fig. 24.16, are used in applications requiring a low solid height and an increased lateral stability or resistance to surging. Conical springs can be designed so that each coil nests wholly or partly within an adjacent coil. Solid height can be as low as one wire diameter. The rate for conical springs usually increases with deflection (see Fig. 24.17) because the number of active coils decreases progressively as the spring approaches solid. By varying the pitch, conical springs can be designed to have a uniform rate. The rate for conical springs is calculated by considering the spring as many springs in series. The rate for

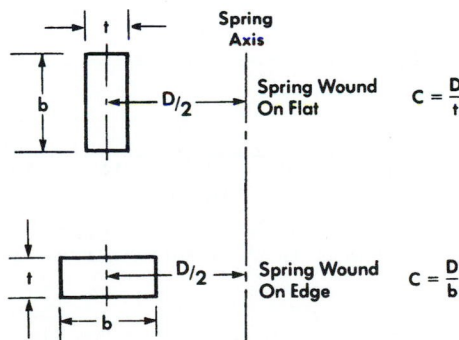

FIGURE 24.12 Rectangular-wire compression spring wound on flat or edge. *(Associated Spring, Barnes Group Inc.)*

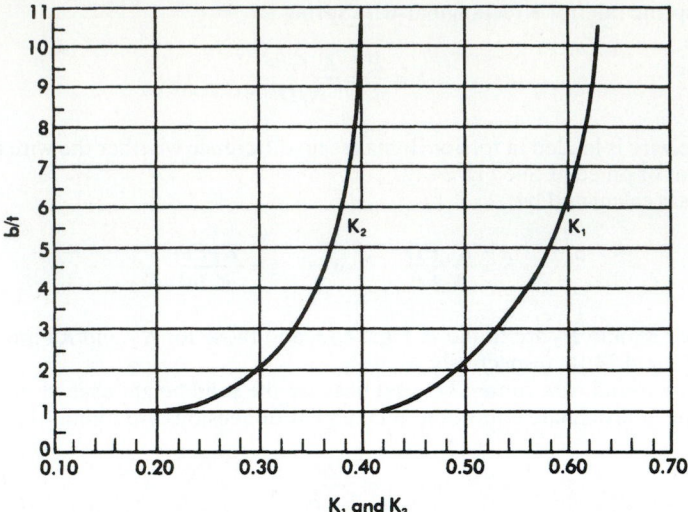

FIGURE 24.13 Constants for rectangular wire in torsion. *(Associated Spring, Barnes Group Inc.)*

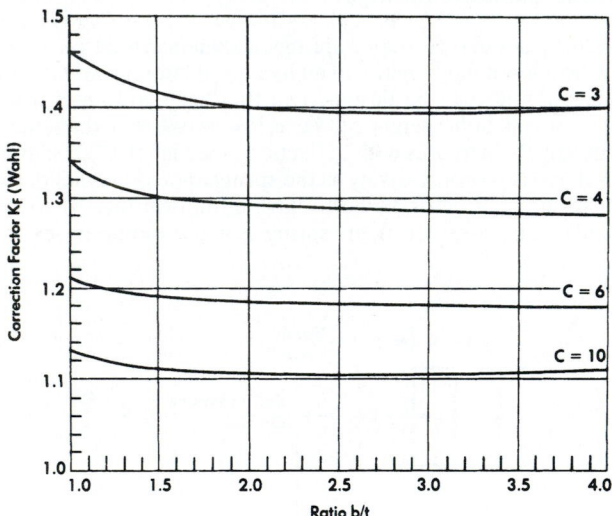

FIGURE 24.14 Stress correction factors for rectangular-wire compression springs wound on flat. *(Associated Spring, Barnes Group Inc.)*

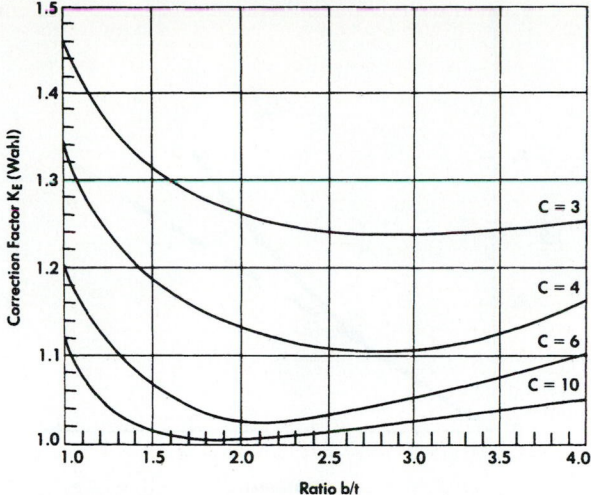

FIGURE 24.15   Stress correction factors for rectangular-wire compression springs wound on edge. *(Associated Spring, Barnes Group Inc.)*

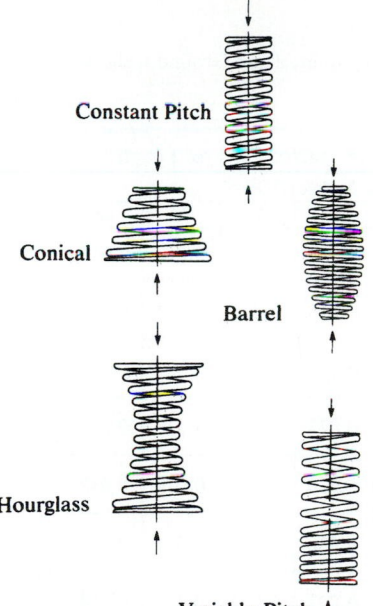

FIGURE 24.16   Various compression-spring body shapes. *(Associated Spring, Barnes Group Inc.)*

each turn or fraction of a turn is calculated by using the standard rate equation. The rate for a complete spring is then determined, given that the spring rate follows the series relationship in Eq. (23.4).

To calculate the highest stress at a given load, the mean diameter of the largest active coil at load is used. The solid height of a uniformly tapered, but not telescoping, spring with squared and ground ends made from round wire can be estimated from

$$L_S = N_a \sqrt{d^2 - u^2} + 2d \qquad (24.18)$$

where u = OD of large end minus OD of small end, divided by $2N_a$.

Barrel- and hourglass-shaped springs are calculated as two conical springs in series.

24.4.8   Commercial Tolerances

Standard commercial tolerances are presented in Tables 24.8, 24.9, and 24.10 for free length, coil diameter, and load tolerances, respectively. These tolerances represent the best tradeoffs between manufacturing costs and performance.

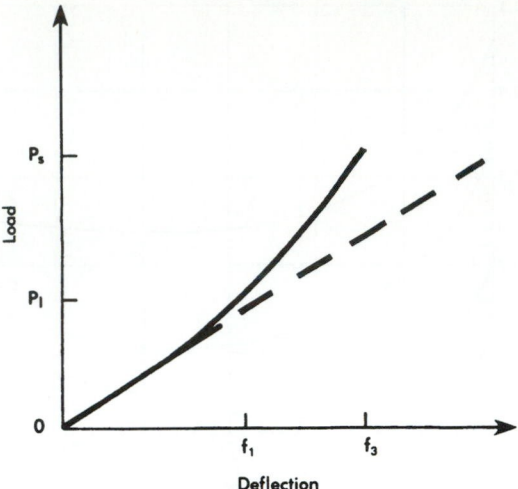

FIGURE 24.17 Typical load-deflection curve for variable-diameter springs (solid line). *(Associated Spring, Barnes Group Inc.)*

TABLE 24.8 Free-Length Tolerances of Squared and Ground Helical Compression Springs

Number of Active coils per mm(in.)	Tolerances: ±mm/mm (in./in.) of Free Length						
	Spring Index (D/d)						
	4	6	8	10	12	14	16
0.02 (0.5)	0.010	0.011	0.012	0.013	0.015	0.016	0.016
0.04 (1)	0.011	0.013	0.015	0.016	0.017	0.018	0.019
0.08 (2)	0.013	0.015	0.017	0.019	0.020	0.022	0.023
0.2 (4)	0.016	0.018	0.021	0.023	0.024	0.026	0.027
0.3 (8)	0.019	0.022	0.024	0.026	0.028	0.030	0.032
0.5 (12)	0.021	0.024	0.027	0.030	0.032	0.034	0.036
0.6 (16)	0.022	0.026	0.029	0.032	0.034	0.036	0.038
0.8 (20)	0.023	0.027	0.031	0.034	0.036	0.038	0.040

For springs less than 12.7 mm (0.500″) long, use the tolerances for 12.7 mm (0.500″). For closed ends not ground, multiply above values by 1.7.

SOURCE: Associated Spring, Barnes Group Inc.

TABLE 24.9 Coil Diameter Tolerances of Helical Compression and Extension Springs

Wire Dia., mm(in.)	Tolerances: ±mm (in.)						
	Spring Index (D/d)						
	4	6	8	10	12	14	16
0.38 (0.015)	0.05 (0.002)	0.05 (0.002)	0.08 (0.003)	0.10 (0.004)	0.13 (0.005)	0.15 (0.006)	0.18 (0.007)
0.58 (0.023)	0.05 (0.002)	0.08 (0.003)	0.10 (0.004)	0.15 (0.006)	0.18 (0.007)	0.20 (0.008)	0.25 (0.010)
0.89 (0.035)	0.05 (0.002)	0.10 (0.004)	0.15 (0.006)	0.18 (0.007)	0.23 (0.009)	0.28 (0.011)	0.33 (0.013)
1.30 (0.051)	0.08 (0.003)	0.13 (0.005)	0.18 (0.007)	0.25 (0.010)	0.30 (0.012)	0.38 (0.015)	0.43 (0.017)
1.93 (0.076)	0.10 (0.004)	0.18 (0.007)	0.25 (0.010)	0.33 (0.013)	0.41 (0.016)	0.48 (0.019)	0.53 (0.021)
2.90 (0.114)	0.15 (0.006)	0.23 (0.009)	0.33 (0.013)	0.46 (0.018)	0.53 (0.021)	0.64 (0.025)	0.74 (0.029)
4.34 (0.171)	0.20 (0.008)	0.30 (0.012)	0.43 (0.017)	0.58 (0.023)	0.71 (0.028)	0.84 (0.033)	0.97 (0.038)
6.35 (0.250)	0.28 (0.011)	0.38 (0.015)	0.53 (0.021)	0.71 (0.028)	0.90 (0.035)	1.07 (0.042)	1.24 (0.049)
9.53 (0.375)	0.41 (0.016)	0.51 (0.020)	0.66 (0.026)	0.94 (0.037)	1.17 (0.046)	1.37 (0.054)	1.63 (0.064)
12.70 (0.500)	0.53 (0.021)	0.76 (0.030)	1.02 (0.040)	1.57 (0.062)	2.03 (0.080)	2.54 (0.100)	3.18 (0.125)

SOURCE: Associated Spring, Barnes Group Inc.

24.5 HELICAL EXTENSION SPRINGS

24.5.1 General

Helical extension springs store energy and exert a pulling force. They are usually made from round wire and are close-wound with initial tension. They have various types of end hooks or loops by which they are attached to the loads.

Like compression springs, extension springs are stressed in torsion in the body coils. The design procedures for the body coil are similar to those discussed in Sec. 24.4 except for the initial tension and the hook stresses.

Most extension springs are made with the body coils held tightly together by a force called *initial tension*. The measure of initial tension is the load required to overcome the internal force and start coil separation.

Extension springs, unlike compression springs, seldom have set removed. Furthermore, they have no solid stop to prevent overloading. For these reasons, the design stresses are normally held to lower values than those for compression springs.

The pulling force exerted by an extension spring is transmitted to the body coils through hooks or loops. Careful attention must be given to the stresses in the hooks. The hook ends must be free of damaging tool marks so that spring performance will not be limited by hook failure.

TABLE 24.10 Load Tolerances of Helical Compression Springs

Tolerances: ± % of Load. Start with Tolerance from Table 24-8 Multiplied by L_F.

Length Tolerance ± mm (in.)	Deflection from Free Length to Load, mm (in.)														
	1.27 (0.050)	2.54 (0.100)	3.81 (0.150)	5.08 (0.200)	6.35 (0.250)	7.62 (0.300)	10.2 (0.400)	12.7 (0.500)	19.1 (0.750)	25.4 (1.00)	38.1 (1.50)	50.8 (2.00)	76.2 (3.00)	102 (4.00)	152 (6.00)
0.13 (0.005)	12.	7.	6.	5.	—	—	—	—	—	—	—	—	—	—	—
0.25 (0.010)	—	12.	8.5	7.	6.5	5.5	5.	—	—	—	—	—	—	—	—
0.51 (0.020)	—	22.	15.5	12.	10.	8.5	7.	6.	5.	—	—	—	—	—	—
0.76 (0.030)	—	—	22.	17.	14.	12.	9.5	8.	6.	5.	—	—	—	—	—
1.0 (0.040)	—	—	—	22.	18.	15.5	12.	10.	7.5	6.	5.	—	—	—	—
1.3 (0.050)	—	—	—	—	22.	19.	14.5	12.	9.	7.	5.5	—	—	—	—
1.5 (0.060)	—	—	—	—	25.	22.	17.	14.	10.	8.	6.	5.	—	—	—
1.8 (0.070)	—	—	—	—	—	25.	19.5	16.	11.	9.	6.5	5.5	—	—	—
2.0 (0.080)	—	—	—	—	—	—	22.	18.	12.5	10.	7.5	6.	5.	—	—
2.3 (0.090)	—	—	—	—	—	—	25.	20.	14.	11.	8.	6.	5.	—	—
2.5 (0.100)	—	—	—	—	—	—	—	22.	15.5	12.	8.5	7.	5.5	—	—
5.1 (0.200)	—	—	—	—	—	—	—	—	—	22.	15.5	12.	8.5	7.	5.5
7.6 (0.300)	—	—	—	—	—	—	—	—	—	—	22.	17.	12.	9.5	7.
10.2 (0.400)	—	—	—	—	—	—	—	—	—	—	—	21.	15.	12.	8.5
12.7 (0.500)	—	—	—	—	—	—	—	—	—	—	—	25.	18.5	14.5	10.5

First load test at not less than 15% of available deflection.
Final load test at not more than 85% of available deflection.

SOURCE: Associated Spring, Barnes Group Inc.

24.5.2 Initial Tension

Initial tension is illustrated in Fig. 24.18. The point of intersection on the ordinate is initial tension P_I. The amount of initial tension is governed by the spring index, material, method of manufacture, and the post-stress-relief heat treatment temperature. Note that a high stress-relief temperature can reduce the initial tension. This is sometimes used as a means to control initial tension in low-stress, low-index springs. It follows that an extension spring requiring no initial tension can be made either by removing the initial tension with heat treatment or by keeping the coils open during coiling. The levels of initial tension obtainable are shown in Fig. 24.19.

24.5.3 Types of Ends

Extension springs require a means of attachment to the system which is to be loaded. A variety of end configurations have been developed over the years. The configurations most commonly used are shown in Fig. 24.20. Loops or hooks longer than recommended will require special setup and are more expensive. Specifying an angular relationship for the loops may also add to the cost. Allow a random relationship of loops whenever possible.

Stresses in the loops are often higher than those in the body coils. In such cases, the loops are the performance limiters, particularly in cyclic applications. Generous bend radii, elimination of tool marks, and a reduced diameter of end coils are methods used to reduce loop stresses. In a full-twist loop, stress reaches a maximum in bending at point A (Fig. 24.21) and a maximum in torsion at point B. The stresses at these locations are complex, but useful approximations are, for bending,

$$S_A = \frac{16K_1DP}{\pi d^3} + \frac{4P}{\pi d^2} \qquad (24.19)$$

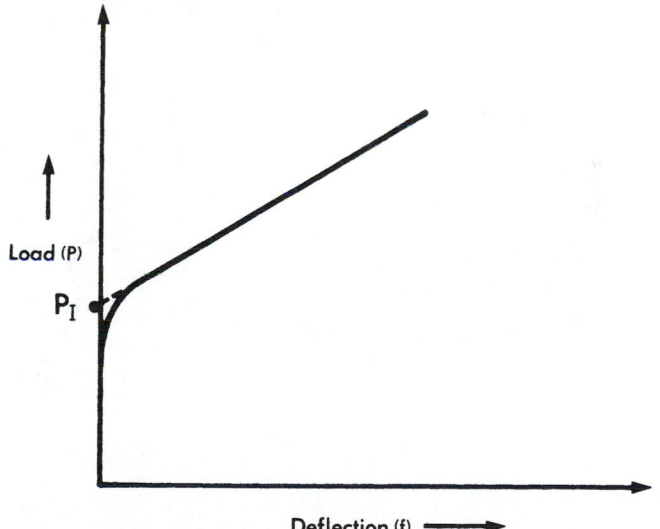

FIGURE 24.18 Load-deflection curve for a helical extension spring with initial tension. *(Associated Spring, Barnes Group Inc.)*

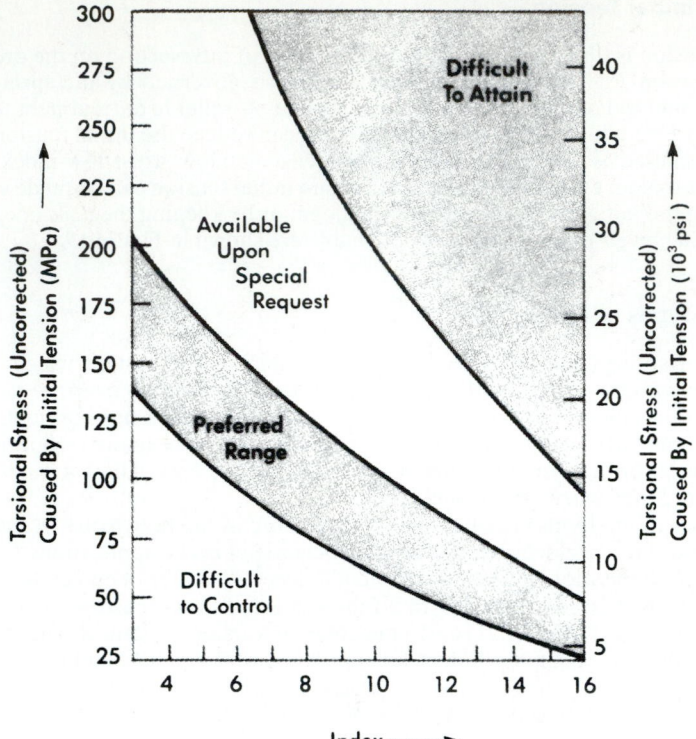

FIGURE 24.19   Torsional stress resulting from initial tension as a function of index in helical extension springs. *(Associated Spring, Barnes Group Inc.)*

where the constants are

$$K_1 = \frac{4C^2 - C_1 - 1}{4C_1(C_1 - 1)} \tag{24.20}$$

and

$$C_1 = \frac{2R_1}{d} \tag{24.21}$$

The torsional stresses are

$$S_B = \frac{8DP}{\pi d^3} \frac{4C_2 - 1}{4C_2 - 4} \tag{24.22}$$

where

$$C_2 = \frac{2R_2}{d} \tag{24.23}$$

General practice is to make C_2 greater than 4.

Type	Configurations	Recommended Length Min.-Max.
Twist Loop or Hook		0.5-1.7 I.D.
Cross Center Loop or Hook		I.D.
Side Loop or Hook		0.9-1.0 I.D.
Extended Hook		1.1 I.D. and up, as required by design
Special Ends		As required by design

FIGURE 24.20 Common end configurations for helical extension springs. Recommended length is distance from last body coil to inside of end. ID is inside diameter of adjacent coil in spring body. (*Associated Spring, Barnes Group Inc.*)

24.31

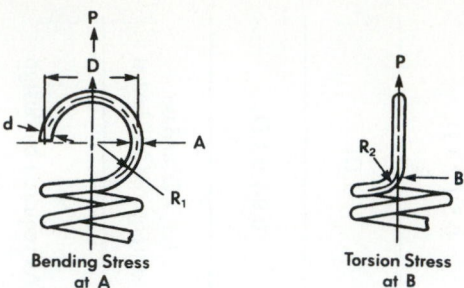

FIGURE 24.21   Location of maximum bending and torsional stresses in twist loops. *(Associated Spring, Barnes Group Inc.)*

24.5.4   Extension Spring Dimensioning

The dimensioning shown in Fig. 24.22 is generally accepted for extension springs. The free length is the distance between the inside surfaces of the loops. The body length is $L_B = d(N+1)$. The loop opening, or gap, can be varied. The number of active coils is equal to the number of coils in the body of the spring. However, with special ends such as threaded plugs or swivel hooks, the number of active coils will be less than the number of body coils.

24.5.5   Design Equations

The design equations are similar to those for compression springs with the exception of initial tension and loop stresses. The rate is given by

$$k = \frac{P - P_I}{f} = \frac{Gd^4}{8D^3 N_a}$$ (24.24)

where P_I is initial tension. Stress is given by

$$S = \frac{K_w 8PD}{\pi d^3}$$ (24.25)

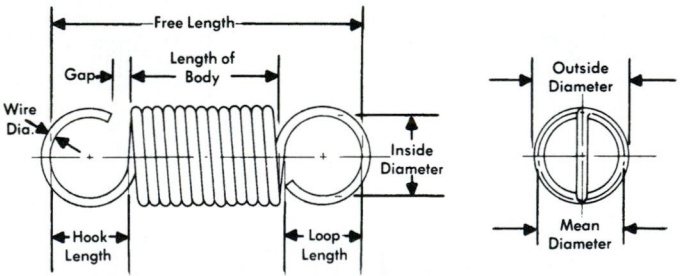

FIGURE 24.22   Typical extension-spring dimensions. *(Associated Spring, Barnes Group Inc.)*

Dynamic considerations discussed previously are generally applicable to extension springs. Natural frequency with one end fixed, in SI units, is

$$n = \frac{560d}{D^2 N_a} \sqrt{\frac{Gg}{\rho}} \qquad (24.26)$$

For steel, this equation becomes

$$n = \frac{176\ 000d}{N_a D^2} \qquad (24.27)$$

where n = frequency in hertz. The corresponding equation in USCS units is

$$n = \frac{d}{18D^2 N_a} \sqrt{\frac{Gg}{\rho}} \qquad (24.28)$$

And for steel we have

$$n = \frac{7000d}{N_a D^2} \qquad (24.29)$$

24.5.6 Choice of Operating Stress—Static

The maximum stresses recommended for extension springs in static applications are given in Table 24.11. Note that extension springs are similar to compression springs without set removed. For body coil stresses in springs that cannot be adequately stress-relieved because of very high initial-tension requirements, use the maximum recommended stress in torsion, given for the end loops.

24.5.7 Choice of Operating Stress—Cyclic

Table 24.12 presents the maximum stresses for extension springs used in cyclic applications. The data are for stress-relieved springs with initial tension in the preferred range.

TABLE 24.11 Maximum Allowable Stresses (K_{w_1} Corrected) for Helical Extension Springs in Static Applications

Materials	Percent of Tensile Strength		
	In Torsion		In Bending
	Body	End	End
Patented, cold-drawn or hardened and tempered carbon and low alloy steels	45–50	40	75
Austenitic stainless steel and nonferrous alloys	35	30	55

This information is based on the following conditions: set not removed and low temperature heat treatment applied.
For springs that require high initial tension, use the same percent of tensile strength as for end.

SOURCE: Associated Spring, Barnes Group Inc.

TABLE 24.12   Maximum Allowable Stresses for ASTM A228 and Type 302 Stainless-Steel Helical Extension Springs in Cyclic Applications

Number of Cycles	Percent of Tensile Strength		
	In Torsion		In Bending
	Body	End	End
10^5	36	34	51
10^6	33	30	47
10^7	30	28	45

This information is based on the following conditions: not shot-peened, no surging and ambient environment with a low temperature heat treatment applied. Stress ratio = 0.

SOURCE: Associated Spring, Barnes Group Inc.

24.5.8   Tolerances

Extension springs do not buckle or require guide pins when they are deflected, but they may vibrate laterally if loaded or unloaded suddenly. Clearance should be allowed in these cases to eliminate the potential for noise or premature failure. The load tolerances are the same as those given for compression springs. Tolerances for free length and for angular relationship of ends are given in Tables 24.13 and 24.14.

24.6   HELICAL TORSION SPRINGS

Helical springs that exert a torque or store rotational energy are known as *torsion springs*. The most frequently used configuration of a torsion spring is the single-body type (Fig. 24.23). Double-bodied springs, known as double-torsion springs, are sometimes used where dictated by restrictive torque, stress, and space requirements. It is often less costly to make a pair of single-torsion springs than a double-torsion type.

TABLE 24.13   Commercial Free-Length Tolerances for Helical Extension Springs with Initial Tension

Spring Free Length (inside hooks) mm (in.)	Tolerance ± mm (in.)
Up to 12.7 (0.500)	0.51 (0.020)
Over 12.7 to 25.4 (0.500 to 1.00)	0.76 (0.030)
Over 25.4 to 50.8 (1.00 to 2.00)	1.0 (0.040)
Over 50.8 to 102 (2.00 to 4.00)	1.5 (0.060)
Over 102 to 203 (4.00 to 8.00)	2.4 (0.093)
Over 203 to 406 (8.00 to 16.0)	4.0 (0.156)
Over 406 to 610 (16.0 to 24.0)	5.5 (0.218)

SOURCE: Associated Spring, Barnes Group Inc.

TABLE 24.14 Tolerances on Angular Relationship of Extension Spring Ends

Angular Tolerance per Coil: ± Degrees									
Index									
4	5	6	7	8	9	10	12	14	16
0.75	0.9	1.1	1.3	1.5	1.7	1.9	2.3	2.6	3

For example, tolerance for a 10-coil spring with an index of 8 is $10 \times \pm1.5 = \pm15°$.

If angular tolerance is greater than ± 45°, or if closer tolerances than indicated must be held, consult with Associated Spring.

SOURCE: Associated Spring, Barnes Group Inc.

Torsion springs are used in spring-loaded hinges, oven doors, clothespins, window shades, ratchets, counterbalances, cameras, door locks, door checks, and many other applications. Torsion springs are almost always mounted on a shaft or arbor with one end fixed. They can be wound either right or left hand.

In most cases the springs are not stress-relieved and are loaded in the direction that winds them up or causes a decrease in body diameter. The residual forming stresses which remain are favorable in that direction. Although it is possible to load a torsion spring in the direction to unwind and enlarge the body coils, ordinarily it is not good design practice and should be avoided. Residual stresses in the unwind direction are unfavorable. Torsion springs which are plated or painted and subsequently baked or are stress-relieved will have essentially no residual stresses and can be loaded in either direction, but at lower stress levels than springs which are not heat-treated.

Correlation of test results between manufacturer and user may be difficult because there are few, if any, standardized torsion-spring testing machines. The springs will have varying degrees of intercoil friction and friction between the mounting arbor and the body coils. Often duplicate test fixtures must be made and test methods coordinated.

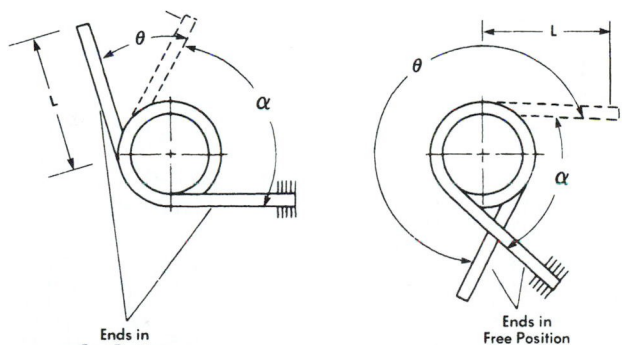

Ends in
Free Position

Ends in
Free Position

FIGURE 24.23 Specifying load and deflection requirements for torsion spring: α = angle between ends; P = load on ends at α; L = moment arm; θ = angular deflection from free position. (*Associated Spring, Barnes Group Inc.*)

Spring ends most commonly used are shown in Fig. 24.24, although the possible variations are unlimited. In considering spring mounting, it must be recognized that for each turn of windup, the overall length L of the spring body will increase as

$$L_1 = d(N_a + 1 + \theta) \qquad (24.30)$$

where θ = deflection in revolutions.

Also note that the body coil diameter will be reduced to

$$D = \frac{D_I N_a}{Na + \theta} \qquad (24.31)$$

where D_I = initial mean coil diameter. Experience indicates that the diameter of the arbor over which the spring operates should be approximately 90 percent of the smallest inside diameter to which the spring is reduced under maximum load. Too large an arbor will interfere with deflection, while too small an arbor will provide too little support. Both conditions lead to unexpectedly early failure. Coil diameter tolerances are given in Table 24.17.

24.6.1 Spring Rate

The spring rate, or moment per turn, is given by

$$k = \frac{M}{\theta} = \frac{Ed^4}{10.8DN_a} \qquad (24.32)$$

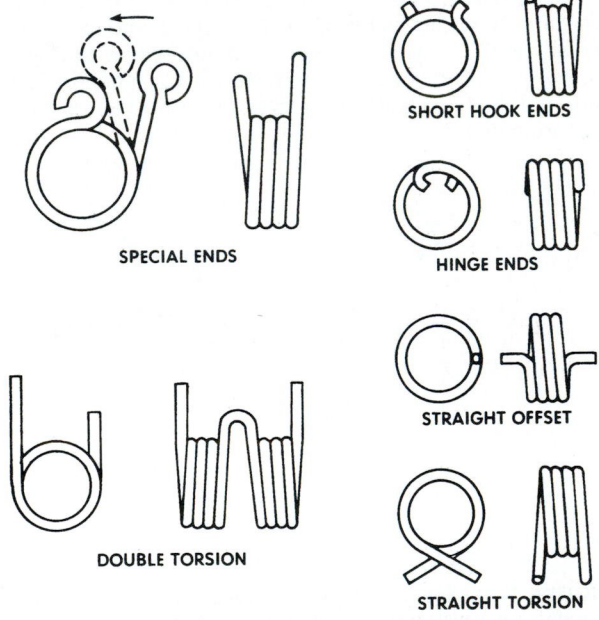

FIGURE 24.24 Common helical torsion-spring end configurations. *(Associated Spring, Barnes Group Inc.)*

The number of coils is equal to the number of body coils plus a contribution from the ends. The effect is more pronounced when the ends are long. The number of equivalent coils in the ends is

$$N_e = \frac{L_1 + L_2}{3\pi D}$$
(24.33)

where L_1 and L_2 = lengths of ends, and so $N_a = N_b + N_e$, where N_b = number of body coils.

The load should be specified at a fixed angular relationship of the spring ends rather than at a specific angular deflection from free or load positions. Helical torsion springs are stressed in bending. Rectangular sections are more efficient than round sections, but round sections are normally used because there is usually a premium cost for rectangular wire.

24.6.2   Stresses

Stress in round-wire torsion springs is given by

$$S = \frac{32 K_B M}{\pi d^3}$$
(24.34)

where K_B = a stress correction factor. Stress is higher on the inner surface of the coil. A useful approximation of this factor is

$$K_B = \frac{4C - 1}{4C - 4}$$
(24.35)

24.6.3   Rectangular-Wire Torsion Springs

When rectangular wire is formed into coils, it approaches a keystone according to the relation

$$b_I = b \, \frac{C + 0.5}{C}$$
(24.36)

where b_I = axial dimension b after keystoning. The radial dimension is always t. The rate equation is

$$k = \frac{M}{\theta} = \frac{E b t^3}{6.6 D N_a}$$
(24.37)

Stress in rectangular-wire torsion springs is given by

$$S = \frac{6 K_B M}{b t^2}$$
(24.38)

where $K_{B_{1D}} = 4C/(4C - 3)$ and b = axial dimension of rectangular cross section. Maximum recommended stresses are given in Table 24.15 for static applications and in Table 24.16 for cyclic applications.

TABLE 24.15 Maximum Recommended Bending Stresses for Helical
Torsion Springs in Static Applications

Material	Percent of Tensile Strength	
	Stress-Relieved (1) (K_B Corrected)	With Favorable Residual Stress (2) (No Correction Factor)
Patented and Cold Drawn	80	100
Hardened and Tempered Carbon and Low Alloy Steels	85	100
Austenitic Stainless Steels and Non-Ferrous Alloys	60	80

(1) Also for springs without residual stresses.
(2) Springs that have not been stress-relieved and which have bodies and
ends loaded in a direction that decreases the radius of curvature.
SOURCE: Associated Spring, Barnes Group Inc.

TABLE 24.16 Maximum Recommended Bending Stresses
(K_B Corrected) for Helical Torsion Springs in Cyclic Applications

Fatigue Life (cycles)	ASTM A228 and Type 302 Stainless Steel		ASTM A230 and A232	
	Not Shot-Peened	Shot-Peened*	Not Shot-Peened	Shot-Peened*
10^5	53	62	55	64
10^6	50	60	53	62

This information is based on the following conditions: no surging,
springs are in the "as-stress-relieved" condition
*Not always possible.

SOURCE: Associated Spring, Barnes Group Inc.

24.6.4 Tolerances

The tolerances for coil diameter and end position are given in Tables 24.17 and
24.18, respectively. Use them as guides.

24.7 BELLEVILLE SPRING WASHER

Belleville washers, also known as *coned-disk springs,* take their name from their
inventor, Julian F. Belleville. They are essentially circular disks formed to a conical
shape, as shown in Fig. 24.25. When load is applied, the disk tends to flatten. This
elastic deformation constitutes the spring action.

TABLE 24.17 Commercial Tolerances for Torsion-Spring Coil Diameters

Wire Diameter mm (in.)	Tolerance: ±mm(in.)						
	Spring Index D/d						
	4	6	8	10	12	14	16
0.38 (0.015)	0.05 (0.002)	0.05 (0.002)	0.05 (0.002)	0.05 (0.002)	0.08 (0.003)	0.08 (0.003)	0.10 (0.004)
0.58 (0.023)	0.05 (0.002)	0.05 (0.002)	0.05 (0.002)	0.08 (0.003)	0.10 (0.004)	0.13 (0.005)	0.15 (0.006)
0.89 (0.035)	0.05 (0.002)	0.05 (0.002)	0.08 (0.003)	0.10 (0.004)	0.15 (0.006)	0.18 (0.007)	0.23 (0.009)
1.30 (0.051)	0.05 (0.002)	0.08 (0.003)	0.13 (0.005)	0.18 (0.007)	0.20 (0.008)	0.25 (0.010)	0.31 (0.012)
1.93 (0.076)	0.08 (0.003)	0.13 (0.005)	0.18 (0.007)	0.23 (0.009)	0.31 (0.012)	0.38 (0.015)	0.46 (0.018)
2.90 (0.114)	0.10 (0.004)	0.18 (0.007)	0.25 (0.010)	0.33 (0.013)	0.46 (0.018)	0.56 (0.022)	0.71 (0.028)
4.37 (0.172)	0.15 (0.006)	0.25 (0.010)	0.33 (0.013)	0.51 (0.020)	0.69 (0.027)	0.86 (0.034)	1.07 (0.042)
6.35 (0.250)	0.20 (0.008)	0.36 (0.014)	0.56 (0.022)	0.76 (0.030)	1.02 (0.040)	1.27 (0.050)	1.52 (0.060)

SOURCE: Associated Spring, Barnes Group Inc.

TABLE 24.18 End-Position Tolerances (for *D/d* Ratios up to and Including 16)

Total Coils	Tolerance: ± Degrees*
Up to 3	8
Over 3–10	10
Over 10–20	15
Over 20–30	20
Over 30	25

*Closer tolerances available
SOURCE: Associated Spring, Barnes Group Inc.

Belleville springs are used in two broad types of applications. First, they are used to provide very high loads with small deflections, as in stripper springs for punch-press dies, recoil mechanisms, and pressure-relief valves. Second, they are used for their special nonlinear load-deflection curves, particularly those with a constant-load portion. In loading a packing seal or a live center for a lathe, or in injection molding machines, Belleville washers can maintain a constant force throughout dimensional changes in the mechanical system resulting from wear, relaxation, or thermal change.

The two types of performance depend on the ratio of height to thickness. Typical load-deflection curves for various height-thickness ratios are shown in Fig. 24.26. Note that the curve for a small *h/t* ratio is nearly a straight line. At *h/t* = 1.41 the curve shows a nearly constant load for approximately the last 50 percent of deflection

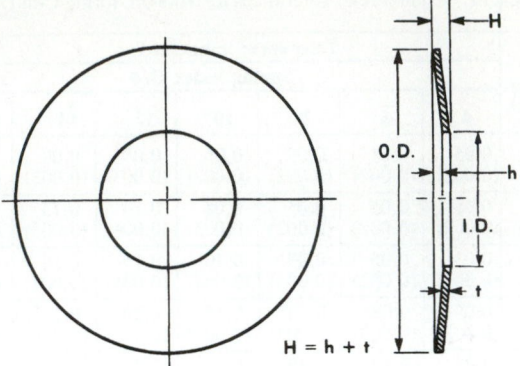

FIGURE 24.25 Belleville washer. *(Associated Spring, Barnes Group Inc.)*

before the flat position. Above $h/t = 1.41$ the load decreases after reaching a peak. When h/t is 2.83 or more, the load will go negative at some point beyond flat and will require some force to be restored to its free position. In other words, the washer will turn inside out.

The design equations given here are complex and may present a difficult challenge to the occasional designer. Use of charts and the equation transpositions presented here have proved helpful. Note that these equations are taken from the mathematical analysis by Almen and Laszlo [24.5]. The symbols used here are those originally used by the authors and may not necessarily agree with those used elsewhere in the text.

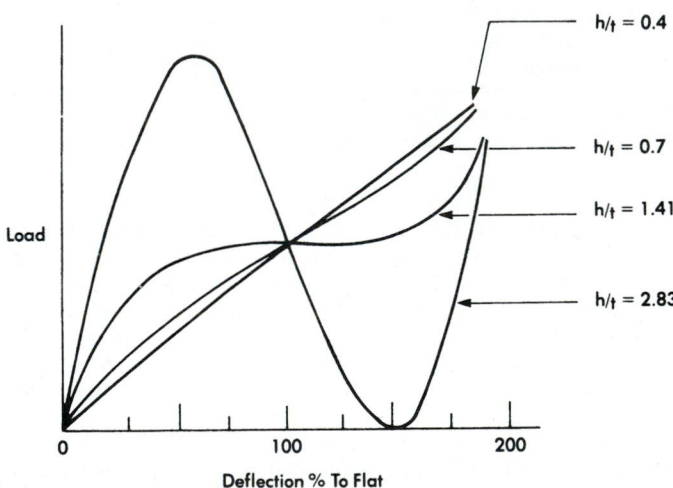

FIGURE 24.26 Load-deflection curves for Belleville washers with various h/t ratios. *(Associated Spring, Barnes Group Inc.)*

24.7.1 Nomenclature

a	OD/2, mm (in)
C_1	Compressive stress constant (see formula and Fig. 24.28)
C_2	Compressive stress constant (see formula and Fig. 24.28)
E	Modulus of elasticity (see Table 24.19), MPa (psi)
f	Deflection, mm (in)
h	Inside height, mm (in)
ID	Inside diameter, mm (in)
M	Constant
OD	Outside diameter, mm (in)
P	Load, N (lb)
P_f	Load at flat position, N (lb)
R	OD/ID
S_c	Compressive stress (Fig. 24.27), MPa (psi)
S_{T_1}	Tensile stress (Fig. 24.27), MPa (psi)
S_{T_2}	Tensile stress (Fig. 24.27), MPa (psi)
t	Thickness, mm (in)
T_1	Tensile stress constant (see formula and Fig. 24.29)
T_2	Tensile stress constant (see formula and Fig. 24.29)
μ	Poisson's ratio (Table 24.19)

24.7.2 Basic Equations

$$P = \frac{Ef}{(1-\mu^2)Ma^2}\left[(h-f)\left(h-\frac{f}{2}\right)t+t^3\right] \qquad (24.39)$$

$$P_F = \frac{Eht^3}{(1-\mu^2)Ma^2} \qquad (24.40)$$

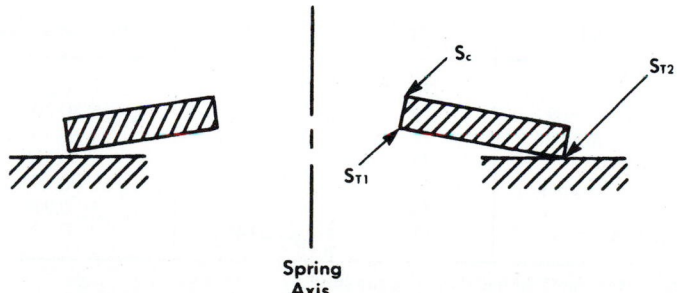

FIGURE 24.27 Highest stressed regions in Belleville washers. *(Associated Spring, Barnes Group Inc.)*

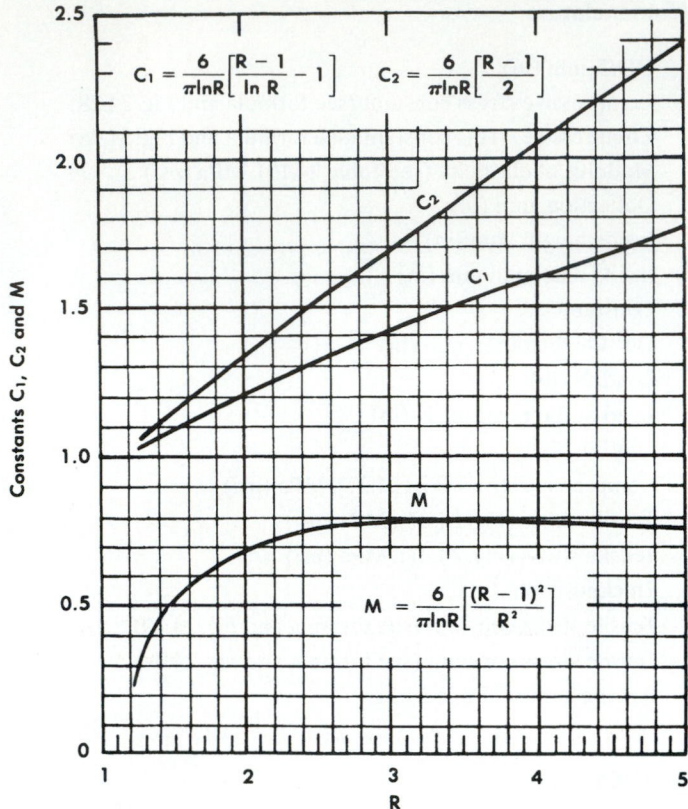

FIGURE 24.28 Compressive stress constants for Belleville washers. *(Associated Spring, Barnes Group Inc.)*

TABLE 24.19 Elastic Constants of Common Spring Materials

Material	Modulus of Elasticity E		Poisson's ratio μ
	Mpsi	**GPa**	
Steel	30	207	0.30
Phosphor bronze	15	103	0.20
17-7 PH stainless	29	200	0.34
302 stainless	28	193	0.30
Beryllium copper	18.5	128	0.33
Inconel	31	214	0.29
Inconel X	31	214	0.29

SOURCE: **Associated Spring, Barnes Group Inc.**

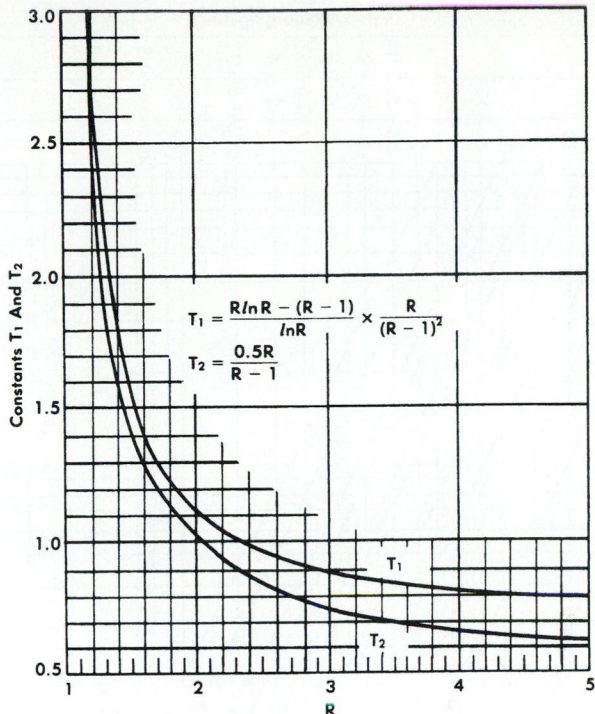

FIGURE 24.29 Tensile stress constants for Belleville washers. *(Associated Spring, Barnes Group Inc.)*

$$S_c = \frac{Ef}{(1-\mu^2)Ma^2}\left[C_1\left(h-\frac{f}{2}\right)+C_2t\right] \qquad (24.41)$$

$$S_{T_1} = \frac{Ef}{(1-\mu^2)Ma^2}\left[C_1\left(h-\frac{f}{2}\right)-C_2t\right] \qquad (24.42)$$

$$S_{T_2} = \frac{Ef}{(1-\mu^2)a^2}\left[T_1\left(h-\frac{f}{2}\right)+T_2t\right] \qquad (24.43)$$

The design approach recommended here depends on first determining the loads and stresses at flat position, as shown in Fig. 24.30. Intermediate loads are determined from the curves in Fig. 24.31.

Figure 24.30 gives the values graphically for compressive stresses S_c at flat position. The stress at intermediate stages is approximately proportional to the deflection. For critical applications involving close tolerances or unusual proportions, stresses should be checked by using the equation before the design is finalized.

The stress level for static applications is evaluated in accordance with Eq. (24.41). This equation has been used most commonly for appraising the design of a Belleville spring because it gives the highest numerical value. It gives the compressive stress at the point shown in Fig. 24.27.

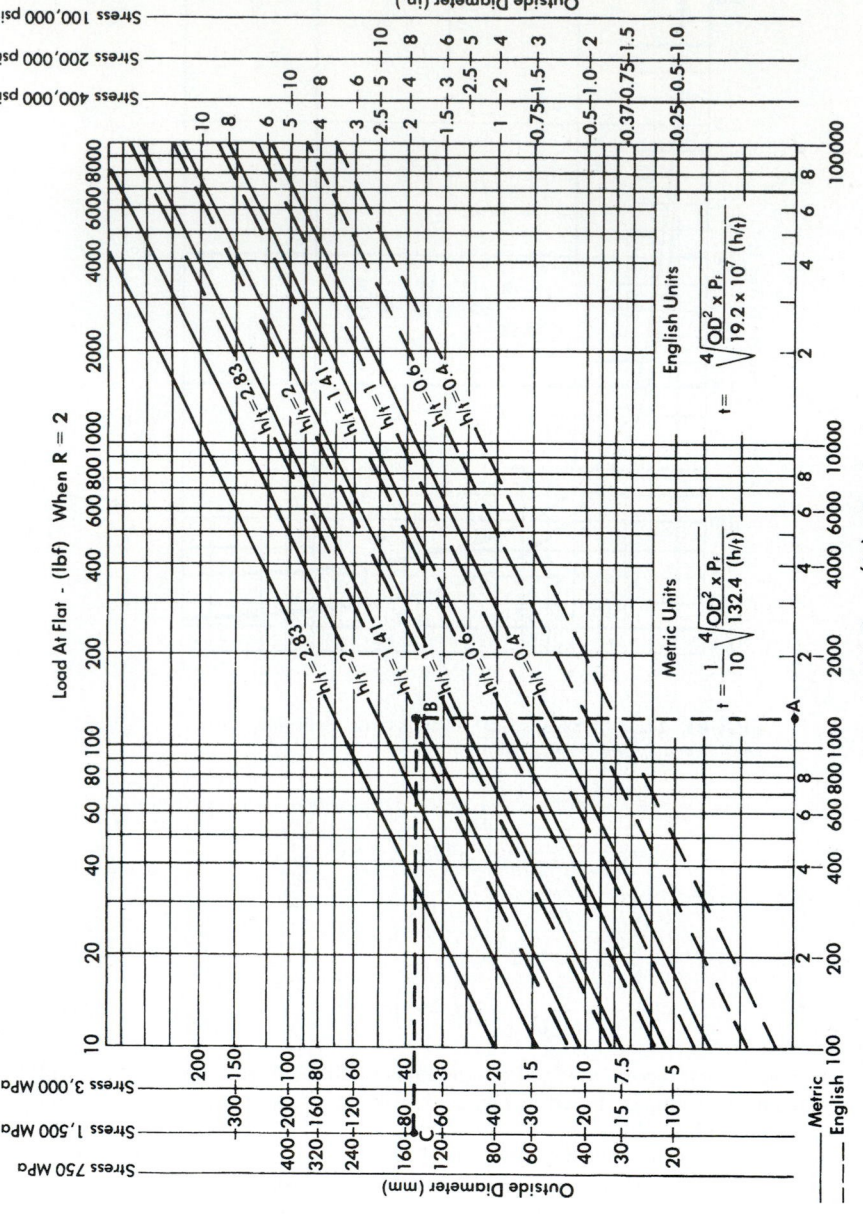

FIGURE 24.30 Loads and compressive stresses S_c for Belleville washers with various outside diameters and h/t ratios. (*Associated Spring, Barnes Group Inc.*)

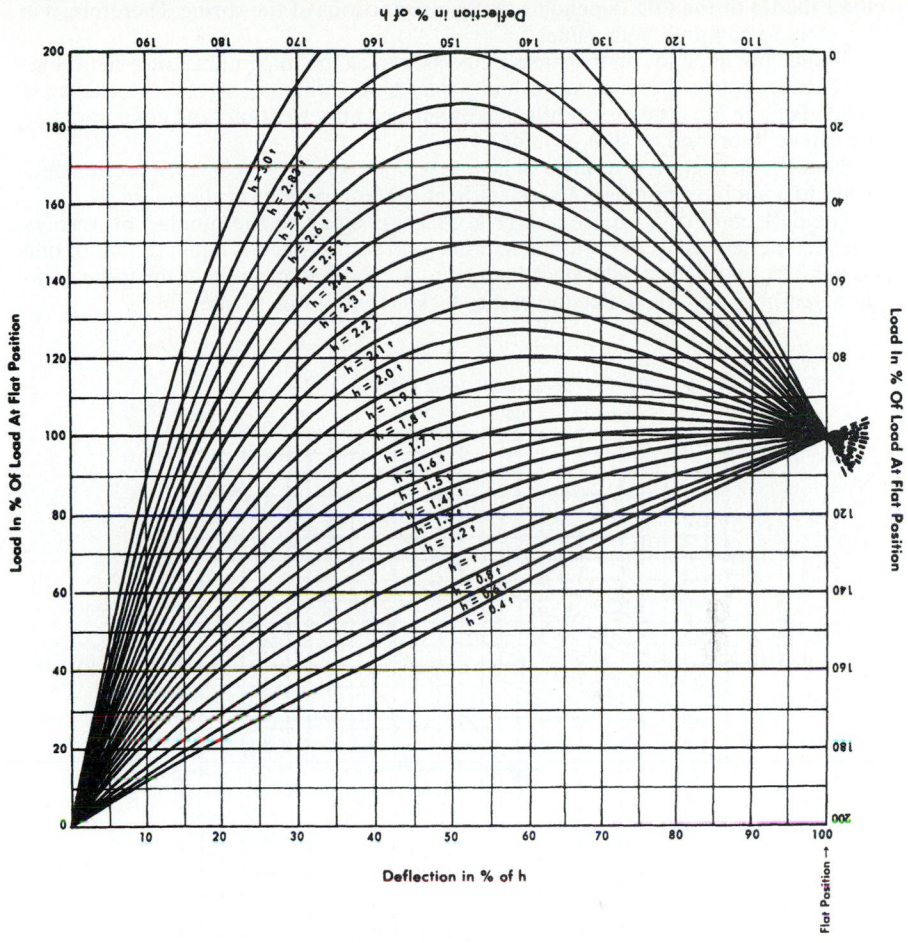

FIGURE 24.31 Load-deflection characteristics for Belleville washers. If a washer is supported and loaded at its edges so that it is deflected beyond the flat position, then the greatest possible deflection can be utilized. Since the load-deflection curve beyond the horizontal position is symmetric with the first part of the curve, this chart has been labeled at the right and top to be read upside down for deflection beyond horizontal. Dotted lines extending beyond the chart indicate continuation of curves beyond flat. *(Associated Spring, Barnes Group Inc.)*

A Belleville spring washer should be designed so that it can be compressed flat by accidental overloading, without setting. This can be accomplished either by using a stress so low that the spring will not set or by forming the spring higher than the design height and removing set by compressing flat or beyond flat (see Table 24.21). The table values should be reduced if the washers are plated or used at elevated temperatures.

For fatigue applications it is necessary to consider the tensile stresses at the points marked S_{T_1} and S_{T_2} in Fig. 24.27. The higher value of the two can occur at either the ID or the OD, depending on the proportions of the spring. Therefore, it is necessary to compute both values.

Fatigue life depends on the stress range as well as the maximum stress value. Figure 24.32 predicts the endurance limits based on either S_{T_1} or S_{T_2}, whichever is higher. Fatigue life is adversely affected by surface imperfections and edge fractures and can be improved by shot peening.

Since the deflection in a single Belleville washer is relatively small, it is often necessary to combine a number of washers. Such a combination is called a *stack*.

The deflection of a series stack (Fig. 24.33) is equal to the number of washers times the deflection of one washer, and the load of the stack is equal to that of one washer. The load of a parallel stack is equal to the load of one washer times the number of washers, and the deflection of the stack is that of one washer.

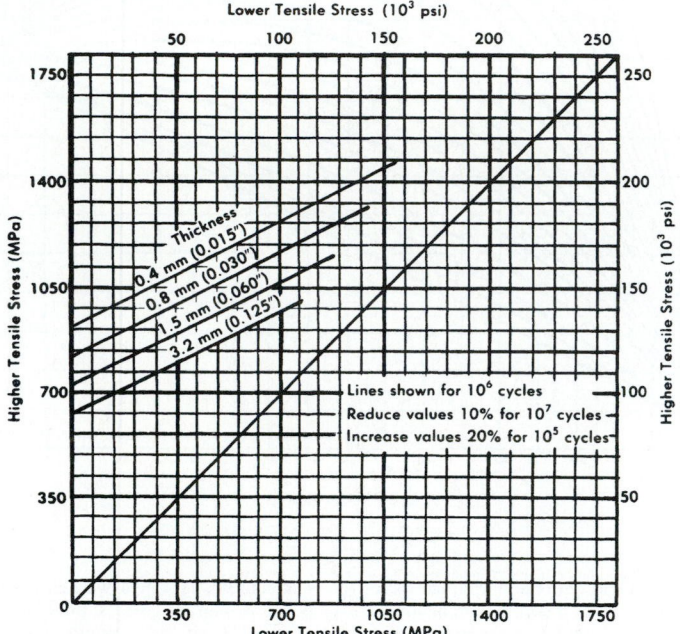

FIGURE 24.32 Modified Goodman diagram for Belleville washers; for carbon and alloy steels at 47 to 49 R_c with set removed, but not shot-peened. (*Associated Spring, Barnes Group Inc.*)

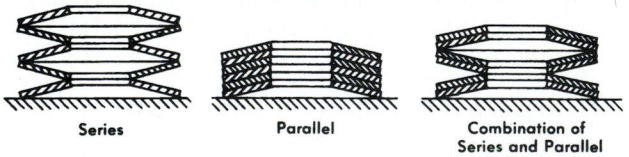

FIGURE 24.33 Stacks of Belleville washers. (*Associated Spring, Barnes Group Inc.*)

Because of production variations in washer parameters, both the foregoing statements carry cautionary notes. In the series stack, springs of the constant-load type ($h/t = 1.41$) may actually have a negative rate in some portion of their deflection range. When such a series stack is deflected, some washers will snap through, producing jumps in the load-deflection curve. To avoid this problem, the h/t ratio in a series stack design should not exceed 1.3.

In the parallel stack, friction between the washers causes a hysteresis loop in the load-deflection curve (Fig. 24.34). The width of the loop increases with each washer added to the stack but may be reduced by adding lubrication as the washers burnish each other during use.

Stacked washers normally require guide pins or sleeves to keep them in proper alignment. These guides should be hardened steel at HRC 48 minimum hardness. Clearance between the washer and the guide pin or sleeve should be about 1.5 percent of the appropriate diameter.

24.7.3 Tolerances

Load tolerances should be specified at test height. For carbon-steel washers with $h/t < 0.25$, use load tolerance of ±15 percent. For washers with $h/t > 0.25$, use ±10 percent. The recommended load tolerance for stainless steel and nonferrous washers is ±15 percent. See Table 24.20 for outside- and inside-diameter tolerances.

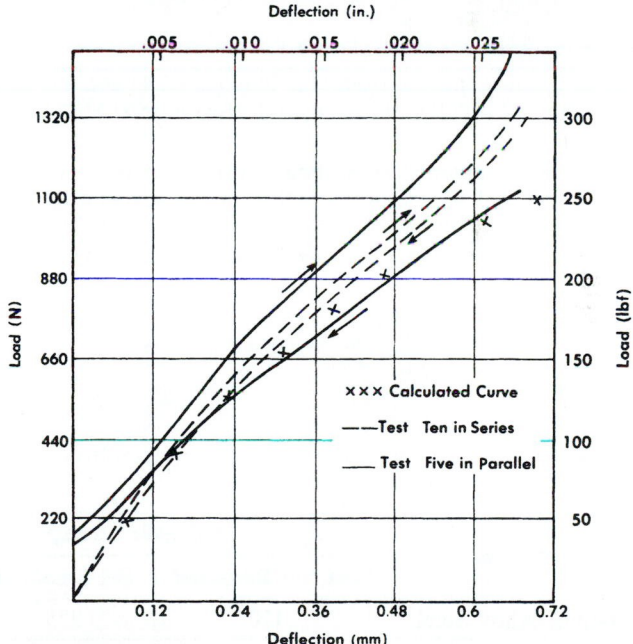

FIGURE 24.34 Hysteresis in stacked Belleville washers. *(Associated Spring, Barnes Group Inc.)*

TABLE 24.20 Belleville Washer Diameter Tolerances

Diameter, mm (in.)	O.D. mm (in.)	I.D. mm (in.)
	+0.00	−0.00
Up to 5 (0.197)	−0.20 (−0.008)	+0.20 (+0.008)
5–10 (0.197–0.394)	−0.25 (−0.010)	+0.25 (+0.010)
10–25 (0.394–0.984)	−0.30 (−0.012)	+0.30 (+0.012)
25–50 (0.984–1.969)	−0.40 (−0.016)	+0.40 (+0.016)
50–100 (1.969–3.937)	−0.50 (−0.020)	+0.50 (+0.020)

Based on R = 2, increased tolerances are required for lower R ratios.
SOURCE: Associated Spring, Barnes Group Inc.

Example.   In a clutch, a minimum pressure of 202 lb (900 N) is required. This pressure must be held nearly constant as the clutch facing wears down 0.31 in (7.9 mm). The washer OD is 2.99 in (76 mm). The material washer OD is 2.99 in (76 mm). The material selected for the application is spring steel HRC 47-50.
   Solution

1. Base the load on a value 10 percent above the minimum load, or 202 + 10 percent = 223 lb (998 N). Assume OD/ID = 2. From Fig. 24.31, select a load-deflection curve which gives approximately constant load between 50 and 100 percent of deflection to flat. Choose the $h/t = 1.41$ curve.

2. From Fig. 24.31, the load at 50 percent of deflection to flat is 88 percent of the flat load.

3. Flat load is $P_F = 223/0.88 = 252$ lb (1125 N).

4. From Fig. 24.30 [follow line AB from 1125 N to $h/t = 1.41$ and line BC to approximately 76-mm (2.99-in) OD], the estimated stress is 1500 MPa [218 kilopounds per square inch (kpsi)].

5. From Table 24.21 maximum stress without set removed is 120 percent of tensile strength. From Fig. 24.3, the tensile strength at HRC 48 will be approximately 239 kpsi (1650 MPa). Yield point without residual stress will be (239 kpsi)(1.20) = 287 kpsi. Therefore 218 kpsi stress is less than the maximum stress of 287 kpsi.

6. Stock thickness is

$$t = \sqrt[4]{\frac{OD^2(P_F)}{19.2(10^7)(h/t)}} = 0.054 \text{ in (1.37 mm)}$$

TABLE 24.21 Maximum Recommended Stress Levels for Belleville Washers in Static Applications

Material	Percent of Tensile Strength	
	Set Not Removed	Set Removed
Carbon or Alloy Steel	120	275
Nonferrous and Austenitic Stainless Steel	95	160

SOURCE: Associated Spring, Barnes Group Inc.

7.
$$h = 1.41t = 1.41(0.054) = 0.076 \text{ in}$$

$$H = h + t = 0.076 + 0.054 = 0.130 \text{ in}$$

8. Refer to Fig. 24.31. The load of 202 lb will be reached at $f_1 = 50$ percent of maximum available deflection. And $f_1 = 0.50(0.076) = 0.038$ in deflection, or the load of 223 lb will be reached at $H_1 = H - f_1 = 0.130 - 0.038 = 0.092$ in height at load. To allow for wear, the spring should be preloaded at $H_2 = H_1 - f(\text{wear}) = 0.092 - 0.032 = 0.060$ in height. This preload corresponds to a deflection $f_2 = H - H_2 = 0.130 - 0.060 = 0.070$ in. Then $f_2/h = 0.070/0.076 = 0.92$, or 92 percent of h.

9. Because 92 percent of h exceeds the recommended 85 percent (the load-deflection curve is not reliable beyond 85 percent deflection when the washer is compressed between flat surfaces), increase the deflection range to 40 to 85 percent. From Fig. 24.31, the load at 40 percent deflection is 78.5 percent, and $P_F = 223/0.785 = 284$ lb. Repeat previous procedures 4, 5, 6, 7, and 8, and find that $100(f_2/h) = 81$ percent of h. The final design is as follows:

Material: AISI 1074

OD = 2.99 in (76 mm)

ID = 1.50 in (38 mm)

$t = 0.055$ in (1.40 mm) nominal

$h = 0.078$ in (1.95 mm) nominal

Tensile stress $S_{T_1} = -29.5$ kpsi (-203 MPa) at $f_2 = 85$ percent of h

Tensile stress $S_{T_2} = 103$ kpsi (710 MPa) at $f_2 = 85$ percent of h

24.8 SPECIAL SPRING WASHERS

Spring washers are being used increasingly in applications where there is a requirement for miniaturization and compactness of design. They are used to absorb vibrations and both side and end play, to distribute loads, and to control end pressure.

Design equations have been developed for determining the spring characteristics of curved, wave, and Belleville washers. There are no special design equations for slotted and finger washers. They are approximated by using Belleville and cantilever equations and then are refined through sampling and testing.

24.8.1 Curved Washers

These springs (Fig. 24.35) exert relatively light thrust loads and are often used to absorb axial end play. The designer must provide space for diametral expansion which occurs as the washer is compressed during loading. Bearing surfaces should be hard, since the washer edges tend to dig in. The spring rate is approximately linear up to 80 percent of the available deflection. Beyond that the rate will be much higher than calculated. Load tolerance should not be specified closer than ±20 percent.

Approximate equations are

$$P = \frac{4fEt^3}{\text{OD}^2(K)} \tag{24.44}$$

and

$$S = \frac{1.5KP}{t^2} \tag{24.45}$$

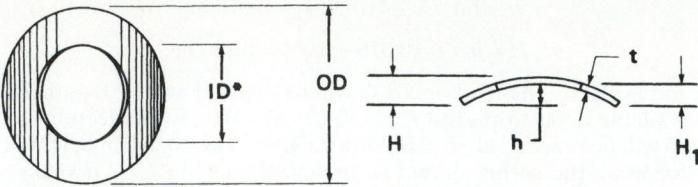

*Long axis of the washer in free position

FIGURE 24.35   Curved washer. *(Associated Spring, Barnes Group Inc.)*

where K is given in Fig. 24.36 and f is 80 percent of h or less.

Maximum recommended stress levels for static operations are given in Table 24.22. Favorable residual stresses can be induced by shot peening and, to a lesser extent, by removing set. The maximum recommended stresses for cyclic applications are given in Table 24.23.

Tensile strengths for carbon steel are obtained from Fig. 24.3.

24.8.2   Wave Washer

These spring washers (Fig. 24.37) are regularly used in thrust loading applications, for small deflections, and for light to medium loads. The rate is linear between 20 and 80 percent of available deflection. Load tolerances should be no less than ±20 percent. In the most commonly used range of sizes, these washers can have three, four, or six waves.

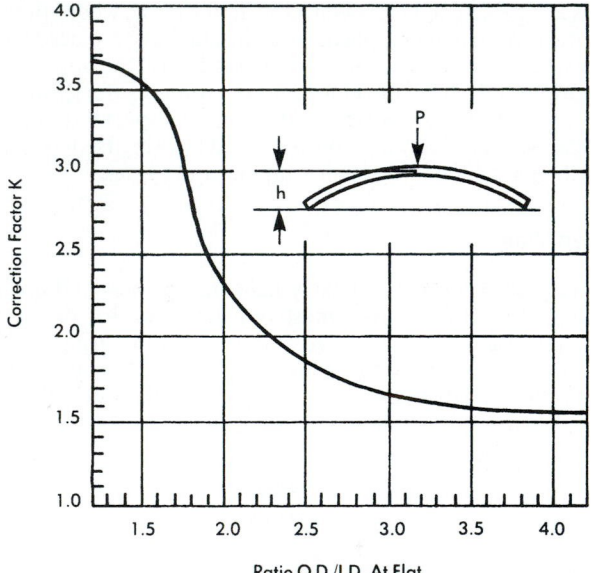

FIGURE 24.36   Empirical correction factor K for curved spring washers. *(Associated Spring, Barnes Group Inc.)*

TABLE 24.22 Maximum Recommended Operating Stress Levels for Special Spring Washers in Static Applications

Material	Percent of Tensile Strength	
	Stress-Relieved	With Favorable Residual Stresses
Steels, Alloy Steels	80	100
Nonferrous Alloys and Austenitic Steel	—	80

Finger washers are not generally supplied with favorable residual stresses.
SOURCE: Associated Spring, Barnes Group Inc.

TABLE 24.23 Maximum Recommended Operating Stress Levels for Steel Curved and Wave Washers in Cyclic Applications

Life (Cycles)	Percent of Tensile Strength
	Maximum Stress
10^4	80
10^5	53
10^6	50

This information is based on the following conditions: ambient environment, free from sharp bends, burrs, and other stress concentrations. AISI 1075
SOURCE: Associated Spring, Barnes Group Inc.

Design equations are

$$\frac{P}{f} = \frac{Ebt^3N^4(\text{OD})}{2.4D^3(\text{ID})} \tag{24.46}$$

and

$$S = \frac{3\pi PD}{4bt^2N^2} \tag{24.47}$$

where $D = \text{OD} - b$. The washer expands in diameter when compressed, according to the formula

$$D' = \sqrt{D^2 + 0.458h^2N^2} \tag{24.48}$$

Maximum recommended stress levels for static applications are given in Table 24.22. Favorable residual stresses are induced by shot peening or removing set. Table 24.23 gives the maximum recommended stress levels for cyclic applications. Figure 24.3 provides tensile strengths for carbon steel.

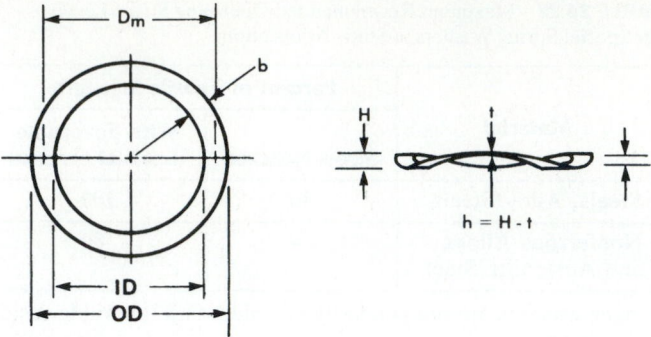

FIGURE 24.37 Typical wave spring washer. *(Associated Spring, Barnes Group Inc.)*

24.8.3 Finger Washers

Finger washers (Fig. 24.38) have both the flexibility of curved washers and the distributed points of loading of wave washers. They are calculated, approximately, as groups of cantilever springs; then samples are made and tested to prove the design. They are most frequently used in static applications such as applying axial load to ball-bearing races to reduce vibration and noise. These washers are not used in cyclic applications because of the shear cuts.

24.8.4 Slotted Washers

These are more flexible than plain dished washers but should be designed to maintain a constant pressure rather than to operate through a deflection range (see Fig. 24.39).

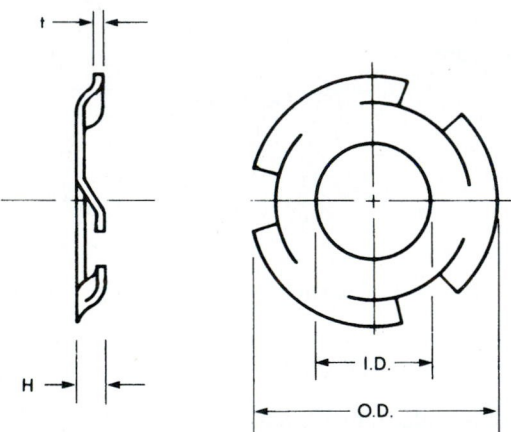

FIGURE 24.38 Finger washer. *(Associated Spring, Barnes Group Inc.)*

24.9 FLAT SPRINGS

24.9.1 Introduction

The classification *flat springs* applies to a wide range of springs made from sheet, strip, or plate material. Exceptions to this classification are power springs and washers. Flat springs may contain bends and forms. Thus the classification refers to the raw material and not to the spring itself.

Flat springs can perform functions beyond normal spring functions. A flat spring may conduct electricity, act as a latch, or hold a part in position. In some flat springs, only a portion of the part may have a spring function.

FIGURE 24.39 Slotted washers. *(Associated Spring, Barnes Group Inc.)*

Most flat springs are custom designs, and the tooling is often a major cost consideration. Flat springs can be cantilever or simple elliptical beams or combinations of both. These two elementary forms are discussed in this section. For a description of the methods used to compute complex flat-spring designs, see [24.6].

Load specification in flat springs is closely connected with the dimensioning of the form of the spring. From the equations it can be seen that the deflection and load vary in proportion to the third power of the material thickness. The important factors in load control are, first, the material thickness and, second, the deflection. Where close load control is required, the material may have to be selected to restricted thickness tolerance, and/or the free shape may be trued.

24.9.2 Cantilever Springs

The basic type of cantilever is a rectangular spring as shown in Fig. 24.40. The maximum bending stress occurs at the clamping point, and the stress is not uniform through the section. This stress is

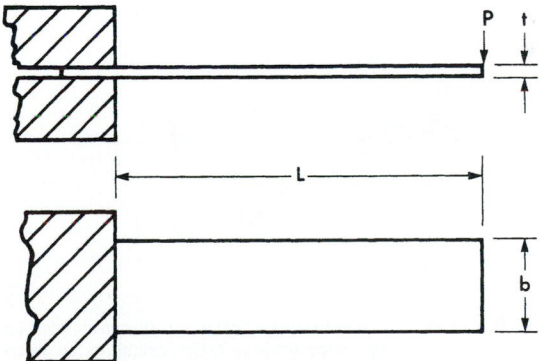

FIGURE 24.40 Rectangular cantilever spring. *(Associated Spring, Barnes Group Inc.)*

$$S = \frac{6PL}{bt^2} \qquad (24.49)$$

The load is given by

$$P = \frac{fEbt^3}{4L^3} \qquad (24.50)$$

These equations are satisfactory when the ratio of deflection to length f/L is less than 0.3. For larger deflections, use the method described in Fig. 24.41.

In cantilever springs with a trapezoidal or triangular configuration (Fig. 24.42), the stress is uniform throughout and is

$$S = \frac{6PL}{b_o t^2} \qquad (24.51)$$

The corresponding load is

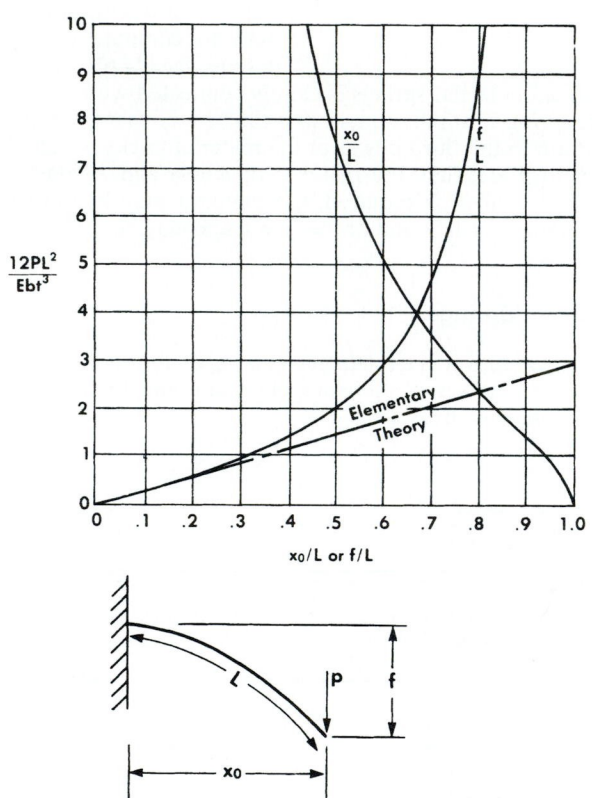

FIGURE 24.41 Calculating large deflection in cantilever beams [24.7]. To utilize this figure for any load P, first calculate the quantity $12PL^3/Ebt^3$. Using this value, from the curves find f/L and x_o/L, where x_o is the moment arm of the load P. Deflection then equals L multiplied by f/L. The maximum stress is reduced in the ratio x_o/L. (*Associated Spring, Barnes Group Inc.*)

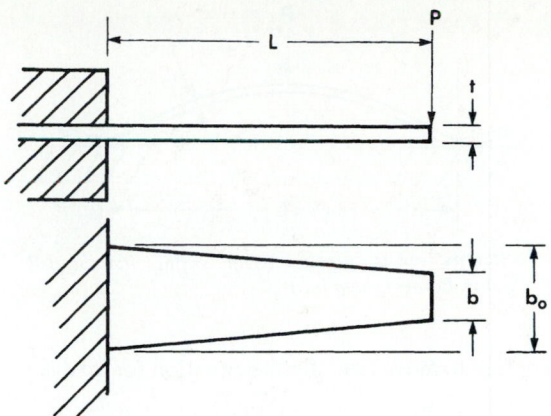

FIGURE 24.42 Trapezoidal cantilever spring. *(Associated Spring, Barnes Group Inc.)*

$$P = \frac{fEb_ot^3}{4L^3K}$$
(24.52)

where K = constant based on the ratio b/b_o (Fig. 24.43). These equations are valid for f/L ratios of less than 0.3.

24.9.3 Simple Beams or Elliptical Springs

Simple beams are usually rectangular and are formed into an arc as in Fig. 24.44. If holes are introduced for clamping purposes, stress will increase at the hole and at the clamping point owing to stress concentration.

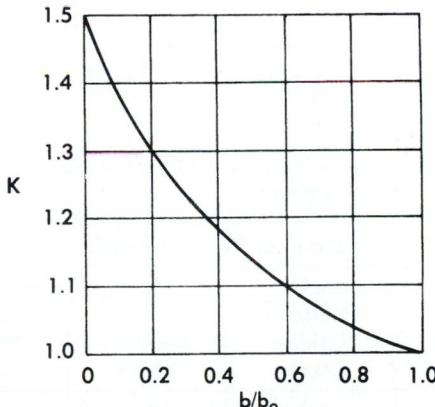

FIGURE 24.43 Correction factor for trapezoidal beam-load equation. *(Associated Spring, Barnes Group Inc.)*

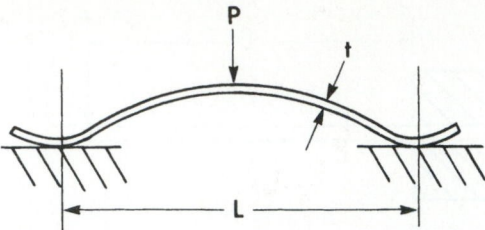

FIGURE 24.44 Simple beam spring. *(Associated Spring, Barnes Group Inc.)*

When ends are free to move laterally, the equation for load is

$$P = \frac{4fEbt^3}{L^3} \tag{24.53}$$

and stress is given by

$$S = \frac{1.5PL}{bt^2} \tag{24.54}$$

These equations apply when the ratio f/L is less than 0.15.

Stress Considerations. The maximum design stresses for cantilevers and simple beams are given in Table 24.24 for static applications and in Table 24.25 for cyclic applications. These recommendations do not apply when holes, sharp corners, notches, or abrupt changes in cross section are incorporated in the design, and should be used for guidance only.

24.10 CONSTANT-FORCE SPRINGS

A constant-force spring is a roll of prestressed material which exerts a nearly constant restraining force to resist uncoiling. Its unique characteristic is *force independent of deflection*. The force required to produce a unit deflection is the same for

TABLE 24.24 Maximum Design Stresses for Cantilever and Simple Beam Springs in Static Applications

Percent of Tensile Stength			
Ferrous Material		Nonferrous Material	
No Residual Stress	Maximum Residual Stress	No Residual Stress	Maximum Residual Stress
80	100	75	80

SOURCE: Associated Spring, Barnes Group Inc.

TABLE 24.25 Maximum Design Stresses for Carbon-Steel Cantilever and Simple Beam Springs in Cyclic Applications

Number of Cycles	Percent of Tensile Strength	
	Not Shot-Peened	Shot-Peened*
10^5	53	62
10^6	50	60
10^7	48	58

*Shot peening is not recommended for thin materials and complex shapes. This information is based on an ambient environment. Stress ratio = 0.
SOURCE: Associated Spring, Barnes Group Inc.

each increment of coil because the radius of curvature of each increment is the same as any other.

Although these springs are not constant-load or constant-torque springs in the precise meaning of those terms, they produce a more nearly constant load over a greater deflection than any other spring design covered here. See Fig. 24.45. Constant-force springs are made of both type 301 stainless steel and ultra-high-strength high-carbon steels, with many of the applications using stainless steel because of its inherent resistance to corrosion.

One of the most severe limitations on the use of constant-force springs is their relatively short operating life. The most efficient use of material will produce a life of about 3000 cycles. Although life of hundreds of thousands of cycles is possible,

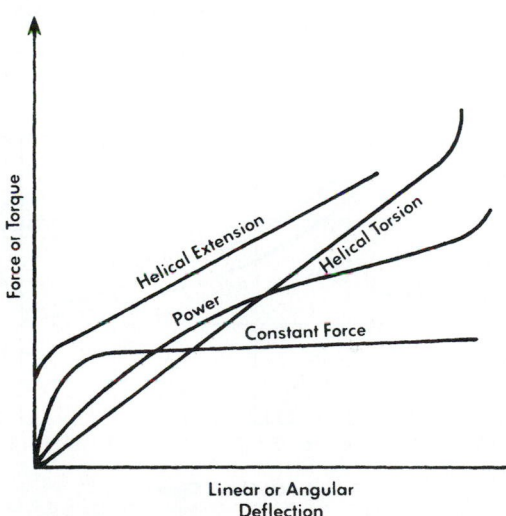

FIGURE 24.45 Load-deflection curves for various spring configurations. *(Associated Spring, Barnes Group Inc.)*

most applications fall into the range of 3000 to 30 000 cycles. Figure 24.46 shows the relationship between stress and fatigue life. These curves are derived from experimentally obtained data.

Some applications involving constant-force and constant-torque springs are simple extension springs, window sash counterbalances, camera motors, toys, machine carriage returns, constant-pressure electric-motor brush springs, space vehicle applications, and retraction devices.

24.10.1 Extension Type

This type of spring is a spiral spring made of strip material wound on the flat with an inherent curvature such that, in repose, each coil wraps tightly on its inner neighbor. In use the strip is extended with the free end loaded and the inner end supported on a drum or arbor. Very long deflections are possible, but the strip becomes unstable in long deflections and must be guided or supported to avoid kinking or snarling on the return stroke.

The rated load is not reached until after an initial deflection of 1.25 times the drum diameter, as shown in Fig. 24.47. Idler pulleys can be used but should be no smaller in diameter than the natural diameter of the coils and should never be used in a direction to cause backbending against the strip curvature.

24.10.2 Design Equations

$$P = \frac{Ebt^3}{6.5D_n^2} \qquad\qquad \text{for } N \le 10 \qquad\qquad (24.55)$$

$$P = \frac{Ebt^3}{6.5D_1}\left(\frac{2}{D_n} - \frac{1}{D_1}\right) \qquad \text{for } N > 10 \qquad\qquad (24.56)$$

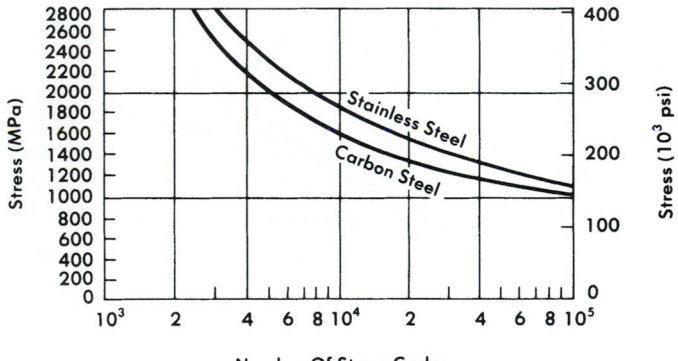

FIGURE 24.46 Maximum bending stress versus number of stress cycles for constant-force springs. These curves are based on no. 1 round-edge strip. *(Associated Spring, Barnes Group Inc.)*

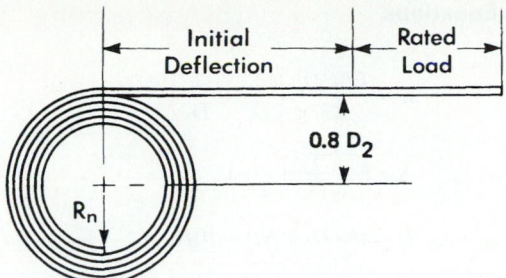

FIGURE 24.47 Typical constant-force extension spring (extension form). *(Associated Spring, Barnes Group Inc.)*

If unknown, let $b/t = 100/1$, $D_2 = 1.2\,D_n$,

$$S = \frac{Et}{D_n} \qquad (24.57)$$

and

$$L = 1.57N(D_1 + D_2) \qquad \text{or} \qquad L \simeq f + 5D_2$$

where N = number of turns
  D_1 = outside coil diameter
  D_2 = drum (arbor) diameter
  D_n = natural diameter
  E = modulus of elasticity

24.10.3 Spring Motor Type

When a constant-force spring is mounted on two drums of different diameters and the spring is backbent onto the larger diameter, the result is a constant-force spring motor. The strip is in repose on the smaller (storage) drum and is backbent onto the larger (output) drum. Torque is taken from the output drum shaft as shown in Fig. 24.48.

Note here that constant torque does not mean constant speed. Constant torque implies uniform acceleration, and the mechanism so driven will continue to speed up unless restrained by a governor mechanism. Load tolerances are normally held within ±10 percent.

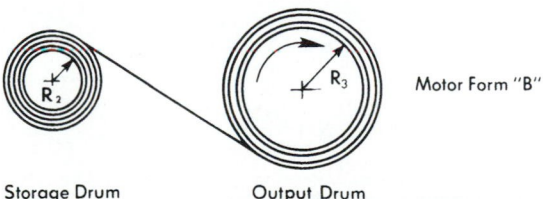

FIGURE 24.48 Typical constant-torque motor spring. *(Associated Spring, Barnes Group Inc.)*

24.10.4 Design Equations

$$M = \frac{Ebt^3D_3}{13}\left(\frac{1}{D_n}+\frac{1}{D_3}\right)^2 \tag{24.58}$$

$$S = Et\left(\frac{1}{D_n}+\frac{1}{D_3}\right) \tag{24.59}$$

$$L = \pi N(D_3 + Nt) + 10D_3 \tag{24.60}$$

$$R_c = R_n\sqrt{4 + \frac{4R_3}{R_n} + \frac{R_n}{R_3} + \left(\frac{R_3}{R_n}\right)^2} \tag{24.61}$$

Design Suggestions.   Let

$$\frac{b}{t}=100 \qquad \frac{D_n}{t}=250 \qquad \frac{D_3}{D_n}=2 \qquad \frac{D_3}{D_2}=1.6$$

where D_n = natural diameter
    R_n = natural radius
    D_2 = storage-drum diameter
    D_3 = output-drum diameter
    R_3 = output-drum radius
    N = number of revolutions
    R_c = minimum center-to-center distance of drums

24.11 TORSION BARS

Torsion bars used as springs are usually straight bars of spring material to which a twisting couple is applied. The stressing mode is torsional. This type of spring is very efficient in its use of material to store energy. The major disadvantage with the torsion bar is that unfavorable stress concentrations occur at the point where the ends are fastened.

Although both round and rectangular bar sections are used, the round section is used more often.

24.11.1 Design Equations: Round Sections

$$\phi = \frac{584ML}{d^4G} \tag{24.62}$$

$$S = \frac{16M}{\pi d^3} \tag{24.63}$$

where ϕ = rotation angle in degrees
    S = shear stress
    L = active length

24.11.2 Design Equations: Rectangular Sections

$$\phi = \frac{57.3ML}{K_1 bt^3 G} \qquad (24.64)$$

$$S = \frac{M}{K_2 bt^2} \qquad (24.65)$$

where factors K_1 and K_2 are taken from Table 24.26.

The assumptions used in deriving these equations are (1) the bar is straight, (2) the bar is solid, and (3) loading is in pure torsion.

Torsion-bar springs are often preset in the direction in which they are loaded by twisting the bar beyond the torsional elastic limit. Care must be taken in the use of a preset bar: It must be loaded in the same direction in which it was preset; otherwise, excessive set will occur.

24.12 POWER SPRINGS

Power springs, also known as clock, motor, or flat coil springs, are made of flat strip material which is wound on an arbor and confined in a case. Power springs store and release rotational energy through either the arbor or the case in which they are retained. They are unique among spring types in that they are almost always stored in a case or housing while unloaded. Figure 24.49 shows typical retainers, a case, and various ends.

24.12.1 Design Considerations

Power springs are stressed in bending, and stress is related to torque by

$$S = \frac{6M}{bt^2} \qquad (24.66)$$

Load-deflection curves for power springs are difficult to predict. As a spring is wound up, material is wound onto the arbor. This material is drawn from that which

TABLE 24.26 Factors for Computing Rectangular Bars in Torsion

b/t	K_1	K_2
1.0	0.140	0.208
1.5	0.196	0.231
2.0	0.229	0.246
2.5	0.249	0.258
3.0	0.263	0.267
5.0	0.291	0.291

SOURCE: A. M. Wahl, *Mechanical Springs*, 2d ed., McGraw-Hill Book Company, New York, 1963.

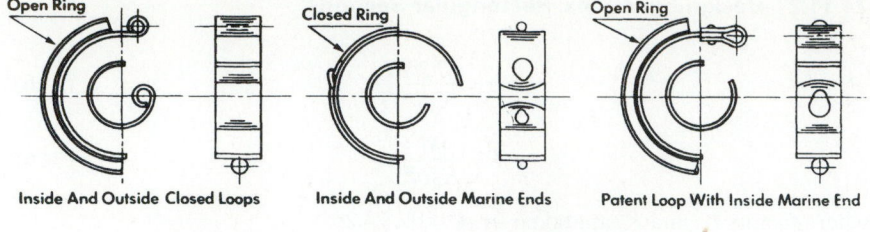

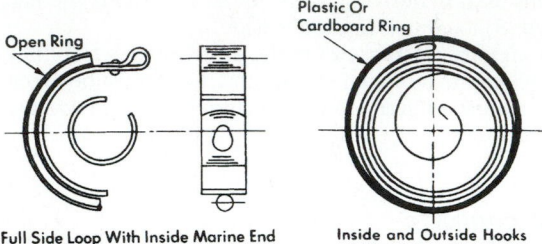

FIGURE 24.49   Typical power spring retainers and ends. *(Associated Spring, Barnes Group Inc.)*

was at rest against the case. Thus, the length of active material is constantly changing, which makes it difficult to develop a workable expression for the spring rate. For these reasons, ratios, tables, and graphical presentations are used to develop the design criteria.

The ratio of arbor diameter to thickness D_a/t is sometimes called the *life factor.* If it is too small, fatigue life will suffer. The life factor is usually maintained from 15 to 25. The ratio of active strip length to thickness L/t determines the flatness of the spring-gradient (torque-revolution) curve. The curve is flatter when L is longer. The usual range of the L/t ratio is from 5000 to 10 000. The ratio of the inside diameter of cup (case or housing) to thickness D_c/t is the *turns factor.* This determines the motion capability of the spring or indicates how much space is available between the arbor and the material lying against the inside of the case.

24.12.2   Design Procedure

In order to design a power spring that will deliver a given torque and number of turns, first determine its maximum torque in the fully wound condition. If a spring is required to deliver a minimum torque of 0.5 N·m for 10 revolutions (r) of windup and 10 r equals 80 percent unwound from solid, then from Fig. 24.50 we see that the torque at that point is 50 percent of the fully wound. Thus the fully wound torque is 1.0 N·m. Table 24.27 shows that a strip of steel 0.58 mm thick and 10 mm wide will provide 1.0 N·m of torque at the fully wound position per 10 mm of strip width.

Figure 24.51 shows that the average maximum solid stress for 0.58-mm-thick stock is about 1820 MPa. At the hardness normally supplied in steel strip for power springs, this is about 95 percent of tensile strength.

In Fig. 24.52, 10 turns relate to a length-to-thickness L/t ratio of 4300. With $t = 0.58$, L equals 2494 mm. Similarly, 4300 L/t relates to a D_c/t ratio of 107. Then $D_c = 62.06$ mm. If

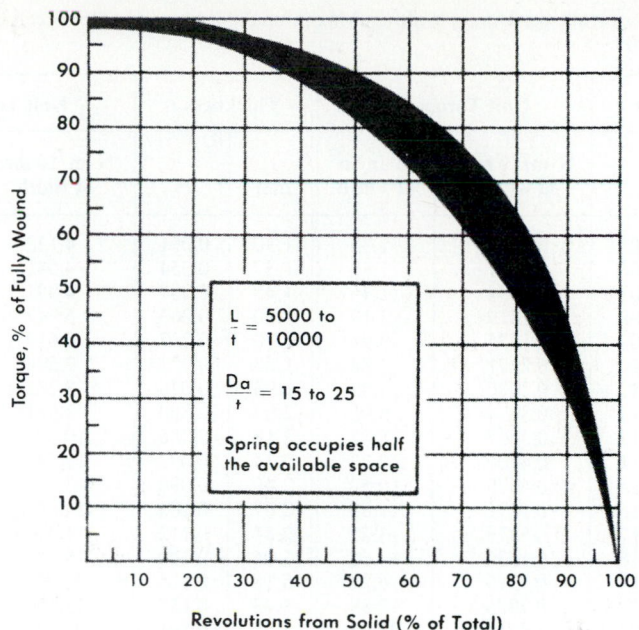

FIGURE 24.50 Typical normalized torque-revolution curve for power springs. *(Associated Spring, Barnes Group Inc.)*

$$L = \frac{D_c^2 - D_a^2}{2.55t} \tag{24.67}$$

then $D_a = \sqrt{D_c^2 - 2.55Lt} = 12.72$ mm and $D_a/t = 22$.

The equation for the number of turns a power spring will deliver, when it occupies half the space between arbor and case, is

$$\theta = \frac{\sqrt{2(D_c^2 + D_a^2)} - (D_c + D_a)}{2.55t} \tag{24.68}$$

In this example $\theta = 10$ r.

Experience shows that highly stressed power springs, made from pretempered AISI 1095 steel with a hardness of HRC 50 to 52 and stressed to 100 percent of tensile strength, could be expected to provide approximately 10 000 full-stroke life cycles. If the maximum stress were 50 percent of tensile strength at full stroke, then a life of about 100 000 cycles could be expected.

The final design is as follows:

$t = 0.023$ in (0.58 mm)
1095 carbon steel, HRC 51, no. 1 round edge
$b = 0.394$ in (10 mm)
$L = 98.188$ in (2494 mm)
$D_a = 0.501$ in (12.72 mm)
$D_c = 2.443$ in (62.06 mm)

TABLE 24.27   Torque per Unit of Width at Maximum Allowable Stress for Steel; L/t Range Is 5000 to 10 000

Thickness t		Unit Torque M		Thickness t		Unit Torque M	
mm	in	N·m/10 mm of width	lb·in/in of width	mm	in	N·m/10 mm of width	lb·in/in of width
0.127	0.005	0.0587	1.32	1.30	0.051	4.132	92.90
0.152	0.006	0.0841	1.89	1.37	0.054	4.541	102.1
0.178	0.007	0.1094	2.46	1.45	0.057	4.991	112.2
0.203	0.008	0.1419	3.19	1.60	0.063	5.947	133.7
0.229	0.009	0.1775	3.99	1.70	0.067	6.619	148.8
0.254	0.010	0.2171	4.88	1.83	0.072	7.504	168.7
0.279	0.011	0.2620	5.89	1.93	0.076	8.282	186.2
0.305	0.012	0.3074	6.91	2.03	0.080	8.981	201.9
0.330	0.013	0.3567	8.02	2.18	0.086	10.37	233.2
0.356	0.014	0.4101	9.22	2.34	0.092	11.74	264.0
0.381	0.015	0.4679	10.52	2.49	0.098	13.12	295
0.406	0.016	0.5271	11.85	2.67	0.105	14.86	334
0.432	0.017	0.5876	13.21	2.84	0.112	16.59	373
0.457	0.018	0.6530	14.68	3.05	0.120	18.82	423
0.483	0.019	0.7215	16.22	3.18	0.125	20.24	455
0.508	0.020	0.7953	17.88	3.43	0.135	23.35	525
0.584	0.023	1.025	23.05	3.58	0.141	25.35	570
0.635	0.025	1.189	26.72	3.76	0.148	27.76	624
0.711	0.028	1.452	32.65	3.96	0.156	30.69	690
0.813	0.032	1.841	41.40	4.11	0.162	33.00	742
0.889	0.035	2.144	48.20	4.50	0.177	39.23	882
1.041	0.041	2.824	63.50	4.75	0.187	43.81	985
1.19	0.047	3.585	80.60				

SOURCE: Associated Spring, Barnes Group Inc.

24.13   HOT-WOUND SPRINGS

24.13.1   Introduction

Springs are usually cold-formed when bar or wire diameters are less than 10 mm (approximately ⅜ in). When the bar diameter exceeds 16 mm (approximately ⅝ in), cold forming becomes impractical and springs are hot-wound.

Hot winding involves heating the steel into the austenitic range, winding hot, quenching to form martensite, and then tempering to the required properties. Although the most common types of hot-wound springs are compression springs for highway, off-highway, and railroad-vehicle suspension applications, torsion and extension springs can also be hot-wound.

24.13.2   Special Design Considerations

Design equations for hot-wound springs are the same as those for cold-formed springs except for the use of an empirical factor K_H which adjusts for effects related to hot-winding springs. Multiply the spring rate by K_H.

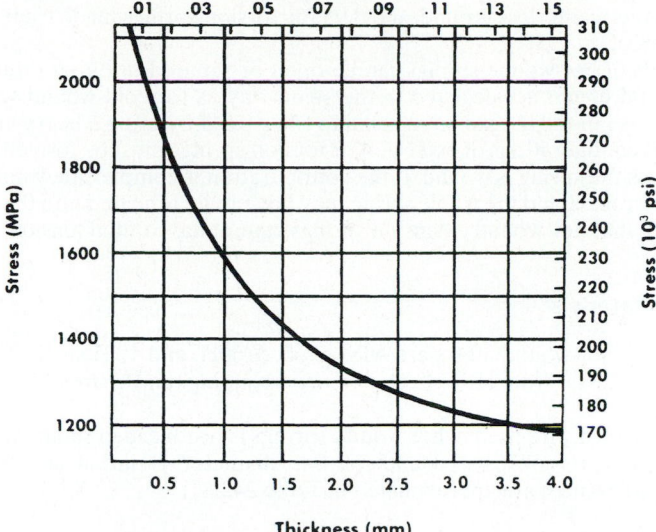

FIGURE 24.51 Average maximum solid stress in carbon-steel power springs. *(Associated Spring, Barnes Group Inc.)*

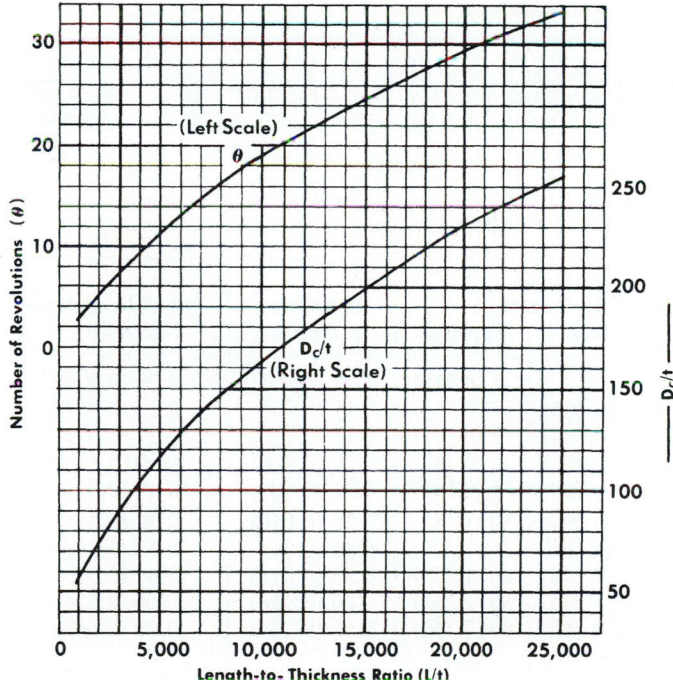

FIGURE 24.52 Relationships among number of revolutions, case diameter, strip length, and thickness for power springs. *(Associated Spring, Barnes Group Inc.)*

The values for factor K_H are 0.91 for springs made from hot-rolled carbon or low-alloy steel, *not* centerless ground; 0.96 for springs made from hot-rolled carbon or low-alloy steel, centerless ground; and 0.95 for torsion springs made from carbon or low-alloy steel.

The ends of hot-wound springs can be open or squared or either ground or not ground. Solid height is calculated in the same way as for cold-wound springs; but when space is limited, L_s can be reduced to $(N_t - 0.5)d$ by using a heavy grind.

The end configurations of extension or torsion springs must be formed hot at the same time as the spring is wound. If the configuration is complex, they may become cool in the process, and the whole spring may have to be reheated into the austenitic range. Note that hot-wound extension springs cannot have initial tension.

24.13.3   Materials

The common hot-wound alloys are AISI 5160, 5160H, and 1095 steels. The normal range of hardness is from HRC 44 to 48. Corresponding tensile strengths are 1430 to 1635 MPa.

The hot-rolled wire used in hot-wound springs is produced in standard sizes. Section 48.3 lists preferred bar diameters. Bar diameter variation and bar out-of-roundness tolerances are approximated in Table 24.28.

24.13.4   Choice of Operating Stress

Static Applications.   The stress is calculated as in cold-wound springs. Use Table 24.29 for set-point information.

Cyclic Applications.   Hot-wound springs made from hot-rolled wire are used in cyclic applications because rolled bars are subject to a variety of characteristic material defects mostly related to the bar surface condition. Therefore Table 24.30 can be

TABLE 24.28   Diameter and Out-of-Roundness Tolerances for Hot-Rolled Carbon-Steel Bars

Diameter mm(in.)		Tolerance ± mm(in.)	Out-of-Roundness mm(in.)
Over	Through		
	8 (0.315)	0.13 (0.005)	0.20 (0.008)
8 (0.315)	10 (0.394)	0.15 (0.006)	0.22 (0.009)
10 (0.394)	15 (0.591)	0.18 (0.007)	0.27 (0.011)
15 (0.591)	20 (0.787)	0.20 (0.008)	0.30 (0.012)
20 (0.787)	25 (0.984)	0.23 (0.009)	0.34 (0.013)
25 (0.984)	30 (1.181)	0.25 (0.010)	0.38 (0.015)
30 (1.181)	35 (1.378)	0.30 (0.012)	0.45 (0.018)
35 (1.378)	40 (1.575)	0.35 (0.014)	0.52 (0.020)
40 (1.575)	60 (2.362)	0.40 (0.016)	0.60 (0.024)
60 (2.362)	80 (3.150)	0.60 (0.024)	0.90 (0.035)

SOURCE: Associated Spring, Barnes Group Inc.

TABLE 24.29 Maximum Allowable Torsional Stress for Hot-Wound
Helical Compression Springs in Static Applications

Before Set Removal	After Set Removal
50% of TS	65–75% of TS

Torsional stress after set removal depends on material size and
amount of set removed.
SOURCE: Associated Spring, Barnes Group Inc.

TABLE 24.30 Maximum Allowable Torsional Stress for Hot-Wound
Helical Compression Springs in Cyclic Applications

Fatigue Life (Cycles)	Percent of Tensile Strength	
	Not Shot-Peened	Shot-Peened
10^5	40	48
10^6	38	46
10^7	35	43

This information is based on **centerless ground** AISI 5160, 5160H
and 1095, HRC 44 to 48, 25 mm (1″) diameter. Set has not been
removed. Conditions are: no surging, room temperature and noncorrosive environment.

$$\text{Stress ratio in fatigue} = \frac{S \text{ Minimum}}{S \text{ Maximum}} = 0.$$

SOURCE: Associated Spring, Barnes Group Inc.

used only for centerless ground alloy bars. Practical manufacturing tolerances for
hot-wound springs can be found in ASTM A125.

REFERENCES

24.1 Leonard G. Johnson, *The Statistical Treatment of Fatigue Experiments,* Elsevier Publishing Co., New York, 1964.

24.2 R. E. Little and E. H. Jebe, *Statistical Design of Fatigue Experiments,* John Wiley & Sons,
New York, 1975.

24.3 A. M. Wahl, *Mechanical Springs,* 2d ed., McGraw-Hill, New York, 1963.

24.4 W. R. Berry, *Spring Design: A Practical Treatment,* Emmot & Co., London, 1961.

24.5 J. O. Almen and A. Laszlo, "The Uniform Section Disc Spring," *ASME Transactions,* vol.
58, no. 4, May 1936, pp. 305–314.

24.6 A. Blake, *Design of Curved Members for Machines,* Industrial Press, New York, 1966.

24.7 K. E. Bisshopp and D. C. Drucker, "Large Deflections of Cantilever Beams," *Quarterly
of Applied Mathematics,* vol. 3, no. 3, 1945, p. 272.

24.8 J. H. Maker, "Steel Springs," *Metals Handbook,* 9th ed., vol. 1, American Society for Metals, Metals Park, Ohio, 1978, pp. 283–313.

24.9 *Design and Manufacture of Volute Springs,* SAE, New York, 1945.

24.10 R. E. Peterson, *Stress Concentration Design Factors,* John Wiley & Sons, New York, 1953, p. 102.

24.11 Karl W. Maier, "Dynamic Loading of Compression Springs," *Product Engineering,* January 1954, pp. 162–167; "Dynamic Loading of Compression Springs," *Product Engineering,* March 1955, pp. 162–172; "Surge Waves in Compression Springs," *Product Engineering,* August 1957, pp. 167–174.

CHAPTER 25

LUBRICATION

A. R. Lansdown, M.Sc., Ph.D.
Director, Swansea Tribology Centre
University College of Swansea
Swansea, United Kingdom

25.1 FUNCTIONS AND TYPES OF LUBRICANT

Whenever relative movement takes place between two surfaces in contact, there will be resistance to movement. This resistance is called the *frictional force,* or simply *friction.* Where this situation exists, it is often desirable to reduce, control, or modify the friction.

Broadly speaking, any process by which the friction in a moving contact is reduced may be described as *lubrication.* Traditionally this description has presented no problems. Friction reduction was obtained by introducing a solid or liquid material, called a *lubricant,* into the contact, so that the surfaces in relative motion were separated by a film of the lubricant. Lubricants consisted of a relatively few types of material, such as natural or mineral oils, graphite, molybdenum disulfide, and talc; and the relationship between lubricants and the process of lubrication was clear and unambiguous.

Recent technological developments have confused this previously clear picture. Friction reduction may now be provided by liquids, solids, or gases or by physical or chemical modification of the surfaces themselves. Alternatively, the sliding components may be manufactured from a material which is itself designed to reduce friction or within which a lubricant has been uniformly or nonuniformly dispersed. Such systems are sometimes described as "unlubricated," but this is clearly a matter of terminology. The system may be unconventionally lubricated, but it is certainly not unlubricated.

On the other hand, lubrication may be used to modify friction but not specifically to reduce it. Certain composite brake materials may incorporate graphite or molybdenum disulfide, whose presence is designed to ensure steady or consistent levels of friction. The additives are clearly lubricants, and it would be pedantic to assert that their use in brake materials is not lubrication.

This introduction is intended only to generate an open-minded approach to the processes of lubrication and to the selection of lubricants. In practice, the vast majority of systems are still lubricated by conventional oils or greases or by equally ancient but less conventional solid lubricants. It is when some aspect of the system makes the use of these simple lubricants difficult or unsatisfactory that the wider interpretation of lubrication may offer solutions. In addition to their primary function of reducing or controlling friction, lubricants are usually expected to reduce wear and perhaps also to reduce heat or corrosion.

In terms of volume, the most important types of lubricant are still the liquids (oils) and semiliquids (greases). Solid lubricants have been rapidly increasing in importance since about 1950, especially for environmental conditions which are too severe for oils and greases. Gases can be used as lubricants in much the same way as liquids, but as is explained later, the low viscosities of gases increase the difficulties of bearing design and construction.

25.2 SELECTION OF LUBRICANT TYPE

A useful first principle in selecting a type of lubrication is to choose the simplest technique which will work satisfactorily. In very many cases this will mean inserting a small quantity of oil or grease in the component on initial assembly; this is almost never replaced or refilled. Typical examples are door locks, hinges, car-window winders, switches, clocks, and watches.

This simple system is likely to be unsatisfactory if the loads or speeds are high or if the service life is long and continuous. Then it becomes necessary to choose the lubricant with care and often to use a replenishment system.

The two main factors in selecting the type of lubricant are the speed and the load. If the speed is high, then the amount of frictional heating tends to be high, and low-viscosity lubricants will give lower viscous friction and better heat transfer. If the loads are high, then low-viscosity lubricants will tend to be expelled from the contact. This situation is summarized in Fig. 25.1.

It is difficult to give precise guidance about the load and speed limits for the various lubricant types, because of the effects of geometry, environment, and variations within each type, but Fig. 25.2 gives some approximate limits.

Some other property of the system will sometimes restrict the choice of lubricant type. For example, in watches or instrument mechanisms, any lubricant type could meet the load and speed requirements, but because of the need for low friction, it is normal to use a very low-viscosity oil. However, for open gears, wire ropes, or chains, the major problem is to prevent the lubricant from being thrown off the moving parts, and

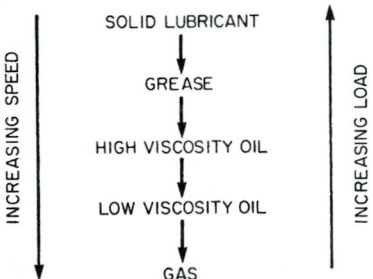

FIGURE 25.1 Effect of speed and load on choice of lubricant type. *(From Ref. [25.1].)*

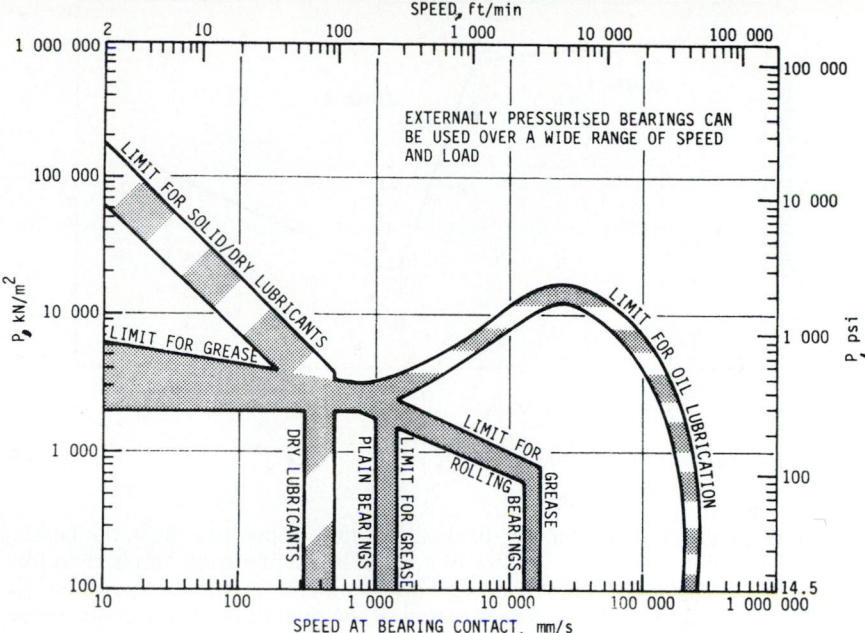

FIGURE 25.2 Speed and load limitations for different types of lubricants. *(From Ref. [25.2].)*

it is necessary to use a "tacky" bituminous oil or grease having special adhesive properties.

In an existing system the geometry may restrict the choice of lubricant type. Thus, an unsealed rolling bearing may have to be lubricated with grease because oil would not be retained in the bearing. But where the lubrication requirements are difficult or particularly important, it will usually be essential to first choose the lubricant type and then design a suitable system for that lubricant. Some very expensive mistakes have been made, even in high technology such as aerospace engineering, where systems that could not be lubricated have been designed and built.

25.3 LIQUID LUBRICANTS: PRINCIPLES AND REQUIREMENTS

The most important single property of a liquid lubricant is its viscosity. Figure 25.3 shows how the viscosity of the lubricant affects the nature and quality of the lubrication. This figure is often called a *Stribeck curve,* although there seems to be some doubt as to whether Stribeck used the diagram in the form shown.

The expression $\eta N/P$ is known as the *Sommerfeld number,* in which η is the lubricant viscosity, N represents the relative speed of movement between the counterfaces of the bearing, and P is the mean pressure or specific load supported by the bearing. Of these three factors, only the viscosity is a property of the lubricant. And if N and P are held constant, the figure shows directly the relationship between the coefficient of friction μ and the lubricant viscosity η.

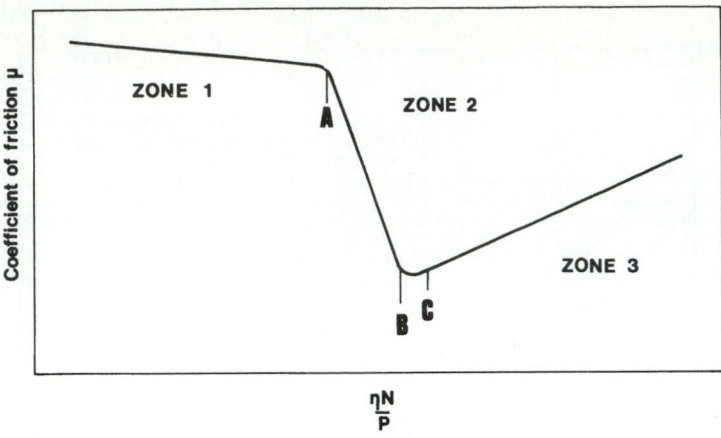

FIGURE 25.3   Effect of viscosity on lubrication.

The graph can be conveniently divided into three zones. In zone 3, the bearing surfaces are fully separated by a thick film of the liquid lubricant. This is, therefore, the zone of *thick-film* or *hydrodynamic lubrication,* and the friction is entirely viscous friction caused by mechanical shearing of the liquid film. There is no contact between the interacting surfaces and therefore virtually no wear.

As the viscosity decreases in zone 3, the thickness of the liquid film also decreases until at point C it is only just sufficient to ensure complete separation of the surfaces. Further reduction in viscosity, and therefore in film thickness, results in occasional contact between asperities on the surfaces. The relatively high friction in asperity contacts offsets the continuing reduction in viscous friction, so that at point B the friction is roughly equal to that at C.

Point C is the ideal point, at which there is zero wear with almost minimum friction, but in practice the design target will be slightly to the right of C, to provide a safety margin.

With further reduction in viscosity from point B, an increasing proportion of the load is carried by asperity contact, and the friction increases rapidly to point A. At this point the whole of the bearing load is being carried by asperity contact, and further viscosity reduction has only a very slight effect on friction.

Zone 1, to the left of point A, is the zone of *boundary lubrication.* In this zone, chemical and physical properties of the lubricant other than its bulk viscosity control the quality of the lubrication; these properties are described in Sec. 25.5.

Zone 2, between points A and B, is the zone of mixed lubrication, in which the load is carried partly by the film of liquid lubricant and partly by asperity interaction. The proportion carried by asperity interaction decreases from 100 percent at A to 0 percent at C.

Strictly speaking, Fig. 25.3 relates to a plain journal bearing, and N usually refers to the rotational speed. Similar patterns arise with other bearing geometries in which some form of hydrodynamic oil film can occur.

The relationship between viscosity and oil-film thickness is given by the Reynolds equation, which can be written as follows:

$$\frac{\partial}{\partial x}\left(h^3 \frac{\partial P}{\partial x}\right) + \frac{\partial}{\partial z}\left(h^3 \frac{\partial P}{\partial z}\right) = \eta \left(6U \frac{\partial h}{\partial x} + 6h \frac{\partial U}{\partial x} + 12V\right)$$

where h = lubricant-film thickness
$\quad\quad P$ = pressure
$\quad\quad x, z$ = coordinates
$\quad\quad U, V$ = speeds in directions x and z

Fuller details of the influence of lubricant viscosity on plain journal bearings are given in Chap. 28.

In nonconformal lubricated systems such as rolling bearings and gears, the relationship between lubricant viscosity and film thickness is complicated by two additional effects: the elastic deformation of the interacting surfaces and the increase in lubricant viscosity as a result of high pressure. The lubrication regime is then known as *elastohydrodynamic* and is described mathematically by various equations.

For roller bearings, a typical equation is the Dowson-Higginson equation:

$$h_{\min} = \frac{2.65(\eta_o U)^{0.7} R^{0.43} \alpha^{0.54}}{E^{0.03} p^{0.13}}$$

where η_o = oil viscosity in entry zone
$\quad\quad R$ = effective radius
$\quad\quad \alpha$ = pressure coefficient of viscosity

Here U represents the speed, p a load parameter, and E a material parameter based on modulus and Poisson's ratio.

For ball bearings, an equivalent equation is the one developed by Archard and Cowking:

$$h_{\min} = \frac{1.4(\eta_o U \alpha)^{0.74} E^{0.074}}{R^{0.74} p^{0.074}}$$

For such nonconformal systems, a diagram similar to Fig. 25.3 has been suggested in which zone 2 represents elastohydrodynamic lubrication. It is difficult to think of a specific system to which the relationship exactly applies, but it may be a useful concept that the lubricant-film thickness and the friction in elastohydrodynamic lubrication bridge the gap between thick-film hydrodynamic lubrication and boundary lubrication.

A form of microelastohydrodynamic lubrication has been suggested as a mechanism for asperity lubrication under boundary conditions (see Sec. 25.5). If this suggestion is valid, the process would probably be present in the zone of mixed lubrication.

Where full-fluid-film lubrication is considered necessary but the viscosity, load, speed, and geometry are not suitable for providing full-fluid-film separation hydrodynamically, the technique of *external pressurization* can be used. Quite simply, this means feeding a fluid into a bearing at high pressure, so that the applied hydrostatic pressure is sufficient to separate the interacting surfaces of the bearing.

Externally pressurized bearings broaden the range of systems in which the benefits of full-fluid-film separation can be obtained and enable many liquids to be used successfully as lubricants which would otherwise be unsuitable. These include aqueous and other low-viscosity process fluids. Remember that the lubricant viscosity considered in Fig. 25.3 and in the various film-thickness equations is the viscosity under the relevant system conditions, especially the temperature. The viscosity of all liquids decreases with increase in temperature, and this and other factors affecting viscosity are considered in Sec. 25.4.

The viscosity and boundary lubrication properties of the lubricant completely define the lubrication performance, but many other properties are important in service. Most of these other properties are related to progressive deterioration of the lubricant; these are described in Sec. 25.6.

25.4 LUBRICANT VISCOSITY

Viscosity of lubricants is defined in two different ways, and unfortunately both definitions are very widely used.

25.4.1 Dynamic or Absolute Viscosity

Dynamic or *absolute viscosity* is the ratio of the shear stress to the resultant shear rate when a fluid flows. In SI units it is measured in pascal-seconds or newton-seconds per square meter, but the centimeter-gram-second (cgs) unit, the centipoise, is more widely accepted, and

$$1 \text{ centipoise (cP)} = 10^{-3} \text{ Pa} \cdot \text{s} = 10^{-3} \text{ N} \cdot \text{s/m}^2$$

The centipoise is the unit of viscosity used in calculations based on the Reynolds equation and the various elastohydrodynamic lubrication equations.

25.4.2 Kinematic Viscosity

The *kinematic viscosity* is equal to the dynamic viscosity divided by the density. The SI unit is square meters per second, but the cgs unit, the centistoke, is more widely accepted, and

$$1 \text{ centistoke (cSt)} = 1 \text{ mm}^2/\text{s}$$

The centistoke is the unit most often quoted by lubricant suppliers and users.

In practice, the difference between kinematic and dynamic viscosities is not often of major importance for lubricating oils, because their densities at operating temperatures usually lie between 0.8 and 1.2. However, for some fluorinated synthetic oils with high densities, and for gases, the difference can be very significant.

The viscosities of most lubricating oils are between 10 and about 600 cSt at the operating temperature, with a median figure of about 90 cSt. Lower viscosities are more applicable for bearings than for gears, as well as where the loads are light, the speeds are high, or the system is fully enclosed. Conversely, higher viscosities are selected for gears and where the speeds are low, the loads are high, or the system is well ventilated. Some typical viscosity ranges at the operating temperatures are shown in Table 25.1.

The variation of oil viscosity with temperature will be very important in some systems, where the operating temperature either varies over a wide range or is very different from the reference temperature for which the oil viscosity is quoted.

The viscosity of any liquid decreases as the temperature increases, but the rate of decrease can vary considerably from one liquid to another. Figure 25.4 shows the

TABLE 25.1   Typical Operating Viscosity Ranges

Lubricant	Viscosity range, cSt
Clocks and instrument oils	5–20
Motor oils	10–50
Roller bearing oils	10–300
Plain bearing oils	20–1500
Medium-speed gear oils	50–150
Hypoid gear oils	50–600
Worm gear oils	200–1000

change of viscosity with temperature for some typical lubricating oils. A graphical presentation of this type is the most useful way to show this information, but it is much more common to quote the viscosity index (VI).

The *viscosity index* defines the viscosity-temperature relationship of an oil on an arbitrary scale in comparison with two standard oils. One of these standard oils has

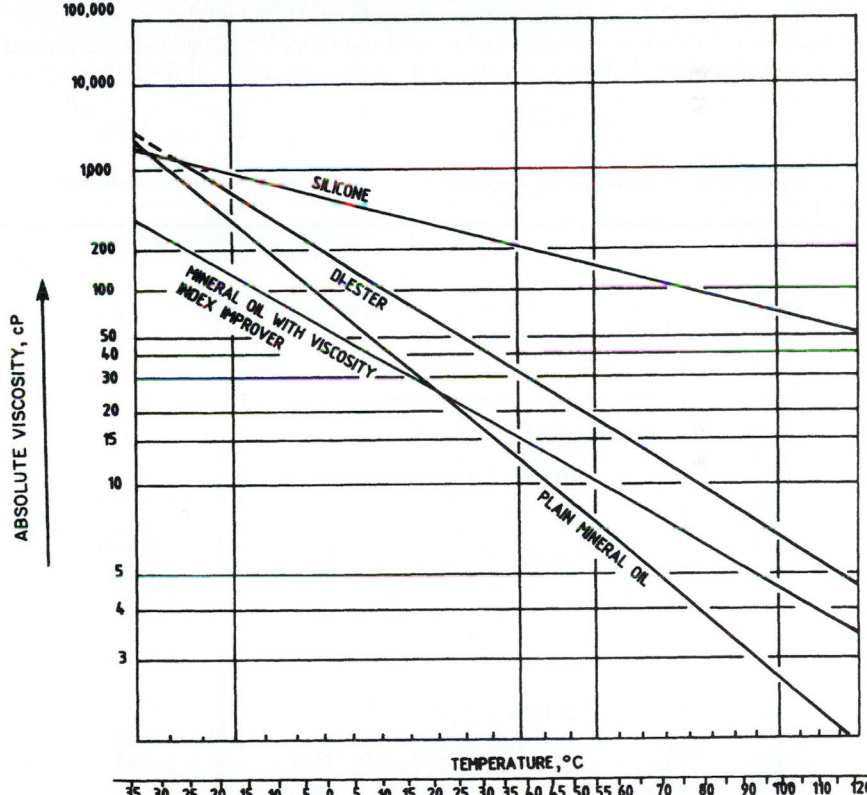

FIGURE 25.4   Variation of viscosity with temperature.

a viscosity index of 0, representing the most rapid change of viscosity with temperature normally found with any mineral oil. The second standard oil has a viscosity index of 100, representing the lowest change of viscosity with temperature found with a mineral oil in the absence of relevant additives.

The equation for the calculation of the viscosity index of an oil sample is

$$VI = \frac{100(L - U)}{L - H}$$

where U = viscosity of sample in centistokes at 40°C, L = viscosity in centistokes at 40°C of oil of 0 VI having the same viscosity at 100°C as the test oil, and H = viscosity at 40°C of oil of 100 VI having the same viscosity at 100°C as the test oil.

Some synthetic oils can have viscosity indices of well over 150 by the above definition, but the applicability of the definition at such high values is doubtful. The viscosity index of an oil can be increased by dissolving in it a quantity (sometimes as high as 20 percent) of a suitable polymer, called a *viscosity index improver.*

The SAE viscosity rating scale is very widely used and is reproduced in Table 25.2. It is possible for an oil to satisfy more than one rating. A mineral oil of high viscosity index could meet the 20W and 30 criteria and would then be called a 20W/30 multigrade oil. More commonly, a VI improved oil could meet the 20W and 50 criteria and would then be called a 20W/50 multigrade oil.

Note that the viscosity measurements used to establish SAE ratings are carried out at low shear rate. At high shear rate in a bearing, the effect of the polymer may

TABLE 25.2 1977 Table of SAE Oil Ratings

SAE no.	Maximum viscosity at −18°C, cP	Viscosity at 100°C, cSt	
		Minimum	Maximum
Engine oils			
5W	1 250	3.8	
10W	2 500	4.1	
20W†	10 000	5.6	
20	.....	5.6	<9.3
30	.....	9.3	<12.5
40	.....	12.5	<16.3
50	.....	16.3	<21.9
Gear oils			
75	3 250		
80	21 600		
90	.....	14	<25
140	.....	25	<43
250	.....	43	

†15W may be used to identify 20W oils which have a maximum viscosity of 5000 cP.

disappear, and a 20W/50 oil at very high shear rate may behave as a thinner oil than a 20W, namely, a 15W or even 10W. In practice, this may not be important, because in a high-speed bearing the viscosity will probably still produce adequate oil-film thickness.

Theoretically the viscosity index is important only where significant temperature variations apply, but in fact there is a tendency to use only high-viscosity-index oils in the manufacture of high-quality lubricant. As a result, a high viscosity index is often considered a criterion of lubricant quality, even where viscosity index as such is of little or no importance.

Before we leave the subject of lubricant viscosity, perhaps some obsolescent viscosity units should be mentioned. These are the *Saybolt viscosity* (SUS) in North America, the *Redwood viscosity* in the United Kingdom, and the *Engler viscosity* in continental Europe. All three are of little practical utility, but have been very widely used, and strenuous efforts have been made by standardizing organizations for many years to replace them entirely by kinematic viscosity.

25.5 *BOUNDARY LUBRICATION*

Boundary lubrication is important where there is significant solid-solid contact between sliding surfaces. To understand boundary lubrication, it is useful to first consider what happens when two metal surfaces slide against each other with no lubricant present.

In an extreme case, where the metal surfaces are not contaminated by an oxide film or any other foreign substance, there will be a tendency for the surfaces to adhere to each other. This tendency will be very strong for some pairs of metals and weaker for others. A few guidelines for common metals are as follows:

1. Identical metals in contact have a strong tendency to adhere.

2. Softer metals have a stronger tendency to adhere than harder metals.

3. Nonmetallic alloying elements tend to reduce adhesion (e.g., carbon in cast iron).

4. Iron and its alloys have a low tendency to adhere to lead, silver, tin, cadmium, and copper and a high tendency to adhere to aluminum, zinc, titanium, and nickel.

Real metal surfaces are usually contaminated, especially by films of their own oxides. Such contaminant films commonly reduce adhesion and thus reduce friction and wear. Oxide films are particularly good lubricants, except for titanium.

Thus friction and wear can usually be reduced by deliberately generating suitable contaminant films on metallic surfaces. Where no liquid lubricant is present, such a process is a type of dry or solid lubrication. Where the film-forming process takes place in a liquid lubricant, it is called boundary lubrication.

Boundary lubricating films can be produced in several ways, which differ in the severity of the film-forming process and in the effectiveness of the resulting film. The mildest film-forming process is adsorption, in which a layer one or more molecules thick is formed on a solid surface by purely physical attraction. Adsorbed films are effective in reducing friction and wear, provided that the resulting film is sufficiently thick. Figure 25.5 shows diagrammatically the way in which adsorption of a long-chain alcohol generates a thick film on a metal surface even when the film is only one molecule thick.

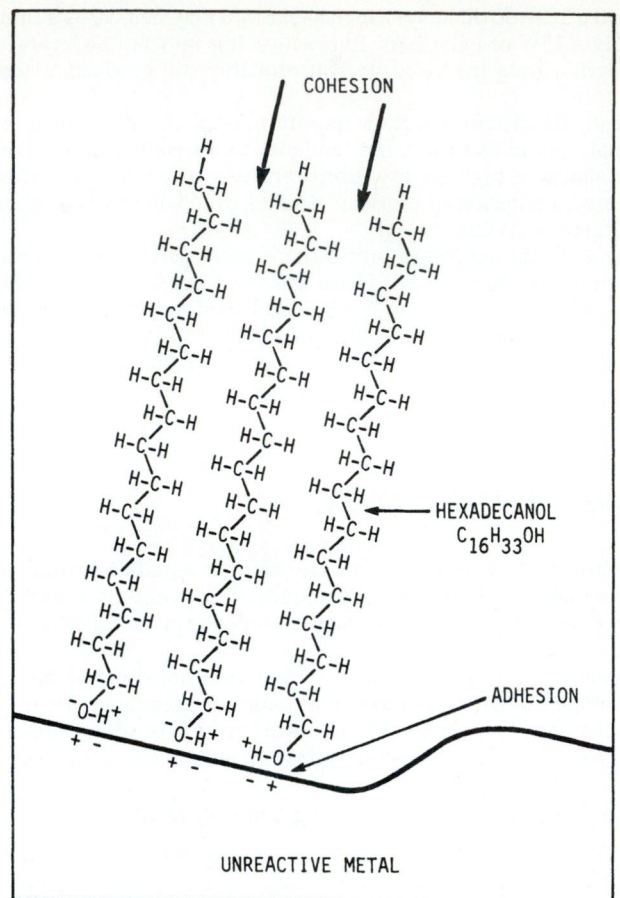

FIGURE 25.5   Representation of adsorption of a long-chain alcohol.
(From Ref. [25.3].)

Mineral oils often contain small amounts of natural compounds which produce useful adsorbed films. These compounds include unsaturated hydrocarbons (olefines) and nonhydrocarbons containing oxygen, nitrogen, or sulfur atoms (known as asphaltenes). Vegetable oils and animal fats also produce strong adsorbed films and may be added in small concentrations to mineral oils for that reason. Other mild boundary additives include long-chain alcohols such as lauryl alcohol and esters such as ethyl stearate or ethyl oleate.

Adsorbed boundary films are removed fairly easily, either mechanically or by increased temperature. A more resistant film is generated by chemisorption, in which a mild reaction takes place between the metal surface and a suitable compound. Typical chemisorbed compounds include aliphatic ("fatty") acids, such as oleic and stearic acids. A chemisorbed film is shown diagrammatically in Fig. 25.6.

Even more resistant films are produced by reaction with the metal surface. The reactive compounds usually contain phosphorus, sulfur, or chlorine and ultimately

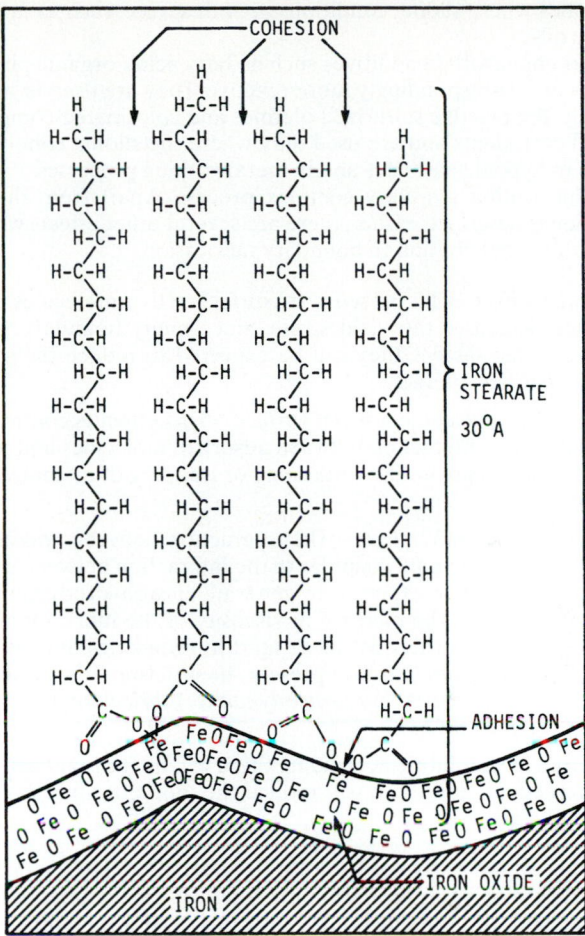

FIGURE 25.6 Representation of chemisorption of a long-chain aliphatic acid. *(From Ref. [25.3].)*

produce films of metal phosphide, sulfide, or chloride on the sliding surface. These reactive additives are known as *extreme-pressure*, or EP, additives.

The processes by which modern boundary lubricant additives generate surface films may be very complex. A single additive such as trixylyl phosphate may be initially adsorbed on the metal surface, then react to form a chemisorbed film of organometallic phosphate, and finally, under severe sliding or heating, react to form metal phosphate or phosphide.

All these boundary lubricant compounds have corresponding disadvantages. As a general rule, they should be used only where the conditions of use require them. The mild, adsorbed compounds have the least undesirable side effects. They are more readily oxidized than the usual mineral-base oils and, as a result, have a higher tendency to produce corrosive acidic compounds and insoluble gums or lacquers. However, these effects are not serious, and mild antiwear additives are widely used in small quantities where sliding conditions are not severe, such as in hydraulic fluids and turbine oils.

The stronger chemisorbed additives such as fatty acids, organic phosphates, and thiophosphates are correspondingly more reactive. They are used in motor oils and gear oils. Finally, the reactive sulfurized olefines and chlorinated compounds are, in fact, controlled corrodents and are used only where the sliding conditions are very severe, such as in hypoid gearboxes and in metalworking processes.

Boundary lubrication is a very complex process. Apart from the direct film-forming techniques described earlier, there are several other effects which probably make an important contribution to boundary lubrication:

1. *The Rehbinder effect*   The presence of surface-active molecules adjacent to a metal surface decreases the yield stress. Since many boundary lubricants are more or less surface-active, they can be expected to reduce the stresses developed when asperities interact.

2. *Viscosity increase adjacent to a metal surface*   This effect is controversial, but it seems probable that interaction between adsorbed molecules and the free ambient oil can result in a greaselike thickening or trapping of oil molecules adjacent to the surface.

3. *Microelastohydrodynamic effects*   The interaction between two asperities sliding past each other in a liquid is similar to the interaction between gear teeth, and in the same way it can be expected to generate elastohydrodynamic lubrication on a microscopic scale. The increase in viscosity of the lubricant and the elastic deformation of the asperities will both tend to reduce friction and wear. However, if the Rehbinder effect is also present, then plastic flow of the asperities is also encouraged. The term *microrheodynamic lubrication* has been used to describe this complex process.

4. *Heating*   Even in well-lubricated sliding there will be transient heating effects at asperity interactions, and these will reduce the modulus and the yield stress at asperity interactions.

Boundary lubrication as a whole is not well understood, but the magnitude of its beneficial effects can be easily seen from the significant reductions in friction, wear, and seizure obtained with suitable liquid lubricants in slow metallic sliding.

25.6   DETERIORATION PROBLEMS

In theory, if the right viscosity and the right boundary properties have been selected, then the lubrication requirements will be met. In practice, there is one further complication—the oil deteriorates. Much of the technology of lubricating oils and additives is concerned with reducing or compensating for deterioration.

The three important types of deterioration are oxidation, thermal decomposition, and contamination. A fourth long-term effect is reaction with other materials in the system, which is considered in terms of compatibility. Oxidation is the most important deterioration process because over a long period, even at normal atmospheric temperature, almost all lubricants show some degree of oxidation.

Petroleum-base oils produced by mild refining techniques oxidize readily above 120°C to produce acidic compounds, sludges, and lacquers. The total oxygen uptake is not high, and this suggests that the trace compounds, such as aromatics and asphaltenes, are reacting, and that possibly in doing so some are acting as oxidation inhibitors for the paraffinic hydrocarbons present. Such mildly refined oils are not much improved by the addition of antioxidants.

More severe refining or hydrogenation produces a more highly paraffinic oil which absorbs oxygen more readily but without producing such harmful oxidation products. More important, however, the oxidation resistance of such highly refined base oils is very considerably improved by the addition of suitable oxidation inhibitors.

Most modern petroleum-base oils are highly refined in order to give consistent products with a wide operating-temperature range. Antioxidants are therefore an important part of the formulation of almost all modern mineral-oil lubricants.

The commonly used antioxidants are amines, hindered phenols, organic phosphites, and organometallic compounds. One particularly important additive is zinc diethyl dithiophosphate, which is a very effective antioxidant and also has useful boundary lubrication and corrosion-inhibition properties.

If no oxygen is present, lubricants can be used at much higher temperatures without breaking down. In other words, their thermal stability is greater than their oxidative stability. This effect can be seen for mineral oils in Table 25.3. To prevent contact of oxygen with the oil, the system must be sealed against the entry of air or purged with an inert gas such as nitrogen. Some critical hydraulic systems, such as those in high-speed aircraft, are operated in this way.

In high-vacuum systems such as spacecraft or electron microscopes, there is no oxygen contact. But in high vacuum an increase in temperature tends to vaporize the

TABLE 25.3 Range of Temperature Limits in Degrees Celsius for Mineral Oils as a Function of Required Life

Oil condition	Life, h				
	1	10	10^2	10^3	10^4
Thermal stability limit; insignificant oxygen present	415 to 435	385 to 405	355 to 375	320 to 340	290 to 310
Limit dependent on amount of oxygen present and presence or absence of catalysts	190 to 415	170 to 385	140 to 355	155 to 320	90 to 290
Limit imposed by oxidation where oxygen supply is unlimited; for oils containing antioxidants	175 to 190	155 to 170	125 to 140	100 to 115	80 to 90
Limit imposed by oxidation where oxygen supply is unlimited; for oils without antioxidants	155 to 165	130 to 140	95 to 110	65 to 80	35 to 50
Lower temperature limit imposed by pour point; varies with oil source, viscosity, treatment, and additives	−65 to 0	−65 to 0	−65 to 0	−65 to 0	−65 to 0

SOURCE: Ref. [25.2].

oil, so that high thermal stability is of little or no value. It follows that oxidative stability is usually much more important than thermal stability.

Compatibility of lubricating oils with other materials in the system is complex, and Table 25.4 lists some of the possible problems and solutions. Compatibility problems with synthetic lubricants are even more complicated; these are considered further in the next section.

25.7 SELECTING THE OIL TYPE

So far most of the information in this chapter has been related to mineral oils. For almost 150 years the availability, good performance, variety, and cheapness of mineral oils have made them the first choice for most applications. They still represent over 90 percent of total lubricant use, but many other liquids are used successfully as lubricants and can provide special features which make them the best choice in particular situations.

Table 25.5 shows the most important types of lubricating oil and their advantages and disadvantages as compared with mineral oils. The natural oils comprise a wide variety of compounds of vegetable or animal origin, consisting mainly of organic esters. They all have better low-friction and boundary lubrication properties than mineral oils, but lower thermal and oxidative stability. Before mineral oils became generally available, natural oils and fats were the most common lubricants, and several are still widely used because their properties make them particularly suitable for special applications, as shown in Table 25.6.

The diesters were the first synthetic lubricating oils to be used in large quantities. Their higher thermal and oxidative stability made them more suitable than mineral

TABLE 25.4 Examples of Compatibility Problems and Possible Solutions

Problem	Solution
1. Attack by mineral oils on natural rubber	Change to nitrile rubber or neoprene
2. Attack by synthetic oils on natural rubber, nitrile, or other rubber	Change to suitable rubber for specific oil, e.g., Viton, resin-cured butyl, or EPR
3. Attack by synthetic oils on plastics or paints	Change to resistant plastics such as PTFE, polyimide, polysulfone, or polyphenylene sulfide
4. Corrosion by dissolved water	Use rust-inhibitor additives such as sulfonates
5. Corrosion by acidic degradation products	Use corrosion inhibitors such as ZDDP, or increase antioxidants to reduce degradation
6. Corrosion by additives of copper alloys or mild steel	Use less powerful EP additives, or change to corrosion-resistant metals
7. Corrosion by synthetic oils	Change to more resistant metals or platings

TABLE 25.5 Advantages and Disadvantages of Main Nonmineral Oils

Oil type	Comparison with mineral oils	
	Advantages	Disadvantages
1. Vegetable oil	Good boundary lubrication; does not cause carburization of steel in metalforming	Decomposes readily to give high viscosity or sludges and lacquers
2. Diesters, hindered esters	Higher temperature stability; high viscosity index	Some attack on rubbers and plastics
3. Polyglycol	Miscibility with water; decomposes without producing solid degradation products	Low maximum temperature
4. Silicones	High temperature stability; resistance to chemicals	Poor boundary lubrication for steel on steel
5. Phosphate ester	Fire resistance; very good boundary lubrication	Attack on rubbers and plastics; poor temperature stability
6. Chlorinated diphenyls	Fire resistance; chemical stability; boundary lubrication	Poor viscosity index; attack on plastics and copper alloys
7. Fluorocarbon	Excellent temperature and chemical stability	Price; poor viscosity index

TABLE 25.6 Some Uses of Natural Oils and Fats

Oil type	Uses
1. Rapeseed oil	a. To reduce friction in plain bearings where oil-film thickness is inadequate by addition of 5% to 10% to mineral oil
	b. In metal forming to give low friction and EP properties without staining or carburizing
	c. Has been used as lubricant in continuous casting
2. Castor oil	a. As low-viscosity hydraulic fluid for compatibility with natural rubber
	b. To give low viscous drag and good boundary lubrication in racing car engines and early aircraft engines
3. Tallow	a. For low friction in metal forming
4. Sperm oil	a. For outstanding boundary lubrication in metal cutting especially in sulfurized form; now virtually obsolete because of whale protection laws

oils for gas-turbine lubrication, and by about 1960 they were almost universally used for aircraft jet engines. For the even more demanding conditions of supersonic jet engines, the more complex ester lubricants such as hindered phenols and triesters were developed.

Phosphate esters and chlorinated diphenyls have very low-flammability characteristics, and this has led to their wide use where critical fire-risk situations occur, such as in aviation and coal mining. Their overall properties are mediocre, but are sufficiently good for use where fire resistance is particularly important.

Other synthetic fluids such as silicones, chlorinated silicones, fluorinated silicones, fluorinated hydrocarbon, and polyphenyl ethers are all used in relatively small quantities for their high-temperature stability, but all are inferior lubricants and very expensive compared with mineral oils.

Several types of water-containing fluid are used in large quantities, and these are listed in Table 25.7. They are used almost entirely to provide either fire resistance or superior cooling.

Mineral oils can be considered as the normal, conventional oils, and alternative types are used only when they can offer some particular advantage over mineral oils. Table 25.8 summarizes the selection of oil type in relation to the special properties required.

It is difficult to give precise high-temperature limits for the use of specific oil types, because the limiting temperature depends on the required life and the amount of degradation which is acceptable. Even for water-containing lubricants, the upper temperature limit may be from 50 to 85°C depending on the required life, the degree of ventilation, and the amount of water loss which is acceptable. Table 25.9 summarizes the temperature limits for a few synthetic oils, but the limits shown should be considered only approximate.

Serious incompatibility problems can occur with lubricating oils, especially with nonmetallic materials such as rubber seals and hoses. Table 25.10 lists some satisfactory and unsatisfactory materials for use with various lubricants.

TABLE 25.7 Some Water-Containing Lubricants

Oil type	Applications
1. Invert emulsions (water in mineral oil)	Used as hydraulic fluids for fire resistance, e.g., in coal mining. Good lubricating properties.
2. Dilute emulsions (5% mineral oil in water)	Used for fire resistance and cheapness where good lubrication properties not needed (e.g., roof jacks in coal mining).
3. "Soluble" oils (about 1% oil in water)	Used for their good cooling properties in metal cutting and grinding operations.
4. Water/Polyglycol	Used for fire resistance where increased viscosity and lack of solid degradation products are required.
5. "Synthetic" Coolants (solutions of boundary additives in water)	Used for excellent cooling and stability in metal cutting operations.

TABLE 25.8 Choice of Oil Type for Specific Properties

Property required	Choice of oil type
1. Wide range of viscosities	Mineral oil; silicone; polyglycol
2. Good boundary lubrication	Natural oil or fat; mineral oil with suitable additives; ester; phosphate ester
3. Long life	Mineral oil; silicone; fluorocarbon; ester; polyphenyl ether
4. High temperature stability	Polyphenyl ether; fluorocarbon; silicone; ester
5. Fire resistance	Emulsions; fluorocarbon; fluorosilicone; chlorinated biphenyl; phosphate ester
6. Cheapness	Emulsions; mineral oil

25.8 LUBRICATING GREASES

Lubricating greases are not simply very viscous oils. They consist of lubricating oils, often of quite low viscosity, which have been thickened by means of finely dispersed solids called thickeners. The effect of the thickeners is to produce a semirigid structure in which the dispersion of thickener particles is stabilized by electric charges. The liquid phase is firmly held by a combination of opposite electric charges, adsorp-

TABLE 25.9 Range of Temperature Limits in Degrees Celsius for Some Synthetic Oils as a Function of the Required Life

Name of lubricant; type of limit	Life, h				
	1	10	10^2	10^3	10^4
Polyphenyl ethers; thermal stability limit	545	520	490	455	425
Polyphenyl ethers; oxidation limit	350	330	305	280	260
Silicones; thermal stability limit	280 to 290	260 to 275	240 to 260	220 to 245	200 to 230
Esters and silicones; oxidation limit	225 to 260	215 to 245	200 to 240	185 to 220	175 to 210
Phosphate esters; thermal and oxidative limit	160	145	130	110	100
Polyphenal ethers; pour-point limit	0	0	0	0	0
Silicones and esters; pour-point limit	−60	−60	−60	−60	−60

SOURCE: Ref. [25.2].

TABLE 25.10 Some Compatible and Incompatible Materials for Different Oil Types

Oil type	Rubbers and plastics	
	Satisfactory	Unsatisfactory
1. Natural oils	Most rubbers, including natural rubber; most plastics	SBR rubber; highly plasticized polyethylene and polypropylene
2. Mineral oil	Nitrile rubber; neofrene; Viton; EPR; most unplasticized plastics	Natural rubber; SBR; highly plasticized plastics; polyurethanes
3. Esters	High nitrile; Viton; nylons; PPS; polyethersulfones	Natural rubber; SBR; low nitrile; polyacrylates; polyurethanes
4. Silicones	High nitrile; Viton; nylons; PPS	Natural rubber; silicone rubber; plasticized plastics
5. Phosphate ester	Resin-cured butyl rubber; EPR; PPS	Most other rubbers; many plastics

tion, and mechanical trapping. As a result, the whole grease behaves as a more or less soft solid, and there is only a very slight tendency for the oil to flow out of the grease.

Greases can probably be made from any type of lubricating oil, but in practice the majority are based on mineral oils, and only a few other base oils are of any real importance. Diesters have been used to produce greases for higher and lower temperatures than greases based on mineral oils are suitable for. Silicones are used for higher temperatures again, and fluorinated hydrocarbons for even higher temperatures; both these types are also used because of their chemical inertness, but the total quantities are relatively small. Phosphate esters have been used for fire resistance, and vegetable oils for compatibility with foodstuffs; but, again, the quantities are very small.

The most commonly used thickeners are soaps, which are salts of organic acids with calcium, sodium, lithium, or aluminum. The soaps take the form of fibrous particles which interlock to give a high level of stiffness at low soap concentrations. Many other substances which have been used as grease thickeners tend to be more spherical and have to be used at higher concentrations than soaps to achieve the same degree of thickening.

Most of the additives used in lubricating oils are also effective in greases. And some, such as the solid lubricants graphite and molybdenum disulfide, are much more effective in greases than in oils.

Table 25.11 lists some of the many different components which may be used in greases. The possible combinations of these components, and their different proportions, lead to an infinite range of grease formulations. In practice, a typical grease consists of a mineral oil in which are dispersed about 10 percent of a soap thickener, about 1 percent of antioxidant, and small amounts of other additives such as corrosion inhibitors, antiwear or extreme-pressure agents, and structure modifiers.

The most important physical characteristic of a grease is its relative hardness or softness, which is called *consistency*. Consistency is assessed by measuring the dis-

TABLE 25.11 Some Components Used in Grease Manufacture

Base oils	Thickeners	Additives
Mineral oils	Sodium soap	Antioxidants
Silicones	Lithium soap	EP additives
Diesters	Aluminum soap	Corrosion inhibitors
Chlorinated silicone	Lithium complex	Metal deactivators
Fluorocarbons	Aluminum complex	Tackiness additives
Phosphate esters	Bentonite clay	Water repellants
	PTFE	Structure modifiers
	Indanthrene dye	

tance in tenths of a millimeter to which a standard metal cone penetrates the grease under a standard load; the result is known as the *penetration.* A widely used classification of greases is that of the American National Lubricating Grease Institute (NLGI), and Table 25.12 shows the relationship between NLGI number and penetration.

TABLE 25.12 NLGI Grease Classification

NLGI number	Worked penetration at 25°C
000	445–475
00	400–430
0	355–385
1	310–340
2	265–295
3	220–250
4	175–205
5	130–160
6	85–115

The consistency of a grease varies with temperature, and there is generally an irregular increase in penetration (softening) as the temperature increases. Eventually a temperature is reached at which the grease is soft enough for a drop to fall away or flow from the bulk of the grease; this is called the *drop point.* The drop point is usually taken to be the maximum temperature at which the grease can be used in service, but several factors confuse this situation:

1. The drop point is measured in a standard apparatus which bears no resemblance to any service equipment, so that the correlation with service use may be poor.
2. Some greases will never give a drop point because chemical decomposition begins before the thickener structure breaks down.
3. A grease may be a satisfactory lubricant above its drop point, although then it will behave like an oil rather than a grease.

4. Some greases can be heated above their drop points and will again form a grease when cooled, although normally the re-formed grease will be markedly inferior in properties.

At high temperature greases will decompose thermally or oxidatively in the same way as lubricating oils. In addition, the grease structure may break down, as explained previously, or the thickener itself may decompose. Table 25.13 depicts the general effects of temperature on lubricating greases.

A grease behaves as an extreme form of non-Newtonian fluid, and its viscous properties change when it is sheared in a feed line or a bearing. Occasionally the viscosity increases with small shear rates, but more commonly the viscosity decreases as the shear rate increases, until eventually the viscosity reaches that of the base oil. For this reason, the viscosity of the base oil may be important if the grease is to be used in high-speed equipment.

The mechanism by which a grease lubricates is more complicated than that for an oil, and it depends partly on the geometry of the system. Some part of the total grease fill distributes itself over the contacting surfaces and is continually sheared in the same way as an oil. This part of the grease performs the lubricating function, giving either hydrodynamic lubrication or boundary lubrication according to the load, speed, and effective viscosity.

The remainder of the grease is swept out of the path of the moving parts and remains almost completely static in the covers of a bearing or the upswept parts of a gearbox. Because of the solid nature of the grease, there is virtually no circulation or exchange between the static, nonlubricating portion and the moving, lubricating portion.

In a plain bearing or a closely fitting gearbox, a high proportion of the grease fill is being continuously sheared at the contacting surfaces. In a roller bearing or a spa-

TABLE 25.13   Temperature Limits in Degrees Celsius for Greases as a Function of Required Life

Grease; type of limit	Life, h				
	1	10	10^2	10^3	10^4
Synthetic greases; oxidation limit with unlimited oxygen present	275 to 285	255 to 265	225 to 240	200 to 225	175 to 200
Synthetic greases; drop-point limit with inorganic thickeners	250	250	250	250	250
Mineral-oil greases; upper limit imposed by drop point depends on thickener; oxidation dependent on amount of oxygen present	80 to 200	80 to 200	80 to 200	80 to 200	80 to 200
Mineral greases; oxidation limit with unlimited oxygen	185 to 200	160 to 175	135 to 150	110 to 125	85 to 100
Mineral greases; lower limit imposed by high torque	−50 to −10	−50 to −10	−50 to −10	−50 to −10	−50 to −10
Synthetic greases; lowest limit imposed by high torque	−70 to −80	−70 to −80	−70 to −80	−70 to −80	−70 to −80

SOURCE: Ref. [25.2].

cious gearbox, a small proportion of the grease is continuously sheared and provides all the lubrication, while the larger proportion is inactive.

If a rolling bearing or gearbox is overfilled with grease, it may be impossible for the surplus to escape from the moving parts. Then a large quantity of grease will be continuously sheared, or "churned," and this causes a buildup of temperature which can severely damage the grease and the components. It is, therefore, important with grease lubrication to leave a void space which is sufficient to accommodate all the surplus grease; in a ball bearing, this could be more than 60 percent of the total space available.

The static grease which is not involved in lubrication may fulfill two useful functions: It may provide a very effective seal against the ingress of dust or other contaminants, and it can prevent loss of base oil from the grease fill. In addition, the static grease may form a reservoir from which to resupply the lubricated surfaces if the lubricating portion of the grease becomes depleted.

If the void space in the system is large, i.e., in a large bearing or gearbox, then usually it is desirable to use a stiffer grease to avoid the surplus grease "slumping" into the moving parts and being continuously churned. The advantages and disadvantages of grease lubrication are summarized in Table 25.14.

The selection of a grease for a specific application depends on five factors: speed, load, size, temperature range, and any grease feed system. For average conditions of speed, load, and size with no feed system, an NLGI no. 2 grease would be the normal choice, and such a grease with a mineral-oil base is sometimes known as a multipurpose grease. The effect of the various factors on selection can then be summarized in a few paragraphs.

1. *Speed* For high speeds, a stiffer grease, NLGI no. 3, should be used except in plain bearings, where no. 2 would usually be hard enough. For lower speeds, a softer grease such as no. 1 or no. 0 should be used.

TABLE 25.14 Advantages and Disadvantages of Grease Lubrication

Advantages
1. Maintain effective lubricant film on surfaces during a shutdown
2. Provide useful squeeze-film lubrication
3. Give effective sealing of rolling bearings
4. Maintain a reserve supply of lubricant in the vicinity of the bearing
5. Reduce contamination problems compared with oil
6. Provide an effective carrier for solid lubricants for antiseize or highly loaded situations

Disadvantages
1. Ineffective cooling
2. Limitations on bearing speed
3. Possible incompatibility with other similar greases
4. Lower oxidation resistance
5. Poorer storage stability

2. *Load*   For high loads, it may be advantageous to use EP additives or molybdenum disulfide. Because higher loads will lead to higher power consumption and therefore higher temperature, a stiffer grease such as no. 3 or a synthetic-base oil may help.

3. *Size*   For large systems, use a stiffer grease, no. 3 or no. 4. For very small systems, use a softer grease, such as no. 1 or no. 0.

4. *Temperature range*   The drop point should be higher than the maximum predicted operating temperature. For sustained operation at higher temperatures, a synthetic-base oil may be necessary. For very high temperatures, about 230°C, one of the very expensive fluorocarbon greases may be required.

5. *Feed systems*   If the grease is to be supplied through a centralized system, usually it is desirable to use one grade softer than would otherwise be chosen (i.e., use a no. 0 instead of a no. 1 or a no. 00 instead of a no. 0). Occasionally a particular grease will be found unsuitable for a centralized feed because separation occurs and the lines become plugged with thickener, but this problem is now becoming less common.

25.9   SOLID LUBRICANTS

Any solid material can act as a solid lubricant provided that it shears readily and smoothly when interposed between sliding surfaces. Some of the wide range of solids which can be used are listed in Table 25.15.

TABLE 25.15   Materials Used as Solid Lubricants

Layer-lattice compounds	
Molybdenum disulfide	Graphite
Tungsten diselenide	Tungsten disulfide
Niobium diselenide	Calcium fluoride
Graphite fluoride	

Polymers	
PTFE	PTFCE
PVF_2	FEP
Acetal	Polyimide
Polyphenylenesulfide	Polysulfones

Metals	
Silver	Gold
Tin	Lead
Barium	Gallium

Other inorganics	
Boron nitride	Molybdenum trioxide

There are many other desirable properties, including the following:

1. Ability to adhere to one or both of the bearing surfaces to ensure retention in the contact area
2. Chemical stability over the required temperature range in the particular environment
3. Sufficient resistance to wear
4. Nontoxicity
5. Easy application
6. Economy

Most of the available materials are eliminated by these requirements, and in practice almost all solid lubrication in engineering is provided by three materials—graphite, molybdenum disulfide, and polytetrafluoroethylene (PTFE).

Solid lubricants can be used in several different forms, such as loose powder, adhering powder, bonded film, or solid block. In the form of a solid block, the material is often called a *dry* bearing material rather than a solid lubricant.

25.9.1 Graphite

Graphite is probably the oldest known of the three main solid lubricants, and it has ceased to be the dominant one since about 1950. It is a grayish black crystalline form of carbon in which the atoms are arranged hexagonally in monatomic layers. The strong chemical bonds between the carbon atoms give strength to the layers, so that they resist bending or fracture and can carry useful loads. The bonds between the layers are relatively weak, and so the layers slide easily over each other and can be easily separated.

When graphite is used as a lubricant, the crystals orient themselves so that the layers are parallel to the bearing surfaces. The layers then adhere fairly well to the bearing surfaces, but slide easily over each other to give low friction.

The low shearing forces, and therefore the low friction, are not an inherent property of the graphite but are strongly influenced by the presence of moisture or certain other adsorbents. If graphite is used in a very dry atmosphere, the crystal layers have quite high interlayer bonding forces, and the friction and wear are high.

The biggest advantage of graphite over molybdenum disulfide and PTFE is its electrical conductivity, and it is almost universally used as a component in electric brushes. Its coefficient of friction varies from 0.05 at high loads to 0.15 at low loads, and these low values are maintained to over 500°C in air.

In block form, graphite has quite high structural integrity. It is commonly used in an impure form as graphitized carbon, in which the degree of crystallization can vary from 30 to over 80 percent of that of crystalline graphite. The frictional and structural properties and abrasiveness vary with the purity and degree of graphitization, and graphite technology is complex.

Graphite can be used in block form, as free powder, or as a coating deposited from dispersion in a liquid. It adheres readily to many solid surfaces, but probably its strength of adhesion is generally lower than that of molybdenum disulfide.

25.9.2 Molybdenum Disulfide

Molybdenum disulfide has also been known as a solid lubricant for centuries, but because it is similar in appearance, it has often been confused with graphite. Its use has increased enormously since about 1950, and for high-technology applications it is now generally preferred to graphite. In crude form, molybdenum disulfide is found naturally, sometimes in very large quantities, as molybdenite, the most common ore of molybdenum.

Like graphite, molybdenum disulfide is a dark gray crystalline material with a hexagonal layer-lattice structure. The bond strengths within the layers are very high, whereas those between layers are very low. The load-carrying capacity normal to the crystal planes is therefore high, and the shear strength parallel to the crystal layers is very low.

Unlike graphite, molybdenum disulfide does not require the presence of adsorbed moisture or other vapors to give low interplanar strength. Its low friction is therefore an inherent property which is maintained in high vacuum and in dry atmospheres.

Molybdenum disulfide starts to oxidize significantly above 350°C in oxygen and 450°C in air, but the main oxidation product is molybdic oxide, which is itself a fair high-temperature lubricant. In high vacuum the disulfide is said to be stable to 1000°C, and it outgasses (evaporates) very slowly, so that it has been widely used in space.

The adhesion to metals and many other solid surfaces is excellent, and durable coatings can be produced on metal surfaces by burnishing (a coating of loose powder is rubbed into the surface to give a very thin, shiny, and strong film). The powder may be applied free or from dispersion in a volatile liquid. Durable coatings can also be obtained by sputtering, but this technique is expensive and is not widely used.

Bonded coatings are widely used, in which molybdenum disulfide powder is incorporated in almost any effective adhesive, including many polymers, natural resins, or molten solids. The performance of the softer bonded coatings is also improved if they are carefully burnished before use. The coefficient of friction of burnished films varies from 0.02 to about 0.12. But for bonded films the friction depends on the nature of the binder and the percentage composition, and it can vary from 0.02 to about 0.3.

Molybdenum disulfide is often added to oils or greases to give high load-carrying capacity, especially at low running speeds. There is also strong evidence that the addition of up to 2 percent to vehicle engine oils produces a small but significant fuel savings without any apparent disadvantages.

At one time molybdenum disulfide suffered considerable criticism, especially for reported corrosion of steels and aluminum. Some of this may have been due to its use in conjunction with graphite. Some was certainly caused by failure to understand that solid lubricants, unlike oils and greases, do not normally protect against corrosion. It is probably fair to say that molybdenum disulfide is now well understood and that, when properly used, it is a very valuable solid lubricant.

25.9.3 Polytetrafluoroethylene

Abbreviated PTFE, polytetrafluoroethylene is a polymer produced from ethylene in which all the hydrogen atoms have been replaced by fluorine atoms. This fluorination produces a material of very high chemical stability and low intermolecular bond strength, while the polymerization of an ethylene-type molecule gives long, straight molecular chains.

The result is a white solid which consists of masses of parallel long-chain molecules that slide easily past one another. This leads to the same sort of low shear strength parallel to the chains which is found in molybdenum disulfide and to a high load-carrying capacity normal to the chains, but significantly lower than that of molybdenum disulfide.

PTFE is often used in the form of solid components, occasionally in bonded coatings, and very rarely as free powder. In addition, it has been used very successfully in composites, and two types are particularly effective.

The coefficient of friction of pure PTFE varies from 0.02 at high load to about 0.1 at low load. It is a rather soft solid, so that its load-carrying capacity is limited and its wear rate is high. It therefore needs reinforcement for use in highly loaded bearings. One successful form of reinforcement is to incorporate the PTFE in the pores of a sintered metal, especially bronze. In one composite, further reinforcement is obtained by dispersing fine particles of lead in the PTFE.

A second, and probably even more successful, form of reinforcement is by means of strengthening fibers. Glass fiber or carbon fiber can be incorporated in solid PTFE components, but the resulting high structural strength is obtained at the cost of an increase in the coefficient of friction to between 0.06 and 0.2. An alternative technique is to interweave PTFE fibers and reinforcing fibers of glass, metal, rayon, or other synthetics. Some of the resulting composites have outstanding strength with low wear rate and low friction.

PTFE can be used in air to about 250°C, but in high vacuum it outgasses slowly, and so it is used in spacecraft only in well-shielded locations.

Because of its high chemical stability, PTFE can be used safely in oxygen systems and in many types of chemical plants. It is nontoxic in almost all situations and is therefore used in the pharmaceutical and food industries, even in situations where low friction is not required, and as the nonstick lubricant in domestic cooking utensils.

25.9.4 Miscellaneous Solid Lubricants

Other solid lubricants are used to a relatively minor degree, in situations where they have specific advantages. They can be classified in three broad categories: inorganics, polymers, and metals.

The inorganics include a number of materials similar to molybdenum disulfide, known generally as the lubricating dichalcogenides. None of these occurs naturally, and the synthetic materials are relatively expensive. Tungsten disulfide has a higher oxidation temperature, and both tungsten disulfide and tungsten diselenide oxidize more slowly than molybdenum disulfide. Niobium diselenide has better electrical conductivity and has been used in electric contact brushes, but in fact molybdenum disulfide composites have been shown to be equally satisfactory.

Other inorganics have been used for their much higher temperature limits, and these include molybdic oxide, boron nitride, graphite fluoride, and calcium fluoride.

The low friction and chemical inertness of PTFE make it difficult to bond to other materials, and two other fluorinated polymers have been recommended for their better bonding behavior: polyvinylfluoride (PVF_2) and polytrifluorochloroethylene (PTFCE). But in both cases the advantages of better bonding and slightly higher structural strength are offset by higher friction.

For higher temperatures, polyimide, polysulfones, and polyphenylene sulfide can be used unlubricated. Other polymers, such as nylons, acetals, and phenolics, are occasionally used unlubricated where sliding speeds are low, but they require lubrication by oil, grease, or water for really useful performance.

Silver, gold, and tin have useful antigalling properties in slow sliding, but metallic coatings are mainly used as lubricants only in high vacuum, where silver, gold, barium, gallium, and lead have all been used successfully.

25.10 GAS LUBRICATION

Gases can be used to provide full-fluid-film lubrication in the same ways as liquids in hydrodynamic and externally pressurized bearings. The physical laws governing behavior are the same for both liquids and gases, but the very low viscosities of gases lead to considerable practical differences in their use, especially in self-pressurizing, or "gas-dynamic," bearings:

1. Operating speeds are much higher to compensate for low viscosity.
2. Surface finish and precision must be better because of the much smaller lubricant-film thickness.
3. Lubricant flow rate is higher for the same pressure differential.
4. Load-carrying capacity is generally low.

As a result, gas bearings tend to be small, high-speed, and lightly loaded, with tight tolerances and high-quality surface finishes. The overall design and manufacturing cost is high, and they are mainly used in high-technology applications.

Any gas or vapor can be used provided that it is chemically stable under the operating conditions and does not attack any of the system materials. If no chemical change takes place, there is no upper temperature limit to the use of a gas, and the viscosity increases as temperature increases.

Air is the most common gas used in gas lubrication. Nitrogen or helium may be used where inertness is important. Otherwise, any gas which is available can be used, especially if it is available at high enough pressure for external pressurization.

Some of the advantages of gas lubrication are high precision, very low friction, cleanliness, and ready availability of lubricant. The greatest potential advantage is the wide temperature range. In theory, it should be possible to design a gas bearing to operate from −250 to +2000°C. The corresponding disadvantages include the demanding design and construction requirements, the low load-carrying capacity, and the need for a very clean gas supply. Examples of important applications of gas bearings are dentists' air-turbine drills, high-precision grinding spindles, and inertial gyroscopes.

25.11 LUBRICANT FEED SYSTEMS

In many lubricated components, no feed system is needed, because the initial lubricant fill is sufficient to last the required life. A feed system becomes necessary when the lubricant must be replaced or replenished, for one of the following reasons:

1. The temperature is too high, so that the lubricant must be removed and replaced by a fresh charge of cooler lubricant.
2. Lubricant becomes depleted by leakage or creepage and must be topped up.
3. Lubricant decomposes and must be replaced with a fresh charge.
4. Lubricant becomes contaminated and must be replaced with clean material.

Where the rate of loss or deterioration is relatively low, it will be sufficient to provide a facility for occasional topping up by means of an oil can or a grease gun, provided that access to the lubricated component is available. Where this occasional manual topping up is not adequate, a lubricant feed system will be needed. It is beyond the scope of this chapter to describe the whole range and design of lubricant feed systems available. It is only possible to give a brief description of the main types and the factors involved in selecting them.

25.11.1 Internal Circulation

One obvious way to reduce oil temperature, slow down the increase in contamination, and increase the life is simply to increase the quantity of oil supplied in the initial fill. This requires an increase in the volume of space available for oil or, in other words, the creation of an oil reservoir or sump adjacent to the lubricated bearings or gears.

Circulation of the oil can be ensured by arranging for the moving parts to dip below the surface of the oil. But they should not be completely submerged because the resulting viscous drag and churning of the oil lead to excessive power consumption and heating. For slow-moving components this problem is not serious, but for high speeds the depth of immersion is critical, and the following guidelines are useful:

1. Gears should be immersed to twice the tooth height. In a vertical train, the oil level should be just below the shaft of the lowest gear.
2. Rolling bearings should be immersed to halfway up to the lowest rollers or balls.
3. Crankshafts should be immersed so that the oil level is just above the big-end bearings at their lowest point.
4. The oil level should be higher for slow operation than for higher-speed systems.

Oil is carried by the partly submerged components to contacting surfaces and is also spread by splashing. Where transfer by these two mechanisms is inadequate, the oil feed can be improved by the use of rings or disks, as shown in Fig. 25.7. Both operate by providing a larger surface with higher peripheral speed to transfer the oil, but they do not cause excessive viscous drag because they are both uniform in shape.

Disks have an advantage over rings in that they can be designed to propel oil axially as well as radially, and this is particularly useful for bearing lubrication. Usually plain bearings cannot be adequately lubricated by partial immersion in oil unless the oil flow is augmented by a ring or a disk. If a weir is incorporated, part of the splashed oil can be trapped and directed to critical locations.

(a)

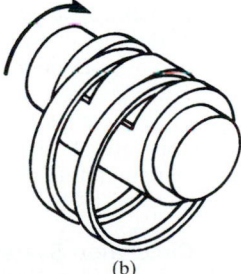
(b)

FIGURE 25.7 Ring and disk lubrication. (*a*) Disk; (*b*) ring. (*From Ref. [25.1].*)

25.11.2 Topping Up Systems

Where the main problem is loss of oil by leakage or creep, it may be sufficient to set up a wick or drip feed to provide a small-scale supply. Wicks or pads consist of porous or permeable materials such as felt which transfer oil by capillary action to the bearing surfaces. The pads may form a path from a reservoir to the bearing or may simply contain a small initial oil fill to increase the quantity available and feed it slowly to the bearing. This latter approach is commonly used in small electric motors.

Drip feeds consist of small reservoirs mounted above the bearing or gears and equipped with a feed tube with some form of flow regulation and usually a sight glass. They can be used to provide a much higher flow rate than a wick.

25.11.3 Centralized Total-Loss Systems

Wick and drip feeds are examples of total-loss systems, in which no attempt is made to collect the oil after feeding it to the bearings or gears. Far more sophisticated total-loss systems can be used to supply oil or grease to a number of separate components.

The requirements of centralized total-loss systems are basically very simple. A typical system consists of a reservoir, a single pump to pressurize a manifold, and a number of controllers, each regulating the feed to a single lubrication point. Alternatively, the flow to several lines may be controlled by a multipiston pump, in which individual single-cylinder piston pumps are operated by cams on a common camshaft immersed in the lubricant. The number of outlets can vary from one to several hundred.

The main advantage of centralized total-loss systems is that they reduce the labor required where a large number of components need relubricating. They are also valuable where the lubrication points are not readily accessible. Their disadvantages are that they do not provide any form of cooling or removal of contaminants, and there is no recovery of used lubricant.

25.11.4 Oil Mist or Fog Systems

One type of total-loss system which has become widely used in recent years is oil mist or fog, in which fine droplets of oil are carried by a stream of air from a reservoir to a bearing or gears. The mist or fog of oil in air is produced by passing the airstream through the reservoir at low speed and low pressure. The oil is usually formulated to have low surface tension. The resulting dispersion of oil droplets is passed through steel, copper, or plastic pipes to the vicinity of the lubricated component. Passage through fine nozzles, or reclassifiers, then increases the linear flow rate to something over 45 meters per second (m/s) and thus causes the oil droplets to coalesce, producing a flow of liquid oil to the lubrication point.

Oil mist and fog systems have two advantages over other total-loss systems in that the oil supply rate can be very low, resulting in a clean system, and the airflow gives a significant amount of cooling.

25.11.5 Oil-Circulation Systems

The most sophisticated centralized systems are those in which the lubricant is collected after use and returned to the reservoir for recirculation. The basic requirements of such a system are a reservoir, a pump, possibly a flow divider or pro-portioner, feed lines, and return lines.

In practice a full-circulation system is likely to be more complex and to include many of or all the following components:

Multiple or divided reservoirs

Heaters

Coolers

Oil-level warning devices

Full-flow filtration to protect the pump

Bypass filtration

Pressure switches and alarms

Water separators

Chip detectors

Sampling points

Sight glass

Circulation systems give a high degree of control over the quality and quantity of oil supplied to each component, enabling the cleanliness and the temperature of the oil to be controlled. Their only disadvantage is their complexity and therefore their cost.

25.12 LUBRICANT STORAGE

The storage of lubricants, like that of any other class of goods, depends first on making rational decisions about the number of varieties and the quantities which it is necessary to store. The special factors which then need to be considered in storing lubricants are as follows:

1. Lubricants are an integral part of the precision components in which they are used. They must therefore be treated as precision components and protected carefully against contamination by dirt, water, or other materials. Never store them in the open.

2. Because most lubricants are liquids, they have no characteristic shapes, and it is very easy to use the wrong lubricant in a machine. This is always undesirable and sometimes catastrophic. So it is important to label lubricant containers carefully and to control their issue and use.

3. Some lubricants deteriorate in storage, and it is important to use supplies in proper rotation and ensure that storage lives are not exceeded.

4. Many lubricants are flammable, and special precautions are necessary to reduce fire risk.

5. Lubricants are slippery, and spillages can cause accidents. Floor gratings and drainage channels should be supplied, and absorbent powders or granules kept available to absorb spilled oil.

6. Oil drums are very convenient and satisfactory containers for storing lubricants, but water can collect in the recessed top and enter even through sealed apertures to contaminate the contents. Drums should always be stored horizontally and never upright.

REFERENCES

25.1 A. R. Lansdown, *Lubrication: A Practical Guide to Lubricant Selection,* Pergamon, New York, 1982.

25.2 M. J. Neale (ed.), *The Tribology Handbook,* Butterworth, London, 1973.

25.3 D. Godfrey, "Boundary Lubrication," in P. M. Ku (ed.), *Interdisciplinary Approach to Friction and Wear,* NASA SP-181, 1968.

CHAPTER 26
GASKETS

Daniel E. Czernik
Director of Product Engineering
Fel-Pro Inc.
Skokie, Illinois

In the field of gaskets and seals, the former are generally associated with sealing mating flanges while the latter are generally associated with sealing reciprocating shafts or moving parts. Some designers refer to gaskets as static seals and consider seals to be dynamic sealing components. This chapter covers gaskets, and Chap. 17 discusses seals.

26.1 DEFINITION

A *gasket* is a material or combination of materials clamped between two separable members of a mechanical joint. Its function is to effect a seal between the members (flanges) and maintain the seal for a prolonged period. The gasket must be capable of sealing mating surfaces, must be impervious and resistant to the medium being sealed, and must be able to withstand the application temperature. Figure 26.1 depicts the nomenclature associated with a gasketed joint.

26.2 STANDARD CLASSIFICATION SYSTEM FOR NONMETALLIC GASKET MATERIALS[†]

This classification system provides a means for specifying or describing pertinent properties of commercial nonmetallic gasket materials. Materials composed of

---

[†] Ref. [26.1] (ANSI/ASTM F104).

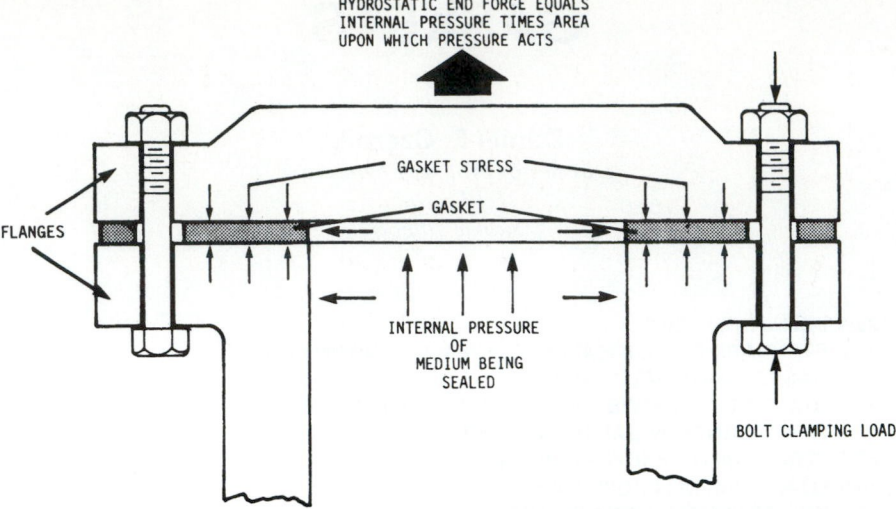

FIGURE 26.1   Nomenclature of a gasketed joint.

asbestos, cork, cellulose, and other organic or inorganic materials in combination with various binders or impregnants are included. Materials normally classified as rubber compounds are not included, since they are covered in ASTM Method D 2000 (SAE J200). Gasket coatings are not covered, since details are intended to be given on engineering drawings or in separate specifications.

This classification is based on the principle that nonmetallic gasket materials can be described in terms of specific physical and mechanical characteristics. Thus, users of gasket materials can, by selecting different combinations of statements, specify different combinations of properties desired in various parts. Suppliers, likewise, can report properties available in their products.

In specifying or describing gasket materials, each *line call-out* shall include the number of this system (minus the date symbol) followed by the letter F and six numerals, for example, ASTM F104 (F125400). Since each numeral of the call-out represents a characteristic (as shown in Table 26.1), six numerals are always required. The numeral 0 is used when the description of any characteristic is not desired. The numeral 9 is used when the description of any characteristic (or related test) is specified by some supplement to this classification system, such as notes on engineering drawings.

26.3 GASKET PROPERTIES, TEST METHODS, AND THEIR SIGNIFICANCE IN GASKETED JOINTS

Table 26.2 lists some of the most significant gasket properties which are associated with creating and maintaining a seal. This table also shows the test method and the significance of each property in a gasket application.

26.4 PERMEABILITY PROPERTIES

For a material to be impervious to a fluid, a sufficient density to eliminate voids which might allow capillary flow of the fluid through the construction must be achieved. This requirement may be met in two ways: by compressing the material to fill the voids and/or by partially or completely filling them during fabrication by means of binders and fillers. Also, for the material to maintain its impermeability for a prolonged time, its constituents must be able to resist degradation and disintegration resulting from chemical attack and temperature of the application [26.2].

Most gasket materials are composed of a fibrous or granular base material, forming a basic matrix or foundation, which is held together or strengthened with a binder. The choice of combinations of binder and base material depends on the compatibility of the components, the conditions of the sealing environment, and the load-bearing properties required for the application.

Some of the major constituents and the properties which are related to impermeability are listed here.

26.4.1 Base Materials—Nonmetallic

Cork and Cork-Rubber. High compressibility allows easy density increase of the material, thus enabling an effective seal at low flange pressures. The temperature limit is approximately 250°F (121°C) for cork and 300°F (149°C) for cork-rubber compositions. Chemical resistance to water, oil, and solvents is good, but resistance to inorganic acids, alkalies, and oxidizing environments is poor. These materials conform well to distorted flanges.

Cellulose Fiber. Cellulose has good chemical resistance to most fluids except strong acids and bases. The temperature limitation is approximately 300°F (149°C). Changes in humidity may result in dimensional changes and/or hardening.

Asbestos Fiber. This material has good heat resistance to 800°F (427°C) and is noncombustible. It is almost chemically inert (crocidolite fibers, commonly known as blue asbestos, resist even inorganic acids) and has very low compressibility. The binder dictates the resistance to temperature and the medium to be sealed.

Nonasbestos Fibers. A number of nonasbestos fibers are being used in gaskets. Some of these are glass, carbon, aramid, and ceramic. These fibers are expensive and are normally used only in small amounts. Temperature limits from 750 to 2400°F (399 to 1316°C) are obtainable. Use of these fillers is an emerging field today, and suppliers should be contacted before these fibers are specified for use.

26.4.2 Binders and Fillers

Rubber. Rubber binders provide varying temperature and chemical resistance depending on the type of rubber used. These rubber and rubberlike materials are used as binders and, in some cases, gaskets:

1. *Natural* This rubber has good mechanical properties and is impervious to water and air. It has uncontrolled swell in petroleum oil and fuel and chlorinated solvents. The temperature limit is 250°F (121°C).

TABLE 26.1   Basic Physical and Mechanical Characteristics

Basic six-digit number	Basic characteristic
First numeral	Type of material (the principal fibrous or particulate reinforcement material from which the gasket is made) shall conform to the first numeral of the basic six-digit number as follows: 0 = not specified 1 = asbestos or other inorganic fibers (type 1) 2 = cork (type 2) 3 = cellulose or other organic fibers (type 3) 4 = fluorocarbon polymer 9 = as specified†

Second numeral	Class of material (method of manufacture or common trade designation) shall conform to the second numeral of the basic six-digit number as follows:
	When first numeral is 1, for second numeral
	    0 = not specified
	    1 = compressed asbestos (class 1)
	    2 = beater addition asbestos (class 2)
	    3 = asbestos paper and millboard (class 3)
	    9 = as specified†
	When first numeral is 2, for second numeral
	    0 = not specified
	    1 = cork composition (class 1)
	    2 = cork and elastomeric (class 2)
	    3 = cork and cellular rubber (class 3)
	    9 = as specified†
	When first numeral is 3, for second numeral
	    0 = not specified
	    1 = untreated fiber—tag, chipboard, vulcanized fiber, etc. (class 1)
	    2 = protein treated (class 2)
	    3 = elastomeric treated (class 3)
	    4 = thermosetting resin treated (class 4)
	    9 = as specified†
	When first numeral is 4, for second numeral
	    0 = not specified
	    1 = sheet PTFE
	    2 = PTFE of expanded structure
	    3 = PTFE filaments, braided or woven
	    4 = PTFE felts
	    5 = filled PTFE
	    9 = as specified†

Third numeral	Compressibility characteristics, determined in accordance with 8.2, shall conform to the percentage indicated by the third numeral of the basic six-digit number (example: 4 = 15 to 25%): 0 = not specified 5 = 20 to 30% 1 = 0 to 10% 6 = 25 to 40% 2 = 5 to 15%‡ 7 = 30 to 50% 3 = 10 to 20% 8 = 40 to 60% 4 = 15 to 25% 9 = as specified†

TABLE 26.1 Basic Physical and Mechanical Characteristics (*Continued*)

Fourth numeral	Thickness increase when immersed in ASTM no. 3 oil, determined in accordance with 8.3, shall conform to the percentage indicated by the fourth numeral of the basic six-digit number (example: 4 = 15 to 30%): 0 = not specified 5 = 20 to 40% 1 = 0 to 15% 6 = 30 to 50 % 2 = 5 to 20% 7 = 40 to 60% 3 = 10 to 25% 8 = 50 to 70% 4 = 15 to 30% 9 = a specified†
Fifth numeral	Weight increase when immersed in ASTM no. 3 oil, determined in accordance with 8.3, shall conform to the percentage indicated by the fifth numeral of the basic six-digit number (example: 4 = 30% maximum): 0 = not specified 5 = 40% max. 1 = 10% max. 6 = 60% max. 2 = 15% max. 7 = 80% max. 3 = 20% max. 8 = 100% max. 4 = 30% max. 9 = as specified†
Sixth numeral	Weight increase when immersed in water, determined in accordance with 8.3, shall conform to the percentage indicated by the sixth numeral of the basic six-digit number (example: 4 = 30% maximum): 0 = not specified 5 = 40% max. 1 = 10% max. 6 = 60% max. 2 = 15% max. 7 = 80% max. 3 = 20% max. 8 = 100% max. 4 = 30% max. 9 = as specified†

†On engineering drawings or other supplement to this classification system. Suppliers of gasket materials should be contacted to find out what line call-out materials are available. Refer to ANSI/ASTM F104 for further details (Ref. [26.1]).

‡From 7 to 17% for type 1, class 1 compressed asbestos sheet.

2. *Styrene/butadiene* This rubber is similar to natural rubber but has slightly improved properties. The temperature limit also is 250°F (121°C).

3. *Butyl* This rubber has excellent resistance to air and water, fair resistance to dilute acids, and poor resistance to oils and solvents. It has a temperature limit of 300°F (149°C).

4. *Nitrile* This rubber has excellent resistance to oils and dilute acids. It has good compression set characteristics and has a temperature limit of 300°F (149°C).

5. *Neoprene* This rubber has good resistance to water, alkalies, nonaromatic oils, and solvents. Its temperature limit is 250°F (121°C).

6. *Ethylene propylene rubber* This rubber has excellent resistance to hot air, water, coolants, and most dilute acids and bases. It swells in petroleum fuels and oils without severe degradation. The temperature limit is 300°F (149°C).

7. *Acrylic* This rubber has excellent resistance to oxidation, heat, and oils. It has poor resistance to low temperature, alkalies, and water. The temperature limit is 450°F (232°C).

TABLE 26.2 Identification, Test Method, and Significance of Various Properties Associated with Gasket Materials

Property	Test method	Significance in gasket applications
Sealability	Fixtures per ASTM F37-62T	Resistance to fluid passage
Heat resistance	Exposure testing at elevated temperatures	Resistance to thermal degradation
Oil and water immersion characteristics	ASTM D-1170	Resistance to fluid attack
Antistick characteristics	Fixture testing at elevated termperatures	Ability to release from flanges after use
Stress vs. compression and spring rates	Various compression test machines	Sealing pressure at various compressions
Compressibility and recovery	ASTM F36-61T	Ability to follow deformation and deflection; indentation characteristics
Creep relaxation and compression set	ASTM F38-62T and D-395-59	Related to torque loss and subsequent loss of sealing pressure
Crush and extrusion characteristics	Compression test machines	Resistance to high loadings and extrusion characteristics at room and elevated temperatures

8. *Silicone*   This rubber has good heat stability and low-temperature flexibility. It is not suitable for high mechanical pressure. Its temperature limit is 600°F (316°C).

9. *Viton*   This rubber has good resistance to oils, fuel, and chlorinated solvents. It also has excellent low-temperature properties. Its temperature limit is 600°F (316°C).

10. *Fluorocarbon*   This rubber has excellent resistance to most fluids, except synthetic lubricants. The temperature limit is 500°F (260°C).

Resins.   These usually possess better chemical resistance than rubber. Temperature limitations depend on whether the resin is thermosetting or thermoplastic.

Tanned Glue and Glycerine.   This combination produces a continuous gel structure throughout the material, allowing sealing at low flange loading. It has good chemical resistance to most oils, fuels, and solvents. It swells in water but is not soluble. The temperature limit is 200°F (93°C). It is used as a saturant in cellulose paper.

Fillers.   In some cases, inert fillers are added to the material composition to aid in filling voids. Some examples are barytes, asbestine, and cork dust.

26.4.3 Reinforcements

Some of the properties of nonmetallic gasket materials can be improved if the gaskets are reinforced with metal or fabric cores. Major improvements in torque retention and blowout resistance are normally seen. Traditionally, perforated or upset metal cores have been used to support gasket facings. A number of designs have been utilized for production. Size of the perforations and their frequency in a given area are the usual specified parameters.

Adhesives have been developed that permit the use of an unbroken metal core to render support to a gasket facing. Laminated composites of this type have certain characteristics that are desired in particular gaskets [26.3].

26.4.4 Metallic Materials

Aluminum. This metal has good conformability and thermal conductivity. Depending on the alloy, aluminum suffers tensile strength loss as a function of temperature. Normally it is recommended up to 800°F (427°C). It is attacked by strong acids and alkalies.

Copper. This metal has good corrosion resistance and heat conductivity. It has ductility and excellent flange conformability. Normally 900°F (482°C) is considered the upper service temperature limit.

Steel. A wide variety of steels—from mild steel to stainless steel—have been used in gasketing. A high clamping load is required. Temperature limits range from 1000 to 2100°F (538 to 1149°C), depending on the alloy.

26.5 LOAD-BEARING PROPERTIES

26.5.1 Conformability and Pressure

Since sealing conditions vary widely depending on the application, it is necessary to vary the load-bearing properties of the gasket elements in accordance with these conditions. Figure 26.2 illustrates stress-compression curves for several gasket components and indicates the difference in the stress-compression properties used for different sealing locations.

Gasket thickness and compressibility must be matched to the rigidity, roughness, and unevenness of the mating flanges. An entire seal can be achieved only if the stress level imposed on the gasket at clampup is adequate for the specific material. Minimum seating stresses for various gasket materials are listed later in this chapter. In addition, the load remaining on the gasket during operation must be high enough to prevent blowout of the gasket. During operation, the hydrostatic end force, which is associated with the internal pressure, tends to unload the gasket. Figure 26.3 is a graphical representation of a gasketed joint depicting the effect of the hydrostatic end force [26.4].

The bolt should be capable of handling the maximum load imposed on it without yielding. The gasket should be capable of sealing at the minimum load resulting on it and should resist blowout at this load level.

Gaskets fabricated from compressible materials should be as thin as possible [26.5]. The gasket should be no thicker than is necessary if it is to conform to the unevenness of the mating flanges. The unevenness is associated with surface finish, flange flatness, and flange warpage during use. It is important to use the gasket's unload curve in considering its ability to conform. Figure 26.4 depicts typical load-compression and unload curves for nonmetallic gaskets.

The unload curve determines the recovery characteristics of the gasket which are required for conformance. Metallic gaskets will show no change in their load and unload curves unless yielding occurs. Load-compression curves are available from gasket suppliers.

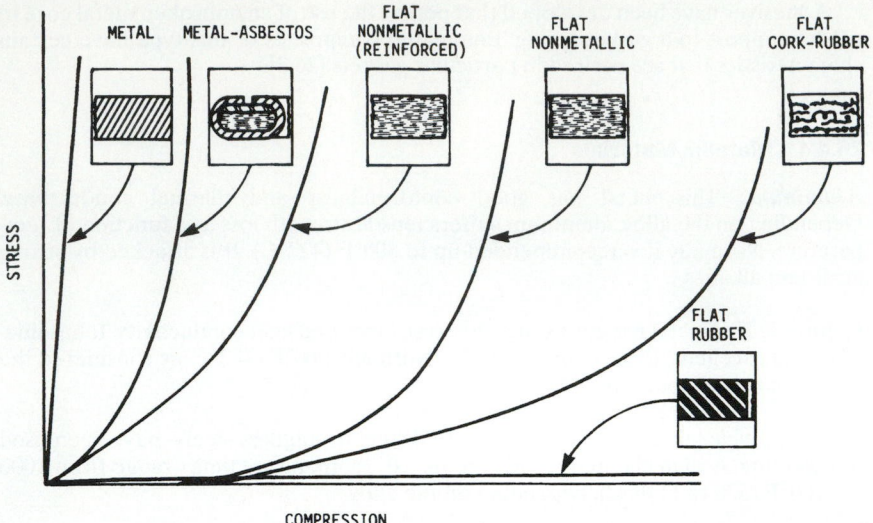

FIGURE 26.2   Stress versus compression for various gasket materials.

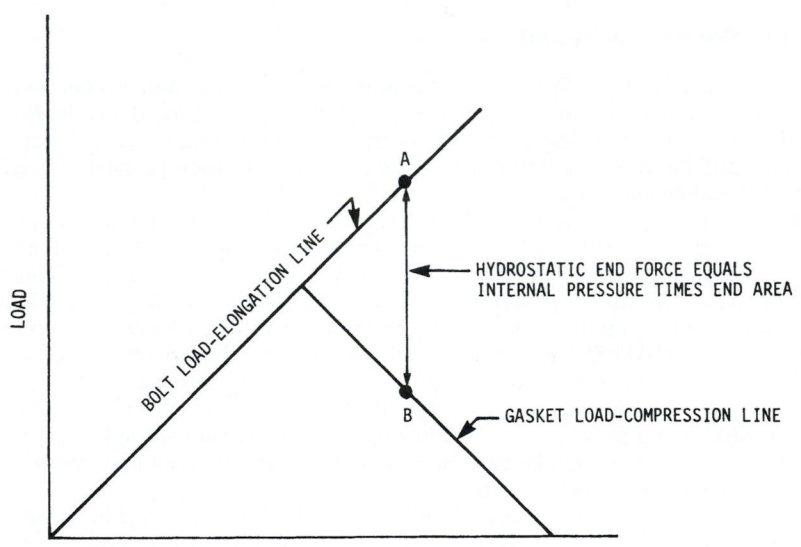

FIGURE 26.3   Graphical representation of a gasketed joint and effect of hydrostatic end force. *A*, Maximum load on gasket; *B*, minimum load on gasket.

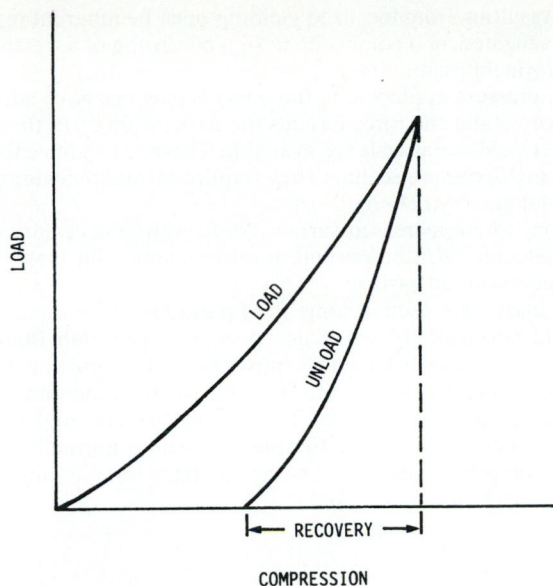

FIGURE 26.4 Load-compression and unload curves for a typical nonmetallic gasket material.

Some advantages of thin gaskets over thick gaskets are

1. Reduced creep relaxation and subsequent torque loss
2. Less distortion of mating flanges
3. Higher resistance to blowout
4. Fewer voids through which sealing media can enter, and so less permeability
5. Lower thickness tolerances
6. Better heat transfer

A common statement in the gasket industry is, "Make the gasket as thin as possible and as thick as necessary."

The following paragraphs describe some of the gasket's design specifications which need to be considered for various applications. A large array of gasket designs and sealing applications are used, and more are coming into use daily. Gaskets are constantly being improved for higher and higher performance.

In high-pressure, clamp load, and temperature applications, a high-spring-rate (stress per unit compression) material is necessary in order to achieve high loading at low compression, thereby sealing the high pressures developed. These applications generally rely on sealing resulting from localized yielding under the unit loading. In addition to the high spring rate, high heat resistance is mandatory. To economically satisfy these conditions, metal is the most commonly used material.

In applications where close tolerances in machining (surface finish and parallelism) are obtainable, a solid steel construction may be used. In those situations where close machining and assembly are not economical, it is necessary to sacrifice some gasket rigidity to allow for conformability. In such cases, conformability exceeding that resulting from localized yielding must be inherent in the design. The metal can be corrugated, or a composite design consisting of asbestos could be used to gain the conformability required.

In very-high-pressure applications, flat gaskets may not have adequate recovery to seal as the hydrostatic end force unseats the gaskets [26.6]. In these cases, various types of self-energized metal seals are available. These seals utilize the internal pressure to achieve high-pressure sealing. They require careful machining of the flanges and have some fatigue restrictions.

In applications where increased surface conformity is necessary and lower temperatures are encountered, asbestos and/or other nonmetallic materials can be used under the limitations noted earlier.

Elastomeric inserts are used in some fluid passages where conformity with sealing surfaces and permeability are major problems and high fluid pressures are encountered. Since the inserts have low spring rates, they must be designed to have appropriate contact areas and restraint in order to effect high unit sealing stresses for withstanding the internal pressures. The inserts also have high degrees of recovery, which allow them to follow high thermal distortions normally associated in the mating flanges. Compression set and heat-aging characteristics must also be considered when elastomeric inserts are used.

26.5.2   Creep and Relaxation

After the initial sealing stress is applied to a gasket, it is necessary to maintain a sufficient sealing stress for the designed life of the unit or equipment. All materials exhibit, in varying degrees, a decrease in applied stress as a function of time, commonly referred to as *stress relaxation.* The reduction of stress on a gasket is actually a combination of two major factors: stress relaxation and creep (compression drift). By definition,

> *Stress relaxation* is a reduction in stress on a specimen under constant strain ($d\sigma/dt; e$ = constant).
>
> *Creep* (compression drift) is a change in strain of a specimen under constant stress ($de/dt; \sigma$ = constant).

In a gasketed joint, stress is applied by tension in a bolt or stud and transmitted as a compressive force to the gasket. After loading, stress relaxation and creep occur in the gasket, causing corresponding lower strain and tension in the bolt. This process continues indefinitely as a function of time. The change in tension of a bolt is related to the often quoted "torque loss" associated with a gasket application. Since the change in stress is due to two primary factors, a more accurate description of the phenomenon would be *creep relaxation,* from now on called *relaxation.*

Bolt elongation, or stretch, is linearly proportional to bolt length. The longer the bolt, the higher the elongation. The higher the elongation, the lower the percentage loss for a given relaxation. Therefore, the bolts should be made as long as possible for best torque retention.

Relaxation in a gasket material may be measured by applying a load on a specimen by means of a strain-gauged bolt-nut-platen arrangement as standardized by ASTM F38-62T. Selection of materials with good relaxation properties will result in the highest retained torque for the application. This results in the highest remaining stress on the gasket, which is desirable for long-term sealing.

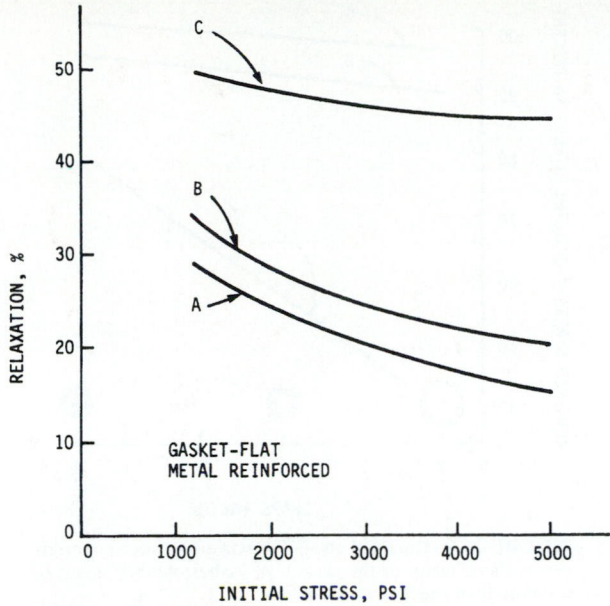

FIGURE 26.5   Relaxation versus stress on a gasket: *A*, 0.030 in–0.035 in thick; *B*, 0.042 in–0.047 in thick; *C*, 0.062 in–0.065 in thick.

The amount of relaxation increases as thickness is increased for a given gasket material. This is another reason why the thinnest gasket that will work should be selected. Figure 26.5 depicts the relaxation characteristics as a function of thickness for a particular gasket design.

Note that as clamping stress is increased, relaxation is decreased. This is the result of more voids being eliminated as the stress level is increased.

26.5.3   Effect of Geometry

The gasket's shape factor has an important effect on its relaxation characteristics. This is particularly true in the case of soft packing materials.

Much of the relaxation of a material may be attributed to the releasing of forces through lateral expansion. Therefore, the greater the area available for lateral expansion, the greater the relaxation. The *shape factor* of a gasket is the ratio of the area of one load face to the area free to bulge. For circular or annular samples, this may be expressed as

$$\text{Shape factor} = \frac{1}{4t}\,(\text{OD} - \text{ID}) \qquad (26.1)$$

where t = thickness of gasket
      OD = outside diameter
      ID = inside diameter

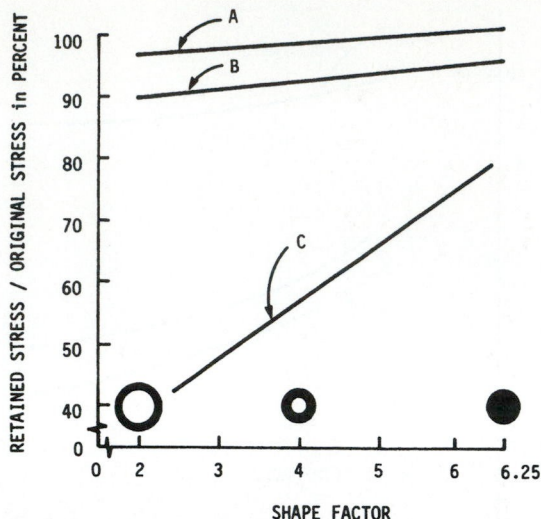

FIGURE 26.6 Retained stress for various gasket materials versus shape factor of the gasket. *A,* Asbestos fiber sheet; *B,* cellulose fiber sheet; *C,* cork-rubber.

As the area free to bulge increases, the shape factor decreases, and the relaxation will increase as the retained stress decreases. Figure 26.6 depicts the effect of shape factor on the gasket's ability to retain stress.

Note that the shape factor decreases with increasing thickness. Therefore, the gasket should be as thin as possible to reduce relaxation. It must be thick enough, however, to permit adequate conformity. The clamp area should be as large as possible, consistent with seating stress requirements. Often designers reduce gasket width, thereby increasing gasket clamping stress to obtain better sealing. Remember, however, that this reduction might decrease the gasket's shape factor, resulting in higher relaxation over time.

26.6 ENVIRONMENTAL CONDITIONS

Many environmental conditions and factors influence the sealing performance of gaskets. Flange design details, in particular, are most important. Design details such as number, size, length, and spacing of clamping bolts; flange thickness and modulus; and surface finish, waviness, and flatness are important factors. Application specifics such as the medium being sealed, as well as the temperatures and pressures involved, also affect the gasket's sealing ability. The material must withstand corrosive attack of the confined medium. In particular, flange bowing is a most common type of problem associated with the sealing of a gasketed joint. The amount of bowing can be reduced by reducing the bolt spacing. For example, if the bolt spacing were cut in half, the bowing would be reduced to one-eighth of its original value [26.7]. Doubling the flange thickness could also reduce bowing to one-eighth of its original value. A method of calculating the minimum stiffness required in a flange is available [26.8].

Different gasket materials and types require different surface finishes for optimum sealing. Soft gaskets such as rubber sheets can seal surface finishes in the vicinity of 500 microinches (μin), whereas some metallic gaskets may require finishes in the range of 32 μin for best sealing. Most gaskets, however, will seal adequately in the surface finish range of 63 to 125 μin, with 90 to 110 μin being preferred. There are two main reasons for the surface finish differences: (1) The gasket must be able to conform to the roughness for surface sealing. (2) It must have adequate bite into the mating flange to create frictional forces to resist radial motion due to the internal pressure, thereby preventing blowout. In addition, elimination of the radial micromotion will result in maintaining the initial clampup sealing condition. Micromotion can result in localized fretting, and a leakage path may be created [26.9].

Because of the complexity that results from the wide variety of environmental conditions, some gaskets for specific applications will have to be designed by trial and error. Understanding Sec. 26.7 will enable a designer to minimize the chance for leaks. Since the factors are so complex, however, adherence to the procedure will not ensure adequate performance in all cases. When inadequate gasket performance occurs, gasket manufacturers should be contacted for assistance.

26.7 GASKET DESIGN AND SELECTION PROCEDURE

26.7.1 Introduction

The first step in the selection of a gasket for sealing in a specific application is to choose a material that is both chemically compatible with the medium being sealed and thermally stable at the operating temperature of the application. The remainder of the selection procedure is associated with the minimum seating stress of the gasket and the internal pressure involved. In these regards, two methods are proposed: the American Society of Mechanical Engineers (ASME) Code method and the simplified method proposed by Whalen.

26.7.2 ASME Code Procedure

The ASME Code for Pressure Vessels, Sec. VIII, Div. 1, App. 2, is the most commonly used design guide for gasketed joints. An important part of this code focuses on two factors: an m factor, called the *gasket material factor,* which is associated with the hydrostatic end force, and a y factor, which is the minimum seating stress associated with particular gasket material. The m factor is essentially a safety factor to increase the clamping load to such an amount that the hydrostatic end force does not unseat the gasket to the point of leakage. The factors were originally determined in 1937, and even though there have been objections to their specific values, these factors have remained essentially unchanged to date. The values are only suggestions and are not mandatory.

This method uses two basic equations for calculating required bolt load, and the larger of the two calculations is used for design. The first equation is associated with W_{m2} and is the required bolt load to initially seat the gasket:

$$W_{m2} = \pi b G y \qquad (26.2)$$

The second equation states that the required bolt operating load must be sufficient to contain the hydrostatic end force and simultaneously maintain adequate compression on the gasket to ensure sealing:

$$W_{m1} = \frac{\pi}{4} G^2 P + 2b\pi GmP \tag{26.3}$$

where W_{m1} = required bolt load for maximum operating or working conditions, lb
$\qquad W_{m2}$ = required initial bolt load at atmospheric temperature conditions without internal pressure, lb
$\qquad G$ = diameter at location of gasket load reaction, generally defined as follows: When $b_0 \leq \frac{1}{4}$ in, G = mean diameter of gasket contact face, in; when $b_0 > \frac{1}{4}$ in, G = outside diameter of gasket contact face less $2b$, in
$\qquad P$ = maximum allowable working pressure, psi
$\qquad b$ = effective gasket or joint-contact-surface seating width, in
$\qquad 2b$ = effective gasket or joint-contact-surface pressure width, in
$\qquad b_0$ = basic gasket seating width per Table 26.4 (the table defines b_0 in terms of flange finish and type of gasket, usually from one-half to one-fourth gasket contact width)
$\qquad m$ = gasket factor per Table 26.3 (the table shows m for different types and thicknesses of gaskets ranging from 0.5 to 6.5)
$\qquad y$ = gasket or joint-contact-surface unit seating load, psi (per Table 26.3, which shows values from 0 to 26 000 psi)

Tables 26.3 and 26.4 are reprints of Tables 2-5-1 and 2-5-2 of the 1980 ASME Code [26.10].

To determine bolt diameter based on required load and a specified torque for the grade of bolt, the following is used:

$$W_b = 0.17DT \qquad \text{(for lubricated bolts)} \tag{26.4}$$

or $\qquad\qquad W_b = 0.2DT \qquad \text{(for unlubricated bolts)} \tag{26.5}$

where W_b = load per bolt, lb
$\qquad D$ = bolt diameter, in
$\qquad T$ = torque for grade of bolt selected, lb $\cdot$ in

Note that W_b is the load per bolt and must be multiplied by the number of bolts to obtain total bolt load.

To determine the bolt diameter based on the required load and the allowable bolt stress for a given grade of bolt, use

$$W_b = \sigma_b A_b \tag{26.6}$$

where W_b = load per bolt, lb
$\qquad \sigma_b$ = allowable bolt stress for grade of bolt selected, psi
$\qquad A_b$ = minimum cross-sectional area of bolt, in^2

26.7.3 Simplified Procedure

A simpler method of calculation has been suggested by Whalen [26.11]. This method is also based on the seating stress σ_g on the gasket, as shown in Table 26.5, and on the hydrostatic end force involved in the application. Basically, Whalen's equations accomplish the same thing as the Code, but they are simplified since they use the full gasket contact width, regardless of the flange width and the surface finish of the sealing faces.

This method is based on the total bolt load F_b being sufficient to

1. Seat the gasket material into the flange surface
2. Prevent the hydrostatic end force from unseating the gasket to the point of leakage

In the first case, Table 26.5 lists a range of seating-stress values. The ranges shown were found in a search of the literature on gasket seating stresses. Gasket suppliers can be contacted to confirm these values.

Table 26.6 depicts various gasket types and comments on them. As a starting point in the design procedure, the mean value of σ_g could be used. Then, depending on the severity of the application and/or the safety factor desired, the upper and lower figures could be utilized.

Two equations are associated with this procedure. The first is

$$F_b = \sigma_g A_g \tag{26.7}$$

where F_b = total bolt load, lb
            σ_g = gasket seating stress, psi (from Table 26.5)
            A_g = gasket contact area, in^2

This equation states that the total bolt load must be sufficient to seat the gasket when the hydrostatic end force is not a major factor. The second equation associated with the hydrostatic end force is

$$F_b = KP_t A_m \tag{26.8}$$

where P_t = test pressure or internal pressure if no test pressure is used
           A_m = hydrostatic area on which internal pressure acts (normally based on gasket's middiameter)
            K = safety factor (from Table 26.7)

The safety factors K from Table 26.7 are based on the joint conditions and operating conditions but not on the gasket type or flange surface finish. They are similar to the m factors in the ASME Code. Equation (26.8) states that the total bolt load must be more than enough to overcome the hydrostatic end force. The middiameter is used in A_m since testing has shown that just prior to leakage, the internal pressure acts up to the middiameter of the gasket.

After the desired gasket has been selected, the minimum seating stress, as given in Table 26.5, is used to calculate the total bolt load required by Eq. (26.7). Then the bolt load required to ensure that the hydrostatic end force does not unseat the gasket is calculated from Eq. (26.8). The total bolt load F_b calculated by Eq. (26.7) must be greater than the bolt load calculated in Eq. (26.8). If it is not, then the gasket design must be changed, the gasket's area must be reduced, or the total bolt load must be increased.

Both the ASME procedure and the simplified procedure are associated with gasketed joints which have rigid, usually cast-iron flanges, have high clamp loads, and generally contain high pressures. A great many gasketed joints have stamped-metal covers and splash or very low fluid pressure. In these cases, the procedures do not

TABLE 26.3 Gasket Materials and Contact Facings[†]
Gasket Factors m *for Operating Conditions and Minimum Design Seating Stress* y

Gasket material	Gasket factor m	Minimum design seating stress y, psi	Sketches	Facing sketch and column to be used from Table 26-4
Self-energizing types (O-rings, metallic, elastomer, other gasket types considered as self-sealing)	0	0		
Elastomers without fabric or high percentage of asbestos fiber:				(1a), (1b), (1c), (1d), (4), (5); column II
Below 75A Shore Durometer	0.50	0		
75A or higher Shore Durometer	1.00	200		
Asbestos with suitable binder for operating conditions:				(1a), (1b), (1c), (1d), (4), (5); column II
⅛ in thick	2.00	1 600		
1/16 in thick	2.75	3 700		
1/32 in thick	3.50	6 500		
Elastomers with cotton fabric insertion	1.25	400		(1a), (1b), (1c), (1d), (4), (5); column II
Elastomers with asbestos fabric insertion (with or without wire reinforcement):				(1a), (1b), (1c), (1d), (4), (5); column II
3-ply	2.25	2 200		
2-ply	2.50	2 900		
1-ply	2.75	3 700		
Vegetable fiber	1.75	1 100		(1a), (1b), (1c), (1d), (4), (5); column II
Spiral wound metal, asbestos-filled:				(1a), (1b); column II
Carbon	2.50	10 000		
Stainless or Monel	3.00	10 000		
Corrugated metal, asbestos inserted or corrugated metal, jacketed asbestos-filled:				(1a), (1b); column II
Soft aluminum	2.50	2 900		
Soft copper or brass	2.75	3 700		
Iron or soft steel	3.00	4 500		
Monel or 4–6% chrome	3.25	5 500		
Stainless steels	3.50	6 500		

TABLE 26.3 Gasket Materials and Contact Facings[†]
Gasket Factors m *for Operating Conditions and Minimum Design Seating Stress* y *(Continued)*

Gasket material	Gasket factor m	Minimum design seating stress y, psi	Sketches	Facing sketch and column to be used from Table 26-4
Corrugated Metal:				
Soft aluminum	2.75	3 700		(1a), (1b), (1c), (1d); column II
Soft copper or brass	3.00	4 500		
Iron or soft steel	3.25	5 500		
Monel or 4–6% chrome	3.50	6 500		
Stainless steels	3.75	7 600		
Flat metal, jacketed asbestos-filled:				
Soft aluminum	3.25	5 500		(1a), (1b), (1c),‡ (1d),‡ (2)‡; column II
Soft copper or brass	3.50	6 500		
Iron or soft steel	3.75	7 600		
Monel or 4–6% chrome	3.50	8 000		
	3.75	9 000		
Stainless steels	3.75	9 000		
Grooved metal:				
Soft aluminum	3.25	5 500		(1a), (1b), (1c), (1d), (2), (3); column II
Soft copper or brass	3.50	6 500		
Iron or soft steel	3.75	7 600		
Monel or 4–6% chrome	3.75	9 000		
Stainless steels	4.25	10 100		
Solid flat metal:				
Soft aluminum	4.00	8 800		(1a), (1b), (1c), (1d), (2), (3), (4), (5); column I
Soft copper or brass	4.75	13 000		
Iron or soft steel	5.50	18 000		
Monel or 4–6% chrome	6.00	21 800		
Stainless steels	6.50	26 000		
Ring joint:				
Iron or soft steel	5.50	18 000		(6); column I
Monel or 4–6% chrome	6.00	21 800		
Stainless steels	6.50	26 000		

†This table gives a list of many commonly used gasket materials and contact facings with suggested design values of *m* and *y* that have generally proved satisfactory in actual service when using effective gasket seating width *b* given in Table 26.4. The design values and other details given in this table are only suggested and are not mandatory.

‡The surface of a gasket having a lap should not be against the nubbin.

TABLE 26.4 Effective Gasket Width[†]

Facing sketch (exaggerated)	Basic gasket seating width b_o	
	Column I	Column II
(1a)	$\dfrac{N}{2}$	$\dfrac{N}{2}$
(1b)‡		
(1c) $w \leq N$	$\dfrac{w+T}{2}\left(\dfrac{w+N}{4} \text{ max.}\right)$	$\dfrac{w+T}{2}\left(\dfrac{w+N}{4} \text{ max.}\right)$
(1d)‡ $w \leq N$		
(2) $\tfrac{1}{64}$" Nubbin $w \leq \dfrac{N}{2}$	$\dfrac{w+N}{4}$	$\dfrac{w+3N}{8}$

	(3) $\tfrac{1}{64}''$ Nubbin	$w \leq \dfrac{N}{2}$	$\dfrac{N}{4}$	$\dfrac{3N}{8}$
(4)‡			$\dfrac{3N}{8}$	$\dfrac{7N}{16}$
(5)‡			$\dfrac{N}{4}$	$\dfrac{3N}{4}$
(6)			$\dfrac{w}{8}$	

S81   Effective gasket seating width b:

$$b = b_o \text{ when } b_o \leq \tfrac{1}{4} \text{ in} \qquad b = 0.5\sqrt{b_o} \text{ when } b_o > \tfrac{1}{4} \text{ in}$$

Location of gasket load reaction:

For $b_O > 1/4$ in.

For $b_O \leq 1/4$ in.

†The gasket factors listed apply only to flanged joints in which the gasket is contained entirely within the inner edges of the bolt holes.
‡Where separations do not exceed $\tfrac{1}{64}$-in-depth and $\tfrac{1}{32}$-in-width spacing, sketches (1b) and (1d) shall be used.

26.19

TABLE 26.5 Minimum Recommended Seating Stresses for Various Gasket Materials

	Material	Gasket type	Minimum seating stress range (S_g), psi†
Nonmetallic	Asbestos fiber sheet $\frac{1}{8}$ in thick $\frac{1}{16}$ in thich $\frac{1}{32}$ in thick	Flat	1400 to 1600 3500 to 3700 6000 to 6500
	Asbestos fiber sheet $\frac{1}{32}$ in thick	Flat with rubber beads	1000 to 1500 lb/in on beads
	Asbestos fiber sheet $\frac{1}{32}$ in thick	Flat with metal grommet	3000 to 4000 lb/in on grommet
	Asbestos fiber sheet $\frac{1}{32}$ in thick	Flat with metal grommet and metal wire	2000 to 3000 lb/in on wire
	Cellulose fiber sheet	Flat	750 to 1100
	Cork composition	Flat	400 to 500
	Cork-rubber	Flat	200 to 300
	Fluorocarbon (TFE) $\frac{1}{8}$ in thick $\frac{1}{16}$ in thick $\frac{1}{32}$ in thick	Flat	1500 to 1700 3500 to 3800 6200 to 6500
	Nonasbestos fiber sheets (glass, carbon, aramid, and ceramics)	Flat	1500 to 3000 depending on composition
	Rubber	Flat	100 to 200
	Rubber with fabric or metal reinforcement	Flat with reinforcement	300 to 500
Metallic	Aluminum	Flat	10 000 to 20 000
	Copper	Flat	15 000 to 45 000 depending on hardness
	Carbon steel	Flat	30 000 to 70 000 depending on alloy and hardness
	Stainless steel	Flat	35 000 to 95 000 depending on alloy and hardness
	Aluminum (soft)	Corrugated	1000 to 3700
	Copper (soft)	Corrugated	2500 to 4500
	Carbon steel (soft)	Corrugated	3500 to 5500
	Stainless steel	Corrugated	6000 to 8000
	Aluminum	Profile	25 000
	Copper	Profile	35 000
	Carbon steel	Profile	55 000
	Stainless steel	Profile	75 000
Jacketed metal-asbestos	Aluminum	Plain	2 500
	Copper	Plain	4 000
	Carbon steel	Plain	6 000
	Stainless steel	Plain	10 000
	Aluminum	Corrugated	2000
	Copper	Corrugated	2500
	Carbon steel	Corrugated	3000
	Stainless steel	Corrugated	4000
	Stainless steel	Spiral-wound	3000 to 30 000

†Stresses in pounds per square inch except where otherwise noted.

TABLE 26.6 Typical Gasket Designs and Descriptions

Type	Cross section	Comments
Flat		Basic form. Available in wide variety of materials. Easily fabricated into different shapes.
Reinforced		Fabric- or metal-reinforced. Improves torque retention and blowout resistance of flat types. Reinforced type can be corrugated.
Flat with rubber beads		Rubber beads located on flat or reinforced material afford high unit sealing pressure and high degree of conformability.
Flat with metal grommet		Metal grommet affords protection to base material from medium and provides high unit sealing stress. Soft metal wires can be put under grommet for higher unit sealing stress.
Plain metal jacket		Basic sandwich type. Filler is compressible. Metal affords protection to filler on one edge and across surfaces.
Corrugated or embossed		Corrugations provide for increased sealing pressure and higher conformability. Primarily circular. Corrugations can be filled with soft filler.
Profile		Multiple sealing surfaces. Seating stress decreases with increase in pitch. Wide varieties of designs are available.
Spiral-wound		Interleaving pattern of metal and filler. Ratio of metal to filler can be varied to meet demands of different applications.

TABLE 26.7 Safety Factors for Gasketed Joints

K factor	When to apply
1.2 to 1.4	For minimum-weight applications where all installation factors (bolt lubrication, tension, parallel seating, etc.) are carefully controlled; ambient to 250°F (121°C) temperature applications; where adequate proof pressure is applied.
1.5 to 2.5	For most normal designs where weight is not a major factor, vibration is moderate and temperatures do not exceed 750°F (399°C). Use high end of range where bolts are not lubricated.
2.6 to 4.0	For cases of extreme fluctuations in pressure, temperature, or vibration; where no test pressure is applied; or where uniform bolt tension is difficult to ensure.

apply, and the compression and stress distribution discussed next should be considered by the designer.

26.8 GASKET COMPRESSION AND STRESS-DISTRIBUTION TESTING

After a gasket has been selected and designed for a particular application, two simple tests can be performed to determine the gasket's compressed thickness and stress distribution. Inadequate compression or nonuniform stress distribution could result in a leaking joint. The tests can be performed to check for these possibilities and permit correction to ensure leaktight joints.

1. *Lead pellet test*   In this test, lead pellets are used to accurately indicate the compressed thicknesses of a gasketed joint. The pellets, commonly called *lead shot,* are available from local gun supply stores. A size approximately twice the thickness of the gasket should be used. Lead solid-core solder can also be used if desired; the size requirements are the same. Pellets or solder are particularly well suited for doing this test, as they exhibit no recovery after compression, whereas the actual gasket material will almost always exhibit some recovery. The degree of nonuniform loading, flange bowing, or distortion will be indicated by the variations in the gasket's compressed thickness.

To begin, the original thickness of the gasket is measured and recorded at uniformly selected points across the gasket. At or near these points, holes are punched or drilled through the gasket. Care should be taken to remove any burrs. The punched holes should be approximately 1½ times the pellet diameter.

Then the gasket is mounted on the flange. A small amount of grease can be put in the punched holes to hold the lead pellets, if required. The pellets are mounted in the grease, and the mating flange is located and torqued to specifications.

Upon careful disassembly of the flange and removal of the pellets, their thicknesses are measured, recorded, and analyzed. Comparison of the pellets' compressed thicknesses to the gasket's stress-compression characteristics permits the desired stress-distribution analysis.

2. *NCR paper test* This test utilizes no-carbon-required (NCR) paper for visual determination of the stress distribution on a gasket. NCR paper is a pressure-sensitive, color-reactive paper. The intensity of color is proportional to the stress imposed on the paper, which is the same as the stress on the gasket.

NCR paper is available from the NCR Corp., the 3M Company, and other paper companies. Various grades are available, but the medium grade is usually chosen. Some papers are only one sheet, whereas others are composed of two sheets. Either type can be used.

To begin, the bolt holes are pierced in a piece of the impression paper. The pierced holes in the paper are made slightly larger than the bolts. The paper is placed on the flange, and the mating flange is assembled per torque specifications.

When you are using the two-piece carbonless paper, make sure to keep the two papers oriented to each other as they were purchased; otherwise, no impression may result. Upon torquing, the impression is made on the paper. The flange is removed, and the impression is inspected for stress distribution. A judgment of the gasket's sealing ability can now be made. Further analysis can be done by calibrating the load versus the color intensity of the paper. Various known stresses can be applied to the paper and the resulting color impressions identified. The impressions can be compared to the test sample, and then the stress on the sample can be determined.

In both the lead pellet and NCR paper tests, gasket manufacturers can be contacted for further interpretation of the results and more detailed analysis.

26.9 INSTALLATION SPECIFICATIONS

An installation is only as good as its gasket; likewise, a gasket is only as good as its installation [26.1]. The following are some recommendations associated with gasket installation:

1. Be sure that mating surfaces are clean and in specification with regard to finish, flatness, and waviness.
2. Check gasket for damage before installing it.
3. Make certain the gasket fits the application.
4. Specify lubricated bolts. Bolt threads and the underside of the bolt head should be lubricated.
5. Specify the torque level.
6. Specify the torquing sequence. In addition to the sequence, two or three stages of torque before reaching the specified level are recommended.

REFERENCES

26.1 ANSI/ASTM F104-79a, *Standard Classification System for Nonmetallic Gasket Materials,* American Society for Testing and Materials.

26.2 D. E. Czernik, J. C. Moerk, Jr., and F. A. Robbins, "The Relationship of a Gasket's Physical Properties to the Sealing Phenomena," SAE paper 650431, May 1965.

26.3 D. E. Czernik, "Recent Developments and New Approaches in Mechanical and Chemical Gasketing," SAE paper 810367, February 1981.

26.4 V. M. Faires, *Design of Machine Elements,* Macmillan, New York, 1955.

26.5 D. J. McDowell, "Choose the Right Gasket Material," *Assembly Engineering,* October 1978.

26.6 H. A. Rothbart, *Mechanical Design and Systems Handbook,* 2d ed., McGraw-Hill, New York, 1985, Sec. 27.4.

26.7 *Armstrong Gasket Design Manual,* Armstrong Cork Co., Lancaster, Pa., 1978.

26.8 J. W. Oren, "Creating Gasket Seals with Rigid Flanges," SAE paper 810362, February 1981.

26.9 D. E. Czernik, "Gasketing the Internal Combustion Engine," SAE paper 800073, February 1980.

26.10 The American Society of Mechanical Engineers, Code for Pressure Vessels, Sec. VIII, Div. 1, App. 2, 1980.

26.11 J. J. Whalen, "How to Select the Right Gasket Material," *Product Engineering,* October 1960.

26.12 D. E. Czernik, "Sealing Today's Engines," *Fleet Maintenance and Specifying,* Irving-Cloud, July 1977.

CHAPTER 27
ROLLING-CONTACT BEARINGS

Charles R. Mischke, Ph.D., P.E.
Professor Emeritus of Mechanical Engineering
Iowa State University
Ames, Iowa

GLOSSARY OF SYMBOLS

a	Exponents; $a = 3$ for ball bearings; $a = {}^{10}\!/_3$ for roller bearings
AF	Application factor
b	Weibull shape parameter
C_s	Static load rating
C_{10}	Basic load rating or basic dynamic load rating
f	Fraction
F	Load
F_a	Axial load
F_{eq}	Equivalent radial load
F_i	ith equivalent radial load
F_r	Radial load
I	Integral
L	Life measure, r or h
L_D	Desired or design life measure
L_R	Rating life measure
L_{10}	Life measure exceeded by 90 percent of bearings tested

n	Design factor
n_D	Desired or design rotative speed, r/min
n_i	Application or design factor at *i*th level
n_R	Rating rotative speed, r/min
R	Reliability
V	Rotation factor; inner ring rotations, $V = 1$; outer ring, $V = 1.20$
x	Life measure in Weibull survival equation
x_0	Weibull guaranteed life parameter
X	Radial factor for equivalent load prediction
Y	Thrust factor for equivalent load prediction
θ	Weibull characteristic life parameter, rotation angle
φ	Period of cyclic variation, rad

27.1 INTRODUCTION

Figures 27.1 to 27.12 illustrate something of the terminology and the wide variety of rolling-contact bearings available to the designer. Catalogs and engineering manuals can be obtained from bearing manufacturers, and these are very comprehensive and of excellent quality. In addition, most manufacturers are anxious to advise designers on specific applications. For this reason the material in this chapter is concerned mostly with providing the designer an independent viewpoint.

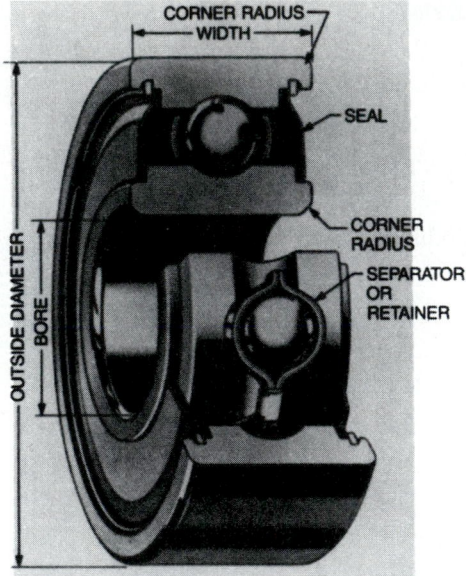

FIGURE 27.1 Photograph of a deep-groove precision ball bearing with metal two-piece cage and dual seals to illustrate rolling-bearing terminology. *(The Barden Corporation.)*

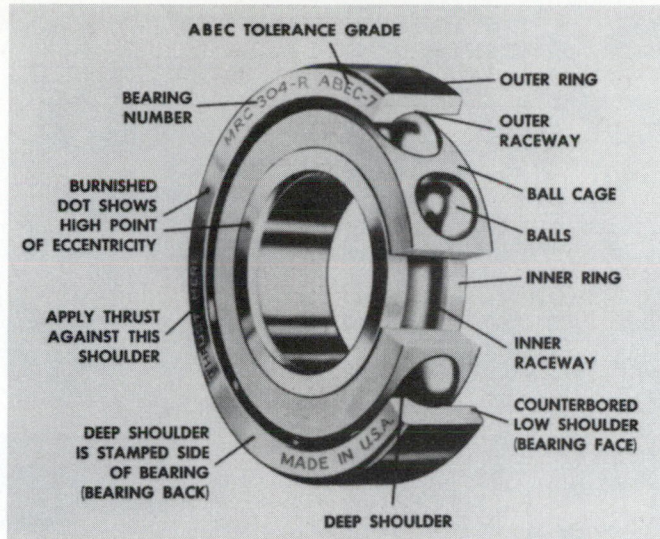

FIGURE 27.2 Photograph of a precision ball bearing of the type generally used in machine-tool applications to illustrate terminology. *(Bearings Division, TRW Industrial Products Group.)*

FIGURE 27.3 Rolling bearing with spherical rolling elements to permit misalignment up to ±3° with an unsealed design. The sealed bearing, shown above, permits misalignment to ±2°. *(McGill Manufacturing Company, Inc.)*

FIGURE 27.4 A heavy-duty cage-guided needle roller bearing with machined race. Note the absence of an inner ring, but standard inner rings can be obtained. *(McGill Manufacturing Company, Inc.)*

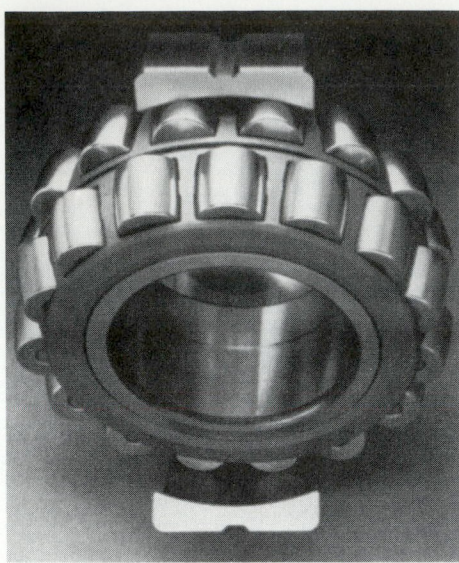

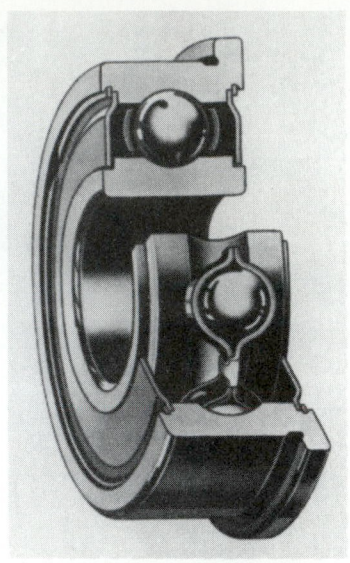

FIGURE 27.5   A spherical roller bearing with two rows of rollers running on a common sphered raceway. These bearings are self-aligning to permit misalignment resulting from either mounting or shaft deflection under load. *(SKF Industries, Inc.)*

FIGURE 27.6   Shielded, flanged, deep-groove ball bearing. Shields serve as dirt barriers; flange facilitates mounting the bearing in a through-bored hole. *(The Barden Corporation.)*

FIGURE 27.7   Ball thrust bearing. *(The Torrington Company.)*

FIGURE 27.8   Spherical roller thrust bearing. *(The Torrington Company.)*

FIGURE 27.9 Tapered-roller thrust bearing. *(The Torrington Company.)*

FIGURE 27.10 Tapered-roller bearing; for axial loads, thrust loads, or combined axial and thrust loads. *(The Timken Company.)*

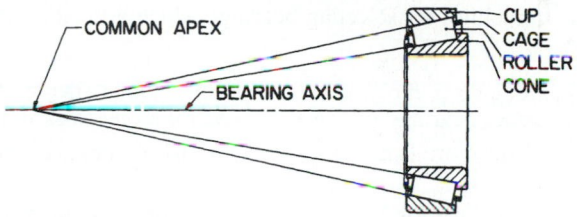

FIGURE 27.11 Basic principle of a tapered-roller bearing with nomenclature. *(The Timken Company.)*

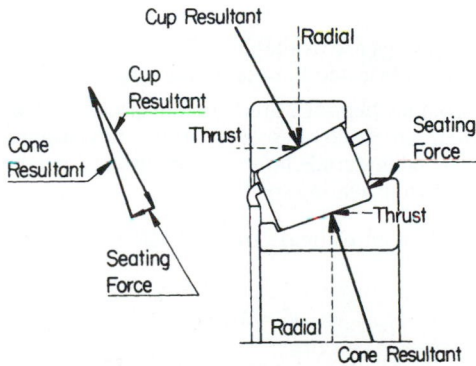

FIGURE 27.12 Force analysis of a Timken bearing. *(The Timken Company.)*

TABLE 27.1 Coefficients of Friction

Bearing type	Coefficient of friction μ
Self-aligning ball	0.0010
Cylindrical roller with flange-guided short rollers	0.0011
Ball thrust	0.0013
Single-row ball	0.0015
Spherical roller	0.0018
Tapered roller	0.0018

SOURCE: Ref. [27.1].

Rolling-contact bearings use balls and rollers to exploit the small coefficients of friction when hard bodies roll on each other. The balls and rollers are kept separated and equally spaced by a separator (cage, or retainer). This device, which is essential to proper bearing functioning, is responsible for additional friction. Table 27.1 gives friction coefficients for several types of bearings [27.1]. Consult a manufacturer's catalog for equations for estimating friction torque as a function of bearing mean diameter, load, basic load rating, and lubrication detail. See also Chap. 25.

Permissible speeds are influenced by bearing size, properties, lubrication detail, and operating temperatures. The speed varies inversely with mean bearing diameter. For additional details, consult any manufacturer's catalog.

Some of the guidelines for selecting bearings, which are valid more often than not, are as follows:

- Ball bearings are the less expensive choice in the smaller sizes and under lighter loads, whereas roller bearings are less expensive for larger sizes and heavier loads.
- Roller bearings are more satisfactory under shock or impact loading than ball bearings.
- Ball-thrust bearings are for pure thrust loading only. At high speeds a deep-groove or angular-contact ball bearing usually will be a better choice, even for pure thrust loads.
- Self-aligning ball bearings and cylindrical roller bearings have very low friction coefficients.
- Deep-groove ball bearings are available with seals built into the bearing so that the bearing can be prelubricated to operate for long periods without attention.
- Although rolling-contact bearings are "standardized" and easily selected from vendor catalogs, there are instances of cooperative development by customer and vendor involving special materials, hollow elements, distorted raceways, and novel applications. Consult your bearing specialist.

It is possible to obtain an estimate of the basic static load rating C_s. For ball bearings,

$$C_s = Mn_b d_b^2 \tag{27.1}$$

For roller bearings,

$$C_s = Mn_r l_e d \tag{27.2}$$

where C_s = basic static loading rating, pounds (lb) [kilonewtons (kN)]
        n_b = number of balls
        n_r = number of rollers
        d_b = ball diameter, inches (in) [millimeters (mm)]
        d = roller diameter, in (mm)
        l_e = length of single-roller contact line, in (mm)

Values of the constant M are listed in Table 27.2.

TABLE 27.2   Value of Constant M for Use in
Eqs. (27.1) and (27.2)

Type of bearing	Constant M	
	C_s, lb	C_s, kN
Radial ball	1.78×10^3	5.11×10^3
Ball thrust	7.10×10^3	20.4×10^3
Radial roller	3.13×10^3	8.99×10^3
Roller thrust	14.2×10^3	40.7×10^3

27.2 *LOAD-LIFE RELATION FOR CONSTANT RELIABILITY*

When proper attention is paid to a rolling-contact bearing so that fatigue of the material is the only cause of failure, then nominally identical bearings exhibit a reliability–life-measure curve, as depicted in Fig. 27.13. The *rating life* is defined as the life measure (revolutions, hours, etc.) which 90 percent of the bearings will equal or exceed. This is also called the L_{10} life or the B_{10} life. When the radial load is adjusted so that the L_{10} life is 1 000 000 revolutions (r), that load is called the *basic load rating* C (SKF Industries, Inc.). The Timken Company rates its bearings at 90 000 000. Whatever the rating basis, the life L can be normalized by dividing by the rating life L_{10}. The *median life* is the life measure equaled or exceeded by half of the bearings. Median life is roughly 5 times rating life.

For steady radial loading, the life at which the first tangible evidence of surface fatigue occurs can be predicted from

$$F^a L = \text{constant} \tag{27.3}$$

where $a = 3$ for ball bearings and $a = {}^{10}\!/_{3}$ for cylindrical and tapered-roller bearings. At constant reliability, the load and life at condition 1 can be related to the load and life at condition 2 by Eq. (27.3). Thus

$$F_1^a L_1 = F_2^a L_2 \tag{27.4}$$

If F_1 is the basic load rating C_{10}, then L_1 is the rating life L_{10}, and so

$$C_{10} = \left(\frac{L}{L_{10}}\right)^{1/a} (F) \tag{27.5}$$

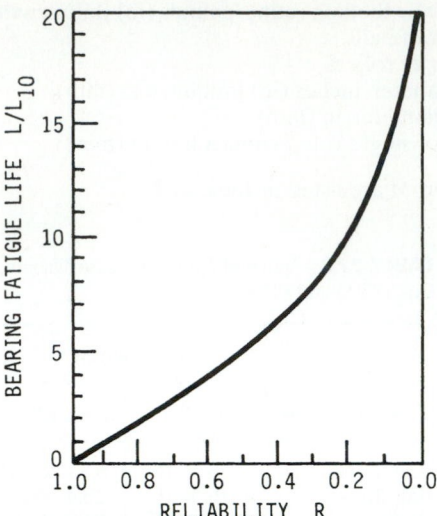

FIGURE 27.13 Survival function representing endurance tests on rolling-contact bearings from data accumulated by SKF Industries, Inc. *(From Ref. [27.2].)*

If L_R is in hours and n_R is in revolutions per minute, then $L_{10} = 60 L_R n_R$. It follows that

$$C_{10} = F_D \left(\frac{L_D n_D}{L_R n_R} \right)^{1/a} \tag{27.6}$$

where the subscript D refers to desired (or design) and the subscript R refers to rating conditions.

27.3 SURVIVAL RELATION AT STEADY LOAD

Figure 27.14 shows how reliability varies as the loading is modified [27.2]. Equation (27.5) allows the ordinate to be expressed as either F/C_{10} or L/L_{10}. Figure 27.14 is based on more than 2500 SKF bearings. If Figs. 27.13 and 27.14 are scaled for recovery of coordinates, then the reliability can be tabulated together with L/L_{10}. Machinery applications use reliabilities exceeding 0.94. An excellent curve fit can be realized by using the three-parameter Weibull distribution (see Table 2.2 and Sec. 2.6). For this distribution the reliability can be expressed as

$$R = \exp \left[-\left(\frac{x - x_0}{\theta - x_0} \right)^b \right] \tag{27.7}$$

where x = life measure, x_0 = Weibull guaranteed life measure, θ = Weibull characteristic life measure, and b = Weibull shape factor. Using the 18 points in Table 27.3 with $x_0 = 0.02$, $\theta = 4.459$, and $b = 1.483$, we see that Eq. (27.7) can be particularized as

$$R = \exp \left[-\left(\frac{L/L_{10} - 0.02}{4.439} \right)^{1.483} \right] \tag{27.8}$$

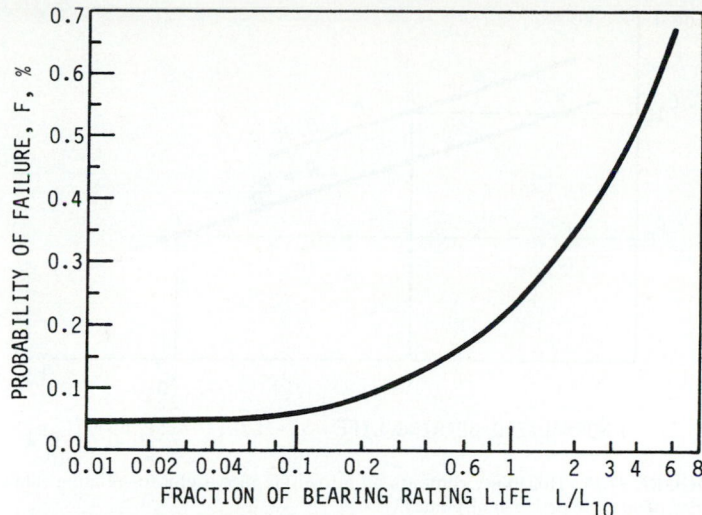

FIGURE 27.14 Survival function at higher reliabilities based on more than 2500 endurance tests by SKF Industries, Inc. *(From Ref. [27.2].)* The three-parameter Weibull constants are $\theta = 4.459$, $b = 1.483$, and $x_0 = 0.02$ when $x = L/L_{10} = Ln/(L_R n_R)$.

For example, for $L/L_{10} = 0.1$, Eq. (27.8) predicts $R = 0.9974$.

27.4 RELATING LOAD, LIFE, AND RELIABILITY GOAL

If Eq. (27.3) is plotted on log-log coordinates, Fig. 27.15 results. The FL loci are rectified, while the parallel loci exhibit different reliabilities. The coordinates of point A are the rating life and the basic load rating. Point D represents the desired (or design) life and the corresponding load. A common problem is to select a bearing which will provide a life L_D while carrying load F_D and exhibit a reliability R_D. Along line BD, constant reliability prevails, and Eq. (27.4) applies:

TABLE 27.3 Survival Equation Points at Higher Reliabilities[†]

Reliability R	Life measure L/L_{10}	Reliability R	Life measure L/L_{10}
0.94	0.67	0.994	0.17
0.95	0.60	0.995	0.15
0.96	0.52	0.996	0.13
0.97	0.435	0.997	0.11
0.975	0.395	0.9975	0.095
0.98	0.35	0.998	0.08
0.985	0.29	0.9985	0.07
0.99	0.23	0.999	0.06
0.992	0.20	0.9995	0.05

[†]Scaled from Ref. [27.2], Fig. 2.

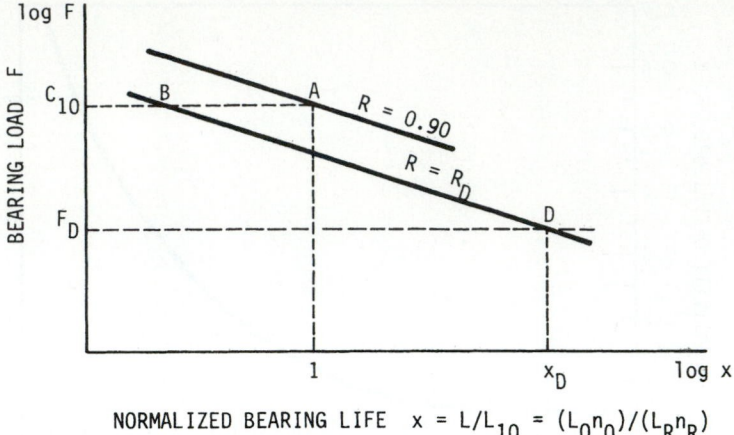

FIGURE 27.15   Reliability contours on a load-life plot useful for relating catalog entry, point A, to design goal, point D.

$$F_B = F_D \left(\frac{x_D}{x_B}\right)^{1/a} \tag{27.9}$$

Along line AB the reliability changes, but the load is constant and Eq. (27.7) applies. Thus

$$R = \exp\left[-\left(\frac{x - x_0}{\theta - x_0}\right)^b\right] \tag{27.10}$$

Now solve this equation for x and particularize it for point B, noting that $R_D = R_B$.

$$x_B = x_0 + (\theta - x_0)\left(\ln \frac{1}{R_D}\right)^{1/b} \tag{27.11}$$

Substituting Eq. (27.11) into Eq. (27.9) yields

$$F_B = C_{10} = F_D\left\{\frac{x_D}{x_0 + (\theta - x_0)[\ln (1/R_D)]^{1/b}}\right\}^{1/a} \tag{27.12}$$

For reliabilities greater than 0.90, which is the usual case, $\ln (1/R) \cong 1 - R$ and Eq. (27.12) simplifies as follows:

$$C_{10} = F_D\left[\frac{x_D}{x_0 + (\theta - x_0)(1 - R)^{1/b}}\right]^{1/a} \tag{27.13}$$

The desired life measure x_D can be expressed most conveniently in millions of revolutions (for SKF).

Example 1.   If a ball bearing must carry a load of 800 lb for 50×10^6 and exhibit a reliability of 0.99, then the basic load rating should equal or exceed

$$C_{10} = 800 \left[\frac{50}{0.02 + (4.439)(1 - 0.99)^{1/1.483}} \right]^{1/3}$$

$$= 4890 \text{ lb}$$

This is the same as 21.80 kN, which corresponds to the capability of a 02 series 35-mm-bore ball bearing. Since selected bearings have different basic load ratings from those required, a solution to Eq. (27.13) for reliability extant after specification is useful:

$$R = 1 - \left[\frac{x_D - x_0(C_{10}/F_D)^a}{(\theta - x_0)(C_{10}/F_D)^a} \right]^b \tag{27.14}$$

Example 2. If the bearing selected for Example 1, a 02 series 50-mm bore, has a basic load rating of 26.9 kN, what is the expected reliability? And $C_{10} = 26.9 \times 10^3)/445 = 6045$ lb. So

$$R = 1 - \left[\frac{50 - 0.02(6045/800)^3}{(4.439)(6045/800)^3} \right]^{1.483} = 0.9966$$

The previous equations can be adjusted to a two-parameter Weibull survival equation by setting x_0 to zero and using appropriate values of θ and b. For bearings rated at a particular speed and time, substitute $L_D n_D/(L_R n_R)$ for x_D.

The survival relationship for Timken tapered-roller bearings is shown graphically in Fig. 27.16, and points scaled from this curve form the basis for Table 27.4. The survival equation turns out to be the two-parameter Weibull relation:

$$R = \exp\left[-\left(\frac{x}{\theta} \right)^b \right] = \exp\left[-\left(\frac{L/L_{10}}{4.890} \right)^{1.4335} \right] \tag{27.15}$$

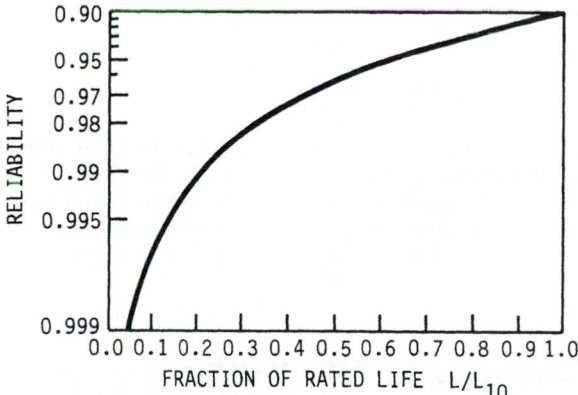

FIGURE 27.16 Survival function at higher reliabilities based on the Timken Company tapered-roller bearings. The curve fit is a two-parameter Weibull function with constants $\theta = 4.48$ and $b = \frac{3}{4}$ ($x_0 = 0$) when $x = Ln/(L_R n_R)$. *(From Ref. [27.3].)*

TABLE 27.4 Survival Equation Points for Tapered-Roller Bearings[†]

Reliability R	Life measure L/L_{10}	Reliability R	Life measure L/L_{10}
0.90	1.00	0.96	0.53
0.91	0.92	0.97	0.43
0.92	0.86	0.98	0.325
0.93	0.78	0.99	0.20
0.94	0.70	0.995	0.13
0.95	0.62	0.999	0.04

[†]Scaled from Fig. 4 of *Engineering Journal*, Sec. 1, The Timken Company, Canton, Ohio, rev. 1978.

The equation corresponding to Eq. (27.13) is

$$C_{10} = F_D \left[\frac{x_D}{\theta(1-R)^{1/b}} \right]^{1/a}$$

$$= F_D \left(\frac{x_D}{\theta} \right)^{1/a} (1-R)^{-1/ab} \tag{27.16}$$

And the equation corresponding to Eq. (27.14) is

$$R = 1 - \left(\frac{x_D}{\theta} \right)^b \left(\frac{C_{10}}{F_D} \right)^{-ab} \tag{27.17}$$

Example 3. A Timken tapered-roller bearing is to be selected to carry a radial load of 4 kN and have a reliability of 0.99 at 1200 hours (h) and a speed of 600 revolutions per minute (r/min). Thus

$$x_D = \frac{L_D n_D}{L_R n_R} = \frac{1200(600)}{3000(500)} = 0.480$$

and

$$C_{10} = 4 \left[\frac{0.48}{4.48(1-0.99)^{1/1.5}} \right]^{3/10} = 5141 \text{ N}$$

Timken bearings are rated in U.S. Customary System (USCS) units or in newtons; therefore, a basic load rating of 5141 N or higher is to be sought.

For any bearings to be specified, check with the manufacturer's engineering manual for survival equation information. This is usually in the form of graphs, nomograms, or equations of available candidates. Check with the manufacturer on cost because production runs materially affect bearing cost.

27.5 COMBINED RADIAL AND THRUST LOADINGS

Ball bearings can resist some thrust loading simultaneously with a radial load. The equivalent radial load is the constant pure radial load which inflicts the same damage on the bearing per revolution as the combination. A common form for weighting the radial load F_r and the axial load F_a is

$$F_e = VXF_r + YF_a \qquad (27.18)$$

where F_e = equivalent radial load. The weighting factors X and Y are given for each bearing type in the manufacturer's engineering manual. The parameter V distinguishes between inner-ring rotation, $V = 1$, and outer-ring rotation, $V = 1.20$. A common form of Eq. (27.18) is

$$F_e = \max(VF_r, X_1VF_r + Y_1F_a, X_2VF_r + Y_2F_a, \ldots) \qquad (27.19)$$

27.6 APPLICATION FACTORS

In machinery applications the peak radial loads on a bearing are different from the nominal or average load owing to a variation in torque or other influences. For a number of situations in which there is a body of measurement and experience, bearing manufacturers tabulate application factors that are used to multiply the average load to properly account for the additional fatigue damage resulting from the fluctuations. Such factors perform the same function as a design factor. In previous equations, F_D is replaced by nF_D or $AF(F_D)$, where AF is the application factor.

27.7 VARIABLE LOADING

At constant reliability the current F^aL product measures progress toward failure. The area under the F^a versus L curve at failure is an index to total damage resulting in failure. The area under the F^aL locus at any time prior to failure is an index to damage so far. If the radial load or equivalent radial load varies during a revolution or several revolutions in a periodic fashion, then the equivalent radial load is related to the instantaneous radial load by

$$F_{eq} = \left(\frac{1}{\phi} \int_0^\phi F^a \, d\theta \right)^{1/a} \qquad (27.20)$$

where ϕ = period of the variation—2π for repetition every revolution, 4π for repetition every second revolution, etc. (see Fig. 27.17).

Example 4. A bearing load is given by $F(\theta) = 1000 \sin\theta$ in pounds force. Estimate the equivalent load by using Simpson's rule,

$$F_{eq} = \left[\frac{1}{\pi} \int_0^\pi (1000 \sin\theta)^{10/3} \, d\theta \right]^{3/10} = 762 \text{ lb}$$

When equivalent loads are applied in a stepwise fashion, the equivalent radial load is expressible by

$$F_{eq} = \left[\sum_{i=1}^k f_i(n_i F_i)^a \right]^{1/a} \qquad (27.21)$$

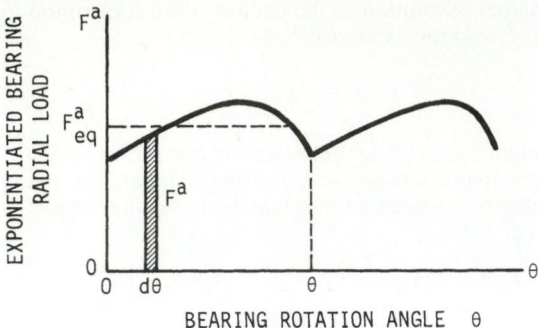

FIGURE 27.17   Equivalent radial load when load varies periodically with angular position.

where f_i = fraction of revolution at load F_i
   n_i = application or design factor
   F_i = ith equivalent radial load
   a = applicable exponent—3 for ball bearings and $^{10}\!/_3$ for roller bearings

Example 5.   A four-step loading cycle is applied to a ball bearing. For one-tenth of the time, the speed is 1000 rpm, $F_r = 800$ lb, and $F_a = 400$ lb; for two-tenths of the time, the speed is 1200 rpm, $F_r = 1000$ lb, and $F_a = 500$ lb; for three-tenths of the time, the speed is 1500 rpm, $F_r = 1500$ lb, and $F_a = 700$ lb; for four-tenths of the time, the speed is 800 rpm, $F_r = 1100$ lb, and $F_a = 500$ lb. For this shallow-angle, angular-contact ball bearing, $X_1 = 1$, $Y_1 = 1.25$, $X_2 = 0.45$, $Y_2 = 1.2$, and $V = 1$. This loading cycle is also depicted in Fig. 27.18.

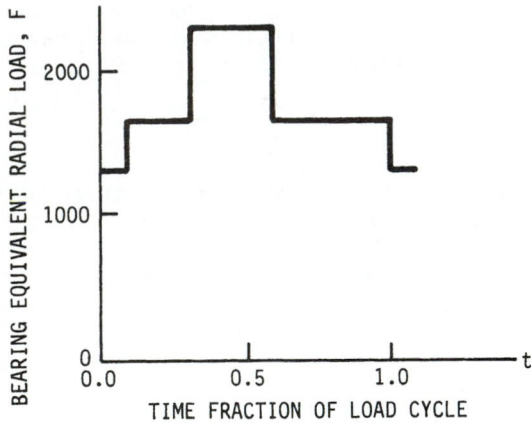

FIGURE 27.18   Loading cycle: one-tenth of time at 1000 rpm, $F_r = 800$, $F_a = 400$; two-tenths of time at 1200 rpm, $F_r = 1000$, $F_a = 500$; three-tenths of time at 1500 rpm, $F_r = 1500$, $F_a = 700$; four-tenths of time at 800 rpm, $F_r = 1100$, $F_a = 500$; $X_1 = 1$, $Y_1 = 1.25$, $X_2 = 0.45$, $Y_2 = 1.2$, $V = 1$.

TABLE 27.5 Tabulation for Example 5

Time fraction	Speed, rpm	Product	Revolution fraction f_i	Radial load F_r	Axial load F_a	Equivalent load F_e	Application factor AF	Product (AF)(F_e)
0.1	1000	100	0.090	800	400	1300	1.1	1430
0.2	1200	240	0.216	1000	500	1625	1.25	2031
0.3	1500	450	0.405	1500	700	2375	1.25	2969
0.4	800	320	0.288	1100	500	1725	1.50	2588
		1110						

27.15

The first step in the solution is to create Table 27.5. The equivalent radial load is

$$F_{eq} = [0.090(1430)^3 + 0.216(2031)^3 + 0.405(2969)^3 + 0.288(2588)^3]^{1/3} = 2604 \text{ lb}$$

Without the use of design factors, the equivalent radial load is

$$F_{eq} = [0.090(1300)^3 + 0.216(1625)^3 + 0.405(2375)^3 + 0.288(1725)^3]^{1/3} = 2002 \text{ lb}$$

The overall design factor is 2604/2002, or 1.30. If this sequence were common in a machinery application, a bearing manufacturer might recommend an application factor of 1.30 for this particular application.

27.8 MISALIGNMENT

The inner ring of a rolling-contact bearing is tightly fitted to the shaft, and the axis of rotation is oriented, as is the shaft centerline. The outer ring is held by some form of housing, and its axis is oriented as demanded by the housing. As the shaft deflects under load, these two axes lie at an angle to each other. This misalignment for very small angles is accommodated in "slack," and no adverse life consequences are exhibited. As soon as the slack is exhausted, the intended deflection is resisted and the bearing experiences unintended loading. Life is reduced below prediction levels. A shaft design which is too limber does not fail, but bearings are replaced with much greater frequency. It is too easy to be critical of bearings when the problem lies in the shaft design.

Figure 27.19 shows the dramatic fractional life reduction owing to misalignment in line-contact bearings [27.4].

If there is misalignment, it should not exceed 0.001 radian (rad) in cylindrical and tapered-roller bearings, 0.0087 rad for spherical ball bearings, or about 0.004 rad for

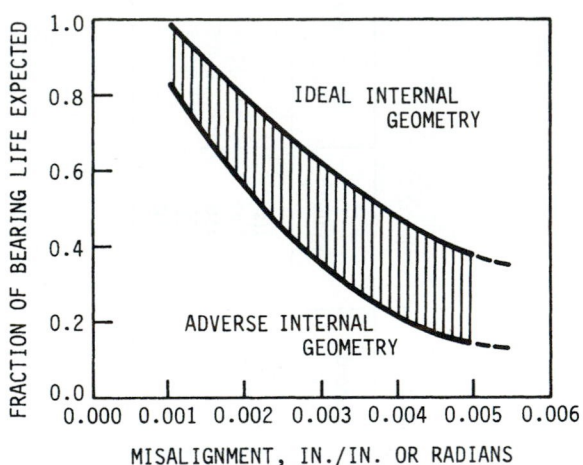

FIGURE 27.19 Fractional bearing life to be expected as a function of misalignment in line-contact bearings. *(From Ref. [27.4], Fig. 11.)*

deep-groove ball bearings. Self-aligning ball or spherical roller bearings are more tolerant of misalignment. The bibliography of Ref. [27.4] is extensive on this subject.

REFERENCES

27.1 *SKF Engineering Data,* SKF Industries, Inc., Philadelphia, 1979.

27.2 T. A. Harris, "Predicting Bearing Reliability," *Machine Design,* vol. 35, no. 1, Jan. 3, 1963, pp. 129–132.

27.3 *Bearing Selection Handbook,* rev. ed., The Timken Company, Canton, Ohio, 1986.

27.4 E. N. Bamberger, T. A. Harris, W. M. Kacmarsky, C. A. Moyer, R. J. Parker, J. J. Sherlock, and E. V. Zaretsky, *Life Adjustment Factors for Ball and Roller Bearings,* ASME, New York, 1971.

CHAPTER 28

JOURNAL BEARINGS

Theo G. Keith, Jr., Ph.D.
Professor and Chairman of Mechanical Engineering
University of Toledo
Toledo, Ohio

LIST OF SYMBOLS

a	Axial-flow land width
a_f	Pad load coefficient
A	Area
b	Circumferential-flow land width
C	Clearance
C^*	Specific heat
D	Diameter
e	Eccentricity
f	Coefficient of friction
F_j	Friction on journal
h	Film thickness
h_0	Minimum film thickness
H	Dimensionless film thickness
J	Mechanical equivalent of heat
k	Permeability
L	Bearing width
M	Rotor mass at bearing
M_j	Frictional torque on journal

n	Number of pads or recesses
N	Revolutions per unit time
p	Pressure
p_a	Ambient pressure
p_0	Short-bearing pressure
p_r	Recess pressure
p_s	Supply pressure
p_∞	Long-bearing pressure
$\bar{p}$	Dimensionless pressure
P	Unit loading
q	Volume flow rate per unit length
q_f	Flow factor
Q	Volume flow rate
Q_s	Side leakage flow rate
R	Radius of journal
R_b	Radius of bearing $= R + C$
s	Stiffness
S	Sommerfeld number $= (\mu N/P)(R/C)^2$
t	Time
t_p	Thickness of porous liner
T	Temperature
u, v, w	Velocity in x, y, z directions, respectively
U	Velocity of journal
W	Load
W_R	Load component directed along line of centers
W_T	Load component normal to line of centers
x, y, z	Rectangular coordinates
X	Dimensionless minimum-film-thickness parameter $= (h_0/R)[P/(2\pi N\mu)]^{1/2}$
Y	Dimensionless frictional torque parameter $= [M_j/(WR)][P/(2\pi N\mu)]^{1/2}$
α	Porous material slip coefficient
β	Included angle of partial bearing, porous bearing parameter
β_1	Angle from line of centers to leading edge of partial bearing
γ	Circumferential-flow parameter
ε	Eccentricity ratio e/c
ζ	Dimensionless axial dimension $= z/(L/2)$
θ	Angular position measured from line of centers
θ_1	Angular position to leading edge of film
θ_2	Angular position to zero pressure in film
θ_3	Angular position to trailing edge of film
θ_{cav}	Angular position to cavitation boundary

Λ	Bearing number $= (6\mu\omega/p_a)(R/C)^2$
λ	Ratio of heat conduction loss to heat generation rate, reduced bearing number $= \Lambda/6$
μ	Dynamic viscosity
ρ	Density
τ	Shear stress
ϕ	Attitude angle
ω	Angular velocity
Ω	Porous bearing parameter

28.1 INTRODUCTION

The design of journal bearings is of considerable importance to the development of rotating machinery. Journal bearings are essential machine components for compressors, pumps, turbines, internal-combustion engines, motors, generators, etc.

In its most basic form (Fig. 28.1), a *journal bearing* consists of a rotatable shaft (the journal) contained within a close-fitting cylindrical sleeve (the bearing). Generally, but not always, the bearing is fixed in a housing. The journal and bearing surfaces are separated by a film of lubricant (liquid or gas) that is supplied to the *clearance* space between the surfaces. The clearance space is generally quite small (on the order of one-thousandth of the journal radius) and has four major functions:

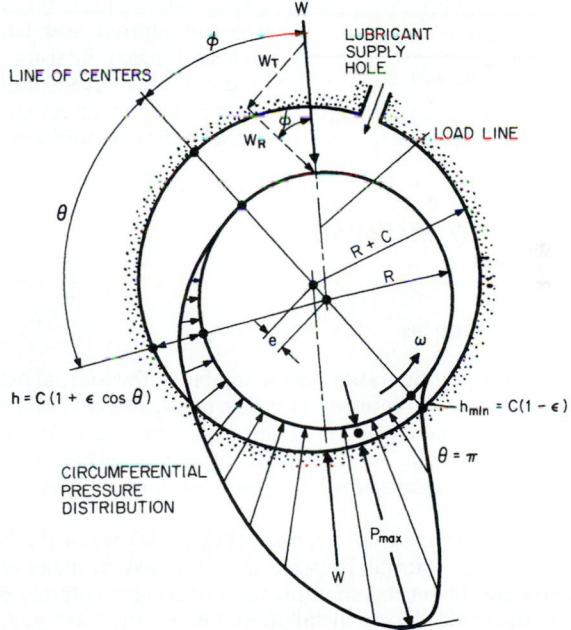

FIGURE 28.1 Journal and bearing notation.

to permit assembly of the journal and bearing, to provide space for the lubricant, to accommodate unavoidable thermal expansions, and to tolerate any shaft misalignment or deflection.

The fundamental purpose of a journal bearing is to provide radial support to a rotating shaft. Under load, the centers of the journal and the bearing are not coincident but are separated by a distance called the *eccentricity*. This eccentric arrangement establishes a converging-wedge geometry which, in conjunction with the relative motion of the journal and the bearing, permits a pressure to be developed by viscous effects within the thin film of lubricant and thus produces a load-carrying capability. However, if the load is too large or the shaft rotation too slow, the wedge-like geometry will not form and solid-to-solid contact can occur.

Journal bearings can operate in any of three lubrication regimes: *thick-film lubrication, thin-film lubrication,* or *boundary lubrication.* Generally, thick-film operation is preferred. Figure 28.2 is a diagram of the three lubrication regimes. Table 28.1 provides some of the characteristics of each regime.

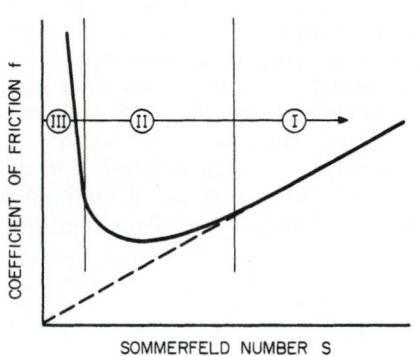

FIGURE 28.2 Three lubrication regimes: I, thick film; II, thin film; III, boundary.

Journal bearings may be classified according to the fluid mechanism that establishes the film load capacity: *Hydrodynamic* journal bearings, also called *self-acting* bearings, depend entirely on the relative motion of the journal and the bearing to produce film pressure for load support. *Hydrostatic* journal bearings, also called *externally pressurized* bearings, achieve load support by the supply of fluid from an external high-pressure source and require no relative motion between journal and bearing surfaces. *Hybrid* journal bearings are designed to use both hydrodynamic and hydrostatic principles to achieve load support between moving surfaces.

28.2 BEARING AND JOURNAL CONFIGURATIONS

28.2.1 Bearing Geometries

A wide range of bearing configurations are available to the journal bearing designer. Figure 28.3 depicts several of these bearings. The configurations range from the very simple plain journal bearing to the very complex tilting-pad bearing. The choice of bearing configuration depends on several factors. Among the more important are cost, load, power loss, dynamic properties, ease of construction, and difficulty of installation.

Journal bearings are termed *full bearings* (Fig. 28.3a) when the bearing surface completely surrounds the journal. Because they are easy to make and do not cost much, full bearings are the most commonly used bearing in rotating machinery. Full bearings become distorted during installation, and so they are generally not perfectly circular.

Journal bearings are called *partial bearings* when the bearing surface extends over only a segment of the circumference, generally 180° or less (Fig. 28.3b). Par-

TABLE 28.1   Characteristics of Lubrication Regimes

Lubrication regime	Contact of bearing surfaces	Range of film thickness, in	Coefficient of friction	Degree of wear	Comments
Thick film	Only during startup or stopping	10^{-3}–10^{-4}	0.01–0.005	None	1. Light-loading high-speed regime 2. Friction coefficient proportional to $\mu N/[W/(LD)]$
Thin film	Intermittent; dependent on surface roughness	10^{-4} to 0.5×10^{-4}	0.005–0.05	Mild	1. High operating temperatures
Boundary	Surface to surface	0.5×10^{-4} to molecular thicknesses	0.05– 0.15	Large	1. Heavy-loading (unit load $>$ 3000 psi) low-speed ($<$ 60 fpm) operating regime 2. Heat generation and friction not dependent on lubricant viscosity

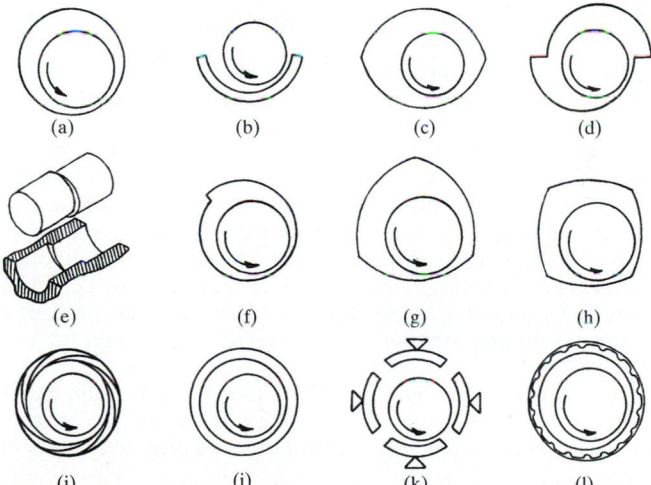

FIGURE 28.3   Journal bearing geometries. (*a*) Full bearing; (*b*) partial bearing; (*c*) elliptical, or lemon, bearing; (*d*) offset bearing; (*e*) rocking journal bearing; (*f*) pressure dam bearing; (*g*) three-lobe bearing; (*h*) four-lobe bearing; (*i*) multileaf bearing; (*j*) floating-ring bearing; (*k*) tilting- or pivoted-pad bearing; (*l*) foil bearing.

tial bearings are used in situations where the load is mainly unidirectional. Partial journal bearings have been found to reduce frictional torque on the journal and provide convenient accessibility, and they do not, in many instances, require strict manufacturing tolerance. Partial journal bearings in which the bearing radius exceeds the journal radius are called *clearance bearings*, whereas partial journal bearings in which the bearing and the journal radii are equal are termed *fitted bearings*.

Geometries in which two circular sectors are employed are called *elliptical*, or *lemon, bearings* (Fig. 28.3c). These bearings are really not elliptical at all but are fabricated by uniting two halves of a circular bearing which have had their mating faces machined so that the bearing has an approximately elliptical appearance. Lemon bearings are probably the most widely used bearing at low and moderate speeds. They are extensively used in turbine applications.

Elliptical bearings in which the two cylindrical halves are laterally displaced along the major axis are termed *offset bearings* (Fig. 28.3d). The relative displacement of the center of each half of the bearing is called the *preset*. When the upper half of the bearing is displaced horizontally in the direction of rotation, the bearing has negative preset. It is found that load capacity increases with preset. Offset bearings have relatively high horizontal stiffness, which helps prevent dynamic instability. Further, offset bearings allow greater lubricant flow and so run cooler.

Novel offset journal bearing designs for reducing power loss and wear in duty cycles which combine nonreversing loading with limited journal angular oscillation or in steady operation with counterrotation of journal and bearing under a constant load have been studied. In these applications, conventional journal bearings are found to develop extremely thin lubricant films, which in turn results in high friction and wear. Figure 28.3e depicts a journal bearing in which both the journal and the bearing are divided axially into segments with offset centerlines. This arrangement produces a dynamic rocking motion which promotes a thicker lubricating film. Accordingly the assembly has been called a *rocking journal bearing*.

When a step is milled from the surface of the bearing (Fig. 28.3f), the resulting bearing is called a *pressure dam*, or *step, bearing*. The purpose of the step is to create additional hydrodynamic pressure on the top of the journal as the lubricant is rotated into the step. In turn, this pressure buildup enhances the load on the journal and therefore diminishes its susceptibility to vibration problems. Pressure dam bearings are very popular in the petrochemical industry.

Bearing geometries consisting of three or more sectors (Fig. 28.3g and h) are termed *lobed*, or *multilobed, bearings*. Generally, bearings with more than three lobes are used only in gas bearing applications. Multilobe bearings act as a number of partial bearings in series. The cost of multilobed bearings is considered moderate.

The *multileaf journal bearing* (Fig. 28.3i) is a variant of a multilobe bearing. It consists of a number of identical circular arcs, or leaves, whose centers are equally spaced around the generating circle. The operating characteristics of a multileaf bearing are practically independent of the direction of loading for bearings with eight or more leaves.

In a *floating-ring journal bearing* (Fig. 28.3j), the lubricating film is divided in two by the addition of a "floating" ring between the journal and the bearing. Floating-ring bearings have lower frictional losses and reduced heat generation and provide better stability.

Hydrodynamic journal bearings may be distinguished as to whether the bearing surface can pivot. The basic advantage of *pivoting*, or *tilting-pad, journal bearings* (Fig. 28.3k) over fixed-pad journal bearings is that they can accommodate, with little loss in performance, any shaft deflection or misalignment.

A *foil journal bearing* (Fig. 28.3*l*) consists of a very thin compliant bearing surface resting atop a series of corrugations. When it is compared to a conventional gas bearing, the foil bearing has a thicker film, higher load capacity, lower power loss, better stability, and superior endurance to high operating temperatures.

28.2.2 Journal Shapes

Although the journal is generally assumed to be perfectly circular, wear effects or poor manufacture can lead to journals with the shapes shown in Fig. 28.4*a, b,* and *c*. In addition, the possibility of developing pressure by grooving the surface of the journal has been investigated. Three grooved patterns that were found to yield good stability characteristics are shown in Fig. 28.4*c, d,* and *e*.

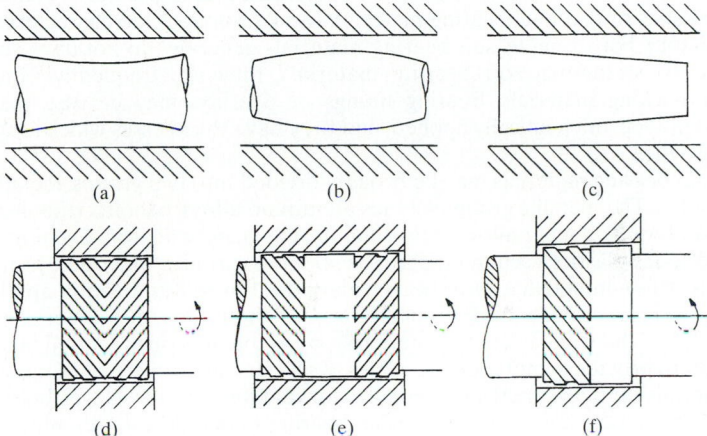

FIGURE 28.4 Journal shapes. (*a*) Hourglass; (*b*) barrel; (*c*) tapered; (*d*) herringbone; (*e*) partly grooved symmetrical pattern: (*f*) partly grooved asymmetrical pattern. *(Parts (d), (e), and (f) are from [28.1].)*

28.3 BEARING MATERIALS AND SELECTION CRITERIA

28.3.1 Bearing Materials

The ideal journal bearing material would have the following characteristics:

1. High compressive strength to withstand the applied radial loading
2. High fatigue strength to endure any cyclic changes in load direction and/or load intensity
3. Compatibility with the journal material to minimize surface scoring and bearing seizure whenever the journal and bearing surfaces come into contact (e.g., during startup)
4. Embedability to permit foreign particles in the lubricant to penetrate the bearing surface to avoid scoring and wear

5. Conformability of surface to tolerate journal misalignment, deflection, or manufacturing inaccuracies

6. High corrosion resistance to withstand chemical attack by the lubricant

7. High thermal conductivity to permit generated heat to be transported from the lubricant film

8. Appropriate coefficient of thermal expansion to avoid differences in thermal expansion of the journal and bearing

9. Low wear to prevent surface destruction, especially under boundary lubrication conditions (i.e., thin-film high-friction lubrication) and thereby lengthen the life of the bearing

Besides all these, the material should be inexpensive, highly available, and easily machined.

To be sure, no single material has been developed that satisfactorily combines all characteristics of the ideal bearing material. In fact, some of the characteristics are contradictory. For example, soft bearing materials generally do not have sufficient strength. To strengthen soft bearing materials, they are frequently bonded to stronger backing materials. Bearing linings or overlays may be cast, electrodeposited, sprayed, or chemically applied, and they have thicknesses which range from 0.01 to 0.5 inch (in).

Journal bearing materials may be broadly divided into two groups: metallics and nonmetallics. The metallic group includes aluminum alloys, babbitts (tin-, lead-, and aluminum-based), copper alloys (brass and bronze), zinc, and iron. The nonmetallic group includes plastics, carbon graphites, cemented carbides, and other proprietary materials. The nonmetallics have been widely used in self-lubrication applications because they can provide low friction and wear without the aid of a lubricant.

Because of the wide diversity of materials available for use in journal bearings, it is difficult to provide comprehensive tables of all relevant properties. Manufacturers and materials suppliers are the best sources for that information. Nevertheless, some physical properties of a variety of journal bearing materials are presented in Table 28.2 [28.2]. Typical applications and useful comments concerning a number of journal bearing alloys are displayed in Table 28.3, while Table 28.4 contains a numerical ranking of the performance characteristics of these alloys.

General information for a variety of self-lubricating materials is given in Table 28.5 [28.3]. Note that the table contains maximum values of the *PV* factor. This factor is the product of the bearing load per unit of projected area and the sliding velocity (i.e., speed in revolutions per minute times the bearing circumference). The *PV* parameter provides an indication of material wear and internal heat generation. Failure in self-lubricated bearings is frequently the direct result of internal overheating.

28.3.2 Bearing Material Selection Criteria

Selection of a bearing material invariably requires a compromise based on particular characteristics regarded by the designer to be of principal importance to the application at hand. DeGee [28.4] has developed a systematic approach for selecting a material for lubricated journal bearings. In this method, certain component criteria are identified within major property groups. Table 28.6 gives one such listing.

Not all the criteria presented in Table 28.6 need be considered. For example, in a particular application, environmental properties may be of no concern because the

TABLE 28.2 Physical Properties of Journal Bearing Materials

Material	Hardness H_B	Tensile strength, kpsi	Modulus of elasticity, Mpsi	Thermal conductivity, Btu/(h·ft·°F)	Coefficient of expansion, μin/(in·°F)	Density, lbm/ft³
Metals						
Lead babbitt	21	10	4.2	14	14	630
Tin babbitt	25	11	7.6	32	13	462
Copper lead	25	8	7.6	170	11	562
Silver	25	23	11	238	10.9	655
Cadmium	35	...	8	53	16.6	537
Aluminum alloy	45	22	10.3	119	13.5	181
Lead bronze	60	34	14	27	9.9	555
Tin bronze	70	45	16	29	10	549
Steel	150	75	30	29	6.4	487
Cast iron	180	35	23	30	5.7	449
Porous metals						
Bronze	40	18	...	17	10.5	399
Iron	50	25	...	16	6.7	381
Aluminum	H55†	15	...	...	...	144
Plastics						
TFE	D60‡	3	0.06	0.10	55	137
Nylon	M79†	11	0.41	0.14	55	71
Phenolic	M100	10	0.5	0.21	12	85
Acetal	M94	10	0.41	0.13	45	89
Polycarbonate	M70	8.5	0.32	0.11	70	75
Filled polyimide	E99†	7.5	...	0.44	22	89
Other nonmetallics						
Rubber	...	...	...	0.09	43	75
Wood	...	1.1	1.8	0.11	2.7	42
Carbon graphite	75§	2	2	10	1.5	106
Cemented tungsten carbide	A91†	130	81	40	3.3	886
Fused aluminum oxide	A85	30	50	1.6	8.2	243

†Rockwell. ‡Shore durometer. §Shore scleroscope.
SOURCE: Ref. [28.2].

TABLE 28.3 Bearing Alloy Material Applications

Material	Nominal composition, % by weight	Applications and remarks
Aluminum, low tin	Al 92 Sn 8	Tin added to improve compatibility; too much tin lowers strength. Has thermal expansion problems in steel housings. Requires hard journals. Good at high temperatures. Used in diesel engines and compressors.
Aluminum, high tin	Al 80 Sn 20	Produced by special working and annealing process so tin content does not greatly reduce strength. Used in automotive engines (crankshafts) and in aircraft equipment.
Babbitt, tin-based	Sn 84 Cu 8 Sb 8	Fatigue strength decreases as thickness increases. Low load capacity, thus usually bonded to one (bimetal) or two (trimetal) backing materials. Good in dirty applications, motors.
Babbitt, lead-based	Pb 75 Sn 10 Sb 15	Antimony (Sb) greater than 15% can cause brittleness. Cheaper than tin-based babbitt. Used in crankshaft bearings, transmission bushings, and electric equipment.
Lead bronze	Cu 70 Pb 25 Sn 5	Good for high-load high-speed applications; can be used with soft journals. Used as bushings in pumps, many home appliances, railroad cars.
Phosphor bronze	Cu 80 Sn 10 Pb 10	General-duty popular bushing; tin added to improve strength. Has high hardness; should be used with harder journals (300 BHN). Good impact resistance; used in lathes, pumps, home appliances.
Copper lead (cast)	Cu 75 Pb 25	Lead in pockets in copper matrix. Lead improves bearing surface but has corrosion problems. Frequently used as lining material on steel-backed bearings. Used in heavy-duty applications.
Copper lead (sintered)	Cu 75 Pb 25	Frequently used with a babbitt overlay in a trimetal bearing. Widely used in heavy-duty (high-temperature high-load) applications.
Silver (oven-plated)		Frequently used with lead indium overlay.

bearing operates in a clean, moderate-temperature environment and is not part of an electric machine.

After the list of criteria has been established, each component criterion is compared with all other criteria, and a graduation mark is allocated from 0, if there is no difference in the criterion, to 3, if there are large differences. For example, compressive strength (A1 in Table 28.6) might receive a 0 when compared with fatigue strength (A2) but receive a 3 when compared to thermal conductivity (B1), and so forth. When all component criteria have been compared with one another and graduation marks assigned, the graduation marks of each criterion are totaled and the sum of all these totals is divided into each amount, to obtain the component criteria weighting factors. The sum of all the weighting factors obviously is unity.

TABLE 28.4 Performance Ratings from 5 (High) to 1 (Low) for Bearing Alloy Materials

Material	Nominal composition, % by weight	Fatigue strength	Corrosion resistance	Seizure resistance	Embedability	Compatibility	Thermal conductivity
Aluminum, low tin	Al 92 Sn 8	2	4	2	1	1	4
Aluminum, high tin	Al 80 Sn 20	2	4	2	2	2	4
Babbitt, tin-based	Sn 84 Cu 8 Sb 8	1	5	4	5	4	3
Babbitt, lead-based	Pb 75 Sn 10 Sb 15	1	3	4	5	4	2
Lead bronze	Cu 70 Pb 25 Sn 5	4	2	3	3	2	3
Phosphor bronze	Cu 80 Sn 10 Pb 10	4	2	3	1	2	3
Copper lead (cast)	Cu 75 Pb 25	3	1	3	3	3	4
Copper lead (sintered)	Cu 60 Pb 40	3	1	3	3	3	4
Silver (over-plated)		5	4	1	1	1	5

TABLE 28.5 General Information on Self-Lubricating Bearing Materials

Material	Maximum load,† kpsi	Maximum speed, fpm	Maximum PV factor,‡ kpsi·fpm	Friction coefficient	Critical temperature, °F	Resistance to humidity	Resistance to chemicals§
Nylon	1.5	200–400	1	0.1–0.4	400	Fair	Good
Acetal	1.5	200–500	1	0.1–0.4	300	Good	Good
Polyimide	10	1000	0.3	0.1–0.3	600	Good	Good
Phenolic	4	1000	0.1	0.9–1.1	300–400	Good	Good
Filled nylon	2	200–400	1	0.1–0.4	400	Fair	Good
Acetal PTFE filled	1.8	800	2.5	0.05–0.15	300	Good	Good
Filled polyimide	10	1000	6	0.1–0.3	600	Good	Good
Reinforced phenolic	4–5	200	4	0.1–0.4	300–400	Good	Good
Filled PTFE	1	500–1000	5–20	0.05–0.25	500	Excellent	Excellent
PTFE	100	50	1–10	0.05–0.25		Excellent	Excellent

†Load on projected area at zero speed.
‡For continuous service.
§At bearing surface.
SOURCE Ref. [28.3].

TABLE 28.6 Journal Bearing Material Selection Criteria

Major property group	Component criteria
A. Mechanical	1. Compressive strength 2. Fatigue strength 3. Conformability (modulus of elasticity)
B. Thermal	1. Thermal conductivity 2. Thermal expansion
C. Chemical	1. Corrosion rate
D. Manufacturing	1. Cost 2. Machinability 3. Availability of material
E. Environmental	1. Behavior under abrasive conditions (embedability) 2. Resistance against electric-discharge pitting 3. Resistance to thermal degradation
F. Tribological	1. Wear rate 2. Coefficient of friction 3. Cavitation erosion resistance

Next the candidate materials are given quality marks for the various component criteria. These marks range from 5 (excellent, or high) to 1 (poor, or low). For instance, tin-based babbitts are known to have only fair (2) fatigue strength, whereas they have excellent (5) resistance to corrosion. The final ranking of the candidate materials is obtained by comparing the sums of the products of all component criteria weighting factors and quality marks.

28.4 PRESSURE EQUATION FOR A LUBRICATING FILM

28.4.1 Reynolds Equation

The differential equation which governs the pressure in a lubricating film is called the *Reynolds equation.* Bearing performance can be evaluated once the solution of this equation is in hand.

To develop the Reynolds equation, consider a portion of the fluid film of a journal bearing (Fig. 28.1). In general, there are three velocity components in the film: u, v, and w. There are three equations of motion (momentum equations), one for each coordinate direction. The collection is known as the *Navier-Stokes equations,* and each equation may be written in the following form:

$$\text{Inertial forces} = \text{pressure forces} + \text{body forces} + \text{viscous forces} \qquad (28.1)$$

The Navier-Stokes equations in their complete form are too involved for analytical solution. They can, however, be reduced, and subsequently solved, by making several simplifying yet plausible assumptions:

1. The flow is laminar.
2. The inertial and body forces are small compared to the pressure and viscous forces.
3. The curvature of the film is negligible; the bearing surfaces are, therefore, nearly parallel.
4. The variation of pressure across the film $\partial P/\partial y$ is negligibly small.
5. The transverse velocity component across the film, v, is small compared to the other velocity components.
6. The velocity gradients across the film dominate over all other velocity gradients.

Application of these assumptions to mathematical versions of Eq. (28.1), and to an integrated version of the conservation of mass (the continuity equation), yields the *Reynolds equation* for a liquid-lubricated bearing:

$$\frac{\partial}{\partial x}\left(\frac{h^3}{\mu}\frac{\partial p}{\partial x}\right) + \frac{\partial}{\partial z}\left(\frac{h^3}{\mu}\frac{\partial p}{\partial z}\right) = 6(U_b + U_j)\frac{\partial h}{\partial x} + 6h\frac{\partial(U_b + U_j)}{\partial x} + 12\frac{\partial h}{\partial t} \qquad (28.2)$$

The first grouping on the right-hand side of Eq. (28.2) is called the *wedge* term and must be negative to generate positive pressures. The third term on the right is called the *squeeze* term, and it will generate positive pressures when $\partial h/\partial t < 0$. The squeeze term vanishes for a steadily loaded bearing. Both the wedge and the squeeze terms vanish for a purely hydrostatic case. If the bearing surface is fixed ($U_b = 0$), if the shaft is rotating with a speed ω (that is, $U_j = U = R\omega$), and if the viscosity of the lubricant is constant, then Eq. (28.2) may be written as

$$\frac{\partial}{\partial x}\left(h^3\frac{\partial p}{\partial x}\right) + \frac{\partial}{\partial z}\left(h^3\frac{\partial p}{\partial z}\right) = 6\mu R\omega\frac{\partial h}{\partial x} + 6\mu h\frac{\partial U}{\partial x} + 12\mu\frac{\partial h}{\partial t} \qquad (28.3)$$

For comparative purposes, it is useful to cast the steady version of Eq. (28.3) into nondimensional form. This can be accomplished by defining the following nondimensional variables:

$$\theta \equiv \frac{x}{R} \qquad \zeta \equiv \frac{z}{L/2} \qquad H \equiv \frac{h}{C}$$

$$\overline{p} \equiv \frac{p}{6\mu(U/R)}\left(\frac{C}{R}\right)^2 = \frac{p}{6\mu\omega}\left(\frac{C}{R}\right)^2$$

Also, since $\omega = 2\pi N$,

$$\overline{p} = \frac{p}{12\pi\mu N}\left(\frac{C}{R}\right)^2$$

Substituting these into Eq. (28.3) yields

$$\frac{\partial}{\partial\theta}\left(H^3\frac{\partial\overline{p}}{\partial\theta}\right) + \left(\frac{D}{L}\right)^2\frac{\partial}{\partial\zeta}\left(H^3\frac{\partial\overline{p}}{\partial\zeta}\right) = \frac{\partial H}{\partial\theta} \qquad (28.4)$$

This equation can be interpreted as

Circumferential pressure flow + axial pressure flow = shear flow

The governing equation of a gas film differs from that of a liquid film by the appearance of the *density* ρ. The steady compressible version of the Reynolds equation for an isoviscous gas can be written

$$\frac{\partial}{\partial x}\left(h^3\rho\,\frac{\partial p}{\partial x}\right) + \frac{\partial}{\partial z}\left(h^3\rho\,\frac{\partial p}{\partial z}\right) = 6\mu R\omega\,\frac{\partial(\rho h)}{\partial x} \qquad (28.5)$$

Since the energy dissipated by frictional forces is very small in normal gas bearing operation, we may assume that the film has a constant temperature and $p = k\rho$; thus, Eq. (28.5) becomes

$$\frac{\partial}{\partial x}\left(h^3\,\frac{\partial p^2}{\partial x}\right) + \frac{\partial}{\partial z}\left(h^3\,\frac{\partial p^2}{\partial z}\right) = 12\,\mu R\omega\,\frac{\partial(ph)}{\partial x} \qquad (28.6)$$

This equation can be written in nondimensional form as

$$\frac{\partial}{\partial\theta}\left(H^3\,\frac{\partial \bar{p}^2}{\partial\theta}\right) + \left(\frac{D}{L}\right)^2 \frac{\partial}{\partial\zeta}\left(H^3\,\frac{\partial \bar{p}^2}{\partial\zeta}\right) = 2\Lambda\,\frac{\partial(\bar{p}H)}{\partial\theta} \qquad (28.7)$$

where

$$\bar{p} \equiv \frac{p}{p_a} \qquad p_a = \text{ambient or supply pressure}$$

$$H \equiv \frac{h}{C} \qquad \zeta = \frac{z}{L/2} \qquad \Lambda \equiv \frac{6\mu\omega}{p_a}\left(\frac{R}{C}\right)^2$$

Here Λ is called the *bearing*, or *compressibility*, *number*.

To solve the Reynolds equation, we require an expression for the *film thickness* h. From the triangle ABC in Fig. 28.4, we may write

$$\overline{AC} = h + R = R_b \cos\xi + e\cos\theta$$

where θ is measured from the line of centers. And since $e \ll R_b$, $\cos\xi \sim 1$, and thus

$$h = R_b - R + e\cos\theta$$

The *radial clearance* is $C = R_b - R$, and the *eccentricity ratio* is $\varepsilon \equiv e/C$; hence

$$h = C(1 + \varepsilon\cos\theta) \qquad (28.8)$$

28.4.2 Boundary Conditions

Three sets of circumferential boundary conditions are commonly applied to the solutions of the Reynolds equation. These boundary conditions have been given the names *Sommerfeld, Gumbel,* and *Swift-Stieber.* Of the three, the Sommerfeld conditions are the easiest to apply, but they yield certain unrealistic results. For example, in a liquid-lubricated bearing, Sommerfeld conditions produce negative pressures in the film. This results in the shaft's being displaced at right angles to the load line as the load is increased. Gumbel conditions are similar to Sommerfeld conditions except that all negative pressures are disregarded. Although this approach leads to more realistic load results, it produces a violation of the conservation of mass. The Swift-Stieber conditions come closest to representing the actual conditions in a film, but they are more difficult to apply. They are widely used in numerical investigations. The three sets of boundary conditions are summarized in Table 28.7.

28.5   JOURNAL BEARING PERFORMANCE

Once the pressure distribution is established, the journal bearing performance may be determined. Performance is generally measured in terms of four quantities: bearing load capacity, frictional losses, lubricant flow requirement, and temperature rise.

28.5.1   Bearing Load Relations

Five parameters are associated with the load capacity of a journal bearing:

1. The radial load component W_R acts along the line of centers (Fig. 28.1) and is computed from

$$W_R = -R \int_{-L/2}^{L/2} \int_{\theta_1}^{\theta_3} p \cos \theta \, d\theta \, dz$$

TABLE 28.7   Typical Boundary Conditions on the Reynolds Equation

Names associated with boundary conditions	Pressure profile	Mathematical expression
Sommerfeld (full Sommerfeld)		$p(\theta_1) = p(\theta_3) = 0$ (zero pressure means ambient or atmospheric pressure) For complete journal bearings: $\theta_1 = 0$, $\theta_3 = 2\pi$ For partial journal bearings: $\theta_1 = \beta_1$, $\theta_3 = \theta_1 + \beta$
Gumbel (half Sommerfeld)		$p(\theta_1) = p(\theta_2) = 0$ $p(\theta_2 \le \theta \le \theta_3) = 0$ For complete journal bearings: $\theta_1 = 0$, $\theta_2 = \pi$, $\theta_3 = 2\pi$
Swift-Stieber (Reynolds)		$p(\theta_2) = p_{cav} =$ atmospheric pressure $\dfrac{\partial p}{\partial \theta}(\theta_2) = 0$ $\theta_2 = \theta_{cav}$ which must be determined

where θ_1 and θ_3 are, respectively, the leading and trailing angular locations of the lubricating film.

2. The tangential load component W_T acts perpendicular to the line of centers:

$$W_T = R \int_{-L/2}^{L/2} \int_{\theta_1}^{\theta_3} p \sin \theta \, d\theta \, dz$$

3. The bearing load W must be supported by the pressure developed within the lubricating film. Generally the load is specified or enters the design via the *unit load P*, which is defined as the load per unit projected area, or

$$P = \frac{W}{LD}$$

Typical values of the unit load are given in Table 28.8.

TABLE 28.8 Range of Unit Loads for Various Applications

Application	Bearing	Unit load range, kpsi
Automotive engines	Main	0.6–0.8
	Crankpin	1.7–2.3
Diesel engines	Main	0.9–2.3
	Crankpin	1.1–2.3
	Wristpin	2.0–2.3
Steam turbines	Main	0.12–0.25
Air compressors	Main	0.14–0.28
	Crankpin	0.28–0.5
Centrifugal pumps	Shaft	0.1–0.18
Electric motors	Shaft	0.12–0.25

4. The *Sommerfeld number S* is a dimensionless parameter that characterizes bearing performance; large S (say, greater than 0.15) indicates a lightly loaded bearing operating at a small eccentricity. The Sommerfeld number may be calculated from

$$S = \frac{\mu U L}{\pi W} \left(\frac{R}{C}\right)^2 = \frac{\mu N}{P} \left(\frac{R}{C}\right)^2$$

5. The *attitude angle* ϕ is the angular distance between the load line and the line of centers (Fig. 28.1). It locates the minimum film thickness as measured from the load line. Because $W_R = W \cos \phi$ and $W_T = W \sin \phi$,

$$\phi = \tan^{-1} \frac{W_T}{W_R}$$

28.5.2 Bearing Friction Relations

Four parameters are involved with the frictional behavior of a journal bearing:

1. The *shear stress* τ, acting on either the shaft or the bearing surface, consists of two terms; one is due to motion of the shaft (pure shear), and the other is due to the circumferential pressure distribution (pressure-induced shear):

$$\tau = \frac{\mu U}{h} \pm \frac{h}{2R}\frac{\partial p}{\partial \theta}$$

The plus sign corresponds to the shear stress on the journal surface; the minus sign, to the shear stress on the bearing surface.

2. The *frictional force* acting on the journal F_j is found by integrating the pure shear stress over the entire surface of the journal and the pressure-induced shear stress up to the trailing edge of the film. This yields

$$F_j = \left(\frac{2\pi}{1-\varepsilon^2}\right)(\mu UL)\left(\frac{R}{C}\right) + \frac{\varepsilon W_T}{2}\left(\frac{C}{R}\right)$$

3. The *friction coefficient* f is the ratio of the journal frictional force to the bearing load:

$$f = \frac{F_j}{W}$$

The *friction variable* is the product $(R/C)(f)$ and so may be written

$$\left(\frac{R}{C}\right)(f) = \left(\frac{2\pi^2}{1-\varepsilon^2}\right)(S) + \frac{\varepsilon \sin \phi}{2}$$

4. The power that must be supplied to the journal to overcome friction is called the *frictional horsepower loss* HP, and it may be computed from

$$HP = C_1 F_j U = C_2 f W R N$$

where C_1 and C_2 depend on the system of units. For F_j in pounds (lb) and U in inches per second (in/s), $C_1 = \frac{1}{6600}$ for W (lb), R (in), and N in revolutions per second (r/s), $C_2 = 2\pi C_1 = 9.51998 \times 10^{-4}$.

The relevant geometry for the execution of these tasks is depicted in Fig. 28.5.

28.5.3 Lubricant Flow Relations

Lubricant flow rates are needed to estimate the capacity of the lubricant supply system and to determine the cooling requirements of the bearing. This involves evaluation of the lubricant flow within the clearance space, the lubricant flow that leaks out the sides of the bearing, and the lubricant flow that is supplied to the bearing.

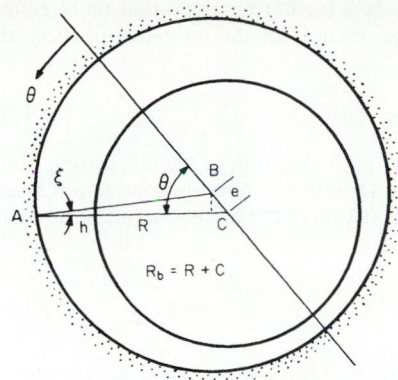

FIGURE 28.5 Film thickness geometry

In general, volume flow rate per unit length is composed of a term due to surface motion (shear flow) and another due to the pressure (pressure-induced flow). Journal bearing circumferential and axial flows per unit length are

$$q_\theta = \frac{Uh}{2} - \frac{h^3}{12\mu R}\frac{\partial p}{\partial \theta} \qquad q_z = -\frac{h^3}{12\mu}\frac{\partial p}{\partial z}$$

$$(28.9)$$

The total flow rates may be found by integrating Eq. (28.9) across the bearing length for the circumferential flow rate and around the bearing circumference for the axial flow rate.

Assuming that lubricant is supplied in the unloaded portion of the bearing (Fig. 28.6), we see that the rate at which lubricant leaks out of the active portion of the film is

$$Q_{sa} = Q_1 - Q_2 \qquad (28.10)$$

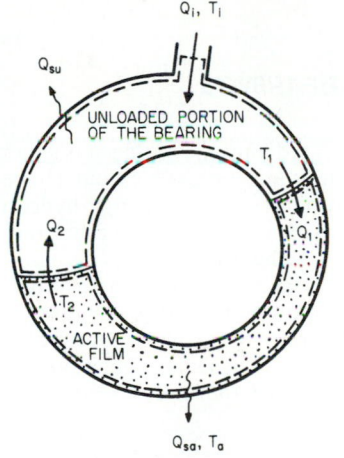

FIGURE 28.6 Control volumes for lubricant flow and heat balances.

where Q_1 = flow into the leading edge of the film and Q_2 = flow out of the trailing edge.

When the input flow rate equals the leakage flow rate, Q_1 is called the *classical rate*. For given values of W, U, and μ, the classical rate is the largest flow that can be carried into the active film by shaft rotation. However, in practice, Q_i, the input flow rate, may be greater (the *flooded condition*) or less (the *starved condition*) than the appropriate value to achieve the classical rate. For flooded conditions, side flow will also occur in the unloaded portion of the bearing, and we may write

$$Q_{su} = Q_2 + Q_i - Q_1$$

$$(28.11)$$

thus, $Q_{su} + Q_{sa} = Q_s = Q_i$

28.5.4 Bearing Thermal Relations

The steady energy balance equation can be simply expressed as

Energy inflow rate – energy outflow rate + energy generation rate = 0

An energy balance may be performed on the unloaded portion of the film (Fig. 28.6). Toward that end, it is assumed that there is complete mixing between the inlet

Q_i and the carryover Q_2 flows so that $T_u = T_1$. It is further assumed that there is no energy generation and negligible heat transfer. Hence, for the unloaded portion of the film,

$$Q_i T_i + Q_2 T_2 = (Q_2 + Q_i)(T_1) \qquad (28.12)$$

Next an energy balance is performed on the active portion of the lubricating film (Fig. 28.6). The energy generation rate is taken to be $F_j U/J$, and the conduction heat loss to the shaft and bearing are taken to be a portion of the heat generation rate, or $\lambda F_j U/J$. Accordingly,

$$\rho Q_1 C^* T_1 - (\rho Q_{sa} C^* T_a + \rho Q_2 C^* T_2) + \frac{(1-\lambda)F_j U}{J} = 0 \qquad (28.13)$$

Combining Eqs. (28.10) to (28.13) and assuming that the side-flow leakage occurs at the average film temperature $T_a = (T_1 + 2\,T_2)/2$, we find that

$$\frac{J\rho C^*(T_a - T_i)}{(1-\lambda)P} = \frac{1 + 2Q_2/Q_i}{2 - Q_{sa}/Q_1}\,\frac{4\pi(R/C)(f)}{Q_1/(RCNL)} \qquad (28.14)$$

This shows that the lubricant temperature rise is $1 - \lambda$ times the rise when conduction is neglected.

28.6 LIQUID-LUBRICATED JOURNAL BEARINGS

In the hydrodynamic operation of a liquid-lubricated journal bearing, it is generally assumed that the lubricant behaves as a continuous incompressible fluid. However, unless the lubricant is admitted to the bearing under relatively high hydrostatic head, the liquid film can experience periodic vaporization which can cause the film to rupture and form unstable pockets, or cavities, within the film. This disruption of the film is called *cavitation,* and it occurs when the pressure within the bearing falls to the vapor pressure of the lubricant. Narrow liquid-lubricated bearings are especially susceptible to this problem. Figure 28.7 illustrates the general film condition in which lubricant is admitted through a lubricating groove at some angular position θ_0. Clearly incomplete films complicate the analysis, and therefore the design, of a liquid-lubricated journal bearing.

28.6.1 L/D Effects on Cylindrical Full Journal Bearings

Long-Length Bearings.   When the length of a bearing is such that $L > 2D$, the axial pressure flow term in the Reynolds equation may be neglected and the bearing performs as if it were infinitely long. Under this condition, the reduced Reynolds equation can be directly integrated. Table 28.9 contains long-bearing results for both Sommerfeld and Gumbel boundary conditions.

Short-Length Bearings.   When the length of a bearing is such that $L < D/4$, the axial pressure flow will dominate over the circumferential flow, and again the Reynolds equation can be readily integrated. Results of such a short-bearing integration with Gumbel boundary conditions are shown in Table 28.10.

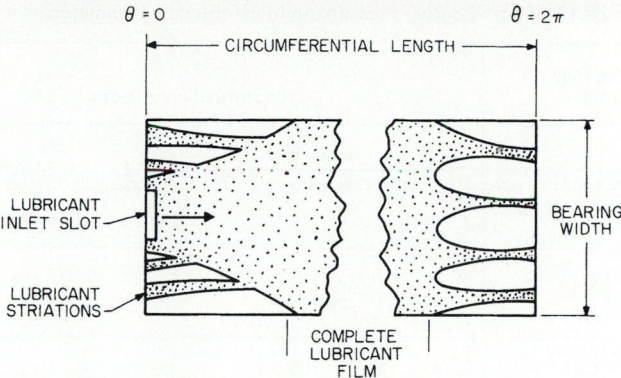

FIGURE 28.7 Diagram of an incomplete fluid film.

TABLE 28.9 Long-Bearing Pressure and Performance Parameters

Performance parameter	Sommerfeld conditions	Gumbel conditions
$\dfrac{p}{12\pi\mu N}\left(\dfrac{C}{R}\right)^2$	$\dfrac{(\varepsilon\sin\theta)(2+\varepsilon\cos\theta)}{(2+\varepsilon^2)(1+\varepsilon\cos\theta)^2}$	$\dfrac{(\varepsilon\sin\theta)(2+\varepsilon\cos\theta)}{(2+\varepsilon^2)(1+\varepsilon\cos\theta)^2}\quad 0<\theta<\pi,$ $0,\ \pi<\theta<2\pi$
$\dfrac{W_R}{3\mu UL}\left(\dfrac{C}{R}\right)^2$	0	$\dfrac{4\varepsilon^2}{(2+\varepsilon^2)(1-\varepsilon^2)}$
$\dfrac{W_T}{3\mu UL}\left(\dfrac{C}{R}\right)^2$	$\dfrac{4\pi\varepsilon}{(2+\varepsilon^2)\sqrt{1-\varepsilon^2}}$	$\dfrac{2\pi\varepsilon}{(2+\varepsilon^2)\sqrt{1-\varepsilon^2}}$
ϕ	$\dfrac{\pi}{2}$	$\tan^{-1}\left(\dfrac{\pi}{2}\dfrac{\sqrt{1-\varepsilon^2}}{\varepsilon}\right)$
S	$\dfrac{(2+\varepsilon^2)\sqrt{1-\varepsilon^2}}{12\pi^2\varepsilon}$	$\dfrac{(2+\varepsilon^2)(1-\varepsilon^2)}{6\pi\varepsilon\sqrt{4\varepsilon^2+\pi^2(1-\varepsilon^2)}}$
$\dfrac{F_j}{\mu UL}\dfrac{C}{R}$	$\dfrac{4\pi(1+2\varepsilon^2)}{(2+\varepsilon^2)\sqrt{1-\varepsilon^2}}$	$\dfrac{\pi(4+5\varepsilon^2)}{(2+\varepsilon^2)\sqrt{1-\varepsilon^2}}$
$\left(\dfrac{R}{C}\right)(f)$	$\dfrac{1+2\varepsilon^2}{3\varepsilon}$	$\dfrac{4+5\varepsilon^2}{6\varepsilon}\sqrt{\dfrac{1-\varepsilon^2}{4\varepsilon^2+\pi(1-\varepsilon^2)}}$
$\dfrac{Q_s}{RCNL}$	0	0

TABLE 28.10 Short-Bearing Pressure and Performance Parameters

Performance parameter	Gumbel conditions
$\dfrac{p}{12\pi\mu N}\left(\dfrac{C}{R}\right)^2$	$-\dfrac{\varepsilon \sin\theta}{2(1+\varepsilon\cos\theta)^3}\left(\dfrac{L}{D}\right)^2(\varsigma^2-1)\qquad 0\le\theta\le\pi,$ $0,\ \pi\le\theta\le2\pi$
$\dfrac{W_R}{3\mu UL}\left(\dfrac{C}{R}\right)^2$	$\dfrac{4\varepsilon^2}{3(1-\varepsilon^2)^2}\left(\dfrac{L}{D}\right)^2$
$\dfrac{W_T}{3\mu UL}\left(\dfrac{C}{R}\right)^2$	$\dfrac{\pi\varepsilon^2}{3(1-\varepsilon^2)^3}\left(\dfrac{L}{D}\right)^2$
$\dfrac{W}{3\mu UL}\left(\dfrac{C}{R}\right)^2$	$\dfrac{\varepsilon\sqrt{\pi^2(1-\varepsilon^2)+16\varepsilon^2}}{3(1-\varepsilon^2)^2}\left(\dfrac{L}{D}\right)^2$
ϕ	$\tan^{-1}\dfrac{\pi(1-\varepsilon^2)^2}{4\varepsilon}$
S	$\dfrac{(1-\varepsilon^2)^2}{\pi\varepsilon\sqrt{\pi^2(1-\varepsilon^2)+16\varepsilon^2}}\left(\dfrac{D}{L}\right)^2$
$\dfrac{F_j}{\mu UL}\dfrac{C}{R}$	$\dfrac{2\pi}{\sqrt{1-\varepsilon^2}}$
$\left(\dfrac{R}{C}\right)(f)$	$\dfrac{(2\pi)(1-\varepsilon^2)^{3/2}}{\varepsilon\sqrt{\pi^2(1-\varepsilon^2)+16\varepsilon^2}}$
$\dfrac{Q_s}{RCNL}$	$2\pi\varepsilon$

Finite-Length Bearings. The *slenderness ratio* L/D for most practical designs ranges between 0.5 and 2.0. Thus, neither the short-bearing theory nor the long-bearing theory is appropriate. Numerous attempts have been made to develop methods which simultaneously account for both length and circumferential effects. Various analytical and numerical methods have been successfully employed. Although such techniques have produced important journal bearing design information, other simplified methods of analysis have been sought. These methods are useful because they do not require specialized analytical knowledge or the availability of large computing facilities. What is more, some of these simple, approximate methods yield results that have been found to be in good agreement with the more exact results. One method is described.

Reason and Narang [28.5] have developed an approximate technique that makes use of both long- and short-bearing theories. The method can be used to accurately design steadily loaded journal bearings on a hand-held calculator.

It was proposed that the film pressure p be written as a harmonic average of the short-bearing pressure p_0 and the long-bearing pressure p_∞, or

$$\frac{1}{p} = \frac{1}{p_0} + \frac{1}{p_\infty} \qquad \text{or} \qquad p = \frac{p_0}{1 + p_0/p_\infty}$$

The pressure and various performance parameters that can be obtained by this combined solution approximation are presented in Table 28.11. Note that several of these parameters are written in terms of two quantities, I_s and I_c. Accurate values of these quantities and the Sommerfeld number are displayed in Table 28.12. With the exception of the entrainment flow, which is increasingly overestimated at large ε and L/D, the predictions of this simple method have been found to be very good.

Example 1. Using the Reason and Narang combined solution approximation, determine the performance of a steadily loaded full journal bearing for the following conditions:

$\mu = 4 \times 10^{-6}$ reyn $\qquad D = 1.5$ in

$N = 1800$ r/min $\qquad L = 1.5$ in

$W = 500$ lbf $\qquad C = 1.5 \times 10^{-3}$

Solution. The unit load is $P = W/(LD) = 222$ pounds per square inch (psi), and the Sommerfeld number is

$$S = \frac{\mu N}{P} \left(\frac{R}{C} \right)^2 = 0.135$$

Entering Table 28.12 at this Sommerfeld number and a slenderness ratio of 1, we find that $\varepsilon = 0.582$, $I_c = 0.2391$, and $I_s = 0.3119$. The bearing performance is computed by evaluating various parameters in Table 28.11. Results are compared in Table 28.13 to values obtained by Shigley and Mischke [28.6] by using design charts.

28.6.2 Design Charts

Design charts have been widely used for convenient presentation of bearing performance data. Separate design graphs are required for every bearing configuration or variation. Use of the charts invariably requires repeated interpolations and extrapolations. Thus, design of journal bearings from these charts is somewhat tedious.

Raimondi-Boyd Charts. The most famous set of design charts was constructed by Raimondi and Boyd [28.7]. They presented 45 charts and 6 tables of numerical information for the design of bearings with slenderness ratios of ¼, ½, and 1 for both partial (60°, 120°, and 180°) and full journal bearings. Consequently, space does not permit all those charts to be presented. Instead a sampling of the charts for bearings with an L/D ratio of 1 is given. Figures 28.8 to 28.13 present graphs of the *minimum-film-thickness variable* h_0/C (note that $h_0/C = 1 - \varepsilon$), the *attitude angle* ϕ (or location of the minimum thickness), the *friction variable* $(R/C)(f)$, the *flow variable* $Q/(RCNL)$, the *flow ratio* Q_s/Q, and the *temperature-rise variable* $J\rho C^* \Delta T/P$. Table 28.14 is a tabular presentation of these data.

TABLE 28.11 Pressure and Performance Parameters of the Combined Solution Approximation

Performance parameter	Equation
$\dfrac{p}{12\pi\mu N}\left(\dfrac{C}{R}\right)^2$	$\dfrac{\dfrac{1}{2}\left(\dfrac{L}{D}\right)^2(1-\zeta^2)\dfrac{\varepsilon\sin\theta}{(1+\varepsilon\cos\theta)^3}}{1+\left(\dfrac{L}{D}\right)^2\dfrac{(2+\varepsilon^2)(1-\zeta^2)}{2(1+\varepsilon\cos\theta)(2+\varepsilon\cos\theta)}}$
$\dfrac{W_R}{3\mu UL}\left(\dfrac{C}{R}\right)^2$	$-2I_c$
$\dfrac{W_T}{3\mu UL}\left(\dfrac{C}{R}\right)^2$	$2I_s$
ϕ	$\tan^{-1}\left(\dfrac{-I_s}{I_c}\right)$
S	$-\dfrac{1}{6\pi\sqrt{I_s^2+I_c^2}}$
$\dfrac{F_j}{\mu UL}\dfrac{C}{R}$	$3\varepsilon I_s+\dfrac{2\pi}{\sqrt{1-\varepsilon^2}}$
$\left(\dfrac{R}{C}\right)(f)$	$6\pi S\left(\dfrac{\varepsilon I_s}{2}+\dfrac{\pi}{3\sqrt{1-\varepsilon^2}}\right)$
$\dfrac{Q_{0,\pi}}{RCNL}$ †	$\pi\left[1+\varepsilon\mp\varepsilon E\left(1-\dfrac{2E}{\sqrt{1+2E}}\tanh^{-1}\dfrac{1}{\sqrt{1+2E}}\right)\left(\dfrac{L}{D}\right)^2\right]$ where $\quad E=\dfrac{(1\pm\varepsilon)(2\pm\varepsilon)}{(2+\varepsilon^2)}\left(\dfrac{D}{L}\right)^2$
$\dfrac{Q_s}{Q_0}$	$1-\dfrac{Q_\pi}{Q_0}$
$\dfrac{J\rho C^*\,\Delta T}{P}$	$\dfrac{1}{1-\tfrac{1}{2}Q_s/Q_0}\dfrac{4\pi(R/C)f}{Q_0/(RCNL)}$

†For Q_0 (flow through maximum film thickness at $\theta=0$) use top signs; for Q_π (flow through minimum film thickness at $\theta=\pi$) use lower signs.

TABLE 28.12 Values of I_s, I_c, and Sommerfeld Number for Various Values of L/D and ε

ε \ L/D	0.25	0.5	0.75	1.0	1.5	2	∞
0.1	0.0032†	0.0120	0.0244	0:0380	0.0636	0.0839	0.1570
	−0.0004	−0.0014	−0.0028	−0.0041	−0.0063	−0.0076	−0.0100
	16.4506	4.3912	2.1601	1.3880	0.8301	0.6297	0.3372
0.2	0.0067	0.0251	0.0505	0.0783	0.1300	0.1705	0.3143
	−0.0017	−0.0062	−0.0118	−0.0174	−0.0259	−0.0312	−0.0408
	7.6750	2.0519	1.0230	0.6614	0.4002	0.3061	0.1674
0.3	0.0109	0.0404	0.0804	0.1236	0.2023	0.2628	0.4727
	−0.0043	−0.0153	−0.0289	−0.0419	−0.0615	−0.0733	−0.0946
	4.5276	1.2280	0.6209	0.4065	0.2509	0.1944	0.1100
0.4	0.0164	0.0597	0.1172	0.1776	0.2847	0.3649	0.6347
	−0.0089	−0.0312	−0.0579	−0.0825	−0.1183	−0.1391	−0.1763
	2.8432	0.7876	0.4058	0.2709	0.1721	0.1359	0.0805
0.5	0.0241	0.0862	0.1656	0.2462	0.3835	0.4831	0.8061
	−0.0174	−0.0591	−0.1065	−0.1484	−0.2065	−0.2391	−0.2962
	1.7848	0.5076	0.2694	0.1845	0.1218	0.0984	0.0618
0.6	0.0363	0.1259	0.2345	0.3306	0.5102	0.6291	0.9983
	−0.0338	−0.1105	−0.1917	−0.2590	−0.3474	−0.3949	−0.4766
	1.0696	0.3167	0.1752	0.1242	0.0859	0.0714	0.0480
0.7	0.0582	0.1927	0.3430	0.4793	0.6878	0.8266	1.2366
	−0.0703	−0.2161	−0.3549	−0.4612	−0.5916	−0.6586	−0.7717
	0.5813	0.1832	0.1075	0.0798	0.0585	0.0502	0.0364
0.8	0.1071	0.3264	0.5425	0.7220	0.9771	1.1380	1.5866
	−0.1732	−0.4797	−0.7283	−0.8987	−0.0941	−1.1891	−0.3467
	0.2605	0.0914	0.0584	0.0460	0.0362	0.0322	0.0255
0.9	0.2761	0.7079	1.0499	1.3002	1.6235	1.8137	2.3083
	−0.6644	−1.4990	−2.0172	−2.3269	−2.6461	−2.7932	−3.0339
	0.0737	0.0320	0.0233	0.0199	0.0171	0.0159	0.0139
0.95	0.6429	1.3712	1.8467	2.1632	2.5455	2.7600	3.2913
	−2.1625	−3.9787	−4.8773	−5.3621	−5.8315	−6.0396	−6.3776
	0.0235	0.0126	0.0102	0.0092	0.0083	0.0080	0.0074
0.99	3.3140	4.9224	5.6905	6.1373	6.6295	6.8881	8.7210
	−22.0703	−28.5960	−30.8608	−31.9219	−32.8642	−33.2602	−33.5520
	0.0024	0.0018	0.0017	0.0016	0.0016	0.0016	0.0015

†The three numbers associated with each ε and L/D pair are, in order from top to bottom, I_s, I_c, and S.

TABLE 28.13 Comparison of Predicted Performance between Two Methods for Example 1

Method \ Parameter	ε	ϕ	$\left(\dfrac{R}{C}\right)(f)$	$\dfrac{Q}{RCNL}$	$\dfrac{Q_s}{Q}$	ΔT
Combined solution approximation	0.582	52.5°	3.508	4.473	0.652	26.6°F
Design charts†	0.58	53.°	3.50	4.28	0.655	26.6°F

†SOURCE: Shigley and Mischke [28.6].

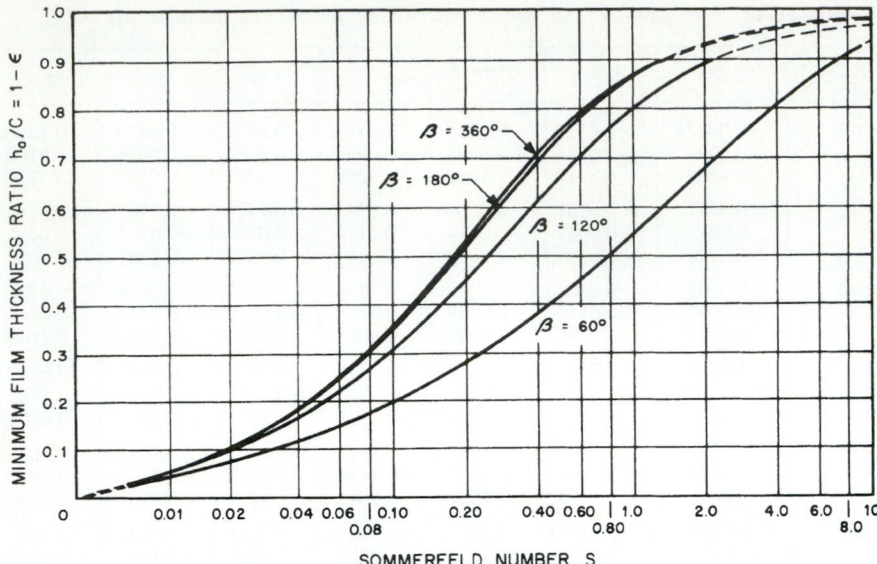

FIGURE 28.8   Minimum film thickness ratio versus Sommerfeld number for full and partial journal bearings, $L/D = 1$, Swift-Stieber boundary conditions. *(From Raimondi and Boyd [28.7].)*

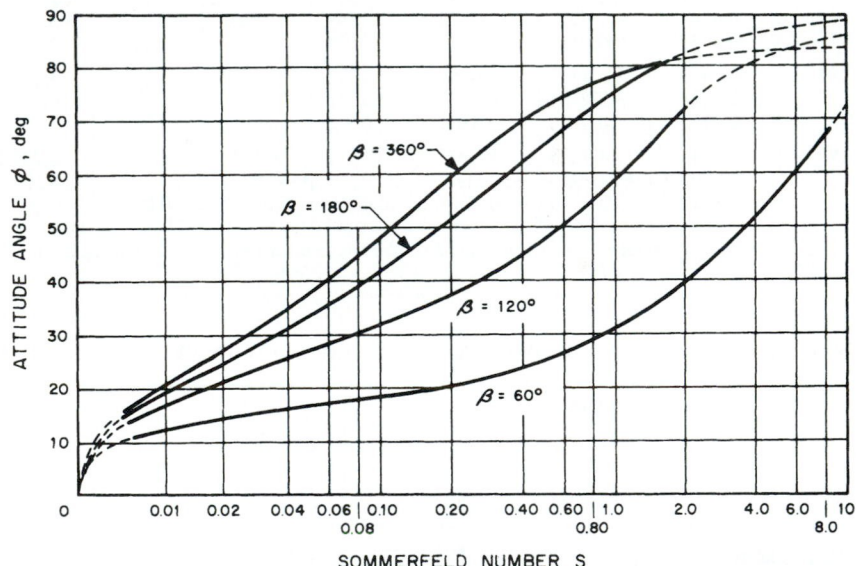

FIGURE 28.9   Attitude angle versus Sommerfeld number for full and partial journal bearings, $L/D = 1$, Swift-Stieber boundary conditions. *(From Raimondi and Boyd [28.7].)*

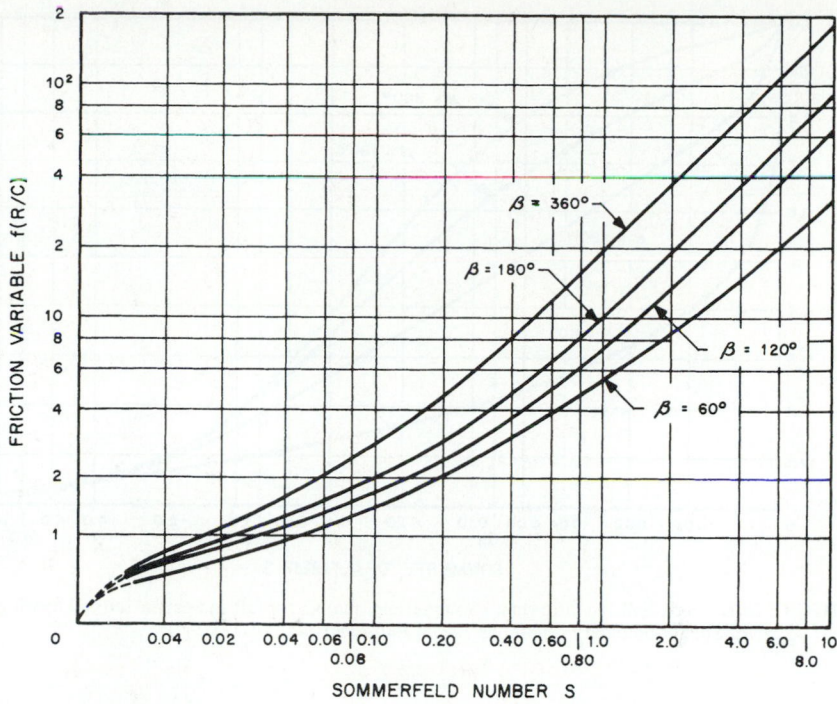

FIGURE 28.10 Friction variable versus Sommerfeld number for full and partial journal bearings, $L/D = 1$, Swift-Stieber boundary conditions. *(From Raimondi and Boyd [28.7].)*

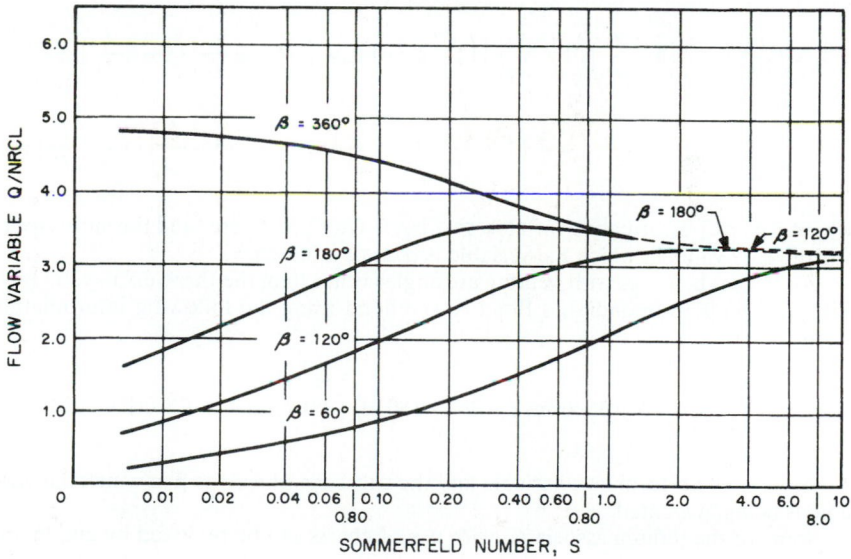

FIGURE 28.11 Flow variable versus Sommerfeld number for full and partial journal bearings, $L/D = 1$, Swift-Stieber boundary conditions. *(From Raimondi and Boyd [28.7].)*

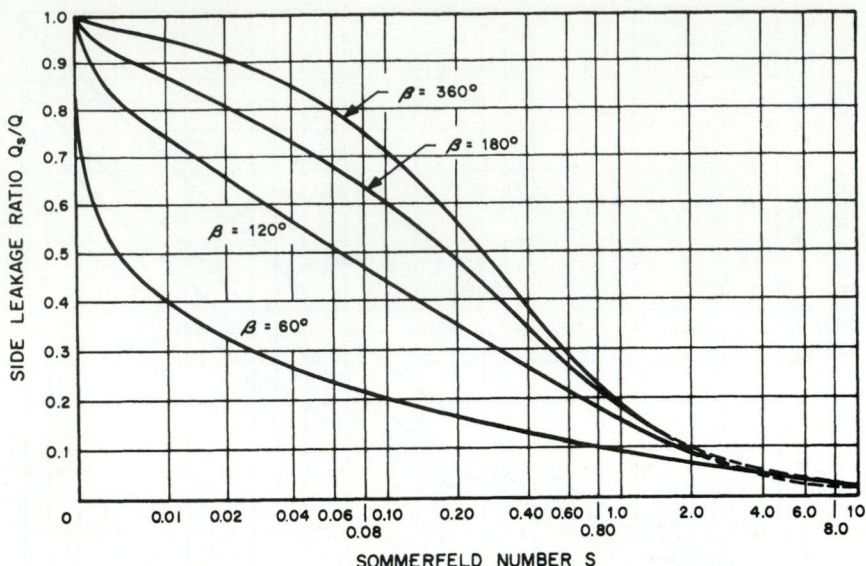

FIGURE 28.12   Side-leakage ratio versus Sommerfeld number for full and partial journal bearings, $L/D = 1$, Swift-Stieber boundary conditions. *(From Raimondi and Boyd [28.7].)*

For slenderness ratios other than the four displayed (∞, 1, ½, and ¼), Raimondi and Boyd suggest the use of the following interpolation formula:

$$y_{L/D} = \left(\frac{D}{L}\right)^3 \left[-\frac{1}{8}\left(1 - \frac{L}{D}\right)\left(1 - 2\frac{L}{D}\right)\left(1 - 4\frac{L}{D}\right)y_\infty + \frac{1}{3}\left(1 - 2\frac{L}{D}\right)\left(1 - 4\frac{L}{D}\right)y_1 \right.$$

$$\left. -\frac{1}{4}\left(1 - \frac{L}{D}\right)\left(1 - 4\frac{L}{D}\right)y_{1/2} + \frac{1}{24}\left(1 - \frac{L}{D}\right)\left(1 - 2\frac{L}{D}\right)y_{1/4} \right]$$

where y = any performance variable, that is, $(R/C)(f)$, h_0/C, etc., and the subscript of y is the L/D value at which the variable is being evaluated.

For partial bearings with bearing arc angles other than the three displayed (180°, 120°, and 60°), Raimondi and Boyd recommend using the following interpolation formula:

$$y_\beta = \frac{1}{7200}\left[(\beta - 120)(\beta - 60)y_{180} - 2(\beta - 180)(\beta - 60)y_{120} + (\beta - 180)(\beta - 120)y_{60}\right]$$

where y = any performance variable and the subscript of y is the β at which the variable is being evaluated.

Some of the tedium associated with use of charts can be removed by employing curve fits of the data. Seireg and Dandage [28.8] have developed approximate equations for the full journal bearing data of the Raimondi and Boyd charts. Table 28.15 gives the coefficients to be used in these curve-fitted equations.

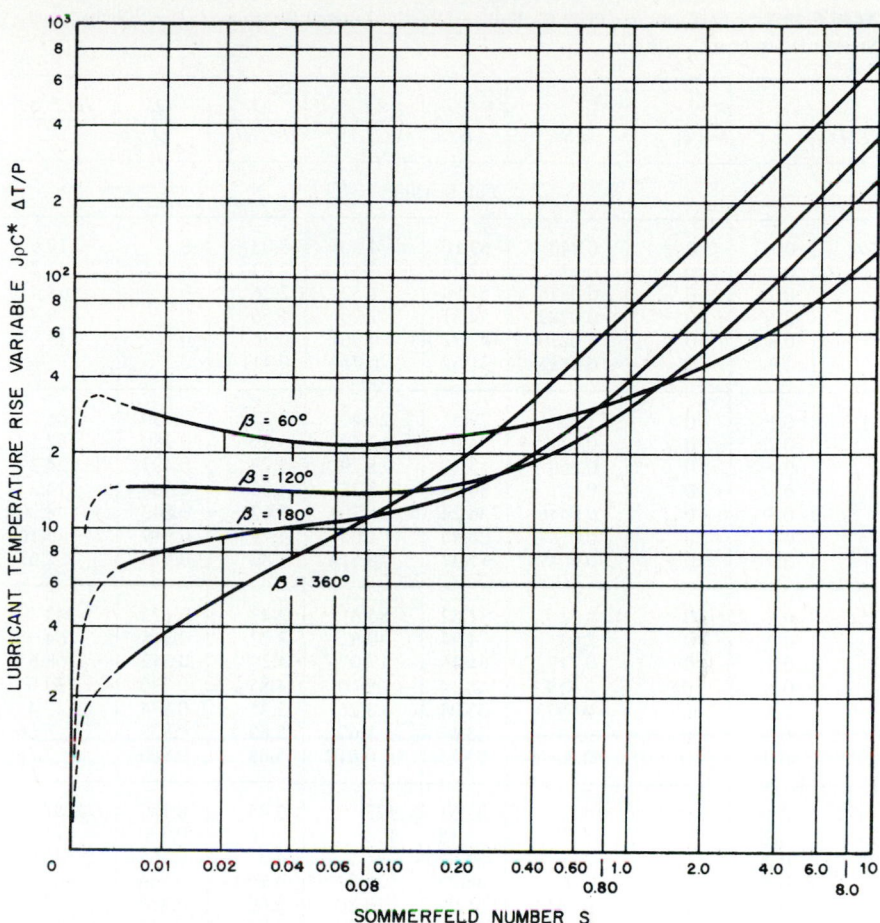

FIGURE 28.13 Lubricant temperature-rise variable versus Sommerfeld number for full and partial journal bearings, $L/D = 1$, Swift-Stieber boundary conditions. *(From Raimondi and Boyd [28.7].)*

Example 2. For the following data

$N = 3600$ r/min $L = 4$ in

$W = 7200$ lbf $C = 6.0 \times 10^3$ in

$D = 6$ in   Lubricant: SAE 20 oil

Inlet temperature $T_i = 110°$F

determine the isoviscous performance of a centrally loaded full journal bearing. The viscosity-temperature relation is contained in Table 28.16.

Solution. Because the viscosity varies with temperature, an iterative procedure is required. By this procedure, a first-guess viscosity is used to determine the film temperature rise. From this an average film temperature is determined, which will permit a second film temperature rise to be determined, and so on, until a converged result is obtained.

TABLE 28.14 Performance Data for Full and Partial Journal Bearings, $L/D = 1$, Swift-Stieber Boundary Conditions

L/D	ε	θ_1	S	ϕ	$\left(\dfrac{R}{C}\right)(f)$	$\dfrac{Q}{RCNL}$	$\dfrac{Q_s}{Q}$	$\dfrac{J\rho C^* \Delta T}{P}$
				Full bearing				
∞	0.1	0	0.240	69.10	4.80	3.03	0	19.9
	0.2	0	0.123	67.26	2.57	2.83	0	11.4
	0.4	0	0.0626	61.94	1.52	2.26	0	8.47
	0.6	0	0.0389	54.31	1.20	1.56	0	9.73
	0.8	0	0.0210	42.22	0.961	0.760	0	15.9
	0.9	0	0.0115	31.62	0.756	0.411	0	
1	0.1	0	1.33	79.5	26.4	3.37	0.150	106
	0.2	0	0.631	74.02	12.8	3.59	0.280	52.1
	0.4	0	0.264	63.10	5.79	3.99	0.497	24.3
	0.6	0	0.121	50.58	3.22	4.33	0.680	14.2
	0.8	0	0.0446	36.24	1.70	4.62	0.842	8.00
	0.9	0	0.0188	26.45	1.05	4.74	0.919	5.16
	0.97	0	0.00474	15.47	0.514	4.82	0.973	2.61
1/2	0.1	0	4.31	81.62	85.6	3.43	0.173	343
	0.2	0	2.03	74.94	40.9	3.72	0.318	164
	0.4	0	0.779	61.45	17.0	4.29	0.552	68.6
	0.6	0	0.319	48.14	8.10	4.85	0.730	33.0
	0.8	0	0.0923	33.31	3.26	5.41	0.874	13.4
	0.9	0	0.0313	23.66	1.60	5.69	0.939	6.66
	0.97	0	0.00609	13.75	0.610	5.88	0.980	2.56
1/4	0.1	0	16.2	82.31	322	3.45	0.180	1287
	0.2	0	7.57	75.18	153	3.76	0.330	611
	0.4	0	2.83	60.86	61.1	4.37	0.567	245
	0.6	0	1.07	46.72	26.7	4.99	0.746	107
	0.8	0	0.261	31.04	8.80	5.60	0.884	35.4
	0.9	0	0.0736	21.85	3.50	5.91	0.945	14.1
	0.97	0	0.0101	12.22	0.922	6.12	0.984	3.73
			Partial bearing, $\beta = 60°$					
∞	0.1	84.00	5.75	65.91	19.7	3.01	0	82.3
	0.2	101.00	2.66	48.91	10.1	2.73	0	46.5
	0.4	118.00	0.931	31.96	4.67	2.07	0	28.4
	0.6	126.80	0.322	23.21	2.40	1.40	0	21.5
	0.8	132.60	0.0755	17.39	1.10	0.722	0	19.2
	0.9	135.06	0.0241	14.94	0.667	0.372	0	22.5
	0.97	139.14	0.00495	10.88	0.372	0.115	0	40.7
1	0.1	82.00	8.52	67.92	29.1	3.07	0.0267	121
	0.2	99.00	3.92	50.96	14.8	2.82	0.0481	67.4
	0.4	116.00	1.34	33.99	6.61	2.22	0.0849	39.1
	0.6	125.50	0.450	24.56	3.29	1.56	0.127	28.2
	0.8	131.60	0.101	18.33	1.42	0.883	0.200	22.5
	0.9	134.67	0.0309	15.33	0.822	0.519	0.287	23.2
	0.97	139.10	0.00584	10.88	0.422	0.226	0.465	30.5

TABLE 28.14 Performance Data for Full and Partial Journal Bearings, $L/D = 1$, Swift-Stieber Boundary Conditions (*Continued*)

L/D	ε	θ_1	S	ϕ	$\left(\dfrac{R}{C}\right)(f)$	$\dfrac{Q}{RCNL}$	$\dfrac{Q_s}{Q}$	$\dfrac{J\rho C^* \Delta T}{P}$
\multicolumn{9}{c}{Partial bearing, $\beta = 60°$ (*Continued*)}
1/2	0.1	81.00	14.2	69.00	48.6	3.11	0.0488	201
	0.2	97.50	6.47	52.60	24.2	2.91	0.0883	109
	0.4	113.00	2.14	37.00	10.3	2.38	0.160	59.4
	0.6	123.00	0.695	26.98	4.93	1.74	0.236	40.3
	0.8	130.40	0.149	19.57	2.02	1.05	0.350	29.4
	0.9	134.09	0.0422	15.91	1.08	0.664	0.464	26.5
	0.97	139.22	0.00704	10.85	0.490	0.329	0.650	27.8
1/4	0.1	78.50	35.8	71.55	121	3.16	0.0666	499
	0.2	91.50	16.0	58.51	58.7	3.04	0.131	260
	0.4	109.00	5.20	41.01	24.5	2.57	0.236	136
	0.6	119.80	1.65	30.14	11.2	1.98	0.346	86.1
	0.8	128.30	0.333	21.70	4.27	1.30	0.496	54.9
	0.9	133.10	0.0844	16.87	2.01	0.894	0.620	41.0
	0.97	139.20	0.0110	10.81	0.713	0.507	0.786	29.1
\multicolumn{9}{c}{Partial bearing, $\beta = 120°$}
∞	0.1	53.300	0.877	66.69	6.02	3.02	0	25.1
	0.2	67.400	0.431	52.60	3.26	2.75	0	14.9
	0.4	81.000	0.181	39.02	1.78	2.13	0	10.5
	0.6	87.300	0.0845	32.67	1.21	1.47	0	10.3
	0.8	93.200	0.0328	26.80	0.853	0.759	0	14.1
	0.9	98.500	0.0147	21.51	0.653	0.388	0	21.2
	0.97	106.15	0.00406	13.86	0.399	0.118	0	42.4
1	0.1	47.500	2.14	72.43	14.5	3.20	0.0876	59.5
	0.2	62.000	1.01	58.25	7.44	3.11	0.157	32.6
	0.4	76.000	0.385	43.98	3.60	2.75	0.272	19.0
	0.6	84.500	0.162	35.65	2.16	2.24	0.384	15.0
	0.8	92.600	0.0531	27.42	1.27	1.57	0.535	13.9
	0.9	98.667	0.0208	21.29	0.855	1.11	0.657	14.4
	0.97	106.50	0.00498	13.49	0.461	0.694	0.812	14.0
1/2	0.1	45.000	5.42	74.99	36.6	3.29	0.124	149
	0.2	56.650	2.51	63.38	18.1	3.32	0.225	77.2
	0.4	72.000	0.914	48.07	8.20	3.15	0.386	40.5
	0.6	81.500	0.354	38.50	4.43	2.80	0.530	27.0
	0.8	92.000	0.0973	28.02	2.17	2.18	0.684	19.0
	0.9	99.000	0.0324	21.02	1.24	1.70	0.787	15.1
	0.97	107.00	0.00631	13.00	0.550	1.19	0.899	10.6
1/4	0.1	43.000	18.4	76.97	124	3.34	0.143	502
	0.2	54.000	8.45	65.97	60.4	3.44	0.260	254
	0.4	68.833	3.04	51.23	26.6	3.42	0.442	125
	0.6	79.600	1.12	40.42	13.5	3.20	0.599	75.8
	0.8	91.560	0.268	28.38	5.65	2.67	0.753	42.7
	0.9	99.400	0.0743	20.55	2.63	2.21	0.846	25.9
	0.97	108.00	0.0105	12.11	0.832	1.69	0.931	11.6

TABLE 28.14 Performance Data for Full and Partial Journal Bearings, $L/D = 1$, Swift-Stieber Boundary Conditions (*Continued*)

L/D	ε	θ_1	S	ϕ	$\left(\dfrac{R}{C}\right)(f)$	$\dfrac{Q}{RCNL}$	$\dfrac{Q_s}{Q}$	$\dfrac{J\rho C^* \Delta T}{P}$
	Partial bearing, $\beta = 180°$								
∞	0.1	17.000	0.347	72.90	3.55	3.04	0	14.7
	0.2	28.600	0.179	61.32	2.01	2.80	0	8.99
	0.4	40.000	0.0898	49.99	1.29	2.20	0	7.34
	0.6	46.900	0.0523	43.15	1.06	1.52	0	8.71
	0.8	56.700	0.0253	33.35	0.859	0.767	0	14.1
	0.9	64.200	0.0128	25.57	0.681	0.380	0	22.5
	0.97	74.650	0.00384	15.43	0.416	0.119	0	44.0
1	0.1	11.500	1.40	78.50	14.1	3.34	0.139	57.0
	0.2	21.000	0.670	68.93	7.15	3.46	0.252	29.7
	0.4	34.167	0.278	58.86	3.61	3.49	0.425	16.5
	0.6	45.000	0.128	44.67	2.28	3.25	0.572	12.4
	0.8	58.000	0.0463	32.33	1.39	2.63	0.721	10.4
	0.9	66.000	0.0193	24.14	0.921	2.14	0.818	9.13
	0.97	75.584	0.00483	14.57	0.483	1.60	0.915	6.96
1/2	0.1	10.000	4.38	79.97	44.0	3.41	0.167	177
	0.2	17.800	2.06	72.14	21.6	3.64	0.302	87.8
	0.4	32.000	0.794	58.01	9.96	3.93	0.506	42.7
	0.6	45.000	0.321	45.01	5.41	3.93	0.665	25.9
	0.8	59.000	0.0921	31.29	2.54	3.56	0.806	15.0
	0.9	67.200	0.0314	22.80	1.38	3.17	0.886	9.80
	0.97	76.500	0.00625	13.63	0.581	2.62	0.951	5.30
1/4	0.1	9.000	16.3	81.40	163	3.44	0.176	653
	0.2	16.300	7.60	73.70	79.4	3.71	0.320	320
	0.4	31.000	2.84	58.99	35.1	4.11	0.534	146
	0.6	45.000	1.08	44.96	17.6	4.25	0.698	79.8
	0.8	59.300	0.263	30.43	6.88	4.07	0.837	36.5
	0.9	68.900	0.0736	21.43	2.99	3.72	0.905	18.4
	0.97	77.680	0.0104	12.28	0.877	3.29	0.961	6.46

SOURCE: Raimondi and Boyd [28.7].

Since $L/D = \frac{2}{3}$, the Raimondi and Boyd charts would require interpolation. Alternatively, the Seireg-Dandage curve-fitted equations are used. The unit load may be immediately computed:

$$P = \frac{W}{LD} = \frac{7200}{(6)(4)} = 300 \text{ psi}$$

The first guess of viscosity is based on the inlet temperature:

$$\mu_1 = 1.36 \times 10^{-8} \exp \frac{1271.6}{110 + 95} = 6.72 \times 10^{-6} \text{ reyn}$$

$$S_1 = \frac{\mu_1 N}{P}\left(\frac{R}{C}\right)^2 = 0.336$$

TABLE 28.15 Seireg-Dandage Curve Fits of Raimondi-Boyd Charts: Variable $= a\left(\dfrac{L}{D}\right)^{b_1} S^{b_2 + b_3(L/D)}$

$$\tfrac{1}{2} \le L/D \le 1$$

Variable		S ≤ 0.15					S > 0.15			
		a	b_1	b_2	b_3		a	b_1	b_2	b_3
$\dfrac{h_0}{C} = 1 - \epsilon$	S ≤ 0.04	2.7258	0.83621	0.75101	0.08113	S ≤ 1	0.91437	0.4538	0.6119	−0.2890
	S > 0.04	1.7176	1.0478	0.4999	0.1868	S > 1	0.89574	0.3895	0.3076	−0.2537
ϕ, rad	......	110.9067	0.60907	0.28856	0.07485	......	74.0225	0.2395	0.3131	−0.2172
$\left(\dfrac{R}{C}\right)(f)$	......	9.9533	−0.4758	0.6705	−0.1124	......	3.5251	−0.2333	−0.1926	0.1149
$\dfrac{Q}{RCNL}$	......	4.1036	−0.306242	−0.024799	−0.009982	......	20.4422	−0.1125	0.8551	0.1014
$\dfrac{J \rho C^* \Delta T}{P}$	......	42.0097	−0.4146	0.6869	−0.1600	......	84.2989	−0.08167	0.85540	0.08787
$\dfrac{p}{p_{max}}$	......	0.79567	0.59651	0.25659	0.04321	......	0.52529	0.2486	0.2335	−0.1870

28.33

TABLE 28.15 Seireg-Dandage Curve Fits of Raimondi-Boyd Charts: Variable $= a\left(\dfrac{L}{D}\right)^{b_1} S^{b_2 + b_3(L/D)}$
(Continued)

Variable		$S \le 0.15$					$S > 0.15$			
		a	b_1	b_2	b_3		a	b_1	b_2	b_3
$\dfrac{h_0}{C} = 1 - \epsilon$	$S \le 0.04$	9.2341	2.0673	0.4286	0.9247	$S \le 1$	1.1674	0.80824	0.48016	−0.02463
	$S > 0.04$	1.1545	0.4637	0.7851	−0.3788	$S > 1$	1.1263	0.7279	0.5117	−0.6581
ϕ, rad	· · · · · ·	112.7756	0.6332	0.2592	0.1336	· · · · · ·	93.6908	0.5313	0.3139	−0.2923
$\left(\dfrac{R}{C}\right)(f)$	· · · · · ·	9.4896	−0.5446	0.7290	−0.2293	· · · · · ·	17.1809	−0.3133	0.7993	0.2887
$\dfrac{Q}{RCNL}$	· · · · · ·	4.5607	−0.153864	−0.005371	−0.48838	· · · · · ·	3.4980	−0.23973	−0.11107	−0.04103
$\dfrac{J\rho C^* \Delta T}{P}$	· · · · · ·	38.7604	−0.5307	0.7322	−0.2505	· · · · · ·	69.4842	−0.3075	0.8003	0.2783
$\dfrac{p}{p_{max}}$	· · · · · ·	0.78635	0.57952	0.23620	0.0840	· · · · · ·	0.64927	0.5037	0.2783	−0.3534

$\tfrac{1}{4} \le L/D \le \tfrac{1}{2}$

SOURCE: Ref. [28.8].

TABLE 28.16 Constants for Use in Viscosity-Temperature Equation for Various Oils

Oil	μ_0†, reyn	b, °F
SAE 10	1.58×10^{-8}	1157.5
SAE 20	1.36×10^{-8}	1271.6
SAE 30	1.41×10^{-8}	1360.9
SAE 40	1.21×10^{-8}	1474.4
SAE 50	1.70×10^{-8}	1509.6
SAE 60	1.87×10^{-8}	1564.0

†$\mu = \mu_0 \exp[b/(T + 95)]$.
SOURCE: Ref. [28.8].

From Table 28.15, the appropriate curve-fitted equation for the temperature rise is

$$\Delta T = 84.2989 \frac{P}{J\rho C^*} \left(\frac{L}{D}\right)^{-0.08167} S^{0.8554 + 0.08787L/D}$$

Taking $\rho = 0.03$ pound mass per cubic inch (lbm/in^3) and $C^* = 0.40$ Btu/(lbm · °F) as representative values for lubricating oil, we obtain

$$\Delta T_1 = 233.33(0.336)^{0.914} = 86.1°\text{F}$$

And so the second estimate of the film mean temperature is

$$T_{a2} = 110 + \frac{86.1}{2} = 153.1°\text{F}$$

Repeated calculations (13 iterations) produce

$$S = 0.176 \qquad \mu = 3.5 \times 10^{-6} \text{ reyn} \qquad \Delta T = 47.7°\text{F}$$

With the Sommerfeld number, the remaining performance parameters are easily calculated.

Connors' Lubricant Supply Charts. The Raimondi-Boyd lubricant flow and temperature rise data are based on the notion that there is no carryover flow into the active film; that is, $Q_2 = 0$ in Eq. (28.14). From an analogous view, these results are applicable to the situation where $Q_i \gg Q_2$. Accordingly, Raimondi-Boyd predictions represent fully flooded bearing conditions and yield the coolest running lubricant temperatures for a given set of operating conditions, not accounting for any heat conduction losses.

To remedy this and thus provide more realistic design information, Connors [28.9] developed design charts which incorporate the influence of lubricant supply rate on the performance of a full journal bearing for $L/D = 1$. Figures 28.14 to 28.16 are plots that can be used over the entire range of flows to determine minimum film thickness, friction, and temperature rise from given values of the Sommerfeld number and the inlet flow variable.

Example 3. Determine the lubricant temperature rise as a function of the inlet flow rate for the following design parameters:

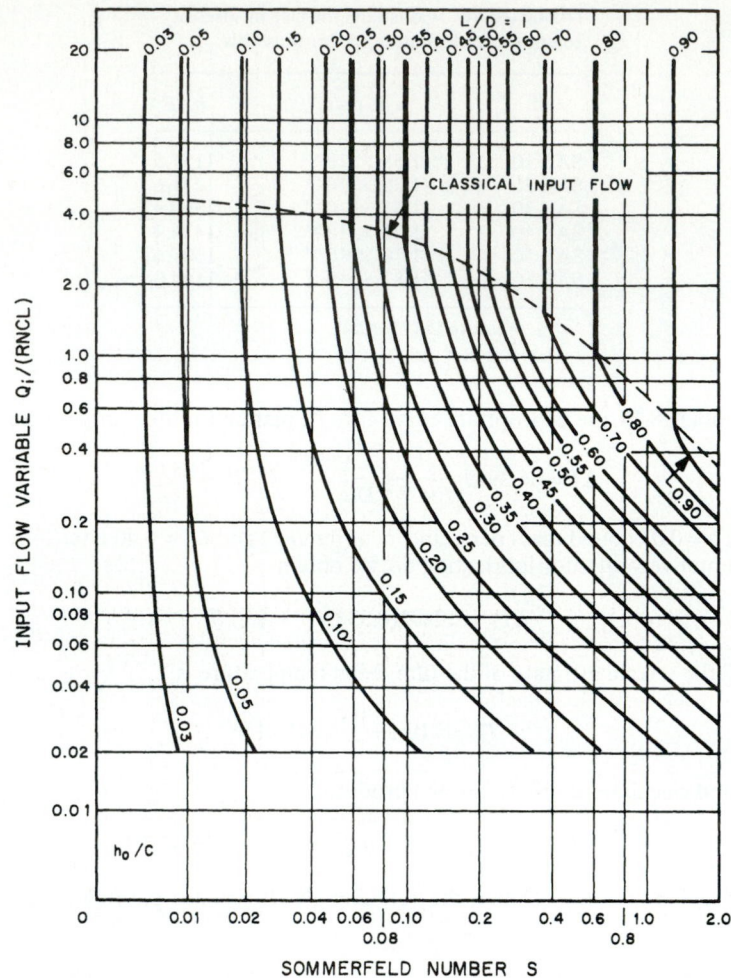

FIGURE 28.14   Inlet flow variable versus Sommerfeld number for parametric values of minimum film thickness ratio; $L/D = 1$, full journal bearing. *(From Connors [28.9].)*

$W = 1500$ lbf	$C = 4 \times 10^{-3}$ in
$N = 1800$ r/min	SAE 30 oil
$D = 4$ in	$T_i = 100°F$
$L = 4$ in	

Solution.   To solve this type of problem, a plot or equation of lubricant viscosity as a function of temperature must be available. The calculation procedure is as follows:

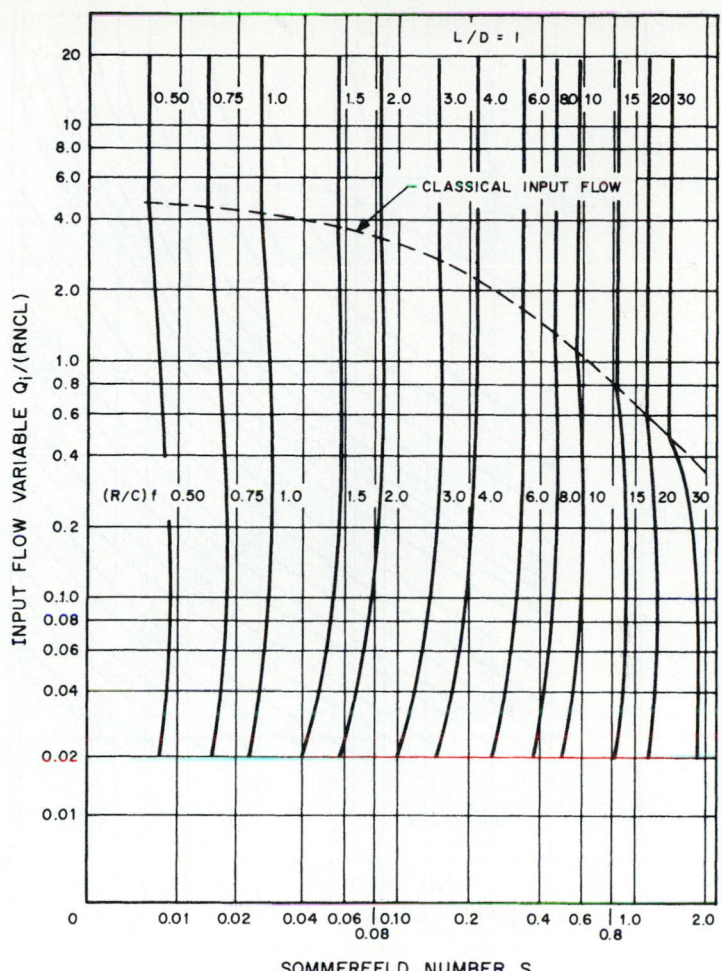

FIGURE 28.15　Inlet flow variable versus Sommerfeld number for parametric values of friction variable; $L/D = 1$, full journal bearing. *(From Connors [28.9].)*

1. Select a value of $Q_i/(RCNL)$.
2. Assume a viscosity value.
3. Compute the Sommerfeld number.
4. Use the $Q_i/(RCNL)$ and S values to find $J\rho C^*(T_a - T_i)/P$ in Fig. 28.16.
5. Calculate the mean film temperature T_a.
6. Increment μ and repeat the process from step 3 until there are sufficient points to establish an intersection with the lubricant's μ versus T data. This intersection represents the operating point for the given $Q_i/(RCNL)$.
7. Increment the input flow variable, and return to step 2.

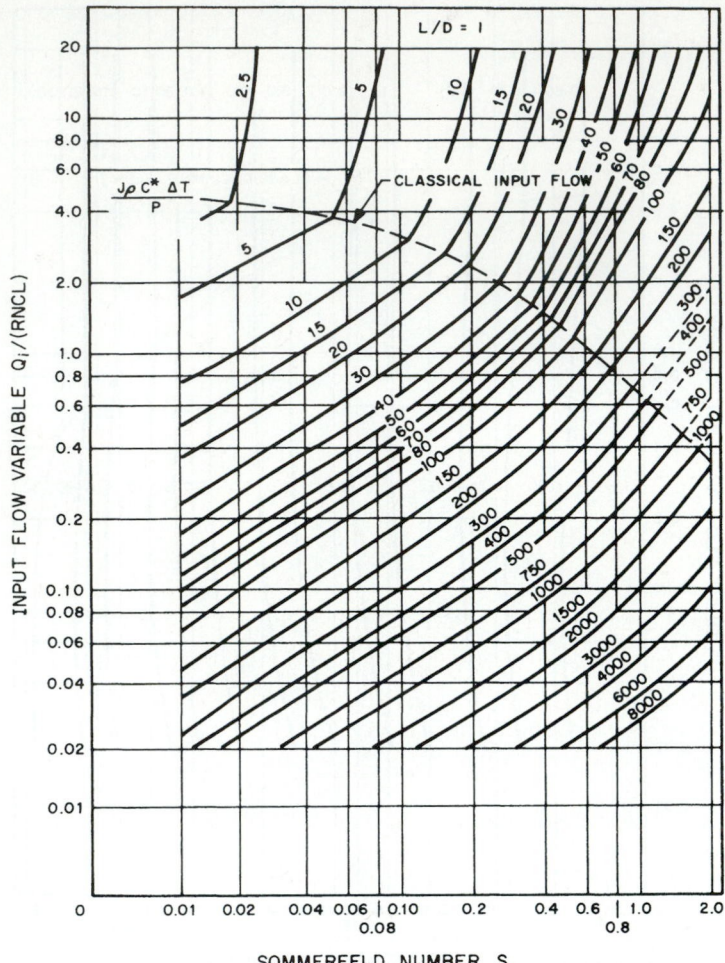

FIGURE 28.16 Inlet flow variable versus Sommerfeld number for parametric values of temperature-rise variable; $L/D = 1$, full journal bearing. *(From Connors [28.9].)*

For $Q_i/(RCNL) = 1$, the following sets of data were obtained by performing this calculation procedure:

$$\mu = \begin{cases} 1.5 \times 10^{-6} \text{ reyn} \\ 3.0 \times 10^{-6} \text{ reyn} \\ 6.0 \times 10^{-6} \text{ reyn} \end{cases} \qquad S = \begin{cases} 0.12 \\ 0.24 \\ 0.48 \end{cases}$$

$$\frac{J\rho C^*(T_a - T_i)}{P} = \begin{cases} 32 \\ 60 \\ 115 \end{cases} \qquad T_a = \begin{cases} 126.8°F \\ 150.2°F \\ 196.2°F \end{cases}$$

Using these and the lubricant μ versus T relation as presented in Table 28.16, we find the operating point to be

$$T_a = 155°F \qquad \mu = 3.2 \times 10^{-6} \text{ reyn}$$

Hence, $S = 0.256$. Also from Fig. 28.14 we obtain $h_0/C = 0.52$, and so $h_0 = 0.00208$. Further, from Fig. 28.15, $(R/C)(f) = 5$, and so $f = 0.01$, which allows us to calculate the power loss to be 0.857 horsepower (hp). Assuming other values of $Q_i/(RCNL)$ permits Fig. 28.17 to be drawn. The Raimondi-Boyd value corresponding to $Q_i \to \infty$ is also presented.

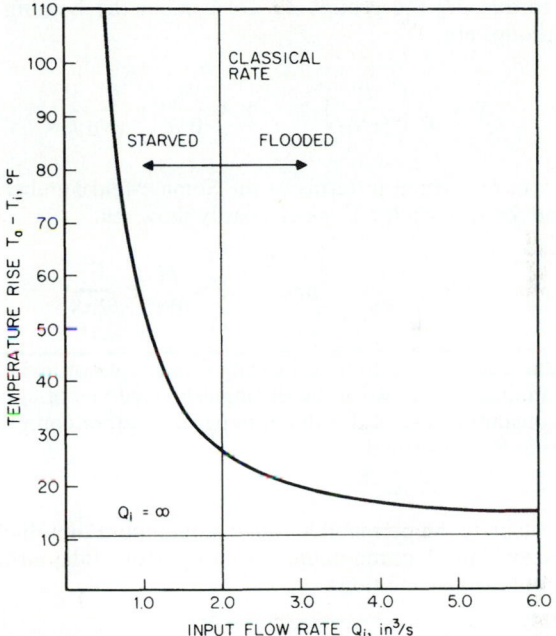

FIGURE 28.17 Lubricant temperature rise versus lubricant input flow rate (Example 3).

In Sec. 28.5.4 it was shown that

$$(T_a - T_i)_{\text{conduction}} = (1 - \lambda)(T_a - T_i)_{\text{no conduction}}$$

where λ = ratio of heat conduction to heat generation rate and is assumed to be a constant. By using this idea, a new operating point for a given $Q_i/(RCNL)$ can be determined. For example, with $\lambda = 0.25$ and $Q_i/(RCNL) = 1$, we find that $T_a = 147°F$, $h_0 = 0.0023$, and HP = 0.960 hp.

28.6.3 Optimization

In designing a journal bearing, a choice must be made among several potential designs for the particular application. Thus the designer must establish an optimum design criterion for the bearing. The design criterion describes the designer's objective, and numerous criteria can be envisioned (e.g., minimizing frictional loss, minimizing the lubricant temperature rise, minimizing the lubricant supply to the bearing, and so forth).

The search for an optimum bearing design is best conducted with the aid of a computer. However, optimum bearing design can also be achieved graphically. Moes and Bosma [28.10] developed a design chart for the full journal bearing which enables the designer to select optimum bearing dimensions. This chart is constructed in terms of two dimensionless groups called X and Y here. The groups include two quantities of primary importance to the bearing designer: minimum film thickness h_0 and frictional torque M_j; the groups do not contain the bearing clearance. The dimensionless groups are

$$X \equiv \frac{h_0}{R}\left(\frac{P}{2\pi N\mu}\right)^{1/2} \qquad Y \equiv \frac{M_j}{WR}\left(\frac{P}{2\pi N\mu}\right)^{1/2} \qquad (28.15)$$

Both X and Y can be written in terms of the Sommerfeld number. Recalling that $h_0 = C(1-\varepsilon)$ and $S = (\mu N/P)(R/C)^2$, we can easily show that

$$X = \frac{1-\varepsilon}{\sqrt{2\pi S}} \qquad \text{and} \qquad Y = \frac{M_j}{WC}\frac{1}{\sqrt{2\pi S}}$$

Figure 28.18 is a plot of full journal bearing design data on the XY plane. In the diagram, two families of curves can be distinguished: curves of constant L/D ratio and curves of constant ε. Use of this diagram permits rather complicated optimization procedures to be performed.

Example 4.   Calculate the permissible range of minimum film thickness and bearing clearance that will produce minimum shaft torque for a full journal bearing operating under the following conditions:

$\mu = 5 \times 10^{-6}$ reyn $D = 4$ in

$N = 1800$ rev/min $L = 3$ in

$W = 1800$ lbf

Solution.   As a first step, we calculate the largest h_0 for the given conditions. This is easily accomplished by locating the coordinates on Fig. 28.18 corresponding to the maximum X for $L/D = \frac{3}{4}$, or

$$\begin{Bmatrix} X \\ Y \\ \varepsilon \end{Bmatrix} = \begin{Bmatrix} 0.385 \\ 4.0 \\ 0.54 \end{Bmatrix} \qquad \text{at } X = X_{max}$$

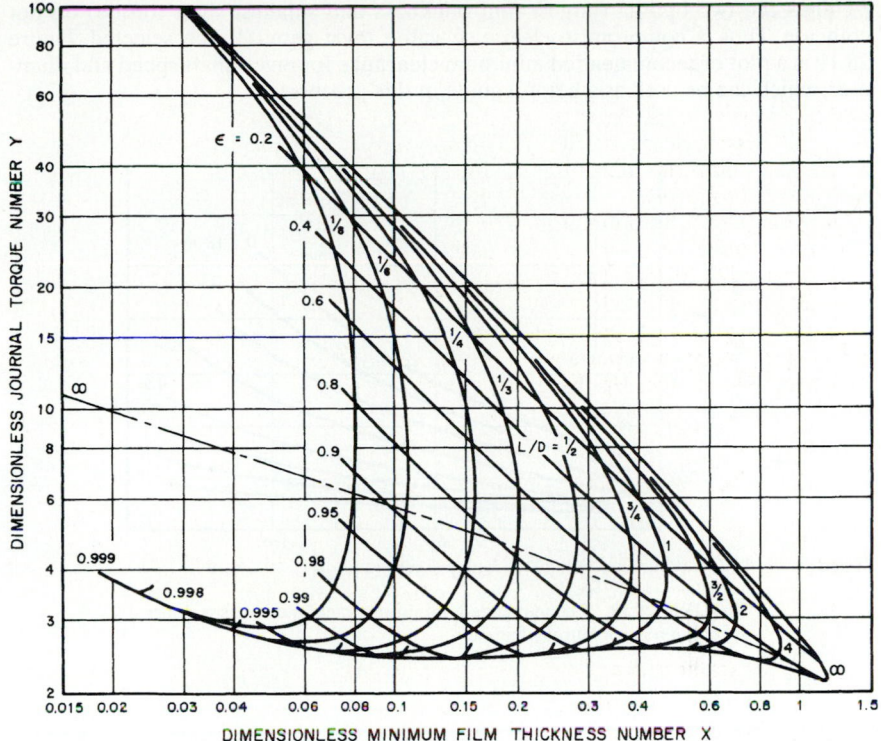

FIGURE 28.18 Optimization chart for full journal bearings. *(From Moes and Bosma [28.10].)*

Thus, from Eq. (28.15),

$$h_0 = 1.93 \times 10^{-3} \text{ in} \qquad \text{and} \qquad M_j = 36.1 \text{ in} \cdot \text{lbf}$$

The clearance is calculated from $C = h_0/(1 - \varepsilon) = 4.2 \times 10^{-3}$ in.
   Next the coordinates on Fig. 28.18 corresponding to the minimum shaft torque (minimum value of Y at $L/D = \frac{3}{4}$) are located at

$$\begin{Bmatrix} X \\ Y \\ \varepsilon \end{Bmatrix} = \begin{Bmatrix} 0.18 \\ 2.5 \\ 0.97 \end{Bmatrix} \qquad \text{at } Y = Y_{\min}$$

Thus, from Eq. (28.15),

$$h_0 = 9.02 \times 10^{-4} \text{ in} \qquad \text{and} \qquad M_j = 22.6 \text{ in} \cdot \text{lbf}$$

The clearance for this case is calculated to be 3.01×10^{-2} in.

Thus the two optima (largest film thickness and smallest shaft torque) do not coincide. Thus, a compromise clearance value must generally be selected. Figure 28.19 is a plot of recommended minimum clearance for given shaft speed and diameter, which can be used as a helpful guide in this process.

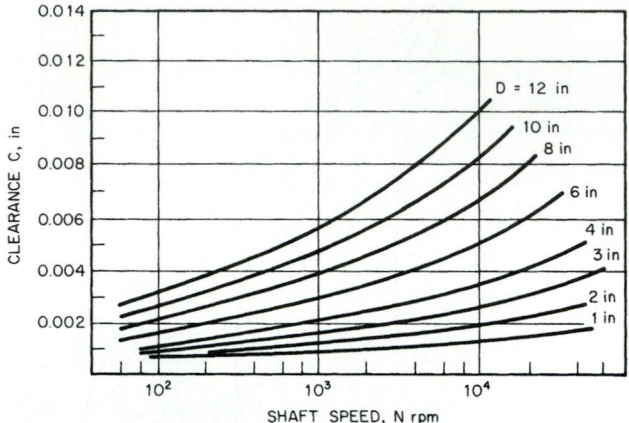

FIGURE 28.19   Recommended minimum clearance versus journal speed for a given journal diameter.

Example 5.   Determine the bearing length and clearance for a full journal bearing operating steadily with minimum friction loss for the following conditions:

$\mu = 17.83 \times 10^{-6}$ $W = 1500$ lbf

$N = 4800$ rev/min $D = 6$ in

and a minimum allowable film thickness of

$$(h_0)_{min} = 4.4 \times 10^{-4}$$

Solution.   Because both X and Y contain the bearing length L, a new set of dimensionless groups not containing L must be developed. This pair is easily seen to be

$$U = (X)\left(\frac{L}{D}\right)^{1/2} = \frac{h_0}{R}\left(\frac{W}{2\pi\mu ND^2}\right)^{1/2} \tag{28.16}$$

$$V = (Y)\left(\frac{L}{D}\right)^{1/2} = \frac{M_j}{WR}\left(\frac{W}{2\pi\mu ND^2}\right)^{1/2} \tag{28.17}$$

For the given information, the smallest value of U (termed U_{min} because the smallest value of h_0 is used) can be directly calculated from Eq. (28.16):

$$U_{min} = 0.10 \qquad \text{Thus,} \qquad X \geq (0.10)\left(\frac{D}{L}\right)^{1/2}$$

Selected L/D values permit an array of X values to be calculated. This in turn permits the minimum values of Y (that is, Y_{min}) to be read from Fig. 28.18. Then corresponding V_{min} values can be calculated from Eq. (28.17). Table 28.17 contains the results of this operation.

TABLE 28.17 Summary of Calculated
Results for Example 5

L/D	X	Y_{min}	V_{min}
1/4	0.2	No solution	1.85
1/3	0.173	3.20	1.85
1/2	0.414	2.43	1.72
3/4	0.115	2.40	2.08
1	0.100	2.40	2.40

Next a cubic equation is fitted to the V_{min} data, and we find

$$V_{min} = 4.057 - 11.927\,\frac{L}{D} + 18.742\left(\frac{L}{D}\right)^{2} - 8.472\left(\frac{L}{D}\right)^{3} \qquad (28.18)$$

The optimum value of L/D for a minimum friction loss can be found by differentiation, which yields the optimum bearing length:

$$L = 0.464D = 2.78 \text{ in}$$

The frictional loss can be calculated by inserting the optimum L/D ratio in the cubic equation:

$$V_{min} = (Y)\left(\frac{L}{D}\right)^{1/2} = 1.712$$

Thus, $Y = 2.513$, and the eccentricity ratio is read from Fig. 28.18 as $\varepsilon \cong 0.98$. Finally the optimum bearing clearance is calculated to be $C = 2.2 \times 10^{-2}$ in.

28.7 GAS-LUBRICATED JOURNAL BEARINGS

Gas-lubricated journal bearings have been employed in a wide variety of modern industrial applications. For example, they are used in dental drills, high-speed machine tools, digital-computer peripheral devices, high-speed turbomachines, and navigational instruments.

Gas bearings produce very little friction even at high speeds; hence, they have low frictional losses and generate minimal amounts of heat. In addition, gas lubricants are chemically very stable over a wide range of temperatures; they neither freeze nor boil; they are nonflammable; and they do not contaminate bearing surfaces. Gas bearings also have low noise characteristics.

On the negative side, gas bearings do not have much load-carrying capacity and have large startup wear. Also, because the gas film thickness is quite small, gas bearings require superior surface finish and manufacture. Great care must be exercised with the journal alignment. Further, thin films offer very little cushion or damping capacity; consequently, gas bearings are prone to certain vibrational instabilities.

28.7.1 Limiting Gas Bearing Solutions

The compressible-flow Reynolds equation, Eq. (28.7), is too complex to permit a general analytical solution, and numerical methods have been used to obtain gas bearing information. However, considerable insight can be gained by considering solutions for limiting cases. Besides the limiting geometric problems of long and short bearings, two other limiting classes of problems are important: gas films with low bearing numbers and gas films with high bearing numbers. The former pertain to low-speed, high-loading applications, and the latter pertain to very-high-speed, light-loading applications. Some bearing performance parameters for gas films with low and high bearing numbers are shown in Table 28.18.

TABLE 28.18 Long Gas Journal Bearing Performance Parameters, Limiting Cases, Long-Bearing Theory

Performance parameter	Low bearing number	Large bearing number
$\dfrac{p}{p_a}$	$\dfrac{1 + \Lambda\varepsilon \sin\theta(2 + \varepsilon\cos\theta)}{(2 + \varepsilon)^2(1 + \varepsilon\cos\theta)^2}$	$\dfrac{\sqrt{1 + 3\varepsilon^2}/2}{1 + \varepsilon\cos\theta}$
$\dfrac{W_R}{p_a DL}$	0	$\dfrac{\pi}{2}\left(1 + \dfrac{3\varepsilon^2}{2}\right)^{1/2}[(1 - \varepsilon^2)^{-1/2} - 1]$
$\dfrac{W_T}{p_a DL}$	$\dfrac{\pi\varepsilon\Lambda}{(2 + \varepsilon^2)\sqrt{1 - \varepsilon^2}}$	0
ϕ	$\dfrac{\pi}{2}$	0
$\dfrac{F_j}{\mu UL}\dfrac{C}{R}$	$\dfrac{4\pi(1 + 2\varepsilon^2)}{(2 + \varepsilon^2)\sqrt{1 - \varepsilon^2}}$	$\dfrac{2\pi}{\sqrt{1 - \varepsilon^2}}$
$\left(\dfrac{R}{C}\right)(f)$	$\dfrac{1 + 2\varepsilon^2}{3\varepsilon}$	$\dfrac{2\mu UR}{C^2 P_a \sqrt{(1 + 3\varepsilon^2/2(1 - \sqrt{1 - \varepsilon^2})}}$

28.7.2 Stability Considerations

The journal bearing of a rotating machine cannot be treated separately. It is part of a complex dynamic system, and its design can influence the dynamic behavior of the entire system. Bearing damping controls vibration amplitude and tolerance to any imbalance and, to a large extent, determines whether the rotor will be dynamically stable. A bearing running in an unstable mode can lose its ability to support load, and rubbing contact can occur between the bearing and the shaft surfaces.

Rotor bearing instabilities are particularly troublesome in gas-lubricated bearings; the film is considerably thinner in gas bearings than in liquid-lubricated bearings and so does not possess the damping capacity. We can differentiate between two forms of dynamic instability: *synchronous whirl* and *half-frequency whirl* (also called fractional-frequency whirl and film whirl).

Synchronous Whirl. A rotating shaft experiences periodic deflection (forced vibration) because of the distribution of load, the method of shaft support, the degree of flexibility of the shaft, and any imbalance within the rotating mass. This deflection causes the journal to orbit within its bearings at the rotational speed. When the frequency of this vibration occurs at the natural frequency (or critical speed) of the system, a resonance condition exists. In this condition, the amplitude of vibration (size of the journal orbit) increases and can cause bearing failure. Because the shaft rotational speed and the critical speed coincide, this form of instability is termed *synchronous whirl*.

Since stable operation occurs on either side of the critical speed, the bearing system must be designed so that critical speeds do not exist in the operating speed range. This can be accomplished by making the critical speeds either very large or very small. Large critical speeds can be established by increasing the bearing stiffness by reducing the bearing clearance. Lower critical speeds can be established by increasing the shaft flexibility. The instability can also be suppressed by mounting the bearing in a flexible housing which introduces additional system damping.

Half-Frequency Whirl. A deflected rotating shaft has its center whirl, or orbit, around the center of the bearing. When the whirl speed Ω is equal to or somewhat less than half the rotational speed of the shaft ω, the hydrodynamic capacity of the bearing to support a load is diminished and, in fact, may fall to zero. With any subsequent rotational speed increase, the whirl amplitude will increase, and the bearing and journal may come into violent contact. This form of self-activated vibrational instability is called *half-frequency* or *half-speed whirl*. Because the whirl speed for gas bearings is generally less than the rotational speed, the instability is also termed *fractional-frequency whirl*. Figure 28.20 is a map for assessing the stability of a full journal bearing.

Half-speed whirl may be suppressed by altering the circumferential symmetry of the bearing. This can be accomplished by providing axial grooves in the bearing surface, by dividing the circumference into a number of sections, or lobes, or by using tilting pads. Also, introducing more damping into the system through the use of flexible housing mounts can be helpful.

28.7.3 Cylindrical Ungrooved Gas Journal Bearings

Complete plain cylindrical journal bearings are generally used in applications that require support of a fixed, unidirectional load. Self-excited whirl instability is a com-

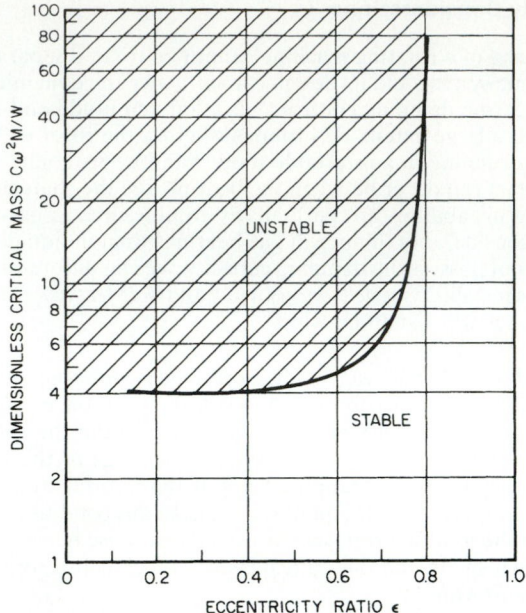

FIGURE 28.20 Bearing stability for full journal bearing, $L/D = 1$.

mon problem with high-speed rotor systems supported by this type of bearing. However, if the load is sufficiently large to have an eccentricity ratio on the order of 0.8, whirl problems are reduced (refer to Fig. 28.20).

Ungrooved cylindrical gas bearings have been analyzed by Raimondi [28.11]. Performance characteristics were developed numerically and presented in the form of various design charts for L/D ratios of ½, 1, and 2. Figures 28.21 to 28.23 are a sampling of these charts for an L/D ratio of 1.

A porous journal bearing is a plain cylindrical journal bearing with a porous liner that is fixed in the bearing housing (Fig. 28.24). An externally pressurized gas is supplied to the outer surface of the liner and flows through the porous material into the bearing clearance space.

The steady performance of a self-acting porous gas bearing of finite length has been determined by Wu [28.12]. Table 28.19 shows a sample of these performance data. The data are applicable for a particular porous linear thickness ratio (0.083) and for particular combinations of the *slip coefficient* α (dependent on the structure of the porous material), the *permeability* k (a physical property of the porous material), and bearing dimensions R and C.

Example 6. For the following conditions

$D = 0.5$ in	$N = 47\,000$ rev/min
$L = 0.5$ in	$\mu = 2.6 \times 10^{-9}$ reyn (air at 80°F)
$C = 2 \times 10^{-4}$ in	$p_a = 40$ psi
	$W = 6.1$ lbf

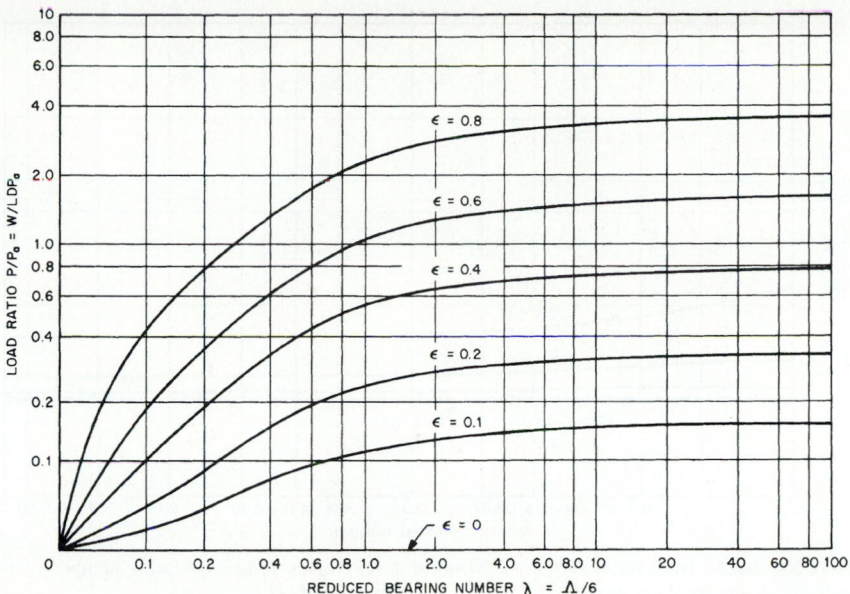

FIGURE 28.21   Load ratio versus reduced bearing number of an ungrooved cylindrical gas bearing; $L/D = 1$. *(From Raimondi [28.11].)*

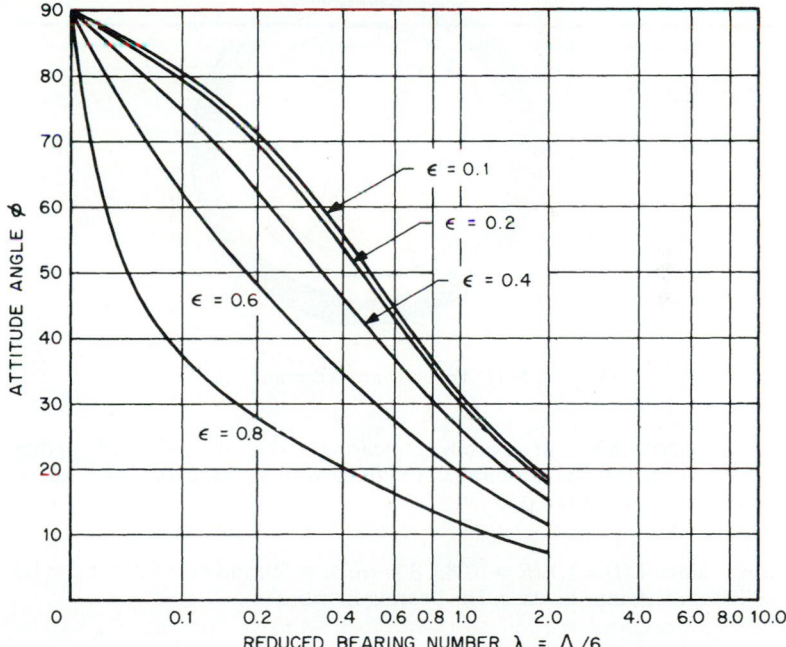

FIGURE 28.22   Attitude angle versus reduced bearing number of an ungrooved cylindrical gas bearing; $L/D = 1$. *(From Raimondi [28.11].)*

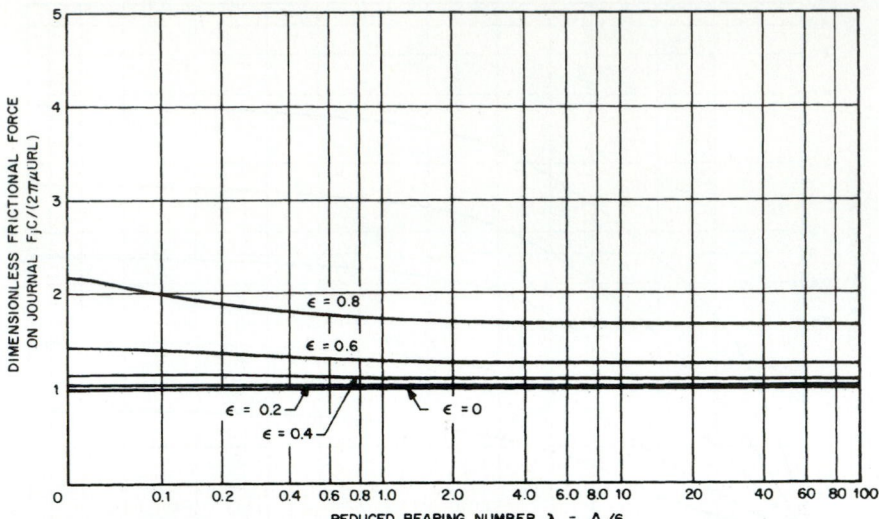

FIGURE 28.23 Dimensionless journal frictional force versus reduced bearing number of an ungrooved cylindrical gas bearing; $L/D = 1$. *(From Raimondi [28.11].)*

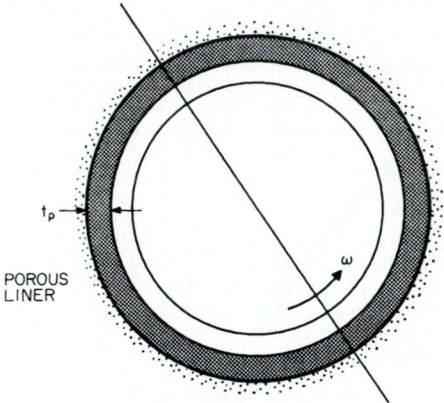

FIGURE 28.24 Porous journal bearing.

compare the performance of a porous gas bearing to that of a solid wall bearing. The porous bearing material is ceramic 0.021 in thick with a permeability $k = 1.33 \times 10^{-12}$ in^2 and a slip coefficient $\alpha = 0.3$.

Solution. Since $L/D = 1$, $t_p/R \cong 0.083$, $\beta = \alpha C/k \cong 50$, and $\Omega = 12kR/C^3 \cong 0.5$, the data of Table 28.19 may be used. The bearing number is

$$\Lambda = \frac{6\mu\omega}{p_a}\left(\frac{R}{C}\right)^2 = 3.0$$

TABLE 28.19 Gas-Lubricated Bearing Performance Data, $L/D = 1$, $t_p/R = 0.083$, $\Omega = 0.5$, $\beta = 50$

ε	Λ	$\dfrac{P}{p_a}$	ϕ, deg	$\left(\dfrac{R}{C}\right)(f)$
0.2	0.6	0.039	80.563	8.1034
	1.2	0.0756	72.109	8.3784
	3.0	0.1594	52.470	9.9199
	6.0	0.2272	34.402	13.8773
	12.0	0.2662	20.290	23.6302
	50.0	0.2959	7.602	88.4725
	100.0	0.3054	5.014	171.9223
	∞	0.3264	0	∞
0.4	0.6	0.0853	77.245	4.0835
	1.2	0.1638	66.670	4.2517
	3.0	0.3441	46.382	5.0012
	6.0	0.5067	30.262	6.7025
	12.0	0.6164	17.806	10.9197
	50.0	0.6982	6.424	39.9776
	100.0	0.7188	4.164	77.6340
	∞	0.7746	0	∞
0.6	0.6	0.1500	70.129	2.8072
	1.2	0.2872	56.903	2.8961
	3.0	0.6102	37.598	3.3003
	6.0	0.9322	24.537	4.2078
	12.0	1.1854	14.488	6.4992
	50.0	1.3774	4.962	23.0667
	100.0	1.4186	3.122	44.7611
	∞	1.5603	0	∞
0.8	0.6	0.2547	61.657	2.2908
	1.2	0.4995	47.468	2.2971
	3.0	1.1050	31.147	2.4980
	6.0	1.7837	20.348	2.9210
	12.0	2.3907	12.272	4.2374
	50.0	2.8736	4.046	14.4249
	100.0	2.9640	2.436	27.9331
	∞	3.4193	0	∞

SOURCE: Ref. [28.12].

The load ratio is

$$\frac{P}{p_a} = \frac{W/(LD)}{p_a} = 0.61$$

Entering Table 28.18 with these values, we find

$$\varepsilon = 0.6 \qquad \phi = 37.598 \qquad \left(\frac{R}{C}\right)(f) = 3.3003$$

For the same values we can find from Figs. 28.21 to 28.23 for the solid wall bearing $\varepsilon = 0.54$, $\phi = 33.7°$, and

$$\frac{F_j C}{2\pi\mu URL} = 1.18$$

From the last value, we may compute for comparative purposes

$$\left(\frac{R}{C}\right)(f) = 3.038$$

Thus the porous bearing operates at a larger eccentricity ratio (which indicates better stability) but has larger frictional loss. On the other hand, for a given ε, the porous bearing has a lower load capacity compared to the solid wall bearing. This loss of load capacity becomes more severe as the eccentricity ratio is increased.

28.7.4 Axially Grooved Gas Journal Bearings

The addition of axial slots, or grooves, in the bearing surface can produce several positive effects. Feed grooves can reduce frictional heating and energy losses. The grooves, in effect, turn the surface into a number of partial arc bearings, which can reduce the tendency for half-frequency whirl instability. What is more, grooving a bearing leads to only a slight increase in the manufacturing costs. Although a variety of grooving arrangements can be envisioned, three or four equally spaced axial grooves are typically employed. Feed grooves of 30° to 60° extent are commonly used. However, bearing stability decreases with increase in the groove size.

Castelli and Pirvics [28.13] presented tabular design data for axially grooved gas bearings. The subtended angle of each groove was taken to be 5°. Figures 28.25 and 28.26 are sample design chart data for a gas bearing with three grooves and an L/D ratio of 2.

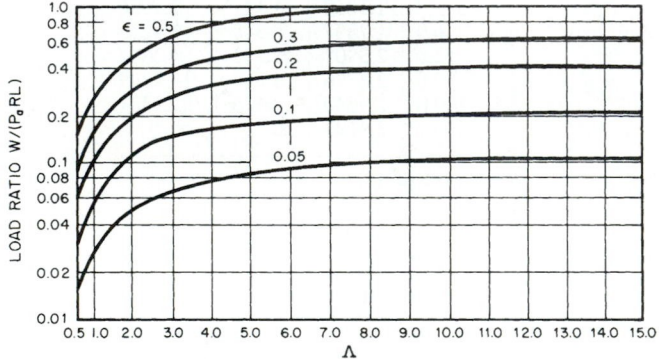

FIGURE 28.25 Load ratio versus compressibility number for an axial-groove gas bearing, $L/D = 2$; three evenly spaced axial grooves—first groove 60° clockwise from load line. *(From Castelli and Pirvics [28.13].)*

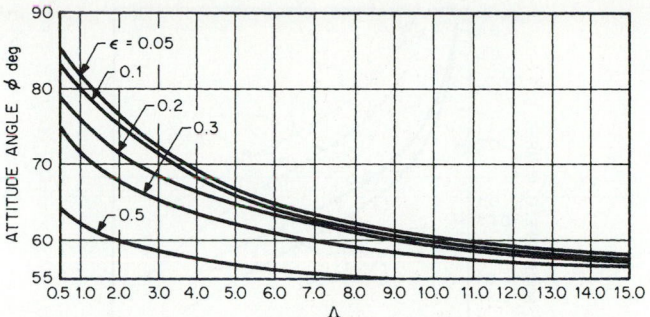

FIGURE 28.26 Attitude angle versus compressibility number for an axial-groove gas bearing, $L/D = 2$; three evenly spaced axial grooves—first groove 60° clockwise from load line. *(From Castelli and Pirvics [28.13].)*

28.7.5 Noncircular Gas Journal Bearings

For low eccentricity ratios, a noncircular gas bearing has superior stability properties but inferior load capacity when it is compared with a circular gas bearing. At higher eccentricity ratios, the load-carrying capacity of the noncircular gas bearing exceeds that of the circular gas bearing. For applications where load capacity is not an issue (e.g., in vertical-shaft cases), use of a noncircular gas bearing minimizes stability concerns that exist at the lower eccentricity ratios.

Pinkus [28.14] has investigated the performance of elliptical and three-lobe gas bearings. It was found that the direction of load application is important to the performance of a noncircular gas bearing. In fact, load-carrying capacity can be optimized by rotating the bearing relative to the load line. The rotation is generally clockwise anywhere from a few degrees to 25°. Table 28.20 presents a comparison of expected eccentricity ratios for given values of S and Λ for circular, elliptical, and three-lobe gas bearings. Higher values of ε required by the noncircular bearings over the ε for the circular bearing at the same Sommerfeld number indicate a lower load capacity. Pinkus also presented envelopes of operation. Figure 28.27 is a plot of Som-

TABLE 28.20 Eccentricity Ratio versus Sommerfeld and Bearing Numbers for Circular, Elliptical, and Three-Lobe Noncircular Gas Bearings

Bearing number Λ	Sommerfeld number S	Eccentricity ratio ε				
		Circular	Elliptical central loading	Elliptical optimum loading	Three-lobe central loading	Three-lobe optimum loading
1	0.365	0.2	0.54	0.525	0.55	0.545
1	0.162	0.4	0.59	0.56	0.615	0.605
1	0.0866	0.6	0.67	0.635	0.70	0.685
1	0.0381	0.8	0.815	0.77	0.84	0.815
3	0.365	0.25	0.57	0.535	0.595	0.57
3	0.162	0.475	0.65	0.59	0.68	0.645
3	0.0860	0.655	0.75	0.685	0.80	0.755

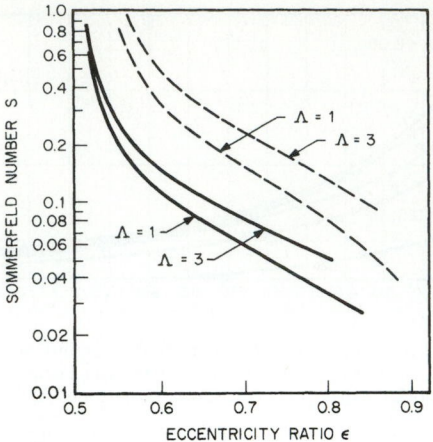

FIGURE 28.27 Sommerfeld number versus eccentricity ratio for a noncentrally loaded elliptical gas bearing, $L/D = 1$. *(From Pinkus [28.14].)*

merfeld number versus the eccentricity ratio for a noncentrally loaded elliptical gas bearing. The upper set of curves represents the largest ε for a given S at a particular Λ, and it would be used in an attempt to avoid stability problems. The lower set of curves represents the optimum loading conditions.

28.8 HYDROSTATIC JOURNAL BEARING DESIGN

Hydrostatic journal bearings (also called *externally pressurized* bearings) offer large radial load-carrying capacities at all rotational speeds (including zero). They exert very little friction and have controllable stiffness. The principal disadvantage of hydrostatic journal bearings is the cost of the pressurized lubricant supply system. These bearings are widely used in the machine-tool industry.

28.8.1 Classification of Bearings and Components

There are basically three types of hydrostatic journal bearings: the single-pad, the multipad, and the multirecess. The various types are depicted in Table 28.21.

The main components of a hydrostatic journal bearing are the pad and the lubricant supply system. The pad consists of a recess, or pocket, region and the surrounding land material. The recess is generally deep compared with the film thickness: typically the depth is taken to be 20 times the radial clearance. In sizing the recess, the land width is reduced as much as practical; however, if the land width is too small, edge effects can become significant. These effects can be avoided if the land width is taken to be greater than 100 times the clearance.

Pressurized lubricant is provided to each bearing recess through the supply system. This system may be as simple as a direct line between the recess and the supply pump. Bearings using such simple systems are termed *noncompensated hydrostatic*

TABLE 28.21 Types of Hydrostatic Journal Bearings

Type		Comments
Single-pad journal bearing		1. Included angle less than 180°. 2. Normally used to support unidirectional loads. 3. May have multiple recesses. 4. Lubricant flow and pad design based on a maximum load applied to a uniform film equaling minimum radial clearance $R_B - R_J$.
Multipad journal bearing		1. Included angle of each pad is less than 180°. 2. Used for rotating, oscillating, or reversing radial loads. 3. Two opposed pads can be used for reversing loads. 4. If angular variation of the load is less than 80°, a two-pad journal bearing may be used. 5. Bearing stiffness is reduced if load is directed between pads. 6. Axial grooves provide cooling capability at high rotational speeds.
Multirecess journal bearing		1. No pressure-reducing grooves between recesses. 2. Usually contains four or more recesses. 3. Used for rotating and reversing loads. 4. Does not require as much flow per recess as multipad type. 5. Circumferential flow from high-pressure pockets to lower-pressure pockets evens pressure distribution around bearing.

bearings and are frequently used to lift highly loaded journals prior to rotation. In that context they are termed *oil lifts*.

In order for a single pump to deliver lubricant to more than one pad or recess, restrictors (or compensating devices) are required in each supply line. Restrictors limit the flow to each pad, thereby permitting all pads to become activated. Without compensating devices, pad imbalance would occur. Restrictors, then, are used to control the operation of a hydrostatic bearing.

There are three common types of restrictors: the capillary tube, the orifice plate, and the flow control valve. Table 28.22 compares these devices. This table was prepared from a comprehensive series of papers on hydrostatic bearing design by Rippel [28.15]. In addition to flow-load control, the resistor affects the stiffness of the bearing. Bearing stiffness is related to the ability of the bearing to tolerate any changes in the applied load.

28.8.2 Design Parameters

For hydrostatic journal bearings at low rotational speeds, the primary design parameters are maximum load, lubricant flow rate, and stiffness. Of secondary importance are considerations of frictional horsepower and lubricant temperature rise.

The load-carrying capacity of a hydrostatic journal bearing is generally written as

$$W = a_f A_p p_r \tag{28.19}$$

where a_f = pad load coefficient, A_p = projected bearing area, and p_r = recess pressure. The pad load coefficient is dimensionless and physically represents the ratio of the average lubricant pressure in the pad to the lubricant pressure supplied to the pad. O'Donoghue and Rowe [28.16] give equations for a_f for both multirecess and multipad bearings.

The lubricant flow rate may be determined from

$$Q = q_f \frac{h^3 p_r}{\mu} \tag{28.20}$$

where q_f = flow factor for a single pad. The flow factor depends on the geometry and land widths of the bearing.

Journal bearing stiffness expresses the ability of the bearing to accommodate any changes in the applied load. Stiffness is proportional to the slope of the bearing load versus film thickness curve, or

$$s = -\frac{dW}{dh} \tag{28.21}$$

Stiffness will depend on the method of flow control (capillary tube, etc.) and the amount of circumferential flow. O'Donoghue and Rowe [28.16] provide a detailed derivation of the journal bearing stiffness and have developed equations for both multipad and multirecess bearings.

28.8.3 Design Procedures

A variety of design procedures are available for hydrostatic bearings. Some are based on experimental findings, whereas others are based on numerical solutions of the Reynolds equation.

Many existing methods incorporate numerous charts and tables which clearly place a limit on the methods; in order to use these procedures, the appropriate reference must be consulted. Alternatively, O'Donoghue and Rowe [28.16] have developed a general approximate method of design that does not require the use of various design charts. The method is strictly valid for thin land bearings, and many of the parameters are conservatively estimated. The following is a condensation of this design procedure for a multirecess bearing:

TABLE 28.22 Comparison of Three Types of Flow Resistors

	Initial cost	Fabrication and installation costs	Reliability	Availability	Ability to remain unclogged	Serviceability	Life	Flow relationship	Restrictions	Notes
Capillary tube	4†	3	5	5	1	4	5	$Q = k_c \dfrac{P_s - P_r}{\mu}$ $k_c = \pi D_c^4/128 L_c$	$L_c > 20 D_c$ $R_e = \dfrac{4\rho Q}{\pi D_c \mu} < 2000$	1. Laminar flow device. 2. Hypodermic needle tubing can generally be used. 3. To prevent clogging $D_c > 0.025$ in.
Sharp-edged orifice	5	2	5	5	3	4	4	$Q = \dfrac{k_0(P_s - P_r)^{1/2}}{}$ $k_0 = \dfrac{C_D \pi D_0^2}{2\sqrt{2\rho}}$	$C_D = f(R_e)$ $R_e = \dfrac{\sqrt{2\rho(P_s - P_r)}\,D_0}{\mu}$ $C_D \sim 0.6$ for $R_e > 15$ and $D_0/D_{line} < 0.1$	1. Turbulent flow device. 2. To prevent clogging, $D_0 > 0.020$ in.
Flow control valve	1	4	2	5	5	2	2	$Q = k_v$ $k_v = $ constant		1. Relatively expensive. 2. Flow is maintained at constant value.

†Rating numbers: 5 is high (best); 1 is low (worst).

28.55

Design Specification

1. Set the maximum load W.
2. Select the number of recesses n (typically $n = 4$, but $n = 6$ for high-precision bearings).
3. Select a pressure ratio p_r/p_s (a design value of 0.5 is recommended).

Bearing Dimensions

1. Calculate the bearing diameter $D = \sqrt{W/50}$ in.
2. Set width $L = D$.
3. Calculate the axial-flow land width a (refer to Fig. 28.28; recommended value: $a = L/4$).

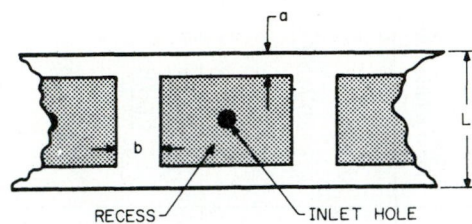

FIGURE 28.28   Hydrostatic pad geometry.

4. Calculate the circumferential land width b [refer to Fig. 28.28; recommended value: $b = \pi D/(4n)$].
5. Calculate the projected pad area $A_p = D(L - a)$.
6. Calculate the recess area for one pad only: $A_r = (\pi D - nb)(L - a)/n$.
7. Calculate the effective frictional area: $A_f = (\pi D L/n) - 0.75A_f$.

Miscellaneous Coefficients and Parameters

1. Establish a design value of film thickness h_d (recommended value: 50 to 10 times larger than the machinery tolerance on h).
2. Calculate the circumferential flow factor $\gamma = na(L - a)/(\pi Db)$; for recommended values of D, L, a, and b: $\gamma = \frac{3}{4}(n/\pi)^2$.
3. Compute the design dimensionless stiffness parameter $\bar{s}$ from Table 28.23 depending on the method of flow control selected.
4. Calculate the flow factor $q_f = \pi D/(6an)$.

Performance Parameters

1. Calculate the minimum supply pressure $p_{s,\min} = 2W/(a_f A_p)$ [recommended: $a_f = \bar{s}$ (capillary tube at design Table 28.23)].

TABLE 28.23 Dimensionless Stiffness at Design Condition for Multirecess Bearing

$$\bar{s} = \frac{C_1}{1 + C_{2\gamma}}$$

Number of recesses n	Capillary tube		Sharp-edged orifice		Flow control valve		Recommended value of γ†
	C_1	C_2	C_1	C_2	C_1	C_2	
3	0.27	0.75	0.36	1.0	0.54	1.5	0.68
4	0.96	0.50	1.28	0.67	1.91	1.01	1.22
5	1.03	0.35	1.38	0.46	2.13	0.69	1.90
6	1.08	0.25	1.43	0.33	2.15	0.50	2.74

†SOURCE: Ref. [28.16].

2. Calculate the stiffness $s = p_s A_p \bar{s}/h_d$ (if this value is too low, either D and/or p_s must be increased).

3. Calculate the minimum running clearance $h_0 = h_d - W/s$.

4. Determine the flow rate $Q = (n q_f p_s h^3 d)/(2\mu)$. A calculated value of viscosity can be used:

$$\mu = \frac{60 P_s h^2 d \sqrt{q_f/(2A_f)}}{\pi D N} \qquad \text{reyn}$$

REFERENCES

28.1 G. G. Hirs, "The Load Capacity and Stability Characteristics of Hydrodynamic Grooved Journal Bearings," *ASLE Transactions,* vol. 8, 1965, pp. 296–305.

28.2 E. R. Booser, "Plain-Bearing Materials," *Machine Design,* June 18, 1970 (bearings reference issue), pp. 14–20.

28.3 V. Hopkins, "Self-Lubricating Bearing Materials—A Review," *Assessment of Lubricant Technology,* ASME, 1972, pp. 21–26.

28.4 A. W. J. DeGee, "Selection of Materials for Lubricated Journal Bearings," *Wear,* vol. 36, 1976, pp. 33–61.

28.5 B. R. Reason and I. P. Narang, "Rapid Design and Performance Evaluation of Steady State Journal Bearings—A Technique Amendable to Programmable Hand Calculators," *ASLE Transactions,* vol. 25, no. 4, 1982, pp. 429–449.

28.6 J. Shigley and C. R. Mischke, *Mechanical Engineering Design,* 5th ed., McGraw-Hill, New York, 1989, pp. 498–507.

28.7 A. A. Raimondi and J. Boyd, "A Solution for the Finite Journal Bearing and Its Application to Analysis and Design," I, II, and III, *ASLE Transactions,* vol. 1, 1958, pp. 159–174, 175–193, and 194–209, respectively.

28.8 A. Seireg and S. Dandage, "Empirical Design Procedure for the Thermodynamic Behavior of Journal Bearings," *ASME Journal of Lubrication Technology,* vol. 104, April 1982, pp. 135–148.

28.9 H. J. Connors, "An Analysis of the Effect of Lubricant Supply Rate on the Performance of the 360° Journal Bearing," *ASLE Transactions,* vol. 5, 1962, pp. 404–417.

28.10 H. Moes and R. Bosma, "Design Charts for Optimum Bearing Configurations: 1—The Full Journal Bearing," *ASME Journal of Lubrication Technology,* vol. 93, April 1971, pp. 302–306.

28.11 A. A. Raimondi, "A Numerical Solution for the Gas Lubricated Full Journal Bearing of Finite Length," *ASLE Transactions,* vol. 4, 1961, pp. 131–155.

28.12 E. R. Wu, "Gas-Lubricated Porous Bearings of Finite Length—Self-Acting Journal Bearings," *ASME Journal of Lubrication Technology,* vol. 101, July 1979, pp. 338–348.

28.13 V. Castelli and J. Pirvics, "Equilibrium Characteristics of Axial-Grooved Gas-Lubricated Bearings," *ASME Journal of Lubrication Technology,* vol. 89, April 1967, pp. 177–196.

28.14 O. Pinkus, "Analysis of Noncircular Gas Journal Bearings," *ASME Journal of Lubrication Technology,* vol. 87, October 1975, pp. 616–619.

28.15 H. C. Rippel, "Design of Hydrostatic Bearings," *Machine Design,* parts 1 to 10, Aug. 1 to Dec. 5, 1963.

28.16 J. P. O'Donoghue and W. B. Rowe, "Hydrostatic Bearing Design," *Tribology,* vol. 2, February 1969, pp. 25–71.

CHAPTER 29
COUPLINGS

Howard B. Schwerdlin
Engineering Manager
Lovejoy, Inc.
Downers Grove, Illinois

GLOSSARY OF SYMBOLS

A	Area or parallel misalignment
b	Bearing spacing
d	Diameter
D	Diameter or distance between equipment
e	Eccentricity
E	Young's modulus or shape factor for maximum allowable stress, psi
F	Force
g	Acceleration due to gravity
h	Height of keyway
I	Second moment of area
J	Polar second moment of area
K_a	U-joint angle correction factor
K_L	U-joint life correction factor
K_s	U-joint speed correction factor
L	Life or length of engagement
ℓ	Length
m	Mass

n	Speed, r/min
N	Number of active elements or bellows convolutions
P	Pressure
PV	Pressure times velocity
r	Radius
R	Operating radius
R_c	Centroidal radius or distance
s	Maximum permissible stroke per convolution for bellows
S	Link length, shape factor, or maximum permissible total bellows stroke
t	Thickness
T	Torque
V	Velocity
w	Width
X	Angular misalignment
Y	Parallel misalignment
α	Rotational position
β	Torsional amplitude
γ	U-joint angle
δ	Deflection or U-joint angle
Δ	Deflection
ζ	Damping ratio
θ	Shaft or joint angle
θ_{eff}	Torsional equivalent angle
τ	Shear stress
ω	Angular velocity

29.1 GENERAL

29.1.1 System Requirements

When selecting a coupling, you have to consider all the system's requirements. It is not enough to know what the driver and load are and how big the shaft is. You must also know how the two halves are assembled and whether there is misalignment, as well as the system's operating range and the operating temperature.

Before you select a coupling, determine the following about the system:

1. *Driver* Type; electric motor, internal-combustion engine, number of cylinders, etc.
2. *Load* Fan, pump, rockcrusher, etc., to determine the inertias.
3. *Nominal torque* T_{kn} Continuous operating torque.
4. *Maximum torque* $T_{\max}$ Peak expected on startup, shutdown, overload, etc.
5. *Vibratory torque* T_{kw} Oscillating torque about the nominal $T_{kn} \pm T_{kw}$.
6. *Number of startups per hour.*

7. *Misalignment* Amount and type of misalignment between the driver and the load: parallel, angular, and/or axial.

8. *Type of mounting* Shaft to shaft, shaft to flywheel, blind fit, etc.

9. *Shaft size* Diameter of the shafts for both the driver and the load.

10. *Operating temperature* General operating temperature and whether the drive is enclosed (unventilated).

11. *Operating speed range* The upper and lower limits of the operating range.

12. *Service factor* A "fudge factor" designed to combine many of the above operating conditions and lump them into one multiplier to oversize the coupling in order to accommodate these parameters. Typical service factors are shown in Table 29.1.

29.1.2 Coupling Characteristics

Once the system requirements have been determined, check the characteristics of the coupling chosen to verify the selection. You should be able to check the following characteristics:

1. *Torque capacity*

2. *Bore size* Minimum and maximum bore

3. *Type of mounting* Mounting configurations available for any given coupling

4. *Maximum speed range*

5. *Misalignment* Degree of misalignment that can be accepted in mounting

6. *Flexible material* Capability of material to withstand heat or oil contamination; torsional stiffness

29.1.3 Selecting the Coupling

The first step is to make a preliminary selection based on the torque transmitted and the shaft dimensions. Then verify that the selection will satisfy the requirements for type of mount, degree of misalignment, operating speed, and operating temperature. Don't forget to check for the possibility of resonance.

Not all systems require all these steps. Smooth operating systems, such as electric motors driving small loads, are seldom subject to severe vibration. The natural frequency probably does not have to be checked.

As a simple guideline for determining system requirements for smooth systems, coupling manufacturers have developed the service factor. The service factor is a rough approximation of the temperature requirements, maximum torque, and natural frequency. It is stated as a multiplier, such as 1.5. To be sure the coupling you have selected is adequate, multiply the nominal torque required for the system by the service factor and select a coupling with that torque rating or better.

The service factor is adequate for some systems. Its drawbacks are that it is imprecise and, in severe applications, does not evaluate all the variables. Also, when you are selecting according to the service factor, be careful not to overspecify, getting more coupling than needed. This is not cost-effective.

Perhaps the most important thing to remember in selecting a coupling is that the coupling manufacturer can make a recommendation for you only based on the

TABLE 29.1   Service Factors and Load Classification for Flexible Couplings[†]

Agitators		Fans (*cont.*)	
Pure liquids	1.0	Cooling towers	2.0
Liquids, variable density	1.0	Forced draft	1.5
Barge puller	2.0	Induced draft w/o damper	2.0
Beaters	1.5	control	
Blowers		Propellor	1.5
Centrifugal	1.0	Induced draft w/damper control	1.25
Lobe	1.25	Feeders	
Vane	1.25	Belt	1.0
Can-filling machinery	1.0	Screw	1.0
Car dumpers	2.5	Reciprocating	2.5
Car pullers	1.5	Generators	
Compressors		Not welding	1.0
Centrifugal	1.0	Welding	2.0
Lobe	1.25	Hoist	1.5
Reciprocating	‡	Hammer mills	2.0
Conveyors, uniformly loaded or fed		Kilns	1.5
Assembly	1.0	Laundry washers, reversing	2.0
Belt	1.0	Line shafting any processing mach.	1.5
Screw	1.0	Lumber machinery	
Bucket	1.25	Barkers	2.0
Live roll, shaker and	3.0	Edger feed	2.0
reciprocating		Live rolls	2.0
Conveyors (heavy-duty), not		Planer	2.0
uniformly fed		Slab conveyor	2.0
Assembly	1.2	Machine tools	
Belt	1.2	Bending roll	2.0
Oven	1.2	Plate planer	2.0
Reciprocating	2.5	Punch press gear driven	2.0
Screw	1.2	Tapping machinery	2.0
Shaker	3.0	Other	
Cranes and hoists		Main drive	1.5
Main hoists	2.0	Aux. drives	1.0
Reversing	2.0	Metal-forming machines	
Skip	2.0	Draw bench carriage	2.0
Trolley drive	2.0	Draw bench main drive	2.0
Bridge drive	2.0	Extruder	2.0
Slope	2.0	Forming machinery	2.0
Crushers		Slitters	1.5
Ore	3.0	Table conveyors	
Stone	3.0	Nonreversing	2.5
Dredges		Reversing	2.5
Cable reels	2.0	Wire drawing	2.0
Conveyors	1.5	Wire winding	1.5
Cutter head drives	2.5	Coilers	1.5
Maneuvering winches	1.5	Mills, rotary type	
Pumps	1.5	Ball	2.0
Elevators		Cement kilns	2.0
Bucket	1.5	Dryers, coolers	2.0
Escalators	1.0	Kilns	2.0
Freight and passenger	2.0	Pebble	2.0
Evaporators	1.0	Rolling	2.0
Fans		Tube	2.0
Centrifugal	1.0	Tumbling	1.5

TABLE 29.1 Service Factors and Load Classification for Flexible Couplings[†] (*Continued*)

Mixers		Pumps (*cont.*)	
Concrete, cont.	1.75	1 cyl., single- or double-acting	2.0
Muller	1.5	2 cyl. single-acting	2.0
Papermills		2 cyl. double-acting	1.75
Agitators (mixers)	1.2	3 or more cyl.	1.5
Barker mech.	2.0	Rubber machinery	
"Barking" drum spur gear	2.5	Mixer	2.5
Beater and pulper	2.0	Rubber calender	2.0
Calenders	1.5	Screens	
Calenders, super	1.5	Air washing	1.0
Converting machines	1.2	Rotary stone or gravel	1.5
Conveyors	1.2	Vibrating	2.5
Dryers	1.5	Water	1.0
Jordans	2.0	Grizzly	2.0
Log haul	2.0	Shredders	1.5
Dresses	2.0	Steering gear	1.0
Reel	1.2	Stokers	1.0
Winder	1.2	Textile machinery	
Printing presses	1.5	Dryers	1.2
Pug mill	1.75	Dyeing mach.	1.2
Pumps		Tumbling barrel	1.75
Centrifugal	1.0	Windlass	2.0
Gear, rotary or vane	1.25	Woodworking machinery	1.0
Reciprocating	[‡]		

[†]The values of the service factors listed are intended only as a general guide. For systems which frequently use the peak torque capacity of the power source, check that this peak torque does not exceed the normal torque capacity of the coupling.

The values of the service factors given are to be used with prime movers such as electric motors, steam turbines, or internal combustion engines having four or more cylinders. For drives involving internal combustion engines of two cylinders, add 0.3 to values; and for a single-cylinder engine add 0.70.

[‡]Consult the manufacturer.

SOURCE: Ref. [29.1].

information you provide. A little time spent selecting the right coupling can save a lot of time and money later.

Selecting a flexible coupling involves more than meeting torque and shaft size requirements. It is also important to understand the functions of a flexible coupling in the system, the operating requirements of the system, and the characteristics of the coupling selected. Flexible couplings serve four main functions in a drive system:

1. They transmit torque and rotation from the drive to the load.

2. They dampen vibration.

3. They accommodate misalignment.

4. They influence the natural frequency of the system.

The torque-handling capacity of a given coupling design defines the basic size of a coupling. The nominal torque T_{kn} is the coupling's continuous load rating under conditions set by the manufacturer. The maximum torque rating T_{max} is the peak torque the coupling can handle on startup, shutdown, running through resonance, and momentary overloads. As defined in the German standards for elastomeric couplings, Ref. [29.2], a coupling should be able to withstand 10^5 cycles of maximum torque at a frequency of not more than 60 per hour. Vibratory torque ($\pm T_{kw}$) is the coupling vibratory rating at 10 hertz (Hz) for elastomeric couplings. The rotary output of the coupling may be uniform (constant velocity) or cyclic (e.g., Hooke's joint).

All drive systems experience some vibration. Vibration can exceed the limits of design, which can cause system failure. Flexible couplings are one method of dampening the amount of vibration from either the driver or the driven equipment.

When a flexible coupling is used, the vibration is transferred to a material which is designed to absorb it rather than transmit it through the entire drive. Soft materials, such as natural rubber, can absorb greater amounts of vibration than stiffer materials, such as Hytrel[†] or steel. As a comparison, the relative vibration damping capabilities of Buna N rubber, Hytrel, and steel are shown in the transmissibility chart of Fig. 29.1.

If a system has misalignment, there are two factors to consider. First, you must use a coupling that can operate between two misaligned shafts. Second, you must be sure that the coupling does not exert excessive forces on the equipment because of misalignment. Perfect alignment between the driver and the load is difficult to obtain and maintain over the life of the system. A cost-effective alternative to precise alignment is a coupling that can accommodate misalignment between two shafts. The amount of misalignment a coupling can accept varies. Steel drive plates, for example, can accept only misalignment equal to their machining tolerances, frequently as little as 0.005 inch (in) parallel. Other couplings can accommodate mis-

---

[†] Hytrel is a trademark of E.I. du Pont de Nemours.

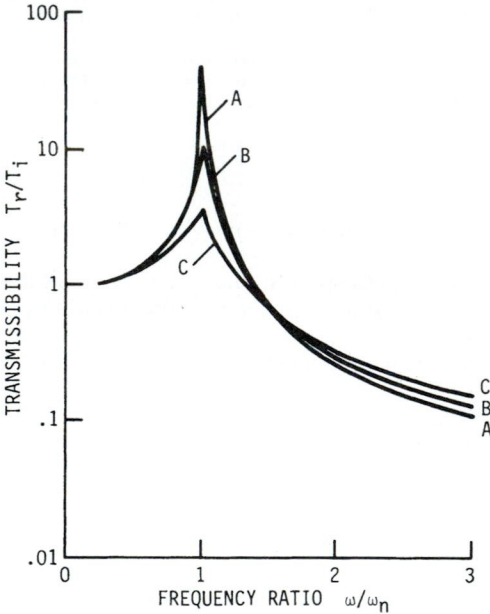

FIGURE 29.1 Effect of damping ratio on torque transmission. A, steel, $\xi = 0.01$; B, Hytrel, $\xi = 0.03$; C, Buna N rubber, $\xi = 0.13$, where T_r is the transmitted torque and T_i the input torque.

alignment up to 45°. The maximum allowable misalignment is a function of the percentage of torque capacity being utilized and the amount of vibratory torque the system is transmitting under perfect alignment.

If there is system misalignment, the material used in the coupling is important. Misalignment may cause radial forces to be exerted on the system. If the radial forces are too great, components such as bearings, seals, and shafts can experience undue stresses and fail prematurely. Different materials exert different radial forces; softer materials typically exert less radial force than stiff materials.

The natural frequency of a system can be altered by changing either the inertia of any of the components or the stiffness of the coupling used. See Chap. 38. Generally, after a system is designed, it is difficult and costly to change the inertia of the components. Therefore, coupling selection is frequently used to alter the natural frequency.

29.2 RIGID COUPLINGS

The solid coupling does not allow for misalignment, except perhaps axial, but enables the addition of one piece of equipment to another. In its simplest form, the rigid coupling is nothing more than a piece of bar stock bored to receive two shafts, as shown in Fig. 29.2. Its torque-handling capacity is limited only by the strength of the material used to make the connection. The coupling is installed on one shaft before the equipment is lined up, and the mating equipment is brought into position without much chance of accurate alignment when the equipment is bolted into position.

The maximum shear stress occurs at the outer radius of the coupling and at the interface of the two bores. This stress can be derived from the torsion formula (see Chap. 49) and is

$$\tau_{max} = \frac{TD_o}{2J} \tag{29.1}$$

where J, the polar second moment of the area, is

$$J = \frac{\pi}{32}(D_o^4 - D_i^4) \tag{29.2}$$

The coupling must be sized so that, typically, the stress given by Eq. (29.1) does not exceed 10 percent of the ultimate tensile strength of the material, as shown in Table 29.2; but see Chap. 12.

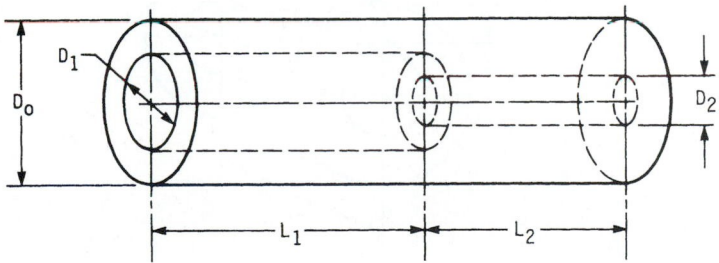

FIGURE 29.2 Schematic view of a rigid coupling.

TABLE 29.2 Maximum Allowable Shear Stress for Some Typical Materials

Material	Stress, psi	Material	Stress, psi
Steel	8000	Powdered iron (Fe-Cu)	4000
Ductile iron (60-45-12)	6000	Aluminum (SAE 380)	4000
Cast iron (Class 40)	4500	Tobin brass	3500

Other factors to consider are the length of engagement into the coupling. The shear stress over the keyway must not exceed the allowable shear stress as given above. Based on Fig. 29.3, the centroidal radius is

$$R_c = \frac{1}{2}\left(\frac{D_o}{2} + \frac{D_i}{2} + h\right) \tag{29.3}$$

The centroid of the bearing area is at radius $(D_i + h)/2$. If the transmitted torque is T, then the compressive force F is $2T/(D_i + h)$. The bearing stress σ_b is

$$\sigma_b = \frac{F}{A} = \frac{4T}{wL(D_i + h)} \tag{29.4}$$

The allowable compressive stress from distortion energy theory of failure is $\sigma_{all} = \tau_{all}/0.577$. Combining this with Eq. (29.4) gives

$$\tau_{all} = \frac{0.577(4)T}{wL(D_i + h)} \tag{29.5}$$

with τ_{all} coming from Table 29.2.

Next, the length of key stock, for keyed shafts, must be examined to keep its shear loading from exceeding the allowable shear stress. Referring to Fig. 29.4, we note that the shear force is $F = T/(D_i/2) = 2T/D_i$. Therefore the average shear stress is

$$\tau = \frac{F}{A} = \frac{2T}{wLD_i} \tag{29.6}$$

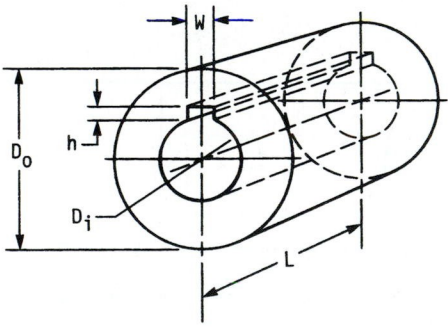

FIGURE 29.3 Portion of coupling showing keyway.

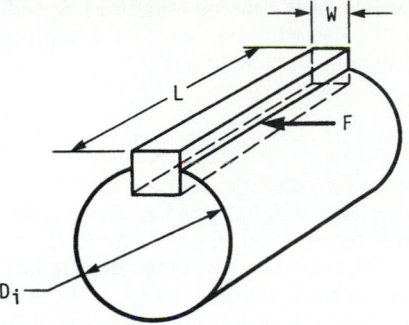

FIGURE 29.4 Portion of shaft showing key.

Both keys must be checked, although experience has shown that small-diameter shafts are more prone to failure of the key and keyway when these precautions are not followed because of their normally smaller key width and length of engagement. As a rule of thumb, the maximum allowable shear stress for some typical materials is shown in Table 29.2.

The *ribbed, hinged,* and *flanged* couplings are shown in Figs. 29.5, 29.6, and 29.7, respectively. These can be analyzed using the same approach as described above.

29.3 FLEXIBLE METALLIC COUPLINGS

29.3.1 Flexible Disk and Link Couplings

In this coupling (Fig. 29.8), misalignment is accommodated by the flexing of steel laminations. Parallel misalignment capacity is virtually zero unless two separated disk packs are used, in which case parallel misalignment is seen in the form of angular misalignment of each pack. This type of coupling can support large imposed radial loads, such as in rolling mills or long, floating shafts. The disk packs can be made from any material and are frequently manufactured from stainless steel for severe service. This coupling requires no lubrication.

The large radial loads imposed by long sections of tubing connecting to widely separated disk packs [up to 20 feet (ft)] are due to the heavy wall section necessary to give the tubing (or shafting) the necessary rigidity to resist whirling due to the

FIGURE 29.5 This ribbed coupling is made of two identical halves, split axially, and bolted together after the shafts have been aligned.

FIGURE 29.6 This hinged coupling is used mostly for light-duty applications. *(CraneVeyor Corp.)*

weight of the tubing (shafting). Specifically, the whirling speed of a uniform tube due to its weight is

$$n_c = \frac{60}{2\pi} \sqrt{\frac{\Delta}{g}} \qquad (29.7)$$

where Δ = static deflection of the tube due to its own weight. See Chap. 50 for deflection formulas, and Chap. 37 for method.

The standard rule of thumb is to keep the critical whirling speed at least 50 percent above the operating speed for subcritical running, or 40 percent below the operating speed for supercritical speeds. This forbidden range of

$$0.6n_c < n_c < 1.4n_c$$

corresponds to the amplification region of a lightly damped resonance curve, as shown in Fig. 29.9. Thus, for a whirling speed of 1800 revolutions per minute (r/min), the operating speed must not be in the range of 1280 to 2700 r/min.

The link coupling in Fig. 29.10 is similar to the metallic disk coupling except that the disk is replaced by links connecting the two shaft hubs. This coupling can be misaligned laterally, considerably more than the disk type. Both the disk and the link type carry torque in tension and compression in alternating arms. Proper bolt torque of the axial bolts holding the links or disks to the hubs is important. Insufficient torque may cause fretting from relative motion between the links or disks. Too much bolt clamping weakens the links or disks at their connecting points as a result of excessive compressive stress.

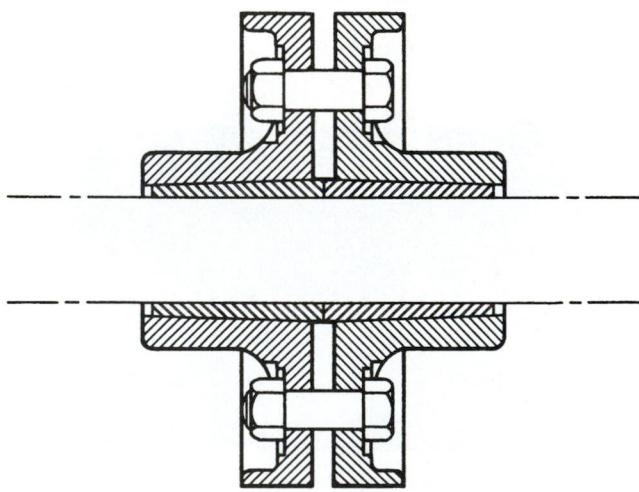

FIGURE 29.7 Schematic view of a flanged sleeve coupling.

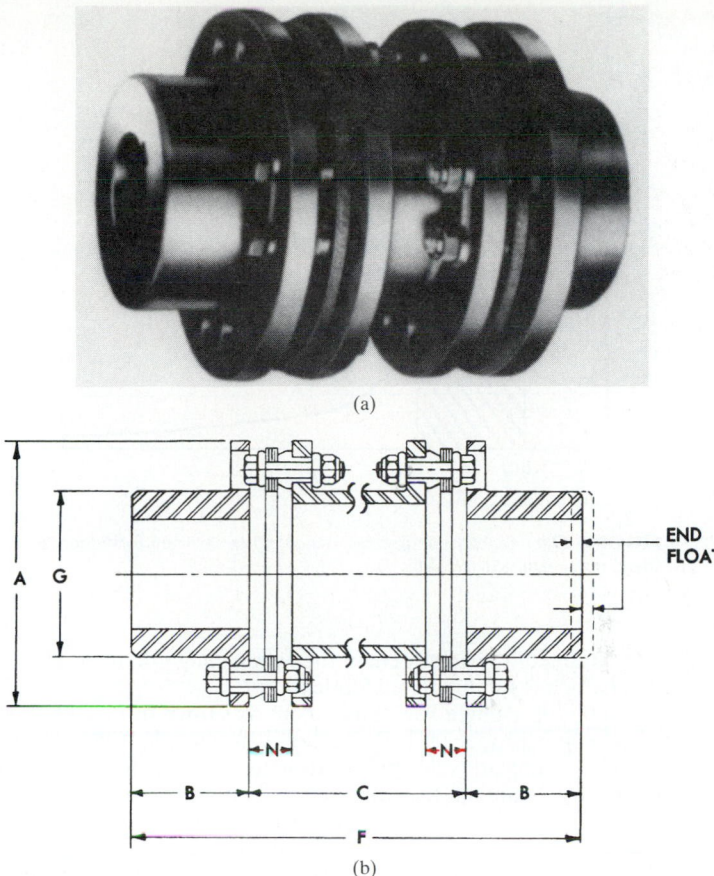

FIGURE 29.8 (*a*) Flexible-disk coupling; (*b*) cross section. (*Rexnord, Inc., Coupling Division.*)

29.3.2 Chain, Grid, and Beam Couplings

The chain coupling of Fig. 29.11 consists of two sprockets joined by an endless double-roller chain or inverted-tooth silent chain. This type of coupling will accommodate small amounts of angular, axial, and radial misalignment, which is provided by clearances between interfacing surfaces of the component parts.

For maximum service life, chain coupling sprockets should have hardened teeth. The coupling should be lubricated and enclosed in a greasetight case. Chain couplings can be assembled by using unhardened sprockets and operated without lubrication or a cover. This can be hazardous and can result in injury to personnel as well as a short coupling service life. This author has seen many such worn-out couplings. The availability of chain couplings is very good worldwide. Most manufacturers publish horsepower ratings to aid in proper coupling selection.

In the grid coupling (Fig. 29.12) the gears are separated by a specific minimum distance that allows for misalignment (Fig. 29.13). Large axial misalignment is

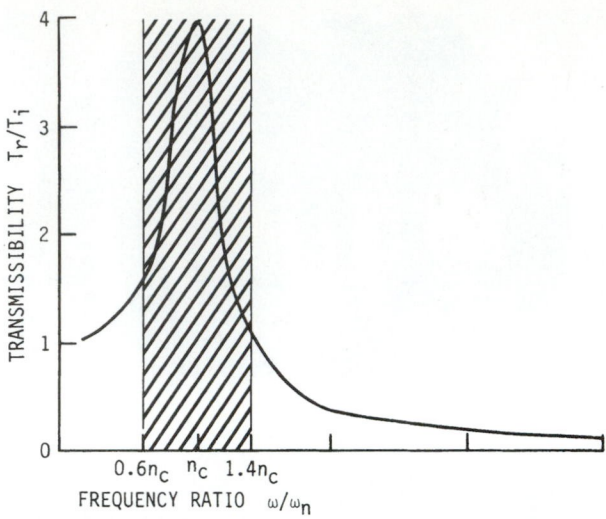

FIGURE 29.9 Lightly damped resonance curve showing forbidden speed range $0.6n_c < n_c < 1.4n_c$.

accomplished by sliding of the gear teeth in the rather long grid area. This coupling requires a guard (which is supplied) and lubrication.

The beam coupling is a single bar of metal cut as shown in Fig. 29.14 so that a continuous helical coil is produced. This coupling is torsionally rigid and free from backlash and provides constant velocity. Two designs exist for this type of coupling, one for maximum misalignment, the other for maximum torque (for the package size where axial movement is not a requirement). This coupling requires no lubrication. Speeds to 25 000 rev/min are possible depending on the coupling size.

29.3.3 Diaphragm and Hydraulic Couplings

As the name suggests, diaphragm couplings are made of a thin diaphragm or multiple thin diaphragms (see Fig. 29.15). Normally the diaphragms are made of metal. The diaphragms may be straight-sided, contoured, tapered, or convoluted; they may have various cutouts in them or take on many other forms. This coupling is connected to one shaft at the periphery [outer diameter (OD)] while the inner diameter (ID) is connected to the shaft or to a spacer piece, which may connect to another

FIGURE 29.10 Link coupling. *(Eaton Corp., Industrial Drives Operation.)*

(a)   (b)

FIGURE 29.11   (*a*) Silent-chain coupling; (*b*) roller-chain coupling. *(Morse Industrial Products, Borg-Warner Corp.)*

diaphragm(s). This coupling is most often used in pairs (two flex elements), which converts parallel misalignment to angular misalignment between two flex elements. Misalignment is accommodated by stretching (straight, contoured, or tapered diaphragms or unrolling convoluted diaphragms) the diaphragm material. This type of coupling requires no lubrication and is considered torsionally rigid.

The hydraulic coupling consists of two sleeves, one with a tapered OD and one with a tapered ID, which slide over one another, as shown in Fig. 29.16. Oil is forced, under pressure, between the two sleeves to allow the outer sleeve to be positioned at a predetermined position on the inner sleeve. The pressure is released, and the outer sleeve firmly compresses the inner sleeve and shafting. In larger couplings, oil is also forced into a piston chamber to force the outer sleeve into position. To remove the coupling, the area between the sleeves is repressurized, and the outer sleeve can be slid away, releasing the coupling.

FIGURE 29.12 Metallic grid coupling with cover removed to show grid detail. *(Falk Corp.)*

29.3.4 Gear Couplings

Double-engagement gear couplings, as shown in Fig. 29.17, transmit more power per unit volume and unit weight than any other flexible coupling design, because of their relatively small OD (compared with other types of similar horsepower). The basic design consists of two gear-type hubs (similar to spur gears) loosely connected by an internal-spline sleeve, which could be one piece or two internal-spline mating flanges bolted together.

Clearance between the mating teeth in the hub and the sleeve allows this type of coupling to absorb angular, parallel,

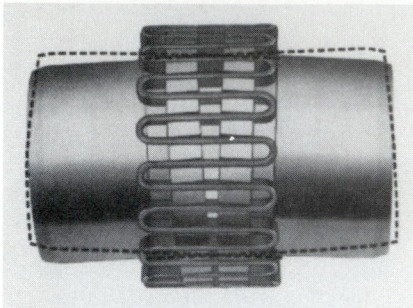

FIGURE 29.13   How the grid coupling accommodates misalignment. *(Falk Corp.)*

FIGURE 29.14   Beam coupling. *(Helical Products Corp.)*

and axial misalignment, as shown in Fig. 29.18. There is no relative rotation between the gear teeth, as in a normal gear set. Various tooth profiles (including crowned and/or barrel-shaped teeth) or changes in pressure angle allow for different misalignment, life, and load capacities. Straight-tooth couplings allow misalignment of 1° per gear mesh; with barrel-shaped teeth on the hub and straight teeth on the sleeve, 6° per mesh can be allowed.

With perfect alignment, all the teeth in the coupling are in contact, and the load is evenly distributed among them. Misalignment concentrates the load on just a few teeth; the number of teeth under load is a function of misalignment and load. The greater the misalignment (angular and parallel), the fewer the number of teeth in contact and the higher the load per tooth. Barrel-shaped teeth distribute the load over a larger area per tooth and may allow a greater number of teeth to be in contact under misaligned conditions, as shown in Fig. 29.19.

The two gear meshes can be separated by large distances, as shown in Fig. 29.20. In this case, two single-engagement couplings are connected by a floating shaft. For

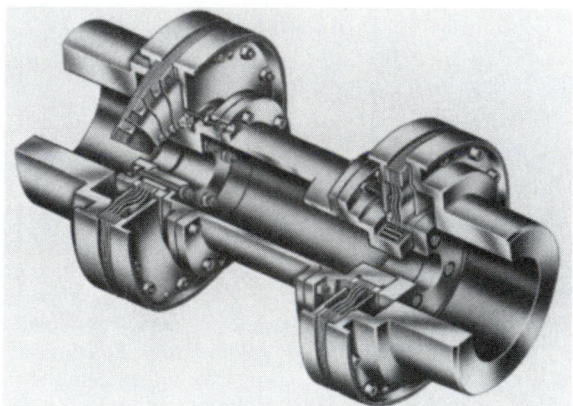

FIGURE 29.15   Cutaway view of diaphragm coupling assembly showing multiple convoluted diaphragms. *(Zurn Industries, Inc., Mechanical Drives Div.)*

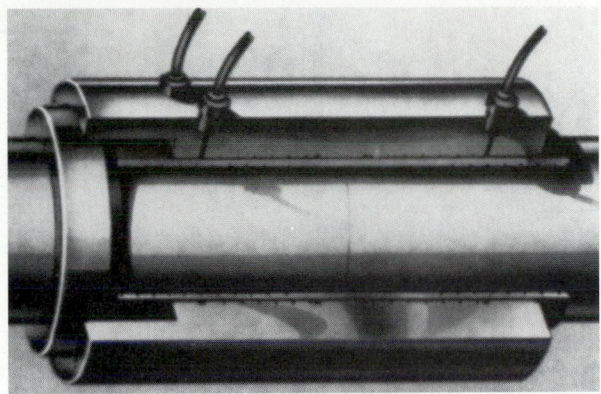

FIGURE 29.16 Hydraulic coupling; cutaway shows oil forced between inner and outer tapered sleeves. Note the oil piston chamber at left. *(SKF Industries.)*

this style coupling, large amounts of parallel misalignment are made possible by converting the angular misalignment capacity per mesh to parallel misalignment.

Parallel misalignment capacity for one single-engagement coupling is virtually nonexistent, however, and these couplings must be used in pairs, as shown in Fig. 29.20, to handle parallel misalignment.

Gear couplings must be lubricated for proper operation. Because of the high contact pressures obtained under misaligned conditions, only extreme-pressure (EP) greases should be used with gear couplings operating at maximum load. At high speeds (over 25 000 rpm), centrifugal effects separate the filler (soap) from the oil in most greases; the filler then collects between the teeth, preventing the oil from lubricating this highly loaded area. To overcome this problem, most high-speed gear couplings use a circulating oil system. The centrifugal effect still separates the fine particles from the oil, even in finely filtered systems. This sludge buildup necessitates cleaning of the teeth at regular intervals to prevent premature coupling failure.

FIGURE 29.17 Cutaway of flange-type gear coupling. *(Dodge Division, Reliance Electric.)*

Gear couplings, while inherently balanced, being machined all over and self-centering, may still require balancing to remove any residual unbalance due to bore runout. The magnitude F of this unbalanced or centrifugal force is

$$F = me\omega^2 \qquad (29.8)$$

where m = mass of the coupling, e = eccentricity, and ω = angular velocity in radians per second. See also Chaps. 37 and 38.

29.3.5 Spring and Flexible Shaft

Flexible shafts are constructed from a casing and a core, which is a series of

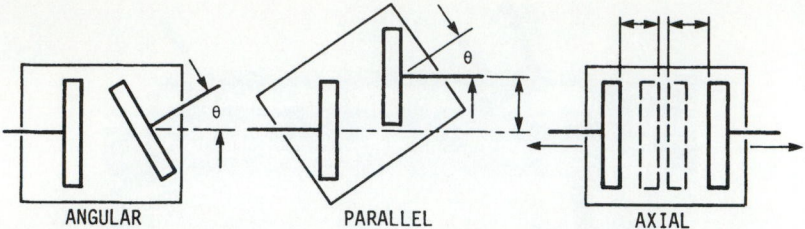

FIGURE 29.18 How double-engagement gear couplings accommodate angular, parallel, and axial misalignment.

multistranded layers of wire successively wrapped about a single central wire. Each wire layer is wound opposite to and at right angles to the layer beneath it to transmit maximum power and retain the greater flexibility.

The casing protects the rotating core from dust and moisture, but does not rotate itself. It is also reinforced to support the core and prevent helixing under torque load. *Helixing* is the tendency for a rope, or wire, to bend back on itself when subjected to torsional stress (Fig. 29.21). The casing also provides a cavity for grease to lubricate the rotating core. The core is attached to the hub on either end and then connected to the equipment.

The power transmission capacity of flexible shafting is limited only by the core construction, minimum radius of curvature of the shafting, and maximum unsupported length.

Flexible shafts are commercially available with ratings up to 1500 pound-inches (lb·in) at 440 r/min. Such a shaft is 1½ in [38 millimeters (mm)] in diameter and has a minimum operating radius R of 24 in (600 mm). In Fig. 29.22, let R be the required

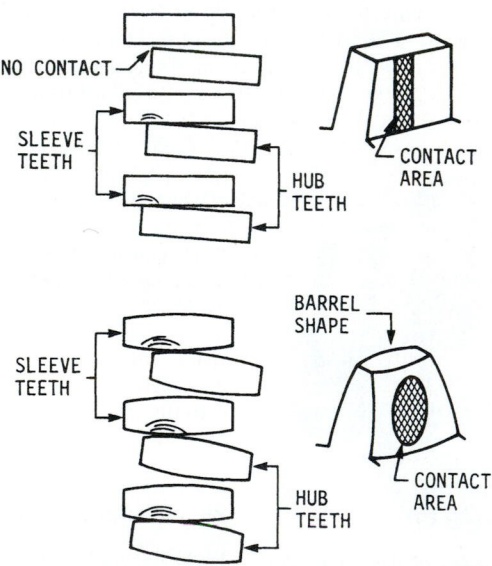

FIGURE 29.19 How change in tooth shape affects load distribution on the teeth of the gear coupling.

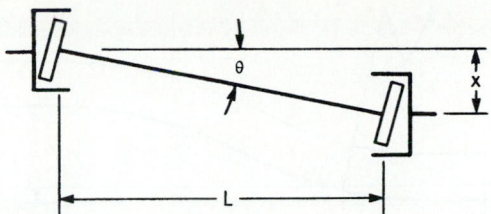

FIGURE 29.20 Diagram shows how parallel misalignment is converted to angular misalignment in each gear coupling mesh. For example, for an extended floating shaft with $L = 12$ in and $\theta = 1°$, the misalignment is $x = L \tan \theta = 0.20$ in. For a standard double-engagement coupling with $L = 2$ in and $\theta = 1°$, $x = 0.03$, which is significantly less.

operating radius corresponding to a misalignment A and a spacing D between equipment. Then the following relations can be derived from Fig. 29.22:

$$R = \frac{D^2 + A^2}{4A} \qquad (29.9)$$

$$A = 2R - (4R^2 - D^2)^{1/2} \qquad (29.10)$$

$$D = [A(4R - A)]^{1/2} \qquad (29.11)$$

$$C_L = \frac{\pi R}{90} \sin^{-1} \frac{D}{2R} \qquad (29.12)$$

where C_L = flexible length between the equipment.

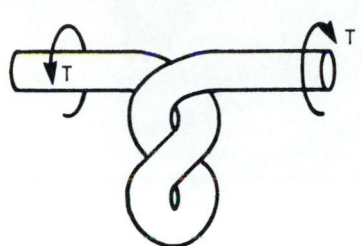

FIGURE 29.21 "Helixed" flexible shaft made of wire or rope.

If the casing is eliminated from the flexible shaft, a flexible coupling is created with much shorter maximum length (owing to lack of support and antihelixing without the casing). Typical commercial availability of this type of coupling is limited to 50 lb·in and 16-in lengths with an 8-in minimum operating radius.

Another similar coupling, the Uniflex,[†] consists of three layers of springs, each with three rectangular wires wound around an open-air core. This coupling (Fig. 29.23) also runs without a casing and has maximum speeds of up to 20 000 rpm, depending on size. It is relatively free from backlash and winds up about 1° at rated torque. This design uses spring elements up to 3 in long with rated torque up to 2000 lb·in at up to 4.5° misalignment.

29.3.6 Bellows Coupling

This type of coupling, shown in Fig. 29.24, consists of an all-metal circular bellows attached to two hubs. The design exhibits zero backlash and constant-velocity operation and is torsionally rigid. However, commercial couplings are typically rated to a maximum of 30 lb·in.

[†] Uniflex is a trademark of Lovejoy, Inc.

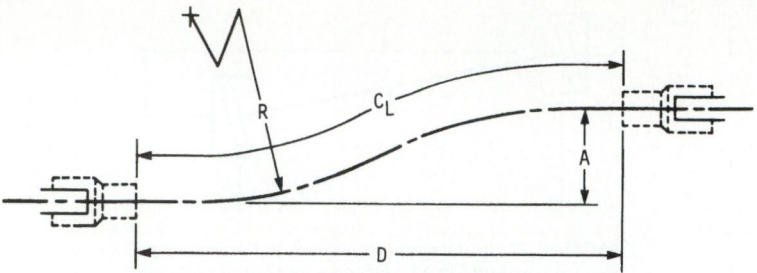

FIGURE 29.22   Maximum parallel misalignment of flexible shafting. *(Stow Manufacturing Company)*

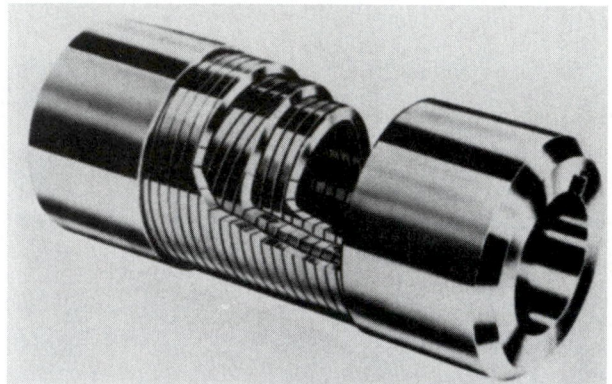

FIGURE 29.23   Uniflex flexible-spring coupling. *(Lovejoy, Inc.)*

The torque rating is obtained from

$$T = \frac{8pkD_o^2}{\ell} \tag{29.13}$$

where T = torque, in·oz
       p = pitch, in
       k = spring rate per convolution, oz/in
       D_o = outside diameter, in
       ℓ = length of a single convolution

The windup of the coupling (angular deflection) is measured in seconds of arc per inch-ounce of torque and is

$$u = \frac{0.08\ell}{(D_o + D_i)^3 t} \tag{29.14}$$

where u = windup, seconds/(in·oz)
       D_i = inside diameter, in
       t = thickness of bellows, in

Equations (29.15) and (29.16), which
follow, apply to the determination of the
life of this type of coupling. This life is
dependent on the flexing motion due to
angular and parallel misalignment, as
shown in Fig. 29.25. The formula gives
the operating misalignment correspond-
ing to 10^5 flexing cycles (5×10^4 r). For
10^8 cycles, derate by 24 percent. Now,

$$X = \frac{71.6NS}{D_o} \qquad (29.15)$$

FIGURE 29.24   A bellows coupling. *(Servometer Corp.)*

$$Y = \frac{N^2 nS}{D_o} \qquad (29.16)$$

where X = angular misalignment in degrees, Y = parallel misalignment, N = number
of bellows convolutions, S = maximum total permissible bellows stroke, D_o = outside
bellows diameter, and n = r/min.

29.4   FLEXIBLE ELASTOMERIC COUPLINGS

In this type of coupling, it is an elastomeric cushioning material—rather than only a
metal part, as in the rigid, flexible metallic, or universal-joint (U-joint) couplings—that
is subjected to the dynamic stresses of the operating system. These flexible elastomeric
couplings need no lubrication and only periodic visual inspection for maintenance.
They are available as compression, precompression, shear, and tension types.

The compression coupling is also known as the jaw, or block, coupling (Fig.
29.26). In this type of coupling, elastomeric cushioning blocks, rollers, or "spiders"
are compressed between alternating pairs of fingers on the two hubs of the coupling.

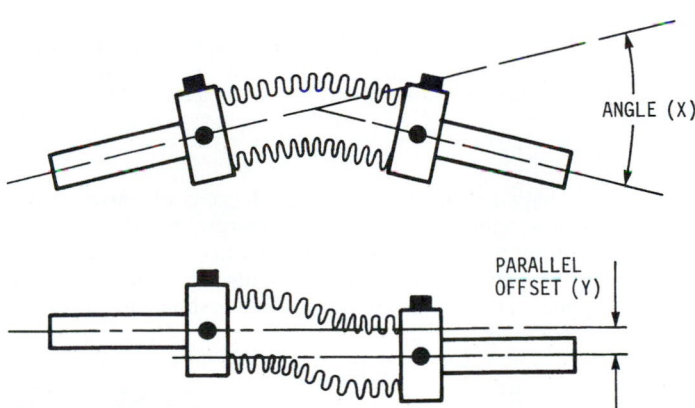

FIGURE 29.25 Bellows coupling with angular misalignment X and parallel
misalignment Y. *(Servometer Corp.)*

FIGURE 29.26 Exploded view of jaw-type compression coupling showing the two hubs and the flexible spider insert. *(Lovejoy Inc.)*

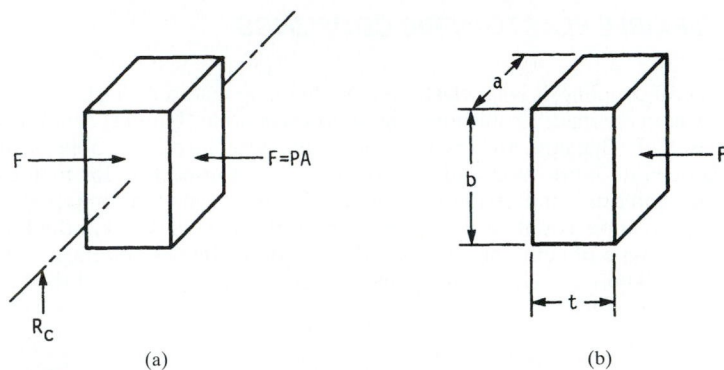

FIGURE 29.27 (*a*) Loading of one cushion at distance R_c from the center of a compression block coupling; (*b*) parameters used to calculate shape factor S for materials such as rubber where load area is ab and force F compresses the material at a thickness t.

The elastomeric elements can be of varying degrees of hardness or different materials to suit load-carrying capacities and temperature and chemical resistance requirements, in addition to torsional stiffness. Overall major dimensions may be altered by changing the number of active elements, element size, and radius on which the load is applied. The loading on one cushion is shown in Fig. 29.27*a*, and the torque rating of the coupling is

$$T = NPAR_c \qquad (29.17)$$

where T = torque, N = number of cushions, P = pressure, and R_c = centroidal radius.

From Eq. (29.17) we can see that large couplings with many active elements are capable of handling very large loads. Typically, the design limit for rubber in compression is 300 pounds per square inch (psi), while a plastic like Hytrel is capable of 1100 psi. The recommended load capacity of some elastomers is dependent on the *shape factor S*, which is the ratio of one load area to the total free area of the cushion. See Fig. 29.27*b*. Thus

$$S = \frac{ab}{2t(a+b)}$$ (29.18)

As this ratio changes from a thin plate (high *S*) to a fat block (low *S*), the maximum allowable stress *E* decreases, and the deformation of the block (compression) increases, as shown in Fig. 29.28.

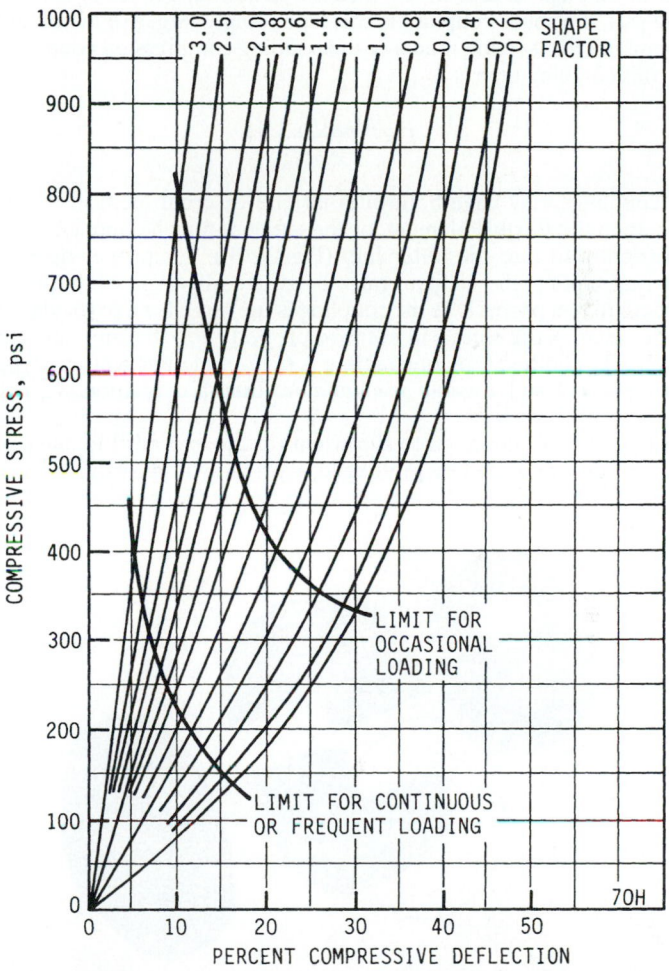

FIGURE 29.28 How the shape factor changes the maximum allowable stress in 70 durometer (Shore A) rubber. (*From Ref. [29.5], p. 77.*)

The basic misalignment capacity of these couplings is determined by the mating tolerances of the two hubs, the elastomeric element, and the particular jaw design. The larger the difference between the cushion and the jaw dimensions, the greater the misalignment that can be accommodated without exerting undue reaction forces on the system. However, this arrangement also leads to noisy operation, pulsating power transmission, and the transmission of shock through the system. A tradeoff between these effects is made in the final design. Alternatively, thick blocks of material can be used and deformed under misalignment. This may cause high restoring forces but decrease noise and pulsation of power owing to loading and unloading of the flexible elements. These couplings exert some axial thrust with increasing torque unless special jaws are used.

The shear unclamped coupling shown in Fig. 29.29 transmits power through shear in the elastomer, which can be rubber or a suitable plastic. Since rubber in shear can be loaded to only 20 percent of the load permitted on rubber in compression (70 versus 350 psi), the shear coupling is correspondingly larger in diameter and has a thicker element cross section than a corresponding compression coupling. The basic rating for this coupling type is

$$T = \frac{\pi t_{max}(r_0^4 - r_i^4)}{2r_0} \tag{29.19}$$

These couplings may have both internal and external teeth in the hubs which mate to teeth in the flexible element, or the element may be bonded to the internal hubs and friction-fit into the outer hub (Fig. 29.30). For proper operation of the toothed-type coupling, the element must be twisted because of torque; this causes the gear teeth to rest properly. If the coupling is too large for a particular application (high service factor), the teeth will not load properly and the coupling will wear out prematurely. The toothed type of coupling is the double-engagement type and may allow more parallel and angular misalignment than a corresponding compression coupling.

The bonded type is designed to slip when torque exceeds the maximum rating; however, the heat generated by prolonged overload will destroy the coupling. The

FIGURE 29.29 Shear unclamped coupling. *(T. B. Wood's Sons, Inc.)*

COUPLINGS   **29.23**

FIGURE 29.30 Bonded type of a shear unclamped coupling. *(Lord Corp., Industrial Products Div.)*

bonded type is torsionally softer than the corresponding toothed type, with windup of 15° to 20° versus 5° to 6° for the toothed type. This coupling is not fail-safe; when the flexible element fails, the driver and driven equipment are no longer connected.

The shear clamped or tire (torus) coupling is designed around a tire-shaped element, as shown in Figs. 29.31 and 29.32. The tire beads are clamped to a hub on either side, and torque is transmitted by shear and tension in the tire body. These couplings may be reinforced or unreinforced (without cords); in either case, the geometry allows very short overall length for a given torque capacity at the expense of diameter, which becomes correspondingly larger. This large diameter can cause some problems resulting from axial pull from the coupling as it is rotated at high speed. Centrifugal forces on the tire cause the two hubs to collapse inward as the speed is increased. The tire shape allows lower restoring force to larger misalignments than the shear unclamped or compression coupling.

The tension coupling shown in Fig. 29.33 is very similar to the link coupling (see Sec. 29.3.1), in that alternate legs (links) are in compression or tension. The links may have reinforcing cords to increase their tensile rating. In this coupling the rating is

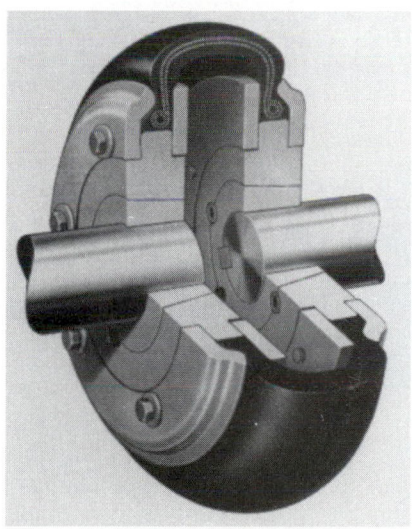

FIGURE 29.31 Tire coupling. *(Dodge Div., Reliance Electric.)*

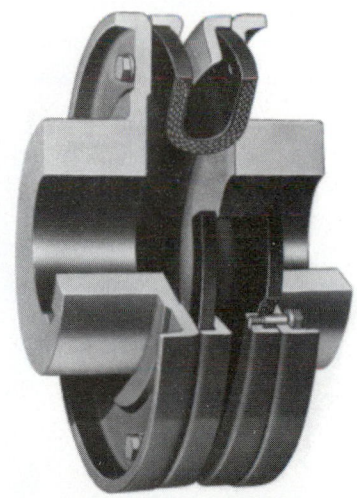

FIGURE 29.32 Torus coupling. *(Falk Corp.)*

FIGURE 29.33   Tension coupling. *(Lovejoy Inc.)*

determined by the compression buckling of the leg pairs or the tensile strength of the other leg pairs, whichever is less. These couplings may be made of plastic, although rubber treated against ozone, when used with reinforcing cords, is also available.

To eliminate the effect of tension on rubber in a design similar to the one in Fig. 29.33, the rubber element is precompressed by assembling the element into a state smaller than the free state. This allows the "trailing leg" (formerly the tension leg) to stretch almost to neutral but never into tension, which would promote early failure.

In the radially restrained type, a tight metal band is slipped around the OD of the element while it is being held in a radially compressed state. After assembly to the mating hub, the band is snipped off, and the assembly bolts hold the element in precompression. Alternatively, radial bolts are used to draw the rubber element into compression. See Fig. 29.34.

Both these designs use fairly thick cross sections for their flexing elements and are quite compliant in all forms of misalignment. Rubber hardness and type can be changed to alter torque capacity, damping, and chemical-temperature resistance.

Alternatively, axial restraint can be used. In this design, the rubber element is installed between two hubs which are set a fixed distance apart, as shown in Fig. 29.35. The rubber element is then installed by means of bolts into this space, which is smaller than the free state of the rubber. This design is also very compliant, but generates an axial force because of the forced axial compression of the element.

FIGURE   29.34   Precompressed   radially   restrained   coupling. *(Lovejoy, Inc.)*

29.5 UNIVERSAL JOINTS AND ROTATING-LINK COUPLINGS

Another class of shaft connectors is composed of the linkage types, which include universal joints (U-joints) and rotating-link couplings. These couplings all rely on translating misalignment to relative rotation between parts of the coupling. The

simplest universal joint is the one for a compass suspension universal joint described by Geronimo Cardano in the sixteenth century and explained mathematically by Robert Hooke in the seventeenth century. The universal joint basically consists of two shaft connections (yokes) which pivot about a pair of centrally located bearings. See Fig. 29.36.

Shafts A and B are inclined to each other at angle θ (shaft centerlines must intersect); when shaft A is rotated, the center member (spider) causes shaft B to rotate by virtue of the connection. The output speed of shaft B is not constant when the shafts are at an angle. Specifically, the output velocity is

FIGURE 29.35 Precompressed axially restrained coupling. *(Koppers Co. Inc., Power Transmission Div.)*

$$\omega_0 = \frac{\omega_i \cos \theta}{1 - \sin^2 \alpha \sin^2 \theta} \qquad (29.20)$$

where α = rotation position angle and θ = joint or shaft angle. In addition, the change in velocity peaks twice during every revolution. This leads to a second-order harmonic excitation frequency for the drive system. Specifically, the torsional magnitude β of this excitation is

$$\beta = \tan^{-1} (\sec \theta)^{1/2} - \tan^{-1} (\cos \theta)^{1/2} \qquad (29.21)$$

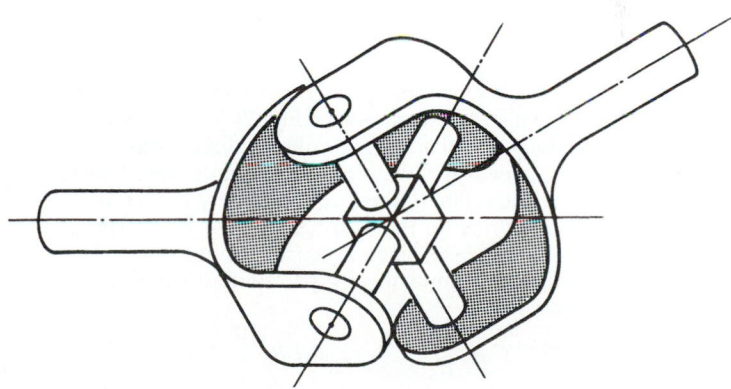

FIGURE 29.36 Basic Hooke's or Cardan universal joint.

The single Hooke's universal joint is not capable of accepting parallel misalignment (shaft centerlines do not meet at pivot member) except for very minor amounts resulting from manufacturing tolerances. Axial misalignment is usually compensated for by using a sliding connection on the input-output shafts or between the two universal joints.

When two universal joints are connected in series, parallel misalignment can be compensated for, because the parallel misalignment is converted to angular in each joint (see Fig. 29.37). By connecting two universal joints in this manner, it is possible to obtain approximately uniform output velocity between A and C. Basically, if joints A and B are aligned such that the pivot pins in their input and output yokes, respectively, are in the same plane, then the nonuniform output of joint A into shaft B is transformed to uniform output in joint C. However, if any angle exists between the joint A input yoke plane and the output yoke plane of joint C, then a nonuniform output will be generated. That nonuniform velocity and angle are, if we refer to Fig. 29.37c,

$$\omega_0 = \frac{\omega_i \cos \theta_{\text{eff}}}{1 - \sin^2 \alpha \sin^2 \theta} \tag{29.22}$$

where θ_{eff} = equivalent torsional angle given by

$$\theta_{\text{eff}} = (\delta^2 - \gamma^2)^{1/2} \tag{29.23}$$

Note that δ and γ are the U-joint angles as given in Fig. 29.37c.

29.5.1 Pin and Block

In its simplest form, the universal joint consists of two yokes and four pivot pins. The pivot pins ride in holes in the yoke and oscillate as the universal joint rotates. All the bearing takes place between the pins and the yoke, without the use of rolling-element bearings. In this type of joint, a thin film of lubricant or surface coating is all that prevents disaster from occurring, namely, the pin freezing in the journal (yoke) because of galling, corrosion, or adhesive transfer of metal. A first approximation to bearing pressure and velocity is as follows (shown in Fig. 29.38):

$$PV = \frac{Tn\theta}{6875 R_c t} \tag{29.24}$$

where T = operating torque, n = operating speed, θ = joint angle in degrees, t = yoke ear thickness, and R_c = the distance, as shown on Fig. 29.38. Since the velocity is actually nonuniform, the average speed of the pin-yoke oscillation has been used. We can see that the product $n\theta$, speed in revolutions per minute times angle in degrees, becomes a convenient unit to relate to the torque capacity. In fact, that is the typical unit used by most manufacturers for their ratings; see Fig. 29.39. These types of rating curves must be used with care, since a practical limit on angle limits the $n\theta$ parameter and the bearing area limits the PV parameter. A typical load-rating formula, given $PV = 1000$ psi·in/s, is

$$T = (0.6875 \times 10^7) \left(\frac{R_c t}{n\theta} \right) \tag{29.25}$$

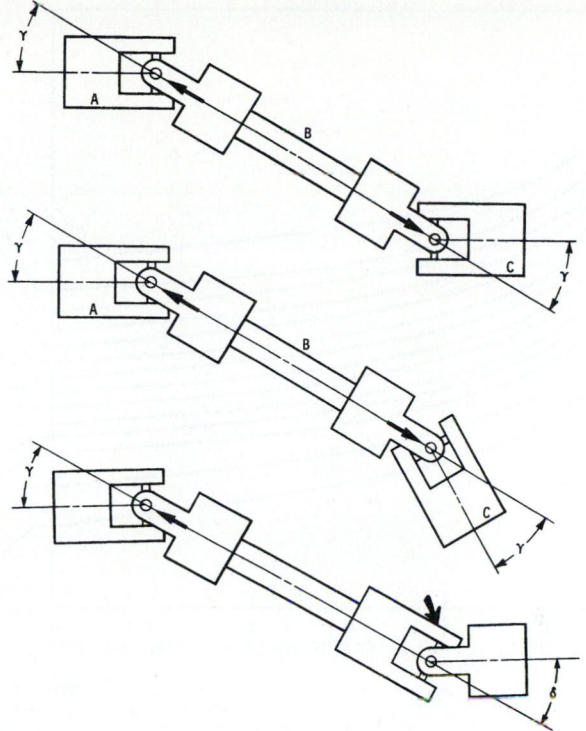

FIGURE 29.37 How two universal joints convert angular misalignment to parallel misalignment. (*a*), (*b*) Correct, yoke ears are aligned and angles are equal. (*c*) Incorrect, yoke ears are not in alignment and angles γ and δ are not equal. A residual angle θ_{eff} results from improper phasing of the two universal joints.

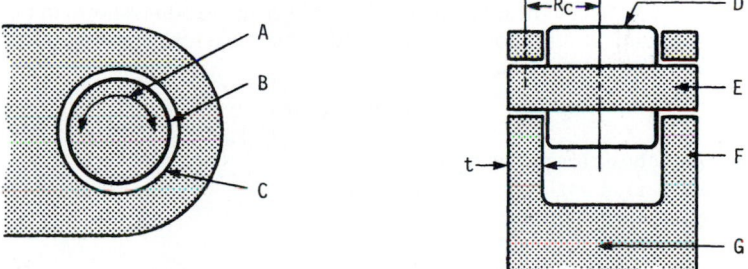

FIGURE 29.38 Plain bearing-type, pin-and-block, universal-joint geometry. A, oscillating pin; B, sliding at yoke interface; C, bore in yoke ear; D, center block; E, bearing pin; F, ear; G, yoke.

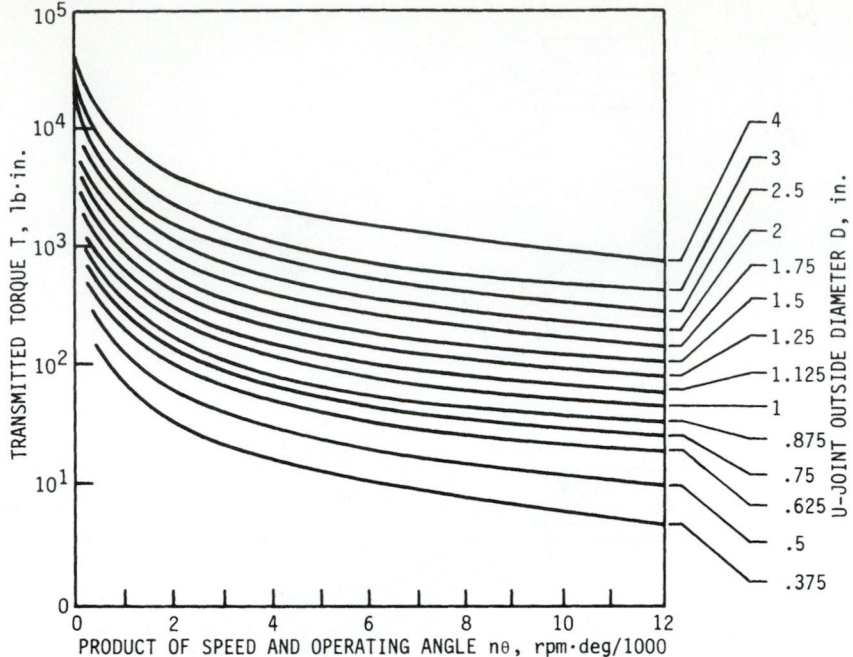

FIGURE 29.39   Torque ratings for pin-and-block universal joint. This chart shows the effect of the speed, operating angle, and OD parameters on the rated torque. These curves are based on a PV value of 1000 psi-in/s.

29.5.2   Needle Bearing

The pin-yoke interface is now changed to a pin and rolling-element bearing. This type of universal joint is also referred to as an automobile type, since it is found in most automobile drivelines. This universal joint also exhibits nonuniform rotational output velocity. A typical needle-bearing universal joint in cross section is shown in Fig. 29.40.

Corresponding to 3000 hours (h) of B_{10} life, the basic torque rating of a needle-bearing universal joint is given by

$$T_R = 4500 K_a K_s d\ell b \tag{29.26}$$

where T_R = basic torque rating, lb·in
K_a = joint angle correction factor
K_s = speed correction factor
d = bearing pin diameter, in
ℓ = length of needle rollers, in
b = bearing spacing, in

The dimensions d, ℓ, and b are shown in Fig. 29.40. The correction factor K_a in Eq. (29.26) is unity when the joint angle γ is 3°. For other angles, use the formula

$$K_a = \frac{1.432}{\gamma^{0.3269}} \tag{29.27}$$

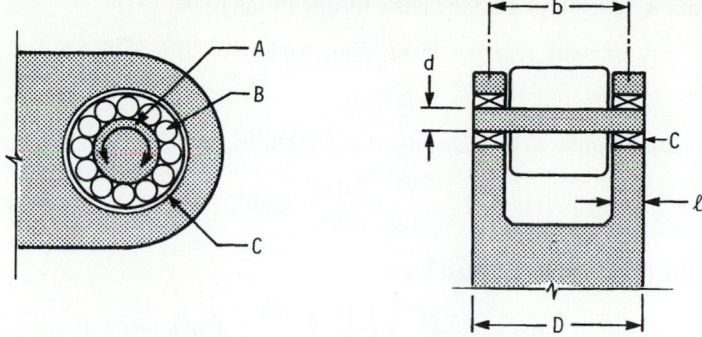

FIGURE 29.40 Geometry of the universal joint with needle bearings. A, oscillating pin; B, rolling motion occurs in the yoke-pin interface; C, needle bearing.

The correction factor K_s in Eq. (29.26) is unity when the speed n is 1000 r/min. For other speeds, use the equation

$$K_s = \frac{9.24}{n^{0.3219}} \qquad (29.28)$$

We also define a life factor K_L as the ratio of the actual transmitted torque to the basic torque rating. Thus

$$K_L = \frac{T}{T_R} \qquad (29.29)$$

For an expected life L of 3000 h, the life factor is unity. For other lives, use the equation

$$L = \left(\frac{15.5}{K_L} \right)^{2.921} \qquad (29.30)$$

where L = expected B_{10} life (see Chap. 27) in hours.

Example 1.   A universal joint has an outside diameter $D = 3$ in, an operating speed $n = 4000$ r/min, and a joint angle $\gamma = 5°$. The needle bearings have a needle length of 0.75 in and a bearing pin diameter $d = 0.875$ in. Find the expected B_{10} life if 2400 lb·in of torque is to be transmitted.
   Solution.   From Fig. 29.40 we find

$$b = D - \ell = 3 - 0.75 = 2.25 \text{ in}$$

Equation (29.27) gives the joint angle correction factor as

$$K_a = \frac{1.432}{\gamma^{0.3269}} = \frac{1.432}{5^{0.3269}} = 0.846$$

The speed correction factor is obtained from Eq. (29.28):

$$K_s = \frac{9.24}{n^{0.3219}} = \frac{9.24}{(4000)^{0.3219}} = 0.640$$

Thus, from Eq. (29.26), we find the basic torque rating to be

$$T_R = 4500 K_a K_s d\ell b = 4500(0.846)(0.640)(0.875)(0.75)(2.25)$$

$$= 3598 \text{ lb} \cdot \text{in}$$

Since the actual torque to be transmitted is 2400 lb·in, the life factor is

$$K_L = \frac{T}{T_R} = \frac{2400}{3598} = 0.667$$

So, from Eq. (29.30), the expected life is

$$L = \left(\frac{15.5}{K_L}\right)^{2.921} = \left(\frac{15.5}{0.667}\right)^{2.921} = 9878 \text{ h}$$

29.5.3 Constant-Velocity Universal Joints

Double Cardan Joint. As mentioned previously, two simple Hooke's (Cardan) universal joints can be connected to give relatively constant output velocity (Fig. 29.37). However, if the universal joints are not exactly at the same angle, a small fluctuation in output velocity will occur. Specifically, the torsional equivalent angle $\theta_{\text{eff}} = \sqrt{\theta_1^2 - \theta_2^2}$ for a universal joint when the bearing planes are in the same plane. When the bearing planes are perpendicular, θ_{eff} becomes

$$\theta_{\text{eff}} = \sqrt{\theta_1^2 + \theta_2^2} \tag{29.31}$$

The torsional equivalent angle is used as if only one Cardan joint were in the system.

Tracta. The Tracta joint operates through the sliding action of two internally connected sliding couplings. This type of joint has a fairly high torque and angle capacity for its diameter; however, because of heat buildup from sliding friction, it cannot be used for any extended time at high loads. It can, however, be used satisfactorily at low angles with high loads.

The C-shaped input-output yokes generally have an opening of less than 180° so as to lock into two grooved couplings, which are, in turn, locked together in a tongue-and-groove fashion. See Fig. 29.41.

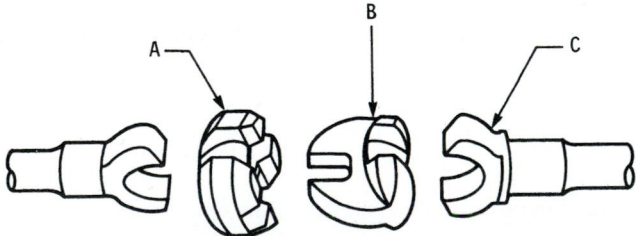

FIGURE 29.41 Exploded view of the Tracta universal joint. A, grooved coupling; B, tongue and groove coupling; C, fork yoke. *(From Ref. [29.6].)*

Rzeppa Universal Joint. This constant-velocity joint uses driving balls and curved axial-groove races; see Fig. 29.42. When the joint is at the same angle θ, the balls on one side of the joint will be farther from the joint centerline than the others. The off-center condition occurs because the groove generation point is displaced to provide camming action necessary for proper ball location and joint operation. Since the outer-race ball grooves act as cam surfaces, frictional locking can occur unless the cam surfaces have a divergent angle of 15° to 17° to prevent this. If the divergent angle were not present, the joint would lock going through zero angle.

Roller and Trunion. If the yoke ears in the simple pin-and-block or needle roller universal joints are elongated to form races, some axial movement is allowed, which decreases greatly the axial force developed by a single rotating universal joint. This joint is not a constant-velocity type. It is also called a *bipot* (see Fig. 29.43).

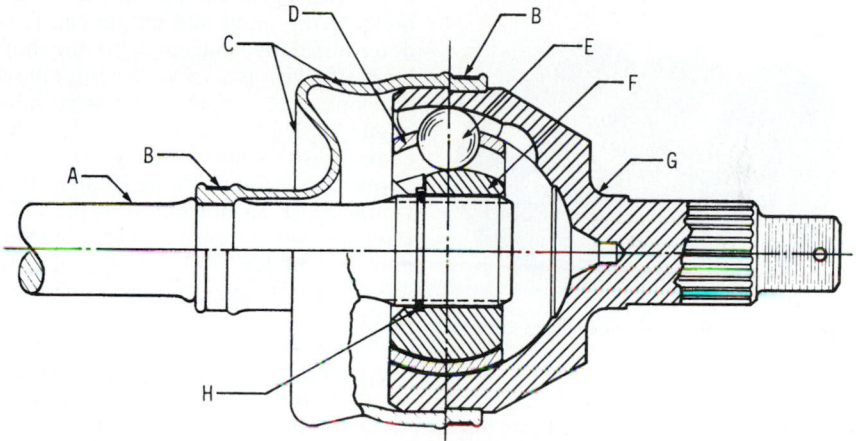

FIGURE 29.42   Rzeppa universal joint in cross section. A, shaft; B, clamp or band; C, boot seal; D, cage; E, ball; F, inner race; G, bell-type outer race; H, retaining ring. *(From Ref. [29.6].)*

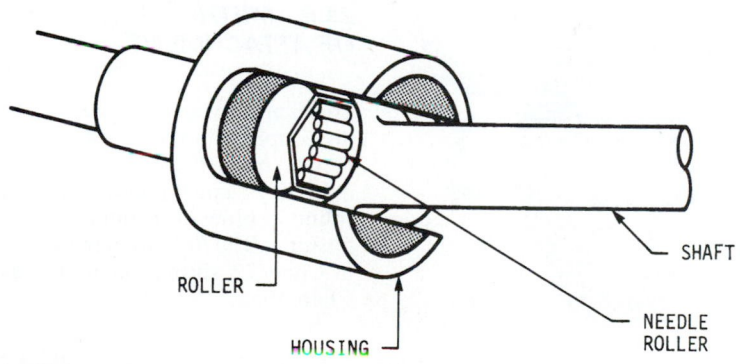

FIGURE 29.43   Roller trunion universal joint.

Tripot.   If three equally spaced rollers are used, the joint will transmit constant velocity because of the cancellation of torque couples within the joint. The center of this joint may be free to float (end motion type), in which case the axial force generated by the joint is low and the joint is self-supporting. In the fixed-center type, the joint center is fixed, which causes the joint to orbit about its center; this limits it to low-speed use, because of inertial effects, and places where this orbital motion can be tolerated (such as in conjunction with self-aligning bearings or other universal joints). See Ref. [29.6].

29.5.4   Rotating-Link Coupling

This coupling consists of three disks connected by two sets of three links, as shown in Fig. 29.44. Typically, one set of links is attached to shafts on the driver and center disks, while the other set of links connects the center to the lower disk. All links are of equal length, and all the shafts are equally spaced on the same bolt circle on each

hub. The maximum parallel displacement of the input and output shafts is determined by and equal to the link length S. With this arrangement, unless the input and output shafts were misaligned at least a small amount S_{min}, the center disk would be free to swing about the center of the shafts. This would cause an unbalance in the coupling and would result in high vibration and limited life. This effect gives the misalignment range of this coupling of S_{min} to S.

FIGURE 29.44   Rotating-link coupling. *(Schmidt Couplings, Inc.)*

The construction of this coupling causes the links to move parallel to one another and results in constant angular velocity. In addition, because three links are equally spaced about each hub and are of equal size and mass, the sum of all link forces is zero, resulting in smooth operation without imposing side loads.

FIGURE 29.45   Pressure bushing. *(FFV Industrial Products.)*

29.6   METHODS OF ATTACHMENT

There are several methods of attaching the coupling hubs to the shafts on equipment. Among these are split-taper bushings, keys and splines, shrink-fit and spline profile distortion, and pressure bushings. The first two items are covered in Chap. 22; shrink fits are discussed in Chap. 19.

Spline profile distortion causes the entire bore area of the coupling hub to deform around the mating shaft. This

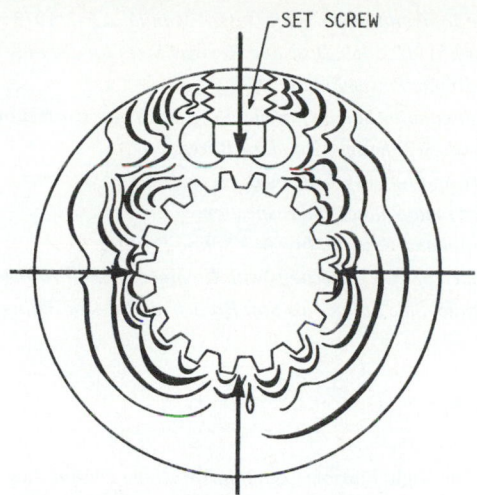

FIGURE 29.46 Drawing of a model of Centaloc spline profile distortion obtained by photoelastic methods showing lines of constant stress. *(Lovejoy, Inc.)*

method of attachment allows a much greater clamping force to be exerted than with a split-taper hub with internal spline. This clamping about the entire spline is important and eliminates fretting corrosion (caused by loose-fitting spline shafts and mating hubs) and point contact obtained with split bushing types, as shown in Fig. 29.46.

Pressure bushings (Fig. 29.45) are similar to the hydraulic coupling except that both the inner and outer surfaces are allowed to deform, compressing onto the shaft and expanding into the coupling hub. The pressure source for this type is usually applied through axial compression of the bladder with a clamping ring.

REFERENCES

29.1 AGMA Standard 514-02. 1971,[†] *Load Classification and Service Factors for Flexible Couplings.*

29.2 DIN 740,[†] *Flexible (Shaft) Couplings; Dimensions, Nominal Torque.*

29.3 AGMA Standard 515.02-1977, *Balance Classification for Flexible Couplings.*

29.4 ISO 1940,[†] *Balance Quality of Rotating Rigid Bodies.* 1973.

29.5 *Handbook of Molded and Extruded Rubber,* Goodyear Tire and Rubber Co., Akron, Ohio, 1969.

[†] Addresses for standards organizations: American Gear Manufacturer's Association (AGMA), 1500 King St., Alexandria, VA 22314; American National Standards Institute (ANSI), 1430 Broadway, New York, NY 10018; Deutsche Industrie Normalische (DIN), Beuth Verlag GMBH, Burggrafensprasse 4-10, 1000 Berlin 30, Germany; International Standards Organization (ISO), 1 Rue de Varembe, Case Postale 56, Ch-1211 Genosa 20, Switzerland; Japanese Industrial Standards (JIS), 1-3-1, Kasumigaseki, Chiyoda-Ku, Tokyo 100, Japan; Society of Automotive Engineers (SAE), 400 Commonwealth Dr., Warrendale, PA 15086. (In the United States, purchase DIN, ISO, and JIS specifications through ANSI.)

29.6 SAE,[†] *Universal Joint and Drive Shaft Design Manual AE-7,* 1979.

29.7 AGMA Standard 511.02-1969, *Bore and Keyway Sizes for Flexible Couplings.*

29.8 SAE J744c, *Hydraulic Power Pumps,* 1970.

29.9 ISO R773, *Rectangular or Square Parallel Keys and Their Corresponding Keyways,* 1969.

29.10 DIN 7154, *ISO-Close Tolerance Standard Bores.*

29.11 JISB 1301,[†] *Bore and Keyway Tolerances.*

29.12 ANSI B92.1,[†] *Involute Splines and Inspection.*

29.13 DIN 5482, *Internal and External Involute Spline Profiles.*

29.14 DIN 5480, *Serrated Shaft Connections with Involute Spline Profiles.*

29.15 SAE J499a, *Parallel Side Splines for Soft Broached Holes in Fittings,* 1975.

BIBLIOGRAPHY

Beercheck, Richard: "The Right Flexible Coupling," *Machine Design,* Aug. 24, 1978, pp. 101–104.

Bigden, J. V., and Ziegler, Carl: "Universal Joints," *Power Transmission Design Handbook 1983–84,* Penton IPC, 1983, pp. C189–192.

Calistrat, Michael: *Flexible Coupling Installation,* Koppers Co. Inc., Baltimore, Md. "Diaphragm Couplings Challenge Disk and Gear Types for High Torque," *Product Engineering,* June 1977, pp. 33–36.

Hitchcox, Alan: "High Performance Couplings Meet Heavy Demands," *Power Transmission Design,* October 1982, pp. 46–48.

McCormick, Doug: "Finding the Right Flexible Coupling," *Design Engineering,* October 1981, pp. 62–66.

Proceedings of the International Conference on Flexible Couplings for High Powers and Speeds, University of Sussex, Brighton, England, July 1972.

Schalitz, August: *Kupplungs-Atlas,* Druckhaus Korntal Fritz Pfitzer, 1975 (German).

Schmidt, Richard: *Unique Couplings for Large Offset Shafts,* Schmidt Couplings, Cincinnati.

Schwerdlin, Howard: "Flexible Couplings," *Power Transmission Design Handbook 1983–84,* Penton IPC, 1983, pp. 123–126.

Schwerdlin, Howard: "Combatting Heat in U-joints," *Machine Design,* July 23, 1981, pp. 83–86.

Wiedenroth, Wolfgang: "Fachgebiete in Jahresübersichten," *VDI-Z,* no. 17, September 1980 (German).

CHAPTER 30
CLUTCHES AND BRAKES

John R. Zimmerman, Ph.D.
Professor of Mechanical and Aerospace Engineering
University of Delaware
Newark, Delaware

GLOSSARY OF SYMBOLS

a	Vehicle deceleration, ft/s^2 (m/s^2)
a	Location of shoe pivot, in (m)
a	Lever arm for larger band force, in (m)
A	Area, in^2 (m^2)
b	Percentage of grade
b	Width of band, shoe, or web, in (m)
c	Lever arm for smaller band force, in (m)
C	Center of pressure
C	Specific heat, Btu/(lbm · °F) [J/(kg · °C)]
d	Inside disk diameter, in (m)
D	Outside disk diameter, in (m)
D	Pitch diameter of gear, in (m)
D_{max}	Maximum roll diameter, in (m)
e	Radius to center of circular brake pad, in (m)
E	Total energy dissipated, ft · lb or Btu (J)
f	Coefficient of friction
f_v	Ventilation factor

F	Actuating force, lb (N)
F_D	Prime-mover factor
F^l	Actuating force on leading shoe, lb (N)
F_L	Load factor
F_n	Normal force, lb (N)
F_S	Starting factor
F_t	Tension force on web, lb (N)
F^t	Actuating force on trailing shoe, lb (N)
g	Local acceleration of gravity, ft/s^2 (m/s^2)
g_c	Gravitational constant, 32.174 lbm $\cdot$ ft/(lb $\cdot$ s^2) [1 kg $\cdot$ m/(N $\cdot$ s^2)]
h	Overall heat transfer coefficient, Btu/(in^2 $\cdot$ s $\cdot$ °F) [W/(m^2 $\cdot$ °C)]
h_c	Convection heat transfer coefficient, Btu/(in^2 $\cdot$ s $\cdot$ °F) [W/(m^2 $\cdot$ °C)]
h_r	Radiation heat transfer coefficient, Btu/(in^2 $\cdot$ s $\cdot$ °F) [W/(m^2 $\cdot$ °C)]
H_{av}	Average rate of heat dissipation, lb $\cdot$ ft/s or Btu/s (W)
I	Mass moment of inertia, lb $\cdot$ in $\cdot$ s^2 (kg $\cdot$ m^2)
I_L	Mass moment of inertia on load side, lb $\cdot$ in $\cdot$ s^2 (kg $\cdot$ m^2)
I_P	Mass moment of inertia on prime-mover side, lb $\cdot$ in $\cdot$ s^2 (kg $\cdot$ m^2)
K_S	Service factor
ℓ	Moment arm of actuating force (drum brake); length of actuating lever in a band brake, in (m)
m	Mass, lbm (kg)
M_a	Moment of actuating force, lb $\cdot$ in (N $\cdot$ m)
M_f	Moment of resultant friction force, lb $\cdot$ in (N $\cdot$ m)
M_n	Moment of resultant normal force, lb $\cdot$ in (N $\cdot$ m)
n	Shaft speed, r/s (Hz) or r/min
N	Number of pairs of friction surfaces in disk clutches or brakes
N	Number of shoes in centrifugal clutch
p	Normal pressure, psi (MPa) or r/min
p_{av}	Average contact pressure, psi (MPa)
p_h	Hydraulic pressure, psi (MPa)
p_{max}	Maximum contact pressure, psi (MPa)
p_{max}^{ℓ}	Maximum contact pressure on leading shoe, psi (MPa)
p_{max}^{t}	Maximum contact pressure on trailing shoe, psi (MPa)
P	Resultant normal force between drum and shoe, lb (N)
P	Power, Btu/s or hp (kW)
P_x	Component of normal force in x direction, lb (N)
P_y	Component of normal force in y direction, lb (N)
P_1	Larger band tension, lb (N)
P_2	Smaller band tension, lb (N)
q	Rate of energy dissipation during clutch slip, Btu/s (W)

Q	Actuating force (band brake), lb (N)
r	Brake drum radius, in (m)
r	Radius to point on disk, in (m)
r_f	Radius to center of pressure, in (m)
R	Tire-rolling radius, in (m)
R	Reaction force (drum brake), lb (N)
R	Radius to rim of centrifugal brake, in (m)
R_e	Effective friction radius, in (m)
R_i	Inside radius, in (m)
R_o	Outside radius, in (m)
s	Total stopping distance, ft (m)
S	Initial tension, lb (n)
S	Stops per hour
t	Web thickness, mils (mm)
t_d	Combined delay time for driver reaction and brake system reaction, s
t_s	Total stopping time, s
T	Torque; nominal torque, lb · ft (N · m)
T_a	Temperature of surrounding air, °F (°C)
T_d	Disk temperature, °F (°C)
ΔT	Temperature rise, °F (°C)
T_{des}	Design torque, lb · ft (N · m)
T_L	Load torque, lb · ft (N · m)
$T_{\max}$	Maximum torque, lb · ft (N · m)
T_P	Prime mover torque, lb · ft (N · m)
V	Rubbing velocity, ft/s (m/s)
V_o	Initial velocity, ft/s (m/s)
V_f	Final velocity, ft/s (m/s)
V_w	Web velocity, ft/s (m/s)
w	Web tension per unit thickness and unit width, lb/(mil · in) [N/(mm · m)]
W	Vehicle weight, lb (N)
α	Cone angle, deg
δ	Multiplier for circular disk brake pads
ε	Angular position of actuation force, deg
θ	Angle of wrap, deg
θ	Angular position, deg
θ_1	Starting position of brake shoe lining, deg
θ_2	Ending position of brake shoe lining, deg
ω	Shaft speed, rad/s
ω_e	Engagement speed, rad/s
ω_o	Initial shaft speed, rad/s

ω_f Final shaft speed, rad/s

Ω_L Initial shaft speed on load side of clutch, rad/s

Ω_P Initial shaft speed on prime-mover side of clutch, rad/s

This chapter begins with an introduction to brakes and clutches, the various types and their applications. The problem of energy dissipation and temperature rise is discussed along with the proper selection of friction materials. Design methods are presented for almost every type of brake and clutch. A discussion of the actuation problems of brakes and clutches, including electromagnetic devices, is also presented.

30.1 TYPES, USES, ADVANTAGES, AND CHARACTERISTICS

30.1.1 Types of Clutches

The characteristic use of a clutch is to connect two shafts rotating at different speeds and bring the output shaft up to the speed of the input shaft smoothly and gradually.

Classifying clutches is done by distinguishing (1) the physical principle used to transmit torque from one member to another and (2) the means by which the members are engaged or by which their relative speed is controlled. Here, we classify clutches as follows:

1. Engagement or actuation method
   a. Mechanical
   b. Pneumatic
   c. Hydraulic
   d. Electrical
   e. Automatic
2. Basic operating principle
   a. Positive contact
      (1) Square jaw
      (2) Spiral jaw
      (3) Toothed
   b. Friction
      (1) Axial
      (2) Radial
      (3) Cone
   c. Overrunning
      (1) Roller
      (2) Sprag
      (3) Wrap-spring
   d. Magnetic
      (1) Magnetic particle
      (2) Hysteresis
      (3) Eddy current
   e. Fluid coupling
      (1) Dry fluid
      (2) Hydraulic

Coupling Methods. *Positive-contact clutches* have interlocking engaging surfaces to form a rigid mechanical junction. Three types of positive-contact clutches are shown in Fig. 30.1.

Frictional clutches are used most frequently. Two opposing surfaces are forced into firm frictional contact. Figures 30.2, 30.3, and 30.4 show *axial, radial,* and *cone* types.

Overrunning clutches are used when two members are to run freely relative to each other in one direction but are to lock in the other. Roller, sprag, and wrap-spring types are shown in Fig. 30.5. In the *roller-ramp clutch* (Fig. 30.5*a*), the members are locked together when the rollers (or balls) ride on a race with a slight cam profile. Eccentric cams are pinched between concentric races in the *sprag-type clutches* (Fig. 30.5*b*). And in the basic *wrap-spring clutch* (Fig. 30.5*c*), the spring's inside diameter is slightly smaller than the outside diameters of the input and output hubs. When the spring is forced over the two hubs, rotation of the input hub in the

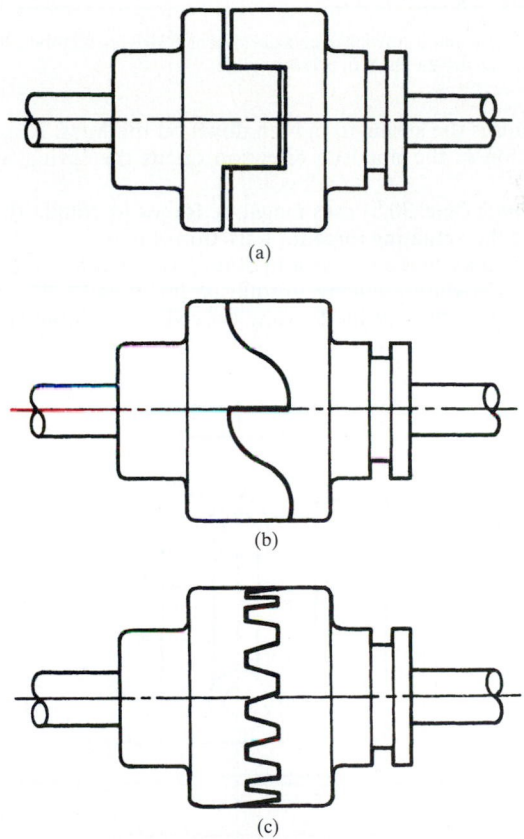

(a)

(b)

(c)

FIGURE 30.1   Positive-contact clutches. (*a*) Square-jaw, the square teeth lock into recesses in the facing plate; (*b*) spiral-jaw, the sloping teeth allow smoother engagement and one way drive; (*c*) toothed-clutch, engagement is made by the radial teeth.

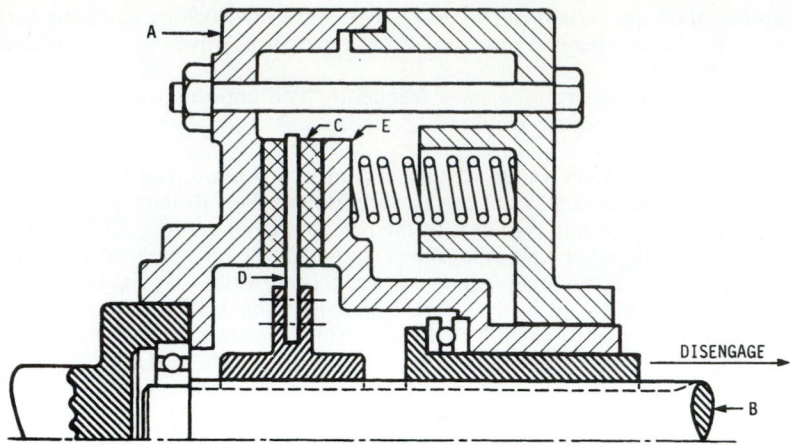

FIGURE 30.2 Schematic drawing of an axial clutch; *A*, driving member; *B*, driven shaft; *C*, friction plates; *D*, driven plate; *E*, pressure plate.

drive direction causes the spring to tighten down on the hubs. Torque is then transmitted. But rotation in the opposite direction opens the spring, and no torque is transmitted.

A *magnetic clutch* (Sec. 30.8) uses magnetic forces to couple the rotating members or to provide the actuating force for a friction clutch.

Fluid couplings may make use of a hydraulic oil or a quantity of heat-treated steel shot. In the *dry-fluid coupling*, torque is developed when the steel shot is thrown centrifugally to the outside housing (keyed to the input shaft) as the input

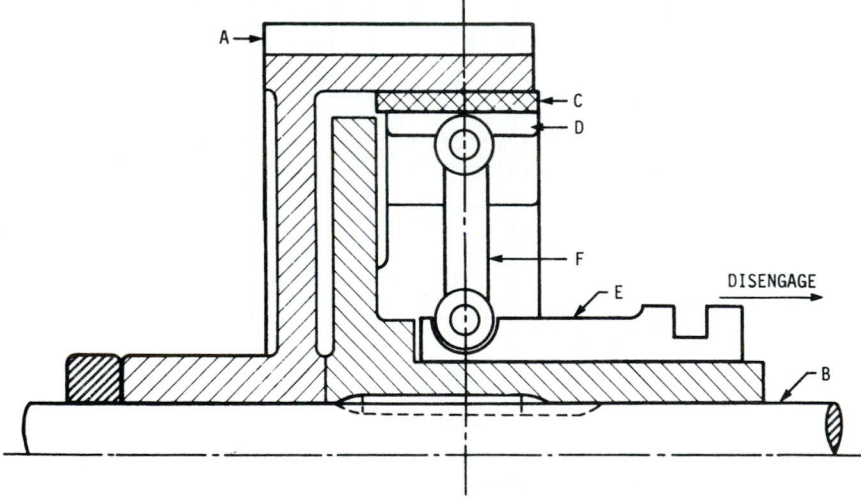

FIGURE 30.3 Schematic drawing of a radial clutch built within a gear; *A*, gear, the driving member; *B*, driven shaft; *C*, friction plate; *D*, pressure plate; *E*, movable sleeve; *F*, toggle link. This type of clutch can also be made within a V-belt sheave.

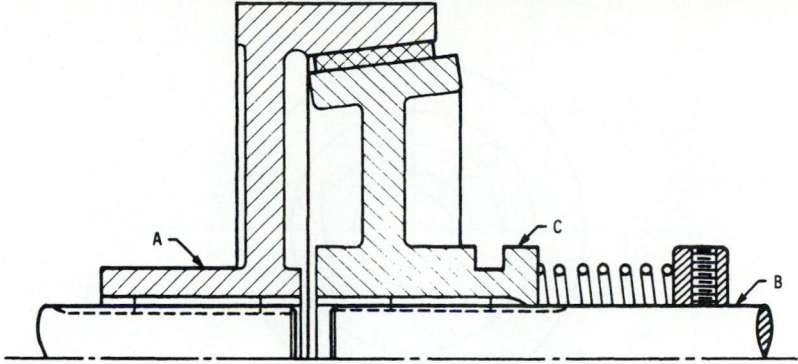

FIGURE 30.4 Schematic drawing of a cone clutch; *A*, driving member; *B*, driven shaft; *C*, movable sleeve.

shaft begins to rotate. At the design speed the shot is solidly packed, and the housing and rotor lock together.

Control Methods. *Mechanical control* is achieved by linkages or by balls or rollers working over cams or wedges. The actuating force can be supplied manually or by solenoid, electric motor, air cylinder, or hydraulic ram.

Electrical control of friction or tooth clutches often involves engaging the clutch electrically and releasing it by spring force. Thus the clutch is *fail-safe:* If power fails, the clutch is disengaged automatically. But where shafts are coupled for much longer periods than they are uncoupled, the opposite arrangement may be used: spring force to engage, electromagnetic force to disengage.

Pneumatic, or *hydraulic, control* is accomplished in several ways. Actuating pistons may be used either to move the actuating linkage or to directly apply a normal force between frictional surfaces. In other designs an inflatable tube or bladder is used to apply the engagement force. Such designs permit close control of torque level.

Automatic control of clutches implies that they react to predetermined conditions rather than simply respond to an external command. Hydraulic couplings and eddy-current clutches both have torque regulated by the slip. *Centrifugal clutches* (Fig. 30.6) use speed to control torque.

30.1.2 Selecting Clutches

A starting point is a selection table constructed by Proctor [30.5] and reproduced here as Table 30.1. Four additional tables in Proctor's article also are useful for preliminary decisions. Designers will have to consult the manufacturers before making final decisions.

30.1.3 Types of Brakes

Physically, brakes and clutches are often nearly indistinguishable. If two shafts initially at different speeds are connected by a device to bring them to the same speed,

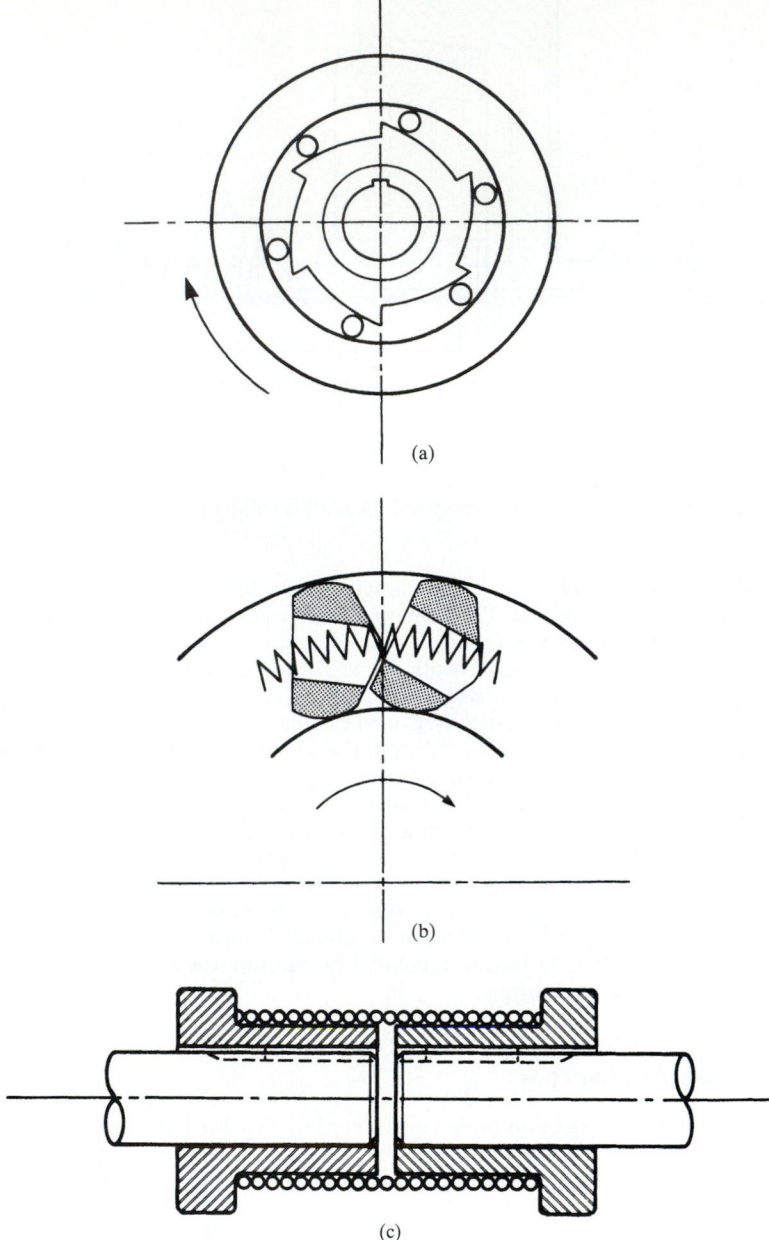

FIGURE 30.5 Overrunning clutches. (*a*) Roller-ramp clutch; springs are often used between the rollers and the stops. (*b*) Portion of a Formsprag clutch. Rockers or *sprags,* acting as cams, are pushed outward by garter springs at both ends of the prismatic sprags. (*c*) Torsion spring winds up when the clutch is in "drive" and grips both hubs. Larger-torque loads can be carried by making the springs of rectangular-section wire.

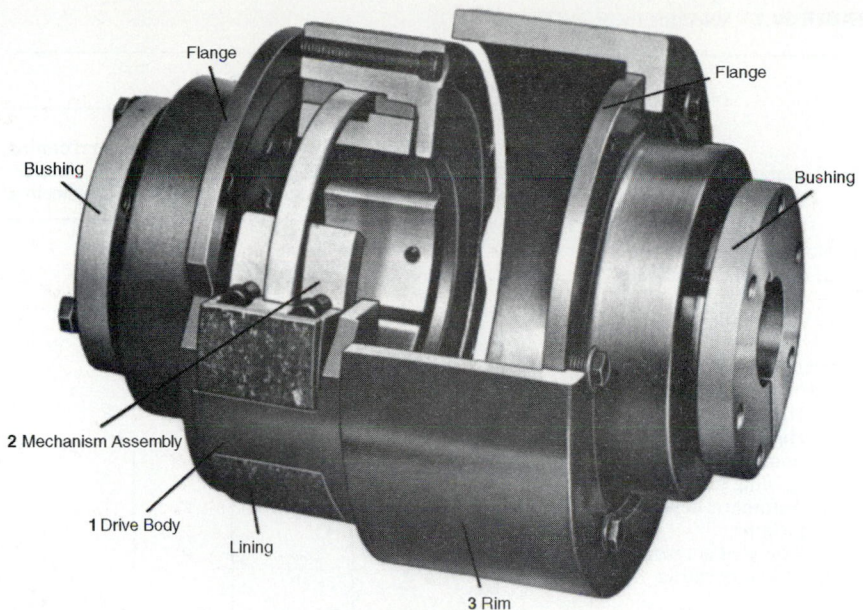

FIGURE 30.6 Centrifugal clutch. *(BLM Automatic Clutch Limited.)*

it is a clutch. If one member is fixed and the torque is used to slow down or stop the rotating member, the device is a brake. A classification scheme for brakes is presented in Fig. 30.7.

Brake Configuration. *Band brakes* can be made *simple* (not self-energizing) or *differential* (self-energizing). In designing a differential band brake (Fig. 30.19), care must be taken to ensure that the brake is not self-locking.

Short-shoe brakes have been used for hoists. Centrifugal brakes employ speed as the actuating signal for short-shoe internal-block brakes and are used in a wide variety of applications.

Drum brakes (Fig. 30.8) are used principally for vehicles, although seldom on the front axles of passenger cars. On the rear axles, drum brakes supply high braking torque for a given hydraulic pressure because one or both of the long shoes can be made self-energizing.

For a *leading shoe,* the friction moment exerted on the shoe by the drum assists in actuating the shoe. The friction moment on a *trailing shoe* opposes the actuating moment. Thus a leading shoe is *self-energizing,* but a trailing shoe is *self-deenergizing.*

The *leading-shoe trailing-shoe* design (Fig. 30.8) provides good braking torque in forward or reverse. The *two-leading-shoe* design has an even higher braking torque in forward, but a much lower braking capacity in reverse. Very high braking torque is available from the *duo-serve* design. Here the friction force on the "leading shoe" assists in actuating the "trailing shoe."

One difficulty with drum brakes is instability. If a brake's output is not sensitive to small changes in the coefficient of friction, the brake is *stable.* But if small changes

TABLE 30.1 Selecting the Right Clutch

Load characteristic or clutch function	General utility	Centrifugal and fluid self-actuating	Continuous slip — Automatic	Continuous slip — Variable	Overrunning or freewheeling
1. No-load start					
a. Manual or externally controlled	✓	...	...	...	✓
b. Automatic	...	✓			
2. Smooth load pickup					
a. Normal load	✓	✓	...	...	✓
b. High-inertia load	...	✓	...	✓	
c. High breakaway load (more than 100% running torque)	...	✓	...	✓	
d. Automatic delayed pickup	...	✓	...	✓	
e. Extended acceleration	...	✓	...	✓	
f. Auxiliary starter	...	...	...	...	✓
3. Running operation					
a. Normal load (no slip at full load, full speed)	✓	✓	✓		
b. Control variable-torque load	...	...	...	✓	
c. Control constant-torque load	...	...	✓	✓	
d. Control constant-tension load	...	...	...	✓	
4. Overload protection and stopping					
a. General protection: transient and infrequent overloads	✓	✓			
b. Limit speed (prevent runaway load)	...	✓	...	...	✓
c. Limit torque	...	✓	✓	...	✓
d. Automatic overload release	✓	...	...	...	✓
e. Dynamic braking	...	✓	...	...	✓
f. Backstopping	...	✓	...	...	✓
5. Intermittent operation					
a. On-off, with driver at speed	✓	...	...	...	✓
b. Inching and jogging	✓				
c. Indexing and load positioning	...	...	...	...	✓
6. Dual-drive and standby operation	...	✓	...	...	✓

SOURCE: Ref. [30.5]

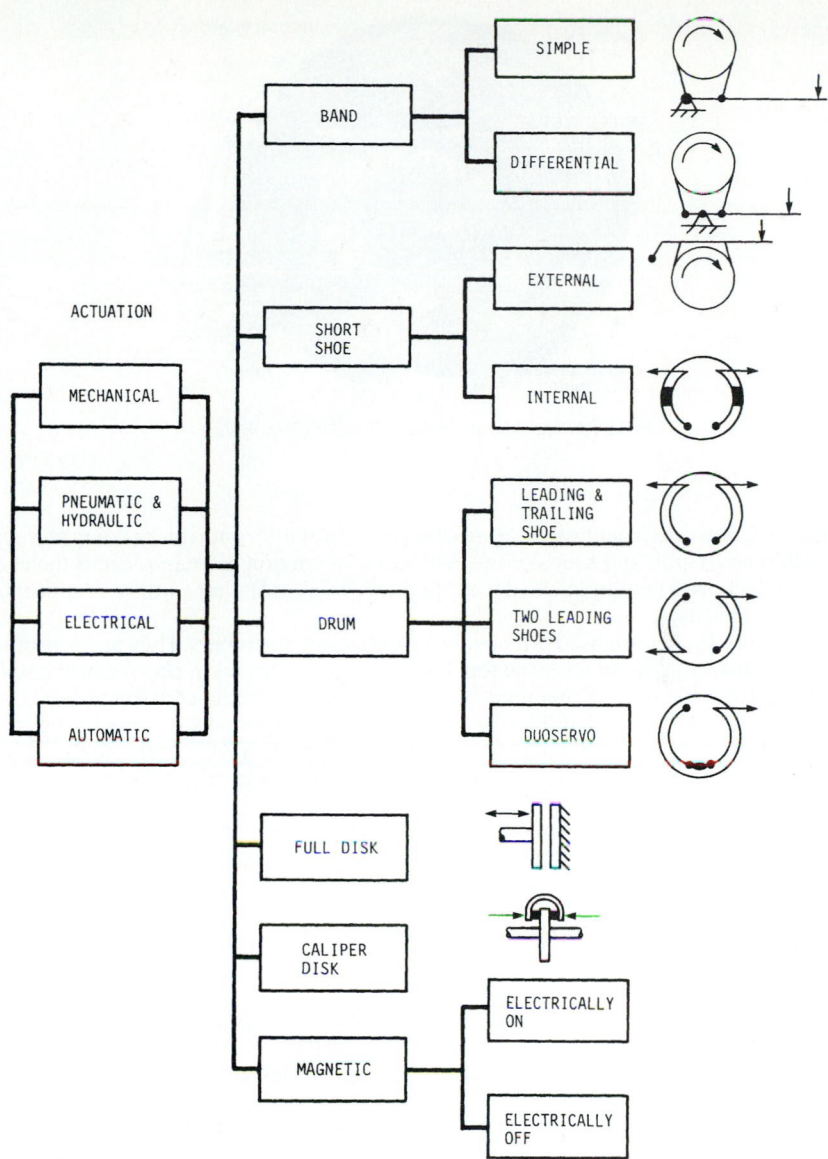

FIGURE 30.7 Classification of brakes.

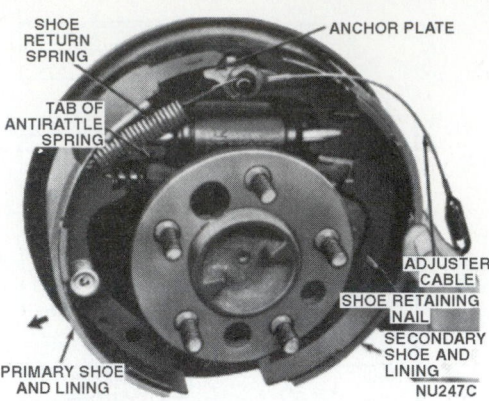

SHOE
RETURN
SPRING

ANCHOR PLATE

TAB OF
ANTIRATTLE
SPRING

ADJUSTER
CABLE
SHOE RETAINING
NAIL
SECONDARY
SHOE AND
LINING
NU247C

PRIMARY SHOE
AND LINING

FIGURE 30.8 Drum brake. *(Chrysler Corporation.)*

in the friction coefficient cause large changes in brake torque, the brake is *unstable.* It will tend to grab and squeal if the coefficient of friction increases. But if the coefficient decreases (say, because of a temperature increase), there will be a noticeable drop in capacity.

Full disk brakes are used principally for industrial machinery. They are very much like full disk clutches in construction. Indeed, they are found in *clutch-brake combination drives* where both members of the drive are full disk in construction.

Caliper disk brakes (Fig. 30.9) are now familiar components of vehicles, but they find applications in industrial equipment as well. The Chrysler brake shown uses a floating caliper. In this design, an automatic mechanism to adjust for pad wear can be incorporated easily.

Generally caliper disk brakes are not self-energizing, although they can be. An advantage of the non-self-energizing disk brake is its great stability; it is relatively insensitive to changes in the coefficient of friction.

Brake Actuation. Four principal actuation methods are shown in the classification chart of Fig. 30.7: mechanical, pneumatic (or hydraulic), electric, and automatic. Sometimes the methods are combined.

The drum brake of Fig. 30.8 and the disk brake of Fig. 30.9 are both *hydraulically operated.* Both are intended for vehicles. In industrial applications, air is often the actuating fluid. The air-tube configuration in Fig. 30.10 can be used for a *pneumatically operated* clutch or brake.

Electrically operated brakes most commonly use electromagnetic forces to actuate a full-disk-friction brake. However, a number of other designs are found (see Sec. 30.8).

Automatically operated brakes are used for both transportation and industrial equipment. Sometimes manual operation is overridden by automatic actuation. Truck brakes are available with spring actuators that engage if the air pressure is lost. The air brake, as originally conceived by George Westinghouse for railroad applications, was of this *fail-safe* design. Electrically, hydraulically, and pneumatically operated brakes can all be designed for automatic operation. Antiskid brakes for automobiles and trucks superimpose on the usual manual control an automatic control that releases braking pressure if lockup and skidding are imminent.

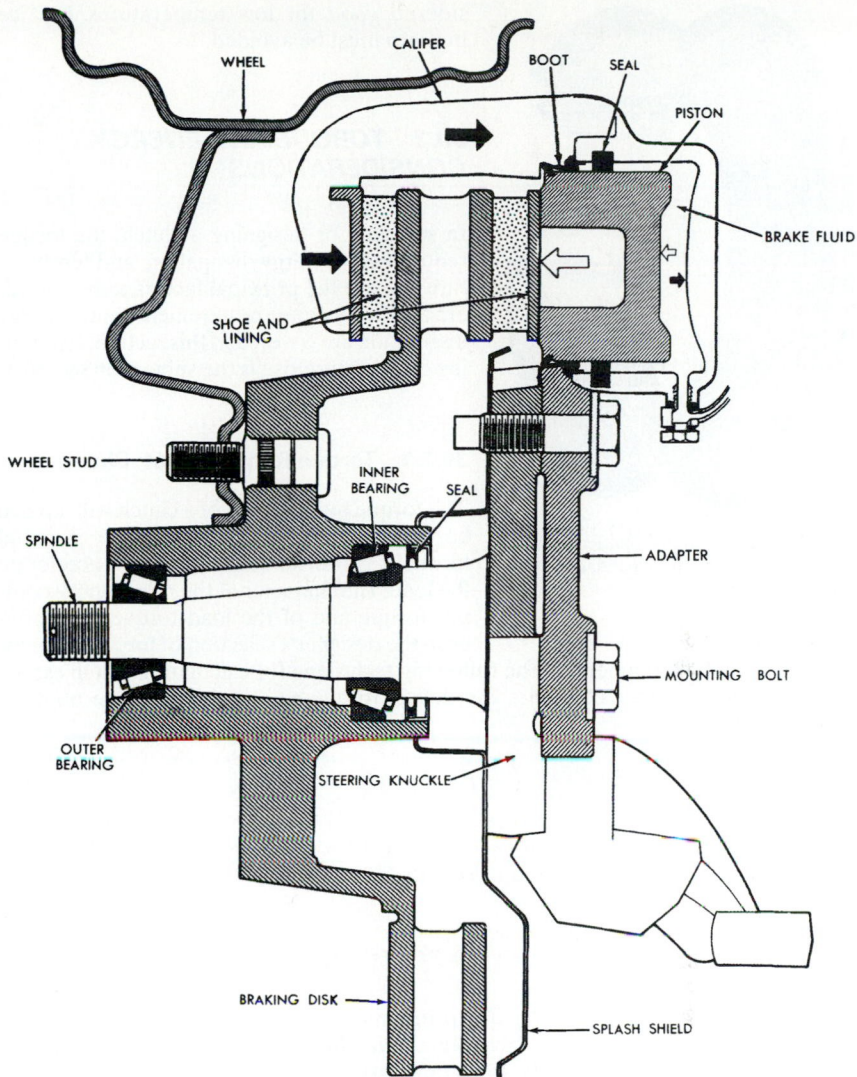

FIGURE 30.9   Automotive disk brake. *(Chrysler Corporation.)*

30.1.4   Selecting a Brake

To help narrow the choice of a brake, Table 30.2 has been provided. Some general indicators are given for performance requirements and environmental conditions. Typical applications are listed as well. The *brake factor* is the ratio of the frictional braking force developed to the actuating force.

Note that temperature considerations have been omitted from the environmental conditions in Table 30.2. For high temperatures, the capacity of all brakes listed is limited by the type of friction material. The performance of all brakes listed is con-

FIGURE 30.10 A pneumatically actuated brake using an expandable tube. *(Eaton Corporation.)*

sidered *good* for low temperatures, but ice buildup must be avoided.

30.2 TORQUE AND ENERGY CONSIDERATIONS

In selecting or designing a clutch, the torque requirement, energy dissipation, and temperature rise are the principal factors to be considered. The torque requirement and energy dissipation are covered in this section. Estimating temperature rise is the subject of Sec. 30.3.

30.2.1 Torque Requirement: Clutches

The torque requirement of a clutch will have to be substantially greater than the nominal torque it is transmitting in order to accelerate the load. The character of the prime-mover output torque and of the load torque also influence the designer's selection of torque capacity.

Gagne [30.4] recommended the following technique for calculating clutch capacity for design purposes. Calculate the design torque as a multiple of the nominal torque T:

$$T_{\text{des}} = K_S T \tag{30.1}$$

where K_S = service factor taking into account the load inertia, the character of the prime mover's output torque, and the character of the load torque. The service factor K_S is

$$K_S = (F_S^2 + F_D^2 + F_L^2 - 2)^{1/2} \tag{30.2}$$

where F_S, F_D, and F_L are the starting, prime-mover, and load factors, respectively. Recommended values for these factors are given in Tables 30.3 to 30.5. Note that if each factor is unity, the service factor is unity also.

But the service factor K_S will usually be greater than unity. Indeed, an old rule of thumb was that the clutch should be designed for a torque capacity at least twice the nominal torque.

Example 1. A multicylinder diesel engine is used to drive an electric generator in a hospital's emergency-power facility. What service factor should be used?

Solution. From Table 30.3, a reasonable selection of values for the torque factors is $F_S = 2$, $F_D = 1.5$, and $F_L = 1.0$. The corresponding service factor is

$$K_S = [2^2 + (1.5)^2 + (1.0)^2 - 2]^{1/2} = 2.29$$

TABLE 30.2 Selecting the Right Brake

| Type of brake | Performance requirements | | | Environmental conditions | | Typical applications |
	Maximum operating temperature	Brake factor	Stability	Wet and humid	Dust and dirt	
Differential band brake	Low	Very high	Very low	Unstable but still effective	Good	Winches, hoists, excavators, tractors, etc.
External drum brake (leading-trailing edge)	Low	Moderate	Moderate	Unstable if humid; poor performance if wet	Good	Mills, elevators, winders
Internal drum brake (leading-trailing edge)	Higher than external brake	Moderate	Moderate	Unstable if humid; completely ineffective if wet	Very good if sealed	Vehicles (rear axles on passenger cars)
Internal drum brake (two leading shoes)	Higher than external brake	High	Low	Unstable if humid; completely ineffective if wet	Very good if sealed	Vehicles (rear axles on passenger cars)
Internal drum brake (duo-servo)	Low	Very high	Low	Unstable if humid; completely ineffective if wet	Very good if sealed	Vehicles (rear axles on passenger cars)
Caliper disk brake	High	Low	High	Good	Poor; should be shielded	Vehicles and industrial machinery
Full disk brake	High	Low	High	Good	Poor; should be shielded	Machine tools and other industrial machinery

SOURCE: Ref. [30.6].

TABLE 30.3 Suggested Values of Torque Starting Factor F_s for Friction Clutches

Type of load	F_S
Free start; no load	1.0
Average inertia load	2.0
High inertia load	3.0

SOURCE: Ref. [30.4].

TABLE 30.4 Suggested Values of Torque Drive (Prime-Mover) Factor F_D for Friction Clutches

Type of drive	F_D
Nonpulsating, such as three-phase motors	1.0
Moderate pulsation—single-phase motors, multicylinder engines, etc.	1.5
Severe pulsation, such as a single-cylinder gas engine	2.0

SOURCE: Ref. [30.4].

TABLE 30.5 Suggested Values of Torque Load Factor F_L for Friction Clutches

Type of load	F_L
Nonpulsating—blowers, centrifugal pumps, generators under steady load, etc.	1.0
Moderate shock, such as a multicylinder pump	1.5–1.75
Severe shock—crane, shovel, single-cylinder compressor, rock crusher, etc.	2.0–3.0

SOURCE: Ref. [30.4].

30.2.2 Equivalent Inertias

Two shafts geared together and rotating at different speeds are shown in Fig. 30.11a. The inertias I_1 and I_2 are each assumed to include the corresponding shaft and gear. For design calculations, it is necessary to have an equivalent inertia for the whole system referred to a single shaft. Figure 30.11b and c shows this. In each case an equivalent inertia has been added to the shaft. So I_2' is the equivalent inertia on shaft 1 of shaft 2 and its hardware. Similarly, I_1' is the equivalent inertia on shaft 2 of shaft 1 and its hardware.

A simple way to find the equivalent inertia is to equate the kinetic energies of the actual and equivalent inertias. Thus, to find the equivalent inertia I_2' referred to shaft 1 (Fig. 30.11b), we write

$$\frac{1}{2}I_2'\omega_1^2 = \frac{1}{2}I_2\omega_2^2$$

Thus

$$I_2' = \left(\frac{\omega_2}{\omega_1}\right)^2 I_2 \qquad (30.3)$$

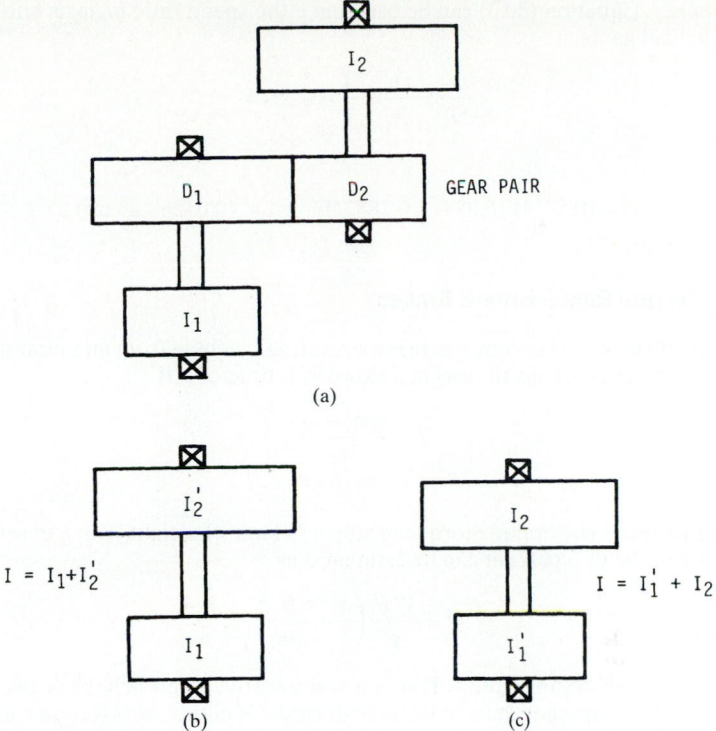

FIGURE 30.11 Equivalent inertia. I_1, I_2 = inertia of input and output shafts, respectively; I' = equivalent inertia. (*a*) Original configuration; (*b*) equivalent system referred to input shaft; (*c*) equivalent system referred to output shaft. For a more extensive treatment of equivalent inertias, see "Suggested Reading" list, Mischke.

Similarly,

$$I_1' = \left(\frac{\omega_1}{\omega_2}\right)^2 I_1 \tag{30.4}$$

In general,

$$I_i' = \left(\frac{\omega_i}{\omega_j}\right)^2 I_i = \left(\frac{n_i}{n_j}\right)^2 I_i \tag{30.5}$$

where the equivalent inertia of the *i*th shaft is referred to the *j*th shaft. Equation (30.5) can be used to reduce a machine with several shafts connected by gears or flexible connectors to a single equivalent shaft.

Example 2. For the two-shaft machine in Fig. 30.11*a*, the inertias are $I_1 = 2.88$ pound-inch-square seconds (lb · in · s²) [0.3254 kilogram-square meters (kg · m²)] and $I_2 = 0.884$ lb · in · s² (0.09988 kg · m²). The pitch diameters of the gears are $D_1 = 4$ in [0.102 meter (m)] and $D_2 = 7$ in (0.178 m). What is the equivalent inertia of shaft 2 referred to shaft 1?

Solution.   Equation (30.3) can be used once the speed ratio ω_2/ω_1 is known. For spur or helical gears,

$$\frac{\omega_2}{\omega_1} = \frac{D_1}{D_2} = \frac{4}{7} = 0.5714$$

Thus

$$I_2' = (0.5714)^2(0.884) = 0.2887 \text{ lb} \cdot \text{in} \cdot \text{s}^2 \ (0.0326 \text{ kg} \cdot \text{m}^2)$$

30.2.3   Torque Requirement: Brakes

Industrial Brakes.   The torque to bring a rotating machine from an initial speed ω_o to a lower one ω_f (perhaps to rest) in a slowdown time of t_S is

$$T = \frac{I(\omega_o - \omega_f)}{t_S} \tag{30.6}$$

Vehicle Brakes.   The braking torque to stop a vehicle of weight W at a deceleration rate a on a grade of b percent can be estimated as

$$T = \frac{WR}{g} \left(\frac{a}{g} + \frac{b}{100} \right) \tag{30.7}$$

Here R is the tire-rolling radius. This is a conservative approach; both tire-rolling resistance and air resistance have been neglected. Of course, this torque capacity T must be allocated to the several brakes in a rational way (for example, in proportion to the weight of the vehicle supported by the corresponding wheel during a panic stop).

For parking-brake capacity, simply set $a = 0$ in Eq. (30.7).

The required acceleration rate a can be determined by setting either a total stopping time t_S or a total stopping distance s:

$$a = \frac{V_o}{t_S - t_d} \tag{30.8}$$

$$a = \frac{V_o^2}{2(s - V_o t_d)} \tag{30.9}$$

In these equations, t_d is the combined delay time (about 1 s for a passenger car) for driver reaction and brake system reaction.

30.2.4   Energy Dissipation: Clutches

A simple model of a clutch connecting a prime mover and a load is shown in Fig. 30.12. The clutch capacity is T, the driving torque provided by the prime mover is T_P, and the load torque is T_L. The inertias I_P and I_L include all rotating masses on their respective sides of the clutch.

If the two sides of the clutch are initially rotating at Ω_P and Ω_L radians per second (rad/s) when the clutch is actuated, the duration of the slip period is

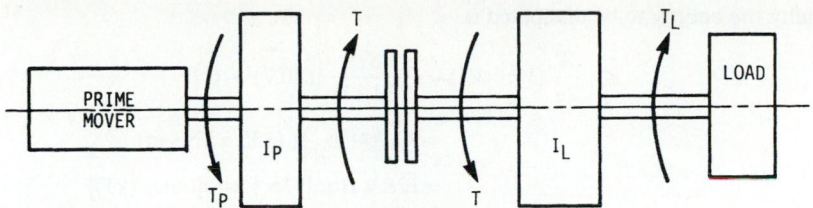

FIGURE 30.12   Abstract model of a machine using a clutch.

$$t_S = \frac{I_P I_L(\Omega_P - \Omega_L)}{T(I_P + I_L) - (I_L T_P + I_P T_L)} \tag{30.10}$$

The rate at which energy is dissipated during the slip period is, at t s from the beginning,

$$q(t) = T\left[\frac{I_L T_P + I_P T_L - T(I_P + I_L)t}{I_P I_L + \Omega_P - \Omega_L}\right] \tag{30.11}$$

And the total energy dissipated in one actuation operation is

$$E = \frac{T(\Omega_P - \Omega_L)^2 I_P I_L}{2T(I_P + I_L) - (I_L T_P + I_P T_L)} \tag{30.12}$$

30.2.5   Energy Dissipation: Brakes

Vehicle Brakes.   When a vehicle of weight W is slowed from an initial velocity V_o to a final velocity V_f, the heat energy that the brakes must dissipate is equal to the change of kinetic energy E:

$$E = \frac{W}{2g}(V_o^2 - V_f^2) \tag{30.13}$$

In dealing with individual brakes, let W be that portion of the vehicle's weight for which the brake is responsible.

Example 3.   A sports car weighing 3185 lb [14.2 kilonewtons (kN)] has 62 percent of its weight on the front axle during an emergency stop. What energy must each of the front-wheel brakes dissipate in braking from 55 miles per hour (mph) [88 kilometers per hour (km/h)] to rest? Local acceleration of gravity is 32.17 feet per second (ft/s) (9.81 m/s).
   Solution.   Each front brake is responsible for a weight of

$$W = 0.5(0.62)(3185) = 987 \text{ lb } (4.39 \text{ kN})$$

The initial velocity is

$$V_o = 55\left(\frac{88}{60}\right) = 80.7 \text{ ft/s } (24.6 \text{ m/s})$$

Finally, the energy to be dissipated is

$$E = \frac{W}{2g}(V_o^2 - V_f^2) = \frac{987}{2(32.17)}[(80.7)^2 - 0^2]$$

$$= 99\,900 \text{ lb} \cdot \text{ft } (135.5 \text{ kN} \cdot \text{m})$$

$$= 128.4 \text{ Btu } [135.5 \text{ kilojoules (kJ)}]$$

Industrial Brakes.   The approach is the same as for vehicular brakes. The heat energy the brake must dissipate equals the change in kinetic energy of the rotating machine:

$$E = \frac{I}{2}(\omega_o{}^2 - \omega_f{}^2) \tag{30.14}$$

where, with n in rev/min,

$$\omega = \frac{2\pi n}{60} \tag{30.15}$$

   In many industrial applications, the brakes are applied frequently. The average rate of heat dissipation is, for S stops per hour,

$$H_{av} = \frac{ES}{3600} \tag{30.16}$$

Tensioning Applications.   In tensioning applications, a continuous application of the brake is required, for example, in unwinding a roll of aluminum foil. The maximum torque occurs at the maximum roll diameter D_{max}. It is

$$T_{max} = \frac{D_{max}F_t}{2} \tag{30.17}$$

The tension F_t is, for material width b and thickness t,

$$F_t = wtb \tag{30.18}$$

Typical values of tension per unit width and per unit thickness for a few materials are given in Table 30.6.

TABLE 30.6   Tension Data for Typical Materials

Material	Unit tension, lb/mil per inch of web width
Aluminum foil	1.00
Cellophane	0.75
Mylar	0.60
Polystyrene	1.00

SOURCE: The Carlson Company, Inc., Wichita, Kansas.

The rate at which heat is generated by the brake friction is

$$H_{av} = F_t V_w \qquad (30.19)$$

Example 4. A printing press is to print on Mylar 0.002 in [0.051 millimeter (mm)] thick. The web velocity is 4000 ft/min (20.3 m/s). The maximum roll diameter is 55 in (1.4 m). The web is 54 in wide. Find the necessary braking torque and the rate at which heat is generated by braking.

   Solution.   For Mylar the unit tension is 0.60 lb/mil per inch (379.2 kN/mm per meter). So the web tension is

$$F = wtb = 0.60(2)(54) = 64.8 \text{ lb } (288 \text{ N})$$

The maximum brake torque is

$$T_{max} = \frac{D_{max} F_t}{2} = \frac{55(64.8)}{2} = 1782 \text{ in} \cdot \text{lb } (201 \text{ N} \cdot \text{m})$$

The rate at which the brake must dissipate heat is, by Eq. (30.19),

$$H_{av} = F_t V_w$$

The web velocity is

$$V_w = \frac{4000}{60} = 66.67 \text{ ft/s } (20.3 \text{ m/s})$$

So H_{av} is

$$H_{av} = 64.8(66.67) = 4320 \text{ ft} \cdot \text{lb/s } [5855 \text{ watts (W)}]$$

$$= 5.55 \text{ Btu/s}$$

30.3 *TEMPERATURE CONSIDERATIONS*

30.3.1 Intermittent Operation: Clutches and Brakes

The temperature rise can be estimated as

$$\Delta T = \frac{E}{Cm} \qquad (30.20)$$

where m [pounds mass (lbm) or kilograms (kg)] = mass of the parts adjacent to the friction surfaces. The specific heat C for steel or cast iron is about 0.12 Btu/(lbm · °F) [500 J/(kg · °C)].

30.3.2 Frequent Operation: Caliper Disk Brakes

The average rate at which heat must be dissipated can be calculated by Eq. (30.16). The disk is capable of dissipating heat by a combination of convection and radiation.

And the convection-heat transfer is sensitive to the velocity of air moving over the disk. The rate at which the disk can dissipate heat is

$$H_{\text{diss}} = hA(T_d - T_a) \tag{30.21}$$

The overall heat transfer coefficient h is

$$h = h_r + f_v h_c \tag{30.22}$$

The heat transfer coefficients for radiation h_r and convection h_c are plotted in Fig. 30.13 against the temperature rise of the disk above the surrounding air. The ventilation factor f_v is plotted against the velocity of the moving air in Fig. 30.14.

Example 5.   An industrial caliper brake is used 19 times per hour on average to stop a machine with a rotating inertia of $I = 328$ lb · in · s^2 (37.06 kg · m^2) from a speed of 315 rev/min. The mean air velocity over the disk will be 30 ft/s (9.14 m/s). What minimum exposed area on the disk is needed to limit the disk's temperature rise to 200°F (111°C)?

Solution.   From Figs. 30.13 and 30.14, $h_r = 3.1 \times 10^{-6}$ Btu/(in^2 · s · °F), $h_c = 2.0 \times 10^{-6}$ Btu/(in^2 · s · °F), and $f_v = 5.25$. The overall heat transfer coefficient is

$$h = 3.1 \times 10^{-6} + 5.25(2.0 \times 10^{-6}) = 13.6 \times 10^{-6} \text{ Btu/(in}^2 \cdot \text{s} \cdot \text{°F)}$$

The energy the brake must dissipate per stop is, by Eq. (30.14),

$$E = \frac{I}{2}(\omega_o^2 - \omega_f^2)$$

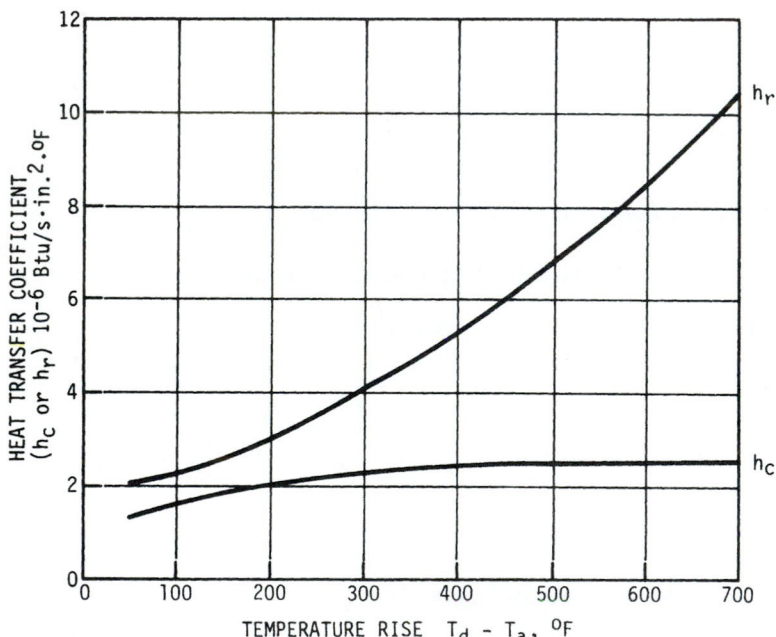

FIGURE 30.13   Heat transfer coefficients in still air. *(Tol-o-matic.)*

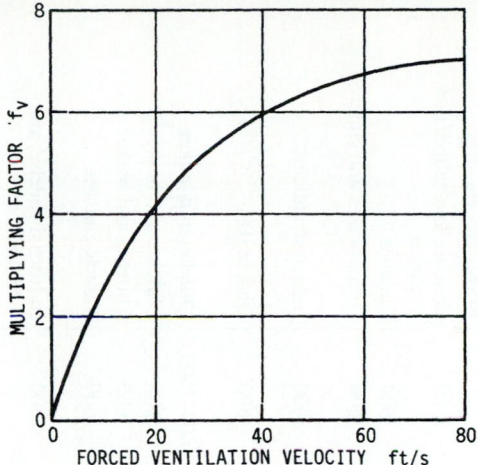

FIGURE 30.14   Ventilation factors. *(Tol-o-matic.)*

Here

$$\omega_o = \frac{2\pi}{60}(315) = 33 \text{ rad/s} \qquad \omega_f = 0$$

Therefore,

$$E = \frac{328}{2}(33^2) = 178.6 \times 10^3 \text{ lb} \cdot \text{in}$$

$$= 19.1 \text{ Btu } (20.150 \text{ kJ})$$

The average rate of energy generation is found by Eq. (30.16):

$$H_{av} = \frac{ES}{3600} = \frac{19.1(19)}{3600} = 0.101 \text{ Btu/s } (107 \text{ W})$$

The disk area needed can be calculated by using Eq. (30.21) and setting $H_{av} = H_{diss} = 0.101$ Btu/s. Thus

$$A = \frac{H_{diss}}{h(T_d - T_a)} = \frac{0.101}{(13.6 \times 10^{-6})(200)} = 37.1 \text{ in}^2 \ (0.0239 \text{ m}^2)$$

30.4   FRICTION MATERIALS

To help in preliminary design, Tables 30.7 and 30.8 have been compiled, principally from data supplied by British and U.S. manufacturers of friction materials. Although these data are representative, they are hardly exhaustive. And they should be used

TABLE 30.7 Characteristics of Friction Materials for Brakes and Clutches

Material	Friction coefficient μ	Maximum pressure p_{max}, psi	Maximum temperature		Maximum velocity V_{max}, ft/min	Applications
			Instantaneous, °F	Continuous, °F		
Cermet	0.32	150	1500	750	3600	Brakes and clutches
Sintered metal (dry)	0.29–0.33	300–400	930–1020	570–660	3600	Clutches and caliper disk brakes
Sintered metal (wet)	0.06–0.08	500	930	570	3600	Clutches
Rigid molded asbestos (dry)	0.35–0.41	100	660–750	350	3600	Drum brakes and clutches
Rigid molded asbestos (wet)	0.06	300	660	350	3600	Industrial clutches
Rigid molded asbestos pads	0.31–0.49	750	930–1380	440–660	4800	Disk brakes
Rigid molded nonasbestos	0.33–0.63	100–150		500–750	4800–7500	Clutches and brakes
Semirigid molded asbestos	0.37–0.41	100	660	300	3600	Clutches and brakes
Flexible molded asbestos	0.39–0.45	100	660–750	300–350	3600	Clutches and brakes
Wound asbestos yarn and wire	0.38	100	660	300	3600	Vehicle clutches
Woven asbestos yarn and wire	0.38	100	500	260	3600	Industrial clutches and brakes
Woven cotton	0.47	100	230	170	3600	Industrial clutches and brakes
Resilient paper (wet)	0.09–0.15	400	300		$PV < 500\,000$ psi·ft/min	Clutches and transmission bands

SOURCES: Ferodo Ltd, Chapel-en-le-frith, England; Scan-pac, Mequon, Wisc.; Raybestos, New York, N.Y. and Stratford, Conn.; Gatke Corp., Chicago, Ill.; General Metals Powder Co, Akron, Ohio; D. A. B. Industries, Troy, Mich.; Friction Products Co., Medina, Ohio.

TABLE 30.8   Area of Friction Material Required for a Given Average Braking Power

Duty cycle	Typical applications	Ratio of area to average braking power, $in^2/(Btu/s)$		
		Band and drum brakes	Plate disk brakes	Caliper disk brakes
Infrequent	Emergency brakes	0.85	2.8	0.28
Intermittent	Elevators, cranes, and winches	2.8	7.1	0.70
Heavy-duty	Excavators, presses	5.6–6.9	13.6	1.41

SOURCES: Refs. [30.6], Sec. A51, and [30.7].

for preliminary design estimates only. A friction materials manufacturer should be consulted both to learn of additional options and to get more authoritative data.

Although Table 30.7 lists maximum recommended values for contact pressure and rubbing velocity, it is not very likely that you can go the limit on both parameters at once. And a careful distinction must be made between the maximum temperature permissible for a short time and the safe temperature level for continuous operation. The temperature limit for continuous operation is much lower than that for a brief temperature peak.

Preliminary design of brakes is aided by calculating the lining area needed for the average rate at which energy has to be dissipated by the brakes (*braking power*). Table 30.8 lists values that are typical of modern design practice. Again, after using these data to make some preliminary design estimates, you will need to contact the manufacturers of the friction materials before making final design decisions.

30.5   TORQUE AND FORCE ANALYSIS OF RIM CLUTCHES AND BRAKES

30.5.1   Long-Shoe   Rim Brake

One shoe of an internal expanding rim brake is shown in Fig. 30.15. Usually there is a second shoe as well. The shoe is pivoted about the fixed point A. It is actuated by a force F which can be provided in a number of ways: mechanically, hydraulically, pneumatically, electromagnetically, or by some combination of these.

The forces on the shoe include the actuating force F, a reaction force R at the pivot, the distributed normal force, and the distributed friction force, the latter two exerted by the drum on the shoe.

For purposes of analysis, the distributed normal and frictional forces on the shoe can be replaced by a resultant normal force P and a resultant frictional force fP. Use of these fictional concentrated forces simplifies the analysis. There is one odd consequence, however. The resultant frictional force fP has to be regarded as acting beyond the surface of the shoe at some point C, the *center of pressure*. Figure 30.16 shows the shoe subjected to this equivalent force system.

Pressure Distribution along Lining.   A first step in developing an equation for the torque capacity of the shoe is to adopt a model for the pressure distribution along

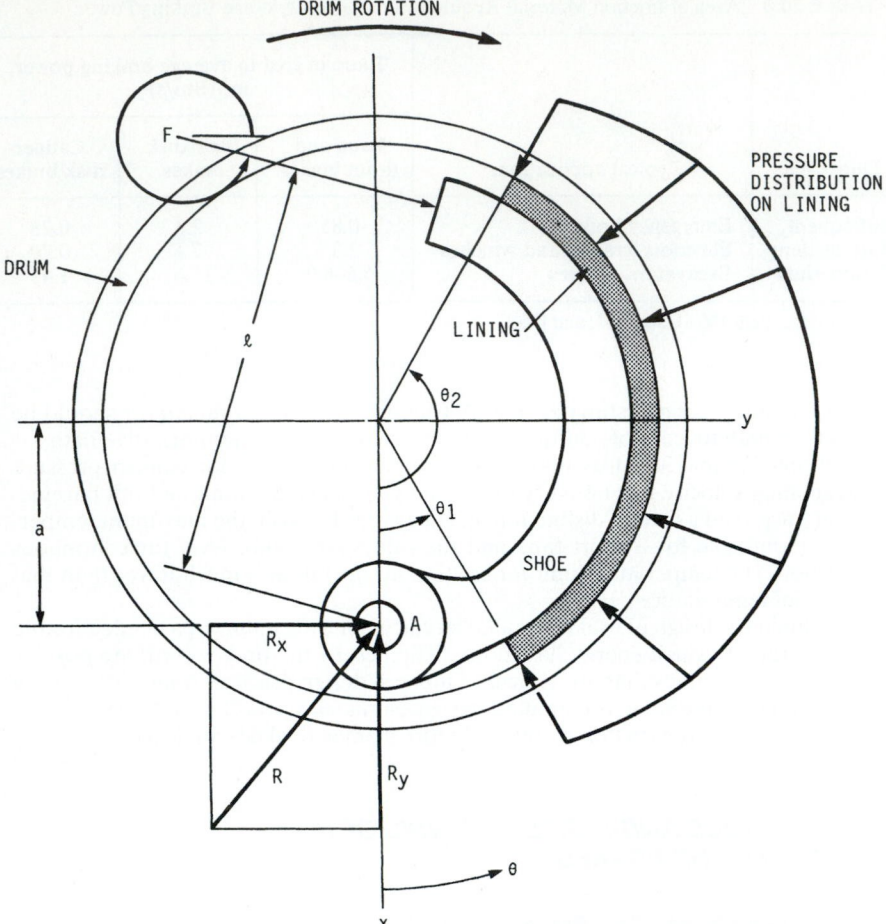

FIGURE 30.15   Forces on an internal long shoe.

the lining of the shoe. Assuming that the lining is elastic and of uniform thickness, but that the drum is perfectly rigid, the expression for the normal pressure p on the lining at any angular position θ is

$$p = \frac{p_{max} \sin \theta}{(\sin \theta)_{max}} \tag{30.23}$$

In the usual case, the shoe straddles the $\theta = 90°$ position and $(\sin \theta)_{max} = 1$.

The variation of contact pressure along the lining is significant. To assume a uniform pressure is to oversimplify. For example, for a shoe extending from $\theta = 30°$ to $\theta = 165°$, the pressure varies from $0.5p_{max}$ at $30°$ to p_{max} at $90°$, to $0.26p_{max}$ at $165°$.

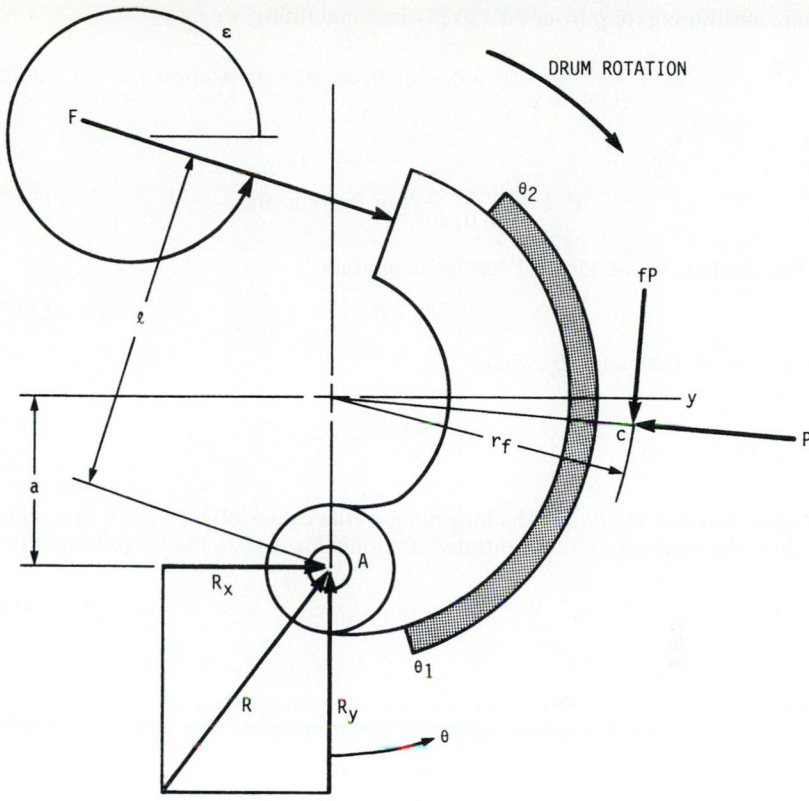

FIGURE 30.16 Equivalent force system on a long internal shoe.

Torque Capacity. With a model for contact pressure variation in hand, the friction torque exerted by the shoe on the drum can be found by a simple integration:

$$T = \int_{\theta_1}^{\theta_2} fpbr^2 \, d\theta \qquad (30.24)$$

where b = width of lining. Substituting for p from Eq. (30.23) and integrating, we get

$$T = \frac{p_{max}}{(\sin \theta)_{max}} fbr^2 (\cos \theta_1 - \cos \theta_2) \qquad (30.25)$$

Normal Force. To determine the actuating force and the pin reaction, it is necessary first to find the normal force P. The components of P are

$$P_x = \int_{\theta_1}^{\theta_2} pbr \, d\theta \sin \theta \qquad (30.26)$$

and

$$P_y = \int_{\theta_1}^{\theta_2} pbr \, d\theta \cos \theta \qquad (30.27)$$

Again, substituting for p from Eq. (30.23) and integrating, we find

$$P_x = \frac{brp_{max}}{2(\sin \theta)_{max}} (\theta_2 - \theta_1 + \sin \theta_1 \cos \theta_1 - \sin \theta_2 \cos \theta_2) \qquad (30.28)$$

and

$$P_y = \frac{brp_{max}}{2(\sin \theta)_{max}} (\cos^2 \theta_1 - \cos^2 \theta_2) \qquad (30.29)$$

The resultant normal force P has the magnitude

$$P = (P_x^2 + P_y^2)^{1/2} \qquad (30.30)$$

and is located at the angle θ_p, where

$$\theta_p = \tan^{-1} \frac{P_x}{P_y} \qquad (30.31)$$

Effective Friction Radius.   The location r_f of the center of pressure C is found by equating the moment of a concentrated frictional force fP to the torque capacity T:

$$T = fPr_f \quad \text{or} \quad r_f = \frac{T}{fP} \qquad (30.32)$$

Brake-Shoe Moments.   The last basic task is to find a relation among actuating force F, normal force P, and the equivalent friction force fP. The moments about the pivot point A are

$$M_a - M_n + M_f = 0 \qquad (30.33)$$

where

$$M_a = F\ell \qquad (30.34)$$

$$M_n = Pa \sin \theta_p \qquad (30.35)$$

$$M_f = P(r_f - a \cos \theta_p) \qquad (30.36)$$

Self-energizing Shoes.   The brake shoe in Fig. 30.16 is said to be *self-energizing*, for the frictional force fP exerts a clockwise moment about point A, thus assisting the actuating force. On vehicle brakes, this would also be called a *leading shoe*. Suppose a second shoe, a *trailing shoe*, were placed to the left of the one shown in Fig. 30.16. For this shoe, the frictional force would exert a counterclockwise moment and oppose the action of the actuating force. Equation (30.33) can be written in a form general enough to apply to both shoes and to external shoes as well:

$$M_a - M_n \pm M_f = 0 \qquad (30.37)$$

Burr [30.2], p. 84, proposes this simple rule for using Eq. (30.37): "If to seat a shoe more firmly against the drum it would have to be rotated in the same sense as the drum's rotation, use the positive sign for the M_f term. Otherwise, use the negative sign."

Pin Reaction. At this point in the analysis, the designer should sketch a free-body diagram of each shoe, showing the components of the actuating force F, the normal force P, and the friction force fP. Then the components of the pin reaction can be found by setting to zero the sum of the force components in each direction (x and y).

Design. The design challenge is to produce a brake with a required torque capacity T. A scale layout will suggest tentative values for the dimensions θ_1, θ_2, a, ℓ, and ε. From the lining manufacturer we learn the upper limit on maximum contact pressure p_{max} and the expected range for values of the frictional coefficient f. The designer must then determine values for lining width b and the actuating force F for each shoe.

Since the friction force assists in seating the shoe for a self-energizing shoe but opposes the actuating moment for a self-deenergizing shoe, a much larger actuating force would be needed to provide as large a contact pressure for a trailing shoe as for a leading shoe. Or if, as is often the case, the same actuating force is used for each shoe, a smaller contact pressure and a smaller torque capacity are achieved for the trailing shoe.

The lining manufacturer will usually specify a likely range of values for the coefficient of friction. It is wise to use a low value in estimating the torque capacity of the shoe.

In checking the design, make sure that a self-energizing shoe is not, in fact, self-locking. For a *self-locking* shoe, the required actuating force is zero or negative. That is, the lightest touch would cause the brake to seize. A brake is self-locking when

$$M_n \le M_f \tag{30.38}$$

As a design rule, make sure that self-locking could occur only if the coefficient of friction were 25 to 50 percent higher than the maximum value cited by the lining manufacturer.

Example 6. Figure 30.17 shows a preliminary layout of an automotive brake with one leading shoe and one trailing shoe. The contact pressure on the lining shall not exceed 1000 kilopascals (kPa). The lining manufacturer lists the coefficient of friction as 0.34 ± 0.02. The brake must be able to provide a braking torque of 550 N · m. Two basic design decisions have already been made: The same actuating force is used on each shoe, and the lining width is the same for each.

Check dimension a to make sure that self-locking will not occur. Determine the lining width b, the actuating force F, and the maximum contact pressure p_{max} for each shoe.

Solution. One way to proceed is to express the braking torque T, the normal moment M_n, and the frictional moment M_f in terms of lining width b and maximum contact pressure p_{max}. Then the design can be completed by equating the actuating force for the two shoes, setting the sum of the braking torques to 550 N · m, and selecting the lining width b so that the maximum contact pressure is within bounds.

1. The dimension a is

$$a = (83^2 + 25^2)^{1/2} = 86.7 \text{ mm} = 0.0867 \text{ m}$$

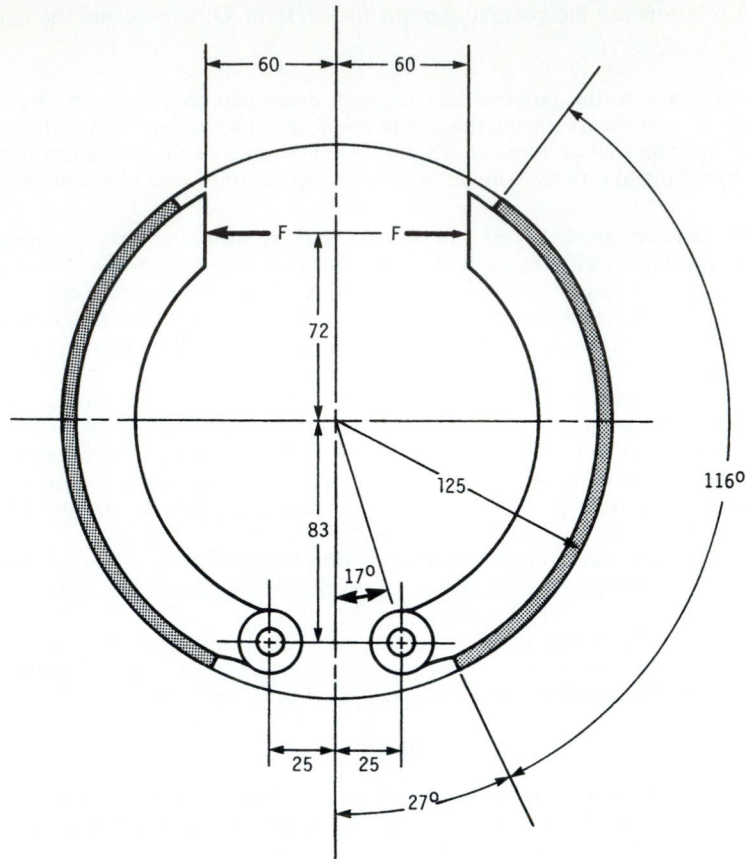

FIGURE 30.17   A leading-shoe trailing-shoe automotive brake.

2. For either shoe the braking torque can be written in terms of the coefficient of friction, the lining width, and the maximum contact pressure. From Eq. (30.25),

$$T = \frac{p_{max}}{(\sin \theta)_{max}} fbr^2 (\cos \theta_1 - \cos \theta_2)$$

$$= \frac{p_{max}}{1} fb(0.125)^2 (\cos 10° - \cos 126°)$$

$$= 0.024\ 57 fbp_{max}$$

The numerical values of T are different for the two shoes, since p_{max} differs.

3. Now the normal force P must be calculated. Equations (30.28), (30.29), (30.30), and (30.31) are used:

$$P_x = \frac{brp_{max}}{2(\sin\theta)_{max}} (\theta_2 - \theta_1 + \sin\theta_1 \cos\theta_1 - \sin\theta_2 \cos\theta_2)$$

$$= \frac{b(0.125)p_{max}}{2(1)} \left(\frac{126° - 10°}{57.296°} + \sin 10° \cos 10° - \sin 126° \cos 126° \right)$$

$$= 0.1671 bp_{max}$$

$$P_y = \frac{brp_{max}}{2(\sin\theta)_{max}} (\cos^2\theta_1 - \cos^2\theta_2)$$

$$= \frac{b(0.125)p_{max}}{2(1)} (\cos^2 10° - \cos^2 126°) = 0.0390 bp_{max}$$

$$P = (P_x^2 + P_y^2)^{1/2} = 0.1716 bp_{max}$$

$$\theta_p = \tan^{-1}\frac{P_x}{P_y} = \tan^{-1}\frac{0.1671}{0.0390} = 76.85°$$

4. The effective friction radius r_f is, by Eq. (30.32),

$$r_f = \frac{T}{fP} = \frac{0.024\,57 fbp_{max}}{f(0.1716 bp_{max})} = 0.1432 \text{ m} = 143.2 \text{ mm}$$

5. The moments about pivot point A are found by Eqs. (30.34) to (30.36):

$$M_a = F\ell = F(0.072 + 0.083) = 0.155F$$

$$M_n = Pa \sin\theta_p = 0.1716 bp_{max}(0.0867)(\sin 76.85°) = 0.014\,49 bp_{max}$$

$$M_f = fP(r_f - a\cos\theta_p)$$

$$= f(0.1716 bp_{max})(0.1432 - 0.0867 \cos 76.85°)$$

$$= 0.021\,19 fbp_{max}$$

6. For the leading (self-energizing) shoe, the proper form of Eq. (30.37) is

$$M_a - M_n + M_f = 0$$

Therefore

$$0.155F^l - 0.014\,49 bp_{max}^l + 0.021\,19 fbp_{max}^l = 0$$

The superscript l has been used to designate the leading shoe.

7. For the trailing shoe, Eq. (30.37) has the form

$$M_a - M_n - M_f = 0$$

Thus

$$0.155F^t - 0.014\,49 bp_{max}^t - 0.021\,19 fbp_{max}^t = 0$$

8. Since the same actuating force is used for each shoe, $F^l = F^t$. After substituting from the two moment equations, we obtain

$$\frac{0.014\,49bp^l_{max} - 0.021\,19fbp^l_{max}}{0.155} = \frac{0.014\,49bp^t_{max} + 0.021\,19fbp^t_{max}}{0.155}$$

After cancellation and substitution of $f = 0.32$ (the most pessimistic assumption),

$$p^l_{max} = 2.759p^t_{max}$$

9. The torque capacities of the shoe must sum to 550 N · m:

$$T^l + T^t = 550$$

$$(0.024\,57)(0.32)b(p^l_{max} + p^t_{max}) = 550$$

Then

$$b(p^l_{max} + p^t_{max}) = 69.95 \text{ kN} \cdot \text{m}$$

10. Enough information has been accumulated to begin to specify the design. Since $p^l_{max} = 1000$ kPa, $p^t_{max} = 1000/2.759 = 362.4$ kPa. Now the distance b can be found. Since

$$b(p^l_{max} + p^t_{max}) = 69.95$$

we have

$$b = \frac{69.95 \times 10^3}{(1000 + 362.4)(10^3)} = 0.0513 \text{ m, or about 51 mm}$$

11. The actuating force F ($F^l = F^t$) can be found from either shoe's moment equation. For the leading shoe,

$$0.155F^l - 0.014\,49bp^l_{max} + 0.021\,19fbp^l_{max} = 0$$

Substituting values of b, f, and p^l_{max} gives $F = 2.55$ kN. Thus $F = F^l = F^t = 2.55$ kN.

12. Now we determine whether the leading shoe is self-locking or too close to self-locking for safety. The moment equation used above in step 11 is used again, but this time with an f value 50 percent higher than the maximum value cited by the lining manufacturer. Use

$$f = 1.5(0.34 + 0.02) = 0.54$$

Then by substituting into the moment equation, the corresponding value of F can be found:

$$0.155F^l - (0.014\,49)(0.0513)(1000) + (0.021\,19)(0.54)(0.0513)(1000) = 0$$

Solving gives $F^l = F^t = 1.001$ kN. Since a large positive force is required to activate the leading shoe even for this very high coefficient of friction, the brake is in no danger of self-locking.

30.5.2 Centrifugal Clutches

The simple centrifugal clutch shown in Fig. 30.18 has a number of shoes which can move radially and against the drum as the input shaft speed increases. A garter spring regulates the engagement speed. At engagement speed the weights contact the drum's inner surface and begin to drive it, and the attached pulley, by means of friction to bring it up to speed.

Design Equations. The normal force between each shoe and the inner circumference of the drum is principally due to centrifugal force. However, the garter spring exerts some inward force. The net normal force [30.1] is

$$F_n = \frac{mr\omega^2}{g_c} - 2S \cos\left(90° - \frac{180°}{N_s}\right) \qquad (30.39)$$

Engagement occurs at the shaft speed ω_e when $F = 0$. The proper initial tension S for the garter spring is found by setting $\omega = \omega_e$ and $F = 0$ in Eq. (30.39). The engagement speed is selected by the designer (for example, about 70 percent of running speed). The required value for S is

$$S = \frac{mr\omega_e^2}{2g_c \cos\left(90° - 180°/N_s\right)} \qquad (30.40)$$

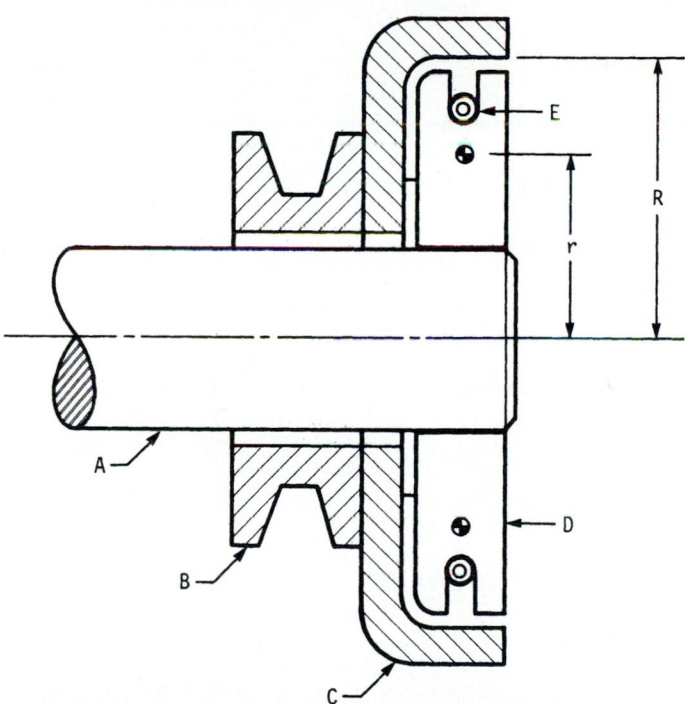

FIGURE 30.18 Free-shoe centrifugal clutch with garter spring to regulate engagement speed. *A*, input shaft; *B*, output pulley; *C*, drum; *D*, weights; *E*, garter spring.

The torque capacity at any shaft speed ω is

$$T = fF_n RN_s \tag{30.41}$$

Example 7. A centrifugal clutch is being considered for an application where the running speed is 3000 rev/min and engagement is to begin at 1000 rev/min. It is planned to use four shoes, each with a mass of 140 grams (g). A coefficient of friction $f = 0.3$ can be achieved. The inner diameter of the drum is 75 mm, and the radius R to the center of gravity of each shoe is 25 mm. (1) What should be the initial tension of the garter spring? (2) What is the normal force on each shoe at running speed? (3) What is the torque capacity at running speed? (4) What is the power capacity at running speed if a service factor of 2.25 is required?
  Solution.

1. The initial tension of the garter spring is found from Eq. (30.40):

$$S = \frac{0.140(0.025)[1000(2\pi/60)]^2}{2(1)\cos(90° - 180°/4)} = 27.1\ \text{N}$$

2. The normal force on each shoe at 3000 rev/min is, by Eq. (30.39),

$$F_n = \frac{(0.140)(0.025)[(2\pi/60)(3000)]^2}{1} - 2(27.1)\cos(90° - 180°/4)$$

$$= 307.1\ \text{N}$$

3. The torque capacity at running speed is

$$T = fF_n RN_s = (0.3)(307.1)\left(\frac{0.075}{2}\right)(4) = 13.8\ \text{N} \cdot \text{m}$$

4. The power capacity uncorrected for service factor is

$$P = \frac{Tn}{9550} = \frac{13.8(3000)}{9550} = 4.34\ \text{kW}$$

  Correcting for service factor, we see the power rating is

$$P_{\text{rating}} = \frac{P}{K_S} = \frac{4.34}{2.25} = 1.93\ \text{kW}$$

30.6 *BAND AND CONE BRAKES AND CLUTCHES*

30.6.1 Band Brakes

A typical design for a band brake is shown in Fig. 30.19. A flexible metal band lined with a friction material contacts the drum over an angle of wrap θ and exerts a braking torque T on the drum. This particular design is self-energizing, since the moment exerted on the lever by force P_1 assists in actuating the brake.

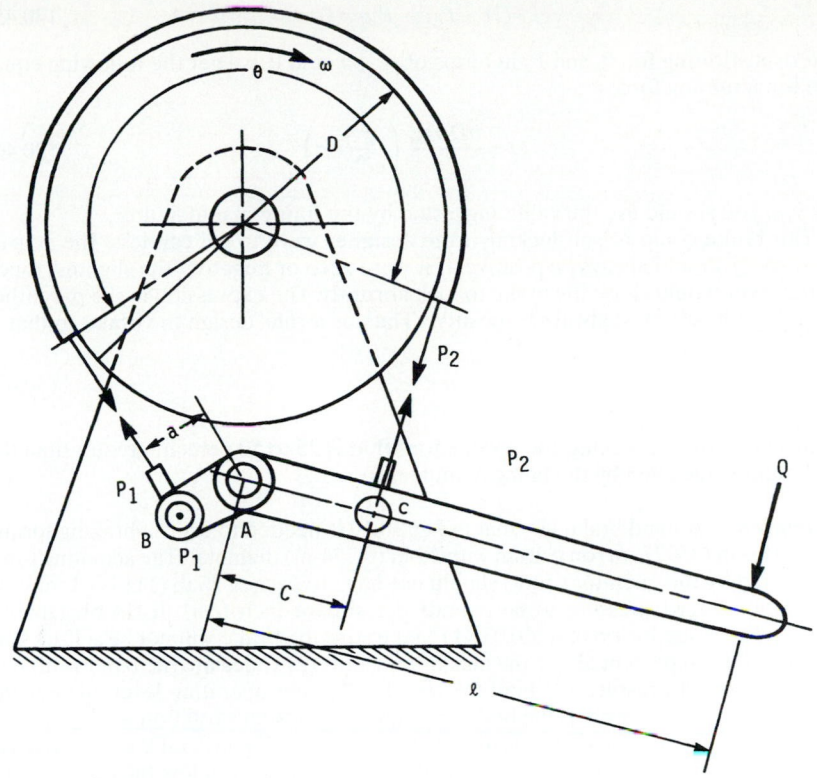

FIGURE 30.19   Forces on a band brake; b = bandwidth.

Four basic relationships are needed for analysis or design. For a band wrapped around a drum, the ratio of tensions is

$$\frac{P_1}{P_2} = e^{f\theta} \tag{30.42}$$

where the notation is indicated in Fig. 30.19. The net torque exerted on the drum by the band is

$$T = (P_1 - P_2)\left(\frac{D}{2}\right) \tag{30.43}$$

The maximum contact pressure between the band and the drum occurs at the more taut P_1 end of the band:

$$p_{\max} = \frac{2P_1}{bD} \tag{30.44}$$

Finally, it is necessary to sum moments on the lever about pivot A to get the relationship involving the actuating force Q.

$$Q\ell - P_2c + P_1a = 0 \tag{30.45}$$

After substituting for P_1 and P_2 in terms of p_{max}, f, and θ, we get the following equation for actuating force:

$$Q = \frac{bDp_{max}}{2\ell}\left(\frac{c}{e^{f\theta}} - a\right) \tag{30.46}$$

For p_{max} you should use the value suggested by the lining manufacturer.

This brake could be self-locking if the designer were to get careless. The actuating force Q should always be positive. If it were zero or negative, the slightest touch on the lever would cause the brake to lock abruptly. The expression in the parentheses in Eq. (30.46) must always be positive. Thus, as a rule, design the brake so that

$$a < \frac{c}{e^{f\theta}} \tag{30.47}$$

In checking for self-locking, use a value for f that is 25 to 50 percent greater than the maximum value cited by the lining manufacturer.

Example 8. A band brake like that in Fig. 30.19 is needed to exert a braking torque of 3100 lb · in (350 N · m) on a drum with 10-in (0.254-m) diameter. The actuating force Q (exerted by the operator's foot) should not have to exceed 25 lb (111 N). Limit the maximum contact pressure to 60 pounds per square inch (psi) [0.414 megapascal (MPa)]. The value for f is 0.31 ± 0.03. (1) Make sure the brake will not be self-locking for an f value 30 percent above the maximum value. (2) Calculate the bandwidth b to limit the contact pressure. (3) Find the length ℓ for the operating lever. (4) For the same actuating force, what is the braking torque if the drum's rotation is reversed?

Solution. A scale layout indicates that $a = 1$ in (0.0254 m) and $\theta = 200°$ are feasible values when the lever's pivot point A is placed directly below the center of the drum. Then $c = 5$ in (0.127 m), corresponding to the drum's radius.

1. First make sure that the brake will not be self-locking. Use $f = 1.3(0.31 + 0.03) = 0.442$ and $\theta = 200°/57.296 = 3.491$ rad. Then, from Eq. (30.47),

$$a < \frac{5}{0.442(3.491)} \quad \text{or} \quad a < 1.069 \text{ in}$$

So the dimension $a = 1$ in will do nicely.

2. Now select a bandwidth b so that p_{max} does not exceed 60 psi (0.414 MPa). By Eqs. (30.42) and (30.43),

$$\frac{P_1}{P_2} = e^{0.28(3.491)} = 2.658$$

$$3100 = (P_1 - P_2)\left(\frac{10}{2}\right)$$

Solving these for P_1 and P_2 gives $P_1 = 994$ lb (4420 N) and $P_2 = 374$ lb (1664 N). So, by Eq. (30.44), the bandwidth is

$$b = \frac{2P_1}{p_{max}D} = \frac{2(994)}{60(10)} = 3.31 \text{ in (say 3.5 in)}$$

3. Next the moment arm length ℓ for the actuating force Q must be found from Eq. (30.45):

$$\ell = \frac{P_2 c - P_1 a}{Q} = \frac{374(5) - 994(1)}{25} = 35 \text{ in}$$

4. Finally, the torque capacity of the brake for the same actuating force, but with the drum rotating counterclockwise, can be evaluated. Since the drum turns counterclockwise, the braking torque must be clockwise. Thus, forces P_1 and P_2 are interchanged. The larger force P_1 is applied at point C, and the force P_2 is applied at point B. Equation (30.45), suitably rewritten, is

$$P_1 c - P_2 a - Q\ell = 0$$

But Eqs. (30.42) and (30.43) do not need to be changed:

$$P_1(5) - P_2(1) - 25(35) = 0$$

or

$$5P_1 - P_2 = 875$$

$$\frac{P_1}{P_2} = e^{f\theta} = 2.658$$

Then $P_1 = 189$ lb and $P_2 = 70$ lb. And the net braking torque is

$$T = (P_1 - P_2)\left(\frac{D}{2}\right) = (189 - 70)\left(\frac{10}{2}\right) = 595 \text{ lb} \cdot \text{in}$$

This is considerably less than the 3100-lb · in capacity for clockwise rotation of the drum.

30.6.2 Cone Brakes and Clutches

Two mating cones kept in contact by an axial force can be used as a clutch, as in Fig. 30.20, or as a brake, as in Fig. 30.21. A small cone angle α produces a wedging action, and a large torque capacity is achieved for a small actuating force. But if the cone angle is too small, it becomes difficult to disengage the cones. A cone angle of 10° to 15° is a reasonable compromise.

Basic Relationships. A cone is shown schematically in Fig. 30.22 with three elementary quantities indicated: area dA, normal contact force dP, and actuating force dF. From these it follows that the elementary torque dT is

$$dT = fr\, dP = frp\frac{2\pi r\, dr}{\sin \alpha} = \frac{2\pi fpr^2\, dr}{\sin \alpha} \qquad (30.48)$$

Similarly, the elementary actuating force dF is

$$dF = dP \sin \alpha = 2\pi pr\, dr \qquad (30.49)$$

The actuating force F and the torque capacity T are found by integrating in Eqs. (30.49) and (30.48) from the inside radius $d/2$ to the outside radius $D/2$:

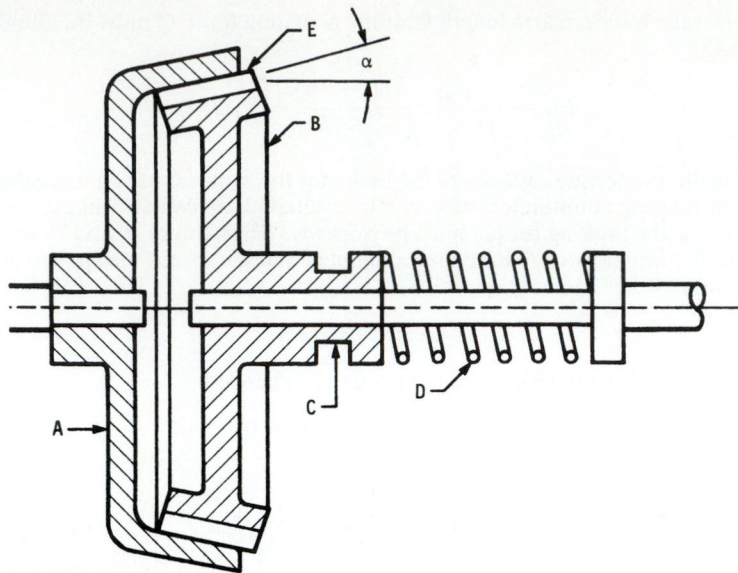

FIGURE 30.20 Cone clutch; *A,* cup; *B,* cone; *C,* shifting groove; *D,* spring; *E,* friction lining; α = cone angle.

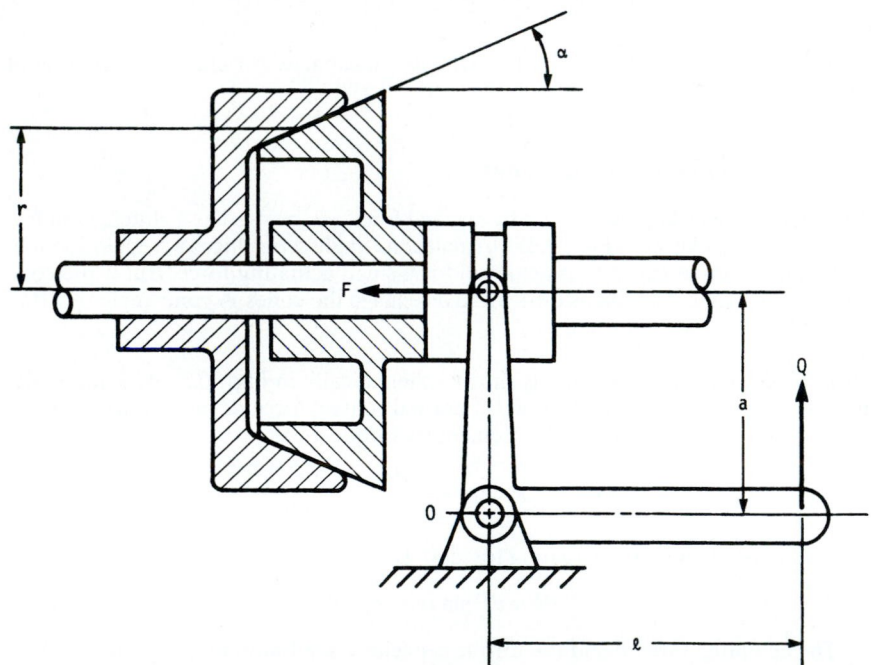

FIGURE 30.21 Cone brake; α = cone angle.

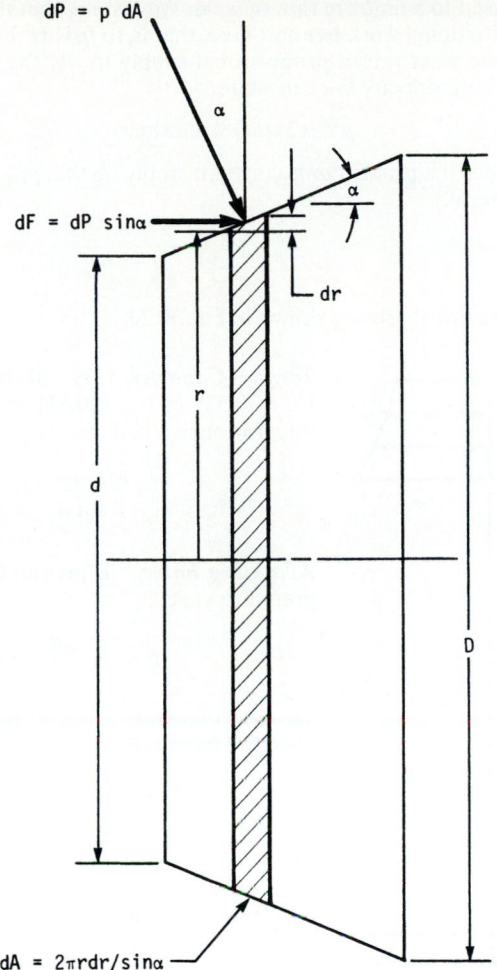

FIGURE 30.22 Elementary quantities on conical surface.

$$F = 2\pi \int_{d/2}^{D/2} pr \, dr \tag{30.50}$$

$$T = \frac{2\pi}{\sin \alpha} \int_{d/2}^{D/2} fpr^2 \, dr \tag{30.51}$$

Before these integrations can be carried out, assumptions have to be made about the way in which contact pressure p and friction coefficient f vary across the active face of the cone. In what follows, the variations of f with pressure and rubbing velocity have been neglected. Only the variation in contact pressure p is used.

Contact-Pressure Distribution.   When the friction surfaces are new, the pressure is fairly uniform across the clutch face. But after an initial wear-in period, the pressure accommodates itself to a uniform rate of wear. We assume that the wear rate is proportional to the frictional work per unit area, that is, to fpV. If the variations in f are neglected, then the wear rate is proportional simply to pV, the product of contact pressure and rubbing velocity. We can write

$$pV = 2\pi r\omega p = \text{constant} \tag{30.52}$$

Thus across the face the product pr is constant, implying that p_{max} occurs at the inner radius $d/2$. In general,

$$p = \frac{p_{max}d}{2r} \tag{30.53}$$

A typical pressure distribution is shown in Fig. 30.23.

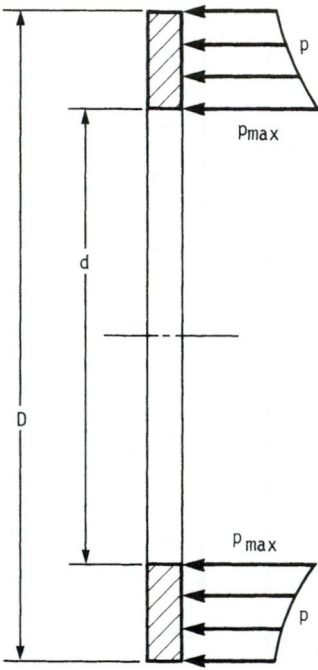

FIGURE 30.23 Contact-pressure distribution on face of cone after wear-in period (for constant f).

Torque Capacity. By substituting for p from Eq. (30.53) in Eq. (30.51) and carrying out the integration, we find the torque capacity T:

$$T = \frac{\pi f p_{max}d}{8 \sin \alpha} (D^2 - d^2) \tag{30.54}$$

Actuating Force. Equation (30.50) can be integrated to yield

$$F = \frac{\pi p_{max}d}{2} (D - d) \tag{30.55}$$

The last two equations can be combined to produce the useful result

$$T = \frac{Ff}{4 \sin \alpha} (D + d) \tag{30.56}$$

The last equation indicates that for the uniform-wear assumption, the mean friction radius is simply the average radius.

Moment Equation for Brake Lever. The axial actuating force for the cone brake shown in Fig. 30.21 can be found by summing moments on the lever about the pivot point O and solving for F. Thus

$$F = \frac{Q\ell}{a} \tag{30.57}$$

30.7 DISK CLUTCHES AND BRAKES

30.7.1 Multidisk Clutches and Brakes

The multidisk clutch in Fig. 30.24 is intended for wet operation using an oil coolant. Similar clutches are built for dry operation. Disk brakes are similar in construction to the disk clutch. In either case, an axial force is applied to the flat surfaces of the elements to produce tangential frictional forces. Typically not more than the outer 40 percent of the radius is used. The ratio of inside to outside diameters may be as high as 0.80/1.

Contact-Pressure Distribution. The reasoning used to establish pressure distribution on the annular clutch or brake plate is the same as that used for cone clutches and brakes. After an initial wear-in period, the pressure distribution accommodates itself to a constant rate of wear across the active portion of the disk (Fig. 30.25). Equations (30.52) and (30.53) apply here also.

Axial Actuating Force. For a given set of dimensions and a permissible contact pressure, the corresponding actuating force is given by Eq. (30.55), which was developed for cone clutches and brakes.

Torque Capacity. For this pressure distribution, the torque capacity is

$$T = \frac{1}{8}[\pi f p_{max} d(D^2 - d^2)N_p] \tag{30.58}$$

Although the torque equation could be derived independently, it can also be derived directly from Eq. (30.54) by setting the cone angle $\alpha = 90°$ and inserting N_p, the num-

FIGURE 30.24 An oil-actuated multiple-disk clutch for enclosed operation in an oil spray or bath. (*Twin Disc, Inc.*)

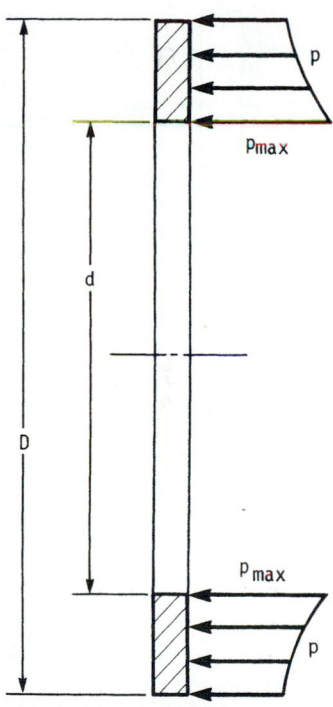

FIGURE 30.25 A friction member of a multiple-disk clutch or brake with the pressure distribution for uniform wear.

ber of pairs of contacting disk faces. Torque capacity can also be expressed in terms of the actuating force F:

$$T = \frac{FfN_p}{4}(D+d) \qquad (30.59)$$

30.7.2 Caliper Disk Brakes

The automotive caliper disk brake shown in Fig. 30.9 is hydraulically operated. Two pads are pressed against opposite sides of the brake disk to provide a braking torque. The principle of operation is shown schematically in Fig. 30.26a.

Usually each pad is nearly the annular shape, illustrated in Fig. 30.26b, but occasionally a circular pad ("puck" or "button") is used (Fig. 30.26c).

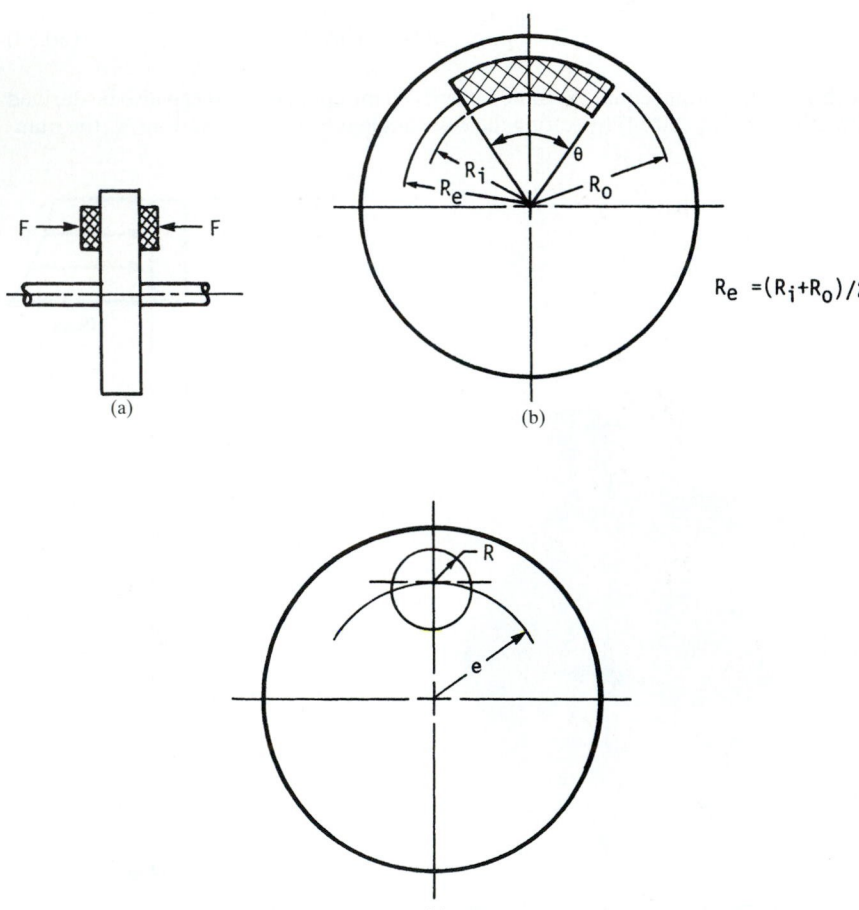

FIGURE 30.26   Caliper disk brake. (a) Principle of operation; (b) annular pad; (c) circular pad.

Torque Capacity. The torque capacity *per pad* is

$$T = fFR_e \tag{30.60}$$

If two pads are used, then the torque capacity is double the value calculated by Eq. (30.60).

This torque equation is quite simple. A friction force fF acting at an effective friction radius R_e produces a braking torque T. The practical issues are (1) the value to use for the effective friction radius R_e and (2) the maximum contact pressure p_{max} developed.

Actuating Force for Annular Pad. Brake designers often assume that the contact pressure does not vary very much over the annular pad. But you may prefer to use the uniform-wear approach as more realistic. Both approaches are given here.

The following two equations are used to calculate the actuating force for the constant-contact-pressure and the uniform-wear approaches, respectively:

$$F = p_{av}\theta \frac{R_o^2 - R_i^2}{2} \tag{30.61}$$

$$F = p_{max}\theta R_i (R_o - R_i) \tag{30.62}$$

The notation is shown in Fig. 30.26a and b.

The relation between average and maximum contact pressures for the uniform-wear approach is

$$\frac{p_{av}}{p_{max}} = \frac{2R_i/R_o}{1 + R_i/R_o} \tag{30.63}$$

In the limit, the average and maximum pressures become equal as the inner radius approaches the outer one. For $R_i/R_o = 0.60$, $p_{av} = 0.75p_{max}$; but for $R_i/R_o = 0.80$, $p_{av} = 0.89p_{max}$.

The uniform-wear assumption is the more conservative approach. For a given actuating force, it implies a smaller torque capacity.

Effective Friction Radius for Annular Pad. If the contact pressure is assumed to be constant over the pad, the effective friction radius is

$$R_e = \frac{2}{3} \frac{R_o^3 - R_i^3}{R_o^2 - R_i^2} \tag{30.64}$$

When the uniform-wear assumption is made, the effective friction radius is simply

$$R_e = \frac{R_i + R_o}{2} \tag{30.65}$$

Circular Pads. Fazekas [30.3] has derived the basic equations for circular pads. The effective friction radius is

$$R_e = \delta e \tag{30.66}$$

where e = radius from the center of the disk to the center of the pad and δ = multiplier found in Table 30.9. Also tabulated in Table 30.9 is the ratio of maximum to average contact pressure.

TABLE 30.9 Design Factors for Caliper Disk Brakes with Circular Pads

R/e	$\delta = R_e/e$	p_{max}/p_{av}
0	1.000	1.000
0.1	0.983	1.093
0.2	0.969	1.212
0.3	0.957	1.367
0.4	0.947	1.578
0.5	0.938	1.875

SOURCE: Ref. [30.3].

The actuating force on each pad is

$$F = \pi R^2 p_{av} \tag{30.67}$$

The torque capacity per pad is found by using Eq. (30.60) after F and R_e have been calculated.

Example 9. A sports car requires disk brakes for the front wheels. It has been decided to use two annular pads per wheel with $R_i = 3.875$ in, $R_o = 5.5$ in, and $\theta = 108°$. The friction material supplier guarantees a coefficient of friction of at least $f = 0.37$. Each pad is actuated by two hydraulic cylinders, each 1.5 in in diameter. Each front-wheel brake provides a braking torque capacity of 13×10^3 lb · in. What hydraulic pressure is needed at the wheel cylinders? What are the average and the maximum contact pressures? Assume uniform wear.
    Solution

1. The torque capacity per pad has to be $(13 \times 10^3)/2 = 6500$ lb · in. Given uniform wear, the effective friction radius is, by Eq. (30.65),

$$R_e = \frac{R_i + R_o}{2} = \frac{3.875 + 5.5}{2} = 4.69 \text{ in}$$

The corresponding actuating force is, by Eq. (30.60),

$$T = \frac{T}{fR_e} = \frac{6500}{0.37(4.69)} = 3750 \text{ lb per pad}$$

or

$$F = 1875 \text{ lb per wheel cylinder}$$

The hydraulic pressure at the wheel cylinder has to be

$$p_h = \frac{F}{A_p} = \frac{4F}{\pi d_p^2} = \frac{4(1875)}{\pi(1.5)^2} = 1060 \text{ psi}$$

2. Equation (30.62) can be used to find the maximum contact pressure:

$$p_{max} = \frac{F}{\theta R_i (R_o - R_i)} = \frac{3750}{(108°/57.296)(3.875)(5.5 - 3.875)} = 316 \text{ psi}$$

For finding the average contact pressure, use Eq. (30.63):

$$p_{av} = p_{max} \frac{2R_i/R_o}{1 + R_i/R_o} = 316 \frac{2(3.875)/5.5}{1 + 3.875/5.5} = 261 \text{ psi}$$

30.8 ELECTROMAGNETIC TYPES

Electromagnetic forces are used in a variety of ways to couple two sides of a clutch or a brake. The use of electrically generated forces implies relatively easy means of automatic control.

30.8.1 Magnetically Actuated Friction Clutches and Brakes

The most common use of electromagnetic forces is to provide the actuating force for a friction brake or clutch. The configuration shown in Fig. 30.27 can be used either as a clutch or as a brake. When power is applied to the coil, the magnet, faced with a friction material, attracts the armature. Torque can be varied by using a potentiometer. If one member is fixed, the device functions as a brake; otherwise, it is a clutch.

The combination clutch-brake in Fig. 30.28 uses a single solenoid coil. When the coil is deenergized, the clutch is disengaged and springs prevent the release of the brake. When the coil is energized, the clutch is engaged and the brake is released. An arrangement of opposing mechanical springs ensures that the input-side clutch is fully engaged for a brief time before the brake is released. This is done by making the springs assisting clutch engagement weaker than those resisting brake release.

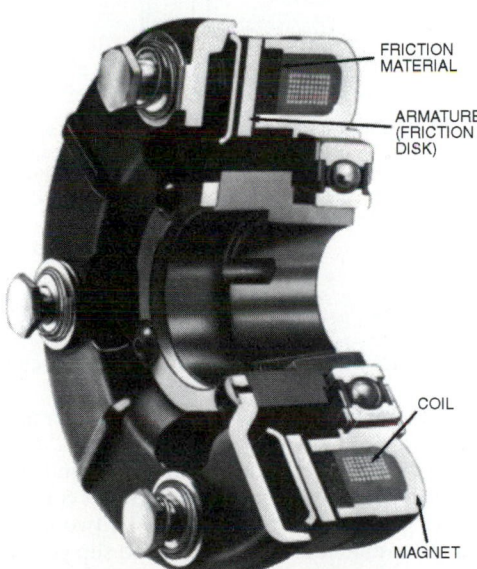

FRICTION
MATERIAL

ARMATURE
(FRICTION
DISK)

COIL

MAGNET

FIGURE 30.27 Electromagnetic friction clutches and brakes. *(Warner Electric Brake and Clutch Co.)*

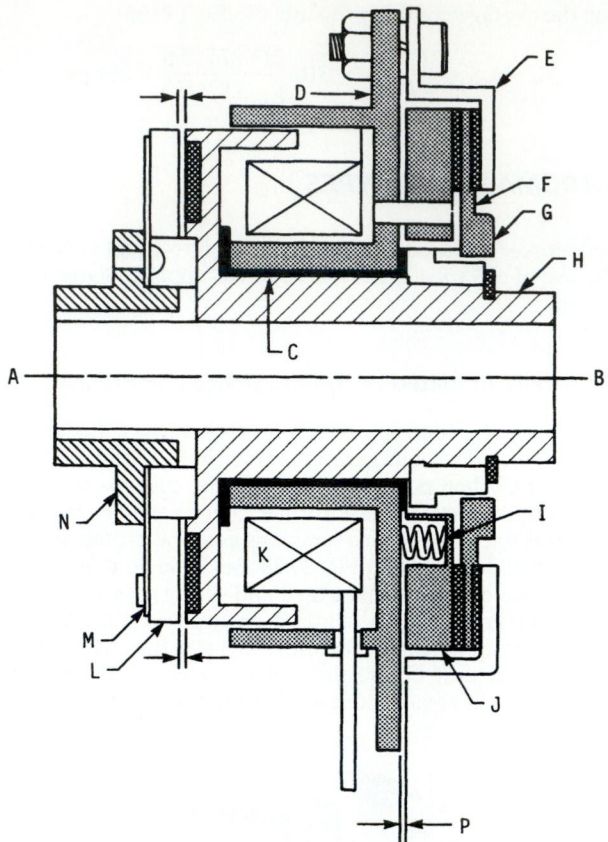

FIGURE 30.28   Clutch-brake transmission. *A,* input; *B,* output; *C,* field coil bearing; *D,* field coil assembly; *E,* pressure cup; *F,* brake plate; *G,* hub spling; *H,* rotor assembly; *I,* brake spring; *J,* brake armature; *K,* field coil; *L,* clutch armature; *M,* clutch spring; *N,* drive plate; *P,* air gap. *(Electroid Corporation.)*

30.8.2   Magnetic Clutches

The operating characteristics of three types of magnetic clutches are shown in Fig. 30.29.

   Magnetic-particle clutches (Fig. 30.29*a*) use an iron powder mixed with a lubricant to partially fill the annular gap between members. When a direct-current (dc) coil induces a magnetic field, the iron particles form chains and provide the means to transmit torque. There is a nearly linear relation between coil current and torque.

   Hysteresis clutches (Fig. 30.29*b*) directly couple the two members as long as the load does not exceed the torque rating. They can also slip continuously to maintain a constant-torque output independent of speed.

   The *eddy-current clutch* (Fig. 30.29*c*) is rather like the hysteresis clutch in construction. Torque is developed if there is slip. The torque is associated with dissipation of eddy currents in the rotor ring's electric resistance.

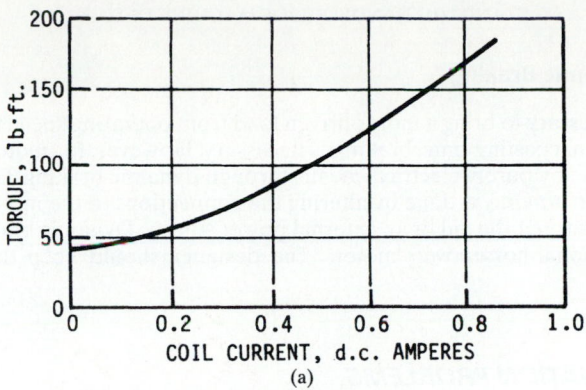

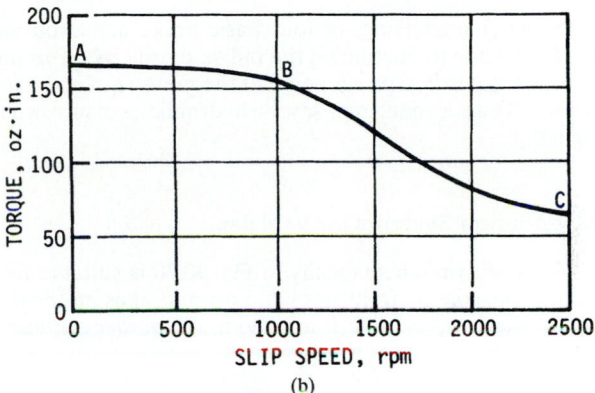

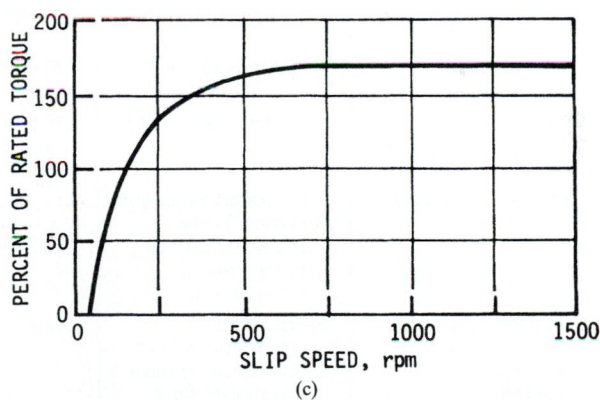

FIGURE 30.29 Torque characteristics of magnetic clutches. (*a*) Magnetic-particle clutch has a characteristic that is independent of slip and increases almost linearly with coil current. (*b*) Hysteresis clutch exhibits almost a constant torque out to the thermal, which begins at *B;* temperature then limits the torque capacity from *B* to *C.* (*c*) Eddy-current clutch exhibits a constant-torque characteristic at rated slip speed. (*From Ref. [30.5].*)

30.8.3 Dynamic Braking

When it is necessary to bring a motor-driven load from operating speed to rest in less than the normal coasting time, braking is necessary. However, for motors, the braking can be done by purely electrical means through dynamic braking. Electric braking, or *dynamic braking,* is done by altering the connections to the motor. It may be done with or without the aid of an external power source. Dynamic braking is available for fractional-horsepower motors. The designer should keep this option in mind.

30.9 ACTUATION PROBLEMS

30.9.1 General

Table 30.10 lists the characteristics of four basic brake actuation methods. Note that many brakes are made to operate on the fail-safe principle. This means that the brake is applied by using strong springs and that the method of actuation releases or holds off the brake. Thus, a reduction, say, in hydraulic pressure would cause the brake to be applied.

30.9.2 Brake Actuation Systems for Vehicles

The hydraulic system shown schematically in Fig. 30.30 is suitable for a passenger vehicle using disk brakes on the front axles and drum brakes on the rear. The pedal force, multiplied by the leverage ratio, is applied to the *master cylinder* to produce a

TABLE 30.10   Characteristics of Various Methods of Actuating Band, Drum, and Disk Brakes

Actuation method	Advantages	Disadvantages	Possible difficulties
Mechanical	Robust; simple; manual operation gives good control	Large leverage needed	Friction losses at pins and pivots
Pneumatic	Large forces available	Compressed air supply needed; brake chambers may be bulky; slow response time	Length of stroke (particularly if diaphragm type)
Hydraulic†	Compact; large forces available; quick response and good control	Special fluid needed; temperatures must not be high enough to vaporize fluid	Seals
Electric‡	Suitable for automatic control; quick response	On/off operation	Air gap

†Used for spot-type disk brakes.
‡Used for spot- and plate-type disk brakes.
SOURCE: Ref. [30.6].

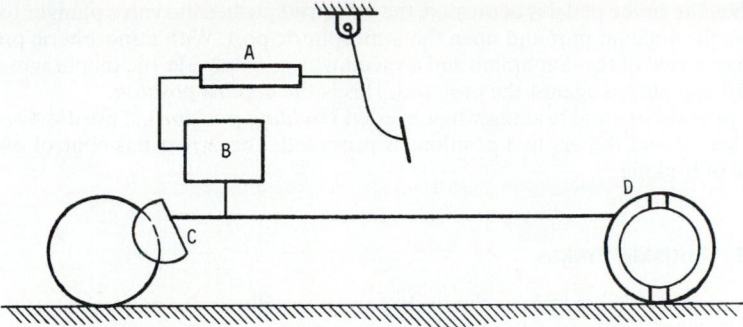

FIGURE 30.30 Hydraulic system with indirect servo. *A,* master cylinder; *B,* slave cylinder and servo; *C,* disk brake; *D,* leading-trailing shoe brake.

hydraulic pressure. This hydraulic pressure can itself be multiplied by a *power-brake assist unit* which uses either the intake manifold's vacuum or a positive hydraulic pressure from the power-steering pump to create a pressure differential.

In a *split system,* the master cylinder has two pistons and two reservoirs, one for the front-wheel brakes, the other for the rear. When the primary piston is pushed forward, the secondary piston is pushed forward as well by the primary piston spring and the buildup of pressure between the two pistons. Thus hydraulic pressure is built up in both systems. The springs return the pistons when the brake-pedal force is removed.

With the split system, one set of brakes can function even if the hydraulic system for the other set of brakes is damaged. For example, if the hydraulic system of the front brakes fails, no hydraulic pressure is built up in the front-brake system, and the secondary piston continues to move until its nose butts against the end of the cylinder. The primary piston continues to move and build up pressure between the primary and secondary pistons to actuate the rear brakes.

Most hydraulic brake systems are equipped with automatic valves:

1. *Pressure-differential valve* to turn on a warning light if either of the hydraulic systems (front or rear) fails.

2. *Proportioning valve* to improve the braking balance between the front and rear brakes and prevent skidding resulting from the rear brakes locking up before the front brakes.

3. *Metering valve* to delay the flow of brake fluid to the front-brake calipers until the system pressure has risen sufficiently. The motives are to overcome the tension of the retracting springs at the rear-brake shoes, expand the shoes, and supply the rear brakes before the front brakes. The metering valve prevents front-brake lockup during light braking on slippery or icy roads.

All three functions may be merged into a *combination valve* mounted near the master cylinder.

A *vacuum-brake booster* functions in three modes: released, applied, and holding. When the brake pedal is released, the engine intake manifold pulls air from the front shell through a check valve. There is a vacuum on both sides of the diaphragm, and the pressures are equal. The diaphragm is held to the rear by its spring. No force is exerted on the master cylinder by the push rod.

When the brake pedal is depressed, the valve rod pushes the valve plunger forward to close the vacuum port and open the atmospheric port. With atmospheric pressure on the rear side of the diaphragm and a vacuum at its front side, the diaphragm moves forward and pushes against the push rod. This is the *applied position.*

To provide gradual braking when needed (*holding position*), a position between the released and the applied positions is provided. The driver has control over the degree of braking.

30.9.3   Antiskid Brakes

In an antiskid braking system, the brakes are normally under manual control; but if wheel lockup is imminent, the antiskid system takes corrective action.

Whenever a rear wheel starts to lock, a wheel sensor detects an abrupt deceleration. A computer then causes the pressure in the rear braking system to decrease slightly, allowing the wheels to accelerate. When the wheel speed approaches its normal level for the vehicle's speed, the wheel cylinder pressure is restored. The antiskid system goes into action repeatedly to prevent wheel lockup and skidding until the vehicle speed drops to about 5 miles per hour (mph).

REFERENCES

30.1 Karl Beach, "Try These Formulas for Centrifugal Clutches," *Product Engineering,* July 9, 1962.

30.2 Arthur H. Burr, *Mechanical Analysis and Design,* Elsevier, New York, 1981.

30.3 G. A. Fazekas, "On Circular Spot Brakes," *Journal of Engineering for Industry, ASME Transactions,* August 1972.

30.4 A. F. Gagne, Jr., "Torque Capacity and Design of Cone and Disk Clutches," *Product Engineering,* December 1953.

30.5 John Proctor, "Selecting Clutches for Mechanical Drives," *Product Engineering,* June 1961.

30.6 M. J. Neale (ed.), *The Tribology Handbook,* Butterworth, London, 1973.

30.7 Ferodo Limited, *Friction Materials for Engineers,* Chapel-en-le-Frith, England, 1968.

SUGGESTED READING

Crouse, William H.: "Automotive Brakes," *Automotive Chassis and Body,* 4th ed., McGraw-Hill, New York, 1971.

Mathews, G. P.: "Art and Science of Braking Heavy Duty Vehicles," special publication SP-251, Society of Automotive Engineers, Warrendale, Pa., 1964.

Mischke, Charles R., *Elements of Mechanical Analysis,* Addison-Wesley, Reading, 1963, pp. 55–65.

Mischke, Charles R., *Mathematical Model Building,* 2d rev. ed., Iowa State University Press, Ames, 1980, pp. 338–344.

Remling, John: *Brakes,* Wiley, New York, 1978.

Wong, J. Y.: *Theory of Ground Vehicles,* Wiley, New York, 1978.

CHAPTER 31
BELT DRIVES

Wolfram Funk, Prof. Dr.-Ing.
Fachbereich Maschinenbau
Fachgebiet Maschinenelemente und Getriebetechnik
Universität der Bundeswehr Hamburg
Hamburg, Germany

NOMENCLATURE

A	Cross section
b	Width
c_β	Angular factor
c_B	Service factor
d_1	Diameter of driving pulley
d_2	Diameter of driven pulley
e	Center distance
E	Modulus of elasticity
F	Force
f_b	Bending frequency
l	Datum length of flexible connector
M	Torque
n	Speed
P	Power
q	Mass per length
r	Radius
s	Belt thickness
t	Pitch
v	Velocity
z	Number

α   Included angle
β   Angle of wrap
ε   Elongation (strain)
μ   Coefficient of friction
η   Efficiency
ψ   Slip
ρ   Specific mass
σ   Stress

Indices

1   Driving
2   Driven
b   Bending
f   Centrifugal
max   Maximum
w   Effective
zul   Allowable
N   Nominal

31.1 GENERAL

Flexible-connector drives are simple devices used to transmit torques and rotational motions from one to another or to several other shafts, which would usually be parallel. Power is transmitted by a flexible element (flexible connector) placed on pulleys, which are mounted on these shafts to reduce peripheral forces. The transmission ratios of torques and speeds at the driving and driven pulleys are determined by the ratio of pulley diameters. Peripheral forces may be transmitted by either frictional (nonpositive) or positive locking of the flexible connector on the pulleys.

Because of their special characteristics, flexible-connector drives have the following advantages and disadvantages as compared with other connector drives:

Advantages:

- Small amount of installation work
- Small amount of maintenance
- High reliability
- High peripheral velocities
- Good adaptability to the individual application
- In some cases, shock- and sound-absorbing
- In some cases, with continuously variable speed (variable-speed belt drive)

Disadvantages:

- Limited power transmission capacity
- Limited transmission ratio per pulley step
- In some cases, synchronous power transmission impossible (slip)
- In some cases, large axle and contact forces required

31.1.1 Classification According to Function

According to function, flexible-connector drives are classified as (1) nonpositive and (2) positive.

Nonpositive flexible-connector drives transmit the peripheral force by means of friction (mechanical force transmission) from the driving pulley to the flexible connector and from there to the driven pulley(s). The transmissible torque depends on the friction coefficient of the flexible connector and the pulleys as well as on the surface pressure on the pulley circumference. The power transmission capacity limit of the drive is reached when the flexible connector starts to slip. By use of wedge-shaped flexible connectors, the surface pressure can be increased, with shaft loads remaining constant, so that greater torques are transmitted. Since nonpositive flexible-connector drives tend to slip, synchronous power transmission is impracticable.

The positive flexible-connector drive transmits the peripheral force by positive locking of transverse elements (teeth) on the connector and the pulleys. The surface pressure required is small. The transmissible torque is limited by the distribution of the total peripheral force to the individual teeth in engagement and by their functional limits. The power transmission capacity limit of the drive is reached when the flexible connector slips. Power transmission is slip-free and synchronous.

31.1.2 Geometry

The dimensions of the different components [pulley diameter, center distance, datum length (pitch length) of the flexible connector] and the operational characteristics (speed ratio, angle of wrap, included angle) are directly interrelated.

Two-Pulley Drives. For the standard two-pulley drive, the geometry is simple (Fig. 31.1). In general, this drive is designed with the center distance and the speed ratio as parameters. The individual characteristics are related as follows: Speed ratio:

$$i = \frac{n_1}{n_2} = \frac{d_2}{d_1} \tag{31.1}$$

Included angle:

$$\sin \alpha = \frac{d_2 - d_1}{2e} = \frac{d_1}{2e} (i - 1) \tag{31.2}$$

Angles of wrap:

$$\beta_1 = 180° - 2\alpha = 180° - 2 \arcsin \frac{d_1}{2e} (i - 1)$$

$$\beta_2 = 180° + 2\alpha = 180° + 2 \arcsin \frac{d_1}{2e} (i - 1)$$

$$\tag{31.3}$$

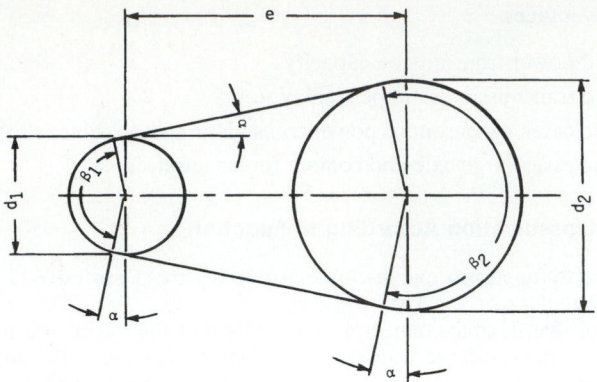

FIGURE 31.1   Two-pulley drive.

Datum length of flexible connector:

$$l = 2e \cos \alpha + \pi \left(d_1 \frac{\beta_1}{360°} + d_2 \frac{\beta_2}{360°} \right)$$

$$= 2e \cos \alpha + \frac{\pi d_1}{360°} \left[180° - 2\alpha + i(180° + 2\alpha) \right] \tag{31.4}$$

Approximate equation:

$$l \approx 2e + 1.57(d_1 + d_2) + \frac{(d_2 - d_1)^2}{4e}$$

$$= 2e + 1.57 d_1(i + 1) + \frac{d_1^2}{4e} (i - 1)^2 \tag{31.5}$$

The minimum diameter allowable for the flexible connector selected is often substituted for the unknown parameter d_1 (driving-pulley diameter) required for the design.

Multiple-Pulley Drives.   For the multiple-pulley drive (one driving pulley, two or more driven pulleys), the geometry is dependent on the arrangement of the pulleys (Fig. 31.2). These drives have the following characteristics: Speed ratios:

$$i_{12} = \frac{n_1}{n_2} = \frac{d_2}{d_1} \qquad i_{13} = \frac{n_1}{n_3} = \frac{d_3}{d_1} \qquad i_{1m} = \frac{n_1}{n_m} = \frac{d_m}{d_1}$$

Included angles:

$$\sin \alpha_{12} = \frac{d_1}{2e_{12}} (i_{12} - 1) \tag{31.6}$$

$$\sin \alpha_{13} = \frac{d_1}{2e_{13}} (i_{13} - 1) \tag{31.7}$$

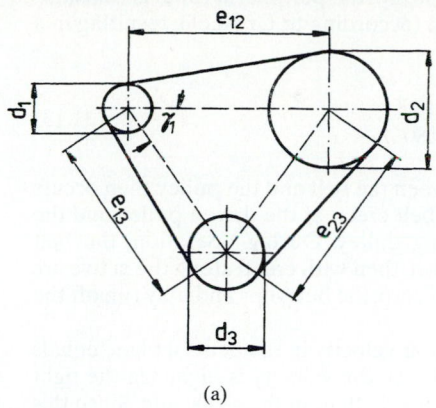

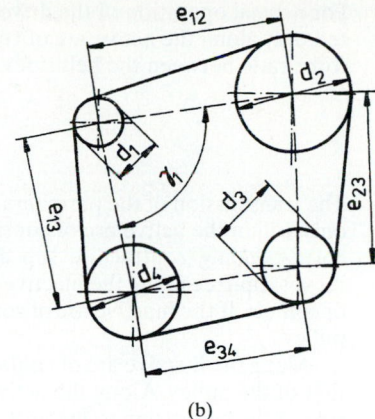

(a)   (b)

FIGURE 31.2 Multiple-pulley drives.

$$\sin \alpha_{1m} = \frac{d_1}{2e_{1m}} \, (i_{1m} - 1) \tag{31.8}$$

$$\sin \alpha_{km} = \frac{d_k}{2e_{km}} \, (i_{km} - 1) \tag{31.9}$$

Angles of wrap:

$$\beta_j = 180° - \alpha_{j,j-1} - \alpha_{j,j+1} - \gamma_j \tag{31.10}$$

where j = index of pulley
γ_j = angle between center distances

$$l = \frac{\beta_1 \pi d_1}{360} + e_{12} \cos \alpha_{12} + \frac{\beta_2 \pi d_2}{360} + e_{23} \cos \alpha_{23} + \cdots$$

$$+ \frac{\beta_k \pi d_k}{360} + e_{km} \cos \alpha_{km} + \frac{\beta_m \pi d_m}{360} + e_{1m} \cos \alpha_{1m} \tag{31.11}$$

31.1.3   Forces in Moving Belt

Friction is employed in transmitting the peripheral forces between the belt and the pulley. The relation of the friction coefficient μ, the arc of contact β, and the belt forces is expressed by Eytelwein's equation. For the extreme case, i.e., slippage along the entire arc of contact, this equation is

$$\frac{F_1'}{F_2'} = \exp \frac{\mu \beta \pi}{180} \tag{31.12}$$

For normal operation of the drive without belt slip, the peripheral force is transmitted only along the active arc of contact $\beta_w < \beta$ (according to Grashof), resulting in a force ratio between the belt sides of

$$\frac{F_1'}{F_2'} = \exp\frac{\mu\beta_w\pi}{180} \qquad (31.13)$$

The transmission of the peripheral force between the belt and the pulley then occurs only within the active arc of contact β_w with belt creep at the driven pulley and the corresponding contraction slip at the driving pulley. During operation, the belt moves slip-free along the inactive arc of contact, then with creep along the active arc of contact. If the inactive arc of contact equals zero, the belt slips and may run off the pulley.

Along the inactive arc of contact, the angular velocity in the neutral plane equals that of the pulley. Along the active arc of contact, the velocity is higher in the tight side of the belt owing to higher tension in that side than in the slack side. Since this velocity difference has to be offset, slip results. This slip leads to a speed difference between the engagement point and the delivery point on each pulley, which amounts up to 2 percent depending on the belt material (modulus of elasticity), and load:

$$\psi = \frac{v_1 - v_2}{v_1} = \frac{(l_2 + \Delta l) - l_2}{l_2 + \Delta l} \approx \Delta\varepsilon = \frac{\sigma_1 - \sigma_2}{E} = \frac{\sigma_n}{E} \qquad (31.14)$$

For practical design purposes, the calculations for a belt drive are usually based on the entire arc of contact β of the smaller pulley (full load), since the active arc of contact is not known, and the belt slips at the smaller pulley first.

$$\frac{F_1'}{F_2'} = m = \exp\frac{\mu\beta\pi}{180} \qquad (31.15)$$

Centrifugal forces acting along the arcs of contact reduce the surface pressure there. As these forces are supported by the free belt sides, they act uniformly along the entire belt:

$$F_f = \rho v^2 A = q v^2 \qquad (31.16)$$

With increasing belt velocity v, constant center distance e, and constant torques, the forces F_1 and F_2 acting along the belt sides as well as the peripheral force (usable force) F_u remain constant, whereas the surface pressure and the usable forces F_1' and F_2' in the belt sides are reduced. Usable forces in belt sides:

$$F_1' = F_1 - F_f = mF_2'$$
$$F_2' = F_2 - F_f = \frac{F_1'}{m} \qquad (31.17)$$

Peripheral force:

$$F_u = F_1' - F_2' = F_1 - F_2 = F_1'\left(1 - \frac{1}{m}\right)$$
$$= F_2'(m - 1) \qquad (31.18)$$

Because

$$F_u = F_2'(m-1) \qquad m = \exp\frac{\mu\beta_w\pi}{180} \tag{31.19}$$

β_w becomes greater, until the belt slips on the pulley with the smaller arc of contact when $\beta_w = \beta$. When $F_f = F_2$, there are no usable forces; that is, $F_2' = F_1' = F_u = 0$. In this case, no torque can be transmitted. If belt velocity v is increased further, the belt runs off the pulley.

The maximum force in the belt sides is given by

$$F_{max} = F_1 = F_2' + F_u + F_f \tag{31.20}$$

With only the centrifugal forces acting, the belt is in equilibrium. They do not act on the pulleys at all. Hence, the shaft load F_w of a belt drive results from only the usable forces F_1' and F_2' in the belt sides (Fig. 31.3):

$$F_w = \sqrt{F_1'^2 + F_2'^2 - 2F_1'F_2'\cos\beta} \tag{31.21}$$

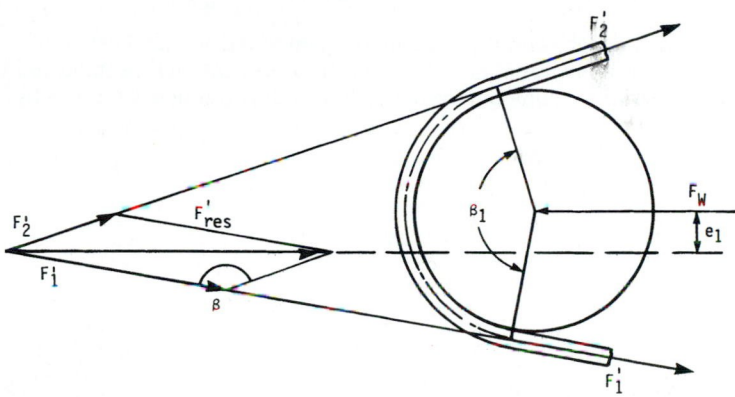

FIGURE 31.3 Equilibrium of forces.

The force rating

$$\Phi = \frac{F_u}{F_w} = (m-1)\sqrt{m^2 + 1 - 2m\cos\beta}$$

defines the minimum shaft tensioning force required for peripheral force production as a function of the friction coefficient μ and the arc of contact β.

The rated output $K = F_u/F_1' = 1 - 1/m$ defines the peripheral force F_u which can be produced by the permissible force F_1' as a function of the friction coefficient μ and the arc of contact β. The reduction in rated output with decreasing arc of contact is defined by an angular factor c_β, based on $\beta = 180°$, that is, a speed ratio of $i = 1$.

The tensions in a homogeneous belt result from the forces acting in the belt and the belt cross section, $A = bs$. For multiple-ply belts, these tensions can be used only as theoretical mean values.

Bending of the belt around the pulley produces the bending stress σ_b. This stress can be calculated from the elongation of the belt fibers with respect to the neutral axis:

$$\Delta l = \beta(r+s) - \beta\left(r + \frac{s}{2}\right) = \beta\frac{s}{2}$$

$$\varepsilon = \frac{\Delta l}{l} = \frac{\beta s/2}{\beta(r+s/2)} = \frac{s}{2r+s} \approx \frac{s}{d} \tag{31.22}$$

$$\sigma_b = \varepsilon E_b \approx \frac{s}{d} E_b \tag{31.23}$$

The strain ε increases with decreasing pulley diameter d. For practical design purposes, σ_b is not taken into consideration, since belt life depends much less on σ_b than on the bending frequency.

The maximum stress is in the tight side of the belt at the beginning and end of the arc of contact, i.e., the points where it passes onto or off the smaller pulley (Fig. 31.4):

$$\sigma_{\max} = \sigma'_1 + \sigma_f + \sigma_b = \frac{F'_1}{A} + \rho v^2 + E_b\frac{s}{d} \tag{31.24}$$

The safety stress depends on the bending frequency and the smallest pulley diameter as well as on the material and the construction of the belt as indicated by the manufacturer. With z = number of pulleys, the bending frequency is given by

$$f_b = \frac{vz}{l} \tag{31.25}$$

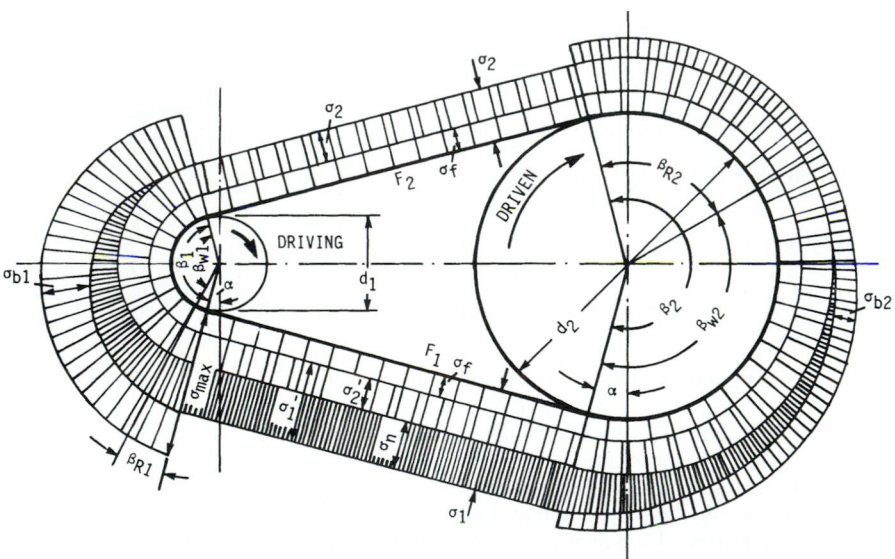

FIGURE 31.4   Stress distribution.

The maximum power transmission capacity of a belt drive can be determined as follows: The power transmission capacity

$$P = F_u v = \sigma_n A v$$

equals zero if the belt velocity v either equals zero or reaches a maximum at which the belt safety stress limit is approached by the centrifugal and bending stresses alone, so that

$$\sigma_1' = \sigma_2' = \sigma_n = 0$$

Then

$$\sigma_{zul} = \sigma_f + \sigma_b = \rho v_{max}^2 + \sigma_b \tag{31.26}$$

from which the maximum belt velocity can be calculated as follows:

$$v_{max} = \sqrt{\frac{\sigma_{zul} - \sigma_b}{\rho}} \tag{31.27}$$

Optimum power transmission is possible only at the optimum belt velocity v_{opt} within the range of $v = 0$ and $v = v_{max}$. It depends on the belt safety stress and is given by

$$v_{opt} = \sqrt{\frac{\sigma_{zul} - \sigma_b}{3\rho}} = \frac{v_{max}}{\sqrt{3}} \tag{31.28}$$

In theory, this equation applies to all flexible connectors, under the assumption of σ_{zul} (belt safety stress) [or F_{zul} (allowable load)] being independent of belt velocity. Since σ_{zul} decreases with increasing belt velocity, though, the stress and power transmission capacity diagrams are as shown in Fig. 31.5.

31.1.4 Arrangement and Tensioning Devices

Because of their good twistability, flexible connectors are suited for drives with pulleys in different planes and nonparallel shafts of equal or opposite directions of rotation. Since the outer fibers of a twisted flat belt or synchronous belt are strained more than the center fibers, stress is higher there, resulting in the reduction of the belt power transmission capacity.

Figures 31.6 and 31.7 show several belt drives with pulleys in different planes. Note that for drives with crossed belts (Fig. 31.6), endless belts have to be used, in order to avoid damage. For half- or quarter-turn belt drives (Fig. 31.7), the side of delivery must lie in the plane of the mating pulley. By the use of step (cone) pulleys, different speed ratios may be obtained (Fig. 31.8). Pulley diameters have to be selected to ensure equal belt lengths on all steps.

The belt rim running onto the larger diameter of a cone pulley (Fig. 31.9) has a higher velocity than the opposite rim. Thus, the following belt portion is skewed and then runs onto a larger diameter. The drive is balanced when the bending moment due to the bending deformation of the belt is compensated by the skew of the belt side running off (Fig. 31.10).

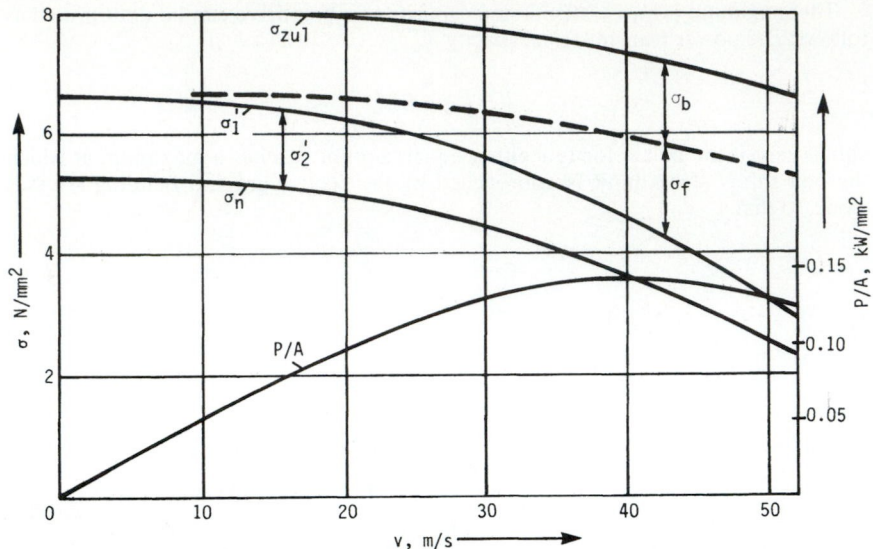

FIGURE 31.5   Stress and power transmission capacity.

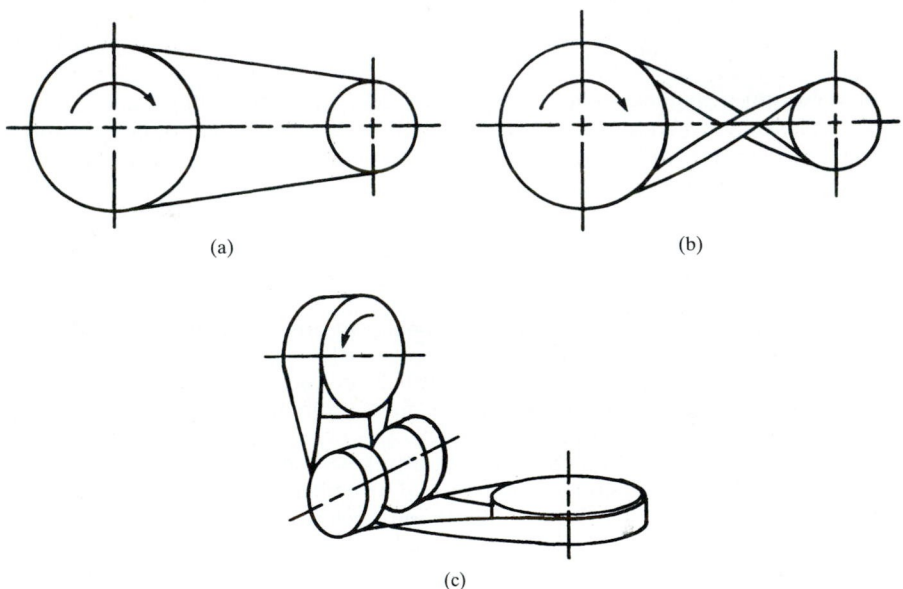

FIGURE 31.6   Examples of crossed belt drives.

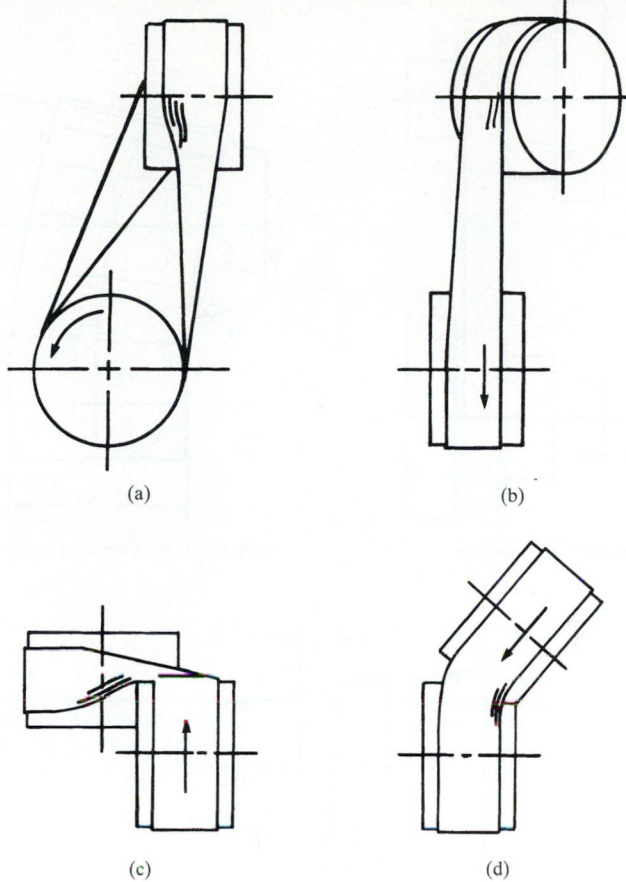

(a)   (b)

(c)   (d)

FIGURE 31.7   Belt drives with pulleys in different planes.

The minimum shaft tensioning force F_w required for nonpositive-type force transmission can be produced as shown in Figs. 31.11 to 31.14.

1. *Pretensioning by belt strain:*   The belt is cut to such a length that it is elastically preloaded when it is placed on the pulleys. Since both the forces F' in the belt sides and the shaft initial tensioning force are reduced by the action of centrifugal forces, σ_f has to be added to the initial tension of the belt to ensure proper transmission of peripheral forces by friction.

2. *Pretensioning by adjustment of center distance:*   The shaft tensioning force F_w is produced by shifting the driving motor on a slide. The belt drive may be preloaded either by adjustment of a threaded spindle or by spring action or weighting.

3. *Pretensioning by means of a belt tightener acting on the slack side:*   The slack side of the belt is provided with a pulley to tighten it—with its own weight or by means of counterweights or by spring action—and increase the arcs of contact on

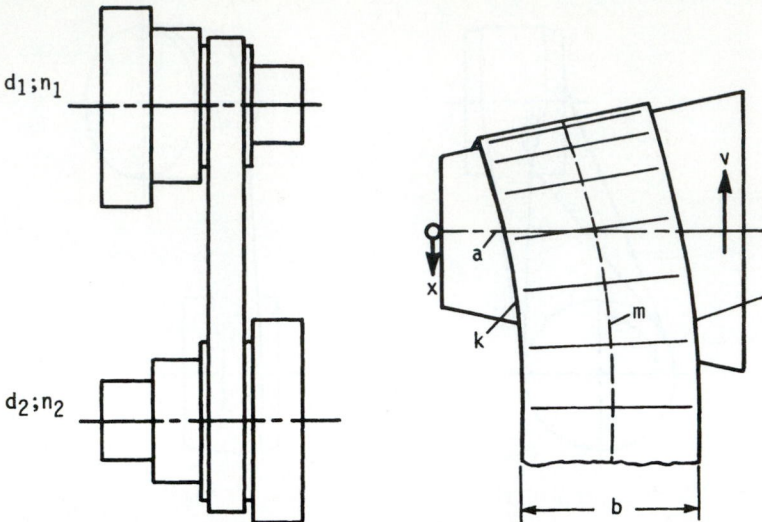

FIGURE 31.8   Step pulleys.   **FIGURE 31.9**   Cone pulley.

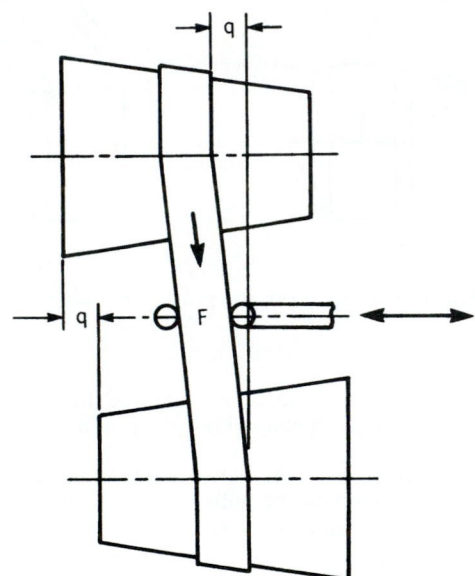

FIGURE 31.10   Cone-pulley drive.

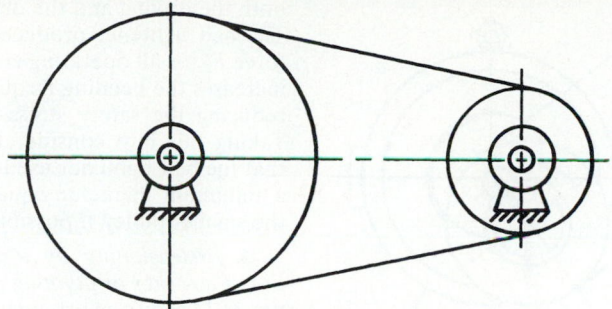

FIGURE 31.11 Pretensioning by belt strain.

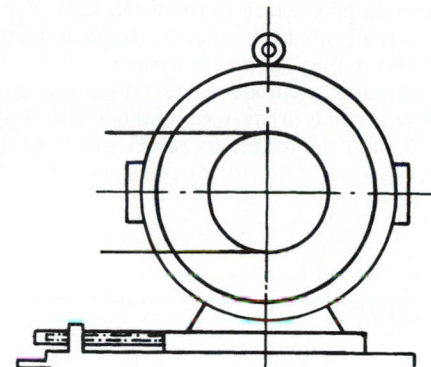

FIGURE 31.12 Pretensioning by adjustment of center distance.

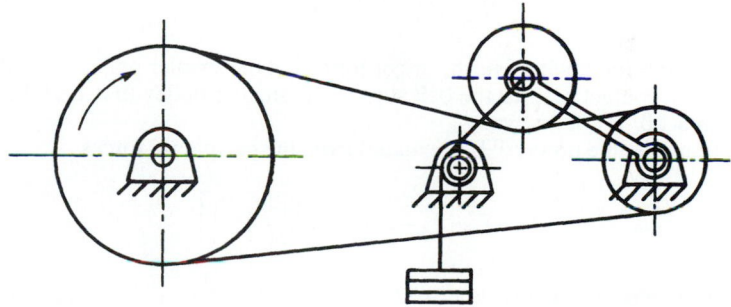

FIGURE 31.13 Pretensioning with a belt tightener.

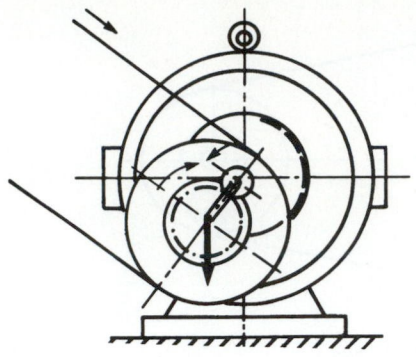

FIGURE 31.14   Pretensioning by torque.

both the driving and the driven pulleys. The belt tightener produces a constant force F_2' for all operating conditions but increases the bending frequency f_b, thus reducing the safety stress of the belt. Taking this into consideration, we see that the belt tightener used should have a minimum diameter equaling that of the smaller pulley, if possible.

4. *Pretensioning by torque making use of a rocker or pivoting pulley:* Figure 31.14 shows an arrangement with an eccentric pulley shaft pivoting the motor pulley shaft. The pulley is driven by a gear assembly. The shaft tensioning force F_w increases almost in proportion to the tooth force F_Z, and with the correct ratio of h_2/h_1, adapts automatically to the specific torque to be transmitted. Belt slipping is impossible with this method of pretensioning.

For all belt pretensioning methods using tensioning devices which can be adjusted during operation of the drive (e.g., methods 2 to 4 described above), the shaft tensioning force F_w and the usable forces F_1' and F_2' in the belt sides are not influenced by centrifugal force. Centrifugal force increases the belt stress by σ_f, though.

31.2   FLAT-BELT DRIVE

Calculations for flat-belt drives are based on Eytelwein's fundamental equation of belt friction:

$$\frac{F_1'}{F_2'} = \exp\frac{\mu\beta\pi}{180} \tag{31.29}$$

For calculations for a belt drive, the arc of contact of the smaller pulley is substituted for the arc of contact β, since the belt slips at the smaller pulley first, and the active arc of contact β_w is not known.

The transmissible power P is calculated from the peripheral forces

$$F_u = F_1' - F_2' = F_2'\left[\exp\left(\frac{\mu\beta\pi}{180}\right) - 1\right] \tag{31.30}$$

and the belt velocity v and is expressed by

$$P_1 = v_1 F_u \qquad P_2 = v_2 F_u \tag{31.31}$$

The belt velocities are

$$v_1 = \pi n_1 d_1 \qquad v_2 = \pi n_2 d_2 \tag{31.32}$$

Taking into consideration a service correction factor c_B (Table 31.1) for peak over-loads due to heavy duty, we can calculate the power rating of the drive from P_1 by

$$P = c_B P_1 \tag{31.33}$$

The speed ratio, which is slightly dependent on the load because of slip, is

$$i = \frac{n_1}{n_2} = \frac{d_2 v_1}{d_1 v_2} = \frac{d_2}{d_1(1 - \sigma_n E)} = \frac{d_2}{d_1(1 - \psi)} \tag{31.34}$$

In calculations, the approximation

$$i \approx \frac{d_2}{d_1} \tag{31.35}$$

may be used.

The efficiency depends on the belt slip only, with the bearing friction and windage being neglected, since the pull of either belt side at both pulleys must be assumed to be equal:

$$\eta = \frac{P_2}{P_1} = \frac{M_2}{iM_1} = 1 - \frac{\sigma_n}{E} = 1 - \psi \tag{31.36}$$

where

$$M_1 = F_u \frac{d_1}{2} \qquad M_2 = F_u \frac{d_2}{2} \tag{31.37}$$

The bending frequency, which is of particular importance for belt life, is calculated by

$$f_b = \frac{vz}{l} \tag{31.38}$$

It should not exceed the limit specified for the particular belt material.

Modern high-performance flat belts are designed as multiple-ply belts. They consist of two or three plies, each serving a special purpose. Leather belts made from hide and Balata belts are no longer used. In some cases, the improvements in power transmission capacity brought about by the development of modern high-performance flat belts are not fully utilized in standard belt-drive applications. The utilization of a drive unit is influenced by the design characteristics on which a flat-belt drive is based. Thus, it is advisable to always consider the whole flat-belt drive unit instead of just the flat belt.

The major components of a multiple-ply belt are the tension ply and the friction ply. The purpose of the tension ply is to absorb the forces resulting from the deformation of the belt by tensioning. The energy stored by tensioning of the belt at minimum elongation is the basis of the power transmission. In addition, the tension ply has to absorb the centrifugal forces acting on the belt during operation.

Since the materials used for the tension ply do not have the required frictional characteristics, a separate, laminated friction ply is used as the second layer. This friction ply, which is adapted to the operating conditions with regard to material and surface finish, transmits the friction forces from the pulley surface finish to the tension ply and vice versa (Fig. 31.15). The tension ply is usually made of highly drawn

TABLE 31.1 Service Factor c_B

		Prime movers		
Application		Alternating-current and three-phase motors with a low starting torque (up to 1.5 times nominal torque); dc shunt motors; internal combustion engines with eight or more cylinders	Alternating-current and three-phase motors with moderate starting torque (1.5 to 2.5 times nominal torque); internal combustion engines with six cylinders	Alternating-current and three-phase motors with high starting torque (above 2.5 times nominal torque); internal combustion engines with four cylinders or fewer
Operating conditions	Examples			
Continuous service, small accelerated masses	Liquid-stirring apparatus, agitators, calenders and drying equipment for paper manufacture, setters, slitters and folders, centrifugal pumps and compressors, fans up to 7.5 kW, light-duty woodworking machinery, sifting plants	1.2	1.4	1.6
Interrupted service without bumps, medium-sized accelerated masses	Agitators and mixers for semifluid media, machine tools (such as grinding, turning, drilling and milling machines), punches, embossing machines, presses, textile machinery, laundry machinery, fans above 7.5 kW, generators and exciters, rotary presses, vibrating screens	1.3	1.5	1.7
Interrupted service with bumps, medium-sized accelerated masses	Elevators and worm conveyors, centrifuges, paper manufacturing machinery such as grinding gear, pumps, shredders, beaters, piston pumps and compressors, blowers, high-power fans	1.5	1.7	1.8
Service with severe bumps, large accelerated masses	Crushers and rolling mills, ball mills, tile-molding machines, compressors and high-capacity pumps, hoists	1.6	1.8	1.9

†The service factor c_B takes into account the type of prime movers and driven machines. Special operating conditions are not taken into account in these values. The factors stated are guide values.

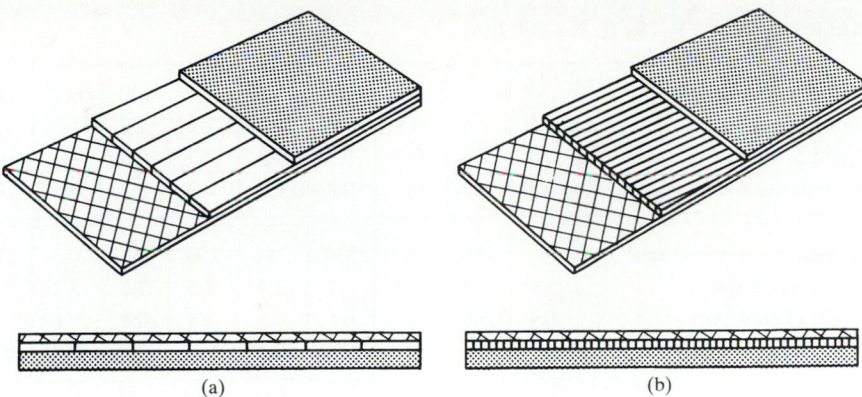

FIGURE 31.15   Multiple-ply belt.

polyamide strips or polyester cord. The friction ply, firmly attached to the tension ply, is made of either synthetic rubber or polyurethane or chrome leather. Table 31.2 shows the most important physical data for high-performance flat belts of the most commonly used tension ply materials, polyamide and polyester. The belts are manufactured in endless form according to the user length requirements or are made endless by heat-cementing the two beveled, feather-edged ends. Table 31.3 shows sizes of pulleys for flat belt drives and tolerances.

Calculations for a high-performance belt drive are usually based on data supplied by the belt manufacturer. Since the latest developments are always taken into consideration in this information, use of the latest manufacturers' data for the calculation is mandatory.

TABLE 31.2   Physical Data of High-Performance Flat Belts

		Tension ply	
Notion	Unit	Polyamid	Polyestercord
Tensile strength	N/mm^2	450–600	700–900
	N/cm	1300–18 000	1300–6600
Elongation at rupture	%	~22	~12–15
Stress at 1% elongation	N/cm	30–400	100–400
Service elongation	%	1.5–3.0	1.0–1.5
Specific nominal peripherical force	N/cm	40–800	100–400
Specific nominal power P_N	kW/cm	≤45	≤60
Maximum belt velocity	m/s	60–80	80–150
Maximum tolerable bending frequency	1/s	80–100	100–250
Elongation slip at nominal peripherical force	%	~0.8–1.0	~0.4–0.6
Attenuation (logarithmic decrement) ϑ		~0.28	~0.25
Efficiency η		0.98–0.99	0.985–0.99
Total thickness a	mm	1.0–8.0	0.8–4.0
Belt width	mm	Max. 1000	Max. 450
Belt length	mm	Unlimited	Max. 12 000

TABLE 31.3   Size of Pulleys for Flat-Belt Drives

Diameter d_1 nominal size	40	50	63	71	80	90	100	112	125
Allowable off size	0.5	0.6	0.8	1	1	1.2	1.2	1.2	1.6
Height of convexity	0.3	0.3	0.3	0.3	0.3	0.3	0.3	0.3	0.4
Tolerance of concentricity	0.2	0.2	0.2	0.2	0.2	0.2	0.2	0.3	0.3
Diameter d_1 nominal size	140	160	180	200	224	250	280	315	355
Allowable off size	1.6	2	2	2	2.5	2.5	3.2	3.2	3.2
Height of convexity	0.4	0.5	0.5	0.6	0.6	0.8	0.8	1	1
Tolerance of concentricity	0.3	0.3	0.4	0.4	0.4	0.4	0.5	0.5	0.5

In addition to the applicable ratings of the various high-performance flat belts, their types, and their configurations, the following data are necessary for the calculations for a single-step flat-belt drive:

1. Type of prime mover (driving assembly), e.g., electric motor, combustion engine, water turbine, etc.; this is important for the determination of the corresponding service-correction factors.
2. Type of machine (driven assembly); this determines corresponding load factors, dependent on acceleration, forces of gravity, changing loads, etc.
3. Power to be transmitted P, in kilowatts.
4. Speed of driving pulley n_1, in revolutions per minute.
5. Diameter of the driving pulley d_1, in millimeters.
6. Speed of the driven pulley n_2, in revolutions per minute.
7. Diameter of the driven pulley d_2, in millimeters.
8. Center distance e, in millimeters.
9. Adjustment range available (of tensioning device).
10. Allowable radial shaft loads of prime mover and driven assembly, loads on which the maximum shaft tensioning force F_w depends.

With the use of the above data and the manufacturer's data, calculations for the flat-belt drive can be made, giving the designer the type of belt, belt width, dynamic and static shaft stresses, and the required elongation, expressed as the percentage of belt strain.

Because of the special characteristics of flat-belt drives with high-performance belts, the determination of the drive data should be based on the following:

1. The belt velocity should be as high as possible (v_{opt}). The higher the belt speed, the smaller the belt width and thus the shaft load.
2. In calculating drives with changing loads or cyclic variations, you should determine to what extent the damping properties of the tension ply materials can be utilized.
3. You have to examine whether the initial tension of the belt or the shaft load can be accurately calculated from belt strain data.

4. The manufacturer can supply belts of all widths and lengths; you should examine, however, whether there are restrictions from the design point of view.

5. Finally, you must examine whether the belt can be manufactured in endless form or has to be assembled open-ended, with the ends being closed by welding after assembly.

The following general guidelines apply to pulley design:

1. The pulleys for open and crossed flat-belt drives are crowned in accordance with ISO R 100 (Table 31.3) in order to align the belt, which tends to move toward the larger pulley diameter.

2. For speed ratios higher than 1/3 ($i > 3$), the smaller pulley may be cylindrical. Spatial belt drives are equipped with cylindrical pulleys.

3. The requirements for smooth running of the belt are as follows: Parallelism of both shafts, smooth pulley faces, static balancing up to belt velocities v of 25 meters per second (m/s), and dynamic balancing for velocities above 25 m/s. When certain aluminum alloys are used, abrasion may occur, reducing friction between belt and pulley to such a degree as to make power transmission impossible.

Flat-belt drives are nonpositive flexible-connector drives used for the transmission of forces and motions between two or more shafts, particularly at greater center distances. This type of drive is superior because of its elasticity, enabling it to absorb shock loads, and its low-noise running. Its disadvantages are the greater forces acting on the shafts and bearings, resulting from the required initial tension, and the unavoidable belt slip.

These properties are decisive for the preferred applications of flat-belt drives, e.g., in machine tools, textile machinery, mixers and grinders, paper machines, gang saws, wire-drawing machines, presses, punches, and compressors. Flat belts with suitable contours may also be used as conveyor belts.

Figure 31.16 shows the drive of a hobbing machine. A high-performance flat belt was used in this case not because of its efficiency or damping properties, but because of the uniformity of rotational transmission from one pulley to the other. Preliminary studies have shown that even slight transmission deviations affect the dimensional accuracy of the tools manufactured on such machines. Belts for this application are subjected to a transmission accuracy test on a special test stand before delivery.

Figure 31.17 shows the tangential belt drive of a textile machine. This drive of a ring spinning frame is typical of a so-called multipoint drive or, in particular, a tangential belt drive. In this machine, a high-performance flat belt of 35-mm width and approximately 82-m length drives a total of 500 spindles on each side of the machine. The total power of 25 to 30 kW per machine side is thus distributed to 500 separate work positions. Depending on the spindle speed, the belt velocity ranges from 25 to 45 m/s. The absolute constancy of the belt operating tension throughout the life of the drive is a necessary prerequisite for this type of application.

31.3   V-BELT DRIVE

V-belt drives are nonpositive drives. The peripheral force F_u is transmitted by frictional forces acting on the flanks of the pulley-and-belt combination (Fig. 31.18). Bottoming of the belt in the groove leads to a reduction of the transmissible peripheral force, to belt slip, and to damages owing to overheating.

FIGURE 31.16   Drive of a hobbing machine. *(Siegling.)*

FIGURE 31.17   Tangential belt drive of a textile machine. *(Siegling.)*

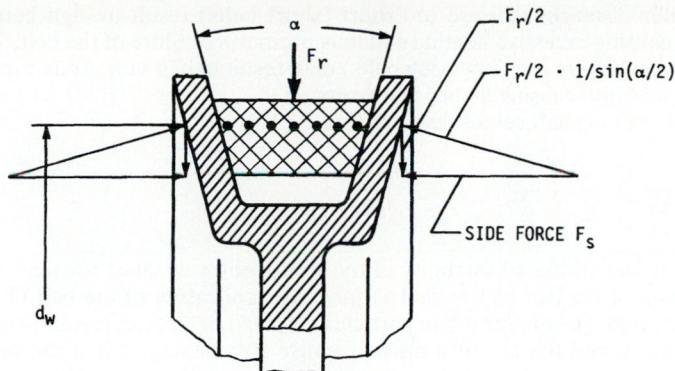

FIGURE 31.18 Section of a V-belt and sheave.

The known form of Eytelwein's equation cannot be used for the calculations for V-belt drives, since the belt creeps in the direction of travel and is simultaneously pulled radially into the groove by the radial component of the belt side force, with friction being reduced by this. Hence

$$\frac{F'_1}{F'_2} > \exp\left[\frac{\mu\beta\pi}{(\sin \alpha/2)180}\right] \qquad (31.39)$$

The arc of contact β of the smaller pulley has to be used in the calculation of a V-belt drive.

In general, a belt angle α on the order of 36° is selected, since $\alpha < 20°$ would cause self-locking. Such a belt would operate with a lot of jerking and little efficiency. By bending around the pulley during operation, the belt is stretched on the outside and compressed on the inside, the belt angle thus being reduced compared to that of the straight belt. The smaller the pulley diameter, the larger the reduction of the belt angle. Since snug fit between belt and groove sides has to be ensured, the groove angle must be adjusted accordingly. Incorrect groove angles will reduce power transmission capacity and belt life.

The calculations for V-belt drives are internationally standardized (ISO R 155). In general, manufacturers' supply data for the calculations for V-belts and other belt types, too, are as follows:

Speed ratio:

$$i = \frac{n_1}{n_2} = \frac{d_{w2}}{d_{w1}} \qquad (31.40)$$

For pulley diameters, standard series have been specified. When these pulley diameters are used, standardized speed ratios will result. Pulley diameters below the minimum values recommended for the belt section in question should not be used because the higher bending stress materially reduces belt life.

Distance between shaft centers:

Recommended lower limit: $e \geq 0.7(d_2 + d_1)$ mm

Recommended upper limit: $e \leq 2(d_2 + d_1)$ mm

$$(31.41)$$

Shaft center distances that are too short (short belts) result in high bending frequencies, causing excessive heating and thus premature failure of the belt. Shaft center distances that are too long (long belts) may result in belt vibrations, especially of the slack side, also causing higher belt stress.

Adjustment of shaft center distance (Fig. 31.19):

$$X \geq 0.03 L_w \qquad \text{mm}$$
$$X \geq 0.015 L_w \qquad \text{mm} \tag{31.42}$$

The importance of the adjustability of the shaft center distance for tensioning and retensioning of the belt (X) as well as for easy application of the belt (Y) is often underestimated. The quantity Y in particular is often neglected, resulting in application problems, and the use of tools may cause belt damage when the belt is first applied.

The transmissible power of the belt is given by

$$P = 2\pi n M \tag{31.43}$$

where M = known torque and n = speed of the corresponding pulley.

Taking into consideration a service-correction factor c_B (Table 31.1), we can calculate the power rating of the drive from P by

$$P = c_B P_1 \tag{31.44}$$

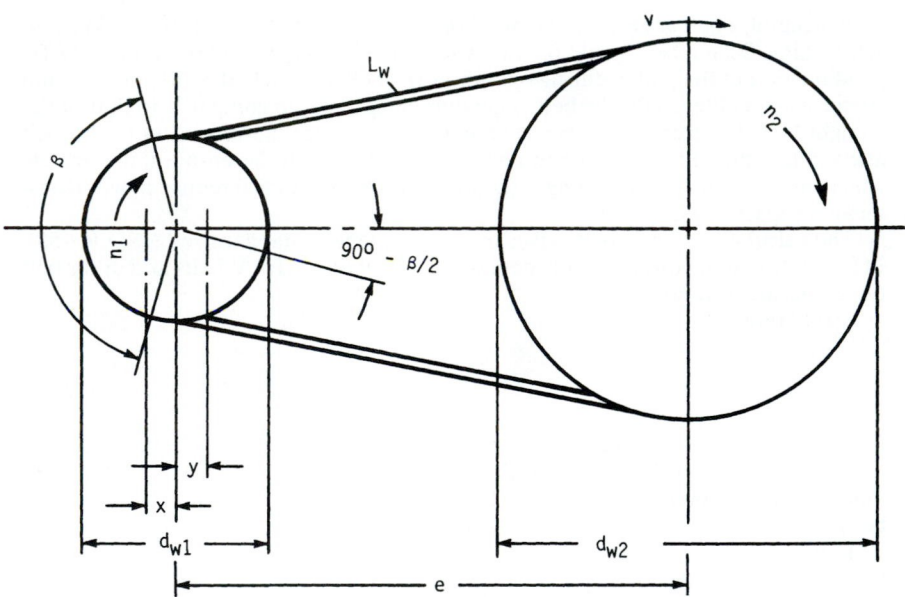

FIGURE 31.19   Adjustment of shaft center distance.

The calculation of belt velocity (peripheral velocity) is not really necessary, as the power rating tables are based on pulley diameter and speed:

$$v = \pi d_1 n_k \approx \pi d_2 n_g \qquad (31.45)$$

For multiple drives, the number of V-belts required is calculated from

$$z = \frac{P c_B}{P_N c_1 c_3} \qquad (31.46)$$

The arc of contact factor c_1, the service factor c_B, and the belt length correction factor c_3 can be found in the manuals of the manufacturers. The power rating P_N of each belt—based on the selected pitch diameter d_{w1} of the smaller pulley, the corresponding speed n_1, and the speed ratio—can be derived from tables. These tables also contain the nominal values of the service-correction factors to be used. When the guidelines mentioned in the design of a V belt are observed, the belt life to be expected is 24 000 hours (h) of operation.

As for flat belts, the bending frequency is

$$f_b = \frac{vz}{l} \qquad (31.47)$$

Normal bending frequencies are

$f_b < 30$ per second for endless standard V-belts

$f_b < 60$ per second for endless narrow-V-belts

The belts have to be pretensioned to limit belt slip to 1 percent. Improperly pretensioned belts have a life substantially shorter than the 24 000 h mentioned. The necessary initial tension will lead to equivalent shaft or bearing loads.

The approximate equation for the calculation of the required average shaft tension force is

$$F_w \cong \frac{aP}{v} \qquad 1.5 \le a \le 2 \qquad (31.48)$$

where P = power to be transmitted and v = belt velocity. More accurate methods may be found in the manufacturers' publications containing a description of the correct initial tension adjustment by force and deflection measurements at the center of the belt side.

The different belt types are distinguished by section dimensions; the configurations, however, are distinguished by belt construction. Figure 31.20 shows the most widely used V-belt configurations.

V-belts consist of the following firmly vulcanized elements:

1. Intermediate ply of high-strength cord (cotton, polyester) for transmission of the rope-pull force
2. Highly elastic belt body (rubber, plastic) for transmission of the peripheral forces between the belt flanks and the cord ply

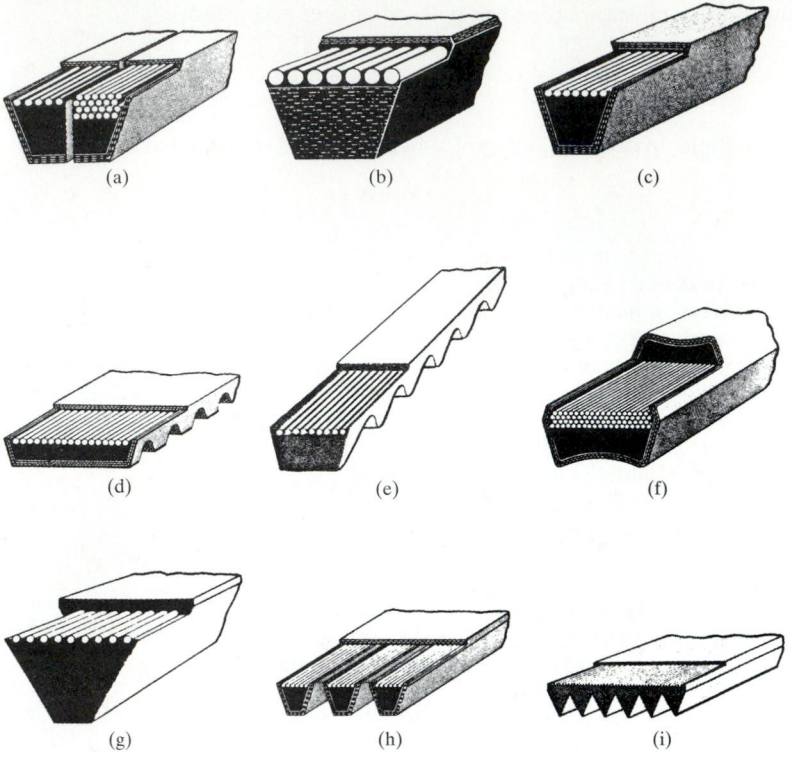

FIGURE 31.20   V-belt configurations.

3. Impregnated wear-resisting fabric coat (textile fibers) for transmission of the friction forces between the belt flanks and the V-belt pulley

Because of its higher modulus of elasticity, the cord ply forms both the neutral axis and the pitch diameter d_w of the V-belt in the groove, i.e., that diameter which is relevant for the transmission of peripheral velocity.

The endless standard V-belts (Fig. 31.20*a*) in accordance with ISO (ISO R 52, R253, R434 and R606) are supplied in specified lengths and cannot be shortened.

Endless narrow-V-belts (Fig. 31.20*c*) in accordance with ISO R 459 and R 460 are most widely used today. Their power transmission capacity is higher than that of standard V-belts of the same pitch width.

FO-type V-belts (Fig. 31.20*b*), made by the Continental Gummi-Werke AG, are distinguished by their basically different construction. They are a promising new development superior to others for the unusually small change of length their tension members undergo when loaded. Transverse short fibers in the rubber filler lead to high lateral stiffness with high flexibility in the direction of belt travel. By grinding the flanks, the traveling accuracy of the FO-type V-belt can be improved systematically. The Continental belts of this type are manufactured as endless standard V-belts of small section dimensions and as endless narrow-V-belts.

The flexibility of the V-belts mentioned thus far may be increased by cross grooves worked into the inner surface of the profile (Fig. 31.20*e*). These grooved belts allow smaller pulley diameters and require less space at slightly reduced power transmission capacity than other belts; however, the grooves are the cause of periodic running in of shock loads and noise.

A V-belt assembly (Fig. 31.20*h*) is composed of up to five standard or narrow-V-belts connected by an elastic cover band which does not rest on the pulleys. The cover band prevents twisting or excessive vibration of individual belts.

The grooves of the V-belt pulleys are standardized (ISO R 52, R 253 for standard V belts and ISO R 459 for narrow-V-belts). The pulleys are castings or weldings or steel-plate parts. In general, pulley diameters below the minimum specified should not be used. The groove flanks must be smooth and clean to ensure a sufficiently long belt life.

Applications of V-belts are practically unlimited. Because of the variety of sections available, the V-belt may be used for fractional-horsepower drives with minimum power transmission capacity in precision machines, phonographic equipment, and domestic appliances; for light drives, e.g., of centrifugal pumps and fans; and for all sizes of drives for general mechanical engineering purposes up to heavy-duty drives, such as rolling mill, rock crusher, excavator, and crane drives.

In general, the power transmission capacity is approximately 700 kW maximum; it may be increased to 1000 kW and in very rare cases to approximately 5000 kW.

Because of the extreme variety of applications in mechanical engineering, general belt-life data cannot be given. Another explanation is that belt life largely depends on the conditions of use. Assembly and operating conditions affect belt life, as do environmental conditions such as oil, dust, and climatic conditions.

The power transmission capacities specified in the relevant standards are based on the very high empiric belt life of 24 000 h, which can be reached only under optimum operating conditions, i.e., no misalignment of the pulleys, no overload, correct belt tension, normal ambient conditions, etc.

One of the particularly important areas of V-belt application is the automotive industry with its large primary equipment production series and substantial spare-part need. The V-belt as a drive unit connects the crankshaft and different accessory units, e.g., generator, cooling-water circulating pump, and fan. The V-belt may also be used for driving air-conditioning equipment or turbosuperchargers. The life of car V-belts is calculated for a total distance of 50 000 to 80 000 kilometers (km).

If higher power transmission capacities are required in mechanical engineering, multiple drives (Fig. 31.21) are used. When used as main drives in situations involving high cost in case of failure, such as mine fan drives and drives used in the metallurgical, glass, and cement industries, multiple drives also meet the demand for safety, since the probability of sudden failure of a complete belt set is very small. The uniformity of initial tensions and speed ratios of all belts of multiple drives is a prerequisite for smooth running and equal distribution of total power to all belts. Differences in speed ratios and initial tensions will lead to a reduction in belt life. Thus the complete set of belts should be replaced, if necessary, not just a single endless belt.

31.4 SYNCHRONOUS-BELT DRIVE

The synchronous belt is a relatively new machine element combining the advantages of positive and nonpositive flexible connectors. In contrast to the nonpositive belt drives, the pulleys and belts of synchronous-belt drives have meshing teeth, allowing synchronous power transmission with angular accuracy (Fig. 31.22). To prevent rid-

FIGURE 31.21 Multiple-V-belt drive of a crusher in mining. *(Arntz-Optibelt-KG.)*

ing off of the belt, at least one of the pulleys has flanges which are slightly beveled to reduce lateral friction.

Because of their special properties, synchronous belts are used wherever synchronous power transmission is asked for and the safety and freedom-from-maintenance requirements are strict.

As for all positive drives, the conformity of belt and pulley pitches is of utmost importance. The advancing belt tooth must correctly mesh with the corresponding pulley groove and remain there until leaving it. To achieve this, the pitch of the deflected belt must correspond exactly with that of the pulley. Hence, the belt must be made of high-strength material experiencing little length change under load; thus, the neutral axis may be assumed to be in the center. The pitch line is situated outside the pulley-tip-circle radius at a distance equaling that of the neutral axis (u value). This value is a largely invariable quantity which depends on the belt construction and must be taken into consideration in the design of the pulleys. Figure 31.22 shows the geometric relations between belt and pulley.

Despite the use of materials with little change of length under load for the tension member (fiber glass, steel cord), the belt pitch is a variable dependent on the strain properties of the belt and the applicable belt pull. Along the arcs of contact there is a step-by-step change in the belt pull with a corresponding change in local belt pitch, so that, in compensating belt and pulley pitch deviation, the teeth are deformed differently and the load differs accordingly. In addition, there is a minor pitch difference between belt and pulley, which is attributable to the production process and has to be compensated by deformation of the belt teeth, too.

The load distribution along the arc of contact as a measure of peripheral force distribution to the individual teeth is thus mainly dependent on the pitch deviation. The pitch deviation within the pretensioned belt (idling) is of particular importance:

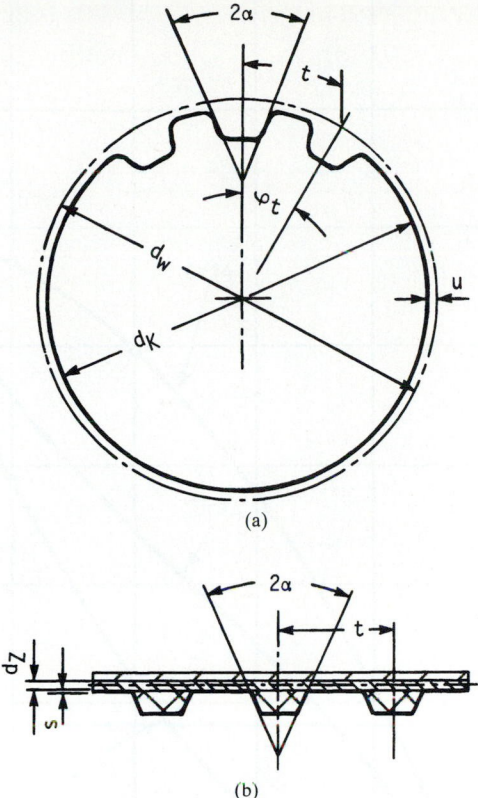

FIGURE 31.22 Synchronous-belt drive. (*a*) Pulley dimensions are d_w = working diameter, d_k = peripheral diameter, t = pitch, φ_t = pitch angle, u = theoretical distance to working diameter, and 2α = angle of tooth; (*b*) belt dimensions are d_z = diameter of tensile member and s = thickness of cover.

- Pitches of the pretensioned belt and the pulleys are exactly the same ($\Delta t = 0$). Load distribution is symmetric. The teeth at the beginning and at the end of the arc of contact carry a higher load than those along the center of the arc.
- The pitch of the pretensioned belt is smaller than the pulley pitch ($\Delta t < 0$). Load distribution is unsymmetric. Power is mainly transmitted at the beginning of the arc of contact (driven pulley) or the end of the arc of contact (driving pulley).
- The pitch of the pretensioned belt is larger than the pulley pitch ($\Delta t > 0$). Load distribution is unsymmetric. Power is mainly transmitted at the end of the arc of contact (driven pulley) or the beginning of the arc of contact (driving pulley).

For the driven pulley Fig. 31.23 shows the pull varying as a function of load-distribution deviations from the beginning of the arc of contact ($n/z = 0$) to the end of the arc of contact ($n/z = 1$).

Synchronous belts are available in different dimensions (pitch, belt width, and tooth geometry) according to the required application and range of capacity. The

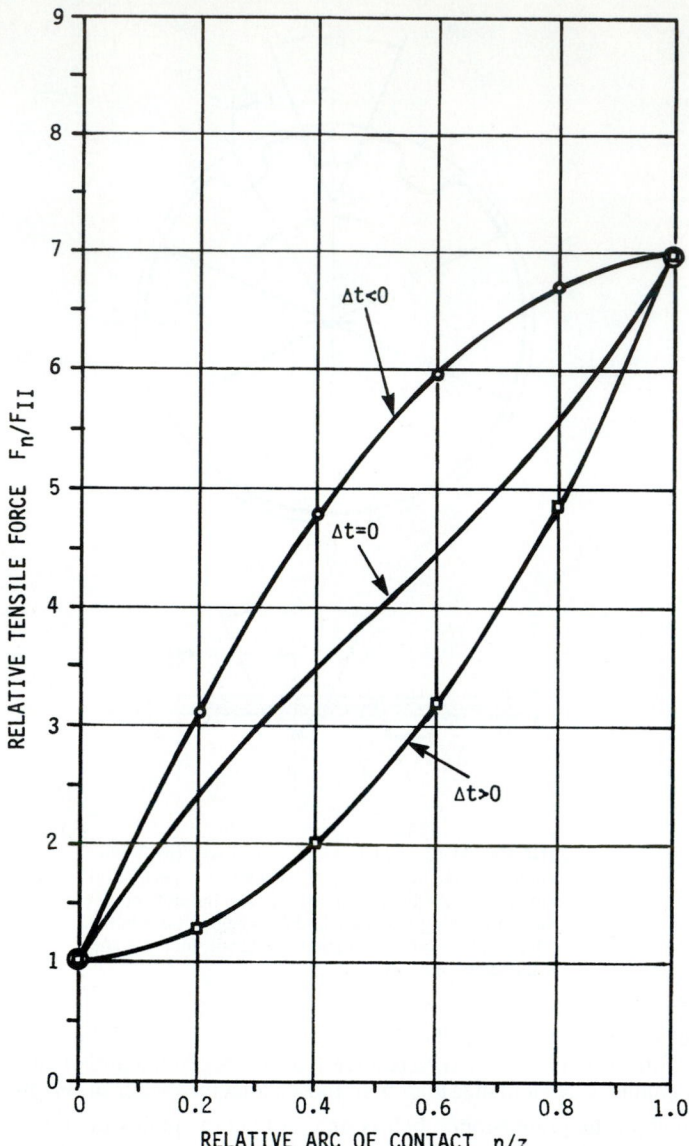

FIGURE 31.23   Theoretical tensile force diagram for the driven pulley.

transmissible power P is determined by the standard capacity P_N of the selected belt and by the operating conditions, expressed by the service-correction factor c_B:

$$P = \frac{P_N}{c_B} \qquad (31.49)$$

The service-correction factors range from 1.0 to 2.10. The accurate value is determined by the type of drive and the type of application as well as the daily period of operation. For special applications, further restrictions may be required; details (meshing factor, acceleration factor, etc.) are contained in the manufacturers' catalogs.

The ranges of power transmission capacity are fixed in accordance with belt pitch. According to ISO 5296, standard belts have the standard pitch values shown in Table 31.4.

The capacities listed in Table 31.4 are nominal values for medium belt velocities; they may vary considerably to the positive or negative depending on the operating conditions. Accurate calculation has to be based on manufacturers' data.

When making calculations for a synchronous-belt drive, you must take into account that the pitch may be changed only in integral numbers, in contrast to non-positive belt drives. The individual design characteristics are as follows:

Speed ratio:

$$i = \frac{z_1}{z_2} = \frac{d_2}{d_1} \qquad (31.50)$$

where z_2 = number of teeth of larger pulley
z_1 = number of teeth of smaller pulley
d_2 = pitch diameter of larger pulley
d_1 = pitch diameter of smaller pulley

Datum length:

$$l = 2e \sin\frac{d}{2} + \frac{t}{2}\left[z_1 + z_2 + \frac{\beta}{90°}(z_2 - z_1)\right] \qquad (31.51)$$

Approximation:

$$l \approx 2e + \frac{t}{2}(z_1 + z_2) + \left(\frac{t}{2\pi}\right)^2 \frac{(z_2 - z_1)^2}{l} \qquad (31.52)$$

where z_1 = number of teeth of smaller pulley
z_2 = number of teeth of larger pulley
t = pitch, in
l = belt datum length = $z_R t$
z_R = number of belt teeth

TABLE 31.4 Standard Pitch Values

	Extra light (XL)	Light (L)	Heavy (H)	Extra heavy (XH)	Double extra heavy (XXH)
Belt pitch, in	$\frac{1}{4}$	$\frac{3}{8}$	$\frac{1}{2}$	$\frac{7}{8}$	$1\frac{1}{4}$
Nominal power, kW	0.15	1.0	10	40	100

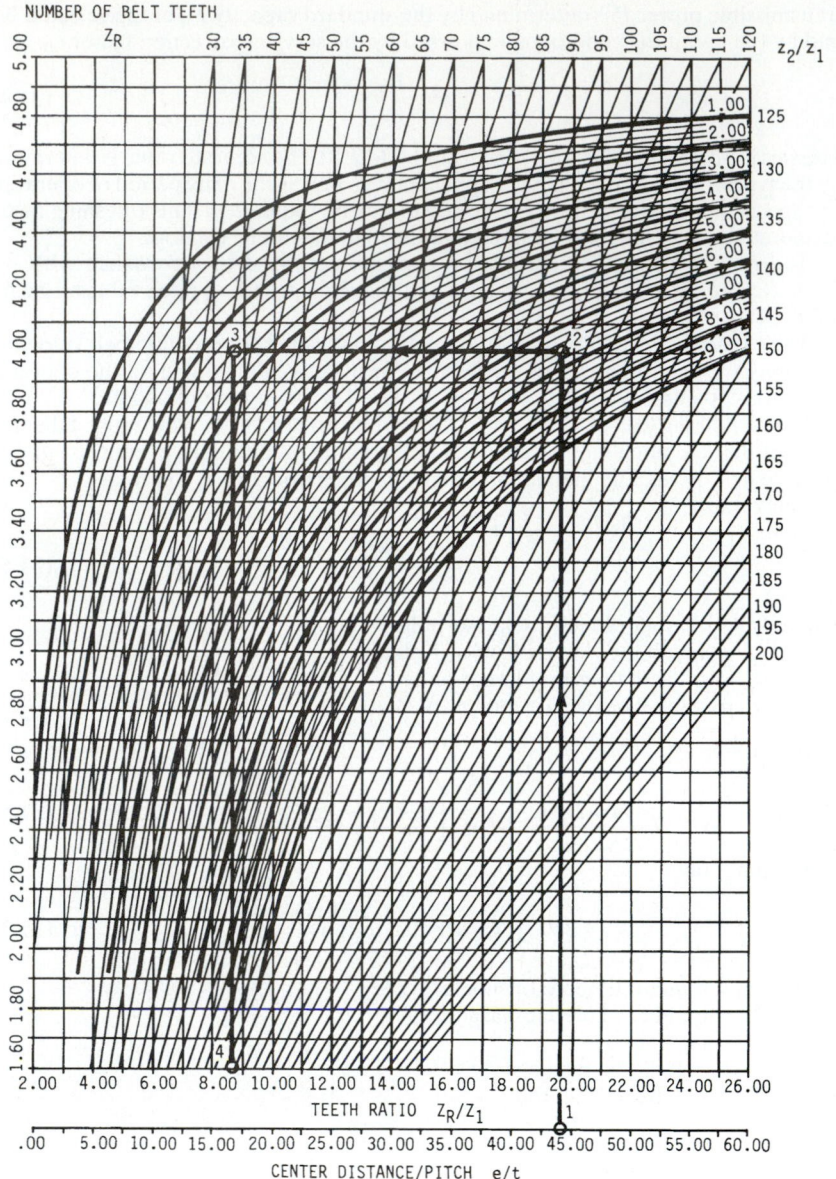

FIGURE 31.24 Determination of center distance of synchronous belts.

The accurate number of pulley and belt teeth for a given center distance may be found from detailed tables included in the manufacturers' catalogs. Guideline values may also be determined by using the nomogram in Fig. 31.24. Enter the nomogram at the ratio of given center distance to pitch (point 1), and proceed vertically to intersection 2 with the straight line indicating the number of teeth of the belt intended for use; then proceed horizontally from this point to intersection 3 with the curve of the given speed ratio. The number of teeth of the smaller pulley can be calculated from the applicable value on the axis of the tooth ratio of the belt and the smaller pulley (point 4), and the number of teeth of the larger pulley from the speed ratio; the results are rounded to the next higher whole numbers.

When calculating the number of teeth or the geometric design data, the designer has to ensure that the number of meshing teeth is at least six to eight. For the transmission of larger peripheral forces, the minimum number of meshing teeth has to be increased in accordance with manufacturers' data.

The transmissible peripheral force F_u is closely related to the effective shaft tensioning force F_w, which is dependent on the initial tensioning force F_A and the operating conditions. As in chain drives, the effective shaft tensioning force F_w increases with increasing load moment. As a guideline, assume the initial tensioning force F_A to be in the range

$$F_u \leq F_A \leq 1.5F_u \tag{31.53}$$

and the belt side-force ratio to be $F_I/F_{II} \cong 5$, where F_I is the tight-side force and F_{II} is the slack-side force.

The permissible initial tensioning force F_A is determined by the construction and the dimensions of the selected belt. Particular attention has to be paid to the fact that the belt should not be strained to such a degree as to prevent proper functioning of the belt-and-pulley combination because of the resultant pitch deviation.

In addition to the standard synchronous belts, according to ISO 5296, other types of synchronous belts with metric pitch but changed tooth configuration or shape are used for special applications. Figure 31.25 shows the different constructions of common synchronous belts.

The teeth of the standard synchronous belt (Fig. 31.25a) are trapezoidal. The belt is made of highly elastic synthetic rubber (polychloroprene) with an intermediate ply (tension member) of high-strength fiber-glass cord with little change of length under load. The pulley side of the belt is made of abrasion-resistant polyamide fabric. The back of the belt is ground to ensure nonpositive driving of belt tighteners or additional pulleys.

The polyurethane synchronous belt, shown in Fig. 31.25b, has a similar tooth shape. This belt is made of a highly elastic and abrasion-resistant plastic (polyurethane) and needs no pulley-side fabric cover to reduce wear. The intermediate ply taking the tension consists of steel cord, and so the belt strain is reduced to a minimum. The belt pitch is metric. Common pitch values are 5 mm for belt type T5, 10 mm for T10, and 20 mm for T20.

Heavy-duty belts are innovations with a tooth shape designed to ensure optimum distribution of the flank pressure over the entire tooth surface. The belt teeth have an enlarged cross-sectional area for the reduction of transverse (shear) stress, their edges being chamfered. Trade names differ among manufacturers (HTD, Super-Torque, etc.). Neither the tooth shape nor the pitch is standardized and both may vary with the product, although the pitch is usually metric. Belt materials and construction are similar to those of standard belts. Common pitch values are 8, 14, and 22 mm.

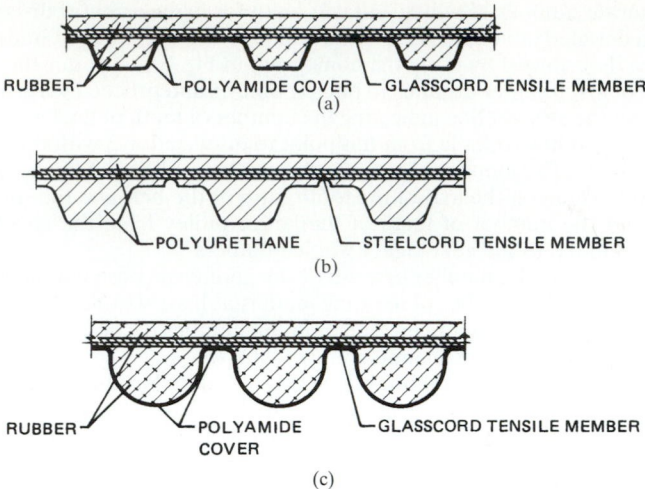

FIGURE 31.25   Construction and differences of timing belts. (*a*) Poly-chloroprene (rubber); (*b*) polyurethane; (*c*) heavy-duty.

In addition to the belt types mentioned above, there are special belts with pitches or tooth shapes adapted to particular applications. There exist also synchronous belts with teeth on the back to provide for a two-side running capability and others with special back configurations, e.g., for transportation purposes. Geometric data for these special belts may be found in the catalogs of the respective manufacturers. Figure 31.26 shows an assortment.

Because of their ability to transmit power synchronously with the required angular accuracy, synchronous belts are used to an ever-increasing extent, with new modes of application being added constantly. Classic applications are as follows:

- Light-duty conveyors (office machinery, food industry)
- Positioning drives (peripheral equipment of electronic data processing systems, machine tools, screening machinery)
- Synchronizing gears [camshaft drives (OHC), textile machinery, paper machinery]

Figures 31.27 to 31.29 show some typical synchronous-belt applications.

To ensure trouble-free operation and a long life of the drive, the following recommendations should be observed:

- The belt should be pretensioned only to the degree necessary to prevent skipping of the belt on starting or braking. Excessive initial tension leads to reduced belt life and in some cases to extreme noise levels.
- To prevent tension member damage (e.g., kinking or breaking), the belt has to be placed on the pulleys without tilting them. To facilitate this, an axial-shifting device should be provided on one of the pulleys.
- To prevent excessive wear of belt sides and noise from lateral contact of belt and pulley flanges, shafts and pulleys have to be aligned with extreme care. Slight lateral contact of belt and pulley flanges is attributable to the manufacturing process

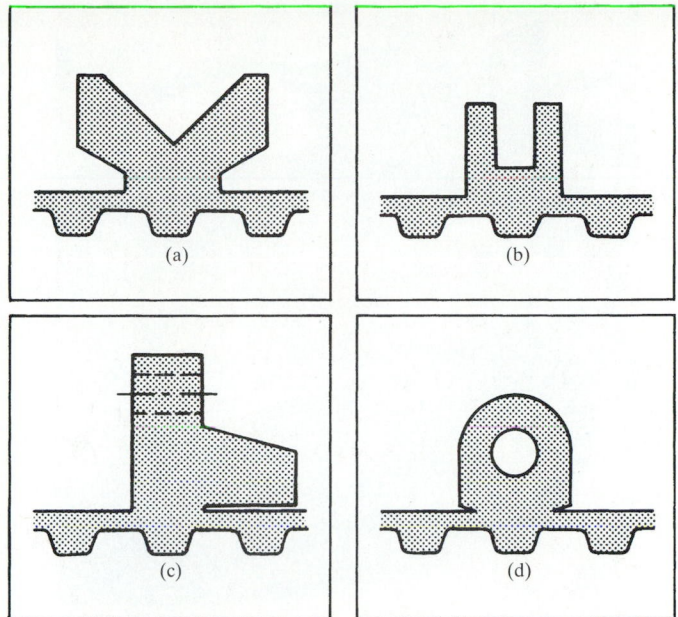

FIGURE 31.26 Synchronous belts with special back configuration.

FIGURE 31.27 Synchrobelt timing belts for sophisticated type-writer drive mechanism. *(Continental Gummi-Werke Aktienge-sellschaft.)*

FIGURE 31.28 Planer machine drive with Synchrobelt timing belt; the driving speed is 30 000 min^{-1}. *(Continental Gummi-Werke Aktiengesellschaft.)*

FIGURE 31.29 Camshaft control with Synchrobelt HTD timing belt. *(Continental Gummi-Werke Aktiengesellschaft.)*

(twisting of tension member and its slight inclination) and does not affect the running characteristics or the life of the belt.

- If feasible, belt tighteners should be arranged to act on the inner surface of the slack side of the belt; they should be toothed, too. The arc of contact should be as small as possible.

31.5 OTHER BELT DRIVES

31.5.1 Special Types

Poly-V-belts are similar to V-belt assemblies; however, their cover band rests on the pulleys, contains the cord, and takes the tension, whereas the individual V-sections, called *ribs,* solely transmit the peripheral force between pulley and cover band. The ribs have no cords (Fig. 31.20*i*) and fill up the pulley grooves almost to the bottom. Thus they are susceptible to foreign matter entering between the belt and the pulley. Because of the relatively rigid cover band resting on the face of the pulley, particularly accurate alignment of the pulleys is required, as it is with flat-belt drives. The belts are remarkably quiet.

V-belts with a belt-groove angle of 60° are cast belts (Fig. 31.20*g*) made of polyurethane with polyester tension members. Because of the higher friction coefficient of polyurethane on steel, the large included angle of 60° is required to prevent self-locking. Because of the manufacturing process, these belts show a substantially greater dimensional accuracy, resulting in particularly quiet operation. They are designed for maximum velocities of up to 50 m/s and are used mainly in machine tools.

Hexagonal (double-V) belts (Fig. 31.20*f*) have a symmetric special profile. The ratio of the maximum width to the height of the section is approximately 1.25. Double-V-belts may be operated in one plane on counterrotating pulleys. The power transmission capacity is approximately equal to that of conventional V-belts of equal width with small pulley diameters. Minimum pulley diameters are also about equal to those of conventional V-belts. Double-V-belts are suited for multiple-shaft drives operated in one plane with counterrotating pulleys. They are especially suited for coupled operation, with a tensioning or coupling shaft riding on the back of the belt and being able to transmit power, too.

Double-V-belts are used for medium-duty drives (combine harvester) as well as for light-duty tools (gardening equipment, rotary sweepers).

Toothed V-belts, which must not be confused with synchronous belts, are available as standard, narrow-, and wide-V-belts with punched or preformed teeth in the belt carcass to increase flexibility. Thus, with only a slight reduction in power transmission capacity, the minimum permissible diameter of the smaller pulley may almost be halved, so that the space requirement of such a belt drive with equal ratio or speed range is reduced substantially. These toothed V-belts are subject to the effects produced by their polygonal shape; i.e., the teeth cause running in of shock loads (which may result in irregular transmission of motion), additional dynamic stress of bearings, and even (in the case of high peripheral velocities) noise.

Round belts are versatile, simple, and reliable connectors for the transmission of small torques and medium velocities. Their main advantage is polydirectional flexibility. There are two different types: homogeneous round belts made of one material (rubber, plastic) and round belts with tension member (Fig. 31.30).

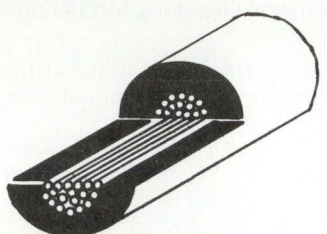

FIGURE 31.30 Continental round section belt.

31.5.2 Variable-Speed Belt Drives

For variable-speed belt drives (with a continuously variable speed ratio), especially wide-V-belts have been developed in toothed and nontoothed configurations. The construction corresponds to that of endless standard V-belts with the exception of the substantially greater belt width. This great belt width (the ratio of the upper width of the belt to the section height $b/h = 2/5$) is required for radial shifting on pulleys, the halves of which can be moved in an axial direction (Fig. 31.20d). These variable-speed belt drives allow for a continuously variable speed ratio within the range of $i_{max}/i_{min} = 4/10$. With larger ratios of belt width to section height, greater speed ranges are possible in general.

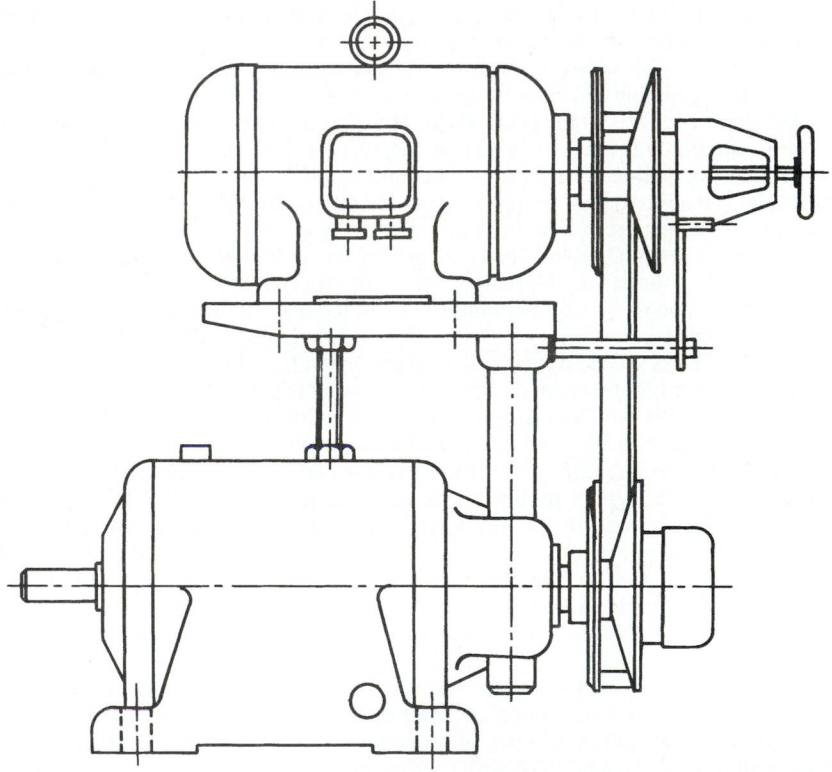

FIGURE 31.31 Variable-speed wide-belt drive.

The power transmission capacity of variable-speed belts is approximately equal to that of standard V-belts of equal section height; with increasing b/h ratio, however, it decreases by up to 20 percent. Depending on the speed range and the design of the variable-speed pulleys, the included angles range from 22° to 34° (ISO recommendation is 26°), with the smaller angles resulting in greater speed ranges but, because of the approach to the self-locking limit, reduced rated outputs.

A variable-speed wide-belt drive combined with an electric motor and a gear train is shown in Fig. 31.31. The top right half of the pulley may be shifted by turning the handwheel. The bottom pulley is an adjusting pulley (Fig. 31.32) which adapts itself automatically to the respective speed ratio.

31.6 COMPARISON OF BELT DRIVES

A flexible connector for a particular application should be selected on the basis of the following considerations:

- *Performance:* The flexible-connector drive has to perform reliably for an adequate period under the given operating conditions (speeds, moments, space

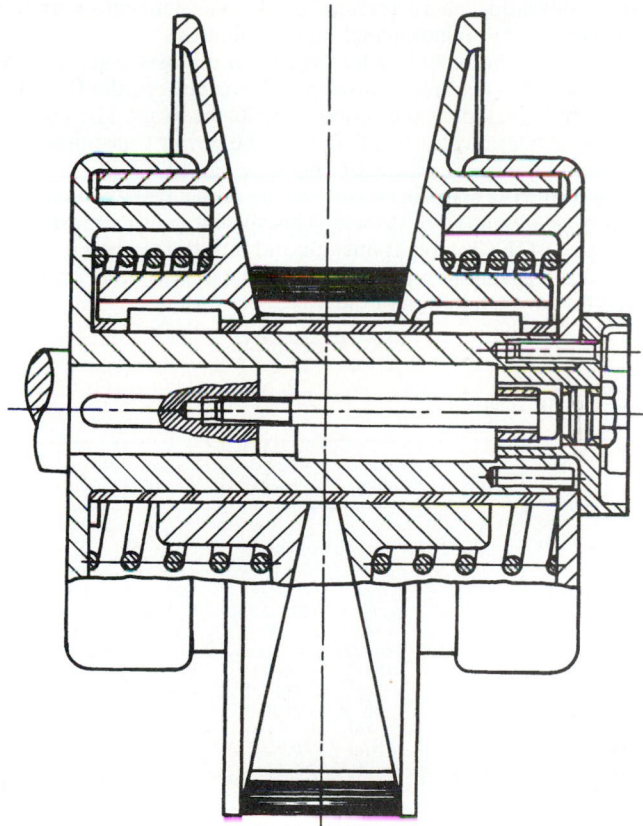

FIGURE 31.32 Adjusting pulley.

requirement, overloads, shaft dislocations, temperatures, and other environmental conditions).

- *Economy:* Those flexible-connector drives suitable under performance criteria are further investigated with respect to their applicability from the economic point of view. The economy of flexible drives is influenced by the following factors:

    1. Production cost of the flexible connector
    2. Subsequent design expenses required because of the selection made (e.g., pulley costs, costs of larger bearings to take up the required initial tensioning force, costs of lubrication facilities and seals required, of special installation arrangements, and of tensioning devices)
    3. Transportation and installation costs, mainly dependent on the space requirement and the weight of the drive
    4. Operation and maintenance costs
    5. Failure risk as well as costs of repair and subsequent costs

The relative importance of these factors varies with the machines used, and the costs of the flexible connector as the drive element are the least important. A drive unit will prevail wherever its special technical properties ensure the most economical drive.

To facilitate the selection of suitable flexible connectors, Table 31.5 compares the most important operating characteristics of flexible connectors and remarks on investment costs as well as maintenance and installation.

Figure 31.33 shows the results of a comparison of costs, when different flexible connectors are used, for a system consisting of two pulleys, the flexible connector, and the driven shaft, including the required roller bearings. The costs were determined following calculations for the drive with the power transmission capacity, the center distance, and the speed of the driving pulley given.

Figure 31.34 shows the specific power transmission capacity of different flexible-connector drives as a function of velocity. This specific power transmission capacity indicates the rating of the flexible connector and thus the space required. The higher the specific power transmission capacity of the flexible connector, the smaller the space required for the installation of the drive.

TABLE 31.5   Properties of Belt Drives

	Flat belts	Toothed belts†	Poly-V belts	Standard V belts	Narrow-V belts	Connected V belts†	Round belts†
Installation cost index	1.1	1.4	1.2	1.4	1.0	ND	ND
Maintenance	Yes	No	Yes	Yes	Yes	Yes	Yes
Power per volume, kW/cm^3	0.8	1.9	1.7	0.7	1.8	ND	ND
Maximum bending frequency, Hz	200	200	100	40	80	40	40
Shaft load	$2F_u\text{–}3F_u$	F_u	$2F_u\text{–}2.5F_u$	$2F_u\text{–}2.5F_u$	$2F_u\text{–}2.5F_u$	$2F_u\text{–}2.5F_u$	$2F_u\text{–}3F_u$
Efficiency, %	98	98	97	95	96	94	95
Diameter transmission	Constant	Constant	Constant		Variable		Constant
Bending rate d/s, dimensionless	15	15–30	5–11	8–14	8–12	8–14	8–10
Girder rate F_u/F_w, dimensionless	0.3–0.4	1	0.4–0.5	0.5–0.6	0.5–0.6	0.4–0.5	0.4
Admissible temperature, °C	−40 to +80	−50 to +120	−50 to +100	−55 to +70	−55 to +70	−55 to +70	−40 to +100
Adjustment of shaft center distance X, mm	$0.02L_w$	NA	$0.014L_w$	$0.02L_w$	$0.02L_w$	$0.02L_w$	$0.03L_w$
Mounting adjustment Y, mm	$0.01L_w$	$0.01L_w$	$0.02L_w$	$0.015L_w$	$0.015L_w$	$0.015L_w$	$0.01L_w$

†NA, not applicable; ND, no data.

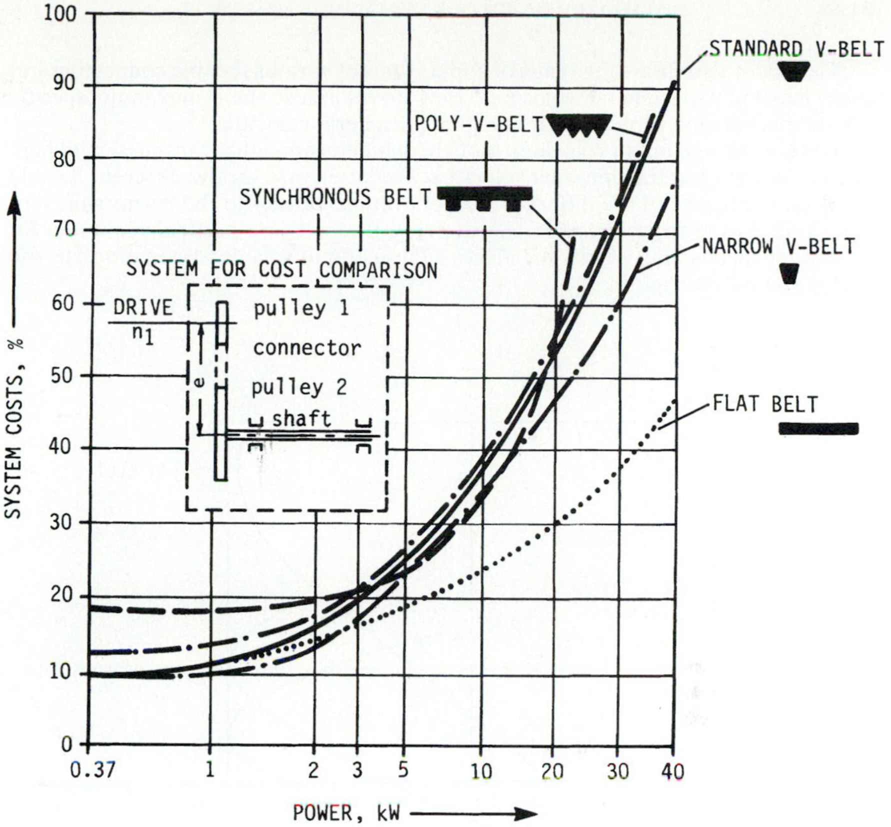

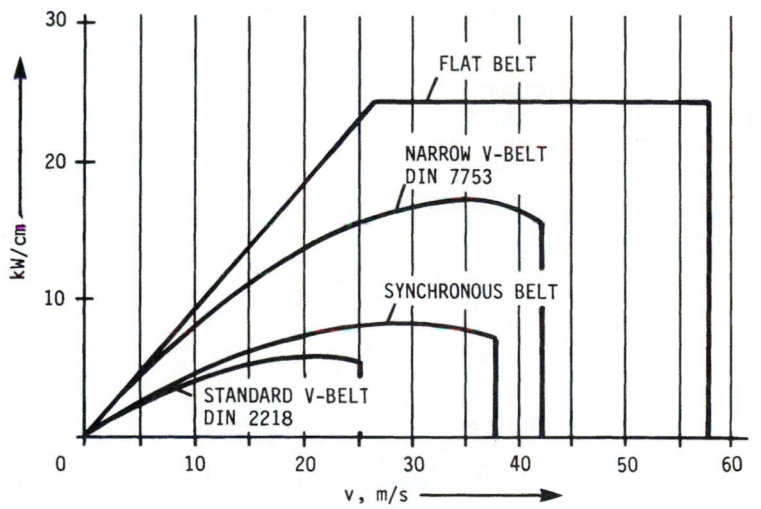

FIGURE 31.33 Comparison of system costs.

FIGURE 31.34 Specific power transmission capacity.

The figure also shows the range of applications of various flexible connectors and their maximum peripheral velocities. Belt drives reach their maximum specific power transmission capacity only at higher peripheral velocities.

It is not the maximum velocities, though, that determine the transmission of high specific powers, but the optimum velocities. The latter vary for the different flexible connectors because of the differing ratio of average density to the permissible tensile strain, as shown in Fig. 31.35. Accordingly, flat belts are particularly suited for maximum speeds and V-belts for most of the medium speeds of common electric motor and piston engines.

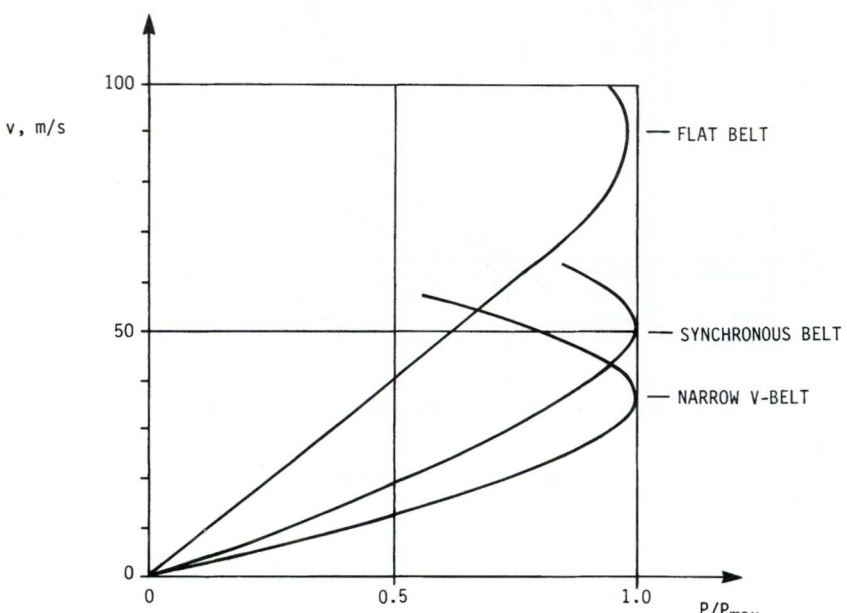

FIGURE 31.35 Optimum velocities.

CHAPTER 32
CHAIN DRIVES

John L. Wright
General Product Manager
Diamond Chain Company
Indianapolis, Indiana

NOTATION

BD	Bottom diameter, in
C	Center distance, in chain pitches
CD	Caliper diameter, in
CCD	Chain clearance diameter, in
D	Roller outside diameter, in
D_p	Gauge pin diameter, in
G	Maximum guide groove diameter, in
H	Maximum chain height, in
HP	Horsepower
K_f	Constant for link plate fatigue
K_r	Constant for roller and bushing impact
L	Chain length, in chain pitches
MHD	Maximum hub or groove diameter, in
MUTS	Minimum ultimate tensile strength, lb
n	Number of chain strands
N	Number of sprocket teeth
N_1	Number of teeth on small sprocket
N_2	Number of teeth on large sprocket

OD	Sprocket outside diameter, in
OGD	Over-gauge diameter, in
P	Chain pitch, in
PD	Sprocket pitch diameter, in
R	Sprocket speed, r/min
T	Thickness of link plate or sidebar, in
W	Chain (roller) width, in

32.1 TYPES, USES, AND CHARACTERISTICS

32.1.1 Chain Drives Compared

Three major types of chain are used for power transmission: roller, engineering steel, and silent. Roller chains are probably the most common and are used in a wide variety of low-speed to high-speed drives. Engineering steel chains are used in many low-speed, high-load drives. Silent chains are mostly used in high-speed drives. Other types of standard chains, and many types of special chains for unique applications, may be found in manufacturers' catalogs.

Chains can span long center distances like belts, and positively transmit speed and torque like gears. For a given ratio and power capacity, chain drives are more compact than belt drives, but less compact than gear drives. Mounting and alignment of chain drives does not need to be as precise as for gear drives. Chain drives can operate at 98 to 99 percent efficiency under ideal conditions. Chain drives are usually less expensive than gear drives and quite competitive with belt drives.

Chain drives can be dangerous. Provide proper guarding to prevent personnel from coming in contact with, or being caught in, a running drive. Any chain can break from unexpected operating conditions. If a chain breaks at speed, it can be thrown off the drive with great force and cause personal injury and property damage. Provide adequate guarding to contain a broken chain or to prevent personnel from entering an area where they might be struck by a broken chain. A broken chain can sometimes release a load and cause personal injury and property damage. Provide an adequate brake or restraint to stop and hold the load in case of a chain breakage.

32.1.2 Roller Chains

Standard Roller Chains. A portion of a typical roller-chain drive is shown in Fig. 32.1. The American National Standards Institute (ANSI) has standardized limiting dimensions, tolerances, and minimum ultimate tensile strength for chains and sprockets of 0.25 to 3.0 in pitch [32.1]. The chain pitch is the distance between successive roller, or bushing, centers, and is the basic dimension for designating roller chains. The standard includes both standard and heavy series chains.

Multiple-Strand Roller Chains. Multiple-strand roller chain consists of two or more parallel strands of chain assembled on common pins. They also are standardized [32.1].

Double-Pitch Roller Chains. Double-pitch roller chains are standardized in Ref. [32.2]. Double-pitch chains have the same pin, bushing, and roller dimensions as cor-

FIGURE 32.1 Typical roller chain on sprocket. *(Diamond Chain Company).*

responding chains in Ref. [32.1], but the pitch of the link plates is twice as long. The standard [32.2] covers chains of 1.0 to 4.0 in pitch.

Nonstandard Roller Chains. Many manufacturers offer high-strength, extra-clearance, sintered metal bushing, sealed-joint, and corrosion-resistant chains for special applications or adverse environments. These chains are not covered by any standard, but most are designed to run on standard sprockets.

Sprockets. Roller-chain sprockets have precisely designed, radiused pockets which smoothly engage the rollers on the chain and positively transmit torque and motion. Driver sprockets receive power from the prime mover and transfer it to the chain. Driven sprockets take power from the chain and transfer it to the selected machinery. Idler sprockets transmit no power; they are used to take up slack chain, increase the amount of chain wrap on another sprocket, guide the chain around other machine members, and reverse the normal direction of rotation of another sprocket.

32.1.3 Engineering Steel Chains

Standard Engineering Steel Chains. The engineering steel chains designated for power transmission are heavy-duty offset sidebar chains. Limiting dimensions, tolerances, and minimum ultimate tensile strength for chains and sprockets of 2.5 to 7.0 in pitch are standardized in Ref. [32.3].

Nonstandard Chains. Some manufacturers offer engineering steel chains in straight-sidebar and multiple-strand versions, and in pitches that are not included in Ref. [32.3]. Although these chains are not standardized, they are listed in manufacturers' catalogs because they are used extensively in special applications.

Sprockets. Machine-cut engineering-steel-chain sprockets look much like roller-chain sprockets, but they have pitch line clearance and undercut bottom diameters to accommodate the dirt and debris in which engineering-class chain drives often operate.

32.1.4 Silent Chain

Standard Silent Chains.   Silent (inverted-tooth) chains are standardized in Ref. [32.3] for pitches of 0.375 to 2.0 in. Silent chain is an assembly of toothed link plates interlaced on common pins. The sprocket engagement side of silent chain looks much like a gear rack. Silent chains are designed to transmit high power at high speeds smoothly and relatively quietly. Silent chains are a good alternative to gear trains where the center distance is too long for one set of gears. The capacity of a given pitch of silent chain varies with its width. Standard widths of silent chain range from 0.5 to 6.0 in for 0.375-in pitch, and from 4.0 to 30.0 in for 2.0-in pitch.

Nonstandard Silent Chains.   Some manufacturers offer silent chains with special rocker-type joints. These chains generally transmit higher horsepower more smoothly and quietly than the standard joint designs. However, they generally require sprockets with special tooth forms.

Sprockets.   Silent-chain sprockets have straight-sided teeth. They are designed to engage the toothed link plates of the chain with mostly rolling and little sliding action.

32.2 ROLLER CHAINS: NOMENCLATURE AND DIMENSIONS

32.2.1 Standard Roller-Chain Nomenclature

Roller Chain.   Roller chain is an assembly of alternating roller links and pin links in which the pins pivot inside the bushings, and the rollers, or bushings, engage the sprocket teeth to positively transmit power, as shown in Fig. 32.1 and the illustration with Table 32.1.

Roller Links.   Roller links are assemblies of two bushings press-fitted into two roller link plates with two rollers free to rotate on the outside of each of the bushings.

Pin Links.   Pin links are assemblies of two pins press-fitted into two pin link plates.

Connecting Links.   Connecting links are pin links in which one of the pin link plates is detachable and is secured either by a spring clip that fits in grooves on the ends of the pins or by cotters that fit in cross-drilled holes through the ends of the pins. Illustrations of connecting links may be found in Ref. [32.1] or [32.4] or in manufacturers' catalogs.

Offset Links.   Offset links are links in which the link plates are bent to accept a bushing in one end and a pin in the other end. The pin may be a press fit in the link plates, or it may be a slip fit in the link plates and be secured by cotters. Illustrations of offset links may be found in Ref. [32.1] or [32.4] or in manufacturers' catalogs.

32.2.2 Roller-Chain Dimensions and Numbering

Standard Chain Dimensions.   The three key dimensions for describing roller chain are pitch, roller diameter, and roller width. The pitch is the distance between adjacent bushing centers. The roller diameter is the outside diameter of the chain rollers.

TABLE 32.1 Roller Chain Dimensions

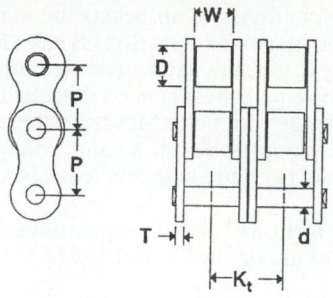

| | | | | | (Dimensions in inches; MUTS in lbf) | | | |
| ANSI chain no. | Chain pitch, P | Roller diameter, D | Roller width, W | Pin diameter, d | Link plate thickness, T | | Transverse pitch, K_t | |
					Std.	Heavy	Std.	Heavy
25	0.250	0.130*	0.125	0.0905	0.030	—	0.252	—
35	0.375	0.200*	0.188	0.141	0.050	—	0.399	—
41**	0.500	0.306	0.250	0.141	0.050	—	—	—
40	0.500	0.312	0.312	0.156	0.060	—	0.566	—
50	0.625	0.400	0.375	0.200	0.080	—	0.713	—
60	0.750	0.469	0.500	0.234	0.094	0.125	0.897	1.028
80	1.000	0.625	0.625	0.312	0.125	0.156	1.153	1.283
100	1.250	0.750	0.750	0.375	0.156	0.187	1.408	1.539
120	1.500	0.875	1.000	0.437	0.187	0.219	1.789	1.924
140	1.750	1.000	1.000	0.500	0.219	0.250	1.924	2.055
160	2.000	1.125	1.250	0.562	0.250	0.281	2.305	2.437
180	2.250	1.406	1.406	0.687	0.281	0.312	2.592	2.723
200	2.500	1.562	1.500	0.781	0.312	0.375	2.817	3.083
240	3.000	1.875	1.875	0.937	0.375	0.500	3.458	3.985

* Bushing diameter. Chain is rollerless.
** Lightweight chain
Illustration courtesy of Diamond Chain Company.

The roller width actually is the inside distance between roller link plates. These and other selected dimensions are shown in Table 32.1.

Ultimate Tensile Strength. The minimum ultimate tensile strength (MUTS) for standard chains is given in Ref. [32.1]. The value is estimated from the equation

$$\text{MUTS} = 12\,500P^2n$$

Chain Numbering. A standard numbering system is described in Ref. [32.1]. The right digit indicates the type of chain: 0 for a standard roller chain, 5 for a rollerless bushing chain, and 1 for a light-duty roller chain. The left one or two digits designate the chain pitch in eighths of an inch; for example, 6 indicates ⅝, or ¾-in pitch. An H immediately following the right digit designates heavy series chain. Multiple-strand chain is designated by a hyphen and one or two digits following the right digit or letter. In Ref. [32.2], 2000 added to the chain number designates a double-pitch chain.

32.2.3 Roller-Chain Sprockets

Definitions and Types.   Four styles of sprockets are standardized in Ref. [32.1]. Style A is a flat plate with no hub extensions. Style B has a hub extension on one side of the plate (flange). Style C has hub extensions on both sides of the flange. The extensions do not have to be equal. Style D has a detachable hub. The style D hub is normally attached to the flange with bolts. Most sprockets have a central bore with a keyway and setscrew to mount them on a shaft. Many other configurations of sprocket hubs and bores may be found in manufacturers' catalogs.

Tooth Form.   The tooth form and profile dimensions for single- and multiple-strand roller-chain sprockets are defined in Ref. [32.1].

Sprocket Diameters.   There are five important sprocket diameters defined in Ref. [32.1]. They are pitch, outside, bottom, caliper, and maximum hub diameters. The equations for those diameters, shown in Fig. 32.2, are

$$\text{PD} = P/\sin(180/N) \qquad \text{OD} = P[0.6\cot(180°/N)]$$

$$\text{BD} = \text{PD} - D \qquad \text{CD} = \text{PD}\cos(90°/N) - D$$

$$\text{MHD} = P[\cot(180°/N) - 1] - 0.030$$

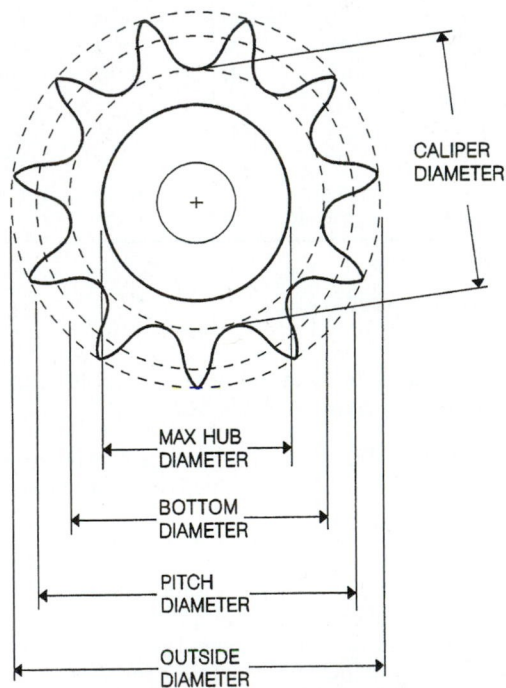

FIGURE 32.2 Roller chain sprocket diameters. *(Diamond Chain Company).*

32.3 SELECTION OF ROLLER-CHAIN DRIVES

32.3.1 General Design Recommendations

The following are only the more important considerations in roller-chain drive design. For more detailed information, consult Ref. [32.5] or manufacturers' catalogs.

Chain Pitch. The most economical drive normally employs the smallest-pitch single-strand chain that will transmit the required power. Small-pitch chains generally are best for lighter loads and higher speeds, whereas large-pitch chains are better for higher loads and lower speeds. The smaller the pitch, the higher the allowable operating speed.

Number of Sprocket Teeth

  Small Sprocket. The small sprocket usually is the driver. The minimum number of teeth on the small sprocket is limited by the effects of chordal action (speed variation), as shown in Fig. 32.3. Lower speeds will tolerate more chordal action than higher speeds. The minimum recommended number of teeth on the small sprocket is

Slow speed	12 teeth
Medium speed	17 teeth
High speed	25 teeth

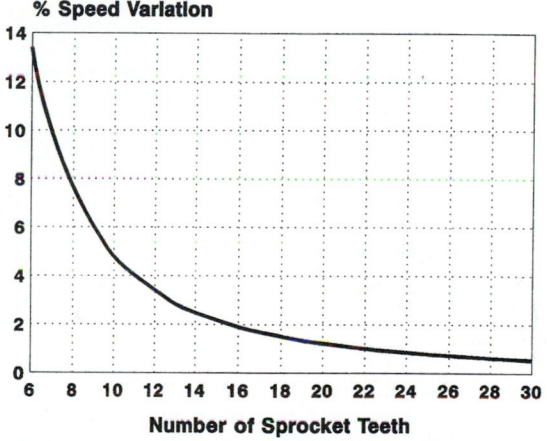

FIGURE 32.3 RC velocity variation versus number of teeth.

  Large Sprocket. The number of teeth on the large sprocket normally should be limited to 120. Larger numbers of teeth are very difficult (expensive) to manufacture. The number of teeth on the large sprocket also limits maximum allowable chain wear elongation. The maximum allowable chain wear elongation, in percent, is $200/N_2$.

Hardened Teeth. The fewer the number of teeth on the sprocket, the higher the tooth loading. Sprocket teeth should be hardened when the number of teeth is less than 25 and any of the following conditions exist:

1. The drive is heavily loaded.
2. The drive runs at high speeds.
3. The drive runs in abrasive conditions.
4. The drive requires extremely long life.

Angle of Wrap.   The minimum recommended angle of wrap on the small sprocket is 120°.

Speed Ratio.   The maximum recommended speed ratio for a single-reduction roller-chain drive is 7:1. Speed ratios up to 10:1 are possible with proper design, but a double reduction is preferred.

Center Distance.   The preferred center distance for a roller-chain drive is 30 to 50 times the chain pitch. At an absolute minimum, the center distance must be at least one-half the sum of the two sprocket outside diameters. A recommended minimum center distance is the pitch diameter of the large sprocket plus one-half the pitch diameter of the small sprocket. The recommended maximum center distance is 80 times the chain pitch.

The center distance should be adjustable to take up chain slack caused by wear. Adjustment of at least 2 pitches is recommended. If a fixed center distance must be used, consult a chain manufacturer.

Chain Length.   Required chain length may be estimated from the following approximate equation:

$$L \doteq 2C + \frac{N_1 + N_2}{2} + \frac{N_2 - N_1}{4\pi^2 C} \tag{32.1}$$

Equation (32.1) will give chain length accurate to within ±½ pitch. If a more precise chain length is required, an equation for the exact chain length may be found in Ref. [32.5] or in manufacturers' literature.

The chain length must be an integral number of pitches. An even number of pitches is preferred. An odd number of pitches requires an offset link, and offset links reduce the chain's capacity.

Wear and Chain Sag.   As a chain wears, it elongates. Roller-chain sprocket teeth are designed to allow the chain to ride higher on the teeth as it wears, to compensate for the elongation. Maximum allowable wear elongation normally is 3 percent. Where timing or smoothness is critical, maximum allowable elongation may be only 1.5 percent. The size of the large sprocket may also limit allowable elongation, as noted earlier.

As a chain elongates from wear, the excess length accumulates as sag in the slack span. In long spans, the sag can become substantial. It is important to design sufficient clearance into the drive to accommodate the expected amount of chain sag. For a drive with an approximately horizontal slack span, the required sag allowance for a particular amount of elongation is shown in Fig. 32.4. The drive centers should be adjusted periodically to maintain sag at 2 to 3 percent of the center distance.

Idlers.   When the center distance is long, the drive centers are near vertical, the center distance is fixed, or machine members obstruct the normal chain path, idler

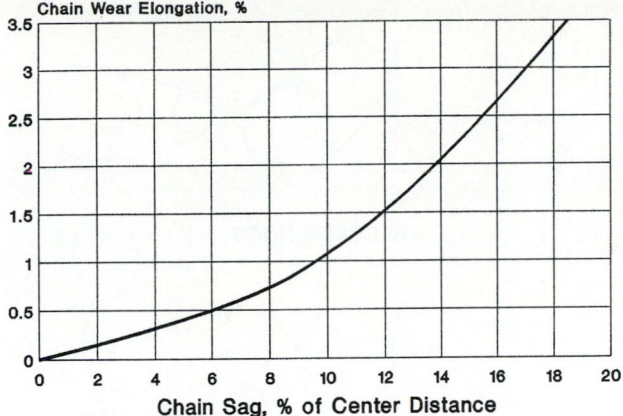

FIGURE 32.4   Chain sag versus center distance.

sprockets may be required. Idler sprockets should engage the chain in the slack span and should not be smaller than the small sprocket. At least 3 teeth on the idler should engage the chain, and there should be at least 3 free pitches of chain between sprocket engagement points.

Multiple-Strand Chain.   Multiple-strand chain may be required when the load and speed are too great for a single-strand chain, or when space restrictions prevent the use of large enough single-strand sprockets.

Drive Arrangements.   A number of recommended, acceptable, and not recommended drive arrangements are shown in Fig. 32.5.

32.3.2   Selection Procedure

Obtain Required Information.   It is very important to obtain all the listed information before making a selection.

1. Source of input power
2. Type of driven equipment
3. Power to be transmitted
4. Speed and size of driver shaft
5. Speed and size of driven shaft
6. Desired center distance and drive arrangement
7. Means of center distance adjustment, if any
8. Available lubrication type
9. Space limitations
10. Adverse environmental conditions

Check for any unusual drive conditions, such as

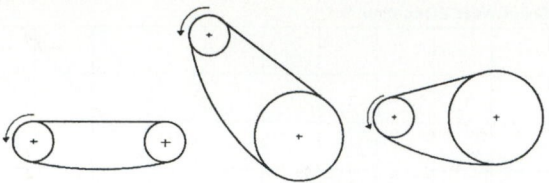

RECOMMENDED

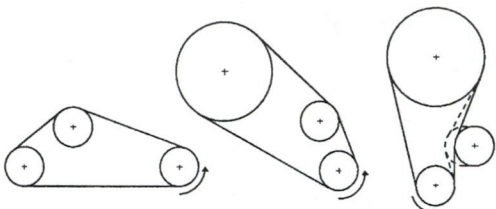

ACCEPTABLE

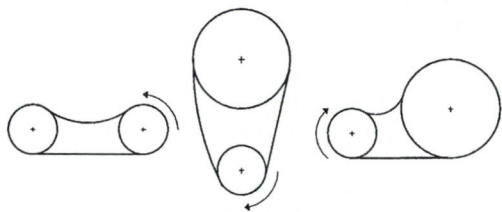

NOT RECOMMENDED

FIGURE 32.5   Drive arrangements.

- Frequent stops and starts
- High starting or inertial loads
- Temperatures above 150°F or below 0°F
- Large cyclic load variations in a single revolution
- Multiple driven shafts

If any of these, or any other unusual drive condition, is found, consult a chain manufacturer for help with the selection.

Determine Service Factor.   The average required power for a drive usually is given. The peak power may be much greater than the average, depending on the power source and the driven equipment. A service factor, obtained from Table 32.2, accounts for the peak loads. The load classification for various types of driven equipment may be found in Ref. [32.1] or [32.5] or in manufacturers' catalogs.

Calculate Design Power.   Obtain the design power by multiplying the average power times the service factor from Table 32.2.

TABLE 32.2 Service Factors for Roller Chain Drives

	Type of input power		
Type of driven load	Internal combustion engine with hydraulic drive	Electric motor or turbine	Internal combustion engine with mechanical drive
Smooth	1.0	1.0	1.2
Moderate shock	1.2	1.3	1.4
Heavy shock	1.4	1.5	1.7

Make Preliminary Chain Selection. Enter the chart, Fig. 32.6, with the design power and the speed of the small sprocket to select a preliminary chain. If no single-strand chain will transmit the design power at the required speed, or if space is restricted, multiple-strand chain may be required. If multiple-strand chain is to be selected, divide the design power by the multiple-strand factor, from Table 32.3, before entering the selection chart. Note that optimally the drive will operate near the peak of the rating curve. If the speed and power are low to moderate and the center distance is long, double-pitch chain may be acceptable. A selection procedure for double-pitch chains is given in Ref. [32.2].

TABLE 32.3 Roller Chain Multiple Strand Factors

Number of strands	Multiple strand factor
2	1.7
3	2.5
4	3.3

Select Small Sprocket. Refer to the horsepower tables in Ref. [32.1], [32.2], or [32.5] or in manufacturers' catalogs to select the small sprocket. Again, if multiple-strand chain is being considered, the design power must be divided by the proper multiple-strand factor from Table 32.3. Several different combinations of chain and sprocket sizes may be satisfactory for a given drive. Study the tables to see if increasing the number of teeth on the small sprocket might allow use of a smaller-pitch chain, or if decreasing the number of teeth on the small sprocket might allow use of a larger-pitch single-strand chain instead of a multiple-strand chain.

Consult sprocket manufacturers' catalogs to ensure that the sprocket bore capacity is adequate for the shaft. If it is not, select a larger sprocket.

Select Large Sprocket. Determine the number of teeth required on the large sprocket by multiplying the number of teeth on the small sprocket by the speed ratio. Ensure that the selected large sprocket will fit within any space restrictions and clear all obstructions. If there is an interference, a smaller-pitch, multiple-strand chain might be needed.

Make Final Chain Selection. Choose the most suitable drive from the alternatives selected earlier. The final choice may be based on economics, performance, effi-

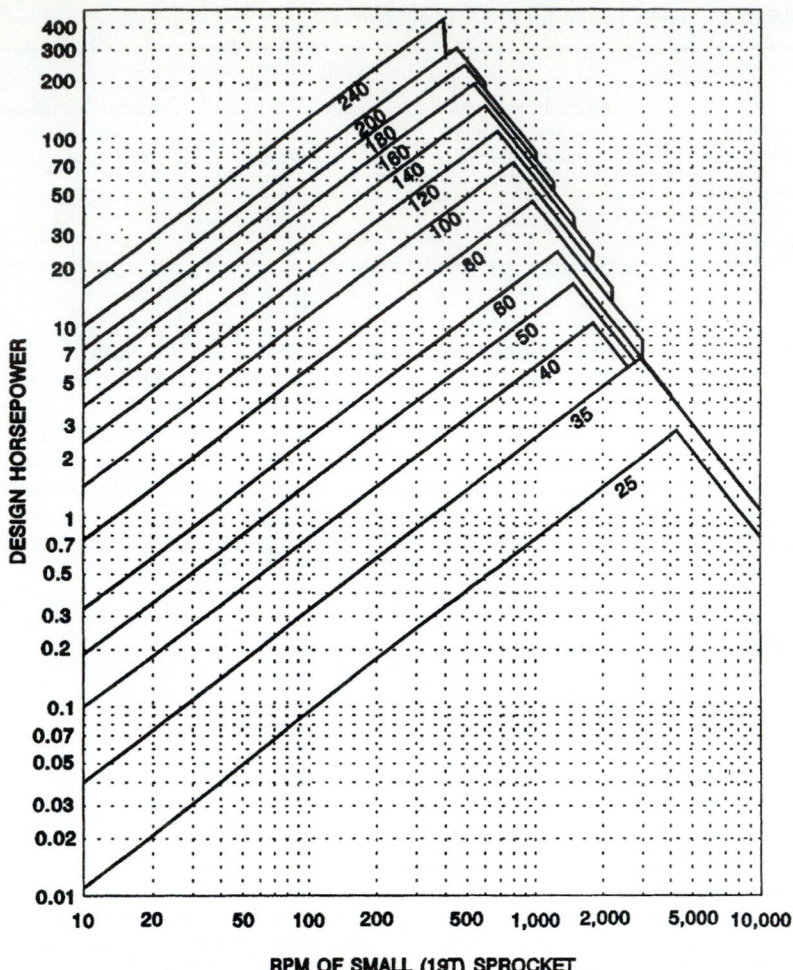

FIGURE 32.6   Roller chain selection chart.

ciency, space utilization, or a number of other considerations. Computer programs are available that automate the preliminary selection process and analyze the alternatives based on parameters provided by the designer.

Calculate Chain Length.   For a two-sprocket drive, the approximate chain length may be estimated by Eq. (32.1). A more accurate chain length may be calculated by equations found in Ref. [32.5] or in manufacturers' catalogs. For three or more sprocket drives, the chain length may be estimated by graphic techniques, geometric layouts, computer programs, or certain CAD packages.

Determine Lubrication Type. The type of lubrication required may be obtained from the horsepower tables in Refs. [32.1] or [32.5], manufacturers' catalogs, or Sec. 32.4. It is very important to provide adequate lubrication to a roller-chain drive. Selecting an inferior type of lubrication can drastically reduce the life of the drive.

32.3.3   Power Ratings of Roller-Chain Drives

Conditions for Ratings. The roller-chain horsepower ratings presented in this section are based on the following conditions:

1. Standard or heavy series chain listed in Ref. [32.1]
2. Service factor of 1
3. Chain length of 100 pitches
4. Use of the recommended lubrication method
5. A two-sprocket drive, driver and driven
6. Sprockets properly aligned on parallel, horizontal shafts and chains
7. A clean, nonabrasive environment
8. Approximately 15 000 hours service life

Horsepower Rating Equations. When operating under the above conditions, the maximum horsepower capacity of standard roller chains is defined by the equations shown. Depending on speed and the number of teeth on the smaller sprocket, the power capacity may be limited by link plate fatigue, roller and bushing impact fatigue, or galling between the pin and the bushing. The power capacity of the chain is the lowest value obtained from the following three equations at the given conditions.

1. Power limited by link plate fatigue:

$$\mathrm{HP}_f = K_f N_1^{1.08} R^{0.9} P^{(3.0 - 0.07P)} \tag{32.2}$$

where $K_f = 0.0022$ for no. 41 chain, and 0.004 for all other numbers.

2. Power limited by roller and bushing impact fatigue:

$$\mathrm{HP}_r = (K_r N_1^{1.5} P^{0.8})/R^{1.5} \tag{32.3}$$

where $K_r = 29\,000$ for nos. 25 and 35 chain, 3400 for no. 41 chain, and 17 000 for nos. 40 through 240 chain.

3. Power limited by galling:

$$\mathrm{HP}_g = (\mathrm{RPN}_1/110.84)(4.413 - 2.073P - 0.0274N_2)$$

$$- [\ln (R/1000)](1.59 \log P + 1.873) \quad (32.4)$$

The loci of these equations are presented in Fig. 32.6.

32.4 LUBRICATION AND WEAR

In all roller and engineering steel chains, and in many silent chains, each pin and bushing joint essentially is a traveling journal bearing. So, it is vital that they receive adequate lubrication to attain full potential wear life. Even silent chains with rocking-type joints are subject to some sliding and fretting, and so they also need good lubrication to obtain optimum wear life.

32.4.1 Purpose of Chain Lubrication

Effective lubrication aids chain performance and life in several ways:

1. By resisting wear between the pin and bushing surfaces
2. By flushing away wear debris and foreign materials
3. By lubricating the chain-sprocket contact surfaces
4. By dissipating heat
5. By cushioning impact loads
6. By retarding rust and corrosion

32.4.2 Lubricant Properties

General Lubricant Characteristics.　　Chain lubrication is usually best achieved by a good grade of nondetergent petroleum-base oil with the following properties:

- Low enough viscosity to penetrate to critical surfaces
- High enough viscosity to maintain an effective lubricating film at prevailing bearing pressures
- Free of contaminants and corrosive substances
- Able to maintain lubricating properties in the full range of operating conditions

Additives that improve film strength, resist foaming, and resist oxidation usually are beneficial, but detergents or additives to improve viscosity index normally are not needed.

Recommended Viscosities.　　The oil must be able to flow into small internal clearances in the chain, and so greases and very high-viscosity oils should not be used. The recommended viscosity for various ambient temperature ranges is shown in Table 32.4.

32.4.3 Application of Lubricant to Chain

Application Location and Flow Direction.　　It is vital to adequately lubricate the pin and bushing surfaces that articulate under load. It also is important to lubricate the surfaces between the roller and the bushing in roller and engineering steel chains. Oil should be applied to the upper link plate edges in the lower chain span just before the chain engages a sprocket. This places the oil where it can pass between the link plate faces and enter the critical bearing area. It also permits

TABLE 32.4 Recommended Oil Viscosity for Various Temperatures

Recommended grade	Temperature, °F
SAE 5	−50 to +50
SAE 10	−20 to +80
SAE 20	+10 to +110
SAE 30	+20 to +130
SAE 40	+30 to +140
SAE 50	+40 to +150

Source: Adapted from Ref. [32.6], p. 8, by courtesy of American Chain Association.

gravity and centrifugal force to aid the flow of oil in the desired direction. The extra oil that spills over the edges of the link plates should be adequate to lubricate the bearing surfaces between the rollers and the bushings in roller and engineering steel chain. It is important to supply oil uniformly across the entire width of silent and multiple-strand roller chains. For more information, see Refs. [32.5] and [32.6].

Flow Rates. When chain drives are transmitting large amounts of power at high speeds, oil-stream lubrication generally is required. The oil stream must cool the chain and carry away wear debris as well as lubricate the drive. A substantial oil flow rate is needed to accomplish all of that. The minimum flow rate for the amount of horsepower transmitted is shown in Table 32.5.

TABLE 32.5 Oil Flow Rates vs. Horsepower

Transmitted horsepower	Minimum flow rate, gal/min
50	0.25
100	0.50
150	0.75
200	1.00
250	1.25
300	1.50
400	2.00
500	2.25
600	3.00
700	3.25
800	3.75
900	4.25
1 000	4.75
1 500	7.00
2 000	10.0

Source: Adapted from Ref. [32.6], p. 12, by courtesy of American Chain Association.

32.4.4 Types of Chain Lubrication

All three types of chain drives—roller, engineering steel, and silent—will work with three types of lubrication system. The type of lubrication system used is dependent on the speed and the amount of power transmitted. The three types of chain drive lubrication systems are

Type 1.   Manual or drip
Type 2.   Oil bath or slinger disk
Type 3.   Oil stream

A description of each type of lubrication follows.

Manual.   Oil is manually applied periodically with a brush or spout can. The time period between applications is often 8 hours, but it may be longer if this is proven adequate for the particular conditions.

Drip.   Oil is dripped between the link plate edges from a lubricator with a reservoir. Rates range from 4 to 20 drops per minute; 10 drops per minute is equal to about one ounce per hour. A distribution pipe is needed to direct oil to all the rows of link plates in multiple-strand chain, and a wick packing in the pipe will ensure uniform distribution of oil to all the holes in the pipe. Windage may misdirect the oil droplets. If that occurs, the lubricator must be relocated.

Oil Bath.   A short section of chain runs through the oil in the sump of a chain casing. The oil level should not be higher than the pitch line of the chain at its lowest point in operation. Long sections of chain running through the oil bath can cause foaming and overheating. If that occurs, slinger disc–type lubrication should be considered.

Slinger Disk.   The chain runs above the oil level while a disk on one shaft picks up oil from the sump and slings it against a collector plate. The oil is then directed into a trough which applies it to the upper edges of the chain link plates in the lower span of the chain. The disk diameter should be sized so that the disk runs at a rim speed of 600 to 8000 ft/min. Slower speeds will not effectively pick up the oil. Higher speeds can cause foaming and overheating.

Oil Stream.   A pump sends a stream or spray of oil under pressure onto the chain. The oil must be applied evenly across the entire width of the chain, and it must be directed onto the lower span from the inside of the chain loop. Excess oil is collected in the sump and returned to the pump reservoir. The oil stream both lubricates and cools the chain when high power is transmitted at high speeds (Table 32.5). The oil may be cooled by radiation from the external surfaces of the reservoir or, if power is very high, by a separate heat exchanger.

32.4.5 Chain Casings

Chain casings provide a reservoir for the oil, contain excess oil slung off the drive, and prevent contaminants from contacting the drive. Chain casings usually are made of sheet metal, and are stiffened by embossed ribs or metal angles. Chain casings generally have doors or panels to allow access to the drive for inspection and maintenance.

Oil-retaining casings have single lap joints and single oil seals at each shaft opening. They are adequate for drip or oil bath types of lubrication. They are relatively inexpensive, but they will allow some oil to escape and some dust and dirt to enter the casing. Oil- and dustproof casings have double lap joints and double oil seals at the shafts. They are strongly recommended for slinger disc and pressure stream types of lubrication. They are more expensive, but they virtually eliminate oil leakage and prevent contaminants from reaching the drive.

Sizing for Clearance.   Sufficient clearance between the chain and the casing is essential if maximum potential wear life from the chain is to be obtained. At least 3 in clearance is needed around the periphery of the chain and ¾ in on each side of the chain. Additional clearance must be provided in the bottom of the casing to accommodate chain sag from wear elongation. The required allowance for chain sag may be obtained from Fig. 32.4.

Sizing for Heat Dissipation.   A chain casing also may need sufficient surface area for heat dissipation. The oil temperature should not exceed 180°F (total of ambient temperature and temperature rise). The temperature rise for a given chain casing may be estimated from Eq. (32.5), which presumes 98 percent chain drive mechanical efficiency.

$$\Delta T_h = \frac{50.9 \text{ HP}}{A K_h} \qquad (32.5)$$

where ΔT_h = temperature rise, °F
          HP = transmitted horsepower
           A = exposed casing area, ft^2
         K_h = overall film coefficient of heat transfer, BTU/(h · ft^2 · °F)
         K_h = 2.0 for still air, 2.7 for normal circulation, and 4.5 for rapid circulation

The equation may be rearranged to obtain the required surface area of the casing:

$$A = \frac{50.9 \text{ HP}}{\Delta T_h K_h} \qquad (32.6)$$

If a casing with the required surface area is too large for the available space, a separate oil cooler may be needed.

This short section gives only the rudiments of designing a chain casing. More information may be found in Refs. [32.5] and [32.6], machine design textbooks, or manufacturers' literature.

32.4.6   Specific Lubrication Recommendations

Roller Chain.   The required type of lubrication is directly related to the speed of a roller-chain drive. The limiting speeds for standard sizes of roller chain are shown in Table 32.6. More detailed information may be found in Refs. [32.5] and [32.6] or in manufacturers' literature.

Engineering Steel Chains.   The required type of lubrication for an engineering steel drive is related to speed, but the relationship is more complex than for a roller-chain drive. The approximate limiting speeds for standard sizes of engineering steel chain, on 12-tooth sprockets only, are shown in Table 32.7. The values in the table are for general guidance only. When designing a drive, consult Ref. [32.5] or manufacturers' literature.

TABLE 32.6 Roller Chain Speed Limits by Lubrication Type

ANSI No.	25	35	40	50	60	80	100	120	140	160	180	200	240
					Limiting speed, fpm								
Type 1	480	350	300	250	215	165	145	125	110	100	90	80	75
Type 2	3250	2650	2200	1900	1750	1475	1250	1170	1050	1000	935	865	790
Type 3	At all speeds higher than for Type 2												

TABLE 32.7 Offset Sidebar Chain Speed Limits by Lubrication Type

ANSI No.	2010	2512	2814	3315	3618	4020	4824	5628
	Limiting speed, rpm of 12-tooth sprocket only							
Type 1	33	33	33	32	29	30	35	38
Type 2	300	200	160	115	100	85	65	N/A
Type 3	At all speeds higher than for Type 2							

Silent Chains.   The required type of lubrication is related to speed for silent-chain drives also, but again, the relationship is more complex than for roller-chain drives. The approximate limiting speeds for standard sizes of silent chain are shown in Table 32.8. The values in the table are for general guidance only. When designing a drive, consult Refs. [32.4] and [32.5] or manufacturers' literature.

32.5 ENGINEERING STEEL CHAINS: NOMENCLATURE AND DIMENSIONS

The engineering steel chains that are specifically designated for power transmission are heavy-duty offset sidebar chains, standardized in Ref. [32.3].

32.5.1 Offset Sidebar Chain Nomenclature

In offset sidebar chain, each link is the same. Each link consists of a pair of offset-bent sidebars with a bushing assembled in one end of the pair of sidebars and a pin assembled in the other end. A roller that is free to rotate is assembled on each bushing. The pin is assembled into, and is free to pivot inside of, the bushing of the adjacent link, as shown in the illustration with Table 32.9.

TABLE 32.8 Silent Chain Speed Limits by Lubrication Type

	21	25	29	35	42	50
	Limiting speed, FPM					
	Number of teeth on small sprocket					
Type 1	1220	1250	1270	1290	1320	1350
Type 2	2400	2500	2650	2800	3000	3200
Type 3	At all speeds higher than for Oil Bath					

TABLE 32.9 Offset Sidebar Chain Dimensions

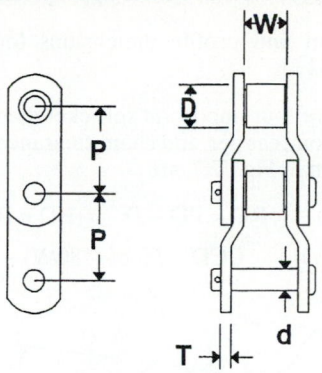

(Dimensions in inches, MUTS and F_w in lbf)

ANSI chain no.	Chain pitch, P	Roller width, W	Roller diameter, D	Pin diameter, d	Sidebar thickness, T	Minimum ultimate tensile strength, MUTS	Maximum allowable working load, F_w
2010	2.500	1.50	1.25	0.625	0.31	57 000	4 650
2512	3.067	1.56	1.62	0.750	0.38	77 000	6 000
2814	3.500	1.50	1.75	0.875	0.50	106 000	7 600
3315	4.073	1.94	1.78	0.938	0.56	124 000	10 000
3618	4.500	2.06	2.25	1.100	0.56	171 000	12 000
4020	5.000	2.75	2.50	1.250	0.62	222 500	17 500
4824	6.000	3.00	3.00	1.500	0.75	287 500	23 600
5628	7.000	3.25	3.25	1.750	0.88	385 000	60 500

Illustration courtesy of Diamond Chain Company.

32.5.2 Offset Sidebar Chain Dimensions and Data

Offset Sidebar Chain Dimensions.   Just as for roller chain, the three key dimensions for describing offset sidebar chain are pitch, roller diameter, and roller width. These, other selected dimensions, minimum ultimate tensile strengths, and maximum working loads for standard offset sidebar chains are shown in Table 32.9.

Offset Sidebar Chain Numbering.   A four-digit numbering system for designating standard offset sidebar chains is given in Ref. [32.3]. The left two digits denote the number of ⅛-in increments in the pitch. The right two digits denote the number of ¹⁄₁₆-in increments in the pin diameter. For example, chain no. 2814 designates a standard offset sidebar chain with 3.5-in pitch and ⅞-in pin diameter.

32.5.3 Sprockets

Machine-cut sprockets for offset sidebar chain look much like sprockets for roller chain. The standard (Ref. [32.3]) defines only tooth form, profile section, and important diameters. It does not define styles of sprockets because a great variety of styles and materials are offered by manufacturers. When designing a drive, consult manufacturers' literature to select sprockets that will be appropriate for the application.

Tooth Form.   The tooth form and profile dimensions for offset-sidebar-chain sprockets are defined in Ref. [32.3].

Sprocket Diameters.   There are four important sprocket diameters defined in Ref. [32.3]. They are pitch, root, bottom, caliper, and chain clearance diameters. The equations for those diameters, shown in Fig. 32.7, are

$$PD = P/\sin (180/N) \qquad RD = PD - D \qquad BD = RD - C_b$$

$$CD = BD[\cos (90/N)] \qquad CCD = P[\cot (180/N) - 0.05] - H$$

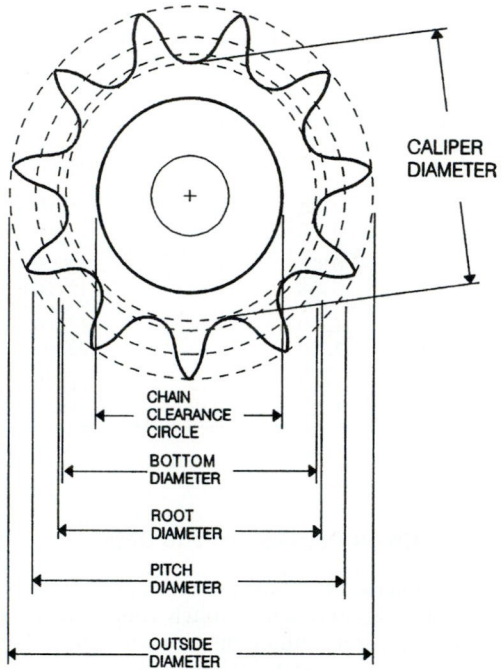

FIGURE 32.7   Engineering steel chain sprocket diameters. *(Diamond Chain Company)*.

32.6   *SELECTION OF OFFSET-SIDEBAR-CHAIN DRIVES*

32.6.1   General Design Recommendations

Some of the general guidelines for offset-sidebar-chain selection are similar to those for roller-chain selection, but many are different. The selection of offset-sidebar-chain drives will be covered fully, even though there may be some repetition.

Chain Pitch. Generally, the smallest-pitch chain that will transmit the required power at the required speed provides the most economical drive.

Number of Sprocket Teeth

Small Sprocket. A small sprocket with 12 teeth is normally recommended. If the speed ratio is high or space is restricted, a small sprocket with 9 teeth is acceptable. If the speed ratio is low, center distance is long, and space is not limiting, a small sprocket with 15 teeth is suggested.

Large Sprocket. The size of the large sprocket is limited by available space, maximum allowable chain wear elongation, and manufacturing feasibility. Large sprockets with more than 66 teeth are not desirable because they limit maximum allowable chain wear elongation to less than 3 percent. Some manufacturers list their maximum-size stock sprockets at 50 teeth or less.

Angle of Wrap. The minimum recommended angle of wrap on the small sprocket is 135°, or three teeth engaging the chain.

Speed Ratio. The maximum recommended speed ratio for a single-reduction drive is 6:1. That is done so as to keep sprocket sizes within reasonable bounds and to obtain reasonably good chain wear life.

Center Distance. The preferred center distance is from 30 to 50 times the chain pitch. The center distance should be adjustable to compensate for chain wear elongation. The minimum adjustment should be equal to at least one chain pitch.

Chain Length. The required chain length may be estimated from Eq. (32.1). This will give chain length accurate to within $\pm\frac{1}{2}$ chain pitch. If a more precise length is required, consult Ref. [32.5] or manufacturers' literature.

Wear and Chain Sag. Maximum allowable wear elongation for offset sidebar chains is determined by component size and hardness. Most offset sidebar chains are designed to provide 5 to 6 percent wear elongation. Wear elongation may be limited by the number of teeth on the large sprocket. For that case, the maximum allowable chain elongation, in percent, is $200/N_2$.

In a drive on approximately horizontal centers, chain wear elongation accumulates as sag in the slack span. The expected amount of sag for various amounts of wear elongation is shown in Fig. 32.4. The drive centers should be adjusted periodically to maintain sag at 2 to 3 percent of horizontal center distance.

Idlers. When the center distance is long, the drive centers are more than 45° from horizontal, or machine members obstruct the normal chain path, an idler sprocket may be required. Idler sprockets should engage the chain in the slack span and should not be smaller than the small sprocket in the drive. At least 3 teeth on the idler should engage the chain, and there should be at least 3 pitches of free chain between engagement points.

Drive Arrangements. A number of recommended, acceptable, and not recommended drive arrangements are shown in Fig. 32.5.

32.6.2 Selection Procedure

Obtain Required Information. Before an offset-sidebar-chain drive is selected, it is essential to obtain all the listed items of information. Note that the first 10 items are the same as for roller chain.

1. Source of input power
2. Type of driven equipment
3. Power to be transmitted
4. Speed and size of driver shaft
5. Speed and size of driven shaft
6. Desired center distance and drive arrangement
7. Means of center distance adjustment, if any
8. Available lubrication type
9. Space limitations
10. Adverse environmental conditions
11. Operating hours per day

In addition, check for any unusual drive conditions, such as

- Higher than listed speeds
- Inadequate lubrication
- Very heavy shock loads
- Corrosive or very abrasive conditions
- More than one driven shaft
- Other than precision-cut sprockets

When any of these, or any other unusual drive condition, is present, consult a chain manufacturer for assistance with the drive selection.

Determine Service Factor. The combined service factor SF is calculated by multiplying the three individual factors from Tables 32.10, 32.11, and 32.12:

$$SF = (SF_1)(SF_2)(SF_3)$$

The load classification for various types of driven equipment may be found in Ref. [32.5] or in manufacturers' literature.

TABLE 32.10 Service Factors for Offset-Sidebar-Chain Drives; Load Type, SF_1

Type of driven load	Type of input power		
	Internal combustion engine with hydraulic drive	Electric motor or turbine	Internal combustion engine with mechanical drive
Smooth	1.0	1.0	1.2
Moderate shock	1.2	1.3	1.4
Heavy shock	1.4	1.5	1.7

TABLE 32.11 Service Factors for Offset-Sidebar-Chain Drives; Environment, SF_2

Environmental conditions	Factor
Relatively clean, moderate temperature	1.0
Moderately dirty, moderate temperature	1.2
Very dirty, abrasive, exposed to weather, mildly corrosive, relatively high temperature	1.4

TABLE 32.12 Service Factors for Offset-Sidebar-Chain Drives; Operating Time, SF_3

Daily operating hours	Factor
8 to 10 hr/day	1.0
10 to 24 hr/day	1.4

Calculate Design Power. Obtain the design power by multiplying the average power times the combined service factor.

Make Preliminary Chain Selection. Enter the chart, Fig. 32.8, with the design power and the speed of the small sprocket to make a preliminary chain selection.

Select Small Sprocket. Refer to the horsepower tables in Ref. [32.5] or manufacturers' catalogs to select the small sprocket. Check the tables to see if increasing the number of teeth on the small sprocket might allow use of a smaller-pitch, more economical chain.

Consult sprocket manufacturers' catalogs to ensure that the sprocket bore capacity will accommodate the shaft. If it will not, a larger sprocket must be selected.

Select Large Sprocket. Determine the number of teeth required on the large sprocket by multiplying the number of teeth on the small sprocket by the speed ratio. Ensure that the selected large sprocket will fit within the available space and clear all obstructions. If there is an interference, a different chain and sprocket combination may have to be selected.

Make Final Chain Selection. Choose the most suitable drive from the alternatives selected earlier. The final choice may be based on economics, performance, efficiency, space utilization, or a number of other considerations. Computer programs are available that automate the preliminary selection process and analyze the alternatives based on parameters provided by the designer.

Calculate Chain Length. Calculate the chain length from Eq. (32.1). More accurate chain length may be calculated using equations in Ref. [32.5] or manufacturers' literature. For drives with more than two sprockets, the chain length may be obtained using graphical techniques, geometric layouts, computer programs, and certain CAD packages.

Determine Lubrication Type. The required lubrication type may be obtained from horsepower tables in Ref. [32.5], manufacturers' literature, or Sec. 32.4. It is very

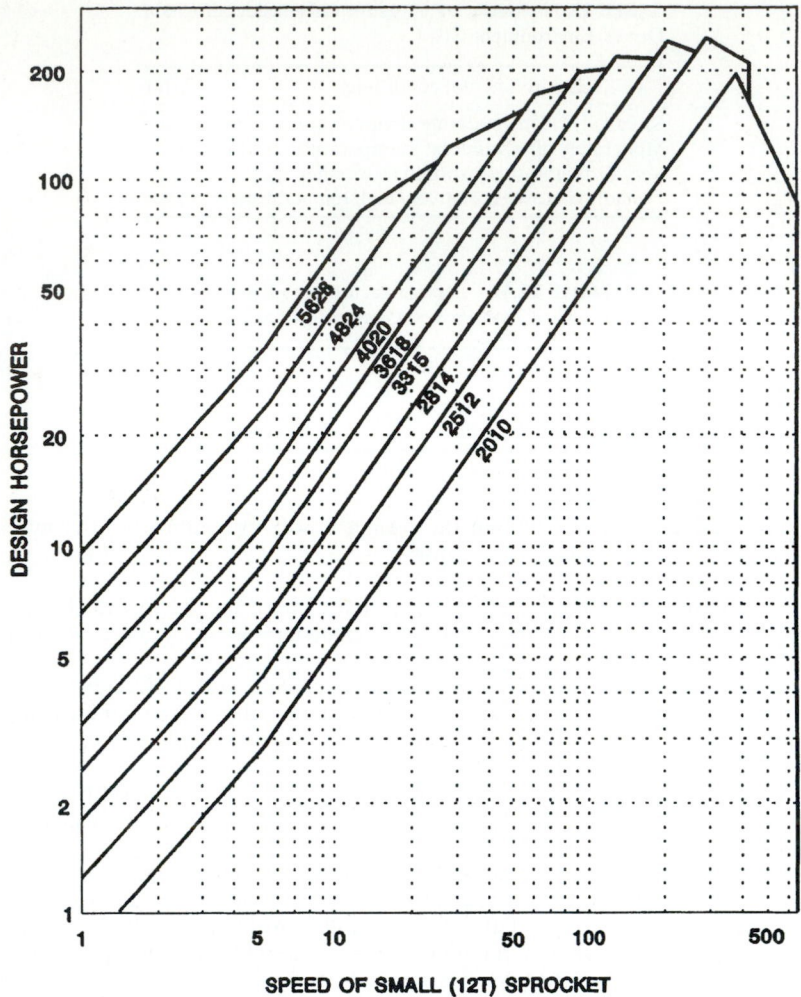

FIGURE 32.8   Engineering steel chain selection chart.

important to provide adequate lubrication to a chain drive. Selecting an inferior type of lubrication can drastically reduce the life of the drive.

Direction of Chain Travel.   The wear life of an offset sidebar chain can be affected by its direction of travel. To obtain greater wear life, in a drive with other than a one-to-one ratio, the narrow or roller end of the links in the taut span should face the smaller sprocket. It may be helpful in many cases for the drive designer to specify the direction in which the chain is to be installed. A complete explanation of this phenomenon may be found in Ref. [32.5].

32.6.3   Horsepower Ratings of Offset Sidebar Chains

Conditions for Ratings.   The offset sidebar chain ratings presented here and in Ref. [32.5] are based on the following conditions:

1. Standard chain listed in Ref. [32.3]
2. Service factor of 1
3. Chain length of 100 pitches
4. Use of the recommended lubrication method
5. A two-sprocket drive, driver and driven
6. Sprockets well aligned on parallel, horizontal shafts
7. Precision-machined tooth sprockets
8. A clean, nonabrasive environment
9. Approximately 15 000 hours service life

Horsepower Ratings.   The ratings for offset sidebar chains on 12-tooth sprockets are shown on the chart in Fig. 32.8. The rating equations for offset sidebar chains are much more complex than those for roller chains, and they are not generally published.

32.7   SILENT CHAINS: NOMENCLATURE AND DIMENSIONS

32.7.1   Silent-Chain Nomenclature

Silent chain is a series of toothed links alternately interlaced on joint components so that the joint between each pitch articulates essentially as shown in the illustration with Table 32.13. The joint components may be any combination of pins, bushings, or specially configured components that cause the toothed links to engage a standard sprocket so that the joint centers lie on the pitch circle. In addition to the toothed links and joint components, silent chain has guide links (not toothed) that run either on the sides of the sprocket or in circumferential grooves in the sprocket to control the chain laterally.

Silent chains do not have discrete strands as roller and engineering steel chains do. Silent chain may be assembled in an almost infinite number of "strands" because of the way the toothed links are alternately interlaced on the joint components, as shown in Fig. 32.9. Silent chain is produced in standard widths, in inches, designated in Ref. [32.4].

32.7.2   Silent-Chain Dimensions and Data

Silent-Chain Dimensions.   The standard for silent chain (Ref. [32.4]) defines the sprockets in more detail than the chains. A silent chain is a standard silent chain if it functions properly on a standard sprocket. The only standardized chain dimensions are those shown in Table 32.13.

TABLE 32.13   Silent Chain Dimensions

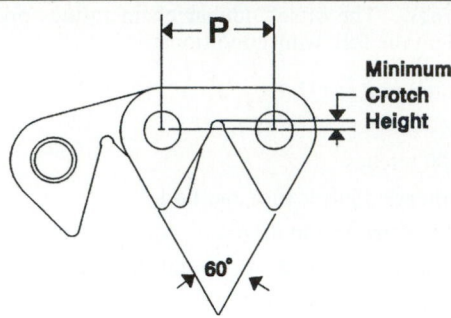

Dimensions in inches

ANSI chain no.	Chain pitch	Min. crotch height	Standard chain widths
SC3	0.375	0.0232	.5, .75, 1, 1.25, 1.5, 1.75, 2, 2.25, 2.5, 3, 4, 5, 6
SC4	0.500	0.0310	.5, .75, 1, 1.25, 1.5, 1.75, 2, 2.25, 2.5, 2.75, 3, 4, 5, 6, 8
SC5	0.625	0.0388	1, 1.25, 1.5, 1.75, 2, 2.5, 3, 4, 5, 6, 7, 8, 10
SC6	0.750	0.0465	1, 1.25, 1.5, 2, 2.5, 3, 3.5, 4, 5, 6, 7, 8, 9, 10, 12
SC8	1.00	0.0620	2, 2.5, 3, 4, 5, 6, 7, 8, 9, 10, 12, 14, 16
SC10	1.25	0.0775	2.5, 3, 4, 5, 6, 7, 8, 9, 10, 12, 14, 16, 18, 20
SC12	1.50	0.0930	3, 4, 5, 6, 7, 8, 9, 10, 12, 14, 16, 18, 20, 22, 24
SC16	2.00	0.1240	4, 5, 6, 7, 8, 10, 12, 14, 16, 18, 20, 22, 24, 30

Illustration courtesy of Diamond Chain Company.

Silent-Chain Numbering.   A standard numbering system is described in Ref. [32.4]. Chain numbers begin with the two letters SC. They are followed by one or two digits that designate the chain pitch in increments of ⅛ in. The pitch designators are then followed by two or three digits that designate the chain width in increments of ¼ in. For example, chain number SC836 is a standard silent chain with a pitch of 1 in and a width of 9 in.

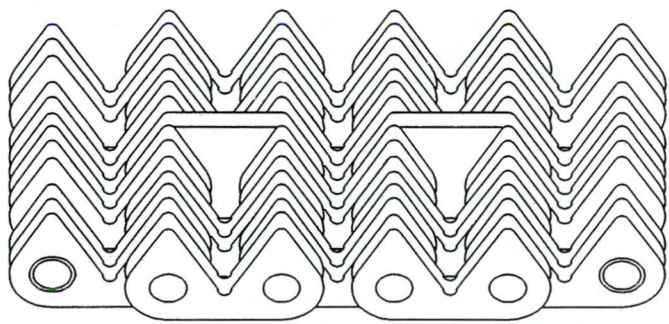

FIGURE 32.9   Typical silent chain. *(Diamond Chain Company).*

32.7.3 Silent-Chain Sprockets

Tooth Form.   The tooth form, profile, and guide groove width and profile dimensions are defined in Ref. [32.4].

Sprocket Diameters.   There are five important sprocket diameters defined in Ref. [32.4]. They are pitch, outside, over-gauge, maximum guide groove, and maximum hub diameters. The equations for those diameters, shown in Fig. 32.10, are

$$PD = P/\sin(180/N) \qquad D_p = 0.625P$$

$$OD \text{ (for rounded teeth)} = P[\cot(180/N) + 0.08]$$

$$OGD \text{ (for even number of teeth)} = PD - 0.125P \csc[30 - (180/N)] + D_p$$

$$OGD \text{ (for odd number of teeth)} = [\cos(90/N)]\{PD - 0.125P \csc[30 - (180/N)]\} + D_p$$

$$G = P[\cot(180/N) - 1.16]$$

$$MHD \text{ (for hobbed teeth)} = P[\cot(180/N) - 1.33]$$

$$MHD \text{ (for straddle-cut teeth)} = P[\cot(180/N) - 1.25]$$

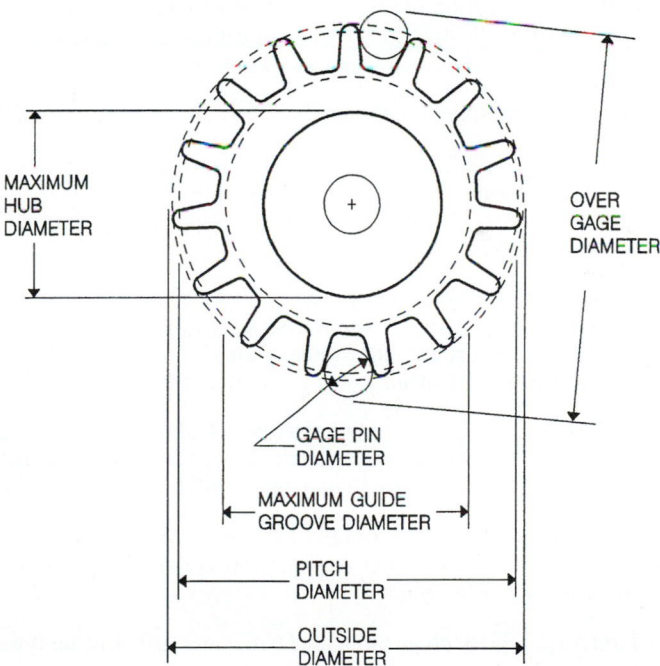

FIGURE 32.10   Silent chain sprocket diameters. *(Diamond Chain Company).*

32.8 SELECTION OF SILENT-CHAIN DRIVES

32.8.1 General Design Recommendations

The following guidelines are only the more important considerations in designing silent-chain drives. For more detailed information, consult Ref. [32.5] or manufacturers' literature.

Chain Pitch.   It is very important to match the silent-chain pitch to the desired shaft speeds. Generally, the smaller the pitch, the greater the allowable speed.

Number of Sprocket Teeth
    Small Sprocket.   The absolute minimum number of teeth that a silent-chain sprocket can have is 12. However, the minimum recommended number of teeth on the small sprocket is much more than that:

Normal operation	21 teeth
Speed increasing (step-up)	23 teeth
High speed (over 1500 ft/min)	25 teeth

    Large Sprocket.   The number of teeth on the large sprocket normally should not exceed 120. The number of teeth on the large sprocket also limits maximum allowable chain wear elongation. The maximum allowable chain wear elongation, in percent, is $200/N_2$.

Hardened Teeth.   The teeth on the small sprocket should be hardened when they are running at moderate to high speeds or transmitting nearly maximum rated power.

Angle of Wrap.   The minimum recommended angle of wrap on the small sprocket is $120°$.

Speed Ratio.   The maximum recommended speed ratio is 8:1 for a single-reduction drive. Speed ratios of up to 12:1 are possible with proper design, but in such situations double-reduction drives are preferred.

Center Distance.   The preferred center distance for a silent-chain drive is 30 to 50 times the pitch. The center distance should be adjustable to take up slack chain caused by wear. The center distance adjustment should be at least 2 pitches. If a fixed center distance must be used, consult a chain manufacturer.

Chain Length.   The required chain length may be calculated from Eq. (32.1). It will give chain length accurate to within ±½ pitch. If a more precise chain length is needed, equations for exact chain length may be found in Ref. [32.5] or in manufacturers' literature.
    Chain length must be an integral number of pitches. An even number of pitches is preferred. An odd number of pitches requires the use of an offset link, and offset links reduce the capacity of the chain.

Wear and Chain Sag.   Silent chain elongates with wear just as other types of chain do. Silent-chain sprocket teeth are designed to allow the chain to run higher on the teeth as it wears, thus compensating for the elongation. The maximum allowable chain wear is limited by the number of teeth on the large sprocket, as noted earlier.

In a drive where the shaft centers are nearly horizontal, chain wear accumulates as sag in the slack span. The amount of sag for various amounts of wear elongation is shown in Fig. 32.4. The drive centers should be adjusted periodically to maintain chain sag at 2 to 3 percent of the center distance.

Idlers.   In drives where the center distance is long, the shaft centers are near vertical, or the center distance is not adjustable, idler sprockets or guide shoes may be needed. Idler sprockets may be used when they engage the inside of the chain loop, but guide shoes with a large radius are advised for contact on the back of the chain. Idlers should have at least 17 teeth. They should have at least 3 teeth engaging the chain, and allow at least 3 pitches of free chain between engagement points. Guide shoes may be of wood, metal, or polymeric material. They should contact the back of the slack span near the small sprocket to increase the angle of wrap on the small sprocket.

Drive Arrangements.   A number of recommended, acceptable, and not recommended drive arrangements are shown in Fig. 32.5. These apply to silent chain as well as roller chain, except that a radiused guide shoe should be used where idler sprockets contact the back side of the chain.

32.8.2   Selection Procedure

Obtain Required Information.   Before a silent-chain drive is selected, it is essential to obtain all the listed items of information. Note that the first 10 items are the same as for roller chain.

1. Source of input power
2. Type of driven equipment
3. Power to be transmitted
4. Speed and size of driver shaft
5. Speed and size of driven shaft
6. Desired center distance and drive arrangement
7. Means of center distance adjustment, if any
8. Available lubrication type
9. Space limitations
10. Adverse environmental conditions
11. Operating hours per day

In addition, check for any unusual drive conditions, such as

- Frequent stops and starts
- High starting or inertial loads
- Multiple driven shafts
- Inferior lubrication type
- Corrosive or abrasive environment

If these or any other unusual drive conditions are present, consult a chain manufacturer for advice on the selection.

Determine Service Factor.   Determine the service factor from Table 32.14. Load classifications may be obtained from Refs. [32.4] and [32.5] or from manufacturers' literature.

TABLE 32.14 Silent Chain Service Factors

Load classification	IC engine with fluid coupling or electric motor		IC engine with mechanical drive		IC engine with torque converter	
	10 hr/day	24 hr/day	10 hr/day	24 hr/day	10 hr/day	24 hr/day
Very smooth	1.0	1.3	1.2	1.5	1.4	1.7
Mild shock	1.2	1.5	1.4	1.7	1.6	1.9
Moderate shock	1.4	1.7	1.6	1.9	1.8	2.1
Heavy shock	1.6	1.9	1.8	2.1	2.0	2.3

Calculate Design Power. Obtain the design power by multiplying the average power times the service factor from Table 32.14.

Make Preliminary Chain Selection. Enter the chart, Fig. 32.11, with design power and the speed of the small sprocket to make a preliminary chain selection. The solid lines on the chart represent the maximum horsepower per inch of width for each silent-chain size when operating on a 21-tooth small sprocket. The dashed lines represent the minimum and maximum recommended speeds for each silent-chain size.

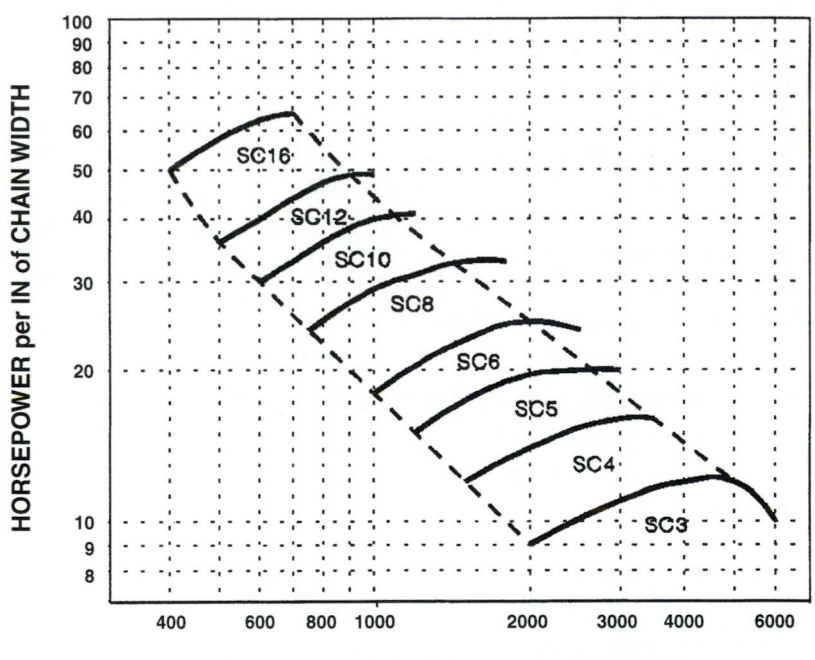

FIGURE 32.11 Silent chain selection chart.

Select Small Sprocket.   Refer to the horsepower tables in Ref. [32.4] or [32.5] or in manufacturers' catalogs to select the small sprocket. Use an odd number of teeth in the small sprocket whenever possible. More than one selection will work well in most situations, so check the tables to see if changing the number of teeth might yield a more economical or more convenient selection.

Consult sprocket manufacturers' catalogs to ensure that the bore in the small sprocket will accommodate the shaft. If it will not, a larger sprocket must be selected.

Select Large Sprocket.   Determine the number of teeth required on the large sprocket by multiplying the number of teeth on the small sprocket by the speed ratio. Ensure that the selected large sprocket will fit within the available space and clear any obstructions. If there is an interference, a different chain and sprocket combination may be necessary.

Make Final Chain Selection.   Choose the most suitable drive from the alternatives selected earlier. The final choice may be based on economics, performance, efficiency, space utilization, or a number of other considerations. Computer programs are available that automate the preliminary selection process and analyze the alternatives based on parameters provided by the designer.

Calculate Chain Length.   Calculate the chain length from Eq. (32.1). More accurate chain lengths may be obtained from equations in Ref. [32.5] or in manufacturers' literature. For drives with three or more sprockets, chain length may be obtained using graphical techniques, geometric layouts, computer programs, and certain CAD packages.

Determine Lubrication Type.   Select the proper type of lubrication from the tables in Ref. [32.4] or [32.5], manufacturers' literature, or Sec. 32.4. It is extremely important to provide adequate lubrication to a silent-chain drive. Selecting an inferior type of lubrication can drastically reduce the life of the drive. In addition, because silent-chain drives generally operate at higher speeds than roller- or engineering steel–chain drives, inadequate lubrication can cause serious damage to the drive or to the machine in which it is used.

32.8.3   Horsepower Ratings of Silent-Chain Drives

Conditions for Ratings.   The silent-chain power ratings presented here are based on the following conditions:

1. Standard chain listed in Ref. [32.4]
2. Service factor of 1
3. Chain length of 100 pitches
4. Use of recommended lubrication type
5. A two-sprocket drive, driver and driven
6. Sprockets properly aligned on parallel, horizontal shafts and chains
7. A clean, nonabrasive environment
8. Approximately 15 000 hours service life

Horsepower Ratings.   Silent-chain power capacity ratings are stated in terms of horsepower per inch of chain width, and they vary with the chain pitch, number of teeth on the small sprocket, and speed of the small sprocket. The ratings consider fatigue strength of the chains, wear resistance of drive components, adequacy of lubrication, and effects of speed and chordal action. The equations for silent-chain ratings are not published, as those for roller-chain ratings are, but complete tables of power ratings for standard silent chains are shown in Ref. [32.4]. Figure 32.11 is a silent-chain quick-selection chart, using an abridged version of the ratings.

REFERENCES

32.1 ANSI B29.1M-1993, "Precision Power Transmission Roller Chains, Attachments, and Sprockets," American Society of Mechanical Engineers, New York, 1993.

32.2 ANSI B29.3M-1994, "Double-Pitch Power Transmission Roller Chains and Sprockets," American Society of Mechanical Engineers, New York, 1994.

32.3 ANSI B29.10M-1981(R1987), "Heavy Duty Offset Sidebar Power Transmission Roller Chains and Sprocket Teeth," American Society of Mechanical Engineers, New York, 1981.

32.4 ANSI B29.2M-1982(1987), "Inverted Tooth (Silent) Chains and Sprockets," American Society of Mechanical Engineers, New York, 1982.

32.5 American Chain Association, *Chains for Power Transmission and Material Handling,* Marcel Dekker, New York, 1982.

32.6 American Chain Association, *Identification, Installation, Lubrication, and Maintenance of Power Transmission Chains,* American Chain Association, Rockville, Md., 1993.

CHAPTER 33
SPUR GEARS

Joseph E. Shigley
Professor Emeritus
The University of Michigan
Ann Arbor, Michigan

33.1 DEFINITIONS

Spur gears are used to transmit rotary motion between parallel shafts. They are cylindrical, and the teeth are straight and parallel to the axis of rotation.

The *pinion* is the smaller of two mating gears; the larger is called the *gear* or the *wheel.*

The *pitch circle, B* in Fig. 33.1, is a theoretical circle upon which all calculations are based. The *operating pitch circles* of a pair of gears in mesh are tangent to each other.

The *circular pitch, p* in Fig. 33.1, is the distance, measured on the theoretical pitch circle, from a point on one tooth to a corresponding point on an adjacent tooth. The circular pitch is measured in inches or in millimeters. Note, in Fig. 33.1, that the circular pitch is the sum of the *tooth thickness t* and the *width of space.*

The *pitch diameter, d* for the pinion and *D* for the gear, is the diameter of the pitch circle; it is measured in inches or in millimeters.

The *module m* is the ratio of the theoretical pitch diameter to the number of teeth *N.* The module is the metric index of tooth sizes and is always given in millimeters.

The *diametral pitch P_d* is the ratio of the number of teeth on a gear to the theoretical pitch diameter. It is the index of tooth size when U.S. customary units are used and is expressed as teeth per inch.

The *addendum a* is the radial distance between the top land *F* and the pitch circle *B* in Fig. 33.1. The *dedendum b* is the radial distance between the pitch circle *B* and the *root circle D* in Fig. 33.1. The *whole depth h_t* is the sum of the addendum and dedendum.

The clearance circle *C* in Fig. 33.1 is tangent to the addendum circle of the mating gear. The distance from the clearance circle to the bottom land is called the *clearance c.*

Backlash is the amount by which the width of a tooth space exceeds the thickness of the engaging tooth measured on the pitch circle.

Undercutting (see distance *u* in Fig. 33.1) occurs under certain conditions when a small number of teeth are used in cutting a gear.

Table 33.1 lists all the relations described above. Additional terminology is shown in Fig. 33.2. Here line *OP* is the *line of centers* connecting the rotation axes of a pair

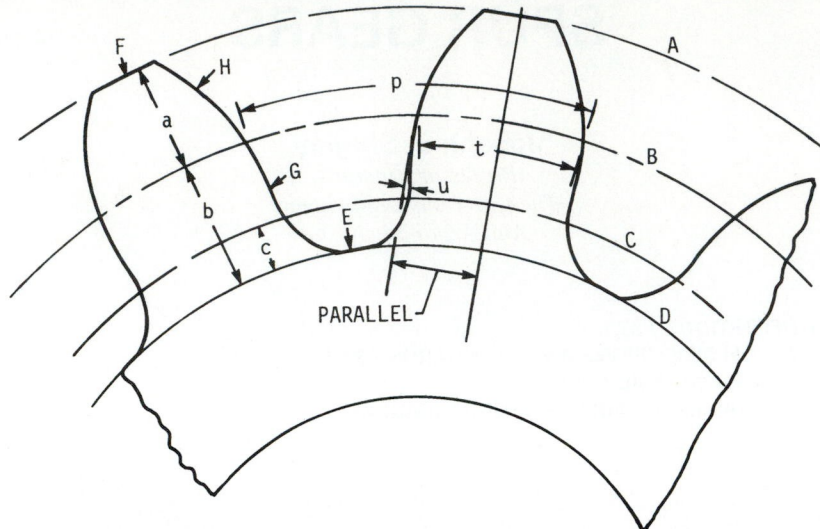

FIGURE 33.1 Terminology of gear teeth. *A*, addendum circle; *B*, pitch circle; *C*, clearance circle; *D*, dedendum circle; *E*, bottom land; *F*, top land; *G*, flank; *H*, face; *a* = addendum distance; *b* = dedendum distance; *c* = clearance distance; *p* = circular pitch; *t* = tooth thickness; *u* = undercut distance.

of meshing gears. Line *E* is the *pressure line,* and the angle φ is the *pressure angle.* The resultant force vector between a pair of operating gears acts along this line.

The pressure line is tangent to both *base circles C* at points *F.* The operating diameters of the pitch circles depend on the center distance used in mounting the gears, but the base circle diameters are constant and depend only on how the tooth forms were generated, because they form the *base* or the starting point of the involute profile.

TABLE 33.1 Basic Formulas for Spur Gears

Quantity desired	Formula	Equation number
Diametral pitch P_d	$P_d = \dfrac{N}{d}$	(33.1)
Module m	$m = \dfrac{d}{N}$	(33.2)
Circular pitch p	$p = \dfrac{\pi d}{N} = \pi m$	(33.3)
Pitch diameter, d or D	$d = \dfrac{N}{P_d} = mN$	(33.4)

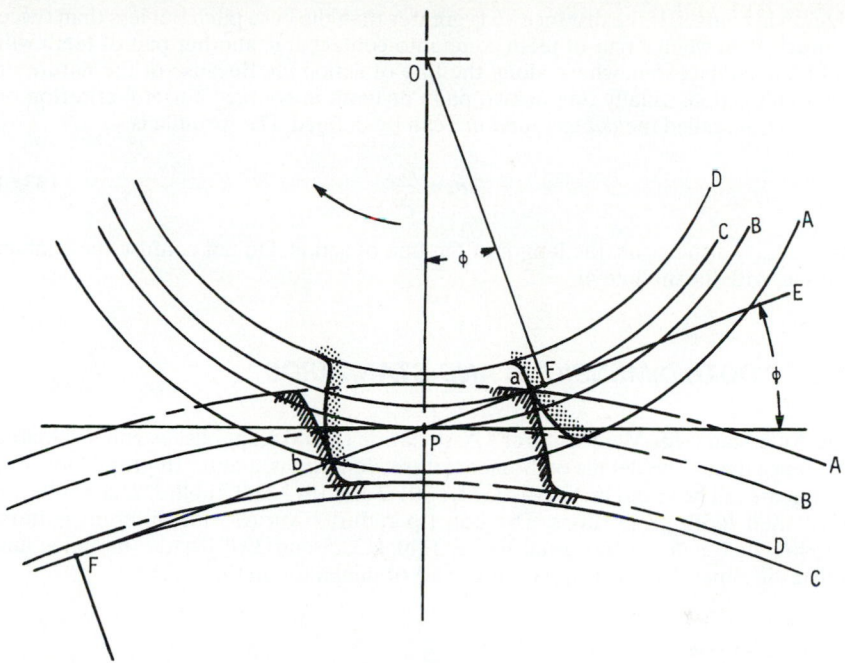

FIGURE 33.2 Layout drawing of a pair of spur gears in mesh. The pinion is the driver and rotates clockwise about the axis at *O*. *A,* addendum circles; *B,* pitch circles; *C,* base circles; *D,* dedendum circles; *E,* pressure line; *F,* tangent points; *P,* pitch point; *a,* initial point of contact; *b,* final point of contact.

Line *aPb* is the *line of action.* Point *a* is the *initial point of contact.* This point is located at the intersection of the addendum circle of the gear with the pressure line. Should point *a* occur on the other side of point *F* on the pinion base circle, the pinion flank would be *undercut* during generation of the profile.

Point *b* of Fig. 33.2 is the *final point of contact.* This point is located at the intersection of the addendum circle of the pinion with the pressure line. For no undercutting of the gear teeth, point *b* must be located between the pitch point *P* and point *F* on the base circle of the gear.

Line *aP* represents the *approach* phase of tooth contact; line *Pb* is the *recess* phase. Tooth contact is a sliding contact throughout the line of action except for an instant at *P* when contact is pure rolling. The nature of the sliding is quite different during the approach action and the recess action; and bevel-gear teeth, for example, are generated to obtain more recess action, thus reducing wear.

Instead of using the theoretical pitch circle as an index of tooth size, the base circle, which is a more fundamental distance, can be used. The result is called the *base pitch* p_b. It is related to the circular pitch p by the equation

$$p_b = p \cos \phi \qquad (33.5)$$

If, in Fig. 33.2, the distance from *a* to *b* exactly equals the base pitch, then, when one pair of teeth are just beginning contact at *a*, the preceding pair will be leaving contact at *b*. Thus, for this special condition, there is never more or less than one pair of teeth in contact. If the distance *ab* is greater than the base pitch but less than twice as much, then when a pair of teeth come into contact at *a*, another pair of teeth will still be in contact somewhere along the line of action *ab*. Because of the nature of this tooth action, usually one or two pairs of teeth in contact, a useful criterion of tooth action, called the *contact ratio m_c*, can be defined. The formula is

$$m_c = \frac{L_{ab}}{p_b} \tag{33.6}$$

where L_{ab} = distance *ab*, the length of the line of action. Do not confuse the contact ratio m_c with the module *m*.

33.2   TOOTH DIMENSIONS AND STANDARDS

The American Gear Manufacturer's Association (AGMA) publishes much valuable reference data.[†] The details on nomenclature, definitions, and tooth proportions for spur gears can be found in ANSI/AGMA 201.2 and 1012-F90. Table 33.2 contains the most used tooth proportions. The hob tip radius r_f varies with different cutters; $0.300/P_d$ or $0.300m$ is the usual value. Tables 33.3 and 33.4 list the modules and pitches in general use. Cutting tools can be obtained for all these sizes.

[†] See Chap. 35 for a special note on AGMA.

TABLE 33.2   Standard and Commonly Used Tooth Systems for Spur Gears

Tooth system	Pressure angle ϕ, deg	Addendum *a*	Dedendum *b*
Full depth	20	$1/P_d$ or $1m$	$1.25/P_d$ or $1.25m$ $1.35/P_d$ or $1.35m$
	$22\frac{1}{2}$	$1/P_d$ or $1m$	$1.25/P_d$ or $1.25m$ $1.35/P_d$ or $1.35m$
	25	$1/P_d$ or $1m$	$1.25/P_d$ or $1.25m$ $1.35/P_d$ or $1.35m$
Stub	20	$0.8/P_d$ or $0.8m$	$1/P_d$ or $1m$

TABLE 33.3   Diametral Pitches in General Use

Coarse pitch	2, $2\frac{1}{4}$, $2\frac{1}{2}$, 3, 4, 6, 8, 10, 12, 16
Fine pitch	20, 24, 32, 40, 48, 64, 96, 120, 150, 200

TABLE 33.4   Modules in General Use

Preferred	1, 1.25, 1.5, 2, 2.5, 3, 4, 5, 6, 8, 10, 12, 16, 20, 25, 32, 40, 50
Next choice	1.125, 1.375, 1.75, 2.25, 2.75, 3.5, 4.5, 5.5, 7, 9, 11, 14, 18, 22, 28, 36, 45

33.3 FORCE ANALYSIS

In Fig. 33.3 a gear, not shown, exerts force W against the pinion at pitch point P. This force is resolved into two components, a radial force W_r, acting to separate the gears, and a tangential component W_t, which is called the *transmitted load*.

Equal and opposite to force W is the shaft reaction F, also shown in Fig. 33.3. Force F and torque T are exerted by the shaft on the pinion. Note that torque T opposes the force couple made up of W_t and F_x separated by the distance $d/2$. Thus

$$T = \frac{W_t d}{2} \tag{33.7}$$

where T = torque, lb · in (N · m)
$\quad\quad\;\; W_t$ = transmitted load, lb (N)
$\quad\quad\;\; d$ = operating pitch diameter, in (m)

The *pitch-line velocity v* is given by

$$v = \frac{\pi d n_P}{12} \;\; \text{ft/min} \quad\quad v = \frac{\pi d n_P}{60} \;\; \text{m/s} \tag{33.8}$$

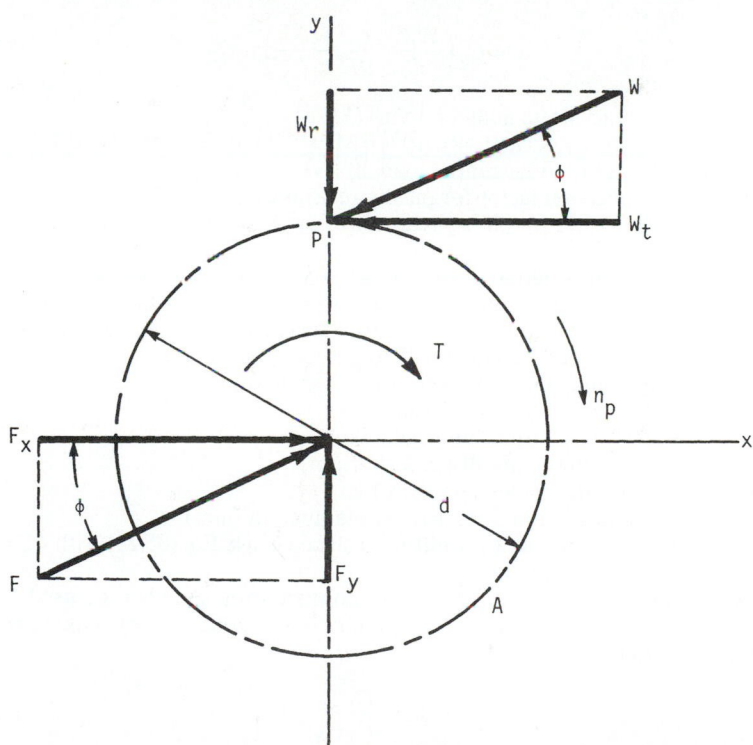

FIGURE 33.3 Force analysis of a pinion. *A*, operating pitch circle; *d*, operating pitch diameter; n_p, pinion speed; ϕ, pressure angle; W_t, transmitted tangential load; W_r, radial tooth load; W, resultant tooth load; T, torque; F, shaft force reaction.

where n_P = pinion speed in revolutions per minute (r/min). The power transmitted is

$$P = \begin{cases} \dfrac{W_t v}{33\,000} & \text{hp} \\[2ex] W_t v & \text{kW} \end{cases} \tag{33.9}$$

33.4 FUNDAMENTAL AGMA RATING FORMULAS[†]

Many of the terms in the formulas that follow require lengthy discussions and considerable space to list their values. This material is considered at length in Chap. 35 and so is omitted here.

33.4.1 Pitting Resistance

The basic formula for *pitting resistance,* or *surface durability,* of gear teeth is

$$s_c = C_p \left(\frac{W_t C_a}{C_v} \frac{C_s}{dF} \frac{C_m C_f}{I} \right)^{1/2} \tag{33.10}$$

where s_c = contact stress number, lb/in^2 (MPa)
C_p = elastic coefficient, (lb/in^2)$^{1/2}$ [(MPa)$^{1/2}$]; see Eq. (35.77) and Table 35.4
W_t = transmitted tangential load, lb (N)
C_a = application factor for pitting resistance; see Table 35.3
C_s = size factor for pitting resistance; use 1.0 or more until values are established
C_m = load distribution factor for pitting resistance; use Tables 33.5 and 33.6
C_f = surface condition factor; use 1.0 or more until values are established
C_v = dynamic factor for pitting resistance; use Fig. 35.4; multiply v in meters per second by 197 to get feet per minute
d = operating pitch diameter of pinion, in (mm)
  = $2C/(m_G + 1.0)$ for external gears
  = $2C/(m_G - 1.0)$ for internal gears
C = operating center distance, in (mm)
m_G = gear ratio (never less than 1.0)
F = net face width of narrowest member, in (mm)
I = geometry factor for pitting resistance; use Eq. (35.24) with $C_\psi = 1.0$

Allowable Contact Stress Number.   The contact stress number s_c, used in Eq. (33.10), is obtained from the *allowable contact stress number s_{ac}* by making several adjustments as follows:

$$s_c \leq s_{ac} \frac{C_L C_H}{C_T C_R} \tag{33.11}$$

[†] See Ref. [35.1].

TABLE 33.5 Load-Distribution Factors C_m and K_m for Spur Gears Having a Face Width of 6 in (150 mm) and Greater[†]

Face-diameter ratio F/d	Contact	C_m, K_m
1 or less	95% face width contact at one-third torque	1.4 at one-third torque
	95% face width contact at full torque	1.1 at full torque
	75% face width contact at one-third torque	1.8 at one-third torque
	95% face width contact at full torque	1.3 at full torque
	35% face width contact at one-third torque	3.0 at one-third torque
	95% face width contact at full torque	1.9 at full torque
	20% face width contact at one-third torque	5.0 at one-third torque
	75% face width contact at full torque	2.5 at full torque
	Teeth are crowned:	
	35% face width contact at one-third torque	2.5 at one-third torque
	85% face width contact at full torque	1.7 at full torque
Over 1 and less than 2	Calculated combined twist and bending of pinion not over 0.001 in (0.025 mm) over entire face:	
	Pinion not over 250 bhn hardness:	
	75% face width contact at one-third torque	2.0 at one-third torque
	95% face width contact at full torque	1.4 at full torque
	30% face width contact at one-third torque	4.0 at one-third torque
	75% face width contact at full torque	3.0 at full torque

[†]For an alternate approach see Eq. (35.21).

SOURCE: ANSI/AGMA 2001-B88.

where s_{ac} = allowable contact stress number, lb/in^2 (MPa); see Fig. 35.40
C_L = life factor for pitting resistance; use Fig. 35.49
C_H = hardness ratio factor; use Figs. 35.47 and 35.48
C_T = temperature factor for pitting resistance; use 1.0 or more, but see Sec. 35.5.1
C_R = reliability factor for pitting resistance; use Table 35.6

TABLE 33.6   Load-Distribution Factors C_m and K_m for Spur Gears

	Face width			
Condition of support	Up to 2 in (50 mm)	6 in (150 mm)	9 in (225 mm)	Over 16 in (400 mm)
Accurate mounting, low bearing clearances, minimum elastic deflection, precision gears	1.3	1.4	1.5	1.8
Less rigid mountings, less accurate gears, contact across full face	1.6	1.7	1.8	2.0
Accuracy and mounting such that less than full-face contact exists	Over 2.0			

SOURCE: ANSI/AGMA 2001-B88. For an alternate approach see Eq. (35.21).

Pitting Resistance Power Rating.   The allowable power rating P_{ac} for pitting resistance is given by

$$P_{ac} = \begin{cases} \dfrac{n_P F}{126\,000} \dfrac{I C_v}{C_s C_m C_f C_a} \left(\dfrac{d s_{ac}}{C_p} \dfrac{C_L C_H}{C_T C_R} \right)^2 & \text{hp} \\[3ex] \dfrac{n_P F}{1.91(10^7)} \dfrac{I C_v}{C_s C_m C_f C_a} \left(\dfrac{d s_{ac}}{C_p} \dfrac{C_L C_H}{C_T C_R} \right)^2 & \text{kW} \end{cases} \tag{33.12}$$

33.4.2   Bending Strength

The basic formula for the bending stress number in a gear tooth is

$$S_t = \begin{cases} \dfrac{W_t K_a}{K_v} \dfrac{P_d}{F} \dfrac{K_s K_m}{J} & \text{lb/in}^2 \\[3ex] \dfrac{W_t K_a}{K_v} \dfrac{1.0}{Fm} \dfrac{K_s K_m}{J} & \text{MPa} \end{cases} \tag{33.13}$$

where
    s_t = bending stress number, lb/in² (MPa)
    K_a = application factor for bending strength; use Table 35.3
    K_s = size factor for bending strength; use 1.0 or more until values are established
    K_m = load distribution factor for bending strength; use Tables 33.5 and 33.6
    K_v = dynamic factor for bending strength; use Fig. 35.4; multiply v in meters per second by 197 to get feet per minute
    J = geometry factor for bending strength; use Eq. (35.46) with $C_\psi = 1.0$ and Figs. 35.11 to 35.22
    m = module, mm
    P_d = nominal diametral pitch, teeth per inch

Allowable Bending Stress Number. The bending stress number s_t in Eq. (33.13) is related to the *allowable bending stress number* s_{at} by

$$s_t \leq \frac{s_{at}K_L}{K_T K_R} \tag{33.14}$$

where s_{at} = allowable bending stress number, lb/in^2 (MPa); use Fig. 35.41
  K_L = life factor for bending strength; use Figs. 35.49 and 35.50
  K_T = temperature factor for bending strength; use 1.0 or more; see Sec. 35.5.1
  K_R = reliability factor for bending strength; use Table 35.6

Bending Strength Power Rating. The allowable power rating P_{at} for bending strength is given by

$$P_{at} = \begin{cases} \dfrac{n_P d K_v}{126\,000 K_a} \dfrac{FJ}{P_d K_s K_m} \dfrac{s_{at}K_L}{K_R K_T} & \text{hp} \\[3ex] \dfrac{n_P d K_v}{1.91(10)^7 K_a} Fm \dfrac{J}{K_s K_m} \dfrac{S_{at}K_L}{K_R K_T} & \text{kW} \end{cases} \tag{33.15}$$

CHAPTER 34
BEVEL AND HYPOID GEARS

Theodore J. Krenzer, M.S.
Director of Research and Development
Gleason Machine Division
Rochester, New York

Robert G. Hotchkiss, B.S.
Director, Gear Technology
Gleason Machine Division
Rochester, New York

34.1 INTRODUCTION

This chapter provides you with information necessary to design a bevel- or hypoid-gear set. It includes guidelines for selecting the type and size of a gear set to suit the application requirements. Equations and graphs are provided for calculating gear-tooth geometry, strength, surface durability, and bearing loads.

Although the text provides sufficient data to design a gear set, reference is also made to appropriate American Gear Manufacturer's Association (AGMA) publications and software available for computer-aided design.

34.2 TERMINOLOGY

34.2.1 Types of Bevel and Hypoid Gears

Straight-bevel gears are the simplest form of bevel gears. The teeth are straight and tapered, and if extended inward, they would pass through the point of intersection of the axes. See Fig. 34.1.

Spiral-bevel gears have teeth that are curved and oblique to their axes. The contact begins at one end of the tooth and progresses to the other. See Fig. 34.2.

Zerol bevel gears have teeth that are in the same general direction as straight-bevel gears and are curved similarly to spiral-bevel gears. See Fig. 34.3.

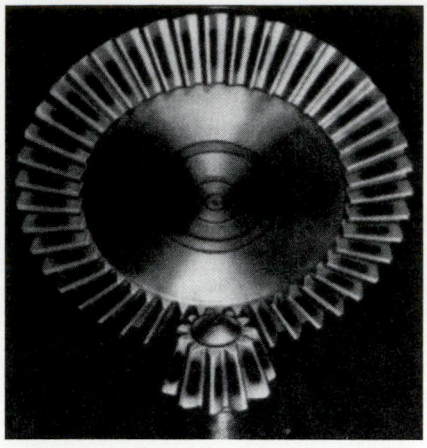

FIGURE 34.1 Straight-bevel set. *(Gleason Machine Division.)*

FIGURE 34.2 Spiral-bevel set. *(Gleason Machine Division.)*

Hypoid gears are similar in appearance to spiral-bevel gears. They differ from spiral-bevel gears in that the axis of the pinion is offset from the axis of the gear. See Fig. 34.4.

34.2.2 Tooth Geometry

The nomenclature used in this chapter relative to bevel and hypoid gears is illustrated in Figs. 34.5, 34.6, and 34.7.

The following terms are used to define the geometry:

Addendum of pinion (gear) a_p (a_G) is the height that the tooth projects above the pitch cone.

Backlash allowance B is the amount by which the circular tooth thicknesses are reduced to provide the necessary backlash in assembly.

Clearance c is the amount by which the dedendum in a given gear exceeds the addendum of its mating gear.

Cone distance, mean A_m is the distance from the apex of the pitch cone to the middle of the face width.

Cone distance, outer A_o is the distance from the apex of the pitch cone to the outer ends of the teeth.

Control gear is the term adopted for bevel gearing in place of the term *master gear,* which implies a gear with all tooth specifications held to close tolerances.

Crown to crossing point on the pinion (gear) x_o (X_o) is the distance in an axial section from the crown to the crossing point, measured in an axial direction.

Cutter radius r_c is the nominal radius of the face-type cutter or cup-shaped grinding wheel that is used to cut or grind the spiral-bevel teeth.

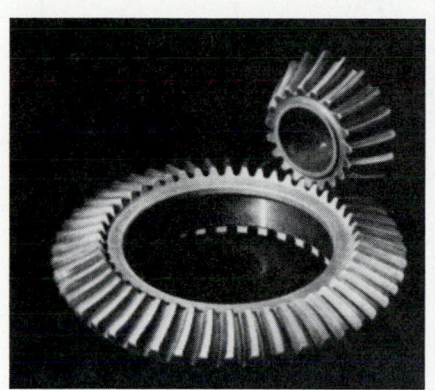

FIGURE 34.3 Zerol bevel set. *(Gleason Machine Division.)*

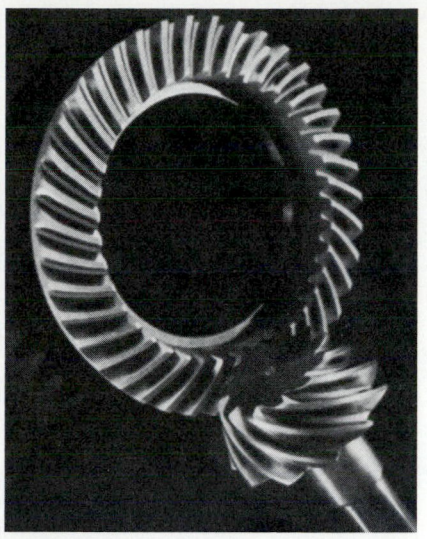

FIGURE 34.4 Hypoid set. *(Gleason Machine Division.)*

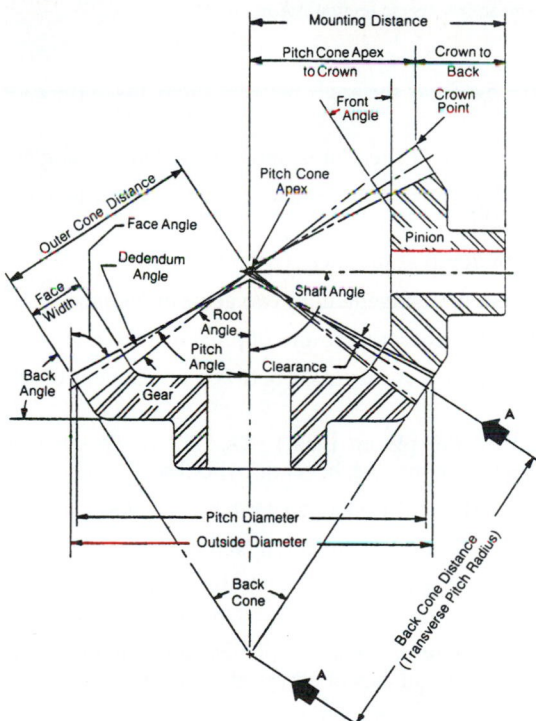

FIGURE 34.5 Bevel-gear nomenclature—axial plane. Section A-A is illustrated in Fig. 34.6.

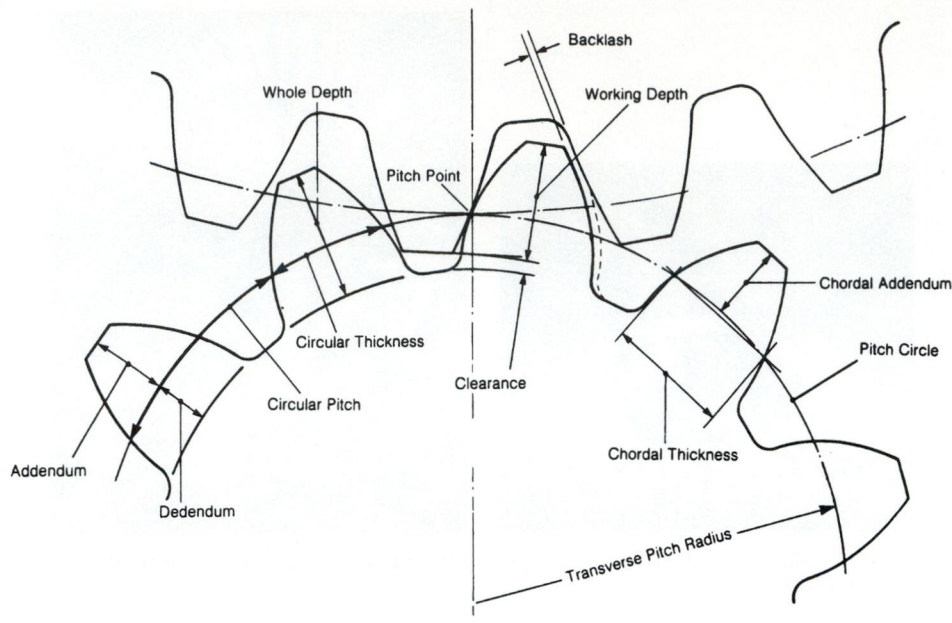

FIGURE 34.6   Bevel-gear nomenclature—mean transverse section AA in Fig. 34.5.

Dedendum angle of pinion (gear) δ_P (δ_G) is the angle between elements of the root cone and pitch cone.

Dedendum angles, sum of $\Sigma\delta$ is the sum of the pinion and gear dedendum angles.

Dedendum of pinion (gear) b_p (b_G) is the depth of the tooth space below the pitch cone.

Depth, mean whole h_m is the tooth depth at midface.

Depth, mean working h is the depth of engagement of two gears at midface.

Diametral pitch P_d is the number of gear teeth per unit of pitch diameter.

Face angle of pinion (gear) blank γ_o (Γ_o) is the angle between an element of the face cone and its axis.

Face apex beyond crossing point on the pinion (gear) G_o (Z_o) is the distance between the face apex and the crossing point on a bevel or hypoid set.

Face width F is the length of the teeth measured along a pitch-cone element.

Factor, mean addendum c_1 is the addendum modification factor.

Front crown to crossing point on the pinion (gear) x_i (X_i) is the distance in an axial section from the front crown to the crossing point, measured in the axial direction.

Hypoid offset E is the distance between two parallel planes, one containing the gear axis and the other containing the pinion axis of a hypoid-gear set.

Number of teeth in pinion (gear) n (N) is the number of teeth contained in the whole circumference of the pitch cone.

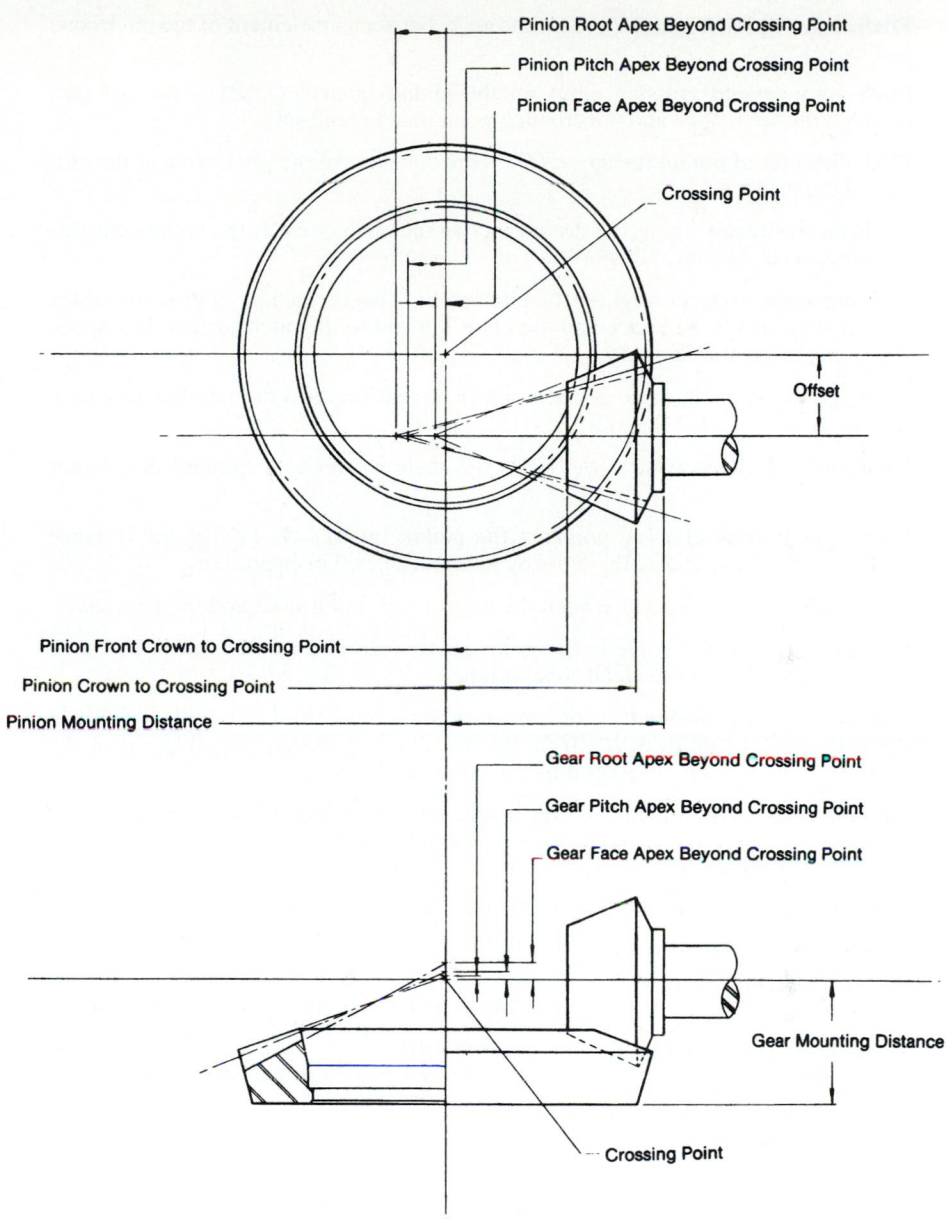

Pinion Root Apex Beyond Crossing Point

Pinion Pitch Apex Beyond Crossing Point

Pinion Face Apex Beyond Crossing Point

Crossing Point

Offset

Pinion Front Crown to Crossing Point

Pinion Crown to Crossing Point

Pinion Mounting Distance

Gear Root Apex Beyond Crossing Point

Gear Pitch Apex Beyond Crossing Point

Gear Face Apex Beyond Crossing Point

Gear Mounting Distance

Crossing Point

FIGURE 34.7   Hypoid gear nomenclature.

Pitch angle of pinion (gear) $\gamma\,(\Gamma)$ is the angle between an element of the pitch cone and its axis.

Pitch apex beyond crossing point on the pinion (gear) $G\,(Z)$ is the distance between the pitch apex and the crossing point on a hypoid set.

Pitch diameter of pinion (gear) $d\,(D)$ is the diameter of the pitch cone at the outside of the blank.

Pitch, mean circular p_m is the distance along the pitch circle at the mean cone distance between corresponding profiles of adjacent teeth.

Pressure angle ϕ is the angle at the pitch point between the line of pressure which is normal to the tooth surface and the plane tangent to the pitch surface. It is specified at the mean cone distance.

Ratio, gear m_G is the ratio of the number of gear teeth to the number of pinion teeth.

Root angle of pinion (gear) $\gamma_R\,(\Gamma_R)$ is the angle between an element of the root cone and its axis.

Root apex beyond crossing point on the pinion (gear) $G_R\,(Z_R)$ is the distance between the root apex and the crossing point on a bevel or hypoid set.

Shaft angle Σ is the angle between the axes of the pinion shaft and the gear shaft.

Spiral angle ψ is the angle between the tooth trace and an element of the pitch cone. It is specified at the mean cone distance.

Spiral-bevel gear, left-hand   is one in which the outer half of a tooth is inclined in the counterclockwise direction from the axial plane through the midpoint of the tooth, as viewed by an observer looking at the face of the gear.

Spiral-bevel gear, right-hand   is one in which the outer half of a tooth is inclined in the clockwise direction from the axial plane through the midpoint of the tooth, as viewed by an observer looking at the face of the gear.

Tangential force W_t is the force applied to a gear tooth at the mean cone distance in a direction tangent to the pitch cone and normal to a pitch-cone element.

Thickness of pinion (gear), mean circular $t\,(T)$ is the length of arc on the pitch cone between the two sides of the tooth at the mean cone distance.

Thickness of pinion (gear), mean normal chordal $t_{nc}\,(T_{nc})$ is the chordal thickness of the pinion tooth at the mean cone distance in a plane normal to the tooth trace.

34.2.3   Calculation Methods

Four methods of blank design are commonly used in the design of bevel and hypoid gears:

1. Standard taper
2. Duplex taper
3. Uniform taper
4. Tilted root-line taper

The taper you select depends in some instances on the manufacturing equipment available for producing the gear set. Therefore, before starting calculations, you should familiarize yourself with the equipment and method used by the gear manufacturer.

34.3 GEAR MANUFACTURING

34.3.1 Methods of Generation

Generation is the basic process in the manufacture of bevel and hypoid gears in that at least one member of every set must be generated. The theory of generation as applied to these gears involves an imaginary generating gear, which can be a crown gear, a mating gear, or some other bevel or hypoid gear. The gear blank or workpiece is positioned so that when it is rolled with the generating gear, the teeth of the workpiece are enveloped by the teeth of the generating gear.

In the actual production of the gear teeth, at least one tooth of the generating gear is described by the motion of the cutting tool or grinding wheel. The tool and its motion are carried on a rotatable machine member called a cradle, the axis of which is identical with the axis of the generating gear. The cradle and the workpiece roll together on their respective axes exactly as would the workpiece and the generating gear.

The lengthwise tooth curve of the generating gear is selected so that it is easily followed with a practical cutting tool and mechanical motion. Figure 34.8 illustrates the representation of a generating gear by a face-mill cutter. Figure 34.9 shows the basic machine elements of a bevel-gear face-mill generator.

Most generating gears are based on one of two fundamental concepts. The first is complementary crown gears, where two gears with 90° pitch angles fit together like mold castings. Each of the crown gears is the generating gear for one member of the mating set. Gears generated in this manner have line contact and are said to be *conjugate* to each other. With the second concept, the teeth of one member are form-cut without generation. This member becomes the generating gear for producing the mating member. Again, gears generated in this manner are conjugate to each other.

34.3.2 Localization of Contact

Any displacement in the nominal running position of either member of a mating conjugate gear set shifts the contact to the edges of the tooth. The result is concentrated loading and irregular motion. To accommodate assembly tolerances and deflections resulting from load, tooth surfaces are relieved in both the lengthwise and profile directions. The resulting localization of the contact pattern is achieved by using a generating setup which is deliberately modified from the conjugate generating gear.

34.3.3 Testing

The smoothness and quietness of operation, the tooth contact pattern, the tooth size, the surface finish, and appreciable runout can be checked in a running test. This is a subjective test. The machine consists of two spindles that can be set at the correct shaft angle, mounting distances, and offset. The gear to be inspected is mounted on

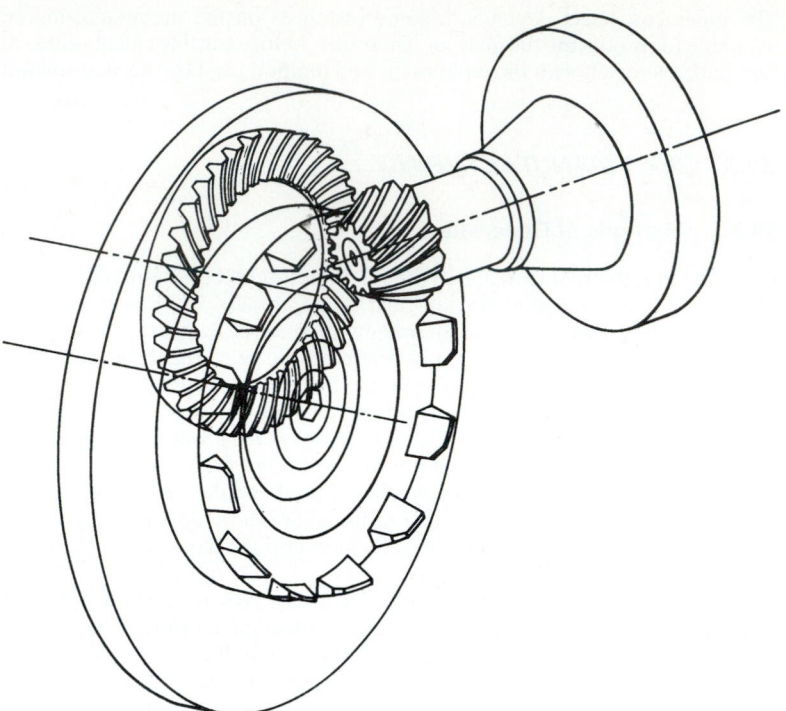

FIGURE 34.8   Imaginary generating gear.

one spindle, and the mating gear or a control gear is mounted on the other spindle. Tooth contact is evaluated by coating the teeth with a gear-marking compound and running the set under light load for a short time. At the same time, the smoothness of operation is observed. Spacing errors and runout are evaluated by noting variations in the contact pattern on the teeth around the blank. Poor surface finish shows up as variations within the marked contact pattern. Tooth size is measured by locking one member and rotating a tooth of the mating member within the slot to determine the backlash.

The contact pattern is shifted lengthwise along the tooth to the inside and outside of the blank by displacing one member along its axis and in the offset direction. The amount of displacement is used as a measure of the set's adjustability.

It is normal practice for tooth spacing and runout to be measured with an additional operation on inspection equipment designed specifically for that purpose. AGMA publication 390.03a specifies allowable tolerances for spacing and runout based on diametral pitch and pitch diameter.

Double- and single-flank test equipment can be used to measure tooth-profile errors, tooth spacing, and runout. The test equipment has transducers on the work spindles, and the output data are in chart form. The output data not only provide a record of the quality of the gear set, but can also be related to gear noise.

Three-dimensional coordinate-measuring machines can be used to compare the actual gear-tooth geometry with theoretical data.

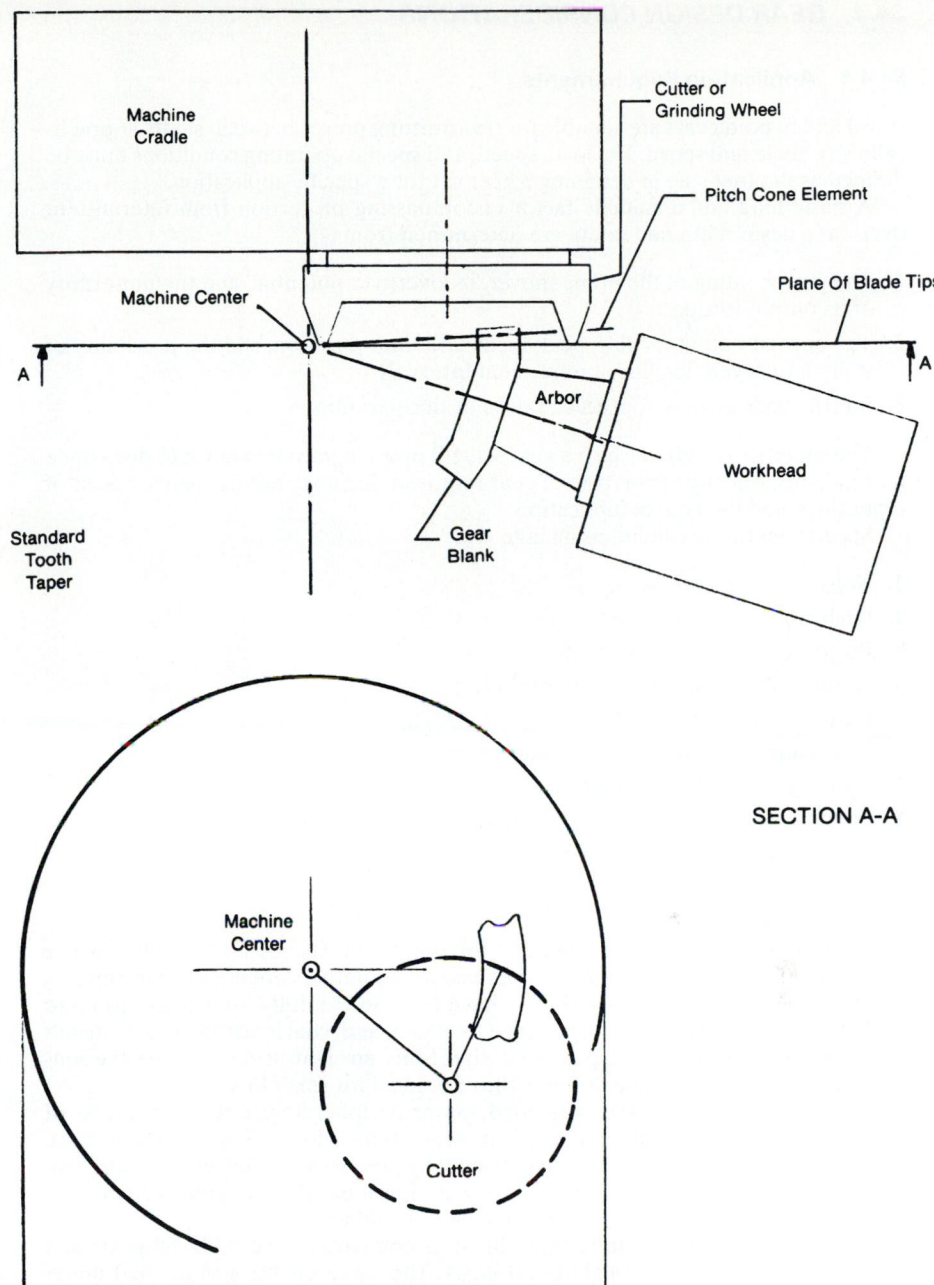

FIGURE 34.9 Basic machine setup of spiral-bevel face-mill generator.

34.4 GEAR DESIGN CONSIDERATIONS

34.4.1 Application Requirements

Bevel and hypoid gears are suitable for transmitting power between shafts at practically any angle and speed. The load, speed, and special operating conditions must be defined as the first step in designing a gear set for a specific application.

A basic load and a suitable factor encompassing protection from intermittent overloads, desired life, and safety are determined from

1. The power rating of the prime mover, its overload potential, and the uniformity of its output torque
2. The normal output loading, peak loads and their duration, and the possibility of stalling or severe loading at infrequent intervals
3. Inertia loads arising from acceleration or deceleration

The speed or speeds at which a gear set will operate must be known to determine inertia loads, velocity factor, type of gear required, accuracy requirements, design of mountings, and the type of lubrication.

Special operating conditions include

1. Noise-level limitations
2. High ambient temperature
3. Presence of corrosive elements
4. Abnormal dust or abrasive atmosphere
5. Extreme, repetitive shock loading or reversing
6. Operating under variable alignment
7. Gearing exposed to weather
8. Other conditions that may affect the operation of the set

34.4.2 Selection of Type of Gear

Straight-bevel gears are recommended for peripheral speeds up to 1000 feet per minute (ft/min) where maximum smoothness and quietness are not of prime importance. However, ground straight bevels have been successfully used at speeds up to 15 000 ft/min. Plain bearings may be used for radial and axial loads and usually result in a more compact and less expensive design. Since straight-bevel gears are the simplest to calculate, set up, and develop, they are ideal for small lots.

Spiral-bevel gears are recommended where peripheral speeds are in excess of 1000 ft/min or 1000 revolutions per minute (r/min). Motion is transmitted more smoothly and quietly than with straight-bevel gears. So spiral-bevel gears are preferred also for some lower-speed applications. Spiral bevels have greater load sharing, resulting from more than one tooth being in contact.

Zerol bevel gears have little axial thrust as compared to spiral-bevel gears and can be used in place of straight-bevel gears. The same qualities as defined under straight bevels apply to Zerol bevels. Because Zerol bevel gears are manufactured on the same equipment as spiral-bevel gears, Zerol bevel gears are preferred by some manufacturers. They are more easily ground because of the availability of bevel grinding equipment.

Hypoid gears are recommended where peripheral speeds are in excess of 1000 ft/min and the ultimate in smoothness and quietness is required. They are somewhat stronger than spiral bevels. Hypoids have lengthwise sliding action, which enhances the lapping operation but makes them slightly less efficient than spiral-bevel gears.

34.4.3 Estimated Gear Size

Figures 34.10 and 34.11 relate size of bevel and hypoid gears to gear torque, which should be taken at a value corresponding to maximum sustained peak or one-half peak, as outlined below.

If the total duration of the peak load exceeds 10 000 000 cycles during the expected life of the gear, use the value of this peak load for estimating gear size. If, however, the total duration of the peak load is less than 10 000 000 cycles, use one-half the peak load or the value of the highest sustained load, whichever is greater.

Given gear torque and the desired gear ratio, the charts give gear pitch diameter. The charts are based on case-hardened steel and should be used as follows:

1. For other materials, multiply the gear pitch diameter by the material factor from Table 34.1.
2. For general industrial gearing, the preliminary gear size is based on surface durability.
3. For straight-bevel gears, multiply the gear pitch diameter by 1.2; for Zerol bevel gears, multiply the gear pitch diameter by 1.3.
4. For high-capacity spiral-bevel and hypoid gears, the preliminary gear size is based on both surface capacity and bending strength. Choose the larger of the gear diameters, based on the durability chart and the strength chart.

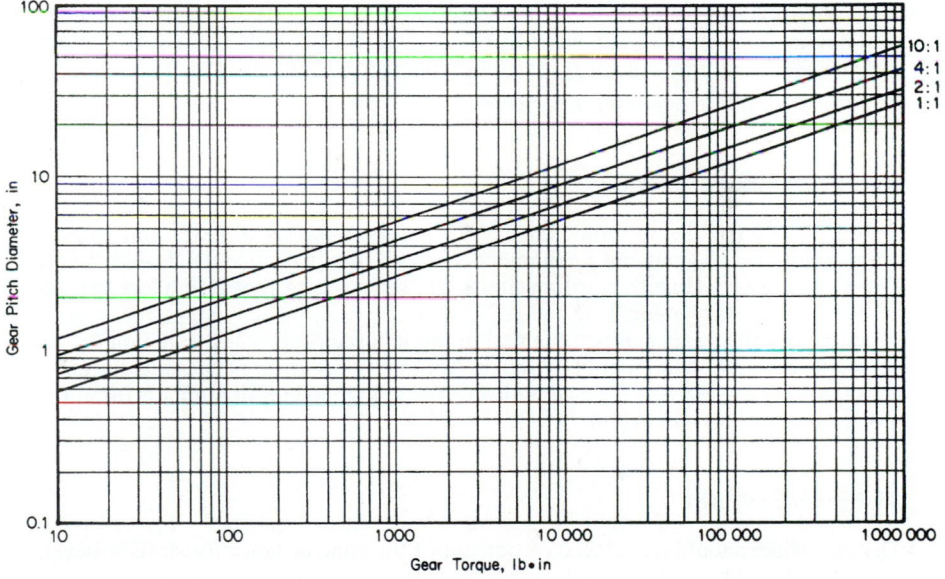

FIGURE 34.10 Gear pitch diameter based on surface durability.

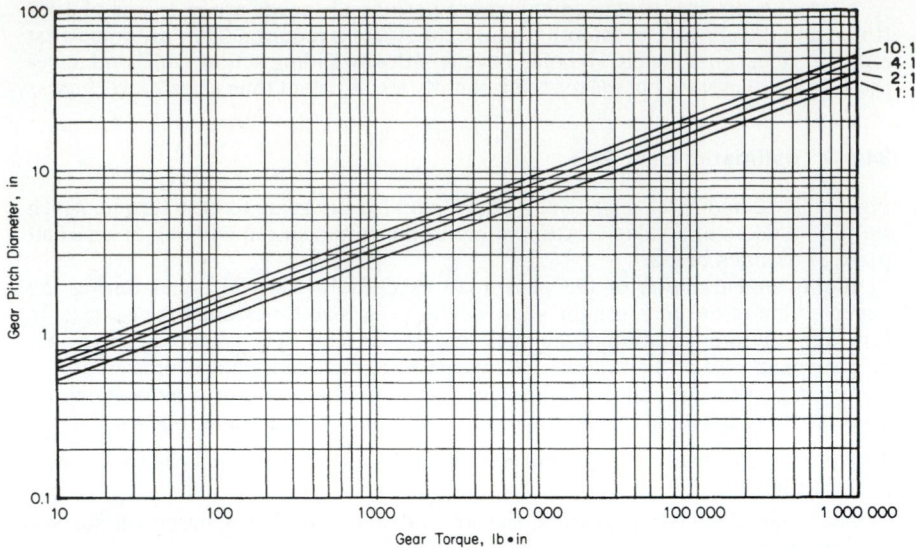

FIGURE 34.11   Gear pitch diameter based on bending strength.

5. For high-capacity ground spiral-bevel and hypoid gears, the gear diameter from the durability chart should be multiplied by 0.80.
6. For hypoid gears, multiply the gear pitch diameter by $D/(D + E)$.
7. Statically loaded gears should be designed for bending strength rather than surface durability. For statically loaded gears subject to vibration, multiply the gear diameter from the strength chart by 0.70. For statically loaded gears not subject to vibration, multiply the gear diameter from the strength chart by 0.60.
8. Estimated pinion diameter is $d = Dn/N$.

34.4.4   Number of Teeth

Figure 34.12 gives the recommended tooth numbers for spiral-bevel and hypoid gears. Figure 34.13 gives the recommended tooth numbers for straight-bevel and Zerol bevel gears. However, within limits, the selection of tooth numbers can be made in an arbitrary manner.

   More uniform gears can be obtained in the lapping process if a common factor between gear and pinion teeth is avoided. Automotive gears are generally designed with fewer pinion teeth. Table 34.2 indicates recommended tooth numbers for automotive spiral-bevel and hypoid drives.

34.4.5   Face Width

The face width should not exceed 30 percent of the cone distance for straight-bevel, spiral-bevel, and hypoid gears and should not exceed 25 percent of the cone distance for Zerol bevel gears. In addition, it is recommended that the face width F be limited to

TABLE 34.1 Material Factors C_M

Gear		Pinion		Material factor C_M
Material	Hardness	Material	Hardness	
Case-hardened steel	58 R_C†	Case-hardened steel	60 R_C†	0.85‡
Case-hardened steel	55 R_C†	Case-hardened steel	55 R_C†	1.00
Flame-hardened steel	50 R_C†	Case-hardened steel	55 R_C†	1.05
Flame-hardened steel	50 R_C†	Flame-hardened steel	50 R_C†	1.05
Oil-hardened steel	375–425 H_B	Oil-hardened steel	375–425 H_B	1.20
Heat-treated steel	250–300 H_B	Case-hardened steel	55 R_C†	1.45
Heat-treated steel	210–245 H_B	Heat-treated steel	245–280 H_B	1.65
Cast iron		Case-hardened steel	55 R_C†	1.95
Cast iron		Flame-hardened steel	50 R_C†	2.00
Cast iron		Annealed steel	160–200 H_B	2.10
Cast iron		Cast iron		3.10

†Minimum values.
‡Gears must be file-hard.

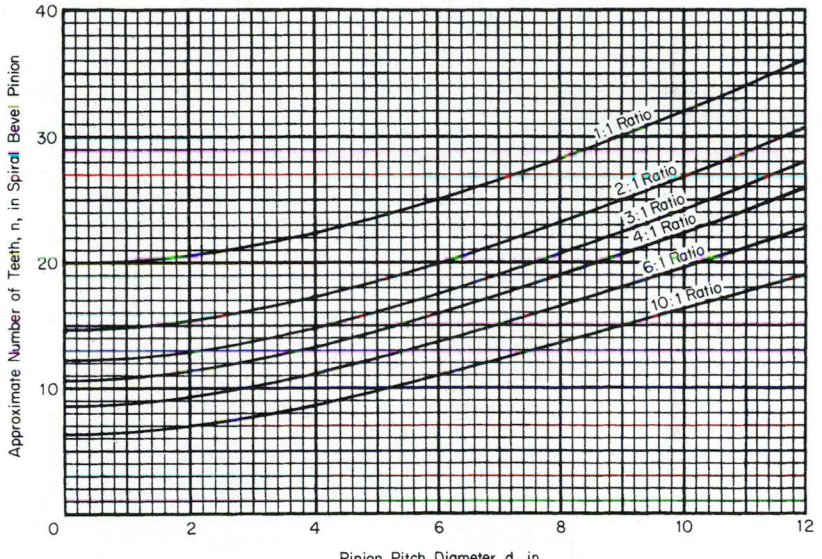

FIGURE 34.12 Recommended tooth numbers for spiral-bevel and hypoid gears.

$$F \le \frac{10}{P_d}$$

The design chart in Fig. 34.14 gives the approximate face width for straight-bevel, spiral-bevel, and hypoid gears. For Zerol bevel gears, the face width given by this chart should be multiplied by 0.83.

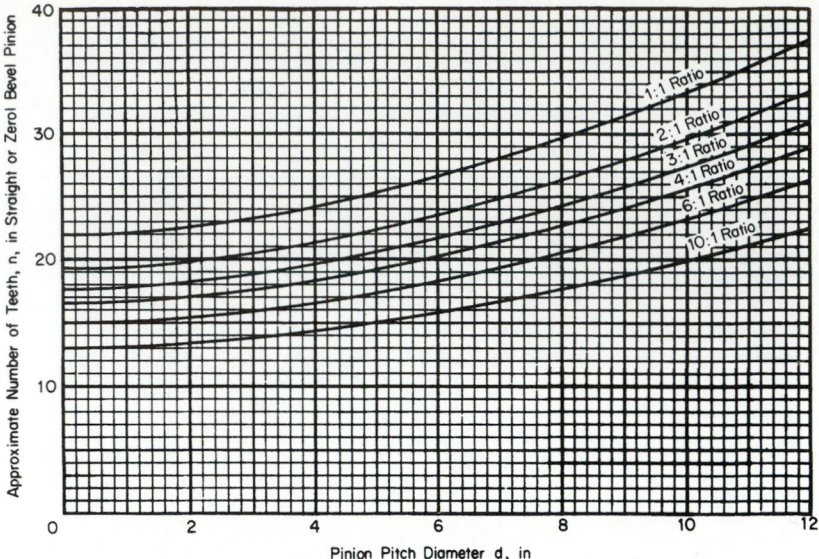

FIGURE 34.13   Recommended tooth numbers for straight- and Zerol-bevel gears.

34.4.6   Diametral Pitch

The diametral pitch is now calculated by dividing the number of teeth in the gear by the gear pitch diameter. Because tooling for bevel gears is not standardized according to pitch, it is not necessary that the diametral pitch be an integer.

TABLE 34.2   Recommended Tooth Numbers for Automotive Applications

Approximate ratio	Preferred no. pinion teeth	Allowable range
1.50/1.75	14	12 to 16
1.75/2.00	13	11 to 15
2.0/2.5	11	10 to 13
2.5/3.0	10	9 to 11
3.0/3.5	10	9 to 11
3.5/4.0	10	9 to 11
4.0/4.5	9	8 to 10
4.5/5.0	8	7 to 9
5.0/6.0	7	6 to 8
6.0/7.5	6	5 to 7
7.5/10.0	5	5 to 6

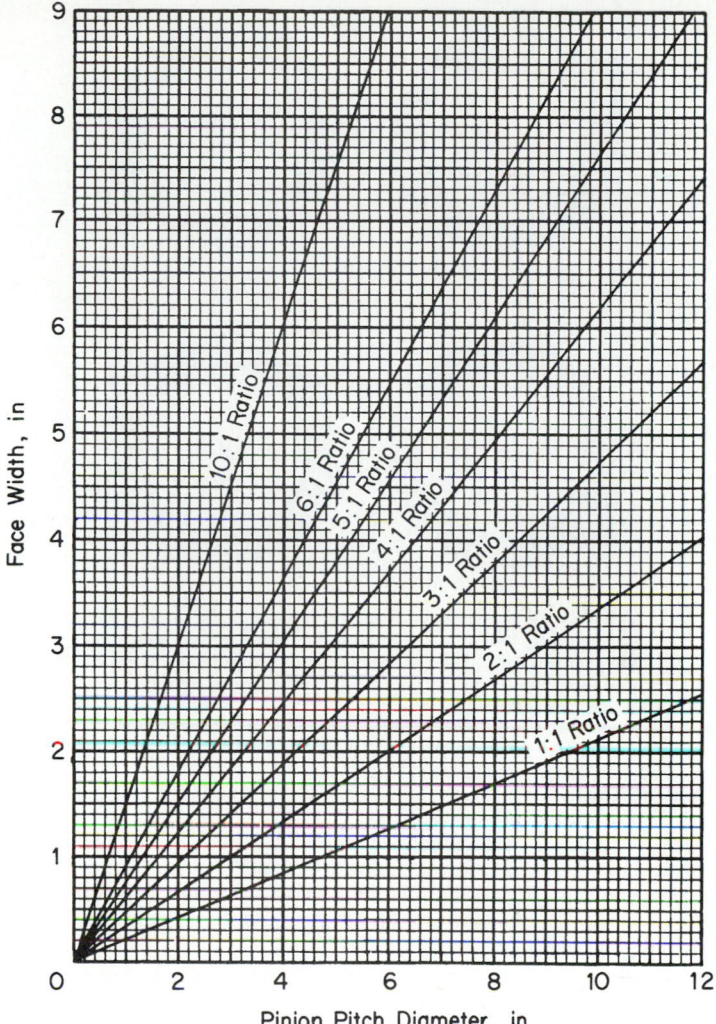

FIGURE 34.14 Face width of spiral-bevel and hypoid gears.

34.4.7 Hypoid Offset

In the design of hypoid gears, the offset is designated as being above or below center. Figure 34.15a and b illustrates the below-center position, and Fig. 34.15c and d illustrates the above-center position. In general, the shaft offset for power drives should not exceed 25 percent of the gear pitch diameter, and on very heavily loaded gears, the offset should be limited to 12.5 percent of the gear pitch diameter.

Hypoid pinions are larger in diameter than the corresponding spiral-bevel pinion. This increase in diameter may be as great as 30 percent, depending on the offset, spiral angle, and gear ratio.

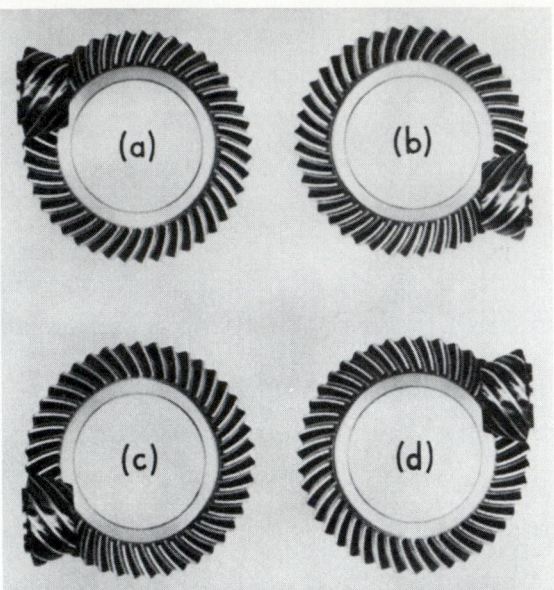

FIGURE 34.15 Hypoid offset. To determine the direction of offset, always look at the gear with the pinion at the right. Thus the gear sets of (*a*) and (*b*) are both offset *below* center; similar reasoning shows that (*c*) and (*d*) are offset *above* center. (*Gleason Machine Division.*)

34.4.8 Spiral Angle

In designing spiral-bevel gears, the spiral angle should be sufficient to give a face-contact ratio of at least 1.25. For maximum smoothness and quietness, the face-contact ratio should be between 1.50 and 2.00. High-speed applications should be designed with a face-contact ratio of 2.00 or higher for best results. Figure 34.16 may be used to assist in the selection of the spiral angle.

For hypoid gears, the desired pinion spiral angle can be calculated by

$$\psi_P = 25 + 5\sqrt{\frac{N}{n}} + 90\,\frac{E}{D}$$

where ψ_P is in degrees.

34.4.9 Pressure Angle

The commonly used pressure angle for bevel gears is 20°, although pressure angles of 22.5° and 25° are used for heavy-duty drives.

In the case of hypoids, the pressure angle is unbalanced on opposite sides of the gear teeth in order to produce equal contact ratios on the two sides. For this reason, the average pressure angle is specified for hypoids. For automotive drives, use 18° or 20°, and for heavy-duty drives, use 22.5° or 25°.

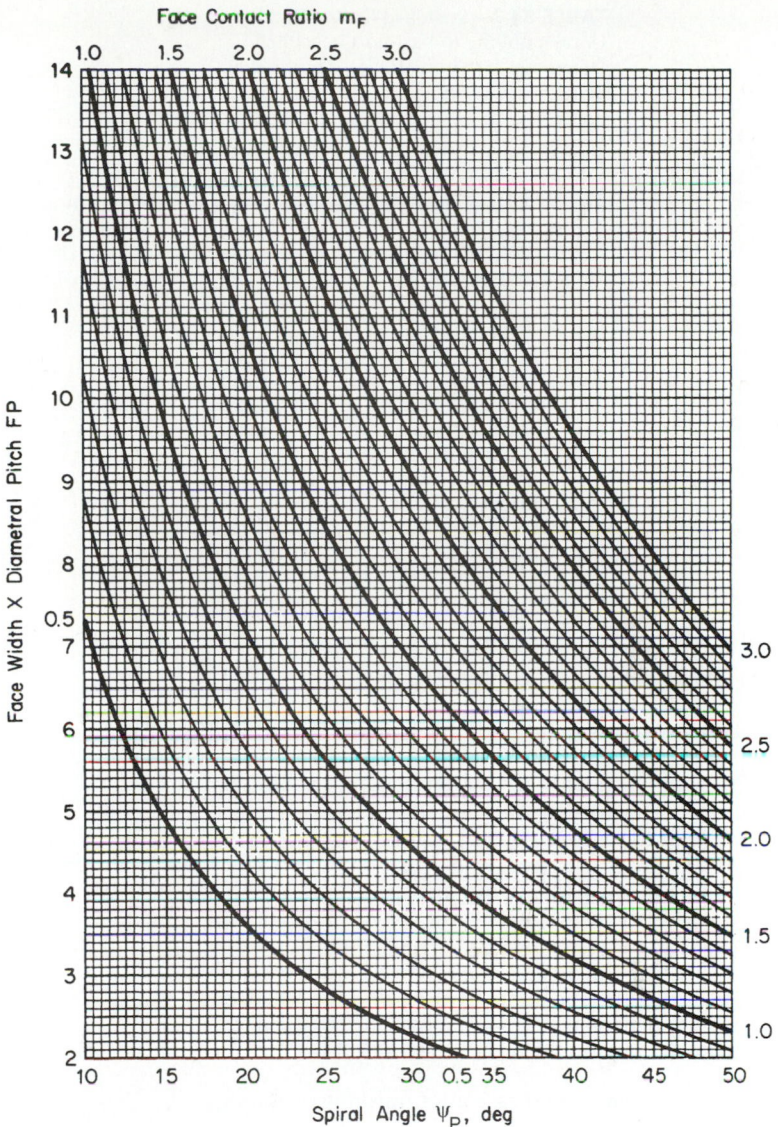

FIGURE 34.16 Selection of spiral angle.

34.4.10 Cutter Diameter

A cutter diameter must be selected for spiral-bevel, Zerol bevel, and hypoid gears. The usual practice is to use a cutter diameter approximately equal to the gear diameter. To increase adjustability of the gear set and obtain maximum strength, a smaller cutter should be used. Cutter diameters are standardized. Therefore, Table 34.3 is included to aid in cutter selection.

TABLE 34.3   Standard Cutter Radii Corresponding to Various Gear Pitch Diameters for Spiral-Bevel, Zerol Bevel, and Hypoid Gears

Pitch diameter D of gear, in	Cutter radius r_c, in (mm)
3.000–5.250	1.750
3.875–6.750	2.250
4.250–7.500	2.500
5.125–9.000	3.000
6.500–11.250	3.750
7.750–13.500	4.500
9.000–15.750	5.250
10.250–18.000	6.000
12.000–21.000	7.000
13.750–24.000	8.000
15.500–27.000	9.000
18.000–31.500	10.500
21.750–60.000	(320)
27.250–75.000	(400)
34.250–100.000	(500)

34.4.11   Materials and Heat Treatment

Through-hardening steels are used when medium wear resistance and medium load-carrying capacity are desired. The following steels are some of those used and listed, beginning with the steel of lowest hardenability: AISI 1045, 1144, 4640, 4150, and 4340. When greater hardenability is required for larger gears, it is sometimes necessary to increase the carbon content of these steels or to select a different steel.

Carburized gears are used when high wear resistance and high load-carrying capacity are required. Carburizing steels used in gears normally have a carbon content of 0.10 to 0.25 percent and should have sufficient alloy content to allow hardening in the section sizes in which they are used. For low-heat-treat distortion, 4620 or 8620 might be used; if high core hardness is desired, 9310 might be used. The following steels have been commonly used and are listed, beginning with the steel of lowest core hardenability: AISI 4620, 8620, 9310, and 4820.

Carburized gears should be specified as follows:

1. Total depth of carburized case after finishing operations
2. Surface hardness
3. Core hardness
4. Maximum case carbon content (optional)

Gears should be quenched from a temperature which will ensure a minimum amount of retained austenite.

Nitriding steels are used in applications which require high wear resistance with minimum distortion in heat treating. The commonly used steels are AISI 4140, 4150, and 4340. If extreme hardness and wear resistance are required, the nitralloy steels can be used. To achieve the desired results in the nitriding operation, all material should be hardened and tempered above the nitriding temperature prior to finish machining. Sharp corners should be avoided on external surfaces.

Nitrided gears should be specified as follows:

1. Total depth of nitrided case after finishing operations
2. Surface hardness
3. Core hardness

Cast iron is used in place of non-heat-treated steel where good wear resistance plus excellent machineability is required. Complicated shapes can be cast more easily from iron than they can be produced by machining from bars or forgings.

34.5 GEAR-TOOTH DIMENSIONS

34.5.1 Calculation of Basic Bevel-Gear-Tooth Dimensions

All bevel-gear-tooth dimensions are calculated in a similar manner. Therefore, straight-bevel, spiral-bevel, and Zerol bevel gears are considered as a group. In Sec. 34.4 we selected

1. Number of pinion teeth n
2. Number of gear teeth N
3. Diametral pitch P_d
4. Shaft angle Σ
5. Face width F
6. Pressure angle ϕ
7. Spiral angle ψ
8. Hand of spiral (pinion), left-hand/right-hand (LH/RH)
9. Cutter radius r_c

The formulas in Table 34.4 are now used to calculate the blank and tooth dimensions.

34.5.2 Tooth Taper

Spiral-bevel- and hypoid-gear blanks are designed by one of four methods—standard taper, duplex taper, tilted root line, or uniform depth.

Standard taper is the case where the root lines of mating members, if extended, would intersect the pitch-cone apex. The tooth depth changes in proportion to the cone distance.

Duplex taper is the case where the root lines are tilted so that the slot width is constant. This condition permits each member of a pair to be finished in one operation, using circular cutters which cut in one slot.

Tilted root-line taper is a compromise between duplex and standard taper. The blanks are designed with duplex taper except when the taper becomes excessive. When the taper is 1.3 times standard taper or greater, 1.3 times standard taper is used.

Uniform-depth taper is the case where the root lines are not tilted. The tooth depth is uniform from the inside to the outside of the blank.

TABLE 34.4 Formulas for Computing Blank and Tooth Dimensions

Item	Item no.	Member	Formula
Pitch diameter	1	Pinion	$d = \dfrac{n}{P_d}$
		Gear	$D = \dfrac{N}{P_d}$
Pitch angle	2	Pinion	$\gamma = \tan^{-1} \dfrac{\sin \Sigma}{N/n + \cos \Sigma}$
		Gear	$\Gamma = \Sigma - \gamma$
Outer cone distance	3	Both	$A_o = \dfrac{0.50D}{\sin \Gamma}$
Mean cone distance	4	Both	$A_m = A_o - 0.5F$
Depth factor k_1	5	Both	Table 34.5
Mean working depth	6	Both	$h = \dfrac{k_1 A_m}{P_d A_o} \cos \psi$
Clearance factor k_2	7	Both	Table 34.6
Clearance	8	Both	$c = k_2 h$
Mean whole depth	9	Both	$h_m = h + c$
Equivalent 90° ratio	10	Both	$m_{90} = \sqrt{\dfrac{N \cos \gamma}{n \cos \Gamma}}$
Mean addendum factor C_1	11	Both	Table 34.7
Mean circular pitch	12	Both	$P_m = \dfrac{\pi A_m}{P_d A_o}$
Mean addendum	13	Pinion	$a_P = h - a_G$
		Gear	$a_G = C_1 h$
Mean dedendum	14	Pinion	$b_P = h_m - a_P$
		Gear	$b_G = h_m - a_G$
Sum of dedendum angles	15	Both	$\Sigma\delta$ (see Sec. 34.5.2)

TABLE 34.4 Formulas for Computing Blank and Tooth Dimensions (*Continued*)

Item	Item no.	Member	Formula
Dedendum angle	16	Pinion	δ_P (see Sec. 34.5.2)
		Gear	δ_G (see Sec. 34.5.2)
Face angle of blank	17	Pinion	$\gamma_o = \gamma + \delta_G$
		Gear	$\Gamma_o = \Gamma + \delta_P$
Root angle of blank	18	Pinion	$\gamma_R = \gamma - \delta_P$
		Gear	$\Gamma_R = \Gamma - \delta_G$
Outer addendum	19	Pinion	$a_{oP} = a_P + 0.5F \tan \delta_G$
		Gear	$a_{oG} = a_G + 0.5F \tan \delta_P$
Outer dedendum	20	Pinion	$b_{oP} = b_P + 0.5F \tan \delta_P$
		Gear	$b_{oG} = b_G + 0.5F \tan \delta_G$
Outer working depth	21	Both	$h_k = a_{oP} + a_{oG}$
Outer whole depth	22	Both	$h_t = a_{oP} + b_{oP}$
Outside diameter	23	Pinion	$d_o = d + 2a_{oP} \cos \gamma$
		Gear	$D_o = D + 2a_{oG} \cos \Gamma$
Pitch apex to crown	24	Pinion	$x_o = A_o \cos \gamma - a_{oP} \sin \gamma$
		Gear	$X_o = A_o \cos \Gamma - a_{oG} \sin \Gamma$
Mean diametral pitch	25	Both	$P_{dm} = P_d \dfrac{A_o}{A_m}$
Mean pitch diameter	26	Pinion	$d_m = \dfrac{n}{P_{dm}}$
		Gear	$D_m = \dfrac{N}{P_{dm}}$
Thickness factor K	27	Both	Fig. 34.17
Mean normal circular thickness	28	Pinion	$t_n = P_m \cos \psi - T_n$
		Gear	$T_n = \dfrac{P_m}{2 \cos \psi} - (a_P - a_G) \tan \phi + \dfrac{K \cos \psi}{P_{dm} \tan \phi}$

TABLE 34.4 Formulas for Computing Blank and Tooth Dimensions (*Concluded*)

Item	Item no.	Member	Formula
Outer normal backlash allowance	29	Both	B (Table 34.8)
Mean normal chordal thickness	30	Pinion	$t_{nc} = t_n - \dfrac{t_n^3}{6d_m^2} - 0.5B\dfrac{A_m}{A_o}\sec\phi$
		Gear	$T_{nc} = T_n - \dfrac{T_n}{6D_m^2} - 0.5B\left(\dfrac{A_m}{A_o}\right)\sec\phi$
Mean chordal addendum	31	Pinion	$a_{cP} = a_P + \dfrac{t_n^2\cos\gamma}{4d_m}$
		Gear	$a_{cG} = a_G + \dfrac{T_n^2\cos\Gamma}{4D_m}$

TABLE 34.5 Depth Factor

Type of gear	No. pinion teeth	Depth factor k_1
Straight bevel	12 and higher	2.000
Spiral bevel	12 and higher	2.000
	11	1.995
	10	1.975
	9	1.940
	8	1.895
	7	1.835
	6	1.765
Zerol bevel	13 and higher	2.000
Hypoid	11 and higher	4.000
	10	3.900
	9	3.8
	8	3.7
	7	3.6
	6	3.5

TABLE 34.6 Clearance Factors

Type of gear	Clearance factor k_2
Straight bevel	0.140
Spiral bevel	0.125
Zerol bevel	0.110
Hypoid	0.150

TABLE 34.7 Mean Addendum Factor

Type of gear	No. pinion teeth	Mean addendum factor C_1
Straight bevel	12 and higher	C_1†
Spiral bevel	12 and higher	C_1†
	11	0.490
	10	0.435
	9	0.380
	8	0.325
	7	0.270
	6	0.215
Zerol bevel	13 and higher	C_1†
Hypoid	21 and higher	C_1†
	9 to 20	0.170
	8	0.150
	7	0.130
	6	0.110

†Use $C_1 = 0.270 + 0.230/(m_{90})^2$.

TABLE 34.8 Minimum Normal Backlash Allowance†

Range of diametral pitch, teeth/in	Allowance, in (for AGMA quality number range)	
	4 to 9	10 to 13
1.00–1.25	0.032	0.024
1.25–1.50	0.027	0.020
1.50–2.00	0.020	0.015
2.00–2.50	0.016	0.012
2.50–3.00	0.013	0.010
3.00–4.00	0.010	0.008
4.00–5.00	0.008	0.006
5.00–6.00	0.006	0.005
6.00–8.00	0.005	0.004
8.00–10.00	0.004	0.003
10.00–12.00	0.003	0.002
12.00–16.00	0.003	0.002
16.00–20.00	0.002	0.001
20.00–25.00	0.002	0.001

†Measured at outer cone in inches.

In many cases, the type of taper depends on the manufacturing method. Before selecting a tooth taper, you should consult with the manufacturer to ensure compatibility between the design and the cutting method.

Straight-bevel gears are usually designed with standard taper. Zerol bevel gears are usually designed with duplex taper.

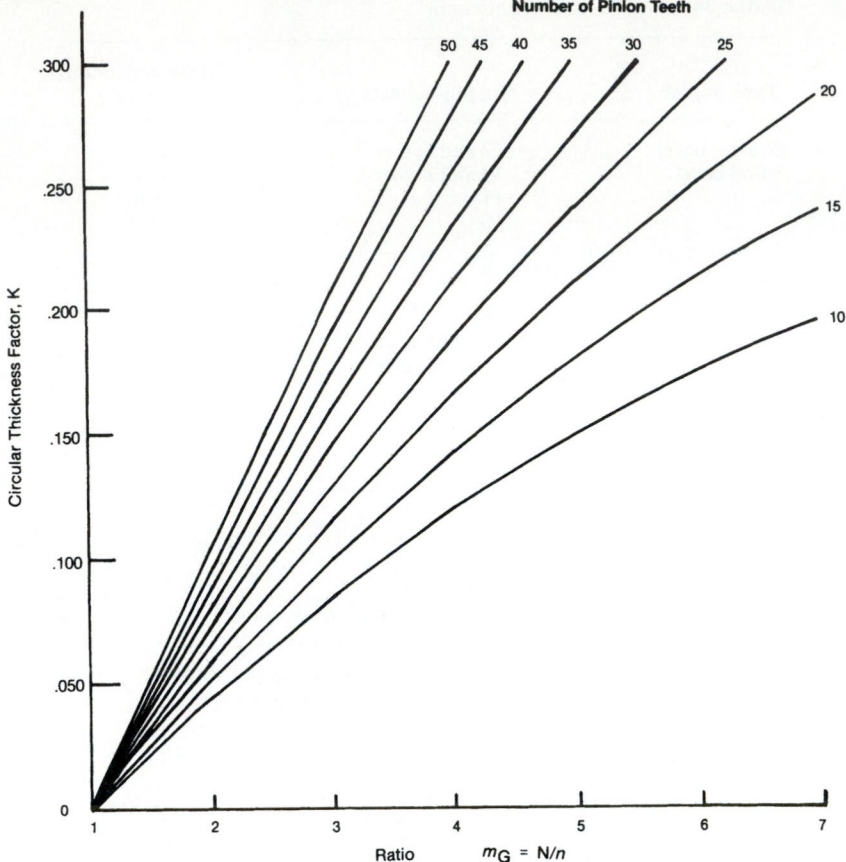

FIGURE 34.17   Circular thickness factor. These curves are plotted from the equation $K = -0.088 + 0.092m_G - 0.004m_G^2 + 0.0016\,(n-30)\,(m_G-1)$.

The formulas used to calculate the sum of dedendum angles and the dedendum angles are shown in Table 34.9.

34.5.3   Hypoid Dimensions

The geometry of hypoid gears is complicated by the offset between the axes of the mating members. Therefore a separate set of calculation formulas is needed.

The starting data are the same as for bevel gears with the following exceptions:

1. Hypoid offset E is required.
2. Pinion spiral angle ψ_P is specified.

The formulas in Table 34.10 are now used to calculate the blank and tooth dimensions.

TABLE 34.9 Formulas for Computing Dedendum Angles and Their Sum

Type of taper	Formula
Standard	$$\Sigma\delta = \tan^{-1}\frac{b_P}{A_m} + \tan^{-1}\frac{b_G}{A_m}$$ $$\delta_P = \tan^{-1}\frac{b_P}{A_m} \qquad \delta_G = \Sigma\delta - \delta_P$$
Duplex	$$\Sigma\delta = \frac{90[1 - (A_m/r_c)\sin\psi]}{(P_d A_o \tan\phi\cos\psi)}$$ $$\delta_P = \frac{a_G}{h}\Sigma\delta \qquad \delta_G = \Sigma\delta - \delta_P$$
Tilted root line	Use $$\Sigma\delta = \frac{90[1 - (A_m/r_c)\sin\psi]}{(P_d A_o \tan\phi\cos\psi)}$$ or $$= 1.3\tan^{-1}\frac{b_P}{A_m} + 1.3\tan^{-1}\frac{b_G}{A_m}$$ whichever is smaller. $$\delta_P = \frac{a_G}{h} \qquad \delta_G = \Sigma\delta - \delta_P$$
Uniform depth	$$\Sigma\delta = 0$$ $$\delta_P = \delta_G = 0$$

34.5.4 AGMA References[†]

The following AGMA standards are helpful in designing bevel and hypoid gears:

AGMA Design Manual for Bevel Gears, 2005
AGMA Rating Standard for Bevel Gears, 2003

These are available through American Gear Manufacturer's Association, 1500 King Street, Suite 201, Alexandria, VA 22314-2730.

34.6 GEAR STRENGTH

Under ideal conditions of operation, bevel and hypoid gears have a tooth contact which utilizes the full working profile of the tooth without load concentration in any

---

[†] The notation and units used in this chapter are the same as those used in the AGMA standards. These may differ in some respects from those used in other chapters of this Handbook.

TABLE 34.10 Formulas for Computing Blank and Tooth Dimensions of Hypoid Gears

Item	No.	Formula
Pitch diameter of gear	1	$D = \dfrac{N}{P_d}$
	2	$m = \dfrac{n}{N}$
	3	$\psi_{Po} = \psi_P$
	4	$\Delta\Sigma = 90 - \Sigma$
	5	$\tan \Gamma_i = \dfrac{\cos \Delta\Sigma}{1.2(m - \sin \Delta\Sigma)}$
	6	$R = 0.5(D - F \sin \Gamma_i)$
	7	$\sin \varepsilon_i' = \dfrac{E}{R} \sin \Gamma_i$
	8	$K_1 = \tan \psi_{Po} \sin \varepsilon_i' + \cos \varepsilon_i'$
	9	$R_{P2} = mRK_1$
	10	$\tan \eta = \dfrac{E}{R(\tan \Gamma_i \cos \Delta\Sigma - \sin \Delta\Sigma) + R_{P2}}$ first trial
	11	$\sin \varepsilon_2 = \dfrac{E - R_{P2} \sin \eta}{R}$
	12	$\tan \gamma_2 = \dfrac{\sin \eta}{\tan \varepsilon_2 \cos \Delta\Sigma} + \tan \Delta\Sigma \cos \eta$
	13	$\sin \varepsilon_2' = \dfrac{\sin \varepsilon_2 \cos \Delta\Sigma}{\cos \gamma_2}$
	14	$\tan \psi_{P2} = \dfrac{K_1 - \cos \varepsilon_2'}{\sin \varepsilon_2'}$
	15	$\Delta K = \sin \varepsilon_2' (\tan \psi_{Po} - \tan \psi_{P2})$
	16	$\dfrac{\Delta R_P}{R} = m(\Delta K)$
	17	$\sin \varepsilon_1 = \sin \varepsilon_2 - \dfrac{\Delta R_P}{R} \sin \eta$
Pinion pitch angle	18	$\tan \gamma = \dfrac{\sin \eta}{\tan \varepsilon_1 \cos \Delta\Sigma} + \tan \Delta\Sigma \cos \eta$
	19	$\sin \varepsilon_1' = \dfrac{\sin \varepsilon_1 \cos \Delta\Sigma}{\cos \gamma_1}$
Pinion spiral angle	20	$\tan \psi_P = \dfrac{K_1 + \Delta K - \cos \varepsilon_1'}{\sin \varepsilon_1'}$

34.26

Item	No.	Formula
Gear spiral angle	21	$\psi_G = \psi_P - \varepsilon_1'$
Gear pitch angle	22	$\tan \Gamma = \dfrac{\sin \varepsilon_1}{\tan \eta \cos \Delta\Sigma} + \cos \varepsilon_1 \tan \Delta\Sigma$
Gear mean cone distance	23	$A_{mG} = \dfrac{R}{\sin \Gamma}$
Pinion mean cone distance	24	$\Delta R_P = R\left(\dfrac{\Delta R_P}{R}\right)$
	25	$A_{mP} = \dfrac{R_{P2} + \Delta R_P}{\sin \gamma}$
	26	$R_P = A_{mP} \sin \gamma$
Limit pressure angle	27	$-\tan \phi_{01} = \dfrac{\tan \gamma \tan \Gamma}{\cos \varepsilon_1'} \times \dfrac{A_{mP} \sin \psi_P - A_{mG} \sin \psi_G}{A_{mP} \tan \gamma + A_{mG} \tan \Gamma}$
	28	$\mathrm{Den} = -\tan \phi_{01}\left(\dfrac{\tan \psi_P}{A_{mP} \tan \gamma} + \dfrac{\tan \psi_G}{A_{mG} \tan \Gamma}\right) + \dfrac{1}{A_{mP} \cos \psi_P}$ $- \dfrac{1}{A_{mG} \cos \psi_G}$
	29	$r_{c1} = \dfrac{\sec \phi_{01}\,(\tan \psi_P - \tan \phi_G)}{\mathrm{Den}}$
	30	$\left|\dfrac{r_c}{r_{c1}} - 1\right| \le 0.01$ Loop back to no. 10 and change η until satisfied.
Gear pitch apex beyond crossing point	31	$Z_P = A_{mP} \tan \gamma \sin \Gamma - \dfrac{E \tan \Delta\Sigma}{\tan \varepsilon_1}$
	32	$Z = \dfrac{R}{\tan \Gamma} - Z_P$
Gear outer cone distance	33	$A_o = \dfrac{0.5D}{\sin \Gamma}$
	34	$\Delta F_o = A_o - A_{mG}$
Depth factor	35	k_1 (see Table 34.5)
Addendum factor	36	C_1 (see Table 34.7)
Mean working depth	37	$h = \dfrac{k_1 R \cos \psi_G}{N}$
Mean addendum	38	$a_P = h - a_G \qquad a_G = C_1 h$

TABLE 34.10 Formulas for Computing Blank and Tooth Dimensions of Hypoid Gears (*Continued*)

Item	No.	Formula
Clearance factor	39	k_2 (see Table 34-6)
Mean dedendum	40	$b_P = b_G + a_G - a_P \qquad b_G = h(1 + k_2 - C_1)$
Clearance	41	$c = k_2 h$
Mean whole depth	42	$h_m = a_G + b_G$
Sum of dedendum angle	43	$\Sigma\delta$ (see Sec. 34.5.2)
Gear dedendum angle	44	δ_G (see Sec. 34.5.2)
Gear addendum angle	45	$\alpha_G = \Sigma\delta - \delta_G$
Gear outer addendum	46	$a_{oG} = a_G + \Delta F_o \sin \alpha_G$
Gear outer dedendum	47	$b_{oG} = b_G + \Delta F_o \sin \delta_G$
Gear whole depth	48	$h_t = a_{oG} + b_{oG}$
Gear working depth	49	$h_k = h_{tG} - c$
Gear root angle	50	$\Gamma_R = \Gamma - \delta_G$
Gear face angle	51	$\Gamma_o = \Gamma + \alpha_G$
Gear outside diameter	52	$D_o = 2a_{oG} \cos \Gamma + D_G$
Gear crown to crossing point	53	$X_o = Z_P + \Delta F_o \cos \Gamma - a_{oG} \sin \Gamma$
Gear root apex beyond crossing point	54	$Z_R = Z + \dfrac{A_{mG} \sin \delta_G - b_G}{\sin \Gamma_R}$
Gear face apex beyond crossing point	55	$Z_o = Z + \dfrac{A_{mG} \sin \alpha_G - a_G}{\sin \Gamma_o}$
	56	$Q_R = \dfrac{A_{mG} \cos \delta_G}{\cos \Gamma_R} - Z$
	57	$Q_o = \dfrac{A_{mG} \cos \alpha_G}{\cos \Gamma_o} - Z$
	58	$\tan \xi_R = \dfrac{E \tan \Delta\Sigma}{Q_R}$

Item	No.	Formula
Gear face apex beyond crossing point (*continued*)	59	$\tan \xi_o = \dfrac{E \tan \Delta\Sigma}{Q_o}$
	60	$\sin (\varepsilon_R + \xi_R) = \dfrac{E \cos \xi_R \tan \Gamma_R}{Q_R}$
	61	$\sin(\varepsilon_o + \xi_o) = \dfrac{E \cos \xi_o \tan \Gamma_o}{Q_o}$
Pinion face angle	62	$\sin \gamma_o = \sin \Delta\Sigma \sin \Gamma_R + \cos \Delta\Sigma \cos \Gamma_R \cos \varepsilon_R$
Pinion root angle	63	$\sin \gamma_R = \sin \Delta\Sigma \sin \Gamma_o + \cos \Delta\Sigma \cos \Gamma_o \cos \varepsilon_o$
Pinion face apex beyond crossing point	64	$G_o = \dfrac{E \sin \varepsilon_R \cos \Gamma_R - Z_R \sin \Gamma_R - c}{\sin \gamma_o}$
Pinion root apex beyond crossing point	65	$G_R = \dfrac{E \sin \varepsilon_o \cos \Gamma_o - Z_o \sin \Gamma_o - c}{\sin \gamma_R}$
	66	$\tan \lambda' = \dfrac{m \sin \varepsilon'_i \cos \Gamma}{\cos \gamma + m \cos \Gamma \cos \varepsilon'_i}$
Pinion addendum angle	67	$\alpha_P = \gamma_o - \gamma$
Pinion dedendum angle	68	$\delta_P = \gamma - \gamma_R$
Pinion whole depth	69	$h_{tP} = \dfrac{(x_o + G_o) \sin \delta_P}{\cos \gamma_o} - \sin \gamma_R (G_R - G_o)$
	70	$\Delta F_i = F - \Delta F_o$
	71	$\Delta F_{oP} = h \sin \varepsilon_R (1 - m)$
	72	$F_{oP} = \dfrac{\Delta F_o \cos \lambda'}{\cos (\varepsilon'_i - \lambda')}$
	73	$F_{iP} = \dfrac{\Delta F_i \cos \lambda'}{\cos (\varepsilon'_i - \lambda')}$
	74	$\Delta B_o = \dfrac{F_o \cos \gamma_o}{\cos \alpha_P} + \Delta F_{oP} - (b_G - c) \sin \gamma$
	75	$\Delta B_i = \dfrac{F \cos \gamma_o}{\cos \alpha_P} + \Delta F_{oP} - (b_G - c) \sin \gamma$
Pinion crown to crossing point	76	$x_o = \dfrac{E}{\tan \varepsilon_1 \cos \Delta\Sigma} \quad R_P \tan \gamma \ | \ \Delta D_o$

TABLE 34.10   Formulas for Computing Blank and Tooth Dimensions of Hypoid Gears (*Concluded*)

Item	No.	Formula
Pinion front crown to crossing point	77	$x_i = \dfrac{E}{\tan \varepsilon_1 \cos \Delta\Sigma} - R_P \tan \gamma - \Delta B_i$
Pinion outside diameter	78	$d_o = 2 \tan \gamma_o \, (x_o + G_o)$
Pinion face width	79	$F_P = \dfrac{x_o - x_i}{\cos \gamma_o}$
Mean circular pitch	80	$p_m = \dfrac{\pi A_{mG}}{P_d A_o}$
Mean diametral pitch	81	$P_{dm} = P_d \dfrac{A_o}{A_{mG}}$
Thickness factor	82	K (see Fig. 34-17)
Mean pitch diameter	83	$d_m = 2A_{mP} \sin \gamma$
	84	$D_m = 2A_{mG} \sin \Gamma$
Mean normal circular thickness	85	$t_n = p_m \cos \psi_G - T_n$
	86	$T_n = 0.5 p_m \cos \psi_G - (a_P - a_G) \tan \phi + \dfrac{K \cos \psi}{P_{dm} \tan \phi}$
Outer normal backlash allowance	87	B (see Table 34-8)
Mean normal chordal thickness	88	$t_{nc} = t_n - \dfrac{t_n^3}{6d_m^2} - 0.5B \sec \phi \left(\dfrac{A_{mg}}{A_o} \right)$
	89	$T_{nc} = T_n - \dfrac{T_n^3}{6D_m^2} - 0.5B \sec \phi \left(\dfrac{A_{mG}}{A_o} \right)$
Mean chordal addendum	90	$a_{cP} = a_P + \dfrac{0.25 t_n^2 \cos \gamma}{d_m}$
	91	$a_{cG} = a_G + \dfrac{0.25 T_N^2 \cos \Gamma}{D_m}$

area. The recommendations and rating formulas which follow are designed for a tooth contact developed to give the correct pattern in the final mountings under full load.

34.6.1 Formulas for Contact and Bending Stress

The basic equation for contact stress in bevel and hypoid gears is

$$S_c = C_p \sqrt{\frac{2T_p C_o}{C_v} \frac{1}{FD^2} \frac{N}{n} \frac{1.2 C_m C_f}{I}} \tag{34.1}$$

and the basic equation for bending stress is

$$S_t = \frac{2T_G K_o}{K_v} \frac{P_d}{FD} \frac{1.2 K_m}{J} \tag{34.2}$$

where
S_t = calculated tensile bending stress at root of gear tooth, pounds per square inch (lb/in^2)
S_c = calculated contact stress at point on tooth where its value will be maximum, lb/in^2
C_p = elastic coefficient of the gear-and-pinion materials combination, $(lb)^{1/2}/in$
T_P, T_G = transmitted torques of pinion and gear, respectively, pound-inches (lb · in)
K_o, C_o = overload factors for strength and durability, respectively
K_v, C_v = dynamic factors for strength and durability, respectively
K_m, C_m = load-distribution factors for strength and durability, respectively
C_f = surface-condition factor for durability
I = geometry factor for durability
J = geometry factor for strength

34.6.2 Explanation of Strength Formulas and Terms

The elastic coefficient for bevel and hypoid gears with localized tooth contact pattern is given by

$$C_p = \sqrt{\frac{3}{2\pi} \frac{1}{(1 - \mu_P^2)/E_P + (1 - \mu_G^2)/E_G}} \tag{34.3}$$

where
μ_P, μ_G = Poisson's ratio for materials of pinion and gear, respectively (use 0.30 for ferrous materials)
E_P, E_G = Young's modulus of elasticity for materials of pinion and gear, respectively (use 30.0×10^6 lb/in^2 for steel)

The overload factor makes allowance for the roughness or smoothness of operation of both the driving and driven units. Use Table 34.11 as a guide in selecting the overload factor.

The dynamic factor reflects the effect of inaccuracies in tooth profile, tooth spacing, and runout on instantaneous tooth loading. For gears manufactured to AGMA class 11 tolerances or higher, a value of 1.0 may be used for dynamic factor. Curve 2 in Fig. 34.18 gives the values of C_v for spiral bevels and hypoids of lower accuracy or for large, planed spiral-bevel gears. Curve 3 gives the values of C_v for bevels of lower accuracy or for large, planed straight-bevel gears.

TABLE 34.11   Overload Factors K_o, C_o[†]

Prime mover	Character of load on driven member		
	Uniform	Medium shock	Heavy shock
Uniform	1.00	1.25	1.75
Medium shock	1.25	1.50	2.00
Heavy shock	1.50	1.75	2.25

†This table is for speed-decreasing drive; for speed-increasing drives add $0.01(N/n)^2$ to the above factors.

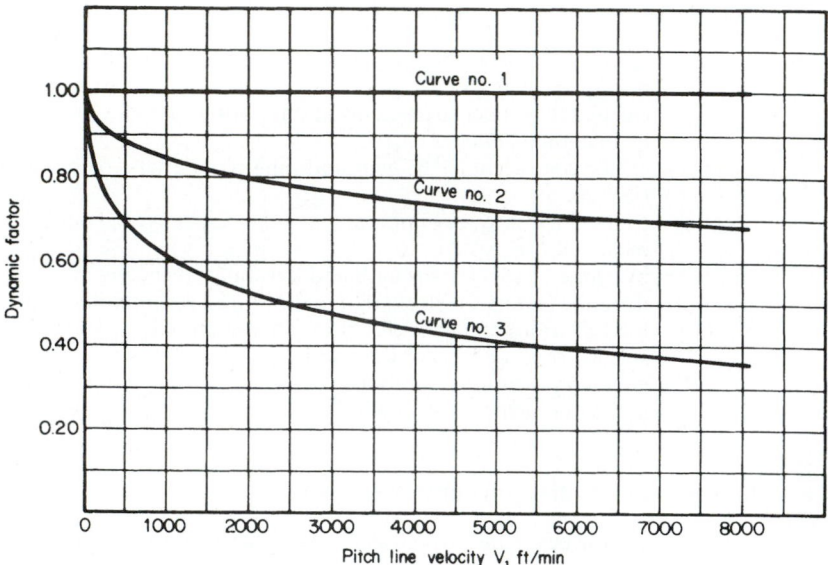

FIGURE 34.18   Dynamic factors K_v and C_v.

The load-distribution factor allows for misalignment of the gear set under operating conditions. This factor is based on the magnitude of the displacements of the gear and pinion from their theoretical correct locations. Use Table 34.12 as a guide in selecting the load-distribution factor.

The surface-condition factor depends on surface finish as affected by cutting, lapping, and grinding. It also depends on surface treatment such as lubrizing. And C_f can be taken as 1.0 provided good gear manufacturing practices are followed.

Use Table 34.13 to locate the charts for the two geometry factors I and J.

The geometry factor for durability I takes into consideration the relative radius of curvature between mating tooth surfaces, load location, load sharing, effective face width, and inertia factor.

The geometry factor for strength J takes into consideration the tooth form factor, load location, load distribution, effective face width, stress correction factor, and inertia factor.

TABLE 34.12 Load-Distribution Factors K_m, C_m

Application	Both members straddle-mounted	One member straddle-mounted	Neither member straddle-mounted
General industrial	1.00–1.10	1.10–1.25	1.25–1.40
Automotive	1.00–1.10	1.10–1.25	
Aircraft	1.00–1.25	1.10–1.40	1.25–1.50

TABLE 34.13 Location of Geometry Factors

Gear type	Pressure angle, ϕ	Shaft angle, Σ	Helix angle, ψ	Figure no.	
				I Factor	J Factor
Straight bevel	20°	90°	0°	34.19	34.20
	25°	90°	0°	34.21	34.22
Spiral bevel	20°	90°	35°	34.23	34.24
	20°	90°	25°	34.25	34.26
	20°	90°	15°	34.27	34.28
	25°	90°	35°	34.29	34.30
	20°	60°	35°	34.31	34.32
	20°	120°	35°	34.33	34.34
	20°†	90°	35°	34.35	34.36
Hypoid	19°	$E/D = 0.10$		34.37	34.38
	19°	$E/D = 0.15$		34.39	34.40
	19°	$E/D = 0.20$		34.41	34.42
	22½°	$E/D = 0.10$		34.43	34.44
	22½°	$E/D = 0.15$		34.45	34.46
	22½°	$E/D = 0.20$		34.47	34.48

† Automotive applications.

Interpolation between charts may be necessary for both the I and J factors.

34.6.3 Allowable Stresses

The maximum allowable stresses are based on the properties of the material. They vary with the material, heat treatment, and surface treatment. Table 34.14 gives nominal values for allowable contact stress on gear teeth for commonly used gear materials and heat treatments. Table 34.15 gives nominal values for allowable bending stress in gear teeth for commonly used gear materials and heat treatments.

Carburized case-hardened gears require a core hardness in the range of 260 to 350 H_B (26 to 37 R_C) and a total case depth in the range shown by Fig. 34.49.

The calculated contact stress S_c times a safety factor should be less than the allowable contact stress S_{ac}. The calculated bending stress S_t times a safety factor should be less than the allowable bending stress S_{at}.

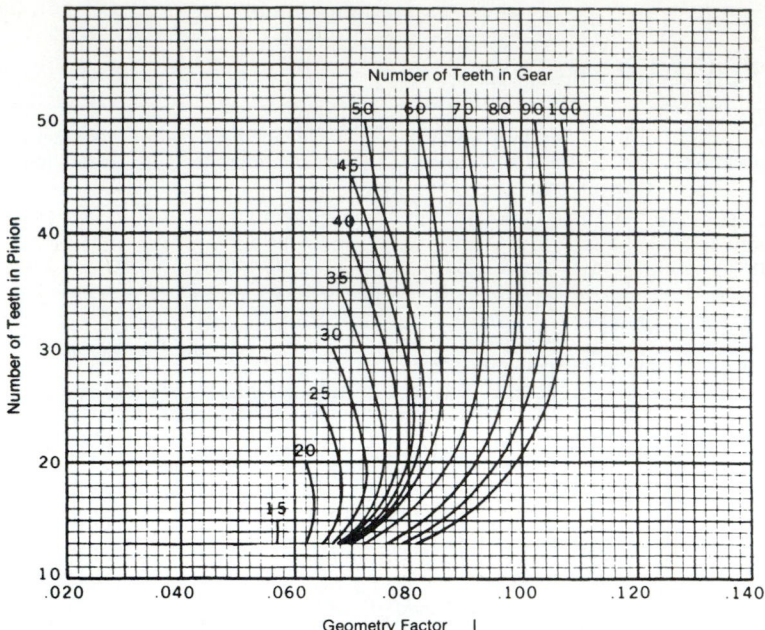

FIGURE 34.19 Geometry factor *I* for durability of straight-bevel gears with 20° pressure angle and 90° shaft angle.

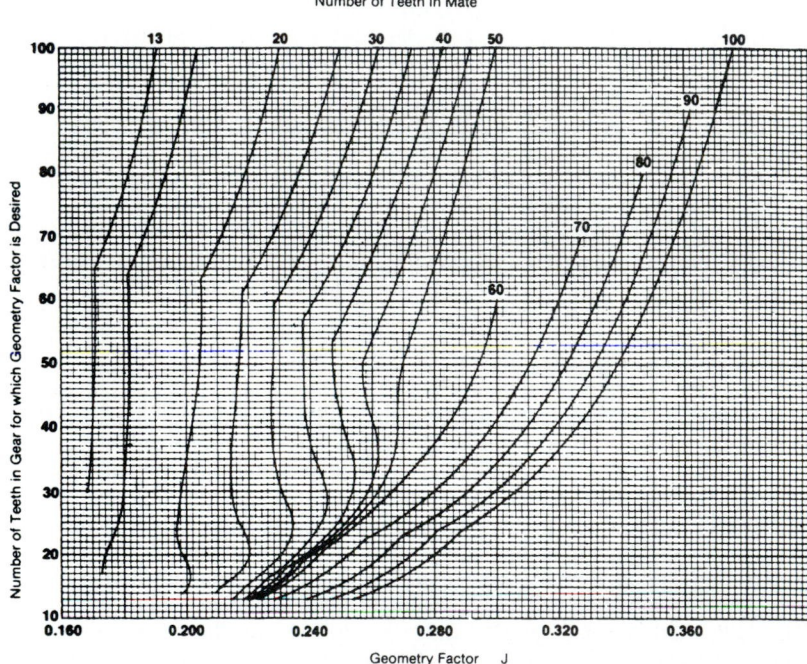

FIGURE 34.20 Geometry factor *J* for strength of straight-bevel gears with 20° pressure angle and 90° shaft angle.

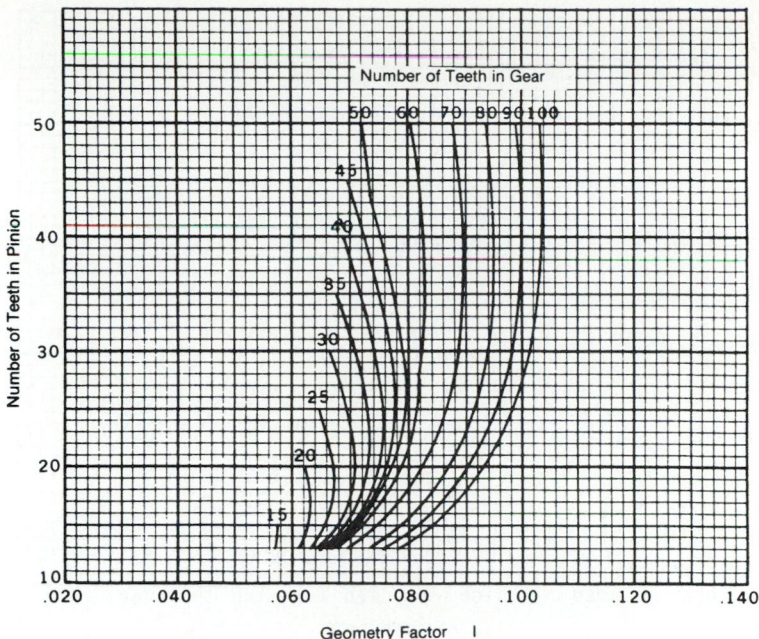

FIGURE 34.21 Geometry factor I for durability of straight-bevel gears with 25° pressure angle and 90° shaft angle.

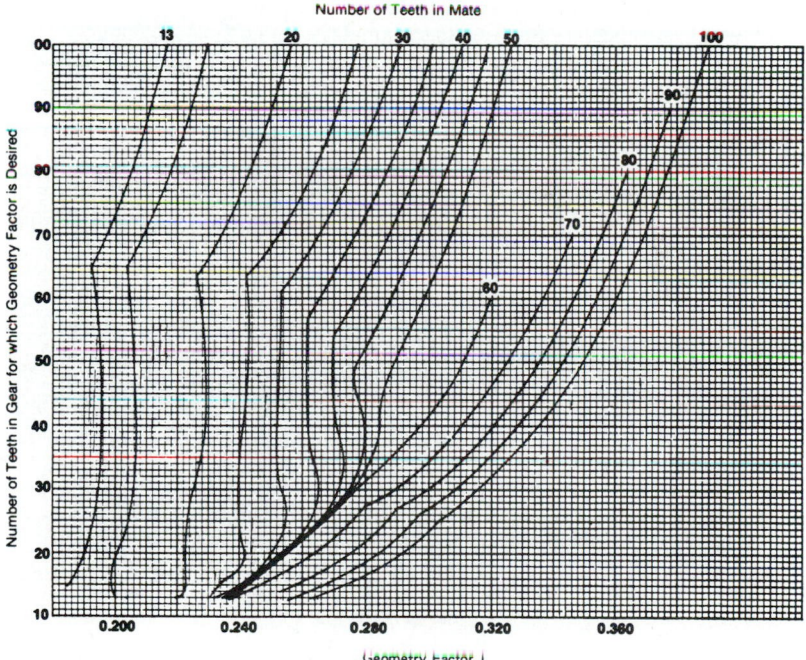

FIGURE 34.22 Geometry factor J for strength of straight-bevel gears with 25° pressure angle and 90° shaft angle.

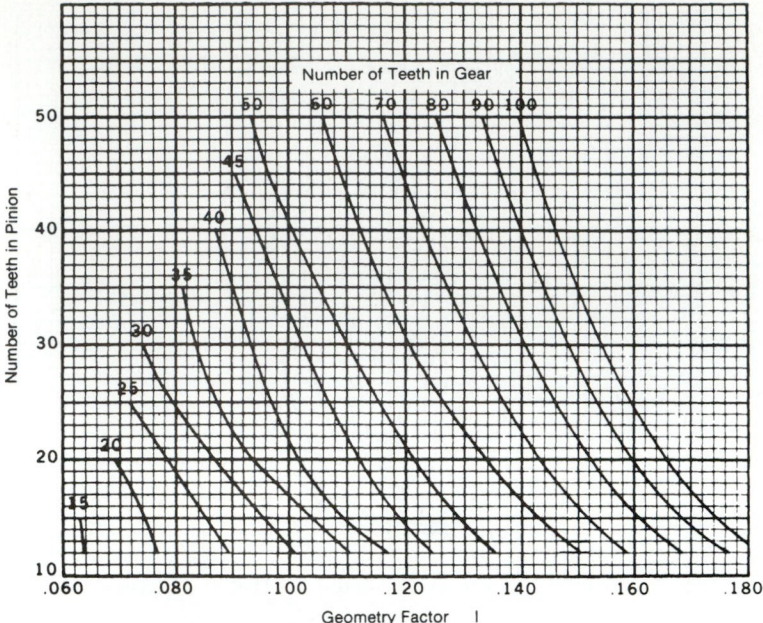

FIGURE 34.23 Geometry factor *I* for durability of spiral-bevel gears with 20° pressure angle, 35° spiral angle, and 90° shaft angle.

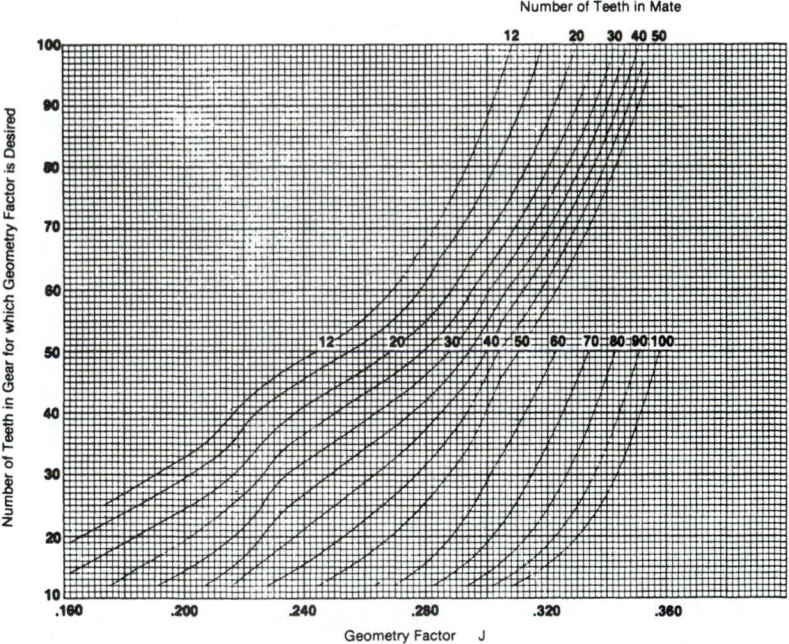

FIGURE 34.24 Geometry factor *J* for strength of spiral-bevel gears with 20° pressure angle, 35° spiral angle, and 90° shaft angle.

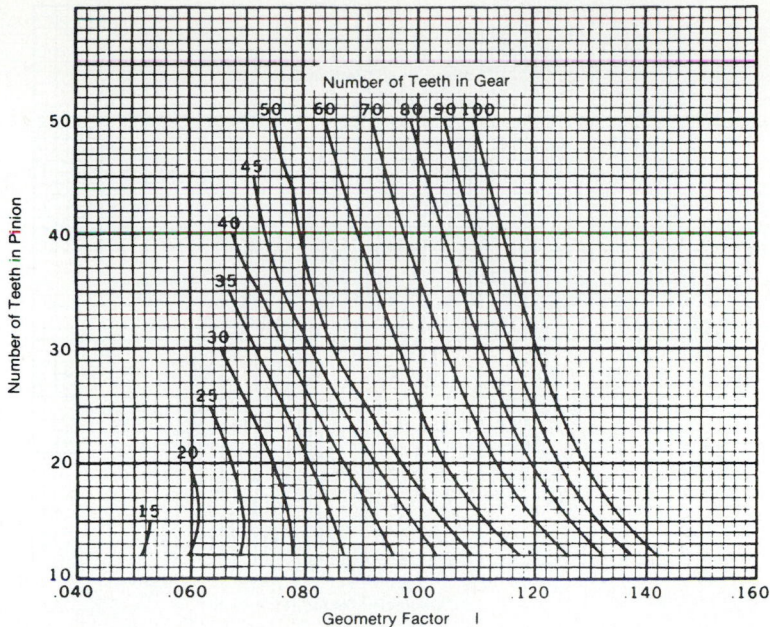

FIGURE 34.25 Geometry factor I for durability of spiral-bevel gears with 20° pressure angle, 25° spiral angle, and 90° shaft angle.

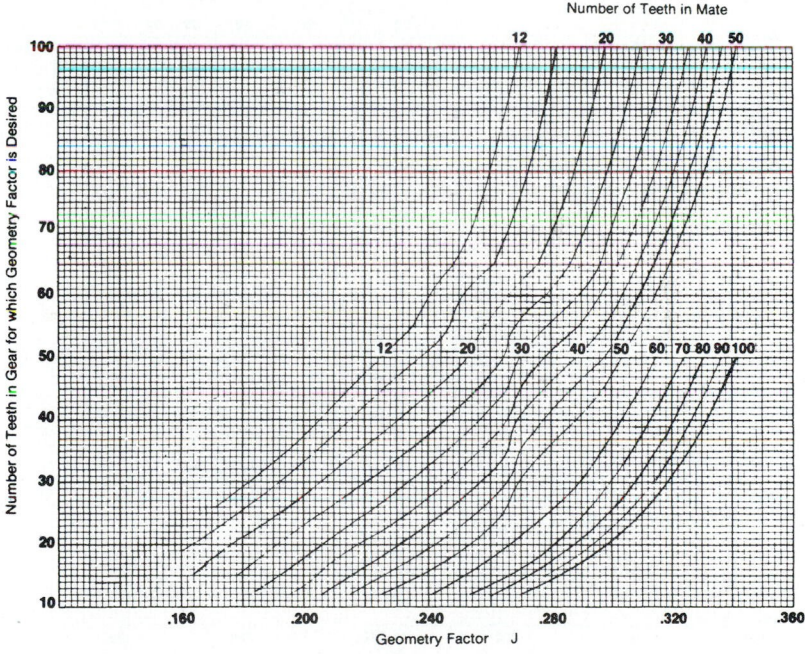

FIGURE 34.26 Geometry factor J for strength of spiral-bevel gears with 20° pressure angle, 25° spiral angle, and 90° shaft angle.

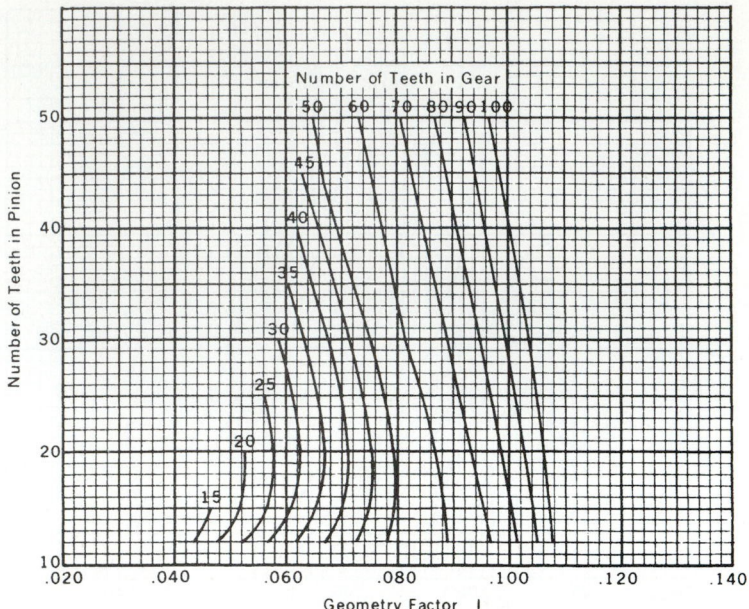

FIGURE 34.27 Geometry factor *I* for durability of spiral-bevel gears with 20° pressure angle, 15° spiral angle, and 90° shaft angle.

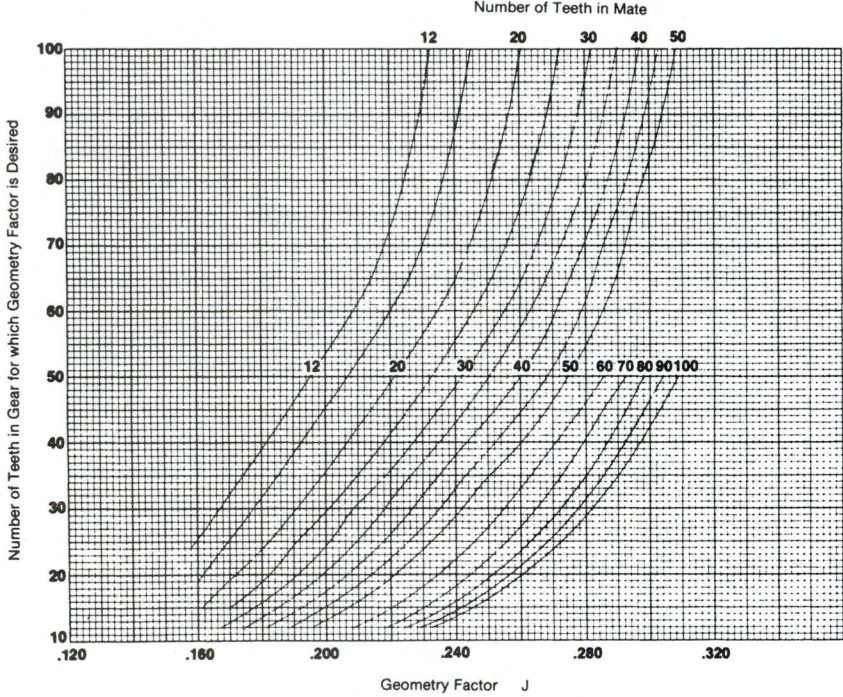

FIGURE 34.28 Geometry factor *J* for strength of spiral-bevel gears with 20° pressure angle, 15° spiral angle, and 90° shaft angle.

34.38

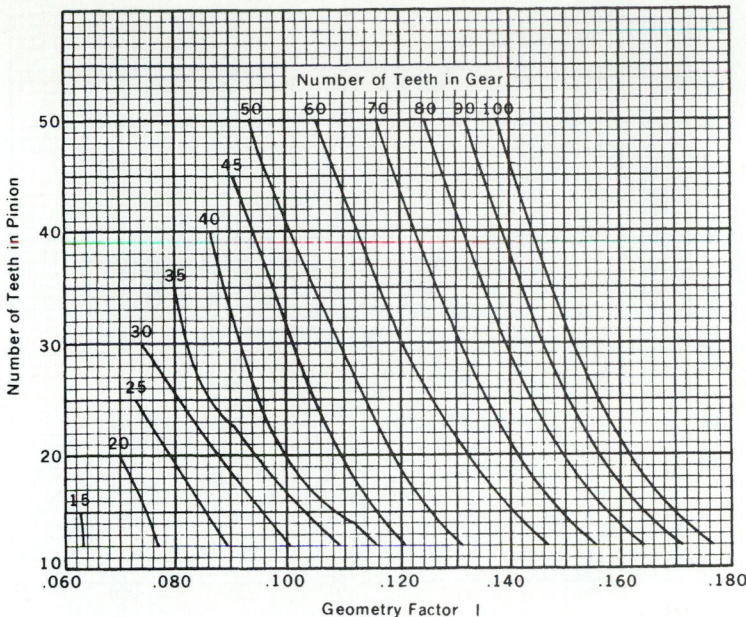

FIGURE 34.29 Geometry factor *I* for durability of spiral-bevel gears with 25° pressure angle, 35° spiral angle, and 90° shaft angle.

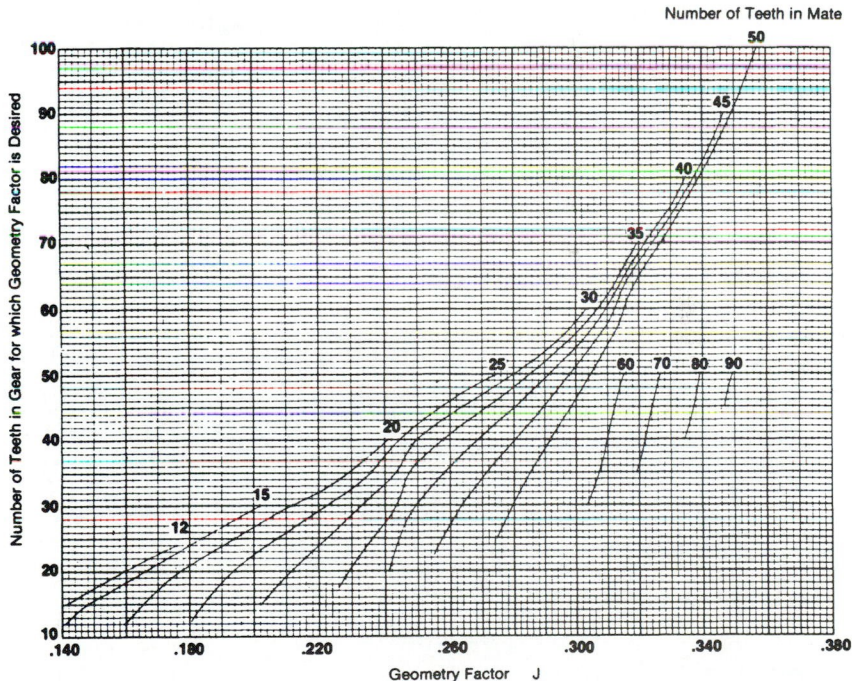

FIGURE 34.30 Geometry factor *J* for strength of spiral bevel gears with 25° pressure angle, 35° spiral angle, and 90° shaft angle.

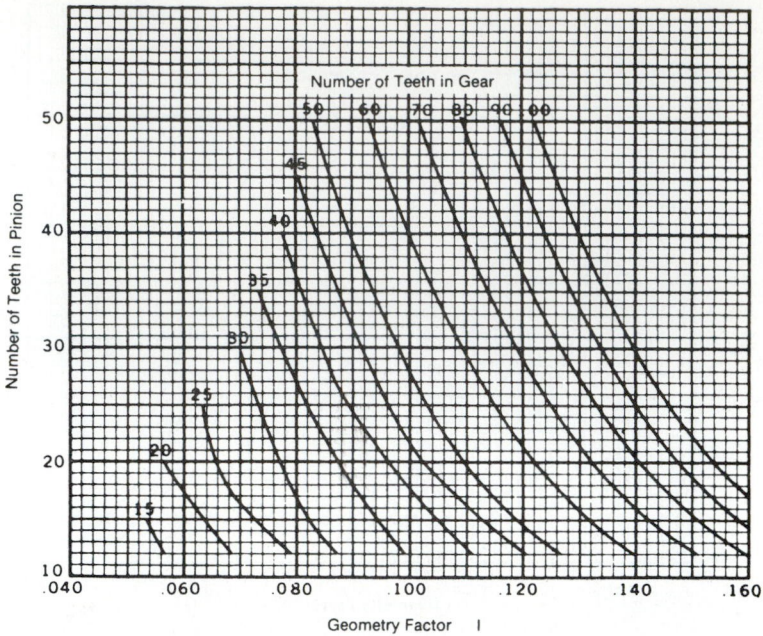

FIGURE 34.31 Geometry factor *I* for durability of spiral-bevel gears with 20° pressure angle, 35° spiral angle, and 60° shaft angle.

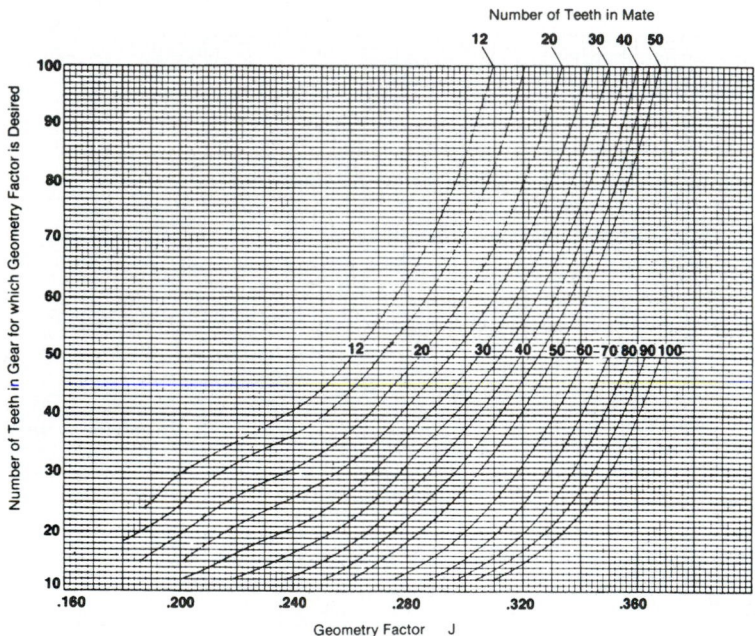

FIGURE 34.32 Geometry factor *J* for strength of spiral-bevel gears with 20° pressure angle, 35° spiral angle, and 60° shaft angle.

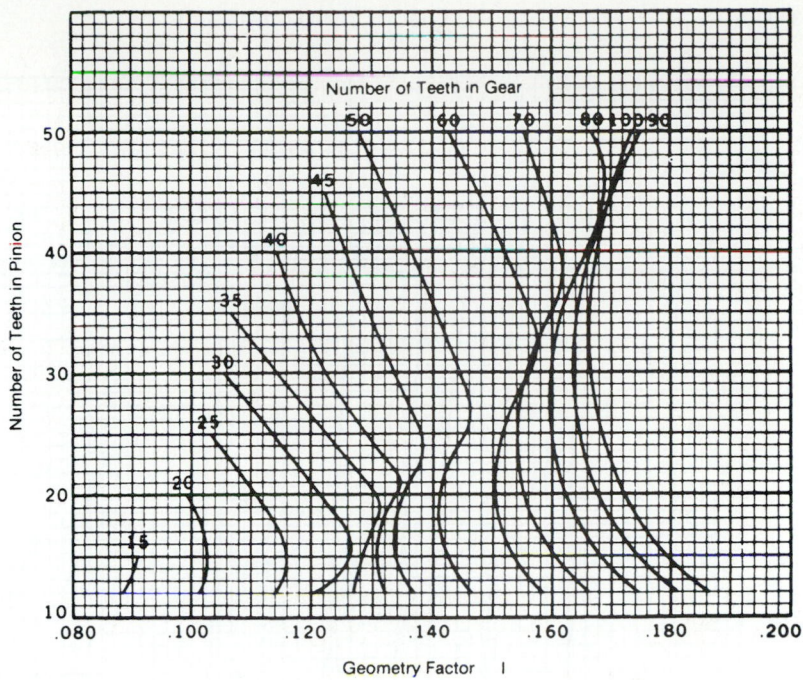

FIGURE 34.33 Geometry factor I for durability of spiral-bevel gears with 20° pressure angle, 35° spiral angle, and 120° shaft angle.

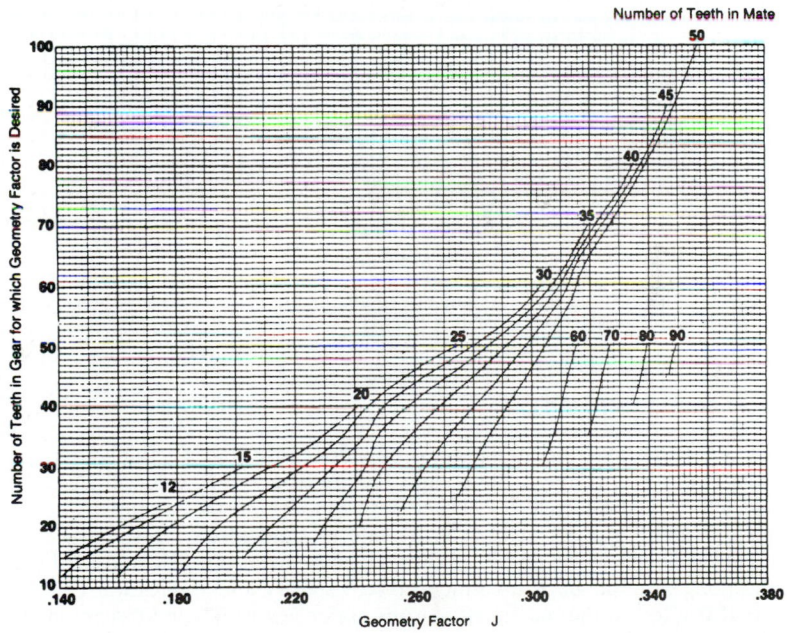

FIGURE 34.34 Geometry factor J for strength of spiral bevel gears with 20° pressure angle, 35° spiral angle, and 120° shaft angle.

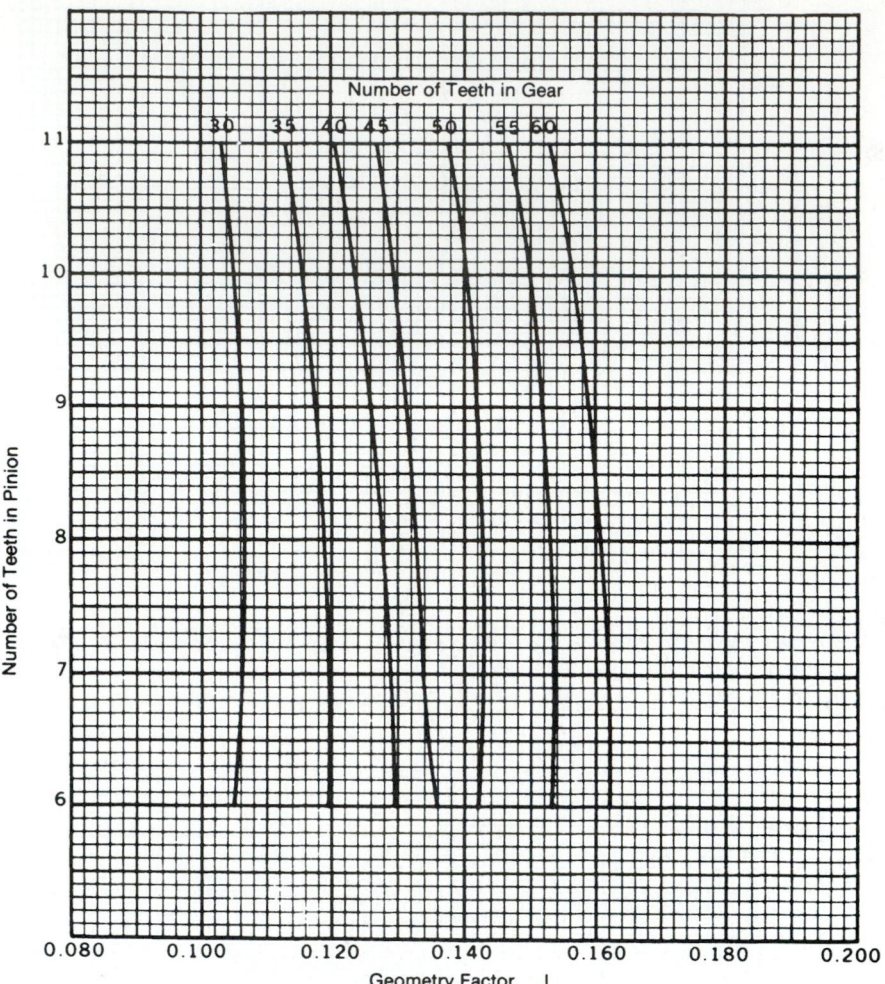

FIGURE 34.35 Geometry factor *I* for durability of automotive spiral-bevel gears with 20° pressure angle, 35° spiral angle, and 90° shaft angle.

34.6.4 Scoring Resistance

Scoring is a temperature-related process in which the surfaces actually tend to weld together. The oil film breaks down, and the tooth surfaces roll and slide on one another, metal against metal. Friction between the surfaces causes heat which reaches the melting point of the tooth material, and scoring results. The factors which could cause scoring are the sliding velocity, surface finish, and load concentrations along with the lubricant temperature, viscosity, and application. But see also Chap. 6. If you follow the recommendations under Sec. 34.7.6 on lubrication and the manufacturer uses acceptable practices in processing the gears, then scoring should not be a problem.

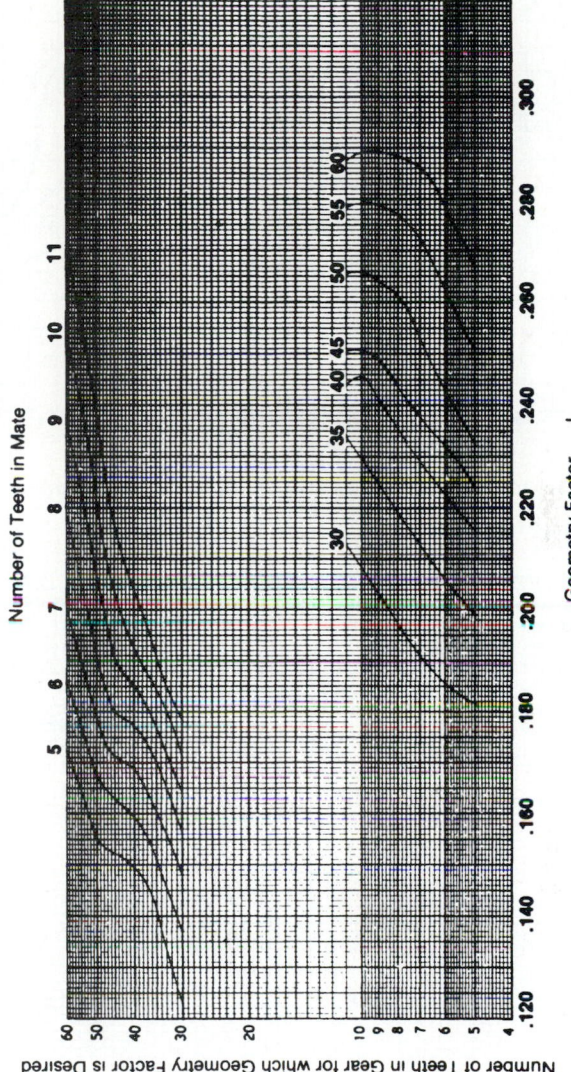

FIGURE 34.36 Geometry factor *J* for strength of automotive spiral-bevel gears with 20° pressure angle, 35° spiral angle, and 90° shaft angle.

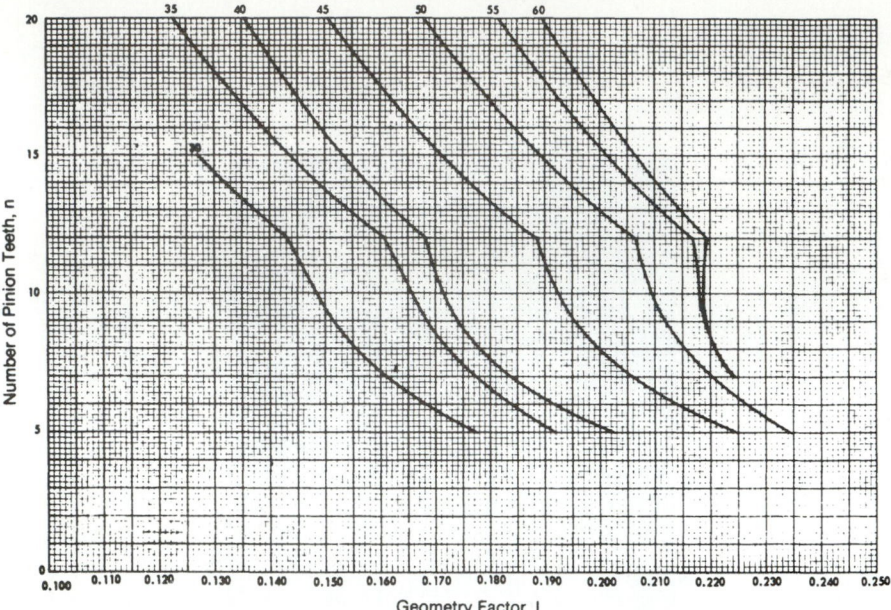

FIGURE 34.37 Geometry factor *I* for hypoid gears with 19° average pressure angle and *E/D* ratio of 0.10.

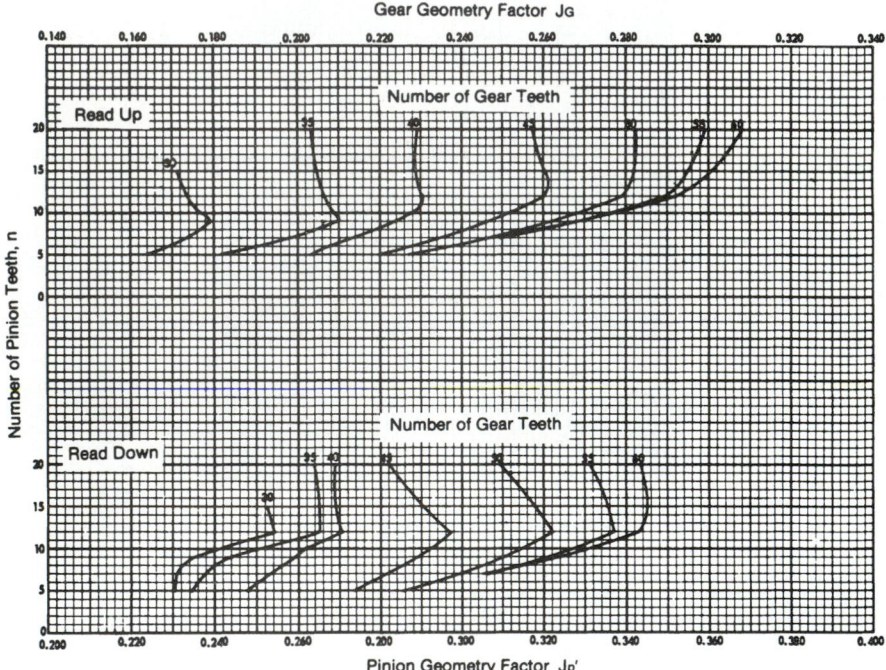

FIGURE 34.38 Geometry factor *J* for strength of hypoid gears with 19° average pressure angle and *E/D* ratio of 0.10.

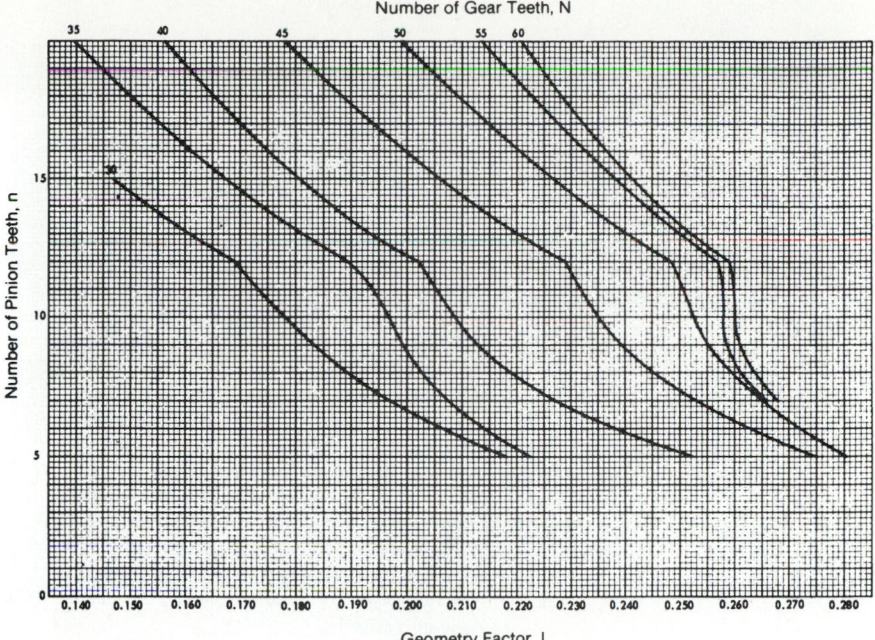

FIGURE 34.39 Geometry factor *I* for durability of hypoid gears with 19° average pressure angle and *E/D* ratio of 0.15.

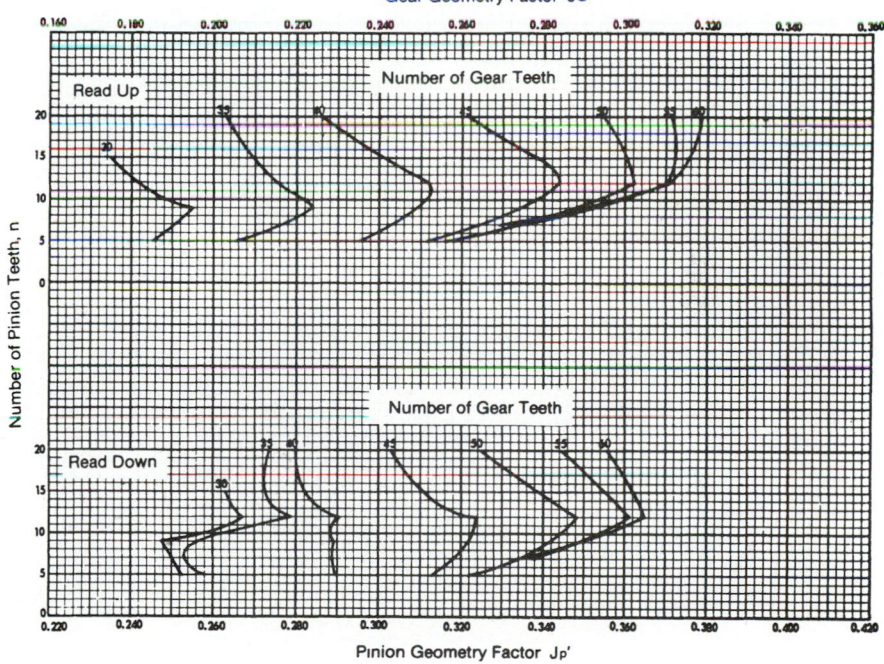

FIGURE 34.40 Geometry factor *J* for strength of hypoid gears with 19° average pressure angle and *E/D* ratio of 0.15.

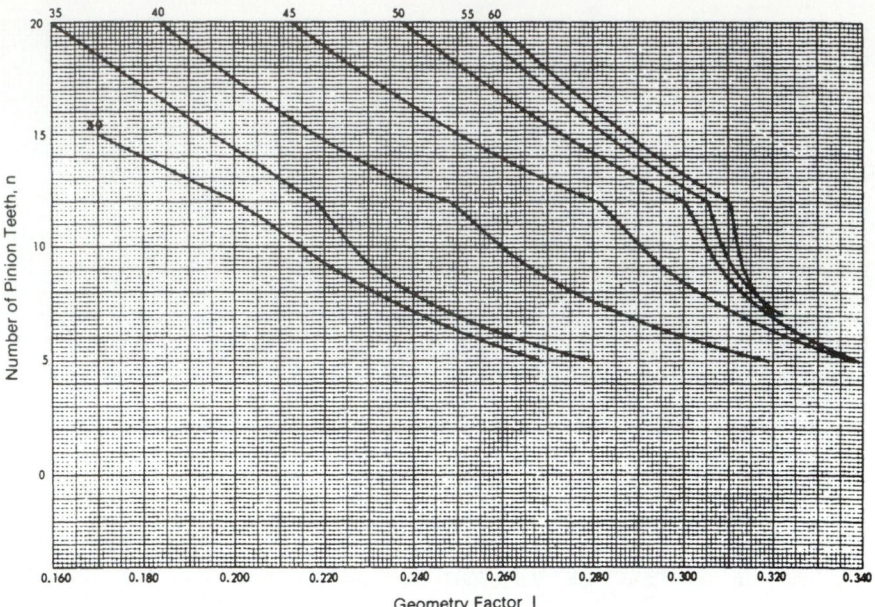

FIGURE 34.41 Geometry factor *I* for durability of hypoid gears with 19° average pressure angle and *E/D* ratio of 0.20.

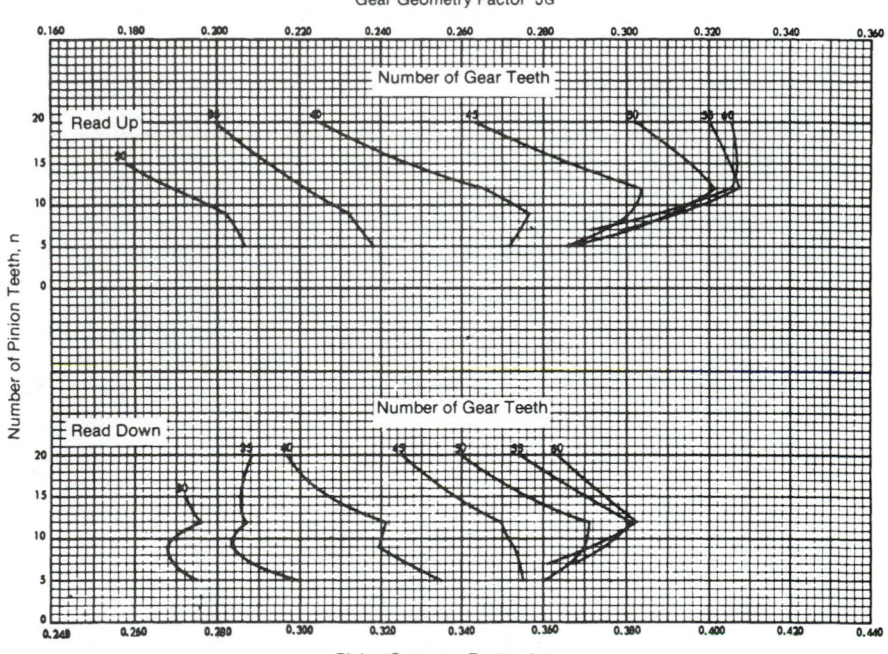

FIGURE 34.42 Geometry factor *J* for strength of hypoid gears with 19° average pressure angle and *E/D* ratio of 0.20.

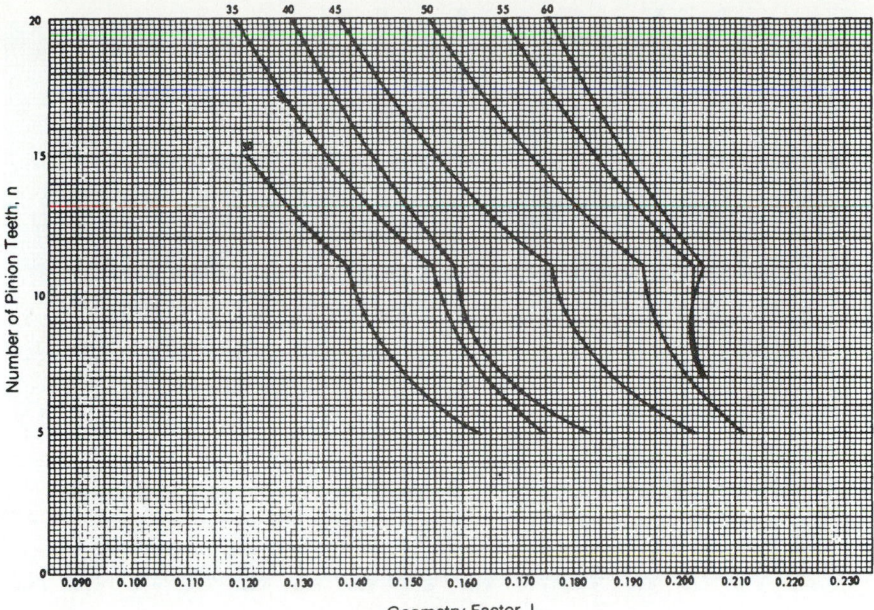

FIGURE 34.43 Geometry factor *I* for durability of hypoid gears with 22½° average pressure angle and *E/D* ratio of 0.10.

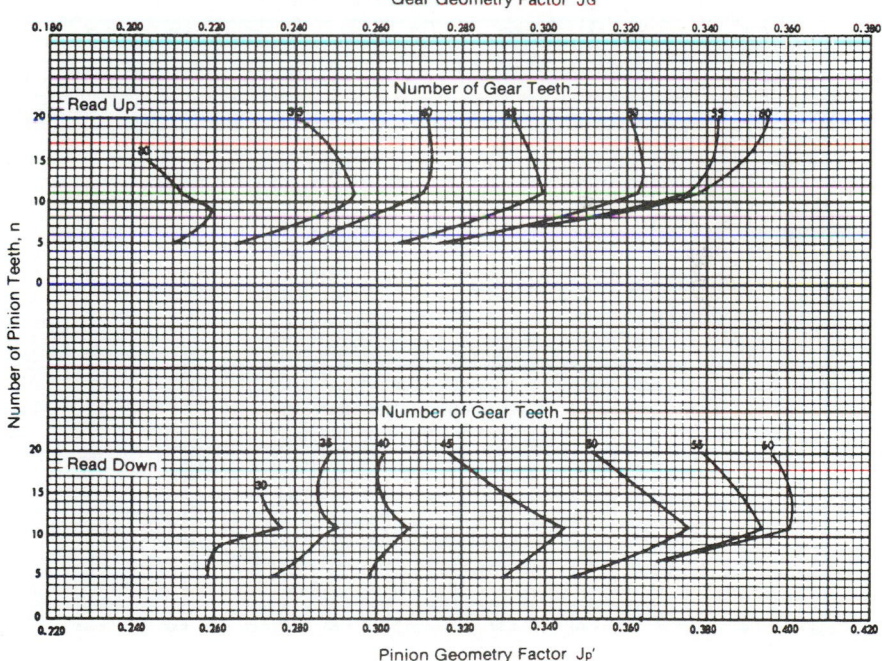

FIGURE 34.44 Geometry factor *J* for strength of hypoid gears with 22½° average pressure angle and *E/D* ratio of 0.10.

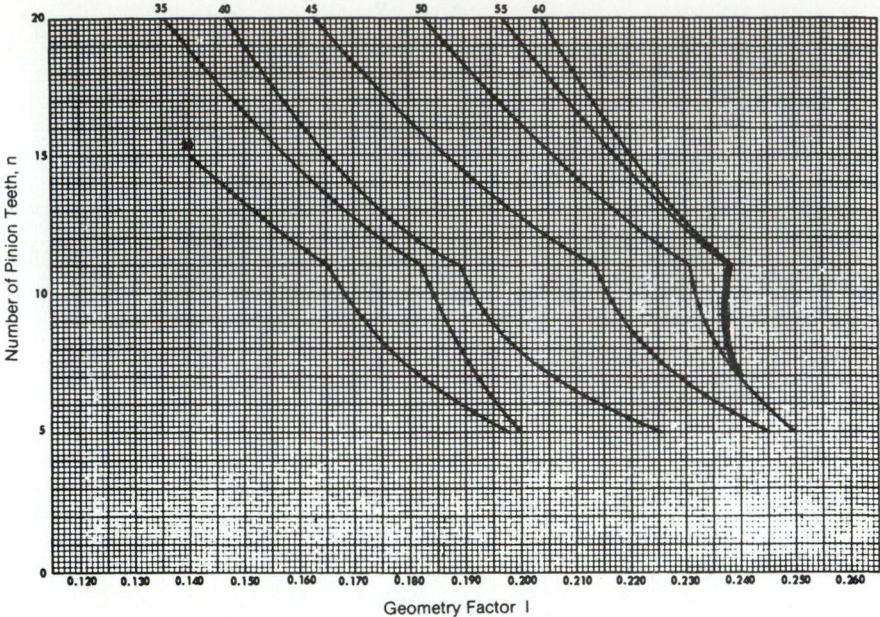

FIGURE 34.45 Geometry factor *I* for durability of hypoid gears with 22½° average pressure angle and *E/D* ratio of 0.15.

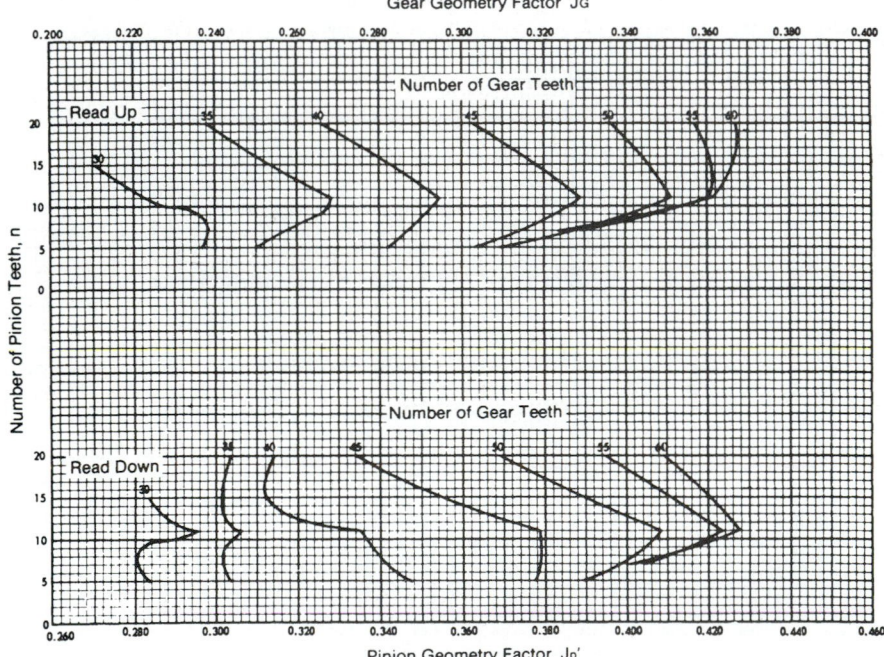

FIGURE 34.46 Geometry factor *J* for strength of hypoid gears with 22½° average pressure angle and *E/D* ratio of 0.15.

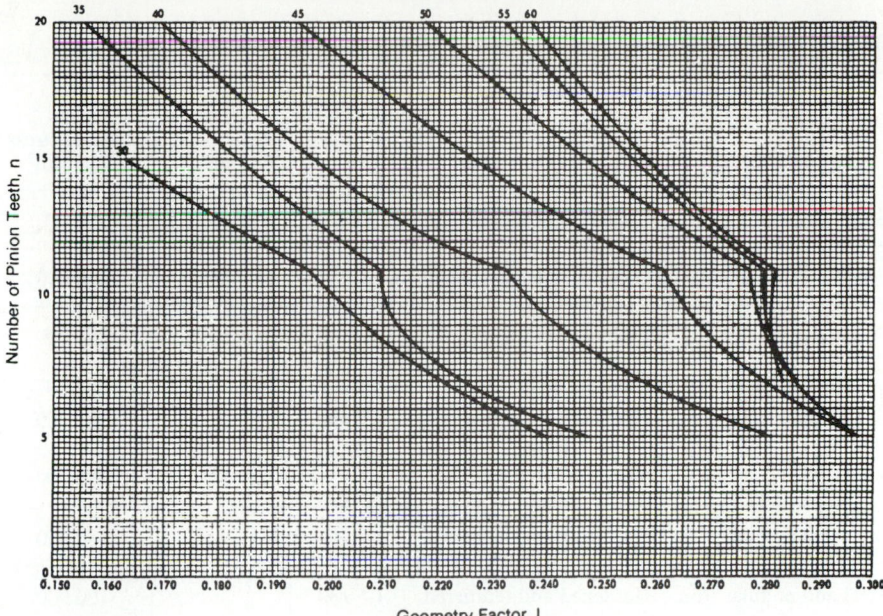

FIGURE 34.47 Geometry factor *I* for durability of hypoid gears with 22½° average pressure angle and *E/D* ratio of 0.20.

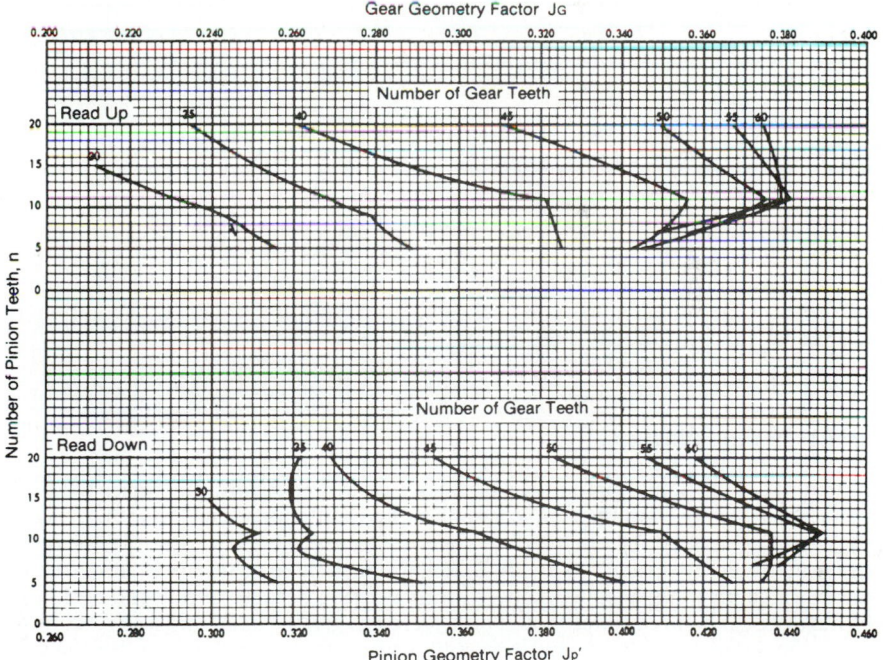

FIGURE 34.48 Geometry factor *J* for strength of hypoid gears with 22½° average pressure angle and *E/D* ratio of 0.20.

TABLE 34.14   Allowable Contact Stress S_{ac}

| Material | Heat treatment | Minimum hardness | | Contact stress S_{ac}, lb/in² |
		Brinell	Rockwell C	
Steel	Carburized (case-hardened)		60	250 000
Steel	Carburized (case-hardened)		55	210 000
Steel	Flame- or induction-hardened	500	50	200 000
Steel and nodular iron	Hardened and tempered	400		180 000
Steel	Nitrided		60	180 000
Steel and nodular iron	Hardened and tempered	300		140 000
Steel and nodular iron	Hardened and tempered	180		100 000
Cast iron	As cast	200		80 000
Cast iron	As cast	175		70 000
Cast iron	As cast			60 000

34.7   DESIGN OF MOUNTINGS

The normal load on the tooth surfaces of bevel and hypoid gears may be resolved into two components: one in the direction along the axis of the gear and the other perpendicular to the axis. The direction and magnitude of the normal load depend on the ratio, pressure angle, spiral angle, hand of spiral, and direction of rotation as well as on whether the gear is the driving or driven member.

34.7.1   Hand of Spiral

In general, a left-hand pinion driving clockwise (viewed from the back) tends to move axially away from the cone center; a right-hand pinion tends to move toward the center because of the oblique direction of the curved teeth. If possible, the hand of spiral should be selected so that both the pinion and the gear tend to move out of mesh, which prevents the possibility of tooth wedging because of reduced backlash. Otherwise, the hand of spiral should be selected to give an axial thrust that tends to move the pinion out of mesh. In a reversible drive, there is no choice unless the pair performs a heavier duty in one direction for a greater part of the time.

TABLE 34.15 Allowable Bending Stress S_{at}

Material	Heat treatment	Surface hardness		Bending stress S_{at}, lb/in²
		Brinell	Rockwell C	
Steel	Carburized (case-hardened)	575–625	55 min.	60 000
Steel	Flame- or induction-hardened (unhardened root fillet)	450–500	50 min.	27 000
Steel	Hardened and tempered	450 min.		50 000
Steel	Hardened and tempered	300 min.		42 000
Steel	Hardened and tempered	180 min.		28 000
Steel	Normalized	140 min.		22 000
Cast iron	As cast	200 min.		13 000
Cast iron	As cast	175 min.		8 500
Cast iron	As cast			5 000

On hypoids when the pinion is below center and to the right (when you are facing the front of the gear), the pinion hand of spiral should always be left-hand. With the pinion above center and to the right, the pinion hand should always be right-hand. See Fig. 34.15.

34.7.2 Tangential Force

The tangential force on a bevel or hypoid gear is given by

$$W_{tG} = \frac{2T_G}{D_m} = \frac{126\,000P}{D_m N} \tag{34.4}$$

where T_G = gear torque, lb · in
P = power, horsepower (hp)
N = speed of gear, r/min

The tangential force on the mating pinion is given by the equation

$$W_{tP} = \frac{W_{tG}\cos\psi_P}{\cos\psi_G} = \frac{2T_P}{d_m} \tag{34.5}$$

where T_P = pinion torque in pound-inches.

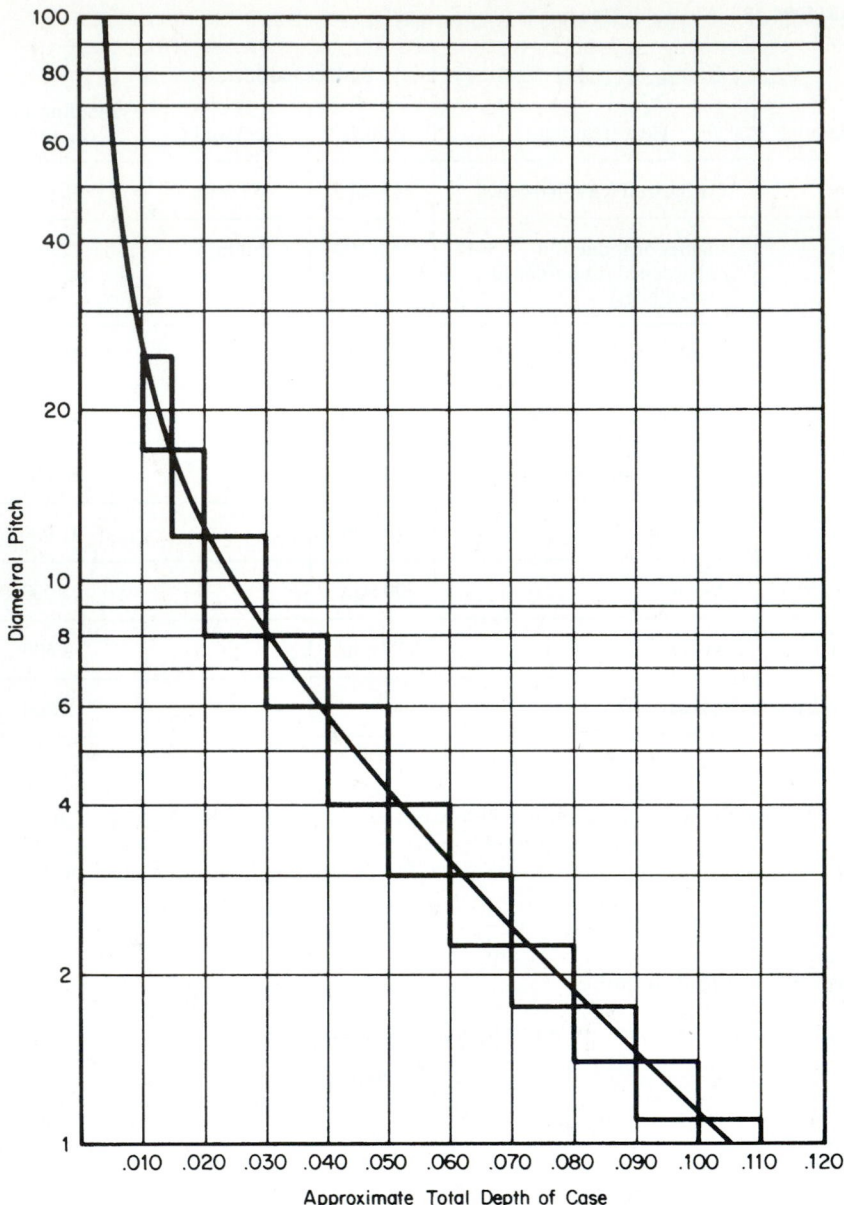

FIGURE 34.49   Diametral pitch versus total case depth. If in doubt, use the greater case depth on ground gears or on short face widths.

34.7.3 Axial Thrust and Radial Separating Forces

The formulas that follow are used to calculate the *axial thrust force* W_x and the *radial separating force* W_R for bevel and hypoid gears. The direction of the pinion (driver) rotation should be viewed from the pinion back.

For a pinion (driver) with a *right-hand (RH) spiral with clockwise (cw) rotation* or a *left-hand (LH) spiral with counterclockwise (ccw) rotation*, the axial and separating force components *acting on the pinion* are, respectively,

$$W_{xP} = W_{tP} \sec \psi_P (\tan \phi \sin \gamma - \sin \psi_P \cos \gamma) \qquad (34.6)$$

$$W_{RP} = W_{tP} \sec \psi_P (\tan \phi \cos \gamma + \sin \psi_P \sin \gamma) \qquad (34.7)$$

For a pinion (driver) with an *LH spiral with cw rotation* or an *RH spiral with ccw rotation*, the force components *acting on the pinion* are, respectively,

$$W_{xP} = W_{tP} \sec \psi_P (\tan \phi \sin \gamma + \sin \psi_P \cos \gamma) \qquad (34.8)$$

$$W_{RP} = W_{tP} \sec \psi_P (\tan \phi \cos \gamma - \sin \psi_P \sin \gamma) \qquad (34.9)$$

For a pinion (driver) with an *RH spiral with cw rotation* or an *LH spiral with ccw rotation*, the force components *acting on the gear* (driven) are, respectively,

$$W_{xG} = W_{tG} \sec \psi_G (\tan \phi \sin \Gamma + \sin \psi_G \cos \Gamma) \qquad (34.10)$$

$$W_{RG} = W_{tG} \sec \psi_G (\tan \phi \cos \Gamma - \sin \psi_G \sin \Gamma) \qquad (34.11)$$

For a pinion (driver) with an *LH spiral and cw rotation* or an *RH spiral with ccw rotation*, the force components *acting on the gear* are, respectively,

$$W_{xG} = W_{tG} \sec \psi_G (\tan \phi \sin \Gamma - \sin \psi_G \cos \Gamma) \qquad (34.12)$$

$$W_{RG} = W_{tG} \sec \psi_G (\tan \phi \cos \Gamma + \sin \psi_G \sin \Gamma) \qquad (34.13)$$

These equations apply to straight-bevel, Zerol bevel, spiral-bevel, and hypoid gears. When you use them for hypoid gears, be sure that the pressure angle corresponds to the driving face of the pinion tooth.

A plus sign for Eqs. (34.6), (34.8), (34.10), and (34.12) indicates that the direction of the axial thrust is *outward,* or away from the cone center. Thus a minus sign indicates that the direction of the axial thrust is *inward,* or toward the cone center.

A plus sign for Eqs. (34.7), (34.9), (34.11), and (34.13) indicates that the direction of the *separating* force is *away* from the mating gear. So a minus sign indicates an *attracting* force *toward* the mating member.

Example. A hypoid-gear set consists of an 11-tooth pinion with LH spiral and ccw rotation driving a 45-tooth gear. Data for the gear are as follows: 4.286 diametral pitch, 8.965-inch (in) mean diameter, 70.03° pitch angle, 31.48° spiral angle, and 30×10^3 lb · in torque. Pinion data are these: 1.500-in offset, 2.905-in mean diameter, concave pressure angle 18.13°, convex pressure angle 21.87°, pitch angle 19.02°, and spiral angle 50°. Determine the force components and their directions for each member of the set.

Solution. From Eq. (34.4) we find the tangential load on the gear to be

$$W_{tG} = \frac{2T_G}{D_m} = \frac{2(30 \times 10^3)}{8.965} = 6693 \text{ lb}$$

Since the pinion has LH spiral angle and rotates ccw, Eqs. (34.10) and (34.11) apply for the gear. Thus

$$W_{xG} = W_{tG} \sec \psi_G (\tan \phi \sin \Gamma + \sin \psi_G \cos \Gamma)$$

$$= 6693 \sec 31.48 (\tan 18.13° \sin 70.03° + \sin 31.48° \cos 70.03°)$$

$$= 3814 \text{ lb}$$

Substituting the same values and angles into Eq. (34.11) gives $W_{RG} = -2974$ lb. Thus the thrust is outward, and the separating force is toward the mating member.

Next we find the tangential load on the pinion from Eq. (34.5):

$$W_{tP} = \frac{W_{tG} \cos \psi_P}{\cos \psi_G} = \frac{6693 \cos 50°}{\cos 31.48°} = 5045 \text{ lb}$$

Equations (34.6) and (34.7) apply to the pinion:

$$W_{xP} = W_{tP} \sec \psi_P (\tan \phi \sin \gamma - \sin \psi_P \cos \gamma)$$

$$= 5045 \sec 50° (\tan 18.13° \sin 19.02° - \sin 50° \cos 19.02°)$$

$$= -4846 \text{ lb}$$

In a similar manner, Eq. (34.7) gives $W_{RP} = 4389$ lb. Thus the axial thrust is inward, and the separating force is away from the gear.

34.7.4 Bearing Loads

The bearings selected must be adequate to support the axial forces W_x for both directions of rotation and for the load conditions on both sides of the teeth.

Radial forces are transmitted indirectly through moment arms to the bearings. The radial bearing loads are derived from the gear separating force, the gear tangential force, and the gear thrust couple, along with the type of mounting and the bearing position.

34.7.5 Types of Mountings

Two types of mountings are generally used: *overhung*, where both bearings are located on the shaft behind the gear, and *straddle*, where one bearing is on either side of the gear. Because of the stiffer configuration, straddle mountings are generally used for highly loaded gears.

34.7.6 Lubrication

The lubrication system for a bevel- or hypoid-gear drive should sufficiently lubricate and adequately cool the gears and bearings. Splash lubrication is generally satisfactory for applications up to peripheral speeds of 2000 ft/min. The oil level should cover the full face of the lowest gear, and the quantity of oil should be sufficient to maintain the oil temperature within recommended limits.

Pressure lubrication is recommended for velocities above 2000 ft/min. The jets should be located to direct the stream to cover the full length of the teeth of both members, preferably close to the mesh point on the leaning side.

Experience has shown that an oil flow of 0.07 to 1.0 gallons per minute (gal/min) per 100 hp will result in an oil temperature rise of approximately 10°F.

Extreme-pressure (EP) lubricants are recommended for hypoid gears and for spiral-bevel gears which are subject to extreme conditions of shock, severe starting conditions, or heavy loads. The lubrication system should be fully protected against contamination by moisture or dirt. For continuous operation at temperatures above 160°F, the lubricants should be approved by the lubricant manufacturer.

In general, for a splash lubrication, an SAE 80 or 90 gear oil should be satisfactory. For a circulating system with an oil spray lubrication, SAE 20 or 30 should be satisfactory. AGMA "Specifications on Lubrication of Enclosed and Open Gearing" is a recommended guide to the type and grade of oil for various operating conditions.

34.7.7 Loaded Contact Check

With highly stressed bevel- and hypoid-gear applications such as aircraft and automotive, it is normal practice to perform a loaded contact check with the gear set assembled in its mountings. A brake load is applied to the output shafts, and the pinion member is rotated slowly at approximately 15 r/min. A marking compound is applied to the pinion and gear teeth to permit observation of the tooth contact pattern at the desired load conditions. The purpose of this test is to evaluate the rigidity of the mountings and ensure that the contact pattern remains within the tooth boundaries under all load conditions. Indicators can be mounted at various positions under load. An analysis of these data can result in modifications of the mounting design or contact pattern to ensure that the contact pattern does not reach the tooth boundaries at operating loads. This will eliminate an edge contact condition which can cause noise or premature failure of the gear teeth.

34.8 COMPUTER-AIDED DESIGN

34.8.1 Computer Timesharing

A computer timesharing service is available to assist you with gear-tooth design, strength calculations, gear-tooth geometry analysis, gear manufacturing, and inspection data for bevel and hypoid gears. Contact

Application Engineering Department
Gleason Machine Division
1000 University Avenue
Rochester, New York 14692

34.8.2 Design Calculating Services

The Gleason Machine Division offers a calculating service which may be used as an alternative to the computer timesharing service mentioned earlier, when you require a computer analysis of the gear-tooth design.

34.8.3 Available Computer Programs

The following computer programs are available from the Gleason Machine Division to assist you with a gear-tooth design analysis:

1. *Dimension Sheet*   Calculation of the basic tooth geometry, contact ratios, stress data, bearing thrust loads, and profile sliding velocities.
2. *Summary*   Calculation of cutting and grinding machine setup data to produce the desired tooth geometry.
3. *Tooth Contact Analysis*   A special analysis program that determines the tooth contact pattern and transmission motion errors based on specified cutting tools and gear-tooth geometry. Figure 34.50 illustrates a typical *tooth contact analysis.*
4. *Undercut Check*   Calculation of the location of undercut lengthwise along the tooth, along with the depth and angle of undercut relative to the tooth profile.
5. *Loaded Tooth Contact Analysis*   An analysis and plot of tooth contact pattern and transmission errors as a function of gear torque. Deflections of the gear mountings may also be considered with this analysis.
6. *Finite-Element Analysis*   Detailed stress data calculated based on a three-dimensional finite-element stress model which considers exact gear-tooth geometry based on cutting tool specifications, machine setup, and generating motions and mounting deflections.

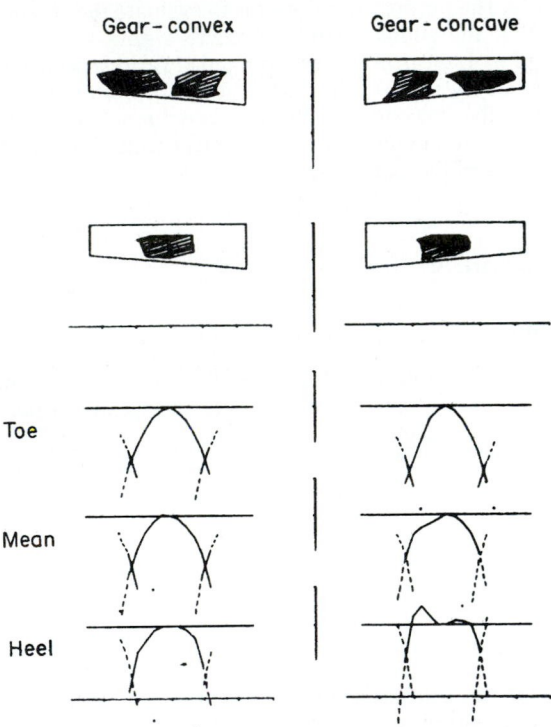

FIGURE 34.50   Typical tooth analysis contact graph.

CHAPTER 35
HELICAL GEARS

Raymond J. Drago, P.E.
Senior Engineer, Advanced Power Train Technology
Boeing Vertol Company
Philadelphia, Pennsylvania

The following is quoted from the Foreword of Ref. [35.1]:

> This AGMA Standard and related publications are based on typical or average data, conditions, or applications. The standards are subject to continual improvement, revision, or withdrawal as dictated by increased experience. Any person who refers to AGMA technical publications should be sure that he has the latest information available from the Association on the subject matter.
>
> Tables or other self-supporting sections may be quoted or extracted in their entirety. Credit line should read: "Extracted from ANSI/AGMA #2001-B88 Fundamental Rating Factors and Calculation Methods for Involute Spur and Helical Gear Teeth, with the permission of the publisher, American Gear Manufacturers Association, 1500 King Street, Alexandria, Virginia 22314."

This reference is cited because numerous American Gear Manufacturer's Association (AGMA) tables and figures are used in this chapter. In each case, the appropriate publication is noted in a footnote or figure caption.

35.1 INTRODUCTION

Helical gearing, in which the teeth are cut at an angle with respect to the axis of rotation, is a later development than spur gearing and has the advantage that the action is smoother and tends to be quieter. In addition, the load transmitted may be somewhat larger, or the life of the gears may be greater for the same loading, than with an equivalent pair of spur gears. Helical gears produce an end thrust along the axis of the shafts in addition to the separating and tangential (driving) loads of spur gears. Where suitable means can be provided to take this thrust, such as thrust collars or ball or tapered-roller bearings, it is no great disadvantage.

Conceptually, helical gears may be thought of as stepped spur gears in which the size of the step becomes infinitely small. For external parallel-axis helical gears to mesh, they must have the same helix angle but be of different hand. An external-internal set will, however, have equal helix angle with the same hand.

Involute profiles are usually employed for helical gears, and the same comments made earlier about spur gears hold true for helical gears.

Although helical gears are most often used in a parallel-axis arrangement, they can also be mounted on nonparallel noncoplanar axes. Under such mounting conditions, they will, however, have limited load capacity.

Although helical gears which are used on crossed axes are identical in geometry and manufacture to those used on parallel axes, their operational characteristics are quite different. For this reason they are discussed separately at the end of this chapter. All the forthcoming discussion therefore applies only to helical gears operating on parallel axes.

35.2 TYPES

Helical gears may take several forms, as shown in Fig. 35.1:

1. Single
2. Double conventional
3. Double staggered
4. Continuous (herringbone)

Single-helix gears are readily manufactured on conventional gear cutting and grinding equipment. If the space between the two rows of a double-helix gear is wide enough, such a gear may also be cut and ground, if necessary, on conventional equipment. Continuous or herringbone gears, however, can be cut only on a special shaping machine (Sykes) and usually cannot be ground at all.

Only single-helix gears may be used in a crossed-axis configuration.

35.3 ADVANTAGES

There are three main reasons why helical rather than straight spur gears are used in a typical application. These are concerned with the noise level, the load capacity, and the manufacturing.

35.3.1 Noise

Helical gears produce less noise than spur gears of equivalent quality because the total contact ratio is increased. Figure 35.2 shows this effect quite dramatically. However, these results are measured at the mesh for a specific test setup; thus, although the trend is accurate, the absolute results are not.

Figure 35.2 also brings out another interesting point. At high values of helix angle, the improvement in noise tends to peak; that is, the curve flattens out. Had data been obtained at still higher levels, the curve would probably drop drastically. This is due to the difficulty in manufacturing and mounting such gears accurately enough to take full advantage of the improvement in contact ratio. These effects at

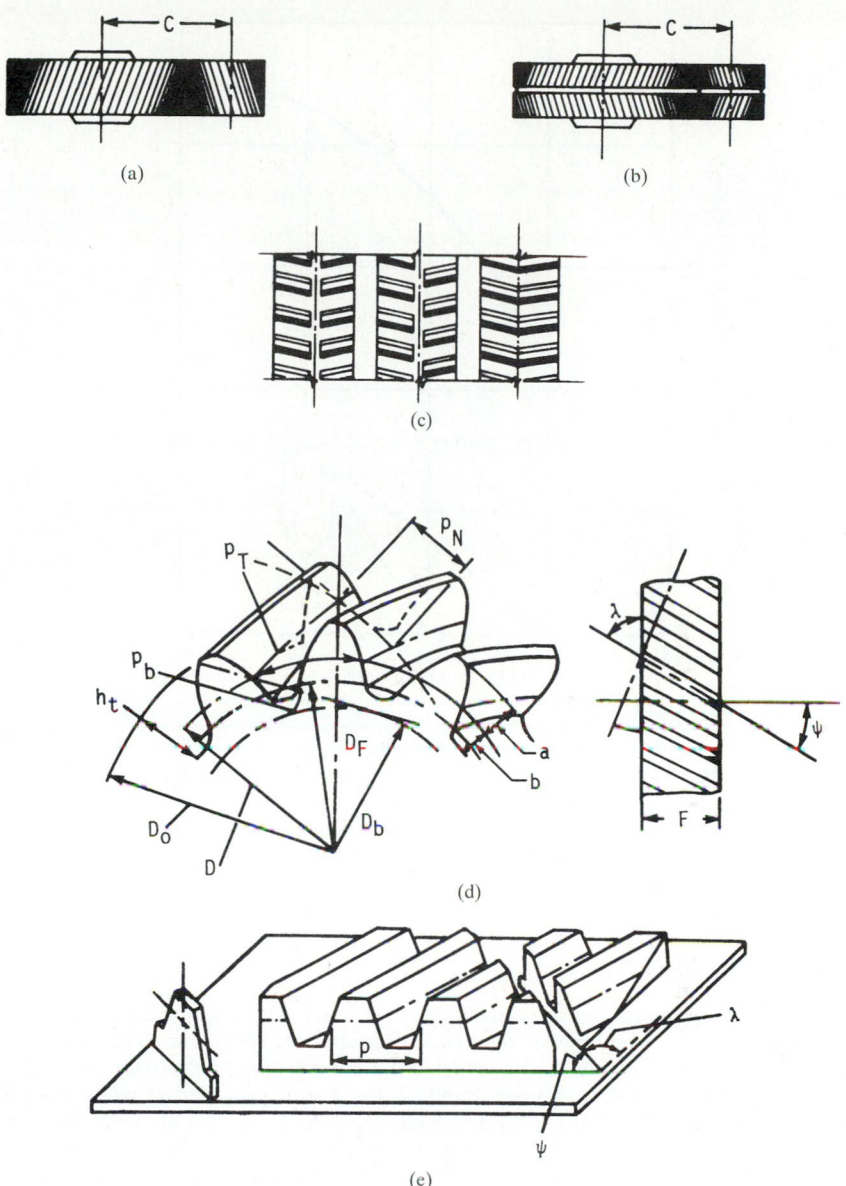

FIGURE 35.1 Terminology of helical gearing. (*a*) Single-helix gear. (*b*) Double-helix gear. (*c*) Types of double-helix gears: left, conventional; center, staggered; right, continous or herringbone. (*d*) Geometry. (*e*) Helical rack.

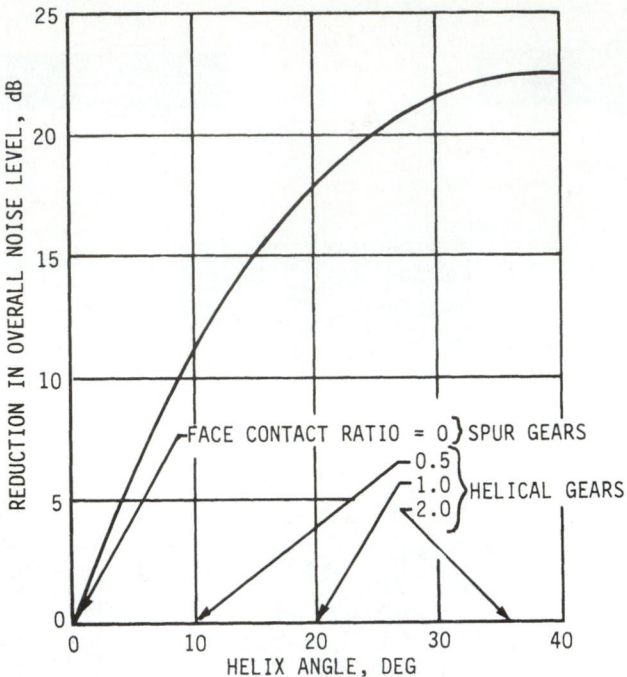

FIGURE 35.2 Effect of face-contact ratio on noise level. Note that increased helix angles lower the noise level.

very high helix angles actually tend to reduce the effective contact ratio, and so noise increases. Since helix angles greater than 45° are seldom used and are generally impractical to manufacture, this phenomenon is of academic interest only.

35.3.2 Load Capacity

As a result of the increased total area of tooth contact available, the load capacity of helical gears is generally higher than that of equivalent spur gears. The reason for this increase is obvious when we consider the contact line comparison which Fig. 35.3 shows. The most critical load condition for a spur gear occurs when a single tooth carries all the load at the highest point of single-tooth contact (Fig. 35.3c). In this case, the total length of the contact line is equal to the face width. In a helical gear, since the contact lines are inclined to the tooth with respect to the face width, the total length of the line of contact is increased (Fig. 35.3b), so that it is greater than the face width. This lowers unit loading and thus increases capacity.

35.3.3 Manufacturing

In the design of a gear system, it is often necessary to use a specific ratio on a specific center distance. Frequently this results in a diametral pitch which is nonstandard. If

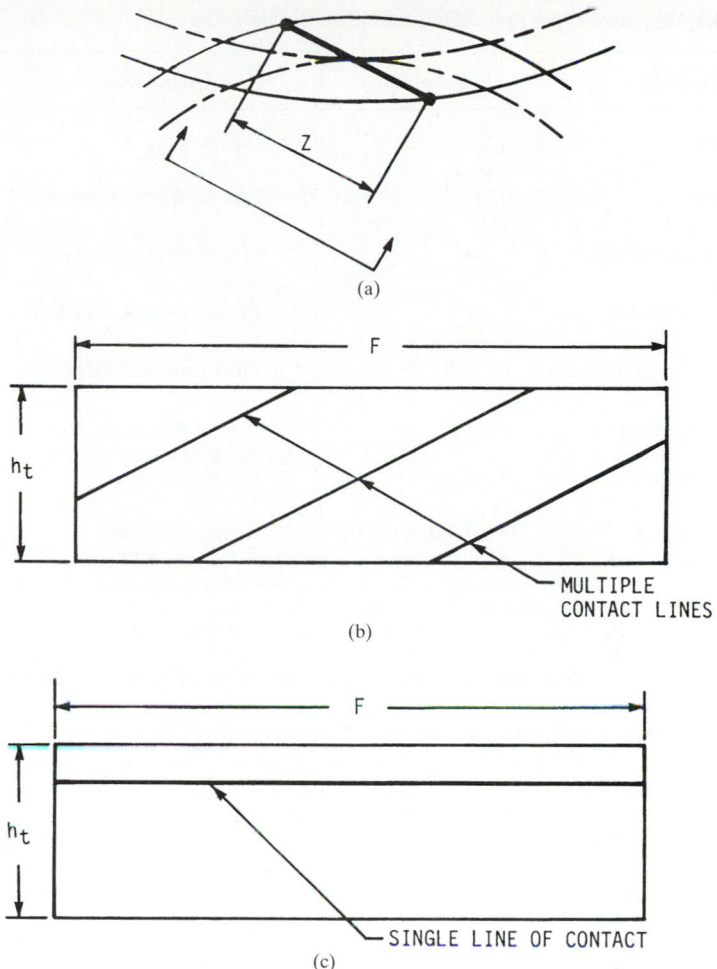

FIGURE 35.3 Comparison of spur and helical contact lines. (*a*) Transverse section; (*b*) helical contact lines; (*c*) spur contact line.

helical gears are employed, a limited number of standard cutters may be used to cut a wide variety of transverse-pitch gears simply by varying the helix angle, thus allowing virtually any center-distance and tooth-number combination to be accommodated.

35.4 GEOMETRY

When considered in the transverse plane (that is, a plane perpendicular to the axis of the gear), all helical-gear geometry is identical to that for spur gears. Standard tooth proportions are usually based on the normal diametral pitch, as shown in Table 35.1.

TABLE 35.1   Standard Tooth Proportions for Helical Gears

Quantity†	Formula	Quantity†	Formula
Addendum	$\dfrac{1.00}{P_N}$	External gears:	
Dedendum	$\dfrac{1.25}{P_N}$	Standard center distance	$\dfrac{D+d}{2}$
Pinion pitch diameter	$\dfrac{N_P}{P_N \cos \psi}$	Gear outside diameter	$D + 2a$
Gear pitch diameter	$\dfrac{N_G}{P_N \cos \psi}$	Pinion outside diameter	$d + 2a$
Normal arc tooth thickness	$\dfrac{\pi}{P_N} - \dfrac{B_N}{2}$	Gear root diameter	$D - 2b$
Pinion base diameter	$d \cos \phi_T$	Pinion root diameter	$d - 2b$
		Internal gears:	
Gear base diameter	$D \cos \phi_T$	Center distance	$\dfrac{D-d}{2}$
Base helix angle	$\tan^{-1}(\tan \psi \cos \phi_T)$	Inside diameter	$d - 2a$
		Root diameter	$D + 2b$

†All dimensions in inches, and angles are in degrees.

It is frequently necessary to convert from the normal plane to the transverse plane and vice versa. Table 35.2 gives the necessary equations. All calculations previously defined for spur gears with respect to transverse or profile-contact ratio, top land, lowest point of contact, true involute form radius, nonstandard center, etc., are valid for helical gears if only a transverse plane section is considered.

For spur gears, the profile-contact ratio (ratio of contact to the base pitch) must be greater than unity for uniform rotary-motion transmission to occur. Helical gears, however, provide an additional overlap along the axial direction; thus their profile-contact ratio need not necessarily be greater than unity. The sum of both the profile-

TABLE 35.2   Conversions between Normal and Transverse Planes

Parameter (normal/ transverse)	Normal to transverse	Transverse to normal
Pressure angle (ϕ_N / ϕ_T)	$\phi_T = \tan^{-1} \dfrac{\tan \phi_N}{\cos \psi}$	$\phi_N = \tan^{-1}(\tan \phi_T \cos \psi)$
Diametral pitch (P_N / P_d)	$P_d = P_N \cos \psi$	$P_N = \dfrac{P_d}{\cos \psi}$
Circular pitch (p_N / p_T)	$P_T = \dfrac{P_N}{\cos \psi}$	$P_N = P_T \cos \psi$
Arc tooth thickness (T_N / T_T)	$T_T = \dfrac{T_N}{\cos \psi}$	$T_N = T_T \cos \psi$
Backlash (B_N / B_T)	$B_T = \dfrac{B_N}{\cos \psi}$	$B_N = B_T \cos \psi$

contact ratio and the axial overlap must, however, be at least unity. The axial overlap, also often called the *face-contact ratio,* is the ratio of the face width to the axial pitch. The face-contact ratio is given by

$$m_F = \frac{P_{do} F \tan \psi_o}{\pi}$$ (35.1)

where P_{do} = operating transverse diametral pitch
ψ_o = helix angle at operating pitch circle
F = face width

Other parameters of interest in the design and analysis of helical gears are the base pitch p_b and the length of the line of action Z, both in the transverse plane. These are

$$p_b = \frac{\pi}{P_d} \cos \phi_T$$ (35.2)

and

$$Z = (r_o^2 - r_b^2)^{1/2} + (R_o^2 - R_b^2)^{1/2} - C_o \sin \phi_o$$ (35.3)

This equation is for an external gear mesh. For an internal gear mesh, the length of the line of action is

$$Z = (R_I^2 - R_b^2)^{1/2} - (r_o^2 - r_b^2)^{1/2} + C_o \sin \phi_o$$ (35.4)

where P_d = transverse diametral pitch as manufactured
ϕ_T = transverse pressure angle as manufactured, degrees (deg)
r_o = effective pinion outside radius, inches (in)
R_o = effective gear outside radius, in
R_I = effective gear inside radius, in
ϕ_o = operating transverse pressure angle, deg
r_b = pinion base radius, in
R_b = gear base radius, in
C_o = operating center distance, in

The operating transverse pressure angle ϕ_o is

$$\phi_o = \cos^{-1} \left(\frac{C}{C_o} \cos \phi_T \right)$$ (35.5)

The manufactured center distance C is simply

$$C = \frac{N_P + N_G}{2 P_d}$$ (35.6)

for external mesh; for internal mesh, the relation is

$$C = \frac{N_G - N_P}{2 P_d}$$ (35.7)

The contact ratio m_P in the transverse plane (profile-contact ratio) is defined as the ratio of the total length of the line of action in the transverse plane Z to the base pitch in the transverse plane p_b. Thus

$$m_P = \frac{Z}{p_b} \tag{35.8}$$

The diametral pitch, pitch diameters, helix angle, and normal pressure angle at the operating pitch circle are required in the load-capacity evaluation of helical gears. These terms are given by

$$P_{do} = \frac{N_P + N_G}{2C_o} \tag{35.9}$$

for external mesh; for internal mesh,

$$P_{do} = \frac{N_G - N_P}{2C_o} \tag{35.10}$$

Also,

$$d = \frac{N_P}{P_{do}} \qquad D = \frac{N_G}{P_{do}} \tag{35.11}$$

$$\psi_B = \tan^{-1}(\tan \psi \cos \phi_T) \tag{35.12}$$

$$\psi_o = \tan^{-1}\frac{\tan \psi_B}{\cos \phi_o} \tag{35.13}$$

$$\phi_{No} = \sin^{-1}(\sin \phi_o \cos \psi_B) \tag{35.14}$$

where P_{do} = operating diametral pitch
ψ_B = base helix angle, deg
ψ_o = helix angle at operating pitch point, deg
ϕ_{No} = operating normal pressure angle, deg
d = operating pinion pitch diameter, in
D = operating gear pitch diameter, in

35.5 LOAD RATING

Reference [35.1] establishes a coherent method for rating external helical and spur gears. The treatment of strength and durability provided here is derived in large part from this source.

Four factors must be considered in the load rating of a helical-gear set: strength, durability, wear resistance, and scoring probability. Although strength and durability must always be considered, wear resistance and scoring evaluations may not be required for every case. We treat each topic in some depth.

35.5.1 Strength and Durability

The strength of a gear tooth is evaluated by calculating the bending stress index number at the root by

$$s_t = \frac{W_t K_a}{K_v} \frac{P_d}{F_E} \frac{K_b K_m}{J} \tag{35.15}$$

where s_t = bending stress index number, pounds per square inch (psi)
  K_a = bending application factor
  F_E = effective face width, in
  K_m = bending load-distribution factor
  K_v = bending dynamic factor
  J = bending geometry factor
  P_d = transverse operating diametral pitch
  K_b = rim thickness factor

The calculated bending stress index number s_t must be within safe operating limits as defined by

$$s_t \leq \frac{s_{at} K_L}{K_T K_R} \tag{35.16}$$

where s_{at} = allowable bending stress index number
  K_L = life factor
  K_T = temperature factor
  K_R = reliability factor

Some of the factors which are used in these equations are similar to those used in the durability equations. Thus we present the basic durability rating equations before discussing the factors:

$$s_c = C_p \sqrt{\frac{W_t C_a}{C_v} \frac{1}{d F_N} \frac{C_m}{I}} \tag{35.17}$$

where s_c = contact stress index number
  C_a = durability application factor
  C_v = durability dynamic factor
  d = operating pinion pitch diameter
  F_N = net face width, in
  C_m = load-distribution factor
  C_p = elastic coefficient
  I = durability geometry factor

The calculated contact stress index number must be within safe operating limits as defined by

$$s_c \leq \frac{s_{ac} C_L C_H}{C_T C_R} \tag{35.18}$$

where s_{ac} = allowable contact stress index number
  C_L = durability life factor
  C_H = hardness ratio factor
  C_T = temperature factor
  C_R = reliability factor

To utilize these equations, each factor must be evaluated. The tangential load W_t is given by

$$W_t = \frac{2T_P}{d} \qquad (35.19)$$

where T_P = pinion torque in inch-pounds (in · lb) and d = pinion operating pitch diameter in inches. If the duty cycle is not uniform but does not vary substantially, then the maximum anticipated load should be used. Similarly, if the gear set is to operate at a combination of very high and very low loads, it should be evaluated at the maximum load. If, however, the loading varies over a well-defined range, then the cumulative fatigue damage for the loading cycle should be evaluated by using Miner's rule. For a good explanation, see Ref. [35.2].

Application Factors C_a and K_a.   The application factor makes the allowances for externally applied loads of unknown nature which are in excess of the nominal tangential load. Such factors can be defined only after considerable field experience has been established. In a *new* design, this consideration places the designer squarely on the horns of a dilemma, since "new" presupposes limited, if any, experience. The values shown in Table 35.3 may be used as a guide if no other basis is available.

TABLE 35.3   Application Factor Guidelines

Power source	Character of load on driven machine		
	Uniform	Moderate shock	Heavy shock
Uniform	1.15	1.25	At least 1.75
Light shock	1.25	1.50	At least 2.00
Medium shock	1.50	1.75	At least 2.50

The application factor should never be set equal to unity except where clear experimental evidence indicates that the loading will be absolutely uniform. Wherever possible, the actual loading to be applied to the system should be defined. One of the most common mistakes made by gear system designers is assuming that the motor (or engine, etc.) "nameplate" rating is also the gear unit rating point.

Dynamic Factors C_v and K_v.   These factors account for internally generated tooth loads which are induced by nonconjugate meshing action. This discontinuous motion occurs as a result of various tooth errors (such as spacing, profile, and runout) and system effects (such as deflections). Other effects, such as system torsional resonances and gear blank resonant responses, may also contribute to the overall dynamic loading experienced by the teeth. The latter effects must, however, be separately evaluated. The effect of tooth accuracy may be determined from Fig. 35.4, which is based on both pitch line velocity and gear quality Q_n as specified in Ref. [35.3]. The pitch line velocity of a gear is

$$v_t = 0.2618nD \qquad (35.20)$$

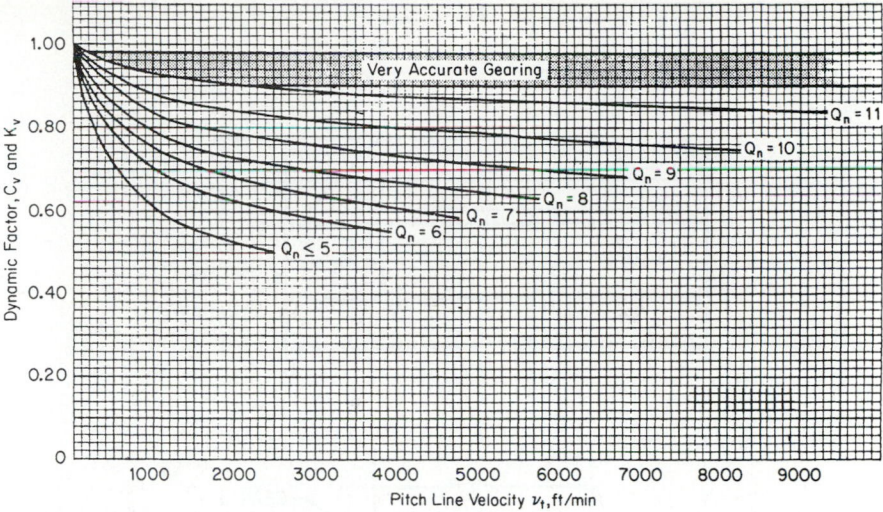

FIGURE 35.4 Dynamic factors C_v and K_v. *(From Ref. [35.1].)*

where v_t = pitch line velocity, feet per minute (ft/min)
              n = gear speed, revolutions per minute (r/min)
              D = gear pitch diameter, in

Effective and Net Face Widths $\mathbf{F_E}$ *and* $\mathbf{F_N}$.   The net minimum face width of the narrowest member should always be used for F_N. In cases where one member has a substantially larger face width than its mate, some advantage may be taken of this fact in the bending stress calculations, but it is unlikely that a very narrow tooth will fully transfer its tooth load across the face width of a much wider gear. At best, the effective face width of a larger-face-width gear mating with a smaller-face-width gear is limited to the minimum face of the smaller member plus some allowance for the extra support provided by the wide face. Figure 35.5 illustrates the definition of net and effective face widths for various cases.

Rim Thickness Factor $\mathbf{K_b}$.   The basic bending stress equations were developed for a single tooth mounted on a rigid support so that it behaves as a short cantilever beam. As the rim which supports the gear tooth becomes thinner, a point is reached at which the rim no longer provides "rigid" support. When this occurs, the bending of the rim itself combines with the tooth bending to yield higher total alternating stresses than would be predicted by the normal equations. Additionally, when a tooth is subjected to fully reversed bending loads, the alternating stress is also increased because of the additive effect of the compressive stress distribution on the normally unloaded side of the tooth, as Fig. 35.6 shows. Both effects are accounted for by the rim thickness factor, as Fig. 35.7 indicates.

    It must be emphasized that the data shown in Fig. 35.7 are based on a limited amount of analytical and experimental (photoelastic and strain-gauge) measurements and thus must be used judiciously. Still, they are the best data available to date and are far better than nothing at all; see Refs. [35.4] and [35.5].

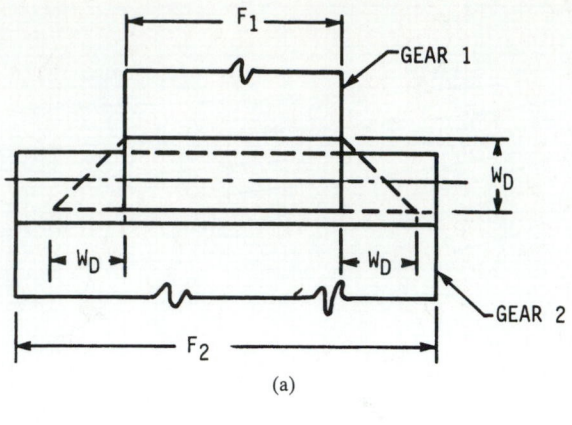

(a)

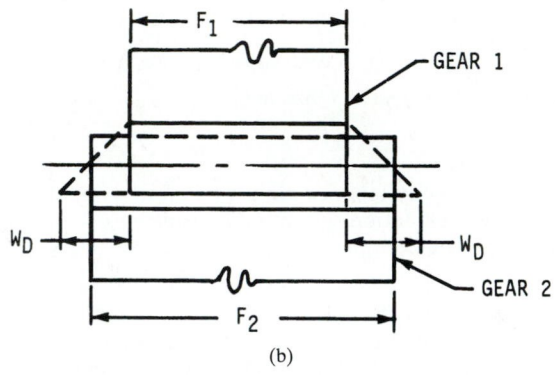

(b)

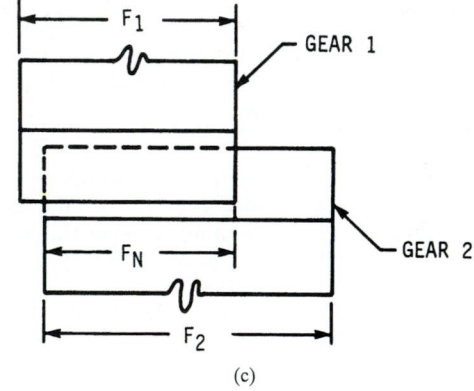

(c)

FIGURE 35.5 Definition of effective face width. (a) $F_{E1} = F_1, F_{E2} = F_1 + 2W_D$; $F_N = F_1$; (b) $F_{E1} = F_1, F_{E2} = F_2, F_N = F_1$; (c) $F_{E1} = F_{E2} = F_N$.

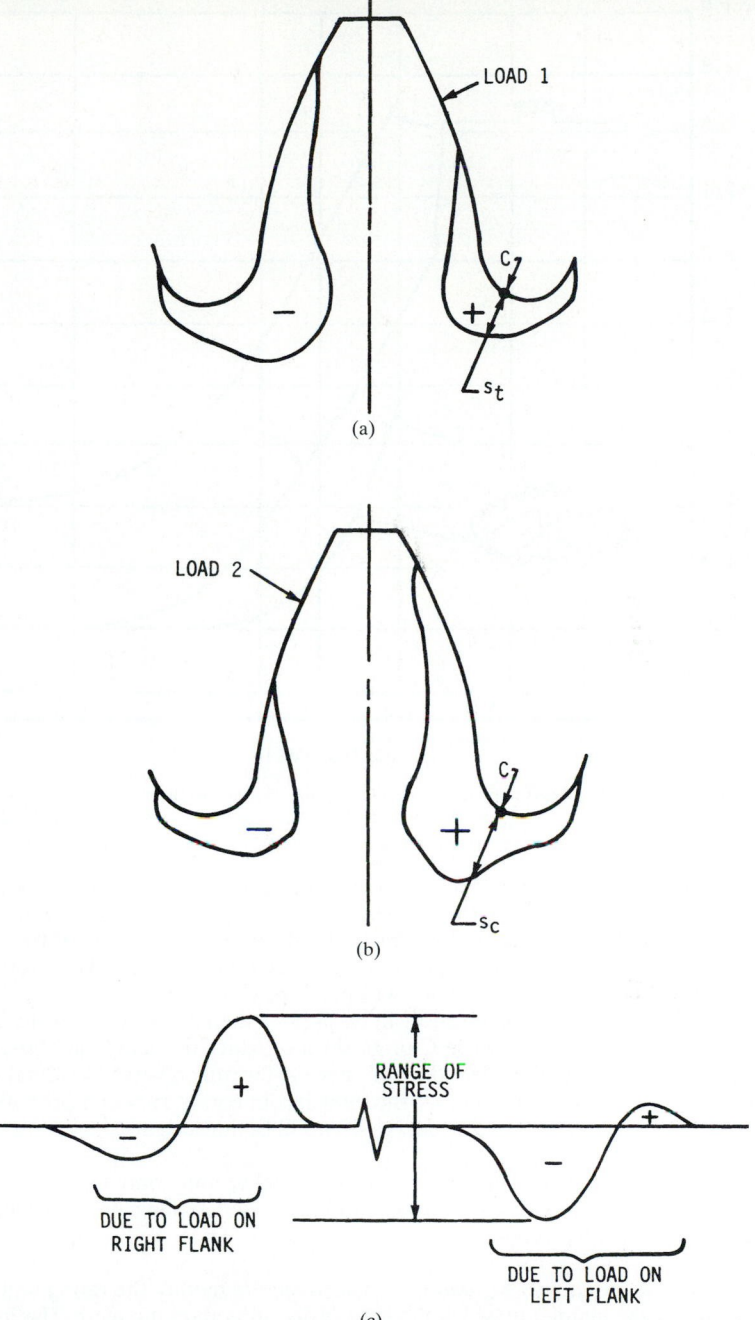

FIGURE 35.6 Stress condition for reversing (as with an idler) loading. (*a*) Load on right flank; (*b*) load on left flank; (*c*) typical waveform for strain gauge at point *C*.

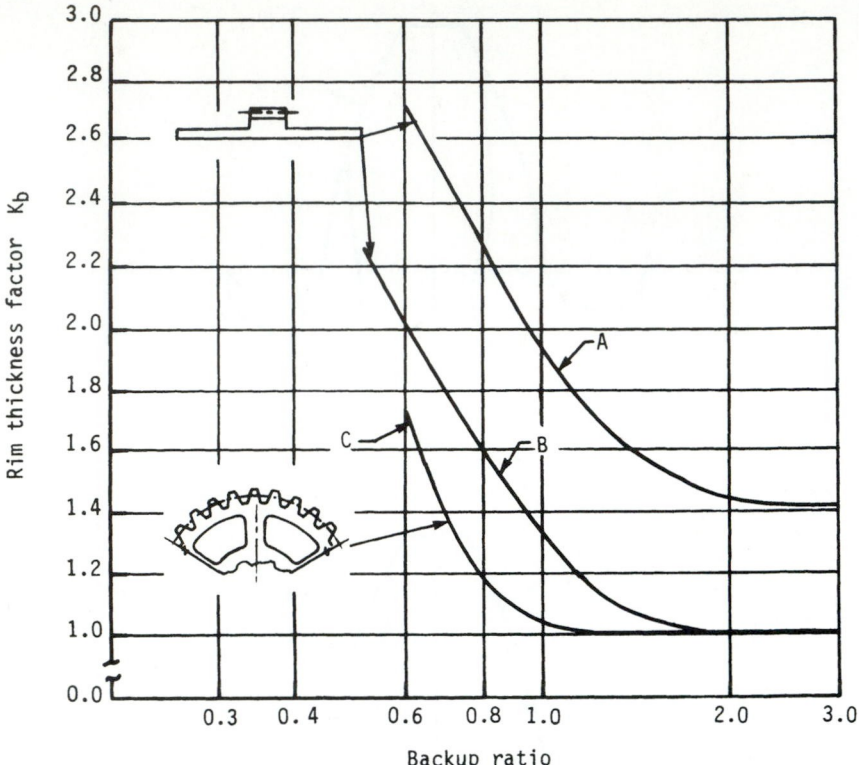

FIGURE 35.7   Rim thickness factor K_b. The *backup ratio* is defined as the ratio of the rim thickness to the tooth height. Curve *A* is fully reversed loading; curves *B* and *C* are unidirectional loading.

For gear blanks which utilize a T-shaped rim and web construction, the web acts as a hard point, if the rim is thin, and stresses will be higher over the web than over the ends of the T. The actual value which should be used for such constructions depends greatly on the relative proportions of the gear face width and the web. If the web spans 70 to 80 percent of the face width, the gear may be considered as having a rigid backup. Thus the backup ratio will be greater than 2.0, and any of the curves shown may be used (that is, curve *C* or *B*, both of which are identical above a 2.0 backup ratio, for unidirectional loading or curve *A* for fully reversed loading). If the proportions are between these limits, the gear lies in a gray area and probably lies somewhere in the range defined by curves *B* and *C*. Some designer discretion should be exercised here.

Finally, note that the rim thickness factor is equal to unity only for unidirectionally loaded, rigid-backup helical gears. For fully reversed loading, its value will be at least 1.4, even if the backup is rigid.

Load-Distribution Factors $\mathbf{K_m}$ ***and*** $\mathbf{C_c}$ These factors modify the rating equations to account for the manner in which the load is distributed on the teeth. The load on a set of gears will never be exactly uniformly distributed. Factors which affect the load distribution include the accuracy of the teeth themselves; the accuracy of the housing which supports the teeth (as it influences the alignment of the gear axes); the deflections of the housing, shafts, and gear blanks (both elastic and thermal); and the internal clearances in the bearings which support the gears, among others.

All these and any other appropriate effects must be evaluated in order to define the total effective alignment error e_t for the gear pair. Once this is accomplished, the load-distribution factor may be calculated.

In some cases it may not be possible to fully define or even estimate the value of e_t. In such cases an empirical approach may be used. We discuss both approaches in some detail.

The empirical approach requires only minimal data, and so it is the simplest to apply. Several conditions must be met, however, prior to using this method:

1. Net face width to pinion pitch diameter ratios must be less than or equal to 2.0. (For double-helix gears, the gap is not included in the face width.)
2. The gear elements are mounted between bearings (not overhung).
3. Face width can be up to 40 in.
4. There must be contact across the full face width of the narrowest member when loaded.
5. Gears are not highly crowned.

The empirical expression for the load-distribution factor is

$$C_m = K_m = 1.0 + C_{mc}(C_{pf}C_{pm} + C_{ma}C_e) \qquad (35.21)$$

where C_{mc} = lead correction factor
C_{pf} = pinion proportion factor
C_{pm} = pinion proportion modifier
C_{ma} = mesh alignment factor
C_e = mesh alignment correction factor

The lead correction factor C_{mc} modifies the peak loading in the presence of slight crowning or lead correction as follows:

$$C_{mc} = \begin{cases} 1.0 & \text{for gear with unmodified leads} \\ 0.8 & \text{for gear with leads properly modified by crowning or lead correction} \end{cases}$$

Figure 35.8 shows the pinion proportion factor C_{pf}, which accounts for deflections due to load. The pinion proportion modifier C_{pm} alters C_{pf} based on the location of the pinion relative to the supporting bearings. Figure 35.9 defines the factors S and S_1. And C_{pm} is defined as follows:

$$C_{pm} = \begin{cases} 1.0 & \text{when } S_1/S < 0.175 \\ 1.1 & \text{when } S_1/S \geq 0.175 \end{cases}$$

The mesh alignment factor C_{ma} accounts for factors other than elastic deformations. Figure 35.10 provides values for this factor for four accuracy groupings. For double-helix gears, this figure should be used with F equal to half of the total face width. The mesh alignment correction factor C_e modifies the mesh alignment factor to allow for the improved alignment which may be obtained when a gear set is adjusted at assembly or when the gears are modified by grinding, skiving, or lapping to more closely match their mates at assembly (in which case, pinion and gear

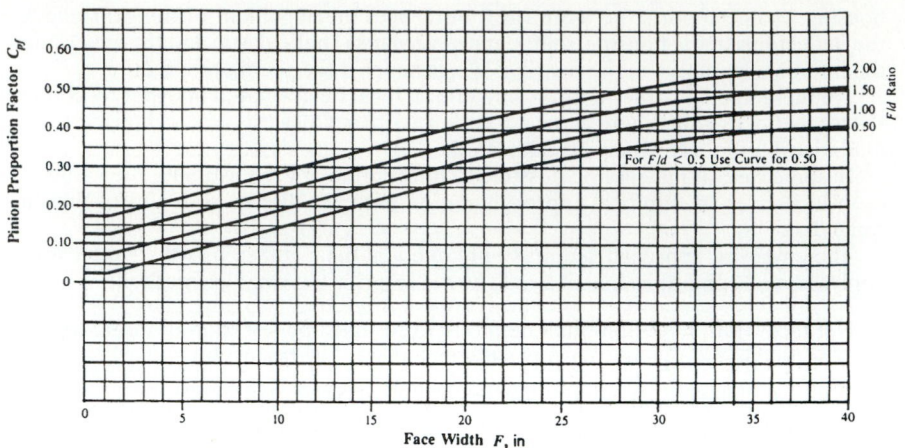

FIGURE 35.8   Pinion proportion factor C_{pf}. *(From Ref. [35.1].)*

become a matched set). Only two values are permissible for C_e—either 1.0 or 0.8, as defined by the following requirements:

$$
C_e = \begin{cases}
0.80 & \text{when the compatibility of the gearing is improved by lapping, grinding, or skiving after trial assembly to improve contact} \\
0.80 & \text{when gearing is adjusted at assembly by shimming support bearings and/or housing to yield uniform contact} \\
1.0 & \text{for all other conditions}
\end{cases}
$$

If enough detailed information is available, a better estimate of the load-distribution factor may be obtained by using a more analytical approach. This method, however, requires that the total alignment error e_t be calculated or estimated. Depending on the contact conditions, one of two expressions is used to calculate the load-distribution factor.

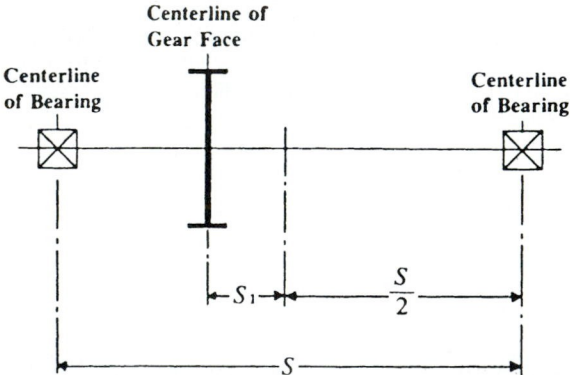

FIGURE 35.9   Definition of distances S and S_1. Bearing span is distance S; pinion offset from midspan is S_1. *(From Ref. [35.1].)*

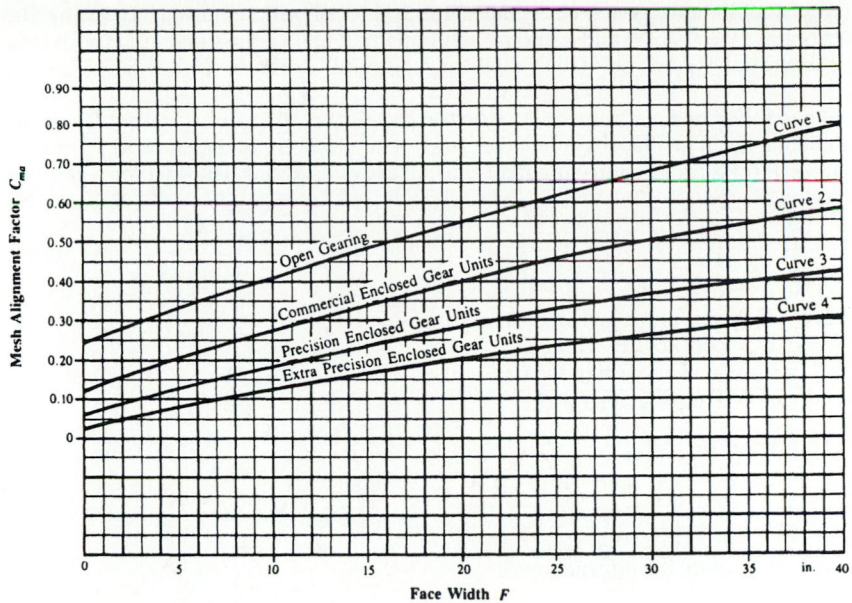

FIGURE 35.10 Mesh alignment factor C_{ma}. For analytical method for determination of C_{ma}, see Eq. (35.21). *(From Ref. [35.1].)*

If the tooth contact pattern at normal operating load essentially covers the entire available tooth face, Eq. (35.22) should be used. If the tooth contact pattern does not cover the entire available tooth face (as would be the case for poorly aligned or high-crowned gears) at normal operating loads, then Eq. (35.23) must be used:

$$C_m = 1.0 + \frac{Ge_tZF}{4W_t p_b} \qquad (35.22)$$

and

$$C_m = \sqrt{\frac{Ge_tZF}{W_t p_b}} \qquad (35.23)$$

where W_t = tangential tooth load, pounds (lb)
G = tooth stiffness constant, (lb/in)/in of face
Z = length of line of contact in transverse plane
e_t = total effective alignment error, in/in
p_b = transverse base pitch, in
F = net face width of narrowest member, in

The value of G will vary with tooth proportions, tooth thickness, and material. For steel gears of standard or close to standard proportions, it is normally in the range of 1.5×10^6 to 2.0×10^6 psi. The higher value should be used for higher-pressure-angle teeth, which are normally stiffer, while the lower value is representative of more flexible teeth. The most conservative approach is to use the higher value in all cases.

For double-helix gears, each half should be analyzed separately by using the appropriate values of F and e_t and by assuming that half of the tangential tooth load is transmitted by each half (the values for p_b, Z, and G remain unchanged).

Geometry Factor **I.**   The geometry factor I evaluates the radii of curvature of the contacting tooth profiles based on the pressure angle, helix, and gear ratio. Effects of modified tooth proportions and load sharing are considered. The I factor is defined as follows:

$$I = \frac{C_c C_x C_\psi^2}{m_N} \tag{35.24}$$

where C_c = curvature factor at operating pitch line
C_x = contact height factor
C_ψ = helical overlap factor
m_N = load-sharing ratio

The curvature factor is

$$C_c = \frac{\cos \phi_o \sin \phi_o}{2} \frac{N_G}{N_G + N_P} \tag{35.25}$$

for external mesh; for internal mesh,

$$C_c = \frac{\cos \phi_o \sin \phi_o}{2} \frac{N_G}{N_G - N_P} \tag{35.26}$$

The contact height factor C_x adjusts the location on the tooth profile at which the critical contact stress occurs (i.e., face-contact ratio > 1.0). The stress is calculated at the mean diameter or the middle of the tooth profile. For low-contact-ratio helical gears (that is, face-contact ratio ≤ 1.0), the stress is calculated at the lowest point of single-tooth contact in the transverse plane and C_x is given by Eq. (35.27):

$$C_x = \frac{R_1 R_2}{R_P R_G} \tag{35.27}$$

where R_P = pinion curvature radius at operating pitch point, in
R_G = gear curvature radius at operating pitch point, in
R_1 = pinion curvature radius at critical contact point, in
R_2 = gear curvature radius at critical contact point, in

The required radii are given by

$$R_P = \frac{d}{2} \sin \phi_o \qquad R_G = \frac{D}{2} \sin \phi_o \tag{35.28}$$

where d = pinion operating pitch diameter, in
D = gear operating pitch diameter, in
ϕ_o = operating pressure angle in transverse plane, deg

$$R_1 = R_P - Z_c \tag{35.29}$$

and

$$R_2 = R_G + Z_c \tag{35.30}$$

for external gears; for internal gears,

$$R_2 = R_G - Z_c \qquad (35.31)$$

where Z_c is the distance along the line of action in the transverse plane to the critical contact point. The value of Z_c is dependent on the transverse contact ratio. For helical gears where the face-contact ratio ≤ 1.0, Z_c is found by using Eq. (35.32). For normal helical gears where the face-contact ratio is > 1.0, Eq. (35.33) is used:

$$Z_c = p_b - 0.5[(d_o^2 - d_b^2)^{1/2} - (d^2 - d_b^2)^{1/2}] \qquad m_F \leq 1.0 \qquad (35.32)$$

and

$$Z_c = 0.5\,[(d^2 - d_b^2)^{1/2} - (d_m^2 - d_b^2)^{1/2}] \qquad m_F > 1.0 \qquad (35.33)$$

where p_b = base pitch, in
  d_o = pinion outside diameter, in
  d_b = pinion base diameter, in
  d_m = pinion mean diameter, in

The pinion mean diameter is defined by Eq. (35.34) or (35.35). For external mesh,

$$d_m = C_o - \frac{D_o - d_o}{2} \qquad (35.34)$$

For internal mesh,

$$d_m = \frac{D_I + d_o}{2} - C_o \qquad (35.35)$$

where D_o = external gear outside diameter and D_I = internal gear inside diameter.

The helical factor C_ψ accounts for the partial helical overlap action which occurs in helical gears with a face-contact ratio $m_F \leq 1.0$. For helical gears with a face-contact ratio > 1.0, C_ψ is set equal to unity; for low-contact helical gears, it is

$$C_\psi = \sqrt{1 - m_F + \frac{C_{xn} Z m_F^2}{C_x F \sin \psi_b}} \qquad (35.36)$$

where Z = total length of line of action in transverse plane, in
  F = net minimum face width, in
  m_F = face-contact ratio
  C_x = contact height factor [Eq. (35.27)]
  C_{xn} = contact height factor for equivalent normal helical gears [Eq. (35.37)]
  ψ_b = base helix angle, deg

The C_{xn} factor is given by

$$C_{xn} = \frac{R_{1n} R_{2n}}{R_P R_G} \qquad (35.37)$$

where R_{1n} = curvature radius at critical point for equivalent normal helical pinion, in
  R_{2n} = curvature radius at critical contact point for equivalent normal helical gear, in

The curvature radii are given by

$$R_{1n} = R_P - Z_c \tag{35.38}$$

$$R_{2n} = R_G + Z \qquad \text{external gears}$$

$$\tag{35.39}$$

$$R_{2n} = R_G - Z_c \qquad \text{internal gears}$$

where Eq. (35.38) applies to external gears and Eq. (35.39) to either, as appropriate. Also, the term Z_c is obtained from Eq. (35.32).

The load-sharing ratio m_N is the ratio of the face width to the minimum total length of the contact lines:

$$m_N = \frac{F}{L_{min}} \tag{35.40}$$

where m_N = load-sharing ratio
F = minimum net face width, in
L_{min} = minimum total length of contact lines, in

The calculation of L_{min} is a rather involved process. For most helical gears which have a face-contact ratio of at least 2.0, a conservative approximation for the load-sharing ratio ratio m_N may be obtained from

$$m_N = \frac{P_N}{0.95Z} \tag{35.41}$$

where p_N = normal circular pitch in inches and Z = length of line of action in the transverse plane in inches. For helical gears with a face-contact ratio of less than 2.0, it is imperative that the actual value of L_{min} be calculated and used in Eq. (35.40). The method for doing this is shown in Eqs. (35.42) through (35.45):

$$L_{min} = \frac{1}{\sin \psi_b} [(P_1 - Q_1) + (P_2 - Q_2) + \cdots$$

$$+ (P_i - Q_i) + \cdots + (P_n - Q_n)] \tag{35.42}$$

where n = limiting number of lines of contact, as given by

$$n = \frac{(Z/\tan \psi_b) + F}{p_x} \tag{35.43}$$

Also, P_i = sum of base pitches in inches. The ith term of P_i is the lesser of

$$ip_x \tan \psi_b \qquad \text{or} \qquad Z \tag{35.44}$$

Finally, Q_i = remainder of base pitches in inches. Its value is

$$Q_i = 0 \qquad \text{if } ip_x \leq F$$

But when $ip_x > F$, then Q_i is the ith term and is the lesser of

$$(ip_x - F) \tan \psi_b \qquad \text{or} \qquad Z \tag{35.45}$$

Geometry Factor J. The bending strength geometry factor is

$$J = \frac{YC_\psi}{K_f m_N} \qquad (35.46)$$

where Y = tooth form factor
K_f = stress correction factor
C_ψ = helical factor
m_N = load-distribution factor

The helical and load-distribution factors were both defined in the discussion of the geometry factor I. The calculation of Y is also a long, tedious process. For helical gears in which load sharing exists among the teeth in contact and for which the face-contact ratio is at least 2.0, the value of Y need not be calculated, since the value for J may be obtained directly from the charts shown in Figs. 35.11 through 35.25 with Eq. (35.47):

$$J = J' Q_{TR} Q_{TT} Q_A Q_H \qquad (35.47)$$

where J' = basic geometry factor
Q_{TR} = tool radius adjustment factor
Q_{TT} = tooth thickness adjustment factor
Q_A = addendum adjustment factor
Q_H = helix-angle adjustment factor

In using these charts, note that the values of addendum, dedendum, and tool-tip radius are given for a 1-normal-pitch gear. Values for any other pitch may be obtained by dividing the factor by the actual normal diametral pitch. For example, if an 8-normal-pitch gear is being considered, the parameters shown on Fig. 35.11 are

$$\text{Addendum } a = \frac{1.0}{8} = 0.125 \text{ in}$$

$$\text{Dedendum } b = \frac{1.35}{8} = 0.168\ 75 \text{ in}$$

$$\text{Tool (hob) tip radius } r_T = \frac{0.42}{8} = 0.0525 \text{ in}$$

The basic geometry factor J' is found from Figs. 35.11 through 35.25. The tool radius adjustment factor Q_{TR} is found from Figs. 35.14 through 35.16 if the edge radius on the tool is other than $0.42/P_d$, which is the standard value used in calculating J'. Similarly, for gears with addenda other than $1.0/P_d$ or tooth thicknesses other than the standard value of $\pi/(2P_d)$, the appropriate factors may be obtained from these charts. In the case of a helical gear, the adjustment factor Q_H is obtained from Figs. 35.23 through 35.25. If a standard helical gear is being considered, Q_{TR}, Q_{TT}, and Q_A remain equal to unity, but Q_H must be found from Figs. 35.23 to 35.25.

These charts are computer-generated and, when properly used, produce quite accurate results. Note that they are also valid for spur gears if Q_H is set equal to unity (that is, enter Figs. 35.23 through 35.25 with 0° helix angle).

The charts shown in Figs. 35.11 through 35.25 assume the use of a standard full-radius hob. Additional charts, still under the assumption that the face-contact ratio is

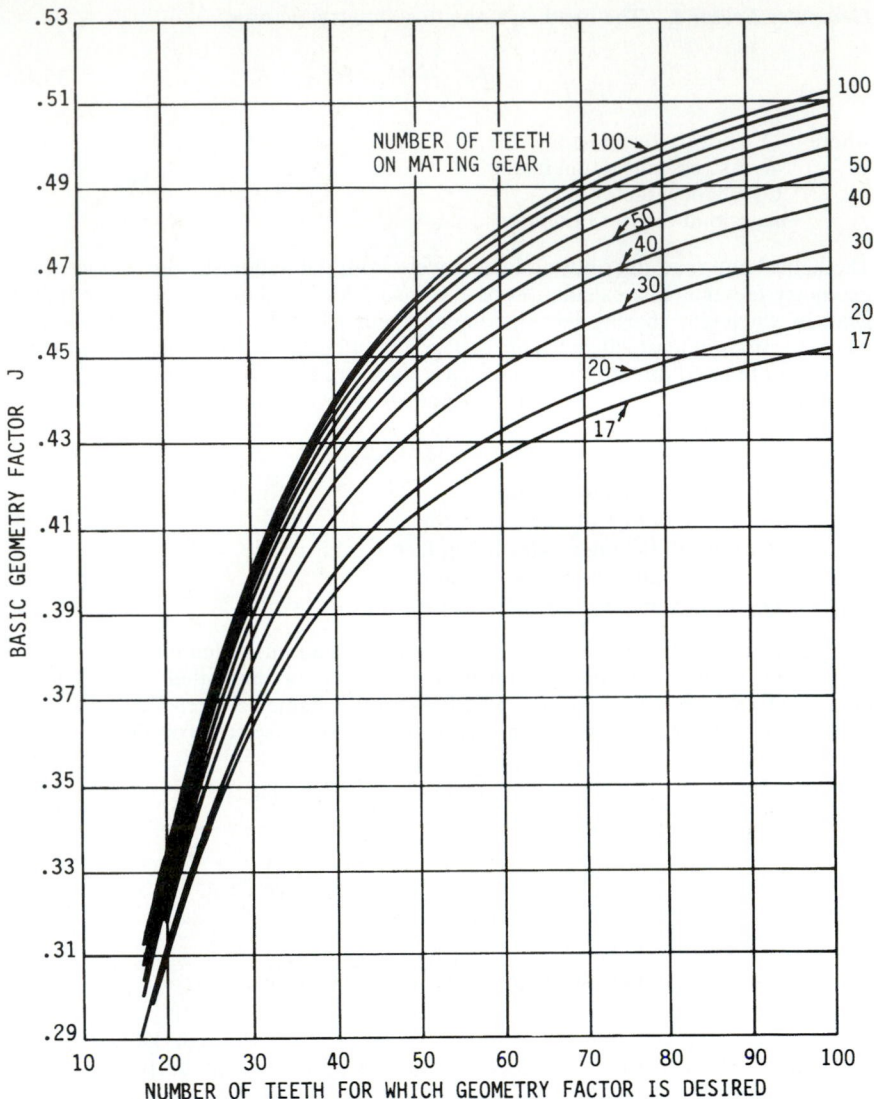

FIGURE 35.11   Basic geometry factors for 20° spur teeth; $\phi_N = 20°$, $a = 1.00$, $b = 1.35$, $r_T = 0.42$, $\Delta t = 0$.

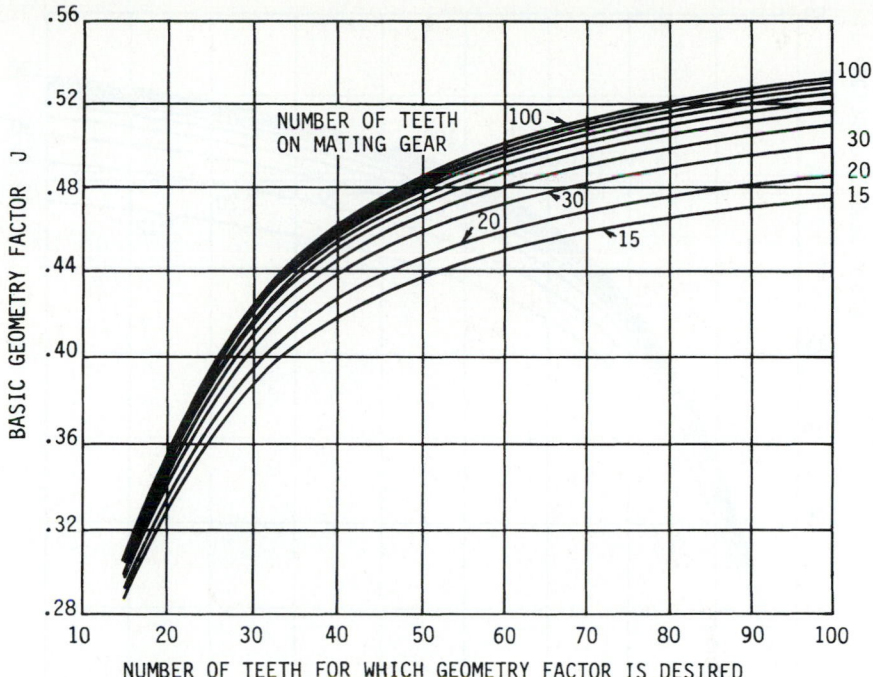

FIGURE 35.12   Basic geometry factors for 22½° spur teeth; $\phi_N = 22\frac{1}{2}°$, $a = 1.00$, $b = 1.35$, $r_T = 0.34$, $\Delta t = 0$.

at least 2.0 for other cutting-tool configurations, are shown in Figs. 35.26 through 35.36.† For these figures,

$$m_N = \frac{P_N}{0.95Z} \tag{35.48}$$

where the value of Z is for an element of indicated number of teeth and a 75-tooth mate. Also, the normal tooth thicknesses of pinion and gear teeth are each reduced 0.024 in, to provide 0.048 in of total backlash corresponding to a normal diametral pitch of unity. Note that these charts are limited to standard addendum, dedendum, and tooth thickness designs.

If the face-contact ratio is less than 2.0, the geometry factor must be calculated in accordance with Eq. (35.46); thus, it will be necessary to define Y and K_f. The definition of Y may be accomplished either by graphical layout or by a numerical iteration procedure. Since this Handbook is likely to be used by the machine designer with an occasional need for gear analysis, rather than by the gear specialist, we present the direct graphical technique. Readers interested in preparing computer codes or calculator routines might wish to consult Ref. [35.6].

The following graphical procedure is abstracted directly from Ref. [35.1] with permission of the publisher, as noted earlier. The Y factor is calculated with the aid

---

† These figures are extracted from AGMA 218.01 with the permission of the AGMA.

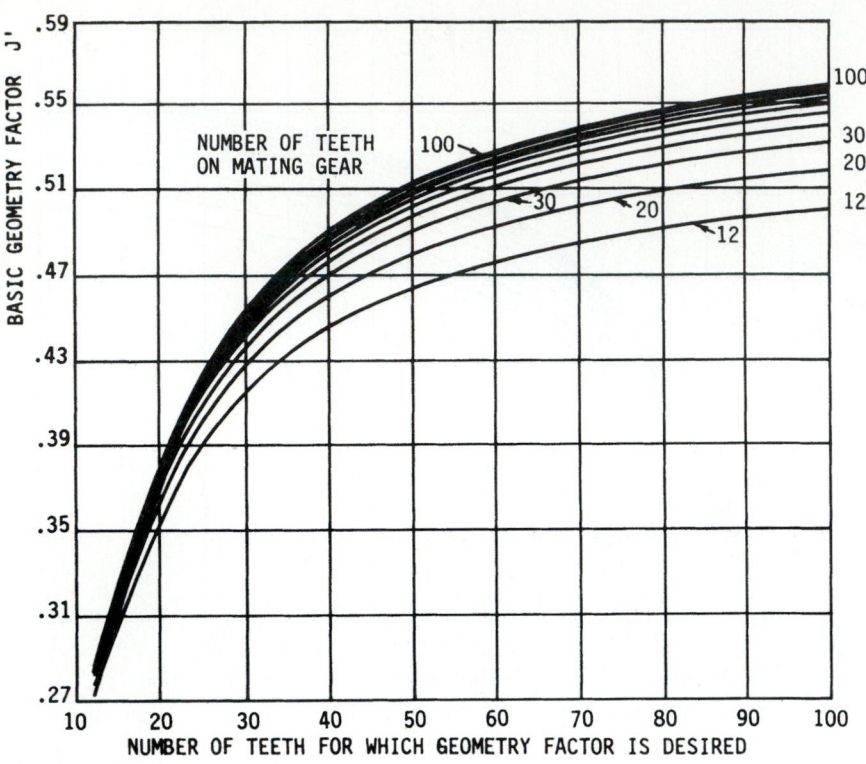

FIGURE 35.13 Basic geometry factors for 25° spur teeth; $\phi_N = 25°$, $a = 1.00$, $b = 1.35$, $r_T = 0.24$, $\Delta t = 0$.

of dimensions obtained from an accurate layout of the tooth profile in the normal plane at a scale of 1 normal diametral pitch. Actually, any scale can be used, but the use of 1 normal diametral pitch is most convenient. Depending on the face-contact ratio, the load is considered to be applied at the highest point of single-tooth contact (HPSTC), Fig. 35.37, or at the tooth tip, Fig. 35.37. The equation is

$$Y = \frac{K_\psi P_s}{[\cos(\phi_L)/\cos(\phi_{No})][(1.5/uC_h) - \tan(\phi_L)/t]}$$ (35.49)

The terms in Eq. (35.49) are defined as follows:

K_ψ = helix-angle factor
ϕ_{No} = normal operating pressure angle [Eq. (35.14)]
ϕ_L = load angle
C_h = helical factor
t = tooth thickness from layout, in
u = radial distance from layout, in
P_s = normal diametral pitch of layout (scale pitch), usually 1.0 in^{-1}

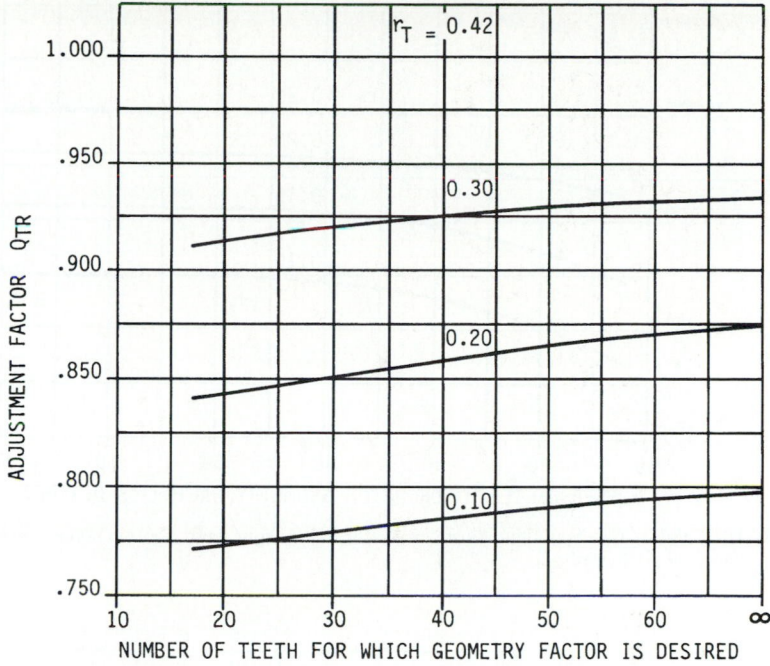

FIGURE 35.14 Tool-tip radius adjustment factor for 20° spur teeth. Tool-tip radius = r_T for a 1-diametral-pitch gear.

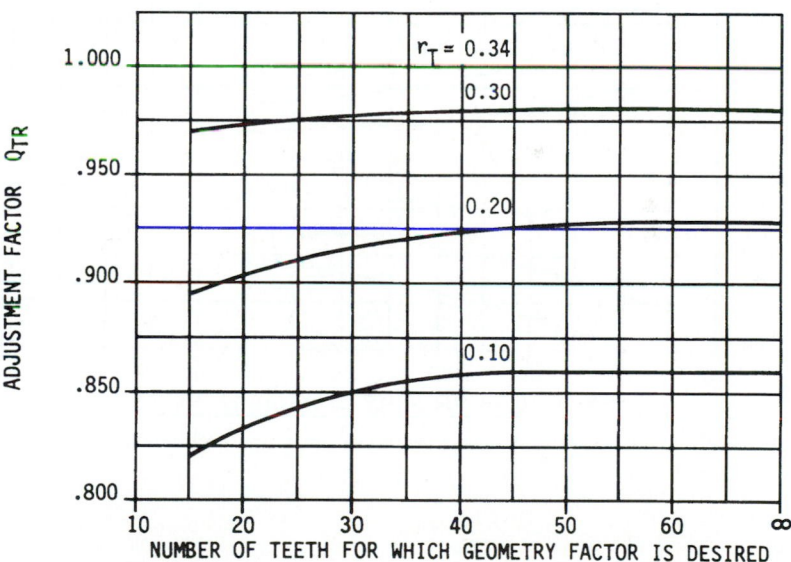

FIGURE 35.15 Tool-tip radius adjustment factor for 22½° spur teeth. Tool-tip radius = r_T for a 1-diametral-pitch gear.

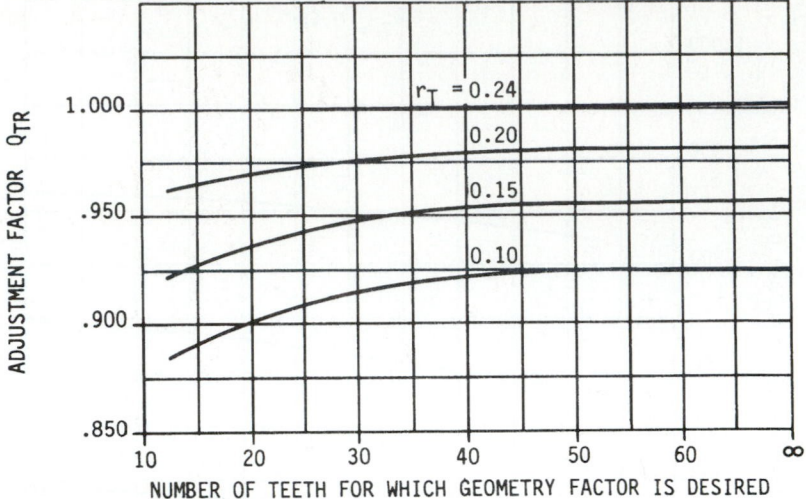

FIGURE 35.16   Tool-tip radius adjustment factor for 25° spur teeth. Tool-tip radius = r_T for a 1-diametral-pitch gear.

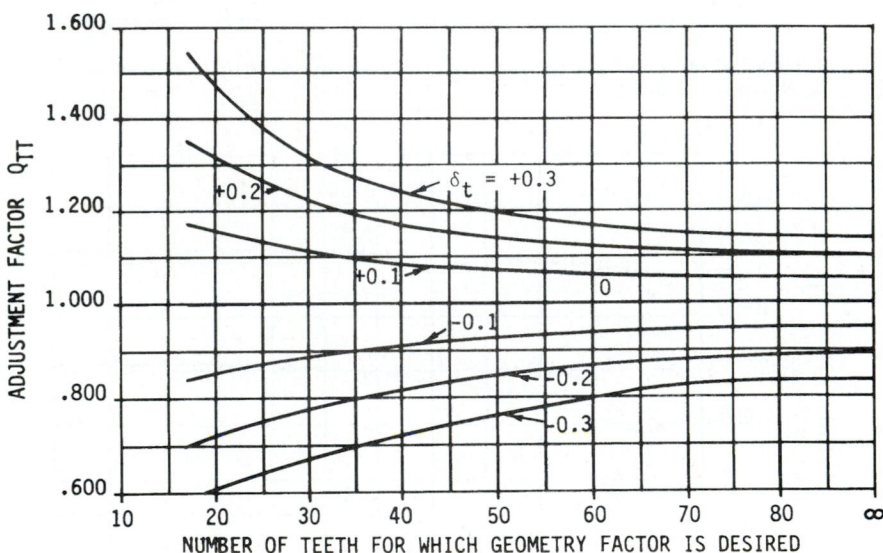

FIGURE 35.17   Tooth thickness adjustment factor Q_{TT} for 20° spur teeth. Tooth thickness modification = δ_t for 1-diametral-pitch gears.

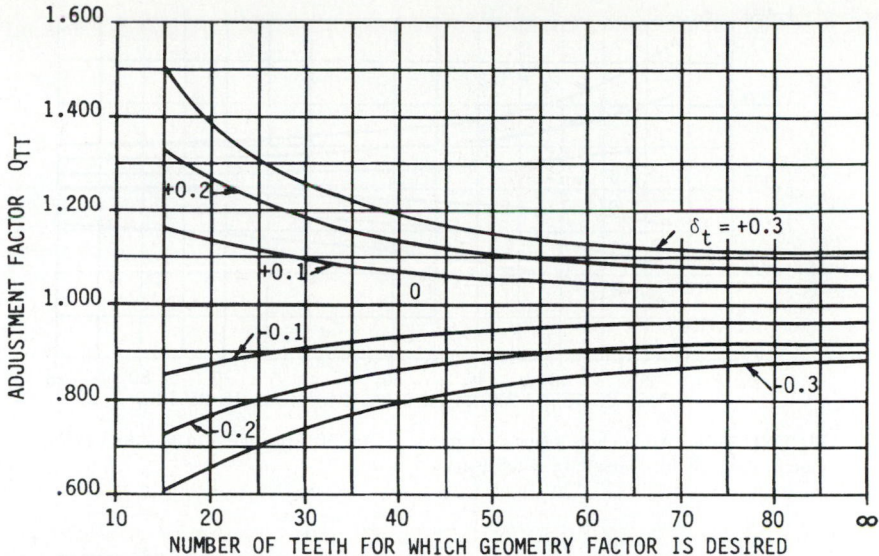

FIGURE 35.18 Tooth thickness adjustment factor Q_{TT} for 22½° spur teeth. Tooth thickness modification = δ_t for 1-diametral-pitch gears.

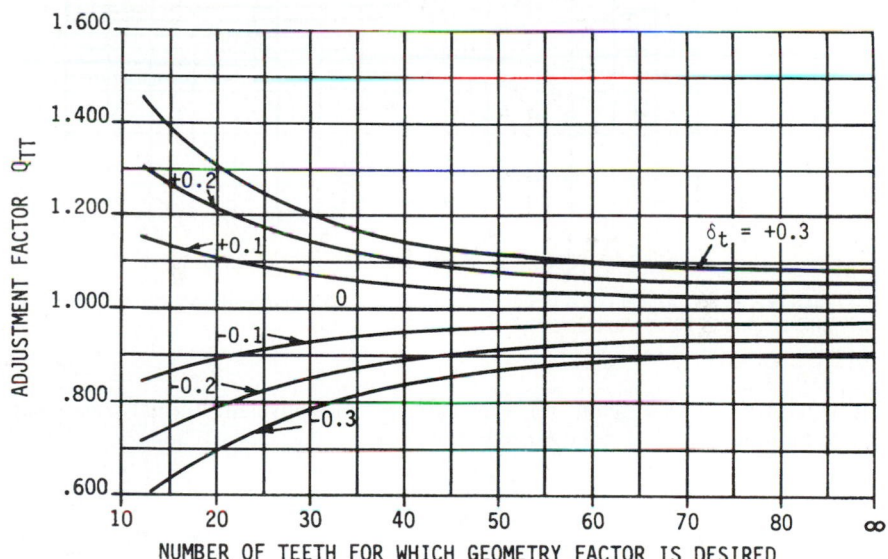

FIGURE 35.19 Tooth thickness adjustment factor Q_{TT} for 25° spur teeth. Tooth thickness modification = δ_t for 1-diametral-pitch gears.

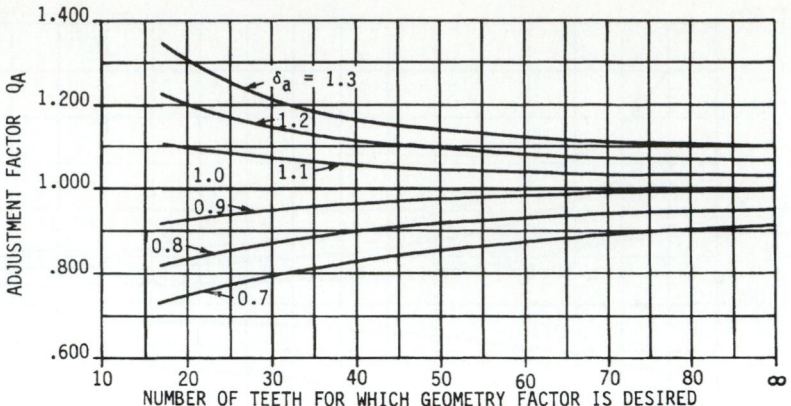

FIGURE 35.20 Addendum adjustment factor Q_A for 20° spur teeth. Addendum factor modification = δ_a for 1-diametral-pitch gears.

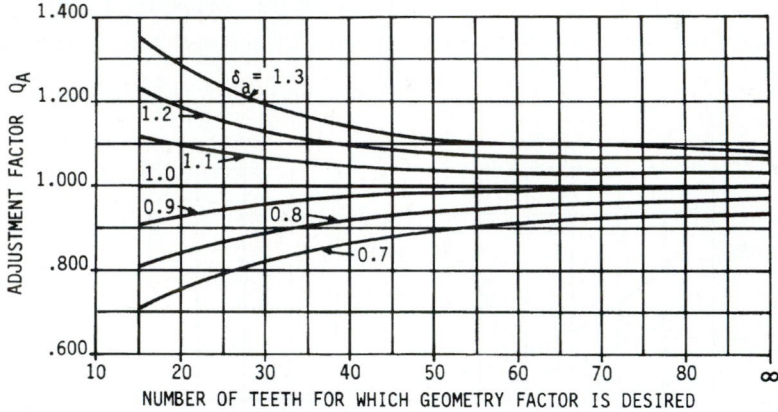

FIGURE 35.21 Addendum adjustment factor Q_A for 22½° spur teeth. Addendum factor modification = δ_a for 1-diametral-pitch gears.

To make the Y factor layout for a helical gear, an equivalent normal-plane gear tooth must be created, as follows:

$$N_e = \frac{N_P}{\cos^3 \psi} \tag{35.50}$$

$$d_c = \frac{dP_{nd}}{\cos^2 \psi} \tag{35.51}$$

$$d_{be} = d_e \cos \phi_n = N_e \cos \phi_c \tag{35.52}$$

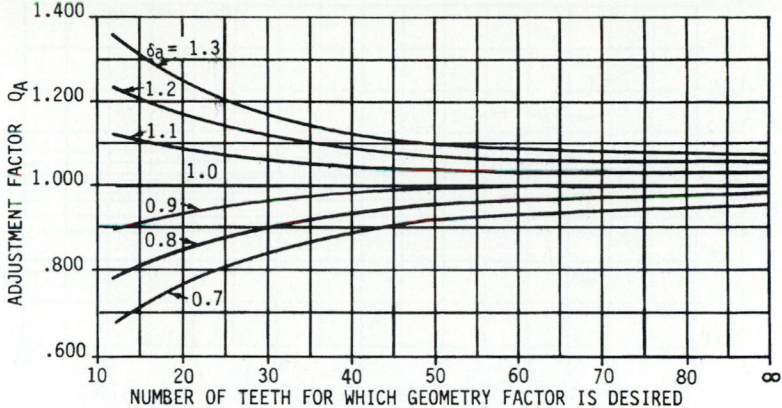

FIGURE 35.22   Addendum adjustment factor Q_A for 25° spur teeth. Addendum factor modification = δ_a for 1-diametral-pitch gears.

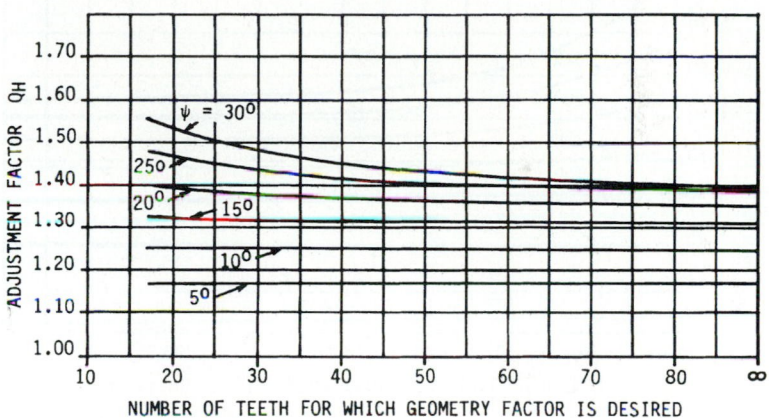

FIGURE 35.23   Helix-angle adjustment factor Q_H for $\phi_N = 20°$.

$$a = \frac{d_o - d}{2}\, P_{nd} \tag{35.53}$$

$$b = \frac{d - d_R}{2}\, P_{nd} \tag{35.54}$$

$$d_{oe} = d_e + 2a \tag{35.55}$$

$$d_{Re} = d_e - 2b \tag{35.56}$$

$$r_1 = \frac{(b - r_{Te})^2}{R_o + b - r_{Te}} \tag{35.57}$$

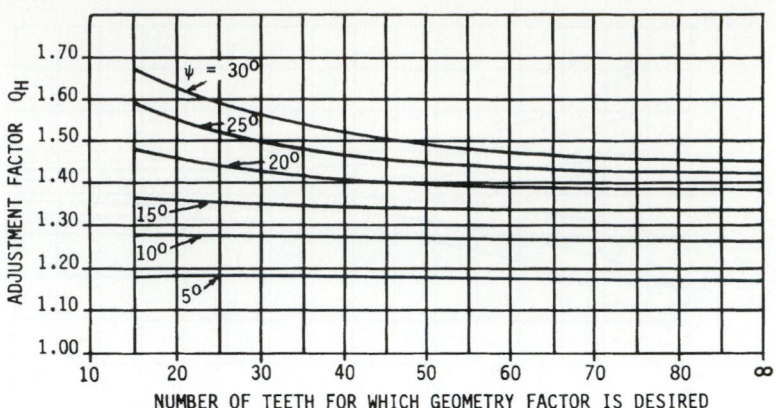

FIGURE 35.24 Helix-angle adjustment factor Q_H for $\phi_N = 22\tfrac{1}{2}°$.

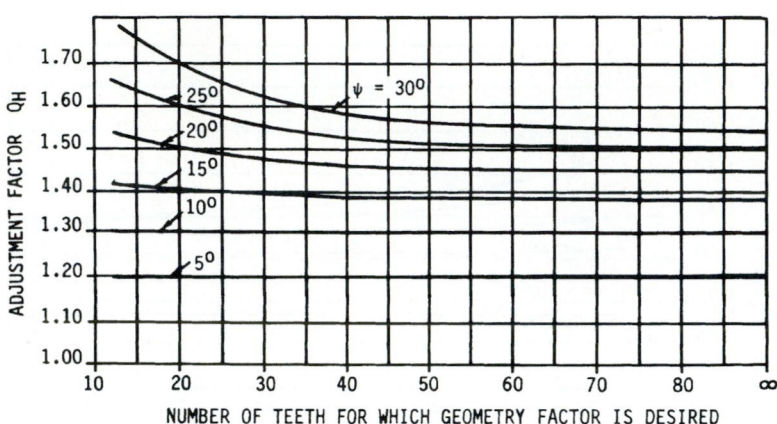

FIGURE 35.25 Helix-angle adjustment factor Q_H for $\phi_N = 25°$.

$$r_f = r_1 + r_{Te} \tag{35.58}$$

$$r_{Te} = r_T P_{nd} \tag{35.59}$$

For a hob or rack-shaped cutting tool,

$$R_o = \frac{d_{se}}{2} \tag{35.60}$$

For a pinion-shaped cutting tool,

$$R_o = \frac{d_{se} D_c P_{nd}}{2(d_{se} + D_c P_{nd})} \tag{35.61}$$

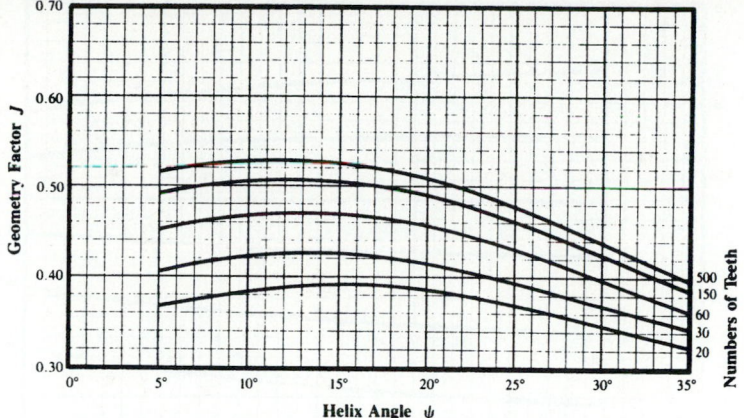

FIGURE 35.26 Geometry factor J for a $14\frac{1}{2}°$ normal-pressure-angle helical gear. These factors are for a standard addendum finishing hob as the final machining operation. See Fig. 35.27a. (*From AGMA 218.01.*)

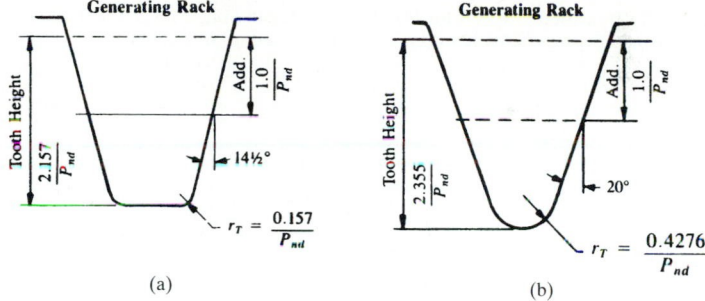

FIGURE 35.27 Generating racks. (*a*) For teeth of Fig. 35.26; (*b*) for teeth of Fig. 35.29. (*From AGMA 218.01.*)

$$D_e = \frac{DP_{nd}}{\cos^2 \psi} \tag{35.62}$$

$$A = \frac{D_o - D}{2} P_{nd} \tag{35.63}$$

$$D_{oe} = D_e + 2A \tag{35.64}$$

$$D_{be} = D_e \cos \phi_n \tag{35.65}$$

$$d_{se} = \frac{N_e}{P_s} \tag{35.66}$$

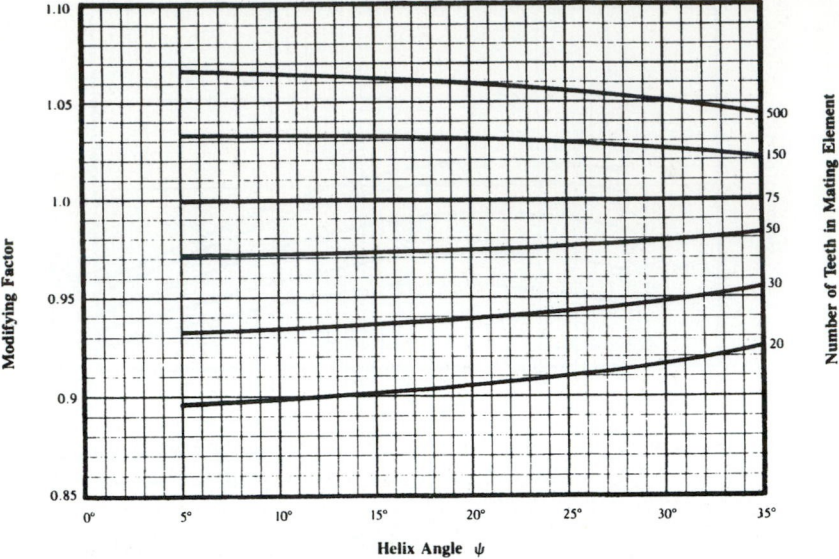

FIGURE 35.28 *J* factor multipliers for 14½° normal-pressure-angle helical gear. These factors can be applied to the *J* factor when other than 75 teeth are used in the mating element. *(From AGMA 218.01.)*

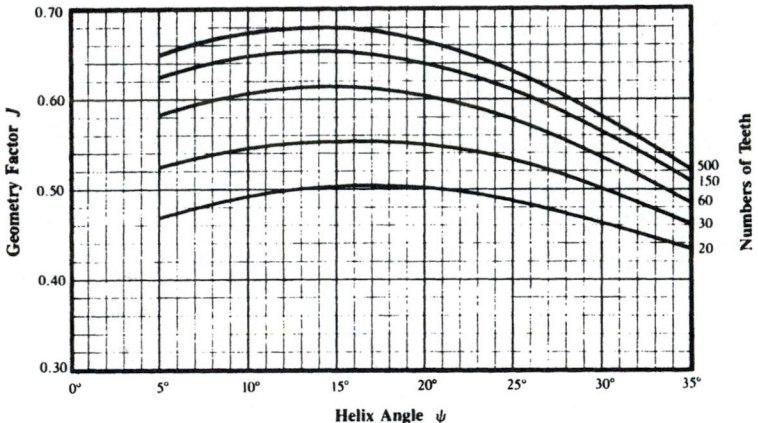

FIGURE 35.29 Geometry factor *J* for a 20° normal-pressure-angle helical gear. These factors are for standard addendum teeth cut with a full-fillet hob. See Fig. 35.27*b*. *(From AGMA 218.01.)*

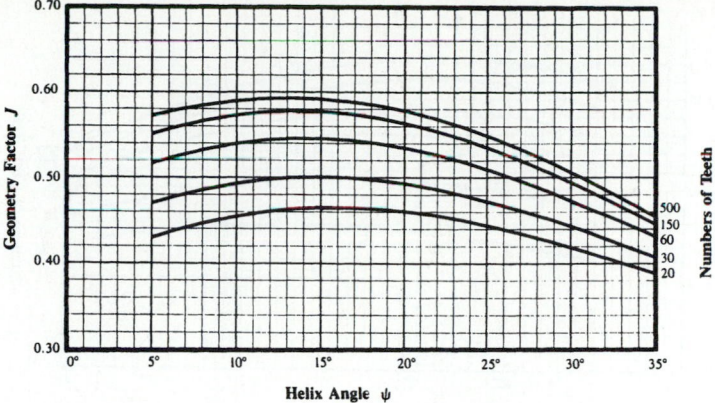

FIGURE 35.30 Geometry factor J for a 20° normal-pressure-angle helical gear. These factors are for standard addendum teeth cut with a finishing hob as the final machining operation. See Fig. 35.31a. *(From AGMA 218.01.)*

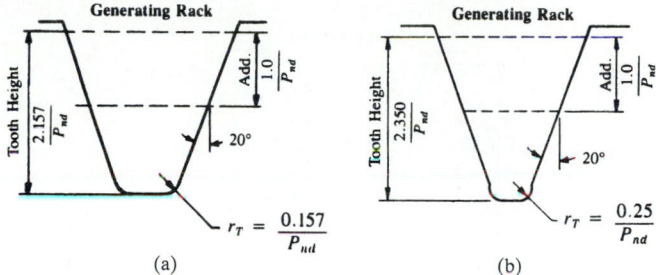

FIGURE 35.31 Generating racks. (a) For teeth of Fig. 35.30; (b) for teeth of Fig. 35.32. *(From AGMA 218.01.)*

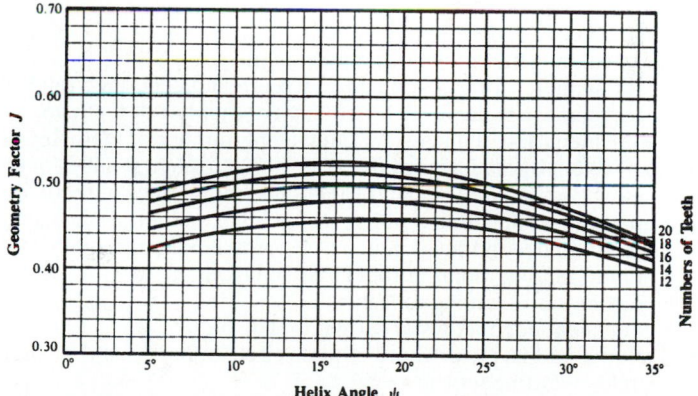

FIGURE 35.32 Geometry factor J for 20° normal-pressure-angle helical gear. These factors are for long-addendum (125 percent of standard) shaved teeth cut with a preshave hob. See Fig. 35.31b. *(From AGMA 218.01.)*

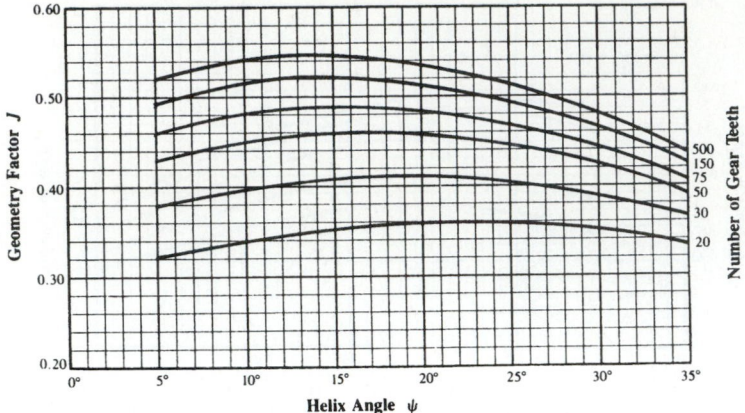

FIGURE 35.33　Geometry factor J for 20° normal-pressure-angle helical gear. These factors are for short-addendum teeth (75 percent of standard) cut with a preshave hob. See Fig. 35.34. *(From AGMA 218.01.)*

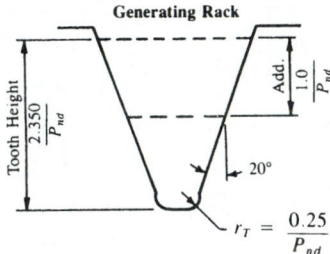

FIGURE 35.34　Generating rack for teeth of Fig. 35.33. *(From AGMA 218.01.)*

where　N_e = equivalent number of pinion teeth
　　　d_{se} = equivalent generating pitch diameter, in
　　　d_R = root diameter for actual number of teeth and generated pitch, in
　　　d_{Re} = equivalent root diameter for equivalent number of teeth, in
　　　d_{be} = equivalent base diameter for equivalent number of teeth, in
　　　d_e = equivalent operating pitch diameter for equivalent number of teeth, in
　　　d_{oe} = equivalent outside diameter for equivalent number of teeth, in
　　　a = operating addendum of pinion at 1 normal diametral pitch, in
　　　b = operating dedendum of pinion at 1 normal diametral pitch, in
　　　r_f = minimum fillet radius at root circle of layout, in
　　　r_T = edge radius of cutting tool, in
　　　r_{Te} = equivalent edge radius of cutting tool, in
　　　R_o = relative radius of curvature of pitch circle of pinion and pitch line or circle of cutting tool, in
　　　D_c = pitch diameter of pinion-shaped cutting tool, in
　　　D_e = equivalent operating pitch diameter of mating gear for equivalent number of teeth, in

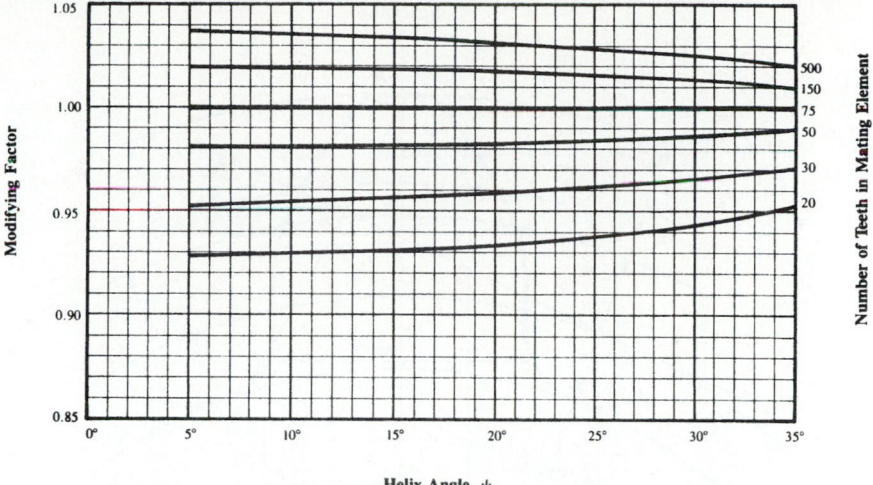

FIGURE 35.35 *J* factor multipliers for 20° normal-pressure-angle helical gears. These factors can be applied to the *J* factor when other than 75 teeth are used in the mating element. *(From Ref. [35.1].)*

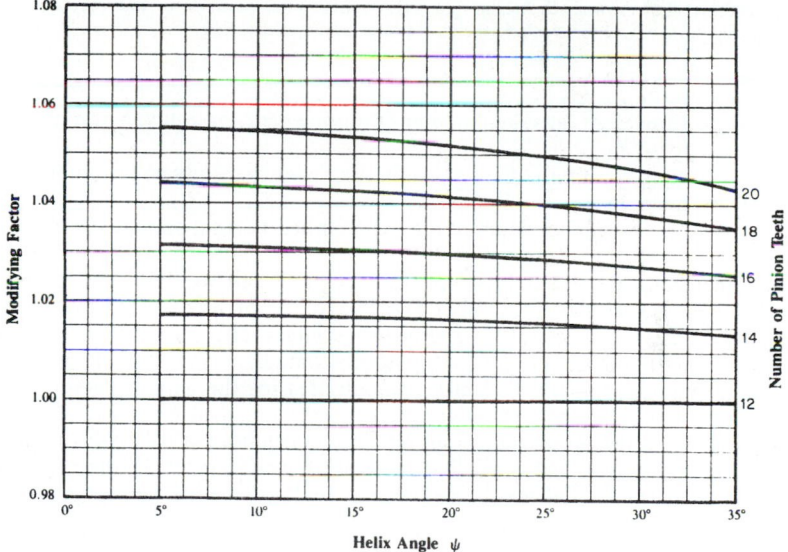

FIGURE 35.36 *J* factor multipliers for 20° normal-pressure-angle helical gears with short addendum (75 percent of standard). These factors can be applied to the *J* factor when other than 75 teeth are used in the mating element. *(From Ref. [35.1].)*

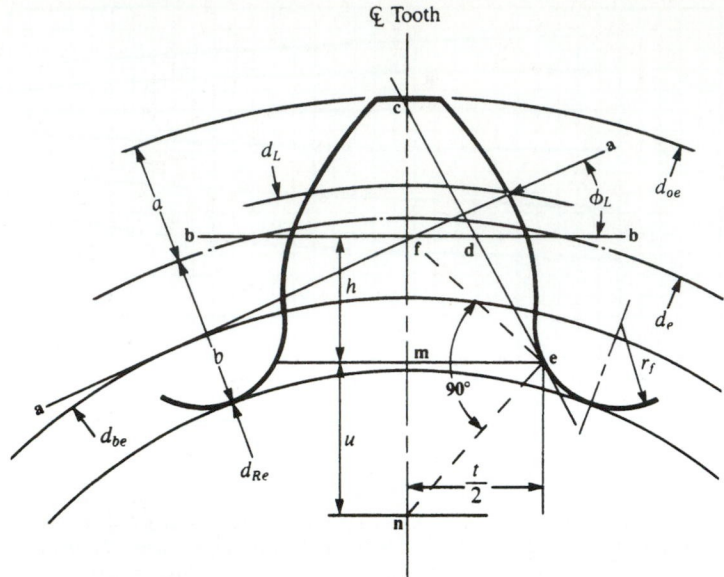

FIGURE 35.37 Tooth form factor with load at highest point of single-tooth contact (HPSTC) shown in the normal plane through the pitch point. Note that r_f occurs at the point where the trochoid meets the root radius. *(From Ref. [35.1].)*

D_{oe} = equivalent outside diameter of mating gear for equivalent number of teeth, in

D_{be} = equivalent base diameter of mating gear for equivalent number of teeth, in

A = operating addendum of mating gear at 1 normal diametral pitch, in

The dimensions defined by Eqs. (35.50) through (35.66) are then used to make a tooth-stress layout, as shown in either Fig. 35.37 or 35.38 as required by the face-contact ratio. That is, helical gears with low face-contact ratio ($m_F \leq 1.0$) are assumed to be loaded at the highest point of single-tooth contact; normal helical gears ($m_F >$ 1.0) use tip loading, and the C_h factor compensates for the actual loading on the oblique line.

To find Y from the above data, a graphical construction, as follows, is required. For low-contact-ratio helical gears (with $m_F \leq 1.0$), using Fig. 35.37, draw a line $\overline{aa}$ through point p, the intersection of diameter d_L with the profile, and tangent to the base diameter d_{be}:

$$d_L = 2 \left\{ \left[\sqrt{\left(\frac{d_e}{2}\right)^2 - \left(\frac{d_{be}}{2}\right)^2} + Z_d \right]^2 + \left(\frac{d_{be}}{2}\right)^2 \right\}^{1/2} \tag{35.67}$$

where Z_d = distance on line of action from highest point of single-tooth contact to pinion operating pitch circle, in inches, and so

$$Z_d = \pi \cos \phi_c - Z_e \tag{35.68}$$

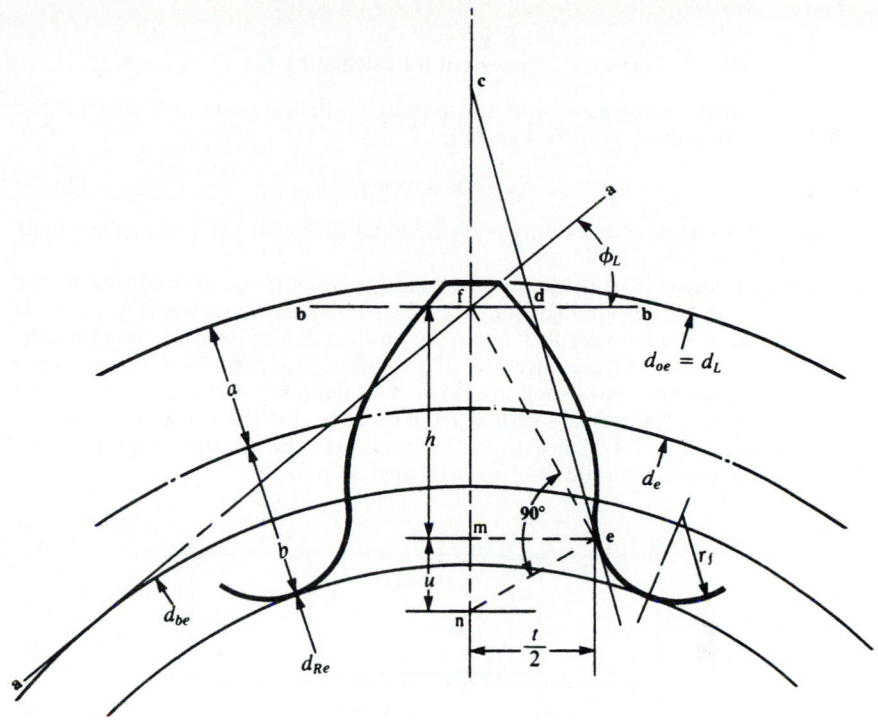

FIGURE 35.38 Tooth form factor layout with load at tooth tip; shown in normal plane through the pitch point. *(From Ref. [35.1].)*

Letting Z_e = distance on line of action from gear outside diameter to pinion operating pitch circle, in inches, we have

$$Z_e = \sqrt{\left(\frac{D_{oe}}{2}\right)^2 - \left(\frac{D_{be}}{2}\right)^2} - \sqrt{\left(\frac{D_e}{2}\right)^2 - \left(\frac{D_{be}}{2}\right)^2} \qquad (35.69)$$

For normal helical gears with $m_F > 1.0$, using Fig. 35.38, we find $D_L = d_{oe}$. Draw a line $\overline{aa}$ through point p, the tip of the tooth profile, and tangent to the base diameter d_{be}. Continue the layout for all gear types as follows:

Through point f, draw a line $\overline{bb}$ perpendicular to the tooth centerline. The included angle between lines $\overline{aa}$ and $\overline{bb}$ is load angle ϕ_L.

Draw line $\overline{cde}$ tangent to the tooth fillet radius r_f at e, intersecting line $\overline{bb}$ at d and the tooth centerline at c so that $\overline{cd} = \overline{de}$.

Draw line $\overline{fe}$.

Through point e, draw a line perpendicular to $\overline{fe}$, intersecting the tooth centerline at n.

Through point e, draw a line $\overline{me}$ perpendicular to the tooth centerline.

Measure the following in inches from the tooth layout:

$$\overline{mn} = u \qquad \overline{me} = \frac{t}{2}$$

and

$$\overline{mf} = h \qquad \text{(required for calculating } K_f\text{)}$$

The helix-angle factor K_ψ is set equal to unity for helical gears with $m_F \le 1.0$, but for helical gears with $m_F > 1.0$, it is given by

$$K_\psi = \cos \psi_o \cos \psi \qquad (35.70)$$

where ψ_o = helix angle at operating pitch diameter [from Eq. (35.13)] and ψ = helix angle at standard pitch diameter.

The helical factor C_h is the ratio of the root bending moment produced by the same intensity of loading applied along the actual oblique contact line (Fig. 35.39). If the face width of one gear is substantially larger than that of its mate, then full buttressing may exist on the wider face gear. If one face is wider than its mate by at least one addendum on *both* sides, then the value of C_h defined below may be increased by 10 percent only. The helical factor is given by either Eq. (35.71) for low-contact-ratio helical gears ($m_F \le 1.0$) or Eq. (35.72) for normal-contact-ratio ($m_F > 1.0$) helicals. These equations are valid only for helix angles up to 30°:

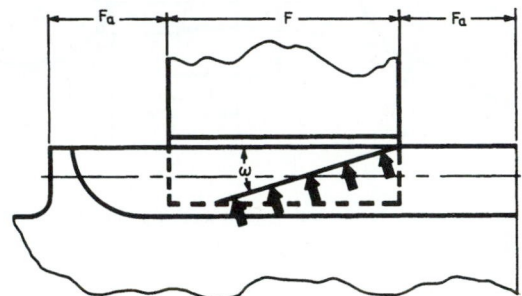

FIGURE 35.39 Oblique contact line. Full buttressing exists when $F_a \ge$ one addendum.

$$C_h = 1.0 \qquad (35.71)$$

$$C_h = \frac{1.0}{1 - [(\omega/100)\,(1 - \omega/100)]^{1/2}} \qquad (35.72)$$

where $\omega = \tan^{-1}(\tan \psi_o \sin \phi_{No})$ = inclination angle, deg
ψ_o = helix angle at operating pitch diameter, deg [Eq. (35.13)]
ϕ_{No} = operating normal pressure angle, deg [Eq. (35.14)]

The tooth form factor Y may now be calculated from Eq. (35.49).

The stress correction factor is the last item which must be calculated prior to finding a value for the bending geometry factor J. Based on photoelastic studies by Dolan and Broghamer, the empirical relations shown in Eqs. (35.73) through (35.76) were developed:

$$K_f = H + \left(\frac{t}{r_f}\right)^L \left(\frac{t}{h}\right)^m \qquad (35.73)$$

$$H = 0.18 - 0.008(\phi_{No} - 20) \tag{35.74}$$

$$L = H - 0.03 \tag{35.75}$$

$$m = 0.45 + 0.010(\phi_{No} - 20) \tag{35.76}$$

Elastic Coefficient C_p. This factor accounts for the elastic properties of various gear materials. It is given by Eq. (35.77). Table 35.4 provides values directly for C_p for various material combinations, for which Poisson's ratio is 0.30.

TABLE 35.4 Values of Elastic Coefficient C_p for Helical Gears with Nonlocalized Contact and for $\upsilon = 0.30$

	Gear material					
Pinion material	Steel	Malleable iron	Nodular iron	Cast iron	Aluminum bronze	Tin bronze
Steel, $E = 30$†	2300	2180	2160	2100	1950	1900
Malleable iron, $E = 25$	2180	2090	2070	2020	1900	1850
Nodular iron, $E = 24$	2160	2070	2050	2000	1880	1830
Cast iron, $E = 22$	2100	2020	2000	1960	1850	1800
Aluminum bronze, $E = 17.5$	1950	1900	1880	1850	1750	1700
Tin bronze, $E = 16$	1900	1850	1830	1800	1700	1650

†Modulus of elasticity E is in megapounds per square inch (Mpsi).

$$C_p = \left\{ \frac{1}{\pi[(1 - \nu_P^2)/E_P + (1 - \nu_G^2)/E_G]} \right\}^{1/2} \tag{35.77}$$

where ν_P, ν_G = Poisson's ratio for pinion and gear, respectively
E_P, E_G = modulus of elasticity for pinion and gear, respectively

Allowable Stresses s_{ac} ***and*** s_{at}. The allowable stresses depend on many factors, such as chemical composition, mechanical properties, residual stresses, hardness, heat treatment, and cleanliness. As a guide, the allowable stresses for helical gears may be obtained from Tables 35.5 and 35.6 or Figs. 35.40 and 35.41. Where a range of values is shown, the lowest values are used for general design. The upper values may be used only when the designer has certified that

1. High-quality material is used.
2. Section size and design allow maximum response to heat treatment.
3. Proper quality control is effected by adequate inspection.
4. Operating experience justifies their use.

Surface-hardened gear teeth require adequate case depth to resist the subsurface shear stresses developed by tooth contact loads and the tooth root fillet tensile stresses. But depths must not be so great as to result in brittle teeth tips and high residual tensile stress in the core.

TABLE 35.5 Allowable Bending Stress Numbers s_{at} and Contact Stress Numbers s_{ac} for a Variety of Materials

AGMA class	Commercial designation	Heat treatment	Minimum hardness		s_{at}, kpsi	s_{ac}, kpsi
			Surface	Core		
		Steel				
A-1 through A-5		Through-hardened and tempered (Fig. 35-40)	180 H_B and less		25–33	85–95
			240 H_B		31–41	105–115
			300 H_B		36–47	120–135
			360 H_B		40–52	145–160
			400 H_B		42–56	155–170
		Flame- or induction-hardened† with type A pattern (Fig. 35-45)	50–54 R_C		45–55	170–190
		Flame- or induction-hardened with type B pattern (Fig. 35-45)			22	
		Carburized† and case-hardened†	55 R_C		55–65	180–200
			60 R_C		55–70	200–225
	AISI 4140	Nitrided†‡	48 R_C	300 H_B	35–45	155–180
	AISI 4340	Nitrided†‡	46 R_C	300 H_B	36–47	150–175
	Nitralloy 135M	Nitrided†‡	60 R_C	300 H_B	38–48	170–195
	2¾% chrome	Nitrided†‡	54–60 R_C	350 H_B	55–65	155–216

ID	Material	Condition	Hardness		
Cast iron					
	20	As cast		5	50–60
	30	As cast	175 H_B	8.5	65–75
	40	As cast	200 H_B	13	75–85
Nodular (ductile) iron					
A-7-a	60-40-18	Annealed,	140 H_B	22–33	77–92
A-7-c	80-55-06	quenched, and	180 H_B	22–33	77–92
A-7-d	100-70-03	tempered	230 H_B	27–40	92–112
A-7-e	120-90-02		270 H_B	31–44	103–126
Malleable iron (pearlitic)					
A-8-c	45007		165 H_B	10	72
A-8-e	50005		180 H_B	13	78
A-8-f	53007		195 H_B	16	83
A-8-i	80002		240 H_B	21	94
Bronze					
Bronze 2	AGMA 2C	Sand-cast	Min. tensile strength 40 kpsi	5.7	30
Al/Br 3	ASTM B-148-52 Alloy 9C	Heat-treated	Min. tensile strength 90 kpsi	23.6	65

† The range of allowable stress numbers indicated corresponds to grade 1 and grade 2 steels. See tables 14-6 and 14-9 to 14-11 of source.
‡ The overload capacity of nitrided gears is low, since the shape of the effective SN curve is flat. The sensitivity to shock should be investigated before proceeding with the design.
SOURCE: **Ref. [35.1]**.

TABLE 35.6　Reliability Factors K_R and C_R

Factor, K_R or C_R	Probabilities, %	
	Success	Failure
1.50	99.99	0.01
1.25	99.90	0.10
1.00	99.00	1.00
0.85	90.00	10.00

The effective case depth for carburized and induction-hardened gears is defined as the depth below the surface at which the Rockwell C hardness has dropped to 50 R_C or to 5 points below the surface hardness, whichever is lower.

The values and ranges shown in Fig. 35.42 have had a long history of successful use for carburized gears and can be used as guides. For gearing in which maximum performance is required, detailed studies must be made of the application, loading, and manufacturing procedures, to obtain desirable gradients of both hardness and internal stress. Furthermore, the method of measuring the case, as well as the allowable tolerance in case depth, should be a matter of agreement between the customer and the manufacturer.

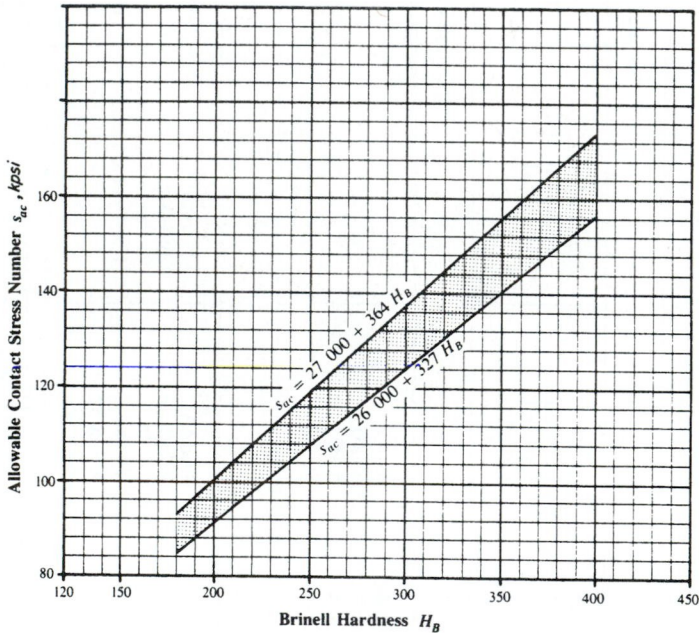

FIGURE 35.40　Allowable contact stress number s_{ac} for steel gears. Lower curve is maximum for grade 1 and upper curve is maximum for grade 2. *(From Ref. [35.1].)*

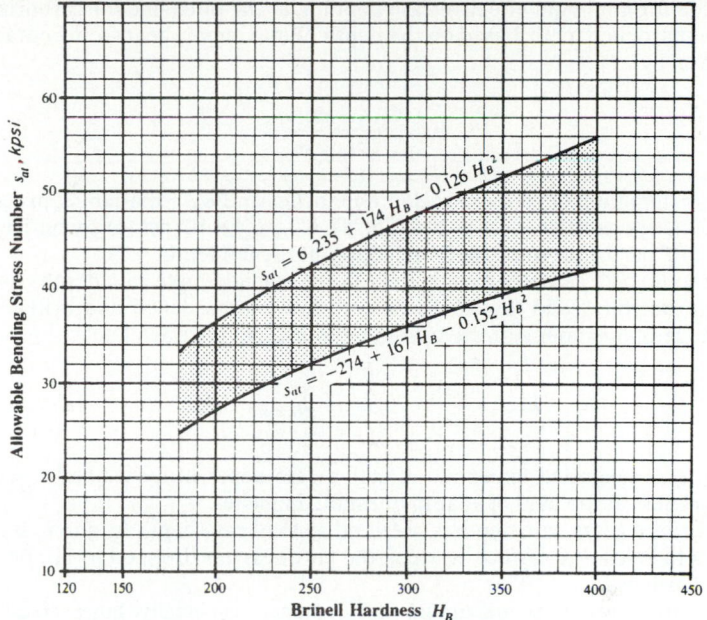

FIGURE 35.41 Allowable bending stress number s_{at} for steel gears. Lower curve is maximum for grade 1 and upper curve is maximum for grade 2. *(From Ref. [35.1].)*

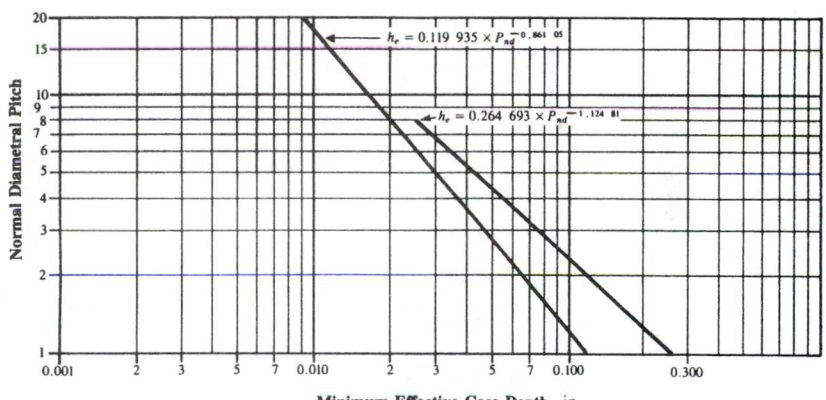

FIGURE 35.42 Effective case depth h_e for carburized gears based on normal diametral pitch. The effective case depth is defined as the depth of case which has a minimum hardness of 50 R_C. The total case depth to core carbon is about $1.5h_e$. The values and ranges shown on the case depth curves are to be used as guides. For gearing in which maximum performance is required, detailed studies must be made of the application, loading, and manufacturing procedures to obtain desirable gradients of both hardness and internal stress. Furthermore, the method of measuring the case as well as the allowable tolerance in case depth should be a matter of agreement between the customer and the manufacturer. *(From Ref. [35.1].)*

A guide for minimum effective case depth h_e at the pitch line for carburized and induction-hardened teeth, based on the depth of maximum shear from contact loading, is given by

$$h_e = \frac{C_G s_c d \sin \phi_o}{U_H \cos \psi_b} \tag{35.78}$$

where h_e = minimum effective case depth in inches and U_H = hardening process factor in pounds per square inch. In Eq. (35.78), $U_H = 6.4 \times 10^6$ psi for carburized teeth and 4.4×10^6 psi for tooth-to-tooth induction-hardened teeth.

You should take care when using Eq. (35.78) that adequate case depths prevail at the tooth root fillet, and that tooth tips are not overhardened and brittle. A suggested value of maximum effective case depth $h_{e,\max}$ at the pitch line is

$$h_{e,\max} = \frac{0.4}{P_d} \qquad \text{or} \qquad h_{e,\max} = 0.56 t_o \tag{35.79}$$

where $h_{e,\max}$ = suggested maximum effective case depth in inches and t_o = normal tooth thickness at top land of gear in question, in inches.

For nitrided gears, case depth is specified as total case depth h_c, and h_c is defined as the depth below the surface at which the hardness has dropped to 110 percent of the core hardness.

For gearing requiring maximum performance, especially large sizes, coarse pitches, and high contact stresses, detailed studies must be made of application, loading, and manufacturing procedures to determine the desirable gradients of hardness, strength, and internal residual stresses throughout the tooth.

A guide for minimum case depth for nitrided teeth, based on the depth of maximum shear from contact loading, is given by

$$h_c = \frac{C_G U_c s_c d \sin \phi_o}{(1.66 \times 10^7)(\cos \psi_b)} \tag{35.80}$$

where h_c = minimum total case depth in inches and U_c = core hardness coefficient, from Fig. 35.43.

If the value of h_c from Eq. (35.80) is less than the value from Fig. 35.44, then the minimum value from Fig. 35.44 should be used. The equation for the lower or left-hand curve in Fig. 35.44 is

$$h_c = (4.328\ 96)(10^{-2}) - P_{nd}(9.681\ 15)(10^{-3}) + P_{nd}^2(1.201\ 85)(10^{-3})$$
$$- P_{nd}^3(6.797\ 21)(10^{-5}) + P_{nd}^4(1.371)(10^{-6}) \tag{35.81}$$

The equation of the right-hand curve is

$$h_c = (6.600\ 90)(10^{-2}) - P_{nd}(1.622\ 24)(10^{-2}) + P_{nd}^2(2.093\ 61)(10^{-3})$$
$$- P_{nd}^3(1.177\ 55)(10^{-4}) + P_{nd}^4(2.331\ 60)(10^{-6}) \tag{35.82}$$

Note that other treatments of the subject of allowable gear-tooth bending recommend that the value obtained from Table 35.6 or Fig. 35.41 be multiplied by 0.70 for teeth subjected to reversed bending. This is not necessary within the context of this analysis, since the rim thickness factor K_b accounts for reversed bending.

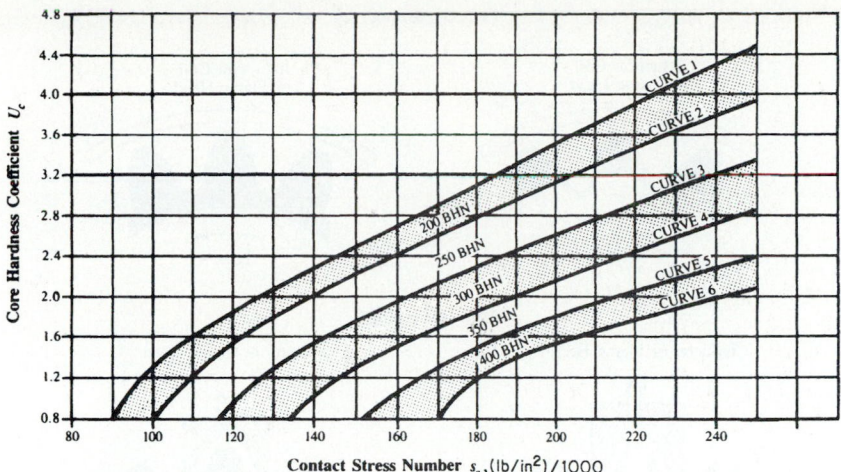

FIGURE 35.43 Core-hardness coefficient U_c as a function of the contact stress number s_c. The upper portion of the core-hardness bands yields heavier case depths and is for general design purposes; use the lower portion of the bands for high-quality material. *(From Ref. [35.1].)*

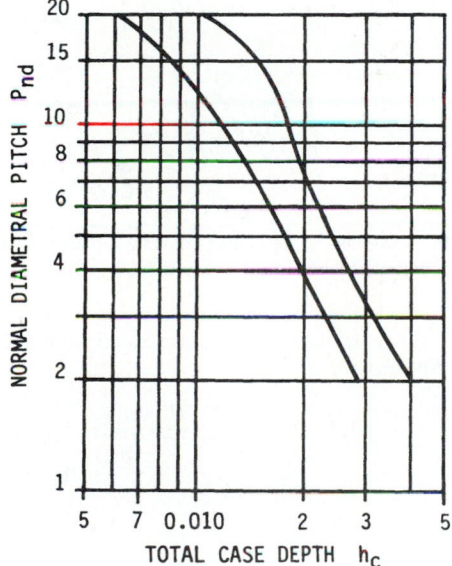

FIGURE 35.44 Minimum total case depth h_c for nitrided gears based on the normal diametral pitch. *(From Ref. [35.1].)*

Spin Hardening

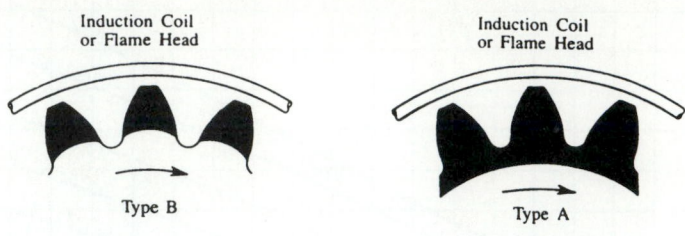

Type B

Type A

Flank Hardening

Type B

Type B

Flank and Root Hardening
Inductor or Flame Head

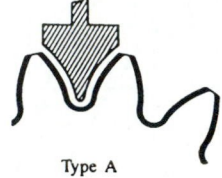

Type A

FIGURE 35.45 Variations in hardening patterns obtainable with flame or induction hardening. *(From Ref. [35.1].)*

For through-hardened gears, the yield stress at maximum peak stress should also be checked as defined by Eq. (35.83):

$$S_{ay}K_y \geq \frac{W_{t,\max}K_a}{K_v} \frac{P_d}{F} \frac{K_m}{K_f}$$ (35.83)

where $W_{t,\max}$ = peak tangential tooth load, lb
    K_a = application factor
    K_v = dynamic factor
    F = minimum net face width, in
    K_m = load-distribution factor
    K_f = stress correction factor
    K_y = yield strength factor
    s_{ay} = allowable yield strength number, psi (from Fig. 35.46)

The yield strength factor should be set equal to 0.50 for conservative practice or to 0.75 for general industrial use.

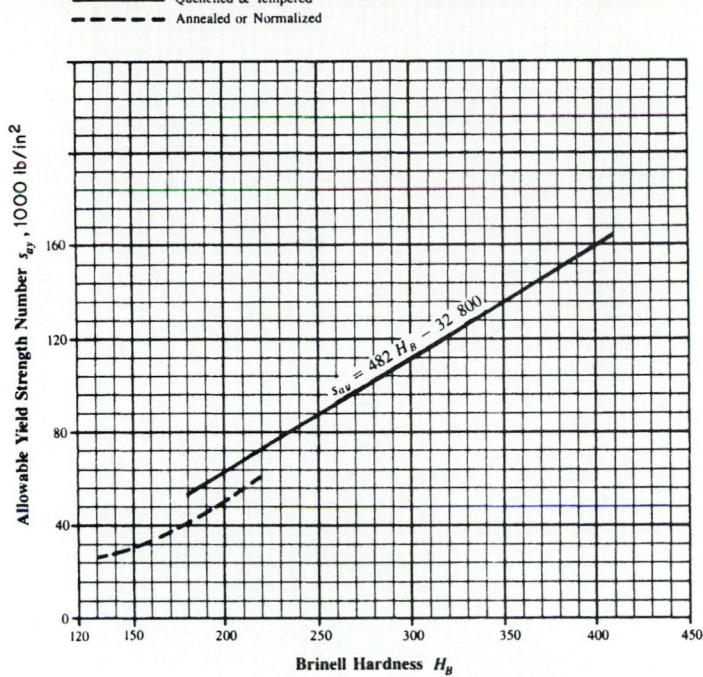

FIGURE 35.46 Allowable yield strength number s_{ay} for steel gears. *(From Ref. [35.1].)*

Hardness Ratio Factor C_H.   It is common practice in using through-hardened gear sets to utilize a higher hardness on the pinion than on the gear. The pinion typically sees many more cycles than the gear; thus a more economical overall design is obtained by balancing the surface durability and wear rate in this manner. Similarly, surface-hardened pinions may be used with through-hardened gears to provide improved overall capacity through the work-hardening effect which a "hard" pinion has on a "soft" gear. The hardness ratio factor adjusts the allowable stresses for this effect.

For through-hardened gear sets, C_H can be found from Fig. 35.47, while Fig. 35.48 provides values for surface-hardened pinions mating with through-hardened gears.

Life Factors K_L ***and*** C_L.   The allowable stresses shown in Tables 35.5 and 35.6 and Figs. 35.40 and 35.41 are based on 10 000 000 load cycles. The life factor adjusts the allowable stresses for design lives other than 10 000 000 cycles. A unity value for the life factor may be used for design lives beyond 10 000 000 cycles only when it is justified by experience with similar designs.

Insufficient specific data are available to define life factors for most materials. For steel gears, however, experience has shown that the curves shown in Figs. 35.49 and 35.50 are valid.

In utilizing these charts, care should be exercised whenever the product of K_L and s_{at} equals or exceeds s_{ay} as shown on Fig. 35.46, since this indicates that localized yielding may occur. For low-speed gears without critical noise vibration or transmis-

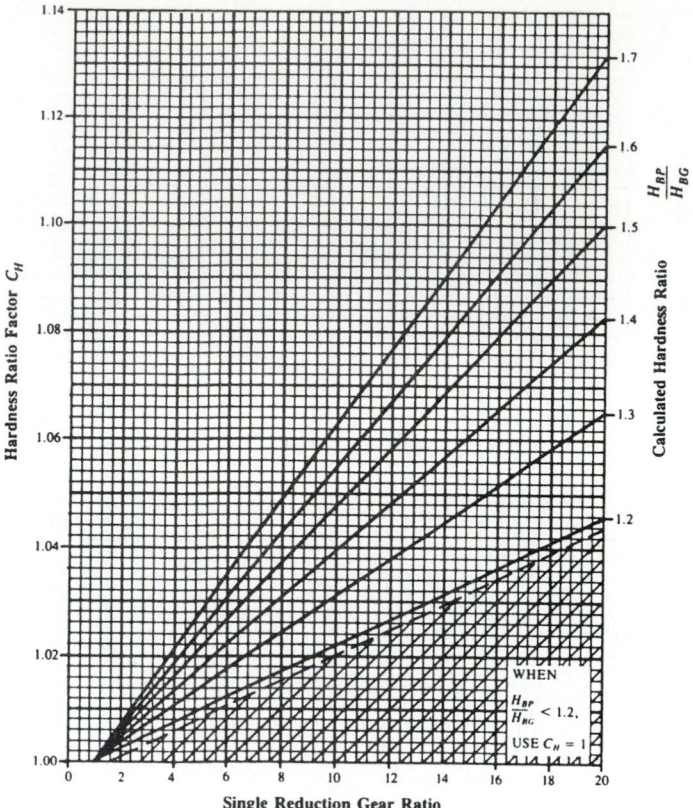

FIGURE 35.47   Hardness ratio factor C_H for through-hardened gears. In this chart, H_{BP} is the Brinell hardness of the pinion, and H_{BG} is the Brinell hardness of the gear. *(From Ref. [35.1].)*

sion accuracy requirements, local yielding may be acceptable, but it should be avoided in general.

Reliability Factors C_R and K_R.   The allowable stress levels are not absolute parameters. Rather, a specific probability of failure is associated with each allowable level. The values shown in Figs. 35.40 and 35.41 and Table 35.5 are based on a 99 percent probability of success (or a 1 percent probability of failure). This means that in a *large* population, at least 99 percent of the gears designed to a particular listed allowable stress will run for at least 10 000 000 cycles without experiencing a failure in the mode (that is, bending or durability) addressed.

In some cases it is desirable to design to higher or lower failure probabilities. Table 35.6 provides values for C_R and K_R which will permit the designer to do so. Before deciding on the reliability factor which is appropriate for a particular design, the analyst should consider what is meant by a "failure." In the case of a durability failure, a failure is said to have occurred when the first pit, or spall, is observed. Obviously a long time will elapse between the occurrence of a durability failure and the time at which the gear will cease to perform its normal power-transmission function.

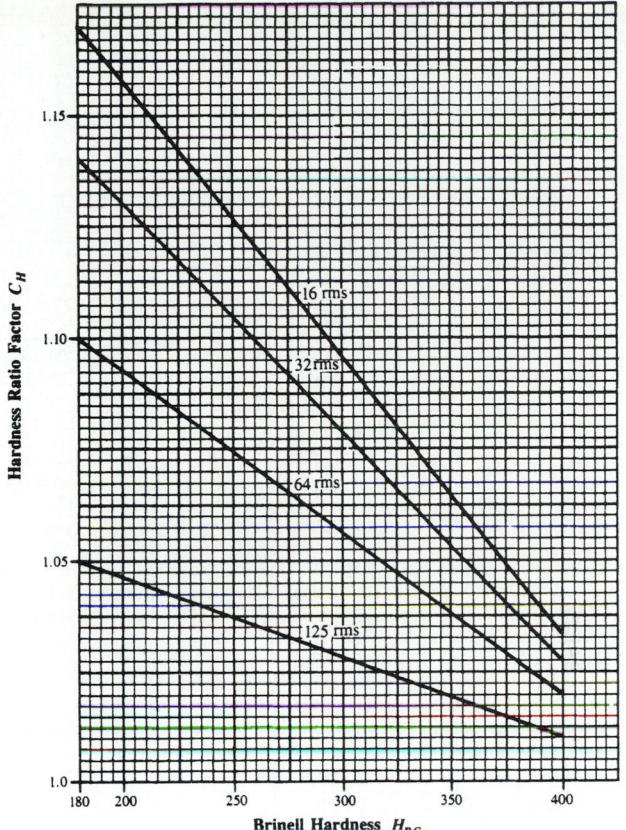

FIGURE 35.48 Hardness ratio factor C_H for surface-hardened teeth. The rms values shown correspond to the surface finish of the pinion f_p in microinches. *(From Ref. [35.1].)*

In the case of a bending failure, the appearance of a crack in the fillet area is the criterion. In most cases, and for most materials, the progression of this crack to the point at which a tooth or a piece of tooth fractures is rather quick. A bending failure will almost always progress to the point where function is lost much more rapidly than a durability failure. For this reason it is sometimes desirable to use a higher value for K_R than for C_R.

Because of the load sharing which occurs on most normal helical gears, a complete fracture of a full single tooth, as often occurs on a spur gear, is not usually the mode of failure on a helical gear. A certain redundancy is built into a helical gear, since initially only a piece of a tooth will normally fracture.

***Temperature Factors* C_T *and* K_T.**   At gear blank operating temperatures below 250°F and above freezing, actual operating temperature has little effect on the allowable stress level for steel gears; thus a temperature factor of unity is used. At higher or lower temperatures, the allowable stress levels are altered considerably. Unfortunately, few hard data are available to define these effects. At very low tem-

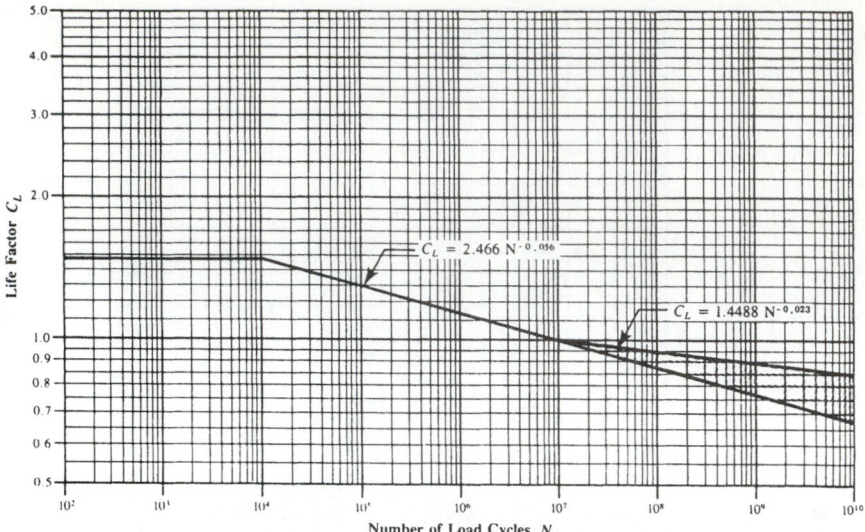

FIGURE 35.49 Pitting-resistance life factor C_L. This curve does not apply where a service factor C_{SF} is used. *Note:* The choice of C_L above 10^7 cycles is influenced by lubrication regime, failure criteria, smoothness of operation required, pitch line velocity, gear material cleanliness, material ductility and fracture toughness, and residual stress. *(From Ref. [35.1].)*

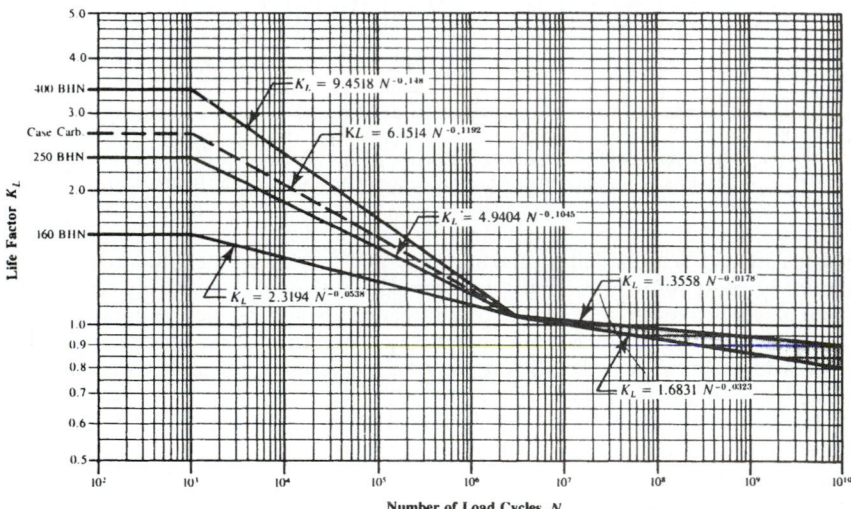

FIGURE 35.50 Bending-strength life factor K_L. This chart does not apply where a service factor K_{SF} is used. *Note:* The choice of K_L above 3×10^6 cycles is influenced by pitch line velocity, gear material cleanliness, residual stress, gear material ductility, and fracture toughness. *(From Ref. [35.1].)*

peratures, the impact resistance and fracture toughness of most materials are reduced; thus special care must be exercised in such designs if nonuniform loading is expected. A temperature factor greater than unity should be used in such cases. Although no specific data are available, a value between 1.25 and 1.50 is recommended for gears which must transmit full power between 0 and −50°F.

At high temperatures, most materials experience a reduction in hardness level. Nonmetallic gears are not ordinarily used at high temperatures; thus our comments are restricted to steel gearing. The temperature factor should be chosen on the basis of the hot hardness curve for the particular material in use. That is, the temperature factor is equal to the allowable stress at room-temperature hardness divided by the allowable stress at the hardness corresponding to the higher temperature. For information related to typical trends, Fig. 35.51 shows the hardness-temperature characteristics for two gear steels (AISI 9310 and VASCO-X2). Two typical bearing steels (M-50 and SAE 52100) are also shown for reference purposes.

Once the strength and durability analyses have been completed, the wear and scoring resistance of the gears must be defined. Wear (see Chap. 6) is usually a concern only for relatively low-speed gears, whereas scoring is a concern only for relatively high-speed gears.

Wear. Gear-tooth wear is a very difficult phenomenon to predict analytically. Fortunately, it is not a major problem for most gear drives operating in the moderate- to

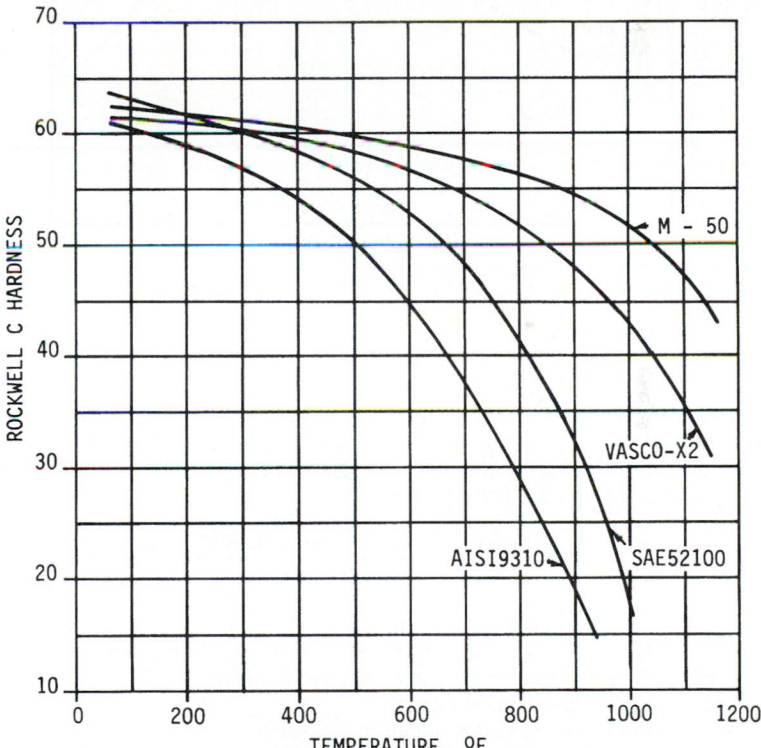

FIGURE 35.51 Hardness as a function of temperature for several steels.

high-speed range. In the case of low-speed gears, however, not only is wear a significant problem, but also it can be the limiting factor in defining the load capacity of the mesh.

In low-speed gear drives, the film which separates the mating tooth surfaces is insufficient to prevent metal-to-metal contact; thus wear occurs. In higher-speed gears, the film becomes somewhat thicker, and gross contact of the mating surfaces is prevented. Indeed, grinding lines are still visible on many aircraft gears after hundreds of hours of operation. The type of surface distress which will occur in a gear set is dependent, to a certain extent, on the pitchline velocity. As shown in Fig. 35.52, wear predominates in the lower-speed range, while scoring rules the upper-speed range. In the midrange, pitting controls the gear life.

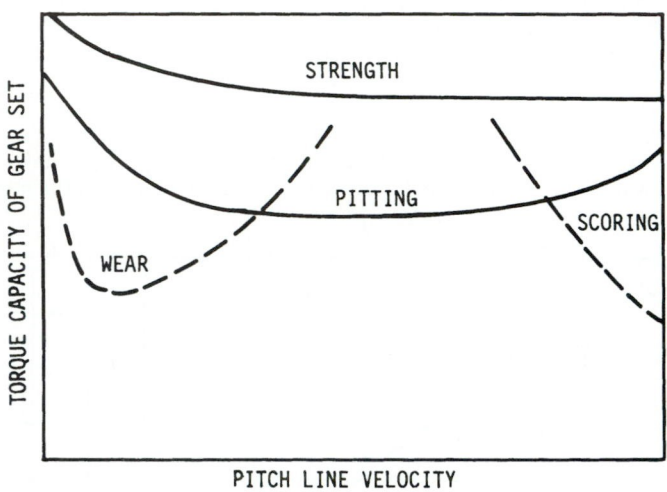

FIGURE 35.52   Gear distress as a function of pitch line velocity.

The elastohydrodynamic (EHD) film thickness can provide some guidance in the evaluation of the wear potential of a gear set. Care must be used in the application of these methods, since the existing data are far from complete and there are many instances of contradictory results. One of the simplest approaches is due to Dowson; see Ref. [35.7] and Chap. 25. The equation is

$$\frac{h}{R'} = \frac{(4.46 \times 10^{-5})(\alpha E')^{0.54}[\mu_o u/(E'R')]^{0.70}}{[w/(E'R')]^{0.13}} \tag{35.84}$$

where h = calculated minimum film thickness, in
   R' = relative radius of curvature in transverse plane at pitch point, in
   α = lubricant pressure-viscosity coefficient, in^2/lb
   E' = effective elastic modulus, psi
   μ_o = sump lubricant viscosity, centipoise (cP)
   u = rolling velocity in transverse plane, inches per second (in/s)
   w = load per unit length of contact, lb/in

Wellauer and Holloway ([35.8]) present a nomograph to compute the film thickness at the pitch point; but this nomograph is quite detailed and is not included here.

The parameter of interest in our discussion is not the film thickness itself, but rather the ratio of the film thickness to the relative surface roughness. This ratio is defined as the *specific film thickness* and is given by

$$\lambda = \frac{h}{S'} \tag{35.85}$$

The relative surface roughness [root-mean-square (rms)] is given by

$$S' = \frac{S_P + S_G}{2} \tag{35.86}$$

Typical values for various gear manufacturing processes are shown in Table 35.7.

TABLE 35.7 Tooth Surface Texture in the As-Finished Condition

Finish method	Surface texture in microinches (rms)	
	Range	Typical
Hobbed	30–80	50
Shaved	10–45	35
Lapped	20–200	93
Lapped and run in	20–100	53
Ground (soft)	5–35	25
Ground (hard)	5–35	15
Honed and polished	4–15	5

Once the specific film thickness has been determined, the probability of surface distress occurring can be determined through the use of Fig. 35.53.

Although the data presented thus far can be quite useful, several factors must be kept in mind in applying them to actual design. Most of the experimental data on which this information is based were obtained from through-hardened gear sets operating with petroleum-based oils. Gears operating with synthetic oils appear able to operate successfully at film thicknesses much less than those predicted by this analysis. The same is true for case-hardened gears of 59 R_C and higher hardness. The results may be further altered by the use of friction modifiers or EP additives in the oil. Finally, wear, of and by itself, is not necessarily a failure. In many cases, wear is an acceptable condition; it is simply monitored until it reaches some predetermined level, at which time the gears are replaced.

Perhaps the most useful application for this analysis is as a comparative, relative rating tool, rather than as an absolute design criterion.

The occurrence of wear is difficult to predict, but the rate of wear is even more so. Equation (35.87) may be useful as a guide in predicting wear, but its accuracy has not been rigorously verified:

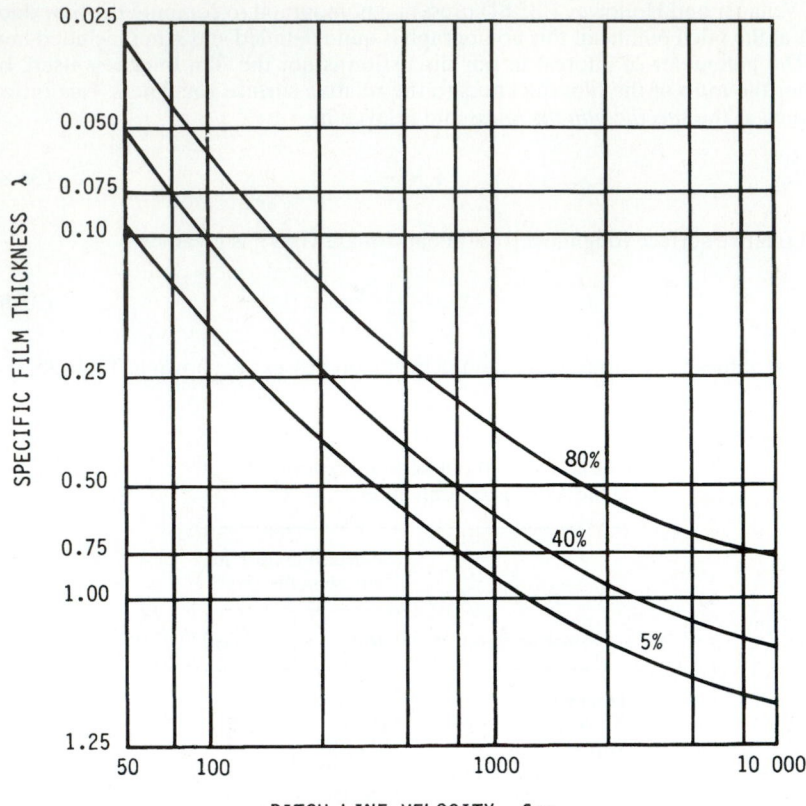

FIGURE 35.53 Surface-distress probability chart as a function of pitch line velocity and specific film thickness. Curves represent 80, 40, and 5 percent probability of distress. The region above the 80 percent line is unsatisfactory; the region below the 5 percent line is good.

$$q = \frac{KW_t\, n_T}{FS_y} \qquad\qquad (35.87)$$

where q = wear, in
        n_T = number of cycles
        S_y = yield strength of gear material, psi
        K = factor from Eq. (35.88)

and

$$3.1 \geq K\lambda^{1.645} \times 10^9 \geq 1.8 \qquad\qquad (35.88)$$

In applying these equations, greater emphasis should be placed on the trend indicated than on the absolute value of the numbers. For example, a new design might be compared with an existing similar design for which the wear characteristics have been established. This could be accomplished by calculating the q value for each by

Eqs. (35.87) and (35.88) and then comparing them, rather than looking at absolute values of either. The ratio of the two q values is far more accurate than the absolute value of either.

Scoring. Very few data concerning the scoring behavior of gears are available in an easily usable form. Scoring is normally a problem for heavily loaded, high-speed steel gears. The exact mechanism by which scoring occurs is not yet fully understood.

At high speeds, the calculated film thickness is often quite large. Yet a wearlike failure mode sometimes occurs. Under high-speed conditions, the sliding motion of one gear tooth on another may create instantaneous conditions of temperature and pressure which destroy the film of oil separating the tooth flanks. When this occurs, the asperities on the surfaces of the mating teeth instantaneously weld. As the gears continue to rotate, these welds break and drag along the tooth flanks, causing scratches, or "score" marks, in the direction of sliding. If the damage which occurs is very slight, it is often referred to as *scuffing* or *frosting*. In some cases, light frosting may heal over and not progress; however, scoring is generally progressively destructive. Though never a catastrophic failure itself, scoring destroys the tooth surface, which leads to accelerated wear, pitting, and spalling. If scoring is allowed to progress unchecked, tooth fracture may ultimately occur.

Note that scoring is not a fatigue phenomenon; that is, its occurrence is not time-dependent. In general, if scoring does not occur within 15 to 25 minutes (min) at a certain operating condition, usually it will not occur *at that condition* at all. Only a change in operating condition, and not the accumulation of cycles, will cause scoring.

A theory known as the *critical-temperature theory,* originally proposed by Harmen Blok, is usually used in the evaluation of scoring hazard for a set of helical gears.

If we consider a simple analogy, the concept of critical temperature will become clear. Consider the old method of making fire by rubbing two sticks together. If the sticks are held together with only light pressure and/or they are rubbed slowly, they will simply wear. If, however, the pressure is increased and the sticks are rubbed more rapidly, then the temperature at the mating surfaces will increase. If the pressure (load) and the rubbing speed (sliding velocity) are progressively increased, eventually the sticks will ignite. At the point of ignition, the sticks have reached their *critical* temperature. Quite obviously, the critical temperature will vary with the type of wood, its moisture content, and other factors.

In a similar manner, as gear-tooth sliding velocity and load are increased, eventually a point will be reached at which the temperature at the conjunction attains a critical value, and then the film separating the tooth flanks will be destroyed. At this point the teeth are in metal-to-metal contact, and instantaneous welding of the surface asperities occurs. The continued rotation of the mesh rips apart these microscopic welds and produces the scored appearance from which this failure derives its name. The critical temperature varies with the type of gear material, surface hardness, surface finish, type and viscosity of oil, additives in the oil, etc. When the film is destroyed, it is sometimes referred to as *flashing;* thus the parameter used to evaluate this condition has come to be known as the *flash* temperature. When the flash temperature reaches its *critical* value, failure by scoring will occur. Note that the flash temperature referred to here is not related in any way to the flash point of the oil; and the oil flash point shown on some manufacturers' specification sheets is in no way related to the allowable flash temperature discussed here.

Many refinements have been made to Blok's original theory, and it is currently accepted as the best method available for evaluating scoring resistance for spur, helical, and bevel gears. Reference [35.9] presents a method of analysis for steel spur and helical gears based on Blok's method. The scoring hazard is evaluated by calculating a flash temperature rise ΔT_{Fi}. The flash temperature rise is added to the gear blank temperature T_B and compared with the allowable tooth flash temperature for the particular material and lubricant combination being used.

The flash temperature rise is given by

$$\Delta T_{Fi} = \left(\frac{W_{ti}C_aC_m}{FC_v}\right)^{0.75}\left(\frac{n_P^{0.5}}{P_d^{0.25}}\right)(\mu Z_{ti})\left(\frac{50}{50-S'}\right) \tag{35.89}$$

where T_{Fi} = flash temperature rise at ith contact point along line of action, °F
W_{ti} = tangential tooth load at ith contact point, lb
F = net minimum face width, in
C_a = application factor
C_m = load-distribution factor
C_v = dynamic factor
n_P = pinion speed, revolutions per minute (r/min)
P_d = transverse diametral pitch
S' = relative surface roughness, Eq. (35.86)
Z_{ti} = scoring geometry factor at ith contact point along line of action

The factors C_a, C_v, and C_m are the same as those used in the durability formula [Eq. (35.17)].

The scoring geometry factor is given by

$$Z_{ti} = \frac{0.2917[\rho_{Pi}^{1/2} - (N_P\rho_{Gi}/N_G)^{1/2}]P_d^{1/4}}{(\cos\phi_i)^{0.75}[\rho_{Pi}\rho_{Gi}/(\rho_{Pi}+\rho_{Gi})]^{0.25}} \tag{35.90}$$

where ρ_{Pi}, ρ_{Gi} = radius of curvature of pinion and gear, respectively, at ith contact point, in
N_P, N_G = tooth numbers of pinion and gear, respectively
P_d = transverse diametral pitch
ϕ_i = pressure angle at ith contact point, deg

The tooth flash temperature is then calculated by

$$T_{Fi} = T_B + \Delta T_{Fi} \tag{35.91}$$

TABLE 35.8 Allowable Flash Temperatures for Some Gear Materials and for Spur and Helical Gears

The surface hardness is 60 R_C for all materials listed.

Gear material	Oil type	Allowable flash temperature, °F
AISI 9310	MIL-L-7808	295
	MIL-L-23699	295
	XAS 2354	335†
VASCO-X2	MIL-L-7808	350
	MIL-L-23699	350
	XAS 2354	375†

†Conservative estimate based on limited current data.

In most cases the blank temperature will be very close to the oil inlet temperature. Thus, unless the actual blank temperature is known, the oil inlet temperature may be used for T_B. Table 35.8 gives allowable values of the total flash temperature.

Equations (35.89) through (35.91) refer to the ith contact point. In utilizing these equations, the entire line of contact should be examined on a point-by-point basis to define the most critical contact point. Depending on the pitch of the tooth, 10 to 25 divisions should be adequate. For hand calculations, this could be quite burdensome. A quick look at the highest and lowest points of single-tooth contact (based on a transverse-plane slice of the helical set) will provide a reasonable approximation.

The range of materials and oils shown in Table 35.8 is limited. Generally, scoring is a problem only in high-speed, high-load applications.

The most likely applications to be affected are aerospace types. This being the case, the material choice is limited to those shown, and usually either MIL-L-23699 or MIL-L-7808 oil is used. Some of the new XAS-2354 oils will provide much improved scoring resistance, but hard data are not presently available.

REFERENCES

35.1 ANSI/AGMA 2001-B88, 1988, "Fundamental Rating Factors and Calculation Methods for Involute Spur and Helical Gear Teeth."

35.2 "Design Guide for Vehicle Spur and Helical Gears," AGMA publ. 170.

35.3 *Gear Handbook,* vol. 1, *Gear Classification, Materials, and Measuring Methods for Unassembled Gears,* AGMA publ. 390.

35.4 Raymond J. Drago, "Results of an Experimental Program Utilized to Verify a New Gear Tooth Strength Analysis," AGMA publ. 229.27, October 1983.

35.5 Raymond J. Drago, "An Improvement in the Conventional Analysis of Gear Tooth Bending Fatigue Strength," AGMA publ. 229.24, October 1982.

35.6 R. Errichello, "An Efficient Algorithm for Obtaining the Gear Strength Geometry Factor on a Programmable Calculator," AGMA publ. 139.03, October 1981.

35.7 D. Dowson, "Elastohydrodynamic Lubrication: Interdisciplinary Approach to the Lubrica-tion of Concentrated Contacts," NASA SP-237, 1970.

35.8 E. J. Wellauer and G. Holloway, "Application of EHD Oil Film Theory to Industrial Gear Drives," ASME paper no. 75PTG-1, 1975.

35.9 "Information Sheet—Gear Scoring Design Guide for Aerospace Spur and Helical Involute Gear Teeth," AGMA publ. 217.

CHAPTER 36
WORM GEARING

K. S. Edwards, Ph.D.
Professor of Mechanical Engineering
University of Texas at El Paso
El Paso, Texas

GLOSSARY OF SYMBOLS

b_G	Dedendum of gear teeth
C	Center distance
d	Worm pitch diameter
d_o	Outside diameter of worm
d_R	Root diameter of worm
D	Pitch diameter of gear in central plane
D_b	Base circle diameter
D_o	Outside diameter of gear
D_t	Throat diameter of gear
f	Length of flat on outside diameter of worm
h_k	Working depth of tooth
h_t	Whole depth of tooth
L	Lead of worm
m_G	Gear ratio $= N_G/N_W$
m_o	Module, millimeters of pitch diameter per tooth (SI use)
m_p	Number of teeth in contact
n_w	Rotational speed of worm, r/min
n_G	Rotational speed of gear, r/min

N_G Number of teeth in gear

N_W Number of threads in worm

p_n Normal circular pitch

p_x Axial circular pitch of worm

P Transverse diametral pitch of gear, teeth per inch of diameter

W Force between worm and gear (various components are derived in the text)

λ Lead angle at center of worm, deg

ϕ_n Normal pressure angle, deg

ϕ_x Axial pressure angle, deg, at center of worm

36.1 INTRODUCTION

Worm gears are used for large speed reduction with concomitant increase in torque. They are limiting cases of helical gears, treated in Chap. 35. The shafts are normally perpendicular, though it is possible to accommodate other angles. Consider the helical-gear pair in Fig. 36.1a with shafts at 90°.

The lead angles of the two gears are described by λ (lead angle is 90° less the helix angle). Since the shafts are perpendicular, $\lambda_1 + \lambda_2 = 90°$. If the lead angle of gear 1 is made small enough, the teeth eventually wrap completely around it, giving the appearance of a screw, as seen in Fig. 36.1b. Evidently this was at some stage taken to resemble a *worm,* and the term has remained. The mating member is called simply the *gear,* sometimes the *wheel.* The helix angle of the gear is equal to the lead angle of the worm (for shafts at 90°).

The worm is always the driver in speed reducers, but occasionally the units are used in reverse fashion for speed increasing. Worm-gear sets are self-locking when the gear cannot drive the worm. This occurs when the tangent of the lead angle is less than the coefficient of friction. The use of this feature in lieu of a brake is not rec-

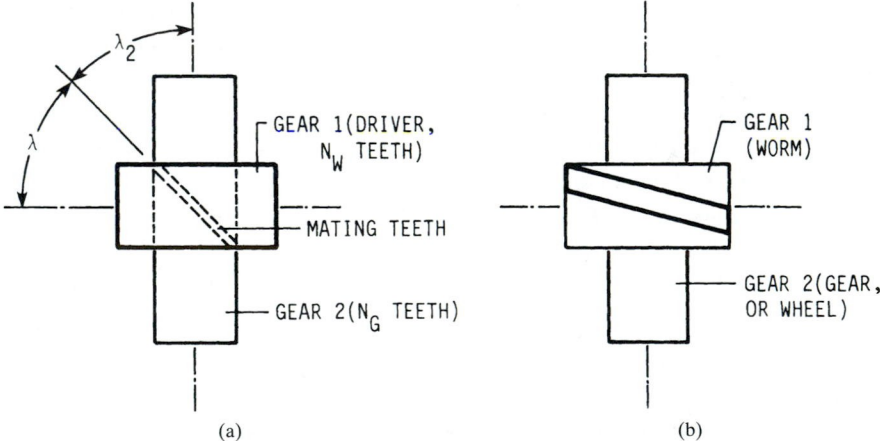

FIGURE 36.1   (*a*) Helical gear pair; (*b*) a small lead angle causes gear one to become a worm.

FIGURE 36.2 Photograph of a worm-gear speed reducer. Notice that the gear partially wraps, or envelopes, the worm. *(Cleveland Worm and Gear Company.)*

ommended, since under running conditions a gear set may not be self-locking at lead angles as small as 2°.

There is only point contact between helical gears as described above. Line contact is obtained in worm gearing by making the gear envelop the worm as in Fig. 36.2; this is termed a *single-enveloping gear set,* and the worm is cylindrical. If the worm and gear envelop each other, the line contact increases as well as the torque that can be transmitted. The result is termed a *double-enveloping gear set.*

The minimum number of teeth in the gear and the reduction ratio determine the number of threads (teeth) for the worm. Generally, 1 to 10 threads are used. In special cases a larger number may be required.

36.2 KINEMATICS

In specifying the pitch of worm-gear sets, it is customary to state the axial pitch p_x of the worm. For 90° shafts this is equal to the transverse circular pitch of the gear. The advance per revolution of the worm, termed the lead L, is

$$L = p_x N_W$$

This and other useful relations result from consideration of the developed pitch cylinder of the worm, seen in Fig. 36.3. From the geometry, the following relations can be found:

$$d = \frac{N_W p_n}{\pi \sin \lambda} \tag{36.1}$$

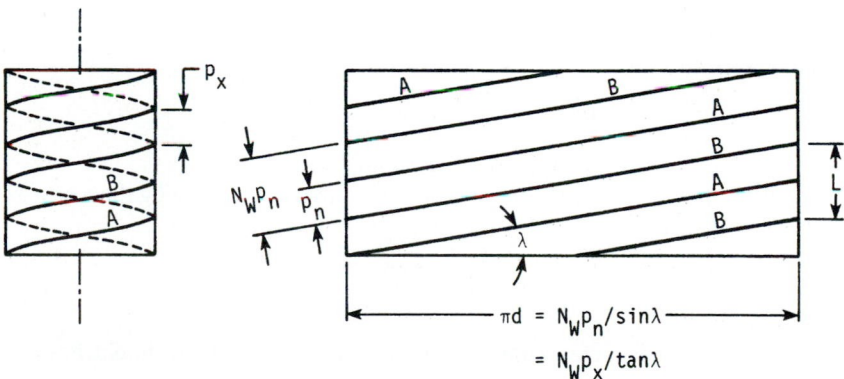

FIGURE 36.3 Developed pitch cylinder of worm.

$$d = \frac{N_W p_x}{\pi \tan \lambda} \tag{36.2}$$

$$\tan \lambda = \frac{L}{\pi d} = \frac{N_W p_x}{\pi d} \tag{36.3}$$

$$p_x = \frac{p_n}{\cos \lambda} \tag{36.4}$$

$$D = \frac{p_x N_G}{\pi} = \frac{N_G p_n}{\pi \cos \lambda} \tag{36.5}$$

From Eqs. (36.1) and (36.5), we find

$$\tan \lambda = \frac{N_W D}{N_G d} = \frac{1}{m_G} \frac{D}{d} \tag{36.6}$$

The center distance C can be derived from the diameters

$$C = \frac{p_n N_W}{2\pi} \left(\frac{m_G}{\cos \lambda} + \frac{1}{\sin \lambda} \right) \tag{36.7}$$

which is sometimes more useful in the form

$$\frac{m_G}{\cos \lambda} + \frac{1}{\sin \lambda} = \begin{cases} \dfrac{2\pi C}{p_n N_W} & \text{U.S. customary units} \\[2ex] \dfrac{2C}{m_o N_W \cos \lambda} & \text{SI units} \\[2ex] \dfrac{2C}{d \sin \lambda} & \text{either} \end{cases} \tag{36.8}$$

For use in the International System (SI), recognize that

$$\text{Diameter} = Nm_o = \frac{Np_x}{\pi}$$

so that the substitution

$$p_x = \pi m_o$$

will convert any of the equations above to SI units.

The pitch diameter of the gear is measured in the plane containing the worm axis and is, as for spur gears,

$$D = \frac{N_G p_x}{\pi} \tag{36.9}$$

The worm pitch diameter is unrelated to the number of teeth. It should, however, be the same as that of the hob used to cut the worm-gear tooth.

36.3 VELOCITY AND FRICTION

Figure 36.4 shows the pitch line velocities of worm and gear. The coefficient of friction between the teeth μ is dependent on the sliding velocity. Representative values of μ are charted in Fig. 36.5. The friction has importance in computing the gear set efficiency, as will be shown.

36.4 FORCE ANALYSIS

If friction is neglected, then the only force exerted by the gear on the worm will be W, perpendicular to the mating tooth surface, shown in Fig. 36.6, and having the three components W^x, W^y, and W^z. From the geometry of the figure,

$$W^x = W \cos \phi_n \sin \lambda$$

$$W^y = W \sin \phi_n \qquad (36.10)$$

$$W^z = W \cos \phi_n \cos \lambda$$

In what follows, the subscripts W and G refer to forces *on* the worm and the gear. The component W^y is the separating, or radial, force for both worm and gear (opposite in direction for the gear). The tangential force is W^x on the worm and W^z on the gear. The axial force is W^z on the worm and W^x on the gear. The gear forces are opposite to the worm forces:

$$W_{W_t} = -W_{G_a} = W^x$$

$$W_{W_r} = -W_{G_r} = W^y \qquad (36.11)$$

$$W_{W_a} = -W_{G_t} = W^z$$

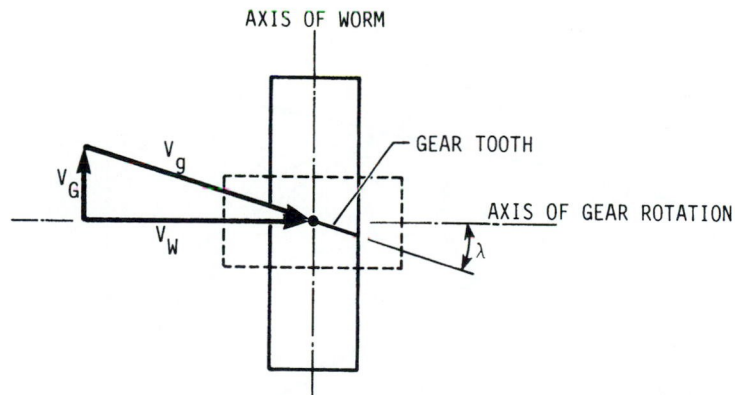

FIGURE 36.4   Velocity components in a worm-gear set. The sliding velocity is $V_S = (V_W^2 + V_G^2)^{1/2} = \dfrac{V_W}{\cos \lambda}$.

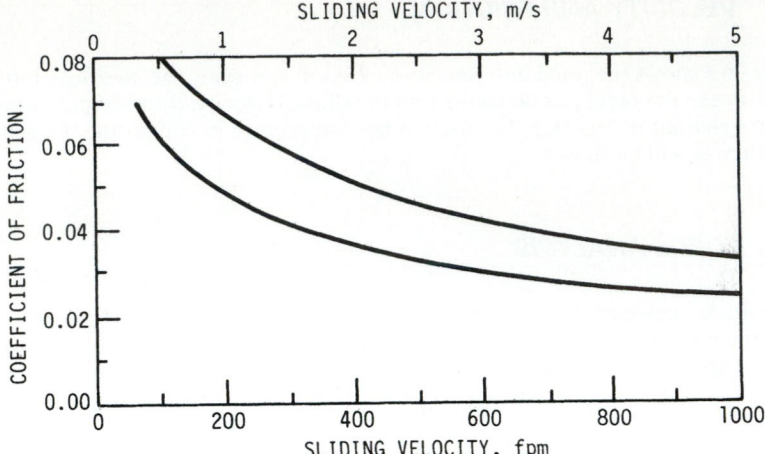

FIGURE 36.5 Approximate coefficients of sliding friction between the worm and gear teeth as a function of the sliding velocity. All values are based on adequate lubrication. The lower curve represents the limit for the very best materials, such as a hardened worm meshing with a bronze gear. Use the upper curve if moderate friction is expected.

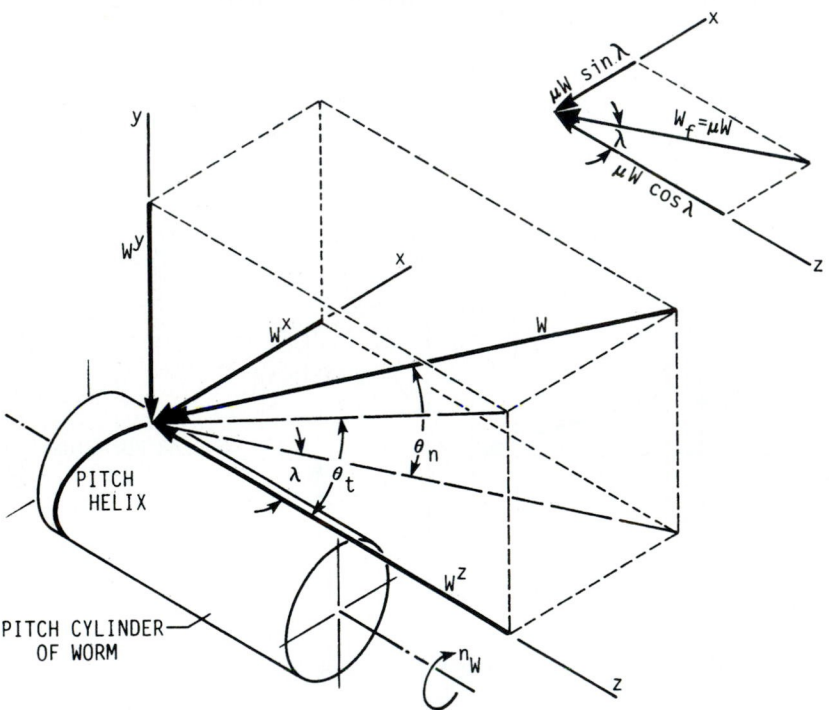

FIGURE 36.6 Forces exerted on worm.

where the subscripts are t for the tangential direction, r for the radial direction, and a for the axial direction. It is worth noting in the above equations that the gear axis is parallel to the x axis and the worm axis is parallel to the z axis. The coordinate system is right-handed.

The force W, which is normal to the profile of the mating teeth, produces a frictional force $W_f = \mu W$, shown in Fig. 36.6, along with its components $\mu W \cos \lambda$ in the negative x direction and $\mu W \sin \lambda$ in the positive z direction. Adding these to the force components developed in Eqs. (36.10) yields

$$W^x = W(\cos \phi_n \sin \lambda + \mu \cos \lambda)$$

$$W^y = W \sin \phi_n \qquad\qquad (36.12)$$

$$W^z = W(\cos \phi_n \cos \lambda - \mu \sin \lambda)$$

Equations (36.11) still apply. Substituting W^z from Eq. (36.12) into the third of Eqs. (36.11) and multiplying by μ, we find the frictional force to be

$$W_f = \mu W = \frac{\mu W_{G_t}}{\mu \sin \lambda - \cos \phi_n \cos \lambda} \qquad (36.13)$$

A relation between the two tangential forces is obtained from the first and third of Eqs. (36.11) with appropriate substitutions from Eqs. (36.12):

$$W_{W_t} = W_{G_t} \frac{\cos \phi_n \sin \lambda + \mu \cos \lambda}{\mu \sin \lambda - \cos \phi_n \cos \lambda} \qquad (36.14)$$

The efficiency can be defined as

$$\eta = \frac{W_{W_t} \text{ (without friction)}}{W_{W_t} \text{ (with friction)}} \qquad (36.15)$$

Since the numerator of this equation is the same as Eq. (36.14) with $\mu = 0$, we have

$$\eta = \frac{\cos \phi_n - \mu \tan \lambda}{\cos \phi_n + \mu \cot \lambda} \qquad (36.16)$$

Table 36.1 shows how η varies with λ, based on a typical value of friction $\mu = 0.05$ and the pressure angles usually used for the ranges of λ indicated. It is clear that small λ should be avoided.

Example 1. A 2-tooth right-hand worm transmits 1 horsepower (hp) at 1200 revolutions per minute (r/min) to a 30-tooth gear. The gear has a transverse diametral pitch of 6 teeth per inch. The worm has a pitch diameter of 2 inches (in). The normal pressure angle is 14½°. The materials and workmanship correspond to the lower of the curves in Fig. 36.5. Required are the axial pitch, center distance, lead, lead angle, and tooth forces.

Solution. The axial pitch is the same as the transverse circular pitch of the gear. Thus

$$p_x = \frac{\pi}{P} = \frac{\pi}{6} = 0.5236 \text{ in}$$

TABLE 36.1 Efficiency of Worm-Gear Sets for $\mu = 0.05$

Normal pressure angle ϕ_n, deg	Lead angle λ, deg	Efficiency η, percent
$14\frac{1}{2}$	1	25.2
	2.5	46.8
	5	62.6
	7.5	71.2
	10	76.8
	15	82.7
20	20	86.0
	25	88.0
	30	89.2

The pitch diameter of the gear is $D = N_G/P = 30/6 = 5$ in. The center distance is thus

$$C = \frac{D+d}{2} = \frac{2+5}{2} = 3.5 \text{ in}$$

The lead is

$$L = p_x N_W = 0.5236(2) = 1.0472 \text{ in}$$

From Eq. (36.3),

$$\lambda = \tan^{-1}\frac{L}{\pi d} = \tan^{-1}\frac{1.0472}{2\pi} = 9.46°$$

The pitch line velocity of the worm, in inches per minute, is

$$V_W = \pi d n_W = \pi(2)(1200) = 7540 \text{ in/min}$$

The speed of the gear is $n_G = 1200(2)/30 = 80$ r/min. The gear pitch line velocity is thus

$$V_G = \pi D n_G = \pi(5)(80) = 1257 \text{ in/min}$$

The sliding velocity is the square root of the sum of the squares of V_W and V_G, or

$$V_S = \frac{V_W}{\cos \lambda} = \frac{7540}{\cos 9.46} = 7644 \text{ in/min}$$

This result is the same as 637 feet per minute (ft/min); we enter Fig. 36.5 and find $\mu = 0.03$.

Proceeding now to the force analysis, we use the horsepower formula to find

$$W_W = \frac{(33\,000)(12)(hp)}{V_W} = \frac{(33\,000)(12)(1)}{7540} = 52.5 \text{ lb}$$

This force is the negative x direction. Using this value in the first of Eqs. (36.12) gives

$$W = \frac{W^x}{\cos \phi_n \sin \lambda + \mu \cos \lambda}$$

$$= \frac{52.5}{\cos 14.5° \sin 9.46° + 0.03 \cos 9.46°} = 278 \text{ lb}$$

From Eqs. (36.12) we find the other components of W to be

$$W^y = W \sin \phi_n = 278 \sin 14.5° = 69.6 \text{ lb}$$

$$W^z = W(\cos \phi_n \cos \lambda - \mu \sin \lambda)$$

$$= 278(\cos 14.5° \cos 9.46° - 0.03 \sin 9.46°)$$

$$= 265 \text{ lb}$$

The components acting on the gear become

$$W_{G_a} = -W^x = 52.5 \text{ lb}$$

$$W_{G_r} = -W^y = 69.6 \text{ lb}$$

$$W_{G_t} = -W^z = -265 \text{ lb}$$

The torque can be obtained by summing moments about the x axis. This gives, in inch-pounds,

$$T = 265(2.5) = 662.5 \text{ in} \cdot \text{lb}$$

It is because of the frictional loss that this output torque is less than the product of the gear ratio and the input torque (778 lb · in).

36.5 STRENGTH AND POWER RATING

Because of the friction between the worm and the gear, power is consumed by the gear set, causing the input and output horsepower to differ by that amount and resulting in a necessity to provide for heat dissipation from the unit. Thus

$$\text{hp(in)} = \text{hp(out)} + \text{hp(friction loss)}$$

This expression can be translated to the gear parameters, resulting in

$$\text{hp(in)} = \frac{W_{G_t} D n_W}{126\,000 m_G} + \frac{V_s W_f}{396\,000} \tag{36.17}$$

The force which can be transmitted W_{G_t} depends on tooth strength and is based on the gear, it being nearly always weaker than the worm (worm tooth strength can be computed by the methods used with screw threads, as in Chap. 20). Based on material strengths, an empirical relation is used. The equation is

$$W_{G_t} = K_s D^{0.8} F_e K_m K_v \tag{36.18}$$

TABLE 36.2   Materials Factor K_s for Cylindrical Worm Gearing[†]

Face width of gear F_G, in	Sand-cast bronze	Static-chill-cast bronze	Centrifugal-cast bronze
Up to 3	700	800	1000
4	665	780	975
5	640	760	940
6	600	720	900
7	570	680	850
8	530	640	800
9	500	600	750

[†]For copper-tin and copper-tin-nickel bronze gears operating with steel worms case-hardened to 58 R_C minimum.

SOURCE: Darle W. Dudley (ed.), *Gear Handbook,* McGraw-Hill, New York, 1962, p. 13–38.

where K_s = materials and size correction factor, values for which are shown in Table 36.2
F_e = effective face width of gear; this is actual face width or two-thirds of worm pitch diameter, whichever is less
K_m = ratio correction factor; values in Table 36.3
K_v = velocity factor (Table 36.4)

Example 2.   A gear catalog lists a 4-pitch, 14½° pressure angle, single-thread hardened steel worm to mate with a 24-tooth sand-cast bronze gear. The gear has a 1½-in face width. The worm has a 0.7854-in lead, 4.767° lead angle, 4½-in face width, 3-in pitch diameter. Find the safe input horsepower.

From Table 36.2, $K_s = 700$. The pitch diameter of the gear is

$$D = \frac{N_G}{P} = \frac{24}{4} = 6 \text{ in}$$

The pitch diameter of the worm is given as 3 in; two-thirds of this is 2 in. Since the face width of the gear is smaller (1.5 in), $F_e = 1.5$ in. Since $m_G = N_G/N_W = 24/1 = $

TABLE 36.3   Ratio Correction Factor K_m

m_G	K_m	m_G	K_m	m_G	K_m
3.0	0.500	8.0	0.724	30.0	0.825
3.5	0.554	9.0	0.744	40.0	0.815
4.0	0.593	10.0	0.760	50.0	0.785
4.5	0.620	12.0	0.783	60.0	0.745
5.0	0.645	14.0	0.799	70.0	0.687
6.0	0.679	16.0	0.809	80.0	0.622
7.0	0.706	20.0	0.820	100.0	0.490

SOURCE: Darle W. Dudley (ed.), *Gear Handbook,* McGraw-Hill, New York, 1962, p. 13–38.

TABLE 36.4   Velocity Factor K_v

Velocity V_S, fpm	K_v	Velocity V_S, fpm	K_v
1	0.649	600	0.340
1.5	0.647	700	0.310
10	0.644	800	0.289
20	0.638	900	0.269
30	0.631	1000	0.258
40	0.625	1200	0.235
60	0.613	1400	0.216
80	0.600	1600	0.200
100	0.588	1800	0.187
150	0.558	2000	0.175
200	0.528	2200	0.165
250	0.500	2400	0.156
300	0.472	2600	0.148
350	0.446	2800	0.140
400	0.421	3000	0.134
450	0.398	4000	0.106
500	0.378	5000	0.089
550	0.358	6000	0.079

SOURCE: Darle W. Dudley (ed.), *Gear Handbook*, McGraw-Hill, New York, 1962, p. 13–39.

24, from Table 36.3, $K_m = 0.823$ by interpolation. The pitch line velocity of the worm is

$$V_W = \pi d n_W = \pi(3)(1800) = 16\ 965 \text{ in/min}$$

The sliding velocity is

$$V_S = \frac{V_W}{\cos \lambda} = \frac{16\ 965}{\cos 4.767°} = 17\ 024 \text{ in/min}$$

Therefore, from Table 36.4, $K_v = 0.215$. The transmitted load is obtained from Eq. (36.18) and is

$$W_{G_t} = K_s d^{0.8} F_e K_m K_v = 700(6^{0.8})(1.5)(0.823)(0.215)$$

$$= 779 \text{ lb}$$

To find the friction load, the coefficient of friction is needed. Converting V_S to feet per minute and using Fig. 36.5, we find $\mu = 0.023$. From Eq. (36.13) we find

$$W_f = \frac{\mu W_{G_t}}{\mu \sin \lambda - \cos \phi_n \cos \lambda}$$

$$= \frac{0.023(779)}{0.023 \sin 4.767° - \cos 14.5° \cos 4.767°}$$

$$= 18.6 \text{ lb}$$

Next, using Eq. (36.17), we find the input horsepower to be

$$\text{hp(in)} = \frac{W_{G_t}Dn_W}{126\,000\,m_G} + \frac{W_f V_S}{396\,000}$$

$$= \frac{779(6)(1800)}{126\,000(24)} + \frac{18.6(17\,024)}{396\,000}$$

$$= 2.78 + 0.80 = 3.58$$

36.6 HEAT DISSIPATION

In the last section we noted that the input and output horsepowers differ by the amount of power resulting from friction between the gear teeth. This difference represents energy input to the gear set unit, which will result in a temperature rise. The capacity of the gear reducer will thus be limited by its heat-dissipating capacity.

The cooling rate for rectangular housings can be estimated from

$$C_1 = \begin{cases} \dfrac{n}{84\,200} + 0.01 & \text{without fan} \\[3mm] \dfrac{n}{51\,600} + 0.01 & \text{with fan} \end{cases} \tag{36.19}$$

where C_1 is the heat dissipated in Btu/(h)(in²)(°F), British thermal units per hour–inch squared–degrees Fahrenheit, and n is the speed of the worm shaft in rotations per minute. Note that the rates depend on whether there is a fan on the worm shaft. The rates are based on the area of the casing surface, which can be estimated from

$$A_c = 43.2C^{1.7} \tag{36.20}$$

where A_c is in square inches.

The temperature rise can be computed by equating the friction horsepower to the heat-dissipation rate. Thus

$$\text{hp(friction)} = \frac{778C_1 A_c \Delta T}{60(33\,000)} \tag{36.21}$$

or

$$\Delta T(°F) = \frac{\text{hp(friction)}(60)(33\,000)}{778C_1 A_c} \tag{36.22}$$

The oil temperature should not exceed 180°F. Clearly the horsepower rating of a gear set may be limited by temperature rather than by gear strength. Both must be checked. Of course, means other than natural radiation and convection can be employed to solve the heat problem.

36.7 DESIGN STANDARDS

The American Gear Manufacturer's Association[†] has issued certain standards relating to worm-gear design. The purpose of these publications, which are the work of broad committees, is to share the experience of the industry and thus to arrive at good standard design practice. The following relate to industrial worm-gear design and are extracted from [36.1] with the permission of the publisher.

Gear sets with axial pitches of ³⁄₁₆ in and larger are termed *coarse-pitch*. Another standard deals with fine-pitch worm gearing, but we do not include these details here. It is not recommended that gear and worm be obtained from separate sources. Utilizing a worm design for which a comparable hob exists will reduce tooling costs.

36.7.1 Number of Teeth of Gear

Center distance influences to a large extent the minimum number of teeth for the gear. Recommended minimums are shown in Table 36.5. The maximum number of teeth selected is governed by high ratios of reduction and considerations of strength and load-carrying capacity.

36.7.2 Number of Threads in Worm

The minimum number of teeth in the gear and the reduction ratio determine the number of threads for the worm. Generally, 1 to 10 threads are used. In special cases, a larger number may be required.

36.7.3 Gear Ratio

Either prime or even gear ratios may be used. However, if the gear teeth are to be generated by a single-tooth "fly cutter," the use of a prime ratio will eliminate the need for indexing the cutter.

[†] American Gear Manufacturer's Association (AGMA), 1500 King Street, Alexandria, VA 22314.

TABLE 36.5 Recommended Minimum Number of Gear Teeth

Center distance, in	Minimum number of teeth[†]
2	20
3	25
5	25
10	29
15	35
20	40
24	45

[†]Lower numbers are permissible for specific applications.

36.7.4 Pitch

It is recommended that pitch be specified in the axial plane of the worm and that it be a simple fraction, to permit accurate factoring for change-gear ratios.

36.7.5 Worm Pitch Diameter

The pitch diameter of the worm for calculation purposes is assumed to be at the mean of the working depth. A worm does not have a true pitch diameter until it is mated with a gear at a specified center distance. If the actual addendum and dedendum of the worm are equal, respectively, to the addendum and dedendum of the gear, then the nominal and actual pitch diameters of the worm are the same. However, it is not essential that this condition exist for satisfactory operation of the gearing.

Although a relatively large variation in worm pitch diameter is permissible, it should be held within certain limits if the power capacity is not to be adversely affected. Therefore, when a worm pitch diameter is selected, the following factors should be considered:

1. Smaller pitch diameters provide higher efficiency and reduce the magnitude of tooth loading.

2. The root diameter which results from selection of a pitch diameter must be sufficiently large to prevent undue deflection and stress under load.

3. Larger worm pitch diameters permit utilization of larger gear face widths, providing higher strength for the gear set.

4. For low ratios, the minimum pitch diameter is governed, to some degree, by the desirability of avoiding too high a lead angle. Generally, the lead is limited to a maximum of 45°. However, lead angles up to 50° are practical.

36.7.6 Gear Pitch Diameter

The selection of an approximate worm pitch diameter permits the determination of a corresponding approximate gear pitch diameter. In the normal case where the addendum and dedendum of the worm are to be equal, respectively, to the addendum and dedendum of the gear, a trial value of gear pitch diameter may be found by subtracting the approximate worm pitch diameter from twice the center distance of the worm and gear. Once the number of teeth for the gear has been selected, it is desirable to arrive at an exact gear pitch diameter by selecting for the gear circular pitch a fraction, which can be conveniently factored into a gear train for processing purposes, and calculating gear pitch diameter from the formula in Table 36.6. Should the actual value of gear pitch diameter differ from the trial value, the worm pitch diameter must be adjusted accordingly through the use of the formula in Table 36.7.

It is not essential that the pitch circle of the gear be at the mean of the working depth. Where there are sufficient teeth in the gear and the pressure angle is high enough to prevent undercutting, the pitch line can be anywhere between the mean of the depth and the throat diameter of the gear, or even outside the throat. This results in a short addendum for the gear teeth and lengthens the angle of recess. It is

TABLE 36.6 Dimensions of the Gear

Quantity	Symbol	Formula
Pitch diameter	D	$\dfrac{N_G p_x}{\pi}$
Throat diameter	D_t	$D + 2a$
Effective face width	F_e	$\sqrt{(d + h_k)^2 - d^2}$

also practical for the gear pitch diameter to be located somewhat below the mean of the working depth.

36.7.7 Worm Thread and Gear-Tooth Proportions

Pressure Angle. Several factors deserve consideration in the selection of the pressure angle. Smaller values of pressure angle decrease the separating forces, extend the line of action, and result in less backlash change with change in center distance. Larger values of pressure angle provide stronger teeth and assist in preventing undercutting of the teeth where lead angles are larger. The recommended pressure angles are listed in Table 36.8. These, used with the system for stubbing teeth (Table 36.9) will avoid undercutting.

TABLE 36.7 Dimensions of the Worm

Quantity	Symbol	Formula
Lead	l	$N_W P_x$
Pitch diameter†	d	$2C - D$
Outside diameter	d_o	$d + 2a$
Minimum face width	f	$2\sqrt{\left(\dfrac{D_t}{2}\right)^2 - \left(\dfrac{D}{2} - a\right)}$
Lead angle	λ	$\tan^{-1}\dfrac{l}{\pi d}$
Normal pitch	p_n	$p_x \cos \lambda$
Normal pressure angle	ϕ_n	See Table 36.8

†Use only where addenda and dedenda of worm and gear are equal.

TABLE 36.8 Recommended Values for the Normal Pressure Angle

Normal pressure angle ϕ_n, deg	Lead angle λ, deg
20	Less than 30
25	30–45

Although its use is discouraged, a 14° normal pressure angle may be used for lead angles up to 17°. A detailed study of gear-tooth action is employed by some designers to utilize pressure angles less than 25° where worm lead angles are above 30°.

Tooth Depth Proportions. The choice of tooth depth proportions is governed, to a great extent, by the need to avoid undercutting of the gear teeth. Commonly used tooth depth proportions for lead angles to, but not including, 30° are listed in Table 36.10. However, other acceptable practices are used by several manufacturers.

TABLE 36.9 System for Stubbing Teeth[†]

Depth, percent	Lead angle λ, deg
90	30–34.99
80	35–39.99
70	40–45

[†]Other systems for stubbing gear teeth such as reducing the depth by 2 percent per degree of lead angle over 30° are also in common use.

TABLE 36.10 Dimensions Common to Both Worm and Gear[†]

Quantity	Symbol	Formula
Addendum	a	$0.3183p_x$
Whole depth	h_t	$0.6866p_x$
Working depth	h_k	$0.6366p_x$
Center distance[‡]	C	$\dfrac{D + d}{2}$

[†]Recommended for lead angles less than 30°. See Table 36.9 for others.
[‡]Nominal, where addenda and dedenda of worm and gear are equal.

Table 36.9 presents a system for stubbing teeth to be used in conjunction with the pressure angles in Table 36.8 for lead angles 30° and above.

Tooth Thickness. The gear-tooth normal thickness preferably should be not less than half the normal pitch at the mean of the working depth. In view of the lower-strength material normally used for the gear, it is the practice of some manufacturers to make the gear tooth appreciably thicker than the worm thread. The extent to which this procedure can be followed is limited by the necessity for providing adequate land thickness at the thread peaks.

Tooth or Thread Forms. The most important detail of the worm thread form is that it must be conjugate to that of the gear tooth. The thread form varies with individual manufacturers' practices and may be anything between the extremes of a straight side and the normal section of an involute helicoid.

36.7.8 Gear Blank Dimensions

Face Width. The effective face width of a worm gear varies with the nominal pitch diameter of the worm and the depth of the thread. The formula for gear face width given in Table 36.6 is based on the maximum effective face width of a worm gear (the length of a tangent to the mean worm diameter) between the points where it is intersected by the outside diameter of the worm. Any additional face width is of very little value and is wasteful of material.

Diameter Increment. This is the amount that is added to the throat diameter of the gear to obtain the outside diameter. The magnitude of this increment is not critical and may vary with manufacturers' practice. Normal practice is to use approximately one addendum. It is general practice to round the outside diameter to the nearest fraction of an inch.

The sharp corners at the point where gear face and outside diameter intersect should be removed by the use of either a chamfer or a radius, as shown in Fig. 36.7.

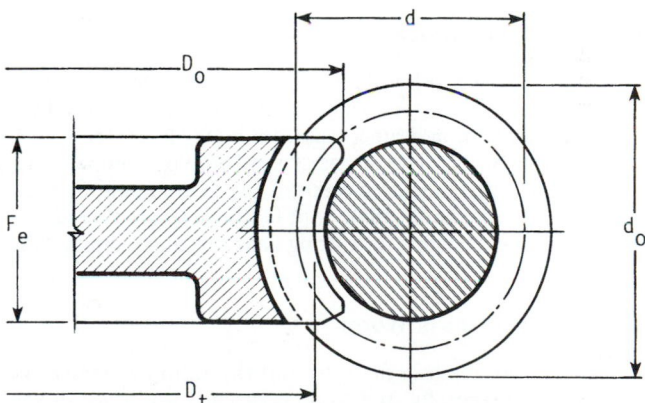

FIGURE 36.7 Section of worm and gear. Note that corners of gear teeth are usually rounded, as shown above the gear centerline; they may, however, be chamfered, as shown below.

Rim thicknesses are generally equal to or slightly greater than the whole depth of the teeth.

36.7.9 Worm Face

The face or length of the worm should be such that it extends beyond the point where contact between its threads and the teeth of the gear begins. Unlike with spur and helical gears, the pressure angle of a worm gear varies along the length of the tooth and becomes quite low on the leaving, or recess, side. This causes contact to occur on the worm almost to the point where the outside diameter of the worm intersects the throat diameter of the gear.

The formula in Table 36.7 provides a conservative value of the worm face width and is based on intersection of worm outside diameter with gear throat diameter.

More exact worm face widths may be determined by detailed calculations or layouts which take into consideration the face width of the gear and fix more definitely the extent of contact along the worm threads.

Good practice includes the breaking or rounding of the sharp edge of the worm threads at the end of the worm face. This procedure is particularly important where the worm face is less than provided for in the formula in Table 36.7.

36.7.10 Bored Worm Blanks

Where it is necessary to use a bored worm, the blank is normally designed with a key seat for driving purposes. The thickness of material between the worm root and the key seat should be at least $0.5h_t$. This is a general recommendation which is governed to some extent by whether the blank is hardened or unhardened. An increase in this amount may be necessary if the blank is hardened, particularly if a case-hardening process is used.

36.8 DOUBLE-ENVELOPING GEAR SETS[†]

36.8.1 Number of Teeth in Gear

The number of teeth for the gear is influenced to a large extent by center distance. The recommended number of teeth for various center distances is listed in Table 36.11. Should special considerations indicate a requirement for fewer teeth, it is advisable to consult a manufacturer of this type of gearing before you complete the design. For multiple-thread worms, the number of teeth in the gear should be within the limits listed in Table 36.11. The maximum number of teeth for single-threaded worms is limited only by the machines available for cutting gear sets and manufacturing tooling.

36.8.2 Number of Threads in Worm

The minimum number of teeth in the gear and the ratio determine the number of threads for the worm. Generally, one to nine threads are used. In special cases, a larger number of threads may be required.

[†] See Ref. [36.2].

TABLE 36.11 Range of Recommended Gear-Tooth Numbers

Center distance, in	No. teeth
2	24–40
3	24–50
4	30–50
8	40–60
15	50–60
20	50–70
24	60–80

36.8.3 Gear Ratio

The gear ratio is the quotient of the number of teeth in the gear and the number of threads in the worm. Either prime or even ratios may be used; however, hob life is improved with even ratios.

36.8.4 Pitch

It is recommended that pitch be specified in the axial section. Pitch is the result of design proportions.

36.8.5 Worm Root Diameter

The recommended root diameter for the worm is

$$d_R = \frac{C^{0.875}}{3} \tag{36.23}$$

It is desirable that the root diameter be not less than that indicated by this formula, even where the worm threads are cut integral with the shaft. For ratios less than 8/1, the worm root diameter may be increased. This increase may vary from zero for an 8/1 ratio to plus 15 percent for a 3/1 ratio.

36.8.6 Worm Pitch Diameter

The pitch diameter of the worm is assumed to be at the mean of the working depth at the center of the worm and is so considered for all calculations. The approximate worm pitch diameter is

$$d = \frac{C^{0.875}}{2.2}$$

and the corresponding root diameter is

$$d_R = d - 2b_G$$

where b_G is the dedendum of gear teeth in inches. Compare this root diameter with that given by Eq. (36.23). If it does not agree, alter the pitch diameter until the root diameter is within the desired limits.

Where horsepower rating is not a factor, there is no limitation regarding pitch diameter of the worm. Where efficiency is not as important as strength or load-carrying capacity, increasing the worm root diameter and gear face width will result in greater capacity.

36.8.7 Base Circle

The base circle may be secured from a layout in the following way. The normal pressure angle is always 20 degrees. The axial pressure angle may be obtained from

$$\phi_x = \tan^{-1} \frac{\tan \phi_n}{\cos \lambda} \tag{36.24}$$

Once the centerline of the worm and gear, the vertical centerline, and the gear pitch circle are laid out, measure along the common worm and gear pitch circle to the right or left of the vertical centerline an amount equal to one-fourth the axial circular pitch p_x. Through the point thus established and at an angle to the vertical centerline equal to the axial pressure angle ϕ_x, extend a line upward. A circle tangent to this line and concentric to the gear axis is the base circle. Adjust this diameter to the nearest 0.01 in. The formula for figuring the base circle diameter is

$$D_b = D \sin \left(\phi_x + \frac{90°}{N_G} \right) \tag{36.25}$$

36.8.8 Tooth Depth Proportions

Formulas for figuring the whole depth, working depth, and dedendum of gear teeth are found in Table 36.12. Note that the working depth is based on the normal circular pitch and so varies for a given axial pitch.

It is common practice in double-enveloping worm gears to proportion the gear tooth and worm thread thickness as follows: The gear tooth thickness is 55 percent of the circular pitch, and the worm thread thickness is 45 percent of the circular pitch. The backlash in the gear set is subtracted from the worm thread thickness. This practice has been followed to secure greater tooth strength in the gear, which is the weaker member.

36.8.9 Tooth or Thread Forms

The thread form is usually straight in the axial section, but any other form may be used. Since there is no rolling action up or down the flanks, the form is unimportant, except that it must be the same on the worm and the hob. The straight-sided tooth in the axial section provides the greatest ease of manufacture and checking, of both the gear sets and the cutting tools.

TABLE 36.12   Recommended Worm Tooth Dimensions

Quantity	Formula
Length of flat on outside diameter of worm, in	$f = \dfrac{p_x}{5.5}$
Whole depth of tooth	$h_t = \dfrac{p_n}{2}$
Working depth of tooth	$h_k = 0.9 h_t$
Dedendum	$b_G = 0.611 h_k$
Normal pressure angle	$\phi_n = 20°$
Axial pressure angle at center of worm	$\phi_x = \tan^{-1} \dfrac{\tan \phi_n}{\cos \lambda_c}$
Lead angle at center of worm	$\lambda_c = \tan^{-1} \dfrac{D}{m_G d}$

36.8.10   Worm Length

The effective length of the worm thread should be the base circle diameter minus $0.10C$ for lead angles up to and including $20°$ and minus $0.20C$ to $0.30C$ for lead angles from 20 to $45°$. The principal reason for altering this length is to secure the proper amount of worm thread overlap. The overlap should be a distance along the worm thread greater than the face width of the gear. The worm thread extending beyond the effective length must be relieved to prevent interference.

The outside diameter of the worm equals the diameter at the tip of the worm thread at the effective length.

A formula for computing the flat on the outside diameter of the worm at the effective length is given in Table 36.12; the worm face equals the effective length plus twice the flat. The worm face angle is generally $45°$.

36.8.11   Gear Blank Dimensions

The face width of the gear should be equal to the root diameter of the worm. Additional face width will not add proportional capacity and is wasteful of material. Where gear sets are to be used at less than their rated horsepower, the face width may be reduced in proportion.

The gear outside diameter may be the point at which the gear face angle intersects the gear throat radius or any desired amount less, except not less than the throat diameter. The gear throat diameter equals the gear pitch diameter plus one working depth.

There are generally three types of gear blanks in use: those having the hub integral, those flanged and counterbored for a bolted spider, and those having a through bore and fastened by setscrews (or bolts) inserted in drilled and tapped holes located half in the joint between the blank and spider. In all designs, the thickness of metal beneath the teeth should be 1¼ to 1½ times the whole depth of the tooth.

36.8.12   Materials[†]

Most of the rating standards are based on the use of worms made from a through-hardened, high-carbon steel heat-treated to 32 to 38 R_C. Where case-hardened worms are employed, somewhat higher ratings may be used.

Many high-strength gear materials (such as aluminum, heat-treated aluminum, and nickel bronzes) are used for slow speeds and heavy loads at higher ratings than shown in [36.3].

REFERENCES

36.1 ANSI/AGMA 6022-C93, "Design of General Industrial Coarse-Pitch Cylindrical Worm-gearing."

36.2 ANSI/AGMA 6030-C87, "Design of Industrial Double-Enveloping Wormgears."

36.3 ANSI/AGMA 6017-E86, "Rating and Application of Single and Multiple Reduction Double-Enveloping Worm and Helical Worm Speed Reducers."

ADDITIONAL REFERENCE

ANSI/AGMA 6034-B92, "Practice for Enclosed Cylindrical Wormgear Speed Reducers and Gearmotors."

[†] See Ref. [36.3].

CHAPTER 37
SHAFTS

Charles R. Mischke, Ph.D., P.E.
Professor Emeritus of Mechanical Engineering
Iowa State University
Ames, Iowa

NOMENCLATURE

a	Distance
A	Area
b	Distance
c_0	Constant
C_1, C_2	Constants
d	Outside diameter of shaft
d_i	Inside diameter of hollow shaft
E	Modulus of elasticity
F	Load
g	Gravitation constant
i	index
I	Second moment of area
J	Polar second area moment
k	Torsional spring rate
K	Transverse shear stress magnification factor
K_f	Fatigue stress concentration factor
ℓ	Span

m	Mass per unit length
M	Bending moment
n	Design factor, factor of safety
p	Shrink-fit pressure
r	Load line slope
R	Bearing reaction
S_a	Strength amplitude ordinate to fatigue locus
S_e	Endurance strength
S_m	Strength steady coordinate to fatigue locus
S_y	Yield strength
S_{ut}	Ultimate tensile strength
T	Torsional or twisting moment
V	Transverse shear force
w_i	Weight of ith segment of shaft
W	Weight of shaft
x	Coordinate
x_a, x_b	Coordinates of bearings
y	Coordinate, deflection
y_0	Constant
z	Coordinate
γ	Weight density
θ	Angle
σ	Normal stress
σ'	Von Mises normal stress
τ	Shear stress
ω	First critical angular frequency

37.1 INTRODUCTION

A shaft is a rotating part used to transmit power, motion, or analogic information. It often carries rotating machine elements (gears, pulleys, cams, etc.) which assist in the transmission. A shaft is a member of a fundamental mechanical pair: the "wheel and axle." Traditional nomenclature includes

Axle   A stationary member supporting rotating parts.
Shaft   A rotating member supporting attached elements.
Spindle   A short shaft or axle.
Head or stud shaft   A shaft integral with a motor or prime mover.
Line shaft   A shaft used to distribute power from one prime mover to many machines.
Jack shaft   A short shaft used for power transmission as an auxiliary shaft between two other shafts (counter shaft, back shaft).

Geometric fidelity is important to many shaft functions. Distortion in a loaded body is unavoidable, and in a shaft design it is controlled so as to preserve function. There are elastic lateral displacements due to bending moment and transverse shear, and there are elastic displacements of an angular nature due to transmitted torque. Fracture due to fatigue and permanent distortion due to yielding destroy function. The tight constraint in shaft design is usually a distortion at a particular location. For example, shaft slope at a bearing centerline should typically be less than 0.001 rad for cylindrical and tapered roller bearings, 0.004 rad for deep-groove ball bearings, and 0.0087 rad for spherical ball bearings (typically). At a gear mesh, the allowable relative slope of two gears with uncrowned teeth can be held to less than 0.0005 rad each. Deflection constraints for involute gears tolerate larger (but not smaller) than theoretical center-to-center distances, with a small increase in pressure angle but with observable increases in backlash. The typical upper bound on center-to-center distance in commercial-quality spur gearing is for diametral pitches up to 10, 0.010 in; for those 11 to 19, 0.005 in; and those for 20 to 50, 0.003 in.

A harsh reality is that a deflection or slope at a shaft section is a function of the geometry and loading *everywhere*. The stress at a shaft section is a function of the local geometry and local bending moment, a simpler problem. Shaft designers often size the shaft to meet the active distortion constraint, then check for strength adequacy. Young's modulus is about the same for most shaft steels, and so adjusting the material and its condition does not significantly undo the distortional adequacy.

Shafts are proportioned so that mounted elements are assembled from one or both ends, which accounts for the stepped cylinder, fat middle aspect. This also efficiently places the most material toward the center. Shaft geometric features may also include chamfers, shoulders, grooves, keyways, splines, tapers, threads, and holes for pins and lubricant access. Shafts may even be hollow, square, etc. The effect of each of these features must be considered when checking shaft performance adequacy.

37.2 *DISTORTION DUE TO BENDING*

Since the most likely active constraint is a slope or a deflection at some shaft section, it is useful to determine the *constant*-diameter shaft that meets the requirement. This establishes in the designer's mind the "heft" of the shaft. Then, as one changes the local diameters and their lengths to accommodate element mounting, the material removed near the bearings has to be replaced in part, but nearer the center. It is a matter of guiding perspective at the outset. Figure 37.1 depicts shafts with a single transverse load F_i or a single point couple M_i which could be applied in either the horizontal or the vertical plane. From [37.1], Tables A-9-6 and A-9-8, expressions for slopes at each bearing can be developed. It follows by superposition that for the left bearing,

$$d = \left(\frac{32n}{3\pi E \ell \Sigma \theta} \left\{ [\Sigma F_i b_i (b_i^2 - \ell^2) + \Sigma M_i (3a_i^2 - 6a_i \ell + 2\ell^2)]_H^2 \right.\right.$$
$$\left.\left. + [\Sigma F_i b_i (b_i^2 - \ell^2) + \Sigma M_i (3a_i^2 - 6a_i \ell + 2\ell^2)]_V^2 \right\}^{1/2} \right)^{1/4} \quad (37.1)$$

and for the right bearing,

$$d = \left(\frac{32n}{3\pi E \ell \Sigma \theta} \left\{ [\Sigma F_i a_i (\ell^2 - a_i^2) + \Sigma M_i (3a_i^2 - \ell^2)]_H^2 \right.\right.$$
$$\left.\left. + [\Sigma F_i a_i (\ell^2 - a_i^2) + \Sigma M_i (3a_i^2 - \ell^2)]_V^2 \right\}^{1/2} \right)^{1/4} \quad (37.2)$$

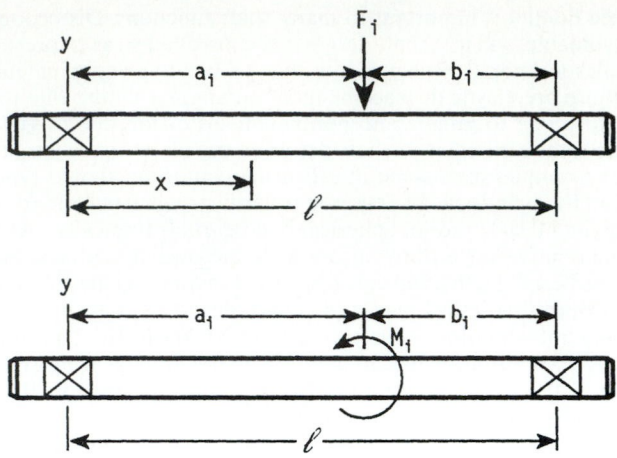

FIGURE 37.1 Simply supported shafts with force F_i and couple M_i applied.

where $\Sigma\theta$ is the absolute value of the allowable slope at the bearing. These equations are an ideal task for the computer, and once programmed interactively, are convenient to use.

Example 1. A shaft is to carry two spur gears between bearings and has loadings as depicted in Fig. 37.2. The bearing at A will be cylindrical roller. The spatial centerline slope is limited to 0.001 rad. Estimate the diameter of the uniform shaft which limits the slope at A with a design factor of 1.5.

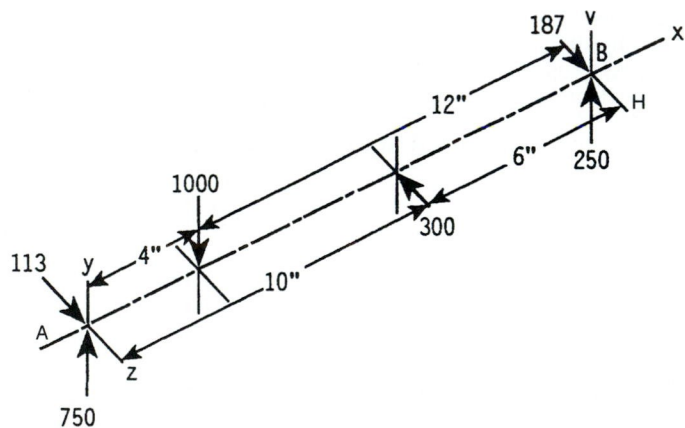

FIGURE 37.2 A shaft carries two spur gears between bearings A and B. The gear loads and reactions are shown.

Solution. Equation (37.1) is used.

$$d = \left(\frac{32n}{3\pi E \ell \Sigma \theta} \left\{ [F_1 b_1 (b_1^2 - \ell^2)]_H^2 + [F_2 b_2 (b_2^2 - \ell^2)]_V^2 \right\}^{1/2} \right)^{1/4}$$

$$= \left(\frac{32(1.5)}{3\pi 30(10)^6\, 16(0.001)} \left\{ [300(6)(6^2 - 16^2)]^2 + [1000(12)(12^2 - 16^2)]^2 \right\}^{1/2} \right)^{1/4}$$

$$= 1.964 \text{ in}$$

Transverse bending due to forces and couples applied to a shaft produces slopes and displacements that the designer needs to control. Bending stresses account for most or all of such distortions. The effects of transverse shear forces will be addressed in Sec. 37.3.

Most bending moment diagrams for shafts are piecewise linear. By integrating once by the trapezoidal rule and a second time using Simpson's rule, one can obtain deflections and slopes that are *exact,* can be developed in tabular form, and are easily programmed for the digital computer. For bending moment diagrams that are piece-wise polynomial, the degree of approximation can be made as close as desired by increasing the number of station points of interest. See [37.1], pp. 103–105, and [37.2]. The method is best understood by studying the tabular form used, as in Table 37.1.

The first column consists of station numbers, which correspond to cross sections along the shaft at which transverse deflection and slope will be evaluated. The mini-mum number of stations consists of those cross sections where M/EI changes in magnitude or slope, namely discontinuities in M (point couples), in E (change of material), and in I (diameter change, such as a shoulder). Optional stations include other locations of interest, including shaft ends. For integration purposes, midstation locations are chosen so that the second integration by Simpson's rule can be exact. The moment column M is dual-entry, displaying the moment as one approaches the station from the left and as one approaches from the right. The distance from the origin to a station x is single-entry. The diameter d column is dual-entry, with the entries differing at a shoulder. The modulus E column is also dual-entry. Usually the shaft is of a single material, and the column need not be filled beyond the first entry. The first integration column is single-entry and is completed by applying the trapezoidal rule to the M/EI column. The second integration column is also single-entry, using the midstation first integration data for the Simpson's rule integration.

TABLE 37.1 Form for Tabulation Method for Shaft Transverse Deflection Due to Bending Moment.

Moment M	Dist. x	Dia. d	Modulus E	$\dfrac{M}{EI}$	$\displaystyle\int_0^x \frac{M}{EI}\, dx$	$\displaystyle\int_0^x \left(\int_0^x \frac{M}{EI}\, dx \right) dx$	Defl. y	Slope $\dfrac{dy}{dx}$
1					0	0		
2								
3								

The deflection entry y is formed from the prediction equation

$$y = \int_0^x \int_0^x \frac{M}{EI}\, dx\, dx + C_1 x + C_2 \tag{37.3}$$

The slope dy/dx column is formed from the prediction equation

$$\frac{dy}{dx} = \int_0^x \frac{M}{EI}\, dx + C_1 \tag{37.4}$$

where the constants C_1 and C_2 are found from

$$C_1 = \frac{\displaystyle\int_0^{x_a}\int_0^{x_a} M/(EI)\, dx\, dx - \int_0^{x_b}\int_0^{x_b} M/(EI)\, dx\, dx}{x_a - x_b} \tag{37.5}$$

$$C_2 = \frac{\displaystyle x_b \int_0^{x_a}\int_0^{x_a} M/(EI)\, dx\, dx - x_a \int_0^{x_b}\int_0^{x_b} M/(EI)\, dx\, dx}{x_a - x_b} \tag{37.6}$$

where x_a and x_b are bearing locations.

This procedure can be repeated for the orthogonal plane if needed, a Pythagorean combination of slope, or deflections, giving the spatial values. This is a good time to plot the end view of the deflected shaft centerline locus in order to see the spatial lay of the loaded shaft.

Given the bending moment diagram and the shaft geometry, the deflection and slope can be found at the station points. If, in examining the deflection column, any entry is too large (in absolute magnitude), find a new diameter d_{new} from

$$d_{new} = d_{old}\left|\frac{n y_{old}}{y_{all}}\right|^{1/4} \tag{37.7}$$

where y_{all} is the allowable deflection and n is the design factor. If any slope is too large in absolute magnitude, find the new diameter from

$$d_{new} = d_{old}\left|\frac{n(dy/dx)_{old}}{(slope)_{all}}\right|^{1/4} \tag{37.8}$$

where $(slope)_{all}$ is the allowable slope. As a result of these calculations, find the largest d_{new}/d_{old} ratio and multiply all diameters by this ratio. The tight constraint will be at its limit, and all others will be loose. Don't be concerned about end journal size, as its influence on deflection is negligible.

Example 2.   A shaft with two loads of 600 and 1000 lbf in the same plane 2 inches (in) inboard of the bearings and 16 in apart is depicted in Fig. 37.3. The loads are from 8-pitch spur gears, and the bearings are cylindrical roller. Establish a geometry of a shaft which will meet distortion constraints, using a design factor of 1.5.

Solution.   The designer begins with identification of a uniform-diameter shaft which will meet the likely constraints of bearing slope. Using Eq. (37.2), expecting the right bearing slope to be controlling,

$$d = \left[\frac{32(1.5)}{3\pi 30(10)^6 16(0.001)}\left|600(2)(16^2 - 2^2) + 1000(14)(16^2 - 14^2)\right|\right]^{1/4}$$

$$= 1.866\ \text{in}$$

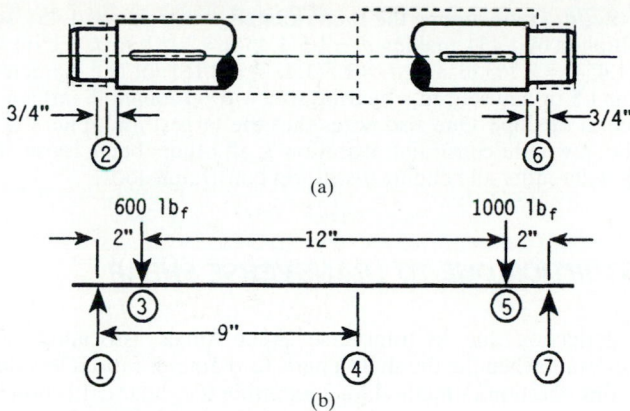

FIGURE 37.3 (*a*) The solid-line shaft detail is the designer's tentative geometry. The dashed lines show shaft sized to meet bending distortion constraints. (*b*) The loading diagram and station numbers.

Based on this, the designer sketches in some tentative shaft geometry as shown in Fig. 37.3*a*. The designer decides to estimate the bearing journal size as 1.5 in, the next diameter as 1.7 in, the diameter beyond a shoulder 9 in from the left bearing as 1.9 in, and the remaining journal as 1.5 in. The next move is to establish the moment diagram and use seven stations to carry out the tabular deflection method by completing Table 37.1. Partial results are shown below.

Station	x, in	Moment M, in · lbf	Diameter d, in	Deflection y, in	Slope dy/dx
1	0	0	1.5	0	−0.787E-03
2	0.75	487.5	1.5/1.7	−0.584E-03	−0.763E-03
3	2	1300	1.7	−0.149E-02	−0.672E-03
4	9	1650	1.7/1.9	−0.337E-02	0.168E-03
5	14	1900	1.9	−0.140E-02	0.630E-03
6	15.25	712.5	1.9/1.5	−0.554E-03	0.715E-03
7	16	0	1.5	0	0.751E-03

The gears are 8 pitch, allowing $0.010/2 = 0.005$ in growth in center-to-center distance, and both y_3 and y_5 have absolute values less than $0.005/1.5 = 0.00333$, so that constraint is loose. The slope constraints of $0.001/1.5$ are violated at stations 1 and 7, so using Eq. (37.8),

$$(d_1)_{new} = 1.5 \left| \frac{1.5(-0.000\ 787)}{0.001} \right|^{1/4} = 1.5(1.042) \text{ in}$$

$$(d_7)_{new} = 1.5 \left| \frac{1.5(0.000\ 751)}{0.001} \right|^{1/4} = 1.5(1.030) \text{ in}$$

and the gear mesh slope constraints are violated at stations 3 and 5, so using Eq. (37.8),

$$(d_3)_{new} = 1.7 \left| \frac{1.5(0.001\ 49)}{0.0005} \right|^{1/4} = 1.7(1.454) \text{ in}$$

$$(d_5)_{new} = 1.9 \left| \frac{1.5(0.001\ 40)}{0.0005} \right|^{1/4} = 1.9(1.432) \text{ in}$$

The largest d_{new}/d_{old} ratio among the four violated constraints is 1.454, so *all* diameters are multiplied by 1.454, making $d_1 = 1.5(1.454) = 2.181$ in, $d_3 = 1.7(1.454) = 2.472$ in, $d_5 = 1.9(1.454) = 2.763$ in, and $d_7 = 1.5(1.454) = 2.181$ in. The diameters d_1 and d_7 can be left at 1.5 or adjusted to a bearing size without tangible influence on transverse deflection or slope. One also notes that the largest multiplier 1.454 is associated with the now tight constraint at station 3, all others being loose. Rounding d_3 and/or d_5 up will render all bending distortion constraints loose.

37.3 DISTORTION DUE TO TRANSVERSE SHEAR

Transverse deflection due to transverse shear forces associated with bending becomes important when the the shaft length-to-diameter ratio is less than 10. It is a short-shaft consideration. A method for estimating the shear deflection is presented in Ref. [37.2]. There are two concerns associated with shear deflection. The first is that it is often forgotten on short shafts. The second is that it is often neglected in formal education, and engineers tend to be uncomfortable with it. Ironically, it is simpler than bending stress deflection.

The loading influence is the familiar shear diagram. The transverse shear force V is piecewise linear, and the single integration required is performed in a tabular method suitable to computer implementation. Table 37.2 shows the form. The left-hand column consists of station numbers which identify cross sections along the shaft at which shear deflection and slope are to be estimated. The minimum number of stations consists of those cross sections where $KV/(AG)$ changes abruptly, namely at discontinuities in transverse shear force V (at loads), in cross-sectional area A (at shoulders), and in torsional modulus G (if the material changes). Optional stations include other locations of interest. There is no need for midstation locations, since the trapezoidal rule will be used for integration, maintaining exactness. The shear force column V is dual-entry, the location x is single-entry, and the diameter d column is dual-entry, as is the torsional modulus G column, if included. The $KV/(AG)$ column is dual-entry, as is the slope dy/dx column.

The single-entry integral column is generated using the trapezoidal rule. The single-entry deflection column y is generated from the prediction equation

$$y = \int_0^x \frac{KV}{AG} \, dx + c_0 x + y_0 \qquad (37.9)$$

TABLE 37.2 Form for Tabulation Method for Shaft Transverse Deflection Due to Transverse Shear.

	Shear V	Dist. x	Dia. d	Modulus G	$\dfrac{KV}{AG}$	$\displaystyle\int_0^x \frac{KV}{AG}\,dx$	Defl. y	Slope $\dfrac{dy}{dx}$	Avg. slope $(dy/dx)_{av.}$
1						0			
2									
3									
4									

The dual-entry slope dy/dx column is generated from the other prediction equation,

$$\frac{dy}{dx} = -\frac{KV}{AG} + c_0 \tag{37.10}$$

where

$$c_0 = \frac{\int_0^{x_a} KV/(AG)\, dx - \int_0^{x_b} KV/(AG)\, dx}{x_a - x_b} \tag{37.11}$$

$$y_0 = \frac{x_a \int_0^{x_b} KV/(AG)\, dx - x_b \int_0^{x_a} KV/(AG)\, dx}{x_a - x_b} \tag{37.12}$$

where x_a and x_b are bearing locations and K is the factor 4/3 for a circular cross section (the peak stress at the centerline is 4/3 the average shear stress on the section). The slope column can have dual entries because Eq. (37.10) contains the discontinuous $KV/(AG)$ term.

Example 3. A uniform 1-in-diameter stainless steel [$G = 10(10)^6$ psi] shaft is loaded as shown in Fig. 37.4 by a 1000-lbf overhung load. Estimate the shear deflection and slope of the shaft centerline at the station locations.
 Solution. Omitting the G column, construct Table 37.3. After the integral column is complete, c_0 and y_0 are given by Eqs. (37.11) and (37.12), respectively:

$$c_0 = \frac{0 - 339.5(10^{-6})}{1 - 11} = 33.95(10^{-6})$$

$$y_0 = \frac{1(339.5)10^{-6} - 11(0)}{1 - 11} = -33.95(10^{-6})$$

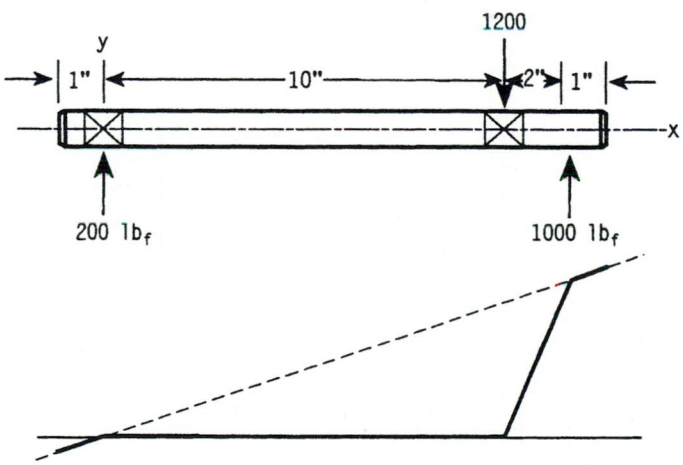

FIGURE 37.4 A short uniform shaft, its loading, and shear deflection.

TABLE 37.3 Transverse Shear Deflection in Shaft of Fig. 37.4

Station	V	x	d	$\dfrac{KV}{AG}$	$\displaystyle\int_0^x \dfrac{KV}{AG}\,dx$	y	dy/dx	$(dy/dx)_{av}$
1	0	0	0	0	0	-33.95E-06	33.95E-06	33.95E-06
	0		1	0			33.95E-06	
2	0	1	1	0	0	0	33.95E-06	16.98E-06
	200			33.95E-06			0	
3	200	11	1	33.95E-06	339.5E-06	0	0	101.9E-06
	-1000		1	-169.8E-06			203.75E-06	
4	-1000	13	1	-169.8E-06	0	407.4E-06	203.75E-06	118.9E-06
	0		1	0			33.95E-06	
5	0	14	1	0	0	441.4E-06	33.95E-06	33.95E-06
	0		0	0			33.95E-06	

The prediction Eqs. (37.9) and (37.10) are

$$y = -\int_0^x \frac{KV}{AG} \, dx + 33.95(10^{-6})x - 33.95(10^{-6})$$

$$\frac{dy}{dx} = 33.95(10^{-6}) - \frac{KV}{AG}$$

and the rest of the table is completed.

A plot of the shear deflection curve is shown under the shaft in Fig. 37.4. Note that it is piecewise linear. The droop of the unloaded overhang is a surprise when the between-the-bearings shaft is straight and undeflected. The discontinuous curve arises from discontinuities in loading V. In reality, V is not discontinuous, but varies rapidly with rounded corners. If a rolling contact bearing is mounted at station 2, the bearing inner race will adopt a compromise angularity between $dy/dx = 33.95(10^{-6})$ and $dy/dx = 0$. This is where the average (midrange) slope $(dy/dx)_{av}$ is useful in estimating the extant slope of the inner race with respect to the outer race of the bearing.

Figure 37.5 shows a short shaft loading in bending. Table 37.4 shows the deflection analysis of Sec. 37.2 for this shaft in columns 3 and 4, the shear deflection analysis of Sec. 37.3 in columns 5 and 6, and their superposition in columns 7 and 8. Figure 37.5 shows the shear deflection at station 7 to be about 28 percent of the bending deflection, and the shear slope at station 9 to be about 15 percent of the bending slope. Both of these locations could involve an active constraint. In the deflection analysis of shafts with length-to-diameter aspect ratios of less than 10, the transverse shear deflections should be included.

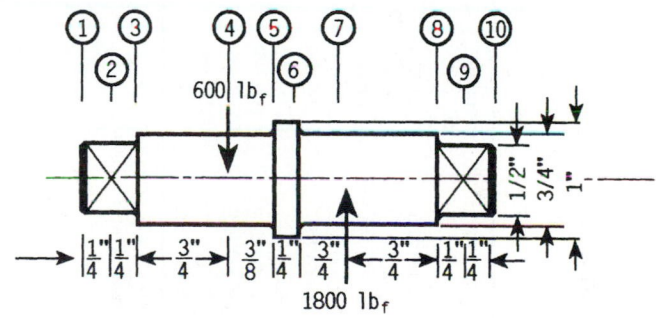

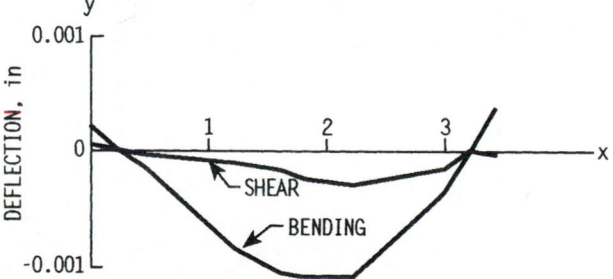

FIGURE 37.5 A short shaft of several diameters, its loading, and the consequential shear and bending deflections.

TABLE 37.4 Deflections of Shaft of Fig. 37.5

Station	x_i	Bending[†] y_i	Bending $(dy/dx)_i$	Shear[‡] y_i	Shear $(dy/dx)_{avi}$	Combined y_i	Combined $(dy/dx)_i$
1		0.000 240	-0.000 959	0.738E-05	-0.295E-04	0.247E-03	-0.988E-03
2	0.250	0	-0.000 959	0	-0.886E-04	0	-0.105E-02
3	0.500	-0.000 234	-0.000 891	-0.369E-04	-0.115E-03	-0.271E-03	-0.101E-02
4	1.250	-0.000 842	-0.000 690	-0.984E-04	-0.116E-03	-0.940E-03	-0.850E-03
5	1.625	-0.001 05	-0.000 408	-0.188E-03	-0.194E-03	-0.124E-02	-0.601E-03
6	1.875	-0.001 14	-0.000 306	-0.225E-03	-0.194E-03	-0.137E-02	-0.500E-03
7	2.250	-0.001 14	0.000 378	-0.315E-03	-0.328E-05	-0.145E-02	0.375E-03
8	3.000	-0.000 403	0.001 38	-0.140E-03	0.397E-03	-0.543E-03	0.178E-02
9	3.250	0	0.001 72	0	0.266E-03	0	0.199E-02
10	3.500	0.000 431	0.001 72	-0.738E-05	-0.295E-04	0.424E-03	0.169E-02

[†] $C_1 = -0.959(10^{-3})$, $C_2 = 0.240(10^{-3})$, Eqs. (37.5) and (37.6).
[‡] $c_0 = -0.295(10^{-4})$, $y_0 = 0.738(10^{-5})$, Eqs. (37.11) and (37.12).

37.12

37.4   DISTORTION DUE TO TORSION

Angular deflection in a right circular cylindrical shaft due to torque T is

$$\theta = \frac{T\ell}{GJ} \qquad \text{rad} \qquad\qquad (37.13)$$

For a stepped shaft of individual cylinder length ℓ_i with torques T_i, the angular deflection is

$$\theta = \Sigma\theta_i = \Sigma \frac{T_i\ell_i}{G_iJ_i} \qquad\qquad (37.14)$$

which becomes $\theta = (T/G)\Sigma(\ell_i/J_i)$ for constant torque through homogeneous material. The torsional stiffness can be defined as $k_i = T_i/\theta_i$, and since $\theta_i = T_i/k_i$ and $\theta = \Sigma\theta_i = \Sigma(T_i/k_i)$, one may write for constant torque $\theta = T\Sigma(1/k_i)$. It follows that

$$\frac{1}{k} = \Sigma \frac{1}{k_i} \qquad\qquad (37.15)$$

The equation $\theta = (T/G)\Sigma\ell_i/J_i$ is not precise, since experimental evidence shows that θ is larger than given by this equation.

The material in a step (shoulder) has a surface free of shear. Some material loafs, so other material is more distressed and distorts more. The existence of keyways, splines, and tapered sections increases angular flexibility also. For quantitative treatment of these realities, see Ref. [37.3], pp. 93–99. When a coupling is keyed or splined to a shaft, that shaft can be considered to twist independently of the coupling for one-third of its hub length.

37.5   SHAFT MATERIALS

Most steels have similar moduli of elasticity, so that the rigidity requirement can be met by geometric decisions, independent of the material choice among steels. Strength to resist loading stresses affects the choice of material. ANSI 1020-1050 steels and 11XX free-machining steels are common choices. Heat treating 1340-50, 3140-50, 4140, 4340, 5140, and 8650 steels produces greater strength. Hardness is a function of size, and the methods of Grossman and Fields and of Crafts and Lamont in Chapter 8 are important to quantitatively relate strength to size and heat-treatment regimen. Carburizing grades 1020, 4320, 4820, and 8620 are chosen for surface-hardening purposes.

Cold-rolled sections are available up to about 3½ in in diameter. Hot-rolled rounds are available up to nearly 6 in. Above this size, forging precedes machining.

When a shaft geometry is created (prior to final machining) by a volume-conservative process (casting or hot or cold forming), then optimality can be pursued by minimizing the material amount if production volume permits. Constraints can be made nearly active at several locations. Many shafts are created for small production runs by machining round stock, and optimality may be achieved by minimizing the amount of material removed from the work piece, which minimizes the machining effort.

37.6   *LOAD-INDUCED STRESSES*

Shafts that transmit power are often loaded in such a way that the torsion which performs the work induces transverse bending forces at gears. If the torsion is stochastic, so is the induced bending due to pitch-line forces. Both the torsion and the bending moment have the same distribution and coefficient of variation. The same is true of a point couple induced at a helical gear.

For ductile shaft materials, distortion energy theory is used, and the array of stresses at a critical location element are combined to form the von Mises stress. If the normal stresses at a point are σ_x, σ_y, σ_z and the associated shear stresses are τ_{xy}, τ_{yz}, τ_{zx}, then the von Mises stress σ' is given by

$$\sigma' = \frac{1}{\sqrt{2}} \left[(\sigma_x - \sigma_y)^2 + (\sigma_y - \sigma_z)^2 + (\sigma_z - \sigma_x)^2 + 6(\tau_{xy}^2 + \tau_{yz}^2 + \tau_{zx}^2) \right]^{1/2} \quad (37.16)$$

In a shaft, the critical location is usually at a surface, and two normal stresses (say σ_y and σ_z) and two shear stresses (say τ_{xz} and τ_{zx}) are zero. Equation (37.16) simplifies to

$$\sigma' = (\sigma_x^2 + 3\tau_{xy}^2)^{1/2} \quad (37.17)$$

The bending stress σ_x is usually expressed as $32K_f M/(\pi d^3)$ and the shear stress τ_{xy} is expressed as $16K_f' T/(\pi d^3)$, or without the stress concentration K_f' if torsion is steady, and so Eq. (37.17) is written as

$$\sigma' = \left[\left(\frac{32K_f M}{d^3} \right)^2 + 3 \left(\frac{16T}{d^3} \right)^2 \right]^{1/2} \quad (37.18)$$

As the shaft rotates and the stress field remains stationary, the bending moment induces a completely reversed stress σ_x on the rotating element in Fig. 37.6. The amplitude component of this stress σ_a' is

$$\sigma_a' = \left| \frac{32K_f M_a}{\pi d^3} \right| \quad (37.19)$$

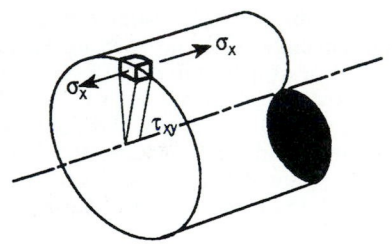

FIGURE 37.6   A stress element at a shaft surface.

The subscript on M_a is to designate the bending moment inducing a completely reversed normal stress on the element as the shaft turns. The bending moment itself may indeed be steady. The steady component of stress σ_m', from Eq. (37.18), is

$$\sigma_m' = \left| \frac{16\sqrt{3}\, T_m}{\pi d^3} \right| \quad (37.20)$$

The stochastic nature of $\mathbf{K}_f$, $\mathbf{M}_a$, and $\mathbf{d}$ controls the nature of σ_a'. Usually the geometric variation in $\mathbf{d}$ involves coefficients of variation of 0.001 or less, and that of $\mathbf{K}_f$ and $\mathbf{M}_a$ is more than an order of magnitude higher, and so $\mathbf{d}$ is usually considered deterministic. The distribution of σ_a' depends on the distributions of $\mathbf{K}_f$ and $\mathbf{M}_a$. When $\mathbf{M}_a$ is lognormal (and since $\mathbf{K}_f$ is robustly lognormal), the distribution of σ_a' is lognormal. When $\mathbf{M}_a$ is not lognormal, then a computer simulation will give the stochastic information on σ_a'.

A press fit induces a surface pressure p and a hoop normal stress of $-p$, so the three orthogonal normal stresses are σ_x, $-p$, and $-p$, and Eq. (37.16) becomes

$$\sigma' = \frac{1}{\sqrt{2}}\{[\sigma_x - (-p)]^2 + [-p - (-p)]^2 + (-p - \sigma_x)^2 + 6\tau_{xy}^2\}^{1/2}$$

$$\sigma' = [(\sigma_x + p)^2 + 3\tau_{xy}^2]^{1/2}$$

The amplitude and steady components of the von Mises stress at a surface element in a press fit are, respectively,

$$\sigma_a' = (\sigma_x^2)^{1/2} = \sigma_x \tag{37.21}$$

$$\sigma_m' = (p^2 + 3\,\tau_{xy}^2)^{1/2} \tag{37.22}$$

On the designer's fatigue diagram, the σ_a', σ_m' coordinates don't necessarily define a point because certain geometric decisions may not yet have been made. In such cases, a locus of possible points which is called the *load line* is established. Often the load line includes the origin, and so the slope together with one point on the line defines the load line. Its slope r is the ratio σ_a'/σ_m'.

37.7 STRENGTH

For the first-quadrant fatigue locus on the designer's fatigue diagram, effective regression models include the 1874 Gerber parabola and the recent ASME-elliptic locus, both of which lie in and among the data. The Gerber parabola is written as

$$\frac{n\sigma_a}{S_e} + \left(\frac{n\sigma_m}{S_{ut}}\right)^2 = 1 \tag{37.23}$$

and the failure locus itself, substituting $n\sigma_a = S_a$ and $n\sigma_m = S_m$ in Eq. (37.23), is expressible as

$$\frac{S_a}{S_e} + \left(\frac{S_m}{S_{ut}}\right)^2 = 1 \tag{37.24}$$

Combining the damaging stress [distortion energy von Mises stress, Eqs. (37.19) and (37.20)] with the strengths in Eq. (37.23) leads to

$$d = \left\{\frac{16nK_fM_a}{\pi S_e}\left[1 + \sqrt{1 + 3\left(\frac{T_mS_e}{K_fM_aS_{ut}}\right)^2}\right]\right\}^{1/3} \tag{37.25}$$

$$\frac{1}{n} = \frac{16K_fM_a}{\pi d^3 S_e}\left[1 + \sqrt{1 + 3\left(\frac{T_mS_e}{K_fM_aS_{ut}}\right)^2}\right] \tag{37.26}$$

Equations (37.25) and (37.26) are called distortion energy–Gerber equations, or D.E.–Gerber equations.

The ASME-elliptic of Ref. [37.4] has a fatigue locus in the first quadrant expressed as

$$\left(\frac{n\sigma_a}{S_e}\right)^2 + \left(\frac{n\sigma_m}{S_y}\right)^2 = 1 \tag{37.27}$$

and the fatigue locus itself is expressed as

$$\left(\frac{S_a}{S_e}\right)^2 + \left(\frac{S_m}{S_y}\right)^2 = 1 \tag{37.28}$$

Combining Eqs. (37.19) and (37.20) with (37.28) gives

$$d = \left\{ \frac{32n}{\pi} \left[\left(\frac{K_f M_a}{S_e}\right)^2 + \frac{3}{4} \left(\frac{T_m}{S_y}\right)^2 \right]^{1/2} \right\}^{1/3} \tag{37.29}$$

$$\frac{1}{n} = \frac{32}{\pi d^3} \left[\left(\frac{K_f M_a}{S_e}\right)^2 + \frac{3}{4} \left(\frac{T_m}{S_y}\right)^2 \right]^{1/2} \tag{37.30}$$

which are called D.E.–elliptic or ASME-elliptic equations.

On the designer's fatigue diagram, the slope of a radial load line r is given by

$$r = \frac{\sigma_a'}{\sigma_m'} = \frac{32 K_f M_a}{\pi d^3} \cdot \frac{\pi d^3}{16\sqrt{3} T_m} = \frac{2}{\sqrt{3}} \frac{K_f M_a}{T_m} \tag{37.31}$$

The expressions for d and n in Eqs. (37.29) and (37.30) are for a threat from fatigue failure. It is also possible on the first revolution to cause local yielding, which changes straightness and strength and involves now-unpredictable loading. The Langer line, $S_a + S_m = S_y$, predicts yielding on the first cycle. The point where the elliptic locus and the Langer line intersect is described by

$$\frac{S_a}{S_e} = \frac{2S_e/S_y}{1 + (S_e/S_y)^2} \tag{37.32}$$

$$\frac{S_m}{S_y} = \frac{1 - (S_e/S_y)^2}{1 + (S_e/S_y)^2} \tag{37.33}$$

The *critical slope* contains this point:

$$r_{crit} = \frac{2(S_e/S_y)^2}{1 - (S_e/S_y)^2} \tag{37.34}$$

If the load line slope r is greater than r_{crit}, then the threat is from fatigue. If r is less than r_{crit}, the threat is from yielding.

For the Gerber fatigue locus, the intersection with the Langer line is described by

$$S_a = \frac{S_{ut}^2 - 2S_e S_y}{2S_e} \left[-1 + \sqrt{1 + \frac{4(S_{ut}^2 - S_y^2)}{(S_{ut}^2/S_e - 2S_y)^2}} \right] \tag{37.35}$$

$$S_m = \frac{S_{ut}^2}{2S_e} \left[1 - \sqrt{1 + \frac{4S_e^2(1 - S_y/S_e)}{S_{ut}^2}} \right] \tag{37.36}$$

and $r_{crit} = S_a/S_m$.

Example 4.   At the critical location on a shaft, the bending moment M_a is 2520 in · lbf and the torque T_m is 6600 in · lbf. The ultimate strength S_{ut} is 80 kpsi, the yield strength S_y is 58 kpsi, and the endurance limit S_e is 31.1 kpsi. The stress concentration factor corrected for notch sensitivity K_f is 1.54. Using an ASME-elliptic fatigue locus, ascertain if the threat is from fatigue or yielding.

Solution. From Eq. (37.31),

$$r = \frac{2(1.54)2520}{\sqrt{3}\ 6600} = 0.679$$

From Eq. (37.34),

$$r_{crit} = \frac{2(31.1/58)^2}{1 - (31.1/58)^2} = 0.807$$

Since $r_{crit} > r$, the primary threat is from fatigue. Using the Gerber fatigue locus, $r_{crit} = S_a/S_m = 26.18/31.8 = 0.823$.

For the distortion energy–Gerber failure locus, the relation for the strength amplitude S_a is given in Eq. (13.34) and C_{Sa} in Eq. (13.35); these quantities are given by Eqs. (13.37) and (13.38), respectively, for the ASME-elliptic failure locus.

37.8   CRITICAL SPEEDS

Critical speeds are associated with uncontrolled large deflections, which occur when inertial loading on a slightly deflected shaft exceeds the restorative ability of the shaft to resist. Shafts must operate well away from such speeds. Rayleigh's equation for the first critical speed of a shaft with transverse inertial loads w_i deflected y_i from the axis of rotation for simple support is given by Ref. [37.6] as

$$\omega = \sqrt{\frac{g\Sigma w_i y_i}{\Sigma w_i y_i^2}} \tag{37.37}$$

where w_i is the inertial load and y_i is the lateral deflection due to w_i and *all other loads.* For the shaft itself, w_i is the inertial load of a shaft section and y_i is the deflection of the center of the shaft section due to all loads. Inclusion of shaft mass when using Eq. (37.37) can be done.

Reference [37.7], p. 266, gives the first critical speed of a uniform simply supported shaft as

$$\omega = \frac{\pi^2}{\ell^2}\sqrt{\frac{EI}{m}} = \frac{\pi^2}{\ell^2}\sqrt{\frac{gEI}{A\gamma}} \tag{37.38}$$

Example 5.   A steel thick-walled tube with 3-in OD and 2-in ID is used as a shaft, simply supported, with a 48-in span. Estimate the first critical speed (*a*) by Eq. (37.38) and (*b*) by Eq. (37.37).

Solution.   (*a*) $A = \pi(3^2 - 2^2)/4 = 3.927$ in^2, $I = \pi(3^4 - 2^4)/64 = 3.19$ in^4, $w = A\gamma = 3.925(0.282) = 1.11$ lbf/in. From Eq. (37.38),

$$\omega = \frac{\pi^2}{48^2}\sqrt{\frac{386(30)(10^6)(3.19)}{3.927(0.282)}} = 782.4\ \text{rad/s} = 7471\ \text{r/min}$$

(b) Divide the shaft into six segments, each 8 in long, and from the equation for the deflection at x of a uniformly loaded, simply supported beam, develop an expression for the deflection at x.

$$y = \frac{wx}{24EI}(2\ell x^2 - x^3 - \ell^3) = \frac{1.11x}{24(30)(10^6)(3.19)}[2(48)x^2 - x^3 - 48^3]$$

$$= 0.483(10^{-9})(x)(96x^2 - x^3 - 48^3)$$

Prepare a table for x, y, and y^2 at the six stations.

x_i	y_i	y_i^2
4	0.000 210 8	4.44(10^{-8})
12	0.000 570 9	32.6(10^{-8})
20	0.000 774 7	60.0(10^{-8})
28	0.000 774 7	60.0(10^{-8})
36	0.000 570 9	32.6(10^{-8})
44	0.000 210 8	4.44(10^{-8})
Σ	0.003 112 8	194(10^{-8})

From Eq. (37.37),

$$\omega = \sqrt{\frac{g\Sigma y_i}{\Sigma y_i^2}} = \sqrt{\frac{386(0.003\ 112\ 8)}{194(10^{-8})}} = 787 \text{ rad/s} = 7515 \text{ r/min}$$

which agrees with the result of part (a), but is slightly higher, as expected, since the static deflection shape was used.

Since most shafts are of variable diameter, Eq. (37.37) will be more useful for estimating the first critical speed, treating simultaneously the contributions of concentrated masses (gears, pulleys, sprokets, cams, etc.) and the distributed shaft mass as well.

Example 6.   Assume that the shaft of Example 2 has been established with its final geometry as shown in Fig. 37.7. The shaft is decomposed into 2-in segments. The weight of each segment is applied as a concentrated force w_i at the segment centroid. Additionally, the left-side gear weighs 30 lbf and the right-side gear weighs 40 lbf. Estimate the first critical speed of the assembly.

Solution.   Bearing in mind the tabular deflection method of Sec. 37.2, twelve stations are established. Also, bending moment diagrams will be superposed.

For the distributed shaft mass load, the shaft weight is estimated as $W = 24.52$ lbf, and it follows that bearing reactions are $R_1' = 11.75$ lbf and $R_2' = 12.77$ lbf. Because each reaction is opposed by a bearing seat weight of 1.772 lbf, the net reactions are $R_1 = 11.75 - 1.722 = 9.98$ lbf and $R_2 = 12.77 - 1.772 = 11.0$ lbf. The bending moments M_i due to shaft segment weights are shown in column 3 of Table 37.5.

For the gears, $R_1 = 31.25$ lbf and $R_2 = 38.75$ lbf, and the resulting bending moments are shown in column 4. The superposition of the moment diagrams for these two sources of bending occurs in column 5. Column 6 displays the shaft segment weights at the station of application. Column 7 shows the concentrated gear weights and their station of application. Column 8 is the superposition of columns 6 and 7. Column 9 is obtained by using the tabular method of Sec. 37.2 and imposing the bending moment diagram of column 5. Columns 10 and 11 are extensions of columns 8 and 9. The sums of columns 10 and 11 are used in Eq. (37.37):

$$\omega = \sqrt{\frac{386(2.348)(10^{-3})}{6.91(10^{-8})}} = 3622 \text{ rad/s} = 34\ 588 \text{ r/min}$$

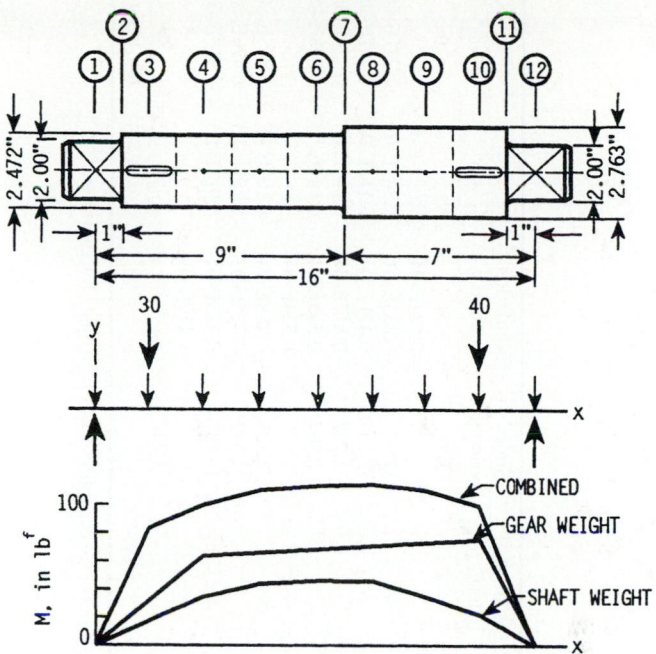

FIGURE 37.7 The final geometry of the shaft of Ex. 37.2. For critical speed estimation, weights of shaft segments and affixed gears generate separate and combined bending moments. The static deflection under such loading found by the tabulation method provides the deflections used in Rayleigh's critical speed equation. See Table 37.5 and Ex. 37.6.

The methods of Secs. 37.2 and 37.3 and this section can be programmed for the digital computer for rapid and convenient use.

37.9 HOLLOW SHAFTS

Advantages accruing to hollow shafting include weight reduction with minor increase in stress (for the same outside diameter), ability to circulate fluids for lubrication or cooling, and the use of thick-walled tubing as shaft stock. However, unbalance must be checked and corrected, and thick-walled tubing may not have enough material in its wall to accommodate the desired external geometry.

For a shaft section with outside diameter d, inside diameter d_i, and $K = d_i/d$, for torsional and bending loading, $d(1 - K^4)^{1/3}$ may be substituted for diameter d in equations such as (37.25), (37.26), (37.29), and (37.30). Equations (37.25) and (37.29) can no longer be solved explicitly for diameter d unless K is known. In cases where it is not known, iterative procedures must be used.

TABLE 37.5 Critical Speed Tabulation for Example 7

Station	x_i	Distributed load M_i	Gear M_i	Super-posed M_i	Shaft section w_i	Concentrated load P_i	Super-posed w_i	Tabulation[†] method y_i	w_iy_i	$w_iy_i^2$
1	0	0	0	0	1.772		1.772	0	0	0
2	1	9.98	31.25	41.23	2.707	30	32.07	-0.122E-04	0.747E-03	1.741E-08
3	2	19.96	62.5	82.46	2.707		2.707	-0.233E-04	0.111E-03	0.457E-08
4	4	34.51	65.0	99.51	2.707		2.707	-0.411E-04	0.140E-03	0.724E-08
5	6	43.64	67.5	111.14	2.707		2.707	-0.517E-04	0.147E-03	0.798E-08
6	8	47.36	70.0	117.36				-0.543E-04		
7	9	46.51	71.25	117.76				-0.524E-04		
8	10	45.66	72.5	118.16	3.382		3.382	-0.488E-04	0.165E-03	0.805E-08
9	12	37.20	75.0	112.2	3.382		3.382	-0.375E-04	0.127E-03	0.476E-08
10	14	21.98	77.5	99.48	3.382	40	43.382	-0.210E-04	0.911E-03	1.913E-08
11	15	10.99	38.75	49.74				-0.110E-04		
12	16	0	0	0	1.772		1.772	0	0	0
									2.348E-03	6.91E-08

[†] Column 9 obtained by tabular method of Sec. 37.2. The constants $C_1 = -0.1249E-04$ and $C_2 = 0$ of prediction Eqs. (37.5) and (37.6) were used.

REFERENCES

37.1 Joseph E. Shigley and Charles R. Mischke, *Mechanical Engineering Design,* 5th ed., McGraw-Hill, New York, 1989.

37.2 Charles R. Mischke, "An Exact Numerical Method for Determining the Bending Deflection and Slope of Stepped Shafts," *Advances in Reliability and Stress Analysis,* Proceedings of the Winter Annual Meeting of A.S.M.E., San Francisco, December 1978, pp. 105–115.

37.3 R. Bruce Hopkins, *Design Analysis of Shafts and Beams,* McGraw-Hill, New York, 1970, pp. 93–99.

37.4 ANSI/ASME B106.1-M-1985, "Design of Transmission Shafting," second printing, March 1986.

37.5 S. Timoshenko, D. H. Young, and W. Weaver, *Vibration Problems in Engineering,* 4th ed., John Wiley & Sons, New York, 1974.

37.6 S. Timoshenko and D. H. Young, *Advanced Dynamics,* McGraw-Hill, New York, 1948, p. 296.

37.7 Charles R. Mischke, *Elements of Mechanical Analysis,* Addison-Wesley, Reading, Mass., 1963.

RECOMMENDED READING

ANSI B17.1, 1967, "Keys and Keyseats."

Mischke, Charles R., "A Probabilistic Model of Size Effect in Fatigue Strength of Rounds in Bending and Torsion," *Transactions of A.S.M.E., Journal of Mechanical Design,* vol. 102, no. 1, January 1980, pp. 32–37.

Peterson, R. E., *Stress Concentration Factors,* John Wiley & Sons, New York, 1974.

Pollard, E. I., "Synchronous Motors . . . , Avoid Torsional Vibration Problems," *Hydrocarbons Processing,* February 1980, pp. 97–102.

Umasankar, G., and C. R. Mischke, "A Simple Numerical Method for Determining the Sensitivity of Bending Deflections of Stepped Shafts to Dimensional Changes," *Transactions of A.S.M.E., Journal of Vibration, Acoustics, Stress and Reliability in Design,* vol. 107, no. 1, January 1985, pp. 141–146.

Umasankar, G., and C. Mischke, "Computer-Aided Design of Power Transmission Shafts Subjected to Size, Strength and Deflection Constraints Using a Nonlinear Programming Technique," *Transactions of A.S.M.E., Journal of Vibration, Acoustics, Stress and Reliability in Design,* vol. 107, no. 1, January 1985, pp. 133–140.

CHAPTER 38
VIBRATION AND CONTROL OF VIBRATION

T. S. Sankar, Ph.D., Eng.
Professor and Chairman
Department of Mechanical Engineering
Concordia University
Montreal, Quebec, Canada

R. B. Bhat, Ph.D.
Associate Professor
Department of Mechanical Engineering
Concordia University
Montreal, Quebec, Canada

38.1 INTRODUCTION

Vibration analysis and control of vibrations are important and integral aspects of every machine design procedure. Establishing an appropriate mathematical model, its analysis, interpretation of the solutions, and incorporation of these results in the design, testing, evaluation, maintenance, and troubleshooting require a sound understanding of the principles of vibration. All the essential materials dealing with various aspects of machine vibrations are presented here in a form suitable for most design applications. Readers are encouraged to consult the references for more details.

38.2 SINGLE-DEGREE-OF-FREEDOM SYSTEMS

38.2.1 Free Vibration

A single-degree-of-freedom system is shown in Fig. 38.1. It consists of a mass m constrained by a spring of *stiffness k,* and a damper with *viscous damping coefficient c.* The stiffness coefficient k is defined as the spring force per unit deflection. The coef-

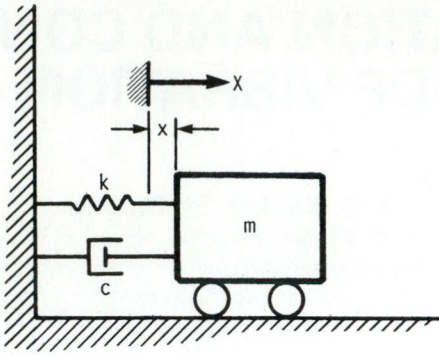

FIGURE 38.1   Representation of a single-degree-of-freedom system.

ficient of viscous damping c is the force provided by the damper opposing the motion per unit velocity.

If the mass is given an initial displacement, it will start vibrating about its equilibrium position. The equation of motion is given by

$$m\ddot{x} + c\dot{x} + kx = 0 \tag{38.1}$$

where x is measured from the equilibrium position and dots above variables represent differentiation with respect to time. By substituting a solution of the form $x = e^{st}$ into Eq. (38.1), the characteristic equation is obtained:

$$ms^2 + cs + k = 0 \tag{38.2}$$

The two roots of the characteristic equation are

$$s = \zeta\omega_n \pm i\omega_n(1 - \zeta^2)^{1/2} \tag{38.3}$$

where $\omega_n = (k/m)^{1/2}$ is undamped *natural frequency*
$\zeta = c/c_c$ is *damping ratio*
$c_c = 2m\omega_n$ is *critical damping coefficient*
$i = \sqrt{-1}$

Depending on the value of ζ, four cases arise.

Undamped System ($\zeta = 0$).   In this case, the two roots of the characteristic equation are

$$s = \pm i\omega_n = \pm i(k/m)^{1/2} \tag{38.4}$$

and the corresponding solution is

$$x = A \cos \omega_n t + B \sin \omega_n t \tag{38.5}$$

where A and B are arbitrary constants depending on the initial conditions of the motion. If the initial displacement is x_0 and the initial velocity is v_0, by substituting these values in Eq. (38.5) it is possible to solve for constants A and B. Accordingly, the solution is

$$x = x_0 \cos \omega_n t + \frac{v_0}{\omega_n} \sin \omega_n t \qquad (38.6)$$

Here, ω_n is the natural frequency of the system in radians per second (rad/s), which is the frequency at which the system executes free vibrations. The *natural frequency* is

$$f_n = \frac{\omega_n}{2\pi} \qquad (38.7)$$

where f_n is in cycles per second, or hertz (Hz). The *period* for one oscillation is

$$\tau = \frac{1}{f_n} = \frac{2\pi}{\omega_n} \qquad (38.8)$$

The solution given in Eq. (38.6) can also be expressed in the form

$$x = X \cos (\omega_n - \theta) \qquad (38.9)$$

where

$$X = \left[x_0^2 + \left(\frac{v_0}{\omega_n} \right)^2 \right]^{1/2} \qquad \theta = \tan^{-1} \frac{v_0}{\omega_n x_0} \qquad (38.10)$$

The motion is harmonic with a *phase angle* θ as given in Eq. (38.9) and is shown graphically in Fig. 38.4.

Underdamped System (0 < ζ < 1). When the system damping is less than the critical damping, the solution is

$$x = [\exp(-\zeta\omega_n t)] \, (A \cos \omega_d t + B \sin \omega_d t) \qquad (38.11)$$

where

$$\omega_d = \omega_n (1 - \zeta^2)^{1/2} \qquad (38.12)$$

is the *damped natural frequency* and A and B are arbitrary constants to be determined from the initial conditions. For an initial amplitude of x_0 and initial velocity v_0,

$$x = [\exp(-\zeta\omega_n t)] \left(x_0 \cos \omega_d t + \frac{\zeta\omega_n x_0 + v_0}{\omega_d} \sin \omega_d t \right) \qquad (38.13)$$

which can be written in the form

$$x = [\exp(-\zeta\omega_n t)] \, X \cos (\omega_d t - \theta)$$

$$X = \left[x_0^2 + \left(\frac{\zeta\omega_n x_0 + v_0}{\omega_d} \right)^2 \right]^{1/2} \qquad (38.14)$$

and

$$\theta = \tan^{-1} \frac{\zeta\omega_n x_0 + v_0}{\omega_d}$$

An underdamped system will execute exponentially decaying oscillations, as shown graphically in Fig. 38.2.

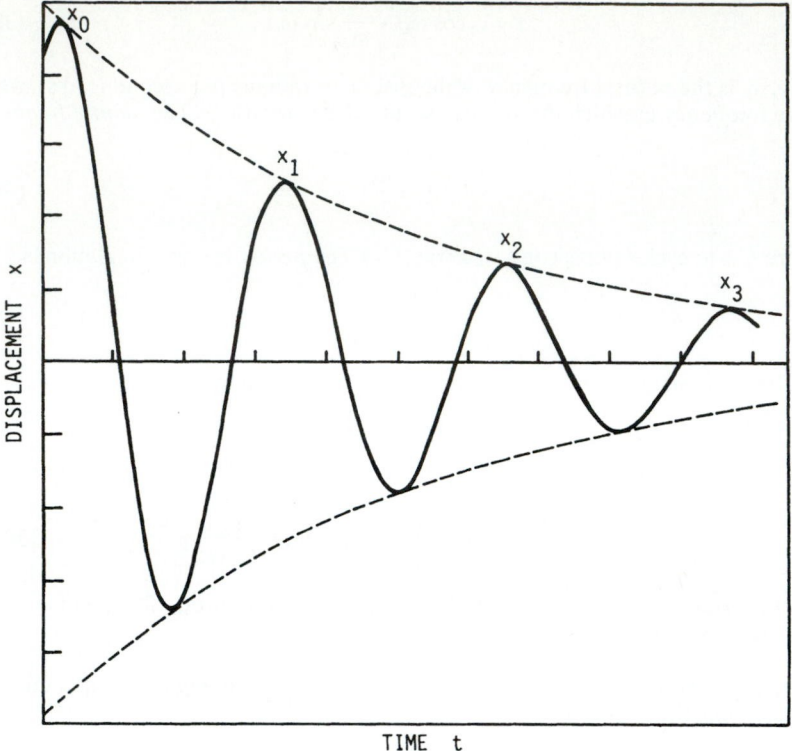

FIGURE 38.2   Free vibration of an underdamped single-degree-of-freedom system.

The successive maxima in Fig. 38.2 occur in a periodic fashion and are marked $X_0, X_1, X_2, \ldots$ The ratio of the maxima separated by n cycles of oscillation may be obtained from Eq. (38.13) as

$$\frac{X_n}{X_0} = \exp{(-n\delta)} \tag{38.15}$$

where

$$\delta = \frac{2\pi\zeta}{(1 - \zeta^2)^{1/2}}$$

is called the *logarithmic decrement* and corresponds to the ratio of two successive maxima in Fig. 38.2. For small values of damping, that is, $\zeta \ll 1$, the logarithmic decrement can be approximated by

$$\delta = 2\pi\zeta \tag{38.16}$$

Using this in Eq. (38.14), we find

$$\frac{X_n}{X_0} = \exp{(-2\pi n\zeta)} \simeq 1 - 2\pi n\zeta \tag{38.17}$$

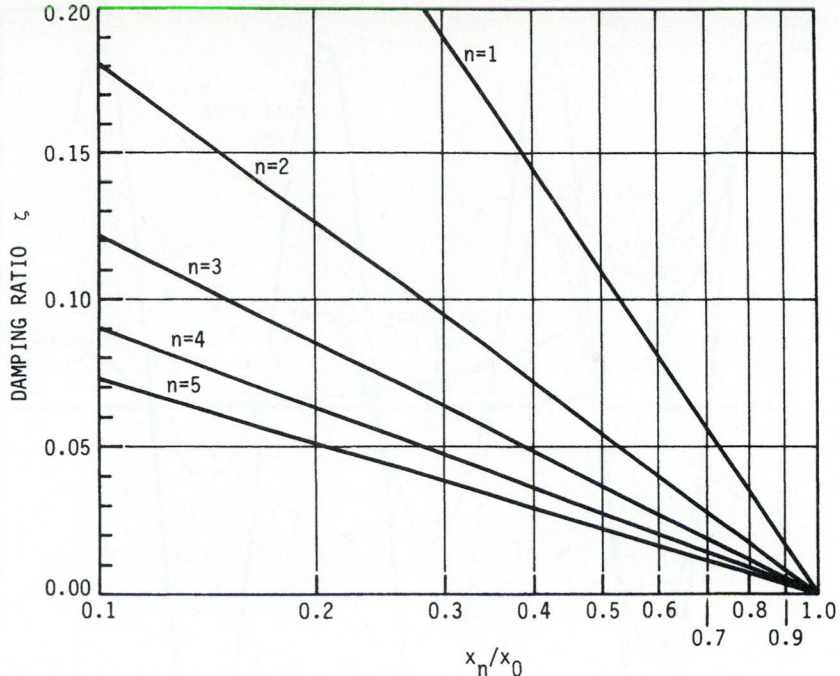

FIGURE 38.3   Variation of the ratio of displacement maxima with damping.

The equivalent viscous damping in a system is measured experimentally by using this principle. The system at rest is given an impact which provides initial velocity to the system and sets it into free vibration. The successive maxima of the ensuing vibration are measured, and by using Eq. (38.17) the damping ratio can be evaluated. The variation of the decaying amplitudes of free vibration with the damping ratio is plotted in Fig. 38.3 for different values of n.

Critically Damped System ($\zeta = 1$).   When the system is critically damped, the roots of the characteristic equation given by Eq. (38.3) are equal and negative real quantities. Hence, the system does not execute oscillatory motion. The solution is of the form

$$x = (A + Bt) \exp(-\omega_n t) \tag{38.18}$$

and after substitution of initial conditions,

$$x = [x_0 + (v_0 + x_0\omega_n)t] \exp(-\omega_n t) \tag{38.19}$$

This motion is shown graphically in Fig. 38.4, which gives the shortest time to rest.

Overdamped System ($\zeta > 1$).   When the damping ratio ζ is greater than unity, there are two distinct negative real roots for the characteristic equation given by Eq. (38.3). The motion in this case is described by

$$x = \exp(-\zeta\omega_n t)\, [A \exp \omega_n t\sqrt{\zeta^2 - 1} + B \exp(-\omega_n t\sqrt{\zeta^2 - 1})] \tag{38.20}$$

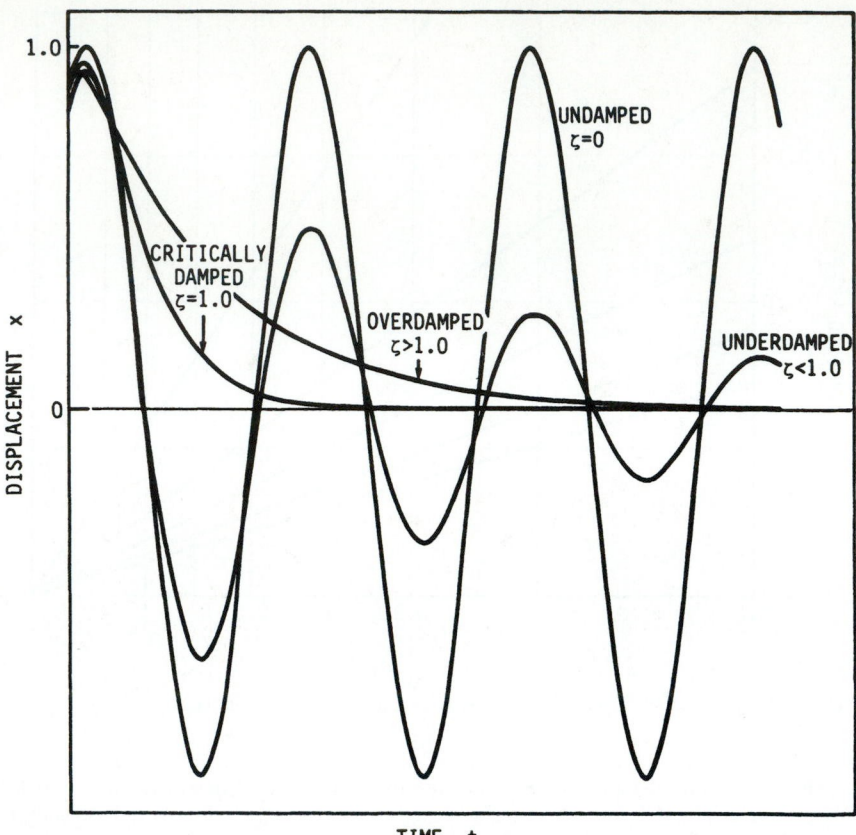

FIGURE 38.4 Free vibration of a single-degree-of-freedom system under different values of damping.

where

$$A = \frac{1}{2}\left(x_0 + \frac{v_0 + \zeta\omega_n x_0}{\omega_n}\right) \qquad B = \frac{1}{2}\left(\frac{x_0 + \zeta\omega_n x_0}{\omega_0}\right)$$

and

$$\omega_0 = \omega_n\sqrt{\zeta^2 - 1}$$

All four types of motion are shown in Fig. 38.4.

If the mass is suspended by a spring and damper as shown in Fig. 38.5, the spring will be stretched by an amount δ_{st}, the static deflection in the equilibrium position. In such a case, the equation of motion is

$$m\ddot{x} + c\dot{x} + k(x + \delta_{st}) = mg \tag{38.21}$$

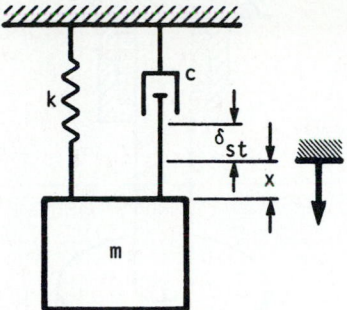

FIGURE 38.5 Model of a single-degree-of-freedom system showing the static deflection due to weight.

Since the force in the spring due to the static equilibrium is equal to the weight, or $k\delta_{st} = mg = W$, the equation of motion reduces to

$$m\ddot{x} + c\dot{x} + kx = 0 \tag{38.22}$$

which is identical to Eq. (38.1). Hence the solution is also similar to that of Eq. (38.1). In view of Eq. (38.21) and since $\omega_n = (k/m)^{1/2}$, the natural frequency can also be obtained by

$$\omega_n = \left(\frac{g}{\delta_{st}}\right)^{1/2} \tag{38.23}$$

An approximate value of the fundamental natural frequency of any complex mechanical system can be obtained by reducing it to a single-degree-of-freedom system. For example, a shaft supporting several disks (wheels) can be reduced to a single-degree-of-freedom system by lumping the masses of all the disks at the center and obtaining the equivalent stiffness of the shaft by using simple flexure theory.

38.2.2 Torsional Systems

Rotating shafts transmitting torque will experience torsional vibrations if the torque is nonuniform, as in the case of an automobile crankshaft.

In rotating shafts involving gears, the transmitted torque will fluctuate because of gear-mounting errors or tooth profile errors, which will result in torsional vibration of the geared shafts.

A single-degree-of-freedom torsional system is shown in Fig. 38.6. It has a massless shaft of torsional stiffness k, a damper with damping coefficient c, and a disk with polar mass moment of inertia J. The torsional stiffness is defined as the resisting torque of the shaft per unit of angular twist, and the damping coefficient is the resisting torque of the damper per unit of angular velocity. Either the damping can be externally applied, or it can be inherent structural damping. The equation of motion of the system in torsion is given

$$J\ddot{\theta} + c\dot{\theta} + k\theta = 0 \tag{38.24}$$

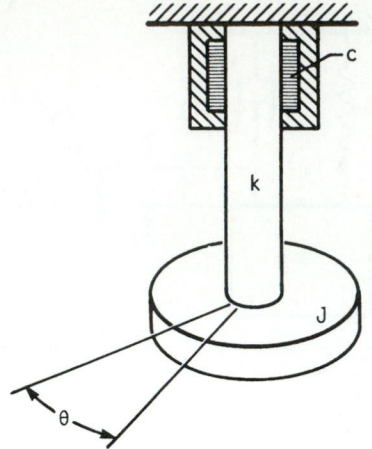

FIGURE 38.6 A representation of a one-freedom torsional system.

Equation (38.24) is in the same form as Eq. (38.1), except that the former deals with moments whereas the latter deals with forces. The solution of Eq. (38.24) will be of the same form as that of Eq. (38.1), except that J replaces m and k and c refer to torsional stiffness and torsional damping coefficient.

38.2.3 Forced Vibration

System Excited at the Mass. A vibrating system with a sinusoidal force acting on the mass is shown in Fig. 38.7. The equation of motion is

$$m\ddot{x} + c\dot{x} + kx = F_0 \sin \omega t \tag{38.25}$$

Assuming that the steady-state response lags behind the force by an angle θ, we see that the solution can be written in the form

$$x_s = X \sin (\omega t - \theta) \tag{38.26}$$

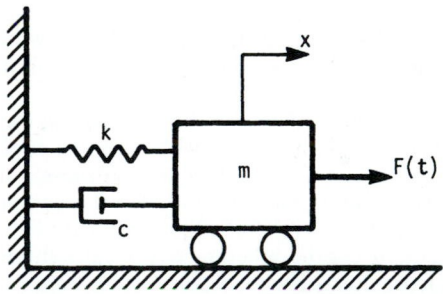

FIGURE 38.7 Oscillating force $F(t)$ applied to the mass.

Substituting in Eq. (38.26), we find that the steady-state solution can be obtained:

$$x_s = \frac{(F_0/k)\sin(\omega t - \theta)}{[(1 - \omega^2/\omega_n^2)^2 + (2\zeta\omega/\omega_n)^2]^{1/2}} \tag{38.27}$$

Using the complementary part of the solution from Eq. (38.19), we see that the complete solution is

$$x = x_s + \exp(-\zeta\omega_n t)\,[A\,\exp(\omega_n t\sqrt{\zeta^2 - 1}) + B\,\exp(-\omega_n t\sqrt{\zeta^2 - 1})] \tag{38.28}$$

If the system is undamped, the response is obtained by substituting $c = 0$ in Eq. (38.25) or $\zeta = 0$ in Eq. (38.28). When the system is undamped, if the exciting frequency coincides with the system natural frequency, say $\omega/\omega_n = 1.0$, the system response will be infinite. If the system is damped, the complementary part of the solution decays exponentially and will be nonexistent after a few cycles of oscillation; subsequently the system response is the steady-state response. At steady state, the nondimensional response amplitude is obtained from Eq. (38.27) as

$$\frac{X}{F_0/k} = \left[\left(\frac{1 - \omega^2}{\omega_n^2}\right)^2 + \left(\frac{2\zeta\omega}{\omega_n}\right)^2\right]^{-1/2} \tag{38.29}$$

and the phase between the response and the force is

$$\theta = \tan^{-1}\frac{2\zeta\omega/\omega_n}{1 - \omega^2/\omega_n^2} \tag{38.30}$$

When the forcing frequency ω coincides with the damped natural frequency ω_d, the response amplitude is given by

$$\frac{X_{\max}}{F_0/k} = \frac{1}{\zeta(4 - 3\zeta^2)^{1/2}} \tag{38.31}$$

The maximum response or resonance occurs when $\omega = \omega_n(1 - 2\zeta^2)^{1/2}$ and is

$$\frac{X_{\max}}{F_0/k} = \frac{1}{2\zeta(1 - \zeta^2)^{1/2}} \tag{38.32}$$

For structures with low damping, ω_d approximately equals ω_n, and the maximum response is

$$\frac{X_{\max}}{F_0/k} = \frac{1}{2\zeta} \tag{38.33}$$

The response amplitude in Eq. (38.29) is plotted against the forcing frequency in Fig. 38.8. The curves start at unity, reach a maximum in the neighborhood of the system natural frequency, and decay to zero at large values of the forcing frequency. The response is larger for a system with low damping, and vice versa, at any given frequency. The phase difference between the response and the excitation as given in Eq. (38.30) is plotted in Fig. 38.9. For smaller forcing frequencies, the response is nearly in phase with the force; and in the neighborhood of the system natural frequency, the response lags behind the force by approximately 90°. At large values of forcing frequencies, the phase is around 180°.

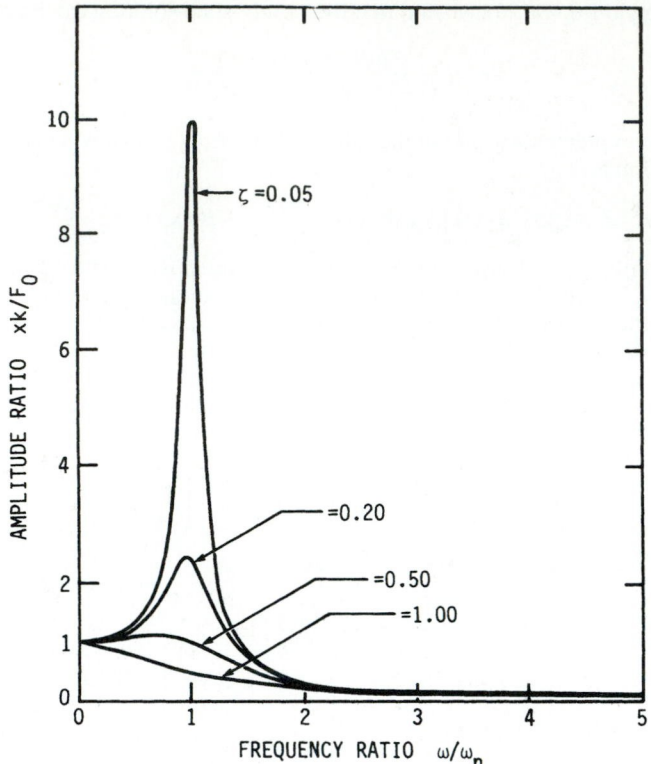

FIGURE 38.8   Displacement-amplitude frequency response due to oscillating force.

Steady-State Velocity and Acceleration Response.   The steady-state velocity response is obtained by differentiating the displacement response, given by Eq. (38.27), with respect to time:

$$\frac{\dot{x}_s}{F_0\omega_n/k} = \frac{\omega/\omega_n}{[(1 - \omega^2/\omega_n^2)^2 + (2\zeta\omega/\omega_n)^2]^{1/2}} \tag{38.34}$$

And the steady-state acceleration response is obtained by further differentiation and is

$$\frac{\ddot{x}_s}{F_0\omega_n^2/k} = \frac{(\omega/\omega_n)^2}{[(1 - \omega^2/\omega_n^2)^2 + (2\zeta\omega/\omega_n)^2]^{1/2}} \tag{38.35}$$

These are shown in Figs. 38.10 and 38.11 and also can be obtained directly from Fig. 38.8 by multiplying the amplitude by ω/ω_n and $(\omega/\omega_n)^2$, respectively.

Force Transmissibility.   The force F_T transmitted to the foundation by a system subjected to an external harmonic excitation is

$$F_T = c\dot{x} + kx \tag{38.36}$$

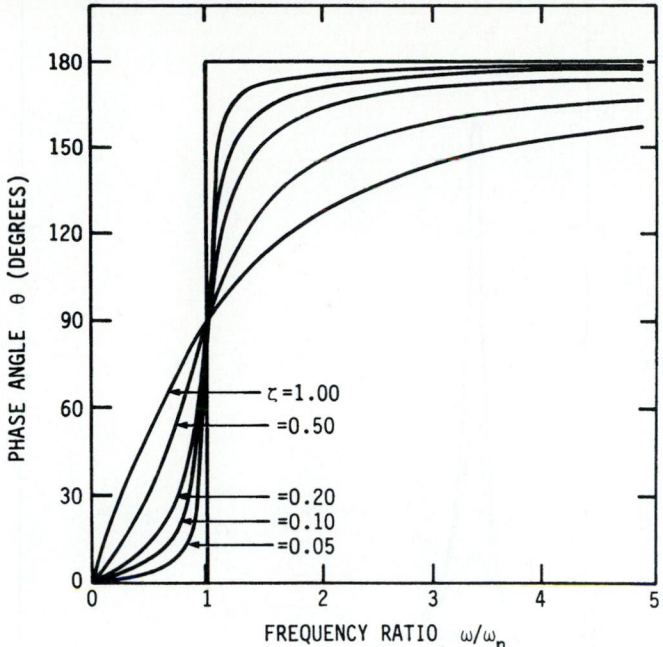

FIGURE 38.9   Phase-angle frequency response for forced motion.

Substituting the system response from Eq. (38.27) into Eq. (38.36) gives

$$\frac{F_T}{F_0} = T \sin(\omega t - \theta) \tag{38.37}$$

where the nondimensional magnitude of the transmitted force T is given by

$$T = \left[\frac{1 + (2\zeta\omega/\omega_n)^2}{(1 - \omega^2/\omega_n^2)^2 + (2\zeta\omega/\omega_n)^2}\right]^{1/2} \tag{38.38}$$

and the phase between F_T and F_0 is given by

$$\theta = \tan^{-1}\frac{2\zeta(\omega/\omega_n)^3}{1 - \omega^2/\omega_n^2 + 4\zeta^2\omega^2/\omega_n^2} \tag{38.39}$$

The *transmissibility* T is shown in Fig. 38.12 versus forcing frequency. At very low forcing frequencies, the transmissibility is close to unity, showing that the applied force is directly transmitted to the foundation. The transmissibility is very large in the vicinity of the system natural frequency, and for high forcing frequencies the transmitted force decreases considerably. The phase variation between the transmitted force and the applied force is shown in Fig. 38.13.

Rotating Imbalance.   When machines with rotating imbalances are mounted on elastic supports, they constitute a vibrating system subjected to excitation from the

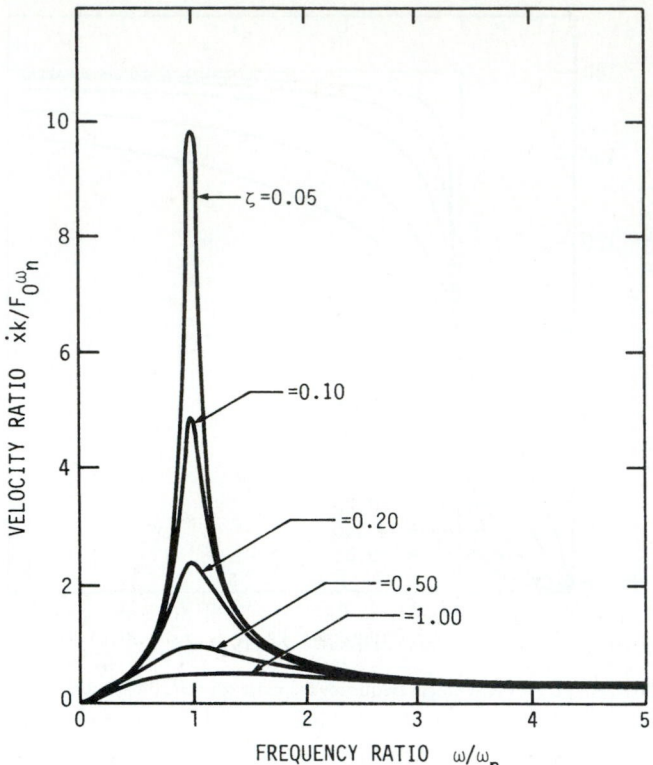

FIGURE 38.10   Velocity frequency response.

rotating imbalance. If the natural frequency of the system coincides with the frequency of rotation of the machine imbalance, it will result in severe vibrations of the machine and the support structure.

Consider a machine of mass M supported as shown in Fig. 38.14. Let the imbalance be a mass m with an eccentricity e and rotating with a frequency ω. Consider the motion x of the mass $M - m$, with x_m as the motion of the unbalanced mass m relative to the machine mass M. The equation of motion is

$$(M - m)\ddot{x} + m(\ddot{x} + \ddot{x}_m) + c\dot{x} + kx = 0 \qquad (38.40)$$

The motion of the unbalanced mass relative to the machine is

$$x_m = e \sin \omega t \qquad (38.41)$$

Substitution in Eq. (38.40) leads to

$$M\ddot{x} + c\dot{x} + kx = me\omega^2 \sin \omega t \qquad (38.42)$$

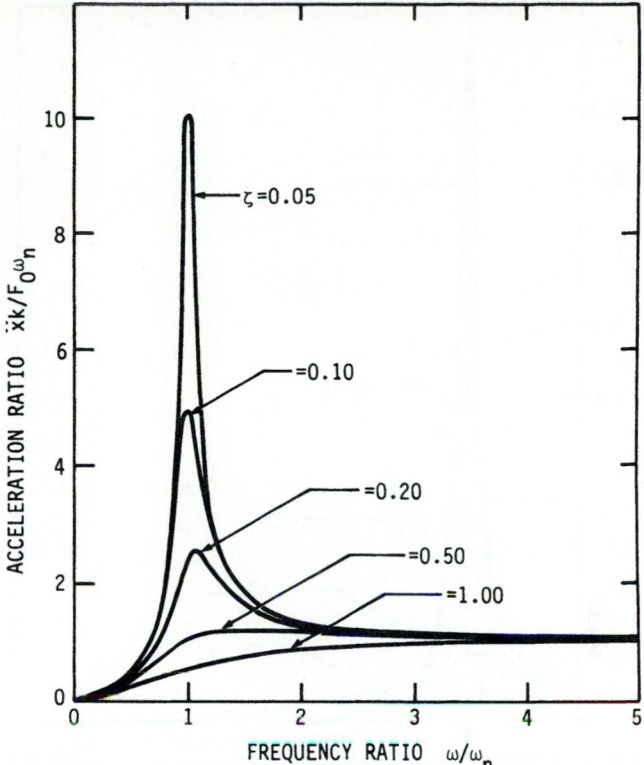

FIGURE 38.11   Acceleration frequency response.

This equation is similar to Eq. (38.25), where the force amplitude F_0 is replaced by $me\omega^2$. Hence, the steady-state solution of Eq. (38.42) is similar in form to Eq. (38.27) and is given nondimensionally as

$$\frac{x}{e}\frac{M}{m} = \frac{(\omega/\omega_n)^2 \sin(\omega t - \theta)}{[(1 - \omega^2/\omega_n^2)^2 + (2\zeta\omega/\omega_n)^2]^{1/2}} \qquad (38.43)$$

where

$$\tan\theta = \frac{2\zeta\omega/\omega_n}{1 - \omega^2/\omega_n^2} \qquad (38.44)$$

Note that since the excitation is proportional to ω^2, the response has an ω^2 term in the numerator and resembles the acceleration response of a system subjected to a force of constant magnitude, given by Eq. (38.35). The complete solution consists of the complementary part of the solution and is

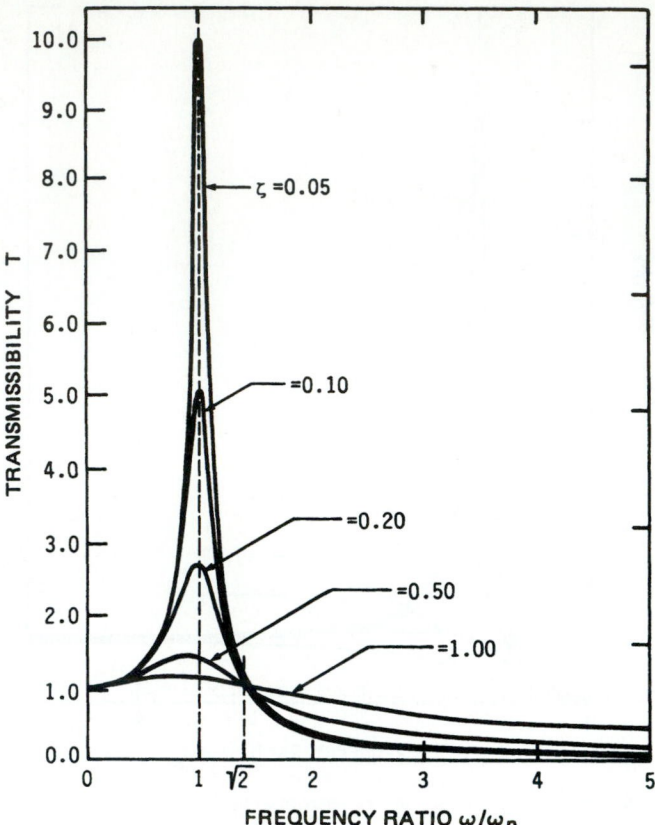

FIGURE 38.12   Transmissibility plot.

$$x = \exp -\zeta\omega_n t \left\{ A \exp \left[(\zeta^2 - 1)^{1/2} \, \omega_n t\right] + B \exp \left[-(\zeta^2 - 1)^{1/2} \, \omega_n t\right] \right\}$$

$$+ \frac{me(\omega/\omega_n)^2 \sin (\omega t - \theta)}{M[(1 - \omega^2/\omega_n^2)^2 + (2\zeta\omega/\omega_n)^2]^{1/2}} \qquad (38.45)$$

System Excited at the Foundation.   When the system is excited at the foundation, as shown in Fig. 38.15, with a certain displacement $u(t) = U_0 \sin \omega t$, the equation of motion can be written as

$$m\ddot{x} + c(\dot{x} - \dot{u}) + k(x - u) = 0 \qquad (38.46)$$

This equation can be written in the form

$$m\ddot{x} + c\dot{x} + kx = cu_0\omega \cos \omega t + ku_0 \sin \omega t \qquad (38.47)$$

$$= F_0 \sin (\omega t + \phi)$$

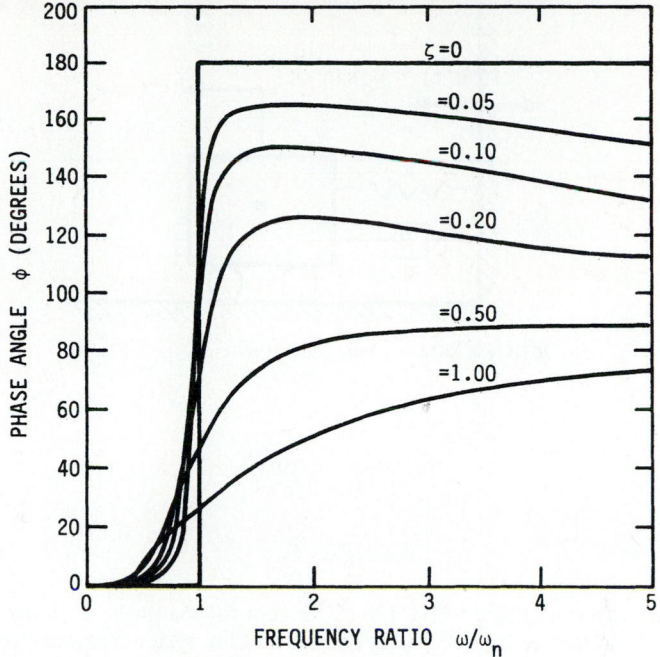

FIGURE 38.13   Phase angle between transmitted and applied forces.

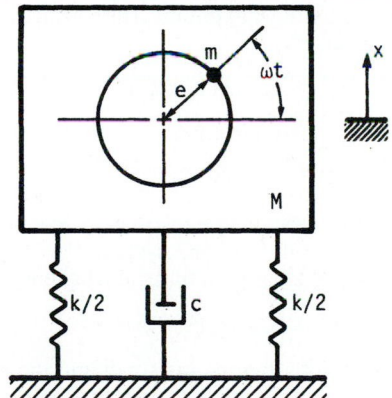

FIGURE 38.14   Dynamic system subject to unbalanced excitation.

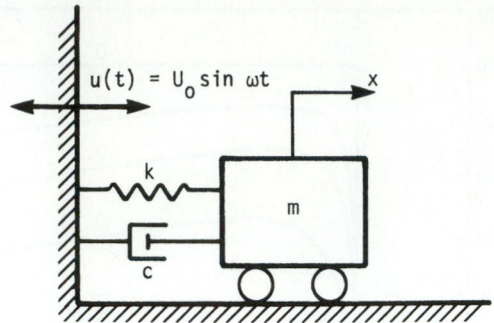

FIGURE 38.15   A base excited system.

where

$$F_0 = u_0 \, (k^2 + c^2\omega^2)^{1/2} \tag{38.48}$$

and

$$\phi = \tan^{-1} \frac{k}{c\omega} \tag{38.49}$$

Equation (38.47) is identical to Eq. (38.25) except for the phase ϕ. Hence the solution is similar to that of Eq. (38.25). If the ratio of the system response to the base displacement is defined as the motion transmissibility, it will have the same form as the force transmissibility given in Eq. (38.38).

Resonance, System Bandwidth, and Q Factor.   A vibrating system is said to be in resonance when the response is maximum. The displacement and acceleration responses are maximum when

$$\omega = \omega_n (1 - 2\zeta^2)^{1/2} \tag{38.50}$$

whereas velocity response is maximum when

$$\omega = \omega_n \tag{38.51}$$

In the case of an undamped system, the response is maximum when $\omega = \omega_n$, where ω_n is the frequency of free vibration of the system. For a damped system, the frequency of free oscillations or the damped natural frequency is given by

$$\omega_d = \omega_n \, (1 - \zeta^2)^{1/2} \tag{38.52}$$

In many mechanical systems, the damping is small and the resonant frequency and the damped natural frequency are approximately the same.

When the system has negligible damping, the frequency response has a sharp peak at resonance; but when the damping is large, the frequency response near resonance will be broad, as shown in Fig. 38.8. A section of the plot for a specific damping value is given in Fig. 38.16.

The Q factor is defined as

$$Q = \frac{1}{2\zeta} = R_{max} \tag{38.53}$$

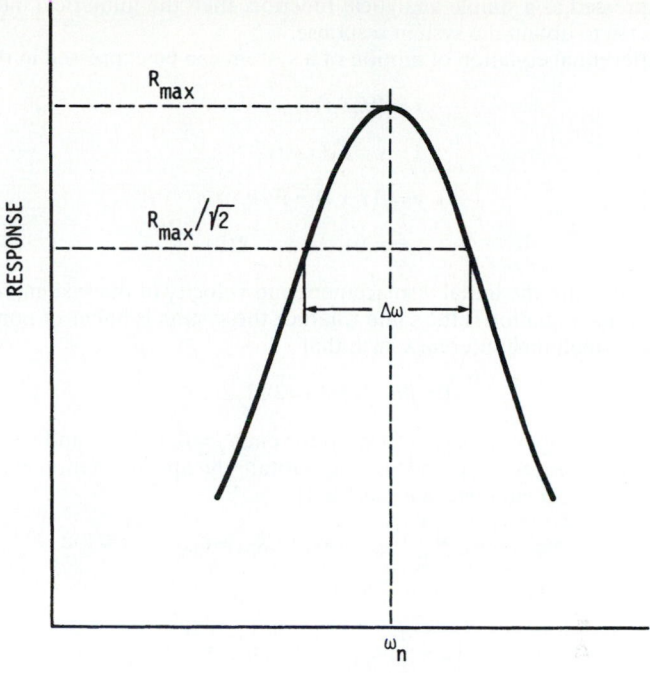

FIGURE 38.16   Resonance, bandwidth, and Q factor.

which is equal to the maximum response in physical systems with low damping. The bandwidth is defined as the width of the response curve measured at the "half-power" points, where the response is $R_{max}/\sqrt{2}$. For physical systems with $\zeta < 0.1$, the bandwidth can be approximated by

$$\Delta\omega = 2\zeta\omega_n = \frac{\omega_n}{Q} \tag{38.54}$$

Forced Vibration of Torsional Systems.   In the torsional system of Fig. 38.3, if the disk is subjected to a sinusoidal external torque, the equation of motion can be written as

$$J\ddot\theta + c\dot\theta + k\theta = T_0 \sin \omega t \tag{38.55}$$

Equation (38.55) has the same form as Eq. (38.25). Hence the solution can be obtained by replacing m by J and F_0 by T_0 and by using torsional stiffness and torsional damping coefficients for k and c, respectively, in the solution of Eq. (38.25).

38.2.4   Numerical Integration of Differential Equations of Motion: Runge-Kutta Method

When the differential equation cannot be integrated in closed form, numerical methods can be employed. If the system is nonlinear or if the system excitation cannot be expressed as a simple analytical function, then the numerical method is the only recourse to obtain the system response.

The differential equation of motion of a system can be expressed in the form

$$\ddot{x} = f(x, \dot{x}, t)$$

or

$$\dot{x} = y = F_1(x, y, t)$$

$$y = f(x, \dot{x}, t) = F_2(x, y, t) \tag{38.56}$$

$$x_0 = x(0) \qquad \dot{x}_0 = \dot{x}(0)$$

where x_0 and $\dot{x}_0$ are the initial displacement and velocity of the system, respectively. The form of the equation is the same whether the system is linear or nonlinear.

Choose a small time interval h such that

$$t_j = jh \qquad \text{for } j = 0, 1, 2, \ldots$$

Let w_{ij} denote an approximation to $x_i\,(t_j)$ for each $j = 0, 1, 2, \ldots$ and $i = 1, 2$. For the initial conditions, set $w_{1,0} = x_0$ and $w_{2,0} = \dot{x}_0$. Obtain the approximation w_{ij+1}, given all the values of the previous steps w_{ij}, as [38.1]

$$w_{i,j+1} = w_{i,j} + \frac{1}{6}\,(k_{1,i} + 2k_{2,i} + 2k_{3,i} + k_{4,i}) \qquad i = 1, 2 \tag{38.57}$$

where

$$k_{1,i} = hF_i(t_j + w_{1,j},\, w_{2,j})$$

$$k_{2,i} = hF_i\!\left(t_j + \frac{h}{2},\, w_{1,j} + \frac{1}{2}\,k_{1,1},\, w_{2i,j} + \frac{1}{2}\,k_{1,2}\right)$$

$$k_{3,i} = hF_i\!\left(t_j + \frac{h}{2},\, w_{1,j} + \frac{1}{2}\,k_{2,1},\, w_{2,j} + \frac{1}{2}\,k_{2,2}\right) \tag{38.58}$$

$$k_{4,i} = hF_i(t_j + h,\, w_{1,i} + k_{3,1},\, w_{2,i} + k_{3,2}) \qquad i = 1, 2$$

Note that $k_{1,1}$ and $k_{1,2}$ must be computed before we can obtain $k_{2,1}$.

Example. Obtain the response of a generator rotor to a short-circuit disturbance given in Fig. 38.17.

The generator shaft may be idealized as a single-degree-of-freedom system in torsion with the following values:

$$\omega_1 = 1737 \text{ cpm} = 28.95(2\pi) \text{ rad/s} = 182 \text{ rad/s}$$

$$J = 8.5428 \text{ lb} \cdot \text{in} \cdot \text{s}^2 \ (25 \text{ kg} \cdot \text{m}^2)$$

$$k = 7.329 \times 10^6 \text{ lb} \cdot \text{in/rad} \ (828\,100 \text{ N} \cdot \text{m/rad})$$

Solution $J\ddot{\theta} + k\theta = f(t) \qquad \ddot{\theta} = \dfrac{k}{J}\,\theta + \dfrac{f(t)}{J}$

Hence, $\dot{\theta} = \phi \qquad \dot{\phi} = \dfrac{-k}{J}\,\theta + \dfrac{f(t)}{J} \tag{38.59}$

where $f(t)$ is tabulated.

Since $\omega_1 = 182$ rad/s, the period $\tau = 2\pi/182 = 0.00345$ s and the time interval h must be chosen to be around 0.005 s. Hence, tabulated values of $f(t)$ must be available for t intervals of 0.005 s, or it has to be interpolated from Fig. 38.17.

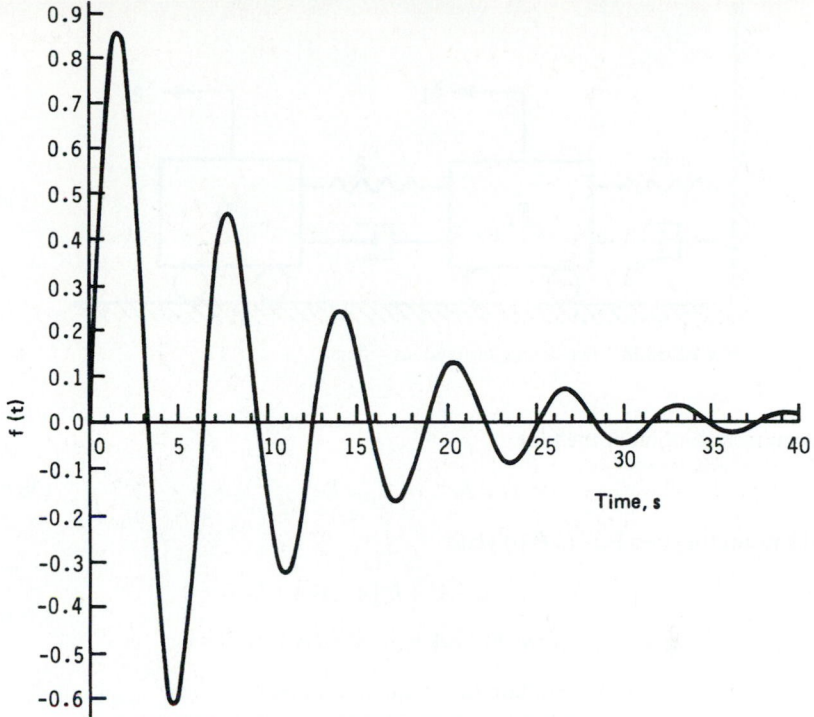

FIGURE 38.17   Short-circuit excitation form.

38.3   SYSTEMS WITH SEVERAL DEGREES OF FREEDOM

Quite often, a single-degree-of-freedom system model does not sufficiently describe the system vibrational behavior. When it is necessary to obtain information regarding the higher natural frequencies of the system, the system must be modeled as a multidegree-of-freedom system. Before discussing a system with several degrees of freedom, we present a system with two degrees of freedom, to give sufficient insight into the interaction between the degrees of freedom of the system. Such interaction can also be used to advantage in controlling the vibration.

38.3.1   System with Two Degrees of Freedom

Free Vibration.   A system with two degrees of freedom is shown in Fig. 38.18. It consists of masses m_1 and m_2, stiffness coefficients k_1 and k_2, and damping coefficients c_1 and c_2. The equations of motion are

$$m_1\ddot{x}_1 + (c_1 + c_2)\dot{x}_1 + (k_1 + k_2)x_1 - c_2\dot{x}_2 - k_2x_2 = 0$$

$$m_2\ddot{x}_2 + c_2\dot{x}_2 + k_2x_2 - c_2\dot{x}_1 - k_2x_1 = 0$$

$$(38.60)$$

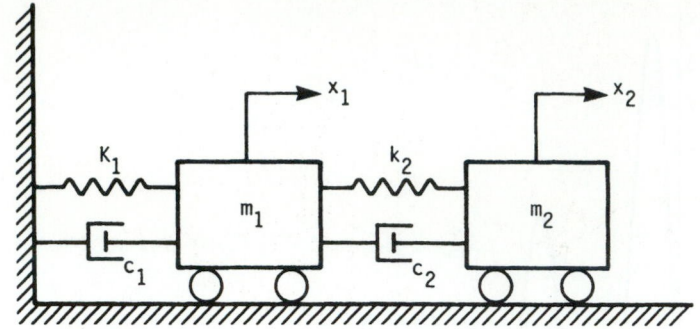

FIGURE 38.18   Two-degree-of-freedom system.

Assuming a solution of the type

$$x_1 = Ae^{st} \qquad x_2 = Be^{st} \qquad (38.61)$$

and substituting into Eqs. (38.60) yield

$$[m_1 s^2 + (c_1 + c_2)s + k_1 + k_2]A - (c_2 s + k_2)B = 0$$
$$- (k_2 + c_2 s)A + (m_2 s^2 + c_2 s + k_2)B = 0 \qquad (38.62)$$

Combining Eqs. (38.62), we obtain the frequency equation

$$[m_1 s^2 + (c_1 + c_2)s + k_1 + k_2](m_2 s^2 + c_2 s + k_2) - (c_2 s + k_2)^2 = 0 \qquad (38.63)$$

This is a fourth-degree polynomial in s, and it has four roots; hence, the complete solution will consist of four constants which can be determined from the four initial conditions x_1, x_2, $\dot{x}_1$, and $\dot{x}_2$. If damping is less than critical, oscillatory motion occurs, and all four roots of Eq. (38.63) are complex with negative real parts, in the form

$$s_{1,2} = -n_1 \pm ip_1 \qquad s_{3,4} = -n_2 \pm ip_2 \qquad (38.64)$$

So the complete solution is

$$x_1 = \exp(-n_1 t)(A_1 \cos p_1 t + A_2 \sin p_1 t)$$
$$+ \exp(-n_2 t)(B_1 \cos p_2 t + B_2 \sin p_2 t)$$
$$x_2 = \exp(-n_1 t)(A_1' \cos p_1 t + A_2' \sin p_1 t) \qquad (38.65)$$
$$+ \exp(-n_2 t)(B_1' \cos p_2 t + B_2' \sin p_2 t)$$

Since the amplitude ratio A/B is determined by Eq. (38.62), there are only four independent constants in Eq. (38.65) which are determined by the initial conditions of the system.

Forced Vibration.   Quite often an auxiliary spring-mass-damper system is added to the main system to reduce the vibration of the main system. The secondary system is called a *dynamic absorber.* Since in such cases the force acts on the main system only, consider a force $P \sin \omega t$ acting on the primary mass m. Referring to Fig. 38.18, we see that the equations of motion are

$$m_1 \ddot{x}_1 + (c_1 + c_2)\dot{x}_1 + (k_1 + k_2)x_1 - c_2\dot{x}_2 - k_2 x_2 = P \sin \omega t$$
$$m_2 \ddot{x}_2 + c_2\dot{x}_2 + k_2 x_2 - c_2\dot{x}_1 - k_2 x_1 = 0 \tag{38.66}$$

Assuming a solution of the type

$$\frac{x_1}{P/k_1} = A_1 \cos \omega t + A_2 \sin \omega t$$

$$\tag{38.67}$$

$$\frac{x_2}{P/k_1} = A_3 \cos \omega t + A_4 \sin \omega t$$

and substituting into Eqs. (38.66), we find that the A_i are given as

$$A_1 = \frac{\omega_1^2[2D_1\omega\zeta_2\omega_2 - D_2(\omega_2^2 - \omega^2)]}{D_1^2 + D_2^2}$$

$$A_2 = \frac{\omega_1^2[D_1(\omega_2^2 - \omega^2) + 2D_2\omega\zeta_2\omega_2]}{D_1^2 + D_2^2}$$

$$\tag{38.68}$$

$$A_3 = \frac{\omega_1^2(2D_1\omega\zeta_2\omega_2 - D_2\omega_2^2)}{D_1^2 + D_2^2}$$

$$A_4 = \frac{\omega_1^2(D_1\omega_2^2 + 2D_2\omega\zeta_2\omega_2)}{D_1^2 + D_2^2}$$

where

$$D_1 = (\omega^2 - \omega_2^2)(\omega^2 - \omega_1^2 - \mu\omega_2^2) - 4\omega^2\zeta_2\omega_2(\zeta_1\omega_1 + \mu\zeta_2\omega_2) - \mu(\omega_2^4 - 4\omega^2\zeta_2^2\omega_2^2)$$
$$D_2 = 2\omega[(\omega_2^2 - \omega^2)(\zeta_1\omega_1 + \mu\zeta_2\omega_2) + \zeta_2\omega_2(\omega_1^2 - \omega^2 + \mu\omega_2^2) - 2\mu\zeta_2\omega_2^3] \tag{38.69}$$

$$\omega_1^2 = \frac{k_1}{m_1} \qquad \omega_2^2 = \frac{k_2}{m_2}$$

$$\zeta_1 = \frac{c_1}{2m_1\omega_1} \qquad \zeta_2 = \frac{c_2}{2m_2\omega_2} \tag{38.70}$$

$$\mu = \frac{m_2}{m_1}$$

Responses may also be written in the form

$$x_1 = B_1 \sin(\omega t - \theta_1) \qquad x_2 = B_2 \sin(\omega t - \theta_2) \tag{38.71}$$

where

$$B_1 = (A_1^2 + A_2^2)^{1/2} \qquad B_2 = (A_3^2 + A_4^2)^{1/2}$$

$$\tan \theta_1 = -\frac{A_1}{A_2} \qquad \tan \theta_2 = -\frac{A_3}{A_4}$$

$$(38.72)$$

Here θ_1 and θ_2 are the phase angles by which the responses of masses m_1 and m_2, respectively, will lag behind the applied force. The response amplitudes B_1 and B_2 are plotted in Figs. 38.19 and 38.20, respectively. The amplitude B_1 has a minimum between ω_1 and ω_2.

The equations of motion for torsional systems with 2 degrees of freedom have the same form as Eqs. (38.60) and (38.66). The solution will also be similar and will exhibit the same characteristics as discussed earlier.

38.3.2 Multidegree-of-Freedom Systems

In many applications, it is necessary to know several higher modes of a vibrating system and evaluate the vibration response. Here, the elastic system has to be treated as one with distributed mass and elasticity. This is possible for simple elements such as beams, plates, or shells of regular geometry. However, when the structural system is complex, it may be modeled as a multidegree-of-freedom discrete

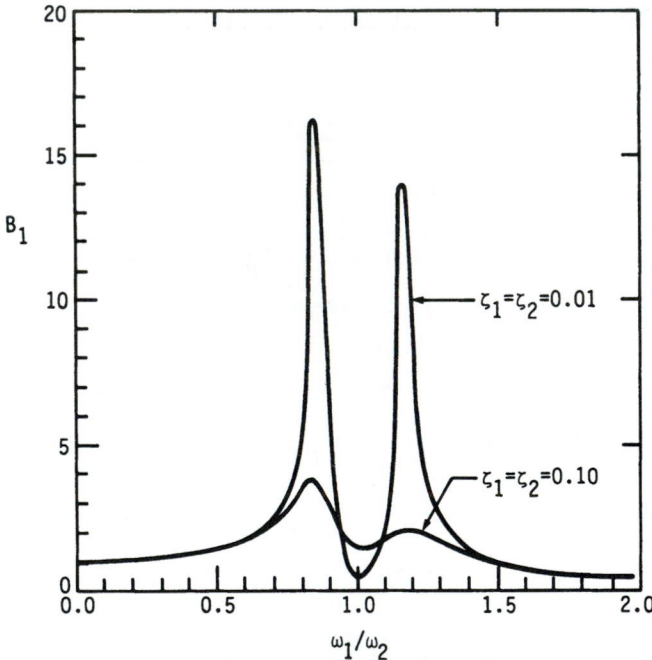

FIGURE 38.19 Amplitude frequency response of the mass of a two-freedom system subject to forced excitation.

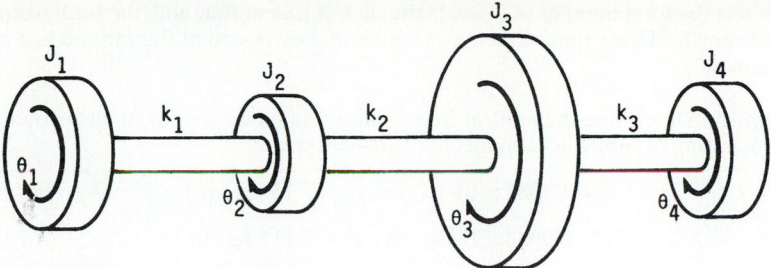

FIGURE 38.20   Torsional system with four freedoms.

system by concentrating its mass and stiffness properties at a number of locations on the structure.

The number of degrees of freedom of a structure is the number of independent coordinates needed to describe the configuration of the structure. In a lumped-mass model, if motion along only one direction is considered, the number of degrees of freedom is equal to the number of masses; and if motion in a plane is of interest, the number of degrees of freedom will equal twice the number of lumped masses.

Holzer Method.   When an undamped torsional system consisting of several disks connected by shafts vibrates freely in one of its natural frequencies, it does not need any external torque to maintain the vibration. In Holzer's method, this fact is used to calculate the natural frequencies and natural modes of a vibrating system. Figure 38.20 shows a torsional system with several disks connected by shafts. In this procedure, an initial value is assumed for the natural frequency, and a unit amplitude is specified at one end. The resulting torques and angular displacements are progressively calculated from disk to disk and carried to the other end. If the resulting torque and displacement at the other end are compatible with boundary conditions, the initial assumed value for the natural frequency is a correct natural frequency; if not, the whole procedure is repeated with another value for the natural frequency until the boundary conditions are satisfied. For a frequency ω and $\theta_1 = 1$, the corresponding inertial torque of the first disk in Fig. 38.20 is

$$T_1 = -J_1 \ddot{\theta}_1 = J_1 \omega^2 \theta_1 \qquad (38.73)$$

This torque is transmitted to disk 2 through shaft 1; hence,

$$T_1 = J_1 \omega^2 \theta_1 = k_1 (\theta_1 - \theta_2) \qquad (38.74)$$

which relates θ_2 and θ_1. The inertial torque of the second disk is $J_2 \omega^2 \theta_2$, and the sum of the inertial torques of disk 1 and disk 2 is transmitted to disk 3 through shaft 2, which gives

$$J_1 \omega^2 \theta_1 + J_2 \omega^2 \theta_2 = k_2 (\theta_2 - \theta_3) \qquad (38.75)$$

Continuing this process, we see the torque at the far end is the combined inertial torques of all the disks and is given by

$$T = \sum_{i=1}^{n} J_i \omega^2 \theta_i \qquad (38.76)$$

where n is the total number of disks. If the disk is free at that end, the total torque T should vanish. Hence, the frequency ω which makes T zero at the far end is a natural frequency.

Example.   Determine the natural frequencies and mode shapes of a torsional system consisting of three disks connected by two shafts.

$$J_1 = 1.7086 \times 10^4 \text{ lb} \cdot \text{in} \cdot \text{s}^2 \qquad (5 \text{ kg} \cdot \text{m}^2)$$

$$J_2 = 3.7588 \times 10^4 \text{ lb} \cdot \text{in} \cdot \text{s}^2 \qquad (11 \text{ kg} \cdot \text{m}^2)$$

$$J_3 = 3.4171 \times 10^4 \text{ lb} \cdot \text{in} \cdot \text{s}^2 \qquad (10 \text{ kg} \cdot \text{m}^2)$$

$$k_1 = 8.8504 \times 10^5 \text{ lb} \cdot \text{in/rad} \qquad (1 \times 10^5 \text{ rad})$$

$$k_2 = 1.7701 \times 10^6 \text{ lb} \cdot \text{in/rad} \qquad (2 \times 10^5 \text{ rad})$$

Solution.   Holzer's procedure can be carried out in a tabulated form as shown in Table 38.1. Two trials are shown in Table 38.1. The calculation can be carried out for more values of ω, and the resulting T_3 can be plotted versus ω, as shown in Fig. 38.21. The frequencies at which $T_3 = 0$ are then the natural frequencies of the system. Better approximation can be obtained by employing the method of false position [38.1]; if $\omega(+)$ and $\omega(-)$ are the frequencies when the torque has corresponding values of $T_3(+)$ and $T_3(-)$, the natural frequency can be obtained by

TABLE 38.1   Holzer's Procedure

Frequency, rad/s	Station						
	1			2			3
ω	θ	$T = J\omega^2\theta$	$\theta = 1 - \dfrac{T}{K}$	$T = T + J\omega^2\theta$	$\theta = \theta - \dfrac{T}{K}$	$T = T + J\omega^2\theta$	
	1	1 1 1	2 1 1	2 1 2 2	3 2 2 2	3 2 3 3	
10	1.0	4 425.2 (500.0)	0.995	1 412.0 (1 594.5)	0.987	22 847.0 (2 581.5)	
20	1.0	17 701.0 (2 000.0)	0.980	55 864.0 (6 312.0)	0.948	89 442.0 (10 106.0)	

$$\omega = \frac{\omega(-)T_3(+) - \omega(+)T_3(-)}{T_3(+) - T_3(-)}$$

The mode shape corresponding to a natural frequency can be obtained by recalculating the values of θ_1, θ_2, and θ_3 in the Holzer table. In the example, the first natural frequency is $\omega_1 = 141.4214$ rad/s, and the corresponding mode shape is

$$\{\theta_1, \theta_2, \theta_3\} = \{1.0, 0.0, -0.5\}$$

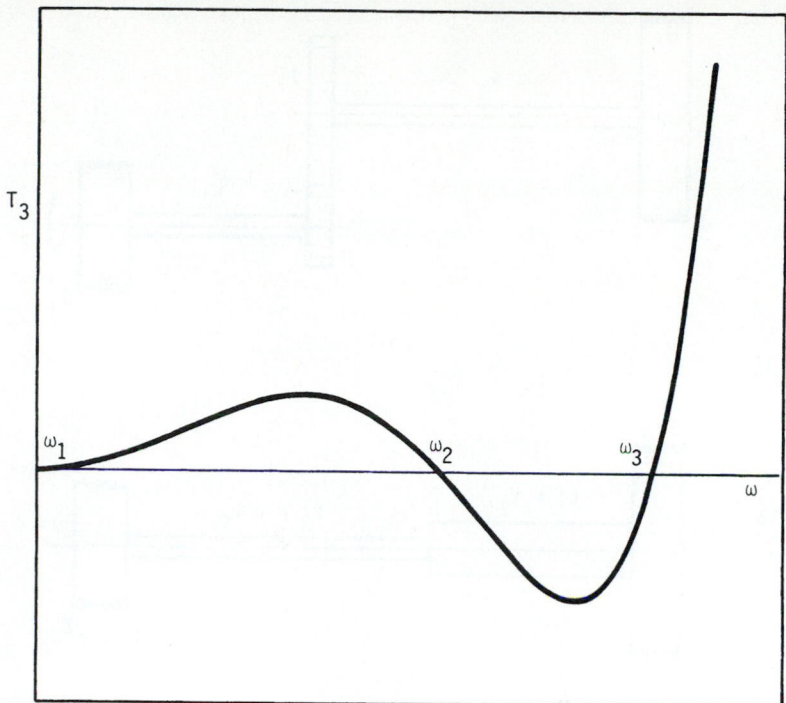

FIGURE 38.21   Variation of end torque with assumed natural frequency.

Geared Systems.   When a shaft transmits torque to another through a gear drive of speed ratio n, it is necessary to reduce the geared torsional system to an equivalent single-shaft system to find its natural frequency. The moments of inertia and the stiffness of the equivalent system are obtained through a consideration of the kinetic and potential energies of the system.

Consider the geared torsional system in Fig. 38.22a. The speed of the second shaft is $\dot{\theta}_2 = n\dot{\theta}_1$. Assuming massless gears, we see that the kinetic energy of the system is

$$T = \frac{1}{2}J_1\dot{\theta}_1^2 + \frac{1}{2}J_2 n^2 \dot{\theta}_1^2 \qquad (38.77)$$

Thus the equivalent mass moment of inertia of disk 2 referred to shaft 1 is $n^2 J_2$. If disks 1 and 2 are clamped and a torque is applied to gear 1, rotating it through an angle θ_1, there will be deformations in both shafts 1 and 2. Gear 2 will rotate through an angle $\theta_2 = n\theta_1$. The potential energy stored in the two shafts is

$$U = \frac{1}{2}k_1\theta_1^2 + \frac{1}{2}k_2 n^2 \theta_1^2 \qquad (38.78)$$

Hence the equivalent stiffness of shaft 2 referred to shaft 1 is $n^2 k_2$. The equivalent torsional system is shown in Fig. 38.22b, where the stiffness and inertia of one side of the system are multiplied by the square of the speed ratio to obtain the corresponding equivalent values for Holzer calculations.

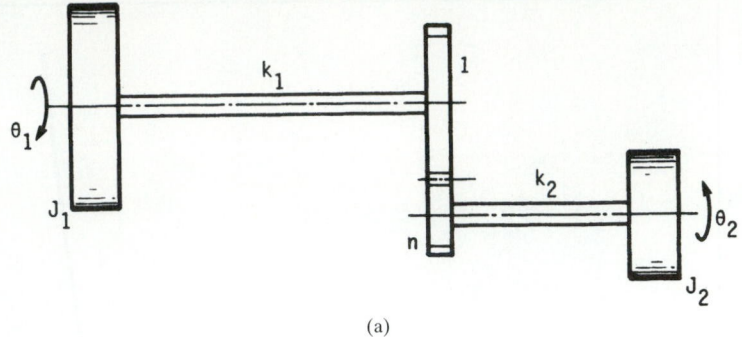

(a)

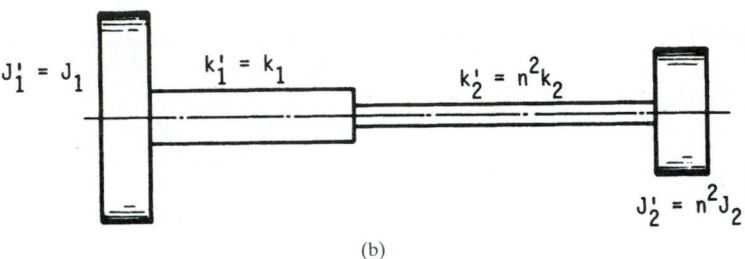

(b)

FIGURE 38.22   (*a*) Geared shaft disk system; (*b*) equivalent torsional system.

38.3.3   Continuous Systems

Engineering structures, in general, have distributed mass and elasticity. Such structures can be treated as multidegree-of-freedom systems by lumping their masses at certain locations and connecting them by representative spring elements. However, it is necessary to consider several such lumped masses and springs to get sufficiently accurate values for the natural frequencies. If only the fundamental natural frequency or the first few natural frequencies are of interest, it is convenient to use some approximate methods based on energy formulations discussed here.

Rayleigh Method.   This method can give the natural frequency of a structure of any specific mode of vibration. A deflection shape satisfying the geometric boundary conditions has to be assumed initially. If the natural frequency of the fundamental mode of vibration is of interest, then a good approximation would be the static deflection shape. For a harmonic motion, the maximum kinetic energy of a structure can be written in the form

$$T_{\max} = \omega^2 C_1 \tag{38.79}$$

where C_1 depends on the assumed deflection shape. The maximum potential energy is of the form

$$U_{max} = C_2 \tag{38.80}$$

Neglecting damping, we see that the maximum kinetic energy must be equal to the maximum potential energy. Hence, the natural frequency is

$$\omega^2 = \frac{C_2}{C_1} \tag{38.81}$$

This estimate will always be higher than the true natural frequency.

Example.   Determine the fundamental natural frequency of a uniform cantilever beam of length L supporting a disk of mass M and diametral mass moment of inertia I_d, as shown in Fig. 38.23. The modulus of elasticity of the beam material is E, the mass moment of inertia of the cross section is I, and the mass per unit length of the beam is m.

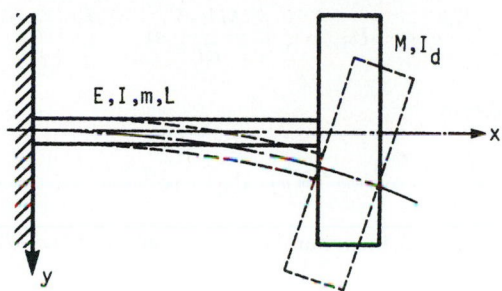

FIGURE 38.23   Cantilever with end mass.

Solution.   The deflection shape may be assumed to be $y(x) = Cx^2$, which satisfies the geometric boundary conditions of zero deflection and zero slope at $x = 0$. The maximum kinetic energy of the structure for harmonic vibration is

$$T_{max} = \frac{1}{2}m\omega^2 \int_0^L y^2(x)\, dx + \frac{1}{2}M\omega^2 y^2(L) + \frac{1}{2}I_d\omega^2 y'^2(L)$$

The strain energy is given by

$$U_{max} = \frac{1}{2}EI \int_0^L y''^2(x)\, dx$$

Substituting $y(x) = Cx^2$ in the above expressions and equating $T_{max} = U_{max}$, we find the natural frequency

$$\omega^2 = \frac{20EI}{mL^4 + 5ML^3 + 20I_dL}$$

In the absence of the disk, $\omega = 4.47 \, (EI/mL^4)^{1/2}$. By comparing this to the exact result $\omega = 3.52 \, (EI/mL^4)^{1/2}$, the error in the approximation is error $= 0.95 \, (EI/mL^4)^{1/2}$, or 26.9 percent.

This error can be reduced by obtaining the strain energy by using a different method. The shear at any section is obtained by integrating the inertial loading from the free end as

$$V(\xi) = \left[\omega^2 \int_{\xi}^{L} m(\xi) y(\xi) \, d\xi \right] + M\omega^2 y(L)$$

$$= \frac{1}{3} \omega^2 mc(L^3 - \xi^3) + M\omega^2 cL^2$$

and the moment at any point x is

$$M(x) = \int_{x}^{L} V(\xi) \, d\xi + I_d y'(L) = \frac{1}{12} \omega^2 mc(3L^4 - 4L^3 x + x^4) + M\omega^2 cL^2(L - x) + 2I_d cL$$

The strain energy is then

$$U_{\max} = \frac{1}{2} \int_{0}^{L} \frac{M^2(x)}{EI} \, dx$$

When the disk is absent,

$$U_{\max} = \frac{\omega^4}{2EI} \frac{m^2 c^2}{144} \frac{312}{135} L^9$$

With $T_{\max}$ and $U_{\max}$ equated, the natural frequency $\omega = 3.53 \, (EI/ML^4)^{1/2}$ has an error of only 0.28 percent.

38.4   VIBRATION ISOLATION

Often machines and components which exhibit vibrations have to be mounted in locations where vibrations may not be desirable. Then the machine has to be isolated properly so that it does not transmit vibrations.

38.4.1   Transmissibility

Active Isolation and Transmissibility.   From Eq. (38.38), the force transmissibility, which is the magnitude of the ratio of the force transmitted to the force applied, is given by

$$T = \left[\frac{1 + (2\zeta\omega/\omega_n)^2}{(1 - \omega^2/\omega_n^2)^2 + (2\zeta\omega/\omega_n)^2} \right]^{1/2} \tag{38.82}$$

Equation (38.82) is plotted in Fig. 38.12 for different values of ζ. All the curves cross at $\omega/\omega_n = \sqrt{2}$. For $\omega/\omega_n > \sqrt{2}$, transmissibility, although below unity, increases with an increase in damping, contrary to normal expectations. At higher frequencies, transmissibility goes to zero.

Since the force amplitude $me\omega^2$ in the case of an unbalanced machine is dependent upon the operating speed of the machine, transmissibility can be defined as

$$T = \frac{F_T}{me\omega^2} = \left(\frac{\omega}{\omega_n}\right)^2 \left[\frac{1 + (2\zeta\omega/\omega_n)^2}{(1 - \omega^2/\omega_n^2)^2 + (2\zeta\omega/\omega_n)^2}\right]^{1/2} \qquad (38.83)$$

where F_T is the amplitude of the transmitted force.

Equation (38.83) is plotted in Fig. 38.24. Transmissibility starts from zero at zero operating frequency, and curves for different damping ratios cross at $\omega/\omega_n = \sqrt{2}$. For higher values of operating speed, transmissibility increases indefinitely with frequency.

Passive Isolation. When a sensitive instrument is isolated from a vibrating foundation, it is called *passive isolation*. Consider the system shown in Fig. 38.15, where the base has a motion $u = U_0 \sin \omega t$. The equation of motion of the system is given in Eq. (38.46).

The ratio of the response and excitation amplitudes is

$$\frac{X}{X_f} = \frac{k + i\omega c}{k - m\omega^2 + i\omega c} \qquad (38.84)$$

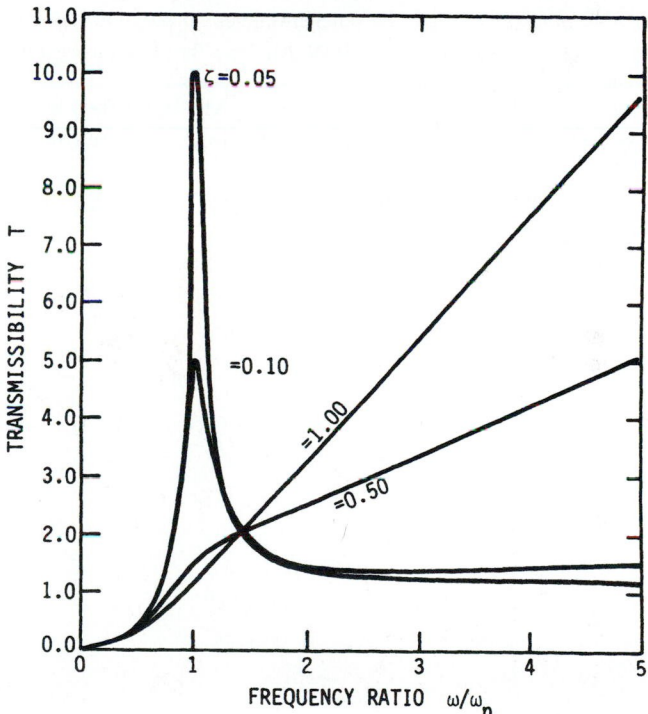

FIGURE 38.24 Transmissibility of a system under unbalanced excitation.

Since we are interested in the motion transmissibility in the case of passive isolation, Eq. (38.84) gives the transmissibility T, which can be put in terms of nondimensional parameters, as

$$T = \frac{X}{U_0} \left[\frac{1 + (2\zeta\omega/\omega_n)^2}{(1 - \omega^2/\omega_n^2)^2 + (2\zeta\omega/\omega_n)^2} \right]^{1/2} \tag{38.85}$$

Equation (38.85) is identical to the force transmissibility in the case of active isolation given in Eq. (38.82).

REFERENCES

38.1 R. L. Burden, J. D. Faires, and A. C. Reynolds, *Numerical Analysis,* 2d ed., Prindle, Weber and Schmidt, Boston, 1981.

38.2 R. B. Bhat, J. S. Rao, and T. S. Sankar, "Optimum Journal Bearing Parameters for Minimum Unbalance Response in Synchronous Whirl," ASME Design Engineering Conference, Paper No. 81-DET-55, 1981. (To be published in *ASME Transactions.*)

38.3 J. P. Den Hartog, *Mechanical Vibrations,* McGraw-Hill, New York, 1962.

38.4 W. T. Thomson, *Theory of Vibration with Applications,* Prentice-Hall, Englewood Cliffs, N.J., 1981.

38.5 T. S. Sankar and G. D. Xistris, "Failure Prediction through the Theory of Stochastic Excursions of Extreme Vibration Amplitudes," *J. Eng. Ind., Trans. ASME,* 1972.

38.6 *NASTRAN Computer Program,* McNeil Schwendler Corporation, Los Angeles.

38.7 *ANSYS Computer Program,* SWANSON Analysis Systems Incorporated, Houston, Pennsylvania.

38.8 *ADINA Computer Program,* ADINA Engineering AB VASTERAS, Sweden.

38.9 *SPAR Structural Analysis System,* Engineering Information Systems, Inc., San Jose, California.

38.10 *PROSSS—Programming Structural Synthesis System,* NASA Langley Research Center, Hampton, Virginia.

CHAPTER 39
A THESAURUS OF MECHANISMS

L. E. Torfason
Profesor of Mechanical Engineering
University of New Brunswick
Fredericton, Canada

GLOSSARY OF SYMBOLS

R	Revolute pair or pin joint
P	Prismatic pair or sliding joint
C	Cylinder pair for joints that allow rotation and sliding along the cylinder axis
G	Spheric pair (globe) for ball joints
S_L	Screw pair with lead L
F	Planar pair (flat) for a joint that maintains two planes in contact

SUMMARY[†]

This chapter is intended to be used as an idea generator. Following the adage that a picture is worth 1000 words, this chapter was assembled with millions of "words" in figures and virtually none using the alphabet. I have taken the liberty of varying dimensions to better show the principle of operation. You should not scale the figures, but follow the regular synthesis procedure to determine the proper dimensions for the application in mind.

In this chapter a new notation is used for the kinematic representation of joints or pairs in a linkage.

[†] Readers will note a difference in the style and character of the figures in this chapter. When this manuscript was received, the illustrations, all conceived and executed by Professor Torfason, were seen to be original and unique. We asked for and received from the publishers special permission to reproduce them exactly as they were drawn—EDS.

COLLATERAL READING

L. J. Kamm, *Designing Cost-Efficient Mechanisms*, McGraw-Hill, New York, 1990.

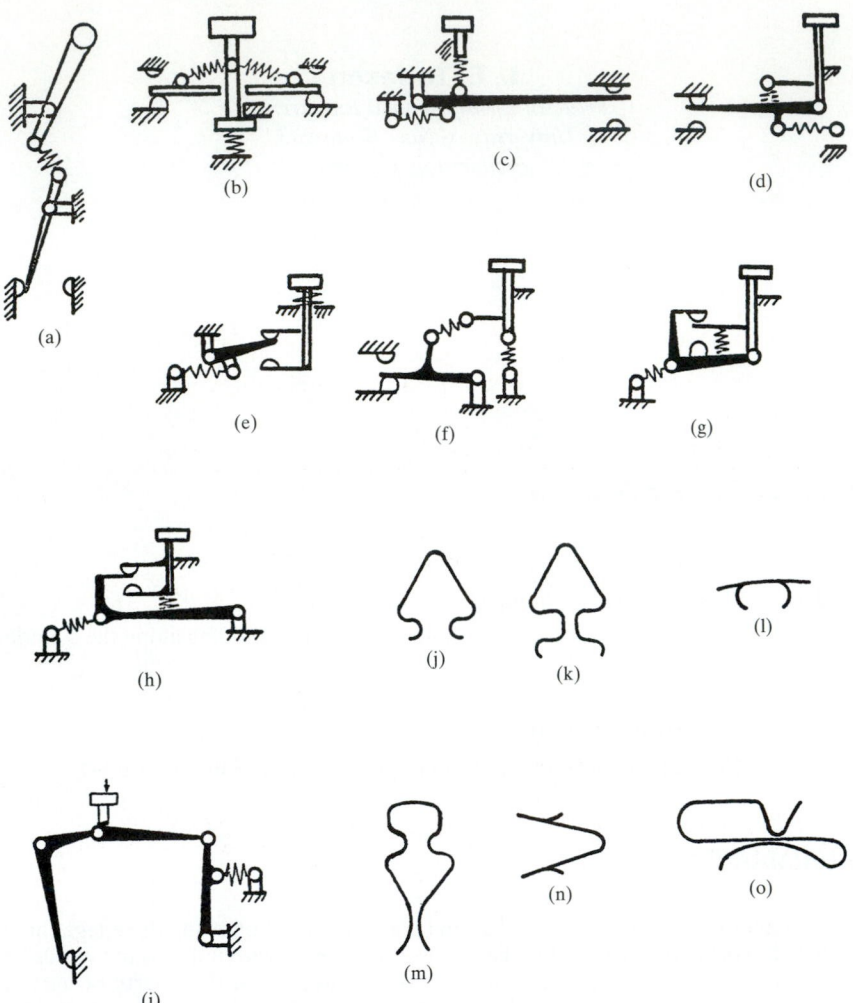

FIGURE 39.1 Snap-action mechanisms. These mechanisms are bistable elements in machines. They are used in switches to quickly make and break electric circuits and for fastening items. (*a*) Snap-action toggle switch; (*b*) to (*h*) seven variations of snap-action switches; (*i*) circuit breaker; (*j*) to (*o*), spring clips.

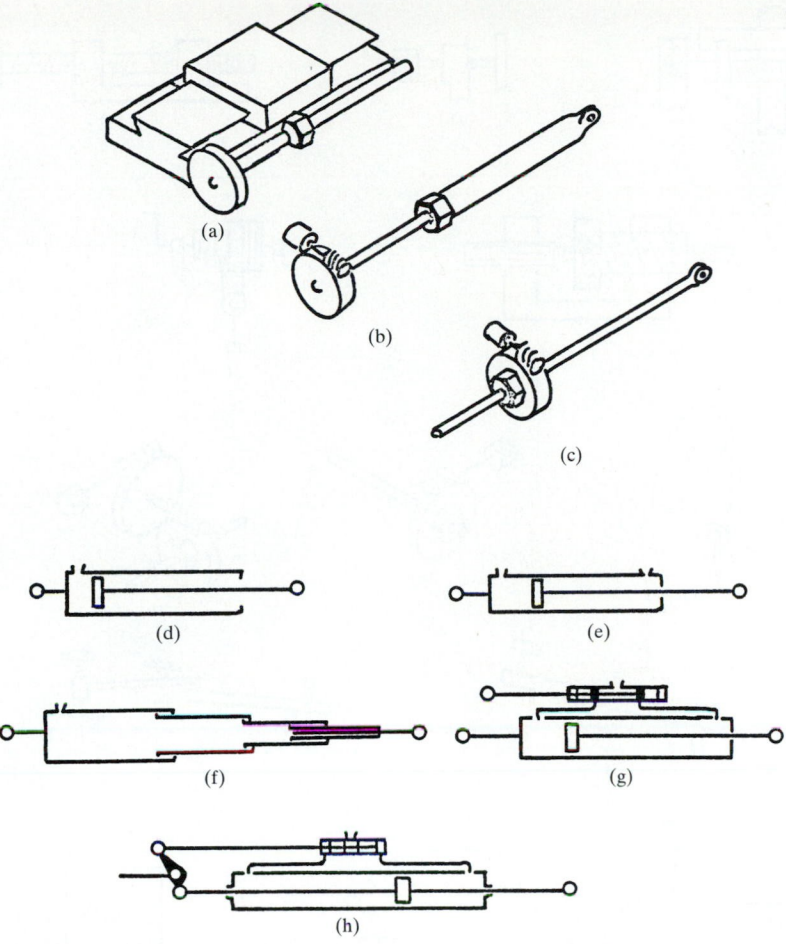

FIGURE 39.2 Linear actuators. These are devices that cause a straight-line displacement between two machine elements. (*a*) Lead screw; (*b*) worm gear with stationary nut; (*c*) worm gear with stationary screw; (*d*) single-acting hydraulic cylinder; (*e*) double-acting hydraulic cylinder; (*f*) telescoping hydraulic cylinder; (*g*) hydraulic cylinder with positional feedback; (*h*) hydraulic cylinder with floating link feedback.

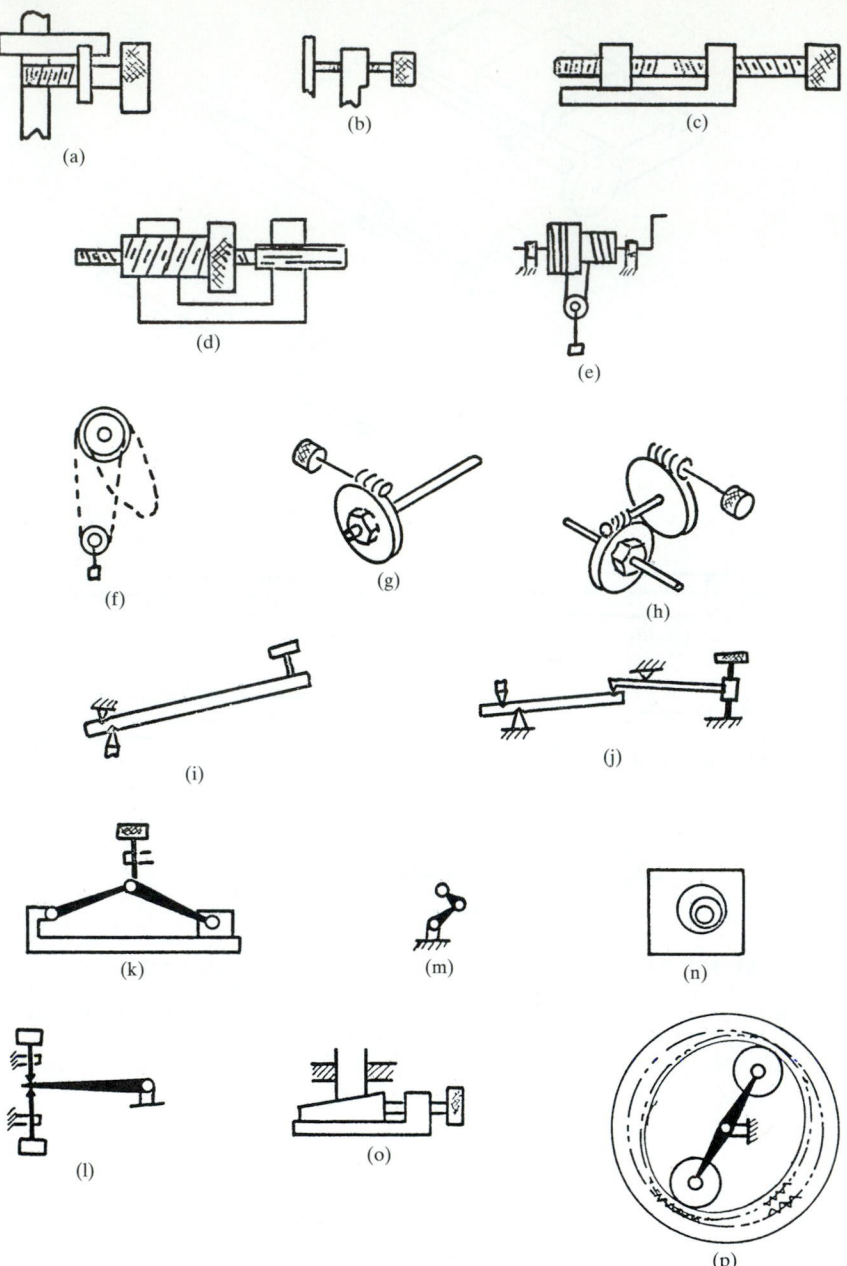

FIGURE 39.3 Fine adjustments I. Fine adjustments for stationary mechanisms are mechanisms that make a small change in the position of a mechanical member. (*a*), (*b*) Screw adjustments; (*c*), (*d*) differential screws; (*e*) Chinese windlass; (*f*) differential hoist; (*g*) worm gear and screw; (*h*) worm gears in series; (*i*) lever; (*j*) levers in series; (*k*) toggle mechanism; (*l*) screws to adjust angular position; (*m*), (*n*) eccentric cranks; (*o*) wedges; (*p*) harmonic drive.

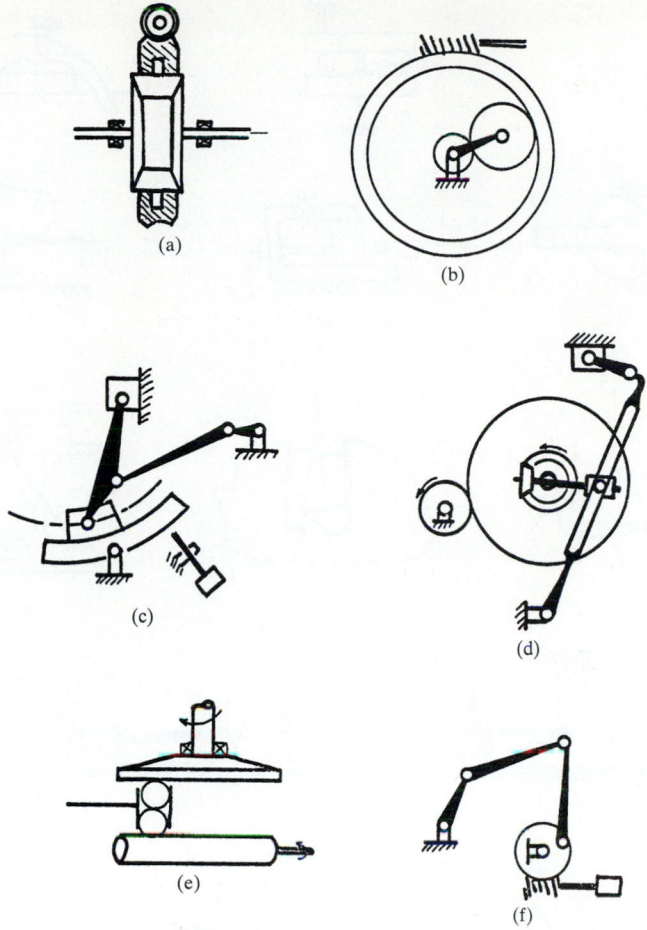

FIGURE 39.4   Fine adjustments II. Fine adjustments for moving mechanisms are adjusting devices which control the motion of linkages such as stroke, etc., while the mechanism is in motion. (*a*), (*b*) Differential gear adjustment; (*c*) adjustable-stroke engine; (*d*) adjustable stroke of shaper mechanism; (*e*) ball and disk speed changer; (*f*) adjusting fixed center of linkage for changing motion properties.

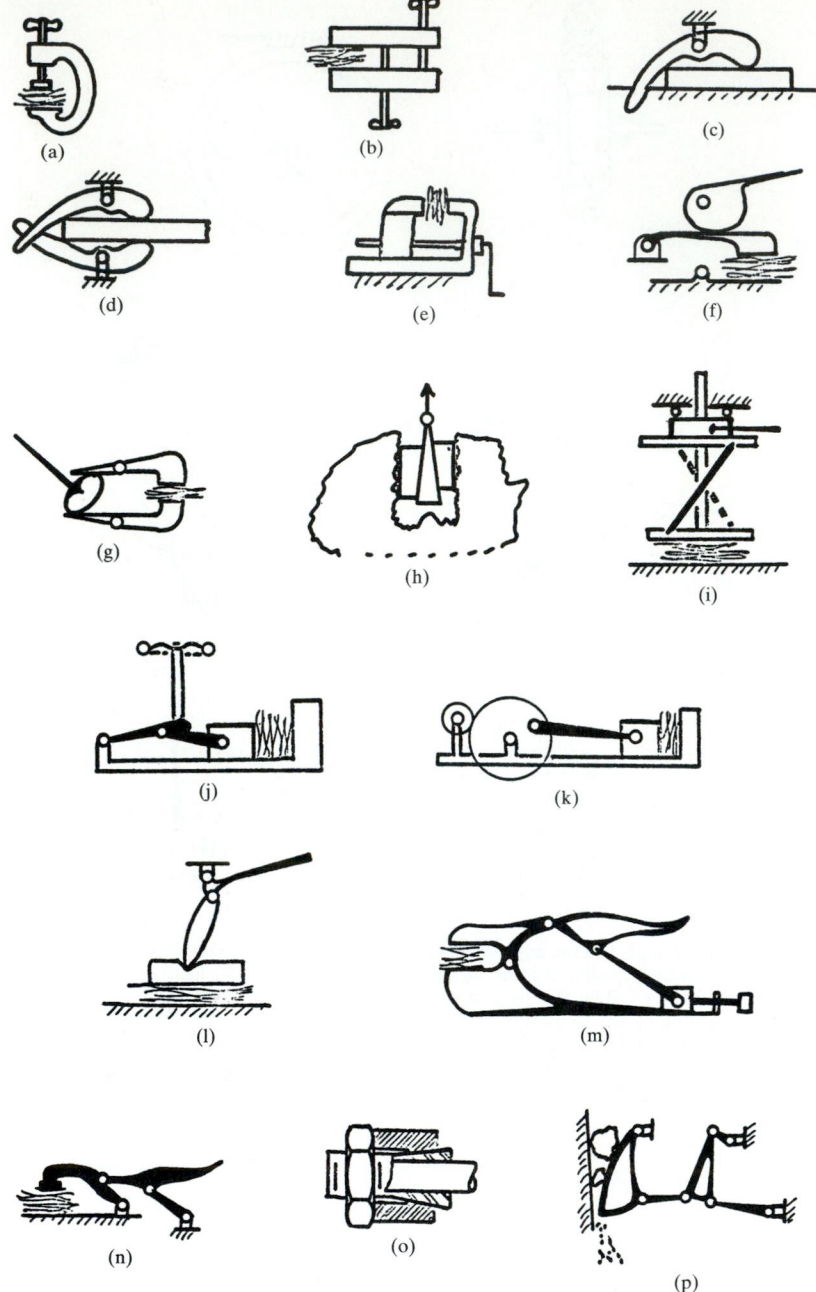

FIGURE 39.5 Clamping mechanisms. These devices are used to hold items for machining operations or to exert great forces for embossing or printing. (*a*) C clamp; (*b*) screw clamp; (*c*) cam clamp; (*d*) double cam clamp; (*e*) vise; (*f*) cam-operated clamp; (*g*) double cam-actuated clamp; (*h*) double wedge; (*i*) to (*l*) toggle press; (*m*) vise grips; (*n*) toggle clamp; (*o*) collet; (*p*) rock crusher.

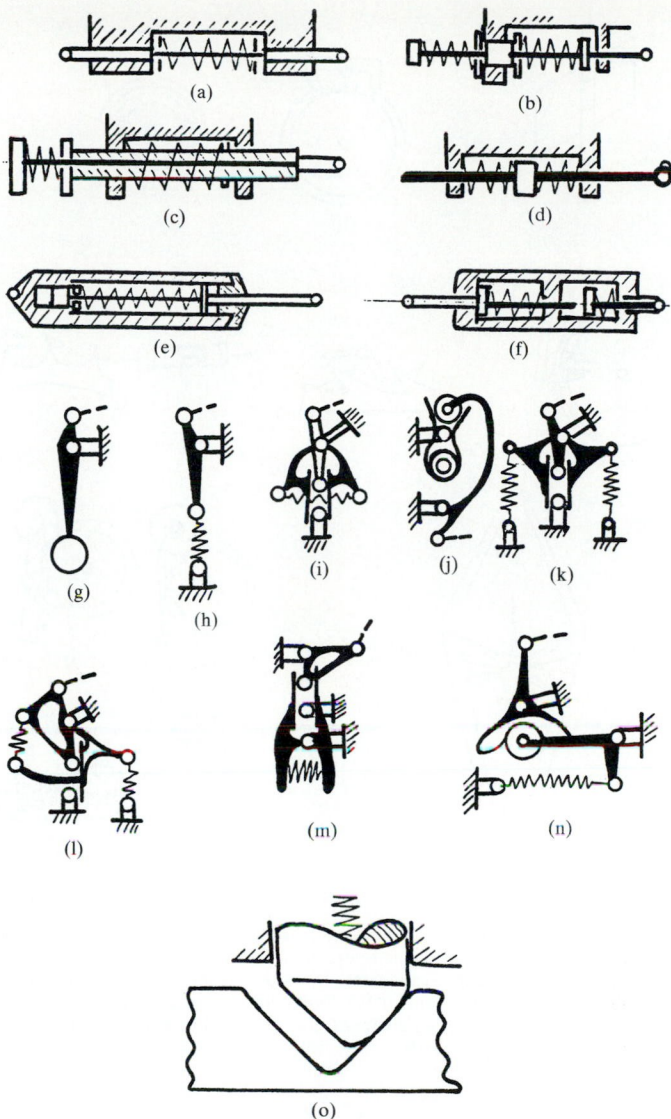

FIGURE 39.6 Locating mechanisms. These are devices which properly position a linkage member when the load is removed. (*a*) to (*f*) Self-centering linear devices; (*g*) to (*n*) self-centering angular devices; (*o*) detent.

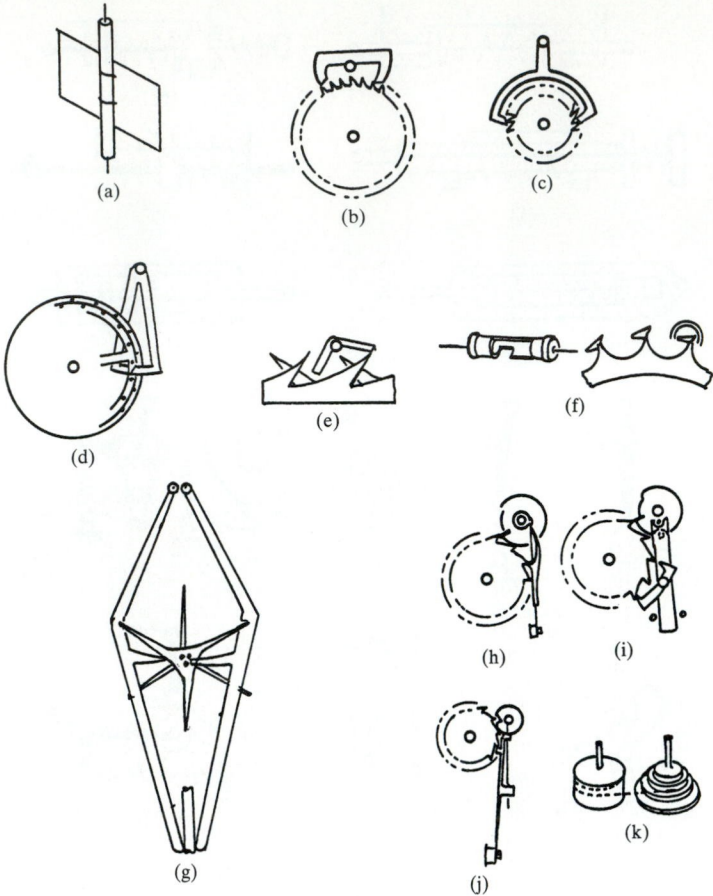

FIGURE 39.7   Escapements. These devices slowly release the potential energy
stored in a spring to control devices such as clocks. (*a*) Paddle wheel; (*b*) recoil
escapement; (*c*) dead-beat escapement; (*d*) stud escapement; (*e*) early anchor
escapement; (*f*) cylinder escapement; (*g*) double three-legged escapement for tower
clocks; (*h*) to (*j*) chronometer escapements; (*k*) fuse used to give uniform torque at
escapement as the spring unwinds.

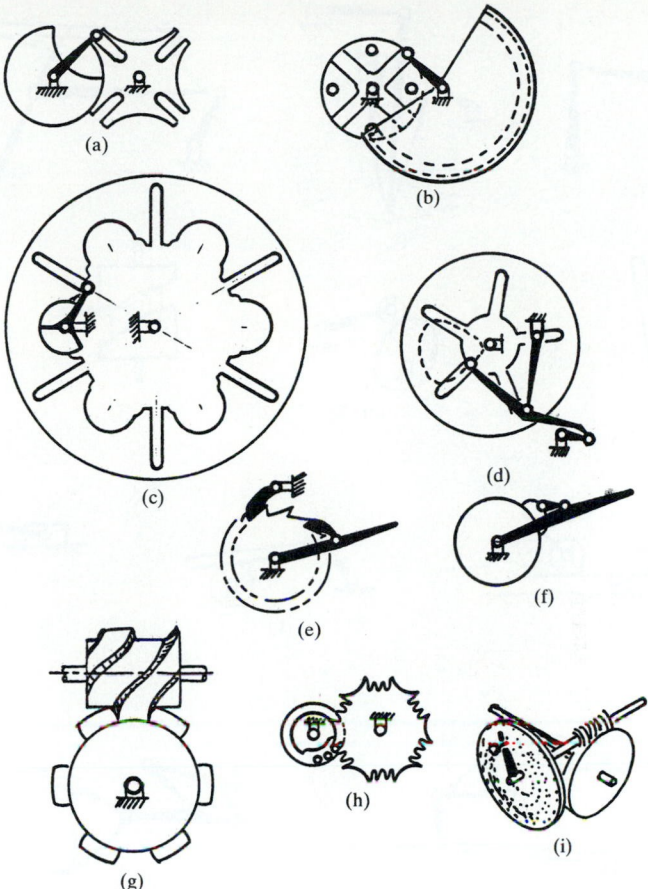

FIGURE 39.8   Indexing mechanisms. These mechanical devices advance a body to a specific position, hold it there for a period, and then advance it again. (*a*) to (*c*) Geneva stops; (*d*) four-bar links used to reduce jerk; (*e*) ratchet mechanism; (*f*) friction ratchet; (*g*) cylindrical cam-stop mechanism; (*h*) pin gearing used in indexing; (*i*) dividing head.

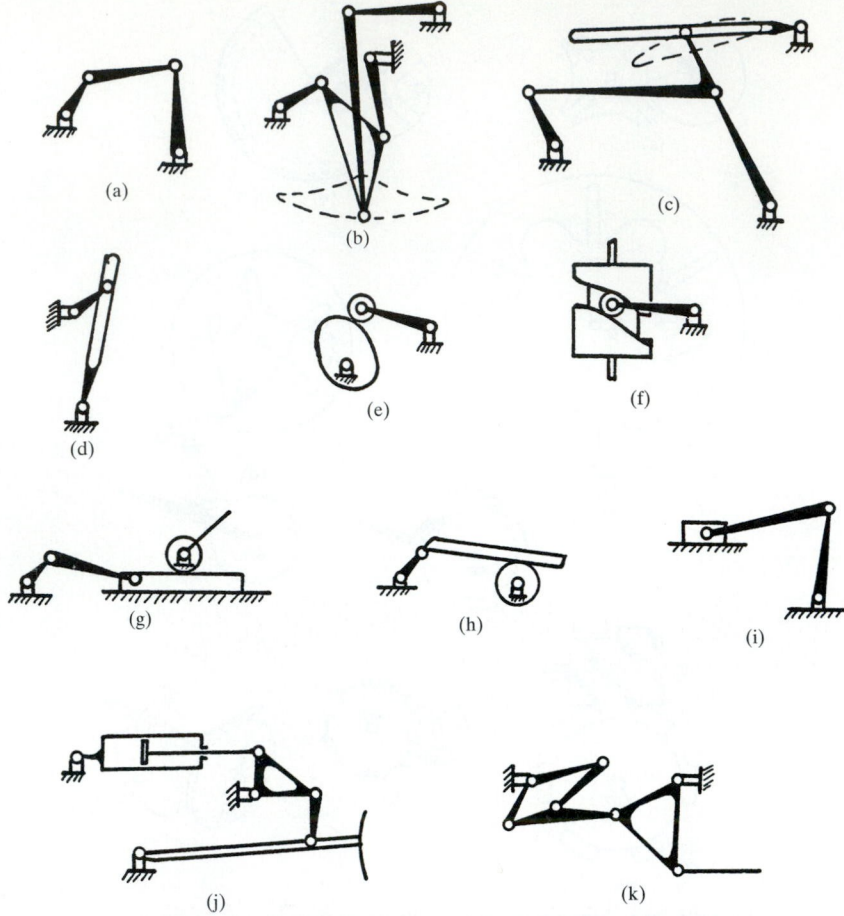

FIGURE 39.9   Oscillating mechanisms I. These mechanisms cause an output to repeatedly swing through a preset angle. (*a*) Four-bar linkage; (*b*) six-bar linkage; (*c*) six-bar linkage with pin in slot; (*d*) inverted slide-crank quick-return linkages; (*e*) radial cam and follower; (*f*) cylindrical cam; (*g*) geared slider crank; (*h*) geared inverted slider crank; (*i*) slider-driven crank; (*j*) bulldozer lift mechanism; (*k*) oscillator of the Corliss valve gear.

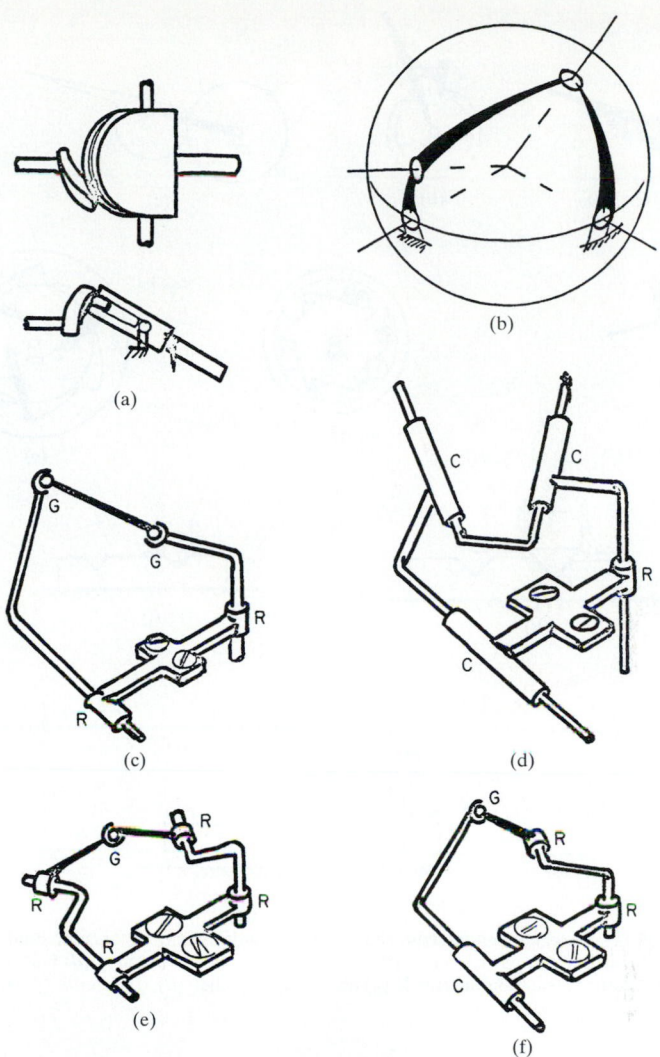

FIGURE 39.10 Oscillating mechanisms II. These all use spatial linkages. (*a*) Spatial pin and yoke; (*b*) spherical four-bar linkage; (*c*) spatial RGGR linkage; (*d*) spatial RCCC; (*e*) spatial RRGRR; (*f*) spatial RRGC.

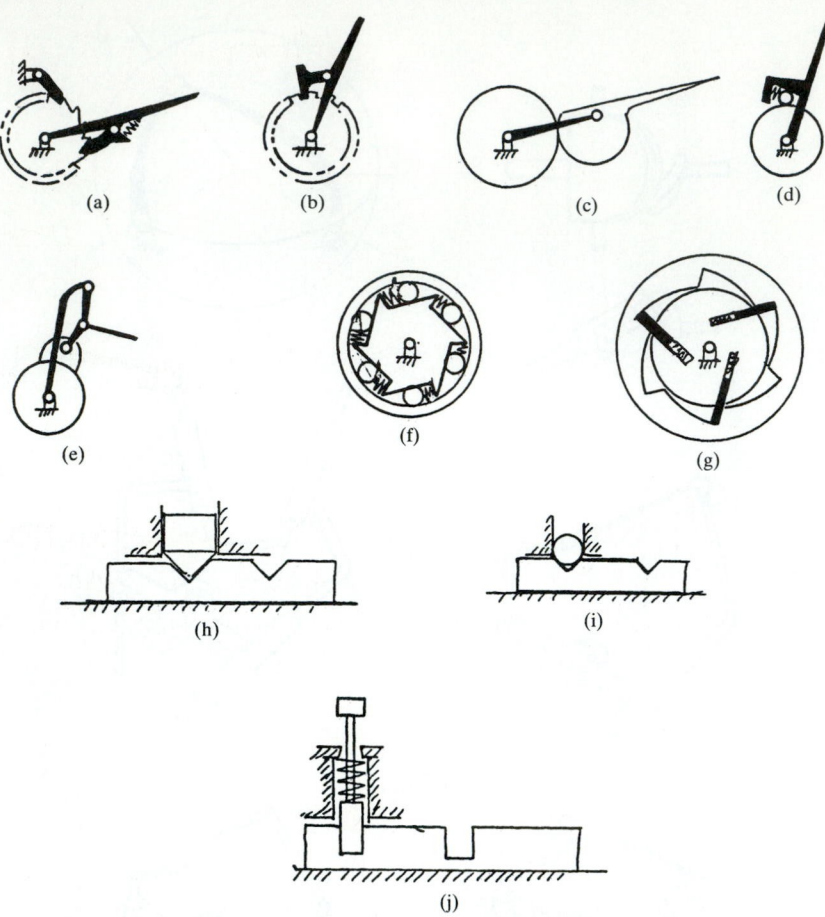

FIGURE 39.11 Ratchets and latches. These are mechanisms that advance or hold a machine member. (*a*) Ratchet and pawl; (*b*) reversible ratchet; (*c*) cam-lock ratchet; (*d*) ball-lock ratchet; (*e*) toggle ratchet; (*f*) overrunning clutch; (*g*) high-torque ratchet; (*h*), (*i*) detents; (*j*) locking bolts.

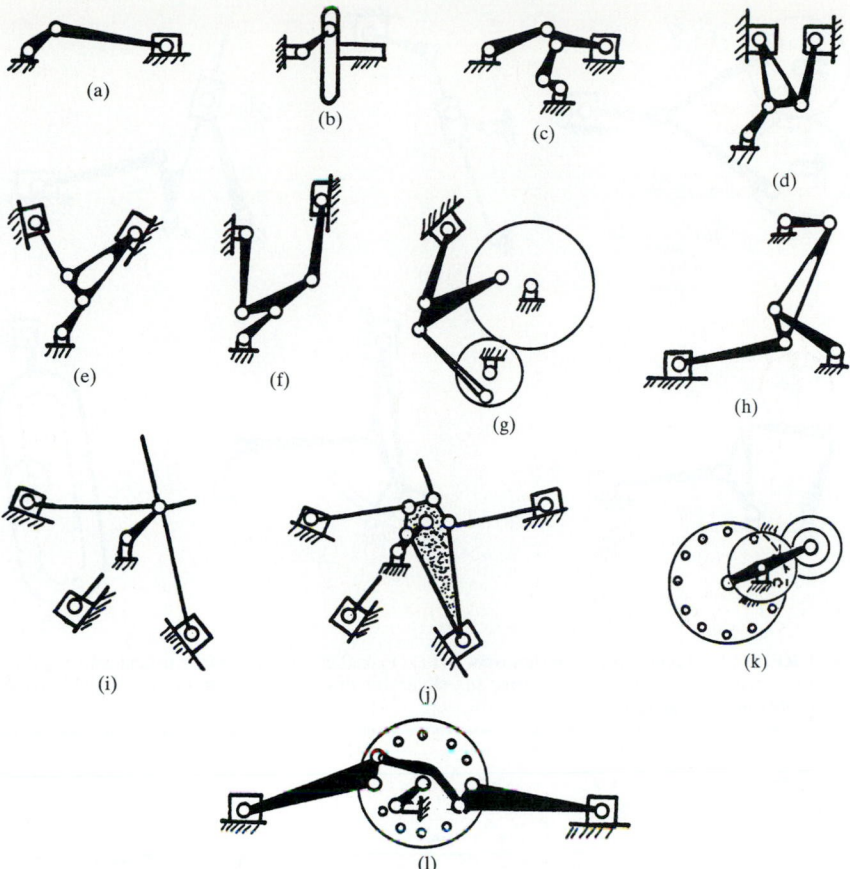

FIGURE 39.12 Reciprocating mechanisms I. These mechanical devices cause a member to translate on a straight line. (*a*) Slider crank; (*b*) Scotch yoke; (*c*) toggle mechanism; (*d*) Zoller engine; (*e*) V engine; (*f*) double-stroke engine; (*g*) geared engine; (*h*) Atkinson gas engine; (*i*) ideal radial engine; (*j*) practical radial engine; (*k*) geared Nordberg radial engine; (*l*) linked Nordberg radial engine.

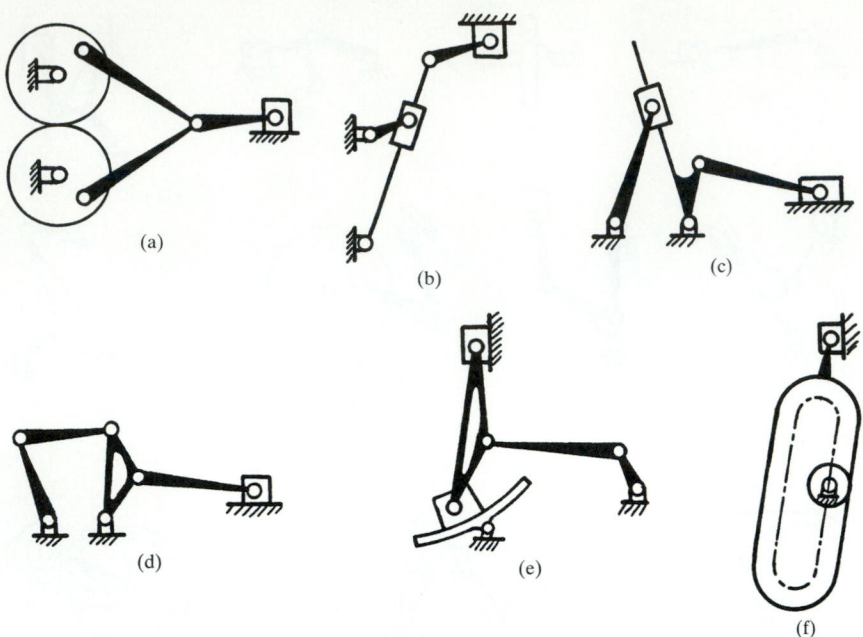

FIGURE 39.13   Reciprocating mechanisms II. (*a*) Geared cranks; (*b*) shaper mechanism; (*c*) slider on Whitworth quick-return mechanisms; (*d*) slider on drag-link mechanism; (*e*) variable-stroke engine; (*f*) gear-driven slider.

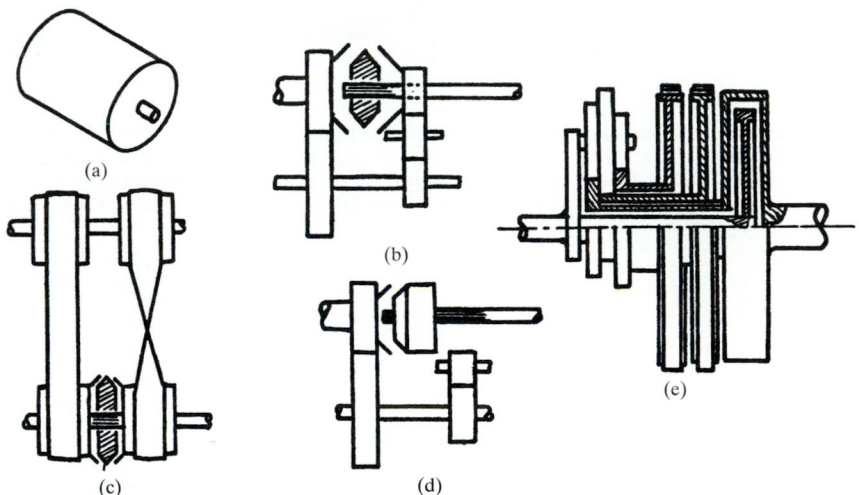

FIGURE 39.14   Reversing mechanism. These mechanical devices change the direction of rotation of the output. (*a*) Reversible prime movers; (*b*) reversing gears; (*c*) reversing belts; (*d*) transmission; (*e*) epicyclic gears as in Model T Ford.

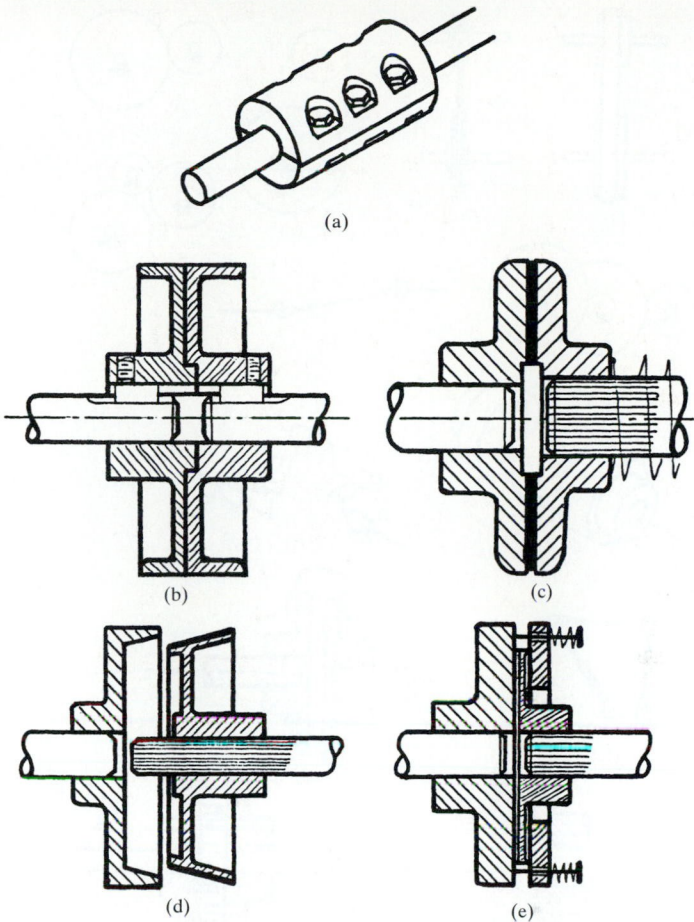

FIGURE 39.15   Couplings and connectors—axial. These are used to connect co-axial shafts. (*a*) Rigid coupling; (*b*) flanged coupling; (*c*) disk clutch; (*d*) cone clutch; (*e*) plate clutch.

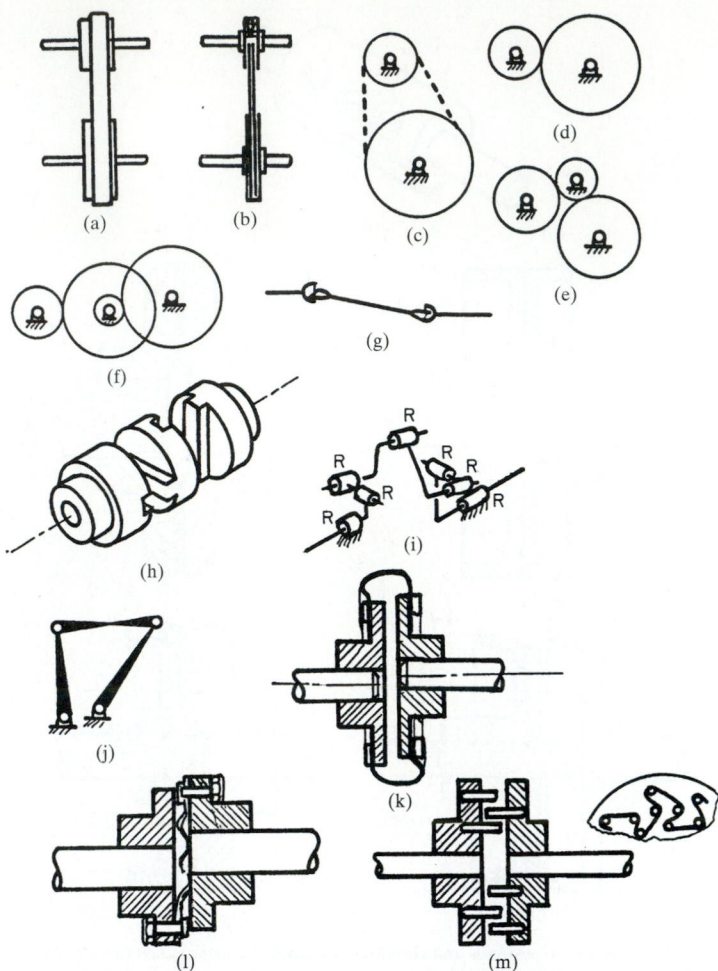

FIGURE 39.16 Couplings and connectors—parallel shafts. (*a*) Flat belt; (*b*) V belt; (*c*) chain; (*d*) to (*f*) gears; (*g*) Hooke joints; (*h*) Oldham coupling; (*i*) Hunt's constant-velocity coupling; (*j*) drag link; (*k*) to (*m*) flexible coupling.

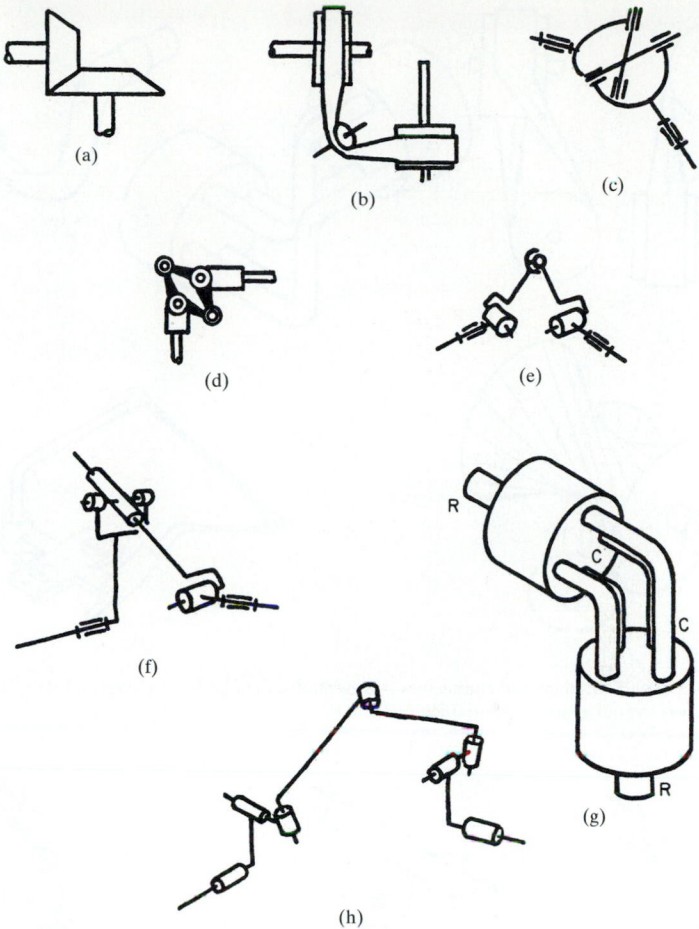

FIGURE 39.17   Couplings and connectors—intersecting shafts. (*a*) Bevel gears; (*b*) flat belts with idlers; (*c*) Hooke joint; (*d*) Hooke's coupling; (*e*) Clemens coupling; (*f*) Rouleaux coupling; (*g*) spatial RCCR; (*h*) Hunt's constant-velocity coupling.

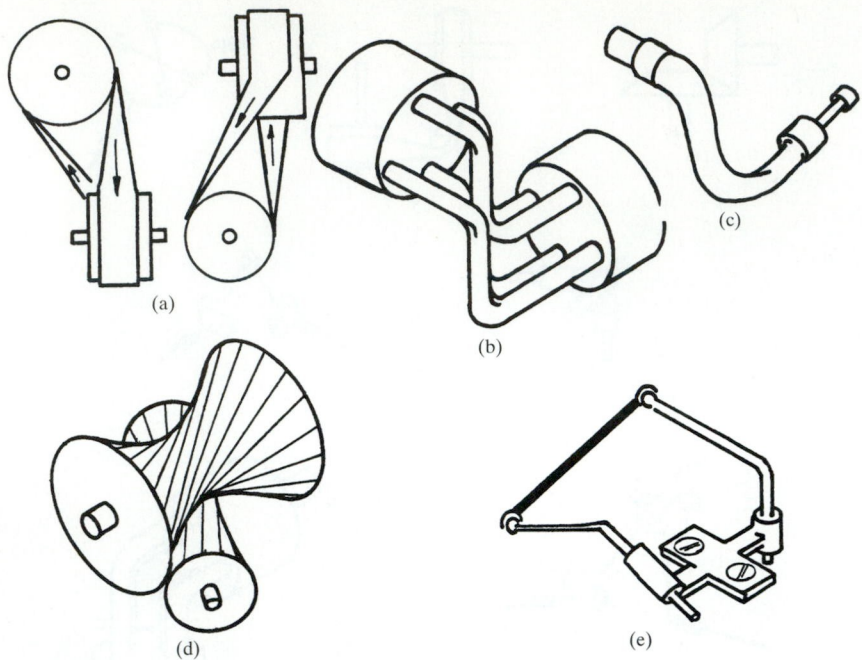

FIGURE 39.18 Couplings and connectors—skew shafts. (*a*) Flat belts; (*b*) spatial RCCR; (*c*) flexible shaft; (*d*) hypoid gears; (*e*) spatial RGGR.

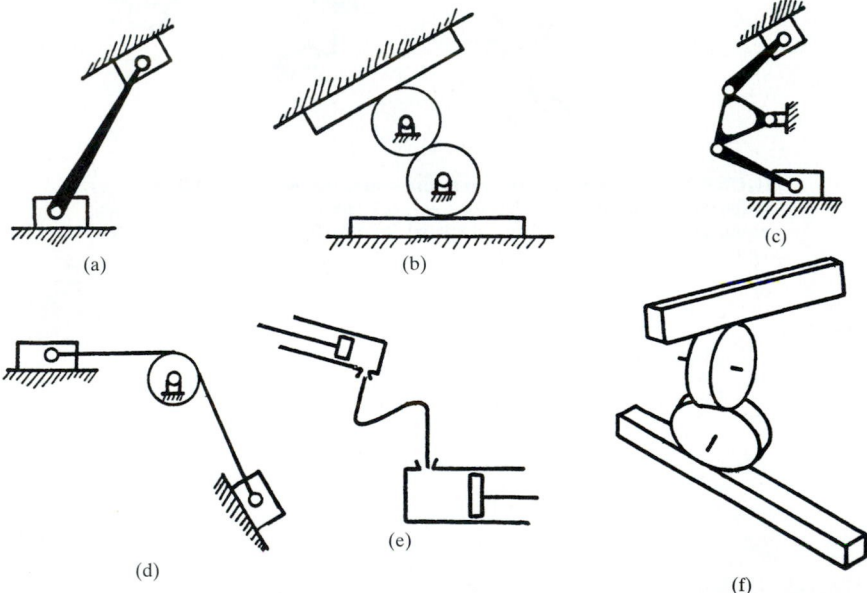

FIGURE 39.19 Slider connectors. These devices connect two or more reciprocating devices. (*a*) Elliptic trammel; (*b*) gears; (*c*) slider-crank-slider; (*d*) cable; (*e*) hydraulic; (*f*) helical gearing.

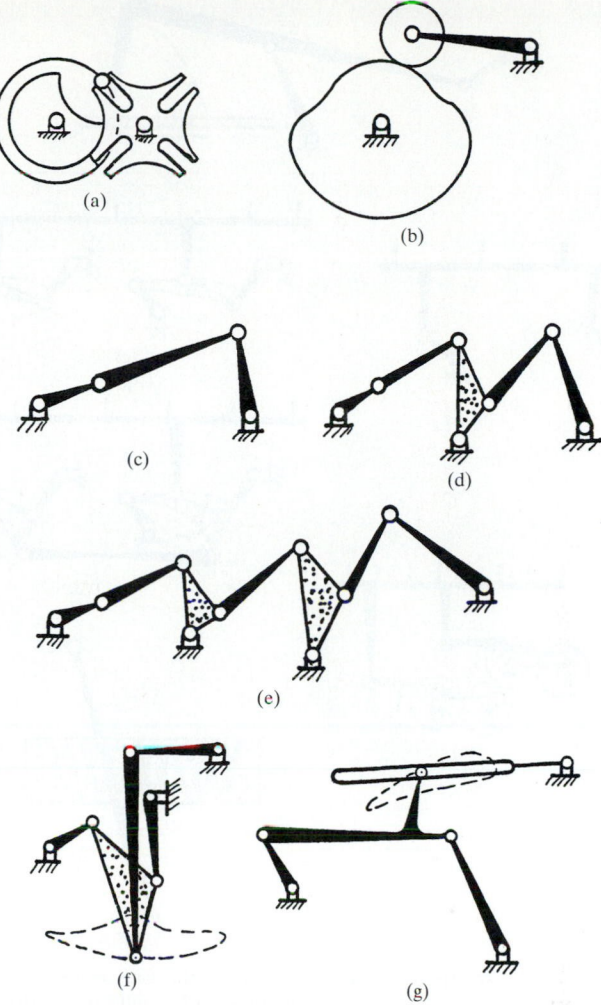

FIGURE 39.20 Stops, pauses, and hesitations. These machine elements cause an output to stop and dwell, to stop and return, to stop and advance, etc. The derivatives of the motion at the stop determine which category the motion fits. (*a*) Geneva stops (this includes all motions in Fig. 39.8); (*b*) cams; (*c*) linkage at extreme limits; (*d*), (*e*) combination of linkages at a limit; (*f*), (*g*) outputs derived from coupler curves.

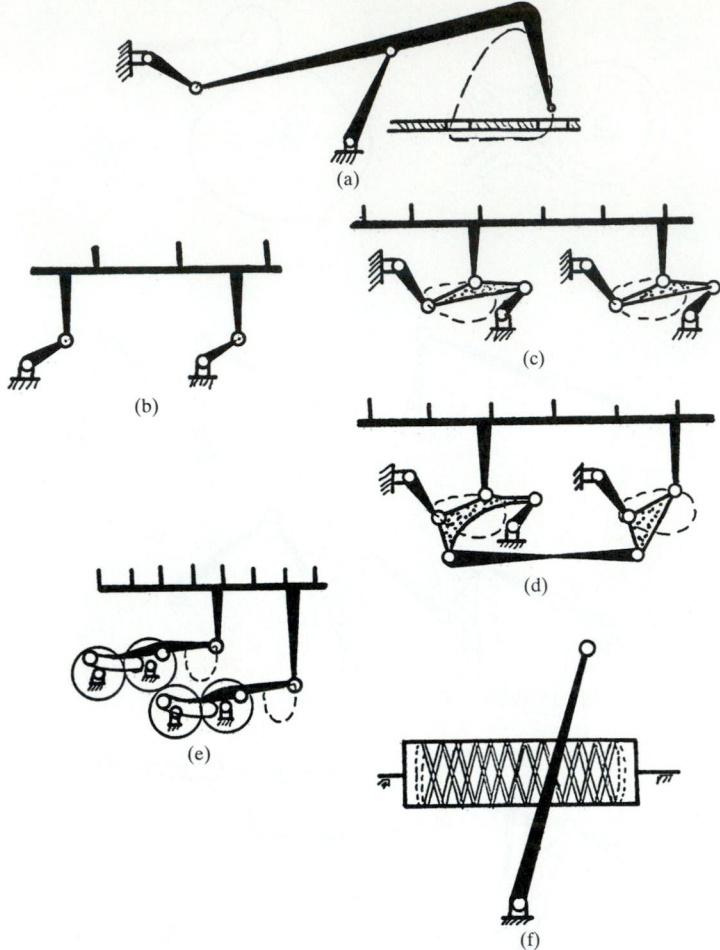

FIGURE 39.21 Transportation devices. These mechanisms move one or more objects a discrete distance in stepped motion. (*a*) Four-bar film advance; (*b*) circular-motion transport; (*c*), (*d*) coupler-curve transport; (*e*) geared linkage transport; (*f*) fishing-reel feed.

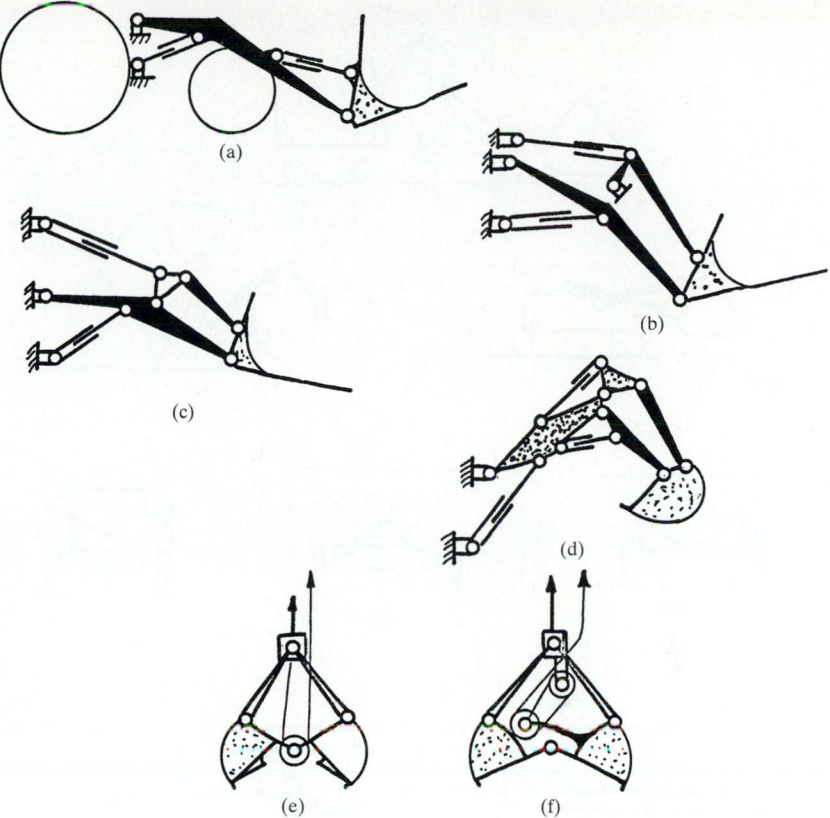

FIGURE 39.22 Loading and unloading mechanisms I. These mechanisms pick up material and transport it to another location. (*a*) to (*c*) Front-end loaders; (*d*) back hoe; (*e*), (*f*) clamshell loaders.

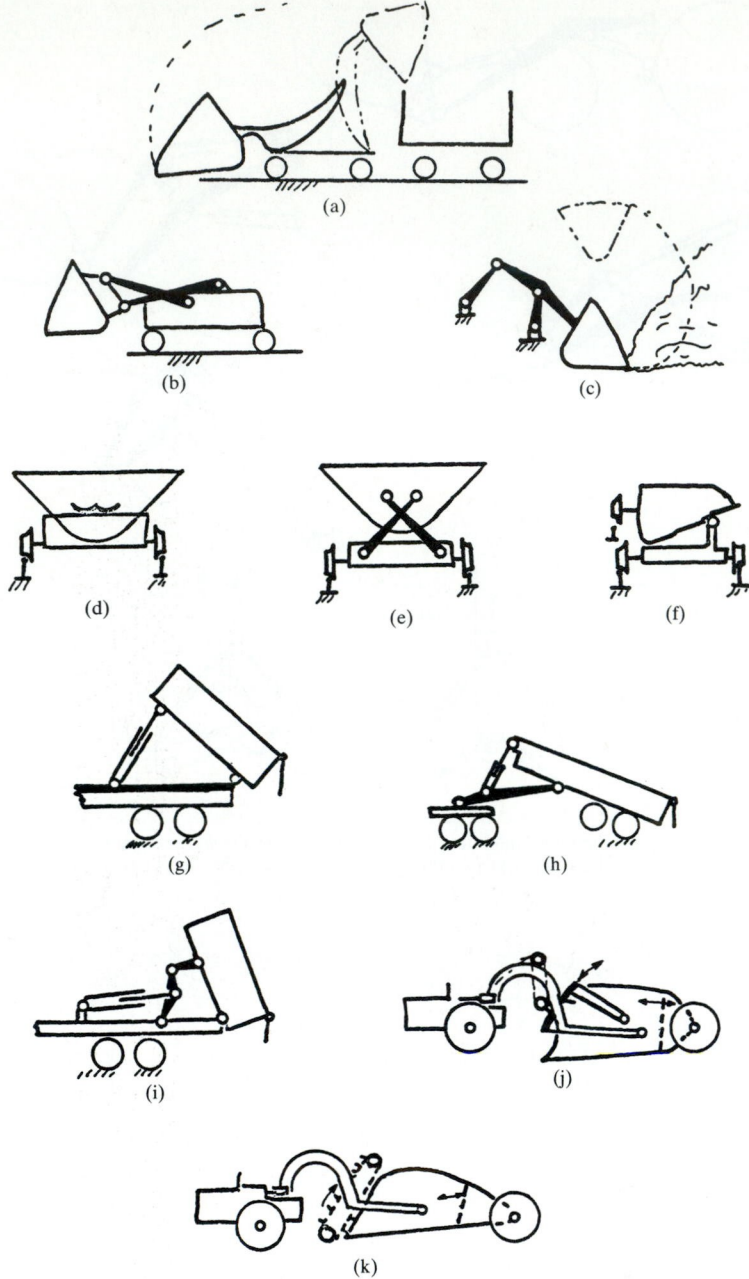

FIGURE 39.23 Loading and unloading mechanisms II. (*a*), (*b*) Mucking machines; (*c*) scooping mechanism; (*d*) to (*f*) dumping mine cars; (*g*) to (*i*) dump trucks; (*j*) motor scraper; (*k*) elevating scraper.

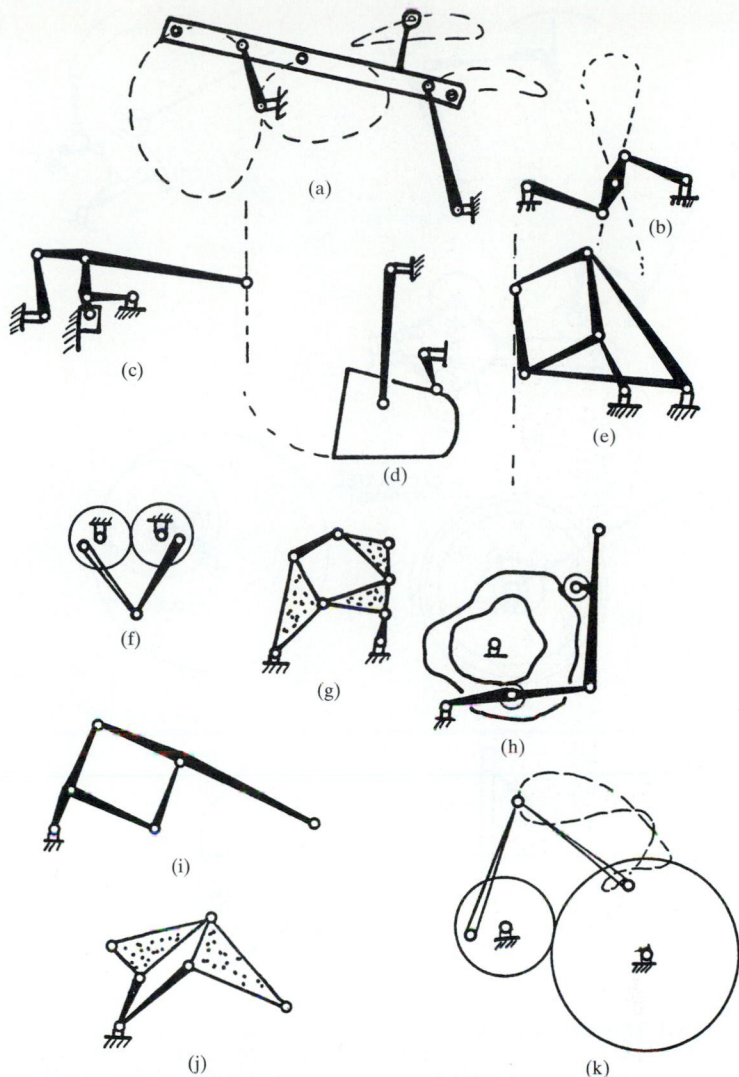

FIGURE 39.24 Path generators. These linkages approximately generate a required curve. (*a*) Four-bar coupler curve; (*b*) Watt straight-line linkage; (*c*) Crosby steam-engine indicator approximates straight line; (*d*) scooping mechanism; (*e*) Peaucellier exact straight-line linkage; (*f*) geared straight-line generators; (*g*) six-bar coupler curve; (*h*) double-cam line generator; (*i*) pantograph; (*j*) Sylvester skew pantograph; (*k*) geared linkage curve generator.

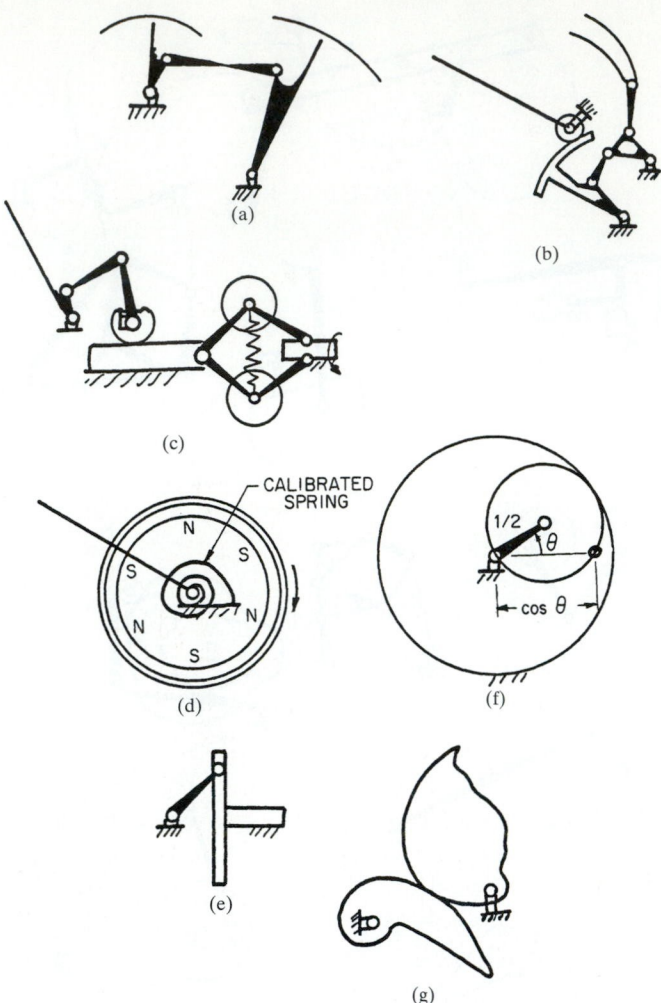

FIGURE 39.25 Function generators. These are mechanical devices in which the output moves as some function of the input $y = f(x)$. (*a*) Four-bar linkage function generator; (*b*) function generator in pressure gauge; (*c*), (*d*) function generator in a speedometer; (*e*) Scotch yoke sine-cosine generator; (*f*) epicyclic sine-cosine generator; (*g*) noncircular gears.

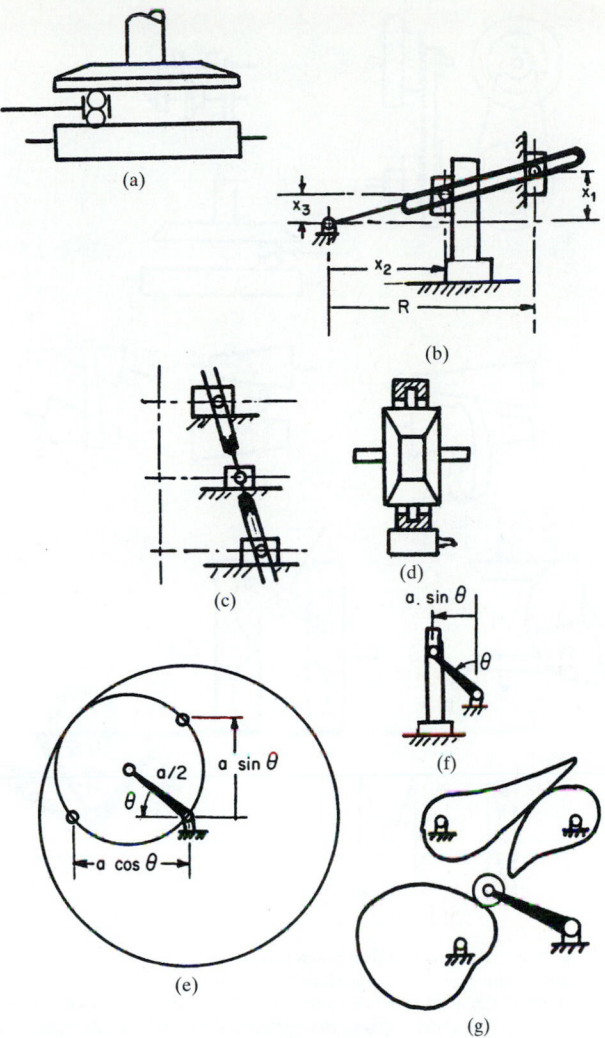

FIGURE 39.26 Computing mechanisms. These devices are used on mechanical computers for performing mathematical operations. (*a*) Ball disk integrator; (*b*) multiplier; (*c*), (*d*) adders; (*e*) epicyclic sine generators; (*f*) Scotch yoke sine generator; (*g*) noncircular gears; (*h*) special-function cams.

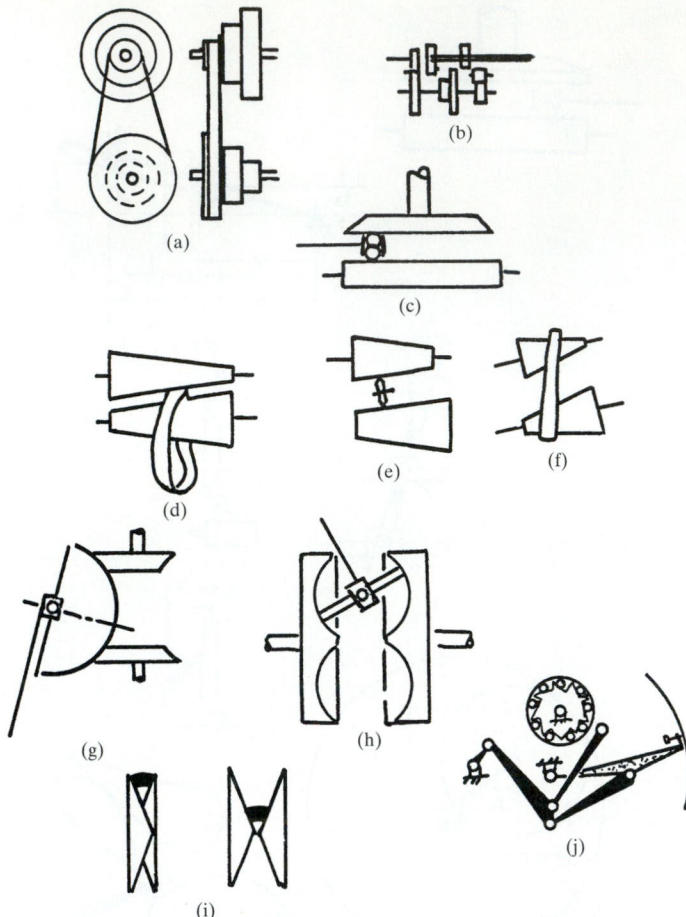

FIGURE 39.27   Speed-changing mechanisms. These devices change the speed of an output shaft while the input shaft has constant speed. (*a*) Stepped pulleys and flat belt; (*b*) geared transmission; (*c*) ball and disk speed changer; (*d*) to (*f*) cone drives; (*g*) sphere drive; (*h*) toroidal drive; (*i*) variable-pitch V belt; (*j*) zero maximum drive.

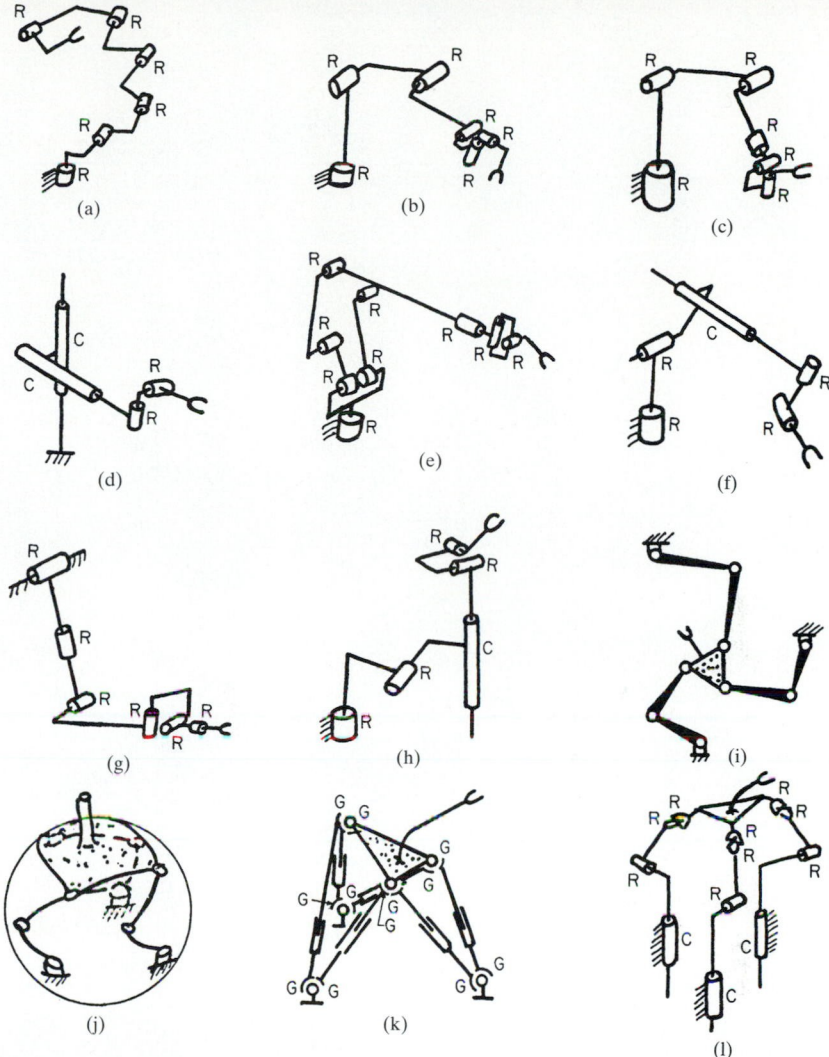

FIGURE 39.28   Robots. These are multidegree-of-freedom devices used for positioning or assembly of items. They usually have some degree of machine intelligence and work under computer control. (*a*) A general 6R robot; (*b*) to (*h*) some forms of existing robots; (*i*) parallel actuation of a planar 3-degrees-of-freedom robot; (*j*) Stewart platform which uses the 3-degrees-of-freedom principle; (*k*) Florida shoulder with parallel actuation; (*l*) general robot with parallel actuation.

CHAPTER 40
CAM MECHANISMS

Andrzej A. Olędzki, D.Sc.[†]
Warsaw Technical University, Poland

SUMMARY

This chapter addresses the design of cam systems in which flexibility is not a consideration. Flexible, high-speed cam systems are too involved for handbook presentation. Therefore only two generic families of motion, trigonometric and polynomial, are discussed. This covers most of the practical problems.

The rules concerning the reciprocating motion of a follower can be adapted to angular motion as well as to three-dimensional cams. Some material concerns circular-arc cams, which are still used in some fine mechanisms. In Sec. 40.3 the equations necessary in establishing basic parameters of the cam are given, and the important problem of accuracy is discussed. Force and torque analysis, return springs, and contact stresses are briefly presented in Secs. 40.4 and 40.5, respectively.

The chapter closes with the logic associated with cam design to assist in creating a computer-aided cam design program.

40.1 CAM MECHANISM TYPES, CHARACTERISTICS, AND MOTIONS

Cam-and-follower mechanisms, as linkages, can be divided into two basic groups:

1. Planar cam mechanisms

2. Spatial cam mechanisms

In a planar cam mechanism, all the points of the moving links describe paths in parallel planes. In a spatial mechanism, that requirement is not fulfilled. The design of mechanisms in the two groups has much in common. Thus the fundamentals of planar cam mechanism design can be easily applied to spatial cam mechanisms, which

---

[†] Prepared while the author was Visiting Professor of Mechanical Engineering, Iowa State University, Ames, Iowa.

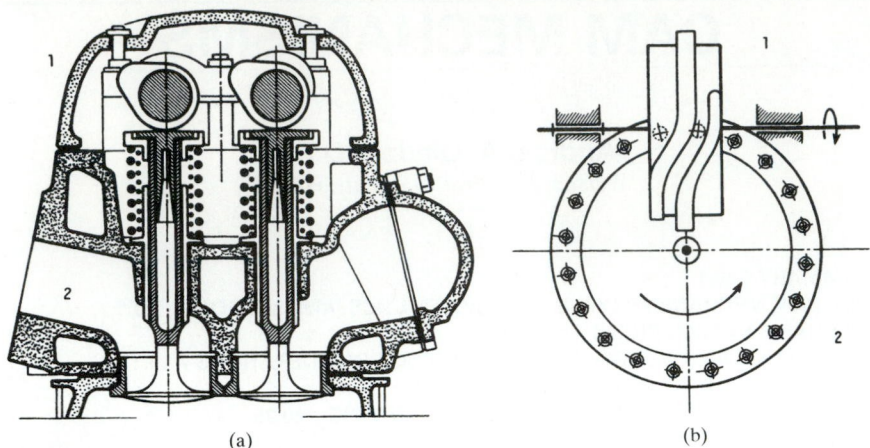

(a)   (b)

FIGURE 40.1   (*a*) Planar cam mechanism of the internal-combustion-engine D-R-D-R type; (*b*) spatial cam mechanism of the 16-mm film projector R-D-R type.

is not the case in linkages. Examples of planar and spatial mechanisms are depicted in Fig. 40.1.

Planar cam systems may be classified in four ways: (1) according to the motion of the follower—reciprocating or oscillating; (2) in terms of the kind of follower surface in contact—for example, knife-edged, flat-faced, curved-shoe, or roller; (3) in terms of the follower motion—such as dwell-rise-dwell-return (D-R-D-R), dwell-rise-return (D-R-R), rise-return-rise (R-R-R), or rise-dwell-rise (R-D-R); and (4) in terms of the constraining of the follower—spring loading (Fig. 40.1*a*) or positive drive (Fig. 40.1*b*).

Plate cams acting with four different reciprocating followers are depicted in Fig. 40.2 and with oscillating followers in Fig. 40.3.

Further classification of reciprocating followers distinguishes whether the centerline of the follower stem is radial, as in Fig. 40.2, or offset, as in Fig. 40.4.

Flexibility of the actual cam systems requires, in addition to the operating speed, some data concerning the dynamic properties of components in order to find discrepancies between rigid and deformable systems. Such data can be obtained from dynamic models. Almost every actual cam system can, with certain simplifications, be modeled by a one-degree-of-freedom system, shown in Fig. 40.5, where m_e

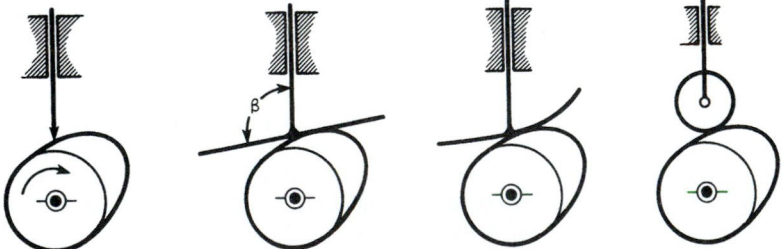

FIGURE 40.2   Plate cams with reciprocating followers.

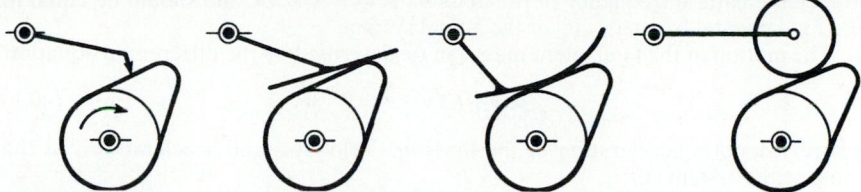

FIGURE 40.3   Plate cams with oscillating followers.

denotes an equivalent mass of the system, k_e equals equivalent stiffness, and s and y denote, respectively, the input (coming from the shape of the cam profile) and the output of the system. The equivalent mass m_e of the system can be calculated from the following equation, based on the assumption that the kinetic energy of that mass equals the kinetic energy of all the links of the mechanism:

$$m_e = \sum_{i=1}^{i=n} \frac{m_i v_i^2}{\dot{s}^2} + \sum_{i=1}^{i=n} \frac{I_i \omega_i^2}{\dot{s}^2}$$

where m_i = mass of link i
   v_i = linear velocity of center of mass of ith link
   I_i = moment of inertia about center of mass for ith link
   ω_i = angular velocity of ith link
   $\dot{s}$ = input velocity

The equivalent stiffness k_e can be found from direct measurements of the actual system (after a known force is applied to the last link in the kinematic chain and the displacement of that link is measured), and/or by assuming that k_e equals the actual stiffness of the most flexible link in the chain. In the latter case, k_e can usually be calculated from data from the drawing, since the most flexible links usually have a simple form (for example, a push rod in the automotive cam of Fig. 40.16c). In such a

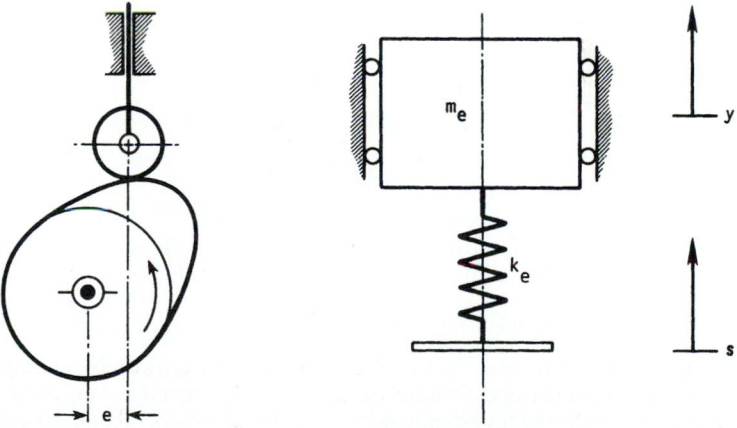

FIGURE 40.4   Plate cam with an offset reciprocating roller follower.

FIGURE 40.5   The one-degree-of-freedom cam system model.

model, the natural frequency of the mass m_e is $\omega_e = \sqrt{k_e/m_e}$ and should be equal to the fundamental frequency ω_n of the actual system.

The motion of the equivalent mass can be described by the differential equation

$$m_e\ddot{y} + k_e(y - s) = 0 \tag{40.1}$$

where $\ddot{y}$ denotes acceleration of the mass m_e. Velocity $\dot{s}$ and acceleration $\ddot{s}$ at the input to the system are

$$\dot{s} = \frac{ds}{dt} = \frac{ds}{d\theta}\frac{d\theta}{dt} = s'\omega \tag{40.2}$$

and

$$\ddot{s} = \frac{d}{dt}s'\omega = \frac{ds'}{dt}\omega + s'\frac{d\omega}{dt} = \frac{ds'}{dt}\frac{d\theta}{dt}\omega + s'\alpha$$

$$\tag{40.3}$$

$$= s''\omega^2 + s'\alpha$$

where θ = angular displacement of cam
        α = angular acceleration of cam
        $s' = ds/d\theta$, the *geometric velocity*
        $s'' = ds'/d\theta = d^2s/d\theta^2$, the *geometric acceleration*

When the cam operates at constant nominal speed $\omega = \omega_0$, $d\omega/dt = \alpha = 0$ and Eq. (40.3) simplifies to

$$\ddot{s} = s''\omega_0^2 \tag{40.4}$$

The same expressions can be used for the actual velocity $\dot{y}$ and for the actual acceleration $\ddot{y}$ at the output of the system. Therefore

$$\dot{y} = y'\omega \tag{40.5}$$

$$\ddot{y} = y''\omega^2 + y'\alpha \tag{40.6}$$

or

$$\ddot{y} = y''\omega_0^2 \qquad \omega = \omega_0 = \text{constant} \tag{40.7}$$

Substituting Eq. (40.7) into Eq. (40.1) and dividing by k_e gives

$$\mu_d y'' + y = s \tag{40.8}$$

where $\mu_d = (m_e/k_e)\omega_0^2$, the *dynamic factor* of the system.

Tesar and Matthew [40.10] classify cam systems by values of μ_d, and their recommendations for the cam designers, depending on the value of μ_d, are as follows:

$\mu_d \cong 10^{-6}$ (for low-speed systems; assume $s = y$)

$\mu_d \cong 10^{-4}$ (for medium-speed systems; use trigonometric, trapezoidal motion specifications, and/or similar ones; synthesize cam at design speed $\omega = \omega_0$, use good manufacturing practices and investigate distortion due to off-speed operations)

$\mu_d \cong 10^{-2}$ (for high-speed systems; use polynomial motion specification and best available manufacturing techniques)

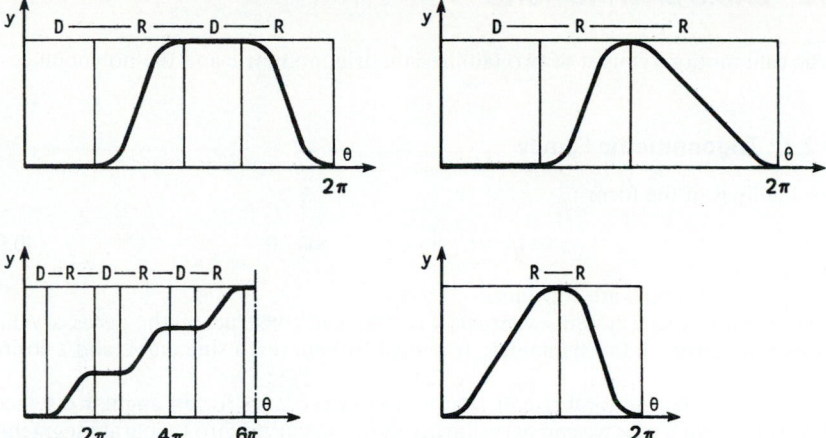

FIGURE 40.6 Types of follower motion.

In all the cases, increasing k_e and reducing m_e are recommended, because it reduces μ_d.

There are two basic phases of the follower motion, *rise* and *return*. They can be combined in different ways, giving types of cams classifiable in terms of the type of follower motion, as in Fig. 40.6.

For positive drives, the symmetric acceleration curves are to be recommended. For cam systems with spring restraint, it is advisable to use unsymmetric curves because they allow smaller springs. Acceleration curves of both the symmetric and unsymmetric types are depicted in Fig. 40.7.

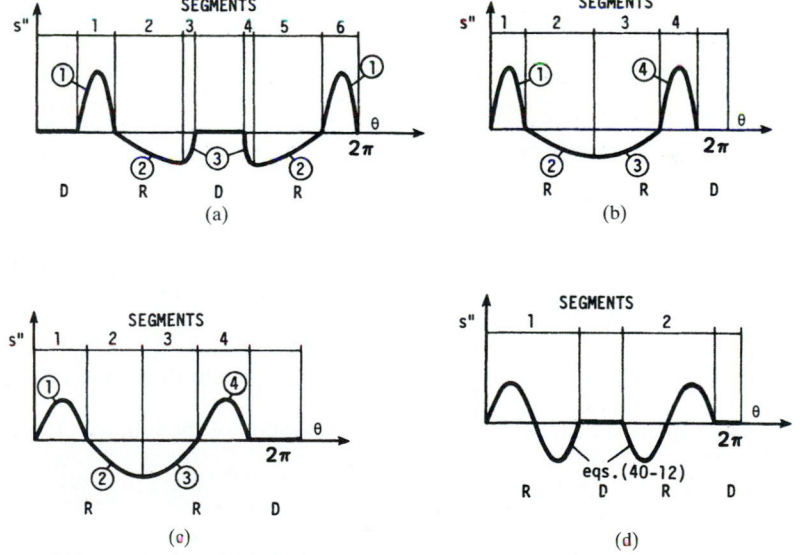

FIGURE 40.7 Acceleration diagrams: (*a*), (*b*) spring loading; (*c*), (*d*) positive drive.

40.2 BASIC CAM MOTIONS

Basic cam motions consist of two families: the trigonometric and the polynomial.

40.2.1 Trigonometric Family

This family is of the form

$$s'' = C_0 + C_1 \sin a\theta + C_2 \cos b\theta \qquad (40.9)$$

where $C_0, C_1, a,$ and b are constants.

For the low-speed systems where $\mu_d < 10^{-4}$, we can construct all the necessary diagrams, symmetric and unsymmetric, from just two curves: a sine curve and a cosine curve.

Assuming that the total rise or return motion s_0 occurs for an angular displacement of the cam $\theta = \beta_0$, we can partition acceleration curves into i separate segments, where $i = 1, 2, 3, \ldots$ with subtended angles $\beta_1, \beta_2, \beta_3, \ldots$ so that $\beta_1 + \beta_2 + \beta_3 + \cdots = \beta_0$. The sum of partial lifts $s_1, s_2, s_3, \ldots$ in the separate segments should be equal to the total rise or return s_0: $s_1 + s_2 + s_3 + \cdots = s_0$. If a dimensionless description θ/β of cam rotation is introduced into a segment, we will have the value of ratio θ/β equal to zero at the beginning of each segment and equal to unity at the end of each segment.

All the separate segments of the acceleration curves can be described by equations of the kind

$$s'' = A \sin \frac{n\pi\theta}{\beta} \qquad n = \frac{1}{2}, 1, 2 \qquad (40.10)$$

or

$$s'' = A \cos \frac{\pi\theta}{2\beta} \qquad (40.11)$$

where A is the maximum or minimum value of the acceleration in the individual segment.

The simplest case is when we have a positive drive with a symmetric acceleration curve (Fig. 40.7d). The complete rise motion can be described by a set of equations

$$s = s_0 \left(\frac{\theta}{\beta} - \frac{1}{2\pi} \sin \frac{2\pi\theta}{\beta} \right) \qquad s'' = \frac{2\pi s_0}{\beta^2} \sin \frac{2\pi\theta}{\beta}$$

$$s' = \frac{s_0}{\beta} \left(1 - \cos \frac{2\pi\theta}{\beta} \right) \qquad s''' = \frac{4\pi^2 s_0}{\beta^3} \cos \frac{2\pi\theta}{\beta} \qquad (40.12)$$

The last term is called *geometric jerk* ($\dddot{s} = \omega^3 s'''$). Traditionally, this motion is called *cycloidal*.

The same equations can be used for the return motion of the follower. It is easy to prove that

$$s_{\text{return}} = s_0 - s_{\text{rise}} \qquad s''_{\text{return}} = -s''_{\text{rise}}$$

$$s'_{\text{return}} = -s'_{\text{rise}} \qquad s'''_{\text{return}} = -s'''_{\text{rise}} \qquad (40.13)$$

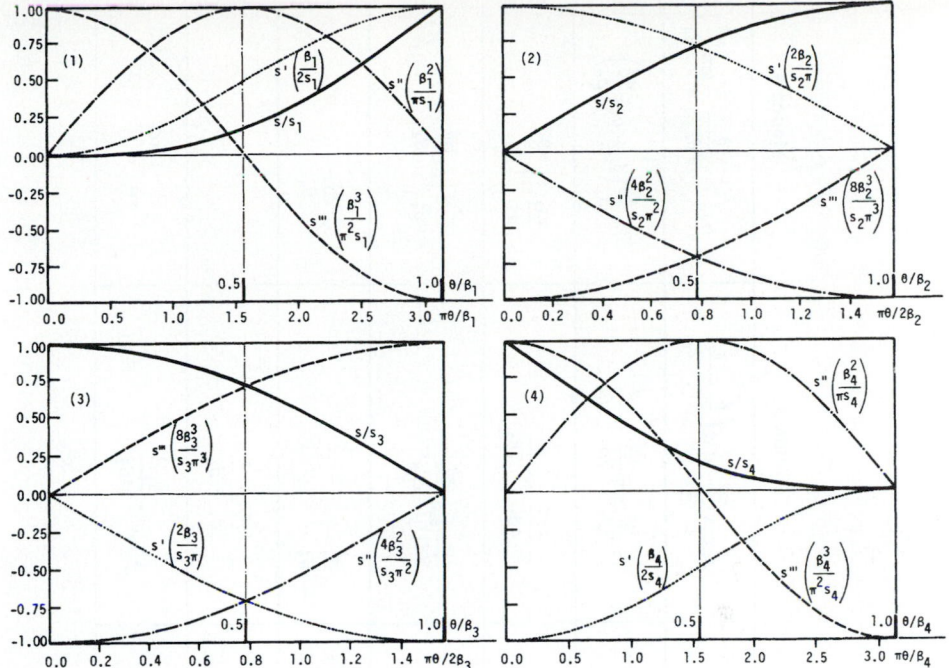

FIGURE 40.8 Trigonometric standard follower motions (according to the equation of Table 40.1, for $c = d = 0$).

All the other acceleration curves, symmetric and unsymmetric, can be constructed from just four trigonometric standard follower motions. They are denoted further by the numbers 1 through 4 (Fig. 40.8). These are displayed in Table 40.1.

Equations in Table 40.1 can be used to represent the different segments of a follower's displacement diagram. Derivatives of displacement diagrams for the adjacent segments should match each other; thus several requirements must be met in order to splice them together to form the motion specification for a complete cam. Motions 1 through 4 have the following applications:

Motion 1 is for the initial part of a rise motion.

Motion 2 is for the end and/or the middle part of a rise motion and the initial part of a return motion. The value c is a constant, equal to zero only in application to the end part of a rise motion.

Motion 3 is for the end part of a rise motion and/or the initial or middle part of a return motion. The value d is a constant, equal to zero only in application to the initial part of a return motion.

Motion 4 is for the end part of a return motion.

The procedure of matching the adjacent segments is best understood through examples.

Example 1. This is an extended version of Example 5-2 from Shigley and Uicker [40.8], p. 229. Determine the motion specifications of a plate cam with reciprocating

TABLE 40.1 Standard Trigonometric Follower Motions

Parameter	Motion 1	Motion 2	Motion 3	Motion 4
s	$\dfrac{s_1}{\pi}\left(\dfrac{\pi\theta}{\beta_1} - \sin\dfrac{\pi\theta}{\beta_1}\right)$	$s_2\sin\dfrac{\pi\theta}{2\beta_2} + c\,\dfrac{\theta}{\beta_2}$	$s_3\cos\dfrac{\pi\theta}{2\beta_3} + d\left(\dfrac{\theta}{\beta_3} - \dfrac{1}{2}\right)$	$s_4\left(1 - \dfrac{\theta}{\beta_4} - \dfrac{1}{\pi}\sin\dfrac{\pi\theta}{\beta_4}\right)$
s'	$\dfrac{s_1}{\beta_1}\left(1 - \cos\dfrac{\pi\theta}{\beta_1}\right)$	$\dfrac{s_2\pi}{2\beta_2}\cos\dfrac{\pi\theta}{2\beta_2} + \dfrac{c}{\beta_2}$	$-\dfrac{s_3\pi}{2\beta_3}\sin\dfrac{\pi\theta}{2\beta_3} + \dfrac{d}{\beta_3}$	$-\dfrac{s_4}{\beta_4}\left(1 + \cos\dfrac{\pi\theta}{\beta_4}\right)$
s''	$\dfrac{\pi s_1}{\beta_1^2}\sin\dfrac{\pi\theta}{\beta_1}$	$-\dfrac{s_2\pi^2}{4\beta_2^2}\sin\dfrac{\pi\theta}{2\beta_2}$	$-\dfrac{s_3\pi^2}{4\beta_3^2}\cos\dfrac{\pi\theta}{2\beta_3}$	$\dfrac{\pi s_4}{\beta_4^2}\sin\dfrac{\pi\theta}{\beta_4}$
s'''	$\dfrac{\pi^2 s_1}{\beta_1^3}\cos\dfrac{\pi\theta}{\beta_1}$	$-\dfrac{s_2\pi^3}{8\beta_2^3}\cos\dfrac{\pi\theta}{2\beta_2}$	$\dfrac{s_3\pi^3}{8\beta_3^3}\sin\dfrac{\pi\theta}{2\beta_3}$	$\dfrac{\pi^2 s_4}{\beta_4^3}\cos\dfrac{\pi\theta}{\beta_4}$
$s'_{i,\,\text{init}}\left(\dfrac{\theta}{\beta_i} = 0\right)$	$\cdots$	$\dfrac{s_2\pi}{2\beta_2}$	$\dfrac{d}{\beta_3}$	$-\dfrac{2s_4}{\beta_4}$
$s'_{i,\,\text{end}}\left(\dfrac{\theta}{\beta_i} = 1\right)$	$\dfrac{2s_1}{\beta_1}$	$\dfrac{c}{\beta_2}$	$-\dfrac{s_3\pi}{2\beta_3} + \dfrac{d}{\beta_3}$	
$s''_{\max};\ s''_{\min}$	$s''_{\max} = \dfrac{\pi s_1}{\beta_1^2}$	$s''_{\min} = -\dfrac{s_2\pi^2}{4\beta_2^2}$	$s''_{\min} = -\dfrac{s_3\pi^2}{4\beta_3^2}$	$s''_{\max} = \dfrac{\pi s_4}{\beta_4^2}$

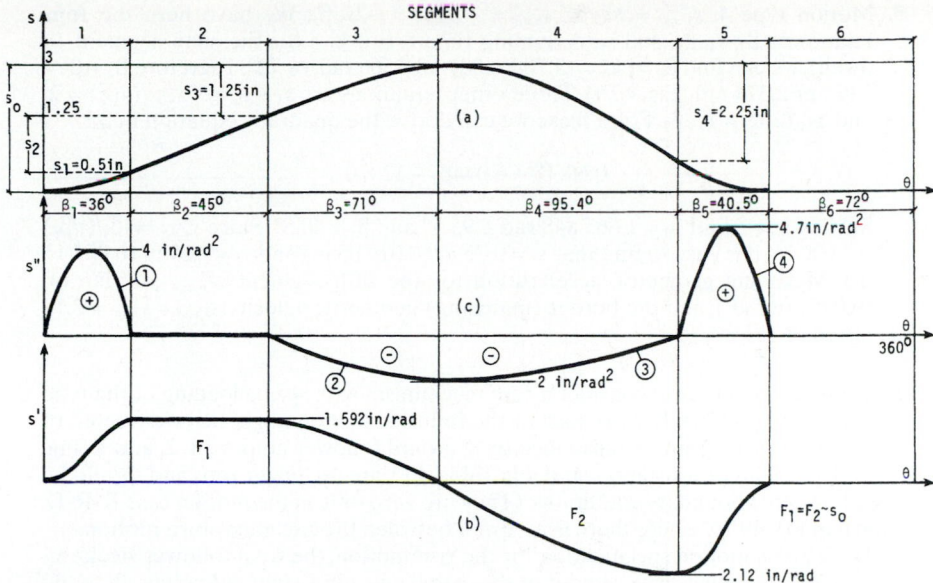

FIGURE 40.9 Example 1: (*a*) displacement diagram, in; (*b*) geometric velocity diagram, in/rad; (*c*) geometric acceleration diagram, in/rad².

follower and return spring for the following requirements: The speed of the cam is constant and equal to 150 r/min. Motion of the follower consists of six segments (Fig. 40.9):

1. Accelerated motion to $\dot{s}_{1,\text{end}} = 25$ in/s (0.635 m/s)
2. Motion with constant velocity 25 in/s, lasting for 1.25 in (0.03175 m) of rise
3. Decelerated motion (segments 1 to 3 describe rise of the follower)
4. Return motion
5. Return motion
6. Dwell, lasting for $t \geq 0.085$ s

The total lift of the follower is 3 in (0.0762 m).

Solution. Angular velocity $\omega = 150\pi/30 = 15.708$ radians per second (rad/s). The cam rotation for 1.25 in of rise is equal to $\beta_2 = 1.25$ in/$s'_2 = 1.25$ in/1.592 in/rad = 0.785 rad = 45°, where $s'_2 = 25/15.708 = 1.592$ in/rad.

The following decisions are quite arbitrary and depend on the designer:

1. Use motion 1; then $s_1 = 0.5$ in, $s''_{\max} = 0.05\pi/\beta_1^2 = 0.5\pi/(0.628)^2 = 4$ in/rad² (0.1016 m/rad²). $s''_{1,\text{end}} = 2(0.5)/\beta_1$; so $\beta_1 = 1/1.592 = 0.628$ rad, or 36°.
2. For the motion with constant velocity, $s'_2 = 1.592$ in/rad (0.4044 m/rad); $s_2 = 1.25$ in.
3. Motion type 2: $s_3 = s_2 = 1.25$ in, $s'_{3,\text{init}} = s_3\pi/(2\beta_3) = 1.592$ in/rad; therefore $\beta_3 = 1.25\pi/[2(1.592)] = 1.233$ rad $\cong 71°$, $s'''_{3,\text{min}} = -(1.25\pi^2)/[4(1.233)^2] = -2$ in/rad². (Points 1 through 3 describe the rise motion of the follower.)
4. Motion type 3: $s''_{4,\text{init}} = s_4\pi^2/(4\beta_r^2) = -2$ in/rad² (the same value as that of $s''_{3,\text{min}}$), $s''_{4,\text{end}} = -\pi s_4/(2\beta_4)$, $s_4 + s_5 = 3$ in.

5. Motion type 4: $s''_{5,\max} = \pi s_5/\beta_5^2$, $s'_{5,\text{init}} = -s'_{4,\text{end}} = -2s_5/\beta_5$. We have here the four unknowns β_4, s_4, β_5, and s_5. Assuming time $t_6 = 0.85$ s for the sixth segment (a dwell), we can find $\beta_6 = \omega t_6 = 15.708(0.08) = 1.2566$ rad, or $72°$. Therefore $\beta_4 + \beta_5 = 136°$, or 2.374 rad (Fig. 40.9). Three other equations are $s_4 + s_5 = 3$, $s_4\pi^2/(4\beta_4^2) = 2$, and $\pi s_4/(2\beta_4) = 2s_5/\beta_5$. From these we can derive the quadratic equation in β_4.

$$0.696\beta_4^2 + 6.044\beta_4 - 12 = 0$$

Solving it, we find $\beta_4 = 1.665\,848$ rad $\cong 95.5°$ and $\beta_5 = 40.5°$. Since $s_4/s_5 = 4\beta_4/(\pi\beta_5)$ $= 3.000\,76$, it is easy to find that $s_5 = 0.75$ in (0.019 05 m) and $s_4 = 2.25$ in (0.057 15 m). Maximum geometric acceleration for the fifth segment $s''_{5,\max} = 4.7$ in/rad² (0.0254 m/rad²), and the border (matching) geometric velocity $s'_{4,\text{end}} = s'_{5,\text{init}} = 2.12$ in/rad (0.253 m/rad).

Example 2. Now let us consider a cam mechanism with spring loading of the type D-R-D-R (Fig. 40.7a). The rise part of the follower motion might be constructed of three segments (1, 2, and 3) described by standard follower motions 1, 2, and 3 (Fig. 40.8). The values of constants c and d in Table 40.1 are no longer zero and should be found from the boundary conditions. (They are zero only in the motion case R-R-D, shown in Fig. 40.7b, where there is no dwell between the rise and return motions.)

For a given motion specification for the rise motion, the total follower stroke s_0, and the total angular displacement of the cam β_0, we have eight unknowns: β_1, s_1, β_2, s_2, β_3, s_3, and constants c and d. The requirements of matching the displacement derivatives will give us only six equations; thus two more must be added to get a unique solution. Two additional equations can be written on the basis of two arbitrary decisions:

1. The maximum value of the acceleration in segment 1, $s''_{1,\max}$ should be greater than that in segment 2 because of spring loading. So $s''_{1,\max} = -as''_{2,\min}$ where $s''_{2,\min}$ is the minimum value of the second-segment acceleration and a is any assumed number, usually greater than 2.

2. The end part of the rise (segment 3), the purpose of which is to avoid a sudden drop in a negative accelerative curve, should have a smaller duration than the basic negative part (segment 2). Therefore we can assume any number b (greater than 5) and write $\beta_2 = b\beta_3$. The following formulas were found after all eight equations for the eight unknowns were solved simultaneously:

$$\beta_1 = \frac{\beta_0}{1 + a + a/b} \qquad\qquad \beta_3 = \beta_0 \frac{a}{a(1+b) + b}$$

$$s_1 = s_0 \frac{\pi b^2}{b^2(\pi + 4a) + 4a(2a+1)} \qquad\qquad s'_2 = s_1 \frac{4a}{\pi}$$

$$s_3 = s_1 \frac{4a}{\pi b^2} \qquad\qquad c = s_1 \frac{8a^2}{\pi b^2}$$

$$d = 2s_3 \qquad\qquad s_0 = s_1 + s'_2 + c + s_3$$

We can assume practical values for a and b (say $a = 2$, $b = 10$) and find from the above equations the set of all the parameters (as functions of s_0 and β_0) necessary to form the motion specification for the rise motion of the follower and the shape of the cam profile. The whole set of parameters is as follows:

$$s_1 = 0.272\ 198s_0 \qquad \beta_1 = 0.312\ 5\beta_0$$

$$s_2' = 0.693\ 147s_0 \qquad c = 0.027\ 726s_0$$

$$\beta_2 = 0.625\beta_0$$

$$s_3 = 0.006\ 931s_0 \qquad d = 2s_3 \quad \text{(always!)}$$

$$\beta_3 = 0.0625\beta_0$$

These can be used for calculations of the table $s = s(\theta)$, which is necessary for manufacturing a cam profile. For such a table, we use as a rule increments of θ equal to about 1° and accuracy of s up to 4×10^{-5} in 1 micrometer (μm). The data of such a table can be easily used for the description of both the return motion of the follower and a cam profile, providing $\beta_0(\text{return}) = \beta_0(\text{rise})$, and the acceleration diagram for the return motion is a mirror image of the acceleration diagram for the rise motion. Table 40.2 can be of assistance in calculating the return portion of the cam profile. The column $s(\text{return})$ is the same as the column $s(\text{rise})$.

TABLE 40.2 Data of Rise Motion Used for Calculations of Return Portion of Cam Profile

Rise		Return	
$\theta(\text{rise})$	$s(\text{rise})$	$\theta(\text{return})$	$s(\text{return})$
0	0	$2\beta_0 + \beta_d - 0$	0
$\vdots$	$\vdots$	$\vdots$	$\vdots$
θ_i	s_i	$2\beta_0 + \beta_d - \theta_i$	s_i
$\vdots$	$\vdots$	$\vdots$	$\vdots$
β_0	s_0	$2\beta_0 + \beta_d - \beta_0$	s_0

The trigonometric acceleration diagram for the positive drive was described at the beginning of this section by Eq. (40.12). The improved diagram (smaller maximum values of acceleration for the same values of s_0 and $\beta_0^\dagger$) can be obtained if we combine sine segments with segments of constant acceleration. Such a diagram, called a *modified trapezoidal acceleration curve,* is shown in Fig. 40.10. Segments 1, 3, 4, and 6 are the sinusoidal type. Sections 2 and 5 are with $s'' = $ constant. It was assumed for that diagram that all the sine segments take one-eighth of the total angular displacement β_0 of the cam during its rise motion. The first half of the motion has three segments. The equations for the first segment are $0 \le \theta/\beta_0 \le \frac{1}{8}$, and so

---

† The maximum acceleration ratio is 4.9/6.28.

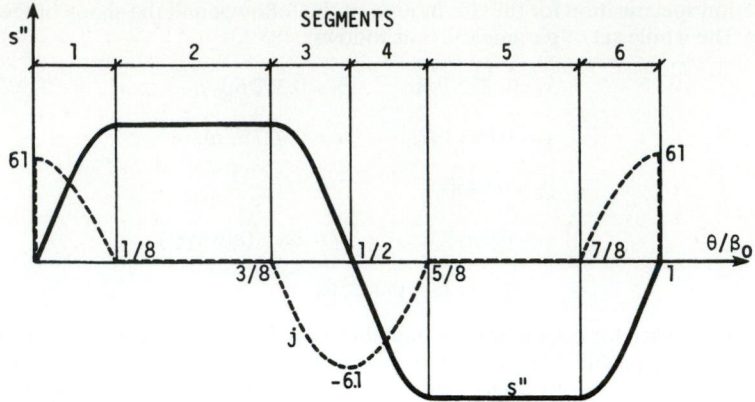

FIGURE 40.10   A modified trapezoidal acceleration diagram.

$$s = \frac{s_0'}{2\pi}\left(\frac{4\pi\theta}{\beta_0} - \sin\frac{4\pi\theta}{\beta_0}\right) \qquad s' = \frac{2s_0'}{\beta_0}\left(1 - \cos\frac{4\pi\theta}{\beta_0}\right)$$

$$ \tag{40.14}$$

$$s'' = 8\pi\frac{s_0'}{\beta_0^2}\sin\frac{4\pi\theta}{\beta_0} \qquad s''' = 32\pi^2\frac{s_0'}{\beta_0^3}\cos\frac{4\pi\theta}{\beta_0}$$

For the second segment, we have $\tfrac{1}{8} \le \theta/\beta_0 \le \tfrac{3}{8}$, and so

$$s = s_0'\left[-\frac{1}{2\pi} + \frac{2\theta}{\beta_0} + 4\pi\left(\frac{\theta}{\beta_0} - \frac{1}{8}\right)^2\right]$$

$$s' = \frac{s_0'}{\beta_0}\left[2 + 8\pi\left(\frac{\theta}{\beta_0} - \frac{1}{8}\right)\right]$$

$$ \tag{40.15}$$

$$s'' = \frac{8\pi s_0'}{\beta_0^2} \qquad s''' = 0$$

The relations for the third segment are $\tfrac{3}{8} \le \theta/\beta_0 \le \tfrac{5}{8}$ ([40.7]); therefore,

$$s = s_0'\left\{-\frac{\pi}{2} + 2(1+\pi)\frac{\theta}{\beta_0} - \frac{1}{2\pi}\sin\left[4\pi\left(\frac{\theta}{\beta_0} - \frac{2}{8}\right)\right]\right\}$$

$$s' = \frac{2s_0'}{\beta_0}\left\{1 + \pi - \cos\left[4\pi\left(\frac{\theta}{\beta_0} - \frac{2}{8}\right)\right]\right\}$$

$$ \tag{40.16}$$

$$s'' = \frac{8\pi s_0'}{\beta_0^2}\sin\left[4\pi\left(\frac{\theta}{\beta_0} - \frac{2}{8}\right)\right]$$

$$s''' = 32\pi^2\frac{s_0'}{\beta_0^3}\cos\left[4\pi\left(\frac{\theta}{\beta_0} - \frac{2}{8}\right)\right]$$

where $s_0' = s_0/(2 + \pi) = 0.194\ 492 s_0$.

Using Eqs. (40.14) through (40.16) for all three segments, we can calculate the s values for the first half of the rise motion, where $\theta/\beta_0 = \tfrac{4}{8}$ and $s = s_0/2$. Since the negative part of the acceleration diagram is a mirror image of the positive part, it is easy to calculate the s values for the second half of the rise motion from the data obtained for the first half. The necessary procedure for that is shown in Table 40.3. The procedure concerns the case with the modified trapezoidal acceleration diagram, but it could be used as well for all the cases with symmetric acceleration diagrams for the rise motion. For the return motion of the follower, when its acceleration diagram is a mirror image of the rise diagram, we can use again the technique shown in Table 40.2. All the calculations can be done simultaneously by the computer after a simple program is written.

TABLE 40.3 Data of First Half of Rise Motion Used for Calculations of Second Half

		$\gamma = \theta/\beta_o$	s	γ	s	
Segments	1	0 $\vdots$ γ_i $\vdots$ $\tfrac{1}{8}$	0 $\vdots$ s_i $\vdots$ $s_1 = s_0'(\pi/2 - 1)/2\pi$	1 $\vdots$ $1 - \gamma_i$ $\vdots$ $\tfrac{7}{8}$	s_0 $\vdots$ $s_0 - s_i$ $\vdots$ $s_0 - s_1$	6
	2	$\tfrac{1}{8}$ $\vdots$ γ_j $\vdots$ $\tfrac{3}{8}$	s_1 $\vdots$ s_j $\vdots$ $s_2 = s_0'(\tfrac{3}{4} - 1/2\pi + \pi/4)$	$\tfrac{7}{8}$ $\vdots$ $1 - \gamma_j$ $\vdots$ $\tfrac{5}{8}$	$s_0 - s_1$ $\vdots$ $s_0 - s_j$ $\vdots$ $s_0 - s_2$	5
	3	$\tfrac{3}{8}$ $\vdots$ γ_k $\vdots$ $\tfrac{4}{8}$	s_2 $\vdots$ s_k $\vdots$ $s_3 = s_0/2$	$\tfrac{5}{8}$ $\vdots$ $1 - \gamma_k$ $\vdots$ $\tfrac{4}{8}$	$s_0 - s_2$ $\vdots$ $s_0 - s_k$ $\vdots$ $s_0 - s_3 = s_0/2$	4

All the trigonometric curves of this section were calculated with finite values of jerk, which is of great importance for the dynamic behavior of the cam mechanism. An example of the jerk diagram is given in Fig. 40.10. The jerk curve j was plotted by using the dimensionless expression

$$ j = \frac{s'''}{s_0/\beta_0^3} \tag{40.17} $$

This form of the jerk description can also be used to compare properties of different acceleration diagrams.

40.2.2 Polynomial Family

The basic polynomial equation is

$$s = C_0 + C_1 \frac{\theta}{\beta_0} + C_2 \left(\frac{\theta}{\beta_0}\right)^2 + C_3 \left(\frac{\theta}{\beta_0}\right)^3 + \cdots \tag{40.18}$$

with constants C_i depending on assumed initial and final conditions.

This family is especially useful in the design of flexible cam systems, where values of the dynamic factor are $\mu_d \geq 10^{-2}$. Dudley (1947) first used polynomials for the synthesis of flexible systems, and his ideal later was improved by Stoddart [40.9] in application to automotive cam gears.

The shape factor s of the cam profile can be found by this method after a priori decisions are made about the motion y of the last link in the kinematic chain. Cams of that kind are called *polydyne cams.*

When flexibility of the system can be neglected, the initial and final conditions ([40.3], [40.4], and [40.8]) might be as follows (positive drive):

1. Initial conditions for full-rise motion are

$$\frac{\theta}{\beta_0} = 0 \qquad s = 0 \qquad s' = 0 \qquad s'' = 0$$

2. Final (end) conditions are

$$\frac{\theta}{\beta_0} = 1 \qquad s = s_0 \qquad s' = 0 \qquad s'' = 0$$

The first and second derivatives of Eq. (40.18) are

$$s' = C_1 + 2C_2 \frac{\theta}{\beta_0} + 3C_3 \left(\frac{\theta}{\beta_0}\right)^2 + 4C_4 \left(\frac{\theta}{\beta_0}\right)^3 + \cdots$$

$$\tag{40.19}$$

$$s'' = 2C_2 + 6C_3 \frac{\theta}{\beta_0} + 12C_4 \left(\frac{\theta}{\beta_0}\right)^2 + \cdots$$

Substituting six initial and final conditions into Eqs. (40.18) and (40.19) and solving them simultaneously for unknowns C_0, C_1, C_2, C_3, C_4, and C_5, we have

$$s = 10s_0 \left[\left(\frac{\theta}{\beta_0}\right)^3 - 1.5 \left(\frac{\theta}{\beta_0}\right)^4 + 0.6 \left(\frac{\theta}{\beta_0}\right)^5 \right]$$

$$s' = 30 \frac{s_0}{\beta_0} \left[\left(\frac{\theta}{\beta_0}\right)^2 - 2 \left(\frac{\theta}{\beta_0}\right)^3 + \left(\frac{\theta}{\beta_0}\right)^4 \right] \tag{40.20}$$

$$s'' = 60 \frac{s_0}{\beta_0^2} \left[\frac{\theta}{\beta_0} - 3 \left(\frac{\theta}{\beta_0}\right)^2 + 2 \left(\frac{\theta}{\beta_0}\right)^3 \right]$$

and for a jerk $s''' = ds''/d\theta$, or

$$s''' = 60 \frac{s_0}{\beta_0^3} \left[1 - 6 \frac{\theta}{\beta_0} + 6 \left(\frac{\theta}{\beta_0}\right)^2 \right]$$

This is called the polynomial 3-4-5, since powers 3, 4, and 5 remain in the displacement equation. It provides a fairly good diagram for the positive drives.

Equations for the full-return polynomial are

$$s(\text{return}) = -s(\text{rise}) + s_0 \qquad s''(\text{return}) = -s'(\text{rise})$$

$$s''(\text{return}) = -s''(\text{rise}) \qquad s'''(\text{return}) = -s'''(\text{rise})$$

(40.21)

All the characteristic curves of the full-rise 3-4-5 polynomial are shown in Fig. 40.11. They were generated by the computer for $s_0 = 1$ displacement unit (inches or centimeters) and $\beta_0 = 1$ rad.

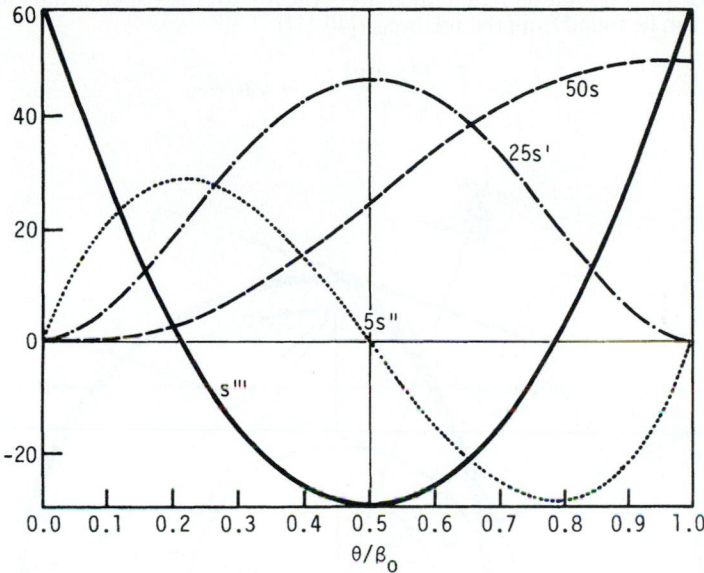

FIGURE 40.11   Full-rise 3-4-5 polynomial motion.

After a proper set of initial and final conditions is established, the basic equation [Eq. (40.18)] can be used for describing any kind of follower motion with an unsymmetric acceleration diagram. Details concerning the necessary procedure can be found in Rothbart [40.7].

40.2.3   Other Cam Motions

The basic cam motions described in the previous sections cover most of the routine needs of the contemporary cam designer. However, sometimes the cost of manufacturing the cam profile may be too high and the dynamic properties of the cam motion may not be severe. This is the case of cams used for generating functions. There is a very effective approach, described by Mischke [40.2], concerning an optimum design of simple eccentric cams. They are very inexpensive, yet can be used even for generating very complicated functions.

The other approach, when we are interested in inexpensive cams, is to use *circular-arc* cams or *tangent* cams. They are still used in low-speed diesel IC engines since the cost of their manufacture is low (compare with Fig. 40.15). An extensive review of these cams can be found in Rothbart [40.7]. They were used quite frequently in the past when the speed of machines was low, but today they are not often recommended because their dynamic characteristics are poor. The only exception can be made for fine- or light-duty mechanisms, such as those of 8- and 16-mm film projectors, where circular-arc cams are still widely used. Those cams are usually of the positive drive kind, where the breadth of the cam is constant. The cam drives a reciprocating follower with two flat working surfaces a fixed distance apart, which contact opposite sides of the cam.

The constant-breadth cam is depicted in Fig. 40.12. For given values of radius ρ, total angle of cam rotation β_0, and total lift of the follower s_0, the basic dimensions of the cam can be found from the relations ([40.11])

$$R_1 = \frac{\rho - b(s_0 + \rho)}{1 - b} \qquad r = R_1 - s_0 \qquad (40.22)$$

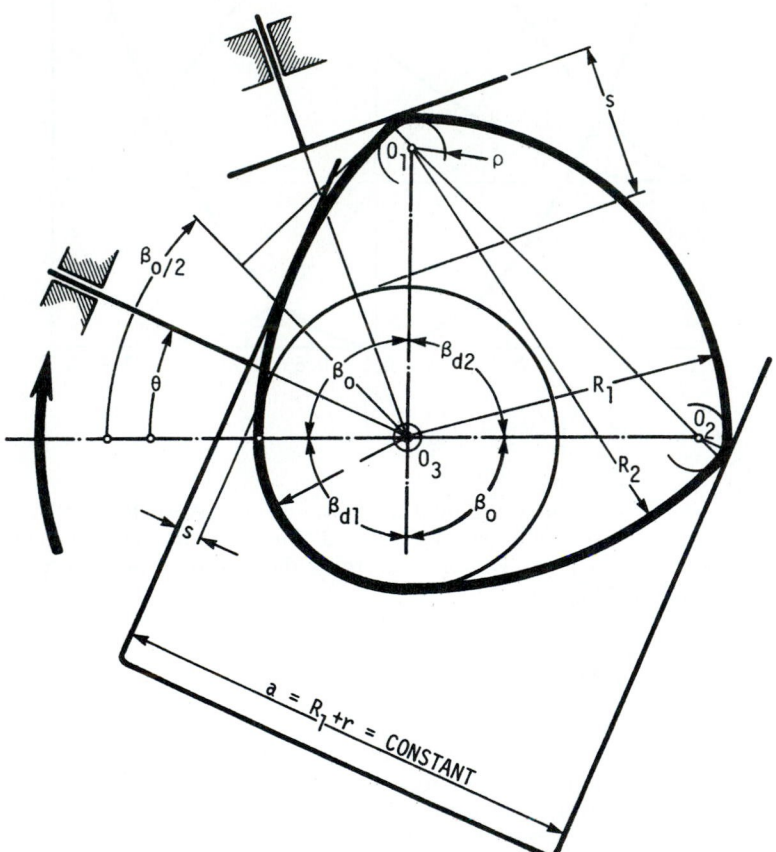

FIGURE 40.12 Constant-breadth circular-arc cam.

where $b = \cos 0.25\beta_0/\cos 0.75\beta_0$. Cam motions for full rise ($0 \geq \theta \geq \beta_0$) are described in Table 40.4. Such cams are symmetric; therefore, $\beta_0(\text{rise}) = \beta_0(\text{return})$, and the two dwells β_{d1} and β_{d2} are the same and equal to $180° - \beta_0$. Table 40.4 can also be used for calculation of full-return motion. Dimensions of the cam (R_1) and maximum values of the acceleration increase with a decrease in β_0. Acceleration diagrams for different values of β_0 are shown in Fig. 40.13.

TABLE 40.4 Basic Equations for a Constant-Breadth Circular-Arc Cam, Using $A = R_1 - \rho$

Parameter	$0 \leq \theta \leq \beta_0/2$	$\beta_0/2 \leq \theta \leq \beta_0$
s	$A(1 - \cos\theta)$	$A\cos(\beta_0 - \theta) - (r - \rho)$
s'	$A\sin\theta$	$A\sin(\beta_0 - \theta)$
s''	$A\cos\theta$	$-A\cos(\beta_0 - \theta)$
s'''	$-A\sin\theta$†	$-A\sin(\beta_0 - \theta)$†

† Both equations are valid, however, only inside the partitions. For $\theta = 0, \beta_0/2$, and $\beta_0, s''' \to \infty$.

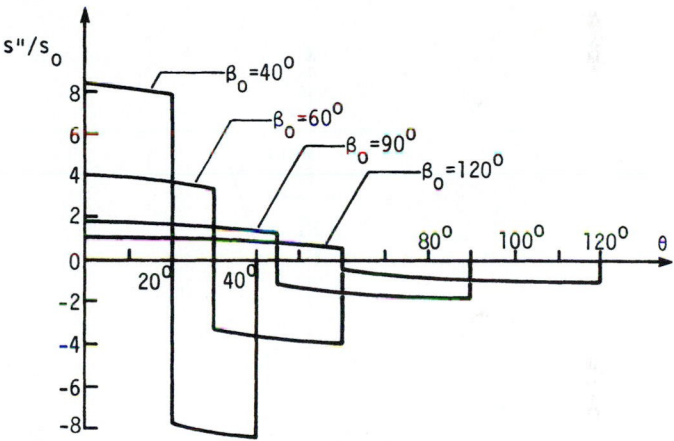

FIGURE 40.13 Acceleration diagrams.

40.3 LAYOUT AND DESIGN; MANUFACTURING CONSIDERATIONS

The cam profile is an inner envelope of the working surface of the follower. After the displacement diagram is determined, the cam layout can be found by using the usual graphical approach or by computer graphics with a rather simple computer program.

In the design of a plate cam with a reciprocating flat-face follower, the geometric parameters necessary for its layout are the prime-circle radius R_0, the minimum width of the follower face F, and the offset e of the follower face. The value R_0 can be found from

$$R_0 > (\rho_{min} - s'' - s)_{max} \qquad (40.23)$$

where ρ_{min} is a minimum value of the radius ρ of the cam-profile curvature. Its value for such practical reasons as contact stresses might be assumed equal to 0.2 to 0.25 in [5 to 6 millimeters (mm)]. Since s is always positive, we should examine that part of the follower acceleration diagram for the rise motion where acceleration is negative.

The face width F can be calculated from

$$F > s'_{max} - s'_{min} \qquad (40.24)$$

To avoid undercutting cams with a roller follower, the radius R_r of the roller must always be smaller than $|\rho|$, where ρ is the radius of curvature.

The pressure angle γ (Fig. 40.14) is an angle between a common normal to both the roller and the cam profile and the direction of the follower motion. This angle can be calculated from

$$\tan \gamma = \frac{s'}{s + R_0 + R_r} \qquad (40.25)$$

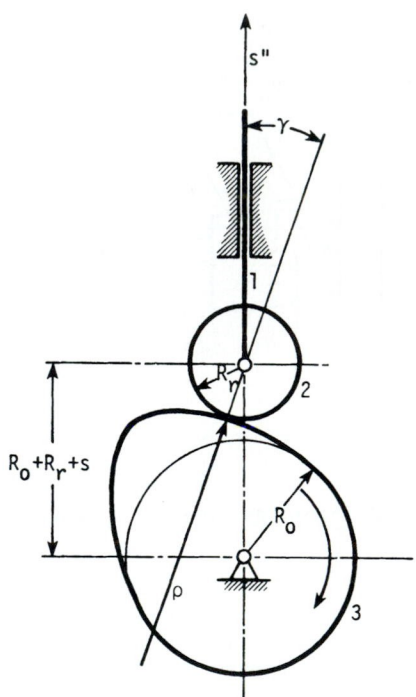

FIGURE 40.14 Cam mechanism with reciprocating roller follower.

It is a common rule of thumb to assume for the preliminary calculation that γ_{max} is not greater than 30° for the reciprocating follower motion (or 45° for the oscillating one). Acceptable values of γ_{max} that can be used without causing difficulties depend, however, on the particular cam mechanism design and should be found for any actual mechanism from the dynamic analysis.

After establishing the value of γ_{max} and R_r in accordance with the preliminary layout of the mechanism, we can find the value of the prime-circle radius R_0 from the equation

$$R_0 \geq \left(\frac{s'}{\tan \gamma_{max}} - s - R_r \right)_{max} \tag{40.26}$$

Now check whether the assumed value of R_r is small enough to avoid undercutting of the cam profile. It can be done ([40.7]) by using Eq. (40.27):

$$R_r \leq \left[\frac{|s'|}{\sin^3 \gamma_{max} \left(\dfrac{1}{\sin^3 \gamma_{max}} + 2 - \dfrac{s''}{|s'| \tan \gamma_{max}} - \dfrac{1}{|s'|} \right)} \right]_{min} - \rho_{min} \tag{40.27}$$

The primary choice of the follower motion should always be guided by a good understanding of the planned manufacturing technique. *Tracer cutting* and *incremental cutting* are two very common methods of cam manufacture. Incremental cutting consists of manufacturing the profile by intermittent cuts based on a table with accurate values of angular cam displacement θ (cam blank) and linear displacement $s(\theta)$ of the follower (cutter). This method is used for making master cams or cams in small numbers. In the tracer control cutting method, the cam surface is milled, shaped, or ground, with the cutter or grinder guided continuously by either a master cam or a computer system. This is the best method for producing large numbers of accurate cam profiles.

In the process of cam and follower manufacturing, several surface imperfections may occur, such as errors, waviness, and roughness. These surface irregularities may induce shock, noise, wear, and vibrations of the cam and follower systems. Imperfections of actual profile cannot exceed an accepted level. Therefore, highly accurate inspection equipment is commonly used in production inspection. Actual displacements of the follower are measured as a function of the cam rotation; then the resulting data can be compared with tabulated theoretical values. By application of the method of finite differences (Sec. 40.3.1), these data can be transformed to actual acceleration curves and compared with theoretical ones. There is, however, a drawback in such a method in that it is based on static measurements.

An example of results obtained from a widely used production inspection method is shown in Fig. 40.15 ([40.5]). Line 1 was obtained from some accurate data from a table of θ values and the corresponding $s(\theta)$ values. Next, two boundary curves were obtained from the basis curve by adding and subtracting 10 percent. This was an arbitrary decision, it being assumed that any acceleration curve contained between such boundaries would be satisfactory. These are shown as upper and lower bounds in Fig. 40.15. The main drawback of the method is that only maximum values of actual acceleration diagrams have been taken into account. It is important to realize that waviness of the real acceleration curve may cause more vibration troubles than will single local surpassing of boundary curves.

A much better method is that of measuring the real acceleration of the follower in an actual cam mechanism at the operating speed of the cam by means of high-quality accelerometers and electronic equipment. To illustrate the importance of proper measurements of the cam profile, we show the results of an investigation of

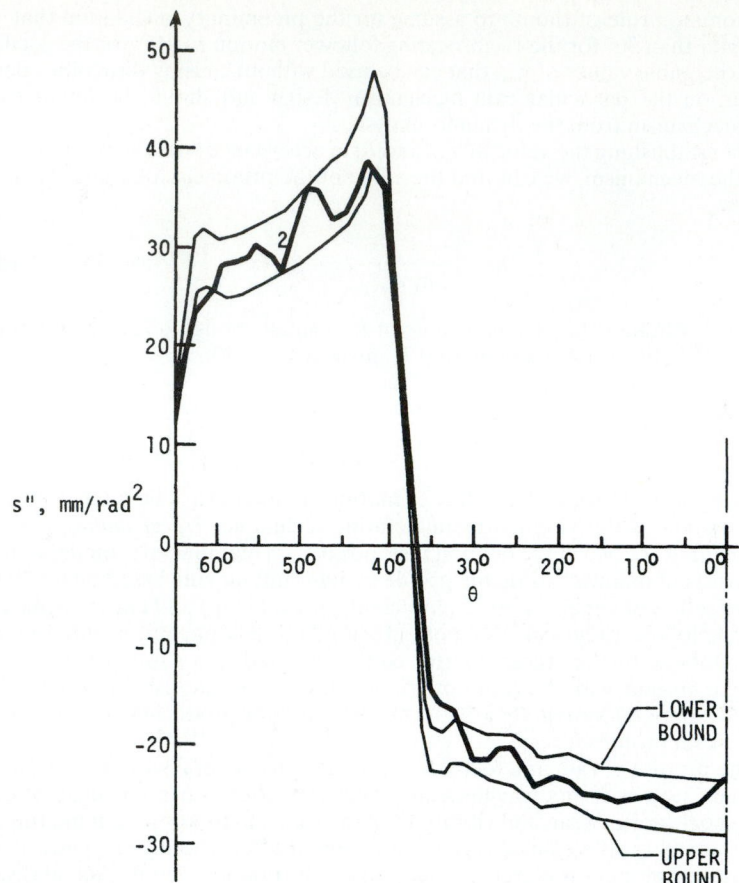

FIGURE 40.15 Example of inspection technique based on acceleration diagram obtained from accurate static measurements of the cam of the Henschel internal combustion engine.

the mechanism used in the Fiat 126 engine ([40.6]). Those results are shown in Figs. 40.16 and 40.17. The acceleration diagram of Fig. 40.16*a* was obtained from designer data by using Eq. (40.29). The diagram plotted as a broken line (1) in Fig. 40.16*b* comes from accurate measurements of a new profile. Here again Eq. (40.29) was applied. The same profile was measured again after 1500 hours (h) of operation, and the acceleration diagram is plotted by a solid line (2) in Fig. 40.16*b*. Comparing curves 1 and 2 of Fig. 40.16*b*, we can see that the wear of the cam smoothed somewhat the waviness of the negative part of the diagram. Accelerations of the follower induced by the same new cam in the actual mechanism (Fig. 40.16*c*) were measured as well by electronic equipment at the design speed, and results of that experiment are presented in Fig. 40.16*d* and *e*. It is obvious from comparison of the diagrams in Fig. 40.16*b* and *d* that the response of the system differs to a considerable extent from the actual input.

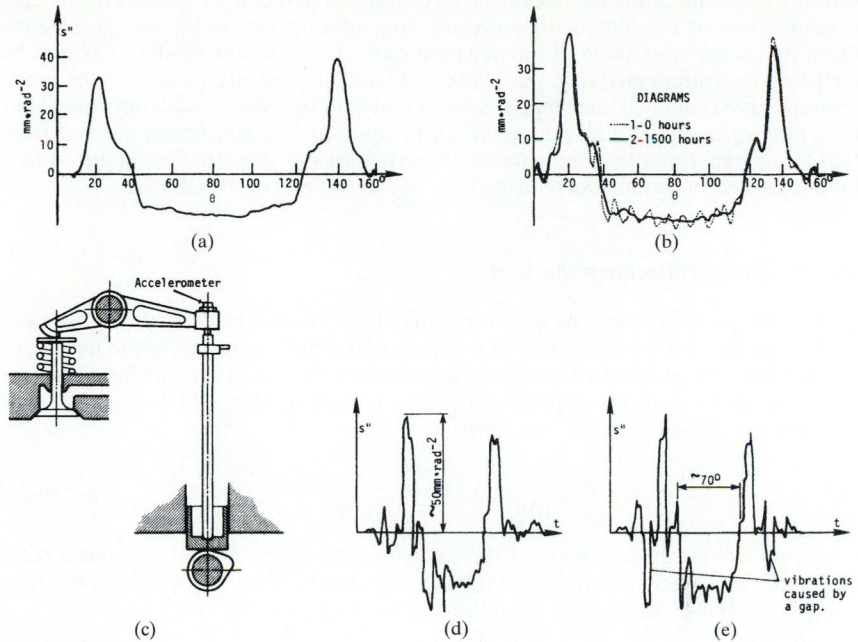

FIGURE 40.16 Comparison between results obtained from static measurements of the Fiat 126 cam profile [(*a*) and (*b*)] and acceleration curves obtained at design speed on the actual engine [(*d*) and (*e*)]. Diagram *d* was obtained for a zero value of backlash and diagram *e* for the factory-recommended 0.2-mm backlash.

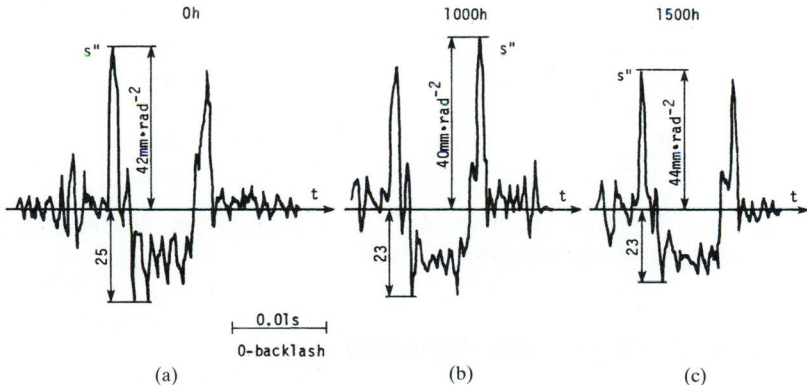

FIGURE 40.17 Changes of acceleration diagram caused by the wear of the cam profile of the Fiat 126.

Eight new cams of the same engine were later used in two separate laboratory stands to find the influence of cam-surface wear on dynamic properties of the cam system. Some of the obtained results are presented in Fig. 40.17. We can observe there that some smoothing of the negative part of the curve (registered as well by statistical measurements) took place after 1500 h. The general character of the acceleration curve remained unchanged, however. (That observation was confirmed later by a Fourier analysis of all the acceleration signals.) The conclusion derived from that single experiment is that dynamic imperfections of the cam system introduced by the process of cam manufacturing may last to the end of the cam life.

40.3.1   Finite-Difference Method

Geometric acceleration of the follower s'' may be estimated by using accurate values of its displacement s from a table of θ versus $s(\theta)$, which comes from the designer's calculations and/or from accurate measurements of the actual cam profile. Denoting as s_{i-1}, s_i, and s_{i+1} three adjacent values of s in such a table, and designating their second finite difference as Δ_i'', we have

$$s_i'' \cong \frac{1}{(\Delta\theta)^2} \Delta_i'' = \frac{s_{i-1} - 2s_i + s_{i+1}}{(\Delta\theta)^2} \qquad (40.28)$$

where $\Delta\theta$ = constant increment of the cam's angular displacement θ. A more accurate value of s_i'' can be found from the *average weighted value* (Oderfeld [40.3]) by using entries of 11 adjacent Δ'' from the table of s versus $s(\theta)$:

$$s_i'' = \frac{1}{(\Delta\theta)^2} \sum_{j=-5}^{i=5} W_j \Delta_{i+j}'' \qquad (40.29)$$

The weights W_j are given in Table 40.5.

TABLE 40.5   Weights Used in the Improved Finite-Difference Method

j	0	± 1	2	± 3	± 4	± 5
W_j	0.31	0.25	0.13	0.015	−0.025	−0.025

An example of an acceleration diagram $s''(\theta)$ of a certain cam obtained using the finite-difference method is presented in Fig. 40.18.

40.4   FORCE AND TORQUE ANALYSIS

A typical approach to dynamic analysis of a rigid cam system can be illustrated by an example of a mechanism with a reciprocating roller follower.[†] A schematic drawing of such a mechanism is depicted in Fig. 40.19a. For the upward motion of the fol-

---

[†] Suggestion of Professor Charles R. Mischke, Iowa State University.

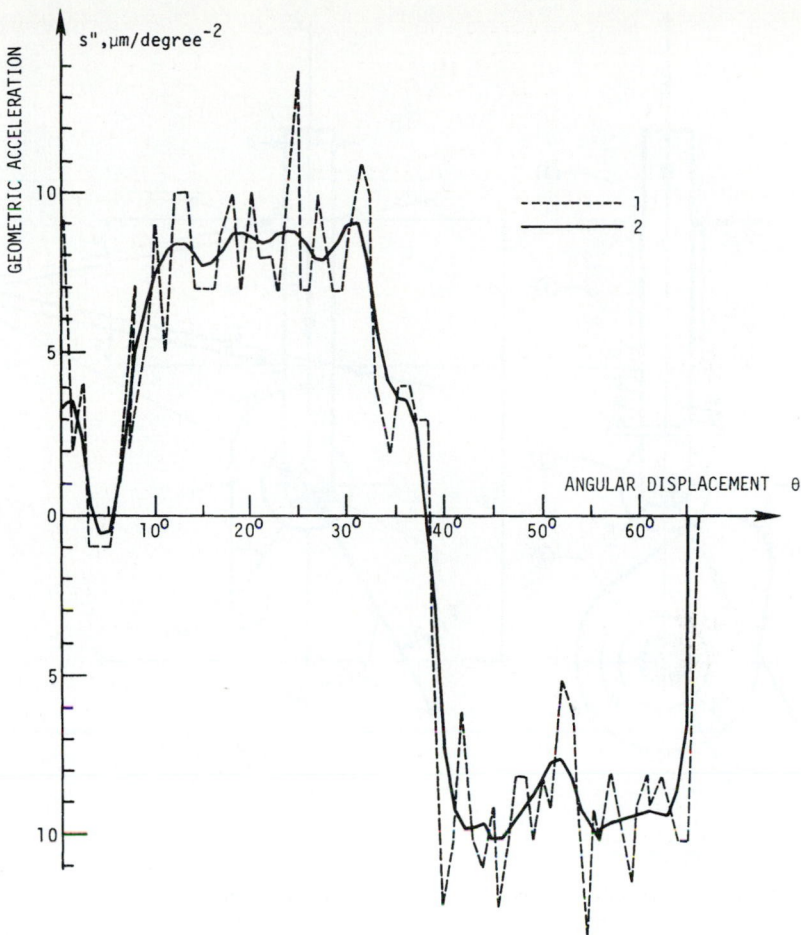

FIGURE 40.18 Acceleration diagrams obtained in the static way. Curve 1 is from Eq. (40.28), curve 2 from Eq. (40.29).

lower, we assume that the follower's stem 4 contacts its guideway at points B and C. As a result of its upward motion, the Coulomb friction at B and C is fully developed and $\tan \psi = \mu$. The free-body diagram of links 3 and 4 is shown in Fig. 40.19b. The cam force F_{23} can be resolved into two components: P_{cr} in the critical-angle (γ_{cr}) direction to sustain motion against friction, and P_y in the y direction to produce accelerated motion or to oppose other forces.

It can be found from the geometry of the follower that

$$\gamma_{cr} = \frac{\pi}{2} - \tan^{-1} \mu \left(\frac{2a - \mu d}{l} - \frac{2s}{l} - 1 \right) \tag{40.30}$$

where $a = l_B - R_0 - R_r$. For $\gamma > \gamma_{cr}$, the cam-follower system is self-locking, and motion is impossible. From the force triangle in Fig. 40.19b and the rule of sines,

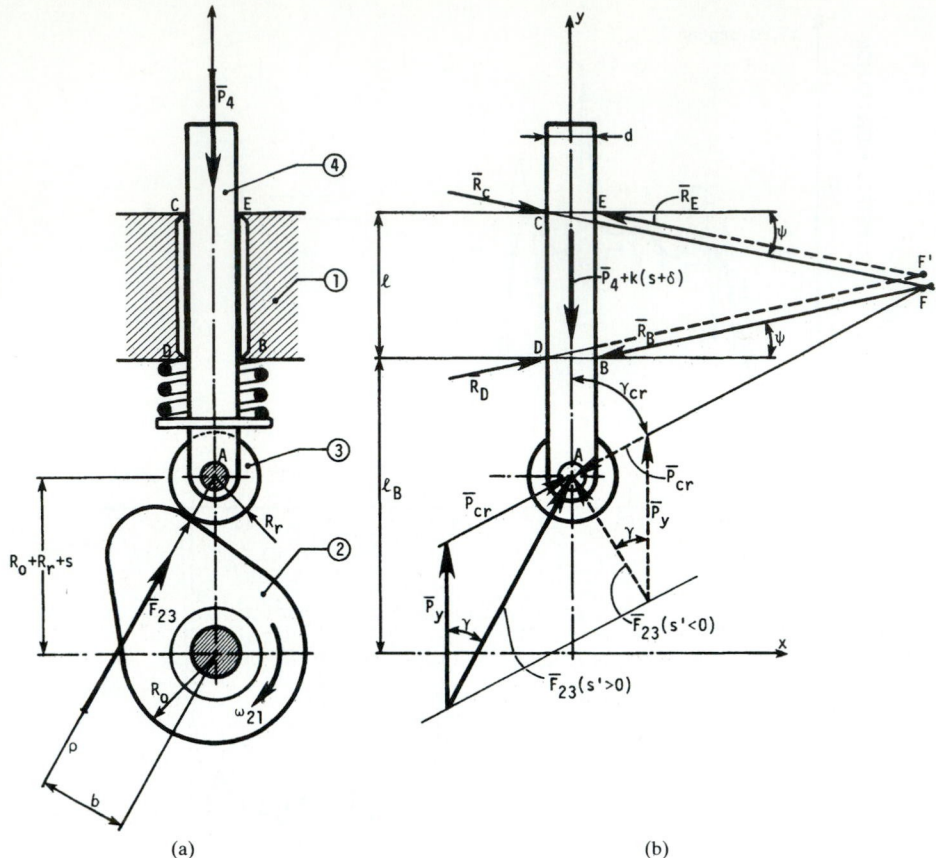

FIGURE 40.19 Force analysis of reciprocating roller-follower cam system.

$$P_y = F_{23} \frac{\sin(\gamma_{cr} - \gamma)}{\sin \gamma_{cr}} \tag{40.31}$$

After finding the vertical component P_y for constant ω_{21} from the force-equilibrium equation, substituting into Eq. (40.31), and solving for F_{23}, we have

$$F_{23} = \frac{\sin \gamma_{cr}(m\omega_{21}^2 s'' + ks + P_4')}{\sin (\gamma_{cr} - \gamma)} \tag{40.32}$$

where $m =$ mass of follower
$k =$ spring rate of retaining spring
$P_4' = P_4 + k\delta$
$\delta =$ preset of spring
$k\delta = P_0$; this force is called *preload* of spring

For $F_{23} = 0$, roller and cam lose their contact. The result is called *jump* ([40.7], [40.8]). Assuming $F_{23} = 0$, we can find the jump speed of the cam from Eq. (40.31). The jump occurs for the upward movement of the follower at

$$\omega_{21} \geq \sqrt{\frac{-ks - P_4'}{ms''}} \qquad (40.33)$$

Since s is always positive, jump may occur only for negative values of s''. To prevent jump, we increase preload P_0 or the spring rate or both. The driving torque is

$$T_{12} = \frac{\sin \gamma_{cr} \sin \gamma}{\sin (\gamma_{cr} - \gamma)} (R_0 + R_r + s)(m\omega_{21}^2 s'' + ks + P_4') \qquad (40.34)$$

We recall that according to Eq. (40.25),

$$\gamma = \tan^{-1} \frac{s'}{s + R_0 + R_r} \qquad (40.35)$$

When motion is downward, the contact point of mating surfaces goes to the right side of the roller, cam force F_{23} changes inclination, and new contact points D and E in the follower's guideway replace old ones (B and C, respectively). The new point of concurrency is now at F'. Since in most practical cases points F and F' almost coincide, we can assume that both the point of concurrency and the line of action of force P_{cr} are unchanged. A new vector P_{cr} (broken line) is rotated by 180° with respect to the old one. It is easy to see in Fig. 40.19b that F_{23} for downward motion, when γ and P_y equal those for the upward motion, is always smaller than F_{23} for upward motion.

40.4.1 Springs

In cam-follower systems, the follower must contact the cam at all times. This is accomplished by a positive drive or a retaining spring. Spring forces should always prevent the previously described jump of the follower for all the operating speeds of the cam. Thus the necessary preload P_0 of the spring and its spring rate k should be chosen for the highest possible velocity of the cam. By plotting inertial and spring forces, we can find values of preload P_0 and spring rate k that will ensure sufficient load margin for the total range of the follower displacement. We use here only the negative portion of the acceleration curve. Since the follower must be held in contact with the cam, even while operating the system with temporary absence of applied forces, that part of the cam-system synthesis may be accomplished without applied forces. At the critical location, where both curves are in closest proximity, the spring force should exceed the inertial force with friction corrections included by not less than 25 to 50 percent.

40.5 CONTACT STRESS AND WEAR: PROGRAMMING

Let us consider the general case of two cylinderlike surfaces in contact. They are represented by a cam and a follower. The radius of curvature of the follower ρ_1 is equal to the radius of the roller R_r for the roller follower, and it goes to infinity for a flat follower. The radius of the cam's curvature ρ_2 can be found from the following equation [compare to Eq. (40.23)]:

$$\rho_2 = R_0 + s'' + s \tag{40.36}$$

Equating $d\rho_2/d\theta = s''' + s'$ to zero, we can find the position of the cam, where a minimum of ρ_2 occurs, and find its value from Eq. (40.36). Assuming perfect alignment of the contacting bodies, we have conditions described by Hertz and can check maximum compressive stress σ_c from his well-known equation

$$\sigma_c = 0.558 \sqrt{\frac{P(1/\rho_1 + 1/\rho_2)}{L[(1 - \mu_1^2)/E_1 + (1 - \mu_2^2)/E_2]}} \tag{40.37}$$

where P = normal load between cam and follower
             L = actual thickness of contacting follower and cam
             μ_1, μ_2 = Poisson's ratios for follower and cam, respectively
             E_1, E_2 = moduli of elasticity of follower and cam, respectively

Some selected data concerning properties of materials for cams and followers are given in [40.7].

Equation (40.37) may be used for finding the minimum permissible value ρ_{min} of the cam's profile radius of curvature. We recall that ρ_{min} was necessary for calculating the values of R_0 and R_r from Eqs. (40.23) and (40.26), respectively.

Rearranging Eq. (40.37) gives, for a roller follower,

$$\rho_{min} \geq \left[3.2\, S_c^2 \frac{L}{P} \left(\frac{1 - \mu_1^2}{E_1} + \frac{1 - \mu_2^2}{E_2} \right) - \frac{1}{R_r} \right]^{-1} \tag{40.38}$$

The same equation holds true for a flat-faced follower, where $1/R_r = 0$.

Using Eq. (40.38), we can easily check to see if the commonly recommended and used value $\rho_{min} = 0.25$ in (6 mm) is justified in the particular design.

Elements of a cam system, as well as of other machine parts, are subject to wear. The proper choice of metal combinations may increase the life of kinematic pairs of the cam system and decrease their wear. Some experience is necessary in choosing materials to fulfill the requirements of satisfactory cam action with low wear over a long period. Designers, as a rule, prefer to make the follower of softer or first-worn-out material, since manufacturing of the follower is less expensive than manufacturing of the cam profile. There are, however, cases where the cam is cheaper and thus is made of softer material.

40.5.1 Programming of Cam Systems

The steps shown in Fig. 40.20 are as follows:

1. Make a preliminary sketch of your cam system, and estimate the dynamic factor $\mu_d = m_e \omega_c^2 / k_e$, according to Eq. (40.8).
2. If your system is a positive drive, go to step 3; otherwise go to step 4.
3. Choose a proper symmetric diagram $s''(\theta)$, and write equations for $s''(\theta), s'(\theta)$, and $s(\theta)$. Write a computer program for θ and $s(\theta)$ with increments of θ equal to 1°.
4. Choose a proper unsymmetric diagram $s''(\theta)$, and proceed as in step 3.
5. Print table of θ and $s(\theta)$.

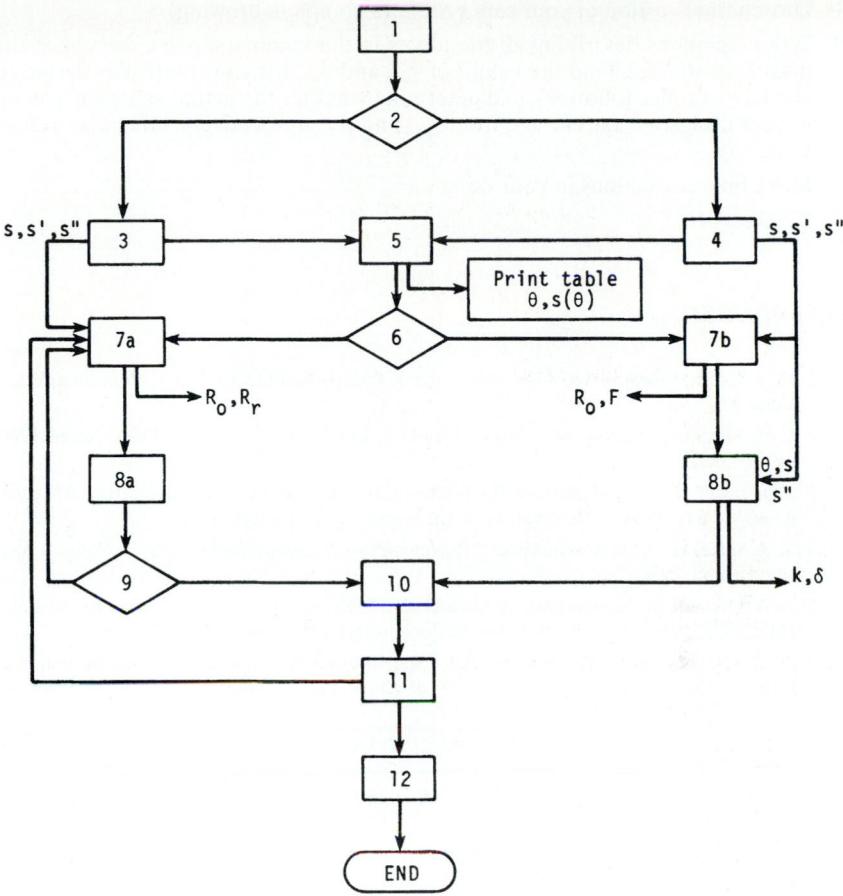

FIGURE 40.20 Programming of cam systems.

6. If you decided to use a roller follower, go to step 7*a;* otherwise go to step 7*b.*

7. *a.* Establish the radius of the roller R_r and the value of maximum pressure angle γ_{max} ($\gamma_{max} \le 30°$ for the reciprocating follower and $\gamma_{max} \le 40°$ for the oscillating one). Find the value of θ for which $s'/\tan \gamma_{max} - s - R_r$ is maximum, by equating $s''/\tan \gamma_{max} - s'$ to zero. Calculate the value of the radius R_0 of the prime circle from Eq. (40.26). Go to step 8*a.*

   b. Find the value of θ for which $\rho_{min} - s'' - s$ is maximum, by setting $-(s''' + s')$ to zero. Calculate the value of the radius R_0 of the prime circle from Eq. (40.23). Calculate the face width F from Eq. (40.24). Go to step 8*b.*

8. *a.* Check undercutting of the cam profile from Eq. (40.27).

   b. Find the spring rate of the retaining spring. Go to step 10.

9. If the assumed value R_r is satisfactory, go to step 10; otherwise, assume a new value of R_r and return to step 7*a.*

10. Draw a final version of your cam system (a complete drawing).

11. Write equations describing all the forces in the kinematic pairs, and check the maximum stresses. Find the values of γ_{max} and ρ_{min} for your particular design (if you have a roller follower), and determine whether the actual values of γ and ρ in your design are far enough from γ_{max} and ρ_{min}, respectively; otherwise, return to step 7*a*.

12. Make final corrections in your design.

13. END.

REFERENCES

40.1 M. P. Koster, *Vibrations of Cam Mechanisms,* Philips Technical Library, Macmillan Press, London, 1974.

40.2 C. R. Mischke, *Mathematical Model Building,* Iowa State University Press, Ames, 1980, pp. 316–322.

40.3 J. Oderfeld, "On Application of the Finite-Difference Method for Kinematics of Mechanisms," *Zastosowania Matematyki IV* (in Polish), 1958, pp. 176–195.

40.4 A. A. Olędzki, "Cam Mechanisms," *Wydawnictwa Naukowo Techniczne* (in Polish), Warsaw, Poland, 1966.

40.5 A. A. Olędzki, M. Krawczyński, I. Siwicki, and F. Źrudelny, "On the Automation of Cam-Profile Control," *IFAC Manufacturing Technology Symposium,* Tokyo, 1977.

40.6 A. A. Olędzki and T. Klimowicz, "An Experimental Attempt to Find the Influence of Cam-Profile Wear on the Kinematics of the Cam-Mechanism," *Archiwum Budowy Maszyn* (in Polish), 1979.

40.7 H. A. Rothbart, *Cams—Design, Dynamics, and Accuracy,* John Wiley & Sons, New York, 1956.

40.8 J. E. Shigley and J. J. Uicker, *Theory of Machines and Mechanisms,* 2d ed., McGraw-Hill, New York, 1995.

40.9 D. A. Stoddart, "Polydyne Cam Design," *Machine Design,* January 1953, p. 124; February 1953, p. 146; March 1953, p. 149.

40.10 D. Tesar and G. K. Matthew, *The Dynamic Synthesis, Analysis, and Design of Modeled Cam Systems,* D. C. Heath, Lexington, Mass., 1976.

CHAPTER 41
LINKAGES

Richard E. Gustavson
Technical Staff Member
The Charles Stark Draper Laboratory, Inc.
Cambridge, Massachusetts

Linkages are mechanical devices that appear very straightforward to both analyze and design. Given proper technique, that is generally the case. The methods described in this chapter reveal the complexity (and, I think, the beauty) of linkages. I have gained significant satisfaction during my 20 years of work with them from both theoretical and functioning hardware standpoints.

41.1 BASIC LINKAGE CONCEPTS

41.1.1 Kinematic Elements

A linkage is composed of rigid-body members, or *links*, connected to one another by rigid kinematic elements, or *pairs*. The nature of those connections as well as the shape of the links determines the kinematic properties of the linkage.

Although many kinematic pairs are conceivable and most do physically exist, only four have general practical use for linkages. In Fig. 41.1, the four cases are seen to include two with 1 degree of freedom ($f = 1$), one with $f = 2$, and one with $f = 3$. Single-degree-of-freedom pairs constitute joints in planar linkages or spatial linkages. The cylindrical and spherical joints are useful only in spatial linkages.

The links which connect these kinematic pairs are usually binary (two connections) but may be tertiary (three connections) or even more. A commonly used tertiary link is the *bell crank* familiar to most machine designers. Since our primary

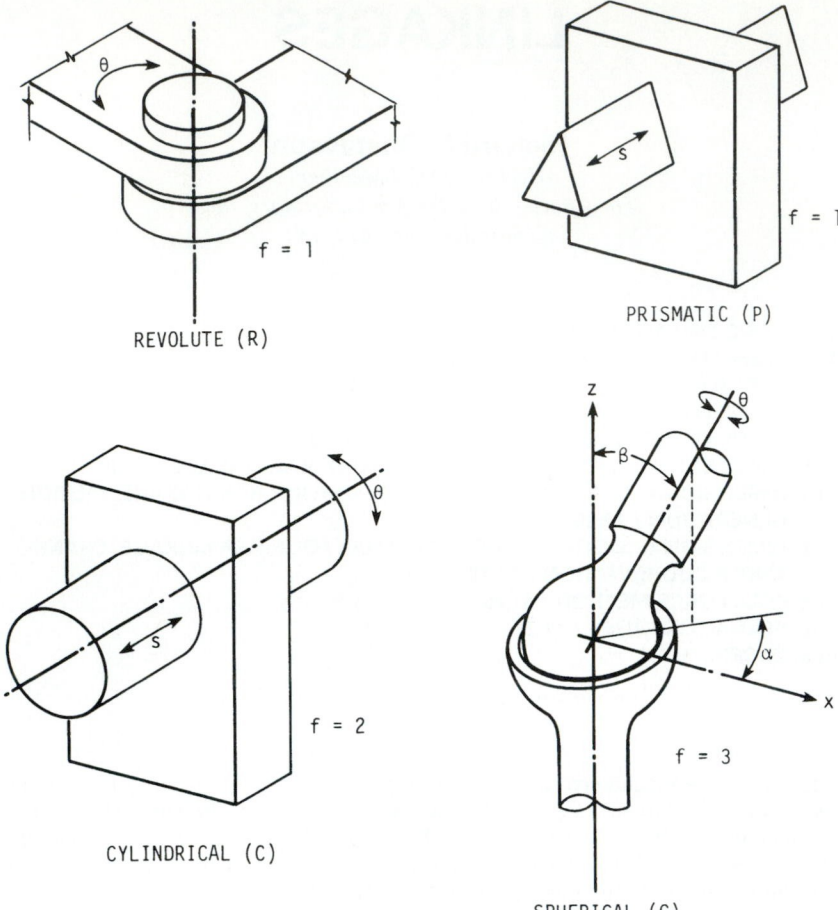

FIGURE 41.1   Kinematic pairs useful in linkage design. The quantity f denotes the number of degrees of freedom.

interest in most linkages is to provide a particular output for a prescribed input, we deal with closed kinematic chains, examples of which are depicted in Fig. 41.2. Considerable work is now under way on robotics, which are basically open chains (see Chap. 47). Here we restrict ourselves to the closed-loop type. Note that many complex linkages can be created by compounding the simple four-bar linkage. This may not always be necessary once the design concepts of this chapter are applied.

41.1.2 Freedom of Motion

The degree of freedom for a mechanism is expressed by the formula

$$F = \lambda(l - j - 1) + \sum_{i=1}^{j} f_i \tag{41.1}$$

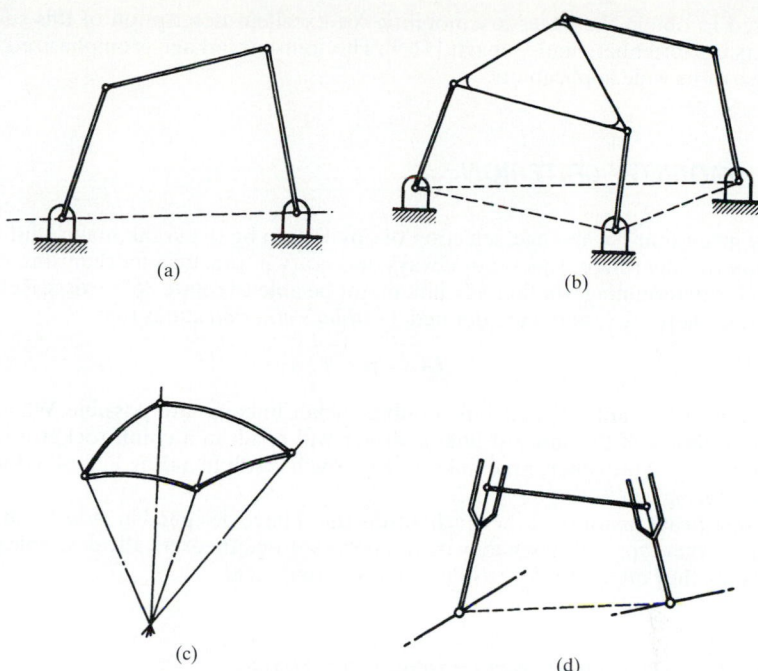

FIGURE 41.2 Closed kinematic chains. (*a*) Planar four-bar linkage; (*b*) planar six-bar linkage; (*c*) spherical four-bar linkage; (*d*) spatial RCCR four-bar linkage.

where l = number of links (fixed link included)
               j = number of joints
              f_i = f of ith joint
              λ = integer
                = 3 for plane, spherical, or particular spatial linkages
                = 6 for most spatial linkages

Since the majority of linkages used in machines are planar, the particular case for plane mechanisms with one degree of freedom is found to be

$$2j - 3l + 4 = 0 \qquad\qquad (41.2)$$

Thus, in a four-bar linkage, there are four joints (either revolute or prismatic). For a six-bar linkage, we need seven such joints. A peculiar special case occurs when a sufficient number of links in a plane linkage are parallel, which leads to such special devices as the pantograph.

Considerable theory has evolved over the years about numerous aspects of linkages. It is often of little help in creating usable designs. Among the best references available are Hartenberg and Denavit [41.9], Hall [41.8], Beyer [41.1], Hain [41.7], Rosenauer and Willis [41.10], Shigley and Uicker [41.11], and Tao [41.12].

41.1.3 Number Synthesis

Before you can dimensionally synthesize a linkage, you may need to use *number synthesis*, which establishes the number of links and the number of joints that are required to obtain the necessary mobility. An excellent description of this subject appears in Hartenberg and Denavit [41.9]. The four-bar linkage is emphasized here because of its wide applicability.

41.2 MOBILITY CRITERION

In any given four-bar linkage, selection of any link to be the crank may result in its inability to fully rotate. This is not always necessary in practical mechanisms. A criterion for determining whether any link might be able to rotate 360° exists. Refer to Fig. 41.3, where l, s, p, and q are defined. *Grubler's criterion* states that

$$l + s < p + q \tag{41.3}$$

If the criterion is not satisfied, only double-rocker linkages are possible. When it is satisfied, choice of the shortest link as driver will result in a crank-rocker linkage; choice of any of the other three links as driver will result in a drag link or a double-rocker mechanism.

A significant majority of the mechanisms that I have designed in industry are the double-rocker type. Although they do not possess some theoretically desirable characteristics, they are useful for various types of equipment.

41.3 ESTABLISHING PRECISION POSITIONS

In designing a mechanism with a certain number of required precision positions, you will be faced with the problem of how to space them. In many practical situations, there will be no choice, since particular conditions must be satisfied.

If you do have a choice, Chebychev spacing should be used to reduce the structural error. Figure 41.4 shows how to space four positions within a prescribed interval [41.9]. I have found that the end-of-interval points can be used instead of those just inside with good results.

41.4 PLANE FOUR-BAR LINKAGE

41.4.1 Basic Parameters

The apparently simple four-bar linkage is actually an incredibly sophisticated device which can perform wonders once proper design techniques are known and used. Figure 41.5 shows the parameters required to define the general case. Such a linkage can be used for three types of motion:

1. *Crank-angle coordination* Motion of driver link b causes prescribed motion of link d.

2. *Path generation* Motion of driver link b causes point C to move along a prescribed path.

3. *Motion generation* Movement of driver link b causes line CD to move in a prescribed planar motion.

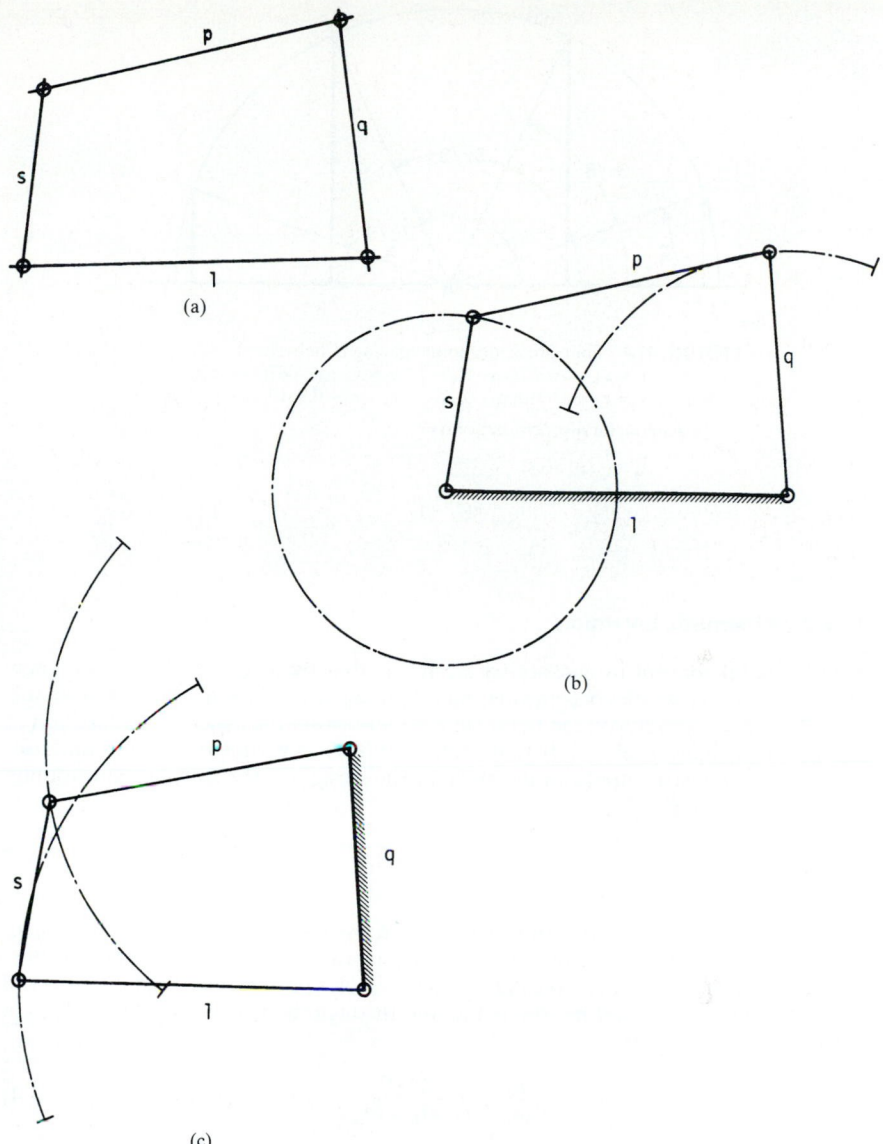

FIGURE 41.3   Mobility characteristics. (*a*) Closed four-link kinematic chain: *l* = longest link, *s* = short-est link, *p, q* = intermediate-length links; (*b*) crank rocker linkage; (*c*) double-rocker linkage.

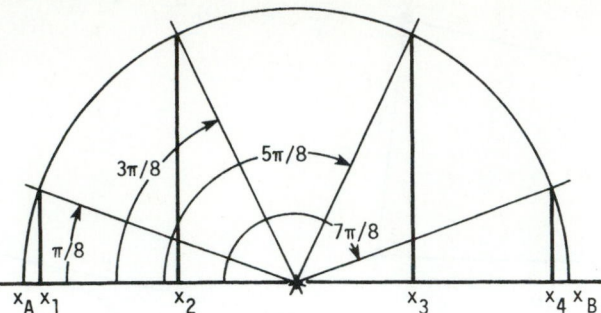

FIGURE 41.4   Four-precision-point spacing (Chebychev)

$$x_1 = x_A + 0.0381(x_B - x_A) \qquad x_2 = x_A + 0.3087(x_B - x_A)$$
$$x_3 = x_A + 0.6913(x_B - x_A) \qquad x_4 = x_A + 0.9619(x_B - x_A)$$

In general, for n precision points

$$x_j = \frac{1}{2}(x_A + x_B)$$
$$-\frac{1}{2}(x_B - x_A)\cos\frac{\pi(2j-1)}{2n} \qquad j = 1, 2, \ldots, n$$

41.4.2   Kinematic Inversion

A very useful concept in mechanism design is that by inverting the motion, new interesting characteristics become evident. By imagining yourself attached to what is actually a moving body, you can determine various properties, such as the location of a joint which connects that body to its neighbor. This technique has been found useful in many industrial applications, such as the design of the four-bar automobile window regulator ([41.6]).

41.4.3   Velocity Ratio

At times the velocity of the output will need to be controlled as well as the corresponding position. When the motion of the input crank and the output crank is coordinated, it is an easy matter to establish the velocity ratio ω_d/ω_b. When you extend line AB in Fig. 41.5 until it intersects the line through the fixed pivots O_A and O_B in a point S, you find that

$$\frac{\omega_d}{\omega_b} = \frac{O_A S}{O_A O_B + O_A S} \tag{41.4}$$

Finding the linear velocity of a point on the coupler is not nearly as straightforward. A very good approximation is to determine the travel distance along the path of the point during a particular motion of the crank.

41.4.4   Torque Ratio

Because of the conservation of energy, the following relationship holds:

$$T_b d\phi = T_d d\psi \tag{41.5}$$

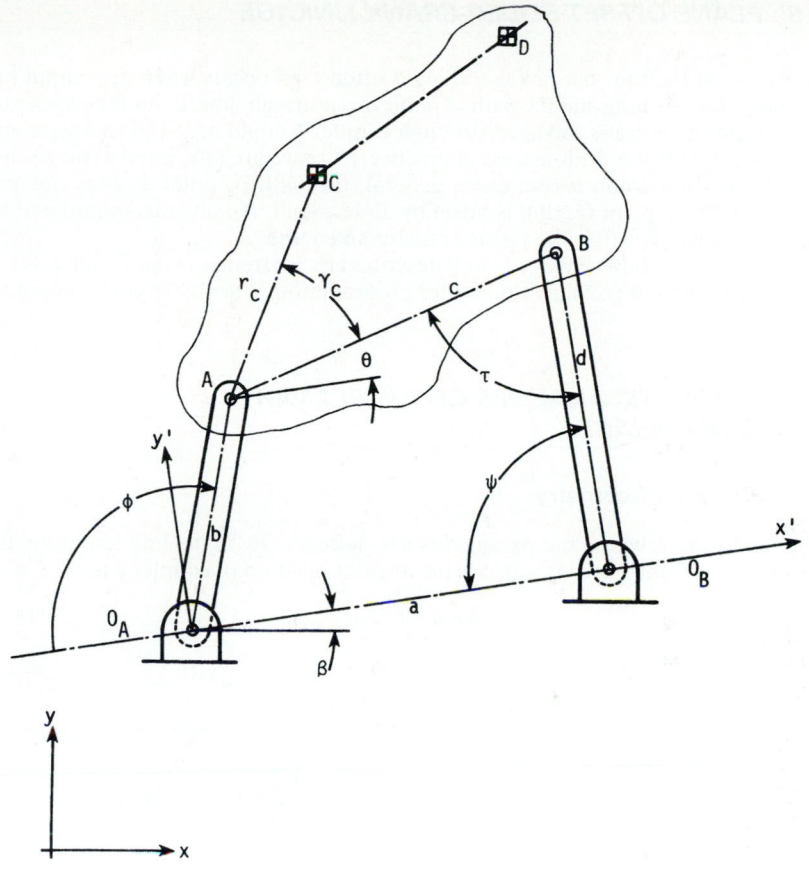

FIGURE 41.5 General four-bar linkage in a plane.

Since both sides of (41.5) can be divided by dt, we have, after some rearranging,

$$n = \frac{T_b}{T_d} = \frac{d\psi/dt}{d\phi/dt} = \frac{\omega_d}{\omega_b} \qquad (41.6)$$

The torque ratio n is thus the inverse of the velocity ratio. Quite a few mechanisms that I have designed have made significant use of torque ratios.

41.4.5 Transmission Angle

For the four-bar linkage of Fig. 41.5, the transmission angle τ occurs between the coupler and the driven link. This angle should be as close to 90° as possible. Useful linkages for motion generation have been created with τ approaching 20°. When a crank rocker is being designed, you should try to keep $45° < \tau < 135°$. Double-rocker or drag link mechanisms usually have other criteria which are more significant than the transmission angle.

41.5   *PLANE OFFSET SLIDER-CRANK LINKAGE*

A variation of the four-bar linkage which is often seen occurs when the output link becomes infinitely long and the path of point B is a straight line. Point B becomes the slider of the slider-crank linkage. Although coupler b could have the characteristics shown in Fig. 41.6, it is seldom used in practice. Here we are interested in the motion of point B while crank a rotates. In general, the path of point B does not pass through the fixed pivot O_A, but is offset by dimension ε. An obvious example of the degenerate case ($\varepsilon \equiv 0$) is the piston crank in an engine.

The synthesis of this linkage is well described by Hartenberg and Denavit [41.9]. I have used the method many times after programming it for the digital computer.

41.6   *KINEMATIC ANALYSIS OF THE PLANAR FOUR-BAR LINKAGE*

41.6.1   Position Geometry

Refer to Fig. 41.7, where the parameters are defined. Given the link lengths a, b, c, and d and the crank position angle ϕ, the angular position of coupler c is

$$\theta = \pi - (\tau + \psi) \qquad (41.7)$$

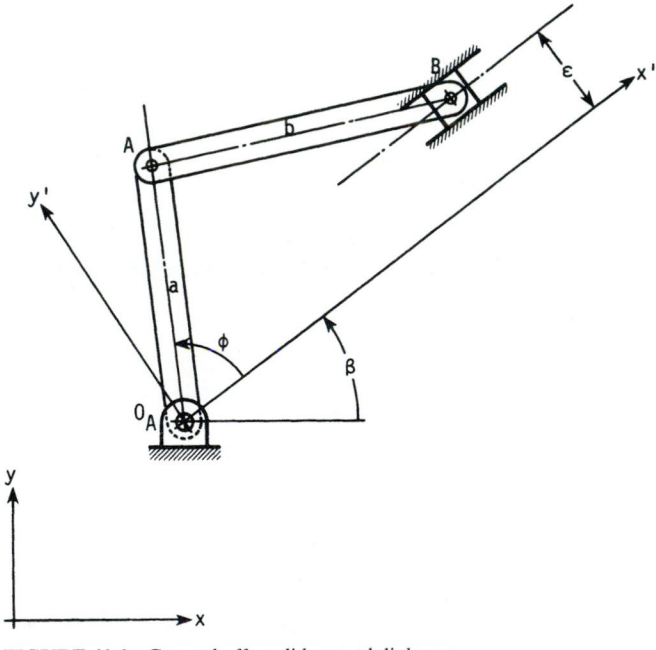

FIGURE 41.6   General offset slider-crank linkage.

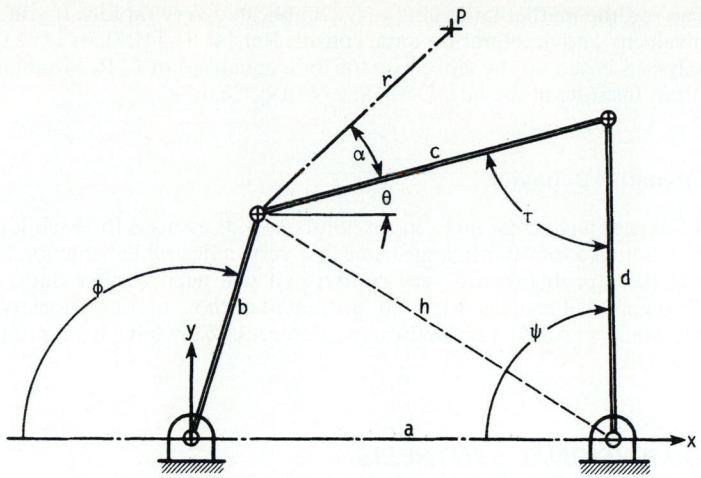

FIGURE 41.7 Parameters for analysis of a four-bar linkage.

The driven link d will be at angle

$$\psi = \cos^{-1} \frac{h^2 + a^2 - b^2}{2hb} + \cos^{-1} \frac{h^2 + d^2 - c^2}{2hd} \tag{41.8}$$

where

$$h^2 = a^2 + b^2 + 2ab \cos \phi \tag{41.9}$$

The transmission angle τ will be

$$\tau = \cos^{-1} \frac{c^2 + d^2 - a^2 - b^2 - 2ab \cos \phi}{2cd} \tag{41.10}$$

A point on coupler P has coordinates

$$P_x = -b \cos \phi + r \cos (\theta + \alpha)$$

$$P_y = b \sin \phi + r \sin (\theta + \alpha) \tag{41.11}$$

41.6.2 Velocity and Acceleration

The velocity of the point on the coupler can be expressed as

$$\frac{dP_x}{dt} = b \frac{d\phi}{dt} \sin \phi - r \frac{d\theta}{dt} \sin (\theta + \alpha)$$

$$\frac{dP_y}{dt} = b \frac{d\phi}{dt} \cos \phi + r \frac{d\theta}{dt} \cos (\theta + \alpha) \tag{41.12}$$

As you can see, the mathematics gets very complicated very rapidly. If you need to establish velocity and acceleration data, consult Ref. [41.1], [41.7], or [41.11]. Computer analysis is based on the closed vector loop equations of C. R. Mischke, developed at Pratt Institute in the late 1950s. See [41.19], Chap. 4.

41.6.3 Dynamic Behavior

Since all linkages have clearances in the joints as well as mass for each link, high-speed operation of a four-bar linkage can cause very undesirable behavior. Methods for solving these problems are very complex. If you need further data, refer to numerous theoretical articles originally presented at the American Society of Mechanical Engineers (ASME) mechanism conferences. Many have been published in ASME journals.

41.7 DIMENSIONAL SYNTHESIS OF THE PLANAR FOUR-BAR LINKAGE: MOTION GENERATION

41.7.1 Two Positions of a Plane

The line A_iB_i defines a plane (Fig. 41.8) which is to be the coupler of the linkage to be designed. When two positions are defined, you can determine a particular point, called the *pole* (in this case P_{12}, since the motion goes from position 1 to position 2). The significance of the pole is that it is the point about which the motion of the body is a simple rotation; the pole is seen to be the intersection of the perpendicular bisectors of A_1A_2 and B_1B_2.

A four-bar linkage can be created by choosing any point on a_1a_2 as O_A and any reasonable point on b_1b_2 as O_B. Note that you do not have a totally arbitrary choice for the fixed pivots, even for this elementary case. There are definite limitations, since the four-bar linkage must produce continuous motion between all positions. When a fully rotating crank is sought, the Grubler criterion must be adhered to. For double-rocker mechanisms, the particular link lengths still have definite criteria to meet. You have to check these for every four-bar linkage that you design.

41.7.2 Three Positions of a Plane

When three positions of a plane are specified by the location of line *CD*, as shown in Fig. 41.9, it is possible to construct the center of a circle through C_1, C_2, and C_3 and through D_1, D_2, and D_3. This is only one of an infinite combination of links that can be attached to the moving body containing line *CD*. If the path of one end of line *CD* lies on a circle, then the other end can describe points on a coupler path which correspond to particular rotation angles of the crank (Fig. 41.10); that is a special case of the motion generation problem.

The general three-position situation describes three poles P_{12}, P_{13}, and P_{23} which form a *pole triangle*. You will find this triangle useful since its interior angles ($\theta_{12}/2$ in Fig. 41.9) define precise geometric relationships between the fixed and moving pivots of links which can be attached to the moving body defined by line *CD*. Examples of this geometry are shown in Fig. 41.11, where you can see that

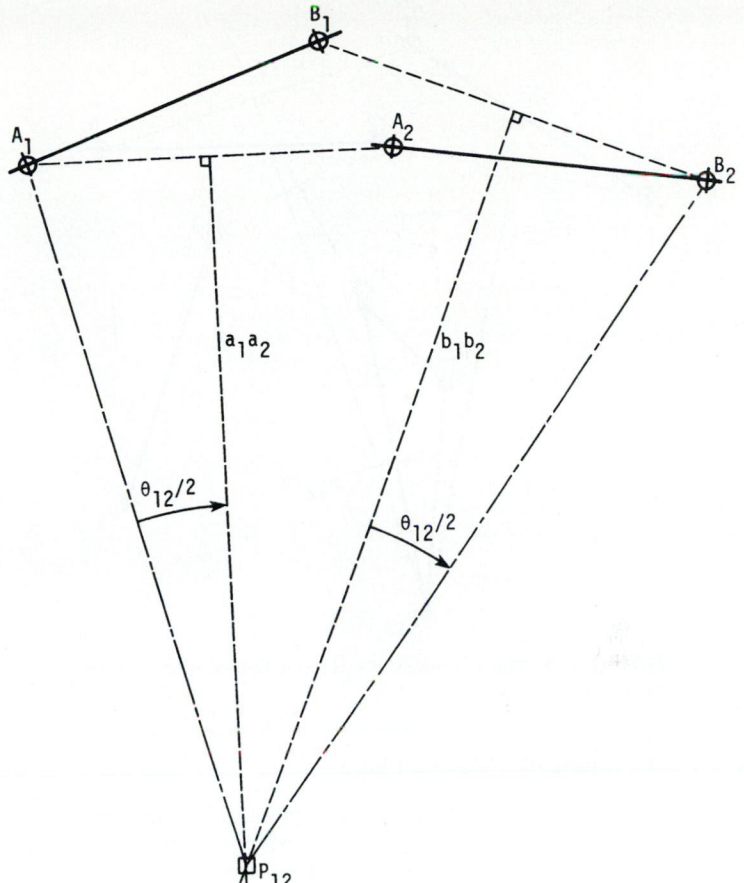

FIGURE 41.8   Two positions of a plane: definition of pole P_{12}.

$$\angle P_{13}P_{12}P_{23} \equiv \angle A_1P_{12}O_A \equiv \angle B_1P_{12}O_B \qquad (41.13)$$

The direction in which these angles are measured is critical. For three positions, you may thus choose the fixed or the moving pivot and use this relationship to establish the location of the corresponding moving or fixed pivot, since it is also true that

$$\angle P_{12}P_{13}P_{23} = \angle A_1P_{13}O_A = \angle B_1P_{13}O_B \qquad (41.14)$$

The intersection of two such lines (Fig. 41.12) is the required pivot point. Note that the lines defined by the pole triangle relationships extend in both directions from the pole; thus a pivot-point angle may appear to be ±180° from that defined within the triangle. This is perfectly valid.

It is important to observe that arbitrary choices for pivot locations are available when three positions, or less, of the moving plane are specified.

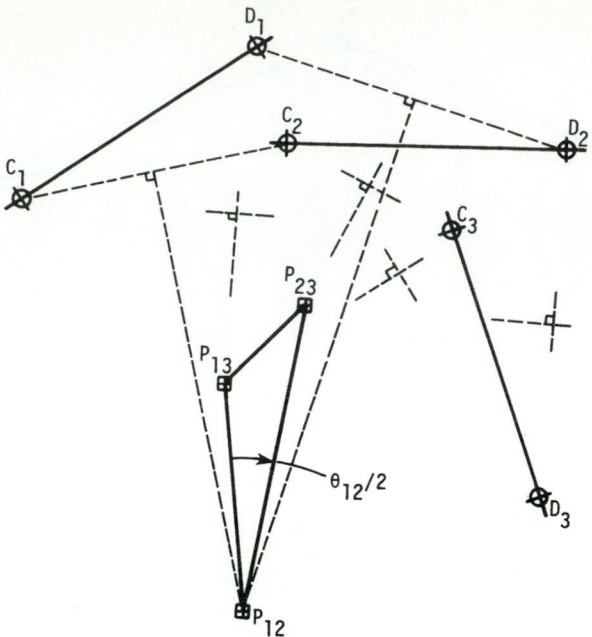

FIGURE 41.9   Three positions of a plane: definition of the pole triangle $P_{12}P_{13}P_{23}$.

41.7.3   Four Positions of a Moving Plane

When four positions are required, appropriate pivot-point locations are precisely defined by theories generated by Professor Burmester in Germany during the 1880s. His work [41.2] is the next step in using the poles of motion. When you define four positions of a moving plane containing line CD as shown in Fig. 41.13, six poles are defined:

$$P_{12} \quad P_{13} \quad P_{14} \quad P_{23} \quad P_{24} \quad P_{34}$$

By selecting opposite poles (P_{12}, P_{34} and P_{13}, P_{24}), you obtain a quadrilateral with significant geometric relationships. For practical purposes, this opposite-pole quadrilateral is best used to establish a locus of points which are the fixed pivots of links that can be attached to the moving body so that it can occupy the four prescribed positions. This locus is known as the *center-point curve* (Fig. 41.14) and can be found as follows:

1. Establish the perpendicular bisector of the two sides $P_{12}P_{24}$ and $P_{13}P_{34}$.
2. Determine points M and M' such that

$$\angle P_{12}MQ_2 \equiv \angle P_{13}M'Q_3$$

3. With M as center and MP_{12} as radius, create circle k. With M' as center and $M'P_{13}$ as radius, create circle k'.
4. The intersections of circles k and k' (shown as c_0 and c_0' in Fig. 41.14) are center points with the particular property that the link whose fixed pivot is c_0 or c_0' has a

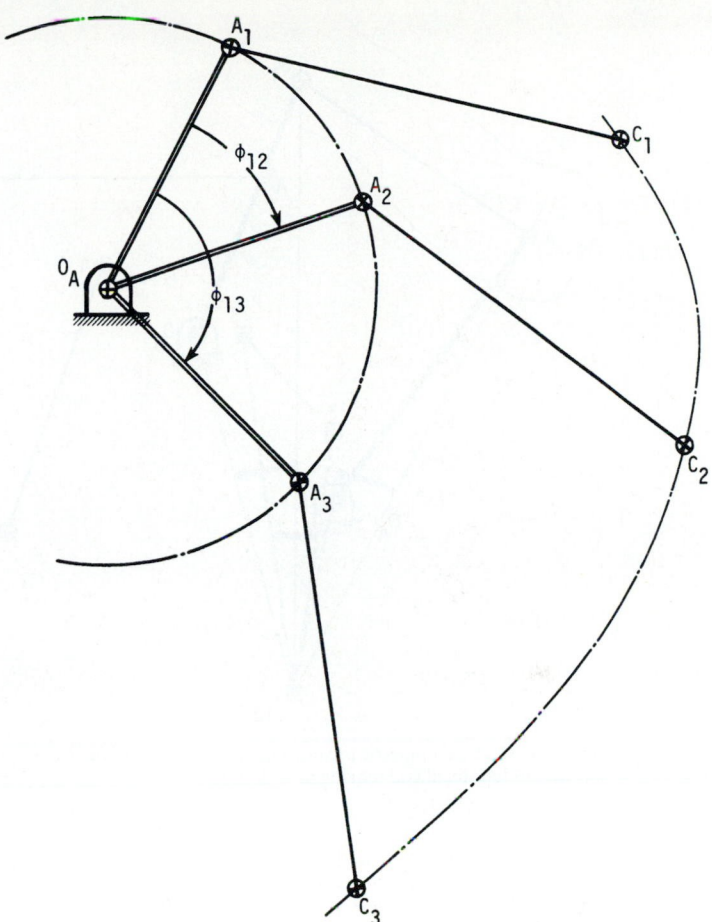

FIGURE 41.10 Path generation as a special case of motion generation.

total rotation angle twice the value defined by the angle(s) in step 2. The magnitude and direction of the link angle ϕ_{14} are defined in the figure.

Note that this construction can produce two, one, or no intersection points. Thus some link rotations are not possible. Depending on how many angles you want to investigate, there will still be plenty of choices. I have found it most convenient to solve the necessary analytic geometry and program it for the digital computer; as many accurate results as desired are easily determined.

Once a center point has been established, the corresponding moving pivot (circle point) can be established. For the first position of the moving body, you need to use the pole triangle $P_{12}P_{13}P_{23}$ angles to establish two lines whose intersection will be the circle point. In Fig. 41.15, the particular angles are

$$\angle P_{13}P_{12}P_{23} \equiv \angle c_1 P_{12} c_0$$

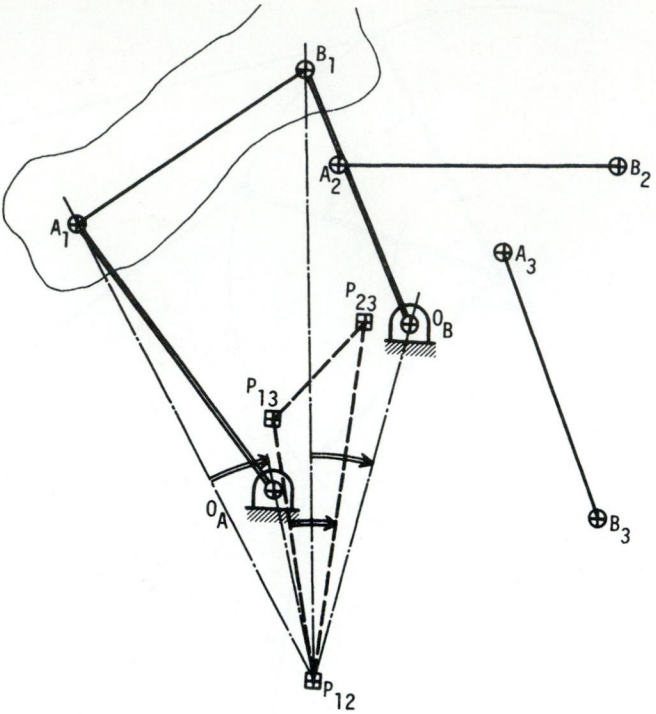

FIGURE 41.11   Geometric relationship between pole triangle angle(s) and location of link fixed and moving pivot points.

and

$$360° - \angle P_{23}P_{13}P_{12} \equiv \angle c_1 P_{13} c_0$$

The second equality could also be written

$$\angle P_{23}P_{13}P_{12} \equiv \angle c_1 P_{13} c_0 \pm 180°$$

A locus of points thus defined can be created as shown in Fig. 41.16. Each point on the circle-point curve corresponds to a particular point on the center-point curve. Some possible links are defined in Fig. 41.16; each has a known first-to-fourth-position rotation angle. Only those links whose length and/or pivot locations are within prescribed limits need to be retained.

The two intermediate positions of the link can be determined by establishing the location of the moving pivot (circle point) in the second and third positions of the moving body. Since the positions lie on the arc with center at the fixed pivot (center point) a and radius aa', it is easy to determine the link rotation angles as

$$\phi_{12} = \angle A_1 O_A A_2 \qquad \phi_{13} = \angle A_1 O_A A_3$$

Linkages need to be actuated or driven by one of the links. Knowing the three rotation angles allows you to choose a drive link which has the desired proportions

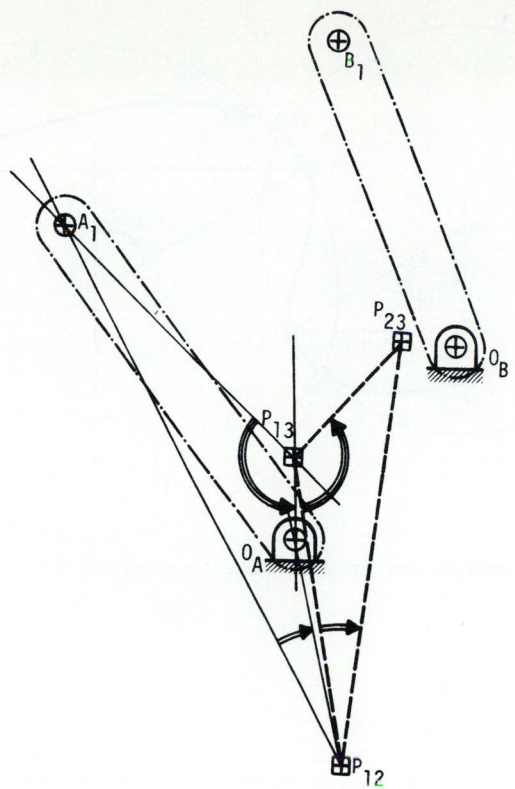

FIGURE 41.12   Determining the moving or fixed pivot by using the pole triangle.

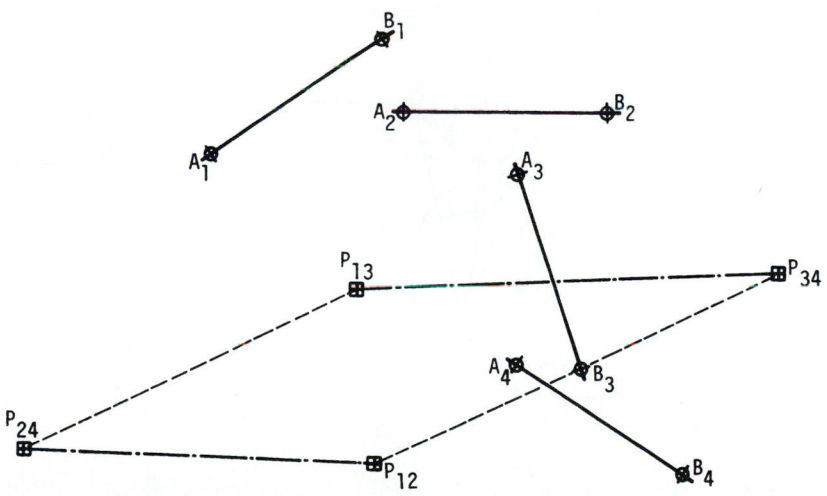

FIGURE 41.13   Four positions of a plane: definition of the opposite-pole quadrilateral formed by lines $P_{13}P_{24}$ and $P_{12}P_{34}$.

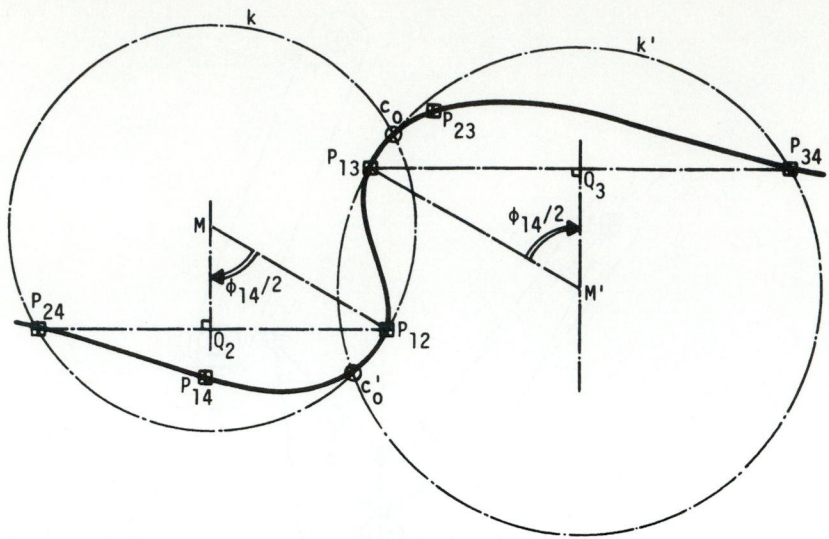

FIGURE 41.14 Determination of points on the center-point curve.

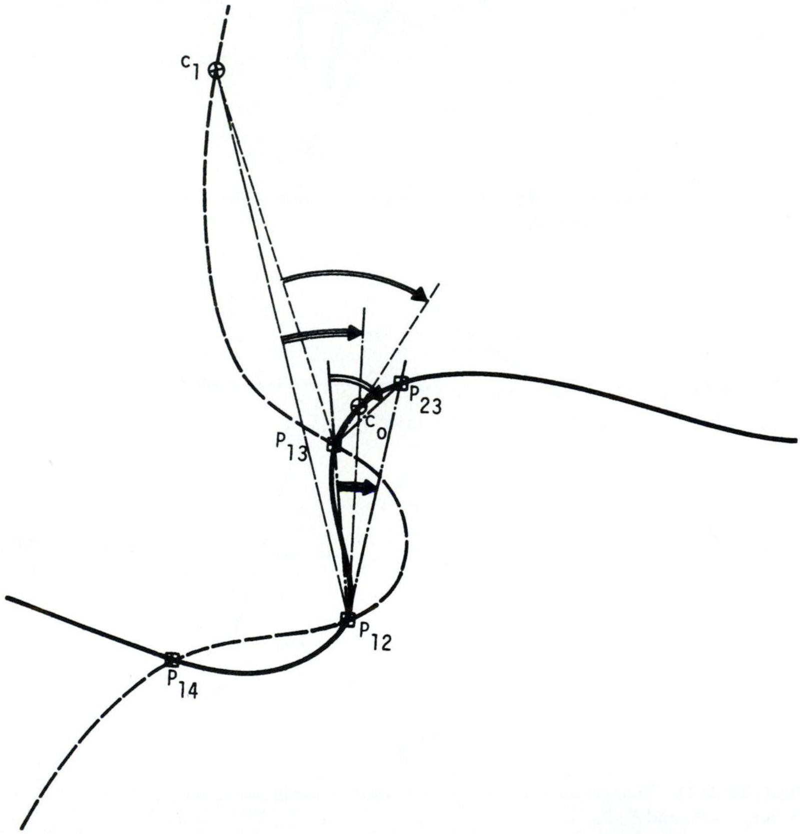

FIGURE 41.15 Determination of a circle point corresponding to a particular center point.

41.16

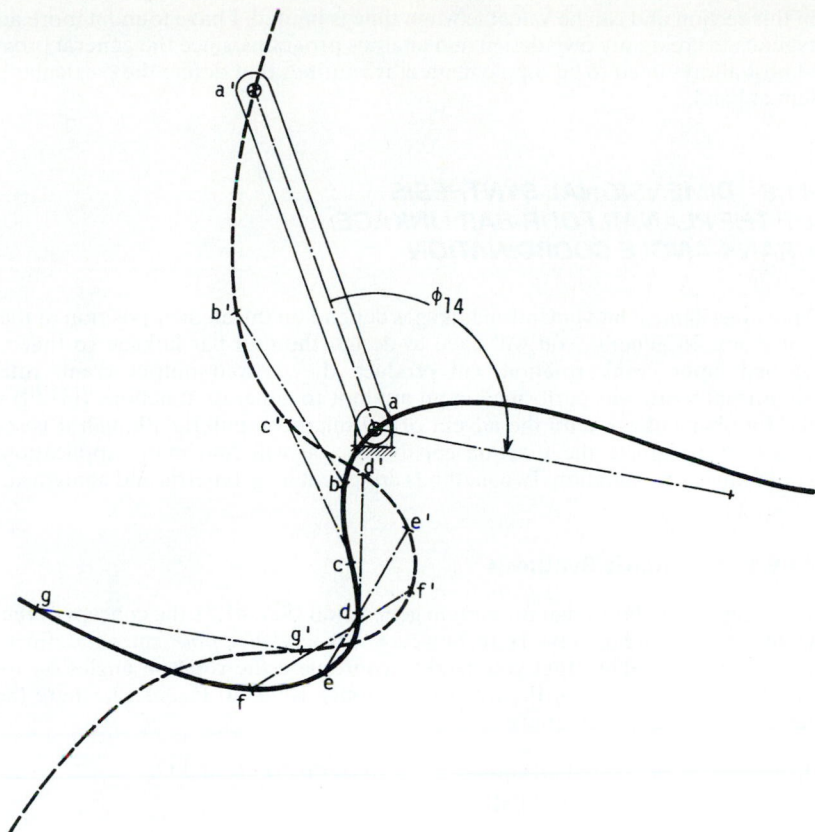

FIGURE 41.16   Some of the links which can be attached to the plane containing *CD*.

of motion. Proper care in selection of the two links will result in a smooth-running four-bar linkage.

41.7.4   Five Positions of a Plane

It would seem desirable to establish as many precision positions as possible. You can choose two sets of four positions (for example, 1235 and 1245) from which the Burmester curves can be created. The intersections (up to six) of those two center-point curves are the only fixed pivots which can be used to guide the moving body through the five positions. Since those pivots and/or link lengths have virtually always been outside the prescribed limits, I never use five-position synthesis.

41.7.5   Available Computer Programs

Two general-purpose planar linkage synthesis programs have been created: KIN-SYN ([41.17]) and LINCAGES ([41.18]). They involve the fundamentals described in this section and can be valuable when time is limited. I have found it more advantageous to create my own design and analysis programs, since the general programs almost always need to be supplemented by routines that define the particular problem at hand.

41.8 DIMENSIONAL SYNTHESIS OF THE PLANAR FOUR-BAR LINKAGE: CRANK-ANGLE COORDINATION

Many mechanical movements in linkages depend on the angular position of the output crank. In general, you will have to design the four-bar linkage so that a prescribed input crank rotation will produce the desired output crank rotation. Significant work was performed in an attempt to generate functions ([41.9]) using the four-bar linkage until the advent of the microcomputer. Although it is seldom necessary to utilize the function capability, you will find many applications for crank-angle coordination. Two methods are possible: geometric and analytical.

41.8.1 Geometric Synthesis

In a manner similar to that for motion generation (Sec. 41.7), the concept of the pole is once again fundamental. Here, however, it is a *relative pole*, since it defines relative motions. Suppose that you need to coordinate the rotation angles ϕ_{12} for the crank (input) and ψ_{12} for the follower (output). Refer to Fig. 41.17, where the following steps have been drawn:

1. Establish convenient locations for the fixed pivots O_A and O_B.
2. Draw an extended fixed link $O_A O_B$.
3. With O_A as vertex, set off a line ℓ at angle $-\phi_{12}/2$ (half rotation angle, opposite direction).

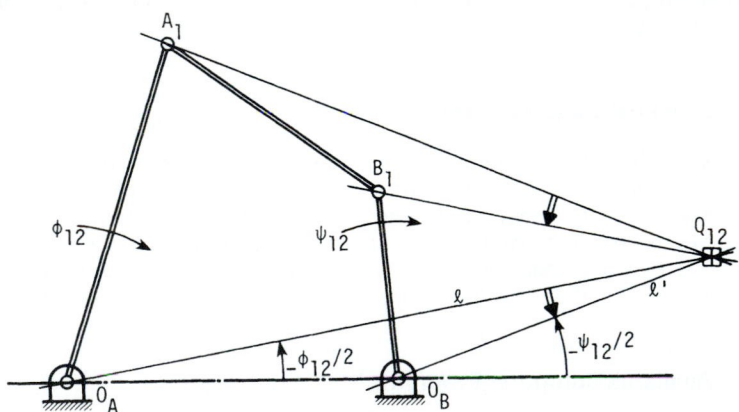

FIGURE 41.17   Crank-angle coordination: definition of relative pole Q_{12}.

· **4.** With O_B as vertex, set off a line ℓ' at angle $-\psi_{12}/2$ (half rotation angle, opposite direction).

5. The intersection of ℓ and ℓ' is the relative pole Q_{12}.

6. Using Q_{12} as the vertex, set off the angle

$$\angle A_1 Q_{12} B_1 = \angle O_A Q_{12} O_B$$

in any convenient location, such as that shown.

When only two positions are required, you may choose A_1 and B_1 anywhere on the respective sides of the angle drawn in step 6. For three positions, two relative poles Q_{12} and Q_{13} are used. You may arbitrarily choose either A_1 or B_1, but the other pivot must be found geometrically. Figure 41.18 shows the necessary constructions.

41.8.2 Analytical Synthesis

Although four-bar linkages had been studied analytically for about 100 years, it was not until 1953 that Ferdinand Freudenstein [41.4] derived the now classic relationship

$$R_1 \cos \phi - R_2 \cos \psi + R_3 = \cos (\phi - \psi) \tag{41.15}$$

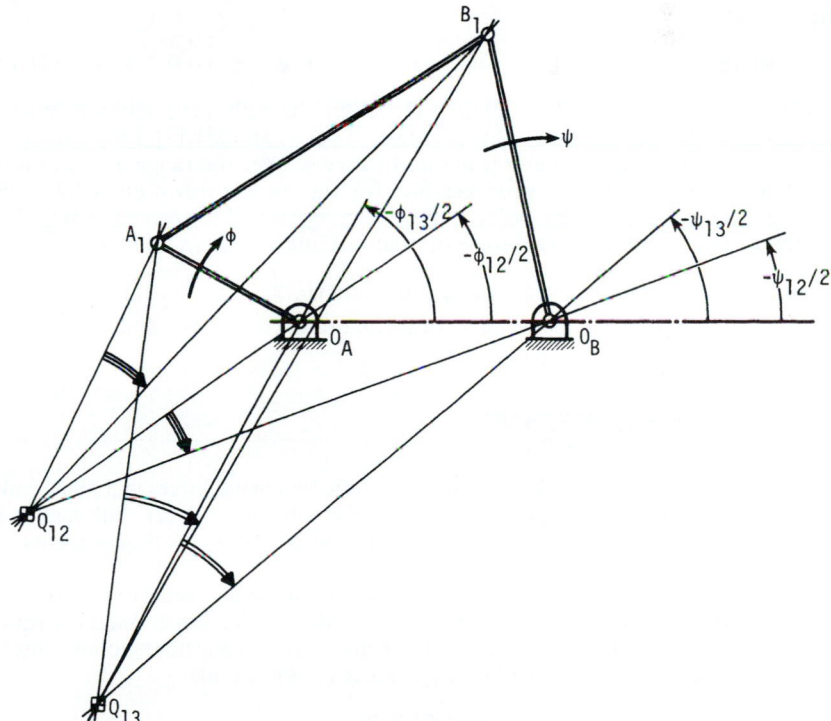

FIGURE 41.18 Geometric construction method for three crank-angle position coordination.

where

$$R_1 = \frac{a}{d} \qquad R_2 = \frac{a}{b} \qquad R_3 = \frac{b^2 - c^2 + d^2 + a^2}{2bd}$$

These link lengths are described in Fig. 41.7. With Eq. (41.15) you can establish a significant variety of linkage requirements. The first derivative of the Freudenstein equation is

$$(R_1 \sin \phi)\left(\frac{d\phi}{dt}\right) - (R_2 \sin \psi)\left(\frac{d\psi}{dt}\right) = \left(\frac{d\phi}{dt} - \frac{d\psi}{dt}\right)\sin(\phi - \psi) \qquad (41.16)$$

which provides a relationship for the velocity or torque ratio. By using the relationship in Sec. 41.4.4, Eq. (41.16) becomes

$$R_1 \sin \phi - nR_2 \sin \psi = (1 - n) \sin(\phi - \psi) \qquad (41.17)$$

where n is the torque or velocity ratio. A further derivative which would deal with accelerations has never been useful to me. If the need arises, see Ref. [41.9].

Since the problem is one of crank-angle coordination, there are potentially five unknowns (R_1, R_2, R_3, ϕ_1, and ψ_1) which you could determine. Combinations of ϕ_{1j}, ψ_{1j}, and n_j may be specified such that a series of equations of the form

$$R_1 \cos(\phi_1 + \phi_{1j}) - R_2 \cos(\psi_1 + \psi_{1j}) + R_3 = \cos(\phi_1 + \phi_{1j} - \psi_1 - \psi_{1j}) \qquad (41.18)$$

and

$$R_1 \sin(\phi_1 + \phi_{1j}) - n_j R_2 \sin(\psi_1 + \psi_{1j}) = (1 - n_j) \sin(\phi_1 + \phi_{1j}) - \psi_1 - \psi_{1j}) \qquad (41.19)$$

can be set up and solved. The nonlinear characteristic makes the solution complicated. Results for certain cases may be found in [41.9] and [41.11]. I have found it most useful in a digital computer program to vary ϕ_1 over the range 0 to π in four simultaneous equations. This produces loci for the moving pivot-point locations which go through the relative poles and are reminiscent of Burmester curves. The two sets of four conditions likely to be of practical interest are as follows:

1. Specify crank rotations ϕ_{12}, ϕ_{13}, ϕ_{14}, ψ_{12}, ψ_{13}, and ψ_{14}.
2. Specify crank rotations and velocity or torque ratios ϕ_{12}, n_1, ψ_{12}, and n_2.

41.9 POLE-FORCE METHOD

An extremely useful scheme for determining static balancing forces in a plane linkage was developed by Hain [41.7] and popularized by Tao [41.12]. Although it is potentially useful for design, I have used it primarily to analyze the requirements for counterbalance springs.

Statically balancing the force on the coupler of a four-bar linkage is a problem often encountered. The solution requires knowledge of the forces and/or torques acting on the four-bar linkage as well as determination of the instantaneous centers. Refer to Fig. 41.19a, in which the following constructions occur:

1. The intersection T_1 of forces F_{ab} and F_{ac} is found.
2. The intersection of the coupler (extended) with force F_{ab} is S_{ab} and with F_{ac} is S_{ac}.

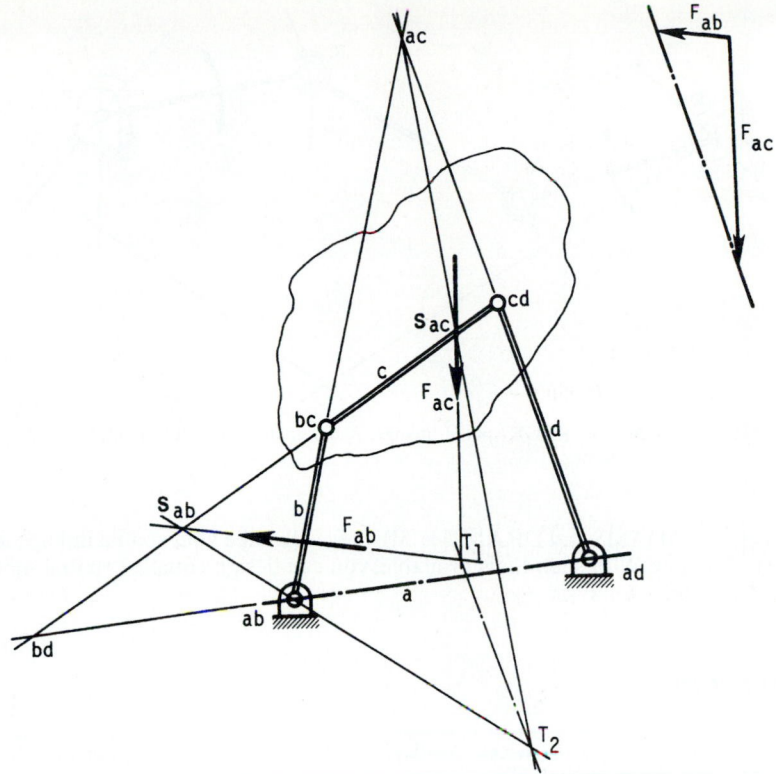

FIGURE 41.19 Pole-force method for balancing a force on the coupler of a four-bar linkage.

3. Determine lines $S_{ab}(ab)$ and $S_{ac}(ac)$; their intersection is T_2.

4. Line T_1T_2 closes the pole-force triangle, which is transferred to Fig. 41.19b.

5. The magnitude of F_{ab} required to balance the coupler force F_{ac} is easily found.

Many other cases, any of which you might encounter in practice, are shown by Tao [41.12].

41.10 SPATIAL LINKAGES

Most practical linkages have motion entirely in a plane or possibly in two parallel planes with duplicated mechanisms such as those in a backhoe or a front loader. Design procedures for some elementary types of spatial four-bar linkage have been created (Refs. [41.9] and [41.11]), principally for the RGGR type (Fig. 41.20).

Three principal mathematical methods for writing the loop-closure equation are vectors ([41.3]), dual-number quaternions ([41.14]), and matrices ([41.13]). These techniques have evolved into general-purpose computer programs such as IMP

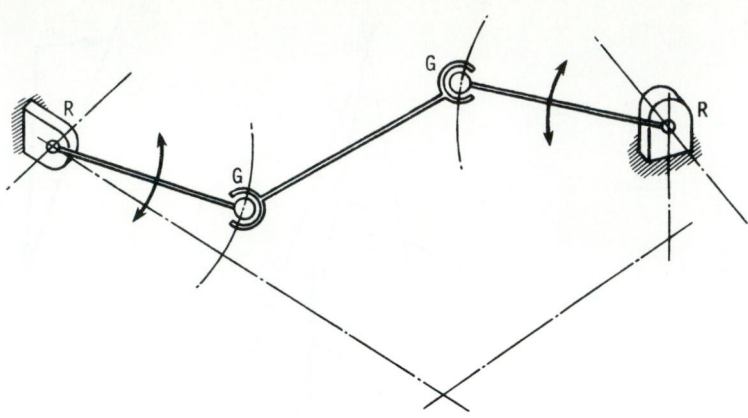

FIGURE 41.20   An RGGR spatial linkage; R designates a revolute joint, G designates a spherical joint.

([41.16]) and ADAMS and DRAM ([41.5]); they will make your spatial linkage analysis much easier. With such tools available, you can design complex spatial mechanisms by iterative analysis.

REFERENCES

41.1 Rudolph A. Beyer, *Kinematic Synthesis of Mechanisms,* Herbert Kuenzel (trans.), McGraw-Hill, New York, 1964.

41.2 Ludwig Burmester, *Lehrbuch der Kinematick* (in German only), A. Felix, Leipzig, 1888.

41.3 Milton A. Chace, "Vector Analysis of Linkages," *J. Eng. Ind.,* ser. B, vol. 55, no. 3, August 1963, pp. 289–297.

41.4 Ferdinand Freudenstein, "Approximate Synthesis of Four-Bar Linkages," *Trans. ASME,* vol. 77, no. 6, 1955, pp. 853–861.

41.5 Ferdinand Freudenstein and George Sandor, "Kinematics of Mechanisms," in Harold A. Rothbart (ed.), *Mechanical Design and Systems Handbook,* 2d ed., McGraw-Hill, New York, 1985.

41.6 Richard E. Gustavson, "Computer-Designed Car-Window Linkage," *Mech. Eng.,* September 1967, pp. 45–51.

41.7 Kurt Hain, *Applied Kinematics,* 2d ed., Herbert Kuenzel, T. P. Goodman, et al. (trans.), McGraw-Hill, New York, 1967.

41.8 Allen S. Hall, Jr., *Kinematics and Linkage Design,* Prentice-Hall, Englewood Cliffs, N.J., 1961.

41.9 Richard S. Hartenberg and Jacques Denavit, *Kinematic Synthesis of Linkages,* McGraw-Hill, New York, 1964.

41.10 N. Rosenauer and A. H. Willis, *Kinematics of Mechanisms,* Dover, New York, 1967.

41.11 Joseph E. Shigley and John J. Uicker, Jr., *Theory of Machines and Mechanisms,* McGraw-Hill, New York, 1980.

41.12 D. C. Tao, *Applied Linkage Synthesis,* Addison-Wesley, Reading, Mass., 1964.

41.13 John J. Uicker, Jr., J. Denavit, and R. S. Hartenberg, "An Iterative Method for the Displacement Analysis of Spatial Linkages," *J. Appl. Mech.*, vol. 31, *ASME Trans.*, vol. 86, ser. E, 1964, pp. 309–314.

41.14 An T. Yang and Ferdinand Freudenstein, "Application of Dual-Number and Quaternion Algebra to the Analysis of Spatial Mechanisms," *J. Appl. Mech. ASME Trans.*, vol. 86, ser. E, 1964, pp. 300–308.

41.15 *ADAMS and DRAM, Automatic Dynamic Analysis of Mechanical Systems and Dynamic Response of Articulated Machinery,* developed by Chace at the University of Michigan.

41.16 *IMP, the Integrated Mechanisms Program,* developed by Uicker at the University of Wisconsin.

41.17 *KINSYN,* primarily for kinematic synthesis, developed by Kaufman while at M.I.T. (he is now at George Washington University).

41.18 *LINCAGES,* for kinematic synthesis and analysis, developed by Erdmann et al. at the University of Minnesota.

41.19 C. R. Mischke, *Elements of Mechanical Analysis,* Addison-Wesley, Reading, Mass., 1963.

A1.23

CHAPTER 42
LOAD-CYCLE ANALYSIS

Russ Henke, P.E.
Russ Henke Associates
Elm Grove, Wisconsin

GLOSSARY OF SYMBOLS

A	Area
F	Force or load
N	Normal force, speed
p	Pressure
Q	Flow rate
t	Time
T	Torque
v	Velocity
$\dot{\theta}$	Angular velocity
μ	Coefficient of friction
ω	Angular velocity

42.1 INTRODUCTION

This chapter deals with the technologies of basic *energy transmission systems* as used by product- and process-oriented industries and the military establishment. Figure 42.1 and Table 42.1 illustrate the essence of these types of systems.

These classes of energy transmission systems can be characterized as follows:

I. Mechanical rotary input in the form of
   A. An input speed N_i which can be constant or variable
   B. An input torque T_i which is variable, responding to the instantaneous demand of the energy transmission system, i.e., the output impedance of the prime mover

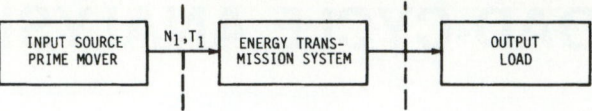

FIGURE 42.1 The energy transmission system is an interface between an input element such as a prime mover and an output element such as a load. The dashed lines indicate interfaces. *(From Ref. [42.1].)*

TABLE 42.1 Various Energy Transfer Systems as Interfaces between Input and Output

Input or source	Energy transmission system	Output or load
AC electric motors DC electric motors Spark-ignition internal-combustion engines Diesel engines Gas turbines Steam turbines Air motors Water motors	1. Electric systems 2. Mechanical systems 3. Fluid power systems	Linear output v_0, F_0 Rotary output N_0, T_0

SOURCE: Ref. [42.1].

II. Mechanical output in two basic forms:
   A. Linear
      1. An output linear velocity v_0 or $\dot{x}$ which can be constant or variable
      2. An output force reaction F_0 which can be constant or variable, responding to the instantaneous changes in load reaction, i.e., the output impedance of the actuator
   B. Rotary
      1. Limited-rotation actuators
         a. An output *rotational* velocity ω or $\dot{\theta}$ which can be constant N_0 or variable θ
         b. An output torque reaction T_0 which can be constant or variable, responding to load changes
      2. Continuous-rotation motors
         a. An output angular velocity which can be constant N_0 (usually as speed N instead of ω_0) or variable θ
         b. An output reaction torque T_0 which can be constant or variable, responding to load changes

In Fig. 42.1 the energy transmission system is an *interface* between an *input* (prime mover) and an *output* (load). The energy transmission system must next be broken down into its *functional sections,* as shown in the block diagram of Fig. 42.2.

42.1.1 Functional Segments

An energy transmission system has three functional sections:

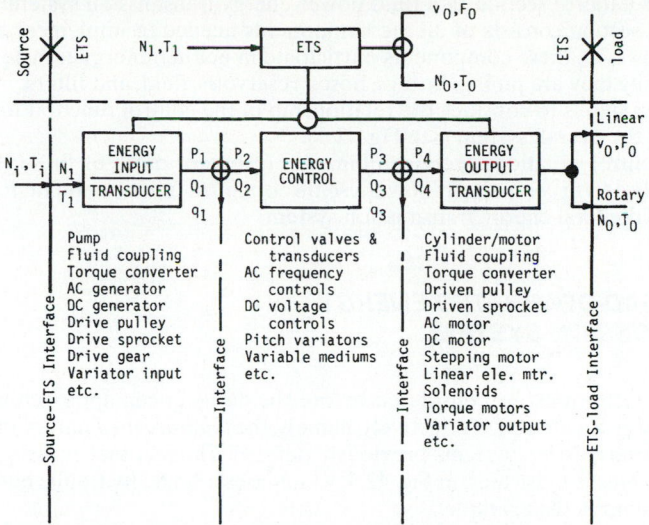

FIGURE 42.2 Block diagram of typical energy transmission system subdivided into its three major categories. *(From Ref. [42.1].)*

I. *Energy input devices* receive the energy from the prime mover across the *source-energy transmission system* interface.
   A. The input variables are the input speed N_i and the input torque T_i.
   B. The output variables are represented by pressure p_1 and flow Q_1.
   C. Typical examples of energy input devices are shown in Fig. 42.2.
II. *Energy output devices* receive the energy transmitted by the energy transmission system, transduce it to mechanical output, and deliver it across the energy transmission system-load interface to the *load*.
   A. Input variables to the energy output devices are p_4 and Q_4.
   B. Output variables from the energy output devices are
      1. Linear output v_0 and F_0.
      2. Rotary output N_0 and T_0.
   C. Typical examples of energy output devices are shown in Fig. 42.2. Commercially there is a wider variety of available energy output devices than of energy input devices.
III. *Energy control devices* receive energy from the energy input devices in the form of input variables p_2 and Q_2. Energy control devices modulate the energy as they transmit it and deliver it in the form of output variables p_3 and Q_3. Note that the intersectional interfaces are shown within the energy transmission systems.
   A. There is an interface between the energy input devices and energy control devices section.
   B. There is an interface between the energy control devices and the energy output devices section of the overall energy transmission system.
   C. Intersectional energy losses (i.e., transmission losses), symbolized by q_e, are shown lumped at a summing point located at the internal interfaces.

There is a fourth section to a fluid power energy transmission system: the *auxiliaries.* This section consists of all the components needed to implement a practical system. However, these components participate in neither energy transfer nor control. Typically they are piping, fittings, hoses, reservoirs, fluid, and filters.

The next step is to consider the relationship of the control function to the other sections of the overall system (see Fig. 42.3).

Most control situations are a combination of two or more of these three basic functions. The term *control* tells how these three control functions relate to the other sections of the total energy transmission system.

42.2 LOAD-DOMINATED ENERGY TRANSMISSION SYSTEM

One more factor must be considered before the designer can approach the subject of fluid power circuit design effectively, namely, the *load-oriented* nature of the kinds of energy transmission systems previously defined. The fact that these systems are *load-dominated* is illustrated in Fig. 42.4, which uses a single hydraulic energy transmission system as the example.

The block diagram in Fig. 42.4 is that of Fig. 42.2. Below the functional representation is a schematic, using International Organization for Standardization (ISO) graphic symbols (see Ref. [42.2]) of the hydraulic system consisting of one pump, one control valve, and either a linear actuator or a continuous-rotation hydraulic motor. The curves below the schematic illustrate the key point: the load domination of the system.

Hydraulic systems, which ordinarily use positive-displacement input-output devices such as pumps, actuators, and motors, transfer energy by means of *potential energy changes in the fluid transfer medium,* that is, by virtue of hydrostatic fluid pressure level differentials Δp. The rate at which the energy is transferred is a function of the flow Q. These two variables—pressure and flow—are essentially independent of each other.

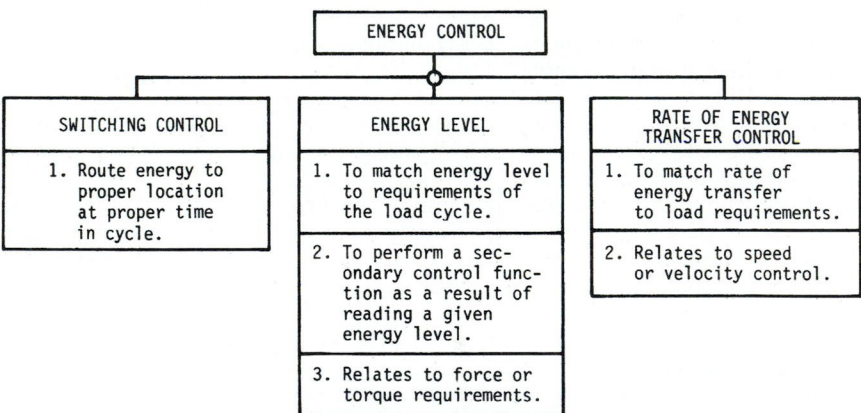

FIGURE 42.3   Block diagram to illustrate the relationship of the control functions to other sections of the overall system. *(From Ref. [42.1].)*

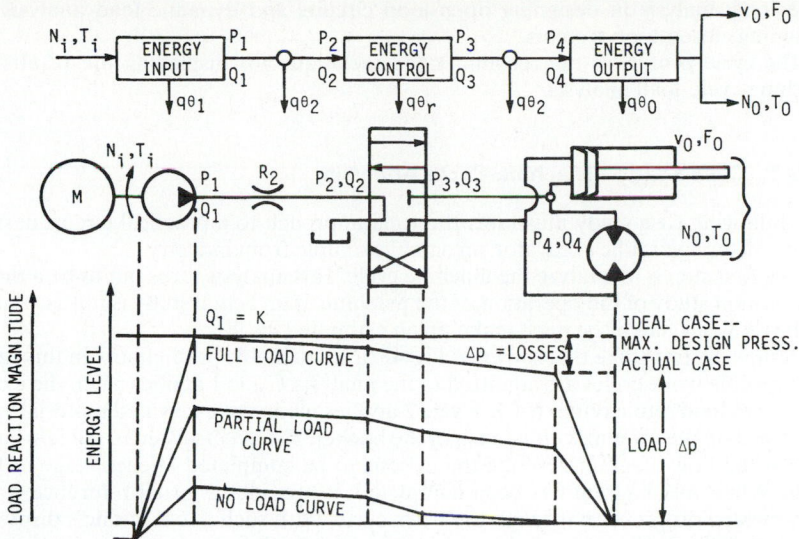

FIGURE 42.4 This diagram illustrates the load-oriented nature of energy transmission systems. *(From Ref. [42.1].)*

There is a common misconception concerning fluid power systems—that the pump generates pressure in the fluid transfer medium. It does not. A positive-displacement pump transfers fluid into a system at a controllable rate *against* an impedance, namely, some resistance to fluid flow. A small part of the resistance emanates from the piping, hoses, fittings, orifices, and other restrictions in the fluid-conducting components. Energy losses due to this part of flow resistance show up as pressure drops—and account for the downward slope of the energy-level curves in Fig. 42.4. The shapes of these curves remain the same for any constant flow in any given system, regardless of overall pressure level. By far the greatest part of resistance to fluid flow comes from the load itself. Pressure (Ref. [42.3], pp. 5–9, 27) is an indication of the potential energy level of the fluid caused by load reaction distributed across the actuator interface area. As load reaction varies, pressure varies accordingly.

As illustrated in Fig. 42.4, when the load reaction is varied, the load curve shifts up and down between the no-load level, representing the summation of pressure differentials $\Sigma \Delta p$ only around the circuit, and the maximum load curve, representing the upper load reaction limit for which the system was designed for safe operation.

Other types of potential energy level transfer systems would exhibit analogous characteristics. Although kinetic energy transfer systems would show different characteristics, the concept would be similar.

42.3 MACHINE-CYCLE ANALYSIS

A complete, quantitative analysis of the machine under consideration is the first requisite for effective fluid power circuit design. It is common practice to use steady-state load analysis in designing open-loop circuits and dynamic load analysis in designing closed-loop systems.

The cycle profile[†] is the recommended technique for displaying the results of machine-cycle-load analysis.

42.3.1 Case Study—Machine-Cycle Analysis

The following case study illustrates the total approach to a practical circuit design problem, the hydraulic excavator, an actual example from industry.

The first step is to analyze the machine cycle. This analysis turns out to be a time-and-motion study of the operation of the machine. If an actual study is not available to the designer, she or he must make up an estimated cycle.

A flow diagram, like that shown in Fig. 42.5, is a valuable preliminary. In this case, two possible work cycles are admitted to the analysis. Cycle 1 applies when the excavator is to load into a dump truck. Cycle 2 applies when the excavated material is to be spread on the ground within reach of the bucket. The events listed in the left-hand side of the flow diagram describe the actions to be completed at each stage of the cycle. When the diagram has been completed, it provides a visual reference for a step-by-step progression through the work cycle. With such a tool at one's disposal, it is very difficult to make a serious error in the cycle plot.

[†] Cycle profile technique is discussed in detail in Ref. [42.1], pp. 16, 26, 29, 40, 41, 249, 348.

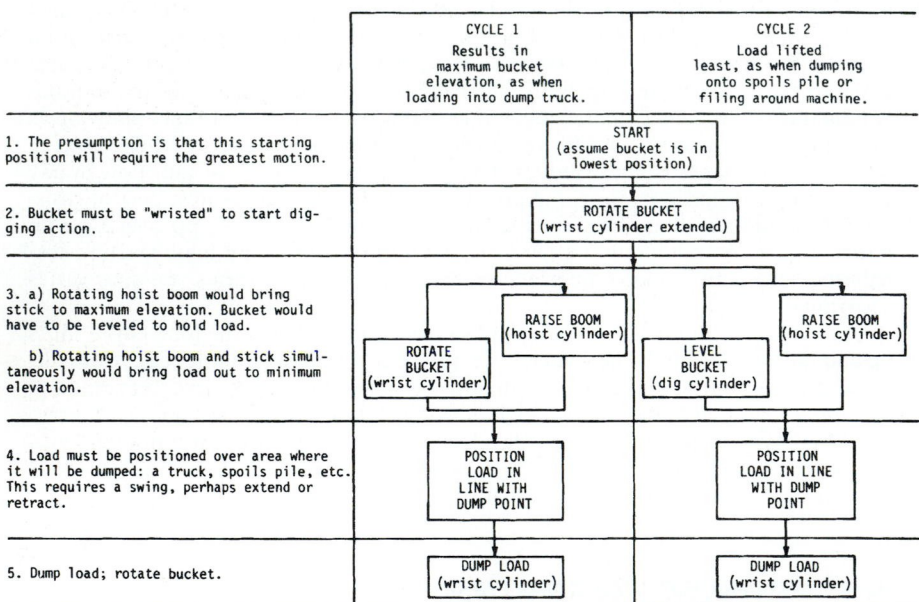

FIGURE 42.5   Flow diagram of the work cycle for a hydraulic excavator. The flow diagram shows primary action only. It is understood that adjustments may be necessary which would necessitate simultaneous operation of the actuators. *(From Ref. [42.1].)*

The next step is to draw a cycle-sequence plot like that shown in Fig. 42.6. The diagrams of machine operations across the top of the sequence plot simplify the communication problem when one is trying to convey the meaning of the plot to persons other than the design group. Each actuator on the machine has been assigned a code letter, from *A* to *J.* These code letters are listed along the left edge of the sequence plot in their order of actuation in the work cycle. The length of the horizontal bar in the diagram indicates the length of time the particular actuator is on. Overlapping of bars indicates that two or more actuators are operating simultaneously. This is an example of how an important point can be brought out graphically by making a sequence plot—in this case, simultaneous operation of motors. It is very difficult to pick up all such instances of overlapping in an intuitive analysis of a circuit.

Now start the load plot, which is the first step in drawing the cycle profile.

42.4 LOAD PLOTS

A separate load plot is required for each actuator in the circuit and for each motion in the cycle.

To better understand the function of a cylinder and its effect on loading, consider the load conditions which occur during a single extension stroke of a cylinder. The question is: What load? General practice has been to work on the basis of the maximum load, either calculated or estimated by the designer. If the engineer is fortunate enough to have an operating system at his or her disposal, a pressure transducer can be used to "look at" the pressure transients that occur as the cylinder is started up and extended. The engineer would see a pressure peak occurring over a short time. This phenomenon is called *breakaway* (Ref. [42.1], pp. 14, 36, 104, 136, 137).

In the time increment $0 < t < dt$, the cylinder must overcome a friction load resistance due to static friction of the total system; this includes external and internal friction. It must also overcome any residual external load applied to the system, for instance, the weight of the arm and bucket, plus any material in the bucket, on the

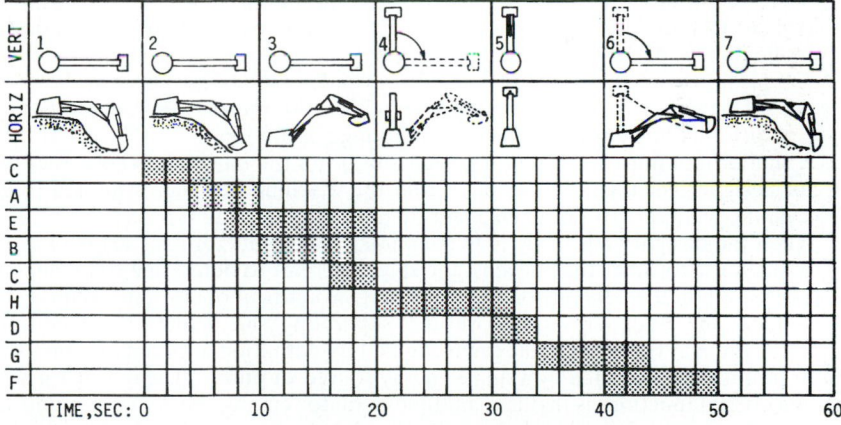

FIGURE 42.6 Cycle-sequence plot. Method 1: maximum bucket elevation. Actuator sequence code is: *A,* extend dig-cylinder rod; *B,* retract dig cylinder; *C,* extend wrist-cylinder rod; *D,* retract wrist cylinder; *E,* extend hoist-cylinder rod; *F,* retract hoist cylinder; *G,* swing clockwise; *H,* swing counterclockwise; *I,* traction forward; *J,* traction reverse. *(From Ref. [42.1].)*

excavator. Note that we have not yet considered acceleration forces, because in time dt the system has not yet started to move. One might say that, so far, the cylinder has simply been taking up the lost motion, or backlash, in the system.

This brings up a fine philosophical point not often recognized. Capacitance in a system, which is due to compressibility of the fluid, compliance, and slip in components, is generally regarded as a negative quantity, that is, one that detracts from the performance of a fluid power system. The consensus seems to be that if capacitance were eliminated, then the system efficiency would be optimized. But consider the following: If the system is at zero velocity at $t = dt$, and if it is accelerated to some velocity v_1 in an infinitesimally small increment of time $+dt$, then as $+dt - dt$ approaches 0, it is necessary for the system to accelerate from zero velocity to some finite velocity in zero time. This would require an infinite acceleration. However, the fact that there is capacitance in the system allows us to transfer energy to the fluid in a finite time interval, thus eliminating the requirement for infinite acceleration. It is quite possible that a fluid power system could not be started if the fluid and the system were perfectly inelastic.

In a practical system, of course, the relief valve also enters into the picture, since it will "crack" and bypass excess fluid every time a cylinder or motor starts up, until the steady-state velocity of the piston matches the flow rate from the pump.

In the time interval $dt < t < \Delta t$, the external portions of the system start to move. Two changes in loading take place:

1. An acceleration force $F = ma$ is introduced in accordance with Newton's second law of motion.

2. The friction force drops from static friction conditions to dynamic friction conditions. Of course, this occurs because the coefficient of static friction is greater than that for kinetic friction.

At the end of the time interval Δt, we note that the piston has reached steady-state velocity $v_{ss} = Q/A_p$. When this occurs, the acceleration force disappears and the steady-state load reduces to components of dynamic friction and external load.

Does this mean that the steady-state load is constant? Certainly not! It is important that the circuit designer recognize this fact, particularly when she or he is dealing with multibranched circuits operating with one pump. In such cases, auxiliary controls, or flow dividers, may be necessary. The designer will not know this unless a complete picture of the load cycle is produced.

A typical load reaction plot for the hoist cylinder on our excavator is shown in Fig. 42.7. This plot must be determined from a layout of the arm and bucket mechanism at different angles during the complete range of motion.

A similar plot must be made for each actuator on the machine. Where does the circuit designer get this information? From the machine designer, who had to go through the analysis in order to engineer the machine in the first place.

When the designer has made individual load-cycle plots for each actuator, then he or she must consider them against the sequence plot to determine simultaneous operation. When the comparison indicates that two critical operations occur at the same time, the designer should consider separating the actuators rather than installing them as branches of the same circuit. (By definition, a circuit, whether single or multibranched, is fed by a single energy source, or pump.) The reason for separating critical functions is that in a multiple-actuator circuit, the actuator requiring the lowest pressure will take *all* the fluid. Besides the inconvenience of not having one of the critical operations occur, there is the danger of dropping load as a result of a change in pressure relationships owing to motion of the system.

The complete load plot for a single actuator might look like Fig. 42.8.

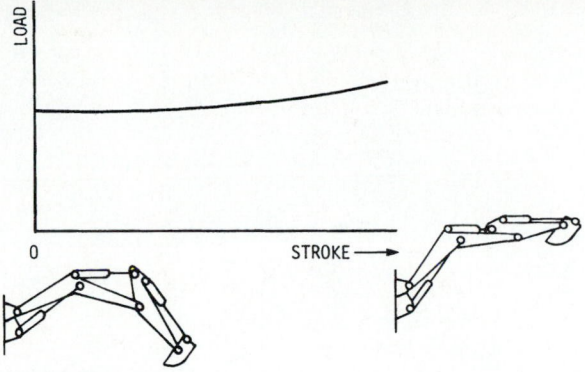

FIGURE 42.7   A load reaction plot for the hoist cylinder. *(From Ref. [42.1].)*

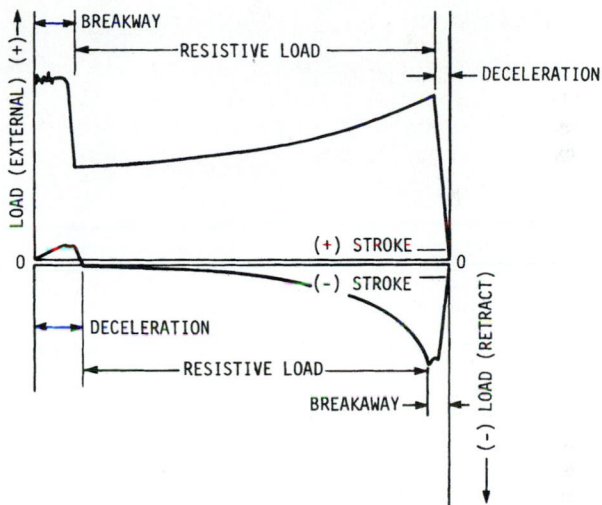

FIGURE 42.8   Load plot for a single actuator. *(From Ref. [42.1].)*

REFERENCES

42.1 Russell W. Henke, P. E., *Fluid Power Systems and Circuits,* Penton/IPC, Cleveland, 1983.

42.2 ANSI standard Y32, 10-1967 (R1979), "Graphic Symbols for Fluid Power Diagrams."

42.3 Russell W. Henke, *Introduction to Fluid Mechanics,* Addison-Wesley, Reading, Mass., 1966.

42.4 William Wolansky, John Negoshian, and Russ Henke, *Fundamentals of Fluid Power,* Houghton Mifflin, Boston, 1977.

CHAPTER 43
FLUID POWER SYSTEMS AND CIRCUIT DESIGN

Russ Henke, P.E.
Russ Henke Associates
Elm Grove, Wisconsin

GLOSSARY OF SYMBOLS

a	Acceleration
A	Area
C	Coefficient
e	Efficiency
F	Force or load
g	Acceleration due to gravity
H	Power
L_N	Normal force component
m	Mass
N	Normal force, speed
p	Pressure
Q	Flow rate

S	Actuator stroke
t	Time
T	Torque
v	Velocity
V	Volume
W	Weight
γ	Specific weight
$\dot{\theta}$	Angular velocity
μ	Coefficient of friction
ω	Angular velocity

Fluid power technology is one of the three primary *energy transmission technologies* used throughout industry, the defense community, agriculture, and all aspects of productive activity in the industrialized world. In addition to fluid power, there are electric and mechanical systems. This chapter deals with the design of hydraulic and pneumatic systems by using the cycle-profile plotting techniques discussed in Chap. 42.

43.1 PRESSURE PLOTS

The pressure plots are the same as the load plots discussed in Sec. 42.4 except for the introduction of a constant reflecting the interface area over which the load is distributed. The equation is

$$p = F_1 \left(\frac{1}{A_p} \right) \tag{43.1}$$

Note that the constant $1/A_p$ is different for each size of cylinder.

Once the load and pressure plots have been completed, the *level* of energy transfer occurring throughout the machine cycle has been defined.

43.2 FLOW PLOTS

The next step is to define the *rate* of energy transfer in the machine. This is a function of the velocities of the various piston rods (or motor shafts, if rotary motors are used).

To determine the required rod velocities, the designer can go back to the sequence chart. The time scale along the bottom of the chart indicates how much time is available to complete each part of the machine cycle. Time (Ref. [43.1], pp. 14, 15) has two basic implications in the design of a cycle:

1. It determines the flow rate requirement relative to the actuator motion pattern.
2. It determines the horsepower requirement of the circuit or branch.

After establishing time increments, the engineer must turn to the machine-element layouts to determine the length of stroke necessary to complete each motion. Next, the steady-state velocity required of the piston can be calculated, allowing for acceleration and deceleration. At this point the designer must make a choice of velocity patterns (Ref. [43.1], pp. 16, 19, 22, 26, 29, 347, 355, 358).

The superimposition of the flow plots for individual actuators has an important implication. If study of the sequence diagram indicates simultaneous operation of two or more cylinders or motors, their flow plots must be superimposed, as shown in Fig. 43.1. Such superimposition of the plots will give the designer an insight into the maximum flow rate required. This affects the selection of the pump or pumps and influences separation of circuit branches.

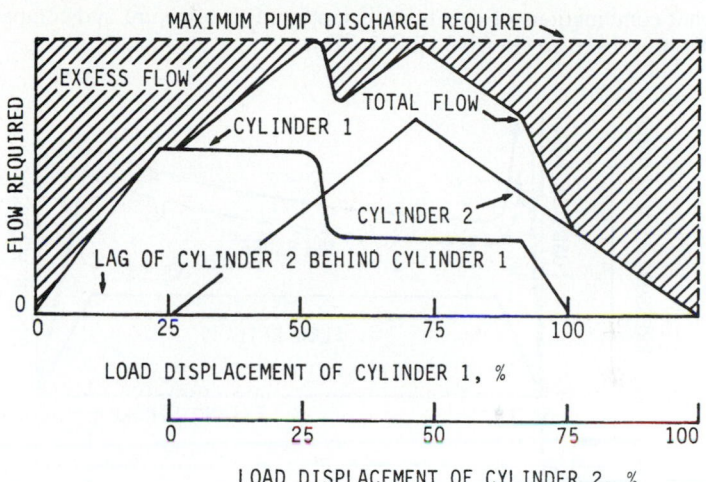

FIGURE 43.1 Flow plot of two actuators superimposed to indicate total flow rates. *(From Ref. [43.1].)*

43.3 POWER PLOTS

With the information developed for the pressure and flow plots, the designer can now make power plots. This is a necessary preliminary to selection of the prime mover. It is particularly useful in pointing out *power peaks* which might otherwise be hidden in averaged calculations. Such peak power demands, occurring when unexpected, could be great enough to stall an undersized prime mover. The fluid horsepower can be calculated from

$$H_f = \frac{pQ}{1714} \tag{43.2}$$

where p is in pounds per square inch (psi) and Q is in gallons per minute (gpm). The input horsepower of the prime mover is then

$$H_i = \frac{H_f}{e_o} \tag{43.3}$$

where e_o = overall efficiency of the pump.

43.4   CYCLE PROFILE

A complete *cycle profile* for a single actuator might look something like that shown in Fig. 43.2. Note that the cycle must be plotted for both directions of motion.

If the circuit designer has intelligently followed the cycle-profile procedure, a complete graphic portrayal of what should happen at any point in the cycle of operation of the machine is the result. The designer should be able to communicate any information necessary to an understanding of the operational capabilities and limitations of the equipment. Even more important, the designer should be able to spot any malfunctions of the machine much more quickly and surely than if she or he had to guess what combinations of events were supposed to transpire and compare them with what was observed (Ref. [43.1], Chap. 26).

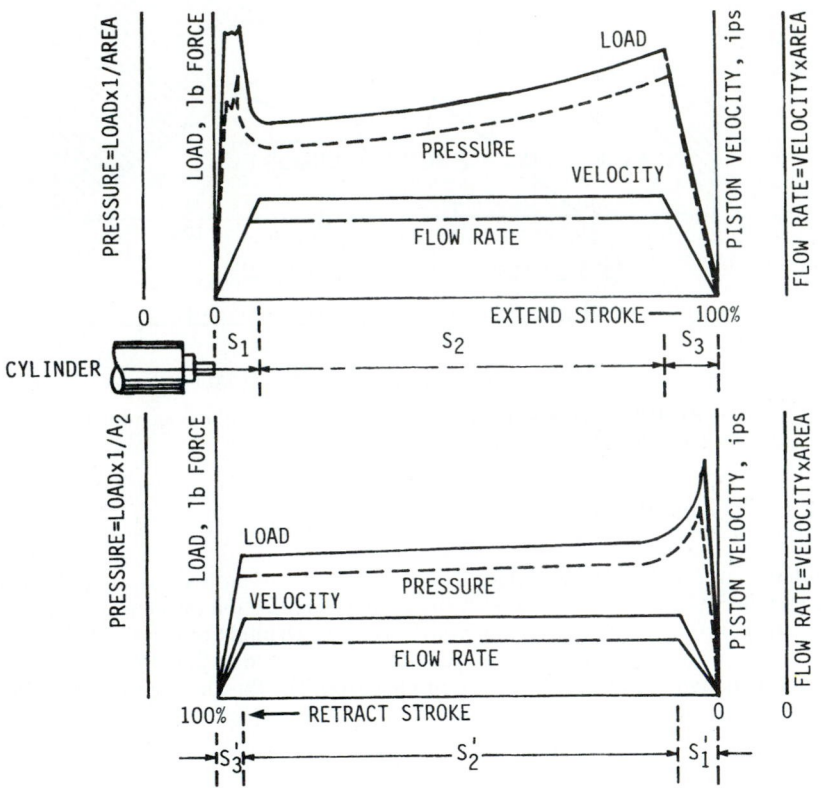

FIGURE 43.2   Cycle profile for a single actuator. *(From Ref. [43.1].)*

43.5   CIRCUIT DESIGN

Fluid power circuits can be thought of as consisting of four sections, as shown in Fig. 43.3. Section I represents energy output, where energy is transferred to the load across the hydromechanical interface. Section II is the control area; fluid switching

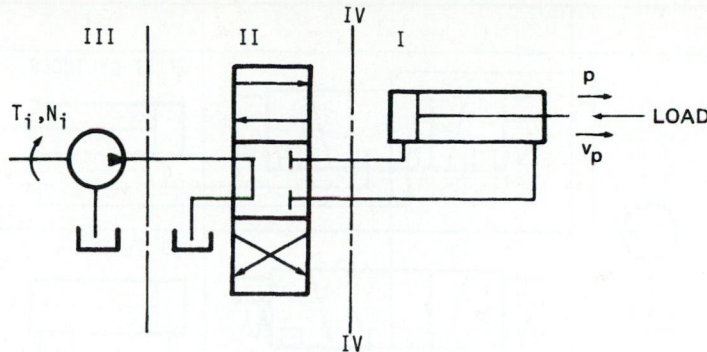

FIGURE 43.3   The four sections of a fluid power circuit. *(From Ref. [43.1].)*

and energy modulation are effected in this section. Section III is the energy input section; this is where the pumps are involved. Section IV is the auxiliaries area; this consists of piping and fittings and all the other components and equipment necessary to make a circuit work, including the fluid.

The pattern of Fig. 43.3 suggests a logical method for solving a circuit design problem. A format like that of Fig. 43.4 is helpful. Divide the sketch sheet into three areas by drawing vertical lines. The left-hand column is reserved for energy input devices, i.e., pumps. The right-hand column is for energy output devices, i.e., motors or actuators. The middle area is for control devices.

Next, sketch the symbols (Ref. [43.5]) for the output components in the right-hand column, in vertical array, as shown in Fig. 43.4. Then divide the page into rows by drawing horizontal lines which separate each actuator from its neighbors. We now have a matrix of sorts, with the columns representing circuit functions (energy output, energy control, and input), and the rows representing machine functions, as typified by the actuators.

The circuit designer can now select functions to match the requirements of the machine functions.

43.6   OPEN-LOOP AND CLOSED-LOOP CIRCUITS

Fluid power systems can be divided into two major groups: open-loop and closed-loop.

In a *closed-loop* system, a *feedback* mechanism continually monitors system output, generating a signal proportional to this output and comparing it to an input or command signal. If the two match, there is no adjustment, and the system continues to operate as programmed. If there is a difference between the input command signal and the feedback signal, the output is adjusted automatically to match command requirements.

There is no feedback mechanism in an *open-loop* system. The performance characteristics of the circuit are determined entirely by the characteristics of the individual components and their interaction in the circuit. A typical open-loop circuit is illustrated in Fig. 43.5a. Most industrial circuits fall into this category.

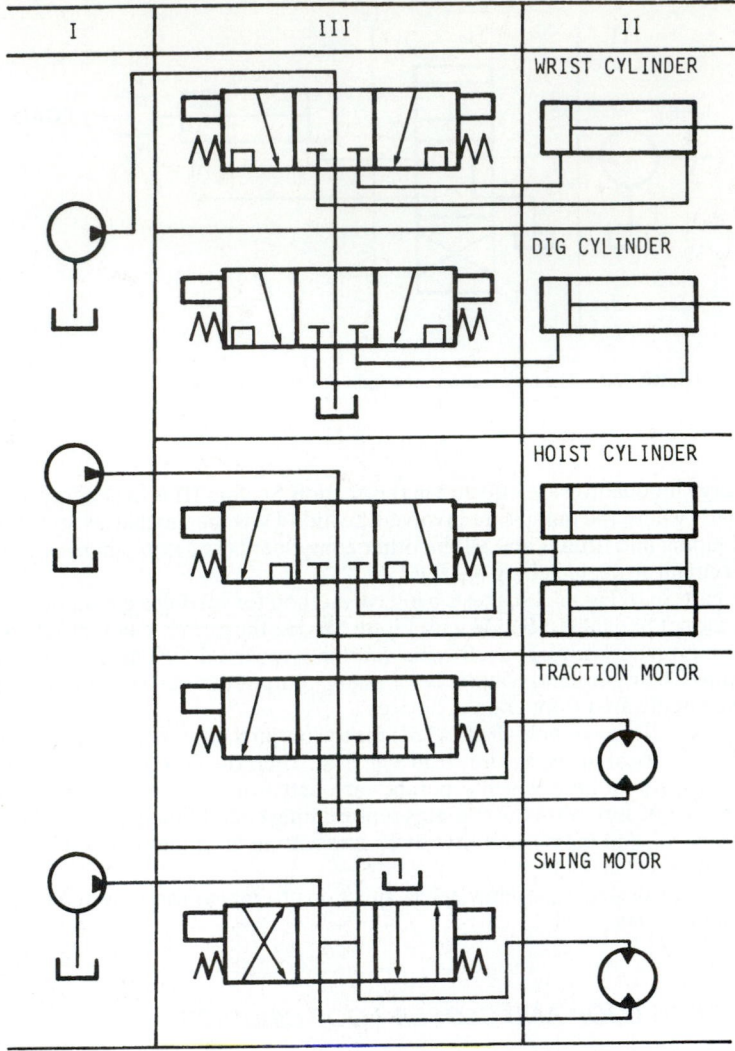

FIGURE 43.4   Graphical layout of a circuit design problem. *(From Ref. [43.1].)*

An *electrohydraulic servo system* is a feedback system in which the output is a mechanical position or function thereof; see Fig. 43.5b.

Open-loop circuits can be grouped by the functions performed or by the control methods.

43.6.1   Functions Performed

Classification of open-loop circuits by function is related to the basic areas of control used in a fluid power system:

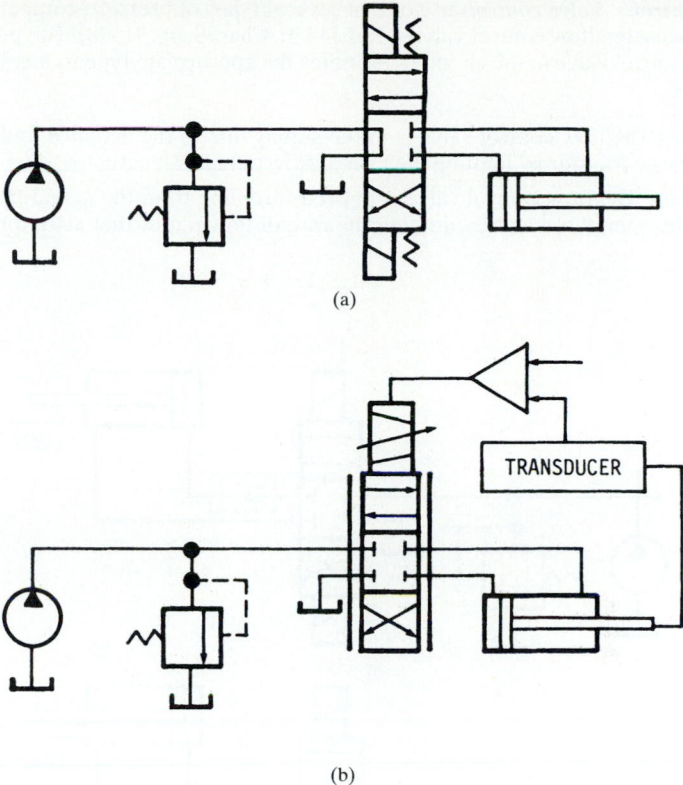

(a)

(b)

FIGURE 43.5 (a) Typical open-loop circuit; (b) typical closed-loop circuit. (*From Ref. [43.1].*)

1. *Directional* controls regulate the distribution of energy (Ref. [43.4], Chap. 12, pp. 79–91, 151–164).

2. *Flow* controls regulate the rate at which energy is transferred by adjusting the flow rate in a circuit or branch of circuit (Ref. [43.4], Chaps. 10, 11, pp. 65–75, 164–168).

3. *Pressure* controls regulate energy transfer by adjusting the pressure level or by using a specific pressure level as a signal to initiate a secondary action (Ref. [43.4], Chaps. 8, 9, pp. 47–60, 143–151).

43.6.2 Control Methods

Directional Control. *Valve controls* make use of one of many types of directional control valves to regulate the distribution of energy throughout the circuit. These valves switch flow streams entering and leaving the valve.

Pump control is limited to reversal of the direction of flow from a variable-displacement reversible pump. *Fluid motor control* is similar to pump control; it uses reversible, variable-displacement motors.

Flow Control. *Valve controls* use one of several types of pressure-compensated or noncompensated flow control valves (Ref. [43.3], Chap. 9, pp. 91–98). The position of the flow control valve in the circuit determines the appropriate type to use. These are as follows:

1. *Meter-in* The flow control valve is in the supply line to the actuator and controls the energy transfer by limiting the rate of flow out of that actuator; see Fig. 43.6a.

2. *Meter-out* The flow control valve is in the return line from the actuator and controls the energy transfer by limiting the rate of flow out of that actuator; see Fig. 43.6b.

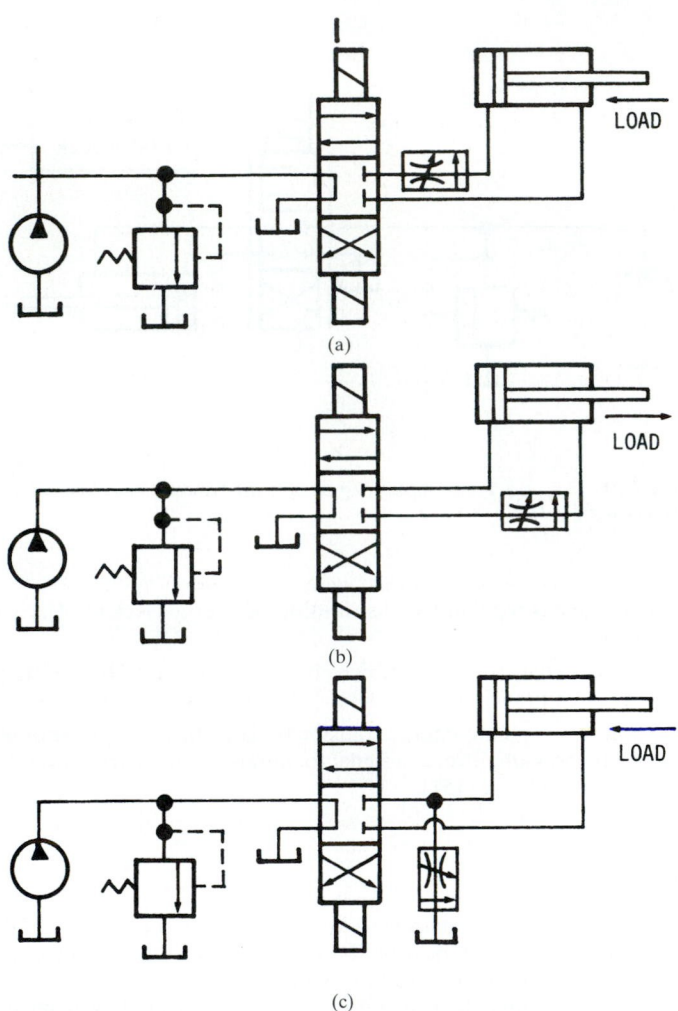

FIGURE 43.6 Valve controls for open-loop circuits. (*a*) Meter-in; (*b*) meter-out; (*c*) bleed-off. (*From Ref. [43.1].*)

3. *Bleed-off* The flow control valve is in parallel with the actuator. It limits the rate of energy transfer to the actuator by controlling the amount of fluid bypassed through the parallel circuit; see Fig. 43.6c.

Pump control involves the use of one of two methods, depending on the type of pump used. *Multiple* pumps provide a step variation in flow (Fig. 43.7a); *variable-displacement* pumps deliver infinitely (from zero to maximum) variable flows (Fig. 43.7b).

Fluid motor controls use techniques similar to pump controls, and this involves the use of multiple motors as in Fig. 43.8a for step variation or variable-displacement motors as in Fig. 43.8b for infinite variation in output speeds.

Pressure Control. *Valve controls* use one or more of six types of pressure control valves:

1. *Relief* valves limit the maximum energy level of the system by limiting the maximum operating pressure; see Fig. 43.9.

2. *Unloading* valves regulate the pressure level by bypassing the supply fluid to the tank at a low energy level. Unloading valves shift when the system pressure reaches a preset level; see Fig. 43.10.

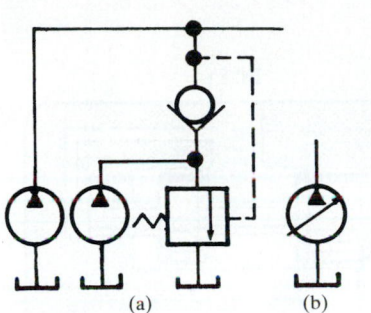

(a)   (b)

FIGURE 43.7 Pump controls for open-loop circuits. (*a*) Multiple pumps; (*b*) variable displacement. (*From Ref. [43.1].*)

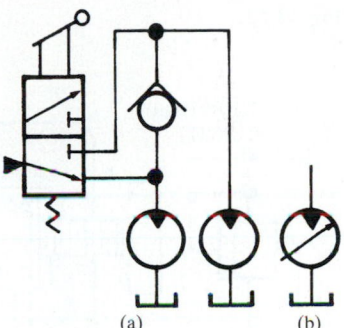

(a)   (b)

FIGURE 43.8 Actuator controls for open-loop circuits. (*a*) Multiple-fluid motors; (*b*) variable displacement. (*From Ref. [43.1].*)

TO SYSTEM

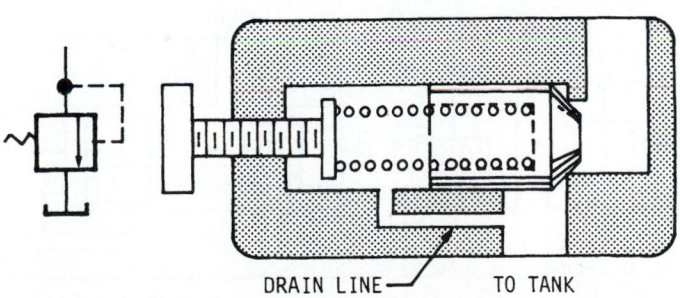

DRAIN LINE   TO TANK

FIGURE 43.9 Pressure relief valve regulates system output fluid pressure. (*From Ref. [43.1].*)

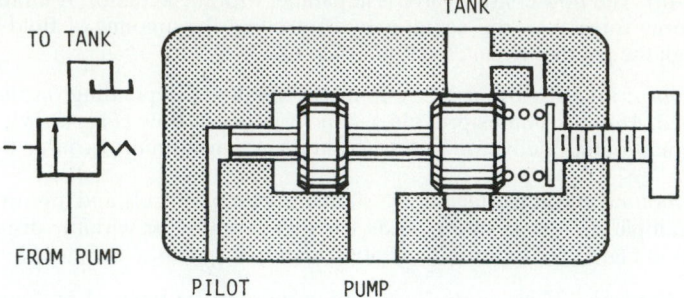

FIGURE 43.10   Pressure unloading valve unloads pump output to the tank at low pressure when high-pressure flow is not required. *(From Ref. [43.1].)*

3. *Sequence* valves react to a pressure signal to divert energy from a primary circuit to a secondary circuit; see Fig. 43.11.

4. *Reducing* valves react to a pressure signal to throttle flow to a secondary circuit, thus delivering energy at a lower level to the secondary than to the primary circuit; see Fig. 43.12.

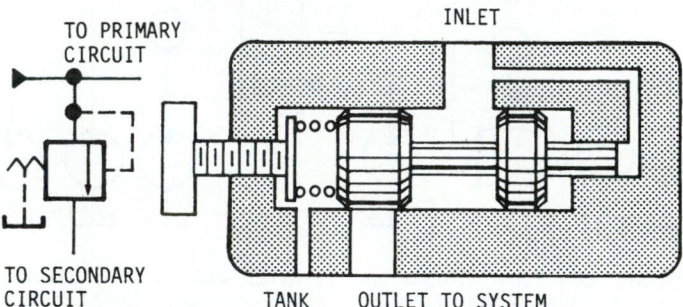

FIGURE 43.11   Sequence valve prevents fluid from entering one branch of a circuit before a preset pressure is reached in the main circuit. *(From Ref. [43.1].)*

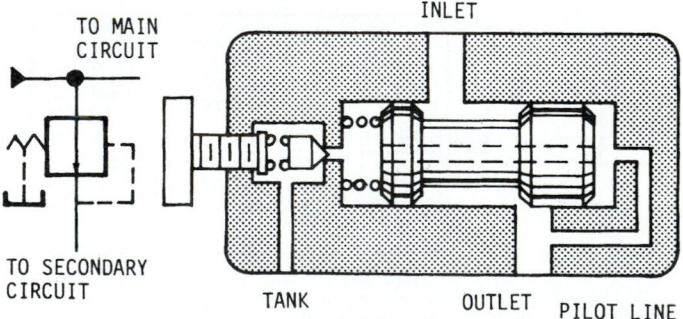

FIGURE 43.12   Pressure-reducing valve allows one branch of a circuit to operate at a lower pressure than the main system. *(From Ref. [43.1].)*

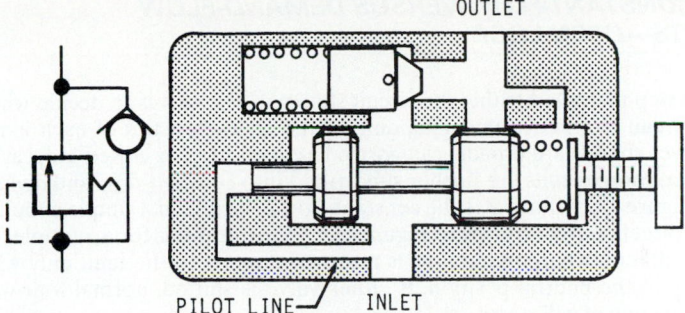

FIGURE 43.13 Counterbalance valve holds fluid pressure in part of a circuit to counterbalance weight on the external force. *(From Ref. [43.1].)*

5. *Counterbalance* valves control the potential energy differential across an actuator by maintaining a preset backpressure in the return line; see Fig. 43.13. Their purpose is to prevent a load from drifting.

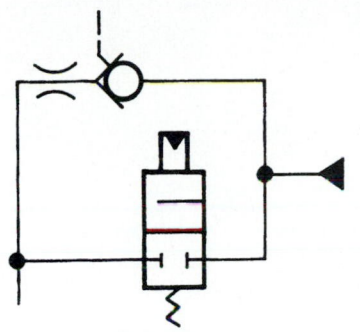

FIGURE 43.14 Decompression valve releases fluid at controlled rate to release energy stored in high-pressure system. *(From Ref. [43.1].)*

6. *Decompression* valves provide controlled release of energy stored in high-pressure systems, because of the elasticity in the system; see Fig. 43.14.

Pump control of pressure fluid in open-loop circuits is generally achieved with pressure-compensated variable-displacement pumps. Energy transfer is controlled by varying the flow from the pump in response to a pressure-level signal across the compensator; see Fig. 43.15.

Rotary actuator control of fluid pressure is not generally used.

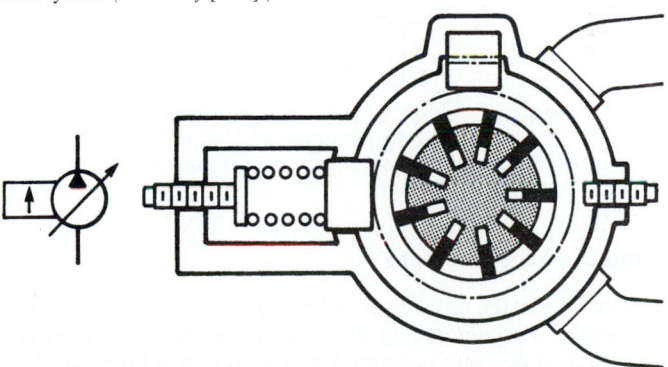

FIGURE 43.15 Pressure-compensated variable-displacement pump. Governor spring loads pump toward full-displacement position. As output pressure rises, it supplies the required force to stroke the cam ring toward deadhead position. *(From Ref. [43.1].)*

43.7 CONSTANT-FLOW VERSUS DEMAND-FLOW CIRCUITS—OPEN LOOP

The next step in implementing the circuit shown in Fig. 43.4 is to decide which basic type of circuit to use. An understanding of the characteristics of each is required. Fluid power circuits are broadly categorized as open-loop or closed-loop, as we have seen. Open-loop circuits are further subdivided into *constant-flow* and *demand-flow* circuits. Figure 43.16 illustrates the constant-flow principle in a simple circuit that has only one directional control valve. Figure 43.17 illustrates this for a multiple- or stack-valve installation where pump flow is returned directly to the tank only when *both* valves are in the neutral position. If either valve is shifted, normal four-way valve directional control will start.

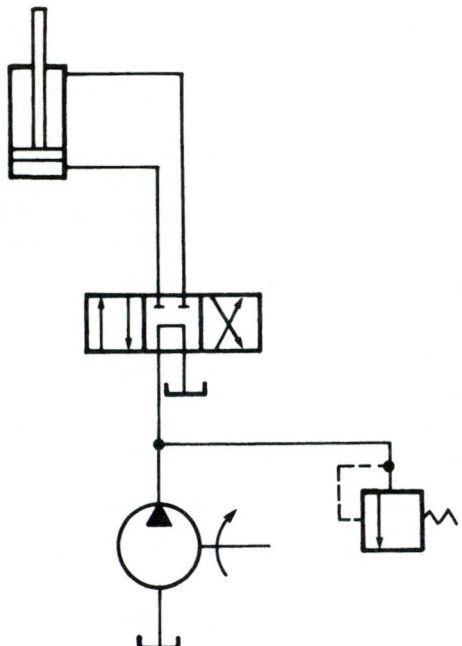

FIGURE 43.16 Typical constant-flow system. When directional control valve is in neutral position, pump output bypasses to tank through tandem center. *(From Ref. [43.1].)*

43.7.1 Pump Discharge Pressure

In constant-flow circuits, the pressure at which the pump discharges fluid is a function of the load resistance encountered by and reflected across the actuator; see Fig. 43.4. The system operating pressure generated by the load is a function of the actuator geometry. If the prime mover can satisfy the energy demand, it will do so. If not, the prime mover will stall; or, as is more likely to happen in actual practice, the relief valve will open to bypass fluid to tank—this wastes energy.

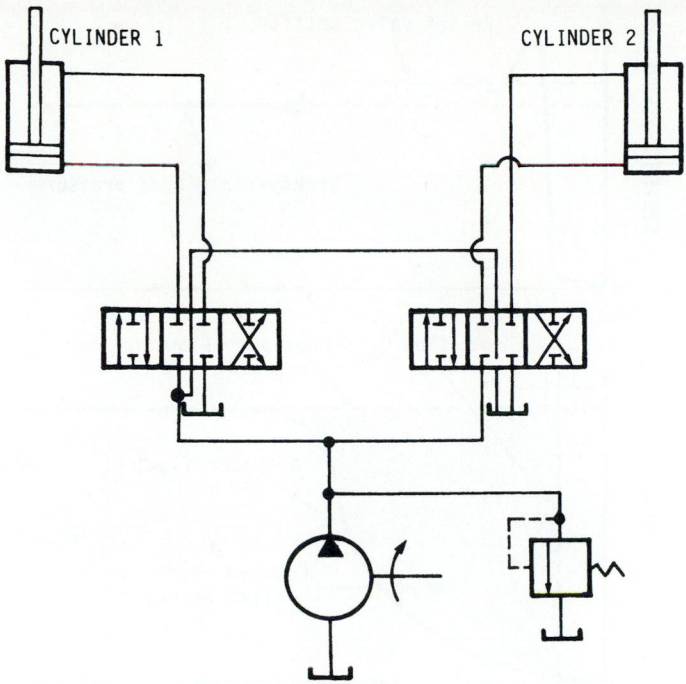

FIGURE 43.17 Constant-flow multiple-valve system. Pump output bypasses to tank only when *both* directional valves are in neutral position. *(From Ref. [43.1].)*

43.7.2 Relation of Pump Discharge to Actuator Speed

In constant-flow circuits, the pump output is *not* determined by the actuator's instantaneous speed requirements. The discharge rate is a function of pump displacement and its speed of rotation. Pump output and actuator displacement jointly determine a steady-state speed, according to the equations

$$v = \frac{Q_p}{A_p} \quad \text{(for a cylinder)} \qquad N_o = \frac{Q_p}{V_m} \quad \text{(for a motor)} \qquad (43.4)$$

where v = velocity
Q_p = pump output
A_p = actuator area
N = output speed
V_m = motor displacement per revolution

Load inertia may preclude rapid acceleration to this steady-state speed; if it does, excess flow from the pump must return to the tank through the relief valve; see Fig. 43.18. At time t_0 the control valve shifts, porting pressure fluid to the actuator. There is a slight time lag caused by such factors as the compressibility of the oil in the system and throttling while the valve spool is shifting. Actuator flow Q_a then increases to full rated output $Q_a = Q_p$; see the vertical dotted line in Fig. 43.18.

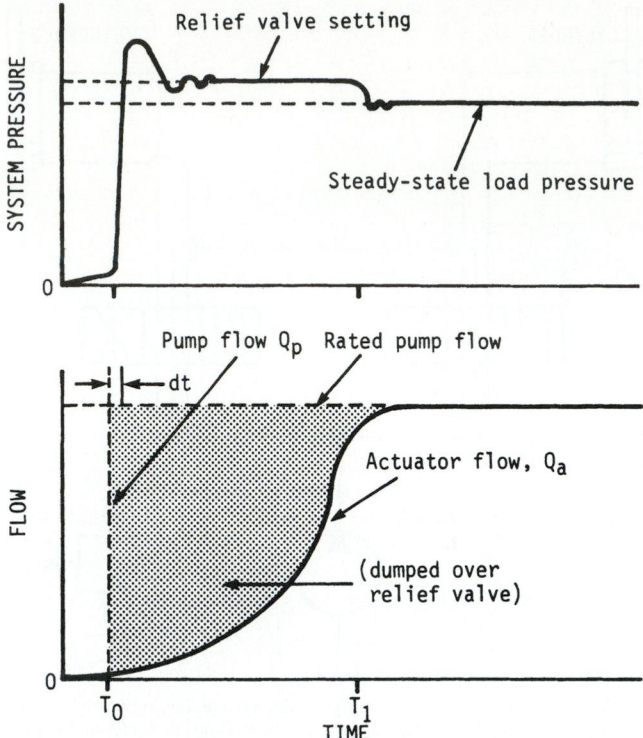

FIGURE 43.18   Should load inertia prevent acceleration to steady-state speed, excess flow from the pump returns to the tank through the relief valve. *(From Ref. [43.1].)*

At time t_0 the actuator velocity is zero. At time $t_0 + dt_0$ (the time that corresponds to full buildup of pump output), the actuator has not yet started to move. Therefore, actuator flow Q_a is zero at that instant. It could be demonstrated mathematically that for these conditions of finite pump output Q_p and zero actuator flow Q_a to coexist, the instantaneous acceleration of the actuator would have to be infinite. An infinitely powerful driving force would be required for this.

If we examine the pressure plot in Fig. 43.18*a* and compare it with the flow plot of Fig. 43.18*b*, we see that the fluid pressure rises rapidly and peaks at some level above the relief valve setting. This level depends on the response time of the relief valve; in addition, it depends on the internal slip in the pump, valve leakage, and actuator slip. Once the relief valve opens, fluid pressure in the system levels out at the relief valve setting.

Now consider the plot of the actuator flow rate Q_a. In any well-designed system consisting of one pump and one actuator, the pump output just matches the actuator input requirement at design speed. Thus, under steady-state conditions,

$$Q_p = Q_a \tag{43.5}$$

At the time dt_0, however, Q_p is equal to rated flow and Q_a is zero.

The actuator and load must accelerate from zero to design velocity. This takes a finite interval of time, from t_0 to t_1. During this time interval, Q_a increases until $Q_a = Q_p$ at time t_1, which is when the actuator reaches design speed. Note that at that time, the system pressure drops to the steady-state design level, and the relief valve closes. The shaded area between the two flow curves (Fig. 43.18b) represents the volume of oil returned to the tank through the relief valve during the acceleration period.

Because this complex sequence of events takes place in a fraction of a second, it is difficult to observe under normal operating conditions. And in most constant-flow circuit applications, it is not even a matter of concern. The designer would analyze this sequence only when dealing with applications that have high performance requirements or when an operating malfunction cannot otherwise be explained. Such a malfunction might occur if the pump's output flow rate and the actuator flow rate were badly matched.

For this reason, the designer must make sure that these two quantities are properly matched, especially when designing multibranch circuits. If a pump must be sized for multibranch circuit operation, as is frequently the case, the designer must choose a pump with a capacity that equals peak flow requirements. Note that the capacity of such a pump exceeds the fluid needs of a single actuator.

Sizing the Actuator. In constant-flow circuits, the designer tries to size actuators to meet speed requirements as a function of pump output. For example, a cylinder might be selected so that

$$\frac{A_p S}{t} = Q_p \qquad (43.6)$$

where
A_p = piston area
S = cylinder stroke
t = time
Q_p = pump flow rate

In some instances this formula may call for a cylinder with a capacity larger than that required for force output alone.

The designer would ordinarily select a fluid motor with a capacity (at desired operating speed) equal to the rated pump output.

$$Q_p = V_a N \qquad (43.7)$$

Unloading the Pump. In a constant-flow circuit, the directional control valve unloads the pump when the valve is in its neutral position. This is an advantage in that auxiliary controls are not required. By unloading the pump, the designer reduces unnecessary energy dissipation during dwell or passive intervals in the cycle, thus minimizing the generation of heat. Care must be taken that the directional control valve selected has enough capacity to bypass the *full* pump output *without* causing excessive pressure drop.

Output Speed Control. One way to control actuator speed in a constant-flow circuit is to restrict flow with a metering or flow control device. The most common metering approach uses one of the many types of flow control valves in combination with one of the basic methods of flow control described previously. Another approach takes advantage of the throttling characteristics of the directional control

valve. This approach is frequently adopted in circuits equipped with manually operated and proportional control valves. The designer must remember that any flow
control method that uses throttling creates energy losses with attendant heat generation.

Application Problem.   Let us consider a typical example of constant-flow analysis.
As previously discussed, a thorough analysis of system objectives is fundamental to
good circuit design. The cycle-profile technique was suggested as one approach to
orderly design. Remember that this approach divides the circuit into sections; see Fig.
43.19. Note that the load is primarily a resistive one. Therefore, under steady-state
conditions,

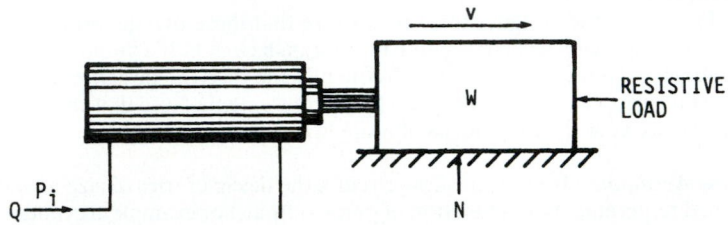

FIGURE 43.19   Example of a typical resistive load system. *(From Ref. [43.1].)*

$$F_a = ma = \frac{W}{g}\left(\frac{v_2 - v_1}{t}\right) \tag{43.8}$$

where F_a = force required to accelerate load
$\quad\quad t$ = time
$\quad\quad v_1$ = initial velocity
$\quad\quad v_2$ = final velocity
$\quad\quad W$ = weight of load, actuator elements, machine-tool carriage, etc.

Since at startup $v_1 = 0$, we have

$$F_a = \frac{Wv_2}{gt}$$

The equation

$$F_R = p_i A_p \tag{43.9}$$

states a relationship among the resistive force F_R required to overcome the resistive
load, the cylinder piston area A_p, and an initial system pressure p_i. However, the equation is not complete because we must consider several other factors: the frictional
component of the resistive load F_f, the breakaway friction F_{fb}, and the running friction F_{fr}.
   The frictional component of the resistive load is given by

$$F_f = \mu N$$

where μ = coefficient of friction and N = normal force. Since

$$N = W + L_N$$

where W = weight of the load, actuator elements, etc., and L_N = normal component of any applied force (such as cable tension or cutter reaction), we may write

$$F_f = \mu(W + L_N)$$

We have noted that breakaway and running friction also enter into the relationship. We must distinguish between these two quantities because the coefficient of friction varies between the static condition μ_s and the dynamic condition μ_d, so that $\mu_s > \mu_d$. The breakaway friction is

$$F_{fb} = \mu_s(W + L_N)$$

and the running friction is

$$F_{fr} = \mu_d(W + L_N)$$

Note that breakaway friction force F_{fb} may vary during the cycle because of a variable normal component force L_N. (There is another frequently neglected component of the total resistance energy requirement which, in some cases, cannot be overlooked—the energy needed to accelerate the mass of oil within the system. We do not discuss this component here, since it is beyond the scope of this presentation. See Ref. [43.1], Chap. 27.)

Thus the total resistive force at *breakaway* is

$$F_{Rb} = p_i A_p + \mu_s(W + L_N) \tag{43.10}$$

and the total resistive force at *running speed* is

$$F_{Rr} = p_i A_p + \mu_d(W + L_N) \tag{43.11}$$

We can now complete the equations for the total load reflected at the actuator. We distinguish three cases as follows:

1. The load at breakaway:

$$F_{ab} = \frac{Wv_2}{gt} + p_i A_p + \mu_s(W + L_N) \tag{43.12}$$

2. The load while running and while the system is accelerating to a constant speed:

$$F_{ar} = \frac{Wv_2}{gt} + p_i A_p + \mu_d(W + L_N) \tag{43.13}$$

3. The load while running at steady-state velocity:

$$F_{ar} = p_i A_p + \mu_d(W + L_N) \tag{43.14}$$

Figure 43.20 shows a typical load-cycle plot for the application. Note that $0 < t < dt$ represents the short interval during which breakaway from zero velocity takes place. This is a transient state, and it would be difficult to plot without use of an analytical instrument such as an oscilloscope.

Qualitatively, however, load components are functions of static friction and the resistive load itself. In the interval $dt < t < \Delta t$ (also of short duration), the load and actuator masses are accelerated. Again, this is a transient state. The components are dynamic friction, resistive load, and the load due to the acceleration of a mass.

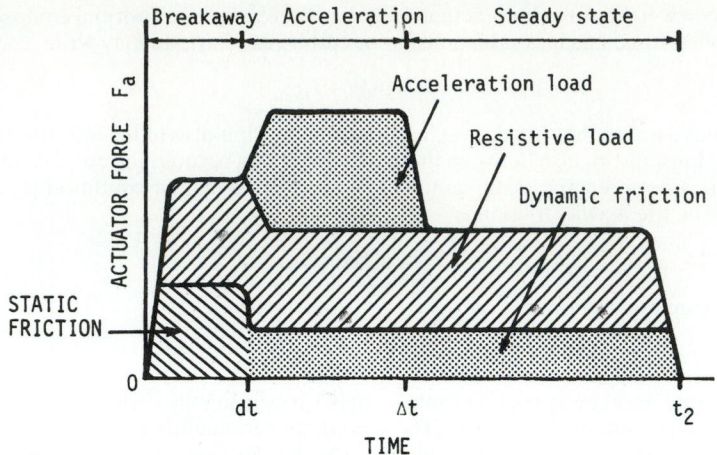

FIGURE 43.20 Typical load-cycle plot for the system illustrated in Fig. 43.19. Note the nonlinear time scale of the abscissa. *(From Ref. [43.1].)*

Beyond $\Delta t < t$, acceleration of the load is essentially a steady-state quantity; at least we usually assume this, even if it is not quite true in actual practice. In this time interval, the important components are resistive load and dynamic friction.

Figure 43.21 shows the shape of a typical system pressure plot determined from the load-cycle plot shown in Fig. 43.20. Note that the highest pressures appear when $0 < dt_1 < \Delta t$. These are the familiar transients, frequently seen on oscilloscopes, caused by the breakaway phenomenon and the superimposition of acceleration forces on normal load resistance. Figure 43.21 also shows that a steady-state pressure p_{ss} is achieved when the load speed corresponds to the actuator design speed.

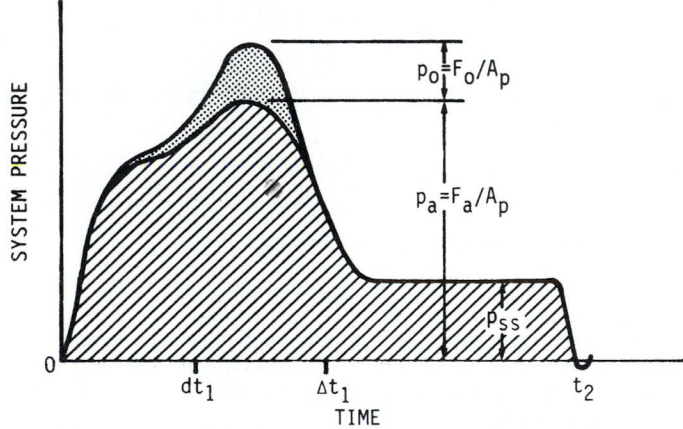

FIGURE 43.21 Shape of typical pressure plot determined from the load-cycle plot shown in Fig. 43.20. Note that the highest pressures appear in the interval $0 < dt_1 < \Delta t_1$. *(From Ref. [43.1].)*

The relief valve setting at this point is ordinarily between p_{ss} and $p_a + p_o$. If Δt_1 is brief in comparison with t_2, the relief valve setting can be close to p_{ss}, because the pressure transient will be so short that the relief valve cannot respond—or even if it could, the relatively small quantity of oil bypassed would not affect circuit operation significantly. If, however, Δt_1 is large in comparison with t_2, the relief valve will have to be set higher.

When interpreting such pressure plots, you should bear in mind that the system will develop a fluid pressure p_o caused by the acceleration of the oil column in the line that connects the pump and the actuator. This pressure is superimposed on the other pressures reflected by the load at the actuator; this relationship develops because p_o is *not* load-reflective and occurs only in the oil in the line to the actuator.

Characteristics of Constant Flow.   In light of this discussion, remember these characteristics of constant-flow circuits:

1. The pump discharge pressure is a function of the load resistance and must build from zero.

2. The pump output is not determined by the actuator speed requirements.

3. Actuators are sized to meet speed requirements as a function of the pump output.

Statements 2 and 3 make sense only if we assume that the pump in a multibranch circuit is sized to accommodate maximum system demand, no matter when or where it occurs. During parts of the cycle when demand is not at peak, the pump output will exceed that required by one actuator. In the simple example of Fig. 43.19, the pump and actuator displacements would have to be matched. Let us analyze the basic circuit of Fig. 43.22 required for the application.

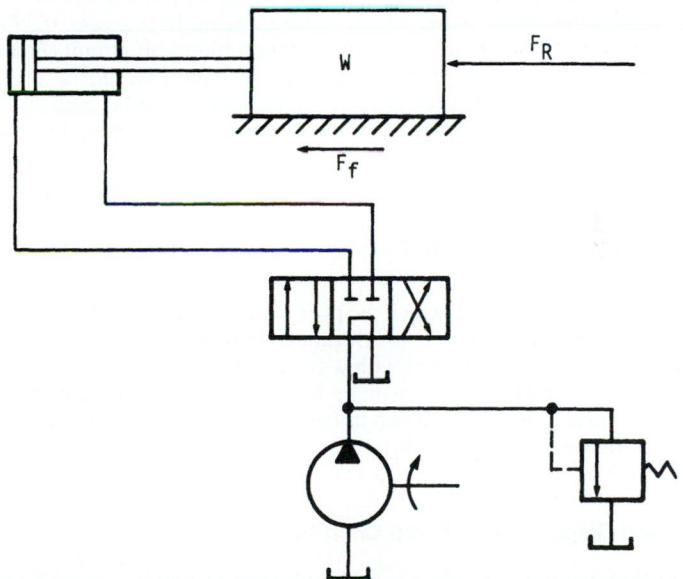

FIGURE 43.22   Basic circuit for application illustrated in Fig. 43.19. *(From Ref. [43.1].)*

Assume that the load resistance of the circuit calls for a piston area A_p for a design pressure p_i and that the load must be moved through a stroke S. The cylinder will displace a volume

$$V_a = A_p S \qquad (43.15)$$

If the job to be done requires that the load be moved in t seconds, then the necessary flow rate to the cylinder is

$$\frac{V_d}{t} = \frac{A_p S}{t} = Q \qquad (43.16)$$

where Q is in cubic inches per second.

(At this point in the analysis, the designer must check the columnar strength of the piston rod, which may turn out to be the critical factor. If a larger rod is needed, the cylinder bore may have to be increased accordingly. Such a change would, in turn, require adjustment of the pump-displacement calculation.)

Assuming that we have satisfactorily calculated the required pump output Q_p, we can complete the input segment of the circuit we are using as an example; see Fig. 43.22. By the very nature of this circuit, we must use the tandem-center four-way valve shown. Also, a constant-flow circuit requires a relief valve with a fixed-displacement pump.

The main functional element still missing from the simple circuit is a method for speed control. Since it was indicated that speed control could be accomplished only by throttling, the manually operated directional control or proportional valve can be used to throttle flow. In all circuits, the pressure drop across the valve has the effect of reducing the pressure available at the actuator. Consequently, the force available to accelerate the load and overcome friction is reduced.

A flow control valve used in a meter-in circuit (Fig. 43.23) would have essentially the same effect, unless it were used in a bleed-off circuit. In this case, the flow to the cylinder would actually be reduced. One could use a bleed-off circuit (Fig. 43.24) for minor speed adjustment, but a valve in a meter-in circuit would be preferred for adjustment over a wide flow range. A meter-out circuit could be used, but since the load in this example is a resistive one, this alternative would have little advantage over a meter-in circuit.

43.8   DEMAND-FLOW CIRCUITS

A closed-center circuit is one in which the port from the pump to the directional control valve is blocked when the valve is in its neutral position (see Fig. 43.25). Typically, demand-flow circuits are equipped with a fixed-displacement pump, an unloading valve, and an accumulator, as shown in Fig. 43.26, or a variable-displacement, pressure-compensated pump, as shown in Fig. 43.27. Closed-center circuits are more accurately characterized as demand-flow circuits.

43.8.1   Fixed-Displacement Pump Circuits

In demand-flow circuits that use a fixed-displacement pump, unloading valve, and accumulator, fluid pressure from the pump is not directly determined by actuator force requirements. As Fig. 43.26 illustrates, the pump charges the accumulator to design pressure when the directional control valve is centered.

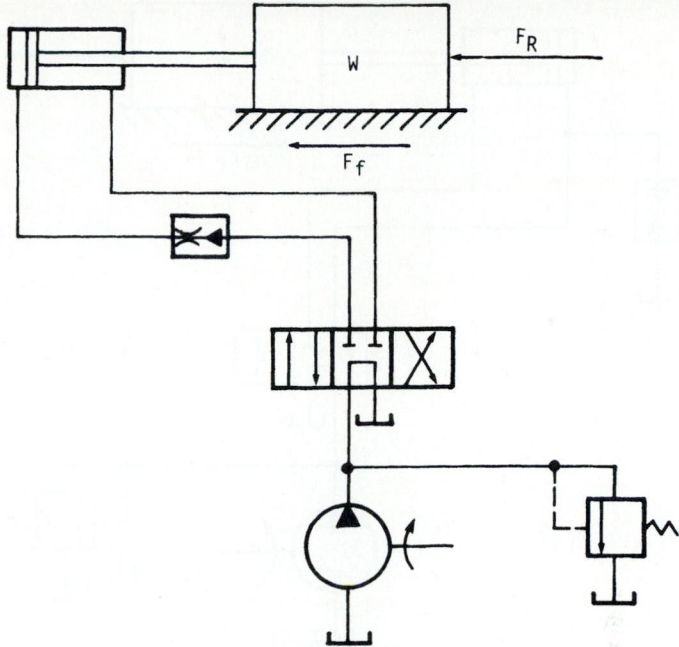

FIGURE 43.23   Flow control valve in meter-in circuit provides speed control for circuit in Fig. 43.22. *(From Ref. [43.1].)*

Maximum design pressure in the circuits is controlled by the spring setting of an unloading valve. When this setting is reached, the valve opens and bypasses oil to the tank, at low pressure. Note that the pilot signal to the relief valve is sensed downstream of a check valve placed between the pump and the accumulator. The check valve prevents the unloading of the accumulator as well as the pump.

When the directional control valve is shifted so that it ports oil to the actuator, the full design pressure (as stored in the accumulator) is immediately available to the system. As the cylinder moves, oil is forced from the accumulator by the compressed gas charge behind the oil. After a time interval, system pressure drops because of the expansion of the gas charge in the accumulator.

At some pressure level for which it has been designed, the unloading valve closes and causes output from the pump to reenter the system rather than bypass to the tank. At this time, the pump will do one of two things:

1. Add its output to that from the accumulator at the lower pressure level.

2. Recharge the accumulator to a higher pressure.

Which event occurs is a function of many other factors.

Some accumulator circuits are designed so that the accumulator supplies all the oil used during the active part of the cycle. It cannot do so at constant pressure, because the pressure of the gas charge drops as the gas expands when the oil flows out of the accumulator. The load cycle must be designed so that the system can still function at

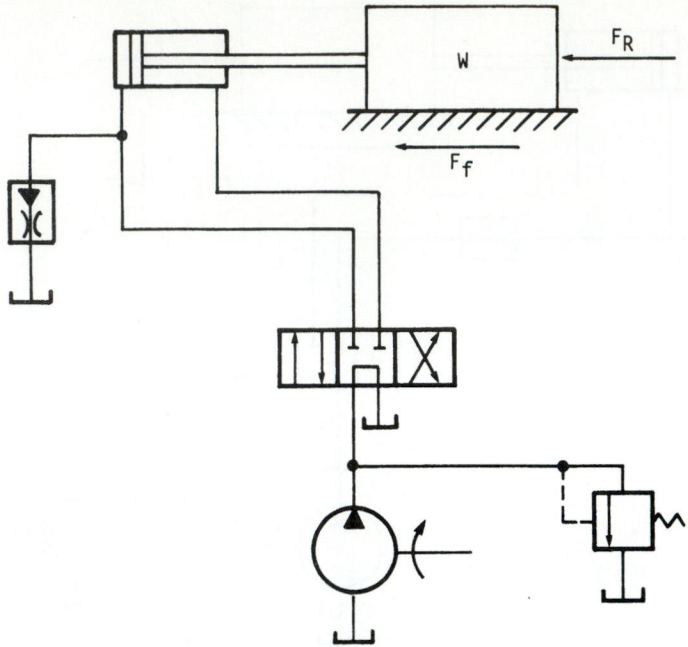

FIGURE 43.24 Bleed-off circuit could be used to control flow to cap end of cylinder, but meter-in circuit is preferred for fine adjustments. *(From Ref. [43.1].)*

the lowest pressure level delivered by the accumulator. This design feature is used where the active, or work, segment of the cycle is rather short and is followed by a relatively long passive, or dwell, segment during which the pump recharges the accumulator. In such circuits, the pump is sized to charge the accumulator during the work-cycle dwell segment; see Fig. 43.28.

Example 1. Assume a circuit similar to that in Fig. 43.26, in which an accumulator supplies 924 cubic inches (in³) of oil to the circuit in 10 seconds (s). What is the required pump output rate if the dwell time between work periods is 50 s?

Solution. The required pump discharge, in gallons per minute, is

$$Q = \frac{924}{50} \, (\text{in}^3/\text{s}) = \frac{18.5(60)}{231} \, (\text{gpm}) = 4.8 \text{ gpm}$$

The horsepower required to drive the pump is

$$H_1 = \frac{pQ}{1714} = \frac{1000(4.8)}{1714} = 2.8 \text{ hp}$$

In circuits that differ from the one illustrated in this example, the accumulator is frequently used to supplement the pump during brief periods when high-rate flows are needed. Thus, if the design calls for a high flow rate for a short time in the active part of the work cycle, the engineer can use a smaller pump in conjunction with an

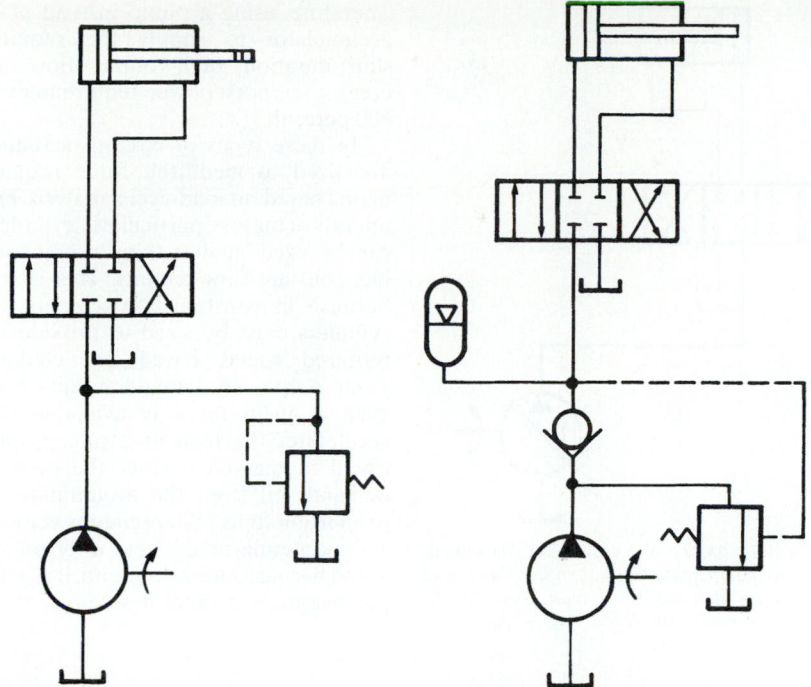

FIGURE 43.25 In a simple demand-flow circuit, the line from the fixed-displacement pump to the valve is blocked when the directional control valve is in the neutral position. *(From Ref. [43.1].)*

FIGURE 43.26 In a simple demand-flow circuit powered by a fixed-displacement pump, an accumulator is added to supply full design pressure immediately. *(From Ref. [43.1].)*

accumulator which it charges during the passive part of the work cycle. When the operator shifts the directional control valve, the accumulator output flow is *added* to the pump flow. Note that the combined flows may exceed several times the output of the pump alone. However, this condition will exist for only a very short time. In designs where a peak flow of short duration may be desirable, this configuration may be much more economical than one that relies on one large pump; see Fig. 43.29.

Example 2. Assume that a pump is used instead of an accumulator to supply the required oil in Example 1. If the operating pressure is 1000 psi, what is the difference in horsepower required to drive the pump in these two examples?

Solution. The total flow to the system is

$$Q_t = Q_p + Q_a = 4.8 + \frac{924}{10}\frac{60}{231} = 28.8 \text{ gpm}$$

The power required to drive the pump in Example 1 was 2.8 hp. In this example, the power required is

$$H_2 = \frac{pQ}{1714} = \frac{1000(28.8)}{1714} = 16.8 \text{ hp}$$

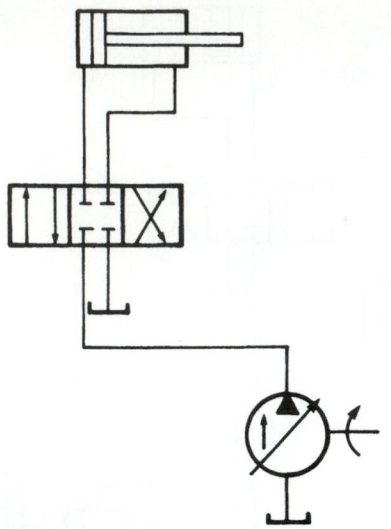

FIGURE 43.27 A variable-displacement, pressure-compensated pump supplies pressure fluid to this demand-flow circuit. Note the absence of a relief valve. *(From Ref. [43.1].)*

Therefore, using a pump instead of an accumulator to supply the required short-duration, high-volume flows increases the horsepower requirement by 600 percent.

In these types of circuits, actuators are sized to meet the force requirements based on load-cycle analysis. Frequently actuators, particularly cylinders, can be sized smaller than in comparable, constant-flow circuits. This is true because in constant-flow circuits, the cylinders must be sized to provide the required speed based on available pump output. In demand circuits, however, a given force is available that accelerates the load at a rate proportional to the mass. Thus, the cylinder demands oil from the accumulator in proportion to its instantaneous velocity. The accumulator delivers only on demand because, unlike a pump, it is not a positive-displacement device.

43.8.2 Pressure-Compensated Pumps

We discussed the simplest form of pressure-compensated pump in a demand-flow circuit in Fig. 43.27. If the demand-flow circuit uses a pressure-compensated pump

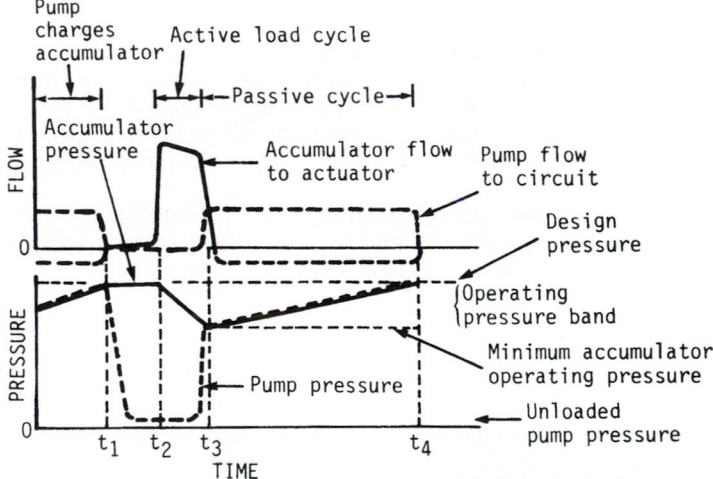

FIGURE 43.28 In systems with cycles that have short work segments and long dwell segments, the pump is sized to charge the accumulator during dwell. *(From Ref. [43.1].)*

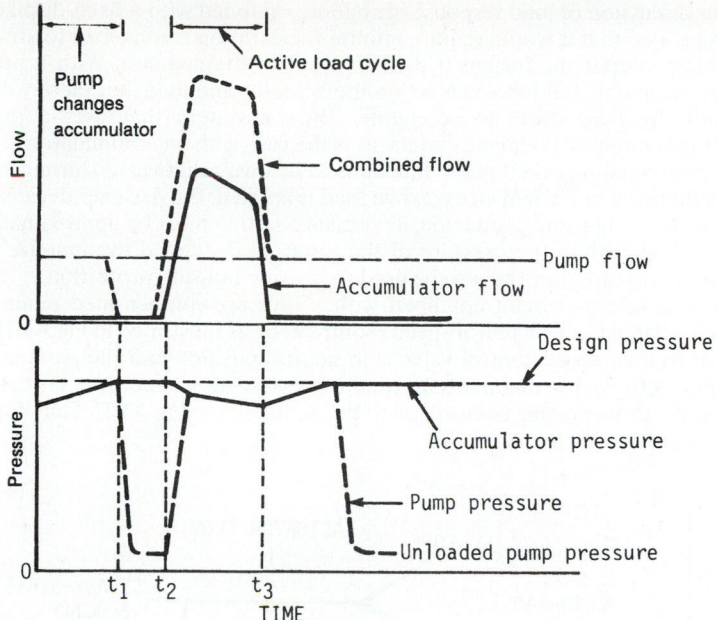

FIGURE 43.29 In systems where short-duration peak flows are needed, an accumulator can often supplement pump output during short periods of high-flow needs. *(From Ref. [43.1].)*

(Ref. [43.1], Chap. 15, pp. 116–122), then the compensator setting determines the maximum circuit pressure; see Fig. 43.30. Pump output is constant until the system reaches a given pressure, called the *cutoff pressure*. At this point the force acting on the compensator begins to exceed the force of the control spring that holds the pump on stroke.

Now, as pressure increases, the pump starts to move off stroke to reduce displacement. The slope of the curve of this decreasing displacement can be controlled by the design of the compensator. Thus, the designer can specify a sharp or gradual cutoff, as required. When fluid pressure in the system reaches the level known as *deadhead pressure,* the pump output flow is zero. The only power consumed by the pump at deadhead is the relatively small amount required to overcome mechanical losses and compensate for internal leakage. The pump maintains full deadhead pressure in the system at this low power input.

As indicated, the compensator setting determines the upper limit of system pressure. Up to its maximum capability, pump output is a function of the ability of the actuator and load to respond to the force exerted on both.

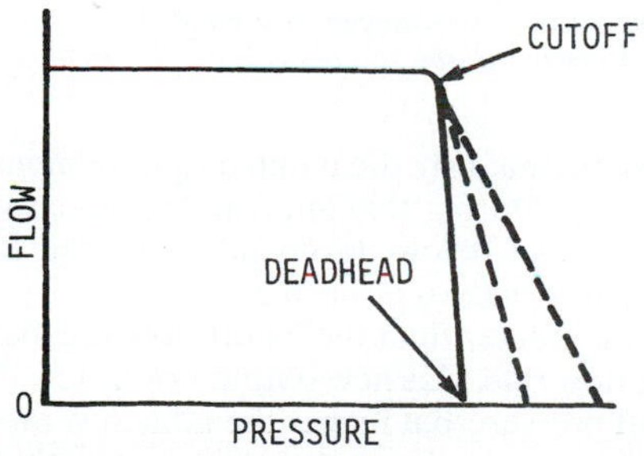

FIGURE 43.30 In a demand-flow circuit equipped with a pressure-compensated, variable-displacement pump, the compensator setting determines maximum circuit pressure. *(From Ref. [43.1].)*

In our discussion of load response in systems equipped with a fixed-displacement pump, we stated that it would require infinite acceleration for the load to absorb the entire pump output the instant it delivers fluid to the actuator. With a pressure-compensated pump, full force can act on the actuator and load, but there will be no flow until the load starts to accelerate. Thus, a system that uses a pressure-compensated pump is a demand system, as is the case with an accumulator.

A pressure-compensated pump functions as its own relief valve, shifting to dead-head conditions if and when an excessive load is applied. If a designer decides to use a relief valve for fail-safe protection, its pressure setting must be approximately 250 to 300 psi higher than the pressure of the pump at deadhead to minimize system instability. A rupture disk can also be used to provide fail-safe protection.

In a demand-flow circuit equipped with a pressure-compensated pump, pump delivery is related to the actuator speed requirement, as illustrated in Fig. 43.31. From time zero to time t_0, the control valve is in neutral position and the pump at dead-head—that is, the pump maintains maximum pressure at zero delivery. At time t_0 the control valve shifts, porting pressure oil to the actuator; see Fig. 43.27. Thus, full deadhead pressure acts on the actuator.

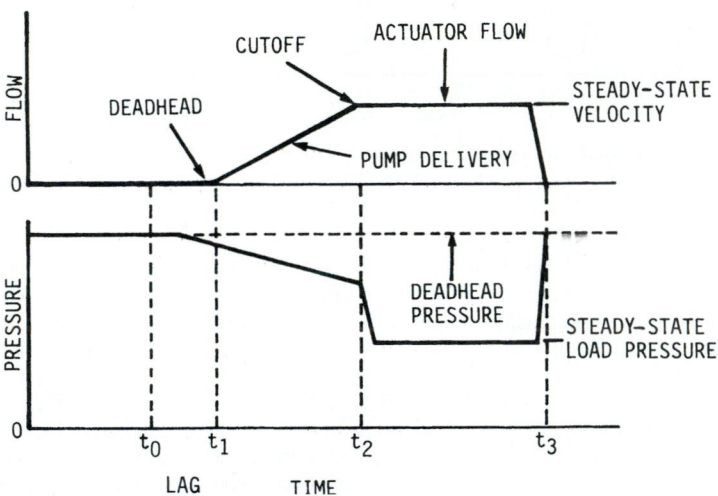

FIGURE 43.31 In a demand-flow circuit with a pressure-compensated pump, the pump output is related to actuator speed requirements. *(From Ref. [43.1].)*

Because the actuator cannot accelerate instantaneously, the pump output remains at zero for a short time $t_0 < t < t_1$; see Fig. 43.31. During this interval, the actuator begins to move. The pressure drops to some level below deadhead—required to accelerate the load level. Simultaneously, the pump moves on stroke.

If the acceleration force requires a pressure greater than the cutoff pressure, the pump will compensate by reducing its output flow rate. This new output flow rate will be lower than that corresponding to the cutoff pressure, but higher than the flow rate corresponding to deadhead pressure. In this sense, a demand-flow circuit with a pressure-compensated pump is a self-regulating system.

Between times t_1 and t_2 in Fig. 43.28, the load accelerates to steady-state speed. By time t_2, the pump has been stroked to full displacement, the load stops accelerating, and the system pressure drops to some value corresponding to the steady-state resistive load. At time t_3, the actuator hits a mechanical stop, or the end of its stroke, and the fluid pressure rises immediately. The pump is destroked, and its output drops to zero; it is at deadhead until the control valve shifts to retract the cylinder. In this circuit no pressurized oil flows over a relief valve; the pump supplies precisely what the system demands.

43.8.3 Flow-Compensated Pumps

The simplest form of a flow-compensated pump is shown schematically in Fig. 43.32a. In these pumps, a control orifice senses the flow rate, with the pressure drop across the orifice being proportional to flow rate (Ref. [43.3], pp. 75ff) according to the equation

$$Q = C_d A_o \sqrt{\frac{2g\,\Delta p}{\gamma}} \qquad (43.17)$$

where C_d = discharge coefficient
A_o = cross-sectional area of orifice opening
g = acceleration due to gravity
Δp = pressure differential
γ = specific weight of fluid

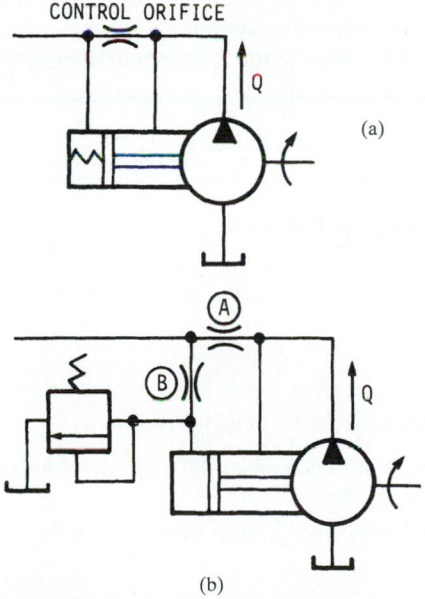

FIGURE 43.32 (a) Schematic of simple flow-compensated pump. (b) Diagram of flow- and pressure-compensating control. Fixed orifice A senses flow; fixed orifice B is in line to compensator. (*From Ref. [43.1].*)

This equation indicates that pressure drop is a function of the square of the flow. The induced pressure drop is felt by the compensator control piston, which adjusts pump output in proportion to flow.

Figure 43.32*b* illustrates a flow rate and pressure-compensating control. This configuration also uses a fixed orifice *A* to sense the flow rate. In addition, it has a second fixed orifice *B* in the line to the spring end of the compensator. A pressure control valve regulates the pressure in the spring-chamber end of the compensator. When pressure in this chamber matches that of the valve setting, the valve opens and bypasses oil to the tank, creating a pressure drop across orifice *B*. Thus the total pressure differential imposed across the compensator piston is the sum of the two pressure drops. This value will exceed the pressure drop induced across orifice *A* by flow alone.

43.9   HYDRAULIC VERSUS PNEUMATIC SYSTEMS

The material discussed previously dealt with fluid power systems, that is, energy transmission systems using a fluid as the transfer medium. The technology covered has been *hydraulics* based on incompressible-fluid liquid transfer media. *Pneumatics* is the second area of fluid power technology in which a compressible fluid—gas—is used as the transfer medium.

Functionally, hydraulics and pneumatics are similar. What has been said about functional design of hydraulic systems is applicable for pneumatic systems as well. The major differences lie in the areas of

1. Hardware used to implement the functions
2. Energy levels usually involved in the applications of each
3. The vastly different characteristics of liquids and gases

Additional information is contained in Ref. [43.1].

43.10   PNEUMATIC CIRCUITS

Pneumatic systems can be classified in two primary functional categories:

1. *Power* systems
2. *Logic* systems

Pneumatic power systems can be further classified as

1. Sequencing circuits, which encompass the majority of industrial applications and are usually open-loop systems
2. *Servo* systems, which are closed-loop, proportional systems

43.11   EFFECT OF FLUID CHARACTERISTICS ON ACTUATOR PERFORMANCE

The performance of the fluid in a hydraulic or pneumatic system is related primarily to compressibility. Figure 43.33 illustrates this point.

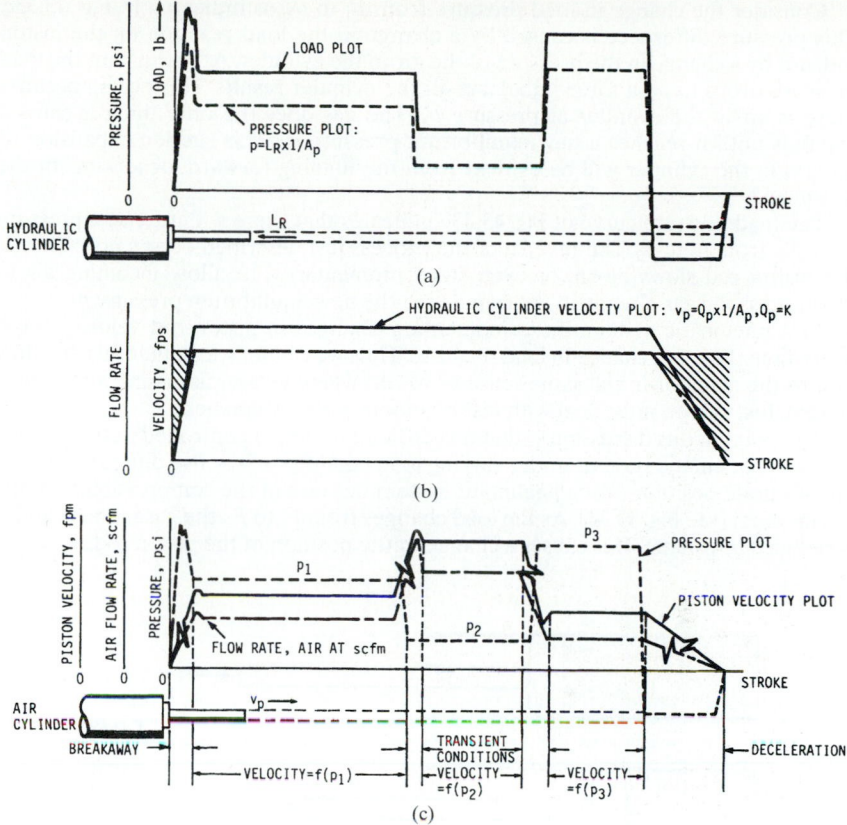

FIGURE 43.33 Performance comparison of hydraulic and pneumatic cylinders.

The load-cycle plot of Fig. 43.33a is identical for both the hydraulic and the pneumatic power system. The pressure plot of Fig. 43.33a is determined from the load plot. If the hydraulic cylinder is driven by a fixed-displacement pump, then the volumetric flow rate to the actuator is constant. This is represented by the horizontal flow curve in Fig. 43.33b. A constant input to the cylinder results in a constant output velocity (of the piston rod). Because of the relative incompressibility of the hydraulic fluid, fairly accurate velocity control is possible.

It is very difficult to determine what the air flow rate to the cylinder really is. For instance, with an initial load pressure of p_1, there is a corresponding initial flow rate of air. The density of the air is a function of the pressure p_1 and the temperature T_1. The source of the compressed air is the central compressor station; the air is delivered through a pressure regulator, which is a throttling device. Since the compressor is assumed to be capable of delivering an unlimited quantity of air to the cylinder, the factors which limit flow rate to the cylinder include load, resistance of the connecting pipes to flow, valve orifices, the regulator, etc. It is important to realize that *all* these factors contribute pressure drops of one sort or another, and that every change in pressure brings about a corresponding change in the volume of the gas.

Consider the change in load pressure from p_1 to p_2, as indicated in Fig. 43.33c. This pressure difference is caused by a change in the load reaction on the piston rod, not by a change in the pressure of the air in the cylinder. At the instant the load pressure drops to p_2, a force imbalance in the cylinder results. This occurs because there is air in the cylinder at pressure p_1. The gas does the only thing it can—it expands until it reaches a new equilibrium pressure p_2. This sudden expansion of the gas in the cylinder will be evident from the lunging forward, or jerking, of the piston rod.

The load pressure curve of Fig. 43.33c indicates that there will next be a pressure increase from p_2 to p_3. The reverse of the process just described occurs now. That is, the piston rod slows down, or even stops momentarily, to allow incoming air to recompress the gas already in the cylinder to the new equilibrium pressure p_3.

In a pneumatic system, these momentary changes in piston-rod velocity occur every time there is a change in load pressure. Thus we cannot speak in terms of a flow rate to the actuator in the same sense as we do when we are discussing a hydraulic system. Instead, we must deal with instantaneous piston velocities.

The point in this discussion is that it is difficult to obtain controlled output velocities with pneumatic systems under varying load conditions. It is also difficult to maintain accurate position with a pneumatic system because of the compressibility of the gas medium (see Fig. 43.34). As the load changes from F_1 to F_2, the gas in the cylinder is reduced in volume. This causes a change in the position of the piston rod.

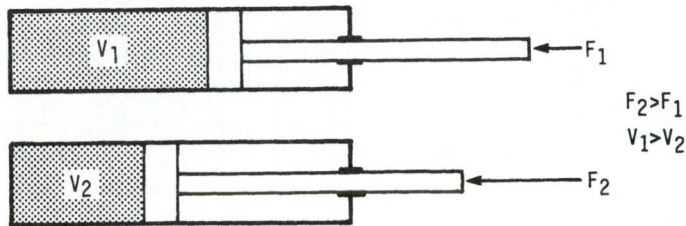

FIGURE 43.34   Explanation of why position control is difficult with pneumatic systems.

It is easy to deduce why hydraulic systems have taken the lead over pneumatic systems in those applications requiring accurate control of position or velocity. Up to the present, pneumatic systems have been used mainly for sequential types of circuits where the end conditions are those of prime importance, i.e., circuits in which the important thing is whether the rod is fully extended or fully retracted. Transfer, clamping, and press circuits are typical of such applications. That the relative importance of pneumatic systems is changing will become apparent in later discussions of logic-circuit design.

Where the economics of the situation dictate the use of a pneumatic power system, yet control requirements are greater than those attainable with a purely pneumatic system, an air-oil system might be used. In air-oil systems (Ref. [43.5], Part II, pp. 149–200), compressed air provides the source of potential energy, and hydraulic oil provides the incompressible-fluid characteristics necessary to achieve the desired degree of control. Figure 43.35 illustrates the use of a tandem air-oil cylinder for such a purpose.

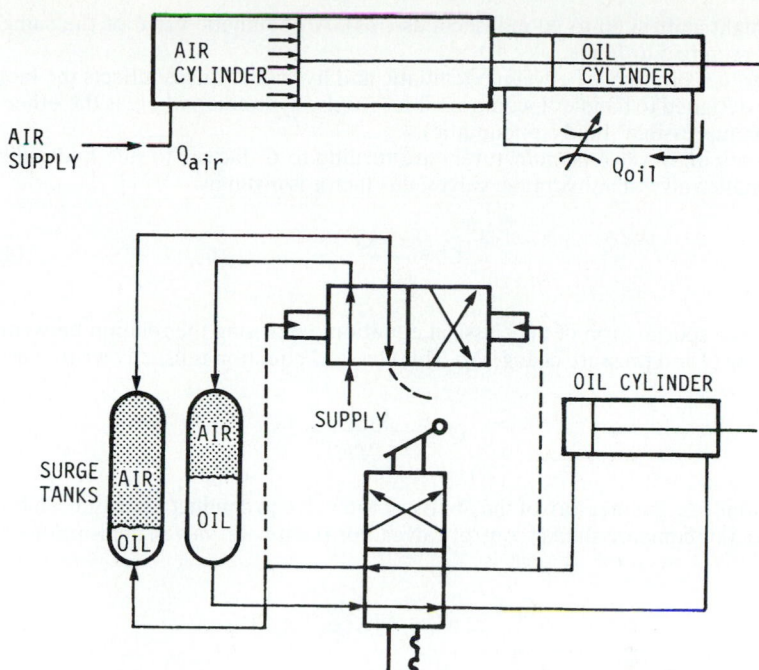

FIGURE 43.35   Example of an air-oil system.

43.12   EFFECT OF FLUID CHARACTERISTICS ON CONTROL-VALVE PERFORMANCE

The compressibility of the gaseous medium makes performance prediction for pneumatic valves more difficult than for their hydraulic counterparts. Pneumatic direction control valves are very similar in function to their hydraulic counterparts. Thus the functional designations—two-way, three-way, four-way, etc.—are applicable to both.

A major difference in the design and construction of pneumatic and hydraulic components reflects the vast difference between them in pressure level. Hydraulic valves, which must operate at pressures from 1000 up to 10 000 psi, are made from heavy castings or bar stock. On the other hand, pneumatic valves seldom encounter pressures over 150 psi and can be die-cast or otherwise fabricated from aluminum, brass, or even zinc alloys.

The types and functions of the valve operators are the same as those previously indicated for hydraulic valves. That is, pneumatic power valves can be operated by solenoids; by pilot controls; by hand, foot, cam, or palm button manual operators, etc. The major difference between hydraulic and pneumatic valve operators reflects the lower operating pressures of the latter. Since lower forces are encountered, the operators are usually smaller.

Pneumatic power valves have shorter response times than hydraulic valves. For example, a pneumatic valve of a given size will probably shift from 3 to 4 times as rapidly as its hydraulic counterpart. A solenoid-operated hydraulic valve of a given size might shift in 30 to 40 milliseconds (ms). A pneumatic valve of the same size might require 5 to 10 ms.

Another difference between pneumatic and hydraulic valves reflects the fact that one is designed to handle "incompressible" fluids (hydraulic), whereas the other handles "compressible" fluids (pneumatic).

Increasingly, valve manufacturers are turning to C_v factors to rate hydraulic and pneumatic valves. For hydraulic valves, this factor is given by

$$C_v = \frac{Q}{(\Delta p / S_G)^{1/2}} \tag{43.18}$$

which is a special form of the classical equation expressing the relation between orifice flow Q and pressure change Δp. This classical equation is usually written as

$$Q = \frac{C_d A_0}{(2g\,\Delta p/\gamma)^{1/2}} \tag{43.19}$$

Essentially C_v is a measure of the ability of the valve to conduct fluid and can be used to select or compare similar control valves. For pneumatic valves, the formula is

$$C_v = \frac{Q}{22.67\{S_G T/[K(p_1 - p_2)]\}^{1/2}} \tag{43.20}$$

where S_G = specific gravity (1 for air)
 T = absolute temperature (460+°F)
 p_1 = absolute inlet pressure (psig + 14.7)
 p_2 = absolute outlet pressure (psig + 14.7)
 K = constant K_1, K_2, or K_3
 K_1 = p_2 when Δp is no more than $0.1p_1$
 K_2 = p_1 when $\Delta p \geq 0.25p_1$
 K_3 = $(Cp_1 + p_2)/2$ when $0.1p_1 < \Delta p < 0.25p_1$
 Q = flow rate in standard cubic feet per minute

The flow coefficient C_v indicates the ability of a valve to conduct a compressible fluid (gas); the pressure p_2 must be greater than $0.53p_1$ to assure subsonic flow through the valve orifice. If Δp across the orifice is greater, flow becomes supersonic and the equation is invalid.

43.13 BASIC PNEUMATIC POWER CIRCUIT

The basic direction control circuit for a single-acting cylinder, shown in Fig. 43.36, illustrates some of the differences between such a circuit and its hydraulic counterpart. Note the absence of an input device, or pump. Most pneumatic circuits use main plant compressors as their source of energy. To provide the input to the circuit, one simply "hooks into" the air manifold at a convenient location. This does not, however, preclude the use of an individual air compressor located at the machine. Conversely, for most hydraulic systems, the use of individual power units does not preclude the use of a central system.

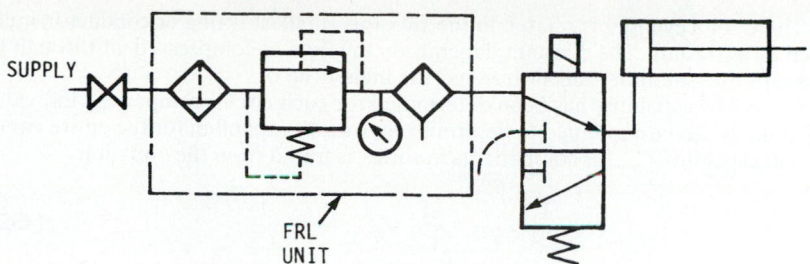

FIGURE 43.36 A pneumatic direction control circuit.

A peculiarity of pneumatic circuits is the use of a *filter-regulator-lubricator* (FRL) unit at the source (the point where the air manifold is tapped). This unit provides clean air at regulated pressure and adds enough lubricant to the air to minimize wear of component parts.

Because of the complexity of handling thermodynamic considerations associated with compressible-fluid flow, much of industrial pneumatics technology has been reduced to empiricism (see Ref. [43.6]).

In theory, all the circuits should work equally well. However, when objectives such as low operating cost, circuit simplicity, energy efficiency, and high productivity are added to the analysis, only one or two of the circuits will actually be feasible.

This winnowing out of impractical designs can be a long and tedious procedure with many pitfalls. A more efficient design procedure (adaptable to the computer) has been developed that focuses on the optimum circuit quickly. Based on experience with the computer technique, a number of guidelines have been developed that help shorten the time from first proposal to final design.

43.13.1 The Basics

The optimization procedure is based on the flow capability (or conductance) of pneumatic devices. Flow capability, the reciprocal of flow resistance, provides a convenient means of evaluating components to optimize operating pressures, valve and line sizes, compressor loads, etc.

Capability is expressed as C_v, a dimensionless number. The National Fluid Power Association (NFPA) and the International Standards Organization (ISO) have prescribed carefully controlled tests for determining C_v values for valves and other fixed-orifice devices.

In general, a C_v value cannot be assigned to a valve; it must be related to a port size—specifically, to the inner diameter (ID) of the smallest conductor or fitting entering the port. For instance, a direct solenoid-actuated valve has a C_v of 3.5 with a ½ NPT port and 3.2 with a ⅜ NPT port.

For conductors, C_v is related to internal diameter, length, and friction factor. Typically, a friction factor of 0.02 is used for smooth-wall conductors and of 0.03 for rough-wall conductors. Ideally, a Reynolds number should be computed to determine the friction factor, but because relatively small conductors are used with compressed air, 0.02 and 0.03 are reasonable values regardless of diameter and air velocity.

Fittings can be converted to equivalent conductor lengths, and the C_v can be calculated in the same manner as for conductors. Some manufacturers of quick connections and fittings publish C_v values for their products.

The C_v of a cylinder is related to the ID of its smallest fitting or conductor, multiplied by a constant. The constant depends on the flow of compressed air through the end caps. The smoother the air passage, the higher the C_v.

Once flow capability has been determined for each circuit component, individual capabilities can be combined to determine an overall capability for the entire circuit. Circuit capability C_{vs}, for components in series, is found from the equation

$$\frac{1}{C_{vs}^2} = \frac{1}{C_{v1}^2} + \frac{1}{C_{v2}^2} + \frac{1}{C_{v3}^2} + \cdots + \frac{1}{C_{vn}^2} \qquad (43.21)$$

For components in parallel, individual flow capabilities are simply added.

Maximum flow possible through a pneumatic circuit is limited by the size of the orifices in the system. The succession of valves, lines, and cylinders obstructs flow, with the restriction being inversely proportional to flow capability. Thus the total system capability is always less than individual component capabilities. With Eq. (43.21) and the instructions for parallel devices, it is not difficult to determine the total flow or system capability.

43.13.2 Analyzing a System

The procedure for determining and combining the flow capabilities of any fixed-orifice pneumatic device provides the means for predicting the response time of an actuator. Furthermore, air usage, productivity, valve size, or a combination of several objectives can be optimized easily by optimizing system C_{vs}.

In a typical analysis, several potential system designs are generated, depending on the design objective. With this information, compromises and tradeoffs can be made until the system is fine-tuned to meet specific needs.

The analysis assesses valve sizes, cylinder sizes, fittings, piping, and operating pressures to minimize energy usage. This analysis is based on several considerations that remain fixed, regardless of design objectives:

- Conductor lengths should be as short as possible.
- Conductor paths from valve to cylinder should be as straight as possible and have as few fittings as possible. A machine may look better cosmetically if the conductors follow a natural contour of the machine, but such piping wastes air.
- Cylinder bore sizes should be selected to handle the expected load plus a reasonable safety factor.
- Cylinder stroke length should be no more than required. A longer stroke than necessary wastes energy.
- Air valves can be oversized without wasting energy.
- Overpressurizing a circuit beyond a certain point does not increase cylinder speed but does waste air.
- If the application calls for two different loads or times for the extension and retraction portions of the cycle, two different pressures should be used.
- Increasing conductor diameter increases C_v but also increases the volume that must be filled and evacuated each cycle. Therefore, each application has an optimum conductor diameter.
- Taking all these points into consideration, required cylinder bore size for the given load conditions and pressures is calculated from the equation

$$D = 1.13 \left(\frac{LS_f}{P_d} \right)^{1/2}$$

where L = cylinder load
         S_f = factor of safety
         P_d = design pressure

The highest load acting on the cylinder is used in this calculation, and the safety factor can range from 15 to 50 percent, depending on the service. Also, the design pressure used in this calculation is 80 percent of the minimum available pressure. The actual cylinder bore specified is the next larger standard bore above that calculated.

Once the required bore size is determined, cylinder performance graphs are used to specify target C_v's for both extension and retraction strokes to meet the load and time requirements.

The required system flow capability can be approximated from cylinder performance graphs such as those shown in Fig. 43.37. This is an important step in the analysis because it provides a check on subsequent steps.

To establish target C_v's for the extension and retraction strokes, plot the intersection of lines from the required stroke time scale and read the C_v values. If the flow capabilities calculated at the end of the analysis are lower than the target C_v's, the cylinder will not extend or retract in the required time, and a larger value of conductor must be selected. If the calculated capacities are higher than the target C_v's, the system will operate satisfactorily.

Determining Conductor Flow Capability. [*]   The conductor size equal to the smallest port size of the selected valve is used to determine the conductor C_v from piping performance graphs, as in Figs. 43.38 and 43.39. The flow capability of pneumatic pipe, tube, and hose depends on the inside diameter and length. Capability can be determined from the two graphs shown in the figures by drawing a horizontal line from the length scale to the conductor size plot and then drawing a vertical line to the capability scale. These graphs can also be used for fittings by converting the fittings to an equivalent conductor length.

The system flow capability is calculated by combining all the component capabilities including the cylinder C_v, which is

$$C_{vc} = 18d^2 \tag{43.23}$$

If the calculated system flow capability is larger than both target C_v's, the system is properly designed. However, if the calculated capability is lower than either target C_v, then a valve with a higher C_v must be selected and the calculations repeated.

The final step in the procedure is to calculate possible extension and retraction times from

$$t_a = \frac{C_{vt}}{C_{vs}t_r} \tag{43.24}$$

---

[*] This discussion of pneumatic power system design is by the courtesy of Numatics Inc., Highland, Michigan, and is from the publication "Practical Air Valve Sizing" by Henry Fleischer, Director of Engineering.

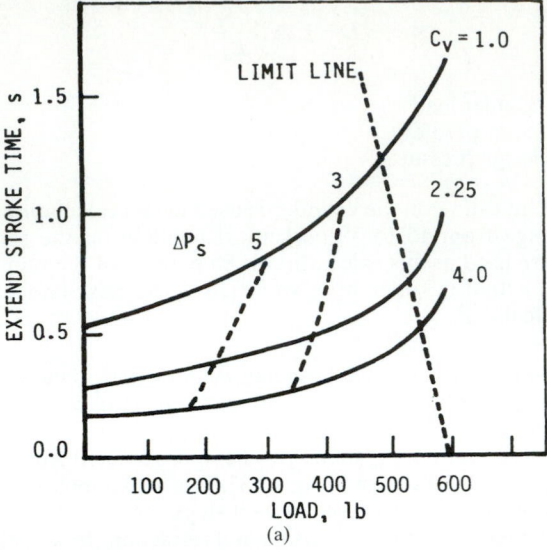

(a)

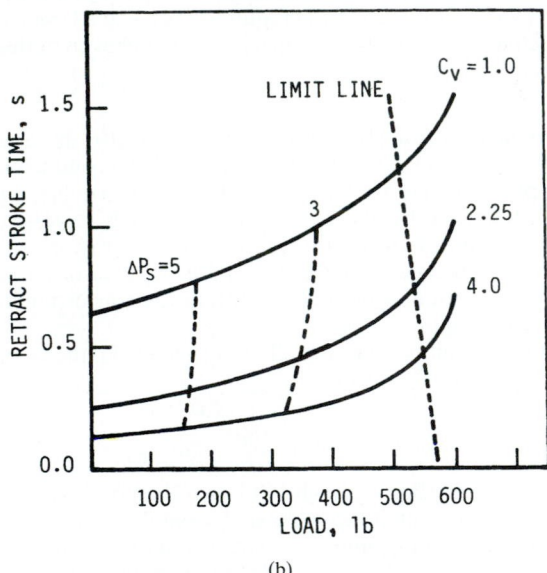

(b)

FIGURE 43.37 Specifying target C_v's using typical cylinder performance charts. Plots shown are for a 4-in-bore, 6-in-stroke, double-acting air cylinder. Supply pressure is 60 psig, initial pressure is 60 psig, and ΔP_s is the mean system pressure drop. (*a*) Extension stroke; (*b*) retraction stroke.

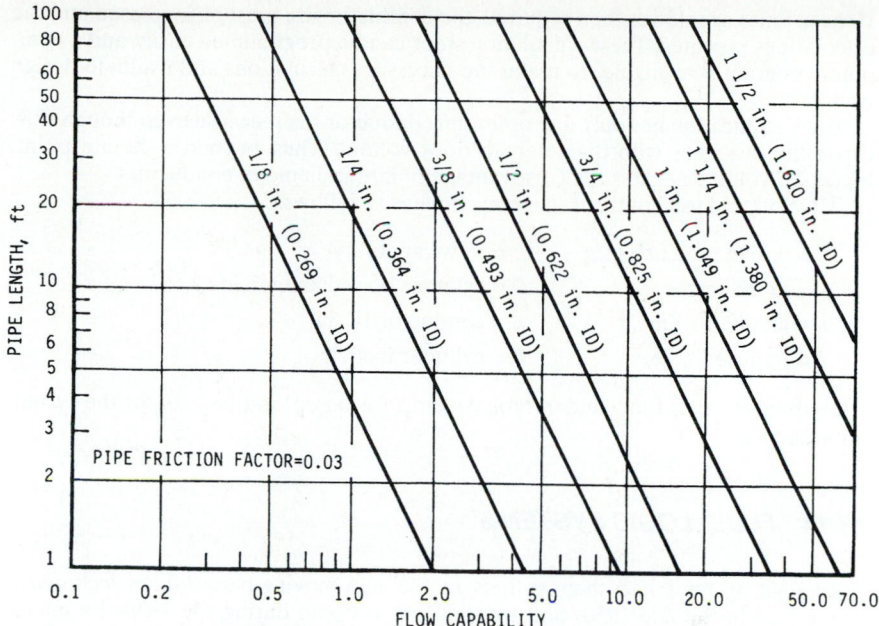

FIGURE 43.38 Determination of conductor flow capability for steel pipe; friction factor = 0.03.

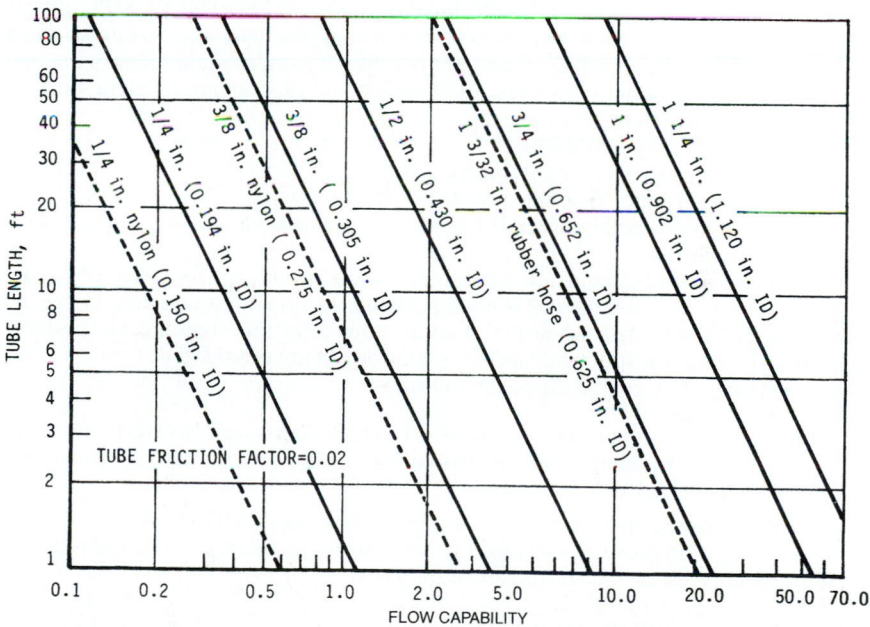

FIGURE 43.39 Determination of conductor flow capability for smooth-wall tubing; tube friction factor = 0.02.

If these times are slower than required, the conductor size must be increased and the calculations repeated. These calculation steps can be programmed easily, and a computer program is available[*] to make the necessary calculations and evaluate design tradeoffs.

Each application has only one optimum conductor size, because even though flow capability increases with diameter, so does volume. Thus, beyond a certain point, increased volume negates the C_v advantage of larger-diameter conductors.

The notation used in the discussion above is as follows:

P_d = design pressure, psig C_v = flow capability

S_f = safety factor D = cylinder bore diameter, in

t_a = attainable time, s d = conductor ID, in

t_r = required time, s L = cylinder load, lb

The subscripts are c for cylinder, s for system, t for target, and $1, \ldots, n$ for the system components.

43.14   FLUID LOGIC SYSTEMS[†]

Fluid logic systems had their genesis in the non-moving-part *fluidics technology* announced in the late 1950s and were all but eclipsed during the 1970s by microelectronics. Today, *fluidics* survives primarily in specialized applications requiring its special characteristics. Typical are medical, food processing, high-temperature, high-radiation, and other environments where all-fluid systems are advantageous.

The residual of two decades of evolution is *moving-part logic* (MPL). MPL utilizes miniature to small-size pneumatic devices which combine logic with power-handling functions. MPL devices include spool, poppet, diaphragm, floating diaphragm, and various proprietary designs of valves. Fluid sensors are used to input system-variable status to the MPL system. Fluid indicators and readouts are also available to provide visual monitoring of system status.

Transistor and microelectronic control technology is having an impact on MPL, as it did on fluidics. Figure 43.40 illustrates the situation wherein MPL and electronic control can be interposed between the *load-sensing* stages and the *fluid power* stage in a typical application.

Pneumatic control circuits are used to provide a prescribed sequence of events. Usually, arriving at that sequence is not a particularly difficult design task. However, problems frequently arise when the analysis stops after this step and "secondary" objectives that help ensure trouble-free operation are not considered.

Typical secondary goals include the following:

• Ensuring that the circuit does not respond to false input signals. Overlooking this detail can cause equipment damage or operator injury. These actions are prevented by including interlocks in the system.

• Providing a circuit that operates over a wide range of conditions. Potential problems arising from pressure variations, contamination, actuating speed, and sensitive adjustments must be recognized and eliminated.

---

[*] From Numatics Inc., Highland, Michigan.

[†] This discussion of fluid logic systems is by the courtesy of ARO Corporation, Bryan, Ohio; Bruce F. McCord, Manager, Control Systems.

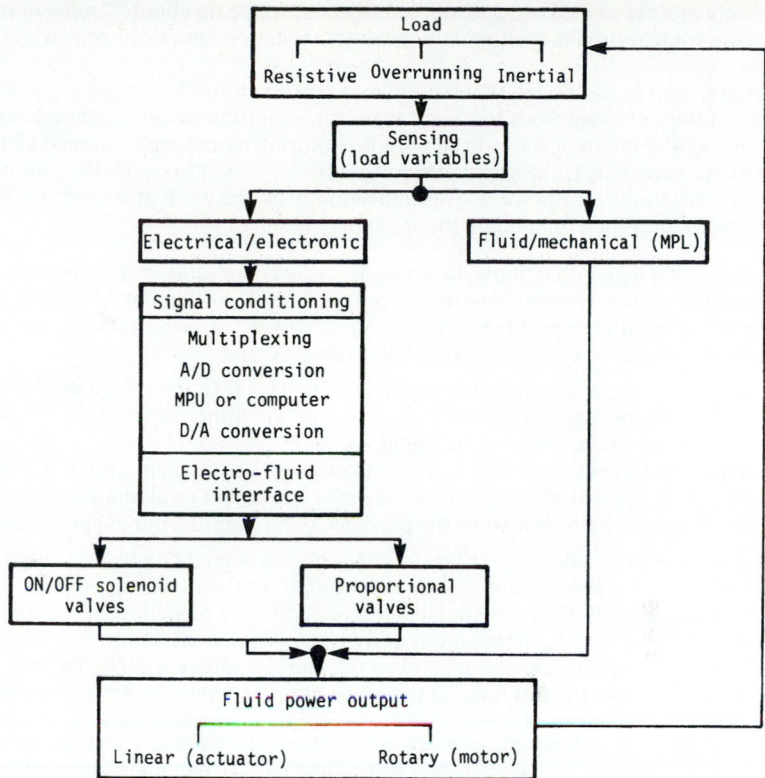

FIGURE 43.40 Comparison between electrical/electronic and MPL control options for a typical industrial application.

- Designing a circuit that assists in troubleshooting and diagnosis of problems. This objective is met in two phases. First, components that can mask problems, such as pulse or single-shot valves, are avoided. Second, display devices, such as pneumatic indicators, are included in the circuit, permitting the cause of failure to be traced quickly.

Meeting all these objectives can be a tall order, but a design technique has been developed that automatically avoids the common pitfalls and meets the design objectives easily. Based on the theory developed for electronic controllers, the method inherently incorporates the interlocks necessary to prevent improper operation. It also produces a circuit that is easy to troubleshoot. Finally, it includes a review procedure that ensures that the system is the least expensive possible.

43.14.1 Circuit Design

Pneumatic controls are used in two types of circuits: combinational and sequential. Combinational circuits monitor an operation and react to certain combinations of inputs regardless of the order of occurrence. Sequential circuits, however, accept inputs only in a prescribed order. Sequential circuits make up about 90 percent of the pneumatic control circuits used, and the method presented here deals only with these circuits.

The first step in the design of a pneumatic circuit is to build a stage register. The purpose of the stage register is to collect signals from input devices (pushbuttons and limit valves) and create a series of maintained output signals, one for each step (or stage) in the sequence. These signals are further manipulated to create the output signals necessary to shift the power valves that actually do the work on a machine. These are the basic rules used in building the stage register:

1. Any input signal that starts an action (or stage) and does not remain on until the end of the cycle is connected to the set port of a flip-flop element. Thus, if an input is not maintained throughout the rest of the cycle (once it is used to start a stage), the flip-flop output can be used as the maintained stage signal.

2. Any input signal that is maintained from the time it starts a stage until the end of the cycle is connected to an AND element input. The other input to the AND element comes from the previous-stage signal.

A maintained signal connected to an AND element acts as an interlock to the previous stage. This prevents the signal from affecting the circuit until that circuit is ready to receive the signal, as indicated by the presence of an output from the previous step.

3. A stage signal is developed for each step in the sequence plus one additional signal, called a *reset*, which returns all circuit elements to their original positions when the cycle is completed. The reset signal is connected to the reset port of the first flip-flop in the circuit. Then the reset output of this first flip-flop is connected to the reset input of the second flip-flop, the second to the third, and so on down the line. This connection interlocks the flip-flops so that they cannot be shifted out of sequence.

The second step in designing a circuit is to convert the maintained stage signals to the on-off signals required to shift the power valves. NOT elements are used to create the power valve signals. A NOT element is similar to a normally passing valve in that a signal applied to the b port is transmitted to the c port until a second signal is applied to the a port. See Fig. 43.41. Thus, to create the signal required by a power valve, the signal from a maintained stage must be applied to the b port of a NOT element. Because this signal is transmitted from the b to the c port, the c port signal becomes the power valve pilot signal.

To remove this signal at the beginning of a later stage, the signal from that stage is applied to the a port of the NOT element. When the a port is pressurized, the output signal from the NOT goes off. This sequence is followed for all power valves in the circuit.

43.14.2 Logic Elements

Pneumatic controls are used in combinational and sequential circuits, as we have learned. Normally, combinational circuits use AND, OR, and NOT elements with an occasional delay or memory (flip-flop) to delay or hold the signals from the other elements (Fig. 43.41). The OR element is sometimes used to add additional functions, such as emergency stops and manual overrides, to a sequential circuit. Also, delay elements are occasionally used to replace limit valves, while amplifiers increase low-pressure signals to the system operating pressure.

In the AND element, both input A and input B must be on for the output to be on. Conversely, if either input is off, the output is off.

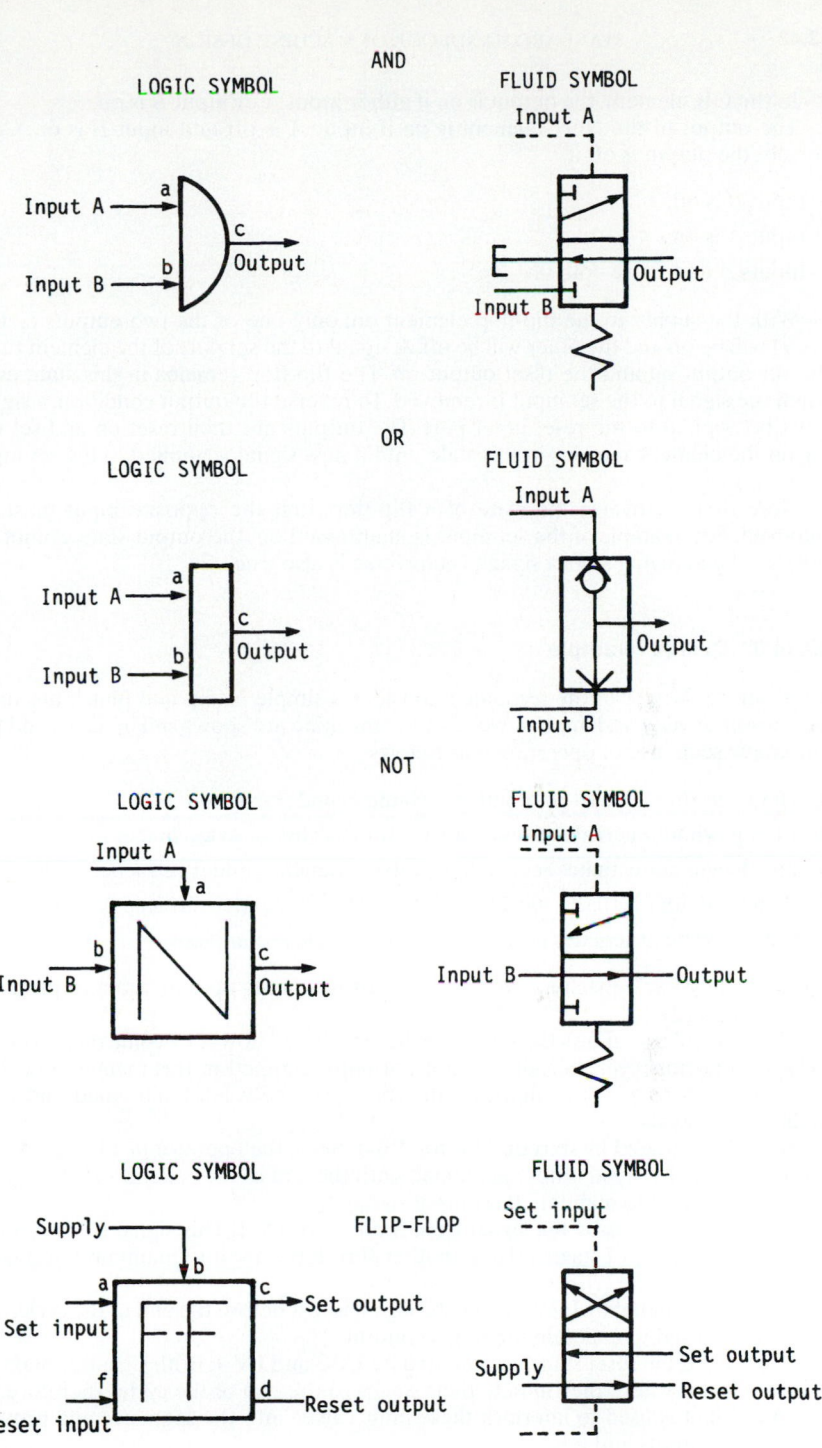

FIGURE 43.41 Logic elements.

43.41

In the OR element, the output is on if either input A or input B is on.

The output of the NOT element is on if input A is off and input B is on. Conversely, the output is off if

- Input B is off.
- Input A is on.
- Inputs A and B are both on.

With the supply to the flip-flop element on, only one of the two outputs (set or reset) will be on and the other will be off. A signal to the set port of the element turns the set output on and the reset output off. The flip-flop remains in this state even when the signal to the set input is removed. To reverse the output condition, a signal must be applied to the reset input port. The outputs are then reset on and set off. Again the element remains in this state until a new signal is applied to the set input port.

Note that to reverse the state of a flip-flop, first the opposite input must be removed. For example, if the set input is maintained on, the output state cannot be reversed by applying a reset signal. The reverse is also true.

43.14.3 Design Example

To illustrate the use of the technique, consider a simple clamp and punch machine. The power devices and input devices for the machine are shown in Fig. 43.42, and the automatic sequence of operation is as follows:

1. Operator presses start pushbutton—clamp cylinder extends.
2. Clamp cylinder actuates a limit valve—punch cylinder extends.
3. Punch cylinder actuates second limit valve—punch cylinder retracts.
4. Punch cylinder actuates third limit valve—clamp cylinder retracts.
5. Clamp cylinder actuates fourth limit valve—cycle is complete.

This is a single-cycle machine, and the operator need only operate the start pushbutton momentarily.

Figure 43.42 also shows the signals to the circuit and how they come on and go off through a normal cycle. Signals are not normally graphed in this manner, but they were in this case to better illustrate the difference between maintained and non-maintained signals.

Stage 1 is initiated by start pushbutton PB-1. Since the operator need not hold the button down, this signal is not maintained until the end of the cycle. Therefore, a flip-flop must be used to maintain this output signal.

Stage 2 is initiated by the signal from limit valve LV-1. This signal is lost shortly after the beginning of stage 4. Thus another flip-flop is used to maintain the stage 2 signal.

Stage 3 is initiated by LV-2. Again, the signal is lost before the end of the cycle, and a flip-flop is used to maintain the stage 3 output.

Stage 4 and the reset stage are initiated by LV-3 and LV-4. Both signals remain on from the point where they initiate their stages to the end of the cycle. Therefore, an AND element is used to interlock these limit valves into the sequence and provide the stage 4 outputs and reset.

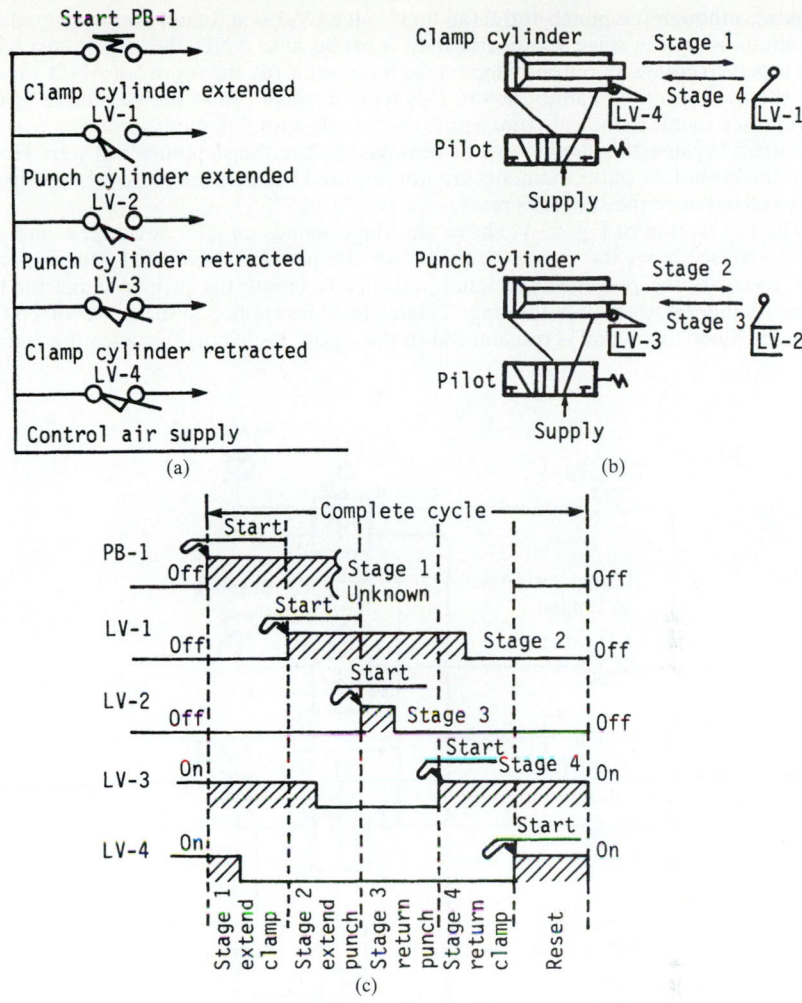

FIGURE 43.42 Basic elements of example circuit. Sketches show the starting positions of (*a*) input and (*b*) power devices for clamp and punch machine. (*c*) The chart shows the sequence of signals through one machine cycle. Such diagrams indicate the type of logic elements needed in the control circuit.

Figure 43.43, the resulting stage register circuit, shows that the stage indicators must come on in numerical order. For example, in the rest condition (all indicators off), the only input this circuit accepts is the input from PB-1. If PB-1 has not been actuated (stage 1 output off) and the clamp extended limit valve, LV-1, is accidentally actuated, then stage 2 output does not go on. The reason is that the reset output coming from the *d* port of flip-flop 1 holds flip-flop 2 in the reset position (pressure at *f* port). Therefore, to initiate stage 2, stage 1 must be actuated first, turning on the stage 1 indicator and releasing the pressure at the reset port of flip-flop 2.

Also, although the punch-retracted limit valve LV-3 is actuated and the signal for the circuit is *on,* the stage 4 indicator light is off because AND element 4 must have two inputs to create an output. Since stage 3 output is off, the *b* port of AND 4 is off and the stage 4 output cannot be on. This type of signal (on at the beginning of the cycle) often seems to indicate that a pulse, or "single-shot," element should be used in the circuit because this signal must be removed before the sequence can start. However, single-shot, or pulse, elements are not required because the effect of the signals is locked out once the system is reset.

The top section of Fig. 43.44 shows the stage signals already developed, and the lower section shows the signals required for the power valves. NOT elements are used to create the power valve signal patterns. To create the proper signal for the clamp-cylinder power valve, the stage 1 signal must be applied to the *b* port of a NOT element. Since this signal is transmitted to the *c* port, the *c* port becomes the clamp-

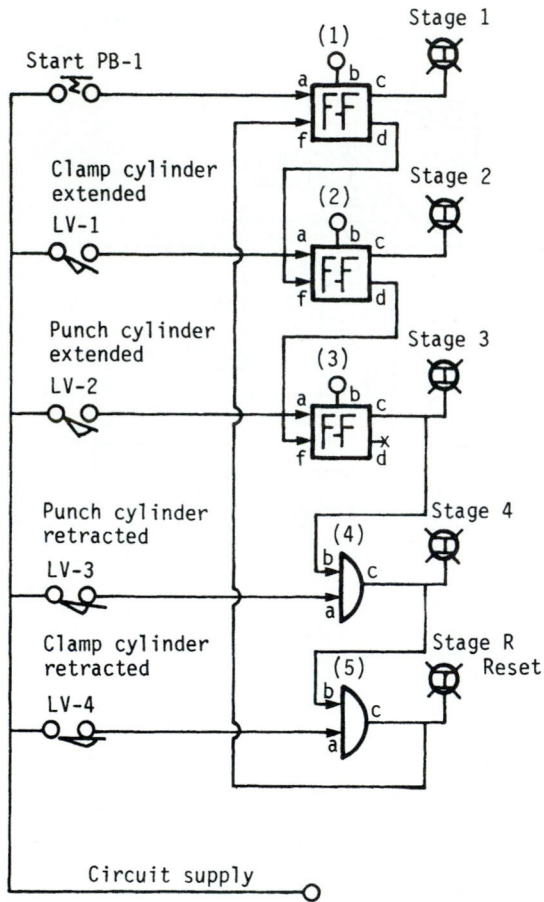

FIGURE 43.43 Stage register for example circuit. Basic control circuit consists of three flip-flops and two AND elements. Indicator lights show when each stage has been completed.

extend pilot signal. To remove this signal at the beginning of stage 4, the stage 4 signal is connected to the *a* port of this NOT element. When the *a* port is pressurized, the output signal from the NOT element goes off. This duplicates the signal pattern shown on the lower portion of Fig. 43.44. The signal for the punch valve is developed by using the same method, applying the stage 2 signal to the *b* port and the stage 3 signal to the *a* port of another NOT element. Figure 43.45 shows the completed circuit.

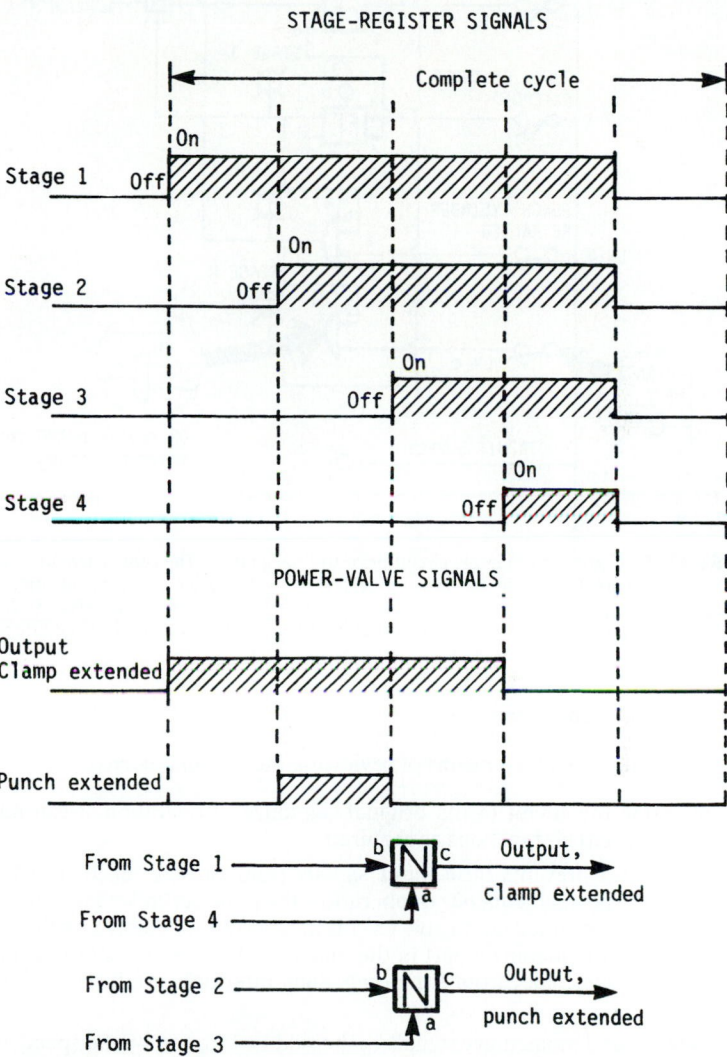

FIGURE 43.44 Circuit signals for example machine. Top portion of graph shows the signals produced by the stage register elements, while bottom portion shows the signals required to actuate the power valves. NOT elements transform the stage register signals to power valve pilot signals.

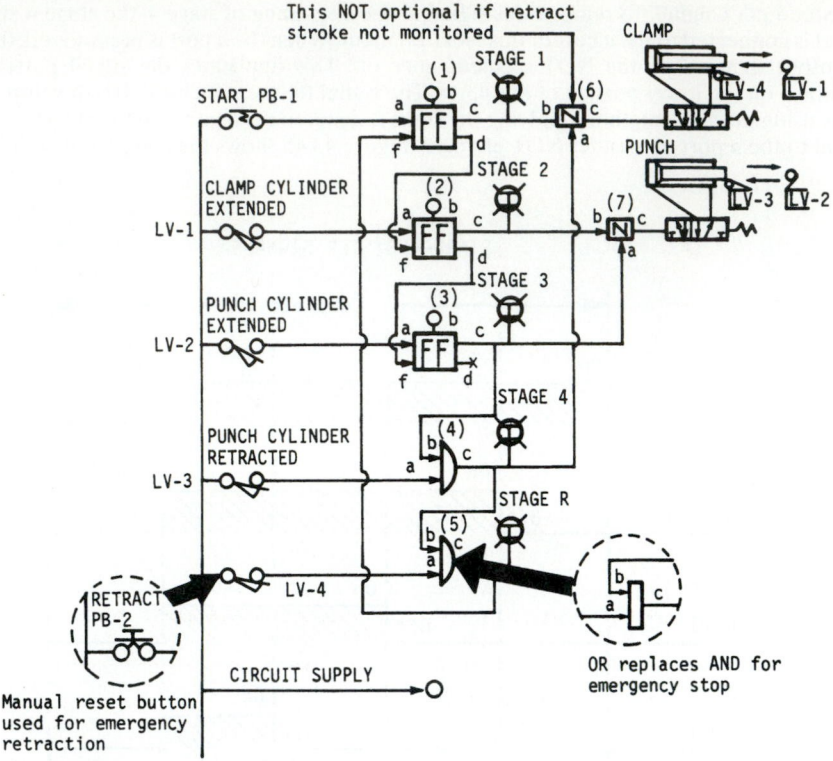

FIGURE 43.45 Completed example circuit prior to final analysis. The control circuit consists of seven logic elements. The circuit can now be evaluated to determine the effect of removing or adding elements. Here, the retraction stroke need not be monitored; therefore AND 5, LV-4, and NOT 6 can be removed. An emergency stop function is provided by adding PB-2 and OR 5.

43.14.4 Circuit Evaluation

The stage register provides a means of reviewing the design objectives.

- By comparing the circuit to the original sequence of operation, it can be determined that the circuit functions as required.
- The circuit also prevents false input signals from causing machine action. For instance, in the normal sequence of operation, the punch cylinder extends when the clamp-cylinder-extended limit valve LV-1 is inadvertently actuated by the operator while loading or unloading a part in the machine; the stage register interlock prevents the punch cylinder from extending. Such interlocks are included automatically.
- Pulse devices and momentary signals that could be affected by the speed at which the limit valves are actuated, length of tubing, pressure changes, leaks, etc., are excluded from the circuit. Therefore, this circuit should operate over as wide a range of conditions as the components themselves can tolerate. Also, no adjustments are required that could affect the operation of this system.

- By including the indicators as shown in Fig. 43.45, the cause of most problems can be traced. Figure 43.46 shows a simple troubleshooting chart that could accompany the machine for use by a repair technician. The bottom three lines of this chart illustrate that the circuit also detects a malfunction in the first three input signals if they happen to remain on. For instance, if a cycle is completed and limit valve 1 is not released, a new cycle cannot be initiated.

This is an important bonus interlock because, on the next cycle, LV-1 would indicate that the clamp was actuated prematurely. However, since the LV-1 signal is applied to the set port of flip-flop 2, it cannot reset and the circuit will not start a new cycle. Instead, the indicators will remain in the position shown in the next-to-last line of Fig. 43.46 and indicate that, in all likelihood, limit valve 1 has failed; if not, a failure in flip-flop 2 is the only other possibility.

By carefully analyzing and removing elements that are not necessary, circuit cost can be reduced considerably. However, this evaluation cannot be made until the circuit has been developed.

If the machine fails to operate and the indicators are in this condition	Follow these steps to locate problem and check in the order listed
1 ○ 2 ○ 3 ○ 4 ○ R ○	Air supply, PB-1, flip-flop-1
✖ ○ ○ ○ ○	Clamp cylinder (should be extended) LV-1, flip-flop-2
✖ ✖ ○ ○ ○	Punch cylinder (should be extended). LV-2, flip-flop-2
✖ ✖ ✖ ○ ○	Punch cylinder (should be retracted) LV-3,AND-4, NOT-7
✖ ✖ ✖ ✖ ○	Clamp cylinder (should be retracted) LV-4, AND-5, NOT-6
✖ ✖ ✖ ✖ ✖	PB-1, flip-flop-1
○ ✖ ✖ ✖ ✖	LV-1, flip-flop-2
○ ○ ✖ ✖ ✖	LV-2, flip-flop-3

FIGURE 43.46 Troubleshooting chart. The chart matches the condition of the indicator lamps to the probable cause of failure. Such a chart should accompany each machine to allow malfunctions to be traced quickly.

REFERENCES

43.1 Russell W. Henke, P. E., *Fluid Power Systems and Circuits*, Penton/IPC, Cleveland, 1983.

43.2 ANSI standard Y32.10-1967 (R1979), "Graphic Symbols for Fluid Power Diagrams."

43.3 Russell W. Henke, *Introduction to Fluid Mechanics,* Addison-Wesley, Reading, Mass., 1966.

43.4 William Wolansky, John Negoshian, and Russ Henke, *Fundamentals of Fluid Power,* Houghton Mifflin, Boston, 1977.

43.5 Russ Henke, *Introduction to Fluid Power Circuits and Systems,* Addison-Wesley, Reading, Mass., 1984.

43.6 Blaine Anderson, *The Analysis and Design of Pneumatic Systems,* John Wiley & Sons, New York, 1967.

CHAPTER 44
CORROSION

Milton G. Wille, Ph.D., P.E.

Professor of Mechanical Engineering
Brigham Young University
Provo, Utah

44.1 INTRODUCTION

Corrosion removal deals with the taking away of mass from the surface of materials by their environment and other forms of environmental attack that weaken or otherwise degrade material properties. The complex nature of corrosion suggests that the designer who is seriously concerned about corrosion review a good readable text such as *Corrosion Engineering* by Fontana and Greene [44.1].

Included in this chapter are many corrosion data for selected environments and materials. It is always hazardous to select one material in preference to another based only on published data because of inconsistencies in measuring corrosion, lack of completeness in documenting environments, variations in test methods, and possible publishing errors. These data do not generally indicate how small variations in temperature or corrosive concentrations might drastically increase or decrease corrosion rates. Furthermore, they do not account for the influence of other associated materials or how combinations of attack mechanisms may drastically alter a given material's behavior. Stray electric currents should be considered along with the various attack mechanisms included in this chapter. Brevity has required simplification and the exclusion of some phenomena and data which may be important in some applications.

The data included in this chapter are but a fraction of those available. *Corrosion Guide* by Rabald [44.2] can be a valuable resource because of its extensive coverage of environments and materials.

Again, all corrosion data included in this chapter or published elsewhere should be used only as a guide for weeding out unsuitable materials or selecting potentially acceptable candidates. Verification of suitability should be based on actual experience or laboratory experimentation. The inclusion or exclusion of data in this chapter should not be interpreted as an endorsement or rejection of any material.

44.2 CORROSION RATES

The vast majority of metal corrosion data in the United States are expressed in terms of surface regression rate *mpy* (mils, or thousands of an inch, per year). Multiply *mpy* by 0.0254 to obtain millimeters per year (mm/yr). To convert to mass-loss rate, multiply the surface regression rate by surface area and material density, using consistent units.

Polymer attack typically involves volume changes, usually increases, caused by liquid absorption; reductions in mechanical properties such as yield strength, tensile strength, flexure strength, and tensile modulus; discoloration; and/or changes in surface texture. Certain plastics are degraded by ultraviolet light, which limits their usefulness in sunlight unless they are pigmented with an opaque substance such as lamp-black carbon.

44.3 METAL ATTACK MECHANISMS

The attack on metals involves oxidation of neutral metal atoms to form positively charged ions which either enter into solution or become part of an oxide layer. This process generates electrons, which must be consumed by other atoms, reducing them, or making them more negatively charged. Conservation of electrons requires that the rate of metal oxidation (corrosion) equal the rate of reduction (absorption of electrons by other atoms).

44.3.1 General Attack

In general attack, oxidation and reduction occur on the same metal surface, with a fairly uniform distribution. Most of the corrosion data in this chapter are for selected materials subject to uniform attack in a given environment.

Once a suitable material is selected, further control of uniform attack can be achieved by coatings, sacrificial anodes (see Galvanic Corrosion), anodic protection (see Passivation), and inhibitors. Coatings are many times multilayered, involving both metallic and polymer layers. Inhibitors are additions to liquid environments that remove corrosives from solution, coat metal surfaces to decrease surface reaction rates, or otherwise alter the aggressiveness of the environment.

Chemically protective metallic coatings for steels are usually zinc (galvanized) or aluminum (aluminized). Aluminized steel is best for elevated temperatures up to 675°C and for severe industrial atmospheres. Both may be deposited by hot dipping, electrochemistry, or arc spraying. Common barrier-type metallic platings are those of chromium and nickel. The Environmental Protection Agency has severely limited or prohibited the use of lead-bearing and cadmium platings and cyanide plating solutions.

Polymer coatings (such as paints) shield metal surfaces from electron-receiving elements, such as oxygen, reducing corrosion attack rates. Under mild conditions, even "decorative paints" can be effective. Under more severe conditions, thicker and tougher films are used which resist the effects of moisture, heat, chlorides, and/or other undesirable chemicals. Acrylics, alkyds, silicones, and silicone-modified alkyds are the most commonly used finishes for industrial equipment, including farm equipment. The silicones have higher heat resistance, making them useful for heaters, engines, boilers, dryers, furnaces, etc.

44.3.2 Galvanic Corrosion and Protection

When two dissimilar metals are electrically connected and both are exposed to the same environment, the more active metal will be attacked at a faster rate than if there had been no electrical connection between the two. Similarly, the less active metal will be protected or suffer less attack because the surface areas of both metals can be used to dissipate the electrons generated by oxidation of the more active metal. The net flow of electrons from the more active to the less active metal increases the attack rate of the more active metal and decreases that of the less active metal.

An *adverse area ratio* is characterized by having a larger surface area of less active metal than that of the more active metal. Cracks in a barrier protective coating (i.e., polymers) applied to the more active metal in a galvanic-couple situation can create an extremely adverse area ratio, resulting in rapid localized attack in the cracks. The standard electromotive force (emf) series of metals (Table 44.1) lists

TABLE 44.1 Standard EMF Series of Metals

Metal–metal ion equilibrium (unit activity)	Electrode potential vs. normal hydrogen electrode at 25°C, V
Au-Au^{3+}	+1.498
Pt-Pt^{2+}	+1.2
Pd-Pd^{2+}	+0.987
Ag-Ag^{+}	+0.799
Hg-Hg_2^{2+}	+0.788
Cu-Cu^{2+}	+0.337
H_2-H^{+}	0.000
Pb-Pb^{2+}	−0.126
Sn-Sn^{2+}	−0.136
Ni-Ni^{2+}	−0.250
Co-Co^{2+}	−0.277
Cd-Cd^{2+}	−0.403
Fe-Fe^{2+}	−0.440
Cr-Cr^{3+}	−0.744
Zn-Zn^{2+}	−0.763
Al-Al^{3+}	−1.662
Mg-Mg^{2+}	−2.363

metals in order of increasing activity, starting with gold (Au), which is the least active. If two of the metals listed were joined in a galvanic couple, the more active one would be attacked and plating or deposition of the less active one would occur. This is based on the fact that solutions contain only unit activity (concentration) of ions of each of the two metals.

The standard EMF series is valid only for pure metals at 25°C and in equilibrium with a solution containing unit activity (concentration) of its own ions. If ion concentrations are greater than unit activity, the potentials are more positive; if less, the opposite is true.

TABLE 44.2   Galvanic Series of Some Commercial Metals and Alloys in Seawater

↑	Platinum
	Gold
Noble or	Graphite
cathodic	Titanium
	Silver
	⎡Chlorimet 3 (62 Ni, 18 Cr, 18 Mo)
	⎣Hastelloy C (62 Ni, 17 Cr, 15 Mo)
	⎡18-8 Mo stainless steel (passive)
	⎢18-8 stainless steel (passive)
	⎣Chromium stainless steel 11-30% Cr (passive)
	⎡Inconel (passive) (80 Ni, 13 Cr, 7 Fe)
	⎣Nickel (passive)
	Silver solder
	⎡Monel (70 Ni, 30 Cu)
	⎢Cupronickels (60-90 Cu, 40-10 Ni)
	⎢Bronzes (Cu-Sn)
	⎢Copper
	⎣Brasses (Cu-Zn)
	⎡Chlorimet 2 (66 Ni, 32 Mo, 1 Fe)
	⎣Hastelloy B (60 Ni, 30 Mo, 6 Fe, 1 Mn)
	⎡Inconel (active)
	⎣Nickel (active)
	Tin
	Lead
	Lead-tin solders
	⎡18-8 Mo stainless steel (active)
	⎣18-8 stainless steel (active)
	Ni-Resist (high Ni cast iron)
	Chromium stainless steel, 13% Cr (active)
	⎡Cast iron
	⎣Steel or iron
	2024 aluminum (4.5 Cu,. 1.5 Mg, 0.6 Mn)
Active or	Cadmium
anodic	Commercially pure aluminum (1100)
↓	Zinc
	Magnesium and magnesium alloys

SOURCE: M. G. Fontana and N. D. Greene, *Corrosion Engineering*, 2d ed., McGraw-Hill, New York, 1978. Used by permission.

The galvanic series (Table 44.2) shows a similar relationship, except that impure metals such as alloys are also included and the medium is seawater. Other media, other concentrations, and other temperatures can further alter the order of the list. Therefore, care should be exercised in applying these data to a given galvanic corrosion situation except as a general, loose guide.

44.3.3 Passivation

Certain common engineering materials, such as iron, nickel, chromium, titanium, and silicon as well as their alloys (i.e., stainless steels), exhibit a characteristic of being able to behave both as a more active and as a less active (passive) material.

Note in the galvanic series (Table 44.2) that several stainless steels are listed twice, once as passive and once as active. Some common metals other than those mentioned also exhibit passivity, but to a lesser extent.

A graphical representation of passivity is shown in Fig. 44.1. The three regions—active, passive, and transpassive—help to explain seemingly inconsistent behavior of active-passive materials under various degrees of attack severity.

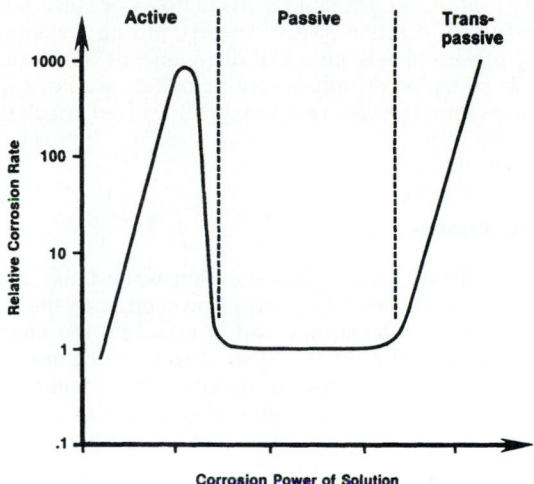

FIGURE 44.1 Corrosion characteristics of an active-passive metal.

There are both advantages and disadvantages to be gained or suffered because of active-passive behavior. In very aggressive environments, a method called anodic protection can be used whereby a *potentiostat* is utilized to electrochemically maintain a passive condition and hence a low rate of corrosion. However, accelerated corrosion test results may be useless because increasing the corrosion power of the medium may cause a shift from a high active corrosion rate to a low passive condition, producing the invalid conclusion that corrosion is not a problem. Another example involves inhibitors which function by maintaining a passive condition. If the concentration of these inhibitors were allowed to decrease, high corrosion could result by passing from a passive to an active condition.

Active-passive materials have a unique advantage in the area of corrosion testing and corrosion rate prediction. *Potentiodynamic polarization* curves can be generated in a matter of hours, which can provide good quantitative insights into corrosion behavior and prediction of corrosion rates in a particular environment. Most other corrosion testing involves months or years of testing to obtain useful results.

44.3.4   Crevice Corrosion and Pitting

Crevice corrosion is related to active-passive materials which are configured such that crevices exist. Mated screw threads, gaskets, packings, and bolted or lapped joints are common examples of crevices. Inside the crevice, oxygen or other corrosives required for passivation have restricted entrance, resulting in reduced concentration as they are consumed by corrosion in the crevice. When the concentration of these corrosives is low enough to fail to maintain passivity, the metal in the crevice becomes active. The large electrically connected, passivated surface outside the crevice completes a galvanic couple with a large adverse-area ratio, providing high attack rates within the crevice. Welding or forming can be used to avoid crevices. However, intergranular corrosion may occur in welded stainless steels (see Sec. 44.3.9).

Pitting is a very localized attack that results in holes, or voids, on a metal surface. Although not restricted to active-passive metals, pitting is commonly related to these. With active-passive metals, pieces of dirt, scale, or other solid particles may rest on the bottom of a pipe or tank where velocities are not sufficient to move them. These particles form crevices, resulting in a localized attack similar to crevice corrosion.

44.3.5 Sacrificial Anodes

Magnesium rods are placed in steel glass-lined hot-water tanks, and zinc is used to coat sheet steel (galvanized steel) to provide protection to the steel against corrosion. As the more active magnesium rod is attacked, the electrons generated are conveyed to the electrically connected steel tank, which needs protection only for regions where cracks or flaws exist in the glass lining. Similarly, for galvanized steel, protection is required only for regions of scratches or where steel edges are exposed.

44.3.6 Stress Corrosion Cracking

In stress corrosion cracking (SCC), most of a metal's surface may show little attack, while fine intergranular or transgranular cracks may penetrate deeply into the surface. There may be a single continuous crack or a multibranched crack, or the entire surface may be covered with a lacy network of cracks. Usually dye penetrants and sectioning are needed to reveal the extent and depth of cracking.

Certain classes of alloys and environments are susceptible to this phenomenon, and usually tensile stresses are involved, with crack penetration rates increasing with increasing tensile stress. The higher the strength condition of a given alloy, the greater seems to be the tendency to suffer SCC. Table 44.3 lists some materials and environments that have been known to produce SCC.

Frequently, a difference in color or texture is noticeable between a stress corrosion crack and an adjacent region of overstress when the fracture is completed by mechanical means. Scanning electron micrographs are frequently useful in identifying SCC.

44.3.7 Selective Leaching

Selective leaching refers to the chemical removal of one metal from an alloy, resulting in a weak, porous structure. Brass sink traps suffer this type of attack by zinc being leached out of the yellow brass, leaving behind a porous structure of reddish copper. Aluminum and silicon bronzes and other alloys are also subjected to selective leaching.

TABLE 44.3 Environments That May Cause Stress Corrosion of Metals and Alloys

Material	Environment	Material	Environment
Aluminum alloys	NaCl-H$_2$O$_2$ solutions NaCl solutions Seawater Air, water vapor Ammonia vapors and solutions	Ordinary steels	NaOH solutions NaOH-Na$_2$SiO$_2$ solutions Calcium, ammonium, and sodium nitrate solutions Mixed acids (H$_2$SO$_4$-HNO$_3$) HCN solutions
Copper alloys	Amines Water, water vapor		Acidic H$_2$S solutions Seawater
Gold alloys	FeCl$_3$ solutions Acetic acid-salt solutions	Stainless steels	Molten Na-Pb alloys Acid chloride solutions such as MgCl$_2$ and BaCl$_2$
Inconel	Caustic soda solutions		NaCl-H$_2$O$_2$ solutions
Lead	Lead acetate solutions		Seawater
Magnesium alloys	NaCl-K$_2$CrO$_4$ solutions Rural and coastal atmospheres Distilled water		H$_2$S HaOH-H$_2$S solutions Condensing steam from chloride waters
Monel	Fused caustic soda Hydrofluoric acid Hydrofluosilicic acid	Titanium alloys	Red fuming nitric acid, seawater, N$_2$O$_4$, methanol-HCl
Nickel	Fused caustic soda		

SOURCE: M. G. Fontana and N. D. Greene, *Corrosion Engineering,* 2d ed., McGraw-Hill, New York, 1978. Used by permission.

44.3.8 Hydrogen Embrittlement

In any electrochemical process where hydrogen ions are reduced, monatomic hydrogen atoms are created prior to their joining in pairs to form diatomic hydrogen gas (H$_2$). Monatomic hydrogen, being small, can diffuse into metals, causing embrittlement. Corrosion of metals by acids, including cleaning by pickling, can produce hydrogen embrittlement. Heating can drive out monatomic hydrogen, reversing the process. If monatomic hydrogen diffuses into voids in a metal, high-pressure pockets of H$_2$ gas are formed which are not eliminated by heating, but rather may form hydrogen blisters.

44.3.9 Intergranular Corrosion

In some alloys, frequently related to prior heating, grain boundaries can experience localized variations in composition that can result in corrosion attack along or immediately adjacent to grain boundaries. The 18-8 stainless steels (such as type 304), when heated in the approximate range of 500 to 790°C, experience the precipitation of chromium carbides in grain boundaries, removing chromium from the regions adjacent to grain boundaries. This process is called *sensitization*. It is theorized that intergranular attack proceeds in the chromium-depleted regions of the grain boundaries, since these lack the protection provided by chromium alloying. When this class of stainless steels is welded, regions a bit removed from the weld axis are heated sufficiently to become sensitized and hence become subject to subsequent intergranular

(continued on page 44.28)

TABLE 44.4 Corrosion Data by Environment and Material[†]

Acetone

Nonmetallics. ABS — satis. **Acetal copolymer**—after 6 mos at 120 F: yld str −19%, tens mod −48%, length +2.1%, weight +4.5%, appearance no change. **Acetal homopolymer**—1 yr at 120 F: tens mod −40%, tens str −7%, length +1.1%, weight +2.6%. **Acrylic**—unsatis in 90% at 100 F. **Butyl rubber**—70 hrs at r.t.: +2% vol change. **Chlorinated polyether**—res at 80 F. **Chlorosulfonated polyethylene rubber**—minor to moderate effect at r.t. **Ethylene-propylene rubbers**—at r.t. after 70 hrs retain 81-83% ten str, vol changes −17 to +4%. **Fluorocarbon (PVF$_2$)**—fair at 70 F, NR at 120 F. **Fluoroelastomer**—severe effect at r.t. **Glass (borosilicate)**—satis at 150 F. **Graphite (impervious)**—res 100% boiling. **Hydrocarbon rubber**—little or no effect at r.t. **Natural rubber**—satis. **Neoprene**—minor to moderate effect at r.t. **Nylon**—satis at 120 F. **Polyacrylate rubber**—after 70 hrs at r.t.: +201% vol change. **Polycarbonate**—not resistant after 6 mos at r.t. **Polyester (glass reinf)**—NR. **Polyethylene (Hi-D)**—unsatis after 1 yr at 70 F. **Polyimide (glass reinf)**—after 7 days exp retains 100% of flex mod and 98% of r.t. flex str. **Polystyrene**—not res. **PVC**—plast and unplast unsatis at 68 F. **PVC-acrylic alloy**—attacked at 73 F. **SBR rubber**—after 70 hrs at r.t.: +18% vol change. **Silicone rubber**—after 7 days at 75 F: ten str −85%, volume +180%. **Styrene-acrylonitrile**—not resistant at 73 F. **Urethane rubber**—severe effect at r.t. **Vinyl ester (glass reinf)**—NR in 100%.

Ammonia

Metals. Aluminum—res. to dry gas even at elevated temp. If moist, attack low for all con. up to 120 F. **Copper and alloys**—generally res. if dry, rapidly attacked if moist. **Iron and steels**—good res. to aqueous and anhydrous sol. **Lead**—res. to dry gas. After 2 days in 1.7% sol. at r.t.: 1.9 mpy under quiet conditions, 1.1 mpy under aerated cond. **Magnesium**—res. to dry gas at r.t.; presence of water vapor may cause attack. **Nickel and alloys**—nickel res. to anhydrous, and to aqueous up to about 1% sol. Nickel alloys generally res., except Ni-Cu. Ni-Cu res. anhydrous ammonia, but readily attacked by aqueous ammonia and ammonium hydroxide. **Stainless steels**—high res. under certain conditions, severely attacked in others, depending on con., temp and pressure. After 2 mos in 99% NH$_3$ vapor at 932 F, 7 to 54 mpy for type 310, 309. more severe attack on 304, 309, 316 and 446 grades. **Tin**—res. to gaseous ammonia, even if heated, but ammonia may become decomposed. Res. liquid ammonia, but readily attacked if sodium in sol.

Nonmetallics. ABS—satis in gas. **Chlorinated polyether**—res to gas at 220 F. **Acrylic**—satis in gas at 100 F. **Chlorosulfonated polyethylene rubber**—minor to moderate effect by anhydrous at r.t. **Fluorocarbon (PVF$_2$)**—exc to 275 F. **Fluorocarbon (TFE, FEP)**—res liquid at 78 F. **Fluoroelastomer**—severe effect by anhydrous at r.t. **Hydrocarbon rubber**—no data, likely to be compatible at r.t. **Neoprene**—little or no effect by anhydrous at r.t. **Nylon**—satis in gas at r.t. **Polyethylene (Hi-D)**—satis after 180 days at 122 F. **PVC**—dry: unplast satis at 140 F; liquid: unplast shows some att or absorp at 68 F and unsatis at 140 F, plast unsatis at 68 F. **Silicone rubber**—no change in vol after 7 days at 75 F. **Urethane rubber**—no data, likely to be compatible with anhydrous.

[†]A footnote on the last page of the table supplies spelled-out forms for the abbreviations used.

Ammonium hydroxide

Metals. Aluminum—low rate of attack in all con. up to 120 F. **Cobalt** —good res. in dilute sol. at r.t.; 0.8 mpy in 5% con. at 77 F under static conditions. **Copper and alloys**— rapidly attacked if more than a few ppm ammonia present, cupronickels being the most resistant. **Irons and steels**—good res.; moderately attacked in hot con. **Lead**—"satisfactory" with liquid or gas at most con. and temps. **Nickel and alloys**—nickel has high res. in very dilute sol., but rapidly attacked in increasing con.; < 1 mpy in 1% sol., over 500 mpy in 13% sol. after 20 hrs at r.t. Aeration may increase res. in low con., but increases attack in high con. Except for Ni-Cu alloys, which are

readily attacked, nickel alloys have high res. in all con. to boil. pt. **Stainless steels**—good res. in all con. up to boil. pt; rapid attack likely above atmospheric boil. pt. **Tin**—0.1 to 0.3 mpy in 1N sol. at 68 F after 24 hrs. **Titanium**—good res.; 0.2 mpy in 5% sol., 0.1 mpy in 28% sol. at r.t. **Tungsten**—good res., only slightly attacked. **Zinc**—12 mpy in quiet (28 mpy for air agitated) 3.4% sol. after 2 days. **Zirconium**—res. in 28% solution, r.t. to 212 F.

Nonmetallics. ABS—satis. **Acetal copolymer**—after 6 mos at 180 F in 10%: yld str −0.3%, tens mod −12%, length· +0.4%, weight +0.74%, discoloration. **Acetal homopolymer**—90 days at 73 F at 10%:

unsatis. **Acrylic**—satis in 30% at 100 F. **Acrylic-PVC alloy**—no change in 10% after 7 days in 10%; res 28% at 73 F. **Alumina (porous)**—res 28% at r.t. **Chlorosulfonated rubber**—little or no effect at 200 F. **Fluorocarbon (PVF₂)** —exc to 275 F. **Fluoroelastomer**— little or no effect at r.t. **Graphite (impervious)**—res in all conc at boiling. **Hydrocarbon rubber**—little or no effect at r.t. **Natural rubber**—satis. **Neoprene**—little or no effect at 158 F. **Nitrile rubber**—rec in 28%. **Nylon**—satisfactory at r.t. **Phenolic** —varies with grade, some show little weight change in 10% and exc appearance after 1 yr. **Polyester (glass reinf)**—rec in 5% to 160 F. **Polyethylene (hi-D)**—satis in 28%

after 90 days at 70 F. **Polyimide (glass reinf)**— 7 days in 10%; retains 81% of flex mod and 77% of r.t. flex str. **Modified polyphenylene oxide**—no effect in 10% after 3 days at 185 F. **Polypropylene**—satis for 30 days at r.t. **Polystyrene**—res conc; heat reduces res. **Silicone rubber**—after 7 days at 75 F in sat'd: ten str −45%. volume +5%. **Styrene-acrylonitrile**—resistant in 30% at 122 F. **Thermoplastic rubber**—satis in 3% after 2 weeks at r.t. **Urethane rubber**—little or no effect at r.t. **Vinyl ester (glass reinf)** —rec in 20% at 150 F, 29% at 100 F.

TABLE 44.4 Corrosion Data by Environment and Material (*Continued*)

Atmosphere — General outdoors except marine

Metals. Aluminum and alloys—high res.; weathering rate is self-limiting, decreasing with time. Alloys tend to acquire light gray patina. In clean atmos away from seacoast, transformation slow, surface may retain some sheen even after many years. Depth of attack ranges from virtually nil in dry rural atmos (Phoenix) to 5 mils max. after 20 yrs in severe industrial atmos (New Kensington, Pa.). **Beryllium**—information limited, but commercially pure grade develops tough, stable, oxide coating which inhibits attack under normal conditions. **Cadmium**—fair to good res.; 0.4 mpy after 1 yr for 0.8 in thk plate in urban indus atmos (N.Y.C.); 0.2 mpy for 3 mos, 0.6 mpy for 9 mos in London. 60 to 90% rusting in severe indus atmos (Altoona, Pa). 4 to 12% in rural (State College, Pa.) after 1 yr. **Carbon steels**—rust rapidly, but rust may be more or less protective depending on steel composition and contaminants in atmos. Rust most protective if surface washed by rain and dries periodically. Plain carbon steel (0.02Cu) attacked to depth of 4 mils after 2 yrs, 13 mils after 10 yrs in severe indus atmos (Pittsburgh). **Cast irons**—fair to good res. depending on type. Austenitic grades generally best; not rust-free, but superior to gray iron and far super-

ior to plain carbon steel. **Chromium** —high res. **Cobalt and alloys**—high res. **Columbium**—high res.; expected to acquire only slight tarnish after 15 yrs in indus atmos. **Copper and alloys**—high res.; copper tarnishes to a brown color which gradually turns black and, after a few yrs, the characteristic green patina starts to form and lasts indefinitely. Some alloys react similarly; but high-zinc brasses and nickel silvers are more resistant to tarnishing than copper. Rate of attack for copper is 0.01 to 0.02 mpy in rural atmos (State College and Phoenix), 0.05 mpy in severe indus atmos (Altoona) after 20 yrs. High-copper alloys (over 70% Cu) have similar res. in above rural areas, somewhat less (0.06 to 0.12 mpy) in Altoona. **Lead**—high res.: 0.01 mpy in rural atmos (State College and Phoenix), 0.01 to 0.02 mpy in urban indus (N.Y.C.) after 20 yrs.; 0.02 to 0.03 mpy in severe indus atmos (Altoona) after 10 yrs. **Low alloy steels**—rust rapidly, but rust may be more or less protective depending on steel composition and contaminants in atmos. Copper structural steel (0.24 Cu) about twice as resistant as plain carbon steel (0.04 Cu) for 0 to 12 yrs in indus atmos (Kearney, N.J.). "High strength low alloy" steels, which include "weathering" grades, at least twice as re-

sistant as copper steel. For 0.2 Cu-0.2 Ni steel, 1.8 mpy after 1 yr and 0.8 mpy after 3 yrs in indus atmos (Bayonne, N.J.). For 5 Ni steel, 1.3 and 0.6 mpy, resp. at same site. **Magnesium**—Good res., may be superior to aluminum in certain atmos. Highly protective oxide film forms upon exposure to atmos. **Molybdenum**—High res.; tarnishes quickly in indus atmos (Bayonne, N.J.) but attacked very slowly (0.03 mpy after 2.2 yrs). **Nickel and alloys** —good to excellent res. Nickel stays bright in clean, dry atmos, tarnishes if relative humidity exceeds about 70%. Tarnishes to faint gray in rural atmos; green corrosion products may form if sheltered from rain. Rate of attack very low in rural areas (State College and Phoenix). Pollutants in severe industrial (Altoona) and urban industrial (N.Y.C.) increase attack markedly. Nickel alloys have high resistance to almost all atmos. 67Ni-33Cu roofing in N.Y.C. shows no measurable loss in thickness after 44 yrs; however, slight pitting (2 to 4 mils) and tarnishing may occur over 20 yrs in Altoona and N.Y.C. **Precious metals** —high res, although some may tarnish under certain conditions. **Stainless steels**—high res for most grades; "300" grades best and will retain brightness for many yrs in

most urban and rural atmos; but slight staining occurs in sulfur-bearing industrial atmos. **Tantalum**—should have high res. **Tin**—high res; corrosion rates (mpy) for 20 yrs: 0.02 in rural atmos (State College and Phoenix); 0.07 in severe indus (Altoona); 0.0008 mpy in an indus —high res; **Titanium and alloys** atmos. **Tungsten**—high res. **Zinc and alloys**—good res; rate of attack after 10 to 20 yrs < 0.01 mpy in dry rural atmos (Phoenix), 0.20 to 0.23 mpy in urban-indus (N.Y.C.) and 0.19 to 0.31 mpy in severe indus (Altoona). Rate of attack is roughly similar whether in form of galvanized steel, die castings or rolled sheet. **Zirconium and alloys**—high res.

Nonmetallics. Acetal copolymer and homopolymer—special UV stabilized and black pigmented grades prevent little loss in prop. **Acrylic**—satis up to 20 yrs. **Epoxy (glass reinf)**—after 1 yr retains 98+% flex str. **Fluorocarbon (PVF₂)**—exc after 8 yrs. **Polyethylene**—not normally res but can be made to produce satis service for 5-20 yrs.

44.10

Metals. Aluminum and alloys—1000, 3000, 5000 and 6000 series alloys have high res. with 5000 grades generally the most suitable. In severe atmos; initial attack may be as high as 4 mpy, but usually tapers off to as low as 0.1 mpy after first yr. After 5 yrs 80 ft and 800 ft from tide at Kure Beach, attack ranged from 0.007 to 0.025 mpy. Most widely used are 5083, 5086, 5154, 5052 and 6061. Many alloys apt to pit but tapers off in time. Above alloys also have good res. in splash zone, where pitting tendency may be less, but attack rate high if pits develop. Corrosion rate higher in meantide than splash zone, but less than if fully immersed. **Beryllium**—information very limited, but believed apt to pit. **Cadmium**—very good res. based on tests at Kure Beach. **Carbon steels**—rapidly attacked in splash zone, rates ranging to 50 mpy, which may be 10 times higher than for same steel submerged. Attack decreases with distance from tide: 47 mpy 80 ft from mean tide, 1.3 to 1.6 mpy 800 ft from tide, at Kure Beach. 2.3 to 2.8 mpy 300 ft from tide at Cristobal, Canal Zone. **Cast irons**—Austenitic cast irons have good res. and plain cast iron is about twice as res. as 0.2% copper steel, based on 7½ yrs exposure at Kure Beach. **Cobalt and alloys**—very good res. 0.1 mpy 80 ft from tide and 0.2 mpy 800 ft from tide after 3 yrs at Kure Beach for cobalt. At same site, 67Co-30Cr-2W, a wear-resistant alloy, lost none of its reflectivity after 1½ yrs.

Columbium—should be res. to attack. **Copper and alloys**—good to high res.; 0.01 to 0.17 mpy for copper, various brasses, and cupronickels exposed for up to 20 yrs at Cristobal, Kure Beach, Key West, La Jolla, Calif. and Sandy Hook, N.J. Rate of attack somewhat higher in tropical zones than in temperate climates. Alloying with aluminum, nickel, zinc tend to increase, silicon and tin decrease, res. over pure copper, but differences slight. In general, alloy with 15% or more zinc susceptible to dezincification, but can be controlled by small additions of arsenic, antimony or phosphorus. Performance in splash zone more similar to that in atmos than in immersion. Generally, alloys having good res. in severe atmos (Cristobal) also good in splash zone. At mean tide, attack about 20 to 60% that for fully immersed. **Lead**—very good res. 0.02 mpy for chemical and antimonial lead after 20 yrs at La Jolla and Sandy Hook; 0.08 mpy after 8 yrs at Cristobal. Even better res. if atmos polluted. **Low alloy steels**—substantially greater res. than plain carbon steels: 0.7 to 0.9 mpy for low alloy steels, 1.8 mpy for copper steel, at Cristobal; in general, total alloy content of 2% seems to give maximum return in performance. At Kure Beach, 800 ft from tide, 0.350 mpy for nickel-copper-molybdenum steel having alloy content of 2%, 0.582 mpy for 1.1%. **Magnesium and alloys**—fair; 1 mpy fairly typical. For AZ31 alloy: 0.94 mpy after 16 yrs

at Cristobal; about 0.9 mpy after 32 mos. at Daytona Beach; 0.57 mpy 80 ft from tide after 4 yrs at Kure Beach. Oxide film which form upon exposure to normal atmos tends to break down in salt-laden atmos, especially salt spray. **Molybdenum and alloys**—high res. in atmos; 0.1 mpy (max pit 2.4 mils) after 7 yrs, 80 ft and 800 ft from tide at Kure Beach. Alloys TZM and Mo30W should behave similarly. **Nickel and alloys**—generally high res.; 0.01 mpy or less for nickel (0.0095 mpy after 7 yrs 80 ft from tide at Kure Beach, 0.0075 mpy and negligible pitting after 16 yrs at Cristobal). Ni-Cu, "Monel 400," will tarnish, but attack rate low (0.014 mpy after 7 yrs at Kure Beach and 16 yrs at Cristobal; other tests show lower rates). Ni-Cr, "Inconel 600": 0.0016 mpy (1.3 mils max pit depth) after 7 yrs 80 ft from tide at Kure Beach. Ni-Cr-Fe, "Incoloy 800" and "825": 0.006 mpy (0.9 and 0.7 mil max pit depths, resp.) after 7 yrs at Kure Beach. Ni-15/22Cr-3/7Mo alloys such as "Hastelloys F" and "G", "Inconel 700" and "718", "Illium R" and "Elgiloy" are even more res. Most res. of all (only titanium alloys have comparable res.) are Ni-16/22Cr-9/18Mo alloys like "Hastelloy C", "C-276" and "X", "Inconel 625", "MP35N" (based on preliminary tests), "Chlorimet 3" and "Rene 41". Res. in splash zone is virtually as good as in atmos, but may be somewhat reduced in tide zone. **Precious metals**—except for silver,

which tarnishes, especially if sulfur compounds present, major noble metals, e.g. platinum, palladium and gold, virtually immune to attack, platinum being the most resistant. **Stainless steel**—good to high res.; austenitic grades generally preferred because of greater res. to staining. Type 304: < 0.1 mpy 800 ft from tide at Kure Beach (somewhat more staining but also negligible attack 80 ft from tide). Type 316 even more resistant. Types 301, 316 and 321 free from pitting and weight loss after 8 yrs at Cristobal. Martensitic grades, typified by 410, may resist after few months, pit on long term (up to 5 mils deep after 8 yrs at Cristobal, but negligible weight loss, e.g., 0.007 mpy). Ferritic 430 subject to partial rusting after about 1 yr at Christobal, but weight loss negligible. Resistance in splash zone also good (austenitic grades again superior). However, subject to some attack in tide zone, e.g., 0.02 mpy for 316 stainless, 0.11 mpy for 304 after 8 yrs in Pacific off Canal Zone. **Tantalum**—should be res. **Tin**—good res.; 0.07 mpy at Sandy Hook, 0.11 mpy at La Jolla, after 20 yrs 0.09 mpy after 10 yrs at Key West. **Titanium and alloys**—excellent res.; immune to crevice attack, pitting and general corrosion at ambient temperatures. Corrosion rate nil for commercially pure titanium after 5 yrs 80 ft and 800 ft from tide at Kure Beach. Also virtually immune to corrosion in splash and tide zones. **Tungsten**—

TABLE 44.4 Corrosion Data by Environment and Material (*Continued*)

Atmosphere — Marine (Continued)

should be res. **Wrought iron**—somewhat similar res. to carbon steel. 1.2 mpy at Halifax, Nova Scotia; 2.2 mpy, Aukland, New Zealand; 4.7 mpy, Plymouth, England; 11 mpy, Colombo, Ceylon; 2.1 to 3.5 mpy, 300 ft from tide at Cristobal. **Zinc and alloys**—good res.; 0.02 to 0.03 mpy at Key West; 0.06 mpy, Sandy Hook; 0.05 to 0.08 mpy, La Jolla, Calif; for various grades of rolled zinc after 10 to 20 yrs. At Kure Beach, rolled zinc contaminated with traces of iron: 0.4 to 0.5 mpy 80 ft from tide (0.3 mpy at 800 ft) after 6 mos.; 0.3 to 0.4 mpy, 80 ft (0.2 at 800 ft) after 1 yr.

Ref. Fink, F. W.; Boyd, W. K.; "The Corrosion of Metals in Marine Environments," DMIC Report 245. Battelle (Columbus), May '70. Published by: Bayer & Co., Col., Ohio.

Carbon tetrachloride

Nonmetallics. ABS—unsatis. **Acetal copolymer**—after 6 mos at 120 F: yld str −11%, ten mod −32%, length +1.2%, weight +5.2%, appearance no change. **Acetal homopolymer**—365 days at 120 F: ten mod −44%, ten str −7%, length −0.3%, weight +5.7%. **Butyl rubber**—70 hrs at r.t.: +214% vol change. **Chlorinated polyether**—res at 80 F. **Chlorosulfonated polyethyl**ene rubber—severe effect at r.t. **Fluorocarbon (PVF₂)**—exc to 275 F. **Fluoroelastomer**—little or no effect at 158 F. **Fluorosilicone rubber**—after 7 days at 75 F: ten str. −45%, volume +20%. **Graphite (impervious)**—res 100% boiling. **Hydrocarbon rubber**—severe effect at r.t. **Nitrile rubber**—severe effect at r.t. **Nylon**—little or no att. **Phenolic**—varies with grade, some show little weight change and exc appearance after 1 yr. **Polycarbonate**—not res after 6 mos at r.t. **Polyester (glass reinf)**—NR. **Polyethylene (hi-D)**—marginal after 7 days at 70 F. **Polyimide (glass reinf)**—after 7 days exp retains 92% of flex mod and 76% of flex str. **Polypropylene**—unsatis after 100 days at 140 F. **Polystyrene**—soluble. **Polysulfone**—7 days at 72 F: weight +0.2%. **PVC-acrylic**: alloy—very slightly whitened after 7 days at 73 F. **Silicone rubber**—after 5 days at 120 F: t.s. −45%, volume +20%. **Styrene-acrylonitrile**—moderately resistant at 73 F. **SBR rubber** —after 70 hrs at r.t.: +207% vol change. **Urethane rubber**—severe effect at 122 F. **Vinyl ester (glass reinf)**—rec at 80 F.

Chlorine

Metals. Aluminum—res. to normal amounts (10 ppm or less) used to treat water. **Carbon steels**—res. to dry, liquid or gaseous at r.t. **Columbium**—little or no attack in wet at 205 F. **Lead**—res. to dry; attacked, but suitable for use if moist up to 230 F; res. to amounts used to treat water. **Magnesium**—res. to dry at r.t., attacked if moist. **Molybdenum** —attacked by wet at r.t. and by dry above 480 F (but little weight loss up to 1470 F). **Nickel and alloys**—res. to dry, liquid or gaseous at r.t., and at elevated temps. under certain conditions. **Precious metals**—gold and palladium rapidly attacked; platinum, rhodium and ruthenium slightly attacked, iridium unaffected by dry or moist at moderate temp. Silver has good res. at r.t. **Stainless steels**—austenitic grades have good res. to dry gas at r.t., severely attacked at high temps. or by wet gas. Rate of attack in dry gas about 10 mpy at 400 F, 60 mpy at 600 F, 400 mpy at 800 F. **Tantalum**—no appreciable attack in wet or dry below 300 F. **Tin**—severely attacked. **Titanium** —exc. res. in moist; < 0.1 mpy at r.t. if more than 0.1% water present; rapid attack if dry (< 0.1% water). **Tungsten**—attacked by dry gas at about 480 F. **Zinc**—res. to dry gas. **Zirconium**—res. to dry gas, attacked if moist. **Nonmetallics. Chlorinated polyether** —res to wet or dry at 80 F. **Chlorosulfonated polyethylene rubber**—dry and wet at r.t.: severe effect. **Fluorocarbon (PVF₂)**—exc in dry and wet to 212 F. **Fluorocarbon (TFE, FEP)**— res at 200 F. **Fluoroelastomer**—dry at 212 F and wet at r.t.: little or no effect. **Graphite (impervious)**—res 100% dry at r.t.: no data, not likely **Hydrocarbon rubber**—dry at r.t.: no data, not likely to be compatible; wet at r.t.: severe effect. **Neoprene**—dry at r.t.: minor to moderate effect; wet at r.t.: severe effect. **Nylon**—unsatis in gas at r.t. **Polyethylene (Hi-D)**—unsatis at 70 F. **Polypropylene**—unsatis in gas and marginal in liquid at 68 F. **PVC** —100% dry: unplast satis at 68 F. Silicon carbide—at 390 F: dry −0.1 mpy, wet +0.1 mpy. **Urethane rubber**—dry and wet at r.t.: no data, not likely to be compatible. **Vinyl ester (glass reinf)**—rec in wet and dry at 210 F.

Metals. Aluminum—generally res. Beryllium—initially attacked, but res. in time. Cast irons—rapidly attacked. Even austenitic grades have poor res., 90 mpy in 5% solution at 60 F. Chromium—good res. in dilute sol. at r.t.; no attack in 10% sol. at 54 F, 7 mpy at 136 F. Copper—moderate res.; 2.2 mpy in 0.2% sol. at 70 F after 5 days. Nickel—good res. in dilute sol. at r.t., 0.8 mpy after 5 days in 2% sol. Moderately attacked in higher con. and temps; 5 mpy in 5% sol. at r.t. after 7 days, 20 mpy at 140 F after 7 days. Aeration increases attack. Silver—good res. Stainless steels—high to moderate res., some pitting may occur. in 10% sol. at 210 to 215 F after 4 hrs: 0.5 mpy for 316, 0.8 mpy for 430, 8 mpy for 302 and 304. 10 mpy for 410. In 60 to 78% sol. at 125 F after 5 wks: 0.1 mpy for 304 and 316. Tin—good res. in dilute, air-free sol.; 0.12 mpy in 0.75% sol. after 9 days. Poor res. in hot, con. or aerated sol. Titanium—high res.; 0.5 mpy in all con. at 212 F. Zinc—attacked. Zirconium—high res.; 0.5 mpy and 0.2 mpy max at 140 and 212 F resp. for all con.

Nonmetallics. Acetal copolymer—after 12 mos at 73 F at 10%: yld str +3%, tens and mod −10%, length +0.2%, weight +1.9%, appearance NC. Acrylic—limited service in 80% at 220 F. Chlorinated polyether—res at 250 F. Chlorosulfonated polyethylene rubber—little or no effect at r.t. Fluorocarbon (PVF_2)—exc to 250 F. Fluoroelastomer—little or no effect at r.t. Graphite (impervious)—res to all conc at boiling. Hydrocarbon rubber—little or no effect at r.t. Neoprene—little or no effect at r.t. Nylon—little or no att to some att in 10% at r.t. Polyethylene (hi-D)—satis after 180 days at 122 F. Polyester (glass reinf)—rec in all conc to 200 F. Polypropylene—satis after 30 days at 140 F. Polystyrene—res to 10%. heat reduces res; slight att in 20%: heat reduces res. Polysulfone—7 days at 72 F at 40%: weight +0.4%. PVC—unplasticized satis at 140 F, plast at 68 F. PVC-acrylic alloy—no change in 10% after 7 days at 73 F. Styrene-acrylonitrile—res in 10% at 122 F. Urethane rubber—little or no effect at r.t. Vinyl ester (glass reinf)—rec at 210 F.

Nonmetallics. ABS—satis. Acetal copolymer—after 6 mos at 180 F: yld str +3%, tens mod −15%, length +0.3%, weight +1%, slight discoloration. Acetal homopolymer—1 yr at 73 F in Lestoil: ten str −4%, weight +0.2%. Ethylene-propylene rubber—at r.t. after 70 hrs ten str is 100-105% of original, vol changes −1 to −2%. Fluorocarbon (TFE, FEP)—res to boiling. Nylon—no att. Phenolic—varies with grade, some show little weight change and exc appearance after 1 yr. Polyester (glass reinf)—rec in sulfonated to 140 F. Polyethylene (hi-D)—satis at 70 F. Polypropylene—satis after 30 days at 140 F. Polysulfone—7 days in Lestoil: weight +0.3%. Styrene-acrylonitrile—resistant at 73 F.

TABLE 44.4 Corrosion Data by Environment and Material (*Continued*)

Ethyl alcohol

Nonmetallics. **ABS**—satis in 50%. **Chlorosulfonated polyethylene rubber**—little or no effect at 200 F. **Fluorocarbon (TFE, (FEP)**—res at 400 F. **Fluoroelastomer**—little or no effect at r.t. **Fluorosilicone rubber**—after 7 days at 75 F: t.s. −30%, volume +5%. **Graphite (impervious)**—res 100% boiling. **Hydrocarbon rubber**—little or no effect at r.t. **Neo**prene— little or no effect at 158 F. **Nitrile rubber**—rec. **Polycarbonate**—res in 96% after 6 mos at r.t. **Polyester (glass reinf)**—rec. **Polypropylene**—satis after 100 days at 140 F. **Polystyrene**—slight att; heat reduces res. **Polysulfide rubber**—exc (0-20% vol swell) for 30 days at 80 F. **PVC**—unplast satis at 68 F, some att or absorp at 140 F, plast satis at 68 F. **PVC-acrylic alloy**—no change in 95% after 7 days at 73 F. **Silicone rubber**—after 7 days at 75 F: t.s. −30%, volume +5%. **Urethane rubber**—severe effect at r.t.

Ethylene glycol

Metals. **Aluminum**—res.; attack may occur if less than 0.01% water present, and at elevated temps. **Cast irons**—gray irons have good res.; austenitic and high-silicon irons even more res. **Copper and alloys**—res. **Magnesium**—res. at r.t. **Nickel and alloys**—res. **Stainless steel**—exc. res.; < 0.1 mpy for 302 and 316 at 70 to 160 F.
Nonmetallics. **ABS**—satis. **Acetal copolymer**—after 6 mos at 180 F at 50%: yld str 0% change, ten mod −18%, length +0.4%, weight +1.3%, slight discoloration. **Acrylic**—satis at 100 F. **Butyl rubber**—70 hrs at 212 F: −1% vol change. **Chlorinated polyether**—res at 220 F. **Chlorosulfonated polyethylene rubber**—little or no effect at 200 F. **Epoxy (glass reinf)**—after 30 days little weight change, retains 100% flex str. **Ethylene-propylene rubber**—at r.t. after 70 hrs t.s. is 87-102% of original, vol change negligible. **Fluorocarbon (PVF₂)**—exc to 275 F. **Fluoroelastomer**—little or no effect at 250 F. **Fluorosilicone rubber**—after 7 days at 180 F: t.s. −5%, volume no change. **Graphite (impervious)**—res all conc at 338 F. **Hydrocarbon rubber**—little or no effect at r.t. **Natural rubber**—satis. **Neoprene**—little or no effect at 158 F. **Nitrile rubber**—rec. **Polyacrylate rubber**—after 70 hrs at 212 F: +37% vol change. **Nylon**—satis at 90% at r.t. **Polyester (glass reinf)**—rec to 200 F. **Polyethylene (Hi-D)**—satis after 180 days at 122 F. **Polypropylene**—satis after 1 yr at 73 F. **Polysulfide rubber**—exc (0-20% vol swell) for 30 days at 80 F. **PVC**—unplast satis at 140 F, plast satis at 68 F. **SBR rubber**—after 70 hrs at 212 F: +4% vol change. **Silicone rubber**—after 7 days at 180 F: t.s. −5%, vol no change. **Styrene-acrylonitrile**—resistant at 73 F. **Urethane rubber**—minor to moderate effect.

Ferric chloride

Metals. Aluminum—attacked. **Cobalt** —negligible attack after 1 day in 2% sol. for 50Co-20Cr-15W-10Ni alloy; 0.2 to 10 mpy for various Co-Cr-W alloys. **Columbium**—high res. No attack in 10% sol. after 1 mo. **Chromium**—high res. in dilute sol. at r.t.; no attack in 10% sol. at 54 F; 16 mpy at 136 F. **Copper and alloys**—moderate to severe attack. **Lead**—rapidly attacked. **Magnesium** —rapidly attacked. **Nickel and alloys** —nickel rapidly attacked except in very dilute sol. Most nickel alloys also readily attacked; however, high-molybdenum and chromium alloys, e.g., 54Ni-16Mo-16Cr, may be exception.

Precious metals—platinum has high res.; 0.01 mpy to 10% sol. at r.t. Hot sol. attacks gold, palladium, platinum and the rhodium-and-iridium-platinum alloys. Iridium, rhodium and ruthenium resist hot sol. **Stainless steels**—most grades apt to pit, some very severely. Molybdenum-bearing grades, e.g., 316, 329, generally most res. **Tantalum**—high res. **Tin**—rapidly attacked. **Titanium**— exc. res.; < 0.5 mpy in 1 to 30% con. at 212 F, 0.7 mpy in 50% con. at 235 F. **Zinc**—attacked. **Zirconium** —exc. res. in very dilute sol. at r.t. and moderately elevated temps. Moderate to severe attack with increas-ing con. and temps. Based on 6-day tests: 0.1 mpy at 95 F, 0.2 mpy (140 F), 0.4 mpy (212 F) in 1% sol., 1 mpy (95 F), 0.7 mpy (140 F), 31 mpy (212 F) in 5% sol., 3.9 mpy (95 F), 5.4 mpy (140 F), 145 mpy (212 F) in 10% sol.

Nonmetallics. Acetal homopolymer— 1 yr at 73 F at 5%: ten str −4%. weight −0.9%. **Chlorinated polyether**—res at 250 F. **Chlorosulfonated polyethylene rubber**—little or no effect at 200 F. **Fluorocarbon (PVF.)** —exc in 50% to 275 F. **Fluoroelastomer**—little or no effect at r.t. **Graphite (impervious)**—res all conc at boiling. **Hydrocarbon rubber**—little or no effect at r.t. **Natural rubber** —ebonite is satis; soft is limited. **Neoprene**—little or no effect at r.t. **Nylon**—unsatis at r.t. **Polyester (glass reinf)**—rec to 200 F. **Polyethylene (hi-D)**—satis at 70 F. **Polypropylene** —satis at 73 F **Polystyrene**—res; heat reduces res. **PVC**—satis at 140 F. **Silicone rubber**—unaffected in 60% after 7 days at 212 F. **Styrene-acrylonitrile**—resistant at 122 F. **Urethane rubber**—little or no effect at r.t. **Vinyl ester (glass reinf)**—rec at 210 F.

Freon

Metals. Aluminum—res. to most types, slight attack in others. **Carbon steel**—should be satisfactory. **Copper and alloys**—res. **Magnesium**— res. if dry at r.t.; attacked if moist and at elevated temps.

Nonmetallics. ABS—varies with type: satis in 12, unsatis in 11. **Chlorinated polyether**—res at 150 F. **Chlorosulfonated polyethylene rubber** —little or no effect by 11, 12, 22, 113, 114 at r.t. **Fluoroelastomer**— no to moderate effect by 11, 12 at r.t.; severe effect by 22 at r.t.; little or no effect by 113, 114 at r.t. **Graphite (impervious)**—res 11, 12 at r.t. **Hydrocarbon rubber**—severe effect by 11, 12, 22, 113, 114 at r.t.; minor to moderate effect by 12 at r.t. **Neoprene**—little or no effect by 12, 22, 113, 114 at r.t. **Nitrile rubber**—rec in 11, 12, 13, 32, 113. **Polyethylene (hi-D)**—satis at 70 F. **Polypropylene**—satis at 73 F. **Polysulfide rubber**—exc (0-20% vol swell) in 11, 12, 13, 32, 112, 114, 115, 218; fair (40-80% vol swell) in 22 and 31; unsatis in 21; all values 30 days, 80 F. **Silicone rubber**—varies with type: in Freon 12 vol changes +45 to +150% after 3 days at 75 F. **Styrene-acrylonitrile** —at r.t. not resistant to 11, res. to 12. **Urethane rubber**—minor-moderate effect by 11 at r.t.; little or no effect by 12, 113 at r.t.; severe effect by 22 at r.t.

44.15

TABLE 44.4 Corrosion Data by Environment and Material (*Continued*)

Gasoline

Metals. Aluminum—high res. to refined and anhydrous, attacked in sour. Copper and alloys—high res. to refined, fair to poor res. in sour. Nickel alloy—"Monel" has high res. to refined, poor res. to sour. Stainless steels—high res. for most grades to refined: in sour, 302, 304, 316 have high res.; 410, 416, 430 fair res. Tin—high res. if moisture and sulfur-free. Nonmetallics. ABS—unsatis. Acetal copolymer—after 6 mos at 120 F: yld str −12%, ten mod −12%, length +0.7%, weight +1.5%, appearance no change. Acetal homopolymer—820 days at 73 F in Texaco: ten mod −17%, ten str −7%, length +0.7%, weight +1.6%. Acrylic—unsatis at 100 F. Chlorinated polyether—res at 220 F. Chlorosulfonated polyethylene rubber—minor to moderate effect at r.t. Fluorocarbon (TFE, FEP)—res at 200 F. Fluoroelastomer—little or no effect at r.t. Graphite (impervious)—res boiling. Hydrocarbon rubber—minor-severe effect at r.t. Neoprene—minor-moderate effect at r.t. Nitrile rubber—rec. Nylon—no att. Polyester (glass reinf)—rec at ambient. Polyethylene (hi-D)—marginal after 365 days at 68 F. Polypropylene—satis after 100 days at 140 F, marginal after 1 yr at 73 F. Polystyrene—not res. Polysulfide rubber—exc (0–20% vol swell) for 30 days at 80 F. Polysulfone—7 days at 72 F: weight +0.1%. PVC-acrylic alloy—no change after 30 days at 73 F except very sl staining. PVC—unplast satis at 140 F. Silicone rubber—after 7 days at 75 F, volume +165%. Styrene-acrylonitrile—resistant at 122 F. Urethane rubber—minor-moderate effect at r.t. Vinyl ester (glass reinf)—rec at 100 F.

Grease

Nonmetallics. Acetal copolymer—after 6 mos at 180 F: yld str +4%, tens mod +3%, length +0.2%, weight −0.03%, appearance NC. Acetal homopolymer—240 days at 200 F in chassis lubricant: ten str 0% change, weight −0.3%. Polypropylene—satis at 68 F. Silicone rubber—after 3 days at 300 F: no change in ten str, volume +20%. Styrene-acrylonitrile—moderately resistant at 73 F.

Nonmetallics. ABS—unsatis in Skydrol-500, 700. **Acetal copolymer**—after 6 mos at 180 F in Lockheed 21: yld str −11%, ten mod −41%, length +1.4%, weight +3.6%, appearance no change. **Acetal homopolymer**—after 310 days at 160 F in Lockheed 21: ten str −23%, length +0.3%, weight +0.9%. **Butyl rubber**—70 hrs at 212 F: +3% vol change. **Chlorosulfonated polyethylene rubber**—severe effect by Skydrol-500 at r.t. **Epoxy (glass reinf)**—after 30 days little weight change, retains 100% flex str. **Ethylene-propylene rubbers**—after 70 hrs at 212 F in Skydrol-500A retain 87-99% of ten str, vol changes −24 to +11%. **Fluoroelastomer**—severe effect by Skydrol-500 at r.t. **Fluorosilicone rubber**—after 7 days at 120 F in Skydrol-500A; ten str −35%, volume +10%. **Hydrocarbon rubber**—little or no effect by Skydrol-500 at 250 F. **Neoprene**—severe effect by Skydrol-500 at r.t. **Nitrile rubber**—rec. **Polyacrylate rubber**—after 70 hrs at 212 F: +116% vol change. **Polyimide (glass reinf)**—after 60 days in Skydrol-500 flex mod is 95% of r.t. value. **Polysulfide rubber**—good (20-40% vol swell) in Skydrol for 30 days at 80 F. **Polysulfone**—3 days at 250 F in Skydrol 500A: dissolves. **Silicone rubber**—after 7 days in Skydrol-500A at 160 F: ten str −10 to −80%, volume +10 to +30%. **Styrene-acrylonitrile**—moderately resistant at 73 F. **SBR rubber**—after 70 hrs at 212 F: +10% vol change. **Urethane rubber**—severe effect by Skydrol-500 at 122 F.

Metals. Aluminum—rapidly attacked. **Beryllium**—rapidly attacked at r.t. **Carbon steels**—rapidly attacked. **Cast irons**—unalloyed and low alloy grades rapidly attacked. High-silicon irons, especially those containing molybdenum, have good res. (up to 5 mpy) under most conditions. **Chromium**—rapidly attacked. **Cobalt alloys**—certain Co-Cr-W and Co-Cr-W-Ni alloys attacked at 9 to 52 mpy in 5% con. at r.t. after 24 hrs; may be more rapidly attacked in 10% to con. sol. at moderately elevated temps. **Columbium**—virtually immune to attack (0 to 0.1 mpy) in 18.37% con. at r.t.; 4 mpy in 37% solution at 230 F. **Copper and alloys**—rapidly attacked in con. sol. and moderately elevated temps. For copper, attack at r.t. may range from 4 mpy in 0.03% sol. to over 160 mpy in 37% con. For more res. alloys, rates may vary from 4 to 32 mpy at r.t. to 256 mpy for con. sol. at higher temps. Aeration accelerates corrosion. **Lead**—attacked at about 10 to 13 mpy in con. < 1%, 20 mpy in 5 to 20% sol. at r.t. Antimonial lead fairly resistant in con. below 18% up to 212 F. **Low alloy steels** —generally similar to carbon steels. **Magnesium**—rapidly attacked. **Molybdenum**—high res. to hot or cold sol.; 1.1 mpy for 1 to 20% con. at 160 F after 2 days, appreciably higher attack for longer periods; 0.3 mpy in con. sol. at 230 F. Presence of oxidizing agents accelerates attack. **Nickel and alloys**—nickel is only moderately attacked (1-10 mpy) in air-free, dilute (up to 10%) sol. at r.t. Rate of attack increases with increasing con.; 70 mpy for 30% air-free sol. at r.t. Substantially greater attack (50 to 90 mpy) in air saturated solutions up to 30% con. at r.t. Certain alloys have better res. (62Ni-

TABLE 44.4 Corrosion Data by Environment and Material (*Continued*)

Hydrochloric acid (Continued)

28Mo considered best, 54Ni-16Mo-16Cr second best). Rates for 62Ni-28Mo: 0.3 to 2 mpy for 2 to 37% con. at r.t.; 6 to 9 mpy for 2 to 15% con. at 150 F; 3 mpy for boil. (214 F) 2% sol. **Precious metals**—generally exc res. in absence of oxidizing agents. No appreciable attack for gold, iridium, osmium, palladium, platinum, rhodium and ruthenium in 36% con. at r.t. after 1 week. At 212 F, iridium, ruthenium and rhodium unattacked; gold and platinum slightly attacked; palladium and osmium attacked at 2 and 6.1 mpy resp. Silver is res. under certain conditions, but subject to attack with increasing con. and temps. **Stainless steels**—rapidly attacked. Higher nickel alloys are less susceptible to attack, but are res. only in very dilute sol. **Tantalum** —high res.; < 1 mpy in 18% solution at 75 F, no attack in 37% solution at 230 F. **Tin**—moderate res., 2 to 14 mpy, in dilute (up to 6%); 2% sol. r.t. Poor res. in high con. or aerated sol. **Titanium**—rapidly attacked; addition of 0.15 Pd to Ti or nitriding improves res. in dilute sol. **Tungsten**—generally res.; no attack in dilute (10%) sol. at r.t., moderate attack (18 mpy) at elevated temps. **Zinc**—rapidly attacked. **Zirconium**—high res.; < 1 mpy in boiling 20% sol. and 37% sol. at 212 F.

Nonmetallics. ABS—in 50% satis after 30 days at r.t., unsatis in 140 F. **Acetal copolymer**—not rec in 10%. **Acetal homopolymer**—90 days at 73 F at 10%: unsatis. **Acrylic**—satis in 40% at 150 F, limited service in 30% at 212 F. **Acrylic-PVC alloy**—no change in 10% after 7 days at 73 F. **Alumina (porous)**—res 35% at 212 F. **Butyl rubber**—70 hrs at r.t. in conc: no change. **Chlorinated polyether**—res to 35% at 250 F. **Chlorosulfonated polyethylene rubber**—varies from little to no effects to 37% at 122 F to minor-moderate effect at 200 F. **Epoxy (glass reinf)**—after 30 days in 10% little weight change, retains 60% flex str. **Ethylene-propylene rubber**—after 70 hrs in 30% at r.t. retains 94-99% of ten str, vol change −1 to +4 %. **Fluoroelastomer**—varies from little or no effect to 37% at 158 F, to minor-moderate effect at 230 F. **Fluorocarbon (TFE, FEP)**—res to 0-100% at boiling. **Fluorocarbon (PVF₂)**—exc in conc to 275 F. **Glass-ceramic**—res in 37% at 194 F. **Graphite (impervious)**—res to all conc at boiling. **Hydrocarbon rubber**—no effect to moderate effect at r.t. **Natural rubber**—satis. **Nitrile rubber**—variable in 20, 30%; not res in 37%. **Nylon**—unsatis in 10% at r.t. **Phenolic**—varies with grade, some show little weight change and exc appearance in 10% after 1 week. **Polyacrylate rubber**—after 70 hrs at r.t. in conc: 4% vol change. **Polyester (glass reinf)**—rec in 37% to 140 F, 20% to 160 F, 10% to 200 F. **Polyethylene (hi-D)**—satis in 37% after 90 days at 70 F. **Modified polyphenylene oxide**—no effect in conc after 3 days at 185 F. **Neoprene**— varies from little or no effect to 37% at r.t. to severe effect at 200 F. **Polysulfide rubber**—varies from exc (0-20% vol swell) in 10%, to good (20-40% vol swell) in 50% to unsatis in 100%; all values for 30 days, 80 F. **Polysulfone**—7 days at 72 F at 20%: weight +0.4%. **Polypropylene**—satis in 35% after 180 days at r.t., marginal after 100 days at 140 F. **Polystyrene**—res 10%, heat reduces res.; slight att in 38%. **PVC**—plasticized and unplast satis in 22% at 140 F; in conc HCl plast and unplast satis at 68 F, plast satis at 140 F, unplast some att at 140 F. **Silicon carbide**— minus 0.1 mpy in 25% at boiling temp. **Silicone rubber**—after 3 days at 150 F, −35%, volume +10%. **Styrene-acrylonitrile**—at 122 F res in 25%, moderately res in 37%. **SBR rubber**—after 70 hrs at r.t. in conc: +3% vol change. **Urethane rubber**—severe effect at r.t. **Vinyl ester (glass reinf)**—rec in 37% at 210 F.

Motor oil

Nonmetallics. ABS—satis. Acetal copolymer—after 6 mos at 180 F: yld str +5%, ten mod 0% change, length −0.1%, weight −0.1%, appearance no change. Acetal homopolymer—1 yr at 160 F: ten mod +2%, ten str +3%, length −0.3%, weight −0.2%. Chlorosulfonated polyethylene rubber—severe effect at r.t. Fluoroelastomer—little or no effect at r.t. Hydrocarbon rubber—severe effect at r.t. Polypropylene—satis after 100 days at 140 F. Polystyrene—slight att; heat does not reduce res. Polysulfide rubber—exc (0-20% vol swell) for 30 days at 80 F. Polysulfonate—7 days at 72 F: weight unchanged. Silicone rubber—after 3 days at 300 F: t.s. −5%, vol no change. Urethane rubber—little or no effect at 158 F.

Nitric acid

Metals. Aluminum—good res. at con. above 82% at r.t. and slightly elevated temps. Beryllium—good res. to con. sol. if cold (violent reaction upon heating). Slow attack in dilute sol. Carbon steels—attacked by dilute and intermediate sol. Cast irons—high-silicon and high-chromium grades generally res.; others rapidly attacked. Chromium—attacked; 12 mpy after 1 day in 32% sol. at 60 F. Cobalt alloys—certain Co-Cr-W and Co-Cr-W-Ni alloys have good res. based on 24 hr tests (0.5 to 6 mpy in boil. 10% sol, 0.5 to 32 mpy in 40% con. at 150 F). Columbium—high res.; no attack in 70% con. at 75 F and 212 F. Copper and alloys—rapidly attacked except by very dilute (0.1% max) sol. Lead—moderately attacked in high con. (48 mpy in 95% sol at r.t.). Severely attacked in con. below 80%. Low alloy steels—generally similar to carbon steels. Magnesium—rapidly attacked. Molybdenum—rapidly attacked by boil. con. sol. and dilute attacked by boil. con. sol. at r.t. Nickel and alloys—nickel rapidly attacked except possibly in very dilute (less than 0.5%) cold sol. Of the alloys, 47Ni-22Cr-17Fe-6Mo and 42Ni-30Fe-22Cr-3Mo are most res. For 42Ni alloy: 0.5 mpy after 1 mo. in white fuming acid at r.t. (43 mpy after 1 wk at 160 F). For 47Ni alloy: 0.1 mpy in 10 to 70% con. at r.t.; 0.1 to 1 mpy at 150 F and 0.4 to 16 mpy at boil., under same conditions. Precious metals—exc. res in 60% and 95% con at r.t. for iridium, platinum, rhodium and ruthenium. These metals also highly res. to 95% concentrations at 212 F. Poor res. for osmium and palladium under these conditions. Gold highly res. to 70% con. at r.t., but subject to some attack in 95% sol. Silver rapidly attacked. Stainless steels—austenitic grades generally have high res.; < 0.1 mpy for 304, 321 and 347 in all con. at r.t., and up to 200 F in dilute sol. These grades have high to moderate res (0.1 to 50 mpy) in 10 to 95% sol from r.t. to boil. pt. (210 to 240 F); rate of attack increasing with con. and temp. Type 430 also has good res.; < 0.1 to 0.4 mpy in con. up to 80% at 0 to about 125 F. In general, all grades are attacked (45 to 400 mpy) by fuming nitric. Tantalum—only slightly attacked by con. sol. at r.t. and mod-

TABLE 44.4 Corrosion Data by Environment and Material (*Continued*)

Nitric acid (Continued)

rately elevated temps.; < 1 mpy in 0% sol at 75 and 185 F. Negligible ttack in red fuming acid at 250 to 00 F. **Tin**—poor res.; 150 mpy in erated or air-free 3% sol. at r.t. itanium—exc res. to all con. including fuming at r.t. and at least up to 0% con. at 95 F. However, under ertain conditions (e.g. less than .34% H2O or more than 6% NO.) 1 fuming nitric, pyrophoric reactions nay occur. Moderate res., 0.2 to 8 py in 5 to 65% sol. at 212 F; < 0 mpy in all con. up to 345 F. **Tungsten**—generally res. in dilute (10%) ol. at r.t. **Zinc**—attacked. **Zirconium**—high res. in con. up to 69.5% at oil. pt.; < 1 mpy at 95 F. in White uming, < 1 mpy at 160 F. in red uming, < 5 mpy at 60 F. **Nonmetallics. ABS**—unsatis in conc; n 20% satis after 30 days at r.t., nsatis at 140 F. **Acetal copolymer**—ot rec in 10%. **Acetal homopolymer** —275 days at 75 F at 10%: unsatis.

Acrylic—lim service at r.t. **Acrylic-PVC alloy**—no change in 70% after 7 days at 73 F. **Alumina (porous)**—res 70% at 212 F. **Butyl rubber**—70 hrs at r.t.; excessive softening in conc. **Chlorinated polyether**—res to 10% at 180 F, 70% at 80 F, 100% not rec. **Chlorosulfonated polyethylene rubber**—little or no effect by 10% at r.t., severe effect by 30% at 158 F. minor-moderate effect by 60-70% at r.t. **Ethylene-propylene rubbers**—after 70 hrs at r.t. in 10% t.s. is 87-109% of original, vol changes −3 to +10%. **Epoxy (glass reinf)**—after 30 days in 10% little weight change, retains 50% flex str. **Fluorocarbon (TFE, FEP)**—res to 0.100% at b.p. **Fluorocarbon (PVF₂)**—exc in conc to 120F. **Fluoroelastomer**—little or no effect by 10-70% at r.t. **Fluorosilicone rubber**—after 7 days at 75 F in 70%; t.s. −40%, volume +5%. **Glass (borosilicate)**—satis at 150 F. **Glass-ceramic**—exc in 70%

at 194 F. 98% at 77 F. **Graphite (impervious)**—res 0.10% at 185 F, 10-20% at 140 F, NR over 20%. **Hydrocarbon rubber**—minor-moderate effect by 10-30% at r.t., severe effect by 30% at 158 F and 60-70% at r.t. **Natural rubber**—not generally rec in 20%. **Neoprene**—minor to moderate effect by 10% at r.t., severe effect by 30%. **Nitrile rubber**—varies in 20, 30%; not rec in 40%. **Nylon**—unsatis in 10% at r.t. **Phenolic**—Varies with grade, some show little weight change and exc appearance in 10% after 1 week. **Polyacrylate rubber**—after 70 days at r.t.: +52% vol change in conc. **Polyester (glass reinf)**—rec in 5% to 140 F. **Polyethylene (hi-D)**—satis in 25% at 70 F, marginal in 50% at 70 F. **Polyimide (glass reinf)**—7 days in 10%: retains 94% of flex mod and 80% of r.t. flex str. **Modified polyphenylene oxide**—no effect in 10% after 3 days at 185 F.

Polypropylene—satis in 75% after 180 days at 68 F, unsatis after 100 days at 100 F. **Polystyrene**—not res to 20%. **Polysulfide rubber**—unsatis in 10% for 30 days at 80 F. **Polysulfone**—7 days at 72 F at 71%: weight +3.8%, surface attacked, discolored. **PVC**—varies from satis at 5% at 140 F to unsatis in 95% at 68 F. **Silicon carbide**—at boiling temp, 0.0 mpy in 50%, −0.2 mpy in 70%. **Silicone rubber**—little change in 10% at 75 F; after 3 days at 150 F in 50% ten str −80%, volume +5%; after 7 days at 75 F in 70% t.s. −40%, volume +5%. **Styrene-acrylonitrile**—at 122 F resistant in 10%, not res in 25%. **SBR rubber**—after 70 hrs in conc at r.t. disintegrates. **Thermoplastic rubber**—after 2 weeks at r.t. 4-5% wt gain, 90+% decrease in ten str. **Urethane rubber**—severe effect by 10-70% at r.t. **Vinyl ester (glass reint)**—rec in 20% at 150 F.

Metals. Aluminum alloys—tend to pit, rate of penetration generally decreasing with increasing oxygen content. Pitting rate highest first yr, tapering off to much lower rate in time. In well-aerated waters pitting not a serious problem for certain alloys; "2000" and "7000" series alloys most susceptible; "5000" alloys relatively immune. Depth of pitting generally increases with increasing ocean depths; even 5000 alloys may show severe pitting at great depths. Alloys most susceptible to pitting also generally most susceptible to crevice attack, and vice versa. Crevice attack also more severe at greater ocean depths than in surface waters. Beryllium—will pit, rate of pitting being most intense during first 2 mos. Carbon steels—chloride ion (sea salt is about 55% chloride) very corrosive to carbon steels, rate of attack increasing with increasing velocity, available oxygen, temperature and pollutants. Certain biofouling, mineral deposits and film formations tend to reduce attack. Rate of attack tends to be less at greater ocean depths than in surface waters. Average penetration rates generally 2 to 5 mpy. Cast iron—res. about half that of carbon steel. Chromium—should be res. in sheet form and local attack apt to be less than for stainless steel. Cobalt and alloys—electrolytic cobalt moderately attacked: 0.7 mpy after 2 yrs; may tend to pit in quite sea water. Co-Cr-Mo alloys have high res. Columbium—should be resistant; no measurable attack after 6 mos. Copper and alloys—for copper: 0.5 to 2 mpy typical in shallow and deep ocean; lower rates reported, e.g. 0.38 mpy after 16 yrs in shallow depths of Pacific off Panama Canal. Single phase brasses, silicon and phosphor bronzes are generally similar. Beryllium-copper slightly more res. Res. of iron-modified alloy 194 seems to be considerably greater than for copper. Brasses vary res.; high zinc grades tend to fail by dezincafication, especially those of two or more phases. Alloys with 15% Zn or less not as apt to fail in this manner. Red brass, Alloy 230, similar to copper (0.5 to 2 mpy) in deep and shallow waters. Arsenic inhibited admiralty brass also resistant (0.6 mpy after 3 yrs in both shallow and deep waters). Aluminum brass (Alloy 687) has very high res.; 0.7 mpy after 5 mos., 0.2 mpy after 3 yrs at 6000 ft. Silicon bronze similar to copper in rate of attack. 5% aluminum bronze has high res. (< 1 mpy after several mos., 0.3 mpy after 3 yrs). 7% aluminum bronze prone to dealuminumization. Cupronickels have high res.; 0.1 to 1.3 mpy after several mos., about 0.8 mpy after 3 yrs. Lead—0.4 to 1.2 mpy at shallow depths (6 mos. to 4 yrs); 0.3 to 1.1 mpy (after 6 mos.) at 2000 to 6000 ft. Low alloy steels—generally similar to carbon steels. Magnesium and alloys—poor res.; highly purified distilled grade attacked at rate of 10 mpy; commercial grade often corrodes at 100 times this rate, largely because of impurity content. Molybdenum—slight attack: 0.3 mpy, in synthetic ocean water at r.t. Attack increases at moderately elevated temps.; 2.1 mpy (140 F), 3.5 mpy (212 F). Nickel and alloys—relatively poor res. for nickel; 5 mpy in rapidly flowing water, greater attack with severe pitting in quiet waters. Ni-Cu "Monel 400" and "K500" have high res. in high velocity waters, but tend to pit in quiet waters. Cupronickels are more resistant than "400." Molybdenum imparts virtual immunity to nickel alloys as evidenced by exc res. of "Hastelloy B" (Ni-28Mo-5Fe) and Ni-Cr-Mo alloys like "Hastelloy C", "Inconel 625", others. "Inconel 600" and "X750" resist well-aerated sea water, but apt to pit. "Inconel 718" (3% Mo) much more res. Overall, "Hastelloy C" and "Inconel 625" are best (based on extensive data): among common metals, their res. to sea water is said to be equalled only by titanium. Precious metals—platinum and gold have exc. res. Palladium also res., but less so than platinum. Stainless steels—In general, all grades susceptible to local attack, highly alloyed grades being most res. Although 304 and 316 will pit, satisfactory service is possible in moderate to high velocity waters. Alloy 20Cb superior to 304 and 316, especially in high velocity waters, but will also pit under low velocity conditions. Ferritic and martensitic grades generally not recommended. 17-4PH can be used effectively under high velocity conditions. Tantalum—tantalum and 90Ta-10W alloy virtually immune at ambient conditions. Tin—tends to pit, severely at times. Titanium alloys— virtually immune to attack at all depths and velocities, including polluted, diluted and hot waters as well as waters containing chlorine, ammonia, hydrogen sulfide gases or excess carbon dioxide. Tungsten—subject only to slight attack at ambient and moderately elevated temps., e.g. 0.3 mpy after 6 months in sea water; 0.2 mpy (95 F), 0.3 mpy (140 F), 0.7 mpy (212 F) in synthetic sea water.

TABLE 44.4 Corrosion Data by Environment and Material (*Continued*)

Sea water (Continued)

Wrought iron—somewhat more res. than carbon steel. **Zinc**—generally attacked at rates of 1 to 2 mpy; higher rates have been reported. **Zirconium**—normally high res.; however, presence of free chlorine will cause attack.

Ref. Fink, F. W.; Boyd, W. K.; "The Corrosion of Metals in Marine Environments". DMIC Report 245, Battelle (Columbus). May '70. Published by: Bayer & Co.. Columbus, Ohio.

Nonmetallics. Chlorosulfonated polyethylene rubber—little or no effect at r.t. **Fluorocarbon (PVF)**—exc to 275 F. **Fluoroelastomers**—little or no effect at r.t. **Glass (borosilicate)**—satis at 150 F. **Hydrocarbon rubber** —little or no effect at r.t. **Neoprene** —little or no effect at r.t. **Polyethylene (hi-D)**—satis at 70 F. **Polypropylene**—satis at 212 F. **PVC**—satis at 140 F. **Urethane rubber**—little or no effect at r.t.

Sodium chloride

Metals. Aluminum—attacked. **Copper**—moderate res., aeration accelerates attack. After 2 days in 1N sol. at r.t.: 2.6 mpy (unagitated), 4.6 mpy (air-agitated). Up to 12 mpy reported after 3 days under strong aeration and agitation in 3% sol. **Chromium**—high res. in dilute sol.; no attack in 10% sol. at 54 F and 136 F. **Lead**—good to moderate res. in dilute sol.; 0.2 to 1.2 mpy in 0.25 to 5.7% con. at 46 F after 200 days. 4 to 5 mpy in 5.5% sol. after 2 days. **Magnesium**—rapidly attacked. **Nickel and alloys**—generally good res. **Tantalum**—res. **Tin**—good res. in very dilute sol.; 0.3 mpy after 1 wk, 0.6 after 1 mo, in 1.3% sol. at 68 F. **Titanium**—exc. res., 0.01 mpy in 3% boil. sol., 0.1 mpy in 29% sol. at 230 F, 0.05 mpy in boil. saturated con. under unaerated conditions. **Zinc**—moderate res. in very dilute sol.; 5 to 10 mpy after 1 mo. in 3.5% sol. at r.t. **Zirconium**—high res. in 3% sol. at r.t.

Nonmetallics. ABS—satis after 30 days at 140 F. **Acetal copolymer**—after 6 mos at 180 F at 10%: yld str +4%, tens mod −10%, length str +4%, weight +0.49%, slight discoloration. **Acrylic**—satis in 30% at 200 F. **Acrylic-PVC alloy**—no change in 10% after 7 days at 73 F. **Butyl rubber**—70 hrs at 212 F: no change in sol'n. **Chlorinated polyether**—res at 250 F. **Chlorosulfonated polyethylene rubber**—little or no effect at r.t. **Epoxy (glass reinf)**—satis in 10%, little weight change, retains 90+% flex str. **Fluorocarbon (PVF₂)**—exc to 275 F. **Fluoroelastomer**—little or no effect at r.t. **Glass (borosilicate)**—satis at 150 F. **Graphite (impervious)**—res all conc at boiling. **Hydrocarbon rubber**—little or no effect at r.t. **Modified polyphenylene oxide**—no effect after 3 days at 185 F. **Neoprene**—little or no effect at r.t. **Nitrile rubber**—rec **Nylon**—satis at r.t. **Phenolic** —varies with grade, some show little weight change and exc appearance in 10% after 1 yr. **Polyacrylate rubber**—after 70 days at 212 F: 2% vol change in sol'n. **Polyester (glass reinf)**—rec to 200 F. **Polypropylene** —satis in 10% after 30 days at 140 F. **Polystyrene**—slight att in 20%. **Polysulfide rubber**—exc (0-20% vol swell) in 10% for 30 days ae 80 F. **PVC**—satis at 140 F. **SBR rubber**—after 70 hrs in sol'n at 212 F: no change. **Silicone rubber**—no vol change in 2% after 7 days at 75 F. **Styrene-acrylonitrile**—resistant at 122 F. **Thermoplastic rubber**—satis in 10% after 2 weeks at r.t. **Urethane rubber** —little or no effect at r.t. **Vinyl ester (glass reinf)**—rec at 210 F.

Metals. Aluminum—rapidly attacked. **Beryllium**—attack not as severe as on many other materials, but use is not recommended for extended periods over 1000F. **Carbon steels**—res. in dilute sol., attacked in hot con. sol. **Cast irons**—moderately res.; 5 mpy for gray cast iron up to 70% con. and up to about 180 F. Austenitic cast irons attacked at rate of < 5 mpy at temps. up to boil. pt. in con. up to 70%. Rate of attack increases rapidly above 70% con. at temps. near boil. **Cobalt**—certain Co-Cr-W and Co-Cr-W-Ni alloys have high res.; negligible attack to 0.6 mpy in 50% con. at 150 F after 1 day. **Columbium**—embrittled by boil. sol. of even low con. **Copper and alloys**—both copper and alloys are moderately attacked in dilute sol. For copper, 14 mpy in 3.9% sol. at 86 F after 2 days under static conditions (20 mpy under air-agitated cond.); lower rates have been reported. Of alloys, cupronickels most resistant. **Iron and carbon steels**—moderate to good res. in very dilute sol.; 1 mpy for mild steel in 50% sol. at 100 F after 7 mos.; rapid attacks in 73% sol. after 4 mos. **Lead**—moderate res. in very dilute unaerated sol.; 9.4 mpy in quiet (47 mpy air-agitated) 3.8% sol. after 2 days at r.t. Poor res. in hot or con. sol. **Moly-**

bdenum—severely attacked in fused at 1000 F. **Nickel and alloys**—exc. res.; 0.01 mpy for nickel, 67Ni-33Cu and 76Ni:16Cr-7Fe in 50% con. at 100 F after 7 mos.; 0.1 mpy for 67Ni:33Cu, 1 mpy for nickel and 76Ni:16Cr-7Fe alloy, in 73% con. at 265 F after 7 mos. In fused 100% con., 0.9 to 2.5 mpy for nickel at 750 to 1075 F. **Precious metals**—silver has high res., even at elevated temps. Gold and platinum metals also have good res. **Stainless steels**—exc. res. for both chromium and chromium-nickel grades in dilute sol. up to moderately elevated temps. High to moderate res. in high con. at moderately elevated temps. Cracking may occur near boil. pt. After 3 to 4 mos.: in 20% con. at 120 to 140 F., < 0.1 mpy for 302, 304, 309, 310; 0.1 mpy for 410, 430. In 72% con. at 245 to 255 F, under moderate aeration, 0.1 mpy for 21Cr-34Ni-0.5Cu alloy, 0.3 mpy for 329, 3.1 mpy for 316, 3.7 mpy for 304, 6 mpy for 410, 32 mpy for 430. In 73% nonaerated sol. at 212 to 248 F, 38 mpy for 302, 45 mpy for 304. **Tantalum**—res. to 5% boil. sol., but attacked in fused at 605 F, severely attacked at 1000 F. **Titanium**—high to moderate res.; 0.8 mpy in 10% boil. sol.: 0.1 mpy in 28% sol. at r.t.; 5 mpy in 40% sol. at 176 F. **Tin**—mol-

ten, attacked at 1000 F. **Tungsten**—slight attack in hot alkaline sol.: severely attacked at 1000 F. **Zinc**—attacked: after 2 days in 3.9% sol. at r.t. 18 mpy under quiet conditions, 35 mpy under air-agitated con. **Zirconium**—res.

Nonmetallics. ABS—satis in 25%. **Acetal copolymer**—after 6 mos at 180 F at 60%: yld str −3%, ten mod −6%, length −0.1%, weight −0.18%, slight discoloration. **Acrylic**—limited service in 10% at 120 F. **Acrylic-PVC alloy**—no change in 10% after 7 days at 73 F. **Alumina (porous)**—edges rounded in 10% at 212 F. **Chlorinated polyether**—res to 70% at 250 F. **Chlorosulfonated polyethylene rubber**—little or no effect by 73% at 280 F. **Ethylene-propylene rubbers**—after 70 hrs at r.t. in 50% t.s. is 99-118% of original, vol changes 0 to −1%. **Fluorocarbon (PVF₂)**—exc in 50% to 275 F. **Fluorocarbon (TFE, FEP)**—res to 0-100% at boiling. **Fluoroelastomer**—little or no effect by 47% at r.t., severe effect by 50% at r.t. **Fluorosilicone rubber**—after 7 days at 75 F in 50%: ten str −10%, volume no change. **Glass (borosilicate)**—satis at 150 F. **Glass-ceramic**—good (less than 20 mpy) in 1% at 194 F, satis (20-25 mpy) in 7% at 194 F. **Graphite (impervious)**—

res 6-67% at boiling, 67-80% at 275 F. **Hydrocarbon rubber**—little or no effect by 20-73%. **Modified polyphenylene oxide**—no effect in conc after 3 days at 185 F. **Neoprene**—little or no effect by 20, 73% at r.t. and 47% at 158 F. **Nitrile rubber**—rec in 50%. **Nylon**—satis at r.t. **Phenolic**—generally poor res in 10% after 1 week. **Polyester (glass reinf)**—rec in 10 and 25% to 130 F. 5% to 160 F, NR in 50%. **Polyethylene (hi-D)**—satis at 70 F. **Polyimide (glass reinf)**—7 days in 10%: retains 93% of flex mod and 82% of r.t. flex str. **Polypropylene**—satis in 60% after 30 days at 140 F. **Polystyrene**—slight att in 1-50%; heat does not reduce res. **Polysulfide rubber**—exc (0-20% vol swell) in 20% for 30 days at 80 F. **Polysulfone**—110 days at 72 F at 5%: weight −0.03%. **PVC**—unplasticized satis at 140 F, plast some att or absorp at 140 F. **Silicon carbide**—+73 mpy in 25% at boiling temp. **Silicone rubber**—after 7 days at 75 F: t.s. −10%, vol no change. **Styrene-acrylonitrile**—resistant in sat at 122 F. **Urethane rubber**—at r.t. little or no effect by 20%, severe effect by 50%. **Vinyl ester (glass reinf)**—rec in 50% at 210 F.

44.23

TABLE 44.4 Corrosion Data by Environment and Material (*Continued*)

Sulfuric acid

Metals. Aluminum—res. attack in very dilute (1% or less) or very high con. (98 to 100%) at r.t. Rapidly attacked at other con. and higher temps. **Beryllium**—rapidly attacked at r.t. **Carbon steels**—moderately res. at con. above 70% (5 to 20 mpy at 75 F, 20 to 50 mpy at 125 F, 50 to 200 mpy at 175 F in static tests; higher rates likely in service); more res. at 100% con. Rapidly attacked by con. below 70%. **Cast irons**—good res. in certain con. and temps. High-silicon irons generally best, followed by austenitic grades. For gray iron < 5 mpy in con. above 65% at r.t., but rapid attack at lower con. **Chromium**—attacked; 28 mpy after 1 day in 17% sol. at 60 F. **Cobalt and alloys**—cobalt has moderate res.; 9 mpy in 5% con. at r.t. under static conditions. Certain Co-Cr-W and Co-Cr-W-Ni alloys suffer negligible attack in high con. (77 to 96%) at r.t., but may be attacked by 25% sol. at moderately elevated temps. (150 F). **Columbium**—virtually immune to attack in 20% con. at 200 F and 40% and 95% con. at 75 F. Slight attack (0.1 mpy) in 98% con. at 75 F. Slight attack (0.1 mpy) in 98% con. at r.t. At high con. (95%) rate of attack increases with temp.; 0.8 mpy at 120 F, 19 mpy at 212 F, 180 mpy at 290 F. **Copper and alloys**—copper has high res.; up to 2 mpy in 10 to 80% con. at r.t.; decreasing temps. increase attack (6 to 15 mpy at 140 F). In 60 to 70% sol., attack moderate (3 to 12 mpy) up to 176 F. Rates roughly similar for Si-Mn bronze and 70Cu-30Ni cupronickel. Aeration increases attack. **Lead**—good res.; < 5 mpy in 5 to 50% con. Attack markedly increased in con. below 5%. Attack up to 50 mpy at 50 to 97% con. up to boil. temp. Antimonial lead superior to chemical lead at high con. **Low alloy steels**—generally similar to carbon steels. **Magnesium**—rapidly attacked. **Molybdenum**—high res.; 0.15 mpy to cold sol. up to about 96% con. Good res. to boil. sol. up to 50% con. Increasing con. and temps. increase attack severely. **Nickel and alloys**—Nickel has moderate res.; 2 to 9 mpy in unaerated, dilute (1 to 20%) sol. at r.t. Aeration increases attack appreciably (50 to 60 mpy for 1% and 5% sol.) Con. sol. more aggressive: 30 and 70 mpy for 70% and 95% con. Attack 10 to 30 mpy for 5 to 48% sol. at 140 to 180 F. Among nickel alloys, Ni-Mo, Ni-Mo-Cr and Ni-Si grades are best overall (< 5 mpy for virtually all con. to about 200 F to 250 F). 67Ni-33Cu also has moderate res. (< 5 mpy) in con. up to 80%. **Precious metals**—gold, iridium, osmium, palladium, platinum, rhodium and ruthenium have exc. res. in 98% con. at r.t. and all but rhodium (moderately attacked) and palladium (excessively attacked) have high res. at 212 F. Iridium and ruthenium also have high res. at 570 F for 7 hrs in 98% sol; gold only slightly attacked (0.7 mpy). Silver res. dilute sol. at r.t. and is only slightly attacked (0.7 mpy) in boil. 10% and 20% sol. **Stainless steels**—several austenitic grades have high res. in aerated sol. at low and moderately elevated temps. In general, increasing con. and temps. and absence of air-accelerate attack. In nitrogen-saturated 5% sol. at 86 F, < 0.1 mpy for 317, 0.6 mpy (316). 1.2 mpy (310 and 321), 9 mpy (301), 12 mpy (347) and 57 mpy (304). 201, 302, 430 rapidly attacked. In aerated dilute sol. (up to about 10%), 304, 310, 316 and 317 have high res.; < 0.1 mpy at temps. of 0 to 160 F. These grades about equally res. at all con. in aerated sol. at temps up to 70 to 125 F. Types 310 and 317 also have high res. in intermediate (20 to 60%) aerated sol. at 125 to 150 F, 310 being somewhat superior. **Tantalum**—high res.; 0 to 0.1 mpy in 20 to 95% con. at 75 to 350 F. Some attack in 95% sol. at higher temp.; 1.5 mpy at 390 F, 29 mpy at 480 F. In fuming acid: 0.3 mpy at 75 F, 9 mpy at 160 F. **Tin**—moderate res., 2 to 10 mpy in dilute (up to 10%), air-free sol. at r.t. Poor res. in high con.; 70 mpy in air-free 20% sol. at r.t. **Titanium**—high res. in very dilute sol. at r.t.; 0.1 mpy in 1% sol. Moderate to poor res. with increasing con. (75 to 80% sol. being most corrosive) and temps.; 4 mpy in 1% sol. at 100 F, 9 mpy in 5% sol. at r.t. and 30 mpy at 100 F, 60 mpy in 40% sol. at r.t., 250 mpy in 50% sol. at 100 F. Addition of 0.15 Pd increases res. in dilute sol., anodizing improves res. in 40% sol. **Zinc**—attacked. **Zirconium**—high res. < 1 mpy in con. up to 70% up to boil. pt. Severe attack in con. above 80%, especially with increasing temp. **Nonmetallics. ABS**—in 50% after 30 days satis at r.t., unsatis at 140 F. **Acetal copolymer**—NR in 30%. **Acetal homopolymer**—316 days at 95 F at 10%; unsatis. **Acrylic**—limited service in 10% at 180 F, 50% at 100 F, unsatis in 50% at 150 F. **Acrylic-PVC alloy**—no change in 30% after 7 days at 73 F. **Alumina (porous)**—res 96% at 212 F. **Butyl rubber**—70 hrs at r.t.: +1% vol change in 50%. **Chlorinated poly-**

ether—res to 80% at 250 F, 90% at 180 F, 96% at 80 F, NR in 98%. **Chlorosulfonated polyethylene rubber** —little or no effect by up to 50% at 250 F, 50-80% at 158 F, 95% at r.t. **Diallyl phthalate**—retains 80% of flex str in 3% after 1 yr. **Epoxy (glass reinf)**—in 3% after 30 days little weight change, retains 75% flex str. **Ethylene-propylene rubber**— after 70 hrs in 98% at r.t. retains 23-80% of ten str, vol change +5 to +8%; in 10% at 212 F t.s. is 96-111% of original, vol change 0 to −2.5%. **Fluorocarbon (TFE, FEP)** —in 0-100% TFE res to 500 F, FEP to 400 F. **Fluoroelastromer**—little or no effect up to 80% at r.t., 60% at 250 F, 90% at 158 F. **Fluorocarbon (PVF$_2$)**—exc in 60% to 230 F, 85% to 150 F. **Glass-ceramic**—exc in 98% at 194 F. **Graphite (impervious)** —res 0-70% at boiling, 70-85% at 338 F, 85-90% at 300 F, 90-93% at 160 F, 93-96% at r.t., NR over 96%. **Hydrocarbon rubber**—at r.t. little or no effect to 50%, severe effect 60-95%. **Natural rubber**—satis in 50%. **Neoprene**—little or no effect up to 50% at 158 F, generally severe effect over 50% at r.t. **Nitrile rubber**—rec in 17, 30%; varies in 42, 56%; not rec in 70%. **Nylon**— unsatis at r.t. **Phenolic**—varies with grade, some show little weight change and exc appearance in 30% after 1 yr. **Polyacrylate rubber**—after 70 days at r.t.: disintegrates in conc. +3% vol change in 50%. **Polyester (glass reinf)**—rec in 70% to 140 F. **Polyethylene (hi-D)** at 70 F satis in 70%, marginal in 95%. **Polyimide (glass reinf)**—7 days in 10%: retains 88% of flex mod and 88% of r.t. flex str. **Modified polyphenylene oxide**—no effect in 90% after 3 days. **Polypropylene**— satis in 97% after 30 days at 140 F. **Polystyrene**—slight att in 10-50%: heat reduces res, NR in conc. **Polysulfide rubber**—varies from exc (0-20% vol swell)in 10%, to fair (40-80% vol swell) in 20%, to unsatis in 50% and 100%; all values 30 days, 80 F. **Polysulfone**—69 days at 72 F at 95%: dissolves. **PVC**—unplast satis to 80% at 140 F, plast satis to 45% at 140 F: check perf at higher conc. **Silicone carbide**—+0.1 mpy in 80% at boiling temp. **Silicone rubber**—after 3 days at 150 F in 50% ten str −35%, volume no change; decomposes in 95% after 7 days at 75 F. **SBR rubber**—after 70 hrs at r.t.: disintegrates in conc. +3% vol change in 50%. **Styrene-acrylonitrile**—at 122 F resistant in 25%, not res in conc. **Thermoplastic rubber**—satis in 10% after 2 weeks at r.t. **Urethane rubber**—severe effect at r.t. **Vinyl ester (glass reinf)**— rec in 70% at 210 F.

TABLE 44.4 Corrosion Data by Environment and Material (*Continued*)

Waters other than sea water

Metals. Aluminum and alloys—high res. in high purity (distilled or deionized) or water vapor up to about 400 F. In general, good res. to most neutral or nearly neutral waters providing waters do not contain compounds other than salts of alkaline earth metals. Acid waters containing chlorides can cause severe pitting; sulfate waters of low pH also aggressive. Compounds of Cu, Pb, Sn, Ni and Co in waters promote pitting attack. **Beryllium**—high res. at ambient temps. in neutral waters, even under static conditions. Under aeration and flowing cond., good res. at moderately elevated temps. Protection required above 500 F. Presence of chloride ions in water markedly increases attack. Sulfate, cupric or ferric ions also increase attack. **Carbon steels**—fresh waters: normally pit in neutral solutions since protection afforded by rust is usually irregular; supply of dissolved oxygen and deposited protective films being most critical factors governing attack. 2 to 5 mpy avg rate of attack in quiet waters free of salts and containing dissolved air. Agitation or aeration usually increase attack; deposition of compounds usually supresses attack. Presence of various salts or other substances may either increase or decrease attack. Boiler water: supply of dissolved oxygen again critical factor; deaeration common corrosion preventative. Mine waters: can cause severe attack. **Cast irons**—generally similar to carbon steels for plain cast irons. **Cobalt**—good res. in distilled water, 0.2 mpy at 77 F under static conditions. Wear resistant alloys undergo little attack in mine and boiler waters at ordinary temps. **Columbium**—good resistance to 500 F in oxygenated water. Res. seems good under both static and dynamic conditions. **Copper and alloys**—copper has good res. to all fresh waters, attack ranging from 0.2 to 1 mpy, sometimes less. Hard waters seldom corrosive, but soft waters, especially with substantial amounts of free carbon dioxide, may be sufficiently corrosive to cause green stains on plumbing fixtures by reacting with soap. Distilled water not very corrosive, but will pick up trace of copper on long standing. Carbonated water much more corrosive, after 20 hrs at r.t. in water saturated with air and carbon dioxide: 2 to 10 mpy in city water, 2 to 6 mpy in distilled water. Some copper alloys, e.g., red brass, better than pure copper for fresh water plumbing. Silicon and phosphor bronzes, cupronickels, cast bronzes and nickel silvers also have high res. Tin-bearing copper alloys, e.g., 88Cu-10Zn-2Sn, most res. to river waters containing acid-mine drainage. **Lead**—not attacked by pure distilled water free of dissolved gases, but aerated distilled water free of carbon dioxide can be corrosive. Also resists non-potable water, except possibly acid mine waters. Soft waters attack lead sufficiently to have discontinued its use for potable soft water systems (toxicity problem). Fresh waters may also be corrosive if containing carbon dioxide or small amounts of organic acids. Small amounts of nitrates in ground water also increase corrosivity. **Low alloy steels**—more or less similar to carbon steels for fresh waters, with variations in attack more likely under short-term submergence, e.g., over long-term no significant difference in attack of copper structural steels and plain carbon steels. In partly stagnant waters corrosion rate of low alloy steel similar to carbon steel, i.e., about $\frac{1}{2}$ rate in aerated waters. Nickel additions may be marginally beneficial, e.g., in Pittsburgh water at 140 to 145 F: 14.6 mpy for mild steel, 13.2 mpy for 1.65Ni steel, 12.2 mpy for 3.61Ni steel, 12.6 mpy for 5.20Ni steel. In simulated reactor boiling water, essentially no difference between carbon steels and steels of up to 5% alloy content. **Magnesium and alloys**—good res. in stagnant distilled water at r.t. Pitting may occur if small amounts of chlorides, heavy metal salts or carbon dioxide present. Agitation or constant replenishment of water may lead to attack, e.g., little attack on AZ31 alloy after 35 days in stagnant distilled water at 125 F; when water continuously replenished to maintain 6.8pH, rate of attack 7 mpy. Corrosion rates in water at different temperatures: MIA—13.2 mpy (95 F), 6.0 mpy (150 F), 13.2 mpy (180 F), 72 mpy (212 F); AZ92A—0.8 mpy (95 F), 16.3 mpy (212 F); AM100A—0.8 mpy (95 F), 26.9 mpy (212 F). **Molybdenum**—tarnished but not attacked by fresh waters up to moderately elevated temps. Poor res. to oxygenated water at 600 F and to water vapor at 1200 F. **Nickel**—high res. to most natural, fresh, distilled, deionized and high purity waters; attack usually < 0.1 mpy. Attack usually < 0.02 mpy in domestic hot water up to 200 F. Res. to carbonated fresh water also good. 0.2 mpy after 10 days at r.t. and 200 psig. Nickel may be attacked in polluted (acid drainage) rivers, e.g., 3.8 mpy in Monongahela River at 5pH (0.3 mpy, 6.5pH), and severely attacked by acid mine waters. **Stainless steels**—high res. to distilled, tap and other fresh waters, including relatively polluted lake and river waters, cold or hot; e.g., after 490 days in Monongahela containing coal mine drainage and spent pickling acid, attack rate < 0.1 mpy for 304, 316, 410 and 430 steels; 2 mpy for 502. In general negligible attack from boiler, high purity and mine waters under most conditions. **Tantalum**—tarnishes in oxygenated water at 500 F. **Tin**—will tarnish, but virtually immune to distilled water and only slightly attacked by carbonated waters. Attacked by drinking waters purified by addition of strong oxidizing agents which produce nascent oxygen. **Titanium and alloys**—high res. even to brackish river waters, and distilled, degassed-distilled or oxygenated-distilled waters at 500 F. **Tungsten**—no attack from cold or hot water. **Wrought iron**—more or less similar to carbon steels, but may be more res. to pitting under certain conditions. **Zinc**—good res., but only in narrow pH range around neutral point. Can provide good galvanic protection to steel in fresh waters. Moderate attack in aerated distilled water up to about 120 F. severe attack from 120 to 200 F. Aeration increases attack in distilled water, especially if trace amounts of carbon dioxide present. Some rates: 2.0 mpy for plain water at 54 F after 2 months; 4.8 mpy in quiet distilled water. **Zirconium and alloys**—good res. to high temperature waters and steam to 900 F. **Nonmetallics. ABS**—satis. **Acetal copolymer**—rec for continuous use at 180 F, retains nearly original tens str in boiling after 22 weeks. **Acetal homopolymer**—not rec for long-term service over 150 F. **Acrylic**—satis after 5 yrs. **Butyl rubber**—disintegrated after 70 hrs at 212 F. **Chlorinated polyether**—res at 250 F. **Chlorosulfonated polyethylene rubber**—little or no effect at 212 F. **Epoxy (glass reinf)**—after 5 yrs immersion minor weight change. **Ethylene-propylene rubber**—after 70 hrs at 212 F t.s. is 76-110% of original, vol change +2%. **Fluoro-**carbon (PVF)—in brine exc to 275 F. **Fluoroelastomer**—little or no effect at 212 F. **Fluorosilicone rubber**—in steam after 1 day at 100 psi: t.s. -20%, volume no change. **Graphite (impervious)**—res boiling. **Hydrocarbon rubber**—little or no effect at 212 F. **Neoprene**—little or no effect at 212 F. **Nitrile rubber**—rec in distilled. **Nylon**—no att in cold, little or no att in hot. **PVC-acrylic alloy**—no change after 7 days at 140 F except slight staining (none at 73 F). **Phenolic**—varies in distilled with grade, some show little weight change and exc appearance after 1 yr. **Polyacrylate rubber**—after 70 days at 212 F: +23% vol change. **Polyester (glass reinf)**—rec to 200 F. **Polyethylene (hi-D)**—satis after 1 yr at 70 F. **Polyimide (glass reinf)**—in boiling after 7 days flex mod is +3% and flex str is +12% over r.t. values. **Polypropylene**—satis in distilled after 160 days at 140 F. **Polystyrene**—in distilled—res; heat reduces res. **Polysulfide rubber**—exc (0-20% vol swell) after 30 days at 80 F. **Polysulfone**—7 days at 72 F: weight +0.6%. 7 days at 210 F: weight +0.9%. **SBR rubber**—after 70 hrs at 212 F: +10% vol change. **Silicone rubber**—no change after 3 days at 212 F. **Styrene-acrylonitrile**—in distilled res at 122 F. **Thermoplastic rubber**—in distilled satis after 2 weeks at r.t. **Urethane rubber**—little or no effect at 212 F.

†Abbreviations used in this table: ABS = acrylonitrile-butadiene-styrene; att = attack; atmos = atmosphere; boil pt = boiling point; con, connc = concentrated; exc = excellent; flex mod = flexural modulus; flex str = flexural strength; indus = industrial; max = maximum; mos = months; mpy = mils per year; plast = plasticized; ppm = parts per million; PVC = polyvinyl chloride; res = resistance; resp = respectively; r.t. = room temperature; satis = satisfactory; SBR = styrene butadiene; sol = soluble; ten mod = tensile modulus; tens str = tensile stress; unsatis = unsatisfactory; unplast = unplasticized; vol = volume; yld str = yield stress or strength; yrs = years.

SOURCE: R. J. Fabian and J. A. Vaccari, eds., "How Materials Stand Up to Corrosion and Chemical Attack," *Materials Engineering*, vol. 73, no. 2, p. 36, 1971. Reprinted with permission of Penton/IPC.

corrosion (weld decay). Heating to higher temperatures, 1060 to 1120°C, followed by water quenching will redissolve the precipitated carbides and keep them in solution. Appropriate alloying changes can reduce carbide precipitation.

44.3.10   Mechanical Contributions

Fluid flow or mechanical rubbing can cause removal of or damage to a protective oxide, increasing the proximity of the bare metal and the attacking medium. This can result in increased attack rates because a stable oxide layer is frequently a corrosion rate limiter. An example is *erosion corrosion* caused by high flow rates of domestic hot water in copper pipes, especially around fittings, which can generate turbulence. Another example involves the press fit of gears, wheels, pulleys, etc., onto shafts that experience elastic torsion, or bending. Small relative motions occur at the contacting surfaces which mechanically break up protective oxide layers. This type of corrosion is known as *fretting corrosion.*

44.4   CORROSION DATA FOR MATERIALS SELECTION[†]

Subject to the limitations mentioned at the beginning of this chapter, the corrosion data in Table 44.4 can be used as a guide in selecting materials for the environments listed. The organization is first by environments and then by materials, metals followed by nonmetals. The abbreviation NR means not recommended.

REFERENCES

44.1 M. G. Fontana and N. D. Greene, *Corrosion Engineering,* 2d ed., McGraw-Hill, New York, 1978.

44.2 E. Rabald, *Corrosion Guide,* 2d ed., rev., Elsevier Scientific Publishing, Amsterdam, 1968.

44.3 R. J. Fabian and J. A. Vaccari (eds.), "How Materials Stand Up to Corrosion and Chemical Attack," *Materials Engineering,* vol. 73, no. 2, February 1971, p. 36.

[†] Reproduced from R. J. Fabian and J. A. Vaccari, eds., *Materials Engineering,* Penton/IPC, Inc., Cleveland, 1971.

CHAPTER 45
NOISE AND ITS CONTROL

David A. Towers, P.E.
Senior Consultant
Harris Miller Miller & Hanson Inc.
Lexington, Massachusetts

Erich K. Bender, Ph.D.
Divisional Vice President

Paul J. Remington, Ph.D.
Principal Engineer

Eric E. Ungar, P.E., Ph.D.
Chief Consulting Engineer
Bolt Beranek and Newman Inc.
Cambridge, Massachusetts

45.1 INTRODUCTION

Noise is unwanted audible sound. Sound is essentially a fluctuating pressure disturbance that may act locally or propagate away from its source.

Extremely intense sound can cause structural damage or equipment malfunctions, but usually its effect on humans is the primary concern. Noise can annoy people, can lead to interference with speech communication, can interfere with the performance of mental and delicate manual tasks, and—if it is intense enough—can cause discomfort, pain, and temporary or permanent hearing damage. Fortunately, many aspects of noise and its effects are well enough understood to permit their consideration in the design process.

Sound can occur and propagate in any gas, liquid, or solid medium, but sound in air is usually of primary interest. Sound may be produced by any phenomenon that can lead to fluctuating pressure disturbances. These phenomena include (1) rapid expansion of gases or injection of fluid volumes, such as from explosions and engine exhausts; (2) repetitive interruptions or modulation of airflows, such as by siren disks or fluctuating valves; (3) turbulence, as present in fluid streams emerging from nozzles or duct grillages; and (4) vibrating solid surfaces. In many practical situations, several noise-generating phenomena may occur simultaneously; for example, an impact press may generate noise not only because of the structural vibrations it produces but also because of the air it expels from between the impacting surfaces.

Sound, being a pressure disturbance, can propagate in the medium in which it is generated. This propagation need not involve flow or net displacement of the medium; only the disturbance and the energy associated with it move away from the source.

Pressure fluctuations in air can induce fluctuations in other media in contact with the air, and vice versa. Therefore, often sound from a given source reaches an observer not only via a direct air path but also via paths that may involve several media. For example, sound radiated from vibrating gears in a housing may propagate from the air in the housing through an oil layer and through the housing wall into the ambient air. In many practical situations, several parallel paths of sound transmission from a given source to a given observer—including some relatively tortuous paths along complex structures—may be similarly important.

It usually is convenient to consider a noise problem from the "source-path-receiver" viewpoint. This approach facilitates accounting for all significant sources (noise generators), receivers (items or persons affected by noise), and paths along which the noise from the sources reaches the receivers. This approach thus encourages evaluation of all relevant facets of the problem.

The remainder of this chapter introduces noise measurement and analysis, noise effects and standards, and noise control techniques relevant to machine design. For treatment of these subjects in greater depth, texts and handbooks on acoustics should be consulted (for example, Refs. [45.1] through [45.5]), as well as the specific references given throughout this chapter.

45.2 NOISE MEASUREMENT AND ANALYSIS

45.2.1 Noise Measures

Sound or noise can be sensed by measurement of *sound pressure,* the variation in air pressure above and below its equilibrium value. The measure most commonly used is the root-mean-square (rms) sound pressure p_{rms}. The rms sound pressure is obtained by squaring the value of the sound pressure disturbance at each instant of time, averaging the squared values over the sample time, and taking the square root of the result.

Because the range of sound pressure amplitude variations that the human ear can detect extends over several factors of 10, a compressed scale based on the logarithm of the mean square pressure is used. The decibel, abbreviated dB, is a measure of this scale. The corresponding noise descriptor is called the *sound pressure level* L_p, defined as

$$L_p = 10 \log \left(\frac{p_{rms}}{p_0} \right)^2 \qquad \text{dB} \qquad (45.1)$$

where p_0 is a reference pressure, standardized as 20 micropascals (µPa) [2.90×10^{-9} pounds per square inch (lb/in^2)]. This very small reference pressure corresponds to 0 dB and represents approximately the weakest sound that can be heard by an average young, alert person with an undamaged hearing mechanism.

Since decibels are logarithmic measures, sound pressure levels cannot be added by ordinary arithmetic. The sound pressure level L_p (total) corresponding to the combination of n sound pressure levels $L_p(i)$ is calculated from[†]

$$L_p \text{ (total)} = 10 \log \left(\sum_{i=1}^{n} 10^{L_p(i)/10} \right) \qquad (45.2)$$

To describe noise adequately, one must measure not only its *amplitude,* which determines the magnitude of the pressure, but also its *frequency,* which determines its pitch. In any sound, the air pressure alternately rises and falls; for repetitive sounds, each time the pressure rises from its minimum value and returns to that value, it completes one cycle. The number of cycles occurring per second is called the *frequency* of the sound; the unit of cycles per second is hertz (Hz). Frequency is observed subjectively as the tone, or pitch, of a sound. The low frequencies (20 to 500 Hz) have a low-pitch, or bass, sound. The midfrequency range, from about 500 to 3000 Hz, is where most speech information is carried. High frequencies, from about 3000 to 20 000 Hz, tend to be prevalent in whistles, jets, and high-speed machines.

The *wavelength of a sound wave* is defined as the distance the wave travels in a stationary medium during one cycle. Wavelength and frequency are related by

$$\lambda = \frac{c}{f} \qquad (45.3)$$

where c = speed of sound, ft/s (m/s)
$\quad\quad\ f$ = frequency, Hz
$\quad\quad\ \lambda$ = wavelength, ft (m).

The speed of sound in gases depends on the temperature, but not on pressure. At 70°F (21°C), for example, the speed of sound in air is 1128 ft/s (344 m/s), and the wavelength of a 1000-Hz sound wave is 1.128 ft (0.3438 m).

The basic properties of a pure-tone (that is, single-frequency) sound wave are summarized in Fig. 45.1. This figure illustrates a time-history graph of the amplitude of a sound. Note that for this sinusoidal wave, the sound pressure amplitude rises from zero to a positive maximum, then falls through zero to a negative maximum, and then returns to zero during one complete cycle. For this type of wave, the rms value is 0.707 times the absolute value of the peak (positive or negative) amplitude.

Noise from common sources, such as machinery, is usually more complex than the pure tone illustrated in Fig. 45.1. In general, noise consists of a combination of many sinusoidal components, all with different frequencies. Description of such noise requires a noise *spectrum,* which is a graph of sound pressure level versus frequency. Frequency analysis (or spectrum analysis) is essential for any comprehensive study of a noise problem for three reasons: (1) people have different hearing sensitivity and different reactions to the various frequency ranges of noise, (2) different noise sources emit differing amounts of noise at different frequencies, and (3) engineering solutions for reducing or controlling noise are different for low- and high-frequency noise.

Although a noise spectrum is useful for purposes of analysis, it is often convenient to use a single-number measure to describe a noise. The most commonly used measure of this type is the *A-weighted sound level,* expressed in units of dBA. From

---

[†] This corresponds to $p_{\text{rms}}^2 \text{ (total)} = \sum_{i=1}^{n} p_{\text{rms}}^2(i)$, where the individual signals are at different frequencies and/or are uncorrelated.

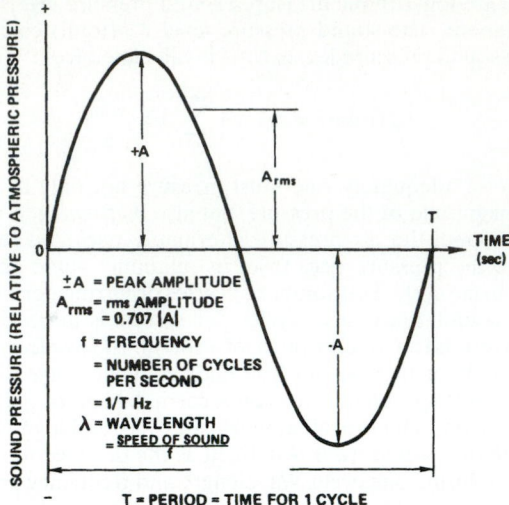

FIGURE 45.1   Basic properties of a sinusoidal (pure-tone) sound wave.

many experiments with human listeners, it was found that human hearing is more sensitive to midrange frequencies than to either low or very high frequencies. This characteristic is taken into account by adjusting, or *weighting,* the various frequency components of a sound in accordance with the sensitivity of human hearing and then combining all the weighted components. The result is a single-number measure of sound level that corresponds approximately to the human subjective perception of the severity of the noise, as well as to its annoyance and hearing damage potential.

Table 45.1 compares representative noise levels for common indoor and outdoor noise sources and environments. The extremes of noise range from 0 dBA (approximate threshold of hearing) to 120 dBA (jet aircraft at 500 ft), although most commonly encountered noise levels fall within the 40- to 100-dBA range.

An understanding of the following subjective perceptions of changes in the A-weighted sound level is useful:

- Changes of 1 dB or less cannot be perceived, except in carefully controlled laboratory experiments.
- A 3-dB increase in A-weighted level generally is just noticeable.
- A 10-dB increase in A-weighted level is perceived as approximately a doubling in loudness, independent of the initial noise level.

All the discussion thus far has been related to sound pressure, since this is the property to which human hearing and microphones respond. However, as discussed later, the magnitude of sound pressure level resulting at a given location and due to a given source depends on the "strength" of the source, on the environment in which the noise source is located, on the distance of the observation location from the source, and sometimes on the direction. Therefore, it is useful in many cases to use a noise measure that describes the intrinsic strength of a given source, that is, its *sound power.* Sound power represents the total sound energy radiated by a source per unit of time and is proportional to the square of the sound pressure at any given location.

TABLE 45.1 Comparison of Various Noise Levels

NOISE LEVEL (dBA)	EXTREMES	HOME APPLIANCES IN ROOMS	SPEECH AT 3 ft	MOTOR VEHICLES AT 50 ft	GENERAL TYPE OF OUTDOOR COMMUNITY ENVIRONMENT	GENERAL TYPE OF INDOOR WORK ENVIRONMENT
120	JET AIRCRAFT AT 500 ft					
110						
100						
90				DIESEL TRUCK (NOT MUFFLED)		HEAVY INDUSTRY
		SHOP TOOLS	SHOUT	DIESEL TRUCK (MUFFLED)		
80						
		BLENDER	LOUD VOICE	AUTOMOBILE AT 70 mph	MAJOR METROPOLIS (DAYTIME)	LIGHT INDUSTRY
70						
		DISHWASHER	NORMAL VOICE	AUTOMOBILE AT 40 mph	URBAN (DAYTIME)	
60						
		AIR CONDITIONER	NORMAL VOICE (BACK TO LISTENER)	AUTOMOBILE AT 20 mph	SUBURBAN (DAYTIME)	OFFICE
50						
		REFRIGERATOR			RURAL (DAYTIME)	
40						
30						
20						
10						
0	THRESHOLD OF HEARING					

As is the case for sound pressure, the range of sound power encountered in acoustics is very large. Thus, a logarithmic (decibel) scale is also used to describe sound power. The *sound power level* L_w is defined as

$$L_w = 10 \log \frac{W}{W_0} \qquad (45.4)$$

where W = source sound power in watts (W) and W_0 = reference sound power, standardized as 10^{-12} W. Sound power level is typically expressed in terms of dB with respect to 10^{-12} W.

45.2.2 Sound Fields

Meaningful measurements must take into account the variation of sound pressure level with position in the vicinity of a noise source. Figure 45.2 illustrates this general relationship and indicates the various sound field regions.

For an ideal nondirectional "point source" in open space, the sound pressure level decreases at the rate of 6 dB per doubling of distance because of spherical spreading of the sound energy. This relation is usually called the *inverse-square law*, because it corresponds to the sound pressure's varying inversely as the square of distance. However, the point-source approximation breaks down at distances very close to the source. At such distances, sound variation is more complex; in this *near field*, the sound pressure level may be either more or less than predicted by the inverse-square law, as shown in Fig. 45.2. The extent of the near field depends on the

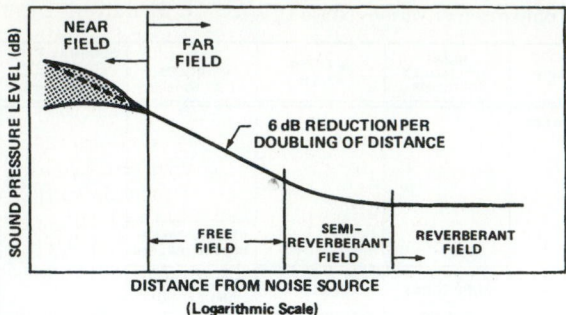

FIGURE 45.2   Sound fields in the vicinity of a noise source.

frequency of the sound, the dimensions of the source, and the phase relations of the various radiating parts of the source. As a rule of thumb, the near field may be assumed to end at a distance about twice the largest dimension of the source (or at 4 times the largest dimension, for sources resting on an acoustically reflective floor). Note that sound pressure levels measured within the near field cannot be used to predict the sound pressure levels at other distances or to evaluate the source sound power level; for these purposes, one must take care to perform measurements in the acoustic *far field* (that is, at distances beyond the near field).

In the acoustic far field, sound pressure levels decrease at a rate of 6 dB per distance doubling, as long as there exists a *free field,* which is, for all practical purposes, a field in which the effects of any air volume boundaries are negligible. Such a free field can be obtained outdoors, in a large room at locations away from the walls, or in an *anechoic chamber.* (In the latter, the walls, floors, and ceiling absorb nearly all sound incident on them.) The extent of the free-field region is characterized in Fig. 45.2 by a line with constant slope.

Sound from a source in any room—but most pronouncedly in a small room with "hard" (i.e., acoustically nonabsorptive) wall, floor, and ceiling surfaces—is reflected many times, so that the total sound at any location is composed of the sound radiated directly from the source (free-field sound) plus all the reflected components. If many reflected sound waves are arriving at an observation point from all directions, the sound field is called *reverberant.* In the reverberant field, the sound pressure level decreases less rapidly with distance than indicated by the inverse-square law, as shown in Fig. 45.2. *Reverberant rooms,* in which sound is uniform throughout, are often used to perform sound measurements that, in effect, average over all directions (for example, for the purpose of evaluating sound power levels of sources).

In practice, noise measurements often must be made in *semireverberant fields,* that is, where the sound propagation characteristics lie somewhere between free-field and reverberant conditions, as indicated by the transition zone in Fig. 45.2. The characteristics of a semireverberant environment are controlled largely by the amount of sound absorption in the room. These characteristics generally need to be evaluated and taken into account in analysis of the measured results.

45.2.3   Measurement Instrumentation

Sound Level Meters.   A sound level meter consists of (1) a transducer (microphone) to convert air pressure fluctuations to an electric signal, (2) an amplifier to raise the electric signal to a usable level, (3) weighting networks to modify the frequency characteristics of the instrument's response, and (4) an indicating device (meter) to display the measured level.

Sound level meters are designated by class, depending on measurement accuracy and tolerances. International Electrotechnical Commission (IEC) standard IEC 651 defines four classes: type 0, laboratory reference; type 1, precision; type 2, general purpose; and type 3, survey. Type 0 sets the most stringent accuracy and tolerance limits, followed by types 1, 2, and 3. Type 1 meters provide sufficient accuracy for field measurements in most cases and are usually selected when cost is not a major consideration. Standards for types 0, 1, and 2 sound level meters are also provided in American National Standards Institute (ANSI) standard S1.4-1983.

The weighting network most commonly used in sound level meters is the *A-weighting* network. Its response, shown in Fig. 45.3, represents the average behavior of human hearing. Measurements made using this network are expressed in A-weighted decibels, abbreviated dBA. Other common weighting networks include the B, C, and D types, used for special purposes. Some sound level meters also include a "linear," or "flat," response, commonly employed when a sound level meter supplies an electrical signal to other instruments.

The indicating meter on a sound level meter displays the sound level in decibels, relative to a standard reference sound pressure (20 µPa, or 2.90×10^{-9} lb/in^2). The speed with which the meter electronics and indicator respond also has been standardized. Most meters include two choices for averaging time: fast, which has a time constant of about ⅛ s, and slow, which has a time constant of about 1 s. The slow response is particularly useful for estimating visually the average value of a sound that fluctuates rapidly. Some sound level meters also have peak-hold and impulse-hold features, which are useful for measuring unsteady or impulsive noises.

Microphones. A microphone is a transducer used to convert air pressure fluctuations to an electric signal. Of the different types currently available, the most commonly used are the condenser, electret, and piezoelectric types. The choice of a particular microphone depends on its intended application and required perfor-

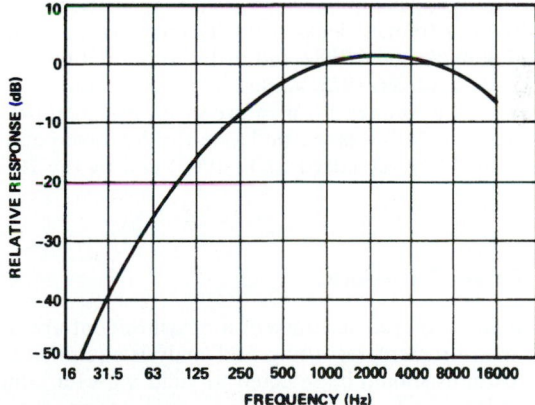

FIGURE 45.3 Frequency response specified for the A-weighting filter of sound level meters *(From ANSI S1.4-1983.)*

mance in terms of stability, precision, directivity, and frequency-response character-istics. Condenser microphones have excellent long-term stability and are insensitive to changes in temperature. However, they are sensitive to moisture. Electret micro-phones vary considerably in their long-term stability and sensitivity to temperature, and so are not as well suited as condenser microphones to measurement environ-ments with large temperature variations. However, they are less sensitive to mois-ture. Piezoelectric microphones are generally more rugged than condenser or electret microphones.

Acoustical Calibrators. An acoustical calibrator is a device that produces a known, stable sound pressure level at the diaphragm of a microphone. The most common calibrators are the pistonphone and loudspeaker.

A pistonphone calibrator produces a known sound pressure level within a closed cavity by means of moving pistons. Calibration is usually restricted to a single fre-quency (typically 250 Hz), and corrections for atmospheric pressure must be applied.

Loudspeaker-type calibrators consist of a battery-operated oscillator and small loudspeaker. In contrast to the pistonphone, some loudspeaker-type calibrators operate over a wide frequency range (125 to 2000 Hz), and the sound pressure level developed is less sensitive to the atmospheric pressure.

Spectrum Analyzers. A *spectrum analyzer* essentially produces a plot of sound pressure level versus frequency. Spectrum analyzers employ electronic filters to sep-arate the frequency components of a sound signal. The range of frequencies covered by an individual filter is called its *bandwidth*. Two basic types of filter sets are used in spectrum analyzers: those that use bands of constant bandwidth (that is, a fixed number of hertz) and those that use bands in which the upper frequency limit of the band is a fixed multiple of the lower frequency limit. Of the latter type, the band-width most commonly used in acoustic analysis covers a frequency range of one octave (that is, a 2-to-1 frequency range); an analyzer having filters with this band-width is called an *octave-band analyzer.* Other analyzers use half octaves ($\sqrt{2}$-to-1 frequency range), one-third octaves ($\sqrt[3]{2}$-to-1 range), or even narrower bands. Nar-rowband filters are often required to determine pure-tone components, such as those resulting from operation of cyclic (reciprocating or rotating) machinery. For narrowband analysis, digital computer-aided real-time analyzers are widely used.

The preferred center frequencies and band limits for spectrum analyzer filters are given in ANSI standard S1.6-1984 and in International Organization for Stan-dardization (ISO) standard 266-1975. Values for octave- and one-third-octave-band filters covering the audio frequency range are given in Table 45.2. Filters that are incorporated in octave-, half-octave-, and one-third-octave-band analyzers have been standardized by ANSI (standard S1.11-1966) and by the IEC (standard 225-1966).

45.2.4 Measurement Procedures

Once the purpose and required accuracy of a measurement are defined, one must select the proper measurement, recording, and analysis equipment.

Microphone positions should be selected to yield a useful sample of the sound field in the area of interest, and the microphone orientations should be chosen on the basis of the frequency-response characteristics of the microphone and of the measurement environment (see microphone manufacturer's instructions). For out-door measurements or for other locations where the air is not calm, the microphone

TABLE 45.2 Center and Approximate Cutoff Frequencies for Octave and One-Third-Octave Frequency Bands Covering the Audio-Frequency Range

Octave, Hz			One-third octave, Hz		
Lower band limit	Center frequency	Upper band limit	Lower band limit	Center frequency	Upper band limit
11	16	22	14.1	16	17.8
			17.8	20	22.4
			22.4	25	28.2
22	31.5	44	28.2	31.5	35.5
			35.5	40	44.7
			44.7	50	56.2
44	63	88	56.2	63	70.8
			70.8	80	89.1
			89.1	100	112
88	125	177	112	125	141
			141	160	178
			178	200	224
177	250	355	224	250	282
			282	315	355
			355	400	447
355	500	710	447	500	562
			562	630	708
			708	800	891
710	1 000	1 420	891	1 000	1 122
			1 122	1 250	1 413
			1 413	1 600	1 778
1 420	2 000	2 840	1 778	2 000	2 239
			2 239	2 500	2 818
			2 818	3 150	3 548
2 840	4 000	5 680	3 548	4 000	4 467
			4 467	5 000	5 623
			5 623	6 300	7 079
5 680	8 000	11 360	7 079	8 000	8 913
			8 913	10 000	11 220
			11 220	12 500	14 130
11 360	16 000	22 720	14 130	16 000	17 780
			17 780	20 000	22 390

SOURCE: ANSI standard S1.6-1984.

should be fitted with a windscreen to avoid extraneous noise generated by air turbulence at the microphone.

Before each set of measurements is made, all equipment should be calibrated according to the manufacturer's instructions. It is also a good idea to measure the electric noise floor (the lower measurement limit) of the instrumentation by replacing the microphone with an equivalent electric impedance (such as a capacitor) or by shielding the microphone from the acoustic background noise.

It is good practice to monitor the output of the sound level meter during the measurements by listening with the aid of a high-quality set of headphones; this permits one to detect electromagnetic pickup, signals due to wind or humidity, or other interference.

For source noise measurements, it is desirable to measure the background noise level (by turning off the noise source) to determine whether the background noise has a significant effect on the measurements. The background noise level should be at least 10 dB below the source noise level, if it is not to affect measured results significantly; otherwise, the measured noise levels must be corrected to obtain the level of the source. Table 45.3 may be used to obtain the appropriate correction.

At the conclusion of each set of measurements, the proper operation and calibration of all equipment should be rechecked, and all pertinent data should be recorded.

45.2.5 Data Evaluation

A set of measured acoustic data usually must be evaluated with regard to the problem of interest. This evaluation often requires conversion or extrapolation of the results. For example, sound pressure level measurements obtained for a machine in an anechoic chamber may need to be used to estimate the sound pressure level of the same machine at a different distance inside an industrial building. Or, one may want to use sound power level data acquired for a noise source in a reverberant room to estimate the sound pressure level at a given distance from the same source located outdoors. Such evaluations may be based on the relation between sound pressure and sound power level, as described below.

For any sound source, the sound pressure level and sound power level are related by

$$L_p = L_w + 10 \log \left(\frac{Q}{4\pi r^2} + \frac{4}{R} \right) + 10.5 \tag{45.5}$$

where L_p = sound pressure level, dB re 20 μPa
L_w = sound power level, dB re 10^{-12} W
Q = directivity factor (dimensionless)
r = distance to observation point from acoustic center of source, ft
R = room constant, ft²

In mks units, this equation converts to

$$L_p = L_w + 10 \log \left(\frac{Q}{4\pi r^2} + \frac{4}{R} \right) \tag{45.6}$$

where r is in meters and R is in square meters.

TABLE 45.3 Correction Factors for Background Noise

Difference between total noise level and background noise level, dB	Correction to be subtracted from total noise level to obtain source noise level, dB
8–10	0.5
6–8	1.0
4.5–6	1.5
4–4.5	2.0
3.5	2.5
3	3.0

The directivity factor Q accounts for the fact that most practical noise sources do not radiate uniformly in all directions. In the case of a nondirectional source (radiating sound uniformly in all directions), $Q = 1$. However, for a source placed on a sound-reflecting surface (for example, a machine on a concrete floor), much of, and sometimes all, the sound that would have been directed downward is reflected upward; here $Q = 2$ for a uniformly radiating source. Similarly, for noise sources located along the edge of a room and at the corner of a room, Q is 4 and 8, respectively.

The acoustic center is the location that would be occupied by a "point source" with the same sound power output as the actual source. For most practical purposes, the acoustic center can be taken as the geometric center of the controlling noise-radiating mechanism. Except for distances r very close to the noise source, errors in the estimation of the location of the acoustic center are not likely to affect the accuracy of the results significantly.

The room constant R is a measure of the sound absorption in a space. In a free field with no sound reflection at all, R is infinite, whereas in a room with no absorption, R is zero. In practice, neither of these extremes exists, and R is calculated as

$$R = \alpha_1 S_1 + \alpha_2 S_2 + \cdots + \alpha_n S_n \tag{45.7}$$

where $\alpha_1, \alpha_2, \ldots, \alpha_n$ are the sound absorption coefficients of materials on various surfaces and $S_1, S_2, \ldots, S_n$ are the areas of various surfaces, in square feet (or meters). The sound absorption coefficient α is a measure of the sound-absorptive property of a material as evaluated by ASTM Method C423, Test for Sound Absorption and Sound Absorption Coefficients by the Reverberation Room Method. And α is defined as the fraction of the randomly incident sound power that is absorbed (or otherwise not reflected) by the material. Table 45.4 (Ref. [45.3]) lists sound absorption coefficients of various construction materials. Note that these coefficients usually vary with frequency, and so does room constant R.

Figure 45.4 gives a graph, obtained from Eqs. (45.5) and (45.6), showing $L_p - L_w$ as a function of $r/\sqrt{Q}$ for various values of room constant R. Thus, given r, Q, and R, the relationship between L_p and L_w can be calculated for each frequency band of interest. Application of these concepts is illustrated in the following example.

Example. Measurements of octave-band sound pressure levels 3 ft from the acoustic center of an air compressor, located inside an anechoic chamber, yielded the results shown in Table 45.5. The compressor is to be installed on the floor at the center of a workroom 60 ft long, 50 ft wide, and 30 ft high. The room has a hard concrete floor, coarse concrete block walls, and a ceiling made of 2-in-thick glass-fiber panels with plastic sheet wrapping and perforated metal facing. What will be the resulting octave-band sound pressure levels and overall A-weighted sound level at a work station located in the workroom 15 ft from the compressor?

The first step is to calculate the octave-band sound power levels for the compressor by using Eq. (45.5). For $r = 3$ ft, $Q = 1$ (spherical radiation), and $R = \infty$ (anechoic chamber), this equation indicates that $L_p - L_w = -10$, and thus $L_w - L_p = +10$ for all frequency bands. The octave-band sound power levels obtained in this way are indicated in Table 45.5.

The next step is to calculate the room constants for the workroom. First, a list is made of the octave-band sound absorption coefficients α for the floor, wall, and ceiling surfaces (see Table 45.5); these values are obtained from Table 45.4 for the materials of the various surfaces. Then these values are multiplied by the respective surface areas (3000 ft² for the floor or ceiling, 6600 ft² for the walls), and the results are summed to yield the room constant R for each octave band.

TABLE 45.4 Sound Absorption Coefficients of Construction Materials

Material	Sound absorption coefficients					
	125 Hz	250 Hz	500 Hz	1000 Hz	2000 Hz	4000 Hz
Ballast or other crushed stone						
3.18-cm (1½-in) screened ballast 15.2 cm (6 in) deep	0.19	0.23	0.43	0.37	0.58	0.62
3.18 cm (1½-in) 30.5 cm (12 in) deep	0.27	0.58	0.48	0.54	0.73	0.63
3.18 cm (1½-in) 45.7 cm (18 in) deep	0.41	0.53	0.64	0.84	0.91	0.63
0.64-cm (¼-in) or less granite aggregate 15.2 cm (6 in) deep	0.22	0.64	0.70	0.79	0.88	0.72
Brick, unglazed	0.03	0.03	0.03	0.04	0.05	0.07
Brick, unglazed, painted	0.01	0.01	0.02	0.02	0.02	0.03
Carpet, heavy, on concrete	0.02	0.06	0.14	0.37	0.60	0.65
Same on 1350-g/m² (40-oz/yd²) hairfelt or foam rubber	0.08	0.24	0.57	0.69	0.71	0.73
Same, with impermeable Latex backing on 1350-g/m² (40-oz/yd²) hairfelt or foam rubber	0.08	0.27	0.39	0.34	0.48	0.63
Concrete block, coarse	0.36	0.44	0.31	0.29	0.39	0.25
Concrete block, painted	0.10	0.05	0.06	0.07	0.09	0.08
Drapes						
Light velour 338 g/m² (10 oz/yd²) hung straight, in contact with wall	0.03	0.04	0.11	0.17	0.24	0.35
Medium velour 475 g/m² (14 oz/yd²) draped to half area	0.07	0.31	0.49	0.75	0.70	0.60
Heavy velour, 610 g/m² (18 oz/yd²) draped to half area	0.14	0.35	0.55	0.72	0.70	0.65

Material	Sound absorption coefficients					
	125 Hz	250 Hz	500 Hz	1000 Hz	2000 Hz	4000 Hz
Fiber-glass boards and blankets						
2.54-cm (1-in) glass wool 24 to 48 kg/m^3 (1.5 to 3.0 lb/ft^3)	0.08	0.25	0.65	0.85	0.80	0.75
5.1-cm (2-in) glass wool 24 to 48 kg/m^3 (1.5 to 3.0 lb/ft^3)	0.17	0.55	0.80	0.90	0.85	0.80
2.54-cm (1-in) glass wool, 2.54-cm (1-in) airspace	0.15	0.55	0.80	0.90	0.85	0.80
5.1-cm (2-in) glass-fiber panels with plastic sheet wrapping and perforated metal facing, as installed	0.33	0.79	0.99	0.91	0.76	0.64
Floors						
Concrete or terrazzo	0.01	0.01	0.015	0.02	0.02	0.02
Linoleum, asphalt, rubber, or cork tile on concrete	0.02	0.03	0.03	0.03	0.03	0.02
Wood	0.15	0.11	0.10	0.07	0.06	0.07
Wood parquet in asphalt on concrete	0.04	0.04	0.07	0.06	0.06	0.07
Glass						
Large panes of heavy plate glass	0.18	0.06	0.04	0.03	0.02	0.02
Ordinary window glass	0.35	0.25	0.18	0.12	0.07	0.04
Gypsum board, 1.27 cm ($\frac{1}{2}$ in), nailed to 5.1 cm × 10.2 cm (2″ × 4″) studs 41 cm (16 in) center to center	0.29	0.10	0.05	0.04	0.07	0.09
Marble or glazed tile	0.01	0.01	0.01	0.01	0.02	0.02
Mineral spray-on materials						
1.27-cm ($\frac{1}{2}$-in) mineral fiber	0.05	0.15	0.45	0.70	0.80	0.80
1.9-cm ($\frac{3}{4}$-in) mineral fiber	0.10	0.30	0.60	0.90	0.90	0.95

Material	Sound absorption coefficients					
	125 Hz	250 Hz	500 Hz	1000 Hz	2000 Hz	4000 Hz
Mineral spray-on materials (*cont.*) 2.5-cm (1-in) mineral fiber	0.16	0.45	0.70	0.90	0.90	0.85
1.27-cm (½-in) mineral fiber on metal lath, 2.54-cm (1-in) airspace	0.25	0.50	0.80	0.90	0.90	0.85
Plaster, gypsum or lime, smooth finish on tile or brick	0.013	0.015	0.02	0.03	0.04	0.05
Plaster, gypsum or lime, rough finish on lath	0.14	0.10	0.06	0.05	0.04	0.03
Same, with smooth finish	0.14	0.10	0.06	0.04	0.04	0.03
Plywood paneling, 1 cm (⅜ in) thick	0.28	0.22	0.17	0.09	0.10	0.11
Water surface, as in a swimming pool	0.008	0.008	0.013	0.015	0.020	0.025

SOURCE: From Harris [45.3]. Used by permission.

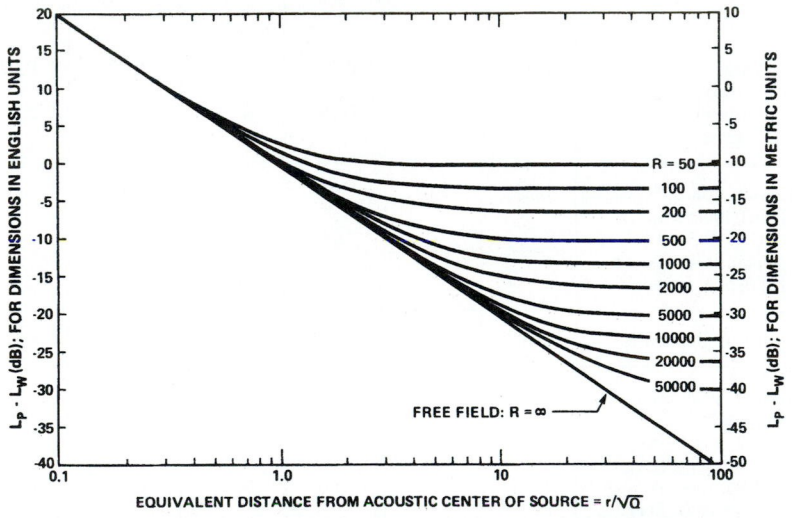

FIGURE 45.4 Relationship between sound pressure level and sound power level as a function of distance and directivity.

TABLE 45.5 Example of Noise Data Evaluation

	Octave-band center frequency, Hz					
	125	250	500	1000	2000	4000
L_p measured at 3 ft in anechoic chamber (dB)	87	86	89	92	92	90
$L_w - L_p$ (dB)	+10	+10	+10	+10	+10	+10
L_w (dB)	97	96	99	102	102	100
α (concrete floor)	0.01	0.01	0.015	0.02	0.02	0.02
α (concrete block walls)	0.36	0.44	0.31	0.29	0.39	0.25
α (glass-fiber panel ceiling)	0.33	0.79	0.99	0.91	0.76	0.64
$S\alpha$ (floor)	30	30	45	60	60	60
$S\alpha$ (walls)	2376	2904	2046	1914	2574	1650
$S\alpha$ (ceiling)	990	2370	2970	2730	2280	1920
R (ft^2)	3396	5304	5061	4344	4914	3630
L_p calculated at 15 ft in workroom (dB)	80	78	81	85	84	83
A-weighting	−16	−9	−3	0	+1	+1
L_p (A-weighted)†	64	69	78	85	85	84

†Overall, A-weighted sound level = 90 dBA.

The final step consists of calculating the octave-band sound pressure levels by using Eq. (45.5). With a directivity factor Q of 2 (for a source on a hard, reflecting surface), a distance r of 15 ft, and the calculated octave-band values of R and L_w, this equation yields the octave-band sound pressure levels L_p in the workroom 15 ft from the compressor given in Table 45.5. Next, the overall A-weighted sound level can be calculated by applying the A-weighting corrections (from Fig. 45.3) to each octave-band level (again, see Table 45.4) and then summing the results logarithmically, by using Eq. (45.2). This calculation yields an overall A-weighted sound level of 90 dBA at the work station.

45.3 NOISE EFFECTS AND STANDARDS

Figure 45.5 summarizes some of the limits and guidelines that can be used for evaluating various effects of noise on people. These effects and standards are discussed below.

45.3.1 Hearing Damage

Extremely intense sounds can produce nearly instantaneous hearing damage, but usually the development of hearing impairment is far more subtle. Persons exposed to high sound levels during part of a day will experience a temporary shift in the threshold of hearing. In other words, they will be unable to hear faint sounds for perhaps a few hours after exposure. After repeated exposure, generally over several

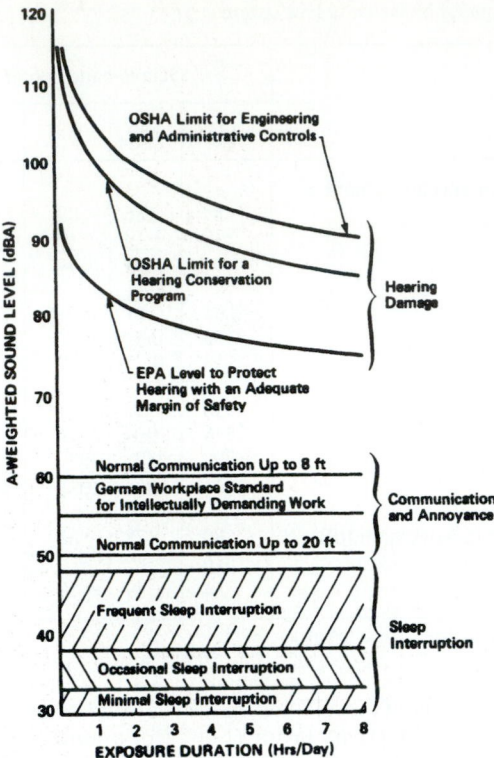

FIGURE 45.5 Noise effects and standards.

years, one's hearing threshold gradually rises and remains permanently high. After sufficiently long exposures to high sound levels, listeners are no longer able to understand normal speech and, in extreme cases, can lose hearing ability almost entirely.

Permanent threshold shifts can be induced by long-term exposures to A-weighted sound levels of about 80 dBA and become increasingly severe with higher levels of exposure [45.3]. Hearing acuity degrades to the greatest extent at frequencies around 4 kHz and less at very high frequencies (6 to 8 kHz) and at low frequencies (0.5 to 2 kHz).

For most U.S. industrial workers, the Occupational Safety and Health Administration (OSHA) established an exposure limit of 90 dBA for 8 hours (h) and uses the "5-dB increase per exposure halving rule" for sounds of lesser durations. That is, 95 dBA is allowed for 4 h, 100 dBA for 2 h, and so forth up to 115 dBA for 15 minutes (min). Exposure to continuous sounds above 115 dBA is not permitted, regardless of the duration. The OSHA exposure limit is given by the top curve in Fig. 45.5.

If workers are exposed to sounds of various durations and levels, a limit applies to the time-weighted average (TWA) sound level or to the daily dosage D. The TWA is calculated from

$$\text{TWA} = 16.61 \log \left(\frac{C_1}{T_1} + \frac{C_2}{T_2} + \cdots + \frac{C_n}{T_n} \right) + 90 \qquad (45.8)$$

where C_i = duration of exposure to a specified level and T_i = total time of exposure permitted at that level according to the foregoing rule. If the TWA exceeds 90 dBA, then the worker is considered to be overexposed. The daily dosage D (as a percentage of full exposure) is computed from

$$D = 100 \left(\frac{C_1}{T_1} + \frac{C_2}{T_2} + \cdots + \frac{C_n}{T_n} \right) \qquad (45.9)$$

If D exceeds 100 percent, then the worker is considered to be overexposed.

OSHA has also established an 85-dBA, 8-h exposure level as the threshold for which a hearing conservation program is required. The same algorithm is used for computing the time-weighted average as Eq. (45.8), except that 90 is replaced with 85. The hearing conservation program requires periodic audiometric testing of exposed workers and provision of hearing protectors for workers exhibiting significant permanent threshold shifts.

The Environmental Protection Agency (EPA) selected a level that would protect "virtually the entire population" against a hearing loss of 5 dB or less ([45.6]). Thus, the EPA recommends that exposure not exceed 70 dBA for 24 h or 75 dBA for 8 h, by the "3-dB rule" (that is, 78 dBA for 4 h, 81 dBA for 2 h, etc.). The EPA-recommended exposure limit is given in Fig. 45.5.

45.3.2 Speech Interference

A useful guide for determining when speech may be understood and the amount of effort required by the speaker is presented in Fig. 45.6 (Ref. [45.7]). Clearly, for the 15- to 20-ft distances common in many homes, schools, or workplaces, background levels should be less than about 50 dBA if communication is to be normal.

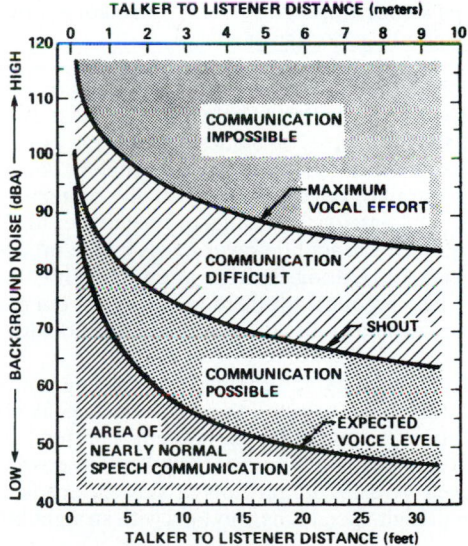

FIGURE 45.6 Quality of speech communication. *(From Miller [45.7].)*

45.3.3 Sleep Interruption

Sound can interrupt sleep in a number of complex ways that depend on the nature of the sound, the stage of sleep in which a person may be, and the individual's susceptibility to disturbance. In spite of these many factors, useful guidelines have been determined for sleep disturbance. A study of the effects of air conditioner noise [45.8] has shown the following reactions to steady noise in sleeping quarters:

Noise level, dBA	Response (complaints)
<33	None
33–38	Occasional
38–48	Frequent
>48	Unlimited

45.4 NOISE CONTROL

Noise control can be incorporated in the design of machinery either by treating the sources of machine noise or by altering the structureborne and/or airborne paths of this noise. This section contains a discussion of these different options for machinery noise control.

45.4.1 Source Control

Controlling noise at the source is often the most cost-effective procedure for machine design. Specific components, operating conditions, and/or geometries can frequently be selected that result in substantially less noise for the same function. This section discusses source noise control for some commonly encountered components and noise-generating mechanisms.

Fans and Blowers.   Fans and blowers move air or other gases by lift forces on rotating fan blades or impeller vanes. This rotating pressure distribution generates some sound, but the fluctuating pressures on blades or impellers usually cause more significant noise. These pressures are generated by the turbulent boundary layers, by irregular vortex shedding at trailing edges, and by spatially and temporally varying inflow.

Where manufacturers' data or the results of special measurements are not available, it is useful to estimate fan sound power levels by the procedure recommended by the American Society of Heating, Refrigeration, and Air Conditioning Engineers (ASHRAE) [45.9], which accounts for the dependence of fan noise on fan type, size, flow rate, and pressure drop as well as on the number of blades and the fan operating point.

Fan source noise control may be achieved by ensuring that the fan is operating efficiently, that is, that the fan is selected to operate at its peak efficiency at the pressure and flow conditions required by the system. For reduced noise, the inflow should be as uniform spatially and as free of turbulence as possible. Noise reductions may often be achieved by using larger fans at slower speeds and fans with swept blades instead of small high-speed fans and fans with straight blades.

Electric Motors.   Electric motor noise is generated primarily by fluctuating magnetic loads, bearings, and cooling fans. Cooling fan noise is the dominant source for most motors.

Noise levels for standard and quieted totally enclosed fan-cooled motors are presented in Fig. 45.7 for a range of horsepower and operating speeds [45.10]. These data clearly show that for a given horsepower rating, sound power levels are distributed within a 10- to 20-dBA range. Standard untreated high-speed motors are invariably the noisiest, whereas quieted motors operating at low speeds are the quietest.

Noise control may be designed into motors through the combined use of high-temperature insulation and low-volume cooling fans. The insulation allows the motor to run hotter than normal, requiring less airflow to dissipate waste heat. The lower airflow and lower heat loss permit the use of smaller, quieter fans.

Noise abatement can also be achieved at the source by operational speed reduction. The data in Fig. 45.7 show that motors built to operate at 1800 instead of 3600 r/min can be as much as 17 dBA quieter and that motors built to operate at 1200 r/min are 2 to 17 dBA quieter than 3600 r/min units.

Gears.   Gear noise is due to the unsteady forces associated with tooth meshing. These forces primarily result from geometric inaccuracies in the gear manufacturing process and from deflections of the teeth under load. The forces result in gear vibration, which is transmitted to the gear housing and often to contiguous structural members, all of which radiate sound. An investigation [45.11] of numerous types of gear sets has shown that radiated sound power ranges from about 2.5×10^{-6} to 10^{-8} times the mechanical power.

Source control of gear noise is best accomplished by selecting high-quality gears and gear boxes. Table 45.6 may be used to obtain a rough estimate of the corresponding reduction in gear noise and the approximate related increase in cost [45.12]. The "maximum conjugacy" tooth form indicated in Table 45.6 incorporates lengthened tooth addenda, circular arc profiles, low-pressure angles, and generous tooth-root radii.

The gear box should be designed to avoid structural resonances corresponding to tooth mesh frequencies, and bearings should be selected to minimize vibration

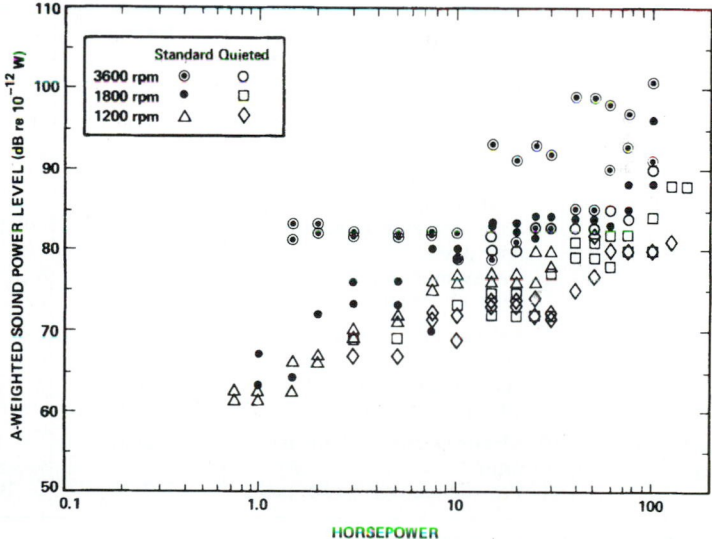

FIGURE 45.7   A-weighted sound power levels of electric motors.

transmission from gear shafts to the housing structure. Preloaded tapered-roller bearings are found to provide 4 to 5 dB of noise reduction for a given gear set when compared with unloaded ball bearings [45.11].

Hydraulic Pumps.   Vane- and piston-type pumps contain oscillating components that result in pump casing vibration and sound radiation. All pumps (vane, piston, and gear) generate fluctuating pressures in working fluids that cause casing vibration and also propagate through inlet and outlet ports. This fluidborne sound radiates through tubing and can excite contiguous structures and other components.

A survey of pump noise levels taken from Ref. [45.13] and the manufacturers' literature is presented in Fig. 45.8. This figure shows how the A-weighted radiated sound power varies with hydraulic horsepower, that is, with the power delivered by the pump (equal to the pressure rise across the pump times the volume flow rate). Sound power levels can be within a large range at any value of hydraulic horsepower. Piston pumps tend to be somewhat noisier than others, and internal gear pumps tend to be somewhat quieter.

Source noise control for hydraulic pumps is best achieved by selecting pumps that have low values of radiated acoustic power, consistent with system requirements. Acoustic data can generally be obtained from manufacturers. If acoustic power levels are to be measured, care should be taken to isolate the pump from contiguous structures, because of the significant fluctuating hydraulic pressure and vibration transmitted through inlet and outlet tubing.

TABLE 45.6   Guidelines for Noise Reduction through Gear Design

Noise reduction principle	Design feature	Approximate reduction in sound pressure level, ΔdBA	Cost increase over cost of hobbed spur gears, %
Increasing total contact ratio	1. Reduce pressure angle to $14\frac{1}{2}°$	0.5–1	0
	2. Lengthen addenda (usually 10 to 25%)	1–2	0–2
	3. Use substantial helix angle (20 to 35%)	3–6	0–5
	4. Use maximum-conjugacy tooth form	6–16	0–5
	5. Shot-peen roots to permit finer pitch	1–3	10–15
	6. Use stronger material to permit finer pitch	1–3	5–20
Improving kinematic accuracy	7. Relieve tooth tips, roots, or both	1–2	0
	8. Shave working surfaces	1–3	20–30
	9. Grind working surfaces	2–4	40–100
Reducing pitch line velocity	10. Maximize face-to-diameter ratio	2–6	10–30
	11. Provide multiple power paths	3–6	100–150

SOURCE: From Roverol [45.12]. Used by permission.

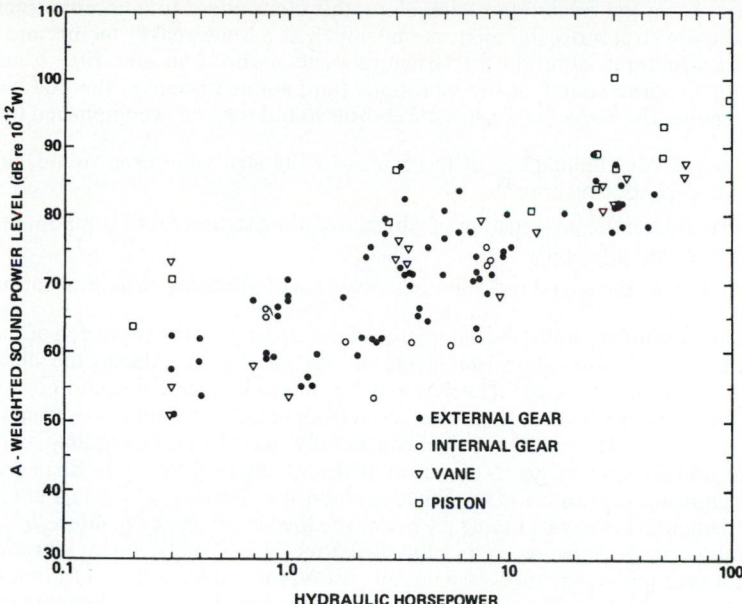

FIGURE 45.8 A-weighted sound power levels of hydraulic pumps.

Other Mechanical Components. Machines are made up of a large variety of mechanical components and mechanisms, each of which should be considered from the perspective of source noise control.

Impact mechanisms excite transient structural vibrations, which subsequently radiate sound. Impact noise can be controlled by seeking alternative mechanisms for performing a given task or by cushioning impacts. Alternative mechanisms could, for example, involve squeezing under hydraulic pressure instead of using impact devices. Cushioning is often accomplished by installing compliant materials at—or just behind—the impact point.

Rolling contact noise from bearings and other devices often results from surface irregularities. Therefore, noise can often be lessened by incorporating precision-ground components and good lubrication. Alternatively, well-lubricated journal bearings, which use an oil film to preclude metal-to-metal contact, may be substituted for ball or roller bearings.

Noise is generated from a wide variety of fluid-handling devices in addition to the pumps, blowers, and fans discussed earlier. Most common are valves and jets, which generate sound by turbulent flow. Since sound levels are strong functions of flow velocity V, source noise control is often best achieved by lowering velocities and by designing structures to reduce turbulence levels where possible.

45.4.2 Path Control

Control of Structureborne Noise. If a structural component is set into vibratory motion—by external mechanical excitation, by an adjacent vibrating structure, by impacts between it and adjacent components, or by incident sound—vibrations can propagate from the excitation region along this component to adjacent structures. Any vibrating structure can radiate sound, much as a loudspeaker membrane does; thus, any structural vibration or "structureborne sound" can give rise to audible noise or "airborne sound" if the vibrations (and sound) occur in the audible frequency range. The reduction of structureborne sound may be accomplished by

1. Reduction (or elimination) of the causes of the structureborne sound, such as vibration-producing impacts

2. Obstruction of the propagation of vibrations along structures to components that radiate sound efficiently

3. Reduction of the sound radiation effectiveness of vibrating structural surfaces

   Shields (secondary walls) that stand free of the structure to be protected or that are connected to it only by soft resilient layers are often useful for reducing the vibrations induced in a structure. Such vibrations may be caused by incident sound fields or by fluctuating pressures, such as those that result from impinging fluid jets or from turbulent flows. To be effective, these shields generally need to be heavy, limp, and well damped, and the resilient layers also need to be soft and well damped—all in order to avoid significant resonances in the shield/resilient-layer/structure system and to keep the fundamental resonance frequency below the lowest frequency of interest.

   Isolation from Sources. Vibration isolators placed between a vibration or impact source and a structure are generally useful for reducing the vibrations transmitted to the structure. These isolators basically consist of resilient elements, such as springs or elastomeric components or cushions, placed between the vibration sources and the structures to be protected. Isolators can be effective only if the stiffnesses of the resilient elements are (1) less than the effective dynamic stiffnesses of the structures to which they are connected, as measured at the attachment points, and (2) low enough that the fundamental resonance of the system, including the isolator, is smaller than the lowest driving frequency by a factor of at least 2.0. In all cases where isolation is used, care must be taken to avoid short-circuiting the isolation by "hard" paths, such as bolts, directly touching structures, or relatively rigid cables or ducts that can transmit vibrations relatively well.

   Discontinuities. The propagation of vibrations along structures can be obstructed by the introduction of discontinuities, such as changes in stiffness or mass, as may be obtained by the addition of ribs or cross braces, by the introduction of abrupt changes in cross section (cut-outs or built-up areas) or in material properties (for example, by adding soft gaskets). Table 45.7 lists equations that may be used to estimate the attenuations provided by various discontinuities in extended beams and plates.

   Added Structural Damping. The attenuation of propagating vibrations may be increased by the addition of structural damping treatments, which convert some of the vibratory energy to heat. Such treatments are useful largely only in frequency regions that encompass structural resonances or at high frequencies, where the structure is three or more wavelengths long. The reduction ΔL in the vibration level (and in the corresponding noise levels) due to adding an amount of damping η_a to an initially present amount η_i may be estimated from[†]

[†] The symbol η, used here with various subscripts, represents the *loss factor,* that is, the ratio of the mechanical energy that is dissipated per radian (or that per cycle, divided by 2π) to the total energy of vibration. This measure of damping is related to other common ones by

$$\eta = 2\frac{c}{c_c} = \frac{\delta}{\pi} = \frac{1}{Q}$$

where c/c_c = ratio of viscous damping coefficient to critical damping coefficient, δ = logarithmic decrement, and Q = amplification at resonance.

$$\Delta L = B \log \left(1 + \frac{\eta_a}{\eta_i} \right) \qquad (45.10)$$

where $B = 20$ for structures vibrating at resonance and $B = 10$ for structures subject to broadband excitation encompassing several resonances [45.14]. Note that the added damping η_a must be considerably greater than the initially present damping η_i if the added damping is to provide a significant reduction in level.

Some increases in the damping of machine structures can usually be obtained by replacing welds with bolted joints, by placing parts of the structures in contact with viscous fluids, or by replacing steel by cast iron or by special-purpose high-damping alloys. However, major damping increases usually can be achieved only by such special means as surrounding the structure with granular materials (such as sand, gravel, or lead shot) or bonding layers of high-damping viscoelastic materials to the structure. These materials typically are plastics or elastomers that are capable of dissipating considerable energy; many types are commercially available.

The increase in damping loss factor η_a that may be expected to be obtained by bonding a viscoelastic layer of thickness h to a structural plate or beam of thickness H may be estimated from

$$\eta_a = \frac{\beta}{1 + \dfrac{EH/(eh)}{3 + 3h/H + 4(h/H)^2}} \qquad (45.11)$$

where β = loss factor of added viscoelastic material
e = elastic modulus of that material[†]
E = elastic modulus of material of basic plate or beam

Considerably increased damping often can be obtained by use of sandwich structures that incorporate high-damping viscoelastic materials. Design relations are available ([45.15]) but involve relatively complex frequency and wavelength dependences.

Reduction of Radiation. Reduction of the sound radiation from structures may be accomplished by minimizing the vibrations of radiating surfaces, as discussed earlier, and by reducing the sound-radiating capabilities of the structures. Because structures can produce sound only by pushing against air as their surfaces move perpendicularly to themselves, their sound-radiating capabilities can be reduced by keeping vibrations parallel to the major planes of structural surfaces, by reducing the areas of vibrating structures, and by providing perforations in structural surfaces.

The decrease in radiated sound level resulting from reduction of the radiating area by the fraction n (or by $100n$ percent) may be estimated from

$$\Delta L = -20 \log (1 - n) \qquad (45.12)$$

However, the level decrease obtained by the introduction of perforations may be greater than that resulting from the corresponding area reduction. Because perforations less than about half an acoustic wavelength apart permit the instantaneous pressure increases produced on one side of a vibrating surface essentially to be can-

---

[†] For most damping materials, β and e vary significantly with frequency and temperature; material selection must reflect these variations.

TABLE 45.7 Vibration Transmission Efficiencies of Structural Discontinuities[†]

	Longitudinal vibrations	Flexural vibrations
	$\lambda_L = \dfrac{\sqrt{E/\rho}}{f}$	$\lambda_F = \sqrt{2\pi R \lambda_L}$
Cross-section change	$\dfrac{1}{\tau} = \dfrac{1}{4}\left(\sqrt{r} + \dfrac{1}{\sqrt{r}}\right)^2$	$\dfrac{1}{\tau} = \left[\dfrac{r^2/2 + \sqrt{r} + 1 + 1/\sqrt{r} + 1/(2r^2)}{r^{5/4} + r^{3/4} + r^{-3/4} + r^{-5/4}}\right]^2$
Change in material	$\dfrac{1}{\tau} = \dfrac{1}{4}\left(\sqrt{\dfrac{E_1\rho_1}{E_2\rho_2}} + \sqrt{\dfrac{E_2\rho_2}{E_1\rho_1}}\right)^2$	$\dfrac{1}{\tau} = \left[\dfrac{a(1+b)^2 + 2b(1+a^2)}{2\sqrt{ab}(1+a)(1+b)}\right]^2$ $a = \left(\dfrac{\rho_2 E_1 R_1^2}{\rho_1 E_2 R_2^2}\right)^{1/4}$ $b = \dfrac{R_2 A_2}{R_1 A_1}\,\sqrt{\dfrac{E_2\rho_2}{E_1\rho_1}}$
Resilient insert	$\dfrac{1}{\tau} = 1 = \left(\dfrac{f}{f_i}\right)^2$ $f_i = \dfrac{k}{\pi A_1\sqrt{E_1\rho_1}}$	$\dfrac{1}{\tau} = 1 + \dfrac{f^3}{f_F^3}$ $f_F^3 = \dfrac{G^2}{2\pi^3\rho_1\sqrt{E_1\rho_1}\,R_1\ell^2}$

Blocking mass

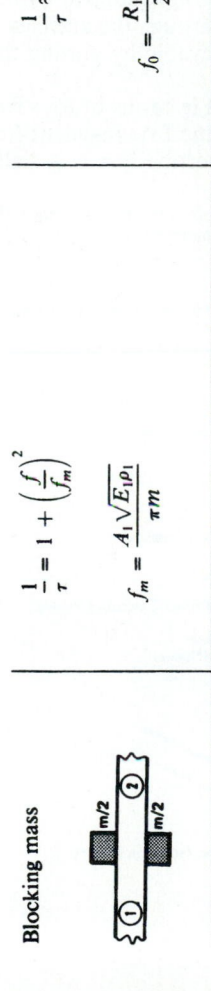

$$\frac{1}{\tau} = 1 + \left(\frac{f}{f_m}\right)^2$$

$$f_m = \frac{A_1\sqrt{E_1\rho_1}}{\pi m}$$

$$\frac{1}{\tau} \approx \begin{cases} 1 & \text{for } f < 0.5f_0 \\ 1 + f/f_B & \text{for } f > 2f_0 \end{cases}$$

$$f_0 = \frac{R_1\sqrt{E_1\rho_1}}{2\pi R_m^2} \qquad f_B = \frac{2\rho_1 A_1 R_1\sqrt{E_1\rho_1}}{\pi m^2}$$

†For structure on both sides of discontinuity assumed infinite—or many wavelengths long and reasonably well damped—and symmetric. (Adapted from Ref. [45.14]. Used by permission.)

τ = vibration transmission efficiency

IL = insertion loss (dB) = 10 log $(1/\tau)$ = 10 log (v_{20}^2/v_2^2)

v_{20} = vibration amplitude on receiving side in absence of discontinuity

v_2 = same, in presence of discontinuity

$r = \begin{cases} h_2/h_1 & \text{for plates} \\ A_2/A_1 & \text{for beams} \end{cases}$ $\begin{array}{l}\lambda_L = \text{wavelength of longitudinal vibrations} \\ \lambda_F = \text{wavelength of flexural vibrations}\end{array}$

h = plate thickness‡

A = cross-sectional area of beam or plate

E = modulus of elasticity $\Big\}$ of plate or beam material‡

ρ = density

R = radius of gyration of beam cross section, for beam‡

  = $h/\sqrt{12}$ for plate‡

G = shear modulus

ℓ = length $\Big\}$ of resilient insert

k = compressional stiffness

m = total mass $\Big\}$ of added masses

R_m = radius of gyration

f = frequency (Hz)

‡Subscripts 1 and 2 refer to structural components on which waves approach and leave discontinuity, respectively.

celed by the simultaneous pressure decreases produced on the other surface,[†] the net sound pressure may be reduced considerably.

Control of Airborne Noise.   *Noise Control with Enclosures.*[‡]   Figure 45.9 is a schematic sketch of a typical complete enclosure. Such an enclosure is essentially a sealed box with stiffened walls. The panels of the walls have damping treatment applied to them to control their resonant vibration, and the interior of the box is covered with absorptive treatment (such as open-cell foam or glass-fiber mat) to prevent the buildup of reverberant sound in the interior. The machine to be quieted is vibration-isolated from the enclosure, so that the machine does not excite the walls of the enclosure, causing them to radiate sound and compromise the noise reduction performance. It is imperative that all openings be carefully sealed or provided with sound-absorbing ducts or mufflers. Even a small leak in a high-performance enclosure can seriously compromise its noise reduction.

A good measure of the performance of an enclosure is its insertion loss (IL), defined as 10 times the logarithm of the ratio of the sound power radiated by the untreated source to the sound power radiated by the source through the enclosure. In other words, the insertion loss is the noise reduction achievable by putting the source inside the enclosure.

The acoustic behavior of an enclosure may be understood in terms of four frequency regions, as indicated in Fig. 45.10. Region I is below the first resonant frequency of the panels of the enclosure. In this region, the insertion loss is initially

---

[†] The perforations need to be configured so as to permit airflow through them with little resistance; the cross-sectional flow area of each opening should be greater than its wetted perimeter area.

[‡] See Refs. [45.16], [45.17], [45.18], and [45.19].

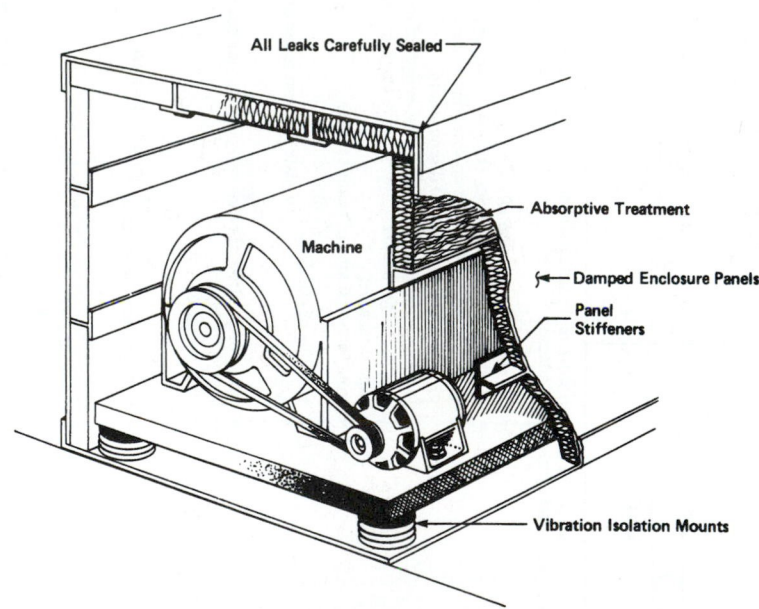

FIGURE 45.9   Cross section of a simple complete enclosure.

constant but dips rapidly near the first panel resonance. This region is often called the *stiffness-controlled* region, since the insertion loss here depends predominantly on the bending stiffness of the panels and increases with increasing stiffness.

Region II lies above the first panel resonance frequency. Here the transmission of sound through the enclosure is controlled by the panel resonances and acoustic standing waves between the enclosed machine and the walls of the enclosure. The acoustic behavior is very complicated. Insertion loss in this region can generally be increased through the use of enclosure panel damping and increased acoustic absorption in the interior of the enclosure.

In region III, the acoustic behavior is much like that in region II, except that the panel resonant modes in region III are so closely spaced in frequency that many resonant panel modes are present in any frequency bandwidth of common interest. In this region, the insertion loss decreases with increases in the transmission coefficient of the panels of the enclosure (that is, the ratio of acoustic power transmitted through the panel to the acoustic power incident on them) and increases with increased average absorption coefficient (or, the ratio of acoustic power absorbed on the surface of the inside of the enclosure to the acoustic power incident there). Since the transmission coefficient of the panels decreases with increases in their mass per unit area and the insertion loss here depends on the mass per unit area, region III is often called the *mass-controlled* region.

In region IV, a pronounced dip occurs at the *coincidence* frequency, or the frequency at which the bending wave speed in the panels equals the acoustic wave speed. At this frequency, sound passes readily through the panels, and the insertion loss can be increased by increasing the mass or damping of the panels. Alternatively, one may change the thickness or material properties of the panels so as to change the coincidence frequency and move it out of the frequency range of interest.

Table 45.8 lists the approximate formulas for estimating the insertion loss in the four frequency regions. The result of using these design formulas is shown by the dashed curves in Fig. 45.10. In region I, the design formula is most accurate well below the first panel resonance f_0. Near f_0 it overestimates the insertion loss. The table

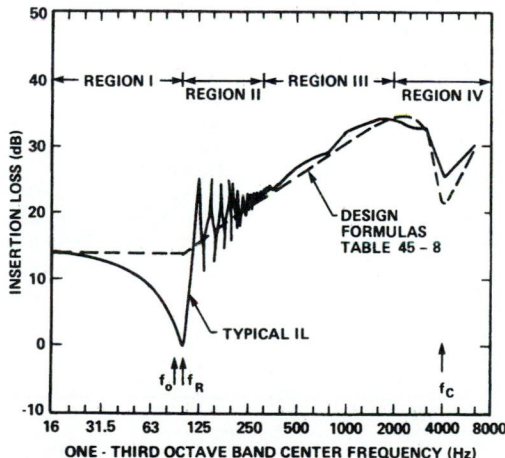

FIGURE 45.10 Insertion loss of a representative enclosure.

includes a formula for estimating the insertion loss at f_R, the frequency where the insertion loss first dips. The frequency f_R is slightly higher than the panel resonance frequency because of the panel stiffening contributed by the air in the enclosure.

As mentioned earlier, the behavior of the insertion loss in region II is quite complicated. An appropriate average value may be obtained by use of the design formula from region III.

For regions III and IV, the same design formulas may be used. The insertion loss here depends on the transmission coefficient τ (fraction of sound transmitted) and the absorption coefficient α (fraction of sound absorbed). Note that if there is no absorption in the enclosure or if α is much less than τ, then the enclosure provides no insertion loss. Absorption is crucial to the functioning of the enclosure. Adequate absorption can be ensured by installing blankets of materials such as glass-fiber mat or open-cell foam on the walls of the enclosure. Typical values of α for some of these materials are given in Table 45.4.

The transmission coefficient can be obtained from Fig. 45.11, where the ratio of τ to the normal incidence mass law transmission coefficient τ_0 is plotted against the ratio of the frequency to the critical frequency. As an aid in using Fig. 45.11, the critical frequency for various panel materials and thicknesses is given in Fig. 45.12. And τ_0 can be calculated by

$$\tau_0 = \frac{1}{1 + [2\pi f m/(2\rho c)]^2} \tag{45.13}$$

where　f = sound frequency
　　　　m = panel mass per unit area
　　　　ρ = mass density of air
　　　　c = speed of sound in air

Sample Calculation.　Calculate the insertion loss of an enclosure with the characteristics given in Fig. 45.13a.

1. Determine the stiffness S of the largest panel from the equation in Table 45.8. For

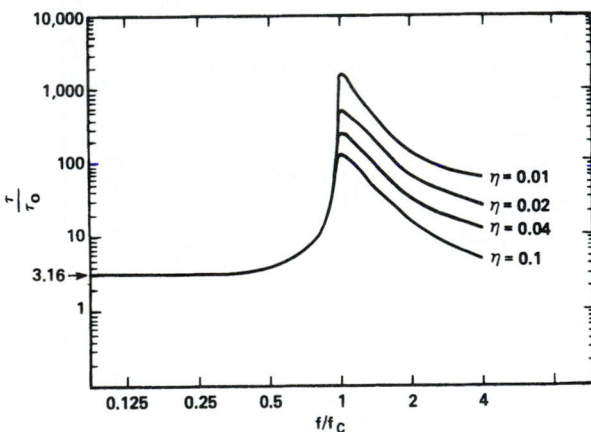

FIGURE 45.11 Transmission coefficient; m is the mass per unit area of the panel; η is the panel loss factor; ρ_0 is the density of air; c_0 is the speed of sound in air; f_c is the critical frequency (see Fig. 45.12).

TABLE 45.8 Enclosure Design Formulas

Region	Frequency range	Design formula	To improve insertion loss, increase:
I. Stiffness-controlled region	$f \ll f_0$	$IL_I = 20 \log\left(1 + \dfrac{Sl}{\rho c^2}\right)$	• Panel stiffness • Enclosure volume
	$f_R \sim f_0\left(1 + \dfrac{\rho c^2}{Sl}\right)^{1/2}$	$IL \sim 10 \log\left\{\left[4\pi\left(\dfrac{f_0 l}{c}\right)\eta + \eta^2 \dfrac{Sl}{\rho c^2}\right]\left(1 + \dfrac{Sl}{\rho c^2}\right)\right\}$	• Panel stiffness • Enclosure volume • Panel damping
II. Resonance-controlled region	$f > f_0$	See region III	• Enclosure absorption • Panel damping
III. Mass-controlled	$\dfrac{f_c}{2} > f \gg f_0$	$IL_{III} = 10 \log \dfrac{A_e \tau + \alpha A_\alpha}{A_e \tau}$	• Panel mass • Enclosure absorption
IV. Critical frequency	$f > \dfrac{f_c}{2}$	$IL_{IV} = IL_{III}$	• Panel mass • Enclosure absorption • Panel damping • f_c by reducing panel thickness

f_0 = first panel resonant frequency: $f_0 = \dfrac{1}{2\pi}\sqrt{\dfrac{S}{m}}$

m = mass per unit area of panels

η = panel loss factor

τ = panel transmission coefficient (see Fig. 45.11)

α = absorption coefficient (see Table 45.4)

A_e = surface area of enclosure

A_α = surface area covered by absorption

f_c = critical frequency (see Fig. 45.12)

f_R = frequency of first dip in insertion loss just above f_0

† S = panel stiffness/unit area
$= \dfrac{Et^3 \pi^4}{12(1 - \mu^2)}\left(\dfrac{1}{a^2} + \dfrac{1}{b^2}\right)^2$ for simply supported panel

t = panel thickness

a, b = panel dimensions

E = panel modulus

μ = Poisson's ratio for panel

l = distance from surface of noise source to enclosure walls

ρ = air density

c = acoustic wave speed

f = frequency, Hz

45.29

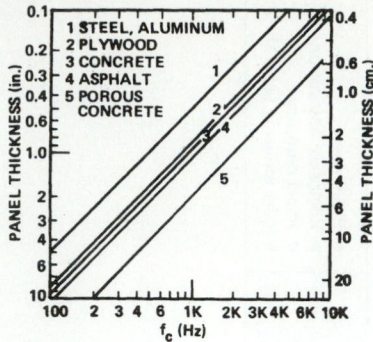

FIGURE 45.12 Critical frequency for panels of various thicknesses and materials. *(From Diehl [45.5].)* Here $f_c = c_0^2/\kappa C_l$, c_0 is the acoustic wave speed; κ is the panel radius of gyration, $\kappa = t/\sqrt{12}$ for the most simple panels; and c_l is the longitudinal wave speed in the panel $c_l = \sqrt{E/\rho_P}$, where E is the modulus and ρ_P is the density of the panel.

$$E = 1.44 \times 10^9 \text{ lb/ft}^2 \qquad t = 0.008 \text{ ft}$$

$$\mu = 0.3 \qquad\qquad a = 1.5 \text{ ft}$$

$$b = 1 \text{ ft}$$

one obtains

$$S = 1.38 \times 10^4 \text{ lb/ft}^3$$

Note that by ignoring the stiffeners in this calculation, one implicitly assumes that the panels are much less stiff than the stiffening beams. Consequently, the stiffening beams must be designed so that if a point load is applied to one of the panels of the enclosure, the deflection of the panel is much greater than that of the frame.

2. Calculate the natural frequency f_0 of that panel from the equation in Table 45.8.

Using S from the above calculation and a surface density of 1.33 lb/ft^2 for 0.1-in-thick aluminum, one obtains

$$m = \rho t = \frac{1.33}{32.2} = 0.0413 \text{ slug/ft}^2$$

and

$$f_0 = \frac{1}{2\pi}\sqrt{\frac{S}{m}} = \frac{1}{2\pi}\sqrt{\frac{1.38 \times 10^4}{0.0413}} = 92 \text{ Hz}$$

Note that the equations for S and f_0 imply simply supported panels. In all likelihood, the true end conditions will tend to stiffen the panel even more, implying that the estimate here of the insertion loss is conservative; i.e., it predicts lower values than those that will occur.

3. Calculate the frequency of the first dip in the insertion loss f_R by using the equation in Table 45.8. For

$$\rho = 0.0023 \text{ slug/ft}^3 \qquad c = 1128 \text{ ft/s} \qquad l = 0.25 \text{ ft}$$

one finds

$$f_R = 125 \text{ Hz}$$

4. Determine the coincidence frequency from curve 1 in Fig. 45.12:

$$f_c = 5000 \text{ Hz}$$

5. Calculate IL for $f < f_R$, using the appropriate equation for region I from Table 45.8:

$$\text{IL}_\text{I} = 7 \text{ dB}$$

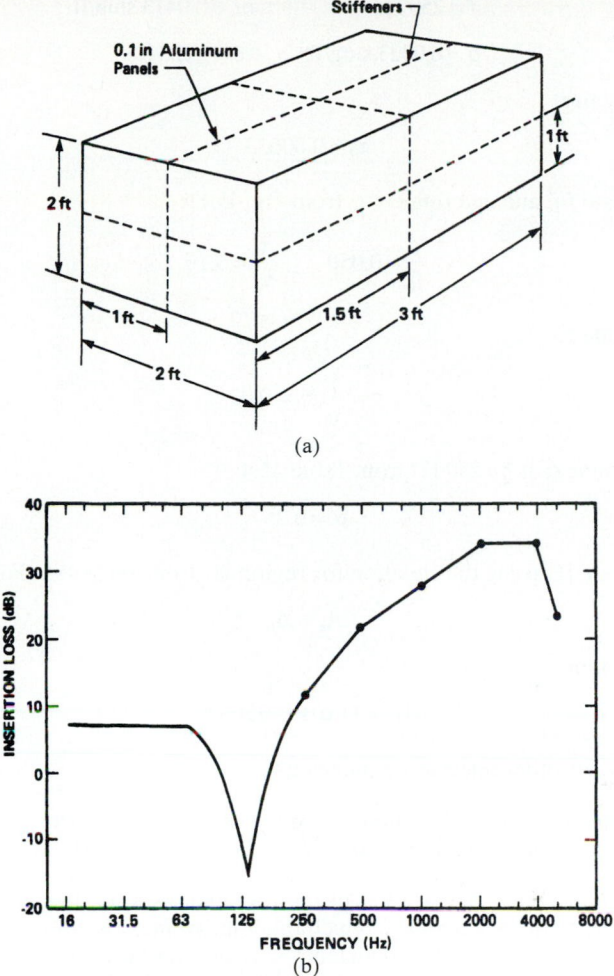

FIGURE 45.13 (*a*) Enclosure characteristics for illustrative calculations. Dimensions, $2 \times 2 \times 3$ ft; material, aluminum of 160-lb/ft^3 density; panel thickness = 0.1 in; panel damping = 0.05 for all frequencies; average spacing between source and panel walls is 3 in; stiffeners: stiffeners in each panel dividing each into four panels of equal size; absorption is 1 in of glass-fiber mat on all interior surfaces. (*b*) Insertion loss calculated for example.

6. Calculate IL at $f = f_R$, using the appropriate equation for region I from Table 45.8:

$$\text{IL} \approx -14.5 \text{ dB}$$

7. Calculate IL for $f > f_R$:
- Select a frequency greater than f_R and calculate τ_0 by using Eq. (45.13). For

$$f = 250 \text{ Hz} \qquad\qquad m = 0.0413 \text{ slug/ft}^2$$
$$\rho = 6.0023 \text{ slug/ft}^3 \qquad c = 1128 \text{ ft/s}$$

one obtains

$$\tau_0 = 0.00633$$

- Calculate f/f_c and determine τ/τ_0 from Fig. 45.11:

$$\frac{f}{f_c} = 0.050 \qquad \frac{\tau}{\tau_0} = 3.16$$

- Calculate τ:

$$\tau = \frac{\tau}{\tau_0}\,\tau_0 = 0.02$$

- Determine α at $f = 250$ Hz from Table 45.4:

$$\alpha = 0.25$$

- Calculate IL, using the equation for region III from Table 46.8. For

$$A_e = A_\alpha$$

one obtains

$$\text{IL} = 11 \text{ dB @ 250 Hz}$$

- Repeat for other selected frequencies.

The results of similar calculations for this example for the octave bands at 500, 1000, 2000, and 4000 Hz, and the coincidence frequency 5000 Hz, are shown in Fig. 45.13b.

The techniques that can be used to improve the insertion loss of an enclosure depend on the frequency region in which the improvement is desired. In the example of Fig. 45.13b, there is a clear deficiency in region I, where the enclosure actually amplifies the noise from the source at 125 Hz. Increasing the insertion loss in this region is usually best accomplished by stiffening of the panels of the enclosure. The approaches that might be employed include

- Increasing the number of stiffeners, so that the effective panel size is reduced
- Increasing the panel thickness
- Using stiffer materials or configurations, such as graphite epoxy or honeycomb composite materials, for instance

One might also damp the panels to reduce the resonant peak in region I, although it usually is difficult to design damping treatments that will be effective at the low frequencies in this region. In regions II, III, and IV, the insertion loss can be increased by decreasing the transmission coefficient or by increasing the absorption coefficient. To increase the absorption coefficient in any region one might, for example,

- Cover more of the interior surfaces of the enclosure with absorptive material
- Select a material (from Table 45.4) that has a higher absorption coefficient in the frequency region of interest
- Use a thicker layer of absorptive material if increased absorption at low frequencies is required

Techniques for decreasing the transmission coefficient depend on the region of interest. In region II, one can

- Damp the panels by gluing on commercially available damping materials
- Increase the panel mass per unit area by using a thicker panel or a panel made from a denser material (e.g., steel instead of aluminum) or by adding layers of limp massive materials, such as leaded vinyl

The above techniques for increasing panel mass are also effective in region III. In this region, increasing panel damping will be effective only insofar as the damping material adds mass to the panel.

In region IV, increased panel mass and increased panel damping both contribute to increasing the enclosure insertion loss. In addition, one may also alter the structure to move the coincidence frequency out of the frequency range of interest by using a thinner or less stiff panel. (Of course, care must be exercised so that such treatments do not degrade the enclosure performance excessively in other frequency regions.)

Partial Enclosures. Where a complete enclosure cannot be used (for example, when openings for cooling air must be provided), mufflers can be employed to prevent noise from escaping from these openings (see Fig. 45.14). With properly designed mufflers, a partial enclosure can be as effective as a complete enclosure. The high-frequency aspect of the corresponding design is discussed in later sections of this chapter. At a certain low frequency, the *Helmholtz resonance frequency,* considerable sound transmission through the enclosure opening can occur. This frequency is given by

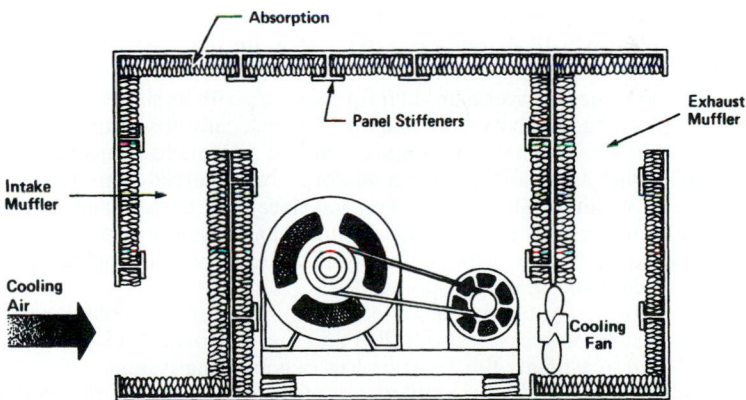

FIGURE 45.14 Schematic sectional view of a partial enclosure with mufflers.

$$f_H = \frac{c}{2\pi} \sqrt{\frac{A}{LV}} \qquad\qquad (45.14)$$

where c = speed of sound
A = open area of duct
L = duct length
V = air volume contained in enclosure (enclosure volume minus volume of machine)

It is important to check that the Helmholtz resonance frequency is low enough that it is outside the frequency range of interest.

If the enclosure in the example above has two ducts 3 in square by 2 ft long and the free volume of the enclosure is 8 ft³, then

$$V = 8 \text{ ft}^3 \qquad A = 0.125 \text{ ft}^2 \qquad \text{(two ducts)}$$

$$L = 2 \text{ ft} \qquad c = 1128 \text{ ft/s}$$

$$f_H = 15 \text{ Hz}$$

This resonance frequency is too low to be of much concern. In fact, this is the case in most practical situations. If it is not, one may increase the enclosure volume or change the duct geometry to reduce f_H.

Noise Control with Mufflers.   Mufflers are special ducts or pipes or openings that allow for the free flow of air or other gases while impeding the transmission of sound.

In internal-combustion engines, these devices are almost always used at the exhaust and are sometimes used at the air inlet. Similarly, if a machine that requires cooling air is placed in an enclosure, the enclosure must be equipped with mufflers through which cooling air can be circulated.

There are two basic muffler types: *reactive* and *resistive*. Reactive mufflers rely on the reflection of acoustic waves at discontinuities (e.g., expansion chambers, side branch resonators) and on the interaction of these waves to reduce the transmission of sound. Mufflers for internal-combustion engines are primarily of this type. Reactive mufflers are essentially tuned devices providing high attenuation in some frequency bands and little attenuation in others.

Resistive, or dissipative, mufflers attenuate sound by the acoustic energy–absorbing action of absorptive material within the muffler. These devices typically provide noise attenuation over a broad frequency range. In its simplest form, a dissipative muffler is a duct with its walls lined with acoustically absorptive material.

Reactive Mufflers.   Reactive mufflers for exhaust and intake silencing of internal-combustion engines are available from a number of commercial firms. These devices come in a great variety of shapes and sizes, and since their performance may be different on different engines, muffler manufacturers have usually measured the insertion loss and backpressure of their mufflers on the particular engine or class of engines for which their use is appropriate.

Techniques for design or analysis of reactive mufflers are too complex for inclusion here. However, both classical ([45.20]) and finite-element ([45.21]) techniques can be employed to estimate the insertion loss of a muffling system.

Resistive Mufflers.   Resistive mufflers are often used in conjunction with enclosures, as discussed earlier. Two geometries are most commonly used: the lined duct and the lined plenum chamber.

The performance of the lined duct is indicated in Fig. 45.15, which shows the attenuation (in decibels) for sound traveling down the duct per unit length of duct of width ℓ, with absorptive lining on only two sides of the duct. If the duct is lined on four sides, one simply adds the attenuation of a duct lined on two sides to the attenuation of a duct lined top and bottom. For a parallel baffle muffler, a special type of lined duct illustrated in Fig. 45.16, the attenuation per unit length can be estimated from Fig. 45.15 by setting ℓ equal to w and t equal to $h/2$.

A lined plenum chamber is shown schematically in Fig. 45.17. Its transmission loss (TL) is given approximately by

$$\text{TL} = -10 \log \left[A_0 \left(\frac{W}{2\pi q^3} + \frac{1}{A_\alpha} \right) \right] \qquad (45.15)$$

where W and q are defined in Fig. 45.17, A_0 = area of the outlet, A_α = surface area covered with absorption, and α = absorption coefficient. The equation yields reasonable estimates, provided that the plenum dimensions are large compared to an acoustic wavelength.

Sample Calculation—Lined Duct.   A parallel baffle muffler is to be constructed in a 1-ft^2 duct, as illustrated in Fig. 45.18. What length of duct will be required to obtain 10-dB attenuation at 1000 Hz?

1. From Fig. 45.18, ℓ is found to be equal to w or 4 in.

2. At 1000 Hz, $\lambda = c/f = 1128/1000 = 1.13$ ft and

$$\frac{\ell}{\lambda} = \frac{4}{12(1.13)} = 0.29$$

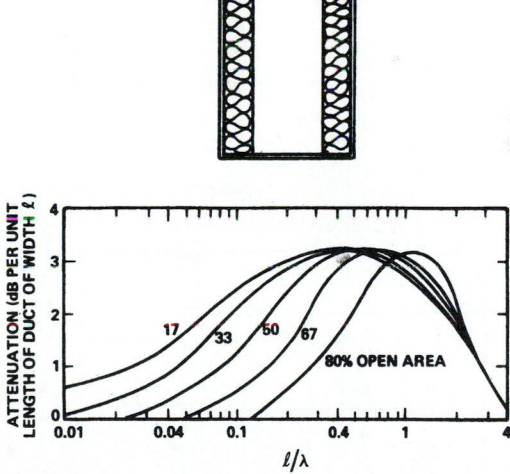

FIGURE 45.15   Attenuation of a sound by a lined duct. (*From Beranek [45.2].*)

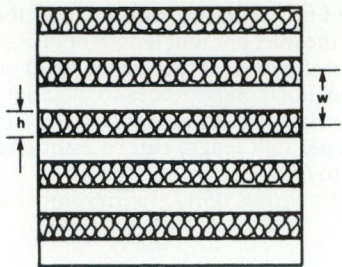

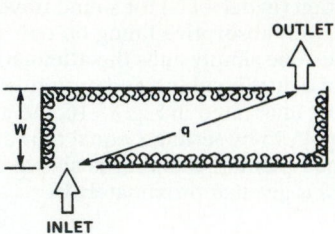

FIGURE 45.16   Parallel baffle muffler.   **FIGURE 45.17**   Plenum chamber.

3. The open area comprises 50 percent. From the corresponding curve in Fig. 45.15, the attenuation is found to be ~ 2.9 dB at $\ell/\lambda = 0.29$.
4. The desired attenuation is 10 dB. Since 2.9 dB is the attenuation every 4 in, the required length is

$$\frac{10}{2.9} \times 4 \approx 14 \text{ in}$$

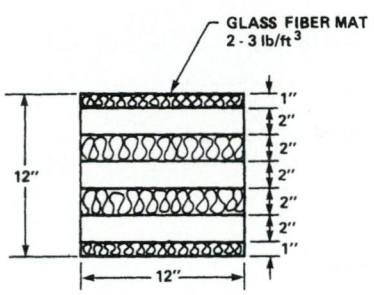

FIGURE 45.18   Parallel baffle muffler (for illustrative calculation).

Sample Calculation—Lined Plenum Chamber. Two plenum chambers are required for cooling air to enter and leave an enclosure. The walls of the enclosure are designed for a transmission loss of 10 dB at 1000 Hz. There is space available to install two plenums of the dimensions shown in Fig. 45.19. If the interior surfaces of each chamber are covered with absorptive material with an absorption coefficient of 0.90 at 1000 Hz and the required outlet area is 1 ft², will the chambers be able to provide adequate transmission loss?

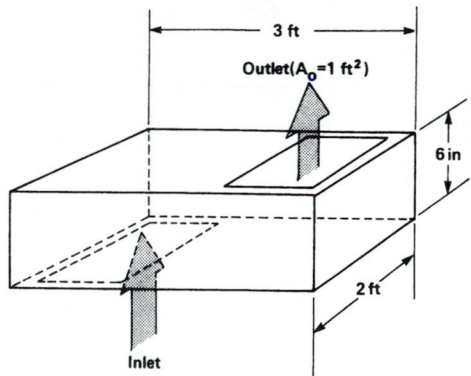

FIGURE 45.19   Plenum chamber for the sample calculation.

1. From Fig. 45.19, the parameters are

$$A_0 = 1 \text{ ft}^2 \qquad W = 0.5 \text{ ft} \qquad \alpha = 0.90$$

$$A_\alpha = 2(3 \times 0.5 + 2 \times 0.5 + 3 \times 2) = 17 \text{ ft}^2$$

$$q \sim [3^2 + (0.5)^2]^{1/2} \approx 3 \text{ ft}$$

2. Substituting these values into Eq. (45.15), one obtains

$$\text{TL} \approx 11 \text{ dB}$$

Therefore, the chamber provides more than the desired 10-dB insertion loss at 1000 Hz.

REFERENCES

45.1 L. E. Kinsler and A. R. Frey, *Fundamentals of Acoustics*, 2d ed., John Wiley & Sons, New York, 1962.

45.2 L. L. Beranek (ed.), *Noise and Vibration Control*, McGraw-Hill, New York, 1971.

45.3 C. M. Harris (ed.), *Handbook of Noise Control*, 2d ed., McGraw-Hill, New York, 1979.

45.4 A. P. G. Peterson and E. E. Gross, Jr., *Handbook of Noise Measurement*, 7th ed., General Radio Company, Concord, Mass., 1972.

45.5 G. M. Diehl, *Machinery Acoustics*, John Wiley & Sons, New York, 1973.

45.6 U.S. Environmental Protection Agency, "Information on Levels of Environmental Noise Requisite to Protect Public Health and Welfare with an Adequate Margin of Safety," EPA Report 550/9-74-004, March 1974.

45.7 J. D. Miller, "Effects of Noise on People," EPA Report NTID300.7, Dec. 31, 1971.

45.8 W. E. Blazier, Jr., "Criteria for Residential Heating and Air Conditioning Systems," *Noise Control*, vol. 5, 1959, pp. 48–53.

45.9 American Society of Heating, Refrigeration, and Air Conditioning Engineers, *ASHRAE Handbook and Product Directory—1980 Systems Volume*, Chapter 35, "Sound and Vibration Control."

45.10 E. K. Bender, "The Economics of Controlling Noise at the Source," Australian Acoustic Society, February 1983.

45.11 H. Opitz, "Noise of Gears," *Phil. Trans. R. Soc. London*, ser. A, vol. 263, 1968, pp. 369–380.

45.12 W. S. Roverol, "Attacking Gear Noise at the Mesh," *Machine Design*, May 10, 1979, pp. 68–71.

45.13 U.S. Army, "Hydraulic System Noise Study," U.S. Army Mobility Equipment Command Report no. FPRC-3M3, Dec. 1973.

45.14 L. Cremer, M. Heckl, and E. E. Ungar, *Structureborne Sound*, Springer-Verlag, New York, 1973.

45.15 E. E. Ungar, "Damping of Panels," Chap. 14 in *Noise and Vibration Control*, L. L. Beranek (ed.), McGraw-Hill, New York, 1971.

45.16 R. S. Jackson, "Performance of Acoustic Hoods at Low Frequencies," *Acoustica*, vol. 12, 1962, pp. 139–152.

45.17 R. S. Jackson, "Some Aspects of the Performance of Acoustic Hoods," *Journal of Sound Vibration*, vol. 3, no. 1, 1966, pp. 82–94.

45.18 T. J. Schultz, "Wrappings, Enclosures and Duct Linings," Chap. 15 in *Noise and Vibration Control*, L. L. Beranek (ed.), McGraw-Hill, New York, 1971.

45.19 J. D. Webb, ed., *Noise Control in Industry,* Sound Research Laboratories Limited, Holbrook Hall, Sudbury, Suffolk, England, 1976.

45.20 M. J. Crocker, "Internal Combustion Engine Exhaust Muffling," *Noise-Con 77 Proc.,* pp. 331–358.

45.21 C. J. Young and M. J. Crocker, "Finite Element Acoustical Analysis of Complex Muffler Systems with and without Wall Vibrations," *Noise Control Engineering,* vol. 9, no. 2, 1977, pp. 86–93.

CHAPTER 46
GEAR TRAINS

Harold L. Johnson, Ph.D.
Associate Professor Emeritus
School of Mechanical Engineering
Georgia Institute of Technology
Atlanta, Georgia

46.1 ORDINARY GEAR TRAINS

Gear trains consist of two or more gears meshed for the purpose of transmitting motion from one axis to another. Ordinary gear trains have axes, relative to the frame, for all gears making up the train. Figure 46.1a shows a simple ordinary train in which there is only one gear for each axis. In Fig. 46.1b, a compound ordinary train is seen to be one in which two or more gears may rotate about a single axis.

The ratio of the angular velocities of a pair of gears is the inverse of their numbers of teeth. The equations for each mesh in the simple train are

$$n_3 = \frac{N_2}{N_3} n_2 \qquad n_4 = \frac{N_3}{N_4} n_3 \qquad n_5 = \frac{N_4}{N_5} n_4 \qquad (46.1)$$

where n is in revolutions per minute (r/min) and N = number of teeth. These equations can be combined to give the velocity ratio of the first gear in the train to the last gear:

$$n_5 = \frac{N_4}{N_5} \frac{N_3}{N_4} \frac{N_2}{N_3} n_2 \qquad (46.2)$$

Note that the tooth numbers in the numerator are those of the driving gears, and the tooth numbers in the denominator belong to the driven gears. Gears 3 and 4 both drive and are, in turn, driven. Thus, they are called *idler gears*. Since their tooth numbers cancel, idler gears do not affect the magnitude of the input-output ratio, but they do change directions of rotation. Note the directional arrows in the figure. Idler gears can also produce a saving of space and money. In Fig. 46.2, the simple train of the previous figure has been repeated. In dotted outline is shown a pair of gears on the same center distance as gears 2 and 5 and having the same input-output ratio as the simple train.

Finally, Eq. (46.2) is simplified to become

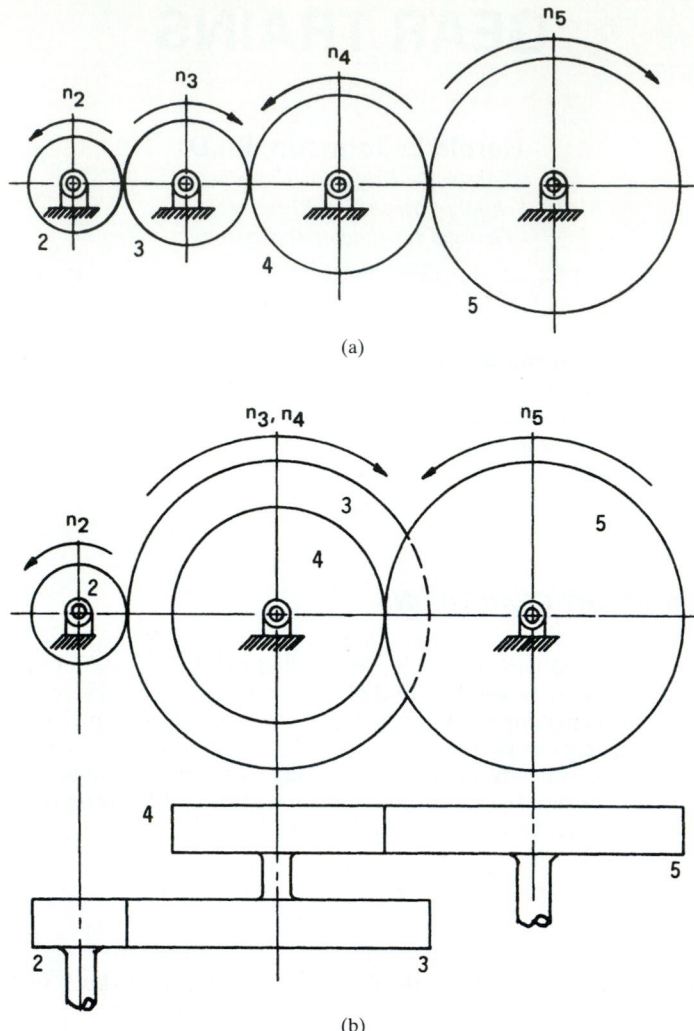

FIGURE 46.1   Ordinary gear trains. (*a*) Simple; (*b*) compound.

$$n_5 = -\frac{N_2}{N_5} n_2 \qquad (46.3)$$

where the minus sign is now introduced to indicate contrarotation of the two gears.

The compound train in Fig. 46.1*b* has the following velocity ratios for the pairs of driver and driven gears:

$$n_3 = -\frac{N_2}{N_3} n_2 \qquad \text{and} \qquad n_5 = -\frac{N_4}{N_5} n_4 \qquad (46.4)$$

and, of course, $n_4 = n_3$. Combining the equations yields

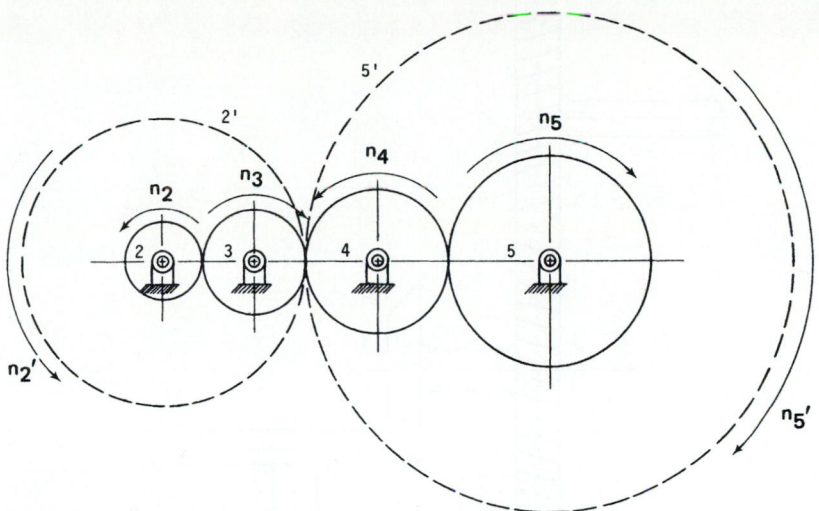

FIGURE 46.2   Gears 2′ and 5′ are required if idler gears are not used.

$$n_5 = \frac{N_2 N_4}{N_3 N_5} n_2 \qquad (46.5)$$

and the thing worthy of note here is that the numbers of teeth of all gears constituting a mesh with a compounded pair are required to determine the velocity ratio
through the system. Compound gear trains have an advantage over simple gear
trains whenever the speed change is large. For example, if a reduction of 12/1 is
required, the final gear in a simple train will have a diameter 12 times that of the
first gear.

46.2   GEAR TYPE SELECTION

The disposition of the axes to be joined by the gear train often suggests the type of
gear to choose. If the axes are parallel, the choices can be spur gears or helical gears.
If the axes intersect, bevel gears can be used. If the axes are nonparallel and nonin-
tersecting, then crossed helicals, worm and gear, or hypoid gears will work. In
Fig. 46.3, a train having various types of gears is shown. Gears 2 and 3, parallel helical gears, have a speed ratio

$$n_3 = -\frac{N_2}{N_3} n_2 \qquad (46.6)$$

Gears 4 and 5, bevel gears, have a speed ratio

$$n_5 = -\frac{N_4}{N_5} n_4 \qquad (46.7)$$

Gears 6 and 7, worm and gear, are considered in a slightly different manner. A worm
is generally spoken of as having threads, one, two, three or more (see Chap. 36). A

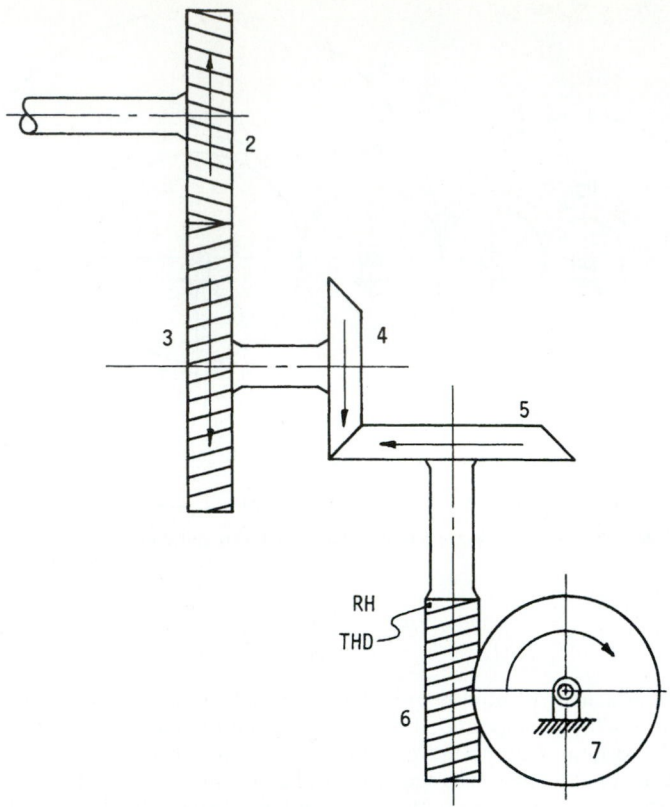

FIGURE 46.3 Various gears used in a train.

worm with one thread would have a lead equal to the pitch of the thread. A worm with two threads would have a lead equal to twice the pitch of the thread. Thus

$$n_7 = \frac{\text{number of threads on 6}}{N_7}\, n_6 \qquad (46.8)$$

Joining Eqs. (46.6), (46.7), and (46.8), we find

$$n_7 = \frac{N_6}{N_7}\frac{N_4}{N_5}\frac{N_2}{N_3}\, n_2 \qquad (46.9)$$

where N_6 represents the number of threads of the worm gear.

To determine the direction of rotation of gear 7, an inversion technique can be used. Fix gear 7 and allow the worm to translate along its axis as it rotates. Here it is necessary to note the hand of the worm, which can be either right or left. In the figure, gear 6 rotates in the same direction as gear 5 and, having a right-hand thread, will move downward (in the drawing). Now, inverting back to the original mechanism, the worm is moved in translation to its proper position, and by doing so, gear 7 is seen to rotate clockwise.

46.3 PLANETARY GEAR TRAINS

Planetary gear trains, also referred to as *epicyclic gear trains,* are those in which one or more gears orbit about the central axis of the train. Thus, they differ from an ordinary train by having a moving axis or axes. Figure 46.4 shows a basic arrangement that is functional by itself or when used as a part of some more complex system. Gear 2 is called a *sun gear,* gear 4 is a *planet,* link 3 is an *arm,* or *planet carrier,* and gear 5 an internal-toothed *ring gear.*

Planetary gear trains are, fundamentally, two-degree-of-freedom systems. Therefore, two inputs are required before they can be uniquely analyzed. Quite frequently a fixed gear is included in the train. Its velocity is zero, but this zero velocity constitutes one of the input values. Any link in the train shown except the planet can serve as an input or an output link. If, for example, the rotations of link 2 and link 5 were the input values, the rotation of the arm would be the output. The term *link* refers to the individual machine elements comprising a mechanism or linkage, and gear trains are included in this broad array of systems. Each link is paired, or joined, with at least two other links by some form of connection, such as pin points, sliding joints, or

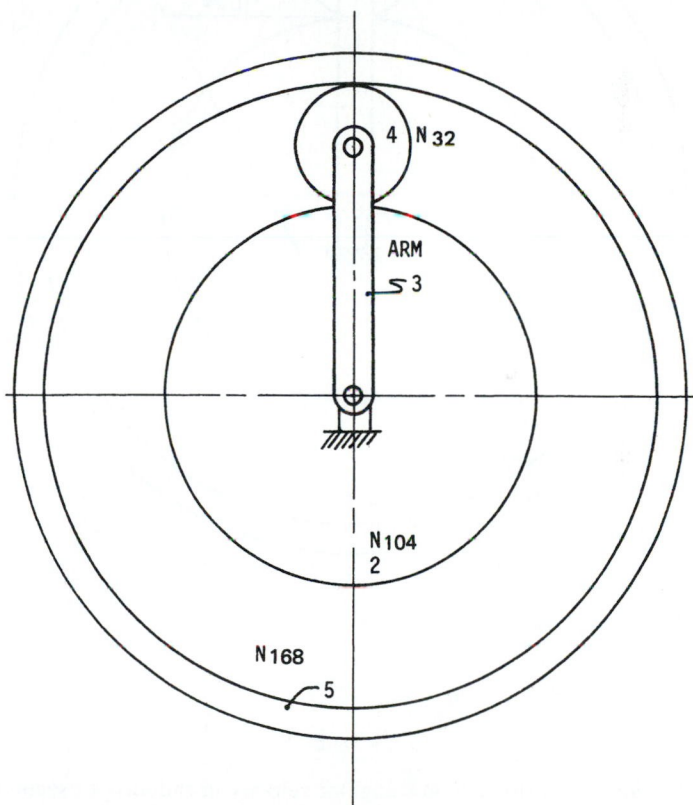

FIGURE 46.4 A basic planetary train.

direct contact, a pairing that is prevalent in cam-and-gear systems. An explanation and an illustration of the *joint* types are found in Refs. [46.1] and [46.2] as well as others (see Chap. 41).

There are several methods for analyzing planetary trains. Among these are instant centers, formula, and tabular methods. By instant centers, as in Ref. [46.3] and on a face view of the train, draw vectors representing the velocities of the instant centers for which input information is known. Then, by simple graphical construction, the velocity of another center can be found and converted to a rotational speed. Figure 46.5 illustrates this technique.

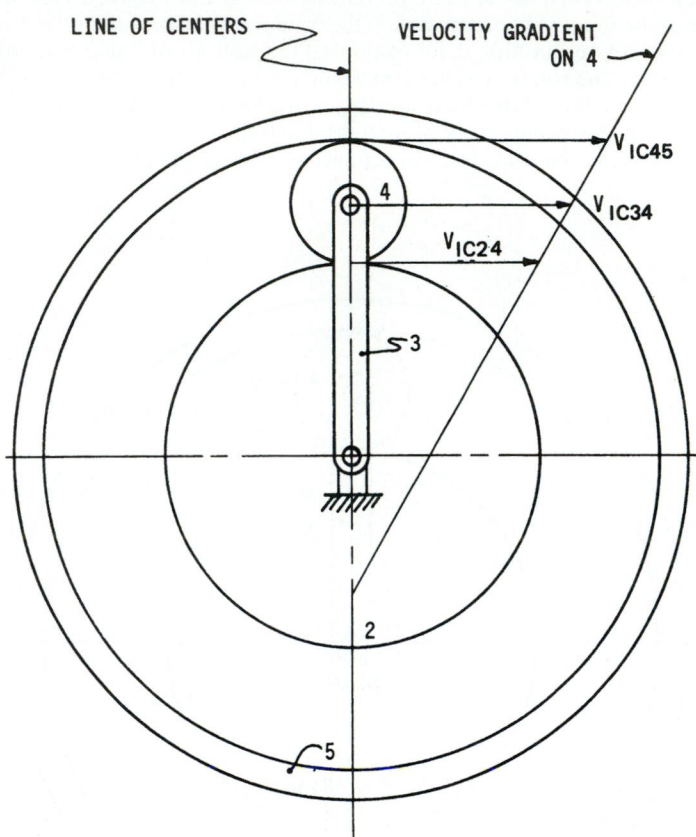

FIGURE 46.5   Instant-centers method of velocity analysis.

Calculate V_{IC24} and V_{IC45} from

$$V = r\omega \tag{46.10}$$

where r = radius dimension and ω = angular velocity in radians per second (rad/s). Draw these vectors to scale in the face view of the train. Then V_{IC24} and V_{IC45} will emanate from their instant-center positions. Now draw a straight line through the termini of the velocity vectors.[†] The velocity of IC34 will be a vector perpendicular to the line of centers and having its terminus on the velocity gradient. Determine ω of link 3 by using Eq. (46.10). Thus,

$$V_{IC24} = r_2\omega_2 \quad \text{and} \quad V_{IC45} = r_5\omega_5$$

Choose a scale and construct the two vectors. Next, draw the gradient line and construct V_{IC34}. Scale its magnitude and determine n_3 according to

$$n_3 = \frac{V_{IC34}}{2\pi r_3} 60 \tag{46.11}$$

where $r_3 =$ radius of the arm and n_3 is in revolutions per minute.

If gear 5 is fixed, then $V_{IC45} = 0$; and using V_{IC24}, connect the terminus of V_{IC24} and IC45 with a straight line, and find V_{IC34} as before. See Fig. 46.6.

[†] This line can be called a *velocity gradient* for link 4.

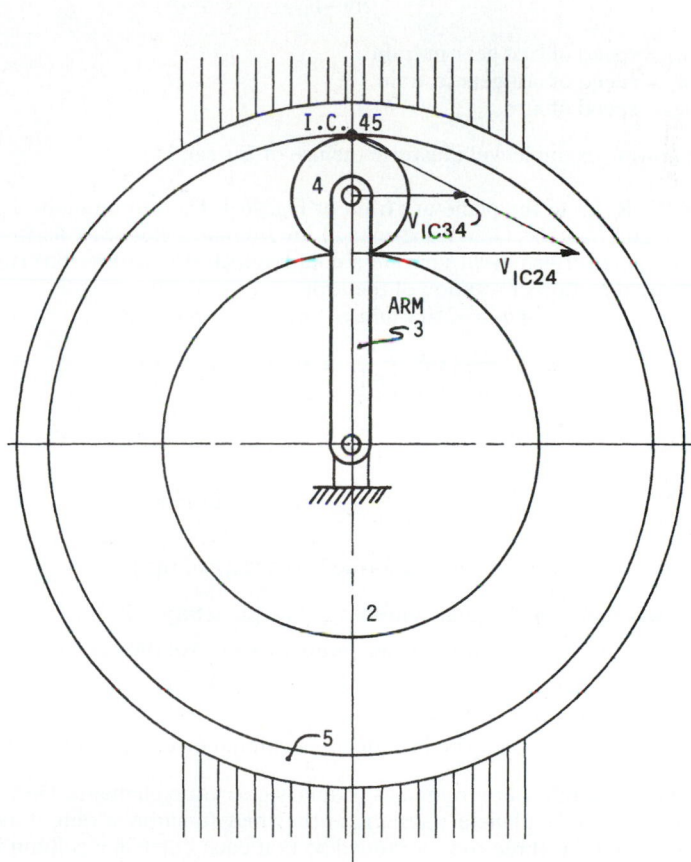

FIGURE 46.6 Gear 5 is fixed.

By formula, the relative-motion equation will establish the velocity of the gears relative to the arm; that is,

$$n_{23} = n_2 - n_3 \tag{46.12}$$

$$n_{53} = n_5 - n_3 \tag{46.13}$$

Then, dividing (46.13) by (46.12), we see that

$$\frac{n_{53}}{n_{23}} = \frac{n_5 - n_3}{n_2 - n_3} \tag{46.14}$$

which represents the ratio of the relative velocity of gear 5 to that of gear 2 with both velocities related to the arm. The right-hand side of the equation is called the *train value*. If the arm should be held fixed, then the ratio of output to input speeds for an ordinary train is obtained.

The equation for train value, which is seen in most references, can be written

$$e = \frac{n_L - n_A}{n_F - n_A} \tag{46.15}$$

where n_F = speed of first gear in train
        n_L = speed of last gear in train
        n_A = speed of arm

The following example will illustrate the use of Eq. (46.15).

Example 1.   Refer to the planetary train of Fig. 46.4. The tooth numbers are N_2 = 104, N_4 = 32, and N_5 = 168. Gear 2 is driven at 250 r/min in a clockwise *negative* direction, and gear 5 is driven at 80 r/min in a counterclockwise *positive* direction. Find the speed and direction of rotation of the arm.

Solution. $n_F = n_2 = -250$ r/min $n_L = n_5 = +80$ r/min

$$e = \left(-\frac{N_2}{N_4}\right)\left(\frac{N_4}{N_5}\right) = \left(-\frac{104}{32}\right)\left(\frac{32}{168}\right) = -\frac{13}{21}$$

In Eq. (46.15),

$$-\frac{13}{21} = \frac{80 - n_3}{-250 - n_3} \qquad n_3 = -46.2 \text{ r/min}$$

By tabular method, a table is first formed according to the following:

1. Include a column for any gear centered on the planetary axis.
2. Do not include a column for any gear whose axis of rotation is fixed and different from the planetary axis.
3. A column for the arm is not necessary.
4. The planet, or planets, may be included in a column or not, as preferred.

Gears which fit rule 2 are treated as ordinary gear train elements. They are used as input motions to the planetary system, or they may function as output motions.

The table contains three rows arranged so that each entry in a column will constitute one term of the relative-motion equation

TABLE 46.1 Solution by Tabulation

Step	Gear 2	Gear 5
1. Gears locked	n_3	n_3
2. Arm fixed	$n_2 - n_3$	$-\dfrac{N_2}{N_4}\left(\dfrac{N_4}{N_5}\right)(n_2 - n_3)$
3. Results	n_2	n_5

$$n_y + n_{xy} = n_x \qquad (46.16)$$

This is best shown by example. Using the planetary train of the previous example, we form Table 46.1, and the equation from the column for gear 5 is

$$n_3 - \frac{N_2 N_4}{N_4 N_5}(n_2 - n_3) = n_5$$

Rearranging and canceling N_4, we find

$$n_3\left(1 + \frac{N_2}{N_5}\right) - n_2\frac{N_2}{N_5} = n_5 \qquad (46.17)$$

This is the characteristic equation of the planetary train, as shown in Fig. 46.4.

Note that three rotational quantities appear—n_3, n_2, and n_5. There must be two input rotations in order to solve for the output. This is easily done when the input rotations and the tooth numbers are inserted. When a positive sense is assigned to counterclockwise and a negative sense to clockwise rotation, the sign of the output rotation indicates its sense of direction.

Note that planet 4 was not included in the table (it could have been); however, gear 4 served its purpose by acting as an idler to change a direction of rotation. This is evidenced by the presence of a negative sign in the second row of the column for gear 5.

A convenient means of representing a planetary train was shown by Levai, Ref. [46.4]. Type A of Fig. 46.7a shows an edge view of the planetary train first seen in Fig. 46.4. It and the other 11 configurations represent all possible variations for a planetary train. The equations in Table 46.2 are the characteristic equations of the 12 types.

An examination of the equations and their corresponding types reveals that certain ones are identical. Types C and D in Fig. 46.7b are identical because of the arrangement of gears. Whereas in type C the meshes of 2 and 4 and of 7 and 8 are external, the input and output meshes are internal in type D. The same relationship can be seen in types G and H in Fig. 46.7c. Certain pair types are alike in equation form but differ in sign. Compare types E and K, F and L in Fig. 46.7b and d, and B and G (or B and H) in Fig. 46.7a and c.

The speed of a planet gear relative to the frame or relative to the arm may be required. If appreciable speeds and forces are involved, this information will facilitate the selection of bearings. Using type A as an example, set up Table 46.3. Row 2 in the column for gear 4 is the speed of gear 4 relative to the arm, and row 3 in the column for gear 4 is its speed relative to the frame.

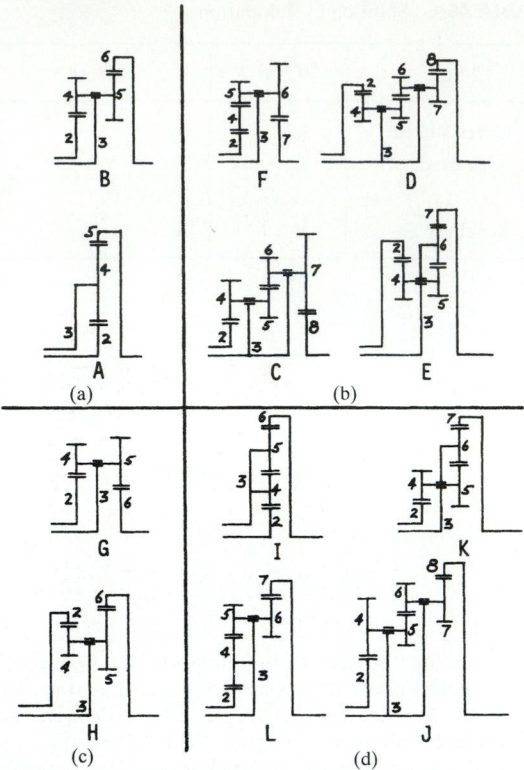

FIGURE 46.7   Twelve variations of planetary trains.

Example 2.   Figure 46.8 shows a planetary gear train with input at gear 2. Also, gear 6' is seen to be part of the frame, in which case its rotation is zero. For $n_2 = 100$ r/min clockwise (negative), find output rotation n_6.

Solution.   Gears 2, 4, 5, and 6 and arm 3 form a type B planetary train:

$$n_3\left(1 + \frac{N_2 N_5'}{N_4 N_6'}\right) - n_2 \frac{N_2 N_6'}{N_4 N_6'} = n_6'$$

Solving for n_3 yields

$$n_3\left(1 + \frac{1}{8}\right) - (-100)\left(\frac{1}{8}\right) = 0$$

$$n_3 = -\frac{100}{9} \text{ r/min}$$

For type G:

$$n_3\left(1 - \frac{N_2 N_5}{N_4 N_6}\right) + \left(n_2 \frac{N_2 N_5}{N_4 N_6}\right) = n_6$$

Then we solve type G for n_6:

TABLE 46.2 Characteristic Equations for 12 Planetary Trains of Fig. 46.7

Type	Equation
A	$n_3\left(1 + \dfrac{N_2}{N_5}\right) - n_2\dfrac{N_2}{N_5} = n_5$
B	$n_3\left(1 + \dfrac{N_2 N_5}{N_4 N_6}\right) - n_2\dfrac{N_2 N_5}{N_4 N_6} = n_6$
C	$n_3\left(1 + \dfrac{N_2 N_5 N_7}{N_4 N_6 N_8}\right) - n_2\dfrac{N_2 N_5 N_7}{N_4 N_6 N_8} = n_8$
D	$n_3\left(1 + \dfrac{N_2 N_5 N_7}{N_4 N_6 N_8}\right) - n_2\dfrac{N_2 N_5 N_7}{N_4 N_6 N_8} = n_8$
E	$n_3\left(1 + \dfrac{N_2 N_5}{N_4 N_7}\right) - n_2\dfrac{N_2 N_5}{N_4 N_7} = n_7$
F	$n_3\left(1 + \dfrac{N_2 N_6}{N_5 N_7}\right) - n_2\dfrac{N_2 N_6}{N_5 N_7} = n_7$
G	$n_3\left(1 - \dfrac{N_2 N_5}{N_4 N_6}\right) + n_2\dfrac{N_2 N_5}{N_4 N_6} = n_6$
H	$n_3\left(1 - \dfrac{N_2 N_5}{N_4 N_6}\right) + n_2\dfrac{N_2 N_5}{N_4 N_6} = n_6$
I	$n_3\left(1 - \dfrac{N_2}{N_6}\right) + n_2\dfrac{N_2}{N_6} = n_6$
J	$n_3\left(1 - \dfrac{N_2 N_5 N_7}{N_4 N_6 N_8}\right) + n_2\dfrac{N_2 N_5 N_7}{N_4 N_6 N_8} = n_8$
K	$n_3\left(1 - \dfrac{N_2 N_5}{N_4 N_7}\right) + n_2\dfrac{N_2 N_5}{N_4 N_7} = n_7$
L	$n_3\left(1 - \dfrac{N_2 N_6}{N_5 N_7}\right) + n_2\dfrac{N_2 N_6}{N_5 N_7} = n_7$

TABLE 46.3 Solution of Type A Train

Step	Gear 2	Gear 4	Gear 5
1. Gears locked	n_3	n_3	n_3
2. Arm fixed	$n_3 - n_2$	$-\dfrac{N_2}{N_4}(n_3 - n_2)$	$-\dfrac{N_2}{N_5}(n_3 - n_2)$
3. Results	n_3	n_4	n_5

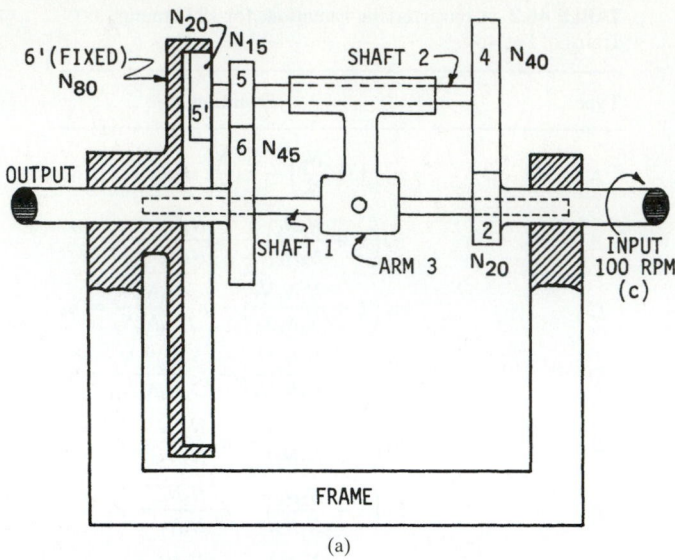

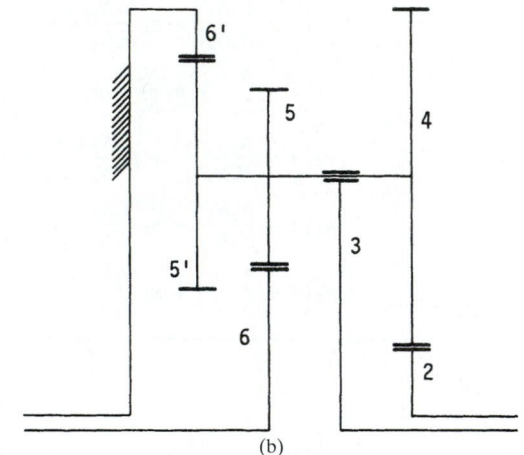

FIGURE 46.8   (*a*) View of a gear train and (*b*) its symbolic notation.

$$-\frac{100}{9}\left(1-\frac{1}{6}\right)+(-100)\left(\frac{1}{6}\right)=n_6$$

$$n_6 = -25.93 \text{ r/min}$$

Example 3.   Figure 46.9 shows a type I planetary train, Ref. [46.2]. Here, if $n_2 = 100$ r/min clockwise and $n_3 = 200$ r/min clockwise, both considered negative, determine n_4, n_5, and n_6.

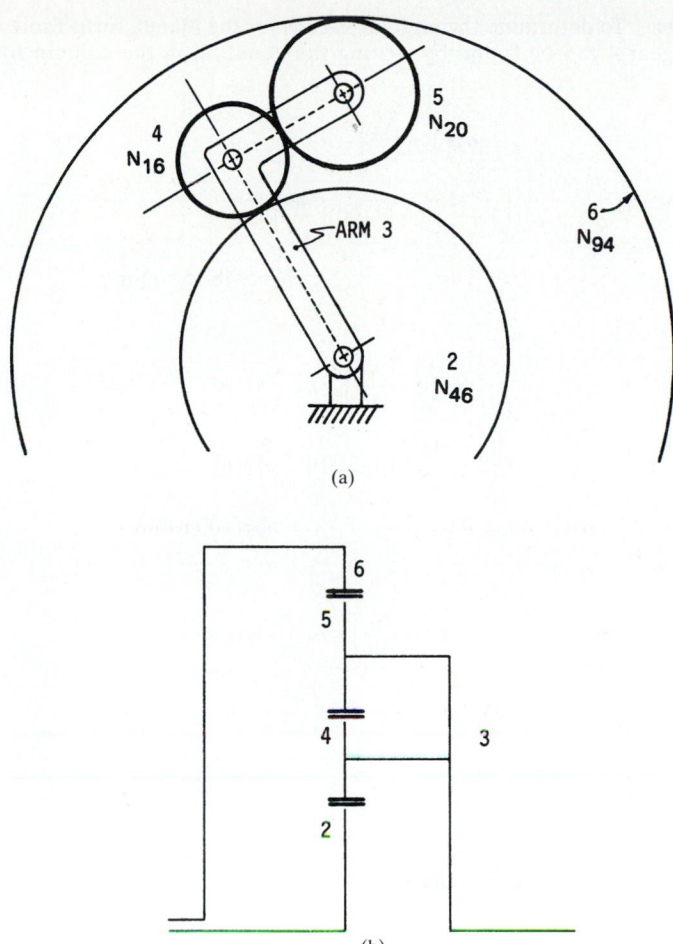

FIGURE 46.9 (*a*) Planetary train; (*b*) symbolic notation.

TABLE 46.4 Solution of Type I Train

Step	Gear 2	Gear 4	Gear 5	Gear 6
1	n_3	n_3	n_3	n_3
2	$n_2 - n_3$	$-\dfrac{N_2}{N_4}(n_2 - n_3)$	$+\dfrac{N_2}{N_5}(n_2 - n_3)$	$+\dfrac{N_2}{N_6}(n_2 - n_3)$
3	n_2	n_4	n_5	n_6

Solution. To determine the angular speeds for the planet, form Table 46.4. The speed of gear 4 can be found by writing the equation in the column for gear 4. Thus,

$$n_3\left(1 + \frac{N_2}{N_4}\right) - n_2\frac{N_2}{N_4} = n_4$$

$$-200\left(1 + \frac{46}{16}\right) - (-100)\left(\frac{46}{16}\right) = n_4$$

$$n_4 = -487.5 \text{ r/min}$$

For gear 5,

$$n_3\left(1 - \frac{N_2}{N_5}\right) + n_2\frac{N_2}{N_5} = n_5$$

$$-200\left(1 - \frac{46}{20}\right) + (-100)\left(\frac{46}{20}\right) = n_5$$

$$n_5 = +30 \text{ r/min}$$

For gear 6,

$$n_3\left(1 - \frac{N_2}{N_6}\right) + n_2\frac{N_2}{N_6} = n_6$$

$$-200\left(1 - \frac{46}{94}\right) + (-100)\left(\frac{46}{94}\right) = n_6$$

$$n_6 = -151 \text{ r/min}$$

46.4 DIFFERENTIAL TRAINS

Differential gear trains are useful as mechanical computing devices. In Fig. 46.10, if ω_4 and ω_6 are input angular velocities and V_A and V_B are the resulting linear velocities of points A and B, respectively, then the velocity of point C on the carrier is

$$V_C = \frac{V_A + V_B}{2} \qquad (46.18)$$

The differential gear train also finds application in the wheel-axle system of an automobile. The planet carrier rotates at the same speed as the wheels when the automobile is traveling in a straight line. When the car goes into a curve, however, the inside wheel rotates at a lesser speed than the outside wheel because of the differential gear action. This prevents tire drag along the road during a turn.

Example 4. See Ref. [46.2], page 329. The tooth numbers for the automotive differential shown in Fig. 46.11 are $N_2 = 17, N_3 = 54, N_4 = 11, N_5 = N_6 = 16$. The drive shaft turns at 1200 r/min. What is the speed of the right wheel if it is jacked up and the left wheel is resting on the road surface?

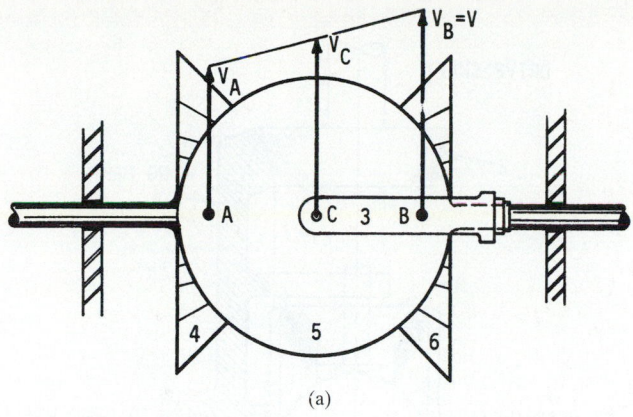

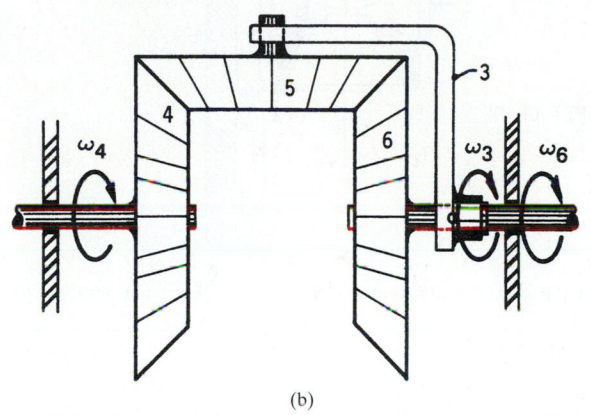

FIGURE 46.10 (*a*) Top and (*b*) front views of a bevel-gear differential used as a mechanical averaging linkage. Point *A* is the pitch point of gears 4 and 5. Point *B* is the pitch point of gears 5 and 6.

Solution. The planet carrier, gear 3, is rotating according to the following equation:

$$n_3 = \frac{N_2}{N_3}\, n_2 = \frac{17}{54}\,(1200) = 377.78 \text{ r/min}$$

Since the r/min of the left wheel is zero, the pitch point of gears 4 and 5 has a linear velocity twice that of the pin which supports the planet. Therefore, the r/min of the right wheel is twice that of the planet, or

$$n_6 = 2n_3 = 755.56 \text{ r/min}$$

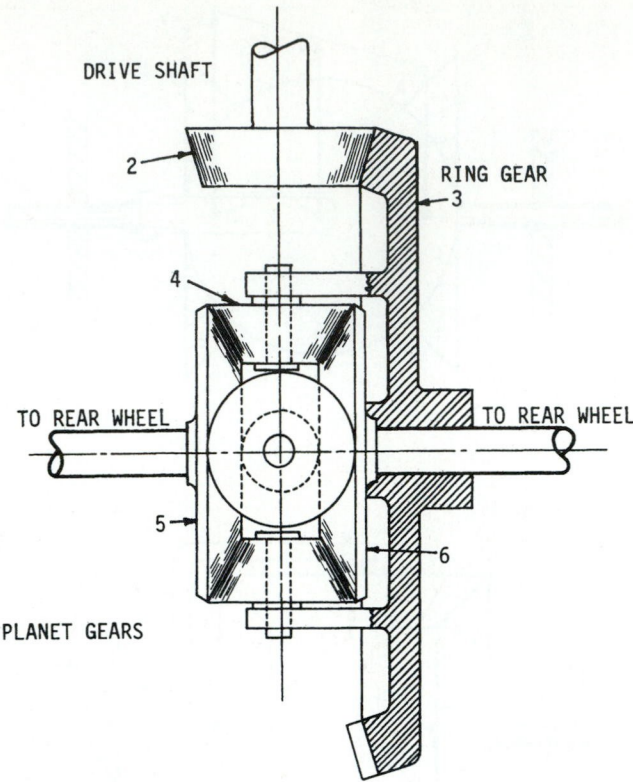

DRIVE SHAFT

2

RING GEAR
3

4

TO REAR WHEEL

TO REAR WHEEL

5

6

PLANET GEARS

FIGURE 46.11   Schematic drawing of a bevel-gear automotive differential.

REFERENCES

46.1 Richard S. Hartenberg and Jacques Denavit, *Kinematic Synthesis of Linkages,* McGraw-Hill, New York, 1964.

46.2 J. E. Shigley and J. J. Uicker, Jr., *Theory of Machines and Mechanisms,* 2d ed., McGraw-Hill, New York, 1995.

46.3 Virgil M. Faires and Robert M. Keown, *Mechanism,* 5th ed., McGraw-Hill, New York, 1960.

46.4 Z. L. Levai, "Theory of Epicyclic Gears and Epicyclic Change-Speed Gears," Technical University of Building, Civil and Transport Engineering, Budapest, 1966.

CHAPTER 47
ROBOTS AND SMART MACHINES

Kenneth J. Waldron, Ph.D.
Professor of Mechanical Engineering
The Ohio State University

47.1 INTRODUCTION

47.1.1 Elements of a Robot System

In recent years, the so-called industrial robot has become a familiar feature of manufacturing plants. This class of machines is, of course, only a part of a much more diverse family of devices characterized by large numbers of degrees of freedom and intelligent controllers. Industrial robots are, however, by far the most numerous, visible, and economically important group of devices in this family. For this reason, much of the material in this chapter is directed at industrial robot design, although an attempt is made to place them in an overall context of intelligent mechanical systems.

Figure 47.1 indicates the hardware subsystems present in a generalized industrial robot system. The manipulator usually has six independently actuated joints, because a body which moves freely in space has six degrees of freedom. Consequently, if the "hand" of the manipulator is to be placed in an arbitrary position and orientation within the manipulator's reach, then the mechanism must have six degrees of freedom. Nevertheless, some industrial robots on the market have as few as four or as many as seven degrees of freedom. Some types of tasks do not require the full six degrees of freedom and can be handled by four- or five-degree-of-freedom robots.

Because the joints usually are not capable of complete rotation but can move only over a restricted angular range, additional degrees of freedom beyond the basic six are often useful for demanding manipulative operations. The joints may be actu-

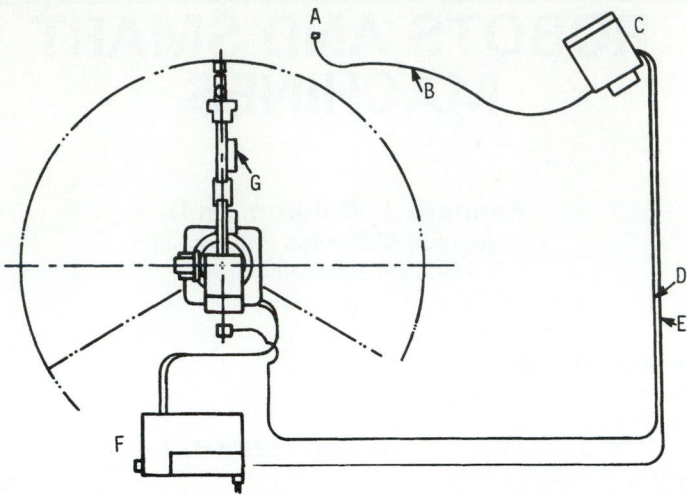

FIGURE 47.1 Components of a generalized industrial robot system: A, teach pendant; B, control to pendant (flexible conduit); C, control cabinet; D, robot to control (flexible conduit); E, power supply to control (flexible conduits); F, power supply; G, industrial robot.

ated electrically, hydraulically, or pneumatically. Each joint is equipped with a position sensor that furnishes the input signal to a servo controller, which positions the joint in response to commands from the central controller.

The hand of the manipulator takes a wide variety of forms depending on the operations for which it is being used. Most often it is a special-purpose tool such as a spot welder, a shielded-arc electrode, a paint sprayer, or a rotary drill. For manipulative or transfer operations it is often a simple gripper, consisting of two jaws which close as a vise on the workpiece.

The power supply varies considerably with the type of actuation used but is always present in some form. High-quality servo systems require well-regulated power regardless of the type of system. Hydraulic systems usually have relatively large and complex power supplies because conversion from electric to hydraulic power is performed locally.

The central controller is a mini- or microcomputer or, more accurately, is a set of software resident in that computer which translates a stored program to a series of position commands to the joint servo controllers to repetitiously generate a series of motions. The controller usually includes permanently stored monitor and operating system firmware. The latter includes a system for programming the machine and may include capability for writing and reading to tape cassettes or floppy disks. This latter capability permits long-term storage and reuse of programs. It also permits a program generated on one machine to be used on other similar machines.

The teach pendant, or programming box, is used in many robots, particularly those of the so-called point-to-point type. The operator uses controls on the teach pendant to place the robot in a series of positions. These positions are recorded and form the data used by the operating program. Additional programming information may be entered by command buttons or by keyboard entry of instructions. Thus, if the robot gripper is to be moved to a specified position and close at that position during the operating cycle, the operator must move the manipulator, using the teach pendant controls, until the gripper is in the desired position. Then the operator enters a command to close the gripper by means of the keyboard and pushes a button on the teach pendant, causing the positions of the joints to be recorded.

Modes of programming and operating industrial robots vary considerably with the class of functions for which each robot is designed. These are discussed further in Sec. 47.2.

47.1.2 Use of the Word *Robot*

The word *robot* has come to be used in a variety of different senses. Its use in naming the rather restricted class of devices introduced above and called *industrial robots* is at variance with the general English use of the word and, indeed, with its technical use outside the industrial robot industry. In common use, a robot is a device which mimics some of or all the characteristics of autonomous intelligence, locomotion, and manipulation found in human beings. In earlier technical use, as in the space program, for example, a robot device was one with a certain amount of local information-processing capability which could execute quite complex operations in response to relatively infrequent supervisory commands. This use was often applied to devices which had no manipulatory or locomotory function at all.

The essential feature of an industrial robot is that it is a *flexibly reprogrammable mechanical device* capable of performing a wide variety of functions. Industrial robots are devices built to perform relatively complex, but nevertheless highly repetitive, operations. They differ from purely mechanical devices, such as cam mechanisms, which also perform complex repetitive operations, in that the constraints which cause them to perform determinate motions are provided by digital data lists via an active actuation system rather than by kinematic constraints. In fact, from the point of view of mechanical design, the definition of the industrial robot is excessively restrictive. It excludes teleoperators (remote handling devices) and more general devices which more closely approach the general concept of a robot.

47.1.3 Externally Constrained Mechanisms

A more useful definition, from the point of view of mechanical design, is that a robot is an *externally constrained mechanism*. An externally constrained mechanism is one which has a relatively large number of degrees of freedom and which performs deterministic motions because additional constraints are provided by means of an active system which interfaces with the mechanism. The active system can be a set of servo actuators controlled by a computer, as in an industrial robot. It can also be an operator manipulating a control harness which is part of the mechanism, as in the case of a passive teleoperator. Many intermediate combinations are also found.

Figure 47.2 shows a purely mechanical teleoperator designed for the handling of radioactive materials. The master and slave arms are kinematically identical, and motion is transmitted between them by means of linkages. The device has six degrees of freedom. The external constraints are provided directly by the operator acting on the master.

The space shuttle manipulator arm is an example of a very sophisticated teleoperator. The operator's movements are not mechanically transmitted to the slave manipulator, as in the device of Fig. 47.2. The master controller geometry, in fact, bears no resemblance to the arm geometry. The controller movements are trans-

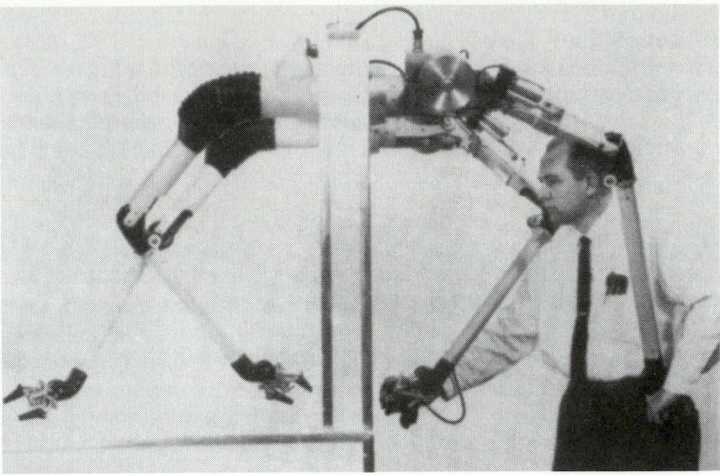

FIGURE 47.2   Mechanical teleoperator system for handling nuclear materials. The motion is transferred by linkages and cables. (*Sargent Industries.*)

formed in a computer to commands to a set of servo actuators which move the arm. Despite being a teleoperator, or master-slave mechanism, this device has a high level of machine intelligence. This is needed not only to transform the movements of the controller to movements of the geometrically dissimilar arm, but also to apply sophisticated control techniques to obtain smooth, stable motion from the highly compliant arm. The device is a teleoperator because the ultimate source of the external constraints is the movements of the operator acting on the controller in real time.

Figure 47.3 shows a typical industrial robot. It has six degrees of freedom. In the operational mode, it is moved by servo actuators in response to commands from a minicomputer which reads a stored list of operator commands. Thus, the essential difference between it and the teleoperator is that although the movement commands still originate from an operator, the operator is not interacting with the device in real time during its normal operation. The important point, then, is that this distinction has little impact on the mechanical design of the devices. They belong to the same class for this purpose.

As larger and larger amounts of processing power are placed on these devices, the distinctions made above among teleoperators, industrial robots, more general robots, and other related devices tend to blur. Figure 47.4 is a scale model of a device which, while very definitely an externally constrained mechanism, fits none of the above categories. It is a vehicle for use in rough terrain conditions. In rugged conditions, legs have significant mechanical advantages over wheels or tracks. It is, however, very much more difficult to actuate the legs efficiently while maintaining the adaptability characteristic of an externally constrained mechanism. The device has a high level of onboard data processing power, allowing it to automatically coordinate the movements of the leg actuators in response to information about the terrain in front of the device received from an optical scanning system, information about the leg loading received from force sensors in the "ankles," and information about the positions of returning legs relative to the ground received from proximity sensors. The operator commands direction and speed but, when cruising, does not directly influence leg movements. This device is certainly not a robot, since it has an opera-

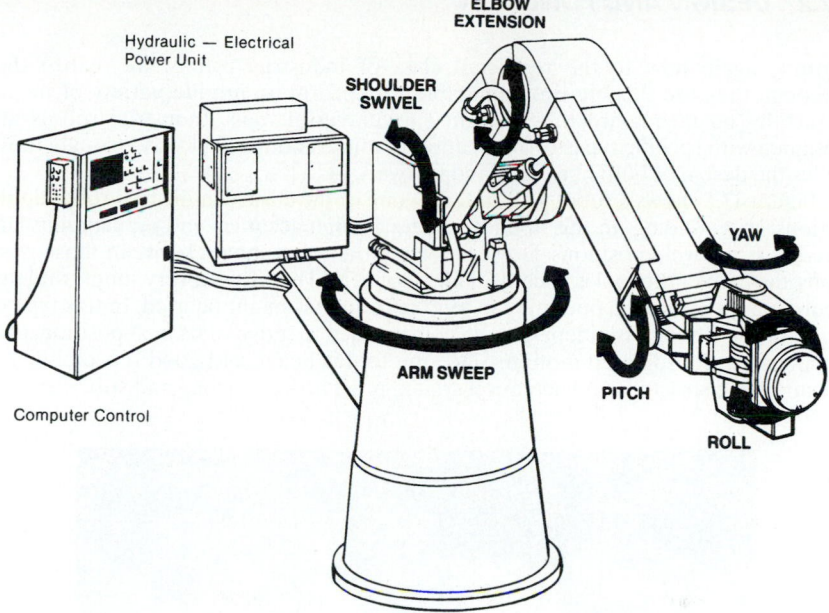

Hydraulic — Electrical
Power Unit

ELBOW
EXTENSION

SHOULDER
SWIVEL

YAW

ARM SWEEP

PITCH

ROLL

Computer Control

FIGURE 47.3 Drawing of an industrial robot indicating the degrees of freedom. (*Cincinnati Milacron.*)

tor on board. Although the operator is the source of some of the information used to provide external constraint, she or he is not the sole source of that information. Thus, it is not a teleoperator either. Nevertheless, the technology used is similar to that used in robots and advanced teleoperators. Once again, the concept of an externally constrained mechanism highlights the essential relationship between the characteristics of these devices.

FIGURE 47.4 The adaptive suspension vehicle (⅒-scale model) This is a vehicle for transportation in very rough terrain conditions which uses legged locomotion. (*The Ohio State University.*)

47.2   DESIGN AND FUNCTION

Turning again now to the restricted class of industrial robots, we realize that although they are flexibly programmable to perform an infinite variety of movements, by no means are they designed as universal tools. Industrial robots are designed with specific types of application in mind, and this fact very strongly influences the design of both hardware and software [47.1].

Figure 47.5 shows a robot well suited to one of the earliest industrial robot applications: spot welding. In this application, fairly high accuracy and repeatability are needed at the weld positions. However, when the robot moves between those positions, the path of the tool is usually of little concern. Therefore, a very simple and fast coordination algorithm operating in joint coordinates might be used. In this type of operation, each joint is independently commanded to move to its next position, producing an uncoordinated motion. Since the tool is heavy and good repeatability is required, the structure and actuation system must be both strong and stiff.

FIGURE 47.5   Heavy-duty industrial robot equipped for spot welding. This is called the NACHI Robot 8000 Series. [*C. Itoh & Co. (America)*.]

A robot suited to seam welding is one of the most important current applications in economic terms. Such robots usually have only five degrees of freedom. Since the tool is a rotationally symmetric electrode, a sixth degree of freedom is unnecessary. Arc-welding robots are, however, used often in conjunction with programmable work tables which, in principle, provide one or two additional degrees of freedom. The use of sliding joints for the first three degrees of freedom is favored in many arc-welding robot designs, even though it leads to a very large structure, because it simplifies the generation of accurate straight lines. These are frequently necessary in the welding of seams. One reason for using programmable tables is to line up the seam to be welded with one of the manipulator slides. Since an arc welder must accurately generate straight lines and curves and must closely control the orientation of the welding head, the type of software suggested above for the spot welder would be quite unsuitable. A relatively sophisticated coordination algorithm based on resolved motion rate control would be preferable. The robot would still be a point-to-point robot. That is, only discrete positions would be taught by the operator. The machine would automatically generate its path between those positions.

When robots are used for spray painting, the loads are light and great accuracy and repeatability are not necessary. Thus, spray-painting robots are often relatively lightly built. The explosive environment in which they operate mandates great care in the use of electric motors. Remote actuation systems, permitting better protection of the motors, are often used. Pneumatic or hydraulic actuation systems may be used to eliminate potential spark sources.

Spray-painting robots are usually taught in a *continuous-path* mode. The operator moves the device through the motion to be taught in real time. Joint positions are sampled at equal time intervals. In playback operation, a simple interpolation algorithm is used between the sampled positions to generate a smooth motion. For this type of algorithm, the geometry of the robot is unimportant. In fact, the computer does not even know what the robot "looks like." Consequently, quite complex geometries are sometimes used for spray-painting robots.

Figure 47.6 shows a robot designed for assembly operations. Actually, the tasks involved in robot assembly are quite diverse. Correspondingly, so-called assembly robots vary from very simple "pick and place" devices up to the most sophisticated

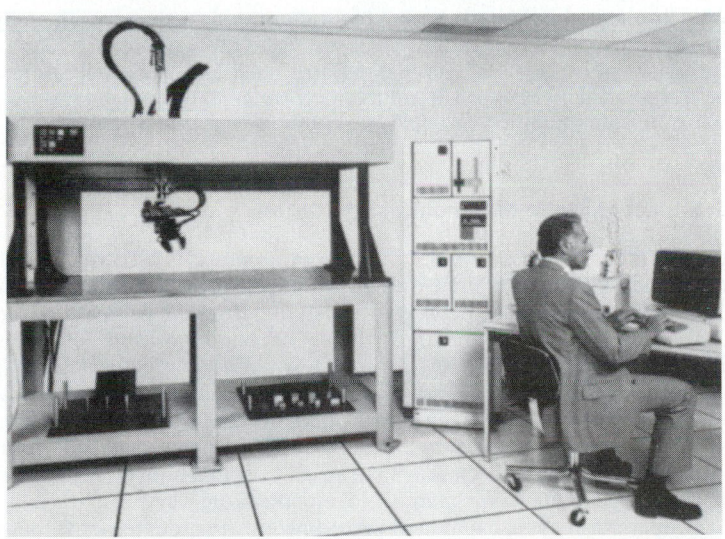

FIGURE 47.6 Robot suitable for assembly operations. Orthogonal slide arrangement gives uniform positioning accuracy and is well adapted to planar transport of parts and unidirectional insertion. Large range of motion in wrist and force-sensing gripper gives good dexterity.

units presently available. Assembly robots tend to be relatively small and geometrically adapted, as in Fig. 47.6, to vertical movements and large movements in the horizontal plane. They may have fewer than six degrees of freedom since many assemblies are designed for all parts to be added from a single direction. The more sophisticated units have more general geometries, sometimes even more than six degrees of freedom, and force sensing in the gripper.

47.3   STRUCTURAL DESIGN

47.3.1   Structural Characteristics

The unique characteristics of externally constrained mechanisms lead to structural design problems which, while certainly not unique to this class of mechanisms, are otherwise relatively uncommon. Most externally coordinated mechanisms have sequential chains of members and joints fixed to a base at one end and loaded at the other. Structurally, this is a cantilever beam, but one which changes geometry.

The overriding structural design constraint for most industrial robots is accuracy and repeatability. For the present generation of industrial robots, which operate without endpoint feedback, this implies high stiffness. Many current-generation robots consequently have massive structures even though their rated load capacity is very modest.

There is a second reason for the importance of stiffness. The servo actuation systems used to operate the joints can be a source of excitation of structural vibration. This is particularly true of the digital servo controllers, which are becoming increasingly popular. The frequency of the update cycle becomes a source of vibration excitation. A rule of thumb in robot design is that the lowest natural frequency of free structural vibrations should be at least 3 to 4 times the servo bandwidth and preferably rather more.

The irregular movements of the manipulator itself also excite vibrations. However, these are transient in nature. If not damped out quickly enough, they may cause problems in fine manipulation operations.

47.3.2   Impact of Mode of Sensing and Control

The need for great stiffness to obtain sufficient accuracy is a consequence of the type of position control used. Since only the joint positions are read and hand position is inferred from them indirectly, errors resulting from the behavior of the intervening structure must be avoided. If, however, hand position relative to the workpiece can be measured directly, a completely different structural design philosophy can be used. Figures 47.7 and 47.8 show two machines which illustrate this point very well. Both are externally constrained mechanisms. The first is a manipulator designed to be used as a self-help device by a quadriplegic (a person wholly or partially paralyzed from the neck down because of a spinal injury). It uses computer coordination to generate coordinated movements from very restricted operator inputs. The operator may be capable of providing only two-degree-of-freedom control movements. For the present purpose, however, the important point is that *endpoint feedback* is available to this device by means of the operator's vision. Despite the fact that the arm is lightly built and quite compliant, there is no problem with accurate positioning.

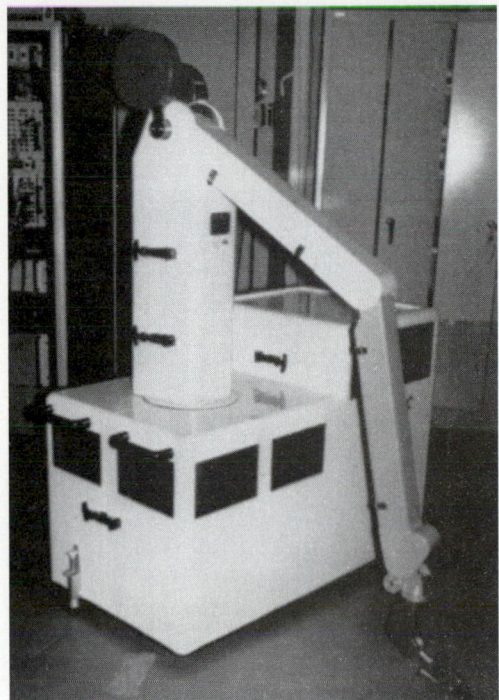

FIGURE 47.7 Manipulator for use by quadriplegic. This device is a teleoperator which operates from very limited inputs and achieves good accuracy despite light, compliant structure and drive because operator's vision provides endpoint feedback. (*Jean Vertut, CEA, Saclay, France.*)

The second device, shown in Fig. 47.8, is a prototype of an industrial robot designed for shearing sheep. Although it is hydraulically actuated, it is also relatively lightly built and is quite compliant. It has eight degrees of freedom. Seven of these move the member carrying the cutter approximately in a series of passes over a computer model of the sheep's body. The remaining joint is a sliding joint which is maintained approximately normal to the surface of the sheep's body and which carries the clippers. It is equipped with a fast, accurate servo actuation system. A contact sensor, which operates by measuring the electric resistance between the clippers and the body of the sheep, and capacitive proximity sensors provide information about the distance of the clippers from the sheep's skin which is used by the sliding-joint servo controller to maintain a constant distance between clippers and skin. Again, the use of direct measurement of the position of the tool relative to the workpiece removes the need for a massive, stiff structure and for fast, accurate response in most joints.

47.3.3 Selection of Structural Sections

Since, in the present generation of industrial robots, strength is unimportant and the lowest vibrational natural frequencies are important, hollow structural sections with

FIGURE 47.8 Prototype sheep-shearing robot. High performance is achieved by means of endpoint feedback from resistive and capacitive sensors mounted in clippers. Robot structure is compliant and has low response except for RAM, which is maintained normal to body of sheep and which responds rapidly. (*Automated Sheep Shearing Project, University of Western Australia, Nedlands.*)

a high degree of symmetry tend to be used. Square sections are particularly popular. Selection of an adequate structural section is not necessarily sufficient to attain adequate stiffness. The compliances of the servo actuators are usually at least comparable to those of the structural members. Because of the complexity of the structure and the mass distribution resulting from the locations of the relatively massive actuators, it is really necessary to use numerical design analysis tools both for determining static deflection and for estimating vibration modes and frequencies. The position in which the arm is fully extended in the horizontal direction with a mass whose weight corresponds to full load lumped at the gripper can usually be regarded as the worst case for both types of analysis.

Even when good structural analysis packages are used, the problem is far from straightforward. Modeling of the manipulator as a solid cantilever with appropriate mass distribution, followed by finite-element analysis, does not usually yield good results for dynamic behavior because of joint compliance. If the joint compliances are high enough for structural compliance to be neglected, better results can be obtained by using one of the mechanism analysis packages which model the members as rigid and the joints as damped springs.

No matter how well the system is modeled, the free-vibration frequencies must be treated with caution. Although vibrational modes can be accurately predicted by good numerical models, modal frequencies are notoriously difficult to estimate accurately. The designer should be prepared to use prototype testing with appropriate adjustments to the structural design to ensure adequate structural performance.

47.3.4   Material Selection

Since strength is not usually a consideration, the materials used in robot construction are quite conventional. The choice is usually based on ease of manufacture. Low-carbon steel has been used in many designs. Some reduction in weight, without losing stiffness and with consequent improvement in vibrational performance, can be achieved by using aluminum. Extruded aluminum tube sections are a convenient structural choice. Glass-fiber-reinforced composites are attractive for the same reason and are appearing in some of the newer designs. Because bending stiffness is the primary requirement, the reinforcing fibers should be laid at a low angle to the section axis. Relatively large tube diameters, or side dimensions for prismatic tubes, with thin walls must be used for the same reason.

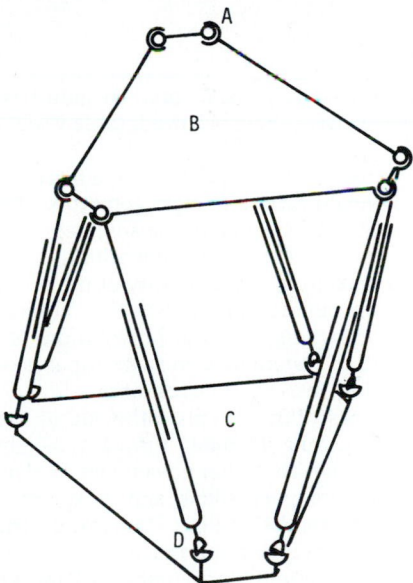

Parallel Structures. As has been noted, the current generation of industrial robots uses series chain structures almost exclusively. There is no basic reason why manipulators must be constructed in this way. The Stewart platform shown in Fig. 47.9 is, equally, a six-degree-of-freedom externally constrained mechanism but has a parallel actuation geometry rather than a series geometry. The Stewart platform has been widely used in aircraft simulators and similar devices, as shown in Fig. 47.10. It functions in a simulator in a very similar manner to an industrial robot. The Stewart platform's vibrational and load-carrying capabilities are certainly superior to those achievable by a comparable series structure. Its motion range is more limited. Nevertheless, it does indicate that there may be something to be gained by thoughtful use of parallel actuation geometries.

FIGURE 47.9 Schematic drawing of the Stewart platform. A, ball-and-socket joint; B, movable platform; C, fixed base; D, Hooke joint. The movable platform is supported on six legs.

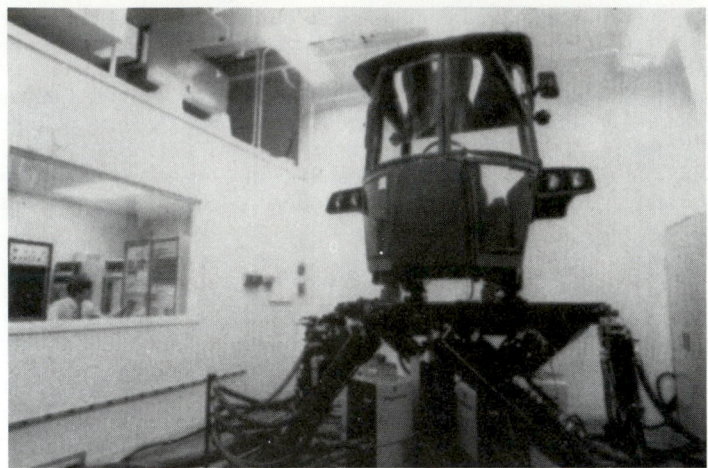

FIGURE 47.10   Stewart platform used to simulate six-degree-of-freedom movement of tractor cab. (*Deere & Company.*)

47.4   ACTUATION AND POWER TRANSMISSION SYSTEMS

47.4.1   Requirements

To date, three different power transmission media have been used in industrial robots: electric, hydraulic, and pneumatic. In each case, there is considerable variety in possible system configurations.

The mechanical requirements for actuators used in robots include high force-to-weight ratio for linear actuators, or torque-to-weight ratio for rotary actuators. This requirement results from the necessity to mount at least some of the actuators outboard on the arm. Joint rates are comparatively low. A typical robot with a maximum reach of about 2 meters (m) will have a maximum hand velocity of the order of 800 inches per minute (in/min) [250 millimeters per second (mm/s)]. This requires a joint rate at the shoulder of 0.2 radian per second (rad/s) or about 2 revolutions per minute (r/min) maximum. Of course, any actuation system used must be capable of being smoothly controlled with at least moderately fast dynamic response.

Electric systems are widely used in industrial robot systems, although, at the present state of the art, their capabilities for providing adequate torque-to-weight ratio with good dynamic response are quite marginal. The other advantages of electric systems include lower cost than hydraulics; compact, silent, and inexpensive power supplies; cleanliness; and ease of routing transmission lines. Because electric systems have trouble meeting the high torque-to-weight requirements of robotic applications, there is a strong trend to use high-technology servomotors. Pancake motors and, more recently, rare-earth motors are very widely used despite their relatively high cost. Rare-earth motors offer an improvement in torque-to-weight ratio by a factor of 2 to 3 over conventional permanent-magnet motors. Current rare-earth designs tend to be long, slender cylinders which present considerable packaging problems to the robot designer. Pancake motors, while not offering such high performance, are short axially with relatively large diameter, as their name implies. This configuration usually presents fewer packaging problems. Other types of electric actuator, such as stepping motors, which might otherwise seem attractive, at present cannot approach the torque-to-weight ratios necessary for robot applications.

Another consequence of the somewhat marginal torque-to-weight performance of electric systems is that they are limited in load-carrying capacity. In the present environment, in which most robot applications are low-load applications, this is not a serious handicap. However, as robot applications diversify, a role for heavy-duty robots will develop. In fact, "heavy-duty" robots have been on the market for a long time. At present *heavy duty* might be defined as anything over a payload of about 150 pounds (lb) [70 kilograms (kg)]. It is no surprise that hydraulic systems dominate this end of the market.

47.4.2 Electric Systems

Electric motors have very low intrinsic torque-to-weight ratios. In compensation, they are capable of operating at very high speeds. Thus, the torque of the motor can be matched to that needed at the joint by means of a speed-reducing and torque-increasing mechanical transmission. Another important feature of electric motors as actuators is that they are highly inefficient at low speeds under load. In particular, they do not hold stationary position well under load. Thus, it is usually advisable to provide solenoid-actuated brakes in electric systems for use when the joint is to hold stationary position. For both these reasons, electric actuators must always be used with mechanical transmissions.

Selection of Transmission. The necessity of a transmission as an electric actuator brings with it a number of problems. Use of very high speed-reduction ratios (which is, in principle, a solution to the torque-to-weight ratio problem) results in very high reflected armature inertias, as viewed from the joint. This results, in turn, in sluggish dynamic response. In addition, because of the discontinuous nature of robot movements, the substantial amounts of kinetic energy which a motor armature stores while running at high speed must be dissipated regularly, usually by conversion to heat. This results in increased system power demand and potential heat-dissipation problems.

Therefore, the selection of transmission ratio and actuator operating speed is a compromise between the increased torque-to-weight ratio achievable with high reduction ratios and the dynamic response and power dissipation problems caused by use of those ratios. Obviously, the actuator and transmission components must be selected or designed on a system basis. The primary design constraint on the system is dynamic response. The torque-to-weight ratio should be maximized consistent with meeting the dynamic response specifications.

The problems just discussed are not the only ones which result from use of a mechanical transmission. The compliance and backlash of the transmission present additional problems. Compliance in the transmission also degrades dynamic response. Thus, this fact should also be considered when the actuation and transmission system is designed to meet specified dynamic response goals. Backlash is often a far more troublesome problem. If joint position is used as the controlled variable of the joint servo system, backlash can lead to instability because the motor position and velocity cannot be inferred from the corresponding joint variables. Use of high reduction ratios exacerbates the problem. Of course, controlling motor shaft position and velocity directly removes this problem, but it introduces errors in joint position. In some successful designs, gravitational loading is used to eliminate backlash during normal operation.

The requirements of low backlash and compliance, relatively high output torque, and large speed reduction in a compact, lightweight package have resulted in wide use of several types of transmission which, while certainly not unknown in other types of mechanical systems, might be regarded as exotic. Harmonic drives, for example, which use a flexible splined intermediate member, have been used in several designs. Their advantages are a large speed-reduction capability in a single stage, with consequent light weight and compactness, and backlash-free operation. Disadvantages are relatively high compliance, a consequence of the flexible spline member, and only modest mechanical efficiency—of the order of 70 percent. Harmonic drives are usually supplied as a set of unpackaged components. They require considerable care in the design of mountings of the drive components and of the support bearings of the input and output shafts to ensure good service life.

Cycloidal drives, now on the market in several types, offer one-stage reduction ratios similar to those of harmonic drives with lower compliance and higher mechanical efficiency. However, robot applications tend to come in at the bottom end of the size range for cycloidal drives, and so they tend to be relatively large and heavy. Nevertheless, they are an attractive alternative in situations in which their weight can be handled.

Gear trains have also been used, both in conventional multistage configurations and in epicyclic trains. Since three or four reductions may be needed to achieve an adequate overall ratio, straight trains tend to be bulky. Backlash may be minimized by use of high-quality, accurately cut gears. It is not usually practical to spring-load multiple-stage gear trains. Gears are a viable option inboard where bulk and weight are not problems. Larger gears than are needed to carry the torque may be used to reduce compliance.

The epicyclic gear heads supplied as matched units by some manufacturers of servomotors tend to have inadequate torque capacity for robot applications. Custom design and manufacture of high-ratio, low-backlash, epicyclic trains is difficult and expensive. Nevertheless, epicyclic speed reducers are capable of filling robot transmission requirements.

The usual low-cost option for large-reduction-ratio transmission, worm gears, is not popular in robot applications. The relatively high backlash and low mechanical efficiency of worm gears are the main reasons. The noncoaxial geometry imposed by a worm reduction often creates packaging problems, which are an important consideration in outboard locations. One feature of worm gear sets is attractive. They can be designed to be self-locking, thus removing the need for brakes.

One of the most elegant solutions to the problem of obtaining a large speed reduction in a compact, lightweight package without backlash and with low compliance is obtained by the use of a ball screw. Ball screws are easily spring-loaded to eliminate backlash by using a preloaded double nut. They have high mechanical efficiency. When a rotary joint is powered in this manner, the system effectively has two speed-reduction stages. The first is in the conversion from rotary to linear motion via the screw. The second is in the conversion from linear back to rotary motion via a lever arm. The result is a very compact and efficient high-ratio reduction. Of course, the lever arm introduces a nonlinear relationship between joint position and motor position and limits the maximum joint rotation range to about 150°. The nonlinearity is of little importance provided the servo system operates on joint position and its time derivatives rather than motor position. The restricted rotational range is adequate for about half of the joints in a typical industrial robot geometry. Comparatively speaking, ball-screw drives are very stiff.

Band mechanisms have also been used in several externally constrained mechanisms. They have been particularly successful in teleoperators for remote handling of nuclear materials. Speed reduction is achieved by means of a block-and-tackle arrangement with metal tape used instead of cable. Figure 47.11 shows the very sophisticated band drives used in a servo-operated teleoperator. Those used in the manipulator of Figure 47.7 are of similar type. Band drives, although very compliant, allow transmission of power from actuators located inboard to the outboard joints. This has the advantage of allowing a lightweight, slender structure. It also allows better shielding of motors during operation in flammable environments such as in spray painting.

FIGURE 47.11   Band drives used in a servo teleoperator for handling nuclear materials. Speed reduction is achieved by use of block-and-tackle sets. (*Jean Vertut, CEA, Saclay, France.*)

Roller-chain drives can also be used to transmit power outboard from actuators mounted inboard. They are much stiffer than band mechanisms. However, it is not possible to use roller chain in block-and-tackle mechanisms, so the direct speed reduction obtainable is very limited. Thus, a speed reducer must be placed between the motor and the drive sprocket of the chain transmission.

Power Control Systems.   Many electric actuation systems use conventional variable-armature-resistance servomotor control. However, the advent of solid-state power-switching devices such as silicon controlled rectifiers (SCRs) has allowed the introduction of phase control and similar modulation techniques. These operate by applying a train of voltage pulses of controllably variable width to the motor. Effectively, the motor responds to the average voltage of this pulse train. These schemes offer substantial improvements in efficiency and, correspondingly, reductions in heat dissipation because they operate on a nondissipative principle. Motor efficiencies in the range from 70 to 90 percent become feasible.

48.4.3 Hydraulic Systems

Hydraulic actuators have the advantage of very high force-to-weight ratio. This allows direct drive of the joints without any intervening transmission. Pressure-regulated, parallel hydraulic circuits perform best at low joint rates and large loads. Hydraulic actuators hold fixed position well under load. Therefore, brakes are not needed. They are worst at high joint rates and low loads. Under those circumstances, large amounts of energy are converted to heat at the control valves. This is the reverse to the pattern of electric drives, which perform best at high joint rates and low loads and worst at low joint rates and high loads. Neither pattern is optimum for manipulator service since a typical operational cycle will include periods both of slow motion or holding position and of rapid motion.

The disadvantages of hydraulic systems are the need for a large, expensive, and noisy power supply; the high cost of servo valves; and the dirt created by oil leaks.

Hydraulic actuators may be linear actuators (hydraulic cylinders), rotary actuators capable of less than 1 revolution (fixed-vane actuators), or rotary actuators capable of continuous rotation (hydraulic motors). Hydraulic cylinders usually give lighter-weight actuation systems than rotary actuators for the same joint torque and motion requirements. However, as was mentioned in relation to ball-screw drives, any crank-type linear-to-rotary drive gives a nonlinear relationship between actuator displacement and joint rotation and is limited to about 150° of joint rotation. Fixed-vane actuators are also lighter than hydraulic motors. The nonlinearity does not cause problems if the control variables of the servo system are joint position and its derivatives, rather than actuator position.

Dynamic response is improved both by locating servo valves so as to minimize the line lengths between them and the actuators they serve and by minimizing actuator volume consistent with load capacity specifications. In particular, connections from the valve to the actuator should be hard-walled tube, not flexible hose. Increasing system operating pressure allows use of smaller actuators. However, since most industrial robot actuators are relatively lightly loaded and losses in small hydraulic actuators are relatively high, moderate supply pressures [1000 to 2000 pounds per square inch (psi)] are favored to permit use of reasonably large actuators.

Cooling Requirements.   In the technology used at present, the actuators are on parallel branches of the circuit. A pressure-regulated supply is used, and the actuators exhaust to a reservoir. A four-way servo control valve acts across both the inlet and the exhaust lines of each actuator. In this configuration, the difference between the pressure drop from supply to reservoir and that needed across the actuator to balance the load is made up by the pressure drops across the two sides of the control valve. Mechanical energy is converted to heat at the valve at a rate which is the product of the flow rate and the sum of the valve pressure drops. Thus, at low loads and high joint rates, large quantities of energy are converted to heat. The heat is carried off in the hydraulic oil. The reservoir and possibly an oil cooler must be sized to allow this heat to be dissipated without undue rise in the oil temperature at the pump inlet. The oil temperature can build up very rapidly if there is insufficient cooling. Excessive temperatures result in breakdown of the oil and damage to the circuit components, particularly the expensive valves. Use of a variable-displacement pump with a pressure-sensing servo loop, instead of the wasteful technique of regulating pressure by a relief valve, helps this situation considerably. Of course, relief valves must be included for safety, but they should remain closed in normal operation.

47.4.4   Pneumatic Systems

Pneumatic actuation is quite extensively used in simple pick-and-place devices with three or four degrees of freedom. These can hardly be called industrial robots, since the control logic needed is minimal. Attempts to power five or six degrees of freedom, and to use rotary inboard joints to achieve reasonably large working volumes from a compact device, quickly run into problems resulting from the high compliance and low force-to-weight ratios of pneumatic actuators. These can be overcome by the use of high-speed air motors. However, as with electric actuators, this requires a mechanical transmission, with its attendant problems.

The components used in pneumatic actuation and control are quite similar to those used in hydraulic systems. However, pneumatic systems are usually supplied from a central compressor rather than from a local power supply, as is used for hydraulic systems. This is practicable by virtue of the much lower viscosity of air as compared to hydraulic fluid.

Pneumatic systems will continue to be important in automated light manufacturing, particularly in simple transfer devices which are lightly loaded and require only a few degrees of freedom. They are also attractive in functions such as spray painting in which special environmental conditions make electric systems less attractive. However, primarily because of the compliance problem, electric or hydraulic systems are more suited to most industrial robot applications.

47.5 SENSING SYSTEMS

47.5.1 Requirements

Although sophisticated robot sensors are under laboratory development, the only sensors used on most industrial robots are joint position sensors. A number of different types of rotary position sensors are available, and several have been used on robots.

A rather wide variation in the accuracy required during the sensing of position at the different joints of a manipulator is usual. A typical industrial robot with a reach of 6 feet (ft) (1.8 m) will be designed to achieve repeatability within 0.05 in (1.3 mm). This requires a resolution at the shoulder joints of 0.0007 rad (2.4 minutes of arc), or 1/9048 revolution. If a digital readout is used, a resolution of 16 384, or 2^{14}, divisions per revolution is needed. At the wrist, the distance to the hand reference point may be 1 ft or less. Thus, the resolution needed is 0.0042 rad (14.3 minutes of arc), or 1/1508 revolution. A digital resolution of 2048, or 2^{11}, divisions per revolution is adequate. Thus, although quite high resolution is needed at the wrist, it is still substantially less than that at the shoulder.

Few position sensors can read 14 bits (2^{14} divisions) per revolution directly. A sensor with lower resolution can be geared up so that it completes several revolutions per joint revolution. Of course, backlash must be avoided during this time. Since the sensor does not load the gear train, it is simple to remove backlash by use of spring-loaded gears. Many electric systems use position sensing on the motor shaft, which accomplishes the same objective.

When gearing up is used to increase resolution in this manner, it is always necessary to start the system from a "home," or reference, position. Otherwise, if the system is started from an arbitrary position, there is no way of knowing how many complete revolutions the sensor has performed since leaving the reference position. This is the reason that many industrial robots must be placed in a home position before automatic operation is begun.

In addition to having high resolution, robot joint position sensors must be highly reliable, since they must read accurately and without noise over many service hours of continuous movement. For this reason, sensors with electric-brush contact, such as potentiometers or commutator-type encoders, should not be used. Several types of position sensor are available which require no mechanical contact and which, for this reason, are much more reliable.

47.5.2 Encoders

Encoders are position sensors which read out in digital form. There are two basic types: incremental encoders and absolute encoders. Absolute encoders read out a binary number which uniquely identifies the joint position within the resolution of the device. Incremental encoders are much simpler devices. Basically, they are pulse generators, generating a voltage pulse every time the joint moves through an angle equal to the encoder resolution. It is necessary for the computer to count the pulses to determine position. Actually, two outputs phase-shifted relative to each other are used. Otherwise, it would be impossible to determine the direction of rotation. Of course, pulses counted during reverse motion must be subtracted from the total which indicates joint position.

Incremental encoders present an additional problem. It is necessary to provide a zero reference position from which to begin the count. This can be done by means of a home position in the same way as with absolute sensors in a multiturn mode. However, it is necessary that joint position be referenced much more accurately for incremental encoders. When multiturn absolute sensors are used, it is necessary only that each joint sensor be on its initial revolution when the manipulator is placed in the home position. Precise location is not required. When incremental encoders are used, each joint must be located to within the resolution of the encoder.

Absolute encoder types available include optical and electromagnetic types as well as the commutator type mentioned earlier as being unsuitable. They are available at accuracies ranging from 6 to 22 bits per revolution (bits/r). High-accuracy encoders are expensive and require great care in mounting. Another consideration is the length of the binary word which the computer can conveniently handle. Since the microprocessors most popular in current new designs use a 16-bit word length, and since 16 bits/r is adequate accuracy, there is little incentive to go to higher accuracies. When 8-bit word length microprocessors are used, gearing up an inexpensive 8-bit/r encoder may be an economically attractive alternative to the use of an expensive 16-bit encoder with the additional complication of reading and manipulating the encoder output as two 8-bit words.

Encoders do not usually read out in natural binary code. Rather, they read out in a binary code known as *gray code*. Figure 47.12 shows the natural binary and gray code representations of the numbers 0 through 15. The reason for using gray code is that only one binary digit changes between any two adjacent gray code divisions. This is not true of natural binary code, and so the use of natural binary code can lead to large errors at the interface between divisions since, because of mechanical or electronic misalignment, the changes in some digits will be sensed momentarily earlier than those of others. Thus, in changing 7 to 8 in natural binary, if the change in the most significant digit is sensed after that of the other three digits, the output may go from 7 to 0 before going to 8. Gray code eliminates that problem. Gray code can be converted to natural binary by applying a logical exclusive-OR operation with the previously converted digit to each binary digit in turn, starting with the most significant digit. An exclusive-OR operation outputs 0 if the digits compared are like and 1 if they are unlike. It is possible to write simple and efficient microprocessor assembly-language routines to perform this conversion.

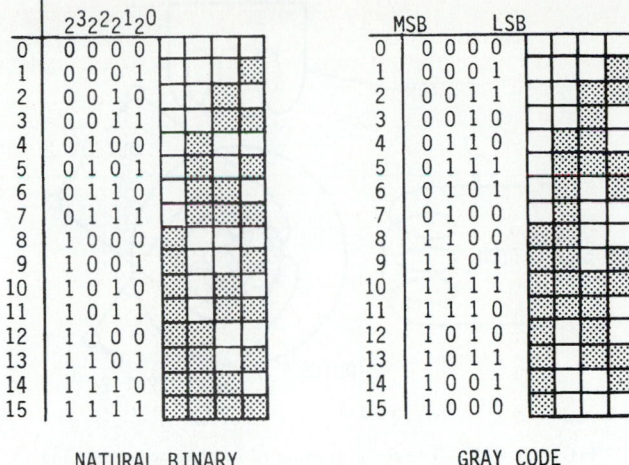

NATURAL BINARY

GRAY CODE

FIGURE 47.12 Comparison of natural binary and gray code representation of numbers 0 through 15. Gray code avoids generation of errors at the transition between divisions.

47.5.3 Resolvers

Resolvers are small, alternating-current (ac), rotating electromagnetic machines related to synchros and linear variable-differential transformers. Figure 47.13 is a schematic diagram showing the two rotor and two stator windings. In both cases, the two windings are arranged to act orthogonally to each other. For simple angle measurement, only one stator winding is excited; the other is short-circuited. If the rotor is at angle θ from its null position, the amplitudes of the alternating electromotive forces excited in the two rotor coils are proportional to $\sin \theta$ and $\cos \theta$, respectively. High-accuracy resolvers are typically designed for 26-volt (V) maximum excitation voltage at 400 hertz (Hz) and have open-circuit output voltage from 0 to 26 V. Resolvers are available with resolution ranging from 6 to 16 bits/r. They are attractive because they do not have mechanical contact and they produce a readily usable analog output voltage.

47.5.4 Force Sensors

Some form of force sensing is highly desirable when grippers are used. The sensor might be simply a contact sensor which indicates whether a workpiece is in the gripper. Such a sensor is easily arranged by mounting a leaf spring along the gripping surface with its free end bearing on a microswitch. By suitably designing the compliance of the spring and cementing strain gauges to it, a readout of gripping pressure is obtained. This is desirable when fragile workpieces are handled. Sensors which will not only measure contact pressure but also locate the point of contact are being developed with solid-state electronic technology. The technology necessary to measure all six components of tool load by means of load cells mounted in either the wrist or the base has been well proved in the laboratory but has not yet found its way into industrial service.

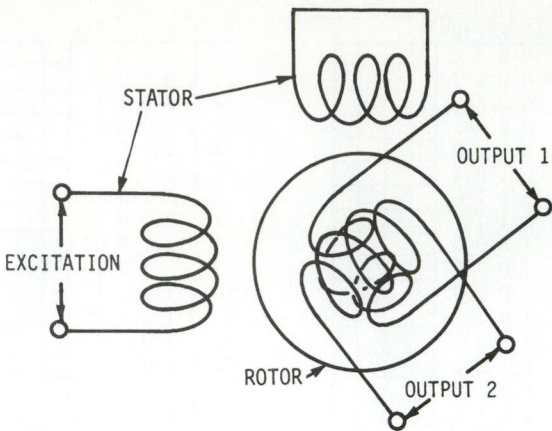

FIGURE 47.13 Schematic drawing of resolver windings. One stator winding is excited at high frequency. Amplitude of electromotive force excited in rotor windings is sinusoidally related to rotor angle.

47.5.5 Robot Vision Systems

Most robot vision systems are based on television camera technology, using either conventional vidicon-tube cameras or array cameras. Constructing a three-dimensional computer model of an object seen in two or three views by television cameras is difficult, is fraught with ambiguity, and requires time-consuming computation. Thus this is not attempted in industrial service. By using highly structured environments, taking advantage of known workpiece geometries, and possibly using structured light sources, it is possible to detect orientations of parts and command a robot to pick them up at the correct position and with the gripper correctly aligned with the part. It is also possible to pick out one type of part from a mixture of several different types. Vidicon-tube cameras give best resolution, but at relatively low frame rates. Array cameras use arrays of photodiodes formed by using integrated-circuit techniques. They are very compact and rugged and read out faster. The best array cameras presently available split a square viewing area into a 512-by-512 array of pixels. This approaches the resolution of an ordinary television set.

47.5.6 Proximity Sensors

Proximity sensing using optical or acoustic techniques is useful when a robot tool is brought into contact with a workpiece. Since the robot is designed to be very stiff and the workpiece is usually quite rigid, the contact force between them builds very rapidly when the robot contacts the workpiece with finite velocity. Even if force sensing is used, the contact force may build to damaging levels before the system can respond. Proximity sensors are short-range, noncontact sensors which allow fine control of tool velocity shortly before contact to avoid severe impacts. Optical systems based on triangulation or simply the intensity of light reflected off the workpiece have been tested. Ultrasonic rangefinders using a sonar-type principle offer an alternative technology.

47.6 COMPUTER HARDWARE AND SOFTWARE ORGANIZATION

47.6.1 Software Organization

Basic Operation. The computer software is divisible into several modules in terms of function. Figure 47.14 shows the relationships of these modules. The monitor is a small program permanently stored in read-only memory (ROM) in the microcomputer. It performs basic traffic direction and boots up the supervisor. That is, when the computer is switched on, the monitor expects to find a program waiting to be read from a mass storage device. It initiates reading of this program, which is, in fact, the remainder of the permanent software. Thus, the software system "pulls itself up by its own bootstraps." This procedure is necessitated by the volatile nature of the devices used for random access memory (RAM). Random access memory is the primary memory of the computer. Information stored in RAM or ROM can be directly addressed by the computer, and so it can be accessed in minimal time. RAM is called *volatile* because the information stored in it is lost if the power is turned off. ROM is nonvolatile. That is why the monitor, at least, must be stored in ROM.

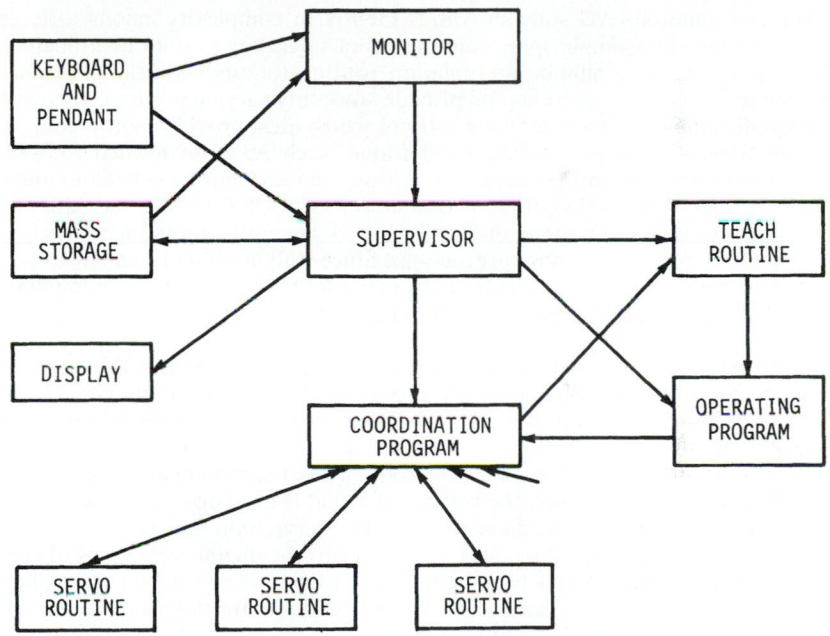

FIGURE 47.14 Computer software architecture typical of industrial robots.

In an industrial robot, storage of the entire permanent system program in ROM is a viable option. ROM is more expensive than RAM, but this configuration eliminates the bootstrap operation. There is a much more important objection to storage of the system program in ROM than cost—the lack of flexibility in terms of updating the system program. It has become customary for robot manufacturers to periodically issue updated versions of the system program which provide additional capabilities. If the system program is stored on a tape cassette or disk, updating requires only substitution of the new cassette or disk for the old. If the system program is on ROM, installation of an update may be much more complicated.

Supervisory Software.   Once the system software has been booted up, the direction of traffic is taken over by the supervisor, which corresponds to the operating system of a computer. The supervisor provides operator access to the computer via keyboard, function buttons, and teach pendant; displays system status; controls reading of operating programs from the mass storage device, or writing of programs into mass storage; controls the power-up and power-down sequences of the manipulator hardware; and contains modules for "teaching" (roughly corresponding to the editor of a data processing computer) and running operating programs.

Coordination Software.   The coordination module is responsible for generating the position commands which are passed to the joint servos to generate smooth motions. It can be viewed as a subroutine which is called by the operating program to generate movements between programmed positions. Likewise, the coordination module is called from the teach pendant when the robot is being programmed to allow the operator to move the manipulator in a smooth and predictable way.

The coordination-level software varies greatly in complexity among different types of robots. In a simple spray-painting robot operating in joint coordinates, it may consist only of a simple interpolation routine to subdivide the differences between recorded joint positions and provide smoothly varying position commands to the joint servos. On the other hand, a robot which must provide coordinated uniform motions with respect to the fixed frame, such as those needed for seam welding, must have a complex coordination program requiring considerable computational power. The function of the coordination software is to take the data positions and the stored program instructions and generate a set of coordinated movements to implement those instructions. Since only a limited number of positions are usually stored, the computer must generate the trajectory the manipulator is to follow between each successive pair of positions.

Servo Software.   The servo programs are exclusively concerned with controlling the joint actuators to produce, as nearly as possible, the joint positions commanded by the coordination module. They are, of course, present only if digital servo controllers are used.

The servo control level closes a control loop around each joint actuator. The sensor is the joint position sensor. The command signal is provided by the next higher level in the hierarchy: the coordination level. The servo loop may be configured as an analog system. This has the advantage over current digital techniques of rapid response. Alternatively, it may be a digital servo loop. Use of a digital servo allows flexible design of the servo characteristics. The capability for tailoring the response characteristics is important in manipulators in which the load and inertia seen by the joint actuator change considerably with arm position.

A digital servo loop must be serviced at frequent intervals. Thus, it is an advantage to have a small microprocessor dedicated to each joint. The microprocessor can perform the servo computations continuously without interference from the demands of other computational functions. However, it is also possible to arrange software for a single central computer which will perform servo loop computations with sufficient frequency by interrupting other computations at appropriate intervals. This arrangement avoids the problem of communication between the central computer and the servo microprocessor.

47.6.2 Microcomputers

Construction. A microcomputer is a computer which uses a microprocessor as its central processing unit. In addition to the microprocessor, the microcomputer contains the RAM and ROM memory chips and interfaces to mass storage, data entry devices, sensors, actuators, and other processors (if used). Usually it consists of one or a few printed-circuit cards on which the integrated-circuit packages are mounted. The cards are plugged into a card cage. A regulated power supply is also needed. Analog-to-digital and digital-to-analog convertors may also be mounted on cards within the computer. Actuator driver circuits, which require relatively high power levels, are usually mounted separately, often adjacent to the actuators they service.

Word Length. Compared with large computers, microcomputers are slow devices. However, the newer forms are very adequate as industrial robot controllers. Their low cost is, of course, a major consideration. The word length of a microprocessor is the number of bits which are grouped for manipulation by the processor. Microprocessors are available with 4-, 8-, 16-, and recently 32-bit word lengths. Most microcomputer controllers for industrial robots now use 16-bit processors. And 8-bit processors are useful as local processors but lack the computing power needed by the central processor. If joint position is read as an 8-bit word, the smallest change in position which can be resolved is more than one degree ($2^8 = 256$). This is quite inadequate for most robotics applications. Of course, it is possible to write software routines in assembly language or machine code to couple two words and perform double-length arithmetic. This strategy results in considerable slowing of computation.

The picture is quite different when a 16-bit word length is provided ($2^{16} = 65\,536$). Thus, computation can be carried out by using numbers with up to five significant decimal digits. At 16-bit accuracy, joint angles are resolved to within 20 seconds of arc, which is more than adequate. Thus, a 16-bit word length meets most computation needs for robotics. One type of computation for which it may not be adequate is computation of trigonometric functions, square roots, etc. As is described later, some manufacturers of recently released microprocessors have developed strategies for dealing with this problem.

Address Register. One of the most important performance parameters of a microprocessor is the length of the address register. This is what determines the number of words of RAM and ROM which can be accessed by the processor. The length of the address register is often greater than the word length used by the processor. Thus, the most commonly used 8-bit microprocessors have 16-bit address registers. A convenient unit of memory is 1000 words, abbreviated to 1K. Actually this usually means $1024 = 2^{10}$ words. A 16-bit address register permits 64K words of memory to be accessed. More accurately, 64K is $2^{16} = 65\,536$. Some modern 16-bit microprocessors have 20-bit address registers, permitting over 1 million words of memory to be directly accessed. The difference in information content between 64K eight-bit words and 2^{20} sixteen-bit words is huge. The cost of the memory modules becomes the most important factor in designing systems capable of handling this much RAM. However, the cost per bit of RAM has been falling rapidly.

47.6.3 Supporting Hardware and Software

Floating-Point Computations. There is much more to selection of a suitable microprocessor than word length, address-register length, and computing-cycle time. Hardware and software support should also be a major consideration. If any of the robot's functions are to be performed in fixed coordinates, floating-point arithmetic, direct and inverse trigonometric functions, and square roots will certainly be required. The availability of a fast software package to provide these capabilities is an important consideration. Several manufacturers of modern 16-bit microprocessors do better than this by producing coprocessors. A *coprocessor* is an additional microprocessor chip which, when used in conjunction with the basic microprocessor, adds additional hard-wired, and therefore very fast, computational operations. Coprocessors to provide the operations listed above are available for at least two types of 16-bit microprocessors.

Interfaces and Programming Aids.   Hardware support also includes interface hardware and microcomputer boards with different configurations. If a distributed processing system is to be used, the availability of a family of microprocessors with compatible characteristics and of shared bus hardware becomes important.

Another class of supporting hardware and software aids in the development of programs. Microcomputers can be programmed directly in machine code, in assembly language, or in a high-level language. Machine code is the actual binary words in which the computer operates. Assembly language is a simple system of mnemonics which allows considerable saving in programming effort with little sacrifice in computational speed. A program called an *assembler,* which is usually part of the monitor, translates the program to machine language. A development system is a computer on which programs can be developed in high-level languages such as PASCAL or C. The development system compiles the program into machine code and then downloads it into the microcomputer's memory. A cross-compiler is a program which allows the same sort of development to be done on a large computer. Likewise, a cross-assembler is a program which allows assembly language programs to be written on a large computer, assembled, and then downloaded to the microcomputer as machine code.

The cost of software development is a large cost component for most industrial robots. For this reason, it is attractive to use development systems or cross-compilers to allow relatively easy programming in high-level languages. The loss in computational speed resulting from this is outweighed by the saving in software development cost. Also, parts of the software in which computational speed is vital can be written as assembly language or machine code macros which are called by the high-level language program. Since coordination software is often very complex, it usually makes sense to write it in a high-level language. Conversely, servo routines should usually run as fast as possible, and so they are usually written in assembly language or machine code.

47.6.4 Computer Structure Options

Architectural Requirements.   The software structure described in Sec. 47.6.1 directly affects the organization of the computer hardware. The supervisor and coordination modules perform systemwide functions and are best incorporated in a single central processor. The servo modules perform local functions and may be handled either in the central processor or in relatively small local processing units. The servo loops must be serviced regularly and at a cycle frequency (20 to 40 Hz) which is high relative to the mechanical natural frequencies of the servo system. Thus, if the servo loops are closed in the central processor, a timed interrupt procedure must be implemented to service the servo loops. A computer is governed by its internal "clock," which is really a very stable, high-frequency oscillator which produces a train of precisely spaced pulses. The timed interrupt procedure counts clock cycles and, after a set number of cycles, interrupts all other processing and initiates running of the servo programs. The interruption is handled in the computer in a manner similar to the interrupts occasioned by swapping in a timesharing computer. The address location of the next instruction of the program which is being interrupted and the contents of the accumulator and other active data registers are stored in a stack so that they may be recalled as soon as servicing of the servo loops is completed.

If local processors are used to service the joint servos, the need for the complication of the timed interrupt procedure is removed. The cost is the introduction of a problem of communication between computers. Actually this is only part of a larger problem of transmission of data between different parts of the system.

Analog-to-Digital and Digital-to-Analog Conversions. Most sensors transmit data in analog form. At some point this must be converted to a digital format to allow the computer to read it. This is done by means of an analog-to-digital converter. Conversely, the driver circuits of actuators require analog input. Thus, the digital output of the computer must be converted to an analog signal by a digital-to-analog converter. Transmission of an analog signal requires only a wire pair, often in the form of a coaxial cable. With the recent advent of relatively inexpensive and compact analog-to-digital converters, considerable flexibility is possible in selecting the appropriate place in the system to perform the conversion. However, the limited number of output lines available on a microcomputer processor board usually requires that digital-to-analog conversion be done in the computer on circuit boards designed for the purpose. The outputs from the digital-to-analog converters are amplified and transmitted as analog signals to the actuator drivers.

Data Transmission. Digital data may be transmitted in either serial or parallel form. In serial data transmission, the bits of each data word are transmitted as a train of pulses. Only a wire pair is needed to do this, but the receiving device must be equipped to identify the beginning of a word and sequentially store the bits. Consequently, serial data transmission is relatively slow.

When parallel data transmission is used, a separate wire is used for each bit of the transmitted word. Thus, to transmit a 16-bit word, a cable with 16 active wires plus a ground is needed. Actually several other wires would be used to carry traffic direction signals. If a number of parallel input channels are to be fed to a microcomputer, a problem arises because of the limited number of input lines provided on the processor board. A multiplexer must be used to sequentially read the parallel input channels and to feed the resulting words to the computer. This, of course, reduces the effective transmission rate by dividing it by the number of incoming channels. Thus, if too many parallel channels must be brought in, the advantage in transmission speed as compared to serial transmission is diminished.

For similar reasons, data output from a central processor to local servo processors must be multiplexed. An alternative technique, which provides more flexibility, is the use of a shared bus. This can be thought of as a segment of memory which can be accessed by both the central processor and one or more of the local processors. This method allows each processor to run, essentially, at its own rate. Shared bus operation requires a high level of compatibility between the processors used. It also requires some means of arbitration when two or more processors seek access to a word of memory simultaneously. If too many devices share a bus, operation is considerably slowed.

Influence of Sensor Type on System Architecture.   Absolute encoders read out position data directly in parallel digital format. It is attractive to feed this directly to local servo processors. This structure largely avoids the problem of multiple parallel inputs to any one processor. Although the coordination software will require joint position values, they need not be updated at high rates. Thus, this need can be serviced by serial or shared bus links.

If absolute encoders are not used, all sensor data are transmitted in analog form. It is then attractive to use a single central processor and to carry data to and from it in analog form. This allows a comparatively simple wiring design.

47.7 CONTROLLER DESIGN

47.7.1 Subdivision

The   control system of an industrial robot is conveniently divided into two levels: coordination and joint servo control. The operations performed by the two levels of the control system are quite different. Joint servo control design, which is briefly discussed in Sec. 47.7.4, is similar to classical analog or digital servo design. Coordination, in contrast, in many ways is unique to industrial robots. The methods used owe more to spatial linkage kinematic theory than to control system theory.

47.7.2 Coordination

As stated earlier, coordination algorithms are those which generate the commanded joint states as functions of time from the data stored when the robot is programmed. In the case of a computer-coordinated teleoperator, the data, rather than being stored, are input by the operator in real time by means of a manual controller. The operation of the coordination algorithms is otherwise very similar.

There are enormous differences in sophistication and complexity among the coordination programs used in different industrial robots. For some types of operation, sophisticated coordination is unnecessary. There is a major division of coordination algorithms into those which operate in *joint coordinates* and those which operate in *base,* or *world, coordinates.* Most industrial robots operate in some type of point-to-point mode. Thus, when the robot is programmed, a series of discrete positions is recorded. The coordination software interpolates motions between those positions. The positions of the robot are recorded by reading the joint position sensors. To compute the position of the hand relative to a fixed frame from the joint positions, a complicated trigonometric transformation must be calculated. In a coordination algorithm which operates in joint coordinates, a simple interpolation scheme is used to command movement of each joint between successive recorded positions without regard to the gross motion of the manipulator which is produced. A typical joint coordinate scheme would command constant angular velocity at each joint with simple linear transitions between the velocities computed before and after a given recorded position. Since little computation is involved, joint coordinate algorithms can be very fast, permitting relatively high operating speeds with simple and inexpensive control computers.

The disadvantage of joint coordinate coordination is that it gives no direct control over the motion of the hand or of the remainder of the manipulator between programmed positions. The path of the hand reference point is a complicated space curve. This is acceptable for applications such as spot welding in which only the programmed positions are important and the path taken between those positions is unimportant. Paths which cause interference with the workpiece or other hardware are easily dealt with by programming intermediate positions that take the machine away from the object it is interfering with.

Joint coordinate coordination algorithms are often used to simulate continuous-path programming. The earliest industrial robots were mostly true continuous-path devices with analog control at both the servo and coordination levels. The positions of the joints were recorded as continuous analog signals on magnetic media. Few devices of this type are now manufactured. Rather, continuous-path operation is simulated by sampling joint positions at regular time intervals during programming. The programmer must lead the machine through the desired motions in real time. A very simple joint coordinate type of interpolation routine is used during automatic operation.

Programming of robots which operate in joint coordinates can be cumbersome. It is necessary to physically place the manipulator in each programmed position. Having the operator control each joint individually, as in a backhoe, turns out to be very clumsy and inefficient. It is better to allow the operator to position the hand directly. Some robots can be placed in a mode in which the joints are unlocked, allowing the arm to be manually manipulated into the desired positions. Others use simulacra. A simulacrum is a light structure with joint geometry identical to that of the manipulator. The joints are completely free and are fitted with position sensors. The operator manually moves the simulacrum to the desired program positions. Strategies such as a "limp" manipulator mode or a simulacrum are vital for programming continuous-path type of operations. Spray painting is a good example of this situation.

When the hand path is to be controlled relative to a fixed reference frame, very much more sophisticated coordination software is necessary. For example, seam-welding robots must generate straight-line paths with high accuracy and constant hand orientation. The type of algorithm used is called a *resolved motion rate control*, or *Jacobian decomposition*, algorithm.

47.7.3 Generation of Specified Hand Trajectories

Coordination algorithms are in use which allow a robot to interpolate a specified hand–reference point path between two taught positions [47.2]. Hand orientation is also coordinated with progress along the reference point path. The point path is usually a straight line but may be a circle or other specified curve. The hand is usually rotated about an axis of constant direction. This type of algorithm depends on repetitive use of a numerical operation called Jacobian decomposition.

An infinite number of forms of Jacobian relationship can be, and are, written in the literature. However, it is possible to write a form which has a readily understandable kinematic meaning. This form also lends itself to simplification by the use of knowledge of the geometry of the manipulator chain. If ω is the angular velocity of the hand, $\dot{\theta}_K$ is the rate of rotation about joint K, and $\mathbf{w}_K$ is a unit vector parallel to joint axis K, then

$$\omega = \sum_{K=1}^{N} \dot{\theta}_K \mathbf{w}_K \qquad (47.1)$$

where N = number of joint axes, usually six. That is, each joint contributes its angular velocity vector to the angular velocity of the hand. Similarly, the velocity $\mathbf{V}$ of the reference point can be expressed as

$$\mathbf{V} = \sum_{K=1}^{N} \dot{\theta}_K \mathbf{w}_K \times \mathbf{r}_K \tag{47.2}$$

where $\mathbf{r}_K$ = any vector from joint axis K to the reference point. And $\dot{\theta}\mathbf{w}_K \times \mathbf{r}_K$ is the velocity the reference point would have if axis K alone were active. These equations may be combined into the form

$$\begin{bmatrix} \omega \\ \mathbf{V} \end{bmatrix} = [\mathbf{J}] \, [\dot{\theta}] \tag{47.3}$$

where $[\dot{\theta}] = [\dot{\theta}_1, \dot{\theta}_2, \ldots, \dot{\theta}_N]^T$. And $[\mathbf{J}]$ is called a *Jacobian* matrix. It is a $6 \times N$ matrix whose Kth column consists of the elements of $\mathbf{w}_K$ followed by the elements of $\mathbf{w}_K \times \mathbf{r}_K$.

In the above, it is assumed that the manipulator joints are rotary joints. If joint K is a sliding joint, the Kth column changes form. The first three elements become 0 (since the joint does not permit rotation). The *second* three elements are now those of $\mathbf{w}_K$.

Equation (47.3) represents six scalar equations which can be solved for the joint rates $\theta_1, \theta_2, \ldots, \theta_N$. Then these are used as the commanded rates for the joint servos in order to produce a nominated angular velocity ω and reference point velocity $\mathbf{V}$ at the hand.

Since Eq. (47.3) must be solved 20 to 30 times per second to produce accurate motion, it is not solved by numerical inversion of $\mathbf{J}$. Rather, the component equations are solved in analytical form to give explicit algebraic expressions for $\theta_1, \theta_2, \ldots, \theta_N$ in terms of the joint variables and the elements of ω and $\mathbf{V}$.

The simplest strategy is to compute the displacement between two reference point positions and the corresponding axis and angle of rotation of the hand and to divide by a suitable time interval to get ω and $\mathbf{V}$. The machine then attempts to displace with constant ω and $\mathbf{V}$. However, since this means that the system is attempting an infinite acceleration and deceleration at the beginning and end of the displacement, it is customary to use transition segments at the beginning and end. These might be simply constant accelerations and decelerations, or they might be more sophisticated. The design of these transitions is strikingly similar in intent and in mathematical form to the design of cam motion programs.

47.7.4 Servo Design

This is a topic for specialized texts on control system synthesis ([47.3], [47.4]). Consequently, in-depth treatment is not attempted here. Rather, we review those aspects which characterize robotic applications.

Joint servos may be implemented as either analog or digital servo loops. Analog servo controllers have the advantage of faster response. Digital servo controllers present great flexibility in tailoring the controller characteristics to the system characteristics. The choice of analog versus digital controllers is also affected by the type of joint instrumentation chosen. Instruments, such as absolute encoders, which produce their output in parallel form, are most conveniently coupled directly to dedicated microprocessor controllers. Instruments such as potentiometers, resolvers, tachometers, etc., which produce analog output, are most easily coupled to analog controllers. However, inexpensive and compact analog-to-digital converters are available, so they may also be coupled to digital controllers with ease.

From the point of view of controller synthesis, the most important feature of the system is the presence of large variations in the inertia to be moved by the actuator as well as, in most cases, in the load. The variation is, in both cases, of the order of 10 to 1. This requires robustly stable controller characteristics.

In the interests of precision, it is attractive to instrument the joint axes with position sensors. This is done in hydraulic systems. In electric systems, however, compliance and backlash in the speed-reducing transmission create stability problems. These are avoided if the position sensor is placed on the motor shaft, but at a cost in accuracy. In many industrial robot designs, gravitational loading takes care of the backlash problem, but power train compliance remains. Deflections due to it are, at least, repeatable, provided the load is repeated.

The physical system design interacts with the design of coordination software and particularly servo design in several ways. The cycle time for updating of the rates commanded by the coordination program is primarily determined by the durations of the velocity transitions which must be tracked. These, in turn, are determined by the accelerations which the joints can achieve under worst-case conditions. The acceleration at a joint required by the software during a transition must be lower than the maximum acceleration which can be achieved. To adequately track a velocity ramp, a sampling interval of about one-tenth the duration of the ramp is needed. Thus, the coordination cycle time should be one-tenth the transition interval or less. The frequency at which the servo controller operates must, in turn, be at least as high as the coordination cycle frequency. It should also be at least 10 times the bandwidth which the actuator and controller hardware would give with an analog controller to achieve performance comparable to an analog system. Thus, a typical robot system which has 0.5-s transitions and 15-Hz analog bandwidth for the joint servos should have a coordination cycle frequency of about 20 Hz and a servo sampling frequency of at least 150 Hz. If a central processor is used to perform both coordination and servo computations, it is convenient to make the servo sampling frequency an even multiple of the coordination frequency.

47.8   GEOMETRIC DESIGN

47.8.1   Structural Subdivision

Most present industrial robot geometries have been developed by trial and error, by using the drawing board and physical models. Lately three-dimensional computer simulation has become important. A theoretical basis is being developed, but some of the concepts involved are difficult. Nevertheless, it is not hard to state some basic principles.

A useful way of thinking about a manipulator structure is to regard the first three members and joints as a *regional structure* which is responsible for transporting the hand to the desired position. The outer three members and joints form an *orientation structure* whose function is to orient the hand. The logic in this is that a rotation of the hand about a "wrist" axis produces a small displacement of the hand reference point (an imaginary point in the working area of the hand fixed with respect to the hand structure). A rotation about a "shoulder" axis produces a large displacement of the reference point. Thus, orientation movements are best performed by using outboard or wrist joints, and translation movements are best performed by using inboard joints. Further, if sliding joints are used, they must be confined to the regional structure, because sliding joints do not permit rotation and so are not appropriate for orientation movements.

It is not, in general, possible to move the reference point to a desired position by using the regional structure and then to rotate the hand to its final position about an axis through the reference point by using the orientation structure without additional movement of the regional structure. That is, translational and rotational movements cannot, in general, be decoupled. Nevertheless, for practical industrial robot or teleoperator geometries, the coupling is weak. We shall see that there are strong geometric reasons for using geometries with these characteristics.

A useful concept for study of manipulator geometry is the *reachable workspace*. It is the space within which the hand reference point may be located. Diagrams of reachable workspace are frequently presented in the literature of industrial robot manufacturers. Figure 47.15 shows a typical example.

47.8.2 Workspace Optimization

Let us consider a manipulator with six revolute joints. Successive joint axes are assumed to be either parallel or normal to one another. For the moment, the further assumption is made that there are no offsets along the joint axes. That is, the common normals to the two joint axes in each of two successive members meet on the axis of the joint connecting those members. This type of geometry is shown in Fig. 47.16. For the moment, it is also assumed that there are no mechanical limits on the rotations of the joints. The members are numbered serially from the base member (0) to the hand (6). The joints are similarly numbered, with joint 1 connecting members 0 and 1. A fixed origin point, 0 in Fig. 47.16, is defined as lying on joint axis 1 at the foot of the common normal of axes 1 and 2. Under these assumptions, the *length* of the manipulator may be defined as the greatest distance from the origin 0 to the reference point R. It is convenient to call the length of the common normal between axis i and axis $i + 1$ a_i. That is, it is the length of the common normal in member i. The length of the manipulator is then

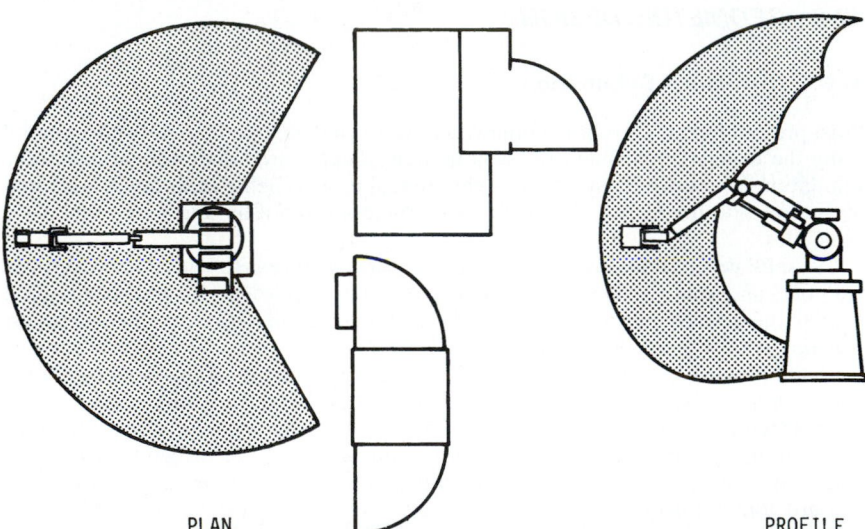

PLAN   PROFILE

FIGURE 47.15 Profiles of reachable workspace of industrial robot. Dimensions can be determined by the grid scale, not shown, which is 6 in (150 mm). (*Cincinnati Milacron.*)

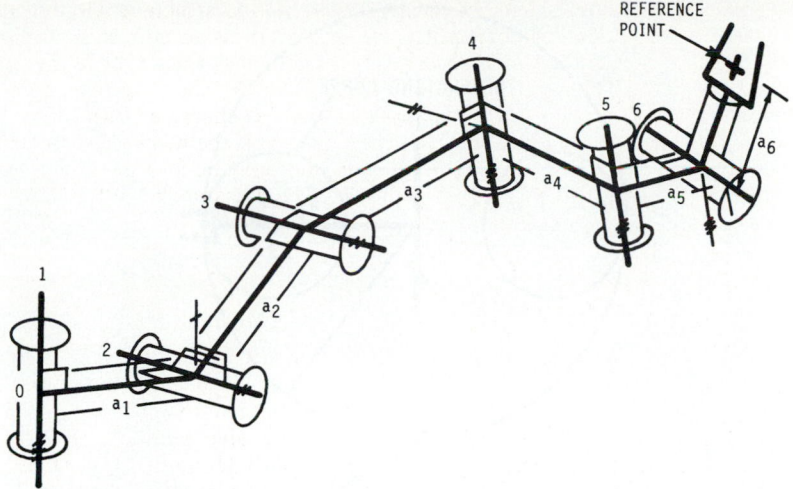

FIGURE 47.16 Manipulator chain with no offsets and with axes parallel or orthogonal. The figure illustrates the notation used in the text.

$$L = \sum_{i=1}^{6} a_i$$

The shape of the workspace boundary is governed by the inboard joints. Figure 47.17 shows the generating curves when $a_1 \neq 0$ and $a_1 = 0$. The first two joints are normal to each other in both cases. In the first case, the workspace is a torus. The generating circle radius is $L - a_1$. The center of that circle moves on a circle about axis 1 of radius a_1. When $a_1 = 0$, the workspace is a sphere of radius L. As a_1 changes, a family of tori is generated, but every member of that family fits inside the sphere of radius L. Thus, for a given length L, the maximum workspace volume is obtained when $a_1 = 0$, that is, when axes 1 and 2 intersect.

If the first two joint axes are not normal to each other but are parallel, with the third normal to the second, the situation is as shown in Fig. 47.18. The right circular torus is distorted into a toroid. Nevertheless, for a given length L, the volume is always less than that obtained when the first two axes intersect at right angles. This is, in fact, true in general for manipulator chains with six rotary joints, even when the restrictions of no joint offsets and of normal or parallel axes are removed. Likewise, the introduction of mechanical motion limits on the joints, while it substantially changes the shape of the workspace generating curve, does not invalidate this general conclusion:

Optimum workspace volume is obtained for a manipulator with all joints rotary joints when the first two joint axes intersect each other at right angles.

The range of orientation achievable by the hand is maximized when axes 4, 5, and 6 are concurrent and when the angles between axes 4 and 5 and between axes 5 and 6 are both right angles. If these conditions are satisfied, if there are no

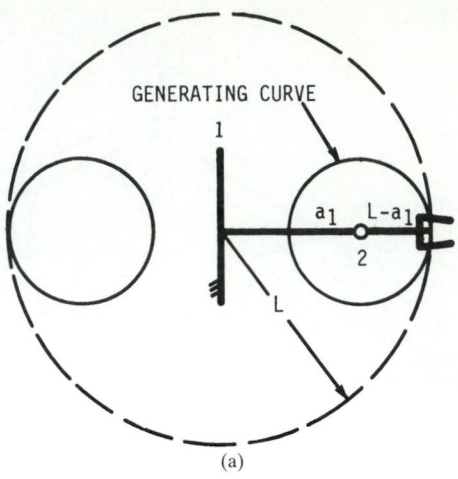

(a)

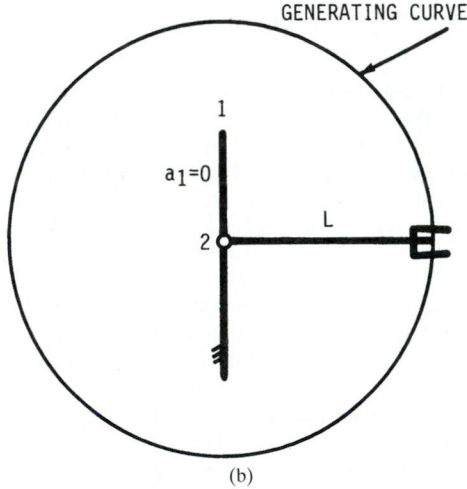

(b)

FIGURE 47.17 Generating curves when first two joint axes are orthogonal. (*a*) Axes do not intersect. (*b*) Axes do intersect; this represents the optimum for a given manipulator length.

mechanical limits on axes 4 and 6, and if axis 5 is capable of more than 180° of rotation, then the hand can be rotated completely about any line through the point of concurrency.

Since the hand reference point should be placed within the gripper area of the hand, it is usually mechanically impossible to locate it at the point of concurrency of

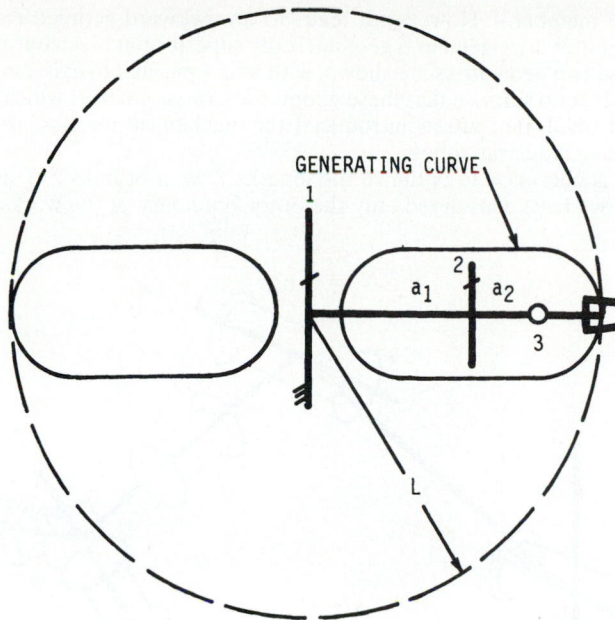

FIGURE 47.18 Generating curve when first two axes are parallel.

joint axes 4, 5, and 6 (if they are concurrent). Thus, although complete rotation about any line through the point of concurrence may be possible, complete rotation about any line through the reference point may not be. Geometric limits (as opposed to mechanical joint motion limits) on hand orientation are minimized by minimizing the distance between the reference point and the point of concurrence. That is, the hand should be as short as possible. Thus, from the point of view of dexterity, the optimum is as follows:

> Joint axes 4 and 5 should meet at right angles, and joint axis 6 should intersect axis 5 at right angles at the same point as axis 4. The hand reference point should be as close as possible to the point of concurrence of axes 4, 5, and 6. Mechanical motion limits should restrict axes 4 and 6 as little as possible. Complete rotation about these axes is desirable. Joint axis 5 should permit 180° of rotation.

If the above rules are obeyed, there is little remaining choice as far as geometric design is concerned. The placement of axis 3 with respect to axes 2 and 4 is all that remains. Axis 3 cannot pass through either the point of concurrency of axes 1 and 2 or the point of concurrency of axes 4, 5, and 6. If it does so, the chain becomes geometrically singular in all positions (the meaning of the term *geometrically singular* is explained in Sec. 47.8.3). Therefore, axis 3 either may intersect axis 2 at right angles at a point different from axis 1 or may be parallel to axis 2. Either arrangement is good from the point of view of ease of control computation. However, the former arrangement is less desirable mechanically, since it leads to large torsional moments on the arm. Likewise, axis 3 either may intersect axis 4 at right angles at a point different from the intersection of axes 4, 5, and 6 or may be parallel to it. The former arrangement is very popular, since it allows the drive for axis 4 to be conveniently placed inside member 4. However, it leads to an awkward geometrically singular position. The latter arrangement is geometrically superior but mechanically less convenient. These two geometries are shown, with axis 3 parallel to axis 2 in both cases, in Fig. 47.19. It is no surprise that these geometries, or geometries which differ from them only in small joint offsets introduced for mechanical reasons, are extremely common among industrial robots.

Finally, it is necessary to optimize the lengths ℓ, m, p of links 2, 3, and 6 in Fig. 47.19. So far we have considered only the outer boundary of the workspace. How-

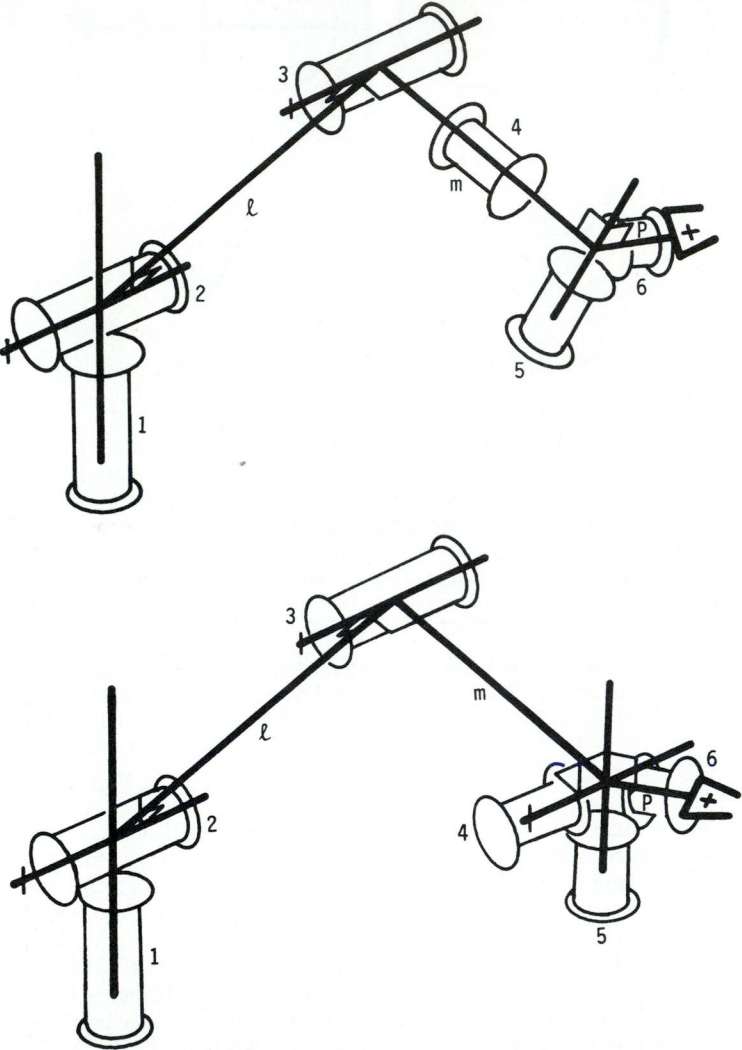

FIGURE 47.19 Two geometries which satisfy optimality conditions of regional structure.

ever, there is also an inner boundary. This is shown in Fig. 47.20 for the two cases $m + p < \ell$ and $m + p > \ell$. Clearly, the geometric optimum is $m + p = \ell$. Since there is usually structure surrounding point O, in practice a geometric capability for reaching O may not be appropriate, and a small positive value of $\ell - m - p$ may be chosen.

The above discussion applies only to geometries in which all joints are rotary joints. If one or more sliding joints are used in the regional structure, the conclusions above are no longer valid. However, a very similar approach to geometric optimization can always be taken. Sliding joints introduce one important difference. Whereas the geometric workspace for a manipulator with only rotary joints is always finite, regardless of the presence or absence of joint motion limits, that of a manipulator with sliding joints is infinite. It is only when mechanical motion limits are placed on the sliding joints that the workspace becomes finite. This is why manipulators with

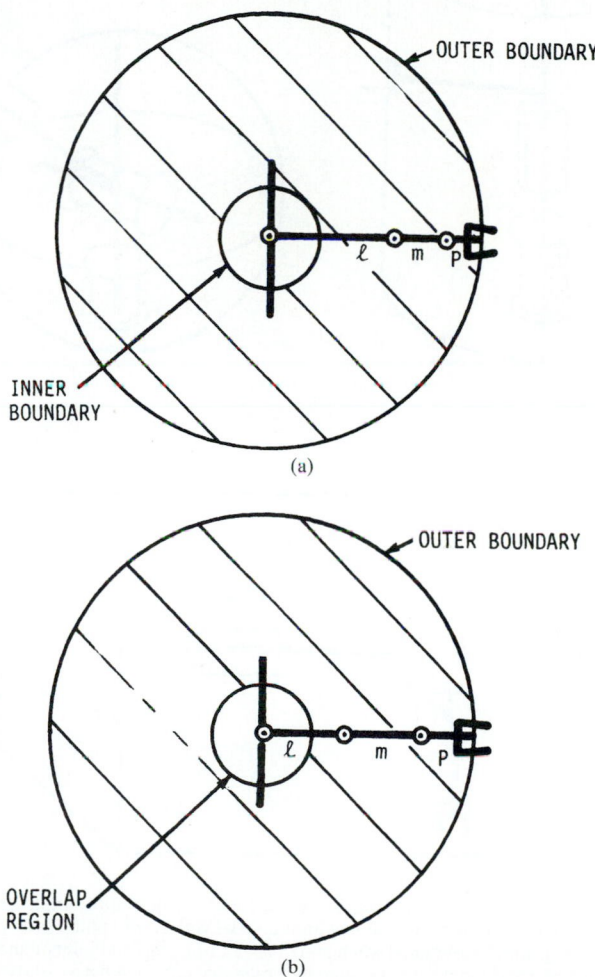

FIGURE 47.20 Relationship of inner and outer workspace boundaries to lengths of segments. (a) $m + p < \ell$; (b) $m + p > \ell$.

orthogonal slides for joints 1 and 2 or joints 1, 2, and 3 can be offered with a variety of workspace volumes. Figure 47.21 shows the workspace geometries obtained when (*a*) joint 2 is a sliding joint coaxial with joint 1; (*b*) joint 3 is a sliding joint and joints 1 and 2 intersect orthogonally; (*c*) joints 1, 2, and 3 are orthogonal sliding joints. All these arrangements can be found on industrial robots in service.

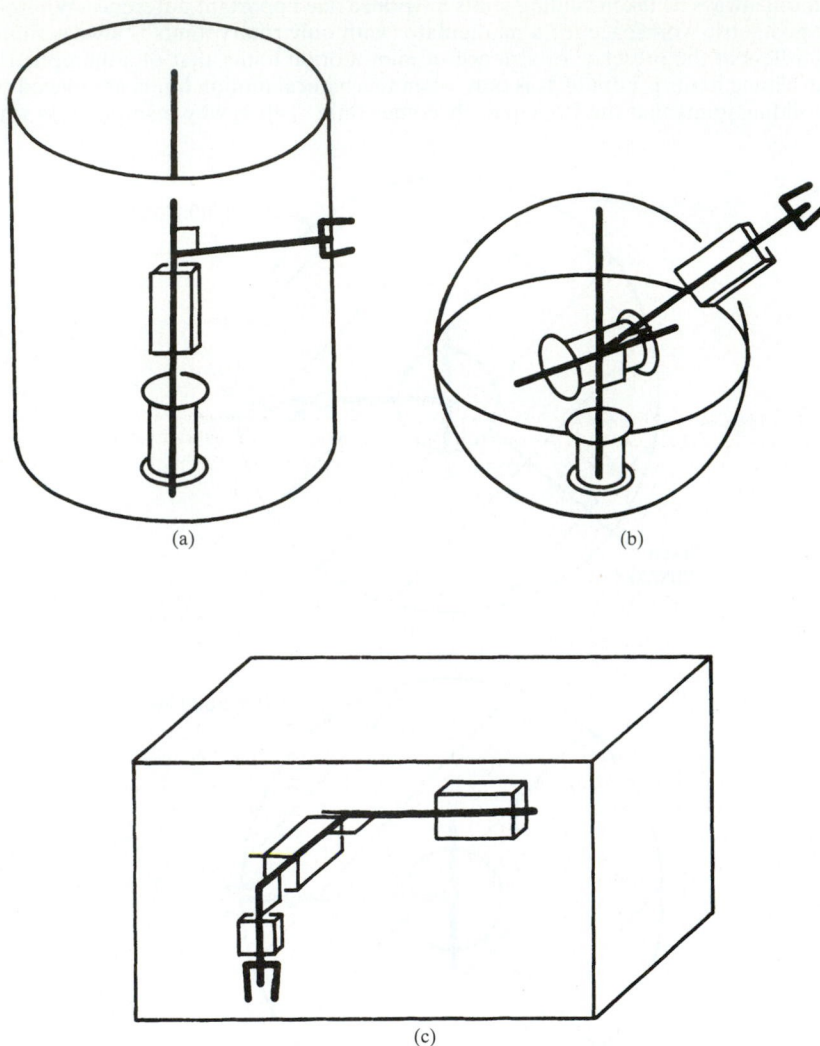

(a)

(b)

(c)

FIGURE 47.21 Some workspace geometries obtained with sliding joints. (*a*) Cylindrical geometry obtained when joint 2 is a sliding joint coaxial with rotary point 1 (or vice versa); (*b*) "spherical polar" geometry obtained when joint 3 is a sliding joint and joints 1 and 2 are orthogonally intersecting rotary joints; (*c*) rectangular geometry obtained when joints 1, 2, and 3 are orthogonal sliding joints.

47.8.3 Singular Positions

A *singular position* is a position in which a manipulator effectively loses a degree of freedom. A very simple example occurs in the roll-pitch-roll wrist configuration shown in Fig. 47.22. In the fully extended position shown in Fig. 47.22b, the axes of joints 4 and 6 are coincident. Thus the motions they contribute at the hand become indistinguishable. Another common example of a singular position occurs in manipulators which have three concurrent wrist joint axes. The singular position occurs if the point of concurrency is placed on the first (most inboard) axis of the manipulator. All positions in which the manipulator is fully extended are singular. This property is often used to trace workspace boundaries.

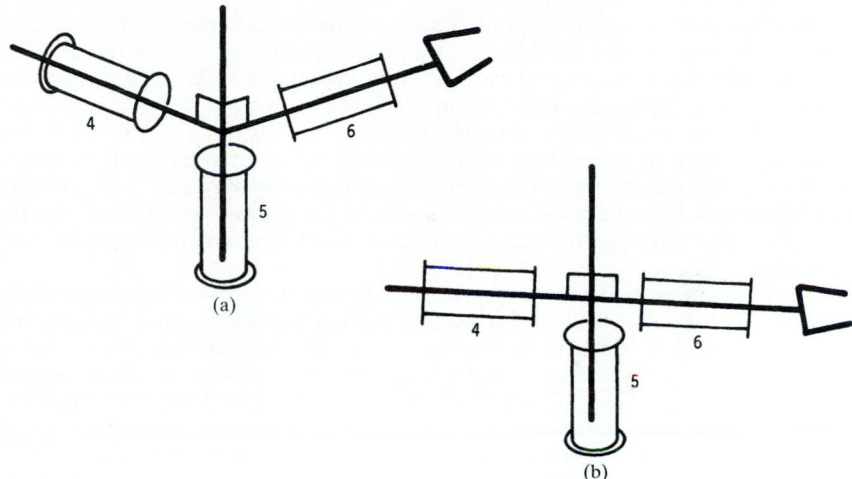

FIGURE 47.22   Geometrically singular position encountered with roll-pitch-roll wrist. (*a*) General position of the wrist; (*b*) singular position.

At a singular position, any coordination algorithm which is set up to control the hand path in *fixed* coordinates will fail. Algorithms which operate in *joint* coordinates are not affected. When the rate decomposition scheme discussed in Sec. 47.7.3 is used, the determinant of the Jacobian matrix becomes zero at a singular position.

Therefore, if coordination in the fixed frame is to be used, it is necessary to take special measures to prevent system failure at singular positions. A straightforward strategy is to provide software joint motion limits which prevent the manipulator from entering singular positions. Of course, there may also be hardware limits on joint motion. These should always be matched by limits in the software, since unexpectedly meeting a mechanical limit will usually result in program failure.

The above strategy is simple to implement and is preferred whenever feasible. It may, however, be unduly limiting. If singular positions remain which the manipulator can assume, it is necessary to make provision to detect the approach to a singular position and to provide a strategy for moving out of it. Temporary suppression of one component of the commanded hand velocity or angular velocity is an effective strategy.

One problem facing the mechanical designer of a manipulator is that of identifying singular positions to which a proposed configuration is subject. This can be done by writing the Jacobian matrix in analytical form, setting its determinant to zero, and expanding to get an algebraic condition on the joint angles which contains all the possible singular positions. This is quite feasible for the simple configurations usual in industrial robot practice, but requires a knowledge of techniques for analytical modeling of manipulator chains. The algebraic techniques involved are the same as those used in constructing coordination algorithms.

47.8.4 Accuracy and Repeatability

There are several different ways of characterizing the accuracy of an industrial robot. Each is useful, but for different purposes. *Positioning error* is the error, in world coordinates, in placing the hand reference point at a specified position. *Orientation error* is the angular error, again in world coordinates, in placing the hand in a specified orientation. *Repeatability* is the error in returning to a specified initial position after an intervening series of moves. It usually refers to the error in positioning the reference point, but angular repeatability may also be of interest. In a typical robot design, both accuracy and repeatability vary drastically over the workspace. For this reason, figures quoted by robot manufacturers can be very misleading.

There are at least four sources of error in industrial robots. One is *manufacturing error*. This includes dimensional and alignment inaccuracy and errors in the zero references of sensors. A second is *structural deflection*. This is predominantly a result of elastic deflection of the members under load. Deflections due to backlash are most conveniently regarded as components of structural deflection. A third is *servo error*, errors due to the limits of resolution of the joint position sensors or to the servo controller characteristics. A fourth is *numerical error*, errors due to the coordination algorithm. Each of these sources of error has different characteristics and affects the system accuracy in different ways.

The "teaching" process, in which a point-to-point robot is physically placed in the positions it is to assume during operation, functions as a calibration process to remove the determinate component of positioning error. The effects of manufacturing error are removed in this manner. The effects of static structural error will also be removed if the machine is "taught" with a simulated load.

Servo errors are random in nature, with range determined predominantly by the resolution of the joint position sensors. However, friction in the joint or actuation system and other effects may also have influence. Repeatability is determined by servo error. The magnitude of positioning error resulting from servo error varies strongly with position. In a six revolute arm with the first two axes intersecting, it is roughly proportional to radial distance from the intersection. In contrast, in a manipulator with orthogonal slides for the first three joints, it is independent of position for the same hand orientation. Orientation error resulting from servo error is more or less independent of position.

Numerical error is not active when the manipulator statically assumes taught positions, because no computation is involved; the system simply reads the recorded joint positions. However, numerical errors influence the accuracy with which a specified trajectory is followed and may, therefore, be significant in operations, such as seam welding, which depend on generation of precise trajectories.

47.9 TOOL DESIGN

47.9.1 Similarities to Fixed Automation

The tooling used on industrial robots ([47.5]) is very similar in character to that used for traditional, fixed automation. Jigs and fixtures, feeders, conveyors, clamping devices, and so on are all used in virtually identical form for both purposes. Although a robot is a flexible tool adaptable to varied tasks, it is usual to design the tooling which is used on it specifically for the task to be performed. Thus, if it is to be used to transfer parts, the gripping tool will be designed around the parts to be handled. This and other tooling is usually designed and constructed by specialist tool manufacturers.

47.9.2 Remote Center Compliance Devices

When a part is to be inserted into a hole during a robot assembly operation, binding due to friction usually causes failure. This can be prevented by installing a remote center compliance device in the wrist of the robot. This places the kinematic center about which the part tends to turn (under the influence of contact forces) inside the hole rather than above it. As a result, the binding problem is usually overcome.

REFERENCES

47.1 B. Roth, "Introduction to Robots," *Design and Application of Small Standardized Components—Data Book 757,* vol. 2, sec. 10, pp. 723–773, Stock Drive Products, New Hyde Park, N.Y. 1983.

47.2 R. P. Paul, *Robot Manipulators,* M.I.T. Press, Cambridge, Mass., 1981.

47.3 J. J. D'Azzo and C. H. Houpis, *Linear Control Systems Analysis and Design,* McGraw-Hill, New York, 1975.

47.4 G. F. Franklin and J. D. Powell, *Digital Control of Dynamic Systems,* Addison-Wesley, Reading, Mass., 1980.

47.5 V. Daniel Hunt, *Industrial Robotics Handbook,* Industrial Press, New York, 1983.

CHAPTER 48

SECTIONS AND SHAPES— TABULAR DATA

Joseph E. Shigley
Professor Emeritus
The University of Michigan
Ann Arbor, Michigan

48.1 CENTROIDS AND CENTER OF GRAVITY

When forces are distributed over a line, an area, or a volume, it is often necessary to determine where the resultant force of such a system acts. To have the same effect, the resultant must act at the centroid of the system. The *centroid* of a system is a point at which a system of distributed forces may be considered concentrated with exactly the same effect.

Figure 48.1 shows four weights W_1, W_2, W_4, and W_5 attached to a straight horizontal rod whose weight W_3 is shown acting at the center of the rod. The centroid of this *weight* or *point group* is located at *G*, which may also be called the *center of gravity* or the *center of mass* of the point group. The total weight of the group is

$$W = W_1 + W_2 + W_3 + W_4 + W_5$$

This weight, when multiplied by the *centroidal distance* $\bar{x}$, must balance or cancel the sum of the individual weights multiplied by their respective distances from the left end. In other words,

$$W\bar{x} = W_1 l_1 + W_2 l_2 + W_3 l_3 + W_4 l_4 + W_5 l_5$$

or

$$\bar{x} = \frac{W_1 l_1 + W_2 l_2 + W_3 l_3 + W_4 l_4 + W_5 l_5}{W_1 + W_2 + W_3 + W_4 + W_5}$$

A similar procedure can be used when the point groups are contained in an area such as Fig. 48.2. The centroid of the group at *G* is now defined by the two centroidal

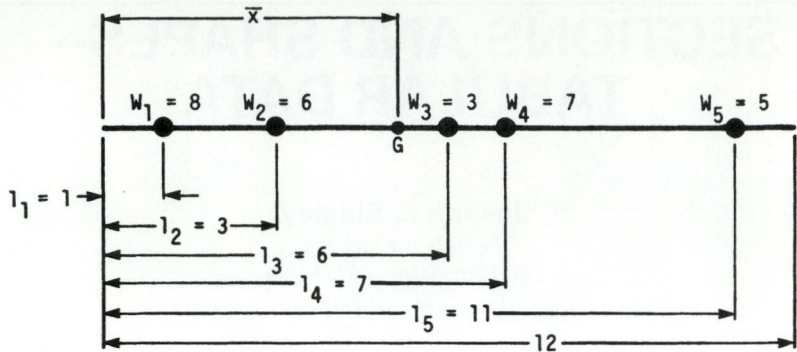

FIGURE 48.1 The centroid of this point group is located at G, a distance of $\bar{x}$ from the left end.

distances $\bar{x}$ and $\bar{y}$, as shown. Using the same procedure as before, we see that these must be given by the equations

$$\bar{x} = \sum_{i=1}^{i=N} A_i x_i \Bigg/ \sum_{i=1}^{i=N} A_i \qquad \bar{y} = \sum_{i=1}^{i=N} A_i y_i \Bigg/ \sum_{i=1}^{i=N} A_i \qquad (48.1)$$

A similar procedure is used to locate the centroids of a group of lines or a group of areas. Area groups are often composed of a combination of circles, rectangles, triangles, and other shapes. The areas and locations of the centroidal axes for many such shapes are listed in Table 48.1. For these, the x_i and y_i of Eqs. (48.1) are taken as the distances to the centroid of each area A_i.

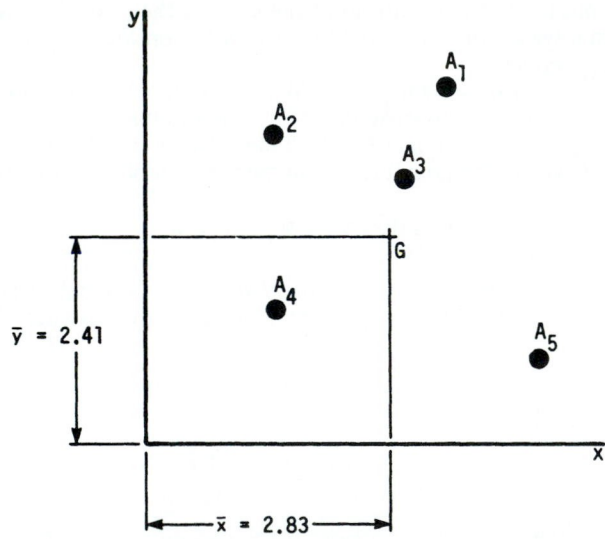

FIGURE 48.2 The weightings and coordinates of the points are designated as A_i (x_i, y_i); they are $A_1 = 0.5(3.5, 4.0)$, $A_2 = 0.5(1.5, 3.5)$, $A_3 = 0.5(3.0, 3.0)$, $A_4 = 0.7(1.5, 1.5)$, and $A_5 = 0.7(4.5, 1.0)$.

TABLE 48.1 Properties of Sections†

1. Rectangle

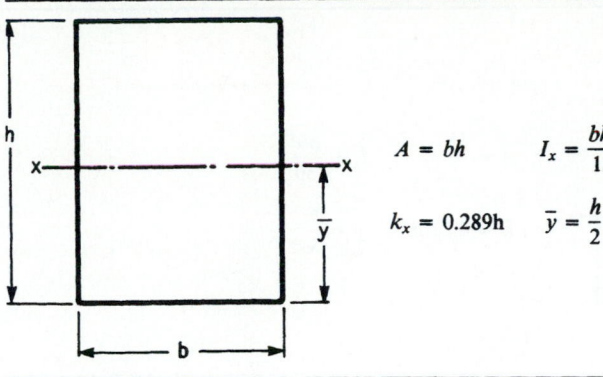

$$A = bh \qquad I_x = \frac{bh^3}{12}$$

$$k_x = 0.289h \qquad \bar{y} = \frac{h}{2}$$

2. Hollow rectangle

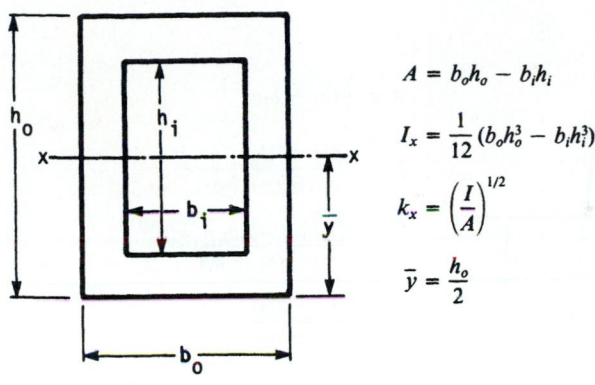

$$A = b_o h_o - b_i h_i$$

$$I_x = \frac{1}{12}(b_o h_o^3 - b_i h_i^3)$$

$$k_x = \left(\frac{I}{A}\right)^{1/2}$$

$$\bar{y} = \frac{h_o}{2}$$

3. Two rectangles

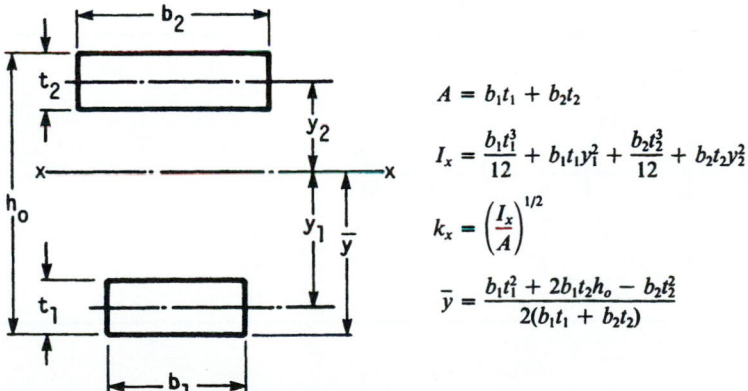

$$A = b_1 t_1 + b_2 t_2$$

$$I_x = \frac{b_1 t_1^3}{12} + b_1 t_1 y_1^2 + \frac{b_2 t_2^3}{12} + b_2 t_2 y_2^2$$

$$k_x = \left(\frac{I_x}{A}\right)^{1/2}$$

$$\bar{y} = \frac{b_1 t_1^2 + 2b_1 t_2 h_o - b_2 t_2^2}{2(b_1 t_1 + b_2 t_2)}$$

†List of symbols: A = area; I = second area moment about principal axis; J_O = second polar area moment with respect to O; k = radius of gyration; and $\bar{x}, \bar{y}$ = centroidal distances.

TABLE 48.1 Properties of Sections (*Continued*)

4. Triangle

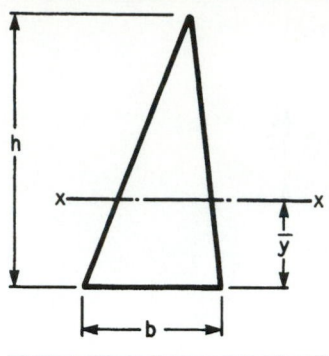

$$A = \frac{bh}{2} \qquad I_x = \frac{bh^3}{36}$$

$$k = 0.236h \qquad \bar{y} = \frac{h}{3}$$

5. Trapezoid

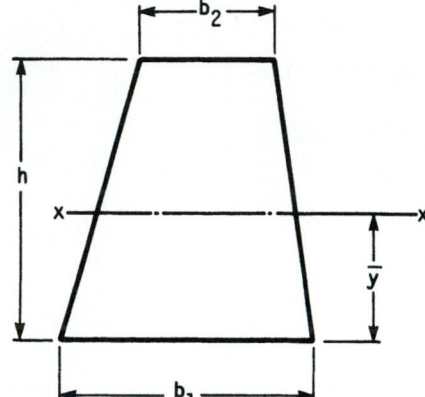

$$A = \frac{h}{2}(b_1 + b_2)$$

$$I_x = \frac{h^3(b_1^2 + 4b_1b_2 + b_2^2)}{36(b_1 + b_2)}$$

$$k_x = \frac{h[2(b_1^2 + 4b_1b_2 + b_2^2)]^{1/2}}{6(b_1 + b_2)}$$

$$\bar{y} = \frac{h(b_1 + 2b_2)}{3(b_1 + b_2)}$$

6. Circle

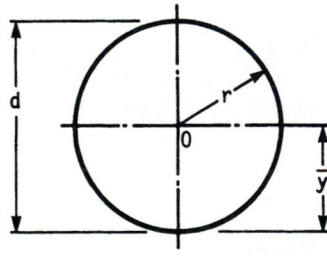

$$A = \pi r^2 = \frac{\pi d^2}{4}$$

$$I = \frac{\pi r^4}{4} = \frac{\pi d^4}{64}$$

$$k = \frac{r}{2} = \frac{d}{4} \qquad \bar{y} = r = \frac{d}{2}$$

TABLE 48.1 Properties of Sections (*Continued*)

---

7. Hollow circle

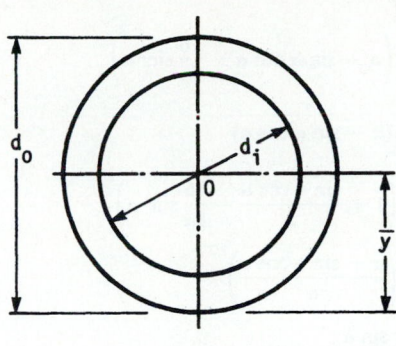

$$A = \frac{\pi}{4}(d_o^2 - d_i^2)$$

$$I = \frac{\pi}{64}(d_o^4 - d_i^4)$$

$$J_O = \frac{\pi}{32}(d_o^4 - d_i^4)$$

$$k = \frac{1}{4}(d_o^2 + d_i^2)^{1/2}$$

$$\bar{y} = \frac{d_o}{2}$$

---

8. Thin ring (annulus)

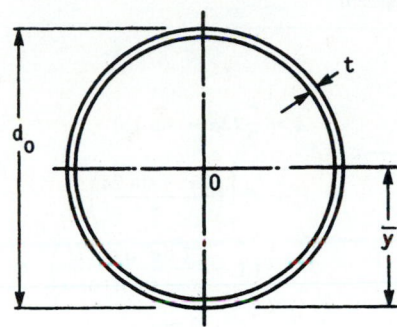

$$A = \pi d_o t \qquad I = \frac{\pi d_o^3 t}{8}$$

$$J_O = \frac{\pi d_o^3 t}{4}$$

$$k = 0.353 d_o \qquad \bar{y} = \frac{d_o}{2}$$

---

9. Semicircle

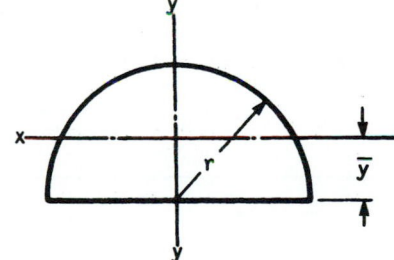

$$A = \frac{\pi r^2}{2} \qquad I_x = 0.1098 r^4$$

$$I_y = \frac{\pi r^4}{8} \qquad k_x = 0.264 r$$

$$k_y = \frac{r}{2} \qquad \bar{y} = 0.424 r$$

TABLE 48.1 Properties of Sections (*Continued*)

10. Circular sector

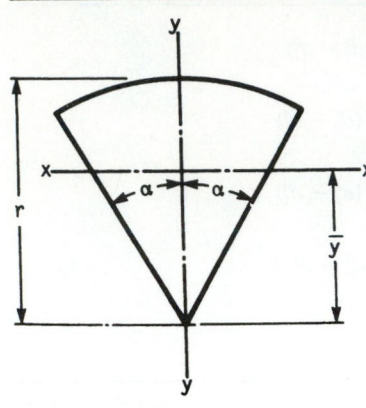

$$A = \alpha r^2$$

$$I_x = \frac{r^4}{4}\left(\alpha + \sin\alpha\cos\alpha - \frac{16}{9\alpha}\sin^2\alpha\right)$$

$$I_y = \frac{r^4}{4}(\alpha - \sin\alpha\cos\alpha)$$

$$k_x = \frac{r}{2}\left(1 + \frac{\sin\alpha\cos\alpha}{\alpha} - \frac{16}{9\alpha}\sin^2\alpha\right)^{1/2}$$

$$k_y = \frac{r}{2}\left(\frac{\alpha - \sin\alpha\cos\alpha}{\alpha}\right)^{1/2}$$

$$\bar{y} = \frac{2r\sin\alpha}{3\alpha}$$

11. Circular segment

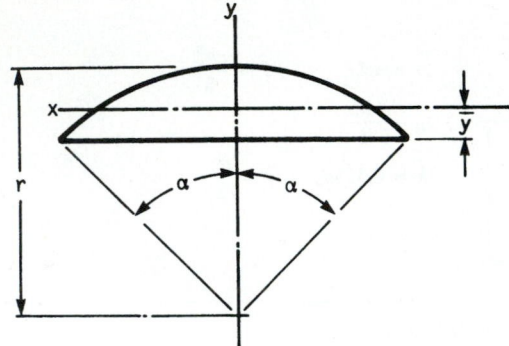

$$A = \frac{r^2}{2}(2\alpha - \sin 2\alpha)$$

$$I_x = r^4\left[\left(\frac{2\alpha - \sin 2\alpha}{8}\right)\right.$$
$$\left(1 + \frac{2\sin^3\alpha\cos\alpha}{\alpha - \sin\alpha\cos\alpha}\right)$$
$$\left. - \frac{8\sin^6\alpha}{9(2\alpha - \sin 2\alpha)}\right]$$

$$k_x = \frac{r}{2}\left[1 + \frac{2\sin^3\alpha\cos\alpha}{\alpha - \sin\alpha\cos\alpha} - \frac{64\sin^6\alpha}{9(2\alpha - \sin 2\alpha)^2}\right]^{1/2}$$

$$\bar{y} = \frac{4r\sin^3\alpha}{6\alpha - 3\sin 2\alpha} - r\cos\alpha$$

TABLE 48.1   Properties of Sections (*Continued*)

12. Parabola

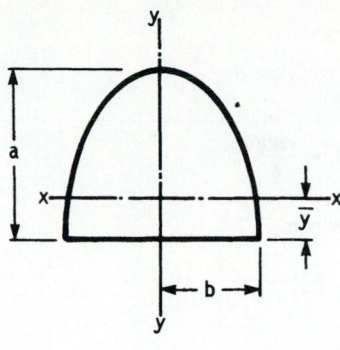

$$A = \frac{4ab}{3} \qquad I_x = \frac{16a^3b}{175}$$

$$I_y = \frac{4ab^3}{15} \qquad \bar{y} = \frac{a}{5}$$

13. Semiparabola

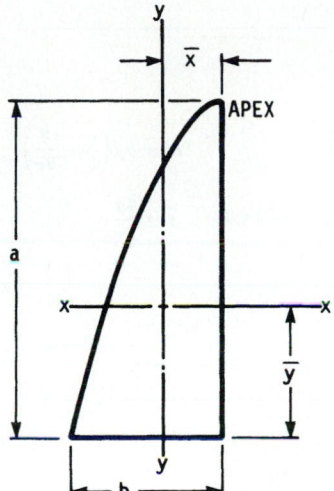

APEX

$$A = \frac{2ab}{3} \qquad I_x = \frac{8a^3b}{175}$$

$$I_y = \frac{19ab^3}{480} \qquad \bar{y} = \frac{2a}{5} \qquad \bar{x} = \frac{3b}{8}$$

TABLE 48.1 Properties of Sections (*Continued*)

14. Ellipse

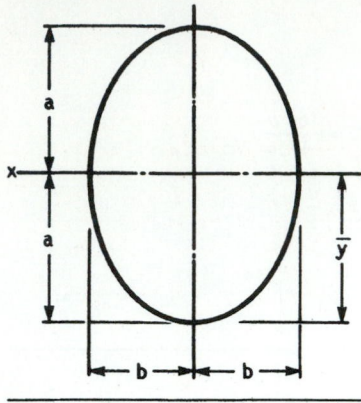

$$A = \pi ab \qquad I_x = \frac{\pi a^3 b}{4}$$

$$k_x = \frac{a}{2} \qquad \bar{y} = a$$

15. Semiellipse

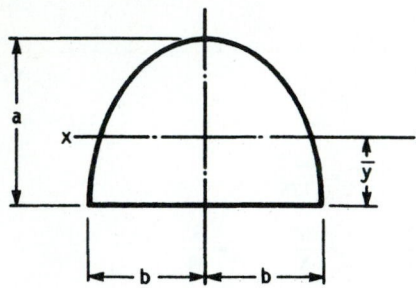

$$A = \frac{\pi ab}{2} \qquad\qquad I_x = a_3 b\left(\frac{\pi}{8} - \frac{8}{9\pi}\right)$$

$$k_x = \frac{b}{6\pi}(9\pi^2 - 64)^{1/2} \qquad \bar{y} = \frac{4a}{3\pi}$$

16. Hollow ellipse

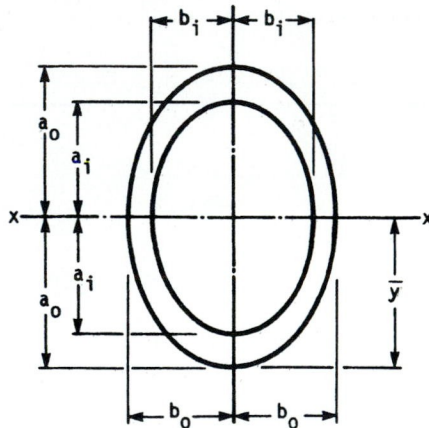

$$A = \pi(a_o b_o - a_i b_i)$$

$$I_x = \frac{\pi(a_o^3 b_o - a_i^3 b_i)}{4}$$

$$k_x = \frac{1}{2}\left(\frac{a_o^3 b_o - a_i^3 b_i}{a_o b_o - a_i b_i}\right)^{1/2} \qquad \bar{y} = a_o$$

TABLE 48.1   Properties of Sections (*Continued*)

17. Regular polygon (*N* sides)

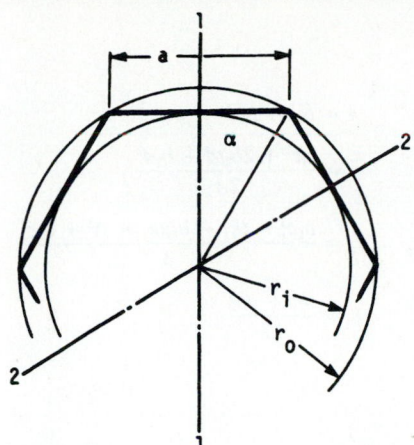

$$A = \frac{N r_o^2 \sin 2\alpha}{2} = N r_i^2 \tan \alpha$$

$$I_1 = \frac{A(6 r_o^2 - a^2)}{24} \qquad I_2 = \frac{A(12 r_i^2 + a^2)}{48}$$

$$k_1 = \left(\frac{6 r_o^2 - a^2}{24}\right)^{1/2} \qquad k_2 = \left(\frac{12 r_i^2 + a^2}{48}\right)^{1/2}$$

18. Angle

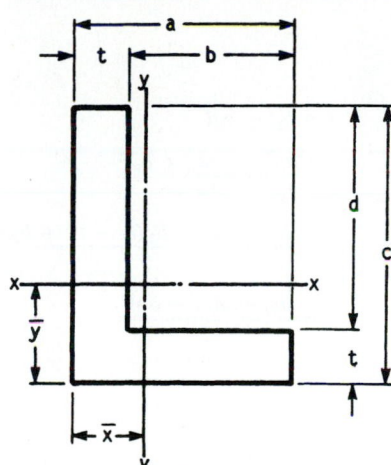

$$A = t(a + d)$$

$$\bar{y} = \frac{c^2 + bt}{2(a + d)} \qquad \bar{x} = \frac{a^2 + dt}{2(a + d)}$$

$$I_x = \frac{1}{3} [t(c - \bar{y})^3 + a\bar{y}^3 - b(\bar{y} - t)^3]$$

$$I_y = \frac{1}{3} [t(a - \bar{x})^3 + c\bar{x}^3 - d(\bar{x} - t)^3]$$

TABLE 48.1 Properties of Sections (*Continued*)

19. T section

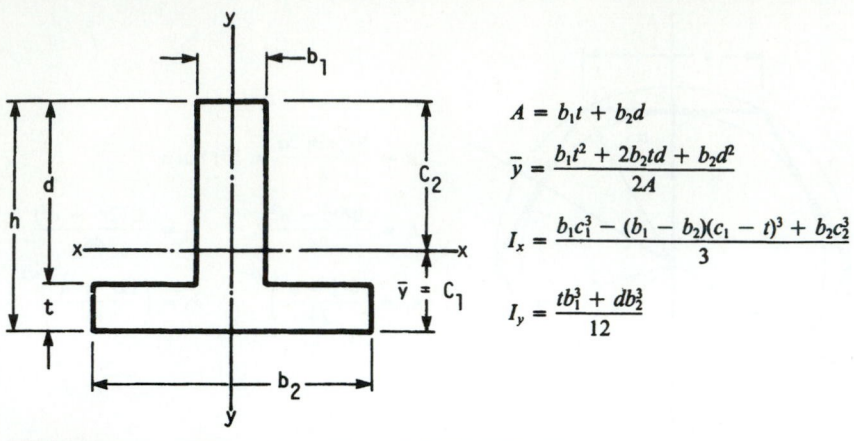

$A = b_1 t + b_2 d$

$$\bar{y} = \frac{b_1 t^2 + 2b_2 t d + b_2 d^2}{2A}$$

$$I_x = \frac{b_1 c_1^3 - (b_1 - b_2)(c_1 - t)^3 + b_2 c_2^3}{3}$$

$$I_y = \frac{t b_1^3 + d b_2^3}{12}$$

20. U Section

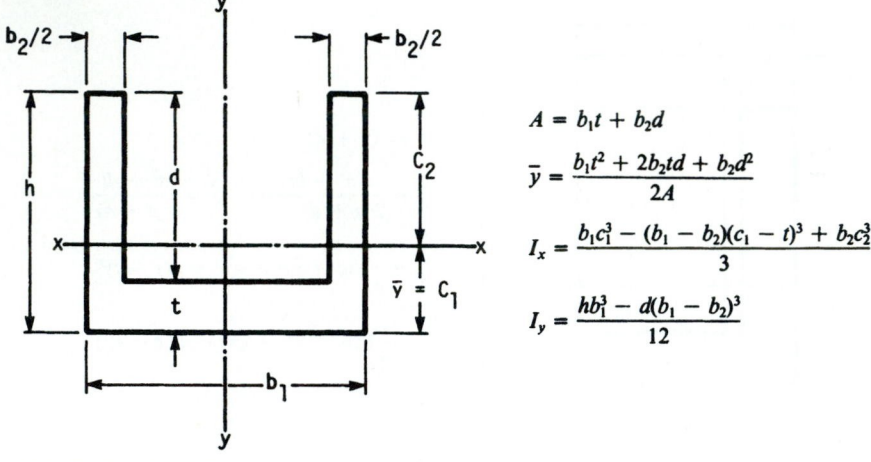

$A = b_1 t + b_2 d$

$$\bar{y} = \frac{b_1 t^2 + 2b_2 t d + b_2 d^2}{2A}$$

$$I_x = \frac{b_1 c_1^3 - (b_1 - b_2)(c_1 - t)^3 + b_2 c_2^3}{3}$$

$$I_y = \frac{h b_1^3 - d(b_1 - b_2)^3}{12}$$

Equations (48.1) can easily be solved on an ordinary calculator using the Σ key twice, once for the denominator and again for the numerator. The equations are also easy to program. Practice these techniques using the data and results in Figs. 48.1, 48.2, and 48.3.

By substituting integration signs for the summation signs in Eqs. (48.1), we get the more general form of the relations as

$$\bar{x} = \frac{\int x'\, dA}{\int dA} \qquad \bar{y} = \frac{\int y'\, dA}{\int dA}$$

These reduce to

$$\bar{x} = \frac{1}{A}\int x'\, dA \qquad \bar{y} = \frac{1}{A}\int y'\, dA \tag{48.2}$$

where x' and y' = coordinate distances to the centroid of the element dA. These equations can be solved by

- Finding expressions for x' and y' and then performing the integration analytically.
- Approximate integration using the routines described in the programming manual of your programmable calculator or computer.
- Using numerical integration routines described in Chap. 4.

48.2   SECOND MOMENTS OF AREAS

The expression $A\bar{x} = \int x'dA$ from Eqs. (48.2) is a *first moment of an area*. A *second moment of an area* is obtained when the element of area is multiplied by the square of a distance to some stated axis. Thus the expressions

$$\int x^2\, dA \qquad \int y^2\, dA \qquad \int r^2\, dA \tag{48.3}$$

are all second moments of areas. Such formulas resemble the equation for *moment of inertia,* which is

$$\int \rho^2\, dm \tag{48.4}$$

where ρ = distance to some axis and dm = an element of mass. Because of the resemblance, Eqs. (48.3) are often called the equations for moment of inertia too, but this is a misnomer because an area cannot have inertia.

We can find the second moment of an area about rectangular axes by using one of the formulas

$$I_x = \int y^2\, dA \qquad I_y = \int x^2\, dA \tag{48.5}$$

Example 1.   Find the second moment of area of the rectangle in Fig. 48.4 about the x axis.

Solution.   Select an element of area dA such that it is everywhere y units from x. Substituting appropriate terms into Eqs. (48.5) gives

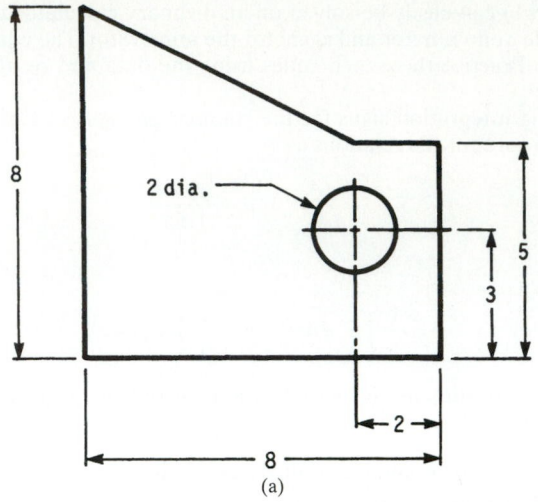

(a)

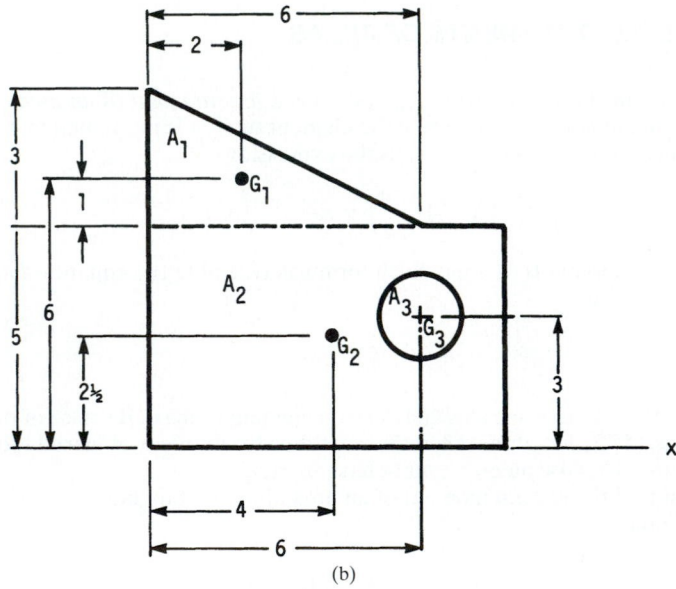

(b)

FIGURE 48.3   A composite shape consisting of a rectangle, a triangle, and a circular hole. The centroidal distances are found to be $\bar{x} = 3.47$ and $\bar{y} = 3.15$.

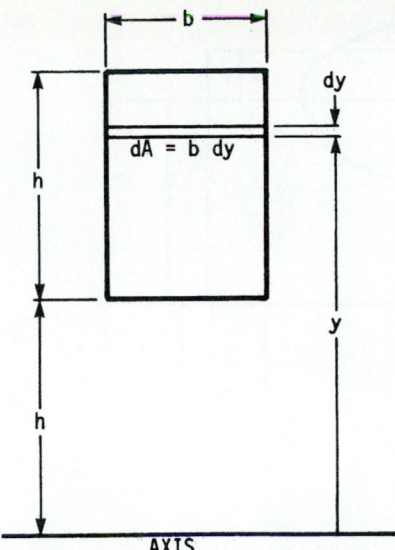

$$I_x = \int y^2 \, dA = \int_h^{2h} y^2 b \, dy = \frac{by^3}{3}\bigg|_h^{2h} = \frac{7bh^3}{3}$$

The *polar second moment of an area* is the second moment taken about an axis *normal to the plane* of an area. The equation is

$$J = \int \rho^2 \, dA \qquad (48.6)$$

Example 2. Find the polar second moment of the area of a circle about its centroidal axis.

Solution. Let the radius of the circle be r. Define a thin elemental ring of thickness $d\rho$ at radius ρ. Then $dA = 2\pi\rho \, d\rho$. We now have

$$J = \int \rho^2 \, dA = \int_0^r \rho^2 (2\pi\rho) \, d\rho = 2\pi \frac{\rho^4}{4}\bigg|_0^r = \frac{\pi r^4}{2}$$

FIGURE 48.4   Second area moment of a rectangle.

48.2.1   Radius of Gyration

If we think of the second moment of an area as the total area times the square of a fictitious distance, then

$$I_x = \int y^2 \, dA = k_x^2 A \qquad \text{or} \qquad k_x = \sqrt{\frac{I_x}{A}} \qquad (48.7)$$

In polar form,

$$J_z = \int \rho^2 \, dA = k_z^2 A \qquad \text{or} \qquad k_z = \sqrt{\frac{J_z}{A}} \qquad (48.8)$$

In each case, k is called the *radius of gyration*.

48.2.2   Transfer Formula

In Fig. 48.5, suppose we know the second moment of the area about x to be I_x. We can find the second moment of the area about some new axis that is parallel to the old using the transfer formula. Thus the second moment of the area in Fig. 48.5 about the x' axis is

$$I' = I_G + d^2 A \qquad (48.9)$$

where I_G = second moment about the centroidal axis and d = transfer distance. Using this formula and the second moments from Table 48.1 makes it possible to compute the second moments of sections made up of a combination of shapes. The procedure has much in common with the example in Fig. 48.3.

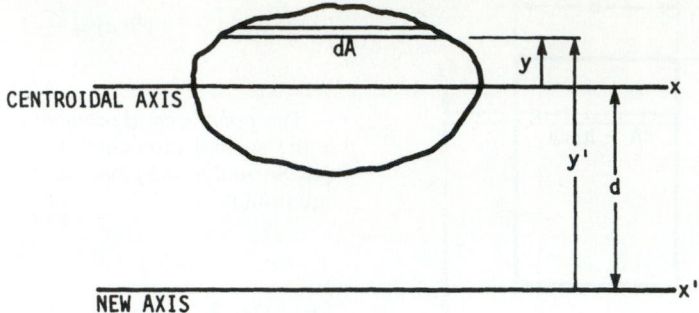

FIGURE 48.5   Use of the transfer formula.

48.2.3 Principal Axes

Sometimes we encounter the integral

$$I_{xy} = \int xy \, dA \qquad\qquad (48.10)$$

which is called the *product moment of an area.* This integral can be either positive or negative because x and y can have positive or negative values.

If one of the axes of an area, say y, is an axis of symmetry, then every element of area dA located by a positive x will have a twin, symmetrically located, having a corresponding negative x. These will sum to zero in the integration, and so the product moment is always zero when either x or y is an axis of symmetry. Since I_{xy} can be either positive or negative, there must be some orientation of rectangular axes where $I_{xy} = 0$. The two axes corresponding to this zero position are called the *principal axes.* If such axes intersect at the centroid of a section, then they are called the *centroidal principal axes.*

48.3 PREFERRED NUMBERS AND SIZES

The recommendations given in this section are not intended to be used as rules for design, since there are none. And even if rules were specified, there would be many occasions when designers would have to deviate from them, because other more pressing considerations may be present.

48.3.1 Preferred Numbers

A set of characteristic values that are to be distributed over a specified range for machines or products can be best obtained using a set of *preferred numbers.* Examples are the horsepower ratings of electric motors, the capacities of presses, or the speeds of a truck transmission. The preferred number system is internationally standardized (ISO3) and is described as the Renard, or R, series. This series is shown in Table 48.2. Some of the interesting characteristics are

TABLE 48.2 Preferred Numbers

First choice R5	Second choice R10	Third choice R20	Fourth choice R40
1			1.06
		1.12	1.18
	1.25		1.32
		1.4	1.5
1.6			1.7
		1.8	1.9
	2		2.12
		2.24	2.36
2.5			2.65
		2.8	3
	3.15		3.35
		3.55	3.75
4			4.25
		4.5	4.75
	5		5.3
		5.6	6
6.3			6.7
		7.1	7.5
	8		8.5
		9	9.5
10			

SOURCE: British standard PD 6481-1977.

1. The series can be applied to any value because it can be increased or decreased by powers of 10.

2. The R10 series contains all the R5 values; the R20 series contains all the R10 values; etc.

3. The preferred numbers can be multiplied, divided, or raised to a power by a single number, and the result, even if slightly rounded, is still a preferred number.

4. The number $3.15 \cong \pi$ in R10 and up means that the diameter, area, and circumference of a circle are also preferred numbers in view of the previous characteristic.

Preferred numbers are based on logarithmic interpolation and are given by

$$x_i = x_s \left(\frac{x_l}{x_s}\right)^{i/n + 1} \tag{48.11}$$

where x_l = largest number
  x_s = smallest number
  x_i = ith interpolated number
  i = interpolation number
  n = number of interpolations

The results should be appropriately rounded.

48.3.2 Preferred Sizes

Table 48.3 provides a list of preferred sizes for linear measurement in SI units. Note that these sizes are not the same as preferred numbers because of the convenience

TABLE 48.3 Preferred Metric Sizes in Millimeters

1st choice	2nd choice	3rd choice	1st choice	2nd choice	3rd choice	1st choice	2nd choice	3rd choice	1st choice	2nd choice	3rd choice
1			10			100			200		
	1.1			11				102			205
1.2			12			105				210	
		1.3			13			108			215
	1.4			14		110			220		
		1.5			15			112			225
1.6			16				115			230	
		1.7			17			118			235
	1.8			18		120			240		
		1.9			19			122			245
2			20				125			250	
		2.1			21			128			255
	2.2			22		130			260		
		2.4			23			132			265
2.5					24		135			270	
		2.6	25					138			275
	2.8				26	140			280		
3				28				142			285
		3.2	30				145			290	
	3.5			32				148			295
		3.8			34	150			300		
4			35					152			305
		4.2			36		155			310	
	4.5			38				158			315
		4.8	40			160			320		
5				42				162			325
		5.2			44		165			330	
	5.5		45					168			335
		5.8			46	170			340		
6				48				172			345
	6.5		50				175			350	
		6.8		52				178			355
	7				54	180			360		
		7.5	55					182			365
8					56		185			370	
		8.5		58				188			375
	9		60			190			380		
		9.5		62				192			385
10					64		195			390	
			65					198			395
					66	200			400†		
				68							
			70								
				72							
					74						
			75								
					76						
				78							
			80								
					82						
				85							
					88						
			90								
					92						
				95							
					98						
			100								

†Continued similarly above 400 mm.
SOURCE: British standard PD 6481-1977.

and simplicity of whole numbers for sizes of things. Preferred sizes in fractions of inches are listed in Table 48.4.

48.4 SIZES AND TOLERANCES OF STEEL SHEETS AND BARS

The dimensions and tolerances of steel products in this section are given in U.S. Customary System (USCS) units. Multiply inches by 25.4 to get the units of millimeters.

48.4.1 Sheet Steel

The *Manufacturer's Standard Gauge* for iron and steel sheets specifies a gauge number based on the weight per square foot. Remember, the gauge size is based on *weight*, not thickness. Steel products having thicknesses of ¼ in and over are called *plates* or *flats*, depending on the width.

The weights and equivalent thicknesses of carbon steel sheets are shown in Table 48.5. Standard widths and lengths available depend on the gauge sizes. Most are also available in coils, but steel warehouses may not stock all sizes.

Tables 48.6 to 48.11 provide the thickness tolerances for various grades of steel sheets. Except as noted, the tables apply to both coils and cut lengths. The width ranges are from *over* the lower limit up to and *including* the upper limit.

48.4.2 Bar Steel

When hot-rolled bars are machined on centers, it is necessary to allow for straightness as well as for the size and out-of-round tolerances in selecting the diameter (see Table 48.12). Tolerances for cold-finished bars are given in Tables 48.13, 48.14, and 48.15.

TABLE 48.4   Preferred Sizes in Fractions of Inches†

$\frac{1}{64}$	$\frac{1}{2}$	$2\frac{1}{4}$	5	$9\frac{1}{2}$	15
$\frac{1}{32}$	$\frac{9}{16}$	$2\frac{1}{2}$	$5\frac{1}{4}$	10	$15\frac{1}{2}$
$\frac{1}{16}$	$\frac{5}{8}$	$2\frac{3}{4}$	$5\frac{1}{2}$	$10\frac{1}{2}$	16
$\frac{3}{32}$	$\frac{11}{16}$	3	$5\frac{3}{4}$	11	$16\frac{1}{2}$
$\frac{1}{8}$	$\frac{3}{4}$	$3\frac{1}{4}$	6	$11\frac{1}{2}$	17
$\frac{5}{32}$	$\frac{7}{8}$	$3\frac{1}{2}$	$6\frac{1}{2}$	12	$17\frac{1}{2}$
$\frac{3}{16}$	1	$3\frac{3}{4}$	7	$12\frac{1}{2}$	18
$\frac{1}{4}$	$1\frac{1}{4}$	4	$7\frac{1}{2}$	13	$18\frac{1}{2}$
$\frac{5}{16}$	$1\frac{1}{2}$	$4\frac{1}{4}$	8	$13\frac{1}{2}$	19
$\frac{3}{8}$	$1\frac{3}{4}$	$4\frac{1}{2}$	$8\frac{1}{2}$	14	$19\frac{1}{2}$
$\frac{7}{16}$	2	$4\frac{3}{4}$	9	$14\frac{1}{2}$	20

†See also ANSI standard Z17.1-1973, Preferred Numbers.

TABLE 48.5 Gauge Sizes of Carbon Steel Sheets

Gauge no.	Thickness, in†	Weight, lb/ft²‡	Gauge no.	Thickness, in†	Weight, lb/ft²‡
7	0.1793	7.500	23	0.0269	1.125
8	0.1644	6.875	24	0.0239	1.000
9	0.1494	6.250	25	0.0209	0.875
10	0.1345	5.625	26	0.0179	0.750
11	0.1196	5.000	27	0.0164	0.6875
12	0.1046	4.375	28	0.0149	0.625
13	0.0897	3.750	29	0.0135	0.5625
14	0.0747	3.125	30	0.0120	0.500
15	0.0673	2.812	31	0.0105	0.4375
16	0.0598	2.500	32	0.0097	0.4062
17	0.0538	2.250	33	0.0090	0.375
18	0.0478	2.000	34	0.0082	0.3438
19	0.0418	1.750	35	0.0075	0.3125
20	0.0359	1.500	36	0.0067	0.2812
21	0.0329	1.375	37	0.0064	0.2656
22	0.0299	1.250	38	0.0060	0.250

†Multiply the thickness in inches by 25.4 to get the thickness in millimeters.
‡Multiply the weight in pounds per square foot by 4.88 to get the mass in kilograms per square meter (SI units).

TABLE 48.6 Thickness Tolerances for Hot-Rolled Carbon Sheets†

Thickness, in	Width, in					
	12–20	20–40	40–48	48–60	60–72	72 up
0.0449–0.0508	5	5	5			
0.0509–0.0567	5	5	6	6	7	
0.0568–0.0709	6	6	6	7	7	
0.0710–0.0971	6	7	7	7	8	8
0.0972–0.1799	7	7	8	8	8	8
0.1800–0.2299	7	8	9			

†Tolerances are plus or minus and in mils (1 mil = 0.001 in). This table applies only to coils.
SOURCE: Ref. [48.1], Sec. 5, Aug. 1979.

TABLE 48.7 Thickness Tolerances for Hot-Rolled Alloy Steel Sheets†

Thickness, in	Width, in					
	24–32	32–40	40–48	48–60	60–72	72–80
0.0568–0.0709	6	6	6	7	7	
0.0710–0.0821	7	7	7	7	8	8
0.0822–0.0971	7	8	8	8	9	9
0.0972–0.1799	8	9	10	10	11	12
0.1800–0.2299	9	9	10			

†Tolerances are plus or minus and in mils (1 mil = 0.001 in).
SOURCE: **Ref** [48.1], **Sec. 5, Aug. 1979.**

TABLE 48.8 Thickness Tolerances for Hot-Rolled High-Strength Steel Sheets†

Thickness, in	Width, in					
	12–15	15–20	20–32	32–40	40–48	48–60
0.0710–0.0821	6	7	7	7	7	7
0.0822–0.0971	6	7	7	8	8	8
0.0972–0.1799	7	8	8	9	10	10
0.1800–0.2299	7	8	9	9	10	

†Tolerances are plus or minus and in mils (1 mil = 0.001 in).
SOURCE: **Ref.** [48.1], **Sec. 5, Aug. 1979**

TABLE 48.9 Thickness Tolerances for Cold-Rolled Carbon Steel Sheets†

Thickness, in	Width, in			
	2–12	12–15	15–72	72 up
0.0142–0.0194	2	2	2	
0.0195–0.0388	3	3	3	3
0.0389–0.0567	4	4	4	4
0.0568–0.0709	5	5	5	5
0.0710–0.0971	5	5	5	6
0.0972–0.1419	....	5	6	7

†Tolerances are plus or minus and in mils (1 mil = 0.001 in).
SOURCE: **Ref** [48.1], **Sec. 5, Aug. 1979.**

TABLE 48.10 Thickness Tolerances for Cold-Rolled Alloy Steel Sheets†

Thickness, in	Width, in					
	24–32	32–40	40–48	48–60	60–70	70–80
0.0195–0.0313	3	3	3	3		
0.0314–0.0508	4	4	4	4	5	
0.0509–0.0567	5	5	5	5	6	
0.0568–0.0709	5	5	5	6	6	
0.0710–0.0821	5	6	6	6	7	7
0.0822–0.0971	6	7	7	8	9	9
0.0972–0.1419	7	8	9	10	10	11
0.1420–0.1799	8	9	10	10	11	12
0.1800–0.2299	8	9	10			

†Tolerances are plus or minus and in mils (1 mil = 0.001 in).
SOURCE: Ref. [48.1], Sec. 5, Aug. 1979.

TABLE 48.11 Thickness Tolerances for Cold-Rolled High-Strength Steel Sheets†

Thickness, in	Width, in					
	2–12	12–15	15–24	24–32	32–40	40–48
0.0142–0.0194	2	2	2	2	2	2
0.0195–0.0388	3	3	3	3	3	3
0.0389–0.0567	4	4	4	4	4	4
0.0568–0.0709	5	5	5	5	5	5
0.0710–0.0971	6	5	5	5	6	6
0.0972–0.1419	....	5	6	6	6	6
0.1419 up	....	6	6	7	7	7

†Tolerances are plus or minus and in mils (1 mil = 0.001 in).
SOURCE: Ref. [48.1], Sec. 5, Aug. 1979.

TABLE 48.12 Machining Allowances for Hot-Rolled Carbon Steel Bars for Turning on Centers†

Diameter, in	Allowance, in	Diameter, in	Allowance, in
To $\frac{7}{8}$	0.025	$2\frac{1}{2}$ to $3\frac{1}{2}$	0.090
$\frac{7}{8}$ to 1	0.028	$3\frac{1}{2}$ to $4\frac{1}{2}$	0.115
1 to $1\frac{1}{8}$	0.031	$4\frac{1}{2}$ to $5\frac{1}{2}$	0.140
$1\frac{1}{8}$ to $1\frac{1}{4}$	0.034	$5\frac{1}{2}$ to $6\frac{1}{2}$	0.165
$1\frac{1}{4}$ to $1\frac{3}{8}$	0.037	$6\frac{1}{2}$ to $8\frac{1}{4}$	0.209
$1\frac{3}{8}$ to $1\frac{1}{2}$	0.040	$8\frac{1}{4}$ to $9\frac{1}{2}$	0.240
$1\frac{1}{2}$ to 2	0.053	$9\frac{1}{2}$ to 10	0.253
2 to $2\frac{1}{2}$	0.065		

†Size range is from over the lower limit up to and including the upper limit; the allowances are on the *radius*.
SOURCE: Ref. [48.1].

TABLE 48.13 Size Tolerances for Cold-Drawn Carbon Steel Bars†

Size and shape, in	Carbon range, percent			
	To 0.28	0.28–0.55	To 0.55‡	Over 0.55§
Rounds:				
To 1½	2	3	4	5
1½ to 2½	3	4	5	6
2½ to 4	4	5	6	7
Hexagons:				
To ¾	2	3	4	6
¾ to 1½	3	4	5	7
1½ to 2½	4	5	6	8
2½ to 3⅛	5	6	7	9
Squares:				
To ¾	2	4	5	7
¾ to 1½	3	5	6	8
1½ to 2½	4	6	7	9
2½ to 4	6	8	9	11
Flats				
To ¾	3	4	6	8
¾ to 1½	4	5	8	10
1½ to 3	5	6	10	12
3 to 4	5	6	10	12
4 to 6	8	10	12	20
Over 6	13	15		

†Includes tolerances for bars that have been annealed, spheroidize annealed, normalized, normalized and tempered, or quenched and tempered before cold finishing. The table *does not* include tolerances for bars that are spheroidize annealed, normalized, normalized and tempered, or quenched and tempered after cold finishing. Size range and carbon range are from over the lower limit up to and including the upper limit. Tolerances are minus and are in mils (1 mil = 0.001 in).

‡Stress relieved or annealed after cold finishing.

§Quenched and tempered or normalized and tempered before cold finishing.

¶These tolerances apply to *both* the widths and thickness of flats.

SOURCE: Ref. [48.1].

TABLE 48.14 Size Tolerances for Cold-Finished, Turned, and Polished Carbon Steel Round Bars†

Diameter, in	Carbon range, percent			
	To 0.28	0.28–0.55	To 0.55‡	Over 0.55§
To 1½	2	3	4	5
1½ to 2½	3	4	5	6
2½ to 4	4	5	6	7
4 to 6	5	6	7	8
6 to 8	6	7	8	9
8 to 9	7	8	9	10
Over 9	8	9	10	11

†Includes tolerances for bars that have been annealed, spheroidize annealed, normalized, normalized and tempered, or quenched and tempered before cold finishing. The table *does not* include tolerances for bars that are spheroidize annealed, normalized, normalized and tempered, or quenched and tempered after cold finishing. Size range and carbon range are from over the lower limit up to and including the upper limit. Tolerances are minus and are in mils (1 mil = 0.001 in).

‡Stress relieved or annealed after cold finishing.

§Quenched and tempered or normalized and tempered before cold finishing.

SOURCE: Ref. [48.1].

TABLE 48.15 Size Tolerances for Ground and Polished Carbon Steel Rounds Prefinished by Cold Drawing or by Turning†

Diameter, in	Prefinish	
	Cold drawn	Turned
To 1½	1	1
1½ to 2½	1.5	1.5
2½ to 3	2	2
3 to 4	3	3
4 to 6		4‡
Over 6		5‡

†Size range is from over the lower limit up to and including the upper limit. Tolerances are minus and in mils (1 mil = 0.001 in).

‡Increase this tolerance by 1 mil if the steels have a sulfur content under 0.08 percent or if they are thermally treated.

SOURCE: Ref. [48.1].

48.4.3 Pipe and Tubing

The outside diameter of pipe having a nominal size of 12 in or smaller is larger than the nominal size. The difference between pipe and tubing is that pipe is intended to be used in piping systems; also, tubing has an outside diameter the same as the nominal size. See Table 48.16 for pipe sizes.

TABLE 48.16 Dimensions and Weights for Threaded and Coupled Pipe

Nominal size, in	Outside diameter, in	Wall thickness, in	Weight,† lb/ft	Weight class	Schedule no.
⅛	0.405	0.068	0.24	STD	40
		0.095	0.32	XS	80
¼	0.540	0.088	0.42	STD	40
		0.119	0.54	XS	80
⅜	0.675	0.091	0.57	STD	40
		0.126	0.74	XS	80

Nominal size, in	Outside diameter, in	Wall thickness, in	Weight,† lb/ft	Weight class	Schedule no.
½	0.840	0.109	0.85	STD	40
		0.147	1.09	XS	80
		0.294	1.72	XXS	
¾	1.050	0.113	1.13	STD	40
		0.154	1.48	XS	80
		0.308	2.44	XXS	
1	1.315	0.133	1.68	STD	40
		0.179	2.18	XS	80
		0.358	3.66	XXS	
1¼	1.660	0.140	2.28	STD	40
		0.191	3.02	XS	80
		0.382	5.22	XXS	
1½	1.900	0.145	2.73	STD	40
		0.200	3.66	XS	80
		0.400	6.41	XXS	
2	2.375	0.154	3.68	STD	40
		0.218	5.07	XS	80
		0.436	9.03	XXS	
2½	2.875	0.203	5.82	STD	40
		0.276	7.73	XS	80
		0.552	13.70	XXS	
3	3.500	0.216	7.62	STD	40
		0.300	10.33	XS	80
		0.600	18.57	XXS	
3½	4.000	0.226	9.20	STD	40
		0.318	12.63	XS	80
4	4.500	0.237	10.89	STD	40
		0.337	15.17	XS	80
		0.674	27.58	XXS	
5	5.563	0.258	14.81	STD	40
		0.375	21.09	XS	80
		0.750	38.61	XXS	
6	6.625	0.280	19.18	STD	40
		0.432	28.89	XS	80
		0.864	53.14	XXS	
8	8.625	0.277	25.55		30
		0.322	29.35	STD	40
		0.500	43.90	XS	80
		0.875	72.44	XXS	
10	10.750	0.279	32.75		
		0.307	35.75		30
		0.365	41.85	STD	40
		0.500	55.82	XS	60
12	12.750	0.330	45.45		30
		0.375	51.15	STD	
		0.500	66.71	XS	

†This is the weight of threaded pipe including the coupling.

SOURCE: ASTM standard A53, Table X3. A greater range of sizes together with SI equivalents is given in ANSI standard B36.10-1979.

TABLE 48.17 Decimal Equivalents of Wire and Sheet-Metal Gauges in Inches

Always specify the name of the gauge when gauge numbers are used.

Gauge Number	American or Brown & Sharpe (nonferrous sheet and rod)	Birmingham or Stubs iron wire (tubing, ferrous strip, flat wire, and spring steel)	United States Standard (ferrous sheet and plate, 480 lb/ft)	Manufacturers Standard (ferrous sheet)	Steel wire or Washburn & Moen (ferrous wire except music wire)	Music wire (music wire)	Stubs steel wire (steel drill rod)	Twist drill (twist drills and drill steel)
7/0	....	....	0.500	....	0.490 0	0.004		
6/0	0.580 0	....	0.468 75	....	0.461 5	0.005		
5/0	0.516 5	....	0.437 5	....	0.430 5	0.006		
4/0	0.460 0	0.454	0.406 25	....	0.393 8	0.007		
3/0	0.409 6	0.425	0.375	....	0.362 5	0.008		
2/0	0.364 8	0.380	0.343 75	....	0.331 0	0.009		
0	0.324 9	0.340	0.312 5	....	0.306 5	0.010	0.227	0.228 0
1	0.289 3	0.300	0.281 25	....	0.283 0	0.011	0.219	0.221 0
2	0.257 6	0.284	0.265 625	....	0.262 5	0.012	0.212	0.213 0
3	0.229 4	0.259	0.25	0.239 1	0.243 7	0.013	0.207	0.209 0
4	0.204 3	0.238	0.234 375	0.224 2	0.225 3	0.014	0.204	0.205 5
5	0.181 9	0.220	0.218 75	0.209 2	0.207 0	0.016	0.201	0.204 0
6	0.162 0	0.203	0.203 125	0.194 3	0.192 0	0.018	0.199	0.201 0
7	0.144 3	0.180	0.187 5	0.179 3	0.177 0	0.020	0.197	0.199 0
8	0.128 5	0.165	0.171 875	0.164 4	0.162 0	0.022	0.194	0.196 0
9	0.114 5	0.148	0.156 25	0.149 5	0.148 3	0.024	0.191	0.193 5
10	0.101 9	0.134	0.140 625	0.134 5	0.135 0	0.026	0.188	0.191 0
11	0.090 74	0.120	0.125	0.119 6	0.120 5	0.029	0.185	0.189 0
12	0.080 81	0.109	0.109 357	0.104 6	0.105 5			

13	0.071 96	0.095	0.093 75	0.089 7	0.091 5	0.031	0.182	0.185 0
14	0.064 08	0.083	0.078 125	0.074 7	0.080 0	0.033	0.180	0.182 0
15	0.057 07	0.072	0.070 312 5	0.067 3	0.072 0	0.035	0.178	0.180 0
16	0.050 82	0.065	0.062 5	0.059 8	0.062 5	0.037	0.175	0.177 0
17	0.045 26	0.058	0.056 25	0.053 8	0.054 0	0.039	0.172	0.173 0
18	0.040 30	0.049	0.05	0.047 8	0.047 5	0.041	0.168	0.169 5
19	0.035 89	0.042	0.043 75	0.041 8	0.041 0	0.043	0.164	0.166 0
20	0.031 96	0.035	0.037 5	0.035 9	0.034 8	0.045	0.161	0.161 0
21	0.028 46	0.032	0.034 375	0.032 9	0.031 7	0.047	0.157	0.159 0
22	0.025 35	0.028	0.031 25	0.029 9	0.028 6	0.049	0.155	0.157 0
23	0.022 57	0.025	0.028 125	0.026 9	0.025 8	0.051	0.153	0.154 0
24	0.020 10	0.022	0.025	0.023 9	0.023 0	0.055	0.151	0.152 0
25	0.017 90	0.020	0.021 875	0.020 9	0.020 4	0.059	0.148	0.149 5
26	0.015 94	0.018	0.018 75	0.017 9	0.018 1	0.063	0.146	0.147 0
27	0.014 20	0.016	0.017 187 5	0.016 4	0.017 3	0.067	0.143	0.144 0
28	0.012 64	0.014	0.015 625	0.014 9	0.016 2	0.071	0.139	0.140 5
29	0.011 26	0.013	0.014 062 5	0.013 5	0.015 0	0.075	0.134	0.136 0
30	0.010 03	0.012	0.012 5	0.012 0	0.014 0	0.080	0.127	0.128 5
31	0.008 928	0.010	0.010 937 5	0.010 5	0.013 2	0.085	0.120	0.120 0
32	0.007 950	0.009	0.010 156 25	0.009 7	0.012 8	0.090	0.115	0.116 0
33	0.007 080	0.008	0.009 375	0.009 0	0.011 8	0.095	0.112	0.113 0
34	0.006 305	0.007	0.008 593 75	0.008 2	0.010 4	.....	0.110	0.111 0
35	0.005 615	0.005	0.007 812 5	0.007 5	0.009 5	.....	0.108	0.110 0
36	0.005 000	0.004	0.007 031 25	0.006 7	0.009 0	.....	0.106	0.106 5
37	0.004 453	.....	0.006 640 625	0.006 4	0.008 5	.....	0.103	0.104 0
38	0.003 965	.....	0.006 25	0.006 0	0.008 0	.....	0.101	0.101 5
39	0.003 531	.....	.....	.....	0.007 5	.....	0.099	0.099 5
40	0.003 145	.....	.....	.....	0.007 0	.....	0.097	0.098 0

SOURCE: Reynolds Metals Co., Richmond, Virginia.

TABLE 48.18 Properties of Square and Rectangular Structural Steel Tubing†

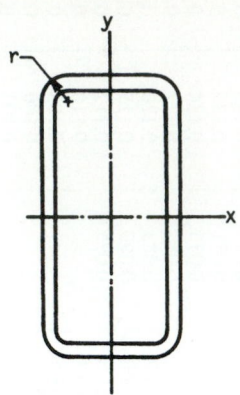

Size, in	Weight, lb/ft	Area A, in²	Radius‡ r, in	I_x, in⁴	I_y, in⁴
$2 \times 2 \times \frac{3}{16}$	4.32	1.27	$\frac{3}{8}$	0.668	
$\frac{1}{4}$	5.41	1.59	$\frac{1}{2}$	0.766	
$3 \times 2 \times \frac{3}{16}$	5.59	1.64	$\frac{3}{8}$	1.24	0.977
$\frac{1}{4}$	7.11	2.09	$\frac{1}{2}$	2.21	1.15
$3 \times 3 \times \frac{3}{16}$	6.87	2.02	$\frac{3}{8}$	2.60	
$\frac{1}{4}$	8.81	2.59	$\frac{1}{2}$	3.16	
$\frac{5}{16}$	10.58	3.11	$\frac{5}{8}$	3.58	
$4 \times 2 \times \frac{3}{16}$	6.87	2.02	$\frac{3}{8}$	3.87	1.29
$\frac{1}{4}$	8.81	2.59	$\frac{1}{2}$	4.69	1.54
$\frac{5}{16}$	10.58	3.11	$\frac{5}{8}$	5.32	1.71
$4 \times 3 \times \frac{3}{16}$	8.15	2.39	$\frac{3}{8}$	5.23	3.34
$\frac{1}{4}$	10.51	3.09	$\frac{1}{2}$	6.45	4.10
$\frac{5}{16}$	12.70	3.73	$\frac{5}{8}$	7.45	4.71
$4 \times 4 \times \frac{3}{16}$	9.42	2.77	$\frac{3}{8}$	6.59	
$\frac{1}{4}$	12.21	3.59	$\frac{1}{2}$	8.22	
$\frac{5}{16}$	14.83	4.36	$\frac{5}{8}$	9.58	
$\frac{3}{8}$	17.27	5.08	$\frac{3}{4}$	10.7	
$\frac{1}{2}$	21.63	6.36	1	12.3	
$5 \times 3 \times \frac{1}{4}$	12.21	3.59	$\frac{1}{2}$	11.3	5.05
$\frac{5}{16}$	14.83	4.36	$\frac{5}{8}$	13.2	5.85
$\frac{3}{8}$	17.27	5.08	$\frac{3}{4}$	14.7	6.48
$\frac{1}{2}$	21.63	6.36	1	16.9	7.33
$5 \times 4 \times \frac{1}{4}$	13.91	4.09	$\frac{1}{2}$	14.1	9.98
$\frac{5}{16}$	16.96	4.98	$\frac{5}{8}$	16.6	11.7
$\frac{3}{8}$	19.82	5.83	$\frac{3}{4}$	18.7	13.2

TABLE 48.18   Properties of Square and Rectangular Structural Steel Tubing† (*Continued*)

Size, in	Weight, lb/ft	Area A, in^2	Radius‡ r, in	I_x, in^4	I_y, in^4
$5 \times 5 \times \frac{1}{4}$	15.62	4.59	$\frac{1}{2}$	16.9	
$\frac{5}{16}$	19.08	5.61	$\frac{5}{8}$	20.1	
$\frac{3}{8}$	22.37	6.58	$\frac{3}{4}$	22.8	
$\frac{1}{2}$	28.43	8.36	1	27.0	
$6 \times 3 \times \frac{1}{4}$	13.91	4.09	$\frac{1}{2}$	17.9	6.00
$\frac{5}{16}$	16.96	4.98	$\frac{5}{8}$	21.1	6.98
$\frac{3}{8}$	19.82	5.83	$\frac{3}{4}$	23.8	7.78
$6 \times 4 \times \frac{1}{4}$	15.62	4.59	$\frac{1}{2}$	22.1	11.7
$\frac{5}{16}$	19.08	5.61	$\frac{5}{8}$	26.2	13.8
$\frac{3}{8}$	22.37	6.58	$\frac{3}{4}$	29.7	15.6
$\frac{1}{2}$	28.43	8.36	1	35.3	18.4
$6 \times 6 \times \frac{1}{4}$	19.02	5.59	$\frac{1}{2}$	30.3	
$\frac{5}{16}$	23.34	6.86	$\frac{5}{8}$	36.3	
$\frac{3}{8}$	27.48	8.08	$\frac{3}{4}$	41.6	
$\frac{1}{2}$	35.24	10.4	1	50.5	
$8 \times 4 \times \frac{5}{16}$	23.34	6.86	$\frac{5}{8}$	53.9	18.1
$\frac{3}{8}$	27.48	8.08	$\frac{3}{4}$	61.9	20.6
$\frac{1}{2}$	35.24	10.4	1	75.1	24.6
$8 \times 6 \times \frac{5}{16}$	27.59	8.11	$\frac{5}{8}$	72.4	46.4
$\frac{3}{8}$	32.58	9.58	$\frac{3}{4}$	83.7	53.5
$\frac{1}{2}$	42.05	12.4	1	103.	65.7
$8 \times 8 \times \frac{5}{16}$	31.84	9.36	$\frac{5}{8}$	90.9	
$\frac{3}{8}$	37.69	11.1	$\frac{3}{4}$	106.	
$\frac{1}{2}$	48.85	14.4	1	131.	
$\frac{5}{8}$	59.32	17.4	$1\frac{1}{4}$	153.	

†Size expressed by outside dimensions and wall thickness; other sizes are available (see Ref. [48.2]).
‡Tolerance is three times the wall thickness.

Seamless mechanical steel tubing is available in a great range of sizes, from about $\frac{3}{16}$ in outside diameter with a wall thickness of no. 24 gauge B and W up to a wall thickness of 1 in and an outside diameter of 12 in or over. Welded tubing is made from strip steel, either hot rolled with a bright finish or cold rolled. Tubing is also available cold drawn and may be obtained with a high-quality inside finish for certain applications.

The wall thickness of tubing is usually specified in gauge sizes or in fractions of an inch when USCS units are used. The tolerances of tubing are generally specified for the outside diameter and the wall thickness. This means that the inside diameter takes all the variation. However, tubing can be ordered using an inside-diameter specification.

TABLE 48.19 Properties of American Standard Channels†

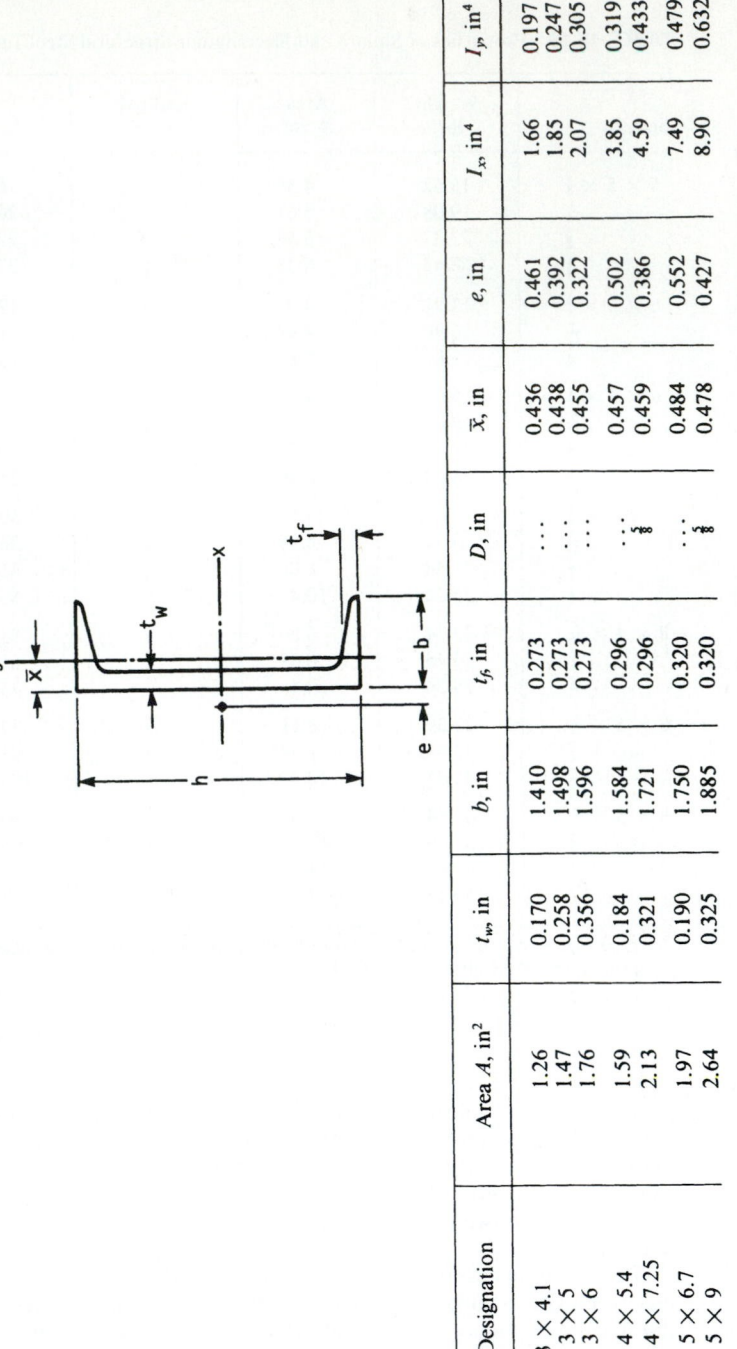

Designation	Area A, in²	t_w, in	b, in	t_f, in	D, in	$\bar{x}$, in	e, in	I_x, in⁴	I_y, in⁴
C 3 × 4.1	1.26	0.170	1.410	0.273	...	0.436	0.461	1.66	0.197
3 × 5	1.47	0.258	1.498	0.273	...	0.438	0.392	1.85	0.247
3 × 6	1.76	0.356	1.596	0.273	...	0.455	0.322	2.07	0.305
C 4 × 5.4	1.59	0.184	1.584	0.296	...	0.457	0.502	3.85	0.319
4 × 7.25	2.13	0.321	1.721	0.296	$\frac{5}{8}$	0.459	0.386	4.59	0.433
C 5 × 6.7	1.97	0.190	1.750	0.320	...	0.484	0.552	7.49	0.479
5 × 9	2.64	0.325	1.885	0.320	$\frac{5}{8}$	0.478	0.427	8.90	0.632

Designation									
C 6 × 8.2	2.40	0.200	1.920	0.343	5/8	0.511	0.599	13.1	0.693
6 × 10.5	3.09	0.314	2.034	0.343	5/8	0.499	0.486	15.2	0.866
6 × 13	3.83	0.437	2.157	0.343	5/8	0.514	0.380	17.4	1.05
C 7 × 9.8	2.87	0.210	2.090	0.366	5/8	0.540	0.647	21.3	0.968
7 × 12.25	3.60	0.314	2.194	0.366	5/8	0.525	0.538	24.2	1.17
7 × 14.75	4.33	0.419	2.299	0.366	5/8	0.532	0.441	27.2	1.38
C 8 × 11.5	3.38	0.220	2.260	0.390	3/4	0.571	0.697	32.6	1.32
8 × 13.75	4.04	0.303	2.343	0.390	3/4	0.553	0.604	36.1	1.53
8 × 18.75	5.51	0.487	2.527	0.390	3/4	0.565	0.431	44.0	1.98
C 9 × 13.4	3.94	0.233	2.433	0.413	3/4	0.601	0.743	47.9	1.76
9 × 15	4.41	0.285	2.485	0.413	3/4	0.586	0.682	51.0	1.93
9 × 20	5.88	0.448	2.648	0.413	3/4	0.583	0.515	60.9	2.42
C 10 × 15.3	4.49	0.240	2.600	0.436	3/4	0.634	0.796	67.4	2.28
10 × 20	5.88	0.379	2.739	0.436	3/4	0.606	0.637	78.9	2.81
10 × 25	7.35	0.526	2.886	0.436	3/4	0.617	0.494	91.2	3.36
10 × 30	8.82	0.673	3.033	0.436	3/4	0.649	0.369	103	3.94
C 12 × 20.7	6.09	0.282	2.942	0.501	7/8	0.698	0.870	129	3.88
12 × 25	7.35	0.387	3.047	0.501	7/8	0.674	0.746	144	4.47
12 × 30	8.82	0.510	3.170	0.501	7/8	0.674	0.618	162	5.14
C 15 × 33.9	9.96	0.400	3.400	0.650	1	0.787	0.896	315	8.13
15 × 40	11.8	0.520	3.520	0.650	1	0.777	0.767	349	9.23
15 × 50	14.7	0.716	3.716	0.650	1	0.798	0.583	404	11.0

†The designation is the channel depth and the unit weight in pounds per foot; D = diameter of maximum flange fastener, and e = location of shear center.

SOURCE: Ref. [48.2]. All the sizes listed here are generally available in aluminum alloys.'or these, the unit weight is obtained by multiplying the area by 0.829.

TABLE 48.20 Properties of Angles†

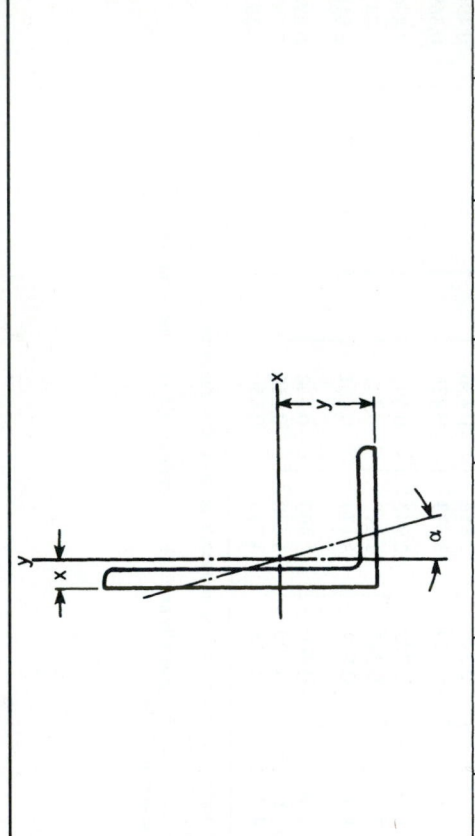

Size, in	w, lb/ft	Area A, in²	y, in	I_x, in⁴	x, in	I_y, in⁴	Tan α
L 2 × 2 × $\frac{1}{8}$	1.65	0.484	0.546	0.190	0.546	0.190	1.000
× $\frac{3}{16}$	2.44	0.715	0.569	0.272	0.569	0.272	1.000
× $\frac{1}{4}$	3.19	0.938	0.592	0.348	0.592	0.348	1.000
× $\frac{5}{16}$	3.92	1.15	0.614	0.416	0.614	0.416	1.000
× $\frac{3}{8}$	4.7	1.36	0.636	0.479	0.636	0.479	1.000
L 2½ × 2 × $\frac{3}{16}$	2.75	0.809	0.764	0.509	0.514	0.291	0.631
× $\frac{1}{4}$	3.62	1.06	0.787	0.654	0.537	0.372	0.626
× $\frac{5}{16}$	4.5	1.31	0.809	0.788	0.559	0.446	0.620
× $\frac{3}{8}$	5.3	1.55	0.831	0.912	0.581	0.514	0.614
L 2½ × 2½ × $\frac{3}{16}$	3.07	0.902	0.694	0.547	⋯	⋯	1.000
× $\frac{1}{4}$	4.10	1.19	0.717	0.703	⋯	⋯	1.000
× $\frac{5}{16}$	5.00	1.46	0.740	0.849	⋯	⋯	1.000
× $\frac{3}{8}$	5.9	1.73	0.762	0.984	⋯	⋯	1.000

L 3 × 2 × $\frac{3}{16}$	3.07	0.902	0.970	0.842	0.470	0.307	0.446
L 3 × 2 × $\frac{1}{4}$	4.1	1.19	0.993	1.09	0.493	0.392	0.440
× $\frac{5}{16}$	5.0	1.46	1.02	1.32	0.516	0.470	0.435
× $\frac{3}{8}$	5.9	1.73	1.04	1.53	0.539	0.543	0.428
L 3 × 2½ × $\frac{3}{16}$	3.39	0.996	0.888	0.907	0.638	0.577	0.688
× $\frac{1}{4}$	4.5	1.31	0.911	1.17	0.661	0.743	0.684
× $\frac{3}{8}$	6.6	1.92	0.956	1.66	0.706	1.04	0.676
L 3 × 3 × $\frac{3}{16}$	3.71	1.09	0.820	0.962	.....	.....	1.000
× $\frac{1}{4}$	4.9	1.44	0.842	1.24	.....	.....	1.000
× $\frac{5}{16}$	6.1	1.78	0.865	1.51	.....	.....	1.000
× $\frac{3}{8}$	7.2	2.11	0.888	1.76	.....	.....	1.000
× $\frac{1}{2}$	9.4	2.75	0.932	2.22	.....	.....	1.000
L 3½ × 2½ × $\frac{1}{4}$	4.9	1.44	1.11	1.80	0.614	0.777	0.506
× $\frac{5}{16}$	6.1	1.78	1.14	2.19	0.637	0.939	0.501
× $\frac{3}{8}$	7.2	2.11	1.16	2.56	0.660	1.09	0.496
L 3½ × 3 × $\frac{1}{4}$	5.4	1.56	1.04	1.91	0.785	1.30	0.727
× $\frac{5}{16}$	6.6	1.93	1.06	2.33	0.808	1.58	0.724
L 3½ × 3 × $\frac{3}{8}$	7.9	2.30	1.08	2.72	0.830	1.85	0.721
L 3½ × 3½ × $\frac{1}{4}$	5.8	1.69	0.968	2.01	.....	.....	1.000
× $\frac{5}{16}$	7.2	2.09	0.990	2.45	.....	.....	1.000
× $\frac{3}{8}$	8.5	2.48	1.01	2.87	.....	.....	1.000
L 4 × 3 × $\frac{1}{4}$	5.8	1.69	1.24	2.77	0.736	1.36	0.558
× $\frac{5}{16}$	7.2	2.09	1.26	3.38	0.759	1.65	0.554
× $\frac{3}{8}$	8.5	2.48	1.28	3.96	0.782	1.92	0.551
× $\frac{1}{2}$	11.1	3.25	1.33	5.05	0.827	2.42	0.543
L 4 × 3½ × $\frac{1}{4}$	6.2	1.81	1.16	2.91	0.909	2.09	0.759
× $\frac{5}{16}$	7.7	2.25	1.18	3.56	0.932	2.55	0.757
× $\frac{3}{8}$	9.1	2.67	1.21	4.18	0.955	2.95	0.755
× $\frac{1}{2}$	11.9	3.50	1.25	5.32	1.00	3.79	0.750

TABLE 48.20 Properties of Angles† (*Continued*)

Size, in	w, lb/ft	Area A, in²	y, in	I_{xx}, in⁴	x, in	I_y, in⁴	Tan α
L 4 × 4 × ¼	6.6	1.94	1.09	3.04	.....	.....	1.000
× 5/16	8.2	2.40	1.12	3.71	.....	.....	1.000
× 3/8	9.8	2.86	1.14	4.36	.....	.....	1.000
× ½	12.8	3.75	1.18	5.56	.....	.....	1.000
L 4 × 4 × 5/8	15.7	4.61	1.23	6.66	.....	.....	1.000
× ¾	18.6	5.44	1.27	7.67	.....	.....	1.000
L 5 × 3 × ¼	6.6	1.94	1.66	5.11	0.657	1.44	0.371
× 5/16	8.2	2.40	1.68	6.26	0.681	1.75	0.368
× 3/8	9.8	2.86	1.70	7.37	0.704	2.04	0.364
× ½	12.8	3.75	1.75	9.45	0.750	2.58	0.357
L 5 × 3½ × 5/16	8.7	2.56	1.59	6.60	0.838	2.72	0.489
× 3/8	10.4	3.05	1.61	7.78	0.861	3.18	0.486
× ½	13.6	4.00	1.66	9.99	0.906	4.05	0.479
× ¾	19.8	5.81	1.75	13.9	0.996	5.55	0.464
L 5 × 5 × 5/16	10.3	3.03	1.37	7.42	.....	.....	1.000
× 3/8	12.3	3.61	1.39	8.74	.....	.....	1.000
× ½	16.2	4.75	1.43	11.3	.....	.....	1.000
× ¾	23.6	6.94	1.52	15.7	.....	.....	1.000
× 7/8	27.2	7.98	1.57	17.8	.....	.....	1.000
L 6 × 3½ × 5/16	9.8	2.87	2.01	10.9	0.763	2.85	0.352
× 3/8	11.7	3.42	2.04	12.9	0.787	3.34	0.350

Size†							
L 6 × 4 × 3/8	12.3	3.61	1.94	13.5	0.941	4.90	0.446
× 1/2	16.2	4.75	1.99	17.4	0.987	6.27	0.440
× 5/8	20.0	5.86	2.03	21.1	1.03	7.52	0.435
× 3/4	23.6	6.94	2.08	24.5	1.08	8.68	0.428
L 6 × 6 × 3/8	14.9	4.36	1.64	15.4	……	……	1.000
× 1/2	19.6	5.75	1.68	19.9	……	……	1.000
× 5/8	24.2	7.11	1.73	24.2	……	……	1.000
× 3/4	28.7	8.44	1.78	28.2	……	……	1.000
× 7/8	33.1	9.73	1.82	31.9	……	……	1.000
× 1	37.4	11.0	1.86	35.5	……	……	1.000
L 7 × 4 × 3/8	13.6	3.98	2.37	20.6	0.870	5.10	0.340
× 1/2	17.9	5.25	2.42	26.7	0.917	6.53	0.335
× 3/4	26.2	7.69	2.51	37.8	1.01	9.05	0.324
L 8 × 4 × 1/2	19.6	5.75	2.86	38.5	0.859	6.74	0.267
× 3/4	28.7	8.44	2.95	54.9	0.953	9.36	0.258
× 1	37.4	11.0	3.05	69.6	1.05	11.6	0.247
L 8 × 6 × 1/2	23.0	6.75	2.47	44.3	1.47	21.7	0.558
× 3/4	33.8	9.94	2.56	63.4	1.56	30.7	0.551
× 1	44.2	13.0	2.65	80.8	1.65	38.8	0.543
L 8 × 8 × 1/2	26.4	7.75	2.19	48.6	……	……	1.000
× 5/8	32.7	9.61	2.23	59.4	……	……	1.000
× 3/4	38.9	11.4	2.28	69.7	……	……	1.000
× 7/8	45.0	13.2	2.32	79.6	……	……	1.000
× 1	51.0	15.0	2.37	89.0	……	……	1.000
× 1 1/8	56.9	16.7	2.41	98.0	……	……	1.000

†Size is the length of each leg and the thickness; unit weight for steel is w. For these, the unit

SOURCE: Ref. [48.2]. Angles up to 6 in inclusive are also available in aluminum alloys. weight is obtained by multiplying the area by 0.829. Sizes in structural steel larger than those listed are available on special order.

TABLE 48.21 Properties of W Shapes†

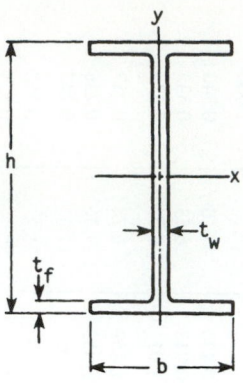

Designation	Area A, in²	h, in	t_w, in	b, in	t_f, in	I_x, in⁴	I_y, in⁴
W 4 × 13	3.83	4.16	0.280	4.060	0.345	11.3	3.86
W 5 × 16	4.68	5.01	0.240	5.000	0.360	21.3	7.51
5 × 19	5.54	5.15	0.270	5.030	0.430	26.2	9.13
W 6 × 9	2.68	5.90	0.170	3.940	0.215	16.4	2.19
6 × 12	3.55	6.03	0.230	4.000	0.280	22.1	2.99
6 × 16	4.74	6.28	0.260	4.030	0.405	32.1	4.43
6 × 15	4.43	5.99	0.230	5.990	0.260	29.1	9.32
6 × 20	5.87	6.20	0.260	6.020	0.365	41.4	13.3
6 × 25	7.34	6.38	0.320	6.080	0.455	53.4	17.1
W 8 × 10	2.96	7.89	0.170	3.940	0.205	30.8	2.09
8 × 13	3.84	7.99	0.230	4.000	0.255	39.6	2.73
8 × 15	4.44	8.11	0.245	4.015	0.315	48.0	3.41
8 × 18	5.26	8.14	0.230	5.25	0.330	61.9	7.97
8 × 21	6.16	8.28	0.250	5.27	0.400	75.3	9.77
8 × 24	7.08	7.93	0.245	6.495	0.400	82.8	18.3
8 × 28	8.25	8.06	0.285	6.535	0.465	98.0	21.7
8 × 31	9.13	8.00	0.285	7.995	0.435	110	37.1
8 × 35	10.3	8.12	0.310	8.020	0.495	127	42.6
8 × 40	11.7	8.25	0.360	8.070	0.560	146	49.1
8 × 48	14.1	8.50	0.400	8.110	0.685	184	60.9
8 × 58	17.1	8.75	0.510	8.220	0.810	228	75.1
8 × 67	19.7	9.00	0.570	8.280	0.935	272	88.6
W 10 × 12	3.54	9.87	0.190	3.960	0.210	53.8	2.18
10 × 15	4.41	9.99	0.230	4.000	0.270	68.9	2.89
10 × 17	4.99	10.11	0.240	4.010	0.330	81.9	3.56
10 × 19	5.62	10.24	0.250	4.020	0.395	96.3	4.29
W 10 × 22	6.49	10.17	0.240	5.75	0.360	118	11.4
10 × 26	7.61	10.33	0.260	5.770	0.440	144	14.1
10 × 30	8.84	10.47	0.300	5.810	0.510	170	16.7
W 10 × 33	9.71	9.73	0.290	7.960	0.435	170	36.6
10 × 39	11.5	9.92	0.315	7.985	0.530	209	45.0
10 × 45	13.3	10.10	0.350	8.020	0.620	248	53.4

†The designation is the nominal depth, and the unit weight for steel is in pounds per foot. Larger sizes are available from W 10 × 49 to W 36 × 300. See Ref. [48.2]. Some of the sizes 8 in and under are available in aluminum alloys which are then called H sections.

TABLE 48.22 Properties of S Shapes†

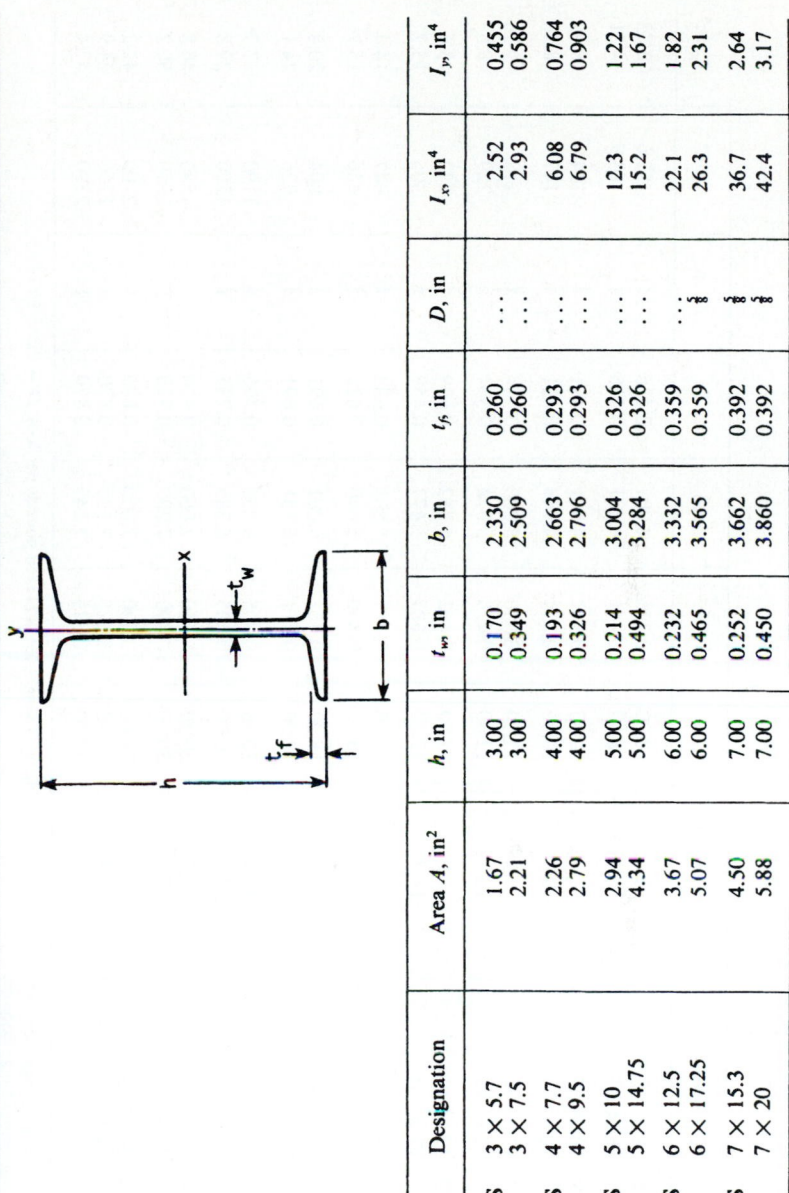

Designation	Area A, in²	h, in	t_w, in	b, in	t_f, in	D, in	I_x, in⁴	I_y, in⁴
S 3 × 5.7	1.67	3.00	0.170	2.330	0.260	...	2.52	0.455
3 × 7.5	2.21	3.00	0.349	2.509	0.260	...	2.93	0.586
S 4 × 7.7	2.26	4.00	0.193	2.663	0.293	...	6.08	0.764
4 × 9.5	2.79	4.00	0.326	2.796	0.293	...	6.79	0.903
S 5 × 10	2.94	5.00	0.214	3.004	0.326	...	12.3	1.22
5 × 14.75	4.34	5.00	0.494	3.284	0.326	...	15.2	1.67
S 6 × 12.5	3.67	6.00	0.232	3.332	0.359	...	22.1	1.82
6 × 17.25	5.07	6.00	0.465	3.565	0.359	5/8	26.3	2.31
S 7 × 15.3	4.50	7.00	0.252	3.662	0.392	5/8	36.7	2.64
7 × 20	5.88	7.00	0.450	3.860	0.392	5/8	42.4	3.17

TABLE 48.22 Properties of S Shapes† (*Continued*)

Designation	Area A, in^2	h, in	t_w, in	b, in	t_f, in	D, in	I_x, in^4	I_y, in^4
S 8 × 18.4	5.41	8.00	0.271	4.001	0.426	3/4	57.6	3.73
8 × 23	6.77	8.00	0.441	4.171	0.426	3/4	64.9	4.31
S 10 × 25.4	7.46	10.00	0.311	4.661	0.491	3/4	124	6.79
10 × 35	10.3	10.00	0.594	4.944	0.491	3/4	147	8.36
S 12 × 31.8	9.35	12.00	0.350	5.000	0.544	3/4	218	9.36
12 × 35	10.3	12.00	0.428	5.078	0.544	3/4	229	9.87
S 12 × 40.8	12.0	12.00	0.462	5.252	0.659	3/4	272	13.6
12 × 50	14.7	12.00	0.687	5.477	0.659	3/4	305	15.7
S 15 × 42.9	12.6	15.00	0.411	5.501	0.622	3/4	447	14.4
15 × 50	14.7	15.00	0.550	5.640	0.622	3/4	486	15.7
S 18 × 54.7	16.1	18.00	0.461	6.001	0.691	7/8	804	20.8
18 × 70	20.6	18.00	0.711	6.251	0.691	7/8	926	24.1
S 20 × 66	19.4	20.00	0.505	6.255	0.795	7/8	1190	27.7
20 × 75	22.0	20.00	0.635	6.385	0.795	7/8	1280	29.8
S 20 × 86	25.3	20.30	0.660	7.060	0.920	1	1580	46.8
20 × 96	28.2	20.30	0.800	7.200	0.920	1	1670	50.2
S 24 × 80	23.5	24.00	0.500	7.000	0.870	1	2100	42.2
24 × 90	26.5	24.00	0.625	7.125	0.870	1	2250	44.9
24 × 100	29.3	24.00	0.745	7.245	0.870	1	2390	47.4

†The designation is the nominal depth and the unit weight for steel is in pounds per foot; D = diameter of maximum flange fastener.

SOURCE: Ref. [48.2]. Many of the sizes in this table up to and including 12 in are also available in aluminum alloys. Multiply the area by 0.829 to get the weight of these shapes.

48.5   WIRE AND SHEET METAL

Gauge sizes of wire and sheet metal of both ferrous and nonferrous materials are tabulated in Table 48.17. The use of SI units is simpler for such products because it is easier to express thicknesses directly in millimeters.

48.6   STRUCTURAL SHAPES

An assortment of various shapes used in structural steel works and their sizes and properties are tabulated in Tables 48.18 to 48.22. These are probably the most useful sizes for machine-design purposes, but other sizes are available or can be obtained on special order. Generally, aluminum shapes are available in a larger range of sizes, especially the smaller ones.

REFERENCES

48.1 *Steel Products Manual,* American Iron and Steel Institute, Washington, D.C.

48.2 *Manual of Steel Construction,* American Institute of Steel Construction, Inc., Chicago, Illinois.

CHAPTER 49
STRESS

Joseph E. Shigley
Professor Emeritus
The University of Michigan
Ann Arbor, Michigan

49.1 DEFINITIONS AND NOTATION

The general two-dimensional stress element in Fig. 49.1a shows two normal stresses σ_x and σ_y, both positive, and two shear stresses τ_{xy} and τ_{yx}, positive also. The element is in static equilibrium, and hence $\tau_{xy} = \tau_{yx}$. The stress state depicted by the figure is called *plane* or *biaxial stress*.

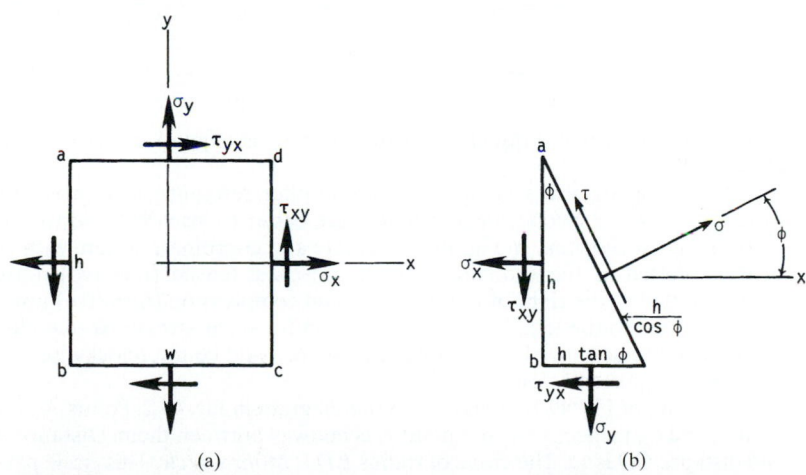

(a)   (b)

FIGURE 49.1   Notation for two-dimensional stress. (*From* Applied Mechanics of Materials, *by Joseph E. Shigley. Copyright © 1976 by McGraw-Hill, Inc. Used with permission of the McGraw-Hill Book Company.*)

Figure 49.1*b* shows an element face whose normal makes an angle ϕ to the *x* axis. It can be shown that the stress components σ and τ acting on this face are given by the equations

$$\sigma = \frac{\sigma_x + \sigma_y}{2} + \frac{\sigma_x - \sigma_y}{2} \cos 2\phi + \tau_{xy} \sin 2\phi \tag{49.1}$$

$$\tau = -\frac{\sigma_x - \sigma_y}{2} \sin 2\phi + \tau_{xy} \cos 2\phi \tag{49.2}$$

It can be shown that when the angle ϕ is varied in Eq. (49.1), the normal stress σ has two extreme values. These are called the *principal stresses*, and they are given by the equation

$$\sigma_1, \sigma_2 = \frac{\sigma_x + \sigma_y}{2} \pm \left[\left(\frac{\sigma_x - \sigma_y}{2} \right)^2 + \tau_{xy}^2 \right]^{1/2} \tag{49.3}$$

The corresponding values of ϕ are called the *principal directions*. These directions can be obtained from

$$2\phi = \tan^{-1} \frac{2\tau_{xy}}{\sigma_x - \sigma_y} \tag{49.4}$$

The shear stresses are always zero when the element is aligned in the principal directions.

It also turns out that the shear stress τ in Eq. (49.2) has two extreme values. These and the angles at which they occur may be found from

$$\tau_1, \tau_2 = \pm \left[\left(\frac{\sigma_x - \sigma_y}{2} \right)^2 + \tau_{xy}^2 \right]^{1/2} \tag{49.5}$$

$$2\phi = \tan^{-1} -\frac{\sigma_x - \sigma_y}{2\tau_{xy}} \tag{49.6}$$

The two normal stresses are equal when the element is aligned in the directions given by Eq. (49.6).

The act of referring stress components to another reference system is called *transformation of stress.* Such transformations are easier to visualize, and to solve, using a *Mohr's circle diagram.* In Fig. 49.2 we create a $\sigma\tau$ coordinate system with normal stresses plotted as the ordinates. On the abscissa, tensile (positive) normal stresses are plotted to the right of the origin *O,* and compression (negative) normal stresses are plotted to the left. The sign convention for shear stresses is that clockwise (cw) shear stresses are plotted *above* the abscissa and counterclockwise (ccw) shear stresses are plotted *below.*

The stress state of Fig. 49.1*a* is shown on the diagram in Fig. 49.2. Points *A* and *C* represent σ_x and σ_y, respectively, and point *E* is midway between them. Distance *AB* is τ_{xy} and distance *CD* is τ_{yx}. The circle of radius *ED* is *Mohr's circle.* This circle passes through the principal stresses at *F* and *G* and through the extremes of the shear stresses at *H* and *I*. It is important to observe that an extreme of the shear stress may *not* be the same as the maximum.

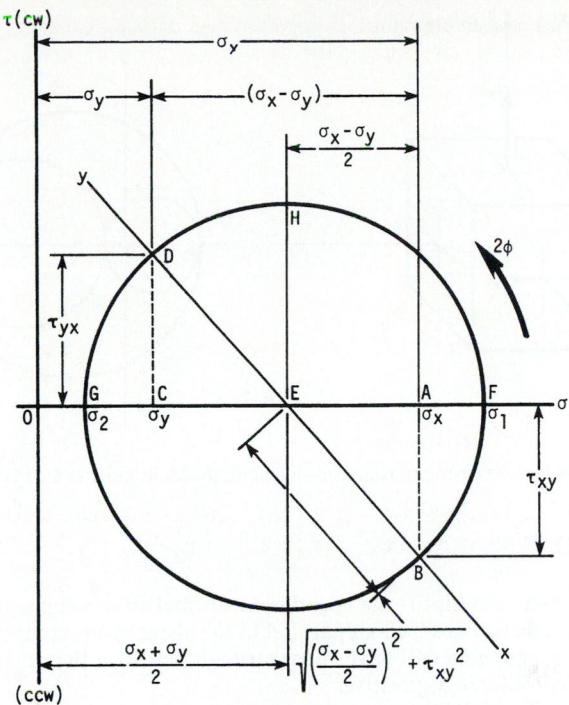

FIGURE 49.2 Mohr's circle diagram for plane stress. (*From Applied Mechanics of Materials, by Joseph E. Shigley. Copyright © 1976 by McGraw-Hill, Inc. Used with permission of the McGraw-Hill Book Company.*)

49.1.1 Programming

To program a Mohr's circle solution, plan on using a rectangular-to-polar conversion subroutine. Now notice, in Fig. 49.2, that $(\sigma_x - \sigma_y)/2$ is the base of a right triangle, τ_{xy} is the ordinate, and the hypotenuse is an extreme of the shear stress. Thus the conversion routine can be used to output both the angle 2ϕ and the extreme value of the shear stress.

As shown in Fig. 49.2, the principal stresses are found by adding and subtracting the extreme value of the shear stress to and from the term $(\sigma_x + \sigma_y)/2$. It is wise to ensure, in your programming, that the angle ϕ indicates the angle *from* the x axis *to* the direction of the stress component of interest; generally, the angle ϕ is considered positive when measured in the ccw direction.

49.2 TRIAXIAL STRESS

The general three-dimensional stress element in Fig. 49.3*a* has three normal stresses σ_x, σ_y, and σ_z, all shown as positive, and six shear-stress components, also shown as positive. The element is in static equilibrium, and hence

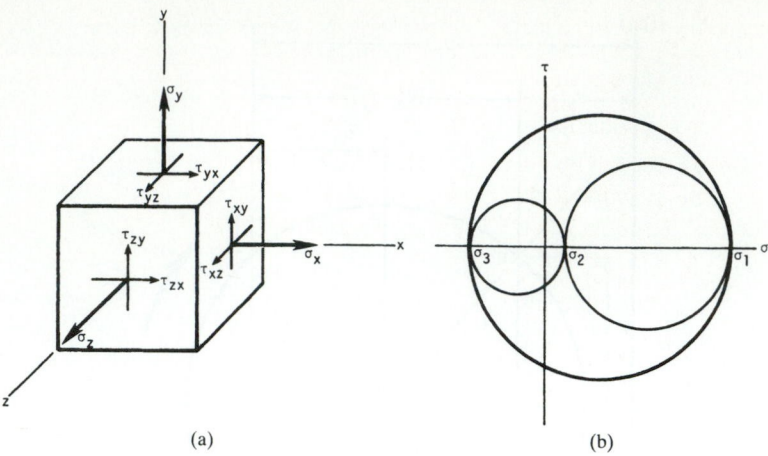

(a)   (b)

FIGURE 49.3   (*a*) General triaxial stress element; (*b*) Mohr's circles for triaxial stress.

$$\tau_{xy} = \tau_{yx} \qquad \tau_{yz} = \tau_{zy} \qquad \tau_{zx} = \tau_{xz}$$

Note that the first subscript is the coordinate normal to the element face, and the second subscript designates the axis parallel to the shear-stress component. The negative faces of the element will have shear stresses acting in the opposite direction; these are also considered as positive.

As shown in Fig. 49.3*b*, there are three principal stresses for triaxial stress states. These three are obtained from a solution of the equation

$$\sigma^3 - (\sigma_x + \sigma_y + \sigma_z)\sigma^2 + (\sigma_x\sigma_y + \sigma_x\sigma_z + \sigma_y\sigma_z - \tau_{xy}^2 - \tau_{yz}^2 - \tau_{zx}^2)\sigma$$

$$- (\sigma_x\sigma_y\sigma_z + 2\tau_{xy}\tau_{yz}\tau_{zx} - \sigma_x\tau_{yz}^2 - \sigma_y\tau_{zx}^2 - \sigma_z\tau_{xy}^2) = 0 \qquad (49.7)$$

In plotting Mohr's circles for triaxial stress, arrange the principal stresses in the order $\sigma_1 > \sigma_2 > \sigma_3$, as in Fig. 49.3*b*. It can be shown that the stress coordinates $\sigma\tau$ for any arbitrarily located plane will always lie on or *inside* the largest circle or on or *outside* the two smaller circles. The figure shows that the maximum shear stress is always

$$\tau_{max} = \frac{\sigma_1 - \sigma_3}{2} \qquad (49.8)$$

when the normal stresses are arranged so that $\sigma_1 > \sigma_2 > \sigma_3$.

49.3   STRESS-STRAIN RELATIONS

The stresses due to loading described as *pure tension, pure compression,* and *pure shear* are

$$\sigma = \frac{F}{A} \qquad \tau = \frac{F}{A} \qquad (49.9)$$

where F is positive for tension and negative for compression and the word *pure* means that there are no other complicating effects. In each case the stress is assumed to be uniform, which requires that

- The member is straight and of a homogeneous material.
- The line of action of the force is through the centroid of the section.
- There is no discontinuity or change in cross section near the stress element.
- In the case of compression, there is no possibility of buckling.

Unit engineering strain ϵ, often called simply *unit strain,* is the elongation or deformation of a member subjected to pure axial loading per unit of original length. Thus

$$\epsilon = \frac{\delta}{l_0} \tag{49.10}$$

where $\quad \delta$ = total strain
$\quad\quad l_0$ = unstressed or original length

Shear strain γ is the change in a right angle of a stress element due to pure shear.

Hooke's law states that, within certain limits, the stress in a material is proportional to the strain which produced it. Materials which regain their original shape and dimensions when a load is removed are called *elastic materials.* Hooke's law is expressed in equation form as

$$\sigma = E\epsilon \quad\quad \tau = G\gamma \tag{49.11}$$

where E = the *modulus of elasticity* and G = the *modulus of ridigity,* also called the *shear modulus of elasticity.*

Poisson demonstrated that, within the range of Hooke's law, a member subjected to uniaxial loading exhibits both an axial strain and a lateral strain. These are related to each other by the equation

$$v = -\frac{\text{lateral strain}}{\text{axial strain}} \tag{49.12}$$

where v is called *Poisson's ratio.*

The three constants given by Eqs. (49.11) and (49.12) are often called *elastic constants.* They have the relationship

$$E = 2G(1 + v) \tag{49.13}$$

By combining Eqs. (49.9), (49.10), and (49.11), it is easy to show that

$$\delta = \frac{Fl}{AE} \tag{49.14}$$

which gives the total deformation of a member subjected to axial tension or compression.

A solid round bar subjected to a *pure* twisting moment or torsion has a shear stress that is zero at the center and maximum at the surface. The appropriate equations are

$$\tau = \frac{T\rho}{J} \qquad \tau_{max} = \frac{Tr}{J} \tag{49.15}$$

where T = torque
ρ = radius to stress element
r = radius of bar
J = second moment of area (polar)

The total angle of twist of such a bar, in radians, is

$$\theta = \frac{Tl}{GJ} \tag{49.16}$$

where l = length of the bar. For the shear stress and angle of twist of other cross sections, see Table 49.1.

49.3.1 Principal Unit Strains

For a bar in uniaxial tension or compression, the principal strains are

$$\epsilon_1 = \frac{\sigma_1}{E} \qquad \epsilon_2 = -\nu\epsilon_1 \qquad \epsilon_3 = -\nu\epsilon_1 \tag{49.17}$$

Notice that the stress state is uniaxial, but the strains are triaxial.
For triaxial stress, the principal strains are

$$\epsilon_1 = \frac{\sigma_1}{E} - \frac{\nu\sigma_2}{E} - \frac{\nu\sigma_3}{E}$$

$$\epsilon_2 = \frac{\sigma_2}{E} - \frac{\nu\sigma_1}{E} - \frac{\nu\sigma_3}{E} \tag{49.18}$$

$$\epsilon_3 = \frac{\sigma_3}{E} - \frac{\nu\sigma_1}{E} - \frac{\nu\sigma_2}{E}$$

These equations can be solved for the principal stresses; the results are

$$\sigma_1 = \frac{E\epsilon_1(1-\nu) + \nu E(\epsilon_2 + \epsilon_3)}{1 - \nu - 2\nu^2}$$

$$\sigma_2 = \frac{E\epsilon_2(1-\nu) + \nu E(\epsilon_1 + \epsilon_3)}{1 - \nu - 2\nu^2} \tag{49.19}$$

$$\sigma_3 = \frac{E\epsilon_3(1-\nu) + \nu E(\epsilon_1 + \epsilon_2)}{1 - \nu - 2\nu^2}$$

The biaxial stress-strain relations can easily be obtained from Eqs. (49.18) and (49.19) by equating one of the principal stresses to zero.

TABLE 49.1 Torsional Stress and Angular Deflection of Various Sections†

Sectional shape	Shape constant	Shear stress
1. Solid round	$K = \dfrac{\pi d^4}{32}$	$\tau_{max} = \dfrac{16T}{\pi d^3}$
2. Round tube	$K = \dfrac{\pi(d_o^4 - d_i^4)}{32}$	$\tau_{max} = \dfrac{16T d_o}{\pi(d_o^4 - d_i^4)}$
3. Square [49.1]	$K = \dfrac{h^4}{7.2}$	$\tau_{max} = \dfrac{4.8T}{h^3}$

4. Square tube, generous fillets [49.2]

$$K = t(h - t)^3$$

$$\tau \cong \frac{T}{2t(h - t)^2}$$

5. Rectangle [49.1]

$h \leq b$

$$K = \frac{bh^3}{A}$$

$$A = 3 + 1.462 \frac{h}{b} + 2.976 \left(\frac{h}{b}\right)^2 - 0.238 \left(\frac{h}{b}\right)^3$$

$$\tau_{max} = \frac{T(3b + 1.8h)}{b^2 h^2}$$

6. Rectangular tube, generous fillets [49.2]

$$K = \frac{2t(b - t)^2(h - t)^2}{b + h - 2t}$$

$$\tau \cong \frac{T}{2t(b - t)(h - t)}$$

Sectional shape	Shape constant	Shear stress
7. Hexagon [49.1]	$K = \dfrac{h^4}{8.8}$	$\tau_{max} = \dfrac{5.7T}{h^3}$

† Deflection is $\theta = Tl/KG$ in rad, where T = torque, l = length, K = shape constant, and G = modulus of rigidity. See [49.2] for additional shapes in torsion.

49.3.2 Plastic Strain

It is important to observe that all the preceding relations are valid only when the material obeys Hooke's law.

Some materials (see Sec. 7.9), when stressed in the plastic region, exhibit a behavior quite similar to that given by Eq. (49.11). For these materials, the appropriate equation is

$$\overline{\sigma} = K\varepsilon^n \qquad (49.20)$$

where $\overline{\sigma}$ = true stress
  K = strength coefficient
  ε = true plastic strain
  n = strain-strengthening exponent

The relations for the true stress and true strain are

$$\overline{\sigma} = \frac{F_i}{A_i} \qquad \varepsilon = \ln \frac{l_i}{l_0} \qquad (49.21)$$

where A_i and l_i are, respectively, the instantaneous values of the area and length of a bar subjected to a load F_i. Note that the areas in Eqs. (49.9) are the original or unstressed areas; the subscript zero was omitted, as is customary. The relations between true and engineering (nominal) stresses and strains are

$$\overline{\sigma} = \sigma \exp \epsilon \qquad \varepsilon = \ln(\epsilon + 1) \qquad (49.22)$$

49.4 FLEXURE

Figure 49.4a shows a member loaded in flexure by a number of forces F and supported by reactions R_1 and R_2 at the ends. At point C a distance x from R_1, we can write

$$\Sigma M_C = \Sigma M_{\text{ext}} + M = 0 \qquad (49.23)$$

where $\Sigma M_{\text{ext}} = -xR_1 + c_1F_1 + c_2F_2$ and is called the *external moment* at section C. The term M, called the *internal* or *resisting moment,* is shown in its positive direction in both parts b and c of Fig. 49.4. Figure 49.5 shows that a positive moment causes the top surface of a beam to be concave. A negative moment causes the top surface to be convex with one or both ends curved downward.

A similar relation can be defined for shear at section C:

$$\Sigma F_y = \Sigma F_{\text{ext}} + V = 0 \qquad (49.24)$$

where $\Sigma F_{\text{ext}} = R_1 - F_1 - F_2$ and is called the *external shear force* at C. The term V, called the *internal shear force,* is shown in its positive direction in both parts b and c of Fig. 49.4.

Figure 49.6 illustrates an application of these relations to obtain a set of shear and moment diagrams.

STRESS **49.11**

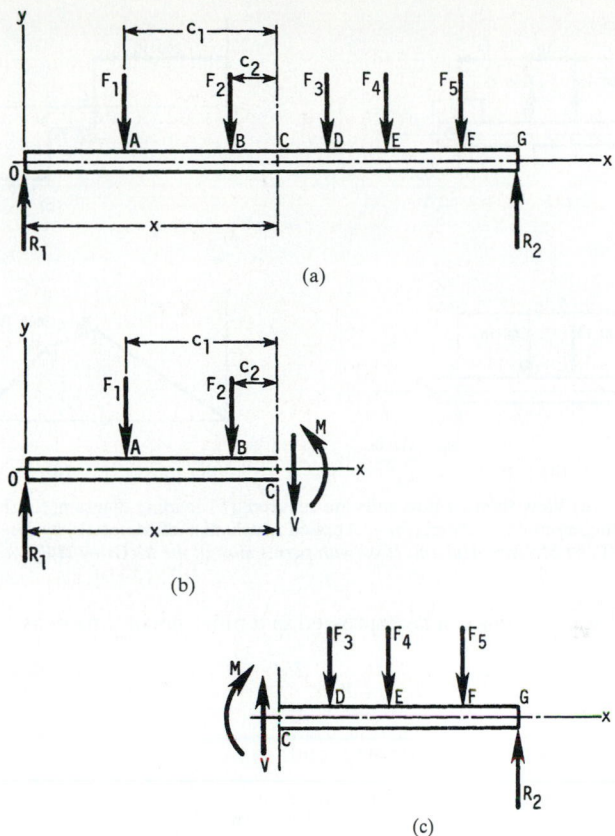

FIGURE 49.4 Shear and moment. (*From* Applied Mechanics of Materials, *by Joseph E. Shigley. Copyright © 1976 by McGraw-Hill, Inc. Used with permission of the McGraw-Hill Book Company.*)

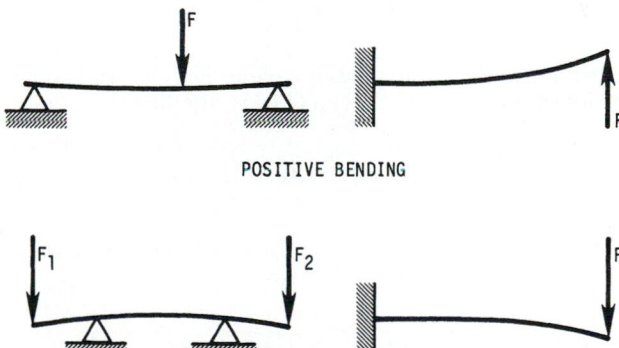

FIGURE 49.5 Sign conventions for bending. (*From* Applied Mechanics of Materials, *by Joseph E. Shigley. Copyright © 1976 by McGraw-Hill, Inc. Used with permission of the McGraw-Hill Book Company.*)

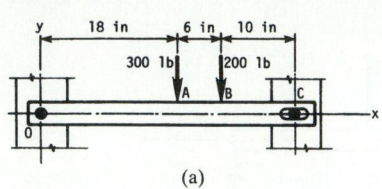

(a)

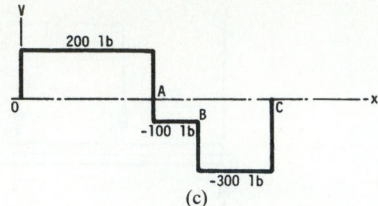

(c)

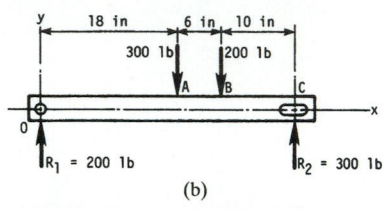

(b)

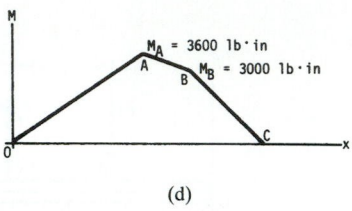

(d)

FIGURE 49.6   (*a*) View showing how ends are secured; (*b*) loading diagram; (*c*) shear-force diagram; (*d*) bending-moment diagram. (*From* Applied Mechanics of Materials, *by Joseph E. Shigley. Copyright © 1976 by McGraw-Hill, Inc. Used with permission of the McGraw-Hill Book Company.*)

The previous relations can be expressed in a more general form as

$$V = \frac{dM}{dx} \tag{49.25}$$

If the flexure is caused by a distributed load,

$$\frac{dV}{dx} = \frac{d^2M}{dx^2} = -w \tag{49.26}$$

where w = a downward-acting load in units of force per unit length. A more general load distribution can be expressed as

$$q = \lim_{\Delta x \to 0} \frac{\Delta F}{\Delta x}$$

where q is called the *load intensity;* thus $q = -w$ in Eq. (49.26). Two useful facts can be learned by integrating Eqs. (49.25) and (49.26). The first is

$$\int_{V_A}^{V_B} dV = \int_{x_A}^{x_B} q \, dx = V_B - V_A \tag{49.27}$$

which states that *the area under the loading function between* x_A *and* x_B *is the same as the change in the shear force from A to B.* Also,

$$\int_{M_A}^{M_B} dM = \int_{x_A}^{x_B} V \, dx = M_B - M_A \tag{49.28}$$

which states that *the area of the shear-force diagram between* x_A *and* x_B *is the same as the change in moment from A to B.*

Figure 49.7 distinguishes between the *neutral axis of a section* and the *neutral axis of a beam,* both of which are often referred to simply as the *neutral axis.* The assumptions used in deriving flexural relations are

- The material is isotropic and homogeneous.
- The member is straight.
- The material obeys Hooke's law.
- The cross section is constant along the length of the member.
- There is an axis of symmetry in the plane of bending (see Fig. 49.7).
- During pure bending (zero shear force), plane cross sections remain plane.

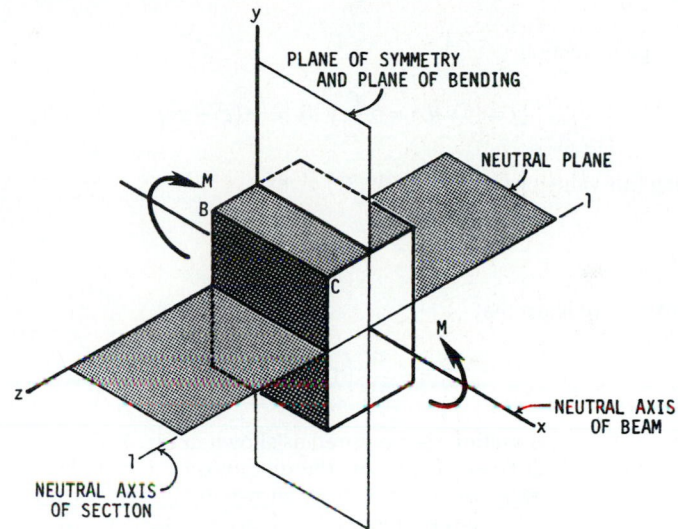

FIGURE 49.7 The meaning of the term *neutral axis.* Note the difference between the *neutral axis of the section* and the *neutral axis of the beam.* (*From Applied Mechanics of Materials, by Joseph E. Shigley. Copyright © 1976 by McGraw-Hill, Inc. Used with permission of the McGraw-Hill Book Company.*)

The *flexural formula* is

$$\sigma_x = -\frac{My}{I} \tag{49.29}$$

for the section of Fig. 49.7. The formula states that a normal compression stress σ_x occurs on a fiber at a distance y from the neutral axis when a *positive moment M* is applied. In Eq. (49.29), I is the *second moment of area.* A number of formulas are listed in Chap. 48.

The maximum flexural stress occurs at $y_{max} = c$ at the outer surface of the beam. This stress is often written in the three forms

$$\sigma = \frac{Mc}{I} \qquad \sigma = \frac{M}{I/c} \qquad \sigma = \frac{M}{Z} \tag{49.30}$$

where Z is called the *section modulus*. Equations (49.30) can also be used for beams having unsymmetrical sections provided that the plane of bending coincides with one of the two principal axes of the section.

When shear forces are present, as in Fig. 49.6c, a member in flexure will also experience shear stresses as given by the equation

$$\tau = \frac{VQ}{Ib} \tag{49.31}$$

where b = section width, and Q = first moment of a vertical face about the neutral axis and is

$$Q = \int_{y_1}^{c} y \, dA \tag{49.32}$$

For a rectangular section,

$$Q = \int_{y_1}^{c} y \, dA = b \int_{y_1}^{c} y \, dy = \frac{b}{2}(c^2 - y_1^2)$$

Substituting this value of Q into Eq. (49.31) gives

$$\tau = \frac{V}{2I}(c^2 - y_1^2)$$

Using $I = Ac^2/3$, we learn that

$$\tau = \frac{3V}{2A}\left(1 - \frac{y_1^2}{c^2}\right) \tag{49.33}$$

The value of b for other sections is measured as shown in Fig. 49.8.

In determining shear stress in a beam, the dimension b is not always measured parallel to the neutral axis. The beam sections shown in Fig. 49.8 show how to measure b in order to compute the static moment Q. It is the tendency of the shaded area to slide relative to the unshaded area which causes the shear stress.

Shear flow q is defined by the equation

$$q = \frac{VQ}{I} \tag{49.34}$$

where q is in force units per unit length of the beam at the section under consideration. So shear flow is simply the shear force per unit length at the section defined by $y = y_1$. When the shear flow is known, the shear stress is determined by the equation

$$\tau = \frac{q}{b} \tag{49.35}$$

49.5 STRESSES DUE TO TEMPERATURE

A *thermal stress* is caused by the existence of a *temperature gradient* in a member. A *temperature stress* is created in a member when it is *constrained* so as to prevent expansion or contraction due to temperature change.

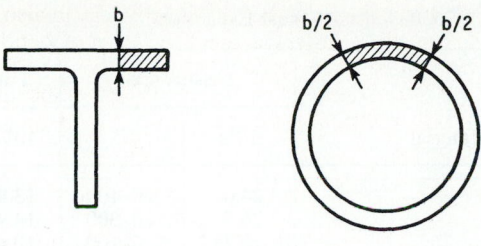

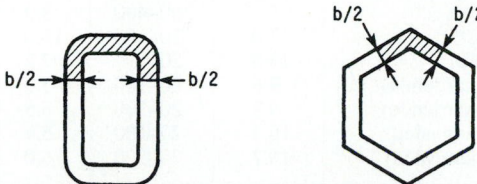

FIGURE 49.8 Correct way to measure dimension *b* to determine shear stress for various sections. (*From* Applied Mechanics of Materials, *by Joseph E. Shigley. Copyright ©* 1976 by McGraw-Hill, Inc. Used with permission of the McGraw-Hill Book Company.)

49.5.1 Temperature Stresses

These stresses are found by assuming that the member is not constrained and then computing the stresses required to cause it to assume its original dimensions. If the temperature of an unrestrained member is uniformly increased, the member expands and the normal strain is

$$\epsilon_x = \epsilon_y = \epsilon_z = \alpha(\Delta T) \tag{49.36}$$

where ΔT = temperature change and α = *coefficient of linear expansion*. The coefficient of linear expansion increases to some extent with temperature. Some mean values for various materials are shown in Table 49.2.

Figure 49.9 illustrates two examples of temperature stresses. For the bar in Fig. 49.9*a*,

$$\sigma_x = -\alpha(\Delta T)E \qquad \sigma_y = \sigma_z = -\nu\sigma_x \tag{49.37}$$

The stresses in the flat plate of Fig. 49.9*b* are

$$\sigma_x = \sigma_y = -\frac{\alpha(\Delta T)E}{1 - \nu} \qquad \sigma_z = -\nu\sigma_x \tag{49.38}$$

TABLE 49.2 Coefficients of Linear Expansion

Material	Celsius scale		Fahrenheit scale	
	$10^6\alpha$	°C	$10^6\alpha$	°F
Aluminum	24.0	20–100	13.4	68–212
Aluminum	26.7	20–300	14.9	68–572
Brass (cast)	18.75	0–100	10.4	32–212
Brass (wire)	19.3	0–100	10.7	32–212
Brass (spring)	19.8	25–300	11.0	77–572
Cast iron	10.6	40	5.9	104
Carbon steel	10.8	40	6.0	104
Carbon steel	11.5	100–200	6.4	212–392
Carbon steel	15	300–400	8.3	572–752
Magnesium (cast)	27.0	20–100	15.0	68–212
Nickel steel (10%)	13.0	20	7.2	68
Stainless steel (hardened)	9.6	20–100	5.3	68–212
Stainless steel (hardened)	9.8	20–200	5.5	68–392
Stainless steel (annealed)	10.3	20–100	5.7	68–212
Stainless steel (annealed)	10.7	20–200	6.0	68–392

49.5.2 Thermal Stresses

Heating of the top surface of the restrained member in Fig. 49.10a causes end moments of

$$M = \frac{\alpha(\Delta T)EI}{h} \qquad (49.39)$$

and maximum bending stresses of

$$\sigma_x = \pm\frac{\alpha(\Delta T)E}{2} \qquad (49.40)$$

with compression of the top surface. If the constraints are removed, the bar will curve to a radius

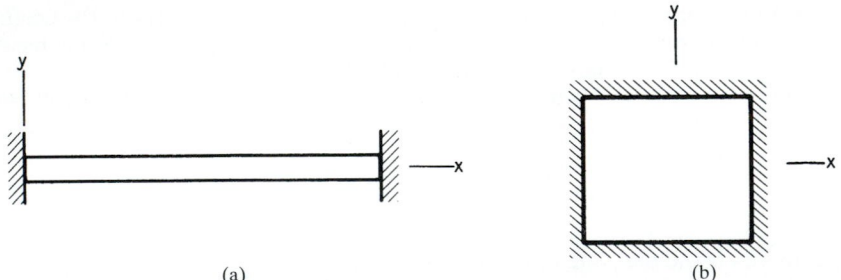

(a)　　　　　　　　　　　　(b)

FIGURE 49.9 Examples of temperature stresses. In each case the temperature rise ΔT is uniform throughout. (a) Straight bar with ends restrained; (b) flat plate with edges restrained.

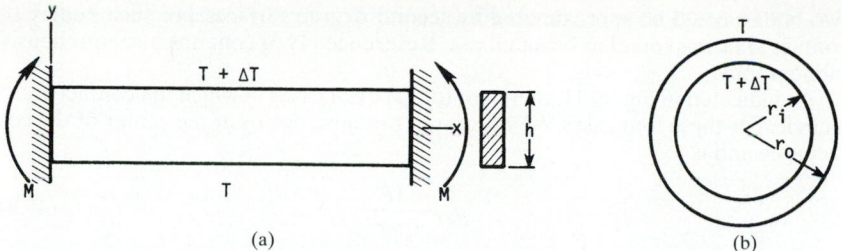

FIGURE 49.10 Examples of thermal stresses. (*a*) Rectangular member with ends restrained (temperature difference between top and bottom results in end moments and bending stresses); (*b*) thick-walled tube has maximum stresses in tangential and longitudinal directions.

$$r = \frac{h}{\alpha(\Delta T)}$$

The thick-walled tube of Fig. 49.10*b* with a hot interior surface has tangential and longitudinal stresses in the outer and inner surfaces of magnitude

$$\sigma_{lo} = \sigma_{to} = \frac{\alpha(\Delta T)E}{2(1-\nu)\,\ln(r_o/r_i)}\left[1 - \frac{2r_i^2\ln(r_o/r_i)}{r_o^2 - r_i^2}\right] \tag{49.41}$$

$$\sigma_{li} = \sigma_{ti} = \frac{-\alpha(\Delta T)E}{2(1-\nu)\,\ln(r_o/r_i)}\left[1 - \frac{2r_o^2\ln(r_o/r_i)}{r_o^2 - r_i^2}\right] \tag{49.42}$$

where the subscripts i and o refer to the inner and outer radii, respectively, and the subscripts t and l refer to the tangential (circumferential) and longitudinal directions. Radial stresses of lesser magnitude will also exist, although not at the inner or outer surfaces.

If the tubing of Fig. 49.10*b* is thin, then the inner and outer stresses are equal, although opposite, and are

$$\sigma_{lo} = \sigma_{to} = \frac{\alpha(\Delta T)E}{2(1-\nu)}$$

$$\sigma_{li} = \sigma_{ti} = -\frac{\alpha(\Delta T)E}{2(1-\nu)} \tag{49.43}$$

at points not too close to the tube ends.

49.6   CONTACT STRESSES

When two elastic bodies having curved surfaces are pressed against each other, the initial point or line of contact changes into area contact, because of the deformation, and a three-dimensional state of stress is induced in both bodies. The shape of the contact area was originally deduced by Hertz, who assumed that the curvature of the two bodies could be approximated by second-degree surfaces. For such bodies, the contact area was found to be an ellipse. Reference [49.3] contains a comprehensive bibliography.

As indicated in Fig. 49.11, there are four special cases in which the contact area is a circle. For these four cases, the maximum pressure occurs at the center of the contact area and is

$$p_o = \frac{3F}{2\pi a^2} \tag{49.44}$$

where a = the radius of the contact area and F = the normal force pressing the two bodies together.

In Fig. 49.11, the x and y axes are in the plane of the contact area and the z axis is normal to this plane. The maximum stresses occur on this axis, they are principal stresses, and their values for all four cases in Fig. 49.11 are

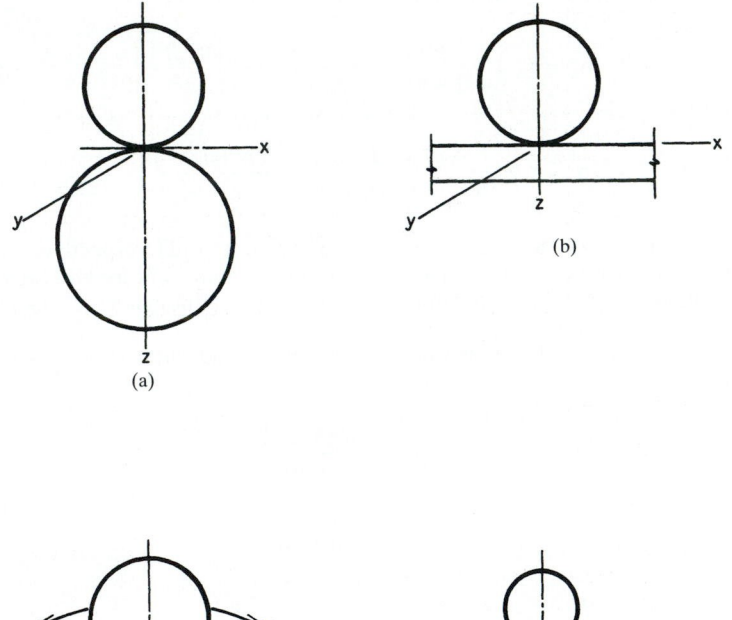

FIGURE 49.11 Contacting bodies having a circular contact area. (*a*) Two spheres; (*b*) sphere and plate; (*c*) sphere and spherical socket; (*d*) crossed cylinders of equal diameters.

$$\sigma_x = \sigma_y = -p_o \left\{ \left(1 - \frac{z}{a} \tan^{-1} \frac{1}{z/a} \right)(1+\nu) - \frac{1}{2(1+z^2/a^2)} \right\} \qquad (49.45)$$

$$\sigma_z = \frac{-p_o}{1 + z^2/a^2} \qquad (49.46)$$

These equations are plotted in Fig. 49.12 together with the two shear stresses τ_{xz} and τ_{yz}. Note that $\tau_{xy} = 0$ because $\sigma_x = \sigma_y$.

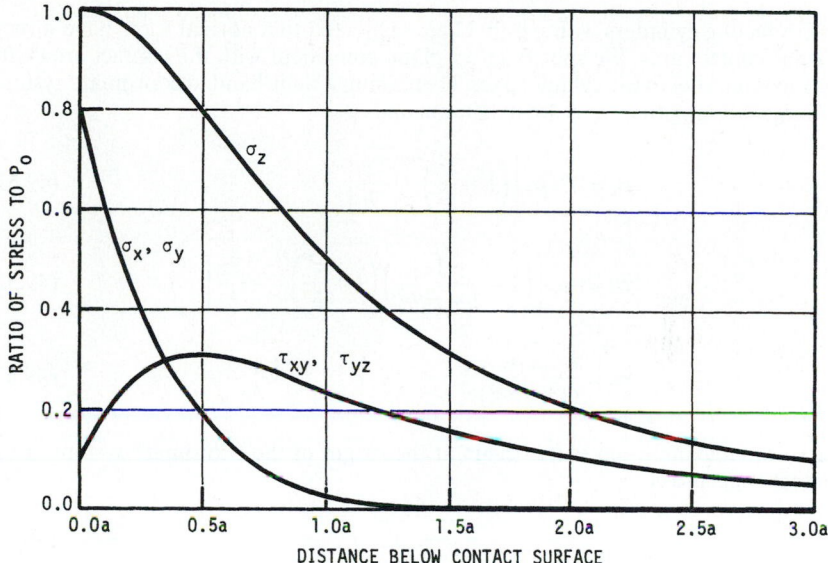

FIGURE 49.12 Magnitude of the stress components on the z axis below the surface as a function of the maximum pressure. Note that the two shear-stress components are maximum slightly below the surface. The chart is based on a Poisson's ratio of 0.30.

The radii a of the contact circles depend on the geometry of the contacting bodies. For two spheres, each having the same diameter d, or for two crossed cylinders, each having the diameter d, and in each case with like materials, the radius is

$$a = \left(\frac{3Fd}{8} \frac{1-\nu^2}{E} \right)^{1/3} \qquad (49.47)$$

where ν and E are the elastic constants.

For two spheres of unlike materials having diameters d_1 and d_2, the radius is

$$a = \left[\frac{3F}{8} \frac{d_1 d_2}{d_1 + d_2} \left(\frac{1-\nu_1^2}{E_1} + \frac{1-\nu_2^2}{E_2} \right) \right]^{1/3} \qquad (49.48)$$

For a sphere of diameter d and a flat plate of unlike materials, the radius is

$$a = \left[\frac{3Fd}{8} \left(\frac{1 - v_1^2}{E_1} + \frac{1 - v_2^2}{E_2} \right) \right]^{1/3} \tag{49.49}$$

For a sphere of diameter d_1 and a spherical socket of diameter d_2 of unlike materials, the radius is

$$a = \left[\frac{3F}{8} \frac{d_1 d_2}{d_2 - d_1} \left(\frac{1 - v_1^2}{E_1} + \frac{1 - v_2^2}{E_2} \right) \right]^{1/3} \tag{49.50}$$

Contacting cylinders with parallel axes subjected to a normal force have a rectangular contact area. We specify an xy plane coincident with the contact area with the x axis parallel to the cylinder axes. Then, using a right-handed coordinate system, the stresses along the z axis are maximum and are

$$\sigma_x = -2vp_o \left[\left(1 + \frac{z^2}{b^2} \right)^{1/2} - \frac{z}{b} \right] \tag{49.51}$$

$$\sigma_y = -p_o \left[\left(2 - \frac{1}{1 + z^2/b^2} \right) \left(1 + \frac{z^2}{b^2} \right)^{1/2} - \frac{2z}{b} \right] \tag{49.52}$$

$$\sigma_z = \frac{-p_o}{(1 + z^2/b^2)^{1/2}} \tag{49.53}$$

where the maximum pressure occurs at the origin of the coordinate system in the contact zone and is

$$p_o = \frac{2F}{\pi b l} \tag{49.54}$$

where l = the length of the contact zone measured parallel to the cylinder axes, and b = the half width. Equations (49.51) to (49.53) give the principal stresses. These equations are plotted in Fig. 49.13. The corresponding shear stresses can be found from a Mohr's circle; they are plotted in Fig. 49.14. Note that the maximum is either τ_{xz} or τ_{yz} depending on the depth below the contact surface.

The half width b depends on the geometry of the contacting cylinders. The following cases arise most frequently: Two cylinders of equal diameter and of the same material have a half width of

$$b = \left(\frac{2Fd}{\pi l} \frac{1 - v^2}{E} \right)^{1/2} \tag{49.55}$$

For two cylinders of unequal diameter and unlike materials, the half width is

$$b = \left[\frac{2F}{\pi l} \frac{d_1 d_2}{d_1 + d_2} \left(\frac{1 - v_1^2}{E_1} + \frac{1 - v_2^2}{E_2} \right) \right]^{1/2} \tag{49.56}$$

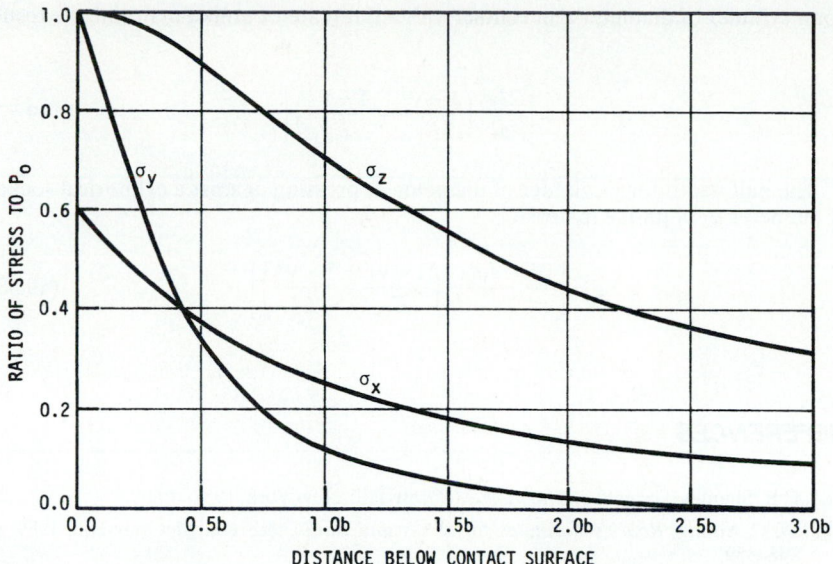

FIGURE 49.13 Magnitude of the principal stresses on the z axis below the surface as a function of the maximum pressure for contacting cylinders. Based on a Poisson's ratio of 0.30.

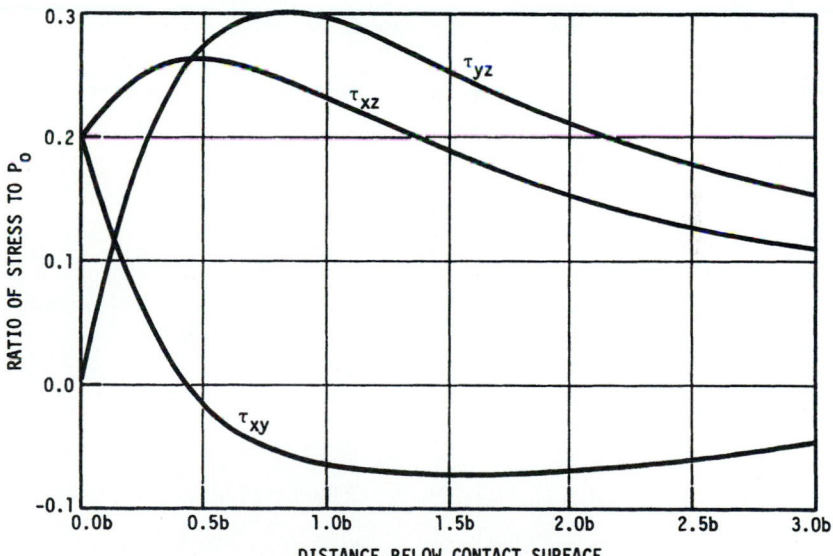

FIGURE 49.14 Magnitude of the three shear stresses computed from Fig. 49.13.

For a cylinder of diameter d in contact with a flat plate of unlike material, the result is

$$b = \left[\frac{2Fd}{\pi l} \left(\frac{1 - v_1^2}{E_1} + \frac{1 - v_2^2}{E_2} \right) \right]^{1/2} \tag{49.57}$$

The half width for a cylinder of diameter d_1 pressing against a cylindrical socket of diameter d_2 of unlike material is

$$b = \left[\frac{2F}{\pi l} \frac{d_1 d_2}{d_2 - d_1} \left(\frac{1 - v_1^2}{E_1} + \frac{1 - v_2^2}{E_2} \right) \right]^{1/2} \tag{49.58}$$

REFERENCES

49.1 F. R. Shanley, *Strength of Materials,* McGraw-Hill, New York, 1957, p. 509.

49.2 W. C. Young, *Roark's Formulas for Stress and Strain,* 6th ed., McGraw-Hill, 1989, p. 348–359.

49.3 J. L. Lubkin, "Contact Problems," in W. Flugge (ed.), *Handbook of Engineering Mechanics,* McGraw-Hill, New York, 1962, pp. 42-10 to 42-12.

CHAPTER 50
DEFLECTION

Joseph E. Shigley
Professor Emeritus
The University of Michigan
Ann Arbor, Michigan

Charles R. Mischke, Ph.D., P.E.
Professor Emeritus of Mechanical Engineering
Iowa State University
Ames, Iowa

GLOSSARY OF SYMBOLS

a	Dimension
A	Area
b	Dimension
C	Constant
D, d	Diameter
E	Young's modulus
F	Force
G	Shear modulus
I	Second moment of area
J	Second polar moment of area
k	Spring rate
K	Constant
ℓ	Length
M	Moment
$M(I)$	Moment relation, $(M/EI)_i$
N	Number

q	Unit load
Q	Fictitious force
R	Support reaction
T	Torque
U	Strain energy
V	Shear force
w	Unit weight
W	Total weight
x	Coordinate
y	Coordinate
δ	Deflection
θ	Slope, torsional deflection
ϕ	An integral
ψ	An integral

50.1 STIFFNESS OR SPRING RATE

The *spring rate* (also called *stiffness* or *scale*) of a body or ensemble of bodies is defined as the partial derivative of force (torque) with respect to colinear displacement (rotation). For a helical tension or compression spring,

$$F = \frac{d^4 G y}{8 D^3 N} \qquad \text{thus} \qquad k = \frac{\partial F}{\partial y} = \frac{d^4 G}{8 D^3 N} \tag{50.1}$$

where D = mean coil diameter
   d = wire diameter
   N = number of active turns

In a round bar subject to torsion,

$$T = \frac{GJ\theta}{\ell} \qquad \text{thus} \qquad k = \frac{\partial T}{\partial \theta} = \frac{GJ}{\ell} \tag{50.2}$$

and the tensile force in an elongating bar of any cross section is

$$F = \frac{AE\delta}{\ell} \qquad \text{thus} \qquad k = \frac{\partial F}{\partial \delta} = \frac{AE}{\ell} \tag{50.3}$$

If k is constant, as in these cases, then displacement is said to be linear with respect to force (torque). For contacting bodies with all four radii of curvature finite, the approach of the bodies is proportional to load to the two-thirds power, making the spring rate proportional to load to the one-third power. In hydrodynamic film bearings, the partial derivative would be evaluated numerically by dividing a small change in load by the displacement in the direction of the load.

50.2 DEFLECTION DUE TO BENDING

The relations involved in the bending of beams are well known and are given here
for reference purposes as follows:

$$\frac{q}{EI} = \frac{d^4y}{dx^4} \tag{50.4}$$

$$\frac{V}{EI} = \frac{d^3y}{dx^3} \tag{50.5}$$

$$\frac{M}{EI} = \frac{d^2y}{dx^2} \tag{50.6}$$

$$\theta = \frac{dy}{dx} \tag{50.7}$$

$$y = f(x) \tag{50.8}$$

These relations are illustrated by the beam of Fig. 50.1. Note that the x axis is *positive* to the right and the y axis is *positive* upward. All quantities—loading, shear
force, support reactions, moment, slope, and deflection—have the same sense as y;
they are positive if upward, negative if downward.

50.3 PROPERTIES OF BEAMS

Table 50.1 lists a number of useful properties of beams having a variety of loadings.
These must all have the same cross section throughout the length, and a linear relation must exist between the force and the deflection. Beams having other loadings
can be solved using two or more sets of these relations and the principle of superposition.

In using Table 50.1, remember that the deflection at the center of a beam with off-
center loads is usually within 2.5 percent of the maximum value.

50.4 COMPUTER ANALYSIS

In this section we will develop a computer method using numerical analysis to determine the slope and deflection of any simply supported beam having a variety of concentrated loads, including point couples, with any number of step changes in cross
section. The method is particularly applicable to stepped shafts where the transverse
bending deflections and neutral-axis slopes are desired at specified points.

The method uses numerical analysis to integrate Eq. (50.6) twice in a marching
method. The first integration uses the trapezoidal rule; the second uses Simpson's
rule (see Sec. 4.6). The procedure gives exact results.

Let us define the two successive integrals as

$$\phi = \int_0^x \frac{M}{EI}\, dx \qquad \psi = \int_0^x \phi\, dx \tag{50.9}$$

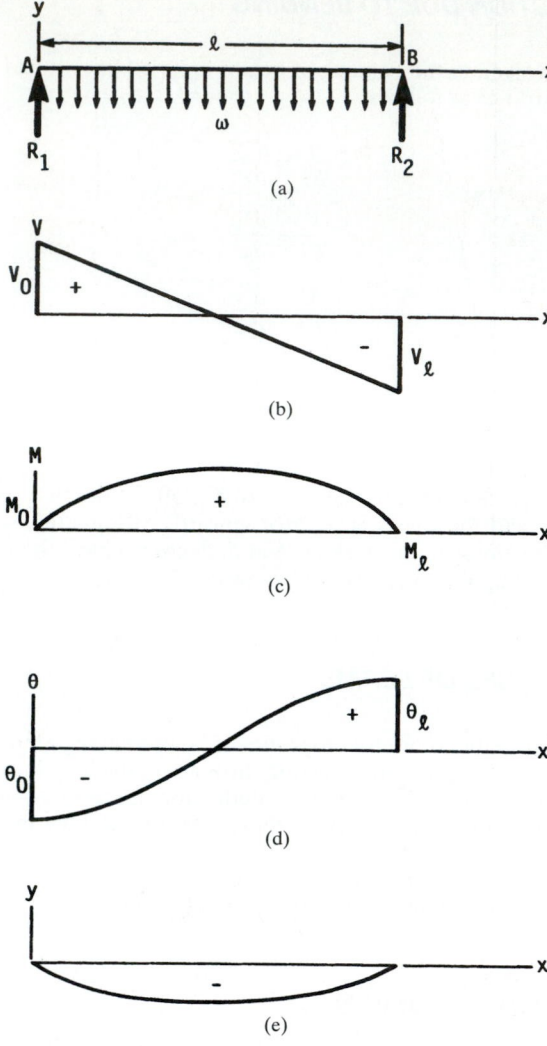

FIGURE 50.1 (*a*) Loading diagram showing beam supported at *A* and *B* with uniform load *w* having units of force per unit length, $R_1 = R_2 = w\ell/2$; (*b*) shear-force diagram showing end conditions; (*c*) moment diagram; (*d*) slope diagram; (*e*) deflection diagram.

But from Eq. (50.7), the slope is

$$\theta = \frac{dy}{dx} = \int_0^x \frac{M}{EI}\, dx + C_1$$

$$= \phi + C_1 \qquad\qquad (a)$$

TABLE 50.1 Properties of Beams

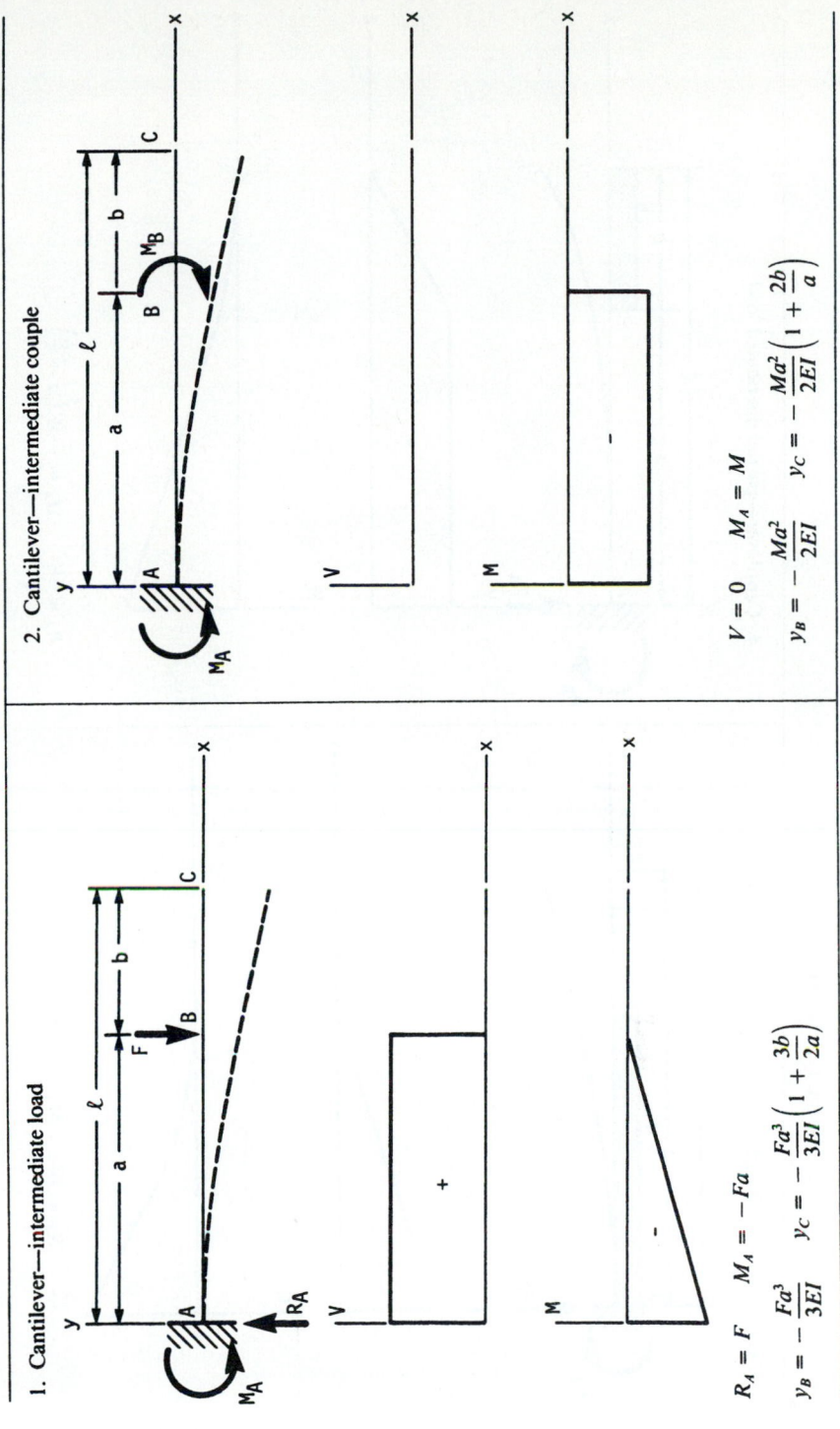

1. Cantilever—intermediate load

$$R_A = F \qquad M_A = -Fa$$

$$y_B = -\frac{Fa^3}{3EI} \qquad y_C = -\frac{Fa^3}{3EI}\left(1 + \frac{3b}{2a}\right)$$

2. Cantilever—intermediate couple

$$V = 0 \qquad M_A = M$$

$$y_B = -\frac{Ma^2}{2EI} \qquad y_C = -\frac{Ma^2}{2EI}\left(1 + \frac{2b}{a}\right)$$

TABLE 50.1 Properties of Beams (*Continued*)

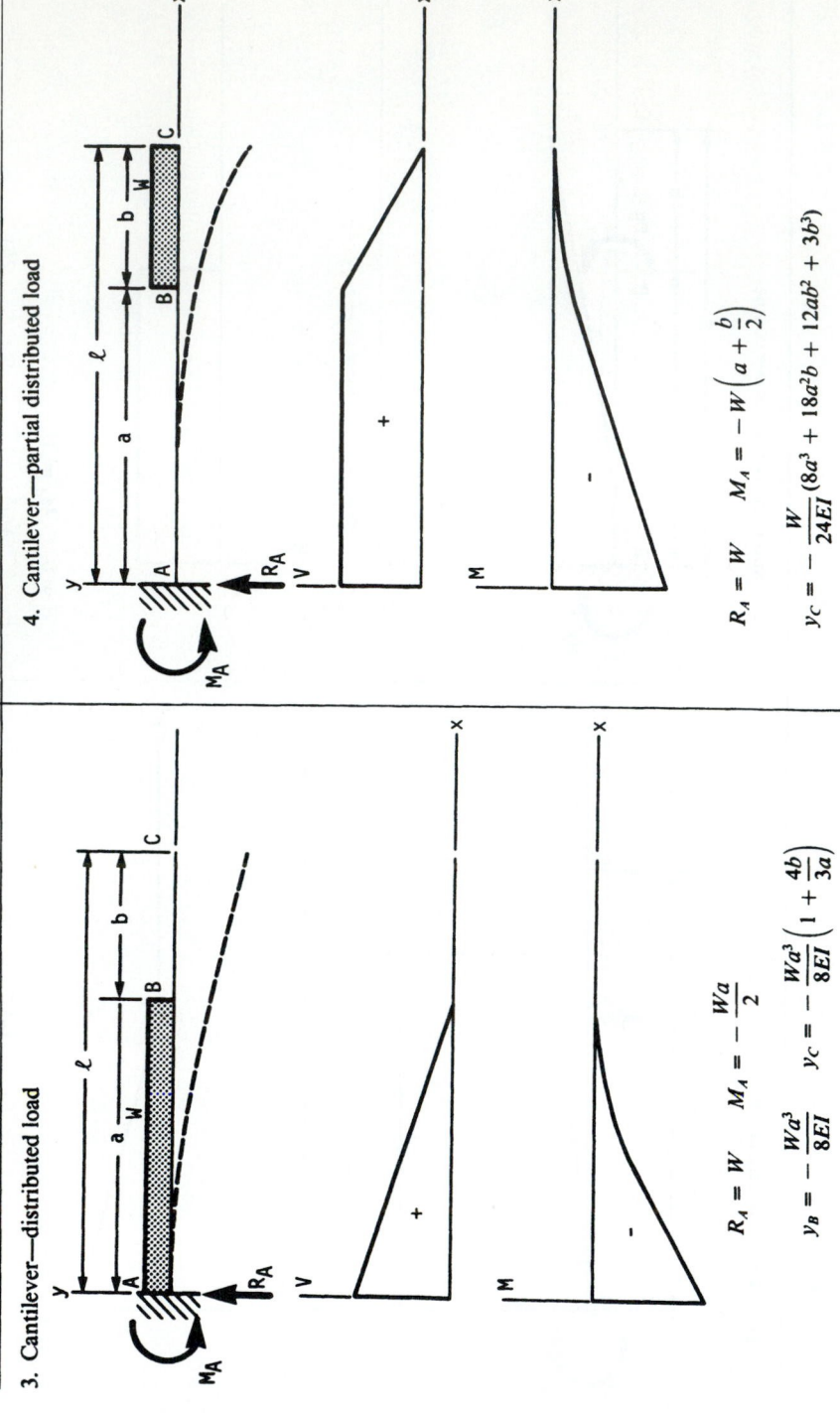

3. Cantilever—distributed load

$$R_A = W \qquad M_A = -\frac{Wa}{2}$$

$$y_B = -\frac{Wa^3}{8EI} \qquad y_C = -\frac{Wa^3}{8EI}\left(1 + \frac{4b}{3a}\right)$$

4. Cantilever—partial distributed load

$$R_A = W \qquad M_A = -W\left(a + \frac{b}{2}\right)$$

$$y_C = -\frac{W}{24EI}(8a^3 + 18a^2b + 12ab^2 + 3b^3)$$

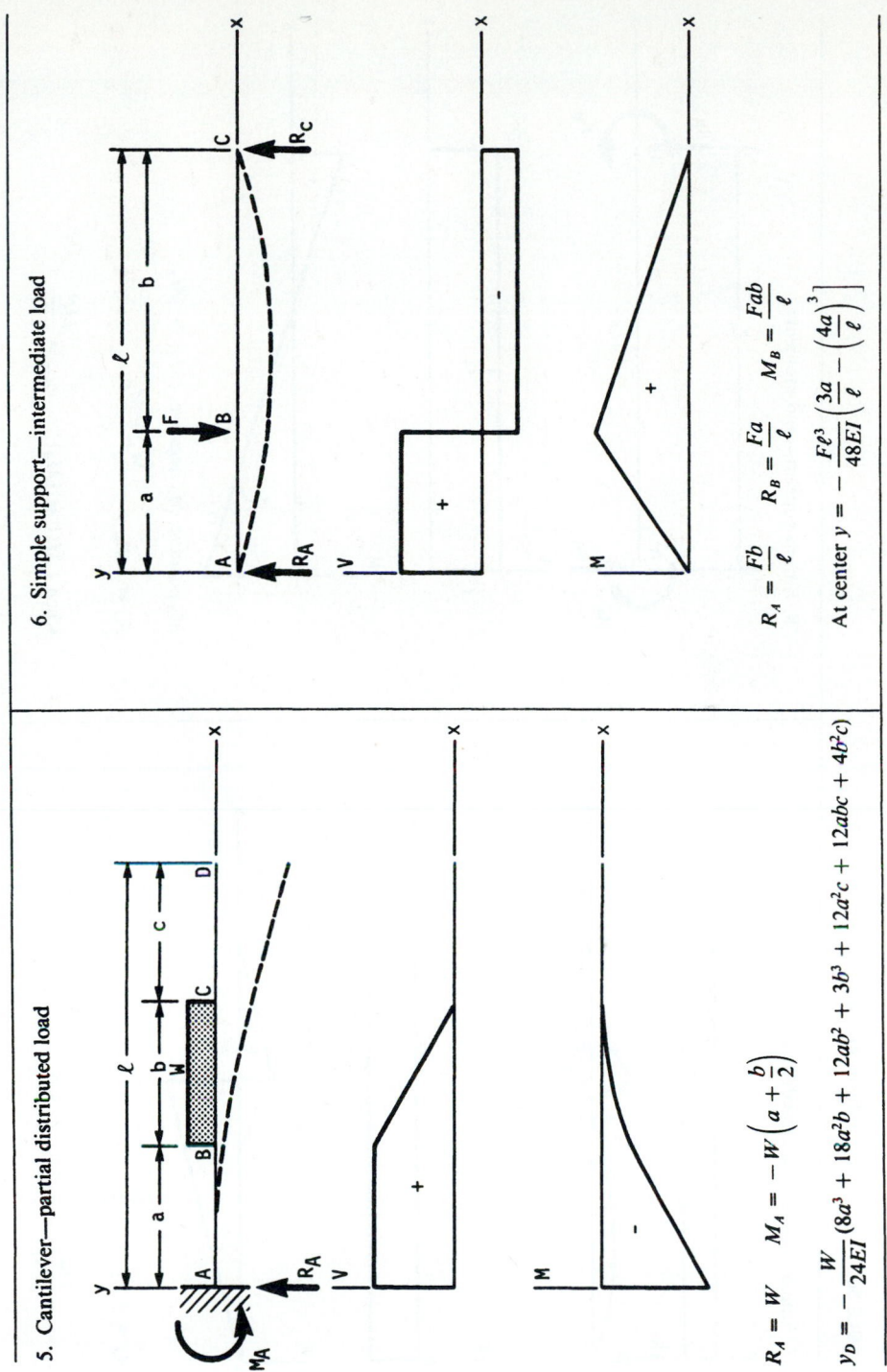

5. Cantilever—partial distributed load

$$R_A = W \qquad M_A = -W\left(a + \frac{b}{2}\right)$$

$$y_D = -\frac{W}{24EI}(8a^3 + 12a^2b + 18a^2b + 12ab^2 + 3b^3 + 12a^2c + 12abc + 4b^2c)$$

6. Simple support—intermediate load

$$R_A = \frac{Fb}{\ell} \qquad R_B = \frac{Fa}{\ell} \qquad M_B = \frac{Fab}{\ell}$$

$$\text{At center } y = -\frac{F\ell^3}{48EI}\left(\frac{3a}{\ell} - \left(\frac{4a}{\ell}\right)^3\right)$$

50.7

TABLE 50.1 Properties of Beams (*Continued*)

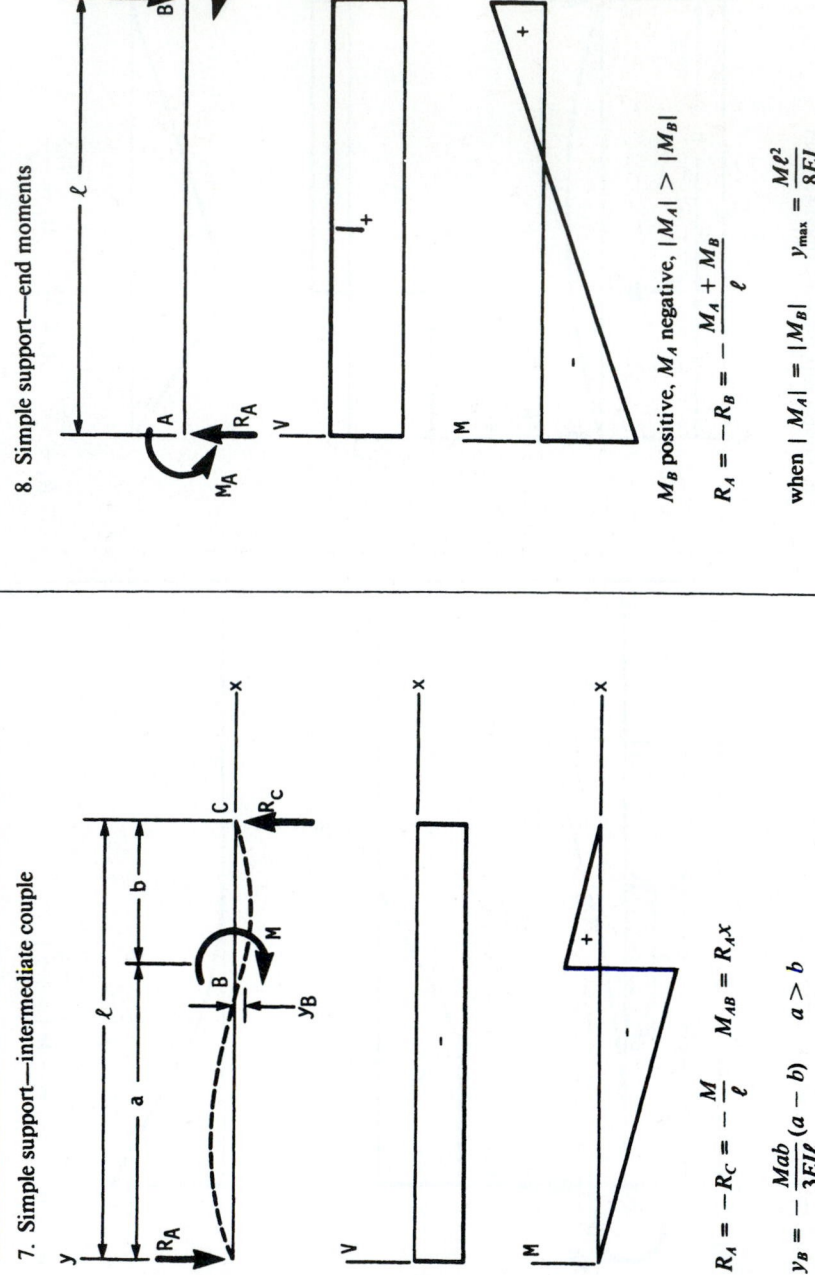

7. Simple support—intermediate couple

$$R_A = -R_C = -\frac{M}{\ell} \quad M_{AB} = R_A x$$

$$y_B = -\frac{Mab}{3EI\ell}(a-b) \quad a > b$$

8. Simple support—end moments

M_B positive, M_A negative, $|M_A| > |M_B|$

$$R_A = -R_B = -\frac{M_A + M_B}{\ell}$$

when $|M_A| = |M_B| \quad y_{max} = \frac{M\ell^2}{8EI}$

50.8

9. Simple support—overhung load

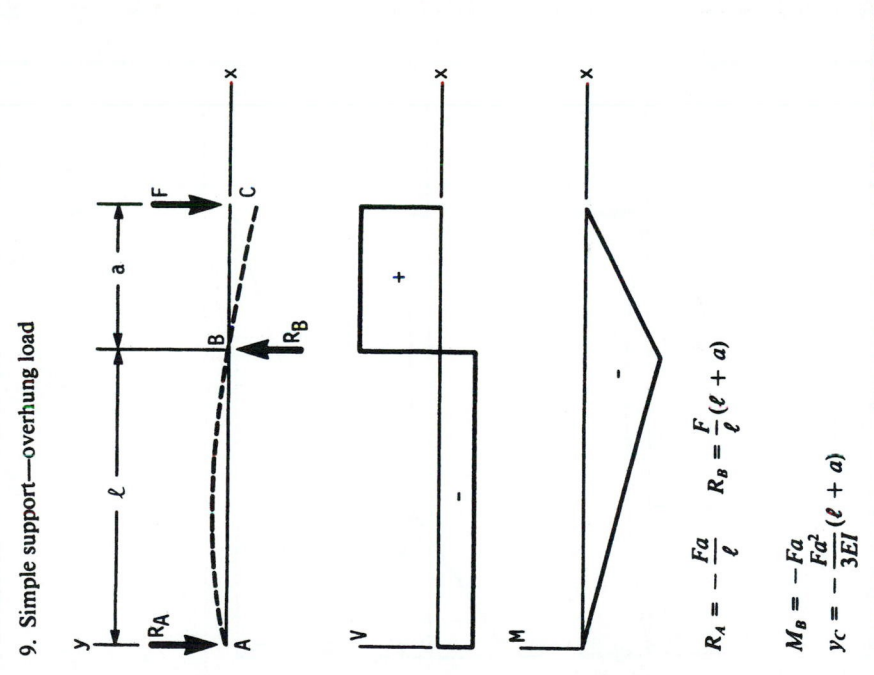

$$R_A = -\frac{Fa}{\ell} \qquad R_B = \frac{F}{\ell}(\ell + a)$$

$$M_B = -Fa$$

$$y_C = -\frac{Fa^2}{3EI}(\ell + a)$$

10. Simple support— uniform loading

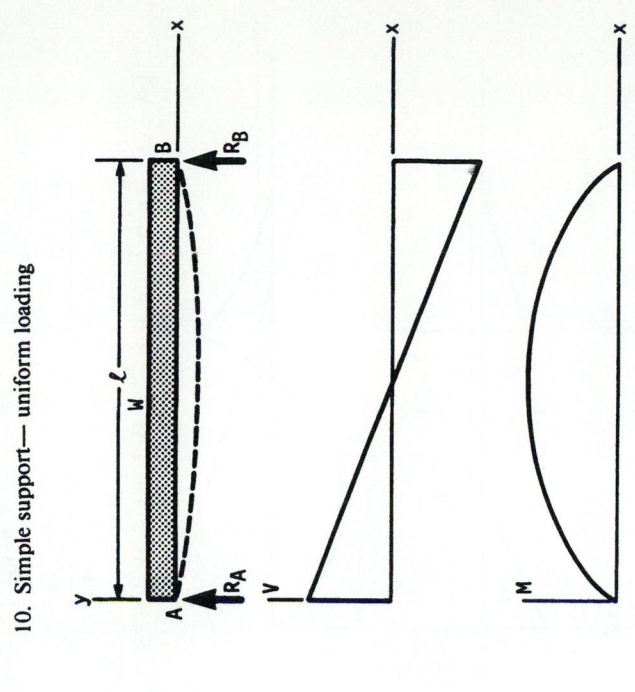
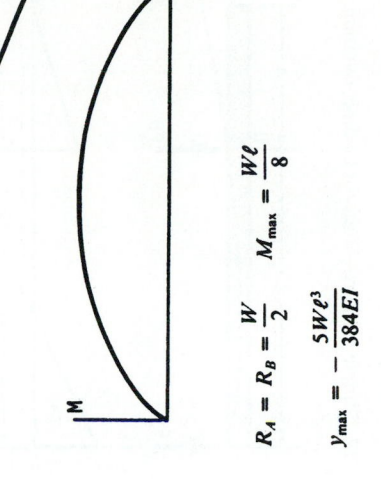

$$R_A = R_B = \frac{W}{2} \qquad M_{max} = \frac{W\ell}{8}$$

$$y_{max} = -\frac{5W\ell^3}{384EI}$$

TABLE 50.1 Properties of Beams (*Continued*)

11. Simple support—partial uniform loading

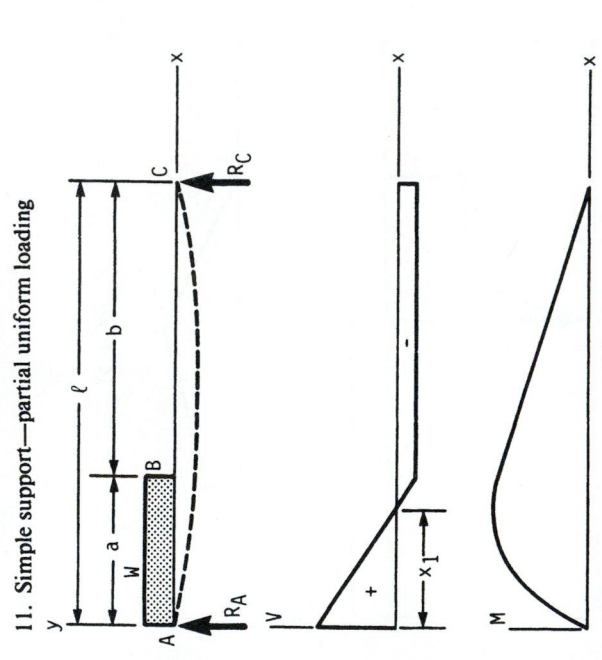

$$R_A = \frac{W}{2\ell}(2\ell - a) \qquad R_B = \frac{Wa}{2\ell} \qquad x_1 = \frac{a}{2\ell}(2\ell - a) \qquad a < \frac{\ell}{2}$$

$$\text{At center } y = -\frac{Wa}{48EI}(a^2 + 2\ell^2)$$

12. Simple support—uniform loading, overhung

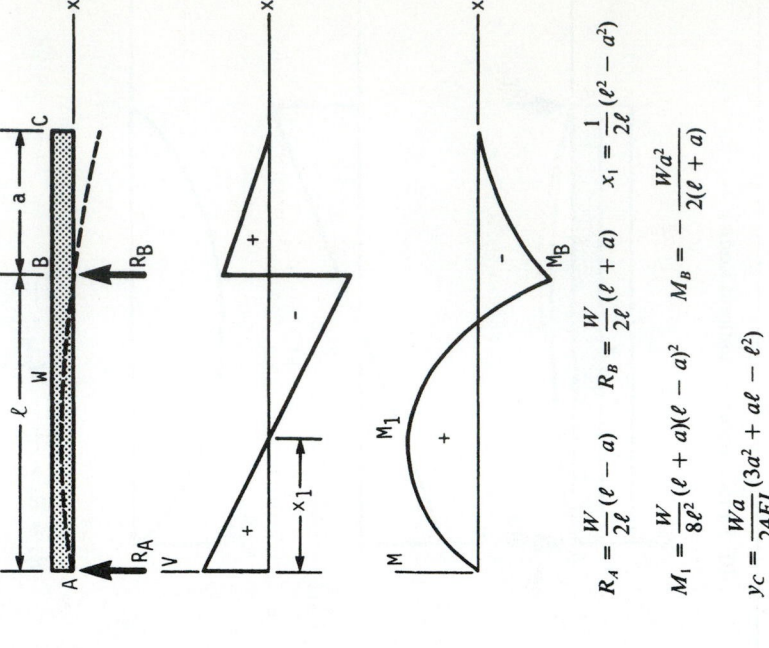

$$R_A = \frac{W}{2\ell}(\ell - a) \qquad R_B = \frac{W}{2\ell}(\ell + a) \qquad x_1 = \frac{1}{2\ell}(\ell^2 - a^2)$$

$$M_1 = \frac{W}{8\ell^2}(\ell + a)(\ell - a)^2 \qquad M_B = -\frac{Wa^2}{2(\ell + a)}$$

$$y_C = \frac{Wa}{24EI}(3a^2 + a\ell - \ell^2)$$

14. Fixed and simple support—intermediate load

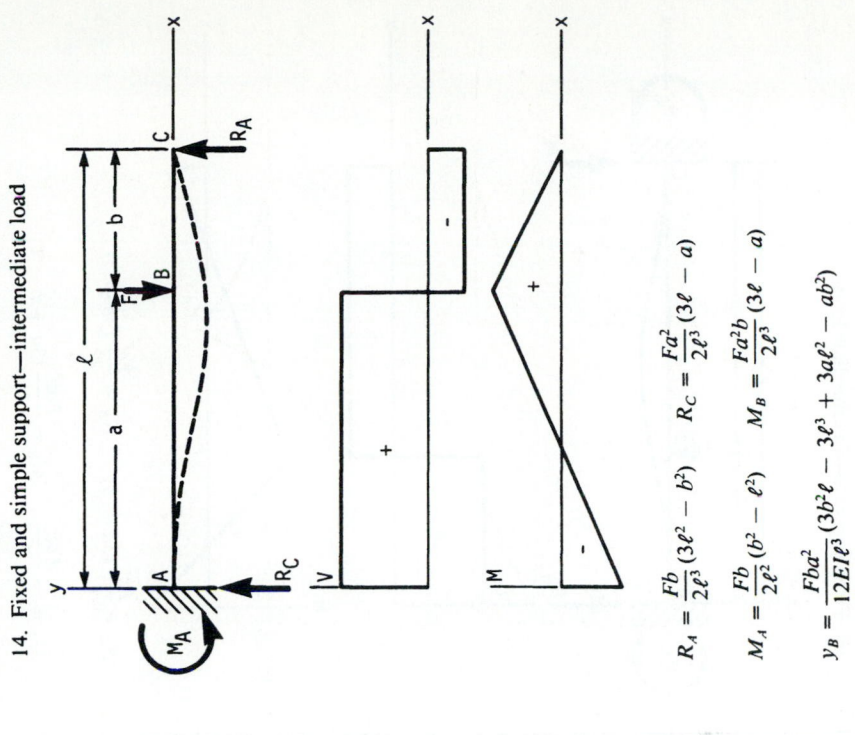

$$R_A = \frac{Fb}{2\ell^3}(3\ell^2 - b^2) \qquad R_C = \frac{Fa^2}{2\ell^3}(3\ell - a)$$

$$M_A = \frac{Fb}{2\ell^2}(b^2 - \ell^2) \qquad M_B = \frac{Fa^2b}{2\ell^3}(3\ell - a)$$

$$y_B = \frac{Fba^2}{12EI\ell^3}(3b^2\ell - 3\ell^3 + 3a\ell^2 - ab^2)$$

13. Simple support—overhung uniform load

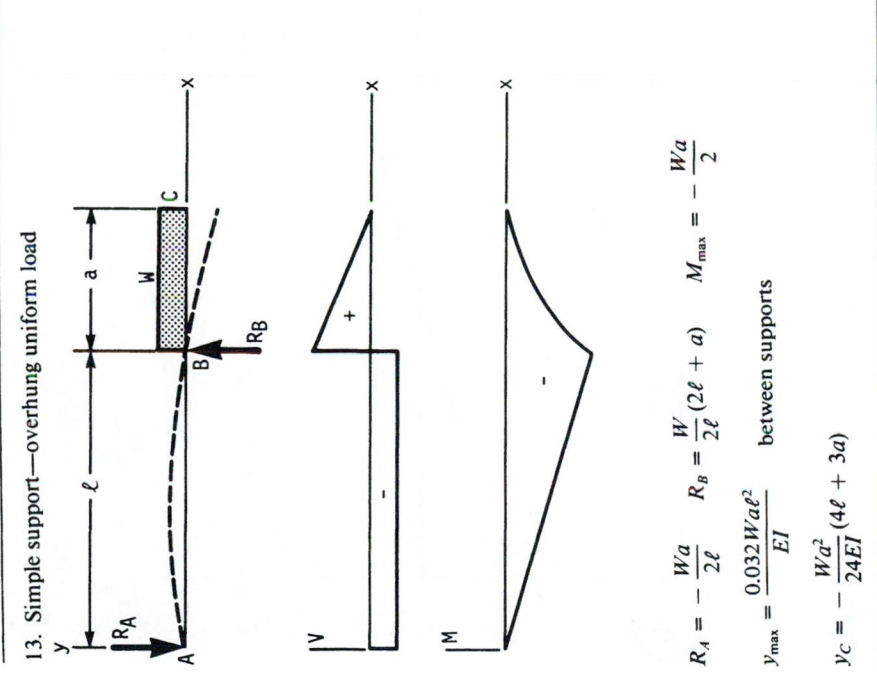

$$R_A = -\frac{Wa}{2\ell} \qquad R_B = \frac{W}{2\ell}(2\ell + a) \qquad M_{max} = -\frac{Wa}{2}$$

between supports

$$y_{max} = \frac{0.032Wa\ell^2}{EI}$$

$$y_C = -\frac{Wa^2}{24EI}(4\ell + 3a)$$

TABLE 50.1 Properties of Beams (*Continued*)

16. Fixed supports—intermediate load

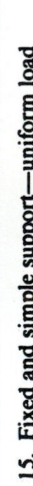

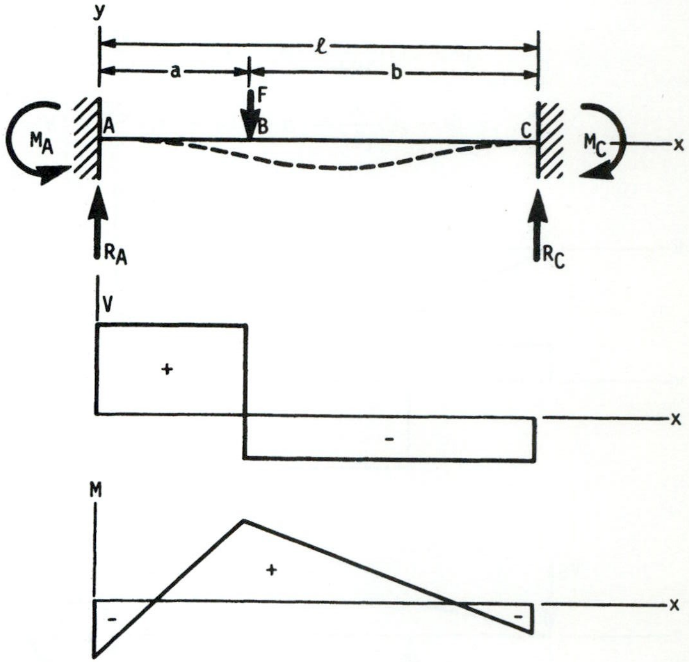

$$R_A = \frac{Fb^2}{\ell^3}(3a+b) \qquad R_B = \frac{Fa^2}{\ell^3}(3b+a)$$

$$M_A = -\frac{Fab^2}{\ell^2} \qquad M_B = \frac{Fab^2}{\ell^3}(3a+b-\ell)$$

$$M_C = -\frac{Fa^2b}{\ell^2} \qquad y_B = \frac{Fa^3b^2}{6EI\ell^3}(3a+b-3\ell)$$

15. Fixed and simple support—uniform load

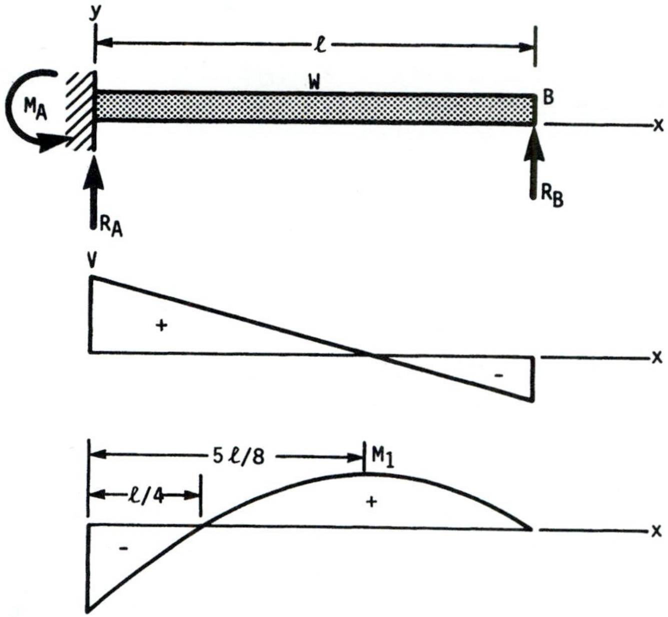

$$R_A = \frac{5W}{8} \qquad R_B = \frac{3W}{8}$$

$$M_A = -\frac{W\ell}{8} \qquad M_1 = \frac{9W\ell}{128}$$

$$y_{max} = -\frac{W\ell^3}{185EI}$$

17. Fixed supports—uniform load

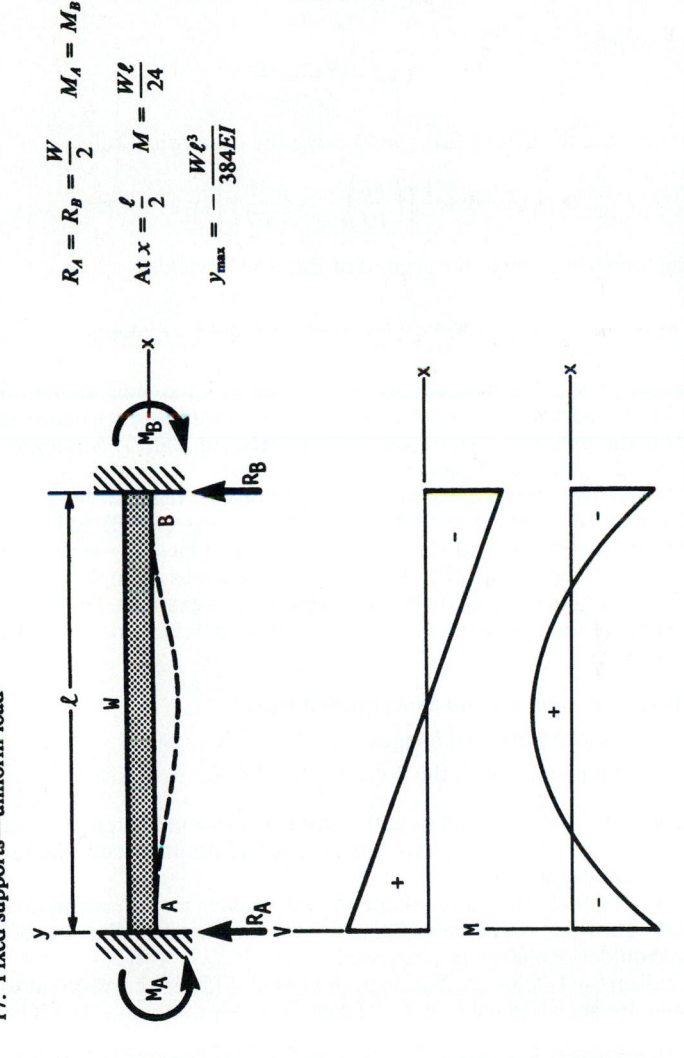

$$R_A = R_B = \frac{W}{2} \qquad M_A = M_B = -\frac{W\ell}{12}$$

$$\text{At } x = \frac{\ell}{2} \qquad M = \frac{W\ell}{24}$$

$$y_{max} = -\frac{W\ell^3}{384EI}$$

A second integration gives

$$y = \psi + C_1 x + C_2 \qquad (b)$$

It is convenient to write Eqs. (a) and (b) as

$$\theta = K(\phi + C_1) \qquad (50.10)$$

$$y = K(\psi + C_1 x + C_2) \qquad (50.11)$$

where K depends on the units used.

Locating supports at $x = a$ and $x = b$ and specifying zero deflection at these supports provides the two conditions for finding C_1 and C_2. The results are

$$C_1 = \frac{\psi_b - \psi_a}{x_a - x_b} \qquad (50.12)$$

$$C_2 = \frac{x_b \psi_a - x_a \psi_b}{x_a - x_b} \qquad (50.13)$$

Now we write the first of Eqs. (50.9) using the trapezoidal rule:

$$\phi_{i+2} = \phi_i + \frac{1}{2}\left[\left(\frac{M}{EI}\right)_{i+2} + \left(\frac{M}{EI}\right)_i\right](x_{i+2} - x_i) \qquad (50.14)$$

Applying Simpson's rule to the second of Eqs. (50.9) yields

$$\psi_{i+4} = \psi_i + \frac{1}{6}(\phi_{i+4} + 4\phi_{i+2} + \phi_i)(x_{i+4} - x_i) \qquad (50.15)$$

As indicated previously, these equations are used in a marching manner. Thus, using Eq. (50.14), we successively compute $\phi_1, \phi_3, \phi_5, \ldots$ beginning at x_1 and ending at x_N, where N is the number of M/EI values. Similarly, Eq. (50.15) is integrated successively to yield $\psi_1, \psi_5, \psi_9, \ldots, \psi_N$.

After these two integrations have been performed, the constants C_1 and C_2 can be found from Eqs. (50.12) and (50.13), and then Eqs. (50.10) and (50.11) can be solved for the deflection and slope. These terms will have the same indices as the integral ψ. See Chap. 37, pp. 37.5–37.8 for a shaft analysis example.

The details of the method are best explained by an example. The shaft of Fig. 50.2 has all points of interest designated by the station letters $A, B, C, \ldots$. These points must include

- Location of all supports and concentrated loads
- Location of cross-sectional changes
- Location of points at which the deflection and slope are desired

Refer now to Table 50.2 and note that coordinates x tabulated in column 2 correspond to each station. Note also the presence of additional x coordinates; these are selected as halfway stations.

Column 4 of Table 50.2 shows that two M/EI values must be computed for each x coordinate. These are needed to account for the fact that M/EI has an abrupt change at every shoulder or change in cross section.

The indices $i = 1, 2, 3, \ldots, N$ in Eqs. (50.14) and (50.15) correspond to the M/EI values and are shown in column 3 of Table 50.2. A program in BASIC is shown in Fig. 50.3. Note that the term M(I) is used for $(M/EI)_i$.

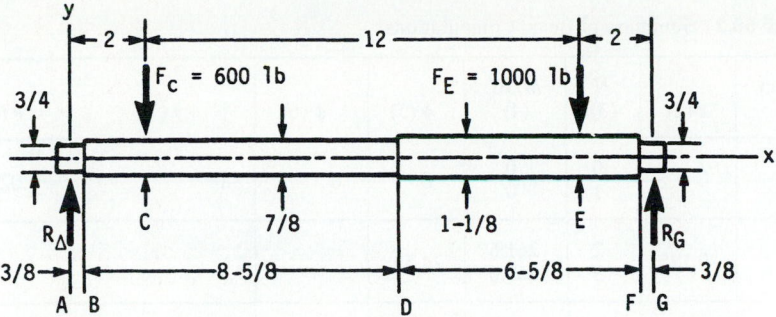

FIGURE 50.2 Simply supported stepped shaft loaded by forces F_C and F_E and supported by bearing reactions R_A and R_G. All dimensions in inches.

50.5   ANALYSIS OF FRAMES

Castigliano's theorem is introduced in Chap. 16, and the energy equations needed for its use are listed in Table 16.2. The method can be used to find the deflection at any point of a frame such as the one shown in Fig. 50.4. For example, the deflection δ_C at C in the direction of F_2 can be found using Eq. (16.2) as

$$\delta_C = \frac{\partial U}{\partial F_2} \tag{50.16}$$

where U = the strain energy stored in the entire frame due to all the forces. If the deflection is desired in another direction or at a point where no force is acting, then a fictitious force Q is added to the system at that point and in the direction in which the deflection is desired. After the partial derivatives have been found, Q is equated to zero, and the remaining terms give the wanted deflection.

The first step in using the method is to make a force analysis of each member of the frame. If Eq. (a) is to be solved, then the numerical values of F_1 and F_2 can be used in the force analysis, but the value of F_2 must *not* be substituted until after each member has been analyzed and the partial derivatives obtained. The following example demonstrates the technique.

Example 1.   Find the downward deflection of point D of the frame shown in Fig. 50.5.
Solution.   A force analysis of the system gives an upward reaction at E of $R_E = 225 + 3F_2$. The reaction at A is downward and is $R_A = 75 - 2F_2$.
The strain energy for member CE is

$$U_{CE} = \frac{R_A^2 \ell}{2AE} \tag{1}$$

The partial deflection is taken with respect to F_2 because the deflection at D in the direction of F_2 is desired. Thus

$$\frac{\partial U_{CE}}{\partial F_2} = \frac{2R_A \ell}{2AE} \frac{\partial R_A}{\partial F_2} \tag{2}$$

TABLE 50.2 Summary of Beam Computations†

Station (1)	x (2)	N (3)	M/EI (4)	φ (5)	ψ (6)	y (7)	θ (8)
A	0	0 1	0 0	0	0	0	−1.028E-02
	0.188	2 3	261.6 261.6	24.59			
B	0.375	4 5	523.2 282.4	98.0	12.27	−3.8444E-03	−1.019E-02
	1.188	6 7	894.6 894.6	576.4			
C	2	8 9	1 506 1 506	1 551	1 083	−0.0195	−8.733E-03
	5.5	10 11	1 708.7 1 708.7	7 177			
D	9	12 13	1 911.4 699.5	13 512	52 149	−4.0408E-02	3.228E-03
	11.5	14 15	752.5 752.5	15 327			
E	14	16 17	805.5 805.5	17 274	128 894	−1.5084E-02	6.990E-03
	14.813	18 19	478.1 478.1	17 796			
F	15.625	20 21	151.0 764.7	18 052	157 741	−2.9488E-03	7.768E-03
	15.813	22 23	382.3 382.3	18 159			
G	16	24 25	0 0	18 195	164 546	0	7.911E-03

†The units are in for x, lb·in for M, Mpsi for E, in^4 for I, in for y, and rad for θ.

```
10 PRINT "YOU MAY USE EITHER U.S. CUSTOMARY UNITS IN THIS PROGRAM"
20 PRINT "OR METRIC UNITS.  IF U.S. CUSTOMARY UNITS ARE USED"
30 PRINT "M IS IN INCH-POUNDS, E IN MPSI, AND I IN INCHES TO"
40 PRINT "THE FOURTH POWER.  IF METRIC UNITS ARE USED, M IS IN"
50 PRINT "NEWTON-METERS, E IN GPA, AND I IN CENTIMETERS TO THE"
60 PRINT "FOURTH POWER."
70 PRINT "WILL YOU USE METRIC UNITS (Y OR N)";U$
80 INPUT U$
90 IF U$= "Y" THEN 100 ELSE 110
100 K = .0001:GOTO 120
110 K = .000001
120 DIM M(65),X(65),PHI(65),PSI(65),Y(65),THETA(65)
130 INPUT "N=";N: FOR I = 1 TO N
140 INPUT "M=";M(I): LPRINT "M("I")="M(I)
150 NEXT I
160 FOR I = 1 TO N STEP 2
170 INPUT "X=";X(I): LPRINT "X("I")="X(I)
180 NEXT I
190 STOP
200 LPRINT "PHI( 1 )="PHI(1)
210 FOR I = 1 TO (N-2) STEP 2
220 PHI(I+2) = PHI(I) + ((M(I+1) + M(I))*(X(I+2) - X(I))*.5)
230 LPRINT "PHI("I+2")="PHI(I+2)
240 NEXT I
250 LPRINT "PSI( 1 )="PSI(1)
260 FOR I = 1 TO (N-4) STEP 4
270 PSI(I+4)=PSI(I)+(((PHI(I+4)+(4*PHI(I+2))+PHI(I))*(X(I+4)-X(I)))/6)
280 LPRINT "PSI("I+4")="PSI(I+4)
290 NEXT I
300 PRINT "SPECIFY VALUES OF X AND PSI AT SUPPORT A"
310 INPUT "X=";A    : LPRINT "X(A)="A
320 INPUT "PSI=";PSIA   : LPRINT "PSI(A)="PSIA
330 LPRINT
340 PRINT "SPECIFY VALUES OF X AND PSI AT SUPPORT B"
350 INPUT "X=";B    : LPRINT "X(B)="B
360 INPUT "PSI=";PSIB   : LPRINT "PSI(B)="PSIB
370 LPRINT
380 C1  = (PSIB - PSIA)/(A-B)
390 LPRINT "C(1)="C1
400 C2 =((B*PSIA)-(A*PSIB))/(A-B)
410 LPRINT "C(2)="C2
420 LPRINT
430 FOR I = 1 TO N STEP 4
440 IF X(I) = A THEN 450 ELSE 460
450 Y(I)=0: GOTO 490
460 IF X(I)=B THEN 470 ELSE 480
470 Y(I) = 0: GOTO 490
480 Y(I) =(PSI(I) + (C1  *X(I)) + C2)*K
490 LPRINT "Y("I")="Y(I)
500 NEXT I
510 LPRINT
520 FOR I = 1 TO N STEP 4
530 THETA(I) =(PHI(I) + C1)*K
540 LPRINT "THETA("I")="THETA(I)
550 NEXT I
560 END
```

FIGURE 50.3 Beam problem programmed in BASIC.

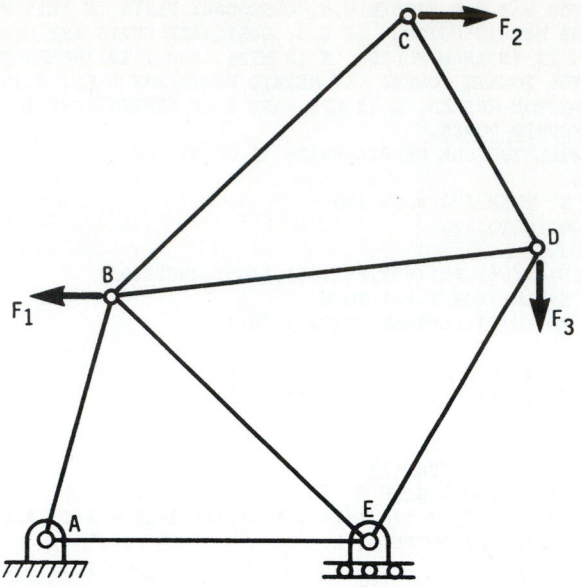

FIGURE 50.4   Frame loaded by three forces.

Also,

$$\frac{\partial R_A}{\partial F_2} = -2$$

Thus Eq. (2) becomes

$$\frac{\partial U_{CE}}{\partial F_2} = \frac{(75 - 2F_2)(30)}{0.2E}(-2) = \frac{37\,500}{E} \tag{3}$$

Note that we were able to substitute the value of F_2 in Eq. (3) because the partial derivative had been taken.

The strain energy stored in member $ABCD$ will have to be computed in three parts because of the change in direction of the bending moment diagram at points B and C. For part AB, the moment is

$$M_{AB} = R_A x = (75 - 2F_2)x$$

The strain energy is

$$U_{AB} = \int_0^6 \frac{M_{AB}^2}{2EI}\,dx \tag{4}$$

Taking the partial derivative with respect to F_2 as before gives

$$\frac{\partial U_{AB}}{\partial F_2} = \int_0^6 \frac{2M_{AB}}{2EI}\frac{\partial M_{AB}}{\partial F_2}\,dx \tag{5}$$

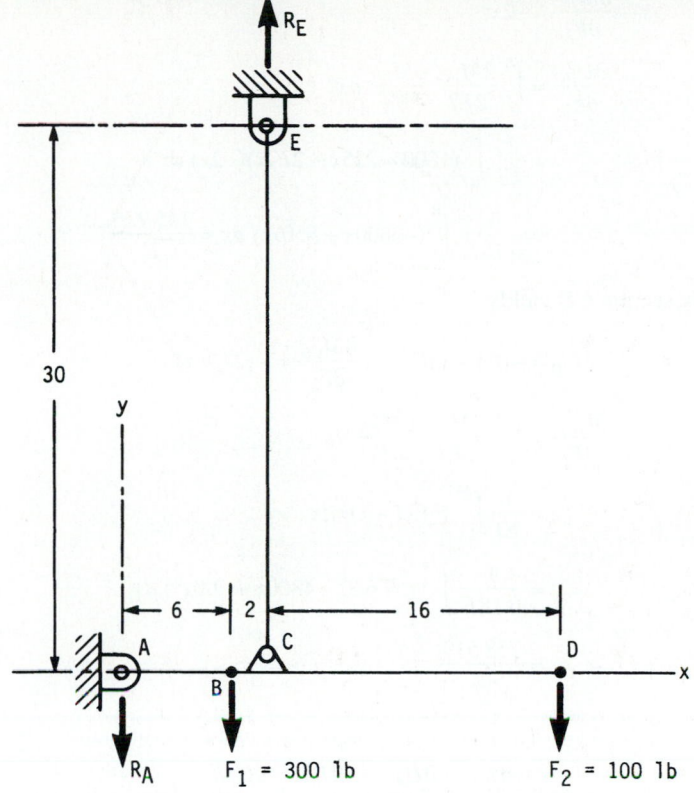

FIGURE 50.5   Frame loaded by two forces. Dimensions in inches: $A_{CE} = 0.20$ in^2; $I_{AD} = 0.18$ in^4; $E = 30 \times 10^6$ psi.

But

$$\frac{\partial M_{AB}}{\partial F_2} = -2x \tag{6}$$

Therefore, Eq. (5) may be written

$$\frac{\partial U_{AB}}{\partial F_2} = \frac{1}{EI} \int_0^6 x(75 - 2F_2)(-2x)\, dx$$

$$= \frac{1}{0.18E} \int_0^6 250x^2\, dx = \frac{100\,000}{E} \tag{7}$$

where the value of F_2 again has been substituted after taking the partial derivative.
   For section BC, we have

$$M_{BC} = R_A x - F_1(x - 6) = 1800 - 225x - 2F_2 x$$

$$\frac{\partial M_{BC}}{\partial F_2} = -2x$$

$$\frac{\partial U_{BC}}{\partial F_2} = \int_6^8 \frac{2M_{BC}}{2EI} \frac{\partial M_{BC}}{\partial F_2} \, dx$$

$$= \frac{1}{EI} \int_6^8 (1800 - 225x - 2F_2x)(-2x) \, dx$$

$$= \frac{1}{0.18E} \int_6^8 (-3600x + 850x^2) \, dx = \frac{145\,926}{E}$$

Finally, section CD yields

$$M_{CD} = -(24-x)F_2 \qquad \frac{\partial M_{CD}}{\partial F_2} = -(24-x)$$

$$\frac{\partial U_{CD}}{\partial F_2} = \int_8^{24} \frac{2M_{CD}}{2EI} \frac{\partial M_{CD}}{\partial F_2} \, dx$$

$$= \frac{1}{EI} \int_8^{24} F_2(24-x)^2 \, dx$$

$$= \frac{1}{0.18E} \int_8^{24} (57\,600 - 4800x + 100x^2) \, dx$$

$$= \frac{758\,519}{E}$$

Then

$$y_D = \frac{\partial U_{CE}}{\partial F_2} + \frac{\partial U_{AB}}{\partial F_2} + \frac{\partial U_{BC}}{\partial F_2} + \frac{\partial U_{CD}}{\partial F_2}$$

$$= \frac{1}{30(10)^6} (37\,500 + 100\,000 + 145\,926 + 758\,519)$$

$$= 0.0347 \text{ in} \qquad \text{(when rounded)}$$

50.5.1 Redundant Members

A frame consisting of one or more redundant members is statically indeterminate because the use of statics is not sufficient to determine all the reactions. In this case, Castigliano's theorem can be used first to determine these reactions and second to determine the desired deflection.

Let R_1, R_2, and R_3 be a set of three indeterminate reactions. The deflection at the supports must be zero, and so Castigliano's theorem can be written three times. Thus

$$\frac{\partial U}{\partial R_1} = 0 \qquad \frac{\partial U}{\partial R_2} = 0 \qquad \frac{\partial U}{\partial R_3} = 0 \qquad (50.17)$$

and so the number of equations to be solved is the same as the number of indeterminate reactions.

In setting up Eqs. (50.17), *do not* substitute the numerical value of the particular force corresponding to the desired deflection. This force symbol must appear in the reaction equations because the partial derivatives must be taken with respect to this force when the deflection is found. The method is illustrated by the following example.

Example 2.   Find the downward deflection at point D of the frame shown in Fig. 50.6.
   Solution.   Choose R_B as the statically indeterminate reaction. A static force analysis then gives the remaining reactions as

$$R_A = R_C = 0.625(F - R_B) \tag{1}$$

The frame consists only of tension members, so the strain energy in each member is

$$U_{AD} = U_{DC} = \frac{R_A^2 \ell_{AD}}{2 A_{AD} E} \qquad U_{BD} = \frac{R_B^2 \ell_{BD}}{2 A_{BD} E} \tag{2}$$

Using Eq. (50.17), we now write

$$0 = \frac{\partial U}{\partial R_B} = \frac{2 R_A \ell_{AD}}{A_{AD} E} \frac{\partial R_A}{\partial R_B} + \frac{R_B \ell_{BD}}{A_{BD} E} \frac{\partial R_B}{\partial R_B} \tag{3}$$

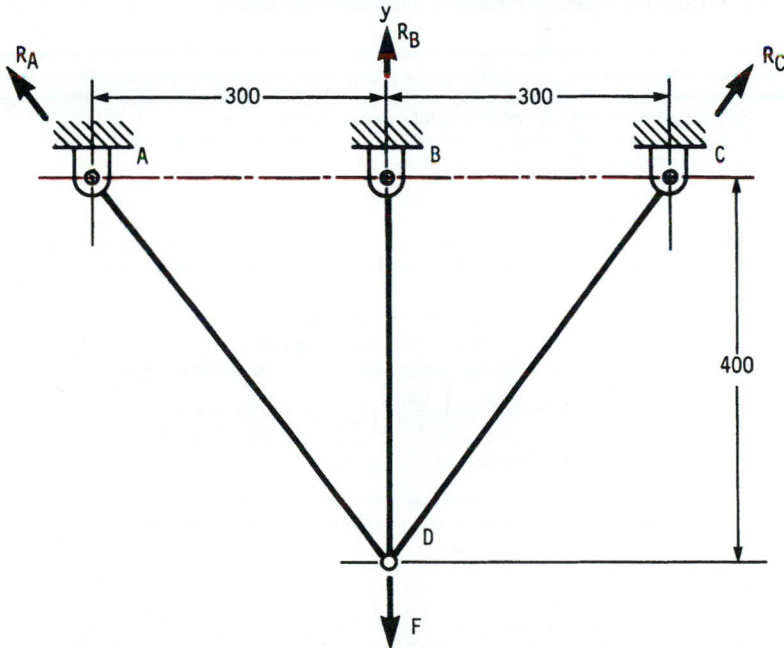

FIGURE 50.6   Frame loaded by a single force. Dimensions in millimeters: $A_{AD} = A_{CD} = 2 \text{ cm}^2$, $A_{BD} = 1.2 \text{ cm}^2$, $E = 207 \text{ GPa}$, $F = 20 \text{ kN}$.

Equation (1) gives $\partial R_A/\partial R_B = -0.625$. Also, $\partial R_B/\partial R_B = 1$. Substituting numerical values in Eq. (3), except for F, gives

$$\frac{2(0.625)(F - R_B)(500)(-0.625)}{2(207)} + \frac{R_B(400)(1)}{1.2(207)} = 0 \tag{4}$$

Solving gives $R_B = 0.369F$. Therefore, from Eq. (1), $R_A = R_C = 0.394F$. This completes the solution of the case of the redundant member. The next problem is to find the deflection at D.

Using Eq. (2), again we write

$$y_D = \frac{\partial U}{\partial F} = \frac{2R_A \ell_{AD}}{A_{AD}E} \frac{\partial R_A}{\partial F} + \frac{R_B \ell_{BD}}{A_{BD}E} \frac{\partial R_B}{\partial F} \tag{5}$$

For use in this equation, we note that $\partial R_A/\partial F = 0.394$ and $\partial R_B/\partial F = 0.369$. Having taken the derivatives, we can now substitute the numerical value of F. Thus Eq. (5) becomes[†]

$$y_D = \left\{ \frac{2[0.394(20)](500)(0.394)}{2(207)} + \frac{[0.369(20)](400)(0.369)}{1.2(207)} \right\} 10^{-2}$$

$$= 0.119 \text{ mm}$$

If care is taken to refrain from substituting numerical values for reactions or forces until after partial derivatives are taken, Castigliano's theorem is applicable to statically indeterminate frames containing redundant members.

[†] In general, when using metric quantities, prefixed units are chosen so as to produce number strings of not more than four members. Thus some preferred units in SI are MPa (N/mm²) for stress, GPa for modulus of elasticity, mm for length, and, say, cm⁴ for second moment of area.

People are sometimes confused when they encounter an equation containing a number of mixed units. Suppose we wish to solve a deflection equation of the form

$$y = \frac{64F\ell^3}{3\pi d^4 E}$$

where $F = 1.30$ kN, $\ell = 300$ mm, $d = 2.5$ cm, and $E = 207$ GPa. Form the equation into two parts, the first containing the numbers and the second containing the prefixes. This converts everything to base units, including the result. Thus,

$$y = \frac{64(1.30)(300)^3}{3\pi(2.5)^4(207)} \frac{\text{(kilo)(milli)}^3}{\text{(centi)}^4\text{(giga)}}$$

Now compute the numerical value of the first part and substitute the prefix values in the second. This gives

$$y = (29.48 \times 10^3) \left[\frac{10^3(10^{-3})^3}{(10^{-2})^4(10^9)} \right] = 29.48 \times 10^{-4} \text{ m}$$

$$= 2.948 \text{ mm}$$

Note that we multiplied the result by 10^3 mm/m to get the answer in millimeters. When this approach is used with Eq. (5), it is found that the result must be multiplied by $(10)^{-2}$ to get y in millimeters.

INDEX

ABOUT THE EDITORS

Joseph Edward Shigley is Late Professor Emeritus of Mechanical Engineering, The University of Michigan. Refer to the Dedication (p. iii).

Charles R. Mischke, Professor Emeritus of Mechanical Engineering, Iowa State University, B.S.M.E. (1947), M.M.E. (1950) Cornell University, Ph.D. (1953) University of Wisconsin. He served on the faculty of mechanical engineering, University of Kansas (1953–1957), as Professor and Chairman of Mechanical Engineering, Pratt Institute (1957–1964), and as Professor of Mechanical Engineering at Iowa State University (1964–1992). His books include *Elements of Mechanical Analysis* (1963), *Introduction to Computer-Aided Design* (1968), and *Introduction to Engineering through Mathematical Model Building* (1980), and he was Coeditor-in-Chief, with J. E. Shigley, of the *Standard Handbook of Machine Design* (1986) and *Mechanical Designers' Workbooks,* eight volumes (1989). He coauthored with J. E. Shigley *Mechanical Engineering Design,* 5th ed. (1989).

He had authored many technical papers on designing to a reliability specification, computer-aided design, and design morphology. He created the CADET (*C*omputer-*A*ugmented *D*esign *E*ngineering *T*echnique) software, and he served on the Reliability, Stress Analysis and Failure Prevention Committee of the American Society of Mechanical Engineers (Reliability Subcommittee chair).

Dr. Mischke's honors and awards include Life Fellow, A.S.M.E. (1993), Centennial Certificate of Recognition of A.S.E.E. (1993), the Ralph Coats Roe Award of A.S.E.E. (1991), The Iowa Legislature Teaching Excellence Award (1991), the Machine Design Award of A.S.M.E. (1990), the Association of American Publishers Award (1987), Outstanding Teaching Award, Iowa State University (1980), the Ralph Teeter Award of the Society of Automotive Engineers (1977), and Alcoa Foundation Professor (1974).

Standard Prefixes for Metric Units

SI prefix*	Symbol	Multiple
exa	E	1 000 000 000 000 000 000 = 10^{18}
peta	P	1 000 000 000 000 000 = 10^{15}
tera	T	1 000 000 000 000 = 10^{12}
giga	G	1 000 000 000 = 10^{9}
mega	M	1 000 000 = 10^{6}
kilo	k	1 000 = 10^{3}
hecto	h	100 = 10^{2}
deka	da	10 = 10^{1}
deci	d	0.1 = 10^{-1}
centi	c	0.01 = 10^{-2}
milli	m	0.001 = 10^{-3}
micro	μ	0.000 001 = 10^{-6}
nano	n	0.000 000 001 = 10^{-9}
pico	p	0.000 000 000 001 = 10^{-12}
femto	f	0.000 000 000 000 001 = 10^{-15}
atto	a	0.000 000 000 000 000 001 = 10^{-18}

*SI = International System of units.

Greek Letters

alpha	A	α, a	nu	N	ν
beta	B	β	xi	Ξ	ξ
gamma	Γ	γ, ∂	omicron	O	o
delta	Δ	δ, ∂	pi	Π	π
epsilon	E	ϵ, ε	rho	P	ρ
zeta	Z	ζ	sigma	Σ	σ, σ
eta	H	η	tau	T	τ
theta	Θ, Θ	θ, ϑ	upsilon	Υ	υ
iota	I	ι	phi	Φ	ϕ, φ
kappa	K	κ	chi	X	χ
lambda	Λ	λ	psi	Ψ	ψ
mu	M	μ	omega	Ω	ω

Conversion Factors A to Convert Input X to Output Y Using the Formula $Y = AX$*

Multiply input X	by factor A	to get output Y	Multiply input X	by factor A	to get output Y
British thermal unit, Btu	1055	joule, J	moment of inertia, lbm·ft²	0.0421	kilogram-meter², kg·m²
Btu/second, Btu/s	1.05	kilowatt, kW	moment of inertia, lbm·in²	293	kilogram-millimeter², kg·mm²
calorie	4.19	joule, J	moment of section (second moment of area), in⁴	41.6	centimeter⁴, cm⁴
centimeter of mercury (0°C)	1.333	kilopascal, kPa	ounce-force, oz	0.278	newton, N
centipoise, cP	0.001	pascal-second, Pa·s	ounce-mass, oz	0.0311	kilogram, kg
degree (angle)	0.0174	radian, rad	pound, lb*	4.45	newton, N
foot, ft	0.305	meter, m	pound-foot, lb·ft	1.36	newton-meter, N·m
foot², ft²	0.0929	meter², m²	pound/foot², lb/ft²	47.9	pascal, Pa
foot/minute, ft/min (fpm)	0.0051	meter/second, m/s	pound-inch, lb·in	0.113	joule, J
foot-pound, ft·lb	1.35	joule, J	lb·in	0.113	newton-meter, N·m
foot-pound/second, ft·lb/s	1.35	watt, W	pound/inch, lb/in	175	newton/meter, N/m
foot/second, ft/s	0.305	meter/second, m/s	pound/inch², psi (lb/in²)	6.89	kilopascal, kPa
gallon (U.S.), gal	3.785	liter, l	pound-mass, lbm	0.454	kilogram, kg
horsepower, hp	0.746	kilowatt, kW	pound-mass/second, lbm/s	0.454	kilogram/second, kg/s
inch, in	0.0254	meter, m	quart (U.S. liquid), qt	946	milliliters, ml
inch, in	25.4	millimeter, mm	section modulus, in³	16.4	centimeter³, cm³
inch², in²	645	millimeter², mm²	slug	14.6	kilogram, kg
inch of mercury (32°F)	3.386	kilopascal, kPa	ton (short 2000 lbm)	907	kilogram, kg
kilopound, kip	4.45	kilonewton, kN	yard, yd	0.914	meter, m
kilopound/inch², kpsi (ksi)	6.89	megapascal, MPa (N/mm²)			
mass, lb·s²/in	175	kilogram, kg			
mile, mi	1.610	kilometer, km			
mile/hour, mi/h	1.61	kilometer/hour, km/h			
mile/hour, mi/h	0.447	meter/second, m/s			

*The U.S. Customary System unit of the pound-force is often abbreviated as lbf to distinguish it from the pound-mass, which is abbreviated as lbm. In most places in this book the pound force is usually written simply as the pound and abbreviated as lb.